PERIODIC TABLE OF THE ELEMENTS

	Main groups		Transition metals										Main groups					
	1A[a] 1	2A 2	3B 3	4B 4	5B 5	6B 6	7B 7	8B 8	8B 9	8B 10	1B 11	2B 12	3A 13	4A 14	5A 15	6A 16	7A 17	8A 18
1	1 **H** 1.00794																	2 **He** 4.002602
2	3 **Li** 6.941	4 **Be** 9.012182											5 **B** 10.811	6 **C** 12.0107	7 **N** 14.0067	8 **O** 15.9994	9 **F** 18.998403	10 **Ne** 20.1797
3	11 **Na** 22.989770	12 **Mg** 24.3050											13 **Al** 26.981538	14 **Si** 28.0855	15 **P** 30.973761	16 **S** 32.065	17 **Cl** 35.453	18 **Ar** 39.948
4	19 **K** 39.0983	20 **Ca** 40.078	21 **Sc** 44.955910	22 **Ti** 47.867	23 **V** 50.9415	24 **Cr** 51.9961	25 **Mn** 54.938049	26 **Fe** 55.845	27 **Co** 58.933200	28 **Ni** 58.6934	29 **Cu** 63.546	30 **Zn** 65.39	31 **Ga** 69.723	32 **Ge** 72.64	33 **As** 74.92160	34 **Se** 78.96	35 **Br** 79.904	36 **Kr** 83.80
5	37 **Rb** 85.4678	38 **Sr** 87.62	39 **Y** 88.90585	40 **Zr** 91.224	41 **Nb** 92.90638	42 **Mo** 95.94	43 **Tc** [98]	44 **Ru** 101.07	45 **Rh** 102.90550	46 **Pd** 106.42	47 **Ag** 107.8682	48 **Cd** 112.411	49 **In** 114.818	50 **Sn** 118.710	51 **Sb** 121.760	52 **Te** 127.60	53 **I** 126.90447	54 **Xe** 131.293
6	55 **Cs** 132.90545	56 **Ba** 137.327	71 **Lu** 174.967	72 **Hf** 178.49	73 **Ta** 180.9479	74 **W** 183.84	75 **Re** 186.207	76 **Os** 190.23	77 **Ir** 192.217	78 **Pt** 195.078	79 **Au** 196.96655	80 **Hg** 200.59	81 **Tl** 204.3833	82 **Pb** 207.2	83 **Bi** 208.98038	84 **Po** [208.98]	85 **At** [209.99]	86 **Rn** [222.02]
7	87 **Fr** [223.02]	88 **Ra** [226.03]	103 **Lr** [262.11]	104 **Rf** [261.11]	105 **Db** [262.11]	106 **Sg** [266.12]	107 **Bh** [264.12]	108 **Hs** [269.13]	109 **Mt** [268.14]	110 [271.15]	111 [272.15]	112 [277]		114 [285]		116 [289]		

*Lanthanide series	57 ***La** 138.9055	58 **Ce** 140.116	59 **Pr** 140.90765	60 **Nd** 144.24	61 **Pm** [145]	62 **Sm** 150.36	63 **Eu** 151.964	64 **Gd** 157.25	65 **Tb** 158.92534	66 **Dy** 162.50	67 **Ho** 164.93032	68 **Er** 167.259	69 **Tm** 168.93421	70 **Yb** 173.04
†Actinide series	89 **†Ac** [227.03]	90 **Th** 232.0381	91 **Pa** 231.03588	92 **U** 238.02891	93 **Np** [237.05]	94 **Pu** [244.06]	95 **Am** [243.06]	96 **Cm** [247.07]	97 **Bk** [247.07]	98 **Cf** [251.08]	99 **Es** [252.08]	100 **Fm** [257.10]	101 **Md** [258.10]	102 **No** [259.10]

[a]The labels on top (1A, 2A, etc.) are common American usage. The labels below these (1, 2, etc.) are those recommended by the International Union of Pure and Applied Chemistry.

The names and symbols for elements 110 and above have not yet been decided.

Atomic weights in brackets are the masses of the longest-lived or most important isotope of radioactive elements.

Further information is available at http://www.shef.ac.uk/chemistry/web-elements/

The production of element 116 was reported in May 1999 by scientists at Lawrence Berkeley National Laboratory.

LIST OF ELEMENTS WITH THEIR SYMBOLS AND ATOMIC WEIGHTS

Element	Symbol	Atomic number	Atomic weight	Element	Symbol	Atomic number	Atomic weight	Element	Symbol	Atomic number	Atomic weight
Actinium	Ac	89	227.03[a]	Helium	He	2	4.002602	Rhenium	Re	75	186.207
Aluminum	Al	13	26.981538	Holmium	Ho	67	164.93032	Rhodium	Rh	45	102.90550
Americium	Am	95	243.06[a]	Hydrogen	H	1	1.00794	Rubidium	Rb	37	85.4678
Antimony	Sb	51	121.760	Indium	In	49	114.818	Ruthenium	Ru	44	101.07
Argon	Ar	18	39.948	Iodine	I	53	126.90447	Rutherfordium	Rf	104	261.11[a]
Arsenic	As	33	74.92160	Iridium	Ir	77	192.217	Samarium	Sm	62	150.36
Astatine	At	85	209.99[a]	Iron	Fe	26	55.845	Scandium	Sc	21	44.955910
Barium	Ba	56	137.327	Krypton	Kr	36	83.80	Seaborgium	Sg	106	266[a]
Berkelium	Bk	97	247.07[a]	Lanthanum	La	57	138.9055	Selenium	Se	34	78.96
Beryllium	Be	4	9.012182	Lawrencium	Lr	103	262.11[a]	Silicon	Si	14	28.0855
Bismuth	Bi	83	208.98038	Lead	Pb	82	207.2	Silver	Ag	47	107.8682
Bohrium	Bh	107	264.12[a]	Lithium	Li	3	6.941	Sodium	Na	11	22.989770
Boron	B	5	10.811	Lutetium	Lu	71	174.967	Strontium	Sr	38	87.62
Bromine	Br	35	79.904	Magnesium	Mg	12	24.3050	Sulfur	S	16	32.065
Cadmium	Cd	48	112.411	Manganese	Mn	25	54.938049	Tantalum	Ta	73	180.9479
Calcium	Ca	20	40.078	Meitnerium	Mt	109	268.14[a]	Technetium	Tc	43	98[a]
Californium	Cf	98	251.08[a]	Mendelevium	Md	101	258.10[a]	Tellurium	Te	52	127.60
Carbon	C	6	12.0107	Mercury	Hg	80	200.59	Terbium	Tb	65	158.92534
Cerium	Ce	58	140.116	Molybdenum	Mo	42	95.94	Thallium	Tl	81	204.3833
Cesium	Cs	55	132.90545	Neodymium	Nd	60	144.24	Thorium	Th	90	232.0381
Chlorine	Cl	17	35.453	Neon	Ne	10	20.1797	Thulium	Tm	69	168.93421
Chromium	Cr	24	51.9961	Neptunium	Np	93	237.05[a]	Tin	Sn	50	118.710
Cobalt	Co	27	58.933200	Nickel	Ni	28	58.6934	Titanium	Ti	22	47.867
Copper	Cu	29	63.546	Niobium	Nb	41	92.90638	Tungsten	W	74	183.84
Curium	Cm	96	247.07[a]	Nitrogen	N	7	14.0067	Uranium	U	92	238.02891
Dubnium	Db	105	262.11[a]	Nobelium	No	102	259.10[a]	Vanadium	V	23	50.9415
Dysprosium	Dy	66	162.50	Osmium	Os	76	190.23	Xenon	Xe	54	131.293
Einsteinium	Es	99	252.08[a]	Oxygen	O	8	15.9994	Ytterbium	Yb	70	173.04
Erbium	Er	68	167.259	Palladium	Pd	46	106.42	Yttrium	Y	39	88.90585
Europium	Eu	63	151.964	Phosphorus	P	15	30.973761	Zinc	Zn	30	65.39
Fermium	Fm	100	257.10[a]	Platinum	Pt	78	195.078	Zirconium	Zr	40	91.224
Fluorine	F	9	18.9984032	Plutonium	Pu	94	244.06[a]	*[b]		110	271.15[a]
Francium	Fr	87	223.02[a]	Polonium	Po	84	208.98[a]	*[b]		111	272.15[a]
Gadolinium	Gd	64	157.25	Potassium	K	19	39.0983	*[b]		112	277[a]
Gallium	Ga	31	69.723	Praseodymium	Pr	59	140.90765	*[b]		114	285[a]
Germanium	Ge	32	72.64	Promethium	Pm	61	145[a]	*[b]		116	289[a]
Gold	Au	79	196.96655	Protactinium	Pa	91	231.03588				
Hafnium	Hf	72	178.49	Radium	Ra	88	226.03[a]				
Hassium	Hs	108	269.13[a]	Radon	Rn	86	222.02[a]				

[a]Massof longest-lived or most important isotope.

[b]The names of elements 110 and above have not yet been decided.

Chemistry

Chemistry

The Central Science

Ninth Edition

Theodore L. Brown
University of Illinois at Urbana-Champaign

H. Eugene LeMay, Jr.
University of Nevada, Reno

Bruce E. Bursten
The Ohio State University

Julia R. Burdge
Florida Atlantic University

PEARSON EDUCATION, INC.
Upper Saddle River, New Jersey 07458

Library of Congress Cataloging-in-Publication Data
Brown, Theodore L.
Chemistry : the central science/Theodore L. Brown, H. Eugene LeMay, Jr., Bruce E. Bursten ; with contributions by Julia R. Burdge.--9th ed.
p. cm.
Includes index.
ISBN 0-13-066997-0
1. Chemistry I. LeMay, H. Eugene (Harold Eugene), 1940- II. Bursten, Bruce Edward. III. Title.

QD31.3 .B76 2003
540--dc21 2001058236

Senior Editor: Nicole Folchetti
Media Editor: Paul Draper
Art Director: Heather Scott
Assistant Art Director: John Christiana
Executive Managing Editor: Kathleen Schiaparelli
Assistant Managing Editor, Science Media: Nicole Bush
Assistant Managing Editor, Science Supplements: Dinah Thong
Development Editor, Text: John Murdzek
Development Editor, Media: Anne Madura
Project Manager: Kristen Kaiser
Media Production Editor: Richard Barnes
Supplements Production Editor: Natasha Wolfe
Art Editor: Thomas Benfatti
Editorial Assistants: Nancy Bauer/Eliana Ortiz
Photo Editor: Debbie Hewitson
Senior Marketing Manager: Steve Sartori
Creative Director: Carole Anson
Director, Creative Services: Paul Belfanti
Manufacturing Manager: Trudy Pisciotti
Assistant Manufacturing Manager: Michael Bell
Editor in Chief, Physical Science: John Challice
Editor in Chief, Development: Ray Mullaney
Vice President ESM Production and Manufacturing: David W. Riccardi
Interior Design: Judith A. Matz-Coniglio
Photo Researcher: Truitt & Marshall
Art Studio: Artworks: Senior Manager: Patty Burns/Production Manager: Ronda Whitson
Manager, Production Technologies: Matthew Haas/Project Coordinator: Connie Long
Illustrators: Royce Copenheaver, Jay McElroy, Daniel Knopsnyder, Mark Landis, Jonathan Derk
Quality Assurance: Stacy Smith, Pamela Taylor, Timothy Nguyen
Contributing Art Studio: Precision Graphics
Cover Art: Ken Eward, Biografx
Cover Designer: Joseph Sengotta
Production Services/Composition: Preparé, Inc.

Pearson Education, Inc.
Upper Saddle River, NJ 07458

Printed in the United States of America
ISBN 0-13-066997-0 (student edition)
10 9 8 7 6

ISBN 0-13-048450-4 (school edition)
10 9 8 7

Pearson Education Ltd.
Pearson Education Australia Pty. Ltd.
Pearson Education Singapore, Pte. Ltd.
Pearson Education North Asia Ltd.
earson Education Canada, Inc.
arson Educacíon de Mexico, S.A. de C.V.
rson Education—Japan
on Education Malaysia, Pte. Ltd.

To our students, whose enthusiasm and curiosity have often inspired us, and whose questions and suggestions have sometimes taught us.

Brief Contents

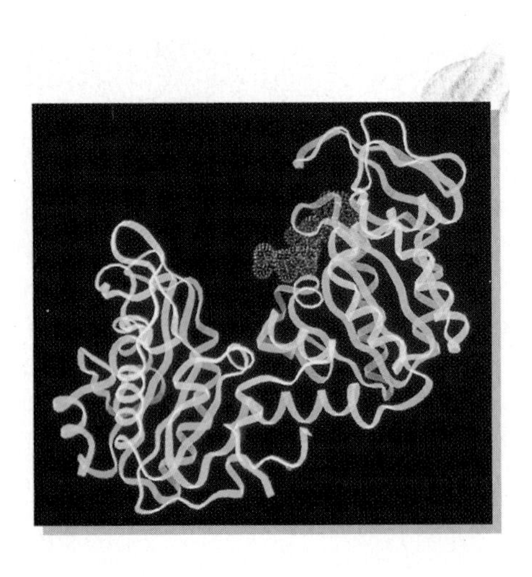

Contents

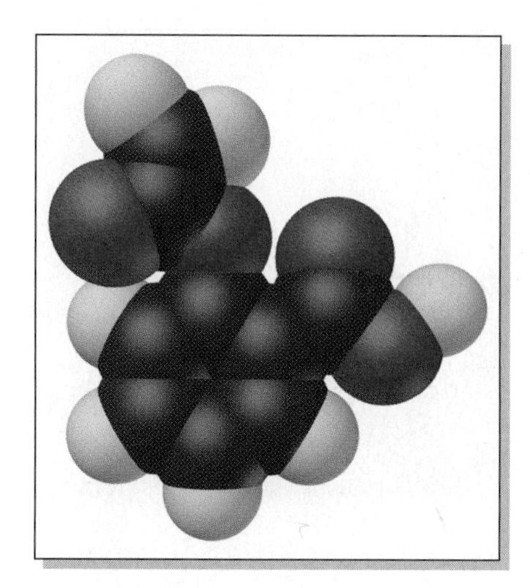

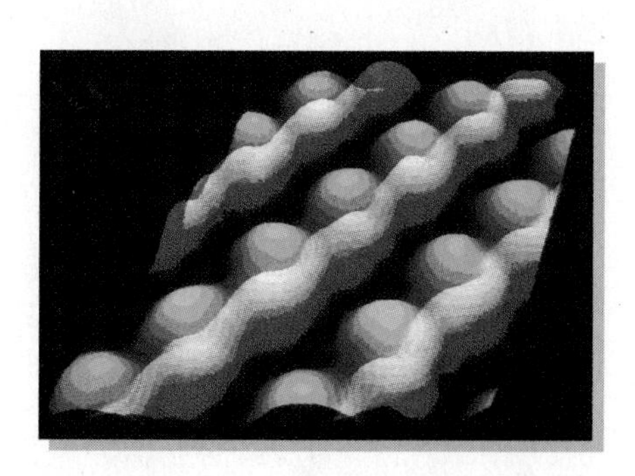

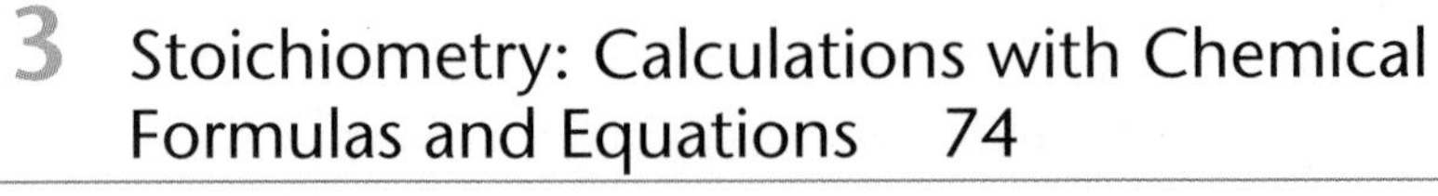

5 Thermochemistry 152

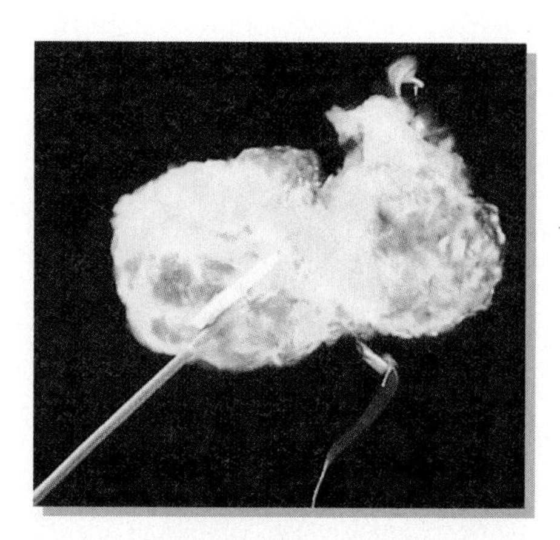

6 Electronic Structure of Atoms 198

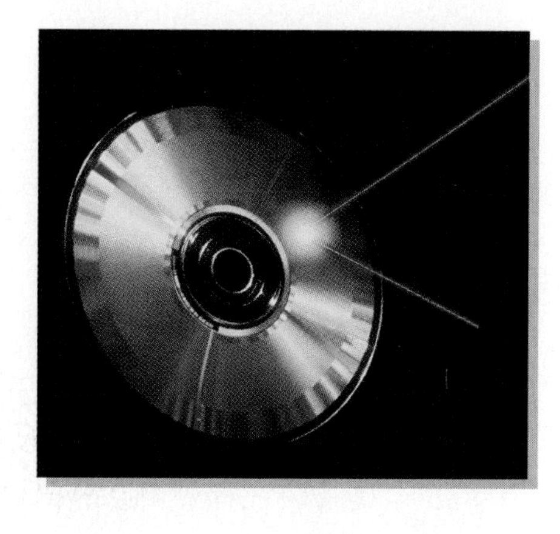

7 Periodic Properties of the Elements 236

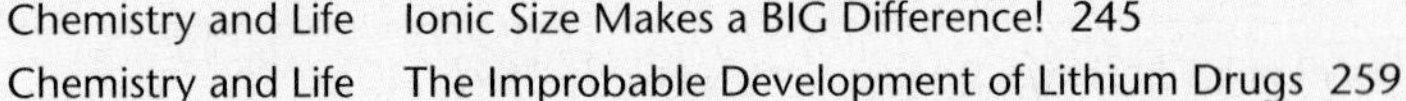

8 Basic Concepts of Chemical Bonding 274

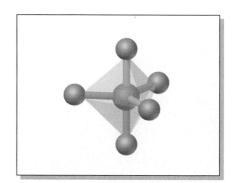

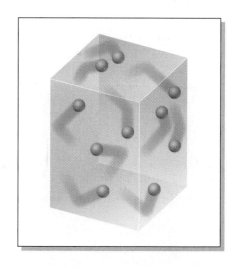

11 Intermolecular Forces, Liquids, and Solids 406

12 Modern Materials 450

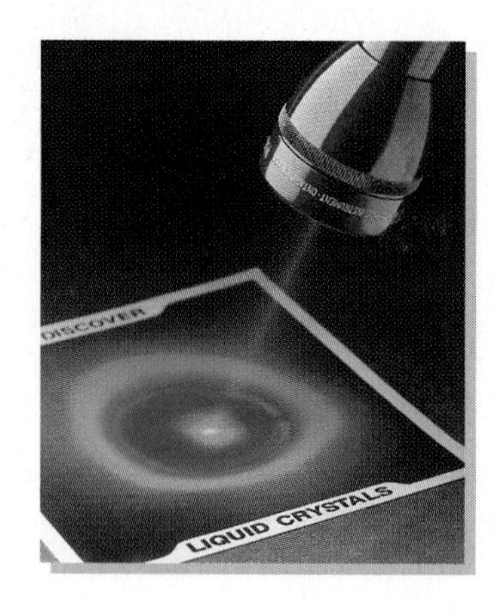

13 Properties of Solutions 484

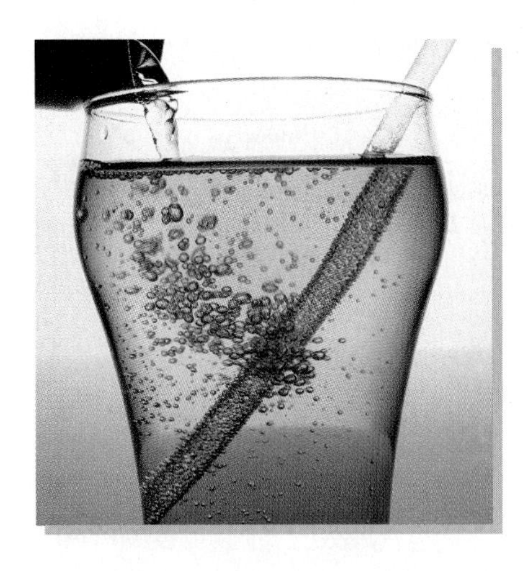

14 Chemical Kinetics 524

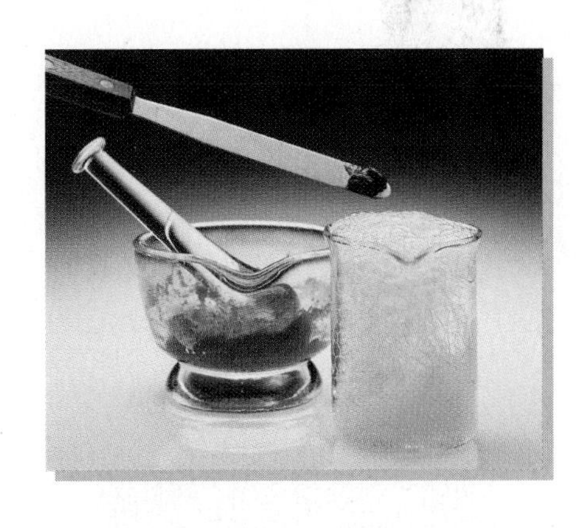

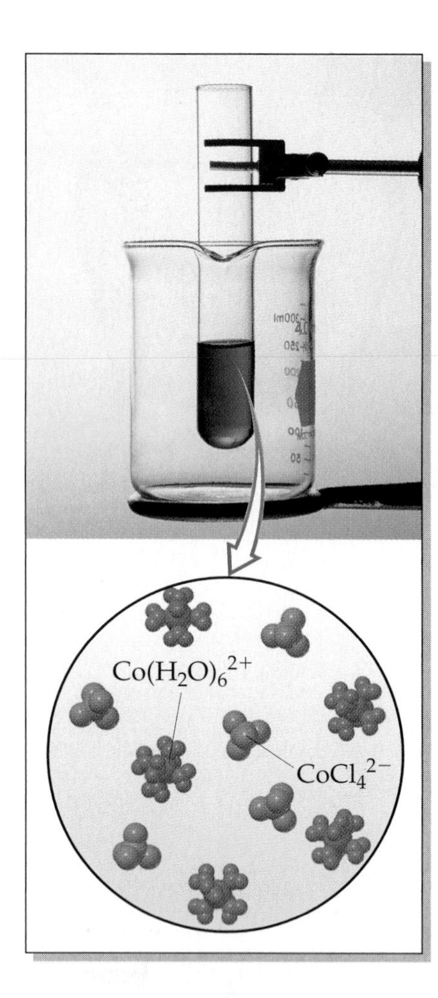

15 Chemical Equilibrium 574

16 Acid-Base Equilibria 612

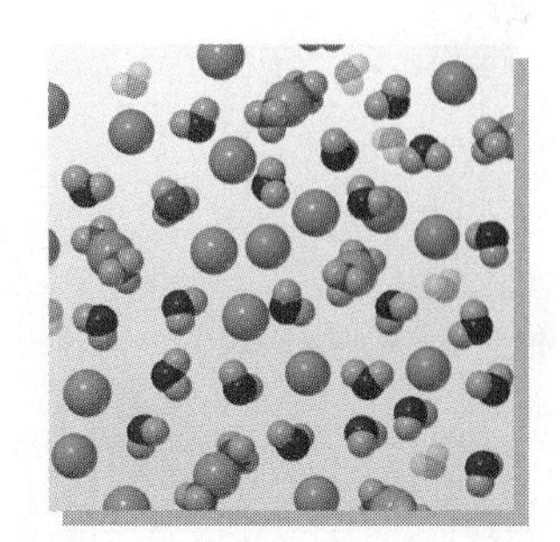

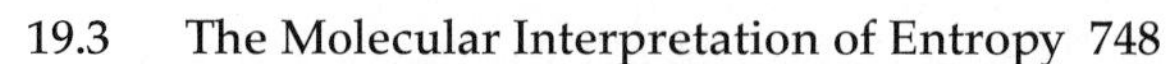

20 Electrochemistry 776

21 Nuclear Chemistry 830

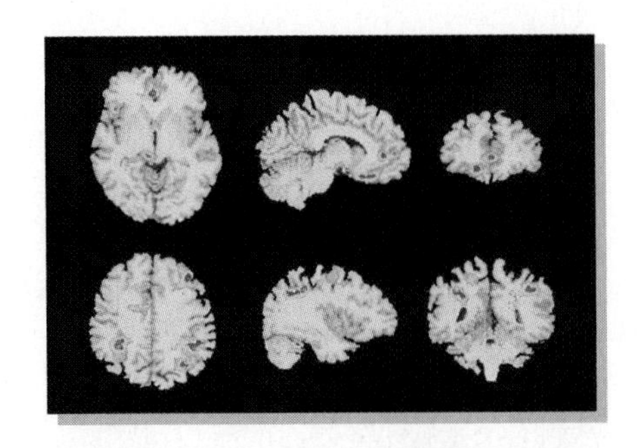

22 Chemistry of the Nonmetals 866

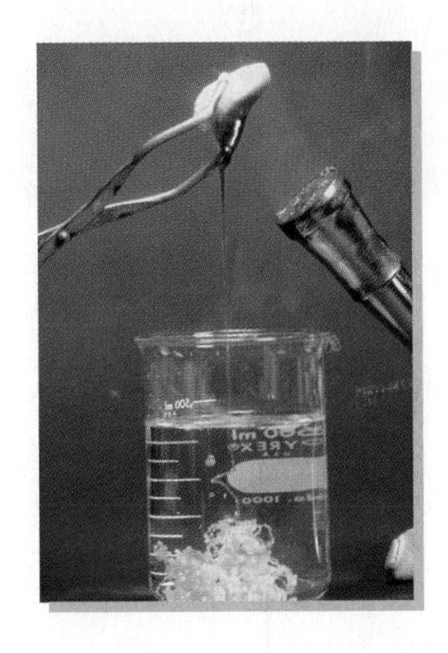

23 Metals and Metallurgy 918

24 Chemistry of Coordination Compounds 948

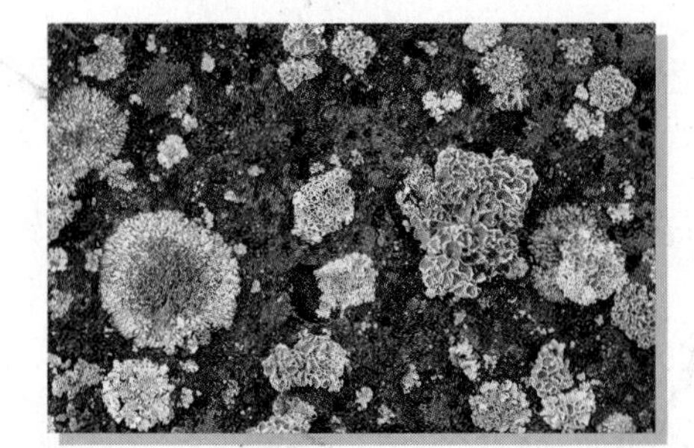

25 The Chemistry of Life: Organic and Biological Chemistry 982

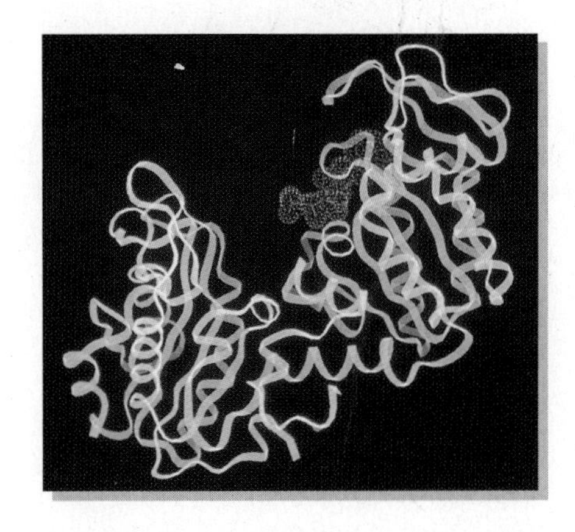

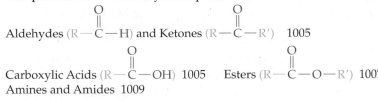

Appendices

Chemical Applications and Essays

Chemistry at Work

A Closer Look

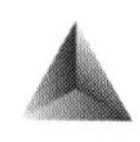

Chemistry and Life

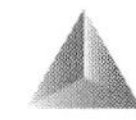

Strategies in Chemistry

Preface

To the Instructor

Philosophy

Throughout the evolution of this text, certain goals have guided our writing efforts. The first is that a text should show students the importance of chemistry in their major areas of study, as well as in their daily lives. We believe that students are more enthusiastic about learning chemistry when they see its importance to their own goals and interests. With this in mind, we have included interesting and significant applications of chemistry. At the same time, the text provides the background in modem chemistry that students need to serve their professional interests, and, as appropriate, to prepare for more advanced chemistry courses.

Second, we want students to see not only that chemistry provides the basis for much of what goes on in our world but also that it is a vital, continually developing science. We have kept the book up to date in terms of new concepts and applications and have tried to convey the excitement of the field.

Third, we feel that if the text is to support your role as teacher effectively, it must be addressed to the students. We have sought to keep our writing clear and interesting and the book attractive and well-illustrated. Furthermore, we have provided numerous in-text study aids for students, including carefully placed descriptions of problem-solving strategies. Together, we have over a hundred years of teaching experience. We hope this is evident in our pacing and choice of examples.

Organization

In the present edition the first five chapters give a largely macroscopic, phenomenological view of chemistry. The basic concepts introduced—such as nomenclature, stoichiometry, and thermochemistry—provide necessary background for many of the laboratory experiments usually performed in general chemistry. We believe that an early introduction to thermochemistry is desirable because so much of our understanding of chemical processes is based on considerations of energy change. Thermochemistry is also important when we come to a discussion of bond enthalpies.

The next four chapters (Chapters 6–9) deal with electronic structure and bonding. The focus then changes to the next level of the organization of matter: the states of matter (Chapters 10 and 11) and solutions (Chapter 13). Also included in this section is an applications chapter on the chemistry of modern materials (Chapter 12), which builds on the student's understanding of chemical bonding and intermolecular interactions.

The next several chapters examine the factors that determine the speed and extent of chemical reactions: kinetics (Chapter 14), equilibria (Chapters 15–17), thermodynamics (Chapter 19), and electrochemistry (Chapter 20). Also in this section is a chapter on environmental chemistry (Chapter 18), in which the concepts developed in preceding chapters are applied to a discussion of the atmosphere and hydrosphere.

After a discussion of nuclear chemistry (Chapter 21), the final chapters survey the chemistry of nonmetals, metals, organic chemistry, and biochemistry (Chapters 22–25). These chapters are developed in a parallel fashion and can be treated in any order.

Our chapter sequence provides a fairly standard organization, but we recognize that not everyone teaches all the topics in exactly the order we have chosen. We have therefore made sure that instructors can make common changes in teaching sequence with no loss in student comprehension. In particular, many instructors prefer to introduce gases (Chapter 10) after stoichiometry or after thermochemistry rather than with states of matter. The chapter on gases has been written to permit this change with *no* disruption in the flow of material. It is also possible to treat the balancing of redox equations (Sections 20.1 and 20.2) earlier, after the introduction of redox reactions in Section 4.4. Finally, some instructors like to cover organic chemistry (Chapter 25) right after bonding (Chapter 9). With the exception of the discussion of stereochemistry (which is introduced in Section 24.3), this, too, is a seamless move.

We have always attempted to introduce students to descriptive organic and inorganic chemistry by integrating examples throughout the text. You will find pertinent and relevant examples of "real" chemistry woven into all the chapters as a means to illustrate principles and applications. Some chapters, of course, more directly address the properties of elements and their compounds, especially Chapters 4, 7, 12, 18, and 22–25. We also incorporate descriptive organic and inorganic chemistry in the end-of-chapter exercises.

Changes in this Edition

Our major goal in the ninth edition has been to strengthen an already strong textbook while retaining its effective and popular style. The traditional strengths of *Chemistry: The Central Science* include its clarity of writing, its scientific accuracy and currency, its strong end-of-chapter exercises, and its consistency in level of coverage. In making changes to this edition, we have tried to be responsive to the feedback we received from the faculty and students who used the eighth edition. Students appreciate the student-friendly style of writing, and we have preserved this style in the ninth edition. Sections that have seemed most difficult to students have in many cases been rewritten and augmented with improved artwork. In order to make the text easier for students to use, we have tried for an even more open, clean design in the layout of the book.

We have also continued to strengthen the art program, to better convey the beauty, excitement, and concepts of chemistry to students. The expanded use of computer-generated molecular art gives students a greater sense of molecular architecture through ball-and-stick and space-filling representations of molecules. In addition, we have added charge distribution maps in selected cases where we believe they can enhance student understanding. We have continued a greater emphasis on three-dimensional representations in the line art. Our goal continues to be to use color and photos to emphasize important points, to focus the student's attention, and to give the text an uncluttered, inviting look.

We still emphasize concept-oriented learning throughout the text. A new feature in this edition is the What's Ahead summary at the opening of each chapter. What's Ahead gives the student a brief overview of the major ideas and relationships that the chapter will cover. We expect that students will begin their study of the chapter with more confidence for having a sense of the direction in which their study will take them. *Concept links* (∞) continue to provide easy-to-see cross-references to pertinent material covered earlier in the text. The essays titled *Strategies in Chemistry*, which provide advice to students on problem solving and "thinking like a chemist," continue to be an important feature. We have added more conceptual exercises to the end-of-chapter exercises. The Integrative Exercises, which give students the opportunity to solve more challenging problems that integrate concepts from the present chapter with those of previous chapters, have also been increased in number.

We have kept the text fresh by keeping it current. References to current events help students relate their studies of chemistry with their everyday life experiences. New essays in our well-received *Chemistry at Work* and *Chemistry and Life* series emphasize world events, scientific discoveries, and medical breakthroughs that have occurred since publication of the eighth edition. We maintain our focus on the positive aspects of chemistry, without neglecting the problems that can arise in an increasingly technological world. Our goal is to help students appreciate the real-world perspective of chemistry and the ways in which chemistry affects their lives.

You'll also find that we've:

- Revised the end-of-chapter Exercises, with particular focus on the black-numbered exercises (those not answered in the Appendix).
- Integrated more conceptual questions into the end-of-chapter material. For the convenience of instructors, these are identified by the annotation in the Annotated Instructor's Edition, but not in the student edition of the text.
- Updated the eMedia Exercises in the end-of-chapter material. These exercises take advantage of the integrated media components and extend student's understanding, using the advantages that interactive, media-rich presentations offer.
- Continued the practice of using a Student Activity icon in the margins to indicate where students can extend understanding of a concept or topic by looking at an activity located on the Web site or the Accelerator CD-ROM.
- Carried the stepwise, Analyze, Plan, Solve, Check, problem-solving strategy into a majority of the Sample Exercises of the book to provide additional guidance in problem solving.
- Added dual-column problem-solving strategies in selected Sample Exercises that outline the process underlying mathematical calculations to teach students how to better perform mathematical calculations.
- Reviewed and revised all chapters based on feedback from reviewers and users. For example, we have:
  - —Added a brief introduction to organic chemistry in Chapter 2.
  - —Improved the presentation of the first law of thermodynamics in Chapter 5.
  - —Expanded the discussion of superconductivity in Chapter 12.
  - —Revised the introductory treatment of equilibrium to eliminate the artificial distinction between equilibrium constants in gas and aqueous phases.
  - —Added a new section on Green Chemistry, which focuses on the environmental impacts of chemical processes.
  - —Improved the treatment of coordination compounds in Chapter 24.

Please see the next pages for more specific details about how the Ninth Edition's integrated learning program will help your students succeed.

Supplements

For the Instructor

- **Annotated Instructor's Edition (with Guide to Print and Media Resources) (0-13-038168-3)** This special instructor's edition provides marginal notes and information for instructors and TAs, including MediaPortfolio and transparency icons, suggested lecture demonstrations, teaching tips, and background references from the chemical education literature for key topics.

- **Solutions to Exercises (0-13-009798-5)** Full solutions to all end-of-chapter exercises in the text are provided. With an instructor's permission, this manual may be made available to students.
- **Instructor's Resource Manual (0-13-009802-7)** This useful guide describes all the different resources available to instructors and shows how to integrate them into your course. Organized by chapter, this manual offers detailed lecture outlines and complete descriptions of all available lecture demonstrations, the animated concept sequences, all video demonstrations, common student misconceptions, and much more.
- **Test Item File (0-13-009792-6)** The Test Item File now provides a selection of more than 3800 test questions, a 25% increase over the previous version.
- **TestGen-EQ (0-13-009793-4)** New testbank software designed with algorithmic questions in mind. This computerized version of the Test Item File includes electronic versions of all 3800 test questions. TestGen-EQ allows you to create and tailor exams to your own needs and includes tools for course management, algorithmic question generation, and administering tests over a local area network.
- **Transparencies (0-13-009794-2)** Two-hundred seventy-six full-color images, more than ever before, are included in an easy-to-use binder. For each transparency, we've made the type even larger for easier viewing in large classrooms.
- ***Central Science Live*—Companion Web site http://www.prenhall.com/brown** The Companion Web site is the focal point for access to the media suite. Instructors can use the Syllabus Manager to administer a date-driven syllabus, including on-line homework assignments or other activities. You can have your students practice their reading comprehension and skills in the Problem Solving Center, peruse the Web for chapter-related resources, or view the Student Activities referred to in the text. If you have adopted the use of the Premium Access Code to the Web site, the eChapters, media-rich presentations that echo the book, are also available to your students. Standard and Premium Access Codes are available; contact your Prentice Hall representative for more information.
- **MediaPortfolio (0-13-009805-1)** An instructor CD/DVD set that contains almost all the art from the text, more than 30 lab demonstration video segments, and more than a 100 animations of core concepts. Using the included MediaPortfolio software, instructors can browse for figures and other media elements by thumbnail and description, as well as search by key word or title. In addition, all of the Student Activities available on the student Accelerator CD are available on the instructor CD/DVD, as well. The images and videos can be cut and pasted, or dragged into your MS PowerPoint® lecture presentation or other documents. The set also contains the Instructor's Resource Manual in MS Word® format, a pre-built PowerPoint Presentation for every chapter, as well as all the responsive media elements specifically developed for *Chemistry: The Central Science, Ninth Edition.*
- **Course Management Options** Prentice Hall provides support for course management systems that are most popular at institutions today, including WebCT®, WebAssign®, BlackBoard®, and Pearson Education's own CourseCompass® (powered by BlackBoard). Course management systems allow complete course administration, including roster and gradebook management, distribution of course materials, setup and maintenance of bulletin boards and announcements, and other tasks.

Prentice Hall can provide the content for a complete chemistry course tailored to *Chemistry: The Central Science, Ninth Edition*, and your course can even include the entire text on-line. In addition to the gallery of animations, we provide quizzing and testing material and a wide range of customizing options. For example, instructors can edit questions, modify/delete/add to the testing database, categorize material by level of difficulty, award different point values for different problems, and give partial credit.

See *http://www.prenhall.com/demo* for a demonstration of our course management options.

- **PH GradeAssist** PH GradeAssist is a new homework and assignment system that allows students unlimited practice with problems that are algorithmically generated and media-enhanced. Instructors can administer quizzes and assignments, control the content and assignment parameters, and receive assignments and view performance statistics with the built-in gradebook.

  In addition, Prentice Hall has partnered with WebAssign, an online system that specializes in the administration of on-line homework. For information on this system, contact your PH representative.

For the Lab

- **Laboratory Experiments *(Nelson/Kemp)* (0-13-009797-7)** This manual includes 41 finely tuned experiments chosen to introduce students to basic lab techniques and to illustrate core chemical principles. It contains pre-lab questions and detachable report sheets. This new edition has been revised to correlate more tightly with the text. Safety and disposal information has also been updated.
- **Annotated Instructor's Edition to *Laboratory Experiments* (0-13-009803-5)** This AIE combines the full student lab manual with front and back appendices covering the proper disposal of chemical waste, safety instructions for the lab, descriptions of standard lab equipment and materials, answers to questions, and more.

For the Student

- *Central Science Live*—The media suite for the Ninth Edition consists of two components that can stand alone or work in concert: the Companion Web site and the Accelerator CD. Access to the materials on the Companion Web site are available through both Standard and Premium Access Codes. Many of the rich media assets available on *Central Science Live* are available on both the CD and on the Web site, and if used together, logging into the Web site provides a rapid, seamless, fully integrated experience for the student.

  ***Central Science Live*—Companion Web site http://www.prenhall.com/brown** Now even more integrated and easier to use, this innovative on-line resource center is designed specifically to support and enhance *Chemistry: The Central Science*, Ninth Edition. Now the front-door for Central Science Live, it features:

  —A Problem-Solving Center, where students have access to more than 2000 additional problems—including algorithmically generated questions and non-multiple-choice questions—all organized by chapter, each with specific hints and detailed feedback. Also included are cumulative quizzes and MCAT review questions.

  —A Visualizing Molecules module, with pre-built 3-D models of molecules discussed in the text that can be manipulated in real time and displayed in different representations.

- —Constantly updated Current Topics Module, linking your students to recently published articles from the lay press, and a Web Resources Center that links your students to other carefully selected, chemistry-related Web sites.
- —A Student Activities module, with hundreds of movies, animations, and interactive simulations that help students discover chemistry. The movies show real chemistry being performed in demonstrations, the animations focus on molecular processes that can't be seen any other way, and the interactive simulations allow students to do experiments and draw conclusions based on simulated experimental results.
- —eChapters, available only on the Premium version of the Web site, which are short synopses of the chapter material, written to include and point to the many Student Activities available in eMedia Chemistry, and including interactive and algorithmically generated self-assessment questions and worked examples. Many students find this an excellent way to preview or review the chapter material in the textbook.
- —An eBook (electronic version of the full textbook) enables students to link directly from Web-based activities and from eChapters to the appropriate sections of the text. This allows students to work through Web exercises without having the actual text in front of them.

***Central Science Live*—Accelerator CD** This book-specific companion to *Chemistry: The Central Science*, Ninth Edition, presents core chemistry content in a dynamic and interactive way. Designed for students, it includes:

- —Over 60 short, narrated animations presenting selected topics that are more easily conveyed in a visual fashion, and over 30 laboratory demonstration video clips showing chemistry in live action.
- —Over 100 Student Activities, responsive activities and simulations that allow students to learn by taking the initiative, changing conditions, adjusting variables, and establishing trends.
- —MediaPortfolio software that allows thumbnail browsing, as well as search capabilities for words and media types, with links to text content.

- **Solutions to Red Exercises (0-13-009799-3)** Full solutions to all of the red-numbered exercises in the text are provided. (Short answers to red exercises are found in the appendix of the text).
- **Solutions to Black Exercises (0-13-009790-X)** Full solutions to all of the black-numbered exercises in the text are provided.
- **Student's Guide (0-13-009795-0)** This book assists students through the text material with chapter overviews, learning objectives, review of key terms, cumulative chapter review quizzes, and self-tests. Included are answers to all *Student's Guide* exercises. Chapter summaries are correlated to those in the Instructor's Resource Manual.
- **Math Review Toolkit (0-13-009801-9)** This free book reinforces the skills necessary to succeed in chemistry. It is keyed specifically to chapters in *Chemistry: The Central Science, Ninth Edition,* and includes additional mathematics review, problem-solving tools and examples, and a section on writing for the laboratory.
- **Lecture Notebook (0-13-038169-1)** This lecture notebook contains the art from the text with notetaking sections to obviate the need for students to spend time re-drawing figures in lecture and instead, concentrate on taking notes.
- **Prentice Hall/*The New York Times* "Themes of the Times"—Chemistry** This innovative program is designed to bring current and relevant applications into the classroom. Adopters of *Chemistry: The Central Science, Ninth Edition,* are eligible to receive these unique "mini-newspapers" that bring together

a collection of the latest and best chemistry articles from the highly respected pages of *The New York Times.* (Updated twice annually.)

- **Prentice Hall Molecular Model Set for General and Organic Chemistry (0-13-955444-0)** This ball-and-stick model kit is designed for use in general chemistry and the student's next course in organic chemistry. It includes trigonal bipyramidal and octahedral atom centers as well as 14 carbon atoms.

To the Student

Chemistry: The Central Science, Ninth Edition, has been written to introduce you to modern chemistry. During the many years that we have been practicing chemists, we have found chemistry to be an exciting intellectual challenge and an extraordinarily rich and varied part of our cultural heritage. We hope that as you advance in your study of chemistry, you will share with us some of that enthusiasm, excitement, and appreciation. We also hope that you will come to realize the importance of chemistry in your everyday life. As authors, we have, in effect, been engaged by your instructor to help you learn chemistry. Based on the comments of students and instructors who have used this book in its previous editions, we believe that we have done that job well. Of course, we expect the text to continue to evolve through future editions. We invite you to write to us to tell us what you like about the book so that we will know where we have helped you most. Also, we would like to learn of any shortcomings, so that we might further improve the book in subsequent editions. Our addresses are given at the end of the Preface.

Advice for Learning and Studying Chemistry

Learning chemistry requires both the assimilation of many new concepts and the development of analytical skills. In this text we have provided you with numerous tools to help you succeed in both. We have provided details of the features of this text in the "walk-through" on pages xxviii–xxxiii. You will find it helpful to examine those features.

As you proceed through your course in chemistry, it is important for you to develop good study habits to help you in the learning process. We offer the following tips for success in your study of chemistry:

Don't fall behind! In your chemistry course, new topics will build on material already presented. If you don't keep up in your reading and problem solving, you will find it much harder to follow the lectures and discussions on current topics. "Cramming" just before an exam has been shown to be an ineffective way to study any subject, chemistry included.

Focus your study. The amount of information you will receive in your chemistry course can sometimes seem overwhelming. It is essential to recognize those concepts and skills that are particularly important. Listen intently to the guidance and emphasis provided by your instructors. Pay attention to the skills stressed in the Sample Exercises and homework assignments. Notice the italicized statements in the text, and study the concepts presented in the chapter summaries.

Keep good lecture notes. Your lecture notes will provide you with a clear and concise record of what your instructor regards as the most important material to learn. Use your lecture notes in conjunction with this text; that's your best way to determine which material to study.

Skim topics in the text before they are covered in lecture. Reviewing a topic before lecture will make it easier for you to take good notes. First read the introduction and Summary, then quickly read through the chapter, skipping Sample Exercises and supplemental sections. Pay attention to the titles of sections and subsections, which give you a feeling for the scope of topics. Try to avoid thinking that you must learn and understand everything right away.

After lecture, carefully read the topics covered in class. You will probably need to read assigned material more than once to master it. As you read, pay attention to the concepts presented and to the application of these concepts in the Sample Exercises. Once you think you understand a Sample Exercise, test your understanding by working the accompanying Practice Exercise. As you progress through the text, you will encounter *Sample Integrative Exercises: Putting Concepts Together*. These are designed to help you see how concepts and methods learned in earlier chapters can be put together with newly learned materials.

Learn the language of chemistry. As you study chemistry, you will encounter many new words. It is important to pay attention to these words and to know their meanings, or the entities to which they refer. Knowing how to identify chemical substances from their names is an important skill; it can help you avoid painful mistakes on examinations.

Attempt all the assigned end-of-chapter exercises. Working the exercises that have been selected by your instructor provides necessary practice in recalling and using the essential ideas of the chapter. You cannot learn merely by observing; you must be a participant. In particular, try to resist checking the Solutions Manual (if you have one) until you have made a sincere effort to solve the exercise yourself. If you really get stuck on an exercise, however, get help from your instructor, your teaching assistant, or from another student. Spending more than 20 minutes on a single exercise is rarely effective unless you know that it is particularly challenging.

Make use of the Web site. Some things are more easily learned by discovery, and others are best shown in three dimensions. Use the Companion Web site to this text to get the most out of your time in chemistry.

The bottom line is to work hard, study effectively, and use the tools that are available to you, including this textbook. We want to help you learn more about the world of chemistry and why it is the *central science*.

Acknowledgments

This book owes its final shape and form to the assistance and hard work of many people. Several colleagues helped us immensely by sharing their insights, reviewing our initial writing efforts, or providing suggestions for improving the text. We would especially like to thank the following:

Ninth Edition Reviewers

John Arnold	University of California, Berkeley
Merrill Blackman (Col.)	US Military Academy
Daeg Scott Brenner	Clark University
Gregory Alan Brewer	Catholic University of America
Gary Buckley	Cameron University
Gene O. Carlisle	Texas A&M University
Dana Chatellier	University of Delaware
William Cleaver	University of Vermont
Elzbieta Cook	University of Calgary
Dwaine Davis	Forsyth Tech Community College
Angel C. deDios	Georgetown University
John Farrar	University of St. Francis
Clark L. Fields	University of Northern Colorado
Jan M. Fleischner	The College of New Jersey
Peter Gold	Penn State University
Michael Greenlief	University of Missouri
John M. Halpin	New York University
Robin Horner	Fayetteville Tech Community College
Roger K. House	Moraine Valley College
William Jensen	South Dakota State University
Siam Kahmis	University of Pittsburgh
John W. Kenney	Eastern New Mexico University
George P. Kreishman	University of Cincinnati
Paul Kreiss	Anne Arundel Community College
David Lehmpuhl	University of Southern Colorado
Gary L. Lyon	Louisiana State University
Albert H. Martin	Moravian College
William A. Meena	Rock Valley College
Massoud Miri	Rochester Institute of Technology
Eric Miller	San Juan College
Mohammad Moharerrzadeh	Bowie State University
Kathleen E. Murphy	Daemen College

Robert T. Paine	University of New Mexico	Richard S. Treptow	Chicago State University
Albert Payton	Broward C. C	Claudia Turro	The Ohio State University
Kim Percell	Cape Fear Community College	Maria Vogt	Bloomfield College
Nancy Peterson	North Central College	Sarah West	University of Notre Dame
James P. Schneider	Portland Community College	Linda M. Wilkes	University of Southern Colorado
Eugene Stevens	Binghamton University	Darren L. Williams	West Texas A&M University
James Symes	Cosumnes River College	Troy D. Wood	SUNY-Buffalo
Edmund Tisko	University of NE at Omaha	David Zax	Cornell University

Ninth Edition Accuracy Checkers

Boyd Beck	Snow College	Robert Paine	Rochester Institute of Technology
B. Edward Cain	Rochester Institute of Technology	Christopher J. Peeples	University of Tulsa
Thomas Edgar Crumm	Indiana University of Pennsylvania	Jimmy R. Rogers	University of Texas at Arlington
Angel deDios	Georgetown University	Iwao Teraoka	Polytechnic University
David Easter	Southwest Texas State University	Richard Treptow	Chicago State University
Jeffrey Madura	Duquesne University	Maria Vogt	Bloomfield College
Hilary L. Maybaum	Think Quest, Inc.		

Special thanks to those who provided valuable feedback to the author and/or publisher:

James Birk	Arizona State University	Roger DeKock	Calvin College
Rik Blumenthal	Auburn University	Friedrich Koknat	Youngstown State University
Daniel T. Haworth	Marquette University	Thomas R. Webb	Auburn University

Previous Edition Reviewers

John J. Alexander	University of Cincinnati	Donald E. Linn, Jr.	Indiana University-Purdue University Indianapolis
Robert Allendoerfer	SUNY-Buffalo	David Lippmann	Southwest Texas State
Boyd R. Beck	Snow College	Ramón López de la Vega	Florida International University
James A. Boiani	College at Geneseo-SUNY	Preston J. MacDougall	Middle Tennessee State University
Kevin L. Bray	Washington State University	Asoka Marasinghe	Moorhead State University
Edward Brown	Lee University	Earl L. Mark	ITT Technical Institute
Donald L. Campbell	University of Wisconsin-Eau Claire	William A. Meena	Rock Valley College
Stanton Ching	Connecticut College	Gordon Miller	Iowa State University
Robert D. Cloney	Fordham University	Massoud (Matt) Miri	Rochester Institute of Technology
Edward Werner Cook	Tunxis Community Technical College	Kathleen E. Murphy	Daemon College
John M. DeKorte	Glendale Community College	Ross Nord	Eastern Michigan University
Roger Frampton	Tidewater Community College	Robert H. Paine	Rochester Institute of Technology
Joe Franek	University of Minnesota	Mary Jane Patterson	Brazosport College
John I. Gelder	Oklahoma State University	Robert C. Pfaff	Saint Joseph's College
Thomas J. Greenbowe	Iowa State University	Jeffrey A. Rahn	Eastern Washington University
Eric P. Grimsrud	Montana State University	Mark G. Rockley	Oklahoma State University
Marie Hankins	University of Southern Indiana	Jimmy Rogers	University of Texas, Arlington
Robert M. Hanson	St. Olaf College	James E. Russo	Whitman College
Gary G. Hoffman	Florida International University	Michael J. Sanger	University of Northern Iowa
Robin Horner	Fayetteville Tech Community College	Jerry L. Sarquis	Miami University
Donald Kleinfelter	University of Tennessee-Knoxville	Gray Scrimgeour	University of Toronto
Manickam Krishnamurthy	Howard University	Richard Treptow	Chicago State University
Brian D. Kybett	University of Regina	Laurence Werbelow	New Mexico Institute of Mining and Technology
William R. Lammela	Nazareth College	Troy D. Wood	SUNY-Buffalo
John T. Landrum	Florida International University		
N. Dale Ledford	University of South Alabama		
Ernestine Lee	Utah State University		

Previous Edition Accuracy Checkers

Leslie Kinsland	University of Louisiana, Lafayette	Robert H. Paine	Rochester Institute of Technology
Albert Martin	Moravian College	Richard Perkins	University of Louisiana, Lafayette
Robert Nelson	Georgia Southern University		

Special thanks to others involved in the review of the previous edition text and its various components:

Pat Amateis	Virginia Polytechnic Institute and State University	Helen Richter	University of Akron
Randy Hall	Louisiana State University	David Shinn	University of Hawaii at Hilo
Daniel T. Haworth	Marquette University	John Vincent	University of Alabama
Neil Kestner	Louisiana State University	Karen Weichelman	University of Louisiana, Lafayette
Barbara Mowery	Yorktown, VA		

We would also like to express our deep gratitude to our colleagues at Prentice Hall who have worked so hard to make this edition possible: Nicole Folchetti, our chemistry editor, who contributed imagination and energy to this edition, and who pulled all the parts together, sometimes in spite of us; Carol Trueheart and Ray Mullaney, our development editorial managers, whose long-term commitments to this book have helped keep it in the forefront; John Challice, Editor in Chief, who continues his record of support and invaluable contributions; Kathleen Schiaparelli, Executive Managing Editor, for her support and encouragement; John Murdzek, our developmental editor, whose good judgement and keen eye have ensured the book's style and quality of presentation; Fran Daniele, who worked with a very difficult schedule to bring us through the production process; Paul Draper, our media editor, and Ann Madura, media development editor, who have continued to enhance the value and range of the book's media materials; Jerry Marshall, our photo researcher, Kristen Kaiser, project manager and Eliana Ortiz, editorial assistant, for their special and valuable contributions to the overall project.

We offer a special thanks to all the students and faculty who gave us comments and suggestions about *Chemistry: The Central Science, Ninth Edition*. You will see many of your suggestions incorporated into the ninth edition.

Finally, we thank our families and friends for their love, support, and patience as we brought this edition to completion.

Theodore L. Brown
School of Chemical Sciences
University of Illinois
Urbana, IL 61801
tlbrown@uiuc.edu

H. Eugene LeMay, Jr.
Department of Chemistry
University of Nevada
Reno, NV 89557
lemay@unr.edu

Bruce E. Bursten
Department of Chemistry
The Ohio State University
Columbus, OH 43210
bursten.1@osu.edu

Julia R. Burdge
Florida Atlantic University
Honors College
Jupiter, Florida 33458
jburdge@fau.edu

A Student's Guide to Using this Text

The following pages walk you through some of the main features of this text and its integrated media components. This learning system was designed with you, the student, in mind. We hope you enjoy your study of Chemistry—*the central science*.

Getting Oriented

▶ What's Ahead ◀

- Our discussion begins with a brief history of the periodic table.
- We will see that many properties of atoms depend on both the net attraction between the nucleus and the outer electrons (due to the *effective nuclear charge*) and on the average distance of those electrons from the nucleus.
- We will examine periodic trends of three key properties of atoms: *atomic size, ionization energy*, (the energy required to remove electrons) and the *electron affinity* (energy associated with adding electrons).
- As part of these discussions, we will also examine the sizes of ions and their electron configurations.

What's Ahead Sections

At the beginning of each chapter, reading the "What's Ahead" sections will give you a sense of direction for studying the chapter and help you to recognize key ideas and relationships of the topics within the chapter.

Problem Solving

Learning effective problem-solving skills is one of your most important goals in this course. To help you solve problems with confidence, the text integrates problem-solving pedagogy.

Worked Solutions

demonstrate the strategy and thought process involved in solving each exercise.

SAMPLE EXERCISE 3.13

Ascorbic acid (vitamin C) contains 40.92% C, 4.58% H, and 54.50% O by mass. What is the empirical formula of ascorbic acid?

Solution

Analyze: We are given the mass percentages of the elements in ascorbic acid and asked for its empirical formula.

Plan: The strategy for determining the empirical formula of a substance from its elemental composition involves the four steps given in Figure 3.11.

Solve: We first assume, for simplicity, that we have exactly 100 g of material (although any number can be used). In 100 g of ascorbic acid, we will have

40.92 g C, 4.58 g H, and 54.50 g O.

Second, we calculate the number of moles of each element in this sample:

$$\text{Moles C} = (40.92 \text{ g C})\left(\frac{1 \text{ mol C}}{12.01 \text{ g C}}\right) = 3.407 \text{ mol C}$$

$$\text{Moles H} = (4.58 \text{ g H})\left(\frac{1 \text{ mol H}}{1.008 \text{ g H}}\right) = 4.54 \text{ mol H}$$

$$\text{Moles O} = (54.50 \text{ g O})\left(\frac{1 \text{ mol O}}{16.00 \text{ g O}}\right) = 3.406 \text{ mol O}$$

Third, we determine the simplest whole-number ratio of moles by dividing each number of moles by the smallest number of moles, 3.406:

$$\text{C: } \frac{3.407}{3.406} = 1.000 \qquad \text{H: } \frac{4.54}{3.406} = 1.33 \qquad \text{O: } \frac{3.406}{3.406} = 1.000$$

The ratio for H is too far from 1 to attribute the difference to experimental error; in fact, it is quite close to $1\frac{1}{3}$. This suggests that if we multiply the ratio by 3, we will obtain whole numbers:

$$\text{C:H:O} = 3(1:1.33:1) = 3:4:3$$

The whole-number mole ratio gives us the subscripts for the empirical formula. Thus, the empirical formula is

$C_3H_4O_3$

Check: It is reassuring that the subscripts are moderately sized whole numbers. Otherwise, we have little by which to judge the reasonableness of our answer.

PRACTICE EXERCISE

A 5.325-g sample of methyl benzoate, a compound used in the manufacture of perfumes, is found to contain 3.758 g of carbon, 0.316 g of hydrogen, and 1.251 g of oxygen. What is the empirical formula of this substance?

Answer: C_4H_4O

Analyze/Plan/Solve/Check Theme

provides a consistent framework for helping you understand what you are being asked to solve, plan how you will solve each problem and check to make sure your answer is correct.

Dual-Column Worked Examples

found in selected sample exercises provide an explanation of the thought process involved in each step of a mathematical calculation to give you a conceptual understanding of the math.

Practice Exercises

include answers but not solutions giving you the opportunity to test your knowledge and get instant feedback

Strategies in Chemistry

Strategies in Chemistry boxes teach you ways to analyze information and organize thoughts, helping to improve your problem-solving and critical-thinking abilities.

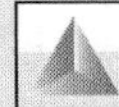

Strategies in Chemistry Pattern Recognition

Someone once said that drinking at the fountain of knowledge in a chemistry course is like drinking from a fire hydrant. Indeed, the pace can sometimes seem brisk. More to the point, however, we can drown in the facts if we don't see the general patterns. The value of recognizing patterns and learning rules and generalizations is that they free us from learning (or trying to memorize) many individual facts. The patterns and rules tie ideas together, so we don't get lost in the details.

Many students struggle with chemistry because they don't see how the topics relate to one another, how ideas connect together. They therefore treat every idea and problem as being unique instead of as an example or application of a general rule, pro... top...

summarize a large body of information. Notice, for example, how atomic structure helps us understand the existence of isotopes (as seen in Table 2.2) and how the periodic table aids us in remembering the charges of ions (as seen in Figure 2.22). You may surprise yourself by observing patterns that are not even explicitly spelled out yet. Perhaps you've even noticed certain trends in chemical formulas. Moving across the periodic table from element 11 (Na) we find that the elements form compounds with F having the following compositions: NaF, MgF_2, and AlF_3. Does this trend continue? Do SiF_4, PF_5, and SF_6 exist? Indeed they do. If you have picked up on trends like this from the scraps of information you've seen so far, then you're ahead

End-of-Chapter Exercises

- The first section of exercises is grouped by topic. They are presented in matched pairs, giving you multiple opportunities to test each concept.
- *Additional Exercises* follow the paired exercises and are not categorized, since many of these exercises draw on multiple concepts from within the chapter.
- *Integrative Exercises* appear at the end of appropriate chapters and connect concepts for the current chapter with those from previous chapters. They help you gain a deeper understanding of how chemistry fits together. In addition, they serve as an overall review of key concepts. Many chapters also contain a *Sample Integrative Exercise* at the end of the chapter to allow you to practice solving problems that encompass more than one concept.
- Answers are provided in the back of the book for red-numbered exercises. More challenging execises are indicated by brackets around the exercise number.
- *eMedia Exercises* are answered by using the movies and simulations available on the student Companion Website. By answering these questions, you will increase your practical understanding of the material.

Exercises

The Nature of Energy

5.1 In what two ways can an object possess energy? How do these two ways differ from one another?

5.2 Suppose you toss a tennis ball upward. **(a)** Does the kinetic energy of the ball increase or decrease as it moves higher? **(b)** What happens to the potential energy of the ball as it moves higher? **(c)** If the same amount of energy were imparted to a ball the same size as a tennis ball, but of twice the mass, how high would it go in comparison to the tennis ball? Explain your answers.

5.3 **(a)** Calculate the kinetic energy in joules of a 45-g golf ball moving at 61 m/s. **(b)** Convert this energy to calories. **(c)** What happens to this energy when the ball lands in a sand trap?

5.4 **(a)** What is the kinetic energy in joules of a 950-lb motorcycle moving at 68 mph? **(b)** By what factor will the kinetic energy change if the speed of the motorcycle is decre... of th...

5.5 In much engineering work it is common to use the British thermal unit (Btu). A Btu is the amount of heat required to raise the temperature of 1 lb of water by 1°F. Calculate the number of joules in a Btu.

5.6 A watt is a measure of power (the rate of energy change) equal to 1 J/s. Calculate the number of joules in a kilowatt-hour.

5.7 An adult person radiates heat to the surroundings at about the same rate as a 100-watt electric incandescent light bulb. What is the total amount of energy in kcal radiated to the surroundings by an adult in 24 hours?

5.8 Describe the source of the energy and the nature of the energy conversions involved when a 100-watt electric lightbulb radiates energy to its surroundings. Compare this with the energy source and energy conversions involved when an adult person radiates energy to the surroundings.

5.9 Suppose that a pellet is shot from an air gun straight up...

SAMPLE INTEGRATIVE EXERCISE 11: Putting Concepts Together

The substance CS_2 has a melting point of −110.8°C and a boiling point of 46.3°C. Its density at 20°C is 1.26 g/cm^3. It is highly inflammable. **(a)** What is the name of this compound? **(b)** If you were going to look up the properties of this substance in the *CRC Handbook of Chemistry and Physics*, would you look under the physical properties of inorganic or organic compounds? Explain. **(c)** How would you classify CS_2( ) as to type of crystalline solid? **(d)** Write a balanced equation for the combustion of this compound in air. (You will have to decide on the most likely oxidation products.) **(e)** The critical temperature and pressure for CS_2 are 552 K and 78 atm, respectively. Compare these values with those for CO_2 (Table 11.5), and discuss the possible origins of the differences. **(f)** Would you expect the density of CS_2 at 40°C to be greater or less than at 20°C? What accounts for the difference?

Solution **(a)** The compound is named carbon disulfide, in analogy with the naming of other binary molecular compounds. (Section 2.8) **(b)** The substance will be listed as an inorganic compound. It contains no carbon–carbon bonds, nor any C—H bonds,

eMedia Exercises

4.105 The **Electrolytes and Non-Electrolytes** movie (*eChapter 4.1*) and the **Aqueous Acids** and **Aqueous Bases** movies (*eChapter 4.3*) illustrate the behavior of various substances in aqueous solution. For each of the seven substances mentioned in the movies, write the chemical equation that corresponds to dissolution in water. (The chemical formula of sugar is $C_{12}H_{22}O_{11}$.) Where appropriate, use the double arrow notation.

4.109 After watching the **Solution Formation from a Solid** movie (*eChapter 4.5*), answer the following questions: **(a)** If we neglect to account for the mass of the weighing paper, how would our calculated concentration differ from the actual concentration of the solution? **(b)** Describe the process of preparing an aqueous solution of known concentration, starting with a solid. **(c)** Why is it necessary to make the solution as described in the movie, rather than

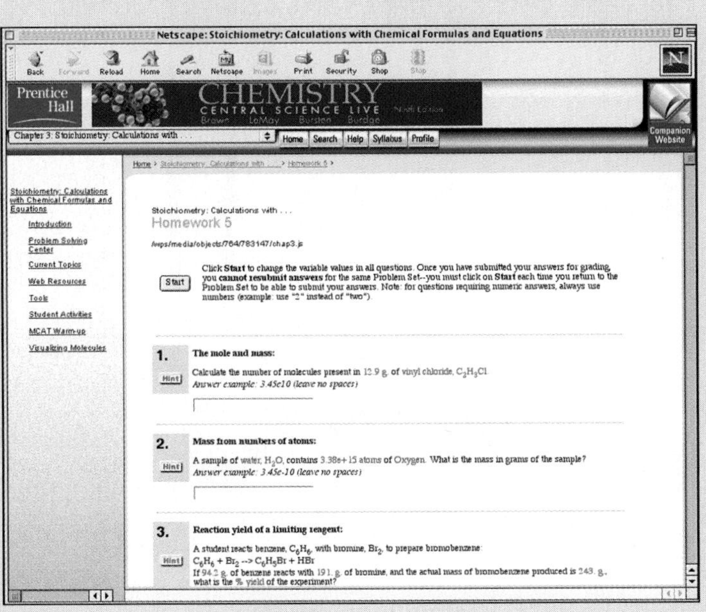

Central Science Live—Companion Website

http://www.prenhall.com/brown

The Brown/LeMay/Bursten Website was designed specifically to support and enhance your study of chemistry. The site provides a Problem-Solving Center where you have access to more than 2500 additional conceptual and quantitative exercises (including algorithmically generated, multiple choice and essay questions). Each problem is categorized by chapter and referenced to the text, and each problem offers hints and specific feedback for incorrect answers.

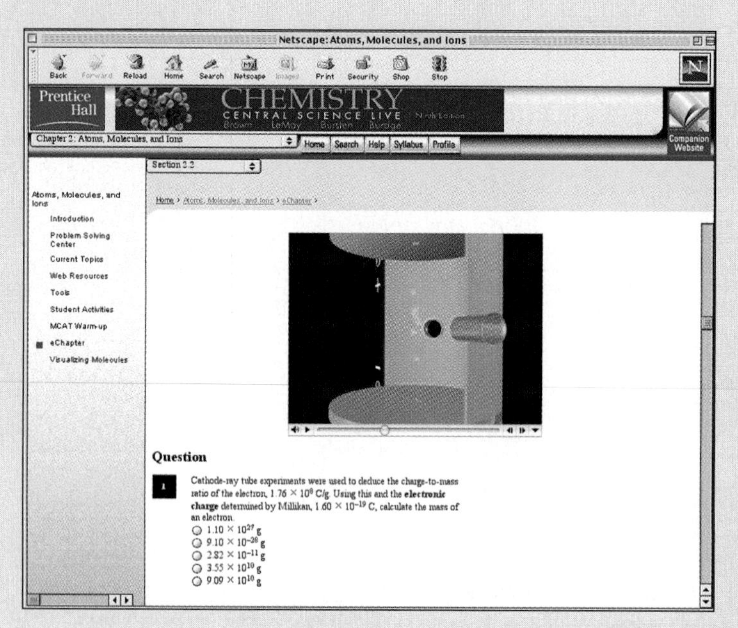

Central Science Live—eChapters

The premium version of the website contains self-assessment questions specifically aimed at the simulations and animations available in eChapters, following the book section by section.

Visualization

One of the challenges facing you in general chemistry is the often abstract nature of the subject. First, chemistry relies on a symbolic language based on chemical formulas and chemical equations. Second, chemistry is based on the behavior of molecules and atoms—particles far too small to see.

This text has been designed expressly to help you better visualize the chemistry you need to learn and succeed in your course. Spend time with the illustrations in the text, they'll help you understand the chemistry concepts being discussed.

Symbolic and Molecular Representations
The careful inclusion of molecular art with chemical formulas helps you see the connection between the symbols you write and the molecules to which they refer.

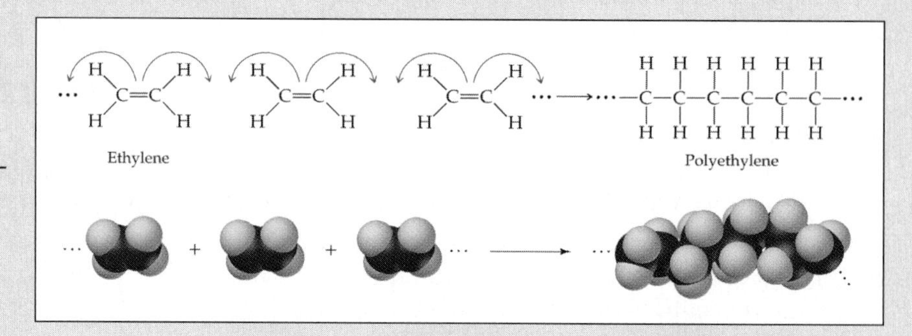

Compound Illustrations
Compound Illustrations combine photographs with molecular art. They give you a better sense of the relationships between the macroscopic properties of matter and its underlying structure at the atomic and molecular levels.

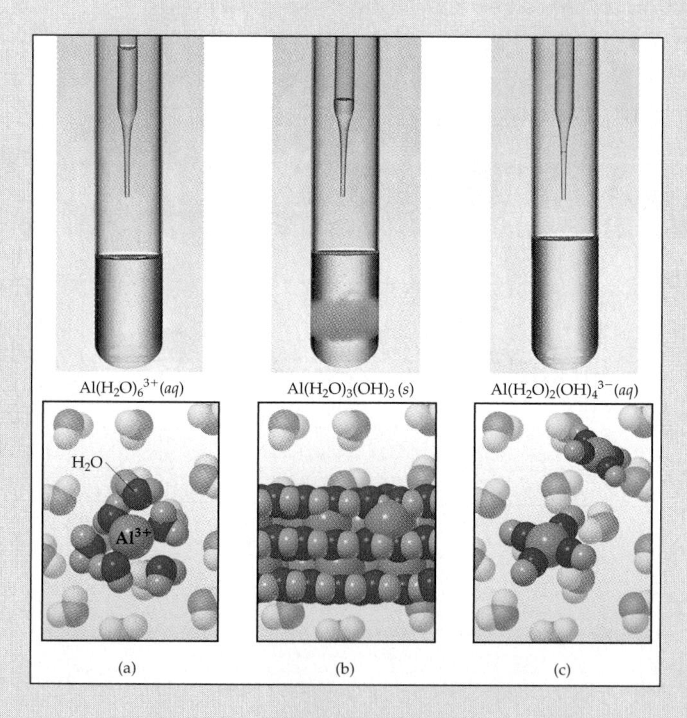

Molecular Illustrations

Computer generated renditions of molecules and materials provide visual representations of matter at the atomic level. These drawings help you visualize molecules in three dimensions and enhance your understanding of molecular architecture.

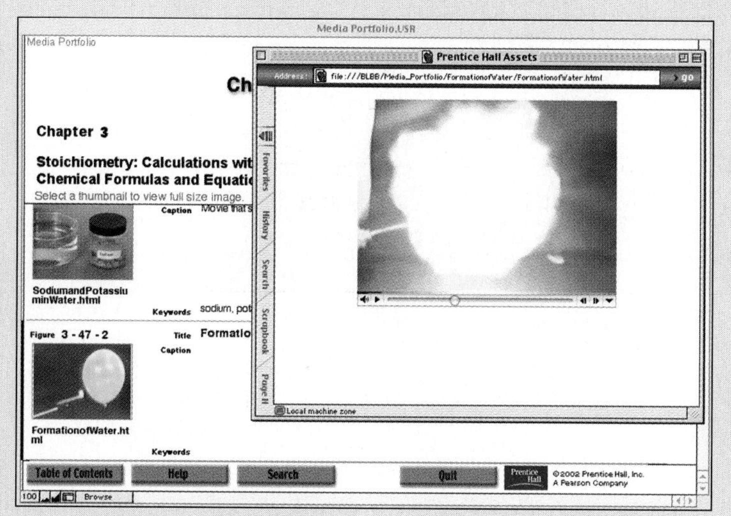

Central Science Live—Accelerator CD-ROM

Though your primary view of the media elements is through the Companion Website, they will load much faster if you also use the Accelerator CD. You can even view the elements offline using a browsing tool on the CD.

The media companion to *Chemistry: The Central Science, Ninth Edition* brings the molecular world to life for you with **more than 90 movies** (including animations written and developed by Ted Brown) as well as numerous 3-D **Chime models**. The Companion Website also features **more than 40 simulations**, where you are guided through virtual experiments to discover and enhance your understanding of chemical concepts. You know to look on the Companion Website when you see this icon

Central Science Live—Companion Website

http://www.prenhall.com/brown

The Brown/LeMay/Bursten website provides a Visualizing Molecules module containing hundreds of molecular models. New to this edition is the Student Activities module, with videos, molecules, animations and activitites. These are the same Student Activities that are also present in the context of eChapters on the premium version of the website.

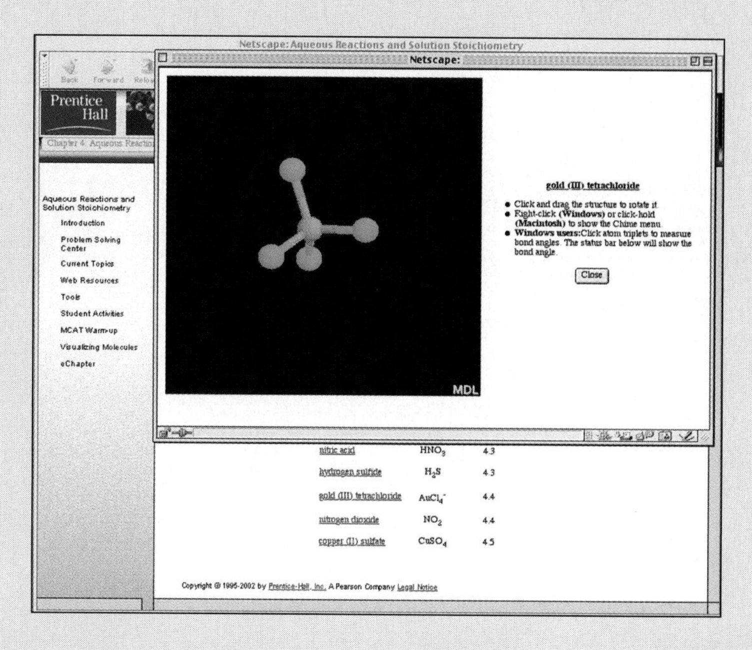

Applications

Why is chemistry "the central science"? In this text and in your course, you will recognize the importance of chemistry for your major areas of study and in the world around you. You will see why chemistry is fundamental to understanding our environment, for new developments in technology, and in our society.

Chemistry and Life and *Chemistry at Work*

Chemistry and Life and *Chemistry at Work* boxes emphasize chemistry's connection to world events, scientific discoveries, and medical breakthroughs. Your instructor may ask you to read these boxes or you may just want to read them on your own. They'll show you a new way of looking at the world around you.

Chemistry and Life Arsenic in Drinking Water

In 2001 the Environmental Protection Agency (EPA) issued a rule reducing the standard for arsenic in public water supplies from 50 ppb (equivalent to 50 $\mu g/L$) to 10 ppb, effective by 2006. Most regions of the United States tend to have low to moderate (2–10 ppb) groundwater arsenic levels (Figure 22.43 ▶). The Western region tends to have higher levels, coming mainly from natural geological sources in the area.

In water the most common forms of arsenic are the arsenate ion and its protonated hydrogen anions (AsO_4^{3-}, $HAsO_4^{2-}$, and $H_2AsO_4^-$), and the arsenite ion and its protonated forms (AsO_3^{3-}, $HAsO_3^{2-}$, $H_2AsO_3^-$, and H_3AsO_3). These species are collectively referred to by the oxidation number of the arsenic as arsenic(V) and arsenic(III), respectively. Arsenic(V) is more prevalent in oxygen-rich (aerobic) surface waters, whereas arsenic(III) is more likely to occur in oxygen-poor (anaerobic) groundwaters. In the pH range from 4 to 10, the arsenic(V) is present primarily as $HAsO_4^{2-}$ and $H_2AsO_4^-$, and the arsenic(III) is present ...sO_3.

...rmining the health effects of ...e different chemistry of the ...ll as the different concentra- ...sponses in different individ- ...ing arsenic levels with the ...indicate a lung and bladder ...evels of arsenic. A 2001 report ...il suggests, for example, that ... ppb arsenic daily have about ... forms of cancer during their ...roximately 3 in 1000. ...r removing arsenic perform ...nic in the form of arsenic(V), ...equire preoxidation of the

Alaska
Hawaii
USGS
EXPLANATION
At least 10% of samples exceed 10 ppm As
At least 10% of samples exceed 5 ppm As
At least 10% of samples exceed 3 ppm As
Fewer than 10% of samples exceed 3 ppm As
Insufficient data

▲ **Figure 22.43** Counties in which at least 10% of the groundwater samples exceeded 10 ppm As are indicated by the darkest color on the scale. As the color of the scale becomes lighter, the scale moves from 10 ppm to 5 ppm to 3 ppm and then cases where fewer than 10% of samples exceed 3 ppm. The white areas are those for which there were insufficient data.

drinking water. Once in the form of arsenic(V), there are a number of possible removal strategies. For example, $Fe_2(SO_4)_3$ could be added to precipitate $FeAsO_4$, which is then removed by filtration. Small utilities in areas where arsenic occurs naturally in groundwater fear that the costs of reducing arsenic even to the 10-ppb level will force them out of business, leaving households dependent on untreated well water.

Chemistry at Work Toward the Plastic Car

Many polymers can be formulated and processed to have sufficient structural strength, rigidity, and heat stability to displace metals, glass, and other materials in a variety of applications. The housings of electric motors and kitchen appliances such as coffee makers and can openers, for example, are now commonly formed from specially formulated polymers. *Engineering polymers* are tailored to particular applications through choice of polymers, blending of polymers, and modifications of processing steps. They generally have lower costs or superior performance over the materials they replace. In addition, shaping and coloring of the individual parts and their assembly to form the final product are often much easier.

The modern automobile provides many examples of the inroads engineering polymers have made in automobile design and construction. Car interiors have long been formed mainly of plastics. Through development of high-performance materials, significant progress has been made in introducing engineering polymers as engine components and car body parts. Figure 12.14 ◀, for example, shows the manifold in a series of Ford V-8 pickup and van engines. Use of an engineering polymer in this application eliminates machining and several assembly steps. The manifold, which is made of nylon, must be stable at high temperatures.

Car body parts can also be formed from engineering polymers. Components formed from engineering polymers usually weigh less than the components they replace, thus improving fuel economy. The fenders of the Volkswagen New Beetle (Figure 12.15 ◀), for example, are made of nylon reinforced with a second polymer, polyphenylene ether (ppe), which has the following structure:

$$-\!\!\left[\!-C_6H_4-O-\right]_n\!\!-$$

Because the polyphenylene ether ... rigid, the ppe confers rigidity and ...

A Closer Look

A Closer Look essays supplement the chapter material by covering high-interest topics in more detail.

A Closer Look Glenn Seaborg and the Story of Seaborgium

Prior to 1940 the periodic table ended at uranium, element number 92. Since that time, no scientist has had a greater effect on the periodic table than Glenn Seaborg (1912–1999). Seaborg (Figure 2.18 ▶) became a faculty member in the chemistry department at the University of California, Berkeley, in 1937. In 1940 he and his colleagues Edwin McMillan, Arthur Wahl, and Joseph Kennedy succeeded in isolating plutonium (Pu) as a product of the reaction of uranium with neutrons. We will talk about reactions of this type, called *nuclear reactions*, in Chapter 21. We will also discuss the key role that plutonium plays in nuclear fission reactions, such as those that occur in nuclear power plants and atomic bombs.

During the period 1944 through 1958, Seaborg and his coworkers also succeeded in identifying the elements with

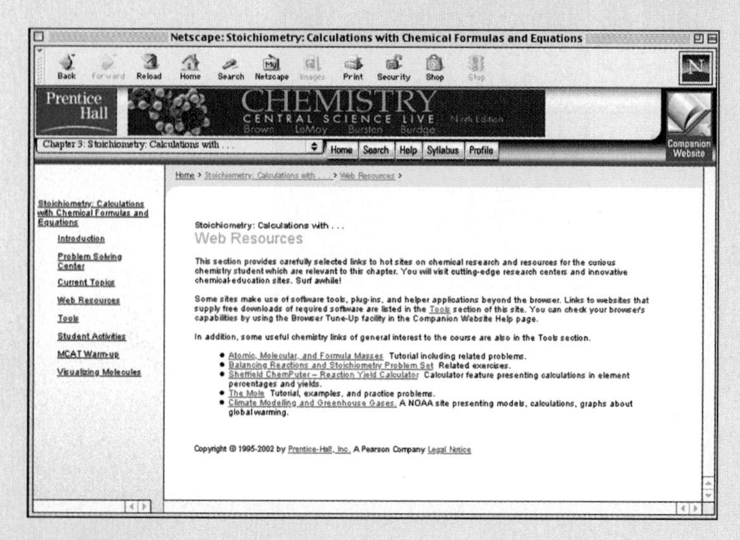

Central Science Live—Companion Website

http://www.prenhall.com/brown

The Brown/LeMay/Bursten website provides a Web Resources Center that links you to selected chemistry-related Websites. The site also connects you to a Current Topics area where we regularly post articles about chemistry from the popular scientific press.

About the Authors

Theodore L. Brown received his Ph.D. from Michigan State University in 1956. Since then, he has been a member of the faculty of the University of Illinois, Urbana-Champaign, where he is now Professor of Chemistry, Emeritus. He served as Vice Chancellor for Research, and Dean, The Graduate College, from 1980 to 1986, and as Founding Director of the Arnold and Mabel Beckman Institute for Advanced Science and Technology from 1987 to 1993. Professor Brown has been an Alfred P. Sloan Foundation Research Fellow and has been awarded a Guggenheim Fellowship. In 1972 he was awarded the American Chemical Society Award for Research in Inorganic Chemistry, and received the American Chemical Society Award for Distinguished Service in the Advancement of Inorganic Chemistry in 1993. He has been elected a Fellow of both the American Association for the Advancement of Science and the American Academy of Arts and Sciences.

H. Eugene LeMay, Jr., received his B.S. degree in Chemistry from Pacific Lutheran University (Washington) and his Ph.D. in Chemistry in 1966 from the University of Illinois (Urbana). He then joined the faculty of the University of Nevada, Reno, where he is currently Professor of Chemistry. He has enjoyed Visiting Professorships at the University of North Carolina at Chapel Hill, at the University College of Wales in Great Britain, and at the University of California, Los Angeles. Professor LeMay is a popular and effective teacher, who has taught thousands of students during more than 35 years of university teaching. Known for the clarity of his lectures and his sense of humor, he has received several teaching awards, including the University Distinguished Teacher of the Year Award (1991) and the first Regents' Teaching Award given by the State of Nevada Board of Regents (1997).

Bruce E. Bursten received his Ph.D. in Chemistry from the University of Wisconsin in 1978. After two years as a National Science Foundation Postdoctoral Fellow at Texas A&M University, he joined the faculty of The Ohio State University, where he is currently Distinguished University Professor. Professor Bursten has been a Camille and Henry Dreyfus Foundation Teacher-Scholar and an Alfred P. Sloan Foundation Research Fellow. At Ohio State he has received the University Distinguished Teaching Award in 1982 and 1996, the Arts and Sciences Student Council Outstanding Teaching Award in 1984, and the University Distinguished Scholar Award in 1990. In addition to his teaching activities, Professor Bursten's research program focuses on compounds of the transition-metal and actinide elements. His research is currently supported by grants from the National Science Foundation and the Department of Energy.

Julia R. Burdge received her B.A. (1987) and M.S. (1990) degrees in Chemistry from the University of South Florida (Tampa), and her Ph.D. in Chemistry from the University of Idaho (Moscow) in 1994. She then joined the faculty of the University of Akron, where she directed the general chemistry program from 1994 to 2001. Professor Burdge implemented the use of new educational technologies and put significant resources in place to enhance the general chemistry curriculum, including a state-of-the-art computer laboratory for use by general chemistry students. She is a well-liked teacher, known for her ability to explain the principles of chemistry in ways that students can understand and appreciate. Professor Burdge recently accepted a position at Florida Atlantic University's new Honors College in Jupiter, Florida, where, in addition to teaching, she will pursue environmental research with undergraduates.

Chapter 1

Introduction: Matter and Measurement

Hubble Space Telescope image of a galaxy that lies 13 million light years from Earth in the southern constellation Circinus. Large amounts of gas and dust occur within the galaxy. Hydrogen gas is the most plentiful, occurring as molecules in cool regions, as atoms in hotter regions and as ions in the hottest regions. The processes that occur within the stars are responsible for creating other chemical elements from hydrogen, which is the simplest element.

HAVE YOU EVER wondered why ice melts and water evaporates? Why leaves turn colors in the fall and how a battery generates electricity? Why keeping foods cold slows their spoilage and how our bodies use food to maintain life? Chemistry supplies answers to these questions and countless others like them. **Chemistry** is the study of the properties of materials and the changes that materials undergo. One of the joys of learning chemistry is seeing how chemical principles operate in all aspects of our lives, from everyday activities like lighting a match to more far-reaching matters like the development of drugs to cure cancer.

You are just beginning the journey of learning chemistry. In a sense, this text is your guide on this journey. Throughout your studies, we hope that you will find this text enjoyable as well as educational. As you study, keep in mind that the chemical facts and concepts you are asked to learn are not ends in themselves, but tools to help you better understand the world around you. This first chapter lays a foundation for our studies by providing an overview of what chemistry is about and dealing with some fundamental concepts of matter and scientific measurements. The list to the right, entitled "What's Ahead," gives a brief overview of some of the ideas that we will be considering in this chapter.

▶ What's Ahead ◀

- We begin our studies by providing a very brief perspective of what chemistry is about and why it is useful to learn chemistry.
- Next we examine some fundamental ways to classify materials, distinguishing between *pure substances* and *mixtures* and noting that there are two fundamentally different kinds of pure substances: *elements* and *compounds*.
- We then consider some of the different kinds of characteristics or *properties* that we use to characterize, identify, and separate substances.
- Many properties rely on quantitative measurements, involving both numbers and units.
- The units of measurement used throughout science are those of the *metric system*, a decimal system of measurement.
- The uncertainties inherent in all measured quantities and those obtained from calculations involving measured quantities are expressed by the number of significant digits or *significant figures* used to report the number.
- Units as well as numbers are carried through calculations, and obtaining correct units for the result of a calculation is an important way to check whether the calculation is correct.

1.1 The Study of Chemistry

Before traveling to an unfamiliar city, you might look at a map to get some sense of where you are heading. Chemistry may be unfamiliar to you, too, so it's useful to get a general idea of what lies ahead before you embark on your journey. In fact, you might even ask why you are taking the trip.

The Molecular Perspective of Chemistry

Chemistry involves studying the properties and behavior of matter. **Matter** is the physical material of the universe; it is anything that has mass and occupies space. This book, your body, the clothes you are wearing, and the air you are breathing are all samples of matter. Not all forms of matter are so common or so familiar, but countless experiments have shown that the tremendous variety of matter in our world is due to combinations of only about 100 very basic or elementary substances called **elements**. As we proceed through this text, we will seek to relate the properties of matter to its composition, that is, to the particular elements it contains.

Chemistry also provides a background to understanding the properties of matter in terms of **atoms**, the almost infinitesimally small building blocks of matter. Each element is composed of a unique kind of atom. We will see that the properties of matter relate not only to the kinds of atoms it contains (*composition*), but also to the arrangements of these atoms (*structure*).

Atoms can combine to form **molecules** in which two or more atoms are joined together in specific shapes. Throughout this text you will see molecules represented using colored spheres to show how their component atoms connect to each other (Figure 1.1 ▼). The color merely provides a convenient way to distinguish between the atoms of different elements. Molecules of ethanol and ethylene glycol, which are depicted in Figure 1.1, differ somewhat in composition. Ethanol contains one red sphere, which represents an oxygen atom, whereas ethylene glycol contains two.

Even apparently minor differences in the composition or structure of molecules can cause profound differences in their properties. Ethanol, also called grain alcohol, is the alcohol in beverages such as beer and wine. Ethylene glycol,

3-D MODEL
Oxygen, Water, Carbon Dioxide, Ethanol, Ethylene Glycol, Aspirin

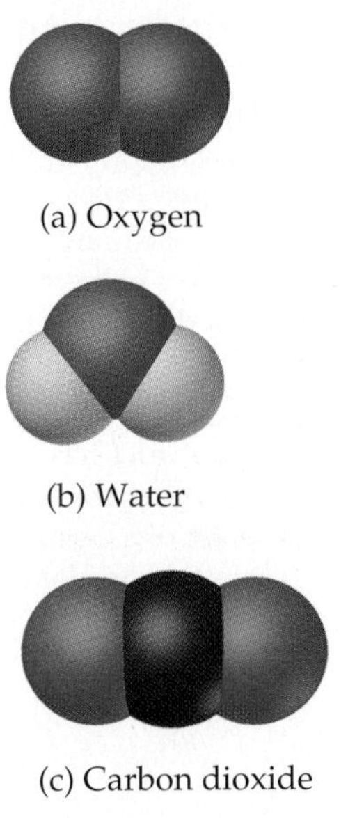

(a) Oxygen

(b) Water

(c) Carbon dioxide

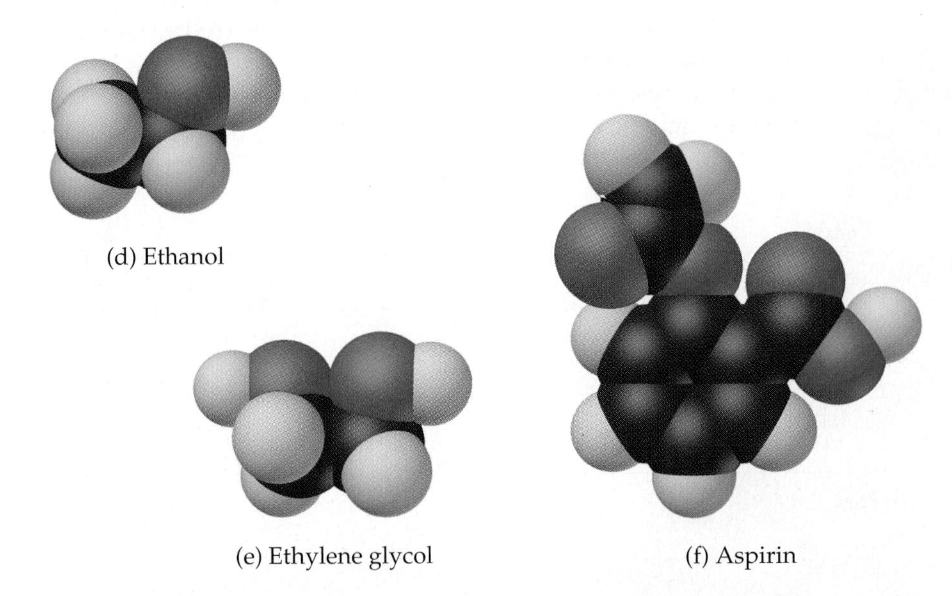

(d) Ethanol

(e) Ethylene glycol

(f) Aspirin

▲ **Figure 1.1** Molecular models. The white, dark gray, and red spheres represent atoms of hydrogen, carbon, and oxygen, respectively.

on the other hand, is a viscous liquid used as automobile antifreeze. The properties of these two substances differ in a great number of ways, including the temperatures at which they freeze and boil. One of the challenges that chemists undertake is to alter molecules in a controlled way, creating new substances with different properties.

Every change in the observable world—from boiling water to the changes that occur as our bodies combat invading viruses—has its basis in the unobservable world of atoms and molecules. Thus, as we proceed with our study of chemistry, we will find ourselves thinking in two realms, the *macroscopic* realm of ordinary-sized objects (*macro* = large) and the submicroscopic realm of atoms. We make our observations in the macroscopic world with our everyday senses—in the laboratory and in our surroundings. In order to understand that world, however, we must visualize how atoms behave.

Why Study Chemistry?

Chemistry provides important understanding of our world and how it works. It is an extremely practical science that greatly impacts our daily living. Indeed, chemistry lies near the heart of many matters of public concern: improvement of health care, conservation of natural resources, protection of the environment, and provision of our everyday needs for food, clothing, and shelter. Using chemistry, we have discovered pharmaceutical chemicals that enhance our health and prolong our lives. We have increased food production through the development of fertilizers and pesticides. We have developed plastics and other materials that are used in almost every facet of our lives. Unfortunately, some chemicals also have the potential to harm our health or the environment. It is in our best interest as educated citizens and consumers to understand the profound effects, both positive and negative, that chemicals have on our lives and to strike an informed balance about their uses.

Most of you are studying chemistry, however, not merely to satisfy your curiosity or to become more informed consumers or citizens, but because it is an essential part of your curriculum. Your major might be biology, engineering, agriculture, geology, or some other field. Why do so many diverse subjects share an essential tie to chemistry? The answer is that chemistry, by its very nature, is the *central science*. Our interactions with the material world raise basic questions about the materials around us. What are their compositions and properties? How do they interact with us and our environment? How, why, and when do they undergo change? These questions are important whether the material is part of high-tech computer chips, an aged pigment used by a Renaissance painter, or the DNA that transmits genetic information in our bodies (Figure 1.2 ▼). Chemistry provides answers to these and countless other questions.

(a)

(b)

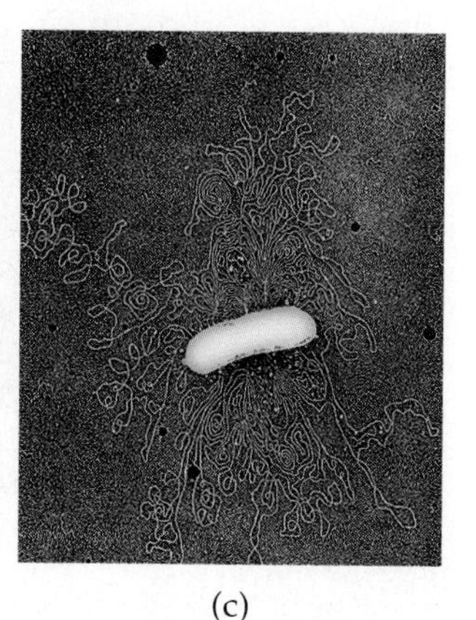
(c)

◀ **Figure 1.2** (a) A microscopic view of a computer chip. (b) A Renaissance painting, *Young Girl Reading*, by Vittore Carpaccio (1472–1526). (c) A long strand of DNA that has spilled out of the damaged cell wall of a bacterium.

By studying chemistry, you will learn to use the powerful language and ideas that have evolved to describe and enhance our understanding of matter. The language of chemistry is a universal scientific language that is widely used in other disciplines. Furthermore, an understanding of the behavior of atoms and molecules provides powerful insights in other areas of modern science, technology, and engineering. For this reason, chemistry will probably play a significant role in your future. You will be better prepared for the future if you increase your understanding of chemical principles, and it is our goal to help you achieve this end.

Chemistry at Work Chemistry and the Chemical Industry

Many people are familiar with common household chemicals such as those shown in Figure 1.3 ▶, but few realize the size and importance of the chemical industry. Worldwide sales of chemicals and related products manufactured in the United States total over $400 billion annually. The chemical industry employs over 10% of all scientists and engineers and is a major contributor to the U.S. economy.

Vast amounts of chemicals are produced each year and serve as raw materials for a variety of uses, including the manufacture of metals, plastics, fertilizers, pharmaceuticals, fuels, paints, adhesives, pesticides, synthetic fibers, microprocessor chips, and numerous other products. Table 1.1 ▼ lists the top ten chemicals produced in the United States. We will discuss many of these substances and their uses as the course progresses.

People who have degrees in chemistry hold a variety of positions in industry, government, and academia. Those who work in the chemical industry find positions as laboratory chemists, carrying out experiments to develop new products (research and development), analyzing materials (quality control), or assisting customers in using products (sales and service). Those with more experience or training may work as managers or company directors. There are also alternate careers that a chemistry degree prepares you for such as teaching, medicine, biomedical research, information science, environmental work, technical sales, work with government regulatory agencies, and patent law.

▲ **Figure 1.3** Many common supermarket products have very simple chemical compositions.

TABLE 1.1 The Top Ten Chemicals Produced by the Chemical Industry in 2000[a]

Rank	Chemical	Formula	2000 Production (billions of pounds)	Principal End Uses
1	Sulfuric acid	H_2SO_4	87	Fertilizers, chemical manufacturing
2	Nitrogen	N_2	81	Fertilizers
3	Oxygen	O_2	55	Steel, welding
4	Ethylene	C_2H_4	55	Plastics, antifreeze
5	Lime	CaO	44	Paper, cement, steel
6	Ammonia	NH_3	36	Fertilizers
7	Propylene	C_3H_6	32	Plastics
8	Phosphoric acid	H_3PO_4	26	Fertilizers
9	Chlorine	Cl_2	26	Bleaches, plastics, water purification
10	Sodium hydroxide	$NaOH$	24	Aluminum production, soap

[a]Most data from *Chemical and Engineering News*, June 25, 2001, pp. 45, 46.

1.2 Classifications of Matter

Let's begin our study of chemistry by examining some fundamental ways in which matter is classified and described. Two principal ways of classifying matter are according to its physical state (as a gas, liquid, or solid) and according to its composition (as an element, compound, or mixture).

States of Matter

A sample of matter can be a gas, a liquid, or a solid. These three forms of matter are called the **states of matter**. The states of matter differ in some of their simple observable properties. A **gas** (also known as *vapor*) has no fixed volume or shape; rather, it conforms to the volume and shape of its container. A gas can be compressed to occupy a smaller volume, or it can expand to occupy a larger one. A **liquid** has a distinct volume independent of its container but has no specific shape: It assumes the shape of the portion of the container that it occupies. A **solid** has both a definite shape and a definite volume: It is rigid. Neither liquids nor solids can be compressed to any appreciable extent.

The properties of the states can be understood on the molecular level (Figure 1.4 ▼). In a gas the molecules are far apart and are moving at high speeds, colliding repeatedly with each other and with the walls of the container. In a liquid the molecules are packed more closely together, but still move rapidly, allowing them to slide over each other; thus, liquids pour easily. In a solid the molecules are held tightly together, usually in definite arrangements, in which the molecules can wiggle only slightly in their otherwise fixed positions. Thus, solids have rigid shapes.

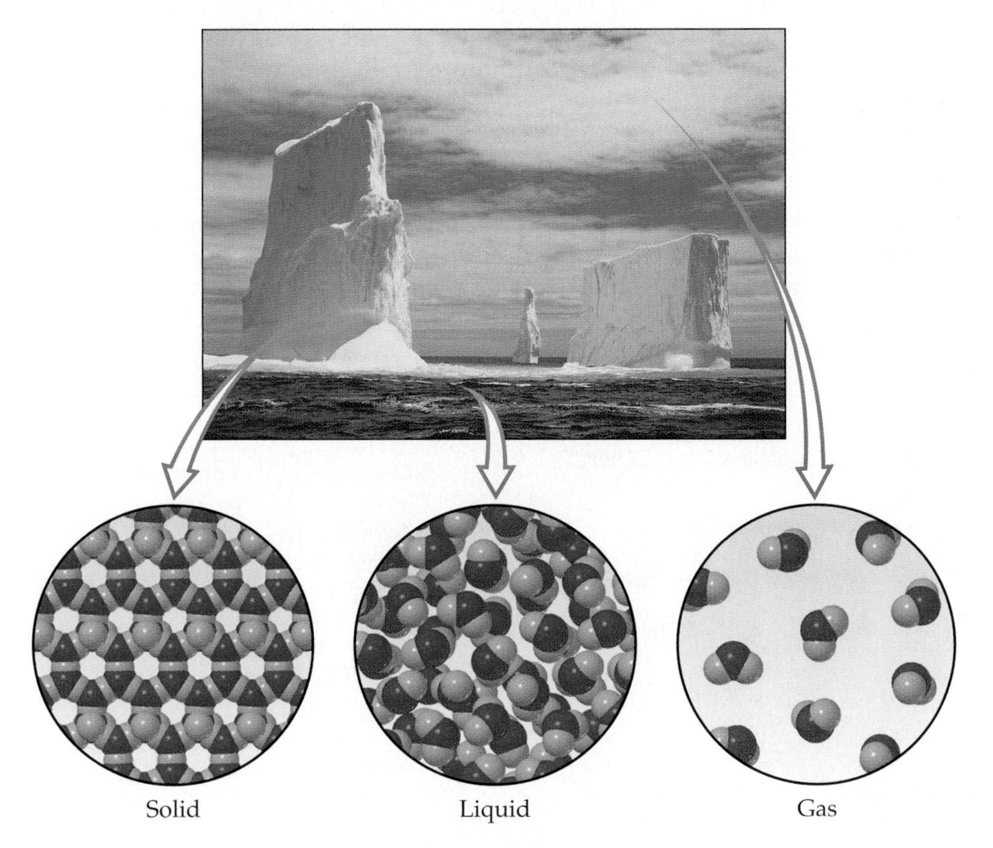

◀ **Figure 1.4** The three physical states of water are water vapor, liquid water, and ice. In this photo we see both the liquid and solid states of water. We cannot see water vapor. What we see when we look at steam or clouds is tiny droplets of liquid water dispersed in the atmosphere. The molecular views show that the molecules in the solid are arranged in a more orderly way than in the liquid. The molecules in the gas are much farther apart than those in the liquid or the solid.

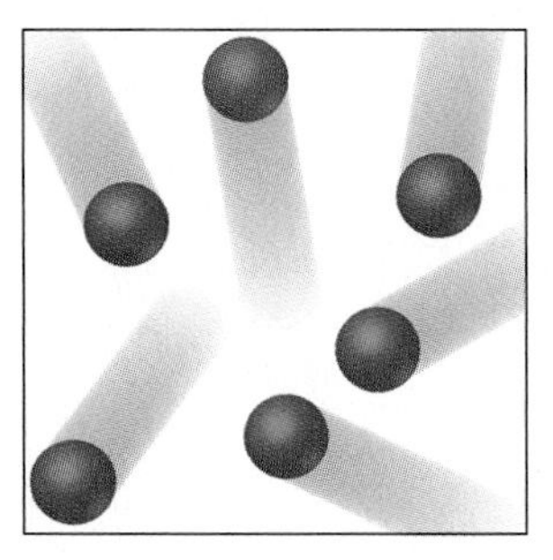

(a) Atoms of an element

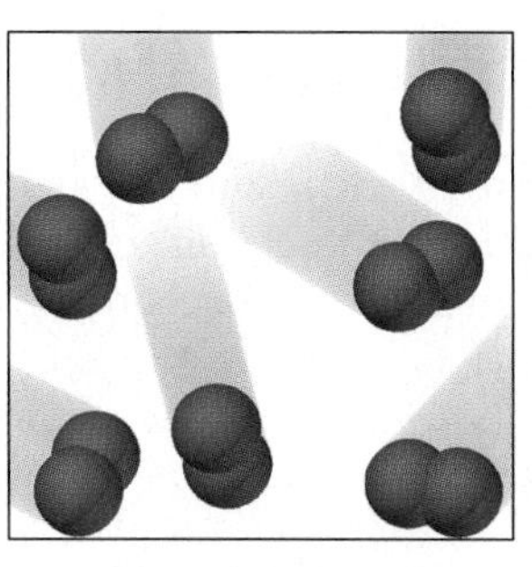

(b) Molecules of an element

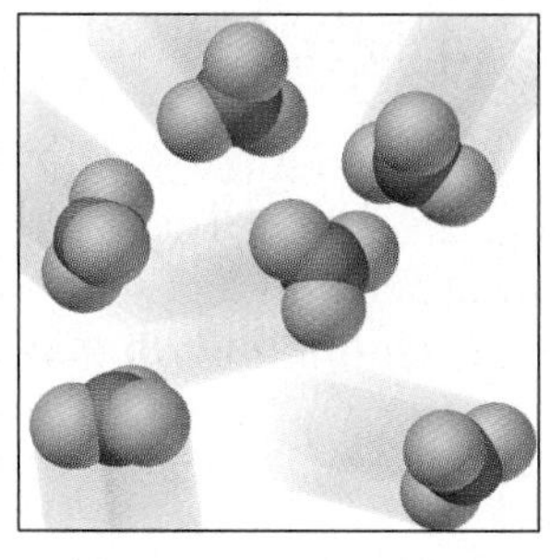

(c) Molecules of a compound

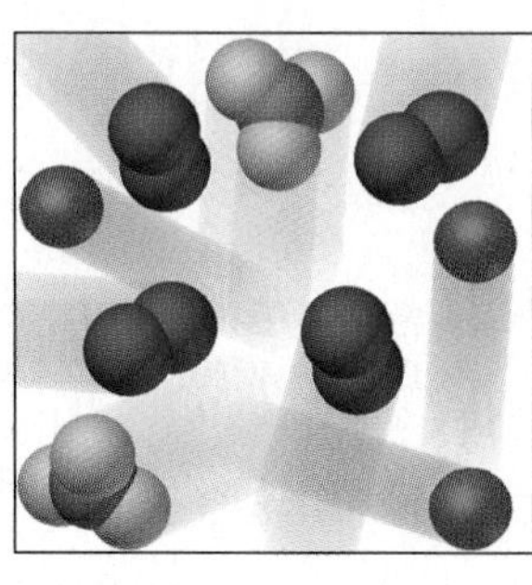

(d) Mixture of elements and a compound

▲ **Figure 1.5** Each element contains a unique kind of atom. Elements might consist of individual atoms, as in (a), or molecules, as in (b). Compounds contain two or more different atoms chemically joined together, as in (c). A mixture contains the individual units of its components, shown in (d) as both atoms and molecules.

Pure Substances

Most forms of matter that we encounter—for example, the air we breathe (a gas), gasoline for cars (a liquid), and the sidewalk on which we walk (a solid)—are not chemically pure. We can, however, resolve, or separate, these kinds of matter into different pure substances. A **pure substance** (usually referred to simply as a *substance*) is matter that has distinct properties and a composition that doesn't vary from sample to sample. Water and ordinary table salt (sodium chloride), the primary components of seawater, are examples of pure substances.

All substances are either elements or compounds. **Elements** cannot be decomposed into simpler substances. On the molecular level, each element is composed of only one kind of atom [Figure 1.5(a and b) ▲]. **Compounds** are substances composed of two or more elements, so they contain two or more kinds of atoms [Figure 1.5(c)]. Water, for example, is a compound composed of two elements, hydrogen and oxygen. Figure 1.5(d) shows a mixture of substances. **Mixtures** are combinations of two or more substances in which each substance retains its own chemical identity.

Elements

At the present time 114 elements are known. These elements vary widely in their abundance, as shown in Figure 1.6 ▼. For example, only five elements account for over 90% of the Earth's crust: oxygen, silicon, aluminum, iron, and calcium. In contrast, just three elements (oxygen, carbon, and hydrogen) account for over 90% of the mass of the human body.

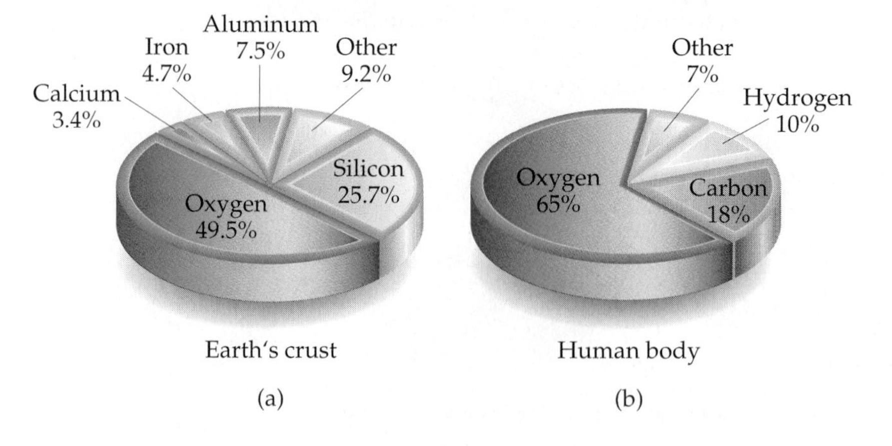

▶ **Figure 1.6** Elements in percent by mass in (a) Earth's crust (including oceans and atmosphere) and (b) the human body.

TABLE 1.2 Some Common Elements and Their Symbols

Carbon	C	Aluminum	Al	Copper	Cu (from *cuprum*)
Fluorine	F	Barium	Ba	Iron	Fe (from *ferrum*)
Hydrogen	H	Calcium	Ca	Lead	Pb (from *plumbum*)
Iodine	I	Chlorine	Cl	Mercury	Hg (from *hydrargyrum*)
Nitrogen	N	Helium	He	Potassium	K (from *kalium*)
Oxygen	O	Magnesium	Mg	Silver	Ag (from *argentum*)
Phosphorus	P	Platinum	Pt	Sodium	Na (from *natrium*)
Sulfur	S	Silicon	Si	Tin	Sn (from *stannum*)

Some of the more familiar elements are listed in Table 1.2 ▲, along with the chemical abbreviations—or chemical *symbols*—used to denote them. All the known elements and their symbols are listed on the front inside cover of this text. The table in which the symbol for each element is enclosed in a box is called the *periodic table*. In the periodic table the elements are arranged in vertical columns so that closely related elements are grouped together. We describe this important tool in more detail in Section 2.5.

The symbol for each element consists of one or two letters, with the first letter capitalized. These symbols are often derived from the English name for the element, but sometimes they are derived from a foreign name instead (last column in Table 1.2). You will need to know these symbols and to learn others as we encounter them in the text.

Compounds

Most elements can interact with other elements to form compounds. Hydrogen gas, for example, burns in oxygen gas to form water. Conversely, water can be decomposed into its component elements by passing an electrical current through it, as shown in Figure 1.7 ▶. Pure water, regardless of its source, consists of 11% hydrogen and 89% oxygen by mass. This macroscopic composition corresponds to the molecular composition, which consists of two hydrogen atoms combined with one oxygen atom. As seen in Table 1.3 ▼, the properties of water bear no resemblance to the properties of its component elements. Hydrogen, oxygen, and water are each unique substances.

The observation that the elemental composition of a pure compound is always the same is known as the **law of constant composition** (or the **law of definite proportions**). It was first put forth by the French chemist Joseph Louis Proust (1754–1826) in about 1800. Although this law has been known for 200 years, the general belief persists among some people that a fundamental difference exists between compounds prepared in the laboratory and the corresponding compounds found in nature. However, a pure compound has the same composition and properties regardless of its source. Both chemists and nature must use the same elements and operate under the same natural laws. When two materials differ in composition and properties, we know that they are composed of different compounds or that they differ in purity.

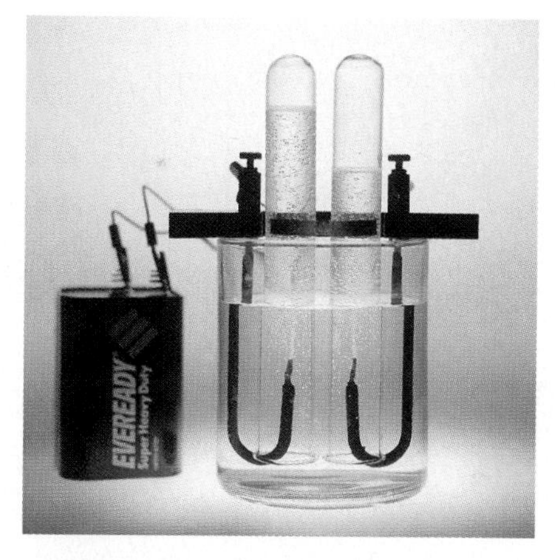

▲ **Figure 1.7** Water decomposes into its component elements, hydrogen and oxygen, when a direct electrical current is passed through it. The volume of hydrogen (on the right) is twice the volume of oxygen (on the left).

ANIMATION
Electrolysis of Water

TABLE 1.3 Comparison of Water, Hydrogen, and Oxygen

	Water	Hydrogen	Oxygen
State[a]	Liquid	Gas	Gas
Normal boiling point	100°C	−253°C	−183°C
Density[a]	1.00 g/mL	0.084 g/L	1.33 g/L
Flammable	No	Yes	No

[a] At room temperature and atmospheric pressure. (See Section 10.2.)

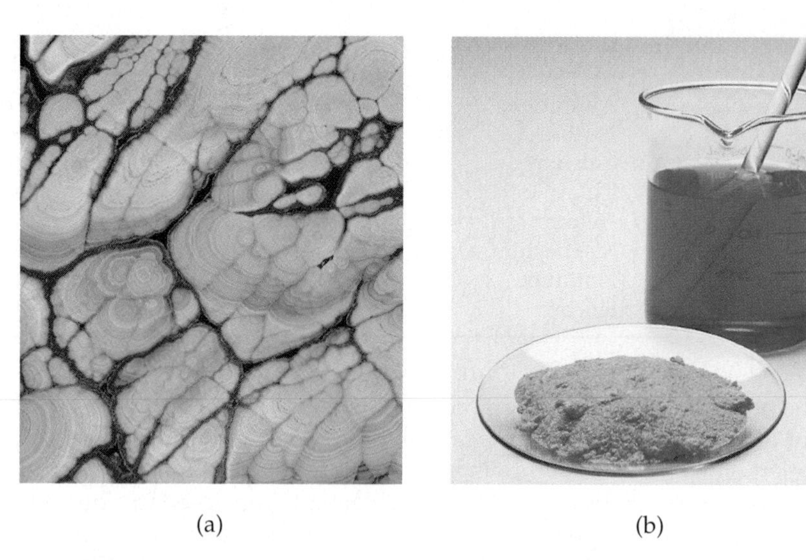

(a) (b)

▶ **Figure 1.8** (a) Many common materials, including rocks, are heterogeneous. This close-up photo is of *malachite*, a copper mineral. (b) Homogeneous mixtures are called solutions. Many substances, including the blue solid shown in this photo (copper sulfate), dissolve in water to form solutions.

Mixtures

Most of the matter we encounter consists of mixtures of different substances. Each substance in a mixture retains its own chemical identity and hence its own properties. Whereas pure substances have fixed compositions, the compositions of mixtures can vary. A cup of sweetened coffee, for example, can contain either a little sugar or a lot. The substances making up a mixture (such as sugar and water) are called *components* of the mixture.

Some mixtures, such as sand, rocks, and wood, do not have the same composition, properties, and appearance throughout the mixture. Such mixtures are *heterogeneous* [Figure 1.8(a) ▲]. Mixtures that are uniform throughout are *homogeneous*. Air is a homogeneous mixture of the gaseous substances nitrogen, oxygen, and smaller amounts of other substances. The nitrogen in air has all the properties that pure nitrogen does because both the pure substance and the mixture contain the same nitrogen molecules. Salt, sugar, and many other substances dissolve in water to form homogeneous mixtures [Figure 1.8(b)]. Homogeneous mixtures are also called **solutions**. Figure 1.9 ▶ summarizes the classification of matter into elements, compounds, and mixtures.

SAMPLE EXERCISE 1.1

"White gold," used in jewelry, contains two elements, gold and palladium. Two different samples of white gold differ in the relative amounts of gold and palladium that they contain. Both are uniform in composition throughout. Without knowing any more about the materials, how would you classify white gold?

Solution Let's use the scheme shown in Figure 1.9. Because the material is uniform throughout, it is homogeneous. Because its composition differs for the two samples, it cannot be a compound. Instead, it must be a homogeneous mixture. Gold and palladium can be said to form a solid solution with one another.

PRACTICE EXERCISE

Aspirin is composed of 60.0% carbon, 4.5% hydrogen, and 35.5% oxygen by mass, regardless of its source. Is aspirin a mixture or a compound?
Answer: a compound because of its constant composition

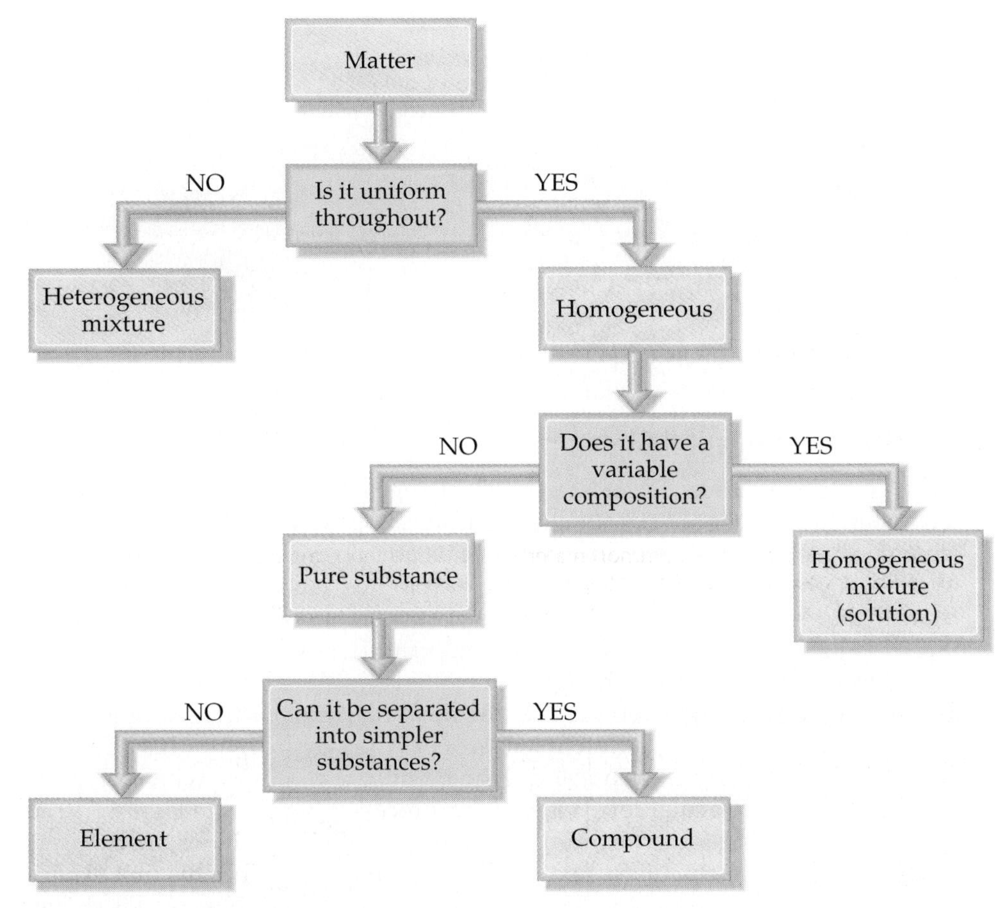

▲ **Figure 1.9** Classification scheme for matter. At the chemical level all matter is classified ultimately as either elements or compounds.

1.3 Properties of Matter

Every substance has a unique set of *properties*—characteristics that allow us to recognize it and to distinguish it from other substances. For example, the properties listed in Table 1.3 allow us to distinguish hydrogen, oxygen, and water from one another. The properties of matter can be categorized as physical or chemical. **Physical properties** can be measured without changing the identity and composition of the substance. These properties include color, odor, density, melting point, boiling point, and hardness. **Chemical properties** describe the way a substance may change or *react* to form other substances. A common chemical property is flammability, the ability of a substance to burn in the presence of oxygen.

Some properties—such as temperature, melting point, and density—do not depend on the amount of the sample being examined. These properties, called **intensive properties**, are particularly useful in chemistry because many can be used to *identify* substances. **Extensive properties** of substances depend on the quantity of the sample and include measurements of mass and volume. Extensive properties relate to the *amount* of substance present.

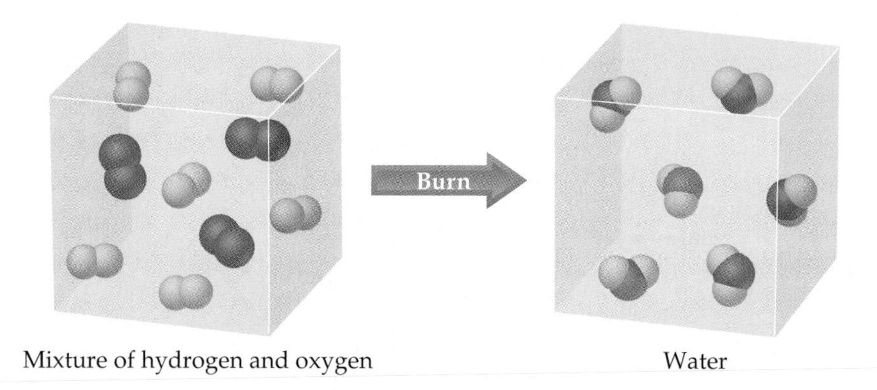

▶ **Figure 1.10** In chemical reactions the chemical identities of substances change. Here, a mixture of hydrogen and oxygen undergoes a chemical change to form water.

Physical and Chemical Changes

As with the properties of a substance, the changes that substances undergo can be classified as either physical or chemical. During **physical changes** a substance changes its physical appearance, but not its composition. The evaporation of water is a physical change. When water evaporates, it changes from the liquid state to the gas state, but it is still composed of water molecules, as depicted earlier in Figure 1.4. All **changes of state** (for example, from liquid to gas or from liquid to solid) are physical changes.

In **chemical changes** (also called **chemical reactions**) a substance is transformed into a chemically different substance. When hydrogen burns in air, for example, it undergoes a chemical change because it combines with oxygen to form water. The molecular-level view of this process is depicted in Figure 1.10 ▲.

Chemical changes can be dramatic. In the account that follows, Ira Remsen, author of a popular chemistry text published in 1901, describes his first experiences with chemical reactions. The chemical reaction that he observed is shown in Figure 1.11 ▼.

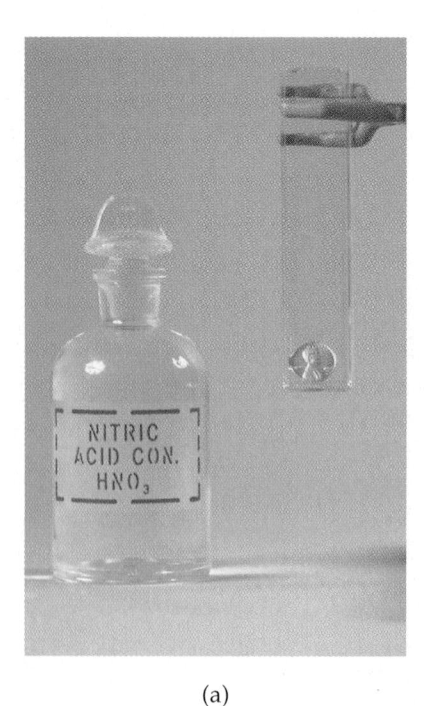

(a)

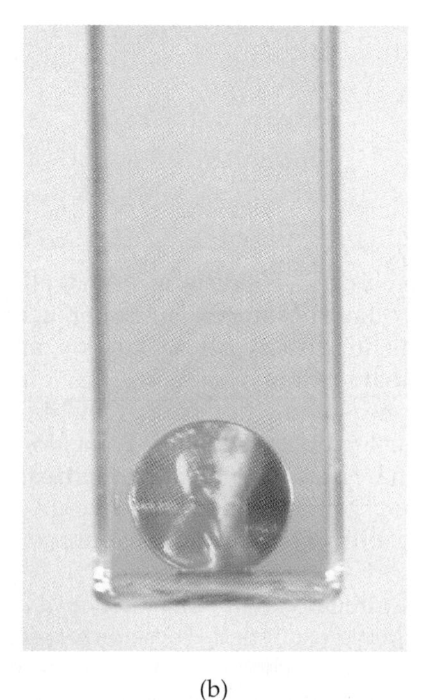

(b)

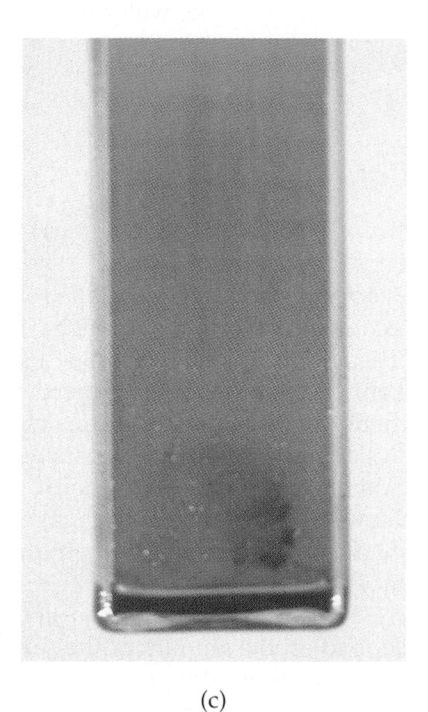

(c)

▲ **Figure 1.11** The chemical reaction between a copper penny and nitric acid. The dissolved copper produces the blue-green solution; the reddish brown gas produced is nitrogen dioxide.

While reading a textbook of chemistry, I came upon the statement "nitric acid acts upon copper," and I determined to see what this meant. Having located some nitric acid, I had only to learn what the words "act upon" meant. In the interest of knowledge I was even willing to sacrifice one of the few copper cents then in my possession. I put one of them on the table, opened a bottle labeled "nitric acid," poured some of the liquid on the copper, and prepared to make an observation. But what was this wonderful thing which I beheld? The cent was already changed, and it was no small change either. A greenish-blue liquid foamed and fumed over the cent and over the table. The air became colored dark red. How could I stop this? I tried by picking the cent up and throwing it out the window. I learned another fact: nitric acid acts upon fingers. The pain led to another unpremeditated experiment. I drew my fingers across my trousers and discovered nitric acid acts upon trousers. That was the most impressive experiment I have ever performed. I tell of it even now with interest. It was a revelation to me. Plainly the only way to learn about such remarkable kinds of action is to see the results, to experiment, to work in the laboratory.

Separation of Mixtures

Because each component of a mixture retains its own properties, we can separate a mixture into its components by taking advantage of the differences in their properties. For example, a heterogeneous mixture of iron filings and gold filings could be sorted individually by color into iron and gold. A less tedious approach would be to use a magnet to attract the iron filings, leaving the gold ones behind. We can also take advantage of an important chemical difference between these two metals: Many acids dissolve iron, but not gold. Thus, if we put our mixture into an appropriate acid, the iron would dissolve and the gold would be left behind. The two could then be separated by *filtration*, a procedure illustrated in Figure 1.12 ▼. We would have to use other chemical reactions, which we will learn about later, to transform the dissolved iron back into metal.

We can separate homogeneous mixtures into their components in similar ways. For example, water has a much lower boiling point than table salt; it is more *volatile*. If we boil a solution of salt and water, the more volatile water evaporates and the salt is left behind. The water vapor is converted back to liquid

(a)

(b)

◀ **Figure 1.12** Separation by filtration. A mixture of a solid and a liquid is poured through a porous medium, in this case filter paper. The liquid passes through the paper while the solid remains on the paper.

MOVIE
Mixtures and Compounds

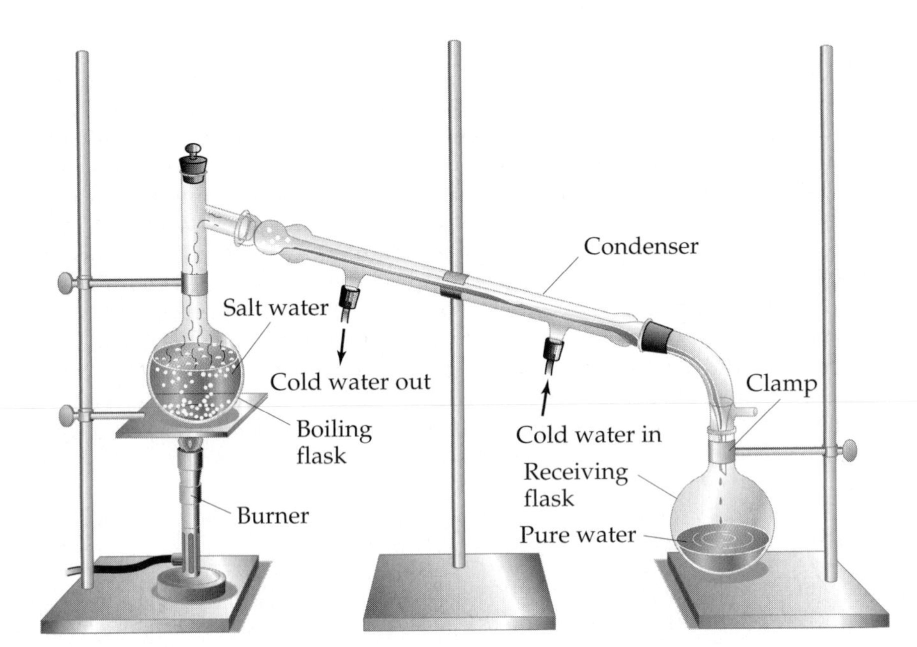

▲ **Figure 1.13** A simple apparatus for the separation of a sodium chloride solution (salt water) into its components. Boiling the solution evaporates the water, which is condensed, then collected in the receiving flask. After all the water has boiled away, pure sodium chloride remains in the boiling flask.

form on the walls of the condenser (Figure 1.13 ▲). This process is called *distillation*.

The differing abilities of substances to adhere to the surfaces of various solids such as paper and starch can also be used to separate mixtures. This is the basis of *chromatography* (literally "the writing of colors"), a technique that can give beautiful and dramatic results. An example of the chromatographic separation of ink is shown in Figure 1.14 ▼.

▲ **Figure 1.14** Separation of ink into components by paper chromatography. (a) Water begins to move up the paper. (b) Water moves past the ink spot, dissolving different components of the ink at different rates. (c) Water has separated the ink into its several different components.

A Closer Look The Scientific Method

Chemistry is an experimental science. The idea of using experiments to understand nature seems like such a natural pattern of thought to us now, but there was a time, before the seventeenth century, when experiments were rarely used. The ancient Greeks, for example, did not rely on experiments to test their ideas.

Although two different scientists rarely approach the same problem in exactly the same way, there are guidelines for the practice of science that have come to be known as the **scientific method**. These guidelines are outlined in Figure 1.15 ▼. We begin by collecting information, or *data*, by observation and experiment. The collection of information, however, is not the ultimate goal. The goal is to find a pattern or sense of order in our observations and to understand the origin of this order.

As we perform our experiments, we may begin to see patterns that lead us to a *tentative explanation*, or **hypothesis**, that guides us in planning further experiments. Eventually, we may be able to tie together a great number of observations in a single statement or equation called a scientific law. A **scientific law** *is a concise verbal statement or a mathematical equation that summarizes a broad variety of observations and experiences*. We tend to think of the laws of nature as the basic rules under which nature operates. However, it is not so much that matter obeys the laws of nature, but rather that the laws of nature describe the behavior of matter.

At many stages of our studies we may propose explanations of why nature behaves in a particular way. If a hypothesis is sufficiently general and is continually effective in predicting facts yet to be observed, it is called a theory or model. A **theory** *is an explanation of the general principles of certain phenomena, with considerable evidence or facts to support it*. For example, Einstein's theory of relativity was a revolutionary new way of thinking about space and time. It was more than just a simple hypothesis, however, because it could be used to make predictions that could be tested experimentally. When these experiments were conducted, the results were generally in agreement with the predictions and were not explainable by the earlier theory of space-time based on Newton's work. Thus, the special theory of relativity was supported, but not proven. Indeed, theories can never be proven to be absolutely correct.

As we proceed through this text, we will rarely have the opportunity to discuss the doubts, conflicts, clashes of personalities, and revolutions of perception that have led to our present ideas. We need to be aware that just because we can spell out the results of science so concisely and neatly in textbooks does not mean that scientific progress is smooth, certain, and predictable. Some of the ideas we have presented in this text took centuries to develop and involved large numbers of scientists. We gain our view of the natural world by standing on the shoulders of the scientists who came before us. Take advantage of this view. As you study, exercise your imagination. Don't be afraid to ask daring questions when they occur to you. You may be fascinated by what you discover!

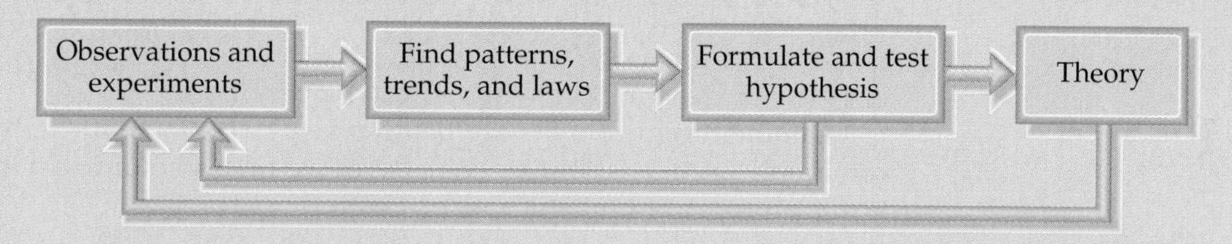

▲ **Figure 1.15** The scientific method is a general approach to problems that involves making observations, seeking patterns in the observations, formulating hypotheses to explain the observations, and testing these hypotheses by further experiments. Those hypotheses that withstand such tests and prove themselves useful in explaining and predicting behavior become known as theories.

1.4 Units of Measurement

Many properties of matter are *quantitative*; that is, they are associated with numbers. When a number represents a measured quantity, the units of that quantity must always be specified. To say that the length of a pencil is 17.5 is meaningless. To say that it is 17.5 centimeters (cm) properly specifies the length. The units used for scientific measurements are those of the **metric system**.

The metric system, which was first developed in France during the late eighteenth century, is used as the system of measurement in most countries throughout the world. The United States has traditionally used the English system, although use of the metric system has become more common in recent years.

▲ **Figure 1.16** Metric measurements are becoming increasingly common in the United States, as exemplified by the volume printed on this container.

For example, the contents of most canned goods and soft drinks in grocery stores are now given in metric as well as in English units as shown in Figure 1.16 ◀.

SI Units

In 1960 an international agreement was reached specifying a particular choice of metric units for use in scientific measurements. These preferred units are called **SI units**, after the French *Système International d'Unités*. The SI system has seven *base units* from which all other units are derived. Table 1.4 ▼ lists these base units and their symbols. In this chapter we will consider the base units for length, mass, and temperature.

TABLE 1.4 SI Base Units

Physical Quantity	Name of Unit	Abbreviation
Mass	Kilogram	kg
Length	Meter	m
Time	Second	s[a]
Temperature	Kelvin	K
Amount of substance	Mole	mol
Electric current	Ampere	A
Luminous intensity	Candela	cd

[a] The abbreviation sec is frequently used.

Prefixes are used to indicate decimal fractions or multiples of various units. For example, the prefix *milli-* represents a 10^{-3} fraction of a unit: A milligram (mg) is 10^{-3} gram (g), a millimeter (mm) is 10^{-3} meter (m), and so forth. Table 1.5 ▼ presents the prefixes commonly encountered in chemistry. In using the SI system and in working problems throughout this text, you must be comfortable using exponential notation. If you are unfamiliar with exponential notation or want to review it, refer to Appendix A.1.

Although non-SI units are being phased out, there are still some that are commonly used by scientists. Whenever we first encounter a non-SI unit in the text, the proper SI unit will also be given.

Length and Mass

The SI base unit of *length* is the meter (m), a distance only slightly longer than a yard. The relations between the English and metric system units that we will use most frequently in this text appear on the back inside cover. We will discuss how to convert English units into metric units, and vice versa, in Section 1.6.

TABLE 1.5 Selected Prefixes Used in the Metric System

Prefix	Abbreviation	Meaning	Example
Giga	G	10^9	1 gigameter (Gm) $= 1 \times 10^9$ m
Mega	M	10^6	1 megameter (Mm) $= 1 \times 10^6$ m
Kilo	k	10^3	1 kilometer (km) $= 1 \times 10^3$ m
Deci	d	10^{-1}	1 decimeter (dm) $= 0.1$ m
Centi	c	10^{-2}	1 centimeter (cm) $= 0.01$ m
Milli	m	10^{-3}	1 millimeter (mm) $= 0.001$ m
Micro	μ[a]	10^{-6}	1 micrometer (μm) $= 1 \times 10^{-6}$ m
Nano	n	10^{-9}	1 nanometer (nm) $= 1 \times 10^{-9}$ m
Pico	p	10^{-12}	1 picometer (pm) $= 1 \times 10^{-12}$ m
Femto	f	10^{-15}	1 femtometer (fm) $= 1 \times 10^{-15}$ m

[a] This is the Greek letter mu (pronounced "mew").

Mass* is a measure of the amount of material in an object. The SI base unit of mass is the kilogram (kg), which is equal to about 2.2 pounds (lb). This base unit is unusual because it uses a prefix, *kilo-*, instead of the word *gram* alone. We obtain other units for mass by adding prefixes to the word *gram*.

SAMPLE EXERCISE 1.2

What is the name given to the unit that equals **(a)** 10^{-9} gram; **(b)** 10^{-6} second; **(c)** 10^{-3} meter?

Solution In each case we can refer to Table 1.5, finding the prefix related to each of the decimal fractions: **(a)** nanogram, ng; **(b)** microsecond, μs; **(c)** millimeter, mm.

PRACTICE EXERCISE

(a) What decimal fraction of a second is a picosecond, ps? **(b)** Express the measurement 6.0×10^3 m using a prefix to replace the power of ten. **(c)** Use standard exponential notation to express 3.76 mg in grams.
Answers: **(a)** 10^{-12} second; **(b)** 6.0 km; **(c)** 3.76×10^{-3} g

Temperature

We sense temperature as a measure of the hotness or coldness of an object. Indeed, temperature determines the direction of heat flow. Heat always flows spontaneously from a substance at higher temperature to one at lower temperature. Thus, we feel the influx of energy when we touch a hot object, and we know that the object is at a higher temperature than our hand.

The temperature scales commonly employed in scientific studies are the Celsius and Kelvin scales. The **Celsius scale** is also the everyday scale of temperature in most countries (Figure 1.17 ▶). It was originally based on the assignment of 0°C to the freezing point of water and 100°C to its boiling point at sea level (Figure 1.18 ▼).

▲ **Figure 1.17** Many countries employ the Celsius temperature scale in everyday use, as illustrated by this Australian stamp.

*Mass and weight are not interchangeable terms and are often incorrectly thought to be the same. The weight of an object is the force that the mass exerts due to gravity. In space, where gravitational forces are very weak, an astronaut can be weightless, but he or she cannot be massless. In fact, the astronaut's mass in space is the same as it is on Earth.

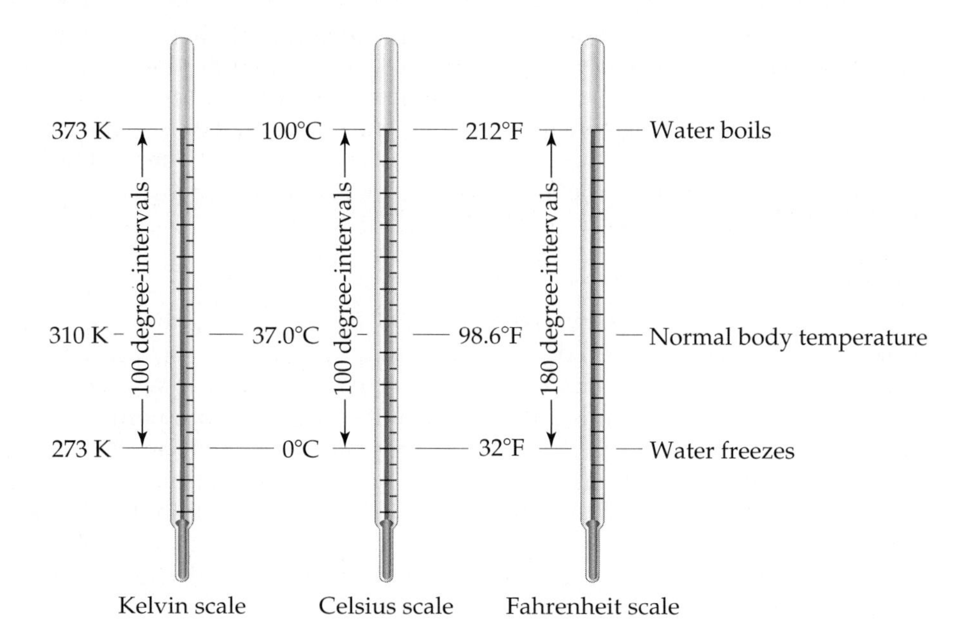

◀ **Figure 1.18** Comparison of the Kelvin, Celsius, and Fahrenheit temperature scales.

The **Kelvin scale** is the SI temperature scale, and the SI unit of temperature is the kelvin (K). Historically, the Kelvin scale was based on the properties of gases; its origins will be considered in Chapter 10. Zero on this scale is the lowest attainable temperature, $-273.15°C$, a temperature referred to as *absolute zero*. Both the Celsius and Kelvin scales have equal-sized units—that is, a kelvin is the same size as a degree Celsius. Thus, the Kelvin and Celsius scales are related as follows:

$$K = °C + 273.15 \qquad [1.1]$$

The freezing point of water, 0°C, is 273.15 K (Figure 1.18). Notice that we do not use a degree sign (°) with temperatures on the Kelvin scale.

The common temperature scale in the United States is the *Fahrenheit scale*, which is not generally used in scientific studies. On the Fahrenheit scale water freezes at 32°F and boils at 212°F. The Fahrenheit and Celsius scales are related as follows:

$$°C = \frac{5}{9}(°F - 32) \quad \text{or} \quad °F = \frac{9}{5}(°C) + 32 \qquad [1.2]$$

SAMPLE EXERCISE 1.3

If a weather forecaster predicts that the temperature for the day will reach 31°C, what is the predicted temperature **(a)** in K; **(b)** in °F?

Solution

(a) Using Equation 1.1, we have $K = 31 + 273 = 304\ K$

(b) Using Equation 1.2, we have $°F = \frac{9}{5}(31) + 32 = 56 + 32 = 88°F$

PRACTICE EXERCISE

Ethylene glycol, the major ingredient in antifreeze, freezes at $-11.5°C$. What is the freezing point in **(a)** K; **(b)** °F?
Answers: **(a)** 261.7 K; **(b)** 11.3°F

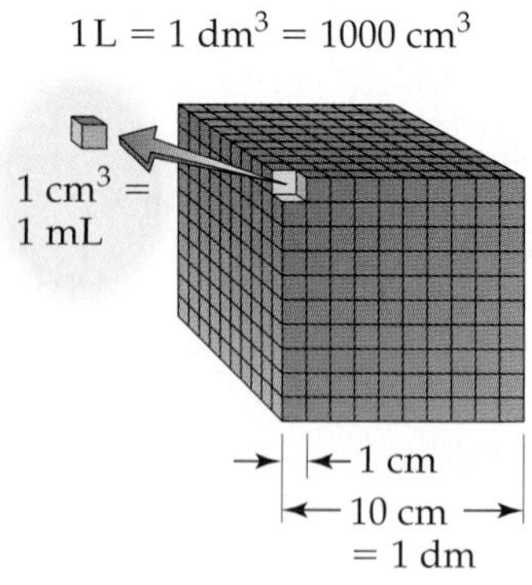

Derived SI Units

The SI base units in Table 1.4 are used to derive the units of other quantities. To do so, we use the defining equation for the quantity, substituting the appropriate base units. For example, speed is defined as the ratio of distance to elapsed time. Thus, the SI unit for speed is the SI unit for distance (length) divided by the SI unit for time, m/s, which we read as "meters per second." We will encounter many derived units, such as those for force, pressure, and energy, later in this text. In this chapter we examine the derived units for volume and density.

$1\ L = 1\ dm^3 = 1000\ cm^3$

$1\ cm^3 = 1\ mL$

1 cm

10 cm = 1 dm

▲ **Figure 1.19** A liter is the same volume as a cubic decimeter, $1\ L = 1\ dm^3$. Each cubic decimeter contains 1000 cubic centimeters, $1\ dm^3 = 1000\ cm^3$. Each cubic centimeter equals a milliliter, $1\ cm^3 = 1\ mL$.

Volume

The *volume* of a cube is given by its length cubed, $(\text{length})^3$. Thus, the basic SI unit of volume is the cubic meter, or m^3, the volume of a cube that is 1 m on each edge. Smaller units, such as cubic centimeters, cm^3 (sometimes written as cc), are frequently used in chemistry. Another unit of volume commonly used in chemistry is the *liter* (L), which equals a cubic decimeter, dm^3, and is slightly larger than a quart. The liter is the first metric unit we have encountered that is *not* an SI unit. There are 1000 milliliters (mL) in a liter (Figure 1.19 ◀), and each milliliter is the same volume as a cubic centimeter: $1\ mL = 1\ cm^3$. The terms *milliliter* and *cubic centimeter* are used interchangeably in expressing volume.

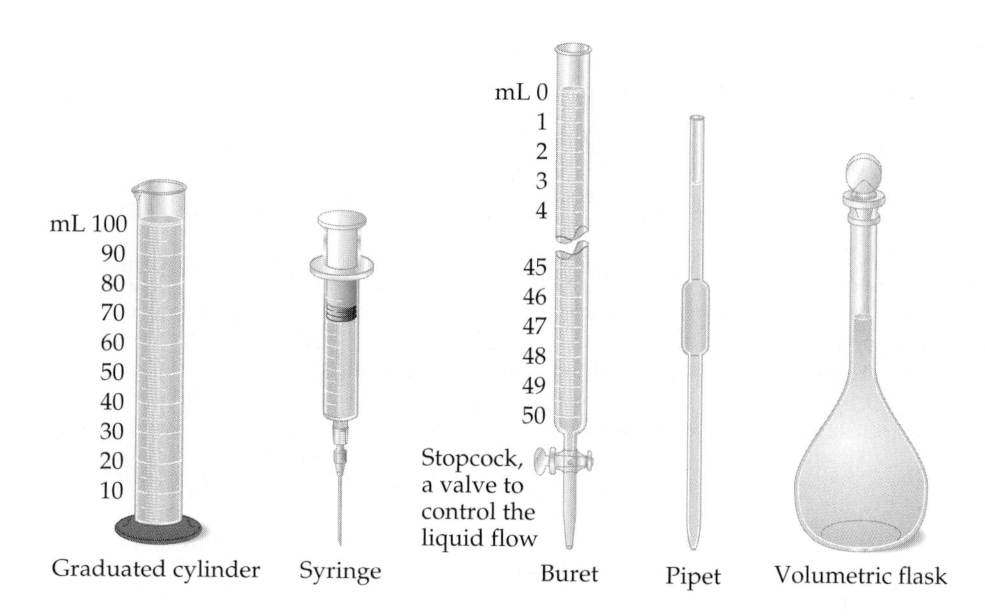

◀ **Figure 1.20** Common devices used in chemistry laboratories for the measurement and delivery of volumes of liquid. The graduated cylinder, syringe, and buret are used to deliver variable volumes of liquid; the pipet is used to deliver a specific volume of liquid. The volumetric flask contains a specific volume of liquid when filled to the mark.

The devices used most frequently in chemistry to measure volume are illustrated in Figure 1.20 ▲. Syringes, burets, and pipets deliver liquids with more accuracy than graduated cylinders. Volumetric flasks are used to contain specific volumes of liquid.

Density

Density is widely used to characterize substances. It is defined as the amount of mass in a unit volume of the substance:

$$\text{Density} = \frac{\text{mass}}{\text{volume}} \qquad [1.3]$$

The densities of solids and liquids are commonly expressed in units of grams per cubic centimeter (g/cm^3) or grams per milliliter (g/mL). The densities of some common substances are listed in Table 1.6 ▼. It is no coincidence that the density of water is 1.00 g/mL; the gram was originally defined as the mass of 1 mL of water at a specific temperature. Because most substances change volume when heated or cooled, densities are temperature dependent. When reporting densities, the temperature should be specified. We usually assume that the temperature is 25°C, close to normal room temperature, if no temperature is reported.

TABLE 1.6 Densities of Some Selected Substances at 25°C

Substance	Density (g/cm^3)
Air	0.001
Balsa wood	0.16
Ethanol	0.79
Water	1.00
Ethylene glycol	1.09
Table sugar	1.59
Table salt	2.16
Iron	7.9
Gold	19.32

Chemistry at Work Chemistry in the News

Chemistry is a very lively, active field of science. Because it is so central to our lives, there are reports on matters of chemical significance in the news nearly every day. Some tell of recent breakthroughs in the development of new pharmaceuticals, materials, and processes. Others deal with environmental and public safety issues. As you study chemistry, we hope you will develop the skills to better understand the impact of chemistry on your life. You need these skills to take part in public discussions and debates about matters related to chemistry that affect your community, the nation, and the world. By way of examples, here are summaries of a few recent stories in which chemistry plays a role.

"Fuel Cells Produce Energy Directly from Hydrocarbons"

The arrival of electric cars, such as the one in Figure 1.21 ▼, as a practical mode of transportation has been delayed for years by problems in finding a suitable energy source. The batteries that are available at reasonable cost are too heavy, and they permit only a limited mileage before needing to be recharged. The fuel cell, in which a chemical reaction is used to furnish electrical energy directly, is an alternative to a battery. Up until now successful fuel cells have required the use of hydrogen as a fuel. Hydrogen is expensive to produce, and storing it presents problems and poses potential dangers.

Recently researchers at the University of Pennsylvania have demonstrated that more convenient, less-expensive, and potentially safer fuels, such as butane and diesel fuel, can be used directly to produce electricity in a newly designed fuel cell. Butane and diesel fuel are composed of hydrocarbons, molecules containing just hydrogen and carbon atoms. The key to the new technology is the development of a new electrode material for the fuel cell, one containing the element copper, which presumably helps catalyze the appropriate electrochemical reactions at the electrode.

Though this new technology appears very promising, you won't be able to place your order for an electric car that incorporates it just yet. Several engineering and cost issues need to be resolved before it can become a commercial reality. Nevertheless, several automobile companies have set their goal to have a fuel-cell powered automobile on the market by 2004 or shortly thereafter.

"Adding Iron to the Ocean Spurs Photosynthesis"

Microscopic plant life—phytoplankton—is scarce in certain parts of the ocean (Figure 1.22 ▼). Several years ago scientists proposed that this scarcity is caused by the lack of plant nutrients, primarily iron. Because phytoplankton take up carbon dioxide in photosynthesis, it was also proposed that relatively small amounts of iron distributed in appropriate regions of the oceans could reduce atmospheric carbon dioxide, thereby reducing global warming. If the phytoplankton sank to the bottom of the ocean when they died, the carbon dioxide would not return to the atmosphere when the microbes decomposed.

Recently, studies have been conducted in which iron was added to surface waters of the southern ocean near Antarctica to study its effect on phytoplankton. Adding iron resulted in a substantial buildup in the amount of phytoplankton and at least a short-term drop in the amount of carbon dioxide in the air immediately above them. These results were consistent with similar experiments performed earlier in the equatorial Pacific Ocean, confirming the hypothesis that iron is the limiting nutrient of these microorganisms in much of the ocean. However, there was no increase in the amount of microbes sinking out of the top layer of ocean water. Thus, this procedure may be of no use for the long-term reduction of atmospheric carbon dioxide.

"Nanotechnology: Hype and Hope"

The past 15 years have witnessed an explosion of relatively inexpensive equipment and techniques for probing and manipulating materials on the nanometer-length scale. These capabilities have led to optimistic forecasts of futuristic nano- technologies including molecular-scale machines and robots that can manipulate matter with atomic precision. Many believe that such futuristic visions are mere hype, while others express the hope that they can be realized.

▲ **Figure 1.21** Cutaway view of car powered by fuel cells.

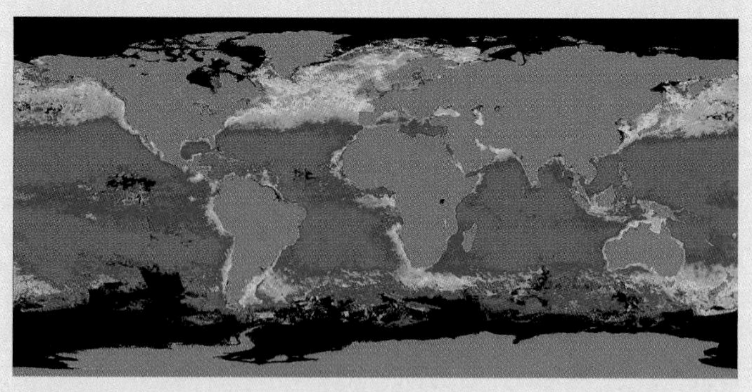

▲ **Figure 1.22** A color-enhanced satellite image of the global ocean, highlighting the distribution and concentration of phytoplankton. The red and orange regions have the greatest concentration, whereas the light blue and dark purple have the least.

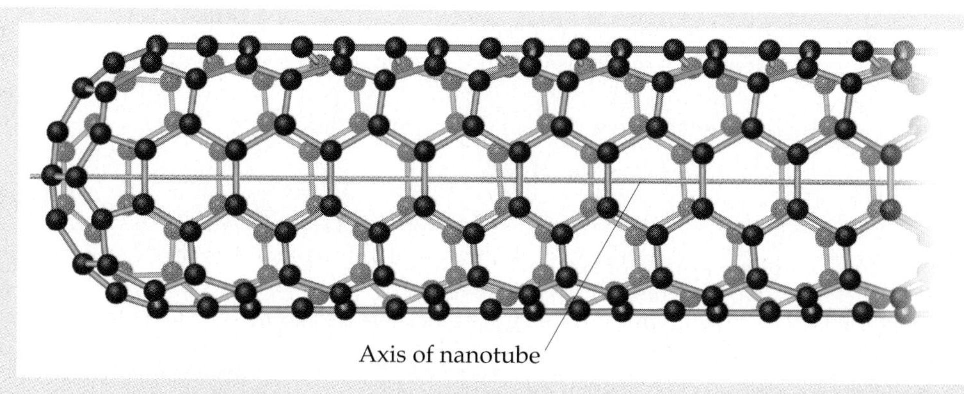

◀ **Figure 1.23** A section of carbon nanotube. Each intersection in the network represents a carbon atom chemically joined to three others.

Nanoscale materials do exhibit chemical and physical properties from different bulk materials. For example, carbon can be made to form tubular structures as shown in Figure 1.23 ▲. These tubes, called nanotubes, resemble a cylindrical roll of chicken wire. When nanotubes are perfectly formed, they conduct electricity like a metal.

Scientists have learned that the electric and optical properties of certain nanometer-size particles can be tuned by adjusting the particle size or shape. Their properties are therefore of interest for applications in optical data-storage devices and ultrafast data communications systems. Although such applications are still years from commercial fruition, they nevertheless offer the promise of dramatically changing not only the size of electronic devices, sensors and many other items, but also the way they are manufactured. It suggests that such devices might be assembled from simpler, smaller components such as molecules and other nanostructures. This approach is similar to the one nature uses to construct complex biological architectures.

"The Search for a Super-aspirin"

Aspirin, introduced in 1899, was one of the first drugs ever developed and is still one of the most widely used. It is estimated that 20 billion aspirin tablets are taken each year in the United States. Originally intended to relieve pain and soothe aching joints and muscles, it has proven to be an immensely complicated medication with unexpected powers and limitations. It has been found to reduce the incidence of heart attacks and is effective in reducing the incidences of Alzheimer's disease and digestive tract cancers. At the same time, however, aspirin attacks the stomach lining, causing bleeding or even ulcers, and it often causes intestinal problems.

One of the ways that aspirin works is by blocking an enzyme (a type of protein) called COX-2, which promotes inflammation, pain, and fever. Unfortunately, it also interferes with the COX-1, a related enzyme that makes hormones essential for the health of the stomach and kidneys. An ideal pain reliever and anti-inflammatory agent would inhibit COX-2 but not interfere with COX-1. The shape of the aspirin molecule is shown in Figure 1.24(a) ▼. Aspirin works by transferring part of its molecule, called the acetyl group, to COX-2, thereby disabling it. A replacement for aspirin must retain this feature of the molecule, which is highlighted in Figure 1.24(a). The replacement should also retain the general shape and size of the aspirin molecule, so that it fits into the space on the enzyme in the same way aspirin does.

One promising variant of the aspirin molecule is shown in Figure 1.24(b). The changed portion consists of a sulfur atom (yellow) followed by a "tail" of carbon atoms (black) and attached hydrogen atoms (white). This molecule is a potent COX-2 inhibitor that does not appear to affect COX-1. This and other "super-aspirin" molecules must pass tests of long-term safety before they can appear on the shelves at the drugstore, but in time they may replace aspirin and the other popular nonsteroidal anti-inflammatory drugs.

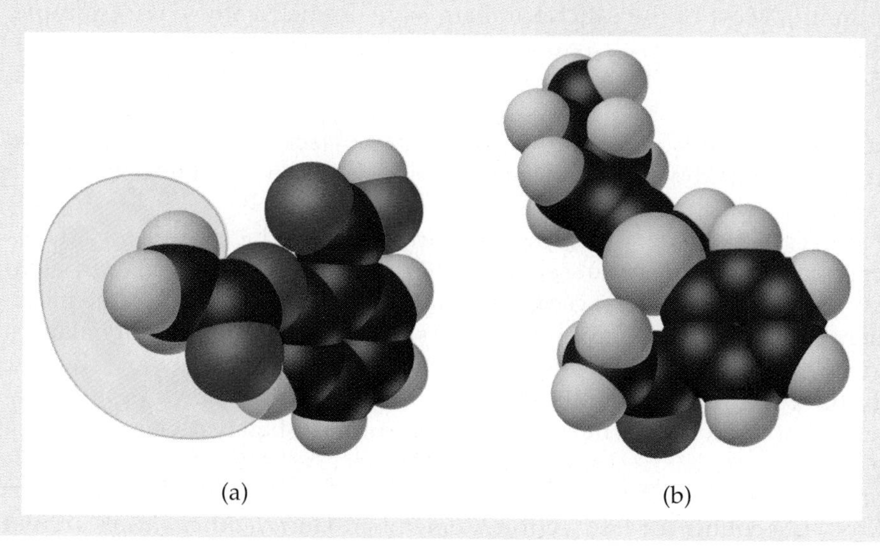

◀ **Figure 1.24** (a) A molecular model of aspirin, the highlighted portion of the molecule is transferred when aspirin deactivates the COX-2 enzyme. (b) Molecular model of a potential new "super-aspirin" whose molecular structure is related to that of aspirin.

The terms *density* and *weight* are sometimes confused. A person who says that iron weighs more than air generally means that iron has a higher density than air; 1 kg of air has the same mass as 1 kg of iron, but the iron occupies a smaller volume, thereby giving it a higher density. If we combine two liquids that do not mix, the less dense one will float on the more dense one.

SAMPLE EXERCISE 1.4

(a) Calculate the density of mercury if 1.00×10^2 g occupies a volume of 7.36 cm^3.
(b) Calculate the volume of 65.0 g of the liquid methanol (wood alcohol) if its density is 0.791 g/mL.
(c) What is the mass in grams of a cube of gold (density = 19.32 g/cm^3) if the length of the cube is 2.00 cm.

Solution

(a) We are given mass and volume, so Equation 1.3 yields

$$\text{Density} = \frac{\text{mass}}{\text{volume}} = \frac{1.00 \times 10^2\ \text{g}}{7.36\ \text{cm}^3} = 13.6\ \text{g/cm}^3$$

(b) Solving Equation 1.3 for volume and then using the given mass and density gives

$$\text{Volume} = \frac{\text{mass}}{\text{density}} = \frac{65.0\ \text{g}}{0.791\ \text{g/mL}} = 82.2\ \text{mL}$$

(c) We can calculate the mass from the volume of the cube and its density. The volume of a cube is given by its length cubed:

$$\text{Volume} = (2.00\ \text{cm})^3 = (2.00)^3\ \text{cm}^3 = 8.00\ \text{cm}^3$$

Solving Equation 1.3 for mass and substituting the volume and density of the cube we have

$$\text{Mass} = \text{volume} \times \text{density} = (8.00\ \text{cm}^3)(19.32\ \text{g/cm}^3) = 155\ \text{g}$$

PRACTICE EXERCISE

(a) Calculate the density of a 374.5-g sample of copper if it has a volume of 41.8 cm^3. **(b)** A student needs 15.0 g of ethanol for an experiment. If the density of the alcohol is 0.789 g/mL, how many milliliters of alcohol are needed? **(c)** What is the mass, in grams, of 25.0 mL of mercury (density = 13.6 g/mL)?
Answers: **(a)** 8.96 g/cm^3; **(b)** 19.0 mL; **(c)** 340 g

1.5 Uncertainty in Measurement

There are two kinds of numbers in scientific work: *exact numbers* (those whose values are known exactly) and *inexact numbers* (those whose values have some uncertainty). Most of the exact numbers have defined values. For example, there are exactly 12 eggs in a dozen, exactly 1000 g in a kilogram, and exactly 2.54 cm in an inch. The number 1 in any conversion factor between units, as in 1 m = 100 cm or 1 kg = 2.2046 lb, is also an exact number. Exact numbers can also result from counting numbers of objects. For example, we can count the exact number of marbles in a jar or the exact number of people in a classroom.

Numbers obtained by measurement are always *inexact*. There are always inherent limitations in the equipment used to measure quantities (equipment errors), and there are differences in how different people make the same measurement (human errors). Suppose that 10 students with 10 different balances are given the same dime to weigh. The 10 measurements will vary slightly. The balances might be calibrated slightly differently, and there might be differences in how each student reads the mass from the balance. Counting very large numbers of objects usually has some associated error as well. Consider, for example, how difficult it is to obtain accurate census information for a city or vote counts for an election. Remember: *Uncertainties always exist in measured quantities.*

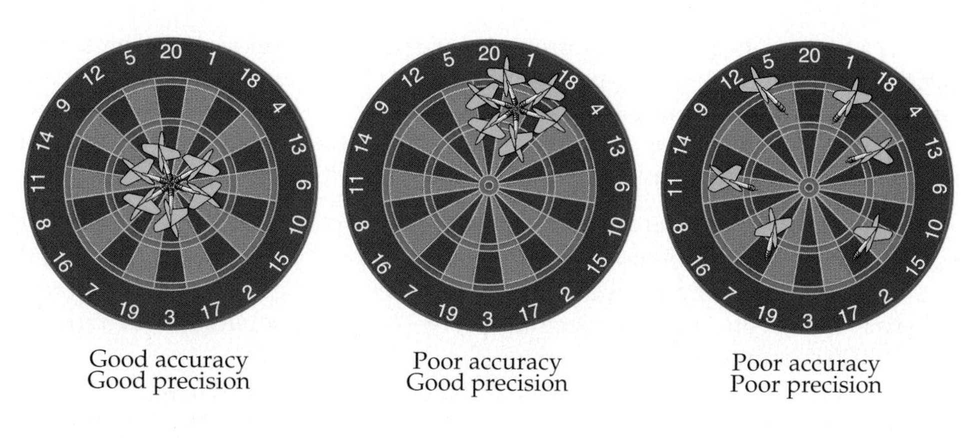

◀ **Figure 1.25** The distribution of darts on a target illustrates the difference between accuracy and precision.

Precision and Accuracy

The terms precision and accuracy are often used in discussing the uncertainties of measured values. **Precision** is a measure of how closely individual measurements agree with one another. **Accuracy** refers to how closely individual measurements agree with the correct, or "true," value. The analogy of darts stuck in a dartboard pictured in Figure 1.25 ▲ illustrates the difference between these two concepts.

In the laboratory we often perform several different "trials" of the same experiment. We gain confidence in the accuracy of our measurements if we obtain nearly the same value each time. Figure 1.25 should remind us, however, that precise measurements can be inaccurate. For example, if a very sensitive balance is poorly calibrated, the masses we measure will be consistently either high or low. They will be inaccurate even if they are precise.

Significant Figures

Suppose you weigh a dime on a balance capable of measuring to the nearest 0.0001 g. You could report the mass as 2.2405 ± 0.0001 g. The ± notation (read "plus or minus") expresses the uncertainty of a measurement. In much scientific work we drop the ± notation with the understanding that an uncertainty of at least one unit exists in the last digit of the measured quantity. That is, *measured quantities are generally reported in such a way that only the last digit is uncertain.*

Figure 1.26 ▶ shows a thermometer with its liquid column between the scale marks. We can read the certain digits from the scale and estimate the uncertain one. From the scale marks, we see that the liquid is between the 25°C and 30°C marks. We might estimate the temperature to be 27°C, being somewhat uncertain of the second digit of our measurement.

All digits of a measured quantity, including the uncertain one, are called **significant figures**. A measured mass reported as 2.2 g has two significant figures, whereas one reported as 2.2405 g has five significant figures. The greater the number of significant figures, the greater is the certainty implied for the measurement.

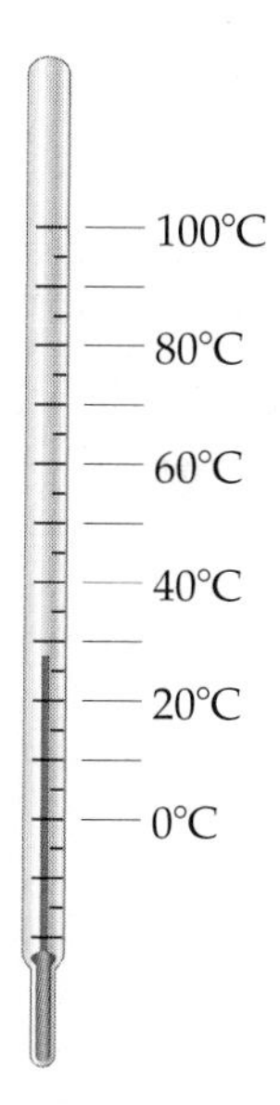

▲ **Figure 1.26** A thermometer whose markings are shown only every 5°C. The temperature is between 25°C and 30°C and is approximately 27°C.

SAMPLE EXERCISE 1.5

What is the difference between 4.0 g and 4.00 g?

Solution Many people would say there is no difference, but a scientist would note the difference in the number of significant figures in the two measurements. The value 4.0 has two significant figures, while 4.00 has three. This implies that the first measurement has more uncertainty. A mass of 4.0 g indicates that the mass is between 3.9 and 4.1 g; the mass is 4.0 ± 0.1 g. A measurement of 4.00 g implies that the mass is between 3.99 and 4.01 g; the mass is 4.00 ± 0.01 g.

PRACTICE EXERCISE

A balance has a precision of ±0.001 g. A sample that weighs about 25 g is weighed on this balance. How many significant figures should be reported for this measurement?
Answer: 5, as in the measurement 24.995 g

In any measurement that is properly reported, all nonzero digits are significant. Zeros, however, can be used either as part of the measured value or merely to locate the decimal point. Thus, zeros may or may not be significant, depending on how they appear in the number. The following guidelines describe the different situations involving zeros:

1. Zeros between nonzero digits are always significant—1005 kg (4 significant figures); 1.03 cm (three significant figures).
2. Zeros at the beginning of a number are never significant; they merely indicate the position of the decimal point—0.02 g (one significant figure); 0.0026 cm (two significant figures).
3. Zeros that fall both at the end of a number and after the decimal point are always significant—0.0200 g (3 significant figures); 3.0 cm (2 significant figures).
4. When a number ends in zeros but contains no decimal point, the zeros may or may not be significant—130 cm (two or three significant figures); 10,300 g (three, four, or five significant figures).

The use of exponential notation (Appendix A) eliminates the potential ambiguity of whether the zeros at the end of a number are significant (rule 4). For example, a mass of 10,300 g can be written in exponential notation showing three, four, or five significant figures:

1.03×10^4 g (three significant figures)

1.030×10^4 g (four significant figures)

1.0300×10^4 g (five significant figures)

In these numbers all the zeros to the right of the decimal point are significant (rules 1 and 3). (All significant figures come before the exponent; the exponential term does not add to the number of significant figures.)

Exact numbers can be treated as if they have an infinite number of significant figures. This rule applies to many definitions between units. Thus, when we say, "There are 12 inches in 1 foot," the number 12 is exact and we need not worry about the number of significant figures in it.

SAMPLE EXERCISE 1.6

How many significant figures are in each of the following numbers (assume that each number is a measured quantity): **(a)** 4.003; **(b)** 6.023×10^{23}; **(c)** 5000?

Solution **(a)** Four; the zeros are significant figures. **(b)** Four; the exponential term does not add to the number of significant figures. **(c)** One, two, three, or four. In this case the ambiguity could have been avoided by using exponential notation. Thus 5×10^3 has only one significant figure, whereas 5.00×10^3 has three.

PRACTICE EXERCISE

How many significant figures are in each of the following measurements: **(a)** 3.549 g; **(b)** 2.3×10^4 cm; **(c)** 0.00134 m^3?
Answers: **(a)** four; **(b)** two; **(c)** three

Significant Figures in Calculations

When carrying measured quantities through calculations, observe these points: (1) The least certain measurement used in a calculation limits the certainty of the calculated quantity. (2) The final answer for any calculation should be reported with only one uncertain digit.

To keep track of significant figures in calculations, we will make frequent use of two rules. The first involves multiplication and division, and the second involves addition and subtraction. *In multiplication and division the result must be reported with the same number of significant figures as the measurement with the fewest significant figures.* When the result contains more than the correct number of significant figures, it must be rounded off.

For example, the area of a rectangle whose measured edge lengths are 6.221 cm and 5.2 cm should be reported as 32 cm^2 even though a calculator shows the product of 6.221 and 5.2 to have more digits:

$$\text{Area} = (6.221\ \text{cm})(5.2\ \text{cm}) = 32.3492\ \text{cm}^2 \Rightarrow \text{round off to } 32\ \text{cm}^2$$

We round off to two significant figures because the least precise number—5.2 cm—has only two significant figures.

In rounding off numbers, look at the leftmost digit to be dropped:

1. If the leftmost digit to be removed is less than 5, the preceding number is left unchanged. Thus, rounding 7.248 to two significant figures gives 7.2.
2. If the leftmost digit to be removed is 5 or greater, the preceding number is increased by 1. Rounding 4.735 to three significant figures gives 4.74, and rounding 2.376 to two significant figures gives 2.4.*

The guidelines used to determine the number of significant figures in addition and subtraction are different from those for multiplication and division. *In addition and subtraction the result can have no more decimal places than the measurement with the fewest number of decimal places.* In the following example the uncertain digits appear in color:

This number limits	20.4	← one decimal place
the number of significant	1.322	← three decimal places
figures in the result ⟶	83	← zero decimal places
	104.722	← round off to 105 (zero decimal places)

SAMPLE EXERCISE 1.7

The width, length, and height of a small box are 15.5 cm, 27.3 cm, and 5.4 cm, respectively. Calculate the volume of the box using the correct number of significant figures in your answer.

Solution The volume of a box is determined by the product of its width, length, and height. In reporting the product, we can show only as many significant figures as given in the dimension with the fewest significant figures, that for the height (two significant figures):

$$\begin{aligned}\text{Volume} &= \text{width} \times \text{length} \times \text{height}\\ &= (15.5\ \text{cm})(27.3\ \text{cm})(5.4\ \text{cm}) = 2285.01\ \text{cm}^3 \Rightarrow 2.3 \times 10^3\ \text{cm}^3\end{aligned}$$

When using a calculator, the display first shows 2285.01, which we must round off to two significant figures. Because the resulting number is 2300, it should be reported in standard exponential notation, 2.3×10^3, to clearly indicate two significant figures. Notice that we round off the result at the end of the calculation.

PRACTICE EXERCISE

It takes 10.5 s for a sprinter to run 100.00 m. Calculate the average speed of the sprinter in meters per second, and express the result to the correct number of significant figures.

Answer: 9.52 m/s (3 significant figures)

*Your instructor may wish you to use a slight variation on the rule when the leftmost digit to be removed is exactly 5, with no following digits or only zeros. One common practice is to round up to the next higher number if that number will be even, and down to the next lower number otherwise. Thus, 4.7350 would be rounded to 4.74, and 4.7450 would also be rounded to 4.74.

SAMPLE EXERCISE 1.8

A gas at 25°C fills a container previously determined to have a volume of $1.05 \times 10^3\ cm^3$. The container plus gas are weighed and found to have a mass of 837.6 g. The container, when emptied of all gas, has a mass of 836.2 g. What is the density of the gas at 25°C?

Solution

To calculate the density we must know both the mass and the volume of the gas. The mass of the gas is just the difference in the masses of the full and empty container:

$$(837.6 - 836.2)\ g = 1.4\ g$$

In subtracting numbers, we determine significant figures by paying attention to decimal places. Thus the mass of the gas, 1.4 g, has only two significant figures, even though the masses from which it is obtained have four.

Using the volume given in the question, $1.05 \times 10^3\ cm^3$, and the definition of density, we have

$$\text{Density} = \frac{\text{mass}}{\text{volume}} = \frac{1.4\ g}{1.05 \times 10^3\ cm^3}$$
$$= 1.3 \times 10^{-3}\ g/cm^3 = 0.0013\ g/cm^3$$

In dividing numbers, we determine the number of significant figures in our result by considering the number of significant figures in each factor. There are two significant figures in our answer, corresponding to the smaller number of significant figures in the two numbers that form the ratio.

PRACTICE EXERCISE

To how many significant figures should the mass of the container be measured (with and without the gas) in Sample Exercise 1.8 in order for the density to be calculated to three significant figures?
Answer: 5 (In order for the difference in the two masses to have three significant figures, there must be two decimal places in the masses of the filled and empty containers.)

When a calculation involves two or more steps and you write down answers for intermediate steps, retain at least one additional digit—past the number of significant figures—for the intermediate answers. This procedure ensures that small errors from rounding at each step do not combine to affect the final result. When using a calculator, you may enter the numbers one after another, rounding only the final answer. Accumulated rounding-off errors may account for small differences between results you obtain and answers given in the text for numerical problems.

1.6 Dimensional Analysis

Throughout the text we use an approach called **dimensional analysis** as an aid in problem solving. In dimensional analysis we carry units through all calculations. Units are multiplied together, divided into each other, or "canceled." Dimensional analysis will help ensure that the solutions to problems yield the proper units. Moreover, dimensional analysis provides a systematic way of solving many numerical problems and of checking our solutions for possible errors.

The key to using dimensional analysis is the correct use of conversion factors to change one unit into another. A **conversion factor** is a fraction whose numerator and denominator are the same quantity expressed in different units. For example, 2.54 cm and 1 in. are the same length, 2.54 cm = 1 in. This relationship allows us to write two conversion factors:

$$\frac{2.54\ cm}{1\ in.} \quad \text{and} \quad \frac{1\ in.}{2.54\ cm}$$

We use the first of these factors to convert inches to centimeters. For example, the length in centimeters of an object that is 8.50 in. long is given by

$$\text{Number of centimeters} = (8.50\ \cancel{\text{in.}})\frac{2.54\ \text{cm}}{1\ \cancel{\text{in.}}} = 21.6\ \text{cm}$$

Desired unit

Given unit

The units of inches in the denominator of the conversion factor cancel the units of inches in the given data (8.50 *inches*). The centimeters in the numerator of the conversion factor become the units of the final answer. Because the numerator and denominator of a conversion factor are equal, multiplying any quantity by a conversion factor is equivalent to multiplying by the number 1 and so does not change the intrinsic value of the quantity. The length 8.50 in. is the same as 21.6 cm.

In general, we begin any conversion by examining the units of the given data and the units we desire. We then ask ourselves what conversion factors we have available to take us from the units of the given quantity to those of the desired one. When we multiply a quantity by a conversion factor, the units multiply and divide as follows:

$$\cancel{\text{Given unit}} \times \frac{\text{desired unit}}{\cancel{\text{given unit}}} = \text{desired unit}$$

If the desired units are not obtained in a calculation, then an error must have been made somewhere. Careful inspection of units often reveals the source of the error.

SAMPLE EXERCISE 1.9

If a woman has a mass of 115 lb, what is her mass in grams? (Use the relationships between units given on the back inside cover of the text.)

Solution Because we want to change from lb to g, we look for a relationship between these units of mass. From the back inside cover we have 1 lb = 453.6 g. In order to cancel pounds and leave grams, we write the conversion factor with grams in the numerator and pounds in the denominator:

$$\text{Mass in grams} = (115\ \cancel{\text{lb}})\left(\frac{453.6\ \text{g}}{1\ \cancel{\text{lb}}}\right) = 5.22 \times 10^4\ \text{g}$$

The answer can be given to only three significant figures, the number of significant figures in 115 lb.

PRACTICE EXERCISE

By using a conversion factor from the back inside cover, determine the length in kilometers of a 500.0-mi automobile race.
Answer: 804.7 km

Strategies in Chemistry **Estimating Answers**

A friend once remarked cynically that calculators let you get the wrong answer more quickly. What he was implying by that remark was that unless you have the correct strategy for solving a problem and have punched in the correct numbers, the answer will be incorrect. If you learn to *estimate* answers, however, you will be able to check whether the answers to your calculations are reasonable.

The idea is to make a rough calculation using numbers that are rounded off in such a way that the arithmetic can be easily performed without a calculator. This approach is often referred to as making a "ball-park" estimate, meaning that while it doesn't give an exact answer, it gives one that is roughly of the right size. By working with units using dimensional analysis and by estimating answers, we can readily check the reasonableness of our answers to calculations.

Using Two or More Conversion Factors

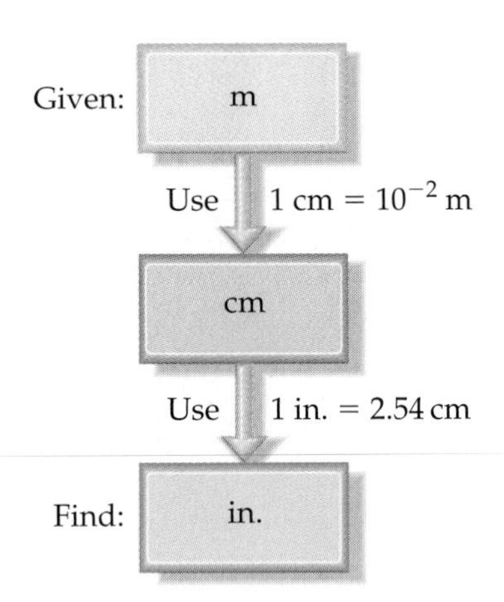

It is often necessary to use more than one conversion factor in the solution of a problem. For example, suppose we want to know the length in inches of an 8.00-m rod. The table on the back inside cover doesn't give the relationship between meters and inches. It *does* give the relationship between centimeters and inches (1 in. = 2.54 cm), though, and from our knowledge of metric prefixes we know that 1 cm = 10^{-2} m. Thus, we can convert step by step, first from meters to centimeters, and then from centimeters to inches as diagrammed in the column.

Combining the given quantity (8.00 m) and the two conversion factors, we have

$$\text{Number of inches} = (8.00\ \cancel{\text{m}})\left(\frac{100\ \cancel{\text{cm}}}{1\ \cancel{\text{m}}}\right)\left(\frac{1\ \text{in.}}{2.54\ \cancel{\text{cm}}}\right) = 315\ \text{in.}$$

The first conversion factor is applied to cancel meters and convert the length to centimeters. Thus, meters are written in the denominator and centimeters in the numerator. The second conversion factor is written to cancel centimeters, so it has centimeters in the denominator and inches, the desired unit, in the numerator.

SAMPLE EXERCISE 1.10

The average speed of a nitrogen molecule in air at 25°C is 515 m/s. Convert this speed to miles per hour.

Solution To go from the given units, m/s, to the desired units, mi/hr, we must convert meters to miles and seconds to hours. From the relationships given on the back inside cover of the book, we find that 1 mi = 1.6093 km. From our knowledge of metric prefixes we know that 1 km = 10^3 m. Thus, we can convert m to km and then convert km to mi. From our knowledge of time we know that 60 s = 1 min and 60 min = 1 hr. Thus, we can convert s to min and then convert min to hr.

Applying first the conversions for distance and then those for time, we can set up one long equation in which unwanted units are canceled:

$$\text{Speed in mi/hr} = \left(515\frac{\cancel{\text{m}}}{\cancel{\text{s}}}\right)\left(\frac{1\ \cancel{\text{km}}}{10^3\ \cancel{\text{m}}}\right)\left(\frac{1\ \text{mi}}{1.6093\ \cancel{\text{km}}}\right)\left(\frac{60\ \cancel{\text{s}}}{1\ \cancel{\text{min}}}\right)\left(\frac{60\ \cancel{\text{min}}}{1\ \text{hr}}\right)$$

$$= 1.15 \times 10^3\ \text{mi/hr}$$

Our answer has the desired units. We can check our calculation using the estimating procedure described in the previous "Strategies" box. The given speed is about 500 m/s. Dividing by 1000 converts m to km, giving 0.5 km/s. Because 1 mi is about 1.6 km, this speed corresponds to 0.5/1.6 = 0.3 mi/s. Multiplying by 60 gives about 0.3 × 60 = 20 mi/min. Multiplying again by 60 gives 20 × 60 = 1200 mi/hr. The approximate solution (about 1200 mi/hr) and the detailed solution (1150 mi/hr) are reasonably close. The answer to the detailed solution has three significant figures, corresponding to the number of significant figures in the given speed in m/s.

PRACTICE EXERCISE

A car travels 28 mi per gallon of gasoline. How many kilometers per liter will it go?
Answer: 12 km/L

Conversions Involving Volume

The conversion factors previously noted convert from one unit of a given measure to another unit of the same measure, such as from length to length. We also have conversion factors that convert from one measure to a different one. The density of a substance, for example, can be treated as a conversion factor between mass and volume. Suppose that we want to know the mass in grams of two cubic inches (2.00 in.^{3}) of gold, which has a density of 19.3 g/cm^3. The density gives us the following factors:

$$\frac{19.3\ \text{g}}{1\ \text{cm}^3} \quad \text{and} \quad \frac{1\ \text{cm}^3}{19.3\ \text{g}}$$

Because the answer we want is a mass in grams, we can see that we will use the first of these factors, which has mass in grams in the numerator. To use this factor, however, we must first convert cubic inches to cubic centimeters. The relationship between in.^{3} and cm^3 is not given on the back inside cover, but the relationship between inches and centimeters is given: 1 in. = 2.54 cm (exactly). Cubing both sides of this equation gives $(1\ \text{in.})^3 = (2.54\ \text{cm})^3$ from which we write the desired conversion factor:

$$\frac{(2.54\ \text{cm})^3}{(1\ \text{in.})^3} = \frac{(2.54)^3\ \text{cm}^3}{(1)^3\ \text{in.}^3} = \frac{16.39\ \text{cm}^3}{1\ \text{in.}^3}$$

Notice that both the numbers and the units are cubed. Also, because 2.54 is an exact number, we can retain as many digits of $(2.54)^3$ as we need. We have used four, one more than the number of digits in the density (19.3 g/cm^3). Applying our conversion factors, we can now solve the problem:

$$\text{Mass in grams} = (2.00\ \cancel{\text{in.}^3})\left(\frac{16.39\ \cancel{\text{cm}^3}}{1\ \cancel{\text{in.}^3}}\right)\left(\frac{19.3\ \text{g}}{1\ \cancel{\text{cm}^3}}\right) = 633\ \text{g}$$

The final answer is reported to three significant figures, the same number of significant figures as is in 2.00 and 19.3.

SAMPLE EXERCISE 1.11

What is the mass in grams of 1.00 gal of water? The density of water is 1.00 g/mL.

Solution Before we begin solving this exercise, we note the following:

1. We are given 1.00 gal of water.
2. We wish to obtain the mass in grams.
3. We have the following conversion factors either given, commonly known, or available on the back inside cover of the text:

$$\frac{1.00\ \text{g water}}{1\ \text{mL water}} \quad \frac{1\ \text{L}}{1000\ \text{mL}} \quad \frac{1\ \text{L}}{1.057\ \text{qt}} \quad \frac{1\ \text{gal}}{4\ \text{qt}}$$

The first of these conversion factors must be used as written (with grams in the numerator) to give the desired result, whereas the last conversion factor must be inverted in order to cancel gallons. The solution is given by

$$\text{Mass in grams} = (1.00\ \cancel{\text{gal}})\left(\frac{4\ \cancel{\text{qt}}}{1\ \cancel{\text{gal}}}\right)\left(\frac{1\ \cancel{\text{L}}}{1.057\ \cancel{\text{qt}}}\right)\left(\frac{1000\ \cancel{\text{mL}}}{1\ \cancel{\text{L}}}\right)\left(\frac{1.00\ \text{g}}{1\ \cancel{\text{mL}}}\right)$$

$$= 3.78 \times 10^3\ \text{g water}$$

The units of our final answer are appropriate, and we've also taken care of our significant figures. We can further check our calculation by the estimation procedure. We can round 1.057 off to 1. Focusing on the numbers that don't equal 1 then gives merely $4 \times 1000 = 4000$ g, in agreement with the detailed calculation.

PRACTICE EXERCISE

(a) Calculate the mass of 1.00 qt of benzene if it has a density of 0.879 g/mL.
(b) If the volume of an object is reported as 5.0 ft^3, what is the volume in cubic meters?
Answers: **(a)** 832 g; **(b)** 0.14 m^3

Strategies in Chemistry The Importance of Practice

If you've ever played a musical instrument or participated in athletics, you know that the keys to success are practice and discipline. You can't learn to play a piano merely by listening to music, and you can't learn how to play basketball merely by watching games on television. Likewise, you can't learn chemistry by merely watching your instructor do it. Simply reading this book, listening to lectures, or reviewing notes will not usually be sufficient when exam time comes around. Your task is not merely to understand how someone else uses chemistry, but to be able to do it yourself. That takes practice on a regular basis, and anything that you have to do on a regular basis requires self-discipline until it becomes a habit.

Throughout the book, we have provided sample exercises in which the solutions are shown in detail. A practice exercise, for which only the answer is given, accompanies each sample exercise. It is important that you use these exercises as learning aids. End-of-chapter exercises provide additional questions to help you understand the material in the chapter. Red numbers indicate exercises for which answers are given at the back of the book. A review of basic mathematics is given in Appendix A.

The practice exercises in this text and the homework assignments given by your instructor provide the minimal practice that you will need to succeed in your chemistry course. Only by working all the assigned problems will you face the full range of difficulty and coverage that your instructor expects you to master for exams. There is no substitute for a determined and perhaps lengthy effort to work problems on your own. If you do get stuck on a problem, however, get help from your instructor, a teaching assistant, a tutor, or a fellow student. Spending an inordinate amount of time on a single exercise is rarely effective unless you know that it is particularly challenging and requires extensive thought and effort.

Summary and Key Terms

Introduction and Section 1.1 **Chemistry** is the study of the composition, structure, properties, and changes of **matter**. The composition of matter relates to the kinds of **elements** it contains. The structure of matter relates to the ways the **atoms** of these elements are arranged. A **molecule** is an entity composed of two or more atoms with the atoms attached to one another in a specific way.

Section 1.2 Matter exists in three physical states, **gas**, **liquid**, and **solid**, which are known as the **states of matter**. There are two kinds of **pure substances**: **elements** and **compounds**. Each element has a single kind of atom and is represented by a chemical symbol consisting of one or two letters, with the first letter capitalized. Compounds are composed of two or more elements joined chemically. The **law of constant composition**, also called the **law of definite proportions**, states that the elemental composition of a pure compound is always the same. Most matter consists of a mixture of substances. **Mixtures** have variable compositions and can be either homogeneous or heterogeneous; homogeneous mixtures are called **solutions**.

Section 1.3 Each substance has a unique set of **physical properties** and **chemical properties** that can be used to identify it. During a **physical change** matter does not change its composition. **Changes of state** are physical changes. In a **chemical change (chemical reaction)** a substance is transformed into a chemically different substance. **Intensive properties** are independent of the amount of matter examined and are used to identify substances. **Extensive properties** relate to the amount of substance present. Differences in physical and chemical properties are used to separate substances.

The **scientific method** is a dynamic process used to answer questions about our physical world. Observations and experiments lead to **scientific laws**, general rules that summarize how nature behaves. Observations also lead to tentative explanations or **hypotheses**. As a hypothesis is tested and refined, a **theory** may be developed.

Section 1.4 Measurements in chemistry are made using the **metric system**. Special emphasis is placed on a particular set of metric units called **SI units**, which are based on the meter, the kilogram, and the second as the basic units of length, **mass**, and time, respectively. The metric system employs a set of prefixes to indicate decimal fractions or multiples of the base units. The SI temperature scale is the **Kelvin scale**, although the **Celsius scale** is frequently used as well. **Density** is an important property that equals mass divided by volume.

Section 1.5 All measured quantities are inexact to some extent. The **precision** of a measurement indicates how closely different measurements of a quantity agree with one another. The **accuracy** of a measurement indicates how well a measurement agrees with the accepted or "true" value. The **significant figures** in a measured quantity include one estimated digit, the last digit of the measurement. The significant figures indicate the extent of the uncertainty of the measurement. Certain rules must be followed so that a calculation involving measured quantities is reported with the appropriate number of significant figures.

Section 1.6 In the **dimensional analysis** approach to problem solving, we keep track of units as we carry measurements through calculations. The units are multiplied together, divided into each other, or canceled like algebraic quantities. Obtaining the proper units for the final result is an important means of checking the method of calculation. When converting units and when carrying out several other types of problems, **conversion factors** can be used. These factors are ratios constructed from valid relations between equivalent quantities.

Exercises

Classification and Properties of Matter

1.1 Classify each of the following as a pure substance or a mixture; if a mixture, indicate whether it is homogeneous or heterogeneous: **(a)** rice pudding; **(b)** seawater; **(c)** magnesium; **(d)** gasoline.

1.2 Classify each of the following as a pure substance or a mixture; if a mixture, indicate whether it is homogeneous or heterogeneous: **(a)** air; **(b)** tomato juice; **(c)** iodine crystals; **(d)** sand.

1.3 Give the chemical symbols for the following elements: **(a)** aluminum; **(b)** sodium; **(c)** bromine; **(d)** copper; **(e)** silicon; **(f)** nitrogen; **(g)** magnesium; **(h)** helium.

1.4 Give the chemical symbol for each of the following elements: **(a)** carbon ; **(b)** potassium; **(c)** chlorine; **(d)** zinc; **(e)** phosphorus; **(f)** argon; **(g)** calcium; **(h)** silver.

1.5 Name the chemical elements represented by the following symbols: **(a)** H; **(b)** Mg; **(c)** Pb; **(d)** Si; **(e)** F; **(f)** Sn; **(g)** Mn; **(h)** As.

1.6 Name each of the following elements: **(a)** Cr; **(b)** I; **(c)** Li; **(d)** Se; **(e)** Pb; **(f)** V; **(g)** Hg; **(h)** Ga.

1.7 A solid white substance A is heated strongly in the absence of air. It decomposes to form a new white substance B and a gas C. The gas has exactly the same properties as the product obtained when carbon is burned in an excess of oxygen. Based on these observations, can we determine whether solids A and B and the gas C are elements or compounds? Explain your conclusions for each substance.

1.8 In 1807 the English chemist Humphry Davy passed an electric current through molten potassium hydroxide and isolated a bright, shiny reactive substance. He claimed the discovery of a new element, which he named potassium. In those days, before the advent of modern instruments, what was the basis on which one could claim that a substance was an element?

1.9 Make a drawing, like that in Figure 1.5, showing a homogeneous mixture of water vapor and argon gas (which occurs as argon atoms).

1.10 Make a drawing, like that in Figure 1.5, showing a heterogeneous mixture of aluminum metal (which is composed of aluminum atoms) and oxygen gas (which is composed of molecules containing two oxygen atoms each).

1.11 In the process of attempting to characterize a substance, a chemist makes the following observations: The substance is a silvery white, lustrous metal. It melts at 649°C and boils at 1105°C. Its density at 20°C is 1.738 g/cm^3. The substance burns in air, producing an intense white light. It reacts with chlorine to give a brittle white solid. The substance can be pounded into thin sheets or drawn into wires. It is a good conductor of electricity. Which of these characteristics are physical properties, and which are chemical properties?

1.12 Read the following description of the element zinc, and indicate which are physical properties and which are chemical properties. Zinc is a silver–gray-colored metal that melts at 420°C. When zinc granules are added to dilute sulfuric acid, hydrogen is given off and the metal dissolves. Zinc has a hardness on the Mohs scale of 2.5 and a density of 7.13 g/cm^3 at 25°C. It reacts slowly with oxygen gas at elevated temperatures to form zinc oxide, ZnO.

1.13 Label each of the following as either a physical process or a chemical process: **(a)** corrosion of aluminum metal; **(b)** melting of ice; **(c)** pulverizing an aspirin; **(d)** digesting a candy bar; **(e)** explosion of nitroglycerin.

1.14 A match is lit and held under a cold piece of metal. The following observations are made: **(a)** The match burns. **(b)** The metal gets warmer. **(c)** Water condenses on the metal. **(d)** Soot (carbon) is deposited on the metal. Which of these occurrences are due to physical changes, and which are due to chemical changes?

1.15 A beaker contains a clear, colorless liquid. If it is water, how could you determine whether it contained dissolved table salt? Do *not* taste it!

1.16 Suggest a method of separating each of the following mixtures into two components: **(a)** sugar and sand; **(b)** iron and sulfur.

Units and Measurement

1.17 What decimal power do the following abbreviations represent **(a)** d; **(b)** c; **(c)** f; **(d)** μ; **(e)** M; **(f)** k; **(g)** n; **(h)** m; **(i)** p?

1.18 Use appropriate metric prefixes to write the following measurements without use of exponents: **(a)** 6.5×10^{-6} m; **(b)** 6.35×10^{-4} L; **(c)** 2.5×10^{-3} L; **(d)** 4.23×10^{-9} m^3; **(e)** 12.5×10^{-8} kg; **(f)** 3.5×10^{-11} s; **(g)** 6.54×10^{9} fs.

1.19 Perform the following conversions: **(a)** 25.5 mg to g; **(b)** 4.0×10^{-10} m to nm; **(c)** 0.575 mm to μm.

1.20 Convert **(a)** 1.48×10^{2} kg to g; **(b)** 0.0023 μm to nm; **(c)** 7.25×10^{-4} s to ms.

1.21 Identify each of the following as measurements of length, area, volume, mass, density, time, or temperature: **(a)** 5 ns; **(b)** 5.5 kg/m^3; **(c)** 0.88 pm; **(d)** 540 km^2; **(e)** 173 K; **(f)** 2 mm^3; **(g)** 23°C.

1.22 What type of quantity (for example, length, volume, density) do the following units indicate: **(a)** mL; **(b)** cm^2; **(c)** mm^3; **(d)** mg/L; **(e)** ps; **(f)** nm; **(g)** K?

1.23 **(a)** A sample of carbon tetrachloride, a liquid once used in dry cleaning, has a mass of 39.73 g and a volume of 25.0 mL at 25°C. What is its density at this temperature? Will carbon tetrachloride float on water? (Materials that are less dense than water will float.) **(b)** The density of platinum is 21.45 g/cm^3 at 20°C. Calculate the mass of 75.00 cm^3 of platinum at this temperature. **(c)** The density of magnesium is 1.738 g/cm^3 at 20°C. What is the volume of 87.50 g of this metal at this temperature?

1.24 **(a)** A cube of osmium metal 1.500 cm on a side has a mass of 76.31 g at 25°C. What is its density in g/cm^3 at this temperature? **(b)** The density of titanium metal is 4.51 g/cm^3 at 25°C. What mass of titanium displaces 65.8 mL of water at 25°C? **(c)** The density of benzene at 15°C is 0.8787 g/mL. Calculate the mass of 0.1500 L of benzene at this temperature.

1.25 **(a)** To identify a liquid substance, a student determined its density. Using a graduated cylinder, she measured out a 45-mL sample of the substance. She then measured the mass of the sample, finding that it weighed 38.5 g. She knew that the substance had to be either isopropyl alcohol (density 0.785 g/mL) or toluene (density 0.866 g/mL). What is the calculated density and the probable identity of the substance? **(b)** An experiment requires 45.0 g of ethylene glycol, a liquid whose density is 1.114 g/mL. Rather than weigh the sample on a balance, a chemist chooses to dispense the liquid using a graduated cylinder. What volume of the liquid should he use? **(c)** A cubic piece of metal measures 5.00 cm on each edge. If the metal is nickel, whose density is 8.90 g/cm^3, what is the mass of the cube?

1.26 **(a)** After the label fell off a bottle containing a clear liquid believed to be benzene, a chemist measured the density of the liquid to verify its identity. A 25.0-mL portion of the liquid had a mass of 21.95 g. A chemistry handbook lists the density of benzene at 15°C as 0.8787 g/mL. Is the calculated density in agreement with the tabulated value? **(b)** An experiment requires 15.0 g of cyclohexane, whose density at 25°C is 0.7781 g/mL. What volume of cyclohexane should be used? **(c)** A spherical ball of lead has a diameter of 5.0 cm. What is the mass of the sphere if lead has a density of 11.34 g/cm^3? (The volume of a sphere is $\left(\frac{4}{3}\right)\pi r^3$.)

[1.27] Gold can be hammered into extremely thin sheets called gold leaf. If a 200-mg piece of gold (density = 19.32 g/cm^3) is hammered into a sheet measuring 2.4 × 1.0 ft, what is the average thickness of the sheet in meters? How might the thickness be expressed without exponential notation, using an appropriate metric prefix?

[1.28] A cylindrical rod formed from silicon is 16.8 cm long and has a mass of 2.17 kg. The density of silicon is 2.33 g/cm^3. What is the diameter of the cylinder? (The volume of a cylinder is given by $\pi r^2 h$, where r is the radius, and h is its length.)

1.29 Make the following conversions: **(a)** 62°F to °C; **(b)** 216.7°C to °F; **(c)** 233°C to K; **(d)** 315 K to °F; **(e)** 2500°F to K.

1.30 **(a)** The temperature on a warm summer day is 87°F. What is the temperature in °C? **(b)** The melting point of sodium bromide (a salt) is 755°C. What is this temperature in °F? **(c)** Toluene freezes at −95°C. What is its freezing point in kelvins and in degrees Fahrenheit? **(d)** Many scientific data are reported at 25°C. What is this temperature in kelvins and in degrees Fahrenheit? **(e)** Neon, the gaseous element used to make electronic signs, has a melting point of −248.6°C and a boiling point of −246.1°C. What are these temperatures in kelvins?

Uncertainty in Measurement

1.31 Indicate which of the following are exact numbers: **(a)** the mass of a paper clip; **(b)** the surface area of a dime; **(c)** the number of inches in a mile; **(d)** the number of ounces in a pound; **(e)** the number of microseconds in a week; **(f)** the number of pages in this book.

1.32 Indicate which of the following are exact numbers: **(a)** the mass of a 32-oz can of coffee; **(b)** the number of students in your chemistry class; **(c)** the temperature of the surface of the sun; **(d)** the mass of a postage stamp; **(e)** the number of milliliters in a cubic meter of water; **(f)** the average height of students in your school.

1.33 What is the length of the pencil in the following figure? How many significant figures are there in this measurement?

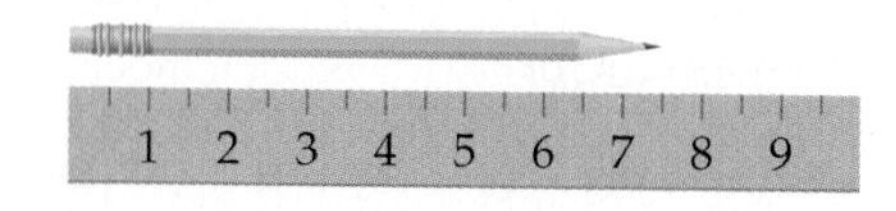

1.34 An oven thermometer with a circular scale is shown. What temperature does the scale indicate? How many significant figures are in the measurement?

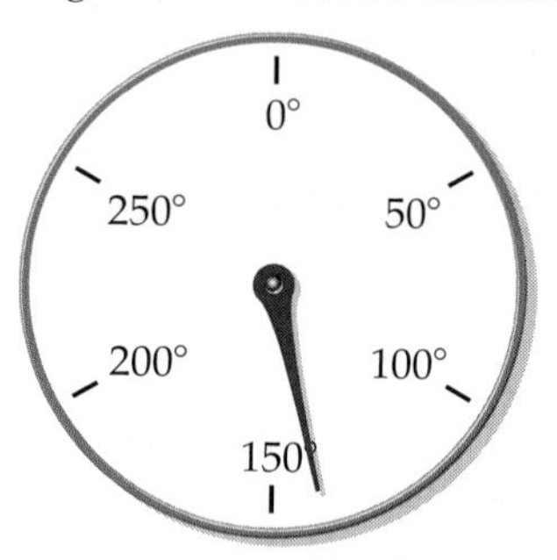

1.35 What is the number of significant figures in each of the following measured quantities? **(a)** 1282 kg; **(b)** 0.00296 s; **(c)** 8.070 mm; **(d)** 0.0105 L; **(e)** 9.7750×10^{-4} cm.

1.36 Indicate the number of significant figures in each of the following measured quantities: **(a)** 5.404×10^2 km; **(b)** 0.0234 m^2; **(c)** 5.500 cm; **(d)** 430.98 K; **(e)** 204.080 g.

1.37 Round each of the following numbers to four significant figures, and express the result in standard exponential notation: **(a)** 300.235800; **(b)** 456,500; **(c)** 0.006543210; **(d)** 0.000957830; **(e)** 50.778×10^3; **(f)** -0.035000.

1.38 Round each of the following numbers to three significant figures, and express the result in standard exponential notation: **(a)** 143,700; **(b)** 0.09750; **(c)** 890,000; **(d)** 6.764×10^4; **(e)** 33,987.22; **(f)** -6.5559.

1.39 Carry out the following operations, and express the answers with the appropriate number of significant figures: **(a)** $21.2405 + 5.80$; **(b)** $13.577 - 21.6$; **(c)** $(5.03 \times 10^{-4})(3.6675)$; **(d)** $0.05770/75.3$.

1.40 Carry out the following operations, and express the answer with the appropriate number of significant figures: **(a)** $320.55 - (6104.5/2.3)$; **(b)** $[(285.3 \times 10^5) - (1.200 \times 10^3)] \times 2.8954$; **(c)** $(0.0045 \times 20{,}000.0) + (2813 \times 12)$; **(d)** $863 \times [1255 - (3.45 \times 108)]$.

Dimensional Analysis

1.41 When you convert units, how do you decide which part of the conversion factor is in the numerator and which is in the denominator?

1.42 Using the information on the back inside cover, write down the conversion factors needed to convert: **(a)** mi to km; **(b)** oz to g; **(c)** qt to L.

1.43 Perform the following conversions: **(a)** 0.076 L to mL; **(b)** 5.0×10^{-8} m to nm; **(c)** 6.88×10^5 ns to s; **(d)** 1.55 kg/m^3 to g/L; (e) 5.850 gal/hr to L/s.

1.44 **(a)** The speed of light in a vacuum is 2.998×10^8 m/s. What is its speed in km/hr? **(b)** The oceans contain approximately 1.35×10^9 km^3 of water. What is this volume in liters? **(c)** An individual suffering from a high cholesterol level in her blood has 232 mg of cholesterol per 100 mL of blood. If the total blood volume of the individual is 5.2 L, how many grams of total blood cholesterol does the individual contain?

1.45 Perform the following conversions: **(a)** 5.00 days to s; **(b)** 0.0550 mi to m; **(c)** $1.89/gal to dollars per liter; **(d)** 0.510 in./ms to km/hr; **(e)** 22.50 gal/min to L/s; **(f)** 0.02500 ft^3 to cm^3.

1.46 Carry out the following conversions: **(a)** 145.7 ft to m; **(b)** 0.570 qt to mL; **(c)** 3.75 μm/s to km/hr; **(d)** 3.977 yd^3 to m^3; **(e)** $2.99/lb to dollars per kg; **(f)** 9.75 lb/ft^3 to g/mL.

1.47 **(a)** How many liters of wine can be held in a wine barrel whose capacity is 31 gal? **(b)** The recommended adult dose of Elixophyllin®, a drug used to treat asthma, is 6 mg/kg of body mass. Calculate the dose in milligrams for a 150-lb person. **(c)** If an automobile is able to travel 254 mi on 11.2 gal of gasoline, what is the gas mileage in km/L? **(d)** A pound of coffee beans yields 50 cups of coffee (4 cups = 1 qt). How many milliliters of coffee can be obtained from 1 g of coffee beans?

1.48 **(a)** If an electric car is capable of going 225 km on a single charge, how many charges will it need to travel from Boston, Massachusetts, to Miami, Florida, a distance of 1486 mi, assuming that the trip begins with a full charge? **(b)** If a migrating loon flies at an average speed of 14 m/s, what is its average speed in mi/hr? **(c)** What is the engine piston displacement in liters of an engine whose displacement is listed as 450 $in.^3$? **(d)** In March 1989, the *Exxon Valdez* ran aground and spilled 240,000 barrels of crude petroleum off the coast of Alaska. One barrel of petroleum is equal to 42 gal. How many liters of petroleum were spilled?

1.49 The density of air at ordinary atmospheric pressure and 25°C is 1.19 g/L. What is the mass, in kilograms, of the air in a room that measures $12.5 \times 15.5 \times 8.0$ ft?

1.50 The concentration of carbon monoxide in an urban apartment is 48 $\mu g/m^3$. What mass of carbon monoxide in grams is present in a room measuring $9.0 \times 14.5 \times 18.8$ ft?

1.51 A copper refinery produces a copper ingot weighing 150 lb. If the copper is drawn into wire whose diameter is 8.25 mm, how many feet of copper can be obtained from the ingot? The density of copper is 8.94 g/cm^3.

1.52 The Morgan silver dollar has a mass of 26.73 g. By law, it was required to contain 90% silver, with the remainder being copper. **(a)** When the coin was minted in the late 1800s, silver was worth $1.18 per troy ounce (31.1 g). At this price, what is the value of the silver in the silver dollar? **(b)** Today, silver sells for $5.30 per troy ounce. How many Morgan silver dollars are required to obtain $25.00 worth of pure silver?

1.53 By using estimation techniques, determine which of the following is the heaviest and which is the lightest: a 5-lb bag of potatoes, a 5-kg bag of sugar, or 1 gal of water (density = 1.0 g/mL)?

1.54 By using estimation techniques, arrange these items in order from shortest to longest: a 57-cm length of string, a 14-in. long shoe, and a 1.1-m length of pipe.

Additional Exercises

1.55 What is meant by the terms composition and structure when referring to matter?

1.56 Classify each of the following as a pure substance, a solution, or a heterogeneous mixture: a gold coin; a cup of coffee; a wood plank. What ambiguities are there in clearly determining the nature of the material from the description given?

1.57 **(a)** What is the difference between a hypothesis and a theory? **(b)** Explain the difference between a theory and a scientific law. Which addresses how matter behaves, and which addresses why it behaves that way?

1.58 A sample of ascorbic acid (vitamin C) is synthesized in the laboratory. It contains 1.50 g of carbon and 2.00 g of oxygen. Another sample of ascorbic acid isolated from citrus fruits contains 6.35 g of carbon. How many grams of oxygen does it contain? Which law are you assuming in answering this question?

1.59 Two students determine the percentage of lead in a sample as a laboratory exercise. The true percentage is

22.52%. The students' results for three determinations are as follows:

1. 22.52, 22.48, 22.54
2. 22.64, 22.58, 22.62

(a) Calculate the average percentage for each set of data, and tell which set is the more accurate based on the average. **(b)** Precision can be judged by examining the average of the deviations from the average value for that data set. (Calculate the average value for each data set, then calculate the average value of the absolute deviations of each measurement from the average.) Which set is more precise?

1.60 Is the use of significant figures in each of the following statements appropriate? Why or why not? **(a)** The 1976 circulation of *Reader's Digest* was 17,887,299. **(b)** There are more than 1.4 million people in the United States who have the surname Brown. **(c)** The average annual rainfall in San Diego, California, is 20.54 cm. **(d)** In Canada, between 1978 and 1992, the prevalence of obesity in men went from 6.8% to 12.0%.

1.61 Neon has a boiling point of −246.1°C. What is this temperature in kelvins? In °F?

1.62 Give the derived SI units for each of the following quantities in base SI units: **(a)** acceleration = distance/time^2; **(b)** force = mass × acceleration; **(c)** work = force × distance; **(d)** pressure = force/area; **(e)** power = work/time.

1.63 A 40-lb container of peat moss measures 14 × 20 × 30 in. A 40-lb container of topsoil has a volume of 1.9 gal. Calculate the average densities of peat moss and topsoil in units of g/cm^3. Would it be correct to say that peat moss is "lighter" than topsoil? Explain.

1.64 Small spheres of equal mass are made of lead (density = 11.3 g/cm^3), silver (10.5 g/cm^3), and aluminum (2.70 g/cm^3). Which sphere has the largest diameter and which has the smallest?

1.65 The liquid substances mercury (density = 13.5 g/mL), water (1.00 g/mL), and cyclohexane (0.778 g/mL) do not form a solution when mixed, but separate in distinct layers. Sketch how the liquids would position themselves in a test tube.

1.66 The annual production of sodium hydroxide in the United States in 1999 was 23.2 billion pounds. **(a)** How many grams of sodium hydroxide were produced in that year? **(b)** The density of sodium hydroxide is 2.130 g/cm^3. How many cubic kilometers were produced?

1.67 **(a)** You are given a bottle that contains 4.59 cm^3 of a metallic solid. The total mass of the bottle and solid is 35.66 g. The empty bottle weighs 14.23 g. What is the density of the solid? **(b)** Mercury is traded by the "flask," a unit that has a mass of 34.5 kg. What is the volume of a flask of mercury if the density of mercury is 13.6 g/mL? **(c)** An undergraduate student has the idea of removing a decorative stone sphere with a radius of 28.9 cm from in front of a campus building. If the density of the stone is 3.52 g/cm^3, what is the mass of the sphere? [The volume of a sphere is $V = (4/3)\pi r^3$.] Is he likely to be able to walk off with it unassisted?

[1.68] A 32.65-g sample of a solid is placed in a flask. Toluene, in which the solid is insoluble, is added to the flask so that the total volume of solid and liquid together is 50.00 mL. The solid and toluene together weigh 58.58 g. The density of toluene at the temperature of the experiment is 0.864 g/mL. What is the density of the solid?

[1.69] Suppose you decide to define your own temperature scale using the freezing point (−11.5°C) and boiling point (197.6°C) of ethylene glycol. If you set the freezing point as 0°G and the boiling point as 100°G, what is the freezing point of water on this new scale?

1.70 Recently, one of the text authors completed a half-marathon, a 13-mi, 192-yd road race, in a time of 1 hr, 44 min, and 18 s. (a) What was the runner's average speed in miles per hour? (b) What was the runner's pace in minutes and seconds per mile?

1.71 The distance from Earth to the Moon is approximately 240,000 mi. **(a)** What is this distance in meters? **(b)** The *Concorde SST* has an air speed of about 2400 km/hr. If the *Concorde* could fly to the Moon, how many seconds would it take?

1.72 The U.S. quarter has a mass of 5.67 g and is approximately 1.55 mm thick. **(a)** How many quarters would have to be stacked to reach 575 ft, the height of the Washington Monument? **(b)** How much would this stack weigh? **(c)** How much money would this stack contain? **(d)** In 1998 the national debt was $4.9 trillion. How many stacks like the one described would be necessary to pay off this debt?

1.73 In the United States water used for irrigation is measured in acre-feet. An acre-foot of water covers an acre to a depth of exactly 1 ft. An acre is 4840 yd^2. An acre-foot is enough water to supply two typical households for 1.00 yr. Desalinated water costs about $2480 per acre-foot. **(a)** How much does desalinated water cost per liter? **(b)** How much would it cost one household per day if it were the only source of water?

[1.74] A cylindrical container of radius r and height h has a volume of $\pi r^2 h$. **(a)** Calculate the volume in cubic centimeters of a cylinder with a radius of 3.55 cm and a height of 75.3 cm. **(b)** Calculate the volume in cubic meters of a cylinder whose height is 22.5 in. and whose diameter is 12.9 in. **(c)** Calculate the mass in kilograms of a volume of mercury equal to the volume of the cylinder in part (b). The density of mercury is 13.6 g/cm^3.

[1.75] A 15.0-cm long cylindrical glass tube, sealed at one end, is filled with ethanol. The mass of ethanol needed to fill the tube is found to be 11.86 g. The density of ethanol is 0.789 g/mL. Calculate the inner diameter of the tube in centimeters.

[1.76] Gold is alloyed (mixed) with other metals to increase its hardness in making jewelry. **(a)** Consider a piece of gold jewelry that weighs 9.85 g and has a volume of 0.675 cm^3. The jewelry contains only gold and silver, which have densities of 19.3 g/cm^3 and 10.5 g/cm^3, respectively. Assuming that the total volume of the jewelry is the sum of the volumes of the gold and silver that it contains, calculate the percentage of gold (by mass) in the jewelry. **(b)** The relative amount of gold in an alloy is commonly expressed in units of carats. Pure gold is 24-carat, and the percentage of gold in an alloy is given as a percentage of this value. For example, an alloy that is 50% gold is 12-carat. State the purity of the gold jewelry in carats.

[1.77] Suppose you are given a sample of a homogeneous liquid. What would you do to determine whether it is a solution or a pure substance?

[1.78] Chromatography (Figure 1.14) is a simple, but reliable, method for separating a mixture into its constituent substances. Suppose you are using chromatography to separate a mixture of two substances. How would you know whether the separation is successful? Can you propose a means of quantifying how good or how poor the separation is?

[1.79] You are assigned the task of separating a desired granular material, with a density of 3.62 g/cm^3, from an undesired granular material that has a density of 2.04 g/cm^3. You want to do this by shaking the mixture in a liquid in which the heavier material will fall to the bottom and the lighter material will float. A solid will float on any liquid that is more dense. Using a handbook of chemistry, find the densities of the following substances: carbon tetrachloride, hexane, benzene, and methylene iodide. Which of these liquids will serve your purpose, assuming no chemical interaction between the liquid and the solids?

[1.80] The concepts of accuracy and precision are not always easy to grasp. Here are two sets of studies: **(a)** The mass of a secondary weight standard is determined by weighing it on a very precise balance under carefully controlled laboratory conditions. The average of 18 different weight measurements is taken as the weight of the standard. **(b)** A group of 10,000 males between the ages of 50 and 55 is surveyed to ascertain a relationship between calorie intake and blood cholesterol level. The survey questionnaire is quite detailed, asking the respondents about what they eat, smoking and drinking habits, and so on. The results are reported as showing that for men of comparable lifestyles, there is a 40% chance of the blood cholesterol level being above 230 for those who consume more than 40 calories per gram of body weight per day, as compared with those who consume less than 30 calories per gram of body weight per day.

Discuss and compare these two studies in terms of the precision and accuracy of the result in each case. How do the two studies differ in nature in ways that affect the accuracy and precision of the results? What makes for high precision and accuracy in any given study? In each of these studies, what factors might not be controlled that could affect the accuracy and precision? What steps can be taken generally to attain higher precision and accuracy?

eMedia Exercises

1.81 Experiment with the **Phases of the Elements** activity (*eChapter 1.2*). **(a)** How many elements are liquids at room temperature and what are they? **(b)** Choose two temperatures—one higher and one lower than room temperature—and determine how many elements are liquids at those temperatures.

1.82 Watch the **Electrolysis of Water** movie (*eChapter 1.2*). **(a)** How can you tell from this experiment that water is a compound and not an element? **(b)** If you were to perform a similar experiment using liquid bromine instead of liquid water in the apparatus, what would you expect to happen?

1.83 The principle that oppositely charged particles attract one another and like charges repel one another is summarized in Coulomb's law. Try some experiments using the **Coulomb's Law** activity (*eChapter 1.3*) to get a feel for the magnitudes of attractive and repulsive forces between charged particles. **(a)** Between which particles is the attractive force the stronger: a particle with a charge of -2 at a distance of 3 Å from a particle with a charge of $+1$; or a particle with a charge of -1 at a distance of 2 Å from a particle with a charge of $+1$? **(b)** Consider a particle with a charge of $+3$ at a distance of 5 Å from a particle with a charge of -3. If there were another negatively charged particle in between the two, what would you expect to happen to the magnitude of the attractive force between them?

1.84 The **Changes of State** movie (*eChapter 1.3*) shows what happens to a solid when it is heated. **(a)** Describe the changes that occur. **(b)** Is the change from solid to liquid a chemical change or a physical change? **(c)** Is the change from liquid to gas a chemical change or a physical change? **(d)** Is enough information given to determine whether the original solid is an element, a compound, or a mixture? Explain.

1.85 **(a)** Use the **Significant Figures** activity (*eChapter 1.5*) to verify your answers to Exercises 1.39 and 1.40. **(b)** Is it possible for the sum of a column of numbers, each containing two significant figures, to have more than two significant figures? Explain. **(c)** How many significant figures should there be in the answer to the following calculation? $(35.2 - 30.1) \times 1.23 =$ ______.

Chapter 2

Atoms, Molecules, and Ions

Microdroplets emerge from the sample nozzle of an electrospray mass spectrometer. The electrospray technique can be used to obtain the mass spectrum of very large molecules, such as proteins.

WE SAW IN Chapter 1 that chemistry is concerned with the properties of materials. The materials in our world exhibit a striking and seemingly infinite variety of properties, including different colors, textures, solubilities, and chemical reactivities. When we see that diamonds are transparent and hard, table salt is brittle and dissolves in water, gold conducts electricity and can be hammered into thin sheets, and nitroglycerin is explosive, we are making observations in the *macroscopic* world, the world of our everyday senses. In chemistry we seek to understand and explain these properties in the *submicroscopic* world, the world of atoms and molecules.

The diversity of chemical behavior results from only about 100 different elements and, thus, only 100 different kinds of atoms. In a sense, the atoms are like the 26 letters of the alphabet that join together in different combinations to form the immense number of words in our language. But how do the atoms combine with each other? What rules govern the ways in which they can combine? How do the properties of a substance relate to the kinds of atoms it contains? Indeed, what is an atom like, and what makes the atoms of one element different from those of another?

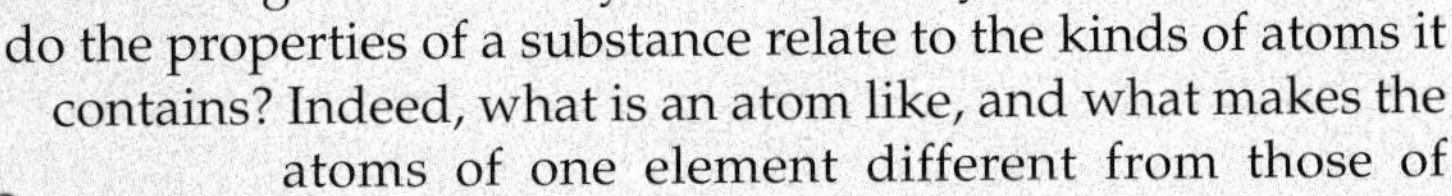

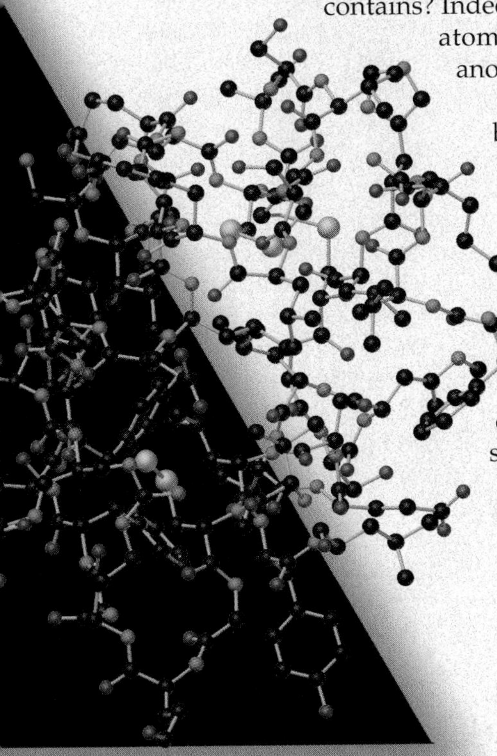

The submicroscopic view of matter forms the basis for understanding why elements and compounds react in the ways they do and why they exhibit specific physical and chemical properties. In this chapter we begin to explore the fascinating world of atoms and molecules. We will examine the basic structure of the atom and briefly discuss the formation of molecules and ions. We will also introduce the systematic procedures used to name compounds. Our discussions in this chapter provide the foundation for exploring chemistry more deeply in later chapters.

▶ What's Ahead ◀

- We begin our discussion by providing a brief history of the notion that *atoms* are the smallest pieces of matter and Dalton's development of an atomic theory.
- Next we look in greater detail at some of the key experiments that led to the discovery of *electrons* and to the *nuclear model* of the atom.
- We then discuss the modern theory of atomic structure, including the ideas of *atomic numbers*, *mass numbers*, and *isotopes*.
- We introduce the concept of *atomic weights* and how they relate to the masses of individual atoms.
- Our discussion of atoms leads to the organization of the elements into the *periodic table*, in which elements are put in order of increasing atomic number and grouped by chemical similarity.
- Our understanding of atoms allows us to discuss the assemblies of atoms called *molecules* and their *molecular formulas* and *empirical formulas*.
- We learn that atoms can gain or lose electrons to form *ions*, and we look at how to use the periodic table to predict the charges on ions and the empirical formulas of *ionic compounds*.
- We will see the systematic way in which substances are named, called *nomenclature*, and how it is applied to inorganic compounds.
- Finally, we introduce some of the basic ideas of *organic chemistry*, which is the chemistry of the element carbon.

2.1 The Atomic Theory of Matter

The world around us is made of many different materials, some living, some inanimate. Moreover, matter often changes from one chemical form to another. In efforts to explain these observations, philosophers from the earliest times have speculated about the nature of the fundamental "stuff" from which the world is made. Democritus (460–370 BC) and other early Greek philosophers thought that the material world must be made up of tiny indivisible particles that they called *atomos*, meaning "indivisible." Later, Plato and Aristotle formulated the notion that there can be no ultimately indivisible particles. The "atomic" view of matter faded for many centuries during which Aristotelean philosophy dominated Western culture.

▲ **Figure 2.1** John Dalton (1766–1844) was the son of a poor English weaver. Dalton began teaching at the age of 12. He spent most of his years in Manchester, where he taught both grammar school and college. His lifelong interest in meteorology led him to study gases and hence to chemistry and eventually to the atomic theory.

The notion of atoms reemerged in Europe during the seventeenth century when scientists tried to explain the properties of gases. Air is composed of something invisible and in constant motion; we can feel the motion of the wind against us, for example. It is natural to think of tiny invisible particles as giving rise to these familiar effects. Isaac Newton, the most famous scientist of his time, favored the idea of atoms. But thinking of atoms in this sense is different from thinking of atoms as the ultimate *chemical* building blocks of nature. As chemists learned to measure the amounts of materials that reacted with one another to make new substances, the ground was laid for a chemical atomic theory. That theory came into being during the period 1803–1807 in the work of an English schoolteacher, John Dalton (Figure 2.1 ◀). Reasoning from a large number of observations, Dalton made the following postulates:

1. Each element is composed of extremely small particles called atoms.
2. All atoms of a given element are identical; the atoms of different elements are different and have different properties (including different masses).
3. Atoms of an element are not changed into different types of atoms by chemical reactions; atoms are neither created nor destroyed in chemical reactions.
4. Compounds are formed when atoms of more than one element combine; a given compound always has the same relative number and kind of atoms.

ACTIVITY
Postulates of Atomic Theory, Multiple Proportions

ANIMATION
Multiple Proportions

According to Dalton's atomic theory, **atoms** are the basic building blocks of matter. They are the smallest particles of an element that retain the chemical identity of the element. ∞ (Section 1.1) As noted in the postulates of Dalton's theory, an element is composed of only one kind of atom, whereas a compound contains atoms of two or more elements.

Dalton's theory explains several simple laws of chemical combination that were known in his time. One of these was the *law of constant composition* (Section 1.2): In a given compound the relative numbers and kinds of atoms are constant. This law is the basis of Dalton's Postulate 4. Another fundamental chemical law was the *law of conservation of mass* (also known as the *law of conservation of matter*): The total mass of materials present after a chemical reaction is the same as the total mass before the reaction. This law is the basis for Postulate 3. Dalton proposed that atoms always retain their identities and that during chemical reactions the atoms rearrange to give new chemical combinations.

A good theory should not only explain the known facts but should also predict new ones. Dalton used his theory to deduce the *law of multiple proportions*: If two elements A and B combine to form more than one compound, the masses of B that can combine with a given mass of A are in the ratio of small whole numbers. We can illustrate this law by considering the substances water and hydrogen peroxide, both of which consist of the elements hydrogen and oxygen. In forming water, 8.0 g of oxygen combines with 1.0 g of hydrogen. In hydrogen peroxide, there are 16.0 g of oxygen per 1.0 g of hydrogen. In other words, the ratio of the mass of oxygen per gram of hydrogen in the two compounds is 2:1. Using the atomic theory, we can conclude that hydrogen peroxide contains twice as many atoms of oxygen per hydrogen atom as does water.

2.2 The Discovery of Atomic Structure

Dalton reached his conclusion about atoms on the basis of chemical observations in the macroscopic world of the laboratory. Neither he nor those who followed him during the century after his work was published had direct evidence for the existence of atoms. Today, however, we can use powerful new instruments to measure the properties of individual atoms and even provide images of them (Figure 2.2 ▶).

As scientists began to develop methods for more detailed probing of the nature of matter, the atom, which was supposed to be indivisible, began to show signs of a more complex structure: We now know that the atom is composed of still smaller **subatomic particles**. Before we summarize the current model of atomic structure, we will briefly consider a few of the landmark discoveries that led to that model. We'll see that the atom is composed in part of electrically charged particles, some with a positive (+) charge and some with a negative (−) charge. As we discuss the development of our current model of the atom, keep in mind a simple statement of the behavior of charged particles with one another: *Particles with the same charge repel one another, whereas particles with unlike charges are attracted to one another.*

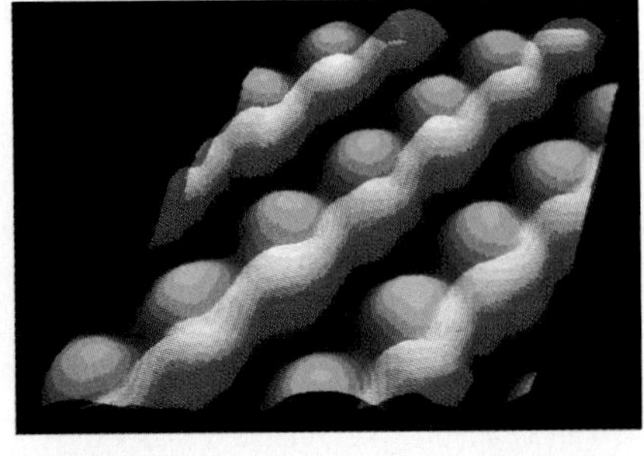

▲ **Figure 2.2** An image of the surface of the semiconductor GaAs (gallium arsenide) as obtained by a technique called tunneling electron microscopy. The color was added to the image by computer to distinguish the gallium atoms (blue spheres) from the arsenic atoms (red spheres).

Cathode Rays and Electrons

In the mid-1800s, scientists began to study electrical discharge through partially evacuated tubes (tubes that had been pumped almost empty of air), such as those shown in Figure 2.3 ▼. A high voltage produces radiation within the tube. This radiation became known as **cathode rays** because it originated from the negative electrode, or cathode. Although the rays themselves could not be seen, their movement could be detected because the rays cause certain materials, including glass, to *fluoresce*, or give off light. (Television picture tubes are cathode-ray tubes; a television picture is the result of fluorescence from the television screen.)

Scientists held differing views about the nature of the cathode rays. It was not initially clear whether the rays were a new form of radiation or rather consisted of an invisible stream of particles. Experiments showed that cathode rays were deflected by electric or magnetic fields, suggesting that the rays carried an electrical charge [Figure 2.3(c)]. The British scientist J. J. Thomson observed many properties of the rays, including the fact that the nature of the rays is the same regardless of the identity of the cathode material and that a metal plate exposed to cathode rays acquires a negative electrical charge. In a paper published in 1897 he summarized his observations and concluded that the cathode rays are streams of negatively charged particles with mass. Thomson's paper is generally accepted as the "discovery" of what became known as the *electron*.

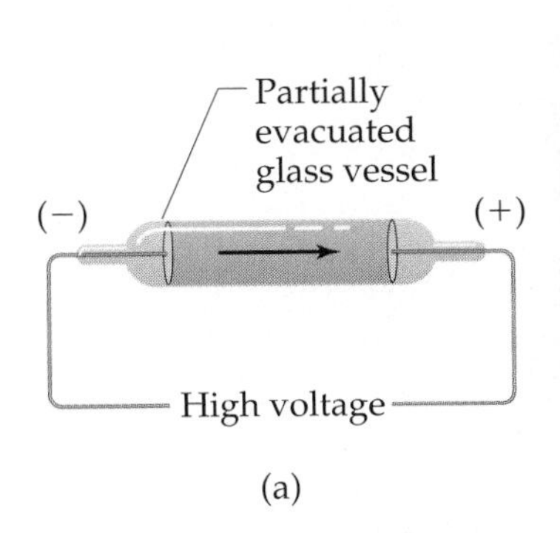

(a)

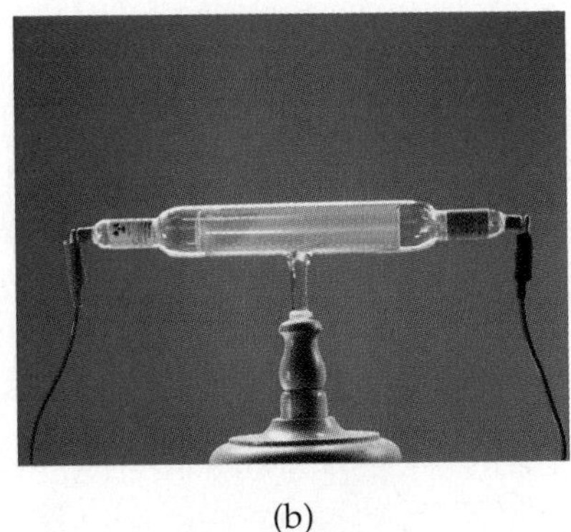

(b)

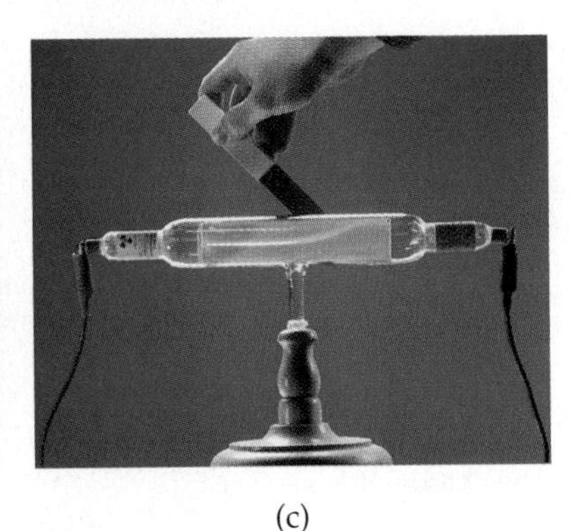

(c)

▲ **Figure 2.3** (a) In a cathode-ray tube, electrons move from the negative electrode (cathode) to the positive electrode (anode). (b) A photo of a cathode-ray tube containing a fluorescent screen to show the path of the cathode rays. (c) The path of the cathode rays is deflected by the presence of a magnet.

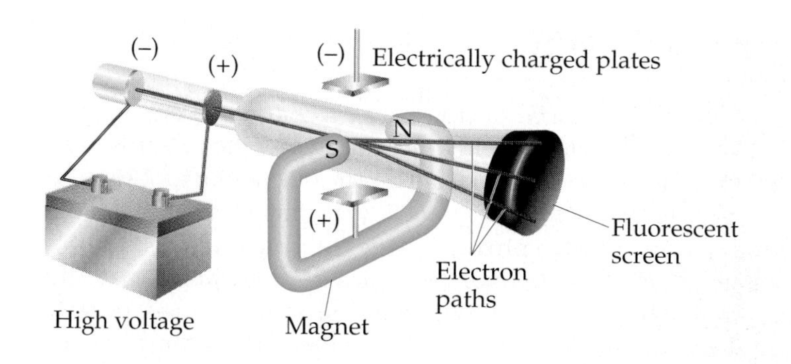

▶ **Figure 2.4** Cathode-ray tube with perpendicular magnetic and electric fields. The cathode rays (electrons) originate from the negative plate on the left and are accelerated toward the positive plate, which has a hole in its center. A beam of electrons passes through the hole and is then deflected by the magnetic and electric fields. The charge-to-mass ratio of the electron can be determined by measuring the effects of the magnetic and electric fields on the direction of the beam.

Thomson constructed a cathode-ray tube with a fluorescent screen, such as that shown in Figure 2.4 ▲, so that he could quantitatively measure the effects of electric and magnetic fields on the thin stream of electrons passing through a hole in the positively charged electrode. These measurements made it possible to calculate a value of 1.76×10^8 coulombs per gram for the ratio of the electron's electrical charge to its mass.*

Once the charge-to-mass ratio of the electron was known, measuring either the charge or the mass of an electron would also yield the value of the other quantity. In 1909 Robert Millikan (1868–1953) of the University of Chicago succeeded in measuring the charge of an electron by performing what is known as the "Millikan oil-drop experiment" (Figure 2.5 ▼). He could then calculate the mass of the electron by using his experimental value for the charge, 1.60×10^{-19} C, and Thomson's charge-to-mass ratio, 1.76×10^8 C/g:

ANIMATION
Millikan Oil Drop Experiment

$$\text{Electron mass} = \frac{1.60 \times 10^{-19}\ \text{C}}{1.76 \times 10^{8}\ \text{C/g}} = 9.10 \times 10^{-28}\ \text{g}$$

Using slightly more accurate values, the presently accepted value for the mass of the electron is 9.10939×10^{-28} g. This mass is about 2000 times smaller than that of hydrogen, the lightest atom.

Radioactivity

In 1896 the French scientist Henri Becquerel (1852–1908) was studying a uranium mineral called *pitchblende*, when he discovered that it spontaneously emits high-

* The coulomb (C) is the SI unit for electrical charge.

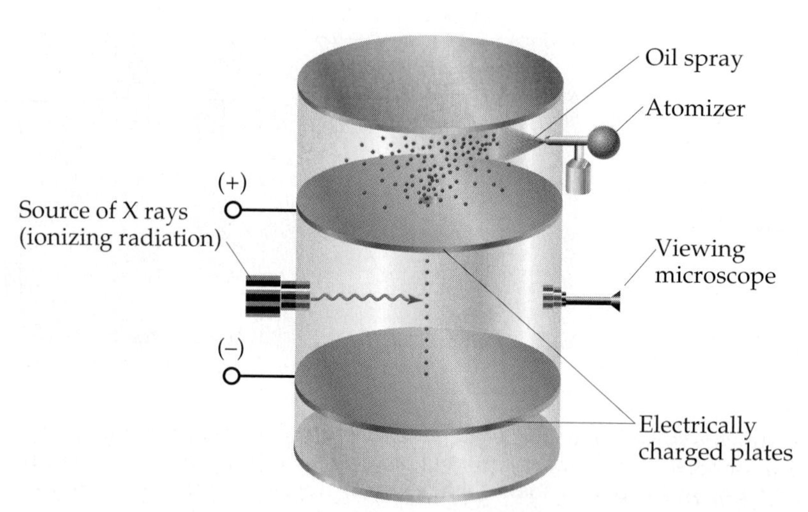

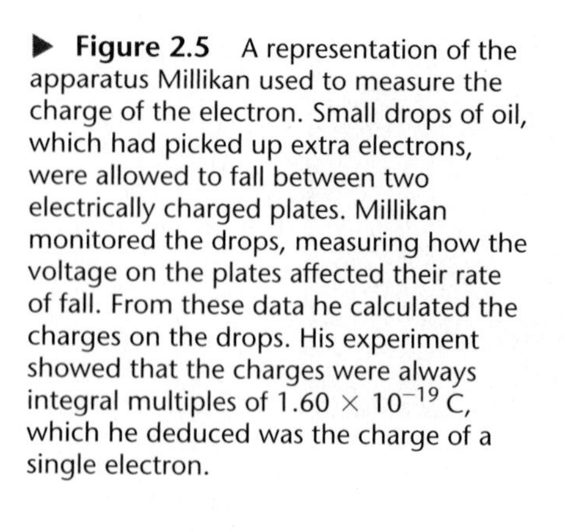

▶ **Figure 2.5** A representation of the apparatus Millikan used to measure the charge of the electron. Small drops of oil, which had picked up extra electrons, were allowed to fall between two electrically charged plates. Millikan monitored the drops, measuring how the voltage on the plates affected their rate of fall. From these data he calculated the charges on the drops. His experiment showed that the charges were always integral multiples of 1.60×10^{-19} C, which he deduced was the charge of a single electron.

◀ **Figure 2.6** Marie Sklodowska Curie (1867–1934). When M. Curie presented her doctoral thesis, it was described as the greatest single contribution of any doctoral thesis in the history of science. Among other things, two new elements, polonium and radium, had been discovered. In 1903 Henri Becquerel, M. Curie, and her husband, Pierre, were jointly awarded the Nobel Prize in physics. In 1911 M. Curie won a second Nobel Prize, this time in chemistry.

ANIMATION
Separation of Alpha, Beta, and Gamma Rays

▲ **Figure 2.7** Ernest Rutherford (1871–1937), whom Einstein called "the second Newton," was born and educated in New Zealand. In 1895 he was the first overseas student ever to be awarded a position at the Cavendish Laboratory at Cambridge University in England, where he worked with J. J. Thomson. In 1898 he joined the faculty of McGill University in Montreal. While at McGill, Rutherford did his research on radioactivity that led to his being awarded the 1908 Nobel Prize in chemistry. In 1907 Rutherford moved back to England to be a faculty member at Manchester University, where in 1910 he performed his famous α-particle scattering experiments that led to the nuclear model of the atom. In 1992 his native New Zealand honored Rutherford by putting his likeness, along with his Nobel Prize medal, on their $100 currency note.

energy radiation. This spontaneous emission of radiation is called **radioactivity**. At Becquerel's suggestion Marie Curie (Figure 2.6 ▲) and her husband, Pierre, began experiments to isolate the radioactive components of the mineral.

Further study of the nature of radioactivity, principally by the British scientist Ernest Rutherford (Figure 2.7 ▶), revealed three types of radiation: alpha (α), beta (β), and gamma (γ) radiation. Each type differs in its response to an electric field, as shown in Figure 2.8 ▼. The paths of both α and β radiation are bent by the electric field, although in opposite directions, whereas γ radiation is unaffected.

Rutherford showed that both α and β rays consist of fast-moving particles, which were called α and β particles. In fact, β particles are high-speed electrons and can be considered the radioactive equivalent of cathode rays. They are therefore attracted to a positively charged plate. The α particles are much more massive than the β particles and have a positive charge. They are thus attracted toward a negative plate. In units of the charge of the electron, β particles have a charge of $1-$, and α particles a charge of $2+$. Rutherford showed further that α particles combine with electrons to form atoms of helium. He thus concluded that an α particle consists of the positively charged core of the helium atom. He further concluded that γ radiation is high-energy radiation similar to X rays; it does not consist of particles and carries no charge. We will discuss radioactivity in greater detail in Chapter 21.

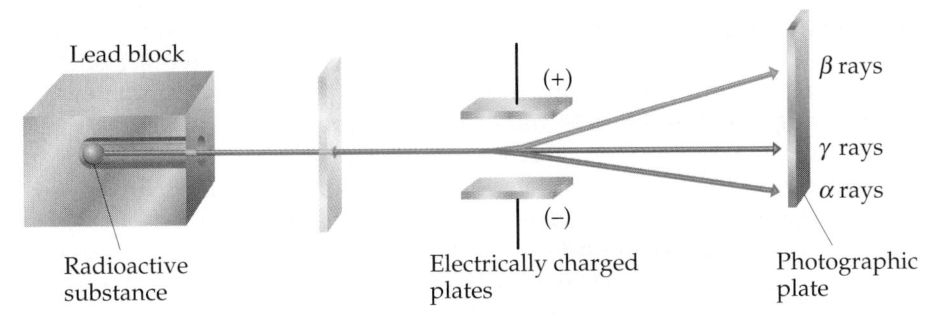

▲ **Figure 2.8** Behavior of alpha (α), beta (β), and gamma (γ) rays in an electric field.

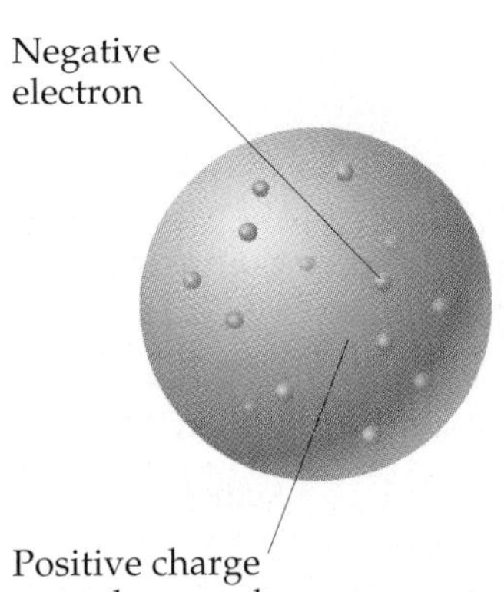

▲ **Figure 2.9** J. J. Thomson's "plum-pudding" model of the atom. He pictured the small electrons to be embedded in the atom much like raisins in a pudding or like seeds in a watermelon. Ernest Rutherford proved this model wrong.

The Nuclear Atom

With the growing evidence that the atom is composed of even smaller particles, attention was given to how the particles fit together. In the early 1900s Thomson reasoned that because electrons comprise only a very small fraction of the mass of an atom, they probably were responsible for an equally small fraction of the atom's size. He proposed that the atom consisted of a uniform positive sphere of matter in which the electrons were embedded, as shown in Figure 2.9 ▶.

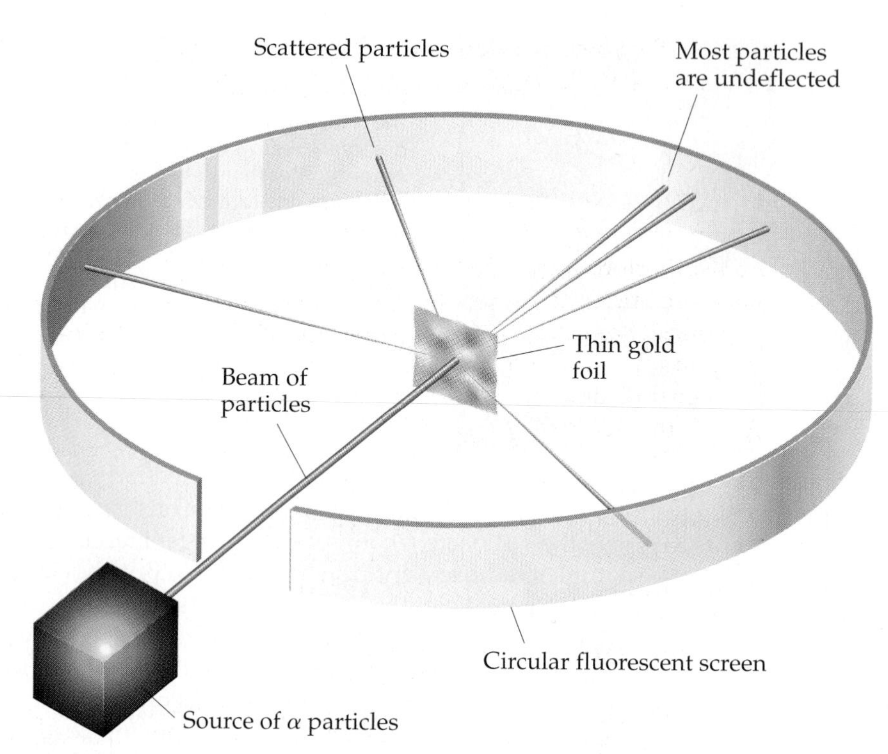

▲ **Figure 2.10** Rutherford's experiment on the scattering of α particles.

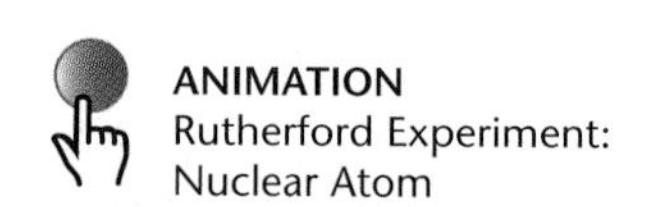

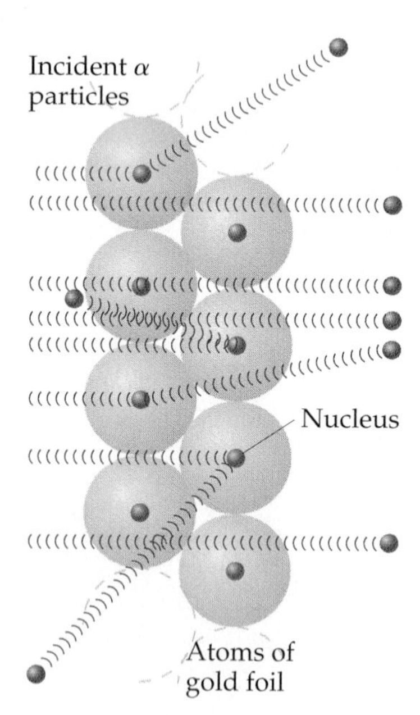

▲ **Figure 2.11** Rutherford's model explaining the scattering of α particles (Figure 2.10). The gold foil is several thousand atoms thick. When an α particle collides with (or passes very close to) a gold nucleus, it is strongly repelled. The less massive α particle is deflected from its path by this repulsive interaction.

This so-called "plum-pudding" model, named after a traditional English dessert, was very short-lived.

In 1910 Rutherford and his coworkers performed an experiment that disproved Thomson's model. Rutherford was studying the angles at which α particles were scattered as they passed through a thin gold foil a few thousand atomic layers in thickness (Figure 2.10 ▲). He and his coworkers discovered that almost all the α particles passed directly through the foil without deflection. A small percentage were found to be slightly deflected, on the order of 1 degree, consistent with Thomson's atomic model. Just for the sake of completeness, Rutherford suggested that Ernest Marsden, an undergraduate student working in the laboratory, look hard for evidence of scattering at large angles. To everyone's complete surprise, a small amount of scattering at large angles was observed. Some particles were even reflected back in the direction from which they had come. The explanation for these results was not immediately obvious, but they were clearly inconsistent with Thomson's "plum-pudding" model.

By 1911 Rutherford was able to explain these observations; he postulated that most of the mass of the atom and all of its positive charge reside in a very small, extremely dense region, which he called the **nucleus**. Most of the total volume of the atom is empty space in which electrons move around the nucleus. In the α-scattering experiment most α particles pass directly through the foil because they do not encounter the minute nucleus; they merely pass through the empty space of the atom. Occasionally, however, an α particle comes into the close vicinity of a gold nucleus. The repulsion between the highly charged gold nucleus and the α particle is strong enough to deflect the less massive α particle, as shown in Figure 2.11 ◄.

Subsequent experimental studies led to the discovery of both positive particles (*protons*) and neutral particles (*neutrons*) in the nucleus. Protons were discovered in 1919 by Rutherford. Neutrons were discovered in 1932 by the British scientist James Chadwick (1891–1972). We examine these particles more closely in Section 2.3.

2.3 The Modern View of Atomic Structure

Since the time of Rutherford, physicists have learned much about the detailed composition of atomic nuclei. In the course of these discoveries the list of particles that make up nuclei has grown long and continues to increase. As chemists, we can take a very simple view of the atom because only three subatomic particles—the **proton**, **neutron**, and **electron**—have a bearing on chemical behavior.

The charge of an electron is -1.602×10^{-19} C, and that of a proton is $+1.602 \times 10^{-19}$ C. The quantity 1.602×10^{-19} C is called the **electronic charge**. For convenience, the charges of atomic and subatomic particles are usually expressed as multiples of this charge rather than in coulombs. Thus, the charge of the electron is $1-$, and that of the proton is $1+$. Neutrons are uncharged and are therefore electrically neutral (which is how they received their name). *Atoms have an equal number of electrons and protons, so they have no net electrical charge.*

Protons and neutrons reside together in the nucleus of the atom, which, as Rutherford proposed, is extremely small. The vast majority of an atom's volume is the space in which the electrons reside. The electrons are attracted to the protons in the nucleus by the force that exists between particles of opposite electrical charge. In later chapters we will see that the strength of the attractive forces between electrons and nuclei can be used to explain many of the differences between different elements.

Atoms have extremely small masses. The mass of the heaviest known atom, for example, is on the order of 4×10^{-22} g. Because it would be cumbersome to express such small masses in grams, we use instead the **atomic mass unit**, or amu.* One amu equals 1.66054×10^{-24} g. The masses of the proton and neutron are very nearly equal, and both are much greater than that of an electron: A proton has a mass of 1.0073 amu, a neutron 1.0087 amu, and an electron 5.486×10^{-4} amu. It would take 1836 electrons to equal the mass of 1 proton, so the nucleus contains most of the mass of an atom. Table 2.1 ▼ summarizes the charges and masses of the subatomic particles. We will have more to say about atomic masses in Section 2.4.

Atoms are also extremely small. Most atoms have diameters between 1×10^{-10} m and 5×10^{-10} m, or 100–500 pm. A convenient, although non-SI, unit of length used to express atomic dimensions is the **angstrom** (Å). One angstrom equals 10^{-10} m. Thus, atoms have diameters on the order of 1–5 Å. The diameter of a chlorine atom, for example, is 200 pm, or 2.0 Å. Both picometers and angstroms are commonly used to express the dimensions of atoms and molecules.

* The SI abbreviation for the atomic mass unit is merely u. We will use the more common abbreviation amu.

TABLE 2.1 Comparison of the Proton, Neutron, and Electron

Particle	Charge	Mass (amu)
Proton	Positive (1+)	1.0073
Neutron	None (neutral)	1.0087
Electron	Negative (1−)	5.486×10^{-4}

Sample Exercise 2.1 illustrates further how very small atoms are compared with more familiar objects.

SAMPLE EXERCISE 2.1

The diameter of a U.S. penny is 19 mm. The diameter of a silver atom, by comparison, is only 2.88 Å. How many silver atoms could be arranged side by side in a straight line across the diameter of a penny?

Solution The unknown is the number of silver (Ag) atoms. We use the relationship 1 Ag atom = 2.88 Å as a conversion factor relating the number of atoms and distance. Thus, we can start with the diameter of the penny, first converting this distance into angstroms and then using the diameter of the Ag atom to convert distance to the number of Ag atoms:

$$\text{Ag atoms} = (19\ \cancel{\text{mm}})\left(\frac{10^{-3}\ \cancel{\text{m}}}{1\ \cancel{\text{mm}}}\right)\left(\frac{1\ \cancel{\text{Å}}}{10^{-10}\ \cancel{\text{m}}}\right)\left(\frac{1\ \text{Ag atom}}{2.88\ \cancel{\text{Å}}}\right) = 6.6 \times 10^{7}\ \text{Ag atoms}$$

That is, 66 million silver atoms could sit side by side across a penny!

PRACTICE EXERCISE

The diameter of a carbon atom is 1.54 Å. **(a)** Express this diameter in picometers. **(b)** How many carbon atoms could be aligned side by side in a straight line across the width of a pencil line that is 0.20 mm wide?
Answers: **(a)** 154 pm; **(b)** 1.3×10^{6} C atoms

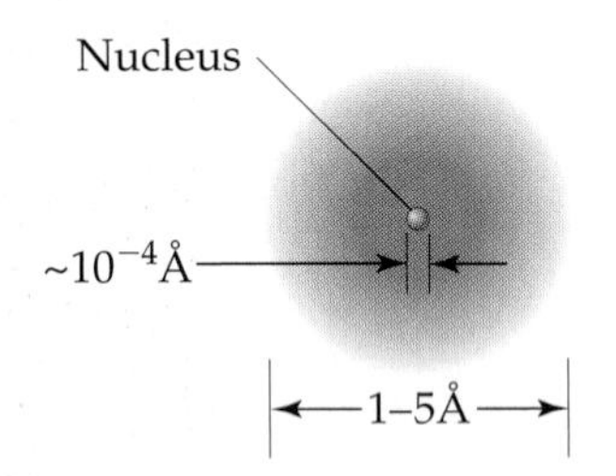

▲ **Figure 2.12** Schematic cross-sectional view through the center of an atom. The nucleus, which contains protons and neutrons, is the location of virtually all the mass of the atom. The rest of the atom is the space in which the light, negatively charged electrons reside.

The diameters of atomic nuclei are on the order of 10^{-4} Å, only a small fraction of the diameter of the atom as a whole. You can appreciate the relative sizes of the atom and its nucleus by imagining that if the atom were as large as a football stadium, the nucleus would be the size of a small marble. Because the tiny nucleus carries most of the mass of the atom in such a small volume, it has an incredible density—on the order of 10^{13}–10^{14} g/cm^3. A matchbox full of material of such density would weigh over 2.5 billion tons! Astrophysicists have suggested that the interior of a collapsed star may approach this density.

An illustration of the atom that incorporates the features we have just discussed is shown in Figure 2.12 ◀. The electrons, which take up most of the volume of the atom, play the major role in chemical reactions. The significance of representing the region containing the electrons as an indistinct cloud will become clear in later chapters when we consider the energies and spatial arrangements of the electrons.

A Closer Look Basic Forces

There are four basic forces, or interactions, known in nature: gravity, electromagnetism, the strong nuclear forces, and the weak nuclear forces. *Gravitational forces* are attractive forces that act between all objects in proportion to their masses. Gravitational forces between atoms or subatomic particles are so small that they are of no chemical significance.

Electromagnetic forces are attractive or repulsive forces that act between electrically charged or magnetic objects. Electric and magnetic forces are intimately related. Electric forces are of fundamental importance in understanding the chemical behavior of atoms. The magnitude of the electric force between two charged particles is given by *Coulomb's law*: $F = kQ_1Q_2/d^2$, where Q_1 and Q_2 are the magnitudes of the charges on the two particles, d is the distance between their centers, and k is a constant determined by the units for Q and d. A negative value for the force indicates attraction, whereas a positive value indicates repulsion.

All nuclei except those of hydrogen atoms contain two or more protons. Because like charges repel, electrical repulsion would cause the protons to fly apart if a stronger attractive force did not keep them together. This force is called the *strong nuclear force*. It acts between subatomic particles, as in the nucleus. At this distance this force is stronger than the electric force, so the nucleus holds together. The *weak nuclear force* is weaker than the electric force but stronger than gravity. We are aware of its existence only because it shows itself in certain types of radioactivity.

Isotopes, Atomic Numbers, and Mass Numbers

What makes an atom of one element different from an atom of another element? *All atoms of an element have the same number of protons in the nucleus.* The specific number of protons is different for different elements. Furthermore, because an atom has no net electrical charge, the number of electrons in it must equal its number of protons. All atoms of the element carbon, for example, have six protons and six electrons. Most carbon atoms also have six neutrons, although some have more and some have less.

Atoms of a given element that differ in the number of neutrons, and consequently in mass, are called **isotopes**. The symbol ${}^{12}_{6}\text{C}$ or simply ${}^{12}\text{C}$ (read "carbon twelve," carbon-12) represents the carbon atom with six protons and six neutrons. The number of protons, which is called the **atomic number**, is shown by the subscript. The atomic number of each element is listed with the name and symbol of the element on the front inside cover of the text. Because all atoms of a given element have the same atomic number, the subscript is redundant and hence is usually omitted. The superscript is called the **mass number**; it is the total number of protons plus neutrons in the atom. Some carbon atoms, for example, contain six protons and eight neutrons and are consequently represented as ${}^{14}\text{C}$ (read "carbon fourteen"). Several isotopes of carbon are listed in Table 2.2 ▼.

We will generally use the notation with subscripts and superscripts only when referring to a particular isotope of an element. An atom of a specific isotope is called a **nuclide**. Thus, an atom of ${}^{14}\text{C}$ is referred to as a ${}^{14}\text{C}$ nuclide.

All atoms are made up of protons, neutrons, and electrons. Because these particles are the same in all atoms, the difference between atoms of distinct elements (gold and oxygen, for example) is due entirely to the difference in the number of subatomic particles in each atom. We can therefore consider an atom to be the smallest sample of an element because breaking an atom into subatomic particles destroys its identity.

SAMPLE EXERCISE 2.2

How many protons, neutrons, and electrons are in an atom of ${}^{197}\text{Au}$?

Solution The superscript 197 is the mass number, the sum of the numbers of protons and neutrons. According to the list of elements given on the front inside cover of this text, gold has an atomic number of 79. Consequently, an atom of ${}^{197}\text{Au}$ has 79 protons, 79 electrons, and $197 - 79 = 118$ neutrons.

PRACTICE EXERCISE

How many protons, neutrons, and electrons are in a ${}^{138}\text{Ba}$ atom?
Answer: 56 protons, 56 electrons, and 82 neutrons

TABLE 2.2 Some of the Isotopes of Carbon[a]

Symbol	Number of Protons	Number of Electrons	Number of Neutrons
${}^{11}\text{C}$	6	6	5
${}^{12}\text{C}$	6	6	6
${}^{13}\text{C}$	6	6	7
${}^{14}\text{C}$	6	6	8

[a] Almost 99% of the carbon found in nature is ${}^{12}\text{C}$.

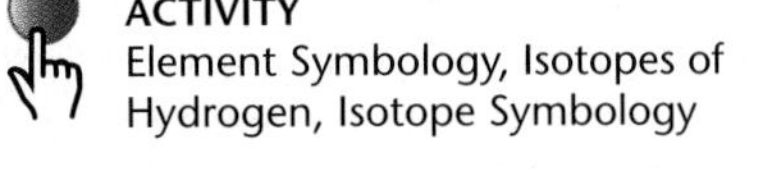

ACTIVITY
Element Symbology, Isotopes of Hydrogen, Isotope Symbology

SAMPLE EXERCISE 2.3

Magnesium has three isotopes, with mass numbers 24, 25, and 26. **(a)** Write the complete chemical symbol for each of them. **(b)** How many neutrons are in a nuclide of each isotope?

Solution **(a)** Magnesium has atomic number 12, so all atoms of magnesium contain 12 protons and 12 electrons. The three isotopes are therefore represented by $^{24}_{12}Mg$, $^{25}_{12}Mg$, and $^{26}_{12}Mg$. **(b)** The number of neutrons in each isotope is the mass number minus the number of protons. The number of neutrons in a nuclide of each isotope is therefore 12, 13, and 14, respectively.

PRACTICE EXERCISE

Give the complete chemical symbol for the nuclide that contains 82 protons, 82 electrons, and 126 neutrons.
Answer: $^{208}_{82}Pb$

2.4 Atomic Weights

Atoms are small pieces of matter, so they have mass. As noted in Section 2.1, a key postulate of Dalton's atomic theory is that mass is conserved during chemical reactions. Much of what we know about chemical reactions and the behavior of substances, therefore, has been derived by accurate measurements of the masses of atoms and molecules (and macroscopic collections of atoms and molecules) that are undergoing change. Chances are that you are already using mass measurements in the laboratory portion of your course in order to monitor changes that occur in chemical reactions. In this section we will discuss the mass scale that is used for atoms and introduce the concept of *atomic weights*. In Section 3.3 we will extend these concepts to show how these atomic masses are used to determine the masses of compounds and *molecular weights*.

The Atomic Mass Scale

Although scientists of the nineteenth century knew nothing about subatomic particles, they were aware that atoms of different elements have different masses. They found, for example, that each 100.0 g of water contains 11.1 g of hydrogen and 88.9 g of oxygen. Thus, water contains $88.9/11.1 = 8$ times as much oxygen, by mass, as hydrogen. Once scientists understood that water contains two hydrogen atoms for each oxygen, they concluded that an oxygen atom must have $2 \times 8 = 16$ times as much mass as a hydrogen atom. Hydrogen, the lightest atom, was arbitrarily assigned a relative mass of 1 (no units), and atomic masses of other elements were at first determined relative to this value. Thus, oxygen was assigned an atomic mass of 16.

Today we can determine the masses of individual atoms with a high degree of accuracy. For example, we know that the ^{1}H atom has a mass of 1.6735×10^{-24} g and the ^{16}O atom has a mass of 2.6560×10^{-23} g. As we noted in Section 2.3, it is convenient to use the *atomic mass unit* (amu) when dealing with these extremely small masses:

$$1 \text{ amu} = 1.66054 \times 10^{-24} \text{ g} \quad \text{and} \quad 1 \text{ g} = 6.02214 \times 10^{23} \text{ amu}$$

The amu is presently defined by assigning a mass of exactly 12 amu to an atom of the ^{12}C isotope of carbon. In these units the mass of the ^{1}H nuclide is 1.0078 amu and that of the ^{16}O nuclide is 15.9949 amu.

Average Atomic Masses

Most elements occur in nature as mixtures of isotopes. We can determine the *average atomic mass* of an element by using the masses of its various isotopes and their relative abundances. Naturally occurring carbon, for example, is composed of 98.93% ^{12}C and 1.07% ^{13}C. The masses of these nuclides are 12 amu (exactly) and 13.00335 amu, respectively. We calculate the average atomic mass of carbon from the fractional abundance of each isotope and the mass of that isotope:

$$(0.9893)(12\text{ amu}) + (0.0107)(13.00335\text{ amu}) = 12.01\text{ amu}$$

The average atomic mass of each element (expressed in amu) is also known as its **atomic weight**. Although the term *average atomic mass* is more proper and the simpler term *atomic mass* is frequently used, the term *atomic weight* is most common. The atomic weights of the elements are listed both in the periodic table and in the table of elements, which are found inside the front cover of this text.

A Closer Look The Mass Spectrometer

The most direct and accurate means for determining atomic and molecular weights is provided by the **mass spectrometer** (Figure 2.13 ▼). A gaseous sample is introduced at *A* and bombarded by a stream of high-energy electrons at *B*. Collisions between the electrons and the atoms or molecules of the gas produce positive ions, mostly with a 1+ charge. These ions are accelerated toward a negatively-charged wire grid (*C*). After they pass through the grid, they encounter two slits that allow only a narrow beam of ions to pass. This beam then passes between the poles of a magnet, which deflects the ions into a curved path, much as electrons are deflected by a magnetic field (Figure 2.4). For ions with the same charge, the extent of deflection depends on mass—the more massive the ion, the less the deflection. The ions are thereby separated according to their masses. By changing the strength of the magnetic field or the accelerating voltage on the negatively charged grid, ions of varying masses can be selected to enter the detector at the end of the instrument.

A graph of the intensity of the signal from the detector versus the mass of the ion is called a *mass spectrum*. The mass spectrum of chlorine atoms, shown in Figure 2.14 ▼, reveals the presence of two isotopes. Analysis of a mass spectrum gives both the masses of the ions reaching the detector and their relative abundances. The abundances are obtained from the intensities of their signals. Knowing the atomic mass and the abundance of each isotope allows us to calculate the average atomic mass of an element, as shown in Sample Exercise 2.4.

Mass spectrometers are used extensively today to identify chemical compounds and analyze mixtures of substances. When a molecule loses electrons, it falls apart, forming an array of positively charged fragments. The mass spectrometer measures the masses of these fragments, producing a chemical "fingerprint" of the molecule and providing clues about how the atoms were connected together in the original molecule. Thus, a chemist might use this technique to determine the molecular structure of a newly synthesized compound or to identify a pollutant in the environment.

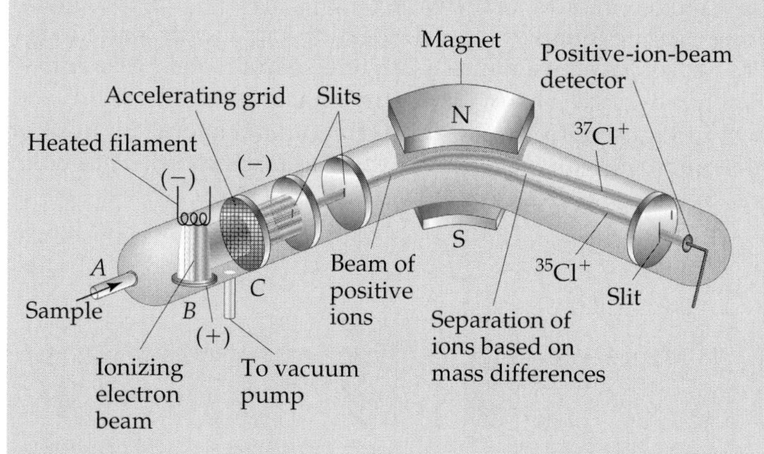

▲ **Figure 2.13** Diagram of a mass spectrometer, tuned to detect $^{35}Cl^+$ ions. The heavier $^{37}Cl^+$ ions are not deflected enough for them to reach the detector.

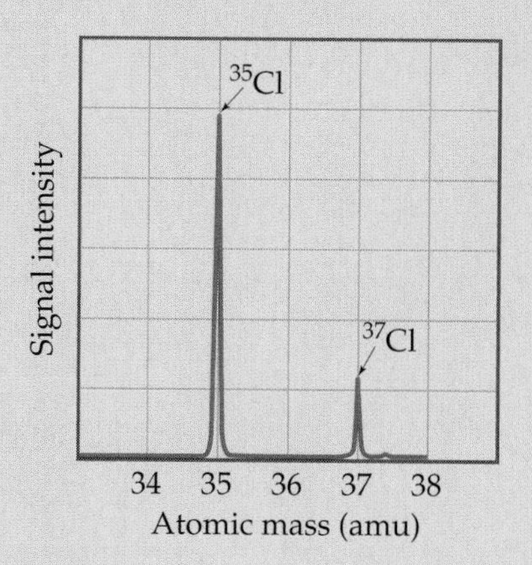

▲ **Figure 2.14** Mass spectrum of atomic chlorine.

SAMPLE EXERCISE 2.4

Naturally occurring chlorine is 75.78% ^{35}Cl, which has an atomic mass of 34.969 amu, and 24.22% ^{37}Cl, which has an atomic mass of 36.966 amu. Calculate the average atomic mass (that is, the atomic weight) of chlorine.

Solution The average atomic mass is found by multiplying the abundance of each isotope by its atomic mass and summing these products. Because 75.78% = 0.7578 and 24.22% = 0.2422, we have

$$\begin{aligned}\text{Average atomic mass} &= (0.7578)(34.969\text{ amu}) + (0.2422)(36.966\text{ amu})\\ &= 26.50\text{ amu} + 8.953\text{ amu}\\ &= 35.45\text{ amu}\end{aligned}$$

This answer makes sense: The average atomic mass of Cl is between the masses of the two isotopes and is closer to the value of ^{35}Cl, which is the more abundant isotope.

PRACTICE EXERCISE

Three isotopes of silicon occur in nature: ^{28}Si (92.23%), which has a mass of 27.97693 amu; ^{29}Si (4.68%), which has a mass of 28.97649 amu; and ^{30}Si (3.09%), which has a mass of 29.97377 amu. Calculate the atomic weight of silicon.
Answer: 28.09 amu

2.5 The Periodic Table

Dalton's atomic theory set the stage for a vigorous growth in chemical experimentation during the early 1800s. As the body of chemical observations grew and the list of known elements expanded, attempts were made to find regular patterns in chemical behavior. These efforts culminated in the development of the periodic table in 1869. We will have much to say about the periodic table in later chapters, but it is so important and useful that you should become acquainted with it now. You will quickly learn that *the periodic table is the most significant tool that chemists use for organizing and remembering chemical facts.*

Many elements show very strong similarities to each other. For example, lithium (Li), sodium (Na), and potassium (K) are all soft, very reactive metals. The elements helium (He), neon (Ne), and argon (Ar) are very nonreactive gases. If the elements are arranged in order of increasing atomic number, their chemical and physical properties are found to show a repeating, or periodic, pattern. For example, each of the soft, reactive metals—lithium, sodium, and potassium—comes immediately after one of the nonreactive gases—helium, neon, and argon—as shown in Figure 2.15 ▼. The arrangement of elements in order of increasing atomic number, with elements having similar properties placed in vertical columns, is known as the **periodic table**. The periodic table is shown

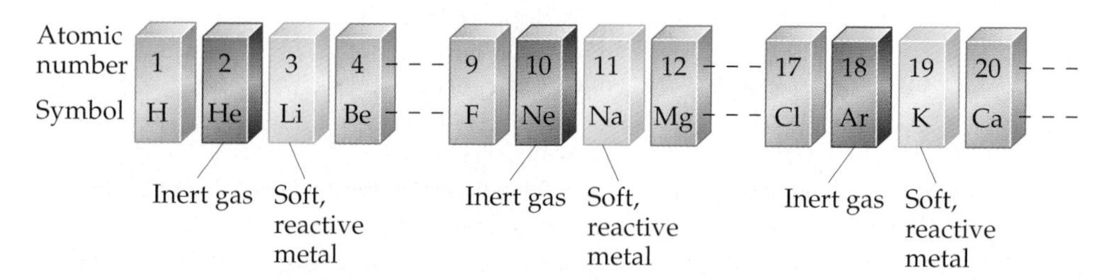

▲ **Figure 2.15** Arranging the elements by atomic number illustrates the periodic, or repeating, pattern in properties that is the basis of the periodic table.

1A 1	2A 2	3B 3	4B 4	5B 5	6B 6	7B 7	8B 8	8B 9	8B 10	1B 11	2B 12	3A 13	4A 14	5A 15	6A 16	7A 17	8A 18
1 H																	2 He
3 Li	4 Be											5 B	6 C	7 N	8 O	9 F	10 Ne
11 Na	12 Mg											13 Al	14 Si	15 P	16 S	17 Cl	18 Ar
19 K	20 Ca	21 Sc	22 Ti	23 V	24 Cr	25 Mn	26 Fe	27 Co	28 Ni	29 Cu	30 Zn	31 Ga	32 Ge	33 As	34 Se	35 Br	36 Kr
37 Rb	38 Sr	39 Y	40 Zr	41 Nb	42 Mo	43 Tc	44 Ru	45 Rh	46 Pd	47 Ag	48 Cd	49 In	50 Sn	51 Sb	52 Te	53 I	54 Xe
55 Cs	56 Ba	71 Lu	72 Hf	73 Ta	74 W	75 Re	76 Os	77 Ir	78 Pt	79 Au	80 Hg	81 Tl	82 Pb	83 Bi	84 Po	85 At	86 Rn
87 Fr	88 Ra	103 Lr	104 Rf	105 Db	106 Sg	107 Bh	108 Hs	109 Mt	110	111	112		114		116		

57 La	58 Ce	59 Pr	60 Nd	61 Pm	62 Sm	63 Eu	64 Gd	65 Tb	66 Dy	67 Ho	68 Er	69 Tm	70 Yb
89 Ac	90 Th	91 Pa	92 U	93 Np	94 Pu	95 Am	96 Cm	97 Bk	98 Cf	99 Es	100 Fm	101 Md	102 No

Metals

Metalloids

Nonmetals

▲ **Figure 2.16** Periodic table of the elements, showing the division of elements into metals, metalloids, and nonmetals.

ACTIVITY
Periodic Table

in Figure 2.16 ▲ and is also given on the front inside cover of the text. For each element in the table, the atomic number and atomic symbol are given, and the atomic weight (average atomic mass) is often given as well, as in the following typical entry for potassium:

19 ← atomic number
K ← atomic symbol
39.0983 ← atomic weight

You may notice slight variations in periodic tables from one book to another or between those in the lecture hall and in the text. These are simply matters of style, or they might concern the particular information included; there are no fundamental differences.

The elements in a column of the periodic table are known as a **group**. The way in which the groups are labeled is somewhat arbitrary, and three different labeling schemes are in common use, two of which are shown in Figure 2.16. The top set of labels, which have A and B designations, is widely used in North America. Roman numerals, rather than Arabic ones, are often employed in this scheme. Group 7A, for example, is often labeled VIIA. Europeans use a similar convention that numbers the columns from 1A through 8A and then from 1B through 8B, thereby giving the label 7B (or VIIB) instead of 7A to the group headed by fluorine (F). In an effort to eliminate this confusion, the International Union of Pure and Applied Chemistry (IUPAC) has proposed a convention that numbers the groups from 1 through 18 with no A or B designations, as shown in the lower set of labels at the top of the table in Figure 2.16. We will use the traditional North American convention.

TABLE 2.3 Names for Some of the Groups in the Periodic Table

Group	Name	Elements
1A	Alkali metals	Li, Na, K, Rb, Cs, Fr
2A	Alkaline earth metals	Be, Mg, Ca, Sr, Ba, Ra
6A	Chalcogens	O, S, Se, Te, Po
7A	Halogens	F, Cl, Br, I, At
8A	Noble gases (or rare gases)	He, Ne, Ar, Kr, Xe, Rn

Elements that belong to the same group often exhibit some similarities in their physical and chemical properties. For example, the "coinage metals"—copper (Cu), silver (Ag), and gold (Au)—all belong to group 1B. As their name suggests, the coinage metals are used throughout the world to make coins. Many other groups in the periodic table also have names, as listed in Table 2.3 ▲.

We will learn in Chapters 6 and 7 that the elements in a group of the periodic table have similar properties because they have the same type of arrangement of electrons at the periphery of their atoms. However, we need not wait until then to make good use of the periodic table; after all, the table was invented by chemists who knew nothing about electrons! We can use the table, as they intended, to correlate the behaviors of elements and to aid in remembering many facts. You will find it helpful to refer to the periodic table frequently when studying the remainder of this chapter.

▲ **Figure 2.17** Some familiar examples of metals and nonmetals. The nonmetals (from bottom left) are sulfur (yellow powder), iodine (dark, shiny crystals), bromine (reddish brown liquid and vapor in glass vial), and three samples of carbon (black charcoal powder, diamonds, and graphite in the pencil lead). The metals are in the form of an aluminum wrench, copper pipe, lead shot, silver coins, and gold nuggets.

All the elements on the left side and in the middle of the periodic table (except for hydrogen) are **metallic elements**, or metals. The majority of elements are metallic. Metals share many characteristic properties, such as luster and high electrical and heat conductivity. All metals, with the exception of mercury (Hg), are solids at room temperature. The metals are separated from the **nonmetallic elements** by a diagonal steplike line that runs from boron (B) to astatine (At), as shown in Figure 2.16. Hydrogen, although on the left side of the periodic table, is a nonmetal. At room temperature some of the nonmetals are gaseous, some are liquid, and some are solid. They generally differ from the metals in appearance (Figure 2.17 ◀) and in other physical properties. Many of the elements that lie along the line that separates metals from nonmetals, such as antimony (Sb), have properties that fall between those of metals and nonmetals. These elements are often referred to as **metalloids**.

SAMPLE EXERCISE 2.5

Which two of the following elements would you expect to show the greatest similarity in chemical and physical properties: B, Ca, F, He, Mg, P?

Solution Elements that are in the same group of the periodic table are most likely to exhibit similar chemical and physical properties. We therefore expect that Ca and Mg should be most alike because they are in the same group (group 2A, the alkaline earth metals).

PRACTICE EXERCISE

Locate Na (sodium) and Br (bromine) on the periodic table. Give the atomic number of each and label each a metal, metalloid, or nonmetal.

Answer: Na, atomic number 11, is a metal; Br, atomic number 35, is a nonmetal.

A Closer Look Glenn Seaborg and the Story of Seaborgium

Prior to 1940 the periodic table ended at uranium, element number 92. Since that time, no scientist has had a greater effect on the periodic table than Glenn Seaborg (1912–1999). Seaborg (Figure 2.18 ▶) became a faculty member in the chemistry department at the University of California, Berkeley, in 1937. In 1940 he and his colleagues Edwin McMillan, Arthur Wahl, and Joseph Kennedy succeeded in isolating plutonium (Pu) as a product of the reaction of uranium with neutrons. We will talk about reactions of this type, called *nuclear reactions*, in Chapter 21. We will also discuss the key role that plutonium plays in nuclear fission reactions, such as those that occur in nuclear power plants and atomic bombs.

During the period 1944 through 1958, Seaborg and his coworkers also succeeded in identifying the elements with atomic numbers 95 through 102 as products of nuclear reactions. All these elements are radioactive and are not found in nature; they can be synthesized only via nuclear reactions. For their efforts in identifying the elements beyond uranium (the *transuranium* elements), McMillan and Seaborg shared the 1951 Nobel Prize in chemistry.

From 1961 to 1971 Seaborg served as the chairman of the U. S. Atomic Energy Commission (now the Department of Energy). In this position he had an important role in establishing international treaties to limit the testing of nuclear weapons. Upon his return to Berkeley he was part of the team that in 1974 first identified element number 106; that discovery was corroborated by another team at Berkeley in 1993. In 1994 to honor Seaborg's many contributions to the discovery of new elements, the American Chemical Society proposed that element number 106 be named "seaborgium," with a proposed symbol of Sg. After several years of controversy about whether an element could be named after a living person, the name seaborgium was officially adopted by the IUPAC in 1997, and Seaborg became the first person to have an element named after him while still alive. The IUPAC also named element 105 "dubnium" (chemical symbol Db) in honor of a nuclear laboratory at Dubna, Russia, that competed with the Berkeley laboratory in the discovery of several new elements.

▲ **Figure 2.18** Glenn Seaborg at Berkeley in 1941 using a Geiger counter to try to detect radiation produced by plutonium. Geiger counters will be discussed in Section 21.5.

2.6 Molecules and Molecular Compounds

The atom is the smallest representative sample of an element, but only the noble-gas elements are normally found in nature as isolated atoms. Most matter is composed of molecules or ions, both of which are formed from atoms. We examine molecules here and ions in Section 2.7.

A **molecule** is an assembly of two or more atoms tightly bound together. The resultant "package" of atoms behaves in many ways as a single, distinct object, just as a television set composed of many parts can be recognized as a single object. We will discuss the forces that hold the atoms together (the chemical bonds) in Chapters 8 and 9.

Molecules and Chemical Formulas

Many elements are found in nature in molecular form; that is, two or more of the same type of atom are bound together. For example, the oxygen normally found in air consists of molecules that contain two oxygen atoms. We represent this molecular form of oxygen by the **chemical formula** O_2 (read "oh two"). The subscript in the formula tells us that two oxygen atoms are present in each molecule.

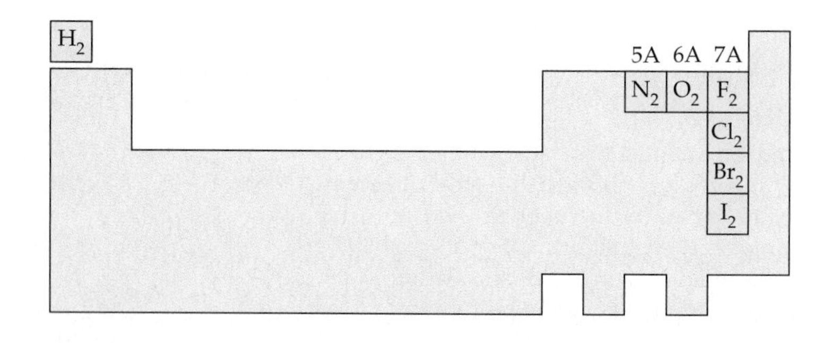

▶ **Figure 2.19** Common elements that exist as diatomic molecules at room temperature.

A molecule that is made up of two atoms is called a **diatomic molecule**. Oxygen also exists in another molecular form known as *ozone*. Molecules of ozone consist of three oxygen atoms, so its chemical formula is O_3. Even though "normal" oxygen (O_2) and ozone are both composed only of oxygen atoms, they exhibit very different chemical and physical properties. For example, O_2 is essential for life, but O_3 is toxic; O_2 is odorless, whereas O_3 has a sharp, pungent smell.

The elements that normally occur as diatomic molecules are hydrogen, oxygen, nitrogen, and the halogens. Their locations in the periodic table are shown in Figure 2.19 ▲. When we speak of the substance hydrogen, we mean H_2 unless we explicitly indicate otherwise. Likewise, when we speak of oxygen, nitrogen, or any of the halogens, we are referring to O_2, N_2, F_2, Cl_2, Br_2, or I_2. Thus, the properties of oxygen and hydrogen listed in Table 1.3 are those of O_2 and H_2. Other, less common forms of these elements behave much differently.

Compounds that are composed of molecules are called **molecular compounds** and contain more than one type of atom. A molecule of water, for example, consists of two hydrogen atoms and one oxygen atom. It is therefore represented by the chemical formula H_2O. Lack of a subscript on the O indicates one atom of O per water molecule. Another compound composed of these same elements (in different relative proportions) is hydrogen peroxide, H_2O_2. The properties of these two compounds are very different.

Several common molecules are shown in Figure 2.20 ◀. Notice how the composition of each compound is given by its chemical formula. Notice also that these substances are composed only of nonmetallic elements. *Most molecular substances that we will encounter contain only nonmetals.*

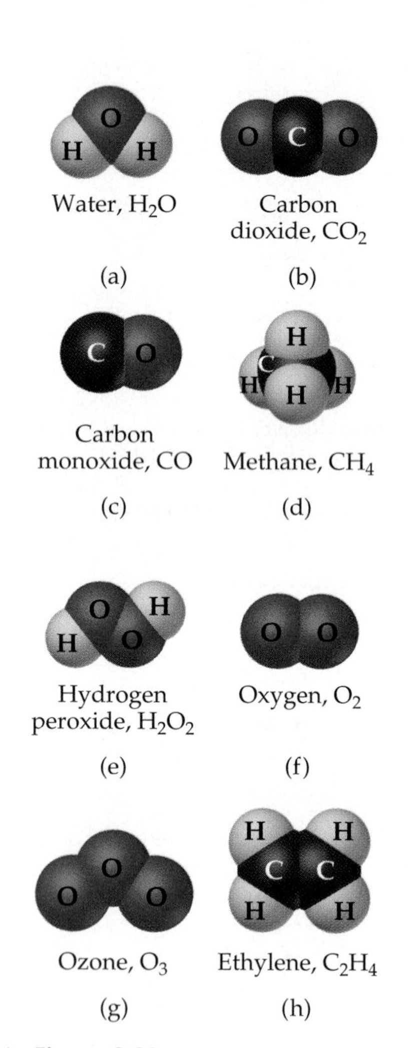

▲ **Figure 2.20** Representation of some common simple molecules.

Molecular and Empirical Formulas

Chemical formulas that indicate the actual numbers and types of atoms in a molecule are called **molecular formulas**. (The formulas in Figure 2.20 are molecular formulas.) Chemical formulas that give only the relative number of atoms of each type in a molecule are called **empirical formulas**. The subscripts in an empirical formula are always the smallest possible whole-number ratios. The molecular formula for hydrogen peroxide is H_2O_2, for example, whereas its empirical formula is HO. The molecular formula for ethylene is C_2H_4, so its empirical formula is CH_2. For many substances, the molecular formula and the empirical formula are identical, as in the case of water, H_2O.

Molecular formulas provide greater information about molecules than empirical formulas. Whenever we know the molecular formula of a compound, we can determine its empirical formula. The converse is not true, however; if we know the empirical formula of a substance, we can't determine its molecular formula unless we have more information. So why do chemists bother with empirical formulas? As we will see in Chapter 3, certain common methods of analyzing substances lead to the empirical formula only. Once the empirical formula is known, however, additional experiments can give the information needed to convert the empirical formula to the molecular one. In addition, there are substances, such as the most common forms of elemental carbon, that don't exist as isolated molecules. For these substances, we must rely on empirical formulas. Thus, carbon is represented by its chemical symbol, C, which is its empirical formula.

SAMPLE EXERCISE 2.6

Write the empirical formulas for the following molecules: **(a)** glucose, a substance also known as blood sugar or dextrose, whose molecular formula is $C_6H_{12}O_6$; **(b)** nitrous oxide, a substance used as an anesthetic and commonly called laughing gas, whose molecular formula is N_2O.

Solution (a) The subscripts of an empirical formula are the smallest whole-number ratios. The smallest ratios are obtained by dividing each subscript by the largest common factor, in this case 6. The resultant empirical formula for glucose is CH_2O.

(b) Because the subscripts in N_2O are already the lowest integral numbers, the empirical formula for nitrous oxide is the same as its molecular formula, N_2O.

PRACTICE EXERCISE

Give the empirical formula for the substance called *diborane*, whose molecular formula is B_2H_6.
Answer: BH_3

Picturing Molecules

The molecular formula of a substance summarizes its composition but does not show how the atoms come together to form the molecule. The **structural formula** of a substance shows which atoms are attached to which within the molecule. For example, the formulas for water, hydrogen peroxide, and methane (CH_4) can be written as follows:

H—O—H (Water)

H—O—O—H (Hydrogen peroxide)

H—C(H)(H)—H (Methane)

Water Hydrogen peroxide Methane

The atoms are represented by their chemical symbols, and lines are used to represent the bonds that hold the atoms together.

A structural formula usually does not depict the actual geometry of the molecule, that is, the actual angles at which atoms are joined together. A structural formula can be written as a *perspective drawing*, however, to give some sense of three-dimensional shape, as shown in Figure 2.21 ▶.

Scientists also rely on various models to help them visualize molecules. *Ball-and-stick models* show atoms as spheres and the bonds as sticks, and they accurately represent the angles at which the atoms are attached to one another within the molecule (Figure 2.21). All atoms may be represented by balls of the same size, or the relative sizes of the balls may reflect the relative sizes of the atoms. Sometimes the chemical symbols of the elements are superimposed on the balls, but often the atoms are identified simply by color.

A *space-filling model* depicts what the molecule would look like if the atoms were scaled up in size (Figure 2.21). These models show the relative sizes of the atoms, but the angles between atoms, which help define their molecular geometry, are often more difficult to see than in ball-and-stick models. As in ball-and-stick models, the identities of the atoms are indicated by their colors, but they may also be labeled with the element's symbol.

H—C(H)(H)—H

Structural formula

Perspective drawing

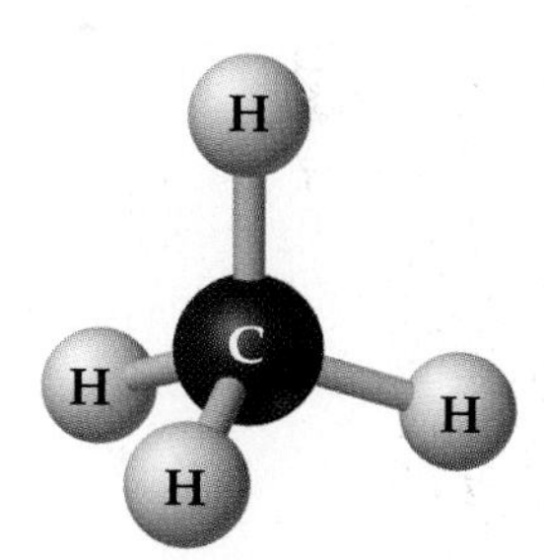

Ball-and-stick model

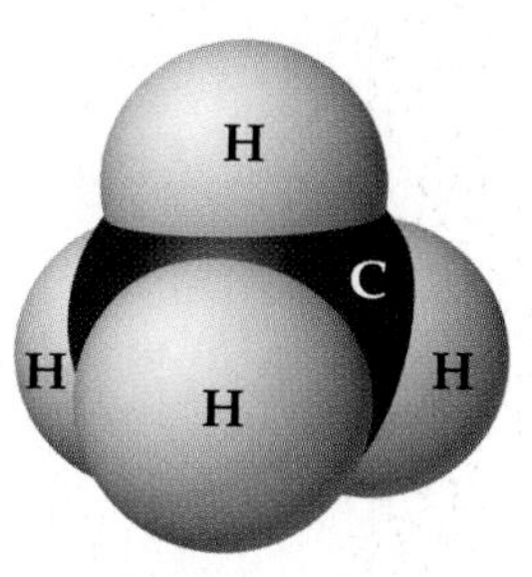

Space-filling model

▲ **Figure 2.21** Some of the ways in which molecules are represented and visualized.

2.7 Ions and Ionic Compounds

The nucleus of an atom is unchanged by ordinary chemical processes, but atoms can readily gain or lose electrons. If electrons are removed from or added to a neutral atom, a charged particle called an **ion** is formed. An ion with a positive charge is called a **cation** (pronounced CAT-ion); a negatively charged ion is called an **anion** (AN-ion). The sodium atom, for example, which has 11 protons and 11 electrons, easily loses one electron. The resulting cation has 11 protons and 10 electrons, so it has a net charge of 1+. The net charge on an ion is represented by a superscript; +, 2+, and 3+ mean a net charge resulting from the loss of one, two, or three electrons, respectively. The superscripts −, 2−, and 3− represent net charges resulting from the gain of one, two, or three electrons, respectively. The formation of the Na^+ ion from an Na atom is shown schematically as follows:

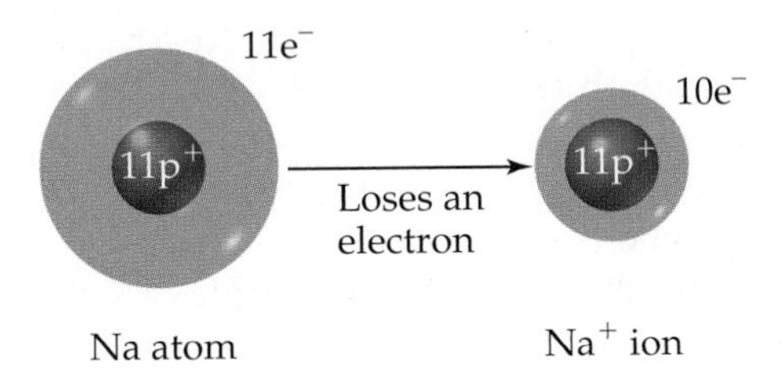

Chlorine, with 17 protons and 17 electrons, often gains an electron in chemical reactions, producing the Cl^- ion.

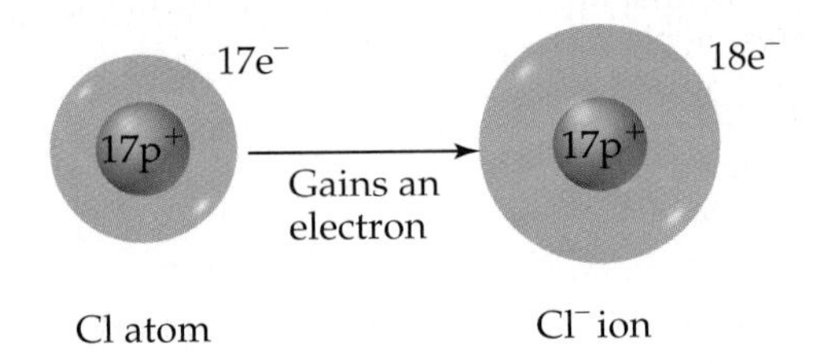

In general, metal atoms tend to lose electrons to form cations, whereas nonmetal atoms tend to gain electrons to form anions.

SAMPLE EXERCISE 2.7

Give the chemical symbols, including mass numbers, for the following ions: **(a)** The ion with 22 protons, 26 neutrons, and 19 electrons; **(b)** the ion of sulfur that has 16 neutrons and 18 electrons.

Solution **(a)** The number of protons (22) is the atomic number of the element, so the element is Ti (titanium). The mass number of this isotope is $22 + 26 = 48$ (the sum of the protons and neutrons). Because the ion has three more protons than electrons, it has a net charge of 3+. Thus, the symbol for the ion is $^{48}Ti^{3+}$.

(b) By referring to a periodic table or table of elements, we see that sulfur (symbol S) has an atomic number of 16. Thus, each atom or ion of sulfur has 16 protons. We are told that the ion also has 16 neutrons, so the mass number of the ion is $16 + 16 = 32$. Because the ion has 16 protons and 18 electrons, its net charge is 2−. Thus, the symbol for the ion is $^{32}S^{2-}$.

In general, we will focus on the net charges of ions and ignore their mass numbers unless the circumstances dictate that we specify a certain isotope.

PRACTICE EXERCISE

How many protons and electrons does the Se^{2-} ion possess?
Answer: 34 protons and 36 electrons

In addition to simple ions such as Na^+ and Cl^-, there are **polyatomic ions** such as NO_3^- (nitrate ion) and SO_4^{2-} (sulfate ion). These ions consist of atoms joined as in a molecule, but they have a net positive or negative charge. We will consider further examples of polyatomic ions in Section 2.8.

The chemical properties of ions are very different from those of the atoms from which they are derived. The difference is like the change from Dr. Jekyll to Mr. Hyde: Although the body may be essentially the same (plus or minus a few electrons), the behavior is much different.

Predicting Ionic Charges

Many atoms gain or lose electrons so as to end up with the same number of electrons as the noble gas closest to them in the periodic table. The members of the noble-gas family are chemically very nonreactive and form very few compounds. We might deduce that this is because their electron arrangements are very stable. Nearby elements can obtain these same stable arrangements by losing or gaining electrons. For example, loss of one electron from an atom of sodium leaves it with the same number of electrons as the neutral neon atom (atomic number 10). Similarly, when chlorine gains an electron, it ends up with 18, the same as argon (atomic number 18). We will use this simple observation to explain the formation of ions in Chapter 8, where we discuss chemical bonding.

SAMPLE EXERCISE 2.8

Predict the charges expected for the most stable ions of barium and oxygen.

Solution We will assume that these elements form ions that have the same number of electrons as the nearest noble-gas atom. From the periodic table, barium has atomic number 56. The nearest noble gas is xenon, atomic number 54. Barium can attain a stable arrangement of 54 electrons by losing two of its electrons, forming the Ba^{2+} cation.

Oxygen has atomic number 8. The nearest noble gas is neon, atomic number 10. Oxygen can attain this stable electron arrangement by gaining two electrons, thereby forming the O^{2-} anion.

PRACTICE EXERCISE

Predict the charge of the most stable ion of aluminum.
Answer: 3+

The periodic table is very useful for remembering the charges of ions, especially those of the elements on the left and right sides of the table. As Figure 2.22 ▼ shows,

1A	2A	Transition metals	3A	4A	5A	6A	7A	8A
H^+							H^-	NOBLE GASES
Li^+					N^{3-}	O^{2-}	F^-	
Na^+	Mg^{2+}		Al^{3+}			S^{2-}	Cl^-	
K^+	Ca^{2+}					Se^{2-}	Br^-	
Rb^+	Sr^{2+}					Te^{2-}	I^-	
Cs^+	Ba^{2+}							

▲ **Figure 2.22** Charges of some common ions found in ionic compounds. Notice that the steplike line that divides metals from nonmetals also separates cations from anions.

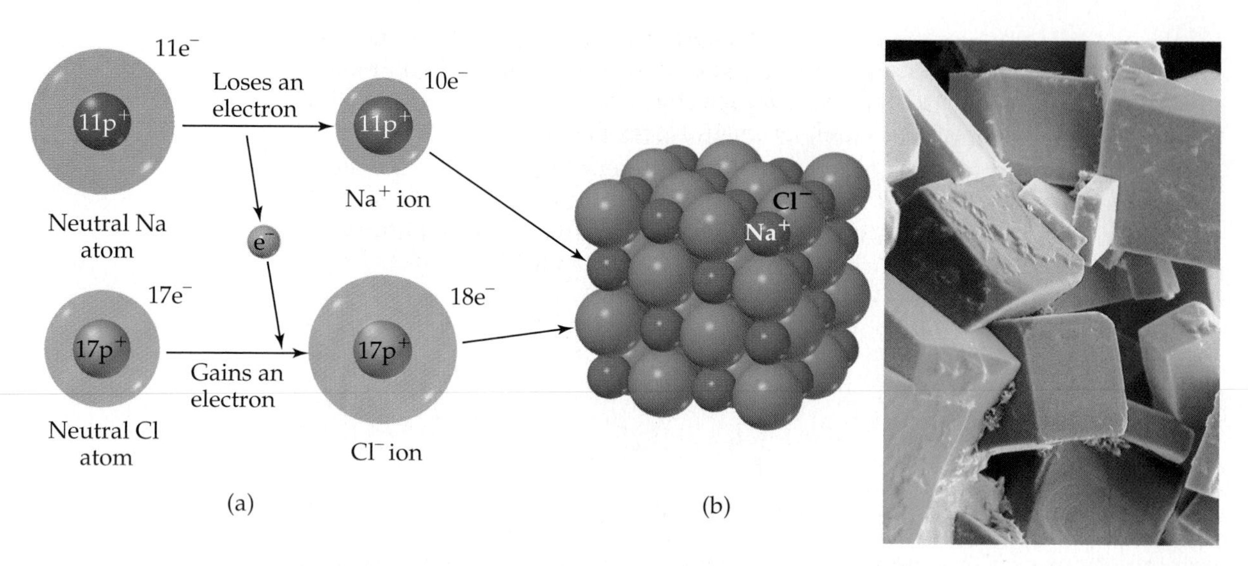

▲ **Figure 2.23** (a) The transfer of an electron from a neutral Na atom to a neutral Cl atom leads to the formation of an Na^+ ion and a Cl^- ion. (b) Arrangement of these ions in solid sodium chloride (NaCl) is pictured at the right.

the charges of these ions relate in a simple way to their positions in the table. On the left side of the table, for example, the group 1A elements (the alkali metals) form 1+ ions, and the group 2A elements (the alkaline earths) form 2+ ions. On the other side of the table the group 7A elements (the halogens) form 1− ions, and the group 6A elements form 2− ions. As we will see later in the text, many of the other groups do not lend themselves to such simple rules.

Ionic Compounds

A great deal of chemical activity involves the transfer of electrons between substances. Ions form when one or more electrons transfer from one neutral atom to another. Figure 2.23 ▲ shows that when elemental sodium is allowed to react with elemental chlorine, an electron transfers from a neutral sodium atom to a neutral chlorine atom. We are left with an Na^+ ion and a Cl^- ion. Objects of opposite charge attract, however, so the Na^+ and the Cl^- ions bind together to form the compound sodium chloride (NaCl), which we know better as common table salt. Sodium chloride is an example of an **ionic compound**, a compound that contains both positively and negatively charged ions.

We can often tell whether a compound is ionic (consisting of ions) or molecular (consisting of molecules) from its composition. In general, cations are metal ions, whereas anions are nonmetal ions. Consequently, *ionic compounds are generally combinations of metals and nonmetals*, as in NaCl. In contrast, *molecular compounds are generally composed of nonmetals only*, as in H_2O.

SAMPLE EXERCISE 2.9

Which of the following compounds would you expect to be ionic: N_2O, Na_2O, $CaCl_2$, SF_4?

Solution We would predict that Na_2O and $CaCl_2$ are ionic compounds because they are composed of a metal combined with a nonmetal. The other two compounds, composed entirely of nonmetals, are predicted (correctly) to be molecular compounds.

PRACTICE EXERCISE

Which of the following compounds are molecular: CBr_4, FeS, P_4O_6, PbF_2?
Answer: CBr_4 and P_4O_6

The ions in ionic compounds are arranged in three-dimensional structures. The arrangement of Na^+ and Cl^- ions in NaCl is shown in Figure 2.23. Because there is no discrete molecule of NaCl, we are able to write only an empirical formula for this substance. In fact, only empirical formulas can be written for most ionic compounds.

We can readily write the empirical formula for an ionic compound if we know the charges of the ions of which it is composed. Chemical compounds are always electrically neutral. Consequently, the ions in an ionic compound always occur in such a ratio that the total positive charge equals the total negative charge. Thus, there is one Na^+ to one Cl^- (giving NaCl), one Ba^{2+} to two Cl^- (giving $BaCl_2$), and so forth.

As you consider these and other examples, you will see that if the charges on the cation and anion are equal, the subscript on each ion will be 1. If the charges are not equal, the charge on one ion (without its sign) will become the subscript on the other ion. For example, the ionic compound formed from Mg (which forms Mg^{2+} ions) and N (which forms N^{3-} ions) is Mg_3N_2:

$$Mg^{2+} \quad N^{3-} \longrightarrow Mg_3N_2$$

Chemistry and Life Elements Required by Living Organisms

Figure 2.24 ▼ shows the elements that are essential for life. Over 97% of the mass of most organisms is due to just six elements—oxygen, carbon, hydrogen, nitrogen, phosphorus, and sulfur. Water (H_2O) is the most common compound in living organisms, accounting for at least 70% of the mass of most cells. Carbon is the most prevalent element (by mass) in the solid components of cells. Carbon atoms are found in a vast variety of organic molecules in which the carbon atoms are bonded to other carbon atoms or to atoms of other elements, principally H, O, N, P, and S. All proteins, for example, contain the following group of atoms that occurs repeatedly within the molecules:

$$\begin{array}{c} \quad\quad O \\ \quad\quad \| \\ -N-C- \\ | \quad\quad \\ R \quad\quad \end{array}$$

(R is either an H atom or a combination of atoms such as CH_3.)

In addition, 23 more elements have been found in various living organisms. Five are ions that are required by all organisms: Ca^{2+}, Cl^-, Mg^{2+}, K^+, and Na^+. Calcium ions, for example, are necessary for the formation of bone and for the transmission of signals in the nervous system, such as those that trigger the contraction of cardiac muscles, causing the heart to beat. Many other elements are needed in only very small quantities, so they are called *trace* elements. For example, trace quantities of copper are required in our diet to aid in the synthesis of hemoglobin.

▼ **Figure 2.24** The elements that are essential for life are indicated by colors. Red denotes the six most abundant elements in living systems (hydrogen, carbon, nitrogen, oxygen, phosphorus, and sulfur). Blue indicates the five next most abundant elements. Green indicates the elements needed in only trace amounts.

1A	2A	3B	4B	5B	6B	7B	8B: 8	9	10	1B	2B	3A	4A	5A	6A	7A	8A
H																	He
Li	Be											B	C	N	O	F	Ne
Na	Mg											Al	Si	P	S	Cl	Ar
K	Ca	Sc	Ti	V	Cr	Mn	Fe	Co	Ni	Cu	Zn	Ga	Ge	As	Se	Br	Kr
Rb	Sr	Y	Zr	Nb	Mo	Tc	Ru	Rh	Pd	Ag	Cd	In	Sn	Sb	Te	I	Xe
Cs	Ba	La	Hf	Ta	W	Re	Os	Ir	Pt	Au	Hg	Tl	Pb	Bi	Po	At	Rn

SAMPLE EXERCISE 2.10

What are the empirical formulas of the compounds formed by **(a)** Al^{3+} and Cl^- ions; **(b)** Al^{3+} and O^{2-} ions; **(c)** Mg^{2+} and NO_3^- ions?

Solution **(a)** Three Cl^- ions are required to balance the charge of one Al^{3+} ion. Thus, the formula is $AlCl_3$.

(b) Two Al^{3+} ions are required to balance the charge of three O^{2-} ions (that is, the total positive charge is 6+, and the total negative charge is 6−). Thus, the formula is Al_2O_3.

(c) Two NO_3^- ions are needed to balance the charge of one Mg^{2+}. Thus the formula is $Mg(NO_3)_2$. In this case the formula for the entire polyatomic ion NO_3^- must be enclosed in parentheses so that it is clear that the subscript 2 applies to all the atoms of that ion.

PRACTICE EXERCISE

Write the empirical formulas for the compounds formed by the following ions: **(a)** Na^+ and PO_4^{3-}; **(b)** Zn^{2+} and SO_4^{2-}; **(c)** Fe^{3+} and CO_3^{2-}.
Answers: **(a)** Na_3PO_4; **(b)** $ZnSO_4$; **(c)** $Fe_2(CO_3)_3$

Strategies in Chemistry Pattern Recognition

Someone once said that drinking at the fountain of knowledge in a chemistry course is like drinking from a fire hydrant. Indeed, the pace can sometimes seem brisk. More to the point, however, we can drown in the facts if we don't see the general patterns. The value of recognizing patterns and learning rules and generalizations is that they free us from learning (or trying to memorize) many individual facts. The patterns and rules tie ideas together, so we don't get lost in the details.

Many students struggle with chemistry because they don't see how the topics relate to one another, how ideas connect together. They therefore treat every idea and problem as being unique instead of as an example or application of a general rule, procedure, or relationship. Begin to notice the structure of the topic. Pay attention to the trends and rules that are given to summarize a large body of information. Notice, for example, how atomic structure helps us understand the existence of isotopes (as seen in Table 2.2) and how the periodic table aids us in remembering the charges of ions (as seen in Figure 2.22). You may surprise yourself by observing patterns that are not even explicitly spelled out yet. Perhaps you've even noticed certain trends in chemical formulas. Moving across the periodic table from element 11 (Na) we find that the elements form compounds with F having the following compositions: NaF, MgF_2, and AlF_3. Does this trend continue? Do SiF_4, PF_5, and SF_6 exist? Indeed they do. If you have picked up on trends like this from the scraps of information you've seen so far, then you're ahead of the game and you've already prepared yourself for some topics we will address in later chapters.

2.8 Naming Inorganic Compounds

To find information about a particular substance, you must know its chemical formula and name. The names and formulas of compounds are essential vocabulary in chemistry. The naming of substances is called **chemical nomenclature** from the Latin words *nomen* (name) and *calare* (to call).

There are now more than 19 million known chemical substances. Naming them all would be a hopelessly complicated task if each had a special name independent of all others. Many important substances that have been known for a long time, such as water (H_2O) and ammonia (NH_3), do have individual, traditional names (so-called "common" names). For most substances, however, we rely on a systematic set of rules that leads to an informative and unique name for each substance, based on its composition.

The rules for chemical nomenclature are based on the division of substances into different categories. The major division is between organic and inorganic compounds. *Organic compounds* contain carbon, usually in combination with hydrogen, oxygen, nitrogen, or sulfur. All others are *inorganic compounds*. Early chemists associated organic compounds with plants and animals, and they associated inorganic compounds with the nonliving portion of our world. Although this distinction between living and nonliving matter is no longer pertinent, the classification between organic and inorganic compounds continues to be useful. In this section we consider the basic rules for naming inorganic compounds. Among inorganic compounds we will consider three categories of substances: ionic compounds, molecular compounds, and acids. We will also introduce the names of some simple organic compounds in Section 2.9.

ACTIVITY
Naming Cations, Naming Anions

Names and Formulas of Ionic Compounds

Recall from Section 2.7 that ionic compounds usually consist of chemical combinations of metals and nonmetals. The metals form the positive ions, and the nonmetals form the negative ions. Let's examine the naming of positive ions, then the naming of negative ones. After that, we will consider how to put the names of the ions together to identify the complete ionic compound.

1. **Positive Ions (Cations)**

(a) *Cations formed from metal atoms have the same name as the metal.*

Na^+ sodium ion Zn^{2+} zinc ion Al^{3+} aluminum ion

Ions formed from a single atom are called *monoatomic ions.*

(b) *If a metal can form cations of differing charges, the positive charge is given by a Roman numeral in parentheses following the name of the metal.*

Fe^{2+} iron(II) ion Cu^+ copper(I) ion

Fe^{3+} iron(III) ion Cu^{2+} copper(II) ion

Ions with different charges exhibit different properties, such as color (Figure 2.25 ▶).

▲ **Figure 2.25** Compounds of ions of the same element but with different charge can be very different in appearance. Both substances shown are complex salts of iron with K^+ and CN^- ions. The one on the left is potassium ferrocyanide, which contains Fe(II) bound to CN^- ions. The one on the right is potassium ferricyanide, which contains Fe(III) bound to CN^- ions. Both substances are used extensively in blueprinting and other dyeing processes.

Most of the metals that have variable charge are *transition metals*, elements that occur in the middle block of elements from group 3B to group 2B in the periodic table. The charges of these ions are indicated by Roman numerals. The common metal ions that do not have variable charges are the ions of group 1A (Li^+, Na^+, K^+, and Cs^+) and group 2A (Mg^{2+}, Ca^{2+}, Sr^{2+}, and Ba^{2+}), as well as Al^{3+} (group 3A) and two transition-metal ions: Ag^+ (group 1B) and Zn^{2+} (group 2B). Charges are not shown explicitly when naming these ions. If there is any doubt in your mind whether a metal forms more than one type of cation, indicate the charge using Roman numerals. It is never wrong to do so, even though it may be unnecessary.

An older method still widely used for distinguishing between two differently charged ions of a metal is to apply the ending *-ous* or *-ic*. These endings represent the lower and higher charged ions, respectively. They are added to the root of the element's Latin name:

Fe^{2+} ferrous ion Cu^+ cuprous ion

Fe^{3+} ferric ion Cu^{2+} cupric ion

Although we will only rarely use these older names in this text, you might encounter them elsewhere.

(c) *Cations formed from nonmetal atoms have names that end in* -ium:

NH_4^+ ammonium ion H_3O^+ hydronium ion

These two ions are the only ions of this kind that we will encounter frequently in the text. They are both *polyatomic* (composed of many atoms). The vast majority of cations are monoatomic metal ions.

TABLE 2.4 Common Cations

Charge	Formula	Name	Formula	Name
1+	H^+	Hydrogen ion	NH_4^+	Ammonium ion
	Li^+	Lithium ion	Cu^+	Copper(I) or cuprous ion
	Na^+	Sodium ion		
	K^+	Potassium ion		
	Cs^+	Cesium ion		
	Ag^+	Silver ion		
2+	Mg^{2+}	Magnesium ion	Co^{2+}	Cobalt(II) or cobaltous ion
	Ca^{2+}	Calcium ion	Cu^{2+}	Copper(II) or cupric ion
	Sr^{2+}	Strontium ion	Fe^{2+}	Iron(II) or ferrous ion
	Ba^{2+}	Barium ion	Mn^{2+}	Manganese(II) or manganous ion
	Zn^{2+}	Zinc ion	Hg_2^{2+}	Mercury(I) or mercurous ion
	Cd^{2+}	Cadmium ion	Hg^{2+}	Mercury(II) or mercuric ion
			Ni^{2+}	Nickel(II) or nickelous ion
			Pb^{2+}	Lead(II) or plumbous ion
			Sn^{2+}	Tin(II) or stannous ion
3+	Al^{3+}	Aluminum ion	Cr^{3+}	Chromium(III) or chromic ion
			Fe^{3+}	Iron(III) or ferric ion

The names and formulas of some of the most common cations are shown in Table 2.4 ▲ and are also included in a table of common ions that is placed in the back inside cover of the text. The ions listed on the left are the monoatomic ions that do not have variable charges. Those listed on the right are either polyatomic cations or cations with variable charges. The Hg_2^{2+} ion is unusual because this metal ion is not monoatomic. It is called the mercury(I) ion because it can be thought of as two Hg^+ ions fused together.

2. Negative Ions (Anions)

(a) *Monoatomic (one-atom) anions have names formed by replacing the ending of the name of the element with* -ide:

H^- hydride ion O^{2-} oxide ion N^{3-} nitride ion

A few simple polyatomic anions also have names ending in *-ide*:

OH^- hydroxide ion CN^- cyanide ion O_2^{2-} peroxide ion

(b) *Polyatomic (many-atom) anions containing oxygen have names ending in* -ate *or* -ite. These anions are called **oxyanions**. The ending *-ate* is used for the most common oxyanion of an element. The ending *-ite* is used for an oxyanion that has the same charge but one less O atom:

NO_3^- nitrate ion SO_4^{2-} sulfate ion

NO_2^- nitrite ion SO_3^{2-} sulfite ion

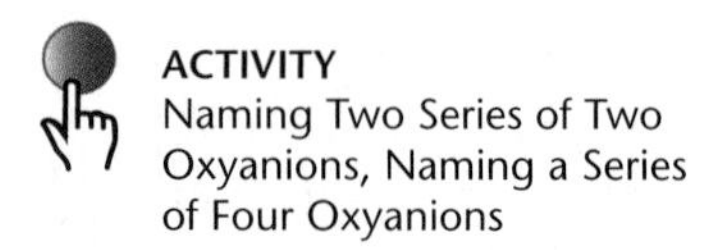
ACTIVITY
Naming Two Series of Two Oxyanions, Naming a Series of Four Oxyanions

Prefixes are used when the series of oxyanions of an element extends to four members, as with the halogens. The prefix *per-* indicates one more O atom than the oxyanion ending in *-ate*; the prefix *hypo-* indicates one less O atom than the oxyanion ending in *-ite*:

ClO_4^- perchlorate ion (one more O atom than chlorate)

$\mathbf{ClO_3^-}$ **chlorate ion**

ClO_2^- chlorite ion (one less O atom than chlorate)

ClO^- hypochlorite ion (one less O atom than chlorite)

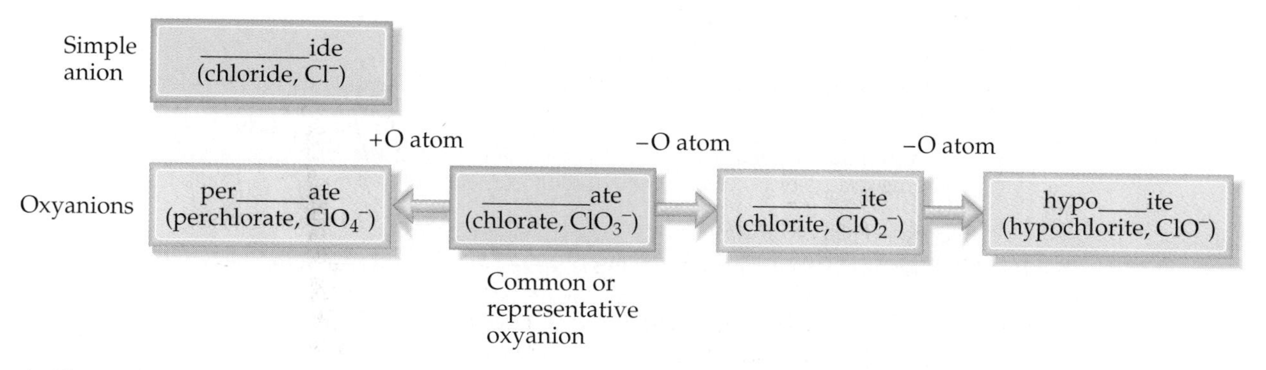

▲ **Figure 2.26** A summary of the procedure for naming anions. The root of the name (such as "chlor" for chlorine) goes in the blank.

If you learn the rules just presented, you only need to know the name for one oxyanion in a series to deduce the names for the other members. These rules are summarized in Figure 2.26 ▲.

(c) *Anions derived by adding H^+ to an oxyanion are named by adding as a prefix the word* hydrogen *or* dihydrogen, *as appropriate*:

CO_3^{2-}	carbonate ion	PO_4^{3-}	phosphate ion
HCO_3^-	hydrogen carbonate ion	$H_2PO_4^-$	dihydrogen phosphate ion

Notice that each H^+ reduces the negative charge of the parent anion by one. An older method for naming some of these ions is to use the prefix *bi-*. Thus, the HCO_3^- ion is commonly called the bicarbonate ion, and HSO_4^- is sometimes called the bisulfate ion.

The names and formulas of the common anions are listed in Table 2.5 ▼ and on the back inside cover of the text. Those whose names end in *-ide* are listed on the left portion of the table, whereas those whose names end in *-ate* are listed on the right. The formulas of the ions whose names end with *-ite* can be derived from those ending in *-ate* by removing an O atom. Notice the location of the monoatomic ions in the periodic table. Those of group 7A always have a 1− charge (F^-, Cl^-, Br^-, and I^-), whereas those of group 6A have a 2− charge (O^{2-} and S^{2-}).

TABLE 2.5 Common Anions

Charge	Formula	Name	Formula	Name
1−	H^-	Hydride ion	$C_2H_3O_2^-$	Acetate ion
	F^-	Fluoride ion	ClO_3^-	Chlorate ion
	Cl^-	Chloride ion	ClO_4^-	Perchlorate ion
	Br^-	Bromide ion	NO_3^-	Nitrate ion
	I^-	Iodide ion	MnO_4^-	Permanganate ion
	CN^-	Cyanide ion		
	OH^-	Hydroxide ion		
2−	O^{2-}	Oxide ion	CO_3^{2-}	Carbonate ion
	O_2^{2-}	Peroxide ion	CrO_4^{2-}	Chromate ion
	S^{2-}	Sulfide ion	$Cr_2O_7^{2-}$	Dichromate ion
			SO_4^{2-}	Sulfate ion
3−	N^{3-}	Nitride ion	PO_4^{3-}	Phosphate ion

SAMPLE EXERCISE 2.11

The formula for the selenate ion is SeO_4^{2-}. Write the formula for the selenite ion.

Solution The ending *-ite* indicates an oxyanion with the same charge but one less O atom than the corresponding oxyanion that ends in *-ate*. Thus, the selenite ion has the same charge but one less oxygen than the selenate ion: SeO_3^{2-}.

PRACTICE EXERCISE

The formula for the bromate ion is BrO_3^-. Write the formula for the hypobromite ion.
Answer: BrO^-

3. Ionic Compounds

Names of ionic compounds consist of the cation name followed by the anion name:

$CaCl_2$	calcium chloride
$Al(NO_3)_3$	aluminum nitrate
$Cu(ClO_4)_2$	copper(II) perchlorate (or cupric perchlorate)

In the chemical formulas for aluminum nitrate and copper(II) perchlorate, parentheses followed by the appropriate subscript are used because the compounds contain two or more polyatomic ions.

SAMPLE EXERCISE 2.12

Name the following compounds: **(a)** K_2SO_4; **(b)** $Ba(OH)_2$; **(c)** $FeCl_3$.

Solution Each compound is ionic and is named using the guidelines we have already discussed. In naming ionic compounds, it is important to recognize polyatomic ions and to determine the charge of cations with variable charge. **(a)** The cation in this compound is K^+, and the anion is SO_4^{2-}. (If you thought the compound contained S^{2-} and O^{2-} ions, you failed to recognize the polyatomic sulfate ion.) Putting together the names of the ions, we have the name of the compound, potassium sulfate. **(b)** In this case the compound is composed of Ba^{2+} and OH^- ions. Ba^{2+} is the barium ion and OH^- is the hydroxide ion. Thus, the compound is called barium hydroxide. **(c)** You must determine the charge of Fe in this compound because iron can have variable charges. Because the compound contains three Cl^- ions, the cation must be Fe^{3+}, which is the iron(III) or ferric ion. The Cl^- ion is the chloride ion. Thus, the compound is iron(III) chloride or ferric chloride.

PRACTICE EXERCISE

Name the following compounds: **(a)** NH_4Br; **(b)** Cr_2O_3; **(c)** $Co(NO_3)_2$.
Answers: **(a)** ammonium bromide; **(b)** chromium(III) oxide; **(c)** cobalt(II) nitrate

SAMPLE EXERCISE 2.13

Write the chemical formulas for the following compounds: **(a)** potassium sulfide; **(b)** calcium hydrogen carbonate; **(c)** nickel(II) perchlorate.

Solution In going from the name of an ionic compound to its chemical formula, you must know the charges of the ions to determine the subscripts. **(a)** The potassium ion is K^+, and the sulfide ion is S^{2-}. Because ionic compounds are electrically neutral, two K^+ ions are required to balance the charge of one S^{2-} ion, giving the empirical formula of the compound, K_2S. **(b)** The calcium ion is Ca^{2+}. The carbonate ion is CO_3^{2-}, so the hydrogen carbonate ion is HCO_3^-. Two HCO_3^- ions are needed to balance the positive charge of Ca^{2+}, giving $Ca(HCO_3)_2$. **(c)** The nickel(II) ion is Ni^{2+}. The perchlorate ion is ClO_4^-. Two ClO_4^- ions are required to balance the charge on one Ni^{2+} ion, giving $Ni(ClO_4)_2$.

PRACTICE EXERCISE

Give the chemical formula for **(a)** magnesium sulfate; **(b)** silver sulfide; **(c)** lead(II) nitrate.
Answers: **(a)** $MgSO_4$; **(b)** Ag_2S; **(c)** $Pb(NO_3)_2$

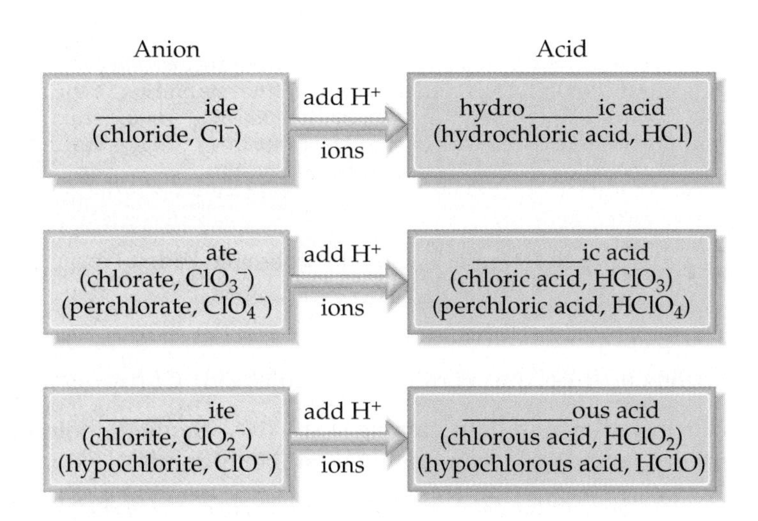

◀ **Figure 2.27** Summary of the way in which anion names and acid names are related. The prefixes *per-* and *hypo-* are retained in going from the anion to the acid.

Names and Formulas of Acids

Acids are an important class of hydrogen-containing compounds and are named a special way. For our present purposes, an *acid* is a substance whose molecules yield hydrogen ions (H^+) when dissolved in water. When we encounter the chemical formula for an acid at this stage of the course, it will be written with H as the first element, as in HCl and H_2SO_4.

We can consider an acid to be composed of an anion connected to enough H^+ ions to totally neutralize or balance the anion's charge. Thus, the SO_4^{2-} ion requires two H^+ ions, forming H_2SO_4. The name of an acid is related to the name of its anion, as summarized in Figure 2.27 ▲.

1. *Acids based on anions whose names end in* -ide. Anions whose names end in *-ide* have associated acids that have the *hydro-* prefix and an *-ic* ending, as in the following examples:

Anion	Corresponding Acid
Cl^- (chloride)	HCl (hydrochloric acid)
S^{2-} (sulfide)	H_2S (hydrosulfuric acid)

2. *Acids based on anions whose names end in* -ate *or* -ite. Anions whose names end in *-ate* have associated acids with an *-ic* ending, whereas anions whose names end in *-ite* have acids with an *-ous* ending. Prefixes in the name of the anion are retained in the name of the acid. These rules are illustrated by the oxyacids of chlorine:

Anion	Corresponding Acid
ClO_4^- (perchlorate)	$HClO_4$ (perchloric acid)
ClO_3^- (chlorate)	$HClO_3$ (chloric acid)
ClO_2^- (chlorite)	$HClO_2$ (chlorous acid)
ClO^- (hypochlorite)	HClO (hypochlorous acid)

SAMPLE EXERCISE 2.14

Name the following acids: **(a)** HCN; **(b)** HNO_3; **(c)** H_2SO_4; **(d)** H_2SO_3.

Solution **(a)** The anion from which this acid is derived is CN^-, the cyanide ion. Because this ion has an *-ide* ending, the acid is given a *hydro-* prefix and an *-ic* ending: hydrocyanic acid. Only water solutions of HCN are referred to as hydrocyanic acid: The pure compound, which is a gas under normal conditions, is called hydrogen cyanide.

Both hydrocyanic acid and hydrogen cyanide are *extremely* toxic. **(b)** Because NO_3^- is the nitrate ion, HNO_3 is called nitric acid (the *-ate* ending of the anion is replaced with an *-ic* ending in naming the acid). **(c)** Because SO_4^{2-} is the sulfate ion, H_2SO_4 is called sulfuric acid. **(d)** Because SO_3^{2-} is the sulfite ion, H_2SO_3 is sulfurous acid (the *-ite* ending of the anion is replaced with an *-ous* ending).

PRACTICE EXERCISE

Give the chemical formulas for **(a)** hydrobromic acid; **(b)** carbonic acid.
Answers: **(a)** HBr; **(b)** H_2CO_3

Names and Formulas of Binary Molecular Compounds

The procedures used for naming *binary* (two-element) molecular compounds are similar to those used for naming ionic compounds:

1. *The name of the element farthest to the left in the periodic table is usually written first.* An exception to this rule occurs in the case of compounds that contain oxygen. Oxygen is always written last except when combined with fluorine.
2. *If both elements are in the same group in the periodic table, the lower one is named first.*
3. *The name of the second element is given an* -ide *ending.*
4. *Greek prefixes (Table 2.6 ◀) are used to indicate the number of atoms of each element.* The prefix *mono-* is never used with the first element. When the prefix ends in *a* or *o* and the name of the second element begins with a vowel (such as *oxide*), the *a* or *o* is often dropped.

TABLE 2.6 Prefixes Used in Naming Binary Compounds Formed Between Nonmetals

Prefix	Meaning
Mono-	1
Di-	2
Tri-	3
Tetra-	4
Penta-	5
Hexa-	6
Hepta-	7
Octa-	8
Nona-	9
Deca-	10

The following examples illustrate these rules:

Cl_2O dichlorine monoxide NF_3 nitrogen trifluoride
N_2O_4 dinitrogen tetroxide P_4S_{10} tetraphosphorus decasulfide

It is important to realize that you cannot predict the formulas of most molecular substances in the same way that you predict the formulas of ionic compounds. That is why we name them using prefixes that explicitly indicate their composition. Compounds that contain hydrogen and one other element are an important exception, however. These compounds can be treated as if they contained H^+ ions. Thus, HCl is hydrogen chloride (this is the name used for the pure compound; water solutions of HCl are called hydrochloric acid). Similarly, H_2S is hydrogen sulfide.

SAMPLE EXERCISE 2.15

Name the following compounds: **(a)** SO_2; **(b)** PCl_5; **(c)** N_2O_3.

Solution The compounds consist entirely of nonmetals, so they are probably molecular rather than ionic. Using the prefixes in Table 2.6, we have **(a)** sulfur dioxide, **(b)** phosphorus pentachloride, and **(c)** dinitrogen trioxide.

PRACTICE EXERCISE

Give the chemical formula for **(a)** silicon tetrabromide; **(b)** disulfur dichloride.
Answers: **(a)** $SiBr_4$; **(b)** S_2Cl_2

2.9 Some Simple Organic Compounds

The study of compounds of carbon is called **organic chemistry**. Compounds that contain carbon and hydrogen, often in combination with oxygen, nitrogen, or other elements, are called *organic compounds*. We will examine organic compounds and organic chemistry in some detail in Chapter 25. You will see a number of organic compounds throughout this text; many of them have practical applications or are relevant to the chemistry of biological systems. Here we present a very brief introduction to some of the simplest organic compounds so as to provide you with a sense of what these molecules look like and how they are named.

Alkanes

Compounds that contain only carbon and hydrogen are called **hydrocarbons**. In the most basic class of hydrocarbons, each carbon atom is bonded to four other atoms. These compounds are called **alkanes**. The three simplest alkanes, which contain one, two, and three carbon atoms, respectively, are methane (CH_4), ethane (C_2H_6), and propane (C_3H_8). The structural formulas of these three alkanes are as follows:

$$\begin{array}{ccc} \begin{array}{c} \mathrm{H} \\ | \\ \mathrm{H-C-H} \\ | \\ \mathrm{H} \end{array} & \begin{array}{c} \mathrm{H\;\;H} \\ |\;\;\;| \\ \mathrm{H-C-C-H} \\ |\;\;\;| \\ \mathrm{H\;\;H} \end{array} & \begin{array}{c} \mathrm{H\;\;H\;\;H} \\ |\;\;\;|\;\;\;| \\ \mathrm{H-C-C-C-H} \\ |\;\;\;|\;\;\;| \\ \mathrm{H\;\;H\;\;H} \end{array} \\ \text{Methane} & \text{Ethane} & \text{Propane} \end{array}$$

Each of the alkanes has a name that ends in *-ane*. Longer alkanes can be made by adding additional carbon atoms to the "skeleton" of the molecule. For alkanes with five or more carbon atoms, the names are derived from prefixes like those in Table 2.6. An alkane with eight carbon atoms, for example, is called *octane* (C_8H_{18}), where the *octa-* prefix for eight is combined with the *-ane* ending for an alkane. Gasoline consists primarily of octanes, as will be discussed in Chapter 25.

Some Derivatives of Alkanes

Other classes of organic compounds are obtained when hydrogen atoms of alkanes are replaced with *functional groups*, which are specific groups of atoms. An **alcohol**, for example, is obtained by replacing an H atom of an alkane with an —OH group. The name of the alcohol is derived from that of the alkane by adding an *-ol* ending:

$$\begin{array}{ccc} \begin{array}{c} \mathrm{H} \\ | \\ \mathrm{H-C-OH} \\ | \\ \mathrm{H} \end{array} & \begin{array}{c} \mathrm{H\;\;H} \\ |\;\;\;| \\ \mathrm{H-C-C-OH} \\ |\;\;\;| \\ \mathrm{H\;\;H} \end{array} & \begin{array}{c} \mathrm{H\;\;H\;\;H} \\ |\;\;\;|\;\;\;| \\ \mathrm{H-C-C-C-OH} \\ |\;\;\;|\;\;\;| \\ \mathrm{H\;\;H\;\;H} \end{array} \\ \text{Methanol} & \text{Ethanol} & \text{1-Propanol} \end{array}$$

Alcohols have properties that are very different from the alkanes from which they are obtained. For example, methane, ethane, and propane are all colorless gases under normal conditions, whereas methanol, ethanol, and propanol are colorless liquids. We will discuss the reasons for these differences in properties in Chapter 11.

The prefix "1" in the name of 1-propanol indicates that the replacement of H with OH has occurred at one of the "outer" carbon atoms rather than the "middle" carbon atom; a different compound called 2-propanol (also known as isopropyl alcohol) is obtained if the OH functional group is attached to the middle carbon atom. Ball-and-stick models of 1-propanol and 2-propanol are presented in Figure 2.28 ▶. As you will learn in Chapter 25, the nomenclature of organic compounds provides ways in which we can unambiguously define which atoms are bonded to one another.

Much of the richness of organic chemistry is possible because compounds with long chains of carbon-carbon bonds are found in nature or can be synthesized. The series of alkanes and alcohols that begins with methane, ethane, and

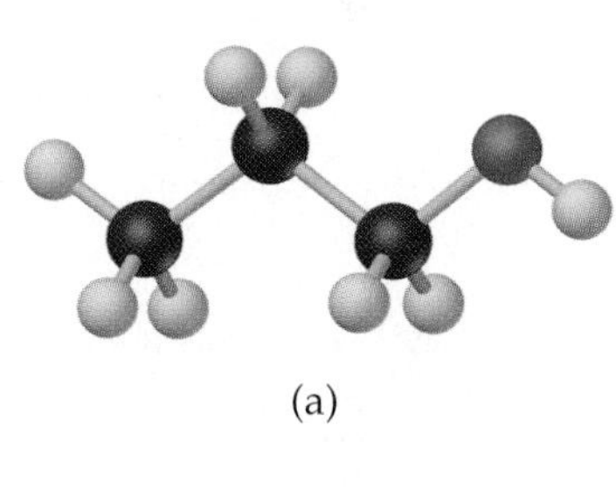

(a)

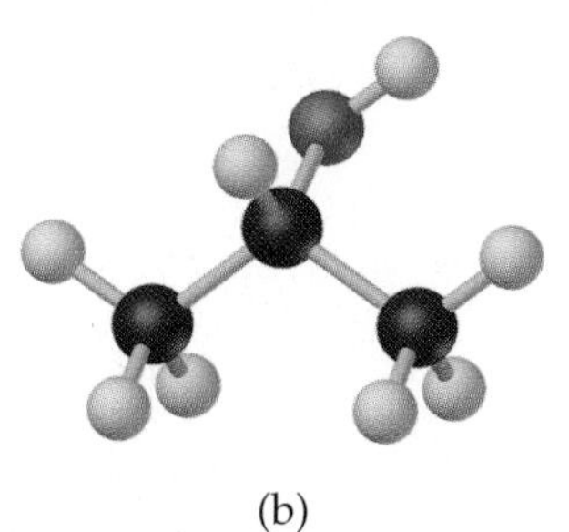

(b)

▲ **Figure 2.28** Ball-and-stick models of the two forms of propanol (C_3H_8O): (a) 1-propanol, in which the OH group is attached to one of the end carbon atoms, and (b) 2-propanol, in which the OH group is attached to the middle carbon atom.

propane can be extended for as long as we desire, in principle. The properties of alkanes and alcohols change as the chains get longer. Octanes, which are alkanes with eight carbon atoms, are liquids under normal conditions. If the alkane series is extended to tens of thousands of carbon atoms, we obtain *polyethylene*, a solid substance that is used to make thousands of plastic products, such as plastic bags, food containers, and laboratory equipment. Polyethylene is an example of a *polymer*, a substance that is made by adding together thousands of smaller molecules. We will discuss polymers in greater detail in Chapter 12.

In all the compounds discussed so far, the carbon atoms in the structural formula are linked to four other atoms by a single line; in later chapters you will learn that the single line represents a *single bond* between the carbon atom and the other atom. Carbon, however, can also form *multiple bonds* to itself and to other atoms, such as oxygen and nitrogen. Multiple bonds greatly change the properties of organic molecules and are one of the main reasons that many of you will take a year-long course dedicated entirely to organic chemistry! Some familiar organic substances that contain double bonds to carbon are shown below. In each case, we have given the proper name of the compound, which is derived from the prefix of an alkane, and the "common" name by which you probably know the substance:

```
H       H          H   O                H   O   H
 \     /           |   ‖                |   ‖   |
  C = C        H — C — C — OH       H — C — C — C — H
 /     \           |                    |       |
H       H          H                    H       H

   Ethene         Ethanoic acid          Propanone
 (ethylene)       (acetic acid)          (acetone)
```

Ethylene is an *unsaturated hydrocarbon*, which is a compound with a carbon-carbon multiple bond. The carbon-carbon double bond makes ethylene much more reactive than alkanes. Acetic acid is a *carboxylic acid*. It is the characteristic component of vinegar. Acetone is a *ketone*. Acetone is a common organic solvent that is used in households as a lacquer and nail-polish remover. Figure 2.29 ▼ shows space-filling models of acetic acid and acetone. You will encounter other organic molecules throughout the text, and you should note the numbers of carbon atoms involved and the types of other atoms to which carbon is bonded. As noted earlier, we will provide a more complete discussion of organic chemistry in Chapter 25.

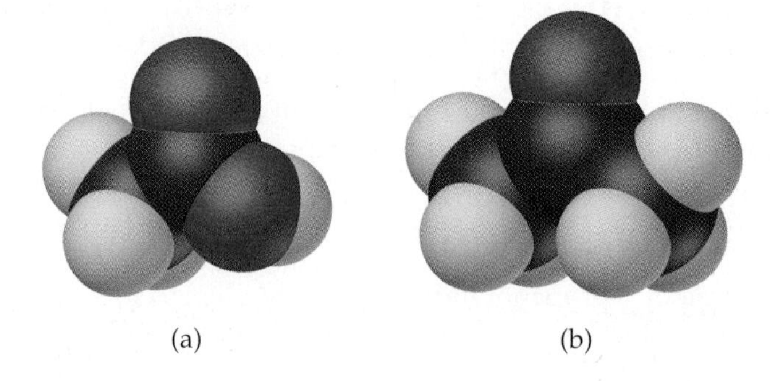

▶ **Figure 2.29** Space-filling models of (a) acetic acid ($HC_2H_3O_2$), and (b) acetone (C_3H_6O).

SAMPLE EXERCISE 2.16

Consider the alkane called *pentane*. **(a)** Assuming that the carbon atoms are in a straight line, write a structural formula for pentane. **(b)** What is the molecular formula for pentane?

Solution **(a)** Alkanes contain only carbon and hydrogen, and each carbon atom is attached to four other atoms. The name pentane contains the prefix *penta-* for five (Table 2.6), so we can assume that pentane contains five carbon atoms bonded in a chain. If we then add enough hydrogen atoms to make four bonds to each carbon atom, we obtain the following structural formula:

```
    H   H   H   H   H
    |   |   |   |   |
H — C — C — C — C — C — H
    |   |   |   |   |
    H   H   H   H   H
```

This form of pentane is often called *n*-pentane, where the *n*- stands for "normal" because all five carbon atoms are in one line in the structural formula.

(b) Once the structural formula is written, we can determine the molecular formula by counting the atoms present. Thus, *n*-pentane has the formula C_5H_{12}.

PRACTICE EXERCISE

Butane is the alkane with four carbon atoms. **(a)** What is the molecular formula of butane? **(b)** What is the name and molecular formula of an alcohol derived from butane?

Answers: **(a)** C_4H_{10}; **(b)** butanol, $C_4H_{10}O$

Summary and Key Terms

Sections 2.1 and 2.2 **Atoms** are the basic building blocks of matter; they are the smallest units of an element that can combine with other elements. Atoms are composed of even smaller particles, called **subatomic particles**. Some of these subatomic particles are charged and follow the usual behavior of charged particles: Particles with the same charge repel one another, whereas particles with unlike charges are attracted to one another. We considered some of the important experiments that led to the discovery and characterization of subatomic particles. Thomson's experiments on the behavior of **cathode rays** in magnetic and electric fields led to the discovery of the electron and allowed its charge-to-mass ratio to be measured; Millikan's oil-drop experiment determined the charge of the electron; Becquerel's discovery of **radioactivity**, the spontaneous emission of radiation by atoms, gave further evidence that the atom has a substructure; and Rutherford's studies of how thin metal foils scatter α-particles showed that the atom has a dense, positively charged **nucleus**.

Section 2.3 Atoms have a nucleus that contains **protons** and **neutrons**; **electrons** move in the space around the nucleus. The magnitude of the charge of the electron, 1.602×10^{-19} C, is called the **electronic charge**. The charges of particles are usually represented as multiples of this charge; thus, an electron has a 1− charge, and a proton has a 1+ charge. The masses of atoms are usually expressed in terms of **atomic mass units** (1 amu = 1.66054×10^{-24} g). The dimensions of atoms are often expressed in units of **angstroms** (1 Å = 10^{-10} m).

Elements can be classified by **atomic number**, the number of protons in the nucleus of an atom. All atoms of a given element have the same atomic number. The **mass number** of an atom is the sum of the numbers of protons and neutrons. Atoms of the same element that differ in mass number are known as **isotopes**. An atom of a specific isotope is called a **nuclide**.

Section 2.4 The atomic mass scale is defined by assigning a mass of exactly 12 amu to a ^{12}C atom. The **atomic weight** (average atomic mass) of an element can be calculated from the relative abundances and masses of that element's isotopes. The **mass spectrometer** provides the most direct and accurate means of experimentally measuring atomic (and molecular) weights.

Section 2.5 The **periodic table** is an arrangement of the elements in order of increasing atomic number. Elements with similar properties are placed in vertical columns. The elements in a column are known as a periodic **group**. The **metallic elements**, which comprise the majority of the elements, dominate the left side and the middle of the table; the **nonmetallic elements** are located on the upper right side. Many of the elements that lie along the line that separates metals from nonmetals are **metalloids**.

Section 2.6 Atoms can combine to form **molecules**. Compounds composed of molecules (**molecular compounds**) usually contain only nonmetallic elements. A molecule that contains two atoms is called a **diatomic molecule**. The composition of a substance is given by its **chemical formula**. A molecular substance can be represented by its **empirical formula**, which gives the relative numbers of atoms of each kind. It is usually represented by its **molecular formula**, however, which gives the actual numbers of each type of atom in a molecule. **Structural formulas** show the order in which the atoms in a molecule are connected. Ball-and-stick models and space-filling models are often used to represent molecules.

Section 2.7 Atoms can either gain or lose electrons, forming charged particles called **ions**. Metals tend to lose electrons, becoming positively charged ions (**cations**). Nonmetals tend to gain electrons, forming negatively charged ions (**anions**). Because **ionic compounds** are electrically neutral, containing both cations and anions, they usually contain both metallic and nonmetallic elements. Atoms that are joined together, as in a molecule, but carry a net charge are called **polyatomic ions**. The chemical formulas used for ionic compounds are empirical formulas, which can be written readily if the charges of the ions are known. The total positive charge of the cations in an ionic compound equals the total negative charge of the anions.

Section 2.8 The set of rules for naming chemical compounds is called **chemical nomenclature**. We studied the systematic rules used for naming three classes of inorganic substances: ionic compounds, acids, and binary molecular compounds. In naming an ionic compound, the cation is named first and then the anion. Cations formed from metal atoms have the same name as the metal. If the metal can form cations of differing charges, the charge is given using Roman numerals. Monatomic anions have names ending in *-ide*. Polyatomic anions containing oxygen and another element **(oxyanions)** have names ending in *-ate* or *-ite*.

Section 2.9 **Organic chemistry** is the study of compounds that contain carbon. The simplest class of organic molecules are the **hydrocarbons**, which contain only carbon and hydrogen. Hydrocarbons in which each carbon atom is attached to four other atoms are called **alkanes**. Alkanes have names that end in *-ane*, such as methane and ethane. Other organic compounds are formed when an H atom of a hydrocarbon is replaced with a functional group. An **alcohol**, for example, is a compound in which an H atom of a hydrocarbon is replaced by an OH functional group. Alcohols have names that end in *-ol*, such as methanol and ethanol. Other organic molecules have multiple bonds between a carbon atom and other atoms.

Exercises

Atomic Theory and the Discovery of Atomic Structure

2.1 How does Dalton's atomic theory account for the fact that when 1.000 g of water is decomposed into its elements, 0.111 g of hydrogen and 0.889 g of oxygen are obtained regardless of the source of the water?

2.2 Hydrogen sulfide is composed of two elements: hydrogen and sulfur. In an experiment, 6.500 g of hydrogen sulfide is fully decomposed into its elements. **(a)** If 0.384 g of hydrogen is obtained in this experiment, how many grams of sulfur must be obtained? **(b)** What fundamental law does this experiment demonstrate? **(c)** How is this law explained by Dalton's atomic theory?

2.3 A chemist finds that 30.82 g of nitrogen will react with 17.60 g, 35.20 g, 70.40 g, or 88.00 g of oxygen to form four different compounds. **(a)** Calculate the mass of oxygen per gram of nitrogen in each compound. **(b)** How do the numbers in part (a) support Dalton's atomic theory?

2.4 In a series of experiments, a chemist prepared three different compounds that contain only iodine and fluorine and determined the mass of each element in each compound:

Compound	Mass of Iodine (g)	Mass of Fluorine (g)
1	4.75	3.56
2	7.64	3.43
3	9.41	9.86

(a) Calculate the mass of fluorine per gram of iodine in each compound. **(b)** How do the numbers in part (a) support the atomic theory?

2.5 Summarize the evidence used by J. J. Thomson to argue that cathode rays consist of negatively charged particles.

2.6 A negatively charged particle is caused to move between two electrically charged plates, as illustrated in Figure 2.8. **(a)** Why does the path of the charged particle bend? **(b)** As the charge on the plates is increased, would you expect the bending to increase, decrease, or stay the same? **(c)** As the mass of the particle is increased while the speed of the particles remains the same, would you expect the bending to increase, decrease, or stay the same? **(d)** An unknown particle is sent through the apparatus. Its path is deflected in the opposite direction from the negatively charged particle, and it is deflected by a smaller magnitude. What can you conclude about this unknown particle?

2.7 **(a)** What is the purpose of the X-ray source in the Millikan oil-drop experiment (Figure 2.5)? **(b)** As shown in Figure 2.5, the positively-charged plate is above the negatively-charged plate. What do you think would be the effect on the rate of oil drops descending if the charges on the plates were reversed (negative above positive)? **(c)** In his original series of experiments, Millikan measured the charge on 58 separate oil drops. Why do you suppose he chose so many drops before reaching his final conclusions?

2.8 Millikan determined the charge on the electron by studying the static charges on oil drops falling in an electric field. A student carried out this experiment using several oil drops for her measurements and calculated the charges on the drops. She obtained the following data:

Droplet	Calculated Charge (C)
A	1.60×10^{-19}
B	3.15×10^{-19}
C	4.81×10^{-19}
D	6.31×10^{-19}

(a) What is the significance of the fact that the droplets carried different charges? **(b)** What conclusion can the student draw from these data regarding the charge of the electron? **(c)** What value (and to how many significant figures) should she report for the electronic charge?

2.9 **(a)** In Figure 2.8, the γ rays are not deflected by an electric field. What can we conclude about γ-radiation from this observation? **(b)** Why are α and β rays deflected in opposite directions by an electric field, as illustrated in Figure 2.8?

2.10 Why is Rutherford's nuclear model of the atom more consistent with the results of his α-particle scattering experiment than Thomson's "plum-pudding" model?

Modern View of Atomic Structure; Atomic Weights

2.11 The radius of an atom of krypton (Kr) is about 1.9 Å. **(a)** Express this distance in nanometers (nm) and in picometers (pm). **(b)** How many krypton atoms would have to be lined up to span 1.0 mm? **(c)** If the atom is assumed to be a sphere, what is the volume in cm^3 of a single Kr atom?

2.12 An atom of rhodium (Rh) has a diameter of about 2.5×10^{-8} cm. **(a)** What is the radius of a rhodium atom in angstroms (Å) and in meters (m). **(b)** How many Rh atoms would have to be placed side by side to span a distance of 6.0 μm **(c)** If the atom is assumed to be a sphere, what is the volume in m^3 of a single Rh atom?

2.13 Answer the following questions without referring to Table 2.1: **(a)** What are the main subatomic particles that make up the atom? **(b)** What is the charge, in units of the electronic charge, of each of the particles? **(c)** Which of the particles is the most massive? Which is the least massive?

2.14 Determine whether each of the following statements is true or false; if false, correct the statement to make it true: **(a)** The nucleus has most of the mass and comprises most of the volume of an atom; **(b)** every atom of a given element has the same number of protons; **(c)** the number of electrons in an atom equals the number of neutrons in the atom; **(d)** the protons in the nucleus of the helium atom are held together by a force called the strong nuclear force.

2.15 How many protons, neutrons, and electrons are in the following atoms: **(a)** ^{28}Si; **(b)** ^{60}Ni; **(c)** ^{85}Rb; **(d)** ^{128}Xe; **(e)** ^{195}Pt; **(f)** ^{238}U?

2.16 Each of the following nuclides is used in medicine. Indicate the number of protons and neutrons in each nuclide: **(a)** phosphorus–32; **(b)** chromium–51; **(c)** cobalt–60; **(d)** technetium–99; **(e)** iodine–131; **(f)** thallium–201.

2.17 Fill in the gaps in the following table, assuming each column represents a neutral atom:

Symbol	^{52}Cr			
Protons		33		77
Neutrons		42	20		
Electrons			20	86	
Mass no.				222	193

2.18 Fill in the gaps in the following table assuming each column represents a neutral atom:

Symbol	^{121}Sb				
Protons		38			94
Neutrons		50	108		
Electrons			74	57	
Mass no.				139	239

2.19 Write the correct symbol, with both superscript and subscript, for each of the following. Use the list of elements on the front inside cover as needed: **(a)** the nuclide of hafnium that contains 107 neutrons; **(b)** the isotope of argon with mass number 40; **(c)** an α particle; **(d)** the isotope of indium with mass number 115; **(e)** the nuclide of silicon that has an equal number of protons and neutrons.

2.20 One way in which the Earth's evolution as a planet can be understood is by measuring the amounts of certain nuclides in rocks. One quantity recently measured is the ratio of ^{129}Xe to ^{130}Xe in some minerals. In what way do these two nuclides differ from one another, and in what respects are they the same?

2.21 **(a)** What isotope is used as the standard in establishing the atomic mass scale? **(b)** The atomic weight of chlorine is reported as 35.5, yet no atom of chlorine has the mass of 35.5 amu. Explain.

2.22 **(a)** What is the mass in amu of a carbon-12 atom? **(b)** Why is the atomic weight of carbon reported as 12.011 in the table of elements and the periodic table in the front inside cover of this text?

2.23 The element lead (Pb) consists of four naturally occurring isotopes with masses 203.97302, 205.97444, 206.97587, and 207.97663 amu. The relative abundances of these four isotopes are 1.4, 24.1, 22.1, and 52.4%, respectively. From these data, calculate the average atomic mass of lead.

2.24 Only two isotopes of copper occur naturally, ^{63}Cu (mass = 62.9296 amu; abundance 69.17%) and ^{65}Cu (mass = 64.9278 amu; abundance 30.83%). Calculate the atomic weight (average atomic mass) of copper.

2.25 **(a)** In what fundamental way is mass spectrometry related to Thomson's cathode-ray experiments (Figure 2.4)? **(b)** What are the labels on the axes of a mass spectrum? **(c)** In order to measure the mass spectrum of an atom, the atom must first lose or gain one or more electrons. Why is this so?

2.26 **(a)** The mass spectrometer in Figure 2.13 has a magnet as one of its components. What is the purpose of the magnet? **(b)** The atomic weight of Cl is 35.5 amu. However, the mass spectrum of Cl (Figure 2.14) does not show a peak at this mass. Explain. **(c)** A mass spectrum of phosphorus (P) atoms shows only a single peak at a mass of 31. What can you conclude from this observation?

2.27 Naturally occurring magnesium has the following isotopic abundances:

Isotope	Abundance	Mass
^{24}Mg	78.99%	23.98504
^{25}Mg	10.00%	24.98584
^{26}Mg	11.01%	25.98259

(a) What is the average atomic mass of Mg? **(b)** Sketch the mass spectrum of Mg.

2.28 Mass spectrometry is more often applied to molecules than to atoms. We will see in Chapter 3 that the *molecular weight* of a molecule is the sum of the atomic weights of the atoms in the molecule. The mass spectrum of H_2 is taken under conditions that prevent decomposition into H atoms. The two naturally occurring isotopes of hydrogen are ^{1}H (mass = 1.00783 amu; abundance 99.9885%) and ^{2}H (mass = 2.01410 amu; abundance 0.0115%). **(a)** How many peaks will the mass spectrum have? **(b)** Give the relative atomic masses of each of these peaks. **(c)** Which peak will be the largest, and which the smallest?

The Periodic Table; Molecules and Ions

2.29 For each of the following elements, write its chemical symbol, locate it in the periodic table, and indicate whether it is a metal, metalloid, or nonmetal: **(a)** silver; **(b)** helium; **(c)** phosphorus; **(d)** cadmium; **(e)** calcium; **(f)** bromine; **(g)** arsenic.

2.30 Locate each of the following elements in the periodic table; indicate whether it is a metal, metalloid, or nonmetal; and give the name of the element: **(a)** Li; **(b)** Sc; **(c)** Ge; **(d)** Yb; **(e)** Mn; **(f)** Au; **(g)** Te.

2.31 For each of the following elements, write its chemical symbol, determine the name of the group to which it belongs (Table 2.3), and indicate whether it is a metal, metalloid, or nonmetal: **(a)** potassium; **(b)** iodine; **(c)** magnesium; **(d)** argon; **(e)** sulfur.

2.32 The elements of group 4A show an interesting change in properties with increasing period. Give the name and chemical symbol of each element in the group, and label it as a nonmetal, metalloid, or metal.

2.33 What can we tell about a compound when we know the empirical formula? What additional information is conveyed by the molecular formula? By the structural formula? Explain in each case.

2.34 Two compounds have the same empirical formula. One substance is a gas, the other is a viscous liquid. How is it possible for two substances with the same empirical formula to have markedly different properties?

2.35 Determine the molecular and empirical formulas of the following: **(a)** The organic solvent *benzene*, which has six carbon atoms and six hydrogen atoms. **(b)** The compound *silicon tetrachloride*, which has a silicon atom and four chlorine atoms and is used in the manufacture of computer chips.

2.36 Write the molecular and empirical formulas of the following: **(a)** the reactive substance *diborane*, which has two boron atoms and six hydrogen atoms; **(b)** the sugar called *glucose*, which has six carbon atoms, twelve hydrogen atoms, and six oxygen atoms.

2.37 How many hydrogen atoms are in each of the following: **(a)** C_2H_5OH; **(b)** $Ca(CH_3COO)_2$; **(c)** $(NH_4)_3PO_4$?

2.38 How many of the indicated atoms are represented by each chemical formula: **(a)** carbon atoms in $C_2H_5COOCH_3$; **(b)** oxygen atoms in $Ca(ClO_3)_2$; **(c)** hydrogen atoms in $(NH_4)_2HPO_4$?

2.39 Write the molecular and structural formulas for the compounds represented by the following molecular models:

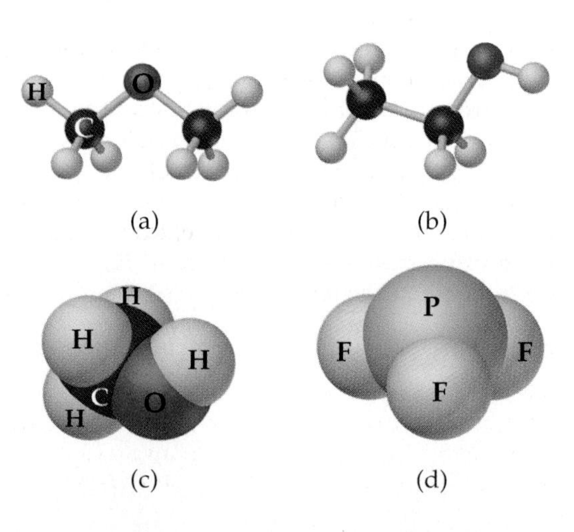

2.40 Write the molecular and structural formulas for the compounds represented by the following models:

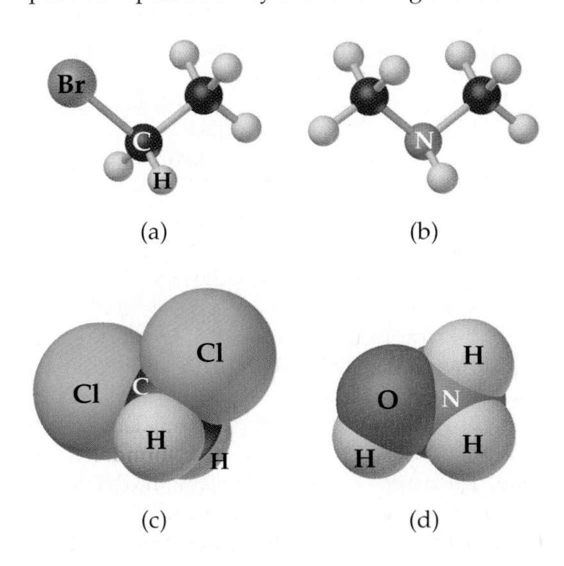

2.41 Write the empirical formula corresponding to each of the following molecular formulas: **(a)** Al_2Br_6; **(b)** C_8H_{10}; **(c)** $C_4H_8O_2$; **(d)** P_4O_{10}; **(e)** $C_6H_4Cl_2$; **(f)** $B_3N_3H_6$.

2.42 From the following list, find the groups of compounds that have the same empirical formula: C_2H_2, N_2O_4, C_2H_4, C_6H_6, NO_2, C_3H_6, C_4H_8.

2.43 Each of the following elements is capable of forming an ion in chemical reactions. By referring to the periodic table, predict the charge of the most stable ion of each: **(a)** Al; **(b)** Ca; **(c)** S; **(d)** I; **(e)** Cs.

2.44 Using the periodic table, predict the charges of the ions of the following elements: **(a)** Sc; **(b)** Sr; **(c)** P; **(d)** K; **(e)** F.

2.45 Using the periodic table to guide you, predict the formula and name of the compound formed by the following elements: **(a)** Ga and F; **(b)** Li and H; **(c)** Al and I; **(d)** K and S.

2.46 The most common charge associated with silver in its compounds is 1+. Indicate the empirical formulas you would expect for compounds formed between Ag and **(a)** iodine; **(b)** sulfur; **(c)** fluorine.

2.47 Predict the empirical formula for the ionic compound formed by **(a)** Ca^{2+} and Br^-; **(b)** NH_4^+ and Cl^-; **(c)** Al^{3+} and $C_2H_3O_2^-$; **(d)** K^+ and SO_4^{2-}; **(e)** Mg^{2+} and PO_4^{3-}.

2.48 Predict the chemical formulas of the compounds formed by the following pairs of ions: **(a)** NH_4^+ and SO_4^{2-}; **(b)** Cu^+ and S^{2-}; **(c)** La^{3+} and F^-; **(d)** Ca^{2+} and PO_4^{3-}; **(e)** Hg_2^{2+} and CO_3^{2-}.

2.49 Predict whether each of the following compounds is molecular or ionic: **(a)** B_2H_6; **(b)** CH_3OH; **(c)** $LiNO_3$; **(d)** Sc_2O_3; **(e)** CsBr; **(f)** NOCl; **(g)** NF_3; **(h)** Ag_2SO_4.

2.50 Which of the following are ionic, and which are molecular? **(a)** PF_5; **(b)** NaI; **(c)** SCl_2; **(d)** $Ca(NO_3)_2$; **(e)** $FeCl_3$; **(f)** LaP; **(g)** $CoCO_3$; **(h)** N_2O_4.

Naming Inorganic Compounds; Organic Molecules

2.51 Give the chemical formula for **(a)** chlorite ion; **(b)** chloride ion; **(c)** chlorate ion; **(d)** perchlorate ion; **(e)** hypochlorite ion.

2.52 Selenium, an element required nutritionally in trace quantities, forms compounds analogous to sulfur. Name the following ions: **(a)** SeO_4^{2-}; **(b)** Se^{2-}; **(c)** HSe^-; **(d)** $HSeO_3^-$.

2.53 Name the following ionic compounds: **(a)** AlF_3; **(b)** $Fe(OH)_2$; **(c)** $Cu(NO_3)_2$; **(d)** $Ba(ClO_4)_2$; **(e)** Li_3PO_4; **(f)** Hg_2S; **(g)** $Ca(C_2H_3O_2)_2$; **(h)** $Cr_2(CO_3)_3$; **(i)** K_2CrO_4; **(j)** $(NH_4)_2SO_4$.

2.54 Name the following ionic compounds: **(a)** Li_2O; **(b)** $Fe_2(CO_3)_3$; **(c)** NaClO; **(d)** $(NH_4)_2SO_3$; **(e)** $Sr(CN)_2$; **(f)** $Cr(OH)_3$; **(g)** $Co(NO_3)_2$; **(h)** NaH_2PO_4; **(i)** $KMnO_4$; **(j)** $Ag_2Cr_2O_7$.

2.55 Write the chemical formulas for the following compounds: **(a)** copper(I) oxide; **(b)** potassium peroxide; **(c)** aluminum hydroxide; **(d)** zinc nitrate; **(e)** mercury(I) bromide; **(f)** iron(III) carbonate; **(g)** sodium hypobromite.

2.56 Give the chemical formula for each of the following ionic compounds: **(a)** potassium dichromate; **(b)** cobalt(II) nitrate; **(c)** chromium(III) acetate; **(d)** sodium hydride; **(e)** calcium hydrogen carbonate; **(f)** barium bromate; **(g)** copper(II) perchlorate.

2.57 Give the name or chemical formula, as appropriate, for each of the following acids: **(a)** $HBrO_3$; **(b)** HBr; **(c)** H_3PO_4; **(d)** hypochlorous acid; **(e)** iodic acid; **(f)** sulfurous acid.

2.58 Provide the name or chemical formula, as appropriate, for each of the following acids: **(a)** hydrobromic acid; **(b)** hydrosulfuric acid; **(c)** nitrous acid; **(d)** H_2CO_3; **(e)** $HClO_3$; **(f)** $HC_2H_3O_2$.

2.59 Give the name or chemical formula, as appropriate, for each of the following molecular substances: **(a)** SF_6; **(b)** IF_5; **(c)** XeO_3; **(d)** dinitrogen tetroxide; **(e)** hydrogen cyanide; **(f)** tetraphosphorus hexasulfide.

2.60 The oxides of nitrogen are very important ingredients in determining urban air pollution. Name each of the following compounds: **(a)** N_2O; **(b)** NO; **(c)** NO_2; **(d)** N_2O_5; **(e)** N_2O_4.

2.61 Write the chemical formula for each substance mentioned in the following word descriptions (use the front inside cover to find the symbols for the elements you don't know). **(a)** Zinc carbonate can be heated to form zinc oxide and carbon dioxide. **(b)** On treatment with hydrofluoric acid, silicon dioxide forms silicon tetrafluoride and water. **(c)** Sulfur dioxide reacts with water to form sulfurous acid. **(d)** The substance hydrogen phosphide, commonly called phosphine, is a toxic gas. **(e)** Perchloric acid reacts with cadmium to form cadmium(II) perchlorate. **(f)** Vanadium(III) bromide is a colored solid.

2.62 Assume that you encounter the following phrases in your reading. What is the chemical formula for each substance mentioned? **(a)** Sodium hydrogen carbonate is used as a deodorant. **(b)** Calcium hypochlorite is used in some bleaching solutions. **(c)** Hydrogen cyanide is a very poisonous gas. **(d)** Magnesium hydroxide is used as a cathartic. **(e)** Tin(II) fluoride has been used as a fluoride additive in toothpastes. **(f)** When cadmium sulfide is treated with sulfuric acid, fumes of hydrogen sulfide are given off.

2.63 **(a)** What is a hydrocarbon? **(b)** Are all hydrocarbons alkanes? **(c)** Write the structural formula for ethane (C_2H_6). **(d)** *n*-Butane is the alkane with four carbon atoms in a line. Write a structural formula for this compound, and determine its molecular and empirical formulas.

2.64 **(a)** What ending is used for the names of alkanes? **(b)** Are all alkanes hydrocarbons? **(c)** Write the structural formula for propane (C_3H_8). **(d)** *n*-Hexane is an alkane with all its carbon atoms in one line. Write a structural formula for this compound and determine its molecular and empirical formulas. (*Hint:* You might need to refer to Table 2.6.)

2.65 **(a)** What is a functional group? **(b)** What functional group characterizes an alcohol? **(c)** With reference to Exercise 2.63, write a structural formula for *n*-butanol, the alcohol derived from *n*-butane by making a substitution on one of the end carbon atoms.

2.66 **(a)** What do ethane, ethanol, and ethylene have in common? **(b)** How does 1-propanol differ from propane? **(c)** Based on the structural formula for ethanoic acid given in the text, propose a structural formula for *propanoic acid*. What is its molecular formula?

Additional Exercises

2.67 Describe a major contribution to science made by each of the following scientists: **(a)** Dalton; **(b)** Thomson; **(c)** Millikan; **(d)** Rutherford.

[2.68] Suppose a scientist repeats the Millikan oil-drop experiment, but reports the charges on the drops using an unusual (and imaginary) unit called the *warmomb* (wa). He obtains the following data for four of the drops:

Droplet	Calculated Charge (wa)
A	3.84×10^{-8}
B	4.80×10^{-8}
C	2.88×10^{-8}
D	8.64×10^{-8}

(a) If all the droplets were the same size, which would fall most slowly through the apparatus? **(b)** From these data, what is the best choice for the charge of the electron in warmombs? **(c)** Based on your answer to part (b), how many electrons are there on each of the droplets? **(d)** What is the conversion factor between warmombs and coulombs?

2.69 What is radioactivity? Indicate whether you agree or disagree with the following statement, and indicate your reasons: Henri Becquerel's discovery of radioactivity shows that the atom is not indivisible, as had been believed for so long.

2.70 How did Rutherford interpret the following observations made during his α-particle scattering experiments? **(a)** Most α particles were not appreciably deflected as they passed through the gold foil. **(b)** A few α particles were deflected at very large angles. **(c)** What differences would you expect if beryllium foil were used instead of gold foil in the α-particle scattering experiment?

[2.71] An α particle is the nucleus of a ^{4}He atom. **(a)** How many protons and neutrons are in an α particle? **(b)** What force holds the protons and neutrons together in the α particle? **(c)** What is the charge on an α particle in units of electronic charge? **(d)** The charge-to-mass ratio of an α particle is 4.8224×10^4 C/g. Based on the charge on the particle, calculate its mass in grams and in amu. **(e)** By using the data in Table 2.1, compare your answer for part (d) with the sum of the masses of the individual subatomic particles. Can you explain the difference in mass? (If not, we will discuss such mass differences further in Chapter 21.)

2.72 The natural abundance of ^{3}He is 0.000137%. **(a)** How many protons, neutrons, and electrons are in an atom of ^{3}He? **(b)** Based on the sum of the masses of their subatomic particles, which is expected to be more massive, an atom of ^{3}He or an atom of ^{3}H (which is also called *tritium*)? **(c)** Based on your answer for part (b), what would need to be the precision of a mass spectrometer that is able to differentiate between peaks due to of $^3He^+$ and $^3H^+$?

2.73 A cube of gold that is 1.00 cm on a side has a mass of 19.3 g. A single gold atom has a mass of 197.0 amu. **(a)** How many gold atoms are in the cube? **(b)** From the information given, estimate the diameter in Å of a single gold atom. **(c)** What assumptions did you make in arriving at your answer for part (b)?

[2.74] The diameter of a rubidium atom is 4.95 Å. We will consider two different ways of placing the atoms on a surface. In arrangement A, all the atoms are lined up with one another. Arrangement B is called a *close-packed* arrangement because the atoms sit in the "depressions" formed by the previous row of atoms:

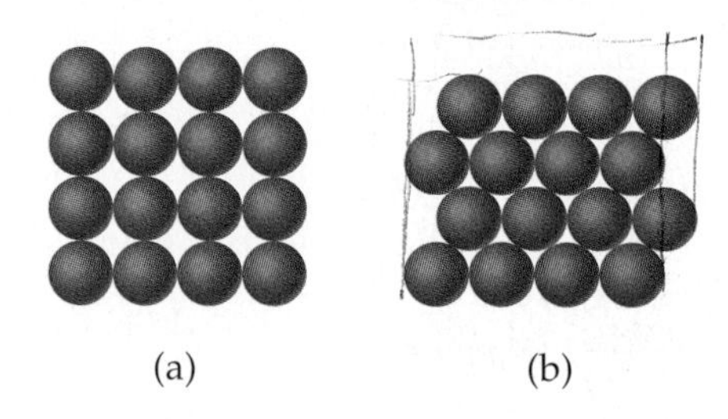

(a) Using arrangement A, how many Rb atoms could be placed on a square surface that is 1.0 cm on a side? **(b)** How many Rb atoms could be placed on a square surface that is 1.0 cm on a side using arrangement B? **(c)** By what factor has the number of atoms on the surface increased in going to arrangement B from arrangement A? If extended to three dimensions, which arrangement would lead to a greater density for Rb metal?

[2.75] **(a)** Assuming the dimensions of the nucleus and atom shown in Figure 2.12, what fraction of the *volume* of the atom is taken up by the nucleus? **(b)** Using the mass of the proton from Table 2.1 and assuming its diameter is 1.0×10^{-15} m, calculate the density of a proton in g/cm^3.

2.76 The element oxygen has three naturally occurring isotopes, with 8, 9, and 10 neutrons in the nucleus, respectively. **(a)** Write the full chemical symbols for these three isotopes. **(b)** Describe the similarities and differences between the three kinds of atoms of oxygen.

2.77 Chemists generally use the term *atomic weight* rather than *average atomic mass*. We state in the text that the latter term is more correct. By considering the units of weight and mass, can you explain why this is so?

2.78 Gallium (Ga) consists of two naturally occurring isotopes with masses of 68.926 and 70.925 amu. **(a)** How many protons and neutrons are in the nucleus of each isotope? Write the complete atomic symbol for each, showing the atomic number and mass number. **(b)** The average atomic mass of Ga is 69.72 amu. Calculate the abundance of each isotope.

[2.79] Using a suitable reference such as the *CRC Handbook of Chemistry and Physics*, look up the following information for nickel: **(a)** the number of known isotopes; **(b)** the atomic masses (in amu) and the natural abundance of the five most abundant isotopes.

[2.80] There are two different isotopes of bromine atoms. Under normal conditions, elemental bromine consists of Br_2 molecules (Figure 2.19), and the mass of a Br_2 molecule is the sum of the masses of the two atoms in the molecule. The mass spectrum of Br_2 consists of three peaks:

Mass (amu)	Relative Size
157.836	0.2569
159.834	0.4999
161.832	0.2431

(a) What is the origin of each peak (of what isotopes does each consist)? **(b)** What is the mass of each isotope? **(c)** Determine the average molecular mass of a Br_2 molecule. **(d)** Determine the average atomic mass of a bromine atom. **(e)** Calculate the abundances of the two isotopes.

2.81 It is common in mass spectrometry to assume that the mass of a cation is the same as that of its parent atom. **(a)** Using data in Table 2.1, determine the number of significant figures that must be reported before the difference in mass of ^1H and $^1H^+$ is significant. **(b)** What percentage of the mass of an ^1H atom does the electron represent?

2.82 *Bronze* is a metallic alloy often used in decorative applications and in sculpture. A typical bronze consists of copper, tin, and zinc, with lesser amounts of phosphorus and lead. Locate each of these five elements in the periodic table, write their symbols, and identify the group of the periodic table to which they belong.

2.83 From the following list of elements—Ar, H, Ga, Al, Ca, Br, Ge, K, O—pick the one that best fits each description; use each element only once: **(a)** an alkali metal; **(b)** an alkaline earth metal; **(c)** a noble gas; **(d)** a halogen; **(e)** a metalloid; **(f)** a nonmetal listed in group 1A; **(g)** a metal that forms a 3+ ion; **(h)** a nonmetal that forms a 2− ion; **(i)** an element that resembles aluminum.

2.84 The first atoms of seaborgium (Sg) were identified in 1974. The longest-lived isotope of Sg has a mass number of 266. **(a)** How many protons, electrons, and neutrons are in a ^{266}Sg nuclide? **(b)** Atoms of Sg are very unstable, and it is therefore difficult to study this element's properties. Based on the position of Sg in the periodic table, what element should it most closely resemble in its chemical properties?

2.85 From the molecular structures shown here, identify the one that corresponds to each of the following species: **(a)** chlorine gas; **(b)** propane, C_3H_8; **(c)** nitrate ion; **(d)** sulfur trioxide; **(e)** methyl chloride, CH_3Cl.

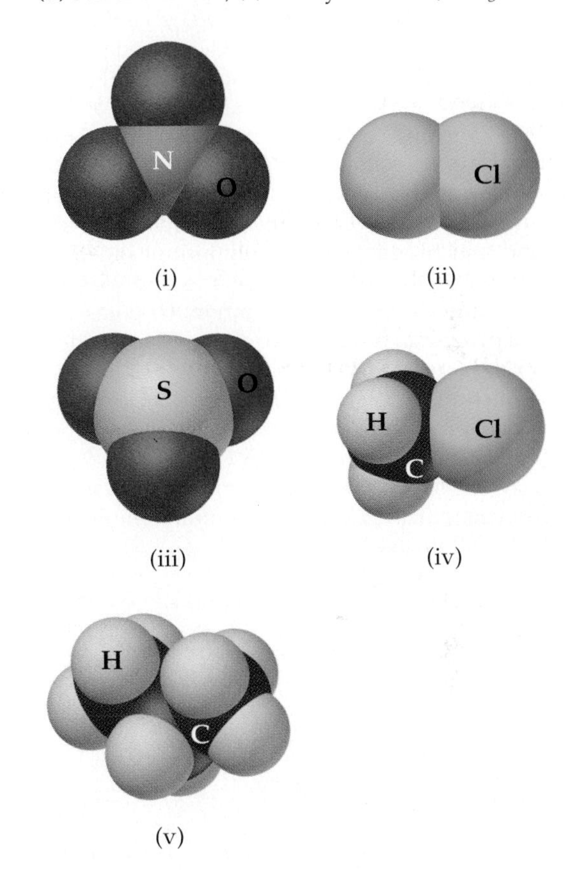

2.86 Fill in the gaps in the following table:

Symbol	$^{102}Ru^{3+}$				Ce
Protons		34	76		
Neutrons		46	116	74	82
Electrons		36		54	
Net charge			2+	1−	3+

2.87 Name each of the following oxides. Assuming that the compounds are ionic, what charge is associated with the metallic element in each case? **(a)** NiO; **(b)** MnO_2; **(c)** Cr_2O_3; **(d)** MoO_3.

2.88 Iodic acid has the molecular formula HIO_3. Write the formulas for the following: **(a)** the iodate anion; **(b)** the periodate anion; **(c)** the hypoiodite anion; **(d)** hypoiodous acid; **(e)** periodic acid.

2.89 Elements in the same group of the periodic table often form oxyanions with the same general formula. The anions are also named in a similar fashion. Based on these observations, suggest a chemical formula or name, as appropriate, for each of the following ions: **(a)** BrO_4^-; **(b)** SeO_3^{2-}; **(c)** arsenate ion; **(d)** hydrogen tellurate ion.

2.90 Give the chemical names of each of the following familiar compounds: **(a)** NaCl (table salt); **(b)** $NaHCO_3$ (baking soda); **(c)** NaOCl (in many bleaches); **(d)** NaOH (caustic soda); **(e)** $(NH_4)_2CO_3$ (smelling salts); **(f)** $CaSO_4$ (plaster of Paris).

2.91 Many familiar substances have common, unsystematic names. For each of the following, give the correct systematic name: **(a)** saltpeter, KNO_3; **(b)** soda ash, Na_2CO_3; **(c)** lime, CaO; **(d)** muriatic acid, HCl; **(e)** Epsom salts, $MgSO_4$; **(f)** milk of magnesia, $Mg(OH)_2$.

2.92 Many ions and compounds have very similar names, and there is great potential for confusing them. Write the correct chemical formulas to distinguish between **(a)** calcium sulfide and calcium hydrogen sulfide; **(b)** hydrobromic acid and bromic acid; **(c)** aluminum nitride and aluminum nitrite; **(d)** iron(II) oxide and iron(III) oxide; **(e)** ammonia and ammonium ion; **(f)** potassium sulfite and potassium bisulfite; **(g)** mercurous chloride and mercuric chloride; **(h)** chloric acid and perchloric acid.

[2.93] Using the *CRC Handbook of Chemistry and Physics*, find the density, melting point, and boiling point for **(a)** PF_3; **(b)** $SiCl_4$; **(c)** ethanol, C_2H_6O.

2.94 *Aromatic hydrocarbons* are hydrocarbons that are derived from benzene (C_6H_6). The structural formula for benzene is the following:

```
          H
          |
  H     C      H
    \ //  \   /
     C      C
     |      ||
     C      C
    / \\  /  \
  H     C      H
        |
        H
```

(a) What is the empirical formula of benzene? **(b)** Is benzene an alkane? Explain your answer briefly. **(c)** The alcohol derived from benzene, called *phenol*, is used as a disinfectant and a topical anesthetic. Propose a structural formula for phenol, and determine its molecular formula.

[2.95] Benzene (C_6H_6, see preceding exercise) contains 0.9226 g of carbon per gram of benzene; the remaining mass is hydrogen. The following table lists the carbon content per gram of substance for several other aromatic hydrocarbons:

Aromatic Hydrocarbon	Grams of Carbon per Gram of Hydrocarbon
Xylene	0.9051
Biphenyl	0.9346
Mesitylene	0.8994
Toluene	0.9125

(a) For benzene, calculate the mass of H that combines with 1 g of C. **(b)** For the hydrocarbons listed in the table, calculate the mass of H that combines with 1 g of C. **(c)** By comparing the results for part (b) to those for part (a), determine small-number ratios of hydrogen atoms per carbon atom for the hydrocarbons in the table. **(d)** Write empirical formulas for the hydrocarbons in the table.

[2.96] The compound *cyclohexane* is an alkane in which six carbon atoms form a ring. The partial structural formula of the compound is as follows:

```
     C
   /   \
  C     C
  |     |
  C     C
   \   /
     C
```

(a) Complete the structural formula for cyclohexane. **(b)** Is the molecular formula for cyclohexane the same as that for *n*-hexane, in which the carbon atoms are in a straight line? If possible, comment on the source of any differences. **(c)** Propose a structural formula for *cyclohexanol*, the alcohol derived from cyclohexane. **(d)** Propose a structural formula for *cyclohexene*, which has one carbon-carbon double bond. Does it have the same molecular formula as cyclohexane?

2.97 The periodic table helps organize the chemical behaviors of the elements. As a class discussion or as a short essay, describe how the table is organized, and mention as many ways as you can think of in which the position of an element in the table relates to the chemical and physical properties of the element.

eMedia Exercises

2.98 **(a)** After watching the **Multiple Proportions** movie (*eChapter 2.1*), show how the oxygen-to-hydrogen mass ratios of H_2O and H_2O_2 illustrate the law of multiple proportions. **(b)** Refer to Exercise 2.3, and sketch molecular models (similar to those in the movie) of the three N- and O-containing compounds in the exercise. **(c)** Do the same thing for Exercise 2.4, sketching models of the I- and F-containing molecules.

2.99 Prior to Rutherford's gold-foil experiment, the mass and positively charged particles of an atom were thought to be evenly distributed throughout the volume of the atom. **(a)** Watch the movie of the **Rutherford Experiment** (*eChapter 2.2*), and describe how the experimental results would have been different if the earlier model had been correct. **(b)** What specific feature of the modern view of atomic structure was illuminated by Rutherford's experiment?

2.100 The **Separation of Radiation** movie (*eChapter 2.2*) shows how three different types of radioactive emissions behave in the presence of an electric field. **(a)** Which of the three types of radiation does not consist of a stream of particles? **(b)** In Exercise 2.9 you explained why α and β rays are deflected in opposite directions. In the movie, the difference in the *magnitude* of deflection of α versus β particles is attributed primarily to a difference in mass. How much more massive are α particles than β particles? What factors other than mass influence the magnitude of deflection?

2.101 Use the **Periodic Table** (*eChapter 2.4*) activity to answer the following questions: **(a)** What element has the largest number of isotopes, and how many isotopes does it have? **(b)** Certain atomic numbers correspond to exceptionally large numbers of isotopes. What does the graph of atomic number versus the number of isotopes suggest regarding the stability of atoms with certain numbers of protons? **(c)** What is the most dense element, and what is its density?

2.102 Give the correct formula and name for the ionic compound formed by each of the indicated combinations: NH_4^+ and Al^{3+} each with Br^-, OH^-, S^{2-}, CO_3^{2-}, NO_3^-, and ClO_4^-. Use the **Ionic Compounds** activity (*eChapter 2.7*) to check your answers.

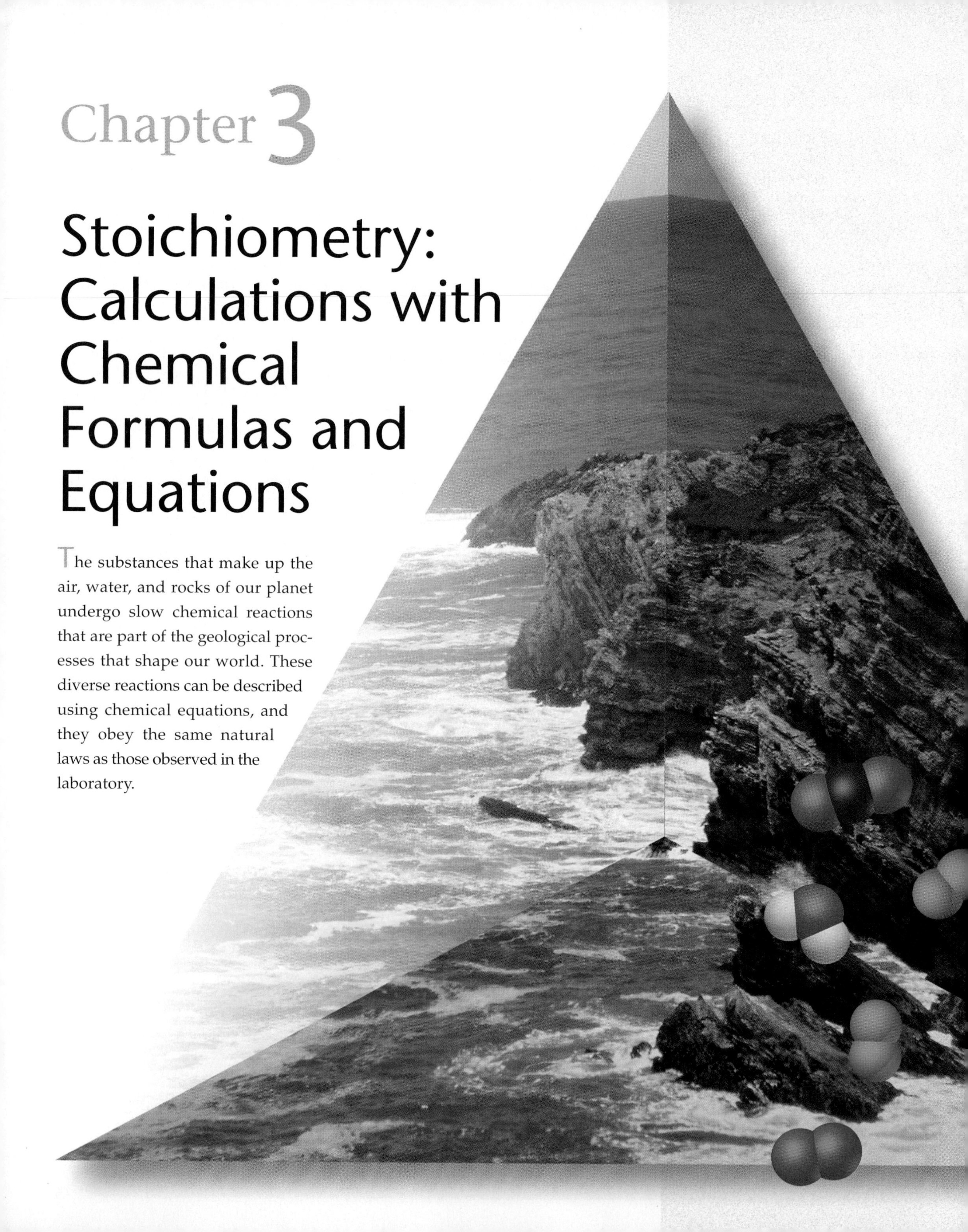

Chapter 3

Stoichiometry: Calculations with Chemical Formulas and Equations

The substances that make up the air, water, and rocks of our planet undergo slow chemical reactions that are part of the geological processes that shape our world. These diverse reactions can be described using chemical equations, and they obey the same natural laws as those observed in the laboratory.

IN CHAPTER 2 we saw that we can represent substances by their chemical formulas. Although chemical formulas are invariably shorter than chemical names, they are not merely abbreviations. Encoded in each chemical formula is important quantitative information about the substance that it represents.

In this chapter we examine several important uses of chemical formulas, as outlined in the "What's Ahead" list. The area of study that we will be examining is known as **stoichiometry** (pronounced stoy-key-OM-uh-tree), a name derived from the Greek *stoicheion* ("element") and *metron* ("measure"). Stoichiometry is an essential tool in chemistry. Such diverse problems as measuring the concentration of ozone in the atmosphere, determining the potential yield of gold from an ore, and assessing different processes for converting coal into gaseous fuels all use aspects of stoichiometry.

Stoichiometry is built on an understanding of atomic masses (Section 2.4) and on a fundamental principle, the **law of conservation of mass**: *The total mass of all substances present after a chemical reaction is the same as the total mass before the reaction.* The French nobleman and scientist Antoine Lavoisier (Figure 3.1 ▶) discovered this important chemical law in the late 1700s. In a chemistry text published in 1789, Lavoisier stated the law in this eloquent way: "We may lay it down as an incontestable axiom that, in all the operations of art and nature, nothing is created; an equal quantity of matter exists both before and after the experiment."

With the advent of the atomic theory, chemists came to understand the basis for the law of conservation of mass: *Atoms are neither created nor destroyed during any chemical reaction.* Thus, the same collection of atoms is present both before and after a reaction. The changes that occur during any reaction merely rearrange the atoms. We begin our discussions in this chapter by examining how chemical formulas and chemical equations are used to represent the rearrangements of atoms that occur in chemical reactions.

▶ What's Ahead ◀

- We begin by considering how we can use chemical formulas to write equations that represent chemical reactions.
- We then use chemical formulas to relate the masses of substances with the numbers of atoms, molecules, or ions they contain, which leads to the crucially important concept of a mole. A mole is 6.022×10^{23} objects (atoms, molecules, ions, or whatever).
- We will apply the mole concept to determine chemical formulas from the masses of each element in a given quantity of a compound.
- We will use the quantitative information inherent in chemical formulas and equations together with the mole concept to predict the amounts of substances consumed and/or produced in chemical reactions.
- A special situation arises when one of the reactants is used up before the others, and the reaction therefore stops leaving some of the excess starting material unreacted.

▲ **Figure 3.1** Antoine Lavoisier (1734–1794) conducted many important studies on combustion reactions. Unfortunately, his career was cut short by the French Revolution. He was a member of the French nobility and a tax collector. He was guillotined in 1794 during the final months of the Reign of Terror. He is now generally considered to be the father of modern chemistry because he conducted carefully controlled experiments and used quantitative measurements.

3.1 Chemical Equations

Chemical reactions are represented in a concise way by **chemical equations**. When hydrogen (H_2) burns, for example, it reacts with oxygen (O_2) in the air to form water (H_2O) (Figure 3.2 ▼). We write the chemical equation for this reaction as follows:

$$2H_2 + O_2 \longrightarrow 2H_2O \quad [3.1]$$

We read the + sign as "reacts with" and the arrow as "produces." The chemical formulas on the left of the arrow represent the starting substances, called **reactants**. The chemical formulas on the right of the arrow represent substances produced in the reaction, called **products**. The numbers in front of the formulas are *coefficients*. (As in algebraic equations, the numeral 1 is usually not written.)

Because atoms are neither created nor destroyed in any reaction, a chemical equation must have an equal number of atoms of each element on each side of the arrow. When this condition is met, the equation is said to be *balanced*. On the right side of Equation 3.1, for example, there are two molecules of H_2O, each composed of two atoms of hydrogen and one atom of oxygen. Thus, $2H_2O$ (read "two molecules of water") contains $2 \times 2 = 4$ H atoms and $2 \times 1 = 2$ O atoms as seen in the art shown in the margin. Because there are also 4 H atoms and 2 O atoms on the left side of the equation, the equation is balanced.

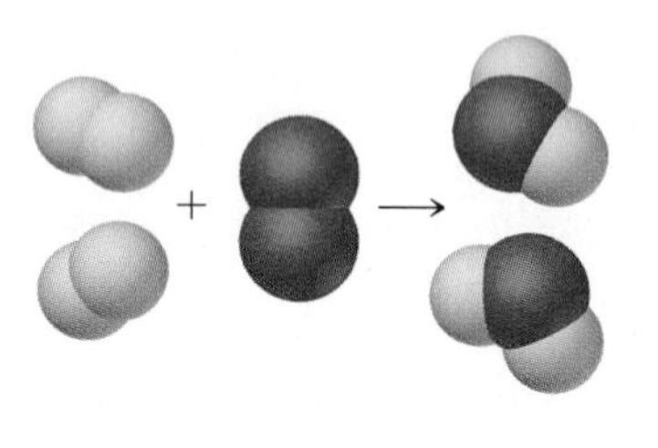

Once we know the formulas of the reactants and products in a reaction, we can write the unbalanced equation. We then balance the equation by determining the coefficients that provide equal numbers of each type of atom on each side of the equation. For most purposes, a balanced equation should contain the smallest possible whole-number coefficients.

In balancing equations, it is important to understand the difference between a coefficient in front of a formula and a subscript in a formula. Refer to Figure 3.3 ▶. Notice that changing a subscript in a formula—from H_2O to H_2O_2, for example—changes the identity of the chemical. The substance H_2O_2, hydrogen peroxide, is quite different from water. *Subscripts should never be changed in balancing an equation.* In contrast, placing a coefficient in front of a formula changes only the *amount* and not the *identity* of the substance. Thus, $2H_2O$ means two molecules of water, $3H_2O$ means three molecules of water, and so forth.

To illustrate the process of balancing equations, consider the reaction that occurs when methane (CH_4), the principal component of natural gas, burns in air to produce carbon dioxide gas (CO_2) and water vapor (H_2O). Both of these products contain oxygen atoms that come from O_2 in the air. We say that combustion in air is "supported by oxygen," meaning that oxygen is a reactant. The unbalanced equation is

$$CH_4 + O_2 \longrightarrow CO_2 + H_2O \quad \text{(unbalanced)} \quad [3.2]$$

▶ **Figure 3.2** Combustion of hydrogen gas. The gas is bubbled through a soap solution forming hydrogen-filled bubbles. As the bubbles float upwards, they are ignited by a candle on a long pole. The orange flame is due to the reaction of the hydrogen with oxygen in the air and results in the formation of water vapor.

ACTIVITY
Reading a Chemical Equation

Chemical symbol	Meaning	Composition
H_2O	One molecule of water:	Two H atoms and one O atom
$2H_2O$	Two molecules of water:	Four H atoms and two O atoms
H_2O_2	One molecule of hydrogen peroxide:	Two H atoms and two O atoms

◀ **Figure 3.3** Illustration of the difference between a subscript in a chemical formula and a coefficient in front of the formula. Notice that the number of atoms of each type (listed under composition) is obtained by multiplying the coefficient and the subscript associated with each element in the formula.

It is usually best to balance first those elements that occur in the fewest chemical formulas on each side of the equation. In our example both C and H appear in only one reactant and, separately, in one product each, so we begin by focusing attention on CH_4. Let's consider first carbon and then hydrogen.

One molecule of CH_4 contains the same number of C atoms (one) as one molecule of CO_2. The coefficients for these substances *must* be the same, therefore, and we choose them both to be 1 as we start the balancing process. However, the reactant CH_4 contains more H atoms (four) than the product H_2O (two). If we place a coefficient 2 in front of H_2O, there will be four H atoms on each side of the equation:

$$CH_4 + O_2 \longrightarrow CO_2 + 2H_2O \quad \text{(unbalanced)} \qquad [3.3]$$

At this stage the products have more total O atoms (four—two from CO_2 and two from $2H_2O$) than the reactants (two). If we place a coefficient 2 in front of O_2, we complete the balancing by making the number of O atoms equal on both sides of the equation:

$$CH_4 + 2O_2 \longrightarrow CO_2 + 2H_2O \quad \text{(balanced)} \qquad [3.4]$$

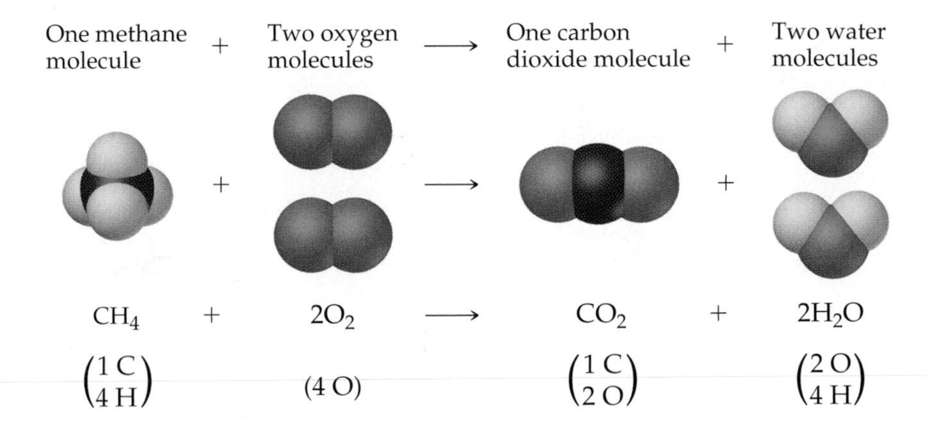

▶ **Figure 3.4** Balanced chemical equation for the combustion of CH_4. The drawings of the molecules involved call attention to the conservation of atoms through the reaction.

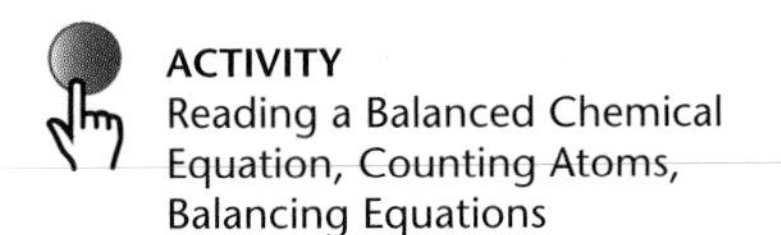

ACTIVITY
Reading a Balanced Chemical Equation, Counting Atoms, Balancing Equations

The molecular view of the balanced equation is shown in Figure 3.4 ▲.

The approach we have taken to balancing Equation 3.4 is largely trial and error. We balance each kind of atom in succession, adjusting coefficients as necessary. This approach works for most chemical equations.

Additional information is often added to the formulas in balanced equations to indicate the physical state of each reactant and product. We use the symbols (*g*), (*l*), (*s*), and (*aq*) for gas, liquid, solid, and aqueous (water) solution, respectively. Thus, Equation 3.4 can be written

$$CH_4(g) + 2O_2(g) \longrightarrow CO_2(g) + 2H_2O(g) \quad [3.5]$$

Sometimes the conditions (such as temperature or pressure) under which the reaction proceeds appear above or below the reaction arrow. The symbol Δ is often placed above the arrow to indicate the addition of heat.

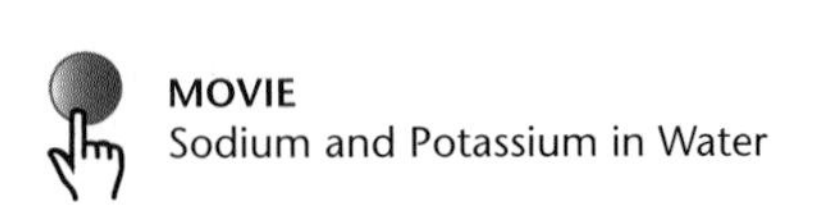

SAMPLE EXERCISE 3.1

Balance the following equation:

$$Na(s) + H_2O(l) \longrightarrow NaOH(aq) + H_2(g)$$

Solution We begin by counting the atoms of each kind on both sides of the arrow. The Na and O atoms are balanced (one Na and one O on each side), but there are two H atoms on the left and three H atoms on the right. To increase the number of H atoms on the left, we place a coefficient 2 in front of H_2O:

$$Na(s) + 2H_2O(l) \longrightarrow NaOH(aq) + H_2(g)$$

This choice is a trial beginning, but it sets us on the correct path. Now that we have $2H_2O$, we must regain the balance in O atoms. We can do so by moving to the other side of the equation and putting a coefficient 2 in front of NaOH:

$$Na(s) + 2H_2O(l) \longrightarrow 2NaOH(aq) + H_2(g)$$

This brings the H atoms into balance, but it requires that we move back to the left and put a coefficient 2 in front of Na to rebalance the Na atoms:

$$2Na(s) + 2H_2O(l) \longrightarrow 2NaOH(aq) + H_2(g)$$

Finally, we check the number of atoms of each element and find that we have two Na atoms, four H atoms, and two O atoms on each side of the equation. The equation is balanced.

PRACTICE EXERCISE

Balance the following equations by providing the missing coefficients:

(a) $\_Fe(s) + \_O_2(g) \longrightarrow \_Fe_2O_3(s)$

(b) $\_C_2H_4(g) + \_O_2(g) \longrightarrow \_CO_2(g) + \_H_2O(g)$

(c) $\_Al(s) + \_HCl(aq) \longrightarrow \_AlCl_3(aq) + \_H_2(g)$

Answers: **(a)** 4, 3, 2; **(b)** 1, 3, 2, 2; **(c)** 2, 6, 2, 3

SAMPLE EXERCISE 3.2

The following diagrams represent a chemical reaction in which the red spheres are oxygen atoms and the blue spheres are nitrogen atoms. **(a)** Write the chemical formulas for the reactants and products. **(b)** Write a balanced equation for the reaction. **(c)** Is the diagram consistent with the law of conservation of mass?

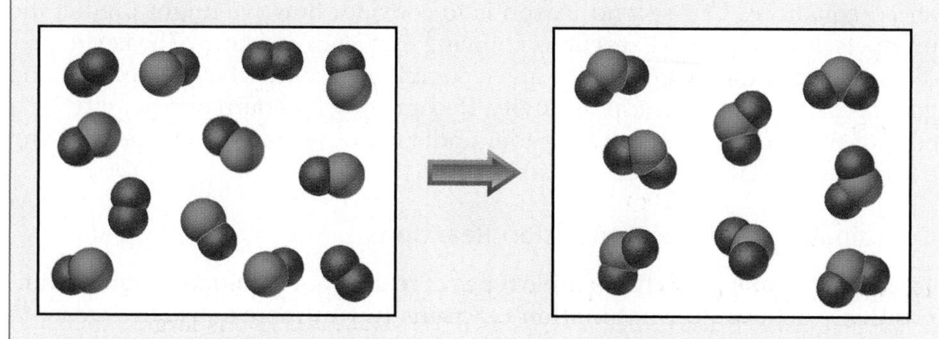

Solution **(a)** The left box, which represents the reactants, contains two kinds of molecules, those composed of two oxygen atoms (O_2) and those composed of one nitrogen atom and one oxygen atom (NO). The right box, which represents the products, contains only molecules composed of one nitrogen atom and two oxygen atoms (NO_2).

(b) The unbalanced chemical equation is

$$O_2 + NO \longrightarrow NO_2 \quad \text{(unbalanced)}$$

In this equation, there are three O atoms on the left side of the arrow and two O atoms on the right side. We can increase the number of O atoms by placing a coefficient 2 on the product side:

$$O_2 + NO \longrightarrow 2NO_2 \quad \text{(unbalanced)}$$

Now there are two N atoms and four O atoms on the right. Placing a coefficient 2 in front of NO brings both the N atoms and O atoms into balance:

$$O_2 + 2NO \longrightarrow 2NO_2 \quad \text{(balanced)}$$

(c) The left box (reactants) contains four O_2 molecules and eight NO molecules. Thus, the molecular ratio is one O_2 for each two NO as required by the balanced equation. The right box (products) contains eight NO_2 molecules. The number of NO_2 molecules on the right equals the number of NO molecules on the left as the balanced equation requires. Counting the atoms, we find eight N atoms in the eight NO molecules in the box on the left. There are also $4 \times 2 = 8$ O atoms in the O_2 molecules and eight O atoms in the NO molecules giving a total of 16 O atoms. In the box on the right, we find eight N atoms and $8 \times 2 = 16$ O atoms in the eight NO_2 molecules. Because there are equal numbers of both N and O atoms in the two boxes, the drawing is consistent with the law of conservation of mass.

PRACTICE EXERCISE

In order to be consistent with the law of conservation of mass, how many NH_3 molecules should be shown in the right box of the following diagram?

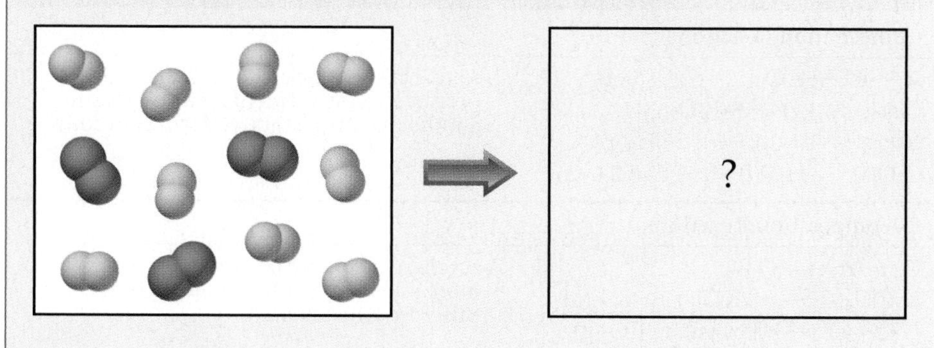

Answer: Six NH_3 molecules

3.2 Some Simple Patterns of Chemical Reactivity

In this section we examine three simple kinds of reactions that we will see frequently throughout this chapter. Our first reason for examining these reactions is merely to become better acquainted with chemical reactions and their balanced equations. Our second reason is to consider how we might predict the products of some of these reactions knowing only their reactants. The key to predicting the products formed by a given combination of reactants is recognizing general patterns of chemical reactivity. Recognizing a pattern of reactivity for a class of substances gives you a broader understanding than merely memorizing a large number of unrelated reactions.

Combination and Decomposition Reactions

Table 3.1 ▼ summarizes two simple types of reactions, combination and decomposition reactions. In **combination reactions** two or more substances react to form one product. There are many examples of such reactions, especially those in which elements combine to form compounds. For example, magnesium metal burns in air with a dazzling brilliance to produce magnesium oxide, as shown in Figure 3.5 ▶:

$$2Mg(s) + O_2(g) \longrightarrow 2MgO(s) \qquad [3.6]$$

This reaction is used to produce the bright flame generated by flares.

When a combination reaction occurs between a metal and a nonmetal, as in Equation 3.6, the product is an ionic solid. Recall that the formula of an ionic compound can be determined from the charges of the ions involved. ∞ (Section 2.7) When magnesium reacts with oxygen, for example, the magnesium loses electrons and forms the magnesium ion, Mg^{2+}. The oxygen gains electrons and forms the oxide ion, O^{2-}. Thus, the reaction product is MgO. You should be able to recognize when a reaction is a combination reaction and to predict the products of a combination reaction in which the reactants are a metal and a nonmetal.

In a **decomposition reaction** one substance undergoes a reaction to produce two or more other substances. Many compounds undergo decomposition reactions when heated. For example, many metal carbonates decompose to form metal oxides and carbon dioxide when heated:

$$CaCO_3(s) \longrightarrow CaO(s) + CO_2(g) \qquad [3.7]$$

The decomposition of $CaCO_3$ is an important commercial process. Limestone or seashells, which are both primarily $CaCO_3$, are heated to prepare CaO, which is known as lime or quicklime. Over 2.0×10^{10} kg (22 million tons) of CaO are used in the United States each year, principally in making glass, in obtaining iron from its ores, and in making mortar to bind bricks.

TABLE 3.1 Combination and Decomposition Reactions

Combination Reactions	
$A + B \longrightarrow C$ $C(s) + O_2(g) \longrightarrow CO_2(g)$ $N_2(g) + 3H_2(g) \longrightarrow 2NH_3(g)$ $CaO(s) + H_2O(l) \longrightarrow Ca(OH)_2(s)$	Two reactants combine to form a single product. Many elements react with one another in this fashion to form compounds.
Decomposition Reactions	
$C \longrightarrow A + B$ $2KClO_3(s) \longrightarrow 2KCl(s) + 3O_2(g)$ $PbCO_3(s) \longrightarrow PbO(s) + CO_2(g)$ $Cu(OH)_2(s) \longrightarrow CuO(s) + H_2O(l)$	A single reactant breaks apart to form two or more substances. Many compounds react this way when heated.

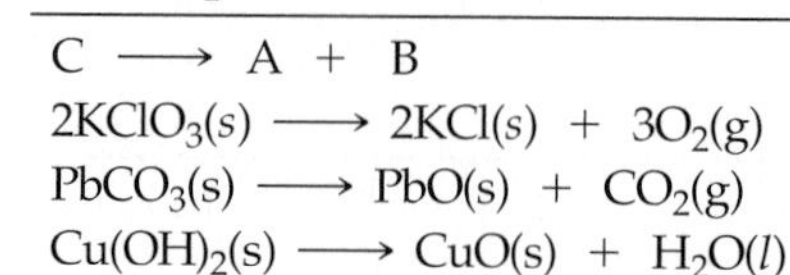

MOVIE
Reactions with Oxygen, Formation of Water

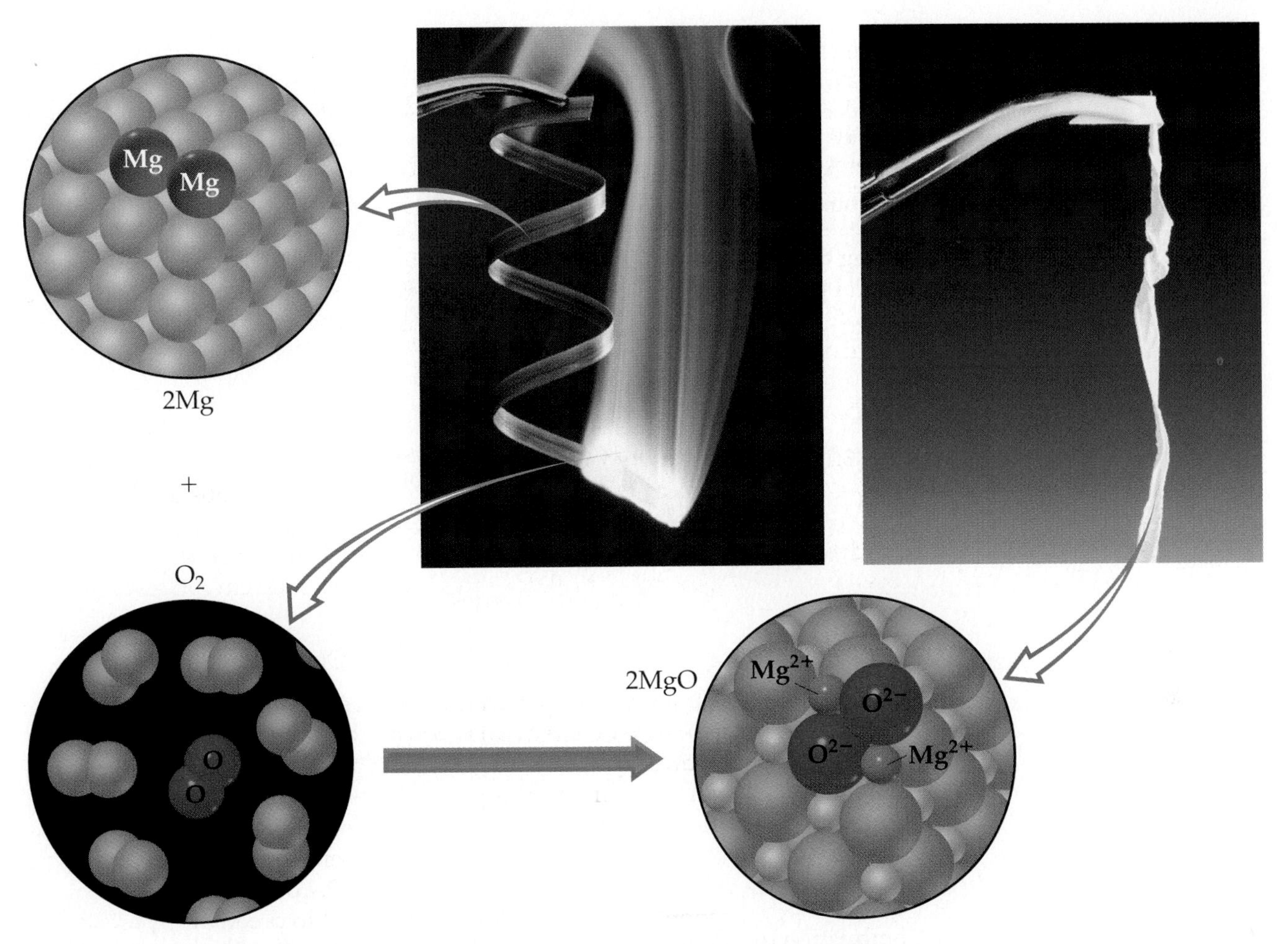

▲ **Figure 3.5** When magnesium metal burns, the Mg atoms react with O_2 molecules from the air to form magnesium oxide, MgO, an ionic solid: $2Mg(s) + O_2(g) \longrightarrow 2MgO(s)$. The photos show what we see in the laboratory. The ribbon of magnesium metal (left) is surrounded by oxygen in the air, and as it burns, an intense flame is produced. At the end of the reaction, a rather fragile ribbon of white solid, MgO, remains. The models show the atomic-level view of the reactants and products.

The decomposition of sodium azide (NaN_3) rapidly releases $N_2(g)$, so this reaction is used to inflate safety air bags in automobiles (Figure 3.6 ▼):

$$2NaN_3(s) \longrightarrow 2Na(s) + 3N_2(g) \qquad [3.8]$$

The system is designed so that an impact ignites a detonator cap, which in turn causes NaN_3 to decompose explosively. A small quantity of NaN_3 (about 100 g) forms a large quantity of gas (about 50 L). We will consider the volumes of gases produced in chemical reactions in Section 10.5.

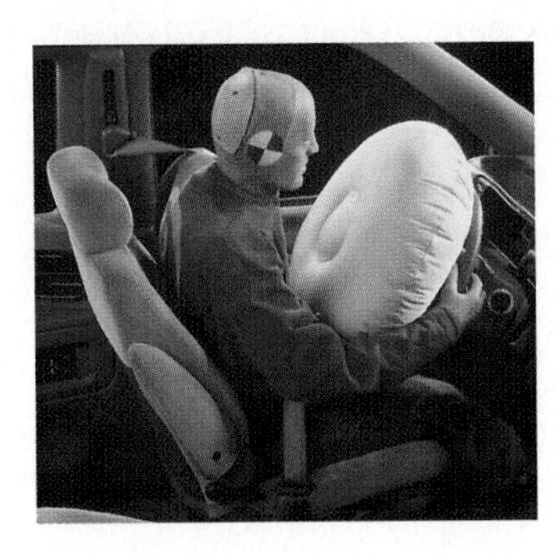

◀ **Figure 3.6** The decomposition of sodium azide, $NaN_3(s)$, is used to inflate automobile air bags. When properly ignited, the NaN_3 decomposes rapidly, forming nitrogen gas, $N_2(g)$, which expands the air bag.

SAMPLE EXERCISE 3.3

Write balanced equations for the following reactions: **(a)** The combination reaction that occurs when lithium metal and fluorine gas react. **(b)** The decomposition reaction that occurs when solid barium carbonate is heated. (Two products form: a solid and a gas.)

Solution (a) The symbol for lithium is Li. With the exception of mercury, all metals are solids at room temperature. Fluorine occurs as a diatomic molecule (see Figure 2.19). Thus, the reactants are Li(*s*) and $F_2(g)$. The product will consist of a metal and a nonmetal, so we expect it to be an ionic solid. Lithium ions have a 1+ charge, Li^+, whereas fluoride ions have a 1− charge, F^-. Thus, the chemical formula for the product is LiF. The balanced chemical equation is

$$2Li(s) + F_2(g) \longrightarrow 2LiF(s)$$

(b) The chemical formula for barium carbonate is $BaCO_3$. As noted in the text, many metal carbonates decompose to form metal oxides and carbon dioxide when heated. In Equation 3.7, for example, $CaCO_3$ decomposes to form CaO and CO_2. Thus, we would expect that $BaCO_3$ decomposes to form BaO and CO_2. Barium and calcium are both in group 2A in the periodic table, moreover, which further suggests they would react in the same way:

$$BaCO_3(s) \longrightarrow BaO(s) + CO_2(g)$$

PRACTICE EXERCISE

Write balanced chemical equations for the following reactions: **(a)** Solid mercury(II) sulfide decomposes into its component elements when heated. **(b)** The surface of aluminum metal undergoes a combination reaction with oxygen in the air.
Answers: **(a)** $HgS(s) \longrightarrow Hg(l) + S(s)$; **(b)** $4Al(s) + 3O_2(g) \longrightarrow 2Al_2O_3(s)$

Combustion in Air

Combustion reactions are rapid reactions that produce a flame. Most of the combustion reactions we observe involve O_2 from air as a reactant. Equation 3.5 and Practice Exercise 3.1(b) illustrate a general class of reactions involving the burning or combustion of hydrocarbon compounds (compounds that contain only carbon and hydrogen, such as CH_4 and C_2H_4). (Section 2.9)

When hydrocarbons are combusted in air, they react with O_2 to form CO_2 and H_2O.* The number of molecules of O_2 required in the reaction and the number of molecules of CO_2 and H_2O formed depend on the composition of the hydrocarbon, which acts as the fuel in the reaction. For example, the combustion of propane (C_3H_8), a gas used for cooking and home heating, is described by the following equation:

$$C_3H_8(g) + 5O_2(g) \longrightarrow 3CO_2(g) + 4H_2O(g) \quad [3.9]$$

The state of the water, $H_2O(g)$ or $H_2O(l)$, depends on the conditions of the reaction. Water vapor, $H_2O(g)$, is formed at high temperature in an open container. The blue flame produced when propane burns is shown in Figure 3.7 ◀.

▲ **Figure 3.7** Propane, C_3H_8, burns in air, producing a blue flame. The liquid propane vaporizes and mixes with air as it escapes through the nozzle.

Combustion of oxygen-containing derivatives of hydrocarbons, such as CH_3OH, also produce CO_2 and H_2O. The simple rule that hydrocarbons and related oxygen-containing derivatives of hydrocarbons form CO_2 and H_2O when they burn in air summarizes the behavior of about 3 million compounds. Many substances that our bodies use as energy sources, such as the sugar glucose

*When there is an insufficient quantity of O_2 present, carbon monoxide (CO) will be produced along with the CO_2. If the amount of O_2 is severely restricted, fine particles of carbon that we call soot will be produced. *Complete* combustion produces CO_2. Unless specifically stated to the contrary, we will take *combustion* to mean *complete combustion*.

($C_6H_{12}O_6$), similarly react in our bodies with O_2 to form CO_2 and H_2O. In our bodies, however, the reactions take place in a series of steps that occur at body temperature. The reactions are then described as *oxidation reactions* rather than combustion reactions.

SAMPLE EXERCISE 3.4

Write the balanced equation for the reaction that occurs when methanol, $CH_3OH(l)$, is burned in air.

Solution When any compound containing C, H, and O is combusted, it reacts with the $O_2(g)$ in air to produce $CO_2(g)$ and $H_2O(g)$. Thus, the unbalanced equation is

$$CH_3OH(l) + O_2(g) \longrightarrow CO_2(g) + H_2O(g)$$

Because CH_3OH has only one C atom, we can start balancing the equation using the coefficient 1 for CO_2. Because CH_3OH has four H atoms, we place a coefficient 2 in front of H_2O to balance the H atoms:

$$CH_3OH(l) + O_2(g) \longrightarrow CO_2(g) + 2H_2O(g)$$

This gives four O atoms among the products and three among the reactants (one in CH_3OH and two in O_2). We can place the fractional coefficient $\frac{3}{2}$ in front of O_2 to give four O atoms among the reactants (there are $\frac{3}{2} \times 2 = 3$ O atoms in $\frac{3}{2}O_2$):

$$CH_3OH(l) + \tfrac{3}{2}O_2(g) \longrightarrow CO_2(g) + 2H_2O(g)$$

Although the equation is now balanced, it is not in its most conventional form because it contains a fractional coefficient. If we multiply each side of the equation by 2, we will remove the fraction and achieve the following balanced equation:

$$2CH_3OH(l) + 3O_2(g) \longrightarrow 2CO_2(g) + 4H_2O(g)$$

PRACTICE EXERCISE

Write the balanced equation for the reaction that occurs when ethanol, $C_2H_5OH(l)$, is burned in air.
Answer: $C_2H_5OH(l) + 3O_2(g) \longrightarrow 2CO_2(g) + 3H_2O(g)$

3.3 Formula Weights

Chemical formulas and chemical equations both have a *quantitative* significance; the subscripts in formulas and the coefficients in equations represent precise quantities. The formula H_2O indicates that a molecule of this substance contains exactly two atoms of hydrogen and one atom of oxygen. Similarly, the balanced chemical equation for the combustion of propane—$C_3H_8(g) + 5O_2(g) \longrightarrow 3CO_2(g) + 4H_2O(g)$, shown in Equation 3.9—indicates that the combustion of one molecule of C_3H_8 requires five molecules of O_2 and produces exactly three molecules of CO_2 and four of H_2O. But how do we relate the numbers of atoms or molecules to the amounts we measure out in the laboratory? Although we cannot directly count atoms or molecules, we can indirectly determine their numbers if we know their masses. Therefore, before we can pursue the quantitative aspects of chemical formulas or equations, we must examine the masses of atoms and molecules, which we do in this section and the next.

Formula and Molecular Weights

The **formula weight** of a substance is the sum of the atomic weights of each atom in its chemical formula. Sulfuric acid (H_2SO_4), for example, has a formula weight of 98.1 amu.*

$$\begin{aligned} \text{FW of } H_2SO_4 &= 2(\text{AW of H})) + (\text{AW of S}) + 4(\text{AW of O}) \\ &= 2(1.0 \text{ amu}) + 32.1 \text{ amu} + 4(16.0 \text{ amu}) \\ &= 98.1 \text{ amu} \end{aligned}$$

We have rounded off the atomic weights to one place beyond the decimal point. We will round off the atomic weights in this way for most problems.

If the chemical formula is merely the chemical symbol of an element, such as Na, then the formula weight equals the atomic weight of the element. If the chemical formula is that of a molecule, then the formula weight is also called the **molecular weight**. The molecular weight of glucose ($C_6H_{12}O_6$), for example, is

$$\text{MW of } C_6H_{12}O_6 = 6(12.0 \text{ amu}) + 12(1.0 \text{ amu}) + 6(16.0 \text{ amu}) = 180.0 \text{ amu}$$

Because ionic substances such as NaCl exist as three-dimensional arrays of ions (Figure 2.23), it is inappropriate to speak of molecules of NaCl. Instead, we speak of *formula units*, represented by the chemical formula of the substance. The formula unit of NaCl consists of one Na^+ ion and one Cl^- ion. Thus, the formula weight of NaCl is the mass of one formula unit:

$$\text{FW of NaCl} = 23.0 \text{ amu} + 35.5 \text{ amu} = 58.5 \text{ amu}$$

SAMPLE EXERCISE 3.5

Calculate the formula weight of **(a)** sucrose, $C_{12}H_{22}O_{11}$ (table sugar), and **(b)** calcium nitrate, $Ca(NO_3)_2$.

Solution **(a)** By adding the weights of the atoms in sucrose, we find it to have a formula weight of 342.0 amu:

$$\begin{aligned} 12 \text{ C atoms} &= 12(12.0 \text{ amu}) = 144.0 \text{ amu} \\ 22 \text{ H atoms} &= 22(1.0 \text{ amu}) = 22.0 \text{ amu} \\ 11 \text{ O atoms} &= 11(16.0 \text{ amu}) = \underline{176.0 \text{ amu}} \\ & \qquad\qquad\qquad\quad 342.0 \text{ amu} \end{aligned}$$

(b) If a chemical formula has parentheses, the subscript outside the parentheses is a multiplier for all atoms inside. Thus, for $Ca(NO_3)_2$, we have

$$\begin{aligned} 1 \text{ Ca atom} &= 1(40.1 \text{ amu}) = 40.1 \text{ amu} \\ 2 \text{ N atoms} &= 2(14.0 \text{ amu}) = 28.0 \text{ amu} \\ 6 \text{ O atoms} &= 6(16.0 \text{ amu}) = \underline{96.0 \text{ amu}} \\ & \qquad\qquad\qquad\quad 164.1 \text{ amu} \end{aligned}$$

PRACTICE EXERCISE

Calculate the formula weight of **(a)** $Al(OH)_3$ and **(b)** CH_3OH.
Answers: **(a)** 78.0 amu; **(b)** 32.0 amu

Percentage Composition from Formulas

Occasionally we must calculate the *percentage composition* of a compound (that is, the percentage by mass contributed by each element in the substance). For example, in order to verify the purity of a compound, we may wish to compare the cal-

*The abbreviation AW is used for atomic weight, FW for formula weight, and MW for molecular weight.

culated percentage composition of the substance with that found experimentally. Calculating percentage composition is a straightforward matter if the chemical formula is known. The calculation depends on the formula weight of the substance, the atomic weight of the element of interest, and the number of atoms of that element in the chemical formula:

$$\% \text{ element} = \frac{(\text{number of atoms of that element})(\text{atomic weight of element})}{\text{formula weight of compound}} \times 100\% \qquad [3.10]$$

SAMPLE EXERCISE 3.6

Calculate the percentage composition of $C_{12}H_{22}O_{11}$.

Solution Let's examine this question using the problem-solving steps in the "Strategies in Chemistry: Problem Solving" essay.
Analyze: Given the chemical formula of a compound, $C_{12}H_{22}O_{11}$, we are asked to calculate its percentage composition, meaning the percent by mass of its component elements (C, H, and O).
Plan: We can use Equation 3.10, relying on a periodic table to obtain the atomic weights of each component element. The atomic weights are first used to determine the formula weight of the compound. (The formula weight of $C_{12}H_{22}O_{11}$, 342.0 amu, was calculated in Sample Exercise 3.5.) We must then do three calculations, one for each element.
Solve: Using Equation 3.10, we have

$$\%C = \frac{(12)(12.0 \text{ amu})}{342.0 \text{ amu}} \times 100\% = 42.1\%$$

$$\%H = \frac{(22)(1.0 \text{ amu})}{342.0 \text{ amu}} \times 100\% = 6.4\%$$

$$\%O = \frac{(11)(16.0 \text{ amu})}{342.0 \text{ amu}} \times 100\% = 51.5\%$$

Check: The percentages of the individual elements must add up to 100%, which they do in this case. We could have used more significant figures for our atomic weights, giving more significant figures for our percentage composition, but we have adhered to our suggested guideline of rounding atomic weights to one digit beyond the decimal point.

PRACTICE EXERCISE

Calculate the percentage of nitrogen, by mass, in $Ca(NO_3)_2$.
Answer: 17.1%

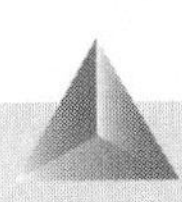

Strategies in Chemistry Problem Solving

The key to success in problem solving is practice. As you practice, you will find that you can improve your skills by following these steps:

Step 1: Analyze the problem. Read the problem carefully for understanding. What does it say? Draw any picture or diagram that will help you visualize the problem. Write down the data you are given. Also, identify the quantity that you need to obtain (the unknown), and write it down.

Step 2: Develop a plan for solving the problem. Consider the possible paths between the given information and the unknown. What principles or equations relate the known data to the unknown? Recognize that some data may not be given explicitly in the problem; you may be expected to know certain quantities (such as Avogadro's number, which we will soon discuss) or look them up in tables (such as atomic weights). Recognize also that your plan may involve a single step or a series of steps with intermediate answers.

Step 3: Solve the problem. Use the known information and suitable equations or relationships to solve for the unknown. Dimensional analysis (Section 1.6) is a very useful tool for solving a great number of problems. Be careful with significant figures, signs, and units.

Step 4: Check the solution. Read the problem again to make sure you have found all the solutions asked for in the problem. Does your answer make sense? That is, is the answer outrageously large or small, or is it in the ballpark? Finally, are the units and significant figures correct?

3.4 The Mole

Even the smallest samples that we deal with in the laboratory contain enormous numbers of atoms, ions, or molecules. For example, a teaspoon of water (about 5 mL) contains 2×10^{23} water molecules, a number so large that it almost defies comprehension. Chemists, therefore, have devised a special counting unit for describing such large numbers of atoms or molecules.

In everyday life we use counting units like dozen (12 objects) and gross (144 objects) to deal with modestly large quantities. In chemistry the unit for dealing with the number of atoms, ions, or molecules in a common-sized sample is the **mole**, abbreviated mol.* A mole is the amount of matter that contains as many objects (atoms, molecules, or whatever objects we are considering) as the number of atoms in exactly 12 g of isotopically pure ^{12}C. From experiments scientists have determined this number to be 6.0221421×10^{23}. Scientists call this number **Avogadro's number**, in honor of Amedeo Avogadro (1776–1856), an Italian scientist. For most purposes we will use 6.02×10^{23} or 6.022×10^{23} for Avogadro's number throughout the text.

A mole of atoms, a mole of molecules, or a mole of anything else all contain Avogadro's number of these objects:

$$1 \text{ mol } ^{12}\text{C atoms} = 6.02 \times 10^{23} \ ^{12}\text{C atoms}$$

$$1 \text{ mol H}_2\text{O molecules} = 6.02 \times 10^{23} \text{ H}_2\text{O molecules}$$

$$1 \text{ mol NO}_3^{-} \text{ ions} = 6.02 \times 10^{23} \text{ NO}_3^{-} \text{ ions}$$

Avogadro's number is so large that it is difficult to imagine. Spreading 6.02×10^{23} marbles over the entire surface of Earth would produce a layer about 3 mi thick. If Avogadro's number of pennies were placed side by side in a straight line, they would encircle Earth 300 trillion (3×10^{14}) times.

SAMPLE EXERCISE 3.7

Without using a calculator, arrange the following samples in order of increasing numbers of carbon atoms: 12 g ^{12}C, 1 mol C_2H_2, 9×10^{23} molecules of CO_2.

Solution

Analyze: We are given amounts of different substances expressed in grams, moles, and number of molecules and asked to arrange these samples in order of increasing numbers of C atoms.

Plan: To determine the number of C atoms in each sample, we must convert g ^{12}C, moles C_2H_2, and molecules CO_2 all to the number of C atoms using the definition of a mole and Avogadro's number.

Solve: A mole is defined as the amount of matter that contains as many as the number of atoms in exactly 12 g of ^{12}C. Thus, 12 g of ^{12}C contains one mole of C atoms (that is, 6.02×10^{23} C atoms). In 1 mol C_2H_2, there are 6×10^{23} C_2H_2 molecules. Because there are two C atoms in each C_2H_2 molecule, this sample contains 12×10^{23} C atoms. Because each CO_2 molecule contains one C atom, the sample of CO_2 contains 9×10^{23} C atoms. Hence, the order is 12 g ^{12}C (6×10^{23} C atoms) $< 9 \times 10^{23}$ CO_2 molecules (9×10^{23} C atoms) $<$ 1 mol C_2H_2 (12×10^{23} C atoms).

PRACTICE EXERCISE

Without using a calculator, arrange the following samples in order of increasing number of O atoms: 1 mol H_2O, 1 mol CO_2, 3×10^{23} molecules O_3.

Answer: 1 mol H_2O $< 3 \times 10^{23}$ molecules O_3 $<$ 1 mol CO_2

---

*The term *mole* comes from the Latin word *moles*, meaning "a mass." The term *molecule* is the diminutive form of this word and means "a small mass."

SAMPLE EXERCISE 3.8

Calculate the number of H atoms in 0.350 mol of $C_6H_{12}O_6$.

Solution

Analyze: We are given both the amount of the substance (0.350 mol) and its chemical formula ($C_6H_{12}O_6$). The unknown is the number of H atoms in this sample.

Plan: Avogadro's number provides the conversion factor between the number of moles of $C_6H_{12}O_6$ and the number of molecules of $C_6H_{12}O_6$. Once we know the number of molecules of $C_6H_{12}O_6$, we can use the chemical formula, which tells us that each molecule of $C_6H_{12}O_6$ contains 12 H atoms. Thus, we convert moles of $C_6H_{12}O_6$ to molecules of $C_6H_{12}O_6$ and then determine the number of atoms of H from the number of molecules of $C_6H_{12}O_6$:

$$\text{mol } C_6H_{12}O_6 \longrightarrow \text{molecules } C_6H_{12}O_6 \longrightarrow \text{atoms H}$$

Solve:

$$\text{H atoms} = (0.350 \cancel{\text{ mol } C_6H_{12}O_6})\left(\frac{6.02 \times 10^{23} \cancel{\text{ molecules}}}{1 \cancel{\text{ mol } C_6H_{12}O_6}}\right)\left(\frac{12 \text{ H atoms}}{1 \cancel{\text{ molecule}}}\right)$$

$$= 2.53 \times 10^{24} \text{ H atoms}$$

Check: The magnitude of our answer is reasonable; it is a large number about the magnitude of Avogadro's number. We can also make the following ballpark calculation: Multiplying $0.35 \times 6 \times 10^{23}$ gives about 2×10^{23} molecules. Multiplying this result by 12 gives $24 \times 10^{23} = 2.4 \times 10^{24}$ H atoms, which agrees with the previous, more detailed calculation. Because we were asked for the number of H atoms, the units of our answer are correct. The given data had three significant figures, so our answer has three significant figures.

PRACTICE EXERCISE

How many oxygen atoms are in **(a)** 0.25 mol $Ca(NO_3)_2$ and **(b)** 1.50 mol of sodium carbonate?

Answers: **(a)** 9.0×10^{23}; **(b)** 2.71×10^{24}

Molar Mass

A dozen is the same number (12), whether we have a dozen eggs or a dozen elephants. Clearly, however, a dozen eggs does not have the same mass as a dozen elephants. Similarly, a mole is always the *same number* (6.02×10^{23}), but a mole of different substances will have *different masses*. Compare, for example, 1 mol of ^{12}C and 1 mol of ^{24}Mg. A single ^{12}C atom has a mass of 12 amu, whereas a single ^{24}Mg atom is twice as massive, 24 amu (to two significant figures). Because a mole always has the same number of particles, a mole of ^{24}Mg must be twice as massive as a mole of ^{12}C. Because a mole of ^{12}C weighs 12 g (by definition), then a mole of ^{24}Mg must weigh 24 g. Thus, the mass of a single atom of an element (in amu) is numerically equal to the mass (in grams) of 1 mol of that element. This statement is true regardless of the element:

$$\text{1 atom of } ^{12}\text{C has a mass of 12 amu} \Rightarrow \text{1 mol } ^{12}\text{C has a mass of 12 g}$$

$$\text{1 atom of Cl has an average mass of 35.5 amu} \Rightarrow \text{1 mol Cl has a mass of 35.5 g}$$

$$\text{1 atom of Au has an average mass of 197 amu} \Rightarrow \text{1 mol Au has a mass of 197 g}$$

Notice that when we are dealing with a particular isotope of an element, we use the mass of that isotope; otherwise we use the atomic weight (the average atomic mass) of the element.

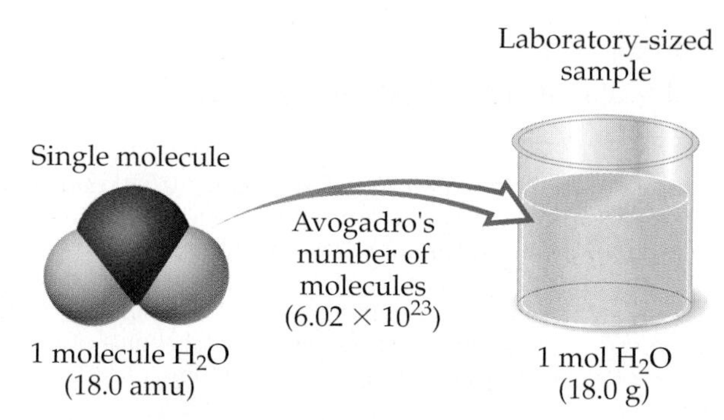

▲ **Figure 3.8** The relationship between a single molecule and its mass and a mole and its mass, using H_2O as an example.

For other kinds of substances, the same numerical relationship exists between the formula weight (in amu) and the mass (in grams) of 1 mol of that substance:

$$1\ H_2O \text{ molecule has a mass of 18.0 amu} \Rightarrow 1 \text{ mol } H_2O \text{ has a mass of 18.0 g}$$

$$1\ NO_3^- \text{ ion has a mass of 62.0 amu} \Rightarrow 1 \text{ mol } NO_3^- \text{ has a mass of 62.0 g}$$

$$1 \text{ NaCl unit has a mass of 58.5 amu} \Rightarrow 1 \text{ mol NaCl has a mass of 58.5 g}$$

Figure 3.8 ▲ illustrates the relationship between the mass of a single molecule of H_2O and that of a mole of H_2O.

The mass in grams of 1 mol of a substance (that is, the mass in grams per mol) is called its **molar mass**. *The molar mass (in g/mol) of any substance is always numerically equal to its formula weight (in amu).* NaCl, for example, has a molar mass of 58.5 g/mol. Further examples of mole relationships are shown in Table 3.2 ▼. Figure 3.9 ◀ shows 1-mol quantities of several common substances.

▲ **Figure 3.9** One mole each of a solid, a liquid, and a gas. One mole of NaCl, the solid, has a mass of 58.45 g. One mole of H_2O, the liquid, has a mass of 18.0 g and occupies a volume of 18.0 mL. One mole of O_2, the gas, has a mass of 32.0 g and occupies a balloon whose diameter is 35 cm.

The entries in Table 3.2 for N and N_2 point out the importance of stating the chemical form of a substance exactly when we use the mole concept. Suppose you read that 1 mol of nitrogen is produced in a particular reaction. You might interpret this statement to mean 1 mol of nitrogen atoms (14.0 g). Unless otherwise stated, however, what is probably meant is 1 mol of nitrogen molecules, N_2 (28.0 g), because N_2 is the usual chemical form of the element. To avoid ambiguity, it is important to state explicitly the chemical form being discussed. Using the chemical formula N_2 avoids ambiguity.

TABLE 3.2 Mole Relationships

Name	Formula	Formula Weight (amu)	Molar Mass (g/mol)	Number and Kind of Particles in One Mole
Atomic nitrogen	N	14.0	14.0	6.022×10^{23} N atoms
Molecular nitrogen	N_2	28.0	28.0	6.022×10^{23} N_2 molecules $2(6.022 \times 10^{23})$ N atoms
Silver	Ag	107.9	107.9	6.022×10^{23} Ag atoms
Silver ions	Ag^+	107.9[a]	107.9	6.022×10^{23} Ag^+ ions
Barium chloride	$BaCl_2$	208.2	208.2	6.022×10^{23} $BaCl_2$ units 6.022×10^{23} Ba^{2+} ions $2(6.022 \times 10^{23})$ Cl^- ions

[a]Recall that the electron has negligible mass; thus, ions and atoms have essentially the same mass.

SAMPLE EXERCISE 3.9

What is the mass in grams of 1.000 mol of glucose, $C_6H_{12}O_6$?

Solution

Analyze: We are given the chemical formula for glucose and asked to determine its molar mass.

Plan: The molar mass of a substance is found by adding the atomic weights of its component atoms.

Solve:

$$\begin{aligned} 6\text{ C atoms} &= 6(12.0) = 72.0\text{ amu} \\ 12\text{ H atoms} &= 12(1.0) = 12.0\text{ amu} \\ 6\text{ O atoms} &= 6(16.0) = \underline{96.0\text{ amu}} \\ & \qquad\qquad\quad 180.0\text{ amu} \end{aligned}$$

Because glucose has a formula weight of 180.0 amu, 1 mol of this substance has a mass of 180.0 g. In other words, $C_6H_{12}O_6$ has a molar mass of 180.0 g/mol.

Check: The magnitude of our answer seems reasonable, and g/mol is the appropriate unit for the molar mass.

Comment: Glucose is sometimes called dextrose. Also known as blood sugar, glucose is found widely in nature, occurring, for example, in honey and fruits. Other types of sugars used as food must be converted into glucose in the stomach or liver before they can be used by the body as energy sources. Because glucose requires no conversion, it is often given intravenously to patients who need immediate nourishment.

PRACTICE EXERCISE

Calculate the molar mass of $Ca(NO_3)_2$.

Answer: 164.1 g/mol

Interconverting Masses, Moles, and Numbers of Particles

Conversions of mass to moles and of moles to mass are frequently encountered in calculations using the mole concept. These calculations are made easy through dimensional analysis, as shown in Sample Exercises 3.10 and 3.11.

SAMPLE EXERCISE 3.10

Calculate the number of moles of glucose ($C_6H_{12}O_6$) in 5.380 g of $C_6H_{12}O_6$.

Solution

Analyze: We are given the number of grams of $C_6H_{12}O_6$ and asked to calculate the number of moles.

Plan: The molar mass of a substance provides the conversion factor for converting grams to moles. The molar mass of $C_6H_{12}O_6$ is 180.0 g/mol (Sample Exercise 3.9).

Solve: Using 1 mol $C_6H_{12}O_6$ = 180.0 g $C_6H_{12}O_6$ to write the appropriate conversion factor, we have

$$\text{Moles } C_6H_{12}O_6 = (5.380\ \cancel{\text{g } C_6H_{12}O_6})\left(\frac{1\text{ mol } C_6H_{12}O_6}{180.0\ \cancel{\text{g } C_6H_{12}O_6}}\right) = 0.02989\text{ mol } C_6H_{12}O_6$$

Check: Because 5.380 g is less than the molar mass, it is reasonable that our answer is less than 1 mol. The units of our answer (mol) are appropriate. The original data had four significant figures, so our answer has four significant figures.

PRACTICE EXERCISE

How many moles of sodium bicarbonate ($NaHCO_3$) are there in 508 g of $NaHCO_3$?

Answer: 6.05 mol $NaHCO_3$

SAMPLE EXERCISE 3.11

Calculate the mass, in grams, of 0.433 mol of calcium nitrate.

Solution

Analyze: We are given the number of moles of calcium nitrate and asked to calculate the mass of the sample in grams.

Plan: In order to convert moles to grams, we need the molar mass, which we can calculate using the chemical formula and atomic weights.

Solve: Because the calcium ion is Ca^{2+} and the nitrate ion is NO_3^-, calcium nitrate is $Ca(NO_3)_2$. Adding the atomic weights of the elements in the compound gives a formula weight of 164.1 amu. Using 1 mol $Ca(NO_3)_2$ = 164.1 g $Ca(NO_3)_2$ to write the appropriate conversion factor, we have

$$\text{Grams } Ca(NO_3)_2 = 0.433 \cancel{\text{mol } Ca(NO_3)_2}\left(\frac{164.1 \text{ g } Ca(NO_3)_2}{1 \cancel{\text{mol } Ca(NO_3)_2}}\right) = 71.1 \text{ g } Ca(NO_3)_2$$

Check: The number of moles is less than 1, so the number of grams must be less than the molar mass, 164.1 g. Using rounded numbers to estimate, we have $0.5 \times 150 = 75$ g. Thus, the magnitude of our answer is reasonable. Both the units (g) and the number of significant figures (3) are correct.

PRACTICE EXERCISE

What is the mass, in grams, of **(a)** 6.33 mol of $NaHCO_3$ and **(b)** 3.0×10^{-5} mol of sulfuric acid?

Answers: **(a)** 532 g; **(b)** 2.9×10^{-3} g

The mole concept provides the bridge between masses and numbers of particles. To illustrate how we can interconvert masses and numbers of particles, let's calculate the number of copper atoms in an old copper penny. Such a penny weighs about 3 g, and we'll assume that it is 100% copper:

$$\text{Cu atoms} = (3 \cancel{\text{g Cu}})\left(\frac{1 \cancel{\text{mol Cu}}}{63.5 \cancel{\text{g Cu}}}\right)\left(\frac{6.02 \times 10^{23} \text{ Cu atoms}}{1 \cancel{\text{mol Cu}}}\right)$$

$$= 3 \times 10^{22} \text{ Cu atoms}$$

Notice how dimensional analysis (Section 1.6) provides a straightforward route from grams to numbers of atoms. The molar mass and Avogadro's number are used as conversion factors to convert grams $\longrightarrow$ moles $\longrightarrow$ atoms. Notice also that our answer is a very large number. Any time you calculate the number of atoms, molecules, or ions in an ordinary sample of matter, you can expect the answer to be very large. In contrast, the number of moles in a sample will usually be much smaller, often less than 1. The general procedure for interconverting mass and number of formula units (atoms, molecules, ions, or whatever is represented by the chemical formula) of a substance is summarized in Figure 3.10 ▼.

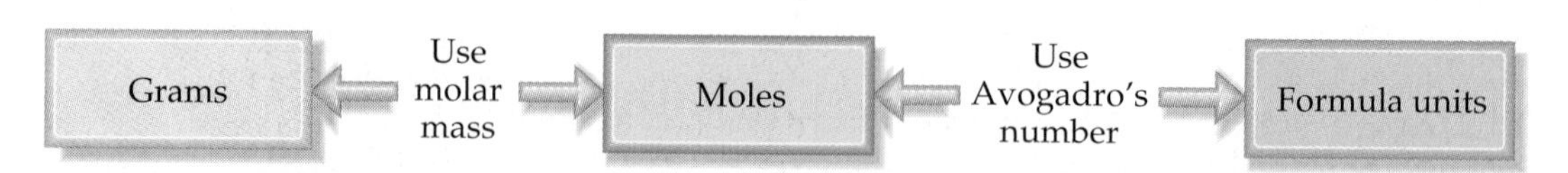

▲ **Figure 3.10** Outline of the procedure used to interconvert the mass of a substance in grams and the number of formula units of that substance. The number of moles of the substance is central to the calculation; thus, the mole concept can be thought of as the bridge between the mass of a substance and the number of formula units.

SAMPLE EXERCISE 3.12

How many glucose molecules are in 5.23 g of $C_6H_{12}O_6$?

Solution

Analyze: We are given the number of grams of glucose and its chemical formula and asked to calculate the number of glucose molecules.

Plan: The strategy for determining the number of molecules in a given quantity of a substance is summarized in Figure 3.10. We must convert 5.23 g $C_6H_{12}O_6$ to moles $C_6H_{12}O_6$, which can then be converted to molecules $C_6H_{12}O_6$. The first conversion uses the molar mass of $C_6H_{12}O_6$: 1 mol $C_6H_{12}O_6$ = 180.0 g $C_6H_{12}O_6$. The second conversion uses Avogadro's number.

Solve:

Molecules $C_6H_{12}O_6$

$$= (5.23\ \cancel{\text{g } C_6H_{12}O_6})\left(\frac{1\ \cancel{\text{mol } C_6H_{12}O_6}}{180.0\ \cancel{\text{g } C_6H_{12}O_6}}\right)\left(\frac{6.023 \times 10^{23}\ \text{molecules } C_6H_{12}O_6}{1\ \cancel{\text{mol } C_6H_{12}O_6}}\right)$$

$$= 1.75 \times 10^{22}\ \text{molecules } C_6H_{12}O_6$$

Check: The magnitude of the answer is reasonable. Because the mass we began with is less than a mole, there should be less than 6.02×10^{23} molecules. We can make a ballpark estimate of the answer: $5/200 = 2.5 \times 10^{-2}$ mol; $2.5 \times 10^{-2} \times 6 \times 10^{23} = 15 \times 10^{21} = 1.5 \times 10^{22}$ molecules. The units (molecules) and significant figures (3) are appropriate.

Comment: If you were also asked for the number of atoms of a particular element, an additional factor would be needed to convert the number of molecules to the number of atoms. For example, there are six O atoms in a molecule of $C_6H_{12}O_6$. Thus, the number of O atoms in the sample is

$$\text{Atoms O} = (1.75 \times 10^{22}\ \cancel{\text{molecules } C_6H_{12}O_6})\left(\frac{6\ \text{atoms O}}{1\ \cancel{\text{molecule } C_6H_{12}O_6}}\right)$$

$$= 1.05 \times 10^{23}\ \text{atoms O}$$

PRACTICE EXERCISE

(a) How many nitric acid molecules are in 4.20 g of HNO_3? **(b)** How many O atoms are in this sample?

Answers: **(a)** 4.01×10^{22} molecules HNO_3; **(b)** 1.20×10^{23} atoms O

3.5 Empirical Formulas from Analyses

The empirical formula for a substance tells us the relative number of atoms of each element it contains. Thus, the formula H_2O indicates that water contains two H atoms for each O atom. This ratio also applies on the molar level; thus, 1 mol of H_2O contains 2 mol of H atoms and 1 mol of O atoms. Conversely, the ratio of the number of moles of each element in a compound gives the subscripts in a compound's empirical formula. Thus, the mole concept provides a way of calculating the empirical formulas of chemical substances, as shown in the following examples.

Mercury forms a compound with chlorine that is 73.9% mercury and 26.1% chlorine by mass. This means that if we had a 100.0-g sample of the solid, it would contain 73.9 g of mercury (Hg) and 26.1 g of chlorine (Cl). (Any size sample can be used in problems of this type, but we will generally use 100.0 g to simplify the calculation of mass from percentage.) Using the atomic weights of the elements to give us molar masses, we then calculate the number of moles of each element in the sample:

$$(73.9 \text{ g Hg})\left(\frac{1 \text{ mol Hg}}{200.6 \text{ g Hg}}\right) = 0.368 \text{ mol Hg}$$

$$(26.1 \text{ g Cl})\left(\frac{1 \text{ mol Cl}}{35.5 \text{ g Cl}}\right) = 0.735 \text{ mol Cl}$$

We then divide the larger number of moles (0.735) by the smaller (0.368) to obtain a Cl:Hg mole ratio of 1.99:1:

$$\frac{\text{moles of Cl}}{\text{moles of Hg}} = \frac{0.735 \text{ mol Cl}}{0.368 \text{ mol Hg}} = \frac{1.99 \text{ mol Cl}}{1 \text{ mol Hg}}$$

Because of experimental errors, the results may not lead to exact integers for the ratios of moles. The number 1.99 is very close to 2, so we can confidently conclude that the empirical formula for the compound is $HgCl_2$. This is the simplest, or empirical, formula because its subscripts are the smallest integers that express the *ratios* of atoms present in the compound. (Section 2.6) The general procedure for determining empirical formulas is outlined in Figure 3.11 ▶.

SAMPLE EXERCISE 3.13

Ascorbic acid (vitamin C) contains 40.92% C, 4.58% H, and 54.50% O by mass. What is the empirical formula of ascorbic acid?

Solution

Analyze: We are given the mass percentages of the elements in ascorbic acid and asked for its empirical formula.

Plan: The strategy for determining the empirical formula of a substance from its elemental composition involves the four steps given in Figure 3.11.

Solve: We first assume, for simplicity, that we have exactly 100 g of material (although any number can be used). In 100 g of ascorbic acid, we will have

40.92 g C, 4.58 g H, and 54.50 g O.

Second, we calculate the number of moles of each element in this sample:

$$\text{Moles C} = (40.92 \text{ g C})\left(\frac{1 \text{ mol C}}{12.01 \text{ g C}}\right) = 3.407 \text{ mol C}$$

$$\text{Moles H} = (4.58 \text{ g H})\left(\frac{1 \text{ mol H}}{1.008 \text{ g H}}\right) = 4.54 \text{ mol H}$$

$$\text{Moles O} = (54.50 \text{ g O})\left(\frac{1 \text{ mol O}}{16.00 \text{ g O}}\right) = 3.406 \text{ mol O}$$

Third, we determine the simplest whole-number ratio of moles by dividing each number of moles by the smallest number of moles, 3.406:

$$\text{C: } \frac{3.407}{3.406} = 1.000 \qquad \text{H: } \frac{4.54}{3.406} = 1.33 \qquad \text{O: } \frac{3.406}{3.406} = 1.000$$

The ratio for H is too far from 1 to attribute the difference to experimental error; in fact, it is quite close to $1\frac{1}{3}$. This suggests that if we multiply the ratio by 3, we will obtain whole numbers:

$$\text{C:H:O} = 3(1:1.33:1) = 3:4:3$$

The whole-number mole ratio gives us the subscripts for the empirical formula. Thus, the empirical formula is

$$C_3H_4O_3$$

Check: It is reassuring that the subscripts are moderately sized whole numbers. Otherwise, we have little by which to judge the reasonableness of our answer.

PRACTICE EXERCISE

A 5.325-g sample of methyl benzoate, a compound used in the manufacture of perfumes, is found to contain 3.758 g of carbon, 0.316 g of hydrogen, and 1.251 g of oxygen. What is the empirical formula of this substance?

Answer: C_4H_4O

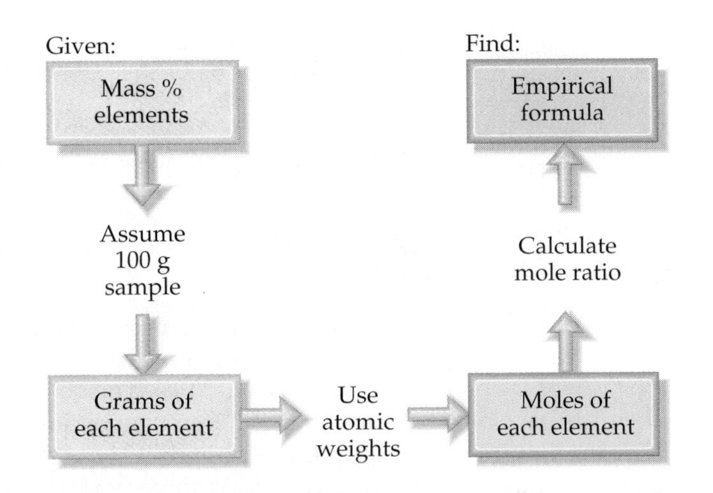

◀ **Figure 3.11** Outline of the procedure used to calculate the empirical formula of a substance from its percentage composition. The procedure is also summarized as "percent to mass, mass to mole, divide by small, multiply 'til whole."

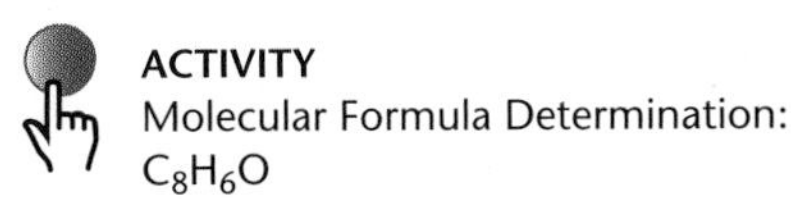

ACTIVITY
Molecular Formula Determination: C_8H_6O

Molecular Formula from Empirical Formula

The formula obtained from percentage compositions is always the empirical formula. We can obtain the molecular formula from the empirical formula if we know the molecular weight of the compound. *The subscripts in the molecular formula of a substance are always a whole-number multiple of the corresponding subscripts in its empirical formula.* (Section 2.6) The multiple is found by comparing the empirical formula weight with the molecular weight. In Sample Exercise 3.13, for example, the empirical formula of ascorbic acid was determined to be $C_3H_4O_3$, giving an empirical formula weight of 3(12.0 amu) + 4(1.0 amu) + 3(16.0 amu) = 88.0 amu. The experimentally determined molecular weight is 176 amu. Thus, the molecule has twice the mass (176/88.0 = 2.00) and must therefore have twice as many atoms of each kind as are given in the empirical formula. Consequently, the subscripts in the empirical formula must be multiplied by 2 to obtain the molecular formula: $C_6H_8O_6$.

SAMPLE EXERCISE 3.14

Mesitylene, a hydrocarbon that occurs in small amounts in crude oil, has an empirical formula of C_3H_4. The experimentally determined molecular weight of this substance is 121 amu. What is the molecular formula of mesitylene?

Solution

Analyze: We are given the empirical formula and molecular weight of mesitylene and asked to determine its molecular formula.

Plan: The subscripts in a molecular formula are whole-number multiples of the subscripts in its empirical formula. To find the appropriate multiple, we must compare the molecular weight with the formula weight of the empirical formula.

Solve: First, we calculate the formula weight of the empirical formula, C_3H_4:

$$3(12.0 \text{ amu}) + 4(1.0 \text{ amu}) = 40.0 \text{ amu}$$

Next, we divide the molecular weight by the empirical formula weight to obtain the factor used to multiply the subscripts in C_3H_4:

$$\frac{\text{molecular weight}}{\text{empirical formula weight}} = \frac{121}{40.0} = 3.02$$

Only whole-number ratios make physical sense because we must be dealing with whole atoms. The 3.02 in this case results from a small experimental error in the molecular weight. We therefore multiply each subscript in the empirical formula by 3 to give the molecular formula: C_9H_{12}.

Check: We can have confidence in the result because dividing the molecular weight by the formula weight yields nearly a whole number.

PRACTICE EXERCISE

Ethylene glycol, the substance used in automobile antifreeze, is composed of 38.7% C, 9.7% H, and 51.6% O by mass. Its molar mass is 62.1 g/mol. **(a)** What is the empirical formula of ethylene glycol? **(b)** What is its molecular formula?
Answers: **(a)** CH_3O; **(b)** $C_2H_6O_2$

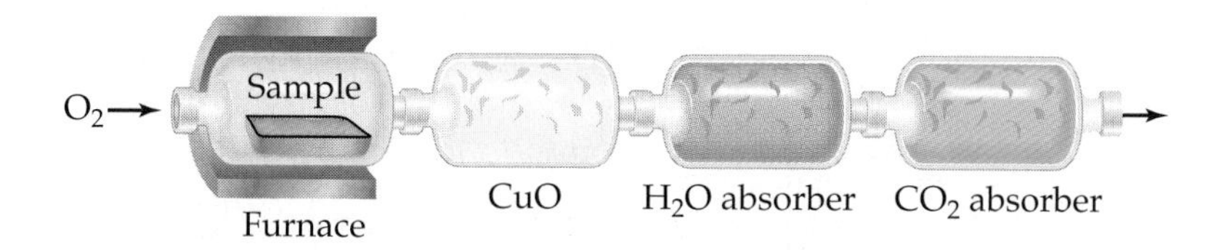

▶ **Figure 3.12** Apparatus to determine percentages of carbon and hydrogen in a compound. Copper oxide helps to oxidize traces of carbon and carbon monoxide to carbon dioxide and to oxidize hydrogen to water.

Combustion Analysis

MOVIE
Reduction of CuO

The empirical formula of a compound is based on experiments that give the number of moles of each element in a sample of the compound. That is why we use the word "empirical," which means "based on observation and experiment." Chemists have devised a number of different experimental techniques to determine the empirical formulas of compounds. One of these is combustion analysis, which is commonly used for compounds containing principally carbon and hydrogen as their component elements.

When a compound containing carbon and hydrogen is completely combusted in an apparatus such as that shown in Figure 3.12 ▲, all the carbon in the compound is converted to CO_2, and all the hydrogen is converted to H_2O. ∞ (Section 3.2) The amounts of CO_2 and H_2O produced are determined by measuring the mass increase in the CO_2 and H_2O absorbers. From the masses of CO_2 and H_2O we can calculate the number of moles of C and H in the original compound and thereby the empirical formula. If a third element is present in the compound, its mass can be determined by subtracting the masses of C and H from the compound's original mass. Sample Exercise 3.15 shows how to determine the empirical formula of a compound containing C, H, and O.

SAMPLE EXERCISE 3.15

Isopropyl alcohol, a substance sold as rubbing alcohol, is composed of C, H, and O. Combustion of 0.255 g of isopropyl alcohol produces 0.561 g CO_2 and 0.306 g H_2O. Determine the empirical formula of isopropyl alcohol.

Solution

Analyze: We are given the quantities of CO_2 and H_2O produced when a given quantity of isopropyl alcohol is combusted. We must use this information to determine the empirical formula for the isopropyl alcohol, a task that requires us to calculate the number of moles of C, H, and O in the sample.

Plan: We can use the mole concept to calculate the number of grams of C present in the CO_2 and the number of grams of H present in the H_2O. These are the quantities of C and H present in the isopropyl alcohol before combustion. The number of grams of O in the compound equals the mass of the isopropyl alcohol minus the sum of the C and H masses. Once we have the number of grams of C, H, and O in the sample, we can then proceed as in Sample Exercise 3.13: Calculate the number of moles of each element, and determine the mole ratio, which gives the subscripts in the empirical formula.

Solve: To calculate the number of grams of C, we first use the molar mass of CO_2, 1 mol CO_2 = 44.0 g CO_2, to convert grams of CO_2 to moles of CO_2. Because there is only 1 C atom in each CO_2 molecule, there is 1 mol of C atoms per mole of CO_2 molecules. This fact allows us to convert the moles of CO_2 to moles of C. Finally, we use the molar mass of C, 1 mol C = 12.0 g C, to convert moles of C to grams of C. Combining the three conversion factors, we have

$$\text{Grams C} = (0.561\ \cancel{\text{g CO}_2})\left(\frac{1\ \cancel{\text{mol CO}_2}}{44.0\ \cancel{\text{g CO}_2}}\right)\left(\frac{1\ \cancel{\text{mol C}}}{1\ \cancel{\text{mol CO}_2}}\right)\left(\frac{12.0\ \text{g C}}{1\ \cancel{\text{mol C}}}\right) = 0.153\ \text{g C}$$

The calculation of the number of grams of H from the grams of H_2O is similar, although we must remember that there are 2 mol of H atoms per 1 mol of H_2O molecules:

$$\text{Grams H} = (0.306\ \cancel{\text{g H}_2\text{O}})\left(\frac{1\ \cancel{\text{mol H}_2\text{O}}}{18.0\ \cancel{\text{g H}_2\text{O}}}\right)\left(\frac{2\ \cancel{\text{mol H}}}{1\ \cancel{\text{mol H}_2\text{O}}}\right)\left(\frac{1.01\ \text{g H}}{1\ \cancel{\text{mol H}}}\right) = 0.0343\ \text{g H}$$

The total mass of the sample, 0.255 g, is the sum of the masses of the C, H, and O. Thus, we can calculate the mass of O as follows:

$$\text{Mass of O} = \text{mass of sample} - (\text{mass of C} + \text{mass of H})$$
$$= 0.255\text{ g} - (0.153\text{ g} + 0.0343\text{ g}) = 0.068\text{ g O}$$

We then calculate the number of moles of C, H, and O in the sample:

$$\text{Moles C} = (0.153\text{ g C})\left(\frac{1\text{ mol C}}{12.0\text{ g C}}\right) = 0.0128\text{ mol C}$$

$$\text{Moles H} = (0.0343\text{ g H})\left(\frac{1\text{ mol H}}{1.01\text{ g H}}\right) = 0.0340\text{ mol H}$$

$$\text{Moles O} = (0.068\text{ g O})\left(\frac{1\text{ mol O}}{16.0\text{ g O}}\right) = 0.0043\text{ mol O}$$

To find the empirical formula, we must compare the relative number of moles of each element in the sample. The relative number of moles of each element is found by dividing each number by the smallest number, 0.0043. The mole ratio of C : H : O so obtained is 2.98 : 7.91 : 1.00. The first two numbers are very close to the whole numbers 3 and 8, giving the empirical formula C_3H_8O.

PRACTICE EXERCISE

(a) Caproic acid, which is responsible for the foul odor of dirty socks, is composed of C, H, and O atoms. Combustion of a 0.225-g sample of this compound produces 0.512 g CO_2 and 0.209 g H_2O. What is the empirical formula of caproic acid? **(b)** Caproic acid has a molar mass of 116 g/mol. What is its molecular formula?
Answers: **(a)** C_3H_6O; **(b)** $C_6H_{12}O_2$

3.6 Quantitative Information from Balanced Equations

The mole concept allows us to use the quantitative information available in a balanced equation on a practical macroscopic level. Consider the following balanced equation:

$$2H_2(g) + O_2(g) \longrightarrow 2H_2O(l) \qquad [3.11]$$

The coefficients tell us that two molecules of H_2 react with each molecule of O_2 to form two molecules of H_2O. It follows that the relative numbers of moles are identical to the relative numbers of molecules:

$2H_2(g)$	+	$O_2(g)$	$\longrightarrow$	$2H_2O(l)$
2 molecules		1 molecule		2 molecules
$2(6.02 \times 10^{23}$ molecules)		6.02×10^{23} molecules		$2(6.02 \times 10^{23}$ molecules)
2 mol		1 mol		2 mol

The coefficients in a balanced chemical equation can be interpreted both as the relative numbers of molecules (or formula units) involved in the reaction and as the relative numbers of moles.

The quantities 2 mol H_2, 1 mol O_2, and 2 mol H_2O, which are given by the coefficients in Equation 3.11, are called *stoichiometrically equivalent quantities*. The relationship between these quantities can be represented as

$$2\text{ mol }H_2 \simeq 1\text{ mol }O_2 \simeq 2\text{ mol }H_2O$$

where the symbol $\simeq$ means "stoichiometrically equivalent to." In other words, Equation 3.11 shows 2 mol of H_2 and 1 mol of O_2 forming 2 mol of H_2O. These stoichiometric relations can be used to convert between quantities of reactants and products in a chemical reaction. For example, the number of moles of H_2O produced from 1.57 mol of O_2 can be calculated as follows:

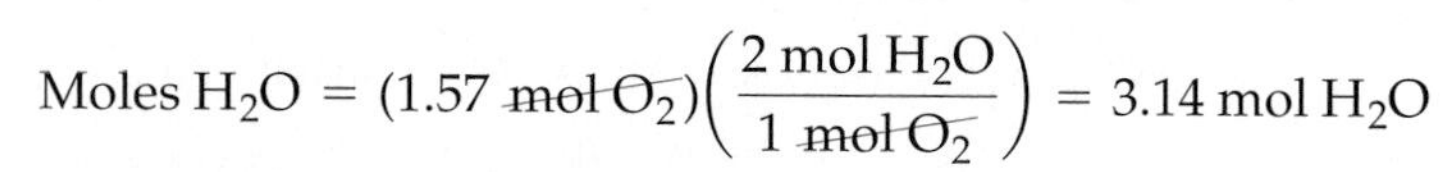

$$\text{Moles } H_2O = (1.57 \text{ mol } O_2)\left(\frac{2 \text{ mol } H_2O}{1 \text{ mol } O_2}\right) = 3.14 \text{ mol } H_2O$$

As an additional example, consider the combustion of butane (C_4H_{10}), the fuel in disposable cigarette lighters:

$$2C_4H_{10}(l) + 13O_2(g) \longrightarrow 8CO_2(g) + 10H_2O(g) \qquad [3.12]$$

Let's calculate the mass of CO_2 produced when 1.00 g of C_4H_{10} is burned. The coefficients in Equation 3.12 tell how the amount of C_4H_{10} consumed is related to the amount of CO_2 produced: 2 mol $C_4H_{10} \simeq$ 8 mol CO_2. In order to use this relationship, however, we must use the molar mass of C_4H_{10} to convert grams of C_4H_{10} to moles of C_4H_{10}. Because 1 mol C_4H_{10} = 58.0 g C_4H_{10}, we have

$$\text{Moles } C_4H_{10} = (1.00 \text{ g } C_4H_{10})\left(\frac{1 \text{ mol } C_4H_{10}}{58.0 \text{ g } C_4H_{10}}\right)$$

$$= 1.72 \times 10^{-2} \text{ mol } C_4H_{10}$$

We can then use the stoichiometric factor from the balanced equation, 2 mol $C_4H_{10} \simeq$ 8 mol CO_2, to calculate moles of CO_2:

$$\text{Moles } CO_2 = (1.72 \times 10^{-2} \text{ mol } C_4H_{10})\left(\frac{8 \text{ mol } CO_2}{2 \text{ mol } C_4H_{10}}\right)$$

$$= 6.88 \times 10^{-2} \text{ mol } CO_2$$

Finally, we can calculate the mass of the CO_2, in grams, using the molar mass of CO_2 (1 mol CO_2 = 44.0 g CO_2):

$$\text{Grams } CO_2 = (6.88 \times 10^{-2} \text{ mol } CO_2)\left(\frac{44.0 \text{ g } CO_2}{1 \text{ mol } CO_2}\right)$$

$$= 3.03 \text{ g } CO_2$$

Thus, the conversion sequence is

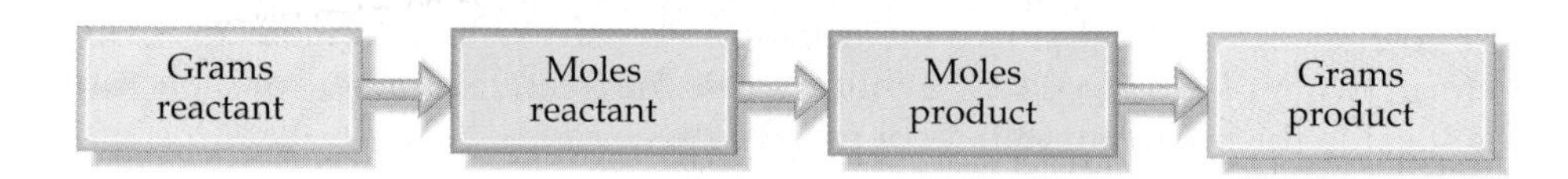

These steps can be combined in a single sequence of factors:

$$\text{Grams } CO_2 = (1.00 \text{ g } C_4H_{10})\left(\frac{1 \text{ mol } C_4H_{10}}{58.0 \text{ g } C_4H_{10}}\right)\left(\frac{8 \text{ mol } CO_2}{2 \text{ mol } C_4H_{10}}\right)\left(\frac{44.0 \text{ g } CO_2}{1 \text{ mol } CO_2}\right)$$

$$= 3.03 \text{ g } CO_2$$

Similarly, we can calculate the amount of O_2 consumed or H_2O produced in this reaction. To calculate the amount of O_2 consumed, we again rely on the coefficients in the balanced equation to give us the appropriate stoichiometric factor: 2 mol $C_4H_{10} \simeq$ 13 mol O_2:

$$\text{Grams } O_2 = (1.00 \text{ g } C_4H_{10})\left(\frac{1 \text{ mol } C_4H_{10}}{58.0 \text{ g } C_4H_{10}}\right)\left(\frac{13 \text{ mol } O_2}{2 \text{ mol } C_4H_{10}}\right)\left(\frac{32.0 \text{ g } O_2}{1 \text{ mol } O_2}\right)$$

$$= 3.59 \text{ g } O_2$$

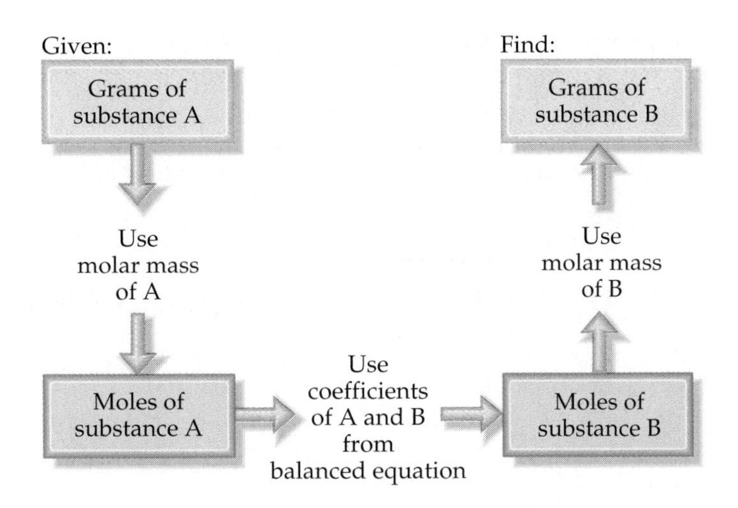

◀ **Figure 3.13** Outline of the procedure used to calculate the number of grams of a reactant consumed or of a product formed in a reaction, starting with the number of grams of one of the other reactants or products.

Figure 3.13 ▲ summarizes the general approach used to calculate the quantities of substances consumed or produced in chemical reactions. The balanced chemical equation provides the relative numbers of moles of reactants and products involved in the reaction.

SAMPLE EXERCISE 3.16

How many grams of water are produced in the oxidation of 1.00 g of glucose, $C_6H_{12}O_6$?

$$C_6H_{12}O_6(s) + 6O_2(g) \longrightarrow 6CO_2(g) + 6H_2O(l)$$

Solution

Analyze: We are given the mass of glucose, a reactant, and are asked to determine the mass of H_2O produced in the given equation.

Plan: The general strategy, as outlined in Figure 3.13, requires three steps. First, the amount of $C_6H_{12}O_6$ must be converted from grams to moles. We can then use the balanced equation, which relates the moles of $C_6H_{12}O_6$ to the moles of H_2O: 1 mol $C_6H_{12}O_6 \simeq$ 6 mol H_2O. Finally, the moles of H_2O must be converted to grams.

Solve: First, use the molar mass of $C_6H_{12}O_6$ to convert from grams $C_6H_{12}O_6$ to moles $C_6H_{12}O_6$:

$$\text{Moles } C_6H_{12}O_6 = (1.00\ \cancel{\text{g } C_6H_{12}O_6})\left(\frac{1 \text{ mol } C_6H_{12}O_6}{180.0\ \cancel{\text{g } C_6H_{12}O_6}}\right)$$

Second, use the balanced equation to convert moles of $C_6H_{12}O_6$ to moles of H_2O:

$$\text{Moles } H_2O = (1.00\ \cancel{\text{g } C_6H_{12}O_6})\left(\frac{1\ \cancel{\text{mol } C_6H_{12}O_6}}{180.0\ \cancel{\text{g } C_6H_{12}O_6}}\right)\left(\frac{6 \text{ mol } H_2O}{1\ \cancel{\text{mol } C_6H_{12}O_6}}\right)$$

Third, use the molar mass of H_2O to convert from moles of H_2O to grams of H_2O:

$$\text{Grams } H_2O = (1.00\ \cancel{\text{g } C_6H_{12}O_6})\left(\frac{1\ \cancel{\text{mol } C_6H_{12}O_6}}{180.0\ \cancel{\text{g } C_6H_{12}O_6}}\right)\left(\frac{6\ \cancel{\text{mol } H_2O}}{1\ \cancel{\text{mol } C_6H_{12}O_6}}\right)\left(\frac{18.0 \text{ g } H_2O}{1\ \cancel{\text{mol } H_2O}}\right)$$

$$= 0.600 \text{ g } H_2O$$

The steps can be summarized in a diagram like that in Figure 3.13:

$$\boxed{1.00 \text{ g } C_6H_{12}O_6} \dashrightarrow \text{no direct calculation} \dashrightarrow \boxed{0.600 \text{ g } H_2O}$$

$$\downarrow \times \left(\frac{1 \text{ mol } C_6H_{12}O_6}{180.0 \text{ g } C_6H_{12}O_6}\right) \qquad\qquad \uparrow \times \left(\frac{18.0 \text{ g } H_2O}{1 \text{ mol } H_2O}\right)$$

$$\boxed{5.56 \times 10^{-3} \text{ mol } C_6H_{12}O_6} \longrightarrow \times \left(\frac{6 \text{ mol } H_2O}{1 \text{ mol } C_6H_{12}O_6}\right) \longrightarrow \boxed{3.33 \times 10^{-2} \text{ mol } H_2O}$$

Check: An estimate of the magnitude of our answer: $18/180 = 0.1$ and $0.1 \times 6 = 0.6$, agrees with the exact calculation. The units, grams H_2O, are correct. The initial data had three significant figures, so three significant figures for the answer is correct.
Comment: An average person ingests 2 L of water daily and eliminates 2.4 L. The difference between 2 L and 2.4 L is produced in the metabolism of foodstuffs, such as in the oxidation of glucose. (*Metabolism* is a general term used to describe all the chemical processes of a living animal or plant.) The desert rat (kangaroo rat), on the other hand, apparently never drinks water. It survives on its metabolic water.

PRACTICE EXERCISE

The decomposition of $KClO_3$ is commonly used to prepare small amounts of O_2 in the laboratory: $2KClO_3(s) \longrightarrow 2KCl(s) + 3O_2(g)$. How many grams of O_2 can be prepared from 4.50 g of $KClO_3$?
Answer: 1.77 g

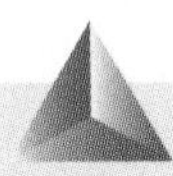

Chemistry at Work CO_2 and the Greenhouse Effect

Coal and petroleum provide the fuels that we use to generate electricity and power our industrial machinery. These fuels are composed primarily of hydrocarbons and other carbon-containing substances. As we have seen, the combustion of 1.00 g of C_4H_{10} produces 3.03 g of CO_2. Similarly, a gallon (3.78 L) of gasoline (density $= 0.70$ g/mL and approximate composition C_8H_{18}) produces about 8 kg (18 lb) of CO_2. Combustion of such fuels releases about 20 billion tons of CO_2 into the atmosphere annually.

Much CO_2 is absorbed into oceans or used by plants in photosynthesis. Nevertheless, we are now generating CO_2 much faster than it is being absorbed. Chemists have monitored atmospheric CO_2 concentrations since 1958. Analysis of air trapped in ice cores taken from Antarctica and Greenland makes it possible to determine the atmospheric levels of CO_2 during the past 160,000 years. These measurements reveal that the level of CO_2 remained fairly constant from the last Ice Age, some 10,000 years ago, until roughly the beginning of the Industrial Revolution, about 300 years ago. Since that time the concentration of CO_2 has increased by about 25% (Figure 3.14 ▼).

Although CO_2 is a minor component of the atmosphere, it plays a significant role by absorbing radiant heat, acting much like the glass of a greenhouse. For this reason, we often refer to CO_2 and other heat-trapping gases as greenhouse gases, and we call the warming caused by these gases the *greenhouse effect*. Some scientists believe that the accumulation of CO_2 and other heat-trapping gases has begun to change the climate of our planet. Other scientists point out that the factors affecting climate are complex and incompletely understood.

We will examine the greenhouse effect more closely in Chapter 18.

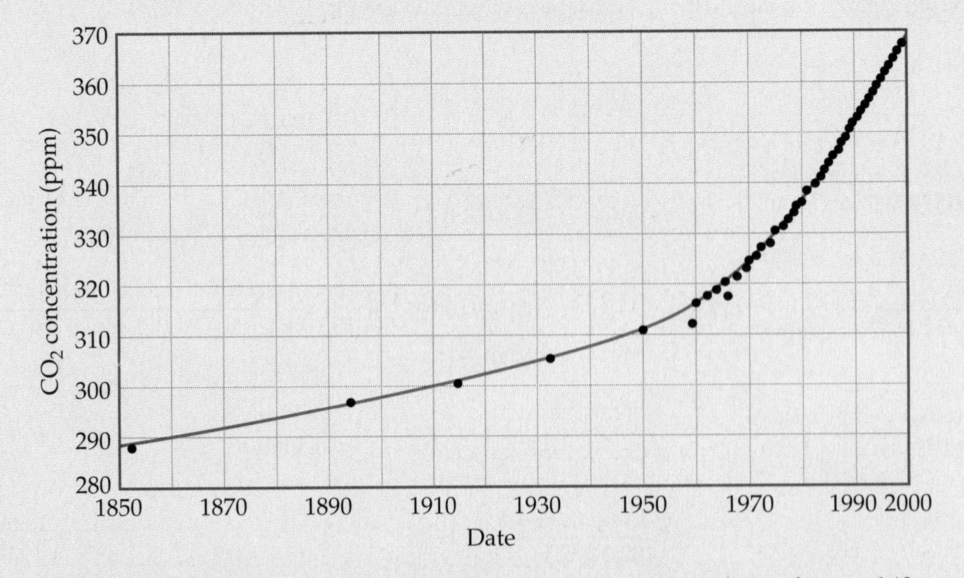

▲ **Figure 3.14** The concentration of atmospheric CO_2 has increased over the past 140 years. Data before 1958 came from analyses of air trapped in bubbles of glacial ice. The concentration in ppm (vertical scale) is the number of molecules of CO_2 per million (10^6) molecules of air.

SAMPLE EXERCISE 3.17

Solid lithium hydroxide is used in space vehicles to remove exhaled carbon dioxide. The lithium hydroxide reacts with gaseous carbon dioxide to form solid lithium carbonate and liquid water. How many grams of carbon dioxide can be absorbed by 1.00 g of lithium hydroxide?

Solution

Analyze: We are given a verbal description of a reaction and asked to calculate the number of grams of carbon dioxide that react with 1.00 g of lithium hydroxide.

Plan: The verbal description of the reaction can be used to write a balanced equation:

$$2LiOH(s) + CO_2(g) \longrightarrow Li_2CO_3(s) + H_2O(l)$$

We are given the grams of LiOH and asked to calculate grams of CO_2. This task can be accomplished by the following set of conversions: grams LiOH $\longrightarrow$ moles LiOH $\longrightarrow$ moles CO_2 $\longrightarrow$ grams CO_2. The conversion from grams of LiOH to moles of LiOH requires the formula weight of LiOH $(6.94 + 16.00 + 1.01 = 23.95)$. The conversion of moles of LiOH to moles of CO_2 is based on the balanced chemical equation: 2 mol LiOH $\simeq$ 1 mol CO_2. To convert the number of moles of CO_2 to grams, we must use the formula weight of CO_2: $12.01 + 2(16.00) = 44.01$.

Solve:

$$(1.00 \text{ g LiOH})\left(\frac{1 \text{ mol LiOH}}{23.95 \text{ g LiOH}}\right)\left(\frac{1 \text{ mol } CO_2}{2 \text{ mol LiOH}}\right)\left(\frac{44.01 \text{ g } CO_2}{1 \text{ mol } CO_2}\right) = 0.919 \text{ g } CO_2$$

Check: Notice that $23.95 \approx 24$; $24 \times 2 = 48$, and 44/48 is slightly less than 1. Thus, the magnitude of the answer is reasonable based on the amount of starting LiOH; the significant figures and units are appropriate, too.

PRACTICE EXERCISE

Propane, C_3H_8, is a common fuel used for cooking and home heating. What mass of O_2 is consumed in the combustion of 1.00 g of propane?

Answer: 3.64 g

3.7 Limiting Reactants

ANIMATION
Limiting Reactant

ACTIVITY
Limiting Reagents

Suppose you wish to make several sandwiches using one slice of cheese and two slices of bread for each sandwich. Using Bd = bread, Ch = cheese, and Bd_2Ch = sandwich, the recipe for making a sandwich can be represented like a chemical equation:

$$2Bd + Ch \longrightarrow Bd_2Ch$$

If you have 10 slices of bread and 7 slices of cheese, you will be able to make only five sandwiches before you run out of bread. You will have two slices of cheese left over. The amount of available bread limits the number of sandwiches.

An analogous situation occurs in chemical reactions when one of the reactants is used up before the others. The reaction stops as soon as any one of the reactants is totally consumed, leaving the excess reactants as leftovers. Suppose, for example, that we have a mixture of 10 mol H_2 and 7 mol O_2, which react to form water:

$$2H_2(g) + O_2(g) \longrightarrow 2H_2O(g)$$

Because 2 mol H_2 $\simeq$ 1 mol O_2, the number of moles of O_2 needed to react with all the H_2 is

$$\text{Moles } O_2 = (10 \text{ mol } H_2)\left(\frac{1 \text{ mol } O_2}{2 \text{ mol } H_2}\right) = 5 \text{ mol } O_2$$

Because 7 mol O_2 was available at the start of the reaction, 7 mol O_2 − 5 mol O_2 = 2 mol O_2 will still be present when all the H_2 is consumed. The example we have considered is depicted on a molecular level in Figure 3.15 ▶.

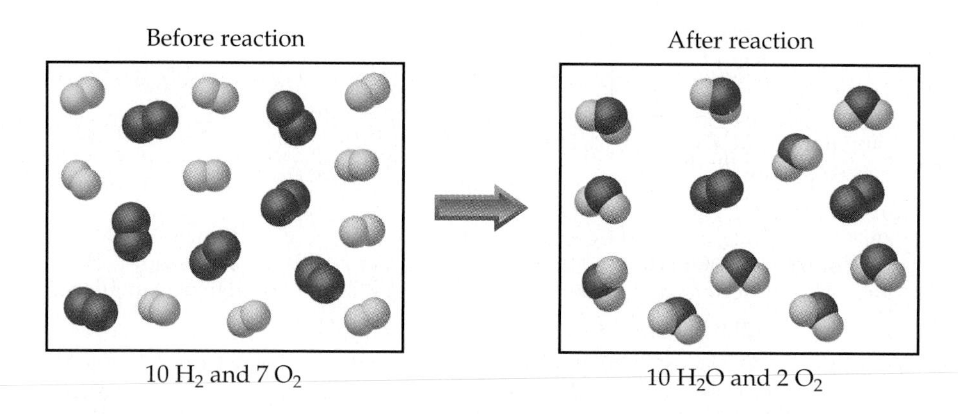

▶ **Figure 3.15** Diagram showing the complete consumption of a limiting reagent in a reaction. Because the H_2 is completely consumed, it is the limiting reagent in this case. Because there is a stoichiometric excess of O_2, some is left over at the end of the reaction.

The reactant that is completely consumed in a reaction is called the **limiting reactant** or **limiting reagent** because it determines, or limits, the amount of product formed. The other reactants are sometimes called *excess reactants* or *excess reagents*. In our example, H_2 is the limiting reactant, which means that once all the H_2 has been consumed, the reaction stops. O_2 is the excess reactant, and some is left over when the reaction stops.

There are no restrictions on the starting amounts of the reactants in any reaction. Indeed, many reactions are carried out using an excess of one reagent. The quantities of reactants consumed and the quantities of products formed, however, are restricted by the quantity of the limiting reactant.

Before we leave our present example, let's summarize the data in a tabular form:

	$2H_2(g)$	$+\ O_2(g)$	$\longrightarrow 2H_2O(g)$
Initial quantities:	10 mol	7 mol	0 mol
Change (reaction):	−10 mol	−5 mol	+10 mol
Final quantities:	0 mol	2 mol	10 mol

The initial amounts of the reactants are what we started with (10 mol H_2 and 7 mol O_2). The second line in the table (change) summarizes the amounts of the reactants consumed and the amount of the product formed in the reaction. These quantities are restricted by the quantity of the limiting reactant and depend on the coefficients in the balanced equation. The mole ratio of $H_2:O_2:H_2O = 10:5:10$ conforms to the ratio of the coefficients in the balanced equation, 2:1:2. The changes are negative for the reactants because they are consumed during the reaction and positive for the product because it is formed during the reaction. Finally, the quantities in the third line of the table (final quantities) depend on the initial quantities and their changes, and these entries are found by adding the entries for the initial quantity and change for each column. There is none of the limiting reactant (H_2) left at the end of the reaction. All that remains is 2 mol O_2 and 10 mol H_2O.

SAMPLE EXERCISE 3.18

The most important commercial process for converting N_2 from the air into nitrogen-containing compounds is based on the reaction of N_2 and H_2 to form ammonia (NH_3):

$$N_2(g) + 3H_2(g) \longrightarrow 2NH_3(g)$$

How many moles of NH_3 can be formed from 3.0 mol of N_2, and 6.0 mol of H_2

Solution
Analyze: We are asked to calculate the number of moles of product, NH_3, given the quantities of each reactant, N_2 and H_2, available in a reaction.
Plan: We are asked for the amount of product formed in a reaction, given the amounts of two reactants, so this is a limiting reactant problem. If we assume that one reactant is completely consumed, we can calculate how much of the second reactant is needed in the reaction. By comparing the calculated quantity with the available amount, we can determine which reactant is limiting. We then proceed with the calculation using the quantity of the limiting reactant.

Solve: The number of moles of H_2 needed for complete consumption of 3.0 mol of N_2 is

$$\text{Moles } H_2 = (3.0 \text{ mol } N_2)\left(\frac{3 \text{ mol } H_2}{1 \text{ mol } N_2}\right) = 9.0 \text{ mol } H_2$$

Because only 6.0 mol H_2 are available, we will run out of H_2 before the N_2 is gone, and H_2 will be the limiting reactant. We use the quantity of the limiting reactant, H_2, to calculate the quantity of NH_3 produced:

$$\text{Moles } NH_3 = (6.0 \text{ mol } H_2)\left(\frac{2 \text{ mol } NH_3}{3 \text{ mol } H_2}\right) = 4.0 \text{ mol } NH_3$$

Comment: The following table summarizes this example:

	$N_2(g)$ +	$3H_2(g)$ ⟶	$2NH_3(g)$
Initial quantities:	3.0 mol	6.0 mol	0 mol
Change (reaction):	−2.0 mol	−6.0 mol	+4.0 mol
Final quantities:	1.0 mol	0 mol	4.0 mol

Notice that we can not only calculate the number of moles of NH_3 formed, but also the number of moles of each of the reactants remaining after the reaction. Notice also that although there are more moles of H_2 present at the beginning of the reaction, it is nevertheless the limiting reactant because of its larger coefficient in the balanced equation.
Check: The summarizing table shows that the mole ratio of reactants used and product formed conforms to the coefficients in the balanced equation, 1:3:2. Also, because H_2 is the limiting reactant, it is completely consumed in the reaction, leaving 0 mol at the end. Because 2.0 mol H_2 has two significant figures, our answer has two significant figures.

PRACTICE EXERCISE

Consider the reaction $2Al(s) + 3Cl_2(g) \longrightarrow 2AlCl_3(s)$. A mixture of 1.50 mol of Al and 3.00 mol of Cl_2 are allowed to react. **(a)** What is the limiting reactant? **(b)** How many moles of $AlCl_3$ are formed? **(c)** How many moles of the excess reactant remain at the end of the reaction?
Answers: **(a)** Al; **(b)** 1.50 mol; **(c)** 0.75 mol Cl_2

SAMPLE EXERCISE 3.19

Consider the following reaction:

$$2Na_3PO_4(aq) + 3Ba(NO_3)_2(aq) \longrightarrow Ba_3(PO_4)_2(s) + 6NaNO_3(aq)$$

Suppose a solution containing 3.50 g of Na_3PO_4 is mixed with a solution containing 6.40 g of $Ba(NO_3)_2$. How many grams of $Ba_3(PO_4)_2$ can be formed?

Solution
Analyze: We are given a chemical reaction and the quantities of two reactants [3.50 g Na_3PO_4 and 6.40 g $Ba(NO_3)_2$]. We are asked to calculate the number of grams of $Ba_3(PO_4)_2$ (one of the products).
Plan: We are asked to calculate the amount of product, given the amounts of two reactants, so this is a limiting reactant problem. Thus, we must first identify the limiting reagent. To do so, we must calculate the number of moles of each reactant and compare their ratio with that required by the balanced equation. We then use the quantity of the limiting reagent to calculate the mass of $Ba_3(PO_4)_2$ that forms.
Solve: From the balanced equation, we have the following stoichiometric relations:

$$2 \text{ mol } Na_3PO_4 \simeq 3 \text{ mol } Ba(NO_3)_2 \simeq 1 \text{ mol } Ba_3(PO_4)_2$$

Using the molar mass of each substance, we can calculate the number of moles of each reactant:

$$\text{Moles } Na_3PO_4 = (3.50 \text{ g } Na_3PO_4)\left(\frac{1 \text{ mol } Na_3PO_4}{164 \text{ g } Na_3PO_4}\right) = 0.0213 \text{ mol } Na_3PO_4$$

$$\text{Moles } Ba(NO_3)_2 = (6.40 \text{ g } Ba(NO_3)_2)\left(\frac{1 \text{ mol } Ba(NO_3)_2}{261 \text{ g } Ba(NO_3)_2}\right) = 0.0245 \text{ mol } Ba(NO_3)_2$$

These calculations show us that there are slightly more moles of $Ba(NO_3)_2$ than moles of Na_3PO_4. The coefficients in the balanced equation indicate, however, that the reaction requires 3 mol $Ba(NO_3)_2$ for each 2 mol Na_3PO_4. [That is, 1.5 times more moles of $Ba(NO_3)_2$ are needed than moles of Na_3PO_4.] Thus, there is insufficient $Ba(NO_3)_2$ to completely consume the Na_3PO_4. That means that $Ba(NO_3)_2$ is the limiting reagent. We therefore use the quantity of $Ba(NO_3)_2$ to calculate the quantity of product formed. We can begin the calculation with the grams of $Ba(NO_3)_2$, but we can save a step by starting with the moles of $Ba(NO_3)_2$ that were calculated previously in the exercise:

$$\text{Grams } Ba_3(PO_4)_2 = (0.0245 \text{ mol } Ba(NO_3)_2)\left(\frac{1 \text{ mol } Ba_3(PO_4)_2}{3 \text{ mol } Ba(NO_3)_2}\right)\left(\frac{602 \text{ g } Ba_3(PO_4)_2}{1 \text{ mol } Ba_3(PO_4)_2}\right)$$

$$= 4.92 \text{ g } Ba_3(PO_4)_2$$

Check: The magnitude of the answer seems reasonable: Starting with the numbers in the two factors on the right, we have $600/3 = 200$; $200 \times 0.025 = 5$. The units are correct, and the number of significant figures (3) corresponds to the number in the quantity of $Ba(NO_3)_2$.

Comment: The quantity of the limiting reagent, $Ba(NO_3)_2$, can also be used to determine the quantity of $NaNO_3$ formed (4.16 g) and the quantity of Na_3PO_4 used (2.67 g). The number of grams of the excess reagent, Na_3PO_4, remaining at the end of the reaction equals the starting amount minus the amount consumed in the reaction, $3.50 \text{ g} - 2.67 \text{ g} = 0.82 \text{ g}$.

PRACTICE EXERCISE

A strip of zinc metal weighing 2.00 g is placed in an aqueous solution containing 2.50 g of silver nitrate, causing the following reaction to occur:

$$Zn(s) + 2AgNO_3(aq) \longrightarrow 2Ag(s) + Zn(NO_3)_2(aq)$$

(a) Which reactant is limiting? **(b)** How many grams of Ag will form? **(c)** How many grams of $Zn(NO_3)_2$ will form? **(d)** How many grams of the excess reactant will be left at the end of the reaction?

Answers: **(a)** $AgNO_3$; **(b)** 1.59 g; **(c)** 1.39 g; **(d)** 1.52 g Zn

Theoretical Yields

The quantity of product that is calculated to form when all of the limiting reactant reacts is called the **theoretical yield**. The amount of product actually obtained in a reaction is called the *actual yield*. The actual yield is almost always less than (and can never be greater than) the theoretical yield. There are many reasons for this difference. Part of the reactants may not react, for example, or they may react in a way different from that desired (side reactions). In addition, it is not always possible to recover all of the reaction product from the reaction mixture. The **percent yield** of a reaction relates the actual yield to the theoretical (calculated) yield:

$$\text{Percent yield} = \frac{\text{actual yield}}{\text{theoretical yield}} \times 100\% \qquad [3.13]$$

In the experiment described in Sample Exercise 3.19, for example, we calculated that 4.92 g of $Ba_3(PO_4)_2$ should form when 3.50 g of Na_3PO_4 is mixed with 6.40 g of $Ba(NO_3)_2$. This is the theoretical yield of $Ba_3(PO_4)_2$ in the reaction. If the actual yield turned out to be 4.70 g, the percent yield would be

$$\frac{4.70 \text{ g}}{4.92 \text{ g}} \times 100\% = 95.5\%$$

SAMPLE EXERCISE 3.20

Adipic acid, $H_2C_6H_8O_4$, is used to produce nylon. It is made commercially by a controlled reaction between cyclohexane (C_6H_{12}) and O_2:

$$2C_6H_{12}(l) + 5O_2(g) \longrightarrow 2H_2C_6H_8O_4(l) + 2H_2O(g)$$

(a) Assume that you carry out this reaction starting with 25.0 g of cyclohexane, and that cyclohexane is the limiting reactant. What is the theoretical yield of adipic acid?

(b) If you obtain 33.5 g of adipic acid from your reaction, what is the percent yield of adipic acid?

Solution

Analyze: We are given a chemical equation and the quantity of one of the reactants (25.0 g of C_6H_{12}). We are asked first to calculate the theoretical yield of a product ($H_2C_6H_8O_4$) and then to calculate its percent yield if only 33.5 g of the substance is actually obtained.

Plan: (a) The theoretical yield is the calculated quantity of adipic acid formed in the reaction. We carry out the following conversions: g $C_6H_{12} \longrightarrow$ mol $C_6H_{12} \longrightarrow$ mol $H_2C_6H_8O_4 \longrightarrow$ g $H_2C_6H_8O_4$. **(b)** Once we have calculated the theoretical yield, we use Equation 3.13 to calculate the percent yield.

Solve:

$$\textbf{(a)}\ \text{Grams } H_2C_6H_8O_4 = (25.0 \cancel{\text{ g } C_6H_{12}})\left(\frac{1 \cancel{\text{ mol } C_6H_{12}}}{84.0 \cancel{\text{ g } C_6H_{12}}}\right) \times \left(\frac{2 \cancel{\text{ mol } H_2C_6H_8O_4}}{2 \cancel{\text{ mol } C_6H_{12}}}\right)\left(\frac{146.0 \text{ g } H_2C_6H_8O_4}{1 \cancel{\text{ mol } H_2C_6H_8O_4}}\right) = 43.5 \text{ g } H_2C_6H_8O_4$$

$$\textbf{(b)}\ \text{Percent yield} = \frac{\text{actual yield}}{\text{theoretical yield}} \times 100\% = \frac{33.5 \text{ g}}{43.5 \text{ g}} \times 100\% = 77.0\%$$

Check: Our answer in **(a)** has the appropriate magnitude, units, and significant figures. In **(b)** the answer is less than 100% as necessary.

PRACTICE EXERCISE

Imagine that you are working on ways to improve the process by which iron ore containing Fe_2O_3 is converted into iron. In your tests you carry out the following reaction on a small scale:

$$Fe_2O_3(s) + 3CO(g) \longrightarrow 2Fe(s) + 3CO_2(g)$$

(a) If you start with 150 g of Fe_2O_3 as the limiting reagent, what is the theoretical yield of Fe? **(b)** If the actual yield of Fe in your test was 87.9 g, what was the percent yield?

Answers: **(a)** 105 g Fe; **(b)** 83.7%

Summary and Key Terms

Introduction and Section 3.1 The study of the quantitative relationships between chemical formulas and chemical equations is known as **stoichiometry**. One of the important concepts of stoichiometry is the **law of conservation of mass**, which states that the total mass of the products of a chemical reaction is the same as the total mass of the reactants. The same numbers of atoms of each type are present before and after a chemical reaction. A balanced **chemical equation** shows equal numbers of atoms of each element on each side of the equation. Equations are balanced by placing coefficients in front of the chemical formulas for the **reactants** and **products** of a reaction, *not* by changing the subscripts in chemical formulas.

Section 3.2 Among the reaction types described in this chapter are (1) **combination reactions**, in which two reactants combine to form one product; (2) **decomposition reactions**, in which a single reactant forms two or more products; and (3) **combustion reactions** in oxygen, in which a hydrocarbon reacts with O_2 to form CO_2 and H_2O.

Section 3.3 Much quantitative information can be determined from chemical formulas and balanced chemical equations by using atomic weights. The **formula weight** of a compound equals the sum of the atomic weights of the atoms in its formula. If the formula is a molecular formula, the formula weight is also called the **molecular weight**. Atomic weights and formula weights can be used to determine the elemental composition of a compound.

Section 3.4 A **mole** of any substance is **Avogadro's number** (6.02×10^{23}) of formula units of that substance. The mass of a mole of atoms, molecules, or ions is the formula weight of that material expressed in grams (the **molar mass**). The mass of a molecule of H_2O, for example, is 18 amu, so the molar mass of H_2O is 18 g/mol.

Section 3.5 The empirical formula of any substance can be determined from its percent composition by calculating the relative number of moles of each atom in 100 g of the substance. If the substance is molecular in nature, its molecular formula can be determined from the empirical formula if the molecular weight is also known.

Sections 3.6 and 3.7 The mole concept can be used to calculate the relative quantities of reactants and products involved in chemical reactions. The coefficients in a balanced equation give the relative numbers of moles of the reactants and products. To calculate the number of grams of a product from the number of grams of a reactant, therefore, first convert grams of reactant to moles of reactant. We then use the coefficients in the balanced equation to convert the number of moles of reactant to moles of product. Finally, we convert moles of product to grams of product.

A **limiting reactant** is completely consumed in a reaction. When it is used up, the reaction stops, thus limiting the quantities of products formed. The **theoretical yield** of a reaction is the quantity of product calculated to form when all of the limiting reagent reacts. The actual yield of a reaction is always less than the theoretical yield. The **percent yield** compares the actual and theoretical yields.

Exercises

Balancing Chemical Equations

3.1 **(a)** What scientific principle or law is used in the process of balancing chemical equations? **(b)** In balancing equations, why shouldn't subscripts in chemical formulas be changed? **(c)** What are the symbols used to represent gases, liquids, solids, and aqueous solutions in chemical equations?

3.2 **(a)** What is the difference between adding a subscript 2 to the end of the formula for CO to give CO_2 and adding a coefficient in front of the formula to give 2CO? **(b)** Is the following chemical equation, as written, consistent with the law of conservation of mass?

$$3Mg(OH)_2(s) + 2H_3PO_4(aq) \longrightarrow Mg_3(PO_4)_2(s) + H_2O(l)$$

Why or why not?

3.3 The reaction between reactant A (blue spheres) and reactant B (red spheres) is shown in the following diagram:

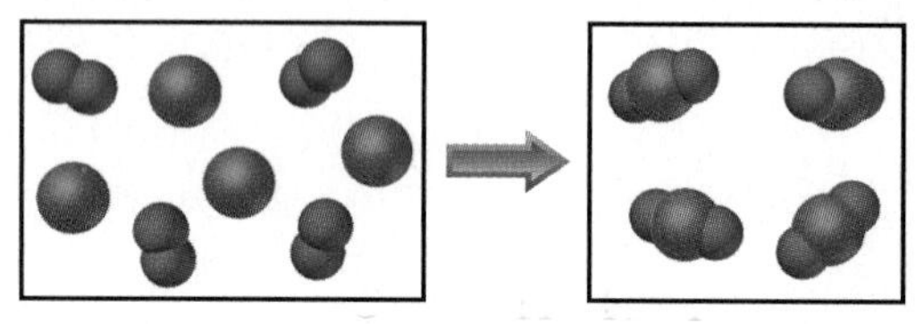

Based on this diagram, which equation best describes the reaction?

(a) $A_2 + B \longrightarrow A_2B$ **(b)** $A_2 + 4B \longrightarrow 2AB_2$
(c) $2A + B_4 \longrightarrow 2AB_2$ **(d)** $A + B_2 \longrightarrow AB_2$

3.4 Under appropriate experimental conditions, H_2 and CO react to form CH_3OH. The drawing represents a sample of H_2. Make a corresponding drawing of the CO needed to react completely with the H_2. How did you arrive at the number of CO molecules to show in your drawing?

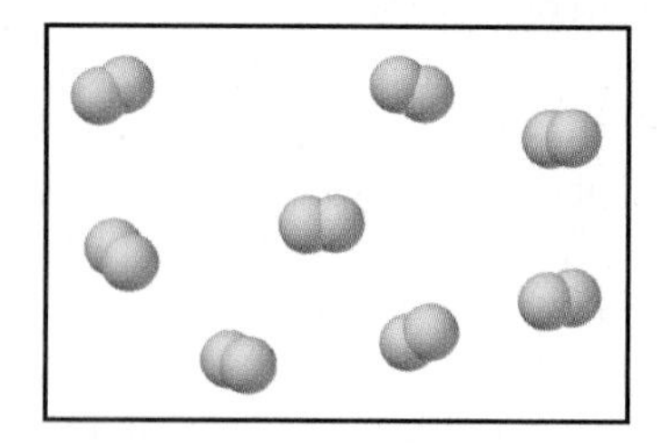

3.5 Balance the following equations:
(a) $SO_2(g) + O_2(g) \longrightarrow SO_3(g)$
(b) $P_2O_5(s) + H_2O(l) \longrightarrow H_3PO_4(aq)$
(c) $CH_4(g) + Cl_2(g) \longrightarrow CCl_4(l) + HCl(g)$
(d) $Al_4C_3(s) + H_2O(l) \longrightarrow Al(OH)_3(s) + CH_4(g)$
(e) $C_4H_{10}O(l) + O_2(g) \longrightarrow CO_2(g) + H_2O(g)$
(f) $Fe(OH)_3(s) + H_2SO_4(aq) \longrightarrow Fe_2(SO_4)_3(aq) + H_2O(l)$
(g) $Mg_3N_2(s) + H_2SO_4(aq) \longrightarrow MgSO_4(aq) + (NH_4)_2SO_4(aq)$

3.6 Balance the following equations:
(a) $Li(s) + N_2(g) \longrightarrow Li_3N(s)$
(b) $TiCl_4(l) + H_2O(l) \longrightarrow TiO_2(s) + HCl(aq)$
(c) $NH_4NO_3(s) \longrightarrow N_2(g) + O_2(g) + H_2O(g)$
(d) $Ca_3P_2(s) + H_2O(l) \longrightarrow Ca(OH)_2(aq) + PH_3(g)$
(e) $Al(OH)_3(s) + HClO_4(aq) \longrightarrow Al(ClO_4)_3(aq) + H_2O(l)$
(f) $AgNO_3(aq) + Na_2SO_4(aq) \longrightarrow Ag_2SO_4(s) + NaNO_3(aq)$
(g) $N_2H_4(g) + N_2O_4(g) \longrightarrow H_2O(g) + N_2(g)$

3.7 Write balanced chemical equations to correspond to each of the following descriptions: **(a)** Solid calcium carbide, CaC_2, reacts with water to form an aqueous solution of calcium hydroxide and acetylene gas, C_2H_2. **(b)** When solid potassium chlorate is heated, it decomposes to form solid potassium chloride and oxygen gas. **(c)** Solid zinc metal reacts with sulfuric acid to form hydrogen gas and an aqueous solution of zinc sulfate. **(d)** When liquid phosphorus trichloride is added to water, it reacts to form aqueous phosphorous acid, $H_3PO_3(aq)$, and aqueous hydrochloric acid. **(e)** When hydrogen sulfide gas is passed over solid hot iron(III) hydroxide, the resultant reaction produces solid iron(III) sulfide and gaseous water.

3.8 Convert these descriptions into balanced equations: **(a)** When sulfur trioxide gas reacts with water, a solution of sulfuric acid forms. **(b)** Boron sulfide, $B_2S_3(s)$, reacts violently with water to form dissolved boric acid, H_3BO_3, and hydrogen sulfide gas. **(c)** Phosphine, $PH_3(g)$, combusts in oxygen gas to form gaseous water and solid tetraphosphorus decoxide. **(d)** When solid mercury(II) nitrate is heated, it decomposes to form solid mercury(II) oxide, gaseous nitrogen dioxide, and oxygen. **(e)** Copper metal reacts with hot concentrated sulfuric acid solution to form aqueous copper(II) sulfate, sulfur dioxide gas, and water.

Patterns of Chemical Reactivity

3.9 **(a)** When the metallic element sodium combines with the nonmetallic element bromine, $Br_2(l)$, how can you determine the chemical formula of the product? How do you know whether the product is a solid, liquid, or gas at room temperature? Write the balanced chemical equation for the reaction. **(b)** When a hydrocarbon burns in air, what reactant besides the hydrocarbon is involved in the reaction? What products are formed? Write a balanced chemical equation for the combustion of benzene, $C_6H_6(l)$, in air.

3.10 **(a)** Determine the chemical formula of the product formed when the metallic element calcium combines with the nonmetallic element oxygen, O_2. Write the balanced chemical equation for the reaction. **(b)** What products form when a compound containing C, H, and O is completely combusted in air? Write a balanced chemical equation for the combustion of acetone, $C_3H_6O(l)$, in air.

3.11 Write a balanced chemical equation for the reaction that occurs when **(a)** $Mg(s)$ reacts with $Cl_2(g)$; **(b)** nickel(II) hydroxide decomposes into nickel(II) oxide and water when heated; **(c)** the hydrocarbon styrene, $C_8H_8(l)$, is combusted in air; **(d)** the gasoline additive MTBE (methyl tertiary-butyl ether), $C_5H_{12}O(l)$, burns in air.

3.12 Write a balanced chemical equation for the reaction that occurs when **(a)** aluminum metal undergoes a combination reaction with $Br_2(l)$; **(b)** strontium carbonate decomposes into strontium oxide and carbon dioxide when heated; **(c)** heptane, $C_7H_{16}(l)$, burns in air; **(d)** dimethylether, $CH_3OCH_3(g)$, is combusted in air.

3.13 Balance the following equations, and indicate whether they are combination, decomposition, or combustion reactions:
(a) $Al(s) + Cl_2(g) \longrightarrow AlCl_3(s)$
(b) $C_2H_4(g) + O_2(g) \longrightarrow CO_2(g) + H_2O(g)$
(c) $Li(s) + N_2(g) \longrightarrow Li_3N(s)$
(d) $PbCO_3(s) \longrightarrow PbO(s) + CO_2(g)$
(e) $C_7H_8O_2(l) + O_2(g) \longrightarrow CO_2(g) + H_2O(g)$

3.14 Balance the following equations, and indicate whether they are combination, decomposition, or combustion reactions:
(a) $C_3H_6(g) + O_2(g) \longrightarrow CO_2(g) + H_2O(g)$
(b) $NH_4NO_3(s) \longrightarrow N_2O(g) + H_2O(g)$
(c) $C_5H_6O(l) + O_2(g) \longrightarrow CO_2(g) + H_2O(g)$
(d) $N_2(g) + H_2(g) \longrightarrow NH_3(g)$
(e) $K_2O(s) + H_2O(l) \longrightarrow KOH(aq)$

Formula Weights

3.15 Determine the formula weights of each of the following compounds: **(a)** H_2S; **(b)** $NiCO_3$; **(c)** $Mg(C_2H_3O_2)_2$; **(d)** $(NH_4)_2SO_4$; **(e)** potassium phosphate; **(f)** iron(III) oxide; **(g)** diphosphorus pentasulfide.

3.16 Determine the formula weights of each of the following compounds: **(a)** nitrous oxide, N_2O, known as laughing gas and used as an anesthetic in dentistry; **(b)** benzoic acid, $HC_7H_5O_2$, a substance used as a food preservative; **(c)** $Mg(OH)_2$, the active ingredient in milk of magnesia; **(d)** urea, $(NH_2)_2CO$, a compound used as a nitrogen fertilizer; **(e)** isopentyl acetate, $CH_3CO_2C_5H_{11}$, responsible for the odor of bananas.

3.17 Calculate the percentage by mass of oxygen in the following compounds: **(a)** SO_2; **(b)** sodium sulfate; **(c)** C_2H_5COOH; **(d)** $Al(NO_3)_3$; **(e)** ammonium nitrate.

3.18 Calculate the percentage by mass of the indicated element in the following compounds: **(a)** carbon in acetylene, C_2H_2, a gas used in welding; **(b)** hydrogen in ammonium sulfate, $(NH_4)_2SO_4$, a substance used as a nitrogen fertilizer; **(c)** oxygen in ascorbic acid, $HC_6H_7O_6$, also known as vitamin C; **(d)** platinum in $PtCl_2(NH_3)_2$, a chemotherapy agent called cisplatin; **(e)** carbon in the female sex hormone estradiol, $C_{18}H_{24}O_2$; **(f)** carbon in capsaicin, $C_{18}H_{27}NO_3$, the compound that gives the hot taste to chili peppers.

3.19 Based on the following structural formulas, calculate the percentage of carbon by mass present in each compound:

```
         H     H
          \   /
           C=C     O
          /   \    ‖
(a) H—C        C—C—H      Benzaldehyde
          \\   //         (almond fragrance)
           C—C
          /   \
         H     H
```

```
        H3CO     H
            \   /
             C=C     O
            /   \    ‖
(b) HO—C         C—C—H     Vanillin
            \\   //        (vanilla flavor)
             C—C
            /   \
           H     H
```

```
           H   H   H       O
           |   |   |       ‖
(c) H3C—C—C—C—O—C—CH3
           |   |   |
         H3C   H   H
       Isopentyl acetate
        (banana flavor)
```

3.20 Calculate the percentage of carbon by mass in each of the compounds represented by the following models:

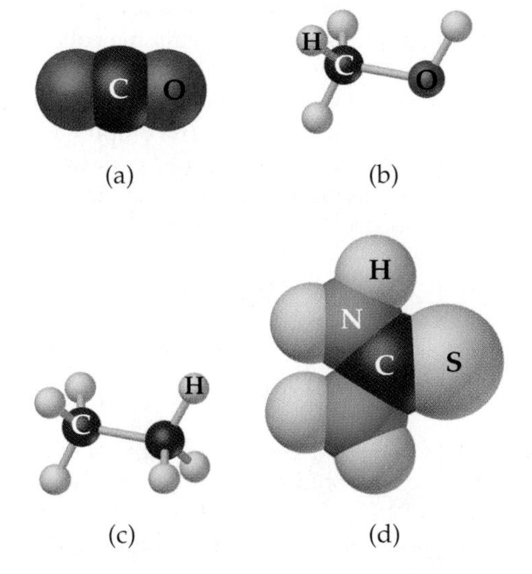

The Mole

3.21 **(a)** What is Avogadro's number, and how is it related to the mole? **(b)** What is the relationship between the formula weight of a substance and its molar mass?

3.22 **(a)** What is the mass, in grams, of a mole of ^{12}C? **(b)** How many carbon atoms are present in a mole of ^{12}C?

3.23 Without doing any detailed calculations (but using a periodic table to give atomic weights), rank the following samples in order of increasing number of atoms: 0.50 mol H_2O; 23 g Na; 6.0×10^{23} N_2 molecules.

3.24 Without doing any detailed calculations (but using a periodic table to give atomic weights), rank the following samples in order of increasing number of atoms: 3.0×10^{23} molecules of H_2O_2; 2.0 mol CH_4; 32 g O_2.

3.25 What is the mass, in kilograms, of an Avogadro's number of Olympic shot-put balls if each one has a mass of 16 lb? How does this compare with the mass of Earth, 5.98×10^{24} kg?

3.26 If Avogadro's number of pennies is divided equally among the 250 million men, women, and children in the United States, how many dollars would each receive? How does this compare with the national debt of the United States, which was $5.5 trillion at the time of the writing of this text?

3.27 Calculate the following quantities:
(a) mass, in grams, of 1.73 mol CaH_2
(b) moles of $Mg(NO_3)_2$ in 3.25 g of this substance
(c) number of molecules in 0.245 mol CH_3OH
(d) number of H atoms in 0.585 mol C_4H_{10}

3.28 Calculate the following quantities:
(a) mass, in grams, of 2.50×10^{-2} mol $MgCl_2$
(b) number of moles of NH_4Cl in 76.5 g of this substance
(c) number of molecules in 0.0772 mol $HCHO_2$
(d) number of NO_3^- ions in 4.88×10^{-3} mol $Al(NO_3)_3$

3.29 **(a)** What is the mass, in grams, of 2.50×10^{-3} mol of aluminum sulfate?
(b) How many moles of chloride ions are in 0.0750 g of aluminum chloride?
(c) What is the mass, in grams, of 7.70×10^{20} molecules of caffeine, $C_8H_{10}N_4O_2$?
(d) What is the molar mass of cholesterol if 0.00105 mol weighs 0.406 g?

3.30 **(a)** What is the mass, in grams, of 0.0714 mol of iron(III) phosphate?
(b) How many moles of ammonium ions are in 4.97 g of ammonium carbonate?
(c) What is the mass, in grams, of 6.52×10^{21} molecules of aspirin, $C_9H_8O_4$?
(d) What is the molar mass of diazepam (Valium®) if 0.05570 mol weighs 15.86 g?

3.31 The molecular formula of allicin, the compound responsible for the characteristic smell of garlic, is $C_6H_{10}OS_2$. **(a)** What is the molar mass of allicin? **(b)** How many moles of allicin are present in 5.00 mg of this substance? **(c)** How many molecules of allicin are in 5.00 mg of this substance? **(d)** How many S atoms are present in 5.00 mg of allicin?

3.32 The molecular formula of aspartame, the artificial sweetener marketed as NutraSweet®, is $C_{14}H_{18}N_2O_5$. **(a)** What is the molar mass of aspartame? **(b)** How many moles of aspartame are present in 1.00 mg of aspartame? **(c)** How many molecules of aspartame are present in 1.00 mg of aspartame? **(d)** How many hydrogen atoms are present in 1.00 mg of aspartame?

3.33 A sample of glucose, $C_6H_{12}O_6$, contains 5.77×10^{20} atoms of carbon. **(a)** How many atoms of hydrogen does it contain? **(b)** How many molecules of glucose does it contain? **(c)** How many moles of glucose does it contain? **(d)** What is the mass of this sample in grams?

3.34 A sample of the male sex hormone testosterone, $C_{19}H_{28}O_2$, contains 3.08×10^{21} atoms of hydrogen. **(a)** How many atoms of carbon does it contain? **(b)** How many molecules of testosterone does it contain? **(c)** How many moles of testosterone does it contain? **(d)** What is the mass of this sample in grams?

3.35 The allowable concentration level of vinyl chloride, C_2H_3Cl, in the atmosphere in a chemical plant is 2.0×10^{-6} g/L. How many moles of vinyl chloride in each liter does this represent? How many molecules per liter?

3.36 At least 25 μg of tetrahydrocannabinol (THC), the active ingredient in marijuana, is required to produce intoxication. The molecular formula of THC is $C_{21}H_{30}O_2$. How many moles of THC does this 25 μg represent? How many molecules?

Empirical Formulas

3.37 The following diagram represents the collection of elements formed by the decomposition of a compound. **(a)** If the blue spheres represent N atoms and the red ones represent O atoms, what was the empirical formula of the original compound? **(b)** Could you draw a diagram representing the molecules of the compound that had been decomposed? Why or why not?

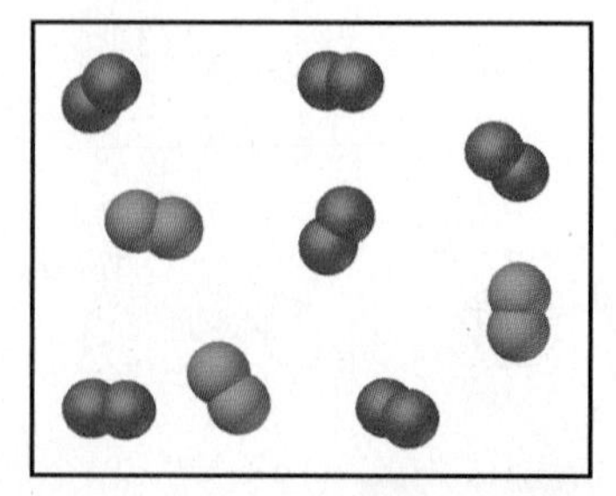

3.38 **(a)** The following diagram represents the collection of CO_2 and H_2O molecules formed by complete combustion of a hydrocarbon. What is the empirical formula of the hydrocarbon? **(b)** Could you draw a diagram representing the oxygen and hydrocarbon molecules that had been combusted? Why or why not?

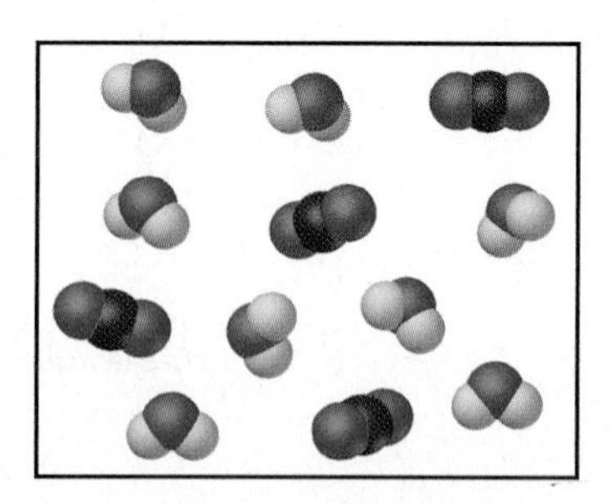

3.39 Give the empirical formula of each of the following compounds if a sample contains **(a)** 0.0130 mol C, 0.0390 mol H, and 0.0065 mol O; **(b)** 11.66 g iron and 5.01 g oxygen; **(c)** 40.0% C, 6.7% H, and 53.3% O by mass.

3.40 Determine the empirical formula of each of the following compounds if a sample contains **(a)** 0.104 mol K, 0.052 mol C, and 0.156 mol O; **(b)** 5.28 g Sn and 3.37 g F; **(c)** 87.5% N and 12.5% H by mass.

3.41 Determine the empirical formulas of the compounds with the following compositions by mass:
(a) 10.4% C, 27.8% S, and 61.7% Cl
(b) 21.7% C, 9.6% O, and 68.7% F
(c) 32.79% Na, 13.02% Al, and 54.19% F

3.42 Determine the empirical formulas of the compounds with the following compositions by mass:
(a) 55.3% K, 14.6% P, and 30.1% O
(b) 24.5% Na, 14.9% Si, and 60.6% F
(c) 62.1% C, 5.21% H, 12.1% N, and 20.7% O

3.43 What is the molecular formula of each of the following compounds?
(a) empirical formula CH_2, molar mass = 84 g/mol
(b) empirical formula NH_2Cl, molar mass = 51.5 g/mol

3.44 What is the molecular formula of each of the following compounds?
(a) empirical formula HCO_2, molar mass = 90.0 g/mol
(b) empirical formula C_2H_4O, molar mass = 88 g/mol

3.45 Determine the empirical and molecular formulas of each of the following substances:
(a) caffeine, a stimulant found in coffee that contains 49.5% C, 5.15% H, 28.9% N, and 16.5% O by mass; molar mass about 195 g/mol
(b) monosodium glutamate (MSG), a flavor enhancer in certain foods that contains 35.51% C, 4.77% H, 37.85% O, 8.29% N, and 13.60% Na; molar mass of 169 g/mol

3.46 Determine the empirical and molecular formulas of each of the following substances:
(a) ibuprofen, a headache remedy that contains 75.69% C, 8.80% H, and 15.51% O by mass; molar mass about 206 g/mol
(b) epinephrine (adrenaline), a hormone secreted into the bloodstream in times of danger or stress: 59.0% C, 7.1% H, 26.2% O, and 7.7% N by mass; MW about 180 amu

3.47 **(a)** Combustion analysis of toluene, a common organic solvent, gives 5.86 mg of CO_2 and 1.37 mg of H_2O. If the compound contains only carbon and hydrogen, what is its empirical formula? **(b)** Menthol, the substance we can smell in mentholated cough drops, is composed of C, H, and O. A 0.1005-g sample of menthol is combusted, producing 0.2829 g of CO_2 and 0.1159 g of H_2O. What is the empirical formula for menthol? If the compound has a molar mass of 156 g/mol, what is its molecular formula?

3.48 **(a)** The characteristic odor of pineapple is due to ethyl butyrate, a compound containing carbon, hydrogen, and oxygen. Combustion of 2.78 mg of ethyl butyrate produces 6.32 mg of CO_2 and 2.58 mg of H_2O. What is the empirical formula of the compound? **(b)** Nicotine, a component of tobacco, is composed of C, H, and N. A 5.250-mg sample of nicotine was combusted, producing 14.242 mg of CO_2 and 4.083 mg of H_2O. What is the empirical formula for nicotine? If the substance has a molar mass of 160 ± 5 g/mol, what is its molecular formula?

3.49 Washing soda, a compound used to prepare hard water for washing laundry, is a hydrate, which means that a certain number of water molecules are included in the solid structure. Its formula can be written as $Na_2CO_3 \cdot xH_2O$, where x is the number of moles of H_2O per mole of Na_2CO_3. When a 2.558-g sample of washing soda is heated at 125°C, all the water of hydration is lost, leaving 0.948 g of Na_2CO_3. What is the value of x?

3.50 Epsom salts, a strong laxative used in veterinary medicine, is a hydrate, which means that a certain number of water molecules are included in the solid structure. The formula for Epsom salts can be written as $MgSO_4 \cdot xH_2O$, where x indicates the number of moles of water per mole of $MgSO_4$. When 5.061 g of this hydrate is heated to 250°C, all the water of hydration is lost, leaving 2.472 g of $MgSO_4$. What is the value of x?

Calculations Based on Chemical Equations

3.51 Why is it essential to use balanced chemical equations when determining the quantity of a product formed from a given quantity of a reactant?

3.52 What parts of balanced chemical equations give information about the relative numbers of moles of reactants and products involved in a reaction?

3.53 The following diagram represents a high-temperature reaction between CH_4 and H_2O. Based on this reaction, how many moles of each product can be obtained starting with 4.0 mol CH_4?

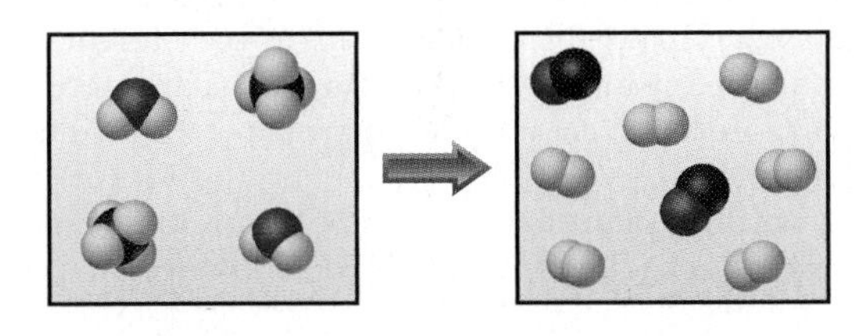

3.54 If 1.5 mol of each of the following compounds is completely combusted in oxygen, which one will produce the largest number of moles of H_2O? Which will produce the least? Explain. C_2H_5OH, C_3H_8, $CH_3CH_2COCH_3$

3.55 Hydrofluoric acid, HF(*aq*), cannot be stored in glass bottles because compounds called silicates in the glass are attacked by the HF(*aq*). Sodium silicate (Na_2SiO_3), for example, reacts as follows:

$$Na_2SiO_3(s) + 8HF(aq) \longrightarrow H_2SiF_6(aq) + 2NaF(aq) + 3H_2O(l)$$

(a) How many moles of HF are needed to react with 0.300 mol of Na_2SiO_3?
(b) How many grams of NaF form when 0.500 mol of HF reacts with excess Na_2SiO_3?
(c) How many grams of Na_2SiO_3 can react with 0.800 g of HF?

3.56 The fermentation of glucose ($C_6H_{12}O_6$) produces ethyl alcohol (C_2H_5OH) and CO_2:

$$C_6H_{12}O_6(aq) \longrightarrow 2C_2H_5OH(aq) + 2CO_2(g)$$

(a) How many moles of CO_2 are produced when 0.400 mol of $C_6H_{12}O_6$ reacts in this fashion?
(b) How many grams of $C_6H_{12}O_6$ are needed to form 7.50 g of C_2H_5OH?
(c) How many grams of CO_2 form when 7.50 g of C_2H_5OH are produced?

3.57 Aluminum sulfide reacts with water to form aluminum hydroxide and hydrogen sulfide. **(a)** Write the balanced chemical equation for this reaction. **(b)** How many grams of aluminum hydroxide are obtained from 10.5 g of aluminum sulfide?

3.58 Calcium hydride reacts with water to form calcium hydroxide and hydrogen gas. **(a)** Write a balanced chemical equation for the reaction. **(b)** How many grams of calcium hydride are needed to form 5.0 g of hydrogen?

3.59 Automotive air bags inflate when sodium azide, NaN_3, rapidly decomposes to its component elements:

$$2NaN_3(s) \longrightarrow 2Na(s) + 3N_2(g)$$

(a) How many moles of N_2 are produced by the decomposition of 2.50 mol of NaN_3?
(b) How many grams of NaN_3 are required to form 6.00 g of nitrogen gas?
(c) How many grams of NaN_3 are required to produce 10.0 ft^3 of nitrogen gas if the gas has a density of 1.25 g/L?

3.60 The complete combustion of octane, C_8H_{18}, a component of gasoline, proceeds as follows:

$$2C_8H_{18}(l) + 25O_2(g) \longrightarrow 16CO_2(g) + 18H_2O(g)$$

(a) How many moles of O_2 are needed to burn 0.750 mol of C_8H_{18}?
(b) How many grams of O_2 are needed to burn 5.00 g of C_8H_{18}?
(c) Octane has a density of 0.692 g/mL at 20°C. How many grams of O_2 are required to burn 1.00 gal of C_8H_{18}?

3.61 A piece of aluminum foil 1.00 cm square and 0.550 mm thick is allowed to react with bromine to form aluminum bromide as shown in the accompanying photo. **(a)** How many moles of aluminum were used? (The density of aluminum is 2.699 g/cm^3.) **(b)** How many grams of aluminum bromide form, assuming that the aluminum reacts completely?

3.62 Detonation of nitroglycerin proceeds as follows:

$$4C_3H_5N_3O_9(l) \longrightarrow 12CO_2(g) + 6N_2(g) + O_2(g) + 10H_2O(g)$$

(a) If a sample containing 3.00 mL of nitroglycerin (density = 1.592 g/mL) is detonated, how many total moles of gas are produced? **(b)** If each mole of gas occupies 55 L under the conditions of the explosion, how many liters of gas are produced? **(c)** How many grams of N_2 are produced in the detonation?

Limiting Reactants; Theoretical Yields

3.63 **(a)** Define the terms limiting reactant and excess reactant. **(b)** Why are the amounts of products formed in a reaction determined only by the amount of the limiting reactant?

3.64 **(a)** Define the terms theoretical yield, actual yield, and percent yield. **(b)** Why is the actual yield in a reaction almost always less than the theoretical yield?

3.65 Nitrogen (N_2) and hydrogen (H_2) react to form ammonia (NH_3). Consider the mixture of N_2 and H_2 shown in the accompanying diagram. The blue spheres represent N, and the white ones represent H. Draw a representation of the product mixture, assuming that the reaction goes to completion. How did you arrive at your representation? What is the limiting reactant in this case?

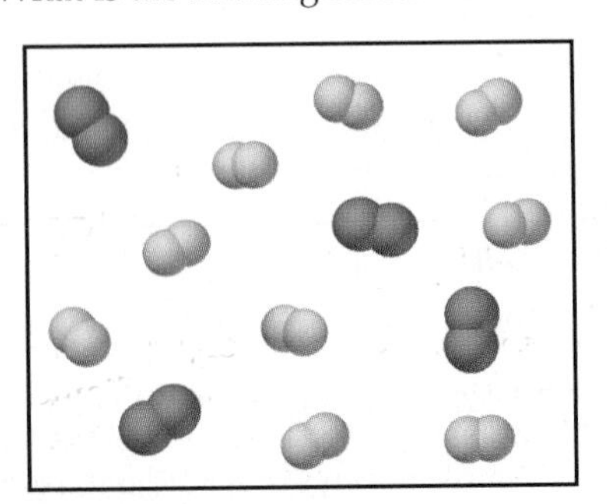

3.66 Nitrogen monoxide and oxygen react to form nitrogen dioxide. Consider the mixture of NO and O_2 shown in the accompanying diagram. The blue spheres represent N, and the red ones represent O. Draw a representation of the product mixture, assuming that the reaction goes to completion. How did you arrive at your representation? What is the limiting reactant in this case?

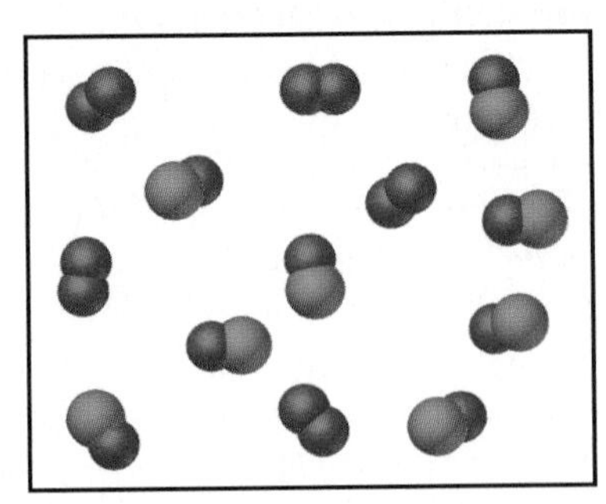

3.67 A manufacturer of bicycles has 4250 wheels, 2755 frames, and 2255 handlebars. **(a)** How many bicycles can be manufactured using these parts? **(b)** How many parts of each kind are left over? **(c)** Which part limits the production of bicycles?

3.68 A bottling plant has 115,350 bottles with a capacity of 355 mL, 122,500 caps, and 39,375 L of beverage. **(a)** How many bottles can be filled and capped? **(b)** How much of each item is left over? **(c)** Which component limits the production?

3.69 Sodium hydroxide reacts with carbon dioxide as follows:

$$2NaOH(s) + CO_2(g) \longrightarrow Na_2CO_3(s) + H_2O(l)$$

Which reagent is the limiting reactant when 1.70 mol NaOH and 1.00 mol CO_2 are allowed to react? How many moles of Na_2CO_3 can be produced? How many moles of the excess reactant remain after the completion of the reaction?

3.70 Aluminum hydroxide reacts with sulfuric acid as follows:

$$2Al(OH)_3(s) + 3H_2SO_4(aq) \longrightarrow Al_2(SO_4)_3(aq) + 6H_2O(l)$$

Which reagent is the limiting reactant when 0.450 mol $Al(OH)_3$ and 0.550 mol H_2SO_4 are allowed to react? How many moles of $Al_2(SO_4)_3$ can form under these conditions? How many moles of the excess reactant remain after the completion of the reaction?

3.71 The fizz produced when an Alka-Seltzer® tablet is dissolved in water is due to the reaction between sodium bicarbonate ($NaHCO_3$) and citric acid ($H_3C_6H_5O_7$):

$$3NaHCO_3(aq) + H_3C_6H_5O_7(aq) \longrightarrow 3CO_2(g) + 3H_2O(l) + Na_3C_6H_5O_7(aq)$$

In a certain experiment 1.00 g of sodium bicarbonate and 1.00 g of citric acid are allowed to react. **(a)** Which is the limiting reactant? **(b)** How many grams of carbon dioxide form? **(c)** How many grams of the excess reactant remain after the limiting reactant is completely consumed?

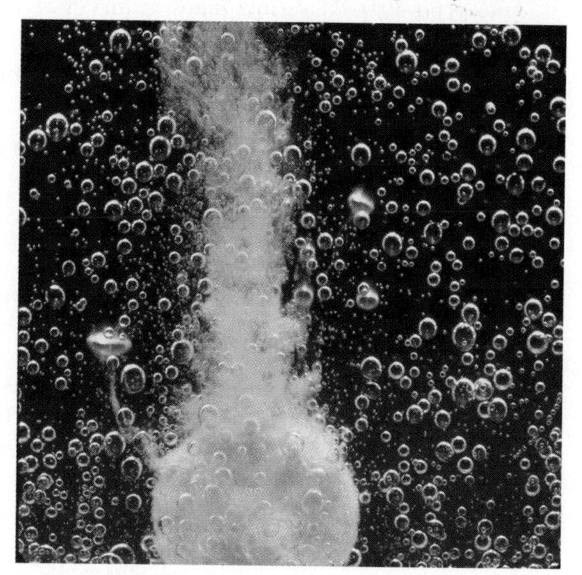

3.72 One of the steps in the commercial process for converting ammonia to nitric acid is the conversion of NH_3 to NO:

$$4NH_3(g) + 5O_2(g) \longrightarrow 4NO(g) + 6H_2O(g)$$

In a certain experiment, 2.25 g of NH_3 reacts with 3.75 g of O_2. **(a)** Which is the limiting reactant? **(b)** How many grams of NO form? **(c)** How many grams of the excess reactant remain after the limiting reactant is completely consumed?

3.73 Solutions of sodium carbonate and silver nitrate react to form solid silver carbonate and a solution of sodium nitrate. A solution containing 6.50 g of sodium carbonate is mixed with one containing 7.00 g of silver nitrate. How many grams of sodium carbonate, silver nitrate, silver carbonate, and sodium nitrate are present after the reaction is complete?

3.74 Solutions of sulfuric acid and lead(II) acetate react to form solid lead(II) sulfate and a solution of acetic acid. If 7.50 g of sulfuric acid and 7.50 g of lead(II) acetate are mixed, calculate the number of grams of sulfuric acid, lead(II) acetate, lead(II) sulfate, and acetic acid present in the mixture after the reaction is complete.

3.75 When benzene (C_6H_6) reacts with bromine (Br_2), bromobenzene (C_6H_5Br) is obtained:

$$C_6H_6 + Br_2 \longrightarrow C_6H_5Br + HBr$$

(a) What is the theoretical yield of bromobenzene in this reaction when 30.0 g of benzene reacts with 65.0 g of bromine? **(b)** If the actual yield of bromobenzene was 56.7 g, what was the percentage yield?

3.76 When ethane (C_2H_6) reacts with chlorine (Cl_2), the main product is C_2H_5Cl, but other products containing Cl such as $C_2H_4Cl_2$ are also obtained in small quantities. The formation of these other products reduces the yield of C_2H_5Cl. **(a)** Assuming that C_2H_6 and Cl_2 react only to form C_2H_5Cl and HCl, calculate the theoretical yield of C_2H_5Cl. **(b)** Calculate the percent yield of C_2H_5Cl if the reaction of 125 g of C_2H_6 with 255 g of Cl_2 produces 206 g of C_2H_5Cl.

3.77 Lithium and nitrogen react to produce lithium nitride:

$$6Li(s) + N_2(g) \longrightarrow 2Li_3N(s)$$

If 5.00 g of each reactant undergo a reaction with a 80.5% yield, how many grams of Li_3N are obtained from the reaction?

3.78 When hydrogen sulfide gas is bubbled into a solution of sodium hydroxide, the reaction forms sodium sulfide and water. How many grams of sodium sulfide are formed if 2.00 g of hydrogen sulfide is bubbled into a solution containing 2.00 g of sodium hydroxide, assuming that the sodium sulfide is made in 92.0% yield?

Additional Exercises

3.79 Write the balanced chemical equation for **(a)** the complete combustion of butyric acid, $HC_4H_7O_2(l)$, a compound produced when butter becomes rancid; **(b)** the decomposition of solid copper(II) hydroxide into solid copper(II) oxide and water vapor; **(c)** the combination reaction between zinc metal and chlorine gas.

3.80 The effectiveness of nitrogen fertilizers depends on both their ability to deliver nitrogen to plants and the amount of nitrogen they can deliver. Four common nitrogen-containing fertilizers are ammonia, ammonium nitrate, ammonium sulfate, and urea [$(NH_2)_2CO$]. Rank these fertilizers in terms of the mass percentage nitrogen they contain.

3.81 **(a)** Diamond is a natural form of pure carbon. How many moles of carbon are in a 1.25-carat diamond (1 carat = 0.200 g)? How many atoms are in this diamond? **(b)** The molecular formula of acetylsalicylic acid (aspirin), one of the most common pain relievers, is $HC_9H_7O_4$. How many moles of $HC_9H_7O_4$ are in a 0.500-g tablet of aspirin? How many molecules of $HC_9H_7O_4$ are in this tablet?

3.82 **(a)** One molecule of the antibiotic known as penicillin G has a mass of 5.342×10^{-21} g. What is the molar mass of penicillin G? **(b)** Hemoglobin, the oxygen-carrying protein in red blood cells, has four iron atoms per molecule and contains 0.340% iron by mass. Calculate the molar mass of hemoglobin.

3.83 Very small crystals composed of 1000 to 100,000 atoms, called quantum dots, are being investigated for use in electronic devices.
(a) Calculate the mass in grams of a quantum dot consisting of 10,000 atoms of silicon.
(b) Assuming that the silicon in the dot has a density of 2.3 g/cm^3, calculate its volume.
(c) Assuming that the dot has the shape of a cube, calculate the length of each edge of the cube.

3.84 Serotonin is a compound that conducts nerve impulses in the brain. It contains 68.2 mass percent C, 6.86 mass percent H, 15.9 mass percent N, and 9.08 mass percent O. Its molar mass is 176 g/mol. Determine its molecular formula.

3.85 The koala dines exclusively on eucalyptus leaves. Its digestive system detoxifies the eucalyptus oil, a poison to other animals. The chief constituent in eucalyptus oil is a substance called eucalyptol, which contains 77.87% C, 11.76% H, and the remainder O. **(a)** What is the empirical formula for this substance? **(b)** A mass spectrum of eucalyptol shows a peak at about 154 amu. What is the molecular formula of the substance?

3.86 Vanillin, the dominant flavoring in vanilla, contains C, H, and O. When 1.05 g of this substance is completely combusted, 2.43 g of CO_2 and 0.50 g of H_2O are produced. What is the empirical formula of vanillin?

[3.87] An organic compound was found to contain only C, H, and Cl. When a 1.50-g sample of the compound was completely combusted in air, 3.52 g of CO_2 was formed. In a separate experiment the chlorine in a 1.00-g sample of the compound was converted to 1.27 g of AgCl. Determine the empirical formula of the compound.

[3.88] An oxybromate compound, $KBrO_x$, where x is unknown, is analyzed and found to contain 52.92% Br. What is the value of x?

[3.89] An element X forms an iodide (XI_3) and a chloride (XCl_3). The iodide is quantitatively converted to the chloride when it is heated in a stream of chlorine:

$$2XI_3 + 3Cl_2 \longrightarrow 2XCl_3 + 3I_2$$

If 0.5000 g of XI_3 is treated, 0.2360 g of XCl_3 is obtained. **(a)** Calculate the atomic weight of the element X. **(b)** Identify the element X.

3.90 A method used by the Environmental Protection Agency (EPA) for determining the concentration of ozone in air is to pass the air sample through a "bubbler" containing sodium iodide, which removes the ozone according to the following equation:

$$O_3(g) + 2NaI(aq) + H_2O(l) \longrightarrow O_2(g) + I_2(s) + 2NaOH(aq)$$

(a) How many moles of sodium iodide are needed to remove 3.8×10^{-5} mol O_3? **(b)** How many grams of sodium iodide are needed to remove 0.550 mg of O_3?

3.91 A chemical plant uses electrical energy to decompose aqueous solutions of NaCl to give Cl_2, H_2, and NaOH:

$$2NaCl(aq) + 2H_2O(l) \longrightarrow 2NaOH(aq) + H_2(g) + Cl_2(g)$$

If the plant produces 1.5×10^6 kg (1500 metric tons) of Cl_2 daily, estimate the quantities of H_2 and NaOH produced.

3.92 The fat stored in the hump of a camel is a source of both energy and water. Calculate the mass of H_2O produced by metabolism of 1.0 kg of fat, assuming the fat consists entirely of tristearin ($C_{57}H_{110}O_6$), a typical animal fat, and assuming that during metabolism, tristearin reacts with O_2 to form only CO_2 and H_2O.

[3.93] When hydrocarbons are burned in a limited amount of air, both CO and CO_2 form. When 0.450 g of a particular hydrocarbon was burned in air, 0.467 g of CO, 0.733 g of CO_2, and 0.450 g of H_2O were formed. **(a)** What is the empirical formula of the compound? **(b)** How many grams of O_2 were used in the reaction? **(c)** How many grams would have been required for complete combustion?

3.94 A mixture of $N_2(g)$ and $H_2(g)$ reacts in a closed container to form ammonia, $NH_3(g)$. The reaction ceases before either reactant has been totally consumed. At this stage 2.0 mol N_2, 2.0 mol H_2, and 2.0 mol NH_3 are present. How many moles of N_2 and H_2 were present originally?

[3.95] A mixture containing $KClO_3$, K_2CO_3, $KHCO_3$, and KCl was heated, producing CO_2, O_2, and H_2O gases according to the following equations:

$$2KClO_3(s) \longrightarrow 2KCl(s) + 3O_2(g)$$
$$2KHCO_3(s) \longrightarrow K_2O(s) + H_2O(g) + 2CO_2(g)$$
$$K_2CO_3(s) \longrightarrow K_2O(s) + CO_2(g)$$

The KCl does not react under the conditions of the reaction. If 100.0 g of the mixture produces 1.80 g of H_2O, 13.20 g of CO_2, and 4.00 g of O_2, what was the composition of the original mixture? (Assume complete decomposition of the mixture.)

3.96 When a mixture of 10.0 g of acetylene (C_2H_2) and 10.0 g of oxygen (O_2) is ignited, the resultant combustion reaction produces CO_2 and H_2O. **(a)** Write the balanced chemical equation for this reaction. **(b)** Which is the limiting reactant? **(c)** How many grams of C_2H_2, O_2, CO_2, and H_2O are present after the reaction is complete?

3.97 Aspirin ($C_9H_8O_4$) is produced from salicylic acid ($C_7H_6O_3$) and acetic anhydride ($C_4H_6O_3$):

$$C_7H_6O_3 + C_4H_6O_3 \longrightarrow C_9H_8O_4 + HC_2H_3O_2$$

(a) How much salicylic acid is required to produce 1.5×10^2 kg of aspirin, assuming that all of the salicylic acid is converted to aspirin? **(b)** How much salicylic acid would be required if only 80% of the salicylic acid is converted to aspirin? **(c)** What is the theoretical yield of aspirin if 185 kg of salicylic acid is allowed to react with 125 kg of acetic anhydride? **(d)** If the situation described in part (c) produces 182 kg of aspirin, what is the percentage yield?

Integrative Exercises

(These exercises require skills from earlier chapters as well as skills from the present chapter.)

3.98 Consider a sample of calcium carbonate in the form of a cube measuring 1.25 in. on each edge. If the sample has a density of 2.71 g/cm^3, how many oxygen atoms does it contain?

3.99 **(a)** You are given a cube of silver metal that measures 1.000 cm on each edge. The density of silver is 10.49 g/cm^3. How many atoms are in this cube? **(b)** Because atoms are spherical, they cannot occupy all of the space of the cube. The silver atoms pack in the solid in such a way that 74% of the volume of the solid is actually filled with the silver atoms. Calculate the volume of a single silver atom. **(c)** Using the volume of a silver atom, and the formula for the volume of a sphere, calculate the radius in angstroms of a silver atom.

3.100 If an automobile travels 125 mi with a gas mileage of 19.5 mi/gal, how many kilograms of CO_2 are produced? Assume that the gasoline is composed of octane, $C_8H_{18}(l)$, whose density is 0.69 g/mL.

[3.101] In 1865 a chemist reported that he had reacted a weighed amount of pure silver with nitric acid and had recovered all the silver as pure silver nitrate. The mass ratio of silver to silver nitrate was found to be 0.634985. Using only this ratio and the presently accepted values for the atomic weights of silver and oxygen, calculate the atomic weight of nitrogen. Compare this calculated atomic weight with the currently accepted value.

[3.102] A particular coal contains 2.5% sulfur by mass. When this coal is burned, the sulfur is converted into sulfur dioxide gas. The sulfur dioxide reacts with calcium oxide to form solid calcium sulfite. **(a)** Write the balanced chemical equation for the reaction. **(b)** If the coal is burned in a power plant that uses 2000 tons of coal per day, what is the daily production of calcium sulfite?

[3.103] Hydrogen cyanide, HCN, is a poisonous gas. The lethal dose is approximately 300 mg HCN per kilogram of air when inhaled. **(a)** Calculate the amount of HCN that gives the lethal dose in a small laboratory room measuring 12 by 15 by 8.0 ft. The density of air at 26°C is 0.00118 g/cm^3. **(b)** If the HCN is formed by reaction of NaCN with an acid such as H_2SO_4, what mass of NaCN gives the lethal dose in the room?

$$2NaCN(s) + H_2SO_4(aq) \longrightarrow Na_2SO_4(aq) + 2HCN(g)$$

(c) HCN forms when synthetic fibers containing Orlon® or Acrilan® burn. Acrilan® has an empirical formula of CH_2CHCN, so HCN is 50.9% of the formula by mass. A rug measures 12 by 15 ft and contains 30 oz of Acrilan® fibers per square yard of carpet. If the rug burns, will a lethal dose of HCN be generated in the room? Assume that the yield of HCN from the fibers is 20% and that the carpet is 50% consumed.

eMedia Exercises

3.104 **(a)** Balance the three reactions available in the **Balancing Equations** activity (*eChapter 3.1*). **(b)** If, in the case of the reduction of Fe_2O_3 you were to multiply each coefficient in the balanced equation by the same number, would the reaction still be balanced? **(c)** Would it still be balanced if you were to square each coefficient? **(d)** What would the coefficients be in the balanced equation if the reduction reaction produced carbon *mon*oxide instead of carbon dioxide?

3.105 Calculate the percentage composition of each compound in Exercise 3.16. Use the **Molecular Weight and Weight Percent** activity (*eChapter 3.3*) to check your answers.

3.106 Consider the reaction of zinc metal with hydrochloric acid shown in the movie **Limiting Reagents** (*eChapter 3.7*). **(a)** What would the limiting reagent have been if 100 mg of Zn had been combined with 2.0×10^{-3} mol of HCl? **(b)** What volume of hydrogen gas would the reaction have produced? **(c)** For the purposes of the limiting reagent experiment shown in the movie, why is it important that the hydrogen gas have a low solubility in water? How would the apparent yield of a reaction be affected if the evolved gas were very soluble in (or reactive with) water?

3.107 Write the balanced equation and predict the masses of products and remaining reactant for each of the following combinations. Use the **Limiting Reagents (Stoichiometry)** (*eChapter 3.7*) simulation to check your answers. **(a)** 50 g $Pb(NO_3)_2$ and 55 g K_2CrO_4 to form $PbCrO_4$ and KNO_3; **(b)** 150 g $FeCl_2$ and 125 g Na_2S to form FeS and NaCl; **(c)** 96 g $Ca(NO_3)_2$ and 62 g Na_2CO_3 to form $CaCO_3$ and $NaNO_3$.

3.108 **(a)** In the reaction between $FeCl_3$ and NaOH to produce $Fe(OH)_3$ and NaCl, what mass of sodium hydroxide would be required to completely consume 50 g of iron(III) chloride? **(b)** Use the **Limiting Reagents (Stoichiometry)** simulation (*eChapter 3.7*) to combine 50 g $FeCl_3$ with as close as you can get to the stoichiometric amount of sodium hydroxide, making sure that you add enough to completely consume the $FeCl_3$. How much sodium hydroxide is left over?

Chapter 4

Aqueous Reactions and Solution Stoichiometry

Bubbles of CO_2 gas form as an Alka-Seltzer® tablet dissolves in water. The CO_2 is produced by the reaction of citric acid, $H_3C_6H_5O_7$, and sodium bicarbonate, $NaHCO_3$, in the tablet.

ALMOST TWO THIRDS of our planet is covered by water, and water is the most abundant substance in our bodies. Because water is so common, we tend to take its unique chemical and physical properties for granted. We will see repeatedly throughout this text, however, that water possesses many unusual properties essential to support life on Earth.

One of the most important properties of water is its ability to dissolve a wide variety of substances. The water in nature, therefore, whether it is the purest drinking water from the tap or water from a clear mountain stream, invariably contains a variety of dissolved substances. Solutions in which water is the dissolving medium are called **aqueous solutions**.

Many of the chemical reactions that take place within us and around us involve substances dissolved in water. Nutrients dissolved in blood are carried to our cells, where they enter into reactions that help keep us alive. Automobile parts rust when they come into frequent contact with aqueous solutions that contain various dissolved substances. Spectacular limestone caves (Figure 4.1 ▶) are formed by the dissolving action of underground water containing carbon dioxide, $CO_2(aq)$:

$$CaCO_3(s) + H_2O(l) + CO_2(aq) \longrightarrow Ca(HCO_3)_2(aq) \quad [4.1]$$

We saw in Chapter 3 a few simple types of chemical reactions and how they are described. In this chapter we continue to examine chemical reactions by focusing on aqueous solutions. A great deal of important chemistry occurs in aqueous solutions, and we need to learn the vocabulary and concepts used to describe and understand this chemistry. In addition, we will extend the concepts of stoichiometry that we learned in Chapter 3 by considering how solution concentrations can be expressed and used.

▶ What's Ahead ◀

- We begin by examining the nature of the substances dissolved in water, whether they exist in water as ions, molecules, or as some mixture of the two. This information is necessary to understand the nature of reactants in aqueous solutions.
- Three major types of chemical processes occur in aqueous solution: precipitation reactions, acid-base reactions, and oxidation-reduction reactions.
- *Precipitation reactions* are those in which soluble reactants yield an insoluble product.
- *Acid-base reactions* are those in which H^+ ions are transferred between reactants.
- *Oxidation-reduction reactions* are those in which electrons are transferred between reactants.
- Reactions between ions can be represented by *ionic equations* that show, for example, how ions can combine to form precipitates, or how they are removed from the solution or changed in some other way.
- After examining the common types of chemical reactions and how they are recognized and described, we consider how the *concentrations* of solutions can be expressed.
- We conclude the chapter by examining how the concepts of stoichiometry and concentration can be used to determine the amounts or concentrations of various substances.

▲ **Figure 4.1** When CO_2 dissolves in water, the resulting solution is slightly acidic. Limestone caves are formed by the dissolving action of this acidic solution acting on $CaCO_3$ in the limestone.

3-D MODEL
Sodium Chloride, Sucrose

ANIMATION
Electrolytes and Nonelectrolytes

4.1 General Properties of Aqueous Solutions

A *solution* is a homogeneous mixture of two or more substances. (Section 1.2) The substance present in greater quantity is usually called the **solvent**. The other substances in the solution are known as the **solutes**; they are said to be dissolved in the solvent. When a small amount of sodium chloride (NaCl) is dissolved in a large quantity of water, for example, the water is the solvent and the sodium chloride is the solute.

Electrolytic Properties

Imagine preparing two aqueous solutions—one by dissolving a teaspoon of table salt (sodium chloride) in a cup of water and the other by dissolving a teaspoon of table sugar (sucrose) in a cup of water. Both solutions are clear and colorless. How do they differ? One way, which might not be immediately obvious, is in their electrical conductivity: The salt solution is a good conductor of electricity, whereas the sugar solution is not.

Whether or not a solution conducts electricity can be determined by using a device such as that shown in Figure 4.2 ▼. To light the bulb, an electric current must flow between the two electrodes that are immersed in the solution. Although water itself is a poor conductor of electricity, the presence of ions causes aqueous solutions to become good conductors. Ions carry electrical charge from one electrode to another, completing the electrical circuit. Thus, the conductivity of NaCl solutions indicates the presence of ions in the solution, and the lack of conductivity of sucrose solutions indicates the absence of ions. When NaCl dissolves in water, the solution contains Na^+ and Cl^- ions, each surrounded by water molecules. When sucrose ($C_{12}H_{22}O_{11}$) dissolves in water, the solution contains only neutral sucrose molecules surrounded by water molecules.

A substance (such as NaCl) whose aqueous solutions contain ions is called an **electrolyte**. A substance (such as $C_{12}H_{22}O_{11}$) that does not form ions in

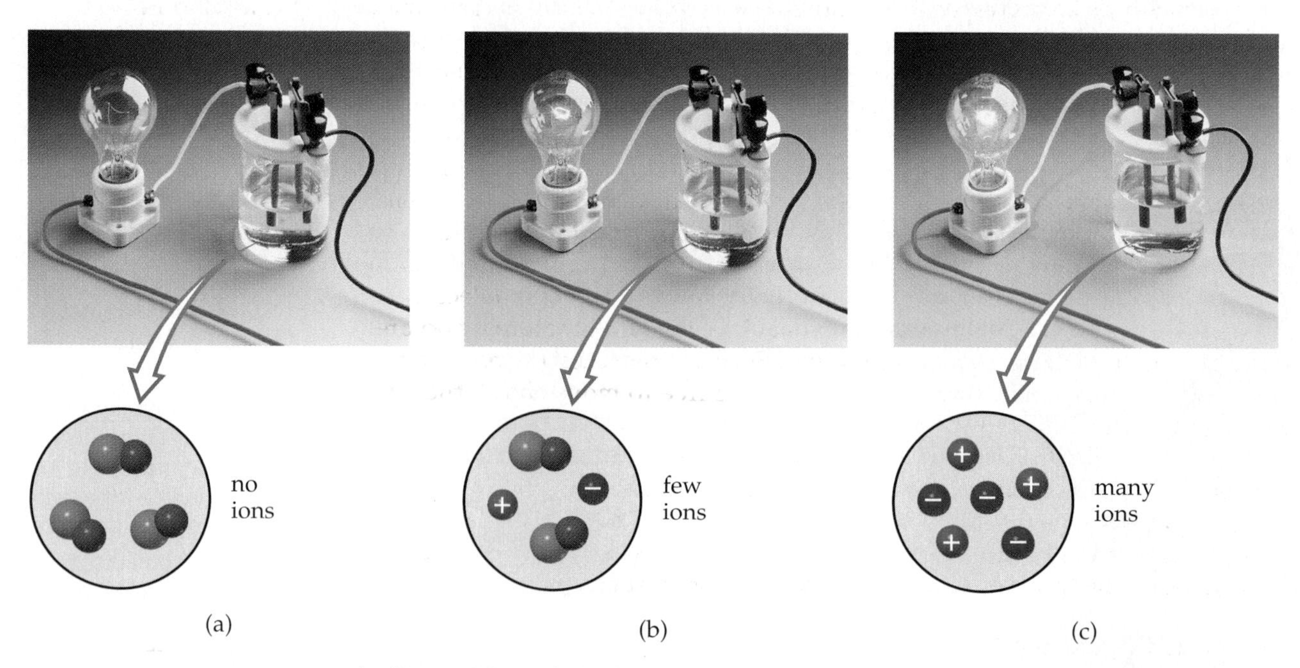

▲ **Figure 4.2** A device for detecting ions in solution. The ability of a solution to conduct electricity depends on the number of ions it contains. (a) A nonelectrolyte solution does not contain ions, and the bulb does not light. (b and c) An electrolyte solution contains ions to serve as charge carriers, causing the bulb to light. If the solution contains a small number of ions, the bulb will be only dimly lit, as in (b). If the solution contains a large number of ions, the bulb will be brightly lit, as in (c).

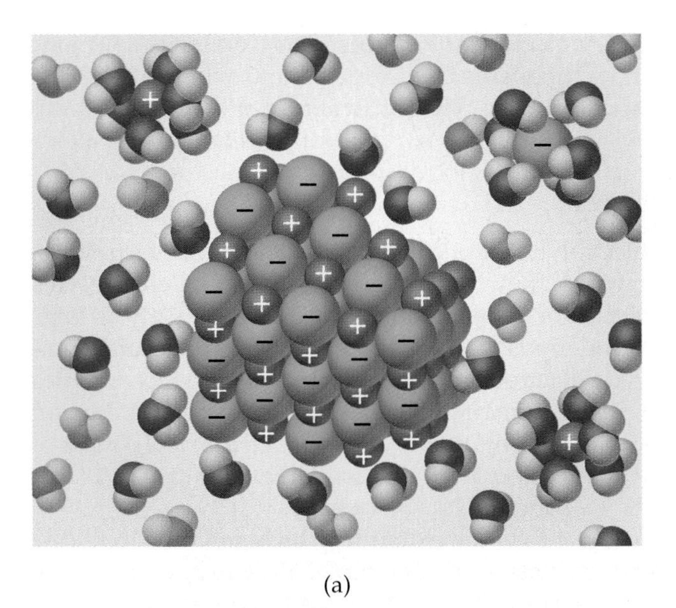

(a)

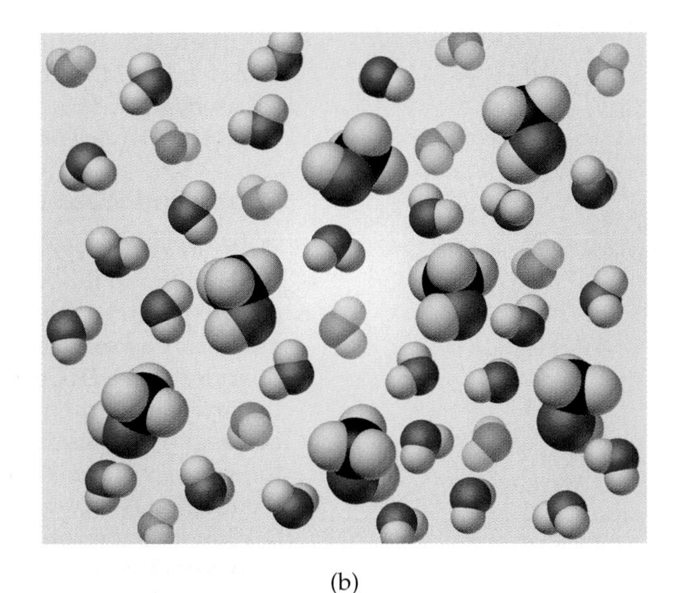

(b)

▲ **Figure 4.3** (a) Dissolution of an ionic compound. When an ionic compound dissolves in water, H_2O molecules separate, surround, and disperse the ions into the liquid. (b) Methanol, CH_3OH, a molecular compound, dissolves without forming ions. The methanol molecules can be found by looking for the black spheres, which represent carbon atoms. In both parts (a) and (b), the water molecules have been moved apart so the solute particles can be seen more clearly.

solution is called a **nonelectrolyte**. The difference between NaCl and $C_{12}H_{22}O_{11}$ arises largely because NaCl is ionic, whereas $C_{12}H_{22}O_{11}$ is molecular.

Ionic Compounds in Water

Recall from Section 2.7 and especially Figure 2.23 that solid NaCl consists of an orderly arrangement of Na^+ and Cl^- ions. When NaCl dissolves in water, each ion separates from the solid structure and disperses throughout the solution as shown in Figure 4.3(a) ▲. The ionic solid *dissociates* into its component ions as it dissolves.

Water is a very effective solvent for ionic compounds. Although water is an electrically neutral molecule, one end of the molecule (the O atom) is rich in electrons and thus possesses a partial negative charge. The other end (the H atoms) has a partial positive charge. Positive ions (cations) are attracted by the negative end of H_2O, and negative ions (anions) are attracted by the positive end. As an ionic compound dissolves, the ions become surrounded by H_2O molecules as shown in Figure 4.3(a). This process helps stabilize the ions in solution and prevents cations and anions from recombining. Furthermore, because the ions and their shells of surrounding water molecules are free to move about, the ions become dispersed uniformly throughout the solution.

We can usually predict the nature of the ions present in a solution of an ionic compound from the chemical name of the substance. Sodium sulfate (Na_2SO_4), for example, dissociates into sodium ions (Na^+) and sulfate ions (SO_4^{2-}). You must remember the formulas and charges of common ions (Tables 2.4 and 2.5) to understand the forms in which ionic compounds exist in aqueous solution.

Molecular Compounds in Water

When a molecular compound dissolves in water, the solution usually consists of intact molecules dispersed throughout the solution. Consequently, most molecular compounds are nonelectrolytes. For example, a solution of methanol (CH_3OH) in water consists entirely of CH_3OH molecules dispersed throughout the water [Figure 4.3(b)].

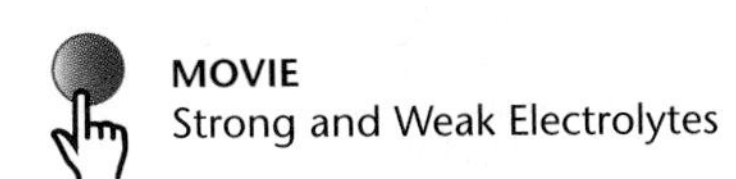

There are, however, a few molecular substances whose aqueous solutions contain ions. Most important of these are acids. For example, when $HCl(g)$ dissolves in water to form hydrochloric acid, $HCl(aq)$, it *ionizes* or breaks apart into $H^+(aq)$ and $Cl^-(aq)$ ions.

Strong and Weak Electrolytes

There are two categories of electrolytes, strong electrolytes and weak electrolytes, which differ in the extent to which they conduct electricity. **Strong electrolytes** are those solutes that exist in solution completely or nearly completely as ions. Essentially all soluble ionic compounds (such as NaCl) and a few molecular compounds (such as HCl) are strong electrolytes. **Weak electrolytes** are those solutes that exist in solution mostly in the form of molecules with only a small fraction in the form of ions. For example, in a solution of acetic acid ($HC_2H_3O_2$) most of the solute is present as $HC_2H_3O_2$ molecules. Only a small fraction (about 1%) of the $HC_2H_3O_2$ is present as $H^+(aq)$ and $C_2H_3O_2^-(aq)$ ions.

We must be careful not to confuse the extent to which an electrolyte dissolves with whether it is strong or weak. For example, $HC_2H_3O_2$ is extremely soluble in water but is a weak electrolyte. $Ba(OH)_2$, on the other hand, is not very soluble, but the amount of the substance that does dissolve dissociates almost completely, so $Ba(OH)_2$ is a strong electrolyte.

When a weak electrolyte such as acetic acid ionizes in solution, we write the reaction in the following manner:

$$HC_2H_3O_2(aq) \rightleftharpoons H^+(aq) + C_2H_3O_2^-(aq) \qquad [4.2]$$

The double arrow means that the reaction is significant in both directions. At any given moment some $HC_2H_3O_2$ molecules are ionizing to form H^+ and $C_2H_3O_2^-$. At the same time, H^+ and $C_2H_3O_2^-$ ions are recombining to form $HC_2H_3O_2$. The balance between these opposing processes determines the relative numbers of ions and neutral molecules. This balance produces a state of **chemical equilibrium** that varies from one weak electrolyte to another. Chemical equilibria are extremely important, and we will devote Chapters 15–17 to examining them in detail.

Chemists use a double arrow to represent the ionization of weak electrolytes and a single arrow to represent the ionization of strong electrolytes. Because HCl is a strong electrolyte, we write the equation for the ionization of HCl as follows:

$$HCl(aq) \longrightarrow H^+(aq) + Cl^-(aq) \qquad [4.3]$$

The single arrow indicates that the H^+ and Cl^- ions have no tendency to recombine in water to form HCl molecules.

In the sections ahead we will begin to look more closely at how we can use the composition of a compound to predict whether it is a strong electrolyte, weak electrolyte, or nonelectrolyte. For the moment, it is important only to remember that *soluble ionic compounds are strong electrolytes*. We identify ionic compounds as being ones composed of metals and nonmetals [such as NaCl, $FeSO_4$, and $Al(NO_3)_3$], or compounds containing the ammonium ion, NH_4^+ [such as NH_4Br and $(NH_4)_2CO_3$].

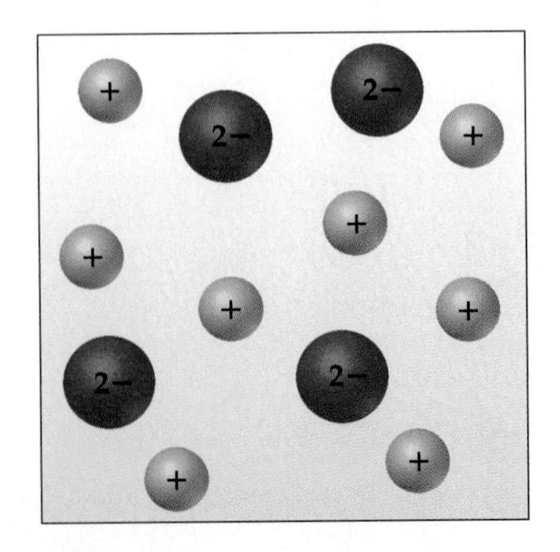

SAMPLE EXERCISE 4.1

The diagram on the left represents an aqueous solution of one of the following compounds: $MgCl_2$, KCl, or K_2SO_4. Which solution does it best represent?

Solution The diagram shows twice as many cations as anions, consistent with the formulation K_2SO_4.

PRACTICE EXERCISE

If you were to draw diagrams (such as that shown on the left) representing aqueous solutions of each of the following ionic compounds, how many anions would you show if the diagram contained six cations? **(a)** $NiSO_4$; **(b)** $Ca(NO_3)_2$; **(c)** Na_3PO_4; **(d)** $Al_2(SO_4)_3$.
Answers: **(a)** 6; **(b)** 12; **(c)** 2; **(d)** 9

4.2 Precipitation Reactions

Figure 4.4 ▼ shows two clear solutions being mixed, one containing lead nitrate [$Pb(NO_3)_2$] and the other containing potassium iodide (KI). The reaction between these two solutes produces an insoluble yellow product. Reactions that result in the formation of an insoluble product are known as **precipitation reactions**. A **precipitate** is an insoluble solid formed by a reaction in solution. In Figure 4.4 the precipitate is lead iodide (PbI_2), a compound that has a very low solubility in water:

$$Pb(NO_3)_2(aq) + 2KI(aq) \longrightarrow PbI_2(s) + 2KNO_3(aq) \quad [4.4]$$

The other product of this reaction, potassium nitrate, remains in solution.

Precipitation reactions occur when *certain pairs of oppositely charged ions attract each other so strongly that they form an insoluble ionic solid.* To predict whether certain combinations of ions form insoluble compounds, we must consider some guidelines or rules concerning the solubilities of common ionic compounds.

▲ **Figure 4.4** The addition of a colorless solution of potassium iodide (KI) to a colorless solution of lead nitrate [$Pb(NO_3)_2$] produces a yellow precipitate of lead iodide (PbI_2) that slowly settles to the bottom of the beaker.

Solubility Guidelines for Ionic Compounds

The **solubility** of a substance is the amount of that substance that can be dissolved in a given quantity of solvent. Only 1.2×10^{-3} mol of PbI_2 dissolves in a liter of water at 25°C. In our discussions any substance with a solubility less than 0.01 mol/L will be referred to as *insoluble*. In those cases the attraction between the oppositely charged ions in the solid is too great for the water molecules to separate them to any significant extent, and the substance remains largely undissolved.

Unfortunately, there are no rules based on simple physical properties such as ionic charge to guide us in predicting whether a particular ionic compound will be soluble or not. Experimental observations, however, have led to guidelines for predicting solubility for ionic compounds. For example, experiments show that all common ionic compounds that contain the nitrate anion, NO_3^-, are soluble in water. Table 4.1 ▼ summarizes the solubility guidelines for common ionic compounds. The table is organized according to the anion in the compound, but it reveals many important facts about cations. Note that *all common ionic compounds of the alkali metal ions (group 1A of the periodic table) and of the ammonium ion* (NH_4^+) *are soluble in water.*

TABLE 4.1 Solubility Guidelines for Common Ionic Compounds in Water

Soluble Ionic Compounds		Important Exceptions
Compounds containing	NO_3^-	None
	$C_2H_3O_2^-$	None
	Cl^-	Compounds of Ag^+, Hg_2^{2+}, and Pb^{2+}
	Br^-	Compounds of Ag^+, Hg_2^{2+}, and Pb^{2+}
	I^-	Compounds of Ag^+, Hg_2^{2+}, and Pb^{2+}
	SO_4^{2-}	Compounds of Sr^{2+}, Ba^{2+}, Hg_2^{2+}, and Pb^{2+}
Insoluble Ionic Compounds		**Important Exceptions**
Compounds containing	S^{2-}	Compounds of NH_4^+, the alkali metal cations, and Ca^{2+}, Sr^{2+}, and Ba^{2+}
	CO_3^{2-}	Compounds of NH_4^+ and the alkali metal cations
	PO_4^{3-}	Compounds of NH_4^+ and the alkali metal cations
	OH^-	Compounds of the alkali metal cations, and Ca^{2+}, Sr^{2+}, and Ba^{2+}

SAMPLE EXERCISE 4.2

Classify the following ionic compounds as soluble or insoluble in water: **(a)** sodium carbonate (Na_2CO_3); **(b)** lead sulfate ($PbSO_4$).

Solution

Analyze: We are given the names and formulas of two ionic compounds and asked to predict whether they are soluble or insoluble in water.

Plan: We can use Table 4.1 to answer the question. Thus, we need to focus on the anion in each compound because the table is organized by anions.

Solve: (a) According to Table 4.1, most carbonates are insoluble, but carbonates of the alkali metal cations (such as sodium ion) are an exception to this rule and are soluble. Thus, Na_2CO_3 is soluble in water.

(b) Table 4.1 indicates that although most sulfates are water soluble, the sulfate of Pb^{2+} is an exception. Thus, $PbSO_4$ is insoluble in water.

PRACTICE EXERCISE

Classify the following compounds as soluble or insoluble in water: **(a)** cobalt(II) hydroxide; **(b)** barium nitrate; **(c)** ammonium phosphate.

Answers: **(a)** insoluble; **(b)** soluble; **(c)** soluble

To predict whether a precipitate forms when we mix aqueous solutions of two strong electrolytes, we must (1) note the ions present in the reactants, (2) consider the possible combinations of the cations and anions, and (3) use Table 4.1 to determine if any of these combinations is insoluble. For example, will a precipitate form when solutions of $Mg(NO_3)_2$ and NaOH are mixed? Because $Mg(NO_3)_2$ and NaOH are both soluble ionic compounds, they are both strong electrolytes. Mixing $Mg(NO_3)_2(aq)$ and $NaOH(aq)$ first produces a solution containing Mg^{2+}, NO_3^-, Na^+, and OH^- ions. Will either of the cations interact with either of the anions to form an insoluble compound? In addition to the reactants, the other possible interactions are Mg^{2+} with OH^- and Na^+ with NO_3^-. From Table 4.1 we see that $Mg(OH)_2$ is insoluble and will thus form a precipitate. $NaNO_3$, however, is soluble, so Na^+ and NO_3^- will remain in solution. The balanced equation for the precipitation reaction is

$$Mg(NO_3)_2(aq) + 2NaOH(aq) \longrightarrow Mg(OH)_2(s) + 2NaNO_3(aq) \qquad [4.5]$$

Exchange (Metathesis) Reactions

Notice in Equation 4.5 that the cations in the two reactants exchange anions—Mg^{2+} ends up with OH^-, and Na^+ ends up with NO_3^-. The chemical formulas of the products are based on the charges of the ions—two OH^- ions are needed to give a neutral compound with Mg^{2+}, and one NO_3^- ion is needed to give a neutral compound with Na^+. ∞ (Section 2.7) It is only after the chemical formulas of the products are determined that the equation can be balanced.

Reactions in which positive ions and negative ions appear to exchange partners conform to the following general equation:

$$AX + BY \longrightarrow AY + BX \qquad [4.6]$$

Example: $AgNO_3(aq) + KCl(aq) \longrightarrow AgCl(s) + KNO_3(aq)$

Such reactions are known as **exchange reactions**, or **metathesis reactions** (meh-TATH-eh-sis, which is the Greek word for "to transpose"). Precipitation reactions conform to this pattern, as do many acid-base reactions, as we will see in Section 4.3.

SAMPLE EXERCISE 4.3

(a) Predict the identity of the precipitate that forms when solutions of $BaCl_2$ and K_2SO_4 are mixed. **(b)** Write the balanced chemical equation for the reaction.

Solution

Analyze: We are given two ionic reactants and asked to predict the insoluble product that they form.

Plan: We need to write down the ions present in the reactants and to exchange the anions between the two cations. Once we have written the chemical formulas for these products, we can use Table 4.1 to determine which is insoluble in water. Knowing the products also allows us to write the equation for the reaction.

Solve: (a) The reactants contain Ba^{2+}, Cl^-, K^+, and SO_4^{2-} ions. If we exchange the anions, we will have $BaSO_4$ and KCl. According to Table 4.1, most compounds of SO_4^{2-} are soluble but those of Ba^{2+} are not. Thus, $BaSO_4$ is insoluble and will precipitate from solution. KCl, on the other hand, is soluble.

(b) From part (a) we know the chemical formulas of the products, $BaSO_4$ and KCl. The balanced equation with phase labels shown is

$$BaCl_2(aq) + K_2SO_4(aq) \longrightarrow BaSO_4(s) + 2KCl(aq)$$

PRACTICE EXERCISE

(a) What compound precipitates when solutions of $Fe_2(SO_4)_3$ and LiOH are mixed? **(b)** Write a balanced equation for the reaction. **(c)** Will a precipitate form when solutions of $Ba(NO_3)_2$ and KOH are mixed?

Answers: **(a)** $Fe(OH)_3$; **(b)** $Fe_2(SO_4)_3(aq) + 6LiOH(aq) \longrightarrow 2Fe(OH)_3(s) + 3Li_2SO_4(aq)$; **(c)** No (both possible products are water soluble)

Ionic Equations

In writing chemical equations for reactions in aqueous solution, it is often useful to indicate explicitly whether the dissolved substances are present predominantly as ions or as molecules. Let's reconsider the precipitation reaction between $Pb(NO_3)_2$ and 2KI, shown previously in Figure 4.4:

$$Pb(NO_3)_2(aq) + 2KI(aq) \longrightarrow PbI_2(s) + 2KNO_3(aq)$$

An equation written in this fashion, showing the complete chemical formulas of the reactants and products, is called a **molecular equation** because it shows the chemical formulas of the reactants and products without indicating their ionic character. Because $Pb(NO_3)_2$, KI, and KNO_3 are all soluble ionic compounds and therefore strong electrolytes, we can write the chemical equation to indicate explicitly the ions that are in the solution:

$$Pb^{2+}(aq) + 2NO_3^-(aq) + 2K^+(aq) + 2I^-(aq) \longrightarrow PbI_2(s) + 2K^+(aq) + 2NO_3^-(aq) \quad [4.7]$$

An equation written in this form, with all soluble strong electrolytes shown as ions, is known as a **complete ionic equation**.

Notice that $K^+(aq)$ and $NO_3^-(aq)$ appear on both sides of Equation 4.7. Ions that appear in identical forms among both the reactants and products of a complete ionic equation are called **spectator ions**. They are present but play no direct role in the reaction. When spectator ions are omitted from the equation (they cancel out like algebraic quantities), we are left with the **net ionic equation:**

$$Pb^{2+}(aq) + 2I^-(aq) \longrightarrow PbI_2(s) \quad [4.8]$$

A net ionic equation includes only the ions and molecules directly involved in the reaction. Charge is conserved in reactions, so the sum of the charges of the ions must be the same on both sides of a balanced net ionic equation. In this case the 2+ charge of the cation and the two 1− charges of the anions add to give zero, the charge of the electrically neutral product. *If every ion in a complete ionic equation is a spectator, then no reaction occurs.*

Net ionic equations are widely used to illustrate the similarities between large numbers of reactions involving electrolytes. For example, Equation 4.8 expresses the essential feature of the precipitation reaction between any strong electrolyte containing Pb^{2+} and any strong electrolyte containing I^-: The $Pb^{2+}(aq)$ and $I^-(aq)$ ions combine to form a precipitate of PbI_2. Thus, a net ionic equation demonstrates that more than one set of reactants can lead to the same net reaction. The complete equation, on the other hand, identifies the actual reactants that participate in a reaction.

Net ionic equations also point out that the chemical behavior of a strong electrolyte solution is due to the various kinds of ions it contains. Aqueous solutions of KI and MgI_2, for example, share many chemical similarities because both contain I^- ions. Each kind of ion has its own chemical characteristics that differ very much from those of its parent atom.

The following steps summarize the procedure for writing net ionic equations:

1. Write a balanced molecular equation for the reaction.
2. Rewrite the equation to show the ions that form in solution when each soluble strong electrolyte dissociates or ionizes into its component ions. *Only strong electrolytes dissolved in aqueous solution are written in ionic form.*
3. Identify and cancel spectator ions.

SAMPLE EXERCISE 4.4

Write the net ionic equation for the precipitation reaction that occurs when solutions of calcium chloride and sodium carbonate are mixed.

Solution

Analyze: Our task is to write a net ionic equation for a precipitation reaction, given the names of the reactants present in solution.

Plan: We first need to write the chemical formulas of the reactants and products and to determine which product is insoluble. Then we write and balance the molecular equation. Next, we write each soluble strong electrolyte as separated ions to obtain the complete ionic equation. Finally, we eliminate the spectator ions to obtain the net ionic equation.

Solve: Calcium chloride is composed of calcium ions, Ca^{2+}, and chloride ions, Cl^-; hence an aqueous solution of the substance is $CaCl_2(aq)$. Sodium carbonate is composed of Na^+ ions and CO_3^{2-} ions; hence an aqueous solution of the compound is $Na_2CO_3(aq)$. In the molecular equations for precipitation reactions, the anions and cations appear to exchange partners. Thus, we put Ca^{2+} and CO_3^{2-} together to give $CaCO_3$ and Na^+ and Cl^- together to give NaCl. According to the solubility guidelines in Table 4.1, $CaCO_3$ is insoluble and NaCl is soluble. The balanced molecular equation is

$$CaCl_2(aq) + Na_2CO_3(aq) \longrightarrow CaCO_3(s) + 2NaCl(aq)$$

In a complete ionic equation, *only* dissolved strong electrolytes (such as soluble ionic compounds) are written as separate ions. As the (*aq*) designations remind us, $CaCl_2$, Na_2CO_3, and NaCl are all dissolved in the solution. Furthermore, they are all strong electrolytes. $CaCO_3$ is an ionic compound, but it is not soluble. We do not write the formula of any insoluble compound as its component ions. Thus, the complete ionic equation is

$$Ca^{2+}(aq) + 2Cl^-(aq) + 2Na^+(aq) + CO_3^{2-}(aq) \longrightarrow CaCO_3(s) + 2Na^+(aq) + 2Cl^-(aq)$$

Cl^- and Na^+ are spectator ions. Canceling them gives the following net ionic equation:

$$Ca^{2+}(aq) + CO_3^{2-}(aq) \longrightarrow CaCO_3(s)$$

Check: We can check our result by confirming that both the elements and the electric charge are balanced. Each side has 1 Ca, 1 C, and 3 O, and the net charge on each side equals 0.

Comment: If none of the ions in an ionic equation is removed from solution or changed in some way, then they all are spectator ions and a reaction does not occur.

PRACTICE EXERCISE

Write the net ionic equation for the precipitation reaction that occurs when aqueous solutions of silver nitrate and potassium phosphate are mixed.

Answer: $3Ag^+(aq) + PO_4^{3-}(aq) \longrightarrow Ag_3PO_4(s)$

▲ **Figure 4.5** Some common acids (left), and bases (right) that are household products.

4.3 Acid-Base Reactions

Many acids and bases are industrial and household substances (Figure 4.5 ▶), and some are important components of biological fluids. Hydrochloric acid, for example, is not only an important industrial chemical but also the main constituent of gastric juice in our stomach. Acids and bases also happen to be common electrolytes.

Acids

Acids are substances that ionize in aqueous solutions to form hydrogen ions, thereby increasing the concentration of $H^+(aq)$ ions. Because a hydrogen atom consists of a proton and an electron, H^+ is simply a proton. Thus, acids are often called proton donors. Molecular models of three common acids, HCl, HNO_3, and $HC_2H_3O_2$, are shown in the margin.

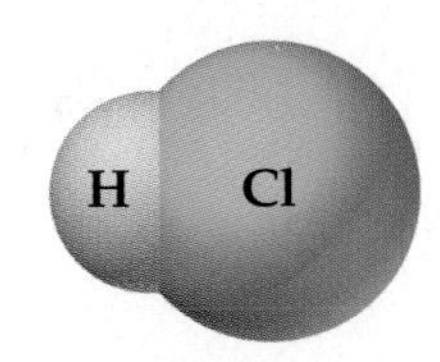

HCl

HNO_3

$HC_2H_3O_2$

Molecules of different acids can ionize to form different numbers of H^+ ions. Both HCl and HNO_3 are *monoprotic* acids, which yield one H^+ per molecule of acid. Sulfuric acid, H_2SO_4, is a *diprotic* acid, one that yields two H^+ per molecule of acid. The ionization of H_2SO_4 and other diprotic acids occurs in two steps:

$$H_2SO_4(aq) \longrightarrow H^+(aq) + HSO_4^-(aq) \quad [4.9]$$

$$HSO_4^-(aq) \rightleftharpoons H^+(aq) + SO_4^{2-}(aq) \quad [4.10]$$

Although H_2SO_4 is a strong electrolyte, only the first ionization is complete. Thus, aqueous solutions of sulfuric acid contain a mixture of $H^+(aq)$, $HSO_4^-(aq)$, and $SO_4^{2-}(aq)$.

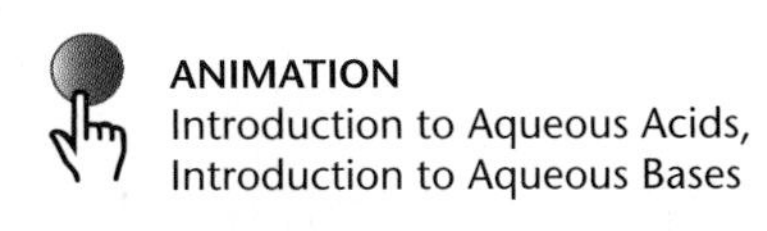

ANIMATION
Introduction to Aqueous Acids, Introduction to Aqueous Bases

Bases

Bases are substances that accept (react with) H^+ ions. Bases produce hydroxide ions (OH^-) when they dissolve in water. Ionic hydroxide compounds such as NaOH, KOH, and $Ca(OH)_2$ are among the most common bases. When dissolved in water, they dissociate into their component ions, introducing OH^- ions into the solution.

Compounds that do not contain OH^- ions can also be bases. For example, ammonia (NH_3) is a common base. When added to water, it accepts an H^+ ion from the water molecule and thereby produces an OH^- ion (Figure 4.6 ◀):

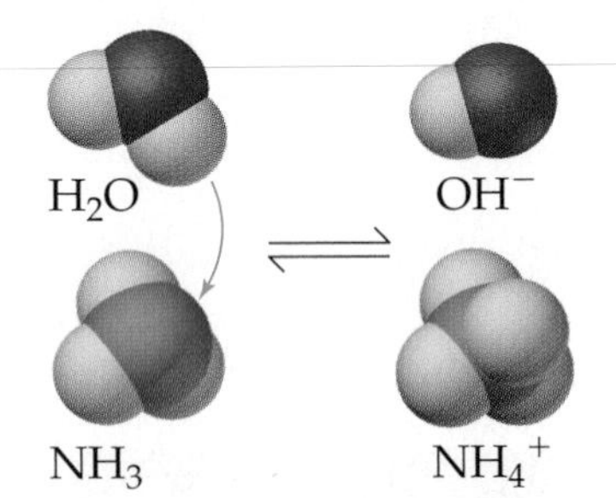

▲ **Figure 4.6** An H_2O molecule acts as a proton donor (acid), and NH_3 as a proton acceptor (base). Only a fraction of the NH_3 reacts with H_2O; NH_3 is a weak electrolyte.

$$NH_3(aq) + H_2O(l) \rightleftharpoons NH_4^+(aq) + OH^-(aq) \quad [4.11]$$

Because only a small fraction of the NH_3 (about 1%) forms NH_4^+ and OH^- ions, ammonia is a weak electrolyte.

Strong and Weak Acids and Bases

Acids and bases that are strong electrolytes (completely ionized in solution) are called **strong acids** and **strong bases**. Those that are weak electrolytes (partly ionized) are called **weak acids** and **weak bases**. Strong acids are more reactive than weak acids when the reactivity depends only on the concentration of $H^+(aq)$. The reactivity of an acid, however, can depend on the anion as well as on $H^+(aq)$. For example, hydrofluoric acid (HF) is a weak acid (only partly ionized in aqueous solution), but it is very reactive and vigorously attacks many substances, including glass. This reactivity is due to the combined action of $H^+(aq)$ and $F^-(aq)$.

Table 4.2 ▼ lists the common strong acids and bases. You should commit these to memory. As you examine this table, notice first that some of the most common acids, such as HCl, HNO_3, and H_2SO_4, are strong. Second, three of the strong acids result from combining a hydrogen atom and a halogen atom. (HF, however, is a weak acid.) Third, the list of strong acids is very short. Most acids are weak. Fourth, the only common strong bases are the hydroxides of Li^+, Na^+, K^+, Rb^+, and Cs^+ (the alkali metals, group 1A) and the hydroxides of Ca^{2+}, Sr^{2+}, and Ba^{2+}

TABLE 4.2 Common Strong Acids and Bases

Strong Acids	Strong Bases
Hydrochloric, HCl	Group 1A metal hydroxides (LiOH, NaOH, KOH, RbOH, CsOH)
Hydrobromic, HBr	Heavy group 2A metal hydroxides [$Ca(OH)_2$, $Sr(OH)_2$, $Ba(OH)_2$]
Hydroiodic, HI	
Chloric, $HClO_3$	
Perchloric, $HClO_4$	
Nitric, HNO_3	
Sulfuric, H_2SO_4	

(the heavy alkaline earths, group 2A). These are the common soluble metal hydroxides. Most other metal hydroxides are insoluble in water. The most common weak base is NH_3, which reacts with water to form OH^- ions (Equation 4.11).

SAMPLE EXERCISE 4.5

The following diagrams represent aqueous solutions of three acids (HX, HY, and HZ) with water molecules omitted for clarity. Rank them from strongest to weakest.

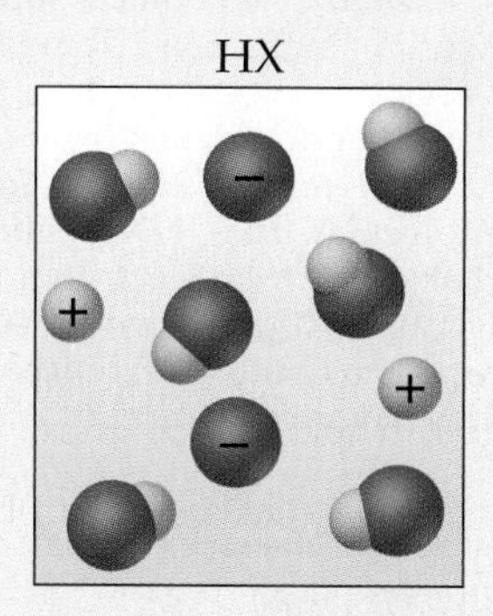

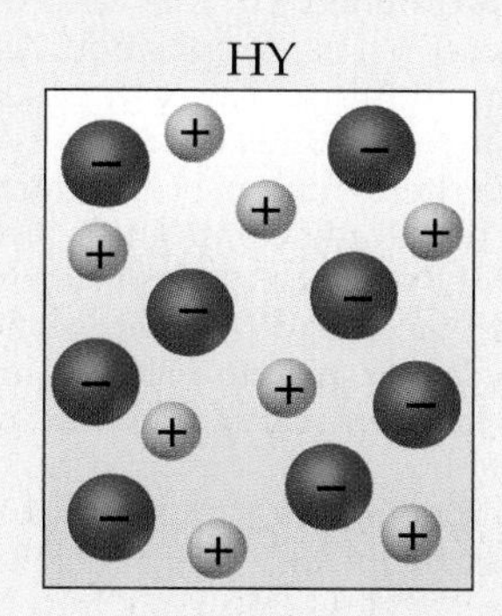

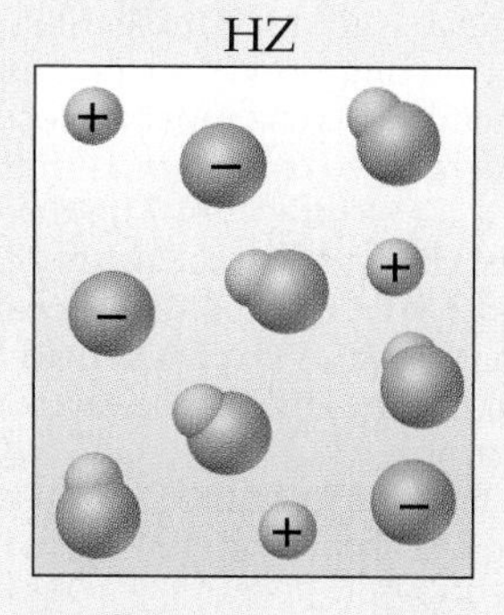

Solution The strongest acid is the one with the most H^+ ions and fewest undissociated acid molecules in solution. Hence, the order is HY > HZ > HX. HY is a strong acid because it is totally ionized (no HY molecules in solution), whereas both HX and HZ are weak acids, whose solutions consist of a mixture of molecules and ions.

PRACTICE EXERCISE

Imagine a diagram showing 10 Na^+ ions and 10 OH^- ions. If this solution were mixed with the one pictured above for HY, what would the diagram look like that represents the solution after any possible reaction? (H^+ ions will react with OH^- ions to form H_2O.)
Answer: The final diagram would show 10 Na^+ ions, 2 OH^- ions, 8 Y^- ions, and 8 H_2O molecules.

Identifying Strong and Weak Electrolytes

ACTIVITY
Strong Acids

If we remember the common strong acids and bases (Table 4.2) and also remember that NH_3 is a weak base, we can make reasonable predictions about the electrolytic behavior of a great number of water-soluble substances. Table 4.3 ▼ summarizes our observations about electrolytes. To classify a soluble substance as a strong electrolyte, weak electrolyte, or nonelectrolyte, we simply work our way down and across this table. We first ask ourselves whether the substance is ionic or molecular. If it is ionic, it is a strong electrolyte. If it is molecular, we ask whether it is an acid. (Does it have H first in the chemical formula?) If it is an acid, we rely on the memorized list from Table 4.2 to determine whether it is a strong or weak electrolyte. If an acid is not listed in Table 4.2, it is probably a weak electrolyte. For example, H_3PO_4, H_2SO_3, and $HC_7H_5O_2$ are not listed in Table 4.2 and are weak acids. NH_3 is the only weak base that we consider in this chapter. (There are compounds called amines that are related to NH_3 and are also molecular bases, but we will not consider them until Chapter 16.) Finally, any molecular substance that we encounter in this chapter that is not an acid or NH_3 is probably a nonelectrolyte.

TABLE 4.3 Summary of the Electrolytic Behavior of Common Soluble Ionic and Molecular Compounds

	Strong Electrolyte	Weak Electrolyte	Nonelectrolyte
Ionic	All	None	None
Molecular	Strong acids (see Table 4.2)	Weak acids (H . . .) Weak bases (NH_3)	All other compounds

SAMPLE EXERCISE 4.6

Classify each of the following dissolved substances as a strong electrolyte, weak electrolyte, or nonelectrolyte: $CaCl_2$, HNO_3, C_2H_5OH (ethanol), $HCHO_2$ (formic acid), KOH.

Solution

Analyze: We are given several chemical formulas and asked to classify each substance as a strong electrolyte, weak electrolyte, or nonelectrolyte.

Plan: The approach we take is outlined in Table 4.3. We can predict whether a substance is ionic or molecular, based on its composition. As we saw in Section 2.7, most ionic compounds we encounter in this text are composed of both a metal and a nonmetal, whereas most molecular compounds are composed only of nonmetals.

Solve: Two compounds fit the criteria for ionic compounds: $CaCl_2$ and KOH. Both are strong electrolytes. The three remaining compounds are molecular. Two, HNO_3 and $HCHO_2$, are acids. Nitric acid, HNO_3, is a common strong acid (a strong electrolyte), as shown in Table 4.2. Because most acids are weak acids, our best guess would be that $HCHO_2$ is a weak acid (weak electrolyte). This is correct. The remaining molecular compound, C_2H_5OH, is neither an acid nor a base, so it is a nonelectrolyte.

Comment: Although C_2H_5OH has an OH group, it is not a metal hydroxide; thus, it is not a base. Rather, it is a member of a class of organic compounds that have C—OH bonds and which are known as alcohols. (Section 2.9)

PRACTICE EXERCISE

Consider solutions in which 0.1 mol of each of the following compounds is dissolved in 1 L of water: $Ca(NO_3)_2$ (calcium nitrate), $C_6H_{12}O_6$ (glucose), $NaC_2H_3O_2$ (sodium acetate), and $HC_2H_3O_2$ (acetic acid). Rank the solutions in order of increasing electrical conductivity, based on the fact that the greater the number of ions in solution, the greater the conductivity.

Answer: $C_6H_{12}O_6$ (nonelectrolyte) $<$ $HC_2H_3O_2$ (weak electrolyte, existing mainly in the form of molecules with few ions) $<$ $NaC_2H_3O_2$ (strong electrolyte that provides two ions, Na^+ and $C_2H_3O_2^-$) $<$ $Ca(NO_3)_2$ (strong electrolyte that provides three ions, Ca^{2+} and $2NO_3^-$)

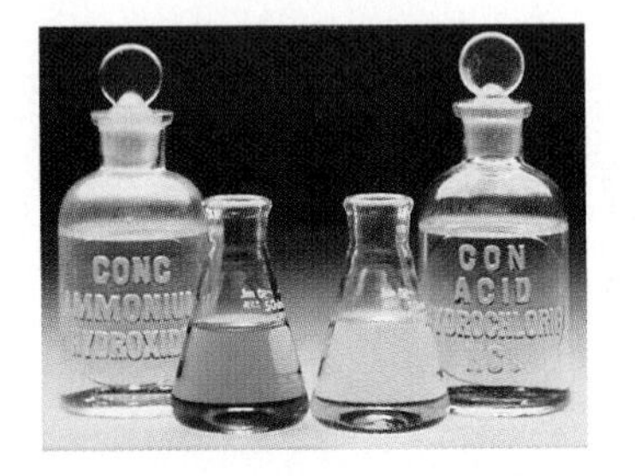

▲ **Figure 4.7** The acid-base indicator bromthymol blue is blue in basic solution and yellow in acidic solution. The left flask shows the indicator in the presence of a base, aqueous ammonia (here labeled as ammonium hydroxide). The right flask shows the indicator in the presence of hydrochloric acid, HCl.

Neutralization Reactions and Salts

Solutions of acids and bases have very different properties. Acids have a sour taste, whereas bases have a bitter taste.* Acids can change the colors of certain dyes in a specific way that differs from the effect of a base (Figure 4.7 ◀). The dye known as litmus, for example, is changed from blue to red by an acid, and from red to blue by a base. In addition, acidic and basic solutions differ in chemical properties in several important ways that we will explore in this chapter and in later chapters.

When a solution of an acid and that of a base are mixed, a **neutralization reaction** occurs. The products of the reaction have none of the characteristic properties of either the acidic or the basic solutions. For example, when hydrochloric acid is mixed with a solution of sodium hydroxide, the following reaction occurs:

$$\underset{\text{(acid)}}{HCl(aq)} + \underset{\text{(base)}}{NaOH(aq)} \longrightarrow \underset{\text{(water)}}{H_2O(l)} + \underset{\text{(salt)}}{NaCl(aq)} \quad [4.12]$$

Water and table salt, NaCl, are the products of the reaction. By analogy to this reaction, the term **salt** has come to mean any ionic compound whose cation comes from a base (for example, Na^+ from NaOH) and whose anion comes from an acid (for example, Cl^- from HCl). In general, *a neutralization reaction between an acid and a metal hydroxide produces water and a salt.*

* Tasting chemical solutions is not a good practice. However, we have all had acids such as ascorbic acid (vitamin C), acetylsalicylic acid (aspirin), and citric acid (in citrus fruits) in our mouths, and we are familiar with their characteristic sour taste. Soaps, which are basic, have the characteristic bitter taste of bases.

(a)

(b)

(c)

▲ **Figure 4.8** (a) Milk of magnesia is a suspension of magnesium hydroxide, $Mg(OH)_2(s)$, in water. (b) The magnesium hydroxide dissolves upon the addition of hydrochloric acid, $HCl(aq)$. (c) The final clear solution contains soluble $MgCl_2(aq)$, shown in Equation 4.15.

Because HCl, NaOH, and NaCl are all soluble strong electrolytes, the complete ionic equation associated with Equation 4.12 is

$$H^+(aq) + Cl^-(aq) + Na^+(aq) + OH^-(aq) \longrightarrow H_2O(l) + Na^+(aq) + Cl^-(aq) \quad [4.13]$$

Therefore, the net ionic equation is

$$H^+(aq) + OH^-(aq) \longrightarrow H_2O(l) \quad [4.14]$$

Equation 4.14 summarizes the essential feature of the neutralization reaction between any strong acid and any strong base: $H^+(aq)$ and $OH^-(aq)$ ions combine to form H_2O.

Figure 4.8 ▲ shows the reaction between hydrochloric acid and another base, $Mg(OH)_2$, which is insoluble in water. A milky white suspension of $Mg(OH)_2$ called milk of magnesia is seen dissolving as the neutralization reaction occurs:

Molecular equation: $Mg(OH)_2(s) + 2HCl(aq) \longrightarrow MgCl_2(aq) + 2H_2O(l)$ [4.15]

Net ionic equation: $Mg(OH)_2(s) + 2H^+(aq) \longrightarrow Mg^{2+}(aq) + 2H_2O(l)$ [4.16]

Notice that the OH^- ions (this time in a solid reactant) and H^+ ions combine to form H_2O. Because the ions exchange partners, neutralization reactions between acids and metal hydroxides are also metathesis reactions.

SAMPLE EXERCISE 4.7

(a) Write a balanced complete chemical equation for the reaction between aqueous solutions of acetic acid ($HC_2H_3O_2$) and barium hydroxide [$Ba(OH)_2$]. **(b)** Write the net ionic equation for this reaction.

Solution

Analyze: We are given the chemical formulas for an acid and a base and asked to write a balanced chemical equation and then a net ionic equation for their neutralization reaction.

Plan: As Equation 4.12 and the italicized statement that follows it indicate, neutralization reactions form two products, H_2O and a salt. We examine the cation of the base and the anion of the acid to determine the composition of the salt.

Solve: (a) The salt will contain the cation of the base (Ba^{2+}) and the anion of the acid ($C_2H_3O_2^-$). Thus, the formula of the salt is $Ba(C_2H_3O_2)_2$. According to the solubility guidelines in Table 4.1, this compound is soluble. The unbalanced equation for the neutralization reaction is

$$HC_2H_3O_2(aq) + Ba(OH)_2(aq) \longrightarrow H_2O(l) + Ba(C_2H_3O_2)_2(aq)$$

To balance the equation, we must provide two molecules of $HC_2H_3O_2$ to furnish the two $C_2H_3O_2^-$ ions and to supply the two H^+ ions needed to combine with the two OH^- ions of the base. The balanced equation is

$$2HC_2H_3O_2(aq) + Ba(OH)_2(aq) \longrightarrow 2H_2O(l) + Ba(C_2H_3O_2)_2(aq)$$

(b) To write the ionic equation, we must determine whether or not each compound in aqueous solution is a strong electrolyte. $HC_2H_3O_2$ is a weak electrolyte (weak acid), $Ba(OH)_2$ is a strong electrolyte, and $Ba(C_2H_3O_2)_2$ is also a strong electrolyte. Thus, the complete ionic equation is

$$2HC_2H_3O_2(aq) + Ba^{2+}(aq) + 2OH^-(aq) \longrightarrow 2H_2O(l) + Ba^{2+}(aq) + 2C_2H_3O_2^-(aq)$$

Eliminating the spectator ions gives

$$2HC_2H_3O_2(aq) + 2OH^-(aq) \longrightarrow 2H_2O(l) + 2C_2H_3O_2^-(aq)$$

Simplifying the coefficients gives the net ionic equation.

$$HC_2H_3O_2(aq) + OH^-(aq) \longrightarrow H_2O(l) + C_2H_3O_2^-(aq)$$

Check: We can determine whether the molecular equation is correctly balanced by counting the number of atoms of each kind on both sides of the arrow. (There are 10 H, 6 O, 4 C, and 1 Ba on each side.) However, it is often easier to check equations by counting groups: There are $2C_2H_3O_2$ groups as well as 1 Ba, and 4 additional H atoms and 2 additional O atoms on each side of the equation. The net ionic equation checks out because the numbers of each kind of element and the net charge are the same on both sides of the equation.

PRACTICE EXERCISE

(a) Write a balanced equation for the reaction of carbonic acid (H_2CO_3) and potassium hydroxide (KOH). **(b)** Write the net ionic equation for this reaction.
Answers: **(a)** $H_2CO_3(aq) + 2KOH(aq) \longrightarrow 2H_2O(l) + K_2CO_3(aq)$; **(b)** $H_2CO_3(aq) + 2OH^-(aq) \longrightarrow 2H_2O(l) + CO_3^{2-}(aq)$; ($H_2CO_3$ is a weak electrolyte, whereas KOH and K_2CO_3 are strong electrolytes.)

Acid-Base Reactions with Gas Formation

There are many bases besides OH^- that react with H^+ to form molecular compounds. Two of these that you might encounter in the laboratory are the sulfide ion and the carbonate ion. Both of these anions react with acids to form gases that have low solubilities in water. Hydrogen sulfide (H_2S), the substance that gives rotten eggs their foul odor, forms when an acid such as $HCl(aq)$ reacts with a metal sulfide such as Na_2S:

Molecular equation: $2HCl(aq) + Na_2S(aq) \longrightarrow H_2S(g) + 2NaCl(aq)$ [4.17]

Net ionic equation: $2H^+(aq) + S^{2-}(aq) \longrightarrow H_2S(g)$ [4.18]

Carbonates and bicarbonates react with acids to form CO_2 gas. Reaction of CO_3^{2-} or HCO_3^- with an acid first gives carbonic acid (H_2CO_3). For example, when hydrochloric acid is added to sodium bicarbonate, the following reaction occurs:

$$HCl(aq) + NaHCO_3(aq) \longrightarrow NaCl(aq) + H_2CO_3(aq) \quad [4.19]$$

Carbonic acid is unstable; if present in solution in sufficient concentrations, it decomposes to form CO_2, which escapes from the solution as a gas.

$$H_2CO_3(aq) \longrightarrow H_2O(l) + CO_2(g) \qquad [4.20]$$

The decomposition of H_2CO_3 produces bubbles of CO_2 gas, as shown in Figure 4.9 ▶. The overall reaction is summarized by the following equations:

Molecular equation:

$$HCl(aq) + NaHCO_3(aq) \longrightarrow NaCl(aq) + H_2O(l) + CO_2(g) \qquad [4.21]$$

Net ionic equation:

$$H^+(aq) + HCO_3^-(aq) \longrightarrow H_2O(l) + CO_2(g) \qquad [4.22]$$

Both $NaHCO_3$ and Na_2CO_3 are used as acid neutralizers in acid spills. The bicarbonate or carbonate salt is added until the fizzing due to the formation of $CO_2(g)$ stops. Sometimes sodium bicarbonate is used as an antacid to soothe an upset stomach. In that case the HCO_3^- reacts with stomach acid to form $CO_2(g)$. The fizz when Alka-Seltzer® tablets are added to water is due to the reaction of sodium bicarbonate and citric acid.

▲ **Figure 4.9** Carbonates react with acids to form carbon dioxide gas. Here $NaHCO_3$ (white solid) reacts with hydrochloric acid; the bubbles contain CO_2.

Chemistry at Work Antacids

The stomach secretes acids to help digest foods. These acids, which include hydrochloric acid, contain about 0.1 mol of H^+ per liter of solution. The stomach and digestive tract are normally protected from the corrosive effects of stomach acid by a mucosal lining. Holes can develop in this lining, however, allowing the acid to attack the underlying tissue, causing painful damage. These holes, known as ulcers, can be caused by the secretion of excess acids or by a weakness in the digestive lining. Recent studies indicate, however, that many ulcers are caused by bacterial infection. Between 10 and 20% of Americans suffer from ulcers at some point in their lives, and many others experience occasional indigestion or heartburn due to digestive acids entering the esophagus.

We can address the problem of excess stomach acid in two simple ways: (1) removing the excess acid, or (2) decreasing the production of acid. Those substances that remove excess acid are called *antacids*, whereas those that decrease the production of acid are called *acid inhibitors*. Figure 4.10 ◀ shows several common, over-the-counter drugs of both types.

Antacids are simple bases that neutralize digestive acids. Their ability to neutralize acids is due to the hydroxide, carbonate, or bicarbonate ions they contain. Table 4.4 ▼ lists the active ingredients in some antacids.

The newer generation of antiulcer drugs, such as Tagamet® and Zantac®, are acid inhibitors. They act on acid-producing cells in the lining of the stomach. Formulations that control acid in this way are now available as over-the-counter drugs.

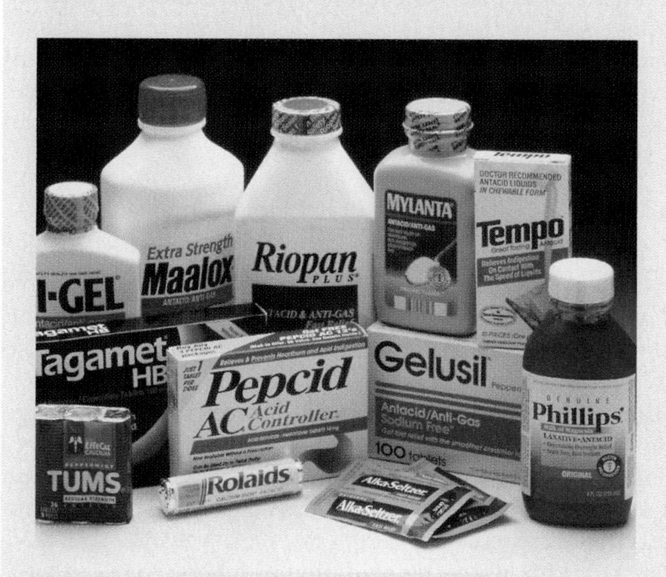

▲ **Figure 4.10** Antacids and acid inhibitors are common, over-the-counter drugs. Tagamet HB® and Pepcid AC® are acid inhibitors, whereas the other products are antacids.

TABLE 4.4 Some Common Antacids

Commercial Name	Acid-Neutralizing Agents
Alka-Seltzer®	$NaHCO_3$
Amphojel®	$Al(OH)_3$
Di-Gel®	$Mg(OH)_2$ and $CaCO_3$
Milk of Magnesia	$Mg(OH)_2$
Maalox®	$Mg(OH)_2$ and $Al(OH)_3$
Mylanta®	$Mg(OH)_2$ and $Al(OH)_3$
Rolaids®	$NaAl(OH)_2CO_3$
Tums®	$CaCO_3$

4.4 Oxidation-Reduction Reactions

In precipitation reactions cations and anions come together to form an insoluble ionic compound. In neutralization reactions H^+ ions and OH^- ions come together to form H_2O molecules. Now let's consider a third important kind of reaction in which electrons are transferred between reactants. Such reactions are called **oxidation-reduction**, or *redox*, **reactions**.

▲ **Figure 4.11** Corrosion at the terminals of a battery, caused by attack of the metal by sulfuric acid from the battery.

ANIMATION
Oxidation-Reduction Reactions: Part I, Oxidation-Reduction Reactions: Part II

Oxidation and Reduction

The corrosion of iron (rusting) and of other metals, such as the corrosion of the terminals of an automobile battery, are familiar processes. What we call *corrosion* is the conversion of a metal into a metal compound by a reaction between the metal and some substance in its environment. Rusting involves the reaction of oxygen with iron in the presence of water. The corrosion shown in Figure 4.11 ◀ results from the reaction of battery acid (H_2SO_4) with the metal clamp.

When a metal corrodes, it loses electrons and forms cations. For example, calcium is vigorously attacked by acids to form calcium ions (Ca^{2+}):

$$Ca(s) + 2H^+(aq) \longrightarrow Ca^{2+}(aq) + H_2(g) \qquad [4.23]$$

When an atom, ion, or molecule has become more positively charged (that is, when it has lost electrons), we say that it has been oxidized. *Loss of electrons by a substance is called* **oxidation**. Thus, Ca, which has no net charge, is *oxidized* (undergoes oxidation) in Equation 4.23, forming Ca^{2+}.

The term oxidation is used because the first reactions of this sort to be studied thoroughly were reactions with oxygen. Many metals react directly with O_2 in air to form metal oxides. In these reactions the metal loses electrons to oxygen, forming an ionic compound of the metal ion and oxide ion. For example, when calcium metal is exposed to air, the bright metallic surface of the metal tarnishes as CaO forms:

$$2Ca(s) + O_2(g) \longrightarrow 2CaO(s) \qquad [4.24]$$

As Ca is oxidized in Equation 4.24, oxygen is transformed from neutral O_2 to two O^{2-} ions. When an atom, ion, or molecule has become more negatively charged (gained electrons), we say that it is *reduced*. *Gain of electrons by a substance is called* **reduction**. When one reactant loses electrons, another reactant must gain them; the oxidation of one substance is always accompanied by the reduction of another as electrons are transferred between them, as shown in Figure 4.12 ◀.

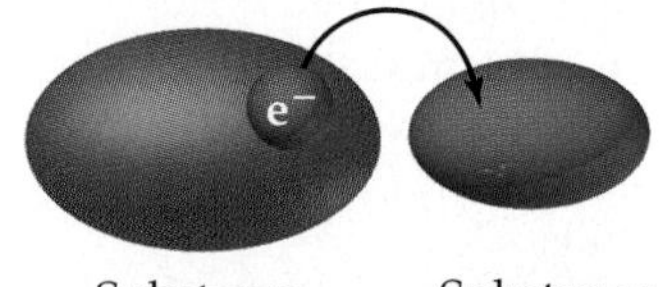

▲ **Figure 4.12** Oxidation is the loss of electrons by a substance; reduction is the gain of electrons by a substance. Oxidation of one substance is always accompanied by reduction of another.

Oxidation Numbers

Before we can properly identify an oxidation-reduction reaction, we must have some way of keeping track of the electrons gained by the substance reduced and those lost by the substance oxidized. The concept of oxidation numbers (also called *oxidation states*) was devised as a simple way of keeping track of electrons in reactions. The **oxidation number** of an atom in a substance is the actual charge of the atom if it is a monoatomic ion; otherwise, it is the hypothetical charge assigned to the atom using a set of rules. Oxidation occurs when there is an increase in oxidation number, whereas reduction occurs when there is a decrease in oxidation number.

We use the following rules for assigning oxidation numbers:

1. *For an atom in its* ***elemental form*** *the oxidation number is always zero.* Thus, each H atom in the H_2 molecule has an oxidation number of 0, and each P atom in the P_4 molecule has an oxidation number of 0.

2. *For any **monoatomic ion** the oxidation number equals the charge on the ion.* Thus, K^+ has an oxidation number of +1, S^{2-} has an oxidation state of −2, and so forth. The alkali metal ions (group 1A) always have a 1+ charge, and therefore the alkali metals always have an oxidation number of +1 in their compounds. Similarly, the alkaline earth metals (group 2A) are always +2, and aluminum (group 3A) is always +3 in their compounds. (In writing oxidation numbers, we will write the sign before the number to distinguish them from the actual electronic charges, which we write with the number first.)
3. ***Nonmetals*** usually have negative oxidation numbers, although they can sometimes be positive:
   (a) *The oxidation number of **oxygen** is usually* −2 in both ionic and molecular compounds. The major exception is in compounds called peroxides, which contain the O_2^{2-} ion, giving each oxygen an oxidation number of −1.
   (b) *The oxidation number of **hydrogen** is +1 when bonded to nonmetals and −1 when bonded to metals.*
   (c) *The oxidation number of **fluorine** is −1 in all compounds.* The other ***halogens*** have an oxidation number of −1 in most binary compounds. When combined with oxygen, as in oxyanions, however, they have positive oxidation states.
4. ***The sum of the oxidation numbers** of all atoms in a neutral compound is zero. The sum of the oxidation numbers in a polyatomic ion equals the charge of the ion.* For example, in the hydronium ion, H_3O^+, the oxidation number of each hydrogen is +1 and that of oxygen is −2. Thus the sum of the oxidation numbers is $3(+1) + (-2) = +1$, which equals the net charge of the ion. This rule is very useful in obtaining the oxidation number of one atom in a compound or ion if you know the oxidation numbers of the other atoms, as illustrated in Sample Exercise 4.8.

SAMPLE EXERCISE 4.8

Determine the oxidation state of sulfur in each of the following: **(a)** H_2S; **(b)** S_8; **(c)** SCl_2; **(d)** Na_2SO_3; **(e)** SO_4^{2-}.

Solution (a) When bonded to a nonmetal, hydrogen has an oxidation number of +1 (rule 3b). Because the H_2S molecule is neutral, the sum of the oxidation numbers must equal zero (rule 4). Letting x equal the oxidation number of S, we have $2(+1) + x = 0$. Thus, S has an oxidation number of −2.

(b) Because this is an elemental form of sulfur, the oxidation number of S is 0 (rule 1).

(c) Because this is a binary compound, we expect chlorine to have an oxidation number of −1 (rule 3c). The sum of the oxidation numbers must equal zero (rule 4). Letting x equal the oxidation number of S, we have $x + 2(-1) = 0$. Consequently, the oxidation number of S must be +2.

(d) Sodium, an alkali metal, always has an oxidation number of +1 in its compounds (rule 2). Oxygen has a common oxidation state of −2 (rule 3a). Letting x equal the oxidation number of S, we have $2(+1) + x + 3(-2) = 0$. Therefore, the oxidation number of S in this compound is +4.

(e) The oxidation state of O is −2 (rule 3a). The sum of the oxidation numbers equals −2, the net charge of the SO_4^{2-} ion (rule 4). Thus we have $x + 4(-2) = -2$. From this relation we conclude that the oxidation number of S in this ion is +6.

These examples illustrate that the oxidation number of a given element depends on the compound in which it occurs. The oxidation numbers of sulfur, as seen in these examples, range from −2 to +6.

PRACTICE EXERCISE

What is the oxidation state of the boldfaced element in each of the following: **(a)** $\mathbf{P}_2O_5$; **(b)** Na**H**; **(c)** $\mathbf{Cr}_2O_7^{2-}$; **(d)** **Sn**Br_4; **(e)** Ba$\mathbf{O}_2$?
Answers: **(a)** +5; **(b)** −1; **(c)** +6; **(d)** +4; **(e)** −1

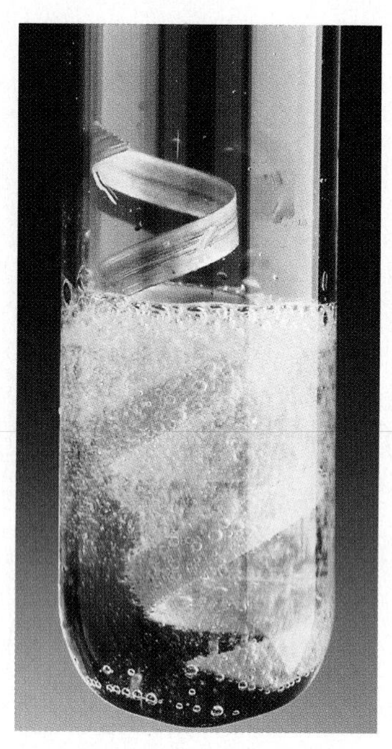

▲ **Figure 4.13** Many metals, such as the magnesium shown here, react with acids to form hydrogen gas. The bubbles are due to the hydrogen gas.

Oxidation of Metals by Acids and Salts

There are many kinds of redox reactions. For example, combustion reactions are redox reactions because elemental oxygen is converted to compounds of oxygen. (Section 3.2) In this chapter we consider the redox reactions between metals and either acids or salts. In Chapter 20 we will examine more complex kinds of redox reactions.

The reaction of a metal with either an acid or a metal salt conforms to the following general pattern:

$$A + BX \longrightarrow AX + B \quad [4.25]$$

Examples:

$$Zn(s) + 2HBr(aq) \longrightarrow ZnBr_2(aq) + H_2(g)$$

$$Mn(s) + Pb(NO_3)_2(aq) \longrightarrow Mn(NO_3)_2(aq) + Pb(s)$$

These reactions are called **displacement reactions** because the ion in solution is displaced or replaced through oxidation of an element.

Many metals undergo displacement reactions with acids, producing salts and hydrogen gas. For example, magnesium metal reacts with hydrochloric acid to form magnesium chloride and hydrogen gas (Figure 4.13 ◀). To show that oxidation and reduction have occurred, the oxidation number for each atom is shown below the chemical equation for this reaction:

$$\underset{0}{Mg(s)} + 2\underset{+1\ -1}{HCl(aq)} \longrightarrow \underset{+2\ -1}{MgCl_2(aq)} + \underset{0}{H_2(g)} \quad [4.26]$$

Notice that the oxidation number of Mg changes from 0 to +2. The increase in the oxidation number indicates that the atom has lost electrons and has therefore been oxidized. The H^+ ion of the acid decreases in oxidation number from +1 to 0, indicating that this ion has gained electrons and has therefore been reduced. The oxidation number of the Cl^- ion remains −1, and it is a spectator ion in the reaction. The net ionic equation is as follows:

$$Mg(s) + 2H^+(aq) \longrightarrow Mg^{2+}(aq) + H_2(g) \quad [4.27]$$

Metals can also be oxidized by aqueous solutions of various salts. Iron metal, for example, is oxidized to Fe^{2+} by aqueous solutions of Ni^{2+}, such as $Ni(NO_3)_2(aq)$:

Molecular equation: $$Fe(s) + Ni(NO_3)_2(aq) \longrightarrow Fe(NO_3)_2(aq) + Ni(s) \quad [4.28]$$

Net ionic equation: $$Fe(s) + Ni^{2+}(aq) \longrightarrow Fe^{2+}(aq) + Ni(s) \quad [4.29]$$

The oxidation of Fe to form Fe^{2+} in this reaction is accompanied by the reduction of Ni^{2+} to Ni. Remember: *Whenever one substance is oxidized, some other substance must be reduced.*

SAMPLE EXERCISE 4.9

Write the balanced molecular and net ionic equations for the reaction of aluminum with hydrobromic acid.

Solution

Analyze: We must write the equation for the redox reaction between a metal and an acid.

Plan: Metals react with acids to form salts and H_2 gas. To write the balanced equation, we must write the chemical formulas for the two reactants and then determine the formula of the salt. The salt is composed of the cation formed by the metal and the anion of the acid.

Solve: The formulas of the given reactants are Al and HBr. The cation formed by Al is Al^{3+}, and the anion from hydrobromic acid is Br^-. Thus, the salt formed in the reaction is $AlBr_3$. Writing the reactants and products and then balancing the equation gives

$$2Al(s) + 6HBr(aq) \longrightarrow 2AlBr_3(aq) + 3H_2(g)$$

Both HBr and $AlBr_3$ are soluble strong electrolytes. Thus, the complete ionic equation is

$$2Al(s) + 6H^+(aq) + 6Br^-(aq) \longrightarrow 2Al^{3+}(aq) + 6Br^-(aq) + 3H_2(g)$$

Because Br^- is a spectator ion, the net ionic equation is

$$2Al(s) + 6H^+(aq) \longrightarrow 2Al^{3+}(aq) + 3H_2(g)$$

Comment: The substance oxidized is the aluminum metal because its oxidation state changes from 0 to +3 in the cation, thereby increasing in oxidation number. The H^+ is reduced because its oxidation state changes from +1 to 0 in H_2.

PRACTICE EXERCISE

(a) Write the balanced molecular and net ionic equations for the reaction between magnesium and cobalt(II) sulfate. **(b)** What is oxidized and what is reduced in the reaction?

Answers: **(a)** $Mg(s) + CoSO_4(aq) \longrightarrow MgSO_4(aq) + Co(s)$; $Mg(s) + Co^{2+}(aq) \longrightarrow Mg^{2+}(aq) + Co(s)$; **(b)** Mg is oxidized and Co^{2+} is reduced.

The Activity Series

Can we predict whether a certain metal will be oxidized either by an acid or by a particular salt? This question is of practical importance as well as chemical interest. According to Equation 4.28, for example, it would be unwise to store a solution of nickel nitrate in an iron container because the solution would dissolve the container. When a metal is oxidized, it appears to be eaten away as it reacts to form various compounds. Extensive oxidation can lead to the failure of metal machinery parts or the deterioration of metal structures.

Different metals vary in the ease with which they are oxidized. Zn is oxidized by aqueous solutions of Cu^{2+}, for example, but Ag is not. Zn, therefore, loses electrons more readily than Ag; that is, Zn is easier to oxidize than Ag.

A list of metals arranged in order of decreasing ease of oxidation is called an **activity series**. Table 4.5 ▼ gives the activity series in aqueous solution for many of the most common metals. Hydrogen is also included in the table. The metals at the top of the table, such as the alkali metals and the alkaline earth metals, are most easily oxidized; that is, they react most readily to form compounds. They

MOVIE
Oxidation-Reduction Chemistry of Tin and Zinc

ACTIVITY
Precipitation, Redox, and Neutralization Reactions

TABLE 4.5 Activity Series of Metals in Aqueous Solution

Metal	Oxidation Reaction
Lithium	$Li(s) \longrightarrow Li^+(aq) + e^-$
Potassium	$K(s) \longrightarrow K^+(aq) + e^-$
Barium	$Ba(s) \longrightarrow Ba^{2+}(aq) + 2e^-$
Calcium	$Ca(s) \longrightarrow Ca^{2+}(aq) + 2e^-$
Sodium	$Na(s) \longrightarrow Na^+(aq) + e^-$
Magnesium	$Mg(s) \longrightarrow Mg^{2+}(aq) + 2e^-$
Aluminum	$Al(s) \longrightarrow Al^{3+}(aq) + 3e^-$
Manganese	$Mn(s) \longrightarrow Mn^{2+}(aq) + 2e^-$
Zinc	$Zn(s) \longrightarrow Zn^{2+}(aq) + 2e^-$
Chromium	$Cr(s) \longrightarrow Cr^{3+}(aq) + 3e^-$
Iron	$Fe(s) \longrightarrow Fe^{2+}(aq) + 2e^-$
Cobalt	$Co(s) \longrightarrow Co^{2+}(aq) + 2e^-$
Nickel	$Ni(s) \longrightarrow Ni^{2+}(aq) + 2e^-$
Tin	$Sn(s) \longrightarrow Sn^{2+}(aq) + 2e^-$
Lead	$Pb(s) \longrightarrow Pb^{2+}(aq) + 2e^-$
Hydrogen	$H_2(g) \longrightarrow 2H^+(aq) + 2e^-$
Copper	$Cu(s) \longrightarrow Cu^{2+}(aq) + 2e^-$
Silver	$Ag(s) \longrightarrow Ag^+(aq) + e^-$
Mercury	$Hg(l) \longrightarrow Hg^{2+}(aq) + 2e^-$
Platinum	$Pt(s) \longrightarrow Pt^{2+}(aq) + 2e^-$
Gold	$Au(s) \longrightarrow Au^{3+}(aq) + 3e^-$

Ease of oxidation increases (upward arrow, from Gold to Lithium)

are called the *active metals*. The metals at the bottom of the activity series, such as the transition elements from groups 8B and 1B, are very stable and form compounds less readily. These metals, which are used to make coins and jewelry, are called *noble metals* because of their low reactivity.

The activity series can be used to predict the outcome of reactions between metals and either metal salts or acids. *Any metal on the list can be oxidized by the ions of elements below it*. For example, copper is above silver in the series. Thus, copper metal will be oxidized by silver ions, as pictured in Figure 4.14 ▼:

$$Cu(s) + 2Ag^{+}(aq) \longrightarrow Cu^{2+}(aq) + 2Ag(s) \quad [4.30]$$

The oxidation of copper to copper ions is accompanied by the reduction of silver ions to silver metal. The silver metal is evident on the surface of the copper wires in Figure 4.14(b) and (c). The copper(II) nitrate produces a blue color in the solution, which is most evident in part (c).

Only those metals above hydrogen in the activity series are able to react with acids to form H_2. For example, Ni reacts with HCl(*aq*) to form H_2:

$$Ni(s) + 2HCl(aq) \longrightarrow NiCl_2(aq) + H_2(g) \quad [4.31]$$

Because elements below hydrogen in the activity series are not oxidized by H^+, Cu does not react with HCl(*aq*). Interestingly, copper does react with nitric acid, as shown previously in Figure 1.11. This reaction, however, is not a simple oxidation of Cu by the H^+ ions of the acid. Instead, the metal is oxidized to Cu^{2+} by the nitrate ion of the acid, accompanied by the formation of brown nitrogen dioxide, $NO_2(g)$:

$$Cu(s) + 4HNO_3(aq) \longrightarrow Cu(NO_3)_2(aq) + 2H_2O(l) + 2NO_2(g) \quad [4.32]$$

What substance is reduced as copper is oxidized in Equation 4.32? In this case the NO_2 results from the reduction of NO_3^-. We will examine reactions of this type in more detail in Chapter 20.

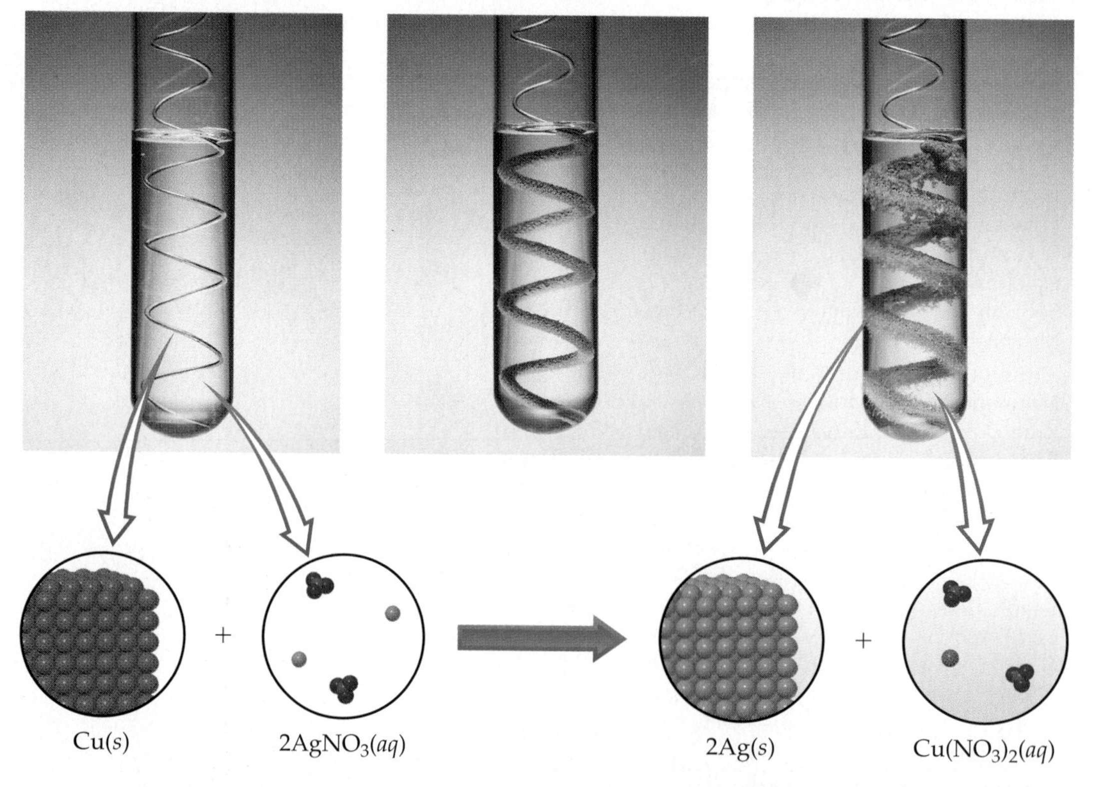

▲ **Figure 4.14** When copper metal is placed in a solution of silver nitrate (a), a redox reaction occurs, forming silver metal and a blue solution of copper(II) nitrate (b and c).

SAMPLE EXERCISE 4.10

Will an aqueous solution of iron(II) chloride oxidize magnesium metal? If so, write the balanced molecular and net ionic equations for the reaction.

Solution

Analyze: We are given two substances—an aqueous salt, $FeCl_2$, and a metal, Mg—and asked if they react with each other.

Plan: A reaction will occur if Mg is above Fe^{2+} in the activity series, Table 4.5. If the reaction occurs, the Fe^{2+} ion in $FeCl_2$ will be reduced to Fe, and the elemental Mg will be oxidized to Mg^{2+}.

Solve: Because Mg is above Fe in the table, the reaction will occur. To write the formula for the salt that is produced in the reaction, we must remember the charges on common ions. Magnesium is always present in compounds as Mg^{2+}; the chloride ion is Cl^-. The magnesium salt formed in the reaction is $MgCl_2$:

$$Mg(s) + FeCl_2(aq) \longrightarrow MgCl_2(aq) + Fe(s)$$

Both $FeCl_2$ and $MgCl_2$ are soluble strong electrolytes and can be written in ionic form. Cl^-, then, is a spectator ion in the reaction. The net ionic equation is

$$Mg(s) + Fe^{2+}(aq) \longrightarrow Mg^{2+}(aq) + Fe(s)$$

The net ionic equation shows that Mg is oxidized and Fe^{2+} is reduced in this reaction.

PRACTICE EXERCISE

Which of the following metals will be oxidized by $Pb(NO_3)_2$: Zn, Cu, Fe?
Answer: Zn and Fe

A Closer Look The Aura of Gold

Gold has been known since the earliest records of human existence. Throughout history people have cherished gold, have fought for it, and have died for it.

The physical and chemical properties of gold serve to make it a special metal. First, its intrinsic beauty and rarity make it precious. Second, gold is soft and can be easily formed into artistic objects, jewelry, and coins (Figure 4.15 ▶). Third, gold is one of the least active metals (Table 4.5). It is not oxidized in air and does not react with water. It is unreactive toward basic solutions and nearly all acidic solutions. As a result, gold can be found in nature as a pure element rather than combined with oxygen or other elements, which accounts for its early discovery.

Many of the early studies of the reactions of gold arose from the practice of alchemy, in which people attempted to turn cheap metals, such as lead, into gold. Alchemists discovered that gold can be dissolved in a 3:1 mixture of concentrated hydrochloric and nitric acids, known as aqua regia ("royal water"). The action of nitric acid on gold is similar to that on copper (Equation 4.32) in that the nitrate ion, rather than H^+, oxidizes the metal to Au^{3+}. The Cl^- ions interact with Au^{3+} to form highly stable $AuCl_4^-$ ions. The net ionic equation for the reaction of gold with aqua regia is

$$Au(s) + NO_3^-(aq) + 4H^+(aq) + 4Cl^-(aq) \longrightarrow AuCl_4^-(aq) + 2H_2O(l) + NO(g)$$

All the gold ever mined would easily fit in a cube 19 m on a side and weighing about 1.1×10^8 kg (125,000 tons). More than 90% of this amount has been produced since the beginning of the California gold rush of 1848. Each year, worldwide production of gold amounts to about 1.8×10^6 kg (2000 tons). By contrast, over 1.5×10^{10} kg (16 million tons) of aluminum are produced annually. Gold is used mainly in jewelry (73%), coins (10%), and electronics (9%). Its use in electronics relies on its excellent conductivity and its corrosion resistance. Gold is used, for example, to plate contacts in electrical switches, relays, and connections. A typical Touch-Tone® telephone contains 33 gold-plated contacts. Gold is also used in computers and other microelectronic devices where fine gold wire is used to link components.

Besides its value for jewelry, currency, and electronics, gold is also important in the health professions. Because of its resistance to corrosion by acids and other substances found in saliva, gold is an ideal metal for dental crowns and caps, which accounts for about 3% of the annual use of the element. The pure metal is too soft to use in dentistry, so it is combined with other metals to form alloys.

◀ **Figure 4.15** Portrait of Pharaoh Tutankhamun (1346–1337 B.C.) made from gold and precious stones. From the inner coffin of the tomb of Tutankhamun.

Strategies in Chemistry Analyzing Chemical Reactions

In this chapter you have been introduced to a great number of chemical reactions. A major difficulty that students face in trying to master material of this sort is gaining a "feel" for what happens when chemicals are allowed to react. In fact, you might marvel at the ease with which your professor or teaching assistant can figure out the results of a chemical reaction. One of our goals in this textbook is to help you become more adept at predicting the outcome of reactions. The key to gaining this "chemical intuition" is understanding how to categorize reactions.

There are so many individual reactions in chemistry that memorizing them all is a futile task. It is far more fruitful to try to use pattern recognition to determine the general category of a reaction, such as metathesis or oxidation-reduction. Thus, when you are faced with the challenge of predicting the outcome of a chemical reaction, ask yourself the following pertinent questions:

- What are the reactants in the reaction?
- Are they electrolytes or nonelectrolytes?
- Are they acids and bases?
- If the reactants are electrolytes, will metathesis produce a precipitate? Water? A gas?
- If metathesis cannot occur, can the reactants possibly engage in an oxidation-reduction reaction? This requires that there be both a reactant that can be oxidized and one that can be reduced.

By asking questions such as these, you should be able to predict what might happen during the reaction. You might not always be correct, but if you keep your wits about you, you will not be far off. As you gain experience with chemical reactions, you will begin to look for reactants that might not be immediately obvious, such as water from the solution or oxygen from the atmosphere.

One of the greatest tools available to us in chemistry is experimentation. If you perform an experiment in which two solutions are mixed, you can make observations that help you understand what is happening. For example, using the information in Table 4.1 to predict whether a precipitate will form is not nearly as exciting as actually seeing the precipitate form, as in Figure 4.4. Careful observations in the laboratory portion of the course will make your lecture material easier to master.

4.5 Concentrations of Solutions

The behavior of solutions often depends not only on the nature of the solutes but also on their concentrations. Scientists use the term **concentration** to designate the amount of solute dissolved in a given quantity of solvent or solution. The concept of concentration is intuitive: The greater the amount of solute dissolved in a certain amount of solvent, the more concentrated the resulting solution. In chemistry we often need to express the concentrations of solutions quantitatively.

Molarity

Molarity (symbol *M*) expresses the concentration of a solution as the number of moles of solute in a liter of solution (soln):

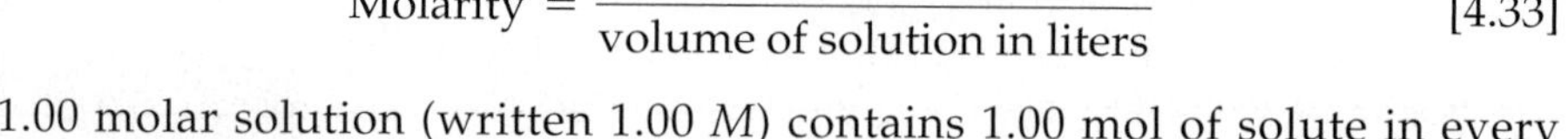

$$\text{Molarity} = \frac{\text{moles solute}}{\text{volume of solution in liters}} \quad [4.33]$$

ANIMATION
Solution Formation from a Solid

A 1.00 molar solution (written 1.00 *M*) contains 1.00 mol of solute in every liter of solution. Figure 4.16 ▶ shows the preparation of 250 mL of a 1.00 *M* solution of $CuSO_4$ by using a volumetric flask that is calibrated to hold exactly 250 mL. First, 0.250 mol of $CuSO_4$ (39.9 g) is weighed out and placed in the volumetric flask. Water is added to dissolve the salt, and the resultant solution is diluted to a total volume of 250 mL. The molarity of the solution is $(0.250 \text{ mol } CuSO_4)/(0.250 \text{ L soln}) = 1.00\ M$.

SAMPLE EXERCISE 4.11

Calculate the molarity of a solution made by dissolving 23.4 g of sodium sulfate (Na_2SO_4) in enough water to form 125 mL of solution.

(a)

(b)

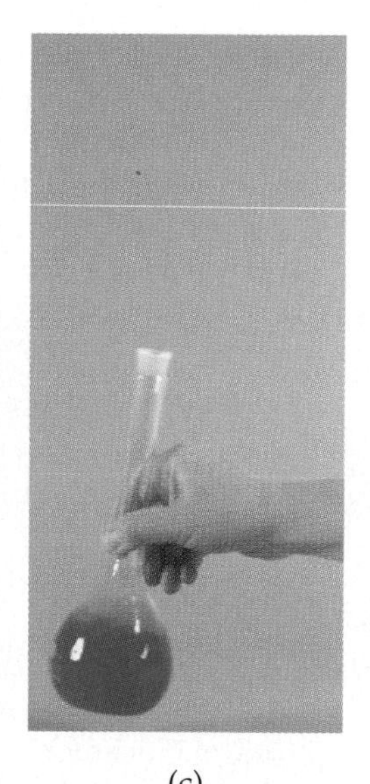

(c)

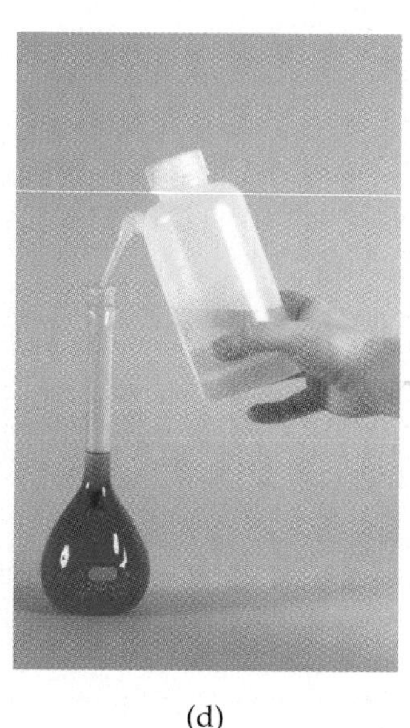

(d)

▲ **Figure 4.16** Procedure for preparation of 0.250 L of 1.00 *M* solution of $CuSO_4$. (a) Weigh out 0.250 mol (39.9 g) of $CuSO_4$ (formula weight = 159.6 amu). (b) Put the $CuSO_4$ (solute) into a 250-mL volumetric flask, and add a small quantity of water. (c) Dissolve the solute by swirling the flask. (d) Add more water until the solution just reaches the calibration mark etched on the neck of the flask. Shake the stoppered flask to ensure complete mixing.

ANIMATION
Dissolution of $KMnO_4$

Solution

Analyze: We are given the number of grams of solute (23.4 g), its chemical formula (Na_2SO_4), and the volume of the solution (125 mL), and we are asked to calculate the molarity of the solution.

Plan: We can calculate molarity using Equation 4.33. To do so, we must convert the number of grams of solute to moles and the volume of the solution from milliliters to liters.

Solve: The number of moles of Na_2SO_4 is obtained from its molar mass.

$$\text{Moles } Na_2SO_4 = (23.4 \text{ g } Na_2SO_4)\left(\frac{1 \text{ mol } Na_2SO_4}{142 \text{ g } Na_2SO_4}\right) = 0.165 \text{ mol } Na_2SO_4$$

Converting the volume of the solution to liters:

$$\text{Liters soln} = (125 \text{ mL})\left(\frac{1 \text{ L}}{1000 \text{ mL}}\right) = 0.125 \text{ L}$$

Thus, the molarity is

$$\text{Molarity} = \frac{0.165 \text{ mol } Na_2SO_4}{0.125 \text{ L soln}} = 1.32\frac{\text{mol } Na_2SO_4}{\text{L soln}} = 1.32\ M$$

Check: Because the numerator is only slightly larger than the denominator, it's reasonable for the answer to be a little over 1 *M*. The units (mol/L) are appropriate for molarity, and three significant figures are appropriate for the answer because each of the initial pieces of data had three significant figures.

PRACTICE EXERCISE

Calculate the molarity of a solution made by dissolving 5.00 g of glucose ($C_6H_{12}O_6$) in sufficient water to form exactly 100 mL of solution.
Answer: 0.278 *M*

Expressing the Concentration of an Electrolyte

When an ionic compound dissolves, the relative concentrations of the ions introduced into the solution depend on the chemical formula of the compound. For example, a 1.0 *M* solution of NaCl is 1.0 *M* in Na^+ ions and 1.0 *M* in Cl^- ions. Similarly, a 1.0 *M* solution of Na_2SO_4 is 2.0 *M* in Na^+ ions and 1.0 *M* in SO_4^{2-} ions. Thus, the concentration of an electrolyte solution can be specified either in terms of the compound used to make the solution (1.0 *M* Na_2SO_4) or in terms of the ions that the solution contains (2.0 *M* Na^+ and 1.0 *M* SO_4^{2-}).

SAMPLE EXERCISE 4.12

What are the molar concentrations of each of the ions present in a 0.025 *M* aqueous solution of calcium nitrate?

Solution

Analyze: We are given the concentration of the ionic compound used to make the solution and asked to determine the concentrations of the ions in the solution.
Plan: We can use the subscripts in the chemical formula of the compound to determine the relative concentrations of the ions.
Solve: Calcium nitrate is composed of calcium ions (Ca^{2+}) and nitrate ions (NO_3^-), so its chemical formula is $Ca(NO_3)_2$. Because there are two NO_3^- ions for each Ca^{2+} ion in the compound, each mole of $Ca(NO_3)_2$ that dissolves dissociates into 1 mol of Ca^{2+} and 2 mol of NO_3^-. Thus, a solution that is 0.025 *M* in $Ca(NO_3)_2$ is 0.025 *M* in Ca^{2+} and $2 \times 0.025\ M = 0.050\ M$ in NO_3^-.
Check: The concentration of NO_3^- ions is twice that of Ca^{2+} ions, as the subscript 2 after the NO_3^- in the chemical formula $Ca(NO_3)_2$ suggests it should be.

PRACTICE EXERCISE

What is the molar concentration of K^+ ions in a 0.015 *M* solution of potassium carbonate?
Answer: 0.030 *M* K^+

Interconverting Molarity, Moles, and Volume

The definition of molarity (Equation 4.33) contains three quantities—molarity, moles solute, and liters of solution. If we know any two of these, we can calculate the third. For example, if we know the molarity of a solution, we can calculate the number of moles of solute in a given volume. Molarity, therefore, is a conversion factor between volume of solution and moles of solute. Calculation of the number of moles of HNO_3 in 2.0 L of 0.200 *M* HNO_3 solution illustrates the conversion of volume to moles:

$$\text{Moles HNO}_3 = (2.0\ \cancel{\text{L soln}})\left(\frac{0.200\ \text{mol HNO}_3}{1\ \cancel{\text{L soln}}}\right)$$

$$= 0.40\ \text{mol HNO}_3$$

Dimensional analysis can be used in this conversion if we express molarity as moles/liter soln. To obtain moles, therefore, we multiply liters and molarity: moles = liters × molarity.

To illustrate the conversion of moles to volume, let's calculate the volume of 0.30 *M* HNO_3 solution required to supply 2.0 mol of HNO_3:

$$\text{Liters soln} = (2.0\ \cancel{\text{mol HNO}_3})\left(\frac{1\ \text{L soln}}{0.30\ \cancel{\text{mol HNO}_3}}\right) = 6.7\ \text{L soln}$$

In this case we must use the reciprocal of molarity in the conversion: liters = moles × 1/*M*.

SAMPLE EXERCISE 4.13

How many grams of Na_2SO_4 are required to make 0.350 L of 0.500 *M* Na_2SO_4?

Solution

Analyze: We are given the volume of the solution (0.350 L), its concentration (0.500 *M*), and the identity of the solute (Na_2SO_4) and asked to calculate the number of grams of the solute in the solution.

Plan: We can use the definition of molarity (Equation 4.33) to determine the number of moles of solute, and then convert moles to grams using the molar mass of the solute:

$$M_{Na_2SO_4} = \frac{\text{moles } Na_2SO_4}{\text{liters soln}}$$

Solve: Calculating the moles of Na_2SO_4 using the molarity and volume of solution gives:

$$\text{Moles } Na_2SO_4 = \text{liters soln} \times M_{Na_2SO_4}$$
$$= (0.350 \cancel{\text{ L soln}})\left(\frac{0.500 \text{ mol } Na_2SO_4}{1 \cancel{\text{ L soln}}}\right)$$
$$= 0.175 \text{ mol } Na_2SO_4$$

Because each mole of Na_2SO_4 weighs 142 g, the required number of grams of Na_2SO_4 is

$$\text{Grams } Na_2SO_4 = (0.175 \cancel{\text{ mol } Na_2SO_4})\left(\frac{142 \text{ g } Na_2SO_4}{1 \cancel{\text{ mol } Na_2SO_4}}\right) = 24.9 \text{ g } Na_2SO_4$$

Check: The magnitude of the answer, the units, and the number of significant figures are all appropriate.

PRACTICE EXERCISE

(a) How many grams of Na_2SO_4 are there in 15 mL of 0.50 *M* Na_2SO_4? **(b)** How many milliliters of 0.50 *M* Na_2SO_4 solution are needed to provide 0.038 mol of this salt?
Answers: **(a)** 1.1 g; **(b)** 76 mL

Dilution

Solutions that are used routinely in the laboratory are often purchased or prepared in concentrated form (called *stock solutions*). Hydrochloric acid, for example, is purchased as a 12 *M* solution (concentrated HCl). Solutions of lower concentrations can then be obtained by adding water, a process called **dilution**.*

To illustrate the preparation of a dilute solution from a concentrated one, suppose we wanted to prepare 250 mL (that is, 0.250 L) of 0.100 *M* $CuSO_4$ solution by diluting a stock solution containing 1.00 *M* $CuSO_4$. When solvent is added to dilute a solution, the number of moles of solute remains unchanged.

$$\text{Moles solute before dilution} = \text{moles solute after dilution} \qquad [4.34]$$

Because we know both the volume and concentration of the dilute solution, we can calculate the number of moles of $CuSO_4$ it contains.

$$\text{Mol } CuSO_4 \text{ in dil soln} = (0.250 \cancel{\text{ L soln}})\left(0.100 \frac{\text{mol } CuSO_4}{\cancel{\text{L soln}}}\right) = 0.0250 \text{ mol } CuSO_4$$

Now we can calculate the volume of the concentrated solution needed to provide 0.0250 mol $CuSO_4$:

$$\text{L of conc soln} = (0.0250 \cancel{\text{ mol } CuSO_4})\left(\frac{1 \text{ L soln}}{1.00 \cancel{\text{ mol } CuSO_4}}\right) = 0.0250 \text{ L}$$

* In diluting a concentrated acid or base, the acid or base should be added to water and then further diluted by adding more water. Adding water directly to concentrated acid or base can cause spattering because of the intense heat generated.

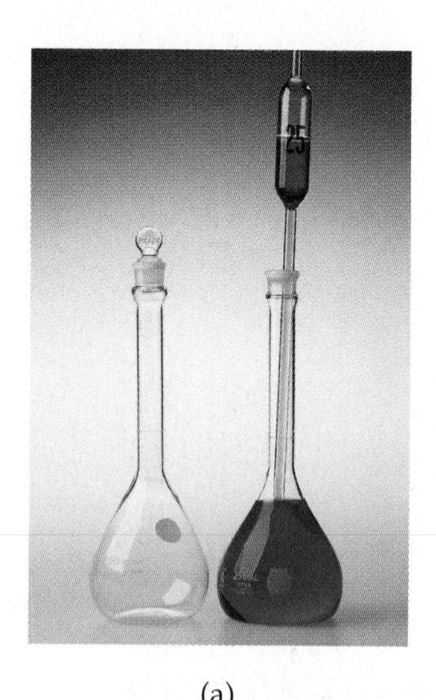

(a)

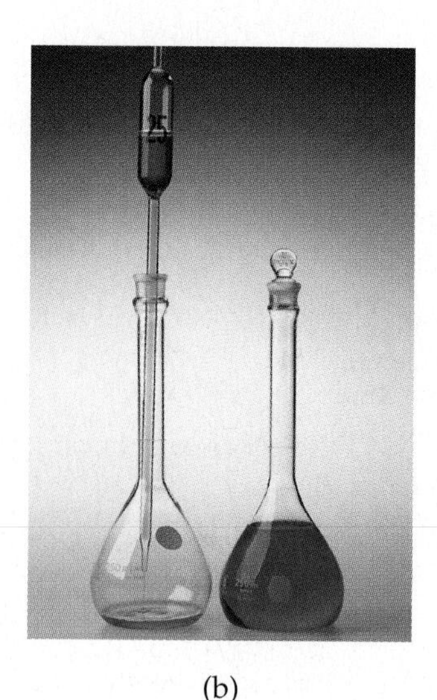

(b)

(c)

▲ **Figure 4.17** Procedure for preparing 250 mL of 0.100 *M* $CuSO_4$ by dilution of 1.00 *M* $CuSO_4$. (a) Draw 25.0 mL of the 1.00 *M* solution into a pipet. (b) Add this to a 250-mL volumetric flask. (c) Add water to dilute the solution to a total volume of 250 mL.

Thus, this dilution is achieved by withdrawing 0.0250 L (that is, 25.0 mL) of the 1.00 *M* solution using a pipet, adding it to a 250-mL volumetric flask, and then diluting it to a final volume of 250 mL, as shown in Figure 4.17 ▲. Notice that the diluted solution is less intensely colored than the concentrated one.

In laboratory situations, calculations of this sort are often made very quickly with a simple equation that can be derived by remembering that the number of moles of solute is the same in both the concentrated and dilute solutions and that moles = molarity × liters:

$$\text{Moles solute in conc soln} = \text{moles solute in dil soln}$$

$$M_{\text{conc}} \times V_{\text{conc}} = M_{\text{dil}} \times V_{\text{dil}} \qquad [4.35]$$

The molarity of the more concentrated stock solution (M_{conc}) is always larger than the molarity of the dilute solution (M_{dil}). Because the volume of the solution increases upon dilution, V_{dil} is always larger than V_{conc}. Although Equation 4.35 is derived in terms of liters, any volume unit can be used so long as that same unit is used on both sides of the equation. For example, in the calculation we did for the $CuSO_4$ solution, we have

$$(1.00\ M)(V_{\text{conc}}) = (0.100\ M)(250\ \text{mL})$$

Solving for V_{conc} gives $V_{\text{conc}} = 25.0$ mL as before.

SAMPLE EXERCISE 4.14

How many milliliters of 3.0 *M* H_2SO_4 are needed to make 450 mL of 0.10 *M* H_2SO_4?

Solution

Analyze: We need to dilute a concentrated solution. We are given the molarity of a more concentrated solution (3.0 *M*) and the volume and molarity of a more dilute one containing the same solute (450 mL of 0.10 *M* solution). We must calculate the volume of the concentrated solution needed to prepare the dilute solution.

Plan: We can calculate the number of moles of solute, H_2SO_4, in the dilute solution and then calculate the volume of the concentrated solution needed to supply this amount of solute. Alternatively, we can directly apply Equation 4.35. Let's compare the two methods.

Solve: Calculating the moles of H_2SO_4 in the dilute solution:

$$\text{Moles } H_2SO_4 \text{ in dilute solution} = (0.450\ \cancel{\text{L soln}})\left(\frac{0.10\ \text{mol } H_2SO_4}{1\ \cancel{\text{L soln}}}\right) = 0.045\ \text{mol } H_2SO_4$$

Calculating the volume of concentrated solution that contains 0.045 mol H_2SO_4:

$$\text{L conc soln} = (0.045 \text{ mol } H_2SO_4)\left(\frac{1 \text{ L soln}}{3.0 \text{ mol } H_2SO_4}\right) = 0.015 \text{ L soln}$$

Converting liters to milliliters gives 15 mL.

If we apply Equation 4.35, we get the same result:

$$(3.0\ M)(V_{conc}) = (0.10\ M)(450 \text{ mL})$$

$$V_{conc} = \frac{(0.10\ M)(450 \text{ mL})}{3.0\ M} = 15 \text{ mL}$$

Either way, we see that if we start with 15 mL of 3.0 *M* H_2SO_4 and dilute it to a total volume of 450 mL, the desired 0.10 *M* solution will be obtained.

Check: The calculated volume seems reasonable because a small volume of concentrated solution is used to prepare a large volume of dilute solution.

PRACTICE EXERCISE

(a) What volume of 2.50 *M* lead nitrate solution contains 0.0500 mol of Pb^{2+}? **(b)** How many milliliters of 5.0 *M* $K_2Cr_2O_7$ solution must be diluted to prepare 250 mL of 0.10 *M* solution? **(c)** If 10.0 mL of a 10.0 *M* stock solution of NaOH is diluted to 250 mL, what is the concentration of the resulting solution?

Answers: **(a)** 0.0200 L = 20.0 mL; **(b)** 5.0 mL; **(c)** 0.40 *M*

4.6 Solution Stoichiometry and Chemical Analysis

Imagine that you have to determine the concentrations of several ions in a sample of lake water. Although many instrumental methods have been developed for such analyses, chemical reactions such as those discussed in this chapter continue to be used. In Chapter 3 we learned that if you know the chemical equation and the amount of one reactant consumed in the reaction, you can calculate the quantities of other reactants and products. In this section we briefly explore such analyses of solutions.

Recall that the coefficients in a balanced equation give the relative number of moles of reactants and products. (Section 3.6) To use this information, we must convert the quantities of substances involved in a reaction into moles. When we are dealing with grams of substances, as we were in Chapter 3, we use the molar mass to achieve this conversion. When we are working with solutions of known molarity, however, we use molarity and volume to determine the number of moles (moles solute = $M \times$ L). Figure 4.18 ▼ summarizes this approach to using stoichiometry.

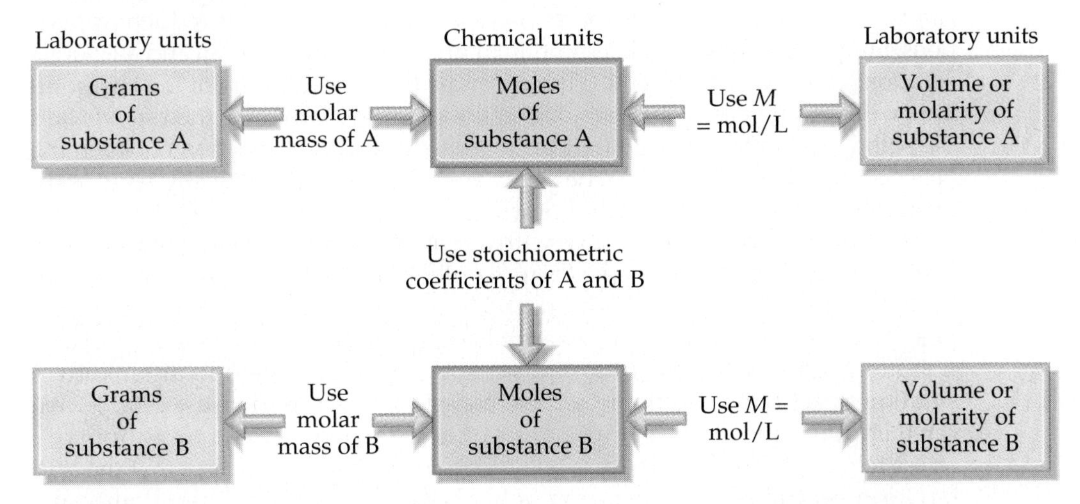

▲ **Figure 4.18** Outline of the procedure used to solve stoichiometry problems that involve measured (laboratory) units of mass, solution concentration (molarity), or volume.

SAMPLE EXERCISE 4.15

How many grams of $Ca(OH)_2$ are needed to neutralize 25.0 mL of 0.100 *M* HNO_3?

Solution

Analyze: The reactants are an acid, HNO_3, and a base, $Ca(OH)_2$. The volume and molarity of HNO_3 are given, and we are asked how many grams of $Ca(OH)_2$ are needed to neutralize this quantity of HNO_3.

Plan: We can use the molarity and volume of the HNO_3 solution to calculate the number of moles of HNO_3. We then use the balanced equation to relate the moles of HNO_3 to moles of $Ca(OH)_2$. Finally, we can convert moles of $Ca(OH)_2$ to grams. These steps can be summarized as follows:

$$L_{HNO_3} \times M_{HNO_3} \Rightarrow \text{mol } HNO_3 \Rightarrow \text{mol } Ca(OH)_2 \Rightarrow \text{g } Ca(OH)_2$$

Solve: The product of the molar concentration of a solution and its volume in liters gives the number of moles of solute:

$$\text{Moles } HNO_3 = L_{HNO_3} \times M_{HNO_3} = (0.0250 \cancel{L})\left(0.100 \frac{\text{mol } HNO_3}{\cancel{L}}\right)$$
$$= 2.50 \times 10^{-3} \text{ mol } HNO_3$$

Because this is an acid-base neutralization reaction, HNO_3 and $Ca(OH)_2$ react to form H_2O and the salt containing Ca^{2+} and NO_3^-:

$$2HNO_3(aq) + Ca(OH)_2(s) \longrightarrow 2H_2O(l) + Ca(NO_3)_2(aq)$$

Thus, 2 mol $HNO_3 \simeq$ 1 mol $Ca(OH)_2$. Therefore,

$$\text{Grams } Ca(OH)_2 = (2.50 \times 10^{-3} \cancel{\text{mol } HNO_3})\left(\frac{1 \cancel{\text{mol } Ca(OH)_2}}{2 \cancel{\text{mol } HNO_3}}\right)\left(\frac{74.1 \text{ g } Ca(OH)_2}{1 \cancel{\text{mol } Ca(OH)_2}}\right)$$
$$= 0.0926 \text{ g } Ca(OH)_2$$

Check: The size of the answer is reasonable. A small volume of dilute acid will require only a small amount of base to neutralize it.

PRACTICE EXERCISE

(a) How many grams of NaOH are needed to neutralize 20.0 mL of 0.150 *M* H_2SO_4 solution? **(b)** How many liters of 0.500 *M* HCl(*aq*) are needed to react completely with 0.100 mol of $Pb(NO_3)_2(aq)$, forming a precipitate of $PbCl_2(s)$?
Answers: **(a)** 0.240 g; **(b)** 0.400 L

Titrations

To determine the concentration of a particular solute in a solution, chemists often carry out a **titration**, which involves combining a sample of the solution with a reagent solution of known concentration, called a **standard solution**. Titrations can be conducted using acid-base, precipitation, or oxidation-reduction reactions. Suppose we have an HCl solution of unknown concentration and an NaOH solution we know to be 0.100 *M*. To determine the concentration of the HCl solution, we take a specific volume of that solution, say 20.00 mL. We then slowly add the standard NaOH solution to it until the neutralization reaction between the HCl and NaOH is complete. The point at which stoichiometrically equivalent quantities are brought together is known as the **equivalence point** of the titration.

In order to titrate an unknown with a standard solution, there must be some way to determine when the equivalence point of the titration has been reached. In acid-base titrations, dyes known as acid-base **indicators** are used for this purpose. For example, the dye known as phenolphthalein is colorless in acidic solution but is pink in basic solution. If we add phenolphthalein to an unknown solution of acid, the solution will be colorless, as seen in Figure 4.19(a) ▶. We can then add standard base from a buret until the solution barely turns from colorless to pink, as seen in Figure 4.19(b). This color change indicates that the acid has been neutralized and the drop of base that caused the solution to become

(a)

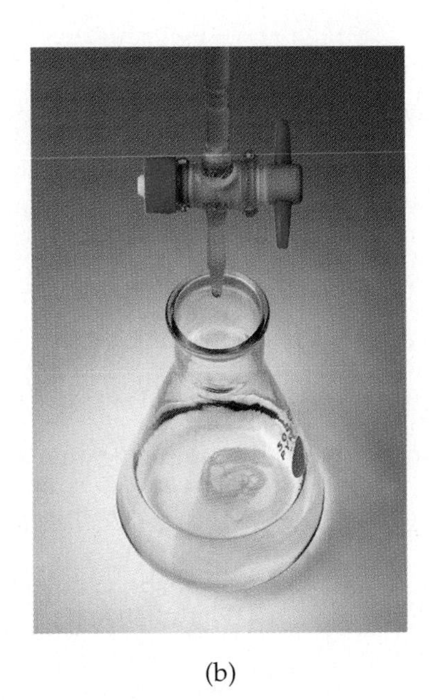

(b)

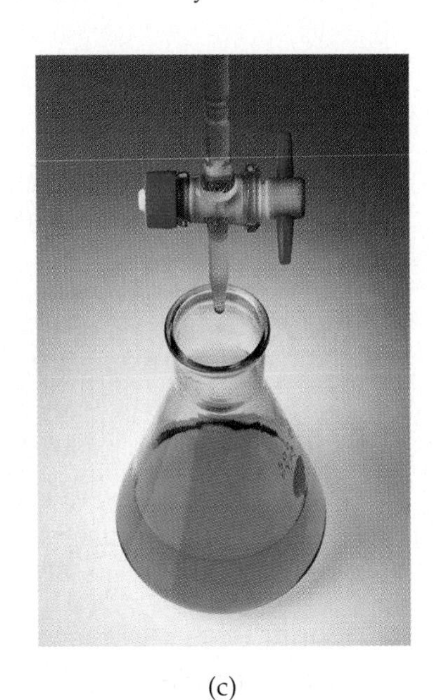

(c)

▲ **Figure 4.19** Change in appearance of a solution containing phenolphthalein indicator as base is added. Before the end point, the solution is colorless (a). As the end point is approached, a pale pink color forms where the base is added (b). At the end point, this pale pink color extends throughout the solution after the mixing. As even more base is added, the intensity of the pink color increases (c).

colored has no acid to react with. The solution therefore becomes basic, and the dye turns pink. The color change signals the *end point* of the titration, which usually coincides very nearly with the equivalence point. Care must be taken to choose indicators whose end points correspond to the equivalence point of the titration. We will consider this matter in Chapter 17. The titration procedure is summarized in Figure 4.20 ▼.

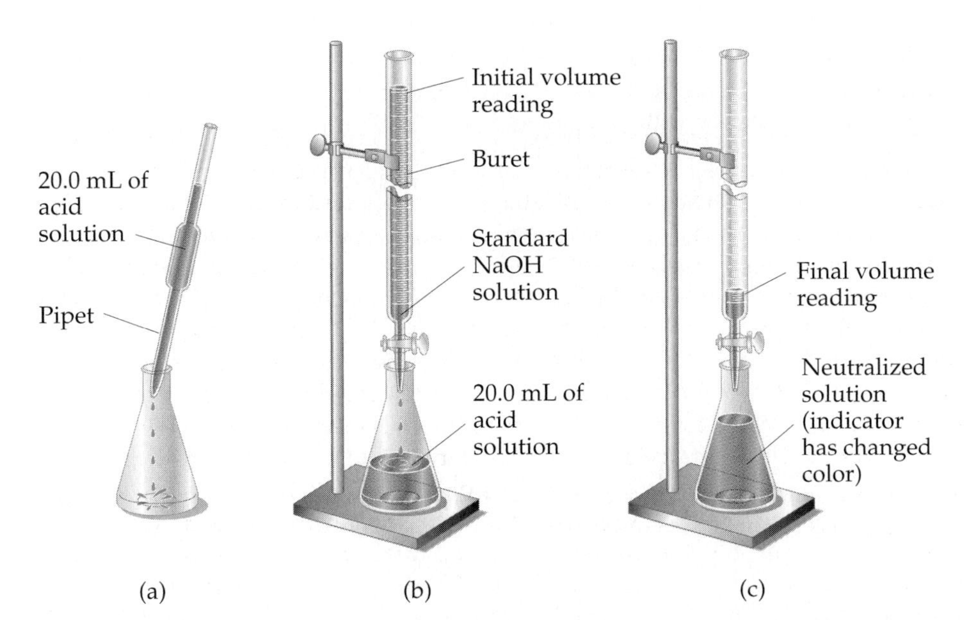

▲ **Figure 4.20** Procedure for titrating an acid against a standardized solution of NaOH. (a) A known quantity of acid is added to a flask. (b) An acid-base indicator is added, and standardized NaOH is added from a buret. (c) End point is signaled by a color change in the indicator.

SAMPLE EXERCISE 4.16

The quantity of Cl^- in a water supply is determined by titrating the sample with Ag^+.

$$Ag^+(aq) + Cl^-(aq) \longrightarrow AgCl(s)$$

(a) How many grams of chloride ion are in a sample of the water if 20.2 mL of 0.100 *M* Ag^+ is needed to react with all the chloride in the sample? **(b)** If the sample has a mass of 10.0 g, what percent Cl^- does it contain?

Solution

Analyze: We are given the volume (20.2 mL) and molarity (0.100 *M*) of a solution of Ag^+ and the chemical equation for reaction of this ion with Cl^-. We are asked first to calculate the number of grams of Cl^- in the sample and, second, to calculate the mass percent of Cl^- in the sample.

(a) Plan: We begin by using the volume and molarity of Ag^+ to calculate the number of moles of Ag^+ used in the titration. We can then use the balanced equation to determine the moles of Cl^- and from that the grams of Cl^-.

Solve:

$$\text{Mol } Ag^+ = (20.2 \text{ mL soln})\left(\frac{1 \text{ L soln}}{1000 \text{ mL soln}}\right)\left(0.100 \frac{\text{mol } Ag^+}{\text{L soln}}\right)$$

$$= 2.02 \times 10^{-3} \text{ mol } Ag^+$$

From the balanced equation we see that 1 mol $Ag^+ \simeq$ 1 mol Cl^-. Using this information and the molar mass of Cl, we have

$$\text{Grams } Cl^- = (2.02 \times 10^{-3} \text{ mol } Ag^+)\left(\frac{1 \text{ mol } Cl^-}{1 \text{ mol } Ag^+}\right)\left(\frac{35.5 \text{ g } Cl^-}{1 \text{ mol } Cl^-}\right) = 7.17 \times 10^{-2} \text{ g } Cl^-$$

(b) Plan: To calculate the percentage of Cl^- in the sample, we compare the number of grams of Cl^- in the sample, 7.17×10^{-2} g, with the original mass of the sample, 10.0 g.

Solve:

$$\%Cl^- = \frac{7.17 \times 10^{-3} \text{ g}}{10.0 \text{ g}} \times 100\% = 0.717\% \ Cl^-$$

Comment: Chloride ion is one of the most common ions in water and sewage. Ocean water contains 1.92% Cl^-. Whether water containing Cl^- tastes salty depends on the other ions present. If the only accompanying ions are Na^+, a salty taste may be detected with as little as 0.03% Cl^-.

PRACTICE EXERCISE

A sample of an iron ore is dissolved in acid, and the iron is converted to Fe^{2+}. The sample is then titrated with 47.20 mL of 0.02240 *M* MnO_4^- solution. The oxidation-reduction reaction that occurs during titration is as follows: $MnO_4^-(aq) + 5Fe^{2+}(aq) + 8H^+(aq) \longrightarrow Mn^{2+}(aq) + 5Fe^{3+}(aq) + 4H_2O(l)$ **(a)** How many moles of MnO_4^- were added to the solution? **(b)** How many moles of Fe^{2+} were in the sample? **(c)** How many grams of iron were in the sample? **(d)** If the sample had a mass of 0.8890 g, what is the percentage of iron in the sample?

Answers: **(a)** 1.057×10^{-3} mol MnO_4^-; **(b)** 5.286×10^{-3} mol Fe^{2+}; **(c)** 0.2952 g; **(d)** 33.21%

SAMPLE EXERCISE 4.17

One commercial method used to peel potatoes is to soak them in a solution of NaOH for a short time, remove them from the NaOH, and spray off the peel. The concentration of NaOH is normally in the range of 3 to 6 *M*. The NaOH is analyzed periodically. In one such analysis, 45.7 mL of 0.500 *M* H_2SO_4 is required to neutralize a 20.0-mL sample of NaOH solution. What is the concentration of the NaOH solution?

Solution

Analyze: We are given the volume (45.7 mL) and molarity (0.500 *M*) of an H_2SO_4 solution that reacts completely with a 20.0-mL sample of NaOH. We are asked to calculate the molarity of the NaOH solution.

Plan: We can use the volume and molarity of the H_2SO_4 to calculate the number of moles of this substance. Then, we can use this quantity and the balanced equation for the reaction to calculate the number of moles of NaOH. Finally, we can use the moles of NaOH and the volume of this solution to calculate molarity.

Solve: The number of moles of H_2SO_4 is given by the product of the volume and molarity of this solution:

$$\text{Moles } H_2SO_4 = (45.7 \text{ mL soln})\left(\frac{1 \text{ L soln}}{1000 \text{ mL soln}}\right)\left(0.500 \frac{\text{mol } H_2SO_4}{\text{L soln}}\right) = 2.28 \times 10^{-2} \text{ mol } H_2SO_4$$

Acids react with metal hydroxides to form water and a salt. Thus, the balanced equation for the neutralization reaction is

$$H_2SO_4(aq) + 2NaOH(aq) \longrightarrow 2H_2O(l) + Na_2SO_4(aq)$$

According to the balanced equation, 1 mol H_2SO_4 $\simeq$ 2 mol NaOH. Therefore,

$$\text{Moles NaOH} = (2.28 \times 10^{-2} \text{ mol } H_2SO_4)\left(\frac{2 \text{ mol NaOH}}{1 \text{ mol } H_2SO_4}\right) = 4.56 \times 10^{-2} \text{ mol NaOH}$$

Knowing the number of moles of NaOH present in 20.0 mL of solution allows us to calculate the molarity of this solution:

$$\text{Molarity NaOH} = \frac{\text{mol NaOH}}{\text{L soln}} = \left(\frac{4.56 \times 10^{-2} \text{ mol NaOH}}{20.0 \text{ mL soln}}\right)\left(\frac{1000 \text{ mL soln}}{1 \text{ L soln}}\right) = 2.28 \frac{\text{mol NaOH}}{\text{L soln}} = 2.28\ M$$

PRACTICE EXERCISE

What is the molarity of an NaOH solution if 48.0 mL is needed to neutralize 35.0 mL of 0.144 *M* H_2SO_4?
Answer: 0.210 *M*

SAMPLE INTEGRATIVE EXERCISE 4: Putting Concepts Together

Note: Integrative exercises require skills from earlier chapters as well as ones from the present chapter.

A sample of 70.5 mg of potassium phosphate is added to 15.0 mL of 0.050 *M* silver nitrate, resulting in the formation of a precipitate. **(a)** Write the molecular equation for the reaction. **(b)** What is the limiting reactant in the reaction? **(c)** Calculate the theoretical yield, in grams, of the precipitate that forms.

Solution (a) Potassium phosphate and silver nitrate are both ionic compounds. Potassium phosphate contains K^+ and PO_4^{3-} ions, so its chemical formula is K_3PO_4. Silver nitrate contains Ag^+ and NO_3^- ions, so its chemical formula is $AgNO_3$. Because both reactants are strong electrolytes, the solution contains K^+, PO_4^{3-}, Ag^+, and NO_3^- ions before the reaction occurs. According to the solubility guidelines in Table 4.1, Ag^+ and PO_4^{3-} form an insoluble compound, so Ag_3PO_4 will precipitate from the solution. In contrast, K^+ and NO_3^- will remain in solution because KNO_3 is water soluble. Thus, the balanced molecular equation for the reaction is

$$K_3PO_4(aq) + 3AgNO_3(aq) \longrightarrow Ag_3PO_4(s) + 3KNO_3(aq)$$

(b) To determine the limiting reactant, we must examine the number of moles of each reactant. (Section 3.7) The number of moles of K_3PO_4 is calculated from the mass of the sample using the molar mass as a conversion factor. (Section 3.4) The molar mass of K_3PO_4 is 3(39.1) + 31.0 + 4(16.0) = 212.3 g/mol. Converting milligrams to grams and then to moles, we have

$$(70.5 \text{ mg } K_3PO_4)\left(\frac{10^{-3} \text{ g } K_3PO_4}{1 \text{ mg } K_3PO_4}\right)\left(\frac{1 \text{ mol } K_3PO_4}{212.3 \text{ g } K_3PO_4}\right) = 3.32 \times 10^{-4} \text{ mol } K_3PO_4$$

We determine the number of moles of $AgNO_3$ from the volume and molarity of the solution. (Section 4.5) Converting milliliters to liters and then to moles, we have

$$(15.0 \text{ mL})\left(\frac{10^{-3} \text{ L}}{1 \text{ mL}}\right)\left(\frac{0.050 \text{ mol } AgNO_3}{\text{L}}\right) = 7.5 \times 10^{-4} \text{ mol } AgNO_3$$

Comparing the amounts of the two reactants, we find that there are $(7.5 \times 10^{-4})/(3.32 \times 10^{-4}) = 2.3$ times as many moles of $AgNO_3$ as there are moles of K_3PO_4. According to the balanced equation, however, 1 mol K_3PO_4 requires 3 mol of $AgNO_3$. Thus, there is insufficient $AgNO_3$ to consume the K_3PO_4, and $AgNO_3$ is the limiting reactant.

(c) The precipitate is Ag_3PO_4, whose molar mass is $3(107.9) + 31.0 + 4(16.0) = 418.7$ g/mol. To calculate the number of grams of Ag_3PO_4 that could be produced in this reaction (the theoretical yield), we use the number of moles of the limiting reactant, converting mol $AgNO_3 \Rightarrow$ mol $Ag_3PO_4 \Rightarrow$ g Ag_3PO_4. We use the coefficients in the balanced equation to convert moles of $AgNO_3$ to moles Ag_3PO_4, and we use the molar mass of Ag_3PO_4 to convert the number of moles of this substance to grams.

$$(7.5 \times 10^{-4}\ \text{mol AgNO}_3)\left(\frac{1\ \text{mol Ag}_3\text{PO}_4}{3\ \text{mol AgNO}_3}\right)\left(\frac{418.7\ \text{g Ag}_3\text{PO}_4}{1\ \text{mol Ag}_3\text{PO}_4}\right) = 0.10\ \text{g Ag}_3\text{PO}_4$$

The answer has only two significant figures because the quantity of $AgNO_3$ is given to only two significant figures.

Summary and Key Terms

Introduction and Section 4.1 Solutions in which water is the dissolving medium are called **aqueous solutions**. The component of the solution that is in the greater quantity is the **solvent**. The other components are **solutes**.

Any substance whose aqueous solution contains ions is called an **electrolyte**. Any substance that forms a solution containing no ions is a **nonelectrolyte**. Those electrolytes that are present in solution entirely as ions are **strong electrolytes**, whereas those that are present partly as ions and partly as molecules are **weak electrolytes**. Ionic compounds dissociate into ions when they dissolve, and they are strong electrolytes. Most molecular compounds are nonelectrolytes, although some are weak electrolytes and a few are strong electrolytes. When representing the ionization of a weak electrolyte in solution, a double arrow is used, indicating that the forward and reverse reactions can achieve a chemical balance called a **chemical equilibrium**.

Section 4.2 **Precipitation reactions** are those in which an insoluble product, called a **precipitate**, forms. Solubility guidelines help determine whether or not an ionic compound will be soluble in water. (The **solubility** of a substance is the amount that dissolves in a given quantity of solvent.) Reactions such as precipitation reactions, in which cations and anions appear to exchange partners, are called **exchange reactions**, or **metathesis reactions**.

Chemical equations can be written to show whether dissolved substances are present in solution predominantly as ions or molecules. When the complete chemical formulas of all reactants and products are used, the equation is called a **molecular equation**. A **complete ionic equation** shows all dissolved strong electrolytes as their component ions. In a **net ionic equation,** those ions that go through the reaction unchanged **(spectator ions)** are omitted.

Section 4.3 Acids and bases are important electrolytes. **Acids** are proton donors; they increase the concentration of $H^+(aq)$ in aqueous solutions to which they are added. **Bases** are proton acceptors; they increase the concentration of $OH^-(aq)$ in aqueous solutions. Those acids and bases that are strong electrolytes are called **strong acids** and **strong bases,** respectively. Those that are weak electrolytes are **weak acids** and **weak bases**. When solutions of acids and bases are mixed, a **neutralization reaction** results. The neutralization reaction between an acid and a metal hydroxide produces water and a **salt**. Gases can also be formed as a result of acid-base reactions. The reaction of a sulfide with an acid forms $H_2S(g)$; the reaction between a carbonate and an acid forms $CO_2(g)$.

Section 4.4 **Oxidation** is the loss of electrons by a substance, whereas **reduction** is the gain of electrons by a substance. **Oxidation numbers** help us keep track of electrons during chemical reactions and are assigned to atoms by using specific rules. The oxidation of an element results in an increase in its oxidation number, whereas reduction is accompanied by a decrease in oxidation number. Oxidation is always accompanied by reduction, giving **oxidation-reduction**, or redox, **reactions**.

Many metals are oxidized by O_2, acids, and salts. The redox reactions between metals and acids and between metals and salts are called **displacement reactions**. The products of these displacement reactions are always an element (H_2 or a metal) and a salt. Comparing such reactions allows us to rank metals according to their ease of oxidation. A list of metals arranged in order of decreasing ease of oxidation is called an **activity series**. Any metal on the list can be oxidized by ions of metals (or H^+) below it in the series.

Section 4.5 The composition of a solution expresses the relative quantities of solvent and solutes that it contains. One of the common ways to express the **concentration** of a solute in a solution is in terms of molarity. The **molarity** of a solution is the number of moles of solute per liter of solution. Molarity makes it possible to interconvert solution volume and number of moles of solute. Solutions of known molarity can be formed either by weighing out the solute and diluting it to a known volume or by the **dilution** of a more concentrated solution of known concentration (a stock solution). Adding solvent to the solution (the process of dilution) decreases the concentration of the solute without changing the number of moles of solute in the solution ($M_{conc} \times V_{conc} = M_{dil} \times V_{dil}$).

Section 4.6 In the process called **titration**, we combine a solution of known concentration (a **standard solution**) with a solution of unknown concentration in order to determine the unknown concentration or the quantity of solute in the unknown. The point in the titration at which stoichiometrically equivalent quantities of reactants are brought together is called the **equivalence point**. An **indicator** can be used to show the end point of the titration, which coincides closely with the equivalence point.

Exercises

Electrolytes

4.1 Although pure water is a poor conductor of electricity, we are cautioned not to operate electrical appliances around water. Why?

4.2 When asked what causes electrolyte solutions to conduct electricity, a student responds that it is due to the movement of electrons through the solution. Is the student correct? If not, what is the correct response?

4.3 When methanol, CH_3OH, is dissolved in water, a nonconducting solution results. When acetic acid, $HC_2H_3O_2$, dissolves in water, the solution is weakly conducting and acidic in nature. Describe what happens upon dissolution in the two cases, and account for the different results.

4.4 We have learned in this chapter that many ionic solids dissolve in water as strong electrolytes, that is, as separated ions in solution. What properties of water facilitate this process?

4.5 Specify how each of the following strong electrolytes ionizes or dissociates into ions upon dissolving in water: **(a)** $ZnCl_2$; **(b)** HNO_3; **(c)** K_2SO_4; **(d)** $Ca(OH)_2$.

4.6 Specify how each of the following strong electrolytes ionizes or dissociates into ions upon dissolving in water: **(a)** MgI_2; **(b)** $Al(NO_3)_3$; **(c)** $HClO_4$; **(d)** $(NH_4)_2SO_4$.

4.7 Aqueous solutions of three different substances, AX, AY, and AZ are represented by the three diagrams below. Identify each substance as a strong electrolyte, weak electrolyte, or nonelectrolyte.

AX

(a)

AY

(b)

AZ

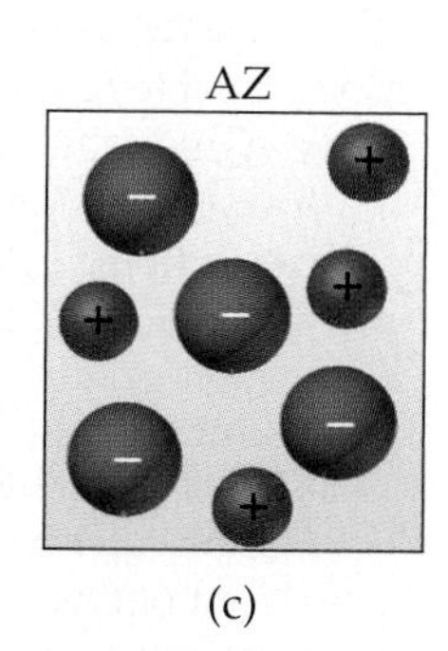

(c)

4.8 The two diagrams represent aqueous solutions of two different substances, AX and BY. Are these substances strong electroytes, weak electrolytes, or nonelectrolytes? Which do you expect to be the better conductor of electricity? Explain.

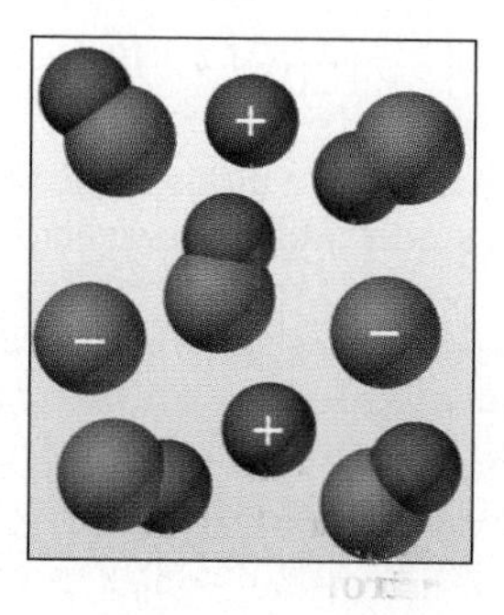

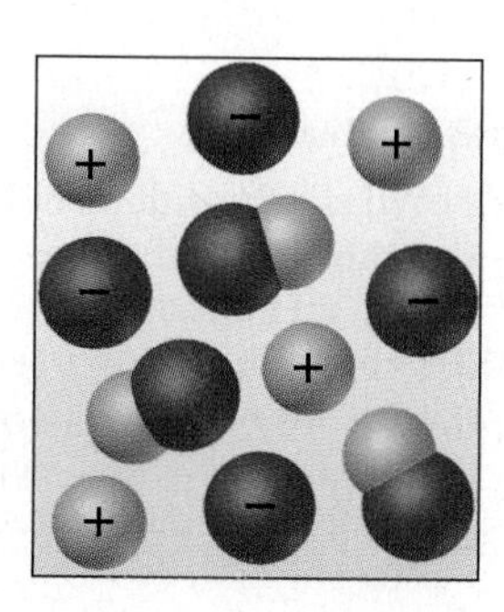

4.9 Formic acid, $HCHO_2$, is a weak electrolyte. What solute particles are present in an aqueous solution of this compound? Write the chemical equation for the ionization of $HCHO_2$.

4.10 Acetone, CH_3COCH_3, is a nonelectrolyte; hypochlorous acid, $HClO$, is a weak electrolyte; and ammonium chloride, NH_4Cl, is a strong electrolyte. **(a)** What are the solute particles present in aqueous solutions of each compound? **(b)** If 0.1 mol of each compound is dissolved in solution, which one contains 0.2 mol of solute particles, which contains 0.1 mol of solute particles, and which contains somewhere between 0.1 and 0.2 mol of solute particles?

Precipitation Reactions and Net Ionic Equations

4.11 Using solubility guidelines, predict whether each of the following compounds is soluble or insoluble in water: **(a)** $NiCl_2$; **(b)** Ag_2S; **(c)** Cs_3PO_4; **(d)** $SrCO_3$; **(e)** $(NH_4)_2SO_4$.

4.12 Predict whether each of the following compounds is soluble in water: **(a)** $Ni(OH)_2$; **(b)** $PbSO_4$; **(c)** $Ba(NO_3)_2$; **(d)** $AlPO_4$; **(e)** $AgC_2H_3O_2$.

4.13 Will precipitation occur when the following solutions are mixed? If so, write a balanced chemical equation for the reaction. **(a)** Na_2CO_3 and $AgNO_3$; **(b)** $NaNO_3$ and $NiSO_4$; **(c)** $FeSO_4$ and $Pb(NO_3)_2$.

4.14 Identify the precipitate (if any) that forms when the following solutions are mixed, and write a balanced equation for each reaction. **(a)** $Sn(NO_3)_2$ and $NaOH$; **(b)** $NaOH$ and K_2SO_4; **(c)** Na_2S and $Cu(C_2H_3O_2)_2$.

4.15 Write the balanced complete ionic equations and net ionic equations for the reactions that occur when each of the following solutions is mixed.
(a) $Na_2CO_3(aq)$ and $MgSO_4(aq)$

(b) $Pb(NO_3)_2(aq)$ and $Na_2S(aq)$
(c) $(NH_4)_3PO_4(aq)$ and $CaCl_2(aq)$

4.16 Write balanced net ionic equations for the reactions that occur in each of the following cases. Identify the spectator ion or ions in each reaction.
(a) $Cr_2(SO_4)_3(aq) + (NH_4)_2CO_3(aq) \longrightarrow$
(b) $AgNO_3(aq) + K_2SO_4(aq) \longrightarrow$
(c) $Pb(NO_3)_2(aq) + KOH(aq) \longrightarrow$

4.17 Separate samples of a solution of an unknown salt are treated with dilute solutions of HBr, H_2SO_4, and NaOH. A precipitate forms only with H_2SO_4. Which of the following cations could the solution contain: K^+; Pb^{2+}; Ba^{2+}?

4.18 Separate samples of a solution of an unknown ionic compound are treated with dilute $AgNO_3$, $Pb(NO_3)_2$, and $BaCl_2$. Precipitates form in all three cases. Which of the following could be the anion of the unknown salt: Br^-; CO_3^{2-}; NO_3^-?

4.19 The labels have fallen off two bottles, one containing $Mg(NO_3)_2$ and the other containing $Pb(NO_3)_2$. You have a bottle of dilute H_2SO_4. How could you use it to test a portion of each solution to identify which solution is which?

4.20 You know that an unlabeled bottle contains one of the following: $AgNO_3$, $CaCl_2$, or $Al_2(SO_4)_3$. A friend suggests that you test a portion of the bottle with $Ba(NO_3)_2$ and then with NaCl. What behavior would you expect when each of these compounds is added to the unlabeled bottle?

Acid-Base Reactions

4.21 What is the difference between: **(a)** a monoprotic acid and a diprotic acid; **(b)** a weak acid and a strong acid; **(c)** an acid and a base?

4.22 Explain the following observations: **(a)** NH_3 contains no OH^- ions, and yet its aqueous solutions are basic; **(b)** HF is called a weak acid, and yet it is very reactive; **(c)** although sulfuric acid is a strong electrolyte, an aqueous solution of H_2SO_4 contains more HSO_4^- ions than SO_4^{2-} ions. Explain.

4.23 Classify each of the following as a strong or weak acid or base: **(a)** $HClO_4$; **(b)** $HClO_2$; **(c)** NH_3; **(d)** $Ba(OH)_2$.

4.24 Classify each of the following as a strong or weak acid or base: **(a)** CsOH **(b)** H_3PO_4; **(c)** $HC_7H_5O_2$ **(d)** H_2SO_4.

4.25 Label each of the following substances as an acid, base, salt, or none of the above. Indicate whether the substance exists in aqueous solution entirely in molecular form, entirely as ions, or as a mixture of molecules and ions. **(a)** HF; **(b)** acetonitrile, CH_3CN; **(c)** $NaClO_4$; **(d)** $Ba(OH)_2$.

4.26 An aqueous solution of an unknown solute is tested with litmus paper and found to be acidic. The solution is weakly conducting compared with a solution of NaCl of the same concentration. Which of the following substances could the unknown be: KOH, NH_3, HNO_3, $KClO_2$, H_3PO_3, CH_3COCH_3 (acetone)?

4.27 Classify each of the following substances as a nonelectrolyte, weak electrolyte, or strong electrolyte in water: **(a)** H_2SO_3; **(b)** C_2H_5OH (ethanol); **(c)** NH_3; **(d)** $KClO_3$; **(e)** $Cu(NO_3)_2$.

4.28 Classify each of the following aqueous solutions as a nonelectrolyte, weak electrolyte, or strong electrolyte: **(a)** HBrO; **(b)** HNO_3; **(c)** KOH; **(d)** CH_3COCH_3 (acetone); **(e)** $CoSO_4$; **(f)** $C_{12}H_{22}O_{11}$ (sucrose).

4.29 Complete and balance the following molecular equations, and then write the net ionic equation for each:
(a) $HBr(aq) + Ca(OH)_2(aq) \longrightarrow$
(b) $Cu(OH)_2(s) + HClO_4(aq) \longrightarrow$
(c) $Al(OH)_3(s) + HNO_3(aq) \longrightarrow$

4.30 Write the balanced molecular and net ionic equations for each of the following neutralization reactions:
(a) Aqueous acetic acid is neutralized by aqueous potassium hydroxide.
(b) Solid chromium(III) hydroxide reacts with nitric acid.
(c) Aqueous hypochlorous acid and aqueous calcium hydroxide react.

4.31 Write balanced molecular and net ionic equations for the following reactions, and identify the gas formed in each: **(a)** solid cadmium sulfide reacts with an aqueous solution of sulfuric acid; **(b)** solid magnesium carbonate reacts with an aqueous solution of perchloric acid.

4.32 Write a balanced molecular equation and a net ionic equation for the reaction that occurs when **(a)** solid $CaCO_3$ reacts with an aqueous solution of nitric acid; **(b)** solid iron(II) sulfide reacts with an aqueous solution of hydrobromic acid.

4.33 Because the oxide ion is basic, metal oxides react readily with acids. **(a)** Write the net ionic equation for the following reaction: $FeO(s) + 2HClO_4(aq) \longrightarrow Fe(ClO_4)_2(aq) + H_2O(l)$. **(b)** Based on the example in part (a), write the net ionic equation for the reaction that occurs between NiO(*s*) and an aqueous solution of nitric acid.

4.34 As K_2O dissolves in water, the oxide ion reacts with water molecules to form hydroxide ions. Write the molecular and net ionic equations for this reaction. Based on the definitions of acid and base, what ion is the base in this reaction? What is the acid? What is the spectator ion in the reaction?

Oxidation-Reduction Reactions

4.35 Define oxidation and reduction in terms of **(a)** electron transfer and **(b)** oxidation numbers.

4.36 Can oxidation occur without accompanying reduction? Explain.

4.37 Where, in general, do the most easily oxidized metals occur in the periodic table? Where do the least easily oxidized metals occur in the periodic table?

4.38 Why are platinum and gold called noble metals? Why are the alkali metals and alkaline earth metals called active metals?

4.39 Determine the oxidation number for the indicated element in each of the following substances: **(a)** S in SO_3; **(b)** C in $COCl_2$; **(c)** Mn in MnO_4^-; **(d)** Br in HBrO; **(e)** As in As_4; **(f)** O in K_2O_2.

4.40 Determine the oxidation number for the indicated element in each of the following compounds: **(a)** Ti in TiO_2; **(b)** Sn in $SnCl_4$; **(c)** C in $C_2O_4^{2-}$; **(d)** N in $(NH_4)_2SO_4$; **(e)** N in HNO_2; **(f)** Cr in $Cr_2O_7^{2-}$.

4.41 Which element is oxidized, and which is reduced in the following reactions?
(a) $Ni(s) + Cl_2(g) \longrightarrow NiCl_2(s)$
(b) $3Fe(NO_3)_2(aq) + 2Al(s) \longrightarrow 3Fe(s) + 2Al(NO_3)_3(aq)$
(c) $Cl_2(aq) + 2NaI(aq) \longrightarrow I_2(aq) + 2NaCl(aq)$
(d) $PbS(s) + 4H_2O_2(aq) \longrightarrow PbSO_4(s) + 4H_2O(l)$

4.42 Which of the following are redox reactions? For those that are, indicate which element is oxidized and which is reduced. For those that are not, indicate whether they are precipitation or acid-base reactions.
(a) $Cu(OH)_2(s) + 2HNO_3(aq) \longrightarrow Cu(NO_3)_2(aq) + 2H_2O(l)$
(b) $Fe_2O_3(s) + 3CO(g) \longrightarrow 2Fe(s) + 3CO_2(g)$
(c) $Sr(NO_3)_2(aq) + H_2SO_4(aq) \longrightarrow SrSO_4(s) + 2HNO_3(aq)$
(d) $4Zn(s) + 10H^+(aq) + 2NO_3^-(aq) \longrightarrow 4Zn^{2+}(aq) + N_2O(g) + 5H_2O(l)$

4.43 Write balanced molecular and net ionic equations for the reactions of **(a)** manganese with sulfuric acid; **(b)** chromium with hydrobromic acid; **(c)** tin with hydrochloric acid; **(d)** aluminum with formic acid, $HCHO_2$.

4.44 Write balanced molecular and net ionic equations for the reactions of **(a)** hydrochloric acid with nickel; **(b)** sulfuric acid with iron; **(c)** hydrobromic acid with magnesium; **(d)** acetic acid, $HC_2H_3O_2$, with zinc.

4.45 Based on the activity series (Table 4.5), what is the outcome of each of the following reactions?
(a) $Al(s) + NiCl_2(aq) \longrightarrow$
(b) $Ag(s) + Pb(NO_3)_2(aq) \longrightarrow$
(c) $Cr(s) + NiSO_4(aq) \longrightarrow$
(d) $Mn(s) + HBr(aq) \longrightarrow$
(e) $H_2(g) + CuCl_2(aq) \longrightarrow$

4.46 Using the activity series (Table 4.5), write balanced chemical equations for the following reactions. If no reaction occurs, simply write NR. **(a)** Iron metal is added to a solution of copper(II) nitrate; **(b)** zinc metal is added to a solution of magnesium sulfate; **(c)** hydrobromic acid is added to tin metal; **(d)** hydrogen gas is bubbled through an aqueous solution of nickel(II) chloride; **(e)** aluminum metal is added to a solution of cobalt(II) sulfate.

4.47 The metal cadmium tends to form Cd^{2+} ions. The following observations are made: (i) When a strip of zinc metal is placed in $CdCl_2(aq)$, cadmium metal is deposited on the strip. (ii) When a strip of cadmium metal is placed in $Ni(NO_3)_2(aq)$, nickel metal is deposited on the strip. **(a)** Write net ionic equations to explain each of the observations made above. **(b)** What can you conclude about the position of cadmium in the activity series? **(c)** What experiments would you need to perform to locate more precisely the position of cadmium in the activity series?

4.48 **(a)** Use the following reactions to prepare an activity series for the halogens: $Br_2(aq) + 2NaI(aq) \longrightarrow 2NaBr(aq) + I_2(aq)$; $Cl_2(aq) + 2NaBr(aq) \longrightarrow 2NaCl(aq) + Br_2(aq)$. **(b)** Relate the positions of the halogens in the periodic table with their locations in this activity series. **(c)** Predict whether a reaction occurs when the following reagents are mixed: $Cl_2(aq)$ and $KI(aq)$; $Br_2(aq)$ and $LiCl(aq)$.

Solution Composition; Molarity

4.49 **(a)** Is the concentration of a solution an intensive or an extensive property? **(b)** What is the difference between 0.50 mol HCl and 0.50 *M* HCl?

4.50 **(a)** Suppose you prepare 500 mL of a 0.10 *M* solution of some salt and then spill some of it. What happens to the concentration of the solution left in the container? **(b)** A certain volume of a 0.50 *M* solution contains 4.5 g of a salt. What mass of the salt is present in the same volume of a 2.50 *M* solution?

4.51 **(a)** Calculate the molarity of a solution that contains 0.0345 mol NH_4Cl in exactly 400 mL of solution. **(b)** How many moles of HNO_3 are present in 35.0 mL of a 2.20 *M* solution of nitric acid? **(c)** How many milliliters of 1.50 *M* KOH solution are needed to provide 0.125 mol of KOH?

4.52 **(a)** Calculate the molarity of a solution made by dissolving 0.145 mol Na_2SO_4 in enough water to form exactly 750 mL of solution. **(b)** How many moles of $KMnO_4$ are present in 125 mL of a 0.0850 *M* solution? **(c)** How many milliliters of 11.6 *M* HCl solution are needed to obtain 0.255 mol of HCl?

4.53 Calculate **(a)** the number of grams of solute in 0.250 L of 0.150 *M* KBr; **(b)** the molar concentration of a solution containing 4.75 g of $Ca(NO_3)_2$ in 0.200 L; **(c)** the volume of 1.50 *M* Na_3PO_4 in milliliters that contains 5.00 g of solute.

4.54 **(a)** How many grams of solute are present in 50.0 mL of 0.850 *M* $K_2Cr_2O_7$? **(b)** If 2.50 g of $(NH_4)_2SO_4$ is dissolved in enough water to form 250 mL of solution, what is the molarity of the solution? **(c)** How many milliliters of 0.387 *M* $CuSO_4$ contain 1.00 g of solute?

4.55 **(a)** Which will have the highest concentration of potassium ion: 0.20 *M* KCl, 0.15 *M* K_2CrO_4, or 0.080 *M* K_3PO_4? **(b)** Which will contain the greater number of moles of potassium ion: 30.0 mL of 0.15 *M* K_2CrO_4 or 25.0 mL of 0.080 *M* K_3PO_4?

4.56 **(a)** Without doing detailed calculations, rank the following solutions in order of increasing concentration of Cl^- ions: 0.10 *M* $CaCl_2$, 0.15 *M* KCl, a solution formed by dissolving 0.10 mol of NaCl in enough water to form 250 mL of solution. **(b)** Which will contain the greater number of moles of chloride ion: 40.0 mL of 0.35 *M* NaCl or 25.0 mL of 0.25 *M* $CaCl_2$?

4.57 Indicate the concentration of each ion or molecule present in the following solutions: **(a)** 0.14 *M* NaOH; **(b)** 0.25 *M* $CaBr_2$; **(c)** 0.25 *M* CH_3OH; **(d)** a mixture of 50.0 mL of 0.10 *M* $KClO_3$ and 25.0 mL of 0.20 *M* Na_2SO_4. Assume the volumes are additive.

4.58 Indicate the concentration of each ion present in the solution formed by mixing: **(a)** 20 mL of 0.100 *M* HCl and 10.0 mL of 0.500 *M* HCl; **(b)** 15.0 mL of 0.300 *M* Na_2SO_4 and 10.0 mL of 0.200 *M* KCl; **(c)** 3.50 g of NaCl in 50.0 mL of 0.500 *M* $CaCl_2$ solution. (Assume that the volumes are additive.)

4.59 **(a)** You have a stock solution of 14.8 *M* NH_3. How many milliliters of this solution should you dilute to make 100.0 mL of 0.250 *M* NH_3? **(b)** If you take a 10.0-mL portion of the stock solution and dilute it to a total volume of 0.250 L, what will be the concentration of the final solution?

4.60 **(a)** How many milliliters of a stock solution of 12.0 *M* HNO_3 would you have to use to prepare 0.500 L of 0.500 *M* HNO_3? **(b)** If you dilute 25.0 mL of the stock solution to a final volume of 0.500 L, what will be the concentration of the diluted solution?

4.61 **(a)** Starting with solid sucrose, $C_{12}H_{22}O_{11}$, describe how you would prepare 125 mL of 0.150 *M* sucrose solution. **(b)** Describe how you would prepare 400.0 mL of 0.100 *M* $C_{12}H_{22}O_{11}$ starting with 2.00 L of 1.50 *M* $C_{12}H_{22}O_{11}$.

4.62 **(a)** How would you prepare 100.0 mL of 0.200 *M* $AgNO_3$ solution starting with pure solute? **(b)** An experiment calls for you to use 250 mL of 1.0 *M* HNO_3 solution. All you have available is a bottle of 6.0 *M* HNO_3. How would you prepare the desired solution?

[4.63] Pure acetic acid, known as glacial acetic acid, is a liquid with a density of 1.049 g/mL at 25°C. Calculate the molarity of a solution of acetic acid made by dissolving 20.00 mL of glacial acetic acid at 25°C in enough water to make 250.0 mL of solution.

[4.64] Glycerol, $C_3H_8O_3$, is a substance used extensively in the manufacture of cosmetics, foodstuffs, antifreeze, and plastics. Glycerol is a water-soluble liquid with a density of 1.2656 g/mL at 15°C. Calculate the molarity of a solution of glycerol made by dissolving 50.000 mL glycerol at 15°C in enough water to make 250.00 mL of solution.

Solution Stoichiometry; Titrations

4.65 What mass of NaCl is needed to precipitate all the silver ions from 20.0 mL of 0.100 *M* $AgNO_3$ solution?

4.66 What mass of NaOH is needed to precipitate all the Fe^{2+} ions from 25.0 mL of 0.500 *M* $Fe(NO_3)_2$ solution?

4.67 **(a)** What volume of 0.115 *M* $HClO_4$ solution is needed to neutralize 50.00 mL of 0.0875 *M* NaOH? **(b)** What volume of 0.128 *M* HCl is needed to neutralize 2.87 g of $Mg(OH)_2$? **(c)** If 25.8 mL of $AgNO_3$ is needed to precipitate all the Cl^- ions in a 785-mg sample of KCl (forming AgCl), what is the molarity of the $AgNO_3$ solution? **(d)** If 45.3 mL of 0.108 *M* HCl solution is needed to neutralize a solution of KOH, how many grams of KOH must be present in the solution?

4.68 **(a)** How many milliliters of 0.120 *M* HCl are needed to completely neutralize 50.0 mL of 0.101 *M* $Ba(OH)_2$ solution? **(b)** How many milliliters of 0.125 *M* H_2SO_4 are needed to neutralize 0.200 g of NaOH? **(c)** If 55.8 mL of $BaCl_2$ solution is needed to precipitate all the sulfate ion in a 752-mg sample of Na_2SO_4, what is the molarity of the solution? **(d)** If 42.7 mL of 0.208 *M* HCl solution is needed to neutralize a solution of $Ca(OH)_2$, how many grams of $Ca(OH)_2$ must be in the solution?

4.69 Some sulfuric acid is spilled on a lab bench. It can be neutralized by sprinkling sodium bicarbonate on it and then mopping up the resultant solution. The sodium bicarbonate reacts with sulfuric acid as follows:

$$2NaHCO_3(s) + H_2SO_4(aq) \longrightarrow Na_2SO_4(aq) + 2H_2O(l) + 2CO_2(g)$$

Sodium bicarbonate is added until the fizzing due to the formation of $CO_2(g)$ stops. If 27 mL of 6.0 *M* H_2SO_4 was spilled, what is the minimum mass of $NaHCO_3$ that must be added to the spill to neutralize the acid?

4.70 The distinctive odor of vinegar is due to acetic acid, $HC_2H_3O_2$. Acetic acid reacts with sodium hydroxide in the following fashion:

$$HC_2H_3O_2(aq) + NaOH(aq) \longrightarrow H_2O(l) + NaC_2H_3O_2(aq)$$

If 2.50 mL of vinegar needs 35.5 mL of 0.102 *M* NaOH to reach the equivalence point in a titration, how many grams of acetic acid are in a 1.00-qt sample of this vinegar?

4.71 A sample of solid $Ca(OH)_2$ is stirred in water at 30°C until the solution contains as much dissolved $Ca(OH)_2$ as it can hold. A 100-mL sample of this solution is withdrawn and titrated with 5.00×10^{-2} *M* HBr. It requires 48.8 mL of the acid solution for neutralization. What is the molarity of the $Ca(OH)_2$ solution? What is the solubility of $Ca(OH)_2$ in water, at 30°C, in grams of $Ca(OH)_2$ per 100 mL of solution?

4.72 In the laboratory 7.52 g of $Sr(NO_3)_2$ is dissolved in enough water to form 0.750 L. A 0.100-L sample is withdrawn from this stock solution and titrated with a 0.0425 *M* solution of Na_2CrO_4. What volume of Na_2CrO_4 solution is needed to precipitate all the $Sr^{2+}(aq)$ as $SrCrO_4$?

4.73 A solution of 100.0 mL of 0.200 *M* KOH is mixed with a solution of 200.0 mL of 0.150 *M* $NiSO_4$. **(a)** Write the balanced chemical equation for the reaction that occurs. **(b)** What precipitate forms? **(c)** What is the limiting reactant? **(d)** How many grams of this precipitate form? **(e)** What is the concentration of each ion that remains in solution?

4.74 A solution is made by mixing 12.0 g of NaOH and 75.0 mL of 0.200 *M* HNO_3. **(a)** Write a balanced equation for the reaction that occurs between the solutes. **(b)** Calculate the concentration of each ion remaining in solution. **(c)** Is the resultant solution acidic or basic?

[4.75] A 0.5895-g sample of impure magnesium hydroxide is dissolved in 100.0 mL of 0.2050 *M* HCl solution. The excess acid then needed 19.85 mL of 0.1020 *M* NaOH for neutralization. Calculate the percent by mass of magnesium hydroxide in the sample, assuming that it is the only substance reacting with the HCl solution.

[4.76] A 1.452-g sample of limestone rock is pulverized and then treated with 25.00 mL of 1.035 *M* HCl solution. The excess acid then required 15.25 mL of 0.1010 *M* NaOH for neutralization. Calculate the percent by mass of calcium carbonate in the rock, assuming that it is the only substance reacting with the HCl solution.

Additional Exercises

4.77 The accompanying photo shows the reaction between a solution of $Cd(NO_3)_2$ and one of Na_2S. What is the identity of the precipitate? What ions remain in solution? Write the net ionic equation for the reaction.

4.78 Suppose you have a solution that might contain any or all of the following cations: Ni^{2+}, Ag^{+}, Sr^{2+}, and Mn^{2+}. Addition of HCl solution causes a precipitate to form. After filtering off the precipitate, H_2SO_4 solution is added to the resultant solution and another precipitate forms. This is filtered off, and a solution of NaOH is added to the resulting solution. No precipitate is observed. Which ions are present in each of the precipitates? Which of the four ions listed above must be absent from the original solution?

4.79 You choose to investigate some of the solubility guidelines for two ions not listed in Table 4.1, the chromate ion (CrO_4^{2-}) and the oxalate ion ($C_2O_4^{2-}$). You are given solutions (A, B, C, D) of four water-soluble salts:

Solution	Solute	Color of Solution
A	Na_2CrO_4	Yellow
B	$(NH_4)_2C_2O_4$	Colorless
C	$AgNO_3$	Colorless
D	$CaCl_2$	Colorless

When these solutions are mixed, the following observations are made:

Expt. Number	Solutions Mixed	Result
1	A + B	No precipitate, yellow solution
2	A + C	Red precipitate forms
3	A + D	No precipitate, yellow solution
4	B + C	White precipitate forms
5	B + D	White precipitate forms
6	C + D	White precipitate forms

(a) Write a net ionic equation for the reaction that occurs in each of the experiments. **(b)** Identify the precipitate formed, if any, in each of the experiments. **(c)** Based on these limited observations, which ion tends to form the more soluble salts, chromate or oxalate?

4.80 Antacids are often used to relieve pain and promote healing in the treatment of mild ulcers. Write balanced net ionic equations for the reactions between the HCl(*aq*) in the stomach and each of the following substances used in various antacids: **(a)** $Al(OH)_3(s)$; **(b)** $Mg(OH)_2(s)$; **(c)** $MgCO_3(s)$; **(d)** $NaAl(CO_3)(OH)_2(s)$; **(e)** $CaCO_3(s)$.

[4.81] Salts of the sulfite ion, SO_3^{2-}, react with acids in a way similar to that of carbonates. **(a)** Predict the chemical formula, and name the weak acid that forms when the sulfite ion reacts with acids. **(b)** The acid formed in part (a) decomposes to form water and an insoluble gas. Predict the molecular formula, and name the gas formed. **(c)** Use a source book such as the *CRC Handbook of Chemistry and Physics* to confirm that the substance in part (b) is a gas under normal room-temperature conditions. **(d)** Write balanced net ionic equations of the reaction of HCl(aq) with (i) $Na_2SO_3(aq)$, (ii) $Ag_2SO_3(s)$, (iii) $KHSO_3(s)$, and (iv) $ZnSO_3(aq)$.

4.82 The commercial production of nitric acid involves the following chemical reactions:

$$4NH_3(g) + 5O_2(g) \longrightarrow 4NO(g) + 6H_2O(g)$$
$$2NO(g) + O_2(g) \longrightarrow 2NO_2(g)$$
$$3NO_2(g) + H_2O(l) \longrightarrow 2HNO_3(aq) + NO(g)$$

(a) Which of these reactions are redox reactions? **(b)** In each redox reaction, identify the element undergoing oxidation and the element undergoing reduction.

4.83 Use Table 4.5 to predict which of the following ions can be reduced to their metal forms by reacting with zinc: **(a)** $Na^{+}(aq)$; **(b)** $Pb^{2+}(aq)$; **(c)** $Mg^{2+}(aq)$; **(d)** $Fe^{2+}(aq)$; **(e)** $Cu^{2+}(aq)$; **(f)** $Al^{3+}(aq)$. Write the balanced net ionic equation for each reaction that occurs.

4.84 Titanium(IV) ion, Ti^{4+}, can be reduced to Ti^{3+} by the careful addition of zinc metal. **(a)** Write the net ionic equation for this process. **(b)** Would it be appropriate to use this reaction as a means of including titanium in the activity series of Table 4.5? Why or why not?

[4.85] Lanthanum metal forms cations with a charge of 3+. Consider the following observations about the chemistry of lanthanum: When lanthanum metal is exposed to air, a white solid (compound A) is formed that contains lanthanum and one other element. When lanthanum metal is added to water, gas bubbles are observed and a different white solid (compound B) is formed. Both A and B dissolve in hydrochloric acid to give a clear solution. When the solution from either A or B is evaporated, a soluble white solid (compound C) remains. If compound C is dissolved in water and sulfuric acid is added, a white precipitate (compound D) forms. **(a)** Propose identities for the substances A, B, C, and D. **(b)** Write net ionic equations for all the reactions described. **(c)** Based on the preceding observations, what can be said about the position of lanthanum in the activity series (Table 4.5)?

4.86 A 25.0-mL sample of 1.00 *M* KBr and a 75.0-mL sample of 0.800 *M* KBr are mixed. The solution is then heated to evaporate water until the total volume is 50.0 mL. What is the molarity of the KBr in the final solution?

4.87 Calculate the molarity of the solution produced by mixing **(a)** 50.0 mL of 0.200 *M* NaCl and 75.0 mL of 0.100 *M* NaCl; **(b)** 24.5 mL of 1.50 *M* NaOH and 25.5 mL of 0.750 *M* NaOH. (Assume that the volumes are additive.)

4.88 Using modern analytical techniques, it is possible to detect sodium ions in concentrations as low as 50 pg/mL. What is this detection limit expressed in: **(a)** molarity of Na^+; **(b)** Na^+ ions per cubic centimeter?

4.89 Hard water contains Ca^{2+}, Mg^{2+}, and Fe^{2+}, which interfere with the action of soap and leave an insoluble coating on the insides of containers and pipes when heated. Water softeners replace these ions with Na^+. If 1.0×10^3 L of hard water contains 0.010 *M* Ca^{2+} and 0.0050 *M* Mg^{2+}, how many moles of Na^+ are needed to replace these ions?

4.90 Tartaric acid, $H_2C_4H_4O_6$, has two acidic hydrogens. The acid is often present in wines and precipitates from solution as the wine ages. A solution containing an unknown concentration of the acid is titrated with NaOH. It requires 22.62 mL of 0.2000 *M* NaOH solution to titrate both acidic protons in 40.00 mL of the tartaric acid solution. Write a balanced net ionic equation for the neutralization reaction, and calculate the molarity of the tartaric acid solution.

4.91 The concentration of hydrogen peroxide in a solution is determined by titrating a 10.0-mL sample of the solution with permanganate ion.

$$2MnO_4^-(aq) + 5H_2O_2(aq) + 6H^+(aq) \longrightarrow 2Mn^{2+}(aq) + 5O_2(g) + 8H_2O(l)$$

If it takes 13.5 mL of 0.109 *M* MnO_4^- solution to reach the equivalence point, what is the molarity of the hydrogen peroxide solution?

[4.92] A solid sample of $Zn(OH)_2$ is added to 0.400 L of 0.500 *M* aqueous HBr. The solution that remains is still acidic. It is then titrated with 0.500 *M* NaOH solution, and it takes 98.5 mL of the NaOH solution to reach the equivalence point. What mass of $Zn(OH)_2$ was added to the HBr solution?

Integrative Exercises

4.93 Calculate the number of sodium ions in 1.00 mL of a 0.0100 *M* solution of sodium phosphate.

4.94 **(a)** By titration, 15.0 mL of 0.1008 *M* sodium hydroxide is needed to neutralize a 0.2053-g sample of an organic acid. What is the molar mass of the acid if it is monoprotic? **(b)** An elemental analysis of the acid indicates that it is composed of 5.89% H, 70.6% C, and 23.5% O by mass. What is its molecular formula?

4.95 A 6.977-g sample of a mixture was analyzed for barium ion by adding a small excess of sulfuric acid to an aqueous solution of the sample. The resultant reaction produced a precipitate of barium sulfate, which was collected by filtration, washed, dried, and weighed. If 0.4123 g of barium sulfate was obtained, what was the mass percentage of barium in the sample?

[4.96] A tanker truck carrying 5.0×10^3 kg of concentrated sulfuric acid solution tips over and spills its load. If the sulfuric acid is 95.0% H_2SO_4 by mass and has a density of 1.84 g/mL, how many kilograms of sodium carbonate must be added to neutralize the acid?

4.97 A sample of 5.53 g of $Mg(OH)_2$ is added to 25.0 mL of 0.200 M HNO_3. **(a)** Write the chemical equation for the reaction that occurs. **(b)** Which is the limiting reactant in the reaction? **(c)** How many moles of $Mg(OH)_2$, HNO_3, and $Mg(NO_3)_2$ are present after the reaction is complete?

4.98 A sample of 1.50 g of lead(II) nitrate is mixed with 125 mL of 0.100 *M* sodium sulfate solution. **(a)** Write the chemical equation for the reaction that occurs. **(b)** Which is the limiting reactant in the reaction? **(c)** What are the concentrations of all ions that remain in solution after the reaction is complete?

4.99 A mixture contains 89.0% NaCl, 1.5% $MgCl_2$, and 8.5% Na_2SO_4 by mass. What is the molarity of Cl^- ions in a solution formed by dissolving 7.50 g of the mixture in enough water to form 500.0 mL of solution?

[4.100] The average concentration of bromide ion in seawater is 65 mg of bromide ion per kg of seawater. What is the molarity of the bromide ion if the density of the seawater is 1.025 g/mL?

[4.101] The mass percentage of chloride ion in a 25.00-mL sample of seawater was determined by titrating the sample with silver nitrate, precipitating silver chloride. It took 42.58 mL of 0.2997 *M* silver nitrate solution to reach the equivalence point in the titration. What is the mass percentage of chloride ion in the seawater if its density is 1.025 g/mL?

4.102 The arsenic in a 1.22-g sample of a pesticide was converted to AsO_4^{3-} by suitable chemical treatment. It was then titrated using Ag^+ to form Ag_3AsO_4 as a precipitate. **(a)** What is the oxidation state of As in AsO_4^{3-}? **(b)** Name Ag_3AsO_4 by analogy to the corresponding compound containing phosphorus in place of arsenic. **(c)** If it took 25.0 mL of 0.102 *M* Ag^+ to reach the equivalence point in this titration, what is the mass percentage of arsenic in the pesticide?

[4.103] A 500-mg tablet of an antacid containing $Mg(OH)_2$, $Al(OH)_3$, and an inert "binder" was dissolved in 50.0 mL of 0.500 *M* HCl. The resulting solution, which was acidic, needed 30.9 mL of 0.255 *M* NaOH for neutralization. **(a)** Calculate the number of moles of OH^- ions in the tablet. **(b)** If the tablet contains 5.0% binder, how many milligrams of $Mg(OH)_2$ and how many of $Al(OH)_3$ does the tablet contain?

[4.104] Federal regulations set an upper limit of 50 parts per million (ppm) of NH_3 in the air in a work environment (that is, 50 molecules of $NH_3(g)$ for every million molecules in the air). Air from a manufacturing operation was drawn through a solution containing 1.00×10^2 mL of 0.0105 *M* HCl. The NH_3 reacts with HCl as follows:

$$NH_3(aq) + HCl(aq) \longrightarrow NH_4Cl(aq)$$

After drawing air through the acid solution for 10.0 min at a rate of 10.0 L/min, the acid was titrated. The remaining acid needed 13.1 mL of 0.0588 *M* NaOH to reach the equivalence point. **(a)** How many grams of NH_3 were drawn into the acid solution? **(b)** How many ppm of NH_3 were in the air? (Air has a density of 1.20 g/L and an average molar mass of 29.0 g/mol under the conditions of the experiment.) **(c)** Is this manufacturer in compliance with regulations?

eMedia Exercises

4.105 The **Electrolytes and Non-Electrolytes** movie (*eChapter 4.1*) and the **Aqueous Acids** and **Aqueous Bases** movies (*eChapter 4.3*) illustrate the behavior of various substances in aqueous solution. For each of the seven substances mentioned in the movies, write the chemical equation that corresponds to dissolution in water. (The chemical formula of sugar is $C_{12}H_{22}O_{11}$.) Where appropriate, use the double arrow notation.

4.106 In the **Strong and Weak Electrolytes** movie (*eChapter 4.1*), the lightbulb glows brightly when the beaker contains aqueous hydrochloric acid, but relatively dimly when the beaker contains aqueous acetic acid. **(a)** For each of the compounds in Exercise 4.3, would you expect an aqueous solution to cause the bulb to light? If so, how brightly? **(b)** Consider the use of aqueous solutions of each of the following compounds in the apparatus shown in the demonstration. For each compound, tell whether you would expect the lightbulb to glow brightly, dimly, or not at all: H_2CO_3; C_2H_5OH; NH_4Cl; CaF_2; and HF.

4.107 **(a)** Use the solubility rules to predict what precipitate, if any, will form as the result of each combination. (i) $Na_2CO_3(aq)$ and $Fe(NO_3)_2(aq)$; (ii) $NH_4NO_3(aq)$ and $K_2SO_4(aq)$; (iii) $AlBr_3(aq)$ and $Fe_2(SO_4)_3(aq)$; (iv) $H_2SO_4(aq)$ and $Pb(NO_3)_2(aq)$; (v) $Na_2S(aq)$ and $(NH_4)_2SO_4(aq)$. Use the **Ionic Compounds** activity (*eChapter 2.8*) to check your answers. **(b)** For each combination that produces a precipitate, write a balanced net ionic equation. **(c)** When $NH_4Cl(aq)$ and $Pb(NO_3)_2(aq)$ are combined, a precipitate forms.What ions are still present in the solution in significant concentration after the precipitation? Explain.

4.108 In the **Redox Chemistry of Tin and Zinc** movie (*eChapter 4.4*), zinc is oxidized by a solution containing tin ions. **(a)** Write the equation corresponding to this redox reaction. **(b)** In addition to the reaction between zinc metal and tin ions, there is another process occurring. Write the net ionic equation corresponding to this process. (Refer to Exercise 4.44.)

4.109 After watching the **Solution Formation from a Solid** movie (*eChapter 4.5*), answer the following questions: **(a)** If we neglect to account for the mass of the weighing paper, how would our calculated concentration differ from the actual concentration of the solution? **(b)** Describe the process of preparing an aqueous solution of known concentration, starting with a solid. **(c)** Why is it necessary to make the solution as described in the movie, rather than simply filling the flask up to the mark with water and then adding the solute? **(d)** Describe how you would prepare the solution in part (a) starting with the concentrated stock solution in the **Solution Formation by Dilution** movie (*eChapter 4.5*).

4.110 Use the **Titration** simulation (*eChapter 4.6*) to determine the concentration of an unknown acid by adding 0.40 *M* NaOH in increments of 1.0 mL. Repeat the titration adding increments of 0.10 mL of base near the end point. Once more, repeat the titration, adding increments of 0.05 mL of base near the end point. If your acid is dilute enough, repeat the titration three more times using 0.10 *M* NaOH in 1.0-mL, 0.50-mL, and 0.05-mL increments. **(a)** Tabulate the acid concentrations that you calculate from your titration data. Are the values all the same? If not, why not? **(b)** Which value do you consider to be most precise and why?

4.112 **(a)** What is the maximum concentration of monoprotic acid that could be titrated in the **Titration** simulation (*eChapter 4.6*) using 0.05 *M* NaOH? **(b)** What is the maximum concentration of a diprotic acid that could be titrated in this simulation using 0.10 *M* base? **(c)** All of the acid-base indicators available in the simulation change color within a pH range of ~4 to 10.5. What effect would the use of an indicator such as metacresol purple have on the experimentally determined value of an unknown acid's concentration? (Metacresol purple changes colors in the pH range of ~1.2 to 2.8.)

Chapter 5

Thermochemistry

Plants utilize solar energy to carry out photosynthesis of carbohydrates, which provide the energy needed for plant growth. Among the molecules produced by plants is ethylene.

MODERN SOCIETY DEPENDS on energy for its existence. Any symptom of an energy shortage—rolling blackouts of electrical power, gasoline shortages, or big increases in the cost of natural gas—are enough to shake people's confidence and roil the markets. Energy is very much a chemical topic. Nearly all of the energy on which we depend is derived from chemical reactions, such as the combustion of fossil fuels, the chemical reactions occurring in batteries, or the formation of biomass through photosynthesis. Think for a moment about some of the chemical processes that we encounter in the course of a typical day: We eat foods to produce the energy needed to maintain our biological functions. We burn fossil fuels (coal, petroleum, natural gas) to produce much of the energy that powers our homes and offices and that moves us from place to place by automobile, plane, or train. We listen to tunes on battery-powered MP3 players.

The relationship between chemical change and energy shows up in various ways. Chemical reactions involving foods and fuels *release* energy. By contrast, the splitting of water into hydrogen and oxygen, illustrated in Figure 1.7, requires an *input* of electrical energy. Similarly, the chemical process we call photosynthesis, which occurs in plant leaves, converts one form of energy, radiant energy from the Sun, to chemical energy. Chemical processes can do more than simply generate heat; they can do work, such as turning an automobile starter, powering a drill, and so on. What we get from all this is that chemical change generally involves energy. If we are to properly understand chemistry, we must also understand the energy changes that accompany chemical change.

The study of energy and its transformations is known as **thermodynamics** (Greek: *thérme-*, "heat"; *dy' namis*, "power"). This area of study began during the Industrial Revolution as the relationships among heat, work, and the energy content of fuels were studied in an effort to maximize the performance of steam engines. Today thermodynamics is enormously important in all areas of science and engineering, as we will see throughout this text. In the last couple of chapters we have examined chemical reactions and their stoichiometry. In this chapter we will examine the relationships between chemical reactions and energy changes involving heat. This aspect of thermodynamics is called **thermochemistry**. We will discuss in detail other aspects of thermodynamics in Chapter 19.

▶ What's Ahead ◀

- We will discuss the nature of *energy* and the forms it takes, notably *kinetic energy*, *potential energy*, thermal energy, and chemical energy.
- In the SI system, the unit of energy is the *joule*, but we often make use of an older, more familiar unit, the calorie.
- Energy is interconvertible from one form to another, although there are limitations and rules on these interconversions. Energy can be employed to accomplish *work*.
- We'll study the *first law of thermodynamics*: Energy cannot be created or destroyed. Energy may be transformed from one form to another or from one part of matter to another, but the total energy of the universe remains constant.
- To explore energy changes, we focus on a particular part of the universe, which we call the *system*. Everything else is called the *surroundings*. The system possesses a certain amount of energy that we can express as the *internal energy*, E. E is called a *state function* because its value depends only on the state of a system now, not on how it came to be in that state.
- A related state function—*enthalpy*, H—is useful because the change in enthalpy, ΔH, measures the quantity of heat energy gained or lost by a system in a process.
- We will also consider how to measure heat changes in chemical processes *(calorimetry)*, how to establish standard values for enthalpy changes in chemical reactions, and how to use them to calculate ΔH values for reactions we can't actually study experimentally.
- We'll examine foods and fuels as sources of energy and discuss some related health and social issues.

(a)

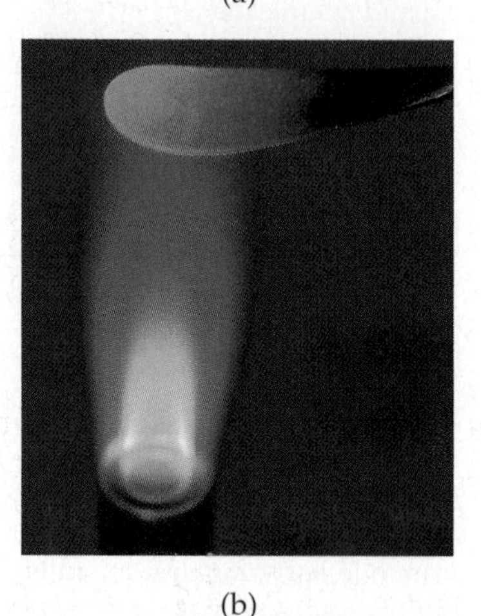

(b)

▲ **Figure 5.1** Energy can be used to achieve two basic types of tasks: (a) Work is energy used to cause an object with mass to move. (b) Heat is energy used to cause the temperature of an object to increase.

5.1 The Nature of Energy

Our discussion of thermodynamics will utilize the concepts of *energy*, *work*, and *heat*. Although these terms are very familiar to us (Figure 5.1 ◀), we will need to develop some precise definitions for our discussion. In particular, we need to examine the ways in which matter can possess energy and how that energy can be transferred from one piece of matter to another.

Kinetic Energy and Potential Energy

Objects, whether they are tennis balls or molecules, can possess **kinetic energy**, the energy of motion. The magnitude of the kinetic energy, E_k, of an object depends on its mass, m, and speed, v:

$$E_k = \frac{1}{2}mv^2 \qquad [5.1]$$

Equation 5.1 shows that the kinetic energy increases as the speed of an object increases. For example, a car moving at 50 miles per hour (mph) has greater kinetic energy than it does at 40 mph. For a given speed, moreover, the kinetic energy increases with increasing mass. Thus, a large sport-utility vehicle traveling at 55 mph has greater kinetic energy than a small sedan traveling at the same speed, because the SUV has greater mass than the sedan. Atoms and molecules have mass and are in motion. They therefore possess kinetic energy, though it is not apparent to us as is the kinetic energy of large-scale objects.

An object can also possess another form of energy, called **potential energy**, by virtue of its *position* relative to other objects. Potential energy arises when there is a force operating on an object. The most familiar force of this kind is gravity. Think of a cyclist poised at the top of a hill, as illustrated in Figure 5.2 ▼. Gravity acts upon her and her bicycle, exerting a force directed toward the center of Earth. At the top of the hill the cyclist and her bicycle possess a certain potential energy by virtue of their elevation. The potential energy is given by the expression mgh, where m is the mass of the object in question (in this case the cyclist and her bicycle), h is the height of the object relative to some reference height, and g is the gravitational constant, 9.8 m/s^2. Once in motion, without any further effort on her part, the cyclist gains speed as the bicycle rolls down the hill. Her potential energy decreases as she moves downward, but the energy does not simply disappear. It is converted to other forms of energy, principally kinetic energy, the energy of motion. In addition, there is friction between the bicycle tires and the pavement and friction of motion through the air, which generate a certain amount of heat. This example illustrates that forms of energy are interconvertible. We will have more to say later about interconversions of energy and the nature of heat.

Gravity is an important kind of force for large objects, such as the cyclist and Earth. Chemistry, however, deals mostly with extremely small objects—atoms

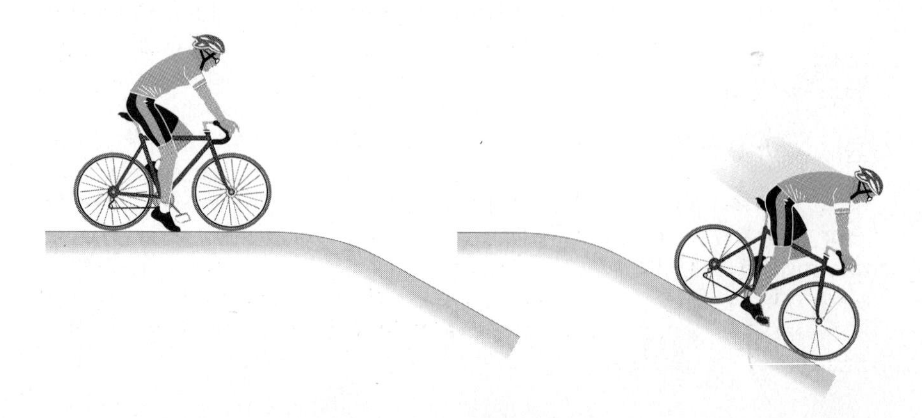

▶ **Figure 5.2** A bicycle at the top of a hill (left) has a high potential energy. Its potential energy relative to the bottom of the hill is mgh, where m is the mass of the cyclist and her bicycle, h is her height relative to the bottom of the hill, and g is the gravitational constant, 9.8 m/s^2. As the bicycle proceeds down the hill (right), the potential energy is converted into kinetic energy, so the potential energy is lower at the bottom than at the top.

and molecules—and gravitational forces play a negligible role in the ways these microscopic objects interact with one another. More important are forces that arise from electrical charges. One of the most important forms of potential energy for our purposes is *electrostatic energy*, which arises from the interactions between charged particles. The electrostatic potential energy, E_{el}, is proportional to the electrical charges on the two interacting objects, Q_1 and Q_2, and inversely proportional to the distance separating them:

$$E_{el} = \frac{\kappa Q_1 Q_2}{d} \quad [5.2]$$

Here κ is simply a constant of proportionality, 8.99×10^9 J-m/C^2 [C is the coulomb, a unit of electrical charge. (Section 2.2)] When Q_1 and Q_2 have the same sign (for example, both are positive), the two charges repel one another, and E_{el} is positive. When they have opposite signs, they attract one another and E_{el} is negative. We will see as we go along that more stable energies are represented by lower or negative values, whereas less stable or repulsive energies are represented by higher or positive values. When dealing with molecular-level objects, the electrical charges Q_1 and Q_2 are typically on the order of magnitude of the charge of the electron (1.60×10^{-19} C).

One of our goals in chemistry is to relate the energy changes that we see in our macroscopic world to the kinetic or potential energy of substances at the atomic or molecular level. Many substances, for example, fuels, release energy when they react. The *chemical energy* of these substances is due to the potential energy stored in the arrangements of the atoms of the substance. Likewise, we will see that the energy a substance possesses because of its temperature (its *thermal energy*) is associated with the kinetic energy of the molecules in the substance. We will soon discuss the transfer of chemical and thermal energy from a reacting substance to its surrounding environment, but first let's revisit the units used to measure energy.

Units of Energy

The SI unit for energy is the **joule** (pronounced "jool"), J, in honor of James Joule (1818–1889), a British scientist who investigated work and heat: 1 J = 1 kg-m^2/s^2. A mass of 2 kg moving at a speed of 1 m/s possesses a kinetic energy of 1 J:

$$E_k = \tfrac{1}{2}mv^2 = \tfrac{1}{2}(2\text{ kg})(1\text{ m/s})^2 = 1\text{ kg-m}^2/\text{s}^2 = 1\text{ J}$$

A joule is not a large amount of energy, and we will often use *kilojoules* (kJ) in discussing the energies associated with chemical reactions.

Traditionally, energy changes accompanying chemical reactions have been expressed in calories, a non-SI unit still widely used in chemistry, biology, and biochemistry. A **calorie** (cal) was originally defined as the amount of energy required to raise the temperature of 1 g of water from 14.5°C to 15.5°C. It is now defined in terms of the joule:

$$1\text{ cal} = 4.184\text{ J (exactly)}$$

A related energy unit used in nutrition is the nutritional *Calorie* (note that this unit is capitalized): 1 Cal = 1000 cal = 1 kcal.

System and Surroundings

When we use thermodynamics to analyze energy changes, we focus our attention on a limited and well-defined part of the universe. The portion we single out for study is called the **system**; everything else is called the **surroundings**. When we study the energy change that accompanies a chemical reaction in the laboratory, the chemicals usually constitute the system. The container and everything beyond it are considered the surroundings. The systems we can most

▲ **Figure 5.3** Hydrogen and oxygen gases in a cylinder. If we are interested only in the properties of the gases, the gases are the system and the cylinder and piston are part of the surroundings.

readily study are called *closed systems*. A closed system can exchange energy but not matter with its surroundings. For example, consider a mixture of hydrogen gas, H_2, and oxygen gas, O_2, in a cylinder, as illustrated in Figure 5.3 ◀. The system in this case is just the hydrogen and oxygen; the cylinder, piston, and everything beyond them (including us) are the surroundings. If the hydrogen and oxygen react to form water, energy is liberated:

$$2H_2(g) + O_2(g) \longrightarrow 2H_2O(g) + \text{energy}$$

Although the chemical form of the hydrogen and oxygen atoms in the system is changed by this reaction, the system has not lost or gained mass; it undergoes no exchange of matter with its surroundings. However, it does exchange energy with its surroundings in the form of *work* and *heat*. These are quantities that we can measure, as we will now discuss.

Transferring Energy: Work and Heat

Figure 5.1 illustrates two of the common ways that we experience energy changes in our everyday lives. In Figure 5.1(a) energy is transferred from the tennis racquet to the ball, changing the direction and speed of the ball's movement. In Figure 5.1(b) energy is transferred in the form of heat. Thus, energy is transferred in two general ways: to cause the motion of an object against a force or to cause a temperature change.

A **force** is any kind of push or pull exerted on an object. As we noted in Figure 5.2, the force of gravity "pulls" a bicycle from the top of a hill to the bottom. Electrostatic force "pulls" unlike charges toward one another or "pushes" like charges apart. Energy used to cause an object to move against a force is called **work**. The work, w, that we do in moving objects against a force equals the product of the force, F, and the distance, d, that the object is moved:

$$w = F \times d \quad [5.3]$$

Thus, we perform work when we lift an object against the force of gravity or when we bring two like charges closer together. If we define the object as the system, then we—as part of the surroundings—are performing work on that system, transferring energy to it.

The other way in which energy is transferred is as heat. **Heat** is the energy transferred from a hotter object to a colder one. A combustion reaction, such as the burning of natural gas illustrated in Figure 5.1(b), releases the chemical energy stored in the molecules of the fuel in the form of heat. ∞ (Section 3.2) The heat raises the temperature of surrounding objects. If we define the reaction taking place as the system and everything else as the surroundings, then energy in the form of heat is transferred from the system to the surroundings.

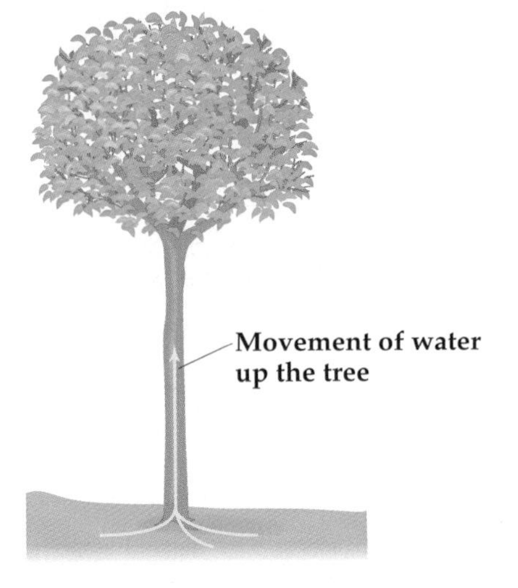

▲ **Figure 5.4** Water flows from ground level into the upper parts of the tree.

SAMPLE EXERCISE 5.1

The movement of water from the ground through the trunk to the upper limbs of a tree, as illustrated in Figure 5.4 ◀, is an important biological process. **(a)** What part of the system, if any, undergoes a change in potential energy? **(b)** Is work done in the process?

Solution

Analyze: The goal here is to associate movements of matter with changes in potential energy and the performance of work.

Plan: We need to identify the parts of Figure 5.4 that change location or that seem to have caused a change in the energy of some other part. Secondly, we must ask whether the change in location involves a change in potential energy. Finally, does this change in potential energy mean that work has been done?

Solve: **(a)** The water changes location as it moves from the ground to the upper part of the tree. It has moved upward, against the force of gravity. This means that the potential energy of the water has changed.

(b) Recall that work is the movement of a mass over a distance against an opposing force. In lifting the groundwater to its upper limbs, the plant does work, just as you would be doing work if you lifted an equivalent amount of water in a container from the ground to some height. How the plant does this work is an interesting subject in its own right.
Check: We have identified a positive change in the potential energy of the water with work performed on the water, which is the correct relationship.

PRACTICE EXERCISE

Which of the following involves the larger change in potential energy? **(a)** A 50-kg object is dropped to the ground from a height of 8 m. **(b)** A 20-kg object is lifted from the ground to a height of 20 m.
Answers: The magnitude of the change in potential energy is the same in each case, $mg\Delta h$.* However, the *sign* of the change is negative in **(a)**, positive in **(b)**.

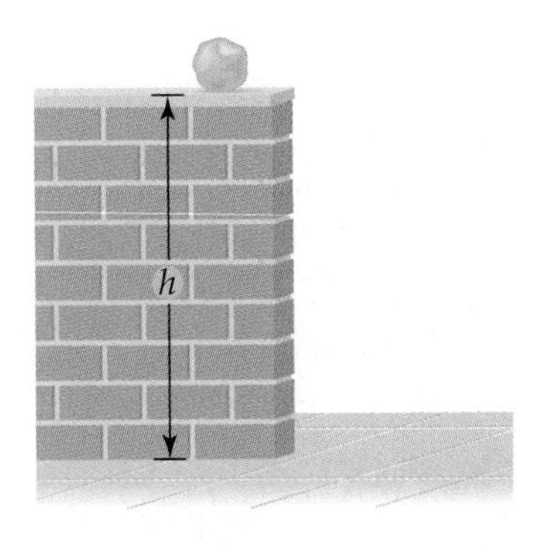

(a)

(b)

(c)

▲ **Figure 5.5** A ball of clay can be used to show energy interconversions. (a) At the top of the wall, the ball has potential energy due to gravity. (b) As the ball falls, its potential energy is converted into kinetic energy. (c) When the ball strikes the ground, some of the kinetic energy is used to do work in squashing the ball; the rest is released as heat.

We can now provide a more precise definition for energy: **Energy** *is the capacity to do work or to transfer heat.* We will end this section with one more example that illustrates some of the concepts of energy that we have covered thus far. Consider a ball of modeling clay, which we will define as the system. If we lift the ball to the top of a wall, as shown in Figure 5.5(a) ▶, we are doing work against the force of gravity. The energy we transfer to the ball by doing work on it increases its potential energy because the ball is now at a greater height. If the ball now rolls off the wall, as in Figure 5.5(b), its downward speed increases as its potential energy is converted to kinetic energy. When the clay ball strikes the ground [Figure 5.5(c)], it stops moving and its kinetic energy goes to zero. Some of the kinetic energy is used to do work in squashing the ball, and the rest is dissipated to the surroundings as heat during the collision with the ground. The "bookkeeping" of the various transfers of energy between the system and the surroundings as work and heat is the focus of Section 5.2.

SAMPLE EXERCISE 5.2

A bowler lifts a 5.4-kg (12-lb) bowling ball from ground level to a height of 1.6 m (5.2 feet) and then drops the ball back to the ground. **(a)** What happens to the potential energy of the bowling ball as it is raised from the ground? **(b)** What quantity of work, in J, is used to raise the ball? **(c)** After the ball is dropped, it gains kinetic energy. If we assume that all of the work in part (b) is converted into kinetic energy at the point of impact with the ground, what is the speed of the ball at the point of impact? (Note: The force due to gravity is $F = m \times g$, where m is the mass of the object and g is the gravitational constant; $g = 9.8\ \text{m/s}^2$.)

Solution

Analyze: We need to relate the potential energy of the bowling ball to its position relative to the ground. We then need to establish the relationship between work and the change in potential energy of the ball. Finally, we need to connect the change in potential energy when the ball is dropped with the kinetic energy attained by the ball.
Plan: We can calculate the work done in lifting the ball by using the relationship $w = F \times d$. We can employ Equation 5.1 to calculate the kinetic energy of the ball at the moment of impact and from that the speed v.
Solve: **(a)** Because the bowling ball is raised to a greater height above the ground, its potential energy increases. There is more energy stored in the ball at greater height than there is at lower height. **(b)** The ball has a mass of 5.4 kg and it is lifted a distance of 1.6 m. To calculate the work performed to raise the ball, we use Equation 5.3 and $F = m \times g$ for the force due to gravity:

$$w = F \times d = m \times g \times d = (5.4\ \text{kg})(9.8\ \text{m/s}^2)(1.6\ \text{m}) = 85\ \text{kg-m}^2/\text{s}^2 = 85\ \text{J}$$

* The symbol Δ is commonly used to denote *change*. For example, a change in height can be represented by Δh.

Thus, the bowler has done 85 J of work to lift the ball to a height of 1.6 m. **(c)** When the ball is dropped, its potential energy is converted to kinetic energy. At the point of impact, we are to assume that the kinetic energy is equal to the work done in part (b), 85 J. Thus, the speed v at impact must have the value such that

$$E_k = \frac{1}{2}mv^2 = 85 \text{ J} = 85 \text{ kg-m}^2/\text{s}^2$$

We can now solve this equation for v:

$$v^2 = \left(\frac{2E_k}{m}\right) = \left(\frac{2(85 \text{ kg-m}^2/\text{s}^2)}{5.4 \text{ kg}}\right) = 31.5 \text{ m}^2/\text{s}^2$$

$$v = \sqrt{31.5 \text{ m}^2/\text{s}^2} = 5.6 \text{ m/s}$$

Check: Work must be done in part (b) to increase the potential energy of the ball, which is in accord with our experiences. The units work out as they should, too, in the calculations of both parts (b) and (c). The work is in units of J and the speed in units of m/s. In part (c) we have carried an additional digit in the intermediate calculation involving the square root, but we report the final value to only two significant figures, as appropriate. A speed of 1 m/s is roughly 2 mph, so the bowling ball has a speed of greater than 10 mph upon impact.

PRACTICE EXERCISE

What is the kinetic energy, in J, of **(a)** an Ar atom moving with a speed of 650 m/s; **(b)** a mole of Ar atoms moving with a speed of 650 m/s?
Answers: **(a)** 1.4×10^{-20} J; **(b)** 8.4×10^3 J

5.2 The First Law of Thermodynamics

We have seen that the potential energy of a system can be converted into kinetic energy, and vice versa. We have also seen that energy can be transferred back and forth between a system and its surroundings in the forms of work and heat. In general, energy can be converted from one form to another, and it can be transferred from one part of the universe to another. Our task is to understand how energy exchanges of heat or work can occur between a system and its surroundings. We begin with one of the most important observations in science, that energy can be neither created nor destroyed. This universal truth, known as the **first law of thermodynamics**, can be summarized by the simple statement: *Energy is conserved.* Any energy that is lost by the system must be gained by the surroundings, and vice versa. To apply the first law quantitatively, let's first define the energy of a system more precisely.

Internal Energy

We will use the first law of thermodynamics to analyze energy changes of chemical systems. To do so, we must consider all the sources of kinetic and potential energy in the system. The **internal energy** of the system is the sum of *all* the kinetic and potential energy of all the components of the system. For the system in Figure 5.3, for example, the internal energy includes the motions of the H_2 and O_2 molecules through space, their rotations and internal vibrations. It also includes the energies of the nuclei of each atom and of the component electrons. We represent the internal energy with the symbol E. We generally don't know the actual numerical value of E. What we can hope to know, however, is ΔE (read "delta E"),* the change in E that accompanies a change in the system.

* Recall that the symbol Δ means "change in."

Imagine that we start with a system with an initial internal energy, $E_{initial}$. The system then undergoes a change, which might involve work being done or heat being transferred. After the change, the final internal energy of the system is E_{final}. We define the *change* in internal energy, ΔE, as the difference between E_{final} and $E_{initial}$:

$$\Delta E = E_{final} - E_{initial} \quad [5.4]$$

We don't really need to know the actual values of E_{final} or $E_{initial}$ for the system. To apply the first law of thermodynamics, we need only the value of ΔE.

Thermodynamic quantities such as ΔE have three parts: a number and a unit that together give the magnitude of the change, and a sign that gives the direction. A *positive* value of ΔE results when $E_{final} > E_{initial}$, indicating the system has gained energy from its surroundings. A *negative* value of ΔE is obtained when $E_{final} < E_{initial}$, indicating the system has lost energy to its surroundings.

In a chemical reaction the initial state of the system refers to the reactants, and the final state refers to the products. When hydrogen and oxygen form water, the system loses energy to the surroundings as heat. Because heat is lost from the system, the internal energy of the products is less than that of the reactants, and ΔE for the process is negative. Thus, the *energy diagram* in Figure 5.6 ▶ shows that the internal energy of the mixture of H_2 and O_2 is greater than that of H_2O.

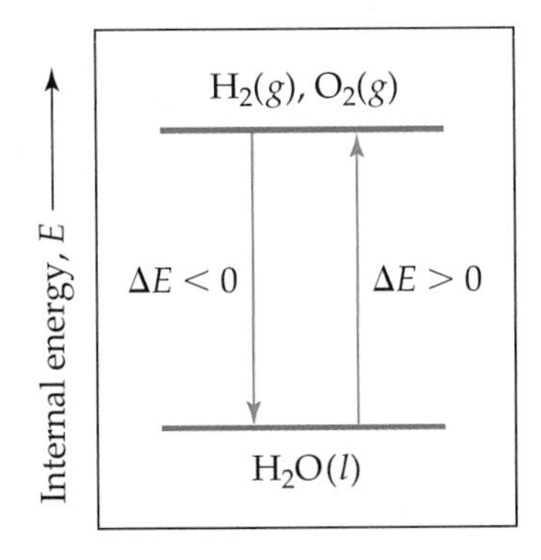

▲ **Figure 5.6** A system composed of $H_2(g)$ and $O_2(g)$ has a greater internal energy than one composed of $H_2O(l)$. The system loses energy ($\Delta E < 0$) when H_2 and O_2 are converted to H_2O. It gains energy ($\Delta E > 0$) when H_2O is decomposed into H_2 and O_2.

Relating ΔE to Heat and Work

As we noted in Section 5.1, any system can exchange energy with its surroundings as heat or as work. The internal energy of a system changes in magnitude as heat is added to or removed from the system or as work is done on it or by it. We can use these ideas to write a very useful algebraic expression of the first law of thermodynamics. When a system undergoes any chemical or physical change, the accompanying change in its internal energy, ΔE, is given by the heat added to or liberated from the system, q, plus the work done on or by the system, w:

$$\Delta E = q + w \quad [5.5]$$

Our everyday experiences tell us that when heat is added to a system or work is done on a system, its internal energy increases. Therefore, when heat is transferred from the surroundings to the system, q has a positive value. Likewise, when work is done on the system by the surroundings, w has a positive value (Figure 5.7 ▼). Conversely, both the heat lost by the system to the surroundings and the

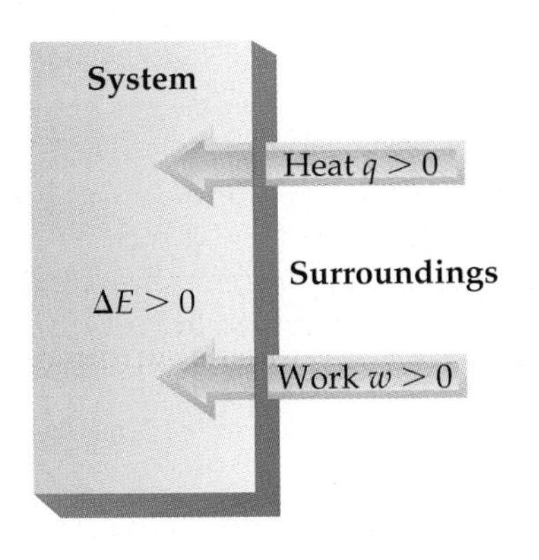

◀ **Figure 5.7** Heat, q, absorbed by the system and work, w, done on the system are both positive quantities. Both serve to increase the internal energy, E, of the system: $\Delta E = q + w$.

TABLE 5.1 Sign Conventions Used and the Relationship Among q, w, and ΔE

Sign Convention for q:	Sign of $\Delta E = q + w$
$q > 0$: Heat is transferred from the surroundings to the system	$q > 0$ and $w > 0$: $\Delta E > 0$
$q < 0$: Heat is transferred from the system to the surroundings	$q > 0$ and $w < 0$: The sign of ΔE depends on the magnitudes of q and w
Sign Convention for w:	$q < 0$ and $w > 0$: The sign of ΔE depends on the magnitudes of q and w
$w > 0$: Work is done by the surroundings on the system	
$w < 0$: Work is done by the system on the surroundings	$q < 0$ and $w < 0$: $\Delta E < 0$

work done by the system on the surroundings have negative values; that is, they lower the internal energy of the system. The relationship between the signs of q and w and the sign of ΔE is presented in Table 5.1 ▲.*

MOVIE
Formation of Water

SAMPLE EXERCISE 5.3

The hydrogen and oxygen gases in the cylinder illustrated in Figure 5.3 are ignited. As the reaction occurs, the system loses 1150 J of heat to the surroundings. The reaction also causes the piston to rise as the hot gases expand. The expanding gas does 480 J of work on the surroundings as it pushes against the atmosphere. What is the change in the internal energy of the system?

Solution Heat is transferred from the system to the surroundings, and work is done by the system on the surroundings. From the sign conventions for q and w (Table 5.1), we see that both q and w are negative: $q = -1150$ J and $w = -480$ J. We can calculate the change in the internal energy, ΔE, by using Equation 5.5:

$$\Delta E = q + w = (-1150\text{ J}) + (-480\text{ J}) = -1630\text{ J}$$

We see that 1630 J of energy has been transferred from the system to the surroundings, some in the form of heat and some in the form of work done on the surroundings

PRACTICE EXERCISE

Calculate the change in the internal energy of the system for a process in which the system absorbs 140 J of heat from the surroundings and does 85 J of work on the surroundings.
Answer: +55 J

Endothermic and Exothermic Processes

When a process occurs in which the system absorbs heat, the process is called **endothermic**. (*Endo-* is a prefix meaning "into.") During an endothermic process, such as the melting of ice, heat flows *into* the system from its surroundings. If we,

* Equation 5.5 is sometimes written $\Delta E = q - w$. When written this way, work done by the system on the surroundings is defined as positive. This convention is used particularly in many engineering applications that focus on the work done by a machine on its surroundings.

(a)

(b)

◀ **Figure 5.8** (a) When ammonium thiocyanate and barium hydroxide octahydrate are mixed at room temperature, an endothermic reaction occurs: $2NH_4SCN(s) + Ba(OH)_2 \cdot 8H_2O(s) \longrightarrow Ba(SCN)_2(aq) + 2NH_3(aq) + 10H_2O(l)$. As a result, the temperature of the system drops from about 20°C to −9°C. (b) The reaction of powdered aluminum with Fe_2O_3 (the thermite reaction) is highly exothermic. The reaction proceeds vigorously to form Al_2O_3 and molten iron: $2Al(s) + Fe_2O_3(s) \longrightarrow Al_2O_3(s) + 2Fe(l)$.

as part of the surroundings, touch a container in which ice is melting, it feels cold to us because heat has passed from our hands to the container.

A process in which the system evolves heat is called **exothermic**. (*Exo-* is a prefix meaning "out of.") During an exothermic process, such as the combustion of gasoline, heat flows *out* of the system and into its surroundings. Figure 5.8 ▲ shows two further examples of chemical reactions, one endothermic and the other highly exothermic. Notice that in the endothermic process shown in Figure 5.8(a) the temperature in the beaker decreases. In this case the "system" is the chemical reactants. The solution in which they are dissolved is part of the surroundings. Heat flows from the solution, as part of the surroundings, into the reactants as products are formed. Thus, the temperature of the solution drops.

MOVIE
Thermite Reaction

ACTIVITY
Dissolution of Ammonium Nitrate

State Functions

Although we usually have no way of knowing the precise value of the internal energy of a system, it does have a fixed value for a given set of conditions. The conditions that influence internal energy include the temperature and pressure. Furthermore, the total internal energy of a system is proportional to the total quantity of matter in the system because energy is an extensive property. (Section 1.3)

Suppose we define our system as 50 g of water at 25°C, as in Figure 5.9 ▼. The system could have arrived at this state by cooling 50 g of water from 100°C

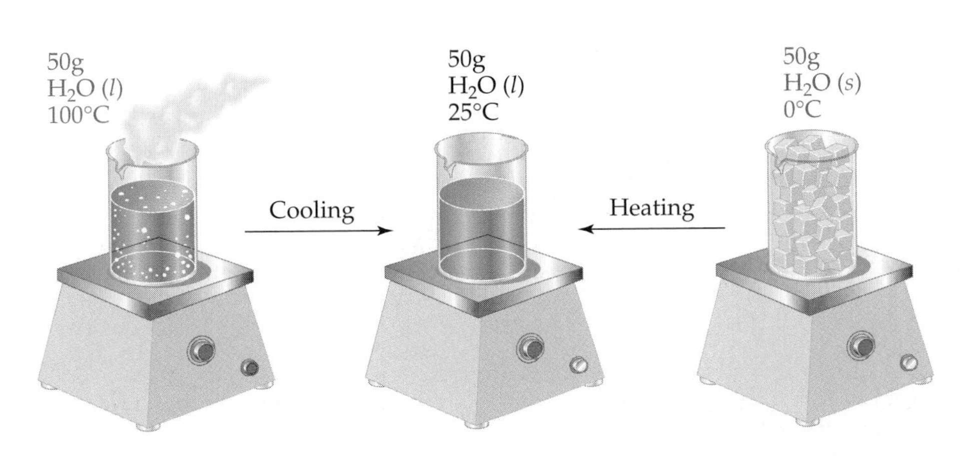

◀ **Figure 5.9** Internal energy, a state function, depends only on the present state of the system and not on the path by which it arrived at that state. The internal energy of 50 g of water at 25°C is the same whether the water is cooled from a higher temperature to 25°C or is obtained by melting 50 g of ice and then warming it to 25°C.

or by melting 50 g of ice and subsequently warming the water to 25°C. The internal energy of the water at 25°C is the same in either case. Internal energy is an example of a **state function**, a property of a system that is determined by specifying its condition or its state (in terms of temperature, pressure, location, and so forth). *The value of a state function depends only on its present condition, not on the particular history of the sample.* Because E is a state function, ΔE depends only on the initial and final states of the system, not on how the change occurs.

An analogy may explain the difference between quantities that are state functions and those that are not. Suppose you are traveling between Chicago and Denver. Chicago is 596 ft above sea level; Denver is 5280 ft above sea level. No matter what route you take, the altitude change will be 4684 ft. The distance you travel, however, will depend on your route. Altitude is analogous to a state function because the change in altitude is independent of the path taken. Distance traveled is not a state function.

Some thermodynamic quantities, such as ΔE, are state functions. Others, such as q and w, are not. Although $\Delta E = q + w$ is a state function, the specific amounts of heat and work produced during a change in the state of the system depend on the way in which the change is carried out, analogous to the choice of travel route between Chicago and Denver. Even though the individual values of q and w are not state functions, their sum *is*; if changing the path from an initial state to a final state increases the value of q, it will also decrease the value of w by exactly the same amount, and so forth.

We can illustrate this principle with the example shown in Figure 5.10 ▼ in which we consider two possible ways of discharging a flashlight battery at constant temperature. If the battery is shorted out by a coil of wire, no work is accomplished because nothing is moved against a force. All the energy is lost from the battery in the form of heat. (The wire coil will get warmer and release heat to the surrounding air.) On the other hand, if the battery is used to make a small motor turn, the discharge of the battery produces work. Some heat will be released as well, although not as much as when the battery is shorted out. The magnitudes of q and w are different for these two cases. If the initial and final states of the battery are identical in both cases, however, then $\Delta E = q + w$ must be the same in both cases because ΔE is a state function.

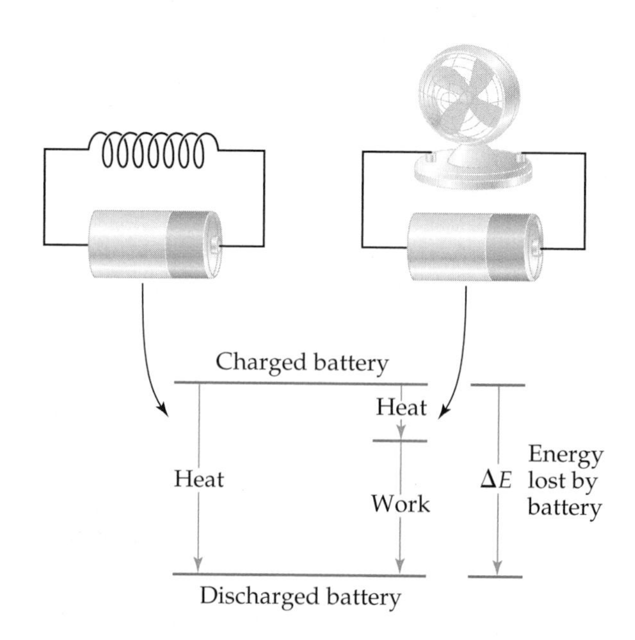

▶ **Figure 5.10** The amounts of heat and work transferred between the system and the surroundings depend on the way in which the system goes from one state to another. (a) A battery shorted out by a wire loses energy to the surroundings as heat; no work is performed. (b) A battery discharged through a motor loses energy as work (to make the fan turn) as well as heat. The value of ΔE is the same for both processes, but the values of q and w are different.

5.3 Enthalpy

Chemical changes can result in the release or absorption of heat, as illustrated in Figure 5.8. They can also cause work to be done, either on the system, or by the system on its surroundings. The relationship between chemical change and electrical work is important, and we will consider it in some detail in Chapter 20, "Electrochemistry." Most commonly, though, the only kind of work produced by chemical change is mechanical work. We usually carry out reactions in the laboratory at constant (atmospheric) pressure. Under these circumstances mechanical work occurs when a gas is produced or consumed in the reaction. Consider, for example, the reaction of zinc metal with hydrochloric acid solution:

$$Zn(s) + 2H^+(aq) \longrightarrow Zn^{2+}(aq) + H_2(g) \quad [5.6]$$

If we carry out this reaction in the laboratory hood in an open beaker, we can see the evolution of hydrogen gas, but it may not be so obvious that work is being done. Still, the hydrogen gas that is being produced must expand against the existing atmosphere. We can see this better by conducting the reaction in a closed vessel at constant pressure, as illustrated in Figure 5.11 ▼. In this apparatus, the piston moves up or down to maintain a constant pressure in the reaction vessel. If we assume for simplicity that the piston is weightless, the pressure in the apparatus is the same as outside, normal atmospheric pressure. As the reaction proceeds, H_2 gas forms, and the piston rises. The gas within the flask is thus doing work on the surroundings by lifting the piston against the force of atmospheric pressure that presses down on it. This kind of work is called **pressure-volume work** (or *P-V* work). When the pressure is constant, as in our example, pressure-volume work is given by

ANIMATION
Changes of State

$$w = -P\,\Delta V \quad [5.7]$$

where ΔV is the change in volume. When the change in volume is positive, as in our example, the work done by the system is negative. That is, work is done *by* the system *on* the surroundings. The "Closer Look" box discusses pressure-volume work in more detail, but all you really need to keep in mind for now is Equation 5.7, which applies to processes occurring at constant pressure. We will take up the properties of gases in more detail in Chapter 10.

The thermodynamic function called **enthalpy** (from the Greek word *enthalpein*, meaning "to warm") accounts for heat flow in chemical changes occurring at constant pressure when no forms of work are performed other than *P-V* work.

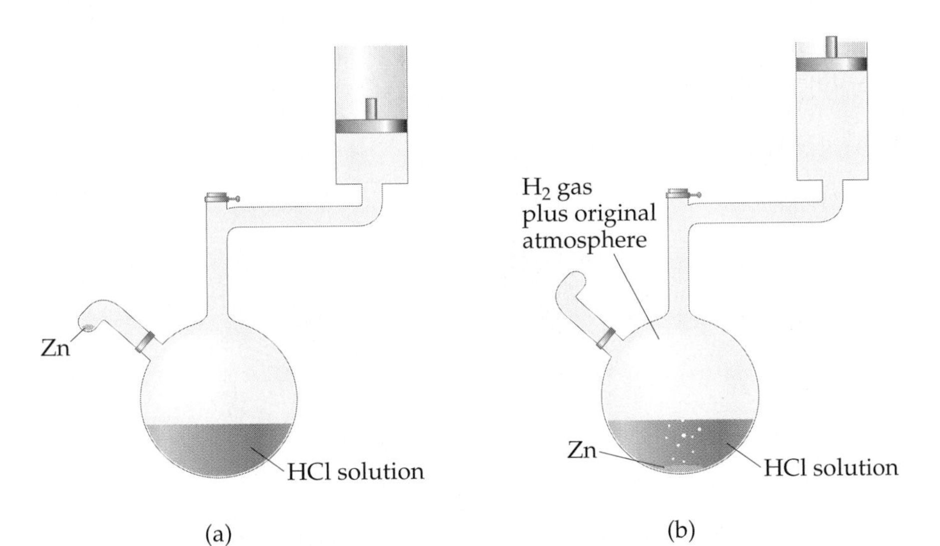

◀ **Figure 5.11** (a) The reaction of zinc metal with hydrochloric acid is conducted at constant pressure. The pressure in the reaction vessel equals atmospheric pressure. (b) When zinc is added to the acid solution, hydrogen gas is evolved. The hydrogen gas does work on the surroundings, raising the piston against atmospheric pressure to maintain constant pressure inside the reaction vessel.

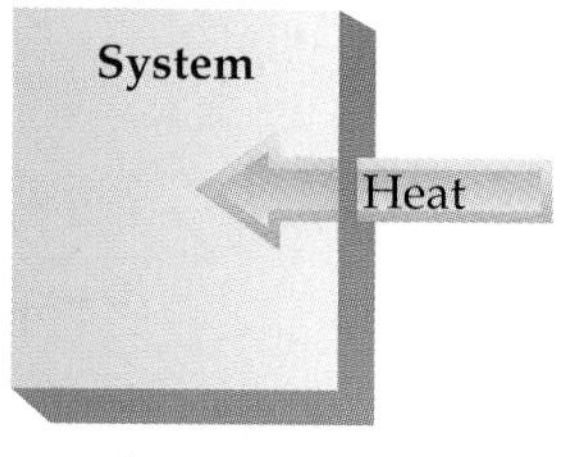

$\Delta H > 0$
Endothermic

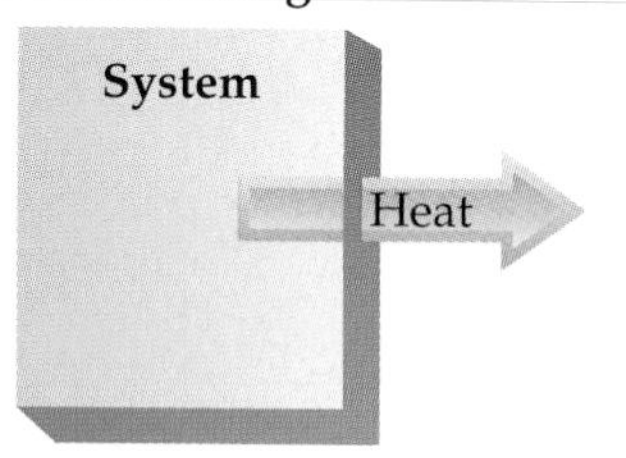

$\Delta H < 0$
Exothermic

▲ **Figure 5.12** (a) If the system absorbs heat, ΔH will be positive ($\Delta H > 0$). (b) If the system loses heat, ΔH will be negative ($\Delta H < 0$).

Enthalpy, which we denote by the symbol H, equals the internal energy plus the product of the pressure and volume of the system:

$$H = E + PV \tag{5.8}$$

Enthalpy is a state function because internal energy, pressure, and volume are all state functions.

Now suppose a change occurs at constant pressure. Then,

$$\begin{aligned}\Delta H &= \Delta(E + PV) \\ &= \Delta E + P\,\Delta V\end{aligned} \tag{5.9}$$

That is, the change in enthalpy is given by the change in internal energy plus the product of the constant pressure times the change in volume. The work of expansion of a gas is given by $w = -P\,\Delta V$, so we can substitute $-w$ for $P\,\Delta V$ in Equation 5.9. From Equation 5.5, moreover, we can substitute $q + w$ for ΔE, yielding for ΔH

$$\Delta H = \Delta E + P\,\Delta V = q_P + w - w = q_P \tag{5.10}$$

where the subscript P on the heat, q, emphasizes changes at constant pressure. The change in enthalpy, therefore, equals the heat gained or lost at constant pressure. Because q_P is something we can measure or readily calculate and because so much of the chemical change of interest to us occurs at constant pressure, enthalpy is a more useful function than internal energy. For most reactions the difference in ΔH and ΔE is small because $P\,\Delta V$ is small.

When ΔH is positive (that is, when q_P is positive), the system has gained heat from the surroundings (Table 5.1), which is an endothermic process. When ΔH is negative, the system has released heat to the surroundings, which is an exothermic process. These cases are diagrammed in Figure 5.12 ◀. Because H is a state function, ΔH (which equals q_P) depends only on the initial and final states of the system, not on how the change occurs. At first glance this statement might seem to contradict our earlier discussion in Section 5.2, in which we said that q is *not* a state function. There is no contradiction, however, because the relationship between ΔH and heat has the special limitation of constant pressure.

A Closer Look Energy, Enthalpy, and P-V Work

In chemistry we are interested mainly in two types of work: electrical work and mechanical work done by expanding gases. We will focus here only on the latter, called pressure–volume, or P-V, work. Expanding gases in the cylinder of an automobile engine do P-V work on the piston, and this work eventually turns the wheels. Expanding gases from an open reaction vessel do P-V work on the atmosphere. This work accomplishes nothing in a practical sense, but we must keep track of all work, useful or not, when monitoring the energy changes of a system.

Consider a gas confined to a cylinder with a movable piston of cross-sectional area A (Figure 5.13 ▶). A downward force, F, acts on the piston. The *pressure, P*, on the gas is the force per area: $P = F/A$. We will assume that the piston is weightless and that the only pressure acting on it is the *atmospheric pressure* due to the weight of Earth's atmosphere, which we will assume to be constant.

Suppose the gas in the cylinder expands, and the piston moves a distance, Δh. From Equation 5.3, the magnitude of the work done by the system equals the distance moved times the force acting on the piston:

$$\text{Magnitude of work} = \text{force} \times \text{distance} = F \times \Delta h \tag{5.11}$$

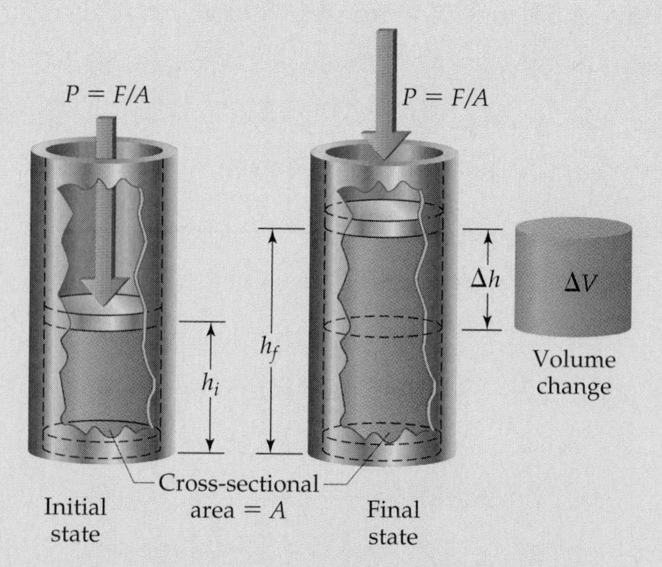

▲ **Figure 5.13** A moving piston does work on the surroundings. The amount of work done is $w = -P\,\Delta V$.

SAMPLE EXERCISE 5.4

Indicate the sign of the enthalpy change, ΔH, in each of the following processes carried out under atmospheric pressure, and indicate whether the process is endothermic or exothermic: **(a)** An ice cube melts; **(b)** 1 g of butane (C_4H_{10}) is combusted in sufficient oxygen to give complete combustion to CO_2 and H_2O; **(c)** a bowling ball is dropped from a height of 8 ft into a bucket of sand.

Solution

Analyze: Our goal is to determine whether ΔH in each case is positive or negative. To accomplish that, we must correctly identify the system.

Plan: We would expect that in each case the process occurs at constant pressure. The change in enthalpy thus equals the amount of heat absorbed or evolved in each process. Processes in which heat is absorbed are endothermic; those in which heat is evolved are exothermic.

Solve: In (a) the water that makes up the ice cube is the system. The ice cube absorbs heat from the surroundings as it melts, so q_P is positive and the process is endothermic. In (b) the system is the 1 g of butane and the oxygen required to combust it. The combustion of butane in oxygen gives off heat, so q_P is negative and the process is exothermic. In (c) the bowling ball is the system. It loses potential energy when it drops from a height of 8 ft into the bucket of sand. Where has the potential energy gone? It first went into kinetic energy of motion, but then the ball came to rest in the sand. In the process of stopping in the sand, the kinetic energy of the ball's motion is converted into heat that is absorbed by the bowling ball's surroundings. Thus, q_P is negative, and the process is exothermic.

PRACTICE EXERCISE

Suppose we confine 1 g of butane and sufficient oxygen to completely combust it in a cylinder like that in Figure 5.13. The cylinder is perfectly insulating, so no heat can escape to the surroundings. A spark initiates combustion of the butane, which forms carbon dioxide and water vapor. If we used this apparatus to measure the enthalpy change in the reaction, would the piston rise, fall, or stay the same?

Answer: The piston must move so as to maintain a constant pressure in the cylinder. Because the products contain more molecules of gas than the reactants, as shown by the balanced equation

$$2C_4H_{10}(g) + 13O_2(g) \longrightarrow 8CO_2(g) + 10H_2O(g)$$

the piston would rise to make room for the additional molecules of gas. Heat is given off, moreover, so the piston would rise to accommodate the expansion of the gases due to the temperature increase.

We can rearrange the definition of pressure, $P = F/A$, to $F = P \times A$. In addition, the volume change, ΔV, resulting from the movement of the piston, is the product of the cross-sectional area of the piston and the distance it moves: $\Delta V = A \times \Delta h$. Substituting into Equation 5.11,

$$\text{Magnitude of work} = F \times \Delta h = P \times A \times \Delta h$$
$$= P \times \Delta V$$

Because the system (the confined gas) is doing work on the surroundings, the sign on the work is negative:

$$w = -P\,\Delta V \qquad [5.12]$$

Now, if P-V work is the only work that can be done, we can substitute Equation 5.12 into Equation 5.5 to give

$$\Delta E = q + w = q - P\,\Delta V \qquad [5.13]$$

When a reaction is carried out in a constant-volume container ($\Delta V = 0$), the heat transferred equals the change in internal energy:

$$\Delta E = q_V \quad \text{(constant volume)} \qquad [5.14]$$

The subscript V indicates that the volume is constant.

Most reactions are run under constant-pressure conditions. In this case Equation 5.13 becomes

$$\Delta E = q_P - P\,\Delta V \quad \text{or}$$
$$q_P = \Delta E + P\,\Delta V \quad \text{(constant pressure)} \qquad [5.15]$$

But we see from Equation 5.9 that the right-hand side of Equation 5.15 is just the enthalpy change under constant pressure conditions.

In summary, the change in internal energy measures the heat gained or lost at constant volume; the change in enthalpy measures the heat gained or lost at constant pressure. The difference between ΔE and ΔH is the amount of P-V work done by the system when the process occurs at constant pressure, $-P\,\Delta V$. The volume change accompanying many reactions is close to zero, which makes $P\,\Delta V$, and therefore the difference between ΔE and ΔH, small. It is generally satisfactory to use ΔH as the measure of energy changes during most chemical processes.

5.4 Enthalpies of Reaction

Because $\Delta H = H_{final} - H_{initial}$, the enthalpy change for a chemical reaction is given by the enthalpy of the products minus the enthalpy of the reactants:

$$\Delta H = H(\text{products}) - H(\text{reactants}) \quad [5.16]$$

ACTIVITY
Enthalpy of Solution

The enthalpy change that accompanies a reaction is called the **enthalpy of reaction** or merely the *heat of reaction* and is sometimes written ΔH_{rxn}, where "rxn" is a commonly used abbreviation for "reaction."

The combustion of hydrogen is shown in Figure 5.14 ▼. When the reaction is controlled so that 2 mol $H_2(g)$ burn to form 2 mol $H_2O(g)$ at a constant pressure, the system releases 483.6 kJ of heat. We can summarize this information as

$$2H_2(g) + O_2(g) \longrightarrow 2H_2O(g) \qquad \Delta H = -483.6 \text{ kJ} \quad [5.17]$$

ΔH is negative, so this reaction is exothermic. Notice that ΔH is reported at the end of the balanced equation, without explicitly mentioning the amounts of chemicals involved. In such cases the coefficients in the balanced equation represent the number of moles of reactants and products producing the associated enthalpy change. Balanced chemical equations that show the associated enthalpy change in this way are called *thermochemical equations*.

The enthalpy change accompanying a reaction can also be represented in an *enthalpy diagram* such as that shown in Figure 5.14(c). Because the combustion of $H_2(g)$ is exothermic, the enthalpy of the products in the reaction is lower than the enthalpy of the reactants. The enthalpy of the system is lower after the reaction because energy has been lost in the form of heat released to the surroundings.

The reaction of hydrogen with oxygen is highly exothermic (ΔH is negative and has a large magnitude), and it occurs rapidly once it starts. It can occur with explosive violence, too, as demonstrated by the disastrous explosions of the German airship *Hindenburg* in 1937 (Figure 5.15 ▶) and the space shuttle *Challenger* in 1986.

The following guidelines are helpful when using thermochemical equations and enthalpy diagrams:

1. *Enthalpy is an extensive property.* The magnitude of ΔH, therefore, is directly proportional to the amount of reactant consumed in the process. For the combustion of methane to form carbon dioxide and liquid water, for

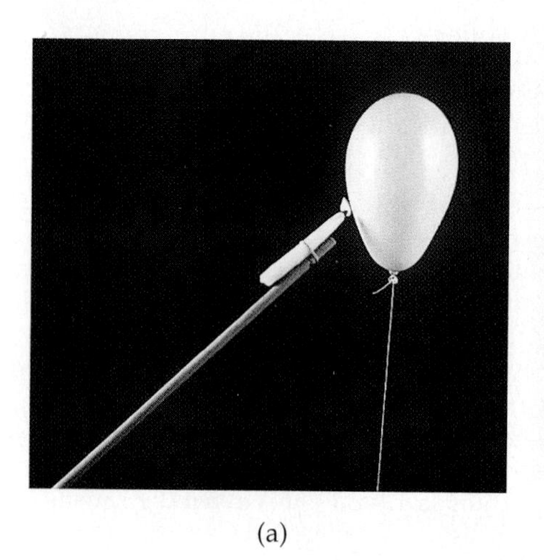

(a)

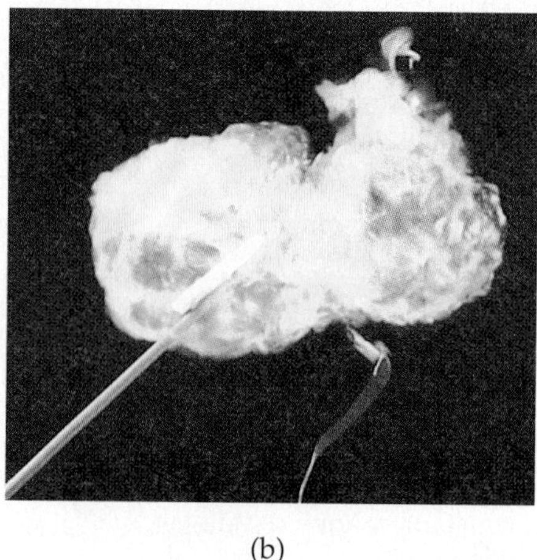

(b)

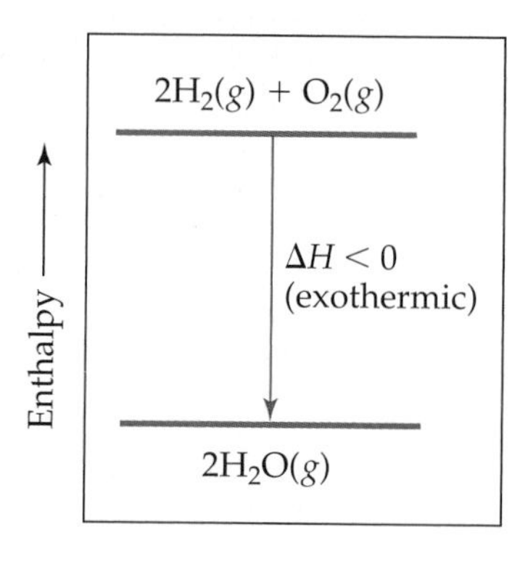

(c)

▲ **Figure 5.14** (a) A candle is held near a balloon filled with hydrogen gas and oxygen gas. (b) The $H_2(g)$ ignites, reacting with $O_2(g)$ to form $H_2O(g)$. The resultant explosion produces the yellow ball of flame. The system gives off heat to its surroundings. (c) The enthalpy diagram for this reaction.

◀ **Figure 5.15** The burning of the hydrogen-filled airship *Hindenburg* in Lakehurst, New Jersey, on May 6, 1937. This photo was taken only 22 seconds after the first explosion occurred. This tragedy led to the discontinuation of hydrogen as a buoyant gas in such craft. Modern-day blimps are filled with helium, which is not as buoyant as hydrogen, but is not flammable.

example, 890 kJ of heat is produced when 1 mol of CH_4 is burned in a constant-pressure system:

$$CH_4(g) + 2O_2(g) \longrightarrow CO_2(g) + 2H_2O(l) \qquad \Delta H = -890 \text{ kJ} \qquad [5.18]$$

ANIMATION
Work of Gas Expansion

Because the combustion of 1 mol of CH_4 with 2 mol of O_2 releases 890 kJ of heat, the combustion of 2 mol of CH_4 with 4 mol of O_2 releases twice as much heat, 1780 kJ.

SAMPLE EXERCISE 5.5

How much heat is released when 4.50 g of methane gas is burned in a constant-pressure system? (Use the information given in Equation 5.18.)

Solution

Analyze: Our goal is to calculate the heat produced when a specific amount of methane gas is combusted.

Plan: According to Equation 5.18, 890 kJ is produced when 1 mol CH_4 is burned at constant pressure ($\Delta H = -890$ kJ). We can treat this information as a stoichiometric relationship: 1 mol $CH_4 \simeq -890$ kJ. To use this relationship, however, we must convert grams of CH_4 to moles of CH_4.

Solve: By adding the atomic weights of C and 4 H, we have 1 mol CH_4 = 16.0 g CH_4. Thus, we can use the appropriate conversion factors to convert grams of CH_4 to moles of CH_4 to kilojoules:

$$\text{Heat} = (4.50 \text{ g } CH_4)\left(\frac{1 \text{ mol } CH_4}{16.0 \text{ g } CH_4}\right)\left(\frac{-890 \text{ kJ}}{1 \text{ mol } CH_4}\right) = -250 \text{ kJ}$$

Check: The negative sign indicates that 250 kJ is released by the system into the surroundings.

PRACTICE EXERCISE

Hydrogen peroxide can decompose to water and oxygen by the following reaction:

$$2H_2O_2(l) \longrightarrow 2H_2O(l) + O_2(g) \qquad \Delta H = -196 \text{ kJ}$$

Calculate the value of q when 5.00 g of $H_2O_2(l)$ decomposes at constant pressure.
Answer: −14.4 kJ

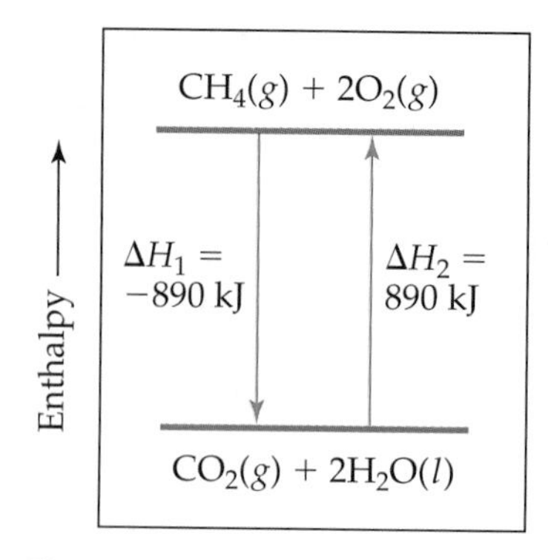

▲ **Figure 5.16** Reversing a reaction changes the sign but not the magnitude of the enthalpy change: $\Delta H_2 = -\Delta H_1$.

2. *The enthalpy change for a reaction is equal in magnitude, but opposite in sign, to ΔH for the reverse reaction.* For example, when Equation 5.18 is reversed, ΔH for the process is +890 kJ:

$$CO_2(g) + 2H_2O(l) \longrightarrow CH_4(g) + 2O_2(g) \qquad \Delta H = 890 \text{ kJ} \qquad [5.19]$$

When we reverse a reaction, we reverse the roles of the products and the reactants; thus, the reactants in a reaction become the products of the reverse reaction, and so forth. From Equation 5.16, we can see that reversing the products and reactants leads to the same magnitude, but a change in sign for ΔH. This relationship is diagrammed for Equations 5.18 and 5.19 in Figure 5.16 ◀.

3. *The enthalpy change for a reaction depends on the state of the reactants and products.* If the product in the combustion of methane (Equation 5.18) were gaseous H_2O instead of liquid H_2O, ΔH would be −802 kJ instead of −890 kJ. Less heat would be available for transfer to the surroundings because the enthalpy of $H_2O(g)$ is greater than that of $H_2O(l)$. One way to see this is to imagine that the product is initially liquid water. The liquid water must be converted to water vapor, and the conversion of 2 mol $H_2O(l)$ to 2 mol $H_2O(g)$ is an endothermic process that absorbs 88 kJ:

$$2H_2O(l) \longrightarrow 2H_2O(g) \qquad \Delta H = +88 \text{ kJ} \qquad [5.20]$$

Thus, it is important to specify the states of the reactants and products in thermochemical equations. In addition, we will generally assume that the reactants and products are both at the same temperature, 25°C, unless otherwise indicated.

There are many situations in which it is valuable to know the enthalpy change associated with a given chemical process. As we will see in the following sections, ΔH_{rxn} can be determined directly by experiment or calculated from the known enthalpy changes of other reactions by invoking the first law of thermodynamics.

Strategies in Chemistry Using Enthalpy as a Guide

If you hold a brick in the air and let it go, it will fall as the force of gravity pulls it toward Earth. A process that is thermodynamically favored to happen, such as a falling brick, is called a *spontaneous* process.

Many chemical processes are thermodynamically favored, or spontaneous, too. By "spontaneous," we don't mean that the reaction will form products without intervention. That can be the case, but often some energy must be imparted to get the process started. The enthalpy change in a reaction gives one indication as to whether it is likely to be spontaneous. The combustion of $H_2(g)$ and $O_2(g)$, for example, is a highly exothermic process:

$$H_2(g) + \tfrac{1}{2}O_2(g) \longrightarrow H_2O(g) \qquad \Delta H = -242 \text{ kJ}$$

Hydrogen gas and oxygen gas can exist together in a volume indefinitely without noticeable reaction occurring, as in Figure 5.14(a). Once initiated, however, energy is rapidly transferred from the system (the reactants) to the surroundings. As the reaction proceeds, large amounts of heat, are released, which greatly increases the temperature of the reactants and the products. The system then loses enthalpy by transferring the heat to the surroundings. (Recall from the first law of thermodynamics that the total energy of the system plus the surroundings will not change; energy is conserved.)

Enthalpy change is not the only consideration in the spontaneity of reactions, however, nor is it a foolproof guide. For example, the melting of ice is an endothermic process:

$$H_2O(s) \longrightarrow H_2O(l) \qquad \Delta H = +6.01 \text{ kJ}$$

Even though this process is endothermic, it is spontaneous at temperatures above the freezing point of water (0°C). The reverse process, the freezing of water to ice, is spontaneous at temperatures below 0°C. Thus, we know that ice at room temperature will melt and that water put into a freezer at −20°C will turn into ice; both of these processes are spontaneous even though they are the reverse of one another. In Chapter 19 we will address the spontaneity of processes more fully. We will see why a process can be spontaneous at one temperature, but not at another, as is the case for the conversion of water to ice.

Despite these complicating factors, however, you should pay attention to the enthalpy changes in reactions. As a general observation, when the enthalpy change is large, it is the dominant factor in determining spontaneity. Thus, reactions for which ΔH is *large* and *negative* tend to be spontaneous. Reactions for which ΔH is *large* and *positive* tend to be spontaneous in the reverse direction. There are a number of ways in which the enthalpy of a reaction can be estimated; from these estimates, the likelihood of the reaction being thermodynamically favorable can be predicted.

5.5 Calorimetry

The value of ΔH can be determined experimentally by measuring the heat flow accompanying a reaction at constant pressure. When heat flows into or out of a substance, the temperature of the substance changes. Experimentally, we can determine the heat flow associated with a chemical reaction by measuring the temperature change it produces. The measurement of heat flow is **calorimetry**; an apparatus used to measure heat flow is a **calorimeter**.

Heat Capacity and Specific Heat

Objects can emit or absorb heat: Red-hot charcoal emits heat in the form of radiant energy; an ice pack absorbs heat when it is placed on a swollen ankle. The emission or absorption of heat causes an object to change temperature. The temperature change experienced by an object when it absorbs a certain amount of energy is determined by its **heat capacity**. The heat capacity of an object is the amount of heat required to raise its temperature by 1 K (or 1°C). The greater the heat capacity, the greater the heat required to produce a given rise in temperature.

For pure substances the heat capacity is usually given for a specified amount of the substance. The heat capacity of 1 mol of a substance is called its **molar heat capacity**. The heat capacity of 1 g of a substance is called its *specific heat capacity*, or merely its **specific heat** (Figure 5.17 ▶). The specific heat of a substance can be determined experimentally by measuring the temperature change, ΔT, that a known mass, m, of the substance undergoes when it gains or loses a specific quantity of heat, q:

$$\text{Specific heat} = \frac{\text{(quantity of heat transferred)}}{\text{(grams of substance)} \times \text{(temperature change)}}$$

$$= \frac{q}{m \times \Delta T} \qquad [5.21]$$

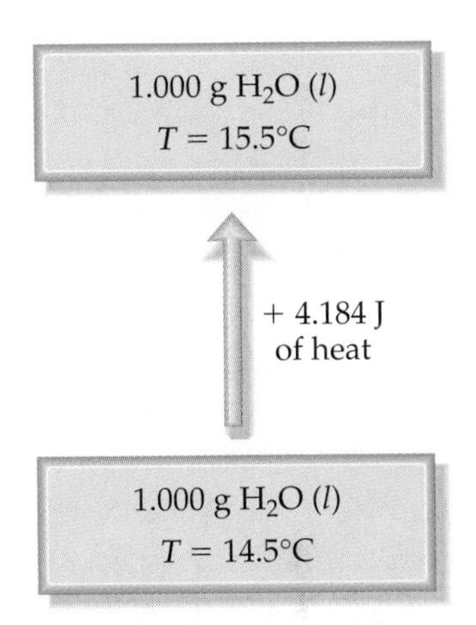

▲ **Figure 5.17** Specific heat indicates the amount of heat that must be added to 1 g of a substance to raise its temperature by 1 K (or 1°C). Specific heats can vary slightly with temperature, so for precise measurements the temperature is specified. For example, the specific heat of $H_2O(l)$ at 14.5°C is 4.184 J/g-K; the addition of 4.184 J of heat raises the temperature to 15.5°C. This amount of energy defines the calorie: 1 cal = 4.184 J.

For example, 209 J is required to increase the temperature of 50.0 g of water by 1.00 K. Thus, the specific heat of water is

$$\text{Specific heat} = \frac{209\ \text{J}}{(50.0\ \text{g})(1.00\ \text{K})} = 4.18\frac{\text{J}}{\text{g-K}}$$

A temperature change in kelvins is equal in magnitude to the temperature change in degrees Celsius: ΔT in K = ΔT in °C. (Section 1.4) When the sample gains heat (positive q), the temperature of the sample increases (positive ΔT).

The specific heats of several substances are listed in Table 5.2 ▼. Notice that the specific heat of liquid water is higher than those of the other substances listed.

TABLE 5.2 Specific Heats of Some Substances at 298 K

Elements		Compounds	
Substance	Specific Heat (J/g-K)	Substance	Specific Heat (J/g-K)
$N_2(g)$	1.04	$H_2O(l)$	4.18
$Al(s)$	0.90	$CH_4(g)$	2.20
$Fe(s)$	0.45	$CO_2(g)$	0.84
$Hg(l)$	0.14	$CaCO_3(s)$	0.82

For example, it is about five times as great as that of aluminum metal. The high specific heat of water affects Earth's climate because it keeps the temperatures of the oceans relatively resistant to change. It also is very important in maintaining a constant temperature in our bodies, as will be discussed in the "Chemistry and Life" box later in this chapter.

We can calculate the quantity of heat that a substance has gained or lost by using its specific heat together with its measured mass and temperature change. Rearranging Equation 5.21, we get

$$q = (\text{specific heat}) \times (\text{grams of substance}) \times \Delta T \qquad [5.22]$$

SAMPLE EXERCISE 5.6

(a) How much heat is needed to warm 250 g of water (about 1 cup) from 22°C (about room temperature) to near its boiling point, 98°C? The specific heat of water is 4.18 J/g-K. **(b)** What is the molar heat capacity of water?

Solution

Analyze: In **(a)** we must find the total quantity of heat needed to warm the sample of water. In **(b)** we must calculate the molar heat capacity.

Plan: We know the total quantity of water and the specific heat (that is, the heat capacity per gram) of water. With this and the total temperature change involved, we can calculate the quantity of heat.

Solve: The water undergoes a temperature change of $\Delta T = 98°C - 22°C = 76°C = 76\text{ K}$. Using Equation 5.22, we have

$$\begin{aligned} q &= (\text{specific heat of } H_2O) \times (\text{grams of } H_2O) \times \Delta T \\ &= (4.18\text{ J/g-K})(250\text{ g})(76\text{ K}) = 7.9 \times 10^4\text{ J} \end{aligned}$$

(b) The molar heat capacity is the heat capacity of 1 mol of substance. Using the atomic weights of hydrogen and oxygen, we have 1 mol H_2O = 18.0 g H_2O. From the specific heat given in part (a), we have

$$\text{Molar heat capacity} = (4.18\text{ J/g-K})\left(\frac{18.0\text{ g}}{1\text{ mol}}\right) = 75.2\text{ J/mol-K}$$

PRACTICE EXERCISE

(a) Large beds of rocks are used in some solar-heated homes to store heat. Assume that the specific heat of the rocks is 0.82 J/g-K. Calculate the quantity of heat absorbed by 50.0 kg of rocks if their temperature increases by 12.0°C. **(b)** What temperature change would these rocks undergo if they emitted 450 kJ of heat?

Answers: **(a)** 4.9×10^5 J; **(b)** 11 K = 11°C decrease

Constant-Pressure Calorimetry

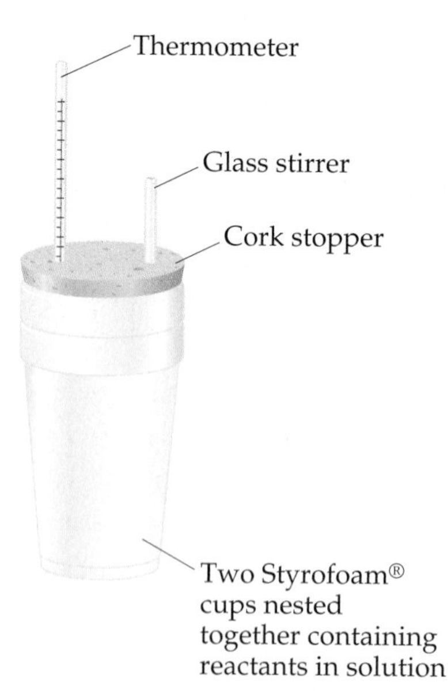

▲ **Figure 5.18** Coffee-cup calorimeter, in which reactions occur at constant pressure.

The techniques and equipment employed in calorimetry depend on the nature of the process being studied. For many reactions, such as those occurring in solution, it is easy to control pressure so that ΔH is measured directly. (Recall that $\Delta H = q_P$.) Although the calorimeters used for highly accurate work are precision instruments, a very simple "coffee-cup" calorimeter, as shown in Figure 5.18 ◀, is often used in general chemistry labs to illustrate the principles of calorimetry. Because the calorimeter is not sealed, the reaction occurs under the essentially constant pressure of the atmosphere.

If we assume that the calorimeter perfectly prevents the gain or loss of heat from the solution to its surroundings, the heat gained by the solution must be produced from the chemical reaction under study. In other words, the heat produced by the reaction, q_{rxn}, is entirely absorbed by the solution; it does not escape the calorimeter. (We also assume that the calorimeter itself does not absorb heat. In the case of the coffee-cup calorimeter, this is a reasonable approximation because the calorimeter has a very low thermal conductivity and heat capacity.) For an exothermic reaction, heat is "lost" by the reaction and "gained" by the solution, so the temperature of the solution rises. The opposite occurs for an endothermic reaction. The heat gained by the solution, q_{soln}, is therefore equal in magnitude and opposite in sign from q_{rxn}: $q_{soln} = -q_{rxn}$. The value of q_{soln} is readily calculated from the mass of the solution, its specific heat, and the temperature change:

$$q_{soln} = (\text{specific heat of solution}) \times (\text{grams of solution}) \times \Delta T = -q_{rxn} \quad [5.23]$$

For dilute aqueous solutions, the specific heat of the solution will be approximately the same as that of water, 4.18 J/g-K.

Equation 5.23 makes it possible to calculate q_{rxn} from the temperature change of the solution in which the reaction occurs. A temperature increase ($\Delta T > 0$) means the reaction is exothermic ($q_{rxn} < 0$).

SAMPLE EXERCISE 5.7

When a student mixes 50 mL of 1.0 *M* HCl and 50 mL of 1.0 *M* NaOH in a coffee-cup calorimeter, the temperature of the resultant solution increases from 21.0°C to 27.5°C. Calculate the enthalpy change for the reaction, assuming that the calorimeter loses only a negligible quantity of heat, that the total volume of the solution is 100 mL, that its density is 1.0 g/mL, and that its specific heat is 4.18 J/g-K.

Solution

Analyze: We need to calculate a heat of reaction per mole, given a temperature increase, the number of moles involved, and enough information to calculate the heat capacity of the system.

Plan: The total heat evolved can be calculated from the temperature change, the volume of the solution, its density, and the specific heat.

Solve: Because the total volume of the solution is 100 mL, its mass is

$$(100 \text{ mL})(1.0 \text{ g/mL}) = 100 \text{ g}$$

The temperature change is

$$27.5°\text{C} - 21.0°\text{C} = 6.5°\text{C} = 6.5 \text{ K}$$

Because the temperature increases, the reaction must be exothermic:

$$q_{rxn} = -(\text{specific heat of solution}) \times (\text{grams of solution}) \times \Delta T$$
$$= -(4.18 \text{ J/g-K})(100 \text{ g})(6.5 \text{ K}) = -2.7 \times 10^3 \text{ J} = -2.7 \text{ kJ}$$

Because the process occurs at constant pressure,

$$\Delta H = q_P = -2.7 \text{ kJ}$$

To express the enthalpy change on a molar basis, we use the fact that the number of moles of HCl and NaOH is given by the product of the respective solution volumes (50 mL = 0.050 L) and concentrations:

$$(0.050 \text{ L})(1.0 \text{ mol/L}) = 0.050 \text{ mol}$$

Thus, the enthalpy change per mole of HCl (or NaOH) is

$$\Delta H = -2.7 \text{ kJ}/0.050 \text{ mol} = -54 \text{ kJ/mol}$$

Check: ΔH is negative (exothermic), which is expected for the reaction of an acid with a base. The molar magnitude of the heat evolved seems reasonable.

PRACTICE EXERCISE

When 50.0 mL of 0.100 *M* $AgNO_3$ and 50.0 mL of 0.100 *M* HCl are mixed in a constant-pressure calorimeter, the temperature of the mixture increases from 22.30°C to 23.11°C. The temperature increase is caused by the following reaction:

$$AgNO_3(aq) + HCl(aq) \longrightarrow AgCl(s) + HNO_3(aq)$$

Calculate ΔH for this reaction, assuming that the combined solution has a mass of 100.0 g and a specific heat of 4.18 J/g-°C.

Answer: −68,000 J/mol = −68 kJ/mol

Bomb Calorimetry (Constant-Volume Calorimetry)

Calorimetry can be used to study the chemical potential energy stored in substances. One of the most important types of reactions studied using calorimetry is combustion, in which a compound (usually an organic compound) reacts completely with excess oxygen. (Section 3.2) Combustion reactions are most conveniently studied using a **bomb calorimeter**, a device shown schematically

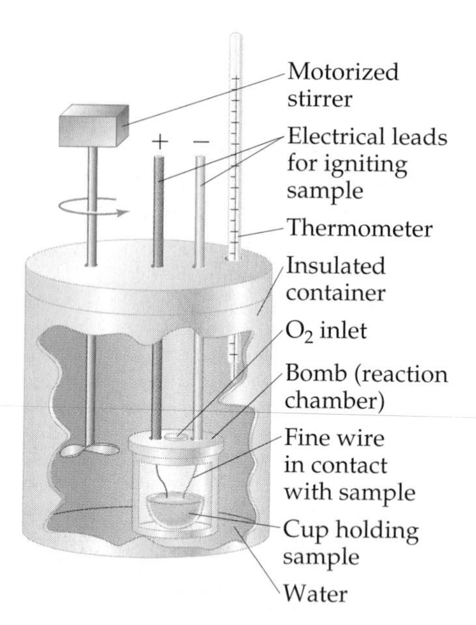

▲ **Figure 5.19** Cutaway view of a bomb calorimeter, in which reactions occur at constant volume.

in Figure 5.19 ◀. The substance to be studied is placed in a small cup within a sealed vessel called a *bomb*. The bomb, which is designed to withstand high pressures, has an inlet valve for adding oxygen and electrical contacts to initiate the combustion. After the sample has been placed in the bomb, the bomb is sealed and pressurized with oxygen. It is then placed in the calorimeter, which is essentially an insulated container, and covered with an accurately measured quantity of water. When all the components within the calorimeter have come to the same temperature, the combustion reaction is initiated by passing an electrical current through a fine wire that is in contact with the sample. When the wire gets sufficiently hot, the sample ignites.

Heat is released when combustion occurs. This heat is absorbed by the calorimeter contents, causing a rise in the temperature of the water. The temperature of the water is very carefully measured before reaction and then after reaction when the contents of the calorimeter have again arrived at a common temperature. The heat evolved in the combustion of the sample is absorbed by its surroundings (the calorimeter contents).

To calculate the heat of combustion from the measured temperature increase in the bomb calorimeter, we must know the heat capacity of the calorimeter, C_{cal}. This quantity is determined by combusting a sample that releases a known quantity of heat and measuring the resulting temperature change. For example, the combustion of exactly 1 g of benzoic acid, $C_7H_6O_2$, in a bomb calorimeter produces 26.38 kJ of heat. Suppose 1.000 g of benzoic acid is combusted in a calorimeter, and it increases the temperature by 4.857°C. The heat capacity of the calorimeter is then given by $C_{cal} = 26.38 \text{ kJ}/4.857°\text{C} = 5.431 \text{ kJ}/°\text{C}$. Once we know the value of C_{cal}, we can measure temperature changes produced by other reactions, and from these we can calculate the heat evolved in the reaction, q_{rxn}:

$$q_{rxn} = -C_{cal} \times \Delta T \quad [5.24]$$

SAMPLE EXERCISE 5.8

Methylhydrazine (CH_6N_2) is commonly used as a liquid rocket fuel. The combustion of methylhydrazine with oxygen produces $N_2(g)$, $CO_2(g)$, and $H_2O(l)$:

$$2CH_6N_2(l) + 5O_2(g) \longrightarrow 2N_2(g) + 2CO_2(g) + 6H_2O(l)$$

When 4.00 g of methylhydrazine is combusted in a bomb calorimeter, the temperature of the calorimeter increases from 25.00°C to 39.50°C. In a separate experiment the heat capacity of the calorimeter is measured to be 7.794 kJ/°C. What is the heat of reaction for the combustion of a mole of CH_6N_2 in this calorimeter?

Solution

Analyze: We are given a temperature change and the total heat capacity of the calorimeter. We are also given the amount of reactant combusted. Our goal is to calculate the enthalpy change per mole for combustion of the reactant.

Plan: We will first calculate the heat evolved for the combustion of the 4.00-g sample. We will then convert this heat to a molar quantity.

Solve: For combustion of the 4.00-g sample of methylhydrazine, the temperature change of the calorimeter is

$$\Delta T = (39.50°\text{C} - 25.00°\text{C}) = 14.50°\text{C}$$

We can use this value and the value for C_{cal} to calculate the heat of reaction (Equation 5.24):

$$q_{rxn} = -C_{cal} \times \Delta T = -(7.794 \text{ kJ}/°\text{C})(14.50°\text{C}) = -113.0 \text{ kJ}$$

We can readily convert this value to the heat of reaction for a mole of CH_6N_2:

$$\left(\frac{-113.0 \text{ kJ}}{4.00 \text{ g } CH_6N_2}\right) \times \left(\frac{46.1 \text{ g } CH_6N_2}{1 \text{ mol } CH_6N_2}\right) = -1.30 \times 10^3 \text{ kJ/mol } CH_6N_2$$

Check: The units cancel properly, and the sign of the answer is negative as it should be for an exothermic reaction.

PRACTICE EXERCISE

A 0.5865-g sample of lactic acid ($HC_3H_5O_3$) is burned in a calorimeter whose heat capacity is 4.812 kJ/°C. The temperature increases from 23.10°C to 24.95°C. Calculate the heat of combustion of **(a)** lactic acid per gram and **(b)** per mole.
Answers: **(a)** −15.2 kJ/g; **(b)** −1370 kJ/mol

Chemistry and Life The Regulation of Human Body Temperature

For most of us, the question "Are you running a fever?" was one of our first introductions to medical diagnosis. Indeed, a deviation in body temperature of only a few degrees indicates that something is amiss. In the laboratory you may have tried to maintain a solution or water bath at a constant temperature, only to find how difficult it can be to keep the solution within a very narrow temperature range. Yet our bodies manage to maintain a near-constant temperature in spite of widely varying weather, levels of physical activity, and periods of high metabolic activity (such as after a meal). How does the human body manage this task, and how does it relate to some of the topics we have discussed in this chapter?

Maintaining a near-constant temperature is one of the primary physiological functions of the human body. Normal body temperature generally ranges from 35.8–37.2°C (96.5–99°F). This very narrow temperature range is essential to proper muscle function and to the control of the rates of the biochemical reactions in the body. You will learn more about the effects of temperature on reaction rates in Chapter 14. The temperature is regulated by a portion of the human brain stem called the *hypothalamus*. The hypothalamus acts as a thermostat for body temperature. When body temperature rises above the high end of the normal range, the hypothalamus triggers mechanisms to lower the temperature. It likewise triggers mechanisms to increase the temperature if body temperature drops too low.

To qualitatively understand how the body's heating and cooling mechanisms operate, we can view the body as a thermodynamic system. The body increases its internal energy content by ingesting foods from the surroundings.
The foods, such as glucose ($C_6H_{12}O_6$), are metabolized—a process that is essentially controlled oxidation to CO_2 and H_2O:

$$C_6H_{12}O_6(s) + 6O_2(g) \longrightarrow 6CO_2(g) + 6H_2O(l) \qquad \Delta H = -2803 \text{ kJ}$$

Roughly 40% of the energy produced is ultimately used to do work in the form of muscle and nerve contractions. The remainder of the energy is released as heat, part of which is used to maintain body temperature. When the body produces too much heat, as in times of heavy physical exertion, it dissipates the excess to the surroundings.

Heat is transferred from the body to its surroundings primarily by *radiation*, *convection*, and *evaporation*. Radiation is the direct loss of heat from the body to cooler surroundings, much as a hot stovetop radiates heat to its surroundings. Convection is heat loss by virtue of heating air that is in contact with the body. The heated air rises and is replaced with cooler air, and the process continues. Warm clothing, which usually consists of insulating layers of material with "dead air" in between, decreases convective heat loss in cold weather. Evaporative cooling occurs when perspiration is generated at the skin surface by the sweat glands. Heat is removed from the body as the perspiration evaporates into the surroundings. Perspiration is predominantly water, so the process involved is the endothermic conversion of liquid water into water vapor:

$$H_2O(l) \longrightarrow H_2O(g) \qquad \Delta H = +44.0 \text{ kJ}$$

The speed with which evaporative cooling occurs decreases as the atmospheric humidity increases, which is why people seem to be more sweaty and uncomfortable on hot and humid days.

When the hypothalamus senses that the body temperature has risen too high, it increases heat loss from the body in two principal ways. First, it increases the flow of blood near the surface of the skin, which allows for increased radiational and convective cooling. The reddish "flushed" appearance of a hot individual is the result of this increased subsurface blood flow. Second, the hypothalamus stimulates the secretion of perspiration from the sweat glands, which increases evaporative cooling. During periods of extreme activity, the amount of liquid secreted as perspiration can be as high as 2–4 liters per hour. As a result, water must be replenished to the body during these periods (Figure 5.20 ▼). If the body loses too much fluid through perspiration, it will no longer be able to cool itself and blood volume decreases, which can lead to *heat exhaustion* or the more serious and potentially fatal *heat stroke*, during which the body temperature can rise to as high as 41–45°C (106–113°F).

When body temperature drops too low, the hypothalamus decreases the blood flow to the surface of the skin, thereby decreasing heat loss. It also triggers small involuntary contractions of the muscles; the biochemical reactions that generate the energy to do this work also generate more heat for the body. When these contractions get large enough—as when the body feels a chill—a *shiver* results. If the body is unable to maintain a temperature above 35°C (95°F), the very dangerous condition called *hypothermia* can result.

The ability of the human body to maintain its temperature by "tuning" the amount of heat it generates and transfers to its surroundings is truly remarkable. If you take courses in human anatomy and physiology, you will see many other applications of thermochemistry and thermodynamics to the ways in which the human body works.

▲ **Figure 5.20** Marathon runners must constantly replenish the water in their bodies that is lost through perspiration.

Because the reactions in a bomb calorimeter are carried out under constant-volume conditions, the heat transferred corresponds to the change in internal energy, ΔE, rather than the change in enthalpy, ΔH (Equation 5.14). For most reactions, however, the difference between ΔE and ΔH is very small. For the reaction discussed in Sample Exercise 5.8, for example, the difference between ΔE and ΔH is only about 1 kJ/mol—a difference of less than 0.1%. It is possible to correct the measured heat changes to obtain ΔH values, and these form the basis of the tables of enthalpy change that we will see in the following sections. However, we need not concern ourselves with how these small corrections are made.

5.6 Hess's Law

Many enthalpies of reaction have been measured and tabulated. In this section and the next we will see that it is often possible to calculate the ΔH for a reaction from the tabulated ΔH values of other reactions. Thus, it is not necessary to make calorimetric measurements for all reactions.

Because enthalpy is a state function, the enthalpy change, ΔH, associated with any chemical process depends only on the amount of matter that undergoes change and on the nature of the initial state of the reactants and the final state of the products. This means that if a particular reaction can be carried out in one step or in a series of steps, the sum of the enthalpy changes associated with the individual steps must be the same as the enthalpy change associated with the one-step process. As an example, the combustion of methane gas, $CH_4(g)$, to form $CO_2(g)$ and liquid water can be thought of as occurring in two steps: (1) the combustion of $CH_4(g)$ to form $CO_2(g)$ and gaseous water, $H_2O(g)$, and (2) the condensation of gaseous water to form liquid water, $H_2O(l)$. The enthalpy change for the overall process is simply the sum of the enthalpy changes for these two steps:

$$CH_4(g) + 2O_2(g) \longrightarrow CO_2(g) + 2H_2O(g) \qquad \Delta H = -802\ \text{kJ}$$

$$\text{(Add)} \qquad 2H_2O(g) \longrightarrow 2H_2O(l) \qquad \Delta H = -88\ \text{kJ}$$

$$CH_4(g) + 2O_2(g) + 2H_2O(g) \longrightarrow CO_2(g) + 2H_2O(l) + 2H_2O(g) \qquad \Delta H = -890\ \text{kJ}$$

The net equation is

$$CH_4(g) + 2O_2(g) \longrightarrow CO_2(g) + 2H_2O(l) \qquad \Delta H = -890\ \text{kJ}$$

To obtain the net equation, the sum of the reactants of the two equations is placed on one side of the arrow and the sum of the products on the other side. Because $2H_2O(g)$ occurs on both sides, it can be canceled like an algebraic quantity that appears on both sides of an equal sign.

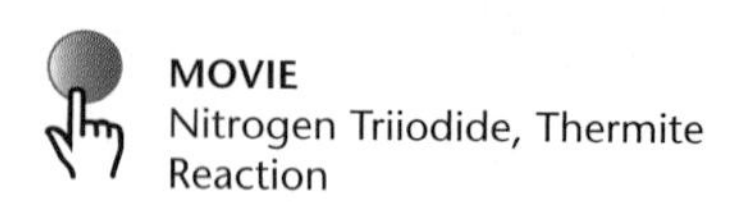

Hess's law states that *if a reaction is carried out in a series of steps, ΔH for the reaction will equal the sum of the enthalpy changes for the individual steps.* The overall enthalpy change for the process is independent of the number of steps or the particular nature of the path by which the reaction is carried out. We can therefore calculate ΔH for any process, as long as we find a route for which ΔH is known for each step. This means that a relatively small number of experimental measurements can be used to calculate ΔH for a vast number of different reactions.

Hess's law provides a useful means of calculating energy changes that are difficult to measure directly. For instance, it is impossible to measure directly the enthalpy of combustion of carbon to form carbon monoxide. Combustion of 1 mol of carbon with 0.5 mol of O_2 produces not only CO, but also CO_2, leaving some carbon unreacted. However, solid carbon and carbon monoxide can both be completely burned in O_2 to produce CO_2. We can use the enthalpy changes of these reactions to calculate the heat of combustion of C to CO, as shown in Sample Exercise 5.9.

SAMPLE EXERCISE 5.9

The enthalpy of combustion of C to CO_2 is −393.5 kJ/mol C, and the enthalpy of combustion of CO to CO_2 is −283.0 kJ/mol CO:

$$(1) \quad C(s) + O_2(g) \longrightarrow CO_2(g) \qquad \Delta H = -393.5 \text{ kJ}$$

$$(2) \quad CO(g) + \tfrac{1}{2}O_2(g) \longrightarrow CO_2(g) \qquad \Delta H = -283.0 \text{ kJ}$$

Using these data, calculate the enthalpy of combustion of C to CO:

$$(3) \quad C(s) + \tfrac{1}{2}O_2(g) \longrightarrow CO(g)$$

Solution

Analyze: We are given two enthalpies of combustion, and our goal is to combine them in such a way as to obtain the enthalpy of combustion of a third process.

Plan: We will manipulate the two equations we have been given so that when added, they yield the desired reaction. At the same time, we employ Hess's law to keep track of the enthalpy changes for the two reactions.

Solve: In order to use equations (1) and (2), we arrange them so that $C(s)$ is on the reactant side and $CO(g)$ is on the product side of the arrow, as in the target reaction, equation (3). Because equation (1) has $C(s)$ as a reactant, we can use that equation just as it is. We need to turn equation (2) around, however, so that $CO(g)$ is a product. Remember that when reactions are turned around, the sign of ΔH is reversed. We arrange the two equations so that they can be added to give the desired equation:

$$C(s) + O_2(g) \longrightarrow CO_2(g) \qquad \Delta H = -393.5 \text{ kJ}$$

$$CO_2(g) \longrightarrow CO(g) + \tfrac{1}{2}O_2(g) \qquad \Delta H = 283.0 \text{ kJ}$$

$$\overline{C(s) + \tfrac{1}{2}O_2(g) \longrightarrow CO(g) \qquad \Delta H = -110.5 \text{ kJ}}$$

When we add the two equations, $CO_2(g)$ appears on both sides of the arrow and therefore cancels out. Likewise, $\tfrac{1}{2}O_2(g)$ is eliminated from each side.

PRACTICE EXERCISE

Carbon occurs in two forms, graphite and diamond. The enthalpy of combustion of graphite is −393.5 kJ/mol and that of diamond is −395.4 kJ/mol:

$$C(\text{graphite}) + O_2(g) \longrightarrow CO_2(g) \qquad \Delta H = -393.5 \text{ kJ}$$

$$C(\text{diamond}) + O_2(g) \longrightarrow CO_2(g) \qquad \Delta H = -395.4 \text{ kJ}$$

Calculate ΔH for the conversion of graphite to diamond:

$$C(\text{graphite}) \longrightarrow C(\text{diamond})$$

Answer: +1.9 kJ

SAMPLE EXERCISE 5.10

Calculate ΔH for the reaction

$$2C(s) + H_2(g) \longrightarrow C_2H_2(g)$$

given the following reactions and their respective enthalpy changes:

$$C_2H_2(g) + \tfrac{5}{2}O_2(g) \longrightarrow 2CO_2(g) + H_2O(l) \qquad \Delta H = -1299.6 \text{ kJ}$$

$$C(s) + O_2(g) \longrightarrow CO_2(g) \qquad \Delta H = -393.5 \text{ kJ}$$

$$H_2(g) + \tfrac{1}{2}O_2(g) \longrightarrow H_2O(l) \qquad \Delta H = -285.8 \text{ kJ}$$

Solution

Analyze: To calculate the enthalpy change for one reaction we must utilize data provided for three other processes. We utilize these data by taking advantage of Hess's law.

Plan: We will sum the three equations or their reverses and multiply each by an appropriate coefficient, so that the net equation is that for the reaction of interest. At the same time, we keep track of the ΔH values, reversing their signs if the reactions are reversed and multiplying them by whatever coefficient is employed on the equation.

Solve: Because the target equation has C_2H_2 as a product, we turn the first equation around; the sign of ΔH is therefore changed. The desired equation has 2C(*s*) as a reactant, so we multiply the second equation and its ΔH by 2. Because the target equation has H_2 as a reactant, we keep the third equation as it is. We then add the three equations and their enthalpy changes in accordance with Hess's law:

$$2CO_2(g) + H_2O(l) \longrightarrow C_2H_2(g) + \tfrac{5}{2}O_2(g) \qquad \Delta H = 1299.6 \text{ kJ}$$
$$2C(s) + 2O_2(g) \longrightarrow 2CO_2(g) \qquad \Delta H = -787.0 \text{ kJ}$$
$$H_2(g) + \tfrac{1}{2}O_2(g) \longrightarrow H_2O(l) \qquad \Delta H = -285.8 \text{ kJ}$$
$$2C(s) + H_2(g) \longrightarrow C_2H_2(g) \qquad \Delta H = 226.8 \text{ kJ}$$

When the equations are added, there are 2 CO_2, $\frac{5}{2}O_2$, and H_2O on both sides of the arrow. These are canceled in writing the net equation.

Check: The procedure must be correct because we obtained the correct net equation. In cases like this you should go back over the numerical manipulations of the ΔH values to ensure that you did not make an inadvertent error with signs.

PRACTICE EXERCISE

Calculate ΔH for the reaction

$$NO(g) + O(g) \longrightarrow NO_2(g)$$

given the following information:

$$NO(g) + O_3(g) \longrightarrow NO_2(g) + O_2(g) \qquad \Delta H = -198.9 \text{ kJ}$$
$$O_3(g) \longrightarrow \tfrac{3}{2}O_2(g) \qquad \Delta H = -142.3 \text{ kJ}$$
$$O_2(g) \longrightarrow 2O(g) \qquad \Delta H = 495.0 \text{ kJ}$$

Answer: −304.1 kJ

In many cases it will turn out that a given reaction could be arrived at by more than one set of stepwise equations. Will the final value of ΔH for a reaction depend on the way in which we break it down to use Hess's law? H is a state function, so *we will always get the same value of* ΔH *for an overall reaction, regardless of how many steps we employ to get to the final products*. For example, consider the reaction of methane (CH_4) and oxygen (O_2) to form CO_2 and H_2O. We can envision the reaction forming CO_2 directly, as we did before, or with the initial formation of CO, which is then combusted to CO_2. These choices are compared in Figure 5.21 ◀. Because H is a state function, both paths *must* produce the same value of ΔH. In the enthalpy diagram, that means $\Delta H_1 = \Delta H_2 + \Delta H_3$.

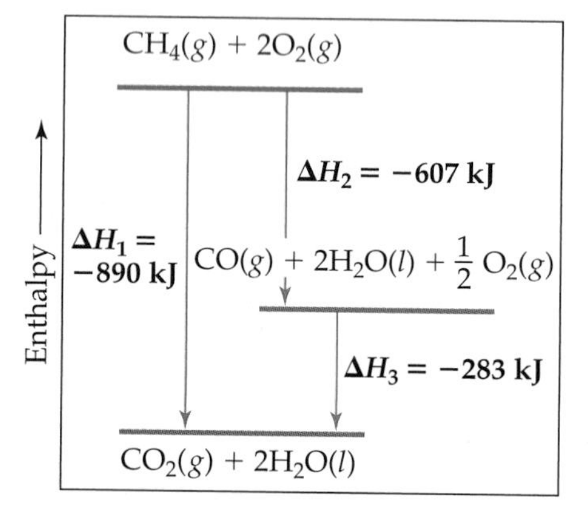

▲ **Figure 5.21** The quantity of heat generated by combustion of 1 mol CH_4 is independent of whether the reaction takes place in one or more steps: $\Delta H_1 = \Delta H_2 + \Delta H_3$.

5.7 Enthalpies of Formation

By using the methods we have just discussed, we can calculate the enthalpy changes for a great many reactions from tabulated ΔH values. Many experimental data are tabulated according to the type of process. For example, extensive tables exist of *enthalpies of vaporization* (ΔH for converting liquids to gases), *enthalpies of fusion* (ΔH for melting solids), *enthalpies of combustion* (ΔH for combusting a substance in oxygen), and so forth. A particularly important process used for tabulating thermochemical data is the formation of a compound from its constituent elements. The enthalpy change associated with this process is called the **enthalpy of formation** (or *heat of formation*) and is labeled ΔH_f, where the subscript f indicates that the substance has been *formed* from its elements.

The magnitude of any enthalpy change depends on the conditions of temperature, pressure, and state (gas, liquid, or solid, crystalline form) of the reactants and products. In order to compare the enthalpies of different reactions, we must define a set of conditions, called a *standard state*, at which most enthalpies are tabulated.

TABLE 5.3 Standard Enthalpies of Formation, ΔH_f°, at 298 K

Substance	Formula	ΔH_f° (kJ/mol)	Substance	Formula	ΔH_f° (kJ/mol)
Acetylene	$C_2H_2(g)$	226.7	Hydrogen chloride	$HCl(g)$	−92.30
Ammonia	$NH_3(g)$	−46.19	Hydrogen fluoride	$HF(g)$	−268.6
Benzene	$C_6H_6(l)$	49.0	Hydrogen iodide	$HI(g)$	25.9
Calcium carbonate	$CaCO_3(s)$	−1207.1	Methane	$CH_4(g)$	−74.8
Calcium oxide	$CaO(s)$	−635.5	Methanol	$CH_3OH(l)$	−238.6
Carbon dioxide	$CO_2(g)$	−393.5	Propane	$C_3H_8(g)$	−103.85
Carbon monoxide	$CO(g)$	−110.5	Silver chloride	$AgCl(s)$	−127.0
Diamond	$C(s)$	1.88	Sodium bicarbonate	$NaHCO_3(s)$	−947.7
Ethane	$C_2H_6(g)$	−84.68	Sodium carbonate	$Na_2CO_3(s)$	−1130.9
Ethanol	$C_2H_5OH(l)$	−277.7	Sodium chloride	$NaCl(s)$	−410.9
Ethylene	$C_2H_4(g)$	52.30	Sucrose	$C_{12}H_{22}O_{11}(s)$	−2221
Glucose	$C_6H_{12}O_6(s)$	−1273	Water	$H_2O(l)$	−285.8
Hydrogen bromide	$HBr(g)$	−36.23	Water vapor	$H_2O(g)$	−241.8

The standard state of a substance is its pure form at atmospheric pressure (1 atm; ∞ Section 10.2) and the temperature of interest, which we usually choose to be 298 K (25°C). The **standard enthalpy** of a reaction is defined as the enthalpy change when all reactants and products are in their standard states. We denote a standard enthalpy as ΔH°, where the superscript ° indicates standard-state conditions.

The **standard enthalpy of formation** of a compound, ΔH_f°, is the change in enthalpy for the reaction that forms 1 mol of the compound from its elements, with all substances in their standard states. We usually report ΔH_f° values at 298 K. If an element exists in more than one form under standard conditions, the most stable form of the element is used for the formation reaction. For example, the standard enthalpy of formation for ethanol, C_2H_5OH, is the enthalpy change for the following reaction:

$$2C(\text{graphite}) + 3H_2(g) + \tfrac{1}{2}O_2(g) \longrightarrow C_2H_5OH(l) \qquad \Delta H_f^\circ = -277.7 \text{ kJ} \quad [5.25]$$

The elemental source of oxygen is O_2, not O or O_3, because O_2 is the stable form of oxygen at 298 K and standard atmospheric pressure. Similarly, the elemental source of carbon is graphite and not diamond, because graphite is more stable (lower energy) at 298 K and standard atmospheric pressure (see Practice Exercise 5.9). Likewise, the most stable form of hydrogen under standard conditions is $H_2(g)$, so this is used as the source of hydrogen in Equation 5.25.

The stoichiometry of formation reactions always indicates that 1 mol of the desired substance is produced, as in Equation 5.25. As a result, enthalpies of formation are reported in kJ/mol of the substance. Several standard enthalpies of formation are given in Table 5.3 ▲. A more complete table is provided in Appendix C. By definition, *the standard enthalpy of formation of the most stable form of any element is zero* because there is no formation reaction needed when the element is already in its standard state. Thus, the values of ΔH_f° for C(graphite), $H_2(g)$, $O_2(g)$, and the standard states of other elements are zero by definition.

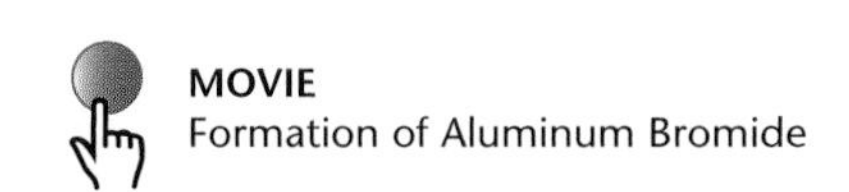

SAMPLE EXERCISE 5.11

For which of the following reactions at 25°C would the enthalpy change represent a standard enthalpy of formation? For those where it does not, what changes would need to be made in the reaction conditions?

(a) $2Na(s) + \tfrac{1}{2}O_2(g) \longrightarrow Na_2O(s)$

(b) $2K(l) + Cl_2(g) \longrightarrow 2KCl(s)$

(c) $C_6H_{12}O_6(s) \longrightarrow 6C(\text{diamond}) + 6H_2(g) + 3O_2(g)$

Solution

Analyze: The standard enthalpy of formation is represented by a reaction in which each reactant is an element in its standard state.

Plan: To solve these problems, we need to examine each equation to determine, first of all, whether the reaction is one in which a substance is formed from the elements. Secondly, we need to determine whether the reactant elements in the reaction are in their standard states at 25°C.

Solve: In (a) Na_2O is formed from the elements sodium and oxygen in their proper states, a solid and gas, respectively. Therefore, the enthalpy change for reaction (a) corresponds to a standard enthalpy of formation.

In (b) potassium is given as a liquid. It must be changed to the solid form, its standard state at room temperature. Furthermore, two moles of product are formed, so the enthalpy change for the reaction as written is twice the standard enthalpy of formation of $KCl(s)$.

Reaction (c) does not form a substance from its elements. Instead, a substance decomposes to its elements, so this reaction must be reversed. Secondly, the element carbon is given as diamond, whereas graphite is the lowest energy solid form of carbon at room temperature and one atmosphere pressure. The equation that correctly represents the enthalpy of formation of glucose from its elements is

$$6C(\text{graphite}) + 6H_2(g) + 3O_2(g) \longrightarrow C_6H_{12}O_6(s)$$

PRACTICE EXERCISE

Write the equation corresponding to the standard enthalpy of formation of liquid carbon tetrachloride (CCl_4).

Answer: $C(s) + 2Cl_2(g) \longrightarrow CCl_4(l)$

Using Enthalpies of Formation to Calculate Enthalpies of Reaction

Tabulations of ΔH_f°, such as those in Table 5.3 and Appendix C, have many important uses. As we will see in this section, we can use Hess's law to calculate the standard enthalpy change for any reaction for which we know the ΔH_f° values for all reactants and products. For example, consider the combustion of propane gas, $C_3H_8(g)$, with oxygen to form $CO_2(g)$ and $H_2O(l)$ under standard conditions:

$$C_3H_8(g) + 5O_2(g) \longrightarrow 3CO_2(g) + 4H_2O(l)$$

We can write this equation as the sum of three formation reactions:

$$C_3H_8(g) \longrightarrow 3C(s) + 4H_2(g) \qquad \Delta H_1 = -\Delta H_f^\circ[C_3H_8(g)] \qquad [5.26]$$

$$3C(s) + 3O_2(g) \longrightarrow 3CO_2(g) \qquad \Delta H_2 = 3\Delta H_f^\circ[CO_2(g)] \qquad [5.27]$$

$$4H_2(g) + 2O_2(g) \longrightarrow 4H_2O(l) \qquad \Delta H_3 = 4\Delta H_f^\circ[H_2O(l)] \qquad [5.28]$$

$$C_3H_8(g) + 5O_2(g) \longrightarrow 3CO_2(g) + 4H_2O(l) \qquad \Delta H_{rxn}^\circ = \Delta H_1 + \Delta H_2 + \Delta H_3 \qquad [5.29]$$

From Hess's law we can write the standard enthalpy change for the overall reaction, Equation 5.29, as the sum of the enthalpy changes for the processes in Equations 5.26 through 5.28. We can then use values from Table 5.3 to compute a numerical value for ΔH_{rxn}°:

$$\begin{aligned} \Delta H_{rxn}^\circ &= \Delta H_1 + \Delta H_2 + \Delta H_3 \\ &= -\Delta H_f^\circ[C_3H_8(g)] + 3\Delta H_f^\circ[CO_2(g)] + 4\Delta H_f^\circ[H_2O(l)] \\ &= -(-103.85 \text{ kJ}) + 3(-393.5 \text{ kJ}) + 4(-285.8 \text{ kJ}) = -2220 \text{ kJ} \end{aligned} \qquad [5.30]$$

Several aspects of this calculation depend on the guidelines we discussed in Section 5.4.

1. Equation 5.26 is the reverse of the formation reaction for $C_3H_8(g)$, so the enthalpy change for this reaction is $-\Delta H_f^\circ[C_3H_8(g)]$.

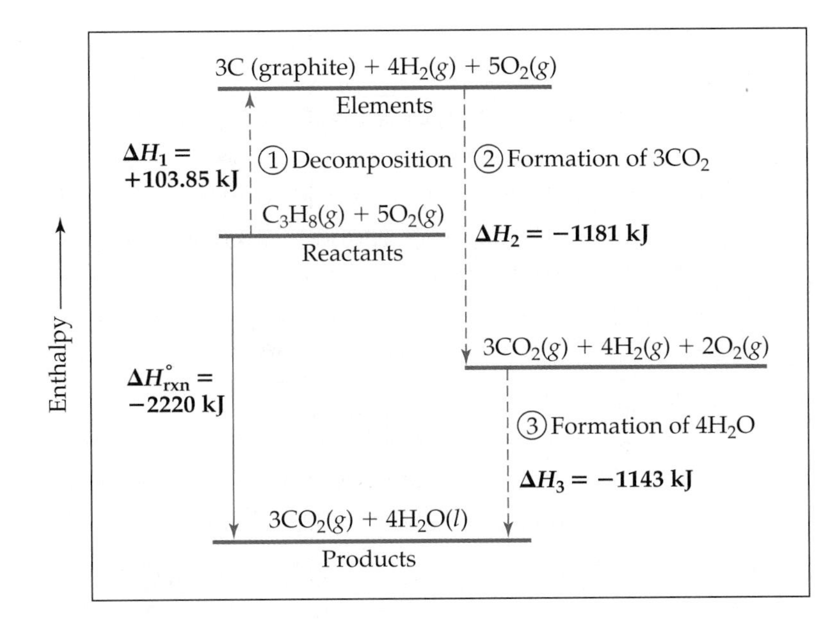

◀ **Figure 5.22** Enthalpy diagram for the combustion of 1 mol of propane gas, $C_3H_8(g)$. The overall reaction is $C_3H_8(g) + 5O_2(g) \longrightarrow 3CO_2(g) + 4H_2O(l)$. We can imagine this reaction as occurring in three steps. First, $C_3H_8(g)$ is decomposed to its elements, so $\Delta H_1 = -\Delta H^\circ_f[C_3H_8(g)]$. Second, 3 mol $CO_2(g)$ are formed, so $\Delta H_2 = 3\Delta H^\circ_f[CO_2(g)]$. Finally, 4 mol $H_2O(l)$ are formed, so $\Delta H_3 = 4\Delta H^\circ_f[H_2O(l)]$. Hess's law tells us that $\Delta H^\circ_{rxn} = \Delta H_1 + \Delta H_2 + \Delta H_3$. This same result is given by Equation 5.30 because $\Delta H^\circ_f[O_2(g)] = 0$.

2. Equation 5.27 is the formation reaction for 3 mol of $CO_2(g)$. Because enthalpy is an extensive property, the enthalpy change for this step is $3\Delta H^\circ_f[CO_2(g)]$. Similarly, the enthalpy change for Equation 5.28 is $4\Delta H^\circ_f[H_2O(l)]$. The reaction specifies that $H_2O(l)$ was produced, so be careful to use the value of ΔH°_f for $H_2O(l)$, not $H_2O(g)$.
3. We assume that the stoichiometric coefficients in the balanced equation represent moles. For Equation 5.29, therefore, the value $\Delta H^\circ_{rxn} = -2220$ kJ represents the enthalpy change for the reaction of 1 mol C_3H_8 and 5 mol O_2 to form 3 mol CO_2 and 4 mol H_2O. The product of the number of moles and the enthalpy change in kJ/mol has the units kJ: (number of moles) × (ΔH°_f in kJ/mol) = kJ. We therefore report ΔH°_{rxn} in kJ.

Figure 5.22 ▲ presents an enthalpy diagram for Equation 5.29, showing how it can be broken into steps involving formation reactions.

We can break down any reaction into formation reactions as we have done here. When we do, we obtain the general result that the standard heat of reaction is the sum of the standard heats of formation of the products minus the standard heats of formation of the reactants:

$$\Delta H^\circ_{rxn} = \sum n\Delta H^\circ_f(\text{products}) - \sum m\Delta H^\circ_f(\text{reactants}) \quad [5.31]$$

The symbol Σ (sigma) means "the sum of," and n and m are the stoichiometric coefficients of the chemical equation. The first term in Equation 5.31 represents the formation reactions of the products, which are written in the "forward" direction, that is, elements reacting to form products. This term is analogous to Equations 5.27 and 5.28 in the previous example. The second term represents the reverse of the formation reactions of the reactants, as in Equation 5.26, which is why the ΔH°_f values have a minus sign in front of them.

SAMPLE EXERCISE 5.12

(a) Calculate the standard enthalpy change for the combustion of 1 mol of benzene, $C_6H_6(l)$, to $CO_2(g)$ and $H_2O(l)$. **(b)** Compare the quantity of heat produced by combustion of 1.00 g propane to that produced by 1.00 g benzene.

Solution (a) We know that a combustion reaction involves $O_2(g)$ as a reactant. Our first step is to write a balanced equation for the combustion reaction of 1 mol $C_6H_6(l)$:

$$C_6H_6(l) + \tfrac{15}{2}O_2(g) \longrightarrow 6CO_2(g) + 3H_2O(l)$$

We can calculate ΔH°_{rxn} for the reaction by using Equation 5.31 and data in Table 5.3. Remember to multiply the ΔH°_f value for each substance in the reaction by that substance's stoichiometric coefficient. Recall also that $\Delta H^\circ_f = 0$ for any element in its most stable form under standard conditions, so $\Delta H^\circ_f[O_2(g)] = 0$:

$$\begin{aligned}\Delta H^\circ_{rxn} &= [6\Delta H^\circ_f(CO_2) + 3\Delta H^\circ_f(H_2O)] - [\Delta H^\circ_f(C_6H_6) + \tfrac{15}{2}\Delta H^\circ_f(O_2)]\\ &= [6(-393.5\text{ kJ}) + 3(-285.8\text{ kJ})] - [(49.0\text{ kJ}) + \tfrac{15}{2}(0\text{ kJ})]\\ &= (-2361 - 857.4 - 49.0)\text{ kJ}\\ &= -3267\text{ kJ}\end{aligned}$$

(b) From the example worked in the text, $\Delta H^\circ_{rxn} = -2220$ kJ for the combustion of 1 mol of propane. In part (a) of this exercise we determined that $\Delta H^\circ_{rxn} = -3267$ kJ for the combustion of 1 mol benzene. To determine the heat of combustion per gram of each substance, we use the molar masses to convert moles to grams:

$$C_3H_8(g): \quad (-2220\text{ kJ/mol})(1\text{ mol}/44.1\text{ g}) = -50.3\text{ kJ/g}$$

$$C_6H_6(l): \quad (-3267\text{ kJ/mol})(1\text{ mol}/78.1\text{ g}) = -41.8\text{ kJ/g}$$

Both propane and benzene are hydrocarbons. As a rule, the energy obtained from the combustion of a gram of hydrocarbon is between 40 and 50 kJ.

PRACTICE EXERCISE

Using the standard enthalpies of formation listed in Table 5.3, calculate the enthalpy change for the combustion of 1 mol of ethanol:

$$C_2H_5OH(l) + 3O_2(g) \longrightarrow 2CO_2(g) + 3H_2O(l)$$

Answer: -1367 kJ

SAMPLE EXERCISE 5.13

The standard enthalpy change for the reaction

$$CaCO_3(s) \longrightarrow CaO(s) + CO_2(g)$$

is 178.1 kJ. From the values for the standard enthalpies of formation of $CaO(s)$ and $CO_2(g)$ given in Table 5.3, calculate the standard enthalpy of formation of $CaCO_3(s)$.

Solution

Analyze: We need to obtain $\Delta H^\circ_f(CaCO_3)$.

Plan: We begin by writing the expression for the standard enthalpy change for the preceding reaction:

$$\Delta H^\circ_{rxn} = [\Delta H^\circ_f(CaO) + \Delta H^\circ_f(CO_2)] - \Delta H^\circ_f(CaCO_3)$$

Solve: Inserting the known values, we have

$$178.1\text{ kJ} = -635.5\text{ kJ} - 393.5\text{ kJ} - \Delta H^\circ_f(CaCO_3)$$

Solving for $\Delta H^\circ_f(CaCO_3)$ gives

$$\Delta H^\circ_f(CaCO_3) = -1207.1\text{ kJ/mol}$$

Check: We expect the enthalpy of formation of a stable solid such as calcium carbonate to be negative, as obtained.

PRACTICE EXERCISE

Given the following standard enthalpy of reaction, use the standard enthalpies of formation in Table 5.3 to calculate the standard enthalpy of formation of $CuO(s)$:

$$CuO(s) + H_2(g) \longrightarrow Cu(s) + H_2O(l) \qquad \Delta H^\circ = -129.7\text{ kJ}$$

Answer: -156.1 kJ/mol

5.8 Foods and Fuels

Most chemical reactions used for the production of heat are combustion reactions. The energy released when 1 g of a material is combusted is often called its **fuel value**. Because fuel values represent the heat *released* in a combustion, fuel values are positive numbers. The fuel value of any food or fuel can be measured by calorimetry.

Foods

Most of the energy our bodies need comes from carbohydrates and fats. The forms of carbohydrate known as starch are decomposed in the intestines into glucose, $C_6H_{12}O_6$. Glucose is soluble in blood, and in the human body it is known as blood sugar. It is transported by the blood to cells, where it reacts with O_2 in a series of steps, eventually producing $CO_2(g)$, $H_2O(l)$, and energy:

$$C_6H_{12}O_6(s) + 6O_2(g) \longrightarrow 6CO_2(g) + 6H_2O(l) \qquad \Delta H^\circ = -2803 \text{ kJ}$$

The breakdown of carbohydrates is rapid, so their energy is quickly supplied to the body. However, the body stores only a very small amount of carbohydrates. The average fuel value of carbohydrates is 17 kJ/g (4 kcal/g).

Like carbohydrates, fats produce CO_2 and H_2O when metabolized and when subjected to combustion in a bomb calorimeter. The reaction of tristearin, $C_{57}H_{110}O_6$, a typical fat, is as follows:

$$2C_{57}H_{110}O_6(s) + 163O_2(g) \longrightarrow 114CO_2(g) + 110H_2O(l) \qquad \Delta H^\circ = -75{,}520 \text{ kJ}$$

The body uses the chemical energy from foods to maintain body temperature (see the "Chemistry and Life" box in Section 5.5), to contract muscles, and to construct and repair tissues. Any excess energy is stored as fats. Fats are well suited to serve as the body's energy reserve for at least two reasons: (1) They are insoluble in water, which facilitates storage in the body, and (2) they produce more energy per gram than either proteins or carbohydrates, which makes them efficient energy sources on a mass basis. The average fuel value of fats is 38 kJ/g (9 kcal/g).

The metabolism of proteins in the body produces less energy than combustion in a calorimeter because the products are different. Proteins contain nitrogen, which is released in the bomb calorimeter as N_2. In the body this nitrogen ends up mainly as urea, $(NH_2)_2CO$. Proteins are used by the body mainly as building materials for organ walls, skin, hair, muscle, and so forth. On average, the metabolism of proteins produces 17 kJ/g(4 kcal/g), the same as for carbohydrates.

The fuel values for a variety of common foods are shown in Table 5.4 ▼. Labels on packaged foods show the amounts of carbohydrate, fat, and protein contained in an average serving, as well as the energy value of the serving (Figure 5.23 ▶). The amount of energy our bodies require varies considerably depending on such factors as weight, age, and muscular activity. About 100 kJ per kilogram of body weight per day is required to keep the body functioning at a minimal level. An average 70-kg (154-lb) person expends about 800 kJ/hr when

▲ **Figure 5.23** Labels of processed foods have information about the quantities of different nutrients in an average serving.

TABLE 5.4 Compositions and Fuel Values of Some Common Foods

	Approximate Composition (% by mass)			Fuel Value	
	Carbohydrate	Fat	Protein	kJ/g	kcal/g(Cal/g)
Carbohydrate	100	–	–	17	4
Fat	–	100	–	38	9
Protein	–	–	100	17	4
Apples	13	0.5	0.4	2.5	0.59
Beer[a]	1.2	–	0.3	1.8	0.42
Bread	52	3	9	12	2.8
Cheese	4	37	28	20	4.7
Eggs	0.7	10	13	6.0	1.4
Fudge	81	11	2	18	4.4
Green beans	7.0	–	1.9	1.5	0.38
Hamburger	–	30	22	15	3.6
Milk (whole)	5.0	4.0	3.3	3.0	0.74
Peanuts	22	39	26	23	5.5

[a]Beers typically contain 3.5% ethanol, which has fuel value.

doing light work, such as slow walking or light gardening. Strenuous activity, such as running, often requires 2000 kJ/hr or more. When the energy content of our food exceeds the energy we expend, our body stores the surplus as fat.

SAMPLE EXERCISE 5.14

A plant such as celery contains carbohydrates in the form of starch and cellulose. These two kinds of carbohydrate have essentially the same fuel values when combusted in a bomb calorimeter. When we consume celery, however, our bodies receive fuel value from the starch only. What can we conclude about the difference between starch and cellulose as foods?

Solution If cellulose does not provide fuel value, we must conclude that it is not converted in the body into CO_2 and H_2O, as starch is. A slight, but critical, difference in the structures of starch and cellulose explains why only starch is broken down into glucose in the body. Cellulose passes through without undergoing significant chemical change. It serves as roughage in the diet, but provides no caloric value.

PRACTICE EXERCISE

The nutritional label on a bottle of canola oil indicates that 10 g of the oil has a fuel value of 86 kcal. A similar label on a bottle of pancake syrup indicates that 60 mL (about 60 g) has a fuel value of 200 kcal. Account for the difference.

Answer: The oil has a fuel value of 8.6 kcal/g, whereas the syrup has a fuel value of about 3.3 kcal/g. The higher fuel value for the canola oil arises because the oil is essentially pure fat, whereas the syrup is a solution of sugars (carbohydrates) in water. The oil has a higher fuel value per gram; in addition, the syrup is diluted by water.

SAMPLE EXERCISE 5.15

(a) A 28-g (1-oz) serving of a popular breakfast cereal served with 120 mL of skim milk provides 8 g protein, 26 g carbohydrates, and 2 g fat. Using the average fuel values of these kinds of substances, estimate the amount of food energy in this serving. **(b)** A person of average weight uses about 100 Cal/mi when running or jogging. How many servings of this cereal provide the fuel value requirements for running 3 mi?

Solution (a) Analyze: The total food value in the serving will be the sum of the food values of the protein, carbohydrates, and fat.
Plan: We are given the mass of protein, carbohydrates, and fat in the serving of cereal. We can use the data in Table 5.4 to convert these masses to their fuel values, which we can sum to get the total food energy.
Solve:

$$(8\text{ g protein})\left(\frac{17\text{ kJ}}{1\text{ g protein}}\right) + (26\text{ g carbohydrate})\left(\frac{17\text{ kJ}}{1\text{ g carbohydrate}}\right) +$$

$$(2\text{ g fat})\left(\frac{38\text{ kJ}}{1\text{ g fat}}\right) = 650\text{ kJ (to two significant figures)}$$

This corresponds to 160 kcal:

$$(650\text{ kJ})\left(\frac{1\text{ kcal}}{4.18\text{ kJ}}\right) = 160\text{ kcal}$$

Recall that the dietary Calorie is equivalent to 1 kcal. Thus, the serving provides 160 Cal.

(b) Analyze: Here we are faced with the reverse problem, calculating the quantity of food that provides a specific amount of caloric food value.
Plan: The problem statement provides a conversion factor between Calories and miles. The answer to part (a) provides us with a conversion factor between servings and Calories.
Solve: We can use these factors in a straightforward dimensional analysis to determine the number of servings needed, rounded to the nearest whole number:

$$\text{Servings} = (3\text{ mi})\left(\frac{100\text{ Cal}}{1\text{ mi}}\right)\left(\frac{1\text{ serving}}{160\text{ Cal}}\right) = 2\text{ servings}$$

PRACTICE EXERCISE

(a) Dry red beans contain 62% carbohydrate, 22% protein, and 1.5% fat. Estimate the fuel value of these beans. **(b)** Very light activity like reading or watching television uses about 7 kJ/min. How many minutes of such activity can be sustained by the energy provided by a serving of chicken noodle soup containing 13 g protein, 15 g carbohydrate, and 5 g fat?
Answers: **(a)** 15 kJ/g; **(b)** 95 min

Fuels

The elemental compositions and fuel values of several common fuels are compared in Table 5.5 ▼. During the complete combustion of fuels, carbon is converted to CO_2 and hydrogen is converted to H_2O, both of which have large negative enthalpies of formation. Consequently, the greater the percentage of carbon and hydrogen in a fuel, the higher its fuel value. Compare, for example, the compositions and fuel values of bituminous coal and wood. The coal has a higher fuel value because of its greater carbon content.

TABLE 5.5 Fuel Values and Compositions of Some Common Fuels

	Approximate Elemental Composition (mass %)			
	C	H	O	Fuel Value (kJ/g)
Wood (pine)	50	6	44	18
Anthracite coal (Pennsylvania)	82	1	2	31
Bituminous coal (Pennsylvania)	77	5	7	32
Charcoal	100	0	0	34
Crude oil (Texas)	85	12	0	45
Gasoline	85	15	0	48
Natural gas	70	23	0	49
Hydrogen	0	100	0	142

In 2000 the United States consumed 1.03×10^{17} kJ of energy. This value corresponds to an average daily energy consumption per person of 1.0×10^{6} kJ, which is roughly 100 times greater than the per capita food-energy needs. We are a very energy-intensive society. Although our population is only about 4.5% of the world's population, the United States accounts for nearly one fourth of the total world energy consumption. Figure 5.24 ▶ illustrates the sources of this energy.

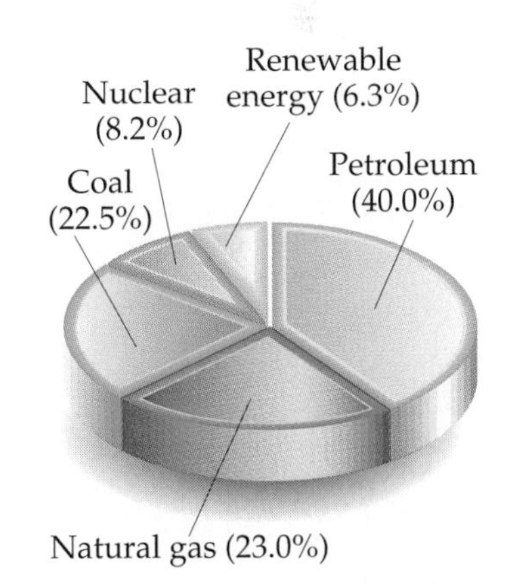

▲ **Figure 5.24** Sources of energy consumed in the United States. In 2000 the United States consumed a total of 1.0×10^{17} kJ of energy.

Coal, petroleum, and natural gas, which are our major sources of energy, are known as **fossil fuels**. All have formed over millions of years from the decomposition of plants and animals and are being depleted far more rapidly than they are being formed. **Natural gas** consists of gaseous hydrocarbons, compounds of hydrogen and carbon. It contains primarily methane (CH_4), with small amounts of ethane (C_2H_6), propane (C_3H_8), and butane (C_4H_{10}). We determined the fuel value of propane in Sample Exercise 5.12. **Petroleum** is a liquid composed of hundreds of compounds. Most of these compounds are hydrocarbons, with the remainder being chiefly organic compounds containing sulfur, nitrogen, or oxygen. **Coal**, which is solid, contains hydrocarbons of high molecular weight as well as compounds containing sulfur, oxygen, or nitrogen. Coal is the most abundant fossil fuel; it constitutes 80% of the fossil fuel reserves of the United States and 90% of those of the world. However, the use of coal presents a number of problems. Coal is a complex mixture of substances, and it contains components that cause air pollution. When coal is combusted, the sulfur it contains is converted mainly to sulfur dioxide, SO_2, a very troublesome air pollutant. Because coal is a solid, recovery from its underground deposits is expensive and often dangerous. Furthermore, coal deposits are not always close to locations of high-energy use, so there are often substantial shipping costs.

One promising way to utilize coal reserves is to use them to produce a mixture of gaseous hydrocarbons called *syngas* (for "*syn*thesis *gas*"). In this process, called *coal gasification*, the coal typically is pulverized and treated with superheated steam. Sulfur-containing compounds, water, and carbon dioxide can be removed from the products, leading to a gaseous mixture of CH_4, H_2, and CO, all of which have high fuel values:

$$\text{coal} + \text{steam} \xrightarrow{\text{conversion}} \text{complex mixture} \xrightarrow{\text{purification}} \text{mixture of } CH_4, H_2, CO \text{ (syngas)}$$

Because it is gaseous, syngas can be easily transported in pipelines. Additionally, because much of the sulfur in coal is removed during the gasification process, combustion of syngas causes less air pollution than burning coal. For these reasons, the economical conversion of coal and petroleum into "cleaner" fuels such as syngas and hydrogen is a very active area of current research in chemistry and engineering.

Other Energy Sources

Nuclear energy is energy that is released in either the splitting or fusion of the nuclei of atoms. Nuclear power is currently used to produce about 22% of the electric power in the United States and comprises about 8% of the total U.S. energy production (Figure 5.24). Nuclear energy is, in principle, free of the polluting emissions that are a major problem in the generation of energy from fossil fuels. However, nuclear power plants produce radioactive waste products, and their use has therefore been fraught with controversy. We will discuss issues related to the production of nuclear energy in Chapter 21.

Fossil fuel and nuclear energy are *nonrenewable* sources of energy; the fuels used are limited resources that we are consuming at a much greater rate than they are regenerated. Eventually these fuels will be expended, although estimates vary greatly as to when this will occur. Because nonrenewable sources of energy will eventually be used up, there is a great deal of research into sources of **renewable energy**, energy sources that are essentially inexhaustible. Renewable energy sources include *solar energy* from the Sun, *wind energy* harnessed by windmills, *geothermal energy* from the heat stored in the mass of Earth, *hydroelectric energy* from flowing rivers, and *biomass energy* from crops, such as trees and corn, and from biological waste matter. Currently, renewable sources provide about 6.3% of the U.S. annual energy consumption, with hydroelectric (3.7%) and biomass (2.9%) sources as the major contributors.

Providing our future energy needs will most certainly depend on developing the technology to harness solar energy with greater efficiency. Solar energy is the world's largest energy source. On a clear day about 1 kJ of solar energy reaches each square meter of Earth's surface every second. The solar energy that falls on only 0.1% of U.S. land area is equivalent to all the energy that this nation currently uses. Harnessing this energy is difficult because it is dilute (it is distributed over a wide area) and it fluctuates with time and weather conditions. The effective use of solar energy will depend on the development of some means of storing the collected energy for use at a later time. Any practical means for doing this will almost certainly involve use of an endothermic chemical process that can be later reversed to release heat. One such reaction is the following:

$$CH_4(g) + H_2O(g) + \text{heat} \rightleftharpoons CO(g) + 3H_2(g)$$

This reaction proceeds in the forward direction at high temperatures, which can be obtained in a solar furnace. The CO and H_2 formed in the reaction could then be stored and allowed to react later, with the heat released being put to useful work.

A survey taken about 20 years ago at Walt Disney's EPCOT Center revealed that nearly 30% of the visitors expected that solar energy would be the principal source of energy in the United States in the year 2000. The future of solar energy

Chemistry at Work The Hybrid Car

The hybrid cars now entering the automobile marketplace nicely illustrate the convertibility of energy from one form to another. Hybrid cars run on either gasoline or electricity. The so-called "full hybrids" are cars capable of running on battery power alone at low speeds. The Honda Insight, Figure 5.25 ▼, is a full hybrid car that achieves 61 miles per gallon in city driving. In the full hybrid cars, an electrical engine is capable of driving the car at lower speeds. The "mild hybrid" cars are best described as electrically assisted gasoline engines. Both General Motors and Ford have announced plans to offer electrically assisted engines for most models, beginning in about 2003.

Full hybrid cars are more efficient than the mild hybrid designs, but they are more costly to produce, and they require more technological advances than the mild hybrid versions. The mild hybrids are likely to be more widely produced and sold within the next several years. Let's consider how they operate and some of the interesting thermodynamic considerations they incorporate.

Figure 5.26 ▼ shows a schematic diagram of the power system for a mild hybrid car. In addition to the 12-volt battery that is standard on conventional autos, the hybrid car carries a 42-volt battery pack. The electrical energy from this battery pack is not employed to move the car directly; an electrical engine capable of doing that, as in the full hybrids, requires from 150 to 300 volts. In the mild hybrid cars the added electrical source is employed to run various auxiliary devices that would otherwise be run off the gasoline engine, such as water pump, power steering, and air systems. To save on energy, when the hybrid car comes to a stop, the engine shuts off. It restarts automatically when the driver presses the accelerator. This feature saves fuel that would otherwise be used to keep the engine idling at traffic lights and other stopping situations.

The idea is that the added electrical system will improve overall fuel efficiency of the car. The added battery, moreover, is not supposed to need recharging from an external power source. Where, then, can the improved fuel efficiency come from? Clearly, if the battery pack is to continue to operate auxiliary devices such as the water pump, it must be recharged. We can think of it this way: The source of the voltage that the battery develops is a chemical reaction. Recharging the battery thus represents a conversion of mechanical energy into chemical potential energy. The recharging occurs in part through the agency of an alternator, which runs off the engine and provides a recharging voltage. In the mild hybrid car, the braking system serves as an additional source of mechanical energy for recharging. When the brakes are applied in a conventional car, the car's kinetic energy is converted through the brake pads in the wheels into heat, so no useful work is done. In the hybrid car, some of the car's kinetic energy is used to recharge the battery when the brakes are applied. Thus, kinetic energy that would otherwise be dissipated as heat is partially converted into useful work. Overall, the mild hybrid cars are expected to yield 10–20% improvements in fuel economy as compared with similar conventional cars.

▲ **Figure 5.25** The Honda Insight, a hybrid car in which both batteries and a gasoline engine provide power to move the car, as well as drive ancillary devices.

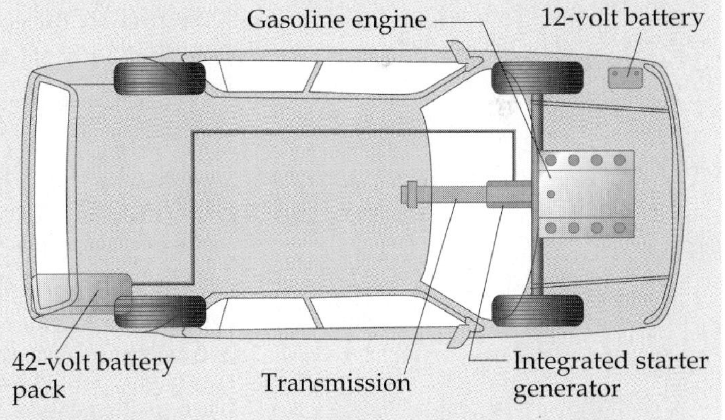

▲ **Figure 5.26** Schematic diagram of a mild hybrid car. The 42-volt battery pack provides energy for operating several auxiliary functions. It is recharged from the engine and through the braking system.

has proven to be a lot like the Sun itself: big and bright, but further away than it seems. Nevertheless, important progress has been made in recent years. Perhaps the most direct way to make use of the Sun's energy is to convert it directly into electricity by use of photovoltaic devices, sometimes called *solar cells*. The efficiencies of solar energy conversion by use of such devices have increased dramatically during the past few years as a result of intensive research efforts. Photovoltaics are vital to the generation of power for the space station. More significant for our Earth-bound concerns, the unit costs of solar panels have been steadily declining, even as their efficiencies have improved dramatically.

As a result, photovoltaics are becoming feasible for large-scale generation of useful energy at Earth's surface. In California, businesses and homes that add solar arrays to their rooftops can receive credit for electricity added directly to the power grid. Now that the year 2000 has come and gone, when do you think solar energy will become a principal source of energy in the United States?

SAMPLE INTEGRATIVE EXERCISE

Trinitroglycerin, $C_3H_5N_3O_9$, (usually referred to simply as nitroglycerin) has been widely used as an explosive. Alfred Nobel used it to make dynamite in 1866. Rather surprisingly, it also is used as a medication, to relieve angina (chest pains resulting from partially blocked arteries to the heart), by dilating the blood vessels. The enthalpy of decomposition at 1 atmosphere pressure of trinitroglycerin to form nitrogen gas, carbon dioxide gas, liquid water, and oxygen gas at 25°C is −1541.4 kJ/mol. **(a)** Write a balanced chemical equation for the decomposition of trinitroglycerin. **(b)** Calculate the standard heat of formation of trinitroglycerin. **(c)** A standard dose of trinitroglycerin for relief of angina is 0.60 mg. Assuming that the sample is eventually completely combusted in the body (not explosively, though!) to nitrogen gas, carbon dioxide gas, and liquid water, what number of calories is released? **(d)** One common form of trinitroglycerin melts at about 3°C. From this information and the formula for the substance, would you expect it to be a molecular or ionic compound? Explain. **(e)** Describe the various conversions of forms of energy when trinitroglycerin is used as an explosive to break rockfaces in highway construction.

Solution **(a)** The general form of the reaction we must balance is

$$C_3H_5N_3O_9(l) \longrightarrow N_2(g) + CO_2(g) + H_2O(l) + O_2(g)$$

We go about balancing in the usual way. To obtain an even number of nitrogen atoms on the left, we multiply the formula for $C_3H_5N_3O_9(s)$ by two. This then gives us 6 mol of $CO_2(g)$ and 5 mol of $H_2O(l)$. Everything is balanced except for oxygen. We have an odd number of oxygen atoms on the right. We can balance the oxygen by adding $\frac{1}{2}$ mol of $O_2(g)$ on the right:

$$2C_3H_5N_3O_9(l) \longrightarrow 3N_2(g) + 6CO_2(g) + 5H_2O(l) + \tfrac{1}{2}O_2(g)$$

We multiply through by two to convert all coefficients to whole numbers:

$$4C_3H_5N_3O_9(l) \longrightarrow 6N_2(g) + 12CO_2(g) + 10H_2O(l) + O_2(g)$$

(b) The heat of formation is the enthalpy change in the balanced chemical equation:

$$3C(s) + \tfrac{3}{2}N_2(g) + \tfrac{5}{2}H_2(g) + \tfrac{9}{2}O_2(g) \longrightarrow C_3H_5N_3O_9(l) \qquad \Delta H_f^\circ = ?$$

We can obtain the value of ΔH_f° by using the equation for the heat of decomposition of trinitroglycerin:

$$4C_3H_5N_3O_9(l) \longrightarrow 6N_2(g) + 12CO_2(g) + 10H_2O(l) + O_2(g)$$

The enthalpy change in this reaction is 4(−1541.4 kJ) = −6155.6 kJ. [We need to multiply by four because there are four moles of $C_3H_5N_3O_9(l)$ in the balanced equation.] This enthalpy change is given by the sum of the heats of formation of the products minus the heats of formation of the reactants, each multiplied by its coefficient in the balanced equation:

$$-6155.6\text{ kJ} = \{6\Delta H_f^\circ(N_2(g)) + 12\Delta H_f^\circ(CO_2(g)) + 10\Delta H_f^\circ(H_2O(l)) + \Delta H_f^\circ(O_2(g))\} - 4\Delta H_f^\circ(C_3H_5N_3O_9(l)).$$

The ΔH_f° values for $N_2(g)$ and $O_2(g)$ are zero, by definition. We look up the values for $H_2O(l)$ and $CO_2(g)$ from Table 5.3 and find that

$$-6155.6\text{ kJ} = 12(-393.5\text{kJ}) + 10(-285.8) - 4\Delta H_f^\circ(C_3H_5N_3O_9(l))$$

$$\Delta H_f^\circ(C_3H_5N_3O_9(l)) = -353.6\text{ kJ/mol}$$

(c) We know that on combustion a mol of $C_3H_5N_3O_9(l)$ yields 1541.4 kJ. We need to calculate the number of moles of $C_3H_5N_3O_9(l)$ in 0.60 mg:

$$0.60 \times 10^{-3}\text{ g } C_3H_5N_3O_9\left(\frac{1\text{ mol } C_3H_5N_3O_9}{227\text{ g } C_3H_5N_3O_9}\right)\left(\frac{1541.4\text{ kJ}}{1\text{ mol } C_3H_5N_3O_9}\right) = 4.1 \times 10^{-3}\text{ kJ}$$

$$= 4.1\text{ J}$$

(d) Because trinitroglycerin melts below room temperature, we expect that it is a molecular compound. With some exceptions, ionic substances are generally hard, crystalline materials that melt at high temperatures. (∞ Sections 2.5 and 2.6) Also, the molecular formula suggests that it is likely to be a molecular substance. All the elements of which it is composed are nonmetals.

(e) The energy stored in trinitroglycerin is chemical potential energy. When the substance reacts explosively in air, it forms substances such as carbon dioxide, water, and nitrogen gas, which are of lower potential energy. In the course of the chemical transformation, energy is released in the form of heat; the gaseous reaction products are very hot. This very high heat energy is transferred to the surroundings; the gases expand against the surroundings, which may be solid materials. Work is done in moving the solid materials and imparting kinetic energy to them. For example, a chunk of rock might be impelled upward. It has been given kinetic energy by transfer of energy from the hot, expanding gases. As the rock rises, its kinetic energy is transformed into potential energy. Eventually, it again acquires kinetic energy as it falls to earth. When it strikes the earth, its kinetic energy is converted largely to thermal energy, though some work may be done on the surroundings as well.

Summary and Key Terms

Introduction and Section 5.1 **Thermodynamics** is the study of energy and its transformations. In this chapter we have focused on **thermochemistry**, the transformations of energy—especially heat—during chemical reactions.

An object can possess energy in two forms: **Kinetic energy** is the energy due to motion of the object, and **potential energy** is the energy that an object possesses by virtue of its position relative to other objects. An electron in motion near a proton, for example, has kinetic energy because of its motion and potential energy because of its electrostatic attraction to the proton. The SI unit of energy is the **joule** (J): $1\ J = 1\ kg\text{-}m^2/s^2$. Another common energy unit is the **calorie** (cal), which was originally defined as the quantity of energy necessary to increase the temperature of 1 g of water by 1°C: 1 cal = 4.184 J.

When we study thermodynamic properties, we define a specific amount of matter as the **system**. Everything outside the system is the **surroundings**. A closed system can exchange energy, but not matter, with the surroundings. Energy can be transferred between the system and the surroundings as work or heat. **Work** is the energy expended to move an object against a **force**. **Heat** is the energy that is transferred from a hotter object to a colder one. In thermodynamics we define **energy** as the capacity to do work or to transfer heat.

Section 5.2 The **internal energy** of a system is the sum of all the kinetic and potential energies of its component parts. The internal energy of a system can change because of energy transferred between the system and the surroundings. The **first law of thermodynamics**, which is also called the law of conservation of energy, states that the change in the internal energy of a system, ΔE, is the sum of the heat, q, transferred into or out of the system and the work, w, done on or by the system: $\Delta E = q + w$. Both q and w have a sign that indicates the direction of energy transfer. When heat is transferred from the surroundings to the system, $q > 0$. Likewise, when the surroundings do work on the system, $w > 0$. In an **endothermic** process the system absorbs heat from the surroundings; in an **exothermic** process the system releases heat to the surroundings.

The internal energy, E, is a **state function**. The value of any state function depends only on the state or condition of the system and not on the details of how it came to be in that state. The temperature of a substance is a state function, too. The heat, q, and the work, w, are not state functions; their values depend on the particular way in which a system changes its state.

Sections 5.3 and 5.4 When a gas is produced or consumed in a chemical reaction occurring at constant pressure, the system may perform **pressure-volume work** against the prevailing pressure. For this reason, we define a new state function called **enthalpy**, H, which is important in thermochemistry. In systems where only pressure-volume work due to gases is involved, the change in the enthalpy of a system, ΔH, equals the heat gained or lost by the system at constant pressure. For an endothermic process, $\Delta H > 0$; for an exothermic process, $\Delta H < 0$.

Every substance has a characteristic enthalpy. In a chemical process, the **enthalpy of reaction** is the enthalpy of the products minus the enthalpy of the reactants: $\Delta H_{rxn} = H(\text{products}) - H(\text{reactants})$. Enthalpies of reaction follow some simple rules: (1) Enthalpy is an extensive property, so the enthalpy of reaction is proportional to the amount of reactant that reacts. (2) Reversing a reaction changes the sign of ΔH. (3) The enthalpy of reaction depends on the physical states of the reactants and products.

Section 5.5 The amount of heat transferred between the system and the surroundings is measured experimentally by **calorimetry**. A **calorimeter** measures the temperature change accompanying a process. The temperature change of a calorimeter depends on its **heat capacity**, the amount of heat required to raise its temperature by 1 K. The heat capacity for 1 mol of a pure substance is called its **molar heat capacity**; for 1 g of the substance, we use the term **specific heat**. Water has a very high specific heat, 4.18 J/g-K. The amount of heat, q, absorbed by a substance is the product of its specific heat, its mass, and its temperature change: $q = (\text{specific heat}) \times (\text{grams of substance}) \times \Delta T$.

If a calorimetry experiment is carried out under a constant pressure, the heat transferred provides a direct measure of the enthalpy change of the reaction. Constant-volume calorimetry is carried out in a vessel of fixed volume called a **bomb calorimeter**. Bomb calorimeters are used to measure the heat evolved in combustion reactions. The heat transferred under constant-volume conditions is equal to ΔE. However, corrections can be applied to ΔE values to yield enthalpies of combustion.

Section 5.6 Because enthalpy is a state function, ΔH depends only on the initial and final states of the system. Thus, the enthalpy change of a process is the same whether the process is carried out in one step or in a series of steps. **Hess's law** states that if a reaction is carried out in a series of steps, ΔH for the reaction will be equal to the sum of the enthalpy changes for the steps. We can therefore calculate ΔH for any process, as long as we can write the process as a series of steps for which ΔH is known.

Section 5.7 The **enthalpy of formation**, ΔH_f, of a substance is the enthalpy change for the reaction in which the substance is formed from its constituent elements. The **standard enthalpy** of a reaction, ΔH°, is the enthalpy change when all reactants and products are at 1 atm pressure and a specific temperature, usually 298 K (25°C). Combining these ideas, the **standard enthalpy of formation**, ΔH_f°, of a substance is the change in enthalpy for the reaction that forms 1 mol of the substance from its elements with all reactants and products at 1 atm pressure and usually 298 K. For any element in its most stable state at 298 K and 1 atm pressure, $\Delta H_f^\circ = 0$. The standard enthalpy change for any reaction can be readily calculated from the standard enthalpies of formation of the reactants and products in the reaction:

$$\Delta H^\circ_{rxn} = \sum n\Delta H_f^\circ(\text{products}) - \sum m\Delta H_f^\circ(\text{reactants})$$

Section 5.8 The **fuel value** of a substance is the heat released when 1 g of the substance is combusted. Different types of foods have different fuel values and differing abilities to be stored in the body. The most common fuels are hydrocarbons that are found as **fossil fuels**, such as **natural gas, petroleum**, and **coal**. Coal is the most abundant fossil fuel, but the sulfur present in most coals causes air pollution. Coal gasification is one possible way to use existing resources as sources of cleaner energy. Sources of **renewable energy** include solar energy, wind energy, biomass, and hydroelectric energy. These energy sources are essentially inexhaustible and will become more important as fossil fuels are depleted.

Exercises

The Nature of Energy

5.1 In what two ways can an object possess energy? How do these two ways differ from one another?

5.2 Suppose you toss a tennis ball upward. **(a)** Does the kinetic energy of the ball increase or decrease as it moves higher? **(b)** What happens to the potential energy of the ball as it moves higher? **(c)** If the same amount of energy were imparted to a ball the same size as a tennis ball, but of twice the mass, how high would it go in comparison to the tennis ball? Explain your answers.

5.3 **(a)** Calculate the kinetic energy in joules of a 45-g golf ball moving at 61 m/s. **(b)** Convert this energy to calories. **(c)** What happens to this energy when the ball lands in a sand trap?

5.4 **(a)** What is the kinetic energy in joules of a 950-lb motorcycle moving at 68 mph? **(b)** By what factor will the kinetic energy change if the speed of the motorcycle is decreased to 34 mph? **(c)** Where does the kinetic energy of the motorcycle go when the rider brakes to a stop?

5.5 In much engineering work it is common to use the British thermal unit (Btu). A Btu is the amount of heat required to raise the temperature of 1 lb of water by 1°F. Calculate the number of joules in a Btu.

5.6 A watt is a measure of power (the rate of energy change) equal to 1 J/s. Calculate the number of joules in a kilowatt-hour.

5.7 An adult person radiates heat to the surroundings at about the same rate as a 100-watt electric incandescent light bulb. What is the total amount of energy in kcal radiated to the surroundings by an adult in 24 hours?

5.8 Describe the source of the energy and the nature of the energy conversions involved when a 100-watt electric lightbulb radiates energy to its surroundings. Compare this with the energy source and energy conversions involved when an adult person radiates energy to the surroundings.

5.9 Suppose that a pellet is shot from an air gun straight up into the air. Why does the pellet eventually stop rising and fall back to Earth, rather than simply moving out into space? In principle, could the pellet ever move on out into space?

5.10 A bowling ball is dropped from a 100-ft-high tower on Earth. Compare the change in potential energy that it undergoes with dropping the same ball from a 100-ft-high tower on the Moon.

5.11 **(a)** What is meant by the term *system* in thermodynamics? **(b)** What is special about a closed system?

5.12 In a thermodynamic study a scientist focuses on the properties of a solution in a flask that is arranged as shown in the illustration. A solution is continuously flowing into the flask at the top and out at the bottom, such that the amount of solution in the flask is constant with time. **(a)** Is the solution in the flask a closed system? Why or why not? **(b)** If it is not a closed system, what could be done with the arrangement in the figure to make a closed system?

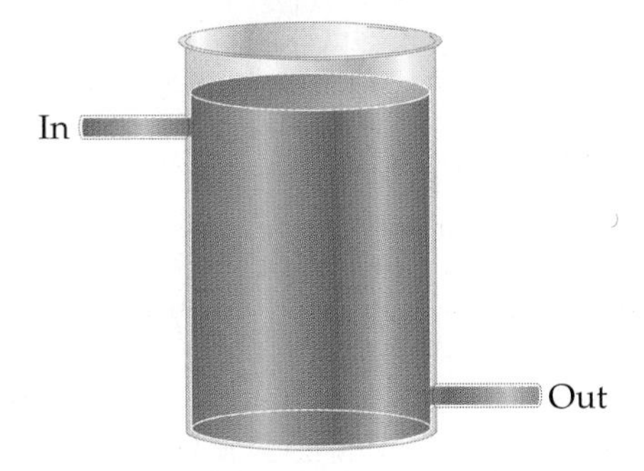

5.13 **(a)** What is work? **(b)** How do we determine the amount of work done, given the force associated with the work?

5.14 **(a)** Not so long ago it was widely believed that heat is not a form of energy. What arguments can you give to convince someone that it is? **(b)** Under what conditions is heat transferred from one object to another?

5.15 Identify the force present, and explain whether work is being performed in the following cases: **(a)** You lift a pencil off the top of a desk. **(b)** A spring is compressed to half its normal length.

5.16 Identify the force present, and explain whether work is done when **(a)** a positively charged particle moves in a circle at a fixed distance from a negatively charged particle; **(b)** an iron nail is pulled off a magnet.

The First Law of Thermodynamics

5.17 **(a)** State the first law of thermodynamics. **(b)** What is meant by the *internal energy* of a system? **(c)** By what means can the internal energy of a system increase?

5.18 **(a)** Write an equation that expresses the first law of thermodynamics. **(b)** In applying the first law, do we need to measure the internal energy of a system? Explain. **(c)** Under what conditions will the quantities q and w be negative numbers?

5.19 Calculate ΔE, and determine whether the process is endothermic or exothermic for the following cases: **(a)** A system releases 113 kJ of heat to the surroundings and does 39 kJ of work on the surroundings; **(b)** $q = 1.62$ kJ and $w = -874$ J; **(c)** the system absorbs 77.5 kJ of heat while doing 63.5 kJ of work on the surroundings.

5.20 For the following processes, calculate the change in internal energy of the system, and determine whether the process is endothermic or exothermic: **(a)** A balloon is heated by adding 900 J of heat. It expands, doing 422 J of work on the atmosphere. **(b)** A 50-g sample of water is cooled from 30°C to 15°C, thereby losing approximately 3140 J of heat. **(c)** A chemical reaction releases 8.65 kJ of heat and does no work on the surroundings.

5.21 The closed box in each of the following illustrations represents a system, and the arrows show the changes to the system in a process. The lengths of the arrows represent the relative magnitudes of q and w. **(a)** Which of these processes is endothermic? **(b)** For which of these processes, if any, is $\Delta E < 0$? **(c)** For which process, if any, is there a net gain in internal energy?

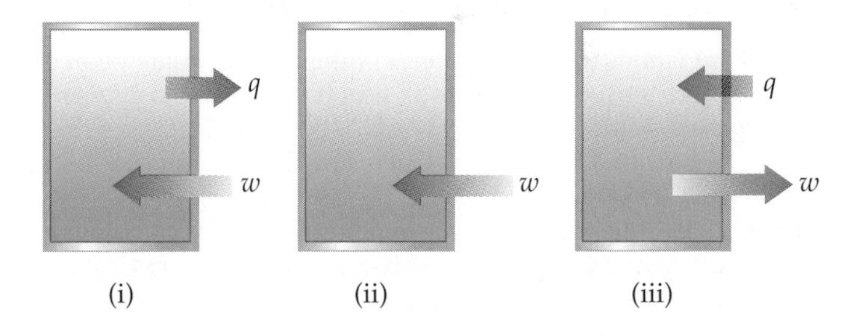

5.22 A system releases heat to its surroundings and has work done on it by the surroundings. **(a)** Sketch a box to represent the system, and use arrows to represent the heat and work transferred. **(b)** Is it possible for ΔE to be positive for this process? Explain. **(c)** Is it possible for ΔE to be negative for this process? Explain.

5.23 A gas is confined to a cylinder fitted with a piston and an electrical heater, as shown in the illustration on the next page. Suppose that current is supplied to the heater so that 100 J of energy are added. Consider two different situations. In case (1) the piston is allowed to move as the energy is added. In case (2) the piston is fixed so that it cannot move. **(a)** In which case does the gas have the higher temperature after addition of the electrical energy? Explain. **(b)** What can you say about the values of q and

w in each of these cases? **(c)** What can you say about the relative values of ΔE for the system (the gas in the cylinder) in the two cases?

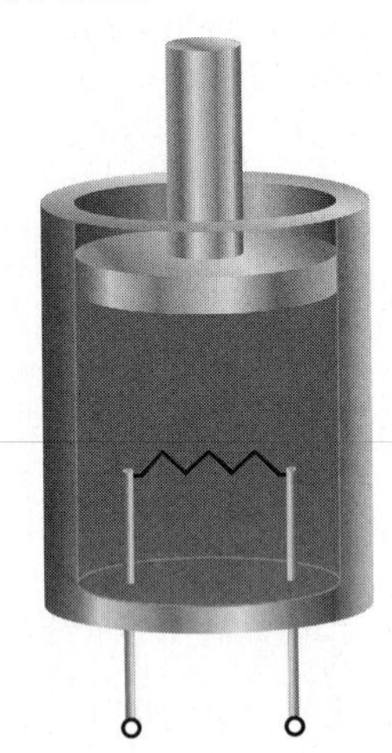

5.24 Consider a system consisting of two oppositely charged spheres hanging by strings and separated by a distance r_1, as shown in the illustration in the next column. Suppose they are separated to a larger distance r_2, by moving them apart along a track. **(a)** What change, if any, has occurred in the potential energy of the system? **(b)** What effect, if any, does this process have on the value of ΔE? **(c)** What can you say about q and w for this process?

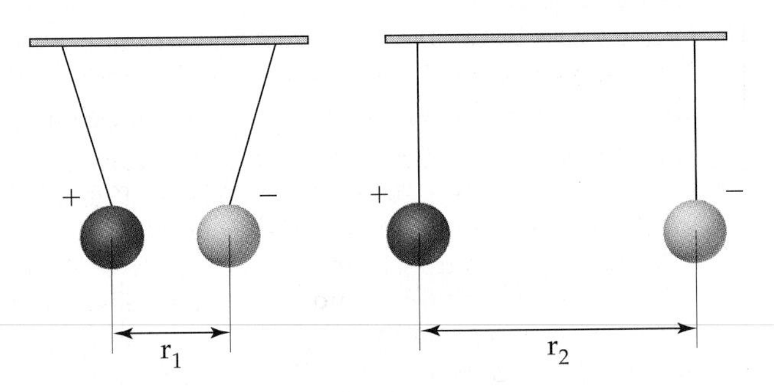

5.25 **(a)** What is meant by the term *state function?* **(b)** Give an example of a quantity that is a state function and one that is not. **(c)** Is temperature a state function? Why or why not?

5.26 Indicate which of the following is independent of the path by which a change occurs: **(a)** the change in potential energy when a book is transferred from table to shelf; **(b)** the heat evolved when a cube of sugar is oxidized to $CO_2(g)$ and $H_2O(g)$; **(c)** the work accomplished in burning a gallon of gasoline.

Enthalpy

5.27 **(a)** Why is the change in enthalpy a meaningful quantity for many chemical processes? **(b)** H is a state function, but q is not a state function. Explain. **(c)** For a given process at constant pressure, ΔH is negative. Is the process endothermic or exothermic?

5.28 **(a)** Under what condition will the enthalpy change of a process equal the amount of heat transferred into or out of the system? **(b)** Enthalpy is said to be a state function. What is it about state functions that makes them particularly useful? **(c)** During a constant-pressure process the system absorbs heat from the surroundings. Does the enthalpy of the system increase or decrease during the process?

5.29 The complete combustion of acetic acid, $HC_2H_3O_2(l)$, to form $H_2O(l)$ and $CO_2(g)$ at constant pressure releases 871.7 kJ of heat per mole of $HC_2H_3O_2$. **(a)** Write a balanced thermochemical equation for this reaction. **(b)** Draw an enthalpy diagram for the reaction.

5.30 The decomposition of zinc carbonate, $ZnCO_3(s)$, into zinc oxide, $ZnO(s)$, and $CO_2(g)$ at constant pressure requires the addition of 71.5 kJ of heat per mole of $ZnCO_3$. **(a)** Write a balanced thermochemical equation for the reaction. **(b)** Draw an enthalpy diagram for the reaction.

5.31 Consider the following reaction, which occurs at room temperature and pressure:

$$2Cl(g) \longrightarrow Cl_2(g) \qquad \Delta H = -243.4 \text{ kJ}$$

Which has the higher enthalpy under these conditions, $2Cl(g)$ or $Cl_2(g)$?

5.32 Without referring to tables, indicate which of the following has the higher enthalpy in each case: **(a)** 1 mol $CO_2(s)$ or 1 mol $CO_2(g)$ at the same temperature; **(b)** 2 mol of hydrogen atoms or 1 mol of H_2; **(c)** 1 mol $H_2(g)$ and 0.5 mol $O_2(g)$ at 25°C or 1 mol $H_2O(g)$ at 25°C; **(d)** 1 mol $N_2(g)$ at 100°C or 1 mol $N_2(g)$ at 300°C.

5.33 Consider the following reaction:

$$2Mg(s) + O_2(g) \longrightarrow 2MgO(s) \qquad \Delta H = -1204 \text{ kJ}$$

(a) Is this reaction exothermic or endothermic? **(b)** Calculate the amount of heat transferred when 2.4 g of $Mg(s)$ reacts at constant pressure. **(c)** How many grams of MgO are produced during an enthalpy change of – 96.0 kJ? **(d)** How many kilojoules of heat are absorbed when 7.50 g of $MgO(s)$ are decomposed into $Mg(s)$ and $O_2(g)$ at constant pressure?

5.34 Consider the following reaction:

$$CH_3OH(g) \longrightarrow CO(g) + 2H_2(g) \qquad \Delta H = +90.7 \text{ kJ}$$

(a) Is heat absorbed or evolved in the course of this reaction? **(b)** Calculate the amount of heat transferred when 1.60 kg of $CH_3OH(g)$ are decomposed by this reaction at constant pressure. **(c)** For a given sample of CH_3OH, the enthalpy change on reaction is 64.7 kJ. How many grams of hydrogen gas are produced? **(d)** What is the value of ΔH for the reverse of the previous reaction? How many kilojoules of heat are released when 32.0 g of $CO(g)$ reacts completely with $H_2(g)$ to form $CH_3OH(g)$ at constant pressure?

5.35 When solutions containing silver ions and chloride ions are mixed, silver chloride precipitates:

$$Ag^+(aq) + Cl^-(aq) \longrightarrow AgCl(s) \qquad \Delta H = -65.5 \text{ kJ}$$

(a) Calculate ΔH for formation of 0.540 mol of AgCl by this reaction. **(b)** Calculate ΔH for the formation of 1.66 g of AgCl. **(c)** Calculate ΔH when 0.188 mmol of AgCl dissolves in water.

5.36 At one time, a common means of forming small quantities of oxygen gas in the laboratory was to heat $KClO_3$:

$$2KClO_3(s) \longrightarrow 2KCl(s) + 3O_2(g) \qquad \Delta H = -89.4 \text{ kJ}$$

For this reaction, calculate ΔH for the formation of **(a)** 4.34 mol of O_2 and **(b)** 200.8 g of KCl. **(c)** The decomposition of $KClO_3$ proceeds spontaneously when it is heated. Do you think that the reverse reaction, the formation of $KClO_3$ from KCl and O_2, is likely to be feasible under ordinary conditions? Explain your answer.

5.37 You are given ΔH for a process that occurs at constant pressure. What additional information is needed to determine ΔE for the process?

5.38 Suppose that the gas phase reaction, $2NO(g) + O_2(g) \longrightarrow 2NO_2(g)$ were carried out in a constant-volume container at constant temperature. Would the measured heat change represent ΔH or ΔE? If there is a difference, which quantity is larger for this reaction? Explain.

5.39 A gas is confined to a cylinder under constant atmospheric pressure, as illustrated in Figure 5.3. When the gas undergoes a particular chemical reaction, it releases 89 kJ of heat to its surroundings and does 36 kJ of P–V work on its surroundings. What are the values of ΔH and ΔE for this process?

5.40 A gas is confined to a cylinder under constant atmospheric pressure, as illustrated in Figure 5.3. When 518 J of heat is added to the gas, it expands and does 127 J of work on the surroundings. What are the values of ΔH and ΔE for this process?

5.41 Consider the combustion of liquid methanol, $CH_3OH(l)$:

$$CH_3OH(l) + \tfrac{3}{2}O_2(g) \longrightarrow CO_2(g) + 2H_2O(l) \qquad \Delta H = -726.5 \text{ kJ}$$

(a) What is the enthalpy change for the reverse reaction? **(b)** Balance the forward reaction with whole-number coefficients. What is ΔH for the reaction represented by this equation? **(c)** Which is more likely to be thermodynamically favored, the forward reaction or the reverse reaction? **(d)** If the reaction were written to produce $H_2O(g)$ instead of $H_2O(l)$, would you expect the magnitude of ΔH to increase, decrease, or stay the same? Explain.

5.42 Consider the decomposition of liquid benzene, $C_6H_6(l)$, to gaseous acetylene, $C_2H_2(g)$:

$$\tfrac{1}{3}C_6H_6(l) \longrightarrow C_2H_2(g) \qquad \Delta H = +210 \text{ kJ}$$

(a) What is the enthalpy change for the reverse reaction? **(b)** What is ΔH for the decomposition of 1 mol of benzene to acetylene? **(c)** Which is more likely to be thermodynamically favored, the forward reaction or the reverse reaction? **(d)** If $C_6H_6(g)$ were consumed instead of $C_6H_6(l)$, would you expect the magnitude of ΔH to increase, decrease, or stay the same? Explain.

Calorimetry

5.43 **(a)** What are the units of heat capacity? **(b)** What are the units of specific heat?

5.44 Two solid objects, A and B, are placed in boiling water and allowed to come to temperature there. Each is then lifted out and placed in separate beakers containing 1000 g water at 10.0°C. Object A increases the water temperature by 3.50°C; B increases the water temperature by 2.60°C. **(a)** Which object has the larger heat capacity? **(b)** What can you say about the specific heats of A and B?

5.45 **(a)** What is the specific heat of liquid water? **(b)** What is the heat capacity of 185 g of liquid water? **(c)** How many kJ of heat are needed to raise the temperature of 10.00 kg of liquid water from 24.6°C to 46.2°C?

5.46 **(a)** What is the molar heat capacity of liquid water? **(b)** What is the heat capacity of 8.42 mol of liquid water? **(c)** How many kJ of heat are needed to raise the temperature of 2.56 kg of water from 44.8°C to 92.0°C?

5.47 The specific heat of copper metal is 0.385 J/g-K. How many J of heat are necessary to raise the temperature of a 1.42-kg block of copper from 25.0°C to 88.5°C?

5.48 The specific heat of toluene (C_7H_8), is 1.13 J/g-K. How many J of heat are needed to raise the temperature of 62.0 g of toluene from 16.3°C to 38.8°C?

5.49 When a 9.55-g sample of solid sodium hydroxide dissolves in 100.0 g of water in a coffee-cup calorimeter (Figure 5.18), the temperature rises from 23.6°C to 47.4°C. Calculate ΔH (in kJ/mol NaOH) for the solution process

$$NaOH(s) \longrightarrow Na^+(aq) + OH^-(aq)$$

Assume that the specific heat of the solution is the same as that of pure water.

5.50 When a 3.88-g sample of solid ammonium nitrate dissolves in 60.0 g of water in a coffee-cup calorimeter (Figure 5.18), the temperature drops from 23.0°C to 18.4°C. Calculate ΔH (in kJ/mol NH_4NO_3) for the solution process

$$NH_4NO_3(s) \longrightarrow NH_4^+(aq) + NO_3^-(aq)$$

Assume that the specific heat of the solution is the same as that of pure water.

5.51 A 2.200-g sample of quinone ($C_6H_4O_2$) is burned in a bomb calorimeter whose total heat capacity is 7.854 kJ/°C. The temperature of the calorimeter increases from 23.44°C to 30.57°C. What is the heat of combustion per gram of quinone? Per mole of quinone?

5.52 A 1.800-g sample of phenol (C_6H_5OH) was burned in a bomb calorimeter whose total heat capacity is 11.66 kJ/°C. The temperature of the calorimeter plus contents increased from 21.36°C to 26.37°C. **(a)** Write a balanced chemical equation for the bomb calorimeter reaction. **(b)** What is the heat of combustion per gram of phenol? Per mole of phenol?

5.53 Under constant-volume conditions the heat of combustion of glucose ($C_6H_{12}O_6$) is 15.57 kJ/g. A 2.500-g sample of glucose is burned in a bomb calorimeter. The temperature of the calorimeter increased from 20.55°C to 23.25°C. **(a)** What is the total heat capacity of the calorimeter? **(b)** If the size of the glucose sample had been exactly twice as large, what would the temperature change of the calorimeter have been?

5.54 Under constant-volume conditions the heat of combustion of benzoic acid ($HC_7H_5O_2$) is 26.38 kJ/g. A 1.640-g sample of benzoic acid is burned in a bomb calorimeter. The temperature of the calorimeter increases from 22.25°C to 27.20°C. **(a)** What is the total heat capacity of the calorimeter? **(b)** A 1.320-g sample of a new organic substance is combusted in the same calorimeter. The temperature of the calorimeter increases from 22.14°C to 26.82°C. What is the heat of combustion per gram of the new substance? **(c)** Suppose that in changing samples, a portion of the water in the calorimeter were lost. In what way, if any, would this change the heat capacity of the calorimeter?

Hess's Law

5.55 State Hess's law. Why is it important to thermochemistry?

5.56 What is the connection between Hess's law and the fact that H is a state function?

5.57 Consider the following hypothetical reactions:

$$A \longrightarrow B \qquad \Delta H = +30 \text{ kJ}$$
$$B \longrightarrow C \qquad \Delta H = +60 \text{ kJ}$$

(a) Use Hess's law to calculate the enthalpy change for the reaction $A \longrightarrow C$. **(b)** Construct an enthalpy diagram for substances A, B, and C, and show how Hess's law applies.

5.58 Suppose you are given the following hypothetical reactions:

$$X \longrightarrow Y \qquad \Delta H = -35 \text{ kJ}$$
$$X \longrightarrow Z \qquad \Delta H = +90 \text{ kJ}$$

(a) Use Hess's law to calculate the enthalpy change for the reaction $Y \longrightarrow Z$. **(b)** Construct an enthalpy diagram for substances X, Y, and Z. **(c)** Would it be valid to do what we have asked in part (a) if the first reaction had been carried out at 25°C and the second at 240°C? Explain.

5.59 Given the enthalpies of reaction

$$P_4(s) + 3O_2(g) \longrightarrow P_4O_6(s) \qquad \Delta H = -1640.1 \text{ kJ}$$
$$P_4(s) + 5O_2(g) \longrightarrow P_4O_{10}(s) \qquad \Delta H = -2940.1 \text{ kJ}$$

calculate the enthalpy change for the reaction

$$P_4O_6(s) + 2O_2(g) \longrightarrow P_4O_{10}(s)$$

5.60 From the heats of reaction

$$2H_2(g) + O_2(g) \longrightarrow 2H_2O(g) \qquad \Delta H = -483.6 \text{ kJ}$$
$$3O_2(g) \longrightarrow 2O_3(g) \qquad \Delta H = +284.6 \text{ kJ}$$

calculate the heat of the reaction

$$3H_2(g) + O_3(g) \longrightarrow 3H_2O(g)$$

5.61 From the enthalpies of reaction

$$H_2(g) + F_2(g) \longrightarrow 2HF(g) \qquad \Delta H = -537 \text{ kJ}$$
$$C(s) + 2F_2(g) \longrightarrow CF_4(g) \qquad \Delta H = -680 \text{ kJ}$$
$$2C(s) + 2H_2(g) \longrightarrow C_2H_4(g) \qquad \Delta H = +52.3 \text{ kJ}$$

calculate ΔH for the reaction of ethylene with F_2:

$$C_2H_4(g) + 6F_2(g) \longrightarrow 2CF_4(g) + 4HF(g)$$

5.62 Given the data

$$N_2(g) + O_2(g) \longrightarrow 2NO(g) \qquad \Delta H = +180.7 \text{ kJ}$$
$$2NO(g) + O_2(g) \longrightarrow 2NO_2(g) \qquad \Delta H = -113.1 \text{ kJ}$$
$$2N_2O(g) \longrightarrow 2N_2(g) + O_2(g) \qquad \Delta H = -163.2 \text{ kJ}$$

use Hess's law to calculate ΔH for the reaction

$$N_2O(g) + NO_2(g) \longrightarrow 3NO(g)$$

Enthalpies of Formation

5.63 **(a)** What is meant by the term *standard conditions*, with reference to enthalpy changes? **(b)** What is meant by the term *enthalpy of formation*? **(c)** What is meant by the term *standard enthalpy of formation*?

5.64 **(a)** Why are tables of standard enthalpies of formation so useful? **(b)** What is the value of the standard enthalpy of formation of an element in its most stable form?

5.65 Suppose it were decided that the standard enthalpies of formation of all elements in their most stable forms should be 100 kJ/mol. Would it still be possible to have standard enthalpies of formation of compounds, as in Table 5.3? If so, would any of the values in Table 5.3 be the same? Explain.

5.66 Using Table 5.3, determine whether the reaction of solid sucrose with liquid water to form solid glucose is an endothermic or exothermic process.

5.67 For each of the following compounds, write a balanced thermochemical equation depicting the formation of 1 mol of the compound from its elements in their standard states and use Appendix C to obtain the value of ΔH_f°: **(a)** $NH_3(g)$; **(b)** $SO_2(g)$; **(c)** $RbClO_3(s)$; **(d)** $NH_4NO_3(s)$.

5.68 Write balanced equations that describe the formation of the following compounds from their elements in their standard states, and use Appendix C to obtain the values of their standard enthalpies of formation: **(a)** $HBr(g)$; **(b)** $AgNO_3(s)$; **(c)** $Hg_2Cl_2(s)$; **(d)** $C_2H_5OH(l)$.

5.69 The following is known as the thermite reaction [(Figure 5.8(b))]:

$$2Al(s) + Fe_2O_3(s) \longrightarrow Al_2O_3(s) + 2Fe(s)$$

This highly exothermic reaction is used for welding massive units, such as propellers for large ships. Using enthalpies of formation in Appendix C, calculate ΔH° for this reaction.

5.70 Many cigarette lighters contain liquid butane, $C_4H_{10}(l)$. Using enthalpies of formation, calculate the quantity of heat produced when 1.0 g of butane is completely combusted in air.

5.71 Using values from Appendix C, calculate the standard enthalpy change for each of the following reactions:
(a) $2SO_2(g) + O_2(g) \longrightarrow 2SO_3(g)$
(b) $Mg(OH)_2(s) \longrightarrow MgO(s) + H_2O(l)$
(c) $4FeO(s) + O_2(g) \longrightarrow 2Fe_2O_3(s)$
(d) $SiCl_4(l) + 2H_2O(l) \longrightarrow SiO_2(s) + 4HCl(g)$

5.72 Using values from Appendix C, calculate the value of $\Delta H°$ for each of the following reactions:
(a) $N_2O_4(g) + 4H_2(g) \longrightarrow N_2(g) + 4H_2O(g)$
(b) $2KOH(s) + CO_2(g) \longrightarrow K_2CO_3(s) + H_2O(g)$
(c) $SO_2(g) + 2H_2S(g) \longrightarrow (\frac{3}{8})S_8(s) + 2H_2O(g)$
(d) $Fe_2O_3(s) + 6HCl(g) \longrightarrow 2FeCl_3(s) + 3H_2O(g)$

5.73 Complete combustion of 1 mol of acetone (C_3H_6O) liberates 1790 kJ:

$$C_3H_6O(l) + 4O_2(g) \longrightarrow 3CO_2(g) + 3H_2O(l) \qquad \Delta H° = -1790 \text{ kJ}$$

Using this information together with data from Appendix C, calculate the enthalpy of formation of acetone.

5.74 Calcium carbide (CaC_2) reacts with water to form acetylene (C_2H_2) and $Ca(OH)_2$. From the following enthalpy of reaction data and data in Appendix C, calculate $\Delta H_f°$ for $CaC_2(s)$:

$$CaC_2(s) + 2H_2O(l) \longrightarrow Ca(OH)_2(s) + C_2H_2(g) \qquad \Delta H° = -127.2 \text{ kJ}$$

5.75 Calculate the standard enthalpy of formation of solid $Mg(OH)_2$, given the following data:

$$2Mg(s) + O_2(g) \longrightarrow 2MgO(s) \qquad \Delta H° = -1203.6 \text{ kJ}$$
$$Mg(OH)_2(s) \longrightarrow MgO(s) + H_2O(l) \qquad \Delta H° = +37.1 \text{ kJ}$$
$$2H_2(g) + O_2(g) \longrightarrow 2H_2O(l) \qquad \Delta H° = -571.7 \text{ kJ}$$

5.76 **(a)** Calculate the standard enthalpy of formation of gaseous diborane (B_2H_6) using the following thermochemical information:

$$4B(s) + 3O_2(g) \longrightarrow 2B_2O_3(s) \qquad \Delta H° = -2509.1 \text{ kJ}$$
$$2H_2(g) + O_2(g) \longrightarrow 2H_2O(l) \qquad \Delta H° = -571.7 \text{ kJ}$$
$$B_2H_6(g) + 3O_2(g) \longrightarrow B_2O_3(s) + 3H_2O(l) \qquad \Delta H° = -2147.5 \text{ kJ}$$

(b) Pentaborane (B_5H_9) is another in a series of boron hydrides. What experiment or experiments would you need to perform to yield the data necessary to calculate the heat of formation of $B_5H_9(l)$? Explain by writing out and summing any applicable chemical reactions.

5.77 Gasoline is composed primarily of hydrocarbons, including many with eight carbon atoms, called *octanes*. One of the cleanest burning octanes is a compound called 2,3,4-trimethylpentane, which has the following structural formula:

```
     CH3  CH3  CH3
      |    |    |
H3C—CH—CH—CH—CH3
```

The complete combustion of 1 mol of this compound to $CO_2(g)$ and $H_2O(g)$ leads to $\Delta H° = -5069$ kJ. **(a)** Write a balanced equation for the combustion of 1 mol of $C_8H_{18}(l)$. **(b)** Write a balanced equation for the formation of $C_8H_{18}(l)$ from its elements. **(c)** By using the information in this problem and data in Table 5.3, calculate $\Delta H_f°$ for 2,3,4-trimethylpentane.

5.78 Naphthalene ($C_{10}H_8$) is a solid aromatic compound often sold as mothballs. The complete combustion of this substance to yield $CO_2(g)$ and $H_2O(l)$ at 25°C yields 5154 kJ/mol. **(a)** Write balanced equations for the formation of naphthalene from the elements and for its combustion. **(b)** Calculate the standard enthalpy of formation of naphthalene.

Foods and Fuels

5.79 **(a)** What is meant by the term *fuel value*? **(b)** What substance is often referred to as *blood sugar*? Why is it significant in the discussion of human foods? **(c)** Which is a greater source of energy as food, 5 g of fat or 9 g of carbohydrate?

5.80 **(a)** Why are fats well suited for energy storage in the human body? **(b)** A particular chip snack food is composed of 12% protein, 14% fat, and the rest carbohydrate. What percentage of the calorie content of this food is fat? **(c)** How many grams of protein provide the same fuel value as 25 g of fat?

5.81 A serving of Campbell's® condensed cream of mushroom soup contains 7 g fat, 9 g carbohydrate, and 1 g protein. Estimate the number of Calories in a serving.

5.82 A pound of plain M&M® candies contains 96 g fat, 320 g carbohydrate, and 21 g protein. What is the fuel value in kJ in a 42-g (about 1.5 oz) serving? How many Calories does it provide?

5.83 The heat of combustion of fructose, $C_6H_{12}O_6$, is −2812 kJ/mol. If a fresh golden delicious apple weighing 4.23 oz (120 g) contains 16.0 g of fructose, what caloric content does the fructose contribute to the apple?

5.84 The heat of combustion of ethanol, $C_2H_5OH(l)$, is −1367 kJ/mol. A batch of sauvignon blanc wine contains 10.6% ethanol by mass. Assuming the density of the wine to be 1.0 g/mL, what caloric content does the alcohol (ethanol) in a 6-oz glass of wine (177 mL) have?

5.85 The standard enthalpies of formation of gaseous propyne (C_3H_4), propylene (C_3H_6), and propane (C_3H_8) are +185.4, +20.4, and −103.8 kJ/mol, respectively. **(a)** Calculate the heat evolved per mole on combustion of each substance to yield $CO_2(g)$ and $H_2O(g)$. **(b)** Calculate the heat evolved on combustion of 1 kg of each substance. **(c)** Which is the most efficient fuel in terms of heat evolved per unit mass?

5.86 It is interesting to compare the "fuel value" of a hydrocarbon in a world where fluorine rather than oxygen is the combustion agent. The enthalpy of formation of $CF_4(g)$ is -679.9 kJ/mol. Which of the following two reactions is the more exothermic?

$$CH_4(g) + 2O_2(g) \longrightarrow CO_2(g) + 2H_2O(g)$$
$$CH_4(g) + 4F_2(g) \longrightarrow CF_4(g) + 4HF(g)$$

Additional Exercises

5.87 At 20°C (approximately room temperature) the average velocity of N_2 molecules in air is 1050 mph. **(a)** What is the average speed in m/s? **(b)** What is the kinetic energy (in J) of an N_2 molecule moving at this speed? **(c)** What is the total kinetic energy of 1 mol of N_2 molecules moving at this speed?

5.88 Suppose an Olympic diver who weighs 52.0 kg executes a straight dive from a 10-m platform. At the apex of the dive, the diver is 10.8 m above the surface of the water. **(a)** What is the potential energy of the diver at the apex of the dive, relative to the surface of the water? (See the caption of Figure 5.5.) **(b)** Assuming that all the potential energy of the diver is converted into kinetic energy at the surface of the water, at what speed in m/s will the diver enter the water? **(c)** Does the diver do work on entering the water? Explain.

5.89 When a mole of Dry Ice®, $CO_2(s)$, is converted to $CO_2(g)$ at atmospheric pressure and -78°C, the heat absorbed by the system exceeds the increase in internal energy of the CO_2. Why is this so? What happens to the remaining energy?

5.90 The air bags that provide protection in autos in the event of an accident expand as a result of a rapid chemical reaction. From the viewpoint of the chemical reactants as the system, what do you expect for the signs of q and w in this process?

[5.91] An aluminum can of a soft drink is placed in a freezer. Later, the can is found to be split open and its contents frozen. Work was done on the can in splitting it open. Where did the energy for this work come from?

5.92 Aside from nuclear reactions, in which matter and energy interconvert to a measurable extent, the classical statement of the first law of thermodynamics can be written as follows: *The energy of the universe is constant*. Is this statement consistent with Equation 5.5? Explain.

[5.93] A sample of gas is contained in a cylinder-and-piston arrangement. It undergoes the change in state shown in the drawing. **(a)** Assume first that the cylinder and piston are perfect thermal insulators that do not allow heat to be transferred. What is the value of q for the state change?

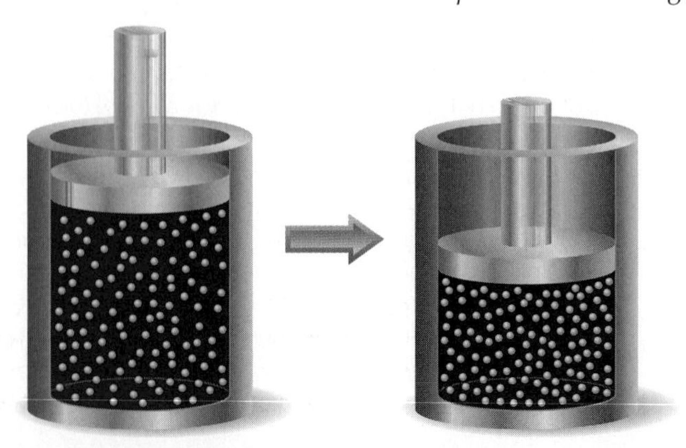

What is the sign of w for the state change? What can be said about ΔE for the state change? **(b)** Now assume that the cylinder and piston are made up of a thermal conductor such as a metal. During the state change, the cylinder gets warmer to the touch. What is the sign of q for the state change in this case? Describe the difference in the state of the system at the end of the process in the two cases. What can you say about the relative values of ΔE?

[5.94] Limestone stalactites and stalagmites are formed in caves by the following reaction:

$$Ca^{2+}(aq) + 2HCO_3^-(aq) \longrightarrow CaCO_3(s) + CO_2(g) + H_2O(l)$$

If 1 mol of $CaCO_3$ forms at 298 K under 1 atm pressure, the reaction performs 2.47 kJ of P-V work, pushing back the atmosphere as the gaseous CO_2 forms. At the same time, 38.95 kJ of heat is absorbed from the environment. What are the values of ΔH and of ΔE for this reaction?

[5.95] Consider the systems shown in Figure 5.10. In one case the battery becomes completely discharged by running the current through a heater, and in the other by running a fan. Both processes occur at constant pressure. In both cases the change in state of the system is the same: The battery goes from being fully charged to being fully discharged. Yet in one case, the heat evolved is large, and in the other it is small. Is the enthalpy change the same in the two cases? If not, how can enthalpy be considered a state function? If it is, what can you say about the relationship between enthalpy change and q in this case, as compared with others that we have considered?

5.96 A house is designed to have passive solar energy features. Brickwork is to be incorporated into the interior of the house to act as a heat absorber. Each brick weighs approximately 1.8 kg. The specific heat of the brick is 0.85 J/g-K. How many bricks must be incorporated into the interior of the house to provide the same total heat capacity as 1.7×10^3 gal of water?

[5.97] A coffee-cup calorimeter of the type shown in Figure 5.18 contains 150.0 g of water at 25.1°C. A 121.0-g block of copper metal is heated to 100.4°C by putting it in a beaker of boiling water. The specific heat of Cu(s) is 0.385 J/g-K. The Cu is added to the calorimeter, and after a time the contents of the cup reach a constant temperature of 30.1°C. **(a)** Determine the amount of heat, in J, lost by the copper block. **(b)** Determine the amount of heat gained by the water. The specific heat of water is 4.18 J/g-K. **(c)** The difference between your answers for (a) and (b) is due to heat loss through the Styrofoam® cups and the heat necessary to raise the temperature of the inner wall of the apparatus. The heat capacity of the calorimeter is the amount of heat necessary to raise the temperature of the apparatus (the cups and the stopper) by 1 K. Calculate the heat capacity of the calorimeter in J/K. **(d)** What would be the final temperature of the system if all the heat lost by the copper block were absorbed by the water in the calorimeter?

[5.98] **(a)** When a 0.235-g sample of benzoic acid is combusted in a bomb calorimeter, the temperature rises 1.642°C. When a 0.265-g sample of caffeine, $C_8H_{10}O_2N_4$, is burned, the temperature rises 1.525°C. Using the value 26.38 kJ/g for the heat of combustion of benzoic acid, calculate the heat of combustion per mole of caffeine at constant volume. **(b)** Assuming that there is an uncertainty of 0.002°C in each temperature reading and that the masses of samples are measured to 0.001 g, what is the estimated uncertainty in the value calculated for the heat of combustion per mole of caffeine?

5.99 A 200-lb man decides to add to his exercise routine by walking up three flights of stairs (45 ft) 20 times per day. He figures that the work required to increase his potential energy in this way will permit him to eat an extra order of French fries, at 245 Cal, without adding to his weight. Is he correct in this assumption?

5.100 Burning methane in oxygen can produce three different carbon-containing products: soot (very fine particles of graphite), $CO(g)$, and $CO_2(g)$. **(a)** Write three balanced equations for the reaction of methane gas with oxygen to produce these three products. In each case assume that $H_2O(l)$ is the only other product. **(b)** Determine the standard enthalpies for the reactions in part (a). **(c)** Why, when the oxygen supply is adequate, is $CO_2(g)$ the predominant carbon-containing product of the combustion of methane?

5.101 From the following data for three prospective fuels, calculate which could provide the most energy per unit volume:

Fuel	Density at 20°C (g/cm^3)	Molar Enthalpy of Combustion (kJ/mol)
Nitroethane, $C_2H_5NO_2(l)$	1.052	−1368
Ethanol, $C_2H_5OH(l)$	0.789	−1367
Methylhydrazine, $CH_6N_2(l)$	0.874	−1305

5.102 The hydrocarbons acetylene (C_2H_2) and benzene (C_6H_6) have the same empirical formula. Benzene is an "aromatic" hydrocarbon, one that is unusually stable because of its structure. **(a)** By using the data in Appendix C, determine the standard enthalpy change for the reaction $3C_2H_2(g) \longrightarrow C_6H_6(l)$. **(b)** Which has greater enthalpy, 3 mol of acetylene gas or 1 mol of liquid benzene? **(c)** Determine the fuel value in kJ/g for acetylene and benzene.

[5.103] Three common hydrocarbons that contain four carbons are listed here, along with their standard enthalpies of formation:

Hydrocarbon	Formula	ΔH_f°(kJ/mol)
1,3-Butadiene	$C_4H_6(g)$	111.9
1-Butene	$C_4H_8(g)$	1.2
n-Butane	$C_4H_{10}(g)$	−124.7

(a) For each of these substances, calculate the molar enthalpy of combustion to $CO_2(g)$ and $H_2O(l)$. **(b)** Calculate the fuel value in kJ/g for each of these compounds. **(c)** For each hydrocarbon, determine the percentage of hydrogen by mass. **(d)** By comparing your answers for parts (b) and (c), propose a relationship between hydrogen content and fuel value in hydrocarbons.

5.104 The two common sugars, glucose ($C_6H_{12}O_6$) and sucrose ($C_{12}H_{22}O_{11}$), are both carbohydrates. Their standard enthalpies of formation are given in Table 5.3. Using these data, **(a)** calculate the molar enthalpy of combustion to $CO_2(g)$ and $H_2O(l)$ for the two sugars; **(b)** calculate the enthalpy of combustion per gram of each sugar; **(c)** determine how your answers to part (b) compare to the average fuel value of carbohydrates discussed in Section 5.8.

[5.105] It is estimated that the net amount of carbon dioxide fixed by photosynthesis on land on Earth is 5.5×10^{16} g/yr of CO_2. All this carbon is converted into glucose. **(a)** Calculate the energy stored by photosynthesis on land per year in kJ. **(b)** Calculate the average rate of conversion of solar energy into plant energy in MW (1 W = 1 J/s). A large nuclear power plant produces about 10^3 MW. The energy of how many such nuclear power plants is equivalent to the solar energy conversion?

[5.106] Ammonia (NH_3) boils at −33°C; at this temperature it has a density of 0.81 g/cm^3. The enthalpy of formation of $NH_3(g)$ is −46.2 kJ/mol, and the enthalpy of vaporization of $NH_3(l)$ is 23.2 kJ/mol. Calculate the enthalpy change when 1 L of liquid NH_3 is burned in air to give $N_2(g)$ and $H_2O(g)$. How does this compare with ΔH for the complete combustion of 1 L of liquid methanol, $CH_3OH(l)$? For $CH_3OH(l)$, the density at 25°C is 0.792 g/cm^3, and ΔH_f° equals −239 kJ/mol.

Integrative Exercises

5.107 Consider the combustion of a single molecule of $CH_4(g)$. **(a)** How much energy, in J, is produced during this reaction? **(b)** A typical X-ray photon has an energy of 8 keV. How does the energy of combustion compare to the energy of the X-ray photon?

5.108 Consider the dissolving of NaCl in water, illustrated in Figure 4.3. Let's say that the system consists of 0.1 mol NaCl and 1 L of water. Considering that the NaCl readily dissolves in the water and that the ions are strongly stabilized by the water molecules, as shown in the figure, is it safe to conclude that the dissolution of NaCl in water results in a lower enthalpy for the system? Explain your response. What experimental evidence would you examine to test this question?

5.109 Consider the following unbalanced oxidation-reduction reactions in aqueous solution:

$$Ag^+(aq) + Li(s) \longrightarrow Ag(s) + Li^+(aq)$$
$$Fe(s) + Na^+(aq) \longrightarrow Fe^{2+}(aq) + Na(s)$$
$$K(s) + H_2O(l) \longrightarrow KOH(aq) + H_2(g)$$

(a) Balance each of the reactions. **(b)** By using data in Appendix C, calculate ΔH° for each of the reactions. **(c)** Based on the values you obtain for ΔH°, which of the reactions would you expect to be favorable? Which would you expect to be unfavorable? **(d)** Use the activity series to predict which of these reactions should occur. (Section 4.4) Are these results in accord with your conclusion in part (c) of this problem?

[5.110] Consider the following acid-neutralization reactions involving the strong base NaOH(*aq*):

$$HNO_3(aq) + NaOH(aq) \longrightarrow NaNO_3(aq) + H_2O(l)$$
$$HCl(aq) + NaOH(aq) \longrightarrow NaCl(aq) + H_2O(l)$$
$$NH_4^+(aq) + NaOH(aq) \longrightarrow NH_3(aq) + Na^+(aq) + H_2O(l)$$

(a) By using data in Appendix C, calculate $\Delta H°$ for each of the reactions. **(b)** As we saw in Section 4.3, nitric acid and hydrochloric acid are strong acids. Write net ionic equations for the neutralization of these acids. **(c)** Compare the values of $\Delta H°$ for the first two reactions. What can you conclude? **(d)** In the third equation $NH_4^+(aq)$ is acting as an acid. Based on the value of $\Delta H°$ for this reaction, do you think it is a strong or a weak acid? Explain.

5.111 Consider two solutions, the first being 50.0 mL of 1.00 *M* $CuSO_4$ and the other 50.0 mL of 2.00 *M* KOH. When the two solutions are mixed in a constant-pressure calorimeter, a precipitate forms and the temperature of the mixture rises from 21.5°C to 27.7°C. **(a)** Before mixing, how many grams of Cu are present in the solution of $CuSO_4$? **(b)** Predict the identity of the precipitate in the reaction. **(c)** Write complete and net ionic equations for the reaction that occurs when the two solutions are mixed. **(d)** From the calorimetric data, calculate ΔH for the reaction that occurs on mixing. Assume that the calorimeter absorbs only a negligible quantity of heat, that the total volume of the solution is 100.0 mL, and that the specific heat and density of the solution after mixing are the same as that of pure water.

5.112 The metathesis reaction between $AgNO_3(aq)$ and NaCl(*aq*) proceeds as follows:

$$AgNO_3(aq) + NaCl(aq) \longrightarrow NaNO_3(aq) + AgCl(s)$$

(a) By using Appendix C, calculate $\Delta H°$ for the net ionic equation of this reaction. **(b)** What would you expect for the value of $\Delta H°$ of the overall molecular equation compared to that for the net ionic equation? Explain. **(c)** Use the results from (a) and (b) along with data in Appendix C to determine the value of ΔH_f° for $AgNO_3(aq)$.

[5.113] A sample of a hydrocarbon is combusted completely in $O_2(g)$ to produce 21.83 g $CO_2(g)$, 4.47 g $H_2O(g)$, and 311 kJ of heat. **(a)** What is the mass of the hydrocarbon sample that was combusted? **(b)** What is the empirical formula of the hydrocarbon? **(c)** Calculate the value of ΔH_f° per empirical-formula unit of the hydrocarbon. **(d)** Do you think that the hydrocarbon is one of those listed in Appendix C? Explain your answer.

5.114 The methane molecule, CH_4, has the geometry shown in Figure 2.21. Imagine a hypothetical process in which the methane molecule is "expanded," by simultaneously extending all four C—H bonds to infinity. We then have the process

$$CH_4(g) \longrightarrow C(g) + 4H(g)$$

(a) Compare this process with the reverse of the reaction that represents the standard enthalpy of formation. **(b)** Calculate the enthalpy change in each case. Which is the more endothermic process? What accounts for the difference in $\Delta H°$ values? **(c)** Suppose that 3.45 g $CH_4(g)$ are reacted with 1.22 g $F_2(g)$, forming $CF_4(g)$ and HF(*g*) as sole products. What is the limiting reagent in this reaction? Assuming that the reaction occurs at constant pressure, what amount of heat is evolved?

eMedia Exercises

5.115 The **Enthalpy of Dissolution** simulation (*eChapter 5.4*) allows you to dissolve five different compounds in water. For each compound, tell whether the process is endothermic or exothermic.

5.116 Using the calculated ΔH from Exercise 5.49, determine how much sodium hydroxide must be dissolved in 100.0 g of water in order for the temperature to rise by 25.0°C. Use the **Enthalpy of Dissolution** simulation (*eChapter 5.4*) to verify your answer.

5.117 **(a)** Which combination (what compound, mass of compound, and mass of water) in the **Enthalpy of Dissolution** simulation produces the largest temperature change? (*eChapter 5.4*) **(b)** Suppose a weighed quantity of a sparingly soluble salt were used for a calorimetry experiment similar to those in the simulation. After measuring the temperature change, you discover that not all of the salt has dissolved. Discuss the errors that would arise in the calculated value of the molar enthalpy of solution as a result of this complication.

5.118 **(a)** Using the heat of combustion given in Exercise 5.54, calculate the temperature increase for the combustion of a 450-mg sample of benzoic acid in a calorimeter with a heat capacity of 420 J/°C that contains 500 g of water. Use the **Calorimetry** simulation (*eChapter 5.5*) to check your prediction. **(b)** Perform calorimetry experiments using the other compounds available in the simulation. For each compound (excluding nitroglycerin), determine the molar heat of combustion. **(c)** Which compound has the highest heat of combustion per mole? **(d)** Which compound has the highest heat of combustion per *gram*?

5.119 **(a)** Using the heat of combustion for sucrose determined in the previous exercise, calculate the amount of heat that would be evolved by the combustion of 228 mg of sucrose. **(b)** What temperature change would be observed if this combustion were carried out with 735 grams of water in the calorimeter? Use the **Calorimetry** simulation (*eChapter 5.5*) to verify your answer.

5.120 Carry out a series of experiments with the **Calorimetry** simulation (*eChapter 5.5*), using 500 grams of water in the calorimeter. Vary the amount of sucrose, using values of 50 mg, 150 mg, 250 mg, and 450 mg. **(a)** Graph your data with mg sucrose on the x-axis and temperature change on the y-axis. **(b)** Is the temperature change proportional to the mass of sucrose combusted?

5.121 Repeat the experiment of the previous exercise with a constant 255 mg of sucrose, but vary the amount of water, using values of 600 g, 700 g, 800 g, and 900 g. **(a)** Again, graph your data with grams of water on the x-axis and temperature change on the y-axis. **(b)** Is the temperature change proportional to the mass of water in the calorimeter? If your answer to part (b) is different from your answer to part (b) of the previous exercise, explain why.

5.122 The enthalpy of combustion of nitroglycerin is -1541 kJ/mol. **(a)** Using this and the data from experiments conducted in the **Calorimetry** simulation (*eChapter 5.5*), determine the molar mass of nitroglycerin. **(b)** Explain how your answer to part (a) would be affected by the use of a poorly insulated calorimeter.

5.123 Compare the **Thermite** and **Formation of Water** movies (*eChapter 5.7*). **(a)** Use the information provided to calculate the $\Delta H°$ for the formation of water, and compare it to the $\Delta H°$ calculated for the thermite reaction in Exercise 5.69. **(b)** Comment on the apparent relationship between the nature of a reaction and the magnitude of its $\Delta H°$.

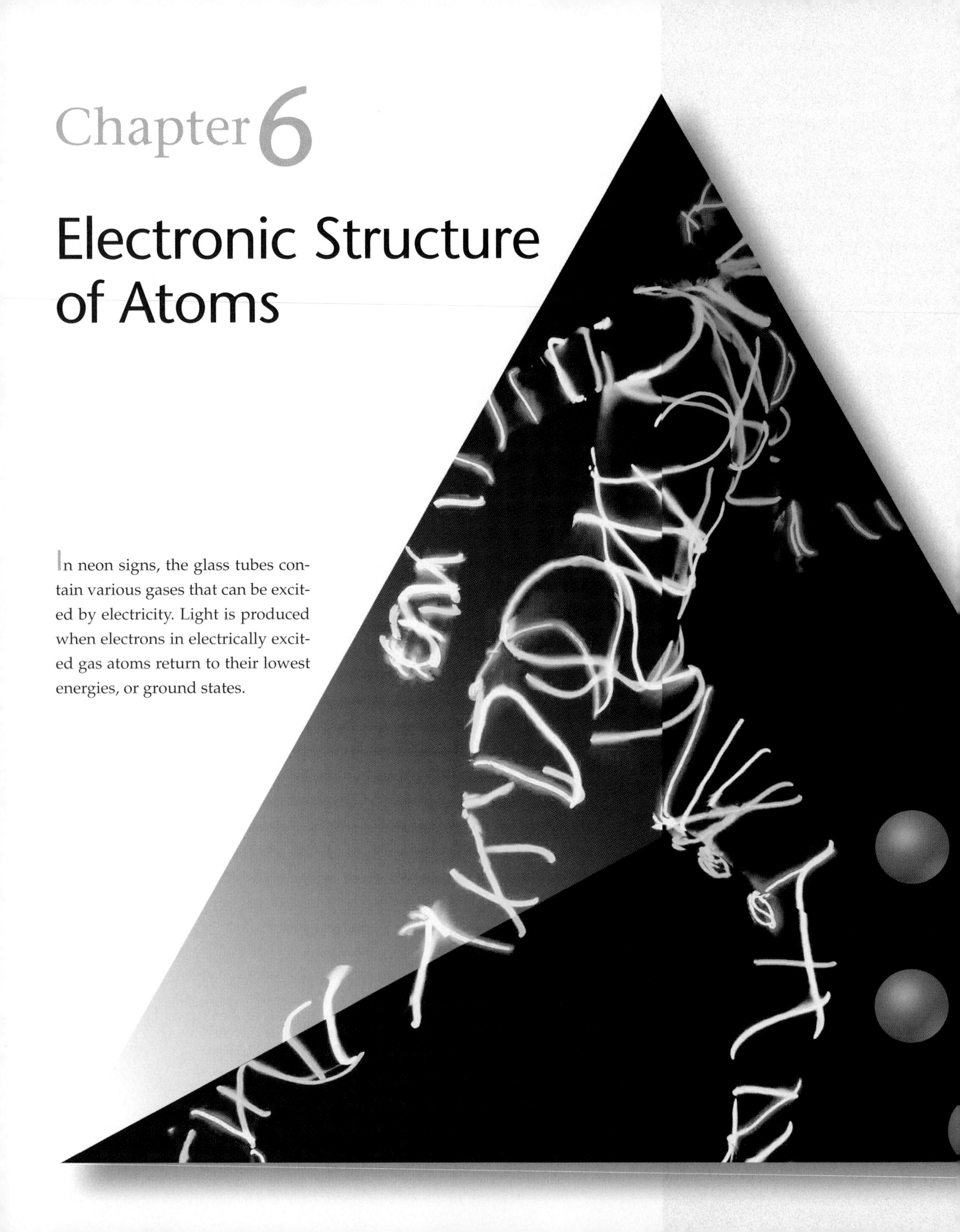

Chapter 6

Electronic Structure of Atoms

In neon signs, the glass tubes contain various gases that can be excited by electricity. Light is produced when electrons in electrically excited gas atoms return to their lowest energies, or ground states.

THE PERIODIC TABLE, discussed in Chapter 2, evolved largely from experimental observations. Elements that exhibit similar properties were placed together in the same column of the table. But what are the fundamental reasons for these similarities? Why, for example, are sodium and potassium both soft, reactive metals? Why are helium and neon both unreactive gases? Why do the halogens all react with hydrogen to form compounds that contain one hydrogen atom and one halogen atom?

When atoms react, it is the electrons that interact. Thus, the key to answering questions like those posed above lies in understanding the behavior of electrons in atoms. The arrangement of electrons in an atom is called its **electronic structure**. The electronic structure of an atom refers not only to the number of electrons that an atom possesses but also to their distribution around the nucleus and to their energies.

As we will see, electrons do not behave like anything we are familiar with in the macroscopic world. Our knowledge of electronic structure is the result of one of the major developments of twentieth-century science, the *quantum theory*. In this chapter we will describe the development of the quantum theory and how it led to a consistent description of the electronic structures of the elements. We will explore some of the tools used in *quantum mechanics*, the new physics that had to be developed to describe atoms correctly. In the chapters that follow, we will see how these concepts are used to explain trends in the periodic table and the formation of bonds between atoms.

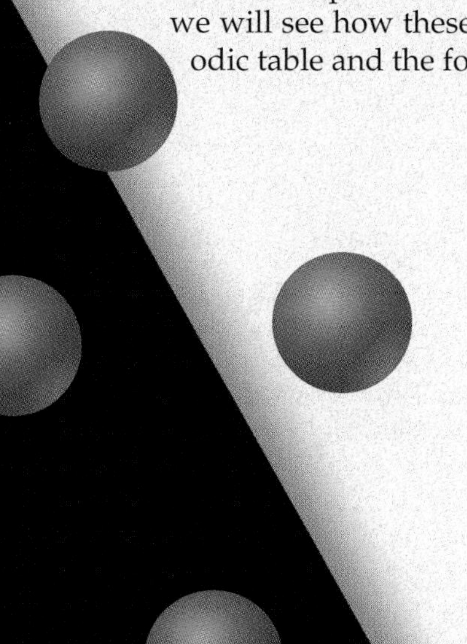

▶ What's Ahead ◀

- Understanding how light (radiant energy, or *electromagnetic radiation*) interacts with matter provides insights into the behavior of electrons in atoms.
- Electromagnetic radiation has wavelike properties characterized by its *wavelength*, *frequency*, and *speed*.
- Studies of the radiation given off by hot objects and of the way in which light striking a metal surface can free electrons indicate that electromagnetic radiation also has particle-like character and can be described in terms of *photons*.
- The fact that atoms give off characteristic colors of light (*line spectra*) provides clues about how electrons are arranged in atoms, leading to two important ideas: Electrons exist only in certain energy levels around nuclei, and energy is involved in moving an electron from one level to another.
- Matter also has wave properties, and it is impossible to determine simultaneously the exact position and exact motion of an electron in an atom (*Heisenberg's uncertainty principle*).
- How electrons are arranged in atoms is described by quantum mechanics in terms of *orbitals*.
- Knowing the energies of orbitals as well as some fundamental characteristics of electrons allows us to determine the ways in which electrons are distributed among various orbitals in an atom (*electron configurations*).
- The electron configuration of an atom is related to the location of the element in the periodic table.

▲ **Figure 6.1** Waves are formed from the boat's movement. The regular variation of the peaks and troughs enables us to sense the motion, or *propagation*, of the waves.

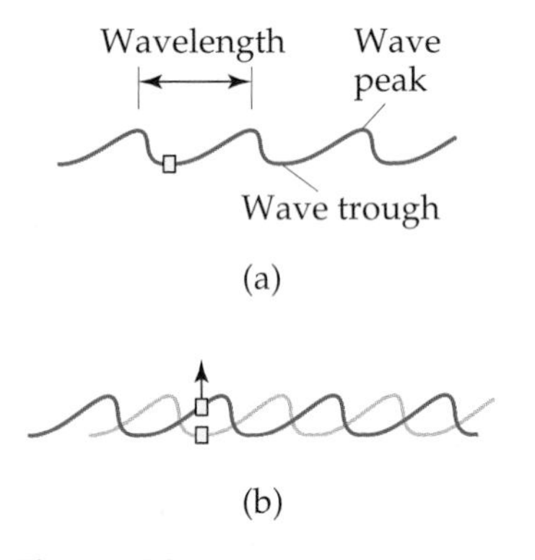

▲ **Figure 6.2** Characteristics of water waves. (a) The distance between corresponding points on each wave is called the *wavelength*. (b) The number of times per second that the cork bobs up and down is called the *frequency*.

6.1 The Wave Nature of Light

Much of our present understanding of the electronic structure of atoms has come from analysis of the light emitted or absorbed by substances. To understand the basis for our current model of electronic structure, therefore, we must first learn more about light. The light that we can see with our eyes, *visible light*, is a type of **electromagnetic radiation**. Because electromagnetic radiation carries energy through space, it is also known as *radiant energy*. There are many types of electromagnetic radiation in addition to visible light. These different forms—such as the radio waves that carry music to our radios, the infrared radiation (heat) from a glowing fireplace, and the X rays used by a dentist—may *seem* very different from one another, yet they share certain fundamental characteristics.

All types of electromagnetic radiation move through a vacuum at a speed of 3.00×10^8 m/s, the *speed of light*. Furthermore, all have wavelike characteristics similar to those of waves that move through water. Water waves are the result of energy imparted to the water, perhaps by the dropping of a stone or the movement of a boat on the water surface (Figure 6.1 ◀). This energy is expressed as the up-and-down movements of the water.

A cross section of a water wave (Figure 6.2 ◀) shows that it is periodic: The pattern of peaks and troughs repeats itself at regular intervals. The distance between successive peaks (or troughs) is called the **wavelength**. The number of complete wavelengths, or *cycles*, that pass a given point each second is the **frequency** of the wave. We can measure the frequency of a water wave by counting the number of times per second that a cork bobbing on the water moves through a complete cycle of upward and downward motion.

The wave characteristics of electromagnetic radiation are due to the periodic oscillations of the intensities of electronic and magnetic forces associated with the radiation. We can assign a frequency and wavelength to these electromagnetic waves, as illustrated in Figure 6.3 ▼. Because all electromagnetic radiation moves at the speed of light, wavelength and frequency are related. If the wavelength is long, there will be fewer cycles of the wave passing a point per second; thus, the frequency will be low. Conversely, for a wave to have a high frequency, the distance between the peaks of the wave must be small (short wavelength). This inverse relationship between the frequency and the wavelength of electromagnetic radiation can be expressed by the following equation:

$$\nu\lambda = c \qquad [6.1]$$

where ν (nu) is the frequency, λ (lambda) is the wavelength, and c is the speed of light.

▼ **Figure 6.3** Radiant energy has wave characteristics; it consists of electromagnetic waves. Notice that the shorter the wavelength, λ, the higher the frequency, ν. The wavelength in (b) is half as long as that in (a), and its frequency is therefore twice as great. The *amplitude* of the wave relates to the intensity of the radiation. It is the maximum extent of the oscillation of a wave. In these diagrams it is measured as the vertical distance from the midline of the wave to its peak. The waves in (a) and (b) have the same amplitude. The wave in (c) has the same frequency as that in (b), but its amplitude is lower.

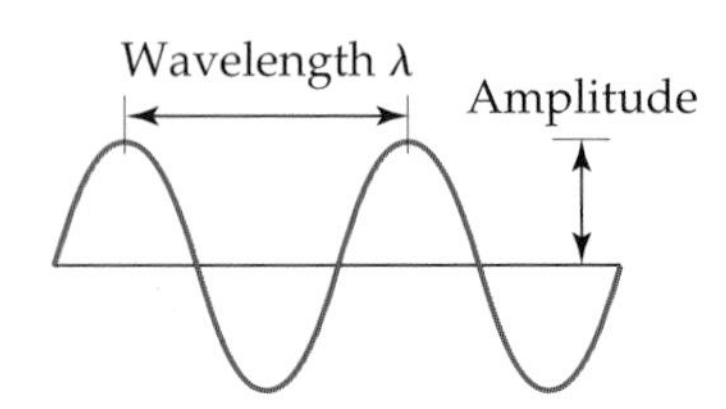

(a) Two complete cycles of wavelength λ

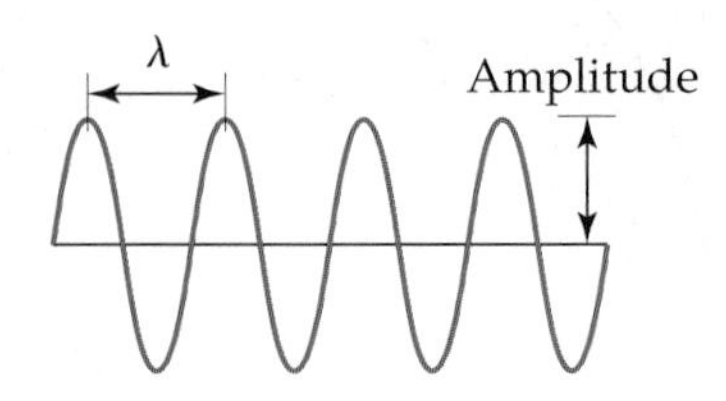

(b) Wavelength half of that in (a); frequency twice as great as in (a)

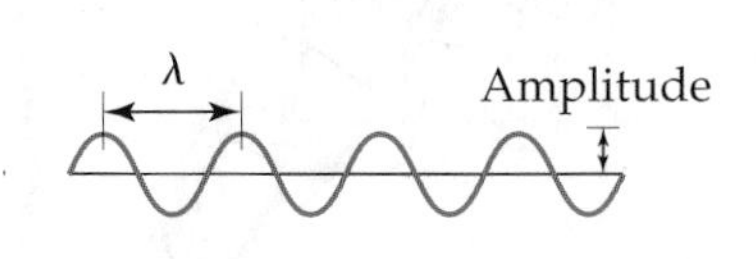

(c) Same frequency as (b), smaller amplitude

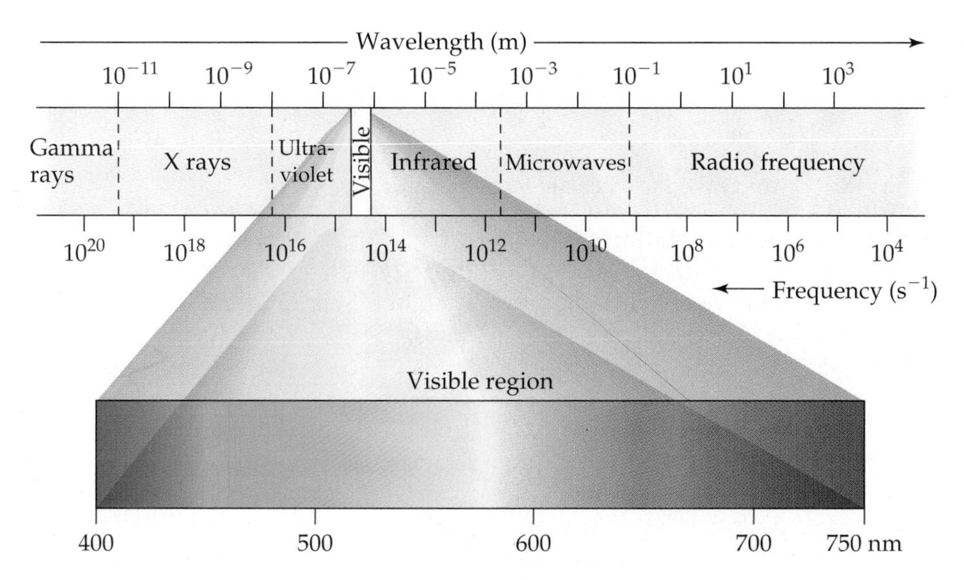

◀ **Figure 6.4** Wavelengths of electromagnetic radiation characteristic of various regions of the electromagnetic spectrum. Notice that color can be expressed quantitatively by wavelength.

Figure 6.4 ▲ shows the various types of electromagnetic radiation arranged in order of increasing wavelength, a display called the *electromagnetic spectrum*. Notice that the wavelengths span an enormous range. The wavelengths of gamma rays are similar to the diameters of atomic nuclei, whereas those of radio waves can be longer than a football field. Notice also that visible light, which corresponds to wavelengths of about 400 to 700 nm, is an extremely small portion of the electromagnetic spectrum. We can see visible light because of chemical reactions that it triggers in our eyes. The unit of length normally chosen to express wavelength depends on the type of radiation, as shown in Table 6.1 ▼.

Frequency is expressed in cycles per second, a unit also called a *hertz* (Hz). Because it is understood that cycles are involved, the units of frequency are normally given simply as "per second," which is denoted by s^{-1} or /s. For example, a frequency of 820 kilohertz (kHz), a typical frequency for an AM radio station, could be written as 820,000 s^{-1}.

SAMPLE EXERCISE 6.1

Two electromagnetic waves are represented in the margin. **(a)** Which wave has the higher frequency? **(b)** If one wave represents visible light and the other represents infrared radiation, which wave is which?

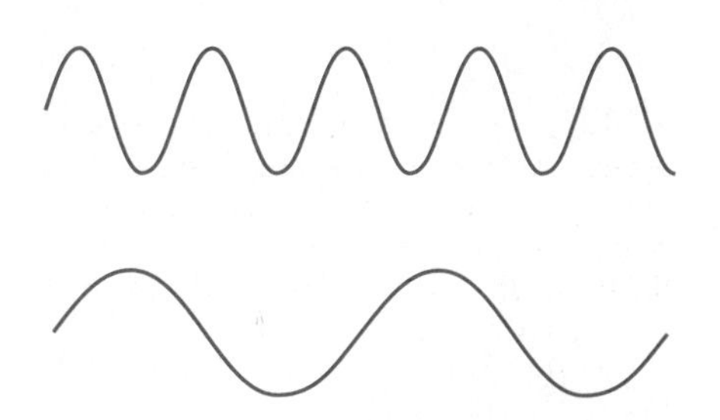

Solution **(a)** The lower wave has a longer wavelength (greater distance between peaks). The longer the wavelength the lower the frequency ($\nu = c/\lambda$). Thus, the lower wave has the lower frequency, and the upper one has the higher frequency.

(b) The electromagnetic spectrum (Figure 6.4) indicates that infrared radiation has a longer wavelength than visible light. Thus, the lower wave would be the infrared radiation.

PRACTICE EXERCISE

If one of the waves in the margin represents blue light and the other red light, which would be which?

Answer: The lower wave has the longer wavelength (smaller frequency) and would be the red light.

TABLE 6.1 Common Wavelength Units for Electromagnetic Radiation

Unit	Symbol	Length (m)	Type of Radiation
Angstrom	Å	10^{-10}	X ray
Nanometer	nm	10^{-9}	Ultraviolet, visible
Micrometer	μm	10^{-6}	Infrared
Millimeter	mm	10^{-3}	Infrared
Centimeter	cm	10^{-2}	Microwave
Meter	m	1	TV, radio

SAMPLE EXERCISE 6.2

The yellow light given off by a sodium vapor lamp used for public lighting has a wavelength of 589 nm. What is the frequency of this radiation?

Solution

Analyze: We are given the wavelength, λ, of the radiation and asked to calculate its frequency, ν.

Plan: The relationship between wavelength and frequency is given by Equation 6.1:

$$\nu\lambda = c$$

We can solve for frequency, ν, because we know both λ and c. (The speed of light, c, is a fundamental constant whose value is given in the text or in the table of fundamental constants on the back inside cover.)

$$c = 3.00 \times 10^8 \text{ m/s}$$

Solve: Solving Equation 6.1 for frequency gives:

$$\nu = c/\lambda$$

When we insert the values for c and λ, we note that the units of length in these two quantities are different. We can convert the wavelength from nanometers to meters, so the units cancel:

$$\nu = \frac{c}{\lambda} = \left(\frac{3.00 \times 10^8 \text{ m/s}}{589 \text{ nm}}\right)\left(\frac{1 \text{ nm}}{10^{-9} \text{ m}}\right) = 5.09 \times 10^{14} \text{ s}^{-1}$$

Check: The high frequency is reasonable because of the short wavelength. The units are proper because frequency has units of "per second" or s^{-1}.

PRACTICE EXERCISE

(a) A laser used in eye surgery to fuse detached retinas produces radiation with a wavelength of 640.0 nm. Calculate the frequency of this radiation. **(b)** An FM radio station broadcasts electromagnetic radiation at a frequency of 103.4 MHz (megahertz; $1 \text{ MHz} = 10^6 \text{ s}^{-1}$). Calculate the wavelength of this radiation.

Answers: **(a)** $4.688 \times 10^{14} \text{ s}^{-1}$; **(b)** 2.901 m

6.2 Quantized Energy and Photons

Although the wave model of light explains many aspects of its behavior, there are several phenomena that it can't explain. Three of these are particularly pertinent to our understanding of how electromagnetic radiation and atoms interact. These three phenomena are (1) the emission of light from hot objects (referred to as *black-body radiation* because the objects studied appear black before heating), (2) the emission of electrons from metal surfaces on which light shines (the *photoelectric effect*), and (3) the emission of light from electronically excited gas atoms (*emission spectra*). We examine the first two here and the third in Section 6.3.

▲ **Figure 6.5** The color and intensity of light emitted by a hot object depends on the temperature of the object. The temperature is highest at the center of this pour of molten steel. As a result, the light emitted from the center is most intense and of shortest wavelength.

Hot Objects and the Quantization of Energy

When solids are heated, they emit radiation, as seen in the red glow of an electric stove burner and the bright white light of a tungsten lightbulb. The wavelength distribution of the radiation depends on temperature, a "red-hot" object being cooler than a "white-hot" one (Figure 6.5 ◀). In the late 1800s a number of physicists were studying this phenomenon, trying to understand the relationship between the temperature and the intensity and wavelengths of the emitted radiation. The prevailing laws of physics could not account for the observations.

In 1900 a German physicist named Max Planck (1858–1947) solved the problem by making a daring assumption: He assumed that energy can be released (or absorbed) by atoms only in discrete "chunks" of some minimum size. Planck gave the name **quantum** (meaning "fixed amount") to the smallest quantity of energy that can be emitted or absorbed as electromagnetic radiation. He proposed that the energy, E, of a single quantum equals a constant times its frequency.

$$E = h\nu \qquad [6.2]$$

The constant h, known as **Planck's constant**, has a value of 6.63×10^{-34} joule-seconds (J-s). According to Planck's theory, energy is always emitted or absorbed by matter in whole-number multiples of $h\nu$ such as $h\nu$, $2h\nu$, $3h\nu$, and so forth. If the quantity of energy emitted by an atom is $3h\nu$, for example, we say that three quanta of energy have been emitted (quanta being the plural of quantum). Furthermore, we say that the allowed energies are quantized; that is, their values are restricted to certain quantities. Planck's revolutionary proposal that energy is quantized was proved correct, and he was awarded the 1918 Nobel Prize in physics for his work on the quantum theory.

If the notion of quantized energies seems strange, it might be helpful to draw an analogy by comparing a ramp and a staircase (Figure 6.6 ▼). As you walk up a ramp, your potential energy increases in a uniform, continuous manner. When you climb a staircase, you can step only *on* individual stairs, not *between* them, so that your potential energy is restricted to certain values and is therefore quantized.

If Planck's quantum theory is correct, why aren't its effects more obvious in our daily lives? Why do energy changes seem continuous rather than quantized or "grainy?" Notice that Planck's constant is an extremely small number. Thus, a quantum of energy, $h\nu$, will be an extremely small amount. Planck's rules regarding the gain or loss of energy are always the same, whether we are concerned with objects on the size scale of our ordinary experience or with microscopic objects. For macroscopic objects, such as humans, the gain or loss of a single quantum of energy goes completely unnoticed. When dealing with matter at the atomic level, however, the impact of quantized energies is far more significant.

The Photoelectric Effect and Photons

A few years after Planck presented his theory, scientists began to see its applicability to a great many experimental observations. It soon became apparent that Planck's theory had within it the seeds of a revolution in the way the physical

(a)

(b)

◀ **Figure 6.6** The potential energy of a person walking up a ramp (a) increases in a uniform, continuous manner, whereas that of a person walking up steps (b) increases in a step-wise, quantized manner.

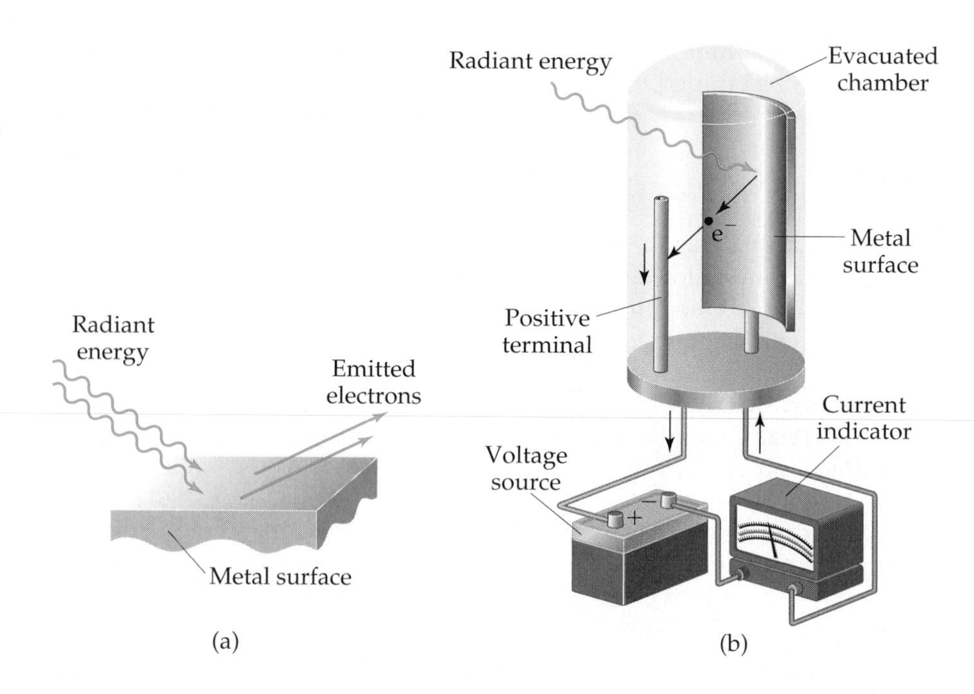

▶ **Figure 6.7** The photoelectric effect. When photons of sufficiently high energy strike a metal surface, electrons are emitted from the metal, as in (a). The photoelectric effect is the basis of the photocell shown in (b). The emitted electrons are drawn toward the positive terminal. As a result, current flows in the circuit. Photocells are used in photographic light meters as well as in numerous other electronic devices.

world is viewed. In 1905 Albert Einstein (1879–1955) used Planck's quantum theory to explain the **photoelectric effect**, which is illustrated in Figure 6.7 ▲. Experiments had shown that light shining on a clean metal surface causes the surface to emit electrons. For each metal, there is a minimum frequency of light below which no electrons are emitted. For example, light with a frequency of $4.60 \times 10^{14}\ s^{-1}$ or greater will cause cesium metal to emit electrons, but light of lower frequency has no effect.

To explain the photoelectric effect, Einstein assumed that the radiant energy striking the metal surface is a stream of tiny energy packets. Each energy packet, called a **photon**, behaves like a tiny particle. Extending Planck's quantum theory, Einstein deduced that each photon must have an energy proportional to the frequency of the light: $E = h\nu$. Thus, radiant energy itself is quantized.

$$\text{Energy of photon} = E = h\nu \qquad [6.3]$$

When a photon strikes the metal, it may literally disappear. When this happens, it may transfer its energy to an electron in the metal. A certain amount of energy is required for an electron to overcome the attractive forces that hold it within the metal. If the photons of the radiation have less energy than this energy threshold, electrons do not acquire sufficient energy to escape from the metal surface, even if the light beam is intense. If the photons have sufficient energy, electrons are emitted. If the photons have more than the minimum energy required to free electrons, the excess appears as the kinetic energy of the emitted electrons.

To better understand what a photon is, imagine that you have a light source that produces radiation with a single wavelength. Further suppose that you could switch the light on and off faster and faster to provide ever-smaller bursts of energy. Einstein's photon theory tells us that you would eventually come to a smallest energy burst, given by $E = h\nu$. This smallest burst of energy consists of a single photon of light.

SAMPLE EXERCISE 6.3

Calculate the energy of one photon of yellow light whose wavelength is 589 nm.

Solution

Analyze: Our task is to calculate the energy, E, of a photon given $\lambda = 589$ nm.

Plan: We can use Equation 6.1 to convert the wavelength to frequency:

$$\nu = c/\lambda$$

We can then use Equation 6.3 to calculate energy:

$$E = h\nu$$

The value of Planck's constant is given both in the text and in the table of physical constants on the back inside cover of the text:

$$h = 6.63 \times 10^{-34}\ \text{J-s}$$

Solve: The frequency, ν, is calculated from the given wavelength, as shown in Sample Exercise 6.2:

$$\nu = c/\lambda = 5.09 \times 10^{14}\ \text{s}^{-1}$$

Thus, we have

$$E = (6.63 \times 10^{-34}\ \text{J-s})(5.09 \times 10^{14}\ \text{s}^{-1}) = 3.37 \times 10^{-19}\ \text{J}$$

Comment: If one photon of radiant energy supplies 3.37×10^{-19} J, then one mole of these photons will supply

$$(6.02 \times 10^{23}\ \text{photons/mol})(3.37 \times 10^{-19}\ \text{J/photon}) = 2.03 \times 10^{5}\ \text{J/mol}$$

This is the magnitude of enthalpies of reactions (Section 5.4), so radiation can break chemical bonds, producing what are called *photochemical reactions*.

PRACTICE EXERCISE

(a) A laser emits light with a frequency of $4.69 \times 10^{14}\ \text{s}^{-1}$. What is the energy of one photon of the radiation from this laser? **(b)** If the laser emits a burst or pulse of energy containing 5.0×10^{17} photons of this radiation, what is the total energy of that pulse? **(c)** If the laser emits 1.3×10^{-2} J of energy during a pulse, how many photons are emitted during the pulse?

Answers: **(a)** 3.11×10^{-19} J; **(b)** 0.16 J; **(c)** 4.2×10^{16} photons

The idea that the energy of light depends on its frequency helps us understand the diverse effects that different kinds of electromagnetic radiation have on matter. For example, the high frequency (short wavelength) of X rays (Figure 6.4) causes photons of this kind to have high energy, sufficient to cause tissue damage and even cancer. Thus, signs are normally posted around X-ray equipment warning us of high-energy radiation.

Although Einstein's theory of light explains the photoelectric effect and a great many other observations, it does pose a dilemma. Is light a wave, or does it consist of particles? The fact is that it possesses properties of both. It behaves macroscopically like a wave, but it consists of a collection of photons. It is when we examine phenomena at the atomic level that we see its particle-like properties. It's as if we move from describing an entire beach and begin to examine the grains of sand from which it is made.

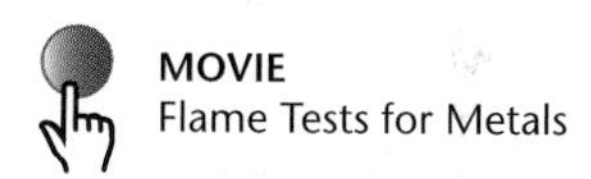

▲ **Figure 6.8** Niels Bohr (right) with Albert Einstein. Bohr (1885–1962) made major contributions to the quantum theory. From 1911 to 1913 he studied in England, working first with J. J. Thomson at Cambridge University and then with Ernest Rutherford at the University of Manchester. He published his quantum theory of the atom in 1914 and was awarded the Nobel Prize in physics in 1922.

6.3 Line Spectra and the Bohr Model

The work of Planck and Einstein paved the way for understanding how electrons are arranged in atoms. In 1913 the Danish physicist Niels Bohr (Figure 6.8 ▶) offered a theoretical explanation of line spectra, another phenomenon that had puzzled scientists in the nineteenth century. Let's first examine this phenomenon and then consider how Bohr used the ideas of Planck and Einstein.

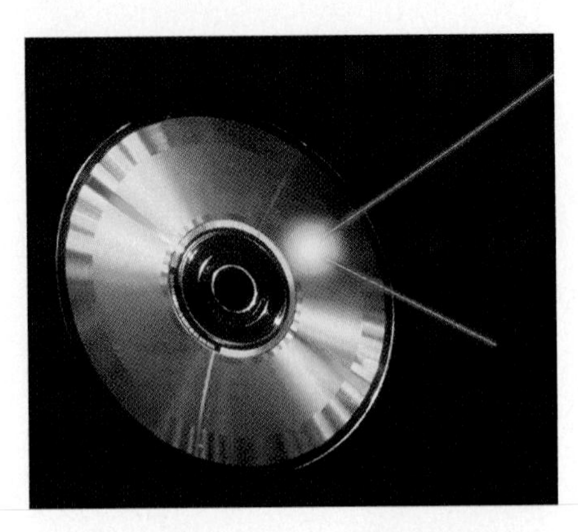

▲ **Figure 6.9** A laser beam reflected from surface of a CD disk. CD players and similar devices use a small laser beam to read the information on the disk.

Line Spectra

A particular source of radiant energy may emit a single wavelength, as in the light from a laser (Figure 6.9 ◀). Radiation composed of a single wavelength is said to be *monochromatic*. However, most common radiation sources, including lightbulbs and stars, produce radiation containing many different wavelengths. When radiation from such sources is separated into its different wavelength components, a **spectrum** is produced. Figure 6.10 ▼ shows how a prism disperses light from a lightbulb. The spectrum so produced consists of a continuous range of colors: Violet merges into blue, blue into green, and so forth, with no blank spots. This rainbow of colors, containing light of all wavelengths, is called a **continuous spectrum**. The most familiar example of a continuous spectrum is the rainbow, produced by the dispersal of sunlight by raindrops or mist.

Not all radiation sources produce a continuous spectrum. When different gases are placed under reduced pressure in a tube and a high voltage is applied, the gases emit different colors of light (Figure 6.11 ▶). The light emitted by neon gas is the familiar red-orange glow of many "neon" lights, whereas sodium vapor emits the yellow light characteristic of some modern streetlights. When light coming from such tubes is passed through a prism, only lines of a few wavelengths are present in the resultant spectra, as shown in Figure 6.12 ▶. The colored lines are separated by black regions, which correspond to wavelengths that are absent in the light. A spectrum containing radiation of only specific wavelengths is called a **line spectrum**.

When scientists first detected the line spectrum of hydrogen in the mid-1800s, they were fascinated by its simplicity. In 1885 a Swiss schoolteacher named Johann Balmer observed that the wavelengths of the four lines of hydrogen shown in Figure 6.12 fit an intriguingly simple formula. Additional lines were found to occur in the ultraviolet and infrared regions. Soon Balmer's equation was extended to a more general one, called the *Rydberg equation*, which allowed the calculation of the wavelengths of all the spectral lines of hydrogen:

$$\frac{1}{\lambda} = (R_H)\left(\frac{1}{n_1^2} - \frac{1}{n_2^2}\right) \qquad [6.4]$$

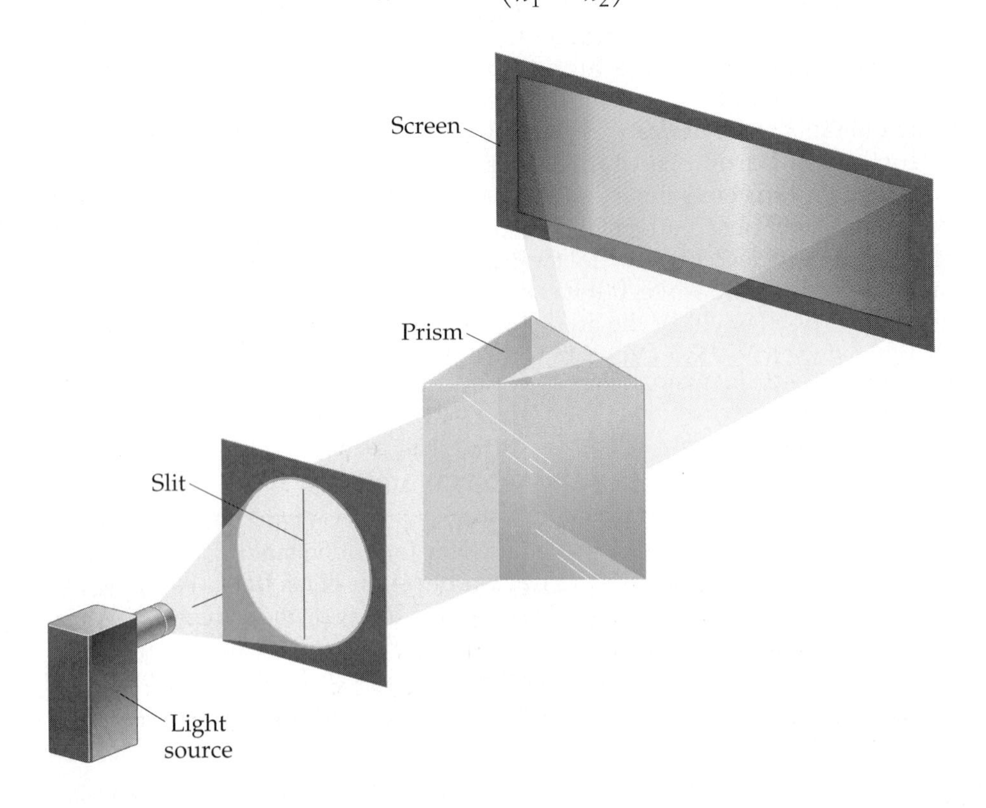

▶ **Figure 6.10** A continuous visible spectrum is produced when a narrow beam of white light is passed through a prism. The white light could be sunlight or light from an incandescent lamp.

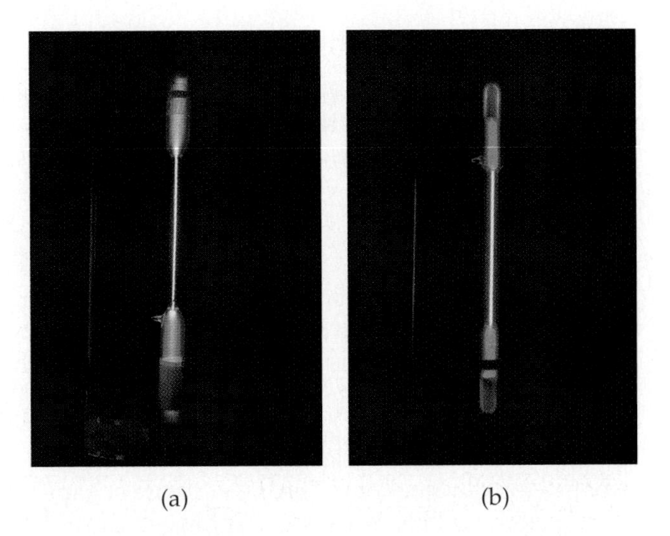

◀ **Figure 6.11** Different gases emit light of different characteristic colors upon excitation in an electrical discharge: (a) hydrogen; (b) neon.

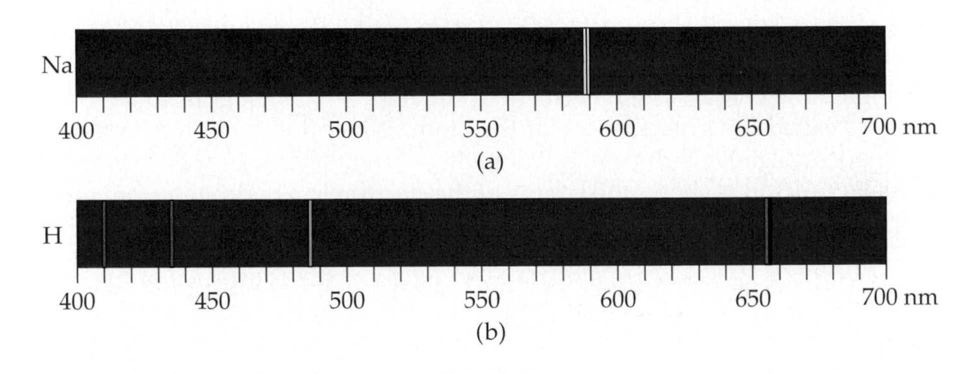

◀ **Figure 6.12** The line spectrum of (a) Na; (b) H.

In this formula λ is the wavelength of a spectral line, R_H is the *Rydberg constant* ($1.096776 \times 10^7\ m^{-1}$), and n_1 and n_2 are positive integers with n_2 being larger than n_1. How could the remarkable simplicity of this equation be explained? It took nearly 30 more years to answer this question, as we will see in the next section.

Bohr's Model

After Rutherford discovered the nuclear nature of the atom (Section 2.2), scientists thought of the atom as a "microscopic solar system" in which electrons orbited the nucleus. To explain the line spectrum of hydrogen, Bohr started by assuming that electrons move in circular orbits around the nucleus. According to classical physics, however, an electrically charged particle (such as an electron) that moves in a circular path should continuously lose energy by the emission of electromagnetic radiation. As the electron loses energy, it should spiral into the nucleus. Bohr approached this problem in much the same way that Planck had approached the problem of the nature of the radiation emitted by hot objects: He assumed that the prevailing laws of physics were inadequate to describe all aspects of atoms. Furthermore, he adopted Planck's idea that energies are quantized.

Bohr based his model on three postulates:

1. Only orbits of certain radii, corresponding to certain definite energies, are permitted for electrons in an atom.
2. An electron in a permitted orbit has a specific energy and is in an "allowed" energy state. An electron in an allowed energy state will not radiate energy and therefore will not spiral into the nucleus.
3. Energy is only emitted or absorbed by an electron as it changes from one allowed energy state to another. This energy is emitted or absorbed as a photon, $E = h\nu$.

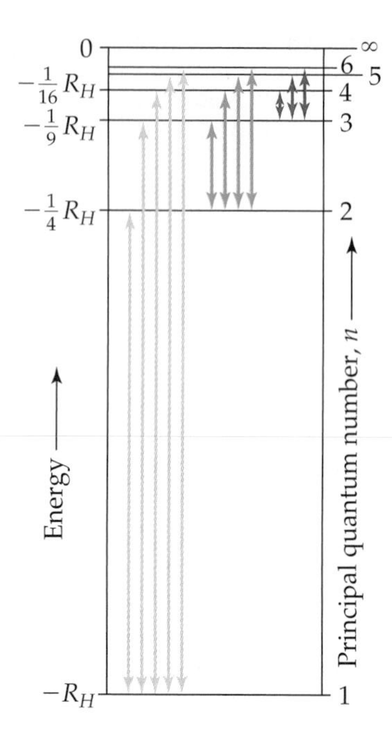

▲ **Figure 6.13** Energy levels in the hydrogen atom from the Bohr model. The arrows refer to the transitions of the electron from one allowed energy state to another. The states shown are those for which $n = 1$ through $n = 6$, and the state for $n = \infty$, for which the energy, E, equals zero.

The Energy States of the Hydrogen Atom

Starting with his three postulates and using classical equations for motion and for interacting electrical charges, Bohr calculated the energies corresponding to each allowed orbit. These energies fit the following formula:

$$E = (-2.18 \times 10^{-18}\,\text{J})\left(\frac{1}{n^2}\right) \qquad [6.5]$$

The integer n, which can have values from 1 to infinity, is called a *quantum number*. Each orbit corresponds to a different value of n, and the radius of the orbit gets larger as n increases. Thus, the first allowed orbit (the one closest to the nucleus) has $n = 1$, the next allowed orbit (the one second closest to the nucleus) has $n = 2$, and so forth.

The energies of the electron of a hydrogen atom given by Equation 6.5 are negative for all values of n. The lower (more negative) the energy is, the more stable the atom will be. The energy is lowest (most negative) for $n = 1$. As n gets larger, the energy becomes successively less negative and therefore increases. We can liken the situation to a ladder in which the rungs are numbered from the bottom rung on up. The higher one climbs the ladder (the greater the value of n), the higher the energy. The lowest energy state ($n = 1$, analogous to the bottom rung) is called the **ground state** of the atom. When the electron is in a higher energy (less negative) orbit—$n = 2$ or higher—the atom is said to be in an **excited state**. Figure 6.13 ◀ shows the energy of the electron in a hydrogen atom for several values of n.

What happens to the orbit radius and the energy as n becomes infinitely large? The radius increases as n^2, so we reach a point at which the electron is completely separated from the nucleus. When $n = \infty$, the energy is zero.

$$E = (-2.18 \times 10^{-18}\,\text{J})\left(\frac{1}{\infty^2}\right) = 0$$

Thus, the state in which the electron is removed from the nucleus is the reference, or zero-energy, state of the hydrogen atom. This zero-energy state is *higher* in energy than the states with negative energies.

In his third postulate, Bohr assumed that the electron could "jump" from one allowed energy state to another by absorbing or emitting photons whose radiant energy corresponds exactly to the energy difference between the two states. Energy must be absorbed for an electron to move to a higher energy state (one with a higher value of n). Conversely, radiant energy is emitted when the electron jumps to a lower energy state (one with a lower value of n). Thus, if the electron jumps from an initial state with energy E_i to a final state with energy E_f, the change in energy is given by the following relationships:

$$\Delta E = E_f - E_i = E_{\text{photon}} = h\nu \qquad [6.6]$$

Bohr's model of the hydrogen atom states, therefore, that only the specific frequencies of light that satisfy Equation 6.6 can be absorbed or emitted by the atom.

Substituting the energy expression in Equation 6.5 into Equation 6.6 and recalling that $\nu = c/\lambda$, we have

$$\Delta E = h\nu = \frac{hc}{\lambda} = (-2.18 \times 10^{-18}\,\text{J})\left(\frac{1}{n_f^2} - \frac{1}{n_i^2}\right) \qquad [6.7]$$

In this equation n_i and n_f are the principal quantum numbers of the initial and final states of the atom, respectively. If n_f is smaller than n_i, the electron moves closer to the nucleus, and ΔE is a negative number, indicating that the atom releases energy. For example, if the electron moves from $n_i = 3$ to $n_f = 1$, we have

$$\Delta E = (-2.18 \times 10^{-18}\,\text{J})\left(\frac{1}{1^2} - \frac{1}{3^2}\right) = (-2.18 \times 10^{-18}\,\text{J})\left(\frac{8}{9}\right) = -1.94 \times 10^{-18}\,\text{J}$$

Knowing the energy for the emitted photon, we can calculate either its frequency or wavelength. For the wavelength, we have

$$\lambda = \frac{c}{\nu} = \frac{hc}{\Delta E} = \frac{(6.63 \times 10^{-34}\,\text{J-s})(3.00 \times 10^{8}\,\text{m/s})}{1.94 \times 10^{-18}\,\text{J}} = 1.03 \times 10^{-7}\,\text{m}$$

We have not included the negative sign of the energy in this calculation because wavelength and frequency are always reported as positive quantities. The direction of energy flow is indicated by saying that a photon of wavelength 1.03×10^{-7} has been *emitted*.

If we solve Equation 6.7 for $1/\lambda$ and exclude the negative sign, we find that this equation derived from Bohr's theory corresponds to the Rydberg equation, Equation 6.4, which was obtained using experimental data:

$$\frac{1}{\lambda} = \frac{2.18 \times 10^{-18}\,\text{J}}{hc}\left(\frac{1}{n_f^2} - \frac{1}{n_i^2}\right)$$

Indeed, the combination of constants, $(2.18 \times 10^{-18}\,\text{J})/hc$, equals the Rydberg constant, R_H, to three significant figures, $1.10 \times 10^{7}\,\text{m}^{-1}$. Thus, the existence of spectral lines can be attributed to the quantized jumps of electrons between energy levels.

SAMPLE EXERCISE 6.4

Using Figure 6.13, predict which of the following electronic transitions produces the longest wavelength spectral line: $n = 2$ to $n = 1$, $n = 3$ to $n = 2$, or $n = 4$ to $n = 3$.

Solution The wavelength increases as frequency decreases ($\lambda = c/v$). Hence the longest wavelength will be associated with the smallest frequency. According to Planck's equation, $E = h\nu$, the lowest frequency is associated with the lowest energy. In Figure 6.13 the shortest line represents the smallest energy change. Thus, the $n = 4$ to $n = 3$ transition produces the longest wavelength (lowest frequency) line.

PRACTICE EXERCISE

Indicate whether each of the following electronic transitions emits energy or requires the absorption of energy: **(a)** $n = 3$ to $n = 1$; **(b)** $n = 2$ to $n = 4$.
Answers: **(a)** emits energy; **(b)** requires absorption of energy

Limitations of the Bohr Model

While the Bohr model offers an explanation for the line spectrum of the hydrogen atom, it cannot explain the spectra of other atoms, except in a rather crude way. Furthermore, there is a problem with describing an electron merely as a small particle circling about the nucleus. As we will see in Section 6.4, the electron exhibits properties of waves, a fact that our model of electronic structure must accommodate. The Bohr model is only an important step along the way toward the development of a more comprehensive model. What is most significant about Bohr's model is that it introduces two important ideas that are also incorporated into our current model: (1) Electrons exist only in certain discrete energy levels, which are described by quantum numbers. (2) Energy is involved in moving an electron from one level to another. In addition, some of the vocabulary associated with our new model goes back to the Bohr model. For example, we still use the idea of ground states and excited states to describe the electronic structures of atoms.

6.4 The Wave Behavior of Matter

In the years following the development of Bohr's model for the hydrogen atom, the dual nature of radiant energy became a familiar concept. Depending on the experimental circumstances, radiation appears to have either a wavelike or a particle-like (photon) character. Louis de Broglie (1892–1987), who was working on his Ph.D. thesis in physics at the Sorbonne in Paris, boldly extended this idea. If radiant energy could, under appropriate conditions, behave as though it were a stream of particles, could matter, under appropriate conditions, possibly show the properties of a wave? Suppose that the electron orbiting the nucleus of a hydrogen atom could be thought of as a wave, with a characteristic wavelength. De Broglie suggested that the electron in its movement about the nucleus has associated with it a particular wavelength. He went on to propose that the characteristic wavelength of the electron or of any other particle depends on its mass, m, and velocity, v.

$$\lambda = \frac{h}{mv} \qquad [6.8]$$

(h is Planck's constant.) The quantity mv for any object is called its **momentum**. De Broglie used the term **matter waves** to describe the wave characteristics of material particles.

Because de Broglie's hypothesis is applicable to all matter, any object of mass m and velocity v would give rise to a characteristic matter wave. However, Equation 6.8 indicates that the wavelength associated with an object of ordinary size, such as a golf ball, is so tiny as to be completely out of the range of any possible observation. This is not so for an electron because its mass is so small, as we see in Sample Exercise 6.5.

SAMPLE EXERCISE 6.5

What is the wavelength of an electron with a velocity of 5.97×10^6 m/s? (The mass of the electron is 9.11×10^{-28} g.)

Solution

Analyze: We are given the mass, m, and velocity, v, of the electron, and we must calculate its de Broglie wavelength, λ.

Plan: The wavelength of a moving particle is given by Equation 6.8, so λ is calculated by merely plugging in the known quantities, h, m, and v. In doing so, however, we must pay attention to units.

Solve: Using the value of Planck's constant, $h = 6.63 \times 10^{-34}$ J-s, and recalling that $1\ \text{J} = 1\ \text{kg-m}^2/\text{s}^2$, we have

$$\lambda = \frac{h}{mv}$$

$$= \frac{(6.63 \times 10^{-34}\ \text{J-s})}{(9.11 \times 10^{-28}\ \text{g})(5.97 \times 10^6\ \text{m/s})}\left(\frac{1\ \text{kg-m}^2/\text{s}^2}{1\ \text{J}}\right)\left(\frac{10^3\ \text{g}}{1\ \text{kg}}\right)$$

$$= 1.22 \times 10^{-10}\ \text{m} = 0.122\ \text{nm}$$

Comment: By comparing this value with the wavelengths of electromagnetic radiation shown in Figure 6.4, we see that the wavelength of this electron is about the same as that of X rays.

PRACTICE EXERCISE

Calculate the velocity of a neutron whose de Broglie wavelength is 500 pm. The mass of a neutron is given in the table on the front inside cover of the text.
Answer: 7.92×10^2 m/s

Within a few years after de Broglie published his theory, the wave properties of the electron were demonstrated experimentally. Electrons were diffracted by crystals, just as X rays are diffracted. Thus, a stream of moving electrons exhibits the same kinds of wave behavior as electromagnetic radiation.

The technique of electron diffraction has been highly developed. In the electron microscope the wave characteristics of electrons are used to obtain pictures of tiny objects. This microscope is an important tool for studying surface phenomena at very high magnifications. Figure 6.14 ▶ is a photograph of an electron microscope image, which demonstrates that tiny particles of matter can behave as waves.

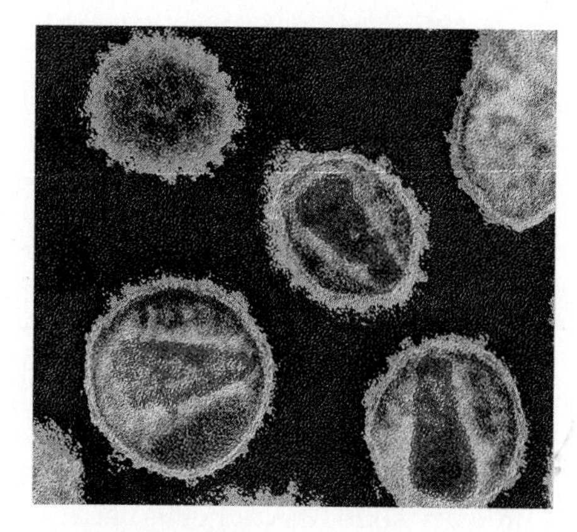

▲ **Figure 6.14** Color enhanced electron microscope image of human immunodeficiency virus (HIV) at a magnification of 240,000. In an electron microscope, the wave behavior of a stream of electrons is utilized in the same way that a conventional microscope uses the wave behavior of a beam of light.

The Uncertainty Principle

The discovery of the wave properties of matter raised some new and interesting questions about classical physics. Consider, for example, a ball rolling down a ramp. By using classical physics, we can calculate its position, direction of motion, and speed at any time, with great accuracy. Can we do the same for an electron that exhibits wave properties? A wave extends in space, and its location is not precisely defined. We might therefore anticipate that it is impossible to determine exactly where an electron is located at a specific time.

The German physicist Werner Heisenberg (Figure 6.15 ▶) concluded that the dual nature of matter places a fundamental limitation on how precisely we can know both the location and the momentum of any object. The limitation becomes important only when we deal with matter at the subatomic level (that is, with masses as small as that of an electron). Heisenberg's principle is called the **uncertainty principle**. When applied to the electrons in an atom, this principle states that it is inherently impossible for us to know simultaneously both the exact momentum of the electron and its exact location in space.

▲ **Figure 6.15** Werner Heisenberg (1901–1976). During his postdoctoral assistantship with Niels Bohr, Heisenberg formulated his famous uncertainty principle. At the age of 25, he became the chair in theoretical physics at the University of Leipzig. At 32 he was one of the youngest scientists to receive the Nobel Prize.

Heisenberg mathematically related the uncertainty of the position (Δx) and the uncertainty in momentum (Δmv) to a quantity involving Planck's constant:

$$\Delta x \cdot \Delta mv \geq \frac{h}{4\pi} \qquad [6.9]$$

A brief calculation illustrates the dramatic implications of the uncertainty principle. The electron has a mass of 9.11×10^{-31} kg and moves at an average speed of about 5×10^{6} m/s in a hydrogen atom. Let's assume that we know the speed to an uncertainty of 1% [that is, an uncertainty of $(0.01)(5 \times 10^{6}\ \text{m/s}) = 5 \times 10^{4}\ \text{m/s}$] and that this is the only important source of uncertainty in the momentum so that $\Delta mv = m\Delta v$. We can then use Equation 6.9 to calculate the uncertainty in the position of the electron:

$$\Delta x \geq \frac{h}{4\pi m \Delta v} = \frac{(6.63 \times 10^{-34}\ \text{J-s})}{4\pi(9.11 \times 10^{-31}\ \text{kg})(5 \times 10^{4}\ \text{m/s})} = 1 \times 10^{-9}\ \text{m}$$

Because the diameter of a hydrogen atom is only about 2×10^{-10} m, the uncertainty is many times greater than the size of the atom. Thus, we have essentially no idea of where the electron is located within the atom. On the other hand, if we were to repeat the calculation with an object of ordinary mass such as a tennis ball, the uncertainty would be so small that it would be inconsequential. In that case, m is large, and Δx is out of the realm of measurement and therefore of no practical consequence.

De Broglie's hypothesis and Heisenberg's uncertainty principle set the stage for a new and more broadly applicable theory of atomic structure. In this new approach, any attempt to define precisely the instantaneous location and momentum of the electron is abandoned. The wave nature of the electron is recognized, and its behavior is described in terms appropriate to waves. The result is a model that precisely describes the energy of the electron while describing its location in terms of probabilities.

A Closer Look Measurement and the Uncertainty Principle

Whenever any measurement is made, some uncertainty exists. Our experience with objects of ordinary dimensions, like balls or trains or laboratory equipment, indicates that the uncertainty of a measurement can be decreased by using more precise instruments. In fact, we might expect that the uncertainty in a measurement can be made indefinitely small. However, the uncertainty principle states that there is an actual limit to the accuracy of measurements. This limit is not a restriction on how well instruments can be made; rather, it is inherent in nature. This limit has no practical consequences when dealing with ordinary-sized objects, but its implications are enormous when dealing with subatomic particles, such as electrons.

To measure an object, we must disturb it, at least a little, with our measuring device. Imagine using a flashlight to locate a large rubber ball in a dark room. You see the ball when the light from the flashlight bounces off the ball and strikes your eyes. When a beam of photons strikes an object of this size, it does not alter its position or momentum to any practical extent. Imagine, however, that you wish to locate an electron by similarly bouncing light off it into some detector. Objects can be located to an accuracy no greater than the wavelength of the radiation used. Thus, if we want an accurate position measurement for an electron, we must use a short wavelength. This means that photons of high energy must be employed. The more energy the photons have, the more momentum they impart to the electron when they strike it, which changes the electron's motion in an unpredictable way. The attempt to measure accurately the electron's position introduces considerable uncertainty in its momentum; the act of measuring the electron's position at one moment makes our knowledge of its future position inaccurate.

Suppose, then, that we use photons of longer wavelength. Because these photons have lower energy, the momentum of the electron is not so appreciably changed during measurement, but its position will be correspondingly less accurately known. This is the essence of the uncertainty principle: *There is an uncertainty in knowing either the position or the momentum of the electron that cannot be reduced beyond a certain minimum level.* The more accurately one is known, the less accurately the other is known. Although we can never know the exact position and momentum of the electron, we can talk about the probability of its being at certain locations in space. In Section 6.5 we introduce a model of the atom that provides the probability of finding electrons of specific energies at certain positions in atoms.

6.5 Quantum Mechanics and Atomic Orbitals

In 1926 the Austrian physicist Erwin Schrödinger (1887–1961) proposed an equation, now known as Schrödinger's wave equation, that incorporates both the wavelike and particle-like behavior of the electron. His work opened a new way of dealing with subatomic particles known as *quantum mechanics* or *wave mechanics*. The application of Schrödinger's equation requires advanced calculus, and we will not be concerned with the details of his approach. We will, however, qualitatively consider the results he obtained, because they give us a powerful new way to view electronic structure. Let's begin by examining the electronic structure of the simplest atom, hydrogen.

Solving Schrödinger's equation leads to a series of mathematical functions called **wave functions** that describe the electron's matter wave. These wave functions are usually represented by the symbol ψ (the Greek lowercase letter *psi*). Although the wave function itself has no direct physical meaning, the square of the wave function, ψ^2, provides information about an electron's location when it is in an allowed energy state.

For the hydrogen atom, the allowed energies are the same as those predicted by the Bohr model. However, the Bohr model assumes that the electron is in a circular orbit of some particular radius about the nucleus. In the quantum mechanical model, the electron's location cannot be described so simply. According to the uncertainty principle, if we know the momentum of the electron with high accuracy, our simultaneous knowledge of its location is very uncertain. Thus, we cannot hope to specify the exact location of an individual electron around the nucleus. Rather, we must be content with a kind of statistical knowledge. In the quantum mechanical model, we therefore speak of the *probability* that the electron will be in a certain region of space at a given instant. As it turns out, the square of the wave function, ψ^2, at a given point in space represents the probability that the electron will be found at that location. For this reason, ψ^2 is called the **probability density**.

One way of representing the probability of finding the electron in various regions of an atom is shown in Figure 6.16 ▶. In this figure the density of the dots represents the probability of finding the electron. The regions with a high density of dots correspond to relatively large values for ψ^2. **Electron density** is another way of expressing probability: Regions where there is a high probability of finding the electron are said to be regions of high electron density. In Section 6.6 we will say more about the ways in which we can represent electron density.

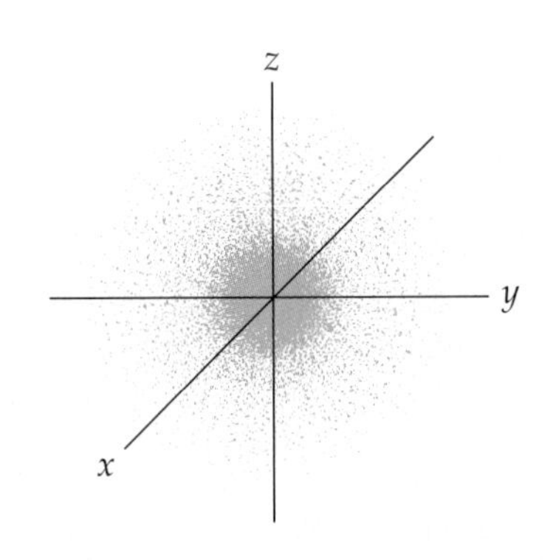

▲ **Figure 6.16** Electron-density distribution in the ground state of the hydrogen atom.

Orbitals and Quantum Numbers

The solution to Schrödinger's equation for the hydrogen atom yields a set of wave functions and corresponding energies. These wave functions are called **orbitals**. Each orbital describes a specific distribution of electron density in space, as given by its probability density. Each orbital therefore has a characteristic energy and shape. For example, the lowest energy orbital in the hydrogen atom has an energy of -2.18×10^{-18} J and the shape illustrated in Figure 6.16. Note that an *orbital* (quantum mechanical model) is not the same as an *orbit* (Bohr model). The quantum mechanical model doesn't refer to orbits because the motion of the electron in an atom cannot be precisely measured or tracked (Heisenberg uncertainty principle).

The Bohr model introduced a single quantum number, n, to describe an orbit. The quantum mechanical model uses three quantum numbers, n, l, and m_l, to describe an orbital. Let's consider what information we obtain from each of these and how they are interrelated.

1. The *principal quantum number*, n, can have positive integral values of 1, 2, 3, and so forth. As n increases, the orbital becomes larger, and the electron spends more time farther from the nucleus. An increase in n also means that the electron has a higher energy and is therefore less tightly bound to the nucleus. For the hydrogen atom, $E_n = -(2.18 \times 10^{-18}\text{ J})(1/n^2)$, as in the Bohr model.
2. The second quantum number—the *azimuthal quantum number*, l—can have integral values from 0 to $n - 1$ for each value of n. This quantum number defines the shape of the orbital. (We will consider these shapes in Section 6.6.) The value of l for a particular orbital is generally designated by the letters s, p, d, and f,* corresponding to l values of 0, 1, 2, and 3, respectively, as summarized here.

Value of l	0	1	2	3
Letter used	s	p	d	f

3. The *magnetic quantum number*, m_l, can have integral values between l and $-l$, including zero. This quantum number describes the orientation of the orbital in space, as we will discuss in Section 6.6.

The collection of orbitals with the same value of n is called an **electron shell**. For example, all the orbitals that have $n = 3$ are said to be in the third shell. Further, the set of orbitals that have the same n and l values is called a **subshell**. Each subshell is designated by a number (the value of n) and a letter (s, p, d, or f, corresponding to the value of l). For example, the orbitals that have $n = 3$ and $l = 2$ are called 3d orbitals and are in the 3d subshell.

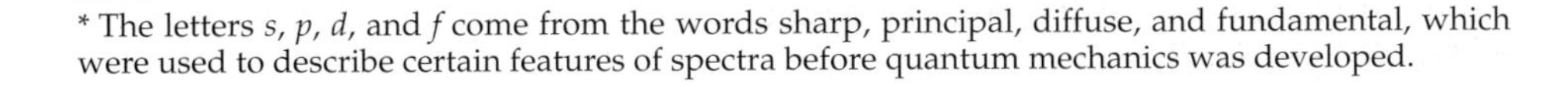
* The letters s, p, d, and f come from the words sharp, principal, diffuse, and fundamental, which were used to describe certain features of spectra before quantum mechanics was developed.

TABLE 6.2 Relationship Among Values of n, l, and m_l Through $n = 4$

n	Possible Values of l	Subshell Designation	Possible Values of m_l	Number of Orbitals in Subshell	Total Number of Orbitals in Shell
1	0	1s	0	1	1
2	0	2s	0	1	
	1	2p	1, 0, −1	3	4
3	0	3s	0	1	
	1	3p	1, 0, −1	3	
	2	3d	2, 1, 0, −1, −2	5	9
4	0	4s	0	1	
	1	4p	1, 0, −1	3	
	2	4d	2, 1, 0, −1, −2	5	
	3	4f	3, 2, 1, 0, −1, −2, −3	7	16

Table 6.2 ▲ summarizes the possible values of the quantum numbers l and m_l for values of n through $n = 4$. The restrictions on the possible values of the quantum numbers give rise to the following very important observations:

1. The shell with principal quantum number n will consist of exactly n subshells. Each subshell corresponds to a different allowed value of l from 0 to $n - 1$. Thus, the first shell ($n = 1$) consists of only one subshell, the 1*s* ($l = 0$); the second shell ($n = 2$) consists of two subshells, the 2*s* ($l = 0$) and 2*p* ($l = 1$); the third shell consists of three subshells, 3*s*, 3*p*, and 3*d*, and so forth.

2. Each subshell consists of a specific number of orbitals. Each orbital corresponds to a different allowed value of m_l. For a given value of l, there are $2l + 1$ allowed values of m_l, ranging from $-l$ to $+l$. Thus, each *s* ($l = 0$) subshell consists of one orbital; each *p* ($l = 1$) subshell consists of three orbitals; each *d* ($l = 2$) subshell consists of five orbitals, and so forth.
3. The total number of orbitals in a shell is n^2, where n is the principal quantum number of the shell. The resulting number of orbitals for the shells—1, 4, 9, 16—is related to a pattern seen in the periodic table: We see that the number of elements in the rows of the periodic table—2, 8, 18, and 32—equal twice these numbers. We will discuss this relationship further in Section 6.9.

Figure 6.17 ▶ shows the relative energies of the hydrogen atom orbitals through $n = 3$. Each box represents an orbital; orbitals of the same subshell, such as the 2*p*, are grouped together. When the electron is in the lowest energy orbital (the 1*s* orbital), the hydrogen atom is said to be in its *ground state*. When the electron is in any other orbital, the atom is in an *excited state*. At ordinary temperatures essentially all hydrogen atoms are in their ground states. The electron can be excited to a higher-energy orbital by absorption of a photon of appropriate energy.

SAMPLE EXERCISE 6.6

(a) Without referring to Table 6.2, predict the number of subshells in the fourth shell, that is, for $n = 4$. **(b)** Give the label for each of these subshells. **(c)** How many orbitals are in each of these subshells?

Solution (a) There are four subshells in the fourth shell, corresponding to the four possible values of l (0, 1, 2, and 3).

(b) These subshells are labeled 4*s*, 4*p*, 4*d*, and 4*f*. The number given in the designation of a subshell is the principal quantum number, n; the following letter designates the value of the azimuthal quantum number, l.

(c) There is one 4*s* orbital (when $l = 0$, there is only one possible value of m_l: 0). There are three 4*p* orbitals (when $l = 1$, there are three possible values of m_l: 1, 0, and −1). There are five 4*d* orbitals (when $l = 2$, there are five allowed values of m_l: 2, 1, 0, −1, −2). There are seven 4*f* orbitals (when $l = 3$, there are seven permitted values of m_l: 3, 2, 1, 0, −1, −2, −3).

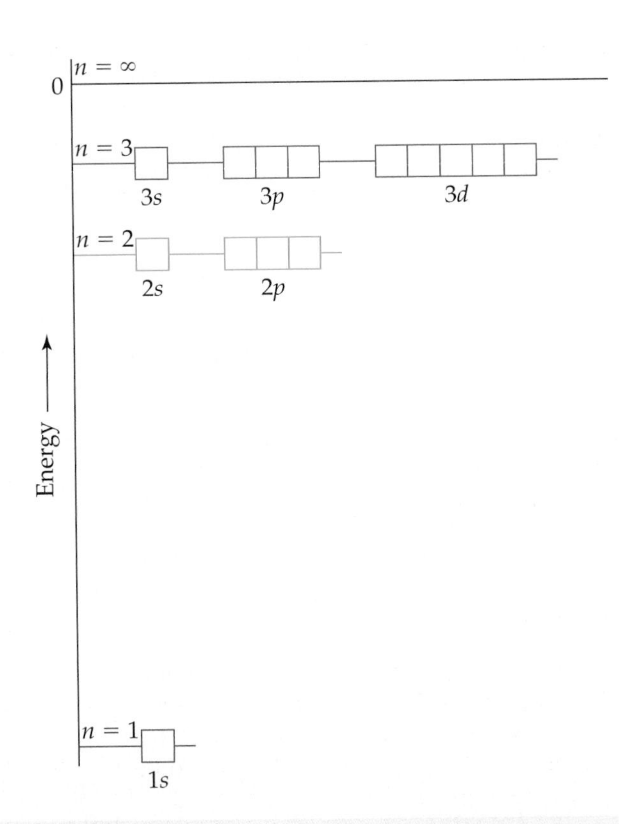

◀ **Figure 6.17** Orbital energy levels in the hydrogen atom. Each box represents an orbital. Note that all orbitals with the same value for the principal quantum number, n, have the same energy. This is true only in one-electron systems.

PRACTICE EXERCISE

(a) What is the designation for the subshell with $n = 5$ and $l = 1$? **(b)** How many orbitals are in this subshell? **(c)** Indicate the values of m_l for each of these orbitals.

Answers: **(a)** 5*p*; **(b)** 3; **(c)** 1, 0, −1

6.6 Representations of Orbitals

In our discussion of orbitals we have so far emphasized their energies. But the wave function also provides information about the electron's location in space when it is in a particular allowed energy state. Let's examine the ways that we can picture the orbitals.

The *s* Orbitals

The lowest energy orbital, the 1*s*, is spherical, as shown in Figure 6.16. Figures of this type, showing electron density, are one of the several ways we use to help us visualize orbitals. This figure indicates that the probability of finding the electron decreases as we move away from the nucleus in any particular direction. When the probability function, ψ^2, for the 1*s* orbital is graphed as a function of the distance from the nucleus, r, it rapidly approaches zero, as shown in Figure 6.18(a) ▶. This effect indicates that the electron, which is drawn toward the nucleus by electrostatic attraction, is unlikely to be found very far from the nucleus.

If we similarly consider the 2*s* and 3*s* orbitals of hydrogen, we find that they are also spherically symmetrical. Indeed, *all s* orbitals are spherically symmetrical. The manner in which the probability function, ψ^2, varies with r for the 2*s* and 3*s* orbitals is shown in Figure 6.18(b) and (c). Notice that for the 2*s* orbital, ψ^2 goes to zero and then increases again in value before finally approaching zero at a larger value of r. The intermediate regions where ψ^2 goes to zero are called **nodes**. The number of nodes increases with increasing value for the principal quantum number, n. The 3*s* orbital possesses two nodes, as illustrated in Figure 6.18(c). Notice also that as n increases, the electron is more and more likely to be located farther from the nucleus. That is, the size of the orbital increases as n increases.

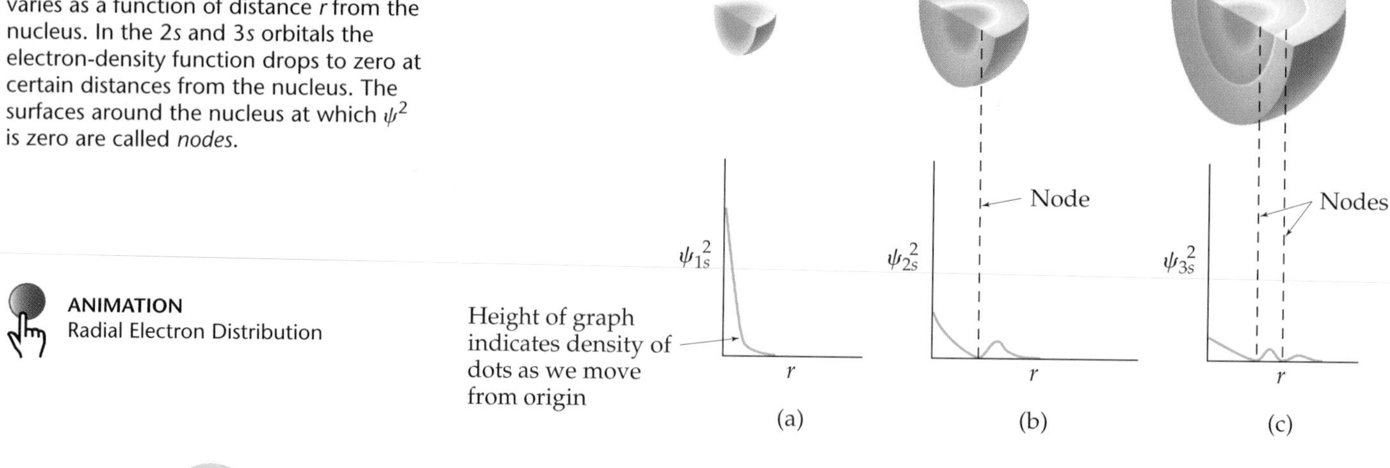

▶ **Figure 6.18** Electron-density distribution in 1*s*, 2*s*, and 3*s* orbitals. The lower part of the figure shows how the electron density, represented by ψ^2, varies as a function of distance *r* from the nucleus. In the 2*s* and 3*s* orbitals the electron-density function drops to zero at certain distances from the nucleus. The surfaces around the nucleus at which ψ^2 is zero are called *nodes*.

ANIMATION
Radial Electron Distribution

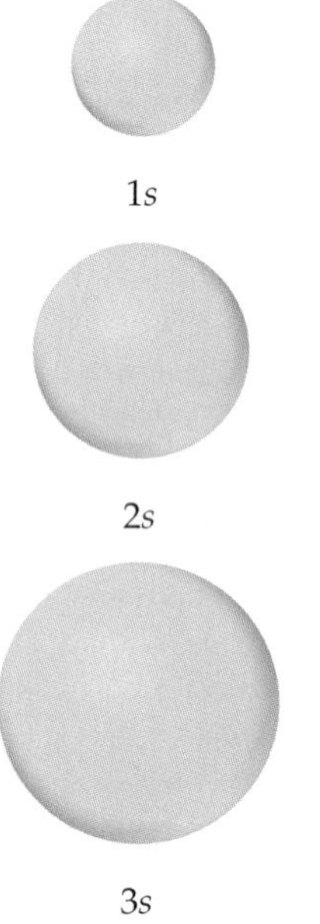

▲ **Figure 6.19** Contour representations of the 1*s*, 2*s*, and 3*s* orbitals. The relative radii of the spheres correspond to a 90% probability of finding the electron within each sphere.

One widely used method of representing orbitals is to display a boundary surface that encloses some substantial portion, say 90%, of the total electron density for the orbital. For the *s* orbitals, these contour representations are merely spheres. The contour or boundary surface representations of the 1*s*, 2*s*, and 3*s* orbitals are shown in Figure 6.19 ◀. They have the same shape, but they differ in size. Although the details of how the electron density varies within the surface are lost in these representations, this is not a serious disadvantage. For more qualitative discussions, the most important features of orbitals are their relative sizes and their shapes. These features are adequately displayed by contour representations.

The *p* Orbitals

The distribution of electron density for a 2*p* orbital is shown in Figure 6.20(a) ▼. As we can see from this figure, the electron density is not distributed in a spherically symmetric fashion as in an *s* orbital. Instead, the electron density is concentrated on two regions on either side of the nucleus, separated by a node at the nucleus. We say that this dumbbell-shaped orbital has two *lobes*. It is useful to recall that we are making no statement of how the electron is moving within the orbital; Figure 6.20(a) portrays the *averaged* distribution of the electron density in a 2*p* orbital.

Each shell beginning with $n = 2$ has three *p* orbitals: Thus, there are three 2*p* orbitals, three 3*p* orbitals, and so forth. The orbitals of a given value of *n* (that is, of a given subshell) have the same size and shape but differ from one another in

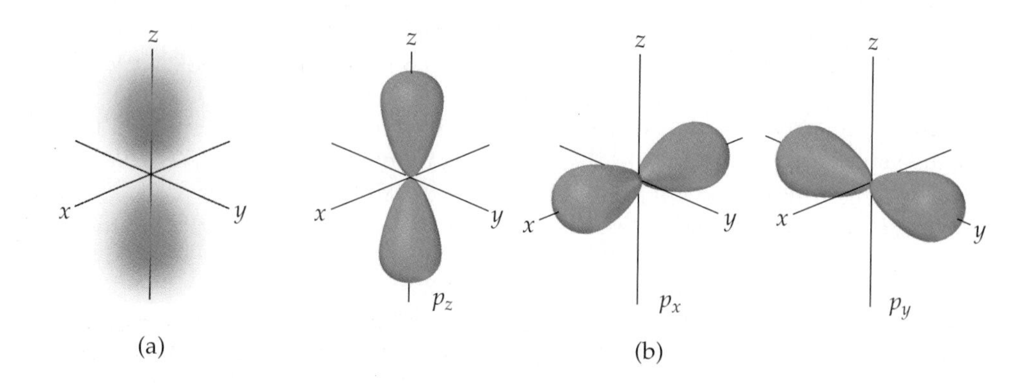

▶ **Figure 6.20** (a) Electron-density distribution of a 2*p* orbital.
(b) Representations of the three *p* orbitals. Note that the subscript on the orbital label indicates the axis along which the orbital lies.

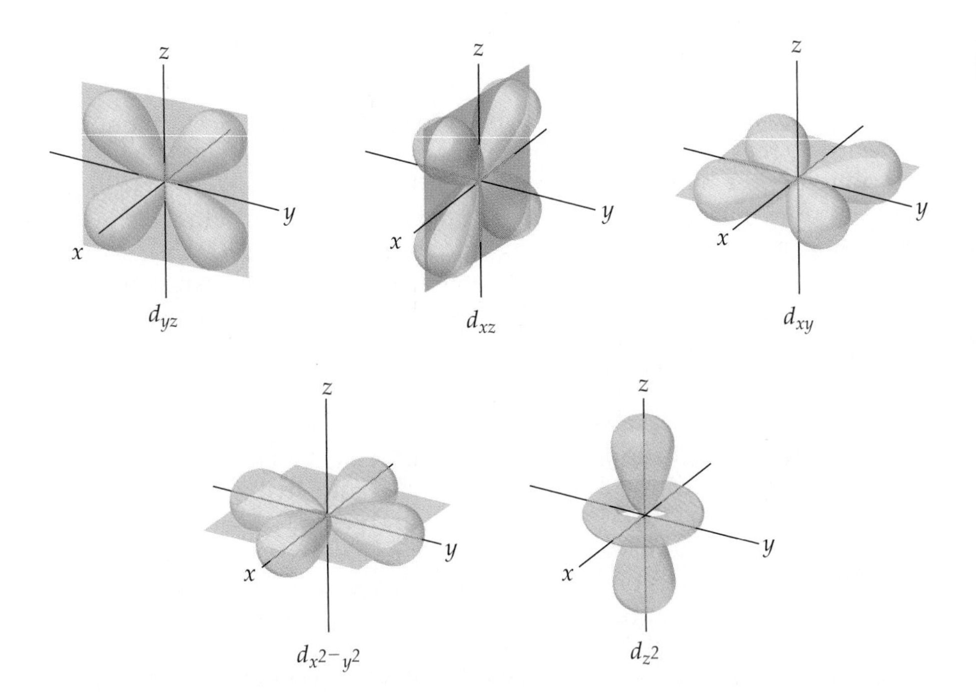

◀ **Figure 6.21** Representations of the five *d* orbitals.

spatial orientation. We usually represent *p* orbitals by drawing the shape and orientation of their wave functions, as shown in Figure 6.20(b). It is convenient to label these as the p_x, p_y, and p_z orbitals. The letter subscript indicates the axis along which the orbital is oriented.* Like *s* orbitals, *p* orbitals increase in size as we move from 2*p* to 3*p* to 4*p*, and so forth.

The *d* and *f* Orbitals

When *n* is 3 or greater, we encounter the *d* orbitals (for which $l = 2$). There are five 3*d* orbitals, five 4*d* orbitals, and so forth. The different *d* orbitals in a given shell have different shapes and orientations in space, as shown in Figure 6.21 ▲. Four of the *d*-orbital contours have "four-leaf clover" shapes, and each lies primarily in a plane. The d_{xy}, d_{xz}, and d_{yz} lie in the *xy*, *xz*, and *yz* planes, respectively, with the lobes oriented between the axes. The lobes of the $d_{x^2-y^2}$ orbital also lie in the *xy* plane, but the lobes lie along the *x* and *y* axes. The d_{z^2} orbital looks very different from the other four: It has two lobes along the *z* axis and a "doughnut" in the *xy* plane. Even though the d_{z^2} orbital looks different, it has the same energy as the other four *d* orbitals. The representations in Figure 6.21 are commonly used for all *d* orbitals, regardless of principal quantum number.

When *n* is 4 or greater, there are seven equivalent *f* orbitals (for which $l = 3$). The shapes of the *f* orbitals are even more complicated than those of the *d* orbitals. We will not present the shapes of the *f* orbitals. As you will see in the next section, however, you must be aware of *f* orbitals as we consider the electronic structure of atoms in the lower part of the periodic table.

In many instances later in the text you will find that knowing the number and shapes of atomic orbitals will help you understand chemistry at the molecular level. You will therefore find it useful to memorize the shapes of the orbitals shown in Figures 6.19 through 6.21.

* We cannot make a simple correspondence between the subscripts (*x*, *y*, and *z*) and the allowed m_l values (1, 0, and −1). To explain why this is so would be beyond the scope of an introductory text.

6.7 Many-Electron Atoms

One of our goals in this chapter has been to determine the electronic structures of atoms. So far we have seen that quantum mechanics leads to a very elegant description of the hydrogen atom. The hydrogen atom, however, has only one electron. How must our description of atomic electronic structure change when we consider atoms with two or more electrons (a *many-electron* atom)? To describe these atoms, we must consider not only the nature of orbitals and their relative energies but also how the electrons populate the available orbitals.

Orbitals and Their Energies

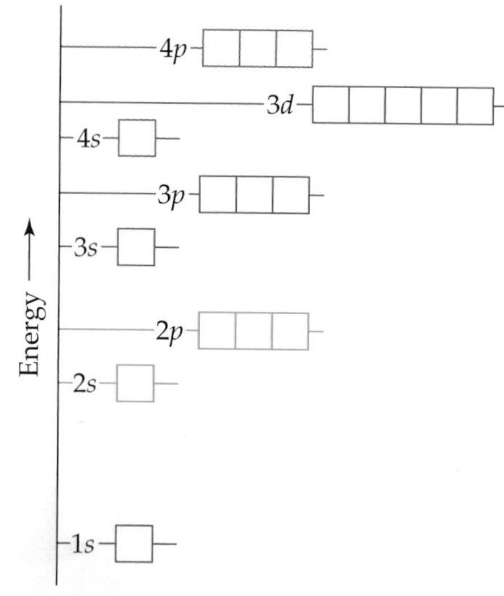

▲ **Figure 6.22** Ordering of orbital energy levels in many-electron atoms, through the 4*p* orbitals. As in Figure 6.17, which shows the orbital energy levels for the hydrogen atom, each box represents an orbital. Note that orbitals in different subshells differ in energy.

The quantum mechanical model would not be very useful if we could not extend what we have learned about hydrogen to other atoms. Fortunately, the atomic orbitals in a many-electron atom are like those of the hydrogen atom. Thus we can continue to designate orbitals as 1*s*, $2p_x$, and so forth. Further, these orbitals have the same general shapes as the corresponding hydrogen orbitals.

Although the shapes of the orbitals for many-electron atoms are the same as those for hydrogen, the presence of more than one electron greatly changes the energies of the orbitals. In hydrogen the energy of an orbital depends only on its principal quantum number, *n* (Figure 6.17); the 3*s*, 3*p*, and 3*d* subshells all have the same energy, for instance. In a many-electron atom, however, the electron-electron repulsions cause the different subshells to be at different energies, as shown in Figure 6.22 ◀. For example, the 2*s* subshell is lower in energy than the 2*p* subshell. To understand why this is so, we must consider the forces between the electrons and how these forces are affected by the shapes of the orbitals. We will, however, forgo this analysis until Chapter 7.

The important idea is this: *In a many-electron atom, for a given value of n, the energy of an orbital increases with increasing value of l.* You can see this illustrated in Figure 6.22. Notice, for example, that the $n = 3$ orbitals (red) increase in energy in the order $s < p < d$. Figure 6.22 is a *qualitative* energy-level diagram; the exact energies and their spacings differ from one atom to another. Notice that all orbitals of a given subshell (such as the 3*d* orbitals) still have the same energy, just as they do in the hydrogen atom. Orbitals with the same energy are said to be **degenerate**.

Electron Spin and the Pauli Exclusion Principle

We have now seen that we can use hydrogen-like orbitals to describe many-electron atoms. What, however, determines the orbitals in which electrons reside? That is, how do the electrons of a many-electron atom populate the available orbitals? To answer this question, we must consider an additional property of the electron.

ACTIVITY
Line Spectrum of Sodium

When scientists studied the line spectra of many-electron atoms in great detail, they noticed a very puzzling feature: Lines that were originally thought to be single were actually closely spaced pairs. This meant, in essence, that there were twice as many energy levels as there were "supposed" to be. In 1925 the Dutch physicists George Uhlenbeck and Samuel Goudsmit proposed a solution to this dilemma. They postulated that electrons have an intrinsic property, called **electron spin**. The electron apparently behaves as if it were a tiny sphere spinning on its own axis.

By now it probably does not surprise you to learn that electron spin is quantized. This observation led to the assignment of a new quantum number for the electron, in addition to *n*, *l*, and m_l that we have already discussed. This new quantum number, the **spin magnetic quantum number**, is denoted m_s (the subscript *s* stands for *spin*). Only two possible values are allowed for m_s, $+\frac{1}{2}$ or $-\frac{1}{2}$, which was first interpreted as indicating the two opposite directions in which the

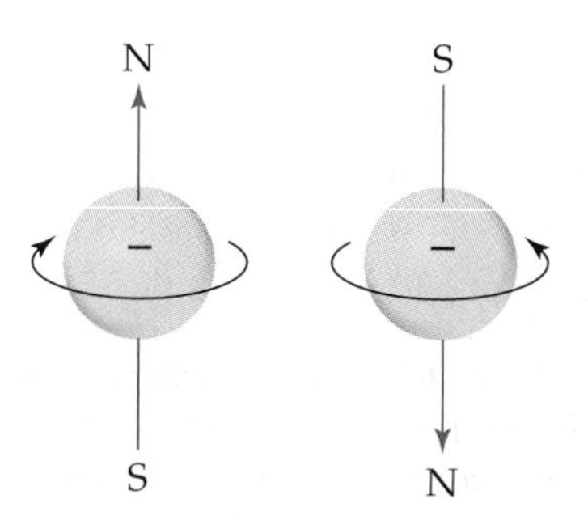

◀ **Figure 6.23** The electron behaves as if it were spinning about an axis through its center, thereby generating a magnetic field whose direction depends on the direction of spin. The two directions for the magnetic field correspond to the two possible values for the spin quantum number, m_s.

electron can spin. A spinning charge produces a magnetic field. The two opposite directions of spin produce oppositely directed magnetic fields, as shown in Figure 6.23 ▲.* These two opposite magnetic fields lead to the splitting of spectral lines into closely spaced pairs.

Electron spin is crucial for understanding the electronic structures of atoms. In 1925 the Austrian-born physicist Wolfgang Pauli (1900–1958) discovered the principle that governs the arrangements of electrons in many-electron atoms. The **Pauli exclusion principle** states that *no two electrons in an atom can have the same set of four quantum numbers n, l, m_l,* and m_s. For a given orbital (1*s*, 2p_z, and so forth), the values of *n*, *l*, and m_l are fixed. Thus, if we want to put more than one electron in an orbital *and* satisfy the Pauli exclusion principle, our only choice is to assign different m_s values to the electrons. Because there are only two such values, we conclude that *an orbital can hold a maximum of two electrons, and they must have opposite spins*. This restriction allows us to index the electrons in an atom, giving their quantum numbers and thereby defining the region in space where each electron is most likely to be found. It also provides the key to one of the great problems in chemistry—understanding the structure of the periodic table of the elements. We will discuss these issues in the next two sections.

* As we discussed earlier, the electron has both particle-like and wavelike properties. Thus, the picture of an electron as a spinning charged sphere is, strictly speaking, just a useful pictorial representation that helps us understand the two directions of magnetic field that an electron can possess.

A Closer Look Experimental Evidence for Electron Spin

Even before electron spin had been proposed, there was experimental evidence that electrons had an additional property that needed explanation. In 1921 Otto Stern and Walter Gerlach succeeded in separating a beam of neutral atoms into two groups by passing them through a nonhomogeneous magnetic field. Their experiment is diagrammed in Figure 6.24 ▶. Let's assume that they used a beam of hydrogen atoms (in actuality, they used silver atoms, which contain just one unpaired electron). We would normally expect that neutral atoms would not be affected by a magnetic field. However, the magnetic field arising from the electron's spin interacts with the magnet's field, deflecting the atom from its straight-line path. As shown in Figure 6.24, the magnetic field splits the beam in two, suggesting that there are two (and only two) equivalent values for the electron's own magnetic field. The Stern–Gerlach experiment could be readily interpreted once it was realized that there are exactly two values for the spin of the electron. These values will produce equal magnetic fields that are opposite in direction.

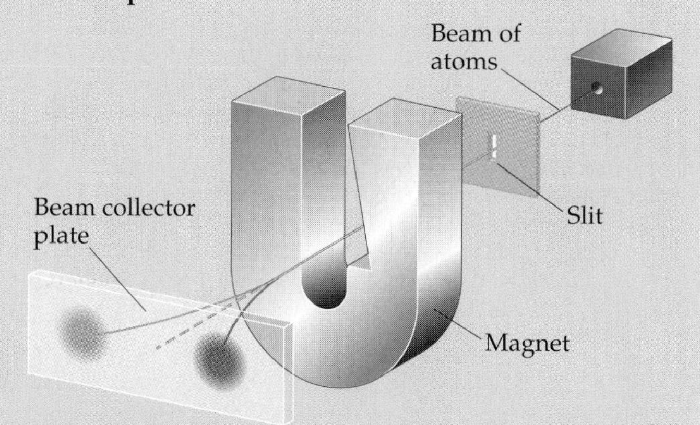

▲ **Figure 6.24** Illustration of the Stern–Gerlach experiment. Atoms in which the electron spin quantum number (m_s) of the unpaired electron is $+\frac{1}{2}$ are deflected in one direction, and those in which m_s is $-\frac{1}{2}$ are deflected in the other.

Chemistry and Life Nuclear Spin and Magnetic Resonance Imaging

A major challenge facing medical diagnosis is seeing inside the human body from the outside. Until recently, this was accomplished primarily by using X rays to image human bones, muscles, and organs. However, there are several drawbacks to using X rays for medical imaging. First, X rays do not give well-resolved images of overlapping physiological structures. Moreover, because damaged or diseased tissue often yields the same image as healthy tissue, X rays frequently fail to detect illness or injuries. Finally, X rays are high-energy radiation that can cause physiological harm, even in low doses.

In the 1980s a new technique called *magnetic resonance imaging* (MRI) moved to the forefront of medical imaging technology. The foundation of MRI is a phenomenon called nuclear magnetic resonance (NMR), which was discovered in the mid-1940s. Today NMR has become one of the most important spectroscopic methods used in chemistry. It is based on the observation that, like electrons, the nuclei of many elements possess an intrinsic spin. Like electron spin, nuclear spin is quantized. For example, the nucleus of ^{1}H (a proton) has two possible magnetic nuclear spin quantum numbers, $+\frac{1}{2}$ and $-\frac{1}{2}$. The hydrogen nucleus is the most common one studied by NMR.

A spinning hydrogen nucleus acts like a tiny magnet. In the absence of external effects, the two spin states have the same energy. However, when the nuclei are placed in an external magnetic field, they can align either parallel or opposed (antiparallel) to the field, depending on their spin. The parallel alignment is lower in energy than the antiparallel one by a certain amount, ΔE (Figure 6.25 ▼). If the nuclei are irradiated with photons with energy equal to ΔE, the spin of the nuclei can be "flipped," that is, excited from the parallel to the antiparallel alignment. Detection of the flipping of nuclei between the two spin states leads to an NMR spectrum. The radiation used in an NMR experiment is in the radiofrequency range, typically 100 to 500 MHz.

Because hydrogen is a major constituent of aqueous body fluids and fatty tissue, the hydrogen nucleus is the most convenient one for study by MRI. In MRI a person's body is placed in a strong magnetic field. By irradiating the body with pulses of radiofrequency radiation and using sophisticated detection techniques, tissue can be imaged at specific depths within the body, giving pictures with spectacular detail (Figure 6.26 ▼). The ability to sample at different depths allows medical technicians to construct a three-dimensional picture of the body.

MRI has none of the disadvantages of X rays. Diseased tissue appears very different from healthy tissue, resolving overlapping structures at different depths in the body is much easier, and the radiofrequency radiation is not harmful to humans in the doses used. The major drawback of MRI is expense: The current cost of a new MRI instrument for clinical applications is over $1.5 million.

▲ **Figure 6.25** Like electron spin, nuclear spin generates a small magnetic field and has two allowed values. In the absence of an external magnetic field (left), the two spin states have the same energy. If an external magnetic field is applied (right), the parallel alignment of the nuclear magnetic field is lower in energy than the antiparallel alignment. The energy difference, ΔE, is in the radiofrequency portion of the electromagnetic spectrum.

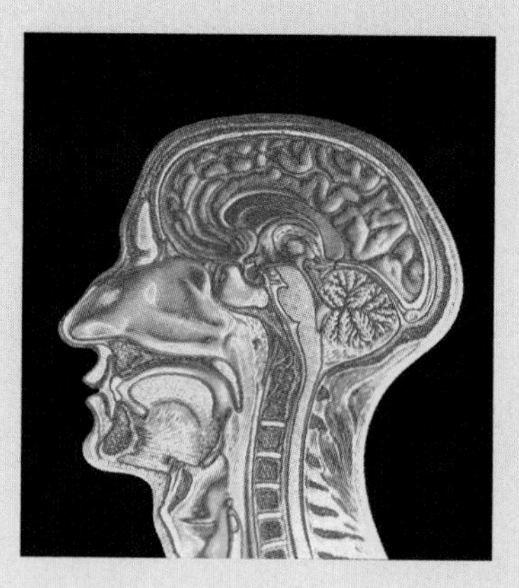

◀ **Figure 6.26** An MRI image of a human head, showing the structures of a normal brain, airways, and facial tissues.

6.8 Electron Configurations

Armed with a knowledge of the relative energies of orbitals and the Pauli exclusion principle, we are now in a position to consider the arrangements of electrons in atoms. The way in which the electrons are distributed among the various orbitals of an atom is called its **electron configuration**. The most stable, or ground state, electron configuration of an atom is that in which the electrons are in the lowest possible energy states. If there were no restrictions on the possible values

ANIMATION
Electron Configurations

for the quantum numbers of the electrons, all the electrons would crowd into the 1s orbital because it is the lowest in energy (Figure 6.22). The Pauli exclusion principle tells us, however, that there can be at most two electrons in any single orbital. Thus, *the orbitals are filled in order of increasing energy, with no more than two electrons per orbital.* For example, consider the lithium atom, which has three electrons. (Recall that the number of electrons in a neutral atom equals its atomic number.) The 1s orbital can accommodate two of the electrons. The third one goes into the next lowest energy orbital, the 2s.

We can summarize any electron configuration by writing the symbol for the occupied subshell and adding a superscript to indicate the number of electrons in that subshell. For example, for lithium we write $1s^2 2s^1$ (read "1s two, 2s one"). We can also show the arrangement of the electrons as

Li	↿⇂	↿
	1s	2s

In this kind of representation, which we will call an *orbital diagram*, each orbital is denoted by a box and each electron by a half arrow. A half arrow pointing upward (↿) represents an electron with a positive spin magnetic quantum number ($m_s = +\frac{1}{2}$), and a half arrow pointing downward (⇂) represents an electron with a negative spin magnetic quantum number ($m_s = -\frac{1}{2}$). This pictorial representation of electron spin is quite convenient. In fact, chemists and physicists often refer to electrons as "spin-up" and "spin-down" rather than specifying the value for m_s.

Electrons having opposite spins are said to be *paired* when they are in the same orbital (↿⇂). An *unpaired electron* is not accompanied by a partner of opposite spin. In the lithium atom the two electrons in the 1s orbital are paired, and the electron in the 2s orbital is unpaired.

Hund's Rule

Consider now how the electron configurations of the elements change as we move from element to element across the periodic table. Hydrogen has one electron, which occupies the 1s orbital in its ground state.

H	↿	: $1s^1$
	1s	

The choice of a spin-up electron here is arbitrary; we could equally well show the ground state with one spin-down electron in the 1s orbital. It is customary, however, to show the unpaired electrons with their spins up.

The next element, helium, has two electrons. Because two electrons with opposite spins can occupy an orbital, both of helium's electrons are in the 1s orbital.

He	↿⇂	: $1s^2$
	1s	

The two electrons present in helium complete the filling of the first shell. This arrangement represents a very stable configuration, as is evidenced by the chemical inertness of helium.

TABLE 6.3 Electron Configurations of Several Lighter Elements

Element	Total Electrons	Orbital Diagram 1s	2s	2p	3s	Electron Configuration
Li	3	↑↓	↑			$1s^22s^1$
Be	4	↑↓	↑↓			$1s^22s^2$
B	5	↑↓	↑↓	↑		$1s^22s^22p^1$
C	6	↑↓	↑↓	↑ ↑		$1s^22s^22p^2$
N	7	↑↓	↑↓	↑ ↑ ↑		$1s^22s^22p^3$
Ne	10	↑↓	↑↓	↑↓ ↑↓ ↑↓		$1s^22s^22p^6$
Na	11	↑↓	↑↓	↑↓ ↑↓ ↑↓	↑	$1s^22s^22p^63s^1$

The electron configurations of lithium and several elements that follow it in the periodic table are shown in Table 6.3 ▲. For the third electron of lithium, the change in principal quantum number represents a large jump in energy and a corresponding jump in the average distance of the electron from the nucleus. It represents the start of a new shell of electrons. As you can see by examining the periodic table, lithium starts a new row of the periodic table. It is the first member of the alkali metals (group 1A).

The element that follows lithium is beryllium; its electron configuration is $1s^22s^2$ (Table 6.3). Boron, atomic number 5, has the electron configuration $1s^22s^22p^1$. The fifth electron must be placed in a $2p$ orbital because the $2s$ orbital is filled. Because all the three $2p$ orbitals are of equal energy, it doesn't matter which $2p$ orbital is occupied.

With the next element, carbon, we encounter a new situation. We know that the sixth electron must go into a $2p$ orbital. However, does this new electron go into the $2p$ orbital that already has one electron, or into one of the others? This question is answered by **Hund's rule**, which states that *for degenerate orbitals, the lowest energy is attained when the number of electrons with the same spin is maximized*. This means that electrons will occupy orbitals singly to the maximum extent possible, with the same spin magnetic quantum number. Electrons arranged in this way are said to have *parallel spins*. For a carbon atom to achieve its lowest energy, therefore, the two $2p$ electrons will have the same spin. In order for this to happen, the electrons must be in different $2p$ orbitals, as shown in Table 6.3. Thus, a carbon atom in its ground state has two unpaired electrons. Similarly, for nitrogen in its ground state, Hund's rule requires that the three $2p$ electrons singly occupy each of the three $2p$ orbitals. This is the only way that all three electrons can have the same spin. For oxygen and fluorine, we place four and five electrons, respectively, in the $2p$ orbitals. To achieve this, we pair up electrons in the $2p$ orbitals, as we will see in Sample Exercise 6.7.

Hund's rule is based in part on the fact that electrons repel one another. By occupying different orbitals, the electrons remain as far as possible from one another, thus minimizing electron-electron repulsions.

SAMPLE EXERCISE 6.7

Draw the orbital diagram representation for the electron configuration of oxygen, atomic number 8. How many unpaired electrons does an oxygen atom possess?

Solution

Analyze and Plan: Because oxygen has an atomic number of 8, the atom has 8 electrons. Figure 6.22 shows the ordering of orbitals. The electrons (represented as arrows), are placed in the orbitals (represented as boxes) beginning with the lowest energy 1*s* orbital. Each orbital can hold a maximum of two electrons (the Pauli exclusion principle). Because the 2*p* orbitals are degenerate, we place one electron in each of these orbitals (spin-up) before pairing any electrons (Hund's rule).

Solve: Two electrons each go into the 1*s* and 2*s* orbitals with their spins paired. This leaves four electrons for the three degenerate 2*p* orbitals. Following Hund's rule, we put one electron into each 2*p* orbital until all three have one each. The fourth electron is then paired up with one of the three electrons already in a 2*p* orbital, so that the representation is

⇅	⇅	⇅ ↑ ↑
1*s*	2*s*	2*p*

The corresponding electron configuration is written $1s^22s^22p^4$. The atom has two unpaired electrons.

PRACTICE EXERCISE

(a) Write the electron configuration of phosphorus, element 15. **(b)** How many unpaired electrons does a phosphorus atom possess?

Answers: **(a)** $1s^22s^22p^63s^23p^3$; **(b)** three

Condensed Electron Configurations

The filling of the 2*p* subshell is complete at neon (Table 6.3), which has a stable configuration with eight electrons (an *octet*) in the outermost shell. The next element, sodium, atomic number 11, marks the beginning of a new row of the periodic table. Sodium has a single 3*s* electron beyond the stable configuration of neon. We can abbreviate the electron configuration of sodium as follows:

$$\text{Na:}\quad [\text{Ne}]3s^1$$

The symbol [Ne] represents the electron configuration of the ten electrons of neon, $1s^22s^22p^6$. Writing the electron configuration in this manner helps focus attention on the outermost electrons of the atom. The outer electrons are the ones largely responsible for the chemical behavior of an element.

In writing the *condensed electron configuration* of an element, the electron configuration of the nearest noble-gas element of lower atomic number is represented by its chemical symbol in brackets. For example, we can write the electron configuration of lithium as

$$\text{Li:}\quad [\text{He}]2s^1$$

We refer to the electrons represented by the symbol for a noble-gas as the noble-gas core of the atom. More usually, these inner-shell electrons are referred to merely as the **core electrons**. The electrons given after the noble gas core are referred to as the outer-shell electrons, or **valence electrons**.

By comparing the electron configuration of lithium with that of sodium, we can appreciate why these two elements are so similar chemically: They have the same type of outer-shell electron configuration. Indeed, all the members of the alkali metal group (1A) have a single *s* electron beyond a noble-gas configuration.

Transition Metals

The noble-gas element argon marks the end of the row started by sodium. The configuration for argon is $1s^22s^22p^63s^23p^6$. The element following argon in the periodic table is potassium (K), atomic number 19. In all its chemical properties, potassium is clearly a member of the alkali metal group. The experimental facts about the properties of potassium leave no doubt that the outermost electron of this element occupies an *s* orbital. But this means that the highest energy electron has *not* gone into a 3*d* orbital, which we might have expected it to do. Here the ordering of energy levels is such that the 4*s* orbital is lower in energy than the 3*d* (Figure 6.22). Hence the condensed electron configuration of potassium is

$$\text{K:}\quad [\text{Ar}]4s^1$$

Following complete filling of the 4*s* orbital (this occurs in the calcium atom), the next set of equivalent orbitals to be filled is the 3*d*. (You will find it helpful as we go along to refer often to the periodic table on the front inside cover.) Beginning with scandium and extending through zinc, electrons are added to the five 3*d* orbitals until they are completely filled. Thus, the fourth row of the periodic table is ten elements wider than the two previous rows. These ten elements are known as **transition elements**, or **transition metals**. Note the position of these elements in the periodic table.

In accordance with Hund's rule, electrons are added to the 3*d* orbitals singly until all five orbitals have one electron each. Additional electrons are then placed in the 3*d* orbitals with spin pairing until the shell is completely filled. The condensed electron configurations and the corresponding orbital diagram representations of two transition elements are as follows:

		4s	3d				
Mn: $[\text{Ar}]4s^2 3d^5$	or [Ar]	↿⇂	↿	↿	↿	↿	↿
Zn: $[\text{Ar}]4s^2 3d^{10}$	or [Ar]	↿⇂	↿⇂	↿⇂	↿⇂	↿⇂	↿⇂

Upon completion of the 3*d* transition series, the 4*p* orbitals begin to be occupied until the completed octet of outer electrons ($4s^2 4p^6$) is reached with krypton (Kr), atomic number 36, another of the noble gases. Rubidium (Rb) marks the beginning of the fifth row. Refer again to the periodic table on the front inside cover. Notice that this row is in every respect like the preceding one, except that the value for n is greater by 1.

The Lanthanides and Actinides

The sixth row of the periodic table begins similarly to the preceding one: one electron in the 6*s* orbital of cesium (Cs) and two electrons in the 6*s* orbital of barium (Ba). Notice however, that the periodic table then has a break, and the subsequent set of elements (elements 57–70) is placed below the main portion of the table. It is at this place that we begin to encounter a new set of orbitals, the 4*f*.

There are seven degenerate 4*f* orbitals, corresponding to the seven allowed values of m_l, ranging from 3 to −3. Thus, it takes 14 electrons to fill the 4*f* orbitals completely. The 14 elements corresponding to the filling of the 4*f* orbitals are known as the **lanthanide** (or rare earth) **elements**. The lanthanide elements are set below the other elements to avoid making the periodic table unduly wide. The properties of the lanthanide elements are all quite similar, and they occur together in nature. For many years it was virtually impossible to separate them from one another.

Because the energies of the 4*f* and 5*d* orbitals are very close, the electron configurations of some of the lanthanides involve 5*d* electrons. For example, the elements lanthanum (La), cerium (Ce) and praseodymium (Pr) have the following electron configurations:

$$\text{La:}\ [\text{Xe}]6s^2 5d^1 \qquad \text{Ce:}\ [\text{Xe}]6s^2 5d^1 4f^1 \qquad \text{Pr:}\ [\text{Xe}]6s^2 4f^3$$

Because La has a single 5*d* electron, it is sometimes placed below yttrium (Y) as the first member of the third series of transition elements, and Ce is then placed as the first member of the lanthanides. Based on their chemistry, however, La can be considered the first element in the lanthanide series. Arranged this way, there are fewer apparent exceptions to the regular filling of the $4f$ orbitals among the subsequent members of the series.

After the lanthanide series, the third transition element series is completed by the filling of the $5d$ orbitals, followed by the filling of the $6p$ orbitals. This brings us to radon (Rn), heaviest of the known noble-gas elements. The final row of the periodic table begins by filling the $7s$ orbitals. The **actinide elements**, of which uranium (U, element 92) and plutonium (Pu, element 94) are the best known, are then built up by completing the $5f$ orbitals. The actinide elements are radioactive, and most of them are not found in nature.

6.9 Electron Configurations and the Periodic Table

Our rather brief survey of electron configurations of the elements has taken us through the periodic table. We have seen that the electron configurations of elements are related to their locations in the periodic table. The periodic table is structured so that elements with the same pattern of outer-shell (valence) electron configuration are arranged in columns. For example, the electron configurations for the elements in groups 2A and 3A are given in Table 6.4 ▶. We see that the 2A elements all have ns^2 outer configurations, while the 3A elements have ns^2np^1 configurations.

Earlier, in Table 6.2, we saw that the total number of orbitals in each shell is equal to n^2: 1, 4, 9, or 16. Because each orbital can hold two electrons, each shell can accommodate up to $2n^2$ electrons: 2, 8, 18, or 32. The structure of the periodic table reflects this orbital structure. The first row has two elements, the second and third rows have eight elements, the fourth and fifth rows have 18 elements, and the sixth row has 32 elements (including the lanthanide metals). Some of the numbers repeat because we reach the end of a row of the periodic table before a shell completely fills. For example, the third row has eight elements, which corresponds to filling the $3s$ and $3p$ orbitals. The remaining orbitals of the third shell, the $3d$ orbitals, do not begin to fill until the fourth row of the periodic table (and after the $4s$ orbital is filled). Likewise, the $4d$ orbitals don't begin to fill until the fifth row of the table, and the $4f$ orbitals don't begin filling until the sixth row.

All these observations are evident in the structure of the periodic table. For this reason, we will emphasize that *the periodic table is your best guide to the order in which orbitals are filled.* You can easily write the electron configuration of an element based on its location in the periodic table. The pattern is summarized in Figure 6.27 ▼. Notice that the elements can be grouped by the *type* of orbital into which the electrons are placed. On the left are *two* columns of elements.

TABLE 6.4 Electron Configurations of the Group 2A and 3A Elements

Group 2A	
Be	[He]$2s^2$
Mg	[Ne]$3s^2$
Ca	[Ar]$4s^2$
Sr	[Kr]$5s^2$
Ba	[Xe]$6s^2$
Ra	[Rn]$7s^2$
Group 3A	
B	[He]$2s^22p^1$
Al	[Ne]$3s^23p^1$
Ga	[Ar]$3d^{10}4s^24p^1$
In	[Kr]$4d^{10}5s^25p^1$
Tl	[Xe]$4f^{14}5d^{10}6s^26p^1$

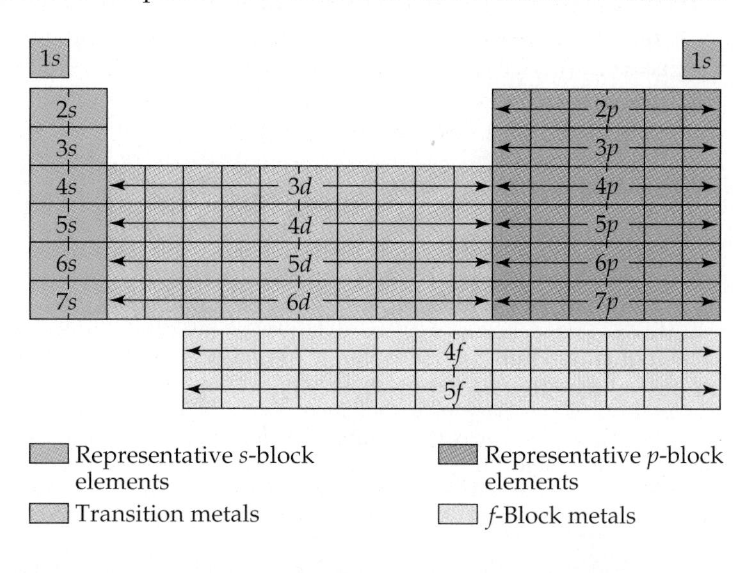

◀ **Figure 6.27** Block diagram of the periodic table showing the groupings of the elements according to the type of orbital being filled with electrons.

These elements, known as the alkali metals (group 1A) and alkaline earth metals (group 2A), are those in which the outer-shell *s* orbitals are being filled. On the right is a block of *six* columns. These are the elements in which the outermost *p* orbitals are being filled. The *s* block and the *p* block of the periodic table contain the **representative** (or **main-group**) **elements**. In the middle of the table is a block of *ten* columns that contain the transition metals. These are the elements in which the *d* orbitals are being filled. Below the main portion of the table are two rows that contain *14* columns. These elements are often referred to as the ***f*-block metals** because they are the ones in which the *f* orbitals are being filled. Recall that the numbers 2, 6, 10, and 14 are precisely the number of electrons that can fill the *s*, *p*, *d*, and *f* subshells, respectively. Recall also that the 1*s* subshell is the first *s* subshell, the 2*p* is the first *p* subshell, the 3*d* is the first *d* subshell, and the 4*f* is the first *f* subshell.

SAMPLE EXERCISE 6.8

What is the characteristic outer-shell electron configuration of the group 7A elements, the halogens?

Solution

Analyze and Plan: We first locate the halogens in the periodic table, write the electron configurations for the first two elements, and then determine the general similarity between them.

Solve: The first member of the halogen group is fluorine, atomic number 9. The abbreviated form of the electron configuration for fluorine is

F: $[He]2s^22p^5$

Similarly, the abbreviated form of the electron configuration for chlorine, the second halogen, is

Cl: $[Ne]3s^23p^5$

From these two examples, we see that the characteristic outer-shell electron configuration of a halogen is ns^2np^5, where n ranges from 2 in the case of fluorine to 6 in the case of astatine.

PRACTICE EXERCISE

What family of elements is characterized by having an ns^2np^2 outer-electron configuration?
Answer: group 4A

SAMPLE EXERCISE 6.9

(a) Write the complete electron configuration for bismuth, element number 83. **(b)** Write the condensed electron configuration for this element, showing the appropriate noble-gas core. **(c)** How many unpaired electrons does each atom of bismuth possess?

Solution **(a)** We write the complete electron configuration by simply moving across the periodic table one row at a time and writing the occupancies of the orbital corresponding to each row (refer to Figure 6.27).

First row	$1s^2$
Second row	$2s^22p^6$
Third row	$3s^23p^6$
Fourth row	$4s^23d^{10}4p^6$
Fifth row	$5s^24d^{10}5p^6$
Sixth row	$6s^24f^{14}5d^{10}6p^3$
Total:	$1s^22s^22p^63s^23p^63d^{10}4s^24p^64d^{10}4f^{14}5s^25p^65d^{10}6s^26p^3$

Note that 3 is the lowest possible value that n may have for a d orbital and that 4 is the lowest possible value of n for an f orbital.

The total of the superscripted numbers should equal the atomic number of bismuth, 83. The electrons may be listed, as shown here, in the order of increasing major quantum number. However, it is equally correct to list the orbitals in an electron configuration in the order in which they are read from the periodic table: $1s^22s^22p^63s^23p^64s^23d^{10}4p^65s^24d^{10}5p^66s^24f^{14}5d^{10}6p^3$.

(b) We can use the periodic table to write the condensed electron configuration of an element. First locate the element of interest (in this case element 83) and then move backward until the first noble gas is encountered (in this case Xe, element 54). Thus the noble gas core is [Xe]. The outer electrons are then read from the periodic table as before. Moving from Xe to Cs, element 55, we find ourselves in the sixth row. Moving across this row to Bi gives us the outer electrons. Thus the abbreviated electron configuration is as follows: $[Xe]6s^2 4f^{14} 5d^{10} 6p^3$ or $[Xe]4f^{14} 5d^{10} 6s^2 6p^3$.

(c) We can see from the abbreviated electron configuration that the only partially occupied subshell is the $6p$. The orbital diagram representation for this subshell is as follows:

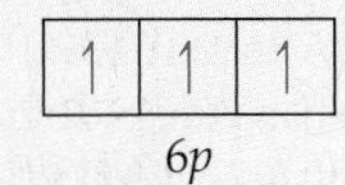

In accordance with Hund's rule, the three $6p$ electrons occupy three $6p$ orbitals singly, with their spins parallel. Thus, there are three unpaired electrons in each atom of bismuth.

PRACTICE EXERCISE

Use the periodic table to write the condensed electron configurations for the following atoms: **(a)** Co (atomic number 27); **(b)** Te (atomic number 52).

Answers: **(a)** $[Ar]4s^2 3d^7$ or $[Ar]3d^7 4s^2$; **(b)** $[Kr]5s^2 4d^{10} 5p^4$ or $[Kr]4d^{10} 5s^2 5p^4$

Figure 6.28 ▼ gives the outer-shell ground-state electron configurations of the elements. You can use this figure to check your answers as you practice writing

Core	1A 1	2A 2	3B 3	4B 4	5B 5	6B 6	7B 7	8	8B 9	10	1B 11	2B 12	3A 13	4A 14	5A 15	6A 16	7A 17	8A 18
	1 H $1s^1$																	2 He $1s^2$
[He]	3 Li $2s^1$	4 Be $2s^2$											5 B $2s^2 2p^1$	6 C $2s^2 2p^2$	7 N $2s^2 2p^3$	8 O $2s^2 2p^4$	9 F $2s^2 2p^5$	10 Ne $2s^2 2p^6$
[Ne]	11 Na $3s^1$	12 Mg $3s^2$											13 Al $3s^2 3p^1$	14 Si $3s^2 3p^2$	15 P $3s^2 3p^3$	16 S $3s^2 3p^4$	17 Cl $3s^2 3p^5$	18 Ar $3s^2 3p^6$
[Ar]	19 K $4s^1$	20 Ca $4s^2$	21 Sc $3d^1 4s^2$	22 Ti $3d^2 4s^2$	23 V $3d^3 4s^2$	24 Cr $3d^5 4s^1$	25 Mn $3d^5 4s^2$	26 Fe $3d^6 4s^2$	27 Co $3d^7 4s^2$	28 Ni $3d^8 4s^2$	29 Cu $3d^{10} 4s^1$	30 Zn $3d^{10} 4s^2$	31 Ga $3d^{10} 4s^2 4p^1$	32 Ge $3d^{10} 4s^2 4p^2$	33 As $3d^{10} 4s^2 4p^3$	34 Se $3d^{10} 4s^2 4p^4$	35 Br $3d^{10} 4s^2 4p^5$	36 Kr $3d^{10} 4s^2 4p^6$
[Kr]	37 Rb $5s^1$	38 Sr $5s^2$	39 Y $4d^1 5s^2$	40 Zr $4d^2 5s^2$	41 Nb $4d^3 5s^2$	42 Mo $4d^5 5s^1$	43 Tc $4d^5 5s^2$	44 Ru $4d^7 5s^1$	45 Rh $4d^8 5s^1$	46 Pd $4d^{10}$	47 Ag $4d^{10} 5s^1$	48 Cd $4d^{10} 5s^2$	49 In $4d^{10} 5s^2 5p^1$	50 Sn $4d^{10} 5s^2 5p^2$	51 Sb $4d^{10} 5s^2 5p^3$	52 Te $4d^{10} 5s^2 5p^4$	53 I $4d^{10} 5s^2 5p^5$	54 Xe $4d^{10} 5s^2 5p^6$
[Xe]	55 Cs $6s^1$	56 Ba $6s^2$	71 Lu $4f^{14} 5d^1 6s^2$	72 Hf $4f^{14} 5d^2 6s^2$	73 Ta $4f^{14} 5d^3 6s^2$	74 W $4f^{14} 5d^4 6s^2$	75 Re $4f^{14} 5d^5 6s^2$	76 Os $4f^{14} 5d^6 6s^2$	77 Ir $4f^{14} 5d^7 6s^2$	78 Pt $4f^{14} 5d^9 6s^1$	79 Au $4f^{14} 5d^{10} 6s^1$	80 Hg $4f^{14} 5d^{10} 6s^2$	81 Tl $4f^{14} 5d^{10} 6s^2 6p^1$	82 Pb $4f^{14} 5d^{10} 6s^2 6p^2$	83 Bi $4f^{14} 5d^{10} 6s^2 6p^3$	84 Po $4f^{14} 5d^{10} 6s^2 6p^4$	85 At $4f^{14} 5d^{10} 6s^2 6p^5$	86 Rn $4f^{14} 5d^{10} 6s^2 6p^6$
[Rn]	87 Fr $7s^1$	88 Ra $7s^2$	103 Lr $5f^{14} 6d^1 7s^2$	104 Rf $5f^{14} 6d^2 7s^2$	105 Db $5f^{14} 6d^3 7s^2$	106 Sg $5f^{14} 6d^4 7s^2$	107 Bh $5f^{14} 6d^5 7s^2$	108 Hs $5f^{14} 6d^6 7s^2$	109 Mt $5f^{14} 6d^7 7s^2$	110	111	112		114		116		

Core	Series														
[Xe]	Lanthanide series	57 La $5d^1 6s^2$	58 Ce $4f^1 5d^1 6s^2$	59 Pr $4f^3 6s^2$	60 Nd $4f^4 6s^2$	61 Pm $4f^5 6s^2$	62 Sm $4f^6 6s^2$	63 Eu $4f^7 6s^2$	64 Gd $4f^7 5d^1 6s^2$	65 Tb $4f^9 6s^2$	66 Dy $4f^{10} 6s^2$	67 Ho $4f^{11} 6s^2$	68 Er $4f^{12} 6s^2$	69 Tm $4f^{13} 6s^2$	70 Yb $4f^{14} 6s^2$
[Rn]	Actinide series	89 Ac $6d^1 7s^2$	90 Th $6d^2 7s^2$	91 Pa $5f^2 6d^1 7s^2$	92 U $5f^3 6d^1 7s^2$	93 Np $5f^4 6d^1 7s^2$	94 Pu $5f^6 7s^2$	95 Am $5f^7 7s^2$	96 Cm $5f^7 6d^1 7s^2$	97 Bk $5f^9 7s^2$	98 Cf $5f^{10} 7s^2$	99 Es $5f^{11} 7s^2$	100 Fm $5f^{12} 7s^2$	101 Md $5f^{13} 7s^2$	102 No $5f^{14} 7s^2$

Metals | Metalloids | Nonmetals

▲ **Figure 6.28** Outer-shell ground-state electron configurations.

electron configurations. We have written these configurations with orbitals listed in order of increasing principal quantum number. As we have seen in Sample Exercise 6.9, the orbitals can also be listed in order of filling, as they would be read off the periodic table.

Anomalous Electron Configurations

If you inspect Figure 6.28 closely, you will see that the electron configurations of certain elements appear to violate the rules we have just discussed. For example, the electron configuration of chromium is $[Ar]3d^54s^1$ rather than $[Ar]3d^44s^2$, as we might have expected. Similarly, the configuration of copper is $[Ar]3d^{10}4s^1$ instead of $[Ar]3d^94s^2$. This anomalous behavior is largely a consequence of the closeness of the 3*d* and 4*s* orbital energies. It frequently occurs when there are enough electrons to lead to precisely half-filled sets of degenerate orbitals (as in chromium) or to a completely filled *d* subshell (as in copper). There are a few similar cases among the heavier transition metals (those with partially filled 4*d* or 5*d* orbitals) and among the *f*-block metals. Although these minor departures from the expected are interesting, they are not of great chemical significance.

SAMPLE INTEGRATIVE EXERCISE 6: Putting Concepts Together

Boron, atomic number 5, occurs naturally as two isotopes, ^{10}B and ^{11}B, with natural abundances of 19.9% and 80.1%, respectively. **(a)** In what ways do the two isotopes differ? Do the electronic configurations of ^{10}B and ^{11}B differ? **(b)** Draw the complete orbital diagram representation for an atom of ^{11}B. Which electrons are the valence electrons (the ones involved in chemical reactions)? **(c)** Indicate three major ways in which the 1*s* and 2*s* electrons in boron differ. **(d)** Elemental boron reacts with fluorine to form BF_3, a gas. Write a balanced chemical equation for the reaction of solid boron with fluorine gas. **(e)** ΔH_f° for $BF_3(g)$ is -1135.6 kJ mol^{-1}. Calculate the standard enthalpy change in the reaction of boron with fluorine. **(f)** When BCl_3, also a gas at room temperature, comes into contact with water, it reacts to form hydrochloric acid and boric acid, H_3BO_3, a very weak acid in water. Write a balanced net ionic equation for this reaction.

Solution **(a)** The two nuclides of boron differ in the number of neutrons in the nucleus. (Sections 2.3 and 2.4) Each of the nuclides contains five protons, but ^{10}B contains five neutrons, whereas ^{11}B contains six neutrons. The two isotopes of boron have identical electron configurations, $1s^22s^22p^1$, because each has five electrons.

(b) The complete orbital diagram is

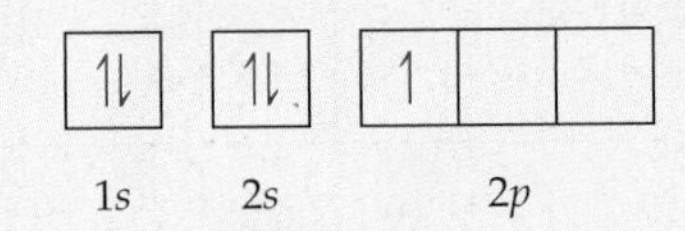

The valence electrons are the outer-shell ones, the $2s^2$ and $2p^1$ electrons. The $1s^2$ electrons constitute the core electrons, which we represent as [He] when we write the condensed electron configuration, $[He]2s^22p^1$.

(c) The 1*s* and 2*s* orbitals are both spherical, but they differ in three important respects: First, the 1*s* orbital is lower in energy than the 2*s* orbital. Second, the average distance of the 2*s* electrons from the nucleus is greater than that of the 1*s* electrons, so the 1*s* orbital is smaller than the 2*s*. Third, the 2*s* orbital has one radial node, whereas the 1*s* orbital has no nodes (Figure 6.18).

(d) The balanced chemical equation is as follows:

$$2B(s) + 3F_2(g) \longrightarrow 2BF_3(g)$$

(e) $\Delta H^\circ = 2(-1135.6) - [0 + 0] = -2271.2$ kJ. The reaction is strongly exothermic.

(f) $BCl_3(g) + 3H_2O(l) \longrightarrow H_3BO_3(aq) + 3H^+(aq) + 3Cl^-(aq)$. Note that because H_3BO_3 is a very weak acid, its chemical formula is written in molecular form, as discussed in Section 4.4.

Summary and Key Terms

Introduction and Section 6.1 The **electronic structure** of an atom describes the energies and arrangement of electrons around the atom. Much of what is known about the electronic structure of atoms was obtained by observing the interaction of light with matter. Visible light and other forms of **electromagnetic radiation** (also known as radiant energy) move through a vacuum at the speed of light, $c = 3.00 \times 10^8$ m/s. Electromagnetic radiation has both electric and magnetic components that vary periodically in wavelike fashion. The wave characteristics of radiant energy allow it to be described in terms of **wavelength**, λ, and **frequency**, ν, which are interrelated: $\lambda\nu = c$.

Section 6.2 Planck proposed that the minimum amount of radiant energy that an object can gain or lose is related to the frequency of the radiation: $E = h\nu$. This smallest quantity is called a **quantum** of energy. The constant h is called **Planck's constant;** $h = 6.63 \times 10^{-34}$ J-s. In the quantum theory, energy is quantized, meaning that it can have only certain allowed values. Einstein used the quantum theory to explain the **photoelectric effect**, the emission of electrons from metal surfaces by light. He proposed that light behaves as if it consisted of quantized energy packets called **photons**. Each photon carries energy $E = h\nu$.

Section 6.3 Dispersion of radiation into its component wavelengths produces a **spectrum**. If the spectrum contains all wavelengths, it is called a **continuous spectrum**; if it contains only certain specific wavelengths, the spectrum is called a **line spectrum**. The radiation emitted by excited hydrogen atoms forms a line spectrum; the frequencies observed in the spectrum follow a simple mathematical relationship that involves small integers.

Bohr proposed a model of the hydrogen atom that explains its line spectrum. In this model the energy of the hydrogen atom depends on the value of a number n, called the quantum number. The value of n must be a positive integer $(1, 2, 3, \ldots)$, and each value of n corresponds to a different specific energy, E_n. The energy of the atom increases as n increases. The lowest energy is achieved for $n = 1$; this is called the **ground state** of the hydrogen atom. Other values of n correspond to **excited states** of the atom. Light is emitted when the electron drops from a higher energy state to a lower energy state; light must be absorbed to excite the electron from a lower energy state to a higher one. The frequency of light emitted or absorbed must be such that $h\nu$ equals the difference in energy between two allowed states of the atom.

Section 6.4 De Broglie proposed that matter, such as electrons, should exhibit wavelike properties; this hypothesis of **matter waves** was proven experimentally by observing the diffraction of electrons. An object has a characteristic wavelength that depends on its **momentum**, mv: $\lambda = h/mv$. Discovery of the wave properties of the electron led to Heisenberg's **uncertainty principle**, which states that there is an inherent limit to the accuracy with which the position and momentum of a particle can be measured simultaneously.

Section 6.5 In the quantum mechanical model of the hydrogen atom, the behavior of the electron is described by mathematical functions called **wave functions**, denoted with the Greek letter ψ. Each allowed wave function has a precisely known energy, but the location of the electron cannot be determined exactly; rather, the probability of its being at a particular point in space is given by the **probability density**, ψ^2. The **electron density** distribution is a map of the probability of finding the electron at all points in space.

The allowed wave functions of the hydrogen atom are called **orbitals**. An orbital is described by a combination of an integer and a letter, corresponding to values of three quantum numbers for the orbital. The principal quantum number, n, is indicated by the integers $1, 2, 3, \ldots$. This quantum number relates most directly to the size and energy of the orbital. The azimuthal quantum number, l, is indicated by the letters s, p, d, f, and so on, corresponding to the values of $0, 1, 2, 3, \ldots$. The l quantum number defines the shape of the orbital. For a given value of n, l can have integer values ranging from 0 to $n - 1$. The magnetic quantum number, m_l, relates to the orientation of the orbital in space. For a given value of l, m_l can have integral values ranging from $-l$ to l. Cartesian labels can be used to label the orientations of the orbitals. For example, the three $3p$ orbitals are designated $3p_x$, $3p_y$, and $3p_z$, with the subscripts indicating the axis along which the orbital is oriented.

An **electron shell** is the set of all orbitals with the same value of n, such as $3s$, $3p$, and $3d$. In the hydrogen atom all the orbitals in an electron shell have the same energy. A **subshell** is the set of one or more orbitals with the same n and l values; for example, $3s$, $3p$, and $3d$ are each subshells of the $n = 3$ shell. There is one orbital in an s subshell, three in a p subshell, five in a d subshell, and seven in an f subshell.

Section 6.6 Contour representations are useful for visualizing the spatial characteristics (shapes) of the orbitals. Represented this way, s orbitals appear as spheres that increase in size as n increases. The wave function for each p orbital has two lobes on opposite sides of the nucleus. They are oriented along the x-, y-, and z-axes. Four of the d orbitals appear as shapes with four lobes around the nucleus; the fifth one, the d_{z^2} orbital, is represented as two lobes along the z-axis and a "doughnut" in the xy plane. Regions in which the wave function is zero are called **nodes**. There is zero probability that the electron will be found at a node.

Section 6.7 In many-electron atoms, different subshells of the same electron shell have different energies. The energy of the subshells increases in the order

$$1s, 2s, 2p, 3s, 3p, 4s, 3d, 4p, \ldots$$

Orbitals within the same subshell are still **degenerate**, meaning they have the same energy.

Electrons have an intrinsic property called **electron spin**, which is quantized. The **spin magnetic quantum number**, m_s, can have two possible values, $+\frac{1}{2}$ and $-\frac{1}{2}$, which can be envisioned as the two directions of an electron spinning about an axis. The **Pauli exclusion principle** states that no two electrons in an atom can have the same values for n, l, m_l, and m_s. This principle places a limit of two on the number of electrons that can occupy any one atomic orbital. These two electrons differ in their value of m_s.

Sections 6.8 and 6.9 The **electron configuration** of an atom describes how the electrons are distributed among the orbitals of the atom. The ground-state electron configurations are generally obtained by placing the electrons in the atomic orbitals of lowest possible energy with the restriction that each orbital can hold no more than two electrons. When electrons occupy a subshell with more than one degenerate orbital, such as the $2p$ subshell, **Hund's rule** states that the lowest energy is attained by maximizing the number of electrons with the same electron spin. For example, in the ground-state electron configuration of carbon, the two $2p$ electrons have the same spin and must occupy two different $2p$ orbitals.

Elements in any given group in the periodic table have the same type of electron arrangements in their outermost shells. For example, the electron configurations of the halogens fluorine and chlorine are $[He]2s^2 2p^5$ and $[Ne]3s^2 3p^5$, respectively. The outer-shell electrons, those that lie outside the orbitals occupied in the next lowest noble-gas element, are called its **valence electrons**, whereas the electrons in the inner shells are called the **core electrons**.

The periodic table is partitioned into different types of elements, based on their electron configurations. Those elements in which the outermost subshell is an s or p subshell are called the **representative** (or **main-group**) **elements**. The alkali metals (group 1A), halogens (group 7A), and noble gases (group 8A) are representative elements. Those elements in which a d subshell is being filled are called the **transition elements** (or **transition metals**). The elements in which the $4f$ subshell is being filled are called the **lanthanide elements**. The **actinide elements** are those in which the $5f$ subshell is being filled. The lanthanide and actinide elements are collectively referred to as the **f-block metals**. These elements are shown as two rows of 14 elements below the main part of the periodic table. The structure of the periodic table, summarized in Figure 6.27, allows us to write the electron configuration of an element from its position in the periodic table.

Exercises

Radiant Energy

6.1 What are the basic SI units for **(a)** the wavelength of light, **(b)** the frequency of light, **(c)** the speed of light?

6.2 **(a)** What is the relationship between the wavelength and the frequency of radiant energy? **(b)** Ozone in the upper atmosphere absorbs energy in the 210–230 nm range of the spectrum. In what region of the electromagnetic spectrum does this radiation occur?

6.3 Label each of the following statements as true or false. For those that are false, correct the statement. **(a)** Visible light is a form of electromagnetic radiation. **(b)** The frequency of radiation increases as the wavelength increases. **(c)** Ultraviolet light has longer wavelengths than visible light. **(d)** Electromagnetic radiation and sound waves travel at the same speed.

6.4 Determine which of the following statements are false, and correct them. **(a)** Electromagnetic radiation is incapable of passing through water. **(b)** Electromagnetic radiation travels through a vacuum at a constant speed, regardless of wavelength. **(c)** Infrared light has higher frequencies than visible light. **(d)** The glow from a fireplace, the energy within a microwave oven, and a foghorn blast are all forms of electromagnetic radiation.

6.5 Arrange the following kinds of electromagnetic radiation in order of increasing wavelength: infrared, green light, red light, radio waves, X rays, ultraviolet light.

6.6 List the following types of electromagnetic radiation in order of increasing wavelength: **(a)** the gamma rays produced by a radioactive nuclide used in medical imaging; **(b)** radiation from an FM radio station at 93.1 MHz on the dial; **(c)** a radio signal from an AM radio station at 680 kHz on the dial; **(d)** the yellow light from sodium vapor streetlights; **(e)** the red light of a light-emitting diode, such as in a calculator display.

6.7 **(a)** What is the frequency of radiation that has a wavelength of 0.452 pm? **(b)** What is the wavelength of radiation that has a frequency of $2.55 \times 10^{16}\ s^{-1}$? **(c)** Would the radiations in part (a) or part (b) be visible to the human eye? **(d)** What distance does electromagnetic radiation travel in 7.50 ms?

6.8 **(a)** What is the frequency of radiation whose wavelength is 589 nm? **(b)** What is the wavelength of radiation that has a frequency of $1.2 \times 10^{13}\ s^{-1}$? **(c)** Would the radiations in part (a) or part (b) be detected by an infrared radiation detector? **(d)** What distance does electromagnetic radiation travel in 10.0 μs?

6.9 Excited mercury atoms emit light strongly at a wavelength of 436 nm. What is the frequency of this radiation? Using Figure 6.4, predict the color associated with this wavelength.

6.10 An argon ion laser emits light at 489 nm. What is the frequency of this radiation? Is this emission in the visible spectrum? If yes, what color is it?

Quantized Energy and Photons

6.11 **(a)** What does it mean when we say energy is quantized? **(b)** Why don't we notice the quantization of energy in everyday activities?

6.12 Einstein's 1905 paper on the photoelectric effect was the first important application of Planck's quantum hypothesis. Describe Planck's original hypothesis, and explain how Einstein made use of it in his theory of the photoelectric effect.

6.13 **(a)** Calculate the smallest increment of energy (a quantum) that can be emitted or absorbed at a wavelength of 812 nm. **(b)** Calculate the energy of a photon of frequency $2.72 \times 10^{13}\ s^{-1}$. **(c)** What wavelength of radiation has photons of energy 7.84×10^{-18} J? In what portion of the electromagnetic spectrum would this radiation be found?

6.14 **(a)** Calculate the smallest increment of energy that can be emitted or absorbed at a wavelength of 3.80 mm. **(b)** Calculate the energy of a photon of frequency 80.5 MHz. **(c)** What frequency of radiation has photons of energy 1.77×10^{-19} J? In what region of the electromagnetic spectrum would this radiation be found?

6.15 **(a)** Calculate and compare the energy of a photon of wavelength 3.3 μm with that of wavelength 0.154 nm. **(b)** Use Figure 6.4 to identify the region of the electromagnetic spectrum to which each belongs.

6.16 An AM radio station broadcasts at 1440 kHz, and its FM partner broadcasts at 94.5 MHz. Calculate and compare the energy of the photons emitted by these two radio stations.

6.17 One type of sunburn occurs on exposure to UV light of wavelength in the vicinity of 325 nm. **(a)** What is the energy of a photon of this wavelength? **(b)** What is the energy of a mole of these photons? **(c)** How many photons are in a 1.00 mJ burst of this radiation?

6.18 The energy from radiation can be used to cause the rupture of chemical bonds. A minimum energy of 495 kJ/mol is required to break the oxygen–oxygen bond in O_2. What is the longest wavelength of radiation that possesses the necessary energy to break the bond? What type of electromagnetic radiation is this?

6.19 A diode laser emits at a wavelength of 987 nm. All of its output energy is absorbed in a detector that measures a total energy of 0.52 J over a period of 32 s. How many photons per second are being emitted by the laser?

6.20 A stellar object is emitting radiation at 1350 nm. If the detector is capturing 8×10^7 photons per second at this wavelength, what is the total energy of the photons detected in one hour?

6.21 Molybdenum metal must absorb radiation with a minimum frequency of $1.09 \times 10^{15}\ s^{-1}$ before it can emit an electron from its surface via the photoelectric effect. **(a)** What is the minimum energy needed to produce this effect? **(b)** What wavelength radiation will provide a photon of this energy? **(c)** If molybdenum is irradiated with light of wavelength of 120 nm, what is the maximum possible kinetic energy of the emitted electrons?

6.22 It requires a photon with a minimum energy of 4.41×10^{-19} J to emit electrons from sodium metal. **(a)** What is the minimum frequency of light necessary to emit electrons from sodium via the photoelectric effect? **(b)** What is the wavelength of this light? **(c)** If sodium is irradiated with light of 439 nm, what is the maximum possible kinetic energy of the emitted electrons? **(d)** What is the maximum number of electrons that can be freed by a burst of light whose total energy is 1.00 μJ?

Bohr's Model; Matter Waves

6.23 Explain how the existence of line spectra is consistent with Bohr's theory of quantized energies for the electron in the hydrogen atom.

6.24 **(a)** In terms of the Bohr theory of the hydrogen atom, what process is occurring when excited hydrogen atoms emit radiant energy of certain wavelengths and only those wavelengths? **(b)** Does a hydrogen atom "expand" or "contract" as it moves from its ground state to an excited state?

6.25 Is energy emitted or absorbed when the following electronic transitions occur in hydrogen? **(a)** from $n = 4$ to $n = 2$; **(b)** from an orbit of radius 2.12 Å to one of radius 8.48 Å; **(c)** an electron adds to the H^+ ion and ends up in the $n = 3$ shell.

6.26 Indicate whether energy is emitted or absorbed when the following electronic transitions occur in hydrogen: **(a)** from $n = 2$ to $n = 6$; **(b)** from an orbit of radius 4.77 Å to one of radius 0.530 Å; **(c)** from the $n = 6$ to the $n = 9$ state.

6.27 Using Equation 6.5, calculate the energy of an electron in the hydrogen atom when $n = 2$, and when $n = 6$. Calculate the wavelength of the radiation released when an electron moves from $n = 6$ to $n = 2$. Is this line in the visible region of the electromagnetic spectrum? If so, what color is it?

6.28 For each of the following electronic transitions in the hydrogen atom, calculate the energy, frequency, and wavelength of the associated radiation, and determine whether the radiation is emitted or absorbed during the transition: **(a)** from $n = 5$ to $n = 1$; **(b)** from $n = 4$ to $n = 2$; **(c)** from $n = 4$ to $n = 6$. Does any of these transitions emit or absorb visible light?

6.29 The visible emission lines observed by Balmer all involved $n_f = 2$. **(a)** Explain why only the lines with $n_f = 2$ were observed in the visible region of the electromagnetic spectrum. **(b)** Calculate the wavelengths of the first three lines in the Balmer series—those for which $n_i = 3, 4$, and 5—and identify these lines in the emission spectrum shown in Figure 6.12.

6.30 The Lyman series of emission lines of the hydrogen atom are those for which $n_f = 1$. **(a)** Determine the region of the electromagnetic spectrum in which the lines of the Lyman series are observed. **(b)** Calculate the wavelengths of the first three lines in the Lyman series—those for which $n_i = 2, 3$, and 4.

[6.31] One of the emission lines of the hydrogen atom has a wavelength of 93.8 nm. **(a)** In what region of the electromagnetic spectrum is this emission found? **(b)** Determine the initial and final values of n associated with this emission.

[6.32] The hydrogen atom can absorb light of wavelength 4055 nm. **(a)** In what region of the electromagnetic spectrum is this absorption found? **(b)** Determine the initial and final values of n associated with this absorption.

6.33 Use the de Broglie relationship to determine the wavelengths of the following objects: **(a)** an 85-kg person skiing at 50 km/hr; **(b)** a 10.0-g bullet fired at 250 m/s; **(c)** a lithium atom moving at 2.5×10^5 m/s.

6.34 Among the elementary subatomic particles of physics is the muon, which decays within a few nanoseconds after formation. The muon has a rest mass 206.8 times that of an electron. Calculate the de Broglie wavelength associated with a muon traveling at a velocity of 8.85×10^5 cm/s.

6.35 Neutron diffraction is an important technique for determining the structures of molecules. Calculate the velocity of a neutron that has a characteristic wavelength of 0.955 Å. (Refer to the back inside cover for the mass of the neutron.)

6.36 The electron microscope has been widely used to obtain highly magnified images of biological and other types of materials. When an electron is accelerated through a particular potential field, it attains a speed of 5.93×10^6 m/s. What is the characteristic wavelength of this electron? Is the wavelength comparable to the size of atoms?

6.37 Using Heisenberg's uncertainty principle, calculate the uncertainty in the position of **(a)** a 1.50-mg mosquito moving at a speed of 1.40 m/s if the speed is known to within ± 0.01 m/s; **(b)** a proton moving at a speed of $(5.00 \pm 0.01) \times 10^4$ m/s. (The mass of a proton is given in the table of fundamental constants in the back inside cover of the text.)

6.38 Calculate the uncertainty in the position of **(a)** an electron moving at a speed of $(3.00 \pm 0.01) \times 10^5$ m/s; **(b)** a neutron moving at this same speed. (The masses of an electron and a neutron are given in the table of fundamental constants in the back inside cover of the text.) **(c)** What are the implications of these calculations to our model of the atom?

Quantum Mechanics and Atomic Orbitals

6.39 According to the Bohr model, an electron in the ground state of a hydrogen atom orbits the nucleus at a specific radius of 0.53 Å. In the quantum mechanical description of the hydrogen atom, the most probable distance of the electron from the nucleus is 0.53 Å. Why are these two statements different?

6.40 **(a)** In the quantum mechanical description of the hydrogen atom, what is the physical significance of the square of the wave function, ψ^2? **(b)** What is meant by the expression "electron density"? **(c)** What is an orbital?

6.41 **(a)** For $n = 4$, what are the possible values of l? **(b)** For $l = 2$, what are the possible values of m_l?

6.42 How many possible values for l and m_l are there when **(a)** $n = 3$; **(b)** $n = 5$?

6.43 Give the numerical values of n and l corresponding to each of the following designations: **(a)** $3p$; **(b)** $2s$; **(c)** $4f$; **(d)** $5d$.

6.44 Give the values for n, l, and m_l for **(a)** each orbital in the $2p$ subshell; **(b)** each orbital in the $5d$ subshell.

6.45 Which of the following represent impossible combinations of n and l: **(a)** $1p$, **(b)** $4s$; **(c)** $5f$; **(d)** $2d$?

6.46 Which of the following are permissible sets of quantum numbers for an electron in a hydrogen atom: **(a)** $n = 2, l = 1, m_l = 1$; **(b)** $n = 1, l = 0, m_l = -1$; **(c)** $n = 4, l = 2, m_l = -2$; **(d)** $n = 3, l = 3, m_l = 0$? For those combinations that are permissible, write the appropriate designation for the subshell to which the orbital belongs (that is, $1s$, and so on).

6.47 Sketch the shape and orientation of the following types of orbitals: **(a)** s; **(b)** p_z; **(c)** d_{xy}.

6.48 Sketch the shape and orientation of the following types of orbitals: **(a)** p_x; **(b)** d_{z^2}; **(c)** $d_{x^2-y^2}$.

6.49 **(a)** What are the similarities and differences between the hydrogen atom $1s$ and $2s$ orbitals? **(b)** In what sense does a $2p$ orbital have directional character? Compare the "directional" characteristics of the p_x and $d_{x^2-y^2}$ orbitals (that is, in what direction or region of space is the electron density concentrated?). **(c)** What can you say about the average distance from the nucleus of an electron in a $2s$ orbital as compared with a $3s$ orbital? **(d)** For the hydrogen atom, list the following orbitals in order of increasing energy (that is, most stable ones first): $4f$, $6s$, $3d$, $1s$, $2p$.

6.50 **(a)** With reference to Figure 6.18, what is the relationship between the number of nodes in an s orbital and the value of the principal quantum number? **(b)** Identify the number of nodes; that is, identify places where the electron density is zero, in the $2p_x$ orbital; in the $3s$ orbital. **(c)** The nodes in s orbitals are spherical surfaces (Figure 6.18). What kind of surface do you expect the nodes to be in the p orbitals (Figure 6.20)? **(d)** For the hydrogen atom, list the following orbitals in order of increasing energy: $3s$, $2s$, $2p$, $5s$, $4d$.

Many-Electron Atoms and Electron Configurations

6.51 For a given value of the principal quantum number, n, how do the energies of the s, p, d, and f subshells vary for **(a)** hydrogen; **(b)** a many-electron atom?

6.52 **(a)** The average distance from the nucleus of a $3s$ electron in a chlorine atom is smaller than that for a $3p$ electron. In light of this fact, which orbital is higher in energy? **(b)** Would you expect it to require more or less energy to remove a $3s$ electron from the chlorine atom, as compared with a $2p$ electron? Explain.

6.53 **(a)** What are the possible values of the electron spin quantum number? **(b)** What piece of experimental equipment can be used to distinguish electrons that have different values of the electron spin quantum number? **(c)** Two electrons in an atom both occupy the 1*s* orbital. What quantity must be different for the two electrons? What principle governs the answer to this question?

6.54 **(a)** State the Pauli exclusion principle in your own words. **(b)** The Pauli exclusion principle is, in an important sense, the key to understanding the periodic table. Explain why.

6.55 What is the maximum number of electrons that can occupy each of the following subshells: **(a)** 3*d*; **(b)** 4*s*; **(c)** 2*p*; **(d)** 5*f*?

6.56 What is the maximum number of electrons in an atom that can have the following quantum numbers: **(a)** $n = 2, m_s = -\frac{1}{2}$; **(b)** $n = 5, l = 3$; **(c)** $n = 4, l = 3, m_l = -3$; **(d)** $n = 4, l = 1, m_l = 1$.

6.57 **(a)** What does each box in an orbital diagram represent? **(b)** What quantity is represented by the direction (either up or down) of the half arrows in an orbital diagram? **(c)** Is Hund's rule needed to write the electron configuration of beryllium? Explain.

6.58 **(a)** What are "outer-shell electrons"? **(b)** What are "unpaired electrons"? **(c)** How many outer-shell electrons does an Si atom possess? How many of these are unpaired?

6.59 Write the condensed electron configurations for the following atoms, using the appropriate noble-gas core abbreviations: **(a)** Cs; **(b)** Ni; **(c)** Se; **(d)** Cd; **(e)** Ac; **(f)** Pb.

6.60 Write the condensed electron configurations for the following atoms: **(a)** Al; **(b)** Sc; **(c)** Co; **(d)** Br; **(e)** Ba **(f)** Re; **(g)** Lu.

6.61 Draw the orbital diagrams for the valence electrons of each of the following elements, and indicate how many unpaired electrons each has: **(a)** S; **(b)** Sr; **(c)** Fe; **(d)** Zr; **(e)** Sb; **(f)** U.

6.62 Using orbital diagrams, determine the number of unpaired electrons in each of the following atoms: **(a)** Ti; **(b)** Ga; **(c)** Rh; **(d)** I; **(e)** Po.

6.63 Identify the specific element that corresponds to each of the following electron configurations: **(a)** $1s^2 2s^2 2p^6 3s^2$; **(b)** $[Ne]3s^2 3p^1$; **(c)** $[Ar]4s^1 3d^5$; **(d)** $[Kr]5s^2 4d^{10} 5p^4$.

6.64 Identify the group of elements that corresponds to each of the following generalized electron configurations:
(a) [noble gas] $ns^2 np^5$
(b) [noble gas] $ns^2(n-1)d^2$
(c) [noble gas] $ns^2(n-1)d^{10}np^1$
(d) [noble gas] $ns^2(n-2)f^6$

6.65 What is wrong with the following electron configurations for atoms in their ground states? **(a)** $1s^2 2s^2 3s^1$; **(b)** $[Ne]2s^2 2p^3$; **(c)** $[Ne]3s^2 3d^5$.

6.66 The following electron configurations represent excited states. Identify the element and write its ground-state condensed electron configuration. **(a)** $1s^2 2s^2 3p^2 4p^1$; **(b)** $[Ar]3d^{10} 4s^1 4p^4 5s^1$; **(c)** $[Kr]4d^6 5s^2 5p^1$.

Additional Exercises

6.67 Consider the two waves shown here, which we will consider to represent two electromagnetic radiations:

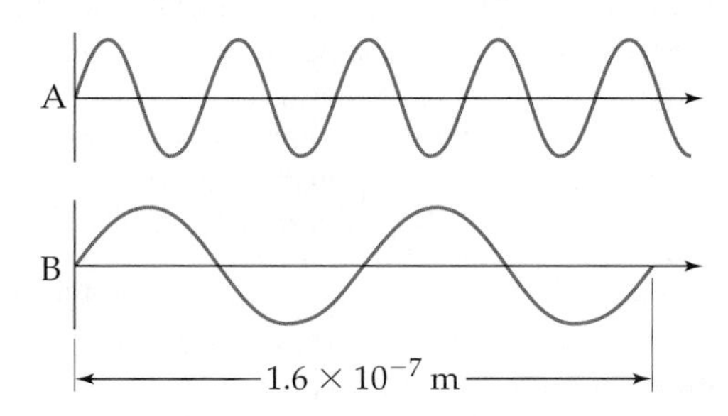

(a) What is the wavelength of wave A? Of wave B?
(b) What is the frequency of wave A? Of wave B?
(c) Identify the regions of the electromagnetic spectrum to which waves A and B belong.

6.68 Certain elements emit light of a specific wavelength when they are burned. Historically, chemists used such emission wavelengths to determine whether specific elements were present in a sample. Some characteristic wavelengths for some of the elements are

Ag	328.1 nm	Fe	372.0 nm
Au	267.6 nm	K	404.7 nm
Ba	455.4 nm	Mg	285.2 nm
Ca	422.7 nm	Na	589.6 nm
Cu	324.8 nm	Ni	341.5 nm

(a) Determine which elements emit radiation in the visible part of the spectrum. **(b)** Which element emits photons of highest energy? Of lowest energy? **(c)** When burned, a sample of an unknown substance is found to emit light of frequency $6.59 \times 10^{14}\ s^{-1}$. Which of these elements is probably in the sample?

6.69 Images of Ganymede, Jupiter's largest moon, were transmitted from *Galileo*, the unmanned spacecraft, when its distance from Earth was 522 million miles. How long did it take for the transmitted signals to travel from the spacecraft to Earth?

6.70 The rays of the Sun that cause tanning and burning are in the ultraviolet portion of the electromagnetic spectrum. These rays are categorized by wavelength: So-called UV-A radiation has wavelengths in the range of 320–380 nm, whereas UV-B radiation has wavelengths in the range of 290–320 nm. **(a)** Calculate the frequency of light that has a wavelength of 320 nm. **(b)** Calculate the energy of a mole of 320-nm photons. **(c)** Which are more energetic, photons of UV-A radiation or photons of UV-B radiation? **(d)** The UV-B radiation from the Sun is considered a greater cause of sunburn in humans than is UV-A radiation. Is this observation consistent with your answer to part (c)?

6.71 The watt is the derived SI unit of power, the measure of energy per unit time: $1\ W = 1\ J/s$. A semiconductor laser in a CD player has an output wavelength of 780 nm and a power level of 0.10 mW. How many photons strike the CD surface during the playing of a CD 69 minutes in length?

6.72 Carotenoids, present in all organisms capable of photosynthesis, extend the range of light absorbed by the organism. They exhibit maximal capacity for absorption of light in the range of 440–470 nm. Calculate the energy represented by absorption of an Avogadro's number of photons of wavelength 455 nm.

[6.73] A photocell, such as the one illustrated in Figure 6.7(b), is a device used to measure the intensity of light. In a certain experiment, when light of wavelength 550 nm is directed on to the photocell, electrons are emitted at the rate of 5.8×10^{-13} C/s. Assume that each photon that impinges on the photocell emits one electron. How many photons per second are striking the photocell? How much energy per second is the photocell absorbing?

6.74 The light-sensitive substance in black-and-white photographic film is AgBr. Photons provide the energy necessary to transfer an electron from Br^- to Ag^+ to produce Ag and Br and thereby darken the film. **(a)** If a minimum energy of 2.00×10^5 J/mol is needed for this process, what is the minimum energy needed by each photon? **(b)** Calculate the wavelength of the light necessary to provide photons of this energy. **(c)** Explain why this film can be handled in a darkroom under red light.

6.75 When the spectrum of light from the Sun is examined in high resolution in an experiment similar to that illustrated in Figure 6.10, dark lines are evident. These are called Fraunhofer lines, after the scientist who studied them extensively in the early nineteenth century. Altogether, about 25,000 lines have been identified in the solar spectrum between 2950 Å and 10,000 Å. The Fraunhofer lines are attributed to absorption of certain wavelengths of the Sun's "white" light by gaseous elements in the Sun's atmosphere. **(a)** Describe the process that causes absorption of specific wavelengths of light from the solar spectrum. **(b)** If a scientist wanted to know which Fraunhofer lines belonged to a given element, say neon, what experiments could she conduct here on Earth to provide data?

[6.76] Bohr's model can be used for hydrogen-like ions—ions that have only one electron, such as He^+ and Li^{2+}. **(a)** Why is the Bohr model applicable to He^+ ions but not to neutral He atoms? **(b)** The ground-state energies of H, He^+, and Li^{2+} are tabulated as follows:

Atom or ion	H	He^+	Li^{2+}
Ground-state energy	-2.18×10^{-18} J	-8.72×10^{-18} J	-1.96×10^{-17} J

By examining these numbers, propose a relationship between the ground-state energy of hydrogen-like systems and the nuclear charge, *Z*. **(c)** Use the relationship you derive in part (b) to predict the ground-state energy of the C^{5+} ion.

6.77 Under appropriate conditions, molybdenum emits X rays that have a characteristic wavelength of 0.711 Å. These X rays are used in diffraction experiments to determine the structures of molecules. How fast would an electron have to be moving in order to have the same wavelength as these X rays?

[6.78] An electron is accelerated through an electric potential to a kinetic energy of 82.4 keV. What is its characteristic wavelength? (*Hint:* Recall that the kinetic energy of a moving object is $E = \frac{1}{2}mv^2$, where m is the mass of the object and v is the speed of the object.)

6.79 Which of the quantum numbers governs **(a)** the shape of an orbital; **(b)** the energy of an orbital; **(c)** the spin properties of the electron; **(d)** the spatial orientation of the orbital?

6.80 Give the subshell designation for each of the following cases: **(a)** $n = 3, l = 1$; **(b)** $n = 6, l = 4$; **(c)** $n = 2, l = 0$; **(d)** $n = 4, l = 3$.

6.81 How many orbitals in an atom can have each of the following designations? **(a)** 3*s*, **(b)** 2*p*; **(c)** 4*d*; **(d)** $n = 3$?

6.82 The "magic numbers" in the periodic table are the atomic numbers of elements with high stability (the noble gases): 2, 10, 18, 36, 54, and 86. In terms of allowed values of orbitals and spin quantum numbers, explain why these electron arrangements correspond to special stability.

[6.83] For non-spherically symmetric orbitals, the contour representations (as in Figures 6.20 and 6.21) suggest where nodal planes exist (that is, where the electron density is zero). For example, the p_x orbital has a node wherever $x = 0$; this equation is satisfied by all points on the yz plane, so this plane is called a nodal plane of the p_x orbital. **(a)** Determine the nodal plane of the p_z orbital. **(b)** What are the two nodal planes of the d_{xy} orbital? **(c)** What are the two nodal planes of the $d_{x^2-y^2}$ orbital?

6.84 Using only a periodic table as a guide, write the condensed electron configurations for the following atoms: **(a)** Se; **(b)** Rh; **(c)** Si; **(d)** Hg; **(e)** Hf.

6.85 Meitnerium, Mt, element 109, named after Lisa Meitner, is a transition metal expected to have the same outer-electron configuration as iridium. By using this observation (and without looking at Figure 6.28), write the electron configuration of meitnerium. Use [Rn] to represent the first 86 electrons of the electron configuration.

6.86 Scientists have speculated that element 126 might have a moderate stability allowing it to be synthesized and characterized. Predict what the condensed electron configuration of this element might be.

Integrative Exercises

[6.87] Microwave ovens use microwave radiation to heat food. The microwaves are absorbed by moisture in the food, which is transferred to other components of the food. As the water becomes hotter, so does the food. Suppose that the microwave radiation has a wavelength of 11.2 cm. How many photons are required to heat 200 mL of coffee from 23°C to 60°C?

6.88 The stratospheric ozone (O_3) layer helps to protect us from harmful ultraviolet radiation. It does so by absorbing ultraviolet light and falling apart into an O_2 molecule and an oxygen atom, a process known as photodissociation.

$$O_3(g) \longrightarrow O_2(g) + O(g)$$

Use the data in Appendix C to calculate the enthalpy change for this reaction. What is the maximum wavelength a photon can have if it is to possess sufficient energy to cause this dissociation? In what portion of the spectrum does this wavelength occur?

6.89 The discovery of hafnium, element number 72, provided a controversial episode in chemistry. G. Urbain, a French chemist, claimed in 1911 to have isolated an element number 72 from a sample of rare earth (elements 58–71) compounds. However, Niels Bohr believed that hafnium was more likely to be found along with zirconium than with the rare earths. D. Coster and G. von Hevesy, working in Bohr's laboratory in Copenhagen, showed in 1922 that element 72 was present in a sample of Norwegian zircon, an ore of zirconium. (The name hafnium comes from the Latin name for Copenhagen, *Hafnia*). **(a)** How would you use electron configuration arguments to justify Bohr's prediction? **(b)** Zirconium, hafnium's neighbor in group 4B, can be produced as a metal by reduction of solid $ZrCl_4$ with molten sodium metal. Write a balanced chemical equation for the reaction. Is this an oxidation-reduction reaction? If yes, what is reduced and what is oxidized? **(c)** Solid zirconium dioxide, ZrO_2, is reacted with chlorine gas in the presence of carbon. The products of the reaction are $ZrCl_4$ and two gases, CO_2 and CO in the ratio 1:2. Write a balanced chemical equation for the reaction. Starting with a 55.4-g sample of ZrO_2, calculate the mass of $ZrCl_4$ formed, assuming that ZrO_2 is the limiting reagent and assuming 100% yield. **(d)** Using their electron configurations, account for the fact that Zr and Hf form chlorides MCl_4 and oxides MO_2.

6.90 **(a)** Account for formation of the following series of oxides in terms of the electron configurations of the elements and the discussion of ionic compounds in Section 2.7: K_2O, CaO, Sc_2O_3, TiO_2, V_2O_5, CrO_3. **(b)** Name these oxides. **(c)** Consider the metal oxides whose enthalpies of formation (in kJ mol^{-1}) are listed here.

Oxide	$K_2O(s)$	$CaO(s)$	$TiO_2(s)$	$V_2O_5(s)$
ΔH_f°	−363.2	−635.1	−938.7	−1550.6

Calculate the enthalpy changes in the following general reaction for each case:

$$M_nO_m(s) + H_2(g) \longrightarrow nM(s) + mH_2O(g)$$

(You will need to write the balanced equation for each case, then compute ΔH°.) **(d)** Based on the data given, estimate a value of ΔH_f° for $Sc_2O_3(s)$.

6.91 The first 25 years of the twentieth century were momentous for the rapid pace of change in scientists' understanding of the nature of matter. **(a)** How did Rutherford's experiments on the scattering of α particles by a gold foil set the stage for Bohr's theory of the hydrogen atom? **(b)** In what ways is de Broglie's hypothesis, as it applies to electrons, consistent with J. J. Thomson's conclusion that the electron has mass? In what sense is it consistent with proposals that preceded Thomson's work, that the cathode rays are a wave phenomenon?

eMedia Exercises

6.92 The **Electromagnetic Spectrum** activity (*eChapter 6.2*) allows you to choose a color in the visible spectrum and see its wavelength, frequency, and energy per photon. **(a)** What is the wavelength range of blue light? **(b)** What are the ranges of its frequency and energy per photon? **(c)** Exercise 6.17 indicates that a type of sunburn is caused by light with wavelength ~325 nm. Would you expect any of the visible wavelengths to cause sunburn? Explain.

6.93 In the **Flame Tests for Metals** movie (*eChapter 6.3*) the characteristic color of the flame is produced by emissions at several visible wavelengths, with the most intense spectral lines dominating the color. For instance, the most intense visible lines in the spectrum of lithium occur at ~671 nm. **(a)** What color is light of this wavelength? **(b)** At what approximate wavelength would you expect to find the most intense lines in the visible spectrum of potassium? **(c)** Based on the movie, how would you expect the intensity of visible lines in the spectrum of potassium to compare to those in the spectrum of lithium? **(d)** Would it be possible to verify the presence of individual metals using flame color if several metal salts were mixed together? If not, explain why not.

6.94 In the **Radial Electron Distribution** movie (*eChapter 6.6*) the radial electron density plots of helium, neon, and argon are all placed on the same graph. **(a)** Explain why the maximum for the helium plot and the first maximum for each of the other two occur at significantly different distances from the nucleus. **(b)** Based on how far the *outermost* maximum for each plot is from the nucleus, predict which pair would have the greater difference between their first ionization energies (the energy required to remove completely an electron from the outermost shell): helium and neon, or neon and argon. Explain your reasoning.

6.95 The electron configuration given in Exercise 6.63(c) is one of a handful of examples of an *s* electron being "stolen" in order to fill or half-fill a *d* subshell. (As seen in the **Electron Configurations** movie (*eChapter 6.7*), there is special stability associated with filled and half-filled subshells.) **(a)** Using the **Electron Configuration** activity (*eChapter 6.8*), identify at least three more examples like this. **(b)** Are there any instances of *both s* electrons being stolen in order to fill a *d* subshell? If so, name the element(s). **(c)** Why is this phenomenon not observed in *p*-block elements? That is, why is chlorine's electron configuration $[Ne]3s^23p^5$ rather than $[Ne]3s^13p^6$?

Chapter 7

Periodic Properties of the Elements

Coral reef in the Red Sea, Egypt. Corals are composed mainly of calcium carbonate, $CaCO_3$. They absorb and release carbon dioxide gas and serve as regulators of the amount of dissolved CO_2 in the oceans.

THE PERIODIC TABLE is the most significant tool that chemists use for organizing and remembering chemical facts. As we saw in Chapter 6, the periodic table arises from the periodic patterns in the electron configurations of the elements. Elements in the same column contain the same number of electrons in their outer-shell orbitals, or **valence orbitals**. For example, O ([He]$2s^22p^4$) and S ([Ne]$3s^23p^4$) are both members of group 6A; the similarity in the occupancies of their valence *s* and *p* orbitals leads to similarities in their properties.

When we compare O and S, however, it is apparent that they exhibit differences as well (Figure 7.1 ▶). One of the major differences between the elements is their electron configurations: The outermost electrons of O are in the second shell, whereas those of S are in the third shell. We will see that electron configurations can be used to explain differences as well as similarities in the properties of elements.

In this chapter we explore how certain properties of elements change as we move across a row or down a column of the periodic table. In many cases the trends within a row or column form patterns that allow us to make predictions about physical and chemical properties.

▶ What's Ahead ◀

- Our discussion begins with a brief history of the periodic table.
- We will see that many properties of atoms depend on both the net attraction between the nucleus and the outer electrons (due to the *effective nuclear charge*) and on the average distance of those electrons from the nucleus.
- We will examine periodic trends of three key properties of atoms: *atomic size, ionization energy* (the energy required to remove electrons), and the *electron affinity* (the energy associated with adding electrons).
- As part of these discussions, we will also examine the sizes of ions and their electron configurations.
- The *metallic character* of an element is demonstrated by the tendency of the element to form cations and by the basicity of its metal oxide.
- We will examine some differences in the physical and chemical properties of metals and nonmetals.
- Finally, we discuss some periodic trends in the chemistry of the active metals (groups 1A and 2A) and of several nonmetals (hydrogen and groups 6A to 8A).

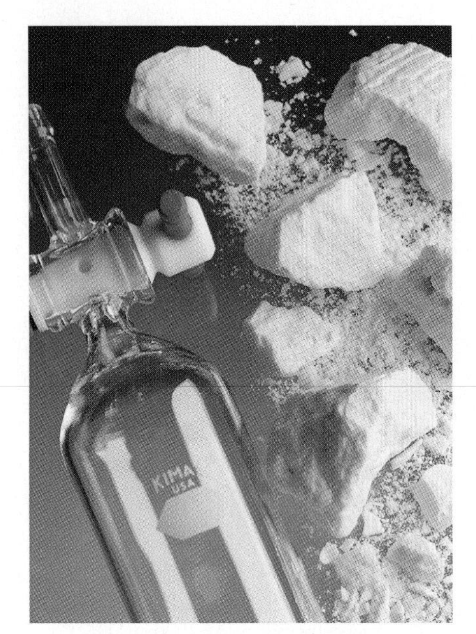

▲ **Figure 7.1** Oxygen and sulfur are both group 6A elements. As such, they have many chemical similarities. However, they also have many differences, including the forms they take as elements at room temperature. Oxygen consists of O_2 molecules that appear as a colorless gas (shown here enclosed in a glass container). In contrast, sulfur consists of S_8 molecules that form a yellow solid.

7.1 Development of the Periodic Table

The discovery of new chemical elements has been an ongoing process since ancient times (Figure 7.2 ▼). Certain elements, such as gold, appear in nature in elemental form and were thus discovered thousands of years ago. In contrast, some elements are radioactive and intrinsically unstable. We know about them only because of twentieth-century technology.

The majority of the elements, although stable, are dispersed widely in nature and are incorporated into numerous compounds. For centuries, therefore, scientists were unaware of their existence. In the early nineteenth century, advances in chemistry made it easier to isolate elements from their compounds. As a result, the number of known elements more than doubled from 31 in 1800 to 63 by 1865.

As the number of known elements increased, scientists began to investigate the possibilities of classifying them in useful ways. In 1869 Dmitri Mendeleev in Russia and Lothar Meyer in Germany published nearly identical classification schemes. Both scientists noted that similar chemical and physical properties recur periodically when the elements are arranged in order of increasing atomic weight. Scientists at that time had no knowledge of atomic numbers. Atomic weights, however, generally increase with increasing atomic number, so both Mendeleev and Meyer fortuitously arranged the elements in proper sequence. The tables of elements advanced by Mendeleev and Meyer were the forerunners of the modern periodic table.

Although Mendeleev and Meyer came to essentially the same conclusion about the periodicity of the properties of the elements, Mendeleev is given credit for advancing his ideas more vigorously and stimulating much new work in chemistry. His insistence that elements with similar characteristics be listed in the same families forced him to leave several blank spaces in his table. For example, both gallium (Ga) and germanium (Ge) were at that time unknown. Mendeleev boldly predicted their existence and properties, referring to them as eka-aluminum and eka-silicon, after the elements they appear under in the periodic table. When these elements were discovered, their properties were found to closely match those predicted by Mendeleev, as illustrated in Table 7.1 ▶.

In 1913, two years after Rutherford proposed the nuclear model of the atom, an English physicist named Henry Moseley (1887–1915) developed the concept

▶ **Figure 7.2** Periodic table showing the dates of discovery of the elements.

H																	He
Li	Be											B	C	N	O	F	Ne
Na	Mg											Al	Si	P	S	Cl	Ar
K	Ca	Sc	Ti	V	Cr	Mn	Fe	Co	Ni	Cu	Zn	Ga	Ge	As	Se	Br	Kr
Rb	Sr	Y	Zr	Nb	Mo	Tc	Ru	Rh	Pd	Ag	Cd	In	Sn	Sb	Te	I	Xe
Cs	Ba	La	Hf	Ta	W	Re	Os	Ir	Pt	Au	Hg	Tl	Pb	Bi	Po	At	Rn
Fr	Ra	Ac	Rf	Db	Sg	Bh	Hs	Mt									

Ce	Pr	Nd	Pm	Sm	Eu	Gd	Tb	Dy	Ho	Er	Tm	Yb	Lu
Th	Pa	U	Np	Pu	Am	Cm	Bk	Cf	Es	Fm	Md	No	Lr

Ancient Times · 1735–1843 · 1894–1918

Middle Ages–1700 · 1843–1886 · 1923–1961 · 1965–

ACTIVITY
Periodic Table

TABLE 7.1 Comparison of the Properties of Eka-Silicon Predicted by Mendeleev with the Observed Properties of Germanium

Property	Mendeleev's Predictions for Eka-Silicon (made in 1871)	Observed Properties of Germanium (discovered in 1886)
Atomic weight	72	72.59
Density (g/cm^3)	5.5	5.35
Specific heat (J/g-K)	0.305	0.309
Melting point (°C)	High	947
Color	Dark gray	Grayish white
Formula of oxide	XO_2	GeO_2
Density of oxide (g/cm^3)	4.7	4.70
Formula of chloride	XCl_4	$GeCl_4$
Boiling point of chloride (°C)	A little under 100	84

of atomic numbers. Moseley determined the frequencies of X rays emitted as different elements were bombarded with high-energy electrons. He found that each element produces X rays of a unique frequency; furthermore, he found that the frequency generally increased as the atomic mass increased. He arranged the X-ray frequencies in order by assigning a unique whole number, called an *atomic number*, to each element. Moseley correctly identified the atomic number as the number of protons in the nucleus of the atom and the number of electrons in the atom. (Section 2.3)

The concept of atomic number clarified some problems in the early version of the periodic table, which was based on atomic weights. For example, the atomic weight of Ar (atomic number 18) is greater than that of K (atomic number 19). However, when the elements are arranged in order of increasing atomic number, rather than increasing atomic weight, Ar and K appear in their correct places in the table. Moseley's studies also made it possible to identify "holes" in the periodic table, which led to the discovery of new elements.

7.2 Effective Nuclear Charge

To understand the properties of atoms, we must be familiar not only with electron configurations, but also with how strongly outer electrons are attracted to the nucleus. Coulomb's law of attraction indicates that the strength of the interaction between two electrical charges depends on the magnitude of the charges and the distance between them. (Section 2.3) Thus, the force of attraction between an electron and the nucleus depends on the magnitude of the net nuclear charge acting on the electron and the average distance between the nucleus and the electron. The force of attraction increases as the nuclear charge increases, and it decreases as the electron moves farther from the nucleus.

In a many-electron atom, each electron is simultaneously attracted to the nucleus and repelled by the other electrons. In general, there are so many electron-electron repulsions that we cannot analyze the situation exactly. We can, however, estimate the energy of each electron by considering how it interacts with the *average* environment created by the nucleus and the other electrons in the atom. This approach allows us to treat each electron individually as if it were moving in the electric field created by the nucleus and the surrounding electron density of the other electrons. This electric field is equivalent to one generated by a charge located at the nucleus, called the **effective nuclear charge**. The effective nuclear charge, Z_{eff}, acting on an electron equals the number of

protons in the nucleus, Z, minus the average number of electrons, S, that are between the nucleus and the electron in question:

$$Z_{eff} = Z - S \quad [7.1]$$

Because S represents an average, it need not be an integer.

Many of an atom's properties are determined by the effective nuclear charge experienced by its outer, or valence, electrons. Any electron density between the nucleus and an outer electron decreases the effective nuclear charge acting on that outer electron. The electron density due to the inner electrons is said to *shield*, or *screen*, the outer electrons from the full charge of the nucleus. Because the core electrons are located mainly between the nucleus and the outer electrons, they are very efficient in shielding the outer electrons. On the other hand, electrons in the same shell hardly shield each other at all from the nucleus. As a result, *the effective nuclear charge experienced by the outer electrons is determined primarily by the difference between the charge on the nucleus and the charge of the core electrons.*

We can crudely estimate the effective nuclear charge using the nuclear charge and the number of core electrons. Magnesium (atomic number 12), for example, has an electron configuration of [Ne]$3s^2$. The nuclear charge of the atom is 12+, and the Ne inner core consists of 10 electrons. Very roughly then, we would expect each outer-shell electron to experience an effective nuclear charge of about $12 - 10 = 2+$ as pictured in a simplified way in Figure 7.3(a) ▼. This calculation underestimates the effective nuclear charge, however, because the outer electrons of an atom have some probability of being inside the core, as shown in Figure 7.3(b). Indeed, more detailed calculations indicate that the effective nuclear charge acting on the outer electrons in Mg is actually 3.3+.

The effective nuclear charge experienced by outer electrons increases as we move from element to element across any row (period) of the table. Although the number of core electrons stays the same as we move across a period, the actual nuclear charge increases. The outer-shell electrons added to counterbalance the

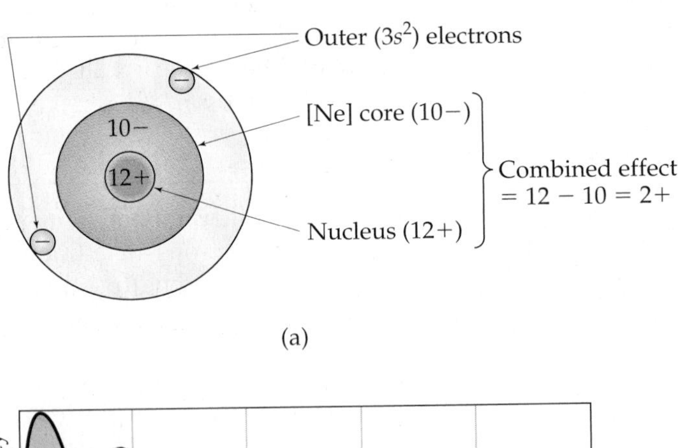

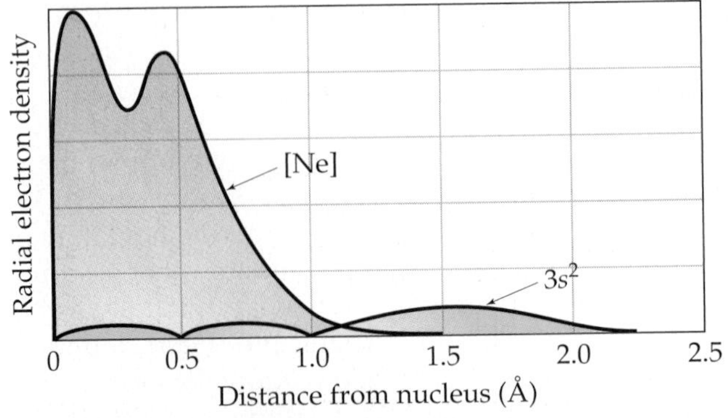

▶ **Figure 7.3** (a) The effective nuclear charge experienced by the valence electrons in magnesium depends mostly on the 12+ charge of the nucleus and the 10− charge of the neon core. If the neon core were totally effective in shielding the valence electrons from the nucleus, each valence electron would experience an effective nuclear charge of 2+. (b) The $3s$ electrons have some probability of being inside the Ne core. As a consequence of this "penetration," the core is not totally effective in screening the $3s$ electrons from the nucleus. Thus, the effective nuclear charge experienced by the $3s$ electrons is greater than 2+.

increasing nuclear charge shield each other very ineffectively. Thus, the effective nuclear charge increases steadily. For example, the inner $1s^2$ electrons of lithium ($1s^2 2s^1$) shield the outer 2*s* electron from the 3+ charged nucleus fairly efficiently. Consequently, the outer electron experiences an effective nuclear charge of roughly $3 - 2 = 1+$. For beryllium ($1s^2 2s^2$) the effective nuclear charge experienced by each outer 2*s* electron is larger; in this case, the inner $1s^2$ electrons are shielding a 4+ nucleus, and each 2*s* electron only partially shields the other from the nucleus. Consequently, the effective nuclear charge experienced by each 2*s* electron is about $4 - 2 = 2+$.

Going down a family, the effective nuclear charge experienced by outer-shell electrons changes far less than it does across a period. For example, we would expect the effective nuclear charge for the outer electrons in lithium and sodium to be about the same, roughly $3 - 2 = 1+$ for lithium and $11 - 10 = 1+$ for sodium. In fact, however, the effective nuclear charge increases slightly as we go down a family because larger electron cores are less able to screen the outer electrons from the nuclear charge. Nevertheless, the slight change in effective nuclear charge that occurs moving down a family is of far less importance than the increase that occurs across a period.

7.3 Sizes of Atoms and Ions

One of the important properties of an atom or ion is its size. We often think of atoms and ions as hard, spherical objects. According to the quantum mechanical model, however, atoms and ions do not have sharply defined boundaries at which the electron distribution becomes zero. (Section 6.5) The edges of atoms and ions are therefore a bit "fuzzy." Nevertheless, we can define their sizes in several different ways based on the distances between atoms in various situations.

Imagine a collection of argon atoms in the gas phase. When the atoms collide with one another in the course of their motions, they ricochet apart—something like billiard balls. This happens because the electron clouds of the colliding atoms cannot penetrate one another to a significant extent. The closest distances separating the nuclei during such collisions determine the *apparent* radii of the argon atoms. We might call this radius the *nonbonding radius* of an atom.

When two atoms are chemically bonded to one another, as in the Cl_2 molecule, there is an attractive interaction between the two atoms leading to a chemical bond. We will discuss the nature of such bonding in Chapter 8. For now, we need only realize that this attractive interaction brings the two atoms closer together than they would be in a nonbonding collision. We can define an atomic radius based on the distances separating the nuclei of atoms when they are chemically bonded to one another. This distance, called the **bonding atomic radius**, is shorter than the nonbonding radius, as illustrated in Figure 7.4 ▶. Space-filling models, such as those in Figure 1.1 or Figure 2.20, use the nonbonding radii (also called *van der Waals radii*) to determine the sizes of the atoms. The bonding atomic radii (also called *covalent radii*) are used to determine the distances between their centers.

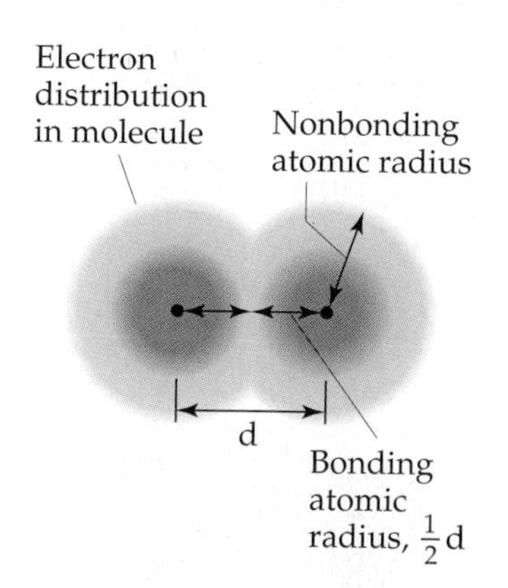

▲ **Figure 7.4** Illustration of the distinction between nonbonding and bonding atomic radius. Values of bonding atomic radii are obtained from measurements of interatomic distances in chemical compounds.

Scientists have developed a variety of methods for measuring the distances separating nuclei in molecules. From observations of these distances in many molecules, each element can be assigned a bonding atomic radius. For example, in the I_2 molecule, the distance separating the iodine nuclei is observed to be 2.66 Å.* We can define the bonding atomic radius of iodine on this basis to

* *Remember:* The angstrom ($1\ \text{Å} = 10^{-10}$ m) is a convenient metric unit for atomic measurements of length. The angstrom is *not* an SI unit. The most commonly used SI unit for such measurements is the picometer ($1\ \text{pm} = 10^{-12}$ m; $1\ \text{Å} = 100$ pm).

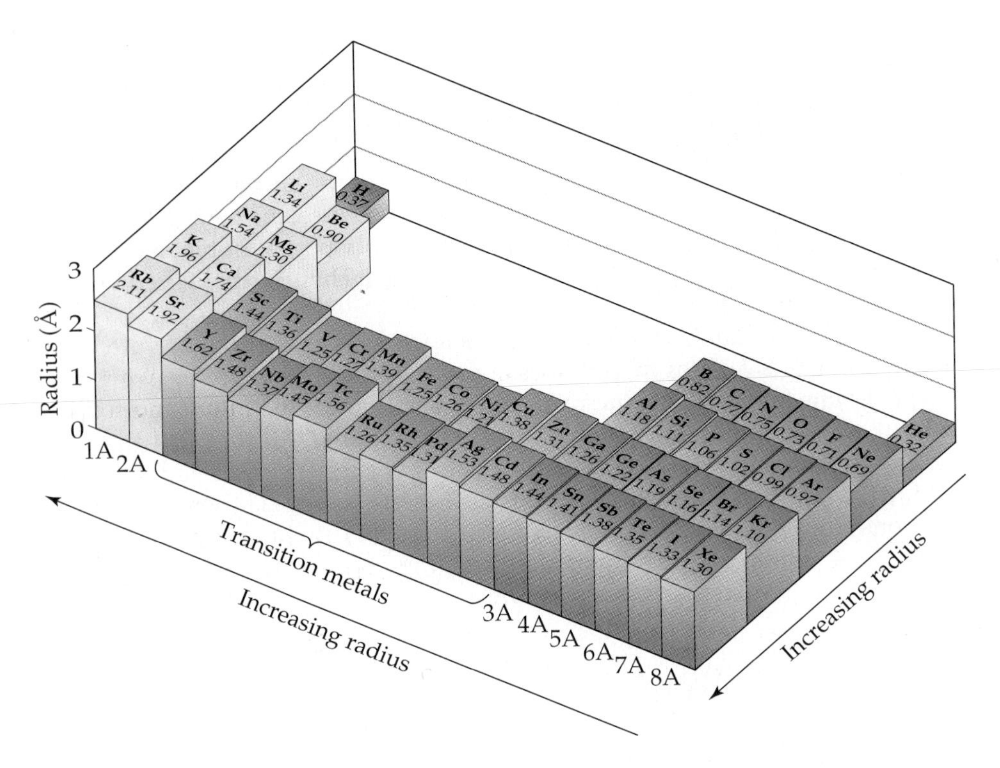

▶ **Figure 7.5** Bonding atomic radii for the first 54 elements of the periodic table. The height of the bar for each element is proportional to its radius, giving a "relief map" view of the radii.

be 1.33 Å. Similarly, the distance separating two adjacent carbon nuclei in diamond, which is a three-dimensional solid network, is 1.54 Å; thus, the bonding atomic radius of carbon is assigned the value 0.77 Å. The radii of other elements can be similarly defined (Figure 7.5 ▲). (For helium and neon, the bonding radii must be estimated because there are no known chemical combinations.)

Atomic radii allow us to estimate the bond lengths between different elements in molecules. For example, the Cl—Cl bond length in Cl_2 is 1.99 Å, so a radius of 0.99 Å is assigned to Cl. In the compound CCl_4 the C—Cl bond length is 1.77 Å, very close to the sum (0.77 + 0.99 Å) of the atomic radii for C and Cl.

SAMPLE EXERCISE 7.1

Natural gas used in home heating and cooking is odorless. Because natural gas leaks pose the danger of explosion or suffocation, various smelly substances are added to the gas to allow detection of a leak. One such substance is methyl mercaptan, CH_3SH, whose structure is shown in the margin. Use Figure 7.5 to predict the lengths of the C—S, C—H, and S—H bonds in this molecule.

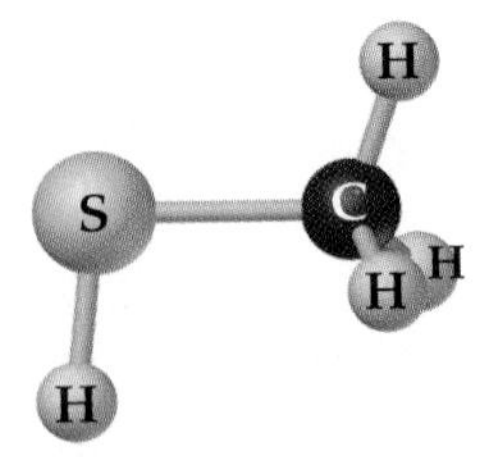

Solution

Analyze and Plan: We are given three specific bonds and the list of atomic radii. We will assume that the bond lengths are the sum of the radii of the atoms involved.

Solve: Using radii for C, S, and H from Figure 7.5, we predict

$$\text{C—S bond length} = \text{radius of C} + \text{radius of S}$$

$$= 0.77\ \text{Å} + 1.02\ \text{Å} = 1.79\ \text{Å}$$

$$\text{C—H bond length} = 0.77\ \text{Å} + 0.37\ \text{Å} = 1.14\ \text{Å}$$

$$\text{S—H bond length} = 1.02\ \text{Å} + 0.37\ \text{Å} = 1.39\ \text{Å}$$

Check: The experimentally determined bond lengths in methyl mercaptan are C—S = 1.82 Å, C—H = 1.10 Å, and S—H = 1.33 Å. (In general, the lengths of bonds involving hydrogen show larger deviations from the values predicted by the sum of the atomic radii than do those bonds involving larger atoms.)

PRACTICE EXERCISE

Using Figure 7.5, predict which will be greater, the P—Br bond length in PBr_3 or the As—Cl bond length in $AsCl_3$.
Answer: the P—Br bond length

Periodic Trends in Atomic Radii

If we examine the "relief map" of atomic radii shown in Figure 7.5, we observe two interesting trends in the data:

1. Within each column (group) the atomic radius tends to increase as we proceed from top to bottom. This trend results primarily from the increase in the principal quantum number (n) of the outer electrons. As we go down a group, the outer electrons spend more time farther from the nucleus, causing the atom to increase in size.
2. Within each row (period) the atomic radius tends to decrease as we move from left to right. The major factor influencing this trend is the increase in the effective nuclear charge (Z_{eff}) as we move across a row. The increasing effective nuclear charge steadily draws the electrons, including the outer ones, closer to the nucleus, causing the radius to decrease.

SAMPLE EXERCISE 7.2

Referring to a periodic table, arrange (as much as possible) the following atoms in order of increasing size: ${}_{15}P$, ${}_{16}S$, ${}_{33}As$, ${}_{34}Se$. (Atomic numbers are given for the elements to help you locate them quickly in the periodic table.)

Solution

Analyze and Plan: We are given the chemical symbols for four elements. We can use their relative positions in the periodic table and the two periodic trends just listed to predict the relative order of their atomic radii.

Solve: Notice that P and S are in the same row of the periodic table, with S to the right of P. Therefore, we expect the radius of S to be smaller than that of P. (Radii decrease as we move from left to right.) Likewise, the radius of Se is expected to be smaller than that of As. We also notice that As is directly below P and that Se is directly below S. We expect, therefore, that the radius of As is greater than that of P and that the radius of Se is greater than that of S (radii increase as we move from top to bottom). From these observations we can conclude that the radii follow the relationships S $<$ P, P $<$ As, S $<$ Se, and Se $<$ As. We can therefore conclude that S has the smallest radius of the four elements and that As has the largest radius.

By using these two general trends, we cannot determine whether P or Se has the larger radius; to go from P to Se in the periodic table, we must move down (radius tends to increase) and to the right (radius tends to decrease). In Figure 7.5 we see that the radius of Se (1.17 Å) is greater than that of P (1.10 Å). If you examine Figure 7.5 carefully, you will discover that for the representative elements the increase in radius upon moving down a column tends to be the greater effect. There are exceptions, however.

Check: From Figure 7.5 we have S (1.02 Å) $<$ P (1.10 Å) $<$ Se (1.17 Å) $<$ As (1.19 Å).

PRACTICE EXERCISE

Arrange the following atoms in order of increasing atomic radius: Na, Be, Mg.
Answer: Be $<$ Mg $<$ Na

Trends in the Sizes of Ions

The sizes of ions are based on the distances between ions in ionic compounds. Like the size of an atom, the size of an ion depends on its nuclear charge, the number of electrons it possesses, and the orbitals in which the outer-shell electrons reside. The formation of a cation vacates the most spatially extended orbitals and also decreases the total electron-electron repulsions. As a consequence,

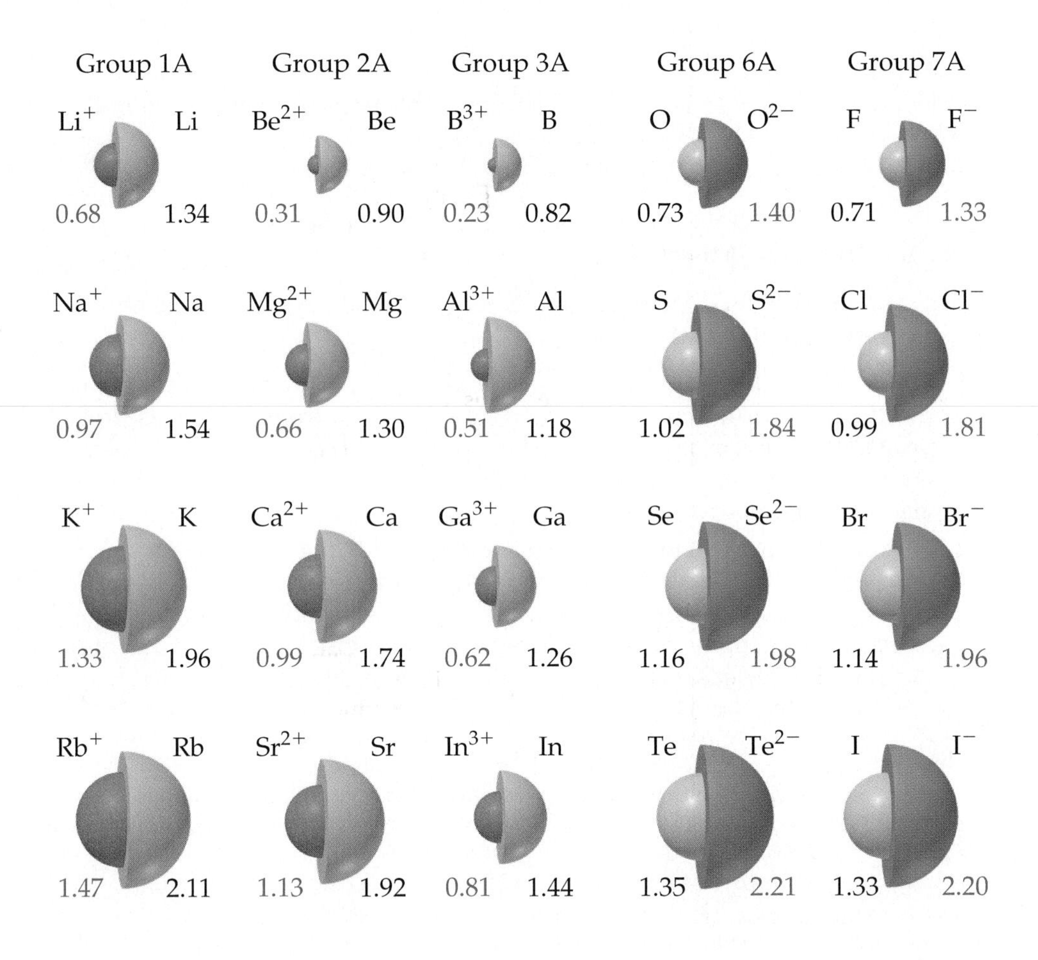

▶ **Figure 7.6** Comparisons of the radii, in Å, of neutral atoms and ions for several of the groups of representative elements. Neutral atoms are shown in gray, cations in red, and anions in blue.

cations are smaller than their parent atoms, as illustrated in Figure 7.6 ▲. The opposite is true of negative ions (anions). When electrons are added to a neutral atom to form an anion, the increased electron-electron repulsions cause the electrons to spread out more in space. Thus, *anions are larger than their parent atoms.*

For ions of the same charge, size increases as we go down a group in the periodic table. This trend is also seen in Figure 7.6. As the principal quantum number of the outer occupied orbital of an ion increases, the size of the ion increases.

SAMPLE EXERCISE 7.3

Arrange these atoms and ions in order of decreasing size: Mg^{2+}, Ca^{2+}, and Ca.

Solution Cations are smaller than their parent atoms, so Ca^{2+} is smaller than the Ca atom. Because Ca is below Mg in group 2A of the periodic table, Ca^{2+} is larger than Mg^{2+}. Consequently, $Ca > Ca^{2+} > Mg^{2+}$.

PRACTICE EXERCISE

Which of the following atoms and ions is largest: S^{2-}, S, O^{2-}?

Answer: S^{2-}

The effect of varying nuclear charge on ionic radii is seen in the variation in radius in an **isoelectronic series** of ions. The term *isoelectronic* means that the ions possess the same number of electrons. For example, each ion in the series O^{2-}, F^-, Na^+, Mg^{2+}, and Al^{3+} has 10 electrons. The nuclear charge in this series increases steadily in the order listed. (Recall that the charge on the nucleus of an atom or monatomic ion is given by the atomic number of the element.) Because the number of electrons remains constant, the radius of the ion decreases with

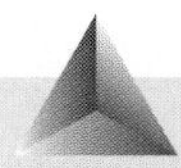

Chemistry and Life Ionic Size Makes a BIG Difference!

Ionic size plays a major role in determining the properties of ions in solution. For example, a small difference in ionic size is often sufficient for one metal ion to be biologically important and another not to be. To illustrate, let's examine some of the biological chemistry of the zinc ion (Zn^{2+}) and compare it with the cadmium ion (Cd^{2+}).

Recall from the "Chemistry and Life" box in Section 2.7 that zinc is needed in our diets in trace amounts. Zinc is an essential part of several enzymes, the proteins that facilitate or regulate the speeds of key biological reactions. For example, one of the most important zinc-containing enzymes is *carbonic anhydrase*. This enzyme is found in red blood cells. Its job is to facilitate the reaction of carbon dioxide (CO_2) with water to form the bicarbonate ion (HCO_3^-):

$$CO_2(aq) + H_2O(l) \longrightarrow HCO_3^-(aq) + H^+(aq) \qquad [7.2]$$

You might be surprised to know that our bodies need an enzyme for such a simple reaction. In the absence of carbonic anhydrase, however, the CO_2 produced in cells when they are oxidizing glucose or other fuels in vigorous exercise would be cleared out much too slowly. About 20% of the CO_2 produced by cell metabolism binds to hemoglobin and is carried to the lungs where it is expelled. About 70% of the CO_2 produced is converted to bicarbonate ion through the action of carbonic anhydrase. When the CO_2 has been converted into bicarbonate ion, it diffuses into the blood plasma and eventually is passed into the lungs in the reverse of Equation 7.2. These processes are illustrated in Figure 7.7 ▶. In the absence of zinc the carbonic anhydrase would be inactive, and serious imbalances would result in the amount of CO_2 present in blood.

Zinc is also found in several other enzymes, including some found in the liver and kidneys. It is obviously an essential element. By contrast, cadmium, zinc's neighbor in group 2B, is extremely toxic to humans. But why are two elements so different? Both occur as 2+ ions, but Zn^{2+} is smaller than Cd^{2+}.

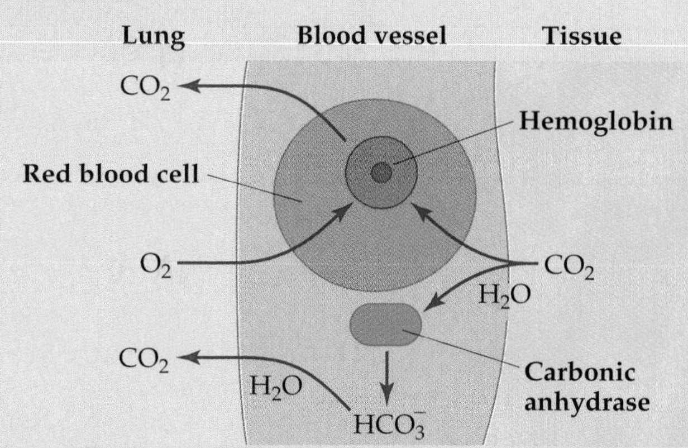

▲ **Figure 7.7** Illustration of the flow of CO_2 from tissues into blood vessels and eventually into the lungs. About 20% of the CO_2 binds to hemoglobin and is released in the lungs. About 70% is converted by carbonic anhydrase into HCO_3^- ion, which remains in the blood plasma until the reverse reaction releases CO_2 into the lungs. Small amounts of CO_2 simply dissolve in the blood plasma and are released in the lungs.

The radius of Zn^{2+} is 0.74 Å, that of Cd^{2+} is 0.95 Å. Can this difference be the cause of such a dramatic reversal of biological properties? The answer is that while size is not the only factor, it is very important. In the carbonic anhydrase enzyme the Zn^{2+} ion is found electrostatically bonded to atoms on the protein, as shown in Figure 7.8 ▼. It turns out that Cd^{2+} binds in this same place preferentially over Zn^{2+}, thus displacing it. When Cd^{2+} is present instead of Zn^{2+}, however, the reaction of CO_2 with water is not facilitated. More seriously, Cd^{2+} inhibits reactions that are essential to the kidney's functioning. Moreover, cadmium is a cumulative poison, so chronic exposure to even very low levels over an extended time leads to poisoning.

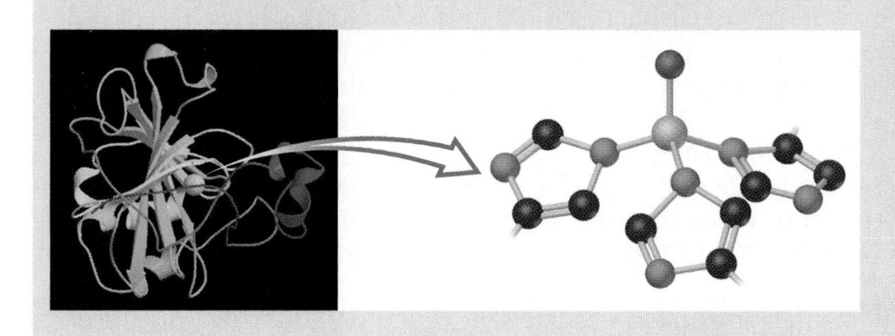

◀ **Figure 7.8** The carbonic anhydrase molecule (left) catalyzes the reaction between CO_2 and the water to form HCO_3^-. The ribbon represents the folding of the protein chain. The "active site" of the enzyme (right) is where the reaction occurs. H atoms are excluded for clarity. Thus, the red sphere represents the oxygen of a water molecule that is bound to the zinc. The water is replaced by CO_2 in the reaction. The bonds coming off the five-member rings attach the active site to the protein.

increasing nuclear charge, as the electrons are more strongly attracted to the nucleus:

—— Increasing nuclear charge ⟶

O^{2-}	F^-	Na^+	Mg^{2+}	Al^{3+}
1.40 Å	1.33 Å	0.97 Å	0.66 Å	0.51 Å

—— Decreasing ionic radius ⟶

Notice the positions of these elements in the periodic table and also their atomic numbers. The nonmetal anions precede the noble gas Ne in the table.

The metal cations follow Ne. Oxygen, the largest ion in this isoelectronic series, has the lowest atomic number, 8. Aluminum, the smallest of these ions, has the highest atomic number, 13.

SAMPLE EXERCISE 7.4

Arrange the ions S^{2-}, Cl^-, K^+, and Ca^{2+} in order of decreasing size.

Solution This is an isoelectronic series of ions, with all ions having 18 electrons. In such a series, size decreases as the nuclear charge (atomic number) of the ion increases. The atomic numbers of the ions are S (16), Cl (17), K (19), and Ca (20). Thus, the ions decrease in size in the order: $S^{2-} > Cl^- > K^+ > Ca^{2+}$.

PRACTICE EXERCISE

Which of the following ions is largest, Rb^+, Sr^{2+}, or Y^{3+}?
Answer: Rb^+

7.4 Ionization Energy

ANIMATION
Gain and Loss of Electrons, Ionization, Energy, Periodic Trends: Ionization Energy

ACTIVITY
Ionization Energy

The ease with which electrons can be removed from an atom is an important indicator of the atom's chemical behavior. The **ionization energy** of an atom or ion is the minimum energy required to remove an electron from the ground state of the isolated gaseous atom or ion. The *first ionization energy*, I_1, is the energy needed to remove the first electron from a neutral atom. For example, the first ionization energy for the sodium atom is the energy required for the following process:

$$Na(g) \longrightarrow Na^+(g) + e^- \quad [7.3]$$

The *second ionization energy*, I_2, is the energy needed to remove the second electron, and so forth, for successive removals of additional electrons. Thus, I_2 for the sodium atom is the energy associated with the following process:

$$Na^+(g) \longrightarrow Na^{2+}(g) + e^- \quad [7.4]$$

The greater the ionization energy, the more difficult it is to remove an electron.

Variations in Successive Ionization Energies

Ionization energies for the elements sodium through argon are listed in Table 7.2 ▼. Notice that the ionization energies for an element increase in magnitude as successive electrons are removed: $I_1 < I_2 < I_3$, and so forth. This trend arises because with each successive removal, an electron is being pulled away from an increasingly more positive ion, requiring increasingly more energy.

A second important feature of Table 7.2 is the sharp increase in ionization energy that occurs when an inner-shell electron is removed. For example, consider silicon, whose electron configuration is $1s^22s^22p^63s^23p^2$ or $[Ne]3s^23p^2$.

TABLE 7.2 Successive Values of Ionization Energies, I, for the Elements Sodium through Argon (kJ/mol)

Element	I_1	I_2	I_3	I_4	I_5	I_6	I_7
Na	496	4560			(inner-shell electrons)		
Mg	738	1450	7730				
Al	578	1820	2750	11,600			
Si	786	1580	3230	4360	16,100		
P	1012	1900	2910	4960	6270	22,200	
S	1000	2250	3360	4560	7010	8500	27,100
Cl	1251	2300	3820	5160	6540	9460	11,000
Ar	1521	2670	3930	5770	7240	8780	12,000

The ionization energies increase steadily from 786 kJ/mol to 4360 kJ/mol for the loss of the four electrons in the outer 3*s* and 3*p* subshells. Removal of the fifth electron, which comes from the 2*p* subshell, requires a great deal more energy: 16,100 kJ/mol. The large increase in energy occurs because the inner-shell 2*p* electron is much closer to the nucleus and experiences a much greater effective nuclear charge than do the valence-shell 3*s* and 3*p* electrons.

Every element exhibits a large increase in ionization energy when electrons are removed from its noble-gas core. This observation supports the idea that only the outermost electrons, those beyond the noble-gas core, are involved in the sharing and transfer of electrons that give rise to chemical bonding and reactions. The inner electrons are too tightly bound to the nucleus to be lost from the atom or even shared with another atom.

SAMPLE EXERCISE 7.5

Three elements are indicated in the periodic table to the right. Based on their locations, predict the one with the largest second ionization energy.

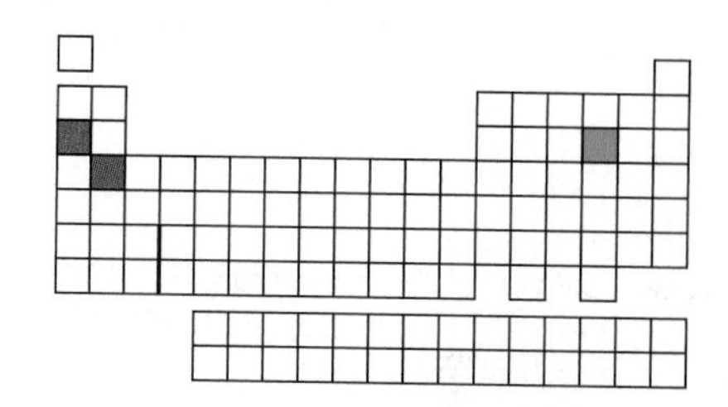

Solution

Analyze and Plan: The locations of the elements in the periodic table allow us to predict the electron configurations of the elements. The greatest ionization energies involve removal of core electrons. Thus, we should look first for an element with only one outer-shell electron.

Solve: The element in group 1A (Na), indicated by the red box, has only one outer electron. The second ionization energy of this element is associated, therefore, with the removal of a core electron. The other elements indicated have two or more outer electrons. Thus, Na has the largest second ionization energy.

Check: If we consult a chemistry handbook, we find the following values for the second ionization energies (I_2) of the respective elements: Ca (1,145 kJ/mol) $<$ S(2,251 kJ/mol) $<$ Na(4,562 kJ/mol).

PRACTICE EXERCISE

Which will have the greater third ionization energy, Ca or S?
Answer: Ca, because the third electron is a core electron

Periodic Trends in First Ionization Energies

We have seen that the ionization energy for a given element increases as we remove successive electrons. What trends do we observe in the ionization energies as we move from one element to another in the periodic table? Figure 7.9 ▼

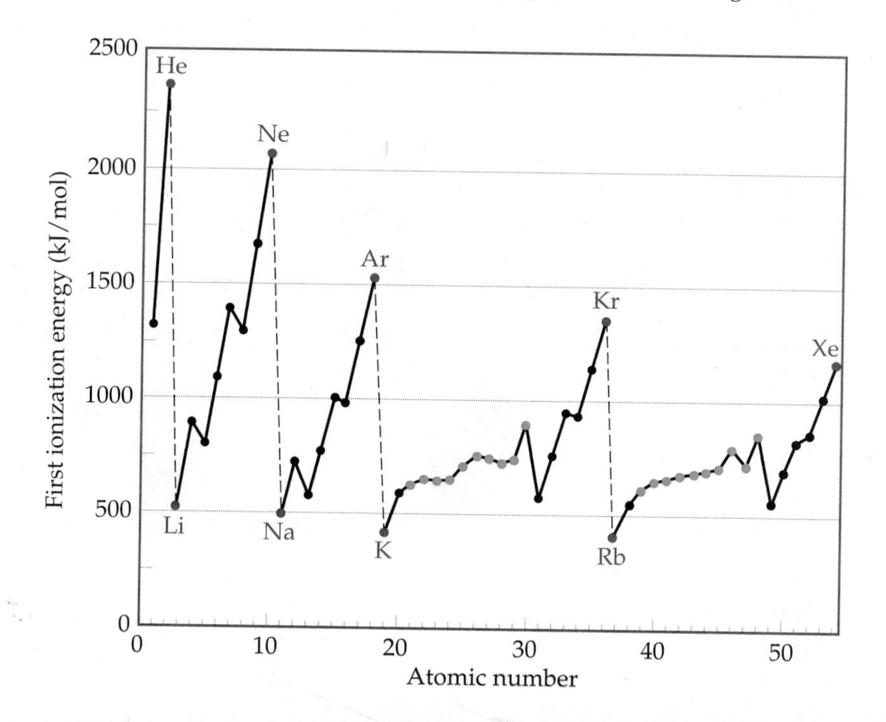

◀ **Figure 7.9** First ionization energy versus atomic number. The red dots mark the beginning of a period (alkali metals), and the blue dots mark the end of a period (noble gases). Green dots are used for the transition metals.

shows a graph of I_1 versus atomic number for the first 54 elements. The important trends are as follows:

1. Within each row I_1 generally increases with increasing atomic number. The alkali metals show the lowest ionization energy in each row, and the noble gases the highest. There are slight irregularities in this trend that we will discuss shortly.
2. Within each group the ionization energy generally decreases with increasing atomic number. For example, the ionization energies of the noble gases follow the order He > Ne > Ar > Kr > Xe.
3. The representative elements show a larger range of values of I_1 than do the transition-metal elements. Generally, the ionization energies of the transition elements increase slowly as we proceed from left to right in a period. The f-block metals, which are not shown in Figure 7.9, also show only a small variation in the values of I_1.

The periodic trends in the first ionization energies of the representative elements are further illustrated in Figure 7.10 ▼.

In general, the smaller atoms have higher ionization energies. The same factors that influence atomic size also influence ionization energies. The energy needed to remove an electron from the outer shell depends on both the effective nuclear charge and the average distance of the electron from the nucleus. Either increasing the effective nuclear charge or decreasing the distance from the nucleus increases the attraction between the electron and the nucleus. As this attraction increases, it becomes harder to remove the electron and, thus, the ionization energy increases. As we move across a period, there is both an increase in effective nuclear charge and a decrease in atomic radius, causing the ionization energy to increase. As we move down a column, however, the atomic radius increases, while the effective nuclear charge changes little. Thus, the attraction between the nucleus and the electron decreases, causing the ionization energy to decrease.

The irregularities within a given row are somewhat more subtle, but are readily explained. For example, the decrease in ionization energy from beryllium

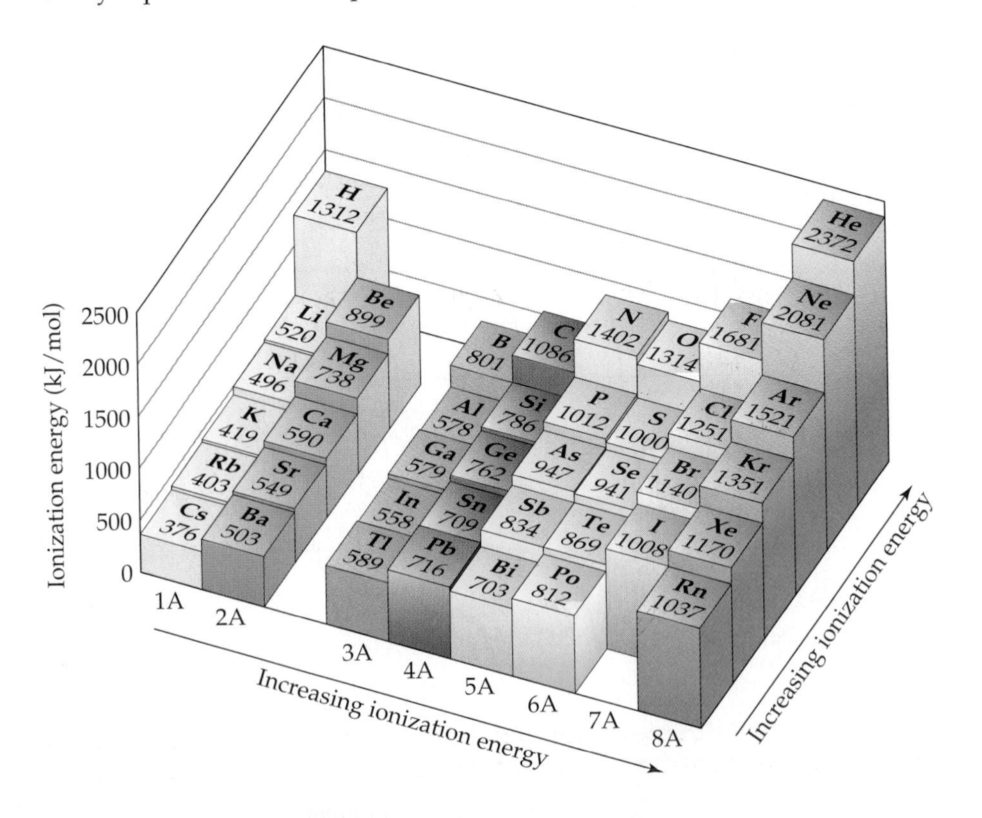

▶ **Figure 7.10** First ionization energies for the representative elements in the first six periods. The ionization energy generally increases from left to right and decreases from top to bottom. The ionization energy of astatine has not been determined.

([He]$2s^2$) to boron ([He]$2s^22p^1$) occurs because the electrons in the filled 2s orbital are more effective at shielding the electrons in the 2p subshell than they are at shielding each other. This is essentially the same reason that in many-electron atoms the 2p orbital is at a higher energy than the 2s (Figure 6.22). The decrease in ionization energy on going from nitrogen ([He]$2s^22p^3$) to oxygen ([He]$2s^22p^4$) is because of repulsion of paired electrons in the p^4 configuration. (Remember that according to Hund's rule, each electron in the p^3 configuration resides in a different p orbital.)

SAMPLE EXERCISE 7.6

Referring to a periodic table, arrange the following atoms in order of increasing first ionization energy: Ne, Na, P, Ar, K.

Solution

Analyze and Plan: We are given the chemical symbols for five elements. In order to rank them according to increasing first ionization energy, we need to locate each element in the periodic table. We can then use their relative positions and the trends in first ionization energies to predict their order.

Solve: The ionization energy increases as we move left to right across a row. It decreases as we move from the top of a group to the bottom. Because Na, P, and Ar are in the same row of the periodic table, we expect I_1 to vary in the order

Na < P < Ar

Because Ne is above Ar in group 8A, we expect Ne to exhibit the greater first ionization energy:

Ar < Ne

Similarly, K is the alkali metal directly below Na in group 1A, so we expect I_1 for K to be less than that of Na:

K < Na

From these observations we conclude that the ionization energies follow the order

K < Na < P < Ar < Ne

Check: The values shown in Figure 7.10 confirm this prediction.

PRACTICE EXERCISE

Based on the trends discussed in this section, predict which of the following atoms—B, Al, C, or Si—has the lowest first ionization energy and which has the highest first ionization energy.
Answer: Al has the lowest and C has the highest

Electron Configurations of Ions

When electrons are removed from an atom to form a cation, they are always removed first from the orbitals with the largest available principal quantum number, n. For example, when one electron is removed from a lithium atom ($1s^22s^1$), it is the $2s^1$ electron that is removed:

$$\text{Li } (1s^22s^1) \Rightarrow \text{Li}^+ (1s^2)$$

Likewise, when two electrons are removed from Fe ([Ar]$3d^64s^2$), the $4s^2$ electrons are the ones removed:

$$\text{Fe } ([\text{Ar}]3d^64s^2) \Rightarrow \text{Fe}^{2+} ([\text{Ar}]3d^6)$$

If an additional electron is removed, forming Fe^{3+}, it now comes from a 3d orbital because all the orbitals with $n = 4$ are now empty:

$$\text{Fe}^{2+} ([\text{Ar}]3d^6) \Rightarrow \text{Fe}^{3+} ([\text{Ar}]3d^5)$$

It may seem odd that the 4s electrons are removed before the 3d electrons in forming transition-metal cations. After all, in writing electron configurations the

4*s* electrons were added before the 3*d*. In writing electron configurations for atoms, however, we are going through an imaginary process in which we move through the periodic table from one element to another. In doing so, we are not only adding an electron, but also a proton to the nucleus, in order to change the identity of the element. In ionization, we do not reverse this process because electrons, but not protons, are removed.

When electrons are added to an atom to form an anion, they are added to an empty or partially filled orbital with the lowest available value of *n*. For example, when an electron is added to a fluorine atom to form the F^- ion, the electron goes into the one remaining vacancy in the 2*p* subshell:

$$\text{F}\ (1s^22s^22p^5) \Rightarrow \text{F}^-\ (1s^22s^22p^6)$$

SAMPLE EXERCISE 7.7

Write the electron configurations for the **(a)** Ca^{2+} ion; **(b)** the Co^{3+} ion; and **(c)** the S^{2-} ion.

Solution

Analyze and Plan: We are asked to write electron configurations for several ions. To do so, we first write the electron configuration of the parent atom. We then remove electrons to form cations or add electrons to form anions. Electrons are first removed from the orbitals with the highest value of *n*. They are added to an empty or partially filled orbitals with the lowest value of *n*.

Solve: (a) Calcium (atomic number 20) has an electron configuration of — Ca: $[Ar]4s^2$

To form a 2+ ion, the two outer electrons must be removed giving an ion that is isoelectronic with Ar: — Ca^{2+}: [Ar]

(b) Cobalt (atomic number 27) has an electron configuration of — Co: $[Ar]3d^74s^2$

To form a 3+ ion, three electrons must be removed. As discussed in the text preceding this Sample Exercise, the 4*s* electrons are removed before the 3*d* electrons. Consequently, the Co^{3+} ion has an electron configuration of — Co^{3+}: $[Ar]3d^6$

(c) Sulfur (atomic number 16) has an electron configuration of — S: $[Ne]3s^23p^4$

To form a 2− ion, two electrons must be added. There is room for two additional electrons in the 3*p* orbitals. Thus, the S^{2-} ion has an electron configuration of — S^{2-}: $[Ne]3s^23p^6 = [Ar]$

PRACTICE EXERCISE

Write the electron configuration for the **(a)** Ga^{3+} ion; **(b)** Cr^{3+} ion; and **(c)** Br^- ion.
Answers: **(a)** $[Ar]3d^{10}$; **(b)** $[Ar]3d^3$; **(c)** $[Ar]3d^{10}4s^24p^6 = [Kr]$

7.5 Electron Affinities

The ionization energy measures the energy changes associated with removing electrons from an atom to form positively charged ions. For example, the first ionization energy of Cl(*g*), 1251 kJ/mol, is the energy change associated with the following process:

$$\textit{Ionization energy:}\quad \underset{[Ne]3s^23p^5}{Cl(g)} \longrightarrow \underset{[Ne]3s^23p^4}{Cl^+(g)} + e^- \qquad \Delta E = 1251\ \text{kJ/mol} \qquad [7.5]$$

The positive value of the ionization energy means that energy must be put into the atom in order to remove the electron.

In addition, most atoms can gain electrons to form negatively charged ions. The energy change that occurs when an electron is added to a gaseous atom is called the **electron affinity** because it measures the attraction, or *affinity*, of the atom for the added electron. For most atoms, energy is released when an electron is added. For example, the addition of an electron to a chlorine atom is accompanied by an energy change of −349 kJ/mol, the negative sign indicating that energy is released during the process. We therefore say that the electron affinity of Cl is −349 kJ/mol:*

$$\textit{Electron affinity:}\quad \underset{[\mathrm{Ne}]3s^23p^5}{\mathrm{Cl}(g)} + e^- \longrightarrow \underset{[\mathrm{Ne}]3s^23p^6}{\mathrm{Cl}^-(g)} \qquad \Delta E = -349\ \mathrm{kJ/mol} \qquad [7.6]$$

It is important to understand the differences between ionization energy and electron affinity: Ionization energy measures the ease with which an atom *loses* an electron, whereas electron affinity measures the ease with which an atom *gains* an electron.

The greater the attraction between a given atom and an added electron, the more negative the atom's electron affinity will be. For some elements, such as the noble gases, the electron affinity has a positive value, meaning that the anion is higher in energy than are the separated atom and electron:

$$\underset{[\mathrm{Ne}]3s^23p^6}{\mathrm{Ar}(g)} + e^- \longrightarrow \underset{[\mathrm{Ne}]3s^23p^64s^1}{\mathrm{Ar}^-(g)} \qquad \Delta E > 0 \qquad [7.7]$$

Because $\Delta E > 0$, the Ar^- ion is unstable and does not form.

ANIMATION
Periodic Trends: Electron Affinity

Figure 7.11 ▼ shows the electron affinities for the representative elements in the first five rows of the periodic table. The electron affinity generally becomes increasingly negative as we proceed in each row toward the halogens. The halogens, which are one electron shy of a filled *p* subshell, have the most negative electron affinities. By gaining an electron, a halogen atom forms a stable negative ion that has a noble-gas configuration (Equation 7.6). The addition of an electron to a noble gas, however, would require that the electron reside in a new, higher-energy subshell (Equation 7.7). Occupying a higher-energy subshell is energetically very unfavorable, so the electron affinity is highly positive. The electron affinities of Be and Mg are positive for the same reason; the added electron would reside in a previously empty *p* subshell that is higher in energy.

---

* Two sign conventions are used for electron affinity. In most introductory texts, including this one, the thermodynamic sign convention is used: A negative sign indicates that the addition of an electron is an exothermic process, as in the electron affinity given for chlorine, −349 kJ/mol. Historically, however, electron affinity has been defined as the energy released when an electron is added to a gaseous atom or ion. Because 349 kJ/mol are *released* when an electron is added to Cl(*g*), the electron affinity by this convention would be +349 kJ/mol.

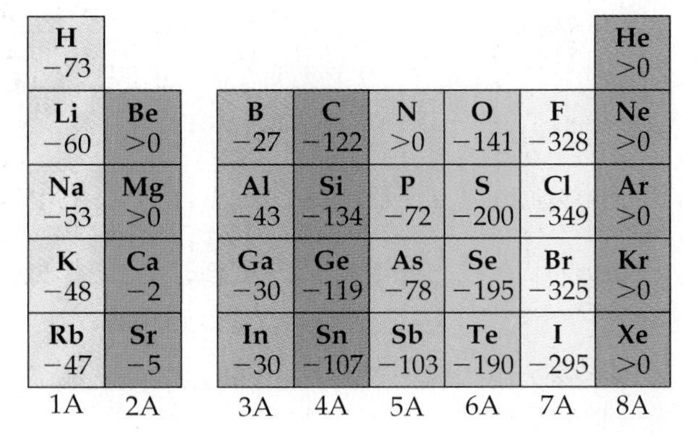

1A	2A	3A	4A	5A	6A	7A	8A
H −73							He >0
Li −60	Be >0	B −27	C −122	N >0	O −141	F −328	Ne >0
Na −53	Mg >0	Al −43	Si −134	P −72	S −200	Cl −349	Ar >0
K −48	Ca −2	Ga −30	Ge −119	As −78	Se −195	Br −325	Kr >0
Rb −47	Sr −5	In −30	Sn −107	Sb −103	Te −190	I −295	Xe >0

◀ **Figure 7.11** Electron affinities in kJ/mol for the representative elements in the first five periods of the periodic table. The more negative the electron affinity, the greater the attraction of the atom for an electron. An electron affinity > 0 indicates that the negative ion is higher in energy than the separated atom and electron.

The electron affinities of the group 5A elements (N, P, As, Sb) are also interesting. Because these elements have half-filled *p* subshells, the added electron must be put in an orbital that is already occupied, resulting in larger electron-electron repulsions. As a result, these elements have electron affinities that are either positive (N) or less negative than their neighbors to the left (P, As, Sb).

Electron affinities do not change greatly as we move down a group. For example, consider the electron affinities of the halogens (Figure 7.11). For F, the added electron goes into a 2*p* orbital, for Cl a 3*p* orbital, for Br a 4*p* orbital, and so forth. As we proceed from F to I, therefore, the average distance of the added electron from the nucleus steadily increases, causing the electron-nucleus attraction to decrease. The orbital that holds the outermost electron is increasingly spread out, however, as we proceed from F to I, thereby reducing the electron-electron repulsions. A lower electron-nucleus attraction is thus counterbalanced by lower electron-electron repulsions.

7.6 Metals, Nonmetals, and Metalloids

The concepts of atomic radii, ionization energies, and electron affinities are properties of individual atoms. With the exception of the noble gases, however, none of the elements exist in nature as individual atoms. To get a broader understanding of the properties of elements, we must also examine periodic trends in properties that involve large collections of atoms.

The elements can be broadly grouped into the categories of metals, nonmetals, and metalloids. (Section 2.5) This classification is shown in Figure 7.12 ▼. Roughly three quarters of the elements are metals, situated in the left and middle portions of the table. The nonmetals are located at the top right corner, and the metalloids lie between the metals and nonmetals. Hydrogen, which is located at the top left corner, is a nonmetal. This is why we set off hydrogen from the remaining group 1A elements. Some of the distinguishing properties of metals and nonmetals are summarized in Table 7.3 ▶.

The more an element exhibits the physical and chemical properties of metals, the greater its **metallic character**. Similarly, we can speak of the *nonmetallic*

Increasing metallic character ←

Increasing metallic character ↓

1A 1	2A 2	3B 3	4B 4	5B 5	6B 6	7B 7	8B 8	8B 9	8B 10	1B 11	2B 12	3A 13	4A 14	5A 15	6A 16	7A 17	8A 18
1 H																	2 He
3 Li	4 Be											5 B	6 C	7 N	8 O	9 F	10 Ne
11 Na	12 Mg											13 Al	14 Si	15 P	16 S	17 Cl	18 Ar
19 K	20 Ca	21 Sc	22 Ti	23 V	24 Cr	25 Mn	26 Fe	27 Co	28 Ni	29 Cu	30 Zn	31 Ga	32 Ge	33 As	34 Se	35 Br	36 Kr
37 Rb	38 Sr	39 Y	40 Zr	41 Nb	42 Mo	43 Tc	44 Ru	45 Rh	46 Pd	47 Ag	48 Cd	49 In	50 Sn	51 Sb	52 Te	53 I	54 Xe
55 Cs	56 Ba	71 Lu	72 Hf	73 Ta	74 W	75 Re	76 Os	77 Ir	78 Pt	79 Au	80 Hg	81 Tl	82 Pb	83 Bi	84 Po	85 At	86 Rn
87 Fr	88 Ra	103 Lr	104 Rf	105 Db	106 Sg	107 Bh	108 Hs	109 Mt	110	111	112		114		116		

57 La	58 Ce	59 Pr	60 Nd	61 Pm	62 Sm	63 Eu	64 Gd	65 Tb	66 Dy	67 Ho	68 Er	69 Tm	70 Yb
89 Ac	90 Th	91 Pa	92 U	93 Np	94 Pu	95 Am	96 Cm	97 Bk	98 Cf	99 Es	100 Fm	101 Md	102 No

Metals

Metalloids

Nonmetals

▲ **Figure 7.12** The periodic table, showing metals, metalloids, nonmetals, and trends in metallic character.

TABLE 7.3 Characteristic Properties of Metals and Nonmetals

Metals	Nonmetals
Have a shiny luster; various colors, although most are silvery	Do not have a luster; various colors
Solids are malleable and ductile	Solids are usually brittle; some are hard, and some are soft
Good conductors of heat and electricity	Poor conductors of heat and electricity
Most metal oxides are ionic solids that are basic	Most nonmetal oxides are molecular substances that form acidic solutions
Tend to form cations in aqueous solution	Tend to form anions or oxyanions in aqueous solution

character of an element. As indicated in Figure 7.12, the metallic character generally increases as we proceed down a column of the periodic table and decreases as we proceed from left to right in a row. Let's now examine the close relationships that exist between electron configurations and the properties of metals, nonmetals, and metalloids.

Metals

Most metallic elements exhibit the shiny luster that we associate with metals (Figure 7.13 ▶). Metals conduct heat and electricity. They are malleable (can be pounded into thin sheets) and ductile (can be drawn into wires). All are solids at room temperature except mercury (melting point $= -39°C$), which is a liquid. Two melt at slightly above room temperature: cesium at 28.4°C and gallium at 29.8°C. At the other extreme, many metals melt at very high temperatures. For example, chromium melts at 1900°C.

Metals tend to have low ionization energies and therefore tend to form positive ions relatively easily. As a result, metals are oxidized (lose electrons) when they undergo chemical reactions. The relative ease of oxidation of common metals was discussed earlier, in Section 4.6. As we noted then, many metals are oxidized by a variety of common substances including O_2 and acids.

▲ **Figure 7.13** Metallic objects are readily recognized by their characteristic luster.

Figure 7.14 ▼ shows the charges of some common ions. As we noted in Section 2.7, the charges of the alkali metals are always 1+ and those of the alkaline earth metals are always 2+ in their compounds. For each of these groups, the outer *s* electrons are easily lost, yielding a noble-gas electron configuration. The charges of the transition metal ions do not follow an obvious pattern. Many transition-metal ions have 2+ charges, but 1+ and 3+ are also encountered. One of the characteristic features of the transition metals is their ability to form more than one positive ion. For example, iron may be 2+ in some compounds and 3+ in others.

1A	2A	Transition metals										3A	4A	5A	6A	7A	8A
H^+																H^-	NOBLE GASES
Li^+														N^{3-}	O^{2-}	F^-	
Na^+	Mg^{2+}											Al^{3+}		P^{3-}	S^{2-}	Cl^-	
K^+	Ca^{2+}				Cr^{3+}	Mn^{2+}	Fe^{2+} Fe^{3+}	Co^{2+}	Ni^{2+}	Cu^+ Cu^{2+}	Zn^{2+}				Se^{2-}	Br^-	
Rb^+	Sr^{2+}									Ag^+	Cd^{2+}		Sn^{2+}		Te^{2-}	I^-	
Cs^+	Ba^{2+}								Pt^{2+}	Au^+ Au^{3+}	Hg_2^{2+} Hg^{2+}		Pb^{2+}	Bi^{3+}			

▲ **Figure 7.14** Charges of some common ions found in ionic compounds. Notice that the steplike line that divides metals from nonmetals also separates cations from anions.

NiO

(a)

(b)

▶ **Figure 7.15** (a) Nickel oxide (NiO), nitric acid (HNO_3), and water. (b) NiO is insoluble in water, but reacts with HNO_3 to give a green solution of $Ni(NO_3)_2$.

Compounds of metals with nonmetals tend to be ionic substances. For example, most metal oxides and halides are ionic solids. To illustrate, the reaction between nickel metal and oxygen produces nickel oxide, an ionic solid containing Ni^{2+} and O^{2-} ions:

$$2Ni(s) + O_2(g) \longrightarrow 2NiO(s) \qquad [7.8]$$

ANIMATION
Periodic Trends: Acid-Base Behavior of Oxides

The oxides are particularly important because of the great abundance of oxygen in our environment.

Most metal oxides are basic. Those that dissolve in water react to form metal hydroxides, as in the following examples:

$$\text{Metal oxide} + \text{water} \longrightarrow \text{metal hydroxide}$$

$$Na_2O(s) + H_2O(l) \longrightarrow 2NaOH(aq) \qquad [7.9]$$

$$CaO(s) + H_2O(l) \longrightarrow Ca(OH)_2(aq) \qquad [7.10]$$

The basicity of metal oxides is due to the oxide ion, which reacts with water according to the following net ionic equation:

$$O^{2-}(aq) + H_2O(l) \rightarrow 2OH^-(aq) \qquad [7.11]$$

Metal oxides also demonstrate their basicity by reacting with acids to form salts and water, as illustrated in Figure 7.15 ▲:

$$\text{Metal oxide} + \text{acid} \longrightarrow \text{salt} + \text{water}$$

$$NiO(s) + 2HCl(aq) \longrightarrow NiCl_2(aq) + H_2O(l) \qquad [7.12]$$

In contrast, we will soon see that nonmetal oxides are acidic, dissolving in water to form acidic solutions and reacting with bases to form salts.

SAMPLE EXERCISE 7.8

(a) Would you expect aluminum oxide to be a solid, liquid, or gas at room temperature? **(b)** Write the balanced chemical equation for the reaction of aluminum oxide with nitric acid.

Solution

Analyze and Plan: We are asked about some of the physical and chemical properties of aluminum oxide, a compound of a metal and nonmetal.

Solve: (a) Because aluminum oxide is the oxide of a metal, we would expect it to be an ionic solid. Indeed it is, and it has a very high melting point, 2072°C.

(b) In its compounds, aluminum has a 3+ charge, Al^{3+}; the oxide ion is O^{2-}. Consequently, the formula of aluminum oxide is Al_2O_3. Metal oxides tend to be basic and therefore to react with acids to form salts and water. In this case the salt is aluminum nitrate, $Al(NO_3)_3$. The balanced chemical equation is

$$Al_2O_3(s) + 6HNO_3(aq) \longrightarrow 2Al(NO_3)_3(aq) + 3H_2O(l)$$

PRACTICE EXERCISE

Write the balanced chemical equation for the reaction between copper(II) oxide and sulfuric acid.

Answer: $CuO(s) + H_2SO_4(aq) \longrightarrow CuSO_4(aq) + H_2O(l)$

Nonmetals

Nonmetals vary greatly in appearance (Figure 7.16 ▶). They are not lustrous and generally are poor conductors of heat and electricity. Their melting points are generally lower than those of metals (although diamond, a form of carbon, melts at 3570°C). Seven nonmetals exist under ordinary conditions as diatomic molecules. Five of them are gases (H_2, N_2, O_2, F_2, and Cl_2), one is a liquid (Br_2), and one is a volatile solid (I_2). The remaining nonmetals are solids that can be hard like diamond or soft like sulfur.

▲ **Figure 7.16** Nonmetals are diverse in their appearances. Shown here are (clockwise from left) carbon as graphite, sulfur, white phosphorus (stored under water), and iodine.

Because of their electron affinities, nonmetals tend to gain electrons when they react with metals. For example, the reaction of aluminum with bromine produces aluminum bromide, an ionic compound containing the aluminum ion, Al^{3+}, and the bromide ion, Br^-:

$$2Al(s) + 3Br_2(l) \longrightarrow 2AlBr_3(s) \quad [7.13]$$

A nonmetal typically will gain enough electrons to fill its outer p subshell completely, giving a noble-gas electron configuration. For example, the bromine atom gains one electron to fill its $4p$ subshell:

$$\text{Br } ([\text{Ar}]4s^23d^{10}4p^5) \Rightarrow \text{Br}^- \ ([\text{Ar}]4s^23d^{10}4p^6)$$

Compounds composed entirely of nonmetals are molecular substances. For example, the oxides, halides, and hydrides of the nonmetals are molecular substances that tend to be gases, liquids, or low-melting solids at room temperature.

Most nonmetal oxides are acidic; those that dissolve in water react to form acids, as in the following examples:

$$\text{Nonmetal oxide} + \text{water} \longrightarrow \text{acid}$$

$$CO_2(g) + H_2O(l) \longrightarrow H_2CO_3(aq) \quad [7.14]$$

$$P_4O_{10}(s) + 6H_2O(l) \longrightarrow 4H_3PO_4(aq) \quad [7.15]$$

The reaction of carbon dioxide with water (Figure 7.17 ▼) accounts for the acidity of carbonated water and, to some extent, rainwater. Because sulfur is present in oil and coal, combustion of these common fuels produces sulfur dioxide and

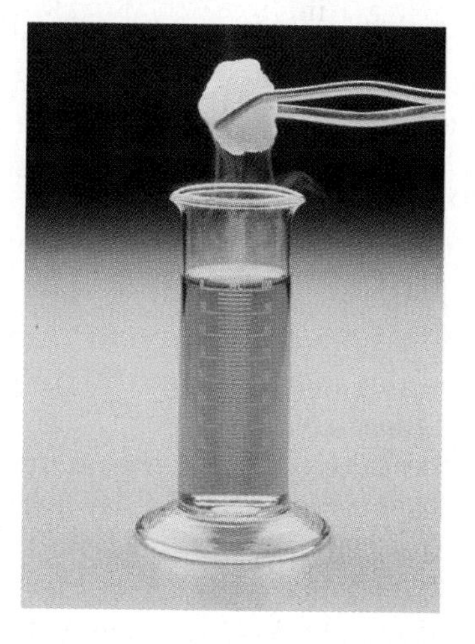

(a)

(b)

◀ **Figure 7.17** The reaction of CO_2 with water. (a) Water has been made slightly basic and contains a few drops of bromthymol blue, an acid-base indicator that is blue in basic solution. (b) When Dry Ice™, $CO_2(s)$, is added, the color changes to yellow, indicating an acidic solution. The mist is due to water droplets condensed from the air by the cold CO_2 gas.

sulfur trioxide. These substances dissolve in water to produce *acid rain*, a major pollution problem in many parts of the world. Like acids, most nonmetal oxides dissolve in basic solutions to form salts:

$$\text{Nonmetal oxide} + \text{base} \longrightarrow \text{salt} + \text{water}$$

$$CO_2(g) + 2NaOH(aq) \longrightarrow Na_2CO_3(aq) + H_2O(l) \quad [7.16]$$

SAMPLE EXERCISE 7.9

Write the balanced chemical equations for the reactions of solid selenium dioxide with **(a)** water, **(b)** aqueous sodium hydroxide.

Solution

Analyze and Plan: We need to write chemical equations for the reaction of a nonmetal oxide first with water then with a base, NaOH. Nonmetal oxides are acidic, reacting with water to form acids and with bases to form salts and water.

Solve: (a) Selenium dioxide is SeO_2. Its reaction with water is like that of carbon dioxide (Equation 7.14):

$$SeO_2(s) + H_2O(l) \longrightarrow H_2SeO_3(aq)$$

(It doesn't matter that SeO_2 is a solid and CO_2 is a gas; the point is that both are water-soluble nonmetal oxides.)

(b) The reaction with sodium hydroxide is like the reaction summarized by Equation 7.16:

$$SeO_2(s) + 2NaOH(aq) \longrightarrow Na_2SeO_3(aq) + H_2O(l)$$

PRACTICE EXERCISE

Write the balanced chemical equation for the reaction of solid tetraphosphorus hexoxide with water.
Answer: $P_4O_6(s) + 6H_2O(l) \longrightarrow 4H_3PO_3(aq)$

▲ **Figure 7.18** Elemental silicon, which is a metalloid. Although it looks metallic, silicon is brittle and is a poor thermal and electrical conductor compared to metals. Large crystals of silicon are sliced into thin wafers for use in integrated circuits.

Metalloids

Metalloids have properties intermediate between those of metals and nonmetals. They may have *some* characteristic metallic properties, but lack others. For example, silicon *looks* like a metal (Figure 7.18 ◀), but it is brittle rather than malleable and is a much poorer conductor of heat and electricity than are metals. Several of the metalloids, most notably silicon, are electrical semiconductors and are the principal elements used in the manufacture of integrated circuits and computer chips.

7.7 Group Trends for the Active Metals

Our discussion of atomic size, ionization energy, electron affinity, and metallic character gives some idea of the way the periodic table can be used to organize and remember facts. Not only do elements in a group possess general similarities, but there are also trends as we move through a group or from one group to another. In this section we will use the periodic table and our knowledge of electron configurations to examine the chemistry of the **alkali metals** (group 1A) and the **alkaline earth metals** (group 2A).

Group 1A: The Alkali Metals

▲ **Figure 7.19** Sodium and the other alkali metals are soft enough to be cut with a knife. The shiny metallic surface quickly tarnishes as the sodium reacts with oxygen in the air.

The alkali metals are soft metallic solids (Figure 7.19 ◀). All have characteristic metallic properties such as a silvery, metallic luster and high thermal and electrical conductivities. The name *alkali* comes from an Arabic word meaning "ashes." Many compounds of sodium and potassium, two alkali metals, were isolated from wood ashes by early chemists.

Sodium and potassium are among the most abundant elements in Earth's crust, in seawater, and in biological systems. We all have sodium ions in our bodies. If we ingest too much, however, it can raise our blood pressure. Potassium is also prevalent in our bodies; a 140-pound person contains about 130 g of potassium, as K^+ ion in intracellular fluids. Plants require potassium for growth and development (Figure 7.20 ▶).

TABLE 7.4 Some Properties of the Alkali Metals

Element	Electron Configuration	Melting Point (°C)	Density (g/cm^3)	Atomic Radius (Å)	I_1 (kJ/mol)
Lithium	$[He]2s^1$	181	0.53	1.34	520
Sodium	$[Ne]3s^1$	98	0.97	1.54	496
Potassium	$[Ar]4s^1$	63	0.86	1.96	419
Rubidium	$[Kr]5s^1$	39	1.53	2.11	403
Cesium	$[Xe]6s^1$	28	1.88	2.60	376

Some of the physical and chemical properties of the alkali metals are given in Table 7.4 ▲. The elements have low densities and melting points, and these properties vary in a fairly regular way with increasing atomic number. We can also see some of the usual trends as we move down the group, such as increasing atomic radius and decreasing first ionization energy. For each row of the periodic table, the alkali metal has the lowest I_1 value (Figure 7.9), which reflects the relative ease with which its outer *s* electron can be removed. As a result, the alkali metals are all very reactive, readily losing one electron to form ions with a 1+ charge. ∞ (Section 4.4)

The alkali metals exist in nature only as compounds. The metals combine directly with most nonmetals. For example, they react with hydrogen to form hydrides, and with sulfur to form sulfides

$$2M(s) + H_2(g) \longrightarrow 2MH(s) \quad [7.17]$$

$$2M(s) + S(s) \longrightarrow M_2S(s) \quad [7.18]$$

(The symbol M in Equations 7.17 and 7.18 represents any one of the alkali metals.) In hydrides of the alkali metals (LiH, NaH, and so forth), hydrogen is present as H^-, called the **hydride ion**. The hydride ion is distinct from the hydrogen ion, H^+, formed when a hydrogen atom loses its electron.

The alkali metals react vigorously with water, producing hydrogen gas and solutions of alkali metal hydroxides:

$$2M(s) + 2H_2O(l) \longrightarrow 2MOH(aq) + H_2(g) \quad [7.19]$$

These reactions are very exothermic. In many cases enough heat is generated to ignite the H_2, producing a fire or explosion (Figure 7.21 ▼). This reaction is most violent for the heavier members of the group, in keeping with their weaker hold on the single valence electron.

▲ **Figure 7.20** Fertilizers applied to these farm fields typically contain large quantities of potassium, phosphorus, and nitrogen to meet the needs of growing plants.

MOVIE
Sodium and Potassium in Water

(a)

(b)

(c)

▲ **Figure 7.21** The alkali metals react vigorously with water. (a) The reaction of lithium is shown by the bubbling of escaping hydrogen gas. (b) The reaction of sodium is more rapid and is so exothermic that the hydrogen gas that is produced burns in air. (c) Potassium reacts almost explosively.

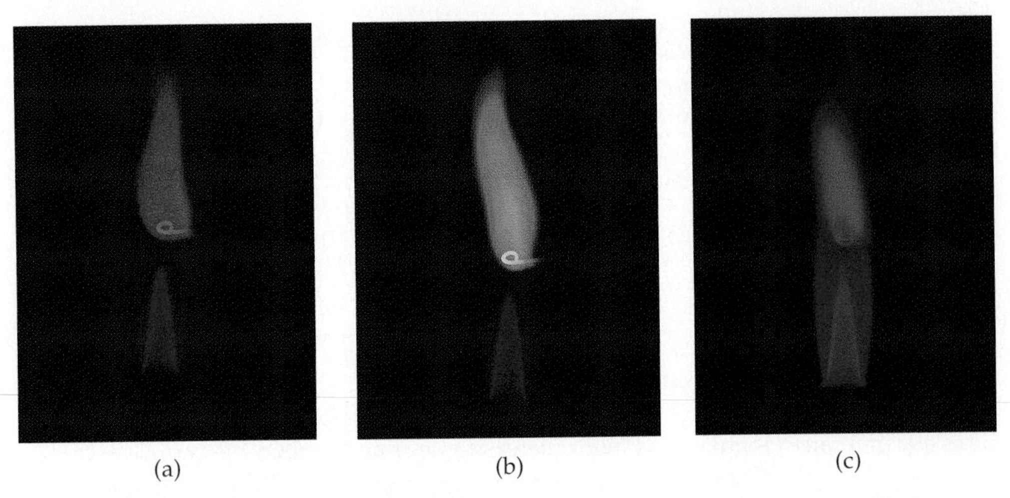

▲ **Figure 7.22** Flame test for (a) Li (crimson red), (b) Na (yellow), and (c) K (lilac).

The reactions between the alkali metals and oxygen are complex. Oxygen usually reacts with metals to form metal oxides, which contain the O^{2-} ion. Indeed, lithium reacts in this manner to form lithium oxide, Li_2O:

$$4Li(s) + O_2(g) \longrightarrow 2Li_2O(s) \quad [7.20]$$

The other alkali metals, however, all react with oxygen to form metal peroxides, which contain the O_2^{2-} ion. For example, sodium forms sodium peroxide, Na_2O_2:

$$2Na(s) + O_2(g) \longrightarrow Na_2O_2(s) \quad [7.21]$$

Surprisingly, potassium, rubidium, and cesium also form MO_2 compounds that contain O_2^-, which we call the superoxide ion. For example, potassium forms potassium superoxide, KO_2:

$$K(s) + O_2(g) \longrightarrow KO_2(s) \quad [7.22]$$

Although alkali metal ions are colorless, they emit characteristic colors when placed in a flame (Figure 7.22 ▲). The alkali metal ions are reduced to gaseous metal atoms in the central region of the flame. The high temperature of the flame electronically excites the valence electron. The atom then emits energy in the form of visible light as it returns to its ground state. Sodium gives a yellow flame because of emission at 589 nm. This wavelength is produced when the excited valence electron drops from the 3*p* subshell to the lower-energy 3*s* subshell. The characteristic yellow emission of sodium is the basis for sodium vapor lamps (Figure 7.23 ▼).

▶ **Figure 7.23** Sodium vapor lamps, which are used for commercial and highway lighting, have a yellow glow due to the emission from excited sodium atoms.

SAMPLE EXERCISE 7.10

Write balanced equations that predict the reactions of cesium metal with **(a)** $Cl_2(g)$; **(b)** $H_2O(l)$; **(c)** $H_2(g)$.

Solution

Analyze and Plan: Cesium is an alkali metal. We therefore expect that its chemistry will be dominated by oxidation of the metal to Cs^+ ions. Further, we recognize that Cs is far down the periodic table, which means it will be among the most active of all metals and will probably react with all three of the substances listed.

Solve: The reaction between Cs and Cl_2 is a simple combination reaction between two elements, one a metal and one a nonmetal, forming the ionic compound CsCl:

$$2Cs(s) + Cl_2(g) \longrightarrow 2CsCl(s)$$

By analogy to Equations 7.19 and 7.17, respectively, we predict the reactions of cesium with water and hydrogen to proceed as follows:

$$2Cs(s) + 2H_2O(l) \longrightarrow 2CsOH(aq) + H_2(g)$$

$$2Cs(s) + H_2(g) \longrightarrow 2CsH(s)$$

In each case cesium forms a Cs^+ ion in its compounds. The chloride (Cl^-), hydroxide (OH^-), and hydride (H^-) ions are all 1− ions, so the final products have 1:1 stoichiometry with Cs^+.

PRACTICE EXERCISE

Write a balanced equation that predicts the products of the reaction between potassium metal and elemental sulfur.

Answer: $2K(s) + S(s) \longrightarrow K_2S(s)$

Chemistry and Life The Improbable Development of Lithium Drugs

The alkali metal ions tend to play a rather unexciting role in most chemical reactions in general chemistry. All salts of the alkali metal ions are soluble, and the ions are spectators in most aqueous reactions (except for those involving the alkali metals in their elemental form, such as in Equation 7.19).

The alkali metal ions play an important role in human physiology, however. Sodium and potassium ions are major components of blood plasma and intracellular fluid, respectively, with average concentrations on the order of 0.1 *M*. These electrolytes serve as vital charge carriers in normal cellular function, and they are two of the principal ions involved in regulation of the heart.

In contrast, the lithium ion (Li^+) has no known function in normal human physiology. Since the discovery of lithium in 1817, however, salts of the element were thought to possess almost mystical healing powers; there were claims that it was an ingredient in ancient "fountain of youth" formulas. In 1927 Mr. C. L. Grigg began marketing a lithium-containing soft drink with the unwieldy name "Bib-Label Lithiated Lemon-Lime Soda." Grigg soon gave his lithiated beverage a much simpler name: Seven-Up® (Figure 7.24 ▶).

◀ **Figure 7.24** The soft drink Seven-Up® originally contained lithium citrate, the lithium salt of citric acid. The lithium was claimed to give the beverage healthful benefits, including "an abundance of energy, enthusiasm, a clear complexion, lustrous hair, and shining eyes!" The lithium was removed from the beverage in the early 1950s, about the same time that the antipsychotic action of Li^+ was discovered.

Because of concerns from the Food and Drug Administration, lithium was removed from Seven-Up® in the early 1950s. At nearly the same time, it was found that the lithium ion has a remarkable therapeutic effect on the mental disorder called *bipolar affective disorder*, or *manic-depressive illness*. Over 1 million Americans suffer from this psychosis, undergoing severe mood swings from deep depression to euphoria. The lithium ion smoothes out these mood swings, allowing the patient to function more effectively in daily life.

The antipsychotic action of Li^+ was discovered by accident in the late 1940s by Australian psychiatrist John Cade. Cade was researching the use of uric acid—a component of urine—to treat manic-depressive illness. He administered the acid to manic laboratory animals in the form of its most soluble salt, lithium urate, and found that many of the manic symptoms seemed to disappear. Later studies showed that uric acid has no role in the therapeutic effects observed; rather, the Li^+ ions were responsible. Because lithium overdose can cause severe side effects in humans, including death, lithium salts were not approved as antipsychotic drugs for humans until 1970. Today Li^+ is usually administered orally in the form of $Li_2CO_3(s)$. Lithium drugs are effective for about 70% of the manic-depressive patients who take it.

In this age of sophisticated drug design and biotechnology, the simple lithium ion is still the most effective treatment of a destructive psychological disorder. Remarkably, in spite of intensive research, scientists still don't fully understand the biochemical action of lithium that leads to its therapeutic effects.

Group 2A: The Alkaline Earth Metals

Like the alkali metals, the group 2A elements are all solids with typical metallic properties, some of which are listed in Table 7.5 ▼. Compared with the alkali metals, the alkaline earth metals are harder and more dense, and they melt at higher temperatures.

The first ionization energies of the alkaline earth elements are low, but not as low as those of the alkali metals. Consequently, the alkaline earths are less reactive than their alkali metal neighbors. As we have noted in Section 7.4, the ease with which the elements lose electrons decreases as we move across the periodic table from left to right and increases as we move down a group. Thus, beryllium and magnesium, the lightest members of the group, are the least reactive.

The trend of increasing reactivity within the group is shown by the behavior of the elements toward water. Beryllium does not react with water or steam, even when heated red-hot. Magnesium does not react with liquid water, but it does react with steam to form magnesium oxide and hydrogen:

$$Mg(s) + H_2O(g) \longrightarrow MgO(s) + H_2(g) \quad [7.23]$$

▲ **Figure 7.25** Calcium metal reacts with water to form hydrogen gas and aqueous calcium hydroxide, $Ca(OH)_2(aq)$.

Calcium and the elements below it react readily with water at room temperature (although more slowly than the alkali metals adjacent to them in the periodic table), as shown in Figure 7.25 ◀:

$$Ca(s) + 2H_2O(l) \longrightarrow Ca(OH)_2(aq) + H_2(g) \quad [7.24]$$

The two preceding reactions illustrate the dominant pattern in the reactivity of the alkaline earth elements—the tendency to lose their two outer *s* electrons and form 2+ ions. For example, magnesium reacts with chlorine at room temperature to form $MgCl_2$, and it burns with dazzling brilliance in air to give MgO (Figure 3.5):

$$Mg(s) + Cl_2(g) \longrightarrow MgCl_2(s) \quad [7.25]$$

$$2Mg(s) + O_2(g) \longrightarrow 2MgO(s) \quad [7.26]$$

In the presence of O_2, magnesium metal is protected from many chemicals by a thin surface coating of water-insoluble MgO. Thus, even though it is high in the activity series (Section 4.4), Mg can be incorporated into lightweight structural alloys used in, for example, automobile wheels. The heavier alkaline earth metals (Ca, Sr, and Ba) are even more reactive toward nonmetals than is magnesium.

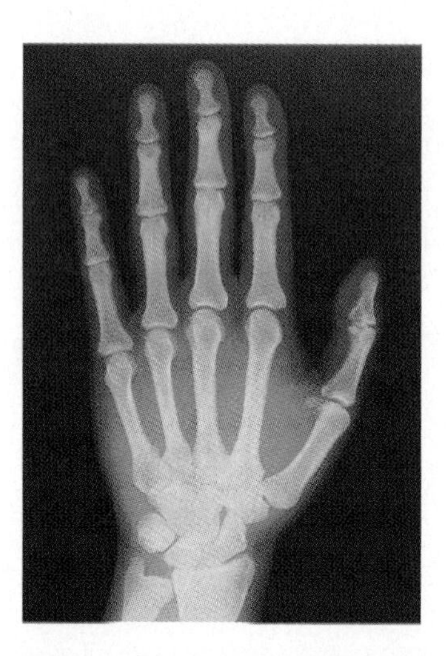

▲ **Figure 7.26** This X-ray photograph shows the bone structure of the human hand. The primary mineral in bone and teeth is hydroxyapatite, $Ca_5(PO_4)_3OH$, in which calcium is present as Ca^{2+}.

The heavier alkaline earth ions give off characteristic colors when strongly heated in a flame. The colored flame produced by calcium is brick red, that of strontium is crimson red, and that of barium is green. Strontium salts produce the brilliant red color in fireworks, and barium salts produce the green color.

Both magnesium and calcium are essential for living organisms (Figure 2.24). Calcium is particularly important for growth and maintenance of bones and teeth (Figure 7.26 ◀). In humans 99% of the calcium is found in the skeletal system.

TABLE 7.5 Some Properties of the Alkaline Earth Metals

Element	Electron Configuration	Melting Point (°C)	Density (g/cm^3)	Atomic Radius (Å)	I_1 (kJ/mol)
Beryllium	[He]$2s^2$	1287	1.85	0.90	899
Magnesium	[Ne]$3s^2$	650	1.74	1.30	738
Calcium	[Ar]$4s^2$	842	1.54	1.74	590
Strontium	[Kr]$5s^2$	777	2.63	1.92	549
Barium	[Xe]$6s^2$	727	3.51	2.15	503

7.8 Group Trends for Selected Nonmetals

Hydrogen

Hydrogen, the first element in the periodic table, has a $1s^1$ electron configuration and is usually placed above the alkali metals. However, it does not truly belong to any particular group. Unlike the alkali metals, hydrogen is a nonmetal that occurs as a colorless diatomic gas, $H_2(g)$, under most conditions. Nevertheless, hydrogen can be metallic at tremendous pressures. The interiors of the planets Jupiter and Saturn, for example, are believed to consist of a rocky core surrounded by a thick shell of metallic hydrogen. The metallic hydrogen is in turn surrounded by a layer of liquid molecular hydrogen with gaseous hydrogen occurring above that near the surface.

Owing to the complete absence of nuclear shielding of its sole electron, the ionization energy of hydrogen, 1312 kJ/mol, is markedly higher than that of the alkali metals. In fact, it is comparable to the I_1 values of other nonmetals, such as oxygen and chlorine. As a result, hydrogen has less tendency to lose an electron than do the alkali metals. Whereas the alkali metals readily lose their valence electron to nonmetals to form ionic compounds, hydrogen shares its electron with nonmetals forming molecular compounds. The reactions between hydrogen and nonmetals can be quite exothermic, as evidenced by the combustion reaction between hydrogen and oxygen to form water (Figure 5.14):

$$2H_2(g) + O_2(g) \longrightarrow 2H_2O(l) \qquad \Delta H^\circ = -571.7 \text{ kJ} \qquad [7.27]$$

We have also seen (Equation 7.17) that hydrogen reacts with active metals to form solid metal hydrides, which contain the hydride ion, H^-. The fact that hydrogen can gain an electron further illustrates that it is not truly a member of the alkali metal family. In fact, it suggests a slight resemblence between hydrogen and the halogens.

In spite of the tendency of hydrogen to form covalent bonds and its ability to even gain electrons, hydrogen can and does lose an electron to form a cation. Indeed, the aqueous chemistry of hydrogen is dominated by the $H^+(aq)$ ion, which we encountered in Chapter 4. We will study this important ion in greater detail in Chapter 16.

Group 6A: The Oxygen Group

As we proceed down group 6A, there is a change from nonmetallic to metallic character. Oxygen, sulfur, and selenium are typical nonmetals. Tellurium has some metallic properties and is classified as a metalloid. Polonium, which is radioactive and quite rare, is a metal. Oxygen is a colorless gas at room temperature; all of the others are solids. Some of the physical properties of the group 6A elements are given in Table 7.6 ▼.

TABLE 7.6 Some Properties of the Group 6A Elements

Element	Electron Configuration	Melting Point (°C)	Density	Atomic Radius (Å)	I_1 (kJ/mol)
Oxygen	$[He]2s^22p^4$	−218	1.43 g/L	0.73	1314
Sulfur	$[Ne]3s^23p^4$	115	1.96 g/cm^3	1.02	1000
Selenium	$[Ar]3d^{10}4s^24p^4$	221	4.82 g/cm^3	1.16	941
Tellurium	$[Kr]4d^{10}5s^25p^4$	450	6.24 g/cm^3	1.35	869
Polonium	$[Xe]4f^{14}5d^{10}6s^26p^4$	254	9.2 g/cm^3	1.9	812

As we saw in Section 2.6, oxygen is encountered in two molecular forms, O_2 and O_3. The O_2 form is the common one. People generally mean O_2 when they say "oxygen," although the name *dioxygen* is more descriptive. The O_3 form is called **ozone**. The two forms of oxygen are examples of *allotropes*. Allotropes are different forms of the same element in the same state. (In this case both forms are gases.) About 21% of dry air consists of O_2 molecules. Ozone, which is toxic and has a pungent odor, is present in very small amounts in the upper atmosphere and in polluted air. It is also formed from O_2 in electrical discharges, such as in lightning storms:

$$3O_2(g) \longrightarrow 2O_3(g) \qquad \Delta H^\circ = 284.6 \text{ kJ} \qquad [7.28]$$

This reaction is endothermic, so O_3 is less stable than O_2.

Oxygen has a great tendency to attract electrons from other elements (to *oxidize* them). Oxygen in combination with metals is almost always present as the oxide ion, O^{2-}. This ion has a noble-gas configuration and is particularly stable. As shown in Equation 7.27, the formation of nonmetal oxides is also often very exothermic and thus energetically favorable.

In our discussion of the alkali metals we noted two less common oxygen anions, namely, the peroxide (O_2^{2-}) and superoxide (O_2^{-}) ions. Compounds of these ions often react with themselves to produce an oxide and O_2. For example, aqueous hydrogen peroxide, H_2O_2, slowly decomposes into water and O_2 at room temperature:

$$2H_2O_2(aq) \longrightarrow 2H_2O(l) + O_2(g) \qquad \Delta H^\circ = -196.1 \text{ kJ} \qquad [7.29]$$

For this reason, bottles of aqueous hydrogen peroxide are topped with caps that are able to release the $O_2(g)$ produced before the pressure inside becomes too great (Figure 7.27 ◀).

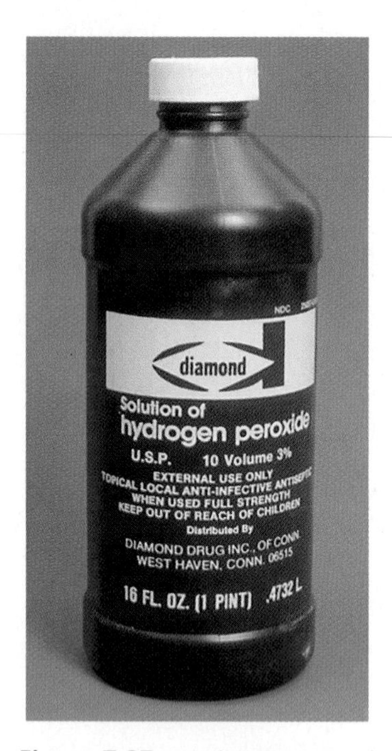

▲ **Figure 7.27** Hydrogen peroxide bottles are topped with caps that allow any excess pressure of $O_2(g)$ to be released from the bottle. Hydrogen peroxide is often stored in dark-colored or opaque bottles to minimize exposure to light, which accelerates its decomposition.

After oxygen, the most important member of group 6A is sulfur. Sulfur also exists in several allotropic forms, the most common and stable of which is the yellow solid with molecular formula S_8. This molecule consists of an eight-membered ring of sulfur atoms, as shown in Figure 7.28 ◀. Even though solid sulfur consists of S_8 rings, we usually write it simply as S(*s*) in chemical equations to simplify the coefficients.

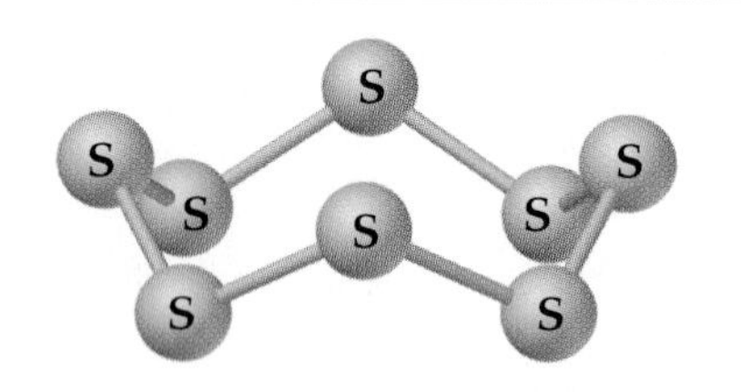

▲ **Figure 7.28** Structure of S_8 molecules as found in the most common allotropic form of sulfur at room temperature.

Like oxygen, sulfur has a tendency to gain electrons from other elements to form sulfides, which contain the S^{2-} ion. In fact, most sulfur in nature is present as metal sulfides. Because sulfur is below oxygen in the periodic table, its tendency to form sulfide anions is not as great as that of oxygen to form oxide ions. As a result, the chemistry of sulfur is more complex than that of oxygen. In fact, sulfur and its compounds (including those in coal and petroleum) can be burned in oxygen. The main product is sulfur dioxide, a major pollutant (Section 18.4):

$$S(s) + O_2(g) \longrightarrow SO_2(g) \qquad [7.30]$$

Group 7A: The Halogens

The group 7A elements are known as the **halogens**, after the Greek words *halos* and *gennao*, meaning "salt formers." Some of the properties of these elements are given in Table 7.7 ▼. Astatine, which is both extremely rare and radioactive, is omitted because many of its properties are not yet known.

TABLE 7.7 Some Properties of the Halogens

Element	Electron Configuration	Melting Point (°C)	Density	Atomic Radius (Å)	I_1 (kJ/mol)
Fluorine	$[He]2s^22p^5$	−220	1.69 g/L	0.71	1681
Chlorine	$[Ne]3s^23p^5$	−102	3.21 g/L	0.99	1251
Bromine	$[Ar]3d^{10}4s^24p^5$	−7.3	3.12 g/cm^3	1.14	1140
Iodine	$[Kr]4d^{10}5s^25p^5$	114	4.93 g/cm^3	1.33	1008

Unlike the group 6A elements, all the halogens are typical nonmetals. Their melting and boiling points increase with increasing atomic number. Fluorine and chlorine are gases at room temperature, bromine is a liquid, and iodine is a solid. Each element consists of diatomic molecules: F_2, Cl_2, Br_2, and I_2. Fluorine gas is pale yellow; chlorine gas is yellow-green; bromine liquid is reddish brown and readily forms a reddish brown vapor; and solid iodine is grayish black and readily forms a violet vapor (Figure 7.29 ▶).

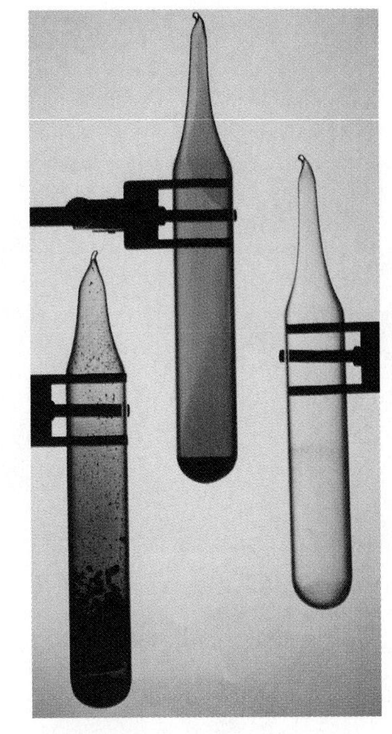

▲ **Figure 7.29** Iodine (I_2), bromine (Br_2), and chlorine (Cl_2) from left to right.

The halogens have highly negative electron affinities (Figure 7.11). Thus, it is not surprising that the chemistry of the halogens is dominated by their tendency to gain electrons from other elements to form halide ions, X^-. (In many equations X is used to indicate any one of the halogen elements.) Fluorine and chlorine are more reactive than bromine and iodine. In fact, fluorine removes electrons from almost any substance with which it comes into contact, including water, and usually does so very exothermically, as in the following examples:

$$2H_2O(l) + 2F_2(g) \longrightarrow 4HF(aq) + O_2(g) \qquad \Delta H = -758.9 \text{ kJ} \qquad [7.31]$$

$$SiO_2(s) + 2F_2(g) \longrightarrow SiF_4(g) + O_2(g) \qquad \Delta H = -704.0 \text{ kJ} \qquad [7.32]$$

As a result, fluorine gas is difficult and dangerous to use in the laboratory, requiring a special apparatus.

Chlorine is the most industrially useful of the halogens. In 2001 total production was 27 billion pounds, making it the eighth most produced chemical in the United States. Unlike fluorine, chlorine reacts slowly with water to form relatively stable aqueous solutions of HCl and HOCl (hypochlorous acid):

$$Cl_2(g) + H_2O(l) \longrightarrow HCl(aq) + HOCl(aq) \qquad [7.33]$$

Chlorine is often added to drinking water and swimming pools, where the HOCl(*aq*) that is generated serves as a disinfectant.

The halogens react directly with most metals to form ionic halides. The halogens also react with hydrogen to form gaseous hydrogen halide compounds:

$$H_2(g) + X_2 \longrightarrow 2HX(g) \qquad [7.34]$$

These compounds are all very soluble in water and dissolve to form the hydrohalic acids. As we discussed in Section 4.3, HCl(*aq*), HBr(*aq*), and HI(*aq*) are strong acids, whereas HF(*aq*) is a weak acid.

Group 8A: The Noble Gases

The group 8A elements, known as the **noble gases**, are all nonmetals that are gases at room temperature. They are all *monoatomic* (that is, they consist of single atoms rather than molecules). Some physical properties of the noble-gas elements are listed in Table 7.8 ▼. The high radioactivity of Rn has inhibited the study of its chemistry.

MOVIE
Physical Properties of the Halogens

TABLE 7.8 Some Properties of the Noble Gases

Element	Electron Configuration	Boiling Point (K)	Density (g/L)	Atomic Radius* (Å)	I_1 (kJ/mol)
Helium	$1s^2$	4.2	0.18	0.32	2372
Neon	$[He]2s^22p^6$	27.1	0.90	0.69	2081
Argon	$[Ne]3s^23p^6$	87.3	1.78	0.97	1521
Krypton	$[Ar]3d^{10}4s^24p^6$	120	3.75	1.10	1351
Xenon	$[Kr]4d^{10}5s^25p^6$	165	5.90	1.30	1170
Radon	$[Xe]4f^{14}5d^{10}6s^26p^6$	211	9.73	—	1037

* Only the heaviest of the noble-gas elements form chemical compounds. Thus, the atomic radii for the lighter noble-gas elements are predicted, estimated values.

The noble gases have completely filled *s* and *p* subshells. All elements of group 8A have large first ionization energies, and we see the expected decrease as we move down the column. Because the noble gases possess such stable electron configurations, they are exceptionally unreactive. In fact, until the early 1960s the elements were called the *inert gases* because they were thought to be incapable of forming chemical compounds. In 1962 Neil Bartlett at the University of British Columbia reasoned that the ionization energy of Xe might be low enough to allow it to form compounds. In order for this to happen, Xe would have to react with a substance with an extremely high ability to remove electrons from other substances, such as fluorine. Bartlett synthesized the first noble-gas compound by combining Xe with the fluorine-containing compound PtF_6. Xenon also reacts directly with $F_2(g)$ to form the molecular compounds XeF_2, XeF_4, and XeF_6 (Figure 7.30 ◀). Krypton has a higher I_1 value than xenon and is therefore less reactive. In fact, only a single stable compound of krypton is known, KrF_2. In 2000, Finnish scientists reported the HArF molecule, which is stable only at low temperatures.

▲ **Figure 7.30** Crystals of XeF_4, which is one of the very few compounds that contain a group 8A element.

SAMPLE INTEGRATIVE EXERCISE 7: Putting Concepts Together

(a) The covalent atomic radii of thallium (Tl) and lead (Pb) are 1.48 Å and 1.47 Å, respectively. Using these values and those in Figure 7.5, predict the covalent atomic radius of the element bismuth (Bi). Explain your answer.

(b) What accounts for the general increase in atomic radius going down the group 5A elements?

(c) A major use of bismuth has been as an ingredient in low-melting metal alloys, such as those used in fire sprinkler systems and in typesetting. The element itself is a brittle white crystalline solid. How do these characteristics fit with the fact that bismuth is in the same periodic group with such nonmetallic elements as nitrogen and phosphorus?

(d) Bi_2O_3 is a basic oxide. Write a balanced chemical equation for its reaction with dilute nitric acid. If 6.77 g of Bi_2O_3 is dissolved in dilute acidic solution to make up 500 mL of solution, what is the molarity of the solution of Bi^{3+} ion?

(e) ^{209}Bi is the heaviest stable isotope of any element. How many protons and neutrons are present in this nucleus?

(f) The density of Bi at 25°C is 9.808 g/cm^3. How many Bi atoms are present in a cube of the element that is 5.00 cm on each edge? How many moles of the element are present?

Solution (a) Note that there is a rather steady decrease in radius for the elements in the row preceding the one we are considering, that is, in the series In–Sn–Sb. It is reasonable to expect a decrease of about 0.02 Å in moving from Pb to Bi, leading to an estimate of 1.45 Å. The tabulated value is 1.46 Å.

(b) The general increase in radius with increasing atomic number in the group 5A elements occurs because additional shells of electrons are being added, with corresponding increases in nuclear charge. The core electrons in each case largely shield the outermost electrons from the nucleus, so the effective nuclear charge is not varying greatly as we go to higher atomic numbers. However, the principal quantum number, *n*, of the outermost electrons is steadily increasing, with a corresponding increase in orbital radius.

(c) The contrast between the properties of bismuth and those of nitrogen and phosphorus illustrates the general rule that there is a trend toward increased metallic character as we move down in a given group. Bismuth, in fact, is a metal. The increased metallic character occurs because the outermost electrons are more readily lost in bonding, a trend that is consistent with lower ionization energy.

(d) Following the procedures described in Section 4.2 for writing molecular and net ionic equations, we have the following:

Molecular equation:

$$Bi_2O_3(s) + 6HNO_3(aq) \longrightarrow 2Bi(NO_3)_3(aq) + 3H_2O(l)$$

Net ionic equation:

$$Bi_2O_3(s) + 6H^+(aq) \longrightarrow 2Bi^{3+}(aq) + 3H_2O(l)$$

In the net ionic equation, nitric acid is a strong acid and $Bi(NO_3)_3$ is a soluble salt, so we need show only the reaction of the solid with the hydrogen ion forming the $Bi^{3+}(aq)$ ion and water.

To calculate the concentration of the solution, we proceed as follows (Section 4.5):

$$\frac{6.77\text{ g }Bi_2O_3}{0.500\text{ L soln}} \times \frac{1\text{ mol }Bi_2O_3}{466.0\text{ g }Bi_2O_3} \times \frac{2\text{ mol }Bi^{3+}}{1\text{ mol }Bi_2O_3} = \frac{0.0581\text{ mol }Bi^{3+}}{\text{L soln}} = 0.0581\ M$$

(e) We can proceed as in Section 2.3. Bismuth is element 83; there are therefore 83 protons in the nucleus. Because the atomic mass number is 209, there are 209 − 83 = 126 neutrons in the nucleus.

(f) We proceed as in Sections 1.4 and 3.4: The volume of the cube is $(5.00)^3\text{ cm}^3 = 125\text{ cm}^3$. Then we have

$$125\text{ cm}^3\text{ Bi} \times \frac{9.780\text{ g Bi}}{1\text{ cm}^3} \times \frac{1\text{ mol Bi}}{209.0\text{ g Bi}} = 5.87\text{ mol Bi}$$

$$5.87\text{ mol Bi} \times \frac{6.022 \times 10^{23}\text{ atom Bi}}{1\text{ mol Bi}} = 3.54 \times 10^{24}\text{ atoms Bi}$$

Summary and Key Terms

Introduction and Section 7.1 The periodic table was first developed by Mendeleev and Meyer on the basis of the similarity in chemical and physical properties exhibited by certain elements. Moseley established that each element has a unique atomic number, which added more order to the periodic table. We now recognize that elements in the same column of the periodic table have the same number of electrons in their **valence orbitals**. This similarity in valence electronic structure leads to the similarities among elements in the same group. The differences among elements in the same group arise because their valence orbitals are in different shells.

Section 7.2 Many properties of atoms are due to the average distance of the outer electrons from the nucleus and to the **effective nuclear charge** experienced by these electrons. The core electrons are very effective in screening the outer electrons from the full charge of the nucleus, whereas electrons in the same shell do not screen each other very effectively at all. As a result, the effective nuclear charge experienced by outer electrons increases as we move left to right across a period.

Section 7.3 The size of an atom can be gauged by its **bonding atomic radius**, based on measurements of the distances separating atoms in their chemical compounds. In general, atomic radii increase as we go down a column in the periodic table and decrease as we proceed left to right across a row.

Cations are smaller than their parent atoms; anions are larger than their parent atoms. For ions of the same charge, size increases going down a column of the periodic table. An **isoelectronic series** is a series of ions that has the same number of electrons. For such a series, size decreases with increasing nuclear charge as the electrons are attracted more strongly to the nucleus.

Section 7.4 The first **ionization energy** of an atom is the minimum energy needed to remove an electron from the atom in the gas phase, forming a cation. The second ionization energy is the energy needed to remove a second electron from the atom, and so forth. Ionization energies show a sharp increase after all the valence electrons have been removed, because of the much higher effective nuclear charge experienced by the core electrons. The first ionization energies of the elements show periodic trends that are opposite those seen for atomic radii, with smaller atoms having higher first ionization energies. Thus, first ionization energies decrease as we go down a column and increase as we proceed left to right across a row.

We can write electron configurations for ions by first writing the electron configuration of the neutral atom and then removing or adding the appropriate number of electrons. Electrons are removed first from the orbitals with the largest value of n. Electrons are added to orbitals with the lowest value of n.

Section 7.5 The **electron affinity** of an element is the energy change upon adding an electron to an atom in the gas phase, forming an anion. A negative electron affinity means that the anion is stable; a positive electron affinity means that the anion will not form readily. In general, electron affinities become more negative as we proceed from left to right across the periodic table. The halogens have the most negative electron affinities. The electron affinities of the noble gases are all positive because the added electron would have to occupy a new, higher-energy subshell.

Section 7.6 The elements can be categorized as metals, nonmetals, and metalloids. Most elements are metals; they occupy the left side and the middle of the periodic table. Nonmetals appear in the upper-right section of the table. Metalloids occupy a narrow band between the metals and nonmetals. The tendency of an element to exhibit the properties of metals, called the **metallic character**, increases as we proceed down a column and decreases as we proceed from left to right across a row.

Metals have a characteristic luster, and they are good conductors of heat and electricity. When metals react with nonmetals, the metal atoms are oxidized to cations and ionic substances are generally formed. Most metal oxides are basic; they react with acids to form salts and water.

Nonmetals lack metallic luster and are poor conductors of heat and electricity. Several are gases at room temperature. Compounds composed entirely of nonmetals are generally molecular. Nonmetals usually form anions in their reactions with metals. Nonmetal oxides are acidic; they react with bases to form salts and water. Metalloids have properties that are intermediate between those of metals and nonmetals.

Section 7.7 The periodic properties of the elements can help us understand the properties of groups of the representative elements. The **alkali metals** (group 1A) are soft metals with low densities and low melting points. They have the lowest ionization energies of the elements. As a result, they are very reactive toward nonmetals, easily losing their outer *s* electron to form 1+ ions. The **alkaline earth metals** (group 2A) are harder and more dense and have higher melting points than the alkali metals. They are also very reactive toward nonmetals, although not as reactive as the alkali metals. The alkaline earth metals readily lose their two outer *s* electrons to form 2+ ions. Both alkali and alkaline earth metals react with hydrogen to form ionic substances that contain the **hydride ion**, H^-.

Section 7.8 Hydrogen is a nonmetal with properties that are distinct from any of the groups of the periodic table. It forms molecular compounds with other nonmetals, such as oxygen and the halogens.

Oxygen and sulfur are the most important elements in group 6A. Oxygen is usually found as a diatomic molecule, O_2. **Ozone**, O_3, is an important allotrope of oxygen. Oxygen has a strong tendency to gain electrons from other elements, thus oxidizing them. In combination with metals, oxygen is usually found as the oxide ion, O^{2-}, although salts of the peroxide ion, O_2^{2-}, and superoxide ion, O_2^-, are sometimes formed. Elemental sulfur is most commonly found as S_8 molecules. In combination with metals, it is most often found as the sulfide ion, S^{2-}.

The **halogens** (group 7A) are nonmetals that exist as diatomic molecules. The halogens have the most negative electron affinities of the elements. Thus their chemistry is dominated by a tendency to form 1− ions, especially in reactions with metals.

The **noble gases** (group 8A) are nonmetals that exist as monoatomic gases. They are very unreactive because they have completely filled *s* and *p* subshells. Only the heaviest noble gases are known to form compounds, and they do so only with very active nonmetals, such as fluorine.

Exercises

Periodic Table; Effective Nuclear Charge

7.1 Why did Mendeleev leave blanks in his early version of the periodic table? How did he predict the properties of the elements that belonged in those blanks?

7.2 **(a)** In the period from about 1800 to about 1865, the atomic weights of many elements were accurately measured. Why was this important to Mendeleev's formulation of the periodic table? **(b)** What property of the atom did Moseley associate with the wavelength of X rays emitted from an element in his experiments? In what ways did this affect the meaning of the periodic table?

7.3 **(a)** What is meant by the term *effective nuclear charge*? **(b)** How does the effective nuclear charge experienced by the valence electrons of an atom vary going from left to right across a period of the periodic table?

7.4 **(a)** How is the concept of effective nuclear charge used to simplify the numerous electron-electron repulsions in a many-electron atom? **(b)** Which experiences a greater effective nuclear charge in a Be atom, the 1*s* electrons or the 2*s* electrons? Explain.

7.5 If each core electron was totally effective in shielding the valence electrons from the full charge of the nucleus and the valence electrons provided no shielding for each other, what would be the effective nuclear charge acting on a valence electron in **(a)** K and **(b)** Br?

7.6 **(a)** If the core electrons were totally effective at shielding the valence electrons from the full charge of the nucleus and the valence electrons provided no shielding for each other, what would be the effective nuclear charge acting on valence electrons in Al? **(b)** Detailed calculations indicate that the effective nuclear charge experienced by the valence electrons is 4.1+. Why is this value larger than that obtained in part (a)?

7.7 Which will experience the greater effective nuclear charge, the electrons in the $n = 3$ shell in Ar or the $n = 3$ shell in Kr? Which will be closer to the nucleus? Explain.

7.8 Arrange the following atoms in order of increasing effective nuclear charge experienced by the electrons in the $n = 3$ electron shell: K, Mg, P, Rh, and Ti. Explain the basis for your order.

Atomic and Ionic Radii

7.9 Because an exact outer boundary cannot be measured or even calculated for an atom, how are atomic radii determined? What is the difference between a bonding radius and a nonbonding radius?

7.10 **(a)** Why does the quantum mechanical description of many-electron atoms make it difficult to define a precise atomic radius? **(b)** When nonbonded atoms come up against one another, what determines how closely the nuclear centers can approach?

7.11 The distance between Au atoms in gold metal is 2.88 Å. What is the atomic radius of a gold atom in this environment? (This radius is called the metallic radius.)

7.12 Based on the radii presented in Figure 7.5, predict the distance between Si atoms in solid silicon.

7.13 Estimate the As—I bond length from the data in Figure 7.5, and compare your value to the experimental As—I bond length in arsenic triiodide, AsI_3, 2.55 Å.

7.14 In the series of group 5A hydrides, of general formula MH_3, the measured bond distances are as follows: P—H, 1.419 Å; As—H, 1.519 Å; Sb—H, 1.707 Å. **(a)** Compare these values with those estimated by use of the atomic radii in Figure 7.5. **(b)** Explain the steady increase in M—H bond distance in this series in terms of the electronic configurations of the M atoms.

7.15 How do the sizes of atoms change as we move **(a)** from left to right across a row in the periodic table **(b)** from top to bottom in a group in the periodic table? **(c)** Arrange the following atoms in order of increasing atomic radius: F, P, S, As.

7.16 **(a)** Among the nonmetallic elements, the change in atomic radius in moving one place left or right in a row is smaller than the change in moving one row up or down. Explain these observations. **(b)** Arrange the following atoms in order of increasing atomic radius: Si, S, Ge, Se.

7.17 Using only the periodic table, arrange each set of atoms in order of increasing radius: **(a)** Ca, Mg, Be; **(b)** Ga, Br, Ge; **(c)** Al, Tl, Si.

7.18 Using only the periodic table, arrange each set of atoms in order of increasing radius: **(a)** Cs, K, Rb; **(b)** In, Te, Sn; **(c)** P, Cl, Sr.

7.19 **(a)** Why are monoatomic cations smaller than their corresponding neutral atoms? **(b)** Why are monoatomic anions larger than their corresponding neutral atoms? **(c)** Why does the size of ions increase as one proceeds down a column in the periodic table?

7.20 Explain the following variations in atomic or ionic radii: **(a)** $I^- > I > I^+$; **(b)** $Ca^{2+} > Mg^{2+} > Be^{2+}$; **(c)** $Fe > Fe^{2+} > Fe^{3+}$.

7.21 Consider a reaction represented by the following spheres:

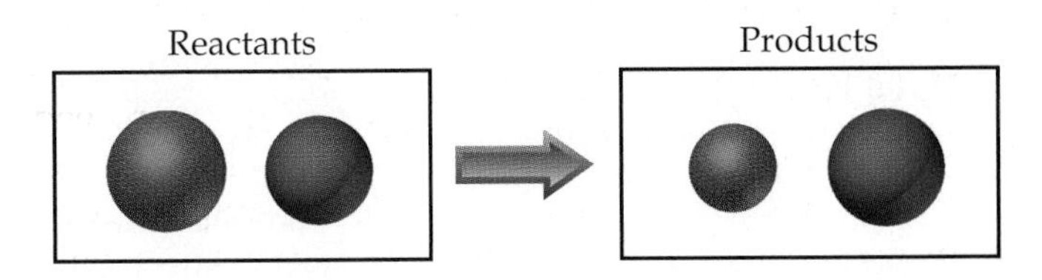

Which sphere represents a metal and which a nonmetal? Explain.

7.22 Consider the following spheres:

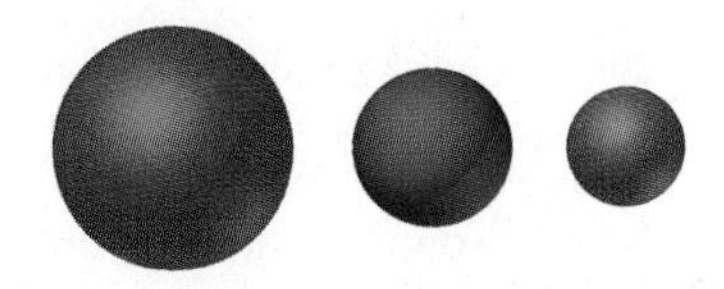

Which one represents Ca, which Ca^{2+}, and which Mg^{2+}?

7.23 **(a)** What is an isoelectronic series? **(b)** Which neutral atom is isoelectronic with each of the following ions: (*i*) Cl^-; (*ii*) Se^{2-}; (*iii*) Mg^{2+}?

7.24 Select the ions or atoms from the following sets that are isoelectronic with each other: **(a)** K^+, Rb^+, Ca^{2+}; **(b)** Cu^+, Ca^{2+}, Sc^{3+}; **(c)** S^{2-}, Se^{2-}, Ar; **(d)** Fe^{2+}, Co^{3+}, Mn^{2+}.

7.25 **(a)** Why do the radii of isoelectronic ions decrease with increasing nuclear charge? **(b)** Which experiences the greatest effective nuclear charge, a 2*p* electron in F^-, a 2*p* electron in Ne, or a 2*p* electron in Na^+?

7.26 Consider S, Cl, and K and their most common ions. **(a)** List the atoms in order of increasing size. **(b)** List the ions in order of increasing size. **(c)** Explain any differences in the orders of the atomic and ionic sizes.

7.27 For each of the following sets of atoms and ions, arrange the members in order of increasing size: **(a)** Se^{2-}, Te^{2-}, Se; **(b)** Co^{3+}, Fe^{2+}, Fe^{3+}; **(c)** Ca, Ti^{4+}, Sc^{3+}; **(d)** Be^{2+}, Na^+, Ne.

7.28 For each of the following statements, provide an explanation: **(a)** Cl^- is larger than Cl; **(b)** S^{2-} is larger than O^{2-}; **(c)** K^+ is larger than Ca^{2+}.

Ionization Energies; Electron Affinities

7.29 Write equations that show the processes that describe the first, second, and third ionization energies of a tellurium atom.

7.30 Write equations that show the process for **(a)** the first two ionization energies of gallium and **(b)** the fourth ionization energy of rhodium.

7.31 **(a)** Why are ionization energies always positive quantities? **(b)** Why does F have a larger first ionization energy than O? **(c)** Why is the second ionization energy of an atom always greater than its first ionization energy?

7.32 **(a)** Why does Li have a larger first ionization energy than Na? **(b)** The difference between the third and fourth ionization energies of scandium is much larger than the difference between the third and fourth ionization energies of titanium. Why? **(c)** Why does Li have a much larger second ionization energy than Be?

7.33 **(a)** What is the general relationship between the size of an atom and its first ionization energy? **(b)** Which element in the periodic table has the largest ionization energy? Which has the smallest?

7.34 **(a)** What is the trend in first ionization energies as one proceeds down the group 7A elements? Explain how this trend relates to the variation in atomic radii. **(b)** What is the trend in first ionization energies as one moves across the fourth period from K to Kr? How does this trend compare with the trend in atomic sizes?

7.35 Based on their positions in the periodic table, predict which atom of the following pairs will have the larger first ionization energy: **(a)** O, Ne; **(b)** Mg, Sr; **(c)** K, Cr; **(d)** Br, Sb; **(e)** Ga, Ge.

7.36 For each of the following pairs, indicate which element has the larger first ionization energy: **(a)** Sr, Cd; **(b)** Si, C; **(c)** In, I; **(d)** Sn, Xe. (In each case use electron configuration and effective nuclear charge to explain your answer.)

7.37 Write the electron configurations for the following ions: **(a)** Sb^{3+}; **(b)** Ga^{+}; **(c)** P^{3-}; **(d)** Cr^{3+}; **(e)** Zn^{2+}; **(f)** Ag^{+}.

7.38 Write electron configurations for the following ions, and determine which have noble-gas configurations: **(a)** Mn^{3+}; **(b)** Se^{2-}; **(c)** Sc^{3+}; **(d)** Ru^{2+}; **(e)** Tl^{+}; **(f)** Au^{+}.

7.39 Write the electron configuration for **(a)** the Co^{2+} ion and **(b)** the In^{+} ion. How many unpaired electrons does each contain?

7.40 Identify the element whose ions have the following electron configurations: **(a)** a 3+ ion with $[Ar]3d^3$; **(b)** a 2+ ion with $[Kr]4d^{10}5s^2$. How many unpaired electrons does each ion contain?

7.41 Write equations, including electron configurations beneath the species involved, that explain the difference between the first ionization energy of Se(*g*) and the electron affinity of Se(*g*).

7.42 While the electron affinity of bromine is a negative quantity, it is positive for Kr. Use the electron configurations of the two elements to explain the difference.

7.43 The electron affinity of lithium is a negative value, whereas the electron affinity of beryllium is a positive value. Use electron configurations to account for this observation.

[7.44] Write an equation for the process that corresponds to the electron affinity of the Mg^{+} ion. Also write the electron configurations of the species involved. What process does this electron affinity equation correspond to? What is the magnitude of the energy change in the process? (*Hint*: The answer is in Table 7.2.)

Properties of Metals and Nonmetals

7.45 How are metallic character and first ionization energy related?

7.46 Arrange the following pure solid elements in order of increasing electrical conductivity: P, Ag, and Sb. Explain the reasoning you used.

7.47 For each of the following pairs, which element will have the greater metallic character: **(a)** Li or Be; **(b)** Li or Na; **(c)** Sn or P; **(d)** Al or B?

7.48 **(a)** What data can you cite from this chapter to support a prediction that the metallic character of the group 5A elements will increase with increasing atomic number? **(b)** Nonmetallic character is the opposite of metallic character—nonmetallic character decreases as metallic character increases. Arrange the following elements in order of increasing nonmetallic character: Se, Ag, Sn, F, and C.

7.49 Predict whether each of the following oxides is ionic or molecular: SO_2, MgO, Li_2O, P_2O_5, Y_2O_3, N_2O, and XeO_3. Explain the reasons for your choices.

7.50 When metal oxides react with water, the oxygen generally ends up as the hydroxide ion, separate from the metal. In contrast, when nonmetallic oxides react with water, the oxygen ends up as part of the nonmetal species. (For example, upon reaction of CO_2 with water, the oxygen remains on the carbon in H_2CO_3.) **(a)** Give two examples each from metals and nonmetals to support these generalizations. **(b)** What connection is there between this contrasting behavior of metal and nonmetal oxides and ionization energies?

7.51 **(a)** What is meant by the terms acidic oxide and basic oxide? **(b)** How can we predict whether an oxide will be acidic or basic based on its composition?

7.52 Arrange the following oxides in order of increasing acidity: CO_2, CaO, Al_2O_3, SO_3, SiO_2, and P_2O_5.

7.53 Write balanced equations for the following reactions: **(a)** barium oxide with water; **(b)** iron(II) oxide with perchloric acid; **(c)** sulfur trioxide with water; **(d)** carbon dioxide with aqueous sodium hydroxide.

7.54 Write balanced equations for the following reactions: **(a)** potassium oxide with water; **(b)** diphosphorus trioxide with water; **(c)** chromium(III) oxide with dilute hydrochloric acid; **(d)** selenium dioxide with aqueous potassium hydroxide.

Group Trends in Metals and Nonmetals

7.55 Compare the elements sodium and magnesium with respect to the following properties: **(a)** electron configuration; **(b)** most common ionic charge; **(c)** first ionization energy; **(d)** reactivity toward water; **(e)** atomic radius. Account for the differences between the two elements.

7.56 **(a)** Compare the electron configurations and atomic radii (see Figure 7.5) of rubidium and silver. In what respects are their electronic configurations similar? Account for the difference in radii of the two elements. **(b)** As with rubidium, silver is most commonly found as the 1+ ion, Ag^+. However, silver is far less reactive. Explain these observations.

7.57 **(a)** Why is calcium generally more reactive than magnesium? **(b)** Why is calcium generally less reactive than potassium?

7.58 **(a)** Why is cesium more reactive toward water than is lithium? **(b)** One of the alkali metals reacts with oxygen to form a solid white substance. When this substance is dissolved in water, the solution gives a positive test for hydrogen peroxide, H_2O_2. When the solution is tested in a burner flame, a lilac-purple flame is produced. What is the likely identity of the metal? **(c)** Write a balanced chemical equation for reaction of the white substance with water.

7.59 Write a balanced equation for the reaction that occurs in each of the following cases: **(a)** Potassium metal burns in an atmosphere of chlorine gas. **(b)** Strontium oxide is added to water. **(c)** A fresh surface of lithium metal is exposed to oxygen gas. **(d)** Sodium metal is reacted with molten sulfur.

7.60 Write a balanced equation for the reaction that occurs in each of the following cases: **(a)** Potassium is added to water. **(b)** Barium is added to water. **(c)** Lithium is heated in nitrogen, forming lithium nitride. **(d)** Magnesium burns in oxygen.

7.61 Use electron configurations to explain why hydrogen exhibits properties similar to those of both Li and F.

7.62 **(a)** As described in Section 7.7, the alkali metals react with hydrogen to form hydrides and react with halogens—for example, fluorine—to form halides. Compare the roles of hydrogen and the halogen in these reactions. In what sense are the forms of hydrogen and halogen in the products alike? **(b)** Write balanced equations for the reaction of fluorine with calcium and for the reaction of hydrogen with calcium. What are the similarities among the products of these reactions?

7.63 Compare the elements fluorine and chlorine with respect to the following properties: **(a)** electron configuration; **(b)** most common ionic charge; **(c)** first ionization energy; **(d)** reactivity toward water; **(e)** electron affinity; **(f)** atomic radius. Account for the differences between the two elements.

7.64 Little is known about the properties of astatine, At, because of its rarity and high radioactivity. Nevertheless, it is possible for us to make many predictions about its properties. **(a)** Do you expect the element to be a gas, liquid, or solid at room temperature? Explain. **(b)** What is the chemical formula of the compound it forms with Na?

7.65 Until the early 1960s the group 8A elements were called the inert gases. Why was this name given? Why is it inappropriate?

7.66 **(a)** Explain the trend in reactivities of the noble gases with fluorine. **(b)** Why is there no comparable pattern of reactivity with chlorine?

7.67 Write a balanced equation for the reaction that occurs in each of the following cases: **(a)** Ozone decomposes to dioxygen. **(b)** Xenon reacts with fluorine. (Write three different equations.) **(c)** Sulfur reacts with hydrogen gas. **(d)** Fluorine reacts with water.

7.68 Write a balanced equation for the reaction that occurs in each of the following cases: **(a)** Chlorine reacts with water. **(b)** Barium metal is heated in an atmosphere of hydrogen gas. **(c)** Lithium reacts with sulfur. **(d)** Fluorine reacts with magnesium metal.

7.69 **(a)** Which would you expect to be a better conductor of electricity, tellurium or iodine? **(b)** How does a molecule of sulfur (in its most common room-temperature form) differ from a molecule of oxygen? **(c)** Why is chlorine generally more reactive than bromine?

7.70 **(a)** Sulfur reacts with fluorine under appropriate conditions to form $SF_4(g)$. Write a balanced chemical equation for the reaction. **(b)** What are the formulas and names of the allotropes of oxygen? **(c)** Why would it not be advisable to store fluorine gas in a silica glass (formed mainly of SiO_2) container?

Additional Exercises

7.71 Consider the stable elements through bismuth ($Z = 83$). In how many instances are the atomic weights of the elements in the reverse order relative to the atomic numbers of the elements? What is the explanation for these cases?

7.72 In 1871 Mendeleev predicted the existence of an element that he called eka-aluminum, which would have the following properties: atomic weight of about 68 amu, density of about 5.9 g/cm^3, low melting point, high boiling point, and oxide with stoichiometry M_2O_3. **(a)** In 1875 the element predicted by Mendeleev was discovered. By what name is the element known? **(b)** Use a reference such as the *CRC Handbook of Chemistry and Physics* or WebElements.com to check the accuracy of Mendeleev's predictions.

7.73 The atoms and ions Na, Mg^+, Al^{2+}, and Si^{3+} are isoelectronic. **(a)** For which of these will the effective nuclear charge acting on the outermost electron be the smallest? **(b)** For which will it be the greatest? **(c)** How do the data in Table 7.2 support your answer?

7.74 **(a)** If the core electrons were totally effective at shielding the valence electrons and the valence electrons provided no shielding for each other, what would be the effective nuclear charge acting on the valence electron in P? **(b)** Detailed calculations indicate that the effective nuclear charge is 5.6+ for the 3*s* electrons and 4.9+ for the 3*p* electrons. Why are the values for the 3*s* and 3*p* electrons different? **(c)** If you remove a single electron from a P atom, which orbital will it come from? Explain.

7.75 As we move across a period of the periodic table, why do the sizes of the transition elements change more gradually than those of the representative elements?

7.76 On the basis of the data in Figure 7.5, predict the bond distances in **(a)** MoF_6; **(b)** SF_6; **(c)** ClF.

7.77 It is possible to produce compounds of the form $GeClH_3$, $GeCl_2H_2$, and $GeCl_3H$. What values do you predict for the Ge—H and Ge—Cl bond lengths in these compounds?

7.78 Nearly all the mass of an atom is in the nucleus, which has a very small radius. When atoms bond together (for example, two fluorine atoms in F_2), why is the distance separating the nuclei so much larger than the radii of the nuclei?

7.79 Note from the following table that the increase in atomic radius in moving from Zr to Hf is smaller than in moving from Y to La. Suggest an explanation for this effect.

Atomic radii (Å)			
Sc	1.44	Ti	1.36
Y	1.62	Zr	1.48
La	1.69	Hf	1.50

7.80 Explain the variation in ionization energies of carbon, as displayed in the following graph:

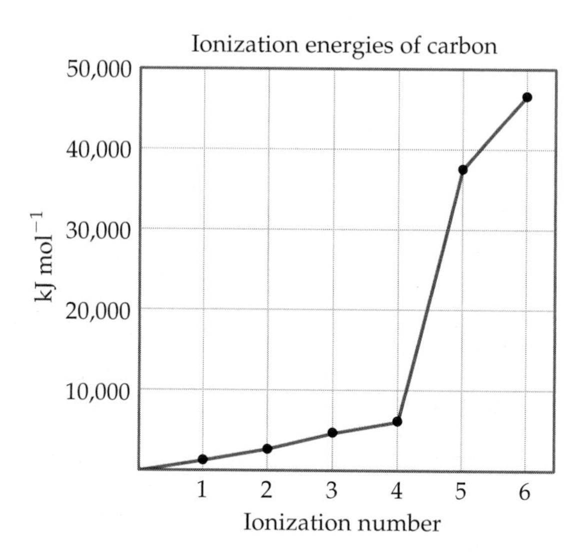

[7.81] Listed here are the atomic and ionic (2+) radii for calcium and zinc:

Radii (Å)			
Ca	1.74	Ca^{2+}	0.99
Zn	1.31	Zn^{2+}	0.74

(a) Explain why the ionic radius in each case is smaller than the atomic radius. **(b)** Why is the atomic radius of calcium larger than that of zinc? **(c)** Suggest a reason why the difference in the ionic radii is much less than the difference in the atomic radii.

7.82 What is the relationship between the ionization energy of an anion with a 1− charge such as F^- and the electron affinity of the neutral atom, F?

7.83 Do you agree with the following statement? "A negative value for the electron affinity of an atom occurs when the outermost electrons only incompletely shield one another from the nucleus." If not, change it to make it more nearly correct in your view. Apply either the statement as given or your revised statement to explain why the electron affinity of bromine is −325 kJ/mol and that for its neighbor Kr is >0.

7.84 Use orbital diagrams to illustrate what happens when an oxygen atom gains two electrons. Why is it extremely difficult to add a third electron to the atom?

[7.85] Use electron configurations to explain the following observations: **(a)** The first ionization energy of phosphorus is greater than that of sulfur. **(b)** The electron affinity of nitrogen is lower (less negative) than those of both carbon and oxygen. **(c)** The second ionization energy of oxygen is greater than that of fluorine. **(d)** The third ionization energy of manganese is greater than those of both chromium and iron.

7.86 The following table gives the electron affinities, in kJ/mol, for the group 1B and group 2B metals:

Cu −119	Zn >0
Ag −126	Cd >0
Au −223	Hg >0

(a) Why are the electron affinities of the group 2B elements greater than zero? **(b)** Why do the electron affinities of the group 1B elements become more negative as we move down the group. (*Hint*: Examine the trends in the electron affinity of other groups as we proceed down the periodic table.)

7.87 Hydrogen is an unusual element because it behaves in some ways like the alkali metal elements and in other ways like a nonmetal. Its properties can be explained in part by its electron configuration and by the values for its ionization energy and electron affinity. **(a)** Explain why the electron affinity of hydrogen is much closer to the values for the alkali elements than for the halogens. **(b)** Is the following statement true? "Hydrogen has the smallest bonding atomic radius of any element that forms chemical compounds." If not, correct it. If it is, explain in terms of electron configurations. **(c)** Explain why the ionization energy of hydrogen is closer to the values for the halogens than for the alkali metals.

[7.88] The first ionization energy of the oxygen molecule is the energy required for the following process:

$$O_2(g) \longrightarrow O_2^+(g) + e^-$$

The energy needed for this process is 1175 kJ/mol, very similar to the first ionization energy of Xe. Would you expect O_2 to react with F_2? If so, suggest a product or products of this reaction.

7.89 Based on your reading of this chapter, arrange the following in order of increasing melting point: K, Br_2, Mg, and O_2. Explain the factors that determine the order.

7.90 Use the electron affinities, ionization energies, and nuclear charges of the atoms to explain the following comparisons: **(a)** Lithium forms Li_2O with oxygen, LiF with fluorine. **(b)** Fluorine has a smaller atomic radius than oxygen. **(c)** Fluorine is a more reactive nonmetal than oxygen.

[7.91] There are certain similarities in properties that exist between the first member of any periodic family and the element located below it and to the right in the periodic table. For example, in some ways Li resembles Mg, Be resembles Al, and so forth. This observation is called the diagonal relationship. Using what we have learned in this chapter, offer a possible explanation for this relationship.

[7.92] The elements at the bottom of groups 1A, 2A, 6A, 7A, and 8A—Fr, Ra, Po, At, and Rn—are all radioactive. As a result, much less is known about their physical and chemical properties than those of the elements above them. Based on what we have learned in this chapter, which of these five elements would you expect **(a)** to have the most metallic character; **(b)** to have the least metallic (that is, the most nonmetallic) character; **(c)** to have the largest first ionization energy; **(d)** to have the smallest first ionization energy; **(e)** to have the greatest (most negative) electron affinity; **(f)** to have the largest atomic radius; **(g)** to resemble least in appearance the element immediately above it; **(h)** to have the highest melting point; **(i)** to react most exothermically with water?

[7.93] A historian discovers a nineteenth-century notebook in which some observations, dated 1822, on a substance thought to be a new element, were recorded. Here are some of the data recorded in the notebook: Ductile, silver-white, metallic-looking. Softer than lead. Unaffected by water. Stable in air. Melting point: 153°C. Density: 7.3 g/cm^3. Electrical conductivity: 20% that of copper. Hardness: About 1% as hard as iron. When 4.20 g of the unknown is heated in an excess of oxygen, 5.08 g of a white solid is formed. The solid could be sublimed by heating to over 800°C. **(a)** Using information in the text and a handbook of chemistry, and making allowances for possible variations in numbers from current values, identify the element reported. **(b)** Write a balanced chemical equation for the reaction with oxygen. **(c)** Judging from Figure 7.2, might this nineteenth-century investigator have been the first to discover a new element?

[7.94] It has been discovered in recent years that many organic compounds that contain chlorine, including dioxins, which had been thought to be entirely of man-made origin, are formed in natural processes. More than 3000 natural organohalogen compounds, most involving chlorine and bromine, are known. These compounds, in which the halogen is attached to carbon, are nearly all nonionic materials. Why are these materials typically not ionic, as are the more abundant inorganic halogen compounds found in nature?

Integrative Exercises

[7.95] Moseley established the concept of atomic number by studying X rays emitted by the elements. The X rays emitted by some of the elements have the following wavelengths:

Element	Wavelength (Å)
Ne	14.610
Ca	3.358
Zn	1.435
Zr	0.786
Sn	0.491

(a) Calculate the frequency, ν, of the X rays emitted by each of the elements, in Hz. **(b)** Using graph paper (or suitable computer software), plot the square root of ν versus the atomic number of the element. What do you observe about the plot? **(c)** Explain how the plot in part (b) allowed Moseley to predict the existence of undiscovered elements. **(d)** Use the result from part (b) to predict the X-ray wavelength emitted by iron. **(e)** A particular element emits X rays with a wavelength of 0.980 Å. What element do you think it is?

[7.96] **(a)** Write the electron configuration for Li, and estimate the effective nuclear charge experienced by the valence electron. **(b)** The energy of an electron in a one-electron atom or ion equals $(-2.18 \times 10^{-18}\ \text{J})\left(\frac{Z^2}{n^2}\right)$, where Z is the nuclear charge and n is the principal quantum number of the electron. Estimate the first ionization energy of Li. **(c)** Compare the result of your calculation with the value reported in Table 7.4, and explain the difference. **(d)** What value of the effective nuclear charge gives the proper value for the ionization energy? Does this agree with your explanation in (c)?

[7.97] One way to measure ionization energies is photoelectron spectroscopy (PES), a technique based on the photoelectric effect. (Section 6.2) In PES, monochromatic light is directed onto a sample, causing electrons to be emitted. The kinetic energy of the emitted electrons is measured. The difference between the energy of the photons and the kinetic energy of the electrons corresponds to the energy needed to remove the electrons (that is, the ionization energy). Suppose that a PES experiment is performed in which mercury vapor is irradiated with ultraviolet light of wavelength 58.4 nm. **(a)** What is the energy of a photon of this light, in eV? **(b)** Write an equation that shows the process corresponding to the first ionization energy of Hg. **(c)** The kinetic energy of the emitted electrons is measured to be 10.75 eV. What is the first ionization energy of Hg, in kJ/mol? **(d)** With reference to Figure 7.10, determine which of the halogen elements has a first ionization energy closest to that of mercury.

7.98 Consider the gas-phase transfer of an electron from a sodium atom to a chlorine atom:

$$Na(g) + Cl(g) \longrightarrow Na^+(g) + Cl^-(g)$$

(a) Write this reaction as the sum of two reactions, one that relates to an ionization energy and one that relates to an electron affinity. **(b)** Use the result from part (a), data in this chapter, and Hess's law to calculate the enthalpy of the above reaction. Is the reaction exothermic or endothermic? **(c)** The reaction between sodium metal and chlorine gas is highly exothermic and produces NaCl(*s*), whose structure was discussed in Section 2.6. Comment on this observation relative to the calculated enthalpy for the aforementioned gas-phase reaction.

[7.99] When magnesium metal is burned in air (Figure 3.6), two products are produced. One is magnesium oxide, MgO. The other is the product of the reaction of Mg with molecular nitrogen, magnesium nitride. When water is added to magnesium nitride, it reacts to form magnesium oxide and ammonia gas. **(a)** Based on the charge of the nitride ion (Table 2.5), predict the formula of magnesium nitride. **(b)** Write a balanced equation for the reaction of magnesium nitride with water. What is the driving force for this reaction? **(c)** In an experiment a piece of magnesium ribbon is burned in air in a crucible. The mass of the mixture of MgO and magnesium nitride after burning is 0.470 g. Water is added to the crucible, further reaction occurs, and the crucible is heated to dryness until the final product is 0.486 g of MgO. What was the mass percentage of magnesium nitride in the mixture obtained after the initial burning? **(d)** Magnesium nitride can also be formed by reaction of the metal with ammonia at high temperature. Write a balanced equation for this reaction. If 6.3 g Mg ribbon reacts with 2.57 g $NH_3(g)$ and the reaction goes to completion, which component is the limiting reactant? What mass of $H_2(g)$ is formed in the reaction? **(e)** The standard enthalpy of formation of solid magnesium nitride is -461.08 kJ mol^{-1}. Calculate the standard enthalpy change for the reaction between magnesium metal and ammonia gas.

7.100 **(a)** The experimental Bi—Br bond length in bismuth tribromide, $BiBr_3$, is 2.63 Å. Based on this value and the data in Figure 7.5, predict the atomic radius of Bi. **(b)** Bismuth tribromide is soluble in acidic solution. It is formed by treating solid bismuth(III) oxide with aqueous hydrobromic acid. Write a balanced chemical equation for this reaction. **(c)** While bismuth(III) oxide is soluble in acidic solutions, it is insoluble in basic solutions such as NaOH(*aq*). On the basis of these properties, is bismuth characterized as a metallic, metalloid, or nonmetallic element? **(d)** Treating bismuth with fluorine gas forms BiF_5. Use the electron configuration of Bi to explain the formation of a compound with this formulation. **(e)** While it is possible to form BiF_5 in the manner just described, pentahalides of bismuth are not known for the other halogens. Explain why the pentahalide might form with fluorine, but not with the other halogens. How does the behavior of bismuth relate to the fact that xenon reacts with fluorine to form compounds, but not with the other halogens?

eMedia Exercises

7.101 The **Periodic Trends: Atomic Radii** movie (*eChapter 7.3*) describes the trends in the sizes of atoms on the periodic table—from left to right and from top to bottom. **(a)** What factors influence atomic radius? **(b)** Based on the factors that influence the atomic radius, explain why the radius of gallium is smaller than that of aluminum.

7.102 The **Electron Gain and Loss** movie (*eChapter 7.3*) illustrates how addition or subtraction of an electron affects the size of an atom. The first ionization of aluminum produces the Al^+ ion, which is smaller than the neutral Al atom. The second ionization of aluminum produces the Al^{2+} ion, which is smaller still. The third ionization of aluminum produces the Al^{3+} ion, and the 3+ cation is even smaller than the 2+ cation. Of the first, second, and third ionizations, which would you expect to cause the biggest change in size? Explain your reasoning.

7.103 According to the information given in the **Periodic Trends: Ionization Energy** movie (*eChapter 7.4*), you might expect fluorine and chlorine to have two of the highest ionization energies among the representative elements. Explain why those two elements are almost always found in nature as ions.

7.104 Although the **Periodic Trends: Ionization Energy** movie (*eChapter 7.4*) depicts clear trends in the magnitudes of first ionization energies, the movie shows that the trend from left to right across the periodic table is not smooth. **(a)** Based on electron configurations, explain why this is so. **(b)** For each pair, predict which one will have the higher first ionization energy: N, O; Be, B; Ca, Ga; P, S.

7.105 **(a)** Using the **Ionization Energy Graph** activity (*eChapter 7.4*), make a graph of first ionization energies versus atomic number for the elements Na through Si. (Plot atomic number on the *x*-axis, I_1 values on the *y*-axis.) **(b)** Using the data in Table 7.2 and the **Ionization Energy Graph** activity, make a graph of I_1 for Na, I_2 for Mg, I_3 for Al, and I_4 for Si. (Plot atomic number on the *x*-axis, the corresponding I value on the *y*-axis.) **(c)** Compare the two graphs. Is the shape of the second graph different from that of the first? If so, explain why.

7.106 **(a)** What happens to the size of the chlorine atom in the **Ionization Energy** movie (*eChapter 7.4*) when the first electron is removed? **(b)** How do you account for this change in size? **(c)** Based on your answer to part (b), how would you expect the size of the chlorine atom to change if an electron were *added* rather than removed?

7.107 **(a)** Based on the information in the **Periodic Trends: Acid/Base Behavior of Oxides** movie (*eChapter 7.6*), which of the following compounds would you expect to form a basic solution with water: NaO; NO; N_2O; K_2O; CO_2? **(b)** Which oxide in each pair would produce the most acidic solution and why? CO_2, CO; NO, N_2O; N_2O_5, N_2O_4.

7.108 Watch the **Sodium and Potassium in Water** movie (*eChapter 7.7*), and note the differences between the two reactions shown. **(a)** What property of the elements involved is responsible for the observed differences? **(b)** How would you expect the reaction of Rb and water to differ from the reactions in the demonstration? Describe what you would expect to happen.

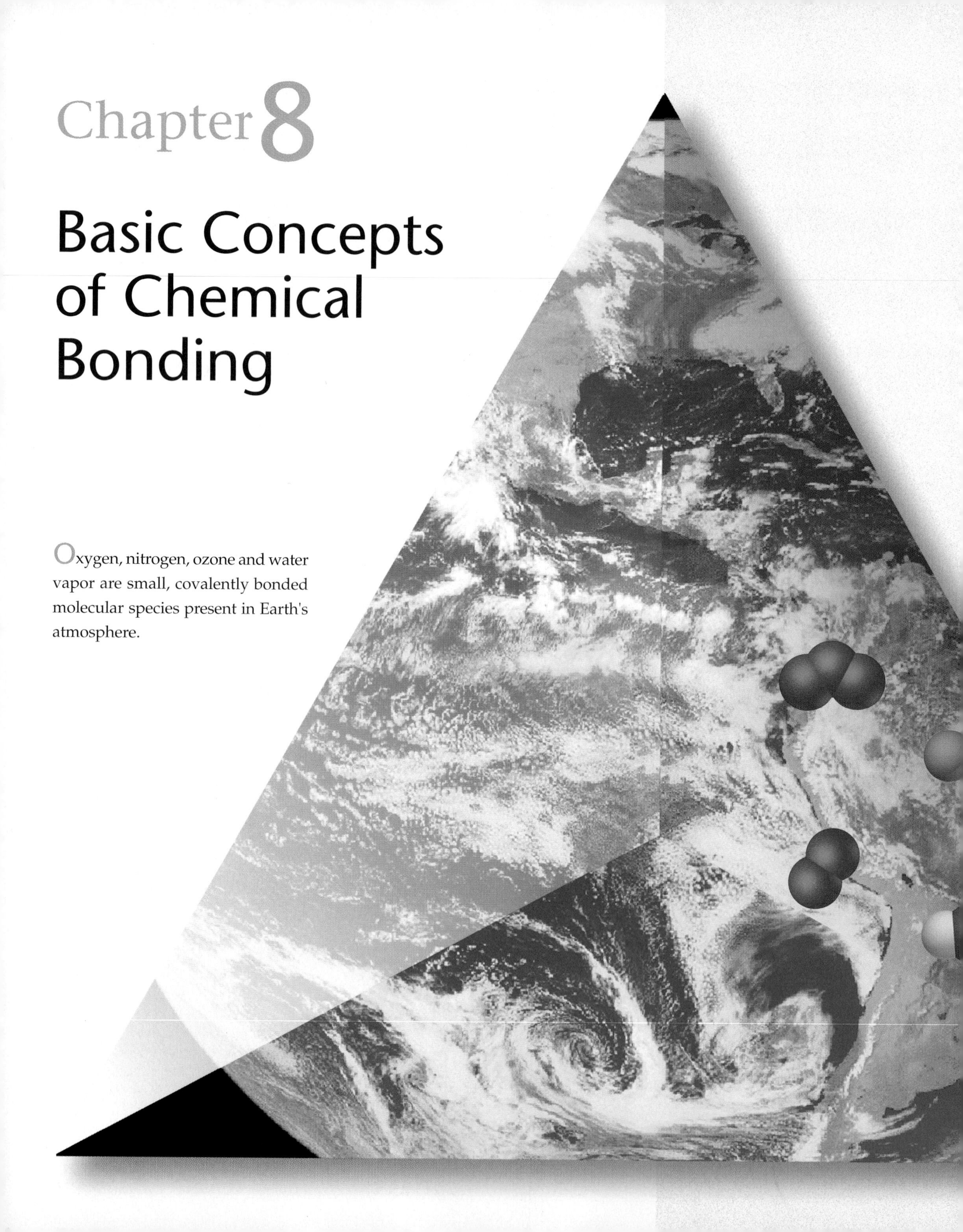

Chapter 8

Basic Concepts of Chemical Bonding

Oxygen, nitrogen, ozone and water vapor are small, covalently bonded molecular species present in Earth's atmosphere.

What's Ahead

- We begin with a brief discussion of chemical bond types and introduce *Lewis symbols*, a way of showing the valence electrons in atoms and ions.
- *Ionic bonding* results from the essentially complete transfer of one or more electrons from one atom to another. We will study the energetics of formation of ionic substances and describe *lattice energy*.
- *Covalent bonding* involves the sharing of one or more electron pairs between atoms, as needed to attain an *octet* of electrons about each atom.
- *Electronegativity* is defined as the ability of an atom to attract electrons to itself in a bonding situation. Electron pairs will in general be shared unequally between atoms of differing electronegativity, leading to *polar covalent bonds*.
- The structures of covalently bonded molecules can be represented by *Lewis structures*, which are extensions of Lewis symbols for atoms.
- *Formal charges* can be assigned to atoms in molecules, by following simple rules.
- It may occur that more than one equivalent Lewis structure can be drawn for a molecule or polyatomic ion. The actual structure in such cases is a blend of two or more contributing Lewis structures, called *resonance structures*.
- *Exceptions to the octet rule* include a relatively few cases in which there are fewer than 8 electrons in the valence-shell orbitals. By contrast, the structures of many compounds of third row and heavier elements can best be described by assuming that the valence-shell orbitals hold more than an octet of electrons.
- The strengths of covalent bonds vary with the number of shared electron pairs as well as other factors. We can use *average bond enthalpy* values to estimate the enthalpies of reaction in cases where thermodynamic data such as heats of formation are unavailable.

ON THE TABLE in most diners you can expect to find two white, crystalline substances: table salt and granulated sugar. In spite of their similarities in appearance, salt and sugar are vastly different kinds of substances. Table salt is sodium chloride, NaCl, which consists of sodium ions, Na^+, and chloride ions, Cl^-. The structure is held together by the attractions between the oppositely charged ions, which we call *ionic bonds*. Granulated sugar, in contrast, does not contain ions at all. It consists of molecules of sucrose, $C_{12}H_{22}O_{11}$, in which attractions called *covalent bonds* hold the atoms together. One consequence of the difference in bonding in salt and sugar is their different behaviors in water: NaCl dissolves in water to yield ions in solution (NaCl is an electrolyte), whereas sucrose dissolves in water to yield aqueous $C_{12}H_{22}O_{11}$ molecules (sucrose is a nonelectrolyte). (Section 4.2)

The properties of substances are determined in large part by the *chemical bonds* that hold their atoms together. What determines the type of bonding in each substance, and just how do the characteristics of these bonds give rise to different physical and chemical properties? The keys to answering the first question are found in the electronic structures of the atoms involved, which we discussed in Chapters 6 and 7. In this chapter and the next we will examine the relationships among electronic structure, chemical bonding forces, and chemical bond type. We'll also see how the properties of ionic and covalent substances arise from the distributions of electronic charge within atoms, ions, and molecules.

8.1 Chemical Bonds, Lewis Symbols, and the Octet Rule

Whenever atoms or ions are strongly attached to one another, we say that there is a **chemical bond** between them. There are three general types of chemical bonds: ionic, covalent, and metallic. Figure 8.1 ◀ shows examples of substances in which we find each of these types of attractive forces.

The term **ionic bond** refers to electrostatic forces that exist between ions of opposite charge. Ions may be formed from atoms by the transfer of one or more electrons from one atom to another. Ionic substances generally result from the interaction of metals on the far left side of the periodic table with nonmetals on the far right side (excluding the noble gases, group 8A). Ionic bonding will be discussed in Section 8.2.

A **covalent bond** results from the sharing of electrons between two atoms. The most familiar examples of covalent bonding are seen in the interactions of nonmetallic elements with one another. We devote much of this chapter and the next to describing and understanding covalent bonds.

Metallic bonds are found in metals such as copper, iron, and aluminum. In the metals each atom is bonded to several neighboring atoms. The bonding electrons are relatively free to move throughout the three-dimensional structure of the metal. Metallic bonds give rise to such typical metallic properties as high electrical conductivity and luster. We will examine these bonds in Chapter 23.

Magnesium oxide

Potassium dichromate Nickel(II) oxide

(a)

Sulfur

Bromine Sucrose

(b)

Magnesium

Gold Copper

(c)

▲ **Figure 8.1** Examples of substances in which (a) ionic, (b) covalent, and (c) metallic bonds are found.

Lewis Symbols

The electrons that are involved in chemical bonding are the *valence electrons*, those residing in the incomplete outer shell of an atom. (Section 6.8) The American chemist G. N. Lewis (1875–1946) suggested a simple way of showing the valence electrons of atoms and tracking them in the course of bond formation, using what are now known as Lewis electron-dot symbols or merely Lewis symbols. The **Lewis symbol** for an element consists of the chemical symbol for the element plus a dot for each valence electron. Sulfur, for example, has the electron configuration [Ne]$3s^2 3p^4$; so its Lewis symbol shows six valence electrons:

The dots are placed on the four sides of the atomic symbol: the top, the bottom, and the left and right sides. Each side can accommodate up to two electrons. All four sides of the symbol are equivalent; the placement of two electrons on one side versus one electron on each side is arbitrary.

The electronic configurations and Lewis symbols for the representative elements of the second and third rows of the periodic table are shown in Table 8.1 ▶. Notice that the number of valence electrons of any representative element is the same as the group number of the element in the periodic table. For example, the Lewis symbols for oxygen and sulfur, members of group 6A, both show six dots.

The Octet Rule

Atoms often gain, lose, or share electrons to achieve the same number of electrons as the noble gas closest to them in the periodic table. The noble gases have very stable electron arrangements, as evidenced by their high ionization energies, low affinity for additional electrons, and general lack of chemical reactivity. (Section 7.8) Because all noble gases (except He) have eight valence electrons, many atoms undergoing reactions also end up with eight valence electrons. This observation has led to a guideline known as the **octet rule**: *Atoms tend to gain, lose, or share electrons until they are surrounded by eight valence electrons.*

An octet of electrons consists of full *s* and *p* subshells on an atom. In terms of Lewis symbols, an octet can be thought of as four pairs of valence electrons

TABLE 8.1 Lewis Symbols

Element	Electron Configuration	Lewis Symbol	Element	Electron Configuration	Lewis Symbol
Li	[He]$2s^1$	Li·	Na	[Ne]$3s^1$	Na·
Be	[He]$2s^2$	·Be·	Mg	[Ne]$3s^2$	·Mg·
B	[He]$2s^22p^1$	·Ḃ·	Al	[Ne]$3s^23p^1$	·Ȧl·
C	[He]$2s^22p^2$	·Ċ̣·	Si	[Ne]$3s^23p^2$	·Ṩi·
N	[He]$2s^22p^3$	·Ṇ̇:	P	[Ne]$3s^23p^3$	·Ṗ̣:
O	[He]$2s^22p^4$	:Ọ̇:	S	[Ne]$3s^23p^4$	:Ṩ:
F	[He]$2s^22p^5$	·F̤̈:	Cl	[Ne]$3s^23p^5$	·C̤̈l:
Ne	[He]$2s^22p^6$	:N̤̈e:	Ar	[Ne]$3s^23p^6$	:Ä̤r:

arranged around the atom, as in the configuration for Ne in Table 8.1. There are many exceptions to the octet rule, but it provides a useful framework for introducing many important concepts of bonding.

ACTIVITY
Lewis Dot Symbols, Octet Rule

ACTIVITY
Lewis Dot Symbols, Octet Rule

MOVIE
Formation of Sodium Chloride

3-D MODEL
Sodium Chloride

8.2 Ionic Bonding

When sodium metal, Na(*s*), is brought into contact with chlorine gas, $Cl_2(g)$, a violent reaction ensues (Figure 8.2 ▼). The product of this very exothermic reaction is sodium chloride, NaCl(*s*).

$$Na(s) + \tfrac{1}{2}Cl_2(g) \longrightarrow NaCl(s) \qquad \Delta H_f^\circ = -410.9 \text{ kJ} \qquad [8.1]$$

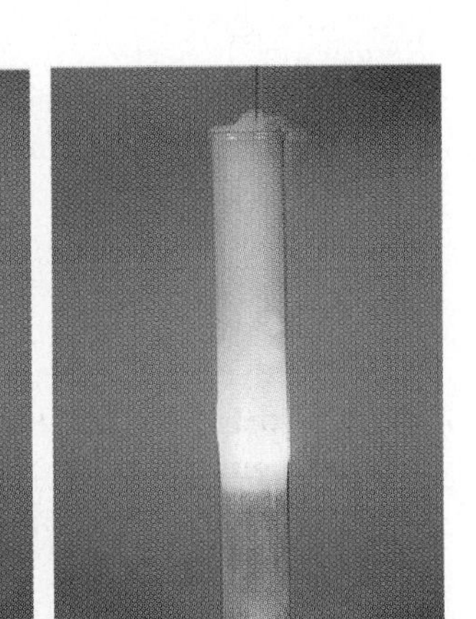

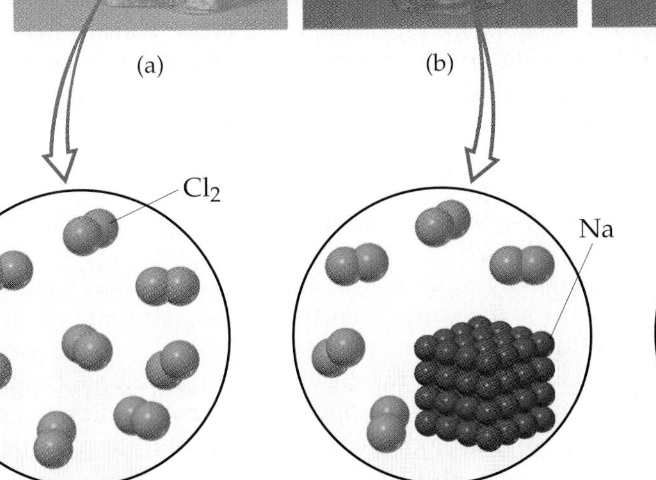

◀ **Figure 8.2** The reaction between sodium metal and chlorine gas to form sodium chloride. (a) A container of chlorine gas (left) and a container of sodium metal (right). (b) Formation of NaCl begins as sodium is added to the chlorine. (c) The reaction a few minutes later. The reaction is strongly exothermic, giving off both heat and light.

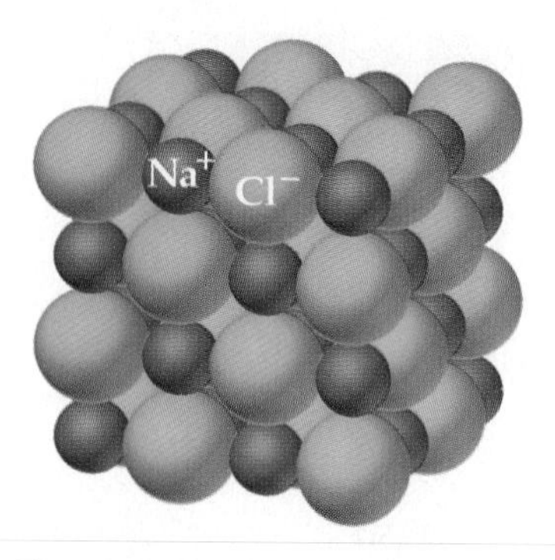

▲ **Figure 8.3** The crystal structure of sodium chloride. Each of the Na^+ ions is surrounded by six Cl^- ions, and each Cl^- ion is surrounded by six Na^+ ions.

Sodium chloride is composed of Na^+ and Cl^- ions, which are arranged in a regular three-dimensional array, as shown in Figure 8.3 ◀.

The formation of Na^+ from Na and Cl^- from Cl_2 indicates that an electron has been lost by a sodium atom and gained by a chlorine atom. Electron transfer to form oppositely charged ions occurs when the atoms involved differ greatly in their attraction for electrons. NaCl is a rather typical ionic compound because it consists of a metal of low ionization energy and a nonmetal with a high affinity for electrons. ∞ (Sections 7.4 and 7.5) Using Lewis electron-dot symbols (and showing a chlorine atom rather than the Cl_2 molecule), we can represent this reaction as follows:

$$\text{Na}\cdot + \cdot\ddot{\underset{\cdot\cdot}{\text{Cl}}}: \longrightarrow \text{Na}^+ + [:\ddot{\underset{\cdot\cdot}{\text{Cl}}}:]^- \qquad [8.2]$$

The arrow indicates the transfer of an electron from the Na atom to the Cl atom. Each ion has an octet of electrons, the octet on Na^+ being the $2s^2 2p^6$ electrons that lie below the single 3*s* valence electron of the Na atom. We've put a bracket around the chloride ion to emphasize that all eight electrons are located exclusively on the Cl^- ion.

Energetics of Ionic Bond Formation

As seen in Figure 8.2, the reaction of sodium with chlorine is *very* exothermic. In fact, Equation 8.1 is the reaction for the formation of NaCl(*s*) from its elements, so that the enthalpy change for the reaction is ΔH_f° for NaCl(*s*). In Appendix C we see that the heat of formation of other ionic substances is also quite negative. What factors make the formation of ionic compounds so exothermic?

In Equation 8.2 we represented the formation of NaCl as the transfer of electrons from Na to Cl. Recall from our discussion of ionization energies, however, that the loss of electrons from an atom is always an endothermic process. ∞ (Section 7.4) Removing an electron from Na(*g*) to form $Na^+(g)$ requires 496 kJ/mol. Conversely, when a nonmetal gains an electron, the process is generally exothermic, as seen from the negative electron affinities of the elements. ∞ (Section 7.5) Adding an electron to Cl(*g*) releases 349 kJ/mol. If the transfer of an electron from one atom to another were the only factor in forming an ionic bond, the overall process would rarely be exothermic. For example, removing an electron from Na(*g*) and adding it to Cl(*g*) is an endothermic process that requires 496 − 349 = 147 kJ/mol. This assumes, however, that the sodium and chlorine atoms are infinitely far apart.

The principal reason that ionic compounds are stable is the attraction between ions of unlike charge. This attraction draws the ions together, releasing energy and causing the ions to form a solid array or lattice such as that shown for NaCl in Figure 8.3. A measure of just how much stabilization results from arranging oppositely charged ions in an ionic solid is given by the **lattice energy**. *The lattice energy is the energy required to completely separate a mole of a solid ionic compound into its gaseous ions.* To get a picture of this process for NaCl, imagine that the structure shown in Figure 8.3 expands from within, so that the distances between the ions increase until the ions are very far apart. This process requires 788 kJ/mol, which is the value of the lattice energy.

$$\text{NaCl}(s) \longrightarrow \text{Na}^+(g) + \text{Cl}^-(g) \qquad \Delta H_{\text{lattice}} = +788 \text{ kJ/mol} \qquad [8.3]$$

The opposite process, therefore, the coming together of $Na(g)^+$ and $Cl(g)^-$ to form NaCl(*s*), is highly exothermic ($\Delta H = -788$ kJ/mol).

Table 8.2 ▶ lists the lattice energies of NaCl and of other ionic compounds. All are large positive values, indicating that the ions are strongly attracted to one another in these solids. The energy released by the attraction between ions of unlike charge more than makes up for the endothermic nature of ionization energies, making the formation of ionic compounds an exothermic process. The strong attractions also cause most ionic materials to be hard and brittle, with high melting points. (NaCl melts at 801°C.)

TABLE 8.2 Lattice Energies for Some Ionic Compounds

Compound	Lattice Energy (kJ/mol)	Compound	Lattice Energy (kJ/mol)
LiF	1030	$MgCl_2$	2326
LiCl	834	$SrCl_2$	2127
LiI	730		
NaF	910	MgO	3795
NaCl	788	CaO	3414
NaBr	732	SrO	3217
NaI	682		
KF	808	ScN	7547
KCl	701		
KBr	671		
CsCl	657		
CsI	600		

The magnitude of the lattice energy of a solid depends on the charges of the ions, their sizes, and their arrangement in the solid. We saw in Chapter 5 (Section 5.1) that the potential energy of two interacting charged particles is given by

$$E_{el} = \kappa \frac{Q_1 Q_2}{d} \qquad [8.4]$$

In this equation Q_1 and Q_2 are the charges on the particles, d is the distance between their centers, and κ is a constant, 8.99×10^9 J-m/C^2. Equation 8.4 indicates that the attractive interaction between two oppositely charged ions increases as the magnitudes of their charges increase and as the distance between their centers decreases. Thus, *for a given arrangement of ions, the lattice energy increases as the charges on the ions increase and as their radii decrease.* The magnitude of lattice energies depends primarily on the ionic charges because ionic radii do not vary over a very wide range.

SAMPLE EXERCISE 8.1

Without consulting Table 8.2, arrange the following ionic compounds in order of increasing lattice energy: NaF, CsI, and CaO.

Solution

Analyze: We need to determine how the distance between ionic centers and magnitude of charge affect lattice energy.

Plan: We will use Equation 8.4 to answer this question.

Solve: NaF consists of Na^+ and F^- ions, CsI of Cs^+ and I^- ions, and CaO of Ca^{2+} and O^{2-} ions. Because the product of the charges, Q_1Q_2, appears in the numerator of Equation 8.4, the lattice energy will increase dramatically when the charges of the ions increase. Thus, we expect the lattice energy of CaO, which has 2+ and 2− ions, to be the greatest of the three.

The ionic charges in NaF and CsI are the same. As a result, the difference in their lattice energies will depend on the difference in the distance between the centers of the ions in their crystals. Because the sizes of ions increase as we go down a group in the periodic table (Section 7.3), we know that Cs^+ is larger than Na^+ and I^- is larger than F^-. Therefore the distance between the Na^+ and F^- ions in NaF should be less than the distance between the Cs^+ and I^- ions in CsI. As a result, the lattice energy of NaF should be greater than that of CsI.

Check: Table 8.2 confirms the order of the lattice energies as CsI $<$ NaF $<$ CaO.

PRACTICE EXERCISE

Which substance would you expect to have the greatest lattice energy, AgCl, CuO, or CrN?

Answer: CrN

Electron Configurations of Ions of the Representative Elements

We began to consider the electron configurations of ions in Section 7.4. In light of our examination of ionic bonding, we will continue with that discussion here. The energetics of ionic bond formation helps explain why many ions tend to have noble-gas electron configurations. For example, sodium readily loses one electron to form Na^+, which has the same electron configuration as Ne:

$$\text{Na} \quad 1s^2 2s^2 2p^6 3s^1 = [\text{Ne}]3s^1$$
$$\text{Na}^+ \quad 1s^2 2s^2 2p^6 = [\text{Ne}]$$

A Closer Look Calculation of Lattice Energies: The Born–Haber Cycle

Lattice energy is a useful concept because it relates directly to the stability of an ionic solid. Unfortunately, the lattice energy cannot be determined directly by experiment. It can, however, be calculated by envisioning the formation of an ionic compound as occurring in a series of well-defined steps. We can then use Hess's law (Section 5.6) to put these steps together in a way that gives us the lattice energy for the compound. By so doing, we construct a **Born–Haber cycle**, a thermochemical cycle named after the German scientists Max Born (1882–1970) and Fritz Haber (1868–1934), who introduced it to analyze the factors contributing to the stability of ionic compounds.

In the Born–Haber cycle for NaCl we consider the formation of NaCl(*s*) from the elements Na(*s*) and $Cl_2(g)$ by two different routes, as shown in Figure 8.4 ▶. The enthalpy change for the direct route (red arrow) is the heat of formation of NaCl(*s*).

$$\text{Na}(s) + \tfrac{1}{2}\text{Cl}_2(g) \longrightarrow \text{NaCl}(s)$$
$$\Delta H_f^\circ[\text{NaCl}(s)] = -411 \text{ kJ} \qquad [8.5]$$

The indirect route consists of five steps, shown by the green arrows in Figure 8.4. First, we generate gaseous atoms of sodium by vaporizing sodium metal. Then, we form gaseous atoms of chlorine by breaking the bonds in the Cl_2 molecules. The enthalpy changes for these processes are available to us as enthalpies of formation (Appendix C):

$$\text{Na}(s) \longrightarrow \text{Na}(g) \qquad \Delta H_f^\circ[\text{Na}(g)] = 108 \text{ kJ} \qquad [8.6]$$
$$\tfrac{1}{2}\text{Cl}_2(g) \longrightarrow \text{Cl}(g) \qquad \Delta H_f^\circ[\text{Cl}(g)] = 122 \text{ kJ} \qquad [8.7]$$

Both of these processes are endothermic; energy is required to generate gaseous sodium and chlorine atoms.

In the next two steps we remove the electron from Na(*g*) to form $Na^+(g)$ and then add the electron to Cl(*g*) to form $Cl^-(g)$. The enthalpy changes for these processes equal the first ionization energy of Na, I_1(Na) and the electron affinity of Cl, denoted *E*(Cl), respectively: (Sections 7.4, 7.5)

$$\text{Na}(g) \longrightarrow \text{Na}^+(g) + e^- \qquad \Delta H = I_1(\text{Na}) = 496 \text{ kJ} \qquad [8.8]$$
$$\text{Cl}(g) + e^- \longrightarrow \text{Cl}^-(g) \qquad \Delta H = E(\text{Cl}) = -349 \text{ kJ} \qquad [8.9]$$

Finally, we combine the gaseous sodium and chloride ions to form solid sodium chloride. Because this process is just the reverse of the lattice energy (breaking a solid into gaseous ions), the enthalpy change is the negative of the lattice energy, the quantity that we want to determine:

$$\text{Na}^+(g) + \text{Cl}^-(g) \longrightarrow \text{NaCl}(s) \qquad \Delta H = -\Delta H_{\text{lattice}} = ? \qquad [8.10]$$

The sum of the five steps in the indirect path gives us NaCl(*s*) from Na(*s*) and $\tfrac{1}{2}Cl_2(g)$. Thus, from Hess's law we know that the sum of the enthalpy changes for these five steps equals that for the direct path, indicated by the red arrow, Equation 8.5:

$$\Delta H_f^\circ[\text{NaCl}(s)] = \Delta H_f^\circ[\text{Na}(g)] + \Delta H_f^\circ[\text{Cl}(g)] + I_1(\text{Na}) + E(\text{Cl}) - \Delta H_{\text{lattice}}$$

$$-411 \text{ kJ} = 108 \text{ kJ} + 122 \text{ kJ} + 496 \text{ kJ} - 349 \text{ kJ} - \Delta H_{\text{lattice}}$$

Solving for $\Delta H_{\text{lattice}}$:

$$\Delta H_{\text{lattice}} = 108 \text{ kJ} + 122 \text{ kJ} + 496 \text{ kJ} - 349 \text{ kJ} + 411 \text{ kJ} = 788 \text{ kJ}$$

Thus, the lattice energy of NaCl is 788 kJ/mol.

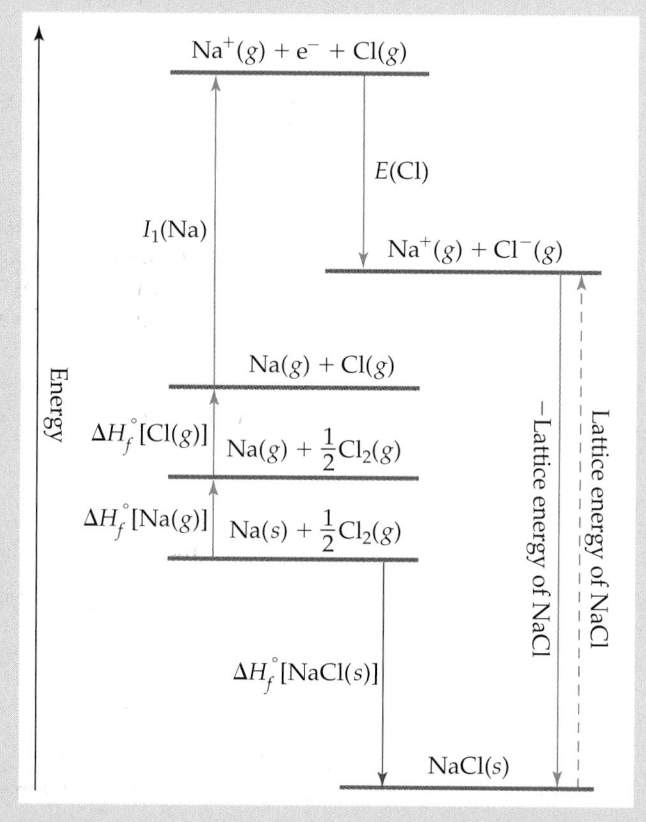

▲ **Figure 8.4** A Born–Haber cycle shows the energetic relationships in the formation of ionic solids from the elements. The enthalpy of formation of NaCl(*s*) from elemental sodium and chlorine (Equation 8.5) is equal to the sum of the energies of several individual steps (Equations 8.6 through 8.10) by Hess's law.

Even though lattice energy increases with increasing ionic charge, we never find ionic compounds that contain Na^{2+} ions. The second electron removed would have to come from the inner shell of the sodium atom, which requires a very large amount of energy. ∞ (Section 7.4) The increase in lattice energy is not enough to compensate for the energy needed to remove an inner-shell electron. Thus, sodium and the other group 1A metals are found in ionic substances only as 1+ ions.

Similarly, the addition of electrons to nonmetals is either exothermic or only slightly endothermic as long as the electrons are being added to the valence shell. Thus, a Cl atom easily adds an electron to form Cl^-, which has the same electron configuration as Ar:

$$\text{Cl} \quad 1s^2 2s^2 2p^6 3s^2 3p^5 = [\text{Ne}]3s^2 3p^5$$

$$\text{Cl}^- \quad 1s^2 2s^2 2p^6 3s^2 3p^6 = [\text{Ne}]3s^2 3p^6 = [\text{Ar}]$$

A second electron would have to be added to the next higher shell of the Cl atom, which is energetically very unfavorable. Therefore, we never observe Cl^{2-} ions in ionic compounds.

By using these concepts, we expect that ionic compounds of the representative metals from groups 1A, 2A, and 3A will contain cations with charges of 1+, 2+, and 3+, respectively. Likewise, ionic compounds of the representative nonmetals of groups 5A, 6A, and 7A usually contain anions of charge 3−, 2−, and 1−, respectively. We very rarely find ionic compounds of the nonmetals from group 4A (C, Si, and Ge). The heaviest elements in group 4A (Sn and Pb) are metals and are usually found as 2+ cations in ionic compounds: Sn^{2+} and Pb^{2+}. This behavior is consistent with the increasing metallic character found as one proceeds down a column in the periodic table. ∞ (Section 7.6)

SAMPLE EXERCISE 8.2

Predict the ion generally formed by each of the following atoms: **(a)** Sr; **(b)** S; **(c)** Al.

Solution In each case we can use the element's position in the periodic table to predict whether it will form a cation or an anion. We can then use its electron configuration to determine the ion that is likely to be formed. **(a)** Strontium is a metal in group 2A and will therefore form a cation. Its electron configuration is $[Kr]5s^2$, so we expect that the two valence electrons can be lost easily to give an Sr^{2+} ion. **(b)** Sulfur is a nonmetal in group 6A and will thus tend to be found as an anion. Its electron configuration ($[Ne]3s^2 3p^4$) is two electrons short of a noble-gas configuration. Thus, we expect that sulfur tends to form S^{2-} ions. **(c)** Aluminum is a metal in group 3A. We therefore expect it to form Al^{3+} ions.

PRACTICE EXERCISE

Predict the charges on the ions formed when magnesium reacts with nitrogen.
Answer: Mg^{2+} and N^{3-}

Transition-Metal Ions

Because ionization energies increase rapidly for each successive electron removed, the lattice energies of ionic compounds are generally large enough to compensate for the loss of up to only three electrons from atoms. Thus, we find cations with charges of 1+, 2+, or 3+ in ionic compounds. Most transition metals, however, have more than three electrons beyond a noble-gas core. Silver, for example, has a $[Kr]4d^{10}5s^1$ electron configuration. Metals of group 1B (Cu, Ag, Au) often occur as 1+ ions (as in CuBr and AgCl). In forming Ag^+, the 5*s* electron is lost, leaving a completely filled 4*d* subshell. As in this example, transition metals generally do not form ions with noble-gas configurations. The octet rule, although useful, is clearly limited in scope.

Recall from our discussion in Section 7.4 that when a positive ion is formed from an atom, electrons are always lost first from the subshell having the largest value of n. Thus, *in forming ions, transition metals lose the valence-shell* s *electrons first, then as many* d *electrons as are required to reach the charge of the ion.* Let's consider Fe, with the electron configuration $[Ar]3d^6 4s^2$. In forming the Fe^{2+} ion, the two $4s$ electrons are lost, leading to an $[Ar]3d^6$ configuration. Removal of an additional electron gives the Fe^{3+} ion, whose electron configuration is $[Ar]3d^5$.

Polyatomic Ions

Let's now briefly reconsider Tables 2.4 and 2.5, which list the common ions. (Section 2.8) Several cations and many common anions are polyatomic. Examples include the ammonium ion, NH_4^+, and the carbonate ion, CO_3^{2-}. In polyatomic ions two or more atoms are bound together by predominantly covalent bonds. They form a stable grouping that carries a charge, either positive or negative. We will examine the covalent bonding forces in these ions in Chapter 9. For now, you need only understand that the group of atoms as a whole acts as a charged species in forming an ionic compound with an ion of opposite charge.

8.3 Covalent Bonding

Ionic substances possess several characteristic properties. They are usually brittle substances with high melting points. They are usually crystalline, meaning that the solids have flat surfaces that make characteristic angles with one another. Ionic crystals often can be cleaved; that is, they break apart along smooth, flat surfaces. These characteristics result from electrostatic forces that maintain the ions in a rigid, well-defined, three-dimensional arrangement such as that shown in Figure 8.3.

The vast majority of chemical substances do not have the characteristics of ionic materials. Most of the substances with which we come in daily contact, such as water, tend to be gases, liquids, or solids with low melting points. Many, such as gasoline, vaporize readily. Many are pliable in their solid forms—for example, plastic bags and paraffin.

For the very large class of substances that do not behave like ionic substances, we need a different model for the bonding between atoms. G. N. Lewis reasoned that atoms might acquire a noble-gas electron configuration by sharing electrons with other atoms. As we noted in Section 8.1, a chemical bond formed by sharing a pair of electrons is called a *covalent bond*.

The hydrogen molecule, H_2, provides the simplest possible example of a covalent bond. When two hydrogen atoms are close to each other, electrostatic interactions occur between them. The two positively charged nuclei and the two negatively charged electrons repel each other, whereas the nuclei and electrons attract each other as shown in Figure 8.5(a) ◀. For the H_2 molecule to exist as a stable entity, the attractive forces must exceed the repulsive ones. But why is this so?

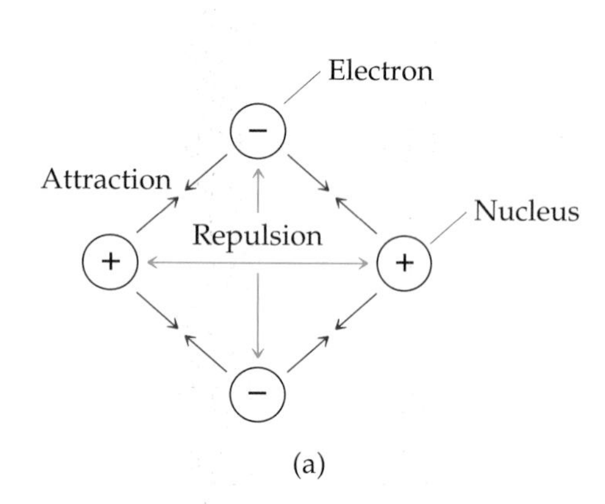

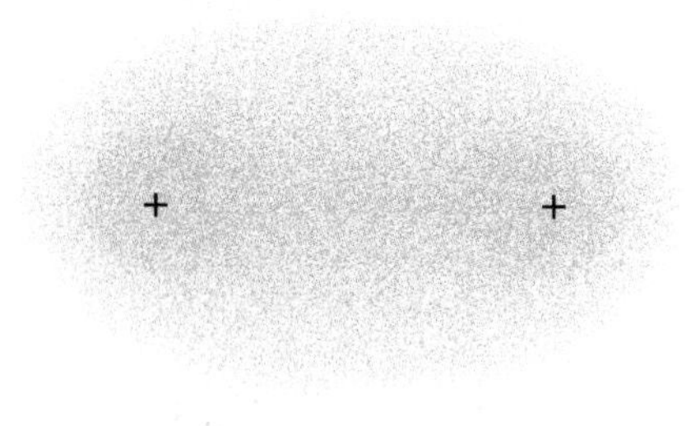

▲ **Figure 8.5** (a) The attractions and repulsions among electrons and nuclei in the hydrogen molecule. (b) Electron distribution in the H_2 molecule. The concentration of electron density between the nuclei leads to a net attractive force that constitutes the covalent bond holding the molecule together.

By using quantum mechanical methods analogous to those employed for atoms (Section 6.5), it is possible to calculate the distribution of electron density in molecules. Such a calculation for H_2 shows that the attractions between the nuclei and the electrons cause electron density to concentrate between the nuclei as shown in Figure 8.5(b). As a result, the overall electrostatic interactions are attractive. Thus, the atoms in H_2 are held together principally because the two nuclei are electrostatically attracted to the concentration of negative charge between them. In essence, the shared pair of electrons in any covalent bond acts as a kind of "glue" to bind atoms together as in the H_2 molecule.

Lewis Structures

The formation of covalent bonds can be represented using Lewis symbols for the constituent atoms. The formation of the H_2 molecule from two H atoms can be represented as

$$\mathrm{H\cdot + \cdot H \longrightarrow (H{:}H)}$$

In this way, each hydrogen atom acquires a second electron, achieving the stable, two electron, noble-gas electron configuration of helium.

The formation of a bond between two Cl atoms to give a Cl_2 molecule can be represented in a similar way:

$$\mathrm{:\ddot{\underset{..}{Cl}}\cdot + \cdot\ddot{\underset{..}{Cl}}: \longrightarrow (:\ddot{\underset{..}{Cl}}{:}\ddot{\underset{..}{Cl}}:)}$$

By sharing the bonding electron pair, each chlorine atom has eight electrons (an octet) in its valence shell. It thus achieves the noble-gas electron configuration of argon.

The structures shown here for H_2 and Cl_2 are called **Lewis structures** (or Lewis electron-dot structures). In writing Lewis structures, we usually show each electron pair shared between atoms as a line, and the unshared electron pairs as dots. Writing them this way, the Lewis structures for H_2 and Cl_2 are shown as follows:

$$\mathrm{H{-}H} \qquad \mathrm{:\ddot{\underset{..}{Cl}}{-}\ddot{\underset{..}{Cl}}:}$$

For the nonmetals, the number of valence electrons on a neutral atom is the same as the group number. Therefore, one might predict that 7A elements, such as F, would form one covalent bond to achieve an octet; 6A elements, such as O, would form two covalent bonds; 5A elements, such as N, would form three covalent bonds; and 4A elements, such as C, would form four covalent bonds. These predictions are borne out in many compounds. For example, consider the simple hydrogen compounds of the nonmetals of the second row of the periodic table:

$$\mathrm{H{-}\ddot{\underset{..}{F}}:} \qquad \mathrm{H{-}\underset{\underset{H}{|}}{\ddot{O}}:} \qquad \mathrm{H{-}\underset{\underset{H}{|}}{\ddot{N}}{-}H} \qquad \mathrm{H{-}\overset{\overset{H}{|}}{\underset{\underset{H}{|}}{C}}{-}H}$$

Thus, the Lewis model succeeds in accounting for the compositions of many compounds of nonmetals, in which covalent bonding predominates.

SAMPLE EXERCISE 8.3

Given the Lewis symbols for the elements nitrogen and fluorine shown in Table 8.1, predict the formula of the stable binary compound formed by reaction of nitrogen with fluorine, and draw its Lewis structure.

Solution

Analyze: The Lewis symbols for nitrogen and fluorine in Table 8.1 reveal that nitrogen has five valence-shell electrons and fluorine has seven.

Plan: We need to find a combination of the two elements that results in an octet of electrons around each atom in the compound. Nitrogen requires three additional electrons to complete its octet, whereas fluorine requires but one. Sharing a pair of electrons between the two elements will result in an octet of electrons for fluorine.

Solve: Nitrogen must share a pair of electrons with three fluorine atoms to complete its octet. Thus, the Lewis structure for the resulting compound, NF_3, is as follows:

$$:\ddot{F}:\ddot{N}:\ddot{F}: \quad \longrightarrow \quad :\ddot{F}-\ddot{N}-\ddot{F}:$$
$$:\ddot{F}: \qquad\qquad\qquad\qquad |\; :\ddot{F}:$$

Check: Each pair of shared electrons is represented as a line. Each of the three fluorine atoms and the central nitrogen atom has an octet of electrons.

PRACTICE EXERCISE

Compare the Lewis symbol for neon with the Lewis structure for methane, CH_4. In what important way are the electron arrangements about neon and carbon alike? In what important respect are they different?

Answer: Both atoms have an octet of electrons about them. However, the electrons about neon are unshared electron pairs, whereas those about carbon are shared with four hydrogen atoms.

Multiple Bonds

The sharing of a pair of electrons constitutes a single covalent bond, generally referred to simply as a **single bond**. In many molecules atoms attain complete octets by sharing more than one pair of electrons between them. When two electron pairs are shared, two lines are drawn, representing a **double bond**. In carbon dioxide, for example, bonding occurs between carbon, with four valence-shell electrons, and oxygen, with six:

$$:\dot{\underset{\cdot}{O}}: + \cdot\dot{\underset{\cdot}{C}}\cdot + :\dot{\underset{\cdot}{O}}: \longrightarrow \ddot{\underset{\cdot\cdot}{O}}::C::\ddot{\underset{\cdot\cdot}{O}} \qquad (\text{or } \ddot{\underset{\cdot\cdot}{O}}{=}C{=}\ddot{\underset{\cdot\cdot}{O}})$$

As the diagram shows, each oxygen acquires an octet of electrons by sharing two electron pairs with carbon. Carbon, on the other hand, acquires an octet of electrons by sharing two pairs with two oxygen atoms.

A **triple bond** corresponds to the sharing of three pairs of electrons, such as in the N_2 molecule:

$$:\dot{\underset{\cdot}{N}}\cdot + \cdot\dot{\underset{\cdot}{N}}: \longrightarrow :N:::N: \qquad (\text{or } :N{\equiv}N:)$$

Because each nitrogen atom possesses five electrons in its valence shell, three electron pairs must be shared to achieve the octet configuration.

The properties of N_2 are in complete accord with its Lewis structure. Nitrogen is a diatomic gas with exceptionally low reactivity that results from the very stable nitrogen-nitrogen bond. Study of the structure of N_2 reveals that the nitrogen atoms are separated by only 1.10 Å. The short N—N bond distance is a result of the triple bond between the atoms. From structure studies of many different substances in which nitrogen atoms share one or two electron pairs, we have learned that the average distance between bonded nitrogen atoms varies with the number of shared electron pairs:

N—N	N=N	N≡N
1.47 Å	1.24 Å	1.10 Å

As a general rule, the distance between bonded atoms decreases as the number of shared electron pairs increases.

8.4 Bond Polarity and Electronegativity

When two identical atoms bond, as in Cl_2 or N_2, the electron pairs must be shared equally. In ionic compounds such as NaCl, on the other hand, there is essentially no sharing of electrons. NaCl is best described as composed of Na^+ and Cl^- ions. The 3*s* electron of the Na atom is, in effect, transferred completely to chlorine. The bonds occurring in most covalent substances fall somewhere between these extremes.

The concept of **bond polarity** helps describe the sharing of electrons between atoms. A **nonpolar covalent bond** is one in which the electrons are shared equally between two atoms. In a **polar covalent bond** one of the atoms exerts a greater attraction for the bonding electrons than the other. If the difference in relative ability to attract electrons is large enough, an ionic bond is formed.

Electronegativity

ANIMATION
Periodic Trends: Electronegativity

We use a quantity called electronegativity to estimate whether a given bond will be nonpolar covalent, polar covalent, or ionic. **Electronegativity** is defined as the ability of an atom *in a molecule* to attract electrons to itself. The greater an atom's electronegativity, the greater is its ability to attract electrons to itself. The electronegativity of an atom in a molecule is related to its ionization energy and electron affinity, which are properties of isolated atoms. The *ionization energy* measures how strongly an atom holds on to its electrons. (Section 7.4) Likewise, the *electron affinity* is a measure of how strongly an atom attracts additional electrons. (Section 7.5) An atom with a very negative electron affinity and high ionization energy will both attract electrons from other atoms and resist having its electrons attracted away; it will be highly electronegative.

Numerical estimates of electronegativity can be based on a variety of properties, not just ionization energy and electron affinity. The first and most widely used electronegativity scale was developed by the American chemist Linus Pauling (1901–1994), who based his scale on thermochemical data. Figure 8.6 ▼ shows

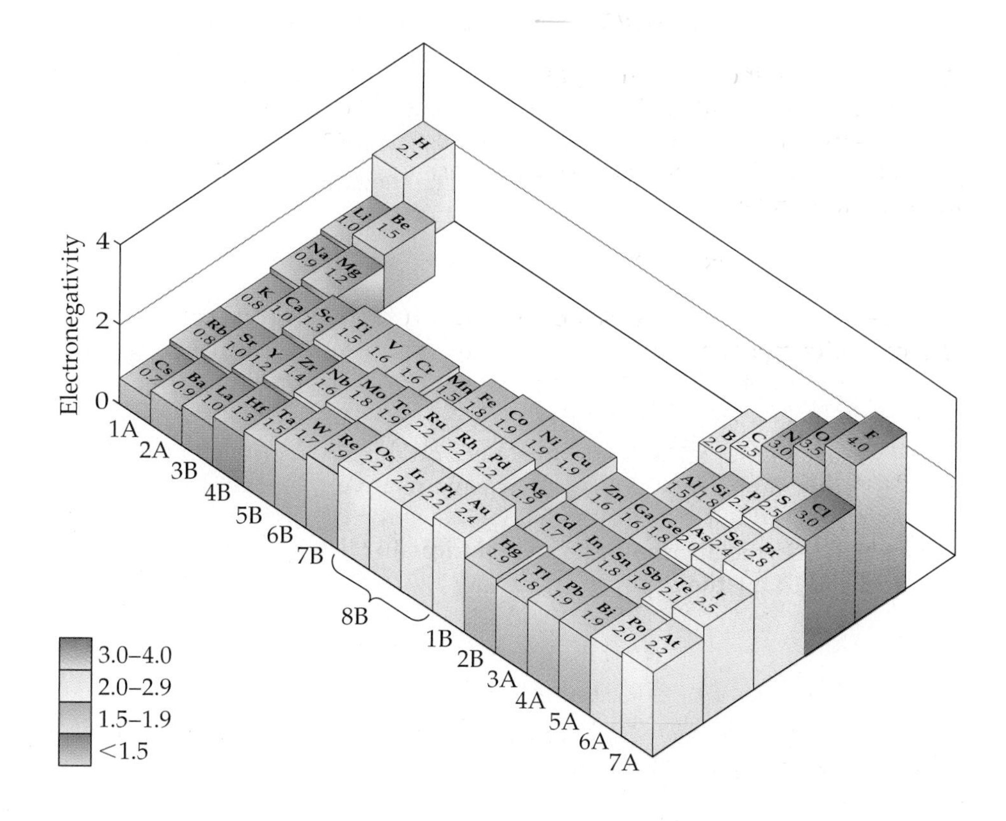

◀ **Figure 8.6** Electronegativities of the elements.

Pauling's electronegativity values for many of the elements. The values are unitless. Fluorine, the most electronegative element, has an electronegativity of 4.0. The least electronegative element, cesium, has an electronegativity of 0.7. The values for all other elements lie between these two extremes.

Within each period there is generally a steady increase in electronegativity from left to right; that is, from the most metallic to the most nonmetallic elements. With some exceptions (especially within the transition metals), electronegativity decreases with increasing atomic number in any one group. This is what we might expect because we know that ionization energies tend to decrease with increasing atomic number in a group and electron affinities don't change very much. You do not need to memorize numerical values for electronegativity. Instead, you should know the periodic trends so that you can predict which of two elements is more electronegative.

Electronegativity and Bond Polarity

We can use the difference in electronegativity between two atoms to gauge the polarity of the bonding between them. Consider these three fluorine-containing compounds:

Compound	F_2	HF	LiF
Electronegativity difference	$4.0 - 4.0 = 0$	$4.0 - 2.1 = 1.9$	$4.0 - 1.0 = 3.0$
Type of bond	Nonpolar covalent	Polar covalent	Ionic

In F_2 the electrons are shared equally between the fluorine atoms, and the covalent bond is *nonpolar*. A nonpolar covalent bond results when the electronegativities of the bonded atoms are equal.

In HF the fluorine atom has a greater electronegativity than the hydrogen atom, so the sharing of electrons is unequal; the bond is *polar*. A polar covalent bond results when the atoms differ in electronegativity. In HF the more electronegative fluorine atom attracts electron density away from the less electronegative hydrogen atom. Thus, some of the electron density around the hydrogen nucleus is pulled toward the fluorine nucleus, leaving a partial positive charge on the hydrogen atom and a partial negative charge on the fluorine atom. We can represent this charge distribution as

$$\overset{\delta+}{\text{H}}-\overset{\delta-}{\text{F}}$$

The $\delta+$ and $\delta-$ (read "delta plus" and "delta minus") symbolize the partial positive and negative charges, respectively. This shift of electron density toward the more electronegative atom can be seen in the results of calculations of electron distributions. Figure 8.7 ▶ shows the electron density distributions in F_2, HF, and LiF, with the regions of space that have relatively higher electron density shown in red, those with a relatively lower electron density shown in blue. You can see that in F_2 the distribution is symmetrical. In HF it is clearly shifted toward fluorine, and in LiF the shift is even greater.*

* The calculation for LiF is for an isolated LiF "molecule." While the bond in this isolated diatomic system is very polar, it is not quite 100% ionic, as is the Li–F bonding in solid lithium fluoride. The solid state promotes a more complete shift of electron density from Li to F because each ion in the solid is surrounded on all sides by ions of opposite charge.

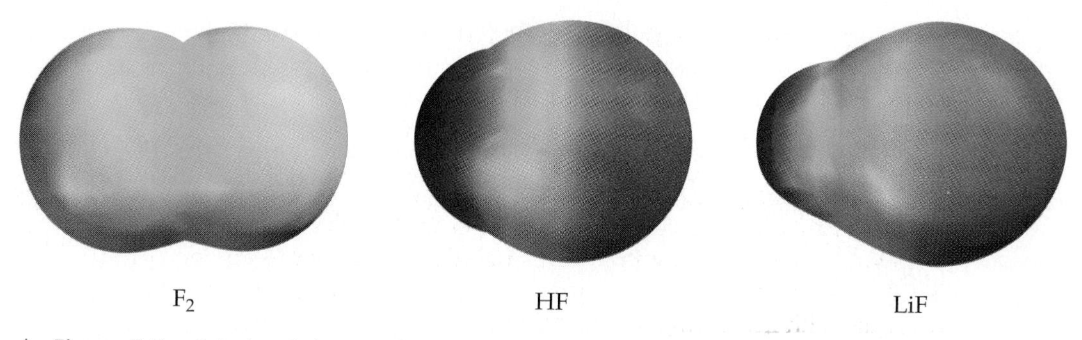

▲ **Figure 8.7** Calculated electron density distributions in F_2, HF, and LiF. The regions of relatively low electron density are shown as blue, those of relatively high electron density are shown as red.

In the three-dimensional structure for LiF, analogous to that shown for NaCl in Figure 8.3, the transfer of electronic charge is essentially complete. The resultant bond is therefore ionic. These examples illustrate, therefore, that *the greater the difference in electronegativity between two atoms, the more polar their bond.* The nonpolar covalent bond lies at one end of a continuum of bond types, and the ionic bond lies at the other end. In between is a broad range of polar covalent bonds, differing in the extent to which there is unequal sharing of electrons.

SAMPLE EXERCISE 8.4

Which bond is more polar: **(a)** B—Cl or C—Cl; **(b)** P—F or P—Cl? Indicate in each case which atom has the partial negative charge.

Solution **(a)** The difference in the electronegativities of chlorine and boron is $3.0 - 2.0 = 1.0$; the difference between chlorine and carbon is $3.0 - 2.5 = 0.5$. Consequently, the B—Cl bond is more polar; the chlorine atom carries the partial negative charge because it has a higher electronegativity. We should be able to reach this same conclusion using periodic trends instead of a table of electronegativities. Because boron is to the left of carbon in the periodic table, we would predict that it has a lower attraction for electrons. Chlorine, being on the right side of the table, has a strong attraction for electrons. The more polar bond will be the one between the atoms having the lowest attraction for electrons (boron) and the highest attraction (chlorine).

(b) Because fluorine is above chlorine in the periodic table, fluorine should be more electronegative. In fact, the electronegativities are F = 4.0, Cl = 3.0. Consequently, the P—F bond will be more polar than the P—Cl bond. You should compare the electronegativity differences for the two bonds to verify this prediction. The fluorine atom carries the partial negative charge.

PRACTICE EXERCISE

Which of the following bonds is most polar: S—Cl, S—Br, Se—Cl, or Se—Br?
Answer: Se—Cl

Dipole Moments

The difference in electronegativity between H and F leads to a polar covalent bond in the HF molecule. As a consequence, there is a concentration of negative charge on the more electronegative F atom, leaving the less electronegative H atom at the positive end of the molecule. A molecule such as HF in which the centers of positive and negative charge do not coincide is said to be a **polar molecule**. Thus, we not only describe bonds as polar and nonpolar, but also entire molecules are described as such.

We can indicate the polarity of the HF molecule in two ways:

$$\overset{\delta+}{\text{H}}-\overset{\delta-}{\text{F}} \quad \text{or} \quad \overset{+\!\!\longrightarrow}{\text{H}-\text{F}}$$

Recall from the preceding section that "δ+" and "δ−" indicate the partial positive and negative charges on the H and F atoms. In the notation on the right, the arrow denotes the shift in electron density toward the fluorine atom. The crossed end of the arrow can be thought of as a plus sign that designates the positive end of the molecule.

ACTIVITY
Molecular Polarity

Polarity helps determine many of the properties of substances that we observe at the macroscopic level, in the laboratory and in everyday life. Polar molecules align themselves with respect to each other and with respect to ions. The negative end of one molecule and the positive end of another attract each other. Polar molecules are likewise attracted to ions. The negative end of a polar molecule is attracted to a positive ion, and the positive end is attracted to a negative ion. These interactions account for many properties of liquids, solids, and solutions, as you will see in Chapters 11, 12, and 13.

How can we quantify the polarity of a molecule such as HF? Whenever two electrical charges of equal magnitude but opposite sign are separated by a distance, a **dipole** is established. The quantitative measure of the magnitude of a dipole is called its **dipole moment**, denoted μ. If two equal and opposite charges, $Q+$ and $Q-$, are separated by a distance r, the magnitude of the dipole moment is the product of Q and r (Figure 8.8 ▼).

$$\mu = Qr \qquad [8.11]$$

The dipole moment will increase in size as the magnitude of charge that is separated increases and as the distance between the charges increases.

Dipole moments of molecules are usually reported in *debyes* (D), a unit that equals 3.34×10^{-30} coulomb-meters (C-m). For molecules, we usually measure charge in units of the electronic charge e, 1.60×10^{-19} C, and distance in units of angstroms, Å. Suppose that two charges, 1+ and 1− (in units of e), are separated by a distance of 1.00 Å. The dipole moment produced is

$$\mu = Qr = (1.60 \times 10^{-19}\ \text{C})(1.00\ \text{Å})\left(\frac{10^{-10}\ \text{m}}{1.00\ \text{Å}}\right)\left(\frac{1\ \text{D}}{3.34 \times 10^{-30}\ \text{C-m}}\right) = 4.79\ \text{D}$$

Measurement of the dipole moments can provide us with valuable information about the charge distributions in molecules, as illustrated in Sample Exercise 8.5.

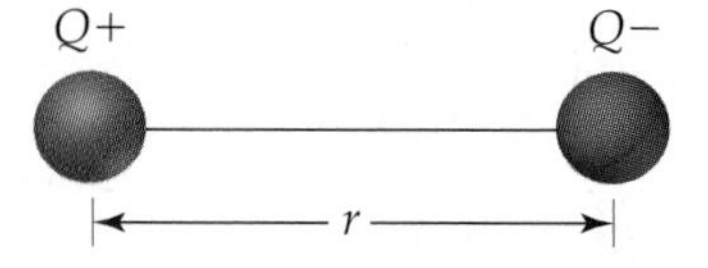

▶ **Figure 8.8** When charges of equal magnitude and opposite sign $Q+$ and $Q-$ are separated by a distance r, a dipole is produced. The size of the dipole is given by the dipole moment, μ, which is the product of the charge separated and the distance of separation between the charge centers: $\mu = Qr$.

SAMPLE EXERCISE 8.5

The distance between the centers of the H and Cl atoms in the HCl molecule (called its *bond length*) is 1.27 Å. **(a)** Calculate the dipole moment, in D, that would result if the charges on the H and Cl atoms were 1+ and 1−, respectively. **(b)** The experimentally measured dipole moment of HCl(*g*) is 1.08 D. What magnitude of charge, in units of e, on the H and Cl atoms would lead to this dipole moment?

Solution

Analyze and Plan: We are asked to calculate the dipole moment of HCl that would result from full charges on each atom, and to use that value to calculate the effective partial charges on H and Cl that would give rise to the observed dipole moment.

Solve: **(a)** The charge on each atom is the electronic charge, e: 1.60×10^{-19} C. The separation is 1.27 Å. The dipole moment is:

$$\mu = Qr = (1.60 \times 10^{-19}\,\text{C})(1.27\,\text{Å})\left(\frac{10^{-10}\,\text{m}}{1\,\text{Å}}\right)\left(\frac{1\,\text{D}}{3.34 \times 10^{-30}\,\text{C-m}}\right) = 6.08\,\text{D}$$

The calculated dipole moment is greater than in the earlier example because the distance between the charges has increased from 1.00 Å to 1.27 Å.

(b) In this instance we know the value of μ, 1.08 D, and the value of r, 1.27 Å, and we want to calculate the value of Q.

$$Q = \frac{\mu}{r} = \frac{(1.08\,\text{D})\left(\dfrac{3.34 \times 10^{-30}\,\text{C-m}}{1\,\text{D}}\right)}{(1.27\,\text{Å})\left(\dfrac{10^{-10}\,\text{m}}{1\,\text{Å}}\right)} = 2.84 \times 10^{-20}\,\text{C}$$

We can readily convert this charge to units of e.

$$\text{Charge in } e = (2.84 \times 10^{-20}\,\text{C})\left(\frac{1\,e}{1.60 \times 10^{-19}\,\text{C}}\right) = 0.178\,e$$

Thus, the experimental dipole moment indicates the following charge separation in the HCl molecule:

$$\overset{0.178+}{\text{H}}-\overset{0.178-}{\text{Cl}}$$

Because the experimental dipole moment is less than that calculated in part (a), the charges on the atoms are less than a full electronic charge. We could have anticipated this because the H—Cl bond is polar covalent rather than ionic.

PRACTICE EXERCISE

The dipole moment of chlorine monofluoride, ClF(g), is 0.88 D. The bond length of the molecule is 1.63 Å. **(a)** Which atom is expected to have a negative charge? **(b)** What is the charge on that atom, in e?
Answers: **(a)** F; **(b)** 0.11−

TABLE 8.3 Bond Lengths, Electronegativity Differences, and Dipole Moments of the Hydrogen Halides

Compound	Bond Length (Å)	Electronegativity Difference	Dipole Moment (D)
HF	0.92	1.9	1.82
HCl	1.27	0.9	1.08
HBr	1.41	0.7	0.82
HI	1.61	0.4	0.44

Table 8.3 ▲ presents the bond lengths and dipole moments of the hydrogen halides. Notice that as we proceed from HF to HI, the electronegativity difference decreases and the bond length increases. The first effect decreases the amount of charge separated and causes the dipole moment to decrease from HF to HI, even though the bond length is increasing. We can "observe" the varying degree of electronic charge shift in these substances from calculations of electron distribution, as shown below. For these molecules, the change in the difference in electronegativity affects the dipole moment more than the bond length.

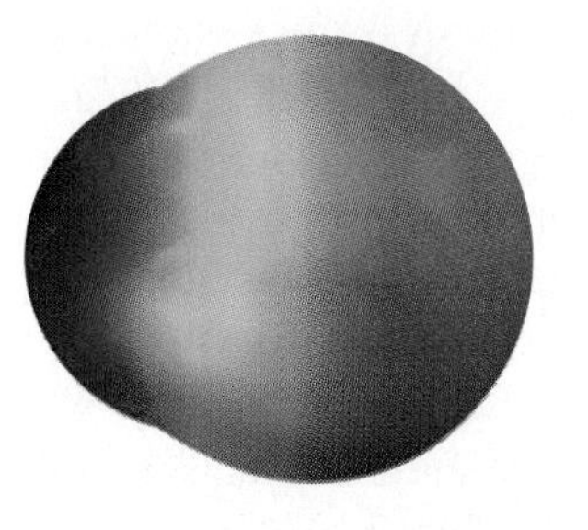

HF

HCl

HBr

HI

Bond Types and Nomenclature

This is a good point for a brief interlude to talk about nomenclature. We saw in Section 2.8 that there are two general approaches to naming binary compounds (compounds composed of two elements): one used for ionic compounds and the other for molecular ones. In both approaches, the name of the less electronegative element is given first. The name of the more electronegative element then follows, modified to have an *-ide* ending. Compounds that are ionic are given names based on their component ions, including the charge of the cation if it is variable. Those that are molecular are named using the prefixes listed in Table 2.6 to indicate the number of atoms of each kind in the substance:

Ionic		*Molecular*	
MgH_2	magnesium hydride	H_2S	dihydrogen sulfide
FeF_2	iron(II) fluoride	OF_2	oxygen difluoride
Mn_2O_3	manganese(III) oxide	Cl_2O_3	dichlorine trioxide

The dividing line between the two approaches, however, is not always clear, and both approaches are often applied to the same substances. TiO_2, for example, which is a commercially important white paint pigment, is sometimes referred to as titanium(IV) oxide but is more commonly called titanium dioxide. The Roman numeral in the first name is the oxidation number of the titanium. (Section 4.4)

One reason for the overlap in the two approaches to nomenclature is that compounds of metals with higher oxidation numbers often behave as though they were molecular rather than ionic. For example, $SnCl_4$ [tin tetrachloride or tin(IV) chloride] is a colorless liquid that freezes at −33°C and boils at 114°C; Mn_2O_7 [dimanganese heptoxide or manganese(VII) oxide] is a green liquid that freezes at 5.9°C. Ionic compounds, on the other hand, are solids at room temperature. When we see the formula of a compound containing a metal with a high oxidation number (above +3), we should not be surprised when it does not exhibit the general properties of ionic compounds.

8.5 Drawing Lewis Structures

Lewis structures can help us understand the bonding in many compounds and are frequently used when discussing the properties of molecules. Drawing Lewis structures is an important skill that you should practice. To do so, you should follow a regular procedure. First we'll outline the procedure, and then we'll go through several examples.

1. *Sum the valence electrons from all atoms.* (Use the periodic table as necessary to help you determine the number of valence electrons in each atom.) For an anion, add an electron for each negative charge. For a cation, subtract an electron for each positive charge. Don't worry about keeping track of which electrons come from which atoms. Only the total number is important.
2. *Write the symbols for the atoms to show which atoms are attached to which, and connect them with a single bond* (a dash, representing two electrons). Chemical formulas are often written in the order in which the atoms are connected in the molecule or ion, as in HCN. When a central atom has a group of other atoms bonded to it, the central atom is usually written first, as in CO_3^{2-} and SF_4. It also helps to remember that the central atom is generally less electronegative than the atoms surrounding it. In other cases, you may need more information before you can draw the Lewis structure.

3. *Complete the octets of the atoms bonded to the central atom.* (Remember, however, that hydrogen can have only two electrons.)
4. *Place any leftover electrons on the central atom*, even if doing so results in more than an octet.
5. *If there are not enough electrons to give the central atom an octet, try multiple bonds.* Use one or more of the unshared pairs of electrons on the atoms bonded to the central atom to form double or triple bonds.

SAMPLE EXERCISE 8.6

Draw the Lewis structure for phosphorus trichloride, PCl_3.

Solution *First*, we sum the valence electrons. Phosphorus (group 5A) has five valence electrons, and each chlorine (group 7A) has seven. The total number of valence-shell electrons is therefore:

$$5 + (3 \times 7) = 26$$

Second, we arrange the atoms to show which atom is connected to which, and we draw a single bond between them. There are various ways the atoms might be arranged. In binary (two-element) compounds, on the other hand, the first element listed in the chemical formula is generally surrounded by the remaining atoms. Thus, we begin with a skeleton structure that shows single bonds between phosphorus and each chlorine:

Cl—P—Cl
    |
   Cl

(It is not crucial to place the atoms in exactly this arrangement.)

Third, complete the octets on the atoms bonded to the central atom. Placing octets around each Cl atom accounts for 24 electrons.

:C̤̈l—P—C̤̈l:
    |
  :C̤̈l:

Fourth, place the remaining two electrons on the central atom, completing the octet around that atom as well.

:C̤̈l—P̈—C̤̈l:
    |
  :C̤̈l:

This structure gives each atom an octet, so we stop at this point. (Remember that in achieving an octet, the bonding electrons are counted for both atoms.)

PRACTICE EXERCISE

(a) How many valence electrons should appear in the Lewis structure for CH_2Cl_2? **(b)** Draw the Lewis structure.

Answers: **(a)** 20; **(b)**

      H
      |
:C̤̈l—C—C̤̈l:
      |
      H

SAMPLE EXERCISE 8.7

Draw the Lewis structure for HCN.

Solution Hydrogen has one valence-shell electron, carbon (group 4A) has four, and nitrogen (group 5A) has five. The total number of valence-shell electrons is therefore $1 + 4 + 5 = 10$. Again, there are various ways we might choose to arrange the atoms.

Because hydrogen can accommodate only one electron pair, it always has only one single bond associated with it in any compound. C—H—N, therefore, is an impossible arrangement. The remaining two possibilities are H—C—N and H—N—C. The first is the arrangement found experimentally. You might have guessed this to be the atomic arrangement because the formula is written with the atoms in this order. Thus we begin with a skeleton structure that shows single bonds between hydrogen, carbon, and nitrogen:

H—C—N

These two bonds account for four electrons. If we then place the remaining six electrons around N to give it an octet, we do not achieve an octet on C:

H—C—N̤̈:

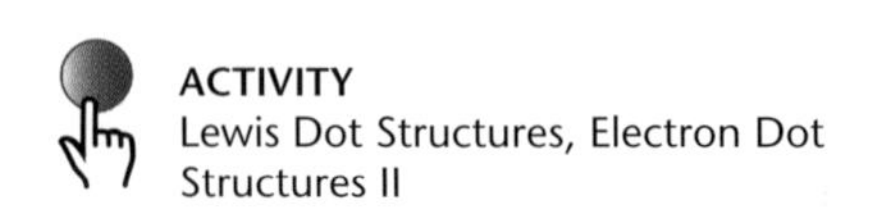
ACTIVITY
Lewis Dot Structures, Electron Dot Structures II

We therefore try a double bond between C and N, using an unshared pair of electrons that we had placed on N. Again, there are fewer than eight electrons on C, so we try a triple bond. This structure gives an octet around both C and N:

H—C=N̤̈: ⟶ H—C≡N:

PRACTICE EXERCISE

Draw the Lewis structure for **(a)** NO^+ ion; **(b)** C_2H_4.

Answers: **(a)** [:N≡O:]$^+$ **(b)** $H_2C{=}CH_2$ (H—C bonds to two H on each C)

SAMPLE EXERCISE 8.8

Draw the Lewis structure for the BrO_3^- ion.

Solution Bromine (group 7A) has seven valence electrons, and oxygen (group 6A) has six. An extra electron is added to account for the ion having a 1− charge. The total number of valence-shell electrons is therefore $7 + (3 \times 6) + 1 = 26$. After putting in the single bonds and distributing the unshared electron pairs, we have

[:Ö—Br̈—Ö: with :Ö: bonded below Br]$^-$

For oxyanions—BrO_3^-, SO_4^{2-}, NO_3^-, CO_3^{2-}, and so forth—the oxygen atoms surround the central nonmetal atoms. Notice here and elsewhere that the Lewis structures of ions are written in brackets with the charge shown outside the bracket at the upper right.

PRACTICE EXERCISE

Draw the Lewis structure for **(a)** ClO_2^- ion; **(b)** PO_4^{3-} ion.

Answers: **(a)** [:Ö—C̈l—Ö:]$^-$ **(b)** [:Ö—P—Ö: with :Ö: above and :Ö: below P]$^{3-}$

Formal Charge

When we draw a Lewis structure, we are describing how the electrons are distributed in a molecule (or ion). In some instances we can draw several different

Lewis structures that all obey the octet rule. How do we decide which one is the most reasonable? One approach is to do some "bookkeeping" of the valence electrons to determine the **formal charge** of each atom in each Lewis structure. The formal charge of an atom is the charge that an atom in a molecule would have if all atoms had the same electronegativity (that is, if all bonding electron pairs were shared equally between atoms.)

To calculate the formal charge on any atom in a Lewis structure, we assign the electrons to the atom as follows:

1. *All* of the unshared (nonbonding) electrons are assigned to the atom on which they are found.
2. *Half* of the bonding electrons are assigned to each atom in the bond.

The formal charge of an atom equals *the number of valence electrons in the isolated atom, minus the number of electrons assigned to the atom in the Lewis structure.*

Let's illustrate these rules by calculating the formal charges on the C and N atoms in the cyanide ion, CN^-, which has the following Lewis structure:

$$[:C{\equiv}N:]^-$$

ANIMATION
Formal Charges

ACTIVITY
Formal Charges

For the C atom, there are 2 nonbonding electrons and 3 electrons from the 6 in the triple bond, for a total of 5. The number of valence electrons on a neutral C atom is 4. Thus, the formal charge on C is $4 - 5 = -1$. For N, there are 2 nonbonding electrons and 3 electrons from the triple bond. Because the number of valence electrons on a neutral N atom is 5, its formal charge is $5 - 5 = 0$. Thus the formal charges on the atoms in the Lewis structure of CN^- are

$$[:\overset{-1}{C}{\equiv}\overset{0}{N}:]^-$$

Notice that the sum of the formal charges equals the overall charge on the ion, 1−. The formal charges on a molecule add to zero, whereas those on an ion add to give the overall charge on the ion.

To see how the formal charge can help to distinguish between alternative Lewis structures, let's consider the CO_2 molecule. As shown in Section 8.3, CO_2 is represented as having two double bonds. The octet rule is also obeyed, however, in a Lewis structure having a single and triple bond. Calculating the formal charge for each atom in these structures, we have

	$\ddot{\underset{..}{O}}$	=C=	$\ddot{\underset{..}{O}}$	$:\ddot{\underset{..}{O}}$	—C≡	O:
valence e^-:	6	4	6	6	4	6
$-(e^-$ assigned to atom):	6	4	6	7	4	5
formal charge:	0	0	0	−1	0	+1

Because CO_2 is a neutral molecule, the formal charges in both Lewis structures add up to zero.

As a general rule, when several Lewis structures are possible, the most stable one will be that in which (1) the atoms bear formal charges closest to zero, and (2) any negative charges reside on the more electronegative atoms. Thus, the first Lewis structure of CO_2 is preferred because the atoms carry no formal charges.

Although the concept of formal charge helps us choose between alternative Lewis structures, *formal charges do not represent real charges on atoms*. Electronegativity differences between atoms are important in determining the actual charge distributions in molecules and ions.

A Closer Look Oxidation Numbers, Formal Charges, and Actual Partial Charges

In Chapter 4 we introduced the rules for assigning *oxidation numbers* to atoms. The concept of electronegativity is the basis of these numbers. The oxidation number of an atom is the charge it would have if its bonds were completely ionic. That is, in determining the oxidation number, all shared electrons are counted with the more electronegative atom. For example, consider the Lewis structure of HCl shown in Figure 8.9(a) ▶. To assign oxidation numbers, the pair of electrons in the covalent bond between the atoms is assigned to the more electronegative Cl atom. This procedure gives Cl eight valence-shell electrons, one more than the neutral atom. Thus, it is assigned an oxidation number of −1. Hydrogen has no valence electrons when they are counted this way, giving it an oxidation number of +1.

In this section we have just considered another way of counting electrons that gives rise to *formal charges*. The formal charge is assigned by completely ignoring electronegativity and assigning the electrons in bonds equally between the bonded atoms. Consider again the HCl molecule, but this time divide the bonding pair of electrons equally between H and Cl as shown in Figure 8.9(b). In this case Cl has seven assigned electrons, the same as that of the neutral Cl atom. Thus, the formal charge of Cl in this compound is 0. Likewise, the formal charge of H is also 0.

Neither the oxidation number nor the formal charge gives an accurate depiction of the actual charges on atoms. Oxidation numbers overstate the role of electronegativity, and formal charges ignore it completely. It seems reasonable that electrons in covalent bonds should be apportioned according to the relative electronegativities of the bonded atoms. From Figure 8.6 we see that Cl has an electronegativity of 3.0 while that of H is 2.1. The more electronegative Cl atom might therefore be expected to have roughly $3.0/(3.0 + 2.1) = 0.59$ of the electrical charge in the bonding pair, whereas the H atom has $2.1/(3.0 + 2.1) = 0.41$ of the charge. Because the bond consists of two electrons, Cl's share is $0.59 \times 2e = 1.18e$, or $0.18e$ more than the neutral Cl atom. This gives rise to a partial charge of 0.18− on Cl and 0.18+ on H.

The dipole moment of HCl gives an experimental measure of the partial charges on each atom. In Sample Exercise 8.6 we saw that the dipole moment of HCl indicates a charge separation with a partial charge of 0.178+ on H and 0.178− on Cl, in surprisingly good agreement with our simple approximation based on electronegativities. Although that type of calculation provides ballpark numbers for the magnitude of charge on atoms, the relationship between electronegativities and charge separation is generally more complicated. As we have already seen, computer programs employing quantum mechanical principles have been developed to calculate the partial charges on atoms, even in complex molecules. Figure 8.9(c) shows a graphical representation of the charge distribution in HCl.

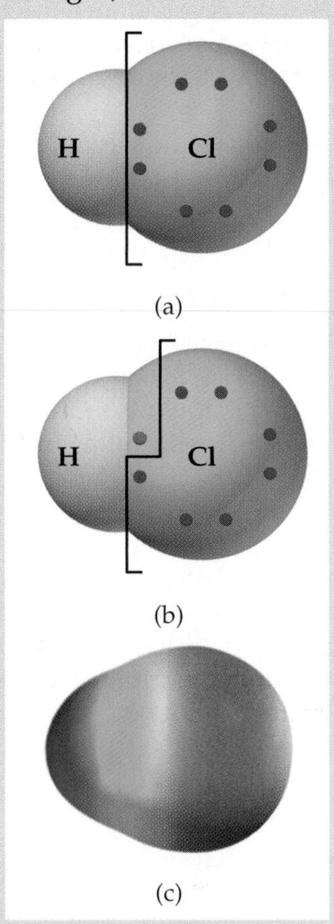

◀ **Figure 8.9** (a) Oxidation numbers are derived by counting all shared electrons with the more electronegative atom (in this case Cl). (b) Formal charges are derived by dividing all shared electrons equally between the bonded atoms. (c) The distribution of electron density on an HCl molecule as calculated by a computer program. Regions of relatively more negative charge are red; those of more positive charge are blue. Negative charge is clearly localized on chlorine.

SAMPLE EXERCISE 8.9

Three possible Lewis structures of the thiocyanate ion, NCS^-, are

$$[:\ddot{\underset{..}{N}}-C\equiv S:]^- \qquad [\ddot{\underset{..}{N}}=C=\ddot{\underset{..}{S}}]^- \qquad [:N\equiv C-\ddot{\underset{..}{S}}:]^-$$

(a) Determine the formal charges of the atoms in each structure. **(b)** Which Lewis structure should be the preferred one?

Solution (a) Neutral N, C, and S atoms have 5, 4, and 6 valence electrons, respectively. By using the rules just discussed, we can determine the following formal charges in the three structures:

$$\overset{-2}{[:\ddot{\underset{..}{N}}}-\overset{0}{C}\equiv\overset{+1}{S}:]^- \qquad \overset{-1}{[\ddot{\underset{..}{N}}}=\overset{0}{C}=\overset{0}{\ddot{\underset{..}{S}}}]^- \qquad \overset{0}{[:N}\equiv\overset{0}{C}-\overset{-1}{\ddot{\underset{..}{S}}}:]^-$$

As they must, the formal charges in all three structures sum to 1−, the overall charge of the ion. **(b)** As discussed in Section 8.4, N is more electronegative than C or S. Therefore, we expect that any negative formal charge will reside on the N atom. Further, we usually choose the Lewis structure that produces the formal charges of smallest magnitude. For these two reasons, the middle structure is the preferred Lewis structure of the NCS^- ion.

PRACTICE EXERCISE

The cyanate ion (NCO^-), like the thiocyanate ion, has three possible Lewis structures. **(a)** Draw these three Lewis structures and assign formal charges to the atoms in each structure. **(b)** Which Lewis structure should be the preferred one?

Answers: **(a)**

(i)	(ii)	(iii)
−2 0 +1	−1 0 0	0 0 −1
$[:\ddot{\underset{..}{N}}-C\equiv O:]^-$	$[\ddot{\underset{..}{N}}=C=\ddot{\underset{..}{O}}]^-$	$[:N\equiv C-\ddot{\underset{..}{O}}:]^-$

(b) Structure (iii), which places a negative charge on oxygen, the most electronegative of the three elements, should be the most important Lewis structure.

8.6 Resonance Structures

We sometimes encounter molecules and ions in which the known arrangement of atoms is not adequately described by a single Lewis structure. Consider ozone, O_3, which consists of bent molecules with both O—O distances the same (Figure 8.10 ▶). Because each oxygen atom contributes 6 valence-shell electrons, the ozone molecule has 18 valence-shell electrons. In writing the Lewis structure, we find that we must have one double bond to attain an octet of electrons about each atom:

$$:\ddot{\underset{..}{O}}-\ddot{O}=\ddot{O}$$

But this structure cannot by itself be correct because it requires that one O—O bond be different from the other, contrary to the observed structure; we would expect the O=O double bond to be shorter than the O—O single bond. In drawing the Lewis structure, however, we could just as easily have put the O=O bond on the left:

$$\ddot{O}=\ddot{O}-\ddot{\underset{..}{O}}:$$

The placement of the atoms in the two alternative Lewis structures for ozone is the same, but the placement of the electrons is different. Lewis structures of this sort are called **resonance structures**. To describe the structure of ozone properly, we write both Lewis structures and indicate that the real molecule is described by an average of the two resonance structures:

$$\ddot{O}=\ddot{O}-\ddot{\underset{..}{O}}: \longleftrightarrow :\ddot{\underset{..}{O}}-\ddot{O}=\ddot{O}$$

The double-headed arrow indicates that the structures shown are resonance structures.

To understand why certain molecules require more than one resonance structure, we can draw an analogy to the mixing of paint (Figure 8.11 ▶). Blue and yellow are both primary colors of paint pigment. An equal blend of blue and yellow pigments produces green pigment. We can't describe green paint in terms of a single primary color, yet it still has its own identity. Green paint does not oscillate

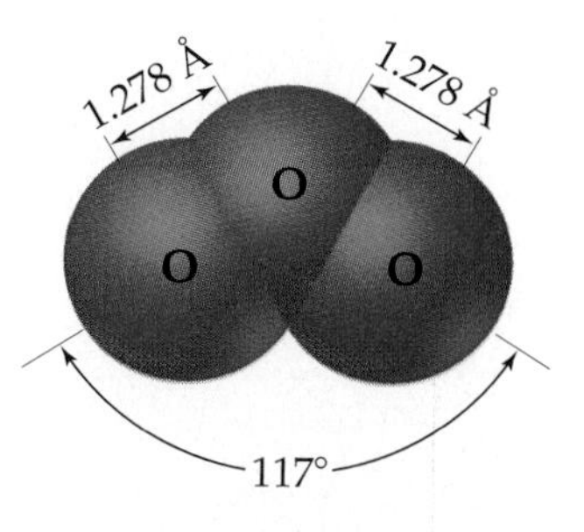

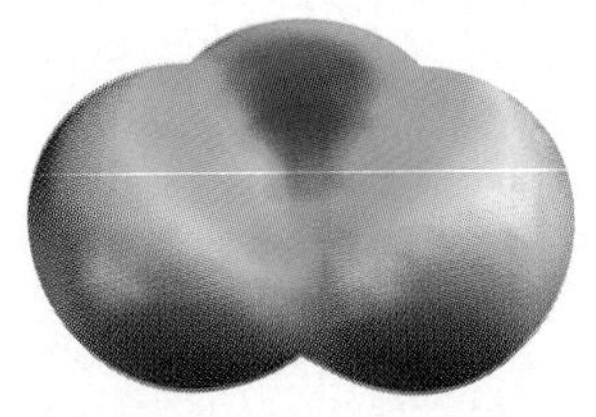

▲ **Figure 8.10** Molecular structure (top) and electron-distribution diagram (bottom) for ozone.

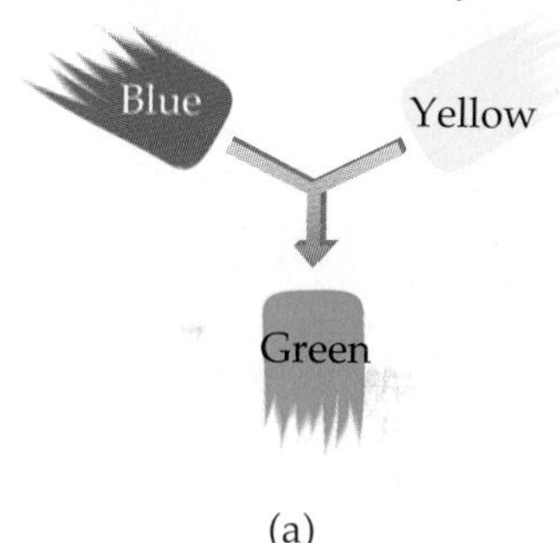

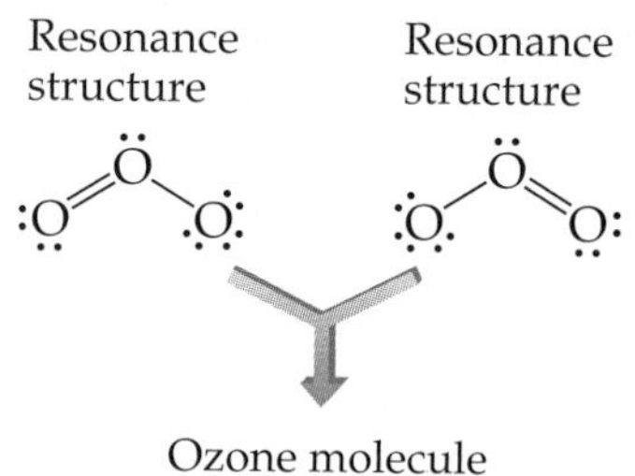

▲ **Figure 8.11** Describing a molecule as a blend of different resonance structures is similar to describing a paint color as a blend of primary colors. (a) Green paint is a blend of blue and yellow. We cannot describe green as a single primary color. (b) The ozone molecule is a blend of two resonance structures. We cannot describe the ozone molecule in terms of a single Lewis structure.

between its two primary colors. It is not blue part of the time and yellow the rest of the time. Similarly, molecules such as ozone cannot be described by a single Lewis structure in which the electrons are "locked into" a particular arrangement. Instead, the true arrangement of the electrons must be considered as a blend of two (or more) Lewis structures. By analogy to the green paint, the molecule has its own identity separate from the individual resonance structures; it does *not* oscillate rapidly between its different resonance structures. For example, the ozone molecule always has two equivalent O—O bonds whose lengths are intermediate between the lengths of oxygen-oxygen single and double bonds. Another way of looking at it is to say that the rules for drawing Lewis structures don't allow us to have a single structure that adequately represents the ozone molecule. For example, there are no rules for drawing half-bonds. But we can draw two equivalent Lewis structures that, when averaged, amount to something very much like what is observed.

As an additional example of resonance structures, consider the nitrate ion, NO_3^-, for which three equivalent Lewis structures can be drawn:

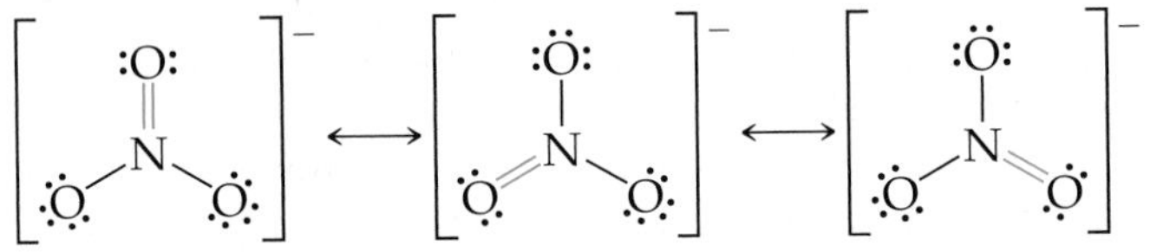

Notice that the arrangement of atoms is the same in each structure; only the placement of electrons differs. In writing resonance structures, the same atoms must be bonded to one another in all structures, so that the only differences are in the arrangements of electrons. All three Lewis structures taken together adequately describe the nitrate ion, which has three equal N—O bond distances. In some instances the Lewis structures may not be equivalent; one or more structures may represent a more stable arrangement than other possibilities. We will encounter examples of this as we proceed.

SAMPLE EXERCISE 8.10

Which is predicted to have the shorter sulfur-oxygen bonds, SO_3 or SO_3^{2-}?

Solution The sulfur atom has 6 valence-shell electrons, as does oxygen. Thus, SO_3 contains 24 valence-shell electrons. In writing the Lewis structure, we see that there are three equivalent resonance structures that can be drawn.

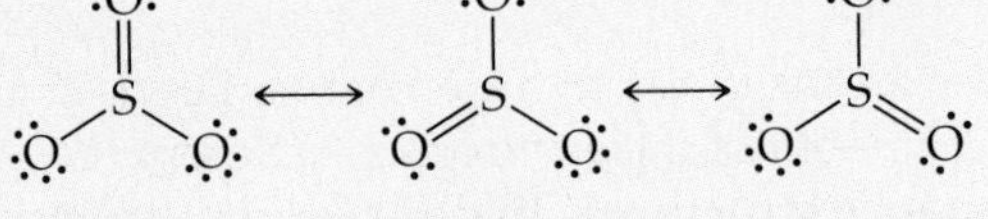

As in the preceding example with NO_3^-, the actual structure of SO_3 is an equal blend of these structures. Thus, each S—O bond distance should be about one third of the way between that of a single and that of a double bond. That is, they should be shorter than single bonds but not as short as double bonds.

The SO_3^{2-} ion has 26 electrons, leading to the following Lewis structure:

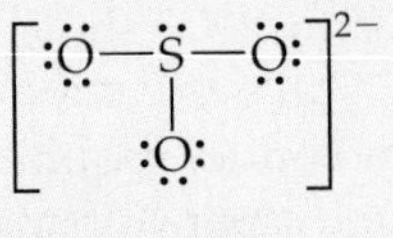

In this case the S—O bonds are all single bonds.

Our analysis of each case suggests that SO_3 should have the shorter S—O bonds and SO_3^{2-} the longer ones. This agrees with experiment; the S—O bond length in SO_3 is 1.42 Å, whereas that in SO_3^{2-} is 1.51 Å.

PRACTICE EXERCISE

Draw two equivalent resonance structures for the formate ion, HCO_2^-.

Answer: $\left[\mathrm{H{-}C({-}\ddot{\underset{..}{O}}{:}){=}\ddot{O}}\right]^- \longleftrightarrow \left[\mathrm{H{-}C({=}\ddot{O}{:}){-}\ddot{\underset{..}{O}}{:}}\right]^-$

Resonance in Benzene

Resonance is an extremely important concept in describing the bonding in organic molecules, particularly in the so-called *aromatic* molecules. Aromatic organic molecules include the hydrocarbon called *benzene*, which has the molecular formula C_6H_6 (Figure 8.12 ▶). The six carbon atoms of benzene are bonded in a hexagonal ring, and one H atom is bonded to each C atom.

We can write two equivalent Lewis structures for benzene, each of which satisfies the octet rule. These two structures are in resonance:

```
   H     H                  H     H
    \   /                    \   /
     C=C                      C—C
    /   \                    //   \\
H—C      C—H   ⟷   H—C       C—H
    \\   //                   \   /
     C—C                      C=C
    /   \                    /   \
   H     H                  H     H
```

Each of these Lewis structures shows three C—C single bonds and three C=C double bonds, but the double bonds are in different places in the two structures. The experimental structure of benzene shows that all six C—C bonds are of equal length, 1.40 Å, intermediate between the values for a C—C single bond (1.54 Å) and a C=C double bond (1.34 Å).

Benzene can be represented by omitting the hydrogen atoms attached to carbon and showing only the carbon-carbon framework with the vertices unlabeled. In this convention the resonance in the benzene molecule is represented as

[hexagon with alternating double bonds] ⟷ [hexagon with alternating double bonds] or [hexagon with circle]

To emphasize the resonance between two Lewis structures, benzene is often represented as a hexagon with a circle in it. This emphasizes that C=C double bonds cannot be assigned to specific edges of the hexagon. Chemists use both representations of benzene interchangeably.

The bonding arrangement in benzene confers special stability. As a result, literally millions of organic compounds contain the six-membered rings characteristic of benzene. Many of these compounds are important in biochemistry, in pharmaceuticals, and in the production of modern materials. We will say more about the bonding in benzene in Chapter 9 and its unusual stability in Chapter 25.

(a)

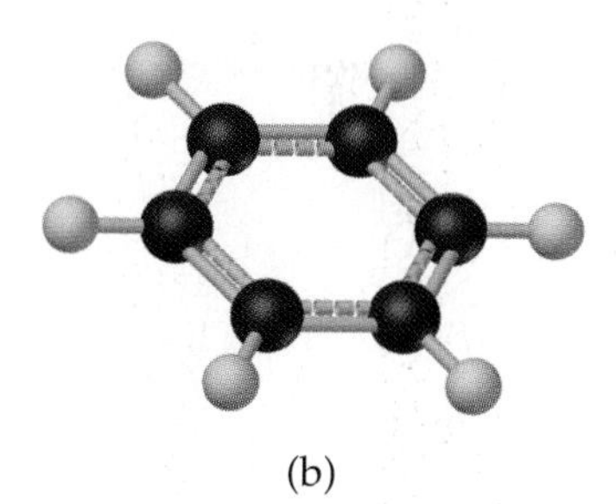

(b)

▲ **Figure 8.12** (a) Benzene is obtained from the distillation of fossil fuels. Over 16 billion pounds of benzene are produced annually in the United States. Benzene is a carcinogen, so its use is closely regulated. (b) The benzene molecule is a regular hexagon of carbon atoms with a hydrogen atom bonded to each one.

8.7 Exceptions to the Octet Rule

The octet rule is so simple and useful in introducing the basic concepts of bonding that you might assume that it is always obeyed. In Section 8.2, however, we noted its limitation in dealing with ionic compounds of the transition metals. The octet rule also fails in many situations involving covalent bonding. These exceptions to the octet rule are of three main types:

1. Molecules with an odd number of electrons
2. Molecules in which an atom has less than an octet
3. Molecules in which an atom has more than an octet

Odd Number of Electrons

In the vast majority of molecules, the number of electrons is even, and complete pairing of electrons occurs. In a few molecules, such as ClO_2, NO, and NO_2, however, the number of electrons is odd. Complete pairing of these electrons is impossible, and an octet around each atom cannot be achieved. For example, NO contains $5 + 6 = 11$ valence electrons. The two most important Lewis structures for this molecule are shown in the margin.

Less than an Octet

A second type of exception occurs when there are fewer than eight electrons around an atom in a molecule or polyatomic ion. This is also a relatively rare situation and is most often encountered in compounds of boron and beryllium. For example, let's consider boron trifluoride, BF_3. If we follow the first four steps of the procedure at the beginning of Section 8.5 for drawing Lewis structures, we obtain the following structure:

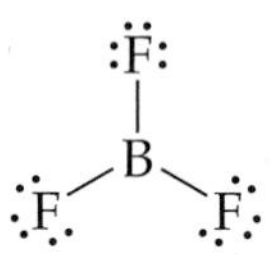

ACTIVITY
Exceptions to the Octet Rule

3-D MODEL
BF_3

There are only six electrons around the boron atom. In this Lewis structure the formal charges on both the B and the F atoms are zero. We could complete the octet around boron by forming a double bond (step 5). In so doing, we see that there are three equivalent resonance structures (the formal charges on each atom are shown in red).

These Lewis structures force a fluorine atom to share additional electrons with the boron atom, which is inconsistent with the high electronegativity of fluorine. In fact, the formal charges tell us that this is an unfavorable situation: The F atom that is involved in the B=F double bond has a formal charge of +1 while the less electronegative B atom has a formal charge of −1. Thus, the Lewis structures in which there is a B—F double bond are less important than the one in which there is less than an octet around boron.

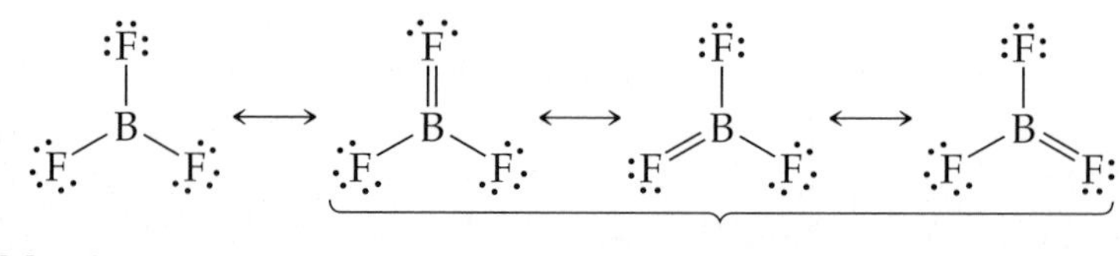

We usually represent BF_3 solely by the leftmost resonance structure in which there are only six valence-shell electrons around boron. The chemical behavior of BF_3 is consistent with this representation. Thus, BF_3 reacts very energetically with molecules having an unshared pair of electrons that can be used to form a bond with boron. For example, it reacts with ammonia, NH_3, to form the compound NH_3BF_3.

3-D MODEL
NH_3BF_3

$$H_3N\colon + BF_3 \longrightarrow H_3N{-}BF_3$$

In this stable compound boron has an octet of electrons.

More than an Octet

The third and largest class of exceptions consists of molecules or ions in which there are more than eight electrons in the valence shell of an atom. When we draw the Lewis structure for PCl_5, for example, we are forced to "expand" the valence shell and place 10 electrons around the central phosphorus atom.

:C̈l:—P(—C̈l:)₄ (Lewis structure of PCl_5)

Other examples of molecules and ions with "expanded" valence shells are SF_4, AsF_6^-, and ICl_4^-. The corresponding molecules with a second-period atom, such as NCl_5 and OF_4, do *not* exist. Let's take a look at why expanded valence shells are observed only for elements in period 3 and beyond in the periodic table.

Elements of the second period have only the $2s$ and $2p$ valence orbitals available for bonding. Because these orbitals can hold a maximum of eight electrons, we never find more than an octet of electrons around elements from the second period. Elements from the third period and beyond, however, have ns, np and unfilled nd orbitals that can be used in bonding. For example, the orbital diagram for the valence shell of a phosphorus atom is as follows:

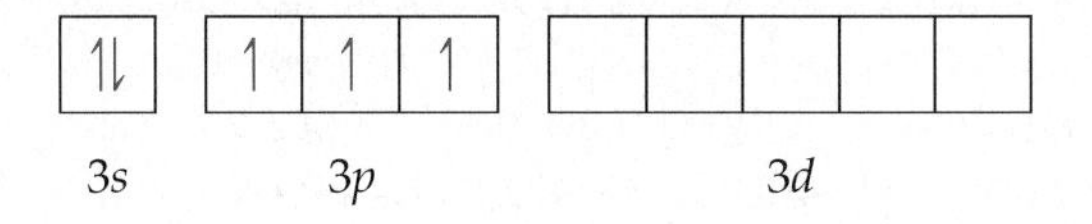

Although third-period elements such as phosphorus often satisfy the octet rule, as in PCl_3, they also often exceed an octet by seeming to use their empty d orbitals to accommodate additional electrons.*

Size also plays an important role in determining whether an atom can accommodate more than eight electrons. The larger the central atom, the larger the number of atoms that can surround it. The occurrences of expanded valence shells therefore increase with increasing size of the central atom. The size of the surrounding atoms is also important. Expanded valence shells occur most often when the central atom is bonded to the smallest and most electronegative atoms, such as F, Cl, and O.

* On the basis of recent theoretical calculations, some chemists have questioned whether valence d orbitals are actually used in the bonding of molecules and ions with expanded valence shells. Nevertheless, the presence of valence d orbitals in period 3 and beyond provides the simplest explanation of this phenomenon, especially within the scope of a general chemistry textbook.

SAMPLE EXERCISE 8.11

Draw the Lewis structure for ICl_4^-.

Solution Iodine (group 7A) has 7 valence electrons; each chlorine (group 7A) also has 7; an extra electron is added to account for the 1− charge of the ion. Therefore, the total number of valence electrons is $7 + 4(7) + 1 = 36$. The I atom is the central atom in the ion. Putting 8 electrons around each Cl atom (including a pair of electrons between I and each Cl to represent the single bonds between these atoms) requires $8 \times 4 = 32$ electrons. We are thus left with $36 - 32 = 4$ electrons to be placed on the larger iodine:

$$\left[\begin{array}{ccc} :\ddot{Cl} & & \ddot{Cl}: \\ & \ddot{\underset{\cdot\cdot}{I}} & \\ :\ddot{Cl} & & \ddot{Cl}: \end{array}\right]^-$$

Iodine has 12 electrons around it, exceeding the common octet of electrons.

PRACTICE EXERCISE

(a) Which of the following atoms is never found with more than an octet of electrons around it: S, C, P, Br? **(b)** Draw the Lewis structure for XeF_2.

Answers: **(a)** C; **(b)** $:\ddot{\underset{\cdot\cdot}{F}}-\ddot{\underset{\cdot\cdot}{Xe}}-\ddot{\underset{\cdot\cdot}{F}}:$

At times you may see Lewis structures written with expanded octets even though structures can be written with an octet. For example, consider the following Lewis structures for the phosphate ion, PO_4^{3-}:

$$\left[\begin{array}{ccc} & :\ddot{O}:^{-1} & \\ & | & \\ :\ddot{\underset{\cdot\cdot}{O}}_{-1}- & \overset{+1}{P} & -\ddot{\underset{\cdot\cdot}{O}}:^{-1} \\ & | & \\ & :\underset{\cdot\cdot}{O}:_{-1} & \end{array}\right]^{3-} \qquad \left[\begin{array}{ccc} & :\ddot{O}:^{-1} & \\ & | & \\ {}^{0}\ddot{\underset{\cdot\cdot}{O}}= & \overset{0}{P} & -\ddot{\underset{\cdot\cdot}{O}}:^{-1} \\ & | & \\ & :\underset{\cdot\cdot}{O}:_{-1} & \end{array}\right]^{3-}$$

The formal charges on the atoms are shown in red. On the left the P atom has an octet; on the right the P atom has an expanded octet of five electron pairs. The structure on the right is often used for PO_4^{3-} because it has smaller formal charges on the atoms. The best representation of PO_4^{3-} is a series of such Lewis structures in resonance with one another. However, theoretical calculations, based on quantum mechanics, suggest that the structure on the left is the best single Lewis structure for the phosphate ion. In general, when choosing between alternative Lewis structures, you should choose one that satisfies the octet rule if it is possible to do so.

8.8 Strengths of Covalent Bonds

The stability of a molecule is related to the strengths of the covalent bonds it contains. The strength of a covalent bond between two atoms is determined by the energy required to break that bond. It is easiest to relate bond strength to the enthalpy change in reactions in which bonds are broken. ∞ (Section 5.4) The **bond enthalpy** is the enthalpy change, ΔH, for the breaking of a particular bond in a mole of gaseous substance. For example, the bond enthalpy for the bond between chlorine atoms in the Cl_2 molecule is the enthalpy change when a mole of Cl_2 is dissociated into chlorine atoms.

$$:\ddot{\underset{\cdot\cdot}{Cl}}-\ddot{\underset{\cdot\cdot}{Cl}}:(g) \longrightarrow 2\,:\ddot{\underset{\cdot\cdot}{Cl}}\cdot(g) \qquad \Delta H = D(\text{Cl}-\text{Cl}) = 242 \text{ kJ}$$

We use the designation D(bond type) to represent bond enthalpies.

It is relatively simple to assign bond enthalpies to bonds in diatomic molecules. The bond enthalpy is just the energy required to break the diatomic molecule into its component atoms. However, for bonds that occur only in polyatomic molecules (such as the C—H bond), we must often utilize average bond enthalpies. For example, the enthalpy change for the following process (called *atomization*) can be used to define an average bond enthalpy for the C—H bond.

$$\mathrm{H{-}\underset{\underset{H}{|}}{\overset{\overset{H}{|}}{C}}{-}H}(g) \longrightarrow \cdot\dot{\underset{\cdot}{\mathrm{C}}}\cdot(g) + 4\,\mathrm{H}\cdot(g) \qquad \Delta H = 1660\ \mathrm{kJ}$$

Because there are four equivalent C—H bonds in methane, the heat of atomization is equal to the total bond enthalpies of the four C—H bonds. Therefore, the average C—H bond enthalpy for CH_4 is $D(\mathrm{C{-}H}) = (1660/4)\mathrm{kJ/mol} = 415\ \mathrm{kJ/mol}$.

The bond enthalpy for a given set of atoms, say C—H, depends on the rest of the molecule of which it is a part. However, the variation from one molecule to another is generally small. This supports the idea that the bonding electron pairs are localized between atoms. If we consider C—H bond enthalpies in many different compounds, we find that the average bond enthalpy is 413 kJ/mol, which compares closely with the 415 kJ/mol value calculated from CH_4.

Table 8.4 ▼ lists several average bond enthalpies. *The bond enthalpy is always a positive quantity*; energy is always required to break chemical bonds. Conversely, energy is always released when a bond forms between two gaseous atoms or molecular fragments. The greater the bond enthalpy, the stronger is the bond.

A molecule with strong chemical bonds generally has less tendency to undergo chemical change than does one with weak bonds. This relationship between strong bonding and chemical stability helps explain the chemical form in which many elements are found in nature. For example, Si—O bonds are among the strongest ones that silicon forms. It should not be surprising, therefore, that SiO_2

TABLE 8.4 Average Bond Enthalpies (kJ/mol)

Single Bonds

Bond	kJ/mol	Bond	kJ/mol	Bond	kJ/mol	Bond	kJ/mol
C—H	413	N—H	391	O—H	463	F—F	155
C—C	348	N—N	163	O—O	146		
C—N	293	N—O	201	O—F	190	Cl—F	253
C—O	358	N—F	272	O—Cl	203	Cl—Cl	242
C—F	485	N—Cl	200	O—I	234		
C—Cl	328	N—Br	243			Br—F	237
C—Br	276			S—H	339	Br—Cl	218
C—I	240	H—H	436	S—F	327	Br—Br	193
C—S	259	H—F	567	S—Cl	253		
		H—Cl	431	S—Br	218	I—Cl	208
Si—H	323	H—Br	366	S—S	266	I—Br	175
Si—Si	226	H—I	299			I—I	151
Si—C	301						
Si—O	368						
Si—Cl	464						

Multiple Bonds

Bond	kJ/mol	Bond	kJ/mol	Bond	kJ/mol
C=C	614	N=N	418	O_2	495
C≡C	839	N≡N	941		
C=N	615	N=O	607	S=O	523
C≡N	891			S=S	418
C=O	799				
C≡O	1072				

and other substances containing Si—O bonds (silicates) are so common; it is estimated that over 90% of Earth's crust is composed of SiO_2 and silicates.

Bond Enthalpies and the Enthalpies of Reactions

We can use the average bond enthalpies in Table 8.4 to estimate the enthalpies of reactions in which bonds are broken and new bonds are formed. This procedure allows us to estimate quickly whether a given reaction will be endothermic ($\Delta H > 0$) or exothermic ($\Delta H < 0$), even if we do not know ΔH_f° for all the chemical species involved. Our strategy for estimating reaction enthalpies is a straightforward application of Hess's law. (Section 5.6) We use the fact that breaking bonds is always an endothermic process (positive ΔH), and bond formation is always exothermic (negative ΔH). We therefore imagine that the reaction occurs in two steps: (1) We supply enough energy to break those bonds in the reactants that are not present in the products. In this step the enthalpy of the system is increased by the sum of the bond enthalpies of the bonds that are broken. (2) We make the bonds in the products that were not present in the reactants. This step will release energy, and it lowers the enthalpy of the system by the sum of the bond enthalpies of the bonds that are formed. The enthalpy of the reaction, ΔH_{rxn}, is estimated as the sum of the bond enthalpies of the bonds broken, minus the sum of the bond enthalpies of the new bonds formed.

$$\Delta H_{rxn} = \Sigma(\text{bond enthalpies of bonds broken}) - \Sigma(\text{bond enthalpies of bonds formed}) \quad [8.12]$$

Consider, for example, the gas-phase reaction between methane (CH_4) and chlorine to produce methyl chloride (CH_3Cl) and hydrogen chloride (HCl):

$$H{-}CH_3(g) + Cl{-}Cl(g) \longrightarrow Cl{-}CH_3(g) + H{-}Cl(g) \qquad \Delta H_{rxn} = ? \quad [8.13]$$

Our two-step procedure is outlined in Figure 8.13 ▼. We note that in the course of this reaction, the following bonds are broken and made:

Bonds broken: 1 mol C—H, 1 mol Cl—Cl
Bonds made: 1 mol C—Cl, 1 mol H—Cl

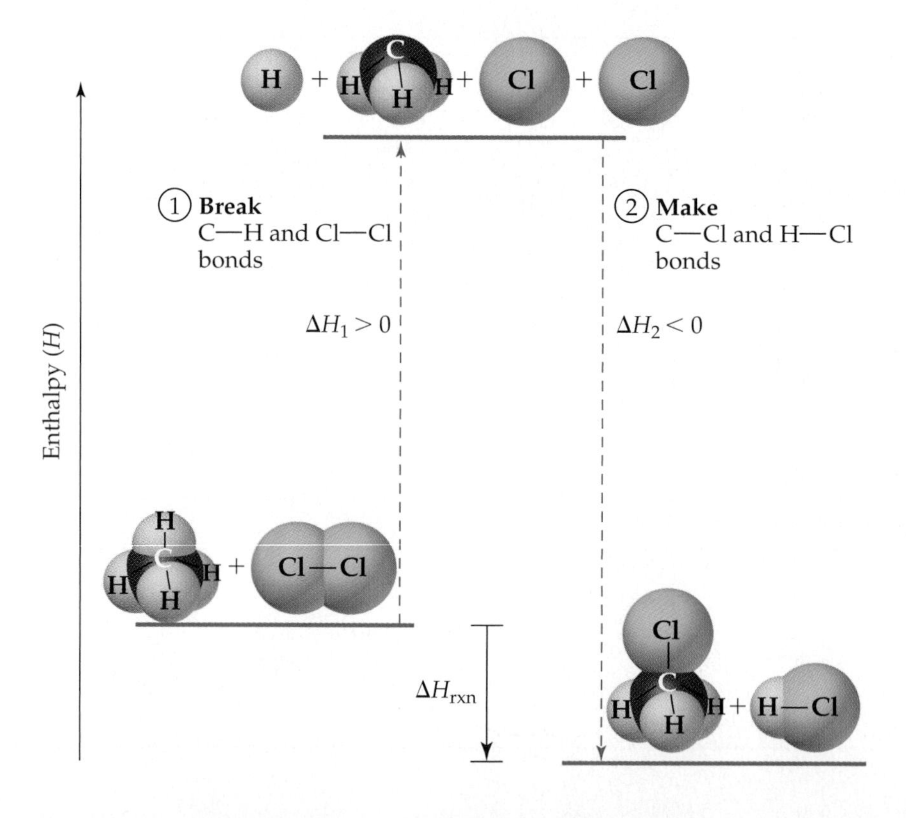

▶ **Figure 8.13** Illustration of the use of average bond enthalpies to estimate ΔH_{rxn} for the reaction in Equation 8.13. Breaking the C—H and Cl—Cl bonds produces a positive enthalpy change (ΔH_1), whereas making the C—Cl and H—Cl bonds causes a negative enthalpy change (ΔH_2). The values of ΔH_1 and ΔH_2 are estimated from the values in Table 8.4. From Hess's law, $\Delta H_{rxn} = \Delta H_1 + \Delta H_2$.

We first supply enough energy to break the C—H and Cl—Cl bonds, which will raise the enthalpy of the system. We then form the C—Cl and H—Cl bonds, which will release energy and lower the enthalpy of the system. By using Equation 8.12 and the data in Table 8.4, we estimate the enthalpy of the reaction as

$$\Delta H_{rxn} = [D(\text{C—H}) + D(\text{Cl—Cl})] - [D(\text{C—Cl}) + D(\text{H—Cl})]$$
$$= (413 \text{ kJ} + 242 \text{ kJ}) - (328 \text{ kJ} + 431 \text{ kJ}) = -104 \text{ kJ}$$

The reaction is exothermic because the bonds in the products (especially the H—Cl bond) are stronger than the bonds in the reactants (especially the Cl—Cl bond).

We usually use bond enthalpies to estimate ΔH_{rxn} only if we do not have the needed ΔH_f° values readily at hand. For the above reaction, we cannot calculate ΔH_{rxn} from ΔH_f° values and Hess's law because the value of ΔH_f° for $CH_3Cl(g)$ is not given in Appendix C. If we obtain the value of ΔH_f° for $CH_3Cl(g)$ from another source (such as the *CRC Handbook of Chemistry and Physics*) and use Equation 5.31, we find that $\Delta H_{rxn} = -99.8$ kJ for the reaction in Equation 8.13. Thus, the use of average bond enthalpies provides a reasonably accurate estimate of the actual reaction enthalpy change.

It is important to remember that bond enthalpies are derived for gaseous molecules and that they are often averaged values. Nonetheless, average bond enthalpies are useful for estimating reaction enthalpies quickly, especially for gas-phase reactions.

SAMPLE EXERCISE 8.12

Using Table 8.4, estimate ΔH for the following reaction (where we explicitly show the bonds involved in the reactants and products):

$$\text{H—}\underset{\text{H}}{\overset{\text{H}}{\text{C}}}\text{—}\underset{\text{H}}{\overset{\text{H}}{\text{C}}}\text{—H}(g) + \tfrac{7}{2}O_2(g) \longrightarrow 2\,\text{O=C=O}(g) + 3\,\text{H—O—H}(g)$$

Solution

Analyze: We are asked to estimate the enthalpy change for a chemical process by using averaged values for chemical bonds that are broken in the reactants and formed in the products.

Plan: Among the reactants, we must break six C—H bonds and a C—C bond in C_2H_6; we also break $\tfrac{7}{2}O_2$ bonds. Among the products, we form four C=O bonds (two in each CO_2) and six O—H bonds (two in each H_2O).

Solve: Using Equation 8.12 and data from Table 8.4, we have

$$\Delta H = 6D(\text{C—H}) + D(\text{C—C}) + \tfrac{7}{2}D(O_2) - 4D(\text{C=O}) - 6D(\text{O—H})$$
$$= 6(413 \text{ kJ}) + 348 \text{ kJ} + \tfrac{7}{2}(495 \text{ kJ}) - 4(799 \text{ kJ}) - 6(463 \text{ kJ})$$
$$= 4558 \text{ kJ} - 5974 \text{ kJ}$$
$$= -1416 \text{ kJ}$$

Check: This estimate can be compared with the value of −1428 kJ calculated from more accurate thermochemical data; the agreement is good.

PRACTICE EXERCISE

Using Table 8.4, estimate ΔH for the following reaction:

$$\text{H—}\underset{\text{H}}{\text{N}}\text{—}\underset{\text{H}}{\text{N}}\text{—H}(g) \longrightarrow \text{N}\equiv\text{N}(g) + 2\,\text{H—H}(g)$$

Answer: −86 kJ

Chemistry at Work Explosives and Alfred Nobel

Enormous amounts of energy can be stored in chemical bonds. Perhaps the most graphic illustration of this fact is seen in certain molecular substances that are used as explosives. Our discussion of bond enthalpies allows us to examine more closely some of the properties of such explosive substances.

An explosive must have the following characteristics: (1) It must decompose very exothermically; (2) the products of its decomposition must be gaseous, so that a tremendous gas pressure accompanies the decomposition; (3) its decomposition must occur very rapidly; and (4) it must be stable enough so that it can be detonated predictably. The combination of the first three effects leads to the violent evolution of heat and gases.

To give the most exothermic reaction, an explosive should have weak chemical bonds and should decompose into molecules with very strong bonds. Looking at bond enthalpies (Table 8.4), the $N \equiv N$, $C \equiv O$, and $C = O$ bonds are among the strongest. Not surprisingly, explosives are usually designed to produce the gaseous products $N_2(g)$, $CO(g)$, and $CO_2(g)$. Water vapor is nearly always produced as well.

Many common explosives are organic molecules that contain nitro (NO_2) or nitrate (NO_3) groups attached to a carbon skeleton. The structures of two of the most familiar explosives, nitroglycerin and trinitrotoluene (TNT), are shown here. TNT contains the six-membered ring characteristic of benzene.

Nitroglycerin

TNT

Nitroglycerin is a pale yellow, oily liquid. It is highly *shock-sensitive*: Merely shaking the liquid can cause its explosive decomposition into nitrogen, carbon dioxide, water, and oxygen gases:

$$4C_3H_5N_3O_9(l) \longrightarrow 6N_2(g) + 12CO_2(g) + 10H_2O(g) + O_2(g)$$

The large bond enthalpies of the N_2 molecules (941 kJ/mol), CO_2 molecules (2×799 kJ/mol), and water molecules (2×463 kJ/mol) make this reaction enormously exothermic. Nitroglycerin is an exceptionally unstable explosive because it is in nearly perfect *explosive balance*: With the exception of a small amount of $O_2(g)$ produced, the only products are N_2, CO_2, and H_2O. Note also that, unlike combustion reactions (Section 3.2), explosions are entirely self-contained. No other reagent, such as $O_2(g)$, is needed for the explosive decomposition.

Because nitroglycerin is so unstable, it is difficult to use as a controllable explosive. The Swedish inventor Alfred Nobel (Figure 8.14 ▼) found that mixing nitroglycerin with an absorbent solid material such as diatomaceous earth or cellulose gives a solid explosive (*dynamite*) that is much safer than liquid nitroglycerin.

▲ **Figure 8.14** Alfred Nobel (1833–1896), the Swedish inventor of dynamite. By many accounts Nobel's discovery that nitroglycerin could be made more stable by absorbing it onto cellulose was an accident. This discovery made Nobel a very wealthy man. He was also a complex and lonely man, however, who never married, was frequently ill, and suffered from chronic depression. He had invented the most powerful military explosive to date, but he strongly supported international peace movements. His will stated that his fortune be used to establish prizes awarding those who "have conferred the greatest benefit on mankind," including the promotion of peace and "fraternity between nations." The Nobel Prize is probably the most coveted award that a scientist, economist, writer, or peace advocate can receive.

TABLE 8.5 Average Bond Lengths for Some Single, Double, and Triple Bonds

Bond	Bond Length (Å)	Bond	Bond Length (Å)
C—C	1.54	N—N	1.47
C=C	1.34	N=N	1.24
C≡C	1.20	N≡N	1.10
C—N	1.43	N—O	1.36
C=N	1.38	N=O	1.22
C≡N	1.16		
		O—O	1.48
C—O	1.43	O=O	1.21
C=O	1.23		
C≡O	1.13		

Bond Enthalpy and Bond Length

Just as we can define an average bond enthalpy, we can also define an average **bond length** for a number of common bond types. The bond length is defined as the distance between the nuclei of the atoms involved in the bond. Some of these are listed in Table 8.5 ▲. Of particular interest is the relationship among bond enthalpy, bond length, and the number of bonds between the atoms. For example, we can use data in Tables 8.4 and 8.5 to compare the bond lengths and bond enthalpies of carbon-carbon single, double, and triple bonds.

C—C	C=C	C≡C
1.54 Å	1.34 Å	1.20 Å
348 kJ/mol	614 kJ/mol	839 kJ/mol

As the number of bonds between the carbon atoms increases, the bond enthalpy increases and the bond length decreases; that is, the carbon atoms are held more closely and more tightly together. In general, *as the number of bonds between two atoms increases, the bond grows shorter and stronger.*

SAMPLE INTEGRATIVE EXERCISE 8: Putting Concepts Together

Phosgene, a substance used in poisonous gas warfare in World War I, is so named because it was first prepared by the action of sunlight on a mixture of carbon monoxide and chlorine gases. Its name comes from the Greek words *phos* (light) and *genes* (born of). Phosgene has the following elemental composition: 12.14% C, 16.17% O, and 71.69% Cl by mass. Its molar mass is 98.9 g/mol. **(a)** Determine the molecular formula of this compound. **(b)** Draw three Lewis structures for the molecule that satisfy the octet rule for each atom. (The Cl and O atoms bond to C.) **(c)** Using formal charges, determine which Lewis structure is the most important one. **(d)** Using average bond enthalpies, estimate ΔH for the formation of gaseous phosgene from $CO(g)$ and $Cl_2(g)$.

Solution (a) The empirical formula of phosgene can be determined from its elemental composition. (Section 3.5) Assuming 100 g of the compound and calculating the number of moles of C, O, and Cl in this sample, we have

$$(12.14 \text{ g C})\left(\frac{1 \text{ mol C}}{12.01 \text{ g C}}\right) = 1.011 \text{ mol C}$$

$$(16.17 \text{ g O})\left(\frac{1 \text{ mol O}}{16.00 \text{ g O}}\right) = 1.011 \text{ mol O}$$

$$(71.69 \text{ g Cl})\left(\frac{1 \text{ mol Cl}}{35.45 \text{ g Cl}}\right) = 2.022 \text{ mol Cl}$$

The ratio of the number of moles of each element, obtained by dividing each number of moles by the smallest quantity, indicates that there is 1 C and 1 O for each 2 Cl in the empirical formula, $COCl_2$.

The molar mass of the empirical formula is 12.01 + 16.00 + 2(35.45) = 98.91 g/mol, the same as the molar mass of the molecule. Thus, $COCl_2$ is the molecular formula.

(b) Carbon has 4 valence electrons, oxygen has 6, and chlorine has 7, giving 4 + 6 + 2(7) = 24 electrons for the Lewis structures. Drawing a Lewis structure with all single bonds does not give the central carbon atom an octet. Using multiple bonds, three structures satisfy the octet rule:

$$:\ddot{\underset{..}{Cl}}-\overset{\overset{:O:}{\|}}{C}-\ddot{\underset{..}{Cl}}: \longleftrightarrow \ddot{\underset{..}{Cl}}=\overset{\overset{:\ddot{O}:}{|}}{C}-\ddot{\underset{..}{Cl}}: \longleftrightarrow :\ddot{\underset{..}{Cl}}-\overset{\overset{:\ddot{O}:}{|}}{C}=\ddot{\underset{..}{Cl}}$$

(c) Calculating the formal charges on each atom gives the following:

$$:\ddot{\underset{..}{Cl}}^{0}-\overset{\overset{:O:^{0}}{\|}}{C}^{0}-\ddot{\underset{..}{Cl}}:^{0} \longleftrightarrow \ddot{\underset{..}{Cl}}^{+1}=\overset{\overset{:\ddot{O}:^{-1}}{|}}{C}^{0}-\ddot{\underset{..}{Cl}}:^{0} \longleftrightarrow :\ddot{\underset{..}{Cl}}^{0}-\overset{\overset{:\ddot{O}:^{-1}}{|}}{C}^{0}=\ddot{\underset{..}{Cl}}^{+1}$$

The first structure is expected to be the most important one because it has the lowest formal charges on each atom. Indeed, the molecule is usually represented by this Lewis structure.

(d) Writing the chemical equation in terms of the Lewis structures of the molecules, we have

$$:C\equiv O: \; + \; :\ddot{\underset{..}{Cl}}-\ddot{\underset{..}{Cl}}: \longrightarrow :\ddot{\underset{..}{Cl}}-\overset{\overset{:O:}{\|}}{C}-\ddot{\underset{..}{Cl}}:$$

Thus, the reaction involves breaking a $C\equiv O$ bond and a $Cl-Cl$ bond and forming a $C=O$ bond and two $C-Cl$ bonds. Using bond enthalpies from Table 8.4, we have

$$\begin{aligned}\Delta H &= D(C\equiv O) + D(Cl-Cl) - D(C=O) - 2D(C-Cl)\\ &= 1072 \text{ kJ} + 242 \text{ kJ} - 799 \text{ kJ} - 2(328 \text{ kJ}) = -141 \text{ kJ}\end{aligned}$$

Summary and Key Terms

Introduction and Section 8.1 In this chapter we have focused on the interactions that lead to the formation of **chemical bonds**. We classify these bonds into three broad groups: **ionic bonds**, which are the electrostatic forces that exist between ions of opposite charge; **covalent bonds**, which result from the sharing of electrons by two atoms; and **metallic bonds**, which bind together the atoms in metals. The formation of bonds involves interactions of the outermost electrons of atoms, their valence electrons. The valence electrons of an atom can be represented by electron-dot symbols, called **Lewis symbols**. The tendencies of atoms to gain, lose, or share their valence electrons often follow the **octet rule**, which can be viewed as an attempt by atoms to achieve a noble-gas electron configuration.

Section 8.2 Ionic bonding results from the complete transfer of electrons from one atom to another, with formation of a three-dimensional lattice of charged particles. The stabilities of ionic substances result from the strong electrostatic attractions between an ion and the surrounding ions of opposite charge. The magnitude of these interactions is measured by the **lattice energy**, which is the energy needed to separate an ionic lattice into gaseous ions. Lattice energy increases with increasing charge on the ions and with decreasing distance between the ions. The **Born–Haber cycle** is a useful thermochemical cycle in which we use Hess's law to calculate the lattice energy as the sum of several steps in the formation of an ionic compound.

An element's position in the periodic table allows us to predict the ion it will tend to form. Metals tend to form cations; nonmetals tend to form anions. We can write electron configurations for ions by first writing the electron configuration of the neutral atom and then removing or adding the appropriate number of electrons.

Section 8.3 A covalent bond results from the sharing of electrons. We can represent the electron distribution in molecules by means of **Lewis structures**, which indicate how many valence electrons are involved in forming bonds and how many remain as unshared electron pairs. The octet rule helps determine how many bonds will be formed between two atoms. The sharing of one pair of electrons produces a **single bond**; the sharing of two or three pairs of electrons between two atoms produces

double or **triple bonds**, respectively. Double and triple bonds are examples of **multiple bonding** between atoms.

Section 8.4 In covalent bonds, the electrons may not necessarily be shared equally between two atoms. **Bond polarity** helps describe unequal sharing of electrons in a bond. In a **nonpolar covalent bond** the electrons in the bond are shared equally by the two atoms; in a **polar covalent bond** one of the atoms exerts a greater attraction for the electrons than the other.

Electronegativity is a numerical measure of the ability of an atom to compete with other atoms for the electrons shared between them. Fluorine is the most electronegative element, meaning it has the greatest ability to attract electrons from other atoms. Electronegativity values range from 0.7 for Cs to 4.0 for F. Electronegativity generally increases from left to right in a row of the periodic table, and decreases going down a column. The difference in the electronegativities of bonded atoms can be used to determine the polarity of a bond. The greater the difference, the more polar is the bond.

A **polar molecule** is one whose centers of positive and negative charge do not coincide. Thus, a polar molecule has a positive side and a negative side. This separation of charge produces a **dipole**, the magnitude of which is given by the **dipole moment**, which is measured in debyes (D). Dipole moments increase with increasing amount of charge separated and increasing distance of separation. Any diatomic molecule X—Y in which X and Y have different electronegativities is a polar molecule.

Sections 8.5 and 8.6 If we know which atoms are connected to one another, we can draw Lewis structures for molecules and ions by a simple procedure. Once we do so, we can determine the **formal charge** of each atom in a Lewis structure, which is the charge that the atom would have if all atoms had the same electronegativity. Most acceptable Lewis structures will have any negative formal charges residing on more electronegative atoms.

Sometimes a single Lewis structure is inadequate to represent a particular molecule (or ion). In such situations, we describe the molecule by using two or more **resonance structures** for the molecule. The molecule is envisioned as a blend of these multiple resonance structures. Resonance structures are important in describing the bonding in the organic molecule benzene, C_6H_6.

Section 8.7 The octet rule is not obeyed in all cases. The exceptions occur when (a) a molecule has an odd number of electrons, (b) it is not possible to complete an octet around an atom without forcing an unfavorable distribution of electrons, or (c) a large atom is surrounded by so many small electronegative atoms that it must have more than an octet of electrons around it. In this last case we envision using the unfilled *d* orbitals of the large atom to "expand" the valence shell of the atom. Expanded octets are observed for atoms in the third row and beyond in the periodic table, for which low-energy *d* orbitals are available.

Section 8.8 The strength of a covalent bond is measured by its **bond enthalpy**, which is the molar enthalpy change upon breaking a particular bond. The strengths of covalent bonds increase with the number of electron pairs shared between two atoms. We can use bond enthalpies to estimate the enthalpy change during chemical reactions in which bonds are broken and new bonds formed. The **bond length** between two bonded atoms is the distance between the two nuclei. The average bond length between two atoms decreases as the number of bonds between the atoms increases.

Exercises

Lewis Symbols and Ionic Bonding

8.1 **(a)** What are valence electrons? **(b)** How many valence electrons does a nitrogen atom possess? **(c)** An atom has the electron configuration $1s^22s^22p^63s^23p^2$. How many valence electrons does the atom have?

8.2 **(a)** What is the octet rule? **(b)** How many electrons must a sulfur atom gain to achieve an octet in its valence shell? **(c)** If an atom has the electron configuration $1s^22s^22p^3$, how many electrons must it gain to achieve an octet?

8.3 Write the electron configuration for phosphorus. Identify a valence electron in this configuration and a nonvalence electron. From the standpoint of chemical reactivity, what is the important difference between them?

8.4 Write the electron configuration for the element scandium, Sc. How many valence electrons does this atom possess? What distinguishes these valence electrons from the other electrons in the atom?

8.5 Write the Lewis symbol for atoms of each of the following elements: **(a)** Ca; **(b)** P; **(c)** Ne **(d)** B.

8.6 What is the Lewis symbol for each of the following atoms or ions: **(a)** Mg; **(b)** As; **(c)** Sc^{3+}; **(d)** Se^{2-}?

8.7 Using Lewis symbols, diagram the reaction between magnesium and oxygen atoms to give the ionic substance MgO.

8.8 Use Lewis symbols to represent the reaction that occurs between Mg and Br atoms.

8.9 In reacting with chlorine, the element potassium loses only one electron per atom, whereas calcium loses two electrons per atom when reacting with chlorine. Explain this fact in terms of energy considerations.

8.10 In reacting with metals, the element bromine accepts one electron to form the Br^- ion. Thus, we have common ionic substances such as KBr or $CaBr_2$. We do not find compounds such as K_2Br, or CaBr. Explain this fact in terms of orbital energies.

8.11 Predict the chemical formula of the ionic compound formed between the following pairs of elements: **(a)** Al and F; **(b)** K and S; **(c)** Y and O; **(d)** Mg and N.

8.12 Which ionic compound is expected to form from combination of the following pairs of elements: **(a)** rubidium and oxygen; **(b)** barium and iodine; **(c)** lithium and oxygen; **(d)** chlorine and magnesium?

8.13 Write the electron configuration for each of the following ions, and determine which ones possess noble-gas configurations: **(a)** Sr^{2+}; **(b)** Ti^{2+}; **(c)** Se^{2-}; **(d)** Ni^{2+}; **(e)** Br^-; **(f)** Mn^{3+}.

8.14 Write electron configurations for the following ions, and determine which have noble-gas configurations: **(a)** Zn^{2+}; **(b)** Te^{2-} **(c)** Se^{3+}; **(d)** Ru^{2+}; **(e)** Tl^+ **(f)** Au^+.

8.15 **(a)** Define the term *lattice energy*. **(b)** Which factors govern the magnitude of the lattice energy of an ionic compound?

8.16 **(a)** The lattice energies of NaF and MgO are given in Table 8.2. Account for the difference in these two quantities. **(b)** Account for the difference in the lattice energies of $MgCl_2$ and $SrCl_2$, which are also listed in that table.

8.17 The ionic substances KF, CaO, and ScN are isoelectronic (they have the same number of electrons). Examine the lattice energies for these substances in Table 8.2, and account for the trends you observe.

8.18 **(a)** Does the lattice energy of an ionic solid increase or decrease (*i*) as the charges of the ions increase; (*ii*) as the sizes of the ions increase? **(b)** Using a periodic table, arrange the following substances according to their expected lattice energies, listing them from lowest lattice energy to the highest: LiCl, NaBr, RbBr, MgO. Compare your list with the data in Table 8.2.

8.19 The lattice energies of KBr and CsCl are nearly equal (Table 8.2). What can you conclude from this observation?

8.20 Explain the following trends in lattice energy: **(a)** $MgO > MgCl_2$; **(b)** $NaCl > RbBr > CsBr$; **(c)** $BaO > KF$.

8.21 Energy is required to remove two electrons from Ca to form Ca^{2+} and is also required to add two electrons to O to form O^{2-}. Why, then, is CaO stable relative to the free elements?

8.22 List the individual steps used in constructing a Born–Haber cycle for the formation of $CaBr_2$ from the elements. Which of these steps would you expect to be exothermic?

[8.23] Use data from Appendix C, Figure 7.11, and Table 7.4 to calculate the lattice energy of RbCl. Is this value greater than or less than the lattice energy of NaCl? Explain.

[8.24] By using data from Appendix C, Figure 7.11, Table 7.5, and the value of the second ionization energy for Ca, 1145 kJ/mol, calculate the lattice energy of $CaCl_2$. Is this value greater than or less than the lattice energy of NaCl? Explain.

Covalent Bonding, Electronegativity, and Bond Polarity

8.25 **(a)** What is meant by the term *covalent bond*? **(b)** Give three examples of covalent bonding. **(c)** A substance XY, formed from two different elements, boils at −33°C. Is XY likely to be a covalent or ionic substance? Explain.

8.26 Which of these elements is unlikely to form covalent bonds: S, H, K, Ar, Si? Explain your choices.

8.27 Using Lewis symbols and Lewis structures, diagram the formation of $SiCl_4$ from Si and Cl atoms.

8.28 Use Lewis symbols and Lewis structures to diagram the formation of NCl_3 from N and Cl atoms.

8.29 **(a)** Construct a Lewis structure for O_2 in which each atom achieves an octet of electrons. **(b)** Explain why it is necessary to form a double bond in the Lewis structure. **(c)** The bond in O_2 is shorter than the O—O bond in compounds that contain an O—O single bond. Explain this observation.

8.30 The C—S bond lengths in carbon disulfide, CS_2, are shorter than would be expected for C—S single bonds. Use a Lewis structure to rationalize this observation.

8.31 **(a)** What is meant by the term electronegativity? **(b)** On the Pauling scale, what is the range of electronegativity values for the elements? **(c)** Which element has the greatest electronegativity? **(d)** Which element has the smallest electronegativity?

8.32 **(a)** What is the trend in electronegativity going from left to right in a row of the periodic table? **(b)** How do electronegativity values generally vary going down a column in the periodic table? **(c)** How do periodic trends in electronegativity relate to those for ionization energy and electron affinity?

8.33 Using only the periodic table as your guide, select the most electronegative atom in each of the following sets: **(a)** P, S, As, Se; **(b)** Be, B, C, Si; **(c)** Zn, Ga, Ge, As; **(d)** Na, Mg, K, Ca.

8.34 By referring only to the periodic table, select **(a)** the most electronegative element in Group 6A; **(b)** the least electronegative element in the group, Al, Si, P; **(c)** the most electronegative element in the group, Ga, P, Cl, Na; **(d)** the element in the group, K, C, Zn, F, that is most likely to form an ionic compound with Ba.

8.35 Which of the following bonds are polar: **(a)** P—O; **(b)** S—F; **(c)** Br—Br; **(d)** O—Cl? Which is the more electronegative atom in each polar bond?

8.36 Arrange the bonds in each of the following sets in order of increasing polarity: **(a)** C—F, O—F, Be—F; **(b)** N—Br, P—Br, O—Br; **(c)** C—S, B—F, N—O.

8.37 **(a)** How does a polar molecule differ from a nonpolar one? **(b)** Atoms X and Y have different electronegativities. Will the diatomic molecule X—Y necessarily be polar? Explain. **(c)** What factors affect the size of the dipole moment of a diatomic molecule?

8.38 Which of the following molecules would you predict to have a nonzero dipole moment? In each case, explain your answer. **(a)** ClF; **(b)** CO; **(c)** CO_2 (a linear molecule); **(d)** H_2O

8.39 From the data in Table 8.3, calculate the effective charges on the H and F atoms of the HF molecule in units of the electronic charge *e*.

8.40 The iodine monobromide molecule, IBr, has a bond length of 2.49 Å and a dipole moment of 1.21 D. **(a)** Which atom of the molecule is expected to have a negative charge? Explain. **(b)** Calculate the effective charges on the I and Br atoms in IBr, in units of the electronic charge *e*.

8.41 Give the name or chemical formula, as appropriate, for each of the following substances, and in each case predict whether the bonding is better described by the ionic-bonding or covalent-bonding model: **(a)** manganese(IV) oxide; **(b)** phosphorus(III) sulfide; **(c)** cobalt(II) oxide; **(d)** Cu_2S; **(e)** ClF_3; **(f)** VF_5.

8.42 Give the name or chemical formula, as appropriate, for each of the following substances, and in each case predict whether the bonding is better described by the ionic-bonding or covalent-bonding model: **(a)** manganese(III) fluoride; **(b)** chromium(VI) oxide; **(c)** arsenic(V) bromide; **(d)** SF_4; **(e)** $MoCl_4$; **(f)** $ScCl_3$.

Lewis Structures; Resonance Structures

8.43 Draw Lewis structures for the following: **(a)** SiH_4; **(b)** CO; **(c)** SF_2; **(d)** H_2SO_4 (H is bonded to O); **(e)** ClO_2^-; **(f)** NH_2OH

8.44 Write Lewis structures for the following: **(a)** H_2CO (both H atoms are bonded to C); **(b)** H_2O_2; **(c)** C_2F_6 (contains a C—C bond) **(d)** AsO_3^{3-}; **(e)** H_2SO_3 (H is bonded to O); **(f)** C_2H_2.

8.45 Write Lewis structures that obey the octet rule for each of the following, and assign formal charges to each atom: **(a)** NO^+; **(b)** $POCl_3$ (P is bonded to three Cls and to the O); **(c)** ClO_4^-; **(d)** $HClO_3$ (H is bonded to O).

8.46 For each of the following molecules or ions of sulfur and oxygen, write a single Lewis structure that obeys the octet rule, and calculate the formal charges on all the atoms: **(a)** SO_2; **(b)** SO_3; **(c)** SO_3^{2-}; **(d)** SO_4^{2-}.

8.47 **(a)** Write one or more appropriate Lewis structures for the nitrite ion, NO_2^-. **(b)** With what compound of oxygen is it isoelectronic? **(c)** What would you predict for the lengths of the bonds in this species relative to N—O single bonds?

8.48 Consider the nitryl cation, NO_2^+. **(a)** Write one or more appropriate Lewis structures for this species. **(b)** Are resonance structures needed to describe the structure? **(c)** With what familiar species is it isoelectronic?

8.49 Predict the ordering of the C—O bond lengths in CO, CO_2, and CO_3^{2-}.

8.50 Based on Lewis structures, predict the ordering of N—O bond lengths in NO^+, NO_2^-, and NO_3^-.

8.51 **(a)** Use the concept of resonance to explain why all six C—C bonds in benzene are equal in length. **(b)** The C—C bond lengths in benzene are shorter than C—C single bonds but longer than C=C double bonds. Use the resonance model to explain this observation.

8.52 Mothballs are composed of naphthalene, $C_{10}H_8$, the structure of which consists of two six-membered rings of carbon that are fused along an edge, as shown in the following incomplete Lewis structure:

(a) Write two different complete Lewis structures for naphthalene. **(b)** The observed C—C bond lengths in the molecule are intermediate between C—C single and C=C double bonds. Explain. **(c)** Represent the resonance in naphthalene in a way analogous to that used to represent it in benzene.

Exceptions to the Octet Rule

8.53 **(a)** State the octet rule. **(b)** Does the octet rule apply to ionic as well as to covalent compounds? Explain, using examples as appropriate.

8.54 Considering the representative nonmetals, what is the relationship between the group number for an element (carbon, for example, belongs to group 14; see the periodic table on the inside front cover) and the number of single covalent bonds that element needs to form to conform to the octet rule?

8.55 What is the most common exception to the octet rule? Give two examples.

8.56 For elements in the third row of the periodic table and beyond, the octet rule is often not obeyed. What factors are usually cited to explain this fact?

8.57 Draw the Lewis structures for each of the following ions or molecules. Identify those that do not obey the octet rule, and explain why they do not. **(a)** CO_3^{2-}; **(b)** BH_3; **(c)** I_3^-; **(d)** GeF_4; **(e)** AsF_6^-.

8.58 Draw the Lewis structures for each of the following molecules or ions. Identify those that do not obey the octet rule, and explain why they do not. **(a)** NO; **(b)** ICl_2^-; **(c)** SO_2; **(d)** BCl_3; **(e)** XeF_4.

8.59 In the vapor phase, $BeCl_2$ exists as a discrete molecule. **(a)** Draw the Lewis structure of this molecule, using only single bonds. Does this Lewis structure satisfy the octet rule? **(b)** What other resonance forms are possible that satisfy the octet rule? **(c)** Using formal charges, select the resonance form from among all the Lewis structures that is most important in describing $BeCl_2$.

8.60 **(a)** Describe the molecule chlorine dioxide, ClO_2, using three possible resonance structures. **(b)** Do any of these resonance structures satisfy the octet rule for every atom in the molecule? Why or why not? **(c)** Using formal charges, select the resonance structure(s) that is (are) most important.

Bond Enthalpies

8.61 Using the bond enthalpies tabulated in Table 8.4, estimate ΔH for each of the following gas-phase reactions:

(a) $$H_2C{=}CH_2 + H{-}O{-}O{-}H \longrightarrow H{-}O{-}CH_2{-}CH_2{-}O{-}H$$

(b) $$H_2C{=}CH_2 + H{-}C{\equiv}N \longrightarrow H{-}CH_2{-}CH_2{-}C{\equiv}N$$

(c) $$2\,Cl{-}N(Cl){-}Cl \longrightarrow N{\equiv}N + 3\,Cl{-}Cl$$

8.62 Using bond enthalpies (Table 8.4), estimate ΔH for the following gas-phase reactions:

(a) $$Br{-}CBr_2{-}H + Cl{-}Cl \longrightarrow Br{-}CBr_2{-}Cl + H{-}Cl$$

(b) $$H{-}S{-}CH_2{-}CH_2{-}S{-}H + H{-}Br \longrightarrow Br{-}CH_2{-}CH_2{-}Br + 2\,H{-}S{-}H$$

(c) $$H{-}NH{-}NH{-}H + Cl{-}Cl \longrightarrow 2\,H{-}NH{-}Cl$$

8.63 Using bond enthalpies (Table 8.4), estimate ΔH for each of the following reactions:

(a) $2NBr_3(g) + 3F_2(g) \longrightarrow 2NF_3(g) + 3Br_2(g)$

(b) $CO(g) + 2H_2(g) \longrightarrow CH_3OH(g)$

(c) $H_2S(g) + 3F_2(g) \longrightarrow SF_4(g) + 2HF(g)$

[8.64] Use bond enthalpies (Table 8.4) to estimate the enthalpy change for each of the following reactions:

(a) $H_2C{=}O(g) + NH_3(g) \longrightarrow H_2C{=}NH + H_2O(g)$

(b) $SiH_3Cl(g) + CH_4(g) \longrightarrow SiH_3CH_3(g) + HCl(g)$.

(c) $8H_2S(g) \longrightarrow 8H_2(g) + S_8(s)$

(See Figure 7.28. Strictly speaking the average bond enthalpy values apply to species in the gas phase. The heat of formation of $S_8(g)$ is 102.3 kJ/mol. Apply the needed correction in order to estimate the enthalpy change for the reaction as shown.)

8.65 Ammonia is produced directly from nitrogen and hydrogen by using the Haber process. The chemical reaction is

$$N_2(g) + 3H_2(g) \longrightarrow 2NH_3(g)$$

(a) Use bond enthalpies (Table 8.4) to estimate the enthalpy change for the reaction and tell whether this reaction is exothermic or endothermic. **(b)** Compare the enthalpy change you calculate in (a) to the true enthalpy change as obtained using ΔH_f° values.

8.66 **(a)** Use bond enthalpies to estimate the enthalpy change for the reaction of hydrogen with ethene:

$$H_2(g) + C_2H_4(g) \longrightarrow C_2H_6(g)$$

(b) Calculate the standard enthalpy change for this reaction using heats of formation. Why does this value differ from that calculated in (a)?

8.67 Given the following bond-dissociation energies, calculate the average bond enthalpy for the Ti—Cl bond.

Reaction	ΔH (kJ/mol)
$TiCl_4(g) \longrightarrow TiCl_3(g) + Cl(g)$	335
$TiCl_3(g) \longrightarrow TiCl_2(g) + Cl(g)$	423
$TiCl_2(g) \longrightarrow TiCl(g) + Cl(g)$	444
$TiCl(g) \longrightarrow Ti(g) + Cl(g)$	519

[8.68] **(a)** Using average bond enthalpy values, predict which of the following reactions will be most exothermic:

(i) $C(g) + 2F_2(g) \longrightarrow CF_4(g)$

(ii) $CO(g) + 3F_2 \longrightarrow CF_4(g) + OF_2(g)$

(iii) $CO_2(g) + 4F_2 \longrightarrow CF_4(g) + 2OF_2(g)$

(b) Explain the trend, if any, that exists between reaction exothermicity and the extent to which the carbon atom is bonded to oxygen.

Additional Exercises

8.69 In each of the following examples of a Lewis symbol, indicate the group in the periodic table in which the element X belongs: **(a)** ·Ẋ·; **(b)** ·X·; **(c)** :Ẋ·

8.70 **(a)** Explain the following trend in lattice energy: BeH_2, 3205 kJ/mol; MgH_2, 2791 kJ/mol; CaH_2, 2410 kJ/mol; SrH_2, 2250 kJ/mol; BaH_2, 2121 kJ/mol. **(b)** The lattice energy of ZnH_2 is 2870 kJ/mol. Based on the data given in part (a), the radius of the Zn^{2+} ion is expected to be closest to that of which group 2A element?

[8.71] From the ionic radii given in Figure 7.6, calculate the potential energy of a K^+ and F^- ion pair that are just touching (the magnitude of the electronic charge is given on the back inside cover). Calculate the energy of a mole of such pairs. How does this value compare with the lattice energy of KF (Table 8.2)? Explain the difference.

[8.72] From Equation 8.4 and the ionic radii given in Figure 7.6, calculate the potential energy of the following pairs of ions. Assume that the ions are separated by a distance equal to the sum of their ionic radii: **(a)** Na^+, Br^-; **(b)** Rb^+, Br^-; **(c)** Sr^{2+}, S^{2-}.

8.73 Based on data in Table 8.2, estimate (within 30 kJ/mol) the lattice energy for each of the following ionic substances: **(a)** LiBr; **(b)** CsBr; **(c)** $CaCl_2$.

8.74 Do you expect the element rhodium, symbol Rh, to have positive ion states in which the metal ion has a noble-gas configuration? Use ionization energies and lattice energies to explain your answer.

[8.75] **(a)** Triazine, $C_3H_3N_3$, is like benzene, except that every other C—H group is replaced by a nitrogen atom. Draw the Lewis structure(s) for this molecule. **(b)** Estimate the carbon-nitrogen bond distances in the ring.

8.76 Which of the following molecules or ions contain polar bonds: **(a)** P_4; **(b)** H_2S; **(c)** NO_2^-; **(d)** S_2^{2-}?

8.77 For the following collection of nonmetallic elements: O, P, Te, I, B, **(a)** which two would form the most polar single bond? **(b)** Which two would form the longest single bond? **(c)** Which two would be likely to form a compound of formula XY_2? **(d)** Which combinations of elements would likely yield a compound of empirical formula X_2Y_3? In each case explain your answer.

[8.78] Using the electronegativities of Cl and F, estimate the partial charges on the atoms in the Cl—F molecule. Using these partial charges and the atomic radii given in Figure 7.6, estimate the dipole moment of the molecule. The measured dipole moment is 0.88 D.

8.79 Calculate the formal charge on the indicated atom in each of the following molecules or ions: **(a)** the central oxygen atom in O_3; **(b)** phosphorus in PF_6^-; **(c)** nitrogen in NO_2; **(d)** iodine in ICl_3; **(e)** chlorine in $HClO_4$ (hydrogen is bonded to O).

8.80 **(a)** Determine the formal charge on the chlorine atom in the hypochlorite ion, ClO^-, and the perchlorate ion, ClO_4^- if the Cl atom has an octet. **(b)** What are the oxidation numbers of chlorine in ClO^- and in ClO_4^-? **(c)** What are the essential differences in the definitions of formal charge and oxidation number that lead to the differences in your answers to parts (a) and (b)?

8.81 The following three Lewis structures can be drawn for N_2O:

:N≡N—Ö: ⟷ :N̈—N≡O: ⟷ :N̈=N=Ö:

(a) Using formal charges, which of these three resonance forms is likely to be the most important? **(b)** The N—N bond length in N_2O is 1.12 Å, slightly longer than a typical N≡N bond and the N—O bond length is 1.19 Å, slightly shorter than a typical N=O bond. (See Table 8.5.) Rationalize these observations in terms of the resonance structures shown previously and your conclusion for (a).

8.82 Although I_3^- is known, F_3^- is not. Using Lewis structures, explain why F_3^- does not form.

8.83 An important reaction for the conversion of natural gas to other useful hydrocarbons is the conversion of methane to ethane.

$$2CH_4(g) \longrightarrow C_2H_6(g) + H_2(g)$$

In practice, this reaction is carried out in the presence of oxygen, which converts the hydrogen produced to water.

$$2CH_4(g) + \tfrac{1}{2}O_2(g) \longrightarrow C_2H_6(g) + H_2O(g)$$

Use bond enthalpies (Table 8.4) to estimate ΔH for these two reactions. Why is the conversion of methane to ethane more favorable when oxygen is used?

8.84 Two compounds are isomers if they have the same chemical formula but a different arrangement of atoms. Use bond enthalpies (Table 8.4) to estimate ΔH for each of the following gas-phase isomerization reactions, and indicate which isomer has the lower enthalpy:

(a) H_3C—CH_2—O—H ⟶ H_3C—O—CH_3

Ethanol ⟶ Dimethyl ether

(b) H—C(—H)—C(—H)—H with a bridging O (three-membered ring) ⟶ H—C(H)(H)—C(=O)—H

Ethylene oxide ⟶ Acetaldehyde

(c) five-membered ring of C atoms with one C=C bond and eight H atoms ⟶ H_2C=CH—CH=CH—CH_3

Cyclopentene ⟶ Pentadiene

(d) H—C(H)(H)—N≡C ⟶ H—C(H)(H)—C≡N

Methyl isocyanide ⟶ Acetonitrile

[8.85] With reference to the "Chemistry at Work" box on explosives, **(a)** use bond enthalpies to estimate the enthalpy change for the explosion of 1.00 g of nitroglycerin. **(b)** Write a balanced equation for the decomposition of TNT. Assume that, upon explosion, TNT decomposes into $N_2(g)$, $CO_2(g)$, $H_2O(g)$, and $C(s)$.

[8.86] The bond lengths of carbon-carbon, carbon-nitrogen, carbon-oxygen, and nitrogen-nitrogen single, double, and triple bonds are listed in Table 8.5. Plot bond enthalpy (Table 8.4) versus bond length for these bonds. What do you conclude about the relationship between bond length and bond enthalpy? What do you conclude about the relative strengths of C—C, C—N, C—O, and N—N bonds?

[8.87] Use the data in Table 8.5 and the following data: S—S distance in S_8 = 2.05 Å; S—O distance in SO_2 = 1.43 Å, to answer the following questions: **(a)** Predict the bond distance in an S—N single bond. **(b)** In an S—O single bond. **(c)** Why is the S—O bond distance in SO_2 considerably shorter than your predicted value for the S—O single bond? **(d)** When elemental sulfur, S_8, is carefully oxidized, a compound S_8O is formed, in which one of the sulfur atoms in the S_8 ring is bonded to an oxygen atom. The S—O distance in this compound is 1.48 Å. In light of this information, write Lewis structures that can account for the observed S—O bond distance. Does the sulfur bearing the oxygen in this compound obey the octet rule?

Integrative Exercises

8.88 The Ti^{2+} ion is isoelectronic with the Ca atom. **(a)** Are there any differences in the electron configurations of Ti^{2+} and Ca? **(b)** With reference to Figure 6.22, comment on the changes in the ordering of the 4*s* and 3*d* subshells in Ca and Ti^{2+}. **(c)** Will Ca and Ti^{2+} have the same number of unpaired electrons? Explain.

[8.89] **(a)** Write the chemical equations that are used in calculating the lattice energy of $SrCl_2(s)$ via a Born–Haber cycle. **(b)** The second ionization energy of Sr(*g*) is 1064 kJ/mol. Use this fact along with data in Appendix C, Figure 7.10, Figure 7.11, and Table 8.2, to calculate ΔH_f° for $SrCl_2(s)$.

[8.90] The electron affinity of oxygen is −141 kJ/mol, corresponding to the reaction

$$O(g) + e^- \longrightarrow O^-(g)$$

The lattice energy of $K_2O(s)$ is 2238 kJ/mol. Use these data along with data in Appendix C and Figure 7.10 to calculate the "second electron affinity" of oxygen, corresponding to the reaction

$$O^-(g) + e^- \longrightarrow O^{2-}(g)$$

8.91 The compound chloral hydrate, known in detective stories as knockout drops, is composed of 14.52% C, 1.83% H, 64.30% Cl, and 19.35% O by mass and has a molar mass of 165.4 g/mol. **(a)** What is the empirical formula of this substance? **(b)** What is the molecular formula of this substance? **(c)** Draw the Lewis structure of the molecule assuming that the Cl atoms bond to a single C atom, and that there is a C—C bond and two C—O bonds in the compound.

[8.92] Acetylene (C_2H_2) and nitrogen (N_2) both contain a triple bond, but they differ greatly in their chemical properties. **(a)** Write the Lewis structures for the two substances. **(b)** By referring to the index, look up the chemical properties of acetylene and nitrogen and compare their reactivities. **(c)** Write balanced chemical equations for the complete oxidation of N_2 to form $N_2O_5(g)$ and of acetylene to form $CO_2(g)$ and $H_2O(g)$. **(d)** Calculate the enthalpy of oxidation per mole of N_2 and C_2H_2 (the enthalpy of formation of $N_2O_5(g)$ is 11.30 kJ/mol). How do these comparative values relate to your response to part (b)? Both N_2 and C_2H_2 possess triple bonds with quite high bond enthalpies (Table 8.4). What aspect of chemical bonding in these molecules or in the oxidation products seems to account for the difference in chemical reactivities?

8.93 Barium azide is composed of 62.04% Ba and 37.96% N. Each azide ion has a net charge of 1−. **(a)** Determine the chemical formula of the azide ion. **(b)** Write three resonance structures for the azide ion. **(c)** Which structure is most important? **(d)** Predict the bond lengths in the azide ion.

[8.94] Under special conditions, sulfur reacts with anhydrous liquid ammonia to form a binary compound of sulfur and nitrogen. The compound is found to consist of 69.6% S and 30.4% N. Measurements of its molecular mass yield a value of 184.3 g mol^{-1}. The compound occasionally detonates on being struck or when heated rapidly. The sulfur and nitrogen atoms of the molecule are joined in a ring. All the bonds in the ring are of the same length. **(a)** Calculate the empirical and molecular formulas for the substance. **(b)** Write Lewis structures for the molecule based on the information you are given. (*Hint*: You should find a relatively small number of dominant Lewis structures.) **(c)** Predict the bond distances between the atoms in the ring (*Note*: The S—S distance in the S_8 ring is 2.05 Å.). **(d)** The enthalpy of formation of the compound is estimated to be 480 kJ mol^{-1}. ΔH_f° of S(*g*) is 222.8 kJ mol^{-1}. Estimate the average bond enthalpy in the compound.

8.95 Use bond enthalpies (Table 8.4), electron affinities (Figure 7.11), and the ionization energy of hydrogen (1312 kJ/mol) to estimate ΔH for the following gas-phase ionization reactions:

(a) $HF(g) \longrightarrow H^+(g) + F^-(g)$

(b) $HCl(g) \longrightarrow H^+(g) + Cl^-(g)$

(c) $HBr(g) \longrightarrow H^+(g) + Br^-(g)$

[8.96] Consider benzene (C_6H_6) in the gas phase. **(a)** Write the reaction for breaking all of the bonds in $C_6H_6(g)$, and use data in Appendix C to determine the enthalpy change for this reaction. **(b)** Write a reaction that corresponds to breaking all of the carbon-carbon bonds in $C_6H_6(g)$. **(c)** By combining your answers to parts (a) and (b) and using the average bond enthalpy for C—H from Table 8.4, calculate the average bond enthalpy for the carbon-carbon bonds in $C_6H_6(g)$. **(d)** Comment on your answer from part (c) as compared to the values for C—C single bonds and C═C double bonds in Table 8.4.

8.97 Average bond enthalpies are generally defined for gas-phase molecules. Many substances are liquids in their standard state. ∞ (Section 5.7) By using appropriate thermochemical data from Appendix C, calculate average bond enthalpies in the liquid state for the following bonds, and compare these values to the gas-phase values given in Table 8.4: **(a)** Br—Br, from $Br_2(l)$; **(b)** C—Cl, from $CCl_4(l)$; **(c)** O—O, from $H_2O_2(l)$ (assume that the O—H bond enthalpy is the same as in the gas phase). **(d)** What can you conclude about the process of breaking bonds in the liquid as compared to the gas phase? Explain the difference in the ΔH values between the two phases.

8.98 The reaction of indium with sulfur leads to three different binary (two-element) compounds, which we will assume to be purely ionic compounds. The three compounds have the following properties:

Compound	Mass % In	Melting Point (°C)
A	87.7	653
B	78.2	692
C	70.5	1050

(a) Determine the empirical formulas of compounds A, B, and C. **(b)** Give the oxidation state of In in each of the three compounds. **(c)** Write the electron configuration for the In ion in each of the three compounds. Do any of these configurations correspond to a noble-gas configuration? **(d)** In which compound is the ionic radius of In expected to be smallest? Explain. **(e)** The melting point of ionic compounds often correlates with the lattice energy. Explain the trends in the melting points of compounds A, B, and C in these terms.

eMedia Exercises

8.99 In Exercise 8.18(a) you assessed the effects of *ionic size* and *ionic charge* on lattice energy separately. When size and charge are *both* significantly different, a more quantitative approach is required to compare lattice energies. Use the ionic radii in Table 8.5 and the **Coulomb's Law** activity (*eChapter 8.2*) to determine which would have the greater lattice energy: CaSe or LiF.

8.100 Covalent bonding can occur when atoms are close enough to share electrons. The **H_2 Bond Formation** movie (*eChapter 8.4*) describes this process for the formation of a single bond between two hydrogen atoms. Bonds can also form this way between nitrogen atoms. Unlike hydrogen atoms, though, nitrogen atoms can form more than one bond. N—N, N=N, and N≡N bonds are all possible. Section 8.3 explained that bond length decreases with increasing number of shared electron pairs between atoms. Thus, the N≡N bond is the shortest of the three. Explain why this is so in terms of the information provided in the movie.

8.102 The **Molecular Polarity** activity (*eChapter 8.5*) allows you to determine the polarity of bonds and of molecules. **(a)** How is it possible for a molecule that contains polar bonds to be nonpolar? **(b)** Is it possible for a molecule containing only nonpolar bonds to be polar? Explain.

8.103 One of the elements that can violate the octet rule is boron. An example of a compound in which boron has fewer than eight electrons around it is BF_3. The **Chime BF_3** molecule (*eChapter 8.7*) illustrates the molecule's most important resonance structure. **(a)** What characteristics of boron allow it to exist with less than a full octet? **(b)** Boron *can* obey the octet rule in substances such as the **Chime $F_3B—NH_3$** molecule (*eChapter 8.7*). Are the formal charges on boron and nitrogen *completely* consistent with the electronegativities of the two elements? (Is it possible to draw a resonance structure in which you think they are *more* consistent?)

8.104 Another element that violates the octet rule is beryllium. The **Chime $BeCl_2$** molecule (*eChapter 8.7*) illustrates this molecule's most important resonance structure. Is it possible for beryllium to accommodate a full octet the way boron does, via the formation of an adduct? Support your answer with a formal charge analysis of such an adduct.

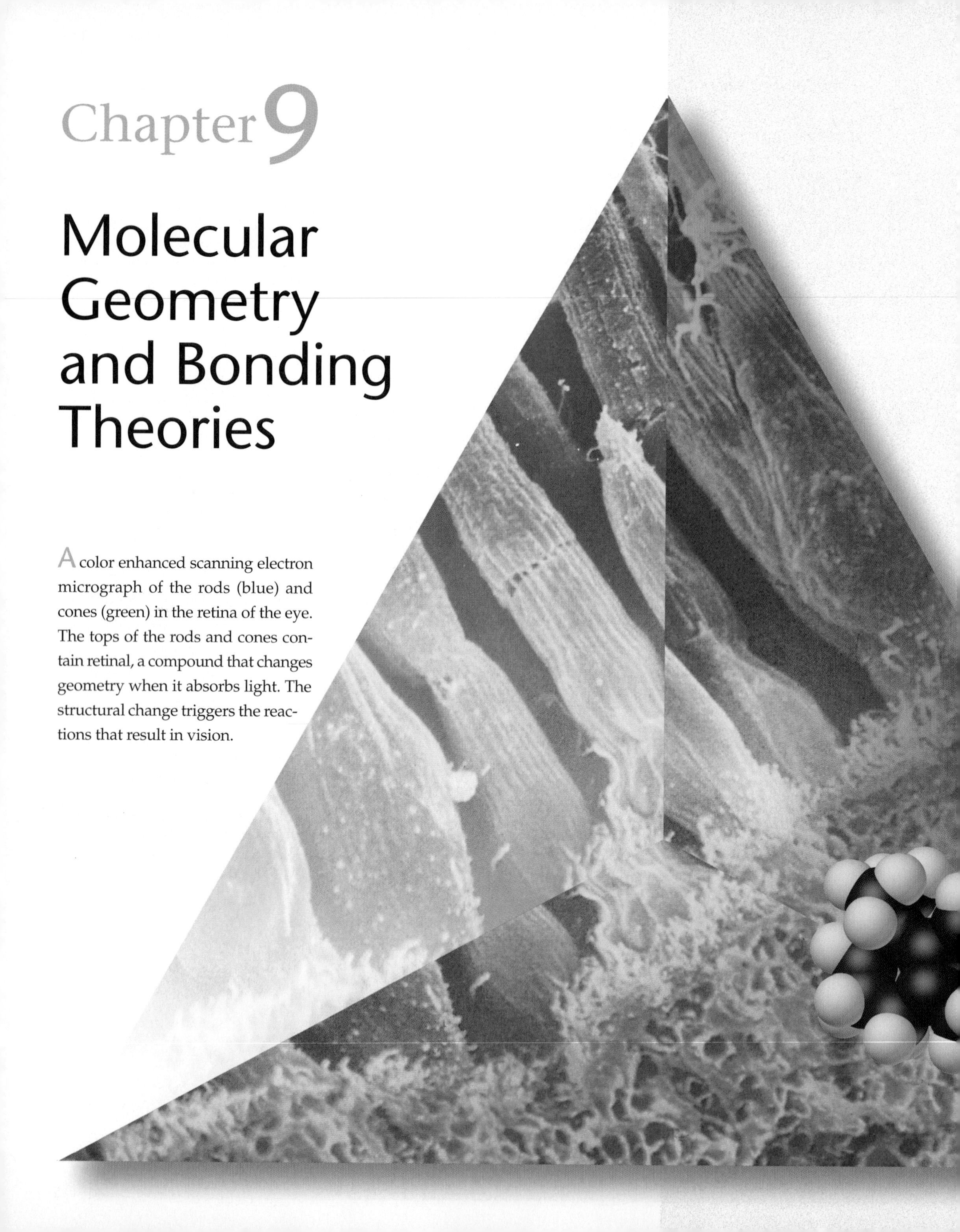

Chapter 9

Molecular Geometry and Bonding Theories

A color enhanced scanning electron micrograph of the rods (blue) and cones (green) in the retina of the eye. The tops of the rods and cones contain retinal, a compound that changes geometry when it absorbs light. The structural change triggers the reactions that result in vision.

WE HAVE SEEN in Chapter 8 that Lewis structures help us understand the compositions of molecules and their covalent bonds. However, they do not show one of the most important aspects of molecules—their overall shapes. Molecules have shapes and sizes that are defined by the angles and distances between the nuclei of their component atoms.

The shape and size of a molecule of a particular substance, together with the strength and polarity of its bonds, largely determine the properties of that substance. Some of the most dramatic examples of the important roles of molecular shape and size are seen in biochemical reactions. For example, a small change in the shape or size of a drug molecule may enhance its effectiveness or reduce its side effects. We saw an example of the influence of molecular shape on drug action in the item "The Search for a Super-aspirin," which was part of the "Chemistry in the News" box in Chapter 1.

The sensations of smell and vision depend in part on molecular shape. When you inhale, molecules in the air are carried past receptor sites in your nose. If the molecules have the right shape and size, they can fit properly on these receptor sites, which transmit impulses to the brain. The brain then identifies these impulses as a particular aroma, such as the aroma of freshly baked bread. The nose is so good at molecular recognition that two substances may produce different sensations of odor even when their molecules differ as subtly as your right hand differs from your left.

Our first goal in this chapter is to learn the relationship between two-dimensional Lewis structures and three-dimensional molecular shapes. Armed with this knowledge, we can then examine more closely the nature of covalent bonds. The lines that are used to depict bonds in Lewis structures provide important clues about the orbitals that molecules use in bonding. By examining these orbitals, we can gain a greater understanding of the behavior of molecules. You will find that the material in this chapter will help you in later discussions of the physical and chemical properties of substances.

▶ What's Ahead ◀

- We begin by discussing how *molecular geometries* are described and by examining some common geometries exhibited by molecules.
- We next consider how molecular shapes can be predicted using a simple model based largely on Lewis structures and the idea of electron-electron repulsions (the *VSEPR model*).
- Being able to predict molecular geometries allows us to predict something about the overall charge distribution in a molecule to determine whether a molecule is *polar* or *nonpolar*.
- We then examine the *valence-bond theory*, a model of molecular bonding that helps us understand why molecules form bonds and why they have the shapes they do.
- In the valence-bond theory, covalent bonds are visualized as resulting from the *overlap* of atomic orbitals, resulting in two general types of bonds, *sigma bonds* and *pi bonds*.
- To account for molecular shape, we consider how the atomic orbitals are reshaped, or *hybridized*, to produce orbitals suitable for bonding within molecules.
- Finally, we discuss the *molecular orbital theory*, a model of chemical bonding that provides additional insight into the electronic structures of molecules.

9.1 Molecular Shapes

In Chapter 8 we used Lewis structures to account for the formulas of covalent compounds. (Section 8.5) Lewis structures, however, do not indicate the shapes of molecules; they simply show the number and types of bonds between atoms. For example, the Lewis structure of CCl_4 tells us only that four Cl atoms are bonded to a central C atom:

$$
\begin{array}{c} :\ddot{\text{Cl}}: \\ | \\ :\ddot{\text{Cl}}-\text{C}-\ddot{\text{Cl}}: \\ | \\ :\ddot{\text{Cl}}: \end{array}
$$

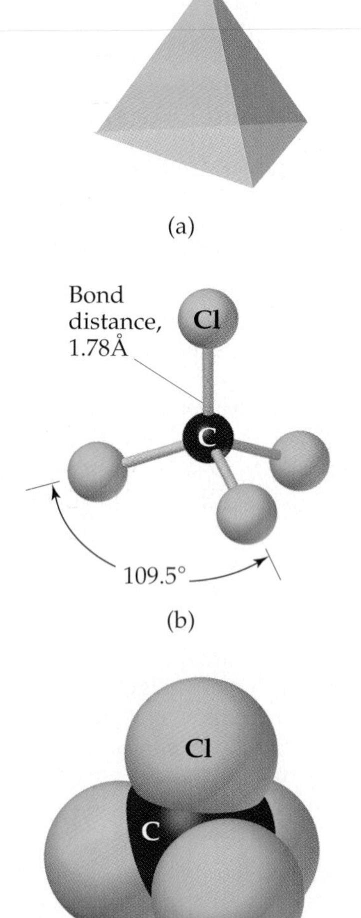

▲ **Figure 9.1** (a) A tetrahedron is an object with four faces and four vertices. Each face is an equilateral triangle. (b) The geometry of the CCl_4 molecule. Each C—Cl bond in the molecule points toward a vertex of a tetrahedron. All of the C—Cl bonds are the same length, and all of the Cl—C—Cl bond angles are the same. This type of drawing of CCl_4 is called a ball-and-stick model. (c) A representation of CCl_4, called a space-filling model. It shows the relative sizes of the atoms, but the geometry is somewhat harder to see.

The Lewis structure is drawn with the atoms in the same plane. In Figure 9.1 ◀, however, the actual three-dimensional arrangement of the atoms shows the Cl atoms at the corners of a *tetrahedron*, a geometric object with four corners and four faces, each of which is an equilateral triangle.

The overall shape of a molecule is determined by its **bond angles**, the angles made by the lines joining the nuclei of the atoms in the molecule. The bond angles of a molecule, together with the bond lengths (Section 8.8), accurately define the shape and size of the molecule. In CCl_4 the bond angles are defined by moving along a bond from a Cl to the central C and then along another bond to another Cl. All six Cl—C—Cl angles have the same value (109.5°, which is characteristic of a tetrahedron). In addition, all four C—Cl bonds are the same length (1.78 Å). Thus, the shape and size of CCl_4 are completely described by stating that the molecule is tetrahedral with C—Cl bonds of length 1.78 Å.

In our discussion of the shapes of molecules we will begin with molecules (and ions) that, like CCl_4, have a single central atom bonded to two or more atoms of the same type. Such molecules conform to the general formula AB_n, in which the central atom A is bonded to n B atoms. Both CO_2 and H_2O are AB_2 molecules, for example, whereas SO_3 and NH_3 are AB_3 molecules, and so on.

The possible shapes of AB_n molecules depend on the value of n. For a given value of n, only a few general shapes are observed. Those commonly found for AB_2 and AB_3 molecules are shown in Figure 9.2 ▶. Thus, an AB_2 molecule must be either linear (bond angle $= 180°$) or bent (bond angle $\neq 180°$). For example, CO_2 is linear and SO_2 is bent. For AB_3 molecules, the two most common shapes place the B atoms at the corners of an equilateral triangle. If the A atom lies in the same plane as the B atoms, the shape is called *trigonal planar*. If the A atom lies above the plane of the B atoms, the shape is called *trigonal pyramidal* (a pyramid with an equilateral triangle as its base). For example, SO_3 is trigonal planar and NF_3 is trigonal pyramidal. Some AB_3 molecules such as ClF_3 exhibit the more unusual *T-shape* shown in Figure 9.2.

The shape for any particular AB_n molecule can usually be derived from one of the five basic geometric structures shown in Figure 9.3 ▶. Starting with a tetrahedron, for example, we can remove atoms successively from the corners as shown in Figure 9.4 ▶. When an atom is removed from one corner of the tetrahedron, the remaining fragment has a trigonal pyramidal geometry such as that found for NF_3. When two atoms are removed, a bent geometry results.

Why do so many AB_n molecules have shapes related to the basic structures in Figure 9.3, and can we predict these shapes? When A is a representative element, one of the elements of the p block of the periodic table, we can answer these questions by using the **valence-shell electron-pair repulsion (VSEPR) model**. Although the name is rather imposing, the model is quite simple, and it has useful predictive capabilities, as we will see in Section 9.2.

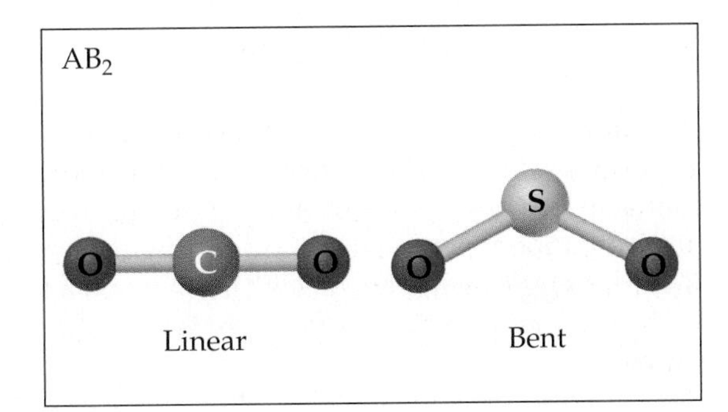

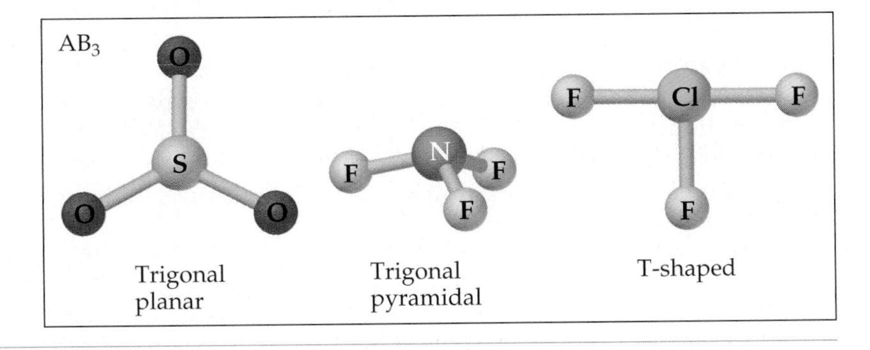

◀ **Figure 9.2** The shapes of some simple AB_2 and AB_3 molecules.

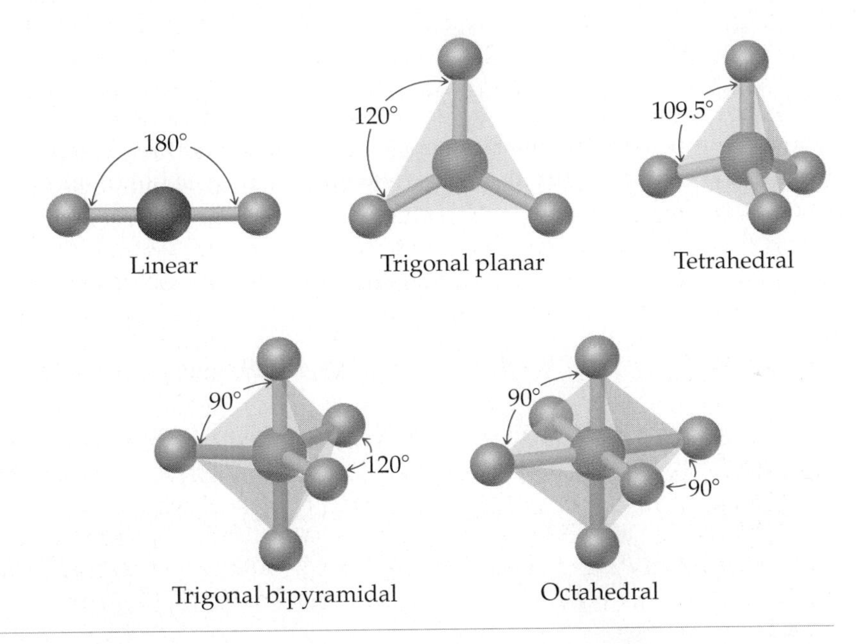

◀ **Figure 9.3** Five fundamental geometries on which the molecular shapes of AB_n molecules are based.

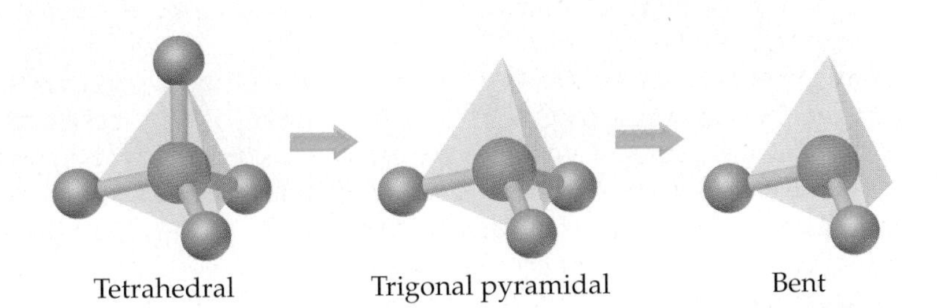

◀ **Figure 9.4** Additional molecular shapes can be obtained by removing corner atoms from the basic geometries shown in Figure 9.3. Here we begin with a tetrahedron and successively remove corners, producing first a trigonal pyramidal geometry and then a bent one, each having ideal bond angles of 109.5°. Molecular shape is meaningful only when there are at least three atoms. If there are only two, they must be arranged next to each other and no special name is given to describe the molecule.

9.2 The VSEPR Model

Imagine tying two identical balloons together at their ends. As shown in Figure 9.5(a) ▼, the balloons naturally orient themselves to point away from each other; that is, they try to "get out of each other's way" as much as possible. If we add a third balloon, the balloons orient themselves toward the vertices of an equilateral triangle as in Figure 9.5(b). If we add a fourth balloon, they adopt a tetrahedral shape [Figure 9.5(c)]. There is an optimum geometry, therefore, for each number of balloons.

In some ways the electrons in molecules behave like the balloons in Figure 9.5. We have seen that a single covalent bond is formed between two atoms when a pair of electrons occupies the space between the atoms. (Section 8.3) A **bonding pair** of electrons thus defines a region in which the electrons will most likely be found. We will refer to such a region as an **electron domain**. Likewise, a **nonbonding pair** (or *lone pair*) of electrons defines an electron domain that is located principally on one atom. For example, the Lewis structure of NH_3 has a total of four electron domains around the central nitrogen atom (three bonding pairs and one nonbonding pair):

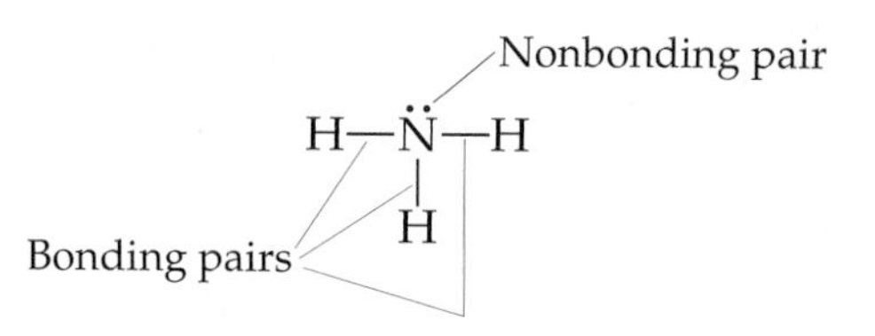

Each multiple bond in a molecule also constitutes a single electron domain. Thus, the following resonance structure for SO_2 has three electron domains around the central sulfur atom (a single bond, a double bond, and a nonbonding pair of electrons):

$$:\ddot{\underset{..}{O}}-\ddot{S}=\ddot{\underset{..}{O}}$$

In general, then, *an electron domain consists of a nonbonding pair, a single bond, or a multiple bond.*

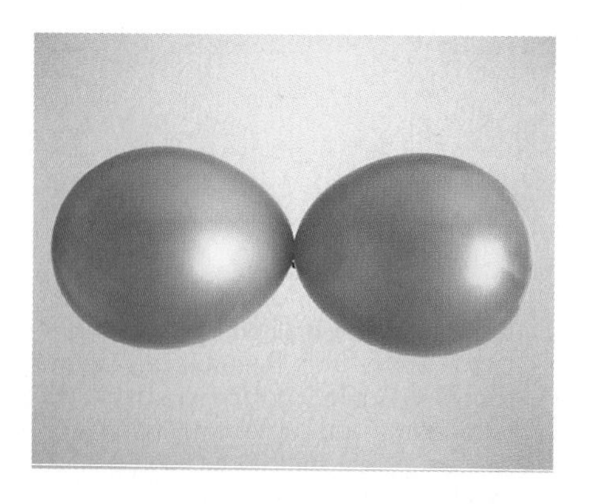

(a)

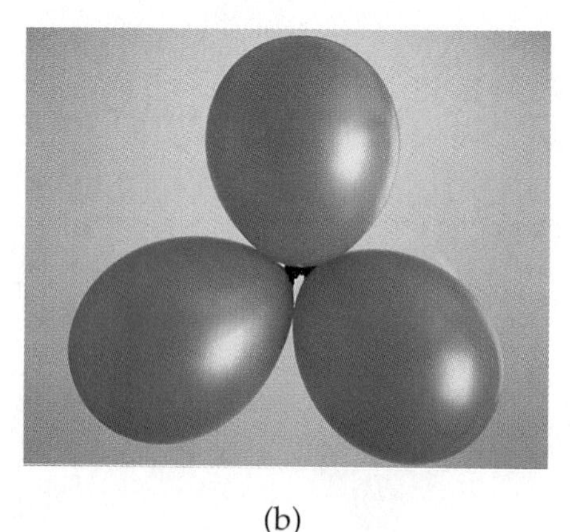

(b)

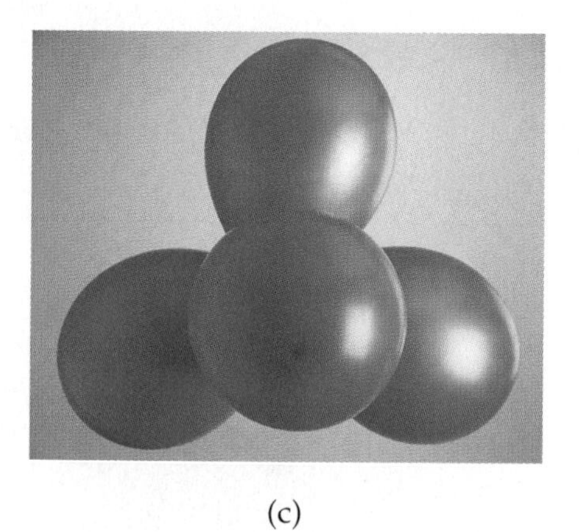

(c)

▲ **Figure 9.5** Balloons tied together at their ends naturally adopt their lowest-energy arrangement. (a) Two balloons adopt a linear arrangement. (b) Three balloons adopt a trigonal-planar arrangement. (c) Four balloons adopt a tetrahedral arrangement.

Because electron domains are negatively charged, they repel one another. Therefore, like the balloons in Figure 9.5, electron domains try to stay out of each other's way. *The best arrangement of a given number of electron domains is the one that minimizes the repulsions among them.* This simple idea is the basis of the VSEPR model. In fact, the analogy between electron domains and balloons is so close that the same preferred geometries are found in both cases. Thus, like the balloons in Figure 9.5, two electron domains are arranged *linearly*, three domains are arranged in a *trigonal-planar* fashion, and four are arranged *tetrahedrally*. These arrangements, together with those for five electron domains (*trigonal bipyramidal*) and six electron domains (*octahedral*), are summarized in Table 9.1 ▼. If you

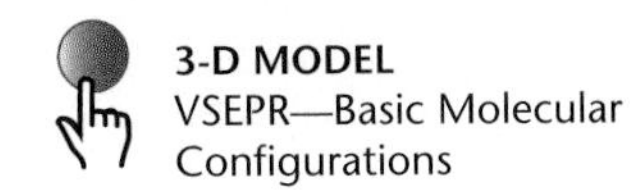

TABLE 9.1 Electron-Domain Geometries as a Function of the Number of Electron Domains

Number of Electron Domains	Arrangement of Electron Domains	Electron-Domain Geometry	Predicted Bond Angles
2	180°	Linear	180°
3	120°	Trigonal planar	120°
4	109.5°	Tetrahedral	109.5°
5	90°, 120°	Trigonal bipyramidal	120° 90°
6	90°, 90°	Octahedral	90°

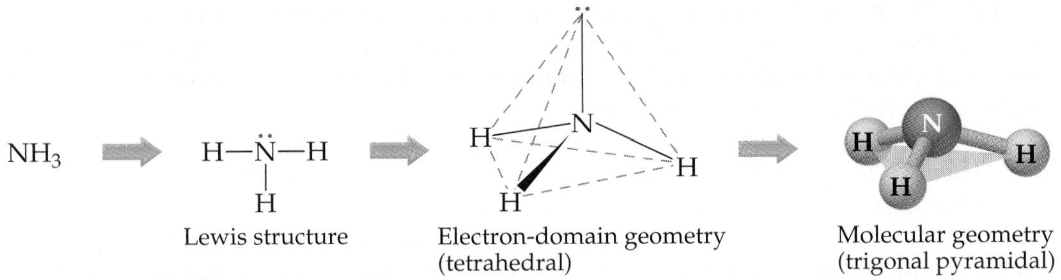

NH_3 ➡ H—N̈—H (Lewis structure) ➡ Electron-domain geometry (tetrahedral) ➡ Molecular geometry (trigonal pyramidal)

▲ **Figure 9.6** The molecular geometry of NH_3 is predicted by first drawing the Lewis structure, then using the VSEPR model to determine the electron-domain geometry, and finally focusing on the atoms themselves to describe the molecular geometry.

compare the geometries in Table 9.1 with those in Figure 9.3, you will see that they are the same. *The shapes of different AB_n molecules or ions depend on the number of electron domains surrounding the central A atom.*

The NH_3 molecule has four electron domains around the nitrogen atom. The repulsions among four electron domains are minimized when the domains point toward the vertices of a tetrahedron (Table 9.1). One of these domains, however, is due to a nonbonding pair of electrons. *Molecular shape describes the arrangement of atoms, not the arrangement of electron domains.* Hence the molecular structure of NH_3 is trigonal pyramidal, as shown in Figure 9.6 ▲. It is the tetrahedral arrangement of the four electron domains, however, that leads us to predict the trigonal-pyramidal molecular geometry.

The arrangement of electron domains about the central atom of an AB_n molecule or ion is called its **electron-domain geometry**. The **molecular geometry** is the arrangement of the atoms in space. In the VSEPR model, we predict the molecular geometry of a molecule or ion from its electron-domain geometry.

To predict the shapes of molecules with the VSEPR model, we use the following steps:

1. Draw the *Lewis structure* of the molecule or ion, and count the total number of electron domains around the central atom. Each nonbonding electron pair, each single bond, each double bond, and each triple bond counts as an electron domain.
2. Determine the *electron-domain geometry* by arranging the total number of electron domains so that the repulsions among them are minimized, as shown in Table 9.1.
3. Use the arrangement of the bonded atoms to determine the *molecular geometry*.

Figure 9.6 shows how these steps are applied to predict the geometry of the NH_3 molecule. Because the trigonal pyramidal structure is based on a tetrahedron, the *ideal bond angles* are 109.5°. As we will soon see, bond angles deviate from the ideal angles when the surrounding atoms and electron domains are not identical.

Let's apply these steps to determine the shape of the CO_2 molecule. We first draw its Lewis structure, which reveals two electron domains (two double bonds) around the central carbon:

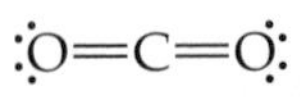

$$:\ddot{O}=C=\ddot{O}:$$

Two electron domains will arrange themselves to give a linear electron-domain geometry (Table 9.1). Because none of the domains is a nonbonding pair of electrons, the molecular geometry is also linear and the O—C—O bond angle is 180°.

Table 9.2 ▼ summarizes the possible molecular geometries when an AB_n molecule has four or fewer electron domains about A. These geometries are important because they include all of the commonly occurring shapes found for molecules or ions that obey the octet rule.

TABLE 9.2 Electron-Domain Geometries and Molecular Shapes for Molecules with Two, Three, and Four Electron Domains Around the Central Atom

Number of Electron Domains	Electron-Domain Geometry	Bonding Domains	Nonbonding Domains	Molecular Geometry	Example
2	Linear	2	0	Linear	O=C=O
3	Trigonal planar	3	0	Trigonal planar	F–B(–F)–F
		2	1	Bent	[O=N–O]$^-$
4	Tetrahedral	4	0	Tetrahedral	H–C(H)(H)H
		3	1	Trigonal pyramidal	N(H)(H)H
		2	2	Bent	O(H)H

SAMPLE EXERCISE 9.1

Use the VSEPR model to predict the molecular geometries of **(a)** O_3; **(b)** $SnCl_3^-$.

Solution

Analyze: We are given the molecular formulas of a molecule and a polyatomic ion, both conforming to the general formula AB_n and both having a central atom from the *p* block of the periodic table.

Plan: To predict the molecular geometries of these species, we first draw their Lewis structures and then count the number of electron domains around the central atom. The number of electron domains gives the electron-domain geometry. We then obtain the molecular geometry from the arrangement of bonding domains.

Solve: (a) We can draw two resonance structures for O_3:

$$:\ddot{\text{O}}-\ddot{\text{O}}=\ddot{\text{O}} \longleftrightarrow \ddot{\text{O}}=\ddot{\text{O}}-\ddot{\text{O}}:$$

Because of resonance, the bonds between the central O atom and the outer O atoms are of equal length. In both resonance structures the central O atom is bonded to the two outer O atoms and has one nonbonding pair. Thus, there are three electron domains about the central O atom. (Remember that a double bond counts as a single electron domain.) For three electron domains, the arrangement is trigonal planar (Table 9.1). Two of the domains are bonding and one is nonbonding, so the molecule has a bent shape with an ideal bond angle of 120°:

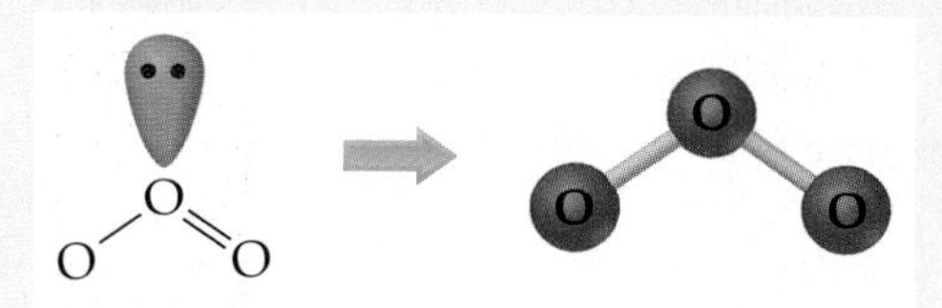

When a molecule exhibits resonance, therefore, any one of the resonance structures can be used to predict the geometry.

(b) The Lewis structure for the $SnCl_3^-$ ion is

$$\left[:\ddot{\text{Cl}}-\ddot{\text{Sn}}-\ddot{\text{Cl}}: \atop \quad | \atop :\ddot{\text{Cl}}:\right]^-$$

The central Sn atom is bonded to the three Cl atoms and has one nonbonding pair. Therefore, the Sn atom has four electron domains around it. The resulting electron-domain geometry is tetrahedral (Table 9.1) with one of the corners occupied by a nonbonding pair of electrons. The molecular geometry is thus trigonal pyramidal:

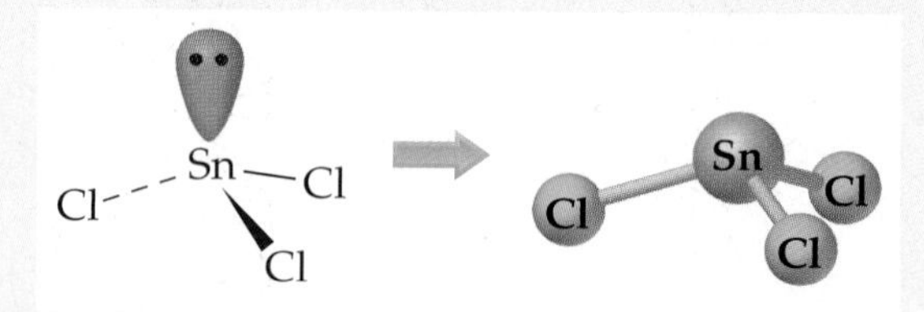

PRACTICE EXERCISE

Predict the electron-domain geometry and the molecular geometry for **(a)** $SeCl_2$; **(b)** CO_3^{2-}.

Answers: **(a)** tetrahedral, bent; **(b)** trigonal planar, trigonal planar

The Effect of Nonbonding Electrons and Multiple Bonds on Bond Angles

We can refine the VSEPR model to predict and explain slight distortions of molecules from the ideal geometries summarized in Table 9.2. For example, consider methane (CH_4), ammonia (NH_3), and water (H_2O). All three have tetrahedral electron-domain geometries, but their bond angles differ slightly:

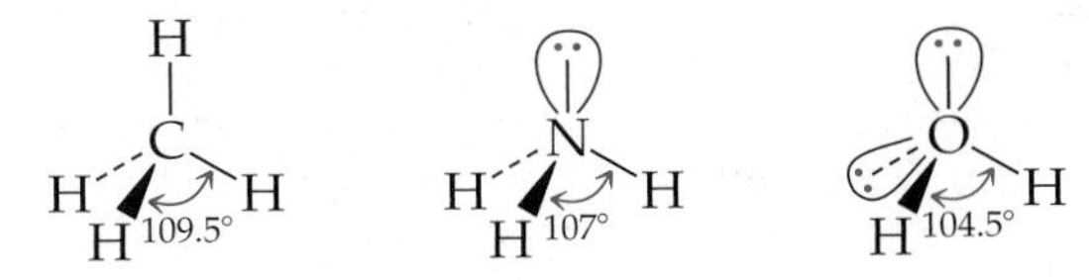

Notice that the bond angles decrease as the number of nonbonding electron pairs increases. A bonding pair of electrons is attracted by both nuclei of the bonded atoms. By contrast, a nonbonding pair is attracted primarily by only one nucleus. Because a nonbonding pair experiences less nuclear attraction, its electron domain is spread out more in space than that for a bonding pair, as shown in Figure 9.7 ▶. As a result, *electron domains for nonbonding electron pairs exert greater repulsive forces on adjacent electron domains and thus tend to compress the bond angles.* Using the analogy in Figure 9.5, we can envision the domains for nonbonding electron pairs as represented by slightly larger and fatter balloons than those for bonding pairs.

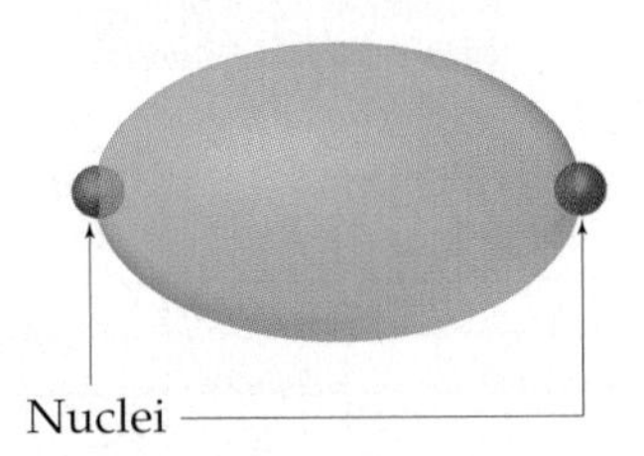

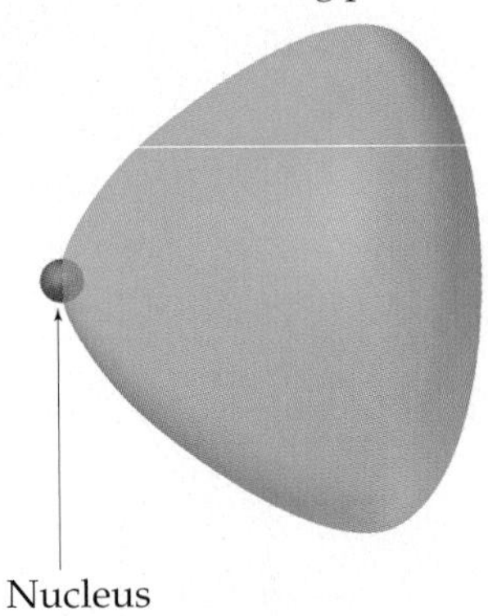

▲ **Figure 9.7** Relative "sizes" of bonding and nonbonding pairs of electrons.

Multiple bonds contain a higher electronic-charge density than single bonds, so multiple bonds also represent larger electron domains ("fatter balloons"). Consider the Lewis structure of *phosgene*, Cl_2CO:

:C̈l:
C=Ö:
:C̈l:

Because the central carbon atom is surrounded by three electron domains, we might expect a trigonal planar geometry with 120°-bond angles. The double bond seems to act much like a nonbonding pair of electrons, however, reducing the Cl—C—Cl bond angle from the ideal angle of 120° to an actual angle of 111°, as shown below.

Cl 124.3°
111.4° C=O
124.3°
Cl

In general, *electron domains for multiple bonds exert a greater repulsive force on adjacent electron domains than single bonds.*

Molecules with Expanded Valence Shells

Our discussion of the VSEPR model thus far has involved molecules with no more than an octet of electrons around the central atom. Recall, however, that when the central atom of a molecule is from the third period of the periodic table and beyond, that atom may have more than four electron pairs around it. (Section 8.7) Molecules with five or six electron domains around the central atom display

TABLE 9.3 Electron-Domain Geometries and Molecular Shapes for Molecules with Five and Six Electron Domains Around the Central Atom

Total Electron Domains	Electron-Domain Geometry	Bonding Domains	Nonbonding Domains	Molecular Geometry	Example
5	 Trigonal bipyramidal	5	0	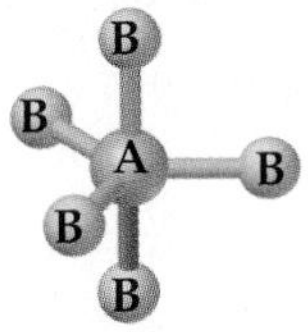 Trigonal bipyramidal	PCl_5
		4	1	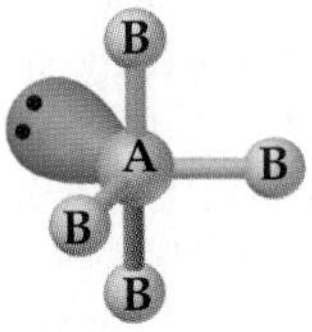 Seesaw	SF_4
		3	2	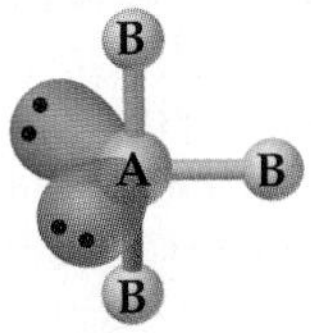 T-shaped	ClF_3
		2	3	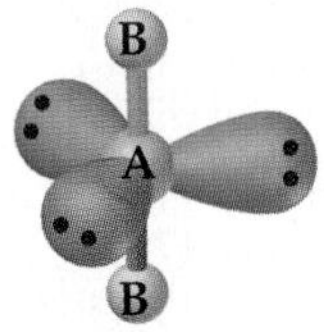 Linear	XeF_2
6	 Octahedral	6	0	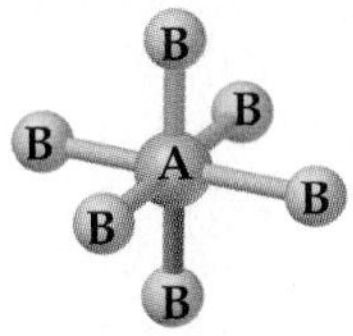 Octahedral	SF_6
		5	1	 Square pyramidal	BrF_5
		4	2	 Square planar	XeF_4

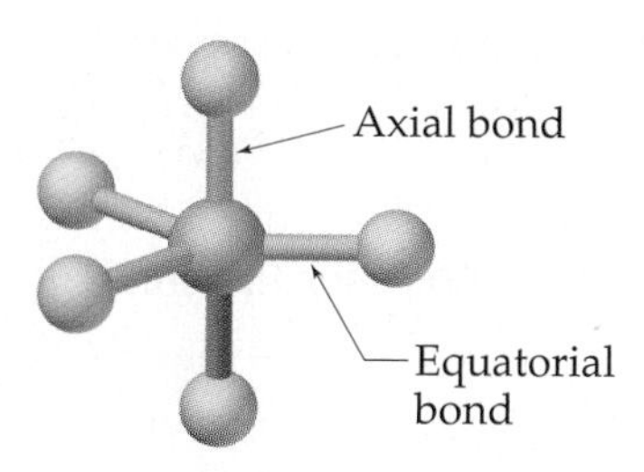

◀ **Figure 9.8** Trigonal bipyramidal arrangement of five electron domains around a central atom. The three *equatorial* electron domains define an equilateral triangle. The two *axial* domains lie above and below the plane of the triangle. If a molecule has nonbonding electron domains, they will occupy the equatorial positions.

a variety of molecular geometries based on the *trigonal bipyramid* (five electron domains) or the *octahedron* (six electron domains), as shown in Table 9.3 ◀.

The most stable electron-domain geometry for five electron domains is the trigonal bipyramid (two trigonal pyramids sharing a base). Unlike the electron-domain geometries we have seen to this point, the electron domains in a trigonal bipyramid can point toward two geometrically distinct types of positions. Two positions are called *axial positions*, and the remaining three are called *equatorial positions* (Figure 9.8 ▲). When pointing toward an axial position, an electron domain is situated 90° from the three equatorial positions. In an equatorial position an electron domain is situated 120° from the other two equatorial positions and 90° from the two axial positions.

Suppose a molecule has five electron domains, one or more of which originates from a nonbonding pair. Will the electron domains from the nonbonding pairs occupy axial or equatorial positions? In order to answer this question, we must determine which location minimizes the repulsions between the electron domains. Repulsions between domains are much greater when they are situated 90° from each other than when they are at 120°. An equatorial domain is 90° from only two other domains (the two axial domains). By contrast, an axial domain is situated 90° from *three* other domains (the three equatorial domains). Hence, an equatorial domain experiences less repulsion than an axial domain. Because the domains from nonbonding pairs exert larger repulsions than those from bonding pairs, they always occupy the equatorial positions in a trigonal bipyramid.

The most stable electron-domain geometry for six electron domains is the *octahedron*. As shown in Figure 9.9 ▼, an octahedron is a polyhedron with six vertices and eight faces, each of which is an equilateral triangle. If an atom has six electron domains around it, that atom can be visualized as being at the center of the octahedron with the electron domains pointing toward the six vertices. All the bond angles in an octahedron are 90°, and all six vertices are equivalent. Therefore, if an atom has five bonding electron domains and one nonbonding domain, we can point the nonbonding domain at any of the six vertices of the octahedron. The result is always a square-pyramidal molecular geometry. When there are two nonbonding electron domains, however, their repulsions are minimized by pointing them toward opposite sides of the octahedron producing a square-planar molecular geometry, as shown in Table 9.3.

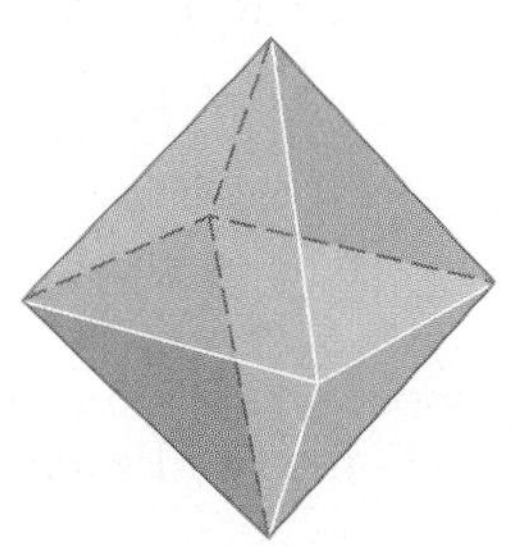

◀ **Figure 9.9** An octahedron is an object with eight faces and six vertices. Each face is an equilateral triangle.

SAMPLE EXERCISE 9.2

Use the VSEPR model to predict the molecular geometry of **(a)** SF_4; **(b)** IF_5.

Solution

Analyze and Plan: The molecules are of the AB_n type with a central atom from the *p* block of the periodic table. Thus, we can predict their structures starting with their Lewis structures and using the VSEPR model.

Solve: (a) The Lewis structure for SF_4 is

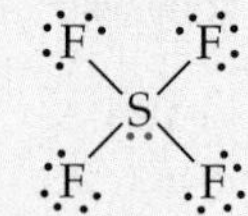

The sulfur has five electron domains around it: four from the S—F bonds and one from the nonbonding pair. Each domain points toward a vertex of a trigonal bipyramid. The domain from the nonbonding pair will point toward an equatorial position. The four bonds point toward the remaining four positions, resulting in a molecular geometry that is described as seesaw-shaped:

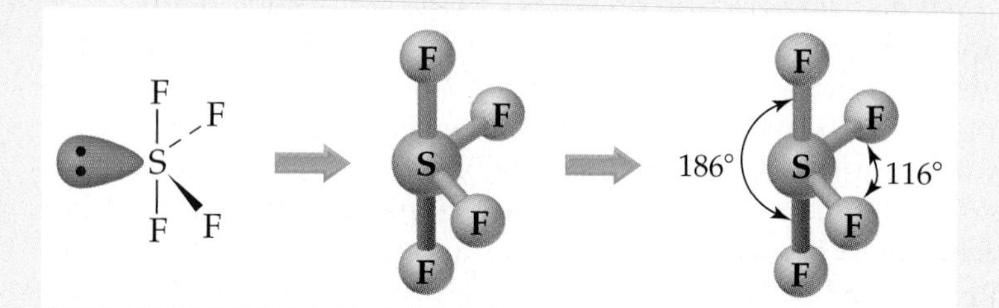

Comment: The experimentally observed structure is shown above on the right, and we can infer that the nonbonding electron domain occupies an equatorial position, as predicted. The axial and equatorial S—F bonds are slightly bent back away from the nonbonding domain, suggesting that the bonding domains are "pushed" by the nonbonding domain, which is larger and has greater repulsion.

(b) The Lewis structure of IF_5 is:

(There are three lone pairs on each of the F atoms, but they are not shown.)

The iodine has a total of six electron domains around it, one of which is from a nonbonding pair. The electron-domain geometry is therefore octahedral, with one position occupied by the nonbonding electron pair.

The resulting molecular geometry is therefore *square pyramidal* (Table 9.3):

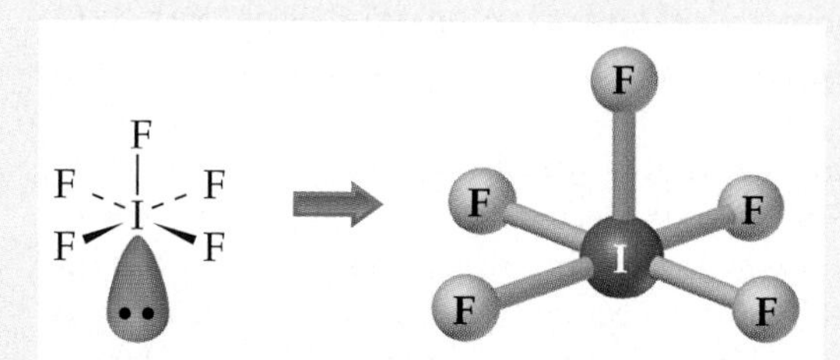

Comment: Because the domain for the nonbonding pair is larger than the other domains, the four F atoms in the base of the pyramid are tipped up slightly toward the F atom on top. Experimentally, it is found that the angle between the base and top F atoms is 82°, smaller than the ideal 90° angle of an octahedron.

PRACTICE EXERCISE

Predict the electron-domain geometry and molecular geometry of **(a)** ClF_3; **(b)** ICl_4^-.
Answers: **(a)** trigonal bipyramidal, T-shaped; **(b)** octahedral, square planar

Shapes of Larger Molecules

Although the molecules and ions whose structures we have thus far considered contain only a single central atom, the VSEPR model can be extended to more complex molecules. Consider the acetic acid molecule, whose Lewis structure is

```
   H  :O:
   |   ||   ..
H—C—C—O—H
   |        ..
   H
```

Acetic acid has three interior atoms, namely, the leftmost C atom, the central C atom, and the rightmost O atom. We can use the VSEPR model to predict the geometry about each of these atoms individually:

```
     H          :O:
     |           ||          ..
  H—C————————C————————O—H
     |                       ..
     H
```

	H—C(H)(H)—	—C(=O)—	—Ö—H
Number of electron domains	4	3	4
Electron-domain geometry	Tetrahedral	Trigonal planar	Tetrahedral
Predicted bond angles	109.5°	120°	109.5°

The leftmost C has four electron domains (all from bonding pairs), so the geometry around that atom is tetrahedral. The central C has three electron domains (counting the double bond as one domain). Thus, the geometry around that atom is trigonal planar. The O atom has four electron domains (two from bonding pairs and two from nonbonding pairs), so its electron-domain geometry is tetrahedral and the molecular geometry around O is bent. The bond angles about the central C atom and the O atom are expected to deviate slightly from the ideal values of 120° and 109.5°, due to the spatial demands of multiple bonds and nonbonding electron pairs. The structure of the acetic acid molecule is shown in Figure 9.10 ▶.

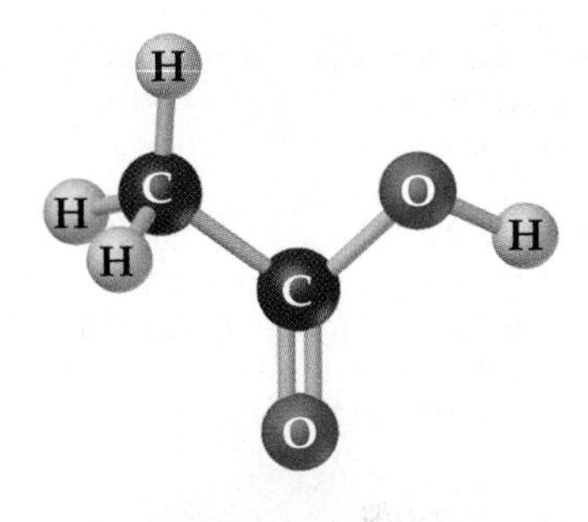

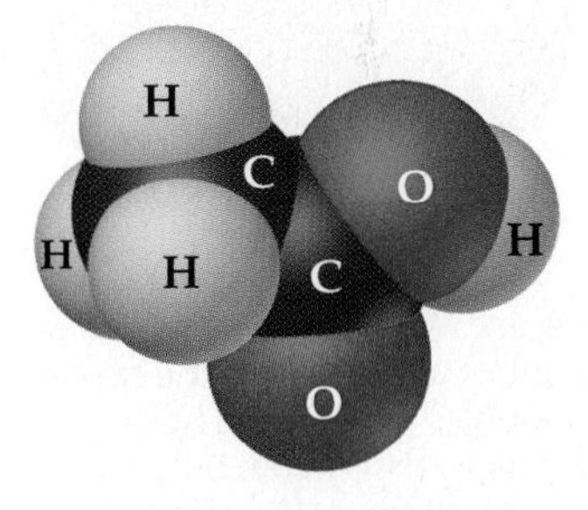

▲ **Figure 9.10** Ball-and-stick (top) and space-filling (bottom) representations of acetic acid, $HC_2H_3O_2$.

SAMPLE EXERCISE 9.3

Eye drops for dry eyes usually contain a water-soluble polymer called *poly(vinyl alcohol)*, which is based on the unstable organic molecule called *vinyl alcohol*:

```
          H   H
   ..     |   |
H—O—C=C—H
   ..
```

Predict the approximate values for the H—O—C and O—C—C bond angles in vinyl alcohol.

Solution

Analyze and Plan: To predict a particular bond angle, we consider the middle atom of the angle and determine the number of electron domains surrounding that atom. The ideal angle corresponds to the electron-domain geometry around the atom. The angle will be compressed somewhat by nonbonding electrons or multiple bonds.

Solve: For the H—O—C bond angle, there are four electron domains around the middle O atom (two bonding and two nonbonding). The electron-domain geometry around O is therefore tetrahedral, which gives an ideal angle of 109.5°. The H—O—C angle will be compressed somewhat by the nonbonding pairs, so we expect this angle to be slightly less than 109.5°.

To predict the O—C—C bond angle, we must examine the leftmost C atom, which is the central atom for this angle. There are three atoms bonded to this C atom and no nonbonding pairs, so it has three electron domains about it. The predicted electron-domain geometry is trigonal planar, resulting in an ideal bond angle of 120°. Because of the larger size of the C=C domain, however, the O—C—C bond angle should be slightly greater than 120°.

PRACTICE EXERCISE

Predict the H—C—H and C—C—C bond angles in the following molecule, called *propyne*:

```
      H
      |
  H—C—C≡C—H
      |
      H
```

Answers: 109.5°, 180°

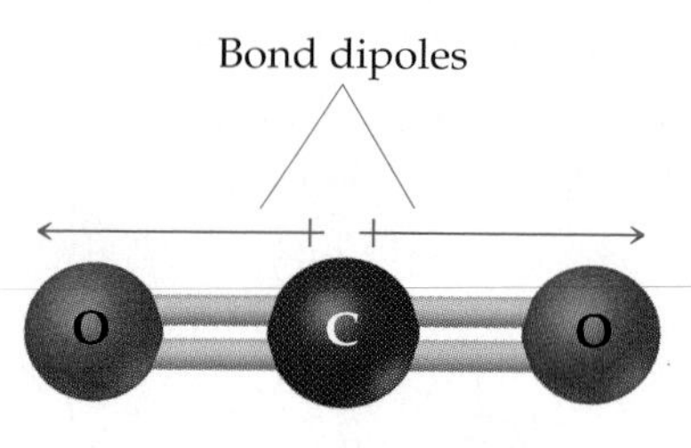

(a)

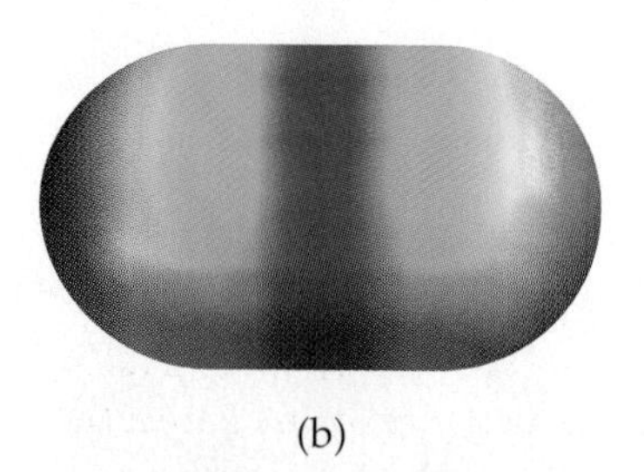

(b)

▲ **Figure 9.11** (a) The overall dipole moment of a molecule is the sum of its bond dipoles. In CO_2 the bond dipoles are equal in magnitude, but exactly oppose each other. The overall dipole moment is zero, therefore, making the molecule nonpolar. (b) The electron-density model shows that the regions of higher electron density (red) are at the ends of the molecule while the region of lower electron density (blue) is at the center.

9.3 Molecular Shape and Molecular Polarity

We now have a better sense of the shapes that molecules adopt and why they do so. We will spend the rest of this chapter looking more closely at the ways in which electrons are shared to form the bonds between atoms in molecules. We will begin by returning to a topic that we first discussed in Section 8.4, namely *bond polarity* and *dipole moments*. Recall that bond polarity is a measure of how equally the electrons in a bond are shared between the two atoms of the bond: As the difference in electronegativity between the two atoms increases, so does the bond polarity. (Section 8.4) We saw that the dipole moment of a diatomic molecule is a quantitative measure of the amount of charge separation in the molecule. The charge separation in molecules has a significant effect on physical and chemical properties. We will see in Chapter 11, for example, how molecular polarity affects boiling points, melting points, and other physical properties.

For a molecule with more than two atoms, *the dipole moment depends on both the polarities of the individual bonds and the geometry of the molecule.* For each bond in the molecule, we can consider the **bond dipole**, which is the dipole moment due only to the two atoms in that bond. Consider the linear CO_2 molecule, for example. As shown in Figure 9.11 ◀, each C=O bond is polar and, because the C=O bonds are identical, the bond dipoles are equal in magnitude. The electron-density model, moreover, shows the regions of high electron density (red) at the ends of the molecule, on the oxygen atoms, and low electron density (blue) in the center, on the carbon atom.

Bond dipoles and dipole moments are *vector* quantities; that is, they have both a magnitude and a direction. The *overall* dipole moment of a polyatomic molecule is the sum of its bond dipoles. Both the magnitudes *and* the directions of the bond dipoles must be considered when summing these vectors. The two bond dipoles in CO_2, although equal in magnitude, are exactly opposite in direction. Adding them together is the same as adding two numbers that are equal in magnitude, but opposite in sign, such as 100 + (−100): The bond dipoles, like the numbers, "cancel" each other. Therefore, the overall dipole moment of CO_2 is zero, even though the individual bonds are polar. Thus, the geometry of the molecule dictates that the overall dipole moment be zero, making CO_2 a *nonpolar* molecule.

Now let's consider H_2O, which is a bent molecule with two polar bonds (Figure 9.12 ▶). Again, both the bonds are identical, so the bond dipoles are equal in magnitude. Because the molecule is bent, however, the bond dipoles do not directly oppose each other and therefore do not cancel each other. Hence, the H_2O molecule has an overall nonzero dipole moment (μ = 1.85 D). Because H_2O has a nonzero dipole moment, it is a *polar* molecule. The oxygen atom carries a partial negative charge, and the hydrogen atoms each have a partial positive charge, as shown by the electron-density model in Figure 9.12(b).

ACTIVITY
Molecular Polarity

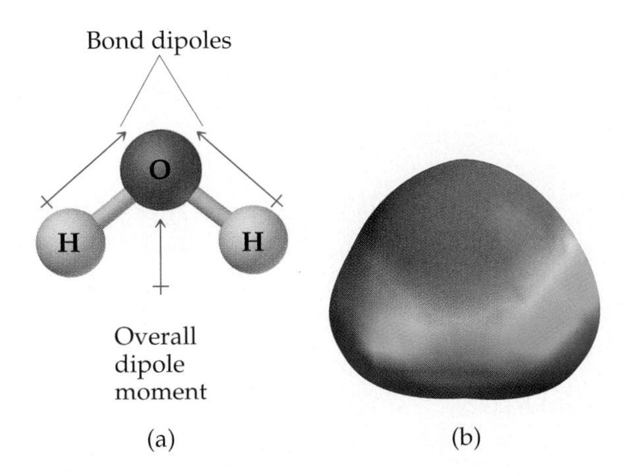

◀ **Figure 9.12** (a) In H_2O the bond dipoles are equal in magnitude, but do not exactly oppose each other. The molecule has a nonzero dipole moment overall, making the molecule polar. (b) The electron-density model shows that one end of the molecule has more electron density (the oxygen end) while the other end has less electron density (the hydrogens).

Figure 9.13 ▼ shows examples of polar and nonpolar molecules, all of which have polar bonds. The molecules in which the central atom is symmetrically surrounded by identical atoms (BF_3 and CCl_4) are nonpolar. For AB_n molecules in which all the B atoms are the same, certain symmetrical shapes—linear (AB_2), trigonal planar (AB_3), tetrahedral and square planar (AB_4), trigonal bipyramidal (AB_5), and octahedral (AB_6)—must lead to nonpolar molecules even though the individual bonds might be polar.

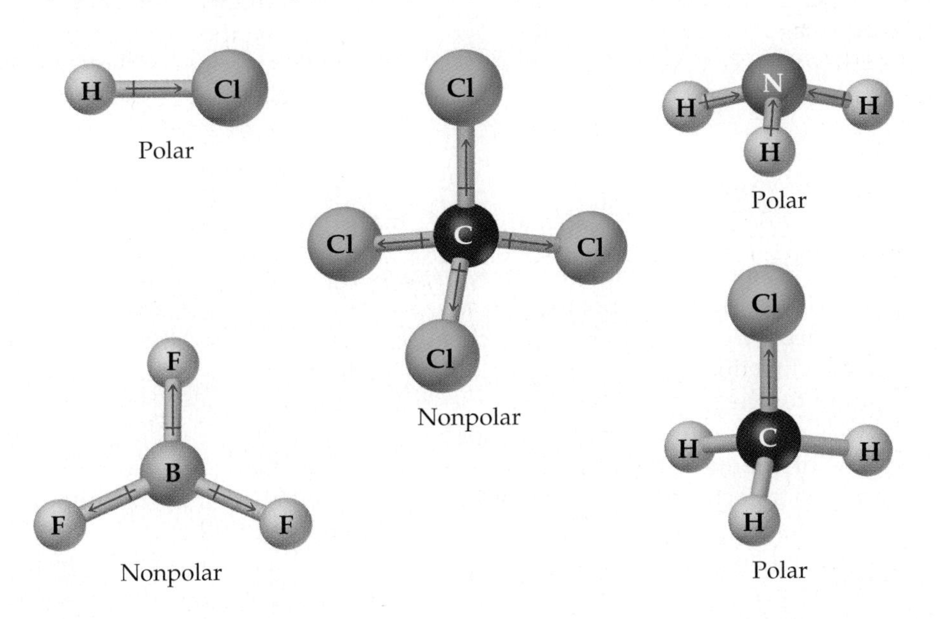

◀ **Figure 9.13** Examples of molecules with polar bonds. Two of these molecules have a zero dipole moment because their bond dipoles cancel one another.

SAMPLE EXERCISE 9.4

Predict whether the following molecules are polar or nonpolar: **(a)** BrCl; **(b)** SO_2; **(c)** SF_6.

Solution

Analyze: We are given the molecular formulas of several substances and asked to predict whether the molecules are polar.

Plan: If the molecule contains only two atoms, it will be polar if the atoms differ in electronegativity. If it contains three or more atoms, its polarity depends on both its molecular geometry and the polarity of its bonds. Thus, we must draw a Lewis structure for each molecule with three or more atoms and determine its molecular geometry. We then use the relative electronegativities of the atoms in each bond to determine the direction of the bond dipoles. Finally, we see if the bond dipoles cancel each other to give a nonpolar molecule or reinforce each other to give a polar one.

Solve: (a) Chlorine is more electronegative than bromine. All diatomic molecules with polar bonds are polar molecules. Consequently, BrCl will be polar, with chlorine carrying the partial negative charge:

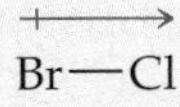

Experimentally, the dipole moment of the molecule is $\mu = 0.57$ D

(b) Because oxygen is more electronegative than sulfur, SO_2 has polar bonds. The following resonance forms for SO_2 can be written:

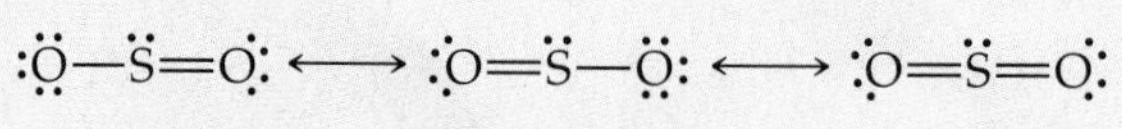

$$:\ddot{O}-\ddot{S}=\ddot{O}: \longleftrightarrow :\ddot{O}=\ddot{S}-\ddot{O}: \longleftrightarrow :\ddot{O}=\ddot{S}=\ddot{O}:$$

For each of these, the VSEPR model predicts a bent geometry. Because the molecule is bent, the bond dipoles do not cancel and the molecule is polar:

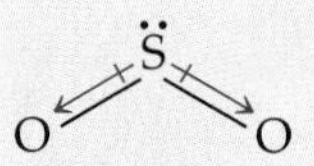

Experimentally, the dipole moment of SO_2 is $\mu = 1.63$ D

(c) Fluorine is more electronegative than sulfur, so the bond dipoles point toward fluorine. The six S—F bonds are arranged octahedrally around the central sulfur:

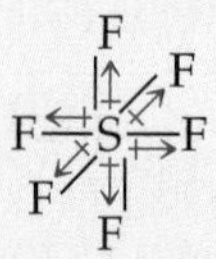

Because the octahedral geometry is symmetrical, the bond dipoles cancel, and the molecule is nonpolar: $\mu = 0$

PRACTICE EXERCISE

Determine whether the following molecules are polar or nonpolar: **(a)** NF_3; **(b)** BCl_3.
Answers: **(a)** polar because polar bonds are arranged in a trigonal-pyramidal geometry; **(b)** nonpolar because polar bonds are arranged in a trigonal-planar geometry.

9.4 Covalent Bonding and Orbital Overlap

The VSEPR model provides a simple means for predicting the shapes of molecules. However, it does not explain why bonds exist between atoms. In developing theories of covalent bonding, chemists have approached the problem from another direction, using quantum mechanics. How can we explain bonding and account for the geometries of molecules using atomic orbitals? The marriage of Lewis's notion of electron-pair bonds to the idea of atomic orbitals leads to a model of chemical bonding called **valence-bond theory**. By extending this approach to include the ways in which atomic orbitals can mix with one another, we obtain a picture that corresponds nicely to the VSEPR model.

In the Lewis theory, covalent bonding occurs when atoms share electrons. Such sharing concentrates electron density between the nuclei. In the valence-bond theory, the buildup of electron density between two nuclei is visualized as occurring when a valence atomic orbital of one atom merges with that of another atom. The orbitals are then said to share a region of space, or to **overlap**. The overlap of orbitals allows two electrons of opposite spin to share the common space between the nuclei, forming a covalent bond.

The approach of two H atoms to form H_2 is depicted in Figure 9.14(a) ▶. Each atom has a single electron in a 1*s* orbital. As the orbitals overlap, electron density is concentrated between the nuclei. Because the electrons in the overlap region are simultaneously attracted to both nuclei, they hold the atoms together, forming a covalent bond.

The idea of orbital overlap producing a covalent bond applies equally well to other molecules. In HCl, for example, chlorine has the electron configuration $[Ne]3s^2 3p^5$. All of the valence orbitals of chlorine are full except one 3*p* orbital, which contains a single electron. This electron pairs up with the single electron of H to form a covalent bond. Figure 9.14(b) shows the overlap of the 3*p* orbital of Cl with the 1*s* orbital of H. Likewise, we can explain the covalent bond in the

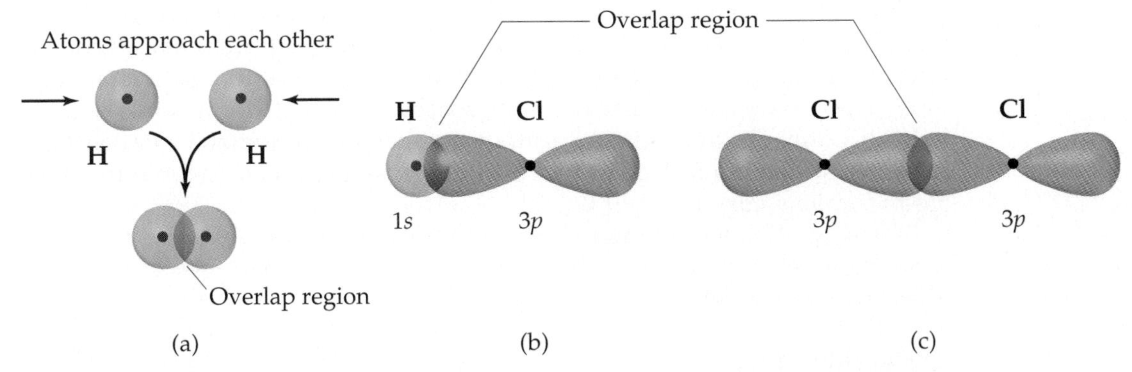

▲ **Figure 9.14** The overlap of orbitals to form covalent bonds. (a) The bond in H_2 results from the overlap of two 1*s* orbitals from two H atoms. (b) The bond in HCl results from the overlap of a 1*s* orbital of H and one of the lobes of a 3*p* orbital of Cl. (c) The bond in Cl_2 results from the overlap of two 3*p* orbitals from two Cl atoms.

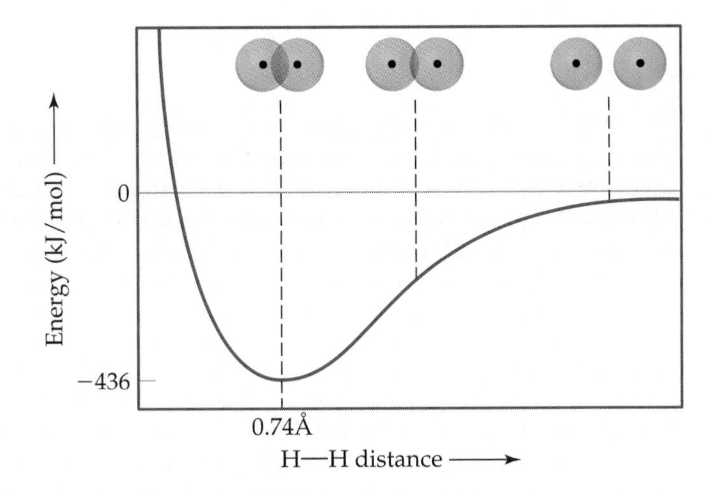

◀ **Figure 9.15** The change in potential energy during the formation of the H_2 molecule. The minimum in the energy, at 0.74 Å, represents the equilibrium bond distance. The energy at that point, −436 kJ/mol, corresponds to the energy change for formation of the H—H bond.

Cl_2 molecule in terms of the overlap of the 3*p* orbital of one atom with the 3*p* orbital of another, as shown in Figure 9.14(c).

There is always an optimum distance between the two bonded nuclei in any covalent bond. Figure 9.15 ▲ shows how the potential energy of the system changes as two H atoms come together to form an H_2 molecule. As the distance between the atoms decreases, the overlap between their 1*s* orbitals increases. Because of the resultant increase in electron density between the nuclei, the potential energy of the system decreases. That is, the strength of the bond increases, as shown by the decrease in the energy on the curve. However, the curve also shows that as the atoms come very close together, the energy increases rapidly. This rapid increase is due mainly to the electrostatic repulsion between the nuclei, which becomes significant at short internuclear distances. The internuclear distance at a minimum of the potential-energy curve corresponds to the observed bond length. Thus, the observed bond length is the distance at which the attractive forces between unlike charges (electrons and nuclei) are balanced by the repulsive forces between like charges (electron-electron and nucleus-nucleus).

9.5 Hybrid Orbitals

Although the idea of orbital overlap allows us to understand the formation of covalent bonds, it is not always easy to extend these ideas to polyatomic molecules. When we apply valence-bond theory to polyatomic molecules, we must explain both the formation of electron-pair bonds *and* the observed geometries of the molecules.

To explain geometries, we often assume that the atomic orbitals on an atom mix to form new orbitals called **hybrid orbitals**. These hybrid orbitals have different shapes than the original atomic orbitals. The process of mixing and thereby changing atomic orbitals as atoms approach each other to form bonds is called **hybridization**. The total number of atomic orbitals on an atom remains constant, however, so the number of hybrid orbitals on an atom equals the number of atomic orbitals mixed.

Let's examine the common types of hybridization. As we do so, notice the connection between the type of hybridization and the five basic electron-domain geometries predicted by the VSEPR model.

sp Hybrid Orbitals

To illustrate the process of hybridization, consider the BeF_2 molecule, which is generated when solid BeF_2 is heated to high temperatures. The Lewis structure of BeF_2 is

:F̤̈—Be—F̤̈:

The VSEPR model correctly predicts that BeF_2 is linear with two identical Be—F bonds, but how can we use valence-bond theory to describe the bonding? The electron configuration of F ($1s^2 2s^2 2p^5$) indicates there is an unpaired electron in a $2p$ orbital. This $2p$ electron can be paired with an unpaired electron from the Be atom to form a polar covalent bond. Which orbitals on the Be atom, however, overlap with those on the F atoms to form the Be—F bonds?

The orbital diagram for a ground-state Be atom is as follows:

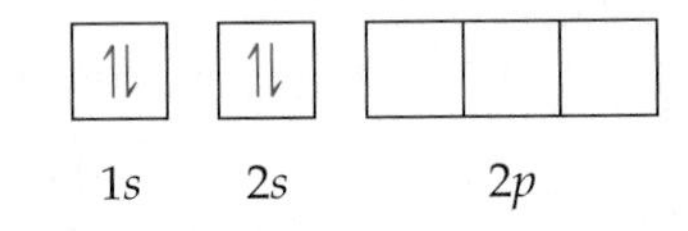

Because it has no unpaired electrons, the Be atom in its ground state is incapable of forming bonds with the fluorine atoms. It could form two bonds, however, by "promoting" one of the $2s$ electrons to a $2p$ orbital:

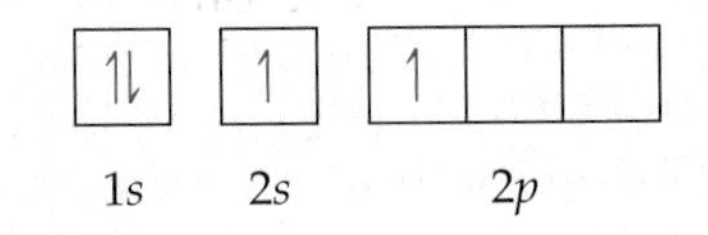

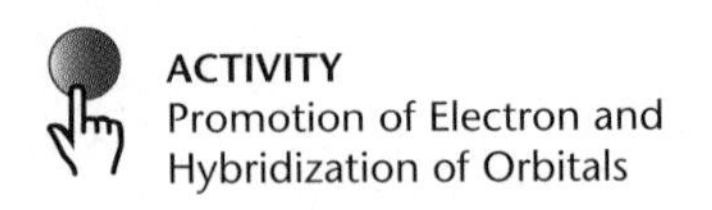

Because the $2p$ orbital is of higher energy than the $2s$, promoting an electron requires energy. The Be atom now has two unpaired electrons and can therefore form two polar covalent bonds with the F atoms. The two bonds would not be identical, however, because a Be $2s$ orbital would be used to form one of the bonds, and a $2p$ orbital would be used for the other. Therefore, although the promotion of an electron allows two Be—F bonds to form, we still haven't explained the structure of BeF_2.

We can solve this dilemma by "mixing" the $2s$ orbital and one of the $2p$ orbitals to generate two new orbitals, as shown in Figure 9.16 ▶. Like *p* orbitals, each of the new orbitals has two lobes. Unlike *p* orbitals, however, one lobe is much larger than the other. The two new orbitals are identical in shape, but their large lobes point in opposite directions. We have created two hybrid orbitals. In this case we have hybridized one *s* and one *p* orbital, so we call each hybrid an *sp* hybrid orbital. *According to the valence-bond model, a linear arrangement of electron domains implies* sp *hybridization.*

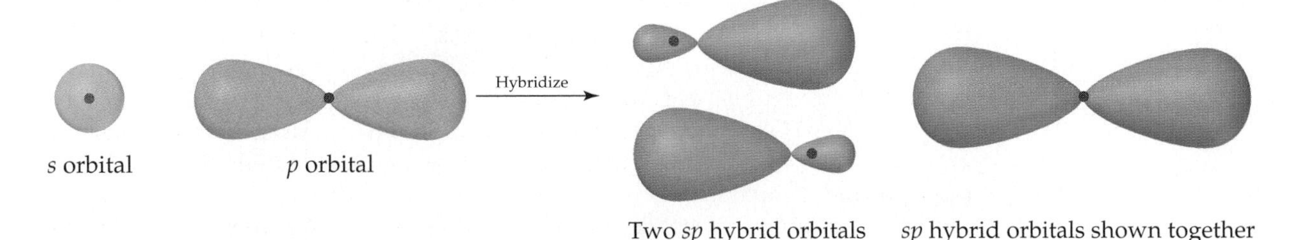

▲ **Figure 9.16** One *s* orbital and one *p* orbital can hybridize to form two equivalent *sp* hybrid orbitals. The two hybrid orbitals have their large lobes pointing in opposite directions, 180° apart.

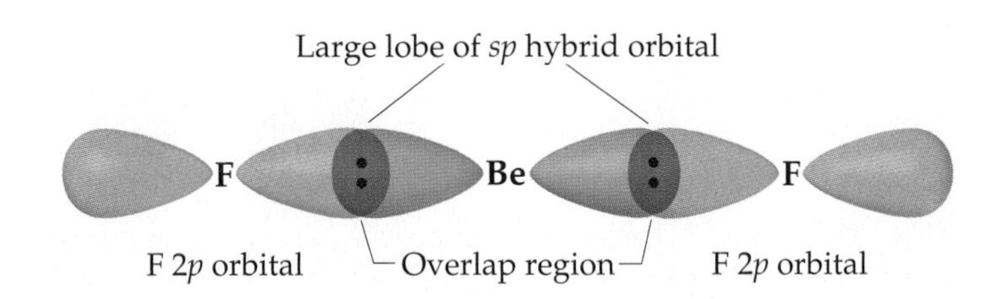

◀ **Figure 9.17** The formation of two equivalent Be—F bonds in BeF_2. Each of the *sp* hybrid orbitals on Be overlaps with a 2*p* orbital on F to form an electron-pair bond.

For the Be atom of BeF_2, we write the orbital diagram for the formation of two *sp* hybrid orbitals as follows:

⇅	1 1	(empty) (empty)
1*s*	*sp*	2*p*

The electrons in the *sp* hybrid orbitals can form shared electron bonds with the two fluorine atoms (Figure 9.17 ▲). Because the *sp* hybrid orbitals are equivalent but point in opposite directions, BeF_2 has two identical bonds and a linear geometry.

The promotion of a 2*s* electron to a 2*p* orbital in Be requires energy. Why, then, do we envision the formation of hybrid orbitals? Hybrid orbitals have one large lobe and can therefore be directed at other atoms better than can unhybridized atomic orbitals. Hence, they can overlap more strongly with the orbitals of other atoms than atomic orbitals, and stronger bonds result. The energy released by the formation of bonds more than offsets the energy that must be expended to promote electrons.

sp^2 and sp^3 Hybrid Orbitals

Whenever we mix a certain number of atomic orbitals, we get the same number of hybrid orbitals. Each of these hybrid orbitals is equivalent to the others, but points in a different direction. Thus, mixing one 2*s* and one 2*p* orbital yields two equivalent *sp* hybrid orbitals that point in opposite directions (Figure 9.16). Other combinations of atomic orbitals can be hybridized to obtain different geometries. In BF_3, for example, a 2*s* electron on the B atom can be promoted to a vacant 2*p* orbital. Mixing the 2*s* and two of the 2*p* orbitals yields three equivalent sp^2 (pronounced "*s*-*p*-two") hybrid orbitals:

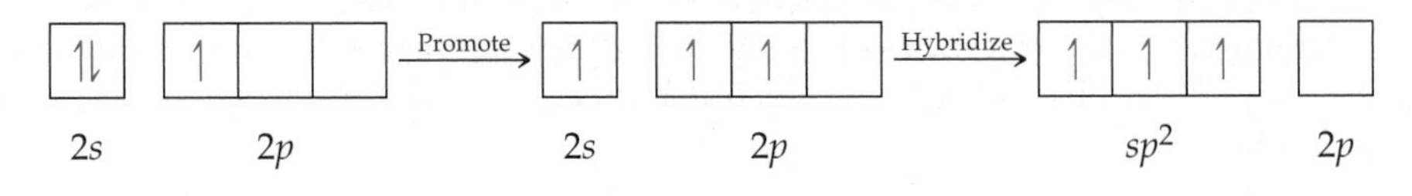

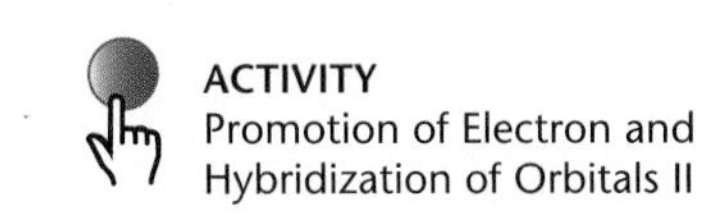

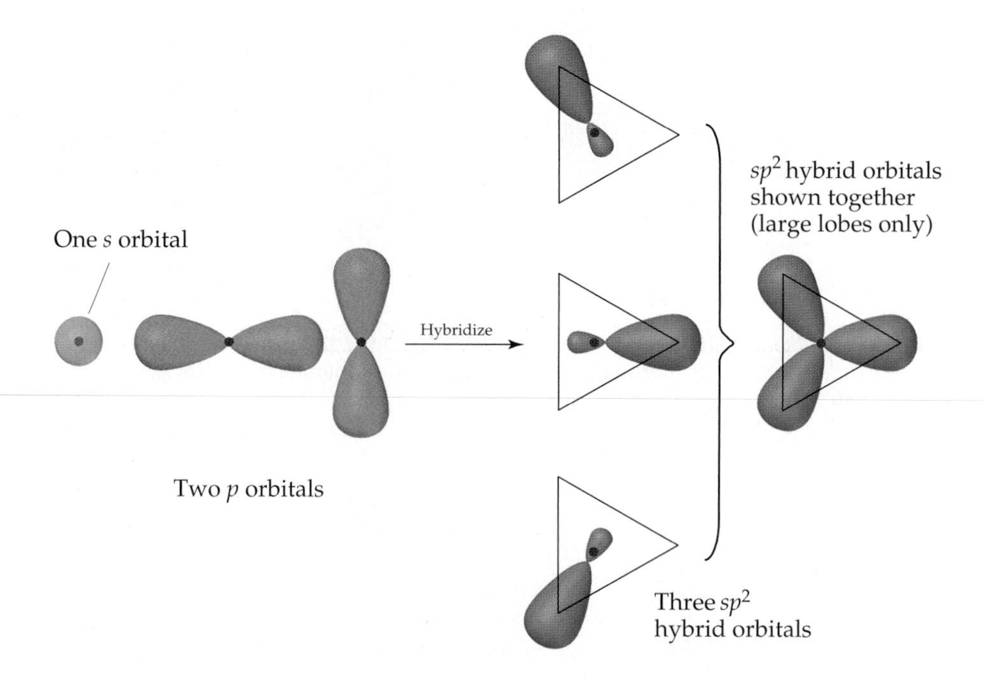

▶ **Figure 9.18** One *s* orbital and two *p* orbitals can hybridize to form three equivalent sp^2 hybrid orbitals. The large lobes of the hybrid orbitals point toward the corners of an equilateral triangle.

The three sp^2 hybrid orbitals lie in the same plane, 120° apart from one another (Figure 9.18 ▲). They are used to make three equivalent bonds with the three fluorine atoms, leading to the trigonal-planar geometry of BF_3. Notice that an unfilled 2*p* orbital remains unhybridized. This unhybridized orbital will be important when we discuss double bonds in Section 9.6.

An *s* orbital can also mix with all three *p* orbitals in the same subshell. For example, the carbon atom in CH_4 forms four equivalent bonds with the four hydrogen atoms. We envision this process as resulting from the mixing of the 2*s* and all three 2*p* atomic orbitals of carbon to create four equivalent sp^3 (pronounced "*s*-*p*-three") hybrid orbitals:

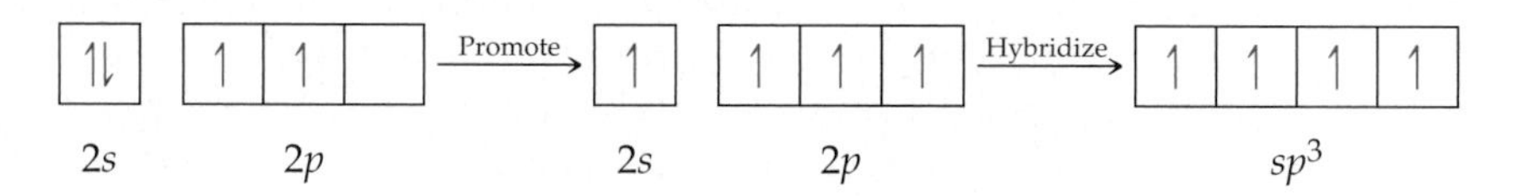

Each of the sp^3 hybrid orbitals has a large lobe that points toward a vertex of a tetrahedron, as shown in Figure 9.19 ▶. These hybrid orbitals can be used to form two-electron bonds by overlap with the atomic orbitals of another atom, such as H. Thus, within valence-bond theory, we can describe the bonding in CH_4 as the overlap of four equivalent sp^3 hybrid orbitals on C with the 1*s* orbitals of the four H atoms to form four equivalent bonds.

The idea of hybridization is used in a similar way to describe the bonding in molecules containing nonbonding pairs of electrons. In H_2O, for example, the electron-domain geometry around the central O atom is approximately tetrahedral. Thus, the four electron pairs can be envisioned as occupying sp^3 hybrid orbitals. Two of these orbitals contain nonbonding pairs of electrons, while the other two are used to form bonds with hydrogen atoms, as shown in Figure 9.20 ▶.

Hybridization Involving *d* Orbitals

Atoms in the third period and beyond can also use *d* orbitals to form hybrid orbitals. Mixing one *s* orbital, three *p* orbitals, and one *d* orbital leads to five sp^3d hybrid

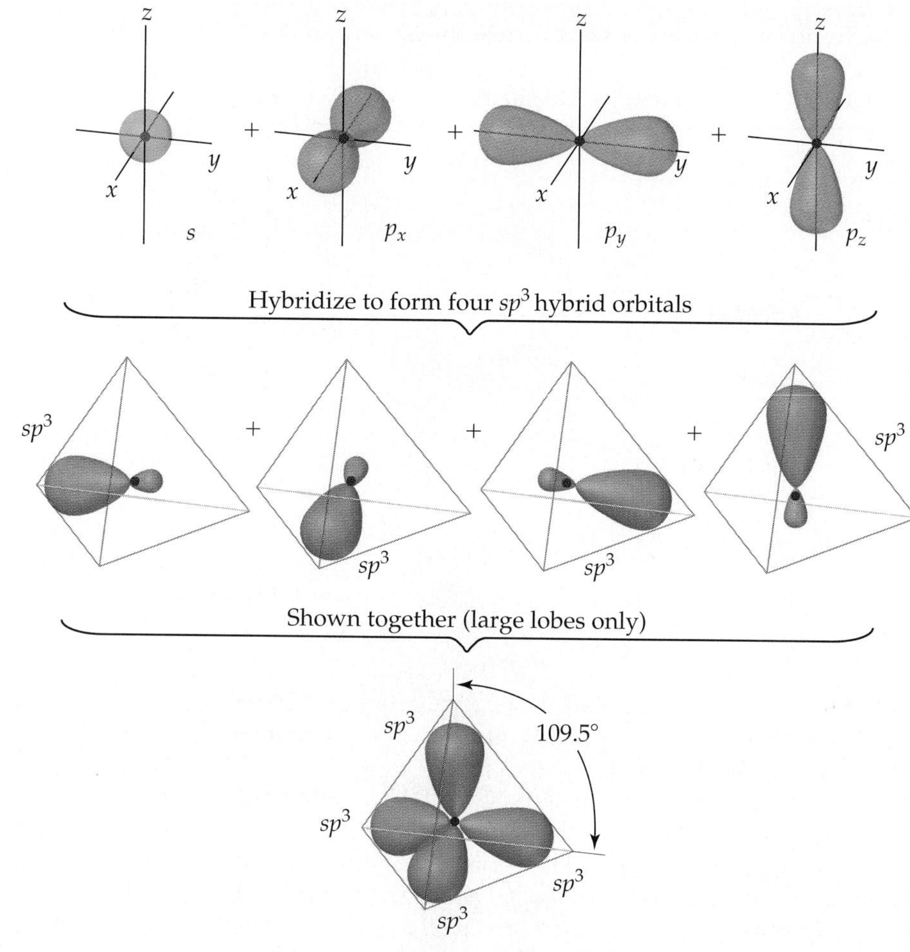

▲ **Figure 9.19** Formation of four sp^3 hybrid orbitals from a set of one *s* orbital and three *p* orbitals.

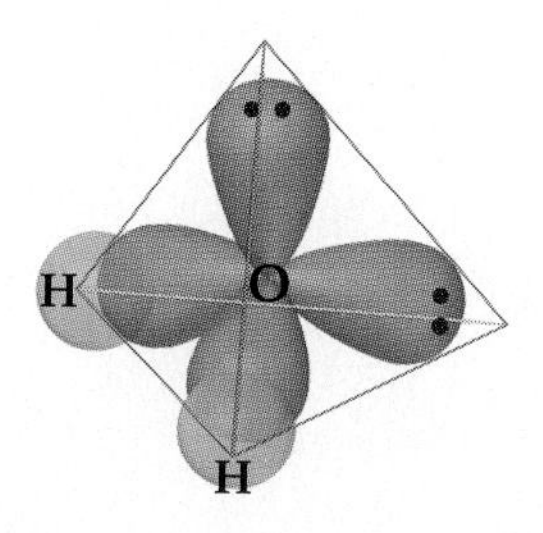

▲ **Figure 9.20** The bonding in H_2O can be envisioned as sp^3 hybridization of the orbitals on O. Two of the four hybrid orbitals overlap with 1*s* orbitals of H to form covalent bonds. The other two hybrid orbitals are occupied by nonbonding pairs of electrons.

ACTIVITY
Promotion of Electron and Hybridization of Orbitals III

ANIMATION
Hybridization

orbitals. These hybrid orbitals are directed toward the vertices of a trigonal bipyramid. The formation of sp^3d hybrids is exemplified by the phosphorus atom in PF_5:

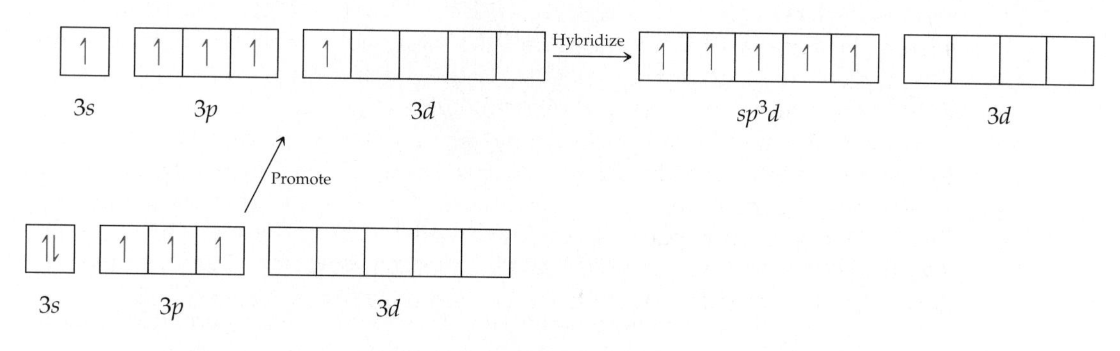

Similarly, mixing one *s* orbital, three *p* orbitals, and two *d* orbitals gives six sp^3d^2 hybrid orbitals, which are directed toward the vertices of an octahedron. The use of *d* orbitals in constructing hybrid orbitals nicely corresponds to the

TABLE 9.4 Geometric Arrangements Characteristic of Hybrid Orbital Sets

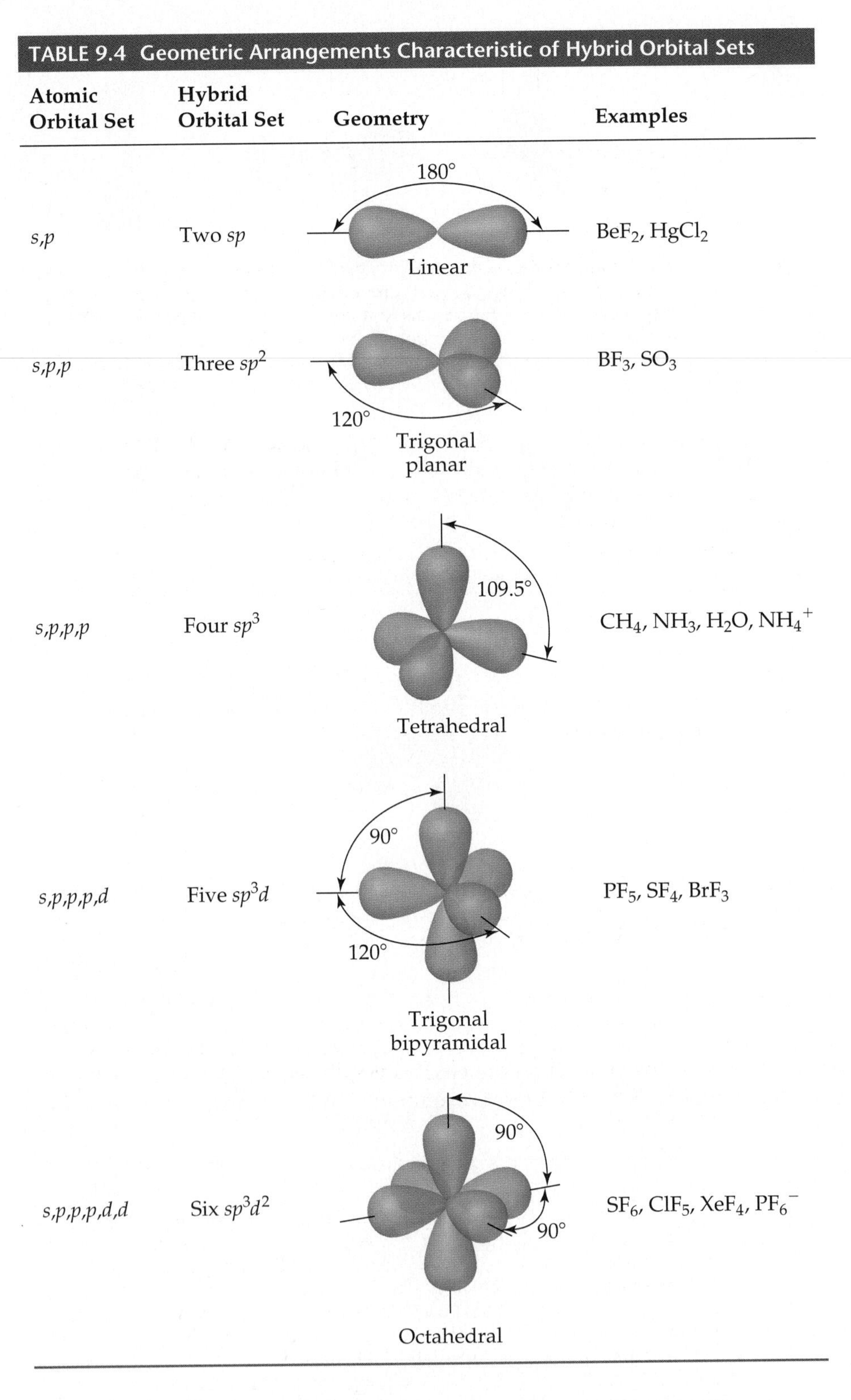

Atomic Orbital Set	Hybrid Orbital Set	Geometry	Examples
s,p	Two *sp*	Linear	BeF_2, $HgCl_2$
s,p,p	Three sp^2	Trigonal planar	BF_3, SO_3
s,p,p,p	Four sp^3	Tetrahedral	CH_4, NH_3, H_2O, NH_4^+
s,p,p,p,d	Five sp^3d	Trigonal bipyramidal	PF_5, SF_4, BrF_3
s,p,p,p,d,d	Six sp^3d^2	Octahedral	SF_6, ClF_5, XeF_4, PF_6^-

notion of an expanded valence shell. (Section 8.7) The geometric arrangements characteristic of hybrid orbitals are summarized in Table 9.4 ▲.

Summary

Hybrid orbitals provide a convenient model for using valence-bond theory to describe covalent bonds in molecules whose geometries conform to the electron-domain geometries predicted by the VSEPR model. The picture of hybrid orbitals

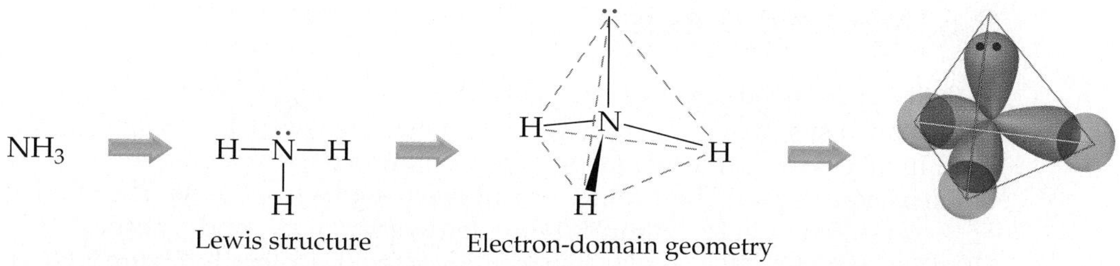

▲ **Figure 9.21** The hybrid orbitals used by N in the NH_3 molecule are predicted by first drawing the Lewis structure, then using the VSEPR model to determine the electron-domain geometry, and then specifying the hybrid orbitals that correspond to that geometry. This is essentially the same procedure as that used to determine molecular structure (Figure 9.6), except the final focus is on the orbitals used to make two-electron bonds and to hold nonbonding pairs.

has limited predictive value; that is, we cannot say in advance that the nitrogen atom in NH_3 uses sp^3 hybrid orbitals. When we know the molecular geometry, however, we can employ hybridization to describe the atomic orbitals used by the central atom in bonding.

The following steps allow us to predict the hybrid orbitals used by an atom in bonding:

1. Draw the *Lewis structure* for the molecule or ion.
2. Determine the electron-domain geometry using the *VSEPR model.*
3. Specify the *hybrid orbitals* needed to accommodate the electron pairs based on their geometric arrangement (Table 9.4).

These steps are illustrated in Figure 9.21 ▲, which shows how the hybridization employed by N in NH_3 is determined.

SAMPLE EXERCISE 9.5

Indicate the hybridization of orbitals employed by the central atom in each of the following: **(a)** NH_2^-; **(b)** SF_4 (see Sample Exercise 9.2).

Solution

Analyze and Plan: To determine the hybrid orbitals used by an atom in bonding, we must know its electron-domain geometry. Thus, we draw the Lewis structure to determine the number of electron domains around the central atom. The hybridization conforms to the number and geometry of electron domains around the central atom as predicted by the VSEPR model.

Solve: (a) The Lewis structure of NH_2^- is as follows:

$$\left[H{:}\ddot{\underset{\cdot\cdot}{N}}{:}H\right]^-$$

Because there are four electron domains around N, the electron-domain geometry is tetrahedral. The hybridization that gives a tetrahedral electron-domain geometry is sp^3 (Table 9.4). Two of the sp^3 hybrid orbitals contain nonbonding pairs of electrons, and the other two are used to make two-electron bonds with the hydrogen atoms.

(b) The Lewis structure and electron-domain geometry of SF_4 are shown in Sample Exercise 9.2. There are five electron domains around S, giving rise to a trigonal bipyramidal electron-domain geometry. With an expanded octet of 10 electrons, a *d* orbital on the sulfur must be used. The trigonal bipyramidal electron-domain geometry corresponds to sp^3d hybridization (Table 9.4). One of the hybrid orbitals that points in an equatorial direction contains a nonbonding pair of electrons; the other four are used to form the S—F bonds.

PRACTICE EXERCISE

Predict the electron-domain geometry and the hybridization of the central atom in **(a)** SO_3^{2-}; **(b)** SF_6.
Answers: **(a)** tetrahedral, sp^3; **(b)** octahedral, sp^3d^2

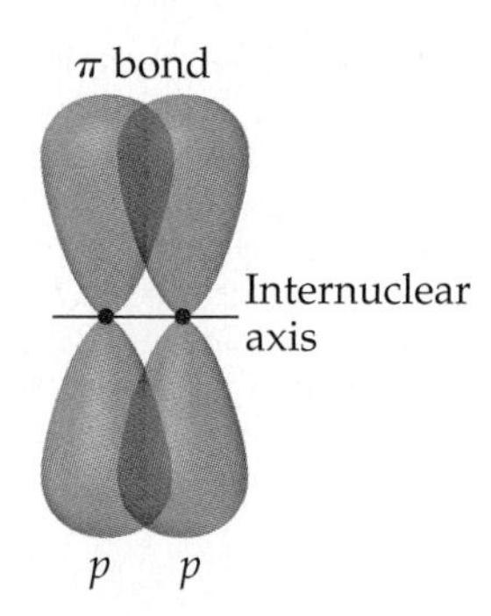

▲ **Figure 9.22** Formation of a π bond by overlap of two *p* orbitals. The two regions of overlap constitute one π bond.

9.6 Multiple Bonds

In the covalent bonds that we have considered thus far, the electron density is concentrated symmetrically about the line connecting the nuclei (the *internuclear axis*). In other words, the line joining the two nuclei passes through the middle of the overlap region. These bonds are all called **sigma (σ) bonds**. The overlap of two *s* orbitals as in H_2 [Figure 9.14(a)], the overlap of an *s* and a *p* orbital as in HCl [Figure 9.14(b)], the overlap between two *p* orbitals as in Cl_2 [Figure 9.14(c)], and the overlap of a *p* orbital with an *sp* hybrid orbital as in BeF_2 (Figure 9.17) are all examples of σ bonds.

To describe multiple bonding, we must consider a second kind of bond that results from the overlap between two *p* orbitals oriented perpendicularly to the internuclear axis (Figure 9.22 ◀). This *side-to-side overlap of* p *orbitals* produces a **pi (π) bond**. A π bond is a covalent bond in which the overlap regions lie above and below the internuclear axis. Unlike a σ bond, in a π bond there is no probability of finding the electron on the internuclear axis. Because the total overlap in the π bonds tends to be less than that in a σ bond, π bonds are generally weaker than σ bonds.

In almost all cases single bonds are σ bonds. A double bond consists of one σ bond and one π bond, and a triple bond consists of one σ bond and two π bonds:

H—H — One σ bond

$H_2C{=}CH_2$ — One σ bond plus one π bond

:N≡N: — One σ bond plus two π bonds

To see how these ideas are used, consider ethylene (C_2H_4), which possesses a C═C double bond. The bond angles in ethylene are all approximately 120° (Figure 9.23 ◀), suggesting that each carbon atom uses sp^2 hybrid orbitals (Figure 9.18) to form σ bonds with the other carbon and with two hydrogens. Because carbon has four valence electrons, after sp^2 hybridization one electron remains in the *unhybridized 2p* orbital:

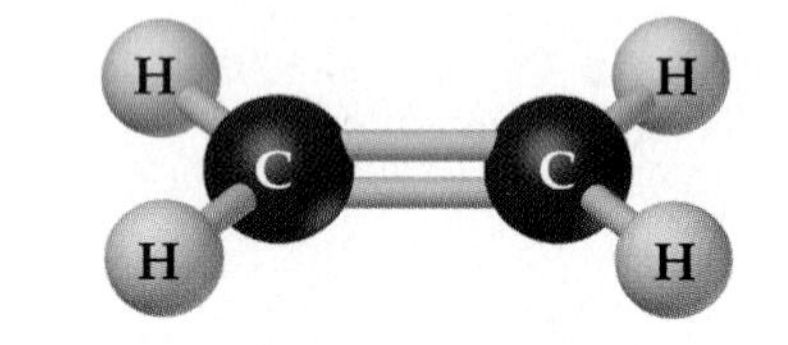

▲ **Figure 9.23** The molecular geometry of ethylene, C_2H_4.

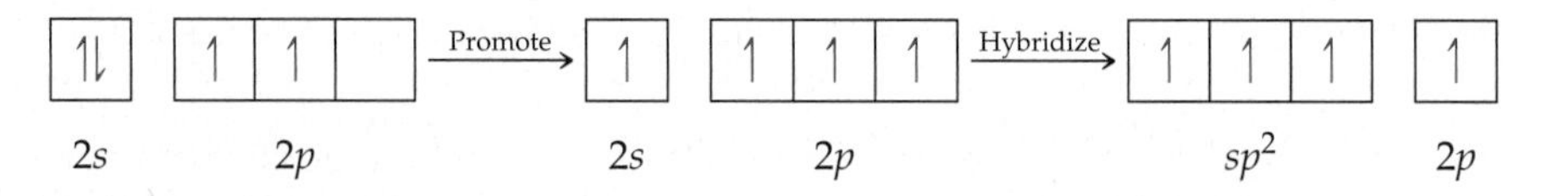

The unhybridized 2*p* orbital is directed perpendicular to the plane that contains the three sp^2 hybrid orbitals.

Each sp^2 hybrid orbital on a carbon atom contains one electron. Figure 9.24 ▼ shows how the four C—H σ bonds are formed by overlap of sp^2 hybrid orbitals on C with the 1*s* orbitals on each H atom. We use eight electrons to form these four electron-pair bonds. The C—C σ bond is formed by the overlap of two sp^2 hybrid orbitals, one on each carbon atom, and requires two more electrons. The C_2H_4 molecule has a total of 12 valence electrons, 10 of which form the five σ bonds.

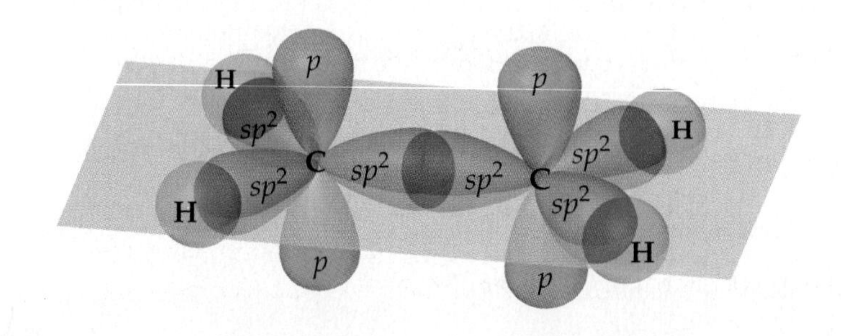

▶ **Figure 9.24** Hybridization of orbitals on carbon in ethylene. The σ bonding framework is formed from sp^2 hybrid orbitals on the carbon atoms. The unhybridized 2*p* orbitals on the C atoms are used to make a π bond.

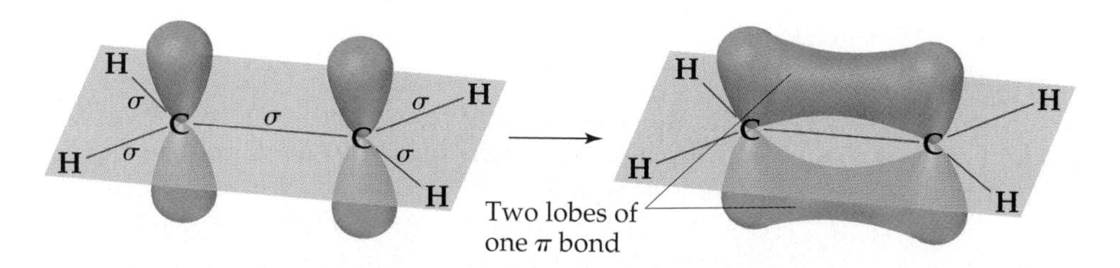

◀ **Figure 9.25** The π bond in ethylene is formed by overlap of the unhybridized 2*p* orbitals on each C atom. The electron density in the π bond is above and below the bond axis, whereas in the σ bonds the electron density lies directly along the bond axes. The two lobes constitute one π bond.

The remaining two valence electrons reside in the unhybridized 2*p* orbitals, one electron on each of the carbon atoms. These 2*p* orbitals can overlap with one another in a side-to-side fashion, as shown in Figure 9.25 ▲. The resultant electron density is concentrated above and below the C—C bond axis, so this is a π bond (Figure 9.22). Thus, the C=C double bond in ethylene consists of one σ bond and one π bond.

Although we cannot experimentally observe a π bond directly (all we can observe are the positions of the atoms), the structure of ethylene provides strong support for its presence. First, the C—C bond length in ethylene (1.34 Å) is much shorter than that in compounds with C—C single bonds (1.54 Å), consistent with the presence of a stronger C=C double bond. Second, all six atoms in C_2H_4 lie in the same plane. Only when the two CH_2 fragments lie in the same plane can the 2*p* orbitals that comprise the π bond achieve a good overlap. If the π bond were absent, there would be no reason for the two CH_2 fragments of ethylene to lie in the same plane. (There would be free rotation about the C—C bond axis.) Because π bonds require that portions of a molecule be planar, they can introduce rigidity into molecules.

Triple bonds can also be explained by using hybrid orbitals. Acetylene, (C_2H_2), for example, is a linear molecule containing a triple bond: H—C≡C—H. The linear geometry suggests that each carbon atom uses *sp* hybrid orbitals to form σ bonds with the other carbon and one hydrogen. Each carbon atom then has two remaining unhybridized 2*p* orbitals at right angles to each other and to the axis of the *sp* hybrid set (Figure 9.26 ▶). These *p* orbitals overlap to form a pair of π bonds. Thus, the triple bond in acetylene consists of one σ bond and two π bonds.

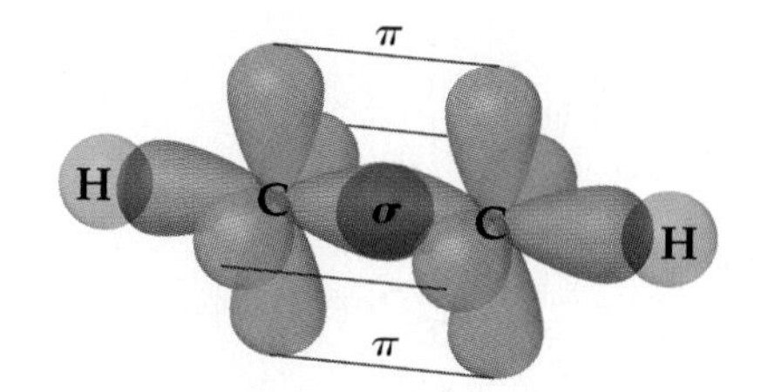

▲ **Figure 9.26** Formation of two π bonds in acetylene, C_2H_2, from the overlap of two sets of unhybridized carbon 2*p* orbitals.

Although it is possible to make π bonds from *d* orbitals, the only π bond we will consider is that formed by the overlap of *p* orbitals. This π bond can form only if unhybridized *p* orbitals are present on the bonded atoms. Therefore, only atoms having *sp* or sp^2 hybridization can be involved in such π bonding. Further, double and triple bonds (and hence π bonds) are more common in molecules with small atoms, especially C, N, and O. Larger atoms, such as S, P, and Si, form π bonds less readily.

SAMPLE EXERCISE 9.6

Formaldehyde has the following Lewis structure:

```
H
 \
  C=Ö:
 /
H
```

Describe how the bonds in formaldehyde are formed in terms of overlaps of appropriate hybridized and unhybridized orbitals.

Solution

Analyze: We are asked to describe the bonding in formaldehyde in terms of orbital overlaps.

Plan: Single bonds will be of the sigma type, whereas double bonds will consist of a σ and a π bond. The ways in which these bonds form can be deduced from the geometry of the molecule, which we predict using the VSEPR model.

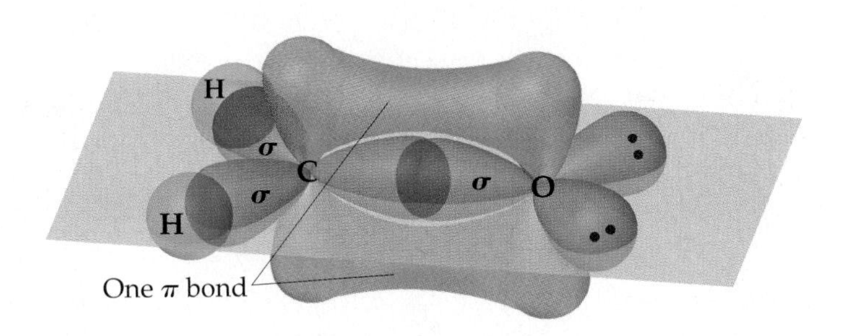

▶ **Figure 9.27** Formation of σ and π bonds in formaldehyde, H_2CO.

Solve: The C atom has three electron domains around it, which suggests a trigonal-planar geometry with bond angles of about 120°. This geometry implies sp^2 hybrid orbitals on C (Table 9.4). These hybrids are used to make the two C—H and one C—O σ bonds to C. There remains an unhybridized $2p$ orbital on carbon, perpendicular to the plane of the three sp^2 hybrids.

The O atom also has three electron domains around it, so we will assume that it has sp^2 hybridization as well. One of these hybrids participates in the C—O σ bond while the other two hybrids hold the two nonbonding electron pairs of the O atom. Like the C atom, therefore, the O atom has an unhybridized $2p$ orbital that is perpendicular to the plane of the molecule. The unhybridized $2p$ orbitals on the C and O atoms overlap to form a C—O π bond, as illustrated in Figure 9.27 ▲.

PRACTICE EXERCISE

Consider the acetonitrile molecule:

```
     H
     |
H—C—C≡N:
     |
     H
```

(a) Predict the bond angles around each carbon atom; **(b)** give the hybridization at each of the carbon atoms; **(c)** determine the total number of σ and π bonds in the molecule.
Answers: **(a)** approximately 109° around the left C and 180° on the right C; **(b)** sp^3, sp; **(c)** five σ bonds and two π bonds

Delocalized π Bonding

In each of the molecules we have discussed thus far in this section, the bonding electrons are *localized*. By this we mean that the σ and π electrons are associated totally with the two atoms that form the bond. In many molecules, however, we cannot adequately describe the bonding as being entirely localized. This situation arises particularly in molecules that have two or more resonance structures involving π bonds.

A molecule that cannot be described with localized π bonds is benzene (C_6H_6), which has the following two resonance structures: (Section 8.6)

⬡ ⟷ ⬡ or ⏣

To describe the bonding in benzene using hybrid orbitals, we first choose a hybridization scheme consistent with the geometry of the molecule. Because each carbon is surrounded by three atoms at 120° angles, the appropriate hybrid set is sp^2. Six localized C—C σ bonds and six localized C—H σ bonds are formed from the sp^2 hybrid orbitals, as shown in Figure 9.28(a) ▶. This leaves a $2p$ orbital on each carbon that is oriented perpendicularly to the plane of the molecule. The situation is very much like that in ethylene, except we now have six carbon $2p$ orbitals arranged in a ring [Figure 9.28(b)]. Each of these p orbitals contributes one electron to the π bonding.

A representation that reflects *both* resonance structures has the six π electrons spread out around the entire ring, as shown in Figure 9.28(c). This figure corresponds to the "circle in a hexagon" drawing that we often use to represent benzene. This model leads to the description of each carbon–carbon bond as having identical

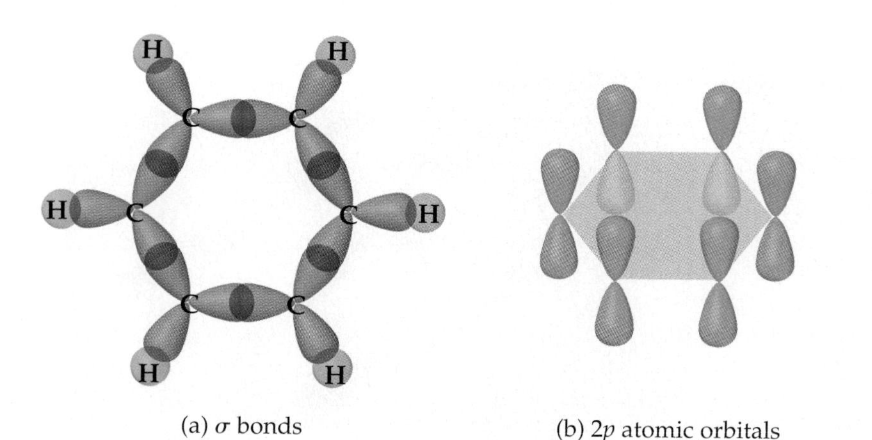

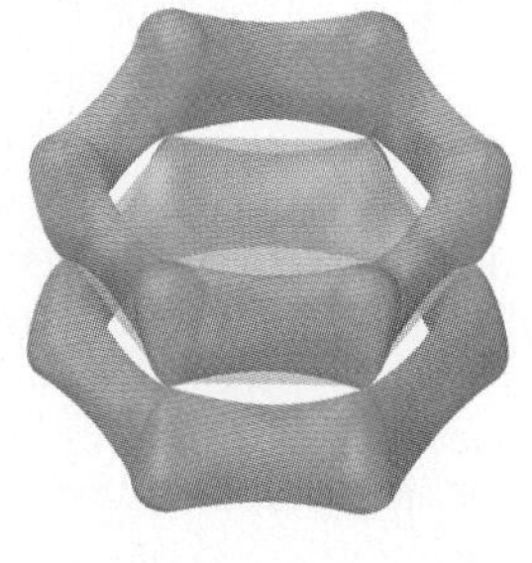

(a) σ bonds (b) $2p$ atomic orbitals (c) Delocalized π bonds

▲ **Figure 9.28** The σ and π bond networks in benzene, C_6H_6. (a) The C—C and C—H σ bonds all lie in the plane of the molecule and are formed by using carbon sp^2 hybrid orbitals. (b) Each carbon atom has an unhybridized $2p$ orbital that lies perpendicular to the molecular plane. (c) The six $2p$ orbitals overlap, forming a π orbital that is smeared out, or delocalized, producing a double-doughnut–shaped electron cloud above and below the plane of the molecule.

bond lengths that are between that of a C—C single bond (1.54 Å) and a C=C double bond (1.34 Å), consistent with the observed bond lengths (1.40 Å).

Because we cannot describe the π bonds in benzene as individual electron-pair bonds between neighboring atoms, we say that the π bonds are **delocalized** among the six carbon atoms. Delocalization of the electrons in its π bonds gives benzene a special stability, as will be discussed in Section 25.4. Delocalization of π bonds is also responsible for the color of many organic molecules. (See the "Chemistry at Work" box on organic dyes at the end of this chapter.) If you take a course in organic chemistry, you will see many examples of how electron delocalization influences the properties of organic molecules.

SAMPLE EXERCISE 9.7

Describe the delocalized π bonding in the nitrate ion, NO_3^-.

Solution

Analyze and Plan: Our first step in describing the bonding in NO_3^- is to construct appropriate Lewis structures. If the resonance structures involve the placement of the double bonds in different locations, that suggests that the π component of the double bonds is actually delocalized in a way suggested by the resonance structures.

Solve: In Section 8.6 we saw that NO_3^- has the following three resonance structures:

In each of these structures the electron-domain geometry at nitrogen is trigonal planar, which implies sp^2 hybridization of the N atom. The sp^2 hybrid orbitals are used to construct the three N—O σ bonds that are present in each of the resonance structures.

The unhybridized $2p$ orbital on the N atom can be used to make π bonds. For any one of the three resonance structures shown, we might imagine the formation of a single localized N—O π bond, formed between the unhybridized $2p$ orbital on N and a $2p$ orbital on one of the O atoms. Because each of the three resonance structures contributes equally to the observed structure of NO_3^-, however, we represent the π bonding as spread out, or delocalized, over the three N—O bonds, as shown in Figure 9.29 ▶.

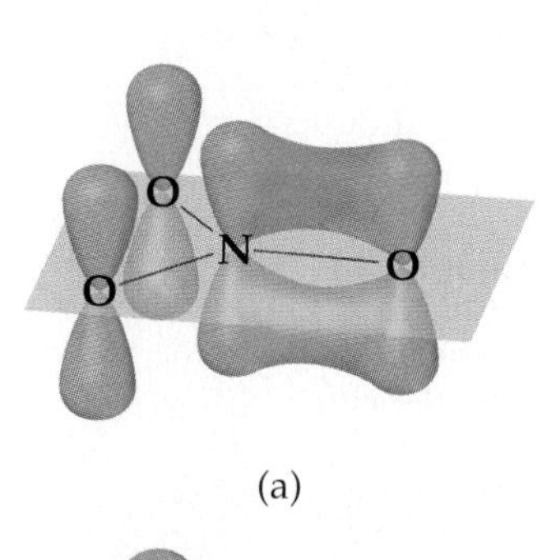

(a)

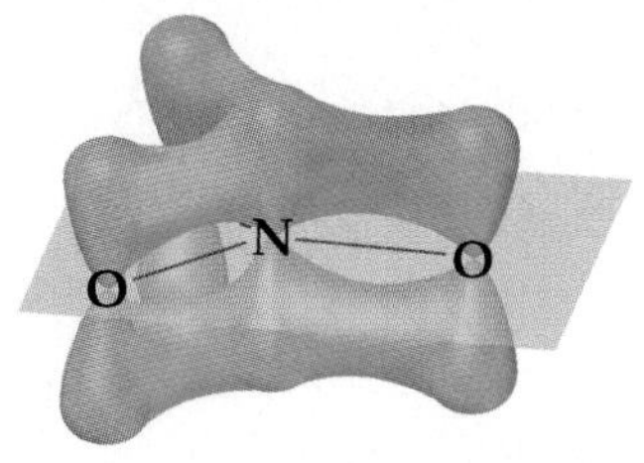

(b)

▲ **Figure 9.29** (a) N—O π bond in one of the resonance structures of NO_3^-. (b) Delocalization of the π bonds in the NO_3^- ion.

PRACTICE EXERCISE

Which of the following molecules or ions will exhibit delocalized bonding? SO_3, SO_3^{2-}, H_2CO, O_3, NH_4^+.

Answer: SO_3 and O_3, as indicated by the presence of two or more resonance structures involving π bonding for each of these molecules

Chemistry and Life The Chemistry of Vision

In recent years scientists have begun to understand the complex chemistry of vision. Vision begins when light is focused by the lens onto the retina, the layer of cells lining the interior of the eyeball. The retina contains *photoreceptor* cells known as rods and cones (Figure 9.30 ▶). The human retina contains about 3 million cones and 100 million rods. The rods are sensitive to dim light and are used in night vision. The cones are sensitive to colors. The tops of the rods and cones contain a molecule called *rhodopsin*. Rhodopsin consists of a protein, called *opsin*, bonded to a reddish purple pigment called *retinal*. Structural changes around a double bond in the retinal portion of the molecule trigger a series of chemical reactions that result in vision.

Double bonds between atoms are stronger than single bonds between the same atoms (Table 8.4). For example, a C═C double bond is stronger [$D(\text{C}{=}\text{C}) = 614$ kJ/mol] than a C—C single bond [$D(\text{C}{-}\text{C}) = 348$ kJ/mol], though not twice as strong. Our recent discussions now allow us to appreciate another aspect of double bonds: the stiffness or rigidity that they introduce into molecules.

Imagine taking the $-CH_2$ group of the ethylene molecule and rotating it relative to the other $-CH_2$ group as shown in Figure 9.31 ▼. This rotation destroys the overlap of *p* orbitals, breaking the π bond, a process that requires considerable energy. Thus, the presence of a double bond restricts the rotation of the bonds in a molecule. In contrast, molecules can rotate almost freely around the bond axis in single (σ) bonds because this motion has no effect on orbital overlap. This rotation allows molecules with single bonds to twist and fold almost as if their atoms were attached by hinges.

▲ **Figure 9.30** A color-enhanced scanning electron micrograph of the rods (yellow) and cones (blue) in the retina of the eye.

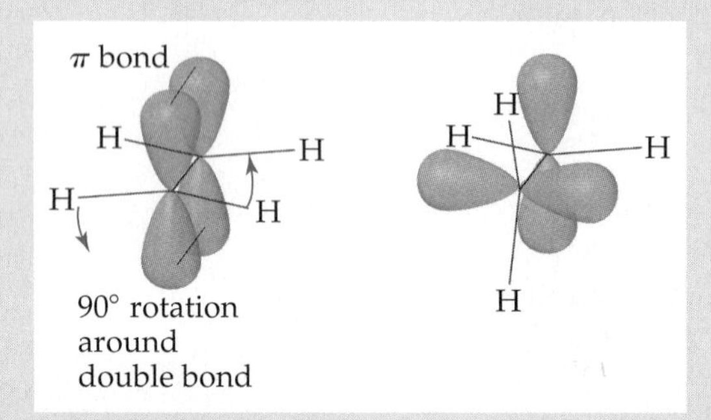

▲ **Figure 9.31** Rotation about a carbon–carbon double bond in ethylene. The overlap of the *p* orbitals that form the π bond is lost in the rotation. For this reason, rotation about double bonds does not occur readily.

Our vision depends on the rigidity of double bonds in retinal. In its normal form, retinal is held rigid by its double bonds, as shown on the left in Figure 9.32 ▼. Light entering the eye is absorbed by rhodopsin, and the energy is used to break the π-bond portion of the indicated double bond. The molecule then rotates around this bond, changing its geometry. The retinal then separates from the opsin, triggering the reactions that produce a nerve impulse that the brain interprets as the sensation of vision. It takes as few as five closely spaced molecules reacting in this fashion to produce the sensation of vision. Thus, only five photons of light are necessary to stimulate the eye.

The retinal slowly reverts to its original form and reattaches to the opsin. The slowness of this process helps explain why intense bright light causes temporary blindness. The light causes all the retinal to separate from opsin, leaving no further molecules to absorb light.

▲ **Figure 9.32** When rhodopsin absorbs visible light, the π component of the double bond shown in red breaks, allowing rotation that produces a change in molecular geometry.

General Conclusions

On the basis of the examples we've seen, we can draw a few helpful conclusions for using the concept of hybrid orbitals to describe molecular structures:

1. Every pair of bonded atoms shares one or more pairs of electrons. In every bond at least one pair of electrons is localized in the space between the atoms in a σ bond. The appropriate set of hybrid orbitals used to form the σ bonds between an atom and its neighbors is determined by the observed geometry of the molecule. The correlation between the set of hybrid orbitals and the geometry about an atom is given in Table 9.4.
2. The electrons in σ bonds are localized in the region between two bonded atoms and do not make a significant contribution to the bonding between any other two atoms.
3. When atoms share more than one pair of electrons, the additional pairs are in π bonds. The centers of charge density in a π bond lie above and below the bond axis.
4. Molecules with two or more resonance structures can have π bonds that extend over more than two bonded atoms. Electrons in π bonds that extend over more than two atoms are delocalized.

9.7 Molecular Orbitals

Valence-bond theory and hybrid orbitals allow us to move in a straightforward way from Lewis structures to rationalizing the observed geometries of molecules in terms of atomic orbitals. For example, we can use this theory to understand why methane has the formula CH_4, how the carbon and hydrogen atomic orbitals are used to form electron-pair bonds, and why the arrangement of the C—H bonds about the central carbon is tetrahedral. This model, however, does not explain all aspects of bonding. It is not successful, for example, in describing the excited states of molecules, which we must understand in order to explain how molecules absorb light, giving them color.

Some aspects of bonding are better explained by another model called **molecular orbital theory**. In Chapter 6 we saw that electrons in atoms can be described by certain wave functions, which we call atomic orbitals. In a similar way molecular orbital theory describes the electrons in molecules by using specific wave functions called **molecular orbitals**. Chemists use the abbreviation **MO** for molecular orbital.

Molecular orbitals have many of the same characteristics as atomic orbitals. For example, an MO can hold a maximum of two electrons (with opposite spins), it has a definite energy, and we can visualize its electron-density distribution by using a contour representation, as we did when we discussed atomic orbitals. Unlike atomic orbitals, however, MOs are associated with the entire molecule, not with a single atom.

The Hydrogen Molecule

To get a sense of the approach taken in MO theory, we will begin with the simplest molecule: the hydrogen molecule, H_2. We will use the two 1*s* atomic orbitals (one on each H atom) to "build" molecular orbitals for the H_2 molecule. *Whenever two atomic orbitals overlap, two molecular orbitals form.* Thus, the overlap of the 1*s* orbitals of two hydrogen atoms to form H_2 produces two MOs (Figure 9.33 ▶).

The lower-energy MO of H_2 concentrates electron density between the two hydrogen nuclei and is called the **bonding molecular orbital**. This sausage-shaped MO results from summing the two atomic orbitals so that the atomic orbital wave functions enhance each other in the bond region. Because an electron in this MO is strongly attracted to both nuclei, the electron is more stable

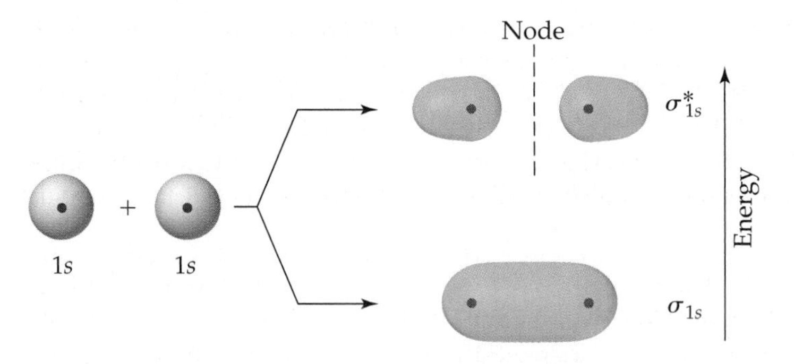

▶ **Figure 9.33** The combination of two H 1*s* atomic orbitals forms two molecular orbitals (MOs) of H_2. In the bonding MO, σ_{1s}, the atomic orbitals combine constructively, leading to a buildup of electron density between the nuclei. In the antibonding MO, σ^*_{1s}, the orbitals combine destructively in the bonding region. Note that the σ^*_{1s} MO has a node between the two nuclei.

(at lower energy) than it is in the 1*s* orbital of an isolated hydrogen atom. Because it concentrates electron density between the nuclei, the bonding MO holds the atoms together in a covalent bond.

The higher-energy MO in Figure 9.33 has very little electron density between the nuclei and is called the **antibonding molecular orbital**. Instead of enhancing each other in the region between the nuclei, the atomic orbitals cancel each other in this region, and the greatest electron density is on opposite sides of the nuclei. Thus, this MO excludes electrons from the very region in which a bond must be formed. An electron in this MO is actually repelled from the bonding region and is therefore less stable (at higher energy) than it is in the 1*s* orbital of a hydrogen atom.

The electron density in both the bonding and the antibonding molecular orbitals of H_2 is centered about the internuclear axis, an imaginary line passing through the two nuclei. MOs of this type are called **sigma (σ) molecular orbitals**. The bonding sigma molecular orbital of H_2 is labeled σ_{1s}, the subscript indicating that the MO is formed from two 1*s* orbitals. The antibonding sigma molecular orbital of H_2 is labeled σ^*_{1s} (read "sigma-star-one-s"), the asterisk denoting that the MO is antibonding.

The interaction between two 1*s* atomic orbitals and the molecular orbitals that result can be represented by an **energy-level diagram** (also called a **molecular orbital diagram**), like those in Figure 9.34 ▼. Such diagrams show the interacting atomic orbitals in the left and right columns and the MOs in the middle column. Note that the bonding molecular orbital, σ_{1s}, is lower in energy than the atomic 1*s* orbitals, whereas the antibonding orbital, σ^*_{1s}, is higher in energy than the 1*s* orbitals. Like atomic orbitals, each MO can accommodate two electrons with their spins paired (Pauli exclusion principle). (Section 6.7)

The molecular orbital diagram of the H_2 molecule is shown in Figure 9.34(a). Each H atom has one electron, so there are two electrons in H_2. These two electrons occupy the lower-energy bonding (σ_{1s}) MO with their spins paired. Electrons occupying a bonding molecular orbital are called *bonding electrons*. Because the σ_{1s} orbital is lower in energy than the isolated 1*s* orbitals, the H_2 molecule is more stable than the two separate H atoms.

▼ **Figure 9.34** Energy-level diagram for (a) the H_2 molecule and (b) the hypothetical He_2 molecule.

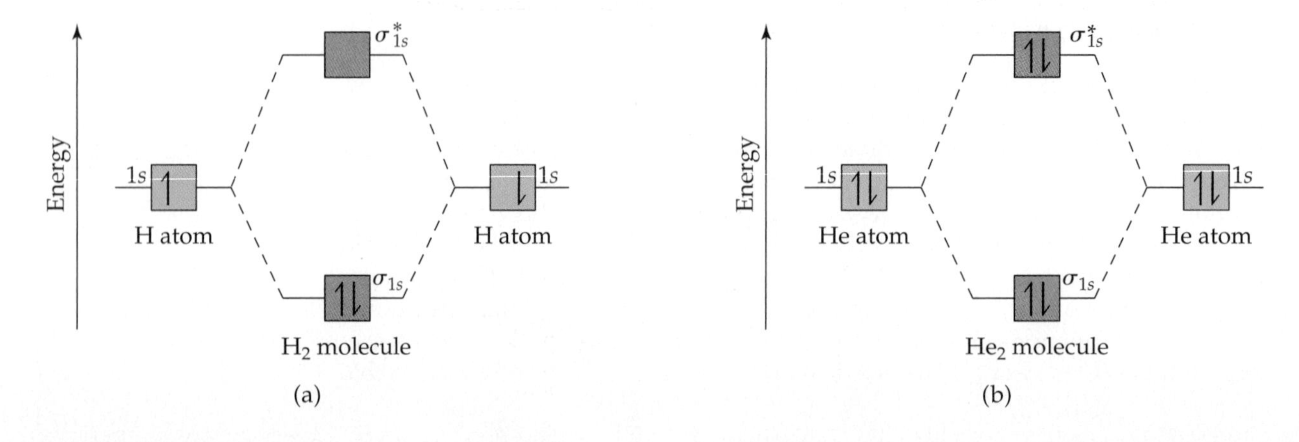

In contrast, the hypothetical He_2 molecule requires four electrons to fill its MO, as in Figure 9.34(b). Because only two electrons can be put in the σ_{1s} orbital, the other two must be placed in the σ^*_{1s} orbital. The energy decrease from the two electrons in the bonding MO is offset by the energy increase from the two electrons in the antibonding MO.* Hence, He_2 is an unstable molecule. Molecular orbital theory correctly predicts that hydrogen forms diatomic molecules but helium does not.

Bond Order

In molecular orbital theory the stability of a covalent bond is related to its **bond order**, defined as follows:

$$\text{Bond order} = \tfrac{1}{2}(\text{no. of bonding electrons} - \text{no. of antibonding electrons})$$

That is, the bond order is half the difference between the number of bonding electrons and the number of antibonding electrons. We take half the difference because we are used to thinking of bonds as pairs of electrons. *A bond order of 1 represents a single bond, a bond order of 2 represents a double bond, and a bond order of 3 represents a triple bond.* Because MO theory also treats molecules with an odd number of electrons, bond orders of 1/2, 3/2, or 5/2 are possible.

Because H_2 has two bonding and zero antibonding electrons [Figure 9.34(a)], it has a bond order of 1. Because He_2 has two bonding and two antibonding electrons [Figure 9.34(b)], it has a bond order of 0. A bond order of 0 means that no bond exists.

SAMPLE EXERCISE 9.8

What is the bond order of the He_2^+ ion? Would you expect this ion to be stable relative to the separated He atom and He^+ ion?

Solution

Analyze and Plan: To determine the bond order, we must determine the number of electrons in the molecule and how these electrons populate the available MOs orbitals. The valence electrons of He are in the 1*s* orbital. Hence, the 1*s* orbitals combine to give an MO diagram like that for H_2.

Solve: The energy-level diagram for the He_2^+ ion is shown in Figure 9.35 ▼. This ion has a total of three electrons. Two are placed in the bonding orbital, the third in the antibonding orbital. Thus, the bond order is

$$\text{Bond order} = \tfrac{1}{2}(2 - 1) = \tfrac{1}{2}$$

Because the bond order is greater than 0, the He_2^+ molecular ion is predicted to be stable relative to the separated He and He^+. Formation of He_2^+ in the gas phase has been demonstrated in laboratory experiments.

PRACTICE EXERCISE

Determine the bond order of the H_2^- ion.
Answer: $\frac{1}{2}$

* In fact, antibonding MOs are slightly more unfavorable than bonding MOs are favorable. Thus, whenever there is an equal number of electrons in bonding and antibonding orbitals, the energy of the molecule is slightly higher than that for the isolated atoms and no bond is formed.

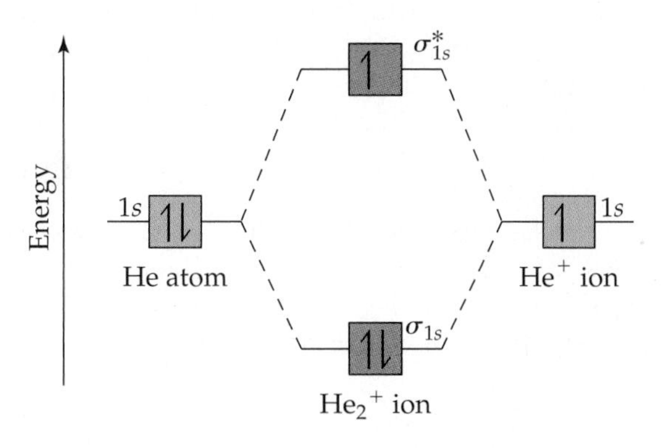

◀ **Figure 9.35** Energy-level diagram for the He_2^+ ion.

9.8 Second-Row Diatomic Molecules

Just as we treated the bonding in H_2 by using molecular orbital theory, we can consider the MO description of other diatomic molecules. Initially we will restrict our discussion to *homonuclear* diatomic molecules (those composed of two identical atoms) of elements in the second row of the periodic table. As we will see, the procedure for determining the distribution of electrons in these molecules closely follows the one we used for H_2.

Second-row atoms have valence 2*s* and 2*p* orbitals, and we need to consider how they interact to form MOs. The following rules summarize some of the guiding principles for the formation of MOs and for how they are populated by electrons:

1. The number of MOs formed equals the number of atomic orbitals combined.
2. Atomic orbitals combine most effectively with other atomic orbitals of similar energy.
3. The effectiveness with which two atomic orbitals combine is proportional to their overlap with one another; that is, as the overlap increases, the bonding MO is lowered in energy, and the antibonding MO is raised in energy.
4. Each MO can accommodate, at most, two electrons, with their spins paired (Pauli exclusion principle).
5. When MOs of the same energy are populated, one electron enters each orbital (with the same spin) before spin pairing occurs (Hund's rule).

Molecular Orbitals for Li_2 and Be_2

Lithium, the first element of the second period, has a $1s^2 2s^1$ electron configuration. When lithium metal is heated above its boiling point (1342°C), Li_2 molecules are found in the vapor phase. The Lewis structure for Li_2 indicates an Li—Li single bond. We will now use MOs to describe the bonding in Li_2.

Because the 1*s* and 2*s* orbitals of Li are so different in energy, we may assume that the 1*s* orbital on one Li atom interacts only with the 1*s* orbital on the other atom (rule 2). Likewise, the 2*s* orbitals interact only with each other. The resulting energy-level diagram is shown in Figure 9.36 ▶. Notice that combining four atomic orbitals produces four MOs (rule 1).

The 1*s* orbitals of Li combine to form σ_{1s} and σ^*_{1s} bonding and antibonding MOs, as they did for H_2. The 2*s* orbitals interact with one another in exactly the same way, producing bonding (σ_{2s}) and antibonding (σ^*_{2s}) MOs. Because the 2*s* orbitals of Li extend farther from the nucleus than the 1*s* orbitals do, the 2*s* orbitals overlap more effectively. As a result, the energy separation between the σ_{2s} and σ^*_{2s} orbitals is greater than that for the 1*s*-based MOs. The 1*s* orbitals of Li are so much lower in energy than the 2*s* orbitals, however, that the σ^*_{1s} antibonding MO is still well below the σ_{2s} bonding MO.

Each Li atom has three electrons, so six electrons must be placed in the MOs of Li_2. As shown in Figure 9.36, these occupy the σ_{1s}, σ^*_{1s}, and σ_{2s} MOs, each with two electrons. There are four electrons in bonding orbitals and two in antibonding orbitals, so the bond order equals 1. The molecule has a single bond, in accord with its Lewis structure.

Because both the σ_{1s} and σ^*_{1s} MOs of Li_2 are completely filled, the 1*s* orbitals contribute almost nothing to the bonding. The single bond in Li_2 is due essentially to the interaction of the valence 2*s* orbitals on the Li atoms. This example illustrates the general rule that *core electrons usually do not contribute significantly to bonding in molecule formation.* The rule is equivalent to using only the valence electrons when drawing Lewis structures. Thus, we need not consider further the 1*s* orbitals while discussing the other second-row diatomic molecules.

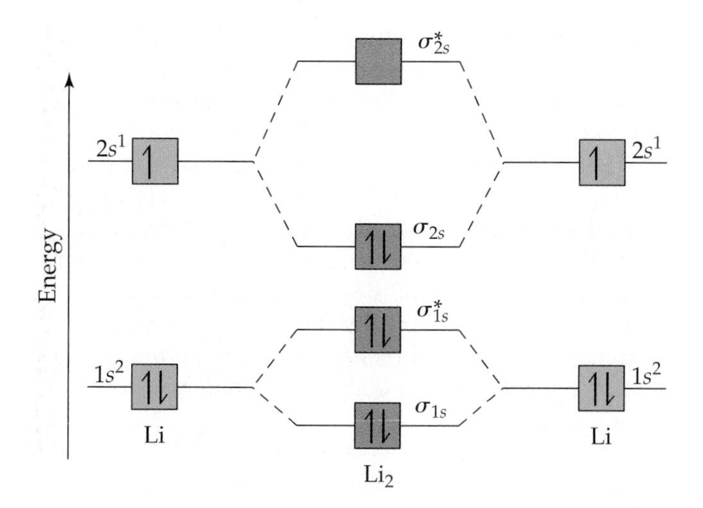

◀ **Figure 9.36** Energy-level diagram for the Li_2 molecule.

The MO description of Be_2 follows readily from the energy-level diagram for Li_2. Each Be atom has four electrons ($1s^22s^2$), so we must place eight electrons in molecular orbitals. Thus, we completely fill the σ_{1s}, $\sigma^*_{1s}\sigma_{2s}$, and σ^*_{2s} MOs. We have an equal number of bonding and antibonding electrons, so the bond order equals 0. Consistent with this analysis, Be_2 does not exist.

Molecular Orbitals from 2*p* Atomic Orbitals

Before we can consider the remaining second-row molecules, we must look at the MOs that result from combining 2*p* atomic orbitals. The interaction between *p* orbitals is shown in Figure 9.37 ▼, where we have arbitrarily chosen the

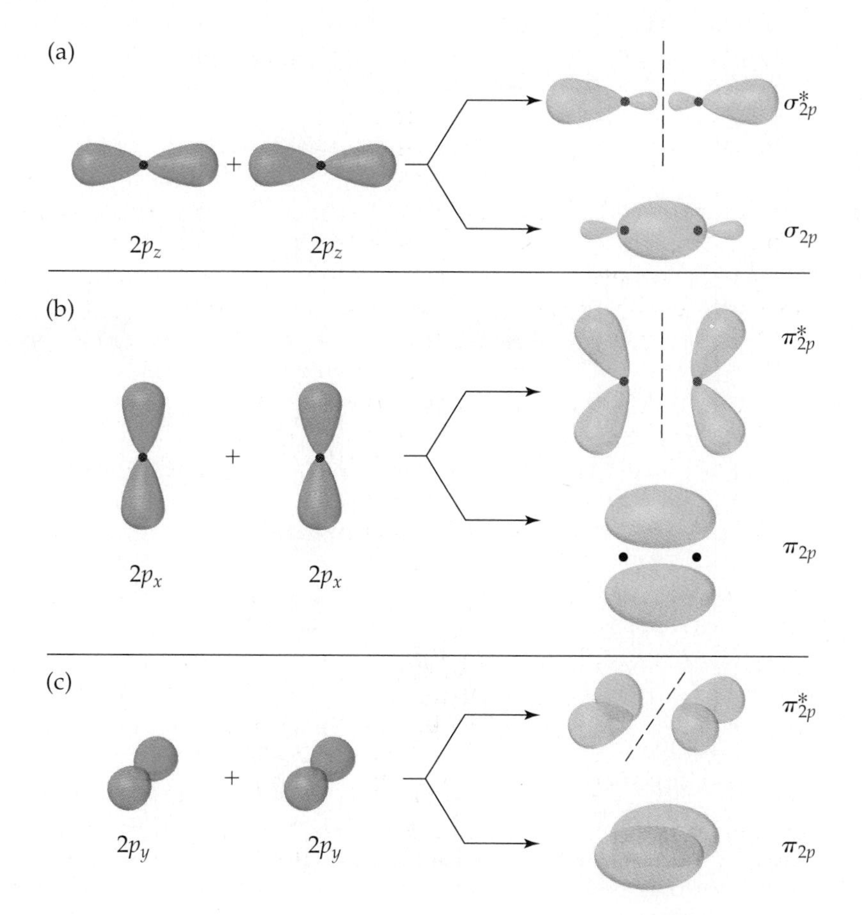

◀ **Figure 9.37** Contour representations of the molecular orbitals formed by the 2*p* orbitals on two atoms. Each time we combine two atomic orbitals, we obtain two MOs: one bonding and one antibonding. In (a) the *p* orbitals overlap "head to head" to form σ and σ^* MOs. In (b) and (c) they overlap "side-to-side" to form π and π^* MOs.

internuclear axis to be the z-axis. The $2p_z$ orbitals face each other in a "head-to-head" fashion. Just as we did for the s orbitals, we can combine the $2p_z$ orbitals in two ways. One combination concentrates electron density between the nuclei and is therefore a bonding molecular orbital. The other combination excludes electron density from the bonding region; it is an antibonding molecular orbital. In each of these MOs the electron density lies along the line through the nuclei, so they are σ molecular orbitals: σ_{2p} and σ^*_{2p}.

The other $2p$ orbitals overlap in a side-to-side fashion and thus concentrate electron density on opposite sides of the line through the nuclei. MOs of this type are called **pi (π) molecular orbitals**. We get one π bonding MO by combining the $2p_x$ atomic orbitals and another from the $2p_y$ atomic orbitals. These two π_{2p} molecular orbitals have the same energy; they are degenerate. Likewise, we get two degenerate π^*_{2p} antibonding MOs.

The $2p_z$ orbitals on two atoms point directly at one another. Hence, the overlap of two $2p_z$ orbitals is greater than that for two $2p_x$ or $2p_y$ orbitals. From rule 3 we therefore expect the σ_{2p} MO to be lower in energy (more stable) than the π_{2p} MOs. Similarly, the σ^*_{2p} MO should be higher in energy (less stable) than the π^*_{2p} MOs.

Electron Configurations for B_2 Through Ne_2

We have thus far independently considered the MOs that result from s orbitals (Figure 9.33) and from p orbitals (Figure 9.37). We can combine these results to construct an energy-level diagram (Figure 9.38 ▼) for homonuclear diatomic molecules of the elements boron through neon, all of which have valence $2s$ and $2p$ atomic orbitals. The following features of the diagram are notable:

1. The $2s$ atomic orbitals are lower in energy than the $2p$ atomic orbitals. (Section 6.7) Consequently, both of the molecular orbitals that result from the $2s$ orbitals, the bonding σ_{2s} and antibonding σ^*_{2s} are lower in energy than the lowest-energy MO that is derived from the $2p$ atomic orbitals.

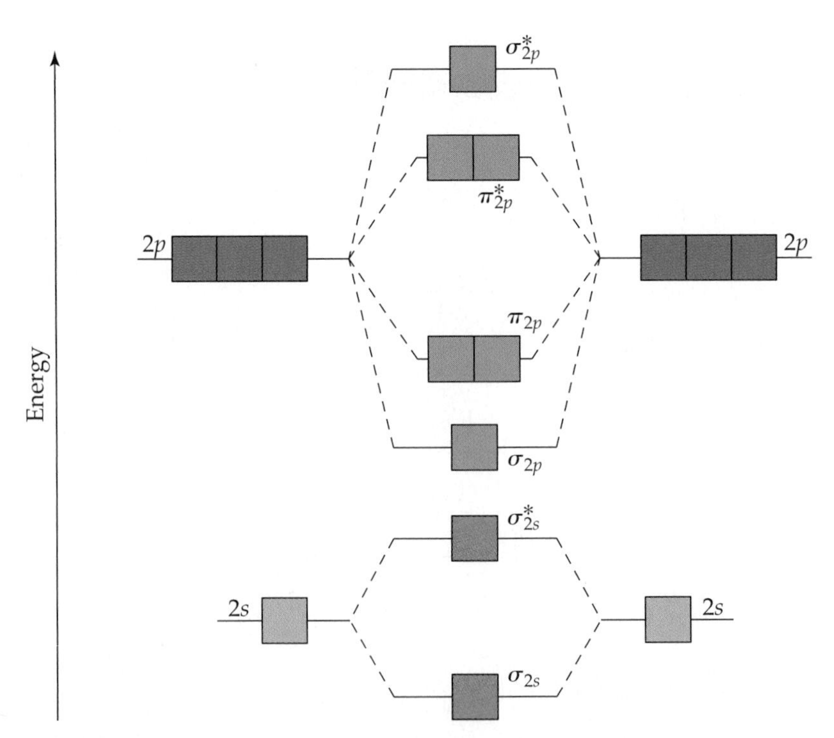

▲ **Figure 9.38** Energy-level diagram for MOs of second-row homonuclear diatomic molecules. The diagram assumes no interaction between the $2s$ atomic orbital on one atom and the $2p$ atomic orbitals on the other atom and experiment shows that it fits only for O_2, F_2, and Ne_2.

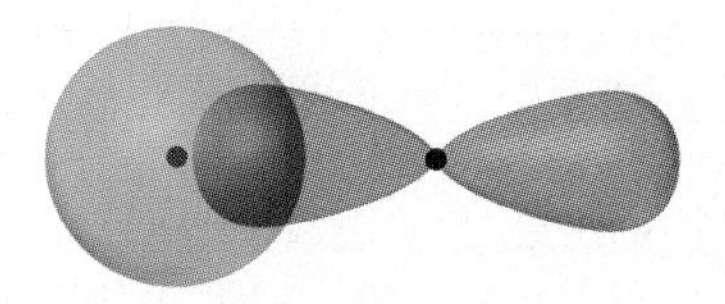

◀ **Figure 9.39** Overlap between a 2*s* orbital on one atom of a diatomic molecule and the $2p_z$ orbital on the other atom. These 2*s*-2*p* interactions can alter the energetic ordering of the MOs of the molecule.

2. The overlap of the two $2p_z$ orbitals is greater than that of the two $2p_x$ or $2p_y$ orbitals. As a result, the bonding σ_{2p} MO is lower in energy than the π_{2p} MOs, and the antibonding σ^*_{2p} MO is higher in energy than the π^*_{2p} MOs.
3. Both the π_{2p} and π^*_{2p} molecular orbitals are *doubly degenerate*; that is, there are two degenerate molecular orbitals of each type.

Before we can add electrons to the energy-level diagram in Figure 9.38, there is one more effect that we must consider. We have constructed the diagram by assuming that there is no interaction between the 2*s* orbital on one atom and the 2*p* orbitals on the other. In fact, such interactions can and do take place. Figure 9.39 ▲ shows the overlap of a 2*s* orbital on one of the atoms with a $2p_z$ orbital on the other. These interactions affect the energies of the σ_{2s} and σ_{2p} molecular orbitals in such a way that these MOs move further apart in energy, the σ_{2s} falling and the σ_{2p} rising in energy (Figure 9.40 ▼). These 2*s*-2*p* interactions are strong enough that the energetic ordering of the MOs can be altered: *For B_2, C_2, and N_2, the σ_{2p} MOs is above the π_{2p} MOs in energy. For O_2, F_2, and Ne_2, the σ_{2p} MO is below the π_{2p} MOs.*

Given the energy ordering of the molecular orbitals, it is a simple matter to determine the electron configurations for the second-row diatomic molecules B_2 through Ne_2. For example, a boron atom has three valence electrons. (Remember that we need to consider the inner-shell 1*s* electrons.) Thus, for B_2 we must place six electrons in MOs. Four of these fully occupy the σ_{2s} and σ^*_{2s} molecular

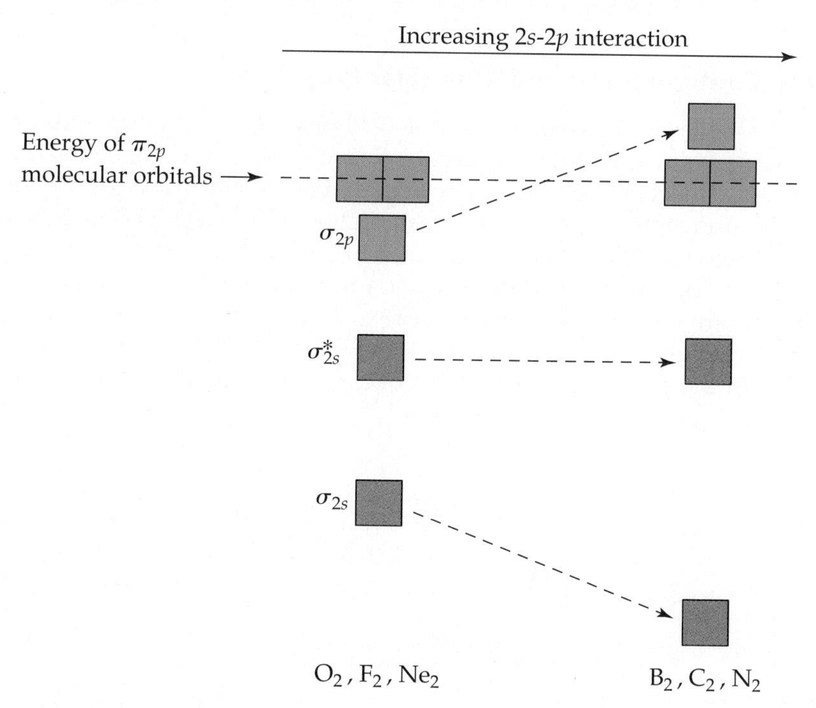

▲ **Figure 9.40** When the 2*s* and 2*p* orbitals interact, the σ_{2s} MO falls in energy and the σ_{2p} MO rises in energy. For O_2, F_2, and Ne_2, the interaction is small, and the σ_{2p} MO remains below the π_{2p} MOs, as in Figure 9.38. For B_2, C_2, and N_2, the 2*s*-2*p* interaction is great enough that the σ_{2p} MO rises above the π_{2p} MOs, as shown on the right.

	Large 2s-2p interaction				Small 2s-2p interaction		
	B_2	C_2	N_2		O_2	F_2	Ne_2
σ^*_{2p}	[ ]	[ ]	[ ]	σ^*_{2p}	[ ]	[ ]	[↑↓]
π^*_{2p}	[ ][ ]	[ ][ ]	[ ][ ]	π^*_{2p}	[↑][↑]	[↑↓][↑↓]	[↑↓][↑↓]
σ_{2p}	[ ]	[ ]	[↑↓]	π_{2p}	[↑↓][↑↓]	[↑↓][↑↓]	[↑↓][↑↓]
π_{2p}	[↑][↑]	[↑↓][↑↓]	[↑↓][↑↓]	σ_{2p}	[↑↓]	[↑↓]	[↑↓]
σ^*_{2s}	[↑↓]	[↑↓]	[↑↓]	σ^*_{2s}	[↑↓]	[↑↓]	[↑↓]
σ_{2s}	[↑↓]	[↑↓]	[↑↓]	σ_{2s}	[↑↓]	[↑↓]	[↑↓]
Bond order	1	2	3		2	1	0
Bond enthalpy (kJ/mol)	290	620	941		495	155	—
Bond length (Å)	1.59	1.31	1.10		1.21	1.43	—
Magnetic behavior	Paramagnetic	Diamagnetic	Diamagnetic		Paramagnetic	Diamagnetic	—

▲ **Figure 9.41** Molecular orbital electron configurations and some experimental data for several second-row diatomic molecules.

orbitals, leading to no net bonding. The last two electrons are put in the π_{2p} bonding MOs; one electron is put in each π_{2p} MO with the same spin. Therefore, B_2 has a bond order of 1. Each time we move to the right in the second row, two more electrons must be placed in the diagram. For example, on moving to C_2, we have two more electrons than in B_2, and these electrons are also placed in the π_{2p} MOs, completely filling them. The electron configurations and bond orders for the diatomic molecules B_2 through Ne_2 are given in Figure 9.41 ▲.

Electron Configurations and Molecular Properties

The behavior of a substance in a magnetic field provides an important insight into the arrangements of its electrons. Molecules with one or more unpaired electrons are attracted into a magnetic field. The more unpaired electrons in a species, the stronger the force of attraction. This type of magnetic behavior is called **paramagnetism**.

Substances with no unpaired electrons are weakly repelled from a magnetic field. This property is called **diamagnetism**. Diamagnetism is a much weaker effect than paramagnetism. A straightforward method for measuring the magnetic properties of a substance, illustrated in Figure 9.42 ▼, involves weighing the substance in the presence and absence of a magnetic field. If the substance is

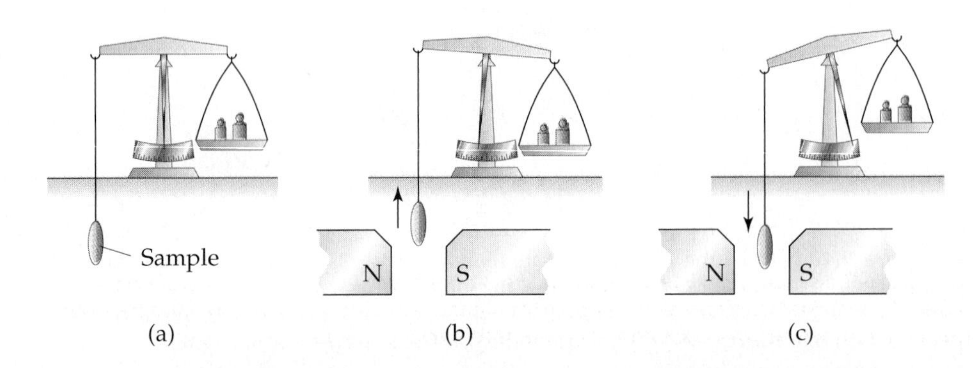

▶ **Figure 9.42** Experiment for determining the magnetic properties of a sample. (a) The sample is first weighed in the absence of a magnetic field. (b) When a field is applied, a diamagnetic sample tends to move out of the field and thus appears to have a lower mass. (c) A paramagnetic sample is drawn into the field and thus appears to gain mass. Paramagnetism is a much stronger effect than diamagnetism.

paramagnetic, it will appear to weigh more in the magnetic field; if it is diamagnetic, it will appear to weigh less. The magnetic behaviors observed for the diatomic molecules of the second-row elements agree with the electron configurations shown in Figure 9.41.

The electron configurations can also be related to the bond distances and bond enthalpies of the molecules. (Section 8.8) As bond orders increase, bond distances decrease and bond enthalpies increase. N_2, for example, whose bond order is 3, has a short bond distance and a large bond enthalpy. The N_2 molecule does not react readily with other substances to form nitrogen compounds. The high bond order of the molecule helps explain its exceptional stability. We should also note, however, that molecules with the same bond orders do *not* have the same bond distances and bond enthalpies. Bond order is only one factor influencing these properties. Other factors include the nuclear charges and the extent of orbital overlap.

Bonding in the dioxygen molecule, O_2, is especially interesting. Its Lewis structure shows a double bond and complete pairing of electrons:

$$\ddot{\underset{..}{O}}=\ddot{\underset{..}{O}}$$

The short O—O bond distance (1.21 Å) and the relatively high bond enthalpy (495 kJ/mol) are in agreement with the presence of a double bond. However, the molecule is found to contain two unpaired electrons. The paramagnetism of O_2 is demonstrated in Figure 9.43 ▼. Although the Lewis structure fails to account for the paramagnetism of O_2, molecular orbital theory correctly predicts that there are two unpaired electrons in the π_{2p}^* orbital of the molecule (Figure 9.41). The MO description also correctly indicates a bond order of 2.

Going from O_2 to F_2, we add two more electrons, completely filling the π_{2p}^* MOs. Thus, F_2 is expected to be diamagnetic and have a F—F single bond, in accord with its Lewis structure. Finally, the addition of two more electrons to make Ne_2 fills all the bonding and antibonding MOs; therefore, the bond order of Ne_2 is zero, and the molecule is not expected to exist.

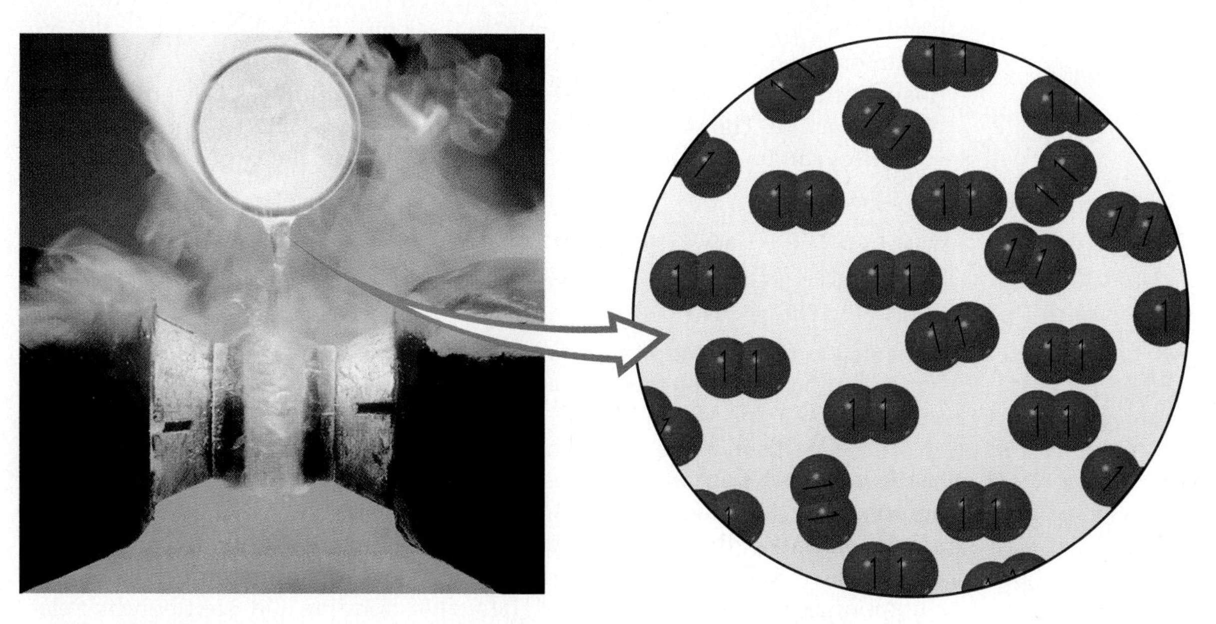

▲ **Figure 9.43** Liquid O_2 being poured between the poles of a magnet. Because each O_2 molecule contains two unpaired electrons, O_2 is paramagnetic. It is therefore attracted into the magnetic field and forms a bridge between the magnetic poles.

SAMPLE EXERCISE 9.9

Predict the following properties of O_2^+: **(a)** number of unpaired electrons; **(b)** bond order; **(c)** bond enthalpy and bond length.

Solution

Analyze and Plan: To determine the requested properties, we must determine the number of electrons in O_2^+ and then write its MO energy diagram. The unpaired electrons are those without a partner of opposite spin. The bond order is one half the difference between the number of bonding and antibonding electrons. After calculating the bond order, we can compare it with similar molecules in Figure 9.41 to estimate the bond enthalpy and bond length.

Solve: (a) The O_2^+ ion has 11 valence electrons, one less than O_2. The electron removed from O_2 to form O_2^+ is one of the two unpaired π^* electrons (see Figure 9.41). Therefore, O_2^+ has just one unpaired electron.

(b) The molecule has eight bonding electrons (the same as O_2) and three antibonding electrons (one less than O_2). Thus, its bond order is

$$\text{Bond order} = \tfrac{1}{2}(8 - 3) = 2\tfrac{1}{2}$$

(c) The bond order of O_2^+ is between that for O_2 (bond order 2) and N_2 (bond order 3). Thus, the bond enthalpy and bond length should be about midway between those for O_2 and N_2, approximately 720 kJ/mol and 1.15 Å, respectively. The observed bond enthalpy and bond length of the ion are 625 kJ/mol and 1.123 Å, respectively.

PRACTICE EXERCISE

Predict the magnetic properties and bond order of **(a)** the peroxide ion, O_2^{2-}; **(b)** the acetylide ion, C_2^{2-}.

Answers: **(a)** diamagnetic, 1;
(b) diamagnetic, 3

Heteronuclear Diatomic Molecules

The NO molecule is a heteronuclear diatomic molecule, one containing two different elements. It has been shown to control several important functions in our bodies. Our bodies use it, for example, to relax muscles, to kill foreign cells, and to reinforce memory. That it plays such important roles in human metabolism was unsuspected before 1987 because NO has an odd number of electrons and is highly reactive. The molecule has 11 valence electrons, and two possible Lewis structures can be drawn. The one with the lower formal charges places the odd electron on the N atom:

$$\dot{\text{N}}{=}\ddot{\text{O}} \longleftrightarrow \overset{-1}{\text{N}}{=}\overset{+1}{\text{O}}$$

Both structures indicate the presence of a double bond, but the experimental bond length (1.15 Å) suggests a higher bond order (Table 8.5). How do we treat NO using the MO model?

If the atoms in a heteronuclear diatomic molecule do not differ too greatly in their electronegativities, the description of their MOs will resemble those for homonuclear diatomics. The MO diagram for NO is shown in Figure 9.44 ▶. The atomic orbitals of the more electronegative O atom are slightly lower in energy than those of N. Nevertheless, the MO energy-level diagram is much like that of a homonuclear diatomic molecule. There are 8 bonding and 3 antibonding electrons, giving a bond order of $\tfrac{1}{2}(8 - 3) = 2\tfrac{1}{2}$, which agrees better with experiment than the Lewis structures do.

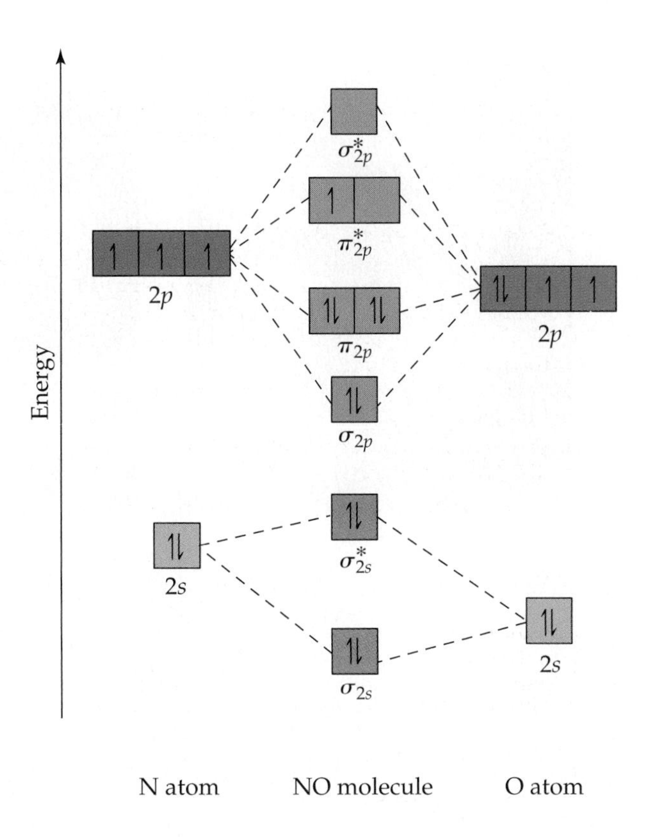

◀ **Figure 9.44** The MO energy diagram for NO.

Chemistry at Work Organic Dyes

The chemistry of color has fascinated people since ancient times. The brilliant colors around you—those of your clothes, the photographs in this book, the foods you eat—are due to the selective absorption of light by chemicals. Light excites electrons in molecules. In a molecular orbital picture, we can envision light exciting an electron from a filled molecular orbital to an empty one at higher energy. Because the MOs have definite energies, only light of the proper wavelengths can excite electrons. The situation is analogous to that of atomic line spectra. ∞ (Section 6.3) If the appropriate wavelength for exciting electrons is in the visible portion of the electromagnetic spectrum, the substance will appear colored: Certain wavelengths of white light are absorbed, others are not. A red traffic light appears red because only red light is transmitted through the lens. The other wavelengths of visible light are absorbed by it.

In using molecular orbital theory to discuss the absorptions of light by molecules, we can focus on two MOs in particular. The *highest occupied molecular orbital* (HOMO) is the MO of highest energy that has electrons in it. The *lowest unoccupied molecular orbital* (LUMO) is the MO of lowest energy that does not have electrons in it. In N_2, for example, the HOMO is the π_{2p} MO and the LUMO is the π^*_{2p} MO (Figure 9.41). The energy difference between the HOMO and the LUMO—known as the HOMO-LUMO gap—is related to the minimum energy needed to excite an electron in the molecule. Colorless or white substances usually have such a large HOMO-LUMO gap that visible light is not energetic enough to excite an electron to the higher level. The minimum energy needed to excite an electron in N_2 corresponds to light with a wavelength of less than 200 nm, which is far into the ultraviolet part of the spectrum (Figure 6.4). As a result, N_2 cannot absorb any visible light and is therefore colorless.

Many rich colors are due to *organic dyes*, organic molecules that strongly absorb selected wavelengths of visible light. Organic dyes are most familiar as the substances that are used to provide vibrant colors to fabrics. They are also used in color photographic film and in new high-tech applications such as the recordable compact discs called CD-R discs (Figure 9.45 ▶). In a CD-R disc a thin layer of a transparent colored organic dye is sandwiched between a reflective surface and a clear, rigid polymer backing. Data are "burned" onto the CD-R disc by use of a laser. When the laser strikes the dye, the dye molecules absorb the light, change structure, and become opaque. The selective production of these opaque "pits" in the CD-R disc gives it the capacity to store data in binary ("clear" or "opaque") form. Because the structure of the dye is irreversibly altered during the writing of data on the disc, data can be written only once to any given part of the disc.

Organic dyes contain extensively delocalized π electrons. The molecules contain atoms that are predominantly sp^2 hybridized, like the carbon atoms in benzene (Figure 9.28). This leaves one unhybridized p orbital on each atom to form π bonds

(continues on next page)

▶ **Figure 9.45** Organic dyes have a variety of useful applications, from the production of colorful fabrics (left) to the production of laser-recordable compact discs (CD-R discs) for computer data storage (right).

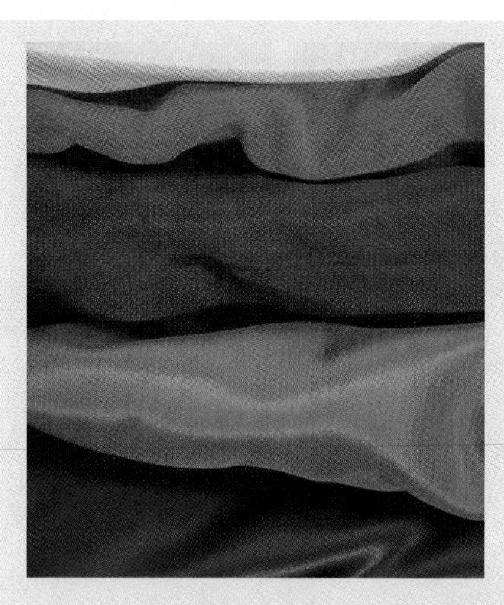

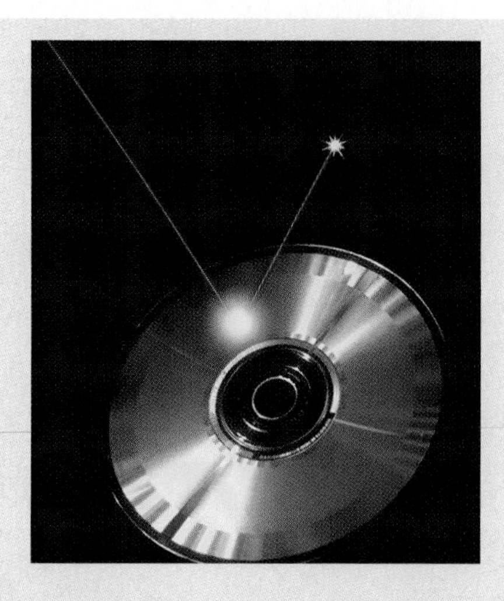

with neighboring atoms. The p orbitals are arranged so that electrons can be delocalized throughout the entire molecule; we say that the π bonds are *conjugated*. The HOMO-LUMO gap in such molecules decreases as the number of conjugated double bonds increases. Butadiene, (C_4H_6), for example, has alternating carbon–carbon double and single bonds:

or

The representation on the right is the shorthand notation that chemists use for organic molecules. There are implicit carbon atoms at the ends of the three straight segments, and there are implicitly sufficient hydrogen atoms to make a total of four bonds at each carbon. Butadiene is planar, so the unhybridized p orbitals on the carbons are pointing in the same direction. The π electrons are delocalized among the four carbon atoms, and the double bonds are said to be *conjugated*.

Because butadiene has only two conjugated double bonds, it still has a fairly large HOMO-LUMO gap. Butadiene absorbs light at 217 nm, still well into the ultraviolet part of the spectrum. It is therefore colorless. If we keep adding new conjugated double bonds, however, the HOMO-LUMO gap keeps shrinking until visible light is absorbed. β-Carotene, for example, is the substance chiefly responsible for the bright orange color of carrots.

Because β-carotene contains 11 conjugated double bonds, its π electrons are very extensively delocalized. It absorbs light of wavelength 500 nm, in the middle of the visible part of the spectrum. The human body converts β-carotene into vitamin A, which in turn is converted into retinal, a component of *rhodopsin*, found in the retinas of your eyes. (See the "Chemistry and Life" box in Section 9.6). The absorption of visible light by rhodopsin is a major reason why "visible" light is indeed visible. Thus, there seems to be a good basis for the maxim that eating carrots is good for your eyesight.

β-carotene

SAMPLE INTEGRATIVE EXERCISE 9: Putting Concepts Together

Elemental sulfur is a yellow solid that consists of S_8 molecules. The structure of the S_8 molecule is a puckered eight-membered ring (Figure 7.28). Heating elemental sulfur to high temperatures produces gaseous S_2 molecules:

$$S_8(s) \longrightarrow 4S_2(g)$$

(a) With respect to electronic structure, which element in the second row of the periodic table is most similar to sulfur? **(b)** Use the VSEPR model to predict the S—S—S bond angles in S_8 and the hybridization at S in S_8. **(c)** Use MO theory to predict the sulfur–sulfur bond order in S_2. Is the molecule expected to be diamagnetic or paramagnetic? **(d)** Use average bond enthalpies (Table 8.4) to estimate the enthalpy change for the reaction just described. Is the reaction exothermic or endothermic?

Solution **(a)** Sulfur is a group 6A element with a $[Ne]3s^2 3p^4$ electron configuration. It is expected to be most similar electronically to oxygen (electron configuration $[He]2s^2 2p^4$), which is immediately above it in the periodic table. (Chapter 7, Introduction) **(b)** The Lewis structure of S_8 is shown below.

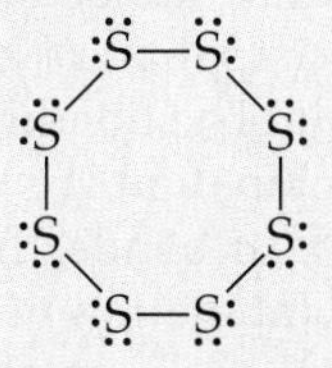

There is a single bond between each pair of sulfur atoms and two nonbonding electron pairs on each S atom. Thus, we see four electron domains around each sulfur atom, and we would expect a tetrahedral electron-domain geometry corresponding to sp^3 hybridization. (Sections 9.2, 9.5) Because of the nonbonding pairs, we would expect the S—S—S angles to be somewhat less than 109.5°, the tetrahedral angle. Experimentally, the S—S—S angle in S_8 is 108°, in good agreement with this prediction. Interestingly, if S_8 were a planar ring (like a stop sign), it would have S—S—S angles of 135°. Instead, the S_8 ring puckers to accommodate the smaller angles dictated by sp^3 hybridization. **(c)** The MOs of S_2 are entirely analogous to those of O_2, although the MOs for S_2 are constructed from the 3*s* and 3*p* atomic orbitals of sulfur. Further, S_2 has the same number of valence electrons as O_2. Thus, by analogy to our discussion of O_2, we would expect S_2 to have a bond order of 2 (a double bond) and to be paramagnetic with two unpaired electrons in the π^*_{3p} molecular orbitals of S_2. (Section 9.8) **(d)** We are considering the reaction in which an S_8 molecule falls apart into four S_2 molecules. From parts (b) and (c), we see that S_8 has S—S single bonds and S_2 has S=S double bonds. During the course of the reaction, therefore, we are breaking eight S—S single bonds and forming four S=S double bonds. We can estimate the enthalpy of the reaction by using Equation 8.12 and the average bond enthalpies in Table 8.4:

$$\Delta H_{rxn} = 8D(\text{S—S}) - 4D(\text{S=S}) = 8(266\text{ kJ}) - 4(418\text{ kJ}) = +456\text{ kJ}$$

Because $\Delta H_{rxn} > 0$, the reaction is endothermic. (Section 5.4) The very positive value of ΔH_{rxn} suggests that high temperatures are required to cause the reaction to occur.

Summary and Key Terms

Introduction and Section 9.1 The three-dimensional shapes and sizes of molecules are determined by their **bond angles** and bond lengths. Molecules with a central atom A surrounded by *n* atoms B, denoted AB_n, adopt a number of different geometric shapes, depending on the value of *n* and on the particular atoms involved. In the overwhelming majority of cases, these geometries are related to five basic shapes (linear, trigonal pyramidal, tetrahedral, trigonal bipyramidal, and octahedral).

Section 9.2 The **valence-shell electron-pair repulsion (VSEPR) model** rationalizes molecular geometries based on the repulsions between **electron domains**, which are regions about a central atom in which electrons are likely to be found. **Bonding pairs** of electrons, which are those involved in making bonds, and **nonbonding pairs** of electrons, also called lone pairs, both create electron domains around an atom. According to the VSEPR model, electron domains orient themselves to minimize electrostatic repulsions; that is, they remain as far apart as possible. Electron domains from nonbonding pairs exert slightly greater repulsions than those from bonding pairs, which leads to certain preferred positions for nonbonding pairs and to the departure of bond angles from idealized values. Electron domains from multiple bonds exert slightly greater repulsions than those from single bonds. The arrangement of electron domains around a central atom is called the **electron-domain geometry**; the arrangement of atoms is called the **molecular geometry**.

Section 9.3 The dipole moment of a polyatomic molecule depends on the vector sum of the dipole moments associated with the individual bonds, called the **bond dipoles**. Certain molecular shapes, such as linear AB_2 and trigonal planar AB_3, assure that the bond dipoles cancel, producing a nonpolar molecule, which is one whose dipole moment is zero. In other shapes, such as bent AB_2 and trigonal pyramidal AB_3, the bond dipoles do not cancel and the molecule will be polar (that is, it will have a nonzero dipole moment).

Section 9.4 **Valence-bond theory** is an extension of Lewis's notion of electron-pair bonds. In valence-bond theory, covalent bonds are formed when atomic orbitals on neighboring atoms **overlap** one another. The overlap region is a favorable one for the two electrons because of their attraction to two nuclei. The greater the overlap between two orbitals, the stronger the bond that is formed.

Section 9.5 To extend the ideas of valence-bond theory to polyatomic molecules, we must envision mixing *s*, *p*, and sometimes *d* orbitals to form **hybrid orbitals**. The process of **hybridization** leads to hybrid atomic orbitals that have a large lobe directed to overlap with orbitals on another atom to make a bond. Hybrid orbitals can also accommodate nonbonding pairs. A particular mode of hybridization can be associated with each of the five common electron-domain geometries (linear = sp; trigonal planar = sp^2; tetrahedral = sp^3; trigonal bipyramidal = sp^3d; and octahedral = sp^3d^2).

Section 9.6 Covalent bonds in which the electron density lies along the line connecting the atoms (the internuclear axis) are called **sigma (σ) bonds**. Bonds can also be formed from the side-to-side overlap of *p* orbitals. Such a bond is called a **pi (π) bond**. A double bond, such as that in C_2H_4, consists of one σ bond and one π bond; a triple bond, such as that in C_2H_2, consists of one σ and two π bonds. The formation of a π bond requires that molecules adopt a specific orientation; the two CH_2 groups in C_2H_4, for example, must lie in the same plane. As a result, the presence of π bonds introduces rigidity into molecules. In molecules that have multiple bonds and more than one resonance structure, such as C_6H_6, the π bonds are **delocalized**; that is, the π bonds are spread among several atoms.

Section 9.7 **Molecular orbital theory** is another model used to describe the bonding in molecules. In this model the electrons exist in allowed energy states called **molecular orbitals (MOs)**. These orbitals can be spread among all the atoms of a molecule. Like an atomic orbital, a molecular orbital has a definite energy and can hold two electrons of opposite spin. The combination of two atomic orbitals leads to the formation of two MOs, one at lower energy and one at higher energy relative to the energy of the atomic orbitals. The lower-energy MO concentrates charge density in the region between the nuclei and is called a **bonding molecular orbital**. The higher-energy MO excludes electrons from the region between the nuclei and is called an **antibonding molecular orbital**. Occupation of bonding molecular orbitals favors bond formation, whereas occupation of antibonding molecular orbitals is unfavorable. The bonding and antibonding MOs formed by the combination of *s* orbitals are **sigma (σ) molecular orbitals**; like σ bonds, they lie on the internuclear axis.

The combination of atomic orbitals and the relative energies of the molecular orbitals are shown by an **energy-level** (or **molecular orbital**) **diagram**. When the appropriate number of electrons are put into the MOs, we can calculate the **bond order** of a bond, which is half the difference between the number of electrons in bonding MOs and the number of electrons in antibonding MOs. A bond order of 1 corresponds to a single bond, and so forth. Bond orders can be fractional numbers.

Section 9.8 Electrons in core orbitals do not contribute to the bonding between atoms, so a molecular orbital description usually needs only to consider electrons in the outermost electron subshells. In order to describe the molecular orbitals of second-row homonuclear diatomic molecules, we need to consider the MOs that can form by the combination of *p* orbitals. The *p* orbitals that point directly at one another can form σ bonding and σ^* antibonding MOs. The *p* orbitals that are oriented perpendicular to the internuclear axis combine to form **pi (π) molecular orbitals**. In diatomic molecules the π molecular orbitals occur as pairs of degenerate (same energy) bonding and antibonding MOs. The σ_{2p} bonding MO is expected to be lower in energy than the π_{2p} bonding MOs because of larger orbital overlap. This ordering is reversed in B_2, C_2, and N_2 because of interaction between the 2*s* and 2*p* atomic orbitals.

The molecular orbital description of second-row diatomic molecules leads to bond orders in accord with the Lewis structures of these molecules. Further, the model predicts correctly that O_2 should exhibit **paramagnetism**, an attraction of a molecule by a magnetic field due to unpaired electrons. Those molecules in which all the electrons are paired exhibit **diamagnetism**, a weak repulsion from a magnetic field.

Exercises

Molecular Shapes; the VSEPR Model

9.1 The molecules BF_3 and SO_3 are both described as trigonal planar. Does this information completely define the bond angles of these molecules?

9.2 Methane (CH_4) and the perchlorate ion (ClO_4^-) are both described as tetrahedral. What does this indicate about their bond angles?

9.3 **(a)** What is meant by the term *electron domain*? **(b)** Explain in what way electron domains behave like the balloons in Figure 9.5. Why do they do so?

9.4 **(a)** How does one determine the number of electron domains in a molecule or ion? **(b)** What is the difference between a *bonding electron domain* and a *nonbonding electron domain*?

9.5 Describe the characteristic electron-domain geometry of each of the following numbers of electron domains about a central atom: **(a)** 3; **(b)** 4; **(c)** 5; **(d)** 6.

9.6 Indicate the number of electron domains about a central atom, given the following angles between them: **(a)** 120°; **(b)** 180°; **(c)** 109.5°; **(d)** 90°.

9.7 What is the difference between the electron-domain geometry and the molecular geometry of a molecule? Use the ammonia molecule as an example in your discussion.

9.8 An AB_3 molecule is described as having a trigonal bipyramidal electron-domain geometry. How many nonbonding domains are on atom A? Explain.

9.9 Give the electron-domain and molecular geometries of a molecule that has the following electron domains on its central atom: **(a)** four bonding domains and no nonbonding domains; **(b)** three bonding domains and two nonbonding domains; **(c)** five bonding domains and one nonbonding domain.

9.10 What are the electron-domain and molecular geometries of a molecule that has the following electron domains on its central atom? **(a)** Three bonding domains and no nonbonding domains; **(b)** three bonding domains and one nonbonding domain; **(c)** two bonding domains and three nonbonding domains.

9.11 Draw the Lewis structure for each of the following molecules or ions, and predict their electron-domain and molecular geometries: **(a)** H_3O^+; **(b)** SCN^-; **(c)** CS_2; **(d)** BrO_3^-; **(e)** SeF_4; **(f)** ICl_4^-.

9.12 Give the electron-domain and molecular geometries for the following molecules and ions: **(a)** N_2O (central N) **(b)** SO_3; **(c)** PCl_3; **(d)** NH_2Cl; **(e)** BrF_5; **(f)** KrF_2.

9.13 The figure that follows shows ball-and-stick drawings of three possible shapes of an AF_3 molecule. **(a)** For each shape, give the electron-domain geometry on which the molecular geometry is based. **(b)** For each shape, how many nonbonding electron domains are there on atom A? **(c)** Which of the following element or elements will lead to an AF_3 molecule with the shape in (ii)? Li, B, N, Al, P, Cl. **(d)** Name an element A that is expected to lead to the AF_3 structure shown in (iii). Explain your reasoning.

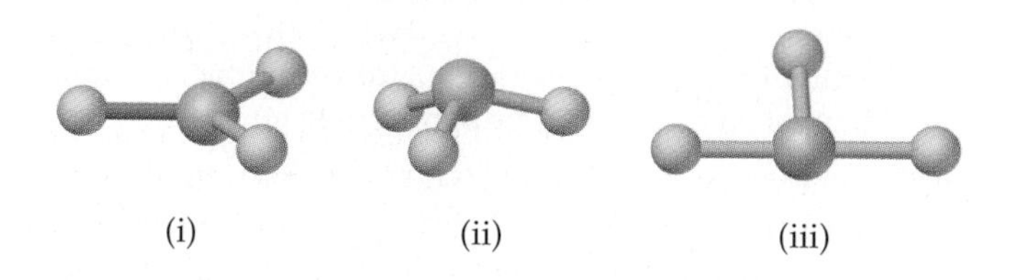

9.14 The figure that follows contains ball-and-stick drawings of three possible shapes of an AF_4 molecule. **(a)** For each shape, give the electron-domain geometry on which the molecular geometry is based. **(b)** For each shape, how many nonbonding electron domains are there on atom A? **(c)** Which of the following element or elements will lead to an AF_4 molecule with the shape in (iii)? Be, C, S, Se, Si, Xe. **(d)** Name an element A that is expected to lead to the AF_4 structure shown in (i).

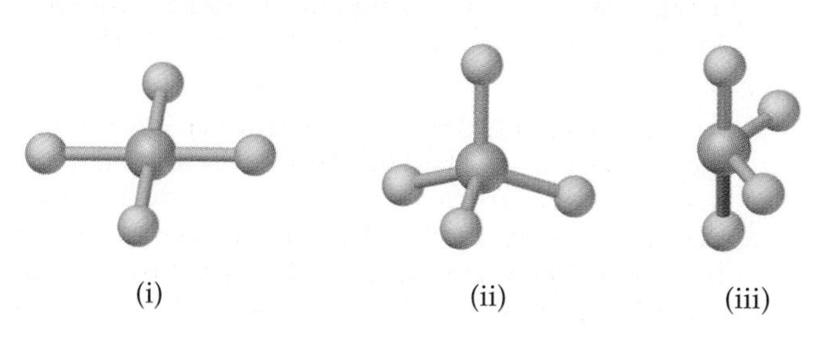

9.15 Give the approximate values for the indicated bond angles in the following molecules:

(a) H—O—Cl—O: (with :O: bonded to Cl; angles 1, 2)

(b) H—C(H)(H)—O—H (angles 3, 4)

(c) H—C≡C—H (angle 5)

(d) H—C(=O)—O—C(H)(H)—H (angles 6, 7, 8)

9.16 Give approximate values for the indicated bond angles in the following molecules:

(a) H—O—N=O (angles 1, 2)

(b) H—C(H)(H)—C(H)=O (angles 3, 4)

(c) H—N(H)—O—H (angles 5, 6)

(d) H—C(H)(H)—C≡N: (angles 7, 8)

9.17 **(a)** Explain why BrF_4^- is square planar whereas BF_4^- is tetrahedral. **(b)** In which of these molecules, CF_4 or SF_4, do you think the actual bond angle is closest to the ideal angle predicted by the VSEPR model? Explain briefly.

9.18 **(a)** Explain why the following ions have different bond angles: ClO_2^- and NO_2^-. Predict the bond angle in each case. **(b)** Given that the spatial requirement of a nonbonding pair of electrons is greater than that of a bonding pair, explain why the XeF_2 molecule is linear and not bent.

9.19 The three species NH_2^-, NH_3, and NH_4^+ have H—N—H bond angles of 105°, 107°, and 109°, respectively. Explain this variation in bond angles.

9.20 Predict the trend in F(axial)—A—F(equatorial) bond angle in the following AF_n molecules: PF_5, SF_4, and ClF_3.

Polarity of Polyatomic Molecules

9.21 Does SO_2 have a dipole moment? If so, in which direction does the net dipole point?

9.22 The H_2O molecule is polar. How does this offer experimental proof that the molecule cannot be linear?

9.23 **(a)** Consider the AF_3 molecules in Exercise 9.13. Which of these will have a nonzero dipole moment? Explain. **(b)** Which of the AF_4 molecules in Exercise 9.14 will have a zero dipole moment?

9.24 **(a)** What conditions must be met if a molecule with polar bonds is nonpolar? **(b)** What geometries will give nonpolar molecules for AB_2, AB_3, and AB_4 geometries?

9.25 Which of the following molecules are polar? BF_3, CO, CF_4, NCl_3, SF_2.

9.26 Predict whether the following molecules are polar or nonpolar: **(a)** IF; **(b)** CS_2; **(c)** SO_3; **(d)** PCl_3; **(e)** SF_6; **(f)** IF_5.

9.27 Dichloroethylene ($C_2H_2Cl_2$) has three forms (isomers), each of which is a different substance:

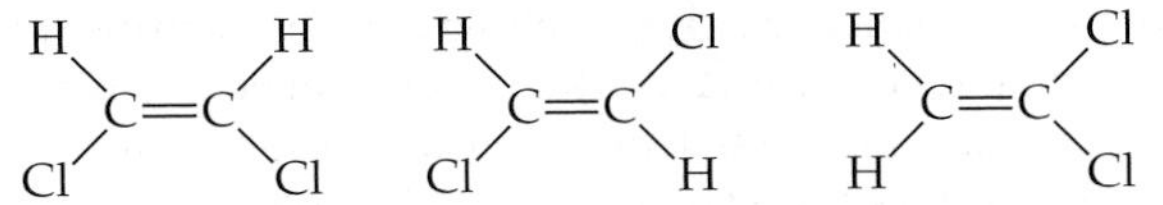

H(Cl)C=C(H)(Cl) H(Cl)C=C(Cl)(H) H(H)C=C(Cl)(Cl)

A pure sample of one of these substances is found experimentally to have a dipole moment of zero. Can we determine which of the three substances was measured?

9.28 Dichlorobenzene, $C_6H_4Cl_2$, exists as three different forms (isomers), called *ortho*, *meta*, and *para*:

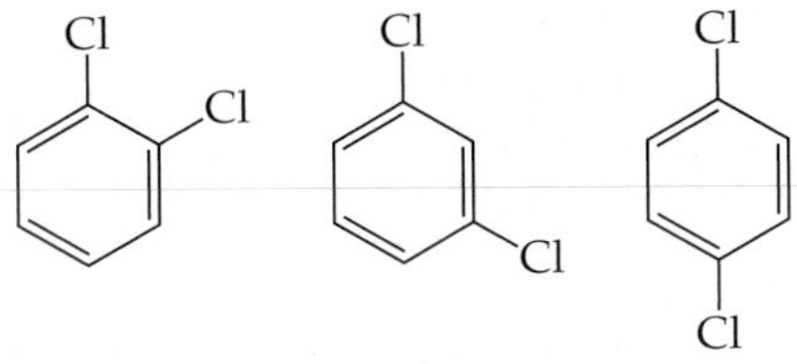

ortho *meta* *para*

Which of these would have a nonzero dipole moment? Explain.

Orbital Overlap; Hybrid Orbitals

9.29 **(a)** What is meant by the term *orbital overlap*? **(b)** What is the significance of overlapping orbitals in valence-bond theory? **(c)** What two fundamental concepts are incorporated in valence-bond theory?

9.30 Draw sketches illustrating the overlap between the following orbitals on two atoms: **(a)** the 2*s* orbital on each; **(b)** the $2p_z$ orbital on each (assume that the atoms are on the *z*-axis); **(c)** the 2*s* orbital on one and the $2p_z$ orbital on the other.

9.31 Indicate the hybridization and bond angles associated with each of the following electron-domain geometries: **(a)** linear; **(b)** tetrahedral; **(c)** trigonal planar; **(d)** octahedral; **(e)** trigonal bipyramidal.

9.32 What is the designation for the hybrid orbitals formed from each of the following combinations of atomic orbitals: **(a)** one *s* and two *p*; **(b)** one *s*, three *p*, and one *d*; **(c)** one *s*, three *p*, and two *d*? What characteristic bond angles are associated with each?

9.33 Draw the Lewis structure for the SO_3^{2-} ion. What is the electron-domain geometry? What is the molecular geometry? Predict the ideal O—S—O bond angle. What hybrid orbitals does S use in bonding?

9.34 What is the maximum number of hybrid orbitals that a carbon atom can form? The minimum number? Explain briefly.

9.35 **(a)** Starting with the orbital diagram of a boron atom, describe the steps needed to construct hybrid orbitals appropriate to describe the bonding in BF_3. **(b)** What is the name given to the hybrid orbitals constructed in (a)? **(c)** On one origin sketch the large lobes of the hybrid orbitals constructed in part (a). **(d)** Are there any valence atomic orbitals of B that are left unhybridized? If so, how are they oriented relative to the hybrid orbitals?

9.36 **(a)** Starting with the orbital diagram of a sulfur atom, describe the steps needed to construct hybrid orbitals appropriate to describe the bonding in SF_2. **(b)** What is the name given to the hybrid orbitals constructed in (a)? **(c)** On one origin sketch the large lobes of the hybrid orbitals constructed in part (a). **(d)** Would the hybridization scheme in part (a) be appropriate for SF_4? Explain.

9.37 Indicate the hybrid orbital set used by the central atom in each of the following molecules and ions: **(a)** BCl_3; **(b)** $AlCl_4^-$; **(c)** CS_2; **(d)** KrF_2; **(e)** PF_6^-.

9.38 What set of hybrid orbitals is used by the central atom in each of the following molecules and ions: **(a)** $SiCl_4$; **(b)** HCN; **(c)** SO_3; **(d)** ICl_2^-; **(e)** BrF_4^-?

Multiple Bonds

9.39 **(a)** Sketch a σ bond that is constructed from *p* orbitals. **(b)** Sketch a π bond that is constructed from *p* orbitals. **(c)** Which is generally the stronger, a σ bond or a π bond? Explain.

9.40 **(a)** If the valence atomic orbitals of an atom are *sp* hybridized, how many unhybridized *p* orbitals remain in the valence shell? How many π bonds can the atom form? **(b)** How many σ and π bonds are generally part of a triple bond? **(c)** How do multiple bonds introduce rigidity into molecules?

9.41 **(a)** Draw Lewis structures for methane (CH_4) and formaldehyde (H_2CO). **(b)** What is the hybridization at the carbon atom in CH_4 and H_2CO? **(c)** The carbon atom in CH_4 cannot participate in multiple bonding, whereas that in H_2CO can. Explain this observation by using the hybridization at the carbon atom.

9.42 The nitrogen atoms in N_2 participate in multiple bonding, whereas those in hydrazine, N_2H_4, do not. How can you explain this observation in light of the hybridization at the nitrogen atoms in the two molecules?

9.43 Acetone, C_3H_6O, is a commonly used organic solvent that is the main component of some nail-polish removers. Its Lewis structure is

```
    H  :O:  H
    |   ||  |
H — C — C — C — H
    |       |
    H       H
```

(a) What is the total number of valence electrons in the acetone molecule? **(b)** How many valence electrons are used to make σ bonds in the molecule? **(c)** How many valence electrons are used to make π bonds in the molecule? **(d)** How many valence electrons remain in nonbonding pairs in the molecule? **(e)** What is the hybridization at the central carbon atom of the molecule?

9.44 The organic molecule *ketene*, C_2H_2O, has the following Lewis structure:

```
H
|
C=C=Ö
|
H
```

(a) What is the hybridization at each of the carbon atoms of the molecule? **(b)** What is the total number of valence electrons in ketene? **(c)** How many of the valence electrons are used to make σ bonds in the molecule? **(d)** How many valence electrons are used to make π bonds? **(e)** How many valence electrons remain in nonbonding pairs in the molecule?

9.45 Consider the Lewis structure for glycine, the simplest amino acid:

```
       H  :O:
       |   ||
H — N — C — C — O — H
    |   |
    H   H
```

(a) What are the approximate bond angles about each of the two carbon atoms, and what are the hybridizations of the orbitals on each of them? **(b)** What are the hybridizations of the orbitals on the two oxygens and the nitrogen atom, and what are the approximate bond angles at the nitrogen? **(c)** What is the total number of σ bonds in the entire molecule, and what is the total number of π bonds?

9.46 The compound with the following Lewis structure is acetylsalicylic acid, better known as aspirin:

```
        H      :O:
        |       ||  2
   H    C       C — O — H
     C     C
  1  |     ||   3       H
     C     C            |
   H    C       O — C — C — H
        |           ||  |
        H          :O:  H
```

(a) What are the approximate values of the bond angles labeled 1, 2, and 3? **(b)** What hybrid orbitals are used about the central atom of each of these angles? **(c)** How many σ bonds are in the molecule?

9.47 **(a)** What is the difference between a localized π bond and a delocalized one? **(b)** How can you determine whether a molecule or ion will exhibit delocalized π bonding? **(c)** Is the π bond in NO_2^- localized or delocalized?

9.48 **(a)** Write a single Lewis structure for SO_3, and determine the hybridization at the S atom. **(b)** Are there other equivalent Lewis structures for the molecule? **(c)** Would you expect SO_3 to exhibit delocalized π bonding? Explain.

Molecular Orbitals

9.49 **(a)** What are the similarities and differences between atomic orbitals and molecular orbitals? **(b)** Why is the bonding molecular orbital of H_2 at lower energy than the electron in a hydrogen atom? **(c)** How many electrons can be placed into each MO of a molecule?

9.50 **(a)** Why is the antibonding molecular orbital of H_2 at higher energy than the electron in a hydrogen atom? **(b)** Does the Pauli exclusion principle (Section 6.7) apply to MOs? Explain. **(c)** If two p orbitals of one atom combine with two p orbitals of another atom, how many MOs result? Explain.

9.51 Consider the H_2^+ ion. **(a)** Sketch the molecular orbitals of the ion, and draw its energy-level diagram. **(b)** How many electrons are there in the H_2^+ ion? **(c)** Write the electron configuration of the ion in terms of its MOs. **(d)** What is the bond order in H_2^+? **(e)** Suppose that the ion is excited by light so that an electron moves from a lower-energy to a higher-energy MO. Would you expect the excited-state H_2^+ ion to be stable or to fall apart? Explain.

9.52 **(a)** Sketch the molecular orbitals of the H_2^- ion, and draw its energy-level diagram. **(b)** Write the electron configuration of the ion in terms of its MOs. **(c)** Calculate the bond order in H_2^-. **(d)** Suppose that the ion is excited by light, so that an electron moves from a lower-energy to a higher-energy molecular orbital. Would you expect the excited-state H_2^- ion to be stable? Explain.

9.53 **(a)** Sketch the σ and σ^* molecular orbitals that can result from the combination of two $2p_z$ atomic orbitals. **(b)** Sketch the π and π^* MOs that result from the combination of two $2p_x$ atomic orbitals. **(c)** Place the MOs from parts (a) and (b) in order of increasing energy, assuming no mixing of 2s and 2p orbitals.

9.54 **(a)** What is the probability of finding an electron on the internuclear axis if the electron occupies a π molecular orbital?

(b) For a homonuclear diatomic molecule, what similarities and differences are there between the π_{2p} MO made from the $2p_x$ atomic orbitals and the π_{2p} MO made from the $2p_y$ atomic orbitals? **(c)** Why are the π_{2p} MOs lower in energy than the π^*_{2p} MOs?

9.55 **(a)** What are the relationships among bond order, bond length, and bond energy? **(b)** According to molecular orbital theory, would either Be_2 or Be_2^+ be expected to exist? Explain.

9.56 Explain the following: **(a)** The *peroxide* ion, O_2^{2-}, has a longer bond than the *superoxide* ion, O_2^-. **(b)** The magnetic properties of B_2 are consistent with the π_{2p} MOs being lower in energy than the σ_{2p} MO.

9.57 **(a)** What is meant by the term *diamagnetism*? **(b)** How does a diamagnetic substance respond to a magnetic field? **(c)** Which of the following ions would you expect to be diamagnetic? N_2^{2-}, O_2^{2-}, Be_2^{2+}, C_2^-.

9.58 **(a)** What is meant by the term *paramagnetism*? **(b)** How can one determine experimentally whether a substance is paramagnetic? **(c)** Which of the following ions would you expect to be paramagnetic? O_2^+, N_2^{2-}, Li_2^+, O_2^{2-}. If the ion is paramagnetic, how many unpaired electrons does it possess?

9.59 Using Figures 9.36 and 9.41 as guides, give the molecular orbital electron configuration for each of the following cations: **(a)** B_2^+; **(b)** Li_2^+; **(c)** N_2^+; **(d)** Ne_2^{2+}. In each case indicate whether the addition of an electron to the ion would increase or decrease the bond order of the species.

9.60 If we assume that the energy-level diagrams for homonuclear diatomic molecules shown in Figure 9.38 can be applied to heteronuclear diatomic molecules and ions, predict the bond order and magnetic behavior of the following: **(a)** CO; **(b)** NO^-; **(c)** OF^+; **(d)** NeF^+.

9.61 Determine the electron configurations for CN^+, CN, and CN^-. Calculate the bond order for each, and indicate which ones are paramagnetic.

9.62 **(a)** The nitric oxide molecule, NO, readily loses one electron to form the NO^+ ion. Why is this consistent with the electronic structure of NO? **(b)** Predict the order of the N—O bond strengths in NO, NO^+, and NO^-, and describe the magnetic properties of each. **(c)** With what neutral homonuclear diatomic molecules are the NO^+ and NO^- ions isoelectronic (same number of electrons)?

[9.63] Consider the molecular orbitals of the P_2 molecule. (Assume that the MOs of diatomics from the third row of the periodic table are analogous to those from the second row.) **(a)** Which valence atomic orbitals of P are used to construct the MOs of P_2? **(b)** The figure that follows shows a sketch of one of the MOs for P_2. What is the label for this MO? **(c)** For the P_2 molecule, how many electrons occupy the MO in the figure? **(d)** Is P_2 expected to be diamagnetic or paramagnetic? Explain.

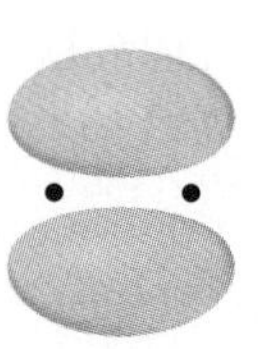

[9.64] The iodine bromide molecule, IBr, is an *interhalogen compound*. Assume that the molecular orbitals of IBr are analogous to the homonuclear diatomic molecule F_2. **(a)** Which valence atomic orbitals of I and of Br are used to construct the MOs of IBr? **(b)** What is the bond order of the IBr molecule? **(c)** One of the MOs of IBr is sketched in the accompanying figure. Why are the atomic orbital contributions to this MO different in size? **(d)** What is the label for the MO? **(e)** For the IBr molecule, how many electrons occupy the MO?

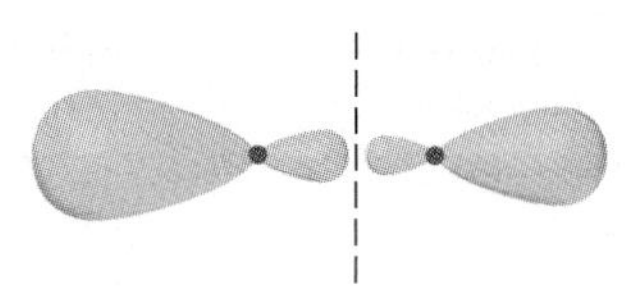

Additional Exercises

9.65 Predict the molecular geometry of **(a)** H_2Se; **(b)** PCl_4^+; **(c)** NO_2^-; **(d)** BrF_3; **(e)** I_3^-.

9.66 **(a)** What is the physical basis for the VSEPR model? **(b)** When applying the VSEPR model, we count a double or triple bond as a single electron domain. Why is this justified?

9.67 The molecules SiF_4, SF_4, and XeF_4 all have molecular formulas of the type AF_4, but the molecules have different molecular geometries. Predict the shape of each molecule, and explain why the shapes differ.

[9.68] The vertices of a tetrahedron correspond to four alternating corners of a cube. By using analytical geometry, demonstrate that the angle made by connecting two of the vertices to a point at the center of the cube is 109.5°, the characteristic angle for tetrahedral molecules.

9.69 From their Lewis structures, determine the number of σ and π bonds in each of the following molecules or ions: **(a)** CO_2; **(b)** thiocyanate ion, NCS^-; **(c)** formaldehyde, H_2CO; **(d)** formic acid, HCO(OH), which has one H and two O atoms attached to C.

9.70 The lactic acid molecule, $CH_3CH(OH)CO(OH)$, gives sour milk its unpleasant, sour taste. Draw the Lewis structure for the molecule, assuming that carbon always forms four bonds in its stable compounds. How many π and how many σ bonds are in the molecule? Which CO bond is shortest in the molecule? What is the hybridization of atomic orbitals around the carbon atom associated with that short bond? What are the approximate bond angles around each carbon atom in the molecule?

9.71 The PF_3 molecule has a dipole moment of 1.03 D, but BF_3 has a dipole moment of zero. How can you explain the difference?

9.72 There are two compounds of the formula $Pt(NH_3)_2Cl_2$:

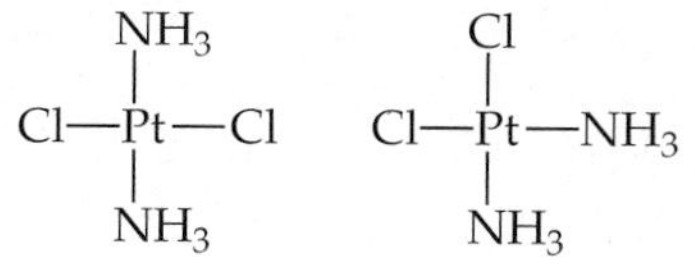

```
   NH3              Cl
    |               |
Cl—Pt—Cl        Cl—Pt—NH3
    |               |
   NH3             NH3
```

The compound on the right, *cisplatin*, is used in cancer therapy. Both compounds have a square-planar geometry. Which compound has a nonzero dipole moment?

[9.73] The O—H bond lengths in the water molecule (H_2O) are 0.96 Å, and the H—O—H angle is 104.5°. The dipole moment of the water molecule is 1.85 D. **(a)** In what directions do the bond dipoles of the O—H bonds point? In what direction does the dipole moment vector of the water molecule point? **(b)** Calculate the magnitude of the bond dipole of the O—H bonds. (*Note*: You will need to use vector addition to do this.) **(c)** Compare your answer from part (b) to the dipole moments of the hydrogen halides (Table 8.3). Is your answer in accord with the relative electronegativity of oxygen?

[9.74] The Lewis structure for allene is

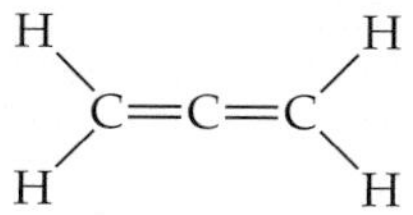

```
H           H
 \         /
  C==C==C
 /         \
H           H
```

Make a sketch of the structure of this molecule that is analogous to Figure 9.27. In addition, answer the following three questions: **(a)** Is the molecule planar? **(b)** Does it have a nonzero dipole moment? **(c)** Would the bonding in allene be described as delocalized? Explain.

[9.75] The reaction of three molecules of fluorine gas with an Xe atom produces the substance xenon hexafluoride, XeF_6:

$$Xe(g) + 3F_2(g) \longrightarrow XeF_6(s)$$

(a) Draw a Lewis structure for XeF_6. **(b)** If you try to use the VSEPR model to predict the molecular geometry of XeF_6, you run into a problem. What is it? **(c)** What could you do to resolve the difficulty in part (b)? **(d)** Suggest a hybridization scheme for the Xe atom in XeF_6. **(e)** The molecule IF_7 has a pentagonal-bipyramidal structure (five equatorial fluorine atoms at the vertices of a regular pentagon and two axial fluorine atoms). Based on the structure of IF_7, suggest a structure for XeF_6.

[9.76] The azide ion, N_3^-, is linear with two N—N bonds of equal length, 1.16 Å. **(a)** Draw a Lewis structure for the azide ion. **(b)** With reference to Table 8.5, is the observed N—N bond length consistent with your Lewis structure? **(c)** What hybridization scheme would you expect at each of the nitrogen atoms in N_3^-? **(d)** Show which hybridized and unhybridized orbitals are involved in the formation of σ and π bonds in N_3^-. **(e)** It is often observed that σ bonds that involve an *sp* hybrid orbital are shorter than those that involve only sp^2 or sp^3 hybrid orbitals. Can you propose a reason for this? Is this observation applicable to the observed bond lengths in N_3^-?

[9.77] In ozone, O_3, the two oxygen atoms on the ends of the molecule are equivalent to one another. **(a)** What is the best choice of hybridization scheme for the atoms of ozone? **(b)** For one of the resonance forms of ozone, which of the orbitals are used to make bonds and which are used to hold nonbonding pairs of electrons? **(c)** Which of the orbitals can be used to delocalize the π electrons? **(d)** How many electrons are delocalized in the π system of ozone?

9.78 Butadiene, C_4H_6, is a planar molecule that has the following carbon–carbon bond lengths:

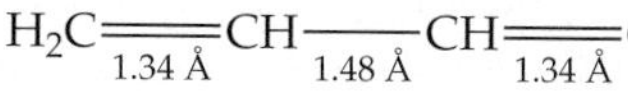

$$H_2C \underset{1.34\ \text{Å}}{=\!=} CH \underset{1.48\ \text{Å}}{—} CH \underset{1.34\ \text{Å}}{=\!=} CH_2$$

(a) Predict the bond angles around each of the carbon atoms, and sketch the molecule. **(b)** Compare the bond lengths to the average bond lengths listed in Table 8.5. Can you explain any differences?

9.79 Using the molecular orbital theory, predict which of the following diatomic molecules is expected to be a stable species: N_2^{2-}, O_2^{2-}, F_2^{2-}.

9.80 Write the electron configuration for the first excited state for N_2 (that is, the state with the highest energy electron moved to the next available energy level). What differences do you expect in the properties of N_2 in its ground state and its first excited state?

[9.81] *Azo dyes* are organic dyes that are used for many applications, such as the coloring of fabrics. Many azo dyes are derivatives of the organic substance *azobenzene*, $C_{12}H_{10}N_2$. A closely related substance is *hydrazobenzene*, $C_{12}H_{12}N_2$. The Lewis structures of these two substances are as follows:

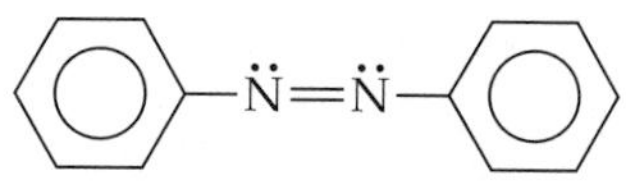

```
(C6H5)—N̈=N̈—(C6H5)          (C6H5)—N̈—N̈—(C6H5)
                                     |   |
                                     H   H
    Azobenzene                  Hydrazobenzene
```

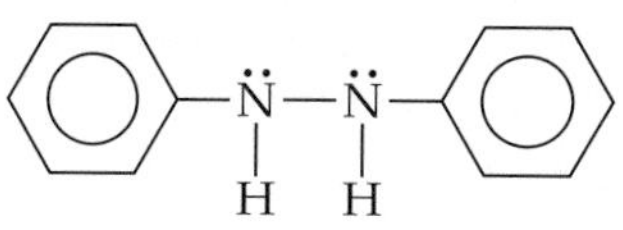

(Recall the shorthand notation used for organic molecules.) **(a)** What is the hybridization at the N atom in each of the substances? **(b)** How many unhybridized atomic orbitals are there on the N and the C atoms in each of the substances? **(c)** Predict the N—N—C angles in each of the substances. **(d)** Azobenzene is said to have greater delocalization of its π electrons than hydrazobenzene. Discuss this statement in light of your answers to (a) and (b). **(e)** All the atoms of azobenzene lie in one plane, whereas those of hydrazobenzene do not. Is this observation consistent with the statement in part (d)? **(f)** Azobenzene is an intense red-orange color, whereas hydrazobenzene is nearly colorless. Discuss this observation with reference to the "Chemistry at Work" box on organic dyes.

Integrative Exercises

9.82 A compound composed of 2.1% H, 29.8% N, and 68.1% O has a molar mass of approximately 50 g/mol. **(a)** What is the molecular formula of the compound? **(b)** What is its Lewis structure if H is bonded to O? **(c)** What is the geometry of the molecule? **(d)** What is the hybridization of the orbitals around the N atom? **(e)** How many σ and how many π bonds are there in the molecule?

9.83 Sulfur tetrafluoride (SF_4) reacts slowly with O_2 to form sulfur tetrafluoride monoxide (OSF_4) according to the following unbalanced reaction:

$$SF_4(g) + O_2(g) \longrightarrow OSF_4(g)$$

The O atom and the four F atoms in OSF_4 are bonded to a central S atom. **(a)** Balance the equation. **(b)** Write a Lewis structure of OSF_4 in which the formal charges of all atoms are zero. **(c)** Use average bond enthalpies (Table 8.4) to estimate the enthalpy of the reaction. Is it endothermic or exothermic? **(d)** Determine the electron-domain geometry of OSF_4, and write two possible molecular geometries for the molecule based on this electron-domain geometry. **(e)** Which of the molecular geometries in part (d) is more likely to be observed for the molecule? Explain.

[9.84] The phosphorus trihalides (PX_3) show the following variation in the bond angle X—P—X: PF_3, 96.3°; PCl_3, 100.3°; PBr_3, 101.0°; PI_3, 102°. The trend is generally attributed to the change in the electronegativity of the halogen. **(a)** Assuming that all electron domains exhibit the same repulsion, what value of the X—P—X angle is predicted by the VSEPR model? **(b)** What is the general trend in the X—P—X angle as the electronegativity increases? **(c)** Using the VSEPR model, explain the observed trend in X—P—X angle as the electronegativity of X changes. **(d)** Based on your answer to part (c), predict the structure of $PBrCl_4$.

[9.85] The molecule 2-butene, C_4H_8, can undergo a geometric change called *cis-trans isomerization*:

$$\underset{cis\text{-2-butene}}{\mathrm{(H_3C)(H)C{=}C(CH_3)(H)}} \longrightarrow \underset{trans\text{-2-butene}}{\mathrm{(H_3C)(H)C{=}C(H)(CH_3)}}$$

As discussed in the "Chemistry and Life" box on the chemistry of vision, such transformations can be induced by light and are the key to human vision. **(a)** What is the hybridization at the two central carbon atoms of 2-butene? **(b)** The isomerization occurs by rotation about the central C—C bond. With reference to Figure 9.31, explain why the π bond between the two central carbon atoms is destroyed halfway through the rotation from *cis*- to *trans*-2-butene. **(c)** Based on average bond enthalpies (Table 8.4), how much energy per molecule must be supplied to break the C—C π bond? **(d)** What is the longest wavelength of light that will provide photons of sufficient energy to break the C—C π bond and cause the isomerization? **(e)** Is the wavelength in your answer to part (d) in the visible portion of the electromagnetic spectrum? Comment on the importance of this result for human vision.

9.86 **(a)** Compare the bond enthalpies (Table 8.4) of the carbon–carbon single, double, and triple bonds to deduce an average π-bond contribution to the enthalpy. What fraction of a single bond does this quantity represent? **(b)** Make a similar comparison of nitrogen–nitrogen bonds. What do you observe? **(c)** Write Lewis structures of N_2H_4, N_2H_2, and N_2, and determine the hybridization around nitrogen in each case. **(d)** Propose a reason for the large difference in your observations of parts (a) and (b).

9.87 Use average bond enthalpies (Table 8.4) to estimate ΔH for the atomization of benzene, C_6H_6:

$$C_6H_6(g) \longrightarrow 6C(g) + 6H(g)$$

Compare the value to that obtained by using ΔH_f° data given in Appendix C and Hess's law. To what do you attribute the large discrepancy in the two values?

[9.88] For both atoms and molecules, ionization energies (Section 7.4) are related to the energies of orbitals: The lower the energy of the orbital, the greater the ionization energy. The first ionization energy of a molecule is therefore a measure of the energy of the highest occupied molecular orbital (HOMO; see "Chemistry at Work" box on organic dyes). The first ionization energies of several diatomic molecules are given in electron-volts in the following table:

Molecule	I_1 (eV)
H_2	15.4
N_2	15.6
O_2	12.1
F_2	15.7

(a) Convert these ionization energies to kJ/mol. **(b)** On the same piece of graph paper, plot I_1 for the H, N, O, and F atoms (Figure 7.10) and I_1 for the molecules listed. **(c)** Do the ionization energies of the molecules follow the same periodic trends as the ionization energies of the atoms? **(d)** Use molecular orbital energy-level diagrams to explain the trends in the ionization energies of the molecules.

[9.89] Many compounds of the transition-metal elements contain direct bonds between metal atoms. We will assume that the *z*-axis is defined as the metal–metal bond axis. **(a)** Which of the 3*d* orbitals (Figure 6.21) can be used to make a σ bond between metal atoms? **(b)** Sketch the σ_{3d} bonding and σ^*_{3d} antibonding MOs. **(c)** Sketch the energy-level diagram for the Sc_2 molecule, assuming that only the 3*d* orbital from part (a) is important. **(d)** What is the bond order in Sc_2?

[9.90] As noted in Section 7.8, the chemistry of astatine (At) is far less developed than that of the other halogen elements. **(a)** In the periodic table the atomic weight of At is written as (210). Why are the parentheses used? How is this nomenclature related to the difficulty of studying astatine? **(b)** Write the complete electron configuration for a neutral At atom. **(c)** Although At_2 is not known, the interhalogen compound AtI has been characterized. Would this compound be expected to have a covalent, polar covalent, or ionic bond? Explain. **(d)** The reaction of AtI with I^- forms the AtI_2^- ion. Use the VSEPR method to predict the geometry of this ion. **(e)** Suppose we construct the molecular orbitals of the unknown At_2 molecule. What bond order is predicted for the molecule? What type of MO is the highest-occupied molecular orbital of the molecule?

eMedia Exercises

9.91 The **VSEPR** movie (*eChapter 9.2*) shows the arrangements of electron domains around a central atom. **(a)** In which of the arrangements are the bond angles not all equal? **(b)** In which of the arrangements are all positions unequal with regard to placement of a single lone pair? **(c)** Are there any electron-domain geometries for which all positions are equal with regard to placement of the first lone pair, but not with regard to placement of the second? Explain.

9.92 In the trigonal-bipyramidal arrangement of electron domains shown in the **VSEPR** movie (*eChapter 9.2*), a single lone pair will occupy an equatorial position rather than an axial position. **(a)** Explain why this is so. **(b)** Is it possible for a molecule with trigonal-bipyramidal electron-domain geometry to have linear molecular geometry? Explain your answer. **(c)** Is it possible for a molecule with trigonal-bipyramidal electron-domain geometry to have a trigonal-planar molecular geometry? If so, how? If not, why not?

9.93 The **VSEPR** tutorial (*eChapter 9.2*) gives examples of electron-domain and molecular geometries for AB_n molecules with and without lone pairs. It also allows you to manipulate and view the molecules from any angle. For each of the examples in the tutorial, give a different example with the same electron-domain and molecular geometries. Write the formula, draw the Lewis dot structure, and give the molecular geometry for each of your examples.

9.94 The **Molecular Polarity** activity (*eChapter 9.3*) allows you to experiment with various central and terminal atoms in AB_2 and AB_3 molecules. For the case of AB_3, where the terminal atoms are all the same element, use geometry to show quantitatively how a molecule with three polar bonds can be nonpolar overall.

9.95 Exercise 9.41 asks about the hybridization at the carbon atom of the molecule H_2CO. The **Hybridization** movies (*eChapter 9.5*) illustrate the formation of sp, sp^2, and sp^3 hybrid orbitals. **(a)** Beginning with the electron configuration of a ground-state carbon atom, diagram and describe the process by which hybridization leads to the formation of the σ bonds in this molecule. **(b)** Using your diagrams from part (a), describe the process by which the π bond forms.

9.96 In Exercise 9.35 you described the steps needed to construct hybrid orbitals to account for the bonding in BF_3. Examine the **Hybridization** movie (*eChapter 9.5*), and use it to explain why promotion and hybridization are necessary to describe the B—F bonds in BF_3.

Chapter 10

Gases

Lightning produces very high temperatures that cause cleavage of both N_2 and O_2, leading to subsequent formation of NO and O_3.

IN THE PAST several chapters we have learned about electronic structures of atoms and about how atoms combine to form molecules and ionic substances. In everyday life, however, we don't have direct experiences with atoms. Instead, we encounter matter as collections of enormous numbers of atoms or molecules that make up gases, liquids, and solids. The fact that matter is actually atomic in nature is not really obvious. Although the idea of atoms extends back to the early Greeks, it took a long time for the concept to gain full acceptance, even among physicists and chemists. Once the atomic nature of matter is understood, however, we can readily understand how atoms and molecules give rise to the properties we observe for matter at the macroscopic level. In this chapter we will focus on gases; in Chapter 11 we will discuss liquids and solids.

In many ways gases are the most easily understood form of matter. Even though different gaseous substances may have very different chemical properties, they behave quite similarly as far as their physical properties are concerned. For example, we live in an atmosphere composed of a mixture of gases that we refer to as air. We breathe air to absorb oxygen, O_2, which supports human life. Air also contains nitrogen, N_2, which has very different chemical properties from oxygen. The atmosphere also contains smaller amounts of other gaseous substances, yet it behaves physically as one gaseous material. The relative simplicity of the gas state affords a good starting point as we seek to understand the properties of matter in terms of its atomic and molecular constitution.

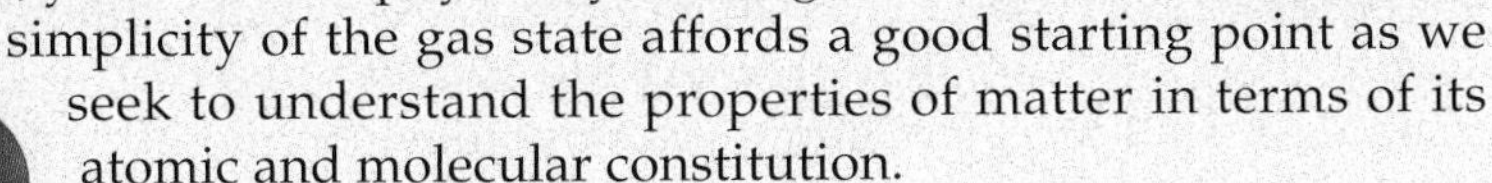

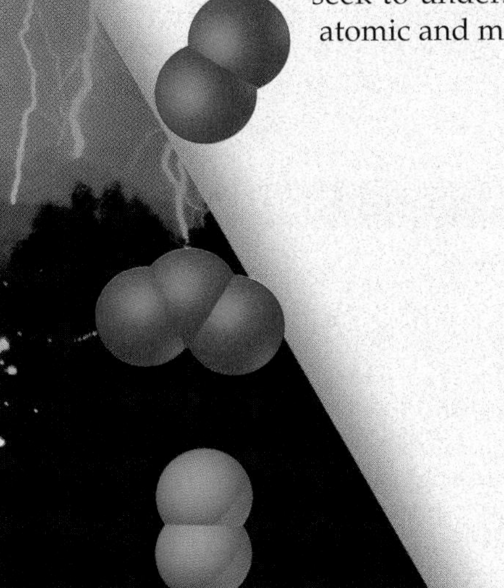

▸ What's Ahead ◂

- We will compare the distinguishing characteristics of gases with those of liquids and solids.
- We'll study gas *pressure*, how it is measured, and the units used to express it, as well as consider Earth's atmosphere and the pressure it exerts.
- The state of a gas can be expressed in terms of volume, pressure, temperature, and quantity of gas. We will examine several empirical relationships that relate these variables to one another. Put together, these empirical relationships yield the *ideal-gas equation*, $PV = nRT$.
- While the ideal-gas equation is not obeyed exactly by any real gas, most gases come quite close to obeying it at temperature and pressure conditions of greatest interest. Consequently, we can use the ideal-gas equation to make many useful calculations.
- In the *kinetic-molecular theory* of gases, the atoms or molecules that make up the gas are assumed to be point masses that move about with an average kinetic energy that is proportional to the gas temperature.
- The kinetic-molecular theory leads to the ideal-gas equation and helps us account for such gas properties as *effusion* through tiny openings and also for *diffusion*.
- Real gases deviate from ideal behavior, primarily because the gas molecules have finite volume and because attractive forces exist between molecules. The *van der Waals equation* gives a more accurate account of real gas behavior at high pressures and low temperatures.

10.1 Characteristics of Gases

In considering the characteristics of gases, there is no more appropriate place to begin than with Earth's atmosphere, vital to all life on the planet. We will consider the atmosphere more fully in Chapter 18. For now, look ahead to Table 18.1, which lists the atmosphere's composition. Notice that air is a complex mixture of many simple substances, either atomic in nature or consisting of small molecules. It consists primarily, however, of N_2 (78%) and O_2 (21%).

In addition to O_2 and N_2, a few other elements (H_2, F_2, Cl_2) exist as gases under ordinary conditions of temperature and pressure. The noble gases (He, Ne, Ar, Kr, and Xe) are all monatomic gases. Many molecular compounds are also gases. Table 10.1 ▼ lists a few of the more common gaseous compounds. Notice that all of these gases are composed entirely of nonmetallic elements. Furthermore, all have simple molecular formulas and, therefore, low molar masses.

3-D MODEL
Hydrogen Cyanide, Hydrogen Chloride, Hydrogen Sulfide, Carbon Monoxide, Carbon Dioxide, Methane, Nitrous Oxide, Nitrogen Dioxide, Ammonia, Sulfur Dioxide

Substances that are liquids or solids under ordinary conditions can usually exist in the gaseous state, too, where they are often referred to as **vapors**. The substance H_2O, for example, can exist as liquid water, solid ice, or water vapor. Under the right conditions, a substance can coexist in all three states of matter, or *phases*, at the same time. A thermos bottle containing a mixture of ice and water at 0°C has some water vapor in the gas phase over the liquid and solid phases.

Gases differ significantly from solids and liquids in several respects. For example, a gas expands spontaneously to fill its container. Consequently, the volume of a gas equals the volume of the container in which it is held. Gases also are highly compressible: When pressure is applied to a gas, its volume readily decreases. Solids and liquids, on the other hand, do not expand to fill their containers, and solids and liquids are not readily compressible.

Gases form homogeneous mixtures with each other regardless of the identities or relative proportions of the component gases. The atmosphere serves as an excellent example. As a further example, when water and gasoline are mixed, the two liquids remain as separate layers. In contrast, the water vapor and gasoline vapors above the liquids form a homogeneous gas mixture. The characteristic properties of gases arise because the individual molecules are relatively far apart. In the air we breathe, for example, the molecules take up only about 0.1% of the total volume, with the rest being empty space. Thus, each molecule behaves largely as though the others weren't present. As a result, different gases behave similarly, even though they are made up of different molecules. In contrast, the individual molecules in a liquid are close together and occupy perhaps 70% of the total space. The attractive forces among the molecules keep the liquid together.

TABLE 10.1 Some Common Compounds That Are Gases At Room Temperature

Formula	Name	Characteristics
HCN	Hydrogen cyanide	Very toxic, slight odor of bitter almonds
H_2S	Hydrogen sulfide	Very toxic, odor of rotten eggs
CO	Carbon monoxide	Toxic, colorless, odorless
CO_2	Carbon dioxide	Colorless, odorless
CH_4	Methane	Colorless, odorless, flammable
C_2H_4	Ethylene	Colorless; ripens fruit
C_3H_8	Propane	Colorless; bottled gas
N_2O	Nitrous oxide	Colorless, sweet odor, laughing gas
NO_2	Nitrogen dioxide	Toxic, red-brown, irritating odor
NH_3	Ammonia	Colorless, pungent odor
SO_2	Sulfur dioxide	Colorless, irritating odor

10.2 Pressure

Among the most readily measured properties of a gas are its temperature, volume, and pressure. It is not surprising, therefore, that many early studies of gases focused on relationships among these properties. We have already discussed volume and temperature. ∞ (Section 1.4) Let's now consider the concept of pressure.

In general terms, **pressure** conveys the idea of a force, a push that tends to move something else in a given direction. Pressure, P, is, in fact, the force, F, that acts on a given area, A.

$$P = \frac{F}{A} \quad [10.1]$$

Gases exert a pressure on any surface with which they are in contact. The gas in an inflated balloon, for example, exerts a pressure on the inside surface of the balloon.

Atmospheric Pressure and the Barometer

You and I, coconuts and nitrogen molecules, all experience an attractive force that pulls toward the center of Earth. When a coconut comes loose from its place at the top of the tree, for example, gravitational attractive force causes it to be accelerated toward Earth, increasing in speed as its potential energy is converted into kinetic energy. ∞ (Section 5.1) The atoms and molecules of the atmosphere experience a gravitational acceleration, too. Because the gas particles have such tiny masses, however, their thermal energies of motion override the gravitational forces, so the atmosphere doesn't just pile up in a thin layer at Earth's surface. Nevertheless, gravity does operate, and it causes the atmosphere as a whole to press down on the surface, creating an atmospheric pressure.

You can demonstrate the existence of atmospheric pressure to yourself with an empty plastic bottle of the sort used to vend water or soft drinks. If you suck on the mouth of the empty bottle, chances are you can cause it to partially cave in. When you break the partial vacuum you have created, the bottle pops out to its original shape. What causes the bottle to cave in when the pressure inside it is reduced, even by the relatively small amount you can manage with your lungs? The atmosphere is exerting a force on the outside of the bottle that is greater than the force within the bottle when some of the gas has been sucked out.

We can calculate the magnitude of the atmospheric pressure as follows: The force, F, exerted by any object is the product of its mass, m, times its acceleration, a; that is, $F = ma$. The acceleration produced by Earth's gravity is $9.8\ \text{m/s}^2$. ∞ (Section 5.1) Now imagine a column of air $1\ \text{m}^2$ in cross section extending through the atmosphere. That column has a mass of roughly 10,000 kg (Figure 10.1 ▼). The force exerted by the column is

$$F = (10{,}000\ \text{kg})(9.8\ \text{m/s}^2) = 1 \times 10^5\ \text{kg-m/s}^2 = 1 \times 10^5\ \text{N}$$

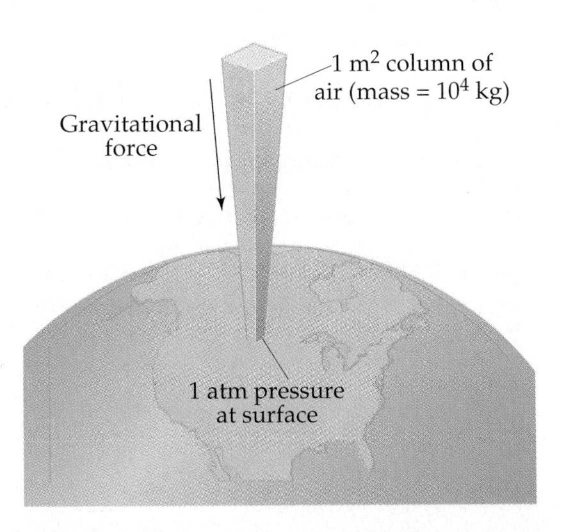

◀ **Figure 10.1** Illustration of the manner in which Earth's atmosphere exerts pressure at the surface of the planet. The mass of a column of atmosphere exactly $1\ \text{m}^2$ in cross-sectional area and extending to the top of the atmosphere exerts a force of 1.01×10^5 N.

The SI unit for force is kg-m/s^2 and is called the *newton* (N): 1 N = 1 kg-m/s^2. The pressure exerted by the column is the force divided by its cross-sectional area, A.

$$P = \frac{F}{A} = \frac{1 \times 10^5\ \text{N}}{1\ \text{m}^2} = 1 \times 10^5\ \text{N/m}^2 = 1 \times 10^5\ \text{Pa} = 1 \times 10^2\ \text{kPa}$$

The SI unit of pressure is N/m^2. It is given the name **pascal** (Pa) after Blaise Pascal (1623–1662), a French mathematician and scientist: 1 Pa = 1 N/m^2. A related unit sometimes used to report pressures is the **bar**, which equals 10^5 Pa. Atmospheric pressure at sea level is about 100 kPa or 1 bar. The actual atmospheric pressure at any location depends on weather conditions and altitude.

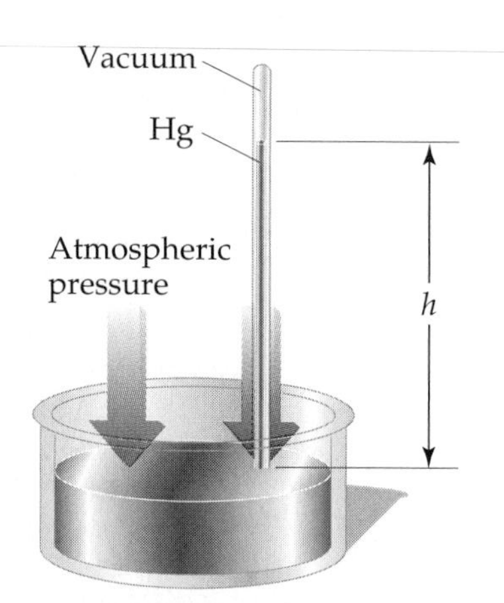

▲ **Figure 10.2** Mercury barometer. The pressure of the atmosphere on the surface of the mercury (represented by the blue arrow) equals the pressure of the column of mercury (red arrow).

In the early part of the seventeenth century it was widely believed that the atmosphere had no weight. Evangelista Torricelli (1608–1647), who was a student of Galileo's, invented the *barometer* (Figure 10.2 ◀) to show that it did. A glass tube more than 760 mm long that is closed at one end is completely filled with mercury and inverted into a dish that contains additional mercury. Care must be taken so that no air gets into the tube. Some of the mercury flows out when the tube is inverted, but a column of mercury remains in the tube. Torricelli argued that the mercury surface in the dish experiences the full force, or weight, of Earth's atmosphere. Because there is no air (and therefore no atmospheric pressure) above the mercury in the tube, the mercury is pushed up the tube until the pressure at the base of the tube due to the mass of the mercury column balances the atmospheric pressure. Thus, the height of the mercury column is a measure of the atmosphere's pressure, so it will change as the atmospheric pressure changes.

Torricelli's proposed explanation met with fierce opposition. Some argued that there could not possibly be a vacuum at the top of the tube. They said, "Nature does not permit a vacuum!" But Torricelli also had his supporters. Blaise Pascal, for example, had one of the barometers carried to the top of Puy de Dome, a volcanic mountain in central France, and compared its readings with a duplicate barometer kept at the foot of the mountain. As the barometer ascended, the height of the mercury column diminished, as expected, because the amount of atmosphere pressing down on the surface decreases as one moves higher. These and experiments by other scientists eventually prevailed, so the idea that the atmosphere has weight became accepted over a period of many years.

Standard atmospheric pressure, which corresponds to the typical pressure at sea level, is the pressure sufficient to support a column of mercury 760 mm high. In SI units, this pressure equals 1.01325×10^5 Pa. Standard atmospheric pressure defines some common non-SI units used to express gas pressures, such as the **atmosphere** (atm) and the *millimeter of mercury* (mm Hg). The latter unit is also called the **torr**, after Torricelli.

$$1\ \text{atm} = 760\ \text{mm Hg} = 760\ \text{torr} = 1.01325 \times 10^5\ \text{Pa} = 101.325\ \text{kPa}$$

Note that the units mm Hg and torr are equivalent: 1 torr = 1 mm Hg.

We will usually express gas pressure in units of atm, Pa (or kPa), or torr, so you should be comfortable converting gas pressures from one set of units to another.

SAMPLE EXERCISE 10.1

(a) Convert 0.357 atm to torr. **(b)** Convert 6.6×10^{-2} torr to atm. **(c)** Convert 147.2 kPa to torr.

Solution

Analyze: In each case we are given the pressure in one unit and asked to convert it to another unit. Our task, therefore, is to choose the appropriate conversion units.

Plan: In solving problems of this type, we can use dimensional analysis.
Solve: (a) We convert atmospheres to torr by using the conversion factor derived from 760 torr = 1 atm.

$$(0.357 \cancel{\text{atm}})\left(\frac{760 \text{ torr}}{1 \cancel{\text{atm}}}\right) = 271 \text{ torr}$$

Note that the units cancel in the required manner.

(b) We use the same relationship as in part (a). To get the appropriate units to cancel, we must use the conversion factor as follows:

$$(6.6 \times 10^{-2} \cancel{\text{torr}})\left(\frac{1 \text{ atm}}{760 \cancel{\text{torr}}}\right) = 8.7 \times 10^{-5} \text{ atm}$$

(c) The relationship 760 torr = 101.325 kPa allows us to write an appropriate conversion factor for this problem:

$$(147.2 \cancel{\text{kPa}})\left(\frac{760 \text{ torr}}{101.325 \cancel{\text{kPa}}}\right) = 1104 \text{ torr}$$

Check: In each case look at the magnitude of the answer and compare it with the starting value. The torr is a much smaller unit than atmosphere, so we expect the *numerical* answer to be larger than the starting quantity in (a), smaller in (b). In (c) the less familiar kPa is involved. The torr is nearly 8 times smaller than the kPa, so the numerical answer in torr should be larger, as obtained.

PRACTICE EXERCISE

In countries that use the metric system, such as Canada, atmospheric pressure in weather reports is given in units of kPa. Convert a pressure of 745 torr to kPa.
Answer: 99.3 kPa

We can use various devices to measure the pressures of enclosed gases. Tire gauges, for example, measure the pressure of air in automobile and bicycle tires. In laboratories we sometimes use a device called a *manometer*. A manometer operates on a principle similar to that of a barometer, as shown in Sample Exercise 10.2.

SAMPLE EXERCISE 10.2

On a certain day the barometer in a laboratory indicates that the atmospheric pressure is 764.7 torr. A sample of gas is placed in a vessel attached to an open-end mercury manometer, shown in Figure 10.3 ▶. A meter stick is used to measure the height of the mercury above the bottom of the manometer. The level of mercury in the open-end arm of the manometer has a measured height of 136.4 mm, and that in the arm that is in contact with the gas has a height of 103.8 mm. What is the pressure of the gas **(a)** in atmospheres; **(b)** in kPa?

Solution

Analyze: We are after the gas pressure in the flask. We know that this pressure must be greater than atmospheric, because the manometer level on the flask side is lower than that on the side open to the atmosphere, as indicated in Figure 10.3.
Plan: We are given the atmospheric pressure (764.7 torr) and the fact that the mercury level in the arm of the manometer that is open to the atmosphere is higher (136.4 mm) than that in the arm attached to the enclosed gas (103.8 mm). We'll use the difference in height between the two arms to obtain the amount by which the pressure of the gas exceeds atmospheric pressure (h in Figure 10.3). In order to use an open-end manometer, we must know the value of the atmospheric pressure. Because a mercury manometer is used, the height difference directly measures the pressure difference in mm Hg or torr.
Solve: (a) The pressure of the gas equals the atmospheric pressure plus the difference in height between the two arms of the manometer:

$$\begin{aligned} P_{gas} &= P_{atm} + (\text{difference in height of arms}) \\ &= 764.7 \text{ torr} + (136.4 \text{ torr} - 103.8 \text{ torr}) \\ &= 797.3 \text{ torr} \end{aligned}$$

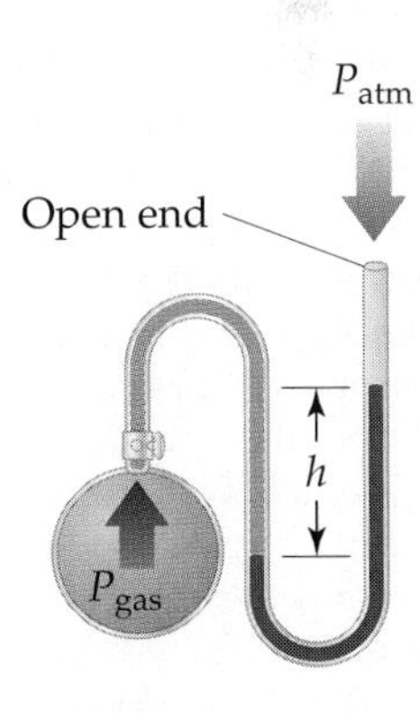

$P_{gas} = P_{atm} + P_h$

▲ **Figure 10.3** A manometer, sometimes employed in the laboratory to measure gas pressures near atmospheric pressure.

ACTIVITY
Manometer

We convert the pressure of the gas to atmospheres:

$$P_{gas} = (797.3 \text{ torr})\left(\frac{1 \text{ atm}}{760 \text{ torr}}\right) = 1.049 \text{ atm}$$

(b) To calculate the pressure in kPa, we employ the conversion factor between atmospheres and kPa:

$$1.049 \text{ atm}\left(\frac{101.3 \text{ kPa}}{1 \text{ atm}}\right) = 106.3 \text{ kPa}$$

Check: The calculated pressure is a bit more than one atmosphere. This makes sense because we anticipated that the pressure in the flask would be greater than the pressure of the atmosphere acting on the manometer, which is a bit greater than one standard atmosphere.

PRACTICE EXERCISE

Convert a pressure of 0.975 atm into Pa and kPa.
Answer: 98.8×10^3 Pa and 98.8 kPa

Chemistry and Life Blood Pressure

The human heart pumps blood to the parts of the body through arteries, and the blood returns to the heart through veins. When your blood pressure is measured, two values are reported, such as 120/80 (120 over 80), which is a normal reading. The first measurement is the *systolic pressure*, the maximum pressure when the heart is pumping. The second is the *diastolic pressure*, the pressure when the heart is in the resting part of its pumping cycle. The units associated with these pressure measurements are torr.

Blood pressure is measured using a pressure gauge attached to a closed, air-filled jacket or cuff that is applied like a tourniquet to the arm (Figure 10.4 ▶). The pressure gauge may be a mercury manometer or some other device. The air pressure in the cuff is increased using a small pump until it is above the systolic pressure and prevents the flow of blood. The air pressure inside the cuff is then slowly reduced until blood just begins to pulse through the artery, as detected by the use of a stethoscope. At this point the pressure in the cuff equals the pressure that the blood exerts inside the arteries. Reading the gauge gives the systolic pressure. The pressure in the cuff is then reduced further until the blood flows freely. The pressure at this point is the diastolic pressure.

Hypertension is the presence of abnormally high blood pressure. The usual criterion for hypertension is a blood pressure greater than 140/90. Hypertension significantly increases the workload on the heart and also places a stress on the walls of the blood vessels throughout the body. These effects increase the risk of aneurysms, heart attacks, and strokes.

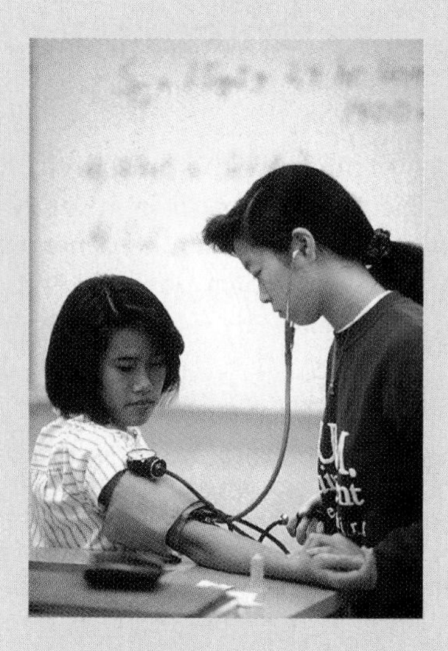

◀ **Figure 10.4** Measuring blood pressure.

10.3 The Gas Laws

Experiments with a large number of gases reveal that four variables are needed to define the physical condition, or *state*, of a gas: temperature, T, pressure, P, volume, V, and the amount of gas, which is usually expressed as the number of moles, n. The equations that express the relationships among T, P, V, and n are known as the *gas laws*.

◀ **Figure 10.5** The volume of gas in this weather balloon will increase as it ascends into the high atmosphere, where the atmospheric pressure is lower than on Earth's surface.

The Pressure-Volume Relationship: Boyle's Law

If the pressure on a balloon is decreased, the balloon expands. That is why weather balloons expand as they rise through the atmosphere (Figure 10.5 ▲). Conversely, when a volume of gas is compressed, the pressure of the gas increases. British chemist Robert Boyle (1627–1691) first investigated the relationship between the pressure of a gas and its volume.

To perform his gas experiments, Boyle used a J-shaped tube like that shown in Figure 10.6 ▼. A quantity of gas is trapped in the tube behind a column of mercury. Boyle changed the pressure on the gas by adding mercury to the tube. He found that the volume of the gas decreased as the pressure increased. For example, doubling the pressure caused the gas volume to decrease to half its original value.

Boyle's law, which summarizes these observations, states that *the volume of a fixed quantity of gas maintained at constant temperature is inversely proportional to the pressure.* When two measurements are inversely proportional, one gets smaller as the other gets larger. Boyle's law can be expressed mathematically as

$$V = \text{constant} \times \frac{1}{P} \quad \text{or} \quad PV = \text{constant} \qquad [10.2]$$

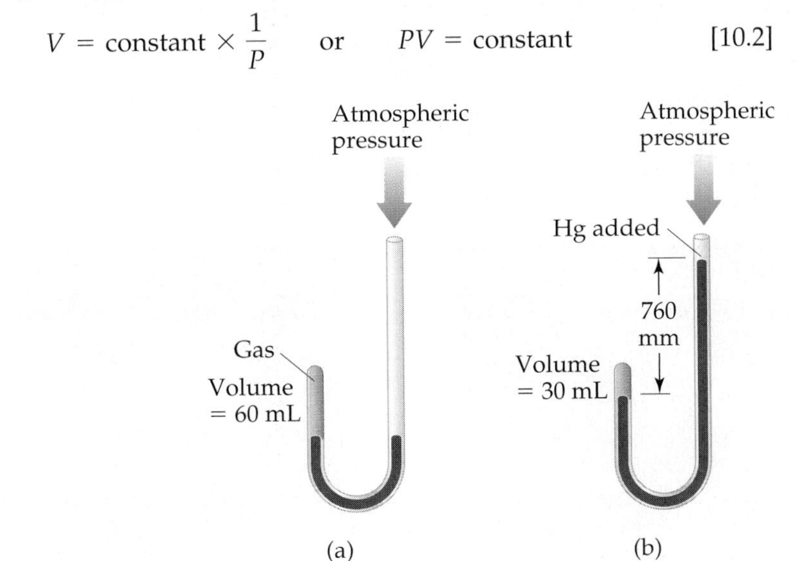

◀ **Figure 10.6** An illustration of Boyle's experiment. In (a) the volume of the gas trapped in the J-tube is 60 mL when the gas pressure is 760 torr. When additional mercury is added, as shown in (b), the trapped gas is compressed. The volume is 30 mL when its total pressure is 1520 torr, corresponding to atmospheric pressure plus the pressure exerted by the 760-mm column of mercury.

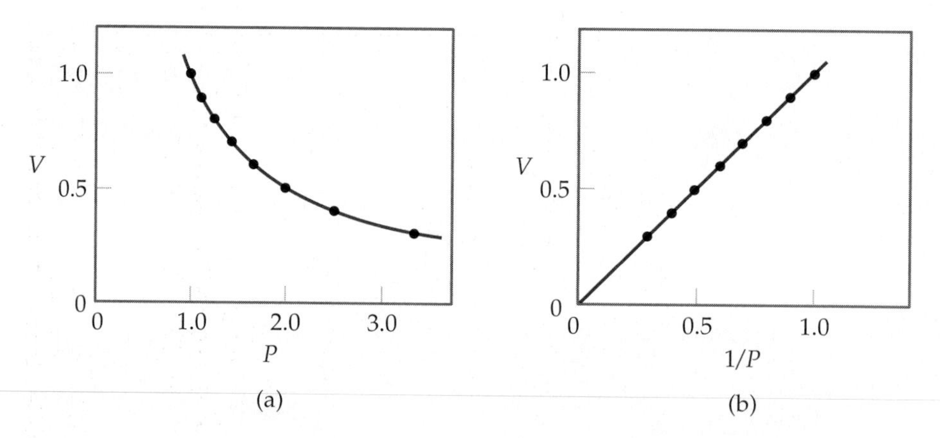

▶ **Figure 10.7** Graphs based on Boyle's law: (a) volume versus pressure; (b) volume versus $1/P$.

The value of the constant depends on the temperature and the amount of gas in the sample. The graph of V versus P in Figure 10.7(a) ▲ shows the type of curve obtained for a given quantity of gas at a fixed temperature. A linear relationship is obtained when V is plotted versus $1/P$ [Figure 10.7(b)].

Simple though it is, Boyle's law occupies a special place in the history of science. Boyle was the first to carry out a series of experiments in which one variable was systematically changed to determine the effect on another variable. The data from the experiments were then employed to establish an empirical relationship, a "law." We apply Boyle's law every time we breathe. The volume of the lungs is governed by the rib cage, which can expand and contract, and the diaphragm, a muscle beneath the lungs. Inhalation occurs when the rib cage expands and the diaphragm moves downward. Both of these actions increase the volume of the lungs, thus decreasing the gas pressure inside the lungs. The atmospheric pressure then forces air into the lungs until the pressure in the lungs equals atmospheric pressure. Exhalation reverses the process: The rib cage contracts and the diaphragm moves up, both of which decrease the volume of the lungs. Air is forced out of the lungs by the resulting increase in pressure.

The Temperature-Volume Relationship: Charles's Law

Hot-air balloons rise because air expands as it is heated. The warm air in the balloon is less dense than the surrounding cool air at the same pressure. This difference in density causes the balloon to ascend. Conversely, a balloon will shrink when the gas in it is cooled, as seen in Figure 10.8 ▼.

The relationship between gas volume and temperature was discovered in 1787 by the French scientist Jacques Charles (1746–1823). Charles found that the volume of a fixed quantity of gas at constant pressure increases linearly with temperature. Some typical data are shown in Figure 10.9 ▶. Notice that the extrapolated (extended) line (which is dashed) passes through −273°C. Note also that the gas is predicted to have zero volume at this temperature. This condition is never realized, however, because all gases liquefy or solidify before reaching this temperature.

▶ **Figure 10.8** As liquid nitrogen (−196°C) is poured over a balloon, the gas in the balloon is cooled and its volume decreases.

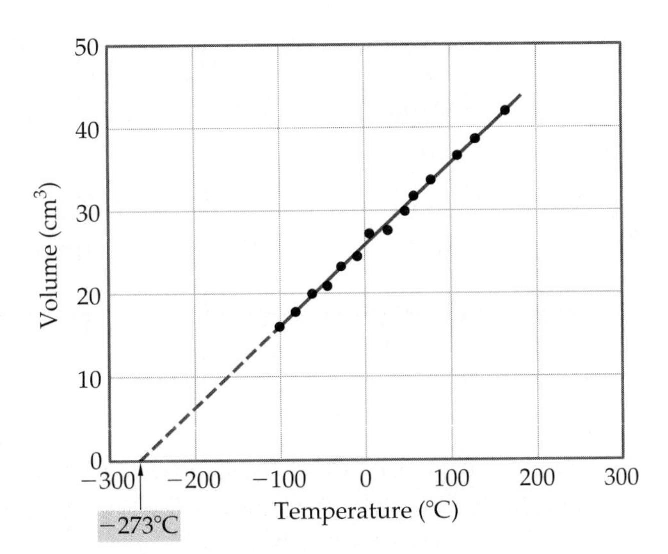

◀ **Figure 10.9** Volume of an enclosed gas as a function of temperature at constant pressure. The dashed line is an extrapolation to temperatures at which the substance is no longer a gas.

In 1848 William Thomson (1824–1907), a British physicist whose title was Lord Kelvin, proposed an absolute-temperature scale, now known as the Kelvin scale. On this scale 0 K, which is called *absolute zero*, equals −273.15°C. (Section 1.4) In terms of the Kelvin scale, **Charles's law** can be stated as follows: *The volume of a fixed amount of gas maintained at constant pressure is directly proportional to its absolute temperature.* Thus, doubling the absolute temperature, say from 200 K to 400 K, causes the gas volume to double. Mathematically, Charles's law takes the following form:

$$V = \text{constant} \times T \qquad \text{or} \qquad \frac{V}{T} = \text{constant} \qquad [10.3]$$

The value of the constant depends on the pressure and amount of gas.

The Quantity-Volume Relationship: Avogadro's Law

As we add gas to a balloon, the balloon expands. The volume of a gas is affected not only by pressure and temperature but also by the amount of gas. The relationship between the quantity of a gas and its volume follows from the work of Joseph Louis Gay-Lussac (1778–1823) and Amadeo Avogadro (1776–1856).

Gay-Lussac is one of those extraordinary figures in the history of science who could truly be called an adventurer. He was interested in lighter-than-air balloons, and in 1804 he made an ascent to 23,000 ft—an exploit that held the altitude record for several decades. To better control lighter-than-air balloons, Gay-Lussac carried out several experiments on the properties of gases. In 1808 he observed the *law of combining volumes*: At a given pressure and temperature, the volumes of gases that react with one another are in the ratios of small whole numbers. For example, two volumes of hydrogen gas react with one volume of oxygen gas to form two volumes of water vapor, as shown in Figure 10.10 ▼.

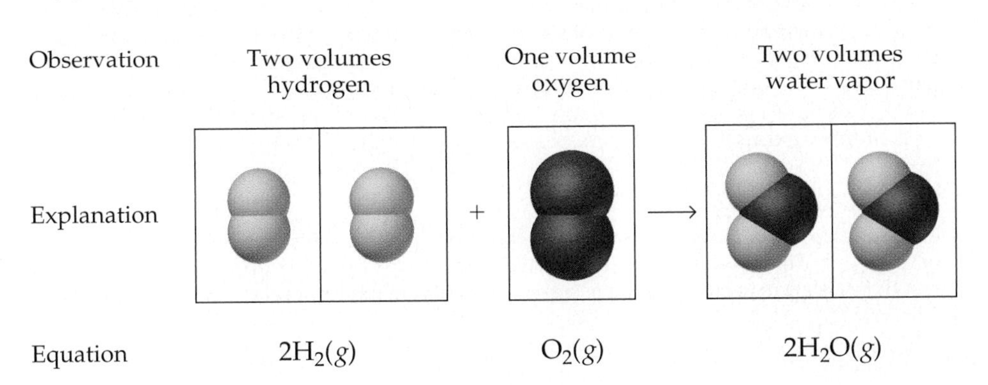

◀ **Figure 10.10** Gay-Lussac's experimental observation of combining volumes shown together with Avogadro's explanation of this phenomenon.

▶ **Figure 10.11** Comparison illustrating Avogadro's hypothesis. Note that helium gas consists of helium atoms. Each gas has the same volume, temperature, and pressure and thus contains the same number of molecules. Because a molecule of one substance differs in mass from a molecule of another, the masses of gas in the three containers differ.

	He	N_2	CH_4
Volume	22.4 L	22.4 L	22.4 L
Pressure	1 atm	1 atm	1 atm
Temperature	0°C	0°C	0°C
Mass of gas	4.00 g	28.0 g	16.0 g
Number of gas molecules	6.02×10^{23}	6.02×10^{23}	6.02×10^{23}

Three years later Amadeo Avogadro (Section 3.4) interpreted Gay-Lussac's observation by proposing what is now known as **Avogadro's hypothesis**: *Equal volumes of gases at the same temperature and pressure contain equal numbers of molecules.* For example, experiments show that 22.4 L of any gas at 0°C and 1 atm contain 6.02×10^{23} gas molecules (that is, 1 mol), as depicted in Figure 10.11 ▲.

Avogadro's law follows from Avogadro's hypothesis: *The volume of a gas maintained at constant temperature and pressure is directly proportional to the number of moles of the gas.* That is,

$$V = \text{constant} \times n \qquad [10.4]$$

Thus, doubling the number of moles of gas will cause the volume to double if T and P remain constant.

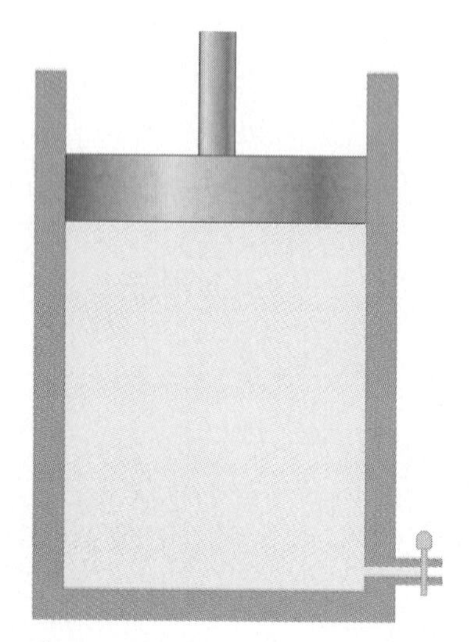

▲ **Figure 10.12** Cylinder with piston and gas inlet valve.

SAMPLE EXERCISE 10.3

Suppose we have a gas confined to a piston as shown in Figure 10.12 ◀. Consider the following changes: **(a)** Heat the gas from 298 K to 360 K, while maintaining the present position of the piston. **(b)** Move the piston to reduce the volume of gas from 1 L to 0.5 L. **(c)** Inject additional gas through the gas inlet valve. Indicate whether each of these changes will:

1. decrease the average distance between molecules
2. increase the pressure of the gas
3. increase the total mass of the gas in the cylinder
4. increase the number of moles of gas present

Solution

Analyze: We need to think how each change in the system affects the gas molecules or the condition in which they exist.

Plan: We'll ask how each of the changes indicated could affect the various quantities (a) through (d).

Solve: (1) Heating the gas while maintaining the position of the piston will cause no change in the number of molecules per unit volume. Thus, the distance between molecules, the total mass of gas molecules and the number of moles of gas remains the same. The increase in temperature will cause the pressure to increase. (2) Moving the piston compresses the same quantity of gas into a smaller volume. The total number of moles of gas, and thus the total mass, remains the same. However, the average distance between molecules must decrease because of the smaller volume in which the gas is confined, so pressure increases. (3) Injecting more gas into the cylinder while keeping the volume and temperature the same will result in more molecules, and thus a greater mass and greater number of moles of gas. The average distance between atoms must decrease because their number per unit volume increases. Correspondingly, the pressure increases.

PRACTICE EXERCISE

CO is oxidized to CO_2 according to the equation, $2CO(g) + O_2(g) \longrightarrow 2CO_2(g)$. If 2 L of $CO(g)$ are mixed with 2 L of $O_2(g)$, what is the resulting total volume of gas after the reaction has gone to completion, assuming no change in temperature or total pressure?
Answer: 3 L

10.4 The Ideal-Gas Equation

In Section 10.3 we examined three historically important gas laws that describe the relationships between the four variables, P, V, T, and n that define the state of a gas. Each law was obtained by holding two variables constant in order to see how the remaining two variables affect each other. We can express each law as a proportionality relationship. Using the symbol $\propto$, which is read "is proportional to," we have

$$\text{Boyle's law:} \quad V \propto \frac{1}{P} \quad (\text{constant } n, T)$$

$$\text{Charles's law:} \quad V \propto T \quad (\text{constant } n, P)$$

$$\text{Avogadro's law:} \quad V \propto n \quad (\text{constant } P, T)$$

We can combine these relationships to make a more general gas law.

$$V \propto \frac{nT}{P}$$

If we call the proportionality constant R, we obtain

$$V = R\left(\frac{nT}{P}\right)$$

Rearranging, we have this relationship in its more familiar form:

$$PV = nRT \qquad [10.5]$$

This equation is known as the **ideal-gas equation**. An **ideal gas** is a hypothetical gas whose pressure, volume, and temperature behavior is completely described by the ideal-gas equation.

The term R in the ideal-gas equation is called the **gas constant**. The value and units of R depend on the units of P, V, n, and T. Temperature must *always* be expressed as an absolute temperature. The quantity of gas, n, is normally expressed in moles. The units chosen for pressure and volume are most often atm and liters, respectively. However, other units can be used. In most countries other than the United States, the SI unit of Pa (or kPa) is most commonly employed. Table 10.2 ▶ shows the numerical value for R in various units. As we saw in the "Closer Look" box on P-V work in Section 5.3, the product PV has the units of energy. Therefore, the units of R can include joules or calories. In working problems with the ideal-gas equation, the units of P, V, n, and T must agree with the units in the gas constant. In this chapter we will most often use the value $R = 0.08206$ L-atm/mol-K (four significant figures) or 0.0821 L-atm/mol-K (three significant figures) whenever we use the ideal-gas equation, consistent with the units of atm for pressure. Use of the value $R = 8.314$ J/mol-K, consistent with the units of Pa for pressure, is also very common.

Suppose we have 1.000 mol of an ideal gas at 1.000 atm and 0.00°C (273.15 K). According to the ideal-gas equation, the volume of the gas is

$$V = \frac{nRT}{P} = \frac{(1.000\ \text{mol})(0.08206\ \text{L-atm/mol-K})(273.15\ \text{K})}{1.000\ \text{atm}} = 22.41\ \text{L}$$

The conditions 0°C and 1 atm are referred to as the **standard temperature and pressure (STP)**. Many properties of gases are tabulated for these conditions. The volume occupied by 1 mol of ideal gas at STP, 22.41 L, is known as the *molar volume* of an ideal gas at STP.

TABLE 10.2 Numerical Values of the Gas Constant, *R*, in Various Units

Units	Numerical Value
L-atm/mol-K	0.08206
J/mol-K[a]	8.314
cal/mol-K	1.987
m^3-Pa/mol-K[a]	8.314
L-torr/mol-K	62.36

[a]SI unit.

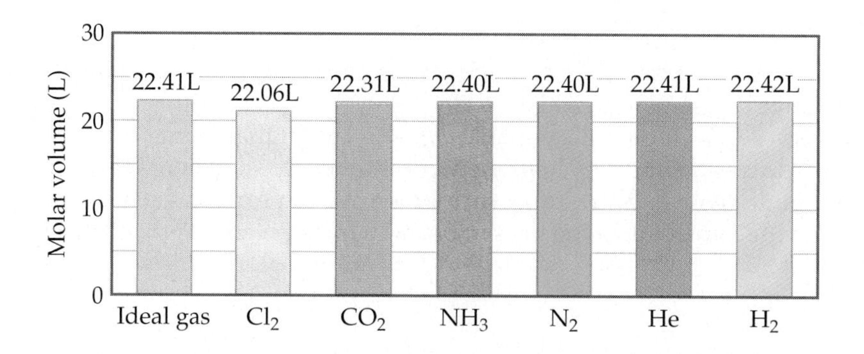

▶ **Figure 10.13** Comparing the molar volumes at STP of the ideal gas with various real gases.

The ideal-gas equation accounts adequately for the properties of most gases under a wide variety of circumstances. It is not exactly correct, however, for any real gas. Thus, the measured volume, V, for given conditions of P, n, and T, might differ from the volume calculated from $PV = nRT$. To illustrate, the measured molar volumes of real gases at STP are compared with the calculated volume of an ideal gas in Figure 10.13 ▲. While these real gases don't match the ideal gas behavior exactly, the differences are so small that we can ignore them for all but the most accurate work. We'll have more to say about the differences between ideal and real gases in Section 10.9.

SAMPLE EXERCISE 10.4

Calcium carbonate, $CaCO_3(s)$, decomposes upon heating to give $CaO(s)$ and $CO_2(g)$. A sample of $CaCO_3$ is decomposed, and the carbon dioxide is collected in a 250-mL flask. After the decomposition is complete, the gas has a pressure of 1.3 atm at a temperature of 31°C. How many moles of CO_2 gas were generated?

Solution

Analyze: We are given the volume (250 mL), pressure (1.3 atm), and temperature (31°C) of a sample of CO_2 gas and asked to calculate the number of moles of CO_2 in the sample.

Plan: Because we are given V, P, and T, we can solve the ideal-gas equation for the unknown quantity, n.

Solve: In analyzing and solving gas-law problems, it is helpful to tabulate the information given in the problems and then to convert the values to units that are consistent with those for R (0.0821 L-atm/mol-K). In this case the given values are

$$P = 1.3 \text{ atm}$$

$$V = 250 \text{ mL} = 0.250 \text{ L}$$

$$T = 31°\text{C} = (31 + 273) \text{ K} = 304 \text{ K}$$

Remember: *Absolute temperature must always be used when the ideal-gas equation is solved.*

We now rearrange the ideal-gas equation (Equation 10.5) to solve for n.

$$n = \frac{PV}{RT}$$

$$n = \frac{(1.3 \text{ atm})(0.250 \text{ L})}{(0.0821 \text{ L-atm/mol-K})(304 \text{ K})} = 0.013 \text{ mol } CO_2$$

Check: Appropriate units cancel, thus ensuring that we have properly rearranged the ideal-gas equation and have converted to the correct units.

PRACTICE EXERCISE

Tennis balls are usually filled with air or N_2 gas to a pressure above atmospheric pressure to increase their "bounce." If a particular tennis ball has a volume of 144 cm^3 and contains 0.33 g of N_2 gas, what is the pressure inside the ball at 24°C?

Answer: 2.0 atm

Strategies in Chemistry Calculations Involving Many Variables

In chemistry and throughout your studies of science and math, you may encounter problems that involve several experimentally measured variables as well as several different physical constants. In this chapter we encounter a variety of problems based on the ideal-gas equation, which consists of four experimental quantities—P, V, n, and T—and one constant, R. Depending on the type of problem, we might need to solve for any of the four quantities.

To avoid any difficulty extracting the necessary information from problems when many variables are involved, we suggest you take the following steps as you analyze, plan, and solve such problems:

1. *Tabulate information.* Read the problems carefully to determine which quantity is the unknown and which quantities are given. Every time you encounter a numerical value, jot it down. In many cases, constructing a table of the given information will be useful.

2. *Convert to consistent units.* As you have already seen, we often use several different units to express the same quantity. Make certain that quantities are converted to the proper units by using the correct conversion factors. In using the ideal-gas equation, for example, we usually use the value of R that has units of L-atm/mol-K. If you are given a pressure in torr, you will need to convert it to atmospheres.

3. *If a single equation relates the variables, rearrange the equation to solve for the unknown.* Make certain that you are comfortable using algebra to solve the equation for the desired variable. In the case of the ideal-gas equation the following algebraic rearrangements will all be used at one time or another:

$$P = \frac{nRT}{V}; \quad V = \frac{nRT}{P}; \quad n = \frac{PV}{RT}; \quad T = \frac{PV}{nR}$$

4. *Use dimensional analysis.* Carry the units through your calculation. Use of dimensional analysis enables you to check that you have solved the equation correctly. If the units of the quantities in the equation cancel properly to give the units of the desired variable, you have probably used the equation correctly.

Sometimes you will not be given values for the necessary variables directly. Rather, you will be given the values of other quantities that can be used to determine the needed variables. For example, suppose you are trying to use the ideal-gas equation to calculate the pressure of a gas. You are given the temperature of the gas, but you are not given explicit values for n and V. However, the problem states that "the sample of gas contains 0.15 mol of gas per liter." We can turn this statement into the expression

$$\frac{n}{V} = 0.15 \text{ mol/L}$$

Solving the ideal-gas equation for pressure yields

$$P = \frac{nRT}{V} = \left(\frac{n}{V}\right)RT$$

Thus, we can solve the equation even though we are not given specific values for n and V. We will examine how to use the density and molar mass of a gas in this fashion in Section 10.5.

As we have continuously stressed, the most important thing you can do to become proficient at solving problems is to practice by using practice exercises and assigned exercises at the end of each chapter. By using systematic procedures, such as those described here, you should be able to minimize difficulties in solving problems involving many variables.

Relating the Ideal-Gas Equation and the Gas Laws

The simple gas laws that we discussed in Section 10.3, such as Boyle's law, are special cases of the ideal-gas equation. For example, when the quantity of gas and the temperature are held constant, n and T have fixed values. Therefore, the product nRT is the product of three constants and must itself be a constant.

$$PV = nRT = \text{constant} \quad \text{or} \quad PV = \text{constant} \qquad [10.6]$$

Thus, we have Boyle's law. We see that if n and T are constant, the individual values of P and V can change, but the product PV must remain constant.

We can use Boyle's law to determine how the volume of a gas changes when its pressure changes. For example, if a metal cylinder holds 50.0 L of O_2 gas at 18.5 atm and 21°C, what volume will the gas occupy if the temperature is maintained at 21°C while the pressure is reduced to 1.00 atm? Because the product PV is a constant when a gas is held at constant n and T, we know that

$$P_1V_1 = P_2V_2 \qquad [10.7]$$

where P_1 and V_1 are initial values and P_2 and V_2 are final values. Dividing both sides of this equation by P_2 gives the final volume, V_2.

$$V_2 = V_1 \times \frac{P_1}{P_2}$$

Substituting the given quantities into this equation gives

$$V_2 = (50.0 \text{ L})\left(\frac{18.5 \text{ atm}}{1.00 \text{ atm}}\right) = 925 \text{ L}$$

The answer is reasonable because gases expand as their pressures are decreased.

In a similar way we can start with the ideal-gas equation and derive relationships between any other two variables, V and T (Charles's law), n and V (Avogadro's law), or P and T. Sample Exercise 10.5 illustrates how these relationships can be derived and used.

SAMPLE EXERCISE 10.5

The gas pressure in an aerosol can is 1.5 atm at 25°C. Assuming that the gas inside obeys the ideal-gas equation, what would the pressure be if the can were heated to 450°C?

Solution

Analyze: We are given the pressure and temperature of the gas at 1.5 atm and 25°C and asked for the pressure at a higher temperature (450°C).

Plan: The volume and number of moles of gas do not change, so we must use a relationship connecting pressure and temperature. Converting temperature to the Kelvin scale and tabulating the given information, we have

	P	T
Initial	1.5 atm	298 K
Final	P_2	723 K

Solve: In order to determine how P and T are related, we start with the ideal-gas equation and isolate the quantities that don't change (n, V, and R) on one side and the variables (P and T) on the other side.

$$\frac{P}{T} = \frac{nR}{V} = \text{constant}$$

Because the quotient P/T is a constant, we can write

$$\frac{P_1}{T_1} = \frac{P_2}{T_2}$$

where the subscripts 1 and 2 represent the initial and final states, respectively. Rearranging to solve for P_2 and substituting the given data gives

$$P_2 = P_1 \times \frac{T_2}{T_1}$$

$$P_2 = (1.5 \text{ atm})\left(\frac{723 \text{ K}}{298 \text{ K}}\right) = 3.6 \text{ atm}$$

Check: This answer is intuitively reasonable—increasing the temperature of a gas increases its pressure. It is evident from this example why aerosol cans carry a warning not to incinerate.

PRACTICE EXERCISE

A large natural-gas storage tank is arranged so that the pressure is maintained at 2.20 atm. On a cold day in December when the temperature is −15°C(4°F), the volume of gas in the tank is 28,500 ft^3. What is the volume of the same quantity of gas on a warm July day when the temperature is 31°C(88°F)?

Answer: 33,600 ft^3

We are often faced with the situation in which P, V, and T all change for a fixed number of moles of gas. Because n is constant under these circumstances, the ideal-gas equation gives

$$\frac{PV}{T} = nR = \text{constant}$$

If we represent the initial and final conditions of pressure, temperature, and volume by subscripts 1 and 2, respectively, we can write

$$\frac{P_1 V_1}{T_1} = \frac{P_2 V_2}{T_2} \qquad [10.8]$$

SAMPLE EXERCISE 10.6

An inflated balloon has a volume of 6.0 L at sea level (1.0 atm) and is allowed to ascend in altitude until the pressure is 0.45 atm. During ascent the temperature of the gas falls from 22°C to −21°C. Calculate the volume of the balloon at its final altitude.

Solution

Analyze: We need to determine a new volume for a gas sample in a situation where both pressure and temperature change.

Plan: Let's again proceed by converting temperature to the Kelvin scale and tabulating the given information.

	P	V	T
Initial	1.0 atm	6.0 L	295 K
Final	0.45 atm	V_2	252 K

Because n is constant, we can use Equation 10.8.

Solve: Rearranging to solve for V_2 gives

$$V_2 = V_1 \times \frac{P_1}{P_2} \times \frac{T_2}{T_1} = (6.0\text{ L})\left(\frac{1.0\text{ atm}}{0.45\text{ atm}}\right)\left(\frac{252\text{ K}}{295\text{ K}}\right) = 11\text{ L}$$

Check: The result appears reasonable. Notice that the calculation involves multiplying the initial volume by a ratio of pressures and a ratio of temperatures. Intuitively we expect that decreasing pressure will cause the volume to increase. Similarly, decreasing temperature should cause the volume to decrease. Note that the difference in pressures is more dramatic than the difference in temperatures. Thus we should expect the effect of the pressure change to predominate in determining the final volume—and it does.

PRACTICE EXERCISE

A 0.50-mol sample of oxygen gas is confined at 0°C in a cylinder with a movable piston, such as that shown in Figure 10.12. The gas has an initial pressure of 1.0 atm. The gas is then compressed by the piston so that its final volume is half the initial volume. The final pressure of the gas is 2.2 atm. What is the final temperature of the gas in degrees Celsius?

Answer: 27°C

10.5 Further Applications of the Ideal-Gas Equation

The ideal-gas equation can be used to define the relationship between the density of a gas and its molar mass, and to calculate the volumes of gases formed or consumed in chemical reactions.

Gas Densities and Molar Mass

The ideal-gas equation has many applications in measuring and calculating gas density. Density has the units of mass per unit volume. We can arrange the gas equation to obtain moles per unit volume.

$$\frac{n}{V} = \frac{P}{RT}$$

Notice that n/V has the units of moles per liter. Suppose that we multiply both sides of this equation by the molar mass, $\mathcal{M}$, which is the number of grams in 1 mol of a substance:

$$\frac{n\mathcal{M}}{V} = \frac{P\mathcal{M}}{RT} \qquad [10.9]$$

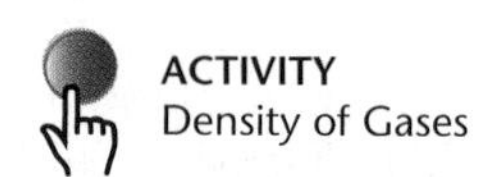

ACTIVITY
Density of Gases

The product of the quantities n/V and $\mathcal{M}$, equals the density in g/L, as seen from their units:

$$\frac{\text{moles}}{\text{liter}} \times \frac{\text{grams}}{\text{mole}} = \frac{\text{grams}}{\text{liter}}$$

Thus, the density, d, of the gas is given by the expression on the right in Equation 10.9:

$$d = \frac{P\mathcal{M}}{RT} \qquad [10.10]$$

▲ **Figure 10.14** The CO_2 gas from a fire extinguisher is denser than air. The CO_2 cools significantly as it emerges from the tank. Water vapor in the air is condensed by the cool CO_2 gas and forms a white fog accompanying the colorless CO_2.

From Equation 10.10, we see that the density of a gas depends on its pressure, molar mass, and temperature. The higher the molar mass and pressure, the more dense the gas; the higher the temperature for a given pressure, the less dense the gas. Although gases form homogeneous mixtures regardless of their identities, a less dense gas will lie above a more dense one in the absence of mixing. For example, CO_2 has a higher molar mass than N_2 or O_2 and is therefore more dense than air. When CO_2 is released from a CO_2 fire extinguisher, as shown in Figure 10.14 ◀, it blankets a fire, preventing O_2 from reaching the combustible material. The fact that a hotter gas is less dense than a cooler one explains why hot air rises. The difference between the densities of hot and cold air is responsible for the lift of hot-air balloons. It is also responsible for many phenomena in weather, such as the formation of large thunderhead clouds during thunderstorms.

SAMPLE EXERCISE 10.7

What is the density of carbon tetrachloride vapor at 714 torr and 125°C?

Solution

Analyze: To find the density, given the temperature and pressure, we need to use Equation 10.10.

Plan: Before we can use Equation 10.10, we need to convert the requisite quantities to the appropriate units. The molar mass of CCl_4 is $12.0 + (4)(35.5) = 154.0$ g/mol. We must convert temperature to the Kelvin scale and pressure to atmospheres.

Solve: Using Equation 10.10, we have

$$d = \frac{(714\ \text{torr})(1\ \text{atm}/760\ \text{torr})(154.0\ \text{g/mol})}{(0.0821\ \text{L-atm/mol-K})(398\ \text{K})} = 4.43\ \text{g/L}$$

Check: If we divide the molar mass (g/mol) by the density (g/L), we end up with L/mol. The numerical value is roughly $154/4.4 = 35$. That is in the right ballpark for the molar volume of a gas heated to 125°C at near atmospheric pressure, so our answer is reasonable.

PRACTICE EXERCISE

The mean molar mass of the atmosphere at the surface of Titan, Saturn's largest moon, is 28.6 g/mol. The surface temperature is 95 K, and the pressure is 1.6 atm. Assuming ideal behavior, calculate the density of Titan's atmosphere.

Answer: 5.9 g/L

Equation 10.10 can be rearranged to solve for the molar mass of a gas:

$$\mathcal{M} = \frac{dRT}{P} \qquad [10.11]$$

Thus, we can use the experimentally measured density of a gas to determine the molar mass of the gas molecules, as shown in Sample Exercise 10.8.

SAMPLE EXERCISE 10.8

A series of measurements are made in order to determine the molar mass of an unknown gas. First, a large flask is evacuated and found to weigh 134.567 g. It is then filled with the gas to a pressure of 735 torr at 31°C and reweighed; its mass is now 137.456 g. Finally, the flask is filled with water at 31°C and found to weigh 1067.9 g. (The density of the water at this temperature is 0.997 g/mL.) Assuming that the ideal-gas equation applies, calculate the molar mass of the unknown gas.

Solution

Analyze: We are given mass information, temperature, and pressure for the gas, and asked to calculate its molar mass.
Plan: We need to use the mass information given to calculate the volume of the container and the mass of the gas within it. From this, we calculate the gas density and then apply Equation 10.11 to calculate the molar mass of the gas.

Solve: The mass of the gas is the difference between the mass of the flask filled with gas and that of the empty (evacuated) flask:

$$137.456 \text{ g} - 134.567 \text{ g} = 2.889 \text{ g}$$

The volume of the gas equals the volume of water that the flask can hold. The volume of water is calculated from its mass and density. The mass of the water is the difference between the masses of the full and empty flask:

$$1067.9 \text{ g} - 134.567 \text{ g} = 933.3 \text{ g}$$

By rearranging the equation for density ($d = m/V$), we have

$$V = \frac{m}{d} = \frac{(933.3 \text{ g})}{(0.997 \text{ g/mL})} = 936 \text{ mL}$$

Knowing the mass of the gas (2.889 g) and its volume (936 mL), we can calculate the density of the gas:

$$2.889 \text{ g}/0.936 \text{ L} = 3.09 \text{ g/L}$$

After converting pressure to atmospheres and temperature to kelvins, we can use Equation 10.11 to calculate the molar mass:

$$\mathcal{M} = \frac{dRT}{P}$$

$$= \frac{(3.09 \text{ g/L})(0.0821 \text{ L-atm/mol-K})(304 \text{ K})}{(735/760) \text{ atm}} = 79.7 \text{ g/mol}$$

Check: The units work out appropriately, and the value of molar mass obtained is reasonable for a substance that is gaseous near room temperature.

PRACTICE EXERCISE

Calculate the average molar mass of dry air if it has a density of 1.17 g/L at 21°C and 740.0 torr.
Answer: 29.0 g/mol

Volumes of Gases in Chemical Reactions

Understanding the properties of gases is important because gases are often reactants or products in chemical reactions. For this reason, we are often faced with calculating the volumes of gases consumed or produced in reactions. We have seen that the coefficients in balanced chemical equations tell us the relative amounts (in moles) of reactants and products in a reaction. The number of moles of a gas, in turn, is related to *P*, *V*, and *T*.

SAMPLE EXERCISE 10.9

The safety air bags in automobiles are inflated by nitrogen gas generated by the rapid decomposition of sodium azide, NaN_3:

$$2NaN_3(s) \longrightarrow 2Na(s) + 3N_2(g)$$

If an air bag has a volume of 36 L and is to be filled with nitrogen gas at a pressure of 1.15 atm at a temperature of 26.0°C, how many grams of NaN_3 must be decomposed?

Solution

Analyze: This is a multistep problem. We are given data for the N_2 gas (volume, pressure, and temperature) and the chemical equation for the reaction by which the N_2 is generated. We must use this information to calculate the number of grams needed to obtain the necessary N_2.

Plan: We need to use the gas data and the ideal-gas equation to calculate the number of moles of N_2 gas that should be formed for the air bag to operate correctly. We can then use the balanced equation to determine the number of moles of NaN_3. Finally, we can convert the moles of NaN_3 to grams.

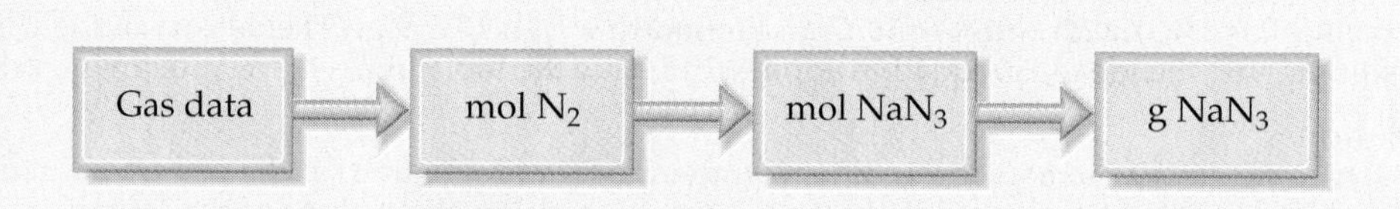

Solve: The number of moles of N_2 is given by

$$n = \frac{PV}{RT} = \frac{(1.15\ \text{atm})(36\ \text{L})}{(0.0821\ \text{L-atm/mol-K})(299\ \text{K})} = 1.7\ \text{mol N}_2$$

From here we use the coefficients in the balanced equation to calculate the number of moles of NaN_3.

$$(1.7\ \text{mol N}_2)\left(\frac{2\ \text{mol NaN}_3}{3\ \text{mol N}_2}\right) = 1.1\ \text{mol NaN}_3$$

Finally, using the molar mass of NaN_3, we convert moles of NaN_3 to grams:

$$(1.1\ \text{mol NaN}_3)\left(\frac{65.0\ \text{g NaN}_3}{1\ \text{mol NaN}_3}\right) = 72\ \text{g NaN}_3$$

Check: The best way to check our approach is to make sure the units cancel properly at each step in the calculation, leaving us with the correct units in the answer, g.

PRACTICE EXERCISE

In the first step in the industrial process for making nitric acid, ammonia reacts with oxygen in the presence of a suitable catalyst to form nitric oxide and water vapor:

$$4NH_3(g) + 5O_2(g) \longrightarrow 4NO(g) + 6H_2O(g)$$

How many liters of $NH_3(g)$ at 850°C and 5.00 atm are required to react with 1.00 mol of $O_2(g)$ in this reaction?
Answer: 14.8 L

Chemistry at Work Gas Pipelines

Most people are quite unaware of the vast network of underground pipelines that undergirds the developed world. Pipelines are used to move massive quantities of liquids and gases over considerable distances. For example, pipelines move natural gas (methane) from huge natural-gas fields in Siberia to Western Europe. Natural gas from Algeria is moved to Italy through a pipeline 120 cm in diameter and 2500 km in length that stretches across the Mediterranean Sea at depths up to 600 m. In the United States the pipeline systems consist of trunk lines, large-diameter pipes for long-distance transport, with branch lines of lower diameter and lower pressure for local transport to and from the trunk lines.

Essentially all substances that are gases at STP are transported commercially by pipeline, including ammonia, carbon dioxide, carbon monoxide, chlorine, ethane, helium, hydrogen, and methane. The largest volume transport by far, though, is natural gas. The methane-rich gas from oil and gas wells is processed to remove particulates, water, and various gaseous impurities such as hydrogen sulfide and carbon dioxide. The gas is then compressed to pressures ranging from 3.5 MPa (35 atm) to 10 MPa (100 atm), depending on the age and diameter of the pipe. (Figure 10.15 ▶) The long-distance pipelines are about 40 cm in diameter and made of steel. Pressure is maintained by large compressor stations along the pipeline, spaced at 50- to 100-mile intervals.

Recall from Figure 5.24 that natural gas is a major source of energy for the United States. To meet this demand, methane must be transported from source wells throughout the United States and Canada to all parts of the nation. The total length of pipeline for natural-gas transport in the United States is about 6×10^5 km, and growing. The United States is divided into seven regions. The total deliverability of natural gas to the seven regions exceeds 2.7×10^{12} L (measured at STP), which is almost 100 billion cubic feet per day! The volume of the pipelines themselves would be entirely inadequate for managing the enormous quantities of natural gas that are placed into and taken out of the system on a continuing basis. For this reason, underground storage facilities, such as salt caverns and other natural formations, are employed to hold large quantities of gas.

▲ **Figure 10.15** A natural gas pipeline relay station.

10.6 Gas Mixtures and Partial Pressures

Thus far we have considered only the behavior of pure gases—those that consist of only one substance in the gaseous state. How do we deal with gases composed of a mixture of two or more different substances? While studying the properties of air, John Dalton (Section 2.1) observed that the *total pressure of a mixture of gases equals the sum of the pressures that each would exert if it were present alone*. The pressure exerted by a particular component of a mixture of gases is called the **partial pressure** of that gas, and Dalton's observation is known as **Dalton's law of partial pressures**.

If we let P_t be the total pressure and P_1, P_2, P_3, and so forth be the partial pressures of the gases in the mixture, we can write Dalton's law as follows:

$$P_t = P_1 + P_2 + P_3 + \cdots \qquad [10.12]$$

This equation implies that each gas in the mixture behaves independently of the others, as we can see by the following analysis. Let n_1, n_2, n_3, and so forth be the number of moles of each of the gases in the mixture and n_t be the total number of moles of gas ($n_t = n_1 + n_2 + n_3 + \cdots$).

If each of the gases obeys the ideal-gas equation, we can write

$$P_1 = n_1\left(\frac{RT}{V}\right), \quad P_2 = n_2\left(\frac{RT}{V}\right), \quad P_3 = n_3\left(\frac{RT}{V}\right), \quad \text{and so forth.}$$

All the gases in the mixture are at the same temperature and occupy the same volume. Therefore, by substituting into Equation 10.12, we obtain

$$P_t = (n_1 + n_2 + n_3 + \cdots)\frac{RT}{V} = n_t\left(\frac{RT}{V}\right) \qquad [10.13]$$

That is, the total pressure at constant temperature and constant volume is determined by the total number of moles of gas present, whether that total represents just one substance or a mixture.

SAMPLE EXERCISE 10.10

A gaseous mixture made from 6.00 g O_2 and 9.00 g CH_4 is placed in a 15.0-L vessel at 0°C. What is the partial pressure of each gas, and what is the total pressure in the vessel?

Solution

Analyze: We need to calculate the pressure for two different gases in the same volume.

Plan: Because each gas behaves independently, we can use the ideal gas equation to calculate the pressure that each would exert if the other were not present.

Solve: We must first convert the mass of each gas to moles:

$$n_{O_2} = (6.00\text{ g }O_2)\left(\frac{1\text{ mol }O_2}{32.0\text{ g }O_2}\right) = 0.188\text{ mol }O_2$$

$$n_{CH_4} = (9.00\text{ g }CH_4)\left(\frac{1\text{ mol }CH_4}{16.0\text{ g }CH_4}\right) = 0.563\text{ mol }CH_4$$

We can now use the ideal-gas equation to calculate the partial pressure of each gas:

$$P_{O_2} = \frac{n_{O_2}RT}{V} = \frac{(0.188\text{ mol})(0.0821\text{ L-atm/mol-K})(273\text{ K})}{15.0\text{ L}} = 0.281\text{ atm}$$

$$P_{CH_4} = \frac{n_{CH_4}RT}{V} = \frac{(0.563\text{ mol})(0.0821\text{ L-atm/mol-K})(273\text{ K})}{15.0\text{ L}} = 0.841\text{ atm}$$

According to Dalton's law (Equation 10.12), the total pressure in the vessel is the sum of the partial pressures:

$$P_t = P_{O_2} + P_{CH_4} = 0.281\text{ atm} + 0.841\text{ atm} = 1.122\text{ atm}$$

Check: Performing rough estimates is good practice, even when you may not feel that you need to do it to check an answer. In this case a pressure of roughly 1 atm seems right for a mixture of about 0.2 mol O_2 (6/32) and a bit more than 0.5 mol CH_4 (9/16), together in a 15-L volume, because one mole of an ideal gas at 1 atm pressure and 0°C occupies about 22 L.

PRACTICE EXERCISE

What is the total pressure exerted by a mixture of 2.00 g of H_2 and 8.00 g of N_2 at 273 K in a 10.0-L vessel?
Answer: 2.86 atm

Partial Pressures and Mole Fractions

Because each gas in a mixture behaves independently, we can relate the amount of a given gas in a mixture to its partial pressure. For an ideal gas, $P = nRT/V$, and so we can write

$$\frac{P_1}{P_t} = \frac{n_1 RT/V}{n_t RT/V} = \frac{n_1}{n_t} \quad [10.14]$$

The ratio n_1/n_t is called the mole fraction of gas 1, which we denote X_1. The **mole fraction**, X, is a dimensionless number that expresses the ratio of the number of moles of one component to the total number of moles in the mixture. We can rearrange Equation 10.14 to give

$$P_1 = \left(\frac{n_1}{n_t}\right)P_t = X_1 P_t \quad [10.15]$$

Thus, the partial pressure of a gas in a mixture is its mole fraction times the total pressure.

The mole fraction of N_2 in air is 0.78 (that is, 78% of the molecules in air are N_2). If the total barometric pressure is 760 torr, then the partial pressure of N_2 is

$$P_{N_2} = (0.78)(760 \text{ torr}) = 590 \text{ torr}$$

This result makes intuitive sense: Because N_2 comprises 78% of the mixture, it contributes 78% of the total pressure.

SAMPLE EXERCISE 10.11

A study of the effects of certain gases on plant growth requires a synthetic atmosphere composed of 1.5 mol percent CO_2, 18.0 mol percent O_2, and 80.5 mol percent Ar. **(a)** Calculate the partial pressure of O_2 in the mixture if the total pressure of the atmosphere is to be 745 torr. **(b)** If this atmosphere is to be held in a 120-L space at 295 K, how many moles of O_2 are needed?

Solution

Analyze: We need to determine the number of moles of O_2 required to make up a synthetic atmosphere, given the percent composition.

Plan: We will calculate the partial pressure of O_2, then go on to calculate the number of moles of O_2 at that pressure needed to occupy 120 L.

Solve: (a) The mole percent is just the mole fraction times 100. Therefore, the mole fraction of O_2 is 0.180. Using Equation 10.15, we have

$$P_{O_2} = (0.180)(745 \text{ torr}) = 134 \text{ torr}$$

(b) Tabulating the given variables and changing them to appropriate units, we have

$$P_{O_2} = (134 \text{ torr})\left(\frac{1 \text{ atm}}{760 \text{ torr}}\right) = 0.176 \text{ atm}$$

$$V = 120 \text{ L}$$

$$n_{O_2} = ?$$

$$R = 0.0821 \frac{\text{L-atm}}{\text{mol-K}}$$

$$T = 295 \text{ K}$$

Solving the ideal-gas equation for n_{O_2}, we have

$$n_{O_2} = P_{O_2}\left(\frac{V}{RT}\right) = (0.176 \text{ atm})\frac{120 \text{ L}}{(0.0821 \text{ L-atm/mol-K})(295 \text{ K})} = 0.872 \text{ mol}$$

Check: The units check out satisfactorily, and the answer seems to be the right order of magnitude.

PRACTICE EXERCISE

From data gathered by *Voyager 1*, scientists have estimated the composition of the atmosphere of Titan, Saturn's largest moon. The total pressure on the surface of Titan is 1220 torr. The atmosphere consists of 82 mol percent N_2, 12 mol percent Ar, and 6.0 mol percent CH_4. Calculate the partial pressure of each of these gases in Titan's atmosphere.
Answer: 1.0×10^3 torr N_2, 1.5×10^2 torr Ar, and 73 torr CH_4

Collecting Gases over Water

An experiment that often comes up in the course of laboratory work involves determining the number of moles of gas collected from a chemical reaction. Sometimes this gas is collected over water. For example, solid potassium chlorate, $KClO_3$, can be decomposed by heating it in a test tube in an arrangement such as that shown in Figure 10.16 ▼. The balanced equation for the reaction is

$$2KClO_3(s) \longrightarrow 2KCl(s) + 3O_2(g) \qquad [10.16]$$

The oxygen gas is collected in a bottle that is initially filled with water and inverted in a water pan.

The volume of gas collected is measured by raising or lowering the bottle as necessary until the water levels inside and outside the bottle are the same. When this condition is met, the pressure inside the bottle is equal to the atmospheric pressure outside. The total pressure inside is the sum of the pressure of gas collected and the pressure of water vapor in equilibrium with liquid water.

$$P_{total} = P_{gas} + P_{H_2O} \qquad [10.17]$$

The pressure exerted by water vapor, P_{H_2O}, at various temperatures is listed in Appendix B.

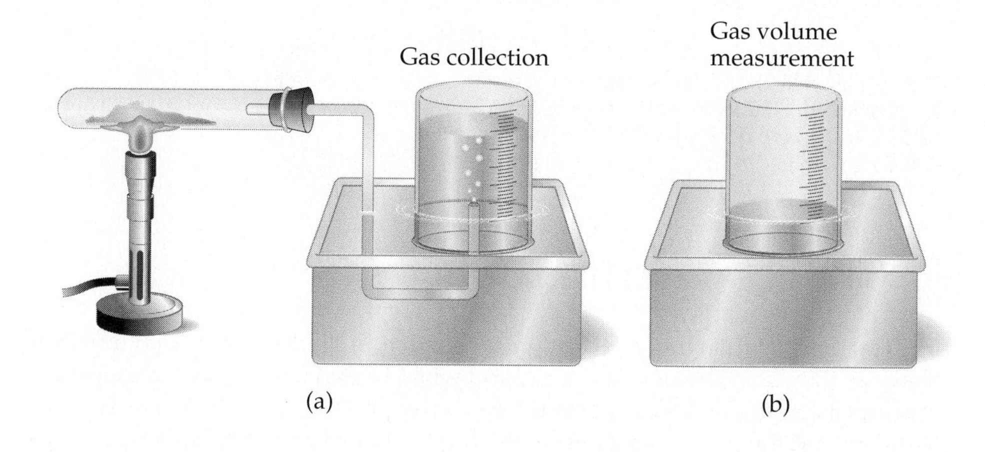

◀ **Figure 10.16** (a) Collection of gas over water. (b) When the gas has been collected, the bottle is raised or lowered so that the heights of the water inside and outside the collection vessel are made equal. The total pressure of the gases inside the vessel is then equal to the atmospheric pressure.

SAMPLE EXERCISE 10.12

A sample of $KClO_3$ is partially decomposed (Equation 10.16), producing O_2 gas that is collected over water as in Figure 10.16. The volume of gas collected is 0.250 L at 26°C and 765 torr total pressure. **(a)** How many moles of O_2 are collected? **(b)** How many grams of $KClO_3$ were decomposed?

Solution

Analyze: We first need to calculate the number of moles of O_2 gas in a container that also contains a second gas (water vapor). Secondly, we need to use the stoichiometry of the reaction to calculate the number of moles of reactant $KClO_3$ decomposed.

(a) Plan: If we tabulate the information presented, we will see that values are given for V and T. In order to calculate the unknown, n_{O_2}, we also must know the pressure of O_2 in the system. We therefore first need to determine the partial pressure of O_2 gas in the mixture of O_2 and H_2O vapor collected over water.
Solve: The partial pressure of the O_2 gas is the difference between the total pressure, 765 torr, and the pressure of the water vapor at 26°C, 25 torr (Appendix B):

$$P_{O_2} = 765 \text{ torr} - 25 \text{ torr} = 740 \text{ torr}$$

We can use the ideal-gas equation to solve for the number of moles of O_2, which gives us

$$n_{O_2} = \frac{P_{O_2}V}{RT} = \frac{(740 \text{ torr})(1 \text{ atm}/760 \text{ torr})(0.250 \text{ L})}{(0.0821 \text{ L-atm/mol-K})(299 \text{ K})} = 9.92 \times 10^{-3} \text{ mol } O_2$$

(b) Plan: We can use the balanced chemical equation to determine the number of moles of $KClO_3$ decomposed from the number of moles of O_2 formed, then convert moles of $KClO_3$ to grams of $KClO_3$.
Solve: From Equation 10.16, we have 2 mol $KClO_3 \simeq$ 3 mol O_2. The molar mass of $KClO_3$ is 122.6 g/mol. Thus, we can convert the moles of O_2 that we found in part (a) to moles of $KClO_3$ and grams of $KClO_3$:

$$(9.92 \times 10^{-3} \text{ mol } O_2)\left(\frac{2 \text{ mol } KClO_3}{3 \text{ mol } O_2}\right)\left(\frac{122.6 \text{ g } KClO_3}{1 \text{ mol } KClO_3}\right) = 0.811 \text{ g } KClO_3$$

Check: As always, we make sure that the units cancel appropriately in the calculations. In addition, the numbers of moles of O_2 and $KClO_3$ seem reasonable, given the small volume of gas collected.
Comment: Many chemical compounds that react with water and water vapor would be degraded by exposure to wet gas. Thus, in research laboratories, gases are often dried by passing wet gas over a substance that absorbs water (a *desiccant*), such as calcium sulfate, $CaSO_4$. Calcium sulfate crystals are sold as a desiccant under the trade name Drierite™.

PRACTICE EXERCISE

Ammonium nitrite, NH_4NO_2, decomposes upon heating to form N_2 gas:

$$NH_4NO_2(s) \longrightarrow N_2(g) + 2H_2O(l)$$

When a sample of NH_4NO_2 is decomposed in a test tube, as in Figure 10.16, 511 mL of N_2 gas is collected over water at 26°C and 745 torr total pressure. How many grams of NH_4NO_2 were decomposed?
Answer: 1.26 g

10.7 Kinetic-Molecular Theory

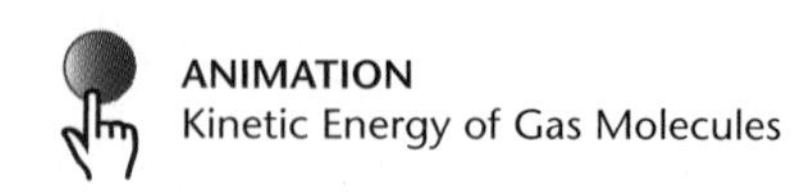

The ideal-gas equation describes *how* gases behave, but it doesn't explain *why* they behave as they do. Why does a gas expand when heated at constant pressure? Or, why does its pressure increase when the gas is compressed at constant temperature? To understand the physical properties of gases, we need a model that helps us picture what happens to gas particles as experimental conditions such as pressure or temperature change. Such a model, known as the **kinetic-molecular theory**, was developed over a period of about 100 years, culminating in 1857 when Rudolf Clausius (1822–1888) published a complete and satisfactory form of the theory.

The kinetic-molecular theory (the theory of moving molecules) is summarized by the following statements:

1. Gases consist of large numbers of molecules that are in continuous, random motion. (The word *molecule* is used here to designate the smallest particle of any gas; some gases, such as the noble gases, consist of individual atoms.)
2. The volume of all the molecules of the gas is negligible compared to the total volume in which the gas is contained.

3. Attractive and repulsive forces between gas molecules are negligible.
4. Energy can be transferred between molecules during collisions, but the *average* kinetic energy of the molecules does not change with time, as long as the temperature of the gas remains constant. In other words, the collisions are perfectly elastic.
5. The average kinetic energy of the molecules is proportional to the absolute temperature. At any given temperature the molecules of all gases have the same average kinetic energy.

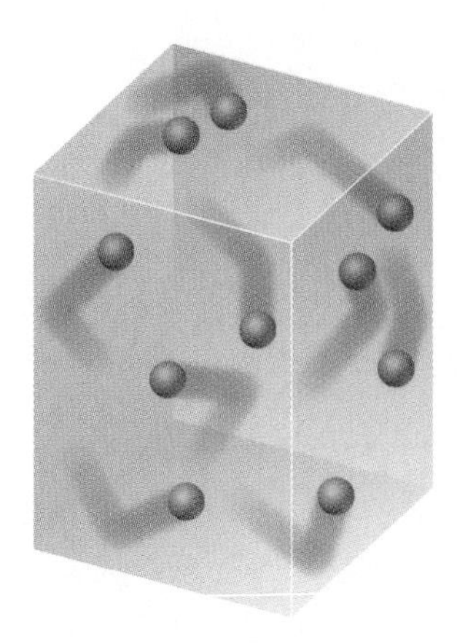

▲ **Figure 10.17** The pressure of a gas is caused by collisions of the gas molecules with the walls of their container.

The kinetic-molecular theory explains both pressure and temperature at the molecular level. The pressure of a gas is caused by collisions of the molecules with the walls of the container, as shown in Figure 10.17 ▶. The magnitude of the pressure is determined both by how often and how forcefully the molecules strike the walls.

The absolute temperature of a gas is a measure of the *average* kinetic energy of its molecules. If two different gases are at the same temperature, their molecules have the same average kinetic energy. If the absolute temperature of a gas is doubled (say from 200 K to 400 K), the average kinetic energy of its molecules doubles. Thus, molecular motion increases with increasing temperature.

Although the molecules in a sample of gas have an *average* kinetic energy and hence an average speed, the individual molecules move at varying speeds. The moving molecules experience frequent collisions with other molecules. Momentum is conserved in each collision, but one of the colliding molecules might be deflected off at high speed while the other is nearly stopped altogether. The result is that the molecules at any instant have a wide range of speeds. Figure 10.18 ▼ illustrates the distribution of molecular speeds for nitrogen gas at 0°C (blue line) and at 100°C (red line). The curve shows us the fraction of molecules moving at each speed. At higher temperatures, a larger fraction of molecules moves at greater speeds; the distribution curve has shifted toward higher speeds and hence toward higher average kinetic energy.

Figure 10.18 also shows the value of the **root-mean-square (rms) speed,** u, of the molecules at each temperature. This quantity is the speed of a molecule possessing average kinetic energy. The rms speed is not quite the same as the average (mean) speed. The difference between the two, however, is small.*

* To illustrate the difference between rms speed and average speed, suppose that we have four objects with speeds of 4.0, 6.0, 10.0, and 12.0 m/s. Their average speed is $\frac{1}{4}(4.0 + 6.0 + 10.0 + 12.0) = 8.0$ m/s. The rms speed, u, however, is the square root of the average squared speeds of the molecules:

$$\sqrt{\tfrac{1}{4}(4.0^2 + 6.0^2 + 10.0^2 + 12.0^2)} = \sqrt{74.0} = 8.6 \text{ m/s}$$

For an ideal gas, the average speed equals $0.921 \times u$. Thus, the average speed is directly proportional to the rms speed, and the two are in fact nearly equal.

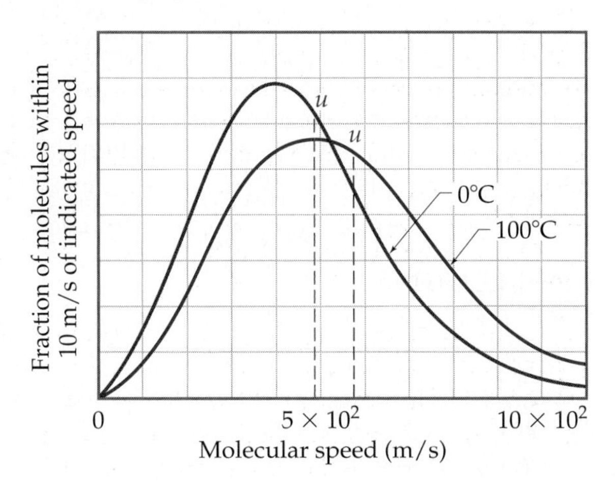

◀ **Figure 10.18** Distribution of molecular speeds for nitrogen at 0°C (blue line) and 100°C (red line).

The rms speed is important because the average kinetic energy of the gas molecules, ϵ, is related directly to u^2:

$$\epsilon = \tfrac{1}{2}mu^2 \qquad [10.18]$$

where m is the mass of the molecule. Mass does not change with temperature. Thus, the increase in the average kinetic energy as the temperature increases implies that the rms speed (and also the average speed) of molecules likewise increases as temperature increases.

Application to the Gas Laws

The empirical observations of gas properties as expressed in the various gas laws are readily understood in terms of the kinetic-molecular theory. The following examples illustrate this point:

1. *Effect of a volume increase at constant temperature*: A constant temperature means that the average kinetic energy of the gas molecules remains unchanged. This in turn means that the rms speed of the molecules, u, is unchanged. If the volume is increased, however, the molecules must move a longer distance between collisions. Consequently, there are fewer collisions per unit time with the container walls, and pressure decreases. Thus, the model accounts in a simple way for Boyle's law.
2. *Effect of a temperature increase at constant volume*: An increase in temperature means an increase in the average kinetic energy of the molecules, and thus an increase in u. If there is no change in volume, there will be more collisions with the walls per unit time. Furthermore, the change in momentum in each collision increases (the molecules strike the walls more forcefully). Hence, the model explains the observed pressure increase.

SAMPLE EXERCISE 10.13

A sample of O_2 gas initially at STP is compressed to a smaller volume at constant temperature. What effect does this change have on **(a)** the average kinetic energy of O_2 molecules; **(b)** the average speed of O_2 molecules; **(c)** the total number of collisions of O_2 molecules with the container walls in a unit time; **(d)** the number of collisions of O_2 molecules with a unit area of container wall per unit time?

Solution

Analyze: We need to apply the concepts of the kinetic-molecular theory to a situation in which a gas is compressed at constant temperature.
Plan: We will determine how each of the quantities in (a)–(d) is affected by the change in pressure at constant volume.
Solve: **(a)** The average kinetic energy of the O_2 molecules is determined only by temperature. The average kinetic energy is unchanged by the compression of O_2 at constant temperature. **(b)** If the average kinetic energy of O_2 molecules doesn't change, the average speed remains constant. **(c)** The total number of collisions with the container walls per unit time must increase because the molecules are moving within a smaller volume but with the same average speed as before. Under these conditions they must encounter a wall more frequently. **(d)** The number of collisions with a unit area of wall increases because the total number of collisions with the walls is higher and the area of the walls is smaller than before.
Check: In a conceptual exercise of this kind, there is no numerical answer to check. We really check our reasoning in the course of solving the problem.

PRACTICE EXERCISE

How is the rms speed of N_2 molecules in a gas sample changed by **(a)** an increase in temperature; **(b)** an increase in volume of the sample; **(c)** mixing with a sample of Ar at the same temperature?
Answers: **(a)** increases; **(b)** no effect; **(c)** no effect

A Closer Look The Ideal-Gas Equation

Beginning with the postulates of the kinetic-molecular theory, it is possible to derive the ideal-gas equation. Rather than proceed through a derivation, let's consider in somewhat qualitative terms how the ideal-gas equation might follow. As we have seen, pressure is force per unit area. (Section 10.2) The total force of the molecular collisions on the walls, and hence the pressure produced by these collisions, depends both on how strongly the molecules strike the walls (impulse imparted per collision) and on the rate at which these collisions occur:

$$P \propto \text{impulse imparted per collision} \times \text{rate of collisions}$$

For a molecule traveling at the rms speed, u, the impulse imparted by a collision with a wall depends on the momentum of the molecule; that is, it depends on the product of its mass and speed, mu. The rate of collisions is proportional to both the number of molecules per unit volume, n/V, and their speed, u. If there are more molecules in a container, there will be more frequent collisions with the container walls. As the molecular speed increases or the volume of the container decreases, the time required for molecules to traverse the distance from one wall to another is reduced, and the molecules collide more frequently with the walls. Thus, we have

$$P \propto mu \times \frac{n}{V} \times u \propto \frac{nmu^2}{V} \qquad [10.19]$$

Because the average kinetic energy, $\frac{1}{2}mu^2$, is proportional to temperature, we have $mu^2 \propto T$. Making this substitution into Equation 10.19 gives

$$P \propto \frac{n(mu^2)}{V} \propto \frac{nT}{V} \qquad [10.20]$$

Let's now convert the proportionality sign to an equal sign by expressing n as the number of moles of gas; we then insert a proportionality constant—R, the molar gas constant:

$$P = \frac{nRT}{V} \qquad [10.21]$$

This expression is the ideal-gas equation.

An eminent Swiss mathematician, Daniel Bernoulli (1700–1782), conceived of a model for gases that was, for all practical purposes, the same as the kinetic theory model. From this model, Bernoulli derived Boyle's law and the ideal-gas equation. His was one of the first examples in science of developing a mathematical model from a set of assumptions, or hypothetical statements. In spite of his eminence, however, Bernoulli's work on this subject was completely ignored, only to be rediscovered a hundred years later by Clausius and others. It was ignored because it conflicted with popular beliefs. For example, his idea that heat is a measure of motional energy was not accepted because it conflicted with the then-popular (and incorrect) caloric theory of heat. Second, Bernoulli was in conflict with Isaac Newton's model for gases (also incorrect). Those idols of the times had to fall before the way was clear for the kinetic-molecular theory. The moral of the story is that science is not a straight road running from here to the "truth." The road is built by humans, so it zigs and zags.

10.8 Molecular Effusion and Diffusion

According to the kinetic-molecular theory, the average kinetic energy of *any* collection of gas molecules, $\frac{1}{2}mu^2$, has a specific value at a given temperature. Thus, a gas composed of light particles, such as He, will have the same average kinetic energy as one composed of much heavier particles, such as Xe, provided the two gases are at the same temperature. The mass, m, of the particles in the lighter gas is smaller than that in the heavier gas. Consequently, the particles of the lighter gas must have a higher rms speed, u, than the particles of the heavier one. The following equation, which expresses this fact quantitatively, can be derived from kinetic-molecular theory:

$$u = \sqrt{\frac{3RT}{\mathcal{M}}} \qquad [10.22]$$

Because the molar mass, $\mathcal{M}$, appears in the denominator, the less massive the gas molecules, the higher the rms speed, u. Figure 10.19 ▶ shows the distribution of molecular speeds for several different gases at 25°C. Notice how the distributions are shifted toward higher speeds for gases of lower molar masses.

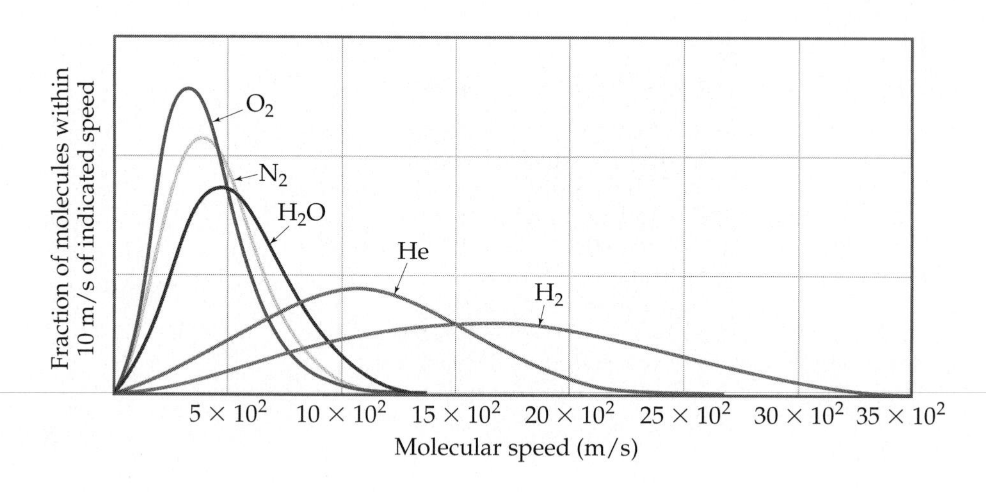

▶ **Figure 10.19** Distribution of molecular speeds for different gases at 25°C.

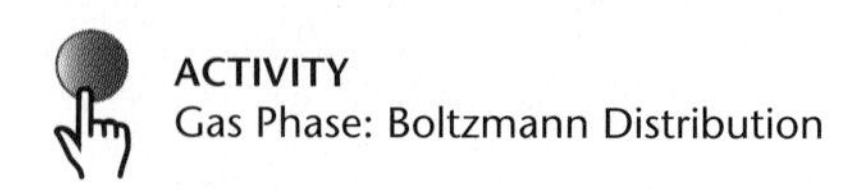

SAMPLE EXERCISE 10.14

Calculate the rms speed, u, of an N_2 molecule at 25°C.

Solution

Analyze: The data given are the identity of the gas and the temperature, the two quantities we need to calculate rms speed.

Plan: We will calculate the rms speed using Equation 10.22.

Solve: In using Equation 10.22, we should convert each quantity to SI units so that all the units are compatible. We will also use R in units of J/mol-K (Table 10.2) in order to make the units cancel correctly.

$$T = 25 + 273 = 298 \text{ K}$$

$$\mathcal{M} = 28.0 \text{ g/mol} = 28.0 \times 10^{-3} \text{ kg/mol}$$

$$R = 8.314 \text{ J/mol-K} = 8.314 \text{ kg-m}^2/\text{s}^2\text{-mol-K}$$

(These units follow from the fact that $1 \text{ J} = 1 \text{ kg-m}^2/\text{s}^2$)

$$u = \sqrt{\frac{3(8.314 \text{ kg-m}^2/\text{s}^2\text{-mol-K})(298 \text{ K})}{28.0 \times 10^{-3} \text{ kg/mol}}} = 5.15 \times 10^2 \text{ m/s}$$

Comment: This corresponds to a speed of 1150 mi/hr. Because the average molecular weight of air molecules is slightly greater than that of N_2, the rms speed of air molecules is a little slower than that for N_2. The speed at which sound propagates through air is about 350 m/s, a value about two-thirds the average rms speed for air molecules.

PRACTICE EXERCISE

What is the rms speed of an He atom at 25°C?
Answer: 1.36×10^3 m/s

The dependence of molecular speeds on mass has several interesting consequences. The first phenomenon is **effusion**, which is the escape of gas molecules through a tiny hole into an evacuated space as shown in Figure 10.20 ◀. The second is **diffusion**, which is the spread of one substance throughout a space or throughout a second substance. For example, the molecules of a perfume diffuse throughout a room.

▲ **Figure 10.20** Effusion of a gas molecule through a pinhole. Molecules escape from their container into the evacuated space only when they happen to hit the hole.

Graham's Law of Effusion

In 1846 Thomas Graham (1805–1869) discovered that the effusion rate of a gas is inversely proportional to the square root of its molar mass. Assume that we have two gases at the same temperature and pressure in containers with identical pinholes. If the rates of effusion of the two substances are r_1 and r_2, and their respective molar masses are $\mathcal{M}_1$ and $\mathcal{M}_2$, **Graham's law** states

$$\frac{r_1}{r_2} = \sqrt{\frac{\mathcal{M}_2}{\mathcal{M}_1}} \qquad [10.23]$$

(a)

(b)

◀ **Figure 10.21** Light atoms or molecules escape through the pores of a balloon faster than heavier ones. (a) Two balloons are filled to the same volume, one with helium and the other with nitrogen. (b) After 48 hr the helium-filled balloon is smaller than the nitrogen-filled one because helium escapes faster than nitrogen.

Equation 10.23 compares the *rates* of effusion of two different gases under identical conditions, and it indicates that the lighter gas effuses more rapidly.

Figure 10.20 illustrates the basis of Graham's law. The only way for a molecule to escape from its container is for it to "hit" the hole. The faster the molecules are moving, the greater is the likelihood that a molecule will hit the hole and effuse. This implies that the rate of effusion is directly proportional to the rms speed of the molecules. Because R and T are constant, we have, from Equation 10.22

$$\frac{r_1}{r_2} = \frac{u_1}{u_2} = \sqrt{\frac{3RT/\mathcal{M}_1}{3RT/\mathcal{M}_2}} = \sqrt{\frac{\mathcal{M}_2}{\mathcal{M}_1}} \qquad [10.24]$$

As expected from Graham's law, helium escapes from containers through tiny pinhole leaks more rapidly than other gases of higher molecular weight (Figure 10.21 ▲).

SAMPLE EXERCISE 10.15

An unknown gas composed of homonuclear diatomic molecules effuses at a rate that is only 0.355 times that of O_2 at the same temperature. What is the identity of the unknown gas?

Solution

Analyze: We are given information related to the relative rate of effusion of an unknown gas, and from that we are asked to find its molar mass. Thus, we need to connect relative rates of effusion to relative molar masses.

Plan: We can use Graham's law of effusion, Equation 10.23, to determine the molar mass of the unknown gas. If we let r_x and $\mathcal{M}_x$ represent the rate of effusion and molar mass of the unknown gas, Equation 10.23 can be written as follows:

$$\frac{r_x}{r_{O_2}} = \sqrt{\frac{\mathcal{M}_{O_2}}{\mathcal{M}_x}}$$

Solve: From the information given,

$$r_x = 0.355 \times r_{O_2}$$

Thus,

$$\frac{r_x}{r_{O_2}} = 0.355 = \sqrt{\frac{32.0\ \text{g/mol}}{\mathcal{M}_x}}$$

We now solve for the unknown molar mass, $\mathcal{M}_x$:

$$\frac{32.0\ \text{g/mol}}{\mathcal{M}_x} = (0.355)^2 = 0.126$$

$$\mathcal{M}_x = \frac{32.0\ \text{g/mol}}{0.126} = 254\ \text{g/mol}$$

Because we are told that the unknown gas is composed of homonuclear diatomic molecules, it must be an element. The molar mass must represent twice the atomic weight of the atoms in the unknown gas. We conclude that the unknown gas is I_2.

PRACTICE EXERCISE

Calculate the ratio of the effusion rates of N_2 and O_2, r_{N_2}/r_{O_2}.
Answer: $r_{N_2}/r_{O_2} = 1.07$

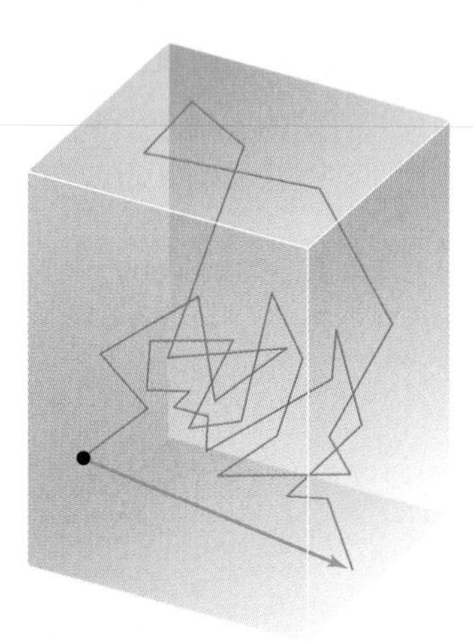

▲ **Figure 10.22** Schematic illustration of the diffusion of a gas molecule. For clarity, no other gas molecules in the container are shown. The path of the molecule of interest begins at the dot. Each short segment of line represents travel between collisions. The blue arrow indicates the net distance traveled by the molecule.

Diffusion and Mean Free Path

Diffusion, like effusion, is faster for light molecules than for heavy ones. In fact, the ratio of rates of diffusion of two gases under identical experimental conditions is approximated by Graham's law, Equation 10.23. Nevertheless, molecular collisions make diffusion more complicated than effusion.

We can see from the horizontal scale in Figure 10.19 that the speeds of molecules are quite high. For example, the average speed of N_2 at room temperature is 515 m/s (1150 mi/hr). In spite of this high speed, if someone opens a vial of perfume at one end of a room, some time elapses—perhaps a few minutes—before the odor is detected at the other end. The diffusion of gases is much slower than molecular speeds due to molecular collisions.* These collisions occur quite frequently for a gas at atmospheric pressure—about 10^{10} times per second for each molecule. Collisions occur because real gas molecules have finite volumes.

Because of molecular collisions, the direction of motion of a gas molecule is constantly changing. Therefore, the diffusion of a molecule from one point to another consists of many short, straight-line segments as collisions buffet it around in random directions, as depicted in Figure 10.22 ◀. First the molecule moves in one direction, then in another; one instant at high speed, the next at low speed.

* The rate at which the perfume moves across the room also depends on how well stirred the air is from temperature gradients and the movement of people. Nevertheless, even with the aid of these factors, it still takes much longer for the molecules to traverse the room than one would expect from the rms speed alone.

Chemistry at Work Gas Separations

The fact that lighter molecules move at higher average speeds than more massive ones has many interesting consequences and applications. For example, the effort to develop the atomic bomb during World War II required scientists to separate the relatively low-abundance uranium isotope ^{235}U (0.7%) from the much more abundant ^{238}U (99.3%). This was accomplished by converting the uranium into a volatile compound, UF_6, which sublimes at 56°C. The gaseous UF_6 was allowed to pass through porous barriers. Because of the lengths of the pores, this is not a simple effusion. Nevertheless, the dependence on molar mass is essentially the same. The slight difference in molar mass between the compounds of the two isotopes caused the molecules to move at slightly different rates:

$$\frac{r_{235}}{r_{238}} = \sqrt{\frac{352.04}{349.03}} = 1.0043$$

Thus, the gas initially appearing on the opposite side of the barrier was very slightly enriched in the lighter molecule. The diffusion process was repeated thousands of times, leading to a nearly complete separation of the two isotopes of uranium.

The rate at which a gas passes through a porous medium is not always determined solely by the molecular mass of the gas molecules. Even weak interactions between the molecules of gas and the molecules of the porous medium will affect the rate. Attractive intermolecular interactions slow the rate at which a gas molecule passes through the narrow passages of the porous medium.

The average distance traveled by a molecule between collisions is called the **mean free path**. The mean free path varies with pressure as the following analogy illustrates. Imagine walking through a shopping mall. When a mall is very crowded (high pressure), the average distance you can walk before bumping into someone else is short (short mean free path). When the mall is empty (low pressure), you can walk a long way (long mean free path) before bumping into someone else. The mean free path for air molecules at sea level is about 60 nm (6×10^{-6} cm). At about 100 km in altitude, where the air density is much lower, the mean free path is about 10 cm, about 1 million times longer than at Earth's surface.

10.9 Real Gases: Deviations from Ideal Behavior

Although the ideal-gas equation is a very useful description of gases, all real gases fail to obey the relationship to some degree. The extent to which a real gas departs from ideal behavior can be seen by rearranging the ideal-gas equation:

$$\frac{PV}{RT} = n \qquad [10.25]$$

For a mole of ideal gas ($n = 1$) the quantity PV/RT equals 1 at all pressures. In Figure 10.23 ▼ PV/RT is plotted as a function of P for 1 mol of several different gases. At high pressures the deviation from ideal behavior ($PV/RT = 1$) is large and is different for each gas. Real gases, therefore, do not behave ideally at high pressure. At lower pressures (usually below 10 atm), however, the deviation from ideal behavior is small, and we can use the ideal-gas equation without generating serious error.

The deviation from ideal behavior also depends on temperature. Figure 10.24 ▼ shows graphs of PV/RT versus P for 1 mol of N_2 at three different temperatures. As temperature increases, the properties of the gas more nearly approach that of the ideal gas. In general, the deviations from ideal behavior increase as temperature decreases, becoming significant near the temperature at which the gas is converted into a liquid.

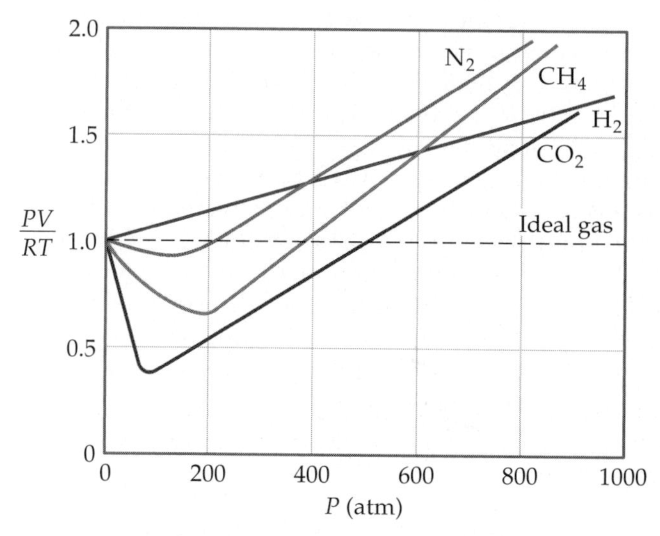

◀ **Figure 10.23** PV/RT versus pressure for 1 mol of several gases at 300 K. The data for CO_2 pertain to a temperature of 313 K because CO_2 liquefies under high pressure at 300 K.

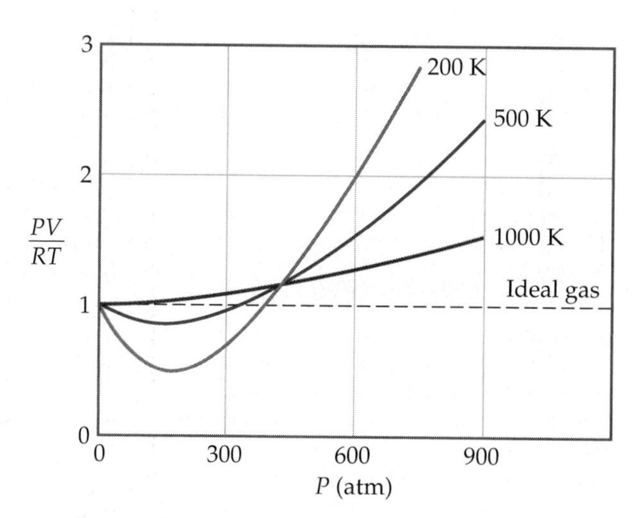

▲ **Figure 10.24** PV/RT versus pressure for 1 mol of nitrogen gas at three different temperatures. As temperature increases, the gas more closely approaches ideal behavior.

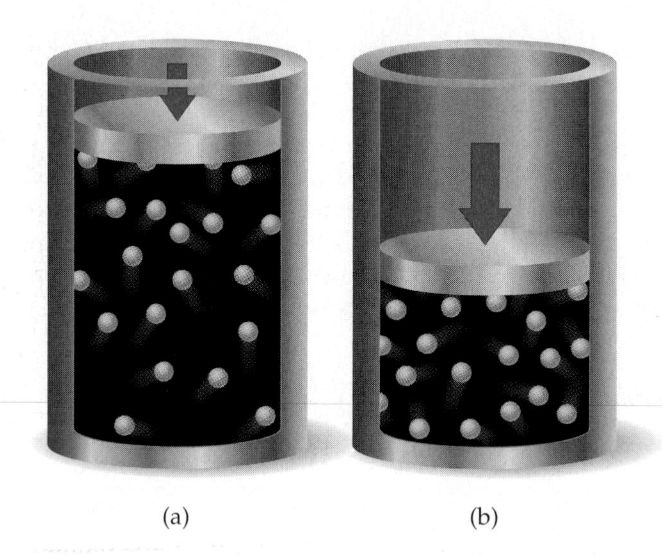

▶ **Figure 10.25** Illustration of the effect of the finite volume of gas molecules on the properties of a real gas at high pressure. In (a), at low pressure, the volume of the gas molecules is small compared with the container volume. In (b), at high pressure, the volume of the gas molecules themselves is a larger fraction of the total space available.

ACTIVITY
Real Gases

The basic assumptions of the kinetic-molecular theory give us insight into why real gases deviate from ideal behavior. The molecules of an ideal gas are assumed to occupy no space and have no attractions for one another. *Real molecules, however, do have finite volumes, and they do attract one another*. As shown in Figure 10.25 ▲, the free, unoccupied space in which molecules can move is somewhat less than the container volume. At relatively low pressures the volume of the gas molecules is negligible, compared with the container volume. Thus, the free volume available to the molecules is essentially the entire volume of the container. As the pressure increases, however, the free space in which the molecules can move becomes a smaller fraction of the container volume. Under these conditions, therefore, gas volumes tend to be slightly greater than those predicted by the ideal-gas equation.

In addition, the attractive forces between molecules come into play at short distances, as when molecules are crowded together at high pressures. Because of these attractive forces, the impact of a given molecule with the wall of the container is lessened. If we could stop the action in a gas, the positions of the molecules might resemble the illustration in Figure 10.26 ◀. The molecule about to make contact with the wall experiences the attractive forces of nearby molecules. These attractions lessen the force with which the molecule hits the wall. As a result, the pressure is less than that of an ideal gas. This effect serves to decrease PV/RT, as seen in Figure 10.23. When the pressure is sufficiently high, these volume effects dominate and PV/RT increases.

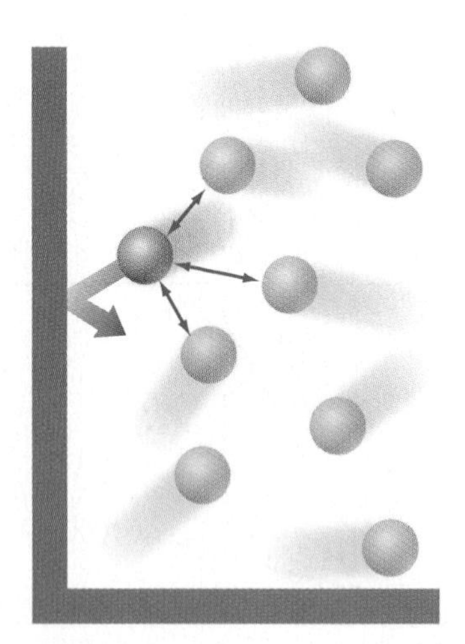

▲ **Figure 10.26** Effect of attractive intermolecular forces on the pressure exerted by a gas on its container walls. The molecule that is about to strike the wall experiences attractive forces from nearby molecules, and its impact on the wall is thereby lessened. The attractive forces become significant only under high-pressure conditions, when the average distance between molecules is small.

Temperature determines how effective attractive forces between gas molecules are. As the gas is cooled, the average kinetic energy decreases, whereas intermolecular attractions remain constant. In a sense, cooling a gas deprives molecules of the energy they need to overcome their mutual attractive influence. The effects of temperature shown in Figure 10.24 illustrate this point very well. As temperature increases, the negative departure of PV/RT from ideal-gas behavior disappears. The difference that remains at high temperature stems mainly from the effect of the finite volumes of the molecules.

The van der Waals Equation

Engineers and scientists who work with gases at high pressures often cannot use the ideal-gas equation to predict the pressure-volume properties of gases because departures from ideal behavior are too large. One useful equation developed to predict the behavior of real gases was proposed by the Dutch scientist Johannes van der Waals (1837–1923).

The ideal-gas equation predicts that the pressure of a gas is

$$P = \frac{nRT}{V} \quad \text{(ideal gas)}$$

Van der Waals recognized that for a real gas, this expression would have to be corrected for the finite volume occupied by the gas molecules and for the attractive forces between the gas molecules. He introduced two constants, a and b, to make these corrections.

$$P = \underbrace{\frac{nRT}{V - nb}}_{\text{Correction for volume of molecules}} - \underbrace{\frac{n^2a}{V^2}}_{\text{Correction for molecular attraction}} \quad [10.26]$$

The volume is decreased by the factor nb, which accounts for the finite volume occupied by the gas molecules (Figure 10.25). The van der Waals constant b is a measure of the actual volume occupied by a mole of gas molecules; b has units of L/mol. The pressure is in turn decreased by the factor n^2a/V^2, which accounts for the attractive forces between the gas molecules (Figure 10.26). The unusual form of this correction results because the attractive forces between pairs of molecules increase as the square of the number of molecules per unit volume $(n/V)^2$. Hence, the van der Waals constant a has units of L^2-atm/mol^2. The magnitude of a reflects how strongly the gas molecules attract each other.

Equation 10.26 is generally rearranged to give the following form of the **van der Waals equation**:

$$\left(P + \frac{n^2a}{V^2}\right)(V - nb) = nRT \quad [10.27]$$

The van der Waals constants a and b are different for each gas. Values of these constants for several gases are listed in Table 10.3 ▼. Note that the values of both a and b generally increase with an increase in mass of the molecule and with an increase in the complexity of its structure. Larger, more massive molecules not only have larger volumes, they also tend to have greater intermolecular attractive forces.

TABLE 10.3 van der Waals Constants for Gas Molecules

Substance	a (L^2-atm/mol^2)	b (L/mol)
He	0.0341	0.02370
Ne	0.211	0.0171
Ar	1.34	0.0322
Kr	2.32	0.0398
Xe	4.19	0.0510
H_2	0.244	0.0266
N_2	1.39	0.0391
O_2	1.36	0.0318
Cl_2	6.49	0.0562
H_2O	5.46	0.0305
CH_4	2.25	0.0428
CO_2	3.59	0.0427
CCl_4	20.4	0.1383

SAMPLE EXERCISE 10.16

If 1.000 mol of an ideal gas were confined to 22.41 L at 0.0°C, it would exert a pressure of 1.000 atm. Use the van der Waals equation and the constants in Table 10.3 to estimate the pressure exerted by 1.000 mol of $Cl_2(g)$ in 22.41 L at 0.0°C.

Solution

Analyze: The quantity we need to solve for is pressure. Because we will use the van der Waals equation, we must identify the appropriate values for the constants that appear there.

Plan: Using Equation 10.26, we have

$$P = \frac{nRT}{V - nb} - \frac{n^2a}{V^2}$$

Solve: Substituting $n = 1.000$ mol, $R = 0.08206$ L-atm/mol-K, $T = 273.2$ K, $V = 22.41$ L, $a = 6.49$ L^2 atm/mol^2, and $b = 0.0562$ L/mol:

$$P = \frac{(1.000 \text{ mol})(0.08206 \text{ L-atm/mol-K})(273.2 \text{ K})}{22.41 \text{ L} - (1.000 \text{ mol})(0.0562 \text{ L/mol})} - \frac{(1.000 \text{ mol})^2(6.49 \text{ L}^2\text{-atm/mol}^2)}{(22.41 \text{ L})^2}$$

$$= 1.003 \text{ atm} - 0.013 \text{ atm} = 0.990 \text{ atm}$$

Check: We expect a pressure not far from 1.000 atm, which would be the value for an ideal gas, so our answer seems very reasonable.

Comment: Notice that the first term, 1.003 atm, is the pressure corrected for molecular volume. This value is higher than the ideal value, 1.000 atm, because the volume in which the molecules are free to move is smaller than the container volume, 22.41 L. Thus, the molecules must collide more frequently with the container walls. The second factor, 0.013 atm, corrects for intermolecular forces. The intermolecular attractions between molecules reduce the pressure to 0.990 atm. We can conclude, therefore, that the intermolecular attractions are the main cause of the slight deviation of $Cl_2(g)$ from ideal behavior under the stated experimental conditions.

PRACTICE EXERCISE

Consider a sample of 1.000 mol of $CO_2(g)$ confined to a volume of 3.000 L at 0.0°C. Calculate the pressure of the gas using **(a)** the ideal-gas equation, and **(b)** the van der Waals equation.

Answers: **(a)** 7.473 atm; **(b)** 7.182 atm

SAMPLE INTEGRATIVE EXERCISE 10: Putting Concepts Together

Cyanogen, a highly toxic gas, is composed of 46.2% C and 53.8% N by mass. At 25°C and 751 torr, 1.05 g of cyanogen occupies 0.500 L. **(a)** What is the molecular formula of cyanogen? **(b)** Predict its molecular structure. **(c)** Predict the polarity of the compound.

Solution

Analyze: We need to determine the molecular formula of a compound from elemental analysis data and data on the properties of the gaseous substance. Thus, we have two separate calculations to do.

(a) Plan: We can use the percentage composition of the compound to calculate its empirical formula. (Section 3.5) Then we can determine the molecular formula by comparing the mass of the empirical formula with the molar mass. (Section 3.5)

Solve: To determine the empirical formula, we assume that we have a 100-g sample of the compound and then calculate the number of moles of each element in the sample:

$$\text{Moles C} = (46.2 \text{ g C})\left(\frac{1 \text{ mol C}}{12.01 \text{ g C}}\right) = 3.85 \text{ mol C}$$

$$\text{Moles N} = (53.8 \text{ g N})\left(\frac{1 \text{ mol N}}{14.01 \text{ g N}}\right) = 3.84 \text{ mol N}$$

Because the ratio of the moles of the two elements is essentially 1 : 1, the empirical formula is CN.

To determine the molar mass of the compound, we use Equation 10.11.

$$\mathcal{M} = \frac{dRT}{P} = \frac{(1.05\ \text{g}/0.500\ \text{L})(0.0821\ \text{L-atm/mol-K})(298\ \text{K})}{(751/760)\text{atm}} = 52.0\ \text{g/mol}$$

The molar mass associated with the empirical formula, CN, is 12.0 + 14.0 = 26.0 g/mol. Dividing the molar mass of the compound by that of its empirical formula gives (52.0 g/mol)/(26.0 g/mol) = 2.00. Thus, the molecule has twice as many atoms of each element as the empirical formula, giving the molecular formula C_2N_2.

(b) Plan: To determine the molecular structure of the molecule, we must first determine its Lewis structure. (Section 8.5) We can then use the VSEPR model to predict the structure. (Section 9.2)

Solve: The molecule has 2(4) + 2(5) = 18 valence-shell electrons. By trial and error, we seek a Lewis structure with 18 valence electrons in which each atom has an octet and in which the formal charges are as low as possible. The following structure meets these criteria:

$$:N{\equiv}C{-}C{\equiv}N:$$

(This structure has zero formal charges on each atom.)

The Lewis structure shows that each atom has two electron domains. (Each nitrogen has a nonbonding pair of electrons and a triple bond, whereas each carbon has a triple bond and a single bond.) Thus the electron-domain geometry around each atom is linear, causing the overall molecule to be linear.

(c) Plan: To determine the polarity of the molecule, we must examine the polarity of the individual bonds and the overall geometry of the molecule.

Solve: Because the molecule is linear, we expect the two dipoles created by the polarity in the carbon–nitrogen bond to cancel each other, leaving the molecule with no dipole moment.

Summary and Key Terms

Section 10.1 Substances that are gases at room temperatures tend to be molecular substances with low molar masses. Air, a mixture composed mainly of N_2 and O_2, is the most common gas we encounter. Some liquids and solids can also exist in the gaseous state, where they are known as **vapors**. Gases are compressible; they mix in all proportions because their component molecules are far apart.

Section 10.2 To describe the state or condition of a gas, we must specify four variables: pressure (P), volume (V), temperature (T), and quantity of gas (n). Volume is usually measured in liters, temperature in kelvins, and quantity of gas in moles. **Pressure** is the force per unit area. It is expressed in SI units as **pascals**, Pa (1 Pa = 1 N/m^2 = 1 kg/m-s^2). A related unit, the **bar**, equals 10^5 Pa. In chemistry, **standard atmospheric pressure** is used to define the **atmosphere** (atm) and the **torr** (also called the millimeter of mercury). One atmosphere pressure equals 101.325 kPa, or 760 torr. A barometer is often used to measure the atmospheric pressure. A manometer can be used to measure the pressure of enclosed gases.

Sections 10.3 and 10.4 Studies have revealed several simple gas laws: For a constant quantity of gas at constant temperature, the volume of the gas is inversely proportional to the pressure (**Boyle's law**). For a fixed quantity of gas at constant pressure, the volume is directly proportional to its absolute temperature (**Charles's law**). Equal volumes of gases at the same temperature and pressure contain equal numbers of molecules (**Avogadro's hypothesis**). For a gas at constant temperature and pressure, the volume of the gas is directly proportional to the number of moles of gas (**Avogadro's law**). Each of these gas laws is a special case of the ideal-gas equation.

The **ideal-gas equation**, $PV = nRT$, is the equation of state for an **ideal gas**. The term R in this equation is the **gas constant**. We can use the ideal-gas equation to calculate variations in one variable when one or more of the others are changed. Most gases at pressures of about 1 atm and temperatures near 273 K and above obey the ideal-gas equation reasonably well. The conditions of 273 K (0°C) and 1 atm are known as the **standard temperature and pressure (STP)**.

Sections 10.5 and 10.6 Using the ideal-gas equation, we can relate the density of a gas to its molar mass: $\mathcal{M} = dRT/P$. We can also use the ideal-gas equation to solve problems involving gases as reactants or products in chemical reactions. In all applications of the ideal-gas equation we must remember to convert temperatures to the absolute-temperature scale (the Kelvin scale).

In gas mixtures the total pressure is the sum of the **partial pressures** that each gas would exert if it were present alone under the same conditions (**Dalton's law of partial pressures**). The partial pressure of a component of a mixture is equal to its mole fraction times the total pressure: $P_1 = X_1 P_t$. The **mole fraction** is the ratio of the moles of one component of a mixture to the total moles of all components. In calculating the quantity of a gas collected over water, correction must be made for the partial pressure of water vapor in the gas mixture.

Section 10.7 The **kinetic-molecular theory** accounts for the properties of an ideal gas in terms of a set of assumptions about the nature of gases. Briefly, these assumptions are: Molecules are in continuous chaotic motion; the volume of gas molecules is negligible compared to the volume of their container; the gas molecules have no attractive forces for one another; their collisions are elastic; and the average kinetic energy of the gas molecules is proportional to the absolute temperature.

The molecules of a gas do not all have the same kinetic energy at a given instant. Their speeds are distributed over a wide range; the distribution varies with the molar mass of the gas and with temperature. The **root-mean-square (rms) speed**, u, varies in proportion to the square root of the absolute temperature and inversely with the square root of the molar mass: $u = \sqrt{3RT/\mathcal{M}}$.

Section 10.8 It follows from kinetic-molecular theory that the rate at which a gas undergoes **effusion** (escapes through a tiny hole into a vacuum) is inversely proportional to the square root of its molar mass (**Graham's law**). The **diffusion** of one gas through the space occupied by a second gas is another phenomenon related to the speeds at which molecules move. Because molecules undergo frequent collisions with one another, the **mean free path**—the mean distance traveled between collisions—is short. Collisions between molecules limit the rate at which a gas molecule can diffuse.

Section 10.9 Departures from ideal behavior increase in magnitude as pressure increases and as temperature decreases. The extent of nonideality of a real gas can be seen by examining the quantity PV/RT for 1 mol of the gas as a function of pressure; for an ideal gas, this quantity is exactly 1 at all pressures. Real gases depart from ideal behavior because the molecules possess finite volume and because the molecules experience attractive forces for one another upon collision. The **van der Waals equation** is an equation of state for gases that modifies the ideal-gas equation to account for intrinsic molecular volume and intermolecular forces.

Exercises

Gas Characteristics; Pressure

10.1 How does a gas differ from a liquid with respect to each of the following properties: **(a)** density; **(b)** compressibility; **(c)** ability to mix with other substances of the same phase to form homogeneous mixtures?

10.2 **(a)** Both a liquid and a gas are moved to larger containers. How does their behavior differ? Explain the difference in molecular terms. **(b)** Although water and carbon tetrachloride, $CCl_4(l)$, do not mix, their vapors form homogeneous mixtures. Explain. **(c)** The densities of gases are generally reported in units of g/L, whereas those for liquids are reported as g/mL. Explain the molecular basis for this difference.

10.3 Consider two people of the same mass standing in a room. One person is standing normally, and the other is standing on one foot. **(a)** Does one person exert a greater force on the floor than the other? **(b)** Does one person exert a greater pressure on the floor than the other?

10.4 The height of the mercury column in a barometer in Denver, elevation 5000 feet, is less than that for a mercury column in Los Angeles, elevation 132 feet. Explain.

10.5 **(a)** How high in meters must a column of water be to exert a pressure equal to that of a 760-mm column of mercury? The density of water is 1.0 g/mL, whereas that of mercury is 13.6 g/mL. **(b)** What is the pressure in atmospheres on the body of a diver if he is 36 ft below the surface of the water when atmospheric pressure at the surface is 0.95 atm?

10.6 The compound 1-iodododecane is a nonvolatile liquid with a density of 1.20 g/mL. The density of mercury is 13.6 g/mL. What do you predict for the height of a barometer column based on 1-iodododecane, when the atmospheric pressure is 752 torr?

10.7 Each of the following statements concerns a mercury barometer such as that shown in Figure 10.2. Identify any incorrect statements, and correct them. **(a)** The tube must be 1 cm^2 in cross-sectional area. **(b)** At equilibrium, the force of gravity per unit area acting on the mercury column at the level of the outside mercury equals the force of gravity per unit area acting on the atmosphere. **(c)** The column of mercury is held up by the vacuum at the top of the column.

10.8 Suppose you make a mercury barometer using a glass tube about 50 cm in length, closed at one end. What would you expect to see if the tube is filled with mercury and inverted in a mercury dish, as in Figure 10.2? Explain.

10.9 The typical atmospheric pressure on top of Mt. Everest (29,028 ft) is about 265 torr. Convert this pressure to **(a)** atm; **(b)** mm Hg; **(c)** pascals; **(d)** bars.

10.10 Perform the following conversions: **(a)** 2.44 atm to torr; **(b)** 682 torr to kilopascals; **(c)** 776 mm Hg to atmospheres; **(d)** 1.456×10^5 Pa to atmospheres; **(e)** 3.44 atm to bars.

10.11 In the United States, barometric pressures are reported in inches of mercury (in. Hg). On a beautiful summer day in Chicago the barometric pressure is 30.45 in. Hg. **(a)** Convert this pressure to torr. **(b)** A meteorologist explains the nice weather by referring to a "high-pressure area." In light of your answer to part (a), explain why this term makes sense.

10.12 **(a)** On Titan, the largest moon of Saturn, the atmospheric pressure is 1.63105 Pa. What is the atmospheric pressure of Titan in atm? **(b)** On Venus the surface atmospheric pressure is about 90 Earth atmospheres. What is the Venusian atmospheric pressure in kilopascals?

10.13 Suppose that a woman weighing 125 lb and wearing high-heeled shoes momentarily places all her weight on the heel of one foot. If the area of the heel is 0.50 in.^{2}, calculate the pressure exerted on the underlying surface in kilopascals.

10.14 A set of bookshelves rests on a hard floor surface on the edges of the two vertical sides of the shelves, each of which has a cross-sectional dimension of 2.2×30 cm. The total mass of the shelves plus the books stacked on them is 262 kg. Calculate the pressure in pascals exerted by the shelf footings on the surface.

10.15 If the atmospheric pressure is 0.975 atm, what is the pressure of the enclosed gas in each of the three cases depicted in the drawing?

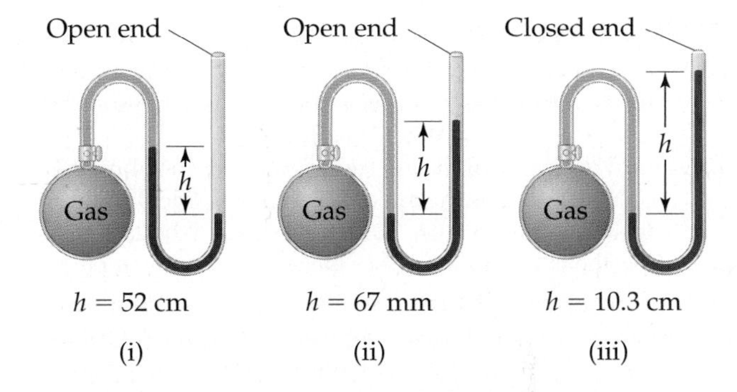

10.16 An open-end manometer containing mercury is connected to a container of gas, as depicted in Sample Exercise 10.2. What is the pressure of the enclosed gas in torr in each of the following situations? **(a)** The mercury in the arm attached to the gas is 13.6 cm higher than in the one open to the atmosphere; atmospheric pressure is 1.05 atm. **(b)** The mercury in the arm attached to the gas is 12 mm lower than in the one open to the atmosphere; atmospheric pressure is 0.988 atm.

The Gas Laws

10.17 Assume that you have a sample of gas in a container with a movable piston such as the one in the drawing. **(a)** Redraw the container to show what it might look like if the temperature of the gas is increased from 300 K to 500 K while the pressure is kept constant. **(b)** Redraw the container to show what it might look like if the pressure on the piston is increased from 1.0 atm to 2.0 atm while the temperature is kept constant.

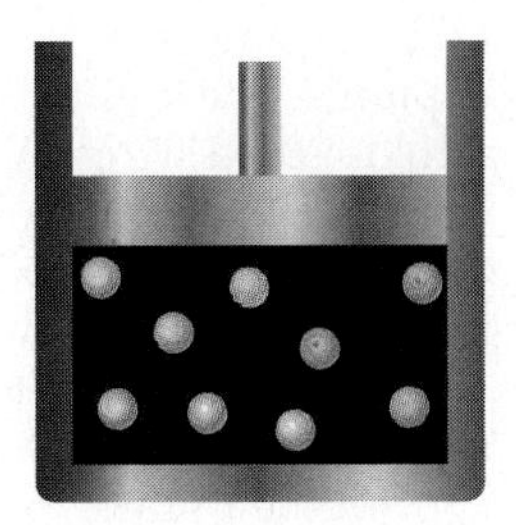

10.18 Assume that you have a cylinder with a movable piston. What would happen to the gas pressure inside the cylinder if you do the following? **(a)** Decrease the volume to one-third the original volume while holding the temperature constant. **(b)** Reduce the Kelvin temperature to half its original value while holding the volume constant. **(c)** Reduce the amount of gas to half while keeping the volume and temperature constant.

10.19 A fixed quantity of gas at 23°C exhibits a pressure of 748 torr and occupies a volume of 10.3 L. **(a)** Use Boyle's law to calculate the volume the gas will occupy at 23°C if the pressure is increased to 1.88 atm. **(b)** Use Charles's law to calculate the volume the gas will occupy if the temperature is increased to 165°C while the pressure is held constant.

10.20 A sample of gas occupies a volume of 1248 ft^3 at 0.988 atm and 28.0°C. **(a)** Calculate the pressure of the gas if its volume is decreased to 978 ft^3 while its temperature is held constant. **(b)** At what temperature in degrees

Celsius is the volume of the gas 1435 ft^3 if the pressure is kept constant?

10.21 **(a)** How is the law of combining volumes explained by Avogadro's hypothesis? **(b)** Consider a 1.0-L flask containing neon gas and a 1.5-L flask containing xenon gas. Both gases are at the same pressure and temperature. According to Avogadro's law, what can be said about the ratio of the number of atoms in the two flasks?

10.22 Nitrogen and hydrogen gases react to form ammonia gas as follows:

$$N_2(g) + 3H_2(g) \longrightarrow 2NH_3(g)$$

At a certain temperature and pressure, 1.2 L of N_2 reacts with 3.6 L of H_2. If all the N_2 and H_2 are consumed, what volume of NH_3, at the same temperature and pressure, will be produced?

The Ideal-Gas Equation

10.23 **(a)** Write the ideal-gas equation, and give the units used for each term in the equation when $R = 0.0821$ L-atm/mol-K. **(b)** What is an ideal gas?

10.24 **(a)** What conditions are represented by the abbreviation STP? **(b)** What is the molar volume of an ideal gas at STP? **(c)** Room temperature is often assumed to be 25°C. Calculate the molar volume of an ideal gas at room temperature.

10.25 Suppose you are given two 1-L flasks and told that one contains a gas of molar mass 30, the other a gas of molar mass 60, both at the same temperature. The pressure in flask A is X atm, and the mass of gas in the flask is 1.2 g. The pressure in flask B is 0.5X atm, and the mass of gas in the flask is 1.2 g. Which flask contains gas of molar mass 30, and which contains the gas of molar mass 60?

10.26 Suppose you are given two flasks at the same temperature, one of volume 2 L and the other of volume 3 L. In the 2-L flask the gas pressure is X atm, and the mass of gas in the flask is 4.8 g. In the 3-L flask the gas pressure is 0.1X, and the mass of gas is 0.36 g. Do the two gases have the same molar mass? If not, which contains the gas of higher molar mass?

10.27 Calculate each of the following quantities for an ideal gas: **(a)** the volume of the gas, in liters, if 2.46 mol has a pressure of 1.28 atm at a temperature of −6°C; **(b)** the absolute temperature of the gas at which 4.79×10^{-2} mol occupies 135 mL at 720 torr; **(c)** the pressure, in atmospheres, if 5.52×10^{-2} mol occupies 413 mL at 88°C; **(d)** the quantity of gas, in moles, if 88.4 L at 54°C has a pressure of 9.84 kPa.

10.28 For an ideal gas, calculate the following quantities: **(a)** the pressure of the gas if 0.215 mol occupies 338 mL at 32°C; **(b)** the temperature (in kelvins) at which 0.0412 mol occupies 3.00 L at 1.05 atm; **(c)** the number of moles in 98.5 L at 236 K and 690 torr; **(d)** the volume occupied by 5.48×10^{-3} mol at 55°C and a pressure of 3.87 kPa.

10.29 The *Hindenburg* was a hydrogen-filled dirigible that exploded in 1937. If the *Hindenburg* held 2.0×10^5 m^3 of hydrogen gas at 23°C and 1.0 atm, what mass of hydrogen was present?

10.30 A neon sign is made of glass tubing whose inside diameter is 4.5 cm and whose length is 5.3 m. If the sign contains neon at a pressure of 2.03 torr at 35°C, how many grams of neon are in the sign? (The volume of a cylinder is $\pi r^2 h$.)

10.31 A scuba diver's tank contains 0.29 kg of O_2 compressed into a volume of 2.3 L. **(a)** Calculate the gas pressure inside the tank at 9°C. **(b)** What volume would this oxygen occupy at 26°C and 0.95 atm?

10.32 An aerosol spray can with a volume of 456 mL contains 3.18 g of propane gas (C_3H_8) as a propellant. **(a)** If the can is at 23°C, what is the pressure in the can? **(b)** What volume would the propane occupy at STP? **(c)** The can says that exposure to temperatures above 130°F may cause the can to burst. What is the pressure in the can at this temperature?

10.33 Chlorine is widely used to purify municipal water supplies and to treat swimming pool waters. Suppose that the volume of a particular sample of Cl_2 gas is 9.22 L at 1124 torr and 24°C. **(a)** How many grams of Cl_2 are in the sample? **(b)** What volume will the Cl_2 occupy at STP? **(c)** At what temperature will the volume be 15.00 L if the pressure is 8.76×10^2 torr? **(d)** At what pressure will the volume equal 6.00 L if the temperature is 58°C?

10.34 Many gases are shipped in high-pressure containers. Consider a steel tank whose volume is 68.0 L and which contains O_2 gas at a pressure of 15,900 kPa at 23°C. **(a)** What mass of O_2 does the tank contain? **(b)** What volume would the gas occupy at STP? **(c)** At what temperature would the pressure in the tank equal 170 atm? **(d)** What would be the pressure of the gas, in kPa, if it were transferred to a container at 24°C whose volume is 52.6 L?

10.35 In an experiment reported in the scientific literature, male cockroaches were made to run at different speeds on a miniature treadmill while their oxygen consumption was measured. In one hour the average cockroach running at 0.08 km/hr consumed 0.8 mL of O_2 at 1 atm pressure and 24°C per gram of insect weight. **(a)** How many moles of O_2 would be consumed in 1 hr by a 5.2-g cockroach moving at this speed? **(b)** This same cockroach is caught by a child and placed in a 1-qt fruit jar with a tight lid. Assuming the same level of continuous activity as in the research, will the cockroach consume more than 20% of the available O_2 in a 48-hr period? (Air is 21 mol percent O_2.)

10.36 After the large eruption of Mount St. Helens in 1980, gas samples from the volcano were taken by sampling the downwind gas plume. The unfiltered gas samples were passed over a gold-coated wire coil to absorb mercury (Hg) present in the gas. The mercury was recovered from the coil by heating it, and then analyzed. In one particular set of experiments scientists found a mercury vapor level of 1800 ng of Hg per cubic meter in the plume, at a gas temperature of 10°C. Calculate **(a)** the partial pressure of Hg vapor in the plume; **(b)** the number of Hg atoms per cubic meter in the gas; **(c)** the total mass of Hg emitted per day by the volcano if the daily plume volume was 1600 km^3.

Further Applications of the Ideal-Gas Equation

10.37 Which gas is most dense at 1.00 atm and 298 K? **(a)** CO_2; **(b)** N_2O; **(c)** Cl_2. Explain.

10.38 Which gas is least dense at 1.00 atm and 298 K? **(a)** SO_3; **(b)** HCl; **(c)** CO_2. Explain

10.39 Which of the following statements best explains why a closed balloon filled with helium gas rises in air?
(a) Helium is an monatomic gas, whereas nearly all the molecules that make up air, such as nitrogen and oxygen, are diatomic.
(b) The average speed of helium atoms is higher than the average speeds of air molecules, and the higher speed of collisions with the balloon walls propels the balloon upward.
(c) Because the helium atoms are of lower mass than the average air molecule, the helium gas is less dense than air. The balloon thus weighs less than the air displaced by its volume.
(d) Because helium has a lower molar mass than the average air molecule, the helium atoms are in faster motion. This means that the temperature of the helium is higher than the air temperature. Hot gases tend to rise.

10.40 Which of the following statements best explains why nitrogen gas at STP is less dense than Xe gas at STP?
(a) Because Xe is a noble gas, there is less tendency for the Xe atoms to repel one another, so they pack more densely in the gas state.
(b) Xe atoms have a higher mass than N_2 molecules. Because both gases at STP have the same number of molecules per unit volume, the Xe gas must be denser.
(c) The Xe atoms are larger than N_2 molecules, and thus take up a larger fraction of the space occupied by the gas.
(d) Because the Xe atoms are much more massive than the N_2 molecules, they move more slowly and thus exert less upward force on the gas container and make the gas appear denser.

10.41 **(a)** Calculate the density of NO_2 gas at 0.970 atm and 35°C. **(b)** Calculate the molar mass of a gas if 2.50 g occupies 0.875 L at 685 torr and 35°C.

10.42 **(a)** Calculate the density of sulfur hexafluoride gas at 455 torr and 32°C. **(b)** Calculate the molar mass of a vapor that has a density of 6.345 g/L at 22°C and 743 torr.

10.43 In the Dumas-bulb technique for determining the molar mass of an unknown liquid, you vaporize the sample of a liquid that boils below 100°C in a boiling-water bath and determine the mass of vapor required to fill the bulb (see drawing). From the following data, calculate the molar mass of the unknown liquid: mass of unknown vapor, 1.012 g; volume of bulb, 354 cm^3; pressure, 742 torr; temperature, 99°C.

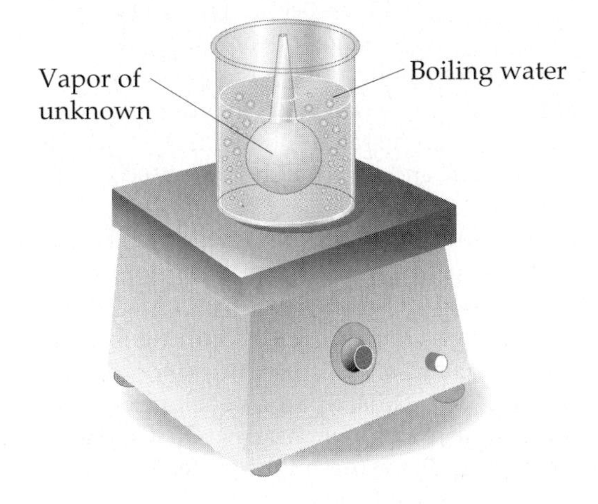

10.44 The molar mass of a volatile substance was determined by the Dumas-bulb method described in Exercise 10.43. The unknown vapor had a mass of 0.963 g; the volume of the bulb was 418 cm^3, pressure 752 torr, and temperature 100°C. Calculate the molar mass of the unknown vapor.

10.45 Magnesium can be used as a "getter" in evacuated enclosures, to react with the last traces of oxygen. (The magnesium is usually heated by passing an electric current through a wire or ribbon of the metal.) If an enclosure of 0.382 L has a partial pressure of O_2 of 3.5×10^{-6} torr at 27°C, what mass of magnesium will react according to the following equation?

$$2Mg(s) + O_2(g) \longrightarrow 2MgO(s)$$

10.46 Calcium hydride, CaH_2, reacts with water to form hydrogen gas:

$$CaH_2(s) + 2H_2O(l) \longrightarrow Ca(OH)_2(aq) + 2H_2(g)$$

This reaction is sometimes used to inflate life rafts, weather balloons, and the like, where a simple, compact means of generating H_2 is desired. How many grams of CaH_2 are needed to generate 64.5 L of H_2 gas if the pressure of H_2 is 814 torr at 32°C?

10.47 Ammonium sulfate, an important fertilizer, can be prepared by the reaction of ammonia with sulfuric acid:

$$2NH_3(g) + H_2SO_4(aq) \longrightarrow (NH_4)_2SO_4(aq)$$

Calculate the volume of $NH_3(g)$ needed at 42°C and 15.6 atm to react with 87 kg of H_2SO_4.

10.48 The metabolic oxidation of glucose, $C_6H_{12}O_6$, in our bodies produces CO_2, which is expelled from our lungs as a gas:

$$C_6H_{12}O_6(aq) + 6O_2(g) \longrightarrow 6CO_2(g) + 6H_2O(l)$$

Calculate the volume of dry CO_2 produced at body temperature (37°C) and 0.970 atm when 24.5 g of glucose is consumed in this reaction.

10.49 Hydrogen gas is produced when zinc reacts with sulfuric acid:

$$Zn(s) + H_2SO_4(aq) \longrightarrow ZnSO_4(aq) + H_2(g)$$

If 159 mL of wet H_2 is collected over water at 24°C and a barometric pressure of 738 torr, how many grams of Zn have been consumed? (The vapor pressure of water is tabulated in Appendix B.)

10.50 Acetylene gas, $C_2H_2(g)$, can be prepared by the reaction of calcium carbide with water:

$$CaC_2(s) + 2H_2O(l) \longrightarrow Ca(OH)_2(s) + C_2H_2(g)$$

Calculate the volume of C_2H_2 that is collected over water at 21°C by reaction of 3.26 g of CaC_2 if the total pressure of the gas is 748 torr? (The vapor pressure of water is tabulated in Appendix B.)

Partial Pressures

10.51 Consider the apparatus shown in the drawing. **(a)** When the stopcock between the two containers is opened and the gases allowed to mix, how does the volume occupied by the N_2 gas change? What is the partial pressure of N_2 after mixing? **(b)** How does the volume of the O_2 gas change when the gases mix? What is the partial pressure of O_2 in the mixture? **(c)** What is the total pressure in the container after the gases mix?

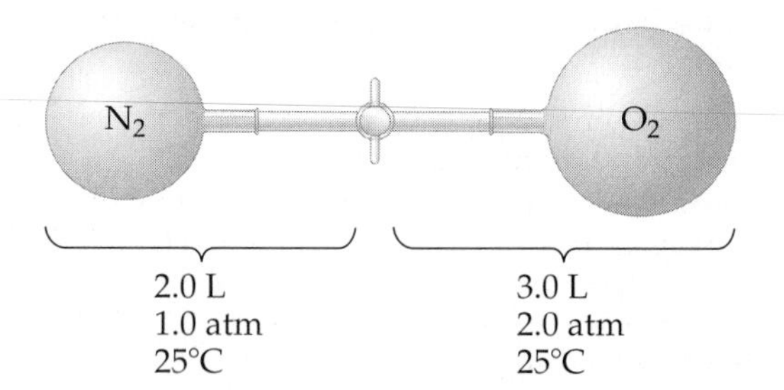

10.52 Consider a mixture of two gases, A and B, confined to a closed vessel. A quantity of a third gas, C, is added to the same vessel at the same temperature. How does the addition of gas C affect the following: **(a)** the partial pressure of gas A; **(b)** the total pressure in the vessel; **(c)** the mole fraction of gas B?

10.53 A mixture containing 0.538 mol He(*g*), 0.315 mol Ne(*g*), and 0.103 mol Ar(*g*) is confined in a 7.00-L vessel at 25°C. **(a)** Calculate the partial pressure of each of the gases in the mixture. **(b)** Calculate the total pressure of the mixture.

10.54 A mixture containing 3.15 g each of $CH_4(g)$, $C_2H_4(g)$, and $C_4H_{10}(g)$ is contained in a 2.00-L flask at a temperature of 64°C. **(a)** Calculate the partial pressure of each of the gases in the mixture. **(b)** Calculate the total pressure of the mixture.

10.55 A mixture of gases contains 0.75 mol N_2, 0.30 mol O_2, and 0.15 mol CO_2. If the total pressure of the mixture is 1.56 atm, what is the partial pressure of each component?

10.56 A mixture of gases contains 12.47 g of N_2, 1.98 g of H_2, and 8.15 g of NH_3. If the total pressure of the mixture is 2.35 atm, what is the partial pressure of each component?

10.57 At an underwater depth of 250 ft, the pressure is 8.38 atm. What should the mole percent of oxygen be in the diving gas for the partial pressure of oxygen in the mixture to be 0.21 atm, the same as in air at 1 atm?

10.58 **(a)** What are the mole fractions of each component in a mixture of 6.55 g of O_2, 4.92 g of N_2, and 1.32 g of H_2? **(b)** What is the partial pressure in atm of each component of this mixture if it is held in a 12.40-L vessel at 15°C?

10.59 A quantity of N_2 gas originally held at 3.80 atm pressure in a 1.00-L container at 26°C is transferred to a 10.0-L container at 20°C. A quantity of O_2 gas originally at 4.75 atm and 26°C in a 5.00-L container is transferred to this same container. What is the total pressure in the new container?

10.60 A sample of 5.25 g of $SO_2(g)$ originally in a 4.00-L vessel at 26°C is transferred to a 13.6-L vessel at 25°C. A sample of 2.35 g $N_2(g)$ originally in a 3.18-L vessel at 20°C is transferred to this same 13.6-L vessel. **(a)** What is the partial pressure of $SO_2(g)$ in the larger container? **(b)** What is the partial pressure of $N_2(g)$ in this vessel? **(c)** What is the total pressure in the vessel?

Kinetic-Molecular Theory; Graham's Law

10.61 What change or changes in the state of a gas bring about each of the following effects? **(a)** The number of impacts per unit time on a given container wall increases. **(b)** The average energy of impact of molecules with the wall of the container decreases. **(c)** The average distance between gas molecules increases. **(d)** The average speed of molecules in the gas mixture is increased.

10.62 Indicate which of the following statements regarding the kinetic-molecular theory of gases are correct. For those that are false, formulate a correct version of the statement. **(a)** The average kinetic energy of a collection of gas molecules at a given temperature is proportional to $\mathcal{M}^{1/2}$. **(b)** The gas molecules are assumed to exert no forces on each other. **(c)** All the molecules of a gas at a given temperature have the same kinetic energy. **(d)** The volume of the gas molecules is negligible in comparison to the total volume in which the gas is contained.

10.63 What property or properties of gases can you point to that support the assumption that most of the volume in a gas is empty space?

10.64 Newton had an incorrect theory of gases in which he assumed that all gas molecules repel one another and the walls of their container. Thus, the molecules of a gas are statically and uniformly distributed, trying to get as far apart as possible from one another and the vessel walls. This repulsion gives rise to pressure. Explain why Charles's law argues for the kinetic-molecular theory and against Newton's model.

10.65 Vessel A contains CO(*g*) at 0°C and 1 atm. Vessel B contains $SO_2(g)$ at 20°C and 0.5 atm. The two vessels have the same volume. **(a)** Which vessel contains more molecules? **(b)** Which contains more mass? **(c)** In which vessel is the average kinetic energy of molecules higher? **(d)** In which vessel is the rms speed of molecules higher?

10.66 Suppose you have two 1-L flasks, one containing N_2 at STP, the other containing CH_4 at STP. How do these systems compare with respect to **(a)** number of molecules; **(b)** density; **(c)** average kinetic energy of the molecules; **(d)** rate of effusion through a pinhole leak?

10.67 **(a)** Place the following gases in order of increasing average molecular speed at 300 K: CO_2, N_2O, HF, F_2, H_2. **(b)** Calculate and compare the rms speeds of H_2 and CO_2 molecules at 300 K.

10.68 **(a)** Place the following gases in order of increasing average molecular speed at 25°C: Ne, HBr, SO_2, NF_3, CO. **(b)** Calculate the rms speed of NF_3 molecules at 25°C.

10.69 Hydrogen has two naturally occurring isotopes, ^{1}H and ^{2}H. Chlorine also has two naturally occurring isotopes, ^{35}Cl and ^{37}Cl. Thus, hydrogen chloride gas consists of four distinct types of molecules: $^{1}H^{35}Cl$, $^{1}H^{37}Cl$, $^{2}H^{35}Cl$, and $^{2}H^{37}Cl$. Place these four molecules in order of increasing rate of effusion.

10.70 As discussed in the "Chemistry at Work" box in Section 10.9, enriched uranium is produced via gaseous diffusion of UF_6. Suppose a process were developed to allow diffusion of gaseous uranium atoms, U(*g*). Calculate the ratio of diffusion rates for ^{235}U and ^{238}U, and compare it to the ratio for UF_6 given in the essay.

10.71 Arsenic(III) sulfide sublimes readily, even below its melting point of 320°C. The molecules of the vapor phase are found to effuse through a tiny hole at 0.28 times the rate of effusion of Ar atoms under the same conditions of temperature and pressure. What is the molecular formula of arsenic(III) sulfide in the gas phase?

10.72 A gas of unknown molecular mass was allowed to effuse through a small opening under constant pressure conditions. It required 105 s for 1.0 L of the gas to effuse. Under identical experimental conditions it required 31 s for 1.0 L of O_2 gas to effuse. Calculate the molar mass of the unknown gas. (Remember that the faster the rate of effusion, the shorter the time required for effusion of 1.0 L; that is, rate and time are inversely proportional.)

Nonideal-Gas Behavior

10.73 **(a)** Under what experimental conditions of temperature and pressure do gases usually behave nonideally? **(b)** What two properties or characteristics of gas molecules cause them to behave nonideally?

10.74 The planet Jupiter has a mass 318 times that of Earth, and its surface temperature is 140 K. Mercury has a mass 0.05 times that of Earth, and its surface temperature is between 600 and 700 K. On which planet is the atmosphere more likely to obey the ideal-gas law? Explain.

10.75 Explain how the function PV/RT can be used to show how gases behave nonideally at high pressures.

10.76 For nearly all real gases, the quantity PV/RT decreases below the value of 1, which characterizes an ideal gas, as pressure on the gas increases. At much higher pressures, however, PV/RT increases and rises above the value of 1. **(a)** Explain the initial drop in value of PV/RT below 1 and the fact that it rises above 1 for still higher pressures. **(b)** The effects we have just noted are smaller for gases at higher temperature. Why is this so?

10.77 Based on their respective van der Waals constants (Table 10.3), is Ar or CO_2 expected to behave more nearly like an ideal gas at high pressures? Explain.

10.78 Briefly explain the significance of the constants a and b in the van der Waals equation.

10.79 Calculate the pressure that CCl_4 will exert at 40°C if 1.00 mol occupies 28.0 L, assuming that **(a)** CCl_4 obeys the ideal-gas equation; **(b)** CCl_4 obeys the van der Waals equation. (Values for the van der Waals constants are given in Table 10.3.)

10.80 It turns out that the van der Waals constant b equals four times the total volume actually occupied by the molecules of a mole of gas. Using this figure, calculate the fraction of the volume in a container actually occupied by Ar atoms **(a)** at STP; **(b)** at 100 atm pressure and 0°C. (Assume for simplicity that the ideal-gas equation still holds.)

Additional Exercises

10.81 Consider the apparatus below, which shows gases in two containers and a third empty container. When the stopcocks are opened and the gases allowed to mix, what is the distribution of atoms in each container, assuming that the containers are of equal volume and ignoring the volume of the tubing connecting them.

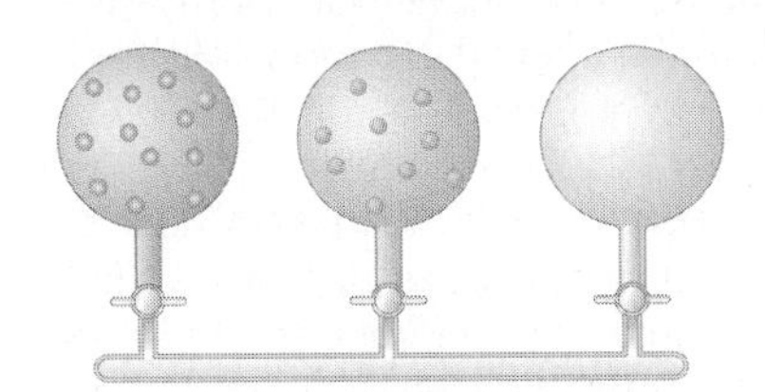

10.82 Suppose the mercury used to make a barometer has a few small droplets of water trapped in it that rise to the top of the mercury in the tube. Will the barometer show the correct atmospheric pressure? Explain.

10.83 A gas bubble with a volume of 1.0 mm^3 originates at the bottom of a lake where the pressure is 3.0 atm. Calculate its volume when the bubble reaches the surface of the lake where the pressure is 695 torr, assuming that the temperature doesn't change.

10.84 To minimize the rate of evaporation of the tungsten filament, 1.4×10^{-5} mol of argon is placed in a 600-cm^3 lightbulb. What is the pressure of argon in the lightbulb at 23°C?

10.85 Propane, C_3H_8, liquefies under modest pressure, allowing a large amount to be stored in a container. **(a)** Calculate the number of moles of propane gas in a 110-L container at 3.00 atm and 27°C. **(b)** Calculate the number of moles of liquid propane that can be stored in the same volume if the density of the liquid is 0.590 g/mL. **(c)** Calculate the ratio of the number of moles of liquid to moles of gas. Discuss this ratio in light of the kinetic-molecular theory of gases.

10.86 What is the total mass (in grams) of O_2 in a room measuring $(10.0 \times 8.0 \times 8.0)$ ft^3 if the air in the room is at STP and contains 20.95% O_2.

10.87 Nickel carbonyl, $Ni(CO)_4$, is one of the most toxic substances known. The present maximum allowable concentration in laboratory air during an 8-hr workday is 1 part in 10^9. Assume 24°C and 1.00 atm pressure. What mass of $Ni(CO)_4$ is allowable in a laboratory that is 54 m^2 in area, with a ceiling height of 3.1 m?

10.88 Consider the arrangement of bulbs shown in the drawing.

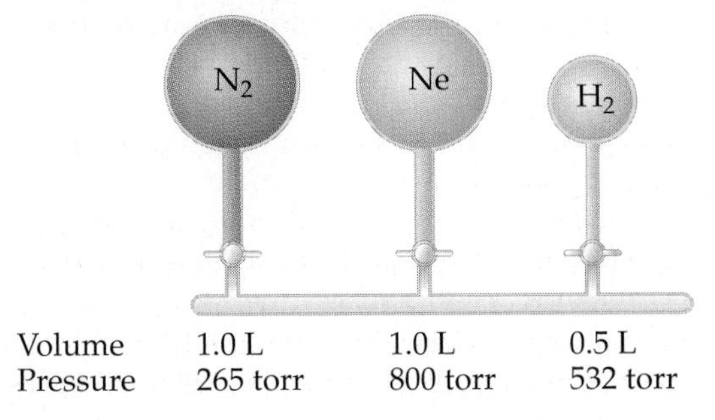

Each of the bulbs contains a gas at the pressure shown. What is the pressure of the system when all the stopcocks are opened, assuming that the temperature remains constant? (We can neglect the volume of the capillary tubing connecting the bulbs.)

10.89 Assume that a single cylinder of an automobile engine has a volume of 524 cm^3. **(a)** If the cylinder is full of air at 74°C and 0.980 atm, how many moles of O_2 are present? (The mole fraction of O_2 in dry air is 0.2095.) **(b)** How many grams of C_8H_{18} could be combusted by this quantity of O_2, assuming complete combustion with formation of CO_2 and H_2O?

[10.90] Ammonia, $NH_3(g)$, and hydrogen chloride, $HCl(g)$, react to form solid ammonium chloride, $NH_4Cl(s)$:

$$NH_3(g) + HCl(g) \longrightarrow NH_4Cl(s)$$

Two 2.00-L flasks at 25°C are connected by a stopcock, as shown in the drawing. One flask contains 5.00 g $NH_3(g)$, and the other contains 5.00 g $HCl(g)$. When the stopcock is opened, the gases react until one is completely consumed. **(a)** Which gas will remain in the system after the reaction is complete? **(b)** What will be the final pressure of the system after the reaction is complete? (Neglect the volume of the ammonium chloride formed.)

10.91 A sample of 1.42 g of helium and an unweighed quantity of O_2 are mixed in a flask at room temperature. The partial pressure of helium in the flask is 42.5 torr, and the partial pressure of oxygen is 158 torr. What is the mass of the oxygen in the container?

[10.92] A gaseous mixture of O_2 and Kr has a density of 1.104 g/L at 435 torr and 300 K. What is the mole percent O_2 in the mixture?

[10.93] A glass vessel fitted with a stopcock has a mass of 337.428 g when evacuated. When filled with Ar, it has a mass of 339.854 g. When evacuated and refilled with a mixture of Ne and Ar, under the same conditions of temperature and pressure, it weighs 339.076 g. What is the mole percent of Ne in the gas mixture?

[10.94] The density of a gas of unknown molar mass was measured as a function of pressure at 0°C, as in the table above. **(a)** Determine a precise molar mass for the gas. (*Hint*: Graph d/P versus P.) **(b)** Why is d/P not a constant as a function of pressure?

Pressure (atm)	1.00	0.666	0.500	0.333	0.250
Density (g/L)	2.3074	1.5263	1.1401	0.7571	0.5660

10.95 Suppose that when Torricelli had his great idea for constructing a mercury manometer, he had rushed into the laboratory and found the following items of glass:

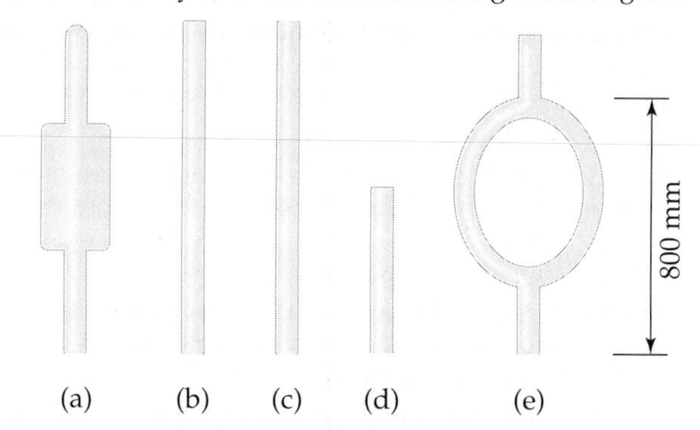

Which of these would have been satisfactory for his use in forming the first manometer? Explain why the unsatisfactory ones would not have worked.

[10.96] Consider the apparatus used in Exercise 10.90. A gas at 1 atm pressure is contained in the left flask, and the right flask is evacuated. When the stopcock is opened, the gas expands to fill both flasks. A very small temperature change is noted when this expansion occurs. Explain how this observation relates to assumption 3 of the kinetic-molecular theory, Section 10.7.

10.97 On a single plot, qualitatively sketch the distribution of molecular speeds for **(a)** Kr(*g*) at 250°C; **(b)** Kr(*g*) at 0°C; **(c)** Ar(*g*) at 0°C.

10.98 Does the effect of intermolecular attraction on the properties of a gas become more significant or less significant if **(a)** the gas is compressed to a smaller volume at constant temperature; **(b)** the temperature of the gas is increased at constant volume.

[10.99] Large amounts of nitrogen gas are used in the manufacture of ammonia, principally for use in fertilizers. Suppose 120.00 kg of $N_2(g)$ is stored in a 1100.0-L metal cylinder at 280°C. **(a)** Calculate the pressure of the gas assuming ideal-gas behavior. **(b)** By using data in Table 10.3, calculate the pressure of the gas according to the van der Waals equation. **(c)** Under the conditions of this problem, which correction dominates, the one for finite volume of gas molecules or the one for attractive interactions?

Integrative Exercises

10.100 Cyclopropane, a gas used with oxygen as a general anesthetic, is composed of 85.7% C and 14.3% H by mass. **(a)** If 1.56 g of cyclopropane has a volume of 1.00 L at 0.984 atm and 50.0°C, what is the molecular formula of cyclopropane? **(b)** Judging from its molecular formula, would you expect cyclopropane to deviate more or less than Ar from ideal-gas behavior at moderately high pressures and room temperature? Explain.

10.101 In the "Chemistry at Work" box on pipelines, Section 10.5, it is mentioned that the total deliverability of natural gas (methane, CH_4) to the various regions of the United States is on the order of 2.7×10^{12} L per day, measured at STP. Calculate the total enthalpy change for combustion of this quantity of methane. (*Note*: Less than this amount of methane is actually combusted daily. Some of the delivered gas is passed through to other regions.)

[10.102] A gas forms when elemental sulfur is heated carefully with AgF. The initial product boils at 15°C. Experiments on several samples yielded a gas density of 0.803 ± 0.010 g/L for the gas at 150 mm pressure and 32°C. When the gas reacts with water, all the fluorine is converted to aqueous HF. Other products are elemental sulfur, S_8, and other sulfur-containing compounds. A 480-mL sample of the dry gas at 126 mm pressure and 28°C, when reacted with 80 mL of water, yielded a 0.081 *M* solution of HF. The initial gaseous product undergoes a transformation over a period of time to a second compound with the same empirical and molecular formula, which boils at −10°C. **(a)** Determine the empirical and molecular formulas of the first compound formed. **(b)** Draw at least two reasonable Lewis structures that represent the initial compound and the one into which it is transformed over time. **(c)** Describe the likely geometries of these compounds, and estimate the single bond distances, given that the S—S bond distance in S_8 is 2.04 Å and the F—F distance in F_2 is 1.43 Å

10.103 Chlorine dioxide gas (ClO_2) is used as a commercial bleaching agent. It bleaches materials by oxidizing them. In the course of these reactions, the ClO_2 is itself reduced. **(a)** What is the Lewis structure for ClO_2? **(b)** Why do you think that ClO_2 is reduced so readily? **(c)** When a ClO_2 molecule gains an electron, the chlorite ion, ClO_2^-, forms. Draw the Lewis structure for ClO_2^-. **(d)** Predict the O—Cl—O bond angle in the ClO_2^- ion. **(e)** One method of preparing ClO_2 is by the reaction of chlorine and sodium chlorite: $Cl_2(g) + 2NaClO_2(s) \longrightarrow 2ClO_2(g) + 2NaCl(s)$. If you allow 10.0 g of $NaClO_2$ to react with 2.00 L of chlorine gas at a pressure of 1.50 atm at 21°C, how many grams of ClO_2 can be prepared?

[10.104] Natural gas is very abundant in many Middle Eastern oil fields. However, the costs of shipping the gas to markets in other parts of the world are high because it is necessary to liquefy the gas, which is mainly methane and thus has a boiling point at atmospheric pressure of −164°C. One possible strategy is to oxidize the methane to methanol, CH_3OH, which has a boiling point of 65°C, and which can therefore be shipped more readily. Suppose that 10.7×10^9 ft^3 of methane at atmospheric pressure and 25°C are oxidized to methanol. **(a)** What volume of methanol is formed if the density of CH_3OH is 0.791 g/L? **(b)** Write balanced chemical equations for the oxidations of methane and methanol to $CO_2(g)$ and $H_2O(l)$. Calculate the total enthalpy change for complete combustion of the 10.7×10^9 ft^3 of methane described above and for complete combustion of the equivalent amount of methanol, as calculated in part (a). **(c)** Methane, when liquefied, has a density of 0.466 g/mL; the density of methanol at 25°C is 0.791 g/L. Compare the enthalpy change upon combustion of a unit volume of liquid methane and liquid methanol. From the standpoint of energy production, which substance has the higher enthalpy of combustion per unit volume?

[10.105] Gaseous iodine pentafluoride, IF_5, can be prepared by the reaction of solid iodine and gaseous fluorine:

$$I_2(s) + 5F_2(g) \longrightarrow 2IF_5(g)$$

A 5.00-L flask containing 10.0 g I_2 is charged with 10.0 g F_2, and the reaction proceeds until one of the reagents is completely consumed. After the reaction is complete, the temperature in the flask is 125°C. **(a)** What is the partial pressure of IF_5 in the flask? **(b)** What is the mole fraction of IF_5 in the flask?

[10.106] A 6.53-g sample of a mixture of magnesium carbonate and calcium carbonate is treated with excess hydrochloric acid. The resulting reaction produces 1.72 L of carbon dioxide gas at 28°C and 743 torr pressure. **(a)** Write balanced chemical equations for the reactions that occur between hydrochloric acid and each component of the mixture. **(b)** Calculate the total number of moles of carbon dioxide that forms from these reactions. **(c)** Assuming that the reactions are complete, calculate the percentage by mass of magnesium carbonate in the mixture.

eMedia Exercises

10.107 Using the **Gas Laws** activity (*eChapter 10.3*), select a mass and a pressure to be held constant, and compare the volumes of N_2 and Xe at various temperatures. Under identical conditions—the same mass at the same pressure and temperature—are the volumes of N_2 and Xe equal? If not, explain why.

10.108 The ***P-V* Relationships** movie (*eChapter 10.3*) illustrates Boyle's law and points out that this law holds only when temperature is constant. **(a)** Reproduce the pressure versus volume graph presented in the movie. **(b)** Using the ideal-gas equation, deduce and superimpose on your graph from part (a) the line you would expect on the *P-V* plot at a temperature higher than the original and at a temperature lower than the original. **(c)** Do the same for the *V* versus $1/P$ plot.

10.109 Automobile air bags are inflated by the explosive decomposition of sodium azide, as shown in the **Air Bags** movie (*eChapter 10.5*). **(a)** If an air bag is to be inflated with 40.0 L of gas, initially at 110°C and 1.05 atm pressure, what mass of sodium azide must be available for decomposition? **(b)** What does the notation O(*s*) stand for in the decomposition reaction? **(c)** Why is it important that the reactants include an oxidant to react with the sodium metal produced by the decomposition?

10.110 Use the **Density of Gases** activity (*eChapter 10.5*) to compare the densities of two different gases at the same pressure and temperature. Explain in terms of kinetic-molecular theory why the molar mass of a gas is necessary to calculate the density of a gas, but not to determine its pressure.

$$d = \frac{P\mathcal{M}}{RT} \qquad P = \frac{nRT}{V}$$

10.111 The relative speeds of helium and neon atoms are shown in the **Kinetic Energy in a Gas** movie (*eChapter 10.7*). **(a)** If the average kinetic energies of both gases are the same at a given temperature, determine how much faster helium atoms are moving (on average) than neon atoms. **(b)** How does the fact that average kinetic energy of a gas is directly proportional to absolute temperature explain Boyle's observation that pressure decreases with increasing volume at constant temperature? **(c)** How does it explain Charles's observation that pressure increases with increasing temperature at constant volume?

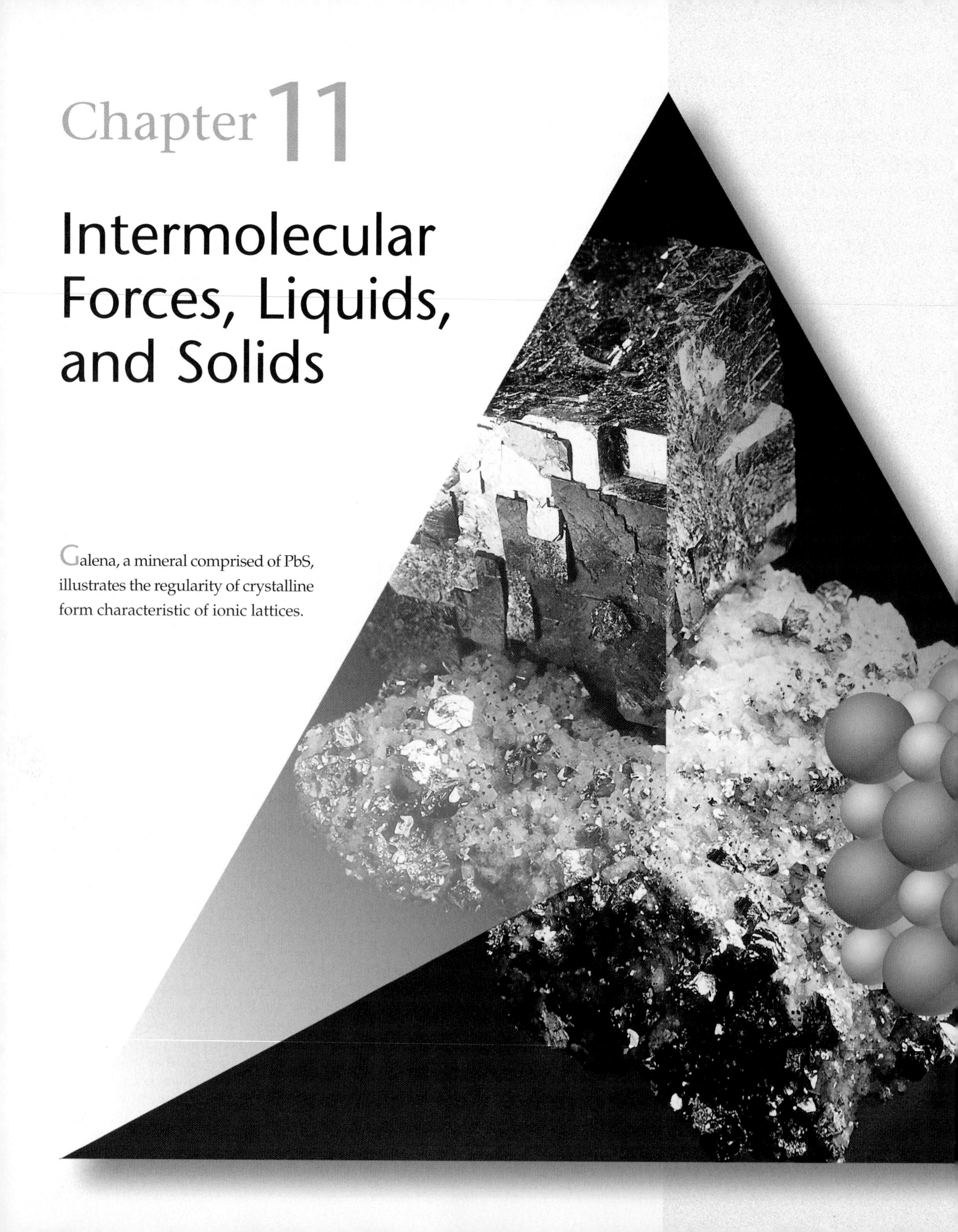

Chapter 11

Intermolecular Forces, Liquids, and Solids

Galena, a mineral comprised of PbS, illustrates the regularity of crystalline form characteristic of ionic lattices.

THE WATER VAPOR—or humidity—in air, the water in a lake, and the ice in a glacier are all forms of the same substance, H_2O. They all have the same chemical properties. Their physical properties differ greatly, however, because the physical properties of a substance depend on its physical state. Some of the characteristic properties of each of the states of matter are listed in Table 11.1 ▶. In Chapter 10 we discussed the gaseous state in some detail. In this chapter we turn our attention to the physical properties of liquids and solids.

Many of the substances that we will consider are molecular. In fact, virtually all substances that are liquids at room temperature are molecular. The forces *within* molecules that give rise to covalent bonding influence molecular shape, bond energies, and many aspects of chemical behavior. The physical properties of molecular liquids and solids, however, are due largely to **intermolecular forces**, the forces that exist *between* molecules. We learned in Section 10.9 that attractions between gas molecules lead to deviations from ideal-gas behavior. But how do these intermolecular attractions arise? By understanding the nature and strength of intermolecular forces, we can begin to relate the composition and structure of molecules to their physical properties.

▶ What's Ahead ◀

- In this chapter we'll study the *intermolecular forces* that exist in gases, liquids, and solids.
- The intermolecular forces between neutral molecules depend on their molecular polarity, size, and shape.
- We'll encounter the *hydrogen bond*, a special kind of intermolecular attractive force in compounds containing O—H, N—H, or F—H bonds.
- *Viscosity*, a measure of resistance to flow, and *surface tension*, a measure of a liquid's resistance to increasing its surface area, are properties characteristic of liquids.
- We'll explore the enthalpy changes that accompany *phase changes*, the transitions of matter between the gaseous, liquid, and solid states.
- We'll examine the *dynamic equilibrium* that exists between a liquid and its gaseous state and introduce the idea of *vapor pressure*.
- In a *phase diagram* the equilibria between the gaseous, liquid, and solid phases are displayed graphically.
- Orderly arrangements of units in three dimensions characterize *crystalline solids*. We'll examine the nature of these solids and how simple objects are most efficiently arranged in three dimensions.
- Solids can be characterized according to both the type of unit that makes up the solid and the attractive forces between the units.

TABLE 11.1 Some Characteristic Properties of the States of Matter

Gas	Assumes both the volume and shape of its container Is compressible Flows readily Diffusion within a gas occurs rapidly
Liquid	Assumes the shape of the portion of the container it occupies Does not expand to fill container Is virtually incompressible Flows readily Diffusion within a liquid occurs slowly
Solid	Retains its own shape and volume Is virtually incompressible Does not flow Diffusion within a solid occurs extremely slowly

11.1 A Molecular Comparison of Liquids and Solids

In Chapter 10 we learned that the physical properties of gases can be understood in terms of the kinetic-molecular theory. Gases consist of a collection of widely separated molecules in constant, chaotic motion. The average kinetic energy of the molecules is much larger than the average energy of the attractions between them. The lack of strong attractive forces between molecules allows a gas to expand to fill its container.

In liquids the intermolecular attractive forces are strong enough to hold molecules close together. Thus, liquids are much denser and far less compressible than gases. Unlike gases, liquids have a definite volume, independent of the size and shape of their container. The attractive forces in liquids are not strong enough, however, to keep the molecules from moving past one another. Thus, liquids can be poured, and they assume the shapes of their containers.

In solids the intermolecular attractive forces are strong enough not only to hold molecules close together, but to virtually lock them in place. Solids, like liquids, are not very compressible because the molecules have little free space between them. Often the molecules take up positions in a highly regular pattern. Solids that possess highly ordered structures are said to be *crystalline*. (The transition from a liquid to a crystalline solid is rather like the change that occurs on a military parade ground when the troops are called to formation.) Because the particles of a solid are not free to undergo long-range movement, solids are rigid. Keep in mind, however, that the units that form the solid, whether ions or molecules, possess thermal energy and vibrate in place. This vibrational motion increases in amplitude as a solid is heated. In fact, the energy may increase to the point that the solid melts or sublimes.

Figure 11.1 ▶ compares the three states of matter. The particles that compose the substance can be individual atoms, as in Ar; molecules, as in H_2O; or ions, as in NaCl. The state of a substance depends largely on the balance between the kinetic energies of the particles and the interparticle energies of attraction. The kinetic energies, which depend on temperature, tend to keep the particles apart and moving. The interparticle attractions tend to draw the particles together. Substances that are gases at room temperature have weaker interparticle attractions than those that are liquids; substances that are liquids have weaker attractions than those that are solids. Because the particles in a solid or liquid are fairly close together compared with those of a gas, we often refer to solids and liquids as *condensed phases*.

We can change a substance from one state to another by heating or cooling, which changes the average kinetic energy of the particles. NaCl, for example, which is a solid at room temperature, melts at 801°C and boils at 1413°C

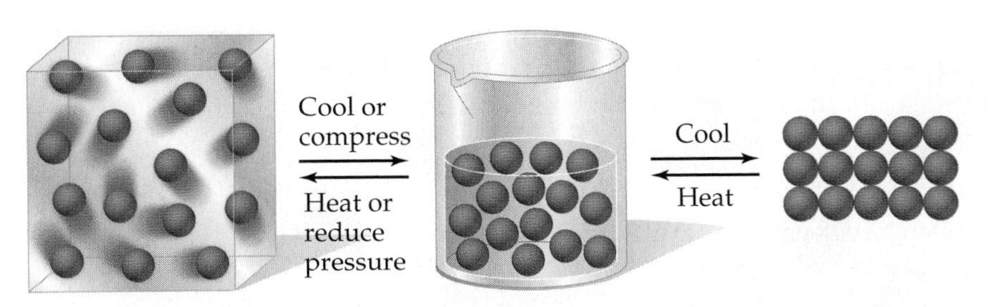

◀ **Figure 11.1** Molecular-level comparison of gases, liquids, and solids. The particles can be atoms, ions, or molecules. The density of particles in the gas phase is exaggerated compared with most real situations.

under 1 atm pressure. N_2O, on the other hand, which is a gas at room temperature, liquefies at −88.5°C and solidifies at −90.8°C under 1 atm pressure.

Increasing the pressure on a substance forces the molecules closer together, which in turn increases the strength of the intermolecular forces of attraction. Propane (C_3H_8) is a gas at room temperature and 1 atm pressure, whereas liquefied propane (LP) gas is a liquid at room temperature because it is stored under much higher pressure.

11.2 Intermolecular Forces

The strengths of intermolecular forces of different substances vary over a wide range, but they are generally much weaker than ionic or covalent bonds (Figure 11.2 ▼). Less energy, therefore, is required to vaporize a liquid or to melt a solid than to break covalent bonds in molecules. For example, only 16 kJ/mol is required to overcome the intermolecular attractions between HCl molecules in liquid HCl in order to vaporize it. In contrast, the energy required to break the covalent bond to dissociate HCl into H and Cl atoms is 431 kJ/mol. Thus, when a molecular substance like HCl changes from solid to liquid to gas, the molecules themselves remain intact.

Many properties of liquids, including their *boiling points*, reflect the strengths of the intermolecular forces. For example, because the forces between HCl molecules are so weak, HCl boils at only −85°C at atmospheric pressure. A liquid boils when bubbles of its vapor form within the liquid. The molecules of a liquid must overcome their attractive forces in order to separate and form a vapor. The stronger the attractive forces, the higher is the temperature at which the liquid boils. Similarly, the *melting points* of solids increase as the strengths of the intermolecular forces increase.

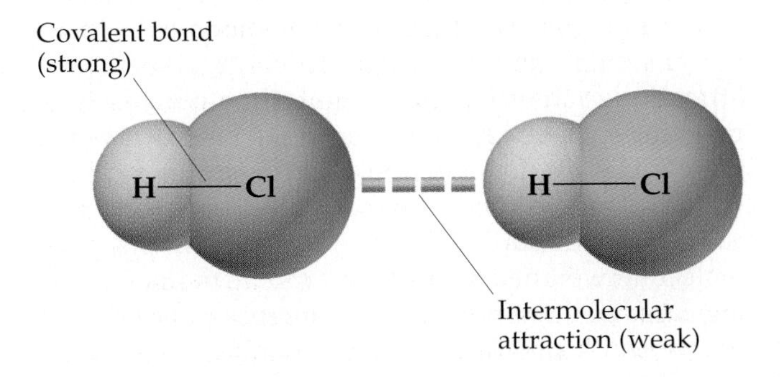

◀ **Figure 11.2** Comparison of a covalent bond (an intramolecular force) and an intermolecular attraction.

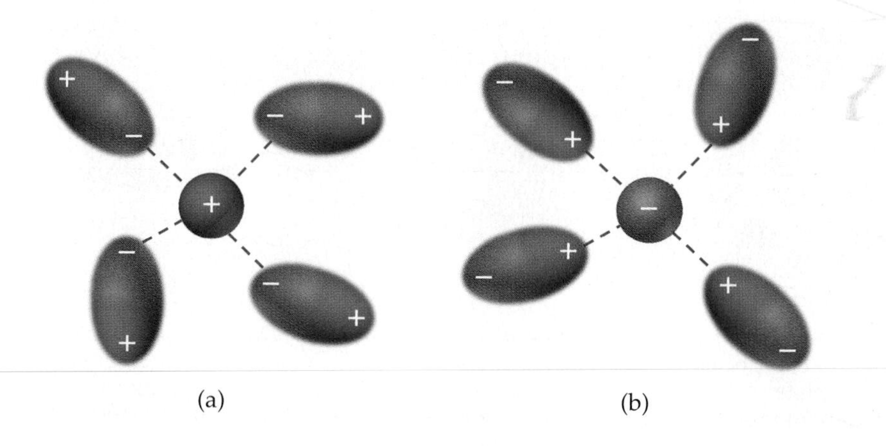

▶ **Figure 11.3** Illustration of the preferred orientation of polar molecules toward ions. The negative end of the polar molecule is oriented toward a cation (a), the positive end toward an anion (b).

Three types of intermolecular attractive forces are known to exist between neutral molecules: dipole-dipole forces, London dispersion forces, and hydrogen-bonding forces. These forces are also called *van der Waals forces* after Johannes van der Waals, who developed the equation for predicting the deviation of gases from ideal behavior. (Section 10.9) Another kind of attractive force, the ion-dipole force, is important in solutions. All four forces are electrostatic in nature, involving attractions between positive and negative species. All tend to be less than 15% as strong as covalent or ionic bonds.

Ion-Dipole Forces

An **ion-dipole force** exists between an ion and the partial charge on the end of a polar molecule. Polar molecules are dipoles; they have a positive end and a negative end. (Section 9.3) HCl is a polar molecule, for example, because the electronegativities of the H and Cl atoms differ.

Positive ions are attracted to the negative end of a dipole, whereas negative ions are attracted to the positive end, as shown in Figure 11.3 ▲. The magnitude of the attraction increases as either the charge of the ion or the magnitude of the dipole moment increases. Ion-dipole forces are especially important for solutions of ionic substances in polar liquids, such as a solution of NaCl in water. (Section 4.1) We will discuss these solutions in more detail in Section 13.1.

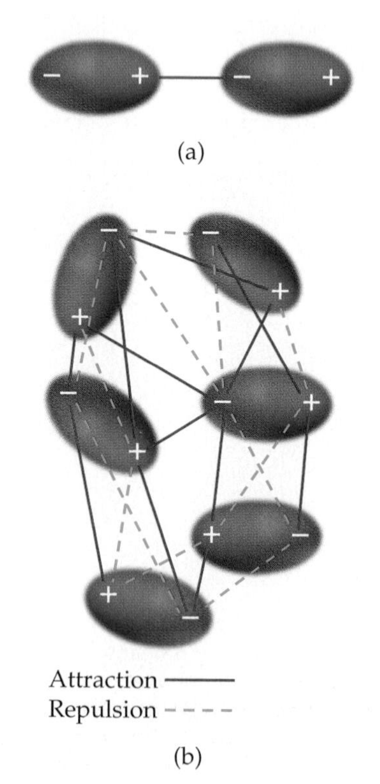

▲ **Figure 11.4** (a) The electrostatic interaction of two polar molecules. (b) The interaction of many dipoles in a condensed state.

Dipole-Dipole Forces

Neutral polar molecules attract each other when the positive end of one molecule is near the negative end of another, as in Figure 11.4(a) ◀. These **dipole-dipole forces** are effective only when polar molecules are very close together, and they are generally weaker than ion-dipole forces.

In liquids polar molecules are free to move with respect to one another. As shown in Figure 11.4(b), they will sometimes be in an orientation that is attractive and sometimes in an orientation that is repulsive. Two molecules that are attracting each other spend more time near each other than do two that are repelling each other. Thus, the overall effect is a net attraction. When we examine various liquids, we find that *for molecules of approximately equal mass and size, the strengths of intermolecular attractions increase with increasing polarity*. We can see this trend in Table 11.2 ▶, which lists several substances with similar molecular weights, but different dipole moments. Notice that the boiling point increases as the dipole moment increases. For dipole-dipole forces to operate, the molecules must be able to get close together in the correct orientation.

TABLE 11.2 Molecular Weights, Dipole Moments, and Boiling Points of Several Simple Organic Substances

Substance	Molecular Weight (amu)	Dipole Moment μ (D)	Boiling Point (K)
Propane, $CH_3CH_2CH_3$	44	0.1	231
Dimethyl ether, CH_3OCH_3	46	1.3	248
Methyl chloride, CH_3Cl	50	1.9	249
Acetaldehyde, CH_3CHO	44	2.7	294
Acetonitrile, CH_3CN	41	3.9	355

For molecules of comparable polarity, therefore, those with smaller molecular volumes generally experience higher dipole-dipole attractive forces.

London Dispersion Forces

There can be no dipole-dipole forces between nonpolar atoms and molecules. There must be some kind of attractive interactions, however, because nonpolar gases can be liquefied. The origin of this attraction was first proposed in 1930 by Fritz London, a German-American physicist. London recognized that the motion of electrons in an atom or molecule can create an *instantaneous* dipole moment.

In a collection of helium atoms, for example, the *average* distribution of the electrons about each nucleus is spherically symmetrical. The atoms are nonpolar and possess no permanent dipole moment. The instantaneous distribution of the electrons, however, can be different from the average distribution. If we could freeze the motion of the electrons in a helium atom at any given instant, both electrons could be on one side of the nucleus. At just that instant, then, the atom would have an instantaneous dipole moment.

Because electrons repel one another, the motions of electrons on one atom influence the motions of electrons on its near neighbors. Thus, the temporary dipole on one atom can induce a similar dipole on an adjacent atom, causing the atoms to be attracted to each other as shown in Figure 11.5 ▼. This attractive interaction is called the **London dispersion force** (or merely the dispersion force). This force, like dipole-dipole forces, is significant only when molecules are very close together.

The ease with which the charge distribution in a molecule can be distorted by an external electric field is called its **polarizability**. We can think of the polarizability of a molecule as a measure of the "squashiness" of its electron cloud; the greater the polarizability of a molecule, the more easily its electron cloud can be distorted to give a momentary dipole. Therefore, more polarizable molecules have stronger London dispersion forces. In general, larger molecules tend to have greater polarizabilities because they have a greater number of electrons and their electrons are farther from the nuclei. The strength of the London dispersion forces, therefore, tends to increase with increasing molecular size. Because molecular size and mass generally parallel each other, *dispersion forces tend to increase in*

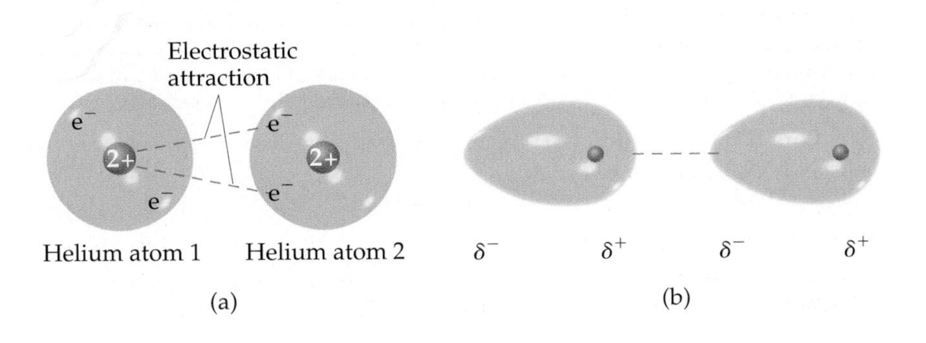

◀ **Figure 11.5** Two schematic representations of the instantaneous dipoles on two adjacent helium atoms, showing the electrostatic attraction between them.

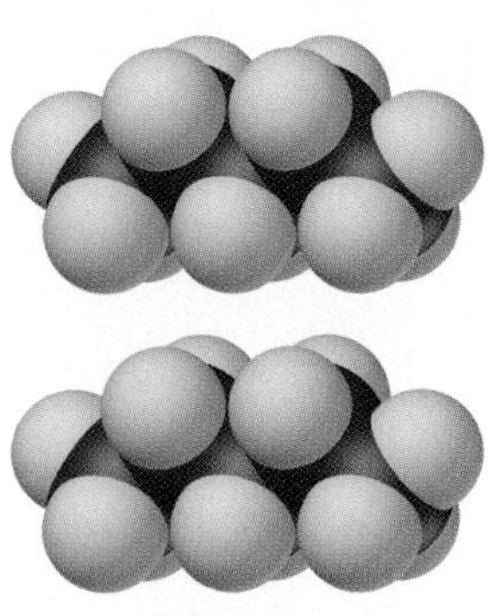

n-Pentane
(bp = 309.4 K)

Neopentane
(bp = 282.7 K)

▲ **Figure 11.6** Molecular shape affects intermolecular attraction. The *n*-pentane molecules make more contact with each other than do the neopentane molecules. Thus, *n*-pentane has the greater intermolecular attractive forces and therefore has the higher boiling point (bp).

TABLE 11.3 Boiling Points of the Halogens and the Noble Gases

Halogen	Molecular Weight (amu)	Boiling Point (K)	Noble Gas	Molecular Weight (amu)	Boiling Point (K)
F_2	38.0	85.1	He	4.0	4.6
Cl_2	71.0	238.6	Ne	20.2	27.3
Br_2	159.8	332.0	Ar	39.9	87.5
I_2	253.8	457.6	Kr	83.8	120.9
			Xe	131.3	166.1

strength with increasing molecular weight. Thus, the boiling points of the halogens and the noble gases increase with increasing molecular weight (Table 11.3 ▲).

The shapes of molecules also influence the magnitudes of dispersion forces. For example, *n*-pentane* and neopentane, illustrated in Figure 11.6 ◀, have the same molecular formula (C_5H_{12}), yet the boiling point of *n*-pentane is 27 K higher than that of neopentane. The difference can be traced to the different shapes of the two molecules. The overall attraction between molecules is greater for *n*-pentane because the molecules can come in contact over the entire length of the long, somewhat cylindrically shaped molecule. Less contact is possible between the more compact and nearly spherical molecules of neopentane.

Dispersion forces operate between all molecules, whether they are polar or nonpolar. Polar molecules experience dipole-dipole interactions, but they also experience dispersion forces at the same time. In fact, dispersion forces between polar molecules commonly contribute more to intermolecular attractions than dipole-dipole forces. In liquid HCl, for example, dispersion forces are estimated to account for more than 80% of the total attraction between molecules; dipole-dipole attractions account for the rest.

When comparing the relative strengths of intermolecular attractions, the following generalizations should be considered:

1. When the molecules have comparable molecular weights and shapes, dispersion forces are approximately equal. In this case differences in the magnitudes of the attractive forces are due to differences in the strengths of dipole-dipole attractions, with the most polar molecules having the strongest attractions.
2. When molecules differ widely in their molecular weights, dispersion forces tend to be decisive. In this case differences in the magnitudes of the attractive forces can usually be associated with differences in molecular weights, with the most massive molecule having the strongest attractions.

The hydrogen bond, which we consider after Sample Exercise 11.1, is one type of intermolecular interaction that is typically stronger than dispersion forces.

SAMPLE EXERCISE 11.1

The dipole moments of acetonitrile, CH_3CN, and methyl iodide, CH_3I, are 3.9 D and 1.62 D, respectively. **(a)** Which of these substances will have the greater dipole-dipole attractions among its molecules? **(b)** Which of these substances will have the greater London dispersion attractions? **(c)** The boiling points of CH_3CN and CH_3I are 354.8 K and 315.6 K, respectively. Which substance has the greatest overall attractive forces?

Solution **(a)** Dipole-dipole attractions increase in magnitude as the dipole moment of the molecule increases. Thus, CH_3CN molecules attract each other by stronger dipole-dipole forces than CH_3I molecules do. **(b)** When molecules differ in their molecular

* The *n* in *n*-pentane is an abbreviation for the word *normal*. A normal hydrocarbon is one in which carbon atoms are arranged in a straight chain. ∞ (Section 2.9)

weights, the more massive molecule generally has the stronger dispersion attractions. In this case CH_3I (142.0 amu) is much more massive than CH_3CN (41.0 amu), so the dispersion forces will be stronger for CH_3I. **(c)** Because CH_3CN has the higher boiling point, we can conclude that more energy is required to overcome attractive forces between CH_3CN molecules. Thus, the total intermolecular attractions are stronger for CH_3CN, suggesting that dipole-dipole forces are decisive when comparing these two substances. Nevertheless, dispersion forces play an important role in determining the properties of CH_3I.

PRACTICE EXERCISE

Of Br_2, Ne, HCl, HBr, and N_2, which is likely to have **(a)** the largest intermolecular dispersion forces; **(b)** the largest dipole-dipole attractive forces?
Answers: **(a)** Br_2; **(b)** HCl

Hydrogen Bonding

Figure 11.7 ▼ shows the boiling points of the simple hydrogen compounds of group 4A and 6A elements. In general, the boiling point increases with increasing molecular weight, owing to increased dispersion forces. The notable exception to this trend is H_2O, whose boiling point is much higher than we would expect on the basis of its molecular weight. The compounds NH_3 and HF also have abnormally high boiling points. In fact, these compounds have many characteristics that distinguish them from other substances of similar molecular weight and polarity. For example, water has a high melting point, a high specific heat, and a high heat of vaporization. Each of these properties indicates that the intermolecular forces between H_2O molecules are abnormally strong.

These strong intermolecular attractions in H_2O result from hydrogen bonding. **Hydrogen bonding** *is a special type of intermolecular attraction between the hydrogen atom in a polar bond (particularly an H—F, H—O, or H—N bond) and an unshared electron pair on a nearby small electronegative ion or atom (usually an F, O, or N atom on another molecule).* For example, a hydrogen bond exists between the H atom in an HF molecule and the F atom of an adjacent HF molecule,

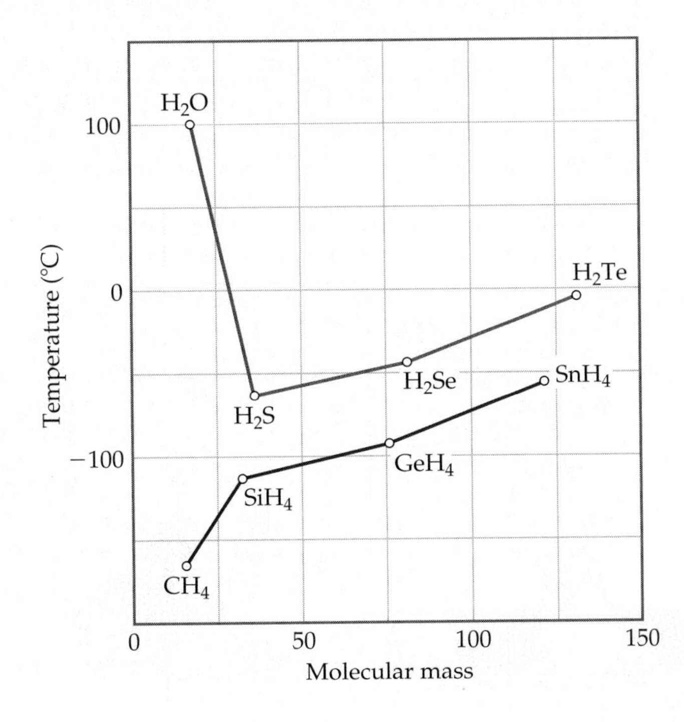

◀ **Figure 11.7** Boiling points of the group 4A (bottom) and 6A (top) hydrides as a function of molecular weight.

H—Ö:···H—Ö:
| |
H   H

H   H
| |
H—N:···H—N:
| |
H   H

H
|
H—N:···H—Ö:
| |
H   H

H
|
H—Ö:···H—N:
| |
H   H

▲ **Figure 11.8** Examples of hydrogen bonding. The solid lines represent covalent bonds; the red dotted lines represent hydrogen bonds.

F—H···F—H (where the dots represent the hydrogen bond between the molecules). Several additional examples are shown in Figure 11.8 ◀.

Hydrogen bonds can be considered unique dipole-dipole attractions. Because F, N, and O are so electronegative, a bond between hydrogen and any of these three elements is quite polar, with hydrogen at the positive end:

$$\overset{\longleftarrow\!\!+}{\text{N—H}} \qquad \overset{\longleftarrow\!\!+}{\text{O—H}} \qquad \overset{\longleftarrow\!\!+}{\text{F—H}}$$

The hydrogen atom has no inner core of electrons. Thus, the positive side of the bond dipole has the concentrated charge of the partially exposed, nearly bare proton of the hydrogen nucleus. This positive charge is attracted to the negative charge of an electronegative atom in a nearby molecule. Because the electron-poor hydrogen is so small, it can approach an electronegative atom very closely and thus interact strongly with it.

The energies of hydrogen bonds vary from about 4 kJ/mol to 25 kJ/mol or so. Thus, they are much weaker than ordinary chemical bonds (see Table 8.4). Nevertheless, because hydrogen bonds are generally stronger than dipole-dipole or dispersion forces, they play important roles in many chemical systems, including those of biological significance. For example, hydrogen bonds help stabilize the structures of proteins, which are key parts of skin, muscles, and other structural components of animal tissues (see Section 25.9). They are also responsible for the way that DNA is able to carry genetic information. ∞ (Section 25.11)

▲ **Figure 11.9** As with most substances, the solid phase of paraffin is denser than the liquid phase, and the solid therefore sinks below the liquid surface (left). In contrast, the solid phase of water, ice, is less dense than its liquid phase (right) causing the ice to float on the water. Richard Megna/Fundamental Photographs.

One of the remarkable consequences of hydrogen bonding is found when the densities of ice and liquid water are compared. In most substances the molecules in the solid are more densely packed than in the liquid. Thus, the solid phase is denser than the liquid phase (Figure 11.9 ◀). By contrast, the density of ice at 0°C (0.917 g/mL) is less than that of liquid water at 0°C (1.00 g/mL), so ice floats on liquid water (Figure 11.9).

The lower density of ice compared to that of water can be understood in terms of hydrogen-bonding interactions between water molecules. In the liquid state each water molecule undergoes continuously changing interactions with its neighbors. Hydrogen bonding is a major component of these interactions. The molecules are as close together as possible, even as their thermal motions keep them in constant motion. When water freezes, however, the molecules assume the ordered, open arrangement shown in Figure 11.10 ▶. This arrangement optimizes the hydrogen bonding interactions between molecules, but it creates a less dense structure for ice compared to that of water: A given mass of ice occupies a greater volume than the same mass of liquid water.

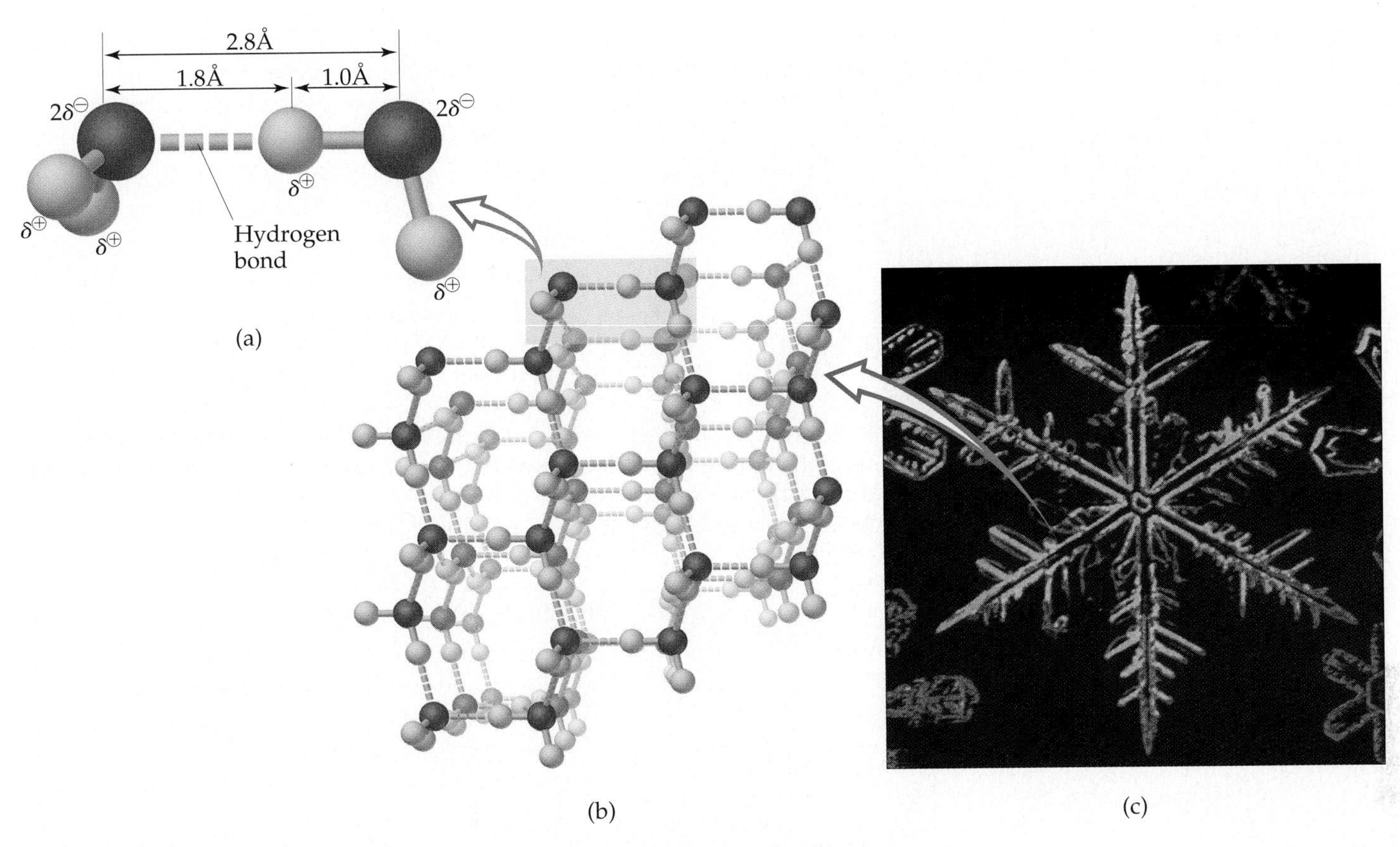

▲ **Figure 11.10** (a) Hydrogen bonding between two water molecules. The distances shown are those found in ice. (b) The arrangement of H_2O molecules in ice. Each hydrogen atom on one H_2O molecule is oriented toward a nonbonding pair of electrons on an adjacent H_2O molecule. As a result, ice has an open, hexagonal arrangement of H_2O molecules. (c) The hexagonal shape is characteristic of snowflakes.

The lower density of ice compared to liquid water profoundly affects life on Earth. Because ice floats (Figure 11.9), it covers the top of the water when a lake freezes in cold weather, thereby insulating the water below. If ice were more dense than water, ice forming at the top of a lake would sink to the bottom, and the lake could freeze solid. Most aquatic life could not survive under these conditions. The expansion of water upon freezing (Figure 11.11 ▼) is also what causes water pipes to break in freezing weather.

◀ **Figure 11.11** Water is one of the few substances that expands upon freezing. The expansion is due to the open structure of ice relative to that of liquid water.

A Closer Look Trends in Hydrogen Bonding

If the hydrogen bond is the result of an electrostatic interaction between the X—H bond dipole and an unshared electron pair on another atom, Y, then the strength of hydrogen bonding should increase as the X—H bond dipole increases. Thus, for the same Y, we would expect hydrogen-bonding strength to increase in the series

$$N{-}H\cdots Y < O{-}H\cdots Y < F{-}H\cdots Y$$

This is indeed true. But what property of Y dictates the strength of the hydrogen bonding? The atom Y must possess an unshared electron pair that attracts the positive end of the X—H dipole. This electron pair must not be too diffuse in space; if the electrons occupy too large a volume, the X—H dipole does not experience a strong directed attraction. For this reason, hydrogen bonding is not very strong unless Y is a small, highly electronegative atom, specifically N, O, or F. Among these three elements, hydrogen bonding is stronger when the electron pair is not attracted *too* strongly to its own nucleus. The electronegativity of Y is a good measure of this aspect. For example, the electronegativity of nitrogen is less than that of oxygen. Nitrogen is thus a better donor of the electron pair to the X—H bond. For a given X—H, hydrogen-bond strength increases in the order

$$X{-}H\cdots F < X{-}H\cdots O < X{-}H\cdots N$$

When X and Y are the same, the energy of hydrogen bonding increases in the order

$$N{-}H\cdots N < O{-}H\cdots O < F{-}H\cdots F$$

When the Y atom carries a negative charge, the electron pair is able to form especially strong hydrogen bonds. The hydrogen bond in the $F{-}H\cdots F^-$ ion is among the strongest known; the reaction

$$F^-(g) + HF(g) \longrightarrow FHF^-(g)$$

has a ΔH value of about -155 kJ/mol.

SAMPLE EXERCISE 11.2

In which of the following substances is hydrogen bonding likely to play an important role in determining physical properties: methane (CH_4), hydrazine (H_2NNH_2), methyl fluoride (CH_3F), or hydrogen sulfide (H_2S)?

Solution All of these compounds contain hydrogen, but hydrogen bonding usually occurs when the hydrogen is directly bonded to N, O, or F. There also needs to be an unshared pair of electrons on an electronegative atom (usually N, O, or F) in a nearby molecule. These criteria eliminate CH_4 and H_2S, which do not contain H bonded to N, O, or F. They also eliminate CH_3F whose Lewis structure shows a central C atom surrounded by three H atoms and an F atom. (Carbon always forms four bonds, whereas hydrogen and fluorine form one each.) Because the molecule contains a C—F bond and not an H—F bond, it does not form hydrogen bonds. In H_2NNH_2, however, we find N—H bonds. Therefore, hydrogen bonds exist between the molecules.

PRACTICE EXERCISE

In which of the following substances is significant hydrogen bonding possible: methylene chloride (CH_2Cl_2), phosphine (PH_3), hydrogen peroxide (HOOH), or acetone (CH_3COCH_3)?
Answer: HOOH

Comparing Intermolecular Forces

We can identify the intermolecular forces that are operative in a substance by considering its composition and structure. Dispersion forces are found in all substances. The strengths of these forces increase with increased molecular weight and depend on molecular shapes. Dipole-dipole forces add to the effect of dispersion forces and are found in polar molecules. Hydrogen bonds, which require H atoms bonded to F, O, or N, also add to the effect of dispersion forces. Hydrogen bonds tend to be the strongest type of intermolecular force. None of these intermolecular forces, however, is as strong as ordinary ionic or covalent bonds. Figure 11.12 ▶ presents a systematic way of identifying the kinds of intermolecular forces in a particular system, including ion-dipole and ion-ion forces.

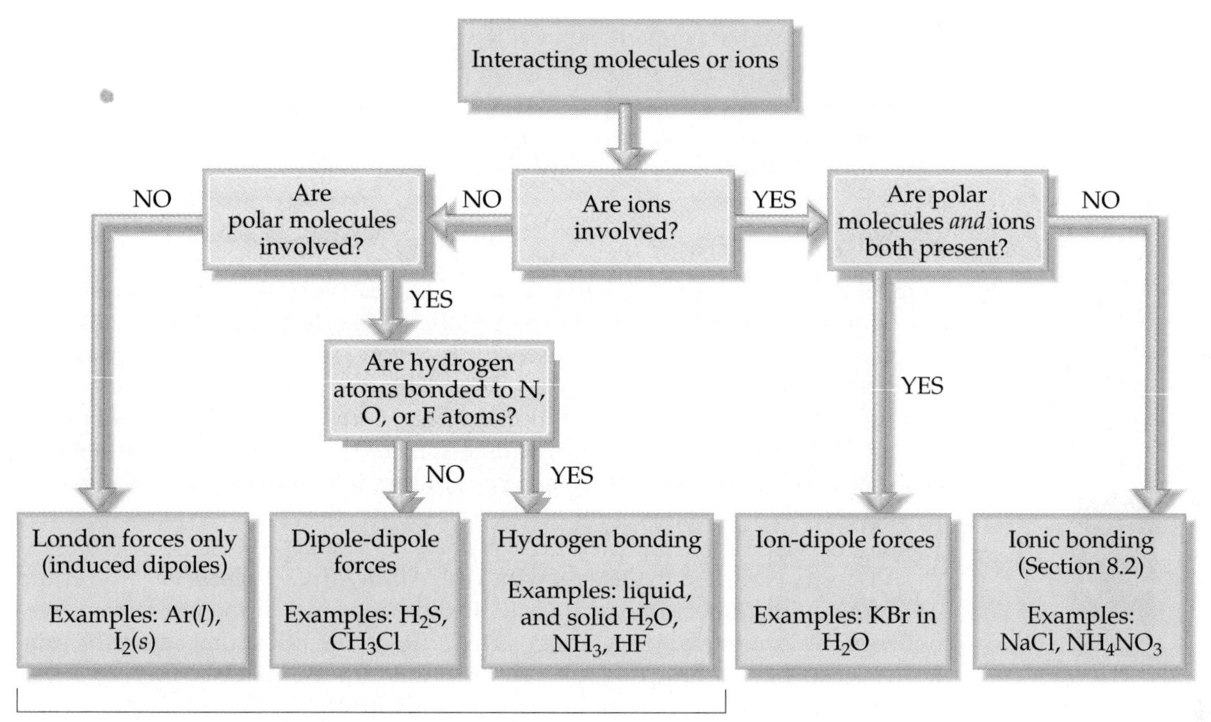

▲ **Figure 11.12** Flowchart for recognizing the major types of intermolecular forces. London dispersion forces occur in all instances. The strengths of the other forces generally increase proceeding from left to right.

SAMPLE EXERCISE 11.3

List the substances $BaCl_2$, H_2, CO, HF, and Ne in order of increasing boiling points.

Solution

Analyze: We need to relate the properties of the listed substances to boiling point.

Plan: The boiling point depends in part on the attractive forces in the liquid. We need to order these according to the relative strengths of the different kinds of forces.

Solve: The attractive forces are stronger for ionic substances than for molecular ones, so $BaCl_2$ should have the highest boiling point. The intermolecular forces of the remaining substances depend on molecular weight, polarity, and hydrogen bonding. The molecular weights are H_2 (2), CO (28), HF (20), and Ne (20). The boiling point of H_2 should be the lowest because it is nonpolar and has the lowest molecular weight. The molecular weights of CO, HF, and Ne are roughly the same. Because HF can hydrogen bond, however, it should have the highest boiling point of the three. Next is CO, which is slightly polar and has the highest molecular weight. Finally, Ne, which is nonpolar, should have the lowest boiling point of these three. The predicted order of boiling points is therefore

$$H_2 < Ne < CO < HF < BaCl_2$$

Check: The actual normal boiling points are H_2 (20 K), Ne (27 K), CO (83 K), HF (293 K), and $BaCl_2$ (1813 K), in agreement with our predictions.

PRACTICE EXERCISE

(a) Identify the intermolecular forces present in the following substances, and **(b)** select the substance with the highest boiling point: CH_3CH_3, CH_3OH, and CH_3CH_2OH.

Answers: **(a)** CH_3CH_3 has only dispersion forces, whereas the other two substances have both dispersion forces and hydrogen bonds; **(b)** CH_3CH_2OH

11.3 Some Properties of Liquids

The intermolecular forces we have just discussed can help us understand many familiar properties of liquids and solids. In this section we examine two important properties of liquids: viscosity and surface tension.

Viscosity

Some liquids, such as molasses and motor oil, flow very slowly; others, such as water and gasoline, flow easily. The resistance of a liquid to flow is called its **viscosity**. The greater a liquid's viscosity, the more slowly it flows. Viscosity can be measured by timing how long it takes a certain amount of the liquid to flow through a thin tube under gravitational force. More viscous liquids take longer (Figure 11.13 ◀). Viscosity can also be determined by measuring the rate at which steel spheres fall through the liquid. The spheres fall more slowly as the viscosity increases. The common unit of viscosity is the *poise*, which equals 1 g/cm-s. Frequently viscosity is reported in centipoise (cP), which is 0.01 poise (P).

▲ **Figure 11.13** The Society of Automotive Engineers (SAE) has established numbers to indicate the viscosity of motor oils. The higher the number, the greater the viscosity at any given temperature. The SAE 40 motor oil on the left is more viscous and flows more slowly than the less viscous SAE 10 oil on the right.

Viscosity is related to the ease with which individual molecules of the liquid can move with respect to one another. It thus depends on the attractive forces between molecules and on whether structural features exist that cause the molecules to become entangled. For a series of related compounds, therefore, viscosity increases with molecular weight, as illustrated in Table 11.4 ▼. For any given substance, viscosity decreases with increasing temperature. Octane, for example, has a viscosity of 0.706 cP at 0°C, and of 0.433 cP at 40°C. At higher temperatures the greater average kinetic energy of the molecules more easily overcomes the attractive forces between molecules.

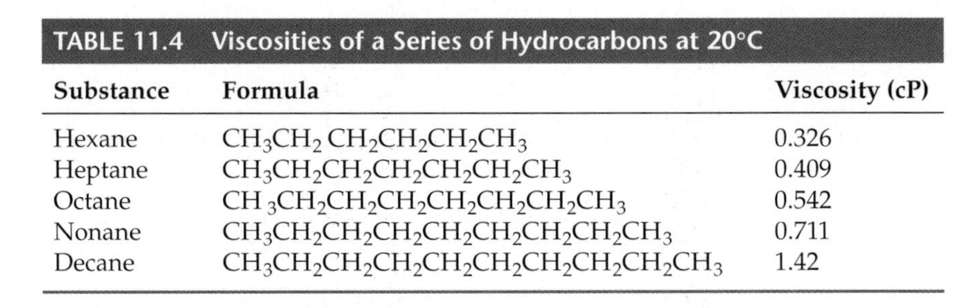

TABLE 11.4 Viscosities of a Series of Hydrocarbons at 20°C

Substance	Formula	Viscosity (cP)
Hexane	$CH_3CH_2CH_2CH_2CH_2CH_3$	0.326
Heptane	$CH_3CH_2CH_2CH_2CH_2CH_2CH_3$	0.409
Octane	$CH_3CH_2CH_2CH_2CH_2CH_2CH_2CH_3$	0.542
Nonane	$CH_3CH_2CH_2CH_2CH_2CH_2CH_2CH_2CH_3$	0.711
Decane	$CH_3CH_2CH_2CH_2CH_2CH_2CH_2CH_2CH_2CH_3$	1.42

Surface Tension

When water is placed on a waxy surface, it "beads up," forming distorted spheres. This behavior is due to an imbalance of intermolecular forces at the surface of the liquid, as shown in Figure 11.14 ▼. Notice that molecules in the inte-

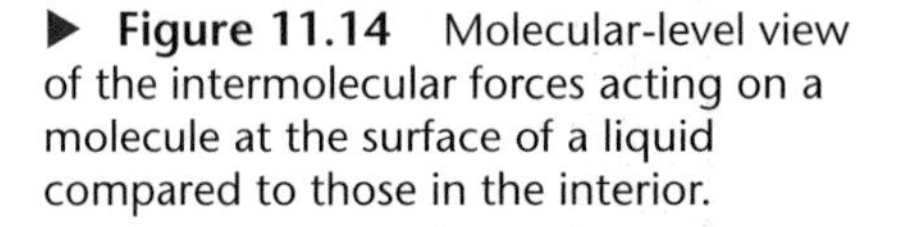

▶ **Figure 11.14** Molecular-level view of the intermolecular forces acting on a molecule at the surface of a liquid compared to those in the interior.

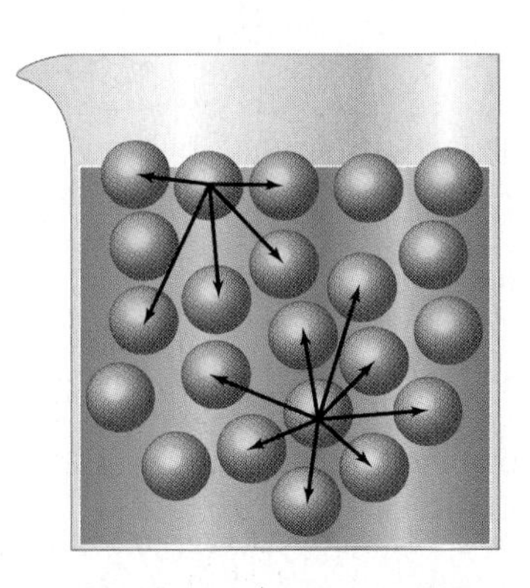

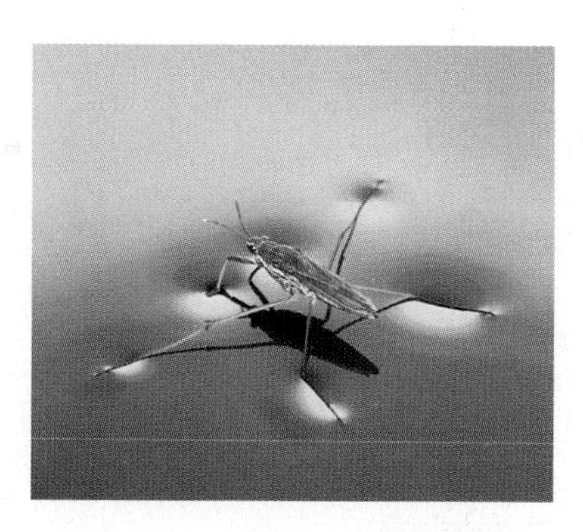

◀ **Figure 11.15** Surface tension permits an insect such as the water strider to "walk" on water.

rior are attracted equally in all directions, whereas those at the surface experience a net inward force. This inward force pulls molecules from the surface into the interior, thereby reducing the surface area. (Spheres have the smallest surface area for their volume.) The inward force also makes the molecules at the surface pack closely together, causing the liquid to behave almost as if it had a skin. This effect permits a carefully placed needle to float on the surface of water and some insects to "walk" on water (Figure 11.15 ▲) even though their densities are greater than that of water.

A measure of the inward forces that must be overcome in order to expand the surface area of a liquid is given by its surface tension. **Surface tension** is the energy required to increase the surface area of a liquid by a unit amount. For example, the surface tension of water at 20°C is $7.29 \times 10^{-2}\ J/m^2$, which means that an energy of 7.29×10^{-2} J must be supplied to increase the surface area of a given amount of water by 1 m^2. Water has a high surface tension because of its strong hydrogen bonds. The surface tension of mercury is even higher ($4.6 \times 10^{-1}\ J/m^2$) because of even stronger metallic bonds between the atoms of mercury.

Intermolecular forces that bind similar molecules to one another, such as the hydrogen bonding in water, are called *cohesive forces*. Intermolecular forces that bind a substance to a surface are called *adhesive forces*. Water placed in a glass tube adheres to the glass because the adhesive forces between the water and glass are even greater than the cohesive forces between water molecules. The curved upper surface, or *meniscus*, of the water is therefore U-shaped (Figure 11.16 ▶). For mercury, however, the meniscus is curved downward where the mercury contacts the glass. In this case the cohesive forces between the mercury atoms are much greater than the adhesive forces between the mercury atoms and the glass.

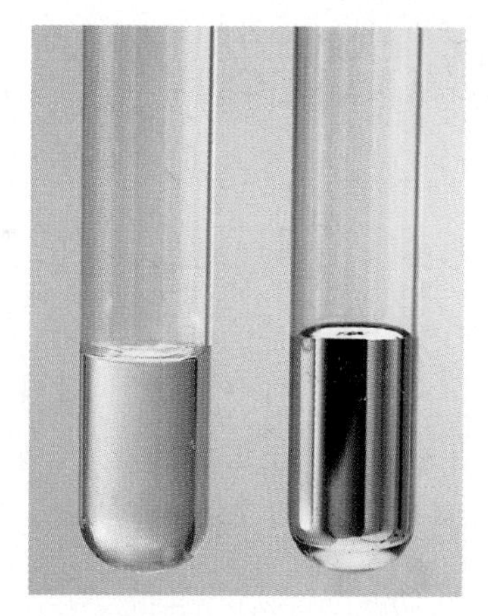

▲ **Figure 11.16** The water meniscus in a glass tube compared with the mercury meniscus in a similar tube.

When a small-diameter glass tube, or capillary, is placed in water, water rises in the tube. The rise of liquids up very narrow tubes is called **capillary action**. The adhesive forces between the liquid and the walls of the tube tend to increase the surface area of the liquid. The surface tension of the liquid tends to reduce the area, thereby pulling the liquid up the tube. The liquid climbs until the adhesive and cohesive forces are balanced by the force of gravity on the liquid. Capillary action helps water and dissolved nutrients move upward through plants.

11.4 Phase Changes

Water left uncovered in a glass for several days evaporates. An ice cube left in a warm room quickly melts. Solid CO_2 (sold as Dry Ice™) *sublimes* at room temperature; that is, it changes directly from the solid to the vapor state. In general, each state of matter can change into either of the other two states. Figure 11.17 ▶ shows the name associated with each of these transformations. These transformations are called **phase changes**, or changes of state.

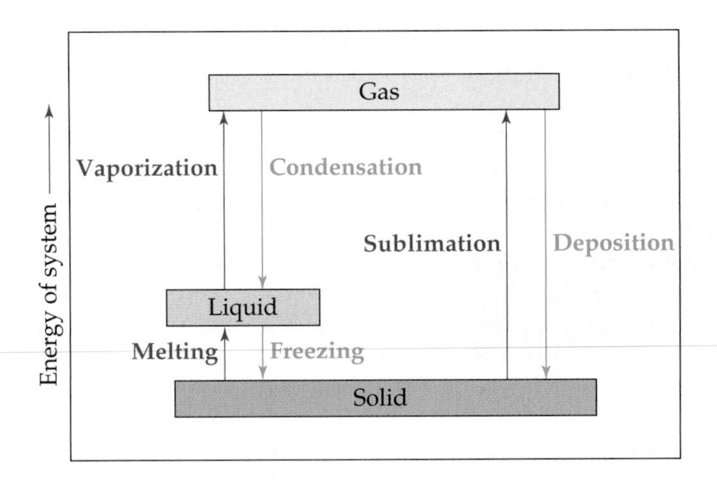

▶ **Figure 11.17** Energy changes accompanying phase changes between the three states of matter, and the names associated with them.

Energy Changes Accompanying Phase Changes

Every phase change is accompanied by a change in the energy of the system. In a solid lattice, for example, the molecules or ions are in more or less fixed positions with respect to one another and closely arranged to minimize the energy of the system. As the temperature of the solid increases, the units of the solid vibrate about their equilibrium positions with increasingly energetic motion. When the solid melts, the units that made up the solid are freed to move with respect to one another, which ordinarily means that their average separations increase. This melting process is called *fusion*. The increased freedom of motion of the molecules or ions comes at a price, measured by the **heat of fusion**, or enthalpy of fusion, denoted ΔH_{fus}. The heat of fusion of ice, for example, is 6.01 kJ/mol.

As the temperature of the liquid phase increases, the molecules of the liquid move about with increasing energy. One measure of this increasing energy is that the concentration of gas-phase molecules over the liquid increases with temperature. These molecules exert a pressure called the vapor pressure. We will explore vapor pressure in Section 11.5. For now we just need to understand that the vapor pressure increases with increasing temperature until it equals the external pressure over the liquid, typically atmospheric pressure. At this point the liquid boils; the molecules of the liquid move into the gaseous state where they are widely separated. The energy required to cause this transition is called the **heat of vaporization**, or enthalpy of vaporization, denoted ΔH_{vap}. For water the heat of vaporization is 40.7 kJ/mol.

Figure 11.18 ▼ shows the comparative values of ΔH_{fus} and ΔH_{vap} for four different substances. ΔH_{vap} values tend to be larger than ΔH_{fus} because in the

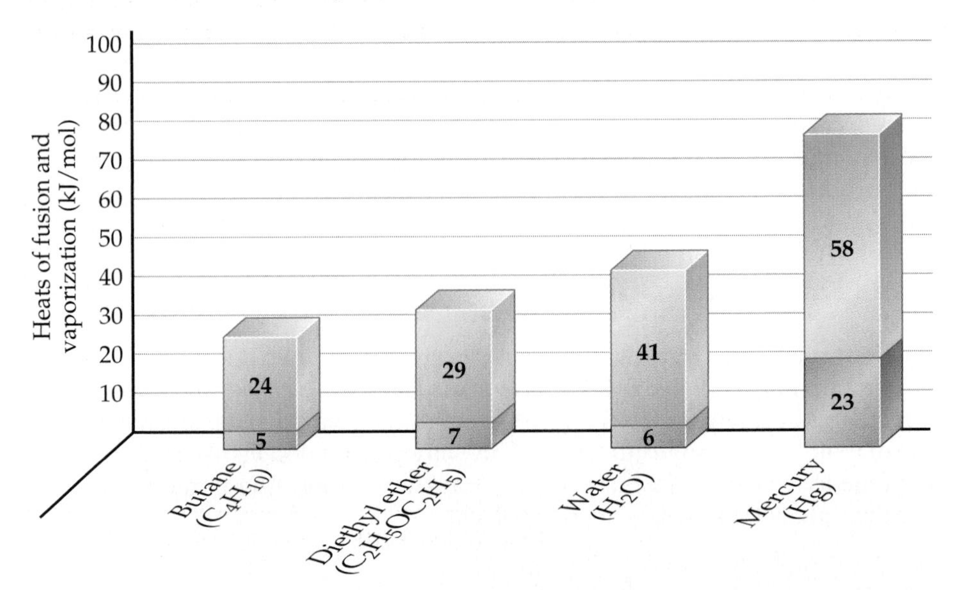

▶ **Figure 11.18** Comparative values of the heats of fusion (violet bars) and vaporization (blue) for several substances.

transition from the liquid to the vapor state the molecules must essentially sever all their intermolecular attractive interactions, whereas in melting, many of these attractive interactions remain.

The molecules of a solid can be transformed directly into the gaseous state. The enthalpy change required for this transition is called the **heat of sublimation**, denoted ΔH_{sub}. For the substances shown in Figure 11.18, ΔH_{sub} is the sum of ΔH_{fus} and ΔH_{vap}. Thus, ΔH_{sub} for water is approximately 47 kJ/mol.

Phase changes of matter show up in important ways in our everyday experiences. We use ice cubes to cool our liquid drinks; the heat of fusion of ice cools the liquid in which the ice is immersed. We feel cool when we step out of a swimming pool or a warm shower because the heat of vaporization is drawn from our bodies as the water evaporates from our skin. Our bodies use the evaporation of water from skin to regulate body temperature, especially when we exercise vigorously in warm weather. A refrigerator also relies on the cooling effects of vaporization. Its mechanism contains an enclosed gas that can be liquefied under pressure. The gas absorbs heat as it is expanded to a chamber where it evaporates, thus cooling the interior of the refrigerator. The vapor is then recycled through a compressor.

What happens to the heat absorbed when the liquid refrigerant vaporizes? According to the first law of thermodynamics (Section 5.2), the heat absorbed by the liquid in vaporizing must be evolved when the reverse process, condensation of the vapor to form the liquid, occurs. As the refrigerator compresses the vapor and liquid is formed, the heat evolved is dissipated through cooling coils in the back of the refrigerator. Just as the heat of condensation is equal in magnitude and opposite in sign from the heat of vaporization, so also the *heat of deposition* is exothermic to the same degree that the heat of sublimation is endothermic; and the *heat of freezing* is exothermic to the same degree that the heat of fusion is endothermic. These relationships, shown in Figure 11.17, are consequences of the first law of thermodynamics.

Heating Curves

What happens when we heat a sample of ice that is initially at −25°C and 1 atm pressure? The addition of heat causes the temperature of the ice to increase. As long as the temperature is below 0°C, the sample remains frozen. When the temperature reaches 0°C, the ice begins to melt. Because melting is an endothermic process, the heat we add at 0°C is used to convert ice to water and the temperature remains constant until all the ice has melted. Once we reach this point, the further addition of heat causes the temperature of the liquid water to increase.

A graph of the temperature of the system versus the amount of heat added is called a *heating curve*. Figure 11.19 ▼ shows a heating curve for transforming ice at −25°C to steam at 125°C under a constant pressure of 1 atm. Heating the

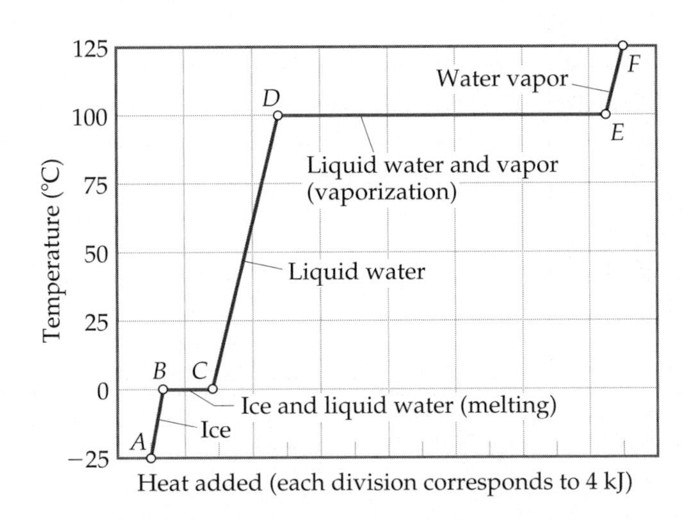

◀ **Figure 11.19** Heating curve for the transformation of 1.00 mol of water from −25°C to 125°C at a constant pressure of 1 atm. Blue lines show the heating of one phase from a lower temperature to a higher one. Red lines show the conversion of one phase to another at constant temperature.

ice from $-25°C$ to $0°C$ is represented by the line segment *AB* in Figure 11.19, while converting the ice at 0°C to water at 0°C is the horizontal segment *BC*. Additional heat increases the temperature of the water until the temperature reaches 100°C (segment *CD*). The heat is then used to convert water to steam at a constant temperature of 100°C (segment *DE*). Once all the water has been converted to steam, the steam is heated to its final temperature of 125°C (segment *EF*).

We can calculate the enthalpy change of the system for each of the segments of the heating curve. In segments *AB*, *CD*, and *EF* we are heating a single phase from one temperature to another. As we saw in Section 5.5, the amount of heat needed to raise the temperature of a substance is given by the product of the specific heat, mass, and temperature change (Equation 5.22). The greater the specific heat of a substance, the more heat we must add to accomplish a certain temperature increase. Because the specific heat of water is greater than that of ice, the slope of segment *CD* is less than that of segment *AB*; we must add more heat to water to achieve a 1°C temperature change than is needed to warm the same quantity of ice by 1°C.

In segments *BC* and *DE* we are converting one phase to another at a constant temperature. The temperature remains constant during these phase changes because the added energy is used to overcome the attractive forces between molecules rather than to increase their average kinetic energy. For segment *BC*, in which ice is converted to water, the enthalpy change can be calculated by using ΔH_{fus}, while for segment *DE* we can use ΔH_{vap}. In Sample Exercise 11.4 we calculate the total enthalpy change for the heating curve in Figure 11.19.

SAMPLE EXERCISE 11.4

Calculate the enthalpy change upon converting 1.00 mol of ice at $-25°C$ to water vapor (steam) at 125°C under a constant pressure of 1 atm. The specific heats of ice, water, and steam are 2.09 J/g-K, 4.18 J/g-K, and 1.84 J/g-K, respectively. For H_2O, $\Delta H_{fus} = 6.01$ kJ/mol, and $\Delta H_{vap} = 40.67$ kJ/mol.

Solution

Analyze: Our goal is to calculate the total heat required to convert 1 mol of ice at $-25°C$ to steam at 125°C.

Plan: We can calculate the enthalpy change for each segment and then sum them to get the total enthalpy change (Hess's law, Section 5.6).

Solve: For segment *AB* in Figure 11.19 we are adding enough heat to ice to increase its temperature by 25°C. A temperature change of 25°C is the same as a temperature change of 25 K, so we can use the specific heat of ice to calculate the enthalpy change during this process:

$$AB: \quad \Delta H = (1.00 \text{ mol})(18.0 \text{ g/mol})(2.09 \text{ J/g-K})(25 \text{ K}) = 940 \text{ J} = 0.94 \text{ kJ}$$

For segment *BC* in Figure 11.19, in which we convert ice to water at 0°C, we can use the molar enthalpy of fusion directly:

$$BC: \quad \Delta H = (1.00 \text{ mol})(6.01 \text{ kJ/mol}) = 6.01 \text{ kJ}$$

The enthalpy changes for segments *CD*, *DE*, and *EF* can be calculated in similar fashion:

$$CD: \quad \Delta H = (1.00 \text{ mol})(18.0 \text{ g/mol})(4.18 \text{ J/g-K})(100 \text{ K}) = 7520 \text{ J} = 7.52 \text{ kJ}$$

$$DE: \quad \Delta H = (1.00 \text{ mol})(40.67 \text{ kJ/mol}) = 40.7 \text{ kJ}$$

$$EF: \quad \Delta H = (1.00 \text{ mol})(18.0 \text{ g/mol})(1.84 \text{ J/g-K})(25 \text{ K}) = 830 \text{ J} = 0.83 \text{ kJ}$$

The total enthalpy change is the sum of the changes of the individual steps:

$$\Delta H = 0.94 \text{ kJ} + 6.01 \text{ kJ} + 7.52 \text{ kJ} + 40.7 \text{ kJ} + 0.83 \text{ kJ} = 56.0 \text{ kJ}$$

Check: The components of the total energy change are reasonable in comparison with the lengths of the horizontal segments of the lines in Figure 11.19. Notice that the largest component is the heat of vaporization.

PRACTICE EXERCISE

What is the enthalpy change during the process in which 100.0 g of water at 50.0°C is cooled to ice at $-30.0°C$? (Use the specific heats and enthalpies for phase changes given in Sample Exercise 11.4.)

Answer: $-20.9 \text{ kJ} - 33.4 \text{ kJ} - 6.27 \text{ kJ} = -60.6 \text{ kJ}$

Cooling a substance has the opposite effect of heating it. Thus, if we start with water vapor and begin to cool it, we would move right to left through the events shown in Figure 11.19. We would first lower the temperature of the vapor ($F \longrightarrow E$), then condense it ($E \longrightarrow D$), and so forth. Sometimes as we remove heat from a liquid, we can temporarily cool it below its freezing point without forming a solid. This phenomenon is called *supercooling*. Supercooling occurs when heat is removed from a liquid so rapidly that the molecules literally have no time to assume the ordered structure of a solid. A supercooled liquid is unstable; particles of dust entering the solution or gentle stirring is often sufficient to cause the substance to solidify quickly.

Critical Temperature and Pressure

A gas liquefies at some point when pressure is applied to it. If we increase the pressure on water vapor at 55°C, for example, then it liquefies when the pressure equals 118 torr, and an equilibrium between the gaseous and liquid phases exists. If the temperature is 110°C, the liquid phase does not form until the pressure is 1075 torr. At 374°C the liquid phase forms only at 1.655×10^5 torr (217.7 atm). Above this temperature, no amount of pressure will cause a distinct liquid phase to form. Instead, as pressure increases, the gas merely becomes steadily more compressed. The highest temperature at which a distinct liquid phase can form is referred to as the **critical temperature**. The **critical pressure** is the pressure required to bring about liquefaction at this critical temperature.

The critical temperatures and pressures are listed for several substances in Table 11.5 ▼. Notice that nonpolar, low molecular weight substances have lower critical temperatures and pressures than those that are polar or of higher molecular weight. As we have already seen, the transition from the gaseous to the liquid state is determined by intermolecular forces. For every gaseous substance, a temperature can be reached at which the motional energies of the molecules are sufficient to overcome the attractive forces that lead to the liquid state, regardless of how closely the molecules are forced to approach by increasing the pressure. Notice that water and ammonia have exceptionally high critical temperatures and pressures. These high values are yet another consequence of strong intermolecular hydrogen-bonding forces.

The critical temperatures and pressures of substances are often of considerable importance to engineers and other people working with gases, because they provide information about the conditions under which gases liquefy. Sometimes we want to liquefy a gas; other times we want to avoid liquefying it. It is useless to try to liquefy a gas by applying pressure if the gas is above its critical temperature. For example, O_2 has a critical temperature of 154.4 K. It must be cooled below this temperature before it can be liquefied by pressure. In contrast, ammonia has a critical temperature of 405.6 K. Thus, it can be liquefied at room temperature (approximately 295 K) by compressing the gas to a sufficient pressure.

TABLE 11.5 Critical Temperatures and Pressures of Selected Substances

Substance	Critical Temperature (K)	Critical Pressure (atm)
Ammonia, NH_3	405.6	111.5
Phosphine, PH_3	324.4	64.5
Argon, Ar	150.9	48
Carbon dioxide, CO_2	304.3	73.0
Nitrogen, N_2	126.1	33.5
Oxygen, O_2	154.4	49.7
Propane, $CH_3CH_2CH_3$	370.0	42.0
Water, H_2O	647.6	217.7
Hydrogen sulfide, H_2S	373.5	88.9

Chemistry at Work Supercritical Fluid Extraction

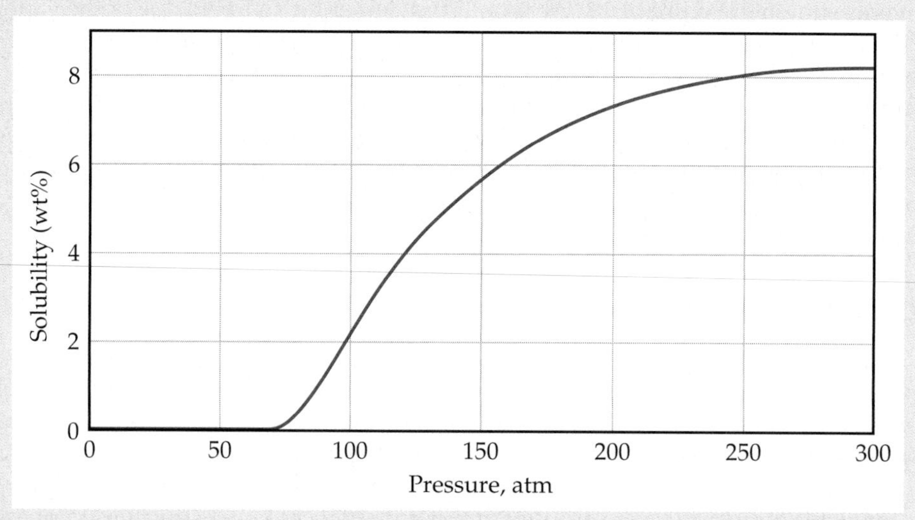

▶ **Figure 11.20** Solubility of naphthalene ($C_{10}H_8$) in supercritical carbon dioxide at 45°C.

At ordinary pressures, a substance above its critical temperature behaves as an ordinary gas. However, as pressure increases up to several hundred atmospheres, its character changes. Like a gas, it still expands to fill the confines of its container, but its density approaches that of a liquid. (For example, the critical temperature of water is 647.6 K, and its critical pressure is 217.7 atm. At this temperature and pressure, the density of water is 0.4 g/mL). A substance at temperatures and pressures higher than its critical temperature and pressure is better considered a *supercritical fluid* rather than a gas.

Like liquids, supercritical fluids can behave as solvents, dissolving a wide range of substances. Using *supercritical fluid extraction*, the components of mixtures can be separated. The solvent power of a supercritical fluid increases as its density increases. Conversely, lowering its density (either by decreasing pressure or increasing temperature) causes the supercritical fluid and the dissolved material to separate. Figure 11.20 ▲ shows the solubility of a typical nonpolar organic solid, naphthalene ($C_{10}H_8$), in supercritical carbon dioxide at 45°C. The solubility of naphthalene is essentially zero below the critical pressure of 73 atm. Solubility increases rapidly, however, with increasing pressure (and, thus, increasing density of the supercritical fluid).

By appropriate manipulation of pressure, supercritical fluid extraction has been used successfully to separate complex mixtures in the chemical, food, pharmaceutical, and energy industries. Supercritical carbon dioxide, for example, is environmentally benign because there are no problems disposing of solvent and there are no toxic residues resulting from the process. In addition, supercritical CO_2 is inexpensive compared to solvents other than water. A process for removing caffeine from green coffee beans by extraction with supercritical CO_2, diagrammed in Figure 11.21 ▶, has been in commercial operation for several years. At the proper temperature and pressure the supercritical CO_2 removes caffeine from the beans by dissolution, but leaves the flavor and aroma components, producing decaffeinated coffee. Other applications of supercritical CO_2 extraction include the extraction of the essential flavor elements from hops for use in brewing, and the isolation of the flavor components of herbs and spices. (See also Section 18.7.)

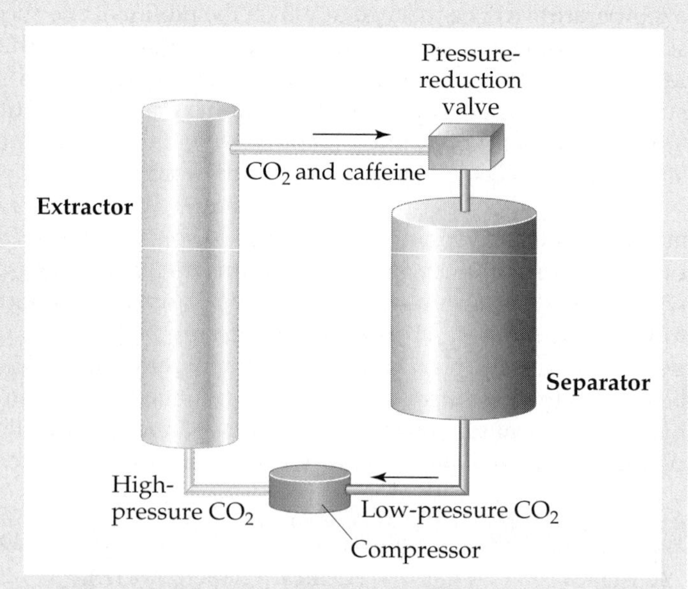

▲ **Figure 11.21** Diagram of a supercritical fluid extraction process. The material to be processed is placed in the extractor. The desired material dissolves in supercritical CO_2 at high pressure, then is precipitated in the separator when the CO_2 pressure is reduced. The carbon dioxide is then recycled through the compressor with a fresh batch of material in the extractor.

11.5 Vapor Pressure

Molecules can escape from the surface of a liquid into the gas phase by vaporization or evaporation. Suppose we conduct an experiment in which we place a quantity of ethanol (C_2H_5OH) in an evacuated, closed container such as that in Figure 11.22 ▼. The ethanol will quickly begin to evaporate. As a result, the pressure exerted by the vapor in the space above the liquid will begin to increase. After a short time the pressure of the vapor will attain a constant value, which we call the **vapor pressure** of the substance.

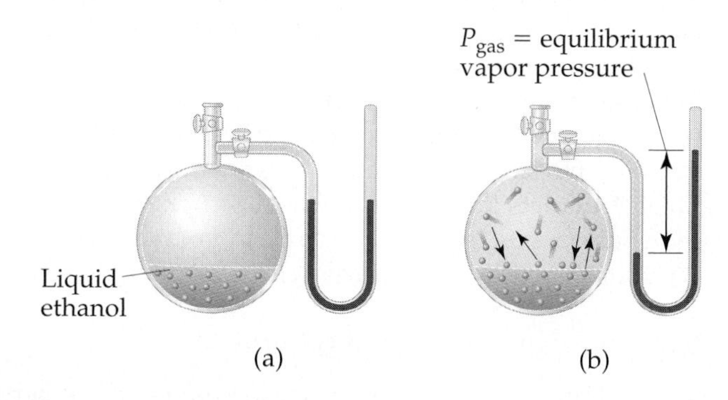

◀ **Figure 11.22** Illustration of the equilibrium vapor pressure over liquid ethanol. In (a) we imagine that no molecules exist in the gas phase; there is zero pressure in the cell. In (b) the rate at which molecules leave the surface equals the rate at which gas molecules pass into the liquid phase. These equal rates produce a stable vapor pressure that does not change as long as the temperature remains constant.

Explaining Vapor Pressure on the Molecular Level

The molecules of a liquid move at various speeds. Figure 11.23 ▼ shows the distribution of kinetic energies of the particles at the surface of a liquid at two temperatures. The distribution curves are like those shown earlier for gases (Figures 10.18 and 10.19). At any instant some of the molecules on the surface of the liquid possess sufficient kinetic energy to overcome the attractive forces of their neighbors and escape into the gas phase. The weaker the attractive forces, the larger is the number of molecules that are able to escape and the higher is the vapor pressure.

At any particular temperature the movement of molecules from the liquid to the gas phase goes on continuously. As the number of gas-phase molecules increases, however, the probability increases that a molecule in the gas phase will strike the liquid surface and be recaptured by the liquid, as shown in Figure 11.22(b). Eventually, the rate at which molecules return to the liquid exactly equals the rate at which they escape. The number of molecules in the gas phase then reaches a steady value, and the pressure of the vapor at this stage becomes constant.

The condition in which two opposing processes are occurring simultaneously at equal rates is called a **dynamic equilibrium**. A liquid and its vapor are in equilibrium when evaporation and condensation occur at equal rates. It may appear that nothing is occurring at equilibrium, because there is no net change in the system. In fact, a great deal is happening; molecules continuously pass from the liquid state to the gas state and from the gas state to the liquid state.

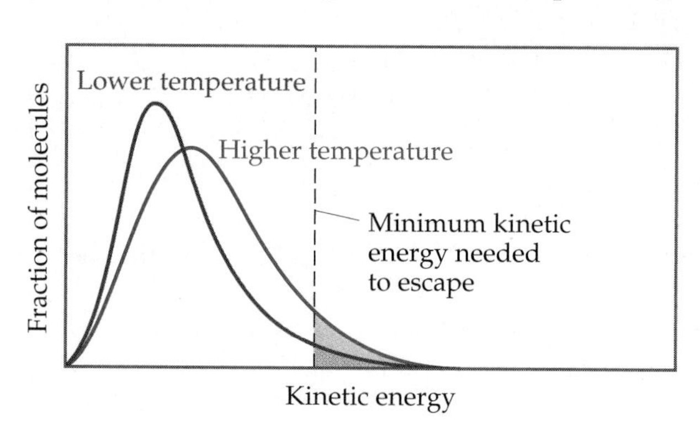

◀ **Figure 11.23** Distribution of kinetic energies of surface molecules of a hypothetical liquid at two temperatures. Only the fastest molecules have sufficient kinetic energy to escape the liquid and enter the vapor, as shown by the shaded areas. The higher the temperature, the larger the fraction of molecules with enough energy to escape.

All equilibria between different states of matter possess this dynamic character. *The vapor pressure of a liquid is the pressure exerted by its vapor when the liquid and vapor states are in dynamic equilibrium.*

Volatility, Vapor Pressure, and Temperature

ANIMATION
Vapor Pressure versus Temperature

ACTIVITY
Equilibrium Vapor Pressure

When vaporization occurs in an open container, as when water evaporates from a bowl, the vapor spreads away from the liquid. Little, if any, is recaptured at the surface of the liquid. Equilibrium never occurs, and the vapor continues to form until the liquid evaporates to dryness. Substances with high vapor pressure (such as gasoline) evaporate more quickly than substances with low vapor pressure (such as motor oil). Liquids that evaporate readily are said to be **volatile**.

Hot water evaporates more quickly than cold water because vapor pressure increases with increasing temperature. We see this effect in Figure 11.23: As the temperature of a liquid is increased, the molecules move more energetically and a greater fraction can therefore escape more readily from their neighbors. Figure 11.24 ▼ depicts the variation in vapor pressure with temperature for four common substances that differ greatly in volatility. Note that the vapor pressure in all cases increases nonlinearly with increasing temperature.

Vapor Pressure and Boiling Point

A liquid boils when its vapor pressure equals the external pressure acting on the surface of the liquid. At this point bubbles of vapor are able to form within the interior of the liquid. The temperature of boiling increases with increasing external pressure. The boiling point of a liquid at 1 atm pressure is called its **normal boiling point**. From Figure 11.24 we see that the normal boiling point of water is 100°C.

The boiling point is important to many processes that involve heating liquids, including cooking. The time required to cook food depends on the temperature. As long as water is present, the maximum temperature of the cooking food is the boiling point of water. Pressure cookers work by allowing steam to escape only when it exceeds a predetermined pressure; the pressure above the water can therefore increase above atmospheric pressure. The higher pressure causes water to boil at a higher temperature, thereby allowing the food to get hotter and to cook more rapidly. The effect of pressure on boiling point also explains why it takes longer to cook food at higher elevations than at sea level. The atmospheric pressure is lower at higher altitudes, so water boils at a lower temperature.

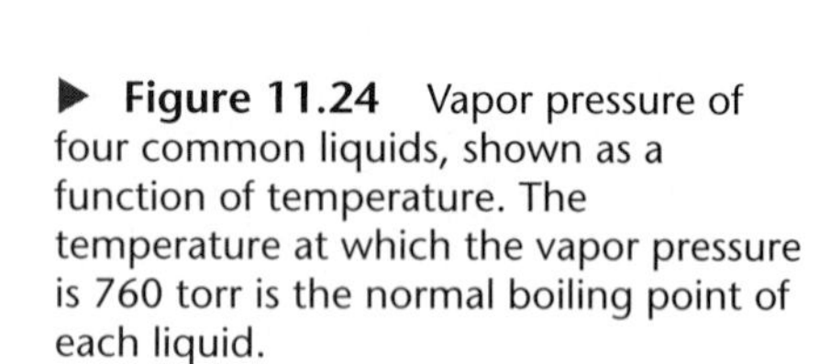

▶ **Figure 11.24** Vapor pressure of four common liquids, shown as a function of temperature. The temperature at which the vapor pressure is 760 torr is the normal boiling point of each liquid.

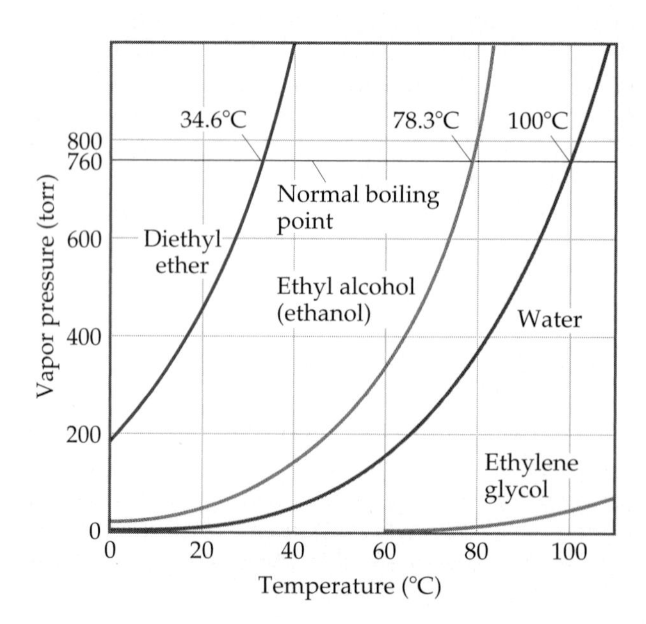

SAMPLE EXERCISE 11.5

Use Figure 11.24 to estimate the boiling point of diethyl ether under an external pressure of 0.80 atm.

Solution The boiling point is the temperature at which the vapor pressure is equal to the external pressure. From Figure 11.24 we see that the boiling point at 0.80 atm is about 27°C, which is close to room temperature. We can make a flask of diethyl ether boil at room temperature by using a vacuum pump to lower the pressure above the liquid to about 0.8 atm (80 kPa).

PRACTICE EXERCISE

At what external pressure will ethanol have a boiling point of 60°C?
Answer: about 0.45 atm

A Closer Look The Clausius–Clapeyron Equation

You might have noticed that the plots of the variation of vapor pressure with temperature shown in Figure 11.24 have a distinct shape: Each curves sharply upward to a higher vapor pressure with increasing temperature. The relationship between vapor pressure and temperature is given by an equation called the *Clausius–Clapeyron equation*:

$$\ln P = \frac{-\Delta H_{vap}}{RT} + C \qquad [11.1]$$

In this equation T is the absolute temperature, R is the gas constant (8.314 J/mol-K), ΔH_{vap} is the molar enthalpy of vaporization, and C is a constant. The Clausius–Clapeyron equation predicts that a graph of ln P versus $1/T$ should give a straight line with a slope equal to $-\Delta H_{vap}/R$. Thus, we can use such a plot to determine the enthalpy of vaporization of a substance as follows:

$$\Delta H_{vap} = -\text{slope} \times R$$

As an example of the application of the Clausius–Clapeyron equation, the vapor-pressure data for ethanol shown in Figure 11.24 are graphed as ln P versus $1/T$ in Figure 11.25 ▶. The data lie on a straight line with a negative slope. We can use the slope of the line to determine ΔH_{vap} for ethanol. We can also extrapolate the line to obtain values for the vapor pressure of ethanol at temperatures above and below the temperature range for which we have data.

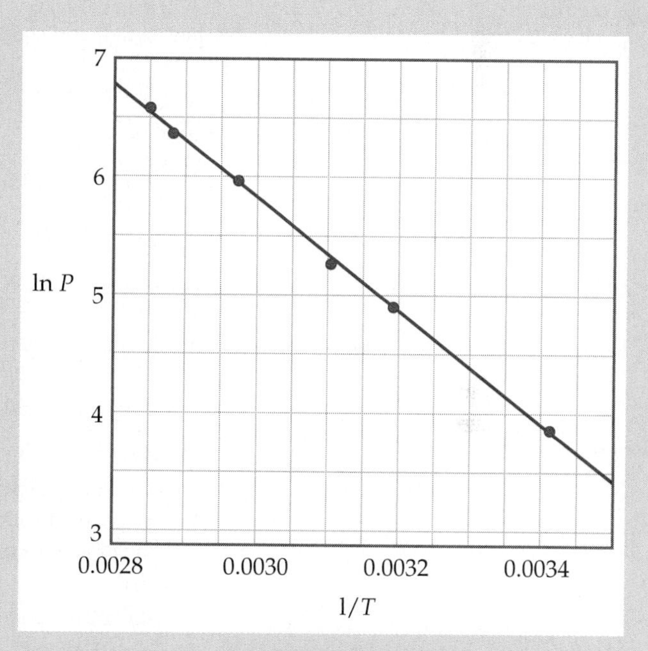

▲ **Figure 11.25** Application of the Clausius–Clapeyron equation, Equation 11.1, to the vapor-pressure-versus-temperature data for ethanol. The slope of the line equals $-\Delta H_{vap}/R$, giving $\Delta H_{vap} = 38.56$ kJ/mol.

11.6 Phase Diagrams

The equilibrium between a liquid and its vapor is not the only dynamic equilibrium that can exist between states of matter. Under appropriate conditions of temperature and pressure a solid can be in equilibrium with its liquid state or even with its vapor state. A **phase diagram** is a graphical way to summarize the conditions under which equilibria exist between the different states of matter. It also allows us to predict the phase of a substance that is stable at any given temperature and pressure.

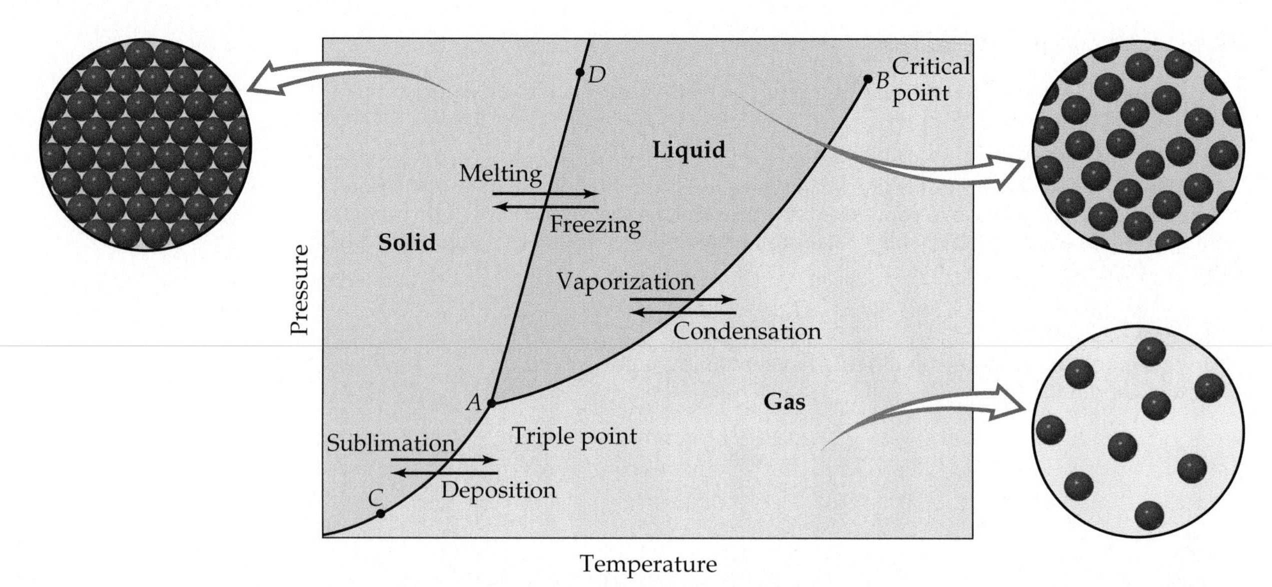

ACTIVITY
Phase Diagram

▲ **Figure 11.26** General shape for a phase diagram of a system exhibiting three phases: gas, liquid, and solid.

The general form of a phase diagram for a substance that exhibits three phases is shown in Figure 11.26 ▲. The diagram is a two-dimensional graph, with pressure and temperature as the axes. The diagram contains three important curves, each of which represents the conditions of temperature and pressure at which the various phases can coexist at equilibrium. The only substance present in the system is the one whose phase diagram is under consideration. The pressure shown in the diagram is the pressure applied to the system or generated by the substance itself. The curves may be described as follows:

1. The line from *A* to *B* is the vapor-pressure curve of the liquid. It represents the equilibrium between the liquid and gas phases. The point on this curve where the vapor pressure is 1 atm is the normal boiling point of the substance. The vapor-pressure curve ends at the *critical point* (*B*), which is at the critical temperature and critical pressure of the substance. Beyond the critical point the liquid and gas phases become indistinguishable.
2. The line *AC* represents the variation in the vapor pressure of the solid as it sublimes at different temperatures.
3. The line from *A* through *D* represents the change in melting point of the solid with increasing pressure. This line usually slopes slightly to the right as pressure increases, because the solid for most substances is denser than the liquid. An increase in pressure usually favors the more compact solid phase; thus, higher temperatures are required to melt the solid at higher pressures. The *melting point* of a substance is identical to its *freezing point*. The two differ only in the direction from which the phase change is approached. The melting point at 1 atm is the **normal melting point**.

Point *A*, where the three curves intersect, is known as the **triple point**. All three phases are in equilibrium at this temperature and pressure. Any other point on the three curves represents an equilibrium between two phases. Any point on the diagram that does not fall on a line corresponds to conditions under which only one phase is present. The gas phase, for example, is stable at low pressures and high temperatures, whereas the solid phase is stable at low temperatures and high pressures. Liquids are stable in the region between the other two.

The Phase Diagrams of H_2O and CO_2

Figure 11.27 ▼ shows the phase diagrams of H_2O and CO_2. The solid-liquid equilibrium (melting point) line of CO_2 follows the typical behavior; its melting point increases with increasing pressure. In contrast, the melting point of H_2O *decreases* with increasing pressure. As seen in Figure 11.11, water is among the very few substances whose liquid form is more compact than its solid form. (Section 11.2)

The triple point of H_2O (0.0098°C and 4.58 torr) is at much lower pressure than that of CO_2 (−56.4°C and 5.11 atm). For CO_2 to exist as a liquid, the pressure must exceed 5.11 atm. Consequently, solid CO_2 does not melt, but sublimes when heated at 1 atm. Thus, CO_2 does not have a normal melting point; instead, it has a normal sublimation point, −78.5°C. Because CO_2 sublimes rather than melts as it absorbs energy at ordinary pressures, solid CO_2 (Dry Ice™) is a convenient coolant. For water (ice) to sublime, however, its vapor pressure must be below 4.58 torr. Food is freeze-dried by placing frozen food in a low-pressure chamber (below 4.58 torr) so that the ice in it sublimes.

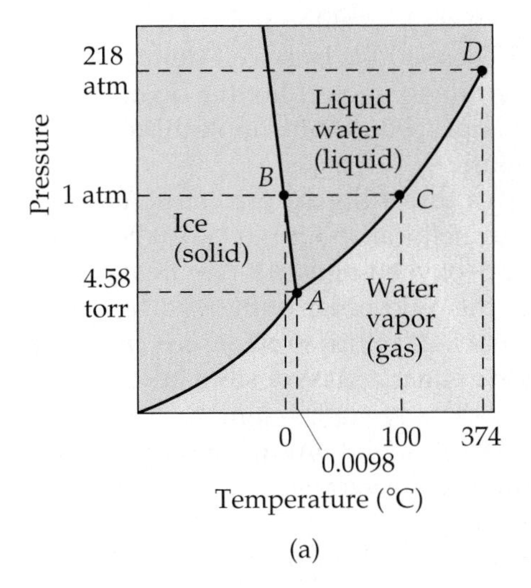

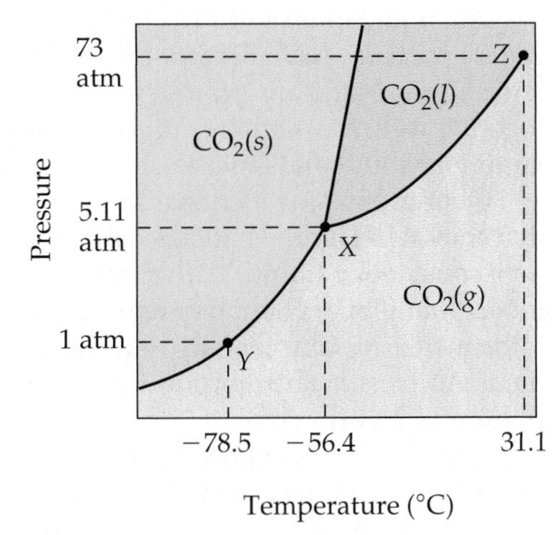

◀ **Figure 11.27** Phase diagram of (a) H_2O and (b) CO_2. The axes are not drawn to scale in either case. In (a), for water, note the triple point *A* (0.0098°C, 4.58 torr), the normal melting (or freezing) point *B* (0°C, 1 atm), the normal boiling point *C* (100°C, 1 atm), and the critical point *D* (374.4°C, 217.7 atm). In (b), for carbon dioxide, note the triple point *X* (−56.4°C, 5.11 atm), the normal sublimation point *Y* (−78.5°C, 1 atm), and the critical point *Z* (31.1°C, 73.0 atm).

SAMPLE EXERCISE 11.6

Referring to Figure 11.28 ▶, describe any changes in the phases present when H_2O is **(a)** kept at 0°C while the pressure is increased from that at point 1 to that at point 5 (vertical line); **(b)** kept at 1.00 atm while the temperature is increased from that at point 6 to that at point 9 (horizontal line).

Solution

Analyze: We are asked to use the phase diagram provided to deduce what phase changes might occur when specific pressure and temperature changes are brought about.

Plan: Trace the path indicated on the phase diagram, and note what phases and phase changes occur.

Solve: (a) At point 1, H_2O exists totally as a vapor. At point 2, a solid-vapor equilibrium exists. Above that pressure, at point 3, all the H_2O is converted to a solid. At point 4, some of the solid melts and an equilibrium between solid and liquid is achieved. At still higher pressures all the H_2O melts, so only the liquid phase is present at point 5. **(b)** At point 6, the H_2O exists entirely as a solid. When the temperature reaches point 4, the solid begins to melt and an equilibrium exists between the solid and liquid phases. At an even higher temperature, point 7, the solid has been converted entirely to a liquid. A liquid-vapor equilibrium exists at point 8. Upon further heating to point 9, the H_2O is converted entirely to the vapor phase.

Check: The indicated phases and phase changes are consistent with our knowledge of the properties of water.

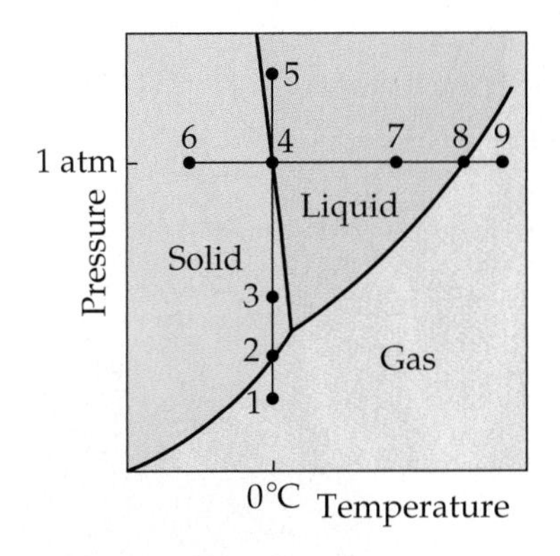

▲ **Figure 11.28** Phase diagram of H_2O.

PRACTICE EXERCISE

Using Figure 11.27(b), describe what happens when the following changes are made in a CO_2 sample initially at 1 atm and −60°C: **(a)** Pressure increases at constant temperature to 60 atm. **(b)** Temperature increases from −60°C to −20°C at constant 60 atm pressure.
Answers: **(a)** $CO_2(g) \longrightarrow CO_2(s)$; **(b)** $CO_2(s) \longrightarrow CO_2(l)$

11.7 Structures of Solids

Throughout the remainder of this chapter we will focus on how the properties of solids relate to their structures and bonding. Solids can be either crystalline or amorphous (noncrystalline). In a **crystalline solid** the atoms, ions, or molecules are ordered in well-defined arrangements. These solids usually have flat surfaces or faces that make definite angles with one another. The orderly stacks of particles that produce these faces also cause the solids to have highly regular shapes (Figure 11.29 ▼). Quartz and diamond are crystalline solids.

An **amorphous solid** (from the Greek words for "without form") is a solid whose particles have no orderly structure. These solids lack well-defined faces and shapes. Many amorphous solids are mixtures of molecules that do not stack together well. Most others are composed of large, complicated molecules. Familiar amorphous solids include rubber and glass.

Quartz (SiO_2) is a crystalline solid with a three-dimensional structure like that shown in Figure 11.30(a) ▶. When quartz melts (at about 1600°C), it becomes a viscous, tacky liquid. Although the silicon–oxygen network remains largely intact, many Si—O bonds are broken, and the rigid order of the quartz is lost. If the melt is rapidly cooled, the atoms are unable to return to an orderly arrangement. As a result, an amorphous solid known as quartz glass or silica glass results [Figure 11.30(b)].

Because the particles of an amorphous solid lack any long-range order, intermolecular forces vary in strength throughout a sample. Thus, amorphous solids do not melt at specific temperatures. Instead, they soften over a temperature range as intermolecular forces of various strengths are overcome. A crystalline solid, in contrast, melts at a specific temperature.

Unit Cells

The characteristic order of crystalline solids allows us to convey a picture of an entire crystal by looking at only a small part of it. We can think of the solid as being built up by stacking together identical building blocks, much as a brick wall is formed by stacking individual "identical" bricks. The repeating unit of a

(a) (b) (c)

▲ **Figure 11.29** Crystalline solids come in a variety of forms and colors: (a) pyrite (fool's gold), (b) fluorite, (c) amethyst.

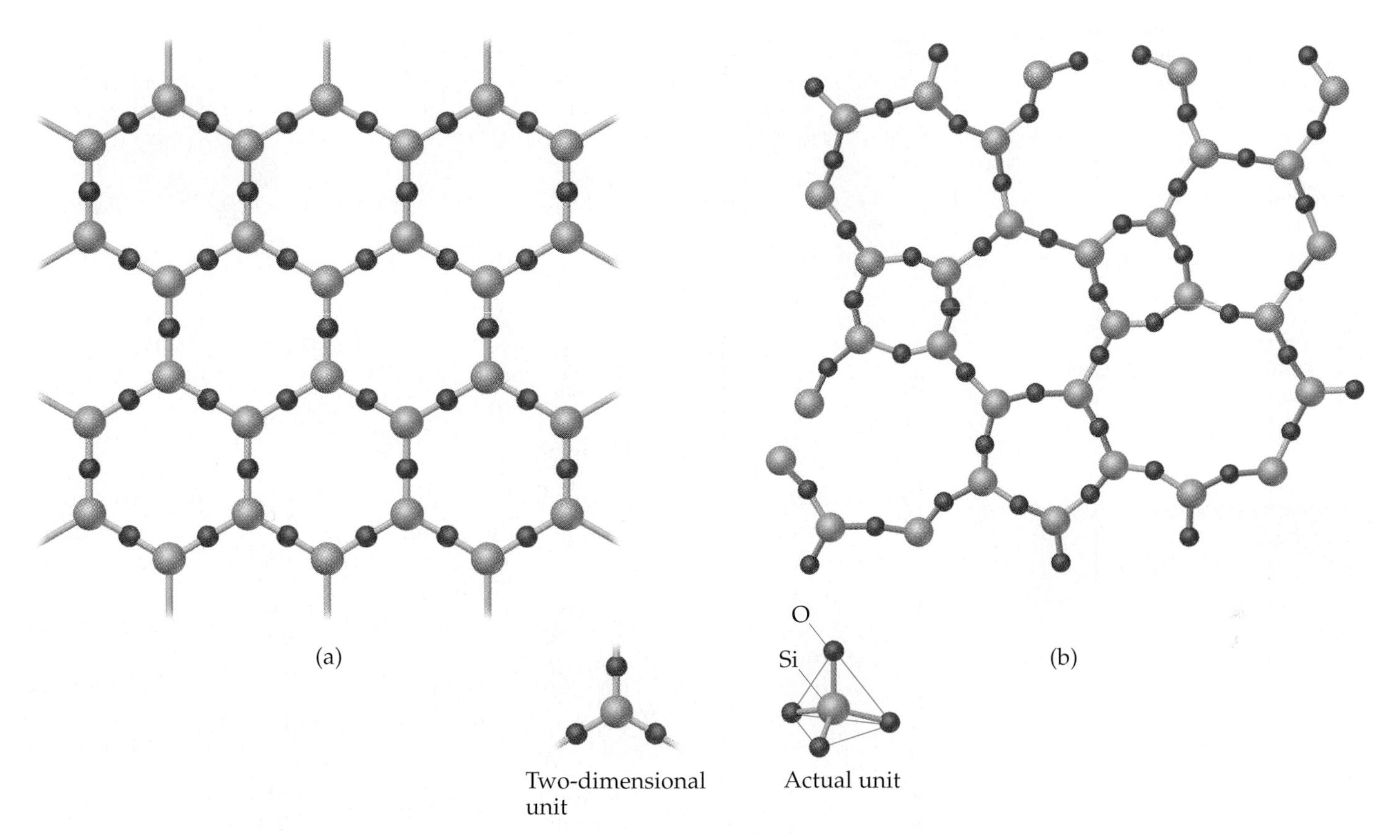

▲ **Figure 11.30** Schematic comparisons of (a) crystalline SiO_2 (quartz) and (b) amorphous SiO_2 (quartz glass). The gray-silver spheres represent silicon atoms; the red spheres represent oxygen atoms. The structure is actually three-dimensional and not planar as drawn. The unit shown as the basic building block (silicon and three oxygens) actually has four oxygens, the fourth coming out of the plane of the paper and capable of bonding to other silicon atoms.

solid, the crystalline "brick," is known as the **unit cell**. A simple two-dimensional example appears in the sheet of wallpaper shown in Figure 11.31 ▶. There are several ways of choosing the repeat pattern, or unit cell, of the design, but the choice is usually the smallest one that shows clearly the symmetry characteristic of the entire pattern.

A crystalline solid can be represented by a three-dimensional array of points, each of which represents an identical environment within the crystal. Such an array of points is called a **crystal lattice**. We can imagine forming the entire crystal structure by arranging the contents of the unit cell repeatedly on the crystal lattice.

Figure 11.32 ▼ shows a crystal lattice and its associated unit cell. In general, unit cells are parallelepipeds (six-sided figures whose faces are parallelograms).

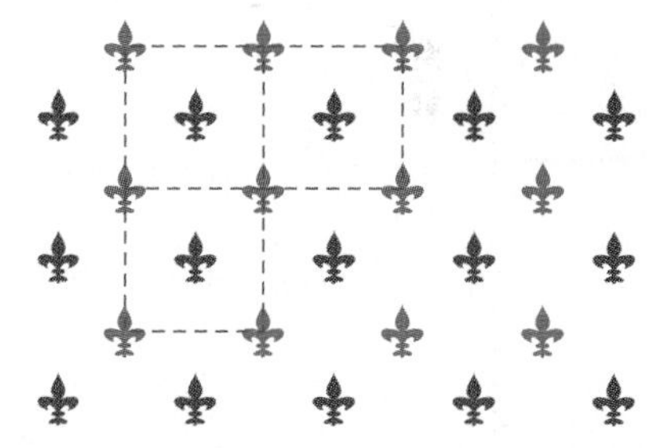

▲ **Figure 11.31** Wallpaper design showing a characteristic repeat pattern. Each dashed blue square denotes a unit cell of the repeat pattern. The unit cell could equally well be selected with red figures at the corners.

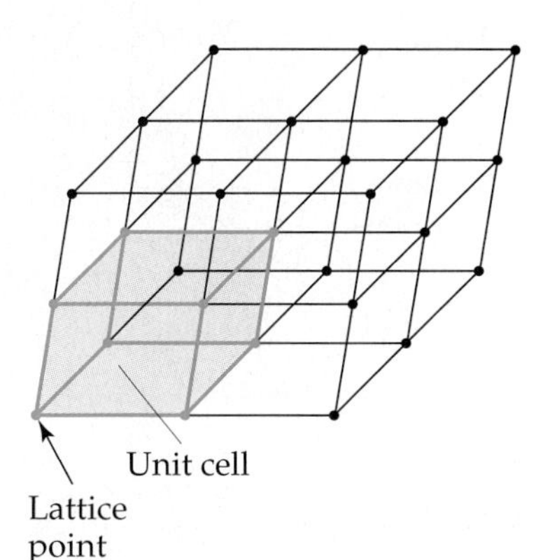

◀ **Figure 11.32** Simple crystal lattice and its associated unit cell.

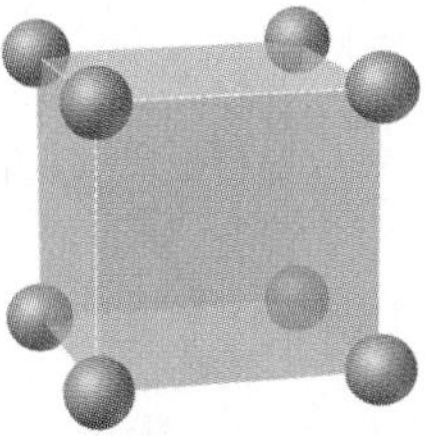

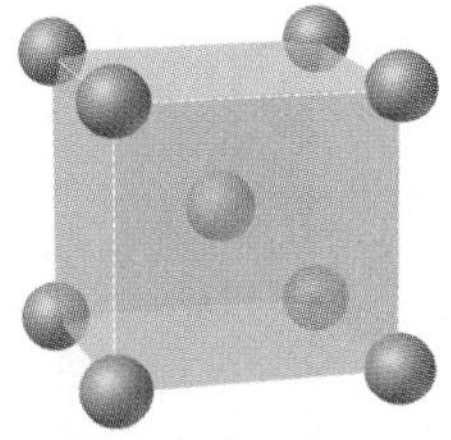
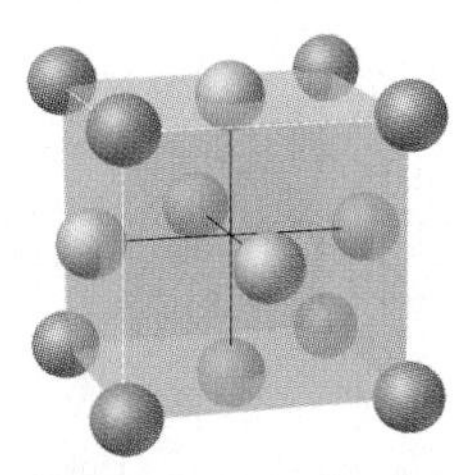

▶ **Figure 11.33** The three types of unit cells found in cubic lattices. For clarity, the corner spheres are red and the body-centered and face-centered ones are yellow. Each sphere represents a lattice point (an identical environment in the solid).

3-D MODEL
Primitive Cubic, Body-Centered Cubic, Face-Centered Cubic

Each unit cell can be described by the lengths of the edges of the cell and by the angles between these edges. The lattices of all crystalline compounds can be described by seven basic types of unit cells. The simplest of these is the cubic unit cell, in which all the sides are equal in length and all the angles are 90°.

There are three kinds of cubic unit cells, as illustrated in Figure 11.33 ▲. When lattice points are at the corners only, the unit cell is called **primitive cubic**. When a lattice point also occurs at the center of the unit cell, the cell is **body-centered cubic**. When the cell has lattice points at the center of each face, as well as at each corner, it is **face-centered cubic**.

The simplest crystal structures are cubic unit cells with only one atom centered at each lattice point. Most metals have such structures. Nickel, for example, has a face-centered cubic unit cell, whereas sodium has a body-centered cubic one. Figure 11.34 ▼ shows how atoms fill the cubic unit cells. Notice that the atoms on the corners and faces do not lie wholly within the unit cell. Instead, these atoms are shared between unit cells. Table 11.6 ◀ summarizes the fraction of an atom that occupies a unit cell when atoms are shared between unit cells.

TABLE 11.6 Fraction of an Atom That Occupies a Unit Cell for Various Positions in the Unit Cell

Position in Unit Cell	Fraction in Unit Cell
Center	1
Face	$\frac{1}{2}$
Edge	$\frac{1}{4}$
Corner	$\frac{1}{8}$

The Crystal Structure of Sodium Chloride

In the crystal structure of NaCl (Figure 11.35 ▶) we can center either the Na^+ ions or the Cl^- ions on the lattice points of a face-centered cubic unit cell. Thus, we can describe the structure as being face-centered cubic.

In Figure 11.35 the Na^+ and Cl^- ions have been moved apart so the symmetry of the structure can be seen more clearly. In this representation no attention is paid to the relative sizes of the ions. The representation in Figure 11.36 ▶, on the other hand, shows the relative sizes of the ions and how they fill the unit cell. Notice that the particles at corners, edges, and faces are shared by other unit cells.

The total cation-to-anion ratio of a unit cell must be the same as that for the entire crystal. Therefore, within the unit cell of NaCl there must be an equal number of Na^+ and Cl^- ions. Similarly, the unit cell for $CaCl_2$ would have one Ca^{2+} for every two Cl^-, and so forth.

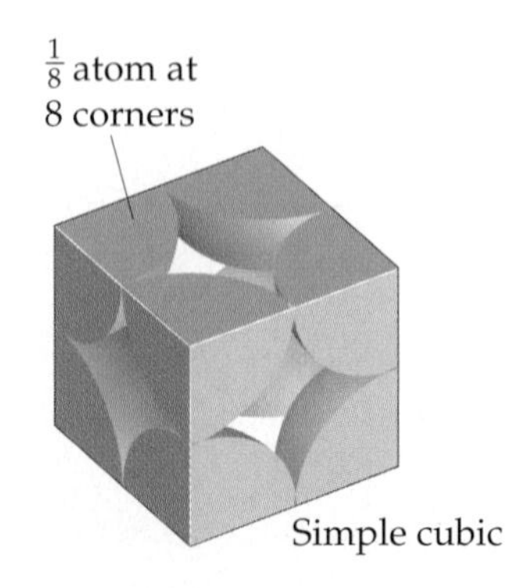

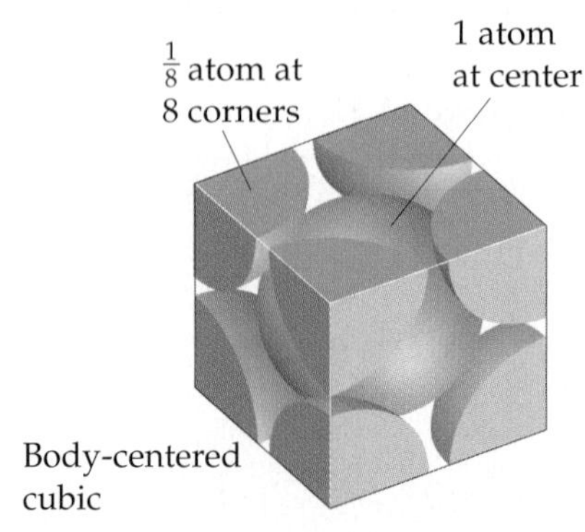

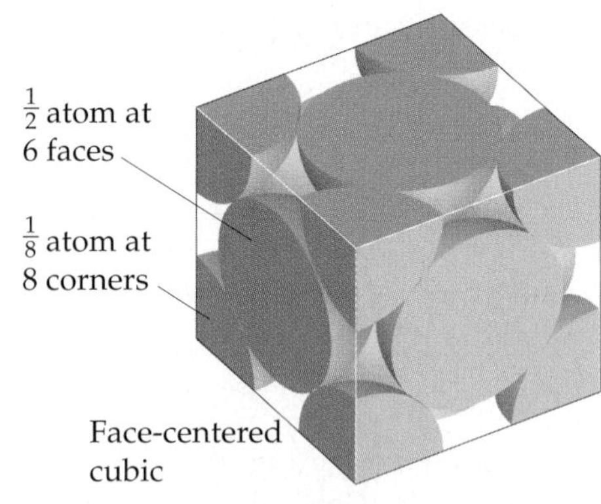

▲ **Figure 11.34** Space-filling view of cubic unit cells. Only the portion of each atom that belongs to the unit cell is shown.

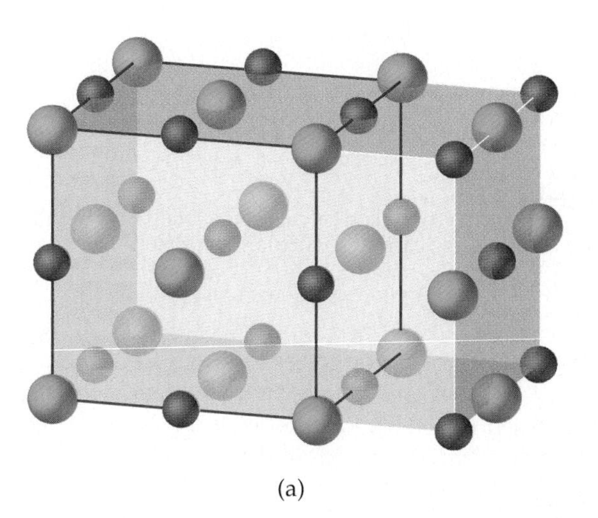

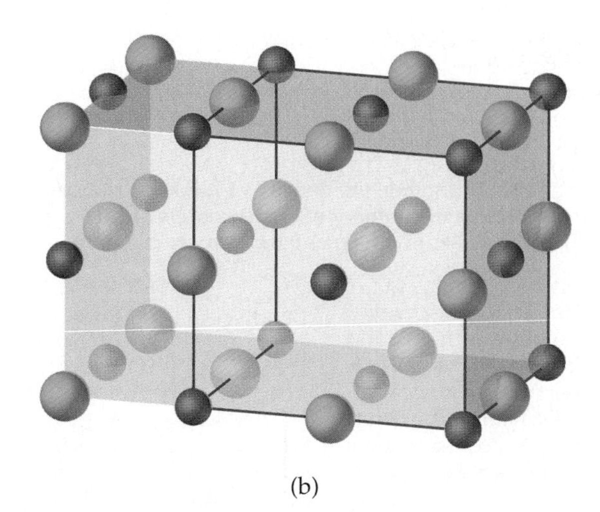

▲ **Figure 11.35** Portion of the crystal lattice of NaCl, illustrating two ways of defining its unit cell. Purple spheres represent Na^+ ions, and green spheres represent Cl^- ions. Red lines define the unit cell. In (a) Cl^- ions are at the corners of the unit cell. In (b) Na^+ ions are at the corners of the unit cell. Both of these choices for the unit cell are acceptable; both have the same volume, and in both cases identical points are arranged in a face-centered cubic fashion.

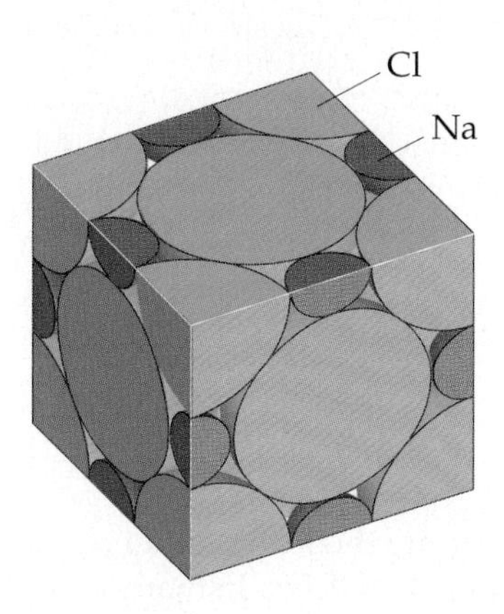

◀ **Figure 11.36** The unit cell of NaCl showing the relative sizes of the Na^+ ions (purple) and Cl^- ions (green). Only portions of most of the ions lie within the boundaries of the single unit cell.

SAMPLE EXERCISE 11.7

Determine the net number of Na^+ and Cl^- ions in the NaCl unit cell (Figure 11.36).

Solution

Analyze: We must sum the various contributing elements to determine the number of Na^+ and Cl^- ions within the unit cell.
Plan: To find the total number of ions of each type, we must identify the different locations within the unit cell and determine the fraction of the ion that lies within the unit cell boundaries.

Solve: There is one fourth of an Na^+ on each edge, a whole Na^+ in the center of the cube (refer also to Figure 11.35), one eighth of a Cl^- on each corner, and one half of a Cl^- on each face. Thus, we have the following:

$$Na^+: \quad \left(\tfrac{1}{4}\ Na^+ \text{ per edge}\right)(12 \text{ edges}) = 3\ Na^+$$

$$(1\ Na^+ \text{ per center})(1 \text{ center}) = 1\ Na^+$$

$$Cl^-: \quad \left(\tfrac{1}{8}\ Cl^- \text{ per corner}\right)(8 \text{ corner}) = 1\ Cl^-$$

$$\left(\tfrac{1}{2}\ Cl^- \text{ per face}\right)(6 \text{ faces}) = 3\ Cl^-$$

Thus, the unit cell contains $4\ Na^+$ and $4\ Cl^-$

Check: This result agrees with the compound's stoichiometry: $1\ Na^+$ for each Cl^-

PRACTICE EXERCISE

The element iron crystallizes in a form called α-iron, which has a body-centered-cubic unit cell. How many iron atoms are in the unit cell?
Answer: two

SAMPLE EXERCISE 11.8

The geometric arrangement of ions in crystals of LiF is the same as that in NaCl. The unit cell of LiF is 4.02 Å on an edge. Calculate the density of LiF.

Solution

Analyze: We are asked to calculate the density of LiF from the size of the unit cell.

Plan: We need to determine the number of formula units of LiF within the unit cell. From that, we can calculate the total mass within the unit cell. Because we know the mass and can calculate the volume of the unit cell, we can then calculate density.

Solve: The arrangement of ions in LiF is the same as that in NaCl, so a unit cell of LiF will contain four Li^+ and four F^- ions (Sample Exercise 11.7). Density measures mass per unit volume. Thus, we can calculate the density of LiF from the mass contained in a unit cell and the volume of the unit cell. The mass contained in one unit cell is:

$$4(6.94\text{ amu}) + 4(19.0\text{ amu}) = 103.8\text{ amu}$$

The volume of a cube of length a on an edge is a^3, so the volume of the unit cell is $(4.02\text{ Å})^3$. We can now calculate the density, converting to the common units of g/cm^3:

$$\text{Density} = \frac{(103.8\text{ amu})}{(4.02\text{ Å})^3}\left(\frac{1\text{ g}}{6.02 \times 10^{23}\text{ amu}}\right)\left(\frac{1\text{ Å}}{10^{-8}\text{ cm}}\right)^3 = 2.65\text{ g/cm}^3$$

Check: This value agrees with that found by simple density measurements 2.640 g/cm^3 at 20°C. The size and contents of the unit cell are therefore consistent with the macroscopic density of the substance.

PRACTICE EXERCISE

The body-centered cubic unit cell of a particular crystalline form of iron is 2.8664 Å on each side. Calculate the density of this form of iron.

Answer: 7.8753 g/cm^3

Close Packing of Spheres

The structures adopted by crystalline solids are those that bring particles in closest contact to maximize the attractive forces between them. In many cases the particles that make up the solids are spherical or approximately so. Such is the case for atoms in metallic solids. It is therefore instructive to consider how equal-sized spheres can pack most efficiently (that is, with the minimum amount of empty space).

The most efficient arrangement of a layer of equal-sized spheres is shown in Figure 11.37(a) ▼. Each sphere is surrounded by six others in the layer. A second layer of spheres can be placed in the depressions on top of the first layer. A third layer can then be added above the second with the spheres sitting in the

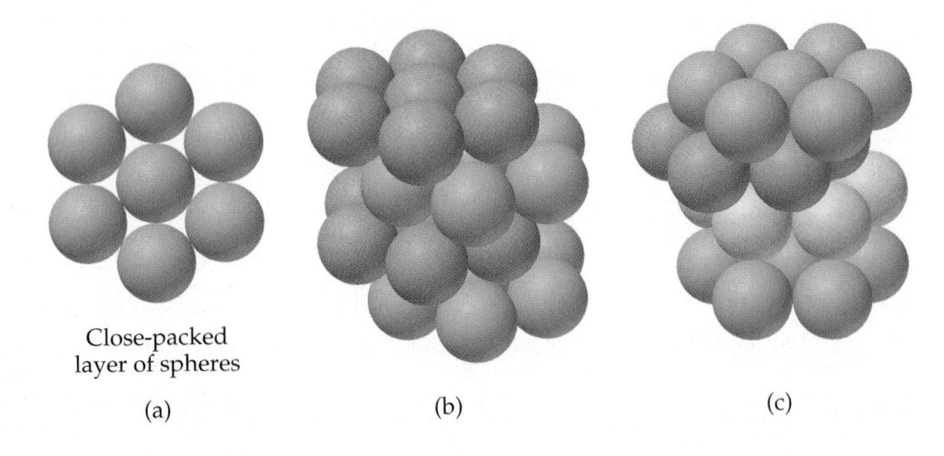

▶ **Figure 11.37** (a) Close packing of a single layer of equal-sized spheres. (b) In the hexagonal close-packed structure the atoms in the third layer lie directly over those in the first layer. The order of layers is ABAB. (c) In the cubic close-packed structure the atoms in the third layer are not over those in the first layer. Instead, they are offset a bit, and it is the fourth layer that lies directly over the first. Thus, the order of layers is ABCA.

depressions of the second layer. However, there are two types of depressions for this third layer, and they result in different structures, as shown in Figure 11.37(b) and (c).

If the spheres of the third layer are placed in line with those of the first layer, as shown in Figure 11.37(b), the structure is known as **hexagonal close packing**. The third layer repeats the first layer, the fourth layer repeats the second layer, and so forth, giving a layer sequence that we denote ABAB.

The spheres of the third layer, however, can be placed so they do not sit above the spheres in the first layer. The resulting structure, shown in Figure 11.37(c), is known as **cubic close packing**. In this case it is the fourth layer that repeats the first layer, and the layer sequence is ABCA. Although it cannot be seen in Figure 11.37(c), the unit cell of the cubic close-packed structure is face-centered cubic.

In both of the close-packed structures, each sphere has 12 equidistant nearest neighbors: six in one plane, three above that plane, and three below. We say that each sphere has a **coordination number** of 12. The coordination number is the number of particles immediately surrounding a particle in the crystal structure. In both types of close packing, 74% of the total volume of the structure is occupied by spheres; 26% is empty space between the spheres. By comparison, each sphere in the body-centered cubic structure has a coordination number of 8, and only 68% of the space is occupied. In the simple cubic structure the coordination number is 6, and only 52% of the space is occupied.

When unequal-sized spheres are packed in a lattice, the larger particles sometimes assume one of the close-packed arrangements, with smaller particles occupying the holes between the large spheres. In Li_2O, for example, the larger oxide ions assume a cubic close-packed structure, and the smaller Li^+ ions occupy small cavities that exist between oxide ions.

11.8 Bonding in Solids

The physical properties of crystalline solids, such as melting point and hardness, depend both on the arrangements of particles and on the attractive forces between them. Table 11.7 ▼ classifies solids according to the types of forces between particles in solids.

TABLE 11.7 Types of Crystalline Solids

Type of Solid	Form of Unit Particles	Forces Between Particles	Properties	Examples
Molecular	Atoms or molecules	London dispersion, dipole-dipole forces, hydrogen bonds	Fairly soft, low to moderately high melting point, poor thermal and electrical conduction	Argon, Ar; methane, CH_4; sucrose, $C_{12}H_{22}O_{11}$; Dry Ice™, CO_2
Covalent-network	Atoms connected in a network of covalent bonds	Covalent bonds	Very hard, very high melting point, often poor thermal and electrical conduction	Diamond, C; quartz, SiO_2
Ionic	Positive and negative ions	Electrostatic attractions	Hard and brittle, high melting point, poor thermal and electrical conduction	Typical salts—for example, NaCl, $Ca(NO_3)_2$
Metallic	Atoms	Metallic bonds	Soft to very hard, low to very high melting point, excellent thermal and electrical conduction, malleable and ductile	All metallic elements—for example, Cu, Fe, Al, Pt

A Closer Look X-Ray Diffraction by Crystals

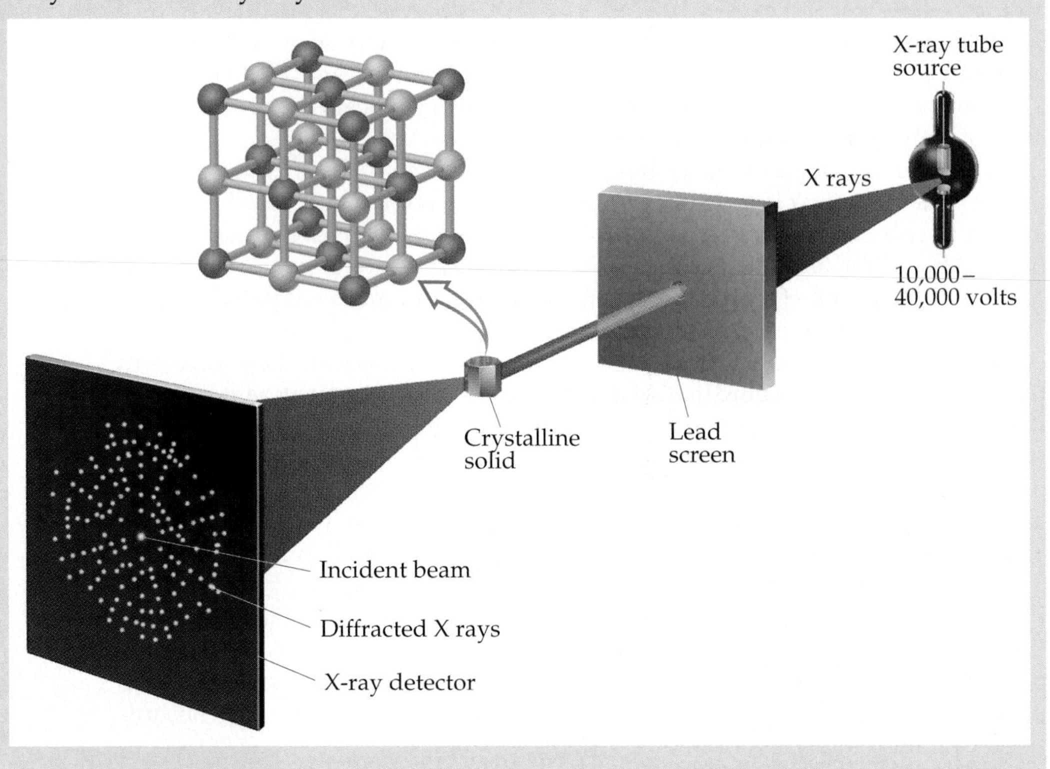

▶ **Figure 11.38** In X-ray crystallography, an X-ray beam is diffracted by a crystal. The diffraction pattern can be recorded as spots where the diffracted X rays strike a detector, which records the positions and intensities of the spots.

When light waves pass through a narrow slit, they are scattered in such a way that the wave seems to spread out. This physical phenomenon is called *diffraction*. When light passes through many evenly spaced narrow slits (a *diffraction grating*), the scattered waves interact to form a series of light and dark bands, known as a diffraction pattern. The most effective diffraction of light occurs when the wavelength of the light and the width of the slits are similar in magnitude.

The spacing of the layers of atoms in solid crystals is usually about 2–20 Å. The wavelengths of X rays are also in this range. Thus, a crystal can serve as an effective diffraction grating for X rays. X-ray diffraction results from the scattering of X rays by a regular arrangement of atoms, molecules, or ions. Much of what we know about crystal structures has been obtained from studies of X-ray diffraction by crystals, a technique known as *X-ray crystallography*. Figure 11.38 ▲ depicts the diffraction of a beam of X rays as it passes through a crystal. The diffracted X rays were formerly detected by a photographic film. Today, crystallographers use an *array detector*, a device analogous to that used in digital cameras, to capture and measure the intensities of the diffracted rays. The diffraction pattern of spots on the detector in Figure 11.38 depends on the particular arrangement of atoms in the crystal. Thus, different types of crystals give rise to different diffraction patterns. In 1913 the English scientists William and Lawrence Bragg (father and son) determined for the first time how the spacing of layers in crystals leads to different X-ray diffraction patterns. By measuring the intensities of the diffracted beams and the angles at which they are diffracted, it is possible to reason backward to the structure that must have given rise to the pattern. One of the most famous X-ray diffraction patterns is the one for crystals of the genetic material DNA (Figure 11.39 ▶), first obtained in the early 1950s. Working from photographs such as this one, Francis Crick, Rosalind Franklin, James Watson, and Maurice Wilkins determined the double-helix structure of DNA, one of the most important discoveries in molecular biology.

Today X-ray crystallography is used extensively to determine the structures of molecules in crystals. The instruments used to measure X-ray diffraction, known as *X-ray diffractometers*, are now computer-controlled, making the collection of diffraction data highly automated. The diffraction pattern of a crystal can be determined very accurately and quickly (sometimes in a matter of hours) even though thousands of diffraction points are measured. Computer programs are then used to analyze the diffraction data and determine the arrangement and structure of the molecules in the crystal.

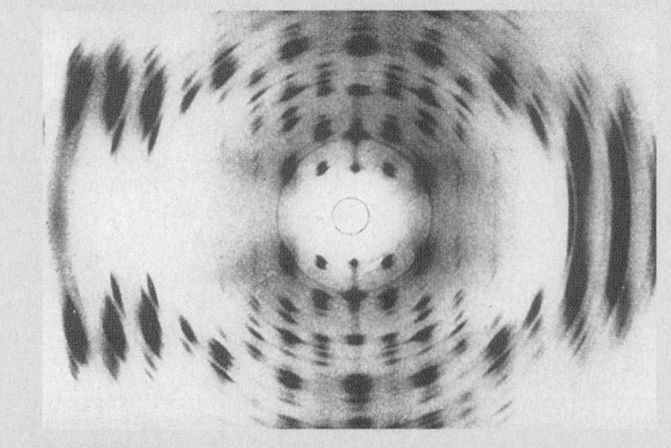

▲ **Figure 11.39** The X-ray diffraction photograph of one form of crystalline DNA, taken in the early 1950s. From the pattern of dark spots, the double-helical shape of the DNA molecule was deduced.

	Benzene	Toluene	Phenol
Melting point (°C)	5	−95	43
Boiling point (°C)	80	111	182

◀ **Figure 11.40** Comparative melting and boiling points for benzene, toluene, and phenol.

Molecular Solids

Molecular solids consist of atoms or molecules held together by intermolecular forces (dipole-dipole forces, London dispersion forces, and hydrogen bonds). Because these forces are weak, molecular solids are soft. Furthermore, they normally have relatively low melting points (usually below 200°C). Most substances that are gases or liquids at room temperature form molecular solids at low temperature. Examples include Ar, H_2O, and CO_2.

The properties of molecular solids depend not only on the strengths of the forces that exist between molecules, but also on the abilities of the molecules to pack efficiently in three dimensions. Benzene (C_6H_6), for example, is a highly symmetrical planar molecule. (Section 8.6) It has a higher melting point than toluene, a compound in which one of the hydrogen atoms of benzene has been replaced by a CH_3 group (Figure 11.40 ▲). The lower symmetry of toluene molecules prevents them from packing as efficiently as benzene molecules. As a result, the intermolecular forces that depend on close contact are not as effective, and the melting point is lower. In contrast, the boiling point of toluene is higher than that of benzene, indicating that the intermolecular attractive forces are larger in liquid toluene than in liquid benzene. Both the melting and boiling points of phenol, another substituted benzene shown in Figure 11.40, are higher than those of benzene because the OH group of phenol can form hydrogen bonds.

Covalent-Network Solids

Covalent-network solids consist of atoms held together in large networks or chains by covalent bonds. Because covalent bonds are much stronger than intermolecular forces, these solids are much harder and have higher melting points than molecular solids. Diamond and graphite, two allotropes of carbon, are covalent-network solids. Other examples include quartz, SiO_2; silicon carbide, SiC; and boron nitride, BN.

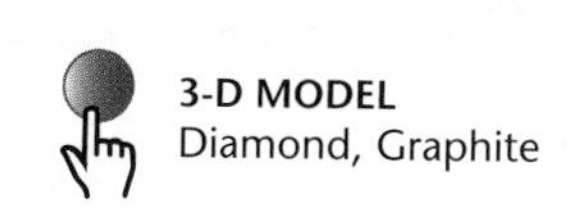

In diamond each carbon atom is bonded to four other carbon atoms as shown in Figure 11.41(a) ▼. This interconnected three-dimensional array of strong

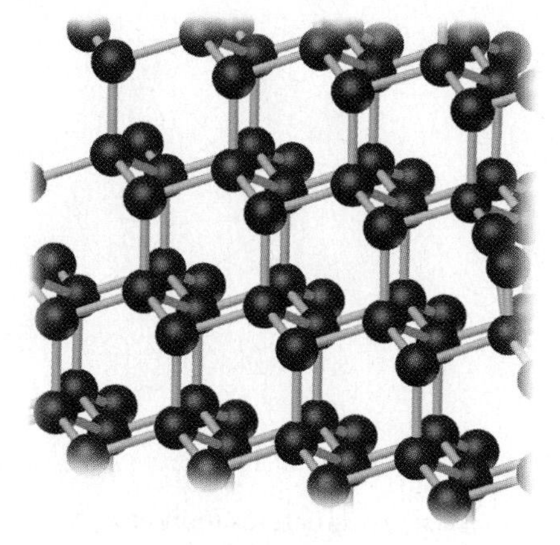

(a) Diamond

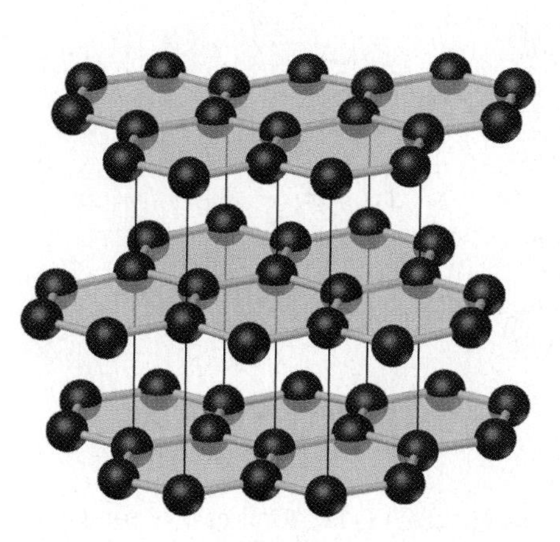

(b) Graphite

◀ **Figure 11.41** Structures of (a) diamond and (b) graphite. The blue color in (b) is added to emphasize the planarity of the carbon layers.

carbon–carbon single bonds contributes to diamond's unusual hardness. Industrial-grade diamonds are employed in the blades of saws for the most demanding cutting jobs. Diamond also has a high melting point, 3550°C.

In graphite the carbon atoms are arranged in layers of interconnected hexagonal rings as shown in Figure 11.41(b). Each carbon atom is bonded to three others in the layer. The distance between adjacent carbon atoms in the plane, 1.42 Å, is very close to the C—C distance in benzene, 1.395 Å. In fact, the bonding resembles that of benzene, with delocalized π bonds extending over the layers. (Section 9.6) Electrons move freely through the delocalized orbitals, making graphite a good conductor of electricity along the layers. (If you have ever taken apart a flashlight battery, you know that the central electrode in the battery is made of graphite.) The layers, which are separated by 3.41 Å, are held together by weak dispersion forces. The layers readily slide past one another when rubbed, giving graphite a greasy feel. Graphite is used as a lubricant and in the "lead" in pencils.

Ionic Solids

Ionic solids consist of ions held together by ionic bonds. (Section 8.2) The strength of an ionic bond depends greatly on the charges of the ions. Thus, NaCl, in which the ions have charges of 1+ and 1−, has a melting point of 801°C, whereas MgO, in which the charges are 2+ and 2−, melts at 2852°C.

The structures of simple ionic solids can be classified as a few basic types. The NaCl structure is a representative example of one type. Other compounds that possess this same structure include LiF, KCl, AgCl, and CaO. Three other common types of crystal structures are shown in Figure 11.42 ▼.

The structure adopted by an ionic solid depends largely on the charges and relative sizes of the ions. In the NaCl structure, for example, the Na^+ ions have a coordination number of 6 because each Na^+ ion is surrounded by six nearest neighbor Cl^- ions. In the CsCl structure [Figure 11.42(a)], by comparison, the Cl^- ions adopt a simple cubic arrangement with each Cs^+ ion surrounded by eight Cl^- ions. The increase in the coordination number as the alkali metal ion changes from Na^+ to Cs^+ is a consequence of the larger size of Cs^+ compared to Na^+.

In the zinc blende (ZnS) structure [Figure 11.42(b)] the S^{2-} ions adopt a face-centered cubic arrangement, with the smaller Zn^{2+} ions arranged so they are each surrounded tetrahedrally by four S^{2-} ions (compare with Figure 11.33). CuCl also adopts this structure.

In the fluorite (CaF_2) structure [Figure 11.42(c)], the Ca^{2+} ions are shown in a face-centered cubic arrangement. As required by the chemical formula of the substance, there are twice as many F^- ions (grey) in the unit cell as there are Ca^{2+} ions. Other compounds that have the fluorite structure include $BaCl_2$ and PbF_2.

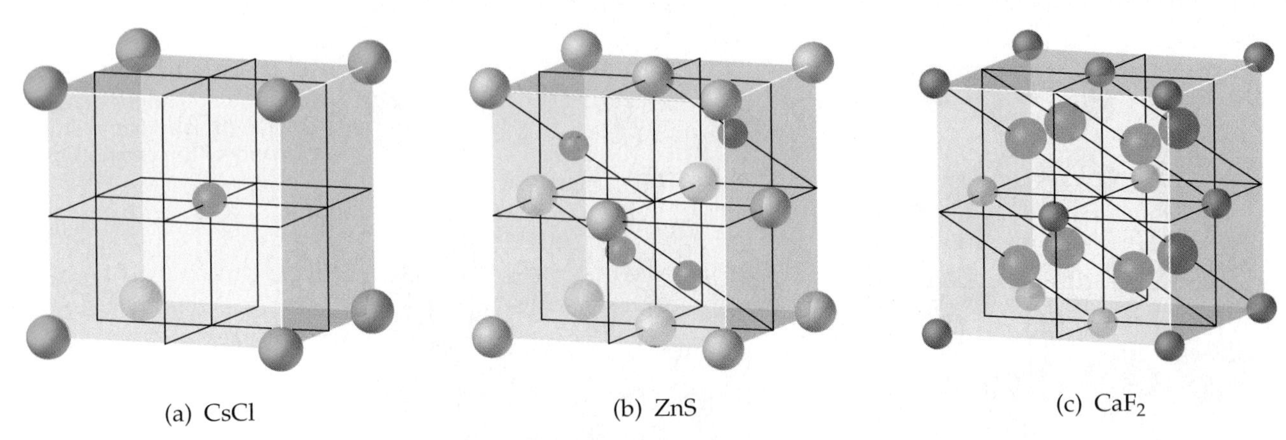

(a) CsCl (b) ZnS (c) CaF_2

▲ **Figure 11.42** Unit cells of some common types of crystal structures found for ionic solids: (a) CsCl; (b) ZnS (zinc blende); (c) CaF_2 (fluorite).

A Closer Look Buckyball

Until the mid-1980s pure solid carbon was thought to exist in only two forms: diamond and graphite, both of which are covalent-network solids. In 1985 a group of researchers led by Richard Smalley and Robert Curl of Rice University in Houston and Harry Kroto of the University of Sussex in England made a startling discovery. They vaporized a sample of graphite with an intense pulse of laser light and used a stream of helium gas to carry the vaporized carbon into a mass spectrometer (see the "Closer Look" box in Section 2.4). The mass spectrum showed peaks corresponding to clusters of carbon atoms, with a particularly strong peak corresponding to molecules composed of 60 carbon atoms, C_{60}.

Because C_{60} clusters were so preferentially formed, the group proposed a radically different form of carbon, namely, C_{60} molecules that were nearly spherical in shape. They proposed that the carbon atoms of C_{60} form a "ball" with 32 faces, of which 12 are pentagons and 20 are hexagons (Figure 11.43 ▼), exactly like a soccer ball. The shape of this molecule is reminiscent of the geodesic dome invented by the U.S. engineer and philosopher R. Buckminster Fuller, so C_{60} was whimsically named "buckminsterfullerene," or "buckyball" for short. Since the discovery of C_{60}, other related molecules of carbon atoms have been discovered. These molecules are now known as fullerenes.

Appreciable amounts of buckyball can be prepared by electrically evaporating graphite in an atmosphere of helium gas. About 14% of the resulting soot consists of C_{60} and a related molecule, C_{70}, which has a more elongated structure. The carbon-rich gases from which C_{60} and C_{70} condense also contain other fullerenes, mostly with more carbon atoms such as C_{76} and C_{84}. The smallest possible fullerene, C_{20}, was first detected in 2000. This small ball-shaped molecule is much more reactive than the larger fullerenes.

Because the fullerenes are composed of individual molecules, they dissolve in various organic solvents, whereas diamond or graphite do not (Figure 11.44 ▼). This solubility permits the fullerenes to be separated from the other components of soot and even from each other. It also allows the study of their reactions in solution. Study of these substances has led to the discovery of some very interesting chemistry. For example, it is possible to place a metal atom inside a buckyball, generating a molecule in which a metal atom is completely enclosed by the carbon sphere. The C_{60} molecules also react with potassium to give K_3C_{60}, which contains a face-centered cubic array of buckyballs with K^+ ions in the cavities between them. This compound is a superconductor at 18 K (Section 12.5), suggesting the possibility that other fullerenes may also have interesting electrical, magnetic, or optical properties. For their discovery and pioneering work with fullerenes, Professors Smalley, Curl, and Kroto were awarded the 1996 Nobel Prize in chemistry.

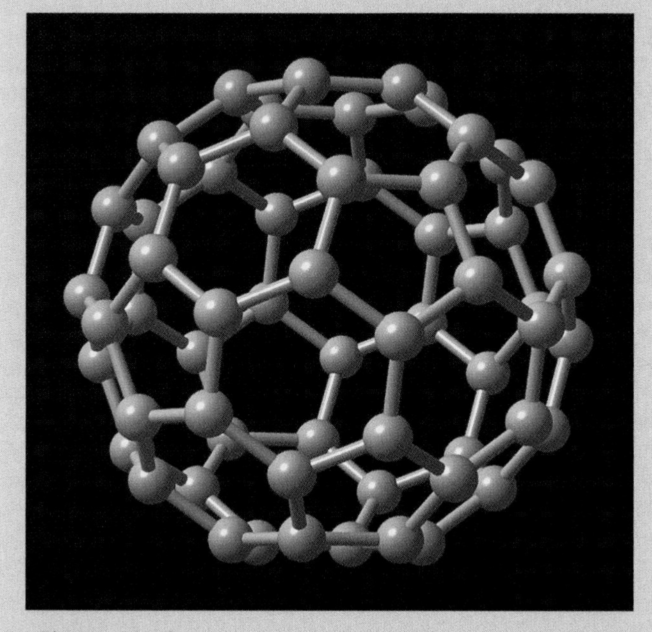

▲ **Figure 11.43** The buckminsterfullerene molecule, C_{60}, has a highly symmetric structure in which the 60 carbon atoms sit at the vertices of a truncated icosahedron—the same geometry as a soccer ball.

▲ **Figure 11.44** Unlike diamond and graphite, the new molecular forms of carbon can be dissolved in organic solvents. The orange solution on the left is a solution of C_{70} in *n*-hexane, which is a colorless liquid. The magenta solution on the right is a solution of buckyball, C_{60}, in *n*-hexane.

Metallic Solids

Metallic solids consist entirely of metal atoms. Metallic solids usually have hexagonal close-packed, cubic close-packed (face-centered cubic), or body-centered cubic structures. Thus, each atom typically has 8 or 12 adjacent atoms.

The bonding in metals is too strong to be due to London dispersion forces, and yet there are not enough valence electrons for ordinary covalent bonds between atoms. The bonding is due to valence electrons that are delocalized throughout the entire solid. In fact, we can visualize the metal as an array of positive ions immersed in a sea of delocalized valence electrons, as shown in Figure 11.45 ▼.

Metals vary greatly in the strength of their bonding, as shown by their wide range of physical properties such as hardness and melting point. In general, however, the strength of the bonding increases as the number of electrons available for bonding increases. Thus, sodium, which has only one valence electron per atom, melts at 97.5°C, whereas chromium, with six electrons beyond the noble-gas core, melts at 1890°C. The mobility of the electrons explains why metals are good conductors of heat and electricity. The bonding and properties of metals will be examined more closely in Chapter 23.

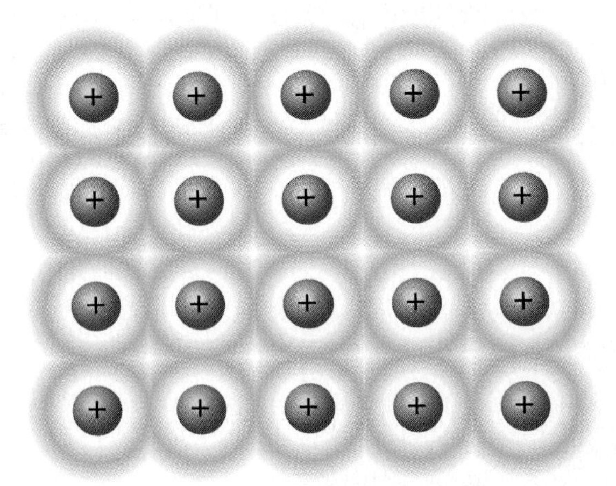

▶ **Figure 11.45** A cross section of a metal. Each sphere represents the nucleus and inner-core electrons of a metal atom. The surrounding colored "fog" represents the mobile sea of electrons that binds the atoms together.

SAMPLE INTEGRATIVE EXERCISE 11: Putting Concepts Together

The substance CS_2 has a melting point of −110.8°C and a boiling point of 46.3°C. Its density at 20°C is 1.26 g/cm^3. It is highly inflammable. **(a)** What is the name of this compound? **(b)** If you were going to look up the properties of this substance in the *CRC Handbook of Chemistry and Physics*, would you look under the physical properties of inorganic or organic compounds? Explain. **(c)** How would you classify $CS_2(s)$ as to type of crystalline solid? **(d)** Write a balanced equation for the combustion of this compound in air. (You will have to decide on the most likely oxidation products.) **(e)** The critical temperature and pressure for CS_2 are 552 K and 78 atm, respectively. Compare these values with those for CO_2 (Table 11.5), and discuss the possible origins of the differences. **(f)** Would you expect the density of CS_2 at 40°C to be greater or less than at 20°C? What accounts for the difference?

Solution (a) The compound is named carbon disulfide, in analogy with the naming of other binary molecular compounds. ∞ (Section 2.8) **(b)** The substance will be listed as an inorganic compound. It contains no carbon–carbon bonds, nor any C—H bonds, which are the usual structural features of organic compounds. **(c)** Because $CS_2(s)$ consists of individual CS_2 molecules, it would be a molecular solid in the classification scheme of Table 11.7. **(d)** The most likely products of the combustion will be CO_2 and SO_2. ∞ (Sections 3.2 and 7.6) Under some conditions SO_3 might be formed, but this would be the less likely outcome. Thus, we have the following equation for combustion:

$$CS_2(l) + 3O_2(g) \longrightarrow CO_2(g) + 2SO_2(g)$$

(e) The critical temperature and pressure of CS_2 (552 K and 78 atm) are both higher than those given for CO_2 in Table 11.5 (304 K and 73 atm). The difference in critical temperatures is especially notable. The higher values for CS_2 arise from the greater London dispersion attractions between the CS_2 molecules compared with CO_2. These greater attractions are due to the larger size of the sulfur compared to oxygen and therefore its greater polarizability. **(f)** The density would be lower at the higher temperature. Density decreases with increasing temperature because the molecules possess higher kinetic energies. Their more energetic movements result in larger average spacings per molecule, which translate into lower densities.

Summary and Key Terms

Introduction and Section 11.1 Substances that are gases or liquids at room temperature are usually composed of molecules. In gases the intermolecular attractive forces are negligible compared to the kinetic energies of the molecules; thus, the molecules are widely separated and undergo constant, chaotic motion. In liquids the **intermolecular forces** are strong enough to keep the molecules in close proximity; nevertheless, the molecules are free to move with respect to one another. In solids the interparticle attractive forces are strong enough to restrain molecular motion and to force the particles to occupy specific locations in a three-dimensional arrangement.

Section 11.2 Three types of intermolecular forces exist between neutral molecules: **dipole-dipole forces, London dispersion forces,** and **hydrogen bonding. Ion-dipole forces** are important in solutions. London dispersion forces operate between all molecules. The relative strengths of the dipole-dipole and dispersion forces depend on the polarity, **polarizability**, size, and shape of the molecule. Dipole-dipole forces increase in strength with increasing polarity. Dispersion forces increase in strength with increasing molecular weight, although molecular shape is also an important factor. Hydrogen bonding occurs in compounds containing O—H, N—H, and F—H bonds. Hydrogen bonds are generally stronger than dipole-dipole or dispersion forces.

Section 11.3 The stronger the intermolecular forces, the greater is the **viscosity**, or resistance to flow, of a liquid. The surface tension of a liquid also increases as intermolecular forces increase in strength. **Surface tension** is a measure of the tendency of a liquid to maintain a minimum surface area. The adhesion of a liquid to the walls of a narrow tube and the cohesion of the liquid account for **capillary action** and the formation of a meniscus at the surface of a liquid.

Section 11.4 A substance may exist in more than one state of matter, or phase. **Phase changes** are transformations from one phase to another. Changes of a solid to liquid (melting), solid to gas (sublimation), and liquid to gas (vaporization) are all endothermic processes. Thus, the **heat of fusion** (melting), the heat of sublimation, and the **heat of vaporization** are all positive quantities. The reverse processes are exothermic. A gas cannot be liquefied by application of pressure if the temperature is above its **critical temperature**. The pressure required to liquefy a gas at its critical temperature is called the **critical pressure**.

Section 11.5 The **vapor pressure** of a liquid indicates the tendency of the liquid to evaporate. The vapor pressure is the partial pressure of the vapor when it is in **dynamic equilibrium** with the liquid. At equilibrium the rate of transfer of molecules from the liquid to the vapor equals the rate of transfer from the vapor to the liquid. The higher the vapor pressure of a liquid, the more readily it evaporates and the more **volatile** it is. Vapor pressure increases nonlinearly with temperature. Boiling occurs when the vapor pressure equals the external pressure. The **normal boiling point** is the temperature at which the vapor pressure equals 1 atm.

Section 11.6 The equilibria between the solid, liquid, and gas phases of a substance as a function of temperature and pressure are displayed on a **phase diagram**. Equilibria between any two phases are indicated by a line. The line through the melting point usually slopes slightly to the right as pressure increases, because the solid is usually more dense than the liquid. The melting point at 1 atm is the **normal melting point**. The point on the diagram at which all three phases coexist in equilibrium is called the **triple point**.

Section 11.7 In a **crystalline solid**, particles are arranged in a regularly repeating pattern. An **amorphous solid** is one whose particles show no such order. The essential structural features of a crystalline solid can be represented by its **unit cell**, the smallest part of the crystal that can, by simple displacement, reproduce the three-dimensional structure. The three-dimensional structures of a crystal can also be represented by its **crystal lattice**. The points in a crystal lattice represent positions in the structure where there are identical environments. The simplest unit cells are cubic. There are three kinds of cubic unit cells: **primitive cubic, body-centered cubic**, and **face-centered cubic**.

Many solids have a close-packed structure in which spherical particles are arranged so as to leave the minimal amount of empty space. Two closely related forms of close packing, **cubic close packing** and **hexagonal close packing**, are possible. In both, each sphere has a **coordination number** of 12.

Section 11.8 The properties of solids depend both on the arrangements of particles and on the attractive forces between them. **Molecular solids**, which consist of atoms or molecules held together by intermolecular forces, are soft and low melting. **Covalent-network solids**, which consist of atoms held together by covalent bonds that extend throughout the solid, are hard and high melting. **Ionic solids** are hard and brittle and have high melting points. **Metallic solids**, which consist of metal cations held together by a "sea" of electrons, exhibit a wide range of properties.

Exercises

Kinetic-Molecular Theory

11.1 List the three states of matter in order of **(a)** increasing molecular disorder, and **(b)** increasing intermolecular attractions.

11.2 List some properties of liquids and solids that reflect the difference in the degree of order in the two states.

11.3 For a given substance, the densities of the liquid and solid phases are usually very similar and very different from the density of the gas. Explain.

11.4 Benzoic acid, C_6H_5COOH, melts at 122°C. The density in the liquid state at 130°C is 1.08 g/cm^3. The density of solid benzoic acid at 15°C is 1.266 g/cm^3. **(a)** In which state is the average distance between molecules the greater? **(b)** Explain the difference in densities at the two temperatures in terms of the kinetic-molecular theory.

11.5 Why does increasing the temperature cause a substance to change in succession from a solid to a liquid to a gas?

11.6 **(a)** Explain why compressing a gas at constant temperature can cause it to liquefy. **(b)** Why are the liquid and solid forms of a substance referred to as *condensed phases*?

Intermolecular Forces

11.7 Which type of intermolecular attractive force operates between **(a)** all molecules; **(b)** polar molecules; **(c)** the hydrogen atom of a polar bond and a nearby small, electronegative atom?

11.8 What type(s) of intermolecular force is (are) common to **(a)** Xe and methanol (CH_3OH); **(b)** CH_3OH and acetonitrile (CH_3CN); **(c)** NH_3 and HF.

11.9 Describe the intermolecular forces that must be overcome to convert each of the following from a liquid to a gas: **(a)** Br_2; **(b)** CH_3OH; **(c)** H_2S.

11.10 What type of intermolecular force accounts for the following differences in each case? **(a)** CH_3OH boils at 65°C, CH_3SH boils at 6°C. **(b)** Xe is liquid at atmospheric pressure and 120 K, whereas Ar is a gas. **(c)** Kr, atomic weight 84, boils at 120.9 K, whereas Cl_2, molecular weight about 71, boils at 238 K. **(d)** Acetone boils at 56°C, whereas 2-methylpropane boils at −12°C.

$$CH_3-\overset{\overset{\displaystyle O}{\|}}{C}-CH_3$$

Acetone

$$CH_3-\overset{\overset{\displaystyle CH_3}{|}}{CH}-CH_3$$

2-Methylpropane

11.11 **(a)** What is meant by the term polarizability? **(b)** Which of the following atoms would you expect to be most polarizable: O, S, Se, or Te? Explain. **(c)** Put the following molecules in order of increasing polarizability: $GeCl_4$, CH_4, $SiCl_4$, SiH_4, and $GeBr_4$. **(d)** Predict the order of boiling points of the substances in part (c).

11.12 **(a)** Why does the strength of dispersion forces increase with increasing polarizability? **(b)** Account for the steady increase in boiling point of the noble-gas elements with increasing atomic weight (Table 11.3). **(c)** What general rule of thumb applies to the relationship between dispersion forces and molecular weight? **(d)** Comment on whether the following statement is correct: "All other factors being the same, dispersion forces between molecules increase with the number of electrons in the molecules."

11.13 Which member of the following pairs has the larger London dispersion forces: **(a)** H_2O or H_2S; **(b)** CO_2 or CO **(c)** CH_4 or CCl_4?

11.14 Which member of the following pairs has the stronger intermolecular dispersion forces: **(a)** Br_2 or O_2; **(b)** CH_3CH_2SH or $CH_3CH_2CH_2SH$; **(c)** $CH_3CH_2CH_2Cl$ or $(CH_3)_2CHCl$?

11.15 Butane and 2-methylpropane, whose space-filling models are shown, are both nonpolar and have the same molecular formula, yet butane has the higher boiling point (−0.5°C compared to −11.7°C). Explain.

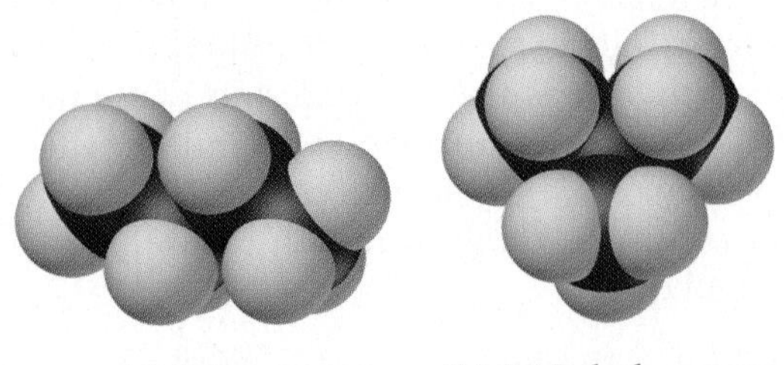

(a) Butane (b) 2-Methylpropane

11.16 Propyl alcohol ($CH_3CH_2CH_2OH$) and isopropyl alcohol [$(CH_3)_2CHOH$], whose space-filling models are shown, have boiling points of 97.2°C and 82.5°C, respectively. Explain why the boiling point of propyl alcohol is higher, even though both have the molecular formula of C_3H_8O?

(a) Propyl alcohol

(b) Isopropyl alcohol

11.17 Which of the following molecules can form hydrogen bonds with other molecules of the same kind: CH_3F, CH_3NH_2, CH_3OH, CH_3Br?

11.18 Ethylene glycol ($HOCH_2CH_2OH$), the major substance in antifreeze, has a normal boiling point of 199°C. By comparison, ethyl alcohol (CH_3CH_2OH) boils at 78°C at atmospheric pressure. Ethylene glycol dimethyl ether ($CH_3OCH_2CH_2OCH_3$) has a normal boiling point of 83°C, and ethyl methyl ether ($CH_3CH_2OCH_3$) has a normal boiling point of 11°C. **(a)** Explain why replacement of a hydrogen on the oxygen by CH_3 generally results in a lower boiling point. **(b)** What are the major factors responsible for the difference in boiling points of the two ethers?

11.19 Rationalize the difference in boiling points between the members of the following pairs of substances: **(a)** HF (20°C) and HCl (−85°C); **(b)** $CHCl_3$ (61°C) and $CHBr_3$ (150°C); **(c)** Br_2 (59°C) and ICl (97°C).

11.20 Identify the types of intermolecular forces present in each of the following substances, and select the substance in each pair that has the higher boiling point: **(a)** C_6H_{14} or C_8H_{18}; **(b)** C_3H_8 or CH_3OCH_3; **(c)** HOOH or HSSH; **(d)** NH_2NH_2 or CH_3CH_3.

11.21 Cite three properties of water that can be attributed to the existence of hydrogen bonding.

11.22 The following quote about ammonia (NH_3) is from a textbook of inorganic chemistry: "It is estimated that 26% of the hydrogen bonding in NH_3 breaks down on melting, 7% on warming from the melting to the boiling point, and the final 67% on transfer to the gas phase at the boiling point." From the standpoint of the kinetic-molecular theory, explain **(a)** why there is a decrease of hydrogen-bonding energy on melting and **(b)** why most of the loss in hydrogen bonding occurs in the transition from the liquid to the vapor state.

Viscosity and Surface Tension

11.23 **(a)** How do the viscosity and surface tension of liquids change as intermolecular forces become stronger? **(b)** How do the viscosity and surface tension of liquids change as temperature increases? Account for these trends.

11.24 **(a)** Distinguish between adhesive forces and cohesive forces. **(b)** Do viscosity and surface tension reflect adhesive forces or cohesive forces of attraction? **(c)** Explain the cause for the U-shaped meniscus formed when water is in a glass tube. **(d)** How is the capacity of paper towels to absorb water related to capillary action?

11.25 Explain the following observations: **(a)** The surface tension of $CHBr_3$ is greater than that of $CHCl_3$. **(b)** As temperature increases, oil flows faster through a narrow tube. **(c)** Raindrops that collect on a waxed automobile hood take on a nearly spherical shape.

11.26 Hydrazine (NH_2NH_2), hydrogen peroxide (HOOH), and water (H_2O) all have exceptionally high surface tensions in comparison with other substances of comparable molecular weights. **(a)** Draw the Lewis structures for these three compounds. **(b)** What structural property do these substances have in common, and how might that account for the high surface tensions?

Changes of State

11.27 Name all the possible phase changes that can occur between different states of matter. Which of these are exothermic, and which are endothermic?

11.28 Name the phase transition in each of the following situations, and indicate whether it is exothermic or endothermic: **(a)** Bromine vapor turns to bromine liquid as it is cooled. **(b)** Crystals of iodine disappear from an evaporating dish as they stand in a fume hood. **(c)** Rubbing alcohol in an open container slowly disappears. **(d)** Molten lava from a volcano turns into solid rock.

11.29 Explain why the heat of fusion of any substance is generally lower than its heat of vaporization.

11.30 Ethyl chloride (C_2H_5Cl) boils at 12°C. When liquid C_2H_5Cl under pressure is sprayed on a room temperature surface in air, the surface is cooled considerably. **(a)** What does this observation tell us about the enthalpy content of $C_2H_5Cl(g)$ as compared with $C_2H_5Cl(l)$? **(b)** In terms of the kinetic-molecular theory, what is the origin of this difference?

11.31 For many years drinking water has been cooled in hot climates by evaporating it from the surfaces of canvas bags or porous clay pots. How many grams of water can be cooled from 35°C to 22°C by the evaporation of 50 g of water? (The heat of vaporization of water in this temperature range is 2.4 kJ/g. The specific heat of water is 4.18 J/g-K.)

11.32 Compounds like CCl_2F_2 are known as chlorofluorocarbons, or CFCs. These compounds were once widely used as refrigerants but are now being replaced by compounds that are believed to be less harmful to the environment. The heat of vaporization of CCl_2F_2 is 289 J/g. What mass of this substance must evaporate in order to freeze 100 g of water initially at 18°C? (The heat of fusion of water is 334 J/g; the specific heat of water is 4.18 J/g-K.)

11.33 Ethanol (C_2H_5OH) melts at −114°C and boils at 78°C. The enthalpy of fusion of ethanol is 5.02 kJ/mol, and its enthalpy of vaporization is 38.56 kJ/mol. The specific heats of solid and liquid ethanol are 0.97 J/g-K and 2.3 J/g-K, respectively. How much heat is required to convert 75.0 g of ethanol at −120°C to the vapor phase at 78°C?

11.34 The fluorocarbon compound $C_2Cl_3F_3$ has a normal boiling point of 47.6°C. The specific heats of $C_2Cl_3F_3(l)$ and $C_2Cl_3F_3(g)$ are 0.91 J/g-K and 0.67 J/g-K, respectively. The heat of vaporization for the compound is 27.49 kJ/mol. Calculate the heat required to convert 25.0 g of $C_2Cl_3F_3$ from a liquid at 5.00°C to a gas at 82.00°C.

11.35 **(a)** What is the significance of the critical pressure of a substance? **(b)** What happens to the critical temperature of a series of compounds as the force of attraction between molecules increases? **(c)** Which of the substances listed in Table 11.5 can be liquefied at the temperature of liquid nitrogen (−196°C)?

11.36 The critical temperatures (K) and pressures (atm) of a series of halogenated methanes are as follows:

Compound	CCl_3F	CCl_2F_2	$CClF_3$	CF_4
Critical Temperature	471	385	302	227
Critical Pressure	43.5	40.6	38.2	37.0

(a) What in general can you say about the variation in intermolecular forces in this series? **(b)** What specific kinds of intermolecular forces are most likely to account for most of the variation in critical parameters in this series?

Vapor Pressure and Boiling Point

11.37 Explain how each of the following affects the vapor pressure of a liquid: **(a)** volume of the liquid; **(b)** surface area; **(c)** intermolecular attractive forces; **(d)** temperature.

11.38 A liquid that has an equilibrium vapor pressure of 130 mm Hg at 25°C is placed into a 1-L vessel like that shown in Figure 11.20. What is the pressure difference shown on the manometer, and what is the composition of the gas in the vessel, under each of the following conditions: **(a)** Two hundred mL of the liquid is introduced into the vessel and frozen at the bottom, then the vessel is evacuated. The vessel is sealed off and the liquid allowed to warm to 25°C. **(b)** Two hundred mL of the liquid is added to the vessel at 25°C under atmospheric pressure, and after a few minutes the vessel is closed off. **(c)** A few mL of the liquid is introduced into the vessel at 25°C while it has a pressure of 1 atm of air in it, without allowing any of the air to escape. After a few minutes there remain a few drops of liquid in the vessel.

11.39 Place the following substances in order of increasing volatility: CH_4, CBr_4, CH_2Cl_2, CH_3Cl, $CHBr_3$, and CH_2Br_2. Explain your answer.

11.40 PCl_3 and $AsCl_3$ are similar substances, with similar geometries and bonding modes. **(a)** Which of these two substances would you expect to be the more volatile at room temperature? **(b)** Which substance would you expect to have the higher boiling point? **(c)** In which substance would the kinetic energies of the molecules be greater at 40°C, a temperature well below either substance's boiling point? **(d)** In which substance would you expect the intermolecular forces to be larger?

11.41 **(a)** Two pans of water are on different burners of a stove. One pan of water is boiling vigorously, while the other is boiling gently. What can be said about the temperature of the water in the two pans? **(b)** A large container of water and a small one are at the same temperature. What can be said about the relative vapor pressures of the water in the two containers?

11.42 Explain the following observations: **(a)** Water evaporates more quickly on a hot, dry day than on a hot, humid day. **(b)** It takes longer to cook hard-boiled eggs at high altitudes than at lower altitudes.

11.43 **(a)** Use the vapor-pressure curve in Figure 11.24 to estimate the boiling point of diethyl ether at 400 torr. **(b)** Use the vapor-pressure table in Appendix B to determine the boiling point of water when the external pressure is 25 torr.

11.44 **(a)** Suppose the pressure inside a pressure cooker reaches 1.2 atm. By using the vapor-pressure table in Appendix B, estimate the temperature at which water will boil in this cooker. **(b)** Use the vapor-pressure curve in Figure 11.24 to estimate the external pressure under which ethyl alcohol will boil at 70°C.

11.45 Mt. Denali in Alaska is the tallest peak in the United States (20,320 ft). **(a)** If the barometric pressure at the top of the mountain is 340 torr, at what temperature will water boil there? Refer to Appendix B. **(b)** If the temperature at the top is 12°C, would a container of diethyl ether experience a pressure greater than the local atmospheric pressure? (See Figure 11.22.)

11.46 Reno, Nevada, is about 4500 ft above sea level. **(a)** If the barometric pressure is 680 mm Hg in Reno, at what temperature will water boil? Refer to Appendix B. **(b)** What can you say about the average kinetic energies of the water molecules at the boiling point in Reno as compared with those at the boiling point in Chicago, where the barometric pressure is 752 mm Hg? If you believe the average kinetic energies are different, explain how it is that boiling of water can occur at two different average kinetic energies of the water molecules.

Phase Diagrams

11.47 On a phase diagram why does the line that separates the gas and liquid phases end rather than go to infinite pressure and temperature?

11.48 **(a)** What is the significance of the triple point in a phase diagram? **(b)** Could you measure the triple point of water by measuring the temperature in a vessel in which water vapor, liquid water, and ice are in equilibrium under one atmosphere of air? Explain.

11.49 Refer to Figure 11.27(a), and describe all the phase changes that would occur in each of the following cases: **(a)** Water vapor originally at 1.0×10^{-3} atm and −0.10°C is slowly compressed at constant temperature until the final pressure is 10 atm. **(b)** Water originally at 100.0°C and 0.50 atm is cooled at constant pressure until the temperature is −10°C.

11.50 Refer to Figure 11.27(b), and describe the phase changes (and the temperatures at which they occur) when CO_2 is heated from −80°C to −20°C at **(a)** a constant pressure of 3 atm; **(b)** a constant pressure of 6 atm.

11.51 The normal melting and boiling points of xenon are −112°C and −107°C, respectively. Its triple point is at −121°C and 282 torr, and its critical point is at 16.6°C and 57.6 atm. **(a)** Sketch the phase diagram for Xe, showing the four points given and indicating the area in which each phase is stable. **(b)** Which is denser, Xe(*s*) or Xe(*l*)? Explain. **(c)** If Xe gas is cooled under an external pressure of 100 torr, will it undergo condensation or deposition? Explain.

11.52 The normal melting and boiling points of O_2 are −218°C and −183°C, respectively. Its triple point is at −219°C and 1.14 torr, and its critical point is at −119°C and 49.8 atm. **(a)** Sketch the phase diagram for O_2, showing the four points given and indicating the area in which each phase is stable. **(b)** Will $O_2(s)$ float on $O_2(l)$? Explain. **(c)** As it is heated, will solid O_2 sublime or melt under a pressure of 1 atm?

Structures of Solids

11.53 How does an amorphous solid differ from a crystalline one? Give an example of an amorphous solid.

11.54 Amorphous silica has a density of about 2.2 g/cm^3, whereas the density of crystalline quartz is 2.65 g/cm^3. Account for this difference in densities.

11.55 What is a unit cell? What properties does it have?

11.56 Perovskite, a mineral composed of Ca, O, and Ti, has the cubic unit cell shown in the drawing. What is the chemical formula of this mineral?

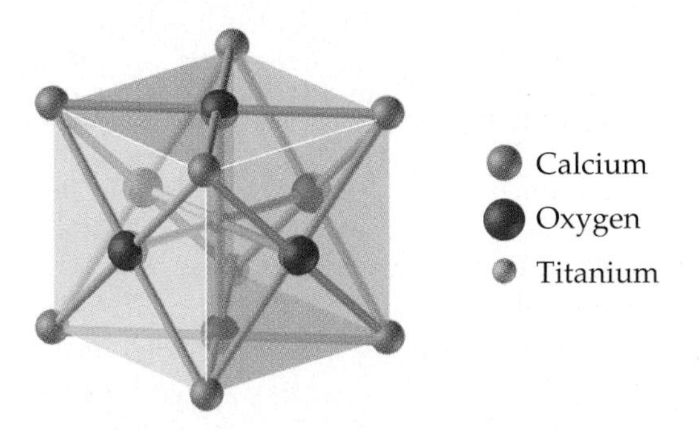

11.57 The elements xenon and gold both have solid-state structures consisting of cubic close-packed arrangements of atoms. Yet Xe melts at −112°C and gold melts at 1064°C. Account for these greatly different melting points.

11.58 Rutile is a mineral composed of Ti and O. Its unit cell, shown in the drawing, contains Ti atoms at each corner and a Ti atom at the center of the cell. Four O atoms are on the opposite faces of the cell, and two are entirely within the cell. **(a)** What is the chemical formula of this mineral? **(b)** What is the nature of the bonding that holds the solid together?

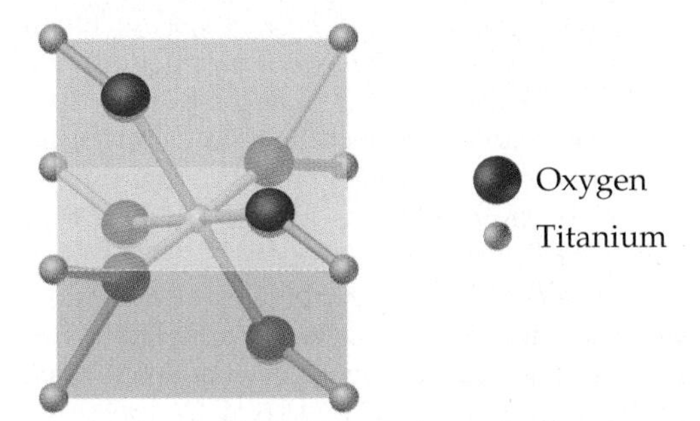

11.59 Iridium crystallizes in a face-centered cubic unit cell that has an edge length of 3.833 Å. The atom in the center of the face is in contact with the corner atoms, as shown in the drawing. **(a)** Calculate the atomic radius of an iridium atom. **(b)** Calculate the density of iridium metal.

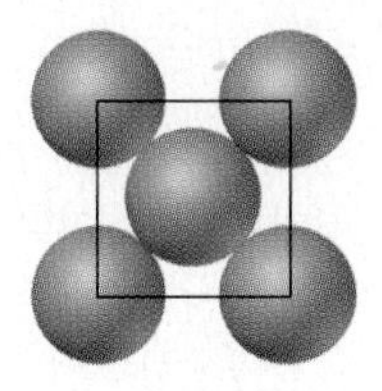

11.60 Aluminum metal crystallizes in a cubic close-packed structure (face-centered cubic cell, Figure 11.34). **(a)** How many aluminum atoms are in a unit cell? **(b)** What is the coordination number of each aluminum atom? **(c)** Assume that the aluminum atoms can be represented as spheres, as shown in the drawing for Exercise 11.59. If each Al atom has a radius of 1.43 Å, what is the length of a side of the unit cell? **(d)** Calculate the density of aluminum metal.

11.61 An element crystallizes in a body-centered cubic lattice. The edge of the unit cell is 2.86 Å, and the density of the crystal is 7.92 g/cm^3. Calculate the atomic weight of the element.

11.62 KCl has the same structure as NaCl. The length of the unit cell is 628 pm. The density of KCl is 1.984 g/cm^3, and its formula mass is 74.55 amu. Using this information, calculate Avogadro's number.

11.63 What is the coordination number of each sphere in **(a)** a three-dimensional, close-packed array of equal-sized spheres; **(b)** a primitive cubic structure; **(c)** a body-centered cubic lattice?

11.64 What is the coordination number of **(a)** Na^+ in the NaCl structure, Figure 11.35; **(b)** Zn^{2+} in the ZnS unit cell, Figure 11.42(b); **(c)** Ca^{2+} in the CaF_2 unit cell, Figure 11.42(c)?

11.65 Clausthalite is a mineral composed of lead selenide (PbSe). The mineral adopts a NaCl-type structure. The density of PbSe at 25°C is 8.27 g/cm^3. Calculate the length of an edge of the PbSe unit cell.

11.66 The mineral oldhamite (CaS) crystallizes in the NaCl type of crystal structure (Figure 11.35). The length of an edge of the unit cell of CaS is 5.689 Å. Calculate the density of CaS.

11.67 The mineral uraninite (UO_2) adopts a fluorite structure [Figure 11.42(c)] in which the length of an edge of the unit cell is 5.468 Å. **(a)** Will the uranium ions be represented by the larger or the smaller spheres in Figure 11.42(c)? Explain. **(b)** Calculate the density of uraninite.

11.68 A particular form of cinnabar (HgS) adopts the zinc blende structure, Figure 11.42(b). The length of the unit cell side is 5.852 Å. **(a)** Calculate the density of HgS in this form. **(b)** The mineral tiemmanite (HgSe) also forms a solid phase with the zinc blende structure. The length of the unit cell side in this mineral is 6.085 Å. What accounts for the larger unit cell length in tiemmanite? **(c)** Which of the two substances has the higher density? How do you account for the difference in densities?

Bonding in Solids

11.69 What kinds of attractive forces exist between particles in **(a)** molecular crystals; **(b)** covalent-network crystals; **(c)** ionic crystals; **(d)** metallic crystals?

11.70 Indicate the type of crystal (molecular, metallic, covalent-network, or ionic) each of the following would form upon solidification: **(a)** $CaCO_3$; **(b)** Pt; **(c)** ZrO_2 (melting point, 2677°C); **(d)** Kr; **(e)** benzene; **(f)** I_2.

11.71 Covalent bonding occurs in both molecular and covalent-network solids. Why do these two kinds of solids differ so greatly in their hardness and melting points?

11.72 Which type (or types) of crystalline solid is characterized by each of the following: **(a)** high mobility of electrons throughout the solid; **(b)** softness, relatively low melting point; **(c)** high melting point and poor electrical conductivity; **(d)** network of covalent bonds; **(e)** charged particles throughout the solid.

11.73 A white substance melts with some decomposition at 730°C. As a solid, it is a nonconductor of electricity, but it dissolves in water to form a conducting solution. Which type of solid (Table 11.7) might the substance be?

11.74 You are given a white substance that sublimes at 3000°C; the solid is a nonconductor of electricity and is insoluble in water. Which type of solid (Table 11.7) might this substance be?

11.75 For each of the following pairs of substances, predict which will have the higher melting point, and indicate why: **(a)** B, BF_3; **(b)** Na, NaCl; **(c)** TiO_2, $TiCl_4$; **(d)** LiF, MgF_2.

11.76 For each of the following pairs of substances, predict which will have the higher melting point and indicate why: **(a)** Ar, Xe; **(b)** SiO_2, CO_2; **(c)** KBr, Br_2; **(d)** C_6Cl_6, C_6H_6.

Additional Exercises

11.77 What are the major differences between intermolecular forces and those that operate within molecules or between ions?

11.78 **(a)** Which of the following substances can exhibit dipole-dipole attractions between its molecules: CO_2, SO_2, H_2, IF, HBr, CCl_4? **(b)** Which of the following substances exhibit hydrogen bonding in their liquid and solid states: CH_3NH_2, CH_3F, PH_3, HCOOH?

11.79 Suppose you have two colorless molecular liquids, one boiling at −84°C, the other at 34°C, and both at atmospheric pressure. Which of the following statements is correct? For those that are not correct, modify the statement so that it is correct. **(a)** The higher boiling liquid has greater total intermolecular forces than the other. **(b)** The lower boiling liquid must consist of nonpolar molecules. **(c)** The lower boiling liquid has a lower molecular weight than the higher boiling liquid. **(d)** The two liquids have identical vapor pressures at their normal boiling points. **(e)** At 34°C, both liquids have vapor pressures of 760 mm Hg.

11.80 Two isomers of the planar compound 1,2-dichloroethylene are shown here, along with their melting and boiling points:

	H(Cl)C=C(Cl)H (cis isomer)	Cl(H)C=C(Cl)H (trans isomer)
Melting point (°C)	−80.5	−49.8
Boiling point (°C)	60.3	47.5

(a) Which of the two isomers will have the stronger dipole-dipole forces? Is this prediction borne out by the data presented here? **(b)** Based on the data presented here, which isomer packs more efficiently in the solid phase?

11.81 In dichloromethane, CH_2Cl_2 ($\mu = 1.60$ D), the dispersion force contribution to the intermolecular attractive forces is about five times larger than the dipole-dipole contribution. Would you expect the relative importance of the two kinds of intermolecular attractive forces to differ **(a)** in dibromomethane ($\mu = 1.43$ D); **(b)** in difluoromethane ($\mu = 1.93$ D)? Explain.

11.82 What molecular-level properties of a liquid are most important in determining **(a)** its ability to flow; **(b)** its tendency to bead up on a surface for which it has no appreciable adhesive forces; **(c)** its boiling point; **(d)** its heat of vaporization.

11.83 As the intermolecular attractive forces between molecules increase in magnitude, do you expect each of the following to increase or decrease in magnitude? **(a)** vapor pressure; **(b)** heat of vaporization; **(c)** boiling point; **(d)** freezing point; **(e)** viscosity; **(f)** surface tension; **(g)** critical temperature.

11.84 When an atom or group of atoms is substituted for an H atom in benzene (C_6H_6), the boiling point changes. Explain the order of the following boiling points: C_6H_6 (80°C), C_6H_5Cl (132°C), C_6H_5Br (156°C), C_6H_5OH (182°C).

11.85 Trimethylamine [$(CH_3)_3N$] boils at 3°C; propylamine ($CH_3CH_2CH_2NH_2$) boils at 49°C. **(a)** What accounts for the difference in boiling points? **(b)** Propylamine is entirely miscible with water; trimethylamine has reasonably high solubility in water. What accounts for these data, considering that isobutane [$(CH_3)_3CH$] is considerably less soluble than trimethylamine?

11.86 Ethylene glycol [$CH_2(OH)CH_2(OH)$] is the major component of antifreeze. It is a slightly viscous liquid, not very volatile at room temperature, with a boiling point of 198°C. Pentane (C_5H_{12}) which has about the same molecular weight, is a nonviscous liquid that is highly volatile at room temperature and whose boiling point is 36.1°C. Explain the differences in the physical properties of the two substances.

[11.87] Using the following list of normal boiling points for a series of hydrocarbons, estimate the normal boiling point for octane, C_8H_{18}: propane (C_3H_8, −42.1°C), butane (C_4H_{10}, −0.5°C), pentane (C_5H_{12}, 36.1°C), hexane (C_6H_{14}, 68.7°C), heptane (C_7H_{16}, 98.4°C). Explain the trend in the boiling points.

11.88 A flask of water is connected to a vacuum pump. A few moments after the pump is turned on, the water begins to boil. After a few minutes, the water begins to freeze. Explain why these processes occur.

[11.89] Notice in Figure 11.24 that there is a pressure reduction valve in the line just before the supercritical CO_2 and dissolved caffeine enter the separator. Use Figure 11.23 to explain the function of this valve in the overall process.

[11.90] The following table gives the vapor pressure of hexafluorobenzene (C_6F_6) as a function of temperature:

Temperature (K)	Vapor Pressure (torr)
280.0	32.42
300.0	92.47
320.0	225.1
330.0	334.4
340.0	482.9

(a) By plotting these data in a suitable fashion, determine whether the Clausius–Clapeyron equation is obeyed. If it is obeyed, use your plot to determine ΔH_{vap} for C_6F_6. **(b)** Use these data to determine the boiling point of the compound.

[11.91] Suppose the vapor pressure of a substance is measured at two different temperatures. **(a)** By using the Clausius–Clapeyron equation, Equation 11.1, derive the following relationship between the vapor pressures, P_1 and P_2, and the absolute temperatures at which they were measured, T_1 and T_2:

$$\ln\frac{P_1}{P_2} = -\frac{\Delta H_{vap}}{R}\left(\frac{1}{T_1} - \frac{1}{T_2}\right)$$

(b) The melting point of potassium is 63.2°C. Molten potassium has a vapor pressure of 10.00 torr at 443°C and a vapor pressure of 400.0 torr at 708°C. Use these data and the equation in part (a) to calculate the heat of vaporization of liquid potassium. **(c)** By using the equation in part (a) and the data given in part (b), calculate the boiling point of potassium. **(d)** Calculate the vapor pressure of liquid potassium at 100°C.

11.92 Gold crystallizes in a face-centered cubic unit cell that has an edge length of 4.078 Å. The atom in the center of the face is in contact with the corner atoms, as shown in the drawing for Exercise 11.59. **(a)** Calculate the apparent radius of a gold atom in this structure. **(b)** Calculate the density of gold metal.

11.93 Consider the cubic unit cells (Figure 11.33) with an atom located at each lattice point. Calculate the net number of atoms in **(a)** a primitive cubic unit cell; **(b)** a body-centered cubic unit cell; **(c)** a face-centered cubic unit cell.

[11.94] The following data present the temperatures at which certain vapor pressures are achieved for dichloromethane (CH_2Cl_2) and methyl iodide (CH_3I):

Vapor Pressure (torr):	10.0	40.0	100.0	400.0
T for CH_2Cl_2 (°C):	−43.3	−22.3	−6.3	24.1
T for CH_3I (°C):	−45.8	−24.2	−7.0	25.3

(a) Which of the two substances is expected to have the greater dipole-dipole forces? Which is expected to have the greater London dispersion forces? Based on your answers, explain why it is difficult to predict which compound would be more volatile. **(b)** Which compound would you expect to have the higher boiling point? Check your answer in a reference book such as the *CRC Handbook of Chemistry and Physics*. **(c)** The order of volatility of these two substances changes as the temperature is increased. What quantity must be different for the two substances in order for this phenomenon to occur? **(d)** Substantiate your answer for part (c) by drawing an appropriate graph.

11.95 In a typical X-ray crystallography experiment, X rays of wavelength $\lambda = 0.71$ Å are generated by bombarding molybdenum metal with an energetic beam of electrons. Why are these X rays more effectively diffracted by crystals than is visible light?

[11.96] **(a)** The density of diamond [Figure 11.41(a)] is 3.5 g/cm^3, and that of graphite [Figure 11.41(b)] is 2.3 g/cm^3. Based on the structure of buckminsterfullerene (Figure 11.43), what would you expect its density to be relative to these other forms of carbon? **(b)** X-ray diffraction studies of buckminsterfullerene show that it has a face-centered cubic lattice of C_{60} molecules. The length of a side of the unit cell is 14.2 Å. Calculate the density of buckminsterfullerene.

Integrative Exercises

11.97 **(a)** At the molecular level, what factor is responsible for the steady increase in viscosity with increasing molecular weight in the hydrocarbon series shown in Table 11.4? **(b)** Although the viscosity varies over a factor of more than two in the series from hexane to nonane, the surface tension at 25°C increases by only about 20% in the same series. How do you account for this? **(c)** *n*-Octyl alcohol, $CH_3CH_2CH_2CH_2CH_2CH_2CH_2CH_2OH$, has a viscosity of 10.1 cP, much higher than nonane, which has about the same molecular weight. What accounts for this difference? How does your answer relate to the difference in normal boiling points for these two substances?

11.98 Acetone, $(CH_3)_2CO$, is widely used as an industrial solvent. **(a)** Draw the Lewis structure for the acetone molecule, and predict the geometry around each carbon atom. **(b)** Is the acetone molecule polar or nonpolar? **(c)** What kinds of intermolecular attractive forces exist between acetone molecules? **(d)** 1-Propanol, $CH_3CH_2CH_2OH$, has a molecular weight that is very similar to that of acetone, yet acetone boils at 56.5°C and 1-propanol boils at 97.2°C. Explain the difference.

11.99 The table shown here lists the molar heats of vaporization for several organic compounds. Use specific examples from this list to illustrate how the heat of vaporization varies with **(a)** molar mass; **(b)** molecular shape; **(c)** molecular polarity; **(d)** hydrogen-bonding interactions. Explain these comparisons in terms of the nature of the intermolecular forces at work. (You may find it helpful to draw out the structural formula for each compound.)

Compound	Heat of Vaporization (kJ/mol)
$CH_3CH_2CH_3$	19.0
$CH_3CH_2CH_2CH_2CH_3$	27.6
$CH_3CHBrCH_3$	31.8
CH_3COCH_3	32.0
$CH_3CH_2CH_2Br$	33.6
$CH_3CH_2CH_2OH$	47.3

11.100 Liquid butane, C_4H_{10}, is stored in cylinders, to be used as a fuel. The normal boiling point of butane is listed as −0.5°C. **(a)** Suppose the tank is standing in the sun and reaches a temperature of 46°C. Would you expect the pressure in the tank to be greater or less than atmospheric pressure? How does the pressure within the tank depend on how much liquid butane is in it? **(b)** Suppose the valve to the tank is opened and a few liters of butane are allowed to escape rapidly. What do you expect would happen to the temperature of the remaining liquid butane in the tank? Explain. **(c)** How much heat must be added to vaporize 155 g of butane if its heat of vaporization is 21.3 kJ/mol? What volume does this much butane occupy at 755 torr and 35°C.

[11.101] Using information in Appendices B and C, calculate the minimum number of grams of $C_3H_8(g)$ that must be combusted to provide the energy necessary to convert 2.50 kg of H_2O from its solid form at −14.0°C to its liquid form at 60.0°C.

11.102 In a certain type of nuclear reactor, liquid sodium metal is employed as a circulating coolant in a closed system, protected from contact with air or water. Much like the coolant that circulates in an automobile engine, the liquid sodium carries heat from the hot reactor core to heat exchangers. **(a)** What properties of the liquid sodium are of special importance in this application? **(b)** The viscosity of liquid sodium varies with temperature as follows:

Temperature (°C)	Viscosity (cP)
100	0.705
200	0.450
300	0.345
600	0.210

What forces within the liquid sodium are likely to be the major contributors to the viscosity? Why does viscosity decrease with increasing temperature?

11.103 The vapor pressure of a volatile liquid can be determined by slowly bubbling a known volume of gas through it at a known temperature and pressure. In an experiment 5.00 L of N_2 gas is passed through 7.2146 g of liquid benzene, C_6H_6, at 26.0°C. The liquid remaining after the experiment weighs 5.1493 g. Assuming that the gas becomes saturated with benzene vapor and that the total gas volume and temperature remain constant, what is the vapor pressure of the benzene in torr?

11.104 The relative humidity of air equals the ratio of the partial pressure of water in the air to the equilibrium vapor pressure of water at the same temperature. If the relative humidity of the air is 45% and its temperature is 23°C, how many molecules of water are present in a room measuring 14 m by 9.0 m by 8.6 m?

[11.105] Use a reference source such as the *CRC Handbook of Chemistry and Physics* to compare the melting and boiling points of the following pairs of inorganic substances: **(a)** W and WF_6; **(b)** SO_2 and SF_4; **(c)** SiO_2 and $SiCl_4$. Account for the major differences observed in terms of likely structures and bonding.

eMedia Exercises

11.106 The **Changes of State** movie (*eChapter 11.4*) shows the heating curve of a substance that melts and then vaporizes. **(a)** Using the information in Figure 11.27(b), sketch the heating curve of carbon dioxide at atmospheric pressure, starting at −100°C and ending at 30°C. **(b)** Sketch the same heating curve as it would appear at a pressure of 10 atm. **(c)** Under what conditions would the heating curve of water resemble the one you sketched for part (a)?

11.107 Using the **Equilibrium Vapor Pressure** simulation (*eChapter 11.5*), compare the vapor pressures of methanol, ethanol, acetic acid, water, and benzene. **(a)** Which compound appears to have the strongest intermolecular forces at 100°C? **(b)** If one compound has a higher equilibrium vapor pressure than another at a particular temperature, will it necessarily have a higher equilibrium vapor pressure at all temperatures? If not, give an example of two compounds whose vapor-pressure curves cross, and at what temperature they have roughly the same vapor pressure.

11.108 Ethanol and acetic acid have very similar molar masses and both exhibit hydrogen bonding. In which of these two compounds is hydrogen bonding a more significant component of the overall intermolecular forces? Support your answer using data from the **Equilibrium Vapor Pressure** simulation (*eChapter 11.5*).

11.109 Water is a substance with some very unusual properties. Based on a comparison of the **water** molecule and the structure of **ice** (*eChapter 11.8*), explain **(a)** why ice is less dense than water, and **(b)** why increasing pressure, within a certain temperature range, causes ice to melt. **(c)** Why is it not possible to convert ice to liquid water by applying pressure at very low temperatures?

11.110 Compare the structure of **ice** (*eChapter 11.8*) with the structure of **diamond** (*eChapter 11.8*). If you rotate the structures just right, you will see distinct similarities, namely sp^3 hybridization (of oxygen and carbon, respectively) and the hexagonal arrangement of atoms. Given this similarity, explain why ice can be melted by applying pressure whereas diamond cannot.

Chapter 12

Modern Materials

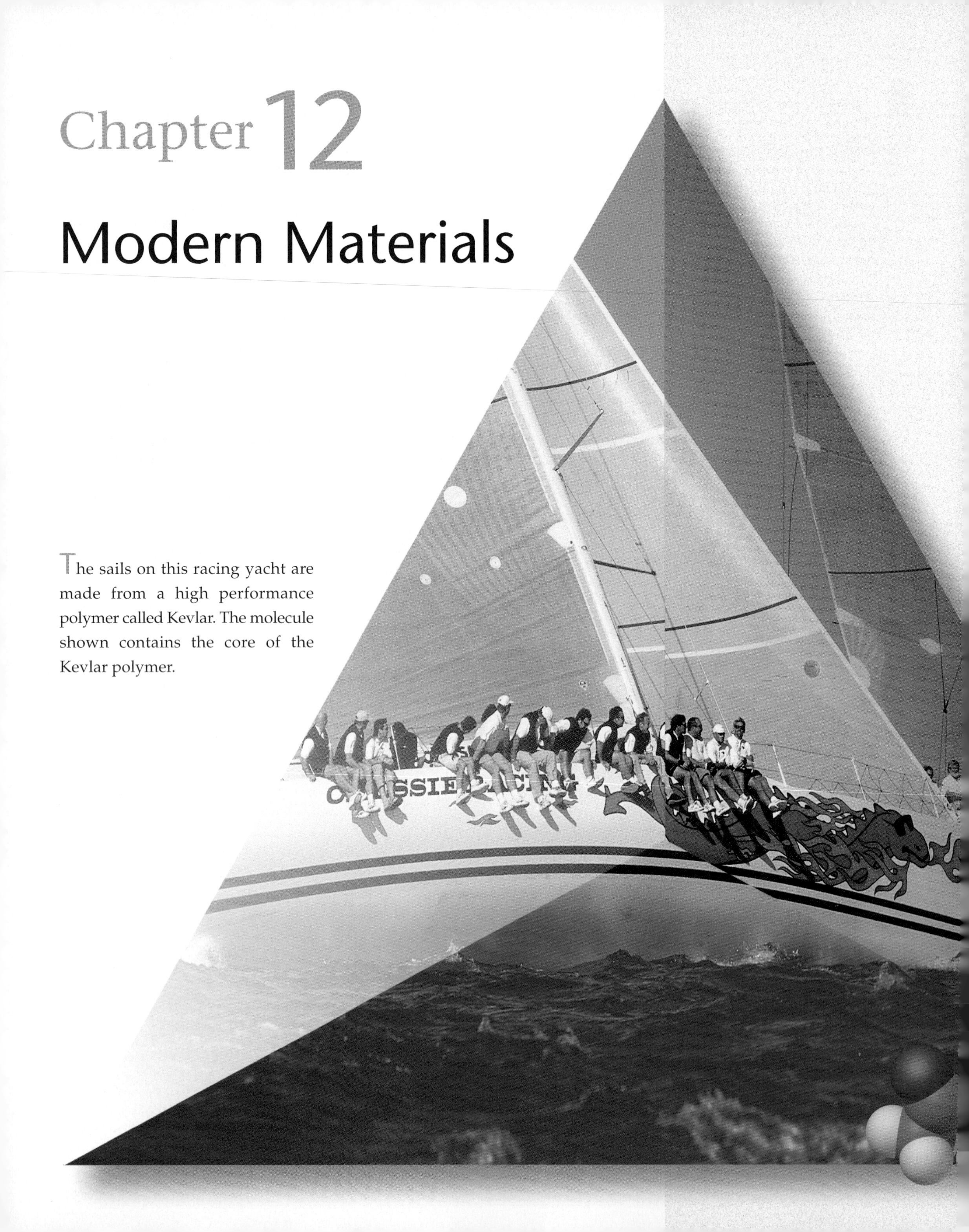

The sails on this racing yacht are made from a high performance polymer called Kevlar. The molecule shown contains the core of the Kevlar polymer.

SINCE THE BEGINNING of the modern era of chemistry in the nineteenth century, one of the important goals of chemical research has been the discovery and development of materials with useful properties. Future advances in technology will depend more than ever on the discovery and development of materials with new, valuable properties. Chemists have contributed to materials science by inventing entirely new substances and developing the means for processing naturally occurring materials to form fibers, films, coatings, adhesives, and substances with special electrical, magnetic, or optical properties. In this chapter we will discuss the properties and applications of several kinds of materials that play important roles in modern society. Our aim is to show how we can understand many special physical or chemical properties by applying the principles we have discussed in earlier chapters.

This chapter demonstrates the important point that *the observable, macroscopic properties of materials are the result of atomic- and molecular-level structures and processes.*

▸ What's Ahead ◂

- We'll learn about the characteristics of *liquid crystals*; what distinguishes them from ordinary liquids, and the molecular structural features that promote liquid crystalline behavior.
- We'll explore synthetic *polymers*, large molecules formed from low molecular weight *monomer* molecules. Polymers are formed most commonly via *addition* or *condensation* polymerization reactions. Their physical and chemical properties make polymers suitable for many practical uses.
- Materials used in biomedical applications (that is, in intimate contact with living organisms) are called *biomaterials*. We'll see that a biomaterial must meet many requirements to be suitable for such applications as heart valves, artificial tissue, or hip replacement components.
- *Ceramics* are nonmetallic inorganic solid materials. They find applications in many areas where hardness, high-temperature stability, and resistance to corrosion are important.
- Many materials at low temperature undergo a phase transition to a state in which they lose all resistance to the flow of electrons. This property is known as *superconductivity*. We will learn some characteristics of the superconducting state, and the types of materials that exhibit superconductivity.
- We'll examine the nature and uses of *thin films*, and some of the methods used to form them.

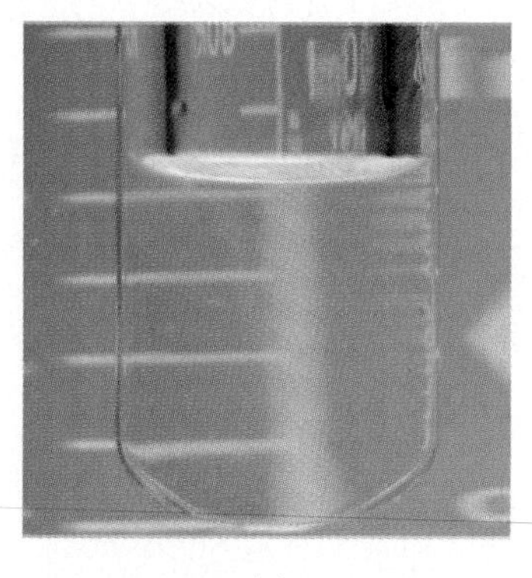

(a)

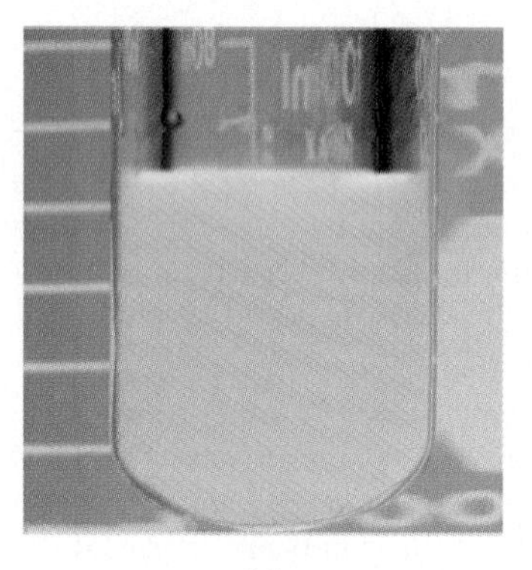

(b)

▲ **Figure 12.1** (a) Molten cholesteryl benzoate at a temperature above 179°C. In this temperature region the substance is a clear liquid. Note that the printing on the surface of the beaker in back of the sample test tube is readable. (b) Cholesteryl benzoate at a temperature between 179°C and its melting point, 145°C. In this temperature interval cholesteryl benzoate exhibits a milky liquid crystalline phase.

12.1 Liquid Crystals

When a solid is heated to its melting point, the added thermal energy overcomes the intermolecular attractions that provide molecular order to the solid. ∞ (Section 11.1) The liquid that forms is characterized by random molecular orientations and considerable molecular motion. Some substances, however, exhibit more complex behavior as their solids are heated.

In 1888 Frederick Reinitzer, an Austrian botanist, discovered that an organic compound he was studying, cholesteryl benzoate, has interesting and unusual properties. When heated, the substance melts at 145°C to form a viscous milky liquid; at 179°C the milky liquid suddenly becomes clear. When the substance is cooled, the reverse processes occur: The clear liquid turns viscous and milky at 179°C (Figure 12.1 ◀), and the milky liquid solidifies at 145°C. Reinitzer's work represents the first systematic report of what we now call a **liquid crystal**.

Instead of passing directly from the solid to the liquid phase when heated, some substances, such as cholesteryl benzoate, pass through an intermediate, liquid crystalline phase that has some of the structure of solids and some of the freedom of motion possessed by liquids. Because of the partial ordering, liquid crystals may be very viscous and possess properties intermediate between those of the solid and liquid phases. The region in which they exhibit these properties is marked by sharp transition temperatures, as in Reinitzer's example.

From the time of their discovery in 1888 until about 30 years ago, liquid crystals were largely a laboratory curiosity. They are now widely used as pressure and temperature sensors and in the displays of electrical devices such as digital watches, calculators, and notebook and handheld computers (Figure 12.2 ▼). Liquid crystals can be used for these applications because the weak intermolecular forces that hold the molecules together in a liquid crystal are easily affected by changes in temperature, pressure, and electromagnetic fields.

Types of Liquid Crystalline Phases

Substances that form liquid crystals are often composed of long, rodlike molecules. In the normal liquid phase, these molecules are oriented randomly

▶ **Figure 12.2** A handheld wireless device with liquid crystal display.

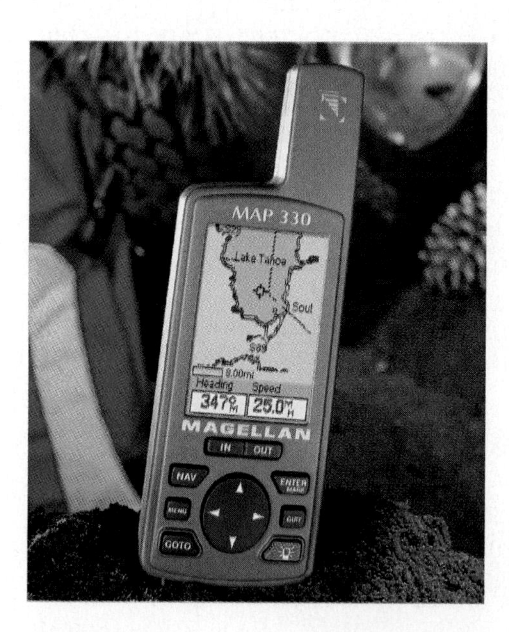

(a) Normal liquid

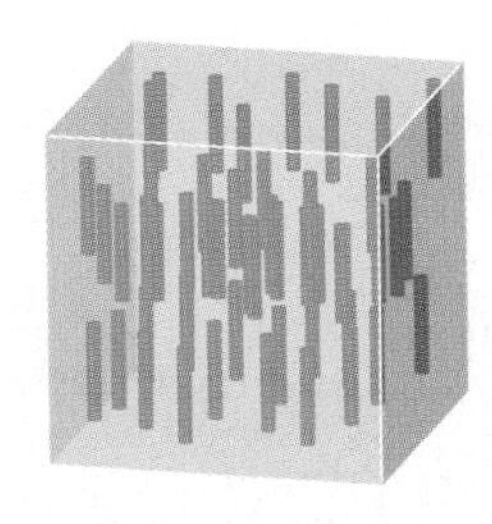
(b) Nematic liquid crystal

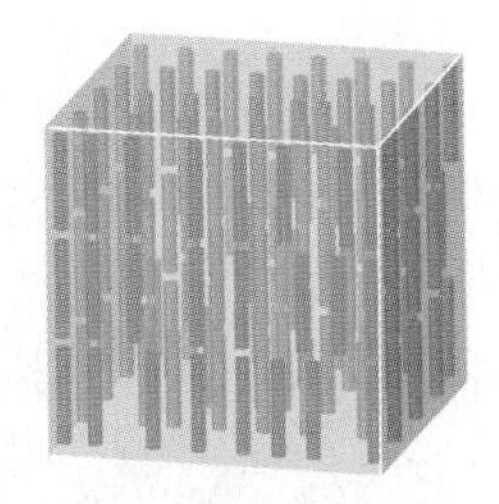
(c) Smectic A liquid crystal

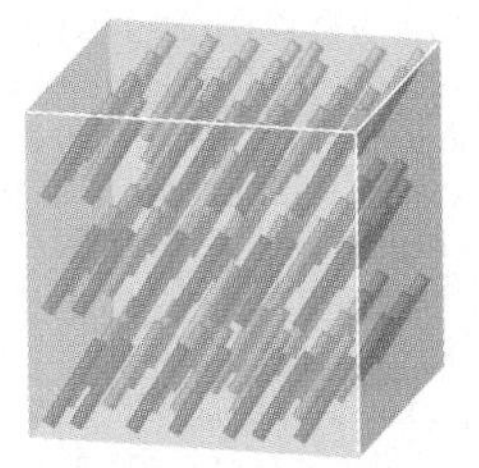
(d) Smectic C liquid crystal

▲ **Figure 12.3** Ordering in liquid crystalline phases, as compared with a normal (non–liquid crystalline) liquid.

[Figure 12.3(a) ▲]. Liquid crystalline phases, by contrast, exhibit some ordering of the molecules. Depending on the nature of the ordering, liquid crystals can be classified as nematic, smectic, or cholesteric.

In the **nematic liquid crystalline phase** the molecules are aligned along their long axes, but there is no ordering with respect to the ends of the molecules [Figure 12.3(b)]. The arrangement of the molecules is like that of a handful of pencils whose ends are not aligned.

In the **smectic liquid crystalline phases** the molecules exhibit additional ordering beyond that of the nematic phase. The smectic phases resemble a handful of pencils whose ends are more nearly aligned. There are different kinds of smectic phases, designated by the letters A, B, C, and so forth. In the smectic A phase the molecules are arranged in layers, with their long axes perpendicular to the layers [Figure 12.3(c)]. Other smectic phases display different types of alignments. For example, in the smectic C phase the molecules are aligned with their long axes tilted relative to the layers in which the molecules are stacked [Figure 12.3(d)].

Two molecules that exhibit liquid crystalline phases are shown in Figure 12.4 ▼. These molecules are fairly long in relation to their thicknesses. The C=N double bond and the benzene rings add stiffness. The flat benzene rings help the molecules stack against one another. In addition, many liquid crystalline molecules contain polar groups; these give rise to dipole-dipole interactions that promote alignments of the molecules. (Section 11.2) Thus, the molecules order themselves quite naturally along their long axes. They can, however, rotate around their axes and slide parallel to one another. In smectic phases the intermolecular forces between the molecules (such as London dispersion forces, dipole-dipole attractions, and hydrogen bonding) limit the ability of the molecules to slide past one another.

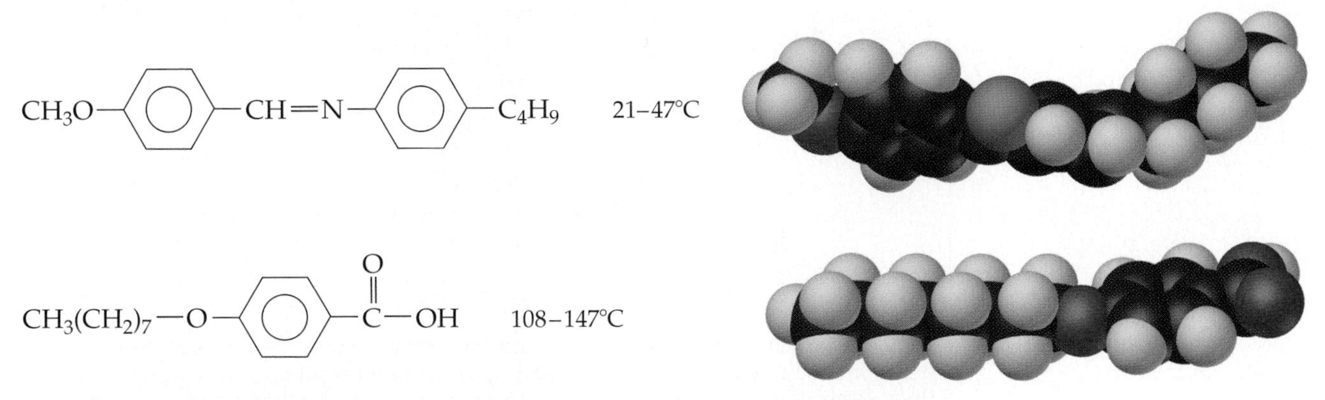

▲ **Figure 12.4** Structures and liquid crystal temperature intervals of two typical liquid crystalline materials. The temperature interval indicates the temperature range in which the substance exhibits liquid crystalline behavior.

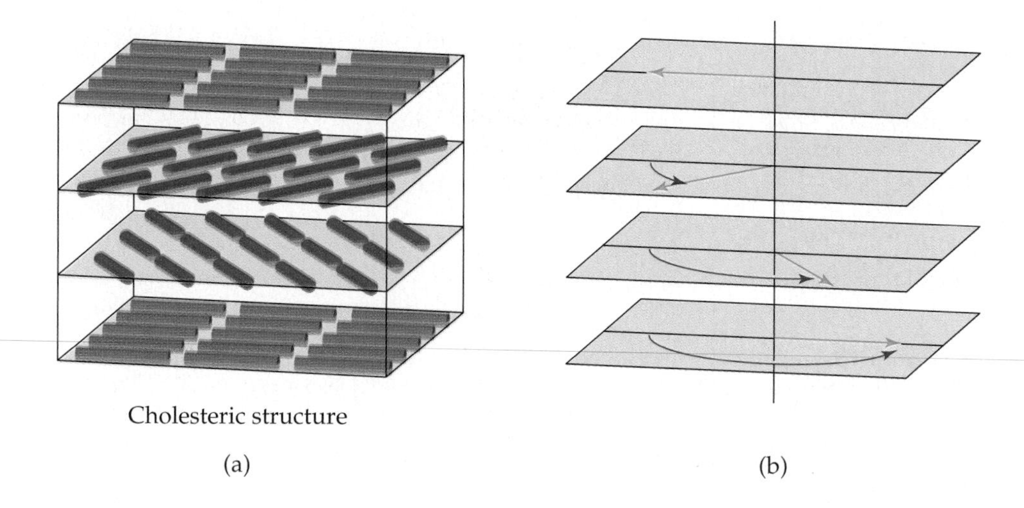

▶ **Figure 12.5** (a) Ordering in a cholesteric liquid crystal. The molecules in successive layers are oriented at a characteristic angle with respect to those in adjacent layers to avoid repulsive interactions. The result is a screwlike axis, as shown in (b).

Figure 12.5 ▲ shows the ordering of the **cholesteric liquid crystalline phase**. The molecules are aligned along their long axes as in nematic liquid crystals, but they are arranged in layers with the molecules in each plane twisted slightly in relation to the molecules in the planes above and below. These liquid crystals are so named because many derivatives of cholesterol adopt this structure. The spiral nature of the molecular ordering produces unusual coloring patterns with visible light. Changes in temperature and pressure change the order and hence the color (Figure 12.6 ◀). Cholesteric liquid crystals have been used to monitor temperature changes in situations where conventional methods are not feasible. For example, they can detect hot spots in microelectronic circuits, which may signal the presence of flaws. They can also be fashioned into thermometers for measuring the skin temperature of infants.

▲ **Figure 12.6** Color change in a cholesteric liquid crystalline material as a function of temperature.

SAMPLE EXERCISE 12.1

Which of the following substances is most likely to exhibit liquid crystalline behavior?

$$CH_3-CH_2-C(CH_3)_2-CH_2-CH_3$$

(i)

$$CH_3CH_2-C_6H_4-N{=}N-C_6H_4-C(=O)-OCH_3$$

(ii)

$$C_6H_5-CH_2-C(=O)-O^-\,Na^+$$

(iii)

Solution

Analyze: We have three molecules of differing molecular structure, and are asked to determine whether any of them are likely to be liquid crystalline substances.

Plan: We need to identify the structural features of each case that might induce liquid crystalline behavior.

Solve: Molecule (i) is not likely to be liquid crystalline because it does not have a long axial structure. Molecule (iii) is ionic; the generally high melting points of ionic materials (Section 8.2) and the absence of a characteristic long axis make it unlikely that this substance will exhibit liquid crystalline behavior. Molecule (ii) possesses the characteristic long axis and the kinds of structural features that are often seen in liquid crystals (Figure 12.5).

PRACTICE EXERCISE

Suggest a reason why the following molecule, decane, does not exhibit liquid crystalline behavior:

$$CH_3CH_2CH_2CH_2CH_2CH_2CH_2CH_2CH_2CH_3$$

Answer: Because rotation can occur about carbon–carbon single bonds, molecules whose backbone consists of C—C single bonds are too flexible; the molecules tend to coil in random ways and thus are not rodlike.

Chemistry at Work Liquid Crystal Displays

Liquid crystals are widely used in electrically controlled liquid crystal display (LCD) devices in watches, calculators, and computer screens, as illustrated in Figure 12.2. These applications are possible because an applied electrical field changes the orientation of liquid crystal molecules and thus affects the optical properties of the device.

LCDs come in a variety of designs, but the structure illustrated in Figure 12.7 ▼ is typical. A thin layer (5–20 μm) of liquid crystalline material is placed between electrically conducting, transparent glass electrodes. Ordinary light passes through a vertical polarizer that permits light in only the vertical plane to pass. Through a special process the liquid crystal molecules are oriented so that the molecules at the front plate are oriented vertically, those at the bottom plate horizontally. The molecules in between vary in orientation in a regular way, as shown in Figure 12.7. Displays of this kind are called "twisted nematic." The plane of polarization of the light is turned by 90° as it passes through the device, and it is thus in the correct orientation to pass through the horizontal polarizer. In a watch display a mirror reflects the light back, and the light retraces its path; the device thus looks bright. When a voltage is applied to the plates, the liquid crystalline molecules align with the voltage [Figure 12.7(b)]. The light rays thus are not properly oriented to pass through the horizontal polarizer, and the device looks dark.

Computer screens employ backlighting rather than reflected light, but the principle is the same. The computer screen is divided into a large number of tiny cells, with the voltages at points on the screen surface controlled by transistors made from thin films of amorphous silicon. Red-green-blue color filters are employed to provide full color. The entire display is refreshed at a frequency of about 60 Hz, so the display can change rapidly with respect to the response time of the human eye. Displays of this kind are remarkable technical achievements, based on a combination of basic scientific discovery and creative engineering.

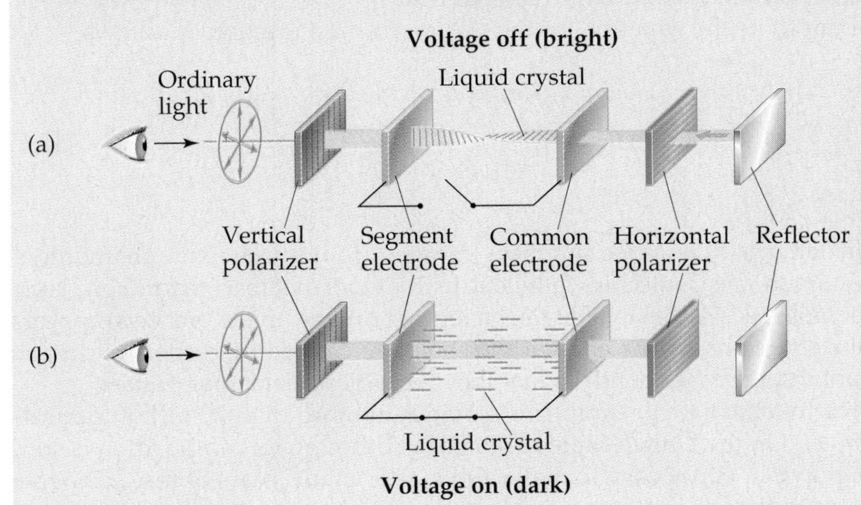

◀ **Figure 12.7** Schematic illustration of the operation of a twisted nematic liquid crystal display (LCD). (a) Ordinary light, which is polarized in all directions, passes through the vertical polarizer. The vertically polarized light then passes into the liquid crystalline layer, where the plane of polarization is rotated 90°. It passes through the horizontal polarizer, is reflected, and retraces its path to give a bright display. (b) When a voltage is applied to the segment electrode that covers the small area, the liquid crystal molecules align along the direction of the light path. Thus, the vertically polarized light is not turned 90° and is unable to pass through the horizontal polarizer. The area encompassed by the small transparent segment electrode therefore appears dark. Digital watches typically have such displays.

12.2 Polymers

Our discussions of chemistry to this point have focused primarily on molecules of fairly low molecular mass. In nature, however, we find many substances of very high molecular mass, running into millions of amu. Starch and cellulose abound in plants; proteins and nucleic acids are found in both plants and animals. In 1827 Jons Jakob Berzelius coined the word **polymer** (from the Greek *polys*, "many," and *meros*, "parts") to denote molecular substances of high molecular mass formed by the *polymerization* (joining together) of **monomers**, molecules with low molecular mass.

For a long time, humans processed naturally occurring polymers, such as wool, leather, and natural rubber, to form useful materials. During the past 60 years or so, chemists have learned to form synthetic polymers by polymerizing monomers through controlled chemical reactions. A great many of these synthetic polymers have a backbone of carbon–carbon bonds because carbon atoms have an exceptional ability to form strong, stable bonds with one another.

Addition Polymerization

The simplest example of a polymerization reaction is the formation of *polyethylene* from ethylene molecules. In this reaction the double bond in each ethylene molecule "opens up," and two of the electrons originally in this bond are used to form new C—C single bonds with two other ethylene molecules:

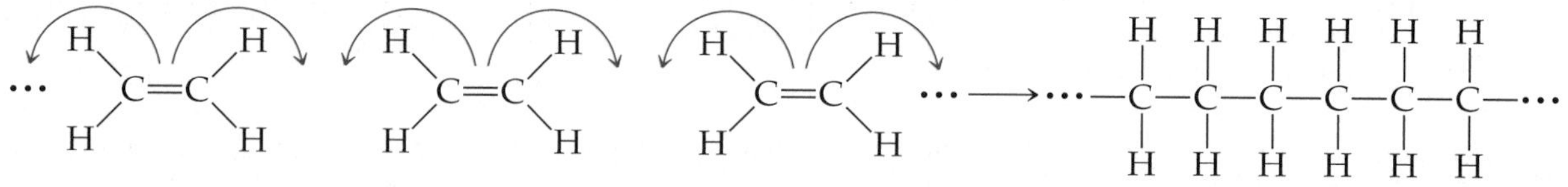

$$\cdots \; \mathrm{H_2C{=}CH_2} \;\; \mathrm{H_2C{=}CH_2} \;\; \mathrm{H_2C{=}CH_2} \; \cdots \longrightarrow \cdots\!-\!\begin{array}{c}\mathrm{H}\\|\\\mathrm{C}\\|\\\mathrm{H}\end{array}\!-\!\begin{array}{c}\mathrm{H}\\|\\\mathrm{C}\\|\\\mathrm{H}\end{array}\!-\!\begin{array}{c}\mathrm{H}\\|\\\mathrm{C}\\|\\\mathrm{H}\end{array}\!-\!\begin{array}{c}\mathrm{H}\\|\\\mathrm{C}\\|\\\mathrm{H}\end{array}\!-\!\begin{array}{c}\mathrm{H}\\|\\\mathrm{C}\\|\\\mathrm{H}\end{array}\!-\!\begin{array}{c}\mathrm{H}\\|\\\mathrm{C}\\|\\\mathrm{H}\end{array}\!-\!\cdots$$

Ethylene

Polyethylene

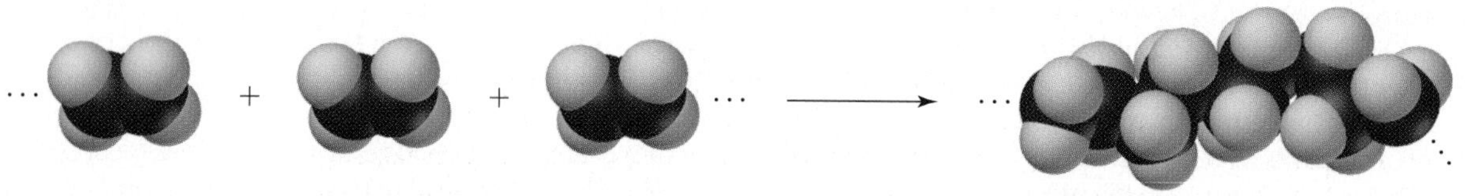

$$\cdots \; + \; + \; \cdots \longrightarrow \cdots$$

Polymerization that occurs through the coupling of monomers using their multiple bonds is called **addition polymerization**.

We can write the equation for the polymerization reaction as follows:

$$n\mathrm{CH_2{=}CH_2} \longrightarrow \left[\begin{array}{c}\mathrm{H}\\|\\\mathrm{C}\\|\\\mathrm{H}\end{array}\!-\!\begin{array}{c}\mathrm{H}\\|\\\mathrm{C}\\|\\\mathrm{H}\end{array}\right]_n \qquad [12.1]$$

Here the letter n is the large number—ranging from hundreds to many thousands—of monomer molecules (ethylene in this case) that react to form one large polymer molecule. Within the polymer a repeat unit (the unit in brackets) appears along the entire chain. The ends of the chain are capped by carbon–hydrogen bonds or by some other bond, so that the end carbons have four bonds.

Polyethylene is a very important material; more than 20 billion pounds are produced in the United States each year. Although its composition is simple, the polymer is not easy to make. Only after many years of research were the right conditions and appropriate catalysts identified for manufacturing a commercially useful polymer. Today many different forms of polyethylene,

TABLE 12.1 Polymers of Commercial Importance

Polymer	Structure	Uses
Addition polymers		
Polyethylene	$\text{+CH}_2\text{—CH}_2\text{+}_n$	Films, packaging, bottles
Polypropylene	$\left[\text{CH}_2\text{—CH(CH}_3\text{)}\right]_n$	Kitchenware, fibers, appliances
Polystyrene	$\left[\text{CH}_2\text{—CH(C}_6\text{H}_5\text{)}\right]_n$	Packaging, disposable food containers, insulation
Polyvinyl chloride	$\left[\text{CH}_2\text{—CH(Cl)}\right]_n$	Pipe fittings, clear film for meat packaging
Condensation polymers		
Polyurethane	$\left[\text{C(=O)—NH—R—NH—C(=O)—O—R'—O}\right]_n$ R, R′ = $\text{—CH}_2\text{—CH}_2\text{—}$ (for example)	"Foam" furniture stuffing, spray-on insulation, automotive parts, footwear, water-protective coatings
Polyethylene terephthalate (a polyester)	$\left[\text{O—CH}_2\text{—CH}_2\text{—O—C(=O)—C}_6\text{H}_4\text{—C(=O)}\right]_n$	Tire cord, magnetic tape, apparel, soft-drink bottles
Nylon 6,6	$\left[\text{NH+CH}_2\text{+}_6\text{NH—C(=O)—(CH}_2\text{)}_4\text{—C(=O)}\right]_n$	Home furnishings, apparel, carpet fibers, fishing line, polymer blends

varying widely in physical properties, are known (refer to the book cover illustration and the accompanying description). Polymers of other chemical compositions provide still greater variety in physical and chemical properties. Table 12.1 ▲ lists several other common polymers that are obtained by addition polymerization.

Condensation Polymerization

A second general kind of reaction used to synthesize commercially important polymers is **condensation polymerization**. In a **condensation reaction** two molecules are joined to form a larger molecule by elimination of a small molecule such as H_2O. For example, an amine (a compound containing the $-NH_2$ group) will react with a carboxylic acid (a compound containing the $-COOH$ group) to form a bond between N and C along with the formation of H_2O.

$$\text{—N(H)—H} + \text{H—O—C(=O)—} \longrightarrow \text{—N(H)—C(=O)—} + \text{H}_2\text{O}$$

+ ⟶ +

Chemistry at Work Recycling Plastics

If you look at the bottom of a plastic container, you are likely to see a recycle symbol containing a number, as seen in Figure 12.8 ▶. The number in the middle of the recycle symbol and the abbreviation below it indicate the kind of polymer from which the container is made, as summarized in Table 12.2 ▼. (The chemical structures of these polymers are shown in Table 12.1.) These symbols make it possible to sort the containers by their composition. In general, the lower the number, the greater the ease with which the material can subsequently be recycled.

TABLE 12.2 Categories Used for Recycling Polymeric Materials in the United States

Number	Abbreviation	Polymer
1	PET	Polyethylene terephthalate
2	HDPE	High-density polyethylene
3	V	Polyvinyl chloride (PVC)
4	LDPE	Low-density polyethylene
5	PP	Polypropylene
6	PS	Polystyrene
7		Others

▲ **Figure 12.8** Plastic containers have symbols on their bottoms, indicating their compositions and suitability for recycling.

MOVIE
Synthesis of Nylon 6,10

Polymers formed from two different monomers are called **copolymers**. In the formation of many nylons a *diamine*, a compound with an $-NH_2$ group at each end, is reacted with a *diacid*, a compound with a $-COOH$ group at each end. For example, nylon 6,6 is formed when a diamine that has six carbon atoms and an amino group on each end is reacted with adipic acid, which also has six carbon atoms.

$$\underset{\text{Diamine}}{n\mathrm{H_2N{+}CH_2{+}_6NH_2}} + \underset{\text{Adipic acid}}{n\mathrm{HO\overset{O}{\overset{\|}{C}}{+}CH_2{+}_4\overset{O}{\overset{\|}{C}}OH}} \longrightarrow \underset{\text{Nylon 6,6}}{\left[\mathrm{NH(CH_2)_6NH{-}\overset{O}{\overset{\|}{C}}(CH_2)_4\overset{O}{\overset{\|}{C}}}\right]_n} + 2n\mathrm{H_2O} \qquad [12.3]$$

A condensation reaction occurs on each end of the diamine and the acid. The components of H_2O are split out, and N—C bonds are formed between molecules. Table 12.1 lists nylon 6,6 and some other common polymers obtained by condensation polymerization. Notice that these polymers have backbones containing N or O atoms as well as C atoms. In Chapter 25 we will see that proteins are also condensation polymers.

Types of Polymers

Plastics are materials that can be formed into various shapes, usually by the application of heat and pressure. **Thermoplastic** materials can be reshaped. For example, plastic milk containers are made from polyethylene of high molecular mass. These containers can be melted down and the polymer recycled for some other use. In contrast, a **thermosetting plastic** is shaped through irreversible chemical processes and therefore cannot be reshaped readily.

An **elastomer** is a material that exhibits rubbery or elastic behavior. When subjected to stretching or bending, it regains its original shape upon removal of the distorting force, provided that it has not been distorted beyond some elastic limit. Some polymers, such as nylon and polyesters, can also be formed into *fibers* that, like hair, are very long compared to their cross-sectional area and are not elastic. These fibers can be woven into fabrics and cords and fashioned into clothing, tire cord, and other useful objects.

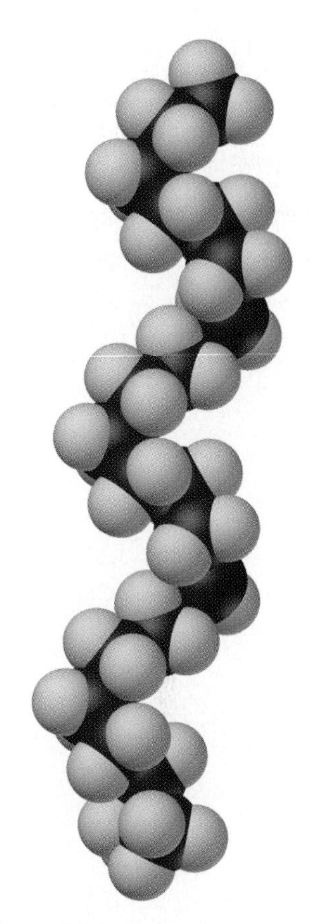

▲ **Figure 12.9** A segment of a polyethylene chain. The segment shown here consists of 20 carbon atoms. In commercial polyethylenes the chain lengths range from about 10^3 to 10^5 CH_2 units. As this illustration implies, the chains are flexible and can coil and twist in random fashion.

Structures and Physical Properties of Polymers

The simple structural formulas given for polyethylene and other polymers are deceptive. Because each carbon atom in polyethylene is surrounded by four bonds, the atoms are arranged in a tetrahedral fashion, so that the chain is not straight as we have depicted it. Furthermore, the atoms are relatively free to rotate around the C—C single bonds. Rather than being straight and rigid, therefore, the chains are very flexible, folding readily (Figure 12.9 ▶). The flexibility in the molecular chains causes the polymer material to be very flexible.

Both synthetic and naturally occurring polymers commonly consist of a collection of macromolecules of different molecular weights. Depending on the conditions of formation, the molecular weights may be distributed over a wide range or be closely clustered around an average value. In part because of this distribution in molecular weights, polymers are largely amorphous (noncrystalline) materials. Rather than exhibiting well-defined crystalline phases with sharp melting points, they soften over a range of temperatures. They may, however, possess short-range order in some regions of the solid, with chains lined up in regular arrays as shown in Figure 12.10 ▶. The extent of such ordering is indicated by the degree of ***crystallinity*** of the polymer. The crystallinity of a polymer can frequently be enhanced by mechanical stretching or pulling to align the chains as the molten polymer is drawn through small holes. Intermolecular forces between the polymer chains hold the chains together in the ordered, crystalline regions, making the polymer denser, harder, less soluble, and more resistant to heat. Table 12.3 ▼ shows how the properties of polyethylene change as the degree of crystallinity increases.

The simple linear structure of polyethylene is conducive to intermolecular interactions that lead to crystallinity. However, the degree of crystallinity in polyethylene strongly depends on the average molecular mass. Polymerization results in a mixture of *macromolecules* (large molecules) with varying n and hence varying molecular masses. So-called low-density polyethylene used in forming films and sheets has an average molecular mass in the range of 10^4 amu and has substantial chain branching. That is, there are side chains off the main chain of the polymer, much like spur lines that branch from a main railway line. These branches inhibit the formation of crystalline regions, reducing the density of the material. High-density polyethylene, used to form bottles, drums, and pipes, has an average molecular mass in the range of 10^6 amu. This form has less branching

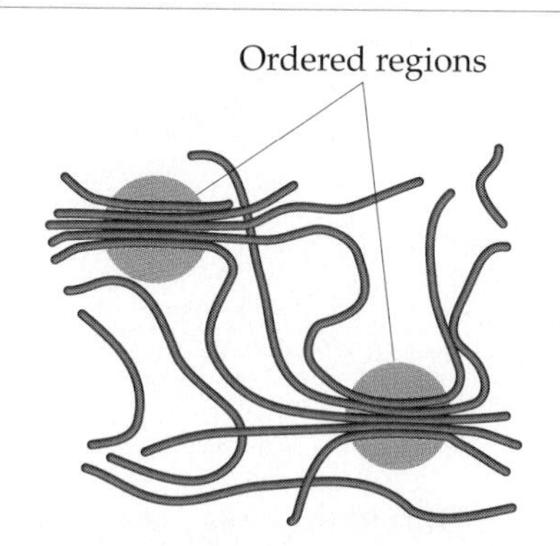

▲ **Figure 12.10** Interactions between polymer chains. In the circled regions, the forces that operate between adjacent polymer-chain segments lead to ordering analogous to the ordering in crystals, though less regular.

TABLE 12.3 Properties of Polyethylene as a Function of Crystallinity

	Crystallinity				
	55%	62%	70%	77%	85%
Melting point (°C)	109	116	125	130	133
Density (g/cm^3)	0.92	0.93	0.94	0.95	0.96
Stiffness*	25	47	75	120	165
Yield stress*	1700	2500	3300	4200	5100

*These test results show that the mechanical strength of the polymer increases with increased crystallinity. The physical units for the stiffness test are psi × 10^{-3} (psi = pounds per square inch); those for the yield stress test are psi. Discussion of the exact meaning and significance of these tests is beyond the scope of this text.

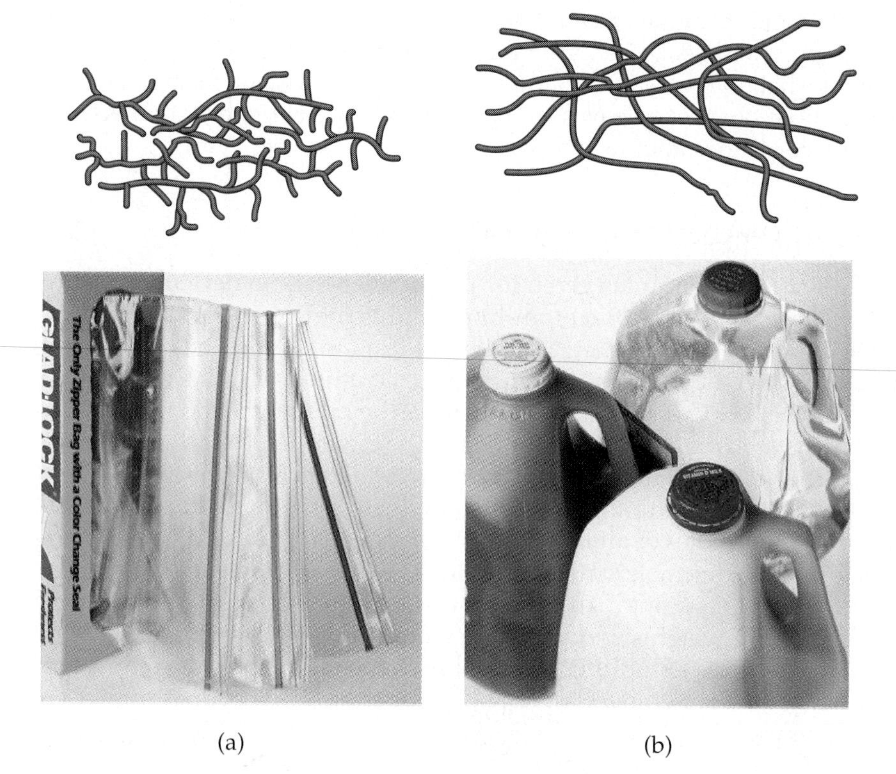

▶ **Figure 12.11** (a) Schematic illustration of the structure of low-density polyethylene (LDPE) and a typical use of LDPE film to form food-storage bags. (b) Schematic illustration of the structure of high-density polyethylene (HDPE) and containers formed from HDPE.

and thus a higher degree of crystallinity. Low-density and high-density polyethylene are illustrated in Figure 12.11 ▲.

Various substances may be added to polymers to provide protection against the effects of sunlight or against degradation by oxidation. For example, manganese(II) salts, such as manganese(II) hypophosphite, $Mn(H_2PO_2)_2$, and copper(I) salts, in concentrations as low as $5 \times 10^{-4}\%$, are added to nylons to provide protection from light and oxidation and help to maintain whiteness. In addition, the physical properties of polymeric materials can be extensively modified by adding substances with lower molecular mass, called *plasticizers*, to reduce the extent of interactions between chains and thus to make the polymer more pliable. Polyvinyl chloride (PVC) (Table 12.1), for example, is a hard, rigid material of high molecular mass that is used to manufacture home drainpipes. When blended with a suitable substance of lower molecular mass, however, it forms a flexible polymer that can be used to make rain boots and doll parts. In some applications the plasticizer in a plastic object may be lost over time because of evaporation. As this happens, the plastic loses its flexibility and becomes subject to cracking.

Cross-linking Polymers

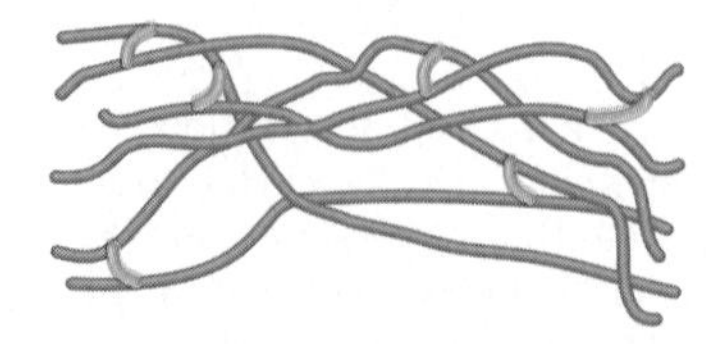

▲ **Figure 12.12** Cross-linking of polymer chains. The cross-linking groups (green) constrain the relative motions of the polymer chains, making the material harder and less flexible.

Polymers can be made stiffer by introducing chemical bonds between the polymer chains, as illustrated in Figure 12.12 ◀. Forming bonds between chains is called **cross-linking**. The greater the number of cross-links in a polymer, the more rigid the material. Whereas thermoplastic materials consist of independent polymer chains, thermosetting ones become cross-linked when heated, and thereby hold their shapes.

An important example of cross-linking is the **vulcanization** of natural rubber, a process discovered by Charles Goodyear in 1839. Natural rubber is formed from a liquid resin derived from the inner bark of the *Hevea brasiliensis* tree. Chemically, it is a polymer of isoprene, C_5H_8.

$$(n+2)\ \underset{\text{Isoprene}}{CH_2{=}C(CH_3){-}CH{=}CH_2} \longrightarrow \underset{\text{Rubber}}{{\sim}CH_2{-}C(CH_3){=}CH{-}CH_2{-}\left[CH_2{-}C(CH_3){=}CH{-}CH_2\right]_n{-}CH_2{-}C(CH_3){=}CH{-}CH_2{\sim}} \quad [12.4]$$

Because rotation about the carbon–carbon double bond does not readily occur, the orientation of the groups bound to the carbons is rigid. In naturally occurring rubber, the chain extensions are on the same side of the double bond, as shown in Equation 12.4. This form is called *cis*-polyisoprene; the prefix *cis*- is derived from a Latin phrase meaning "on this side."

Natural rubber is not a useful plastic because it is too soft and too chemically reactive. Goodyear accidentally discovered that adding sulfur to rubber and then heating the mixture makes the rubber harder and reduces its susceptibility to oxidation or other chemical attack. The sulfur changes rubber into a thermosetting polymer by cross-linking the polymer chains through reactions at some of the double bonds, as shown schematically in Figure 12.13 ▼. Cross-linking of about 5% of the double bonds creates a flexible, resilient rubber. When the rubber is stretched, the cross-links help prevent the chains from slipping, so that the rubber retains its elasticity.

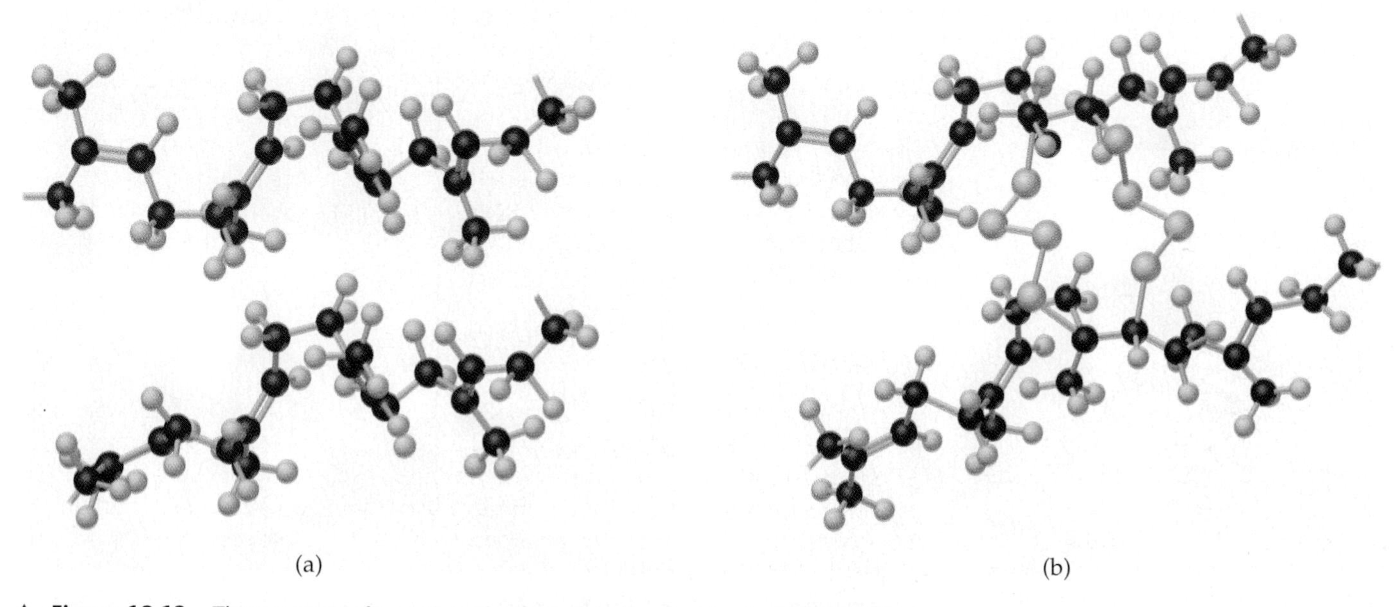

▲ **Figure 12.13** The structure of a section of polymeric natural rubber is shown in (a). There are carbon–carbon double bonds at regular intervals along the chain, as shown in Equation 12.4. (b) Chains of four sulfur atoms have been added across two polymer chains, by the opening of a carbon–carbon double bond on each chain.

Chemistry at Work Toward the Plastic Car

Many polymers can be formulated and processed to have sufficient structural strength, rigidity, and heat stability to displace metals, glass, and other materials in a variety of applications. The housings of electric motors and kitchen appliances such as coffee makers and can openers, for example, are now commonly formed from specially formulated polymers. *Engineering polymers* are tailored to particular applications through choice of polymers, blending of polymers, and modifications of processing steps. They generally have lower costs or superior performance over the materials they replace. In addition, shaping and coloring of the individual parts and their assembly to form the final product are often much easier.

The modern automobile provides many examples of the inroads engineering polymers have made in automobile design and construction. Car interiors have long been formed mainly of plastics. Through development of high-performance materials, significant progress has been made in introducing engineering polymers as engine components and car body parts. Figure 12.14 ◀, for example, shows the manifold in a series of Ford V-8 pickup and van engines. Use of an engineering polymer in this application eliminates machining and several assembly steps. The manifold, which is made of nylon, must be stable at high temperatures.

Car body parts can also be formed from engineering polymers. Components formed from engineering polymers usually weigh less than the components they replace, thus improving fuel economy. The fenders of the Volkswagen New Beetle (Figure 12.15 ◀), for example, are made of nylon reinforced with a second polymer, polyphenylene ether (ppe), which has the following structure:

$$-C_6H_4-O-\left[C_6H_4-O\right]_n-C_6H_4-O-$$

Because the polyphenylene ether polymer is linear and rather rigid, the ppe confers rigidity and shape retention.

A big advantage of most engineering polymers over metals is that they eliminate the need for costly corrosion protection steps in manufacture. In addition, some engineering polymer formulations permit manufacturing with the color built in, as it were, thus eliminating painting steps (Figure 12.16 ▼).

▲ **Figure 12.14** The intake manifold for a series of Ford Motor Company V-8 engines is formed from nylon.

▲ **Figure 12.15** The fenders of this New Beetle are made of General Electric Noryl GTX, a composite of nylon and polyphenylene ether.

▲ **Figure 12.16** This experimental car has body panels that are made of a polycarbonate-polybutylene plastic.

SAMPLE EXERCISE 12.2

If we assume that there are four sulfur atoms per cross-link connection, what mass of sulfur per gram of isoprene, C_5H_8, is required to establish a cross-link as illustrated in Figure 12.13 with 5% of the isoprene units in rubber?

Solution

Analyze: We are asked to calculate the mass of sulfur required per gram of isoprene.
Plan: We need to evaluate the ratio of sulfur atoms to isoprene units based on Figure 12.13, then scale the required mass of sulfur to take account of the 5% cross-linking.
Solve: We can see from the figure that each cross-link involves eight sulfur atoms for every two isoprene units; this means that the ratio of S to C_5H_8 is four. Thus, with 5% (0.05) of the isoprene units cross-linked, we have

$$(1.0 \text{ g } C_5H_8)\left(\frac{1 \text{ mol } C_5H_8}{68.1 \text{ g } C_5H_8}\right)\left(\frac{4 \text{ mol S}}{1 \text{ mol } C_5H_8}\right)\left(\frac{32.1 \text{ g S}}{1 \text{ mol S}}\right)(0.05) = 0.09 \text{ g S}$$

PRACTICE EXERCISE

How would you expect the properties of rubber to vary as the percentage of sulfur in the vulcanized product increases? Explain.
Answer: The rubber would be harder and less flexible as the percentage of sulfur increases, due to an increased degree of cross-linking, which covalently bonds the polymer chains together.

Cross-linking is also found in the polymer formed by melamine and formaldehyde, illustrated in Figure 12.17 ▼. Although it is not obvious from this two-dimensional drawing, the cross-linking extends in three dimensions, creating a hard, rigid, chemically stable material. This class of thermosetting polymers is found in products such as dinnerware, coatings, and Formica™ tabletops.

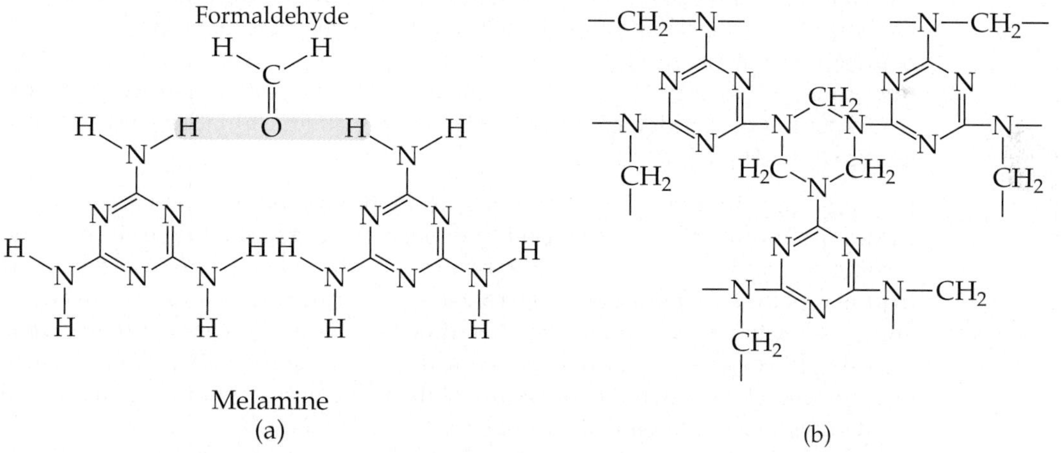

Formaldehyde

Melamine
(a)

(b)

▲ **Figure 12.17** (a) A condensation reaction between formaldehyde and two melamine molecules begins polymer formation. (b) The structure of the melamine-formaldehyde polymer, a highly cross-linked material.

12.3 Biomaterials

For our discussion here, a **biomaterial** is any material that has a biomedical application. The material might have a therapeutic use, for example, in the treatment of an injury or a disease. Or it might have a diagnostic use, as part of a system for identifying a disease or for monitoring a quantity such as the glucose level in blood. Whether the use is therapeutic or diagnostic, the biomaterial is in contact with biological fluids, and this material must have properties that meet the demands of that application. For example, a polymer employed to form a disposable contact lens must be soft and have an easily wetted surface, whereas the polymer used to fill a dental cavity must be hard and wear-resistant.

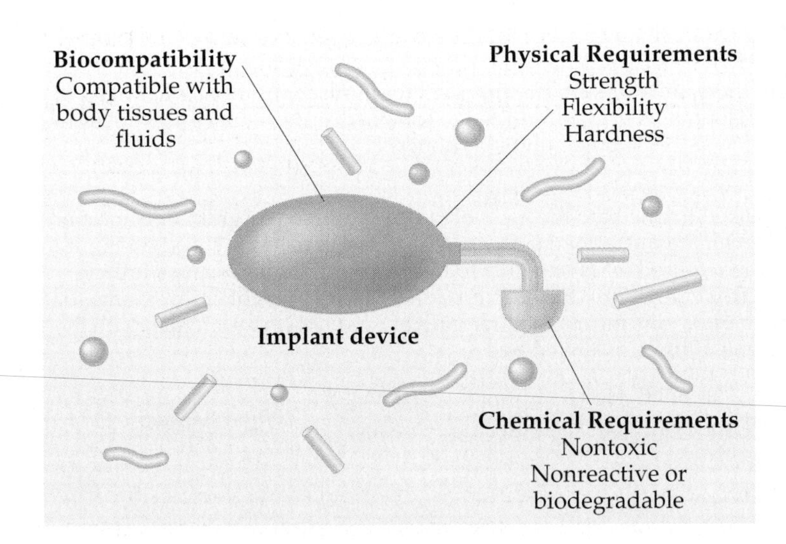

▶ **Figure 12.18** Schematic illustration of a human-made device implanted in a biological system. To function successfully, the device must be biocompatible with its surroundings and meet the necessary physical and chemical requirements, some of which are listed for illustrative purposes.

Characteristics of Biomaterials

The most important characteristics that influence the choice of a biomaterial are biocompatibility, physical requirements, and chemical requirements, as illustrated in Figure 12.18 ▲.

Biocompatibility Living systems, particularly the higher animals, have a complex set of protections against invasions of other organisms. Our bodies have an extraordinary ability to determine whether an object is the body's own material or whether it is foreign. Any substance that is foreign to the body has the potential to generate a response from our immune system. Molecular-sized objects are bound by antibodies and rejected, whereas larger objects induce an inflammatory reaction around them. Some materials are more **biocompatible**; that is, they are more readily integrated into the body without inflammatory reactions. The most important determining factors are the chemical nature and physical texture of an object's surface.

Physical Requirements A biomaterial is often required to meet very exacting demands. Tubing that is to be used to replace a defective blood vessel must be flexible and must not collapse under bending or other distortion. Materials used in joint replacements must be wear-resistant. An artificial heart valve must open and close 70 to 80 times a minute, day after day, for many years. If the valve is to have a life expectancy of twenty years, this means about 750 million cycles of opening and closing! Unlike the failure of the valve in a car engine, failure of the heart valve may have fatal consequences for its owner.

Chemical Requirements Biomaterials must be *medical grade*, which means that they must be approved for use in any particular medical application. Any ingredients present in a medical-grade biomaterial must remain innocuous over the lifetime of the application. Polymers are important biomaterials, but most commercial polymeric materials contain contaminants such as unreacted monomers, traces of the catalyst used to effect the polymerization, fillers or plasticizers, and antioxidants or other stabilizers. The small amounts of foreign materials present in a polymer used as a milk carton (Figure 12.11) pose no threat in that application, but they might if the same plastic material were implanted in the body over a long period.

Polymeric Biomaterials

The degree to which the body accepts the foreign polymer is determined by the nature of the atomic groups along the chain and the possibilities for interactions with the body's own molecules. Our bodies are composed largely of biopolymers such as proteins, polysaccharides (sugars), and polynucleotides (RNA, DNA). We will learn more about these molecules in Chapter 25. For now, we can simply note that the body's own biopolymers have complex structures, with polar groups along the polymer chain. Proteins, for example, are long strings of amino acids that have formed a condensation polymer. The protein chain has the following structure:

$$[-\underset{\mathrm{R}}{\underset{|}{\mathrm{CH}}}-\overset{\mathrm{O}}{\overset{\|}{\mathrm{C}}}-\underset{\mathrm{H}}{\underset{|}{\mathrm{N}}}-\underset{\mathrm{R}}{\underset{|}{\mathrm{CH}}}-\overset{\mathrm{O}}{\overset{\|}{\mathrm{C}}}-\underset{\mathrm{H}}{\underset{|}{\mathrm{N}}}-\underset{\mathrm{R}}{\underset{|}{\mathrm{CH}}}-\overset{\mathrm{O}}{\overset{\|}{\mathrm{C}}}-\underset{\mathrm{H}}{\underset{|}{\mathrm{N}}}-\underset{\mathrm{R}}{\underset{|}{\mathrm{CH}}}-\overset{\mathrm{O}}{\overset{\|}{\mathrm{C}}}-\mathrm{N}-]$$

where the R groups vary along the chain [$-CH_3$, $-CH(CH_3)_2$, and so on]. There are 20 different amino acids present in most proteins. By contrast, human-made polymers are simpler, being formed from a single repeating unit or perhaps two different repeating units, as described in Section 12.2. This difference in complexity is one of the reasons that synthetic polymers are identified by the body as foreign objects. Another reason is that there may be few or no polar groups along the chain that can interact with the body's aqueous medium. (Section 11.2)

We learned in Section 12.2 that polymers can be characterized by their physical properties. Elastomers are used as biomaterials in flexible tubing over leads for implanted heart pacemakers and as catheters (tubes implanted into the body to administer a drug or to drain fluids). Thermoplastics, such as polyethylene or polyesters, are employed as membranes in blood dialysis machines and as replacements for blood vessels. Thermoset plastics find limited but important uses. Because they are hard, inflexible, and somewhat brittle, they are most often used in dental devices or in orthopedic applications, such as in joint replacements. To fill a cavity, for example, the dentist may pack some material into the cavity, then shine an ultraviolet lamp on it. The light initiates a photochemical reaction that forms a hard, highly cross-linked thermoset polymer.

Examples of Biomaterial Applications

We can best appreciate the kinds of problems encountered in using biomaterials by considering a few specific cases.

Heart Replacement and Repairs The term *cardiovascular* pertains to the heart, blood, and blood vessels. The heart is, of course, an absolutely essential organ. A heart that fails completely may be replaced by a donor organ. About 60,000 people suffer terminal heart failure each year in the United States, yet only about 2500 donor hearts become available for transplant. Many attempts have been made—and continue to be made—to produce an artificial heart that can serve over a long period of time as a replacement for the natural organ. We will not devote attention to these efforts, except to note that recent results are quite promising.

It often happens that only a part of the heart, such as the aortic valve, fails and needs replacement. Repair could be made by using foreign tissue (for example, a pig heart valve) or a mechanical heart valve implant to replace a diseased one. About 250,000 valve replacement procedures are performed annually worldwide. In the United States about 45% of the procedures involve a mechanical valve. The most widely used valve is shown in Figure 12.19 ▶. It has two semicircular discs that move to allow blood to flow in the desired direction as the heart pumps, then they fall back together to form a seal against backflow.

▲ **Figure 12.19** A bileaflet disc heart valve, known as the St. Jude valve, which was named for the medical center at which it was developed. The surfaces of the valve are coated with pyrolytic carbon. The valve is secured to the surrounding tissue via a Dacron™ sewing ring. SJM is a registered trademark of St. Jude Medical, Inc.

It is vital to minimize fluid disturbance as the blood passes through artificial devices. Surface roughness in a device causes *hemolysis*, the breakdown of red cells. Furthermore, surface roughness can serve as a site for invading bacteria to adhere and colonize. Finally, rough surfaces also promote coagulation of the blood, which forms a thrombus, or blood clot. Thus, even though we may have a perfectly fine piece of machinery from a mechanical point of view, the heart valve may not be suitable as a long-term implant. To minimize blood clots, the discs in the heart valve must have a smooth, chemically inert surface.

A second challenge in the use of a heart valve implant is to fix it in place. As shown in Figure 12.19, the retainer ring that forms the body of the valve is covered with a mesh of woven fabric. The material of choice is Dacron™, du Pont's trade name for the fiber formed from polyethylene terephthalate (Table 12.1). The mesh acts as a lattice on which the body's tissues can grow, so that the valve becomes incorporated into its surroundings. Tissue grows through the polyester mesh, whereas it does not do so on many other plastics. Apparently the polar, oxygen-containing functional groups along the polyester chain afford attractive interactions to facilitate tissue growth.

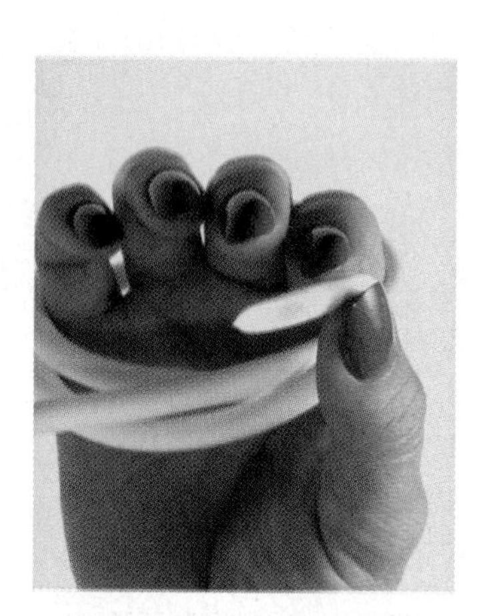

▲ **Figure 12.20** A Dacron™ vascular graft.

Vascular Grafts A vascular graft is a replacement for a segment of diseased arteries. Where possible, diseased blood vessels are replaced with vessels taken from the patient's own body. When this is not feasible, artificial materials must be used. Dacron™ is used as replacement for large-diameter arteries around the heart. For this purpose, it is fabricated into a crimped, woven, tubular form, as shown in Figure 12.20 ◀. The tubing is crimped to enable it to bend without serious decrease in its cross-sectional area. The graft must integrate with surrounding tissue after it has been put in place. It must therefore have an open structure, with pores on the order of 10 μm in diameter. During the healing process blood capillaries grow into the graft, and new tissues form throughout it. Similarly, polytetrafluoroethylene $[—(CF_2CF_2)_n—]$ is used for the smaller diameter vascular grafts in the limbs.

Ideally, the inside surface of the graft would become lined with the same sorts of cells that line the native artery, but this does not occur with materials currently available. Instead, the inside surface of the tubing is recognized as being foreign to the blood. Platelets, which are circulating components of the blood, normally serve the function of sealing up wounds in the blood vessel walls. Unfortunately, they attach to foreign surfaces and cause blood coagulation. The search for a more biocompatible lining for grafts is an area of active research, but at present there is a continuing risk of blood clots. Excess tissue growth at the intersection of the graft with the native artery is also a frequent problem. Because of the possibility of blood clot formation, patients who have received artificial heart valves or vascular grafts are generally required to take anticoagulation drugs on a continuing basis.

Artificial Tissue The treatment of patients who have lost extensive skin tissue—for example, burn patients or those with skin ulcers—is one of the most difficult problems in therapeutic medicine. Today, laboratory-grown skin can be employed to replace grafts in such patients. Ideally, the "artificial" tissue would be grown from cells taken from the patient. When this is not possible, for example, with burn victims, the tissue cells come from another source. When the graft skin is not formed from the patient's own cells, drugs that suppress the body's natural immune defense system must be used, or steps must be taken to modify the new cell line to prevent rejection of the tissue.

The challenge in growing artificial tissue is to get the cells to organize themselves as they would in a living system. The first step in accomplishing this objective is to provide a suitable scaffold for the cells to grow on, one that will keep them in contact with each other and allow them to organize. Such a scaffold must be biocompatible; cells must adhere to the scaffold and differentiate (that is, develop into cells of different types) as the culture grows. The scaffolding must also be mechanically strong and biodegradable.

The most successful scaffolds have been lactic acid—glycolic acid copolymers. Formation of the copolymer via a condensation reaction is shown in Equation 12.5:

$$n\mathrm{HOCH_2\overset{O}{\overset{\|}{C}}{-}OH} + n\mathrm{HO\underset{CH_3}{\underset{|}{C}}H\overset{O}{\overset{\|}{C}}{-}OH} \longrightarrow \mathrm{{-}O{-}CH_2\overset{O}{\overset{\|}{C}}{-}O{+}\underset{CH_3}{\underset{|}{C}}H\overset{O}{\overset{\|}{C}}{-}O{-}CH_2\overset{O}{\overset{\|}{C}}{-}O{+}_n\underset{CH_3}{\underset{|}{C}}H\overset{O}{\overset{\|}{C}}{-}} \quad [12.5]$$

Glycolic acid Lactic acid Copolymer

The copolymer has an abundance of polar carbon–oxygen bonds along the chain, affording many opportunities for hydrogen-bonding interactions. The ester linkages formed in the condensation reaction are susceptible to hydrolysis, which is just the reverse reaction. When the artificial tissue is deployed in the body, the underlying copolymer scaffold slowly hydrolyzes away as the tissue cells continue to develop and merge with adjacent tissue. An example of a graft skin product is shown in Figure 12.21 ▶.

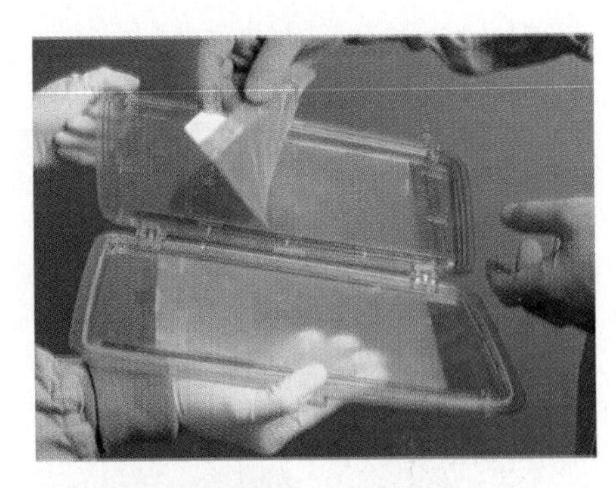

▲ **Figure 12.21** Artificial skin prepared for use as a skin graft.

Hip Replacements More than 750,000 surgeries involving replacement of natural joints with artificial joints are carried out each year. About 200,000 of these are total hip-joint replacements. Most of these replacements are driven by debilitating arthritis. A hip-joint replacement is designed to provide joint mobility and structural support. It must be stable under considerable load, have high fatigue resistance, abrasion resistance, and biocompatibility.

Figure 12.22 ▶ shows the components of a typical modern hip-joint replacement. It includes a metal ball, formed from a hard, corrosion-resistant metal alloy, usually cobalt-chromium. The highly polished ball is attached to a titanium alloy stem that fits into the femur, which has been cut flat and drilled. This lower part of the artificial joint can be held in place in the femur by cementing it, using a cement that forms a hard, tough thermoset polymer. Alternatively, the component fitting into the femur can be coated with a porous layer that promotes bone growth and integration of the implant into the host bone. The disadvantage of the latter approach is that it takes several weeks for bone growth to occur to an appreciable extent, during which time no weight can be applied to the joint. The advantage of this technique compared with cementing the replacement in place is that it leaves more of the femur intact and may have a longer life expectancy.

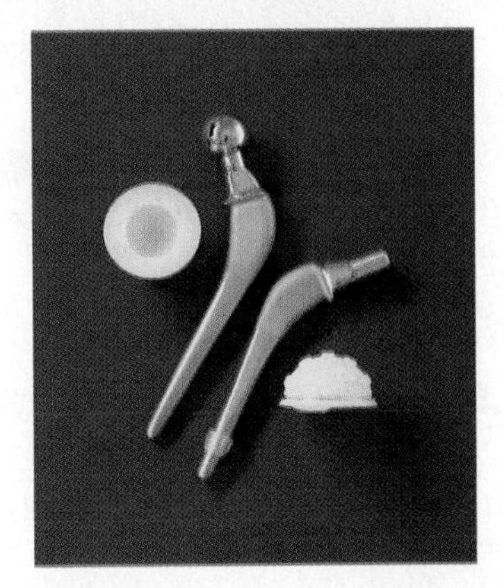

▲ **Figure 12.22** The components of a modern hip-joint replacement. The metal ball is composed of a corrosion-resistant alloy. The cup, which receives the ball, is lined with high molecular weight polyethylene.

The part of the pelvis that accommodates the head of the femur is known as the *acetabulum*. In the artificial joint this component consists of a cup lined with tough, ultrahigh molecular weight polyethylene (Figure 12.22). This material is designed to maintain its shape over a long time and under various loads. Abrasion at the interface between cup and ball must be minimized because any small particles formed during wear can stimulate an inflammatory response. Teflon™ was used in earlier versions of the cup, but it was found not to have adequate abrasion resistance.

12.4 Ceramics

Ceramics are inorganic, nonmetallic, solid materials. They can be crystalline or noncrystalline. Noncrystalline ceramics include glass and a few other materials with amorphous structures. Ceramics can possess a covalent-network structure, ionic bonding, or some combination of the two. ∞ (Section 11.8, Table 11.6) They are normally hard and brittle and are stable to very high temperatures. Ceramic materials include familiar objects such as pottery, china, cement, roof tiles, refractory bricks used in furnaces, and the insulators in spark plugs.

TABLE 12.4 Properties of Some Ceramic and Selected Nonceramic Materials

Material	Melting Point (°C)	Density (g/cm^3)	Hardness (Mohs)[a]	Modulus of Elasticity[b]	Coefficient of Thermal Expansion[c]
Alumina, Al_2O_3	2050	3.8	9	34	8.1
Silicon carbide, SiC	2800	3.2	9	65	4.3
Zirconia, ZrO_2	2660	5.6	8	24	6.6
Beryllia, BeO	2550	3.0	9	40	10.4
Mild steel	1370	7.9	5	17	15
Aluminum	660	2.7	3	7	24

[a] The Mohs scale is a logarithmic scale based on the relative ability of a material to scratch another softer material. Diamond, the hardest material, is assigned a value of 10.

[b] A measure of the stiffness of a material when subjected to a load (MPa $\times 10^4$). The larger the number, the stiffer the material.

[c] In units of ($K^{-1} \times 10^{-6}$). The larger the number, the greater the size change upon heating or cooling.

▲ **Figure 12.23** A variety of ceramic parts made from silicon nitride, Si_3N_4. These ceramic components can replace metal parts in engines or can be used in other applications where high temperatures and wear are involved. (Used with permission from Kyocera Industrial Ceramics Corporation.)

3-D MODEL
Silicon Carbide

Ceramic materials come in a variety of chemical forms, including *silicates* (silica, SiO_2, with metal oxides), *oxides* (oxygen and metals), *carbides* (carbon and metals), *nitrides* (nitrogen and metals), and *aluminates* (alumina, Al_2O_3, with metal oxides). Although most ceramic materials contain metal ions, some do not. Table 12.4 ▲ lists a few ceramic materials and contrasts their properties with those of two common metals.

Ceramics are highly resistant to heat, corrosion, and wear, do not readily deform under stress, and are less dense than the metals used for high-temperature applications. Some ceramics used in aircraft, missiles, and spacecraft weigh only 40% as much as the metal components they replace (Figure 12.23 ◀). In spite of these many advantages, the use of ceramics as engineering materials has been limited because they are extremely brittle. Whereas a metal component might suffer a dent when struck, a ceramic part typically shatters because the bonding prevents the atoms from sliding over one another. Ceramic components are also difficult to manufacture free of defects. Indeed, high fabrication costs and uncertain component reliability are barriers that must be overcome before ceramics are more widely used to replace metals and other structural materials. Attention has therefore focused in recent years on the processing of ceramic materials, as well as on the formation of composite ceramic materials and the development of thin ceramic coatings on conventional materials.

Processing of Ceramics

Ceramic parts often develop random, undetectable microcracks and voids (hollow spaces) during processing. These defects are more susceptible to stress than the rest of the ceramic; thus, they are generally the origin of cracking and fractures. To "toughen" a ceramic—to increase its resistance to fracture—scientists frequently produce very pure uniform particles of the ceramic material that are less than a μm (10^{-6} m) in diameter. These are then *sintered* (heated at high temperature under pressure so that the individual particles bond together) to form the desired object.

The **sol-gel process** is an important method of forming extremely fine particles of uniform size. A typical sol-gel procedure begins with a metal alkoxide. An alkoxide contains organic groups bonded to a metal atom through oxygen atoms. Alkoxides are produced when the metal reacts with an alcohol, which is an organic compound containing an OH group bonded to carbon. To illustrate this process, we will use titanium as the metal and ethanol, CH_3CH_2OH, as the alcohol.

$$Ti(s) + 4CH_3CH_2OH(l) \longrightarrow Ti(OCH_2CH_3)_4(s) + 2H_2(g) \quad [12.6]$$

The alkoxide product, $Ti(OCH_2CH_3)_4$, is dissolved in an appropriate alcohol solvent. Water is then added, and it reacts with the alkoxide to form Ti—OH groups and to regenerate ethanol.

$$Ti(OCH_2CH_3)_4(soln) + 4H_2O(l) \longrightarrow Ti(OH)_4(s) + 4CH_3CH_2OH(l) \quad [12.7]$$

The reaction with ethanol is used, even though the ethanol is simply regenerated, because the direct reaction of $Ti(s)$ with $H_2O(l)$ leads to a complex mixture of titanium oxides and titanium hydroxides. The intermediate formation of $Ti(OC_2H_5)_4(s)$ ensures that a uniform suspension of $Ti(OH)_4$ will be formed. The $Ti(OH)_4$ is present at this stage as a *sol*, a suspension of extremely small particles. The acidity or basicity of the sol is adjusted to split out water from between two of the Ti—OH bonds.

$$(HO)_3Ti{-}O{-}H(s) + H{-}O{-}Ti(OH)_3(s) \longrightarrow (HO)_3Ti{-}O{-}Ti(OH)_3(s) + H_2O(l) \quad [12.8]$$

This is another example of a condensation reaction. (Section 12.2) Condensation also occurs at some of the other OH groups bonded to the central titanium atom, producing a three-dimensional network. The resultant material, called a *gel*, is a suspension of extremely small particles with the consistency of gelatin. When this material is heated carefully at 200°C to 500°C, all the liquid is removed, and the gel is converted to a finely divided metal oxide powder with particles in the range of 0.003 to 0.1 μm in diameter. Figure 12.24 ▶ shows SiO_2 particles, formed into remarkably uniform spheres by a precipitation process similar to the sol-gel process.

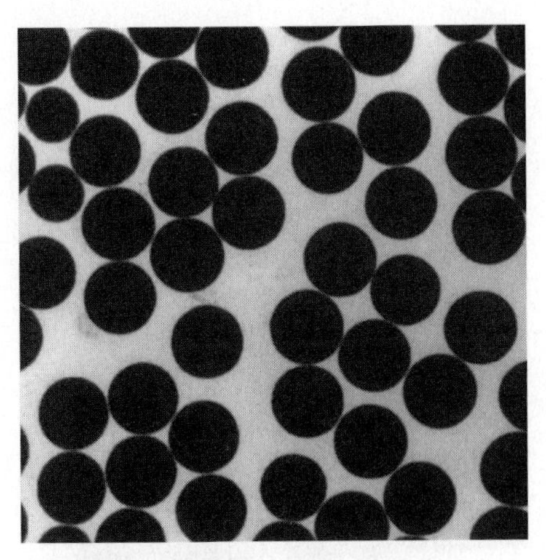

▲ **Figure 12.24** Uniformly sized spheres of amorphous silica, SiO_2, formed by precipitation from a methanol solution of $Si(OCH_3)_4$ upon addition of water and ammonia. The average diameter is 550 nm.

To form a ceramic object with a complex three-dimensional shape, the finely divided ceramic powder, possibly mixed with other powders, is compacted under pressure and then sintered at high temperature. The temperatures required are about 1650°C for alumina, 1700°C for zirconium oxide, and 2050°C for silicon carbide. During sintering the ceramic particles coalesce without actually melting (compare the sintering temperatures with the melting points listed in Table 12.4).

Ceramic Composites

Ceramic objects are much tougher when they are formed from a composite, a complex mixture of two or more materials. The most effective composites are formed by addition of *ceramic fibers* to a ceramic material. Thus, the composite consists of a ceramic matrix containing embedded fibers of a ceramic material, which may or may not be of the same chemical composition as the matrix. By definition, a fiber has a length at least 100 times its diameter. Fibers typically have great strength with respect to loads applied along the long axis. When they are embedded in a matrix, they strengthen it by resisting deformations that exert a stress on the fiber along its long axis.

The formation of ceramic fibers is illustrated by silicon carbide (SiC), or carborundum. The first step in the production of SiC fibers is the synthesis of a polymer, polydimethylsilane.

$$-\underset{CH_3}{\overset{CH_3}{Si}}-\left[\underset{CH_3}{\overset{CH_3}{Si}}\right]_n-\underset{CH_3}{\overset{CH_3}{Si}}-$$

When this polymer is heated to about 400°C, it converts to a material that has alternating carbon and silicon atoms along the chain.

$$-\overset{\displaystyle H}{\underset{\displaystyle CH_3}{Si}}-\left[-CH_2-\overset{\displaystyle H}{\underset{\displaystyle CH_3}{Si}}-\right]_n-CH_2-\overset{\displaystyle H}{\underset{\displaystyle CH_3}{Si}}-$$

Fibers formed from this polymer are then heated slowly to about 1200°C in a nitrogen atmosphere to drive off all the hydrogen and all carbon atoms other than those that directly link the silicon atoms. The final product is a ceramic material of composition SiC, in the form of fibers ranging in diameter from 10 to 15 μm. By similar procedures, beginning with an appropriate organic polymer, ceramic fibers of other compositions such as boron nitride (BN) can be fabricated. When the ceramic fibers are added to a ceramic material that is then processed at high temperature, the resulting product has a much higher resistance to catastrophic crack failure.

Applications of Ceramics

Ceramics, particularly new ceramic composites, are widely used in the cutting-tool industry. For example, alumina reinforced with silicon carbide whiskers (extremely fine fibers) is used to cut and machine cast iron and harder nickel-based alloys. Ceramic materials are also used in grinding wheels and as abrasives because of their exceptional hardness (Table 12.4). Silicon carbide is the most widely used abrasive.

Ceramic materials play an important role in the electronics industry. Semiconductor integrated circuits are typically mounted on a ceramic substrate, usually alumina. Some ceramics, notably quartz (crystalline SiO_2), are *piezoelectric*, which means that they generate an electrical potential when subjected to mechanical stress. This property enables us to use piezoelectric materials to control frequencies in electronic circuits, as in quartz watches and ultrasonic generators.

▲ **Figure 12.25** A worker applying thermally insulating ceramic tiles to the body of the Space Shuttle orbiter.

Ceramic materials are used in the manufacture of ceramic tiles for the surfaces of the Space Shuttle, to protect against overheating on reentry into Earth's atmosphere (Figure 12.25 ◀). The tiles are made of short, high-purity silica fibers reinforced with aluminum borosilicate fibers. The material is formed into blocks, sintered at over 1300°C, and then cut into tiles. The tiles have a density of only 0.2 g/cm^3, yet they are able to keep the shuttle's aluminum skin below 180°C while sustaining a surface temperature of up to 1250°C.

12.5 Superconductivity

In 1911 the Dutch physicist H. Kamerlingh Onnes discovered that when mercury is cooled below 4.2 K, it loses all resistance to the flow of an electrical current. Since that discovery, scientists have found that many substances exhibit this "frictionless" flow of electrons. This property has become known as **superconductivity.** Substances that exhibit superconductivity do so only when cooled below a particular temperature, called the **superconducting transition temperature, T_c**. The observed values of T_c are generally very low. Table 12.5 ▶ lists several major discoveries of superconducting materials. Some are notable for their relatively high value of T_c, others for the fact that a material of that composition could be superconducting at all.

TABLE 12.5 Superconducting Materials: Dates of Discovery and Transition Temperatures

Substance	Discovery Date	T_c (K)
Hg	1911	4.0
Nb_3Sn	1954	18.0
$SrTiO_3$	1966	0.3
Nb_3Ge	1973	22.3
$BaPb_{1-x}Bi_xO_3$	1975	13.0
$La(Ba)_2CuO_4$	1986	35.0
$YBa_2Cu_3O_7$	1987	95.0
$BiSrCaCu_2O_x$	1988	100.0
$Tl_2Ba_2Ca_2Cu_3O_{10}$	1988	125.0
$HgBa_2Ca_2Cu_3O_{8+x}$	1993	133.0
Cs_3C_{60}	1995	40
MgB_2	2001	39

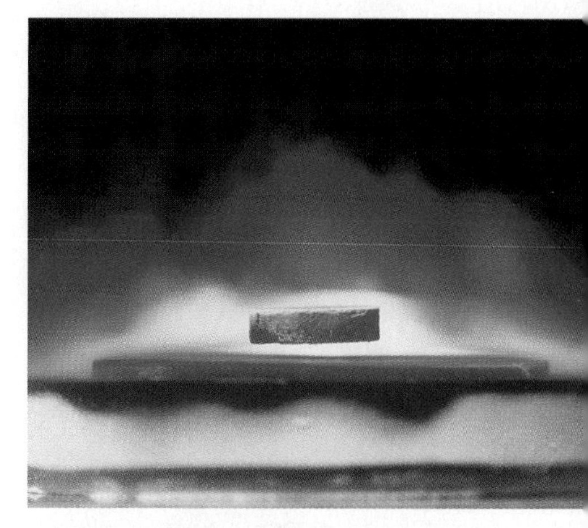

▲ **Figure 12.26** A small permanent magnet is levitated by its interaction with a ceramic superconductor that is cooled to liquid nitrogen temperature, 77 K. The magnet floats in space because the superconductor excludes magnetic field lines, a property known as the Meissner effect.

Superconductivity has tremendous economic potential. If electrical power lines or the conductors in a variety of electrical devices were capable of conducting current without resistance, enormous amounts of energy could be saved. Further, many devices that are not now feasible, including smaller, faster computer chips, could be built. In addition, superconducting materials exhibit a property, called the *Meissner effect* (Figure 12.26 ▶), in which they exclude all magnetic fields from their volume. The Meissner effect could potentially be used to build magnetically levitated high-speed trains. Because superconductivity appears in most materials only at very low temperatures, however, applications of the phenomenon to date have been limited. One important use is in the windings of the large magnets that form the fields needed in magnetic resonance imaging (MRI) instruments, used in medical imaging (Figure 12.27 ▶). The magnet windings, typically formed from Nb_3Sn, must be kept cooled with liquid helium, which boils at about 4 K. The cost of liquid helium is a significant factor in the cost of using MRI.

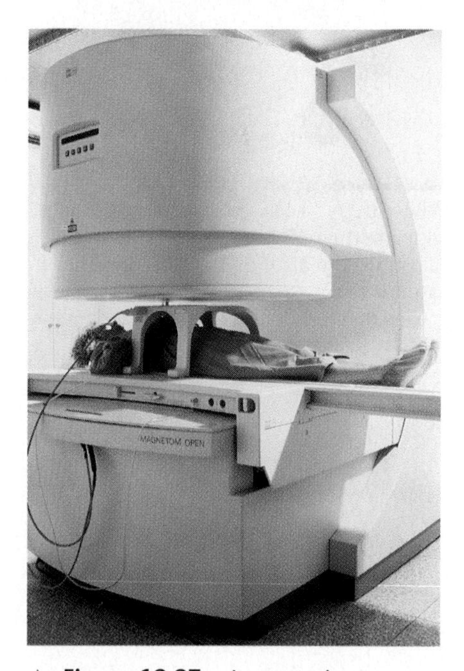

▲ **Figure 12.27** A magnetic resonance imaging (MRI) instrument used in medical diagnosis. The magnetic field needed for the procedure is generated by current flowing in superconducting wires, which must be kept below their superconducting transition temperature, T_c, of 18 K. This requires liquid He as a coolant.

Superconducting Ceramic Oxides

Before the 1980s the highest value that had been observed for T_c was about 23 K for a niobium-germanium compound (Table 12.5). In 1986, however, J. G. Bednorz and K. A. Müller, working at the IBM research laboratories in Zürich, Switzerland, discovered superconductivity above 30 K in a ceramic oxide containing lanthanum, barium, and copper. This material represents the first **superconducting ceramic**. That discovery, for which Bednorz and Müller received the Nobel Prize in 1987, set off a flurry of research activity all over the world. Before the end of 1986 scientists had verified the onset of superconductivity at 95 K in yttrium-barium-copper oxide, $YBa_2Cu_3O_7$. The highest temperature observed to date for onset of zero resistance at 1 atm pressure is 133 K, which was achieved in another complex copper oxide, $HgBa_2Ca_2Cu_3O_{8+x}$, where x represents a slight excess of oxygen.

The discovery of so-called **high-temperature** (high-T_c) **superconductivity** is of great significance. Many applications of superconductivity will become feasible only with the development of usable high-temperature superconductors, because the cost of maintaining extremely low temperatures is very high. The only readily available safe coolant at temperatures below 77 K is liquid helium, which costs about \$2.50 per liter. For materials that undergo the superconducting transition at temperatures well above 77 K, however, liquid nitrogen, which costs only about \$0.05 per liter, can be used. Alternatively, mechanical cooling devices might be feasible in some applications. One of the most widely studied ceramic superconductors is $YBa_2Cu_3O_7$, whose structure is shown in Figure 12.28 ◀. The unit cell is defined by the lines; a few oxygen atoms that lie outside the unit cell are also shown to illustrate the arrangement of oxygens about each copper atom. Extensive work on modifying this and related copper oxide superconductors by introducing other atoms, called dopants, in various atomic positions indicates that the conductivity and superconductivity take place in the copper-oxygen planes. At temperatures above T_c the electrical conductivity parallel to the copper-oxygen planes is 10^4 times greater than in the perpendicular direction. The Cu^{2+} ions have an $[Ar]3d^9$ electron configuration with a single electron in the $3d_{x^2-y^2}$ orbital. Although the mechanism of conduction and superconduction is not yet well understood, it is believed to be important that the lobes of the $3d_{x^2-y^2}$ orbital point toward the neighboring O^{2-} ions.

The new superconducting ceramic materials have immense promise, but a great deal of research is needed before they can be widely applied on a practical basis. At present it is difficult to mold ceramics, which are brittle materials, into useful shapes like tapes or wires on a large scale. In addition, the attainable current densities (the current that can be carried by a wire of a certain cross-sectional area) are still not high enough for many applications. A related problem is the tendency of ceramics to interact with their environment, particularly with water and carbon dioxide. For example, the reaction of $YBa_2Cu_3O_7$ with atmospheric water liberates O_2 and forms $Ba(OH)_2$, $Y_2BaCu_3O_5$, and CuO. Because these materials are so reactive, they must be protected against long-term exposure to the atmosphere. In spite of these limitations, high-temperature superconducting materials have found applications in a few situations already. (See "Chemistry at Work" box.)

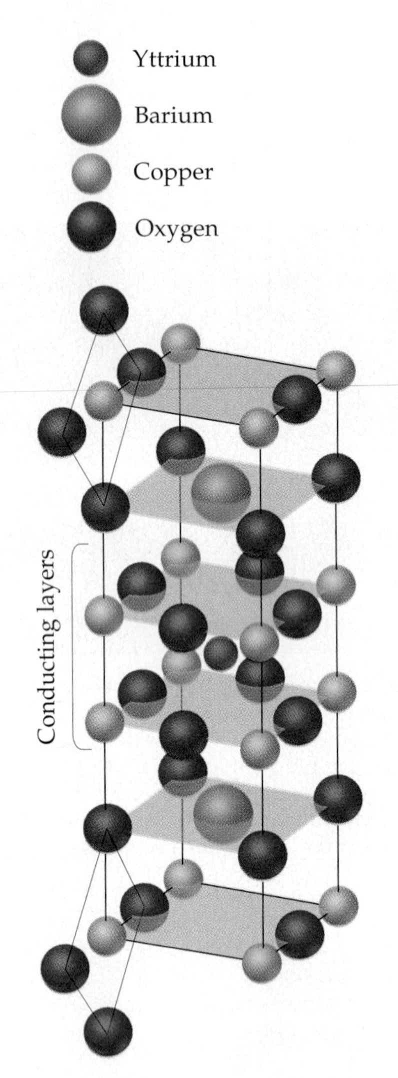

▲ **Figure 12.28** Unit cell of $YBa_2Cu_3O_7$. A few oxygen atoms that fall outside the unit cell are also shown to illustrate the arrangement of oxygen atoms about each copper atom. The unit cell is defined by the lines that describe a rectangular box.

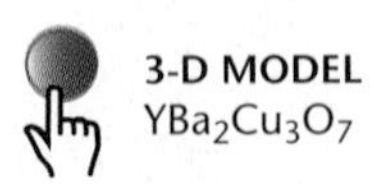

3-D MODEL
$YBa_2Cu_3O_7$

New Superconductors

It is still not clear what makes any particular material a superconductor. Superconductivity in metals and metal alloys, such as Nb_3Sn, is pretty well accounted for by BCS theory, named after its inventors, John Bardeen, Leon Cooper, and Robert Schrieffer. However, after years of research there is still no satisfactory theory of superconductivity in ceramic materials. Because it seems that superconductivity might arise in many different kinds of materials, much empirical research is devoted to the search for new classes of superconductors. As noted in Table 12.5, it was discovered only recently that C_{60} (see "Closer Look" box on page 439) reacted with an alkali metal, which converts it into an electrically conducting material, exhibits a superconducting transition at temperatures up to about 40 K. Even more recently, the simple binary compound magnesium diboride, MgB_2, has been found to become superconducting at 39 K. This is a very surprising and potentially quite important result. MgB_2, which is a conductor of electricity somewhat like graphite, is a relatively cheap material. Other, related compounds in the same family might prove to have higher superconducting transitions. The field of superconductivity has a great deal of promise, but scientists estimate that new discoveries will not be translated into major practical applications for several years. In time, however, new superconductors will likely become part of our everyday lives.

Chemistry at Work Cell Phone Tower Range

Cell phone towers increasingly dot both the rural and urban landscapes (Figure 12.29 ▼), yet it can be difficult to maintain contact with a tower in the midst of a telephone conversation. The cell phone communicates with the system by receiving a signal from the tower's transmitter and transmitting signals back to it. Although the tower's transmitter may be quite powerful, the cell phone has very limited power. As the distance from the tower increases or intervening structures interfere, the cell phone signal becomes too weak to be discerned in the midst of general electronic noise.

The amplifiers in the tower's receiver have electronic filters that discriminate between the desired incoming signal and other electronic signals. The sharper the filter, the higher the discrimination between one channel and another, and therefore, the higher the capability of detecting the desired signal with clarity. Filters can be made from a high-temperature superconducting oxide that, when cooled below T_c, provide much sharper filtering than conventional filters. By incorporating such filters in the receivers at cell phone towers, the range of the tower can be extended up to a factor of two, which saves construction costs and improves the reliability of communication.

Superconducting technology is now being deployed in PC-sized boxes located in cell phone base stations (the little building at the foot of the tower). The filters are formed from a ceramic oxide, typically $YBa_2Cu_3O_7$ or $Tl_2Ba_2CaCu_2O_8$. The requisite cooling is provided by a mechanical cooling device, basically a small refrigeration unit capable of cooling the filter below its T_c (Figure 12.30 ▼)

◀ **Figure 12.29** Wireless communications tower.

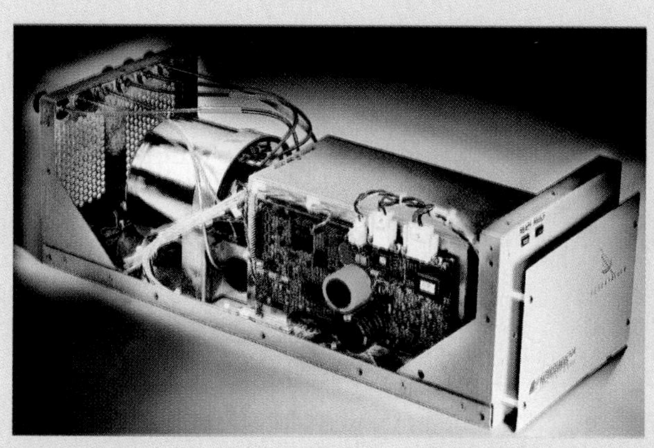

▲ **Figure 12.30** A view of the cryogenic receiver system employing a superconducting filter and a cryo-cooled low noise amplifer (LNA). The cylindrical object to the left is the cryogenic refrigerator used to keep the filter and LNA at a temperature below their T_c values.

12.6 Thin Films

Thin films were first used for decorative purposes. In the seventh century, artists learned how to paint a pattern on a ceramic object with a silver salt solution and then heat the painted object to decompose the salt, leaving a thin film of metallic silver. Thin films are used today for decorative or protective purposes: to form conductors, resistors, and other types of films in microelectronic circuits; to form photovoltaic devices for conversion of solar energy to electricity; and for many other applications (Figure 12.31 ▶). A thin film might be made of any kind of material, including metals, metal oxides, or organic substances.

The term **thin film** does not have a precise definition. In general, it refers to films with thickness ranging from 0.1 μm to about 300 μm. It does not normally refer to coatings such as paint and varnish, which are typically much thicker. For a thin film to be useful, it should possess all or most of the following properties: (a) It should be chemically stable in the environment in which it is to be used; (b) it should adhere well to the surface it covers (the substrate); (c) it should have a uniform thickness; (d) it should be chemically pure or of controlled chemical composition; and (e) it should have a low density of imperfections. In addition to these general characteristics, special properties might be required for certain applications. The film might need to be an insulator or a semiconductor, for example, or to possess special optical or magnetic properties.

▲ **Figure 12.31** The glass panels forming the outer walls of this building have a thin film of metal, which reflects a significant fraction of the outdoor light. The reflective glass provides privacy, reduces interior glare, and reduces the cooling load on the building in hot weather.

A thin film must adhere to its underlying substrate if it is to perform usefully. Because the film is inherently fragile, it must depend on the substrate for structural support. To attain that support, the film must be bound to the substrate by strong forces. The bonding forces may be chemical; that is, a chemical reaction at the interface can connect the film to the underlying material. When a metal oxide is deposited on glass, for example, the oxide lattices of the metal oxide and the glass blend at the interface, forming a thin zone of intermediate composition. In these cases the bonding energies between the film and the substrate are of the same magnitudes as chemical bonds, in the range of 250 to 400 kJ/mol. In some cases, however, the bonding between the film and the substrate is based solely on intermolecular van der Waals and electrostatic forces, as might be the case when an organic polymer film is deposited on a metal surface. The energies that bind the film to the substrate in such cases might be in the range of only 50 to 100 kJ/mol, and the films are not as robust.

Uses of Thin Films

▲ **Figure 12.32** The lenses of these binoculars are coated with a ceramic thin film to reduce reflection and to protect the softer glass against scratching.

Thin films are used in microelectronics as conductors, resistors, and capacitors. They are widely used as optical coatings on lenses (Figure 12.32 ◀) to reduce the amount of light reflected from the lens surface and to protect the lens. Thin metallic films have been used for a long time as protective coatings on metals. They are usually deposited from solutions by the use of electrical currents, as in silver plating and "chrome" plating. (We will defer discussion of electrochemical methods for forming films until Chapter 20.) Metal tool surfaces are coated with ceramic thin films to increase their hardness. Although it is not evident to the consumer, nearly every glass bottle purchased is coated with one or more thin films. The films are applied to the glass to reduce scratching and abrasion and to increase lubricity, that is, the ease with which bottles can slide by one another. The most common thin film for this application is tin(IV) oxide, SnO_2.

Formation of Thin Films

Thin films are formed by a variety of techniques, including vacuum deposition, sputtering, and chemical-vapor deposition.

Vacuum deposition is used to form thin films of substances that can be vaporized or evaporated without destroying their chemical identities. These substances include metals, metal alloys, and simple inorganic compounds such as oxides, sulfides, fluorides, and chlorides. For example, optical lenses are coated with inorganic materials such as MgF_2, Al_2O_3, and SiO_2. The material to be deposited as a thin film is heated—either electrically or by electron bombardment—in a high-vacuum chamber with a pressure of 10^{-5} torr or less. The vaporized molecules travel in a line of sight path to the point of deposition. To obtain a film of uniform thickness, all parts of the surface to be coated must be equally accessible to the vapor phase from which the thin-film material is deposited. Sometimes this uniformity is obtained by rotating the piece to be coated.

Sputtering involves the use of a high voltage to remove material from a source, or target. Atoms removed from the target are carried through the ionized gas within the chamber and deposited on the substrate. The target surface is the negative electrode, or cathode, in the circuit; the substrate may be attached to the positive electrode, or anode. Figure 12.33 ▶ depicts this process. The chamber contains an inert gas such as argon that is ionized in the high-voltage field. The positively charged ions are accelerated toward the target surface, which they strike with sufficient energy to dislodge atoms of the target material. Many of these atoms are accelerated toward the substrate surface. On striking it, they form a thin film.

The sputtered atoms have a lot of energy. The initial atoms striking the surface may penetrate several atomic layers into the substrate, which helps to ensure

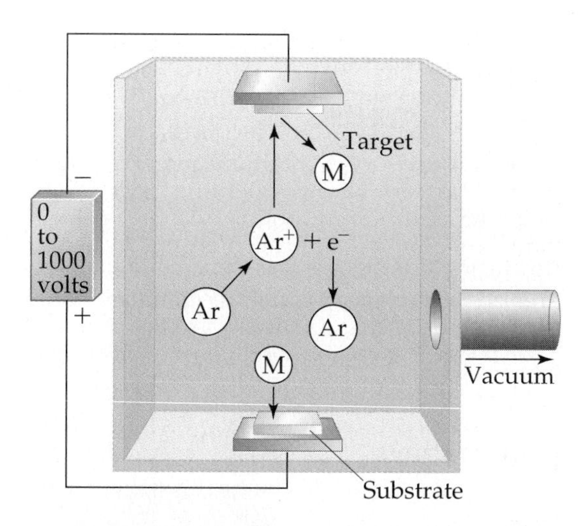

◀ **Figure 12.33** Schematic illustration of a sputtering apparatus. The high voltage applied across the gas results in ionization of Ar atoms. The Ar^+ ions are accelerated toward the negatively charged target. Upon impact they knock atoms of M from the surface. The M atoms travel in all directions with high kinetic energy, and some of them strike the substrate, forming a coating.

good adhesion of the thin-film layer to the substrate. An additional advantage of sputtering is that it is possible to change the target material from which the sputtered atoms arise without disturbing the system, so that multilayer thin films can be formed.

Sputtering is widely used to form thin films of such elements as silicon, titanium, niobium, tungsten, aluminum, gold, and silver. It is also employed to form thin films of refractory materials such as carbides, borides, and nitrides on metal tool surfaces, to form soft lubricating films such as molybdenum disulfide and to apply energy-conserving coatings on architectural glass (Figure 12.31).

In **chemical-vapor deposition**, the surface is coated with a volatile, stable chemical compound at a temperature below the melting point of the surface. The compound then undergoes some form of chemical reaction to form a stable, adherent coat. For example, titanium tetrabromide is evaporated and the gaseous $TiBr_4$ is mixed with hydrogen. The mixture is then passed over a surface heated to about 1300°C. The heated substrate is ordinarily a ceramic such as silica or alumina. The metal halide undergoes reaction with hydrogen to form a coating of titanium metal.

$$TiBr_4(g) + 2H_2(g) \longrightarrow Ti(s) + 4HBr(g) \qquad [12.9]$$

Similarly, it is possible to make silicon films by decomposing $SiCl_4$ in the presence of H_2 at 1100°C to 1200°C.

$$SiCl_4(g) + 2H_2(g) \longrightarrow Si(s) + 4HCl(g) \qquad [12.10]$$

Films of silica, SiO_2, are formed by decomposing $SiCl_4$ in the presence of both H_2 and CO_2 at 600°C to 900°C.

$$SiCl_4(g) + 2H_2(g) + 2CO_2(g) \longrightarrow SiO_2(s) + 4HCl(g) + 2CO(g) \qquad [12.11]$$

Films of ditungsten carbide, W_2C, (Figure 12.34 ▶) can be formed by decomposing WF_6 in the presence of benzene (C_6H_6) and hydrogen gas.

$$12WF_6(g) + C_6H_6(g) + 33H_2(g) \longrightarrow 6W_2C(s) + 72HF(g) \qquad [12.12]$$

▲ **Figure 12.34** The tip of this masonry drill bit has been coated with a thin film of tungsten carbide to impart hardness and resistance against wear.

SAMPLE INTEGRATIVE EXERCISE 12: Putting Concepts Together

Films of silicon nitride, Si_3N_4, can be formed by decomposing silane (SiH_4) in the presence of ammonia, at 900°C to 1100°C, as shown in Equation [12.13]:

$$3SiH_4(g) + 4NH_3(g) \longrightarrow Si_3N_4(s) + 12H_2(g) \qquad [12.13]$$

(a) What type of substance is Si_3N_4? What characteristic properties would you expect it to have? What type of bonding? **(b)** A proposal is made to replace the ammonia in Equation 12.13 with dinitrogen (N_2) because the hydrogen of the ammonia simply comes off anyway. Write a balanced equation for the reaction for formation of Si_3N_4 from reaction of SiH_4 with N_2. **(c)** Is the replacement of NH_3 by N_2 likely to be a good idea? Explain.

Solution **(a)** Si_3N_4 is a ceramic material. We can think of it as analogous to silicon carbide, a very hard covalent-network solid (Section 11.8). It should have high-melting and boiling points and be very hard (Table 12.4). Because Si and N are both nonmetals, the bonding between them should be polar covalent. **(b)** $3SiH_4(g) + 2N_2(g) \longrightarrow Si_3N_4(s) + 6H_2(g)$ **(c)** It would not be a good idea to try to replace NH_3 by N_2. Because the bond between nitrogen atoms in N_2 is a very strong triple bond, (Sections 8.9 and 9.8) it is unlikely that the N_2 bond would be reactive under the reaction conditions to which SiH_4 could be subjected.

Chemistry at Work Diamond Coatings

Besides being one of the hardest known substances, diamond is also highly resistant to corrosion and is stable to exceptionally high temperatures. At present, commercial diamonds are widely used to strengthen cutting and grinding tools. These diamonds are embedded in the tools and are not intimately and uniformly part of the material.

Scientists have recently developed procedures for applying an ultrathin layer of polycrystalline diamond on many materials. These thin films impart the hardness and durability of diamond to a variety of materials, such as glass, paper, plastics, metals, and semiconductor devices. Imagine scratchproof glass; cutting tools that virtually never need sharpening; surfaces that are chemical-resistant; temperature sensors that operate at high temperatures in harsh environments. Because diamond is compatible with biological tissue, it can be used to coat prosthetic materials and biosensors. When deposited on silicon, polycrystalline diamond thin films can serve as high-temperature sensors and be used in electronic devices such as flat-screen displays (Figure 12.35 ▶). In some of these applications the film must be doped during deposition with another element, such as boron, to create a semiconductor.

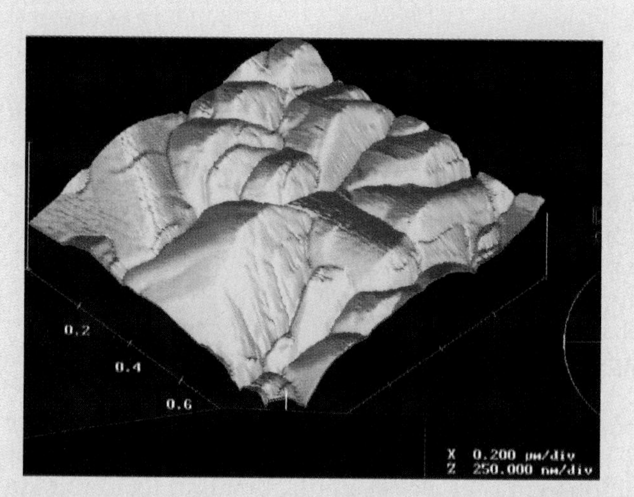

▲ **Figure 12.35** Atomic force microscope image of thin diamond films deposited on silicon. The image is about 1 μm along each edge in the plane of the films. The polycrystalline nature of the film is evident. The mean surface roughness (mean distance between peaks and troughs) is on the order of 30 nm.

One procedure for generating diamond films involves exposing a mixture of methane gas (CH_4) and hydrogen gas (H_2) to intense microwave radiation in the presence of the object to be coated. Typically, total gas pressure is about 50 torr, and H_2 is present in large excess. Under appropriate conditions the CH_4 decomposes, depositing a thin film of diamond. The H_2 dissociates into atomic hydrogen, which reacts faster with graphite than with diamond, effectively removing graphite from the growing film.

Summary and Key Terms

Introduction and Section 12.1 In this chapter we consider five classes of materials: liquid crystals, polymers, biomaterials, ceramics, and thin films. A **liquid crystal** is a substance that exhibits one or more ordered phases at a temperature above the melting point of the solid. In a **nematic liquid crystalline phase** the molecules are aligned along a common direction, but the ends of the molecules are not lined up. In a **smectic liquid crystalline phase** the ends of the molecules are lined up, so that the molecules form sheets. Nematic and smectic liquid crystalline phases are generally composed of molecules with fairly rigid, elongated shapes, with polar groups along the molecules to help retain relative alignment through dipole-dipole interactions. The **cholesteric liquid crystalline phase** is composed of molecules that align as in nematic liquid crystalline phases but with each molecule twisted with respect to its neighbors, to form a helical structure.

Section 12.2 **Polymers** are molecules of high molecular mass formed by joining together large numbers of small molecules, called **monomers**. In an **addition polymerization** reaction, the molecules form new linkages by opening existing bonds. Polyethylene forms, for example, when the carbon-carbon double bonds of ethylene open up. In a **condensation polymerization reaction**, the monomers are joined by splitting out a small molecule from between them. The various kinds of nylon are formed, for example, by removing a water molecule from between an amine and a carboxylic acid. A polymer formed from two different monomers is called a **copolymer**.

Plastics are materials that can be formed into various shapes, usually by application of heat and pressure. **Thermoplastic** polymers can be reshaped, perhaps through heating, whereas **thermosetting plastics** are formed into objects through an irreversible chemical process and cannot readily be reshaped. An **elastomer** is a material that exhibits elastic behavior; that is, it returns to its original shape following stretching or bending.

Polymers are largely amorphous, but some materials possess a degree of **crystallinity**. For a given chemical composition, the crystallinity depends on the molecular mass and the degree of branching along the main polymer chain. High-density polyethylene, for example, with little side-chain branching and a high molecular mass, has a higher degree of crystallinity than low-density polyethylene, which has a lower molecular mass and a relatively high degree of branching. Polymer properties are also strongly affected by **cross-linking**, in which short chains of atoms connect the long polymer chains. Rubber is cross-linked by short chains of sulfur atoms in the process called **vulcanization**.

Section 12.3 A **biomaterial** is any material that has a biomedical application. Biomaterials are typically in contact with body tissues and fluids. They must be **biocompatible**, which means that they are not toxic, nor do they cause an inflammatory reaction. They must meet physical requirements, such as long-term reliability, strength, and flexibility or hardness, depending on the application. They must also meet chemical requirements of nonreactivity in the biological environment, or of biodegradability. Biomaterials are commonly polymers with special properties matched to the application.

Biomaterials have many cardiovascular applications (that is, pertaining to the heart, blood, and blood vessels). Heart valve implants are often mechanical devices. The presentation of a smooth surface is important, to reduce blood clotting and the loss of red blood cells. Vascular grafts are commonly constructed of Dacron™ a polyester material that integrates with surrounding tissues. Artificial tissue is grown on polymeric scaffolding, which holds the cells in place and promotes cell differentiation. Joint replacement devices, such as a hip or knee replacement, involve moving parts that can integrate with surrounding bone to develop strength. Metal ball joints fit into cups formed from very high molecular weight polymeric materials such as polyethylene.

Section 12.4 **Ceramics** are inorganic solids with generally high thermal stability, usually formed through three-dimensional network bonding. The bonding in ceramics may be either covalent or ionic, and ceramics may be crystalline or amorphous. The processing of ceramics generally begins with formation of very small, uniformly sized particles through the **sol-gel process**. The small particles are then compressed and heated at high temperature. They coalesce through a process known as sintering. Ceramics can be made tougher and less subject to cracking by forming **composites**, in which ceramic fibers are added to the ceramic material before processing.

Section 12.5 **Superconductivity** involves a material that is capable of conducting an electrical current without any apparent resistance when cooled below its **superconducting transition temperature**, T_c. Since discovery of the phenomenon in 1911, the number of known superconducting materials has steadily increased. Until recently, however, all observed T_c values were below about 25 K. An important recent development is the discovery of **high-temperature superconductivity** in certain complex oxides. **Superconducting ceramics** such as $YBa_2Cu_3O_7$ are capable of superconductivity at temperatures higher than that for any nonceramic superconductor. Still other classes of compounds have been shown recently to have relatively high T_c values.

Section 12.6 A **thin film** is a very thin layer of a substance, covering an underlying substrate. Thin films can be formed by **vacuum deposition**, in which a material is vaporized or evaporated onto a surface; by **sputtering**, in which a high voltage is used to generate energetic atoms of the material to be deposited; or via **chemical-vapor deposition**, in which a chemical reaction involving a vapor-phase substance occurs on a surface, thereby forming a stable, adherent coat.

Exercises

Liquid Crystals

12.1 In what ways are a nematic liquid crystalline phase and an ordinary liquid the same, and in what ways do their physical properties differ?

12.2 In contrast to ordinary liquids, liquid crystals are said to possess "order." What does this mean?

12.3 Describe what is occurring at the molecular level as a substance passes from the solid to the nematic liquid crystalline to the isotropic (normal) liquid phase upon heating.

12.4 What observations made by Reinitzer on cholesteryl benzoate suggested that this substance possesses a liquid crystalline phase?

12.5 The molecules shown in Figure 12.4 possess polar groups (that is, groupings of atoms that give rise to sizable dipole moments within the molecules). How might the presence of polar groups enhance the tendency toward liquid crystal formation?

12.6 Liquid crystalline phases tend to be more viscous than the isotropic, or normal, liquid phase of the same substance. Why?

12.7 The smectic liquid crystalline phase can be said to be more highly ordered than the nematic. In what sense is this true?

12.8 One of the more effective liquid crystalline substances employed in LCDs is this molecule.

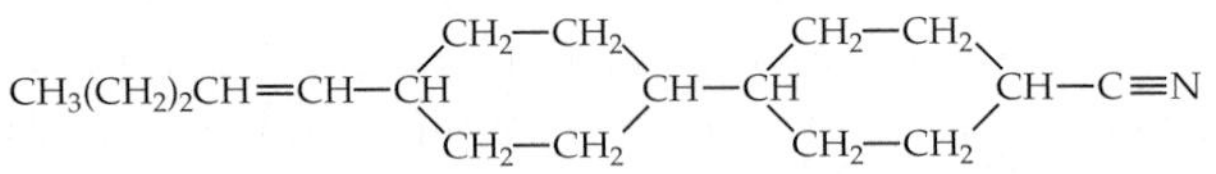

$$CH_3(CH_2)_2CH{=}CH{-}CH\begin{matrix}\diagup CH_2{-}CH_2 \diagdown \\ \diagdown CH_2{-}CH_2 \diagup\end{matrix}CH{-}CH\begin{matrix}\diagup CH_2{-}CH_2 \diagdown \\ \diagdown CH_2{-}CH_2 \diagup\end{matrix}CH{-}C{\equiv}N$$

By comparing this structure with the structural formulas and models shown in Figure 12.4, describe the features of the molecule that promote its liquid crystalline behavior.

12.9 Describe how a cholesteric liquid crystal phase differs from a nematic phase

12.10 It often happens that a substance possessing a smectic liquid crystalline phase just above the melting point passes into a nematic liquid crystalline phase at a higher temperature. Account for this type of behavior in terms of the ideas developed in Chapter 11 relating molecular energies to temperature.

Polymers

12.11 The structure of decane is shown in Sample Exercise 12.1. Decane is not considered to be a polymer, whereas polyethylene is. What is the distinction?

12.12 What is a monomer? Give three examples of monomers, taken from the examples given in this chapter.

12.13 An ester is a compound formed by a condensation reaction between a carboxylic acid and an alcohol. Use the index to find the discussion of esters in Chapter 25, and give an example of a reaction forming an ester. How might this kind of reaction be extended to form a polymer (a polyester)?

12.14 Write a chemical equation for formation of a polymer via a condensation reaction from the monomers succinic acid ($HOOCCH_2CH_2COOH$) and ethylenediamine ($H_2NCH_2CH_2NH_2$).

12.15 Draw the structure of the monomer(s) employed to form each of the following polymers shown in Table 12.1: **(a)** polyvinyl chloride; **(b)** nylon 6,6; **(c)** polyethylene terephthalate.

12.16 Write the chemical equation that represents the formation of **(a)** polychloroprene from chloroprene

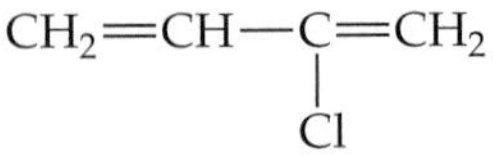

$$CH_2{=}CH{-}\underset{\displaystyle Cl}{\underset{|}{C}}{=}CH_2$$

(Polychloroprene is used in highway-pavement seals, expansion joints, conveyor belts, and wire and cable jackets.); **(b)** polyacrylonitrile from acrylonitrile

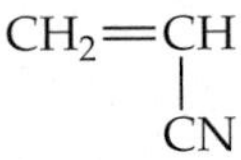

$$CH_2{=}\underset{\displaystyle CN}{\underset{|}{CH}}$$

(Polyacrylonitrile is used in home furnishings, craft yarns, clothing, and many other items.)

12.17 The nylon Nomex™, a condensation polymer, has the following structure:

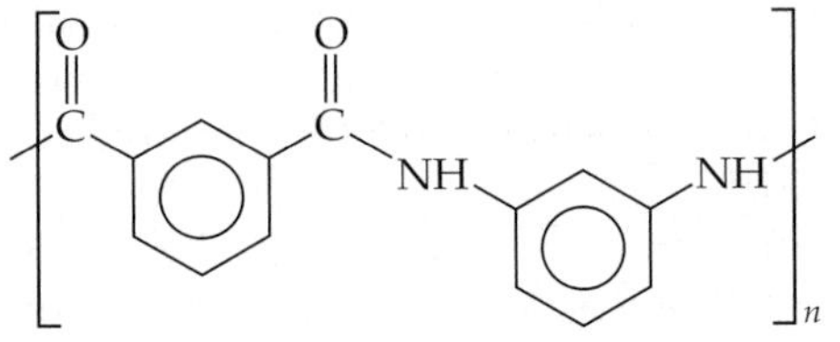

$$\left[\overset{\displaystyle O}{\overset{\|}{C}} - C_6H_4 - \overset{\displaystyle O}{\overset{\|}{C}} - NH - C_6H_4 - NH \right]_n$$

Draw the structures of the two monomers that yield Nomex™.

12.18 Proteins are polymers formed by condensation reactions of amino acids, which have the general structure

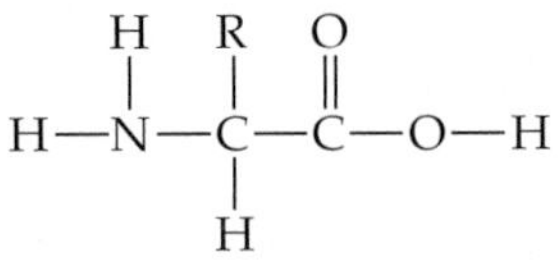

$$H{-}\overset{\displaystyle H}{\overset{|}{N}}{-}\underset{\displaystyle H}{\underset{|}{\overset{\displaystyle R}{\overset{|}{C}}}}{-}\overset{\displaystyle O}{\overset{\|}{C}}{-}O{-}H$$

In this structure, R represents $-H$, $-CH_3$ or another group of atoms. Draw the general structure for a polyamino acid polymer formed by condensation polymerization of the molecule shown here.

12.19 In addition to the condensation of dicarboxylic acids with diamines, as shown in Equation 12.3, nylons can also be formed by the condensation reactions of aminocarboxylic acids with themselves. Nylon 4, for example, is formed by the polycondensation of 4-aminobutyric acid ($NH_2CH_2CH_2CH_2COOH$). Write a chemical equation to show the formation of nylon 4 from this monomer.

12.20 Kevlar™, a high-performance polymer, has the following structure:

$$\left[-\overset{\overset{O}{\|}}{C}-C_6H_4-\overset{\overset{O}{\|}}{C}-NH-C_6H_4-NH- \right]_n$$

Write the structural formulas for the two substances that are condensed to form this polymer.

12.21 What molecular features make a polymer flexible? Explain how cross-linking affects the chemical and physical properties of the polymer.

12.22 What molecular structural features cause high-density polyethylene to be denser than low-density polyethylene?

12.23 Are high molecular masses and a high degree of crystallinity always desirable properties of a polymer? Explain.

12.24 Briefly describe each of the following: **(a)** elastomer; **(b)** thermoplastic; **(c)** thermosetting plastic; **(d)** plasticizer.

Biomaterials

12.25 Neoprene is a polymer of chlorobutadiene:

$$CH_2{=}\underset{\underset{H}{|}}{C}-\underset{\underset{Cl}{|}}{C}{=}CH_2$$

The polymer can be used to form flexible tubing that is resistant to chemical attack from a variety of chemical reagents. Suppose it is proposed to use neoprene tubing as a coating for the wires running to the heart from an implanted pacemaker. What questions would you ask to determine whether it might be suitable for such an application?

12.26 On the basis of the structure shown in Table 12.1 for polystyrene and polyurethane, which of these two classes of polymer would you expect to form the most effective interface with biological systems? Explain.

12.27 Patients who receive vascular grafts formed from polymer material such as Dacron™ are required to take anti-coagulation drugs on a continuing basis to prevent blood clots. Why? What advances in such vascular implants are needed to make this precaution unnecessary?

12.28 Several years ago a biomedical company produced and marketed a new, efficient heart valve implant. It was later withdrawn from the market, however, because patients using it suffered from severe loss of red blood cells. Describe what properties of the valve could have been responsible for this result.

12.29 Skin cells from the body do not differentiate when they are simply placed in a tissue culture medium; that is, they do not organize into the structure of skin, with different layers and different cell types. What is needed to cause such differentiation to occur? Indicate the most important requirements on any material used.

12.30 If you were going to attempt to grow skin cells in a medium that affords an appropriate scaffolding for the cells and you had only two fabrics available, one made from polystyrene and the other from polyethyleneterephthalate (Table 12.1), which would you choose for your experiments? Explain.

Ceramics

12.31 Metals, such as Al or Fe, and many plastics are recyclable. With the exception of many glasses, such as bottle glass, ceramic materials in general are not recyclable. What characteristics of ceramics make them less readily recyclable?

12.32 You have two solid objects, one formed from a ceramic material and the other from a metal, that look pretty much alike. When they are dropped from a height of ten feet onto a concrete surface, one shatters into a variety of pieces, some with sharp edges. The others suffers a big dent and partially splits in two. **(a)** Which one is likely to be the ceramic, and which the metal? **(b)** Why do the two materials differ in their behavior?

12.33 Why is the formation of very small, uniformly sized and shaped particles important for many applications of ceramic materials?

12.34 Describe the general chemical steps in a sol-gel process, beginning with $Zr(s)$ and $CH_3CH_2OH(l)$. Indicate whether each step is an oxidation-reduction reaction (refer to Section 4.4), condensation reaction, or other process.

12.35 Steel reinforcing rods are needed when concrete is used as a highway bed or in construction of a building. Describe the analogy between this practice and the formation of ceramic composites. What does this analogy suggest about the optimal shape and size of the added composite material? Why is this the optimal shape?

12.36 The hardnesses of several substances according to a scale known as the Knoop value are as follows:

Substance	Knoop Value
Ag	60
$CaCO_3$	135
MgO	370
Soda-lime glass	530
Cr	935
ZrB_2	1550
Al_2O_3	2100
TaC	2000

Which of the materials in this list would you classify as a ceramic? What were your criteria for making this classification? Does classification as a ceramic correlate with Knoop hardness? If you think it does, is hardness alone a sufficient criterion to determine whether a substance is a ceramic? Explain.

12.37 Silicon carbide, SiC, has the three-dimensional structure shown in the figure.

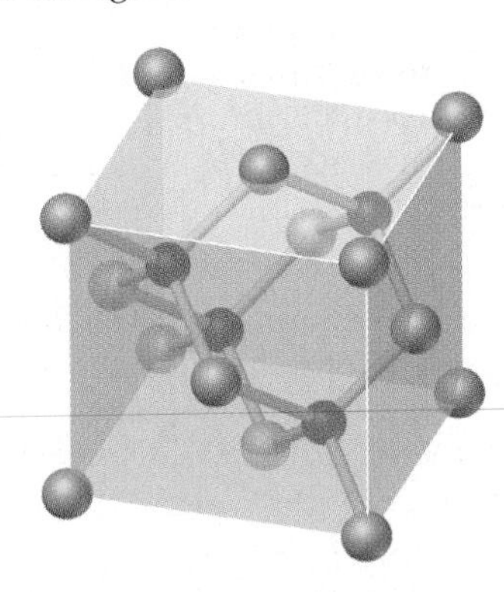

Describe how the bonding and structure of SiC lead to its great thermal stability (to 2700°C) and exceptional hardness.

12.38 What structural characteristics do the four ceramic substances listed in Table 12.4 have in common, and how do these account for the properties listed there, particularly the high melting points and hardnesses?

Superconductivity

12.39 To what does the term *superconductivity* refer? Why might superconductive materials be of value?

12.40 Distinguish between an excellent metallic conductor of electricity (such as silver) and a superconducting substance (such as Nb_3Sn) below its superconducting transition temperature.

12.41 The following graph shows the resistivity of MgB_2 as a function of temperature in the region from about 4 K to 100 K. What is the significance of the sharp drop in resistivity below 40 K?

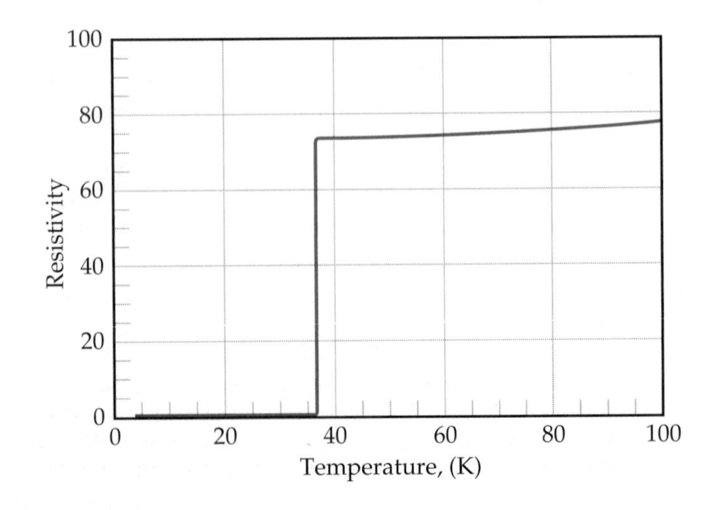

12.42 **(a)** What is the superconducting transition temperature, T_c? **(b)** The discovery by Müller and Bednorz of superconductivity in a copper oxide ceramic at 35 K set off a frantic scramble among physicists and chemists to find materials that exhibit superconductivity at higher temperatures. What is the significance of achieving T_c values above 77 K?

12.43 Superconducting ceramic materials have some limitations as electrical conductors when compared with an ordinary conductor such as copper wire. What are some of these?

12.44 Why would a substance that is a nonconductor not be a good candidate for superconductivity?

Thin Films

12.45 One of the requirements in most applications of thin films is that they must adhere strongly to the underlying substrate. Of the forces discussed in the text, which might lead to adherence of the film?

12.46 List the characteristics that a thin film should possess if it is to have useful application.

12.47 What properties of the thin film illustrated in Figure 12.31 are important to its function in this particular application?

12.48 List the major methods employed for forming thin films. Which of these involves a net chemical change in proceeding from the starting materials to the applied film?

Additional Exercises

12.49 What properties of the typical nematic liquid crystalline molecule are likely to cause it to reorient when it is placed in an electrical field that is perpendicular to the direction of orientation of the molecules?

12.50 Teflon™ is a polymer formed by the polymerization of $F_2C{=}CF_2$. Draw the structure of a section of this polymer. What type of polymerization reaction is required to form it?

12.51 Classify each of the following as a ceramic, polymer, or liquid crystal.

(a) $\left[-CH_2-C(CH_3)(COOCH_3)-\right]_n$ (b) $LiNbO_3$

(c) SiC (d) $\left[-Si(CH_3)_2-\right]_n$

(e) $CH_3O-C_6H_4-N{=}N(O)-C_6H_4-OCH_3$

12.52 The temperature range over which a liquid possesses liquid crystalline behavior is rather narrow (for examples, see Figure 12.4). Why?

12.53 Ceramics are generally brittle, subject to crack failure, and stable to high temperatures. In contrast, plastics are generally deformable under stress and have limited thermal stability. Discuss these differences in terms of the structures and bonding in the two classes of materials.

12.54 A watch with a liquid crystal display does not function properly when it is exposed to low temperatures during a trip to Antarctica. Explain why the LCD might not function well at low temperature.

12.55 Suppose that a liquid crystalline material such as cholesteryl benzoate is warmed to well above its liquid crystalline range and then cooled. On cooling, the sample unexpectedly remains clear until it reaches a temperature just below the melting point, at which time it solidifies. What explanation can you give for this behavior?

[12.56] By using the index of the text to find information about metallic bonding, compare the structures of ceramics such as Al_2O_3 and others listed in Table 12.4 with those of metals, and explain why the ceramic materials are much harder.

12.57 The lining of the acetabular cup that accepts the ball joint in a hip replacement is composed of ultrahigh molecular weight polyethylene. Based on the discussion of polyethylene polymers in Section 12.2, predict some of the general properties of this material. Which properties are of particular importance in this application? What requirements are there on the friction between the metal ball and the polymer surface?

12.58 Write balanced equations to describe: **(a)** formation of a silicon carbide whisker by the two-step thermal decomposition of poly(dimethylsilane) (two equations); **(b)** formation of a thin film of niobium by thermal decomposition of $NbBr_5$ on a hot surface under an H_2 atmosphere; **(c)** formation of $Si(OCH_2CH_3)_4$ by reaction of $SiCl_4$ with ethyl alcohol; **(d)** polymerization of styrene (shown below) to form polystyrene.

$C_6H_5-CH{=}CH_2$

12.59 Hydrogen bonding between polyamide chains plays an important role in determining the properties of a nylon such as nylon 6,6 (Table 12.1). Draw the structural formulas for two adjacent chains of nylon 6,6, and show where hydrogen bonding interactions could occur between them.

[12.60] A particular liquid crystalline substance has the phase diagram shown in the figure. By analogy with the phase diagram for a non–liquid crystalline substance (Section 11.6), identify the phase present in each area.

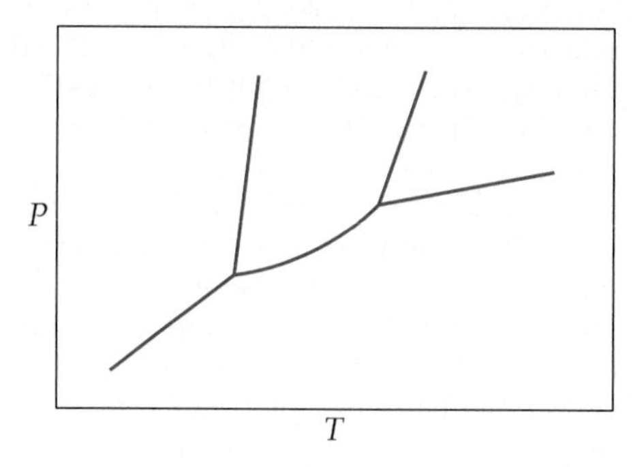

[12.61] In fabricating microelectronics circuits, a ceramic conductor such as $TiSi_2$ is employed to connect various regions of a transistor with the outside world, notably aluminum wires. The $TiSi_2$ is deposited as a thin film via chemical vapor deposition, in which $TiCl_4(g)$ and $SiH_4(g)$ are reacted at the Si surface. Write a balanced equation for the reaction, assuming that the other products are H_2 and HCl. Why might $TiSi_2$ behave better as a conducting interconnect on Si than a metal such as Cu?

[12.62] Indicate the nature of the film formed by thermal decomposition on a heated surface of each of the following: **(a)** SiH_4 with H_2 as carrier gas, in the presence of CO_2 (CO is a product); **(b)** $TiCl_4$ in the presence of water vapor; **(c)** $GeCl_4$ in the presence of H_2 as carrier gas.

Integrative Exercises

[12.63] Whereas thin films of metal oxides can be formed by vacuum deposition, a general rule is that inorganic compounds having anionic components with names ending in *-ite* or *-ate* cannot be successfully vacuum deposited. Why is this so?

[12.64] Employing the bond enthalpy values listed in Table 8.4, estimate the molar enthalpy change occurring upon **(a)** polymerization of ethylene; **(b)** formation of nylon 6,6; **(c)** formation of polyethylene terephthalate (PET).

[12.65] Although polyethylene can twist and turn in random ways, as illustrated in Figure 12.9, the most stable form is a linear one with the carbon backbone oriented as shown in the figure below:

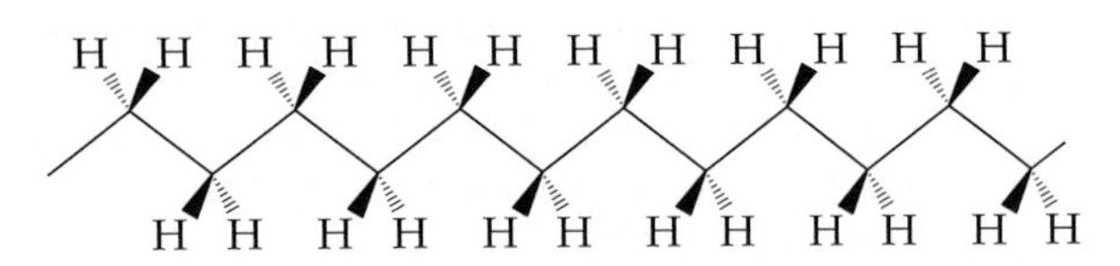

(a) What is the hybridization of orbitals at each carbon atom? What angles do you expect between the bonds?
(b) The solid wedges in the figure indicate bonds from carbon that come out of the plane of the page, the dotted wedges indicate bonds that lie behind the plane of the page. Now imagine that the polymer is polypropylene rather than polyethylene. Draw structures for polypropylene in which (i) the CH_3 groups all lie on the same side of the plane of the paper (this form is called isotactic polypropylene); (ii) the CH_3 groups lie on alternating sides of the plane (syndiotactic polypropylene); or (iii) the CH_3 groups are randomly distributed on either side (atactic polypropylene). Which of these forms would you expect to have the highest crystallinity and melting point, and which the lowest? Explain in terms of intermolecular interactions and molecular shapes.
(c) Polypropylene fibers have been employed in athletic wear. The product is said to be superior to cotton or polyester clothing in "wicking" moisture away from the body through the fabric to the outside. Explain the difference between polypropylene and polyester or cotton (which has many –OH groups along the molecular chain), in terms of intermolecular interactions with water.

[12.66] In the superconducting ceramic $YBa_2Cu_3O_7$, what is the average oxidation state of copper, assuming that Y and Ba are in their expected oxidation states? Yttrium can be replaced with a rare-earth element such as La, and Ba can be replaced with other similar elements without fundamentally changing the superconducting properties of the material. However, general replacement of copper by any other element leads to a loss of superconductivity. In what respects is the electronic structure of copper different from that of the other two metallic elements in this compound?

[12.67] In an experiment in which a diamond thin film was formed by microwave decomposition of CH_4 in the presence of H_2, the total gas pressure was 90 torr at a temperature of 850°C. The ratio of H_2 to CH_4 was 40. **(a)** Compute the average molar mass of the gas and its density under the experimental conditions. **(b)** Assuming that the effective surface area to be covered with the film is 1 cm^2, that the film thickness is 2.5 μm, and that the efficiency of converting the CH_4 to diamond on decomposition is 0.5%, what volume of the gas is consumed in forming the film? (The density of diamond is 3.51 g/cm^3.)

[12.68] A sample of the superconducting oxide, $HgBa_2Ca_2Cu_3O_{8+x}$ is found to contain 14.99% oxygen by mass. **(a)** Assuming that all other elements are present in the ratios represented by the formula, what is the value of x in the formula? **(b)** Which of the metallic elements in the compound is (or are) most likely to have a noninteger average charges? Explain your answer. **(c)** Which of the metallic ions in the substance is likely to have the largest ionic radius? Which will have the smallest?

[12.69] **(a)** In the polydimethylsilane polymer shown on page 469, which bonds have the lowest average bond enthalpy? **(b)** In the thermal conversion of polydimethylsilane to the polymeric substance shown on page 470, which bonds are most likely to rupture first on heating the material? **(c)** Following this first step of bond rupture, what next step occurs to yield the product polymer shown? Employing the values of average bond enthalpy in Table 8.4, estimate the overall enthalpy change in this final step.

[12.70] Consider *para*-azoxyanisole, which is a nematic liquid crystal in the temperature range of 21°C to 47°C:

$$CH_3O-C_6H_4-CH{=}N(\rightarrow O)-C_6H_4-OCH_3$$

(a) Write out the Lewis structure for this molecule, showing all lone-pair electrons as well as bonds. (b) Describe the hybrid orbitals employed by each of the two nitrogens. What bond angles do you anticipate about the nitrogen atom that is bonded to oxygen? (c) Replacing one of the $-OCH_3$ groups in *para*-azoxyanisole by a $-CH_2CH_2CH_2CH_3$ group causes the melting point of the substance to drop; the liquid crystal range changes to 19°C to 76°C. Explain why this substitution produces the observed changes in properties. (d) How would you expect the density of *para*-azoxyanisole to change upon melting at 117°C? Upon passing from the nematic to the isotropic liquid state at 137°C? Explain.

eMedia Exercises

12.71 The **Synthesis of Nylon 610** movie (*eChapter 12.2*) shows the production of a polymer from two organic compounds. The newly formed material is removed from the beaker and wound around a glass rod. If the nylon were not removed, would the same amount be produced? Explain your answer.

12.72 Some of the properties of **Silicon Carbide** (*eChapter 12.4*) are given in Table 12.4. **(a)** In what ways does silicon carbide resemble diamond? **(b)** In what way is silicon carbide different from most ceramics?

12.73 The **1, 2, 3 Superconductor** (*eChapter 12.5*), $YBa_2Cu_3O_7$, was one of the first materials shown to exhibit superconductivity at "high" temperature (95 K). What properties of this material prevent its use in many applications where superconductivity would be desirable?

Chapter 13

Properties of Solutions

A wave breaking. Ocean water is a complex aqueous solution of many dissolved substances, of which sodium chloride is highest in concentration.

MOST OF THE materials that we encounter in everyday life are mixtures. Many mixtures are homogeneous; that is, their components are uniformly intermingled on a molecular level. Homogeneous mixtures are called *solutions*. ∞∞ (Sections 1.2 and 4.1)

Examples of solutions abound in the world around us. The air we breathe is a solution of several gases. Brass is a solid solution of zinc in copper. The fluids that run through our bodies are solutions, carrying a great variety of essential nutrients, salts, and other materials.

Solutions may be gases, liquids, or solids (Table 13.1 ▶). Each of the substances in a solution is called a *component* of the solution. As we saw in Chapter 4, the *solvent* is normally the component present in greatest amount. Other components are called *solutes*. Because liquid solutions are the most common, we will focus our attention on them in this chapter. Our primary goal is to examine the physical properties of solutions, comparing them with the properties of their components. We will be particularly concerned with aqueous solutions of ionic substances because of their central importance in chemistry and in our daily lives.

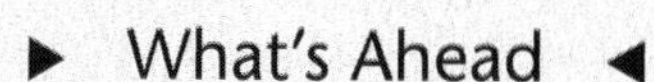

▶ What's Ahead ◀

- We begin by considering what happens at a molecular level when a substance dissolves, paying particular attention to the role of *intermolecular forces* in the process.
- We next examine the changes in *energy* and in *disorder* that accompany the solution process.
- We'll see that in *saturated solutions* the dissolved and undissolved solute are in *equilibrium*.
- The amount of solute in a saturated solution defines its *solubility*, the extent to which a particular solute dissolves in a particular solvent.
- Solubility depends on the nature of the solute and solvent, which determine the identities of the intermolecular forces between and within them. Pressure affects the solubility of gaseous solutes. Temperature also affects solubility.
- Because many physical properties of solutions depend on their concentration, we examine several common ways of expressing concentration.
- Physical properties of solutions that depend only on concentration and not on the identity of the solute are known as *colligative properties*. They include the extent to which the solute lowers the vapor pressure, increases the boiling point, and decreases the freezing point of the solvent. The osmotic pressure of a solution is also a colligative property.
- We close the chapter by investigating *colloids*, mixtures in which particles larger than molecular size are dispersed in another component.

TABLE 13.1 Examples of Solutions

State of Solution	State of Solvent	State of Solute	Example
Gas	Gas	Gas	Air
Liquid	Liquid	Gas	Oxygen in water
Liquid	Liquid	Liquid	Alcohol in water
Liquid	Liquid	Solid	Salt in water
Solid	Solid	Gas	Hydrogen in palladium
Solid	Solid	Liquid	Mercury in silver
Solid	Solid	Solid	Silver in gold

13.1 The Solution Process

A solution is formed when one substance disperses uniformly throughout another. With the exception of gas mixtures, all solutions involve substances in a condensed phase. We learned in Chapter 11 that the molecules or ions of substances in the liquid and solid states experience intermolecular attractive forces that hold them together. Intermolecular forces also operate between solute particles and solvent molecules.

Any of the various kinds of intermolecular forces that we discussed in Chapter 11 can operate between solute and solvent particles in a solution. Ion-dipole forces, for example, dominate in solutions of ionic substances in water. Dispersion forces, on the other hand, dominate when a nonpolar substance like C_6H_{14} dissolves in another nonpolar one like CCl_4. Indeed, a major factor determining whether a solution forms is the relative strengths of intermolecular forces between and among the solute and solvent particles.

Solutions form when the attractive forces between solute and solvent particles are comparable in magnitude with those that exist between the solute particles themselves or between the solvent particles themselves. For example, the ionic substance NaCl dissolves readily in water because the attractive interactions between the ions and the polar H_2O molecules overcome the lattice energy of NaCl(*s*). Let's examine this solution process more closely, paying attention to these attractive forces.

When NaCl is added to water (Figure 13.1 ▼), the water molecules orient themselves on the surface of the NaCl crystals. The positive end of the water

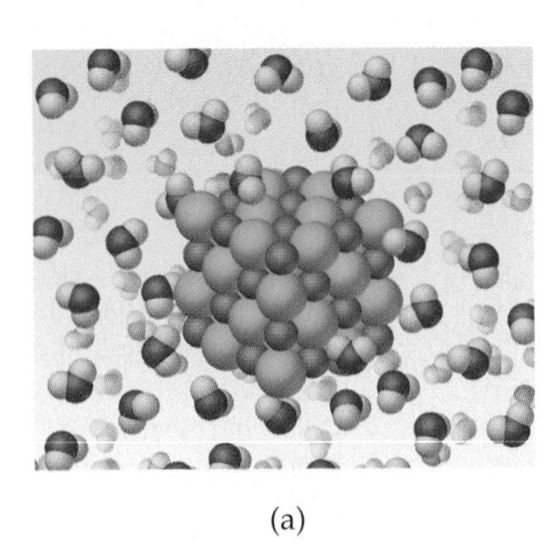

(a)

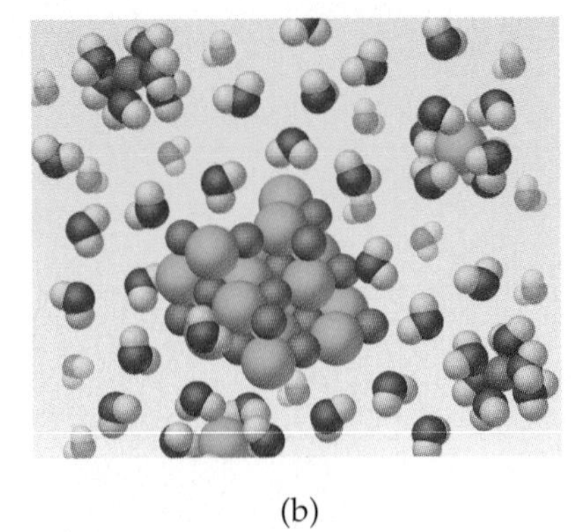

(b)

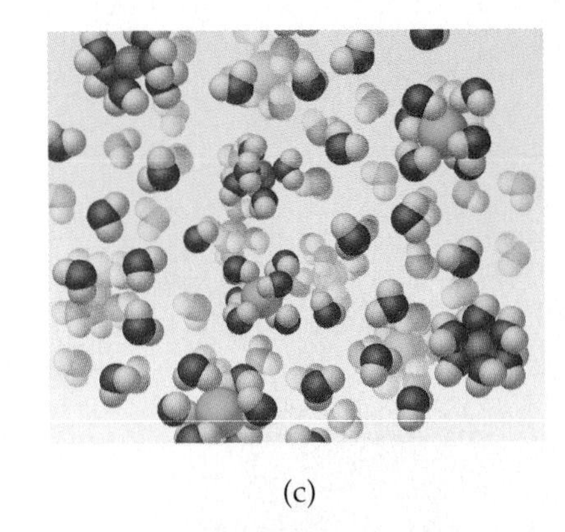

(c)

▲ **Figure 13.1** A schematic illustration of the solution process of an ionic solid in water. (a) The solid substance is hydrated by water molecules, with the oxygen atoms of the water molecules oriented toward the cations and the hydrogens oriented toward the anions. (b, c) As the solution process proceeds, the individual ions are removed from the solid surface and become completely separate hydrated species in solution.

dipole is oriented toward the Cl^- ions, and the negative end of the water dipole is oriented toward the Na^+ ions. The ion-dipole attractions between the ions and water molecules are sufficiently strong to pull the ions from their positions in the crystal.

Once separated from the crystal, the Na^+ and Cl^- ions are surrounded by water molecules, as shown in Figure 13.1(b and c) and Figure 13.2 ▶. Such interactions like this between solute and solvent molecules are known as **solvation**. When the solvent is water, the interactions are known as **hydration**.

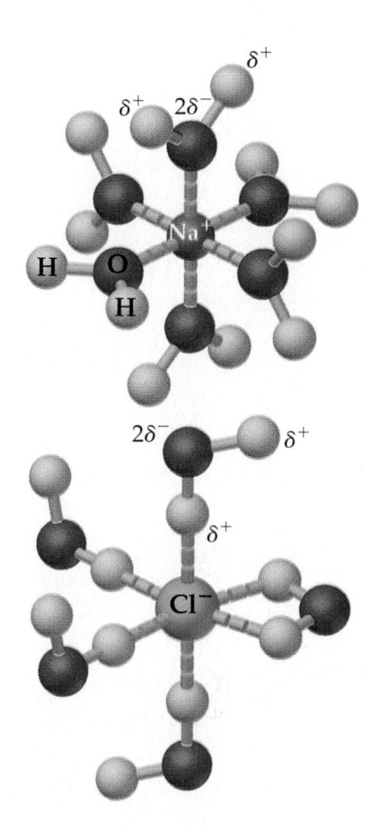

▲ **Figure 13.2** Hydrated Na^+ and Cl^- ions. The negative ends of the water dipole point toward the positive ion, and the positive ends point toward the negative ion.

Energy Changes and Solution Formation

Sodium chloride dissolves in water because the water molecules have a sufficient attraction for the Na^+ and Cl^- ions to overcome the attraction of these two ions for one another in the crystal. To form an aqueous solution of NaCl, water molecules must also separate from one another to form spaces in the solvent that will be occupied by the Na^+ and Cl^- ions. Thus, we can think of the overall energetics of solution formation as having three components, illustrated schematically in Figure 13.3 ▼. The overall enthalpy change in forming a solution, ΔH_{soln}, is the sum of three terms:

$$\Delta H_{soln} = \Delta H_1 + \Delta H_2 + \Delta H_3 \quad [13.1]$$

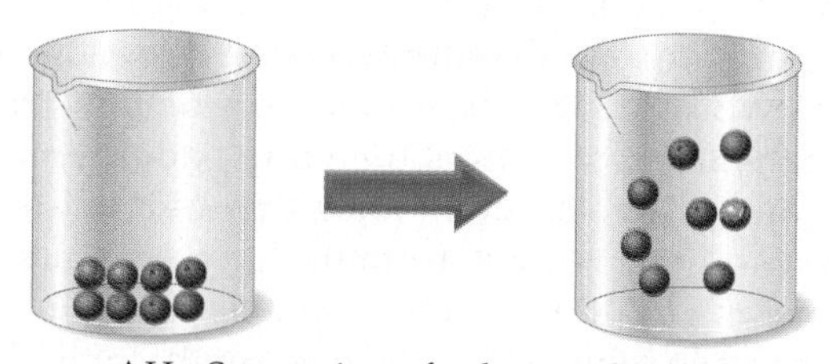

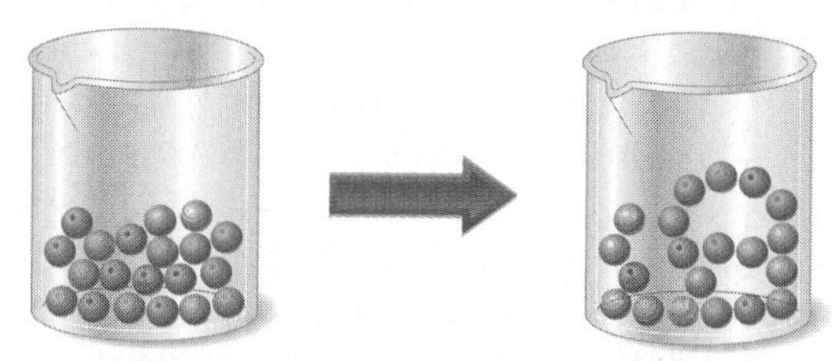

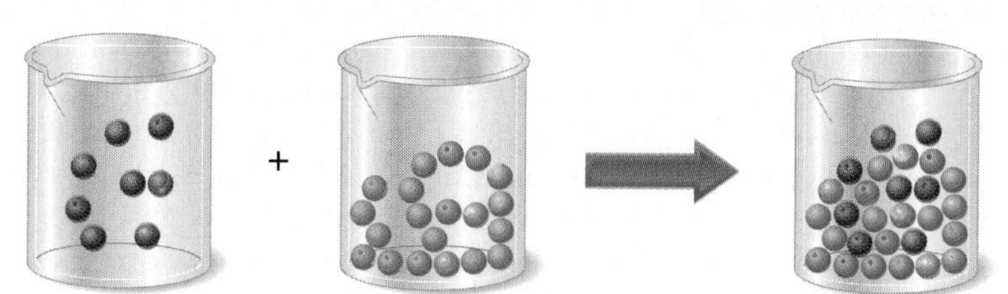

▲ **Figure 13.3** Depiction of the three enthalpic contributions to the overall heat of solution of a solute. As noted in the text, ΔH_1 and ΔH_2 represent endothermic processes, requiring an input of energy, whereas ΔH_3 represents an exothermic process.

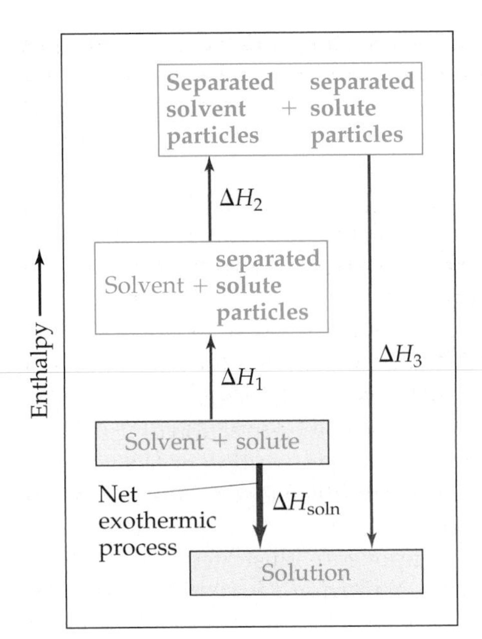

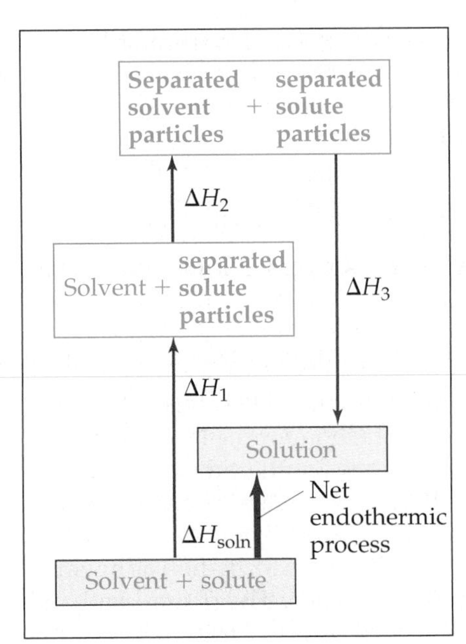

ACTIVITY
Enthalpy of Solution

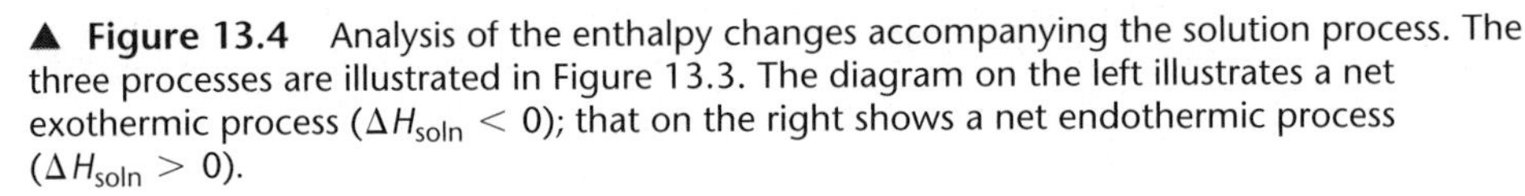

▲ **Figure 13.4** Analysis of the enthalpy changes accompanying the solution process. The three processes are illustrated in Figure 13.3. The diagram on the left illustrates a net exothermic process ($\Delta H_{soln} < 0$); that on the right shows a net endothermic process ($\Delta H_{soln} > 0$).

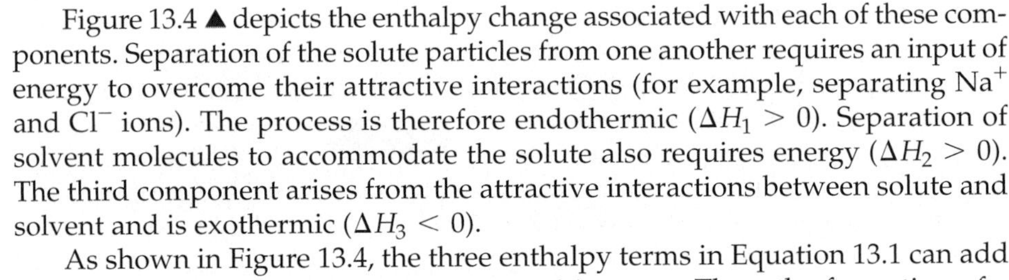

Figure 13.4 ▲ depicts the enthalpy change associated with each of these components. Separation of the solute particles from one another requires an input of energy to overcome their attractive interactions (for example, separating Na^+ and Cl^- ions). The process is therefore endothermic ($\Delta H_1 > 0$). Separation of solvent molecules to accommodate the solute also requires energy ($\Delta H_2 > 0$). The third component arises from the attractive interactions between solute and solvent and is exothermic ($\Delta H_3 < 0$).

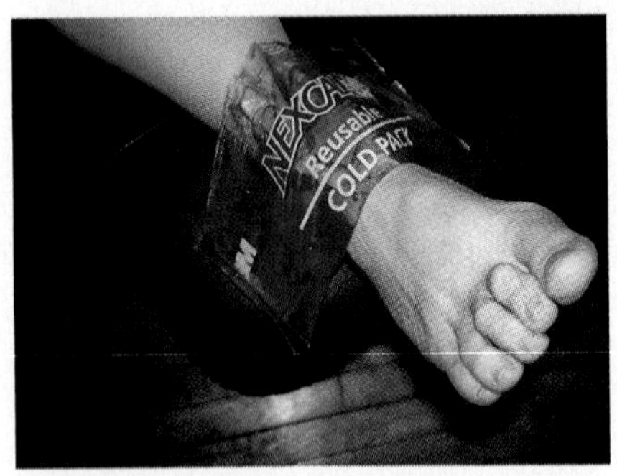

▲ **Figure 13.5** An instant ice pack, containing ammonium nitrate, used to treat athletic injuries. To activate the pack, the container is kneaded, breaking the seal separating solid NH_4NO_3 from water. The heat of solution of NH_4NO_3 is positive, so the temperature of the solution decreases.

As shown in Figure 13.4, the three enthalpy terms in Equation 13.1 can add together to give either a negative or a positive sum. Thus, the formation of a solution can be either exothermic or endothermic. For example, when magnesium sulfate, $MgSO_4$, is added to water, the resultant solution gets quite warm: $\Delta H_{soln} = -91.2$ kJ/mol. In contrast, the dissolution of ammonium nitrate (NH_4NO_3) is endothermic: $\Delta H_{soln} = 26.4$ kJ/mol. These particular substances have been used to make the instant heat packs and ice packs that are used to treat athletic injuries (Figure 13.5 ◀). The packs consist of a pouch of water and a dry chemical, $MgSO_4$ for hot packs and NH_4NO_3 for cold packs. When the pack is squeezed, the seal separating the solid from the water is broken and a solution forms, either increasing or decreasing the temperature.

In Chapter 5 we learned that the enthalpy change in a process can provide information about the extent to which a process will occur. ∞ (Section 5.4) Processes that are exothermic tend to proceed spontaneously. A solution will not form if ΔH_{soln} is too endothermic. The solvent-solute interaction must be strong enough to make ΔH_3 comparable in magnitude to $\Delta H_1 + \Delta H_2$. This is why ionic solutes like NaCl do not dissolve in nonpolar liquids such as gasoline. The nonpolar hydrocarbon molecules of the gasoline would experience only weak attractive interactions with the ions, and these interactions would not compensate for the energies required to separate the ions from one another.

By similar reasoning, a polar liquid such as water does not form solutions with a nonpolar liquid such as octane (C_8H_{18}). The water molecules experience strong hydrogen-bonding interactions with one another. ∞ (Section 11.2)

These attractive forces must be overcome to disperse the water molecules throughout the nonpolar liquid. The energy required to separate the H_2O molecules is not recovered in the form of attractive interactions between H_2O and C_8H_{18} molecules.

Solution Formation, Spontaneity, and Disorder

When carbon tetrachloride (CCl_4) and hexane (C_6H_{14}) are mixed, they dissolve in one another in all proportions. Both substances are nonpolar, and they have similar boiling points (77°C for CCl_4 and 69°C for C_6H_{14}). It is therefore reasonable to suppose that the magnitudes of the attractive forces (London dispersion forces) among molecules in the two substances and in their solution are comparable. When the two are mixed, dissolving occurs spontaneously; that is, it occurs without any extra input of energy from outside the system. Two distinct factors are involved in processes that occur spontaneously. The most obvious is energy; the other is disorder.

If you let go of a book, it falls to the floor because of gravity. At its initial height, it has a higher potential energy than it has when it is on the floor. Unless it is restrained, the book falls and so loses energy. This fact leads us to the first basic principle identifying spontaneous processes and the direction they take: *Processes in which the energy content of the system decreases tend to occur spontaneously*. Spontaneous processes tend to be exothermic. (Section 5.4, "Strategies in Chemistry: Using Enthalpy as a Guide") Change tends to occur in the direction that leads to a lower energy or lower enthalpy for the system.

Some processes, however, do not result in lower energy for a system or may even be endothermic and still occur spontaneously. For example, NH_4NO_3 readily dissolves in water, even though the solution process is endothermic. All such processes are characterized by an increase in the disorder, or randomness, of the system. The mixing of CCl_4 and C_6H_{14} provides another simple example. Suppose that we could suddenly remove a barrier that separates 500 mL of CCl_4 from 500 mL of C_6H_{14}, as in Figure 13.6(a) ▶. Before the barrier is removed, each liquid occupies a volume of 500 mL. All the CCl_4 molecules are in the 500 mL to the left of the barrier and all the C_6H_{14} molecules are in the 500 mL to the right. When equilibrium has been established after removing the barrier, the two liquids together occupy a volume of about 1000 mL. Formation of a homogeneous solution has increased disorder, or randomness, because the molecules of each substance are now mixed and distributed in a volume twice as large as that which they occupied before mixing. The amount of disorder in the system is given by a thermodynamic quantity called **entropy**. This example illustrates our second basic principle: *Processes in which the disorder (entropy) of the system increases tend to occur spontaneously*.

When molecules of different types are brought together, mixing and hence an increase in disorder occur spontaneously unless the molecules are restrained by sufficiently strong intermolecular forces or by physical barriers. Thus, gases spontaneously mix and expand unless restrained by their containers; in this case intermolecular forces are too weak to restrain the molecules. However, because strong bonds hold sodium and chloride ions together, sodium chloride does not spontaneously dissolve in gasoline.

We will discuss spontaneous processes again in Chapter 19. At that time we will consider the balance between the tendencies toward lower enthalpy and toward increased entropy in greater detail. For the moment we need to be aware that the solution process involves two factors: a change in enthalpy and a change in entropy. In most cases *formation of solutions is favored by the increase in entropy that accompanies mixing*. Consequently, a solution will form unless solute-solute or solvent–solvent interactions are too strong relative to the solute-solvent interactions.

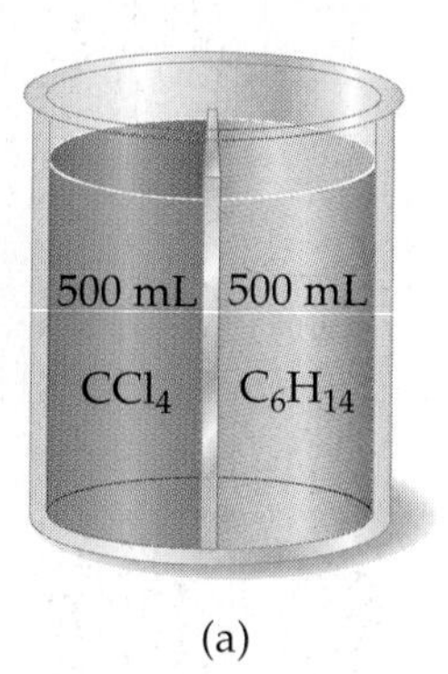

(a)

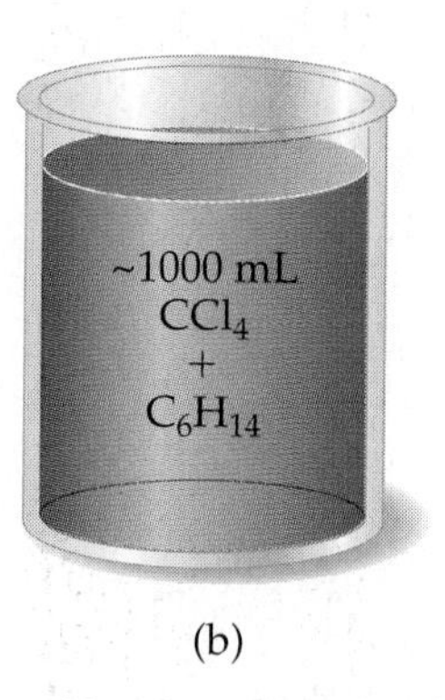

(b)

▲ **Figure 13.6** Formation of a homogeneous solution between CCl_4 and C_6H_{14} upon removal of a barrier separating the two liquids. The solution in (b) is more disordered, or random, in character than the separate liquids before solution formation (a).

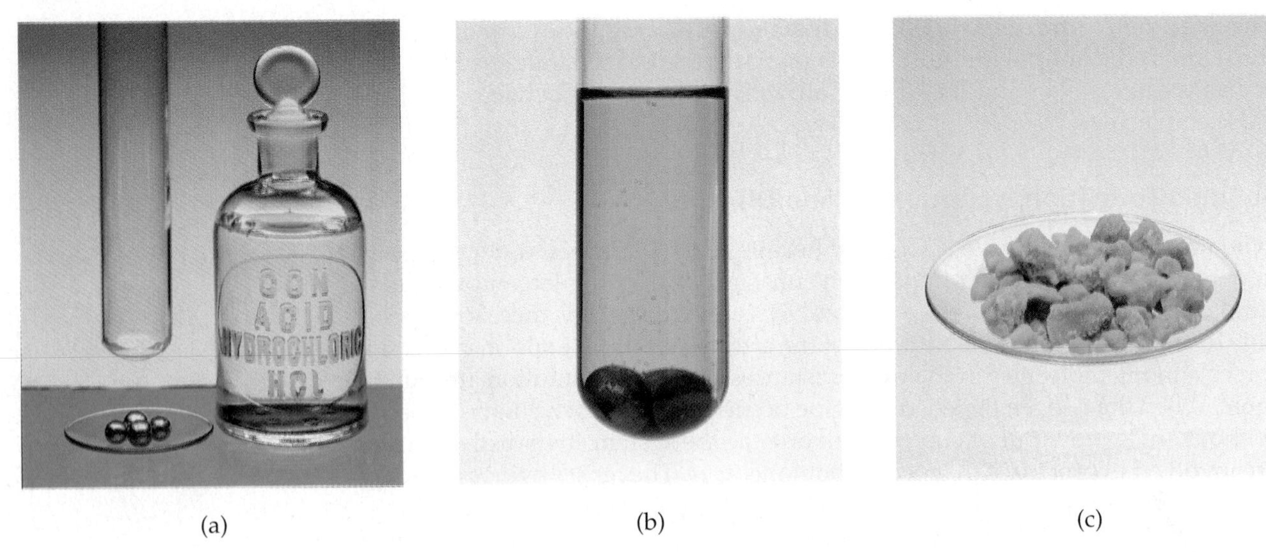

(a) (b) (c)

▲ **Figure 13.7** (a) Nickel metal and hydrochloric acid. (b) Nickel reacts slowly with hydrochloric acid, forming $NiCl_2(aq)$ and $H_2(g)$. (c) $NiCl_2 \cdot 6H_2O$ is obtained when the solution from (b) is evaporated to dryness.

Solution Formation and Chemical Reactions

In all our discussions of solutions, we must be careful to distinguish the physical process of solution formation from chemical reactions that lead to a solution. For example, nickel metal is dissolved on contact with hydrochloric acid solution because the following chemical reaction occurs:

$$Ni(s) + 2HCl(aq) \longrightarrow NiCl_2(aq) + H_2(g) \quad [13.2]$$

In this instance the chemical form of the substance being dissolved is changed from Ni to $NiCl_2$. If the solution is evaporated to dryness, $NiCl_2 \cdot 6H_2O(s)$, not $Ni(s)$, is recovered (Figure 13.7 ▲). When $NaCl(s)$ is dissolved in water, on the other hand, no chemical reaction occurs. If the solution is evaporated to dryness, NaCl is recovered. Our focus throughout this chapter is on solutions from which the solute can be recovered unchanged from the solution.

A Closer Look Hydrates

Frequently, hydrated ions remain in crystalline salts that are obtained by evaporation of water from aqueous solutions. Common examples include $FeCl_3 \cdot 6H_2O$ [iron(III) chloride hexahydrate] and $CuSO_4 \cdot 5H_2O$ [copper(II) sulfate pentahydrate]. The $FeCl_3 \cdot 6H_2O$ consists of $Fe(H_2O)_6{}^{3+}$ and Cl^- ions; the $CuSO_4 \cdot 5H_2O$ consists of $Cu(H_2O)_4{}^{2+}$ and $SO_4(H_2O)^{2-}$ ions. Water molecules can also occur in positions in the crystal lattice that are not specifically associated with either a cation or an anion. $BaCl_2 \cdot 2H_2O$ (barium chloride dihydrate) is an example. Compounds such as $FeCl_3 \cdot 6H_2O$, $CuSO_4 \cdot 5H_2O$, and $BaCl_2 \cdot 2H_2O$, which contain a salt and water combined in definite proportions, are known as *hydrates*; the water associated with them is called *water of hydration*. Figure 13.8 ▶ shows an example of a hydrate and the corresponding anhydrous (water-free) substance.

▲ **Figure 13.8** Samples of hydrated cobalt(II) chloride, $CoCl_2 \cdot 6H_2O$ (left), and anhydrous $CoCl_2$ (right).

13.2 Saturated Solutions and Solubility

As a solid solute begins to dissolve in a solvent, the concentration of solute particles in solution increases, and so do their chances of colliding with the surface of the solid (Figure 13.9 ▶). Such a collision may result in the solute particle becoming reattached to the solid. This process, which is the opposite of the solution process, is called **crystallization**. Thus, two opposing processes occur in a solution in contact with undissolved solute. This situation is represented in Equation 13.3 by use of a double arrow:

$$\text{Solute} + \text{solvent} \underset{\text{crystallize}}{\overset{\text{dissolve}}{\rightleftharpoons}} \text{solution} \qquad [13.3]$$

When the rates of these opposing processes become equal, no further net increase in the amount of solute in solution occurs. A dynamic equilibrium is established similar to the one between evaporation and condensation discussed in Section 11.5.

A solution that is in equilibrium with undissolved solute is **saturated**. Additional solute will not dissolve if added to a saturated solution. The amount of solute needed to form a saturated solution in a given quantity of solvent is known as the **solubility** of that solute. For example, the solubility of NaCl in water at 0°C is 35.7 g per 100 mL of water. This is the maximum amount of NaCl that can be dissolved in water to give a stable equilibrium solution at that temperature.

If we dissolve less solute than that needed to form a saturated solution, the solution is **unsaturated**. Thus, a solution containing only 10.0 g of NaCl per 100 mL of water at 0°C is unsaturated because it has the capacity to dissolve more solute.

Under suitable conditions it is sometimes possible to form solutions that contain a greater amount of solute than that needed to form a saturated solution. Such solutions are **supersaturated**. For example, considerably more sodium acetate ($NaC_2H_3O_2$) can dissolve in water at high temperatures than at low temperatures. When a saturated solution of sodium acetate is made at a high temperature and then slowly cooled, all of the solute may remain dissolved even though the solubility decreases as the temperature is reduced. Because the solute in a supersaturated solution is present in a concentration higher than the equilibrium concentration, supersaturated solutions are unstable. Supersaturated solutions result for much the same reason as supercooled liquids (Section 11.4): In order for crystallization to occur, the molecules or ions of solute must arrange themselves properly to form crystals. Addition of a small crystal of the solute (a seed crystal) provides a template for crystallization of the excess solute, leading to a saturated solution in contact with excess solid (Figure 13.10 ▼).

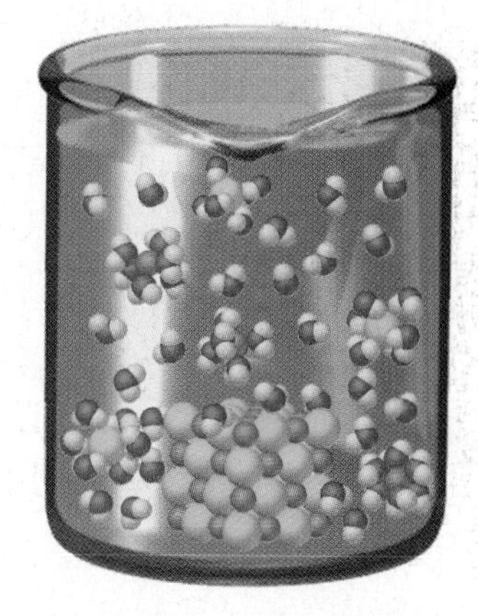

▲ **Figure 13.9** A solution in which excess ionic solute is present. Ions on the surface of the solute are continually passing into the solution as hydrated species, while hydrated ions from the solution are deposited on the surfaces of the solute. At equilibrium in a saturated solution, the two processes occur at equal rates.

(a)

(b)

(c)

▲ **Figure 13.10** Sodium acetate readily forms supersaturated solutions in water. (a) When a seed crystal of $NaC_2H_3O_2$ is added, excess $NaC_2H_3O_2$ crystallizes from the solution, as shown in (b) and (c).

13.3 Factors Affecting Solubility

The extent to which one substance dissolves in another depends on the nature of both the solute and the solvent. It also depends on temperature and, at least for gases, on pressure. Let's consider these factors more closely.

Solute–Solvent Interactions

One factor determining solubility is the natural tendency of substances to mix (the tendency of systems to move toward disorder). If this were all that was involved, however, we would expect substances to be completely soluble in one another. This is clearly not the case. So what other factors are involved? As we saw in Section 13.1, the relative forces of attraction among the solute and solvent molecules also play very important roles in the solution process.

Although the tendency toward disorder and the various interactions among solute and solvent particles are all involved in determining the solubilities, considerable insight can often be gained by focusing on the interaction between the solute and solvent. The data in Table 13.2 ◀ show, for example, that the solubilities of various simple gases in water increase with increasing molecular mass or polarity. The attractive forces between the gas and solvent molecules are mainly of the London dispersion type, which increase with increasing size and mass of the gas molecules. (Section 11.2) Thus, the data indicate that the solubilities of gases in water increase as the attraction between the solute (gas) and solvent (water) increases. In general, when other factors are comparable, *the stronger the attractions between solute and solvent molecules, the greater the solubility.*

TABLE 13.2 Solubilities of Gases in Water at 20°C, with 1 atm Gas Pressure

Gas	Solubility (*M*)
N_2	0.69×10^{-3}
CO	1.04×10^{-3}
O_2	1.38×10^{-3}
Ar	1.50×10^{-3}
Kr	2.79×10^{-3}

As a result of favorable dipole-dipole attractions between solvent molecules and solute molecules, *polar liquids tend to dissolve readily in polar solvents.* Water is not only polar, but also able to form hydrogen bonds. (Section 11.2) Thus, polar molecules, and especially those that can form hydrogen bonds with water molecules, tend to be soluble in water. For example, acetone, a polar molecule whose structural formula is shown below, mixes in all proportions with water. Acetone has a strongly polar C=O bond and pairs of nonbonding electrons on the O atom that can form hydrogen bonds with water.

$$CH_3\overset{\overset{\displaystyle :\!O\!:}{\|}}{C}CH_3$$

Acetone

Pairs of liquids such as acetone and water that mix in all proportions are **miscible**, whereas those that do not dissolve in one another are **immiscible**. Gasoline, which is a mixture of hydrocarbons, is immiscible in water. Hydrocarbons are nonpolar substances because of several factors: The C—C bonds are nonpolar, the C—H bonds are nearly nonpolar, and the shapes of the molecules are symmetrical enough to cancel much of the weak C—H bond dipoles. The attraction between the polar water molecules and the nonpolar hydrocarbon molecules is not sufficiently strong to allow the formation of a solution. *Nonpolar liquids tend to be insoluble in polar liquids.* As a result, hexane (C_6H_{14}) does not dissolve in water.

The series of compounds in Table 13.3 ▶ demonstrates that polar liquids tend to dissolve in other polar liquids and nonpolar liquids in nonpolar ones.

TABLE 13.3 Solubilities of Some Alcohols in Water and in Hexane

Alcohol	Solubility in H_2O[a]	Solubility in C_6H_{14}
CH_3OH (methanol)	∞	0.12
CH_3CH_2OH (ethanol)	∞	∞
$CH_3CH_2CH_2OH$ (propanol)	∞	∞
$CH_3CH_2CH_2CH_2OH$ (butanol)	0.11	∞
$CH_3CH_2CH_2CH_2CH_2OH$ (pentanol)	0.030	∞
$CH_3CH_2CH_2CH_2CH_2CH_2OH$ (hexanol)	0.0058	∞
$CH_3CH_2CH_2CH_2CH_2CH_2CH_2OH$ (heptanol)	0.0008	∞

[a]Expressed in mol alcohol/100 g solvent at 20°C. The infinity symbol indicates that the alcohol is completely miscible with the solvent.

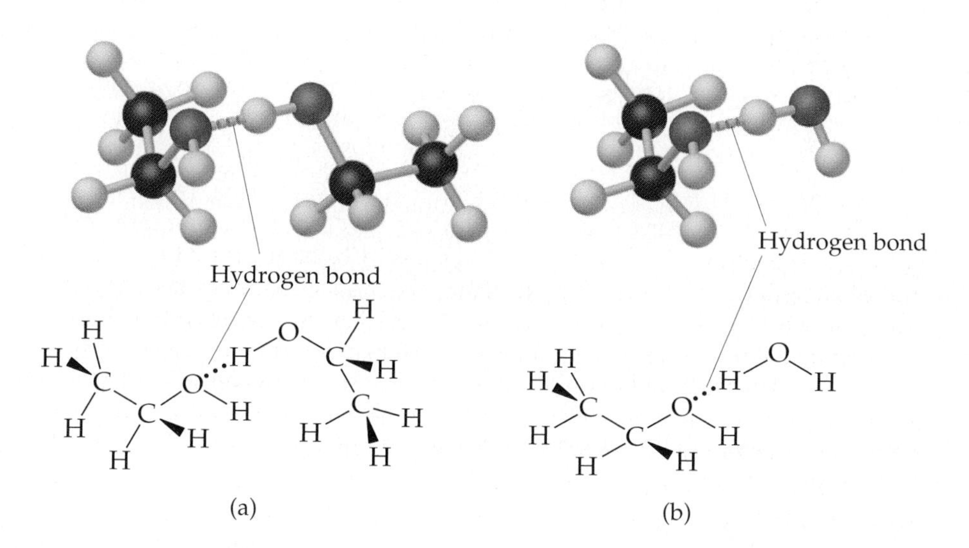

◀ **Figure 13.11** Hydrogen-bonding interactions between ethanol molecules (a) and between ethanol and water (b).

These organic compounds all contain the OH group attached to a C atom. Organic compounds with this molecular feature are called *alcohols*. The O—H bond is not only polar, but is also able to form hydrogen bonds. For example, CH_3CH_2OH molecules can form hydrogen bonds with water molecules as well as with each other (Figure 13.11 ▲). As a result, the solute-solute, solvent-solvent, and solute-solvent forces are not appreciably different within a mixture of CH_3CH_2OH and H_2O. There is no significant change in the environment of the molecules as they are mixed. Therefore, the increase in disorder accompanying mixing plays a significant role in formation of the solution. Ethanol (CH_3CH_2OH), therefore, is completely miscible with water.

The number of carbon atoms in an alcohol affects its solubility in water. As the length of the carbon chain increases, the polar OH group becomes an ever smaller part of the molecule, and the molecule behaves more like a hydrocarbon. The solubility of the alcohol in water decreases correspondingly. On the other hand, the solubility of the alcohol in a nonpolar solvent like hexane (C_6H_{14}) increases as the nonpolar hydrocarbon chain increases in length.

One way to enhance the solubility of a substance in water is to increase the number of polar groups it contains. For example, increasing the number of OH groups along a carbon chain of a solute increases the extent of hydrogen bonding between that solute and water, thereby increasing solubility. Glucose ($C_6H_{12}O_6$) has five OH groups on a six-carbon framework, which makes the molecule very soluble in water (83 g dissolve in 100 mL of water at 17.5°C). The glucose molecule is shown in Figure 13.12 ▶.

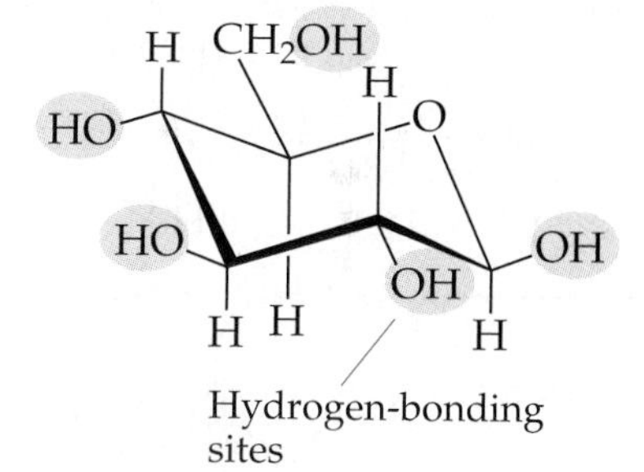

▲ **Figure 13.12** Structure of glucose. Note that OH groups capable of hydrogen bonding with water are prominent on the "surface" of the molecule.

Examination of different combinations of solvents and solutes such as those considered in the preceding paragraphs has led to an important generalization: *Substances with similar intermolecular attractive forces tend to be soluble in one another.* This generalization is often simply stated as "*like dissolves like.*" Nonpolar substances are more likely to be soluble in nonpolar solvents; ionic and polar solutes are more likely to be soluble in polar solvents. Network solids like diamond and quartz are not soluble in either polar or nonpolar solvents because of the strong bonding forces within the solid.

Chemistry and Life Fat- and Water-Soluble Vitamins

Vitamins have unique chemical structures, that affect their solubilities in different parts of the human body. Vitamins B and C are water soluble, for example, whereas vitamins A, D, E, and K are soluble in nonpolar solvents and in the fatty tissue of the body (which is nonpolar). Because of their water solubility, vitamins B and C are not stored to any appreciable extent in the body, and so foods containing these vitamins should be included in the daily diet. In contrast, the fat-soluble vitamins are stored in sufficient quantities to keep vitamin-deficiency diseases from appearing even after a person has subsisted for a long period on a vitamin-deficient diet.

The different solubility patterns of the water-soluble vitamins and the fat-soluble ones can be rationalized in terms of the structures of the molecules. The chemical structures of vitamin A (retinol) and of vitamin C (ascorbic acid) are shown in Figure 13.13 ▼. Notice that the vitamin A molecule is an alcohol with a very long carbon chain. Because the OH group is such a small part of the molecule, the molecule resembles the long-chain alcohols listed in Table 13.3. This vitamin is nearly nonpolar. In contrast, the vitamin C molecule is smaller and has more OH groups that can form hydrogen bonds with water. It is somewhat like glucose, which was discussed earlier. It is a more polar substance.

The Procter & Gamble Company introduced a no-calorie fat substitute called *olestra*™ in 1998. This substance, which is formed by the combination of a sugar molecule with fatty acids, is stable at high temperatures and thus can be used instead of vegetable oil in the preparation of potato chips, tortilla chips, and related products. Although it tastes like a vegetable oil, it passes through the human digestive system without being metabolized and thus contributes no calories to the diet. However, its use has generated some controversy. Because olestra consists of large fatlike molecules, it absorbs fat-soluble vitamins (such as A, D, E, and K). It absorbs other nutrients (such as carotenes) and carries them through the digestive system and out of the body. Critics worry that although olestra-containing foods are fortified with the vitamins that might be lost, the long-term consequences of regular olestra consumption could cause dietary problems.

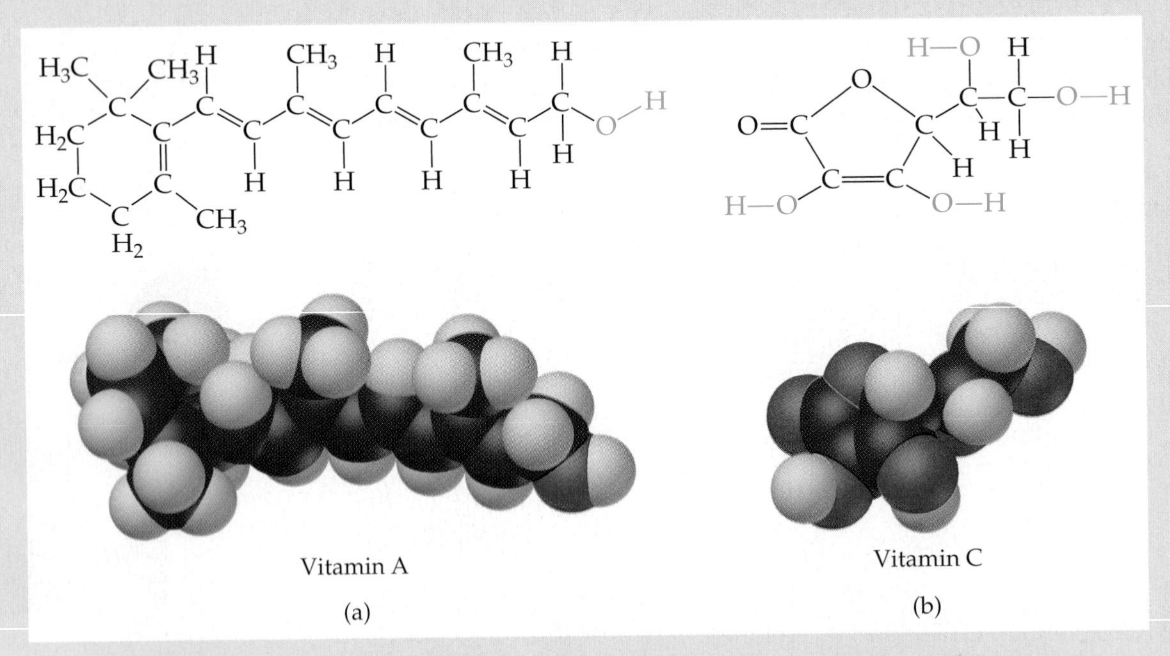

▲ **Figure 13.13** (a) The molecular structure of vitamin A, a fat-soluble vitamin. The molecule is composed largely of carbon–carbon and carbon–hydrogen bonds, so it is nearly nonpolar. (b) The molecular structure of vitamin C, a water-soluble vitamin. Notice the OH groups and the other oxygen atoms in the molecule that can interact with water molecules by hydrogen bonding.

SAMPLE EXERCISE 13.1

Predict whether each of the following substances is more likely to dissolve in carbon tetrachloride (CCl_4) or in water: C_7H_{16}, Na_2SO_4, HCl, and I_2.

Solution

Analyze: We are given two solvents, one that is nonpolar (CCl_4) and the other that is polar (H_2O), and asked to determine which will be the best solvent for each solute listed.

Plan: By examining the formulas of the solutes, we can predict whether they are ionic or molecular. For those that are molecular, we can predict whether they are polar or nonpolar. We can then apply the idea that the nonpolar solvent will be best for the nonpolar solutes, whereas the polar solvent will be best for the ionic and polar solutes.

Solve: C_7H_{16} is a hydrocarbon, so it is molecular and nonpolar. Na_2SO_4, a compound containing a metal and nonmetals, is ionic; HCl, a diatomic molecule containing two nonmetals that differ in electronegativity, is polar; and I_2, a diatomic molecule with atoms of equal electronegativity, is nonpolar. We would therefore predict that C_7H_{16} and I_2 would be more soluble in the nonpolar CCl_4 than in polar H_2O, whereas water would be the better solvent for Na_2SO_4 and HCl.

PRACTICE EXERCISE

Arrange the following substances in order of increasing solubility in water:

```
    H  H  H  H  H              H  H  H  H  H
    |  |  |  |  |              |  |  |  |  |
 H—C—C—C—C—C—H          HO—C—C—C—C—C—OH
    |  |  |  |  |              |  |  |  |  |
    H  H  H  H  H              H  H  H  H  H

    H  H  H  H  H              H  H  H  H  H
    |  |  |  |  |              |  |  |  |  |
 H—C—C—C—C—C—OH          H—C—C—C—C—C—Cl
    |  |  |  |  |              |  |  |  |  |
    H  H  H  H  H              H  H  H  H  H
```

Answer: $C_5H_{12} < C_5H_{11}Cl < C_5H_{11}OH < C_5H_{10}(OH)_2$ (in order of increasing polarity and hydrogen-bonding ability)

Pressure Effects

The solubilities of solids and liquids are not appreciably affected by pressure, whereas the solubility of a gas in any solvent is increased as the pressure over the solvent increases. We can understand the effect of pressure on the solubility of a gas by considering the dynamic equilibrium illustrated in Figure 13.14 ▶. Suppose that we have a gaseous substance distributed between the gas and solution phases. When equilibrium is established, the rate at which gas molecules enter the solution equals the rate at which solute molecules escape from the solution to enter the gas phase. The small arrows in Figure 13.14(a) represent the rates of these opposing processes. Now suppose that we exert added pressure on the piston and compress the gas above the solution, as shown in Figure 13.14(b). If we reduced the volume to half its original value, the pressure of the gas would increase to about twice its original value. The rate at which gas molecules strike the surface to enter the solution phase would therefore increase. As a result, the solubility of the gas in the solution would increase until an equilibrium is again established; that is, solubility increases until the rate at which gas molecules enter the solution equals the rate at which solute molecules escape from the solution. Thus, *the solubility of the gas increases in direct proportion to its partial pressure above the solution.*

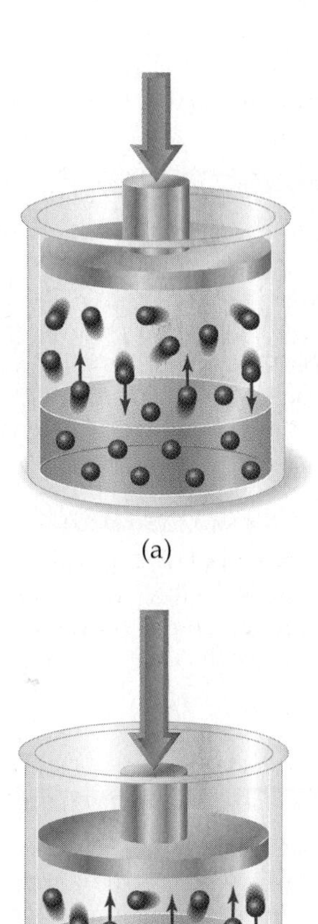

▲ **Figure 13.14** Effect of pressure on the solubility of a gas. When the pressure is increased, as in (b), the rate at which gas molecules enter the solution increases. The concentration of solute molecules at equilibrium increases in proportion to the pressure.

The relationship between pressure and the solubility of a gas is expressed by a simple equation known as **Henry's law**:

$$S_g = kP_g \qquad [13.4]$$

Here, S_g is the solubility of the gas in the solution phase (usually expressed as molarity), P_g is the partial pressure of the gas over the solution, and k is a proportionality constant known as the *Henry's law constant*. The Henry's law constant is different for each solute-solvent pair. It also varies with temperature. As an example, the solubility of N_2 gas in water at 25°C and 0.78 atm pressure is 5.3×10^{-4} *M*. The Henry's law constant for N_2 in water at 25°C is thus given by $(5.3 \times 10^{-4}$ mol/L)/0.78 atm = 6.8×10^{-4} mol/L-atm. If the partial pressure of N_2 is doubled, Henry's law predicts that the solubility in water at 25°C will also double, to 1.06×10^{-3} *M*.

ANIMATION
Henry's Law

Bottlers use the effect of pressure on solubility in producing carbonated beverages such as beer and many soft drinks. These are bottled under a carbon dioxide pressure greater than 1 atm. When the bottles are opened to the air, the partial pressure of CO_2 above the solution decreases. Hence, the solubility of CO_2 decreases, and CO_2 bubbles out of the solution (Figure 13.15 ◀).

▲ **Figure 13.15** CO_2 bubbles out of solution when a carbonated beverage is opened, because the CO_2 partial pressure above the solution is reduced.

SAMPLE EXERCISE 13.2

Calculate the concentration of CO_2 in a soft drink that is bottled with a partial pressure of CO_2 of 4.0 atm over the liquid at 25°C. The Henry's law constant for CO_2 in water at this temperature is 3.1×10^{-2} mol/L-atm.

Solution

Analyze and Plan: We are given the partial pressure of CO_2, P_{CO_2}, and the Henry's law constant, k. With this information, we can use Henry's law, Equation 13.4, to calculate the solubility, S_{CO_2}.

Solve: $S_{CO_2} = kP_{CO_2} = (3.1 \times 10^{-2}\ \text{mol/L-atm})(4.0\ \text{atm}) = 0.12\ \text{mol/L} = 0.12\ M$

Check: The units are correct for solubility, and the answer has two significant figures consistent with both the partial pressure of CO_2 and the value of Henry's constant.

Chemistry and Life Blood Gases and Deep-Sea Diving

Because the solubility of gases increases with increasing pressure, divers who breathe compressed air (Figure 13.16 ▶) must be concerned about the solubility of gases in their blood. Although the gases are not very soluble at sea level, their solubilities can become appreciable deep underwater where their partial pressures are greater. Thus, deep-sea divers must ascend slowly to prevent dissolved gases from being released rapidly from blood and other fluids in the body. These bubbles affect nerve impulses and give rise to the affliction known as decompression sickness, or "the bends," which is painful and can be fatal. Nitrogen is the main problem because it has the highest partial pressure in air and because it can be removed only through the respiratory system. Oxygen, in contrast, is consumed in metabolism.

Deep-sea divers sometimes substitute helium for nitrogen in the air that they breathe, because helium has a much lower solubility in biological fluids than N_2. For example, divers working at a depth of 100 ft experience a pressure of about 4 atm. At this pressure a mixture of 95% helium and 5% oxygen will give an oxygen partial pressure of about 0.2 atm, which is the partial pressure of oxygen in normal air at 1 atm. If the oxygen partial pressure becomes too great, the urge to breathe is reduced, CO_2 is not removed from the body and CO_2 poisoning occurs. At excessive concentrations in the body carbon dioxide acts as a neurotoxin, interfering with nerve conduction and transmission.

▲ **Figure 13.16** Divers who use compressed gases must be concerned about the solubility of gases in their blood.

PRACTICE EXERCISE

Calculate the concentration of CO_2 in a soft drink after the bottle is opened and equilibrates at 25°C under a CO_2 partial pressure of 3.0×10^{-4} atm.
Answer: 9.3×10^{-6} *M*

Temperature Effects

The solubility of most solid solutes in water increases as the temperature of the solution increases. Figure 13.17 ▼ shows this effect for several ionic substances in water. There are exceptions to this rule, however, as seen for $Ce_2(SO_4)_3$ whose solubility curve slopes downward with increasing temperature.

In contrast to solid solutes, *the solubility of gases in water decreases with increasing temperature* (Figure 13.18 ▼). If a glass of cold tap water is warmed, bubbles of air are seen on the inside of the glass. Similarly, carbonated beverages go flat as they are allowed to warm; as the temperature of the solution increases, the solubility of CO_2 decreases and $CO_2(g)$ escapes from the solution. The decreased solubility of O_2 in water as temperature increases is one of the effects of *thermal pollution* of lakes and streams. The effect is particularly serious in deep lakes because warm water is less dense than cold water. It therefore tends to remain on top of cold water, at the surface. This situation impedes the dissolving of oxygen into the deeper layers, thus stifling the respiration of all aquatic life needing oxygen. Fish may suffocate and die under these conditions.

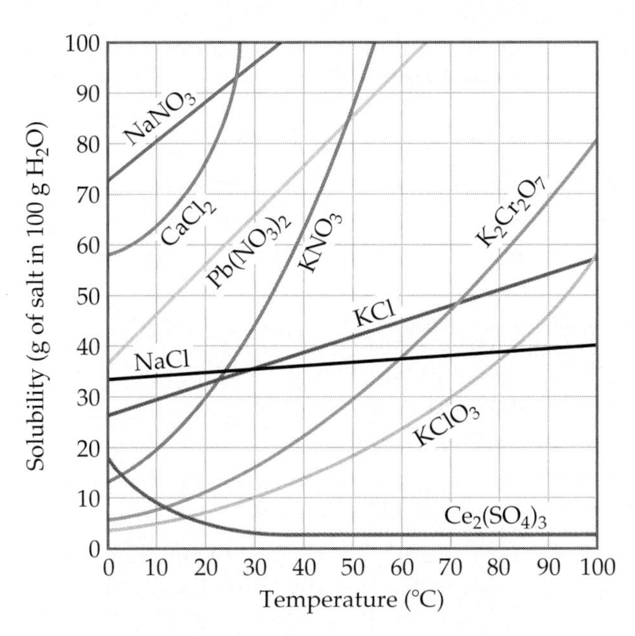

▲ **Figure 13.17** Solubilities of several ionic compounds in water as a function of temperature.

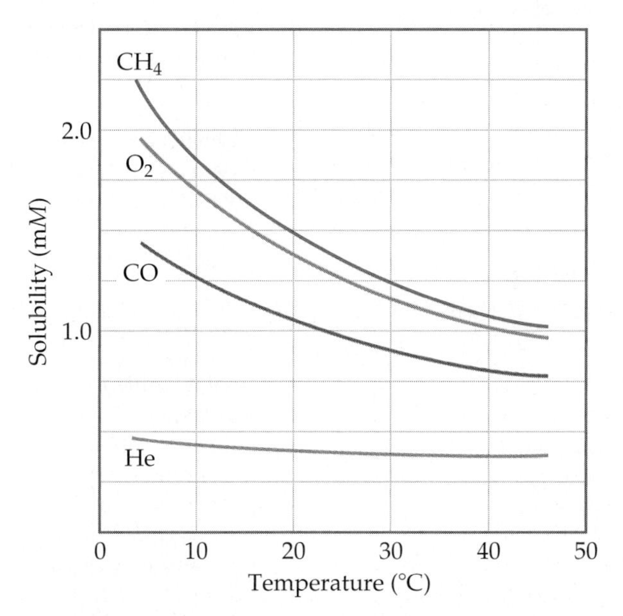

▲ **Figure 13.18** Solubilities of several gases in water as a function of temperature. Note that solubilities are in units of millimoles per liter (mmol/L), for a constant total pressure of 1 atm in the gas phase.

13.4 Ways of Expressing Concentration

The concentration of a solution can be expressed either qualitatively or quantitatively. The terms *dilute* and *concentrated* are used to describe a solution qualitatively. A solution with a relatively small concentration of solute is said to be dilute; one with a large concentration is said to be concentrated. We use several different ways to express concentration in quantitative terms, and we examine four of these in this section: mass percentage, mole fraction, molarity, and molality.

Mass Percentage, ppm, and ppb

One of the simplest quantitative expressions of concentration is the **mass percentage** of a component in a solution, given by

$$\text{Mass \% of component} = \frac{\text{mass of component in soln}}{\text{total mass of soln}} \times 100 \qquad [13.5]$$

where we have abbreviated *solution* as "soln." Thus, a solution of hydrochloric acid that is 36% HCl by mass contains 36 g of HCl for each 100 g of solution.

We often express the concentrations of very dilute solution in **parts per million (ppm)**, defined as

$$\text{ppm of component} = \frac{\text{mass of component in soln}}{\text{total mass of soln}} \times 10^6 \qquad [13.6]$$

A solution whose solute concentration is 1 ppm contains 1 g of solute for each million (10^6) grams of solution or, equivalently, 1 mg of solute per kilogram of solution. Because the density of water is 1 g/mL, 1 kg of a dilute aqueous solution will have a volume very close to 1 L. Thus, 1 ppm also corresponds to 1 mg of solute per liter of solution. The acceptable maximum concentrations of toxic or carcinogenic substances are often expressed in ppm. For example, the maximum allowable concentration of arsenic in drinking water in the United States is 0.010 ppm; that is, 0.010 mg of arsenic per liter of water.

For solutions that are even more dilute, **parts per billion (ppb)** is used. A concentration of 1 ppb represents 1 g of solute per billion (10^9) grams of solution, or 1 microgram (μg) of solute per liter of solution. Thus, the allowable concentration of arsenic in water can be expressed as 10 ppb.

SAMPLE EXERCISE 13.3

(a) A solution is made by dissolving 13.5 g of glucose ($C_6H_{12}O_6$) in 0.100 kg of water. What is the mass percentage of solute in this solution? **(b)** A 2.5-g sample of groundwater was found to contain 5.4 μg of Zn^{2+}. What is the concentration of Zn^{2+} in parts per million?

Solution (a) Analyze and Plan: We are given the number of grams of solute (13.5 g) and the number of grams of solvent (0.100 kg = 100 g). We calculate the mass percentage by using Equation 13.5. The mass of the solution is the sum of the mass of solute (glucose) and the mass of solvent (water).

Solve:

$$\text{Mass \% of glucose} = \frac{\text{mass glucose}}{\text{mass soln}} \times 100 = \frac{13.5\text{ g}}{13.5\text{ g} + 100\text{ g}} \times 100 = 11.9\%$$

Comment: The mass percentage of water in this solution is (100 − 11.9)% = 88.1%.

(b) Analyze and Plan: We are given the number of micrograms of solute. Because 1 μg is 1×10^{-6} g, 5.4 μg = 5.4×10^{-6} g. We calculate the parts per million using Equation 13.6.

Solve:

$$\text{ppm} = \frac{\text{mass of solute}}{\text{mass of soln}} \times 10^6 = \frac{5.4 \times 10^{-6}\text{ g}}{2.5\text{ g}} \times 10^6 = 2.2\text{ ppm}$$

PRACTICE EXERCISE

(a) Calculate the mass percentage of NaCl in a solution containing 1.50 g of NaCl in 50.0 g of water. **(b)** A commercial bleaching solution contains 3.62 mass % sodium hypochlorite, NaOCl. What is the mass of NaOCl in a bottle containing 2500 g of bleaching solution?

Answers: **(a)** 2.91%; **(b)** 90.5 g of NaOCl

Mole Fraction, Molarity, and Molality

Concentration expressions are often based on the number of moles of one or more components of the solution. The three most commonly used are mole fraction, molarity, and molality.

Recall from Section 10.6 that the *mole fraction* of a component of a solution is given by

$$\text{Mole fraction of component} = \frac{\text{moles of component}}{\text{total moles of all components}} \quad [13.7]$$

The symbol X is commonly used for mole fraction, with a subscript to indicate the component of interest. For example, the mole fraction of HCl in a hydrochloric acid solution is represented as X_{HCl}. Thus a solution containing 1.00 mol of HCl (36.5 g) and 8.00 mol of water (144 g) has a mole fraction of HCl of $X_{HCl} = (1.00 \text{ mol})/(1.00 \text{ mol} + 8.00 \text{ mol}) = 0.111$. Mole fractions have no units because the units in the numerator and the denominator cancel. The sum of the mole fractions of all components of a solution must equal 1. Thus, in the aqueous HCl solution, $X_{H_2O} = 1.000 - 0.111 = 0.889$. Mole fractions are very useful when dealing with gases as we saw in Section 10.6, but have limited use when dealing with liquid solutions.

Recall from Section 4.5 that the *molarity* (M) of a solute in a solution is defined as

$$\text{Molarity} = \frac{\text{moles solute}}{\text{liters soln}} \quad [13.8]$$

For example, if you dissolve 0.500 mol of Na_2CO_3 in enough water to form 0.250 L of solution, then the solution has a concentration of (0.500 mol)/(0.250 L) = 2.00 M in Na_2CO_3. Molarity is especially useful for relating the volume of a solution to the quantity of solute it contains, as we saw in our discussions of titrations. (Section 4.6)

The **molality** of a solution, denoted m, is a unit that we haven't encountered in previous chapters. This concentration unit equals the number of moles of solute per kilogram of solvent:

$$\text{Molality} = \frac{\text{moles solute}}{\text{kilograms of solvent}} \quad [13.9]$$

Thus, if you form a solution by mixing 0.200 mol of NaOH (40.0 g) and 0.500 kg of water (500 g), the concentration of the solution is (0.200 mol)/(0.500 kg) = 0.400 m (that is, 0.400 molal) in NaOH.

The definitions of molarity and molality are similar enough that they can be easily confused. Molarity depends on the *volume* of *solution*, whereas molality depends on the *mass* of *solvent*. When water is the solvent, the molality and molarity of dilute solutions are numerically about the same because 1 kg of solvent is nearly the same as 1 kg of solution, and 1 kg of the solution has a volume of about 1 L.

The molality of a given solution does not vary with temperature because masses do not vary with temperature. Molarity, however, changes with temperature because the expansion or contraction of the solution changes its volume. Thus molality is often the concentration unit of choice when a solution is to be used over a range of temperatures.

SAMPLE EXERCISE 13.4

A solution is made by dissolving 4.35 g glucose ($C_6H_{12}O_6$) in 25.0 mL of water. Calculate the molality of glucose in the solution.

Solution

Analyze and Plan: To calculate the molality, we must determine the number of moles of solute (glucose) and the number of kilograms of the solvent (water). We use the molar mass of $C_6H_{12}O_6$ to convert grams to moles. We use the density of water to convert milliliters to kilograms. The molality equals the number of moles of solute divided by the number of kilograms of solvent (Equation 13.9).

Solve: Use the molar mass of glucose, 180.2 g/mol, to convert grams to moles:

$$\text{Mol } C_6H_{12}O_6 = (4.35 \text{ g } C_6H_{12}O_6)\left(\frac{1 \text{ mol } C_6H_{12}O_6}{180.2 \text{ g } C_6H_{12}O_6}\right) = 0.0241 \text{ mol } C_6H_{12}O_6$$

Because water has a density of 1.00 g/mL, the mass of the solvent is (25.0 mL)(1.00 g/mL) = 25.0 g = 0.0250 kg. Finally, use Equation 13.9 to obtain the molality:

$$\text{Molality of } C_6H_{12}O_6 = \frac{0.0241 \text{ mol } C_6H_{12}O_6}{0.0250 \text{ kg } H_2O} = 0.964\ m$$

PRACTICE EXERCISE

What is the molality of a solution made by dissolving 36.5 g of naphthalene ($C_{10}H_8$) in 425 g of toluene (C_7H_8)?

Answer: 0.670 *m*

Conversion of Concentration Units

Sometimes the concentration of a given solution needs to be known in several different concentration units. It is possible to interconvert concentration units as shown in Sample Exercises 13.5 and 13.6.

SAMPLE EXERCISE 13.5

A solution of hydrochloric acid contains 36% HCl by mass. **(a)** Calculate the mole fraction of HCl in the solution. **(b)** Calculate the molality of HCl in the solution.

Solution

Analyze and Plan: In converting concentration units based on the mass or moles of solute and solvent (mass percentage, mole fraction, and molality), it is useful to assume a certain total mass of solution. Let's assume that there is exactly 100 g of solution. Because the solution is 36% HCl, it contains 36 g of HCl and (100 − 36) g = 64 g of H_2O. We must convert grams of solute (HCl) to moles in order to calculate either mole fraction or molality. We must convert grams of solvent (H_2O) to moles to calculate mole fractions, and to kilograms to calculate molality.

Solve: (a) To calculate the mole fraction of HCl, we convert the masses of HCl and H_2O to moles and then use Equation 13.7:

$$\text{Moles HCl} = (36 \text{ g HCl})\left(\frac{1 \text{ mol HCl}}{36.5 \text{ g HCl}}\right) = 0.99 \text{ mol HCl}$$

$$\text{Moles } H_2O = (64 \text{ g } H_2O)\left(\frac{1 \text{ mol } H_2O}{18 \text{ g } H_2O}\right) = 3.6 \text{ mol } H_2O$$

$$X_{HCl} = \frac{\text{moles HCl}}{\text{moles } H_2O + \text{moles HCl}} = \frac{0.99}{3.6 + 0.99} = \frac{0.99}{4.6} = 0.22$$

(b) To calculate the molality of HCl in the solution, we use Equation 13.9. We calculated the number of moles of HCl in part (a), and the mass of solvent is 64 g = 0.064 kg:

$$\text{Molality of HCl} = \frac{0.99 \text{ mol HCl}}{0.064 \text{ kg } H_2O} = 15\ m$$

PRACTICE EXERCISE

A commercial bleach solution contains 3.62 mass % NaOCl in water. Calculate **(a)** the molality and **(b)** the mole fraction of NaOCl in the solution.
Answers: **(a)** 0.505 *m*; **(b)** 9.00×10^{-3}

In order to interconvert molality and molarity, we need to know the density of the solution. Figure 13.19 ▼ outlines the calculation of the molarity and molality of a solution from the mass of solute and the mass of solvent. The mass of the solution is the sum of masses of the solvent and solute. The volume of the solution can be calculated from its mass and density.

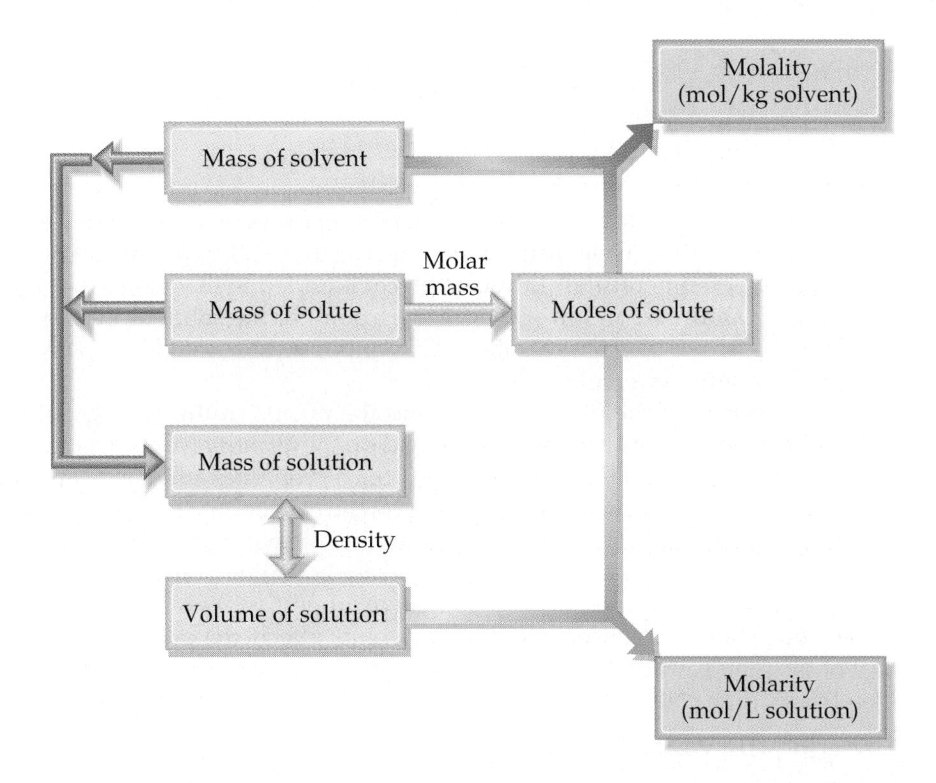

◀ **Figure 13.19** Diagram summarizing the calculation of molality and molarity from the mass of the solute, the mass of the solvent, and the density of the solution.

SAMPLE EXERCISE 13.6

A solution contains 5.0 g of toluene (C_7H_8) and 225 g of benzene and has a density of 0.876 g/mL. Calculate the molarity of the solution.

Solution

Analyze and Plan: The molarity of a solution is the number of moles of solute divided by the number of liters of solution (Equation 13.8). The number of moles of solute (C_7H_8) is calculated from the number of grams and the molar mass. The volume of the solution is obtained from the mass of the solution (mass of solute + mass of solvent = 5.0 g + 225 g = 230 g) and its density.

Solve: The number of moles of solute is

$$\text{Moles } C_7H_8 = (5.0 \text{ g } C_7H_8)\left(\frac{1 \text{ mol } C_7H_8}{92 \text{ g } C_7H_8}\right) = 0.054 \text{ mol}$$

The density of the solution is used to convert the mass of the solution to its volume:

$$\text{Milliliters soln} = (230 \text{ g})\left(\frac{1 \text{ mL}}{0.876 \text{ g}}\right) = 263 \text{ mL}$$

Molarity is moles of solute per liter of solution:

$$\text{Molarity} = \left(\frac{\text{moles } C_7H_8}{\text{liter soln}}\right) = \left(\frac{0.054 \text{ mol } C_7H_8}{263 \text{ mL soln}}\right)\left(\frac{1000 \text{ mL soln}}{1 \text{ L soln}}\right) = 0.21\ M$$

Check: The magnitude of our answer is reasonable. Rounding moles to 0.05 and liters to 0.25 gives a molarity of

$$(0.05 \text{ mol})/(0.25 \text{ L}) = 0.2\ M$$

The units for our answer (mol/L) are correct, and the answer has two significant figures, corresponding to the number of significant figures in the mass of solute (2).

Comment: Because the mass of the solvent (0.225 kg) and the volume of the solution (0.263 L) are similar in magnitude, the molarity and molality are also similar in magnitude:

$$(0.054 \text{ mol } C_7H_8)/(0.225 \text{ kg solvent}) = 0.24\ m$$

PRACTICE EXERCISE

A solution containing equal masses of glycerol ($C_3H_8O_3$) and water has a density of 1.10 g/mL. Calculate **(a)** the molality of glycerol; **(b)** the mole fraction of glycerol; **(c)** the molarity of glycerol in the solution.
Answers: **(a)** 10.9 *m*; **(b)** $X_{C_3H_8O_3} = 0.163$; **(c)** 5.97 *M*

13.5 Colligative Properties

Some physical properties of solutions differ in important ways from those of the pure solvent. For example, pure water freezes at 0°C, but aqueous solutions freeze at lower temperatures. Ethylene glycol is added to the water in radiators of cars as an antifreeze to lower the freezing point of the solution. It also raises the boiling point of the solution above that of pure water, making it possible to operate the engine at a higher temperature.

The lowering of the freezing point and the raising of the boiling point are physical properties of solutions that depend on the *quantity* (concentration), but not the *kind* or *identity* of the solute particles. Such properties are called **colligative properties**. (*Colligative* means "depending on the collection"; colligative properties depend on the collective effect of the number of solute particles.) In addition to the decrease in freezing point and the increase in boiling point, vapor-pressure reduction and osmotic pressure are colligative properties. As we examine each of these, notice how the concentration of the solute affects the property relative to that of the pure solvent.

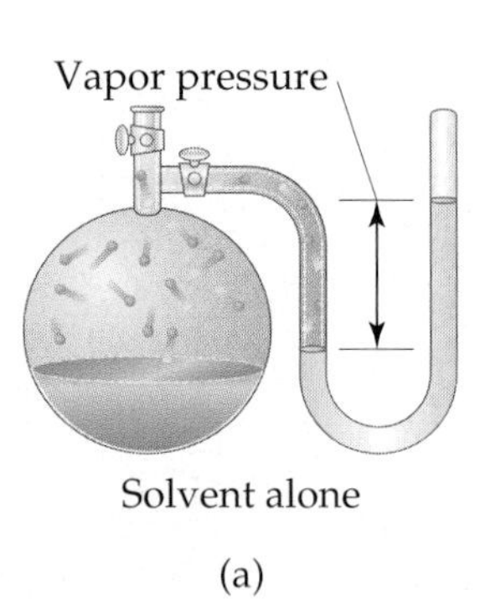

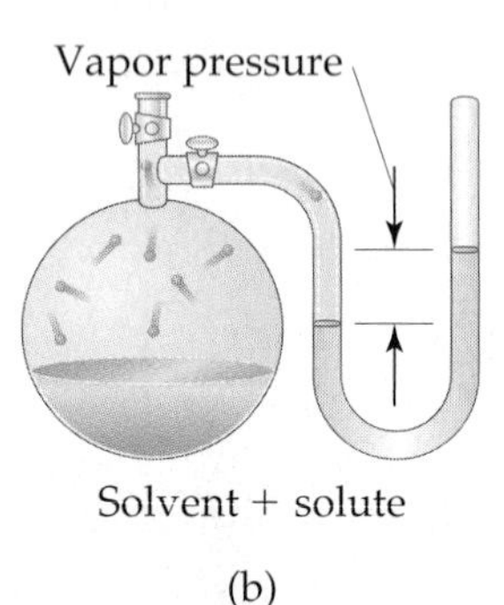

▲ **Figure 13.20** The vapor pressure over a solution formed by a volatile solvent and a nonvolatile solute (b) is lower than that of the solvent alone (a). The extent of the decrease in the vapor pressure upon addition of the solute depends on the concentration of the solute.

Lowering the Vapor Pressure

We learned in Section 11.5 that a liquid in a closed container will establish an equilibrium with its vapor. When that equilibrium is reached, the pressure exerted by the vapor is called the *vapor pressure*. A substance that has no measurable vapor pressure is *nonvolatile*, whereas one that exhibits a vapor pressure is *volatile*.

When we compare the vapor pressures of various solvents with those of their solutions, we find that adding a nonvolatile solute to a solvent always lowers the vapor pressure. This effect is illustrated in Figure 13.20 ◀. The extent to which a nonvolatile solute lowers the vapor pressure is proportional to its concentration. This relationship is expressed by **Raoult's law**, which states that the partial pressure exerted by solvent vapor above a solution, P_A, equals the product of the mole fraction of the solvent in the solution, X_A, times the vapor pressure of the pure solvent:

$$P_A = X_A P_A^\circ \qquad [13.10]$$

For example, the vapor pressure of water is 17.5 torr at 20°C. Imagine holding the temperature constant while adding glucose ($C_6H_{12}O_6$) to the water so that the resulting solution has $X_{H_2O} = 0.800$ and $X_{C_6H_{12}O_6} = 0.200$. According to Equation 13.10, the vapor pressure of water over the solution will be 80.0% of that of pure water:

$$P_{H_2O} = (0.800)(17.5 \text{ torr}) = 14.0 \text{ torr}$$

ACTIVITY
Boiling Point Elevation and Freezing Point Depression

In other words, the presence of the nonvolatile solute lowers the vapor pressure of the volatile solvent by 17.5 torr − 14.0 torr = 3.5 torr.

Raoult's law predicts that when we increase the mole fraction of nonvolatile solute particles in a solution, the vapor pressure over the solution will be reduced. In fact, the reduction in vapor pressure depends on the total concentration of solute particles, regardless of whether they are molecules or ions. Remember that vapor-pressure lowering is a colligative property, so it depends on the concentration of solute particles and not on their kind. In our applications of Raoult's law, however, we will limit ourselves to solutes that are not only nonvolatile but nonelectrolytes as well. We consider the effects of volatile substances on vapor pressure in the "Closer Look" box in this section, and we will consider the effects of electrolytes in our discussions of freezing points and boiling points.

A Closer Look Ideal Solutions with Two or More Volatile Components

Solutions sometimes have two or more volatile components. Gasoline, for example, is a complex solution containing several volatile substances. To gain some understanding of such mixtures, consider an ideal solution containing two components, A and B. The partial pressures of A and B vapors above the solution are given by Raoult's law:

$$P_A = X_A P_A^\circ \quad \text{and} \quad P_B = X_B P_B^\circ$$

The total vapor pressure over the solution is the sum of the partial pressures of each volatile component:

$$P_{total} = P_A + P_B = X_A P_A^\circ + X_B P_B^\circ$$

Consider, for example, a mixture of benzene (C_6H_6) and toluene (C_7H_8) containing 1.0 mol of benzene and 2.0 mol of toluene ($X_{ben} = 0.33$ and $X_{tol} = 0.67$). At 20°C the vapor pressures of the pure substances are

$$\text{Benzene:} \quad P_{ben}^\circ = 75 \text{ torr}$$

$$\text{Toluene:} \quad P_{tol}^\circ = 22 \text{ torr}$$

Thus, the partial pressures of benzene and toluene above the solution are

$$P_{ben} = (0.33)(75 \text{ torr}) = 25 \text{ torr}$$

$$P_{tol} = (0.67)(22 \text{ torr}) = 15 \text{ torr}$$

The total vapor pressure is

$$P_{total} = 25 \text{ torr} + 15 \text{ torr} = 40 \text{ torr}$$

The vapor, therefore, is richer in benzene, the more volatile component. The mole fraction of benzene in the vapor is given by the ratio of its vapor pressure to the total pressure (Equation 10.15):

$$X_{ben} \text{ in vapor} = \frac{P_{ben}}{P_{total}} = \frac{25 \text{ torr}}{40 \text{ torr}} = 0.63$$

Although benzene constitutes only 33% of the molecules in the solution, it makes up 63% of the molecules in the vapor.

When ideal solutions are in equilibrium with their vapor, the more volatile component of the mixture will be relatively richer in the vapor. This fact forms the basis of *distillation*, a technique used to separate (or partially separate) mixtures containing volatile components. Distillation is the procedure by which a moonshiner obtains whiskey using a still and by which petrochemical plants achieve the separation of crude petroleum into gasoline, diesel fuel, lubricating oil, and so forth (Figure 13.21 ▼). It is also used routinely on a small scale in the laboratory. A specially designed *fractional distillation* apparatus can achieve in a single operation a degree of separation that would be equivalent to several successive simple distillations.

▲ **Figure 13.21** Industrial distillation towers, in which the components of a volatile organic mixture are separated according to boiling-point range.

SAMPLE EXERCISE 13.7

Glycerin ($C_3H_8O_3$) is a nonvolatile nonelectrolyte with a density of 1.26 g/mL at 25°C. Calculate the vapor pressure at 25°C of a solution made by adding 50.0 mL of glycerin to 500.0 mL of water. The vapor pressure of pure water at 25°C is 23.8 torr (Appendix B).

Solution

Analyze and Plan: Use Raoult's law (Equation 13.10) to determine the vapor pressure of a solution. The mole fraction of the solvent in the solution, X_A, is the ratio of the number of moles of solvent (H_2O) to total solution (moles $C_3H_8O_3$ + moles H_2O).

Solve: To calculate the mole fraction of water in the solution, we must determine the number of moles of $C_3H_8O_3$ and H_2O:

$$\text{Moles } C_3H_8O_3 = (50.0 \text{ mL } C_3H_8O_3)\left(\frac{1.26 \text{ g } C_3H_8O_3}{1 \text{ mL } C_3H_8O_3}\right)\left(\frac{1 \text{ mol } C_3H_8O_3}{92.1 \text{ g } C_3H_8O_3}\right) = 0.684 \text{ mol}$$

$$\text{Moles } H_2O = (500.0 \text{ mL } H_2O)\left(\frac{1.00 \text{ g } H_2O}{1 \text{ mL } H_2O}\right)\left(\frac{1 \text{ mol } H_2O}{18.0 \text{ g } H_2O}\right) = 27.8 \text{ mol}$$

$$X_{H_2O} = \frac{\text{mol } H_2O}{\text{mol } H_2O + \text{mol } C_3H_8O_3} = \frac{27.8}{27.8 + 0.684} = 0.976$$

We now use Raoult's law to calculate the vapor pressure of water for the solution:

$$P_{H_2O} = X_{H_2O}P^{\circ}_{H_2O} = (0.976)(23.8 \text{ torr}) = 23.2 \text{ torr}$$

The vapor pressure of the solution has been lowered by 0.6 torr relative to that of pure water.

PRACTICE EXERCISE

The vapor pressure of pure water at 110°C is 1070 torr. A solution of ethylene glycol and water has a vapor pressure of 1.00 atm at 110°C. Assuming that Raoult's law is obeyed, what is the mole fraction of ethylene glycol in the solution?
Answer: 0.290

An ideal gas obeys the ideal-gas equation (Section 10.4), and an **ideal solution** obeys Raoult's law. Real solutions best approximate ideal behavior when the solute concentration is low and when the solute and solvent have similar molecular sizes and similar types of intermolecular attractions.

Many solutions do not obey Raoult's law exactly: They are not ideal solutions. If the intermolecular forces between solvent and solute are weaker than those between solvent and solvent and between solute and solute, then the solvent vapor pressure tends to be greater than predicted by Raoult's law. Conversely, when the interactions between solute and solvent are exceptionally strong, as might be the case when hydrogen bonding exists, the solvent vapor pressure is lower than Raoult's law predicts. Although you should be aware that these departures from ideal solution occur, we will ignore them for the remainder of this chapter.

Boiling-Point Elevation

In Sections 11.5 and 11.6 we examined the vapor pressures of pure substances and how they can be used to construct phase diagrams. How will the phase diagram of a solution, and hence its boiling and freezing points, differ from those of the pure solvent? The addition of a nonvolatile solute lowers the vapor pressure of the solution. Thus, as shown in Figure 13.22 ▶, the vapor-pressure curve of the solution (blue line) will be shifted downward relative to the vapor-pressure curve of the pure liquid (black line); at any given temperature, the vapor pressure of the solution is lower than that of the pure liquid. Recall that the normal boiling point of a liquid is the temperature at which its vapor pressure equals 1 atm. (Section 11.5) At the normal boiling point of the pure liquid, the vapor pressure of the solution will be less than 1 atm (Figure 13.22). Therefore, a higher

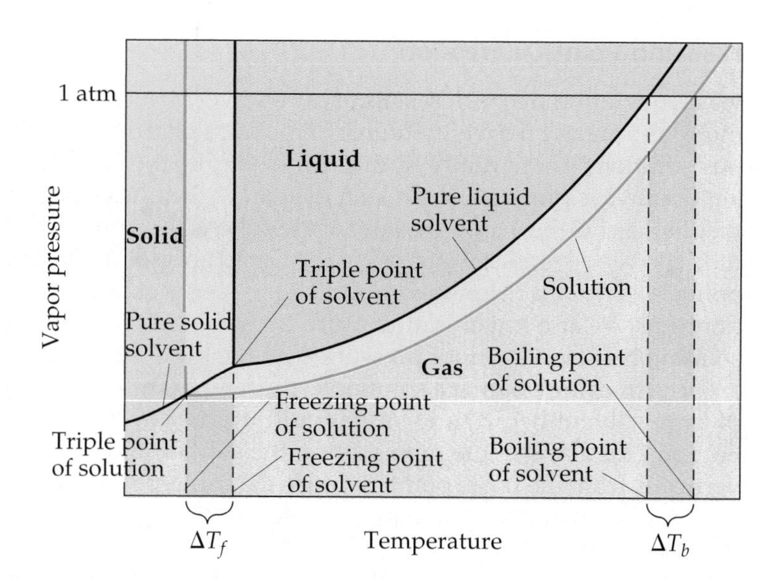

◀ **Figure 13.22** Phase diagrams for a pure solvent and for a solution of a nonvolatile solute. The vapor pressure of the solid solvent is unaffected by the presence of solute if the solid freezes out without containing a significant concentration of solute, as is usually the case.

temperature is required to attain a vapor pressure of 1 atm. Thus, *the boiling point of the solution is higher than that of the pure liquid.*

The increase in boiling point relative to that of the pure solvent, ΔT_b, is directly proportional to the number of solute particles per mole of solvent molecules. We know that molality expresses the number of moles of solute per 1000 g of solvent, which represents a fixed number of moles of solvent. Thus, ΔT_b is proportional to molality:

$$\Delta T_b = K_b m \qquad [13.11]$$

The magnitude of K_b, which is called the **molal boiling-point-elevation constant**, depends only on the solvent. Some typical values for several common solvents are given in Table 13.4 ▼.

For water, K_b is 0.51°C/m; therefore, a 1-m aqueous solution of sucrose or any other aqueous solution that is 1 m in nonvolatile solute particles will boil 0.51°C higher than pure water. The boiling-point elevation is proportional to the concentration of solute particles, regardless of whether the particles are molecules or ions. When NaCl dissolves in water, 2 mol of solute particles (1 mol of Na^+ and 1 mol of Cl^-) are formed for each mole of NaCl that dissolves. Therefore, a 1-m aqueous solution of NaCl is 1 m in Na^+ and 1 m in Cl^-, making it 2 m in total solute particles. As a result, the boiling-point elevation of a 1-m aqueous solution of NaCl is approximately $(2\ m)(0.51°C/m) = 1°C$, twice as large as a 1-m solution of a nonelectrolyte such as sucrose. Thus, to properly predict the effect of a particular solute on the boiling point (or any other colligative property), it is important to know whether the solute is an electrolyte or a nonelectrolyte. ∞ (Sections 4.1 and 4.3)

TABLE 13.4 Molal Boiling-Point-Elevation and Freezing-Point-Depression Constants

Solvent	Normal Boiling Point (°C)	K_b (°C/m)	Normal Freezing Point (°C)	K_f (°C/m)
Water, H_2O	100.0	0.51	0.0	1.86
Benzene, C_6H_6	80.1	2.53	5.5	5.12
Ethanol, C_2H_5OH	78.4	1.22	−114.6	1.99
Carbon tetrachloride, CCl_4	76.8	5.02	−22.3	29.8
Chloroform, $CHCl_3$	61.2	3.63	−63.5	4.68

Freezing-Point Depression

When a solution freezes, crystals of pure solvent usually separate out; the solute molecules are not normally soluble in the solid phase of the solvent. When aqueous solutions are partially frozen, for example, the solid that separates out is almost always pure ice. As a result, the part of the phase diagram in Figure 13.22 that represents the vapor pressure of the solid is the same as that for the pure liquid. The vapor-pressure curves for the liquid and solid phases meet at the triple point. (Section 11.6) In Figure 13.22 we see that the triple point of the solution must be at a lower temperature than that in the pure liquid because the solution has a lower vapor pressure than the pure liquid.

The freezing point of a solution is the temperature at which the first crystals of pure solvent begin to form in equilibrium with the solution. Recall from Section 11.6 that the line representing the solid-liquid equilibrium rises nearly vertically from the triple point. Because the triple-point temperature of the solution is lower than that of the pure liquid, *the freezing point of the solution is lower than that of the pure liquid.*

Like the boiling-point elevation, the decrease in freezing point, ΔT_f, is directly proportional to the molality of the solute:

$$\Delta T_f = K_f m \qquad [13.12]$$

The values of K_f, the **molal freezing-point-depression constant**, for several common solvents are given in Table 13.4. For water, K_f is 1.86°C/m; therefore, a 1-m aqueous solution of sucrose, or any other aqueous solution that is 1 m in nonvolatile solute particles (such as 0.5 m NaCl), will freeze 1.86°C lower than pure water. The freezing-point lowering caused by solutes explains the use of antifreeze in cars (Sample Exercise 13.8) and the use of calcium chloride ($CaCl_2$) to melt ice on roads during winter.

SAMPLE EXERCISE 13.8

Automotive antifreeze consists of ethylene glycol ($C_2H_6O_2$), a nonvolatile nonelectrolyte. Calculate the boiling point and freezing point of a 25.0 mass % solution of ethylene glycol in water.

Solution

Analyze and Plan: In order to calculate the boiling-point elevation and the freezing-point depression using Equations 13.11 and 13.12, we must express the concentration of the solution as molality. Let's assume for convenience that we have 1000 g of solution. Because the solution is 25.0 mass % ethylene glycol, the masses of ethylene glycol and water in the solution are 250 and 750 g, respectively. Using these quantities, we can calculate the molality of the solution, which we use with the molal boiling-point-elevation and freezing-point-depression constants (Table 13.4) to calculate ΔT_b and ΔT_f. We add ΔT_b to the boiling point and subtract ΔT_f from the freezing point of the solvent to obtain the boiling point and freezing point of the solution.

Solve: The molality of the solution is calculated as follows:

$$\text{Molality} = \frac{\text{moles } C_2H_6O_2}{\text{kilograms } H_2O} = \left(\frac{250 \text{ g } C_2H_6O_2}{750 \text{ g } H_2O}\right)\left(\frac{1 \text{ mol } C_2H_6O_2}{62.1 \text{ g } C_2H_6O_2}\right)\left(\frac{1000 \text{ g } H_2O}{1 \text{ kg } H_2O}\right) = 5.37\ m$$

We can now use Equations 13.11 and 13.12 to calculate the changes in the boiling and freezing points:

$$\Delta T_b = K_b m = (0.51°\text{C}/m)(5.37\ m) = 2.7°\text{C}$$

$$\Delta T_f = K_f m = (1.86°\text{C}/m)(5.37\ m) = 10.0°\text{C}$$

Hence, the boiling and freezing points of the solution are

$$\text{Boiling point} = (\text{normal bp of solvent}) + \Delta T_b = 100.0°\text{C} + 2.7°\text{C} = 102.7°\text{C}$$

$$\text{Freezing point} = (\text{normal fp of solvent}) - \Delta T_f = 0.0°\text{C} - 10.0°\text{C} = -10.0°\text{C}$$

Comment: Notice that the solution is a liquid over a larger temperature range than the pure solvent.

PRACTICE EXERCISE

Calculate the freezing point of a solution containing 0.600 kg of $CHCl_3$ and 42.0 g of eucalyptol ($C_{10}H_{18}O$), a fragrant substance found in the leaves of eucalyptus trees. (See Table 13.4.)
Answer: −65.6°C

SAMPLE EXERCISE 13.9

List the following aqueous solutions in order of their expected freezing points: 0.050 m $CaCl_2$; 0.15 m NaCl; 0.10 m HCl; 0.050 m $HC_2H_3O_2$; 0.10 m $C_{12}H_{22}O_{11}$.

Solution

Analyze and Plan: The lowest freezing point will correspond to the solution with the greatest concentration of solute particles. To determine the total concentration of solute particles in each case, we must determine whether the substance is a nonelectrolyte or an electrolyte and consider the number of ions formed when it ionizes.

Solve: $CaCl_2$, NaCl, and HCl are strong electrolytes, $HC_2H_3O_2$ is a weak electrolyte, and $C_{12}H_{22}O_{11}$ is a nonelectrolyte. The molality of each solution in total particles is as follows:

$$0.050\ m\ CaCl_2 \Rightarrow 0.050\ m \text{ in } Ca^{2+} \text{ and } 0.10\ m \text{ in } Cl^- \Rightarrow 0.15\ m \text{ in particles}$$

$$0.15\ m\ NaCl \Rightarrow 0.15\ m\ Na^+ \text{ and } 0.15\ m \text{ in } Cl^- \Rightarrow 0.30\ m \text{ in particles}$$

$$0.10\ m\ HCl \Rightarrow 0.10\ m\ H^+ \text{ and } 0.10\ m \text{ in } Cl^- \Rightarrow 0.20\ m \text{ in particles}$$

$$0.050\ m\ HC_2H_3O_2 \Rightarrow \text{weak electrolyte} \Rightarrow \text{between } 0.050\ m \text{ and } 0.10\ m \text{ in particles}$$

$$0.10\ m\ C_{12}H_{22}O_{11} \Rightarrow \text{nonelectrolyte} \Rightarrow 0.10\ m \text{ in particles}$$

Because the freezing points depend on the total molality of particles in solution, the expected ordering is 0.15 m NaCl (lowest freezing point), 0.10 m HCl, 0.050 m $CaCl_2$, 0.10 m $C_{12}H_{22}O_{11}$, and 0.050 m $HC_2H_3O_2$ (highest freezing point).

PRACTICE EXERCISE

Which of the following solutes will produce the largest increase in boiling point upon addition of 1 kg of water: 1 mol $Co(NO_3)_2$, 2 mol of KCl, 3 mol of ethylene glycol ($C_2H_6O_2$)?

Answer: 2 mol of KCl because it contains the highest concentration of particles, 2 m K^+ and 2 m Cl^-, giving 4 m in total

Osmosis

Certain materials, including many membranes in biological systems and synthetic substances such as cellophane, are *semipermeable*. When in contact with a solution, they allow some molecules to pass through their network of tiny pores, but not others. They often allow small solvent molecules such as water to pass through but block larger solute molecules or ions.

Consider a situation in which only solvent molecules are able to pass through a membrane. If such a membrane is placed between two solutions of different concentration, solvent molecules move in both directions through the membrane. The concentration of *solvent* is higher in the solution containing less solute, however, so the rate with which solvent passes from the less concentrated to the more concentrated solution is greater than the rate in the opposite direction. Thus, there is a net movement of solvent molecules from the less concentrated solution into the more concentrated one. In this process, called **osmosis**, *the net movement of solvent is always toward the solution with the higher solute concentration*.

Figure 13.23(a) ▶ shows two solutions separated by a semipermeable membrane. Solvent moves through the membrane from right to left, as if the solutions were driven to attain equal concentrations. As a result, the liquid levels in the two arms become unequal. Eventually, the pressure difference resulting from the unequal heights of the liquid in the two arms becomes so large that the net flow of solvent ceases, as shown in Figure 13.23(b). Alternatively, we may apply

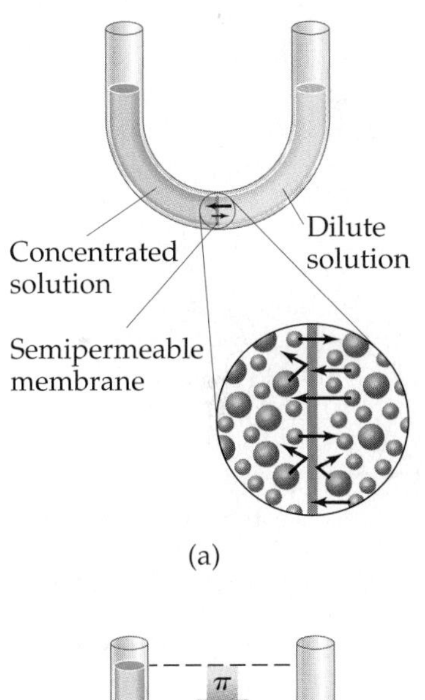

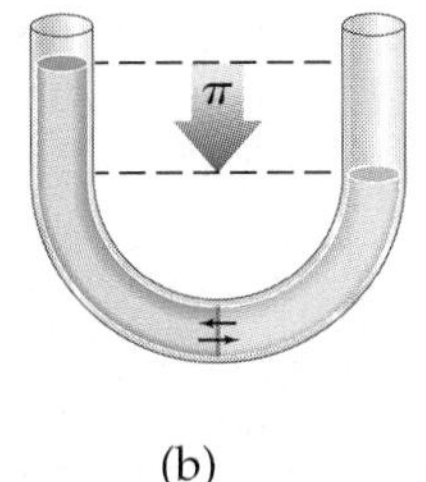

▲ **Figure 13.23** Osmosis: (a) net movement of a solvent from the pure solvent or a solution with low solute concentration to a solution with high solute concentration; (b) osmosis stops when the column of solution on the left becomes high enough to exert sufficient pressure at the membrane to counter the net movement of solvent. At this point the solution on the left has become more dilute, but there still exists a difference in concentrations between the two solutions.

▶ **Figure 13.24** Applied pressure on the left arm of the apparatus stops net movement of solvent from the right side of the semipermeable membrane. This applied pressure is the osmotic pressure of the solution.

Applied pressure, π, stops net movement of solvent.

Solution

Pure solvent

Semipermeable membrane

pressure to the left arm of the apparatus, as shown in Figure 13.24 ▲, to halt the net flow of solvent. The pressure required to prevent osmosis is the **osmotic pressure**, π, of the solution. The osmotic pressure obeys a law similar in form to the ideal-gas law, $\pi V = nRT$, where V is the volume of the solution, n is the number of moles of solute, R is the ideal-gas constant, and T is the temperature on the Kelvin scale. From this equation, we can write

$$\pi = \left(\frac{n}{V}\right)RT = MRT \qquad [13.13]$$

where M is the molarity of the solution.

If two solutions of identical osmotic pressure are separated by a semipermeable membrane, no osmosis will occur. The two solutions are *isotonic*. If one solution is of lower osmotic pressure, it is *hypotonic* with respect to the more concentrated solution. The more concentrated solution is *hypertonic* with respect to the dilute solution.

Osmosis plays a very important role in living systems. The membranes of red blood cells, for example, are semipermeable. Placing a red blood cell in a solution that is hypertonic relative to the intracellular solution (the solution within the cells) causes water to move out of the cell, as shown in Figure 13.25 ▼. This causes the cell to shrivel, a process called *crenation*. Placing the cell in a solution that is hypotonic relative to the intracellular fluid causes water to move into the cell. This causes the cell to rupture, a process called *hemolysis*. People who need body fluids or nutrients replaced but cannot be fed orally are given solutions by intravenous (IV) infusion, which feeds nutrients directly into the veins. To prevent crenation or hemolysis of red blood cells, the IV solutions must be isotonic with the intracellular fluids of the cells.

▶ **Figure 13.25** Osmosis through the semipermeable membrane of a red blood cell: (a) crenation caused by movement of water from the cell; (b) hemolysis caused by movement of water into the cell.

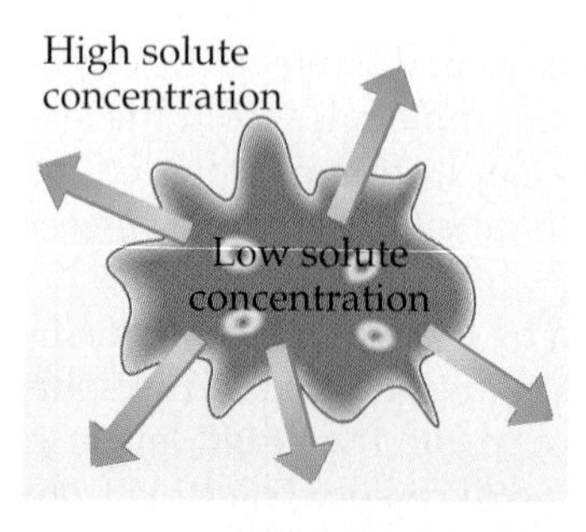

(a)

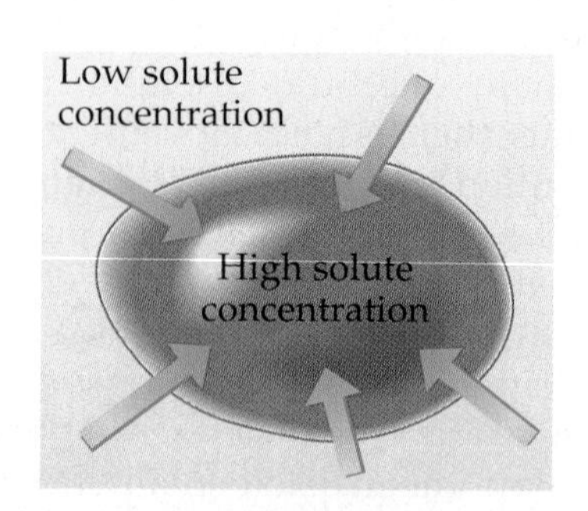

(b)

SAMPLE EXERCISE 13.10

The average osmotic pressure of blood is 7.7 atm at 25°C. What concentration of glucose ($C_6H_{12}O_6$) will be isotonic with blood?

Solution

Analyze and Plan: Because we are given the osmotic pressure and temperature, we can solve for the concentration using Equation 13.13.

Solve:

$$\pi = MRT$$

$$M = \frac{\pi}{RT} = \frac{7.7 \text{ atm}}{\left(0.0821 \dfrac{\text{L-atm}}{\text{mol-K}}\right)(298 \text{ K})} = 0.31\ M$$

Comment: In clinical situations the concentrations of solutions are generally expressed as mass percentages. The mass percentage of a 0.31 M solution of glucose is 5.3%. The concentration of NaCl that is isotonic with blood is 0.16 M because NaCl ionizes to form two particles, Na^+ and Cl^- (a 0.155 M solution of NaCl is 0.310 M in particles). A 0.16 M solution of NaCl is 0.9 mass % in NaCl. This kind of solution is known as a physiological saline solution.

PRACTICE EXERCISE

What is the osmotic pressure at 20°C of a 0.0020 M sucrose ($C_{12}H_{22}O_{11}$) solution?
Answer: 0.048 atm, or 37 torr

There are many interesting examples of osmosis. A cucumber placed in concentrated brine loses water via osmosis and shrivels into a pickle. If a carrot that has become limp because of water loss to the atmosphere is placed in water, the water moves into the carrot through osmosis, making it firm once again. People who eat a lot of salty food retain water in tissue cells and intercellular space because of osmosis. The resultant swelling or puffiness is called *edema*. Water moves from soil into plant roots and subsequently into the upper portions of the plant, at least in part because of osmosis. Bacteria on salted meat or candied fruit lose water through osmosis, shrivel, and die, thus preserving the food.

In osmosis water moves from an area of high water concentration (low solute concentration) into an area of low water concentration (high solute concentration). The movement of a substance from an area where its concentration is high to an area where it is low is spontaneous. Biological cells transport not only water, but also other select materials through their membrane walls. This permits nutrients to enter and waste materials to exit. In some cases substances must be moved from an area of low concentration to one of high concentration. This movement—called *active transport*—is not spontaneous, so cells must expend energy to do it.

Determination of Molar Mass

The colligative properties of solutions provide a useful means of experimentally determining molar mass. Any of the four colligative properties can be used, as shown in Sample Exercises 13.11 and 13.12.

ACTIVITY
Determination of Molar Mass

SAMPLE EXERCISE 13.11

A solution of an unknown nonvolatile nonelectrolyte was prepared by dissolving 0.250 g of the substance in 40.0 g of CCl_4. The boiling point of the resultant solution was 0.357°C higher than that of the pure solvent. Calculate the molar mass of the solute.

Solution

Analyze and Plan: We are given the boiling-point elevation of the solution, $\Delta T_b = 0.357°C$, and Table 13.4 gives K_b for the solvent (CCl_4), $K_b = 5.02°C/m$. Thus, we can use Equation 13.11, $\Delta T_b = K_b m$, to calculate the molality of the solution. Then we can use molality and the quantity of solvent (40.0 g CCl_4) to calculate the number of moles of solute. Finally, the molar mass of the solute equals the number of grams per mole, so we divide the number of grams of solute (0.250 g) by the number of moles we have just calculated.

Solve: From Equation 13.11 we have

$$\text{Molality} = \frac{\Delta T_b}{K_b} = \frac{0.357°C}{5.02°C/m} = 0.0711\ m$$

Thus, the solution contains 0.0711 mol of solute per kilogram of solvent. The solution was prepared using 40.0 g = 0.0400 kg of solvent (CCl_4). The number of moles of solute in the solution is therefore

$$(0.0400 \text{ kg } CCl_4)\left(0.0711 \frac{\text{mol solute}}{\text{kg } CCl_4}\right) = 2.84 \times 10^{-3} \text{ mol solute}$$

The molar mass of the solute is the number of grams per mole of the substance:

$$\text{Molar mass} = \frac{0.250 \text{ g}}{2.84 \times 10^{-3} \text{ mol}} = 88.0 \text{ g/mol}$$

PRACTICE EXERCISE

Camphor ($C_{10}H_{16}O$) melts at 179.8°C, and it has a particularly large freezing-point-depression constant, $K_f = 40.0°C/m$. When 0.186 g of an organic substance of unknown molar mass is dissolved in 22.01 g of liquid camphor, the freezing point of the mixture is found to be 176.7°C. What is the molar mass of the solute?

Answer: 110 g/mol

SAMPLE EXERCISE 13.12

The osmotic pressure of an aqueous solution of a certain protein was measured in order to determine its molar mass. The solution contained 3.50 mg of protein dissolved in sufficient water to form 5.00 mL of solution. The osmotic pressure of the solution at 25°C was found to be 1.54 torr. Calculate the molar mass of the protein.

Solution

Analyze and Plan: The temperature ($T = 25°C$) and osmotic pressure ($\pi = 1.54$ torr) are given and we know the value of R, so we can use Equation 13.13 to calculate the molarity of the solution, M. In doing so, we must convert temperature from °C to K and the osmotic pressure from torr to atm. We then use the molarity and the volume of the solution (5.00 mL) to determine the number of moles of solute. Finally, we obtain the molar mass by dividing the mass of the solute (3.50 mg) by the number of moles of solute.

Solve: Solving Equation 13.13 for molarity gives

$$\text{Molarity} = \frac{\pi}{RT} = \frac{(1.54 \text{ torr})\left(\frac{1 \text{ atm}}{760 \text{ torr}}\right)}{\left(0.0821 \frac{\text{L-atm}}{\text{mol-K}}\right)(298 \text{ K})} = 8.28 \times 10^{-5} \frac{\text{mol}}{\text{L}}$$

Because the volume of the solution is $5.00 \text{ mL} = 5.00 \times 10^{-3}$ L, the number of moles of protein must be

$$\text{Moles} = (8.28 \times 10^{-5} \text{ mol/L})(5.00 \times 10^{-3} \text{ L}) = 4.14 \times 10^{-7} \text{ mol}$$

The molar mass is the number of grams per mole of the substance. The sample has a mass of $3.50 \text{ mg} = 3.50 \times 10^{-3}$ g. The molar mass is the number of grams divided by the number of moles:

$$\text{Molar mass} = \frac{\text{grams}}{\text{moles}} = \frac{3.50 \times 10^{-3} \text{ g}}{4.14 \times 10^{-7} \text{ mol}} = 8.45 \times 10^{3} \text{ g/mol}$$

Comment: Because small pressures can be measured easily and accurately, osmotic pressure measurements provide an excellent way to determine the molar masses of large molecules.

PRACTICE EXERCISE

A sample of 2.05 g of the plastic polystyrene was dissolved in enough toluene to form 0.100 L of solution. The osmotic pressure of this solution was found to be 1.21 kPa at 25°C. Calculate the molar mass of the polystyrene.

Answer: 4.20×10^{4} g/mol

A Closer Look Colligative Properties of Electrolyte Solutions

The colligative properties of solutions depend on the total concentration of solute particles, regardless of whether the particles are ions or molecules. Thus, we would expect a 0.100 *m* solution of NaCl to have a freezing-point depression of $(0.200\ m)(1.86°C/m) = 0.372°C$ because it is 0.100 *m* in $Na^+(aq)$ and 0.100 *m* in $Cl^-(aq)$. The measured freezing-point depression is only 0.348°C, however, and the situation is similar for other strong electrolytes. A 0.100 *m* solution of KCl, for example, freezes at −0.344°C.

The difference between the expected and observed colligative properties for strong electrolytes is due to electrostatic attractions between ions. As the ions move about in solution, ions of opposite charge collide and "stick together" for brief moments. While they are together, they behave as a single particle called an *ion pair* (Figure 13.26 ▶). The number of independent particles is thereby reduced, causing a reduction in the freezing-point depression (as well as in the boiling-point elevation, the vapor-pressure reduction, and the osmotic pressure).

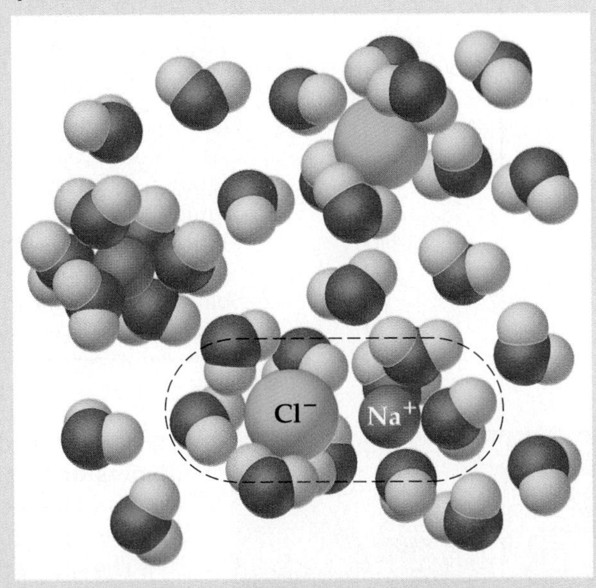

▲ **Figure 13.26** A solution of NaCl contains not only separated $Na^+(aq)$ and $Cl^-(aq)$ ions, but ion pairs as well. Ion pairing becomes more prevalent as the solution concentration increases.

One measure of the extent to which electrolytes dissociate is the *van't Hoff factor, i*. This factor is the ratio of the actual value of a colligative property to the value calculated assuming the substance to be a nonelectrolyte. Using the freezing-point depression, for example, we have

$$i = \frac{\Delta T_f(\text{measured})}{\Delta T_f(\text{calculated for nonelectrolyte})} \qquad [13.14]$$

The ideal value of *i* can be determined for a salt from the number of ions per formula unit. For NaCl, for example, the ideal van't Hoff factor is 2 because NaCl consists of one Na^+ and one Cl^- per formula unit; for K_2SO_4 it is 3 because K_2SO_4 consists of two K^+ and one SO_4^{2-}. In the absence of any information about the actual value of *i* for a solution, we will use the ideal value in calculations.

Table 13.5 ▼ gives the observed van't Hoff factors for several substances at different dilutions. Two trends are evident in these data. First, dilution affects the value of *i* for electrolytes; the more dilute the solution, the more closely *i* approaches the ideal or limiting value. Thus, the extent of ion pairing in electrolyte solutions decreases upon dilution. Second, the lower the charges on the ions, the less *i* departs from the limiting value because the extent of ion pairing decreases as the ionic charges decrease. Both trends are consistent with simple electrostatics: The force of interaction between charged particles decreases as their separation increases and as their charges decrease.

TABLE 13.5 van't Hoff Factors for Several Substances at 25°C

	Concentration			
Compound	0.100 *m*	0.0100 *m*	0.00100 *m*	Limiting Value
Sucrose	1.00	1.00	1.00	1.00
NaCl	1.87	1.94	1.97	2.00
K_2SO_4	2.32	2.70	2.84	3.00
$MgSO_4$	1.21	1.53	1.82	2.00

13.6 Colloids

When finely divided clay particles are dispersed throughout water, they eventually settle out of the water due to gravity. The dispersed clay particles are much larger than molecules and consist of many thousands or even millions of atoms. In contrast, the dispersed particles of a solution are of molecular size. Between these extremes lie dispersed particles that are larger than molecules, but not so large that the components of the mixture separate under the influence of gravity.

TABLE 13.6 Types of Colloids

Phase of Colloid	Dispersing (solventlike) Substance	Dispersed (solutelike) Substance	Colloid Type	Example
Gas	Gas	Gas	—	None (all are solutions)
Gas	Gas	Liquid	Aerosol	Fog
Gas	Gas	Solid	Aerosol	Smoke
Liquid	Liquid	Gas	Foam	Whipped cream
Liquid	Liquid	Liquid	Emulsion	Milk
Liquid	Liquid	Solid	Sol	Paint
Solid	Solid	Gas	Solid foam	Marshmallow
Solid	Solid	Liquid	Solid emulsion	Butter
Solid	Solid	Solid	Solid sol	Ruby glass

▲ **Figure 13.27** Illustration of the Tyndall effect. The vessel on the left contains a colloidal suspension; that on the right, a solution. The path of the beam through the colloidal suspension is seen because the light is scattered by the colloidal particles. Light is not scattered by the individual solute molecules in the solution.

These intermediate types of dispersions or suspensions are called **colloidal dispersions**, or simply **colloids**. Colloids form the dividing line between solutions and heterogeneous mixtures. Like solutions, colloids can be gases, liquids, or solids. Examples of each are listed in Table 13.6 ▲.

The size of the dispersed particle is used to classify a mixture as a colloid. Colloid particles range in diameter from approximately 10 to 2000 Å. Solute particles are smaller. The colloid particle may consist of many atoms, ions, or molecules, or it may even be a single giant molecule. The hemoglobin molecule, for example, which carries oxygen in blood, has molecular dimensions of 65 Å × 55 Å × 50 Å and a molecular weight of 64,500 amu.

Although colloid particles may be so small that the dispersion appears uniform even under a microscope, they are large enough to scatter light very effectively. Consequently, most colloids appear cloudy or opaque unless they are very dilute. (Homogenized milk is a colloid.) Furthermore, because they scatter light, a light beam can be seen as it passes through a colloidal suspension, as shown in Figure 13.27 ◀. This scattering of light by colloidal particles, known as the **Tyndall effect**, makes it possible to see the light beam of an automobile on a dusty dirt road or the sunlight coming through a forest canopy [Figure 13.28(a) ▼]. Not all wavelengths are scattered to the same extent. As a result, brilliant red sunsets are seen when the sun is near the horizon and the air contains dust, smoke, or other particles of colloidal size [Figure 13.28(b)].

▶ **Figure 13.28** (a) Scattering of sunlight by colloidal particles in the misty air of a forest. (b) The scattering of light by smoke or dust particles produces a rich red sunset.

(a)

(b)

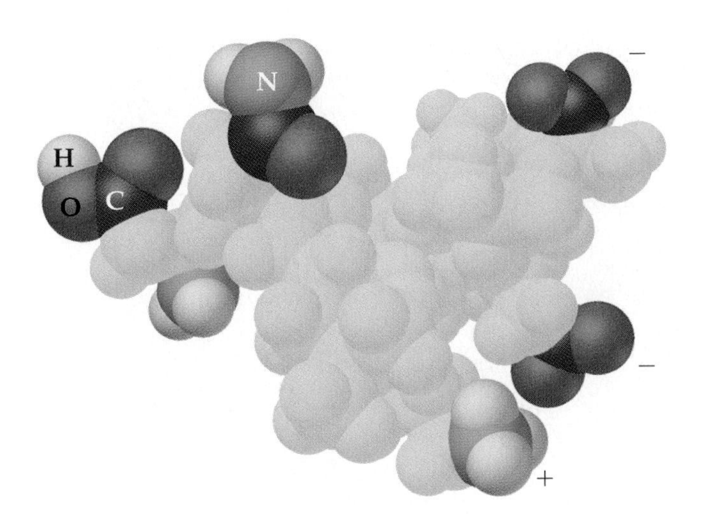

◀ **Figure 13.29** Examples of hydrophilic groups on the surface of a giant molecule (macromolecule) that help to keep the molecule suspended in water.

Hydrophilic and Hydrophobic Colloids

The most important colloids are those in which the dispersing medium is water. These colloids may be **hydrophilic** (water loving) or **hydrophobic** (water fearing). Hydrophilic colloids are most like the solutions that we have previously examined. In the human body the extremely large molecules that make up such important substances as enzymes and antibodies are kept in suspension by interaction with surrounding water molecules. The molecules fold in such a way that the hydrophobic groups are away from the water molecules, on the "inside" of the folded molecule, while the hydrophilic, polar groups are on the surface, interacting with the water molecules. These hydrophilic groups generally contain oxygen or nitrogen. Some examples are shown in Figure 13.29 ▲.

Hydrophobic colloids can be prepared in water only if they are stabilized in some way. Otherwise, their natural lack of affinity for water causes them to separate from the water. Hydrophobic colloids can be stabilized by adsorption of ions on their surface, as shown in Figure 13.30 ▼. (*Adsorption* means to adhere to a surface. It differs from *absorption*, which means to pass into the interior, as when a sponge absorbs water.) These adsorbed ions can interact with water, thereby stabilizing the colloid. At the same time, the mutual repulsion between colloid particles with adsorbed ions of the same charge keeps the particles from colliding and getting larger.

Hydrophobic colloids can also be stabilized by the presence of hydrophilic groups on their surfaces. Small droplets of oil are hydrophobic, for example, so they do not remain suspended in water. Instead, they separate, forming an oil slick on the surface of the water. Sodium stearate (whose structure is in the margin), or any similar substance having one end that is hydrophilic (polar, or

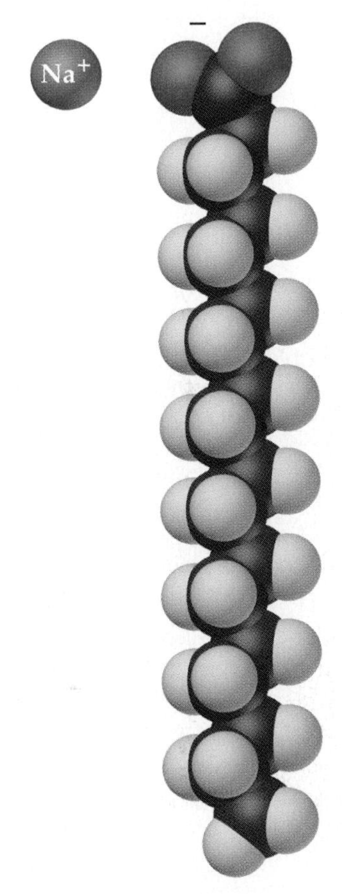

Sodium stearate

$$\underbrace{CH_3(CH_2)_{16}}_{\text{Hydrophobic end}}-\underbrace{\overset{\overset{\displaystyle O}{\|}}{C}O^-\,Na^+}_{\text{Hydrophilic end}}$$

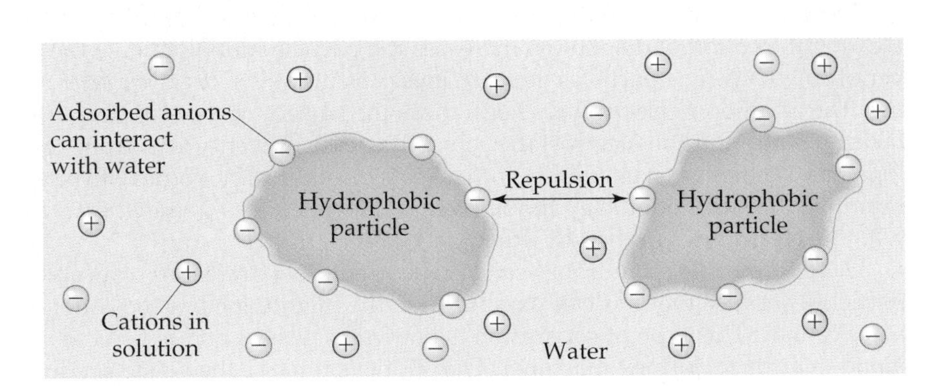

◀ **Figure 13.30** Schematic illustration of the stabilization of a hydrophobic colloid in water by adsorbed ions.

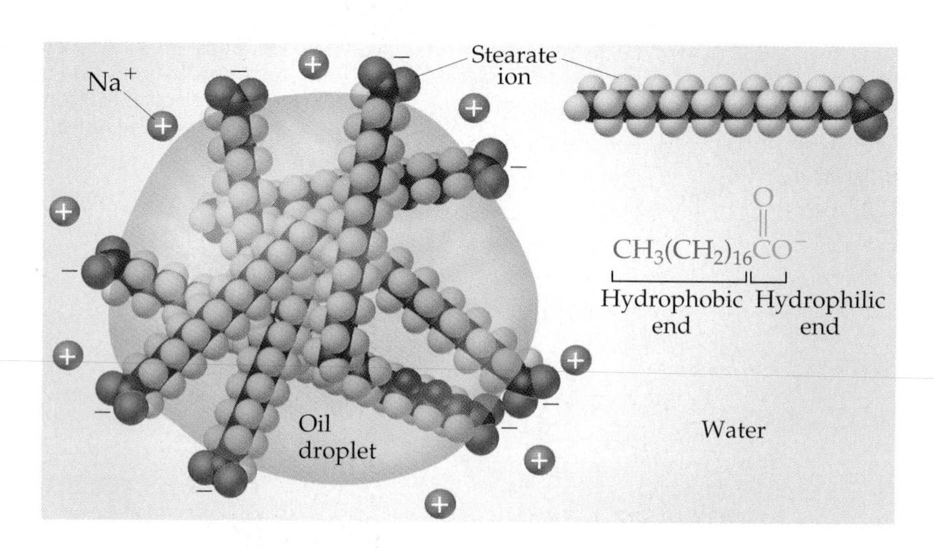

▶ **Figure 13.31** Stabilization of an emulsion of oil in water by stearate ions.

3-D MODEL
Sodium Stearate

charged) and one that is hydrophobic (nonpolar) will stabilize a suspension of oil in water. Stabilization results from the interaction of the hydrophobic ends of the stearate ions with the oil droplet and the hydrophilic ends with the water as shown in Figure 13.31 ▲.

The stabilization of colloids has an interesting application in our own digestive system. When fats in our diet reach the small intestine, a hormone causes the gallbladder to excrete a fluid called bile. Among the components of bile are compounds that have chemical structures similar to sodium stearate; that is, they have a hydrophilic (polar) end and a hydrophobic (nonpolar) end. These compounds emulsify the fats present in the intestine and thus permit digestion and absorption of fat-soluble vitamins through the intestinal wall. The term *emulsify* means "to form an emulsion," a suspension of one liquid in another (Table 13.6). A substance that aids in the formation of an emulsion is called an emulsifying agent. If you read the labels on foods and other materials, you will find that a variety of chemicals are used as emulsifying agents. These chemicals typically have a hydrophilic end and a hydrophobic end.

Removal of Colloidal Particles

Colloidal particles frequently must be removed from a dispersing medium, as in the removal of smoke from stacks or butterfat from milk. Because colloidal particles are so small, they cannot be separated by simple filtration. Instead, the colloidal particles must be enlarged in a process called *coagulation*. The resultant larger particles can then be separated by filtration or merely by allowing them to settle out of the dispersing medium.

Heating or adding an electrolyte to the mixture may bring about coagulation. Heating the colloidal dispersion increases the particle motion and so the number of collisions. The particles increase in size as they stick together after colliding. The addition of electrolytes neutralizes the surface charges of the particles, thereby removing the electrostatic repulsions that prevent them from coming together. Wherever rivers empty into oceans or other salty bodies of water, for example, the suspended clay in the river is deposited as a delta when it mixes with the electrolytes in the salt water.

Semipermeable membranes can also be used to separate ions from colloidal particles, because the ions can pass through the membrane but the colloid particles cannot. This type of separation is known as *dialysis* and is used to purify blood in artificial kidney machines. Our kidneys remove the waste products of metabolism from blood. In a kidney machine, blood is circulated through a dialyzing tube immersed in a washing solution. The washing solution is isotonic in ions that must be retained by the blood, but is lacking the waste products. Wastes therefore dialyze out of the blood, but the ions do not.

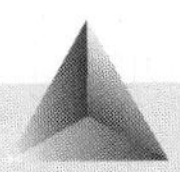

Chemistry and Life Sickle-Cell Anemia

Our blood contains a complex protein called *hemoglobin* that carries oxygen from our lungs to other parts of our body. In the genetic disease known as *sickle-cell anemia*, hemoglobin molecules are abnormal and have a lower solubility, especially in their unoxygenated form. Consequently, as much as 85% of the hemoglobin in red blood cells crystallizes from solution.

The reason for the insolubility of hemoglobin in sickle-cell anemia can be traced to a structural change in one part of an amino acid side chain. Normal hemoglobin molecules have an amino acid in their makeup that has the following side chain protruding from the main body of the molecule:

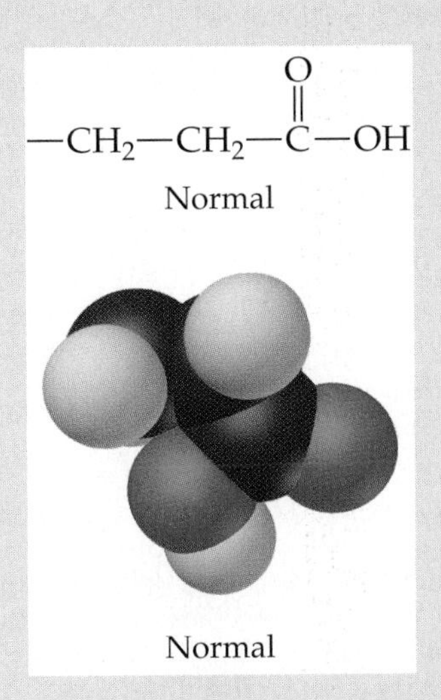

This side chain terminates in a polar group, which contributes to the solubility of the hemoglobin molecule in water. In the hemoglobin molecules of persons suffering from sickle-cell anemia, the side chain is of a different type:

This abnormal group of atoms is nonpolar (hydrophobic), and its presence leads to the aggregation of this defective form of hemoglobin into particles too large to remain suspended in biological fluids. It also causes the cells to distort into a sickle shape, as shown in Figure 13.32 ▼. The sickled cells clog the capillaries, causing severe pain and the gradual deterioration of the vital organs. The disease is hereditary, and if both parents carry the defective genes, it is likely that their children will possess only abnormal hemoglobin.

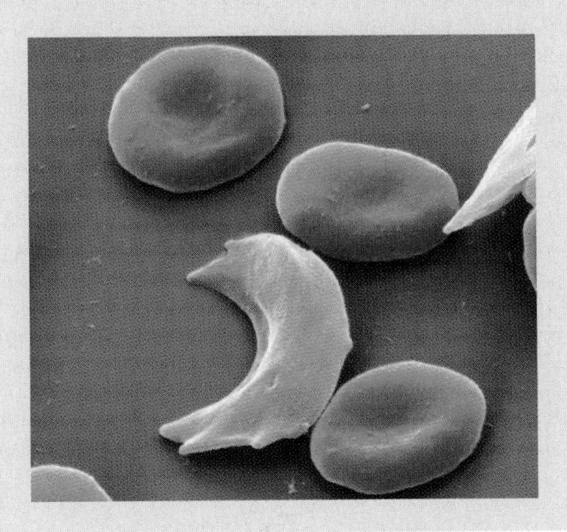

◀ **Figure 13.32** Electron micrograph showing normal red blood cells and sickled red blood cells.

SAMPLE INTEGRATIVE EXERCISE 13: Putting Concepts Together

A 0.100-L solution is made by dissolving 0.441 g of $CaCl_2(s)$ in water. **(a)** Calculate the osmotic pressure of this solution at 27°C, assuming that it is completely dissociated into its component ions. **(b)** The measured osmotic pressure of this solution is 2.56 atm at 27°C. Explain why it is less than the value calculated in (a), and calculate the van't Hoff factor, *i*, for the solute in this solution. (See the "Closer Look" box in Section 13.5 on page 511.) **(c)** The enthalpy of solution for $CaCl_2$ is $\Delta H = -81.3$ kJ/mol. If the final temperature of the solution was 27.0°C, what was its initial temperature? (Assume that the density of the solution is 1.00 g/mL, that its specific heat is 4.18 J/g-K, and that the solution loses no heat to its surroundings.)

Solution (a) The osmotic pressure is given by Equation 13.13, $\pi = MRT$. We know the temperature, $T = 27°C = 300$ K, and the gas constant, $R = 0.0821$ L-atm/mol-K. We can calculate the molarity of the solution from the mass of $CaCl_2$ and the volume of the solution:

$$\text{Molarity} = \left(\frac{0.441\text{ g } CaCl_2}{0.100\text{ L}}\right)\left(\frac{1\text{ mol } CaCl_2}{111.0\text{ g } CaCl_2}\right) = 0.0397\text{ mol } CaCl_2/\text{L}$$

Soluble ionic compounds are strong electrolytes. (Sections 4.1 and 4.3) Thus, $CaCl_2$ consists of metal cations (Ca^{2+}) and nonmetal anions (Cl^-). When completely dissociated, each $CaCl_2$ unit forms three ions (one Ca^{2+} and two Cl^-). Hence the total concentration of ions in the solution is $(3)(0.0397\ M) = 0.119\ M$, and the osmotic pressure is

$$\pi = MRT = (0.119\text{ mol/L})(0.0821\text{ L-atm/mol-K})(300\text{ K}) = 2.93\text{ atm}$$

(b) The actual values of colligative properties of electrolytes are less than those calculated, because the electrostatic interactions between ions limit their independent movements. In this case the van't Hoff factor, which measures the extent to which electrolytes actually dissociate into ions, is given by

$$i = \frac{\pi(\text{measured})}{\pi(\text{calculated for nonelectrolyte})}$$

$$= \frac{2.56\text{ atm}}{(0.0397\text{ mol/L})(0.0821\text{ L-atm/mol-K})(300\text{ K})} = 2.62$$

Thus, the solution behaves as if the $CaCl_2$ has dissociated into 2.62 particles instead of the ideal 3.

(c) If the solution is 0.0397 *M* in $CaCl_2$ and has a total volume of 0.100 L, the number of moles of solute is $(0.100\text{ L})(0.0397\text{ mol/L}) = 0.00397$ mol. Hence the quantity of heat generated in forming the solution is $(0.00397\text{ mol})(-81.3\text{ kJ/mol}) = -0.323$ kJ. The solution absorbs this heat, causing its temperature to increase. The relationship between temperature change and heat is given by Equation 5.19:

$$q = (\text{specific heat})(\text{grams})(\Delta T)$$

The heat absorbed by the solution is $q = +0.323$ kJ $= 323$ J. The mass of the 0.100 L of solution is $(100\text{ mL})(1.00\text{ g/mL}) = 100$ g (to 3 significant figures). Thus the temperature change is

$$\Delta T = \frac{q}{(\text{specific heat of solution})(\text{grams of solution})}$$

$$= \frac{323\text{ J}}{(4.18\text{ J/g-K})(100\text{ g})} = 0.773\text{ K}$$

A kelvin has the same size as a degree Celsius. (Section 1.4) Because the solution temperature increases by 0.773°C, the initial temperature was 27.0°C − 0.773°C = 26.2°C.

Summary and Key Terms

Section 13.1 Solutions form when one substance disperses uniformly throughout another. The attractive interaction of solvent molecules with solute is called **solvation**. When the solvent is water, the interaction is called **hydration**. The dissolution of ionic substances in water is promoted by hydration of the separated ions by the polar water molecules.

The overall enthalpy change upon solution formation may be either positive or negative. Solution formation is favored both by a negative enthalpy change (exothermic process) and by a positive **entropy** change (the increase in disorder) of the system.

Section 13.2 The equilibrium between a saturated solution and undissolved solute is dynamic; the process of solution and the reverse process, **crystallization**, occur simultaneously. In a solution in equilibrium with undissolved solute, the two processes occur at equal rates, giving a **saturated** solution. If there is less solute present than is needed to saturate the solution, the solution is **unsaturated**. When more than the equilibrium concentration of solute is present, the solution is **supersaturated**. This is an unstable condition, and separation of some solute from the solution will occur if the process is initiated with a solute seed crystal. The amount of solute needed to form a saturated solution at any particular temperature is the **solubility** of that solute at that temperature.

Section 13.3 The solubility of one substance in another depends on the natural tendency of systems to become more disordered and on the relative intermolecular solute-solute and solvent-solvent energies compared with solute-solvent interactions. Polar and ionic solutes tend to dissolve in polar solvents, and nonpolar solutes tend to dissolve in nonpolar solvents ("like dissolves like"). Liquids that mix in all proportions are **miscible**; those that do not dissolve significantly in one another are **immiscible**. Hydrogen-bonding interactions between solute and solvent often play an important role in determining solubility; for example, ethanol and water, whose molecules form hydrogen bonds with each other, are miscible.

The solubilities of gases in a liquid are generally proportional to the pressure of the gas over the solution, as expressed by **Henry's law**: $S_g = kP_g$. The solubilities of most solid solutes in water increase as the temperature of the solution increases. In contrast, the solubilities of gases in water generally decrease with increasing temperature.

Section 13.4 Concentrations of solutions can be expressed quantitatively by several different measures, including **mass percentage** [(mass solute/mass solution) $\times 10^2$], **parts per million (ppm)** [(mass solute/mass solution) $\times 10^6$], **parts per billion (ppb)** [(mass solute/mass solution) $\times 10^9$], and mole fraction [mol solute/(mol solute + mol solvent)]. Molarity, M, is defined as moles of solute per liter of solution; **molality**, m, is defined as moles of solute per kg of solvent. Molarity can be converted to these other concentration units if the density of the solution is known.

Section 13.5 A physical property of a solution that depends on the concentration of solute particles present, regardless of the nature of the solute, is a **colligative property**. Colligative properties include vapor-pressure lowering, freezing-point lowering, boiling-point elevation, and osmotic pressure. The lowering of vapor pressure is expressed by **Raoult's law**. An **ideal solution** obeys Raoult's law. Differences in solvent-solute as compared with solvent-solvent and solute-solute intermolecular forces cause many solutions to depart from ideal behavior.

A solution containing a nonvolatile solute possesses a higher boiling point than the pure solvent. The **molal boiling-point-elevation constant**, K_b, represents the increase in boiling point for a 1 m solution of solute particles as compared with the pure solvent. Similarly, the **molal freezing-point-depression constant**, K_f, measures the lowering of the freezing point of a solution for a 1 m solution of solute particles. The temperature changes are given by the equations $\Delta T_b = K_b m$ and $\Delta T_f = K_f m$. When NaCl dissolves in water, two moles of solute particles are formed for each mole of dissolved salt. The boiling point or freezing point is thus elevated or depressed, respectively, approximately twice as much as that of a nonelectrolyte solution of the same concentration. Similar considerations apply to other strong electrolytes.

Osmosis is the movement of solvent molecules through a semipermeable membrane from a less concentrated to a more concentrated solution. This net movement of solvent generates an **osmotic pressure**, π, which can be measured in units of gas pressure, such as atm. The osmotic pressure of a solution as compared with pure solvent is proportional to the solution molarity: $\pi = MRT$. Osmosis is a very important process in living systems, in which cell walls act as semipermeable membranes, permitting the passage of water, but restricting the passage of ionic and macromolecular components.

Section 13.6 Particles that are large on the molecular scale but are still small enough to remain suspended indefinitely in a solvent system form **colloids**, or **colloidal dispersions**. Colloids, which are intermediate between solutions and heterogeneous mixtures, have many practical applications. One useful physical property of colloids, the scattering of visible light, is referred to as the **Tyndall effect**. Aqueous colloids are classified as **hydrophilic** or **hydrophobic**. Hydrophilic colloids are common in living organisms, in which large molecular aggregates (enzymes, antibodies) remain suspended because they have many polar, or charged, atomic groups on their surfaces that interact with water. Hydrophobic colloids, such as small droplets of oil, may remain in suspension through adsorption of charged particles on their surfaces.

Exercises

The Solution Process

13.1 In general, the attractive intermolecular forces between solvent and solute particles must be comparable or greater than solute-solute interactions in order for significant solubility to occur. Explain this statement in terms of the overall energetics of solution formation.

13.2 **(a)** Considering the energetics of solute-solute, solvent-solvent, and solute–solvent interactions, explain why NaCl dissolves in water, but not in benzene (C_6H_6). **(b)** Why do ionic substances with higher lattice energies tend to be less soluble in water than those with lower lattice energies? **(c)** What factors cause a cation to be strongly hydrated?

13.3 Indicate the type of solute-solvent interaction (Section 11.2) that should be most important in each of the following solutions: **(a)** CCl_4 in benzene (C_6H_6); **(b)** $CaCl_2$ in water; **(c)** propanol ($CH_3CH_2CH_2OH$) in water; **(d)** HCl in acetonitrile (CH_3CN).

13.4 Rank the solvent–solute interaction in the following solutions from weakest to strongest, and indicate the principal type of interaction in each case: **(a)** KCl in water; **(b)** CH_2Cl_2 in benzene (C_6H_6); **(c)** methanol (CH_3OH) in water.

13.5 **(a)** In Equation 13.1, which of the energy terms for dissolving an ionic solid would correspond to the lattice energy? **(b)** Which energy terms in this equation are always exothermic?

13.6 The schematic diagram of the solution process as the net sum of three steps in Figure 13.4 does not show the relative magnitudes of the three components because these will vary from case to case. For the dissolution of NH_4NO_3 in water, which of the three enthalpy changes would you expect to be much smaller than the other two? Explain.

13.7 When two nonpolar organic liquids such as hexane (C_6H_{14}) and heptane (C_7H_{16}) are mixed, the enthalpy change that occurs is generally quite small. **(a)** Use the energy diagram in Figure 13.4 to explain why. **(b)** Given that $\Delta H_{soln} < 0$, explain why hexane and heptane spontaneously form a solution.

13.8 The enthalpy of solution of KBr in water is about +19.8 kJ/mol. Nevertheless, the solubility of KBr in water is relatively high. Why does the solution process occur even though it is endothermic?

Saturated Solutions; Factors Affecting Solubility

13.9 The solubility of $Cr(NO_3)_3 \cdot 9H_2O$ in water is 208 g per 100 g of water at 15°C. A solution of $Cr(NO_3)_3 \cdot 9H_2O$ in water at 35°C is formed by dissolving 324 g in 100 g water. When this solution is slowly cooled to 15°C, no precipitate forms. **(a)** What term describes this solution? **(b)** What action might you take to initiate crystallization? Use molecular-level processes to explain how your suggested procedure works.

13.10 The solubility of $MnSO_4 \cdot H_2O$ in water at 20°C is 70 g per 100 mL of water. **(a)** Is a 1.22 *M* solution of $MnSO_4 \cdot H_2O$ in water at 20°C saturated, supersaturated, or unsaturated? **(b)** Given a solution of $MnSO_4 \cdot H_2O$ of unknown concentration, what experiment could you perform to determine whether the new solution is saturated, supersaturated, or unsaturated?

13.11 By referring to Figure 13.17, determine whether the addition of 40.0 g of each of the following ionic solids to 100 g of water at 40°C will lead to a saturated solution: **(a)** $NaNO_3$; **(b)** KCl; **(c)** $K_2Cr_2O_7$; **(d)** $Pb(NO_3)_2$.

13.12 By referring to Figure 13.17, determine the mass of each of the following salts required to form a saturated solution in 250 g water at 30°C: **(a)** $KClO_3$; **(b)** $Pb(NO_3)_2$; **(c)** $Ce_2(SO_4)_3$.

13.13 Water and glycerol, $CH_2(OH)CH(OH)CH_2OH$, are miscible in all proportions. What does this mean? How does the OH group of the alcohol contribute to this miscibility?

13.14 Oil and water are immiscible. What does this mean? Explain in terms of the structural features of their respective molecules and the forces between them.

13.15 Consider a series of carboxylic acids whose general formula is $CH_3(CH_2)_nCOOH$. How would you expect the solubility of these compounds in water and in hexane to change as *n* increases? Explain.

13.16 **(a)** Would you expect stearic acid, $CH_3(CH_2)_{16}COOH$, to be more soluble in water or carbon tetrachloride? Explain. **(b)** Which would you expect to be more soluble in water, cyclohexane or dioxane? Explain.

Dioxane   Cyclohexane

13.17 Which of the following in each pair is likely to be the more soluble in water: **(a)** CCl_4 or $CaCl_2$; **(b)** benzene (C_6H_6) or phenol (C_6H_5OH)? Explain your answer in each case.

13.18 Which of the following in each pair is likely to be the more soluble in hexane (C_6H_{14}): **(a)** cyclohexane (C_6H_{12}) or glucose ($C_6H_{12}O_6$) (Figure 13.12); **(b)** propionic acid (CH_3CH_2COOH) or sodium propionate (CH_3CH_2COONa); **(c)** HCl or ethyl chloride (CH_3CH_2Cl)? Explain in each case.

13.19 **(a)** Explain why carbonated beverages must be stored in sealed containers. **(b)** Once the beverage has been opened, why does it maintain some carbonation when refrigerated?

13.20 Explain why pressure affects the solubility of O_2 in water, but not the solubility of NaCl in water.

13.21 The Henry's law constant for helium gas in water at 30°C is 3.7×10^{-4} M/atm; the constant for N_2 at 30°C is 6.0×10^{-4} M/atm. If the two gases are each present at 1.5 atm pressure, calculate the solubility of each gas.

13.22 The partial pressure of O_2 in air at sea level is 0.21 atm. Using the data in Table 13.2, together with Henry's law, calculate the molar concentration of O_2 in the surface water of a mountain lake saturated with air at 20°C and an atmospheric pressure of 665 torr.

Concentrations of Solutions

13.23 **(a)** Calculate the mass percentage of Na_2SO_4 in a solution containing 11.7 g Na_2SO_4 in 443 g water. **(b)** An ore contains 5.95 g of silver per ton of ore. What is the concentration of silver in ppm?

13.24 **(a)** What is the mass percentage of iodine (I_2) in a solution containing 0.045 mol I_2 in 115 g of CCl_4? **(b)** Seawater contains 0.0079 g Sr^{2+} per kilogram of water. What is the concentration of Sr^{2+} measured in ppm?

13.25 A solution is made containing 7.5 g CH_3OH in 245 g H_2O. Calculate **(a)** the mole fraction of CH_3OH; **(b)** the mass percent of CH_3OH; **(c)** the molality of CH_3OH.

13.26 A solution is made containing 25.5 g phenol (C_6H_5OH) in 495 g ethanol (CH_3CH_2OH). Calculate **(a)** the mole fraction of phenol; **(b)** the mass percent of phenol; **(c)** the molality of phenol.

13.27 Calculate the molarity of the following aqueous solutions: **(a)** 10.5 g $Mg(NO_3)_2$ in 250.0 mL of solution; **(b)** 22.4 g $LiClO_4 \cdot 3H_2O$ in 125 mL of solution; **(c)** 25.0 mL of 3.50 M HNO_3 diluted to 0.250 L.

13.28 What is the molarity of each of the following solutions: **(a)** 15.0 g $Al_2(SO_4)_3$ in 0.350 L solution; **(b)** 5.25 g $Mn(NO_3)_2 \cdot 2H_2O$ in 175 mL of solution; **(c)** 35.0 mL of 9.00 M H_2SO_4 diluted to 0.500 L?

13.29 Calculate the molality of each of the following solutions: **(a)** 10.5 g benzene (C_6H_6) dissolved in 18.5 g carbon tetrachloride (CCl_4); **(b)** 4.15 g NaCl dissolved in 0.250 L of water.

13.30 **(a)** What is the molality of a solution formed by dissolving 1.50 mol of KCl in 16.0 mol of water? **(b)** How many grams of sulfur (S_8) must be dissolved in 100.0 g naphthalene ($C_{10}H_8$) to make a 0.12 m solution?

13.31 A sulfuric acid solution containing 571.6 g of H_2SO_4 per liter of solution has a density of 1.329 g/cm^3. Calculate **(a)** the mass percentage; **(b)** the mole fraction; **(c)** the molality; **(d)** the molarity of H_2SO_4 in this solution.

13.32 Ascorbic acid, (vitamin C, $C_6H_8O_6$) is a water-soluble vitamin. A solution containing 80.5 g of ascorbic acid dissolved in 210 g of water has a density of 1.22 g/mL at 55°C. Calculate **(a)** the mass percentage; **(b)** the mole fraction; **(c)** the molality; **(d)** the molarity of ascorbic acid in this solution.

13.33 The density of acetonitrile (CH_3CN) is 0.786 g/mL, and the density of methanol (CH_3OH) is 0.791 g/mL. A solution is made by dissolving 15.0 mL CH_3OH in 90.0 mL CH_3CN. **(a)** What is the mole fraction of methanol in the solution? **(b)** What is the molality of the solution? **(c)** Assuming that the volumes are additive, what is the molarity of CH_3OH in the solution?

13.34 The density of toluene (C_7H_8) is 0.867 g/mL, and the density of thiophene (C_4H_4S) is 1.065 g/mL. A solution is made by dissolving 10.0 g of thiophene in 250.0 mL of toluene. **(a)** Calculate the mole fraction of thiophene in the solution. **(b)** Calculate the molality of thiophene in the solution. **(c)** Assuming that the volumes of the solute and solvent are additive, what is the molarity of thiophene in the solution?

13.35 Calculate the number of moles of solute present in each of the following aqueous solutions: **(a)** 255 mL of 0.250 M $CaBr_2$; **(b)** 50.0 g of 0.150 m KCl; **(c)** 50.0 g of a solution that is 2.50% glucose ($C_6H_{12}O_6$) by mass.

13.36 Calculate the number of moles of solute present in each of the following solutions: **(a)** 245 mL of 1.50 M $HNO_3(aq)$; **(b)** 50.0 mg of an aqueous solution that is 1.25 m NaCl; **(c)** 75.0 g of an aqueous solution that is 1.50% sucrose ($C_{12}H_{22}O_{11}$) by mass.

13.37 Describe how you would prepare each of the following aqueous solutions, starting with solid KBr: **(a)** 0.75 L of 1.5×10^{-2} M KBr; **(b)** 125 g of 0.180 m KBr; **(c)** 1.85 L of a solution that is 12.0% KBr by mass (the density of the solution is 1.10 g/mL); **(d)** a 0.150 M solution of KBr that contains just enough KBr to precipitate 16.0 g of AgBr from a solution containing 0.480 mol of $AgNO_3$.

13.38 Describe how you would prepare each of the following aqueous solutions: **(a)** 1.50 L of 0.110 M $(NH_4)_2SO_4$ solution, starting with solid $(NH_4)_2SO_4$; **(b)** 120 g of a solution that is 0.65 m in Na_2CO_3, starting with the solid solute; **(c)** 1.20 L of a solution that is 15.0% of $Pb(NO_3)_2$ by mass (the density of the solution is 1.16 g/mL), starting with solid solute; **(d)** a 0.50 M solution of HCl that would just neutralize 5.5 g of $Ba(OH)_2$, starting with 6.0 M HCl.

13.39 Commercial concentrated aqueous ammonia is 28% NH_3 by mass and has a density of 0.90 g/mL. What is the molarity of this solution?

13.40 Commercial aqueous nitric acid has a density of 1.42 g/mL and is 16 M. Calculate the percent HNO_3 by mass in the solution.

13.41 Propylene glycol, $C_3H_6(OH)_2$, is sometimes used in automobile antifreeze solutions. If an aqueous solution has a mole fraction $X_{C_3H_6(OH)_2} = 0.100$, calculate **(a)** the percent propylene glycol by mass; **(b)** the molality of the propylene glycol in the solution.

13.42 Caffeine ($C_8H_{10}N_4O_2$) is a stimulant found in coffee and tea. If a solution of caffeine in chloroform ($CHCl_3$) as a solvent has a concentration of 0.0750 m, calculate **(a)** the percent caffeine by mass; **(b)** the mole fraction of caffeine.

Colligative Properties

13.43 List four properties of a solution that depend on the concentration but not the type of particle or particles present as solute. Write the mathematical expression that describes how each of these properties depends on concentration.

13.44 How does increasing the concentration of a nonvolatile solute in water affect the following properties: **(a)** vapor pressure; **(b)** freezing point; **(c)** boiling point; **(d)** osmotic pressure?

13.45 **(a)** What is an *ideal solution*? **(b)** The vapor pressure of pure water at 60°C is 149 torr. The vapor pressure of water over a solution at 60°C containing equal numbers of moles of water and ethylene glycol (a nonvolatile solute) is 67 torr. Is the solution ideal according to Raoult's law? Explain.

13.46 Consider two solutions, one formed by adding 10 g of glucose ($C_6H_{12}O_6$) to 1 L of water and another formed by adding 10 g of sucrose ($C_{12}H_{22}O_{11}$) to 1 L of water. Are the vapor pressures over the two solutions the same? Why or why not?

13.47 **(a)** Calculate the vapor pressure of water above a solution prepared by adding 15.0 g of lactose ($C_{12}H_{22}O_{11}$) to 100.0 g of water at 338 K. (Vapor-pressure data for water are given in Appendix B.) **(b)** Calculate the mass of propylene glycol ($C_3H_8O_2$) that must be added to 0.500 kg of water to reduce the vapor pressure by 4.60 torr at 40°C.

13.48 **(a)** Calculate the vapor pressure of water above a solution prepared by dissolving 35.0 g of glycerin ($C_3H_8O_3$) in 125 g of water at 343 K. (The vapor pressure of water is given in Appendix B.) **(b)** Calculate the mass of ethylene glycol ($C_2H_6O_2$) that must be added to 1.00 kg of ethanol (C_2H_5OH) to reduce its vapor pressure by 10.0 torr at 35°C. The vapor pressure of pure ethanol at 35°C is 1.00×10^2 torr.

[13.49] At 63.5°C the vapor pressure of H_2O is 175 torr, and that of ethanol (C_2H_5OH) is 400 torr. A solution is made by mixing equal masses of H_2O and C_2H_5OH. **(a)** What is the mole fraction of ethanol in the solution? **(b)** Assuming ideal-solution behavior, what is the vapor pressure of the solution at 63.5°C? **(c)** What is the mole fraction of ethanol in the vapor above the solution?

[13.50] At 20°C the vapor pressure of benzene (C_6H_6) is 75 torr, and that of toluene (C_7H_8) is 22 torr. Assume that benzene and toluene form an ideal solution. **(a)** What is the composition in mole fractions of a solution that has a vapor pressure of 35 torr at 20°C? **(b)** What is the mole fraction of benzene in the vapor above the solution described in part (a)?

13.51 **(a)** Why does a 0.10 m aqueous solution of NaCl have a higher boiling point than a 0.10 m aqueous solution of $C_6H_{12}O_6$? **(b)** Calculate the boiling point of each solution.

13.52 Arrange the following aqueous solutions in order of increasing boiling point: a 10% solution of glucose ($C_6H_{12}O_6$), a 10% solution of sucrose ($C_{12}H_{22}O_{11}$), and a 10% solution of sodium nitrate ($NaNO_3$).

13.53 List the following aqueous solutions in order of decreasing freezing point: 0.040 m glycerin ($C_3H_8O_3$); 0.020 m KBr; 0.030 m phenol (C_6H_5OH).

13.54 List the following aqueous solutions in the order of increasing boiling point: 0.120 M glucose; 0.050 M LiBr; 0.050 M $Zn(NO_3)_2$.

13.55 Using data from Table 13.4, calculate the freezing and boiling points of each of the following solutions: **(a)** 0.35 m glycerol ($C_3H_8O_3$) in ethanol; **(b)** 1.58 mol of naphthalene ($C_{10}H_8$) in 14.2 mol of chloroform; **(c)** 5.13 g KBr and 6.85 g glucose ($C_6H_{12}O_6$) in 255 g of water.

13.56 Using data from Table 13.4, calculate the freezing and boiling points of each of the following solutions: **(a)** 0.40 m glucose in ethanol; **(b)** 20.0 g of $C_{10}H_{22}$ in 455 g $CHCl_3$; **(c)** 0.45 mol ethylene glycol and 0.15 mol KBr in 150 g H_2O.

13.57 What is the osmotic pressure of a solution formed by dissolving 50.0 mg of aspirin ($C_9H_8O_4$) in 0.250 L of water at 25°C?

13.58 Seawater contains 3.4 g of salts for every liter of solution. Assuming that the solute consists entirely of NaCl (over 90% is), calculate the osmotic pressure of seawater at 20°C.

13.59 Adrenaline is the hormone that triggers the release of extra glucose molecules in times of stress or emergency. A solution of 0.64 g of adrenaline in 36.0 g of CCl_4 elevates the boiling point by 0.49°C. What is the molar mass of adrenaline?

13.60 Lauryl alcohol is obtained from coconut oil and is used to make detergents. A solution of 5.00 g of lauryl alcohol in 0.100 kg of benzene freezes at 4.1°C. What is the molar mass of lauryl alcohol?

13.61 Lysozyme is an enzyme that breaks bacterial cell walls. A solution containing 0.150 g of this enzyme in 210 mL of solution has an osmotic pressure of 0.953 torr at 25°C. What is the molar mass of lysozyme?

13.62 A dilute aqueous solution of an organic compound soluble in water is formed by dissolving 2.35 g of the compound in water to form 0.250 L solution. The resulting solution has an osmotic pressure of 0.605 atm at 25°C. Assuming that the organic compound is a nonelectrolyte, what is its molar mass?

[13.63] The osmotic pressure of 0.010 M aqueous solution of $CaCl_2$ is found to be 0.674 atm at 25°C. **(a)** Calculate the van't Hoff factor, i, for the solution. **(b)** How would you expect the value of i to change as the solution becomes more concentrated? Explain.

[13.64] Based on the data given in Table 13.5, which solution would give the larger freezing-point lowering, a 0.030 m solution of NaCl or a 0.020 m solution of K_2SO_4? How do you explain the departure from ideal behavior, and the differences observed between the two salts?

Colloids

13.65 **(a)** Why is there no colloid in which both the dispersed substance and the dispersing substance are gases? **(b)** Michael Faraday first prepared ruby-red colloids of gold particles in water that were stable for indefinite times. To the unaided eye these brightly colored colloids are not distinguishable from solutions. How could you determine whether a given colored preparation is a solution or colloid?

13.66 **(a)** Many proteins that remain homogeneously distributed in aqueous medium have molecular masses in the range of 30,000 amu and larger. In what sense is it appropriate to consider such suspensions to be colloids rather than solutions? Explain. **(b)** What general name is given to a colloidal dispersion of one liquid in another? What is an emulsifying agent?

13.67 Indicate whether each of the following is a hydrophilic or a hydrophobic colloid: **(a)** butterfat in homogenized milk; **(b)** hemoglobin in blood; **(c)** vegetable oil in a salad dressing.

13.68 Explain how each of the following factors helps determine the stability or instability of a colloidal dispersion: **(a)** particulate mass; **(b)** hydrophobic character; **(c)** charges on colloidal particles.

13.69 What is the factor that most commonly prevents colloidal particles from coalescing into larger aggregates? How can colloids be coagulated?

13.70 Explain how **(a)** a soap such as sodium stearate stabilizes a colloidal dispersion of oil droplets in water; **(b)** milk curdles upon addition of an acid.

Additional Exercises

13.71 Butylated hydroxytoluene (BHT) has the following molecular structure:

BHT

It is widely used as a preservative in a variety of foods, including dried cereals. Based on its structure, would you expect BHT to be more soluble in water or in hexane (C_6H_{14})? Explain.

13.72 A saturated solution of sucrose ($C_{12}H_{22}O_{11}$) is made by dissolving excess table sugar in a flask of water. There are 50 g of undissolved sucrose crystals at the bottom of the flask, in contact with the saturated solution. The flask is stoppered and set aside. A year later a single large crystal of mass 50 g is at the bottom of the flask. Explain how this experiment provides evidence for a dynamic equilibrium between the saturated solution and the undissolved solute.

13.73 Fish need at least 4 ppm dissolved O_2 for survival. **(a)** What is this concentration in mol/L? **(b)** What partial pressure of O_2 above the water is needed to obtain this concentration at 10°C? (The Henry's law constant for O_2 at this temperature is 1.71×10^{-3} mol/L-atm.)

13.74 The presence of the radioactive gas radon (Rn) in well water obtained from aquifers that lie in rock deposits presents a possible health hazard in parts of the United States. A sample consisting of various gases contains 3.5×10^{-6} mole fraction of radon. This gas at a total pressure of 32 atm is shaken with water at 30°C. Assume that the solubility of radon in water with 1 atm pressure of the gas over the solution at 30°C is 7.27×10^{-3} *M*. Calculate the molar concentration of radon in the water.

13.75 Glucose makes up about 0.10% by mass of human blood. Calculate the concentration in **(a)** ppm; **(b)** molality. What further information would you need to determine the molarity of the solution?

13.76 Which has the higher molarity of K^+ ions, an aqueous solution that is 15 ppm in KBr or one that is 12 ppm in KCl?

13.77 A 32.0% by weight solution of propanol ($CH_3CH_2CH_2OH$) in water has a density at 20°C of 0.945 g/mL. What are the molarity and molality of the solution?

13.78 Acetonitrile (CH_3CN) is a polar organic solvent that dissolves a wide range of solutes, including many salts. The density of a 1.80 *M* acetonitrile solution of LiBr is 0.826 g/cm^3. Calculate the concentration of the solution in **(a)** molality; **(b)** mole fraction of LiBr; **(c)** mass percentage of CH_3CN.

13.79 Sodium metal dissolves in liquid mercury to form a solution called a sodium amalgam. The densities of Na(*s*) and Hg(*l*) are 0.97 g/cm^3 and 13.6 g/cm^3, respectively. A sodium amalgam is made by dissolving 1.0 cm^3 Na(*s*) in 20.0 cm^3 Hg(*l*). Assume that the final volume of the solution is 21.0 cm^3. **(a)** Calculate the molality of Na in the solution. **(b)** Calculate the molarity of Na in the solution. **(c)** For dilute aqueous solutions, the molality and molarity are generally nearly equal in value. Is that the case for the sodium amalgam described here? Explain.

13.80 A "canned heat" product used to warm chafing dishes consists of a homogeneous mixture of ethanol (C_2H_5OH) and paraffin that has an average formula of $C_{24}H_{50}$. What mass of C_2H_5OH should be added to 620 kg of the paraffin in formulating the mixture if the vapor pressure of ethanol at 35°C over the mixture is to be 8 torr? The vapor pressure of pure ethanol at 35°C is 100 torr.

[13.81] Two beakers are placed in a sealed box at 25°C. One beaker contains 20.0 mL of a 0.060 *M* aqueous solution of a nonvolatile nonelectrolye. The other beaker contains 20.0 mL of a 0.040 *M* aqueous solution of NaCl. The water vapor from the two solutions reach equilibrium. **(a)** In which beaker does the solution level rise and in which one does it fall? **(b)** What are the volumes in the two beakers when equilibrium is attained, assuming ideal behavior?

13.82 Calculate the freezing point of a 0.100 *m* aqueous solution of K_2SO_4, **(a)** ignoring interionic attractions, and **(b)** taking interionic attractions into consideration by using the van't Hoff factor (Table 13.5).

13.83 The cooling system of a car is filled with a solution formed by mixing equal volumes of water (density = 1.00 g/mL) and ethylene glycol, $C_2H_6O_2$ (density = 1.12 g/mL). Estimate the freezing point and boiling point of the mixture.

13.84 When 10.0 g of mercuric nitrate, $Hg(NO_3)_2$, is dissolved in 1.00 kg of water, the freezing point of the solution is −0.162°C. When 10.0 g of mercuric chloride ($HgCl_2$) is dissolved in 1.00 kg of water, the solution freezes at −0.0685°C. Use these data to determine which is the stronger electrolyte, $Hg(NO_3)_2$ or $HgCl_2$.

13.85 Carbon disulfide (CS_2) boils at 46.30°C and has a density of 1.261 g/mL. **(a)** When 0.250 mol of a nondissociating solute is dissolved in 400.0 mL of CS_2, the solution boils at 47.46°C. What is the molal boiling-point-elevation constant for CS_2? **(b)** When 5.39 g of a nondissociating unknown is dissolved in 50.0 mL of CS_2, the solution boils at 47.08°C. What is the molecular weight of the unknown?

[13.86] A 40.0% by weight solution of KSCN in water at 20°C has a density of 1.22 g/mL. **(a)** What is the mole fraction of KSCN in the solution, and what are the molarity and molality? **(b)** Given the calculated mole fraction of salt in the solution, comment on the total number of water molecules available to hydrate each anion and cation. What ion pairing (if any) would you expect to find in the solution? Would you expect the colligative properties of such a solution to be those predicted by the formulas given in this chapter? Explain.

[13.87] A mixture of solid NaCl and solid sucrose ($C_{12}H_{22}O_{11}$) has an unknown composition. When 15.0 g of the mixture is dissolved in enough water to make 500 mL of solution, the solution exhibits an osmotic pressure of 6.41 atm at 25°C. Determine the mass percentage of NaCl in the mixture.

[13.88] A lithium salt used in lubricating grease has the formula $LiC_nH_{2n+1}O_2$. The salt is soluble in water to the extent of 0.036 g per 100 g of water at 25°C. The osmotic pressure of this solution is found to be 57.1 torr. Assuming that molality and molarity in such a dilute solution are the same and that the lithium salt is completely dissociated in the solution, determine an appropriate value of n in the formula for the salt.

Integrative Exercises

13.89 Fluorocarbons (compounds that contain both carbon and fluorine) were, until recently, used as refrigerants. The compounds listed in the following table are all gases at 25°C, and their solubilities in water at 25°C and 1 atm fluorocarbon pressure are given as mass percentages:

Fluorocarbon	Solubility (mass %)
CF_4	0.0015
$CClF_3$	0.009
CCl_2F_2	0.028
$CHClF_2$	0.30

(a) For each fluorocarbon, calculate the molality of a saturated solution. **(b)** Explain why the molarity of each of the solutions should be very close numerically to the molality. **(c)** Based on their molecular structures, account for the differences in solubility of the four fluorocarbons. **(d)** Calculate the Henry's law constant at 25°C for $CHClF_2$, and compare its magnitude to that for N_2 (6.8×10^{-4} mol/L-atm). Can you account for the difference in magnitude?

[13.90] At ordinary body temperature (37°C) the solubility of N_2 in water in contact with air at ordinary atmospheric pressure (1.0 atm) is 0.015 g/L. Air is approximately 78 mol % N_2. Calculate the number of moles of N_2 dissolved per liter of blood, which is essentially an aqueous solution. At a depth of 100 ft, the pressure is 4.0 atm. What is the solubility of N_2 from air in blood at this pressure? If the diver suddenly surfaces, how many milliliters of N_2 gas, in the form of tiny bubbles, are released into the bloodstream from each liter of blood?

[13.91] Consider the following values for enthalpy of vaporization (kJ/mol) of several organic substances:

Acetaldehyde	$CH_3C(=O)-H$	30.4
Ethylene oxide	$H_2C-O-CH_2$ (ring)	28.5
Acetone	$CH_3C(=O)CH_3$	32.0
Cyclopropane	$H_2C-CH_2-CH_2$ (ring)	24.7

(a) Use variations in the intermolecular forces operating in these organic liquids to account for their variations in heats of vaporization. **(b)** How would you expect the solubilities of these substances to vary in hexane as solvent? In ethanol? Use intermolecular forces, including hydrogen-bonding interactions where applicable, to explain your responses.

[13.92] The enthalpies of solution of hydrated salts are generally more positive than those of anhydrous materials. For example, ΔH of solution for KOH is −57.3 kJ/mol, whereas that for $KOH \cdot H_2O$ is −14.6 kJ/mol. Similarly, ΔH_{soln} for $NaClO_4$ is + 13.8 kJ/mol, whereas that for $NaClO_4 \cdot H_2O$ is +22.5 kJ/mol. Use the enthalpy contributions to the solution process depicted in Figure 13.4 to explain this effect.

[13.93] A textbook on chemical thermodynamics states, "The heat of solution represents the difference between the lattice energy of the crystalline solid and the solvation energy of the gaseous ions." **(a)** Draw a simple energy diagram to illustrate this statement. **(b)** A salt such as NaBr is insoluble in most polar nonaqueous solvents such as acetonitrile (CH_3CN) or nitromethane (CH_3NO_2), but salts of large cations, such as tetramethylammonium bromide ($(CH_3)_4NBr$), are generally more soluble. Use the thermochemical cycle you drew in part (a) and the factors that determine the lattice energy (Section 8.2) to explain this fact.

13.94 **(a)** A sample of hydrogen gas is generated in a closed container by reacting 2.050 g of zinc metal with 15.0 mL of 1.00 M sulfuric acid. Write the balanced equation for the reaction, and calculate the number of moles of hydrogen formed, assuming that the reaction is complete. **(b)** The volume over the solution is 122 mL. Calculate the partial pressure of the hydrogen gas in this volume at 25°C ignoring any solubility of the gas in the solution. **(c)** The Henry's law constant for hydrogen in water at 25°C is 7.8×10^{-4} mol/L-atm. Estimate the number of moles of hydrogen gas that remain dissolved in the solution. What fraction of the gas molecules in the system is dissolved in the solution? Was it reasonable to ignore any dissolved hydrogen in part (b)?

[13.95] The following table presents the solubilities of several gases in water at 25°C under a total pressure of gas and water vapor of 1 atm:

Gas	Solubility (m*M*)
CH_4 (methane)	1.3
C_2H_6 (ethane)	1.8
C_2H_4 (ethylene)	4.7
N_2	0.6
O_2	1.2
NO	1.9
H_2S	99
SO_2	1476

(a) What volume of $CH_4(g)$ under standard conditions of temperature and pressure is contained in 4.0 L of a saturated solution at 25°C? **(b)** Explain the variation in solubility among the hydrocarbons listed (the first three compounds), based on their molecular structures and intermolecular forces. **(c)** Compare the solubilities of O_2, N_2, and NO, and account for the variations based on molecular structures and intermolecular forces. **(d)** Account for the much larger values observed for H_2S and SO_2 as compared with the other gases listed. **(e)** Find several pairs of substances with the same or nearly the same molecular masses (for example, C_2H_4 and N_2), and use intermolecular interactions to explain the differences in their solubilities.

13.96 Hexabarbital, used in medicine as a sedative and intravenous anesthetic, is composed of 61.00% C, 6.83% H, 11.86% N, and 20.32% O by mass. A sample of 2.505 mg in 10.00 mL of solution has an osmotic pressure of 19.7 torr at 25°C. What is the molecular formula of hexabarbital?

[13.97] When 0.55 g of pure benzoic acid ($C_7H_6O_2$) is dissolved in 32.0 g of benzene, the freezing point of the solution is 0.36°C lower than the freezing point value of 5.5°C for the pure solvent. **(a)** Calculate the molecular weight of benzoic acid in benzene. **(b)** Use the structure of the solute to account for the observed value:

O
C
O—H

[13.98] At 35°C the vapor pressure of acetone, $(CH_3)_2CO$, is 360 torr, and that of chloroform, $CHCl_3$, is 300 torr. Acetone and chloroform can form weak hydrogen bonds between one another as follows:

Cl
Cl—C—H···O=C
Cl
CH_3
CH_3

A solution composed of an equal number of moles of acetone and chloroform has a vapor pressure of 250 torr at 35°C. **(a)** What would be the vapor pressure of the solution if it exhibited ideal behavior? **(b)** Use the existence of hydrogen bonds between acetone and chloroform molecules to explain the deviation from ideal behavior. **(c)** Based on the behavior of the solution, predict whether the mixing of acetone and chloroform is an exothermic ($\Delta H_{soln} < 0$) or endothermic ($\Delta H_{soln} > 0$) process.

eMedia Exercises

13.99 **(a)** Of the four solids available in the **Enthalpy of Solution** simulation (*eChapter 13.1*), which would become more soluble and which would become less soluble with increasing temperature? **(b)** Comparing sodium chloride and ammonium nitrate, which would you expect to exhibit the greater change in solubility (in g/L) for a given temperature increase? Explain your reasoning.

13.100 The **Enthalpy of Solution** simulation (*eChapter 13.1*) deals specifically with dissolution of solids. Describe the solution process in terms of three distinct steps. Using the three steps, explain why gases, in contrast to most solids, become *less* soluble as temperature increases.

13.101 The **Enthalpy of Solution** simulation (*eChapter 13.1*) allows you to compare the overall energy changes for the dissolution of various salts. Compare the dissolution of sodium hydroxide with that of sodium chloride. **(a)** How do the energy changes for the overall processes compare? **(b)** Discuss the difference in magnitude of each of the three steps for dissolution of the two compounds.

13.102 The **Henry's Law** movie (*eChapter 13.3*) illustrates the relationship between the pressure of a gas above a liquid and the solubility of the gas in the liquid. Sample Exercise 13.2 determines the concentration of CO_2 in a carbonated beverage bottled under 4.0 atm of CO_2. **(a)** If the mole fraction of CO_2 in the air is 0.000355, how many times more concentrated is the CO_2 in the unopened bottle than it is in a glass of the beverage that has gone completely flat? **(b)** Use the kinetic-molecular theory of gases to explain why the solubility of a gas is proportional to its pressure over the liquid.

13.103 The **Colligative Properties** simulation (*eChapter 13.5*) shows how the boiling point and freezing point of water are affected by the addition of various solutes. **(a)** If you were to dissolve 10 g sodium chloride in 500 g water, what would be the boiling point of the solution? **(b)** If you were to heat the solution in part (a), and continue heating it as it boiled, what would happen to the temperature over time? **(c)** In what way is your answer to part (b) contrary to what you learned in Chapter 11 (Section 11.4) regarding constant temperature during phase changes? Explain this apparent contradiciton.

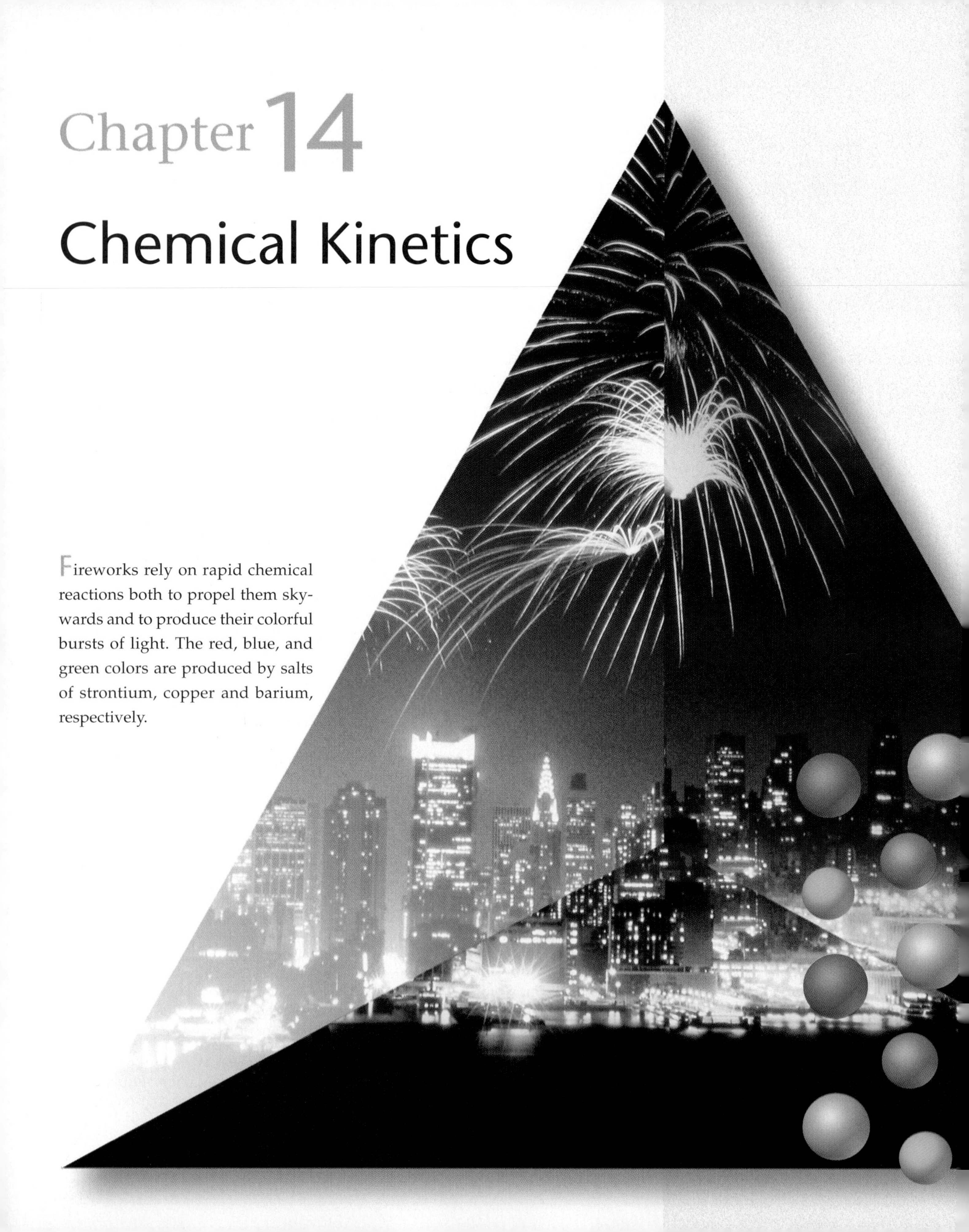

Chapter 14

Chemical Kinetics

Fireworks rely on rapid chemical reactions both to propel them skywards and to produce their colorful bursts of light. The red, blue, and green colors are produced by salts of strontium, copper and barium, respectively.

CHEMISTRY IS, BY its very nature, concerned with change. Chemical reactions convert substances with well-defined properties into other materials with different properties. Much of our study of chemical reactions is concerned with the formation of new substances from a given set of reactants. However, it is equally important to understand how rapidly chemical reactions occur. The rates of reactions span an enormous range, from those that are complete within fractions of seconds, such as certain explosions, to those that take thousands or even millions of years, such as the formation of diamonds or other minerals in Earth's crust (Figure 14.1 ▶).

The area of chemistry that is concerned with the speeds, or rates, of reactions is called **chemical kinetics**. Chemical kinetics is a subject of broad importance. It relates, for example, to how quickly a medicine is able to work, to whether the formation and depletion of ozone in the upper atmosphere are in balance, and to industrial problems such as the development of catalysts to synthesize new materials. Our goal in this chapter is not only to understand how to determine the rates at which reactions occur, but also to consider the factors that control these rates. For example, what factors determine how rapidly food spoils? How does one design a fast-setting material for dental fillings? What determines the rate at which steel rusts? What controls the rate at which fuel burns in an automobile engine? Although we won't address these specific questions directly, we will see that the rates of all chemical reactions are subject to the same basic principles.

▶ What's Ahead ◀

- There are four experimental variables that affect reaction rates: concentration, physical states of reactants, temperature, and catalysts. These factors can be understood in terms of the collisions among reactant molecules that lead to reaction.
- We consider how we express *reaction rates* and how the rates of appearance of reactants and the rates of disappearance of products are related to the stoichiometry of the reaction.
- We then examine how the effect of concentration on rate is expressed quantitatively by *rate laws* and how rate laws can be determined experimentally.
- Rate equations can be written to express how concentrations change with time. We will look at such rate equations for two simple kinds of rate laws.
- We next consider the effect of temperature on rate and the fact that reactions require a minimum input of energy called the *activation energy* in order to occur.
- We then examine the *mechanisms* of reactions, the step-by-step molecular pathways leading from reactants to products.
- The chapter ends with a discussion of how *catalysts* speed reaction rates, including a discussion of biological catalysts, called *enzymes*.

▲ **Figure 14.1** The rates of chemical reactions span a range of time scales. For example, explosions are rapid, occurring in seconds or fractions of seconds; cooking can take minutes or hours; corrosion can take years; and the weathering of rocks takes place over thousands or even millions of years.

14.1 Factors that Affect Reaction Rates

Before we examine the quantitative aspects of chemical kinetics, such as how rates are measured, let's examine the key factors that influence the rates of reactions. Because reactions involve the breaking and forming of bonds, their speeds depend on the nature of the reactants themselves. There are, however, four factors that allow us to change the rates at which particular reactions occur:

1. *The physical state of the reactants.* Reactants must come together in order to react. The more readily molecules collide with each other, the more rapidly they react. Most of the reactions we consider are homogeneous, involving either gases or liquid solutions. When reactants are in different phases, such as when one is a gas and another a solid, the reaction is limited to their area of contact. Thus, reactions that involve solids tend to proceed faster if the surface area of the solid is increased. For example, a medicine in the form of a tablet will dissolve in the stomach and enter the bloodstream more slowly than the same medicine in the form of a fine powder.
2. *The concentrations of the reactants.* Most chemical reactions proceed faster if the concentration of one or more of the reactants is increased. For example, steel wool burns with difficulty in air, which contains 20% O_2, but bursts into a brilliant white flame in pure oxygen (Figure 14.2 ▼). As concentration increases, the frequency with which molecules collide increases, leading to increased rates.

(a)

(b)

▶ **Figure 14.2** (a) When heated in air, steel wool glows red-hot, but oxidizes slowly. (b) When the red-hot steel wool is placed in an atmosphere of pure oxygen, it burns vigorously, forming Fe_2O_3 at a much faster rate. The different behaviors are due to the different concentrations of O_2 in the two environments.

3. *The temperature at which the reaction occurs.* The rates of chemical reactions increase as temperature is increased. It is for this reason that we refrigerate perishable foods such as milk. The bacterial reactions that lead to the spoiling of milk proceed much more rapidly at room temperature than they do at the lower temperatures of a refrigerator. Increasing temperature increases the kinetic energies of molecules. (Section 10.7) As molecules move more rapidly, they collide more frequently and also with higher energy, leading to increased rates.
4. *The presence of a catalyst.* Catalysts are agents that increase reaction rates without being used up. They affect the kinds of collisions (the mechanism) that lead to reaction. Catalysts play a crucial role in our lives. The physiology of most living species depends on *enzymes*, protein molecules that act as catalysts, increasing the rates of selected biochemical reactions.

On a molecular level, reaction rates depend on the frequency of collisions between molecules. The greater the frequency of collisions, the greater the rate of reaction. In order for a collision to lead to reaction, however, it must occur with sufficient energy to stretch bonds to a critical length and with suitable orientation for new bonds to form in the proper locations. We will consider these factors as we proceed through this chapter.

14.2 Reaction Rates

The *speed* of an event is defined as the *change* that occurs in a given interval of *time*: Whenever we talk about speed, we necessarily bring in the notion of time. For example, the speed of a car is expressed as the change in the car's position over a certain period of time. The units of this speed are usually miles per hour (mi/hr)—that is, the quantity that is changing (position, measured in miles) divided by a time interval (hours).

Similarly, the speed of a chemical reaction—its **reaction rate**—is the change in the concentration of reactants or products per unit time. Thus, the units for reaction rate are usually molarity per second (M/s)—that is, the change in concentration (measured in molarity) divided by a time interval (seconds).

Let's consider a simple hypothetical reaction, $A \longrightarrow B$, depicted in Figure 14.3 ▼. Each red sphere represents 0.01 mol of A, and each blue sphere represents 0.01 mol of B. Let's assume that the container has a volume of 1.00 L. At the beginning of the reaction there is 1.00 mol A, so the concentration is 1.00 mol/L = 1.00 M. After 20 s the concentration of A has fallen to 0.54 M, whereas that of B has risen to 0.46 M. The sum of the concentrations is still 1.00 M because one mole of B is produced for each mole of A that reacts. After 40 s the concentration of A is 0.30 M and that of B is 0.70 M.

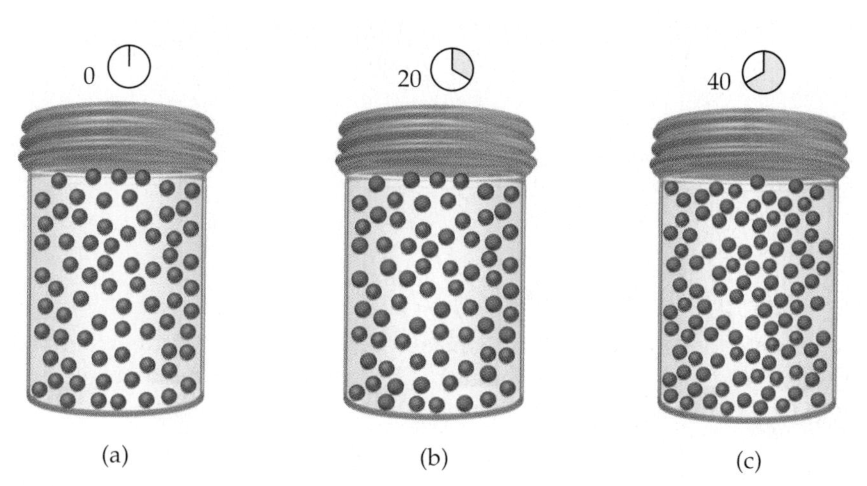

◀ **Figure 14.3** Progress of a hypothetical reaction $A \longrightarrow B$, starting with 1.00 mol A. Each red sphere represents 0.01 mol A, each blue sphere represents 0.01 mol B, and the vessel has a volume of 1.00 L. (a) At time zero the vessel contains 1.00 mol A (100 red spheres) and 0 mol B (0 blue spheres). (b) After 20 s the vessel contains 0.54 mol A and 0.46 mol B. (c) After 40 s the vessel contains 0.30 mol A and 0.70 mol B.

The rate of this reaction can be expressed either as the rate of disappearance of reactant A or as the rate of appearance of product B. The *average* rate of appearance of B over a particular time interval is given by the change in concentration of B divided by the change in time:

$$\text{Average rate with respect to B} = \frac{\text{change in concentration of B}}{\text{change in time}}$$

$$= \frac{[\text{B}] \text{ at } t_2 - [\text{B}] \text{ at } t_1}{t_2 - t_1} = \frac{\Delta[\text{B}]}{\Delta t} \qquad [14.1]$$

We use brackets around a chemical formula, as in [B], to indicate the concentration of the substance in molarity. The Greek letter delta, Δ, is read "change in," and it is always equal to the final quantity minus the initial quantity. (Section 5.2) The average rate of appearance of B over the 20-s interval from the beginning of the reaction ($t_1 = 0$ s to $t_2 = 20$ s) is given by

$$\text{Average rate} = \frac{0.46\ M - 0.00\ M}{20\text{ s} - 0\text{ s}} = 2.3 \times 10^{-2}\ M/\text{s}$$

We could equally well express the rate of the reaction with respect to the change of concentration of the reactant, A. In this case we would be describing the rate of disappearance of A, which we express as

$$\text{Average rate with respect to A} = -\frac{\Delta[\text{A}]}{\Delta t} \qquad [14.2]$$

Notice the minus sign in this equation. By convention, rates are always expressed as positive quantities. Because [A] is decreasing with time, Δ[A] is a negative number. We use the negative sign to convert the negative Δ[A] to a positive rate. Because one molecule of A is consumed for every molecule of B that forms, the average rate of disappearance of A equals the average rate of appearance of B, as the following calculation shows:

$$\text{Average rate} = -\frac{\Delta[\text{A}]}{\Delta t} = -\frac{0.54\ M - 1.00\ M}{20\text{ s} - 0\text{ s}} = 2.3 \times 10^{-2}\ M/\text{s}$$

SAMPLE EXERCISE 14.1

For the reaction pictured in Figure 14.3, calculate the average rate of disappearance of A over the time interval from 20 s to 40 s.

Solution:

Analyze and Plan: The average rate is given by the change in concentration, Δ[A], divided by the corresponding change in time, Δt. Because A is a reactant, a minus sign is used in the calculation to make the rate a positive quantity.

Solve:

$$\text{Average rate} = -\frac{\Delta[\text{A}]}{\Delta t} = -\frac{0.30\ M - 0.54\ M}{40\text{ s} - 20\text{ s}} = 1.2 \times 10^{-2}\ M/\text{s}$$

PRACTICE EXERCISE

For the reaction pictured in Figure 14.3, calculate the average rate of appearance of B over the time interval from 0 to 40 s.

Answer: $1.8 \times 10^{-2}\ M/\text{s}$

Change of Rate with Time

Now, let's consider an actual chemical reaction, one that occurs when butyl chloride (C_4H_9Cl) is placed in water. The products formed are butyl alcohol (C_4H_9OH) and hydrochloric acid:

$$C_4H_9Cl(aq) + H_2O(l) \longrightarrow C_4H_9OH(aq) + HCl(aq) \qquad [14.3]$$

TABLE 14.1 Rate Data for Reaction of C_4H_9Cl with Water

Time, t (s)	$[C_4H_9Cl]$ (M)	Average Rate (M/s)
0.0	0.1000	
		1.9×10^{-4}
50.0	0.0905	
		1.7×10^{-4}
100.0	0.0820	
		1.6×10^{-4}
150.0	0.0741	
		1.4×10^{-4}
200.0	0.0671	
		1.22×10^{-4}
300.0	0.0549	
		1.01×10^{-4}
400.0	0.0448	
		0.80×10^{-4}
500.0	0.0368	
		0.560×10^{-4}
800.0	0.0200	
10,000	0	

Suppose that we prepare a 0.1000 M aqueous solution of C_4H_9Cl and then measure the concentration of C_4H_9Cl at various times after time zero, collecting the data shown in the first two columns of Table 14.1 ▲. We can use these data to calculate the average rate of disappearance of C_4H_9Cl over the intervals between measurements, and these rates are given in the third column. Notice that the average rate decreases over each 50-s interval for the first several measurements and continues to decrease over even larger intervals through the remaining measurements. It is typical for rates to decrease as a reaction proceeds, because the concentration of reactants decreases. The change in rate as the reaction proceeds is also seen in a graph of the concentration of C_4H_9Cl versus time (Figure 14.4 ▼). Notice how the steepness of the curve decreases with time, indicating a decreasing rate of reaction.

The graph shown in Figure 14.4 is particularly useful because it allows us to evaluate the **instantaneous rate**, the rate at a particular moment in the reaction. The instantaneous rate is determined from the slope (or tangent) of this curve at the point of interest. We have drawn two tangents in Figure 14.4, one at $t = 0$ and the other at $t = 600$ s. The slopes of these tangents give the instantaneous

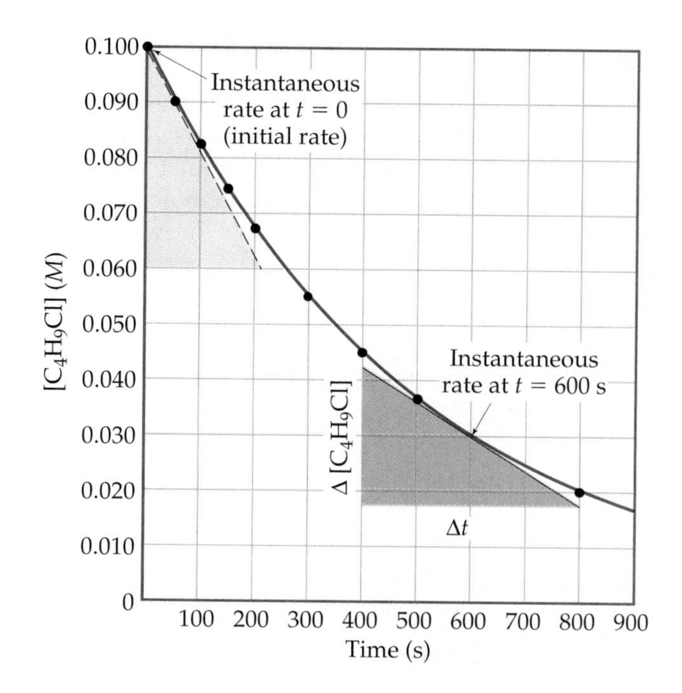

◀ **Figure 14.4** Concentration of butyl chloride (C_4H_9Cl) as a function of time. The dots represent the experimental data from the first two columns of Table 14.2, and the red curve is drawn to smoothly connect the data points. Lines are drawn that are tangent to the curve at $t = 0$ and $t = 600$ s. The slope of each of these tangents is defined as the vertical change divided by the horizontal change (that is, $\Delta[C_4H_9Cl]/\Delta t$). The reaction rate at any time is related to the slope of the tangent to the curve at that time. Because C_4H_9Cl is disappearing, the rate is equal to the negative of the slope.

rates at these times.* For example, to determine the instantaneous rate at 600 s, we draw the tangent to the curve at this time, then construct horizontal and vertical lines to form the right triangle shown. The slope is the ratio of the height of the vertical side to the length of the horizontal side:

$$\text{Instantaneous rate} = \frac{\Delta[C_4H_9Cl]}{\Delta t} = -\frac{(0.017 - 0.042)\,M}{(800 - 400)\,\text{s}}$$

$$= 6.2 \times 10^{-5}\,M/\text{s}$$

In what follows, the term "rate" means "instantaneous rate," unless indicated otherwise. The instantaneous rate at $t = 0$ is called the *initial rate* of the reaction.

To understand better the difference between average rate and instantaneous rate, imagine that you have just driven 98 mi in 2.0 hr. Your average speed is 49 mi/hr, whereas your instantaneous speed at any moment is the speedometer reading at that time.

SAMPLE EXERCISE 14.2

Using Figure 14.4, calculate the instantaneous rate of disappearance of C_4H_9Cl at $t = 0$ (the initial rate).

Solution

Analyze and Plan: To obtain the instantaneous rate at $t = 0$, we must determine the slope of the curve at $t = 0$. The tangent is drawn on the graph. The slope of this straight line equals the change in the vertical axis divided by the corresponding change in the horizontal axis (that is, change in molarity over change in time).

Solve: The straight line falls from $[C_4H_9Cl] = 0.100\,M$ to $0.060\,M$ in the time change from 0 s to 200 s, as indicated by the yellow triangle shown in Figure 14.4. Thus, the initial rate is

$$\text{Rate} = -\frac{\Delta[C_4H_9Cl]}{\Delta t} = -\frac{(0.060 - 0.100)\,M}{(200 - 0)\,\text{s}} = 2.0 \times 10^{-4}\,M/\text{s}$$

PRACTICE EXERCISE

Using Figure 14.4, determine the instantaneous rate of disappearance of C_4H_9Cl at $t = 300$ s.

Answer: $1.1 \times 10^{-4}\,M/\text{s}$

Reaction Rates and Stoichiometry

During our earlier discussion of the hypothetical reaction, $A \longrightarrow B$, we saw that the stoichiometry requires that the rate of disappearance of A equals the rate of appearance of B. Likewise, the stoichiometry of Equation 14.3 indicates that one mole of C_4H_9OH is produced for each mole of C_4H_9Cl consumed. Therefore, the rate of appearance of C_4H_9OH equals the rate of disappearance of C_4H_9Cl:

$$\text{Rate} = -\frac{\Delta[C_4H_9Cl]}{\Delta t} = \frac{\Delta[C_4H_9OH]}{\Delta t}$$

What happens when the stoichiometric relationships are not one to one? For example, consider the following reaction:

$$2HI(g) \longrightarrow H_2(g) + I_2(g)$$

* You may wish to review briefly the idea of graphical determination of slopes by referring to Appendix A. If you are familiar with calculus, you may recognize that the average rate approaches the instantaneous rate as the time interval approaches zero. This limit, in the notation of calculus, is represented as $-d[C_4H_9Cl]/dt$.

We can measure the rate of disappearance of HI or the rate of appearance of either H_2 or I_2. Because 2 mol of HI disappear for each mole of H_2 or I_2 that forms, the rate of disappearance of HI is twice the rate of appearance of either H_2 or I_2. To equate the rates, we must therefore divide the rate of disappearance of HI by 2 (its coefficient in the balanced chemical equation):

$$\text{Rate} = -\frac{1}{2}\frac{\Delta[\text{HI}]}{\Delta t} = \frac{\Delta[\text{H}_2]}{\Delta t} = \frac{\Delta[\text{I}_2]}{\Delta t}$$

In general, for the reaction

$$a\text{A} + b\text{B} \longrightarrow c\text{C} + d\text{D}$$

the rate is given by

$$\text{Rate} = -\frac{1}{a}\frac{\Delta[\text{A}]}{\Delta t} = -\frac{1}{b}\frac{\Delta[\text{B}]}{\Delta t} = \frac{1}{c}\frac{\Delta[\text{C}]}{\Delta t} = \frac{1}{d}\frac{\Delta[\text{D}]}{\Delta t} \qquad [14.4]$$

When we speak of the rate of a reaction without specifying a particular reactant or product, we will mean it in this sense.*

SAMPLE EXERCISE 14.3

(a) How is the rate of disappearance of ozone related to the rate of appearance of oxygen in the following equation: $2O_3(g) \longrightarrow 3O_2(g)$? **(b)** If the rate of appearance of O_2, $\Delta[O_2]/\Delta t$, is 6.0×10^{-5} M/s at a particular instant, what is the value of the rate of disappearance of O_3, $-\Delta[O_3]/\Delta t$, at this same time?

Solution

Analyze and Plan: The relative rates with respect to different reactants and products depend on the coefficients in the balanced chemical equation. Equation 14.4 indicates the general relationships.

Solve: (a) Using the coefficients in the balanced equation and the relationship given by Equation 14.4, we have

$$\text{Rate} = -\frac{1}{2}\frac{\Delta[\text{O}_3]}{\Delta t} = \frac{1}{3}\frac{\Delta[\text{O}_2]}{\Delta t}$$

(b) Solving the equation from part (a) for the rate of disappearance of O_3, $-\Delta[O_3]/\Delta t$, we have

$$-\frac{\Delta[\text{O}_3]}{\Delta t} = \frac{2}{3}\frac{\Delta[\text{O}_2]}{\Delta t} = \frac{2}{3}(6.0 \times 10^{-5}\ M/\text{s}) = 4.0 \times 10^{-5}\ M/\text{s}$$

Check: We can directly apply a stoichiometric factor to convert the rate with respect to O_2 to the rate with respect to O_3:

$$-\frac{\Delta[\text{O}_3]}{\Delta t} = \left(6.0 \times 10^{-5}\frac{\text{mol O}_2/\text{L}}{\text{s}}\right)\left(\frac{2\text{ mol O}_3}{3\text{ mol O}_2}\right) = 4.0 \times 10^{-5}\frac{\text{mol O}_3/\text{L}}{\text{s}}$$

$$= 4.0 \times 10^{-5}\ M/\text{s}$$

PRACTICE EXERCISE

The decomposition of N_2O_5 proceeds according to the following equation:

$$2\text{N}_2\text{O}_5(g) \longrightarrow 4\text{NO}_2(g) + \text{O}_2(g)$$

If the rate of decomposition of N_2O_5 at a particular instant in a reaction vessel is 4.2×10^{-7} M/s, what is the rate of appearance of **(a)** NO_2; **(b)** O_2?

Answers: **(a)** 8.4×10^{-7} M/s; **(b)** 2.1×10^{-7} M/s

ACTIVITY
Decomposition of N_2O_5

---

* Equation 14.4 does not hold true if substances other than C and D are formed in significant amounts during the course of the reaction. For example, sometimes intermediate substances build in concentration before forming the final products. In that case the relationship between the rate of disappearance of reactants and the rate of appearance of products will not be given by Equation 14.4. All reactions whose rates we consider in this chapter obey Equation 14.4.

A Closer Look Using Spectroscopic Methods to Measure Reaction Rates

A variety of techniques can be used to monitor the concentration of a reactant or product during a reaction. Spectroscopic methods, which rely on of the ability of substances to absorb (or emit) electromagnetic radiation, are some of the most useful. Spectroscopic kinetic studies are often performed with the reaction mixture in the sample compartment of the spectrometer. The spectrometer is set to measure the light absorbed at a wavelength characteristic of one of the reactants or products. In the decomposition of $HI(g)$ into $H_2(g)$ and $I_2(g)$, for example, both HI and H_2 are colorless, whereas I_2 is violet. During the course of the reaction, the color increases in intensity as I_2 forms. Thus, visible light of appropriate wavelength can be used to monitor the reaction.

Figure 14.5 ▼ shows the basic components of a spectrometer. The spectrometer measures the amount of light absorbed by the sample by comparing the intensity of the light emitted from the light source with the intensity of the light that emerges from the sample. As the concentration of I_2 increases and its color becomes more intense, the amount of light absorbed by the reaction mixture increases, causing less light to reach the detector.

Beer's law relates the amount of light being absorbed to the concentration of the substance absorbing the light:

$$A = abc \qquad [14.5]$$

In this equation A is the measured absorbance, a is the molar absorptivity constant (a characteristic of the substance being monitored), b is the path length through which the radiation must pass, and c is the molar concentration of the absorbing substance. Thus, the concentration is directly proportional to absorbance.

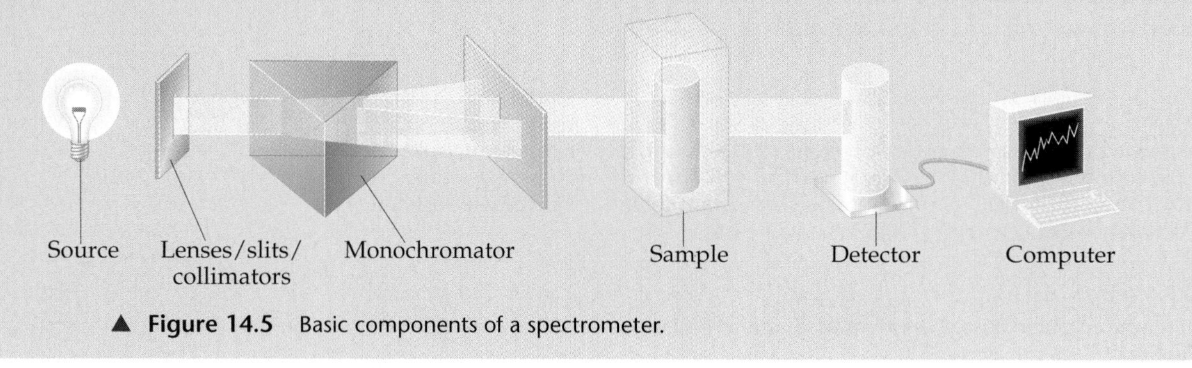

▲ **Figure 14.5** Basic components of a spectrometer.

14.3 Concentration and Rate

One way of studying the effect of concentration on reaction rate is to determine the way in which the rate at the beginning of a reaction (the initial rate) depends on the starting concentrations. To illustrate this approach, consider the following reaction:

$$NH_4^+(aq) + NO_2^-(aq) \longrightarrow N_2(g) + 2H_2O(l)$$

We might study the rate of this reaction by measuring the concentration of NH_4^+ or NO_2^- as a function of time or by measuring the volume of N_2 collected. Because the stoichiometric coefficients on NH_4^+, NO_2^-, and N_2 are all the same, all of these rates will be equal.

Once we determine the initial reaction rate for various starting concentrations of NH_4^+ and NO_2^-, we can tabulate the data as shown in Table 14.2 ▼.

TABLE 14.2 Rate Data for the Reaction of Ammonium and Nitrite Ions in Water at 25°C

Experiment Number	Initial NH_4^+ Concentration (M)	Initial NO_2^- Concentration (M)	Observed Initial Rate (M/s)
1	0.0100	0.200	5.4×10^{-7}
2	0.0200	0.200	10.8×10^{-7}
3	0.0400	0.200	21.5×10^{-7}
4	0.0600	0.200	32.3×10^{-7}
5	0.200	0.0202	10.8×10^{-7}
6	0.200	0.0404	21.6×10^{-7}
7	0.200	0.0606	32.4×10^{-7}
8	0.200	0.0808	43.3×10^{-7}

These data indicate that changing either $[NH_4^+]$ or $[NO_2^-]$ changes the reaction rate. If we double $[NH_4^+]$ while holding $[NO_2^-]$ constant, the rate doubles (compare experiments 1 and 2). If $[NH_4^+]$ is increased by a factor of 4 (compare experiments 1 and 3), the rate changes by a factor of 4, and so forth. These results indicate that the rate is proportional to $[NH_4^+]$ raised to the first power. When $[NO_2^-]$ is similarly varied while $[NH_4^+]$ is held constant, the rate is affected in the same manner. We conclude that the rate is also directly proportional to the concentration of NO_2^-. We can express the overall concentration dependence as follows:

$$\text{Rate} = k[NH_4^+][NO_2^-] \qquad [14.6]$$

An equation such as Equation 14.6, which shows how the rate depends on the concentrations of reactants, is called a **rate law**. For a general reaction,

$$aA + bB \longrightarrow cC + dD$$

the rate law generally has the form

$$\text{Rate} = k[A]^m[B]^n \qquad [14.7]$$

The constant k in the rate law is called the **rate constant**. The magnitude of k changes with temperature and therefore determines how temperature affects rate, as we will see in Section 14.5. The exponents m and n are typically small whole numbers (usually 0, 1, or 2). We will consider these exponents more closely very shortly.

If we know the rate law for a reaction and its rate for a set of reactant concentrations, we can calculate the value of the rate constant, k. For example, using the data in Table 14.2 and the results from experiment 1, we can substitute into Equation 14.6:

$$5.4 \times 10^{-7}\ M/\text{s} = k(0.0100\ M)(0.200\ M)$$

Solving for k gives

$$k = \frac{5.4 \times 10^{-7}\ M/\text{s}}{(0.0100\ M)(0.200\ M)} = 2.7 \times 10^{-4}\ M^{-1}\,\text{s}^{-1}$$

You may wish to verify that this same value of k is obtained using any of the other experimental results given in Table 14.2.

Once we have both the rate law and the value of the rate constant for a reaction, we can calculate the rate of reaction for any set of concentrations. For example, using Equation 14.6 and $k = 2.7 \times 10^{-4}\ M^{-1}\text{s}^{-1}$, we can calculate the rate for $[NH_4^+] = 0.100\ M$ and $[NO_2^-] = 0.100\ M$:

$$\text{Rate} = (2.7 \times 10^{-4}\ M^{-1}\text{s}^{-1})(0.100\ M)(0.100\ M) = 2.7 \times 10^{-6}\ M/\text{s}$$

Exponents in the Rate Law

The rate laws for most reactions have the general form

$$\text{Rate} = k[\text{reactant 1}]^m[\text{reactant 2}]^n \ldots \qquad [14.8]$$

The exponents m and n in a rate law are called **reaction orders**. For example, consider again the rate law for the reaction of NH_4^+ with NO_2^-:

$$\text{Rate} = k[NH_4^+][NO_2^-]$$

Because the exponent of $[NH_4^+]$ is one, the rate is *first order* in NH_4^+. The rate is also first order in NO_2^-. (The exponent "1" is not shown explicitly in rate laws.) The **overall reaction order** is the sum of the orders with respect to each reactant in the rate law. Thus, the rate law has an overall reaction order of $1 + 1 = 2$, and the reaction is *second order overall*.

The exponents in a rate law indicate how the rate is affected by the concentration of each reactant. Because the rate of the reaction of NH_4^+ with NO_2^- depends on $[NH_4^+]$ raised to the first power, the rate doubles when $[NH_4^+]$ doubles. Doubling $[NO_2^-]$ likewise doubles the rate. If a rate law is second order with respect to a reactant, $[A]^2$, then doubling the concentration of that substance will cause the reaction rate to quadruple ($[2]^2 = 4$).

The following are some additional examples of rate laws:

$$2N_2O_5(g) \longrightarrow 4NO_2(g) + O_2(g) \qquad \text{Rate} = k[N_2O_5] \qquad [14.9]$$

$$CHCl_3(g) + Cl_2(g) \longrightarrow CCl_4(g) + HCl(g) \qquad \text{Rate} = k[CHCl_3][Cl_2]^{1/2} \qquad [14.10]$$

$$H_2(g) + I_2(g) \longrightarrow 2HI(g) \qquad \text{Rate} = k[H_2][I_2] \qquad [14.11]$$

Although the exponents in a rate law are sometimes the same as the coefficients in the balanced equation, this is not necessarily the case, as seen in Equations 14.9 and 14.10. *The values of these exponents must be determined experimentally.* In most rate laws, reaction orders are 0, 1, or 2. However, we also occasionally encounter rate laws in which the reaction order is fractional (such as Equation 14.10) or even negative.

SAMPLE EXERCISE 14.4

Consider a reaction $A + B \longrightarrow C$ for which rate $= k[A][B]^2$. Each of the following boxes represents a reaction mixture in which A is shown as red spheres and B as blue ones. Rank these mixtures in order of increasing rate of reaction.

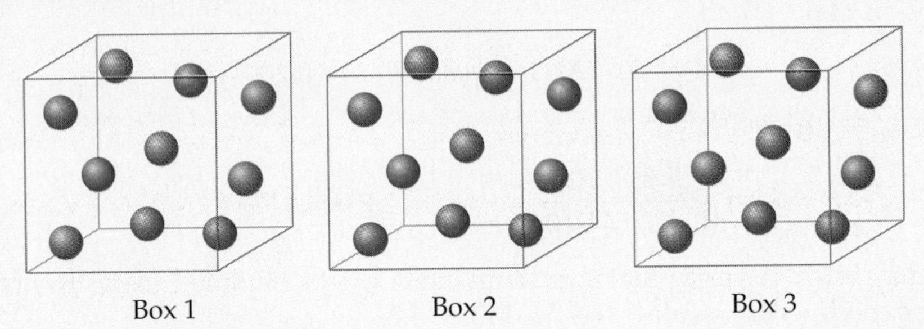

Solution

Analyze and Plan: Each of the boxes has 10 spheres. The rate law indicates that in this case [B] has a greater influence on rate than [A] because it has a higher reaction order. Hence the mixture with the highest concentration of B (most blue spheres) should react fastest.

Solve: The rates vary in the order $2 < 1 < 3$.

Check: If we put the number of spheres of each kind into the rate law we obtain the following results:

$$\text{Box 1—Rate} = k(5)(5)^2 = 125k$$

$$\text{Box 2—Rate} = k(7)(3)^2 = 63k$$

$$\text{Box 3—Rate} = k(3)(7)^2 = 147k$$

These calculations confirm the order $2 < 1 < 3$.

PRACTICE EXERCISE

Assuming that rate $= k[A][B]$, rank the mixtures represented in order of increasing rate.

Answer: $2 = 3 < 1$

Units of Rate Constants

The units of the rate constant depend on the overall reaction order of the rate law. In a reaction that is second order overall, for example, the units of the rate constant must satisfy the equation:

$$\text{Units of rate} = (\text{units of rate constant})(\text{units of concentration})^2$$

Hence, in our usual units of concentration and time,

$$\text{Units of rate constant} = \frac{\text{units of rate}}{(\text{units of concentration})^2} = \frac{M/\text{s}}{M^2} = M^{-1}\text{s}^{-1}.$$

SAMPLE EXERCISE 14.5

(a) What are the overall reaction orders for the reactions described in Equations 14.9 and 14.10? **(b)** What are the usual units of the rate constant for the rate law for Equation 14.9?

Solution

Analyze and Plan: The overall reaction order is the sum of the exponents in the rate law. The units for the rate constant, k, are found by using the normal units for rate (M/s) and concentration (M) in the rate law.

Solve: (a) The rate of the reaction in Equation 14.9 is first order in N_2O_5 and first order overall. The reaction in Equation 14.10 is first order in $CHCl_3$ and one-half order in Cl_2. The overall reaction order is three halves.

(b) For the rate law for Equation 14.9 we have

$$\text{Units of rate} = (\text{units of rate constant})(\text{units of concentration})$$

So

$$\text{Units of rate constant} = \frac{\text{units of rate}}{\text{units of concentration}} = \frac{M/\text{s}}{M} = \text{s}^{-1}.$$

Notice that the units of the rate constant for the first-order reaction are different from those for the second-order reaction discussed previously.

PRACTICE EXERCISE

(a) What is the reaction order of the reactant H_2 in Equation 14.11? **(b)** What are the units of the rate constant for Equation 14.11?
Answers: **(a)** 1; **(b)** $M^{-1}\text{s}^{-1}$

Using Initial Rates to Determine Rate Laws

The rate law for any chemical reaction must be determined experimentally; it cannot be predicted by merely looking at the chemical equation. We often determine the rate law for a reaction by the same method we applied to the data in Table 14.2: We observe the effect of changing the initial concentrations of the reactants on the initial rate of the reaction.

In most reactions the exponents in the rate law are 0, 1, or 2. If a reaction is zero order in a particular reactant, changing its concentration will have no effect on rate (as long as some of the reactant is present) because any concentration raised to the zero power equals 1. On the other hand, we have seen that when a reaction is first order in a reactant, changes in the concentration of that reactant will produce proportional changes in the rate. Thus, doubling the concentration will double the rate, and so forth. Finally, when the rate law is second order in a particular reactant, doubling its concentration increases the rate by a factor of $2^2 = 4$, tripling its concentration causes the rate to increase by a factor of $3^2 = 9$, and so forth.

In working with rate laws, it is important to realize that the *rate* of a reaction depends on concentration, but the *rate constant* does not. As we will see later in this chapter, the rate constant (and hence the reaction rate) is affected by temperature and by the presence of a catalyst.

SAMPLE EXERCISE 14.6

The initial rate of a reaction $A + B \longrightarrow C$ was measured for several different starting concentrations of A and B, and the results are as follows:

Experiment Number	[A] (M)	[B] (M)	Initial Rate (M/s)
1	0.100	0.100	4.0×10^{-5}
2	0.100	0.200	4.0×10^{-5}
3	0.200	0.100	16.0×10^{-5}

Using these data, determine **(a)** the rate law for the reaction; **(b)** the magnitude of the rate constant; **(c)** the rate of the reaction when $[A] = 0.050\ M$ and $[B] = 0.100\ M$.

Solution

Analyze: We are given a table of data that relates concentrations of reactants with initial rates of reaction and asked to determine (a) the rate law, (b) the rate constant, and (c) the rate of reaction for a set of concentrations not listed in the table.

Plan: (a) We assume that the rate law has the following form: $\text{Rate} = k[A]^m[B]^n$, so we must use the given data to deduce the reaction orders m and n. We do so by determining how changes in the concentration change the rate. **(b)** Once we know m and n, we can use the rate law and one of the sets of data to determine the rate constant k. **(c)** Now that we know both the rate constant and the reaction orders, we can use the rate law with the given concentrations to calculate rate.

Solve: (a) As we move from experiment 1 to experiment 2, [A] is held constant and [B] is doubled. Thus, this pair of experiments shows how [B] affects the rate, allowing us to deduce the order of the rate law with respect to B. Because the rate remains the same when [B] is doubled, the concentration of B has no effect on the reaction rate. The rate law is therefore zero order in B (that is, $n = 0$).

In experiments 1 and 3, [B] is held constant so they show how [A] affects rate. Holding [B] constant while doubling [A] increases the rate fourfold. This result indicates that rate is proportional to $[A]^2$ (that is, the reaction is second order in A). Hence the rate law is

$$\text{Rate} = k[A]^2[B]^0 = k[A]^2$$

This rate law could be reached in a more formal way by taking the ratio of the rates from two experiments:

$$\frac{\text{Rate 2}}{\text{Rate 1}} = \frac{4.0 \times 10^{-5}\ M/\text{s}}{4.0 \times 10^{-5}\ M/\text{s}} = 1$$

Using the rate law, then, we have

$$1 = \frac{\text{rate 2}}{\text{rate 1}} = \frac{\cancel{k[0.100\ M]^m}[0.200\ M]^n}{\cancel{k[0.100\ M]^m}[0.100\ M]^n} = \frac{[0.200]^n}{[0.100]^n} = 2^n$$

2^n equals 1 only if $n = 0$. We can deduce the value of m in a similar fashion:

$$\frac{\text{Rate 3}}{\text{Rate 1}} = \frac{16.0 \times 10^{-5}\ M/\text{s}}{4.0 \times 10^{-5}\ M/\text{s}} = 4$$

Using the rate law gives

$$4 = \frac{\text{rate 3}}{\text{rate 1}} = \frac{k[0.200\ M]^m\cancel{[0.100\ M]^n}}{k[0.100\ M]^m\cancel{[0.100\ M]^n}} = \frac{[0.200]^m}{[0.100]^m} = 2^m$$

$2^m = 4$, so $m = 2$.

(b) Using the rate law and the data from experiment 1, we have

$$k = \frac{\text{rate}}{[A]^2} = \frac{4.0 \times 10^{-5}\ M/s}{(0.100\ M)^2} = 4.0 \times 10^{-3}\ M^{-1}s^{-1}$$

(c) Using the rate law from part (a) and the rate constant from part (b), we have

$$\text{Rate} = k[A]^2 = (4.0 \times 10^{-3}\ M^{-1}s^{-1})(0.050\ M)^2 = 1.0 \times 10^{-5}\ M/s$$

Because [B] is not part of the rate law, it is irrelevant to the rate, provided that there is at least some B present to react with A.

Check: A good way to check our rate law is to use the concentrations in experiment 2 or 3 and see if we can correctly calculate the rate. Using data from experiment 3, we have

$$\text{Rate} = k[A]^2 = (4.0 \times 10^{-3}\ M^{-1}s^{-1})(0.200\ M)^2 = 1.60 \times 10^{-4}\ M/s$$

Thus, the rate law correctly reproduces the data, giving both the correct number and the correct units for the rate.

PRACTICE EXERCISE

The following data were measured for the reaction of nitric oxide with hydrogen:

$$2NO(g) + 2H_2(g) \longrightarrow N_2(g) + 2H_2O(g)$$

Experiment Number	[NO] (M)	$[H_2]$ (M)	Initial Rate (M/s)
1	0.10	0.10	1.23×10^{-3}
2	0.10	0.20	2.46×10^{-3}
3	0.20	0.10	4.92×10^{-3}

(a) Determine the rate law for this reaction. **(b)** Calculate the rate constant. **(c)** Calculate the rate when [NO] = 0.050 M and $[H_2]$ = 0.150 M.
Answers: **(a)** rate = $k[NO]^2[H_2]$; **(b)** $k = 1.2\ M^{-2}s^{-1}$; rate = $4.5 \times 10^{-4}\ M/s$

14.4 The Change of Concentration with Time

A rate law tells us how the rate of a reaction changes at a particular temperature as we change reactant concentrations. Rate laws can be converted into equations that tell us what the concentrations of the reactants or products are at any time during the course of a reaction. The mathematics required involves calculus. We don't expect you to be able to perform the calculus operations; however, you should be able to use the resulting equations. We will apply this conversion to two of the simplest rate laws: those that are first order overall and those that are second order overall.

First-Order Reactions

A **first-order reaction** is one whose rate depends on the concentration of a single reactant raised to the first power. For a reaction of the type A $\longrightarrow$ products, the rate law may be first order:

$$\text{Rate} = -\frac{\Delta[A]}{\Delta t} = k[A]$$

Using an operation from calculus called integration, this relationship can be transformed into an equation that relates the concentration of A at the start of the reaction, $[A]_0$, to its concentration at any other time t, $[A]_t$:

$$\ln[A]_t - \ln[A]_0 = -kt \qquad \text{or} \qquad \ln\frac{[A]_t}{[A]_0} = -kt \qquad [14.12]$$

The function "ln" is the natural logarithm (Appendix A.2).* Equation 14.12 can also be rearranged and written as follows:

$$\ln[A]_t = -kt + \ln[A]_0 \qquad [14.13]$$

Equations 14.12 and 14.13 can be used with any concentration units, so long as the units are the same for both $[A]_t$ and $[A]_0$.

For a first-order reaction, Equation 14.12 or 14.13 can be used in several ways. Given any three of the following quantities, we can solve for the fourth: k, t, $[A]_0$, and $[A]_t$. Thus, these equations can be used, for example, to determine (1) the concentration of a reactant remaining at any time after the reaction has started, (2) the time required for a given fraction of a sample to react, or (3) the time required for a reactant concentration to fall to a certain level.

SAMPLE EXERCISE 14.7

The first-order rate constant for the decomposition of a certain insecticide in water at 12°C is $1.45\ \text{yr}^{-1}$. A quantity of this insecticide is washed into a lake on June 1, leading to a concentration of $5.0 \times 10^{-7}\ \text{g/cm}^3$ of water. Assume that the average temperature of the lake is 12°C. **(a)** What is the concentration of the insecticide on June 1 of the following year? **(b)** How long will it take for the concentration of the insecticide to drop to $3.0 \times 10^{-7}\ \text{g/cm}^3$?

Solution

Analyze and Plan: We are told that the rate is first order, and we are asked about concentrations and times, so Equation 14.13 can be used. In (a) we are given $k = 1.45\ \text{yr}^{-1}$, $t = 1.00$ yr, and $[\text{insecticide}]_0 = 5.0 \times 10^{-7}\ \text{g/cm}^3$, so Equation 14.13 can be solved for $[\text{insecticide}]_t$. In (b) we have $k = 1.45\ \text{yr}^{-1}$, $[\text{insecticide}]_0 = 5.0 \times 10^{-7}\ \text{g/cm}^3$, and $[\text{insecticide}]_t = 3.0 \times 10^{-7}\ \text{g/cm}^3$, so we can solve for t.

Solve: (a) Substituting the known quantities into Equation 14.13, we have

$$\ln[\text{insecticide}]_{t=1\ \text{yr}} = -(1.45\ \text{yr}^{-1})(1.00\ \text{yr}) + \ln(5.0 \times 10^{-7})$$

We use the ln function on a calculator to evaluate the second term on the right, giving

$$\ln[\text{insecticide}]_{t=1\ \text{yr}} = -1.45 + (-14.51) = -15.96$$

To obtain $[\text{insecticide}]_{t=1\ \text{yr}}$ we use the inverse natural logarithm, or e^x, function on the calculator:

$$[\text{insecticide}]_{t=1\ \text{yr}} = e^{-15.96} = 1.2 \times 10^{-7}\ \text{g/cm}^3$$

Note that the concentration units for $[A]_t$ and $[A]_0$ must be the same.

(b) Again substituting into Equation 14.13, with $[\text{insecticide}]_t = 3.0 \times 10^{-7}\ \text{g/cm}^3$, gives

$$\ln(3.0 \times 10^{-7}) = -(1.45\ \text{yr}^{-1})(t) + \ln(5.0 \times 10^{-7})$$

Solving for t gives

$$t = -[\ln(3.0 \times 10^{-7}) - \ln(5.0 \times 10^{-7})]/1.45\ \text{yr}^{-1}$$
$$= -(-15.02 + 14.51)/1.45\ \text{yr}^{-1} = 0.35\ \text{yr}$$

Check: In part (a) the concentration remaining after 1.00 yr (that is, $1.2 \times 10^{-7}\ \text{g/cm}^3$) is less than the original concentration ($5.0 \times 10^{-7}\ \text{g/cm}^3$), as it should be. In (b) the given concentration ($3.0 \times 10^{-7}\text{g/cm}^3$) is greater than that remaining after 1.00 yr, indicating that the time must be less than a year. Thus, $t = 0.35$ yr is a reasonable answer.

PRACTICE EXERCISE

The decomposition of dimethyl ether, $(CH_3)_2O$, at 510°C is a first-order process with a rate constant of $6.8 \times 10^{-4}\ \text{s}^{-1}$:

$$(CH_3)_2O(g) \longrightarrow CH_4(g) + H_2(g) + CO(g)$$

If the initial pressure of $(CH_3)_2O$ is 135 torr, what is its partial pressure after 1420 s?
Answer: 51 torr

---

* In terms of base-10, or common, logarithms, Equation 14.12 can be written as

$$\log [A]_t - \log [A]_0 = -\frac{kt}{2.303} \qquad \text{or} \qquad \log\frac{[A]_t}{[A]_0} = -\frac{kt}{2.303}$$

The factor 2.303 arises from the conversion of natural logarithms to base-10 logarithms.

Equation 14.13 can be used to verify whether a reaction is first order and to determine its rate constant. This equation has the form of the general equation for a straight line, $y = mx + b$, in which m is the slope and b is the y-intercept of the line (Appendix A.4):

$$\begin{array}{ccccc} \ln[A]_t & = & -k \cdot t & + & \ln[A]_0 \\ \updownarrow & & \updownarrow \; \updownarrow & & \updownarrow \\ y & = & m \cdot x & + & b \end{array}$$

For a first-order reaction, therefore, a graph of $\ln[A]_t$ versus time gives a straight line with a slope of $-k$ and a y-intercept of $\ln[A]_0$. A reaction that is not first order will not yield a straight line.

As an example, consider the conversion of methyl isonitrile (CH_3NC) to acetonitrile (CH_3CN) (Figure 14.6 ▶). Because experiments show that the reaction is first order, we can write the rate equation:

$$\ln[CH_3NC]_t = -kt + \ln[CH_3NC]_0$$

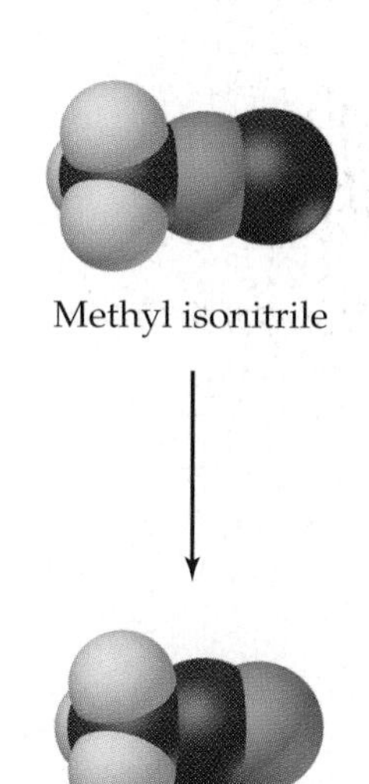

▲ **Figure 14.6** The transformation of methyl isonitrile (CH_3NC) to acetonitrile (CH_3CN) is a first-order process. Methyl isonitrile and acetonitrile are isomers, molecules that have the same atoms arranged differently. This reaction is called an isomerization reaction.

Figure 14.7(a) ▼ shows how the partial pressure of methyl isonitrile varies with time as it rearranges in the gas phase at 198.9°C. We can use pressure as a unit of concentration for a gas because, from the ideal-gas law the pressure is directly proportional to the number of moles per unit volume. Figure 14.7(b) shows a plot of the natural logarithm of the pressure versus time, a plot that yields a straight line. The slope of this line is $-5.1 \times 10^{-5} s^{-1}$. (You should verify this for yourself, remembering that your result may vary slightly from ours because of inaccuracies associated with reading the graph.) Because the slope of the line equals $-k$, the rate constant for this reaction equals $5.1 \times 10^{-5} s^{-1}$.

Second-Order Reactions

A **second-order reaction** is one whose rate depends on the reactant concentration raised to the second power or on the concentrations of two different reactants, each raised to the first power. For simplicity, let's consider reactions of the type A $\longrightarrow$ products or A + B $\longrightarrow$ products that are second order in just one reactant, A:

$$\text{Rate} = -\frac{\Delta[A]}{\Delta t} = k[A]^2$$

With the use of calculus, this rate law can be used to derive the following equation:

$$\frac{1}{[A]_t} = kt + \frac{1}{[A]_0} \qquad [14.14]$$

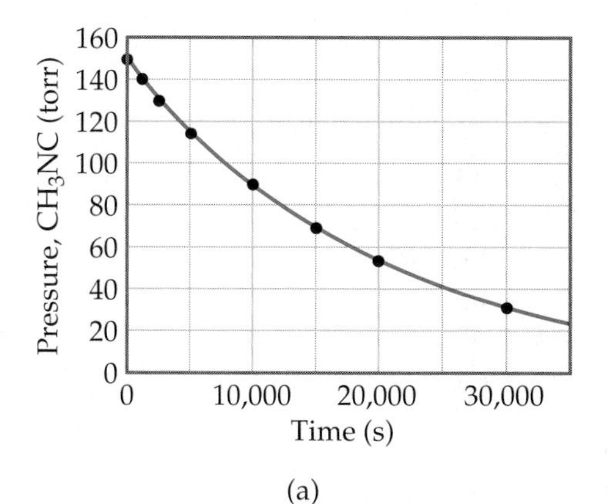

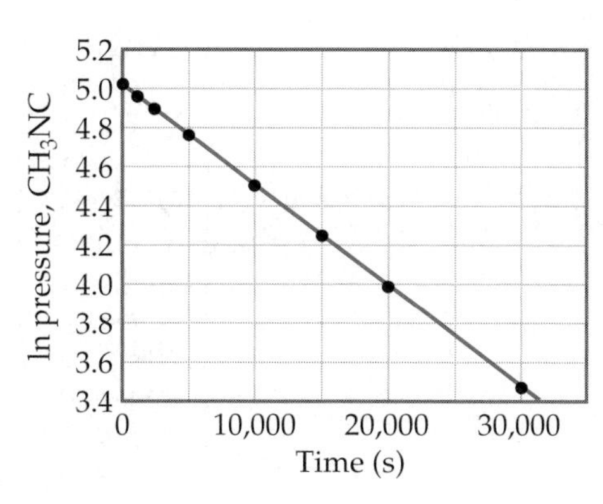

◀ **Figure 14.7** (a) Variation in the partial pressure of methyl isonitrile (CH_3NC) with time at 198.9°C during the reaction $CH_3NC \longrightarrow CH_3CN$. (b) A plot of the natural logarithm of the CH_3NC pressure as a function of time.

This equation, like Equation 14.13, has four variables, k, t, $[A]_0$, and $[A]_t$, and any one of these can be calculated knowing the other three. Equation 14.14 also has the form of a straight line ($y = mx + b$). If the reaction is second order, a plot of $1/[A]_t$ versus t will yield a straight line with a slope equal to k and a y-intercept equal to $1/[A]_0$. One way to distinguish between first- and second-order rate laws is to graph both $\ln[A]_t$ and $1/[A]_t$ against t. If the $\ln[A]_t$ plot is linear, the reaction is first order; if the $1/[A]_t$ plot is linear, the reaction is second order.

SAMPLE EXERCISE 14.8

The following data were obtained for the gas-phase decomposition of nitrogen dioxide at 300°C, $NO_2(g) \longrightarrow NO(g) + \frac{1}{2}O_2(g)$:

Time (s)	$[NO_2]$ (M)
0.0	0.01000
50.0	0.00787
100.0	0.00649
200.0	0.00481
300.0	0.00380

Is the reaction first or second order in NO_2?

Solution

Analyze and Plan: To test whether the reaction is first or second order, we can plot $\ln[NO_2]$ and $1/[NO_2]$ against time. One or the other will be linear, indicating whether the reaction is first or second order.

Solve: In order to graph $\ln[NO_2]$ and $1/[NO_2]$ against time, we will first prepare the following table from the data given:

Time (s)	$[NO_2]$ (M)	$\ln[NO_2]$	$1/[NO_2]$
0.0	0.01000	−4.610	100
50.0	0.00787	−4.845	127
100.0	0.00649	−5.038	154
200.0	0.00481	−5.337	208
300.0	0.00380	−5.573	263

As Figure 14.8 ▼ shows, only the plot of $1/[NO_2]$ versus time is linear. Thus, the reaction obeys a second-order rate law: Rate $= k[NO_2]^2$. From the slope of this straight-line graph, we determine that $k = 0.543\ M^{-1}s^{-1}$ for the disappearance of NO_2.

PRACTICE EXERCISE

Consider again the decomposition of NO_2 discussed in the Sample Exercise. The reaction is second order in NO_2 with $k = 0.543\ M^{-1}\ s^{-1}$. If the initial concentration of NO_2 in a closed vessel is 0.0500 M, what is the remaining concentration after 0.500 hr?

Answer: $1.00 \times 10^{-3} M$

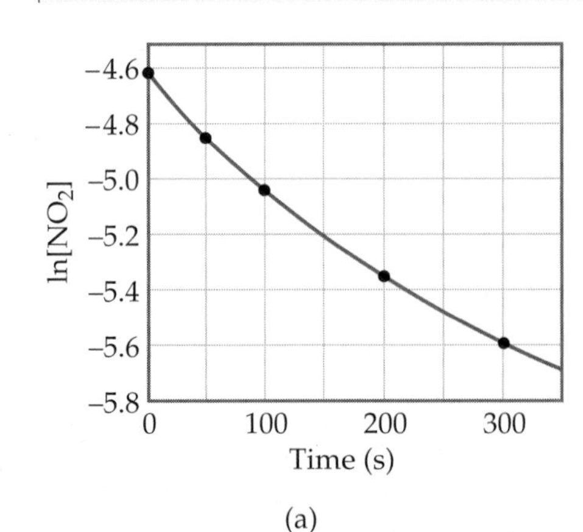

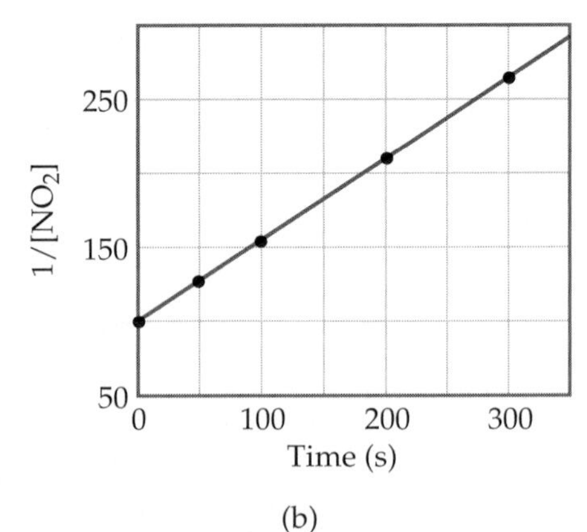

▶ **Figure 14.8** Plots of the kinetic data for the reaction $NO_2(g) \longrightarrow NO(g) + \frac{1}{2}O_2(g)$ at 300°C. A plot of $\ln[NO_2]$ versus time (a) is not linear, indicating that the reaction is not first order in NO_2. A plot of $1/[NO_2]$ versus time (b) is linear, however, indicating that the reaction is second order in NO_2.

Half-Life

The **half-life** of a reaction, $t_{1/2}$, is the time required for the concentration of a reactant to drop to one half of its initial value, $[A]_{t_{1/2}} = \frac{1}{2}[A]_0$. The half-life is a convenient way to describe how fast a reaction occurs, especially if it is a first-order process. A fast reaction will have a short half-life.

We can determine the half-life of a first-order reaction by substituting $[A]_{t_{1/2}}$ into Equation 14.12:

$$\ln \frac{\frac{1}{2}[A]_0}{[A]_0} = -kt_{1/2}$$

$$\ln \tfrac{1}{2} = -kt_{1/2}$$

$$t_{1/2} = -\frac{\ln \frac{1}{2}}{k} = \frac{0.693}{k} \qquad [14.15]$$

Notice that $t_{1/2}$ for a first-order rate law depends only on k. Thus, the half-life of a first-order reaction is unaffected by the initial concentration of reactant. Consequently, the half-life remains constant throughout the reaction. If, for example, the concentration of the reactant is 0.120 M at some moment in the reaction, it will be $\frac{1}{2}(0.120\ M) = 0.060\ M$ after one half-life. After one more half-life passes, the concentration will drop to 0.030 M, and so on.

The change in concentration over time for the first-order rearrangement of methyl isonitrile at 198.9°C is graphed in Figure 14.9 ▼. The first half-life is shown at 13,300 s (that is, 3.69 hr). At a time 13,300 s later, the isonitrile concentration has decreased to one-half of one-half, or one-fourth the original concentration. *In a first-order reaction, the concentration of the reactant decreases by $\frac{1}{2}$ in each of a series of regularly spaced time intervals, namely,* $t_{1/2}$. The concept of half-life is widely used in describing radioactive decay, a first-order process that we will discuss in detail in Section 21.4.

In contrast to the behavior of first-order reactions, the half-life for second-order and other reactions depends on reactant concentrations and therefore changes as the reaction progresses. Using Equation 14.14, we find that the half-life of a second-order reaction is

$$t_{1/2} = \frac{1}{k[A]_0} \qquad [14.16]$$

Notice that this half-life depends on the initial concentration of reactant.

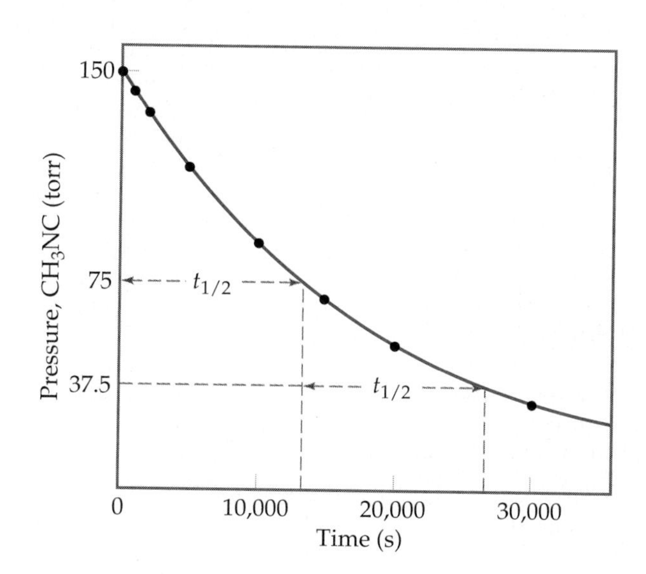

◀ **Figure 14.9** Pressure of methyl isonitrile as a function of time. Two successive half-lives of the isomerization reaction (Figure 14.6) are shown.

SAMPLE EXERCISE 14.9

From Figure 14.4, estimate the half-life of the reaction of C_4H_9Cl with water.

Solution

Analyze and Plan: To estimate a half-life of a first-order reaction from a graph, we can select a concentration and then determine the time required for the concentration to decrease by half.

Solve: From the figure, we see that the initial value of $[C_4H_9Cl]$ is 0.100 *M*. The half-life for this first-order reaction is the time required for $[C_4H_9Cl]$ to decrease to 0.050 *M*, which we can read off the graph. This point occurs at approximately 340 s.

Check: At the end of the second half-life, which should occur at 680 s, the concentration should have decreased by yet another factor of 2, to 0.025 *M*. Inspection of the graph shows that this is indeed the case.

PRACTICE EXERCISE

Using Equation 14.15, calculate $t_{1/2}$ for the decomposition of the insecticide described in Sample Exercise 14.7.

Answer: 0.478 yr = 1.51×10^7 s

Chemistry at Work Methyl Bromide in the Atmosphere

Several small molecules containing carbon–chlorine or carbon–bromine bonds, when present in the stratosphere, are capable of reacting with ozone (O_3) and thus contributing to the destruction of the ozone layer. (The nature of the stratospheric ozone layer and its importance for Earth's ecosystems is discussed in Section 18.3.)

Whether a halogen-containing molecule contributes significantly to destruction of the ozone layer depends in part on its concentrations near Earth's surface and on its average lifetime in the atmosphere. It takes quite a long time for molecules formed at Earth's surface to diffuse through the lower atmosphere (called the troposphere) and move into the stratosphere, where the ozone layer is located (Figure 14.10 ▼). Decomposition in the lower atmosphere competes with diffusion into the stratosphere.

The much-discussed chlorofluorocarbons, or CFCs, contribute to the destruction of the ozone layer because they have long lifetimes in the troposphere. Thus, they persist long enough so that a substantial fraction of the molecules formed at the surface find their way to the stratosphere. Another simple molecule that has the potential to destroy the stratospheric ozone layer is methyl bromide (CH_3Br). This substance has a wide range of uses, including antifungal treatment of plant seeds, and has therefore been produced in large quantity (about 150 million pounds per year). In the stratosphere, the C—Br bond is broken through absorption of short wavelength radiation. (Section 18.2) The Br atoms catalyze decomposition of O_3.

Methyl bromide is removed from the lower atmosphere by a variety of mechanisms, including a slow reaction with ocean water:

$$CH_3Br(g) + H_2O(l) \longrightarrow CH_3OH(aq) + HBr(aq) \quad [14.17]$$

To determine the potential importance of CH_3Br in destruction of the ozone layer, it is important to know how rapidly Equation 14.17 and all other mechanisms together remove CH_3Br from the atmosphere before it can diffuse into the stratosphere. Scientists have recently carried out research to estimate the average lifetime of CH_3Br in Earth's atmosphere. Such an estimate is difficult to make. It cannot be done in laboratory-based experiments because the conditions that exist all over the planet are too complex to be simulated in the laboratory. Instead, scientists gathered nearly 4000 samples of the atmosphere during aircraft flights all over the Pacific Ocean and analyzed them for the presence of several trace organic substances, including methyl bromide. From a detailed analysis of the patterns of concentrations, it was possible for them to estimate that the *atmospheric residence time* for CH_3Br is 0.8 ± 0.1 yr.

The atmospheric residence time equals the half-life for CH_3Br in the lower atmosphere, assuming that it decomposes by a first-order process. That is, a collection of CH_3Br molecules present at any given time will, on average, be 50% decomposed after 0.8 years, 75% decomposed after 1.6 years, and so on. A residence time of 0.8 years, while comparatively short, is still sufficiently long so that CH_3Br contributes significantly to the destruction of the ozone layer. In 1997 an international agreement was reached to phase out use of methyl bromide in the developed countries worldwide by 2005.

Stratosphere
Diffusion toward stratosphere
Troposphere
Decomposition
Surface

▲ **Figure 14.10** Distribution and fate of methyl bromide (CH_3Br) in the atmosphere. Some CH_3Br is removed from the atmosphere by decomposition, and some diffuses upward into the stratosphere, where it contributes to destruction of the ozone layer. The relative rates of decomposition and diffusion determine how extensively methyl bromide is involved in destruction of the ozone layer.

14.5 Temperature and Rate

The rates of most chemical reactions increase as the temperature rises. For example, dough rises faster at room temperature than when refrigerated, and plants grow more rapidly in warm weather than in cold. We can literally see the effect of temperature on reaction rate by observing a chemiluminescent reaction (one that produces light). The characteristic glow of fireflies is a familiar example of chemiluminescence. Another is the light produced by Cyalume® light sticks, which contain chemicals that produce chemiluminescence when mixed. As seen in Figure 14.11 ▼, these light sticks produce a brighter light at higher temperature. The amount of light produced is greater because the rate of the reaction is faster at the higher temperature. Although the light stick glows more brightly initially, its luminescence also dies out more rapidly.

How is this experimentally observed temperature effect reflected in the rate expression? The faster rate at higher temperature is due to an increase in the rate constant with increasing temperature. For example, let's reconsider the first-order reaction $CH_3NC \longrightarrow CH_3CN$ (Figure 14.6). Figure 14.12 ▼ shows the rate constant for this reaction as a function of temperature. The rate constant, and hence the rate of the reaction, increases rapidly with temperature, approximately doubling for each 10°C rise.

The Collision Model

We have seen that reaction rates are affected both by the concentrations of reactants and by temperature. The **collision model**, which is based on the kinetic-molecular theory (Section 10.7), accounts for both of these effects at the molecular level. The central idea of the collision model is that molecules must collide to react. The greater the number of collisions occurring per second, the greater the reaction rate. As the concentration of reactant molecules increases, therefore, the number of collisions increases, leading to an increase in reaction rate. According to the kinetic-molecular theory of gases, furthermore, increasing the temperature increases molecular speeds. As molecules move faster, they collide more forcefully (with more energy) and more frequently, increasing reaction rates.

For reaction to occur, though, more is required than simply a collision. For most reactions, only a tiny fraction of the collisions leads to reaction. For example,

▲ **Figure 14.11** Temperature affects the rate of the chemiluminescence reaction in Cyalume® light sticks. The light stick in hot water (left) glows more brightly than the one in cold water (right); the reaction is initially faster and produces a brighter light at the higher temperature.

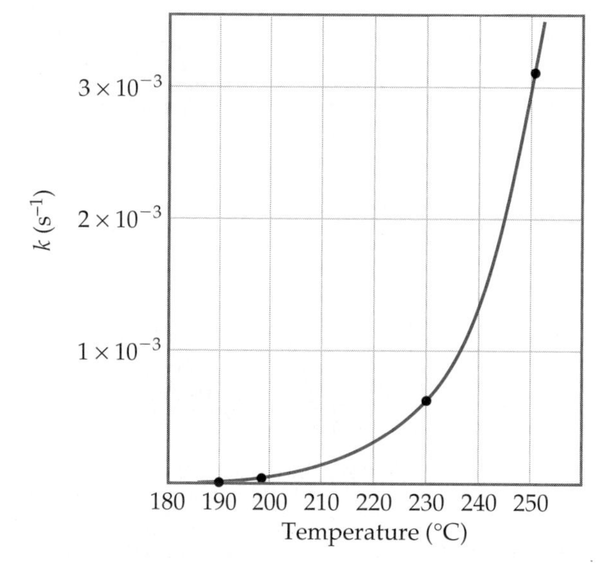

▲ **Figure 14.12** Variation in the first-order rate constant for the rearrangement of methyl isonitrile as a function of temperature. (The four points indicated are used in connection with Sample Exercise 14.11.)

in a mixture of H_2 and I_2 at ordinary temperatures and pressures, each molecule undergoes about 10^{10} collisions per second. If every collision between H_2 and I_2 resulted in the formation of HI, the reaction would be over in much less than a second. Instead, at room temperature the reaction proceeds very slowly. Only about one in every 10^{13} collisions produces a reaction. What keeps the reaction from occurring more rapidly?

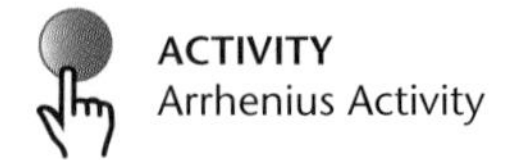

The Orientation Factor

In most reactions, molecules must be oriented in a certain way during collisions in order for a reaction to occur. The relative orientations of the molecules during their collisions determine whether the atoms are suitably positioned to form new bonds. For example, consider the reaction of Cl atoms with NOCl:

$$Cl + NOCl \longrightarrow NO + Cl_2$$

The reaction will take place if the collision brings Cl atoms together to form Cl_2, as shown in Figure 14.13(a) ▼. In contrast, the collision shown in Figure 14.13(b) will be ineffective and will not yield products. Indeed, a great many collisions do not lead to reaction, merely because the molecules are not suitably oriented. There is, however, another factor that is usually even more important in determining whether particular collisions result in reaction.

Activation Energy

In 1888 the Swedish chemist Svante Arrhenius suggested that molecules must possess a certain minimum amount of energy in order to react. According to the collision model, this energy comes from the kinetic energies of the colliding molecules. Upon collision, the kinetic energy of the molecules can be used to stretch, bend, and ultimately break bonds, leading to chemical reactions. If molecules are moving too slowly, with too little kinetic energy, they merely bounce off one another without changing. In order to react, colliding molecules must have a total kinetic energy equal to or greater than some minimum value. The minimum energy required to initiate a chemical reaction is called the **activation energy**, E_a. The value of E_a varies from reaction to reaction.

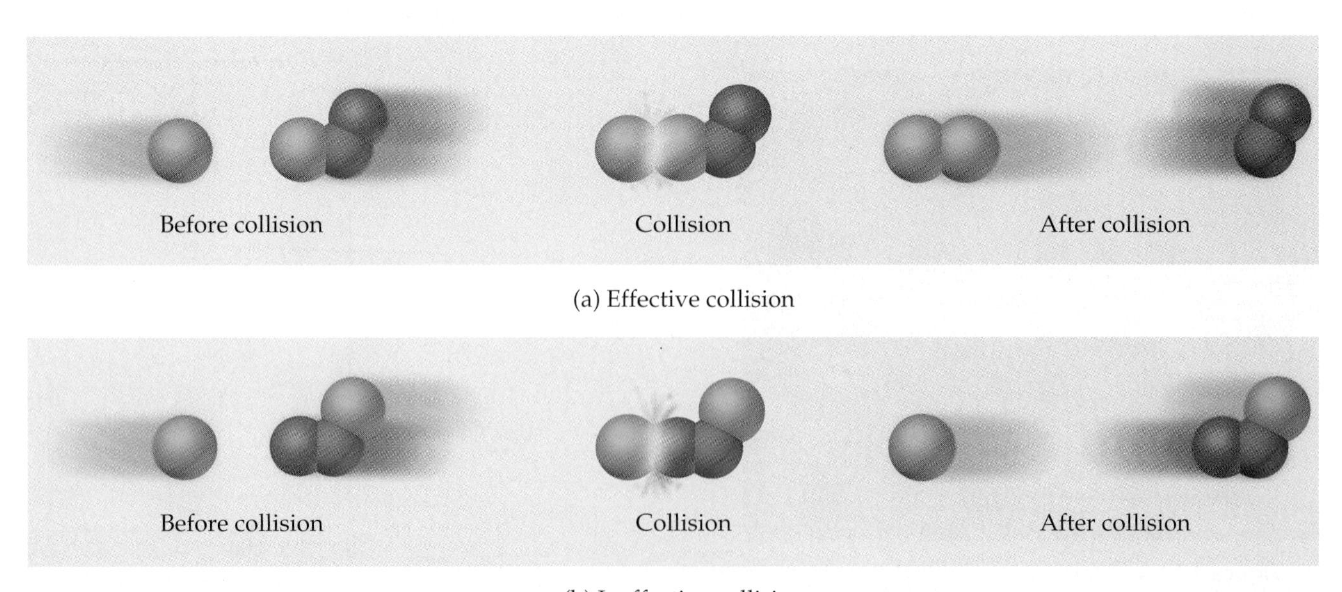

▲ **Figure 14.13** Two possible ways that Cl atoms and NOCl molecules can collide. (a) If molecules are oriented properly, a sufficiently energetic collision will lead to reaction. (b) If the orientation of the colliding molecules is wrong, no reaction occurs.

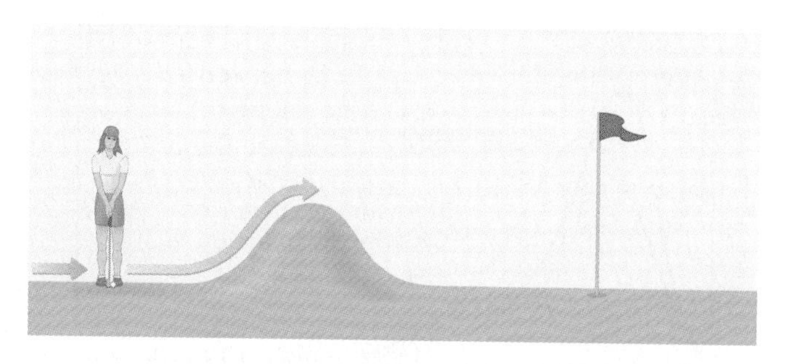

◀ **Figure 14.14** To move the golf ball to the vicinity of the cup, the player must impart enough kinetic energy to the ball to enable it to surmount the barrier represented by the hill. This situation is analogous to a chemical reaction, in which molecules must gain enough energy through collisions to enable them to overcome the barrier to chemical reaction.

The situation during reactions is rather like that shown in Figure 14.14 ▲. The player on the putting green needs to move her ball over the hill to the vicinity of the cup. To do this, she must impart enough kinetic energy with the putter to move the ball to the top of the hill. If she doesn't impart enough energy, the ball will roll partway up the hill and then back down. In the same way, molecules may require a certain minimum energy to break existing bonds during a chemical reaction. In the rearrangement of methyl isonitrile to acetonitrile, for example, we might imagine passing through an intermediate state in which the N≡C portion of the molecule is sitting sideways:

$$H_3C{-}N{\equiv}C{:} \longrightarrow \left[H_3C \cdots \overset{\ddot{C}}{\underset{\ddot{N}}{|||}} \right] \longrightarrow H_3C{-}C{\equiv}N{:}$$

The change in the energy of the molecule during the reaction is shown in Figure 14.15 ▼. The diagram shows that energy must be supplied to stretch the bond between the H_3C group and the N≡C group so as to allow the N≡C group to rotate. After the N≡C group has twisted sufficiently, the C—C bond begins to form and the energy of the molecule drops. Thus, the barrier represents the energy necessary to force the molecule through the relatively unstable intermediate state to the final product. The energy difference between that of the starting molecule and the highest energy along the reaction pathway is the activation energy, E_a. The particular arrangement of atoms at the top of the barrier is called the **activated complex**, or **transition state**.

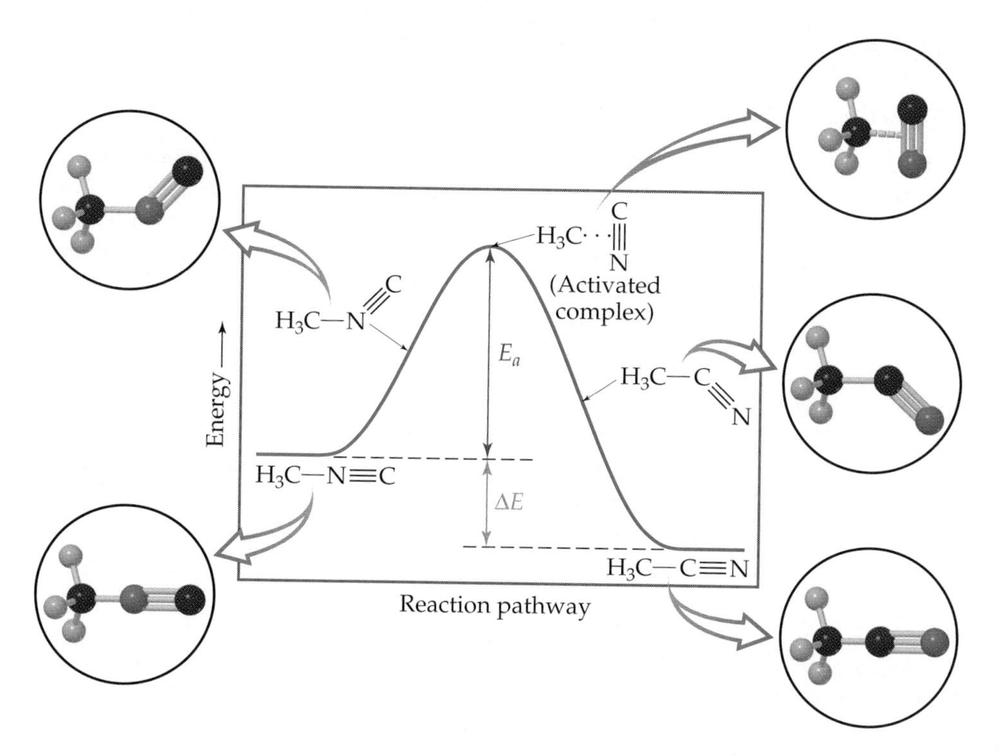

◀ **Figure 14.15** Energy profile for the rearrangement (isomerization) of methyl isonitrile. The molecule must surmount the activation-energy barrier before it can form the product, acetonitrile.

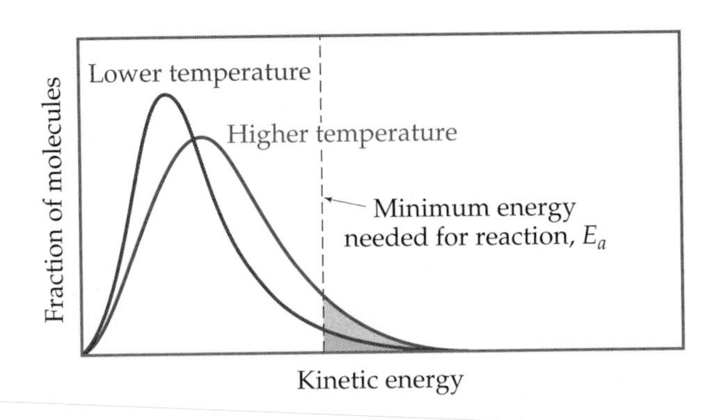

▶ **Figure 14.16** Distribution of kinetic energies in a sample of gas molecules at two different temperatures. At the higher temperature, a larger number of molecules has higher energy. Thus, a larger fraction at any one instant will have more than the minimum energy required for reaction.

The conversion of $H_3C—N\equiv C$ to $H_3C—C\equiv N$ is exothermic. Figure 14.15 therefore shows the product as having a lower energy than the reactant. The energy change for the reaction, ΔE, has no effect on the rate of the reaction. The rate depends on the magnitude of E_a; generally, the lower E_a is, the faster the reaction. Notice that the reverse reaction is endothermic. The activation barrier for the reverse reaction is equal to the sum of ΔE and E_a for the forward reaction.

How does any particular methyl isonitrile molecule acquire sufficient energy to overcome the activation barrier? It does so through collisions with other molecules. Recall from the kinetic-molecular theory of gases that, at any given instant, gas molecules are distributed in energy over a wide range. (Section 10.7) Figure 14.16 ▲ shows the distribution of kinetic energies for two different temperatures, comparing them with the minimum energy needed for reaction, E_a. At the higher temperature a much greater fraction of the molecules has kinetic energy greater than E_a, which leads to a much greater rate of reaction.

The fraction of molecules that has an energy equal to or greater than E_a is given by the expression

$$f = e^{-E_a/RT} \qquad [14.18]$$

In this equation R is the gas constant (8.314 J/mol-K) and T is absolute temperature. To get an idea of the magnitude of f, let's suppose that E_a is 100 kJ/mol, a value typical of many reactions, and that T is 300 K, around room temperature. The calculated value of f is 3.8×10^{-18}, an extremely small number! At 310 K, the fraction is $f = 1.4 \times 10^{-17}$. Thus, a 10-degree increase in temperature produces a 3.7-fold increase in the fraction of molecules possessing at least 100 kJ/mol of energy.

The Arrhenius Equation

Arrhenius noted that for most reactions the increase in rate with increasing temperature is nonlinear, as shown in Figure 14.12. He found that most reaction-rate data obeyed an equation based on three factors: (a) the fraction of molecules possessing an energy of E_a or greater, (b) the number of collisions occurring per second, and (c) the fraction of collisions that have the appropriate orientation. These three factors are incorporated into the **Arrhenius equation**:

$$k = Ae^{-E_a/RT} \qquad [14.19]$$

In this equation k is the rate constant, E_a is the activation energy, R is the gas constant (8.314 J/mol-K), and T is the absolute temperature. The **frequency factor**, A, is constant, or nearly so, as temperature is varied. It is related to the frequency of collisions and the probability that the collisions are favorably oriented for reaction.* As the magnitude of E_a increases, k decreases because the fraction of molecules that possess the required energy is smaller. Thus, *reaction rates decrease as* E_a *increases.*

* Because the frequency of collisions increases with temperature, A also has some temperature dependence, but it is small compared to the exponential term. Therefore, A is considered approximately constant.

SAMPLE EXERCISE 14.10

Consider a series of reactions having the following energy profiles:

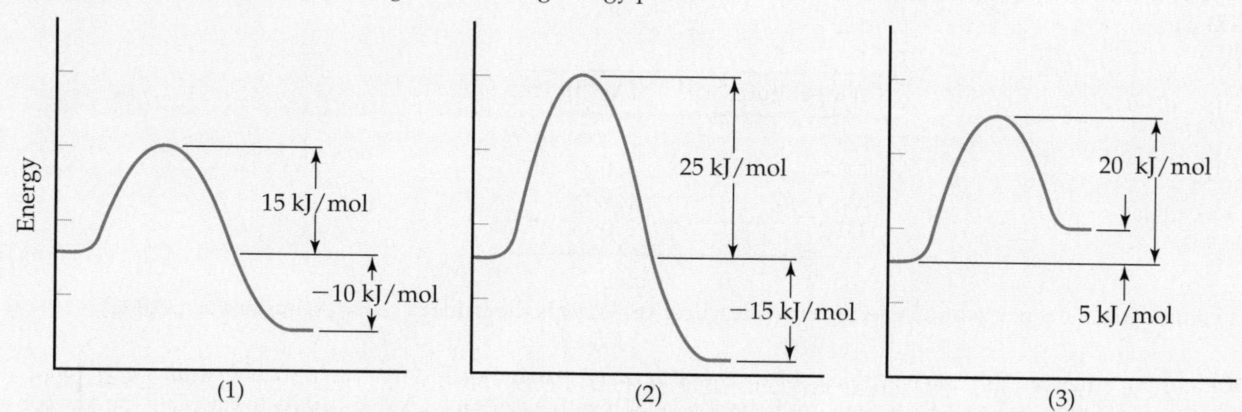

Assuming that all three reactions have nearly the same frequency factors, rank the reactions from slowest to fastest.

Solution

The lower the activation energy, the faster the reaction. The value of ΔE does not affect the rate. Hence the order is (2) < (3) < (1).

PRACTICE EXERCISE

Imagine that these reactions are reversed. Rank these reverse reactions from slowest to fastest.

Answer: (2) < (1) < (3) because E_a values are 40, 25, and 15 kJ/mol, respectively

Determining the Activation Energy

Taking the natural log of both sides of Equation 14.19, we have

$$\ln k = -\frac{E_a}{RT} + \ln A \qquad [14.20]$$

Equation 14.20 has the form of a straight line; it predicts that a graph of $\ln k$ versus $1/T$ will be a line with a slope equal to $-E_a/R$ and a y-intercept equal to $\ln A$. Thus, the activation energy can be determined by measuring k at a series of temperatures, graphing $\ln k$ versus $1/T$, and then calculating E_a from the slope of the resultant line.

We can also use Equation 14.20 to evaluate E_a in a nongraphical way if we know the rate constant of a reaction at two or more temperatures. For example, suppose that at two different temperatures, T_1 and T_2, a reaction has rate constants k_1 and k_2. For each condition, we have

$$\ln k_1 = -\frac{E_a}{RT_1} + \ln A \quad \text{and} \quad \ln k_2 = -\frac{E_a}{RT_2} + \ln A$$

Subtracting $\ln k_2$ from $\ln k_1$ gives

$$\ln k_1 - \ln k_2 = \left(-\frac{E_a}{RT_1} + \ln A\right) - \left(-\frac{E_a}{RT_2} + \ln A\right)$$

Simplifying this equation and rearranging it gives

$$\ln \frac{k_1}{k_2} = \frac{E_a}{R}\left(\frac{1}{T_2} - \frac{1}{T_1}\right) \qquad [14.21]$$

Equation 14.21 provides a convenient way to calculate the rate constant, k_1, at some temperature, T_1, when we know the activation energy and the rate constant, k_2, at some other temperature, T_2.

SAMPLE EXERCISE 14.11

The following table shows the rate constants for the rearrangement of methyl isonitrile at various temperatures (these are the data in Figure 14.12):

Temperature (°C)	k (s^{-1})
189.7	2.52×10^{-5}
198.9	5.25×10^{-5}
230.3	6.30×10^{-4}
251.2	3.16×10^{-3}

(a) From these data, calculate the activation energy for the reaction. **(b)** What is the value of the rate constant at 430.0 K?

Solution

Analyze and Plan: We are given the rate constants, k, measured at several temperatures. We can obtain E_a from the slope of a graph of $\ln k$ versus $1/T$. Once we know E_a, we can use Equation 14.21 together with the given rate data to calculate the rate constant at 430.0 K.

Solve: (a) We must first convert the temperatures from degrees Celsius to kelvins. We then take the inverse of each temperature, $1/T$, and the natural log of each rate constant, $\ln k$. This gives us the following table:

T (K)	$1/T$ (K^{-1})	$\ln k$
462.9	2.160×10^{-3}	−10.589
472.1	2.118×10^{-3}	−9.855
503.5	1.986×10^{-3}	−7.370
524.4	1.907×10^{-3}	−5.757

A graph of $\ln k$ versus $1/T$ results in a straight line, as shown in Figure 14.17 ▶. The slope of the line is obtained by choosing two well-separated points, as shown, and using the coordinates of each:

$$\text{Slope} = \frac{\Delta y}{\Delta x} = \frac{-6.6 - (-10.4)}{0.00195 - 0.00215} = -1.9 \times 10^4$$

Because logarithms have no units, the numerator in this equation is dimensionless. The denominator has the units of $1/T$, namely, K^{-1}. Thus, the overall units for the slope are K. The slope equals $-E_a/R$. We use the value for the molar gas constant R in units of J/mol-K (Table 10.2). We thus obtain

$$\text{Slope} = -\frac{E_a}{R}$$

$$E_a = -(\text{slope})(R) = -(-1.9 \times 10^4\ \text{K})\left(8.31 \frac{\text{J}}{\text{mol-K}}\right)\left(\frac{1\ \text{kJ}}{1000\ \text{J}}\right)$$

$$= 1.6 \times 10^2\ \text{kJ/mol} = 160\ \text{kJ/mol}$$

We report the activation energy to only two significant figures because we are limited by the precision with which we can read the graph in Figure 14.17.

(b) To determine the rate constant, k_1, at $T_1 = 430.0$ K, we can use Equation 14.21 with $E_a = 160$ kJ/mol, and one of the rate constants and temperatures from the given data, such as $k_2 = 2.52 \times 10^{-5} s^{-1}$ and $T_2 = 462.9$ K:

$$\ln\left(\frac{k_1}{2.52 \times 10^{-5} s^{-1}}\right) = \left(\frac{160\ \text{kJ/mol}}{8.31\ \text{J/mol-K}}\right)\left(\frac{1}{462.9\ \text{K}} - \frac{1}{430.0\ \text{K}}\right)\left(\frac{1000\ \text{J}}{1\ \text{kJ}}\right) = -3.18$$

Thus,

$$\frac{k_1}{2.52 \times 10^{-5} s^{-1}} = e^{-3.18} = 4.15 \times 10^{-2}$$

$$k_1 = (4.15 \times 10^{-2})(2.52 \times 10^{-5} s^{-1}) = 1.0 \times 10^{-6} s^{-1}$$

Note that the units of k_1 are the same as those of k_2.

PRACTICE EXERCISE

Using the data in Sample Exercise 14.11, calculate the rate constant for the rearrangement of methyl isonitrile at 280°C.

Answer: $2.2 \times 10^{-2} s^{-1}$

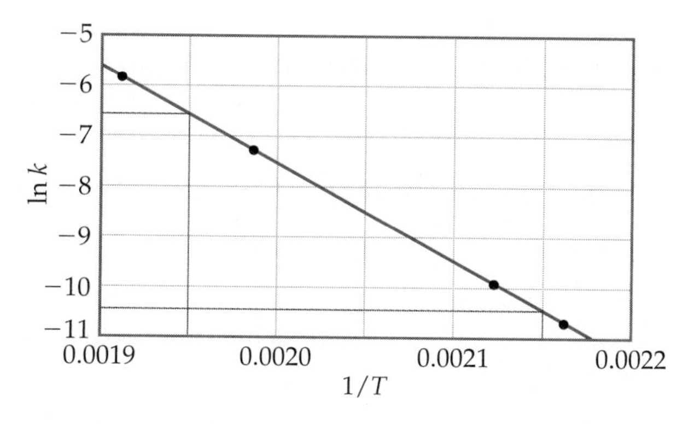

◀ **Figure 14.17** The natural logarithm of the rate constant for the rearrangement of methyl isonitrile as a function of $1/T$. The linear relationship is predicted by the Arrhenius equation.

14.6 Reaction Mechanisms

A balanced equation for a chemical reaction indicates the substances present at the start of the reaction and those produced as the reaction proceeds. It provides no information, however, about how the reaction occurs. The process by which a reaction occurs is called the **reaction mechanism**. At the most sophisticated level, a reaction mechanism will describe in great detail the order in which bonds are broken and formed and the changes in relative positions of the atoms in the course of the reaction. We will begin with more rudimentary descriptions of how reactions occur, considering further the nature of the collisions leading to reaction.

Elementary Steps

We have seen that reactions take place as a result of collisions between reacting molecules. For example, the collisions between molecules of methyl isonitrile (CH_3NC) can provide the energy to allow the CH_3NC to rearrange:

$$H_3C{-}N{\equiv}C: \longrightarrow \left[H_3C\cdots \begin{matrix} \ddot{C} \\ ||| \\ \ddot{N} \end{matrix} \right] \longrightarrow H_3C{-}C{\equiv}N:$$

Similarly, the reaction of NO and O_3 to form NO_2 and O_2 appears to occur as a result of a single collision involving suitably oriented and sufficiently energetic NO and O_3 molecules:

$$NO(g) + O_3(g) \longrightarrow NO_2(g) + O_2(g) \qquad [14.22]$$

Both of these processes occur in a single event or step and are called **elementary steps** (or elementary processes).

The number of molecules that participate as reactants in an elementary step defines the **molecularity** of the step. If a single molecule is involved, the reaction is **unimolecular**. The rearrangement of methyl isonitrile is a unimolecular process. Elementary steps involving the collision of two reactant molecules are **bimolecular**. The reaction between NO and O_3 (Equation 14.22) is bimolecular. Elementary steps involving the simultaneous collision of three molecules are **termolecular**. Termolecular steps are far less probable than unimolecular or bimolecular processes and are rarely encountered. The chance that four or more molecules will collide simultaneously with any regularity is even more remote; consequently, such collisions are never proposed as part of a reaction mechanism.

Multistep Mechanisms

The net change represented by a balanced chemical equation often occurs by a *multistep mechanism*, which consists of a sequence of elementary steps. For example, consider the reaction of NO_2 and CO:

$$NO_2(g) + CO(g) \longrightarrow NO(g) + CO_2(g) \qquad [14.23]$$

Below 225°C, this reaction appears to proceed in two elementary steps, each of which is bimolecular. First, two NO_2 molecules collide, and an oxygen atom is transferred from one to the other. The resultant NO_3 then collides with a CO molecule and transfers an oxygen atom to it:

$$NO_2(g) + NO_2(g) \longrightarrow NO_3(g) + NO(g)$$

$$NO_3(g) + CO(g) \longrightarrow NO_2(g) + CO_2(g)$$

The elementary steps in a multistep mechanism must always add to give the chemical equation of the overall process. In the present example the sum of the elementary steps is

$$2NO_2(g) + NO_3(g) + CO(g) \longrightarrow NO_2(g) + NO_3(g) + NO(g) + CO_2(g)$$

Simplifying this equation by eliminating substances that appear on both sides of the arrow gives Equation 14.23, the net equation for the process. Because NO_3 is neither a reactant nor a product in the overall reaction—it is formed in one elementary step and consumed in the next—it is called an **intermediate**. Multistep mechanisms involve one or more intermediates.

SAMPLE EXERCISE 14.12

It has been proposed that the conversion of ozone into O_2 proceeds via two elementary steps:

$$O_3(g) \longrightarrow O_2(g) + O(g)$$

$$O_3(g) + O(g) \longrightarrow 2O_2(g)$$

(a) Describe the molecularity of each step in this mechanism. **(b)** Write the equation for the overall reaction. **(c)** Identify the intermediate(s).

Solution

Analyze and Plan: The molecularity of each step depends on the number of reactant molecules in that step. The overall equation is the sum of the equations for the elementary steps. The intermediate is a substance formed in one step and used in another and therefore not part of the equation for the overall reaction.

Solve: (a) The first elementary step involves a single reactant and is consequently unimolecular. The second step, which involves two reactant molecules, is bimolecular.

(b) Adding the two elementary steps gives

$$2O_3(g) + O(g) \longrightarrow 3O_2(g) + O(g)$$

Because $O(g)$ appears in equal amounts on both sides of the equation, it can be eliminated to give the net equation for the chemical process:

$$2O_3(g) \longrightarrow 3O_2(g)$$

(c) The intermediate is $O(g)$. It is neither an original reactant nor a final product, but is formed in the first step and consumed in the second.

PRACTICE EXERCISE

For the reaction

$$Mo(CO)_6 + P(CH_3)_3 \longrightarrow Mo(CO)_5P(CH_3)_3 + CO$$

the proposed mechanism is

$$Mo(CO)_6 \longrightarrow Mo(CO)_5 + CO$$

$$Mo(CO)_5 + P(CH_3)_3 \longrightarrow Mo(CO)_5P(CH_3)_3$$

(a) Is the proposed mechanism consistent with the equation for the overall reaction? **(b)** Identify the intermediate(s).

Answers: **(a)** Yes, the two equations add to yield the equation for the reaction; **(b)** $Mo(CO)_5$

Rate Laws for Elementary Steps

In Section 14.3 we stressed that rate laws must be determined experimentally; they cannot be predicted from the coefficients of balanced chemical equations. We are now in a position to understand why this is so: Every reaction is made up of a series of one or more elementary steps, and the rate laws and relative speeds of these steps will dictate the overall rate law. Indeed, the rate law for a reaction can be determined from its mechanism, as we will see shortly. Thus, our next challenge in kinetics is to arrive at reaction mechanisms that lead to rate laws that are consistent with those observed experimentally. We will start by examining the rate laws of elementary steps.

Elementary steps are significant in a very important way: *If we know that a reaction is an elementary step, then we know its rate law.* The rate law of any elementary step is based directly on its molecularity. For example, consider the general unimolecular process.

$$\text{A} \longrightarrow \text{products}$$

As the number of A molecules increases, the number that decompose in a given interval of time will increase proportionally. Thus, the rate of a unimolecular process will be first order:

$$\text{Rate} = k[\text{A}]$$

In the case of bimolecular elementary steps, the rate law is second order, as in the following example:

$$\text{A} + \text{B} \longrightarrow \text{products} \qquad \text{Rate} = k[\text{A}][\text{B}]$$

The second-order rate law follows directly from the collision theory. If we double the concentration of A, the number of collisions between molecules of A and B will double; likewise, if we double [B], the number of collisions will double. Therefore, the rate law will be first order in both [A] and [B], and second order overall.

The rate laws for all feasible elementary steps are given in Table 14.3 ▼. Notice how the rate law for each kind of elementary step follows directly from the molecularity of that step. It is important to remember, however, that we cannot tell by merely looking at a balanced chemical equation whether the reaction involves one or several elementary steps.

TABLE 14.3 Elementary Steps and Their Rate Laws

Molecularity	Elementary Step	Rate Law
*Uni*molecular	$\text{A} \longrightarrow \text{products}$	$\text{Rate} = k[\text{A}]$
*Bi*molecular	$\text{A} + \text{A} \longrightarrow \text{products}$	$\text{Rate} = k[\text{A}]^2$
*Bi*molecular	$\text{A} + \text{B} \longrightarrow \text{products}$	$\text{Rate} = k[\text{A}][\text{B}]$
*Ter*molecular	$\text{A} + \text{A} + \text{A} \longrightarrow \text{products}$	$\text{Rate} = k[\text{A}]^3$
*Ter*molecular	$\text{A} + \text{A} + \text{B} \longrightarrow \text{products}$	$\text{Rate} = k[\text{A}]^2[\text{B}]$
*Ter*molecular	$\text{A} + \text{B} + \text{C} \longrightarrow \text{products}$	$\text{Rate} = k[\text{A}][\text{B}][\text{C}]$

SAMPLE EXERCISE 14.13

If the following reaction occurs in a single elementary step, predict the rate law:

$$H_2(g) + Br_2(g) \longrightarrow 2HBr(g)$$

Solution

Analyze and Plan: Because we are assuming that the reaction occurs as a single elementary step, we are able to write the rate law using the coefficients for the reactants in the equation as the reaction orders.

Solve: If this reaction involves a single elementary step, that step is bimolecular, involving one molecule of H_2 with one molecule of Br_2. The rate law would therefore be first order in each reactant and second order overall:

$$\text{Rate} = k[H_2][Br_2]$$

Comment: Experimental studies of this reaction show that it has a very different rate law:

$$\text{Rate} = k[H_2][Br_2]^{1/2}$$

Because the experimental rate law differs from the one obtained by assuming a single elementary step, we can conclude that the mechanism must involve more than one elementary step.

PRACTICE EXERCISE

Consider the following reaction: $2NO(g) + Br_2(g) \longrightarrow 2NOBr(g)$. **(a)** Write the rate law for the reaction, assuming it involves a single elementary step. **(b)** Is a single-step mechanism likely for this reaction?
Answers: **(a)** rate $= k[NO]^2[Br_2]$; **(b)** no, because termolecular reactions are very rare

Rate Laws for Multistep Mechanisms

As with the reaction in Sample Exercise 14.13, most chemical reactions occur by mechanisms that involve more than one elementary step. Each step has its own rate constant and activation energy. Often one of the steps is much slower than the others. The overall rate of a reaction cannot exceed the rate of the slowest elementary step of its mechanism. Because the slow step limits the overall reaction rate, it is called the **rate-determining step** (or *rate-limiting step*).

To understand the concept of a rate-determining step, consider a toll road with two toll plazas (Figure 14.18 ▼). We will measure the rate at which cars exit the toll road. Cars enter the toll road at point 1 and pass through toll plaza A. They then pass an intermediate point 2 before passing through toll plaza B. Upon exiting, they pass point 3. We can therefore envision this trip along the toll road as occurring in two elementary steps:

Step 1: Point 1 $\longrightarrow$ point 2 (through plaza A)
Step 2: Point 2 $\longrightarrow$ point 3 (through plaza B)
Overall: Point 1 $\longrightarrow$ point 3 (through plazas A and B)

Now suppose that several of the gates at toll plaza A are malfunctioning, so that traffic backs up behind it [Figure 14.18(a)]. The rate at which cars can get to point 3 is limited by the rate at which they can get through the traffic jam at plaza A. Thus, step 1 is the rate-determining step of the journey along the toll road. If, however, traffic flows quickly through plaza A, but gets backed up at plaza B [Figure 14.18(b)], there will be a buildup of cars in the intermediate region between the plazas. In this case step 2 is rate determining: The rate at which cars can travel the toll road is limited by the rate at which they can pass through plaza B.

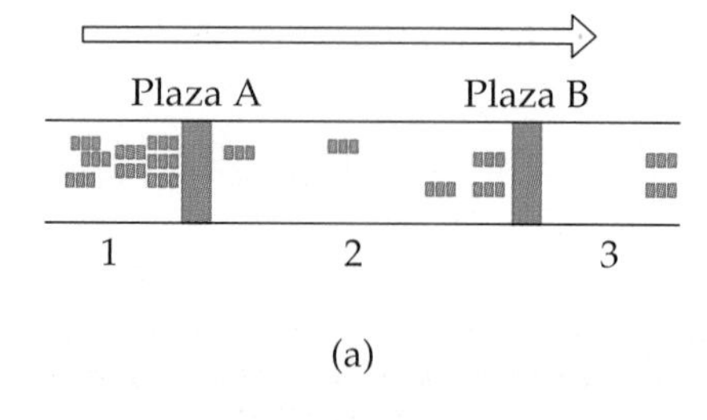

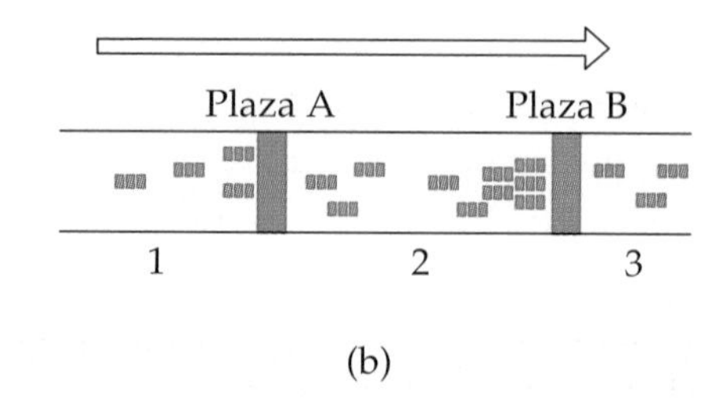

▶ **Figure 14.18** The flow of traffic on a toll road is limited by the flow of traffic through the slowest toll plaza. As cars pass from point 1 to point 3, they pass through plazas A and B. In (a) the rate at which cars can reach point 3 is limited by how quickly they can get through plaza A; getting from point 1 to point 2 is the rate-determining step. In (b) getting from point 2 to point 3 is the rate-determining step.

In the same way, the slowest step in a multistep reaction determines the overall rate. By analogy to Figure 14.18(a), the rate of a faster step following the rate-determining step does not affect the overall rate. If the slow step is not the first one, as in Figure 14.18(b), the faster preceding steps produce intermediate products that accumulate before being consumed in the slow step. In either case *the rate-determining step governs the rate law for the overall reaction.*

Below 225°C, it is found experimentally that the rate law for the reaction of NO_2 and CO to produce NO and CO_2 (Equation 14.23) is second order in NO_2 and zero order in CO: Rate $= k[NO_2]^2$. Can we propose a reaction mechanism that is consistent with this rate law? Consider the following two-step mechanism:*

$$\begin{array}{lll} \text{Step 1:} & NO_2(g) + NO_2(g) \xrightarrow{k_1} NO_3(g) + NO(g) & \text{(slow)} \\ \text{Step 2:} & NO_3(g) + CO(g) \xrightarrow{k_2} NO_2(g) + CO_2(g) & \text{(fast)} \\ \hline \text{Overall:} & NO_2(g) + CO(g) \longrightarrow NO(g) + CO_2(g) & \end{array}$$

Step 2 is much faster than step 1; that is, $k_2 \gg k_1$. The intermediate $NO_3(g)$ is slowly produced in step 1 and is immediately consumed in step 2.

Because step 1 is slow and step 2 is fast, step 1 is rate determining. Thus, the rate of the overall reaction equals the rate of step 1, and the rate law of the overall reaction equals the rate law of step 1. Step 1 is a bimolecular process that has the rate law

$$\text{Rate} = k_1[NO_2]^2$$

Thus, the rate law predicted by this mechanism agrees with the one observed experimentally.

Could we propose a one-step mechanism for the preceding reaction? We might suppose that the overall reaction is a single bimolecular elementary process that involves the collision of a molecule of NO_2 with one of CO. However, the rate law predicted by this mechanism would be

$$\text{Rate} = k[NO_2][CO]$$

Because this mechanism predicts a rate law different from that observed experimentally, we can rule it out.

Mechanisms with an Initial Fast Step

It is difficult to derive the rate law for a mechanism in which an intermediate is a reactant in the rate-determining step. This situation arises in multistep mechanisms when the first step is *not* rate determining. Let's consider one example: the gas-phase reaction of nitric oxide (NO) with bromine (Br_2).

$$2NO(g) + Br_2(g) \longrightarrow 2NOBr(g) \qquad [14.24]$$

The experimentally determined rate law for this reaction is second order in NO and first order in Br_2:

$$\text{Rate} = k[NO]^2[Br_2] \qquad [14.25]$$

We seek a reaction mechanism that is consistent with this rate law. One possibility is that the reaction occurs in a single termolecular step:

$$NO(g) + NO(g) + Br_2(g) \longrightarrow 2NOBr(g) \qquad \text{Rate} = k[NO]^2[Br_2] \qquad [14.26]$$

* The subscript on the rate constant identifies the elementary step involved. Thus, k_1 is the rate constant for step 1, k_2 is the rate constant for step 2, and so forth. A negative subscript refers to the rate constant for the reverse of an elementary step. For example, k_{-1} is the rate constant for the reverse of the first step.

As noted in Practice Exercise 14.13, this does not seem likely because termolecular processes are so rare.

Let's consider an alternative mechanism that does not invoke termolecular steps:

$$\text{Step 1:} \quad NO(g) + Br_2(g) \underset{k_{-1}}{\overset{k_1}{\rightleftharpoons}} NOBr_2(g) \quad \text{(fast)}$$

$$\text{Step 2:} \quad NOBr_2(g) + NO(g) \xrightarrow{k_2} 2NOBr(g) \quad \text{(slow)}$$

In this mechanism step 1 actually involves two processes: a forward reaction and its reverse.

Because step 2 is the slow, rate-determining step, the rate of the overall reaction is governed by the rate law for that step:

$$\text{Rate} = k_2[NOBr_2][NO] \quad [14.27]$$

However, $NOBr_2$ is an intermediate generated in step 1. Intermediates are usually unstable molecules that have a low, unknown concentration. Thus, our rate law depends on the unknown concentration of an intermediate.

Fortunately, with the aid of some assumptions, we can express the concentration of the intermediate ($NOBr_2$) in terms of the concentrations of the starting reactants (NO and Br_2). We first assume that $NOBr_2$ is intrinsically unstable and that it does not accumulate to a significant extent in the reaction mixture. There are two ways for $NOBr_2$ to be consumed once it is formed: It can either react with NO to form NOBr or fall back apart into NO and Br_2. The first of these possibilities is step 2, a slow process. The second is the reverse of step 1, a unimolecular process:

$$NOBr_2(g) \xrightarrow{k_{-1}} NO(g) + Br_2(g) \quad [14.28]$$

Because step 2 is slow, we assume that most of the $NOBr_2$ falls apart according to Equation 14.28. Thus, we have both the forward and reverse reactions of step 1 occurring much faster than step 2. Because they occur rapidly with respect to the reaction in step 2, the forward and reverse processes of step 1 establish an equilibrium. We have seen examples of dynamic equilibrium before, in the equilibrium between a liquid and its vapor (Section 11.5) and between a solid solute and its solution. (Section 13.3) As in any dynamic equilibrium, the rates of the forward and reverse reactions are equal. Thus, we can equate the rate expression for the forward reaction in step 1 with the rate expression for the reverse reaction:

$$\underbrace{k_1[NO][Br_2]}_{\text{Rate of forward reaction}} = \underbrace{k_{-1}[NOBr_2]}_{\text{Rate of reverse reaction}}$$

Solving for $[NOBr_2]$, we have

$$[NOBr_2] = \frac{k_1}{k_{-1}}[NO][Br_2]$$

Substituting this relationship into the rate law for the rate-determining step (Equation 14.27), we have

$$\text{Rate} = k_2\frac{k_1}{k_{-1}}[NO][Br_2][NO] = k[NO]^2[Br_2]$$

This is consistent with the experimental rate law (Equation 14.25). The experimental rate constant, k, equals k_2k_1/k_{-1}. This mechanism, which involves only unimolecular and bimolecular processes, is far more probable than the single termolecular step (Equation 14.26).

In general, *whenever a fast step precedes a slow one, we can solve for the concentration of an intermediate by assuming that an equilibrium is established in the fast step.*

SAMPLE EXERCISE 14.14

Show that the following mechanism for Equation 14.24 also produces a rate law consistent with the experimentally observed one:

$$\text{Step 1:} \quad NO(g) + NO(g) \underset{k_{-1}}{\overset{k_1}{\rightleftharpoons}} N_2O_2(g) \quad \text{(fast equilibrium)}$$

$$\text{Step 2:} \quad N_2O_2(g) + Br_2(g) \xrightarrow{k_2} 2NOBr(g) \quad \text{(slow)}$$

Solution

Analyze and Plan: The rate law of the slow elementary step in a mechanism determines the rate law found experimentally for the overall reaction. Thus, we first write the rate law based on the molecularity of the slow step. In this case the slow step involves the intermediate N_2O_2 as a reactant. Experimental rate laws, however, do not contain the concentrations of intermediates, but are expressed in terms of the concentrations of starting substances. Thus, we must relate the concentration of N_2O_2 to the concentration of NO by assuming that an equilibrium is established in the first step.

Solve: The second step is rate determining, so the overall rate is

$$\text{Rate} = k_2[N_2O_2][Br_2]$$

We solve for the concentration of the intermediate N_2O_2 by assuming that an equilibrium is established in step 1; thus, the rates of the forward and reverse reactions in step 1 are equal:

$$k_1[NO]^2 = k_{-1}[N_2O_2]$$

$$[N_2O_2] = \frac{k_1}{k_{-1}}[NO]^2$$

Substituting this expression into the rate expression gives

$$\text{Rate} = k_2\frac{k_1}{k_{-1}}[NO]^2[Br_2] = k[NO]^2[Br_2]$$

Thus, this mechanism also yields a rate law consistent with the experimental one.

PRACTICE EXERCISE

The first step of a mechanism involving the reaction of bromine is

$$Br_2(g) \underset{k_{-1}}{\overset{k_1}{\rightleftharpoons}} 2Br(g) \quad \text{(fast, equilibrium)}$$

What is the expression relating the concentration of $Br(g)$ to that of $Br_2(g)$?

Answer: $[Br] = \left(\frac{k_1}{k_{-1}}[Br_2]\right)^{1/2}$

14.7 Catalysis

A **catalyst** is a substance that changes the speed of a chemical reaction without undergoing a permanent chemical change itself in the process. Catalysts are very common; most reactions in the body, the atmosphere, the oceans, or in industrial chemistry occur with the help of catalysts.

In your laboratory work you may have carried out the reaction in which oxygen is produced by heating potassium chlorate ($KClO_3$):

$$2KClO_3(s) \xrightarrow{\Delta} 2KCl(s) + 3O_2(g)$$

In the absence of a catalyst, $KClO_3$ does not readily decompose in this manner, even on strong heating. However, mixing black manganese dioxide (MnO_2) with the $KClO_3$ before heating causes the reaction to occur much more readily. The MnO_2 can be recovered largely unchanged from this reaction, and so the overall chemical process is clearly still the same. Thus, MnO_2 acts as a catalyst for the decomposition of $KClO_3$.

Much industrial chemical research is devoted to the search for new and more effective catalysts for reactions of commercial importance. Extensive research efforts also are devoted to finding means of inhibiting or removing certain catalysts that promote undesirable reactions, such as those that corrode metals, age our bodies, and cause tooth decay.

Homogeneous Catalysis

A catalyst that is present in the same phase as the reacting molecules is a **homogeneous catalyst**. Examples abound both in solution and in the gas phase. Consider, for example, the decomposition of aqueous hydrogen peroxide, $H_2O_2(aq)$, into water and oxygen:

$$2H_2O_2(aq) \longrightarrow 2H_2O(l) + O_2(g) \qquad [14.29]$$

In the absence of a catalyst, this reaction occurs extremely slowly. Many different substances are capable of catalyzing the reaction, including bromide ion, $Br^-(aq)$, as shown in Figure 14.19(a) ▼. The bromide ion reacts with hydrogen peroxide in acidic solution, forming aqueous bromine and water:

$$2Br^-(aq) + H_2O_2(aq) + 2H^+ \longrightarrow Br_2(aq) + 2H_2O(l) \qquad [14.30]$$

The brown color observed in Figure 14.19(b) indicates the formation of $Br_2(aq)$. If this were the complete reaction, bromide ion would not be a catalyst because it undergoes chemical change during the reaction. However, hydrogen peroxide also reacts with the $Br_2(aq)$ generated in Equation 14.30:

$$Br_2(aq) + H_2O_2(aq) \longrightarrow 2Br^-(aq) + 2H^+(aq) + O_2(g) \qquad [14.31]$$

The bubbling evident in Figure 14.19(b) is due to the formation of $O_2(g)$. The sum of Equations 14.30 and 14.31 is just Equation 14.29:

$$2H_2O_2(aq) \longrightarrow 2H_2O(l) + O_2(g)$$

MOVIE
Catalysis

When H_2O_2 has been totally decomposed, we are left with a colorless solution of $Br^-(aq)$, as seen in Figure 14.19(c). Bromide ion, therefore, is indeed a catalyst of the reaction because it speeds the overall reaction without itself undergoing any net change. In contrast, Br_2 is an intermediate because it is first formed (Equation 14.30) and then consumed (Equation 14.31).

On the basis of the Arrhenius equation (Equation 14.19), the rate constant (k) is determined by the activation energy (E_a) and the frequency factor (A). A catalyst may affect the rate of reaction by altering the value of either E_a or A.

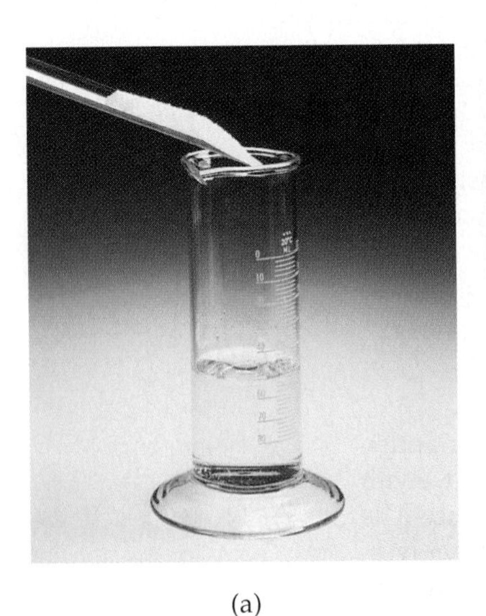
(a)

(b)

(c)

▲ **Figure 14.19** (a) In the absence of a catalyst, $H_2O_2(aq)$ decomposes very slowly. (b) Shortly after the addition of a small amount of $NaBr(aq)$ to $H_2O_2(aq)$, the solution turns brown because Br_2 is generated (Equation 14.30). The buildup of Br_2 leads to rapid evolution of $O_2(g)$, according to Equation 14.31. (c) After all of the H_2O_2 has decomposed, a colorless solution of $NaBr(aq)$ remains. Thus, NaBr has catalyzed the reaction even though it is not consumed during the reaction.

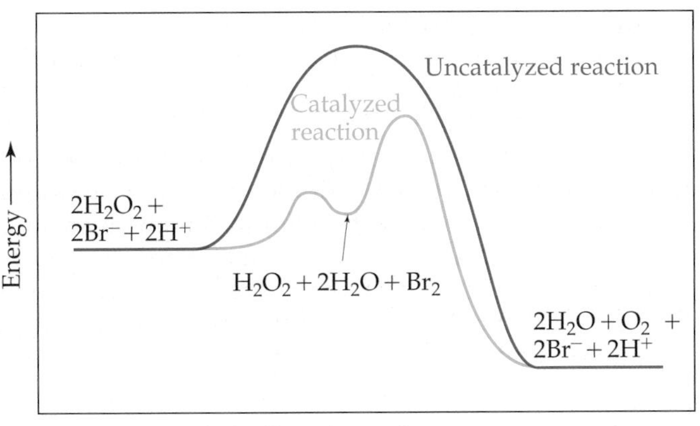

◀ **Figure 14.20** Energy profiles for the uncatalyzed decomposition of hydrogen peroxide and for the reaction as catalyzed by Br^-. The catalyzed reaction involves two successive steps, each of which has a lower activation energy than the uncatalyzed reaction. Notice that the energies of reactants and products are unchanged by the catalyst.

The most dramatic catalytic effects come from lowering E_a. As a general rule, *a catalyst lowers the overall activation energy for a chemical reaction.*

A catalyst usually lowers the overall activation energy for a reaction by providing a completely different mechanism for the reaction. The examples given previously involve a reversible, cyclic reaction of the catalyst with the reactants. In the decomposition of hydrogen peroxide, for example, two successive reactions of H_2O_2, with bromide and then with bromine, take place. Because these two reactions together serve as a catalytic pathway for hydrogen peroxide decomposition, *both* of them must have significantly lower activation energies than the uncatalyzed decomposition, as shown schematically in Figure 14.20 ▲.

Heterogeneous Catalysis

A **heterogeneous catalyst** exists in a different phase from the reactant molecules, usually as a solid in contact with either gaseous reactants or with reactants in a liquid solution. Many industrially important reactions are catalyzed by the surfaces of solids. For example, hydrocarbon molecules are rearranged to form gasoline with the aid of what are called "cracking" catalysts (see the "Chemistry at Work" box in Section 25.3). Heterogeneous catalysts are often composed of metals or metal oxides. Because the catalyzed reaction occurs on the surface, special methods are often used to prepare catalysts so that they have very large surface areas.

The initial step in heterogeneous catalysis is usually **adsorption** of reactants. *Adsorption* refers to the binding of molecules to a surface, whereas *absorption* refers to the uptake of molecules into the interior of another substance. ∞ (Section 13.6) Adsorption occurs because the atoms or ions at the surface of a solid are extremely reactive. Unlike their counterparts in the interior of the substance, they have unfulfilled valence requirements. The unused bonding capability of surface atoms or ions may be used to bond molecules from the gas or solution phase to the surface of the solid. In practice, not all the atoms or ions of the surface are reactive; various impurities may be adsorbed at the surface, and these may occupy many potential reaction sites and block further reaction. The places where reacting molecules may become adsorbed are called **active sites**. The number of active sites per unit amount of catalyst depends on the nature of the catalyst, on its method of preparation, and on its treatment before use.

ANIMATION
Surface Reactions—Hydrogenation

The reaction of hydrogen gas with ethylene gas to form ethane gas provides an example of heterogeneous catalysis:

$$\underset{\text{Ethylene}}{C_2H_4(g)} + H_2(g) \longrightarrow \underset{\text{Ethane}}{C_2H_6(g)} \qquad \Delta H^\circ = -137 \text{ kJ/mol} \qquad [14.32]$$

Even though this reaction is exothermic, it occurs very slowly in the absence of a catalyst. In the presence of a finely powdered metal, however, such as nickel, palladium, or platinum, the reaction occurs rather easily at room temperature. The mechanism by which the reaction occurs is diagrammed in Figure 14.21 ▼. Both ethylene and hydrogen are adsorbed at active sites on the metal surface [Figure 14.21(a)]. Upon adsorption the H—H bond of H_2 breaks, leaving two H atoms that are bonded to the metal surface, as shown in Figure 14.21(b). The hydrogen atoms are relatively free to move about the surface. When a hydrogen encounters an adsorbed ethylene molecule, it can form a σ bond to one of the carbon atoms, effectively destroying the C—C π bond and leaving an *ethyl group* (C_2H_5) bonded to the surface via a metal-to-carbon σ bond [Figure 14.21(c)]. This σ bond is relatively weak, so when the other carbon atom also encounters a hydrogen atom, a sixth C—H σ bond is readily formed, and an ethane molecule is released from the metal surface [Figure 14.21(d)]. The active site is ready to adsorb another ethylene molecule and thus begin the cycle again.

We can understand the role of the catalyst in this process by considering the bond enthalpies involved. (Section 8.8) In the course of the reaction, the H—H σ bond and the C—C π bond must be broken, and to do so requires the input of energy, which we can liken to the activation energy of the reaction. The formation of the new C—H σ bonds *releases* an even greater amount of energy, making the reaction exothermic. When H_2 and C_2H_4 are bonded to the surface of the catalyst, less energy is required to break the bonds, lowering the activation energy of the reaction.

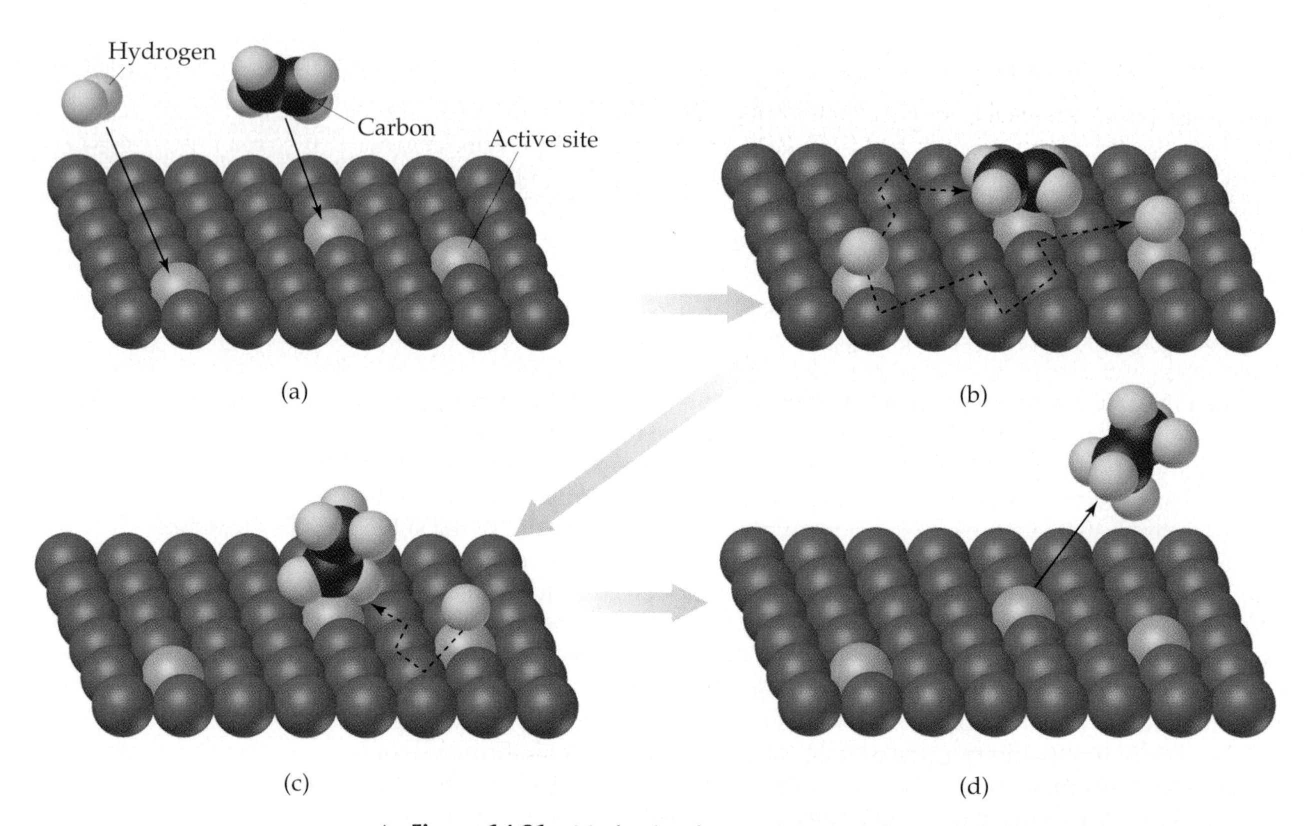

▲ **Figure 14.21** Mechanism for reaction of ethylene with hydrogen on a catalytic surface. (a) The hydrogen and ethylene are adsorbed at the metal surface. (b) The H—H bond is broken to give adsorbed hydrogen atoms. (c) These migrate to the adsorbed ethylene and bond to the carbon atoms. (d) As C—H bonds are formed, the adsorption of the molecule to the metal surface is decreased and ethane is released.

Chemistry at Work Catalytic Converters

Heterogeneous catalysis plays a major role in the fight against urban air pollution. Two components of automobile exhausts that help form photochemical smog are nitrogen oxides and unburned hydrocarbons of various types (Section 18.4). In addition, automobile exhausts may contain considerable quantities of carbon monoxide. Even with the most careful attention to engine design, it is impossible under normal driving conditions to reduce the quantity of these pollutants to an acceptable level in the exhaust gases. It is therefore necessary to remove them from the exhaust before they are vented to the air. This removal is accomplished in the *catalytic converter*.

The catalytic converter, which is part of the exhaust system, must perform two distinct functions: (1) oxidation of CO and unburned hydrocarbons (C_xH_y) to carbon dioxide and water, and (2) reduction of nitrogen oxides to nitrogen gas:

$$CO, C_xH_y \xrightarrow{O_2} CO_2 + H_2O$$

$$NO, NO_2 \longrightarrow N_2$$

These two functions require two distinctly different catalysts, so the development of a successful catalyst system is a difficult challenge. The catalysts must be effective over a wide range of operating temperatures. They must continue to be active in spite of the fact that various components of the exhaust can block the active sites of the catalyst. They must be sufficiently rugged to withstand exhaust gas turbulence and the mechanical shocks of driving under various conditions for thousands of miles.

Catalysts that promote the combustion of CO and hydrocarbons are, in general, the transition-metal oxides and noble metals such as platinum. A mixture of two different metal oxides, CuO and Cr_2O_3, might be used, for example. These materials are supported on a structure (Figure 14.22 ▶) that allows the best possible contact between the flowing exhaust gas and the catalyst surface. Either bead or honeycomb structures made from alumina (Al_2O_3) and impregnated with the catalyst may be employed. Such catalysts operate by first adsorbing oxygen gas, also present in the exhaust gas. This adsorption weakens the O—O bond in O_2, so that oxygen atoms are available for reaction with adsorbed CO to form CO_2. Hydrocarbon oxidation probably proceeds somewhat similarly, with the hydrocarbons first being adsorbed by rupture of a C—H bond.

The most effective catalysts for reduction of NO to yield N_2 and O_2 are transition-metal oxides and noble metals, the same kinds of materials that catalyze the oxidation of CO and hydrocarbons. The catalysts that are most effective in one reaction, however, are usually much less effective in the other. It is therefore necessary to have two different catalytic components.

Catalytic converters are remarkably efficient heterogeneous catalysts. The automotive exhaust gases are in contact with the catalyst for only 100 to 400 ms. In this very short time, 96% of the hydrocarbons and CO are converted to CO_2 and H_2O, and the emission of nitrogen oxides is reduced by 76%.

There are costs as well as benefits associated with the use of catalytic converters. Some of the metals used in the converters are very expensive. Catalytic converters currently account for about 35% of the platinum, 65% of the palladium, and 95% of the rhodium used annually. All of these metals, which come mainly from Russia and South Africa, are far more expensive than gold.

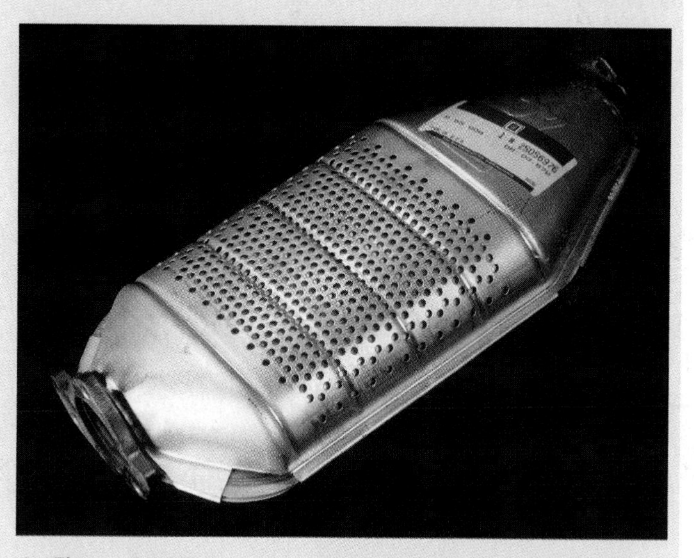

▲ **Figure 14.22** A stainless steel catalytic converter used in automobiles. The converter contains catalysts that promote the conversion of exhaust gases into CO_2, H_2O, and N_2.

Enzymes

Many of the most interesting and important examples of catalysis involve reactions within living systems. The human body is characterized by an extremely complex system of interrelated chemical reactions. All these reactions must occur at carefully controlled rates in order to maintain life. A large number of marvelously efficient biological catalysts known as **enzymes** are necessary for many of these reactions to occur at suitable rates. Most enzymes are large protein molecules with molecular weights ranging from about 10,000 to about 1 million amu. They are very selective in the reactions that they catalyze, and some are absolutely specific, operating for only one substance in only one reaction. The decomposition of hydrogen peroxide, for example, is an important biological process. Because hydrogen

▶ **Figure 14.23** Ground-up beef liver causes hydrogen peroxide to decompose rapidly into water and oxygen. The decomposition is catalyzed by the enzyme *catalase*. Grinding the liver breaks open the cells, so that the reaction takes place more rapidly. The frothing is due to escape of oxygen gas from the reaction mixture.

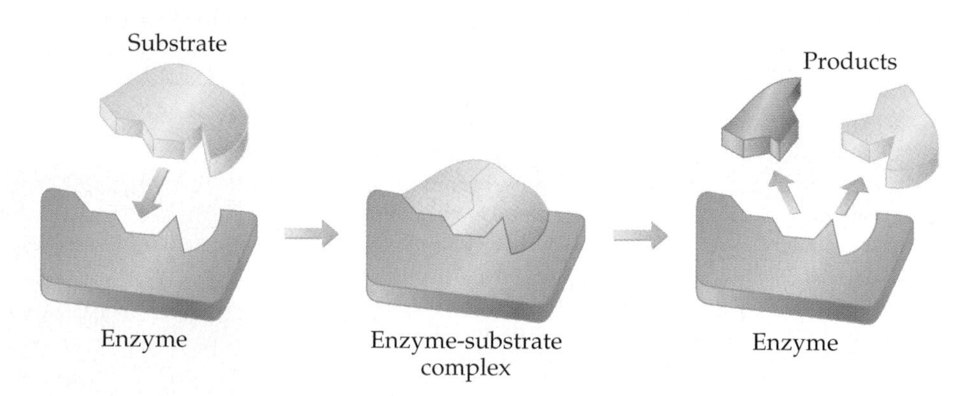

▶ **Figure 14.24** The lock-and-key model for enzyme action. The correct substrate is recognized by its ability to fit the active site of the enzyme, forming the enzyme-substrate complex. After the reaction of the substrate is complete, the products separate from the enzyme.

peroxide is strongly oxidizing, it can be physiologically harmful. For this reason, the blood and livers of mammals contain an enzyme, *catalase*, which catalyzes the decomposition of hydrogen peroxide into water and oxygen (Equation 14.29). Figure 14.23 ▲ shows the dramatic acceleration of this chemical reaction by the catalase in beef liver.

Although an enzyme is a large molecule, the reaction is catalyzed at a very specific location in the enzyme, called the active site. The substances that undergo reaction at this site are called **substrates**. A simple explanation for the specificity of enzymes is provided by the **lock-and-key model**, illustrated in Figure 14.24 ▲. The substrate is pictured as fitting neatly into a special place on the enzyme (the active site), much like a specific key fitting into a lock. The active site is created by coiling and folding of the long protein molecule to form a space, something like a pocket, into which the substrate molecule fits. Figure 14.25 ◀ shows a model of the enzyme *lysozyme* with and without a bound substrate molecule.

The combination of the enzyme and the substrate is called the *enzyme-substrate complex*. Although Figure 14.24 shows both the active site and its complementary substrate as having rigid shapes, there is often a fair amount of flexibility in the active site. Thus, the active site may change shape as it binds the substrate. The binding between the substrate and the active site involves intermolecular forces such as dipole-dipole attractions, hydrogen bonds, and London dispersion forces.

As the substrate molecules enter the active site, they are somehow activated, so that they are capable of extremely rapid reaction. This activation may result from the withdrawal or donation of electron density at a particular bond by the enzyme. In addition, in the process of fitting into the active site, the substrate molecule may be distorted and thus made more reactive. Once the reaction occurs, the products then depart, allowing another substrate molecule to enter.

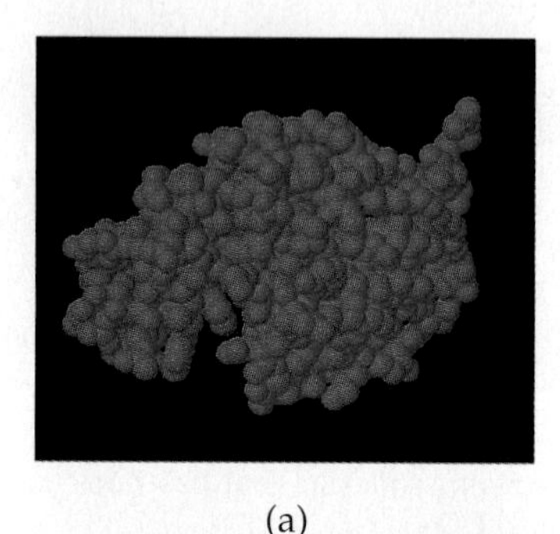

(a)

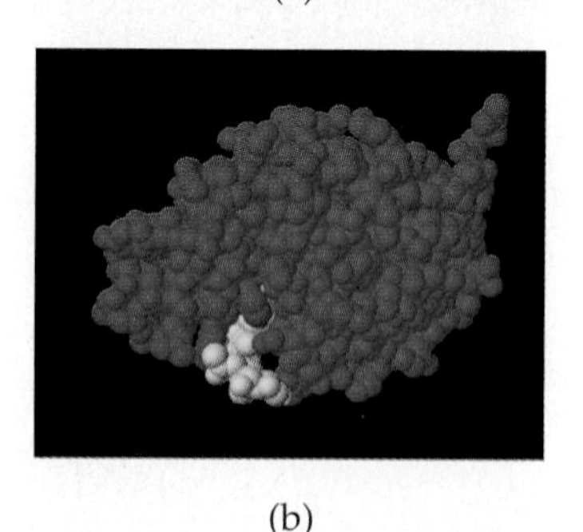

(b)

▲ **Figure 14.25** (a) A molecular model of the enzyme *lysozyme*. Note the characteristic cleft, which is the location of the active site. (b) Lysozyme with a bound substrate molecule.

The activity of an enzyme is destroyed if some molecule in the solution is able to bind strongly to the active site and block the entry of the substrate. Such substances are called *enzyme inhibitors*. Nerve poisons and certain toxic metal ions such as lead and mercury are believed to act in this way to inhibit enzyme activity. Some other poisons act by attaching elsewhere on the enzyme, thereby distorting the active site so that the substrate no longer fits.

Enzymes are enormously more efficient than ordinary nonbiochemical catalysts. The number of individual catalyzed reaction events occurring at a particular active site, called the *turnover number*, is generally in the range of 10^3 to 10^7 per second. Such large turnover numbers correspond to very low activation energies.

Chemistry and Life Nitrogen Fixation and Nitrogenase

Nitrogen is one of the most essential elements in living organisms. It is found in many compounds that are vital to life, including proteins, nucleic acids, vitamins, and hormones. Plants use very simple nitrogen-containing compounds, especially NH_3, NH_4^+, and NO_3^-, as starting materials from which such complex, biologically necessary compounds are formed. Animals are unable to synthesize the complex nitrogen compounds they require from the simple substances used by plants. Instead, they rely on more complicated precursors present in vitamin- and protein-rich foods.

Nitrogen is continually cycling through this biological arena in various forms, as shown in the simplified nitrogen cycle in Figure 14.26 ▼. For example, certain microorganisms convert the nitrogen in animal waste and dead plants and animals into molecular nitrogen, $N_2(g)$, which returns to the atmosphere. In order for the food chain to be sustained, there must be a means of reincorporating this atmospheric N_2 in a form that plants can utilize. The process of converting N_2 into compounds that plants can use is called *nitrogen fixation*. Fixing nitrogen is difficult; N_2 is an exceptionally unreactive molecule, in large part because of its very strong N≡N triple bond. ∞∞ (Section 8.3) Some fixed nitrogen results from the action of lightning on the atmosphere, and some is produced industrially using a process we will discuss in Chapter 15. About 60% of fixed nitrogen, however, is a consequence of the action of a remarkable and complex enzyme called *nitrogenase*. This enzyme is *not* present in humans or other animals; rather, it is found in bacteria that live in the root nodules of certain plants such as the legumes, clover and alfalfa.

Nitrogenase converts N_2 into NH_3, a process that, in the absence of a catalyst, has a very large activation energy. This process is a *reduction* of nitrogen—during the reaction, its oxidation state is reduced from 0 in N_2 to -3 in NH_3. The mechanism by which nitrogenase reduces N_2 is not fully understood. Like many other enzymes, including catalase, the active site of nitrogenase contains transition-metal atoms; such enzymes are called *metalloenzymes*. Because transition metals can readily change oxidation state, metalloenzymes are especially useful for effecting transformations in which substrates are either oxidized or reduced.

It has been known for nearly 20 years that a portion of nitrogenase contains iron and molybdenum atoms. This portion, called the *FeMo-cofactor*, is thought to serve as the active

continues on next page

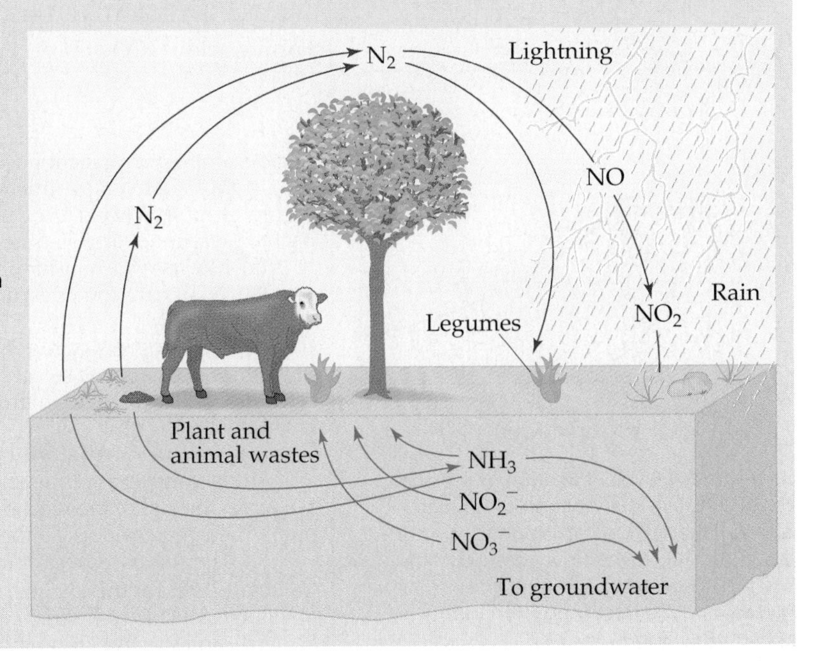

▶ **Figure 14.26** Simplified picture of the nitrogen cycle. The compounds of nitrogen in the soil are water-soluble species, such as NH_3, NO_2^-, and NO_3^-, which can be washed out of the soil by groundwater. These nitrogen compounds are converted into biomolecules by plants and are incorporated into animals that eat the plants. Animal waste and dead plants and animals are attacked by certain bacteria that release N_2 to the atmosphere. Atmospheric N_2 is fixed in the soil predominantly by the action of certain plants that contain the enzyme nitrogenase, thereby completing the cycle.

site of the enzyme. The FeMo-cofactor of nitrogenase is a striking cluster of seven Fe atoms and one Mo atom, all linked by sulfur atoms (Figure 14.27 ▼). Current research on nitrogenase is exploring the possibility that the N_2 molecule can enter the "pocket" inside the FeMo-cofactor, at which point the transformation of nitrogen to ammonia begins.

It is one of the wonders of life that simple bacteria can contain beautifully complex and vitally important enzymes such as nitrogenase. Because of this enzyme, nitrogen is continually cycled between its comparatively inert role in the atmosphere and its critical role in living organisms; without it, life as we know it could not exist on Earth.

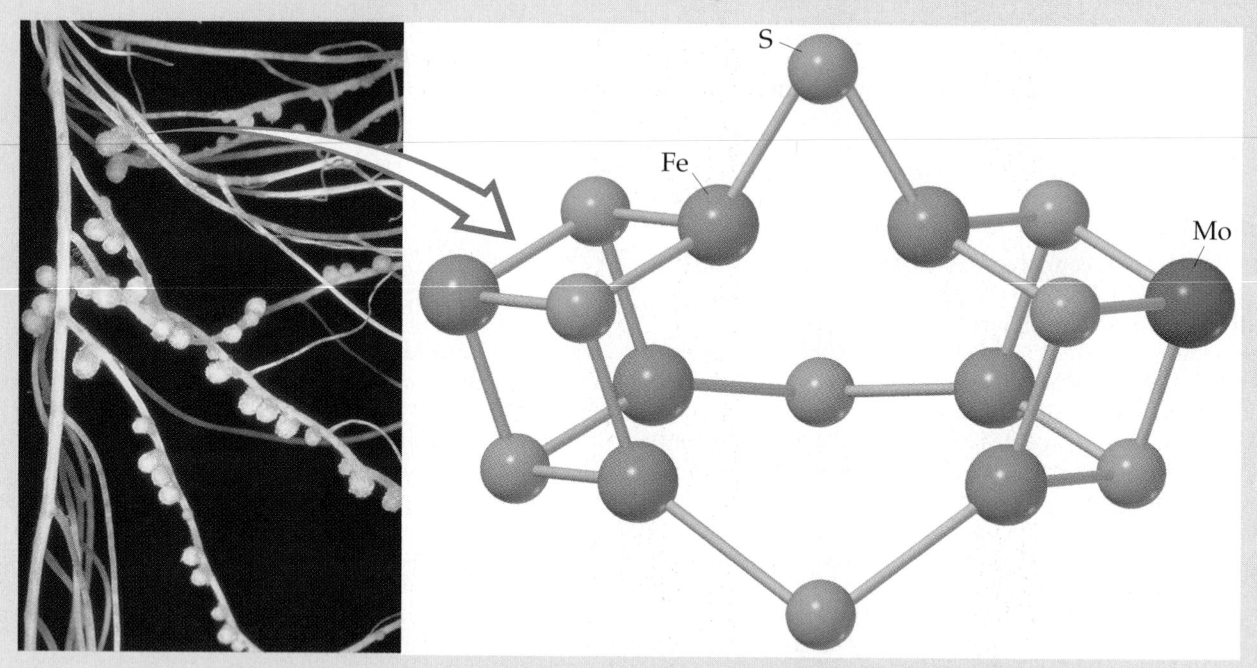

▲ **Figure 14.27** Representation of the FeMo-cofactor of nitrogenase, as determined by X-ray crystallography. Nitrogenase is found in nodules in the roots of certain plants, such as the white clover roots shown at the left. The cofactor, which is thought to be the active site of the enzyme, contains seven Fe atoms and one Mo atom, linked by sulfur atoms. The molecules on the outside of the cofactor connect it to the rest of the protein.

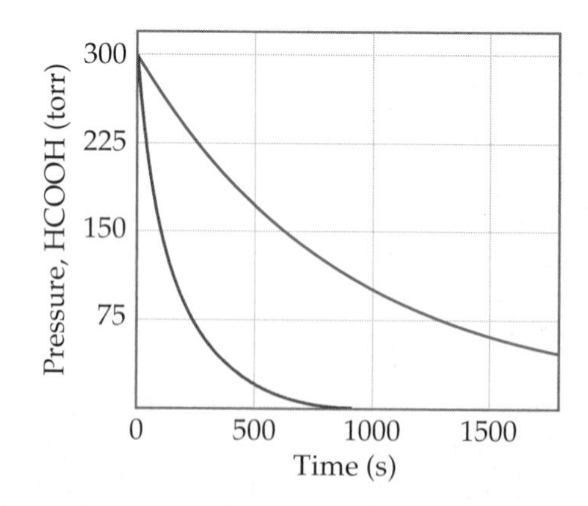

▲ **Figure 14.28** Variation in pressure of HCOOH(g) as a function of time at 838 K. The red line corresponds to decomposition when only gaseous HCOOH is present. The blue line corresponds to decomposition in the presence of added ZnO(s).

SAMPLE INTEGRATIVE EXERCISE 14: Putting Concepts Together

Formic acid (HCOOH) decomposes in the gas phase at elevated temperatures as follows:

$$HCOOH(g) \longrightarrow CO_2(g) + H_2(g)$$

The decomposition reaction is determined to be first order. A graph of the partial pressure of HCOOH versus time for decomposition at 838 K is shown as the red curve in Figure 14.28 ◀. When a small amount of solid ZnO is added to the reaction chamber, the partial pressure of acid versus time varies as shown by the blue curve in Figure 14.28.

(a) Estimate the half-life and first-order rate constant for formic acid decomposition.
(b) What can you conclude from the effect of added ZnO on the decomposition of formic acid?
(c) The progress of the reaction was followed by measuring the partial pressure of formic acid vapor at selected times. Suppose that, instead, we had plotted the concentration of formic acid in units of mol/L. What effect would this have had on the calculated value of k?
(d) The pressure of formic acid vapor at the start of the reaction is 3.00×10^2 torr. Assuming constant temperature and ideal-gas behavior, what is the pressure in the system at the end of the reaction? If the volume of the reaction chamber is 436 cm^3, how many moles of gas occupy the reaction chamber at the end of the reaction?
(e) The standard heat of formation of formic acid vapor is $\Delta H_f^\circ = -378.6$ kJ/mol. Calculate ΔH° for the overall reaction. Assuming that the activation energy (E_a) for the reaction is 184 kJ/mol, sketch an approximate energy profile for the reaction, and label E_a, ΔH°, and the transition state.

Solution (a) The initial pressure of HCOOH is 3.00×10^2 torr. On the graph we move to the level at which the partial pressure of HCOOH is 150 torr, half the initial value. This corresponds to a time of about 6.60×10^2 s, which is therefore the half-life. The first-order rate constant is given by Equation 14.15: $k = 0.693/t_{1/2} = 0.693/660 \text{ s} = 1.05 \times 10^{-3} \text{ s}^{-1}$.

(b) The reaction proceeds much more rapidly in the presence of solid ZnO, so the surface of the oxide must be acting as a catalyst for the decomposition of the acid. This is an example of heterogeneous catalysis.

(c) If we had graphed the concentration of formic acid in units of moles per liter, we would still have determined that the half-life for decomposition is 660 seconds, and we would have computed the same value for k. Because the units for k are s^{-1}, the value for k is independent of the units used for concentration.

(d) According to the stoichiometry of the reaction, two moles of product are formed for each mole of reactant. When reaction is completed, therefore, the pressure will be 600 torr, just twice the initial pressure, assuming ideal-gas behavior (Section 10.6). (Because we are working at quite high temperature and fairly low gas pressure, assuming ideal-gas behavior is reasonable.) The number of moles of gas present can be calculated using the ideal-gas equation (Section 10.4):

$$n = \frac{PV}{RT} = \frac{(600/760) \text{ atm } (0.436 \text{ L})}{(0.0821 \text{ L-atm/mol-K})(838 \text{ K})} = 5.00 \times 10^{-3} \text{ moles}$$

(e) We first calculate the overall change in energy, $\Delta H°$ (Section 5.7 and Appendix C), as in

$$\begin{aligned} \Delta H° &= \Delta H_f°(CO_2(g)) + \Delta H_f°(H_2(g)) - \Delta H_f°(HCOOH(g)) \\ &= -393.5 \text{ kJ/mol} + 0 - (-378.6 \text{ kJ/mol}) \\ &= -14.9 \text{ kJ/mol} \end{aligned}$$

From this and the given value for E_a, we can draw an approximate energy profile for the reaction, in analogy to Figure 14.15.

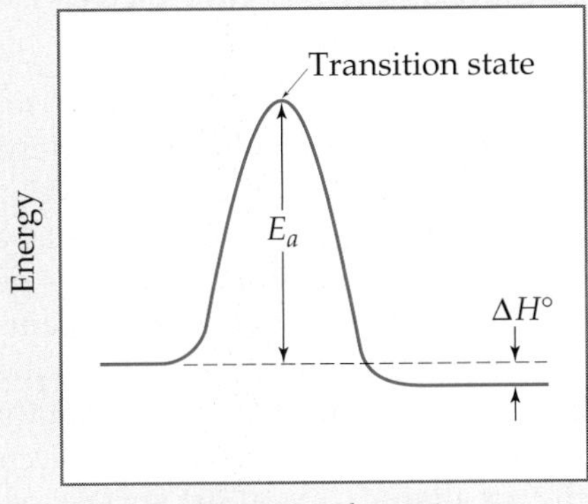

Summary and Key Terms

Introduction and Section 14.1 In this chapter we explored **chemical kinetics**, the area of chemistry that studies the rates of chemical reactions and the factors that affect them, namely, concentration, temperature, and catalysts.

Section 14.2 **Reaction rates** are usually expressed as changes in concentration per unit time: Typically, for reactions in solution, rates are given in units of molarity per second (M/s). For most reactions a plot of molarity versus time shows that the rate slows down as the reaction proceeds. The **instantaneous rate** is the slope of a line drawn tangent to the concentration-versus-time curve at a specific time. Rates can be written in terms of the appearance of products or the disappearance of reactants; the stoichiometry of the reaction dictates the relationship between rates of appearance and disappearance. Spectroscopy is one technique that can be used to monitor the course of a reaction. According to **Beer's law**, the absorption of electromagnetic radiation by a substance at a particular wavelength is directly proportional to its concentration.

Section 14.3 The quantitative relationship between rate and concentration is expressed by a **rate law**, which usually has the following form:

$$\text{Rate} = k[\text{reactant } 1]^m[\text{reactant } 2]^n \ldots$$

The constant k in the rate law is called the **rate constant**; the exponents m, n, and so forth are called **reaction orders** for the reactants. The sum of the reaction orders gives the **overall reaction order**. Reaction orders must be determined experimentally. The units of the rate constant depend on the overall reaction order. For a reaction in which the overall reaction order is 1, k has units of s^{-1}; for one in which the overall reaction order is 2, k has units of $M^{-1}s^{-1}$.

Section 14.4 Rate laws can be used to determine the concentrations of reactants or products at any time during a reaction. In a **first-order reaction** the rate is proportional to the concentration of a single reactant raised to the first power: Rate $= k[A]$. In such cases $\ln[A]_t = -kt + \ln[A]_0$, where $[A]_t$ is the concentration of reactant A at time t, k is the rate constant, and $[A]_0$ is the initial concentration of A. Thus, for a first-order reaction, a graph of ln[A] versus time yields a straight line of slope $-k$.

A **second-order reaction** is one for which the overall reaction order is 2. If a second-order rate law depends on the concentration of only one reactant, then rate $= k[A]^2$, and the time dependence of [A] is given by the following relationship: $1/[A]_t = 1/[A]_0 + kt$. In this case a graph of $1/[A]_t$ versus time yields a straight line.

The **half-life** of a reaction, $t_{1/2}$, is the time required for the concentration of a reactant to drop to one half of its original value. For a first-order reaction, the half-life depends only on the rate constant and not on the initial concentration: $t_{1/2} = 0.693/k$. The half-life of a second-order reaction depends on both the rate constant and the initial concentration of A: $t_{1/2} = 1/k[A]_0$.

Section 14.5 The **collision model**, which assumes that reactions occur as a result of collisions between molecules, helps explain why the magnitudes of rate constants increase with increasing temperature. The greater the kinetic energy of the colliding molecules, the greater the energy of collision. The minimum energy required for a reaction to occur is called the **activation energy**, E_a. A collision with energy E_a or greater can cause the atoms of the colliding molecules to reach the **activated complex** (or **transition state**), which is the highest energy arrangement in the pathway from reactants to products. Even if a collision is energetic enough, it may not lead to reaction; the reactants must also be correctly oriented relative to one another in order for a collision to be effective.

Because the kinetic energy of molecules depends on temperature, the rate constant of a reaction is very dependent on temperature. The relationship between k and temperature is given by the **Arrhenius equation**: $k = Ae^{-E_a/RT}$. The term A is called the **frequency factor**; it relates to the number of collisions that are favorably oriented for reaction. The Arrhenius equation is often used in logarithmic form: $\ln k = \ln A - E_a/RT$. Thus, a graph of ln k versus $1/T$ yields a straight line with slope $-E_a/R$.

Section 14.6 A **reaction mechanism** details the individual steps that occur in the course of a reaction. Each of these steps, called **elementary steps**, has a well-defined rate law that depends on the number of molecules (the **molecularity**) of the step. Elementary steps are defined as either **unimolecular**, **bimolecular**, or **termolecular**, depending on whether one, two, or three reactant molecules are involved, respectively. Termolecular elementary steps are very rare. Unimolecular, bimolecular, and termolecular steps follow rate laws that are first order overall, second order overall, and third order overall, respectively. An elementary step may produce an **intermediate**, a product that is consumed in a later elementary step and therefore does not appear in the overall stoichiometry of the reaction.

If a mechanism has several elementary steps, the overall rate is determined by the slowest elementary step, called the **rate-determining step**. A fast elementary step that follows the rate-determining step will have no effect on the rate law of the reaction. A fast step that precedes the rate-determining step often creates an equilibrium that involves an intermediate. For a mechanism to be valid, the rate law predicted by the mechanism must be the same as that observed experimentally.

Section 14.7 A **catalyst** is a substance that increases the rate of a reaction without undergoing a net chemical change itself. It does so by providing a different mechanism for the reaction, one that has a lower activation energy. A **homogeneous catalyst** is one that is in the same phase as the reactants. A **heterogeneous catalyst** has a different phase from the reactants. Finely divided metals are often used as heterogeneous catalysts for solution- and gas-phase reactions. Reacting molecules can undergo binding, or **adsorption**, at the surface of the catalyst. The sites on the catalyst at which reaction occurs are called **active sites**. The adsorption of a reactant at an active site makes bond breaking easier, lowering the activation energy. Catalysis in living organisms is achieved by **enzymes**, large protein molecules that usually catalyze a very specific reaction. The specific reactant molecules involved in an enzymatic reaction are called **substrates**. In the **lock-and-key model** for enzyme catalysis, substrate molecules bind very specifically to the active site of the enzyme, after which they can undergo reaction.

Exercises

Reaction Rates

14.1 **(a)** What is meant by the term *reaction rate*? **(b)** Name three factors that can affect the rate of a chemical reaction. **(c)** What information is necessary to relate the rate of disappearance of reactants to the rate of appearance of products?

14.2 **(a)** What are the units usually used to express the rates of reactions occurring in solution? **(b)** From your everyday experience, give two examples of the effects of temperature on the rates of reactions. **(c)** What is the difference between average rate and instantaneous rate?

14.3 Consider the following hypothetical aqueous reaction: $A(aq) \longrightarrow B(aq)$. A flask is charged with 0.065 mol of A in a total volume of 100.0 mL. The following data are collected:

Time (min)	0	10	20	30	40
Moles of A	0.065	0.051	0.042	0.036	0.031

(a) Calculate the number of moles of B at each time in the table, assuming that there are no molecules of B at time zero. **(b)** Calculate the average rate of disappearance of A for each 10-min interval, in units of M/s. **(c)** Between $t = 10$ min and $t = 30$ min, what is the average rate of appearance of B in units of M/s? Assume that the volume of the solution is constant.

14.4 A flask is charged with 0.100 mol of A and allowed to react to form B according to the hypothetical gas-phase reaction $A(g) \longrightarrow B(g)$. The following data are collected:

Time (s)	0	40	80	120	160
Moles of A	0.100	0.067	0.045	0.030	0.020

(a) Calculate the number of moles of B at each time in the table. **(b)** Calculate the average rate of disappearance of A for each 40-s interval, in units of mol/s. **(c)** What additional information would be needed to calculate the rate in units of concentration per time?

14.5 The isomerization of methyl isonitrile (CH_3NC) to acetonitrile (CH_3CN) was studied in the gas phase at 215°C, and the following data were obtained:

Time (s)	$[CH_3NC]$ (M)
0	0.0165
2,000	0.0110
5,000	0.00591
8,000	0.00314
12,000	0.00137
15,000	0.00074

Calculate the average rate of reaction, in M/s, for the time interval between each measurement.

14.6 The rate of disappearance of HCl was measured for the following reaction:

$$CH_3OH(aq) + HCl(aq) \longrightarrow CH_3Cl(aq) + H_2O(l)$$

The following data were collected:

Time (min)	[HCl] (M)
0.0	1.85
54.0	1.58
107.0	1.36
215.0	1.02
430.0	0.580

Calculate the average rate of reaction, in M/s, for the time interval between each measurement.

14.7 Using the data provided in Exercise 14.5, graph $[CH_3NC]$ versus time. Use the graph to determine the instantaneous rates in M/s at $t = 5000$ and $t = 8000$ s.

14.8 Using the data provided in Exercise 14.6, graph [HCl] versus time. Use the graph to determine the instantaneous rates in M/min and M/s at $t = 75.0$ and $t = 250$ min.

14.9 For each of the following gas-phase reactions, indicate how the rate of disappearance of each reactant is related to the rate of appearance of each product:
(a) $H_2O_2(g) \longrightarrow H_2(g) + O_2(g)$
(b) $2N_2O(g) \longrightarrow 2N_2(g) + O_2(g)$
(c) $N_2(g) + 3H_2(g) \longrightarrow 2NH_3(g)$

14.10 For each of the following gas-phase reactions, write the rate expression in terms of the appearance of each product or disappearance of each reactant:
(a) $2HBr(g) \longrightarrow H_2(g) + Br_2(g)$
(b) $2SO_2(g) + O_2(g) \longrightarrow 2SO_3(g)$
(c) $2NO(g) + 2H_2(g) \longrightarrow N_2(g) + 2H_2O(g)$

14.11 **(a)** Consider the combustion of $H_2(g)$: $2H_2(g) + O_2(g) \longrightarrow 2H_2O(g)$. If hydrogen is burning at the rate of 0.85 mol/s, what is the rate of consumption of oxygen? What is the rate of formation of water vapor? **(b)** The reaction $2NO(g) + Cl_2(g) \longrightarrow 2NOCl(g)$ is carried out in a closed vessel. If the partial pressure of NO is decreasing at the rate of 23 torr/min, what is the rate of change of the total pressure of the vessel?

14.12 **(a)** Consider the combustion of ethylene, $C_2H_4(g) + 3O_2(g) \longrightarrow 2CO_2(g) + 2H_2O(g)$. If the concentration of C_2H_4 is decreasing at the rate of 0.23 M/s, what are the rates of change in the concentrations of CO_2 and H_2O? **(b)** The rate of decrease in N_2H_4 partial pressure in a closed reaction vessel from the reaction $N_2H_4(g) + H_2(g) \longrightarrow 2NH_3(g)$ is 45 torr/hr. What are the rates of change of NH_3 partial pressure and total pressure in the vessel?

Rate Laws

14.13 A reaction $A + B \longrightarrow C$ obeys the following rate law: Rate $= k[A]^2[B]$. **(a)** If [A] is doubled, how will the rate change? Will the rate constant change? Explain. **(b)** What are the reaction orders for A and B? What is the overall reaction order? **(c)** What are the units of the rate constant?

14.14 A reaction $A + B \longrightarrow C$ obeys the following rate law: Rate $= k[B]^2$. **(a)** If [A] is doubled, will the rate change? Will the rate constant change? Explain. **(b)** What are the reaction orders for A and B? What is the overall reaction order? **(c)** What are the units of the rate constant?

14.15 The decomposition of N_2O_5 in carbon tetrachloride proceeds as follows: $2N_2O_5 \longrightarrow 4NO_2 + O_2$. The rate law is first order in N_2O_5. At 64°C the rate constant is $4.82 \times 10^{-3}\ s^{-1}$. **(a)** Write the rate law for the reaction. **(b)** What is the rate of reaction when $[N_2O_5] = 0.0240\ M$? **(c)** What happens to the rate when the concentration of N_2O_5 is doubled to 0.0480 M?

14.16 Consider the following reaction:

$$2NO(g) + 2H_2(g) \longrightarrow N_2(g) + 2H_2O(g)$$

(a) The rate law for this reaction is first order in H_2 and second order in NO. Write the rate law. **(b)** If the rate constant for this reaction at 1000 K is $6.0 \times 10^4\ M^{-2}\ s^{-1}$, what is the reaction rate when [NO] = 0.050 *M* and $[H_2]$ = 0.010 *M*? **(c)** What is the reaction rate at 1000 K when the concentration of NO is doubled, to 0.10 *M*, while the concentration of H_2 is 0.010 *M*?

14.17 Consider the following reaction:

$$CH_3Br(aq) + OH^-(aq) \longrightarrow CH_3OH(aq) + Br^-(aq)$$

The rate law for this reaction is first order in CH_3Br and first order in OH^-. When $[CH_3Br]$ is 5.0×10^{-3} *M* and $[OH^-]$ is 0.050 *M*, the reaction rate at 298 K is 0.0432 *M*/s. **(a)** What is the value of the rate constant? **(b)** What are the units of the rate constant? **(c)** What would happen to the rate if the concentration of OH^- were tripled?

14.18 The reaction between ethyl bromide (C_2H_5Br) and hydroxide ion in ethyl alcohol at 330 K, $C_2H_5Br(alc) + OH^-(alc) \longrightarrow C_2H_5OH(l) + Br^-(alc)$, is first order each in ethyl bromide and hydroxide ion. When $[C_2H_5Br]$ is 0.0477 *M* and $[OH^-]$ is 0.100 *M*, the rate of disappearance of ethyl bromide is 1.7×10^{-7} *M*/s. **(a)** What is the value of the rate constant? **(b)** What are the units of the rate constant? **(c)** How would the rate of disappearance of ethyl bromide change if the solution were diluted by adding an equal volume of pure ethyl alcohol to the solution?

14.19 You determine that the rate, law for a reaction $A \longrightarrow B + C$ has the form, rate = $k[A]^x$. What is the value of x if **(a)** the rate triples when [A] is tripled? **(b)** the rate increases eightfold when [A] is doubled? **(c)** there is no change in rate when [A] is tripled?

14.20 **(a)** You and your lab partner are studying the rate of a reaction, $A + B \longrightarrow C$. You make measurements of the initial rate under the following conditions:

1	[A] = 1.0 *M*	[B] = 1.0 *M*
2	[A] = 2.0 *M*	[B] = 1.0 *M*

(a) What reactant concentration could you use for experiment 3 in order to determine the rate law, assuming that the rate law is of the form, rate = $k[A]^x[B]^y$? **(b)** For a reaction of the form, $A + B + C \longrightarrow$ products, the following observations are made: Doubling the concentration of A doubles the rate, tripling the concentration of B has no effect on the rate, and tripling the concentration of C increases the rate by a factor of 9. By what factor will the rate change if the concentrations of A, B, and C are all halved?

14.21 The iodide ion reacts with hypochlorite ion (the active ingredient in chlorine bleaches) in the following way: $OCl^- + I^- \longrightarrow OI^- + Cl^-$. This rapid reaction gives the following rate data:

$[OCl^-]$, *M*	$[I^-]$, *M*	Rate, *M*/s
1.5×10^{-3}	1.5×10^{-3}	1.36×10^4
3.0×10^{-3}	1.5×10^{-3}	2.72×10^4
1.5×10^{-3}	3.0×10^{-3}	2.72×10^4

(a) Write the rate law for this reaction. **(b)** Calculate the rate constant. **(c)** Calculate the rate when $[OCl^-] = 1.0 \times 10^{-3}$ *M* and $[I^-] = 5.0 \times 10^{-4}$ *M*.

14.22 The reaction $2ClO_2(aq) + 2OH^-(aq) \longrightarrow ClO_3^-(aq) + ClO_2^-(aq) + H_2O(l)$ was studied with the following results:

Experiment	$[ClO_2]$, *M*	$[OH^-]$, *M*	Rate, *M*/s
1	0.060	0.030	0.0248
2	0.020	0.030	0.00276
3	0.020	0.090	0.00828

(a) Determine the rate law for the reaction. **(b)** Calculate the rate constant. **(c)** Calculate the rate when $[ClO_2]$ = 0.010 *M* and $[OH^-]$ = 0.015 *M*.

14.23 The following data were collected for the rate of disappearance of NO in the reaction $2NO(g) + O_2(g) \longrightarrow 2NO_2(g)$:

Experiment	[NO] (*M*)	$[O_2]$ (*M*)	Initial Rate (*M*/s)
1	0.0126	0.0125	1.41×10^{-2}
2	0.0252	0.0250	1.13×10^{-1}
3	0.0252	0.0125	5.64×10^{-2}

(a) What is the rate law for the reaction? **(b)** What are the units of the rate constant? **(c)** What is the average value of the rate constant calculated from the three data sets?

14.24 The following data were measured for the reaction $BF_3(g) + NH_3(g) \longrightarrow F_3BNH_3(g)$:

Experiment	$[BF_3]$ (*M*)	$[NH_3]$ (*M*)	Initial Rate (*M*/s)
1	0.250	0.250	0.2130
2	0.250	0.125	0.1065
3	0.200	0.100	0.0682
4	0.350	0.100	0.1193
5	0.175	0.100	0.0596

(a) What is the rate law for the reaction? **(b)** What is the overall order of the reaction? **(c)** What is the value of the rate constant for the reaction?

[14.25] Consider the gas-phase reaction between nitric oxide and bromine at 273°C: $2NO(g) + Br_2(g) \longrightarrow 2NOBr(g)$. The following data for the initial rate of appearance of NOBr were obtained:

Experiment	[NO] (*M*)	$[Br_2]$ (*M*)	Initial Rate (*M*/s)
1	0.10	0.20	24
2	0.25	0.20	150
3	0.10	0.50	60
4	0.35	0.50	735

(a) Determine the rate law. **(b)** Calculate the average value of the rate constant for the appearance of NOBr from the four data sets. **(c)** How is the rate of appearance of NOBr related to the rate of disappearance of Br_2? **(d)** What is the rate of disappearance of Br_2 when [NO] = 0.075 *M* and $[Br_2]$ = 0.25 *M*?

[14.26] Consider the reaction of peroxydisulfate ion ($S_2O_8^{2-}$) with iodide ion (I^-) in aqueous solution:

$$S_2O_8^{2-}(aq) + 3I^-(aq) \longrightarrow 2SO_4^{2-}(aq) + I_3^-(aq)$$

At a particular temperature the rate of disappearance of $S_2O_8^{2-}$ varies with reactant concentrations in the following manner:

Experiment	$[S_2O_8^{2-}]$ (*M*)	$[I^-]$ (*M*)	Initial Rate (*M*/s)
1	0.018	0.036	2.6×10^{-6}
2	0.027	0.036	3.9×10^{-6}
3	0.036	0.054	7.8×10^{-6}
4	0.050	0.072	1.4×10^{-5}

(a) Determine the rate law for the reaction. **(b)** What is the average value of the rate constant for the disappearance of $S_2O_8^{2-}$ based on the four sets of data? **(c)** How is the rate of disappearance of $S_2O_8^{2-}$ related to the rate of disappearance of I^-? **(d)** What is the rate of disappearance of I^- when $[S_2O_8^{2-}] = 0.015\ M$ and $[I^-] = 0.040\ M$?

Change of Concentration with Time

14.27 **(a)** Define the following symbols that are encountered in rate equations: $[A]_0$, $t_{1/2}$, $[A]_t$, k **(b)** What quantity, when graphed versus time, will yield a straight line for a first-order reaction?

14.28 **(a)** For a second-order reaction, what quantity, when graphed versus time, will yield a straight line? **(b)** How do the half-lives of first-order and second-order reactions differ?

14.29 **(a)** The gas-phase decomposition of SO_2Cl_2, $SO_2Cl_2(g) \longrightarrow SO_2(g) + Cl_2(g)$, is first order in SO_2Cl_2. At 600 K the half-life for this process is 2.3×10^5 s. What is the rate constant at this temperature? **(b)** At 320°C the rate constant is $2.2 \times 10^{-5}\ s^{-1}$. What is the half-life at this temperature?

14.30 **(a)** The decomposition of H_2O_2, $H_2O_2(aq) \longrightarrow H_2O(l) + \frac{1}{2}O_2(g)$, is a first-order reaction. At a particular temperature near room temperature, the rate constant equals $7.0 \times 10^{-4}\ s^{-1}$. Calculate the half-life at this temperature. **(b)** At 415°C, $(CH_2)_2O$ decomposes in the gas phase, $(CH_2)_2O(g) \longrightarrow CH_4(g) + CO(g)$. If the reaction is first order with a half-life of 56.3 min at this temperature, calculate the rate constant in s^{-1}.

14.31 As described in Exercise 14.29, the decomposition of sulfuryl chloride (SO_2Cl_2) is a first-order process. The rate constant for the decomposition at 660 K is $4.5 \times 10^{-2}\ s^{-1}$. **(a)** If we begin with an initial SO_2Cl_2 pressure of 375 torr, what is the pressure of this substance after 65 s? **(b)** At what time will the pressure of SO_2Cl_2 decline to one-tenth its initial value?

14.32 The first-order rate constant for the decomposition of N_2O_5, $N_2O_5(g) \longrightarrow 2NO_2(g) + O_2(g)$, at 70°C is $6.82 \times 10^{-3}\ s^{-1}$. Suppose we start with 0.0250 mol of $N_2O_5(g)$ in a volume of 2.0 L. **(a)** How many moles of N_2O_5 will remain after 2.5 min? **(b)** How many minutes will it take for the quantity of N_2O_5 to drop to 0.010 mol? **(c)** What is the half-life of N_2O_5 at 70°C?

14.33 The reaction

$$SO_2Cl_2(g) \longrightarrow SO_2(g) + Cl_2(g)$$

is first order in SO_2Cl_2. Using the following kinetic data, determine the magnitude of the first-order rate constant:

Time (s)	Pressure SO_2Cl_2 (atm)
0	1.000
2,500	0.947
5,000	0.895
7,500	0.848
10,000	0.803

14.34 From the following data for the first-order gas-phase isomerization of CH_3NC at 215°C, calculate the first-order rate constant and half-life for the reaction:

Time (s)	Pressure CH_3NC (torr)
0	502
2,000	335
5,000	180
8,000	95.5
12,000	41.7
15,000	22.4

14.35 Consider the data presented in Exercise 14.3. **(a)** By using appropriate graphs, determine whether the reaction is first order or second order. **(b)** What is the value of the rate constant for the reaction? **(c)** What is the half-life for the reaction?

14.36 Consider the data presented in Exercise 14.4. **(a)** Determine whether the reaction is first order or second order. **(b)** What is the value of the rate constant? **(c)** What is the half-life?

14.37 The gas-phase decomposition of NO_2, $NO_2(g) \longrightarrow NO(g) + O_2(g)$, is studied at 383°C, giving the following data:

Time (s)	$[NO_2]$ (*M*)
0.0	0.100
5.0	0.017
10.0	0.0090
15.0	0.0062
20.0	0.0047

(a) Is the reaction first order or second order with respect to the concentration of NO_2? **(b)** What is the value of the rate constant?

14.38 Sucrose ($C_{12}H_{22}O_{11}$), which is commonly known as table sugar, reacts in dilute acid solutions to form two simpler sugars, glucose and fructose, both of which have the formula $C_6H_{12}O_6$:

$$C_{12}H_{22}O_{11}(aq) + H_2O(l) \longrightarrow 2C_6H_{12}O_6(aq)$$

At 23°C and in 0.5 *M* HCl, the following data were obtained for the disappearance of sucrose:

Time (min)	$[C_{12}H_{22}O_{11}]$ (*M*)
0	0.316
39	0.274
80	0.238
140	0.190
210	0.146

(a) Is the reaction first order or second order with respect to $[C_{12}H_{22}O_{11}]$? **(b)** What is the value of the rate constant?

Temperature and Rate

14.39 **(a)** What is the central idea of the collision model? **(b)** What factors determine whether a collision between two molecules will lead to a chemical reaction? **(c)** According to the collision model, why does temperature affect the value of the rate constant?

14.40 **(a)** Explain the rate of a unimolecular (that is, one-molecule) reaction, such as the isomerization of methyl isonitrile (Figure 14.6), in terms of the collision model. **(b)** In a reaction of the form, $A(g) + B(g) \longrightarrow$ products, are all collisions of A with B that are sufficiently energetic likely to lead to reaction? Explain. **(c)** How does the kinetic-molecular theory help us understand the temperature dependence of chemical reactions?

14.41 Calculate the fraction of atoms in a sample of argon gas at 400 K that have an energy of 10.0 kJ or greater.

14.42 **(a)** The activation energy for the isomerization of methyl isonitrile (Figure 14.6) is 160 kJ/mol. Calculate the fraction of methyl isonitrile molecules that have an energy of 160.0 kJ or greater at 500 K. **(b)** Calculate this fraction for a temperature of 510 K. What is the ratio of the fraction at 510 K to that at 500 K?

14.43 For the elementary process, $N_2O_5(g) \longrightarrow NO_2(g) + NO_3(g)$, the activation energy (E_a) and overall ΔE are 154 kJ/mol and 136 kJ/mol, respectively. **(a)** Sketch the energy profile for this reaction, and label E_a and ΔE. **(b)** What is the activation energy for the reverse reaction?

14.44 The gas-phase reaction, $Cl(g) + HBr(g) \longrightarrow HCl(g) + Br(g)$, has an overall enthalpy change of −66 kJ. The activation energy for the reaction is 7 kJ. **(a)** Sketch the energy profile for the reaction, and label E_a and ΔE. **(b)** What is the activation energy for the reverse reaction?

14.45 Based on their activation energies and energy changes and assuming that all collision factors are the same, which of the following reactions would be fastest and which would be slowest? **(a)** $E_a = 45$ kJ/mol; $\Delta E = -25$ kJ/mol; **(b)** $E_a = 35$ kJ/mol; $\Delta E = -10$ kJ/mol; **(c)** $E_a = 55$ kJ/mol; $\Delta E = 10$ kJ/mol.

14.46 Which of the reactions in Exercise 14.45 will be fastest in the reverse direction? Which will be slowest?

14.47 A certain first-order reaction has a rate constant of $2.75 \times 10^{-2}\ s^{-1}$ at 20°C. What is the value of k at 60°C if **(a)** $E_a = 75.5$ kJ/mol; **(b)** $E_a = 105$ kJ/mol?

14.48 Two first-order reactions have the same rate constant at 30°C. Reaction A has an activation energy of 45.5 kJ/mol; reaction B has an activation energy of 25.2 kJ/mol. Calculate the ratio of rate constants, k_A/k_B, at 60°C.

14.49 The rate of the reaction

$$CH_3COOC_2H_5(aq) + OH^-(aq) \longrightarrow CH_3COO^-(aq) + C_2H_5OH(aq)$$

was measured at several temperatures, and the following data were collected:

Temperature (°C)	$k\ (M^{-1}\ s^{-1})$
15	0.0521
25	0.101
35	0.184
45	0.332

Using these data, graph ln k versus $1/T$. Using your graph, determine the value of E_a.

14.50 The temperature dependence of the rate constant for the reaction

$$CO(g) + NO_2(g) \longrightarrow CO_2(g) + NO(g)$$

is tabulated as follows:

Temperature (K)	$k\ (M^{-1}\ s^{-1})$
600	0.028
650	0.22
700	1.3
750	6.0
800	23

Calculate E_a and A.

[14.51] The activation energy of a certain reaction is 65.7 kJ/mol. How many times faster will the reaction occur at 50°C than at 0°C?

[14.52] The following is a quote from an article in the August 18, 1998, issue of *The New York Times* about the breakdown of cellulose and starch: "A drop of 18 degrees Fahrenheit [from 77°F to 59°F] lowers the reaction rate six times; a 36-degree drop [from 77°F to 41°F] produces a fortyfold decrease in the rate." **(a)** Calculate activation energies for the breakdown process based on the two estimates of the effect of temperature on rate. Are the values consistent? **(b)** Assuming the value of E_a calculated from the 36-degree drop, and assuming that the rate of breakdown is first order with a half-life at 25°C of 2.7 years, calculate the half-life for breakdown at a temperature of −15°C.

Reaction Mechanisms

14.53 **(a)** What is meant by the term *elementary step?* **(b)** What is the difference between a *unimolecular* and a *bimolecular* elementary step? **(c)** What is a *reaction mechanism*?

14.54 **(a)** What is meant by the term *molecularity?* **(b)** Why are termolecular elementary steps so rare? **(c)** What is an *intermediate* in a mechanism?

14.55 What is the molecularity of each of the following elementary processes? Write the rate law for each.
(a) $Cl_2(g) \longrightarrow 2Cl(g)$
(b) $OCl^-(g) + H_2O(g) \longrightarrow HOCl(g) + OH^-(g)$
(c) $NO(g) + Cl_2(g) \longrightarrow NOCl_2(g)$

14.56 What is the molecularity of each of the following elementary processes? Write the rate law for each.
(a) $2NO(g) \longrightarrow N_2O_2(g)$
(b) $H_2C{-}CH_2(g)$ (ring closed by CH_2) $\longrightarrow CH_2{=}CH{-}CH_3(g)$
(c) $SO_3(g) \longrightarrow SO_2(g) + O(g)$

14.57 Based on the following reaction profile, how many intermediates are formed in the reaction A $\longrightarrow$ C. How many transition states are there? Which step is the fastest? Is the reaction A $\longrightarrow$ C exothermic or endothermic?

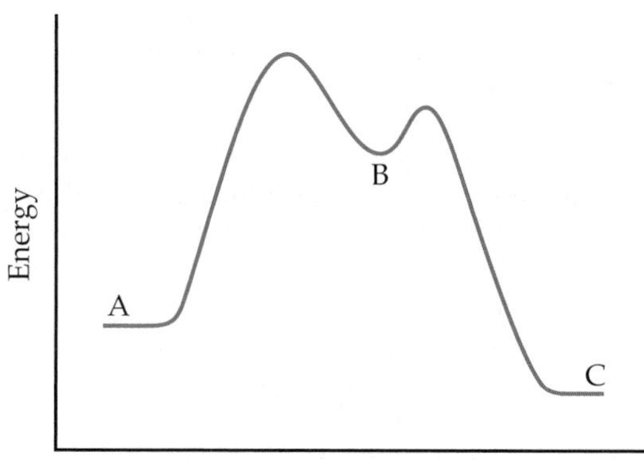

14.58 Based on the following reaction profile, how many intermediates are formed in the reaction A $\longrightarrow$ D? How many transition states are there? Which step is the fastest? Is the reaction A $\longrightarrow$ D exothermic or endothermic?

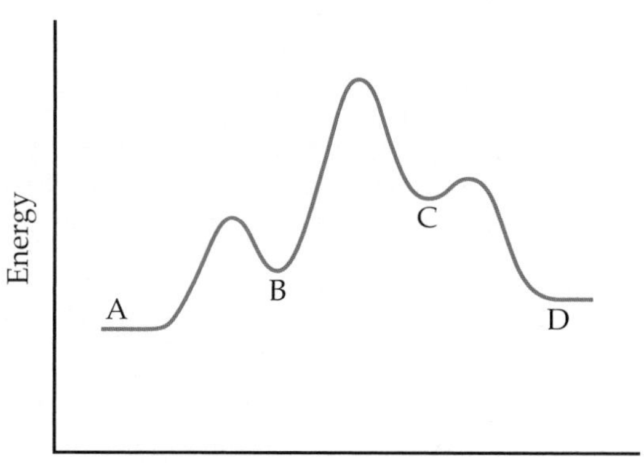

14.59 The following mechanism has been proposed for the gas-phase reaction of H_2 with ICl:

$$H_2(g) + ICl(g) \longrightarrow HI(g) + HCl(g)$$
$$HI(g) + ICl(g) \longrightarrow I_2(g) + HCl(g)$$

(a) Write the balanced equation for the overall reaction. **(b)** Identify any intermediates in the mechanism. **(c)** Write rate laws for each elementary step in the mechanism. **(d)** If the first step is slow and the second one is fast, what rate law do you expect to be observed for the overall reaction?

14.60 The following mechanism has been proposed for the reaction of NO with H_2 to form N_2O and H_2O:

$$NO(g) + NO(g) \longrightarrow N_2O_2(g)$$
$$N_2O_2(g) + H_2(g) \longrightarrow N_2O(g) + H_2O(g)$$

(a) Show that the elementary steps of the proposed mechanism add to provide a balanced equation for the reaction. **(b)** Write a rate law for each elementary step in the mechanism. **(c)** Identify any intermediates in the mechanism. **(d)** The observed rate law is rate $= k[NO]^2[H_2]$. If the proposed mechanism is correct, what can we conclude about the relative speeds of the first and second steps?

14.61 The reaction $2NO(g) + Cl_2(g) \longrightarrow 2NOCl(g)$ obeys the rate law, rate $= k[NO]^2[Cl_2]$. The following mechanism has been proposed for this reaction:

$$NO(g) + Cl_2(g) \longrightarrow NOCl_2(g)$$
$$NOCl_2(g) + NO(g) \longrightarrow 2NOCl(g)$$

(a) What would the rate law be if the first step were rate determining? **(b)** Based on the observed rate law, what can we conclude about the relative rates of the two steps?

14.62 You have studied the gas-phase oxidation of HBr by O_2:

$$4HBr(g) + O_2(g) \longrightarrow 2H_2O(g) + 2Br_2(g)$$

You find the reaction to be first order with respect to HBr and first order with respect to O_2. You propose the following mechanism:

$$HBr(g) + O_2(g) \longrightarrow HOOBr(g)$$
$$HOOBr(g) + HBr(g) \longrightarrow 2HOBr(g)$$
$$HOBr(g) + HBr(g) \longrightarrow H_2O(g) + Br_2(g)$$

(a) Indicate how the elementary steps add to give the overall reaction. (*Hint*: You will need to multiply the coefficients of one of the equations by 2.) **(b)** Based on the rate law, which step is rate-determining? **(c)** What are the intermediates in this mechanism? **(d)** If you are unable to detect HOBr or HOOBr among the products, does this disprove your mechanism?

Catalysis

14.63 **(a)** What part of the energy profile of a reaction is affected by a catalyst? **(b)** What is the difference between a homogeneous and a heterogeneous catalyst?

14.64 **(a)** Most heterogeneous catalysts of importance are extremely finely divided solid materials. Why is particle size important? **(b)** What role does adsorption play in the action of a heterogeneous catalyst?

14.65 The oxidation of SO_2 to SO_3 is catalyzed by NO_2. The reaction proceeds as follows:

$$NO_2(g) + SO_2(g) \longrightarrow NO(g) + SO_3(g)$$
$$2NO(g) + O_2(g) \longrightarrow 2NO_2(g)$$

(a) Show that the two reactions can be summed to give the overall oxidation of SO_2 by O_2 to give SO_3. **(b)** Why do we consider NO_2 a catalyst and not an intermediate in this reaction? **(c)** Is this an example of homogeneous catalysis or heterogeneous catalysis?

14.66 NO catalyzes the decomposition of N_2O, possibly by the following mechanism:

$$NO(g) + N_2O(g) \longrightarrow N_2(g) + NO_2(g)$$
$$2NO_2(g) \longrightarrow 2NO(g) + O_2(g)$$

(a) What is the chemical equation for the overall reaction? Show how the two steps can be added to give the overall equation. **(b)** Why is NO considered a catalyst and not an intermediate? **(c)** If experiments show that during the decomposition of N_2O, NO_2 does not accumulate in measurable quantities, does this rule out the proposed mechanism? If you think not, suggest what might be going on.

14.67 Many metallic catalysts, particularly the precious-metal ones, are often deposited as very thin films on a substance of high surface area per unit mass, such as alumina (Al_2O_3) or silica (SiO_2). Why is this an effective way of utilizing the catalyst material?

14.68 **(a)** If you were going to build a system to check the effectiveness of automobile catalytic converters on cars, what substances would you want to look for in the car exhaust? **(b)** Automobile catalytic converters have to work at high temperatures, as hot exhaust gases stream through them. In what ways could this be an advantage? In what ways a disadvantage? **(c)** Why is the rate of flow of exhaust gases over a catalytic converter important?

14.69 When D_2 reacts with ethylene (C_2H_4) in the presence of a finely divided catalyst, ethane with two deuteriums, $CH_2D{-}CH_2D$, is formed. (Deuterium, D, is an isotope of hydrogen of mass 2.) Very little ethane forms in which two deuteriums are bound to one carbon (for example, $CH_3{-}CHD_2$). Use the sequence of steps involved in the reaction to explain why this is so.

14.70 Heterogeneous catalysts that perform hydrogenation reactions, as illustrated in Figure 14.21, are subject to poisoning, which shuts down their catalytic ability. Compounds of sulfur are often poisons. Suggest a mechanism by which such compounds might act as poisons.

14.71 **(a)** Explain the importance of enzymes in biological systems. **(b)** What chemical transformations are catalyzed (*i*) by the enzyme catalase; (*ii*) by nitrogenase?

14.72 There are literally thousands of enzymes at work in complex living systems such as human beings. What properties of the enzymes give rise to their ability to distinguish one substrate from another?

[14.73] The activation energy of an uncatalyzed reaction is 95 kJ/mol. The addition of a catalyst lowers the activation energy to 55 kJ/mol. Assuming that the collision factor remains the same, by what factor will the catalyst increase the rate of the reaction at **(a)** 25°C; **(b)** 125°C?

[14.74] Suppose that a certain biologically important reaction is quite slow at physiological temperature (37°C) in the absence of a catalyst. Assuming that the collision factor remains the same, by how much must an enzyme lower the activation energy of the reaction in order to achieve a 1×10^5-fold increase in the reaction rate?

Additional Exercises

14.75 Hydrogen sulfide (H_2S) is a common and troublesome pollutant in industrial wastewaters. One way to remove H_2S is to treat the water with chlorine, in which case the following reaction occurs:

$$H_2S(aq) + Cl_2(aq) \longrightarrow S(s) + 2H^+(aq) + 2Cl^-(aq)$$

The rate of this reaction is first order in each reactant. The rate constant for the disappearance of H_2S at 28°C is $3.5 \times 10^{-2}\ M^{-1}s^{-1}$. If at a given time the concentration of H_2S is $1.6 \times 10^{-4}\ M$ and that of Cl_2 is 0.070 M, what is the rate of formation of Cl^-?

14.76 The reaction $2NO(g) + O_2(g) \longrightarrow 2NO_2(g)$ is second order in NO and first order in O_2. When [NO] = 0.040 M and $[O_2]$ = 0.035 M, the observed rate of disappearance of NO is $9.3 \times 10^{-5}\ M/s$. **(a)** What is the rate of disappearance of O_2 at this moment? **(b)** What is the value of the rate constant? **(c)** What are the units of the rate constant? **(d)** What would happen to the rate if the concentration of NO were increased by a factor of 1.8?

14.77 For the reaction of iodide ion with hypochorite ion, $I^-(aq) + OCl^-(aq) \longrightarrow OI^-(aq) + Cl^-(aq)$, the reaction is found to be first order each in iodide and hypochlorite ions, and inversely proportional to the concentration of hydroxide ion present in the solution. **(a)** Write the rate law for the reaction. **(b)** By what factor will the rate change if the concentration of iodide ion is tripled? **(c)** By what factor will the rate change if the hydroxide ion concentration is doubled?

14.78 Consider the following reaction between mercury(II) chloride and oxalate ion:

$$2HgCl_2(aq) + C_2O_4^{2-}(aq) \longrightarrow 2Cl^-(aq) + 2CO_2(g) + Hg_2Cl_2(s)$$

The initial rate of this reaction was determined for several concentrations of $HgCl_2$ and $C_2O_4^{2-}$, and the following rate data were obtained for the rate of disappearance of $C_2O_4^{2-}$:

Experiment	$[HgCl_2]$ (M)	$[C_2O_4^{2-}]$ (M)	Rate (M/s)
1	0.164	0.15	3.2×10^{-5}
2	0.164	0.45	2.9×10^{-4}
3	0.082	0.45	1.4×10^{-4}
4	0.246	0.15	4.8×10^{-5}

(a) What is the rate law for this reaction? **(b)** What is the value of the rate constant? **(c)** What is the reaction rate when the concentration of $HgCl_2$ is 0.12 M and that of $C_2O_4^{2-}$ is 0.10 M if the temperature is the same as that used to obtain the data shown?

14.79 The reaction $2NO_2 \longrightarrow 2NO + O_2$ has the rate constant $k = 0.63\ M^{-1}s^{-1}$. Based on the units for k, is the reaction first or second order in NO_2? If the initial concentration of NO_2 is 0.100 M, how would you determine how long it would take for the concentration to decrease to 0.025 M?

14.80 Urea (NH_2CONH_2) is the end product in protein metabolism in animals. The decomposition of urea in 0.1 M HCl occurs according to the reaction

$$NH_2CONH_2(aq) + H^+(aq) + 2H_2O(l) \longrightarrow 2NH_4^+(aq) + HCO_3^-(aq)$$

The reaction is first order in urea and first order overall. When $[NH_2CONH_2]$ = 0.200 M, the rate at 61.05°C is $8.56 \times 10^{-5}\ M/s$. **(a)** What is the value for the rate constant, k? **(b)** What is the concentration of urea in this solution after 5.00×10^3 s if the starting concentration is 0.500 M? **(c)** What is the half-life for this reaction at 61.05°C?

[14.81] The rate of a first-order reaction is followed by spectroscopy, monitoring the absorption of a colored reactant. The reaction occurs in a 1.00-cm sample cell and the only colored species in the reaction has a molar absorptivity constant of $5.60 \times 10^3\ cm^{-1}M^{-1}$. **(a)** Calculate the initial concentration of the colored reactant if the absorbance is 0.605 at the beginning of the reaction. **(b)** The absorbance falls to 0.250 within 30.0 min. Calculate the rate constant in units of s^{-1}. **(c)** Calculate the half-life of the reaction. **(d)** How long does it take for the absorbance to fall to 0.100?

14.82 Cyclopentadiene (C_5H_6) reacts with itself to form dicyclopentadiene ($C_{10}H_{12}$). A 0.0400 *M* solution of C_5H_6 was monitored as a function of time as the reaction $2C_5H_6 \longrightarrow C_{10}H_{12}$ proceeded. The following data were collected:

Time (s)	$[C_5H_6]$ (*M*)
0.0	0.0400
50.0	0.0300
100.0	0.0240
150.0	0.0200
200.0	0.0174

Plot $[C_5H_6]$ versus time, $\ln[C_5H_6]$ versus time, and $1/[C_5H_6]$ versus time. What is the order of the reaction? What is the value of the rate constant?

14.83 **(a)** Two reactions have identical values for E_a. Does this ensure that they will have the same rate constant if run at the same temperature? Explain. **(b)** Two similar reactions have the same rate constant at 25°C, but at 35°C one of the reactions has a higher rate constant than the other. Account for these observations.

14.84 The first-order rate constant for reaction of a particular organic compound with water varies with temperature as follows:

Temperature (K)	Rate Constant (s^{-1})
300	3.2×10^{-11}
320	1.0×10^{-9}
340	3.0×10^{-8}
355	2.4×10^{-7}

From these data calculate the activation energy in units of kJ/mol.

14.85 The decomposition of hydrogen peroxide is catalyzed by iodide ion. The catalyzed reaction is thought to proceed by a two-step mechanism:

$$H_2O_2(aq) + I^-(aq) \longrightarrow H_2O(l) + IO^-(aq) \quad \text{(slow)}$$

$$IO^-(aq) + H_2O_2(aq) \longrightarrow H_2O(l) + O_2(g) + I^-(aq) \quad \text{(fast)}$$

(a) Assuming that the first step of the mechanism is rate-determining, predict the rate law for the overall process. **(b)** Write the chemical equation for the overall process. **(c)** Identify the intermediate, if any, in the mechanism.

14.86 Using Figure 14.20 as your basis, draw the energy profile for the bromide ion-catalyzed decomposition of hydrogen peroxide. **(a)** Label the curve with the activation energies for reactions [14.30] and [14.31]. **(b)** Notice from Figure 14.19(b) that when $Br^-(aq)$ is added initially, Br_2 accumulates to some extent during the reaction. What does this tell us about the relative rates of reactions [14.30] and [14.31]?

[14.87] The following mechanism has been proposed for the gas-phase reaction of chloroform ($CHCl_3$) and chlorine:

Step 1: $$Cl_2(g) \underset{k_{-1}}{\overset{k_1}{\rightleftharpoons}} 2Cl(g) \quad \text{(fast)}$$

Step 2: $$Cl(g) + CHCl_3(g) \xrightarrow{k_2} HCl(g) + CCl_3(g) \quad \text{(slow)}$$

Step 3: $$Cl(g) + CCl_3(g) \xrightarrow{k_3} CCl_4(g) \quad \text{(fast)}$$

(a) What is the overall reaction? **(b)** What are the intermediates in the mechanism? **(c)** What is the molecularity of each of the elementary steps? **(d)** What is the rate-determining step? **(e)** What is the rate law predicted by this mechanism? (*Hint*: The overall reaction order is not an integer.)

[14.88] In a hydrocarbon solution, the gold compound $(CH_3)_3AuPH_3$ decomposes into ethane (C_2H_6) and a different gold compound, $(CH_3)AuPH_3$. The following mechanism has been proposed for the decomposition of $(CH_3)_3AuPH_3$:

Step 1: $$(CH_3)_3AuPH_3 \underset{k_{-1}}{\overset{k_1}{\rightleftharpoons}} (CH_3)_3Au + PH_3 \quad \text{(fast)}$$

Step 2: $$(CH_3)_3Au \xrightarrow{k_2} C_2H_6 + (CH_3)Au \quad \text{(slow)}$$

Step 3: $$(CH_3)Au + PH_3 \xrightarrow{k_3} (CH_3)AuPH_3 \quad \text{(fast)}$$

(a) What is the overall reaction? **(b)** What are the intermediates in the mechanism? **(c)** What is the molecularity of each of the elementary steps? **(d)** What is the rate-determining step? **(e)** What is the rate law predicted by this mechanism? **(f)** What would be the effect on the reaction rate of adding PH_3 to the solution of $(CH_3)_3AuPH_3$?

14.89 One of the many remarkable enzymes in the human body is carbonic anhydrase, which catalyzes the interconversion of carbonic acid with carbon dioxide and water. If it were not for this enzyme, the body could not rid itself rapidly enough of the CO_2 accumulated by cell metabolism. The enzyme catalyzes the dehydration (release to air) of up to 10^7 CO_2 molecules per second. Which components of this description correspond to the terms *enzyme*, *substrate*, and *turnover number*?

14.90 The enzyme *invertase* catalyzes the conversion of sucrose, a disaccharide, to invert sugar, a mixture of glucose and fructose. When the concentration of invertase is 4.2×10^{-7} *M* and the concentration of sucrose is 0.0077 *M*, invert sugar is formed at the rate of 1.5×10^{-4} *M*/s. When the sucrose concentration is doubled, the rate of formation of invert sugar is doubled also. **(a)** Assuming that the enzyme-substrate model is operative, is the fraction of enzyme tied up as a complex large or small? Explain. **(b)** Addition of inositol, another sugar, decreases the rate of formation of invert sugar. Suggest a mechanism by which this occurs.

Integrative Exercises

14.91 Dinitrogen pentoxide (N_2O_5) decomposes in chloroform as a solvent to yield NO_2 and O_2. The decomposition is first order with a rate constant at 45°C of $1.0 \times 10^{-5}\ s^{-1}$. Calculate the partial pressure of O_2 produced from 1.00 L of 0.600 *M* N_2O_5 solution at 45°C over a period of 20.0 hr if the gas is collected in a 10.0-L container. (Assume that the products do not dissolve in chloroform.)

[14.92] The reaction between ethyl iodide and hydroxide ion in ethanol (C_2H_5OH) solution, $C_2H_5I(alc) + OH^-(alc) \longrightarrow C_2H_5OH(l) + I^-(alc)$, has an activation energy of 86.8 kJ/mol and a frequency factor of $2.10 \times 10^{11}\ M^{-1}s^{-1}$. **(a)** Predict the rate constant for the reaction at 35°C. **(b)** A solution of KOH in ethanol is made up by dissolving 0.335 g KOH in ethanol to form 250.0 mL of solution. Similarly, 1.453 g of C_2H_5I is dissolved in ethanol to form 250.0 mL of solution. Equal volumes of the two solutions are mixed. Assuming the reaction is first order in each reactant, what is the initial rate at 35°C? **(c)** Which reagent in the reaction is limiting, assuming the reaction proceeds to completion?

14.93 Zinc metal dissolves in hydrochloric acid according to the reaction

$$Zn(s) + 2HCl(aq) \longrightarrow ZnCl_2(aq) + H_2(g)$$

Suppose you are asked to study the kinetics of this reaction by monitoring the rate of production of $H_2(g)$. **(a)** By using a reaction flask, a manometer, and any other common laboratory equipment, design an experimental apparatus that would allow you to monitor the partial pressure of $H_2(g)$ produced as a function of time. **(b)** Explain how you would use the apparatus to determine the rate law of the reaction. **(c)** Explain how you would use the apparatus to determine the reaction order for $[H^+]$ for the reaction. **(d)** How could you use the apparatus to determine the activation energy of the reaction? **(e)** Explain how you would use the apparatus to determine the effects of changing the form of Zn(*s*) from metal strips to granules.

14.94 The gas-phase reaction of NO with F_2 to form NOF and F has an activation energy of $E_a = 6.3$ kJ/mol and a frequency factor of $A = 6.0 \times 10^8\ M^{-1}s^{-1}$. The reaction is believed to be bimolecular:

$$NO(g) + F_2(g) \longrightarrow NOF(g) + F(g)$$

(a) Calculate the rate constant at 100°C. **(b)** Draw the Lewis structures for the NO and the NOF molecules, given that the chemical formula for NOF is misleading because the nitrogen atom is actually the central atom in the molecule. **(c)** Predict the structure for the NOF molecule. **(d)** Draw a possible transition state for the formation of NOF, using dashed lines to indicate the weak bonds that are beginning to form. **(e)** Suggest a reason for the low activation energy for the reaction.

14.95 The mechanism for the oxidation of HBr by O_2 to form $2H_2O$ and Br_2 is shown in Exercise 14.62. **(a)** Calculate the overall standard enthalpy change for the reaction process. **(b)** HBr does not react with O_2 at a measurable rate at room temperature under ordinary conditions. What can you infer from this about the magnitude of the activation energy for the rate-determining step? **(c)** Draw a plausible Lewis structure for the intermediate HOOBr. To what familiar compound of hydrogen and oxygen does it appear similar?

14.96 Enzymes, the catalysts of biological systems, are high molecular weight protein materials. The active site of the enzyme is formed by the three-dimensional arrangement of the protein in solution. When heated in solution, proteins undergo *denaturation*, a process in which the three-dimensional structure of the protein unravels or at least partly does so. The accompanying graph shows the variation with temperature of the activity of a typical enzyme. The activity increases with temperature to a point above the usual operating region for the enzyme, then declines rapidly with further temperature increases. What role does denaturation play in determining the shape of this curve? How does your explanation fit in with the lock-and-key model of enzyme action?

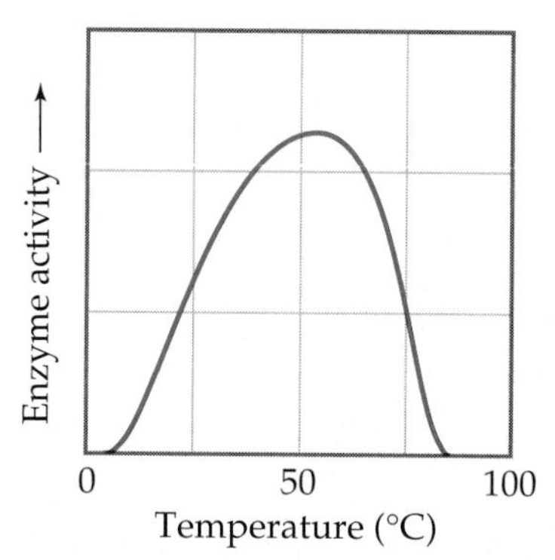

[14.97] Metals often form several cations with different charges. Cerium, for example, forms Ce^{3+} and Ce^{4+} ions, and thallium forms Tl^+ and Tl^{3+} ions. Cerium and thallium ions react as follows:

$$2Ce^{4+}(aq) + Tl^+(aq) \longrightarrow 2Ce^{3+}(aq) + Tl^{3+}(aq)$$

This reaction is very slow and is thought to occur in a single elementary step. The reaction is catalyzed by the addition of $Mn^{2+}(aq)$, according to the following mechanism:

$$Ce^{4+}(aq) + Mn^{2+}(aq) \longrightarrow Ce^{3+}(aq) + Mn^{3+}(aq)$$
$$Ce^{4+}(aq) + Mn^{3+}(aq) \longrightarrow Ce^{3+}(aq) + Mn^{4+}(aq)$$
$$Mn^{4+}(aq) + Tl^+(aq) \longrightarrow Mn^{2+}(aq) + Tl^{3+}(aq)$$

(a) Write the rate law for the uncatalyzed reaction. **(b)** What is unusual about the uncatalyzed reaction? Why might it be a slow reaction? **(c)** The rate for the catalyzed reaction is first order in $[Ce^{4+}]$ and first order in $[Mn^{2+}]$. Based on this rate law, which of the steps in the catalyzed mechanism is rate-determining? **(d)** Use the available oxidation states of Mn to comment on its special suitability to catalyze this reaction.

[14.98] The rates of many atmospheric reactions are accelerated by the absorption of light by one of the reactants. For example, consider the reaction between methane and chlorine to produce methyl chloride and hydrogen chloride:

Reaction 1: $CH_4(g) + Cl_2(g) \longrightarrow CH_3Cl(g) + HCl(g)$

This reaction is very slow in the absence of light. However, $Cl_2(g)$ can absorb light to form Cl atoms:

Reaction 2: $Cl_2(g) + h\nu \longrightarrow 2Cl(g)$

Once the Cl atoms are generated, they can catalyze the reaction of CH_4 and Cl_2, according to the following proposed mechanism:

Reaction 3: $CH_4(g) + Cl(g) \longrightarrow CH_3(g) + HCl(g)$

Reaction 4: $CH_3(g) + Cl_2(g) \longrightarrow CH_3Cl(g) + Cl(g)$

The enthalpy changes and activation energies for these two reactions are tabulated as follows:

Reaction	ΔH°_{rxn} (kJ/mol)	E_a (kJ/mol)
3	+4	17
4	−109	4

(a) By using the bond enthalpy for Cl_2 (Table 8.4), determine the longest wavelength of light that is energetic enough to cause reaction 2 to occur. In which portion of the electromagnetic spectrum is this light found? **(b)** By using the data tabulated here, sketch a quantitative energy profile for the catalyzed reaction represented by reactions 3 and 4. **(c)** By using bond enthalpies, estimate where the reactants, $CH_4(g) + Cl_2(g)$, should be placed on your diagram in part (b). Use this result to estimate the value of E_a for the reaction $CH_4(g) + Cl_2(g) \longrightarrow CH_3(g) + HCl(g) + Cl(g)$. **(d)** The species $Cl(g)$ and $CH_3(g)$ in reactions 3 and 4 are radicals, atoms, or molecules with unpaired electrons. Draw a Lewis structure of CH_3, and verify that it is a radical. **(e)** The sequence of reactions 3 and 4 comprise a radical chain mechanism. Why do you think this is called a "chain reaction"? Propose a reaction that will terminate the chain reaction.

eMedia Exercises

14.99 The **Rates of Reaction** simulation (*eChapter 14.2*) allows you to adjust activation energy, overall energy change, temperature, and starting concentration of a reactant to assess the effect of each variable on initial reaction rate. **(a)** If you want to determine the rate law for the reaction using a series of experiments, which variable(s) *must* you vary, which variable(s) should you *not* vary, and for which variable(s) does it not make any difference? **(b)** Determine the rate law for the reaction. What is the overall order of the reaction? **(c)** Determine the value of the rate constant.

14.100 **(a)** Using data from the **Rates of Reaction** simulation (*eChapter 14.2*), select a temperature (T_1) and an activation energy (E_a) and determine the value of the rate constant (k_1). **(b)** Using the E_a and (T_1) from part (a), predict the value of the rate constant (k_2) at a higher temperature, $T_2 = 2T_1$. **(c)** Again using E_a from part (a) and the temperature from either part (a) or part (b), predict the value of the rate constant (k_3) at a temperature (T_3) that is *lower* than T_1. Use the simulation to check your answers for parts (a), (b), and (c).

14.101 The **First-Order Process** movie (*eChapter 14.3*) illustrates the change in concentration over time for a process that is first order overall. **(a)** Determine the rate constant for the process shown. **(b)** Assuming that the first-order process is a chemical reaction, what would normally happen to the value of the half-life if the temperature were increased? **(c)** Many enzyme-catalyzed reactions are first order in the concentration of substrate. However, increasing the temperature does not always increase the rate. Why might this be so?

14.102 Using the numerical value of the rate constant you calculated for the reaction in the previous question and using the same "unit" concentration as in the **First-Order Process** movie (*eChapter 14.3*), generate a curve for a second-order reaction with Equation 14.16. Reproduce the first-order curve, superimpose the second-order curve onto it, and compare them. What can you say about the relative rates of disappearance of the starting material initially and after a significant period of time has passed?

14.103 In Exercise 14.69, you explained why the combination of ethylene and D_2 in the presence of a catalyst predominantly produces $CH_2D—CH_2D$. A small amount of product has more than one D bound to a carbon, as in $CH_2D—CHD_2$. After watching the **Surface Reaction-Hydrogenation** movie (*eChapter 14.6*), propose a mechanism by which product with more than one D per carbon atom could arise.

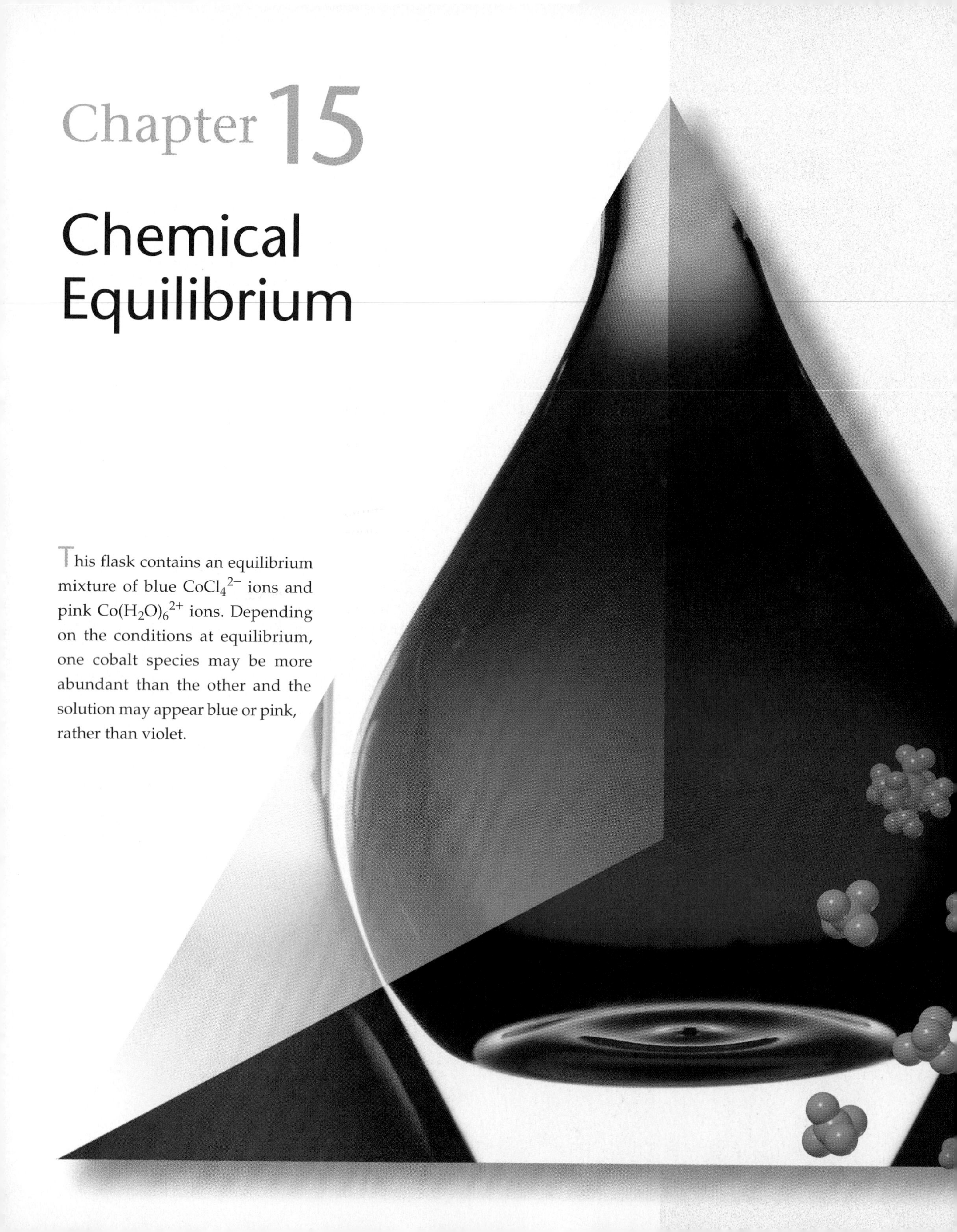

Chapter 15

Chemical Equilibrium

This flask contains an equilibrium mixture of blue $CoCl_4^{2-}$ ions and pink $Co(H_2O)_6^{2+}$ ions. Depending on the conditions at equilibrium, one cobalt species may be more abundant than the other and the solution may appear blue or pink, rather than violet.

IN THE LABORATORY portion of your chemistry course you have had the opportunity to observe a number of chemical reactions. In some cases you have been asked to calculate the amounts of products formed assuming that the reactions go to completion, meaning that the limiting reactants are used up. In fact, many reactions do not go to completion but rather approach an equilibrium state in which both reactants and products are present. Thus, after a certain amount of time, these reactions appear to "stop"—colors stop changing, gases stop evolving, and so forth—before the reaction is complete, leading to a mixture of reactants and products.

As an example, consider N_2O_4 and NO_2 (Figure 15.1 ▶), which readily interconvert. When pure frozen N_2O_4 is warmed above its boiling point (21.2°C), the gas in the sealed tube turns progressively darker as colorless N_2O_4 gas dissociates into brown NO_2 gas (Figure 15.2 ▶):

$$\underset{\text{Colorless}}{N_2O_4(g)} \longrightarrow \underset{\text{Brown}}{2NO_2(g)}$$

Eventually the color change stops even though there is still N_2O_4 in the tube. We are left with a mixture of N_2O_4 and NO_2 in which the concentrations of the gases no longer change.

The condition in which the concentrations of all reactants and products in a closed system cease to change with time is called **chemical equilibrium**. *Chemical equilibrium occurs when opposing reactions are proceeding at equal rates:* The rate at which the products are formed from the reactants equals the rate at which the reactants are formed from the products. For equilibrium to occur, neither reactants nor products can escape from the system.

We have already encountered several equilibrium processes. For example, the vapor above a liquid is in equilibrium with the liquid phase. ∞ (Section 11.5) The rate at which molecules escape from the liquid into the gas phase equals the rate at which molecules in the gas phase strike the surface and become part of the liquid. Similarly, in a saturated solution of sodium chloride the solid sodium chloride is in equilibrium with the ions dispersed in water. ∞ (Section 13.2) The rate at which ions leave the solid surface equals the rate at which other ions are removed from the liquid to become part of the solid.

▶ What's Ahead ◀

- We begin by examining the concept of equilibrium and defining the *equilibrium constant*.
- We learn how to write *equilibrium-constant expressions* for homogeneous reactions and how to interpret the magnitude of an equilibrium constant.
- We then learn how to write equilibrium-constant expressions for heterogeneous reactions.
- The value of an equilibrium constant can be calculated using equilibrium concentrations of reactants and products.
- Equilibrium constants can be used to predict the equilibrium concentrations of reactants and products and to determine the direction in which a reaction must proceed in order to achieve equilibrium.
- The chapter concludes with a discussion of *Le Châtelier's principle*, which predicts how a system at equilibrium responds to changes in concentration, volume, pressure, and temperature.

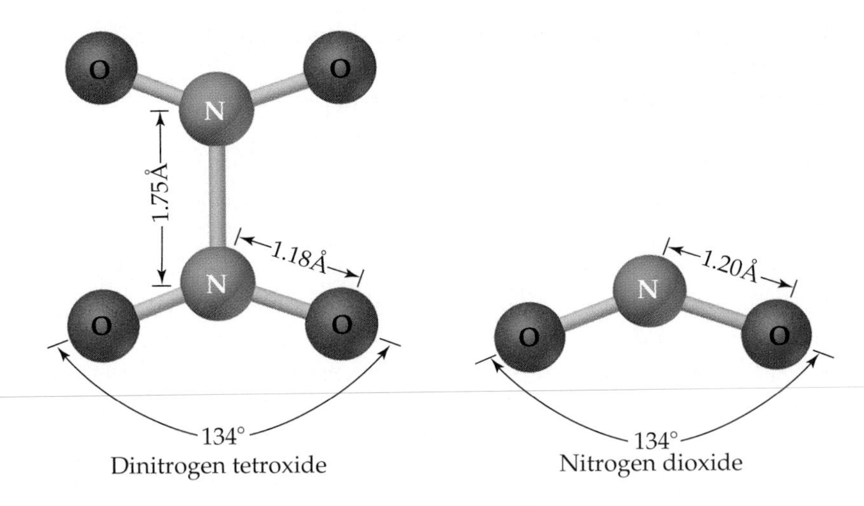

▶ **Figure 15.1** Structures of the N_2O_4 and NO_2 molecules. Both substances are gases at room temperature and atmospheric pressure. Dinitrogen tetroxide (N_2O_4) is colorless, whereas nitrogen dioxide (NO_2) is brown. The molecules readily interconvert: $N_2O_4(g) \rightleftharpoons 2NO_2(g)$.

Both of these examples involve a pair of opposing processes. At equilibrium these opposing processes are occurring at the same rate.

Chemical equilibria explain a great many natural phenomena, and they play important roles in many industrial processes. In this and the next two chapters we will explore chemical equilibria in some detail. Here we will learn how to express

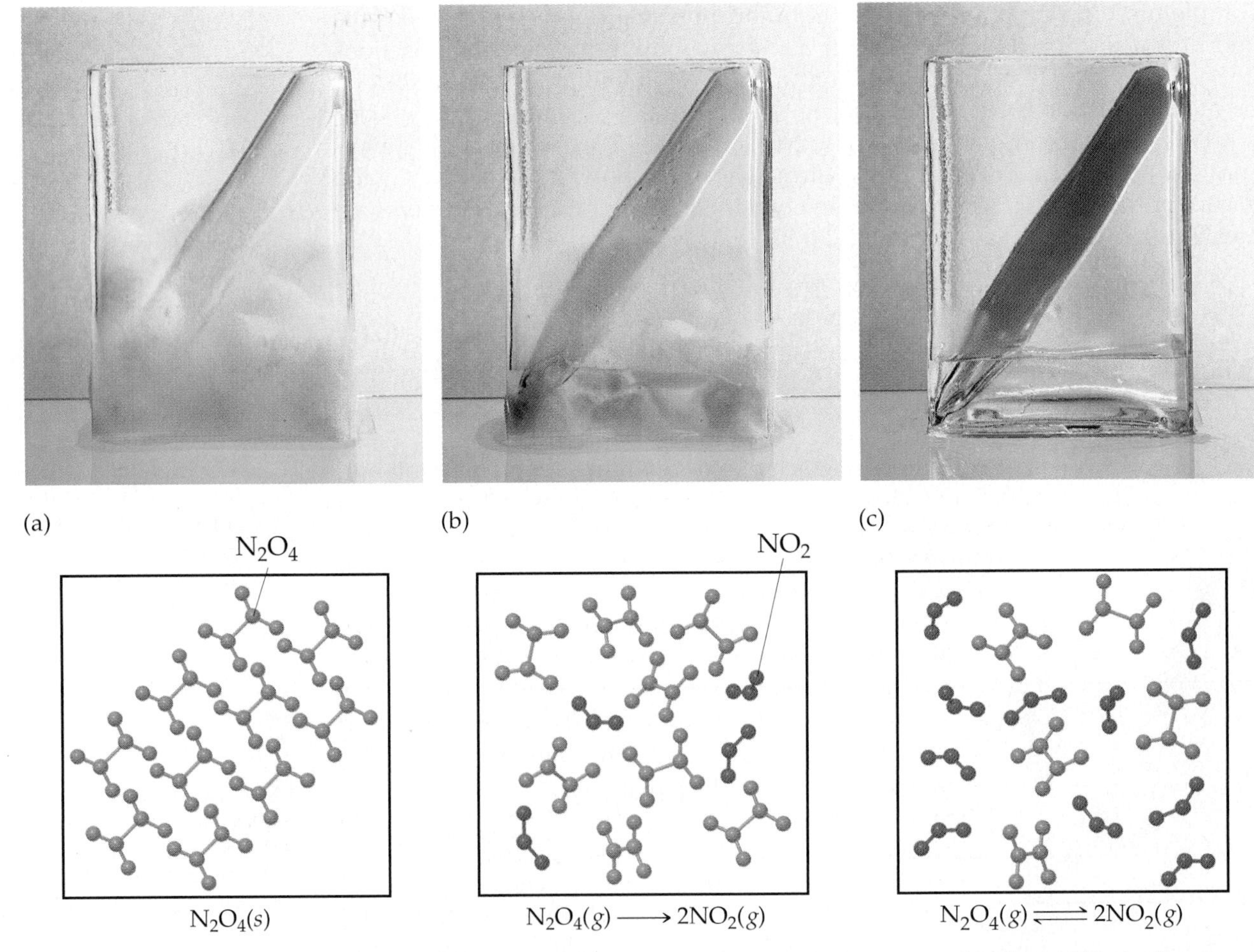

▲ **Figure 15.2** Establishing an equilibrium between N_2O_4 (grey) and NO_2 (red). (a) Frozen N_2O_4 is nearly colorless. (b) As N_2O_4 is warmed above its boiling point, it starts to dissociate into brown NO_2 gas. (c) Eventually the color stops changing as $N_2O_4(g)$ and $NO_2(g)$ reach partial pressures at which they are interconverting at the same rate. The two gases are in equilibrium.

the equilibrium position of a reaction in quantitative terms, and we will study the factors that determine the relative concentrations of reactants and products at equilibrium. We begin by exploring the relationship between the rates of opposing reactions and how this relationship leads to chemical equilibrium.

15.1 The Concept of Equilibrium

Let's begin by looking at some gas-phase reactions and then broaden our discussion to include solids, liquids, and aqueous solutions. As we consider these topics, we will find that we frequently need to express the relative concentrations of the reactants and products present in various equilibrium mixtures. For gases we will express concentrations as partial pressures (in atmospheres). (Section 10.6) For solutes in solution we will use molarities. Now, let's consider the equilibrium state.

At equilibrium the rate at which products are produced from reactants equals the rate at which reactants are produced from products. We can use some of the concepts developed in Chapter 14 to illustrate how equilibrium is reached. Let's imagine that we have a simple gas-phase reaction, $A(g) \longrightarrow B(g)$, and that both this reaction and its reverse, $B(g) \longrightarrow A(g)$, are elementary processes. As we learned in Section 14.6, the rates of these unimolecular reactions are

$$\text{Forward reaction:} \quad A \longrightarrow B \quad \text{Rate} = k_f[A] \qquad [15.1]$$

$$\text{Reverse reaction:} \quad B \longrightarrow A \quad \text{Rate} = k_r[B] \qquad [15.2]$$

where k_f and k_r are the rate constants for the forward and reverse reactions, respectively. For gaseous substances we can use the ideal-gas equation (Section 10.4) to convert between concentration (in molarity, M) and pressure (in atm):

$$PV = nRT, \quad \text{so} \quad M = (n/V) = (P/RT)$$

For substances A and B, therefore,

$$[A] = (P_A/RT) \quad \text{and} \quad [B] = (P_B/RT)$$

And the rates for the forward and reverse reactions can, thus, be expressed as

$$\text{Forward reaction:} \quad \text{Rate} = k_f\frac{P_A}{RT} \qquad [15.3]$$

$$\text{Reverse reaction:} \quad \text{Rate} = k_r\frac{P_B}{RT} \qquad [15.4]$$

Now, let's suppose that we start with pure compound A in a closed container. As A reacts to form compound B, the partial pressure of A decreases while the partial pressure of B increases [Figure 15.3(a) ▼]. As P_A decreases, the rate of the forward reaction decreases, as shown in Figure 15.3(b). Likewise, as P_B

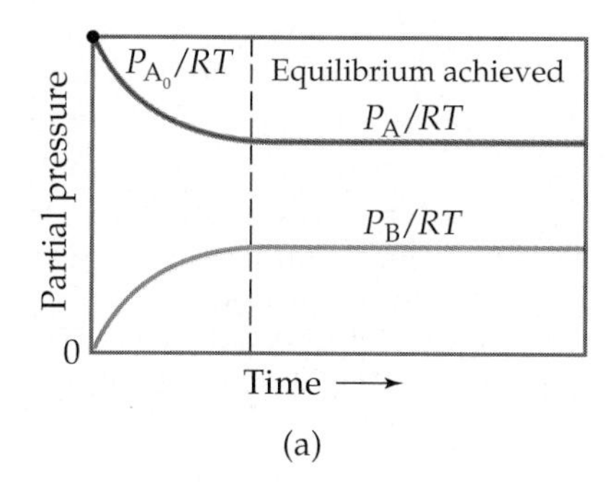

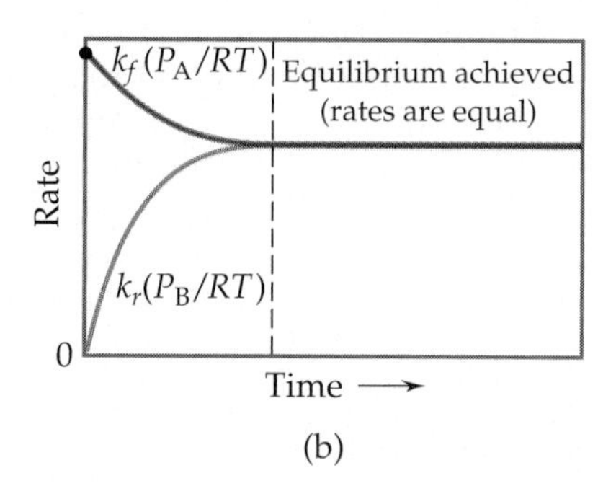

◀ **Figure 15.3** Achieving chemical equilibrium for the reaction $A \rightleftharpoons B$. (a) The reaction of pure compound A, with initial partial pressure P_{A_0}. After a time the partial pressures of A and B do not change. The reason is that (b) the rates of the forward reaction, $k_f(P_A/RT)$, and the reverse reaction, $k_r(P_B/RT)$, have become equal.

increases, the rate of the reverse reaction increases. Eventually the reaction reaches a point at which the forward and reverse rates are the same [Figure 15.3(b)]; compounds A and B are in equilibrium. At equilibrium, therefore,

$$k_f \frac{P_A}{RT} = k_r \frac{P_B}{RT}$$

Rearranging this equation and canceling RT terms gives

$$\frac{(P_B/RT)}{(P_A/RT)} = \frac{P_B}{P_A} = \frac{k_f}{k_r} = \text{a constant} \quad [15.5]$$

The quotient of two constants such as k_f and k_r is itself a constant. Thus, at equilibrium the ratio of the partial pressures of A and B equals a constant as shown in Equation 15.5. (We will consider this constant in Section 15.2.) It makes no difference whether we start with A or B, or even with some mixture of the two. At equilibrium the ratio equals a specific value. Thus, there is an important constraint on the proportions of A and B at equilibrium.

ACTIVITY
Chemical Equilibrium

Once equilibrium is established, the partial pressures of A and B no longer change [Figure 15.3(a)]. The result is an *equilibrium mixture* of A and B. Just because the composition of the equilibrium mixture remains constant with time does not mean, however, that A and B stop reacting. On the contrary, the equilibrium is dynamic. (Section 11.5) Compound A is still converted to compound B, and B to A, but at equilibrium both processes occur at the same rate, so there is no *net* change in their amounts. To indicate that the reaction proceeds in both the forward and reverse directions, we use a double arrow:

$$A \rightleftharpoons B$$

This example illustrates that opposing reactions naturally lead to an equilibrium situation. In order to examine equilibrium for a real chemical system, we will focus on an extremely important chemical reaction—the synthesis of ammonia from nitrogen and hydrogen:

$$N_2(g) + 3H_2(g) \rightleftharpoons 2NH_3(g) \quad [15.6]$$

This reaction is the basis for the **Haber process** for synthesizing ammonia.

15.2 The Equilibrium Constant

The Haber process combines N_2 and H_2 in a high-pressure tank at a total pressure of several hundred atmospheres, in the presence of a catalyst, and at a temperature of several hundred degrees Celsius. The two gases react to form ammonia under these conditions, but the reaction does not lead to complete consumption of the N_2 and H_2. Rather, at some point the reaction appears to stop, with all three components of the reaction mixture present at the same time.

The manner in which the concentrations of H_2, N_2, and NH_3 vary with time is shown in Figure 15.6(a) ▶. The situation is analogous to the one shown in Figure 15.3(a). The relative amounts of N_2, H_2, and NH_3 present at equilibrium do not depend on the amount of catalyst present. However, they do depend on the relative amounts of H_2 and N_2 at the beginning of the reaction. Furthermore, if only ammonia is placed in the tank under the same reaction conditions, an equilibrium mixture of N_2, H_2, and NH_3 will form. The variations in partial pressures as a function of time for this situation are shown in Figure 15.6(b). At equilibrium the relative partial pressures of H_2, N_2, and NH_3 are the same, regardless of whether the starting mixture was a 3:1 molar ratio of H_2 and N_2 or pure NH_3. *The equilibrium condition can be reached from either direction.*

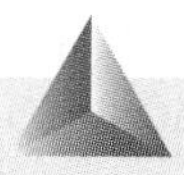

Chemistry at Work The Haber Process

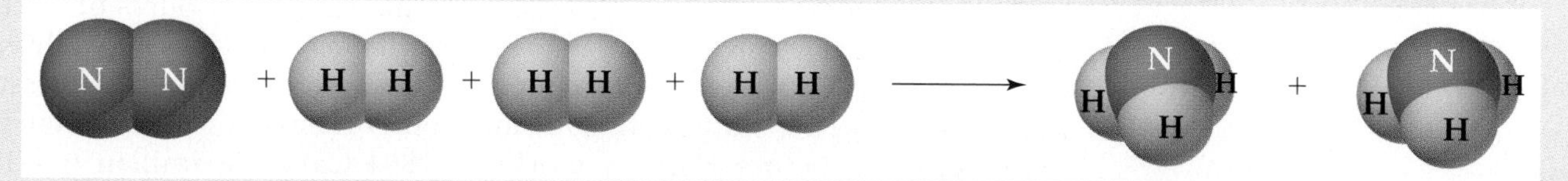

▲ **Figure 15.4** The Haber process is used to convert $N_2(g)$ and $H_2(g)$ to $NH_3(g)$, a process that, although exothermic, requires breaking the very strong triple bond in N_2.

We presented a "Chemistry and Life" box in Section 14.7 that discussed *nitrogen fixation*, the processes that convert N_2 gas into ammonia, which can then be incorporated into living organisms. We learned that the enzyme nitrogenase is responsible for generating most of the fixed nitrogen essential for plant growth. However, the quantity of food required to feed the ever-increasing human population far exceeds that provided by nitrogen-fixing plants, so human agriculture requires substantial amounts of ammonia-based fertilizers that can be applied directly to croplands. Thus, of all the chemical reactions that humans have learned to carry out and control for their own purposes, the synthesis of ammonia from hydrogen and atmospheric nitrogen is one of the most important.

In 1912 the German chemist Fritz Haber (1868–1934) developed a process for synthesizing ammonia directly from nitrogen and hydrogen (Figure 15.4 ▲). The process is sometimes called the *Haber–Bosch process* to also honor Karl Bosch, the engineer who developed the equipment for the industrial production of ammonia. The engineering needed to implement the Haber process requires the use of temperatures and pressures (approximately 500°C and 200 atm) that were difficult to achieve at that time.

The Haber process provides a historically interesting example of the complex impact of chemistry on our lives. At the start of World War I, in 1914, Germany depended on nitrate deposits in Chile for the nitrogen-containing compounds needed to manufacture explosives. During the war the Allied naval blockade of South America cut off this supply. However, by fixing nitrogen from air, Germany was able to continue to produce explosives. Experts have estimated that World War I would have ended before 1918 had it not been for the Haber process.

From these unhappy beginnings as a major factor in international warfare, the Haber process has become the world's principal source of fixed nitrogen. The same process that prolonged World War I has enabled scientists to manufacture fertilizers that have increased crop yields, thereby saving millions of people from starvation. About 40 billion pounds of ammonia are manufactured annually in the United States, mostly by the Haber process. The ammonia can be applied directly to the soil as fertilizer (Figure 15.5 ▼). It can also be converted into ammonium salts—for example, ammonium sulfate, $(NH_4)_2SO_4$, or ammonium hydrogen phosphate, $(NH_4)_2HPO_4$—which, in turn, are used as fertilizers.

Haber was a patriotic German who gave enthusiastic support to his nation's war effort. He served as chief of Germany's Chemical Warfare Service during World War I and developed the use of chlorine as a poison-gas weapon. Consequently, the decision to award him the Nobel Prize for chemistry in 1918 was the subject of considerable controversy and criticism. The ultimate irony, however, came in 1933 when Haber was expelled from Germany because he was Jewish.

▲ **Figure 15.5** Liquid ammonia, produced by the Haber process, can be added directly to the soil as a fertilizer. Agricultural use is the largest single application of manufactured NH_3.

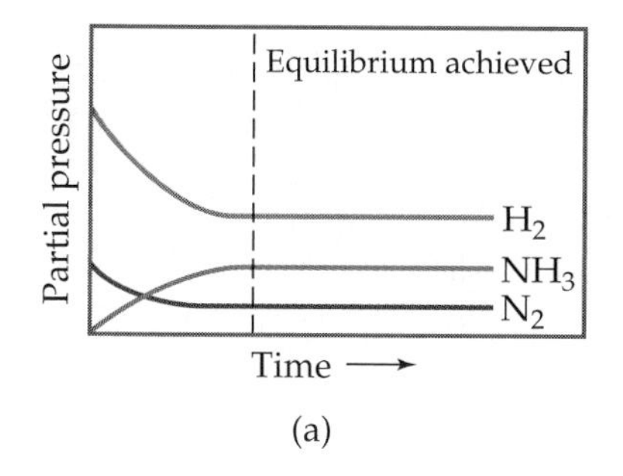

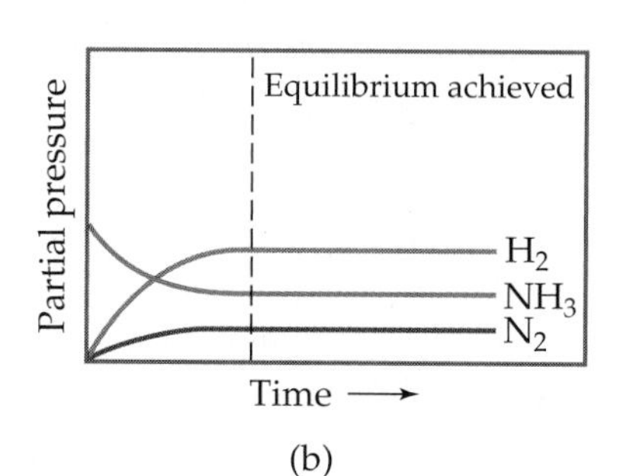

◀ **Figure 15.6** Variation in partial pressures in the approach to equilibrium for $N_2 + 3H_2 \rightleftharpoons 2NH_3$. (a) The equilibrium is approached beginning with H_2 and N_2 in the ratio 3 : 1. (b) The equilibrium is approached beginning with NH_3.

Earlier we saw that when the reaction $A(g) \rightleftharpoons B(g)$ reaches equilibrium, the ratio of the partial pressures of A and B has a constant value (Equation 15.5). A similar relationship governs the partial pressures of N_2, H_2, and NH_3 at equilibrium. If we were to change systematically the relative amounts of the three gases in the starting mixture and then analyze the gas mixtures at equilibrium, we could determine the relationship among the equilibrium partial pressures. Chemists carried out studies of this kind on other chemical systems in the nineteenth century, before Haber's work. In 1864 Cato Maximilian Guldberg (1836–1902) and Peter Waage (1833–1900) postulated their **law of mass action**, which expresses the relationship between the concentrations (expressed as partial pressures for gases and as molarities for solutions) of the reactants and products present at equilibrium in any reaction. Suppose we have the following general equilibrium equation:

$$aA + bB \rightleftharpoons cC + dD \qquad [15.7]$$

where A, B, C, and D are the chemical species involved, and a, b, c, and d are their coefficients in the balanced chemical equation. According to the law of mass action, the equilibrium condition is expressed by the following equation when all of the reactants and products are in the gas phase:

$$K_{eq} = \frac{(P_C)^c(P_D)^d}{(P_A)^a(P_B)^b} \qquad [15.8]$$

When the reactants and products are all in solution, the equilibrium condition is expressed by the same sort of equation but with concentrations in molarity.

$$K_{eq} = \frac{[C]^c[D]^d}{[A]^a[B]^b} \qquad [15.9]$$

We call this relationship the **equilibrium-constant expression** (or merely the equilibrium expression) for the reaction. The constant K_{eq}, which we call the **equilibrium constant**, is the numerical value obtained when we substitute actual equilibrium partial pressures or molar concentrations into the equilibrium-constant expression.

In general, the numerator of the equilibrium-constant expression is the product of the concentrations (expressed as partial pressures for gaseous species and as molarities for aqueous species) of all substances on the product side of the equilibrium equation, each raised to a power equal to its coefficient in the balanced equation. The denominator is similarly derived from the reactant side of the equilibrium equation. For the reaction $A(g) \rightleftharpoons B(g)$, the equilibrium expression is $K_{eq} = P_B/P_A$, in accord with Equation 15.5. For the Haber process (Equation 15.6), the equilibrium-constant expression is

$$K_{eq} = \frac{(P_{NH_3})^2}{P_{N_2}(P_{H_2})^3}$$

Note that once we know the balanced chemical equation for an equilibrium, we can write the equilibrium-constant expression even if we don't know the reaction mechanism. *The equilibrium-constant expression depends only on the stoichiometry of the reaction, not on its mechanism.*

The value of the equilibrium constant at any given temperature does not depend on the initial amounts of reactants and products. It also does not matter whether other substances are present, as long as they do not react with a reactant or a product. The value of the equilibrium constant varies only with temperature.

We can illustrate how the law of mass action was discovered empirically, by considering the gas-phase equilibrium between dinitrogen tetroxide and nitrogen dioxide:

$$N_2O_4(g) \rightleftharpoons 2NO_2(g) \qquad [15.10]$$

TABLE 15.1 Initial and Equilibrium Partial Pressures (*P*) of N_2O_4 and NO_2 at 100°C

Experiment	Initial N_2O_4 Partial Pressure (atm)	Initial NO_2 Partial Pressure (atm)	Equilibrium N_2O_4 Partial Pressure (atm)	Equilibrium NO_2 Partial Pressure (atm)	K_{eq}
1	0.0	0.612	0.0429	0.526	6.45
2	0.0	0.919	0.0857	0.744	6.46
3	0.0	1.22	0.138	0.944	6.46
4	0.612	0.0	0.138	0.944	6.46

Figure 15.2 shows this equilibrium being reached after starting with pure N_2O_4. Because NO_2 is a dark brown gas and N_2O_4 is colorless, the amount of NO_2 in the mixture can be determined by measuring the intensity of the brown color of the gas mixture.

The equilibrium expression for Equation 15.10 is

$$K_{eq} = \frac{(P_{NO_2})^2}{P_{N_2O_4}} \qquad [15.11]$$

How can we determine the numerical value for K_{eq} and verify that it is constant regardless of the starting amounts of NO_2 and N_2O_4? We could perform experiments in which we start with several sealed tubes containing different partial pressures of NO_2 and N_2O_4, as summarized in Table 15.1 ▲. The tubes are kept at 100°C until no further change in the color of the gas is noted. We then analyze the mixtures and determine the equilibrium partial pressures of NO_2 and N_2O_4, as shown in Table 15.1.

To evaluate the equilibrium constant, K_{eq}, the equilibrium partial pressures are inserted into the equilibrium-constant expression (Equation 15.11). For example, using the first set of data, $P_{NO_2} = 0.526$ atm and $P_{N_2O_4} = 0.0429$ atm:

$$K_{eq} = \frac{(P_{NO_2})^2}{P_{N_2O_4}} = \frac{(0.526)^2}{0.0429} = 6.45$$

Proceeding in the same way, the values of K_{eq} for the other samples were calculated, as listed in Table 15.1. Note that the value for K_{eq} is constant ($K_{eq} = 6.46$, within the limits of experimental error) even though the initial partial pressures vary. Furthermore, the results of experiment 4 show that equilibrium can be achieved beginning with N_2O_4 as well as with NO_2. That is, equilibrium can be approached from either direction. Figure 15.7 ▼ shows how both experiments 3 and 4 result in the same equilibrium mixture even though one begins with 1.22 atm NO_2 and the other with 0.612 atm N_2O_4.

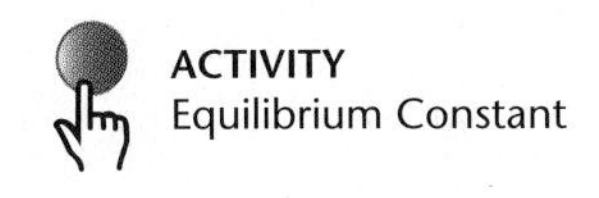

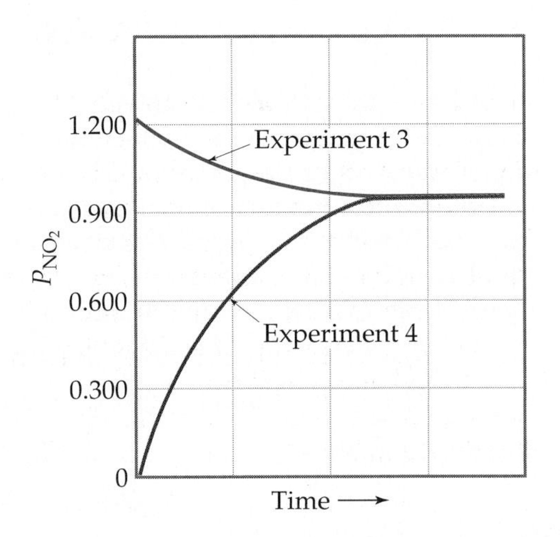

◀ **Figure 15.7** As seen in Table 15.1, the same equilibrium mixture is produced starting with either 1.22 atm NO_2 (experiment 3) or 0.612 atm N_2O_4 (experiment 4).

SAMPLE EXERCISE 15.1

Write the equilibrium expression for K_{eq} for the following reactions:
(a) $2O_3(g) \rightleftharpoons 3O_2(g)$
(b) $2NO(g) + Cl_2(g) \rightleftharpoons 2NOCl(g)$
(c) $Ag^+(aq) + 2NH_3(aq) \rightleftharpoons Ag(NH_3)_2^+(aq)$

Solution

Analyze: We are given three equations and are asked to write an equilibrium-constant expression for each.

Plan: Using the law of mass action, we write each expression as a quotient. The quotient has the concentration terms of the products multiplied together and each raised to the power of its stoichiometric coefficient in the balanced chemical equation, divided by the concentration terms of the reactants multiplied together and each raised to the power of its stoichiometric coefficient.

Solve: **(a)** For a gas-phase reaction, we use the partial pressures of the products and reactants for the concentration terms:

$$K_{eq} = \frac{(P_{O_2})^3}{(P_{O_3})^2}$$

(b) Similarly, we have

$$K_{eq} = \frac{(P_{NOCl})^2}{(P_{NO})^2 P_{Cl_2}}$$

(c) For an aqueous reaction, we use the molar concentrations of the products and reactants for the concentration terms.

$$K_{eq} = \frac{[Ag(NH_3)_2^+]}{[Ag^+][NH_3]^2}$$

PRACTICE EXERCISE

Write the equilibrium-constant expression for **(a)** $H_2(g) + I_2(g) \rightleftharpoons 2HI(g)$, **(b)** $Cd^{2+}(aq) + 4Br^-(aq) \rightleftharpoons CdBr_4^{2-}(aq)$.

Answers: **(a)** $K_{eq} = \frac{(P_{HI})^2}{P_{H_2}P_{I_2}}$; **(b)** $K_{eq} = \frac{[CdBr_4^{2-}]}{[Cd^{2+}][Br^-]^4}$

The Magnitude of Equilibrium Constants

Equilibrium constants can be very large or very small. The magnitude of the constant provides us with important information about the composition of an equilibrium mixture. For example, consider the reaction of carbon monoxide and chlorine gases at 100°C to form phosgene ($COCl_2$), a toxic gas that is used in the manufacture of certain polymers and insecticides.

$$CO(g) + Cl_2(g) \rightleftharpoons COCl_2(g) \qquad K_{eq} = \frac{P_{COCl_2}}{P_{CO}P_{Cl_2}} = 1.49 \times 10^8$$

In order for the equilibrium constant to be so large, the numerator of the equilibrium-constant expression must be much larger than the denominator. Thus, the equilibrium concentration of $COCl_2$ must be much greater than that of CO or Cl_2; an equilibrium mixture of the three gases is essentially pure $COCl_2$. We say that this equilibrium *lies to the right* (that is, toward the product side). Likewise, a very small equilibrium constant indicates that the equilibrium mixture contains mostly reactants. We then say that the equilibrium *lies to the left*. In general,

$K_{eq} \gg 1$: Equilibrium lies to the right; products predominate.
$K_{eq} \ll 1$: Equilibrium lies to the left; reactants predominate.

These situations are summarized in Figure 15.8 ◀.

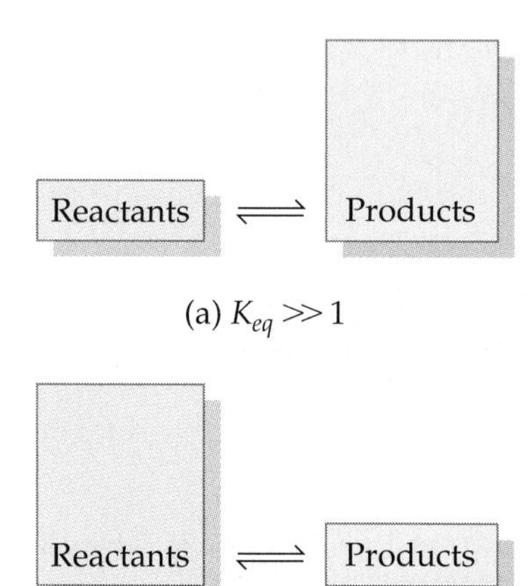

▲ **Figure 15.8** The equilibrium expression has products in the numerator and reactants in the denominator. (a) When $K_{eq} \gg 1$, there are more products than reactants at equilibrium, and the equilibrium is said to lie to the right. (b) When $K_{eq} \ll 1$, there are more reactants than products at equilibrium, and the equilibrium is said to lie to the left.

SAMPLE EXERCISE 15.2

The reaction of N_2 with O_2 to form NO might be considered a means of "fixing" nitrogen.

$$N_2(g) + O_2(g) \rightleftharpoons 2NO(g)$$

The value for the equilibrium constant for this reaction at 25°C is $K_{eq} = 1 \times 10^{-30}$. Describe the feasibility of this reaction for nitrogen fixation.

Solution

Analyze: We are asked to comment on the utility of a reaction based upon the magnitude of its equilibrium constant.

Plan: We consider the magnitude of the equilibrium constant to determine whether or not this reaction is feasible for the production of the desired species.

Solve: Because K_{eq} is so small, very little NO will form at 25°C. The equilibrium lies to the left, favoring the reactants. Consequently, this reaction is an extremely poor choice for nitrogen fixation, at least at 25°C.

PRACTICE EXERCISE

The equilibrium constant for the reaction $H_2(g) + I_2(g) \rightleftharpoons 2HI(g)$ varies with temperature as follows: $K_{eq} = 794$ at 298 K; $K_{eq} = 54$ at 700 K. Is the formation of HI favored more at the higher or lower temperature?

Answer: It is favored at the lower temperature because K_{eq} is larger

The Direction of the Chemical Equation and K_{eq}

Because an equilibrium can be approached from either direction, the direction in which we write the chemical equation for an equilibrium is arbitrary. For example, we have seen that we can represent the N_2O_4–NO_2 equilibrium as

$$N_2O_4(g) \rightleftharpoons 2NO_2(g)$$

For this equation we can write

$$K_{eq} = \frac{(P_{NO_2})^2}{P_{N_2O_4}} = 6.46 \qquad \text{(at 100°C)} \qquad [15.12]$$

We could equally well consider this same equilibrium in terms of the reverse reaction:

$$2NO_2(g) \rightleftharpoons N_2O_4(g)$$

The equilibrium expression is then given by

$$K_{eq} = \frac{P_{N_2O_4}}{(P_{NO_2})^2} = 0.155 \qquad \text{(at 100°C)} \qquad [15.13]$$

Equation 15.13 is just the reciprocal of Equation 15.12. *The equilibrium-constant expression for a reaction written in one direction is the reciprocal of the one for the reaction written in the reverse direction.* Consequently, the numerical value of the equilibrium constant for the reaction written in one direction is the reciprocal of that for the reverse reaction. Both expressions are equally valid, but it is meaningless to say that the equilibrium constant for the equilibrium between NO_2 and N_2O_4 is 6.46 or 0.155 unless we indicate how the equilibrium reaction is written and also specify the temperature.

SAMPLE EXERCISE 15.3

(a) Write the expression for K_{eq} for the following reaction:

$$2NO(g) \rightleftharpoons N_2(g) + O_2(g)$$

(b) Using information in Sample Exercise 15.2, determine the value of this equilibrium constant at 25°C.

Solution

Analyze: We are asked to write an equilibrium-constant expression and determine the value of the equilibrium constant for a gaseous equilibrium.

Plan: As before, we write the equilibrium constant as a quotient of products over reactants, each raised to a power equal to its coefficient in the balanced equation. We can determine the value of the equilibrium constant by relating the equilibrium-constant expression we write for this equation to the equilibrium-constant expression in Sample Exercise 15.2.

Solve: (a) Writing products over reactants, we have

$$K_{eq} = \frac{P_{N_2}P_{O_2}}{(P_{NO})^2}$$

(b) The reaction is just the reverse of the one given in Sample Exercise 15.2. Thus, both the equilibrium-constant expression and the numerical value of the equilibrium constant are the reciprocals of those for the reaction in Sample Exercise 15.2.

$$K_{eq} = \frac{P_{N_2}P_{O_2}}{(P_{NO})^2} = \frac{1}{1 \times 10^{-30}} = 1 \times 10^{30}$$

Regardless of the way we express the equilibrium among NO, N_2, and O_2, at 25°C it lies on the side that favors N_2 and O_2.

PRACTICE EXERCISE

For the formation of NH_3 from N_2 and H_2, $N_2(g) + 3H_2(g) \rightleftharpoons 2NH_3(g)$, $K_{eq} = 4.34 \times 10^{-3}$ at 300°C. What is the value of K_{eq} for the reverse reaction?

Answer: 2.30×10^2

Other Ways to Manipulate Chemical Equations and K_{eq} Values

Just as the K_{eq} values of forward and reverse reactions are reciprocals of one another, equilibrium constants of reactions related in other ways are also related. For example, if we were to multiply the original N_2O_4–NO_2 equilibrium by 2, we would have

$$2N_2O_4(g) \rightleftharpoons 4NO_2(g)$$

The equilibrium-constant expression for this equation is

$$K_{eq} = \frac{(P_{NO_2})^4}{(P_{N_2O_4})^2}$$

which is simply the equilibrium-constant expression for the original equation, given in Equation 15.11, squared. Because the new equilibrium-constant expression equals the original expression squared, the new equilibrium constant equals the original constant squared, in this case $6.46^2 = 41.7$ (at 100°C).

Sometimes, as in problems that utilize Hess's law (Section 5.6), we must use equations made up of two or more steps. We obtain the net equation by adding the individual equations and canceling identical terms. Consider the following two reactions, their equilibrium-constant expressions, and their equilibrium constants at 100°C:

$$2NOBr(g) \rightleftharpoons 2NO(g) + Br_2(g) \qquad K_{eq} = \frac{(P_{NO})^2 P_{Br_2}}{(P_{NOBr})^2} \qquad K_{eq} = 0.42$$

$$Br_2(g) + Cl_2(g) \rightleftharpoons 2BrCl(g) \qquad K_{eq} = \frac{(P_{BrCl})^2}{P_{Br_2}P_{Cl_2}} \qquad K_{eq} = 7.2$$

The sum of these two equations is

$$2NOBr(g) + Cl_2(g) \rightleftharpoons 2NO(g) + 2BrCl(g)$$

and the equilibrium-constant expression for the net equation is

$$K_{eq} = \frac{(P_{NO})^2(P_{BrCl})^2}{(P_{NOBr})^2 P_{Cl_2}}$$

This equilibrium-constant expression is the product of the expressions for the individual steps. Because the net equation's equilibrium-constant expression is the product of two expressions, its equilibrium constant is the product of the two individual equilibrium constants: $0.42 \times 7.2 = 3.0$.

To summarize:

1. The equilibrium constant of a reaction in the reverse direction is the inverse of the equilibrium constant of the reaction in the forward direction.
2. The equilibrium constant of a reaction that has been multiplied by a number is the equilibrium constant raised to a power equal to that number.
3. The equilibrium constant for a net reaction made up of two or more steps is the product of the equilibrium constants for the individual steps.

SAMPLE EXERCISE 15.4

Given the following information,

$$HF(aq) \rightleftharpoons H^+(aq) + F^-(aq) \qquad K_{eq} = 6.8 \times 10^{-4}$$

$$H_2C_2O_4(aq) \rightleftharpoons 2H^+(aq) + C_2O_4^{2-}(aq) \qquad K_{eq} = 3.8 \times 10^{-6}$$

determine the value of the equilibrium constant for the following reaction:

$$2HF(aq) + C_2O_4^{2-}(aq) \rightleftharpoons 2F^-(aq) + H_2C_2O_4(aq)$$

Solution

Analyze: We are given two equilibrium equations and the corresponding equilibrium constants and are asked to determine the equilibrium constant for a third equation, which is related to the first two.

Plan: We cannot simply add the first two equations to get the third. Instead, we need to determine how to manipulate the equations in order to come up with the steps that will add to give us the desired equation.

Solve: If we multiply the first equation by 2 and make the corresponding change to its equilibrium constant (raising to the power 2), we get

$$2HF(aq) \rightleftharpoons 2H^+(aq) + 2F^-(aq) \qquad K_{eq} = 4.6 \times 10^{-7}$$

Reversing the second equation and again making the corresponding change to its equilibrium constant (taking the reciprocal) gives

$$2H^+(aq) + C_2O_4^{2-}(aq) \rightleftharpoons H_2C_2O_4(aq) \qquad K_{eq} = 2.6 \times 10^5$$

Now we have two equations that sum to give the net equation, and we can multiply the individual K_{eq} values to get the desired equilibrium constant.

$$2HF(aq) \rightleftharpoons 2H^+(aq) + 2F^-(aq)$$

$$2H^+(aq) + C_2O_4^{2-}(aq) \rightleftharpoons H_2C_2O_4(aq)$$

$$2HF(aq) + C_2O_4^{2-}(aq) \rightleftharpoons 2F^-(aq) + H_2C_2O_4(aq)$$

$$K_{eq} = (4.6 \times 10^{-7})(2.6 \times 10^5) = 0.12$$

PRACTICE EXERCISE

Given the following information at 700 K, $K_{eq} = 54.0$ for the reaction $H_2(g) + I_2(g) \rightleftharpoons 2HI(g)$, and $K_{eq} = 1.04 \times 10^{-4}$ for the reaction $N_2(g) + 3H_2(g) \rightleftharpoons 2NH_3(g)$, determine the value of the equilibrium constant for the reaction $2NH_3(g) + 3I_2(g) \rightleftharpoons 6HI(g) + N_2(g)$ at 700 K.

Answer: 1.51×10^9

Units of Equilibrium Constants

You may have noticed that none of the values of K_{eq} has any units associated with it, even though the pressures and concentrations that go into the equilibrium expressions have units of atmospheres and moles-per-liter, respectively. This happens because the values we enter into the equilibrium expression are actually *ratios* of pressure to a reference pressure, P_{ref}, or molar concentration to a reference concentration, M_{ref}. The reference pressure, for partial pressures expressed in atmospheres, is 1 atm. The reference concentration, for concentrations expressed in molarity, is 1 M. For example, consider the equilibrium $N_2O_4(g) \rightleftharpoons 2NO_2(g)$. The equilibrium-constant expression for this equation is given in Equation 15.11. If we were simply to substitute partial pressures into the expression, we would obtain a K_{eq} value with units of atm. By first dividing each partial pressure by the reference pressure (1 atm), we eliminate the units of atmospheres and get an equilibrium constant that is dimensionless.

$$K_{eq} = \frac{(P_{NO_2}/P_{ref})^2}{(P_{N_2O_4}/P_{ref})} \qquad [15.14]$$

The same is true for aqueous equilibria. Simply substituting molar concentrations into the equilibrium-constant expression for the formation of $Ag(NH_3)_2{}^+(aq)$ from $Ag^+(aq)$ and $NH_3(aq)$ (Sample Exercise 15.1) would yield an equilibrium constant with units of M^{-2}. First dividing each concentration by 1 M eliminates the units, however, and gives us the desired, dimensionless equilibrium constant. Because each quantity used in an equilibrium-constant expression is rendered dimensionless through division by a reference quantity, we can use partial pressures and molar concentrations in the same expression, where appropriate, as we will see in Section 15.3. For the remainder of the text, we will not show the division by the reference pressure or concentration explicitly because it does not change the numerical value of the result.

15.3 Heterogeneous Equilibria

Many equilibria, such as the hydrogen-nitrogen-ammonia system, involve substances all in the same phase. Such equilibria are called **homogeneous equilibria**. In other cases the substances in equilibrium are in different phases, giving rise to **heterogeneous equilibria**. As an example, consider the dissolution of lead(II) chloride ($PbCl_2$).

$$PbCl_2(s) \rightleftharpoons Pb^{2+}(aq) + 2Cl^-(aq) \qquad [15.15]$$

This system consists of a solid in equilibrium with two aqueous species. If we write the equilibrium-constant expression for this process in the usual way, we obtain

$$K_{eq} = \frac{[Pb^{2+}][Cl^-]^2}{[PbCl_2]} \qquad [15.16]$$

This example presents a problem we have not encountered previously: How do we express the concentration of a solid substance? Although it is possible to express the concentration of a solid in terms of moles per unit volume, it turns out to be unnecessary to do so for the purpose of writing equilibrium-constant expressions. Recall from the last section that what actually gets substituted into an equilibrium expression is a ratio of concentration (or partial pressure) to a reference value. The same is true for pure substances, whether they are solids or liquids, and the reference concentration for any pure substance is simply that of the pure substance itself. Therefore, division by the reference concentration for any pure solid or pure liquid simply yields 1, making it unnecessary for solids and liquids to appear in the equilibrium-constant expression. Thus, *if a pure solid or a pure liquid is involved in a heterogeneous equilibrium, its concentration is not included in the equilibrium-constant expression for the reaction.* When the solvent is involved in the equilibrium, its concentration is also excluded from the equilibrium-constant expression, provided the concentrations of reactants and products are low, so that the solvent is essentially a pure substance. The partial pressures of gases and the molar concentrations of substances in solution, however, are included in equilibrium-constant expressions because these quantities can vary. To summarize:

1. Partial pressures of gases are substituted into the equilibrium-constant expression.
2. Molar concentrations of dissolved species are substituted into the equilibrium-constant expression.
3. Pure solids, pure liquids, and solvents are not included in the equilibrium-constant expression.

Applying these guidelines to the decomposition of calcium carbonate,

$$CaCO_3(s) \rightleftharpoons CaO(s) + CO_2(g) \qquad [15.17]$$

gives the following equilibrium-constant expression:

$$K_{eq} = P_{CO_2} \qquad [15.18]$$

Equation 15.18 tells us that, at a given temperature, an equilibrium among $CaCO_3$, CaO, and CO_2 will always lead to the same partial pressure of CO_2 as long as all three components are present. As shown in Figure 15.9 ▶, we would have the same pressure of CO_2 regardless of the relative amounts of CaO and $CaCO_3$. *Note that even though they do not appear in the equilibrium-constant expression, the pure solids and liquids participating in the reaction must be present at equilibrium.* Applying the guidelines to an equilibrium involving water as a solvent,

$$H_2O(l) + CO_3^{2-}(aq) \rightleftharpoons OH^-(aq) + HCO_3^-(aq) \qquad [15.19]$$

gives the following equilibrium-constant expression:

$$K_{eq} = \frac{[OH^-][HCO_3^-]}{[CO_3^{2-}]} \qquad [15.20]$$

▲ **Figure 15.9** The decomposition of $CaCO_3$ is a heterogeneous equilibrium. At the same temperature the equilibrium pressure of CO_2 is the same in the two bell jars even though the relative amounts of pure $CaCO_3$ and CaO differ greatly.

SAMPLE EXERCISE 15.5

Write the equilibrium-constant expressions for each of the following reactions:

(a) $CO_2(g) + H_2(g) \rightleftharpoons CO(g) + H_2O(l)$

(b) $SnO_2(s) + 2CO(g) \rightleftharpoons Sn(s) + 2CO_2(g)$

(c) $Sn(s) + 2H^+(aq) \rightleftharpoons Sn^{2+}(aq) + H_2(g)$

Solution

Analyze: We are given three chemical equations, all for heterogeneous equilibria, and are asked to write the corresponding equilibrium-constant expressions.

Plan: We employ the law of mass action, remembering to omit any pure solids, pure liquids, and solvents from the expressions.

Solve: (a) The equilibrium-constant expression is

$$K_{eq} = \frac{P_{CO}}{P_{CO_2}P_{H_2}}$$

Because H_2O appears in the reaction as a pure liquid, its concentration does not appear in the equilibrium-constant expression.

(b) The equilibrium-constant expression is

$$K_{eq} = \frac{(P_{CO_2})^2}{(P_{CO})^2}$$

Because SnO_2 and Sn are both pure solids, their concentrations do not appear in the equilibrium-constant expression.

(c) The equilibrium-constant expression is

$$K_{eq} = \frac{[Sn^{2+}]P_{H_2}}{[H^+]^2}$$

Because Sn is a pure solid, its concentration does not appear in the equilibrium-constant expression. Note that both molarities and partial pressures appear in the same expression.

PRACTICE EXERCISE

Write the equilibrium-constant expressions for the reactions **(a)** $3Fe(s) + 4H_2O(g) \rightleftharpoons Fe_3O_4(s) + 4H_2(g)$, and **(b)** $Cr(s) + 3Ag^+(aq) \rightleftharpoons Cr^{3+}(aq) + 3Ag(s)$.

Answers: **(a)** $K_{eq} = (P_{H_2})^4/(P_{H_2O})^4$; **(b)** $K_{eq} = [Cr^{3+}]/[Ag^+]^3$

SAMPLE EXERCISE 15.6

Each of the following mixtures was placed in a closed container and allowed to stand. Which of these mixtures is capable of attaining the equilibrium expressed by Equation 15.17: **(a)** pure $CaCO_3$; **(b)** CaO and a pressure of CO_2 greater than the value of K_{eq}; **(c)** some $CaCO_3$ and a pressure of CO_2 greater than the value of K_{eq}; **(d)** $CaCO_3$ and CaO?

Solution

Analyze: We are asked which of several combinations of species can establish an equilibrium between calcium carbonate and its decomposition products, calcium oxide and carbon dioxide: $CaCO_3(s) \rightleftharpoons CaO(s) + CO_2(g)$.

Plan: In order for equilibrium to be achieved, it must be possible for both forward and reverse processes to occur. In order for the forward process to occur, there must be some calcium carbonate present. In order for the reverse process to occur, there must be both calcium oxide and carbon dioxide. In both cases, the necessary compounds may be present initially, or they may be formed by reaction of the other species.

Solve: Equilibrium can be reached in all cases except (c) as long as sufficient quantities of solids are present. In (a) $CaCO_3$ simply decomposes, forming CaO(*s*) and $CO_2(g)$ until the equilibrium pressure of CO_2 is attained. There must be enough $CaCO_3$, however, to allow the CO_2 pressure to reach equilibrium. In (b) CO_2 combines with the CaO present until its pressure decreases to the equilibrium value. There is no CaO present in (c), so equilibrium can't be attained because there is no way in which the CO_2 pressure can decrease to its equilibrium value (which would require some of the CO_2 to react with CaO). In (d) the situation is essentially the same as in (a): $CaCO_3$ decomposes until equilibrium is attained. The presence of CaO initially makes no difference.

PRACTICE EXERCISE

Which one of the following substances—$H_2(g)$, $H_2O(g)$, $O_2(g)$—when added to $Fe_3O_4(s)$ in a closed container will allow equilibrium to be established in the reaction $3Fe(s) + 4H_2O(g) \rightleftharpoons Fe_3O_4(s) + 4H_2(g)$?

Answer: only $H_2(g)$

15.4 Calculating Equilibrium Constants

One of the first tasks confronting Haber when he approached the problem of ammonia synthesis was finding the magnitude of the equilibrium constant for the synthesis of NH_3 at various temperatures. If the value of K_{eq} for Equation 15.6 were very small, the amount of NH_3 in an equilibrium mixture would be small relative to the amounts of N_2 and H_2. That is, if the equilibrium lies too far to the left, it would be impossible to develop a satisfactory synthesis of ammonia.

Haber and his coworkers therefore evaluated the equilibrium constants for this reaction at various temperatures. The method they employed is analogous to that described in constructing Table 15.1: They started with various mixtures of N_2, H_2, and NH_3, allowed the mixtures to achieve equilibrium at a specific temperature, and measured the concentrations of all three gases at equilibrium. Because the equilibrium partial pressures of all products and reactants were known, the equilibrium constant could be calculated directly from the equilibrium-constant expression.

SAMPLE EXERCISE 15.7

A mixture of hydrogen and nitrogen in a reaction vessel is allowed to attain equilibrium at 472°C. The equilibrium mixture of gases was analyzed and found to contain 7.38 atm H_2, 2.46 atm N_2, and 0.166 atm NH_3. From these data calculate the equilibrium constant, K_{eq}, for

$$N_2(g) + 3H_2(g) \rightleftharpoons 2NH_3(g)$$

Solution

Analyze: We are given a balanced equation and equilibrium partial pressures and are asked to calculate the value of the equilibrium constant.

Plan: Using the balanced equation, we write the equilibrium-constant expression. We then substitute the equilibrium partial pressures into the expression and solve for K_{eq}.

Solve:

$$K_{eq} = \frac{(P_{NH_3})^2}{P_{N_2}(P_{H_2})^3} = \frac{(0.166)^2}{(2.46)(7.38)^3} = 2.79 \times 10^{-5}$$

PRACTICE EXERCISE

An aqueous solution of acetic acid is found to have the following equilibrium concentrations at 25°C: $[HC_2H_3O_2] = 1.65 \times 10^{-2}\ M$; $[H^+] = 5.44 \times 10^{-4}\ M$; and $[C_2H_3O_2^-] = 5.44 \times 10^{-4}\ M$. Calculate the equilibrium constant, K_{eq}, for the ionization of acetic acid at 25°C. (Section 4.3)

Answer: 1.79×10^{-5}

We often don't know the equilibrium concentrations of all chemical species in an equilibrium. If we know the equilibrium concentration of at least one species, however, we can generally use the stoichiometry of the reaction to deduce the equilibrium concentrations of the other species in the chemical equation. We will use the following procedure to do this:

1. Tabulate the known initial and equilibrium concentrations of all species in the equilibrium-constant expression.
2. For those species for which both the initial and equilibrium concentrations are known, calculate the change in concentration that occurs as the system reaches equilibrium.
3. Use the stoichiometry of the reaction (that is, use the coefficients in the balanced chemical equation) to calculate the changes in concentration for all the other species in the equilibrium.
4. From the initial concentrations and the changes in concentration, calculate the equilibrium concentrations. These are used to evaluate the equilibrium constant.

We illustrate the procedure in Sample Exercise 15.8.

SAMPLE EXERCISE 15.8

Enough ammonia is dissolved in 5.00 liters of water at 25°C to produce a solution that is 0.0124 *M* in ammonia. The solution is then allowed to come to equilibrium. Analysis of the equilibrium mixture shows that the concentration of OH^- is $4.64 \times 10^{-4}\ M$. Calculate K_{eq} at 25°C for the reaction

$$NH_3(aq) + H_2O(l) \rightleftharpoons NH_4^+(aq) + OH^-(aq)$$

Solution

Analyze: We are given a starting concentration of ammonia and an equilibrium concentration of one of its dissociation products, and we are asked to determine the value of the equilibrium constant for the dissociation of ammonia in water.

Plan: We construct a table to find equilibrium concentrations of all species and use the equilibrium concentrations to calculate the equilibrium constant.

Solve: First, we tabulate the known initial and equilibrium concentrations of all the species in the equilibrium-constant expression. We also provide space in our table for listing the changes in concentrations. As shown, it is convenient to use the chemical equation as the heading for the table. Note that there are no entries in the column beneath water because water is a solvent and does not appear in the equilibrium-constant expression:

	$NH_3(aq)$	+ $H_2O(l)$ ⇌	$NH_4^+(aq)$ +	$OH^-(aq)$
Initial	0.0124 M		0 M	0 M
Change				
Equilibrium				4.64×10^{-4} M

Second, we calculate the change in concentration of OH^-, using the initial and equilibrium values. The change is the difference between the equilibrium and initial values, 4.64×10^{-4} M.

Third, we use the stoichiometry of the reaction to calculate the changes in the other species. The balanced chemical equation indicates that for each mol of OH^- formed, 1 mol of NH_3 must be consumed. Thus, the amount of NH_3 consumed is also 4.64×10^{-4} M. The same line of reasoning gives us the amount of NH_4^+ produced, which is also 4.64×10^{-4} M.

Fourth, we calculate the equilibrium concentrations, using the initial concentrations and the changes. The equilibrium concentration of NH_3 is the initial concentration minus that consumed:

$$[NH_3] = 0.0124\ M - 4.64 \times 10^{-4}\ M = 0.01194\ M \text{ (carrying one extra significant figure)}$$

Likewise, the equilibrium concentration of NH_4^+ is

$$[NH_4^+] = 0\ M + 4.64 \times 10^{-4}\ M = 4.64 \times 10^{-4}\ M$$

The completed table now looks like the following:

	$NH_3(aq)$	+ $H_2O(l)$ ⇌	$NH_4^+(aq)$ +	$OH^-(aq)$
Initial	0.0124 M		0 M	0 M
Change	-4.64×10^{-4} M		$+4.64 \times 10^{-4}$ M	$+4.64 \times 10^{-4}$ M
Equilibrium	0.0119 M		4.64×10^{-4} M	4.64×10^{-4} M

Finally, now that we know the equilibrium concentration of each reactant and product, we can use the equilibrium-constant expression to calculate the equilibrium constant.

$$K_{eq} = \frac{[NH_4^+][OH^-]}{[NH_3]} = \frac{(4.64 \times 10^{-4})^2}{(0.0119)} = 1.81 \times 10^{-5}$$

Comment: The same method can be applied to gaseous equilibrium problems. Partial pressures are simply used as table entries in place of molar concentrations.

PRACTICE EXERCISE

Sulfur trioxide decomposes at high temperature in a sealed container: $2SO_3(g) \rightleftharpoons 2SO_2(g) + O_2(g)$. Initially the vessel is charged at 1000 K with $SO_3(g)$ at a partial pressure of 0.500 atm. At equilibrium the SO_3 partial pressure is 0.200 atm. Calculate the value of K_{eq} at 1000 K.
Answer: 0.338

15.5 Applications of Equilibrium Constants

ACTIVITY
Using an Equilibrium Table I, Using an Equilibrium Table II

We have seen that the magnitude of K_{eq} indicates the extent to which a reaction will proceed. If K_{eq} is very large, the reaction will tend to proceed far to the right; if K_{eq} is very small (that is, much less than 1), the equilibrium mixture will contain mainly reactants. The equilibrium constant also allows us to (1) predict the direction in which a reaction mixture will proceed to achieve equilibrium and (2) calculate the concentrations of reactants and products when equilibrium has been reached.

Predicting the Direction of Reaction

Suppose we place a mixture of 2.00 mol of H_2, 1.00 mol of N_2, and 2.00 mol of NH_3 in a 1.00-L container at 472°C. Will N_2 and H_2 react to form more NH_3? In this instance we must first calculate the starting partial pressure of each species, using the ideal-gas equation.

$$P_{H_2} = \frac{n_{H_2}RT}{V} = \frac{(2.00 \text{ mol})(0.0821 \text{ L-atm/mol-K})(745 \text{ K})}{1.00 \text{ L}} = 122 \text{ atm}$$

$$P_{N_2} = \frac{n_{N_2}RT}{V} = \frac{(1.00 \text{ mol})(0.0821 \text{ L-atm/mol-K})(745 \text{ K})}{1.00 \text{ L}} = 61.2 \text{ atm}$$

$$P_{NH_3} = \frac{n_{NH_3}RT}{V} = \frac{(2.00 \text{ mol})(0.0821 \text{ L-atm/mol-K})(745 \text{ K})}{1.00 \text{ L}} = 122 \text{ atm}$$

If we insert the starting partial pressures of N_2, H_2, and NH_3 into the equilibrium-constant expression, we have

$$\frac{(P_{NH_3})^2}{P_{N_2}(P_{H_2})^3} = \frac{(122)^2}{(61.2)(122)^3} = 1.34 \times 10^{-4}$$

According to Sample Exercise 15.7, $K_{eq} = 2.79 \times 10^{-5}$ at this temperature. Therefore, the quotient $P^2_{NH_3}/P_{N_2}P^3_{H_2}$ will need to decrease from 1.34×10^{-4} to 2.79×10^{-5} for the system to achieve equilibrium. This change can happen only if the partial pressure of NH_3 decreases and those of N_2 and H_2 increase. Thus, the reaction proceeds toward equilibrium via the formation of N_2 and H_2 from NH_3; that is, the reaction proceeds from right to left.

When we substitute reactant and product partial pressures or concentrations into an equilibrium-constant expression, the result is known as the **reaction quotient** and is represented by the letter Q. *The reaction quotient will equal the equilibrium constant, K_{eq}, only if the system is at equilibrium: $Q = K_{eq}$ only at equilibrium.* We have seen that when $Q > K_{eq}$, substances on the right side of the chemical equation will react to form substances on the left; the reaction moves from right to left in approaching equilibrium. Conversely, if $Q < K_{eq}$, the reaction will achieve equilibrium by forming more products; it moves from left to right. These relationships are summarized in Figure 15.10 ▶.

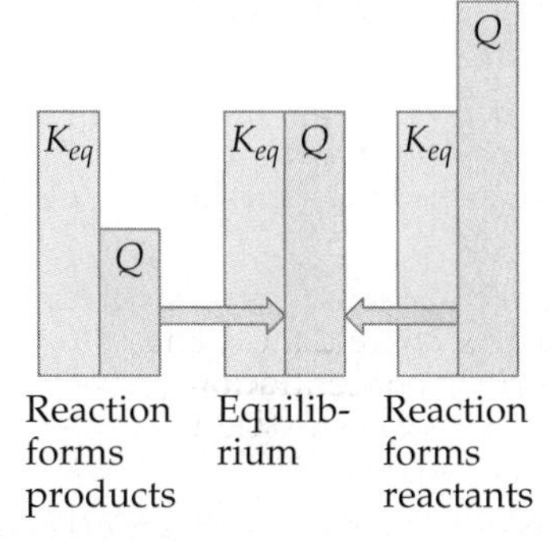

▲ **Figure 15.10** The relative magnitudes of the reaction quotient Q and the equilibrium constant K_{eq} indicate how the reaction mixture changes as it moves toward equilibrium. If Q is smaller than K_{eq}, the reaction proceeds from left to right until $Q = K_{eq}$. When $Q = K_{eq}$, the reaction is at equilibrium and has no tendency to change. If Q is larger than K_{eq}, the reaction proceeds from right to left until $Q = K_{eq}$.

SAMPLE EXERCISE 15.9

At 448°C the equilibrium constant, K_{eq}, for the reaction

$$H_2(g) + I_2(g) \rightleftharpoons 2HI(g)$$

is 51. Predict how the reaction will proceed to reach equilibrium at 448°C if we start with 2.0×10^{-2} mol of HI, 1.0×10^{-2} mol of H_2, and 3.0×10^{-2} mol of I_2 in a 2.00-L container.

Solution

Analyze: We are given a volume and initial molar amounts of the species in a reaction and are asked to determine in which direction the reaction must proceed to achieve equilibrium.

Plan: We can determine the starting partial pressures of all species from the information given. We can then substitute the starting partial pressures into the equilibrium-constant expression to calculate the reaction quotient, Q. Comparing the magnitudes of the equilibrium constant, which is given, and the reaction quotient will tell us in which direction the reaction will proceed.

Solve: The initial partial pressures are

$$P_{HI} = \frac{n_{HI}RT}{V} = \frac{(2.0 \times 10^{-2} \text{ mol})(0.0821 \text{ L-atm/mol-K})(721 \text{ K})}{2.00 \text{ L}} = 0.592 \text{ atm}$$

$$P_{H_2} = \frac{n_{H_2}RT}{V} = \frac{(1.0 \times 10^{-2} \text{ mol})(0.0821 \text{ L-atm/mol-K})(721 \text{ K})}{2.00 \text{ L}} = 0.296 \text{ atm}$$

$$P_{I_2} = \frac{n_{I_2}RT}{V} = \frac{(3.0 \times 10^{-2} \text{ mol})(0.0821 \text{ L-atm/mol-K})(721 \text{ K})}{2.00 \text{ L}} = 0.888 \text{ atm}$$

The reaction quotient is

$$Q = \frac{(P_{HI})^2}{P_{H_2}P_{I_2}} = \frac{(0.592)^2}{(0.296)(0.888)} = 1.3$$

Because $Q < K_{eq}$, the partial pressure of HI must increase and those of H_2 and I_2 must decrease to reach equilibrium; the reaction will proceed from left to right.

PRACTICE EXERCISE

At 1000 K the value of K_{eq} for the reaction $2SO_3(g) \rightleftharpoons 2SO_2(g) + O_2(g)$ is 0.338. Calculate the value for Q, and predict the direction in which the reaction will proceed toward equilibrium if the initial partial pressures of reactants are $P_{SO_3} = 0.16$ atm; $P_{SO_2} = 0.41$ atm; $P_{O_2} = 2.5$ atm.
Answer: $Q = 16$; $Q > K_{eq}$, so the reaction will proceed from right to left, forming more SO_3.

Calculating Equilibrium Concentrations

Chemists frequently need to calculate the amounts of reactants and products present at equilibrium. Our approach in solving problems of this type is similar to the one we used for evaluating equilibrium constants: We tabulate the initial partial pressures or concentrations, the changes therein, and the final equilibrium partial pressures or concentrations. Usually we end up using the equilibrium-constant expression to derive an equation that must be solved for an unknown quantity, as demonstrated in Sample Exercise 15.10.

SAMPLE EXERCISE 15.10

For the Haber process, $N_2(g) + 3H_2(g) \rightleftharpoons 2NH_3(g)$, $K_{eq} = 1.45 \times 10^{-5}$ at 500°C. In an equilibrium mixture of the three gases at 500°C, the partial pressure of H_2 is 0.928 atm and that of N_2 is 0.432 atm. What is the partial pressure of NH_3 in this equilibrium mixture?

Solution

Analyze: We are given an equilibrium constant and the equilibrium partial pressures of the reactants in the equation, and are asked to calculate the equilibrium partial pressure of the product.

Plan: We can set the equilibrium constant, which is given, equal to the equilibrium-constant expression and substitute in the partial pressures that are known. Then we can solve for the only unknown in the equation.

Solve: Because the mixture is in equilibrium, we need not worry about initial concentrations. We tabulate the equilibrium pressures as follows:

	$N_2(g)$ +	$3H_2(g)$ $\rightleftharpoons$	$2NH_3(g)$
Equilibrium pressure (atm):	0.432	0.928	x

Because we do not know the equilibrium pressure of NH_3, we represent it with a variable, x. At equilibrium the pressures must satisfy the equilibrium-constant expression:

$$K_{eq} = \frac{(P_{NH_3})^2}{P_{N_2}(P_{H_2})^3} = \frac{x^2}{(0.432)(0.928)^3} = 1.45 \times 10^{-5}$$

We now rearrange the equation to solve for x:

$$x^2 = (1.45 \times 10^{-5})(0.432)(0.928)^3 = 5.01 \times 10^{-6}$$

$$x = \sqrt{5.01 \times 10^{-6}} = 2.24 \times 10^{-3} \text{ atm} = P_{NH_3}$$

Comment: We can always check our answer by using it to recalculate the value of the equilibrium constant:

$$K_{eq} = \frac{(2.24 \times 10^{-3})^2}{(0.432)(0.928)^3} = 1.45 \times 10^{-5}$$

PRACTICE EXERCISE

At 500 K the reaction $PCl_5(g) \rightleftharpoons PCl_3(g) + Cl_2(g)$ has $K_{eq} = 0.497$. In an equilibrium mixture at 500 K, the partial pressure of PCl_5 is 0.860 atm and that of PCl_3 is 0.350 atm. What is the partial pressure of Cl_2 in the equilibrium mixture?
Answer: 1.22 atm

In many situations we will know the value of the equilibrium constant and the initial amounts of all species. We must then solve for the equilibrium amounts. This usually entails treating the change in partial pressure or concentration as a variable as equilibrium is achieved. The stoichiometry of the reaction gives us the relationship between the changes in the amounts of all the reactants and products, as illustrated in Sample Exercise 15.11.

SAMPLE EXERCISE 15.11

A 1.000-L flask is filled with 1.000 mol of H_2 and 2.000 mol of I_2 at 448°C. The value of the equilibrium constant, K_{eq}, for the reaction

$$H_2(g) + I_2(g) \rightleftharpoons 2HI(g)$$

at 448°C is 50.5. What are the partial pressures of H_2, I_2, and HI in the flask at equilibrium?

Solution

Analyze: We are given a volume, an equilibrium constant, and starting mole amounts of reactants for an equilibrium and are asked to calculate the equilibrium partial pressures of all species.

Plan: In this case, unlike Sample Exercise 15.10, we are not given any of the equilibrium partial pressures. We must develop some relationships that relate the initial partial pressures to those at equilibrium. The procedure is similar in many regards to that outlined in Sample Exercise 15.8.

Solve: First, we calculate the initial partial pressures of H_2 and I_2:

$$P_{H_2} = \frac{n_{H_2}RT}{V} = \frac{(1.000 \text{ mol})(0.0821 \text{ L-atm/mol-K})(721 \text{ K})}{1.00 \text{ L}} = 59.19 \text{ atm}$$

$$P_{I_2} = \frac{n_{I_2}RT}{V} = \frac{(2.000 \text{ mol})(0.0821 \text{ L-atm/mol-K})(721 \text{ K})}{1.00 \text{ L}} = 118.4 \text{ atm}$$

Second, we construct a table in which we tabulate the initial partial pressures:

	$H_2(g)$ +	$I_2(g)$ $\rightleftharpoons$	$2HI(g)$
Initial	59.19 atm	118.4 atm	0 atm
Change			
Equilibrium			

Third, we use the stoichiometry of the chemical equation to determine the changes in partial pressures that occur as the reaction proceeds to equilibrium. The partial pressures of H_2 and I_2 will decrease as equilibrium is established, and that of HI will increase. Let's represent the change in partial pressure of H_2 by the variable x. The balanced chemical equation tells us the relationship between the changes in the partial pressures of the three gases:

For each x atm of H_2 that reacts, x atm of I_2 are also consumed, and $2x$ atm of HI are produced.

Fourth, we use the initial partial pressures and the changes in partial pressures, as dictated by stoichiometry, to express the equilibrium partial pressures. With all of our entries we now have the following table:

	$H_2(g)$ +	$I_2(g)$ $\rightleftharpoons$	$2HI(g)$
Initial	59.19 atm	118.4 atm	0 atm
Change	$-x$ atm	$-x$ atm	$+2x$ atm
Equilibrium	$59.19 - x$ atm	$118.4 - x$ atm	$2x$ atm

Fifth, we substitute the equilibrium partial pressures into the equilibrium-constant expression and solve for the single unknown, x:

$$K_{eq} = \frac{(P_{HI})^2}{P_{H_2}P_{I_2}} = \frac{(2x)^2}{(59.19 - x)(118.4 - x)} = 50.5$$

If you have an equation-solving calculator, you can solve this equation directly for x. If not, expand this expression to obtain a quadratic equation in x:

$$4x^2 = 50.5(x^2 - 177.6x + 7.01 \times 10^3)$$
$$46.5x^2 - 8.97 \times 10^3 x + 3.54 \times 10^5 = 0$$

Solving the quadratic equation (Appendix A.3) leads to two solutions for x:

$$x = \frac{-(-8.97 \times 10^3) \pm \sqrt{(-8.97 \times 10^3)^2 - 4(46.5)(3.54 \times 10^5)}}{2(46.5)} = 137.6 \quad \text{or} \quad 55.3$$

When we substitute the first of these solutions, $x = 137.6$, into the expressions for the equilibrium partial pressures, we find *negative* partial pressures of H_2 and I_2. A negative partial pressure is not chemically meaningful, so we reject this solution. We use the other solution, $x = 55.3$, to find the equilibrium partial pressures:

$$P_{H_2} = 59.19 - x = 3.9 \text{ atm}$$
$$P_{I_2} = 118.4 - x = 63.1 \text{ atm}$$
$$P_{HI} = 2x = 110.6 \text{ atm}$$

Check: We can check our solution by putting these numbers into the equilibrium-constant expression:

$$K_{eq} = \frac{(P_{HI})^2}{P_{H_2}P_{I_2}} = \frac{(110.6)^2}{(3.9)(63.1)} = 50$$

Comment: Whenever we use the quadratic equation to solve an equilibrium problem, one of the solutions will not be chemically meaningful and will be rejected.

PRACTICE EXERCISE

For the equilibrium, $PCl_5(g) \rightleftharpoons PCl_3(g) + Cl_2(g)$, the equilibrium constant, K_{eq}, has the value 0.497 at 500 K. A gas cylinder at 500 K is charged with $PCl_5(g)$ at an initial pressure of 1.66 atm. What are the equilibrium pressures of PCl_5, PCl_3, and Cl_2 at this temperature?

Answer: $P_{PCl_5} = 0.967$ atm; $P_{PCl_3} = P_{Cl_2} = 0.693$ atm

15.6 Le Châtelier's Principle

In developing his process for making ammonia from N_2 and H_2, Haber sought the factors that might be varied to increase the yield of NH_3. Using the values of the equilibrium constant at various temperatures, he calculated the equilibrium amounts of NH_3 formed under a variety of conditions. Some of his results are shown in Figure 15.11 ▶. Notice that the percent of NH_3 present at equilibrium decreases with increasing temperature and increases with increasing pressure. We can understand these effects in terms of a principle first put forward by Henri-Louis Le Châtelier* (1850–1936), a French industrial chemist. **Le Châtelier's principle** can be stated as follows: *If a system at equilibrium is disturbed by a change in temperature, pressure, or the concentration of one of the components, the system will shift its equilibrium position so as to counteract the effect of the disturbance.*

In this section we will use Le Châtelier's principle to make qualitative predictions about the response of a system at equilibrium to various changes in external conditions. We will consider three ways that a chemical equilibrium can be disturbed: (1) adding or removing a reactant or product, (2) changing the pressure, and (3) changing the temperature.

Change in Reactant or Product Concentrations

A system at equilibrium is in a dynamic state; the forward and reverse processes are occurring at equal rates, and the system is in a state of balance. Altering the conditions of the system may disturb the state of balance. If this occurs, the

---

* Pronounced "le-SHOT-lee-ay."

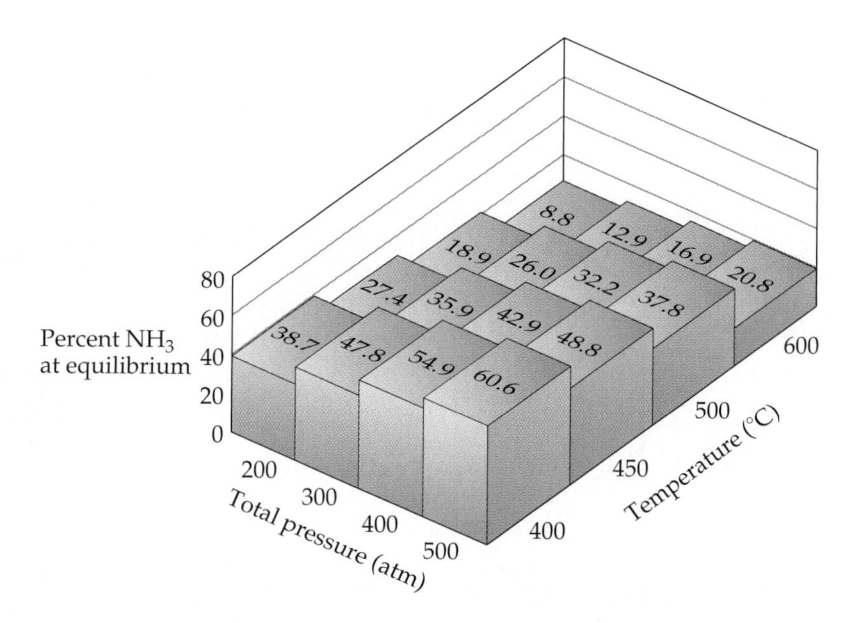

◀ **Figure 15.11** Plot of the effects of temperature and total pressure on the percentage of ammonia present in an equilibrium mixture of N_2, H_2, and NH_3. Each mixture was produced by starting with a 3 : 1 molar mixture of H_2 and N_2.

equilibrium shifts until a new state of balance is attained. Le Châtelier's principle states that the shift will be in the direction that minimizes or reduces the effect of the change. Therefore, *if a chemical system is at equilibrium and we add a substance (either a reactant or a product), the reaction will shift so as to reestablish equilibrium by consuming part of the added substance. Conversely, removing a substance will cause the reaction to move in the direction that forms more of that substance.*

As an example, consider an equilibrium mixture of N_2, H_2, and NH_3:

$$N_2(g) + 3H_2(g) \rightleftharpoons 2NH_3(g)$$

Adding H_2 would cause the system to shift so as to reduce the newly increased concentration of H_2. This can occur only by consuming H_2 and simultaneously consuming N_2 to form more NH_3. This situation is illustrated in Figure 15.12 ▼. Adding more N_2 to the equilibrium mixture would likewise cause the direction of the reaction to shift toward forming more NH_3. Removing NH_3 would also cause a shift toward producing more NH_3, whereas *adding* NH_3 to the system at equilibrium would cause the concentrations to shift in the direction that reduces the greater NH_3 concentration; that is, some of the added ammonia would decompose to form N_2 and H_2.

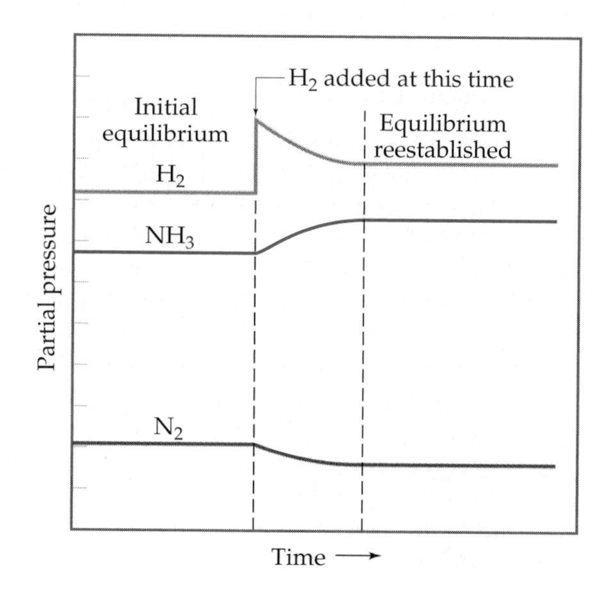

◀ **Figure 15.12** When H_2 is added to an equilibrium mixture of N_2, H_2, and NH_3, a portion of the H_2 reacts with N_2 to form NH_3, thereby establishing a new equilibrium position.

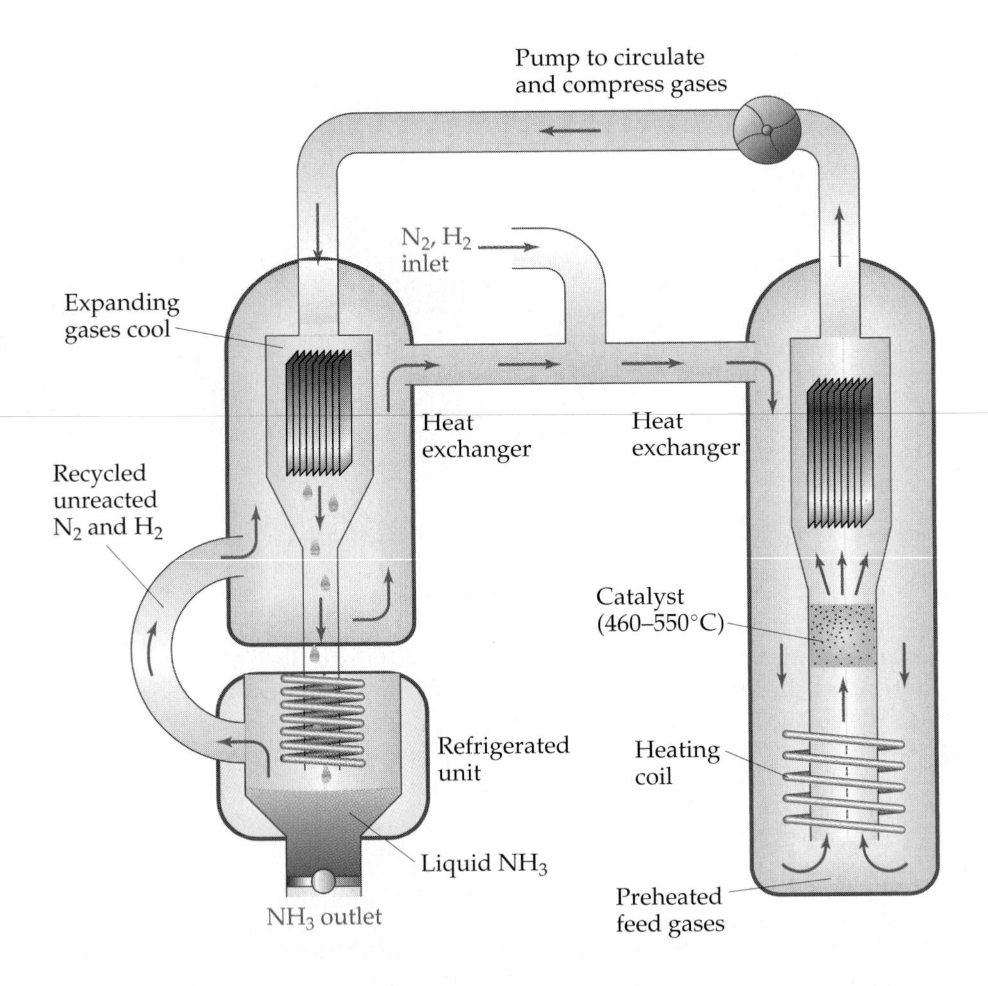

▶ **Figure 15.13** Schematic diagram summarizing the industrial production of ammonia. Incoming N_2 and H_2 gases are heated to approximately 500°C and passed over a catalyst. The resultant gas mixture is allowed to expand and cool, causing NH_3 to liquefy. Unreacted N_2 and H_2 gases are recycled.

In the Haber reaction, removing NH_3 from an equilibrium mixture of N_2, H_2, and NH_3 causes the reaction to shift from left to right to form more NH_3. If the NH_3 can be removed continuously, the yield of NH_3 from the Haber reaction can be increased dramatically. In the industrial production of ammonia the NH_3 is continuously removed by selectively liquefying it; the boiling point of NH_3 (−33°C) is much higher than that of N_2 (−196°C) and H_2 (−253°C). The liquid NH_3 is removed, and the N_2 and H_2 are recycled to form more NH_3, as diagrammed in Figure 15.13 ▲. By continually removing the product, the reaction is driven essentially to completion.

Effects of Volume and Pressure Changes

If a system is at equilibrium and its volume is decreased, thereby increasing its total pressure, Le Châtelier's principle indicates that the system will respond by shifting its equilibrium position to reduce the pressure. A system can reduce its pressure by reducing the total number of gas molecules (fewer molecules of gas exert a lower pressure). Thus, at constant temperature, *reducing the volume of a gaseous equilibrium mixture causes the system to shift in the direction that reduces the number of moles of gas*. Conversely, increasing the volume causes a shift in the direction that produces more gas molecules.

For example, let's consider the equilibrium $N_2O_4(g) \rightleftharpoons 2NO_2(g)$, which we saw in Figure 15.2. What happens if the total pressure of an equilibrium mixture is increased by decreasing the volume as shown in the sequential photos in Figure 15.14 ▶? According to Le Châtelier's principle, we expect the equilibrium will shift to the side that reduces the total number of moles of gas, which is the reactant side in this case. We therefore expect NO_2 to be converted into N_2O_4 as

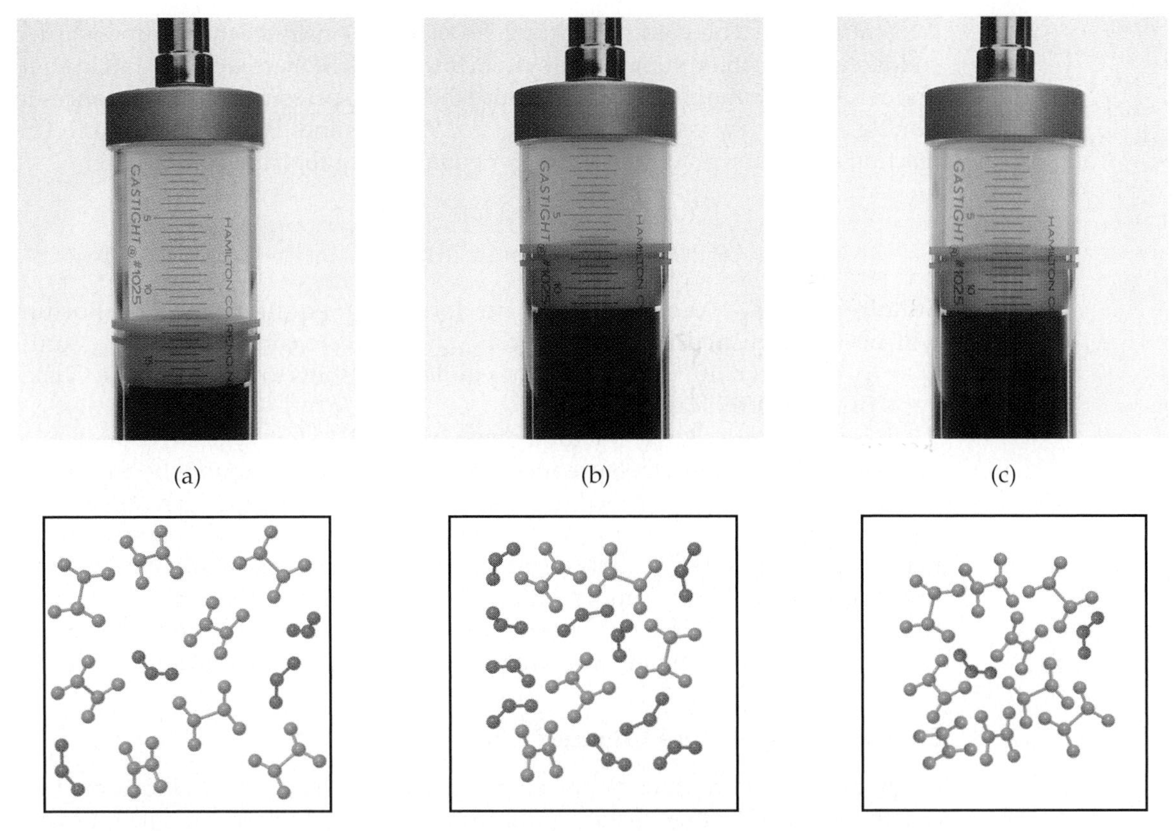

▲ **Figure 15.14** (a) An equilibrium mixture of brown $NO_2(g)$ (red) and colorless $N_2O_4(g)$ (grey) held in a gas-tight syringe. (b) The volume and hence the pressure are changed by moving the plunger. Compression of the mixture temporarily increases its temperature, making the equilibrium shift toward production of NO_2 and causing the mixture to darken. (c) When the mixture has returned to its original temperature, the color is as light as that in (a) because the formation of $N_2O_4(g)$ is favored by the pressure increase.

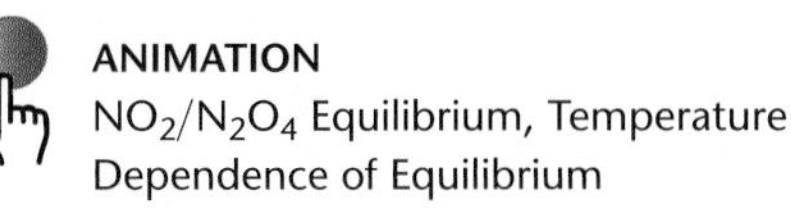

equilibrium is reestablished. Initially, compression of the gas mixture causes an increase in temperature. We will soon see that because the conversion of N_2O_4 to NO_2 is endothermic, $\Delta H = +58.02$ kJ, (Section 5.4), the temperature increase favors the formation of NO_2. Indeed, we see that the gas mixture darkens initially, indicating that the equilibrium has shifted to the right. However, as the temperature drops back to its original value, the equilibrium shifts back toward N_2O_4, causing the color of the gas mixture to lighten. It is important to note that the color of the gas mixture in Figure 15.14(c), after compression and subsequent cooling, is as light in color as that in Figure 15.14(a). This is because the pressure increase, as expected, caused the equilibrium to shift in favor of colorless N_2O_4.

For the reaction $N_2(g) + 3H_2(g) \rightleftharpoons 2NH_3(g)$, there are 2 mol of gas on the right side of the chemical equation ($2NH_3$) and 4 mol of gas on the left ($1N_2 + 3H_2$). Consequently, an increase in pressure (decrease in volume) leads to the formation of more NH_3, as indicated in Figure 15.11; the reaction shifts toward the side with fewer gas molecules. In the case of the reaction $H_2(g) + I_2(g) \rightleftharpoons 2HI(g)$, the number of moles of gaseous products (two) equals the number of moles of gaseous reactants; therefore, changing the pressure will not influence the position of the equilibrium.

Keep in mind that pressure-volume changes do *not* change the value of K_{eq} as long as the temperature remains constant. Rather, they change the partial pressures of the gaseous substances. In Sample Exercise 15.7 we calculated K_{eq} for an equilibrium mixture at 472°C that contained 7.38 atm H_2, 2.46 atm N_2, and

0.166 atm NH_3. The value of K_{eq} is 2.79×10^{-5}. Consider what happens if we suddenly reduce the volume of the system by one half. If there were no shift in equilibrium, this volume change would cause the partial pressures of all substances to double, giving $P_{H_2} = 14.76$ atm, $P_{N_2} = 4.92$ atm, and $P_{NH_3} = 0.332$ atm. The reaction quotient would then no longer equal the equilibrium constant.

$$Q = \frac{(P_{NH_3})^2}{P_{N_2}(P_{H_2})^3} = \frac{(0.332)^2}{(4.92)(14.76)^3} = 6.97 \times 10^{-6} \neq K_{eq}$$

Because $Q < K_{eq}$, the system is no longer at equilibrium. Equilibrium will be reestablished by increasing P_{NH_3} and decreasing P_{N_2} and P_{H_2} until $Q = K_{eq} = 2.79 \times 10^{-5}$. Therefore, the equilibrium shifts to the right as Le Châtelier's principle predicts.

It is possible to change the total pressure of the system without changing its volume. For example, pressure increases if additional amounts of any of the reacting components are added to the system. We have already seen how to deal with a change in concentration of a reactant or product. The total pressure within the reaction vessel might also be increased by adding a gas that is not involved in the equilibrium. For example, argon might be added to the ammonia equilibrium system. The argon would not alter the partial pressures of any of the reacting components and therefore would not cause a shift in equilibrium.

Effect of Temperature Changes

Changes in concentrations or partial pressures cause shifts in equilibrium without changing the value of the equilibrium constant. In contrast, almost every equilibrium constant changes in value as the temperature changes. For example, consider the following equilibrium, which is established when cobalt(II) chloride ($CoCl_2$) is dissolved in hydrochloric acid, $HCl(aq)$:

$$\underset{\text{Pale pink}}{Co(H_2O)_6^{2+}(aq)} + 4Cl^-(aq) \rightleftharpoons \underset{\text{Deep blue}}{CoCl_4^{2-}(aq)} + 6H_2O(l) \qquad \Delta H > 0 \qquad [15.21]$$

The formation of $CoCl_4^{2-}$ from $Co(H_2O)_6^{2+}$ is an endothermic process. We will discuss the significance of this enthalpy change shortly. Because $Co(H_2O)_6^{2+}$ is pink and $CoCl_4^{2-}$ is blue, the position of this equilibrium is readily apparent from the color of the solution. Figure 15.15(a) ▶ shows a room-temperature solution of $CoCl_2$ in $HCl(aq)$. Both $Co(H_2O)_6^{2+}$ and $CoCl_4^{2-}$ are present in significant amounts in the solution; the violet color results from the presence of both the pink and blue ions. When the solution is heated [Figure 15.15(b)], it becomes intensely blue in color, indicating that the equilibrium has shifted to form more $CoCl_4^{2-}$. Cooling the solution, as in Figure 15.15(c), leads to a more pink solution, indicating that the equilibrium has shifted to produce more $Co(H_2O)_6^{2+}$. How can we explain the dependence of this equilibrium on temperature?

We can deduce the rules for the temperature dependence of the equilibrium constant by applying Le Châtelier's principle. A simple way to do this is to treat heat as if it were a chemical reagent. In an endothermic reaction we can consider heat as a reactant, whereas in an exothermic reaction we can consider heat as a product.

$$\text{Endothermic:} \quad \text{Reactants} + \textit{heat} \rightleftharpoons \text{products}$$

$$\text{Exothermic:} \quad \text{Reactants} \rightleftharpoons \text{products} + \textit{heat}$$

When the temperature is increased, it is as if we have added a reactant, or a product, to the system at equilibrium. The equilibrium shifts in the direction that consumes the excess reactant (or product), namely heat. In an endothermic reaction, such as Equation 15.21, heat is absorbed as reactants are converted to products; thus,

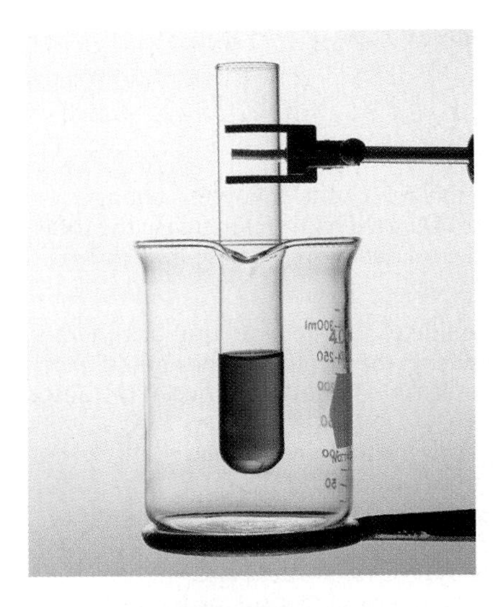

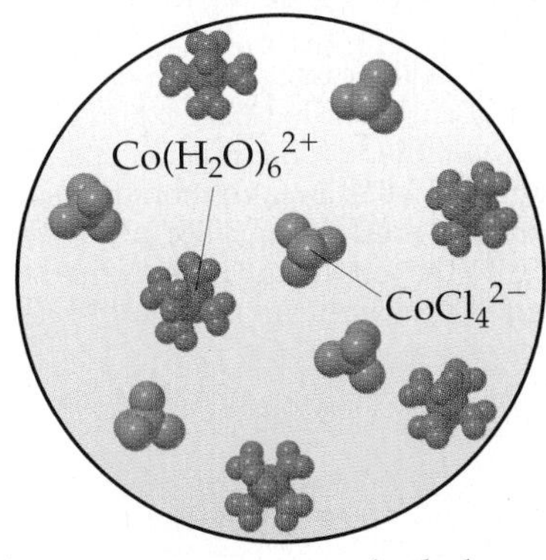

(a) At room temperature both the pink $Co(H_2O)_6^{2+}$ and blue $CoCl_4^{2-}$ ions are present in significant amounts, giving a violet color to the solution.

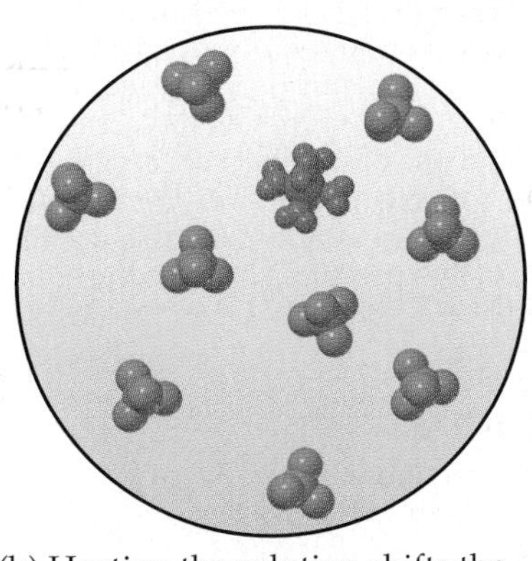

(b) Heating the solution shifts the equilibrium to the right, forming more blue $CoCl_4^{2-}$.

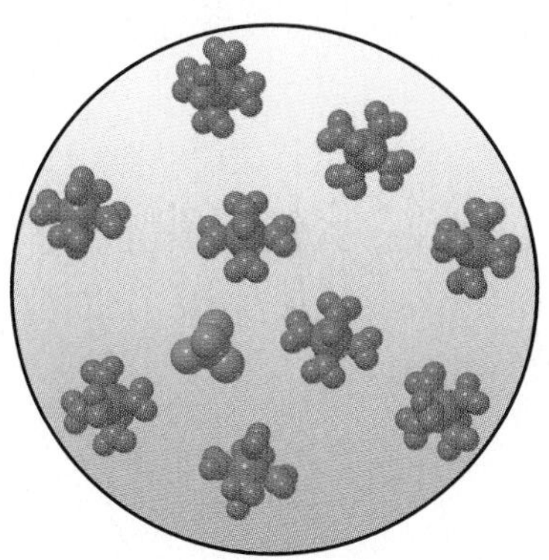

(c) Cooling the solution shifts the equilibrium to the left, toward pink $Co(H_2O)_6^{2+}$.

▲ **Figure 15.15** The effect of temperature on the equilibrium $Co(H_2O)_6^{2+}(aq) + 4Cl^-(aq) \rightleftharpoons CoCl_4^{2-}(aq) + 6H_2O(l)$. (a) At room temperature both the pink $Co(H_2O)_6^{2+}$ and blue $CoCl_4^{2-}$ ions are present in significant amounts, giving a violet color to the solution. (b) Heating the solution shifts the equilibrium to the right, forming more blue $CoCl_4^{2-}$. (c) Cooling the solution shifts the equilibrium to the left, toward pink $Co(H_2O)_6^{2+}$.

increasing the temperature causes the equilibrium to shift to the right, in the direction of products, and K_{eq} increases. For Equation 15.21, increasing the temperature leads to the formation of more $CoCl_4^{2-}$, as observed in Figure 15.15(b). In an exothermic reaction the opposite occurs. Heat is absorbed as products are converted to reactants, so the equilibrium shifts to the left and K_{eq} decreases. We can summarize these results as follows:

Endothermic: Increasing T results in an increase in K_{eq}.

Exothermic: Increasing T results in a decrease in K_{eq}.

Cooling a reaction has the opposite effect of heating it. As we lower the temperature, the equilibrium shifts to the side that produces heat. Thus, cooling an endothermic reaction shifts the equilibrium to the left, decreasing K_{eq}. We observed this effect in Figure 15.15(c). Cooling an exothermic reaction shifts the equilibrium to the right, increasing K_{eq}.

SAMPLE EXERCISE 15.12

Consider the following equilibrium:

$$N_2O_4(g) \rightleftharpoons 2NO_2(g) \qquad \Delta H^\circ = 58.0 \text{ kJ}$$

In what direction will the equilibrium shift when each of the following changes is made to a system at equilibrium: **(a)** add N_2O_4; **(b)** remove NO_2; **(c)** increase the total pressure by adding $N_2(g)$; **(d)** increase the volume; **(e)** decrease the temperature?

Solution

Analyze: We are given a series of changes to be made to a system at equilibrium and are asked to predict what effect each change will have on the position of the equilibrium.

Plan: Le Châtelier's principle can be used to determine the effects of each of these changes.

Solve: (a) The system will adjust to decrease the concentration of the added N_2O_4, so the equilibrium shifts to the right, in the direction of products.

(b) The system will adjust to the removal of NO_2 by shifting to the side that produces more NO_2; thus, the equilibrium shifts to the right.

(c) Adding N_2 will increase the total pressure of the system, but N_2 is not involved in the reaction. The partial pressures of NO_2 and N_2O_4 are therefore unchanged, and there is no shift in the position of the equilibrium.

(d) If the volume is increased, the system will shift in the direction that occupies a larger volume (more gas molecules); thus, the equilibrium shifts to the right. (This is the opposite of the effect observed in Figure 15.14 where the volume was decreased.)

(e) The reaction is endothermic, so we can imagine heat as a reagent on the reactant side of the equation. Decreasing the temperature will shift the equilibrium in the direction that produces heat, so the equilibrium shifts to the left, toward the formation of more N_2O_4. Note that only this last change also affects the value of the equilibrium constant, K_{eq}.

PRACTICE EXERCISE

For the reaction

$$PCl_5(g) \rightleftharpoons PCl_3(g) + Cl_2(g) \qquad \Delta H^\circ = 87.9 \text{ kJ}$$

in what direction will the equilibrium shift when **(a)** $Cl_2(g)$ is removed; **(b)** the temperature is decreased; **(c)** the volume of the reaction system is increased; **(d)** $PCl_3(g)$ is added?

Answers: **(a)** right; **(b)** left; **(c)** right; **(d)** left

SAMPLE EXERCISE 15.13

Using the standard heat of formation data in Appendix C, determine the standard enthalpy change for the reaction

$$N_2(g) + 3H_2(g) \rightleftharpoons 2NH_3(g)$$

Determine how the equilibrium constant for this reaction should change with temperature.

Solution

Analyze: We are asked to determine the standard enthalpy change of a reaction and to determine how the equilibrium constant for the reaction should vary with changing temperature.

Plan: We can use standard enthalpies of formation to calculate ΔH° for the reaction. We can then use Le Châtelier's principle to determine what effect temperature will have on the equilibrium constant. Recall that the standard enthalpy change for a reaction is given by the sum of the standard molar enthalpies of formation of the products, each multiplied by its coefficient in the balanced chemical equation, less the same quantities for the reactants. At 25°C, ΔH_f° for $NH_3(g)$ is −46.19 kJ/mol. The ΔH_f° values for $H_2(g)$ and $N_2(g)$ are zero by definition, because the enthalpies of formation of the elements in their normal states at 25°C are defined as zero (Section 5.7). Because 2 mol of NH_3 is formed, the total enthalpy change is

$$(2 \text{ mol})(-46.19 \text{ kJ/mol}) - 0 = -92.38 \text{ kJ}$$

The reaction in the forward direction is exothermic, so we can consider heat a product of the reaction. An increase in temperature causes the reaction to shift in the direction of less NH_3 and more N_2 and H_2. This effect is seen in the values for K_{eq} presented in Table 15.2 ▶. Notice that K_{eq} changes very markedly with change in temperature and that it is larger at lower temperatures. This is a matter of great practical importance. To form ammonia at a reasonable rate requires higher temperatures. At higher temperatures, however, the equilibrium constant is smaller, and so the percentage conversion to ammonia is smaller. To compensate for this, higher pressures are needed because high pressure favors ammonia formation.

PRACTICE EXERCISE

Using the thermodynamic data in Appendix C, determine the enthalpy change for the reaction

$$2POCl_3(g) \rightleftharpoons 2PCl_3(g) + O_2(g)$$

Use this result to determine how the equilibrium constant for the reaction should change with temperature.
Answer: $\Delta H^\circ = 508$ kJ; the equilibrium constant for the reaction will increase with increasing temperature.

TABLE 15.2 Variation in K_{eq} for the Equilibrium $N_2 + 3H_2 \rightleftharpoons 2NH_3$ as a Function of Temperature

Temperature (°C)	K_{eq}
300	4.34×10^{-3}
400	1.64×10^{-4}
450	4.51×10^{-5}
500	1.45×10^{-5}
550	5.38×10^{-6}
600	2.25×10^{-6}

The Effect of Catalysts

What happens if we add a catalyst to a chemical system that is at equilibrium? As shown in Figure 15.16 ▼, a catalyst lowers the activation barrier between the reactants and products. The activation energy of the forward reaction is lowered to the same extent as that for the reverse reaction. The catalyst thereby increases the rates of both the forward and reverse reactions. As a result, *a catalyst increases the rate at which equilibrium is achieved, but it does not change the composition of the equilibrium mixture.* The value of the equilibrium constant for a reaction is not affected by the presence of a catalyst.

The rate at which a reaction approaches equilibrium is an important practical consideration. As an example, let's again consider the synthesis of ammonia from N_2 and H_2. In designing a process for ammonia synthesis, Haber had to deal with a rapid decrease in the equilibrium constant with increasing temperature, as shown in Table 15.2. At temperatures sufficiently high to give a satisfactory reaction rate, the amount of ammonia formed was too small. The solution to this dilemma was to develop a catalyst that would produce a reasonably rapid approach to equilibrium at a sufficiently low temperature, so that the equilibrium constant was still reasonably large. The development of a suitable catalyst thus became the focus of Haber's research efforts.

After trying different substances to see which would be most effective, Haber finally settled on iron mixed with metal oxides. Variants of the original catalyst

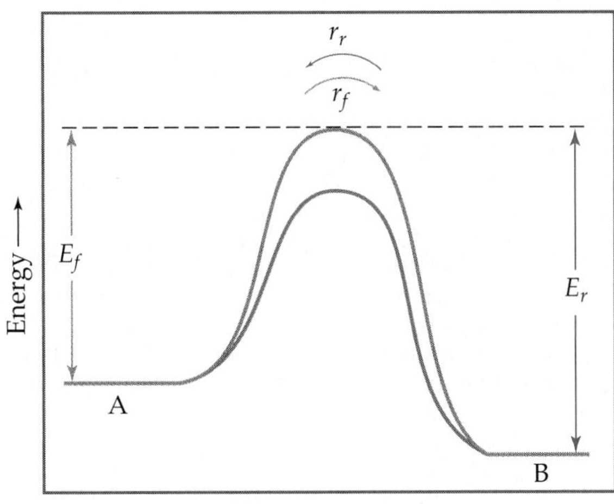

◀ **Figure 15.16** Schematic illustration of chemical equilibrium in the reaction $A \rightleftharpoons B$. At equilibrium the forward reaction rate, r_f, equals the reverse reaction rate, r_r. The violet curve represents the path over the transition state in the absence of a catalyst. A catalyst lowers the energy of the transition state, as shown by the green curve. Thus, the activation energy is lowered for both the forward and the reverse reactions. As a result, the rates of forward and reverse reactions in the catalyzed reaction are increased.

formulations are still used. These catalysts make it possible to obtain a reasonably rapid approach to equilibrium at temperatures around 400°C to 500°C and with gas pressures of 200 to 600 atm. The high pressures are needed to obtain a satisfactory degree of conversion at equilibrium. You can see from Figure 15.11 that if an improved catalyst could be found—one that would lead to sufficiently rapid reaction at temperatures lower than 400°C to 500°C—it would be possible to obtain the same degree of equilibrium conversion at much lower pressures. This would result in great savings in the cost of equipment for ammonia synthesis. In view of the growing need for nitrogen as fertilizer, the fixation of nitrogen is a process of ever-increasing importance.

SAMPLE INTEGRATIVE EXERCISE 15: Putting Concepts Together

At temperatures near 800°C, steam passed over hot coke (a form of carbon obtained from coal) reacts to form CO and H_2:

$$C(s) + H_2O(g) \rightleftharpoons CO(g) + H_2(g)$$

The mixture of gases that results is an important industrial fuel called *water gas*. **(a)** At 800°C the equilibrium constant for this reaction is $K_{eq} = 14.1$. What are the equilibrium partial pressures of H_2O, CO, and H_2 in the equilibrium mixture at this temperature if we start with solid carbon and 0.100 mol of H_2O in a 1.00-L vessel? **(b)** What is the minimum amount of carbon required to achieve equilibrium under these conditions? **(c)** What is the total pressure in the vessel at equilibrium? **(d)** At 25°C the value of K_{eq} for this reaction is 1.7×10^{-21}. Is the reaction exothermic or endothermic? **(e)** To produce the maximum amount of CO and H_2 at equilibrium, should the pressure of the system be increased or decreased?

Solution **(a)** To determine the equilibrium partial pressures, we proceed as in Sample Exercise 15.11, first determining the starting partial pressure of hydrogen.

$$P_{H_2O} = \frac{n_{H_2O}RT}{V} = \frac{(0.100 \text{ mol})(0.0821 \text{ L-atm/mol-K})(1073 \text{ K})}{1.00 \text{ L}} = 8.81 \text{ atm}$$

We then construct a table of starting partial pressures and their changes as equilibrium is achieved:

	$C(s)$ +	$H_2O(g)$ $\rightleftharpoons$	$CO(g)$ +	$H_2(g)$
Initial		8.81 atm	0 atm	0 atm
Change		$-x$	$+x$	$+x$
Equilibrium		$8.81 - x$ atm	x atm	x atm

There are no entries in the table under $C(s)$ because it does not appear in the equilibrium-constant expression. Substituting the equilibrium partial pressures of the other species into the equilibrium-constant expression for the reaction gives

$$K_{eq} = \frac{P_{CO}P_{H_2}}{P_{H_2O}} = \frac{(x)(x)}{(8.81 - x)} = 14.1$$

Multiplying through by the denominator gives a quadratic equation in x.

$$x^2 = (14.1)(8.81 - x)$$

$$x^2 + 14.1x - 124.22 = 0$$

Solving this equation for x using the quadratic formula yields $x = 6.14$ atm. Hence the equilibrium partial pressures are $P_{CO} = x = 6.14$ atm, $P_{H_2} = x = 6.14$ atm, and $P_{H_2O} = (8.81 - x) = 2.67$ atm.

(b) Part (a) shows that $x = 6.14$ atm of H_2O must react in order for the system to achieve equilibrium. We can use the ideal-gas equation to convert this partial pressure into a mole amount.

$$n = \frac{PV}{RT} = \frac{(6.14 \text{ atm})(1.00 \text{ L})}{(0.0821 \text{ L-atm/mol-K})(1073 \text{ K})} = 0.0697 \text{ mol}$$

Thus, 0.0697 mol of H_2O and the same amount of C must react to achieve equilibrium. As a result, there must be at least 0.0697 mol of C (0.836 g C) present among the reactants at the start of the reaction.

(c) The total pressure in the vessel at equilibrium is simply the sum of the equilibrium partial pressures:

$$P_{total} = P_{H_2O} + P_{CO} + P_{H_2} = 2.67 \text{ atm} + 6.14 \text{ atm} + 6.14 \text{ atm} = 14.95 \text{ atm}$$

(d) In discussing Le Châtelier's principle, we saw that endothermic reactions exhibit an increase in K_{eq} with increasing temperature. Because the equilibrium constant for this reaction increases as temperature increases, the reaction must be endothermic. From the enthalpies of formation given in Appendix C, we can verify our prediction by calculating the enthalpy change for the reaction, $\Delta H° = \Delta H_f°(CO) + \Delta H_f°(H_2) - \Delta H_f(C) - \Delta H_f°(H_2O) = +131.3$ kJ. The positive sign for $\Delta H°$ indicates that the reaction is endothermic.

(e) According to Le Châtelier's principle, a decrease in the pressure causes a gaseous equilibrium to shift toward the side of the equation with the greater number of moles of gas. In this case there are two moles of gas on the product side and only one on the reactant side. Therefore, the pressure should be reduced to maximize the yield of the CO and H_2.

Chemistry at Work Controlling Nitric Oxide Emissions

The formation of NO from N_2 and O_2 provides another interesting example of the practical importance of changes in the equilibrium constant and reaction rate with temperature. The equilibrium equation and the standard enthalpy change for the reaction are

$$\tfrac{1}{2}N_2(g) + \tfrac{1}{2}O_2(g) \rightleftharpoons NO(g) \qquad \Delta H° = 90.4 \text{ kJ} \qquad [15.22]$$

The reaction is endothermic; that is, heat is absorbed when NO is formed from the elements. By applying Le Châtelier's principle, we deduce that an increase in temperature will shift the equilibrium in the direction of more NO. The equilibrium constant, K_{eq}, for formation of 1 mol of NO from the elements at 300 K is only about 10^{-15}. In contrast, at a much higher temperature of about 2400 K the equilibrium constant is 10^{13} times as large, about 0.05. The manner in which K_{eq} for Equation 15.22 varies with temperature is shown in Figure 15.17 ▶.

This graph helps to explain why NO is a pollution problem. In the cylinder of a modern high-compression auto engine the temperatures during the fuel-burning part of the cycle may be on the order of 2400 K. Also, there is a fairly large excess of air in the cylinder. These conditions favor the formation of some NO. After the combustion, however, the gases are quickly cooled. As the temperature drops, the equilibrium in Equation 15.22 shifts strongly to the left (that is, in the direction of N_2 and O_2). The lower temperatures also mean that the rate of the reaction is decreased, however, so the NO formed at high temperatures is essentially "frozen" in that form as the gas cools.

The gases exhausting from the cylinder are still quite hot, perhaps 1200 K. At this temperature, as shown in Figure 15.17, the equilibrium constant for formation of NO is much smaller. However, the rate of conversion of NO to N_2 and O_2 is too slow to permit much loss of NO before the gases are cooled still further.

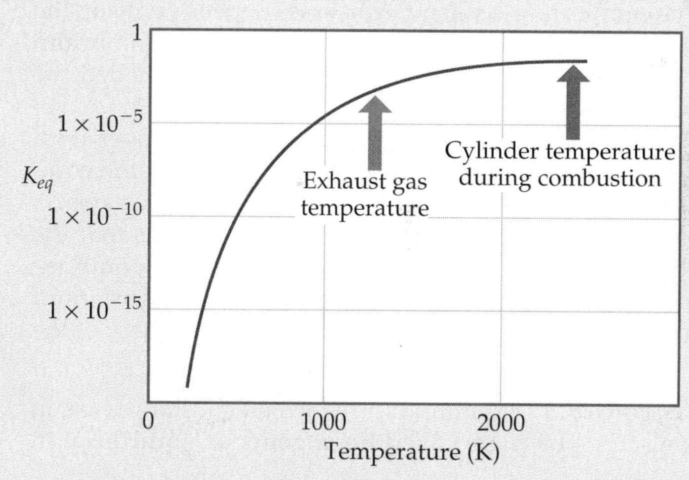

▲ **Figure 15.17** Variation in the equilibrium constant for the reaction $\tfrac{1}{2}N_2(g) + \tfrac{1}{2}O_2(g) \rightleftharpoons NO(g)$ as a function of temperature. It is necessary to use a log scale for K_{eq} because the values of K_{eq} vary over such a large range.

As discussed in the "Chemistry at Work" box in Section 14.7, one of the goals of automotive catalytic converters is to achieve the rapid conversion of NO to N_2 and O_2 at the temperature of the exhaust gas. Some catalysts for this reaction have been developed that are reasonably effective under the grueling conditions found in automotive exhaust systems. Nevertheless, scientists and engineers are continually searching for new materials that provide even more effective catalysis of the decomposition of nitrogen oxides.

Summary and Key Terms

Introduction and Section 15.1 A chemical reaction can achieve a state in which the forward and reverse processes are occurring at the same rate. This condition is called **chemical equilibrium,** and it results in the formation of an equilibrium mixture of the reactants and products of the reaction. The composition of an equilibrium mixture does not change with time. An equilibrium that is used throughout this chapter is the reaction of $N_2(g)$ with $H_2(g)$ to form $NH_3(g)$: $N_2(g) + 3H_2(g) \rightleftharpoons 2NH_3(g)$. This reaction is the basis of the **Haber process** for the production of ammonia.

Section 15.2 The relationship between the concentrations of the reactants and products of a system at equilibrium is given by the **law of mass action.** For a general equilibrium equation of the form $aA + bB \rightleftharpoons cC + dD$, the **equilibrium-constant expression** is written as follows:

$$K_{eq} = \frac{(P_C)^c(P_D)^d}{(P_A)^a(P_B)^b} \text{ for a gas-phase equilibrium or}$$

$$K_{eq} = \frac{[C]^c[D]^d}{[A]^a[B]^b} \text{ for an aqueous equilibrium}$$

The equilibrium-constant expression depends only on the stoichiometry of the reaction. For a system at equilibrium at a given temperature, K_{eq} will be a constant called the **equilibrium constant**.

The value of the equilibrium constant changes with temperature. A large value of K_{eq} indicates that the equilibrium mixture contains more products than reactants. A small value for the equilibrium constant means that the equilibrium lies toward the reactant side. The equilibrium-constant expression and the equilibrium constant of the reverse of a reaction are the reciprocals of those of the forward reaction.

Section 15.3 Equilibria for which all substances are in the same phase are called **homogeneous equilibria**; in **heterogeneous equilibria** two or more phases are present. Because the concentrations of pure solids and liquids are constant, these substances are left out of the equilibrium-constant expression for a heterogeneous equilibrium.

Section 15.4 If the concentrations of all species in an equilibrium are known, the equilibrium-constant expression can be used to calculate the value of the equilibrium constant. The changes in the concentrations of reactants and products on the way to achieving equilibrium are governed by the stoichiometry of the reaction.

Section 15.5 The **reaction quotient,** Q, is found by substituting reactant and product partial pressures or concentrations into the equilibrium-constant expression. If the system is at equilibrium, $Q = K_{eq}$. If $Q \neq K_{eq}$, however, the system is not at equilibrium. When $Q < K_{eq}$, the reaction will move toward equilibrium by forming more products (the reaction moves from left to right); when $Q > K_{eq}$, the reaction will proceed from right to left. Knowing the value of K_{eq} makes it possible to calculate the equilibrium amounts of reactants and products, often by the solution of an equation in which the unknown is the change in a partial pressure or concentration.

Section 15.6 **Le Châtelier's principle** states that if a system at equilibrium is disturbed, the equilibrium will shift to minimize the disturbing influence. By this principle, if a reactant or product is added to a system at equilibrium, the equilibrium will shift to consume the added substance. The effects of removing reactants or products and of changing the pressure or volume of a reaction can be similarly deduced. The enthalpy change for a reaction indicates how an increase in temperature affects the equilibrium: For an endothermic reaction, an increase in temperature shifts the equilibrium to the right; for an exothermic reaction a temperature increase shifts the equilibrium to the left. Catalysts affect the speed at which equilibrium is reached but do not affect the magnitude of K_{eq}.

Exercises

The Concept of Equilibrium; Equilibrium-Constant Expressions

15.1 The following diagrams represent a hypothetical reaction of $A \longrightarrow B$, with A represented by red spheres and B represented by blue spheres. The sequence from left to right represents the system as time passes. Do the diagrams indicate that the system reaches an equilibrium state? Explain.

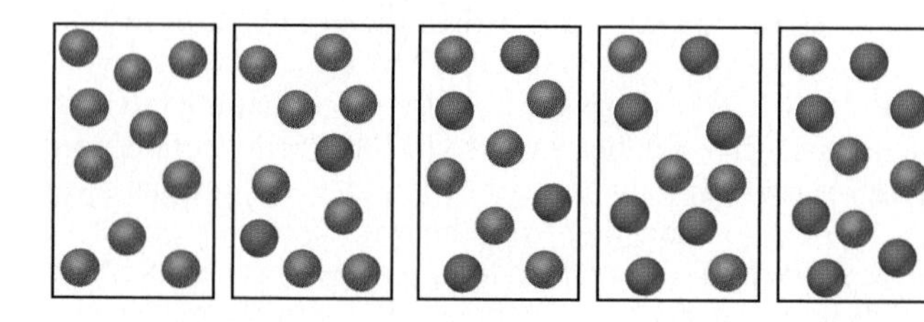

15.2 Explain what is incorrect about the following statements: **(a)** At equilibrium no more reactants are transformed into products. **(b)** At equilibrium the rate constant for the forward reaction equals that of the reverse reaction. **(c)** At equilibrium there are equal amounts of reactants and products.

15.3 Suppose that the gas-phase reactions A $\longrightarrow$ B and B $\longrightarrow$ A are both elementary processes with rate constants of $4.2 \times 10^{-3}\ s^{-1}$ and $1.5 \times 10^{-1}\ s^{-1}$, respectively. **(a)** What is the value of the equilibrium constant for the equilibrium $A(g) \rightleftharpoons B(g)$? **(b)** Which is greater at equilibrium, the partial pressure of A or the partial pressure of B? Explain.

15.4 Consider the reaction $A + B \rightleftharpoons C + D$. We will assume that both the forward and reverse reactions are elementary processes and that the value of the equilibrium constant is very large. **(a)** Which species predominate at equilibrium, reactants or products? **(b)** Which reaction has the larger rate constant, the forward or the reverse? Explain.

15.5 **(a)** What is the *law of mass action*? Illustrate the law by using the reaction $NO(g) + Br_2(g) \rightleftharpoons NOBr_2(g)$. **(b)** What is the difference between the *equilibrium-constant expression* and the *equilibrium constant* for a given equilibrium? **(c)** Describe an experiment that could be used to determine the value of the equilibrium constant for the reaction in part (a).

15.6 **(a)** The mechanism for a certain reaction, $A + B \rightleftharpoons C + D$, is unknown. Is it still possible to apply the law of mass action to the reaction? Explain. **(b)** Write the chemical reaction involved in the *Haber process*. Why is this reaction important to humanity? **(c)** Write the equilibrium-constant expression for the reaction in part (b).

15.7 Write the expression for K_{eq} for the following reactions. In each case indicate whether the reaction is homogeneous or heterogeneous.

(a) $3NO(g) \rightleftharpoons N_2O(g) + NO_2(g)$

(b) $CH_4(g) + 2H_2S(g) \rightleftharpoons CS_2(g) + 4H_2(g)$

(c) $Ni(CO)_4(g) \rightleftharpoons Ni(s) + 4CO(g)$

(d) $HF(aq) \rightleftharpoons H^+(aq) + F^-(aq)$

(e) $2Ag(s) + Zn^{2+}(aq) \rightleftharpoons 2Ag^+(aq) + Zn(s)$

15.8 Write the expressions for K_{eq} for the following reactions. In each case indicate whether the reaction is homogeneous or heterogeneous.

(a) $N_2(g) + O_2(g) \rightleftharpoons 2NO(g)$

(b) $Ti(s) + 2Cl_2(g) \rightleftharpoons TiCl_4(l)$

(c) $2C_2H_4(g) + 2H_2O(g) \rightleftharpoons 2C_2H_6(g) + O_2(g)$

(d) $Co(s) + 2H^+(aq) \rightleftharpoons Co^{2+}(aq) + H_2(g)$

(e) $NH_3(aq) + H_2O(l) \rightleftharpoons NH_4^+(aq) + OH^-(aq)$

15.9 When the following reactions come to equilibrium, does the equilibrium mixture contain mostly reactants or mostly products?

(a) $N_2(g) + O_2(g) \rightleftharpoons 2NO(g)$; $K_{eq} = 1.5 \times 10^{-10}$

(b) $2SO_2(g) + O_2(g) \rightleftharpoons 2SO_3(g)$; $K_{eq} = 2.5 \times 10^9$

15.10 Which of the following reactions lies to the right, favoring the formation of products, and which lies to the left, favoring formation of reactants?

(a) $2NO(g) + O_2(g) \rightleftharpoons 2NO_2(g)$; $K_{eq} = 5.0 \times 10^{12}$

(b) $2HBr(g) \rightleftharpoons H_2(g) + Br_2(g)$; $K_{eq} = 5.8 \times 10^{-18}$

15.11 The equilibrium constant for the reaction

$$2SO_3(g) \rightleftharpoons 2SO_2(g) + O_2(g)$$

is $K_{eq} = 2.4 \times 10^{-3}$ at 200°C. **(a)** Calculate K_{eq} for $2SO_2(g) + O_2(g) \rightleftharpoons 2SO_3(g)$. **(b)** Does the equilibrium favor SO_2 and O_2, or does it favor SO_3 at this temperature?

15.12 The equilibrium constant for the reaction

$$2NO(g) + Br_2(g) \rightleftharpoons 2NOBr(g)$$

is $K_{eq} = 1.3 \times 10^{-2}$ at 1000 K. **(a)** Calculate K_{eq} for $2NOBr(g) \rightleftharpoons 2NO(g) + Br_2(g)$ **(b)** At this temperature does the equilibrium favor NO and Br_2, or does it favor NOBr?

15.13 At 700°C, $K_{eq} = 0.112$ for the reaction

$$SO_2(g) + \tfrac{1}{2}O_2(g) \rightleftharpoons SO_3(g)$$

(a) What is the value of K_{eq} for the reaction $SO_3(g) \rightleftharpoons SO_2(g) + \frac{1}{2}O_2(g)$? **(b)** What is the value of K_{eq} for the reaction $2SO_2(g) + O_2(g) \rightleftharpoons 2SO_3(g)$? **(c)** What is the value of K_{eq} for the reaction $2SO_3(g) \rightleftharpoons 2SO_2(g) + O_2(g)$?

15.14 Consider the following equilibrium, for which $K_{eq} = 0.0752$ at 480°C:

$$2Cl_2(g) + 2H_2O(g) \rightleftharpoons 4HCl(g) + O_2(g)$$

(a) What is the value of K_{eq} for the reaction $4HCl(g) + O_2(g) \rightleftharpoons 2Cl_2(g) + 2H_2O(g)$? **(b)** What is the value of K_{eq} for the reaction $Cl_2(g) + H_2O(g) \rightleftharpoons 2HCl(g) + \frac{1}{2}O_2(g)$? **(c)** What is the value of K_{eq} for the reaction $2HCl(g) + \frac{1}{2}O_2(g) \rightleftharpoons Cl_2(g) + H_2O(g)$?

15.15 The equilibrium constant for the reaction

$$HClO_2(aq) \rightleftharpoons H^+(aq) + ClO_2^-(aq)$$

is $K_{eq} = 1.1 \times 10^{-2}$ at 25°C. **(a)** What is the value of K_{eq} for the reaction $\frac{1}{2}HClO_2(aq) \rightleftharpoons \frac{1}{2}H^+(aq) + \frac{1}{2}ClO_2^-(aq)$? **(b)** What is the value of K_{eq} for the reaction $2HClO_2(aq) \rightleftharpoons 2H^+(aq) + 2ClO_2^-(aq)$? **(c)** What is the value of K_{eq} for the reaction $2H^+(aq) + 2ClO_2^-(aq) \rightleftharpoons 2HClO_2(aq)$?

15.16 Consider the reactions $A(aq) + B(aq) \rightleftharpoons C(aq)$ and $C(aq) + D(aq) \rightleftharpoons E(aq) + A(aq)$ for which the equilibrium constants at 100°C are $K_{eq} = 1.9 \times 10^{-4}$ and $K_{eq} = 8.5 \times 10^2$, respectively. What is the value of K_{eq} for the reaction $B(aq) + D(aq) \rightleftharpoons E(aq)$?

15.17 Mercury(I) oxide decomposes into elemental mercury and elemental oxygen: $2Hg_2O(s) \rightleftharpoons 4Hg(l) + O_2(g)$. **(a)** Write an expression for K_{eq} that includes all the reactants and products. **(b)** Explain why we normally exclude pure solids and liquids from equilibrium-constant expressions. **(c)** Write an expression for K_{eq} that excludes the pure solid and pure liquid from the equilibrium expression.

15.18 Consider the equilibrium $Na_2O(s) + SO_2(g) \rightleftharpoons Na_2SO_3(s)$. **(a)** Write an expression for K_{eq} that includes all the reactants and products. **(b)** Explain why we normally exclude pure solids and liquids from equilibrium-constant expressions. **(c)** Write an expression for K_{eq} that excludes the pure solids from the equilibrium expression.

Calculating Equilibrium Constants

15.19 Gaseous hydrogen iodide is placed in a closed container at 425°C, where it partially decomposes to hydrogen and iodine: $2HI(g) \rightleftharpoons H_2(g) + I_2(g)$. At equilibrium it is found that $P_{HI} = 0.202$ atm, $P_{H_2} = 0.0274$ atm, and $P_{I_2} = 0.0274$ atm. What is the value of K_{eq} at this temperature?

15.20 Methanol (CH_3OH) is produced commercially by the catalyzed reaction of carbon monoxide and hydrogen: $CO(g) + 2H_2(g) \rightleftharpoons CH_3OH(g)$. An equilibrium mixture in a 2.00-L vessel is found to contain 0.0406 mol CH_3OH, 0.170 mol CO, and 0.302 mol H_2 at 500 K. Calculate K_{eq} at this temperature.

15.21 At 500 K the following equilibrium is established: $2NO(g) + Cl_2(g) \rightleftharpoons 2NOCl(g)$. An equilibrium mixture of the three gases has partial pressures of 0.095 atm, 0.171 atm, and 0.28 atm for NO, Cl_2, and NOCl, respectively. Calculate K_{eq} for this reaction at 500 K.

15.22 Phosphorus trichloride gas and chlorine gas react to form phosphorus pentachloride gas: $PCl_3(g) + Cl_2(g) \rightleftharpoons PCl_5(g)$. A gas vessel is charged with a mixture of $PCl_3(g)$ and $Cl_2(g)$, which is allowed to equilibrate at 450 K. At equilibrium the partial pressures of the three gases are $P_{PCl_3} = 0.124$ atm, $P_{Cl_2} = 0.157$ atm, and $P_{PCl_5} = 1.30$ atm. **(a)** What is the value of K_{eq} at this temperature? **(b)** Does the equilibrium favor reactants or products?

15.23 A mixture of 0.10 mol of NO, 0.050 mol of H_2, and 0.10 mol of H_2O is placed in a 1.0-L vessel at 300 K. The following equilibrium is established:

$$2NO(g) + 2H_2(g) \rightleftharpoons N_2(g) + 2H_2O(g)$$

At equilibrium $P_{NO} = 1.53$ atm. **(a)** Calculate the equilibrium partial pressures of H_2, N_2, and H_2O. **(b)** Calculate K_{eq}.

15.24 A mixture of 1.374 g of H_2 and 70.31 g of Br_2 is heated in a 2.00-L vessel at 700 K. These substances react as follows:

$$H_2(g) + Br_2(g) \rightleftharpoons 2HBr(g)$$

At equilibrium the vessel is found to contain 0.566 g of H_2. **(a)** Calculate the equilibrium partial pressures of H_2, Br_2, and HBr. **(b)** Calculate K_{eq}.

15.25 A mixture of 0.2000 mol of CO_2, 0.1000 mol of H_2, and 0.1600 mol of H_2O is placed in a 2.000-L vessel. The following equilibrium is established at 500 K:

$$CO_2(g) + H_2(g) \rightleftharpoons CO(g) + H_2O(g)$$

At equilibrium $P_{H_2O} = 3.51$ atm. **(a)** Calculate the equilibrium partial pressures of CO_2, H_2, and CO. **(b)** Calculate K_{eq} for the reaction.

15.26 A flask is charged with 1.500 atm of $N_2O_4(g)$ and 1.00 atm $NO_2(g)$ at 25°C. The equilibrium reaction is given in Equation 15.10. After equilibrium is reached, the partial pressure of NO_2 is 0.512 atm. **(a)** What is the equilibrium partial pressure of N_2O_4? **(b)** Calculate the value of K_{eq} for the reaction.

Applications of Equilibrium Constants

15.27 **(a)** How does a reaction quotient differ from an equilibrium constant? **(b)** If $Q < K_{eq}$, in which direction will a reaction proceed in order to reach equilibrium? **(c)** What condition must be satisfied so that $Q = K_{eq}$?

15.28 **(a)** How is a reaction quotient used to determine whether a system is at equilibrium? **(b)** If $Q > K_{eq}$, how must the reaction proceed to reach equilibrium? **(c)** At the start of a certain reaction, only reactants are present; no products have been formed. What is the value of Q at this point in the reaction?

15.29 At 100°C the equilibrium constant for the reaction $COCl_2(g) \rightleftharpoons CO(g) + Cl_2(g)$ has the value of $K_{eq} = 6.71 \times 10^{-9}$. Are the following mixtures of $COCl_2$, CO, and Cl_2 at 100°C at equilibrium? If not, indicate the direction that the reaction must proceed to achieve equilibrium. **(a)** $P_{COCl_2} = 6.12 \times 10^{-2}$ atm, $P_{CO} = 1.01 \times 10^{-4}$ atm, $P_{Cl_2} = 2.03 \times 10^{-4}$ atm; **(b)** $P_{COCl_2} = 1.38$ atm, $P_{CO} = 3.37 \times 10^{-6}$ atm, $P_{Cl_2} = 6.89 \times 10^{-5}$ atm; **(c)** $P_{COCl_2} = 3.06 \times 10^{-1}$ atm, $P_{CO} = P_{Cl_2} = 4.53 \times 10^{-5}$ atm.

15.30 As shown in Table 15.2, K_{eq} for the equilibrium

$$N_2(g) + 3H_2(g) \rightleftharpoons 2NH_3(g)$$

is 4.51×10^{-5} at 450°C. For each of the mixtures listed here, indicate whether the mixture is at equilibrium at 450°C. If it is not at equilibrium, indicate the direction (toward product or toward reactants) in which the mixture must shift to achieve equilibrium. **(a)** 105 atm NH_3, 35 atm N_2, 495 atm H_2; **(b)** 35 atm NH_3, 595 atm H_2, no N_2; **(c)** 26 atm NH_3, 42 atm H_2, 202 atm N_2; **(d)** 105 atm NH_3, 55 atm H_2, 5.0 atm N_2.

15.31 At 100°C, $K_{eq} = 2.39$ for the following reaction:

$$SO_2Cl_2(g) \rightleftharpoons SO_2(g) + Cl_2(g)$$

In an equilibrium mixture of the three gases the partial pressures of SO_2Cl_2 and SO_2 are 3.31 atm and 1.59 atm, respectively. What is the partial pressure of Cl_2 in the equilibrium mixture?

15.32 At 900 K the following reaction has $K_{eq} = 0.345$:

$$2SO_2(g) + O_2(g) \rightleftharpoons 2SO_3(g)$$

In an equilibrium mixture the partial pressures of SO_2 and O_2 are 0.165 atm and 0.755 atm, respectively. What is the equilibrium partial pressure of SO_3 in the mixture?

15.33 **(a)** At 1285°C the equilibrium constant for the reaction $Br_2(g) \rightleftharpoons 2Br(g)$ is $K_{eq} = 0.133$. A 0.200-L vessel containing an equilibrium mixture of the gases has 0.245 g $Br_2(g)$ in it. What is the mass of Br(*g*) in the vessel? **(b)** For the reaction $H_2(g) + I_2(g) \rightleftharpoons 2HI(g)$, $K_{eq} = 55.3$ at 700 K. In a 2.00-L flask containing an equilibrium mixture of the three gases, there are 0.056 g H_2 and 4.36 g I_2. What is the mass of HI in the flask?

15.34 **(a)** At 800 K the equilibrium constant for $I_2(g) \rightleftharpoons 2I(g)$ is $K_{eq} = 2.04 \times 10^{-3}$. If an equilibrium mixture in a 10.0-L vessel contains 3.22×10^{-2} g of I(*g*), how many grams of I_2 are in the mixture? **(b)** For $2SO_2(g) + O_2(g) \rightleftharpoons 2SO_3(g)$, $K_{eq} = 3.0 \times 10^4$ at 700 K. In a 2.00-L vessel the equilibrium mixture contains 2.65 g of SO_3 and 1.08 g of O_2. How many grams of SO_2 are in the vessel?

15.35 At 2000°C the equilibrium constant for the reaction

$$2NO(g) \rightleftharpoons N_2(g) + O_2(g)$$

is $K_{eq} = 2.4 \times 10^3$. If the initial partial pressure of NO is 37.3 atm, what are the equilibrium partial pressures of NO, N_2, and O_2?

15.36 For the equilibrium

$$Br_2(g) + Cl_2(g) \rightleftharpoons 2BrCl(g)$$

at 400 K, $K_{eq} = 7.0$. If 0.30 mol of Br_2 and 0.30 mol Cl_2 are introduced into a 1.0-L container at 400 K, what will be the equilibrium partial pressure of BrCl?

15.37 At 373 K, $K_{eq} = 0.416$ for the equilibrium

$$2NOBr(g) \rightleftharpoons 2NO(g) + Br_2(g)$$

If the pressures of NOBr(*g*) and NO(*g*) are equal, what is the equilibrium pressure of $Br_2(g)$?

15.38 At 21.8°C, $K_{eq} = 7.0 \times 10^{-2}$ for the equilibrium

$$NH_4HS(s) \rightleftharpoons NH_3(g) + H_2S(g)$$

Calculate the equilibrium partial pressures of NH_3 and H_2S if a sample of solid NH_4HS is placed in a closed vessel and decomposes until equilibrium is reached.

15.39 At 80°C, $K_{eq} = 5.42 \times 10^{-2}$ for the following reaction:

$$PH_3BCl_3(s) \rightleftharpoons PH_3(g) + BCl_3(g)$$

(a) Calculate the equilibrium partial pressures of PH_3 and BCl_3 if a solid sample of PH_3BCl_3 is placed in a closed vessel and decomposes until equilibrium is reached. **(b)** If the flask has a volume of 0.500 L, what is the minimum mass of $PH_3BCl_3(s)$ that must be added to the flask to achieve equilibrium?

15.40 Consider the following reaction:

$$CaSO_4(s) \rightleftharpoons Ca^{2+}(aq) + SO_4^{2-}(aq)$$

At 25°C the equilibrium constant is $K_{eq} = 2.4 \times 10^{-5}$ for this reaction. **(a)** If excess $CaSO_4(s)$ is mixed with water at 25°C to produce a saturated solution of $CaSO_4$, what are the equilibrium concentrations of Ca^{2+} and SO_4^{2-}? **(b)** If the resulting solution has a volume of 3.0 L, what is the minimum mass of $CaSO_4(s)$ needed to achieve equilibrium?

15.41 For the reaction $I_2(g) + Br_2(g) \rightleftharpoons 2IBr(g)$, $K_{eq} = 280$ at 150°C. Suppose that 0.500 mol IBr in a 1.00-L flask is allowed to reach equilibrium at 150°C. What are the equilibrium partial pressures of IBr, I_2, and Br_2?

15.42 At 25°C the reaction

$$CaCrO_4(s) \rightleftharpoons Ca^{2+}(aq) + CrO_4^{2-}(aq)$$

has an equilibrium constant $K_{eq} = 7.1 \times 10^{-4}$. What are the equilibrium concentrations of Ca^{2+} and CrO_4^{2-} in a saturated solution of $CaCrO_4$?

Le Châtelier's Principle

15.43 Consider the following equilibrium, for which $\Delta H < 0$:

$$2SO_2(g) + O_2(g) \rightleftharpoons 2SO_3(g)$$

How will each of the following changes affect an equilibrium mixture of the three gases? **(a)** $O_2(g)$ is added to the system; **(b)** the reaction mixture is heated; **(c)** the volume of the reaction vessel is doubled; **(d)** a catalyst is added to the mixture; **(e)** the total pressure of the system is increased by adding a noble gas; **(f)** $SO_3(g)$ is removed from the system.

15.44 For the following reaction, $\Delta H° = 2816$ kJ:

$$6CO_2(g) + 6H_2O(l) \rightleftharpoons C_6H_{12}O_6(s) + 6O_2(g)$$

How is the equilibrium yield of $C_6H_{12}O_6$ affected by **(a)** increasing P_{CO_2}; **(b)** increasing temperature; **(c)** removing CO_2; **(d)** decreasing the total pressure; **(e)** removing part of the $C_6H_{12}O_6$; **(f)** adding a catalyst?

15.45 How do the following changes affect the value of the equilibrium constant for a gas phase exothermic reaction: **(a)** removal of a reactant or product; **(b)** decrease in the volume; **(c)** decrease in the temperature; **(d)** addition of a catalyst?

15.46 For a certain gas-phase reaction, the fraction of products in an equilibrium mixture is increased by increasing the temperature and increasing the volume of the reaction vessel. **(a)** What can you conclude about the reaction from the influence of temperature on the equilibrium? **(b)** What can you conclude from the influence of increasing the volume?

15.47 Consider the following equilibrium between oxides of nitrogen:

$$3NO(g) \rightleftharpoons NO_2(g) + N_2O(g)$$

(a) Use data in Appendix C to calculate $\Delta H°$ for this reaction. **(b)** Will the equilibrium constant for the reaction increase or decrease with increasing temperature? Explain. **(c)** At constant temperature would a change in the volume of the container affect the fraction of products in the equilibrium mixture?

15.48 Methanol (CH_3OH) can be made by the reaction of CO with H_2:

$$CO(g) + 2H_2(g) \rightleftharpoons CH_3OH(g)$$

(a) Use thermochemical data in Appendix C to calculate $\Delta H°$ for this reaction. **(b)** In order to maximize the equilibrium yield of methanol, would you use a high or low temperature? **(c)** In order to maximize the equilibrium yield of methanol, would you use a high or low pressure?

Additional Exercises

15.49 Both the forward and reverse reactions of the following equilibrium are believed to be elementary steps:

$$CO(g) + Cl_2(g) \rightleftharpoons COCl(g) + Cl(g)$$

At 25°C the rate constants for the forward and reverse reactions are $1.4 \times 10^{-28}\ M^{-1}\ s^{-1}$ and $9.3 \times 10^{10}\ M^{-1}\ s^{-1}$, respectively. **(a)** What is the value for the equilibrium constant at 25°C? **(b)** Are reactants or products more plentiful at equilibrium?

15.50 A mixture of CH_4 and H_2O is passed over a nickel catalyst at 1000 K. The emerging gas is collected in a 5.00-L flask and is found to contain 8.62 g of CO, 2.60 g of H_2, 43.0 g of CH_4, and 48.4 g of H_2O. Assuming that equilibrium has been reached, calculate K_{eq} for the reaction.

15.51 When 2.00 mol of SO_2Cl_2 is placed in a 2.00-L flask at 303 K, 56% of the SO_2Cl_2 decomposes to SO_2 and Cl_2:

$$SO_2Cl_2(g) \rightleftharpoons SO_2(g) + Cl_2(g)$$

Calculate K_{eq} for this reaction at this temperature.

15.52 A mixture of H_2, S, and H_2S is held in a 1.0-L vessel at 90°C until the following equilibrium is achieved:

$$H_2(g) + S(s) \rightleftharpoons H_2S(g)$$

At equilibrium the mixture contains 0.46 g of H_2S and 0.40 g H_2. **(a)** Write the equilibrium-constant expression for this reaction. **(b)** What is the value of K_{eq} for the reaction at this temperature? **(c)** Why can we ignore the amount of S when doing the calculation in part (b)?

15.53 A sample of nitrosyl bromide (NOBr) decomposes according to the following equation:

$$2NOBr(g) \rightleftharpoons 2NO(g) + Br_2(g)$$

An equilibrium mixture in a 5.00-L vessel at 100°C contains 3.22 g of NOBr, 3.08 g of NO, and 4.19 g of Br_2. **(a)** Calculate K_{eq}. **(b)** What is the total pressure exerted by the mixture of gases?

15.54 Consider the hypothetical reaction $A(g) \rightleftharpoons 2B(g)$. A flask is charged with 0.55 atm of pure A, after which it is allowed to reach equilibrium at 0°C. At equilibrium the partial pressure of A is 0.36 atm. **(a)** What is the total pressure in the flask at equilibrium? **(b)** What is the value of K_{eq}?

15.55 As shown in Table 15.2, the equilibrium constant for the reaction $N_2(g) + 3H_2(g) \rightleftharpoons 2NH_3(g)$ is $K_{eq} = 4.34 \times 10^{-3}$ at 300°C. Pure NH_3 is placed in a 1.00-L flask and allowed to reach equilibrium at this temperature. There are 1.05 g NH_3 in the equilibrium mixture. **(a)** What are the masses of N_2 and H_2 in the equilibrium mixture? **(b)** What was the initial mass of ammonia placed in the vessel? **(c)** What is the total pressure in the vessel?

15.56 For the equilibrium

$$2IBr(g) \rightleftharpoons I_2(g) + Br_2(g)$$

$K_{eq} = 8.5 \times 10^{-3}$ at 150°C. If 0.025 mol of IBr is placed in a 2.0-L container, what is the partial pressure of this substance after equilibrium is reached?

15.57 For the equilibrium

$$PH_3BCl_3(s) \rightleftharpoons PH_3(g) + BCl_3(g)$$

$K_{eq} = 0.052$ at 60°C. Some solid PH_3BCl_3 is added to a closed 0.500-L vessel at 60°C; the vessel is then charged with 0.0128 mol of $BCl_3(g)$. What is the equilibrium partial pressure of PH_3?

[15.58] Solid NH_4HS is introduced into an evacuated flask at 24°C. The following reaction takes place:

$$NH_4HS(s) \rightleftharpoons NH_3(g) + H_2S(g)$$

At equilibrium the total pressure (for NH_3 and H_2S taken together) is 0.614 atm. What is K_{eq} for this equilibrium at 24°C?

[15.59] A 0.831-g sample of SO_3 is placed in a 1.00-L container and heated to 1100 K. The SO_3 decomposes to SO_2 and O_2:

$$2SO_3(g) \rightleftharpoons 2SO_2(g) + O_2(g)$$

At equilibrium the total pressure in the container is 1.300 atm. Find the value of K_{eq} for this reaction at 1100 K.

15.60 Nitric oxide (NO) reacts readily with chlorine gas as follows:

$$2NO(g) + Cl_2(g) \rightleftharpoons 2NOCl(g)$$

At 700 K the equilibrium constant, K_{eq}, for this reaction is 0.26. Predict the behavior of each of the following mixtures at this temperature: **(a)** $P_{NO} = 0.15$ atm, $P_{Cl_2} = 0.31$ atm, and $P_{NOCl} = 0.11$ atm; **(b)** $P_{NO} = 0.12$ atm, $P_{Cl_2} = 0.10$ atm, and $P_{NOCl} = 0.050$ atm; **(c)** $P_{NO} = 0.15$ atm, $P_{Cl_2} = 0.20$ atm, and $P_{NOCl} = 5.10 \times 10^{-3}$ atm.

15.61 At 900°C, $K_{eq} = 0.0108$ for the reaction

$$CaCO_3(s) \rightleftharpoons CaO(s) + CO_2(g)$$

A mixture of $CaCO_3$, CaO, and CO_2 is placed in a 10.0-L vessel at 900°C. For the following mixtures, will the amount of $CaCO_3$ increase, decrease, or remain the same as the system approaches equilibrium?
(a) 15.0 g $CaCO_3$, 15.0 g CaO, and 4.25 g CO_2
(b) 2.50 g $CaCO_3$, 25.0 g CaO, and 5.66 g CO_2
(c) 30.5 g $CaCO_3$, 25.5 g CaO, and 6.48 g CO_2

15.62 Nickel carbonyl, $Ni(CO)_4$, is an extremely toxic liquid with a low boiling point. Nickel carbonyl results from the reaction of nickel metal with carbon monoxide. For temperatures above the boiling point (42.2°C) of $Ni(CO)_4$, the reaction is

$$Ni(s) + 4CO(g) \rightleftharpoons Ni(CO)_4(g)$$

Nickel that is more than 99.9% pure can be produced by the *carbonyl process*: Impure nickel combines with CO at 50°C to produce $Ni(CO)_4(g)$. The $Ni(CO)_4$ is then heated to 200°C, causing it to decompose back into Ni(*s*) and CO(*g*). **(a)** Write the equilibrium-constant expression for the formation of $Ni(CO)_4$. **(b)** Given the temperatures used for the steps in the carbonyl process, do you think this reaction is endothermic or exothermic? **(c)** In the early days of automobiles, nickel-plated exhaust pipes were used. Even though the equilibrium constant for the formation of $Ni(CO)_4$ is very small at the temperature of automotive exhaust gases, the exhaust pipes quickly corroded. Explain why this occurred.

15.63 NiO is to be reduced to nickel metal in an industrial process by use of the reaction

$$NiO(s) + CO(g) \rightleftharpoons Ni(s) + CO_2(g)$$

At 1600 K the equilibrium constant for the reaction is $K_{eq} = 6.0 \times 10^2$. If a CO pressure of 150 torr is to be employed in the furnace and total pressure never exceeds 760 torr, will reduction occur?

[15.64] At 700 K the equilibrium constant for the reaction

$$CCl_4(g) \rightleftharpoons C(s) + 2Cl_2(g)$$

is $K_{eq} = 0.76$. A flask is charged with 2.00 atm of CCl_4, which then reaches equilibrium at 700 K. **(a)** What fraction of the CCl_4 is converted into C and Cl_2? **(b)** What are the partial pressures of CCl_4 and Cl_2 at equilibrium?

[15.65] The reaction $PCl_3(g) + Cl_2(g) \rightleftharpoons PCl_5(g)$ has $K_{eq} = 0.0870$ at 300°C. A flask is charged with 0.50 atm PCl_3, 0.50 atm Cl_2, and 0.20 atm PCl_5 at this temperature. **(a)** Use the reaction quotient to determine the direction the reaction must proceed in order to reach equilibrium. **(b)** Calculate the equilibrium partial pressures of the gases. **(c)** What effect will increasing the volume of the system have on the mole fraction of Cl_2 in the equilibrium mixture? **(d)** The reaction is exothermic. What effect will increasing the temperature of the system have on the mole fraction of Cl_2 in the equilibrium mixture?

[15.66] An equilibrium mixture of H_2, I_2, and HI at 458°C contains 0.112 mol H_2, 0.112 mol I_2, and 0.775 mol HI in a 5.00-L vessel. What are the equilibrium partial pressures when equilibrium is reestablished following the addition of 0.100 mol of HI?

[15.67] Consider the hypothetical reaction $A(g) + 2B(g) \rightleftharpoons 2C(g)$, for which $K_{eq} = 0.25$ at some temperature. A 1.00-L reaction vessel is loaded with 1.00 mol of compound C, which is allowed to reach equilibrium. Let the variable x represent the number of mol/L of compound A present at equilibrium. **(a)** In terms of x, what are the equilibrium concentrations of compounds B and C? **(b)** What limits must be placed on the value of x so that all concentrations are positive? **(c)** By putting the equilibrium concentrations (in terms of x) into the equilibrium-constant expression, derive an equation that can be solved for x. **(d)** The equation from part (c) is a cubic equation (one that has the form $ax^3 + bx^2 + cx + d = 0$). In general, cubic equations cannot be solved in closed form. However, you can estimate the solution by plotting the cubic equation in the allowed range of x that you specified in part (b). The point at which the cubic equation crosses the x-axis is the solution. **(e)** From the plot in part (d), estimate the equilibrium concentrations of A, B, and C. *Hint:* You can check the accuracy of your answer by substituting these concentrations into the equilibrium expression.

15.68 At 1200 K the approximate temperature of automobile exhaust gases (Figure 15.17), K_{eq} for the reaction

$$2CO_2(g) \rightleftharpoons 2CO(g) + O_2(g)$$

is about 1×10^{-13}. Assuming that the exhaust gas (total pressure 1 atm) contains 0.2% CO, 12% CO_2, and 3% O_2 by volume, is the system at equilibrium with respect to the above reaction? Based on your conclusion, would the CO concentration in the exhaust be decreased or increased by a catalyst that speeds up the reaction above?

15.69 Suppose that you worked at the U. S. Patent Office and a patent application came across your desk claiming that a newly developed catalyst was much superior to the Haber catalyst for ammonia synthesis because the catalyst led to much greater equilibrium conversion of N_2 and H_2 into NH_3 than the Haber catalyst under the same conditions. What would be your response?

Integrative Exercises

15.70 Consider the following equilibria in aqueous solution:
(i) $Na(s) + Ag^+(aq) \rightleftharpoons Na^+(aq) + Ag(s)$
(ii) $3Hg(l) + 2Al^{3+}(aq) \rightleftharpoons 3Hg^{2+}(aq) + 2Al(s)$
(iii) $Zn(s) + 2H^+(aq) \rightleftharpoons Zn^{2+}(aq) + H_2(g)$
(a) For each of the reactions, write the equilibrium-constant expression for K_{eq}. **(b)** By using information provided in Table 4.5, predict whether K_{eq} is large ($K_{eq} \gg 1$) or small ($K_{eq} \ll 1$). Explain your reasoning. **(c)** At 25°C the reaction $Cd(s) + Fe^{2+}(aq) \rightleftharpoons Cd^{2+}(aq) + Fe(s)$ has $K_{eq} = 6 \times 10^{-2}$. If Cd were added to Table 4.5, would you expect it to be above or below iron? Explain.

15.71 Silver chloride, AgCl(*s*), is an insoluble strong electrolyte. **(a)** Write the equation for the dissolution of AgCl(*s*) in $H_2O(l)$. **(b)** Write the expression for K_{eq} for the reaction in part (a). **(c)** Based on the thermochemical data in Appendix C and Le Châtelier's principle, predict whether the solubility of AgCl in H_2O increases or decreases with increasing temperature.

15.72 The hypothetical reaction $A + B \rightleftharpoons C$ occurs in the forward direction in a single step. The energy profile of the reaction is shown in the drawing. **(a)** Is the forward or reverse reaction faster at equilibrium? **(b)** Would you expect the equilibrium to favor reactants or products? **(c)** In general, how would a catalyst affect the energy profile shown? **(d)** How would a catalyst affect the ratio of the rate constants for the forward and reverse reactions? **(e)** How would you expect the equilibrium constant of the reaction to change with increasing temperature?

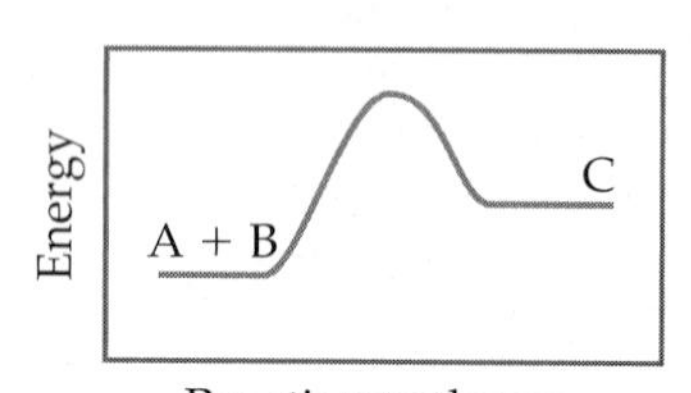

[15.73] Consider the equilibrium $A \rightleftharpoons B$ discussed in Equations 15.1 through 15.5. Assume that the only effect of a catalyst on the reaction is to lower the activation energies of the forward and reverse reactions, as shown in Figure 15.16. Using the Arrhenius equation (Section 14.5), prove that the equilibrium constant is the same for the catalyzed reaction as for the uncatalyzed one.

[15.74] At 25°C the reaction

$$NH_4HS(s) \rightleftharpoons NH_3(g) + H_2S(g)$$

has $K_{eq} = 0.120$. A 5.00-L flask is charged with 0.300 g of pure $H_2S(g)$ at 25°C. Solid NH_4HS is then added until there is excess unreacted solid remaining. **(a)** What is the initial pressure of $H_2S(g)$ in the flask? **(b)** Why does no reaction occur until NH_4HS is added? **(c)** What are the partial pressures of NH_3 and H_2S at equilibrium? **(d)** What is the mole fraction of H_2S in the gas mixture at equilibrium? **(e)** What is the minimum mass, in grams, of NH_4HS that must be added to the flask to achieve equilibrium?

[15.75] Write the equilibrium-constant expression for the following equilibrium:

$$C(s) + CO_2(g) \rightleftharpoons 2CO(g)$$

The table shows the relative mole percentages of $CO_2(g)$ and $CO(g)$ at a total pressure of 1 atm for several temperatures. Calculate the value of K_{eq} at each temperature. Is the reaction exothermic or endothermic? Explain.

Temperature (°C)	CO_2 (mol %)	CO (mol %)
850	6.23	93.77
950	1.32	98.68
1050	0.37	99.63
1200	0.06	99.94

15.76 In Section 11.5 we defined the vapor pressure of a liquid in terms of an equilibrium. **(a)** Write the equation representing the equilibrium between liquid water and water vapor, and the corresponding expression for K_{eq}. **(b)** By using data in Appendix B, give the value of K_{eq} for this reaction at 30°C. **(c)** What is the value of K_{eq} for any liquid in equilibrium with its vapor at the normal boiling point of the liquid?

[15.77] Polyvinyl chloride (PVC) is one of the most commercially important polymers (Table 12.1). PVC is made by addition polymerization of vinyl chloride (C_2H_3Cl). Vinyl chloride is synthesized from ethylene (C_2H_4) in a two-step process involving the following equilibria:

Equilibrium 1: $C_2H_4(g) + Cl_2(g) \rightleftharpoons C_2H_4Cl_2(g)$

Equilibrium 2: $C_2H_4Cl_2(g) \rightleftharpoons C_2H_3Cl(g) + HCl(g)$

The product of Equilibrium 1 is 1,2-dichloroethane, a compound in which one Cl atom is bonded to each C atom. **(a)** Draw Lewis structures for $C_2H_4Cl_2$ and C_2H_3Cl. What are the C—C bond orders in these two compounds? **(b)** Use average bond enthalpies (Table 8.4) to estimate the enthalpy changes in the two equilibria. **(c)** How would the yield of $C_2H_4Cl_2$ in Equilibrium 1 vary with temperature and volume? **(d)** How would the yield of C_2H_3Cl in Equilibrium 2 vary with temperature and volume? **(e)** Look up the normal boiling points of 1,2-dichloroethane and vinyl chloride in a sourcebook, such as the *CRC Handbook of Chemistry and Physics*. Based on these data, propose a reactor design (analogous to Figure 15.13) that could be used to maximize the amount of C_2H_3Cl produced by using the two equilibria.

eMedia Exercises

15.78 You can choose starting concentrations for reactants in the **Chemical Equilibrium** simulation (*eChapter 15.1*). **(a)** Write the equilibrium-constant expression for the reaction in the simulation. **(b)** Calculate the value of the equilibrium constant. **(c)** Does this reaction lie to the right or to the left?

15.79 Using the equilibrium constant that you calculated in the previous question for the reaction in the **Chemical Equilibrium** simulation (*eChapter 15.1*), predict what the equilibrium concentrations of reactants and products will be if you start with 0.0007 M $Fe^{3+}(aq)$ and 0.0004 M $SCN^-(aq)$. Use the simulation to check your answer.

15.80 The significance of an equilibrium constant's magnitude is illustrated in the **Equilibrium Constant** simulation (*eChapter 15.2*). Select the $A \rightleftharpoons B$ reaction and carry out several experiments with varying starting concentrations and varying equilibrium constants. Because reactants are favored when K_{eq} is very small, is it possible to enter a very small value for K_{eq}, start with only A (no B), and still have only A at equilibrium? Explain.

15.81 In Exercise 15.26 you calculated K_{eq} for the reaction $N_2O_4(g) \rightleftharpoons 2NO_2(g)$. **(a)** Which of the reactions in the **Equilibrium Constant** simulation (*eChapter 15.2*) best represents the equilibrium between $N_2O_4(g)$ and $2NO_2(g)$? **(b)** Using the simulation, enter the K_{eq} value from Exercise 15.26 and experiment with various starting concentrations of reactant and products. Can you select starting concentrations of reactants and products such that no change in concentrations is observed? Explain.

15.82 Consider the $A \rightleftharpoons B$ and $A \rightleftharpoons 2B$ reactions in the **Equilibrium Constant** simulation (*eChapter 15.2*). Assume that A and B are both gases; $A \rightleftharpoons B$ is an exothermic reaction; and $A \rightleftharpoons 2B$ is an endothermic reaction. For both reactions, state the effect that each of the following changes would have on the value of K_{eq}, the concentration of B at equilibrium, and the rate of the reaction: **(a)** increased temperature; **(b)** increased pressure; and **(c)** addition of a catalyst.

Chapter 16

Acid-Base Equilibria

Oranges and other citrus fruits contain citric acid and ascorbic acid (more commonly known as vitamin C). These acids give citrus fruits their characteristic sour flavors.

▶ What's Ahead ◀

- We start by reviewing the definitions of *acid* and *base* presented in Chapter 4 and learn that these are the *Arrhenius* definitions.
- We then learn the more general *Brønsted–Lowry* definitions for acid and base. A Brønsted–Lowry acid is a *proton donor* and a Brønsted–Lowry base is a *proton acceptor*.
- The *conjugate base* of a Brønsted–Lowry acid is what remains after the acid has donated a proton. Similarly, the *conjugate acid* of a Brønsted–Lowry base is the species that results when the base accepts a proton. Two such species that differ from each other only by the presence or absence of a proton together are known as a *conjugate acid-base pair*.
- *Autoionization* of water produces small concentrations of hydronium and hydroxide ions in pure water. The *equilibrium constant* for autoionization, K_w, defines the relationship between H_3O^+ and OH^- concentrations in aqueous solutions.
- The pH scale is used to describe the acidity or basicity of a solution.
- *Strong* acids and bases are those that ionize or dissociate completely in aqueous solution, whereas *weak* acids and bases ionize only partially.
- We learn that the ionization of a weak acid in water is an equilibrium process with an equilibrium constant K_a, which can be used to calculate the pH of a weak acid solution.
- Likewise, ionization of a weak base in water is an equilibrium process with equilibrium constant K_b, which we can use to calculate the pH of a weak base solution.
- There is a constant relationship, $K_a \times K_b = K_w$, between the K_a and K_b of any conjugate *acid-base* pair. This relationship can be used to determine the pH of a salt solution.
- We continue by exploring the relationship between chemical structure and acid-base behavior.
- Finally, we learn the *Lewis* definitions of acid and base. A Lewis acid is an *electron acceptor*, and a Lewis base is an *electron donor*. The Lewis definitions are more general and inclusive than either Arrhenius or Brønsted–Lowry.

ACIDS AND BASES are important in numerous chemical processes that occur around us, from industrial processes to biological ones, from reactions in the laboratory to those in our environment. The time required for a metal object immersed in water to corrode, the ability of an aquatic environment to support fish and plant life, the fate of pollutants washed out of the air by rain, and even the rates of reactions that maintain our lives all critically depend upon the acidity or basicity of solutions. Indeed, an enormous amount of chemistry can be understood in terms of acid-base reactions.

We have encountered acids and bases many times in earlier discussions. For example, a portion of Chapter 4 focused on their reactions. But what makes a substance behave as an acid or as a base? In this chapter we reexamine acids and bases, taking a closer look at how they are identified and characterized. In doing so, we will consider their behavior not only in terms of their structure and bonding but also in terms of the chemical equilibria in which they participate.

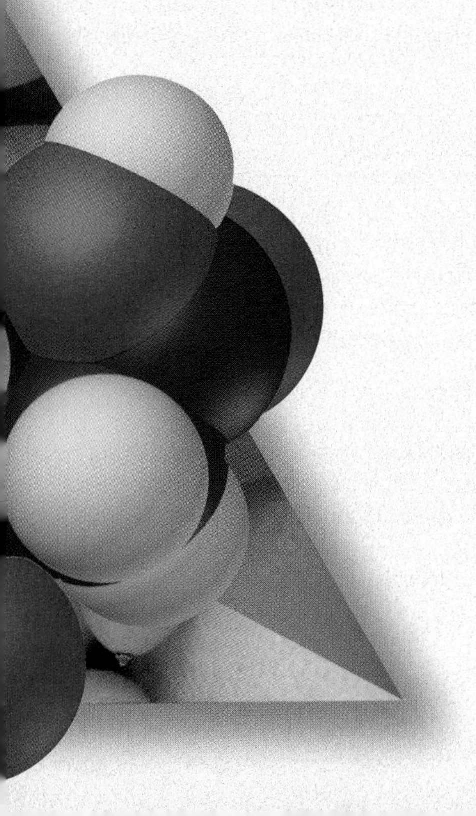

16.1 Acids and Bases: A Brief Review

ANIMATION
Introduction to Aqueous Acids, Introduction to Aqueous Bases

From the earliest days of experimental chemistry, scientists have recognized acids and bases by their characteristic properties. Acids have a sour taste (for example, citric acid in lemon juice) and cause certain dyes to change color (for example, litmus turns red on contact with acids). Indeed, the word *acid* comes from the Latin word *acidus*, meaning sour or tart. Bases, in contrast, have a bitter taste and feel slippery (soap is a good example). The word *base* comes from an old English meaning of the word, which is "to bring low." (We still use the word *debase* in this sense, meaning to lower the value of something.) When bases are added to acids, they lower the amount of acid. Indeed, when acids and bases are mixed in certain proportions, their characteristic properties disappear altogether. (Section 4.3)

Historically, chemists have sought to relate the properties of acids and bases to their compositions and molecular structures. By 1830 it was evident that all acids contain hydrogen but not all hydrogen-containing substances are acids. In the 1880s the Swedish chemist Svante Arrhenius (1859–1927) linked acid behavior with the presence of H^+ ions and base behavior with the presence of OH^- ions in aqueous solution. He defined acids as substances that produce H^+ ions in water, and bases as substances that produce OH^- ions in water. Indeed, the properties of aqueous solutions of acids, such as sour taste, are due to $H^+(aq)$, whereas the properties of aqueous solutions of bases are due to $OH^-(aq)$. Over time the Arrhenius concept of acids and bases came to be stated in the following way: *Acids are substances that, when dissolved in water, increase the concentration of H^+ ions*. Likewise, *bases are substances that, when dissolved in water, increase the concentration of OH^- ions*.

Hydrogen chloride is an Arrhenius acid. Hydrogen chloride gas is highly soluble in water because of its chemical reaction with water, which produces hydrated H^+ and Cl^- ions:

$$HCl(g) \xrightarrow{H_2O} H^+(aq) + Cl^-(aq) \qquad [16.1]$$

The aqueous solution of HCl is known as hydrochloric acid. Concentrated hydrochloric acid is about 37% HCl by mass and is 12 *M* in HCl.

Sodium hydroxide is an Arrhenius base. Because NaOH is an ionic compound, it dissociates into Na^+ and OH^- ions when it dissolves in water, thereby releasing OH^- ions into the solution.

16.2 Brønsted–Lowry Acids and Bases

The Arrhenius concept of acids and bases, while useful, has limitations. For one thing, it is restricted to aqueous solutions. In 1923 the Danish chemist Johannes Brønsted (1879–1947) and the English chemist Thomas Lowry (1874–1936) proposed a more general definition of acids and bases. Their concept is based on the fact that acid-base reactions involve the transfer of H^+ ions from one substance to another.

The H^+ Ion in Water

In Equation 16.1 hydrogen chloride is shown ionizing in water to form $H^+(aq)$. *An H^+ ion is simply a proton with no surrounding valence electron*. This small, positively charged particle interacts strongly with the nonbonding electron pairs of water molecules to form hydrated hydrogen ions. For example, the interaction of a proton with one water molecule forms the **hydronium ion**, $H_3O^+(aq)$:

$$H^+ + :\ddot{O}{-}H \longrightarrow \left[H{-}\ddot{O}{-}H \right]^+ \qquad [16.2]$$
(with an H bonded below each O)

The formation of hydronium ions is one of the complex features of the interaction of the H^+ ion with liquid water. In fact, the H_3O^+ ion can form hydrogen bonds to additional H_2O molecules to generate larger clusters of hydrated hydrogen ions, such as $H_5O_2^+$ and $H_9O_4^+$ (Figure 16.1 ▶).

Chemists use $H^+(aq)$ and $H_3O^+(aq)$ interchangeably to represent the same thing—namely the hydrated proton that is responsible for the characteristic properties of aqueous solutions of acids. We often use the $H^+(aq)$ ion for simplicity and convenience, as we did in Equation 16.1. The $H_3O^+(aq)$ ion, however, more closely represents reality.

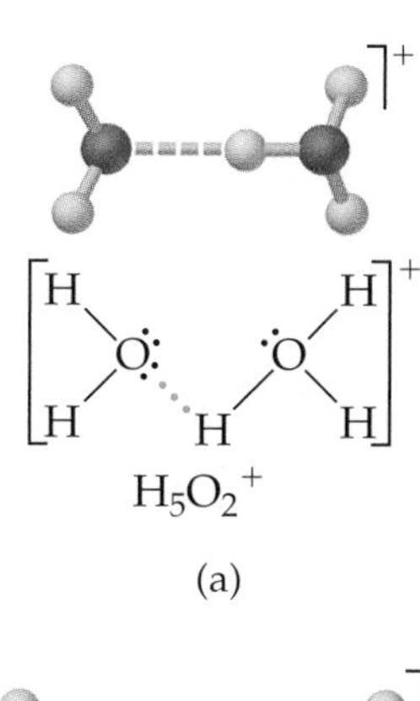

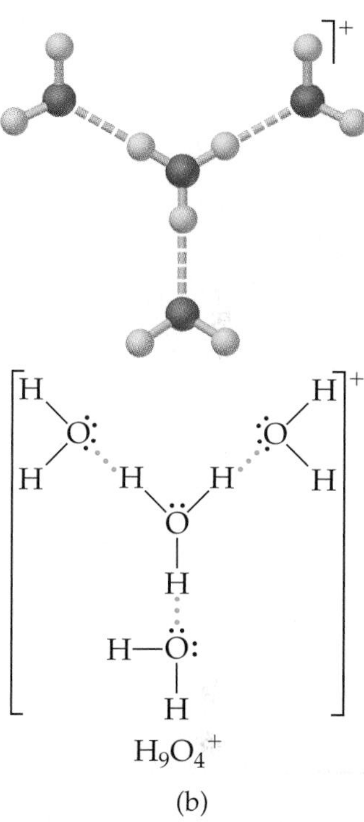

▲ **Figure 16.1** Lewis structures and molecular models for $H_5O_2^+$ and $H_9O_4^+$. There is good experimental evidence for the existence of both these species.

Proton-Transfer Reactions

When we closely examine the reaction that occurs when HCl dissolves in water, we find that the HCl molecule actually transfers an H^+ ion (a proton) to a water molecule as depicted in Figure 16.2 ▼. Thus, we can represent the reaction as occurring between an HCl molecule and a water molecule to form hydronium and chloride ions:

$$HCl(g) + H_2O(l) \longrightarrow H_3O^+(aq) + Cl^-(aq) \qquad [16.3]$$

Brønsted and Lowry proposed defining acids and bases in terms of their ability to transfer protons. According to their definition, *an acid is a substance (molecule or ion) that can donate a proton to another substance.* Likewise, *a base is a substance that can accept a proton.* Thus, when HCl dissolves in water (Equation 16.3), HCl acts as a **Brønsted–Lowry acid** (it donates a proton to H_2O), and H_2O acts as a **Brønsted–Lowry base** (it accepts a proton from HCl).

Because the emphasis in the Brønsted–Lowry concept is on proton transfer, the concept also applies to reactions that do not occur in aqueous solution. In the reaction between HCl and NH_3, for example, a proton is transferred from the acid HCl to the base NH_3:

$$:\ddot{\underset{..}{Cl}}-H + :\underset{H}{\overset{H}{\underset{|}{\overset{|}{N}}}}-H \longrightarrow :\ddot{\underset{..}{Cl}}:^- + \left[H-\underset{H}{\overset{H}{\underset{|}{\overset{|}{N}}}}-H\right]^+ \qquad [16.4]$$

This reaction can occur in the gas phase. The hazy film that forms on the windows of general chemistry laboratories and on glassware in the lab is largely solid NH_4Cl formed by the gas-phase reaction of HCl and NH_3 (Figure 16.3 ▼).

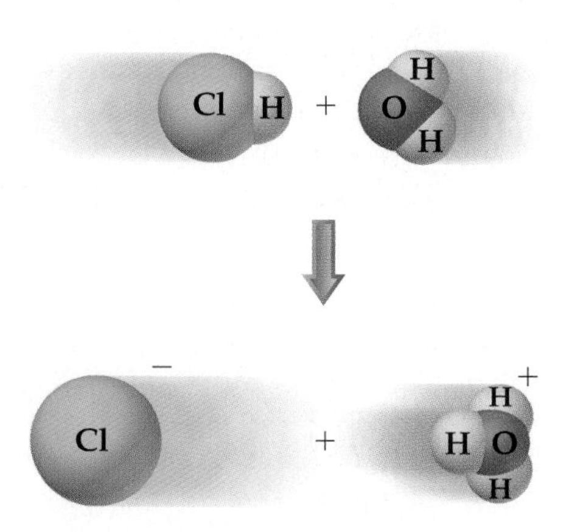

▲ **Figure 16.2** When a proton is transferred from HCl to H_2O, HCl acts as the Brønsted–Lowry acid and H_2O acts as the Brønsted–Lowry base.

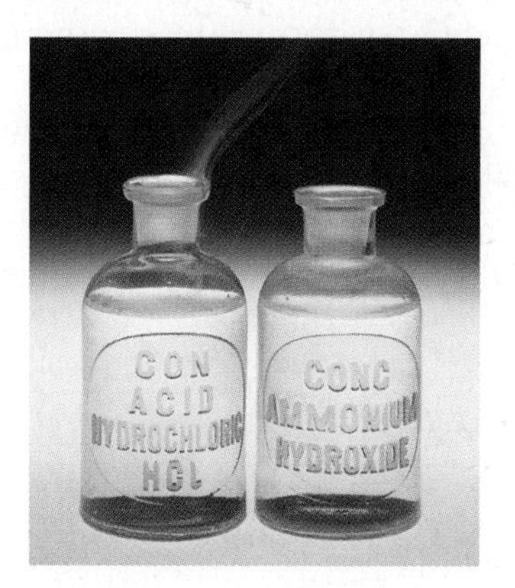

▲ **Figure 16.3** The $HCl(g)$ escaping from concentrated hydrochloric acid and the $NH_3(g)$ escaping from aqueous ammonia (here labeled ammonium hydroxide) combine to form a white fog of $NH_4Cl(s)$.

Let's consider another example that compares the relationship between the Arrhenius definitions and the Brønsted–Lowry definitions of acids and bases—an aqueous solution of ammonia, in which the following equilibrium occurs:

$$NH_3(aq) + H_2O(l) \rightleftharpoons NH_4^+(aq) + OH^-(aq) \quad [16.5]$$

Ammonia is an Arrhenius base because adding it to water leads to an increase in the concentration of $OH^-(aq)$. It is a Brønsted–Lowry base because it accepts a proton from H_2O. The H_2O molecule in Equation 16.5 acts as a Brønsted–Lowry acid because it donates a proton to the NH_3 molecule.

An acid and a base always work together to transfer a proton. In other words, a substance can function as an acid only if another substance simultaneously behaves as a base. To be a Brønsted–Lowry acid, a molecule or ion must have a hydrogen atom that it can lose as an H^+ ion. To be a Brønsted–Lowry base, a molecule or ion must have a nonbonding pair of electrons that it can use to bind the H^+ ion.

Some substances can act as an acid in one reaction and as a base in another. For example, H_2O is a Brønsted–Lowry base in its reaction with HCl (Equation 16.3) and a Brønsted–Lowry acid in its reaction with NH_3 (Equation 16.5). A substance that is capable of acting as either an acid or a base is called **amphoteric**. An amphoteric substance acts as a base when combined with something more strongly acidic than itself, and as an acid when combined with something more strongly basic than itself.

Conjugate Acid-Base Pairs

In any acid-base equilibrium both the forward reaction (to the right) and the reverse reaction (to the left) involve proton transfers. For example, consider the reaction of an acid, which we will denote HX, with water.

$$HX(aq) + H_2O(l) \rightleftharpoons X^-(aq) + H_3O^+(aq) \quad [16.6]$$

In the forward reaction HX donates a proton to H_2O. Therefore, HX is the Brønsted–Lowry acid, and H_2O is the Brønsted–Lowry base. In the reverse reaction the H_3O^+ ion donates a proton to the X^- ion, so H_3O^+ is the acid and X^- is the base. When the acid HX donates a proton, it leaves behind a substance, X^-, which can act as a base. Likewise, when H_2O acts as a base, it generates H_3O^+, which can act as an acid.

An acid and a base such as HX and X^- that differ only in the presence or absence of a proton are called a **conjugate acid-base pair**.* Every acid has a **conjugate base**, formed by removing a proton from the acid. For example, OH^- is the conjugate base of H_2O, and X^- is the conjugate base of HX. Similarly, every base has associated with it a **conjugate acid**, formed by adding a proton to the base. Thus, H_3O^+ is the conjugate acid of H_2O, and HX is the conjugate acid of X^-.

In any acid-base (proton-transfer) reaction we can identify two sets of conjugate acid-base pairs. For example, consider the reaction between nitrous acid (HNO_2) and water:

remove H^+

$$HNO_2(aq) + H_2O(l) \rightleftharpoons NO_2^-(aq) + H_3O^+(aq) \quad [16.7]$$

Acid | Base | Conjugate base | Conjugate acid

add H^+

* The word *conjugate* means "joined together as a pair."

Likewise, for the reaction between NH_3 and H_2O (Equation 16.5), we have

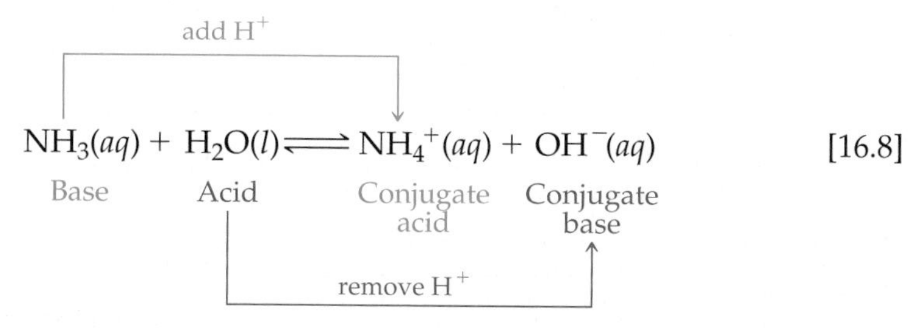

$$NH_3(aq) + H_2O(l) \rightleftharpoons NH_4^+(aq) + OH^-(aq) \quad [16.8]$$

SAMPLE EXERCISE 16.1

(a) What is the conjugate base of each of the following acids: $HClO_4$; H_2S; PH_4^+; HCO_3^-?

(b) What is the conjugate acid of each of the following bases: CN^-; SO_4^{2-}; H_2O; HCO_3^-?

Solution

Analyze: We are asked to give the conjugate base for each of a series of species and to give the conjugate acid for each of another series of species.

Plan: The conjugate base of a substance is simply the parent substance minus one proton, and the conjugate acid of a substance is the parent substance plus one proton.

Solve: (a) $HClO_4$ less one proton (H^+) is ClO_4^-. The other conjugate bases are HS^-, PH_3, and CO_3^{2-}. **(b)** CN^- plus one proton (H^+) is HCN. The other conjugate acids are HSO_4^-, H_3O^+, and H_2CO_3.

Notice that the hydrogen carbonate ion (HCO_3^-) is amphoteric: It can act as either an acid or a base.

PRACTICE EXERCISE

Write the formula for the conjugate acid of each of the following: HSO_3^-; F^-; PO_4^{3-}; CO.

Answers: H_2SO_3; HF; HPO_4^{2-}; HCO^+

SAMPLE EXERCISE 16.2

The hydrogen sulfite ion (HSO_3^-) is amphoteric. **(a)** Write an equation for the reaction of HSO_3^- with water, in which the ion acts as an acid. **(b)** Write an equation for the reaction of HSO_3^- with water, in which the ion acts as a base. In both cases identify the conjugate acid-base pairs.

Solution

Analyze and Plan: We are asked to write two equations representing reactions between HSO_3^- and water, one in which HSO_3^- should donate a proton to water, thereby acting as a Brønsted–Lowry acid, and one in which HSO_3^- should accept a proton from water, thereby acting as a base. We are also asked to identify the conjugate pairs in each equation.

Solve: (a)

$$HSO_3^-(aq) + H_2O(l) \rightleftharpoons SO_3^{2-}(aq) + H_3O^+(aq)$$

The conjugate pairs in this equation are HSO_3^- (acid) and SO_3^{2-} (conjugate base); and H_2O (base) and H_3O^+ (conjugate acid).

(b)

$$HSO_3^-(aq) + H_2O(l) \rightleftharpoons H_2SO_3(aq) + OH^-(aq)$$

The conjugate pairs in this equation are H_2O (acid) and OH^- (conjugate base); and HSO_3^- (base) and H_2SO_3 (conjugate acid).

PRACTICE EXERCISE

When lithium oxide (Li_2O) is dissolved in water, the solution turns basic from the reaction of the oxide ion (O^{2-}) with water. Write the reaction that occurs, and identify the conjugate acid-base pairs.

Answer: $O^{2-}(aq) + H_2O(l) \rightleftharpoons OH^-(aq) + OH^-(aq)$. OH^- is the conjugate acid of the base O^{2-}. OH^- is also the conjugate base of the acid H_2O.

Relative Strengths of Acids and Bases

Some acids are better proton donors than others; likewise, some bases are better proton acceptors than others. If we arrange acids in order of their ability to donate a proton, we find that the more readily a substance gives up a proton, the less readily its conjugate base accepts a proton. Similarly, the more readily a base accepts a proton, the less readily its conjugate acid gives up a proton. In other words, *the stronger an acid, the weaker is its conjugate base; the stronger a base, the weaker is its conjugate acid*. Thus, if we know something about the strength of an acid (its ability to donate protons), we also know something about the strength of its conjugate base (its ability to accept protons).

The inverse relationship between the strengths of acids and the strengths of their conjugate bases is illustrated in Figure 16.4 ▼. Here we have grouped acids and bases into three broad categories based on their behavior in water.

1. The *strong acids* completely transfer their protons to water, leaving no undissociated molecules in solution. (Section 4.3) Their conjugate bases have a negligible tendency to be protonated (to abstract protons) in aqueous solution.
2. The *weak acids* only partly dissociate in aqueous solution and therefore exist in the solution as a mixture of acid molecules and their constituent ions. The conjugate bases of weak acids show a slight ability to remove protons from water. (The conjugate bases of weak acids are weak bases.)
3. The substances with *negligible acidity* are those such as CH_4 that contain hydrogen but do not demonstrate any acidic behavior in water. Their conjugate bases are strong bases, reacting completely, abstracting protons from water molecules to form OH^- ions.

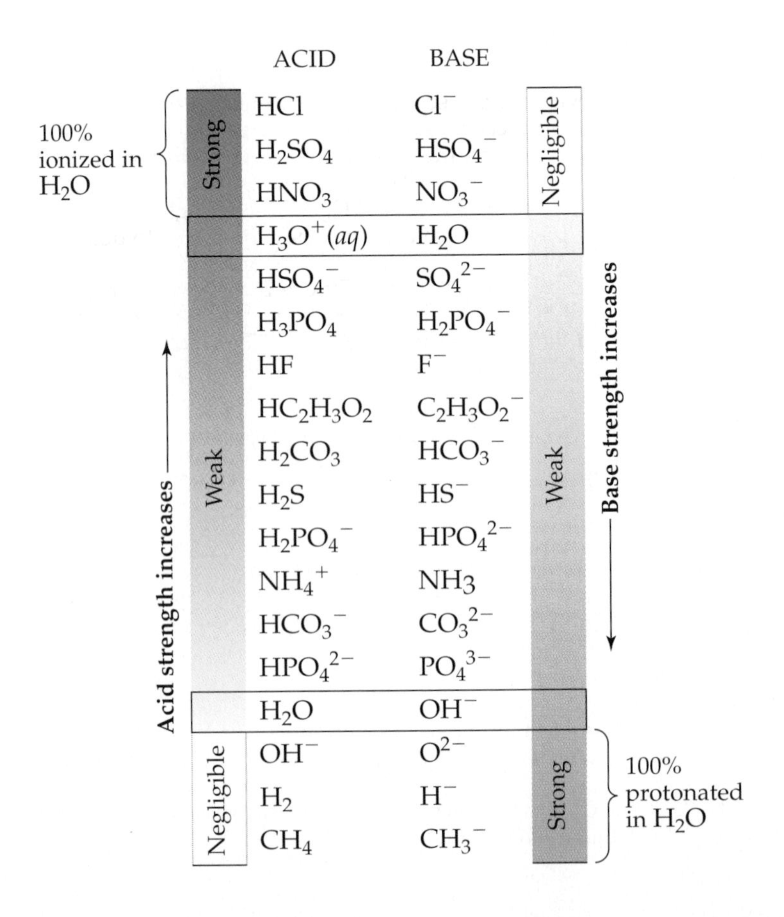

▶ **Figure 16.4** Relative strengths of some common conjugate acid-base pairs, which are listed opposite one another in the two columns.

We can think of proton-transfer reactions as being governed by the relative abilities of two bases to abstract protons. For example, consider the proton transfer that occurs when an acid HX dissolves in water:

$$HX(aq) + H_2O(l) \rightleftharpoons H_3O^+(aq) + X^-(aq) \quad [16.9]$$

If H_2O (the base in the forward reaction) is a stronger base than X^- (the conjugate base of HX), then H_2O will abstract the proton from HX to produce H_3O^+ and X^-. As a result, the equilibrium will lie to the right. This describes the behavior of a strong acid in water. For example, when HCl dissolves in water, the solution consists almost entirely of H_3O^+ and Cl^- ions with a negligible concentration of HCl molecules.

$$HCl(g) + H_2O(l) \longrightarrow H_3O^+(aq) + Cl^-(aq) \quad [16.10]$$

H_2O is a stronger base than Cl^- (Figure 16.4), so H_2O acquires the proton to become the hydronium ion.

When X^- is a stronger base than H_2O, the equilibrium will lie to the left. This situation occurs when HX is a weak acid. For example, an aqueous solution of acetic acid ($HC_2H_3O_2$) consists mainly of $HC_2H_3O_2$ molecules with only a relatively few H_3O^+ and $C_2H_3O_2^-$ ions.

$$HC_2H_3O_2(aq) + H_2O(l) \rightleftharpoons H_3O^+(aq) + C_2H_3O_2^-(aq) \quad [16.11]$$

$C_2H_3O_2^-$ is a stronger base than H_2O (Figure 16.4) and therefore abstracts the proton from H_3O^+. From these examples we conclude that *in every acid-base reaction the position of the equilibrium favors transfer of the proton to the stronger base.*

SAMPLE EXERCISE 16.3

For the following proton-transfer reaction, use Figure 16.4 to predict whether the equilibrium lies predominantly to the left or to the right:

$$HSO_4^-(aq) + CO_3^{2-}(aq) \rightleftharpoons SO_4^{2-}(aq) + HCO_3^-(aq)$$

Solution

Analyze: We are asked to predict whether the equilibrium shown lies to the right, favoring products, or to the left, favoring reactants.

Plan: This is a proton-transfer reaction, and the position of the equilibrium will favor the proton going to the stronger of two bases. The two bases in the equation are CO_3^{2-}, the base in the forward reaction as written, and SO_4^{2-}, the conjugate base of HSO_4^-. We can find the relative positions of these two bases in Figure 16.4 to determine which is the stronger base.

Solve: CO_3^{2-} appears lower in the right-hand column in Figure 16.4 and is therefore a stronger base than SO_4^{2-}. CO_3^{2-}, therefore, will get the proton preferentially to become HCO_3^-, while SO_4^{2-} will remain mostly unprotonated. The resulting equilibrium will lie to the right, favoring products.

$$\underset{\text{Acid}}{HSO_4^-(aq)} + \underset{\text{Base}}{CO_3^{2-}(aq)} \rightleftharpoons \underset{\text{Conjugate base}}{SO_4^{2-}(aq)} + \underset{\text{Conjugate acid}}{HCO_3^-(aq)}$$

Comment: Of the two acids in the equation, HSO_4^- and HCO_3^-, the stronger one gives up a proton while the weaker one retains its proton. Thus, the equilibrium favors the direction in which the proton moves from the stronger acid and becomes bonded to the stronger base. In other words, the reaction favors consumption of the stronger acid and stronger base and formation of the weaker acid and weaker base.

PRACTICE EXERCISE

For each of the following reactions, use Figure 16.4 to predict whether the equilibrium lies predominantly to the left or to the right:

(a) $HPO_4^{2-}(aq) + H_2O(l) \rightleftharpoons H_2PO_4^-(aq) + OH^-(aq)$

(b) $NH_4^+(aq) + OH^-(aq) \rightleftharpoons NH_3(aq) + H_2O(l)$

Answers: **(a)** left; **(b)** right

16.3 The Autoionization of Water

One of the most important chemical properties of water is its ability to act as either a Brønsted acid or a Brønsted base, depending on the circumstances. In the presence of an acid, water acts as a proton acceptor; in the presence of a base, water acts as a proton donor. In fact, one water molecule can donate a proton to another water molecule:

$$\mathrm{H{-}\ddot{\underset{\large |}{\underset{H}{O}}}{:}} + \mathrm{H{-}\ddot{\underset{\large |}{\underset{H}{O}}}{:}} \rightleftharpoons \left[\mathrm{H{-}\ddot{\underset{\large |}{\underset{H}{O}}}{-}H}\right]^+ + \mathrm{:\ddot{\underset{\cdot\cdot}{O}}{-}H^-} \quad [16.12]$$

We call this process the **autoionization** of water. No individual molecule remains ionized for long; the reactions are extremely rapid in both directions. At room temperature only about two out of every 10^9 molecules is ionized at any given instant. Thus, pure water consists almost entirely of H_2O molecules and is an extremely poor conductor of electricity. Nevertheless, the autoionization of water is very important, as we will soon see.

The Ion Product of Water

Because the autoionization of water (Equation 16.12) is an equilibrium process, we can write the following equilibrium-constant expression for it:

$$K_{eq} = [H_3O^+][OH^-] \quad [16.13]$$

ACTIVITY
K_w Activity

Because this equilibrium-constant expression refers specifically to the autoionization of water, we use the symbol K_w to denote the equilibrium constant, which we call the **ion-product constant** for water. At 25°C, K_w equals 1.0×10^{-14}. Thus, we have

$$K_w = [H_3O^+][OH^-] = 1.0 \times 10^{-14} \text{ (at 25°C)} \quad [16.14]$$

Because we use $H^+(aq)$ and $H_3O^+(aq)$ interchangeably to represent the hydrated proton, the autoionization reaction for water can also be written as

$$H_2O(l) \rightleftharpoons H^+(aq) + OH^-(aq) \quad [16.15]$$

Likewise, the expression for K_w can be written in terms of either H_3O^+ or H^+, and K_w has the same value in either case:

$$K_w = [H_3O^+][OH^-] = [H^+][OH^-] = 1.0 \times 10^{-14} \text{ (at 25°C)} \quad [16.16]$$

This equilibrium-constant expression and the value of K_w at 25°C are extremely important, and you should commit them to memory.

What makes Equation 16.16 particularly useful is that it is not only applicable to pure water but also to any aqueous solution. Although the equilibrium between $H^+(aq)$ and $OH^-(aq)$ as well as other ionic equilibria are affected somewhat by the presence of additional ions in solution, it is customary to ignore these ionic effects except in work requiring exceptional accuracy. Thus, Equation 16.16 is taken to be valid for any dilute aqueous solution, and it can be used to calculate either $[H^+]$ (if $[OH^-]$ is known) or $[OH^-]$ (if $[H^+]$ is known).

A solution in which $[H^+] = [OH^-]$ is said to be *neutral*. In most solutions H^+ and OH^- concentrations are not equal. As the concentration of one of these ions increases, the concentration of the other must decrease, so that the product of their concentrations equals 1.0×10^{-14}. In acidic solutions $[H^+]$ exceeds $[OH^-]$. In basic solutions $[OH^-]$ exceeds $[H^+]$.

SAMPLE EXERCISE 16.4

Calculate the values of $[H^+]$ and $[OH^-]$ in a neutral solution at 25°C.

Solution

Analyze: We are asked to determine the concentrations of hydronium and hydroxide ions in a neutral solution at 25°C.

Plan: We will use Equation 16.16 and the fact that, by definition, $[H^+] = [OH^-]$ in a neutral solution.

Solve: We will represent the concentration of $[H^+]$ and $[OH^-]$ in neutral solution with x. This gives

$$[H^+][OH^-] = (x)(x) = 1.0 \times 10^{-14}$$

$$x^2 = 1.0 \times 10^{-14}$$

$$x = 1.0 \times 10^{-7}\,M = [H^+] = [OH^-]$$

In an acid solution $[H^+]$ is greater than $1.0 \times 10^{-7}\,M$; in a basic solution $[H^+]$ is less than $1.0 \times 10^{-7}\,M$.

PRACTICE EXERCISE

Indicate whether solutions with each of the following ion concentrations is neutral, acidic, or basic: **(a)** $[H^+] = 4 \times 10^{-9}\,M$; **(b)** $[OH^-] = 1 \times 10^{-7}\,M$; **(c)** $[OH^-] = 7 \times 10^{-13}\,M$.

Answers: **(a)** basic; **(b)** neutral; **(c)** acidic

SAMPLE EXERCISE 16.5

Calculate the concentration of $H^+(aq)$ in **(a)** a solution in which $[OH^-]$ is 0.010 M; **(b)** a solution in which $[OH^-]$ is $1.8 \times 10^{-9}\,M$. *Note:* In this problem and all that follow, we assume, unless stated otherwise, that the temperature is 25°C.

Solution

Analyze: We are asked to calculate the hydronium ion concentration in an aqueous solution where the hydroxide concentration is known.

Plan: We can use the equilibrium-constant expression for the autoionization of water and the value of K_w to solve for each unknown concentration.

Solve: (a) Using Equation 16.16, we have

$$[H^+][OH^-] = 1.0 \times 10^{-14}$$

$$[H^+] = \frac{1.0 \times 10^{-14}}{[OH^-]} = \frac{1.0 \times 10^{-14}}{0.010} = 1.0 \times 10^{-12}\,M$$

This solution is basic because $[OH^-] > [H^+]$.

(b) In this instance

$$[H^+] = \frac{1.0 \times 10^{-14}}{[OH^-]} = \frac{1.0 \times 10^{-14}}{1.8 \times 10^{-9}} = 5.6 \times 10^{-6}\,M$$

This solution is acidic because $[H^+] > [OH^-]$.

PRACTICE EXERCISE

Calculate the concentration of $OH^-(aq)$ in a solution in which **(a)** $[H^+] = 2 \times 10^{-6}\,M$; **(b)** $[H^+] = [OH^-]$; **(c)** $[H^+] = 100 \times [OH^-]$.

Answers: **(a)** $5 \times 10^{-9}\,M$; **(b)** $1.0 \times 10^{-7}\,M$; **(c)** $1.0 \times 10^{-8}\,M$

16.4 The pH Scale

The molar concentration of $H^+(aq)$ in an aqueous solution is usually very small. For convenience, we therefore usually express $[H^+]$ in terms of **pH**, which is the negative logarithm in base 10 of $[H^+]$.*

$$pH = -\log[H^+] \qquad [16.17]$$

* Because $[H^+]$ and $[H_3O^+]$ are used interchangeably, you might see pH defined as $-\log[H_3O^+]$.

If you need to review the use of logs, see Appendix A.

We can use Equation 16.17 to calculate the pH of a neutral solution at 25°C (that is, one in which $[H^+] = 1.0 \times 10^{-7}\,M$):

$$pH = -\log(1.0 \times 10^{-7}) = -(-7.00) = 7.00$$

The pH of a neutral solution is 7.00 at 25°C.

What happens to the pH of a solution as we make the solution acidic? An acidic solution is one in which $[H^+] > 1.0 \times 10^{-7}\,M$. Because of the negative sign in Equation 16.17, *the pH decreases as* $[H^+]$ *increases*. For example, the pH of an acidic solution in which $[H^+] = 1.0 \times 10^{-3}\,M$ is

$$pH = -\log(1.0 \times 10^{-3}) = -(-3.00) = 3.00$$

At 25°C the pH of an acidic solution is less than 7.00.

We can also calculate the pH of a basic solution, one in which $[OH^-] > 1.0 \times 10^{-7}\,M$. Suppose $[OH^-] = 2.0 \times 10^{-3}\,M$. We can use Equation 16.16 to calculate $[H^+]$ for this solution, and Equation 16.17 to calculate the pH:

$$[H^+] = \frac{K_w}{[OH^-]} = \frac{1.0 \times 10^{-14}}{2.0 \times 10^{-3}} = 5.0 \times 10^{-12}\,M$$

$$pH = -\log(5.0 \times 10^{-12}) = 11.30$$

At 25°C the pH of a basic solution is greater than 7.00. The relationships among $[H^+]$, $[OH^-]$, and pH are summarized in Table 16.1 ▼.

The pH values characteristic of several familiar solutions are shown in Figure 16.5 ▶. Notice that a change in $[H^+]$ by a factor of 10 causes the pH to change by 1. Thus, a solution of pH 6 has 10 times the concentration of $H^+(aq)$ as a solution of pH 7.

You might think that when $[H^+]$ is very small, as it is for some of the examples shown in Figure 16.5, it would be unimportant. Nothing is further from the truth. If $[H^+]$ is part of a kinetic rate law, then changing its concentration will change the rate. ∞ (Section 14.3) Thus, if the rate law is first order in $[H^+]$, doubling its concentration will double the rate even if the change is merely from $1 \times 10^{-7}\,M$ to $2 \times 10^{-7}\,M$. In biological systems many reactions involve proton transfers and have rates that depend on $[H^+]$. Because the speeds of these reactions are crucial, the pH of biological fluids must be maintained within narrow limits. For example, human blood has a normal pH range of 7.35 to 7.45. Illness and even death can result if the pH varies much from this narrow range.

A convenient way to estimate pH is to use the "benchmark" H^+ concentrations in Figure 16.5, those for which $[H^+]$ equals 1×10^{-x}, where x is a whole number from 0 to 14. When $[H^+]$ is one of these benchmark concentrations, the pH is simply the corresponding pH value, x. When $[H^+] = 1 \times 10^{-4}$, for example, the pH is simply 4. When $[H^+]$ falls between two benchmark concentrations, the pH will fall between the two corresponding pH values. Consider a solution that is 0.050 M in H^+. Because 0.050 (that is, 5.0×10^{-2}) is greater than 1.0×10^{-2} and less than 1.0×10^{-1}, we estimate the pH to be between 2.00 and 1.00. Using Equation 16.17 to calculate the pH gives 1.30.

TABLE 16.1 Relationships Among $[H^+]$, $[OH^-]$, and pH at 25°C

Solution Type	$[H^+]$ (M)	$[OH^-]$ (M)	pH Value
Acidic	$> 1.0 \times 10^{-7}$	$< 1.0 \times 10^{-7}$	< 7.00
Neutral	$= 1.0 \times 10^{-7}$	$= 1.0 \times 10^{-7}$	$= 7.00$
Basic	$< 1.0 \times 10^{-7}$	$> 1.0 \times 10^{-7}$	> 7.00

$[H^+]$ (M)	pH	pOH	$[OH^-]$ (M)
$1\ (1\times10^{-0})$	0.0	14.0	1×10^{-14}
1×10^{-1}	1.0	13.0	1×10^{-13}
1×10^{-2}	2.0	12.0	1×10^{-12}
1×10^{-3}	3.0	11.0	1×10^{-11}
1×10^{-4}	4.0	10.0	1×10^{-10}
1×10^{-5}	5.0	9.0	1×10^{-9}
1×10^{-6}	6.0	8.0	1×10^{-8}
1×10^{-7}	7.0	7.0	1×10^{-7}
1×10^{-8}	8.0	6.0	1×10^{-6}
1×10^{-9}	9.0	5.0	1×10^{-5}
1×10^{-10}	10.0	4.0	1×10^{-4}
1×10^{-11}	11.0	3.0	1×10^{-3}
1×10^{-12}	12.0	2.0	1×10^{-2}
1×10^{-13}	13.0	1.0	1×10^{-1}
1×10^{-14}	14.0	0.0	$1\ (1\times10^{-0})$

More acidic
Gastric juice
Lemon juice
Cola, vinegar
Wine
Tomatoes
Banana
Black coffee
Rain
Saliva
Milk
Human blood, tears
Egg white, seawater
Baking soda
Borax
Milk of magnesia
Lime water
Household ammonia
Household bleach
NaOH, 0.1 M
More basic

◀ **Figure 16.5** H^+ concentrations and pH values of some common substances at 25°C. pH and pOH can be estimated using the benchmark concentrations of H^+ and OH^-.

ACTIVITY
pH Estimation

SAMPLE EXERCISE 16.6

Calculate the pH values for the two solutions described in Sample Exercise 16.5.

Solution

Analyze: We are asked to determine the pH of aqueous solutions for which we have already calculated $[H^+]$.

Plan: We can use the benchmarks in Figure 16.5 to determine the pH for part **(a)** and to estimate pH for part **(b)**. We can then use Equation 16.17 to calculate pH for part **(b)**.

Solve: (a) In the first instance we found $[H^+]$ to be $1.0 \times 10^{-12}\ M$. Although we can use Equation 16.17 to determine the pH, 1.0×10^{-12} is one of the benchmarks in Figure 16.5, so the pH can be determined without any formal calculation.

$$pH = -\log(1.0 \times 10^{-12}) = -(-12.00) = 12.00$$

The rule for using significant figures with logs is that *the number of decimal places in the log equals the number of significant figures in the original number* (see Appendix A). Because $\mathbf{1.0} \times 10^{-12}$ has two significant figures, the pH has two decimal places, 12.**00**.

(b) For the second solution, $[H^+] = 5.6 \times 10^{-6}\ M$. Before performing the calculation, it is helpful to estimate the pH. To do so, we note that $[H^+]$ lies between 1×10^{-6} and 1×10^{-5}.

$$1 \times 10^{-6} < 5.6 \times 10^{-6} < 1 \times 10^{-5}$$

Thus, we expect the pH to lie between 6.0 and 5.0. We use Equation 16.17 to calculate the pH.

$$pH = -\log 5.6 \times 10^{-6}\ M = 5.25$$

Check: After calculating a pH, it is useful to compare it to your prior estimate. In this case the pH, as we predicted, falls between 6 and 5. Had the calculated pH and the estimate not agreed, we should have reconsidered our calculation or estimate, or both. Note that although $[H^+]$ lies halfway between the two benchmark concentrations, the calculated pH does not lie halfway between the two corresponding pH values. This is because the pH scale is logarithmic rather than linear.

PRACTICE EXERCISE

(a) In a sample of lemon juice $[H^+]$ is $3.8 \times 10^{-4}\ M$. What is the pH? **(b)** A commonly available window-cleaning solution has a $[H^+]$ of $5.3 \times 10^{-9}\ M$. What is the pH?
Answers: **(a)** 3.42; **(b)** 8.28

SAMPLE EXERCISE 16.7

A sample of freshly pressed apple juice has a pH of 3.76. Calculate $[H^+]$.

Solution

Analyze and Plan: We need to calculate $[H^+]$ from pH. We will use Equation 16.17 for the calculation, but first, we will use the benchmarks in Figure 16.5 to estimate $[H^+]$.
Solve: Because the pH is between 3.0 and 4.0, we know that $[H^+]$ will be between 1×10^{-3} and $1 \times 10^{-4}\ M$. From Equation 16.17, we have

$$\text{pH} = -\log[H^+] = 3.76$$

Thus,

$$\log[H^+] = -3.76$$

To find $[H^+]$ we need to determine the *antilog* of −3.76. Scientific calculators have an antilog function (sometimes labeled INV log or 10^x) that allows us to perform the calculation:

$$[H^+] = \text{antilog}\,(-3.76) = 10^{-3.76} = 1.7 \times 10^{-4}\ M$$

Consult the user's manual for your calculator to find out how to perform the antilog operation. The number of significant figures in $[H^+]$ is two because the number of decimal places in the pH is two.
Check: Our calculated $[H^+]$ falls within the estimated range.

PRACTICE EXERCISE

A solution formed by dissolving an antacid tablet has a pH of 9.18. Calculate $[H^+]$.
Answer: $[H^+] = 6.6 \times 10^{-10}\ M$

Other "p" Scales

The negative log is also a convenient way of expressing the magnitudes of other small quantities. We use the convention that the negative log of a quantity is labeled p (quantity). For example, one can express the concentration of OH^- as pOH:

$$\text{pOH} = -\log[OH^-] \qquad [16.18]$$

By taking the negative log of both sides of Equation 16.16,

$$-\log[H^+] + (-\log[OH^-]) = -\log K_w \qquad [16.19]$$

we obtain the following useful expression:

$$\text{pH} + \text{pOH} = 14 \ \ (\text{at } 25°\text{C}) \qquad [16.20]$$

We will see in Section 16.8 that p scales are also useful when working with equilibrium constants.

Measuring pH

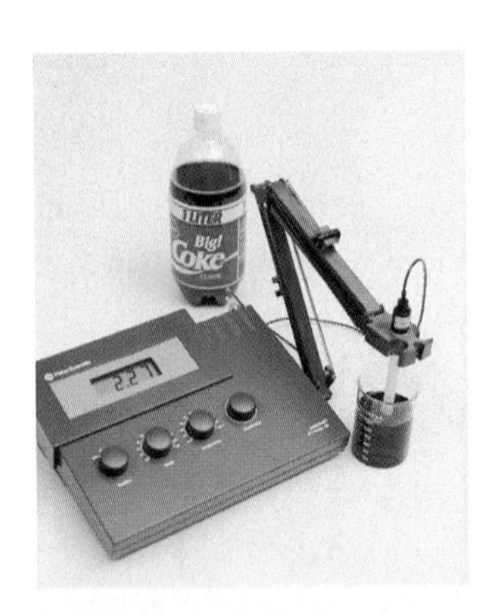

▲ **Figure 16.6** A digital pH meter.

The pH of a solution can be measured quickly and accurately with a *pH meter* (Figure 16.6 ◄). A complete understanding of how this important device works requires a knowledge of electrochemistry, a subject we take up in Chapter 20. In brief, a pH meter consists of a pair of electrodes connected to a meter capable of measuring small voltages, on the order of millivolts. A voltage, which varies with the pH, is generated when the electrodes are placed in a solution. This voltage is read by the meter, which is calibrated to give pH.

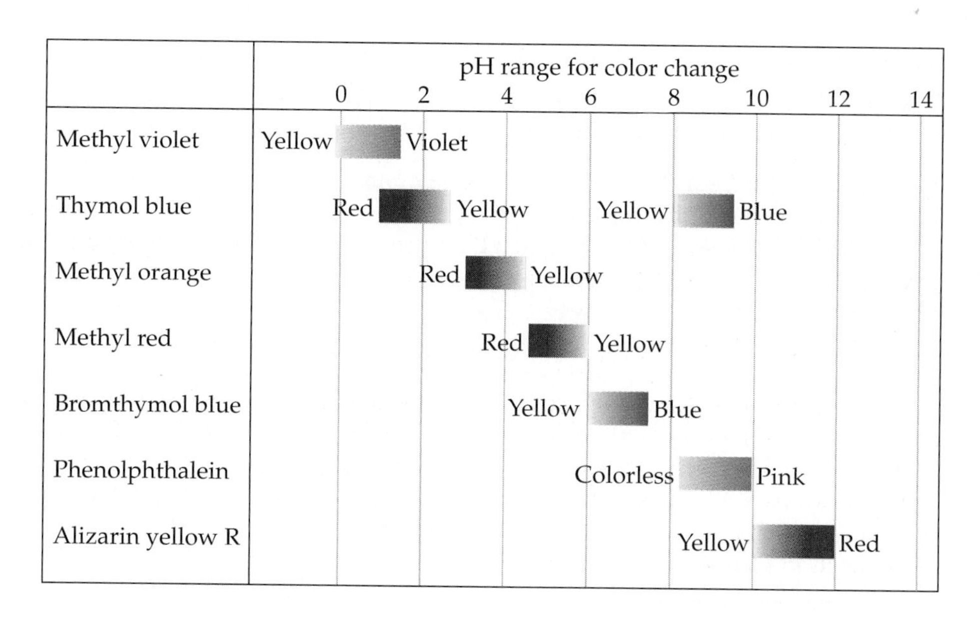

◀ **Figure 16.7** The pH ranges for the color changes of some common acid-base indicators. Most indicators have a useful range of about 2 pH units.

The electrodes used with pH meters come in many shapes and sizes, depending on their intended use. Electrodes have even been developed that are so small that they can be inserted into single living cells in order to monitor the pH of the cell medium. Pocket-size pH meters are also available for use in environmental studies, in monitoring industrial effluents, and in agricultural work.

Although less precise, acid-base indicators can be used to measure pH. An acid-base indicator is a colored substance that itself can exist in either an acid or a base form. The two forms have different colors. Thus, the indicator turns one color in an acid and another color in a base. If you know the pH at which the indicator turns from one form to the other, you can determine whether a solution has a higher or lower pH than this value. Litmus, for example, changes color in the vicinity of pH 7. The color change, however, is not very sharp. Red litmus indicates a pH of about 5 or lower, and blue litmus indicates a pH of about 8 or higher.

Some of the more common indicators are listed in Figure 16.7 ▲. Methyl orange, for example, changes color over the pH interval from 3.1 to 4.4. Below pH 3.1 it is in the acid form, which is red. In the interval between 3.1 and 4.4 it is gradually converted to its basic form, which has a yellow color. By pH 4.4 the conversion is complete, and the solution is yellow. Paper tape that is impregnated with several indicators and comes complete with a comparator color scale is widely used for approximate determinations of pH.

16.5 Strong Acids and Bases

The chemistry of an aqueous solution often depends critically on the pH of the solution. It is therefore important to examine how the pH of solutions relate to the concentrations of acids and bases. The simplest cases are those involving strong acids and strong bases. Strong acids and bases are *strong electrolytes*, existing in aqueous solution entirely as ions. There are relatively few common strong acids and bases, and we listed these substances in Table 4.2.

Strong Acids

The seven most common strong acids include six monoprotic acids (HCl, HBr, HI, HNO_3, $HClO_3$, and $HClO_4$), and one diprotic acid (H_2SO_4). Nitric acid (HNO_3) exemplifies the behavior of the monoprotic strong acids. For all practical purposes, an aqueous solution of HNO_3 consists entirely of H_3O^+ and NO_3^- ions.

$$HNO_3(aq) + H_2O(l) \longrightarrow H_3O^+(aq) + NO_3^-(aq) \text{ (complete ionization)} \quad [16.21]$$

We have not used equilibrium arrows for Equation 16.21 because the reaction lies entirely to the right, the side with the ions. (Section 4.1) As noted in Section 16.3, we use $H_3O^+(aq)$ and $H^+(aq)$ interchangeably to represent the hydrated proton in water. Thus, we often simplify the equations for the ionization reactions of acids as follows:

$$HNO_3(aq) \longrightarrow H^+(aq) + NO_3^-(aq)$$

In an aqueous solution of a strong acid, the acid is normally the only significant source of H^+ ions.* As a result, calculating the pH of a solution of a strong monoprotic acid is straightforward because $[H^+]$ equals the original concentration of acid. In a 0.20 M solution of $HNO_3(aq)$, for example, $[H^+] = [NO_3^-] = 0.20\ M$. The situation with the diprotic acid H_2SO_4 is more complex, as we will see in Section 16.6.

SAMPLE EXERCISE 16.8

What is the pH of a 0.040 M solution of $HClO_4$?

Solution

Analyze and Plan: We are asked to calculate the pH of a 0.040 M solution of $HClO_4$. Because $HClO_4$ is a strong acid, it is completely ionized, giving $[H^+] = [ClO_4^-] = 0.040\ M$. Because $[H^+]$ lies between benchmarks 1×10^{-2} and 1×10^{-1} in Figure 16.5, we estimate that the pH will be between 2.0 and 1.0.

Solve: The pH of the solution is given by

$$pH = -\log(0.040) = 1.40.$$

Check: Our calculated pH falls within the estimated range.

PRACTICE EXERCISE

An aqueous solution of HNO_3 has a pH of 2.34. What is the concentration of the acid?

Answer: 0.0046 M

Strong Bases

There are relatively few common strong bases. The most common soluble strong bases are the ionic hydroxides of the alkali metals (group 1A) and the heavier alkaline earth metals (group 2A), such as NaOH, KOH, and $Ca(OH)_2$. These compounds completely dissociate into ions in aqueous solution. Thus, a solution labeled 0.30 M NaOH consists of 0.30 M $Na^+(aq)$ and 0.30 M $OH^-(aq)$; there is essentially no undissociated NaOH.

Because these strong bases dissociate entirely into ions in aqueous solution, calculating the pH of their solutions is also straightforward, as shown in Sample Exercise 16.9.

SAMPLE EXERCISE 16.9

What is the pH of **(a)** a 0.028 M solution of NaOH; **(b)** a 0.0011 M solution of $Ca(OH)_2$?

Solution

Analyze: We're asked to calculate the pH of two solutions given the concentration of strong base for each.

Plan: We can calculate each pH by two equivalent methods. First, we could use Equation 16.16 to calculate $[H^+]$ and then use Equation 16.17 to calculate the pH. Alternatively, we could use $[OH^-]$ to calculate pOH and then use Equation 16.20 to calculate the pH.

* If the concentration of the acid is 10^{-6} M or less, we also need to consider H^+ ions that result from the autoionization of H_2O. Normally, the concentration of H^+ from H_2O is so small that it can be neglected.

Solve: (a) NaOH dissociates in water to give one OH^- ion per formula unit. Therefore, the OH^- concentration for the solution in (a) equals the stated concentration of NaOH, namely 0.028 *M*.

Method 1:

$$[H^+] = \frac{1.0 \times 10^{-14}}{0.028} = 3.57 \times 10^{-13}\ M \qquad pH = -\log(3.57 \times 10^{-13}) = 12.45$$

Method 2:

$$pOH = -\log(0.028) = 1.55 \qquad pH = 14.00 - pOH = 12.45$$

(b) $Ca(OH)_2$ is a strong base that dissociates in water to give two OH^- ions per formula unit. Thus, the concentration of $OH^-(aq)$ for the solution in part **(b)** is $2 \times (0.0011\ M) = 0.0022\ M$.

Method 1:

$$[H^+] = \frac{1.0 \times 10^{-14}}{0.0022} = 4.55 \times 10^{-12}\ M \qquad pH = -\log(4.55 \times 10^{-12}) = 11.34$$

Method 2:

$$pOH = -\log(0.0022) = 2.66 \qquad pH = 14.00 - pOH = 11.34$$

PRACTICE EXERCISE

What is the concentration of a solution of **(a)** KOH for which the pH is 11.89; **(b)** $Ca(OH)_2$ for which the pH is 11.68?
Answers: **(a)** $7.8 \times 10^{-3}\ M$; **(b)** $2.4 \times 10^{-3}\ M$

Although all the hydroxides of the alkali metals (group 1A) are strong electrolytes, LiOH, RbOH, and CsOH are not commonly encountered in the laboratory. The hydroxides of the heavier alkaline earth metals, $Ca(OH)_2$, $Sr(OH)_2$, and $Ba(OH)_2$, are also strong electrolytes. They have limited solubilities, however, so they are used only when high solubility is not critical.

Strongly basic solutions are also created by certain substances that react with water to form $OH^-(aq)$. The most common of these contain the oxide ion. Ionic metal oxides, especially Na_2O and CaO, are often used in industry when a strong base is needed. Each mole of O^{2-} reacts with water to form 2 mol of OH^-, leaving virtually no O^{2-} remaining in the solution:

$$O^{2-}(aq) + H_2O(l) \longrightarrow 2OH^-(aq) \qquad [16.22]$$

Thus, a solution formed by dissolving 0.010 mol of $Na_2O(s)$ in enough water to form 1.0 L of solution will have $[OH^-] = 0.020\ M$ and a pH of 12.30.

Ionic hydrides and nitrides also react with H_2O to form OH^-:

$$H^-(aq) + H_2O(l) \longrightarrow H_2(g) + OH^-(aq) \qquad [16.23]$$

$$N^{3-}(aq) + 3H_2O(l) \longrightarrow NH_3(aq) + 3OH^-(aq) \qquad [16.24]$$

Because the anions O^{2-}, H^-, and N^{3-} are stronger bases than OH^- (the conjugate base of H_2O), they are able to remove a proton from H_2O.

16.6 Weak Acids

Most acidic substances are weak acids and are therefore only partially ionized in aqueous solution. We can use the equilibrium constant for the ionization reaction to express the extent to which a weak acid ionizes. If we represent a general weak acid as HA, we can write the equation for its ionization reaction in either of the following ways, depending on whether the hydrated proton is represented as $H_3O^+(aq)$ or $H^+(aq)$:

$$HA(aq) + H_2O(l) \rightleftharpoons H_3O^+(aq) + A^-(aq) \qquad [16.25]$$

$$\text{or} \qquad HA(aq) \rightleftharpoons H^+(aq) + A^-(aq) \qquad [16.26]$$

Because $[H_2O]$ is the solvent, it is omitted from the equilibrium-constant expression, which can be written as either

$$K_{eq} = \frac{[H_3O^+][A^-]}{[HA]} \quad \text{or} \quad K_{eq} = \frac{[H^+][A^-]}{[HA]}$$

As we did for the ion-product constant for the autoionization of water, we change the subscript on this equilibrium constant to denote the type of equation to which it corresponds.

$$K_a = \frac{[H_3O^+][A^-]}{[HA]} \quad \text{or} \quad K_a = \frac{[H^+][A^-]}{[HA]} \qquad [16.27]$$

The subscript a on K_a denotes that it is an equilibrium constant for the ionization of an acid, so K_a is called the **acid-dissociation constant**.

Table 16.2 ▼ shows the names, structures, and K_a values for several weak acids. A more complete listing is given in Appendix D. Many weak acids are organic compounds composed entirely of carbon, hydrogen, and oxygen. These compounds usually contain some hydrogen atoms bonded to carbon atoms and some bonded to oxygen atoms. In almost all cases the hydrogen atoms bonded to carbon do not ionize in water; instead, the acidic behavior of these compounds is due to the hydrogen atoms attached to oxygen atoms.

The magnitude of K_a indicates the tendency of the acid to ionize in water: *The larger the value of K_a, the stronger the acid.* Hydrofluoric acid (HF), for example, is the strongest acid listed in Table 16.2, and phenol (HOC_6H_5) is the weakest. Notice that K_a is typically less than 10^{-3}.

TABLE 16.2 Some Weak Acids in Water at 25°C*

Acid	Structural Formula	Conjugate Base	Equilibrium Reaction	K_a
Hydrofluoric (HF)	H—F	F^-	$HF(aq) + H_2O(l) \rightleftharpoons H_3O^+(aq) + F^-(aq)$	6.8×10^{-4}
Nitrous (HNO_2)	H—O—N=O	NO_2^-	$HNO_2(aq) + H_2O(l) \rightleftharpoons H_3O^+(aq) + NO_2^-(aq)$	4.5×10^{-4}
Benzoic ($HC_7H_5O_2$)	H—O—C(=O)—C₆H₅ ring	$C_7H_5O_2^-$	$HC_7H_5O_2(aq) + H_2O(l) \rightleftharpoons H_3O^+(aq) + C_7H_5O_2^-(aq)$	6.3×10^{-5}
Acetic ($HC_2H_3O_2$)	H—O—C(=O)—CH₃ (C bonded to three H)	$C_2H_3O_2^-$	$HC_2H_3O_2(aq) + H_2O(l) \rightleftharpoons H_3O^+(aq) + C_2H_3O_2^-(aq)$	1.8×10^{-5}
Hypochlorous (HClO)	H—O—Cl	ClO^-	$HClO(aq) + H_2O(l) \rightleftharpoons H_3O^+(aq) + ClO^-(aq)$	3.0×10^{-8}
Hydrocyanic (HCN)	H—C≡N	CN^-	$HCN(aq) + H_2O(l) \rightleftharpoons H_3O^+(aq) + CN^-(aq)$	4.9×10^{-10}
Phenol (HC_6H_5O)	H—O—C₆H₅ ring	$C_6H_5O^-$	$HC_6H_5O(aq) + H_2O(l) \rightleftharpoons H_3O^+(aq) + C_6H_5O^-(aq)$	1.3×10^{-10}

*The proton that ionizes is shown in blue.

Calculating K_a from pH

In order to calculate either the K_a value for a weak acid or the pH of its solutions, we will use many of the skills for solving equilibrium problems that we developed in Section 15.5. In many cases the small magnitude of K_a allows us to use approximations to simplify the problem. In doing these calculations, it is important to realize that proton-transfer reactions are generally very rapid. As a result, the measured or calculated pH of a solution always represents an equilibrium condition.

SAMPLE EXERCISE 16.10

A student prepared a 0.10 *M* solution of formic acid ($HCHO_2$) and measured its pH using a pH meter of the type illustrated in Figure 16.6. The pH at 25°C was found to be 2.38. **(a)** Calculate K_a for formic acid at this temperature. **(b)** What percentage of the acid is ionized in this 0.10 *M* solution?

Solution

Analyze: We are given the molar concentration of an aqueous solution of weak acid and the pH of the solution at 25°C, and we are asked to determine the value of K_a for the acid and the percentage of the acid that is ionized.

Plan: Although we are dealing specifically with the ionization of a weak acid, this problem is very similar to the equilibrium problems we encountered in Chapter 15. We can solve it using the method first outlined in Sample Exercise 15.8, starting with the chemical reaction and a tabulation of initial and equilibrium concentrations.

Solve: (a) The first step in solving any equilibrium problem is to write the equation for the equilibrium reaction. The ionization equilibrium for formic acid can be written as follows:

$$HCHO_2(aq) \rightleftharpoons H^+(aq) + CHO_2^-(aq)$$

The equilibrium-constant expression is

$$K_a = \frac{[H^+][CHO_2^-]}{[HCHO_2]}$$

From the measured pH we can calculate $[H^+]$:

$$pH = -\log[H^+] = 2.38$$

$$\log[H^+] = -2.38$$

$$[H^+] = 10^{-2.38} = 4.2 \times 10^{-3}\ M$$

We can do a little accounting to determine the concentrations of the species involved in the equilibrium. We imagine that the solution is initially 0.10 *M* in $HCHO_2$ molecules. We then consider the ionization of the acid into H^+ and CHO_2^-. For each $HCHO_2$ molecule that ionizes, one H^+ ion and one CHO_2^- ion are produced in solution. Because the pH measurement indicates that $[H^+] = 4.2 \times 10^{-3}\ M$ at equilibrium, we can construct the following table:

	$HCHO_2(aq)$	$\rightleftharpoons$ $H^+(aq)$	+ $CHO_2^-(aq)$
Initial	0.10 *M*	0	0
Change	$-4.2 \times 10^{-3}\ M$	$+4.2 \times 10^{-3}\ M$	$+4.2 \times 10^{-3}\ M$
Equilibrium	$(0.10 - 4.2 \times 10^{-3})\ M$	$4.2 \times 10^{-3}\ M$	$4.2 \times 10^{-3}\ M$

Notice that we have neglected the very small concentration of $H^+(aq)$ due to the autoionization of H_2O. Notice also that the amount of $HCHO_2$ that ionizes is very small compared with the initial concentration of the acid. To the number of significant figures we are using, the subtraction yields 0.10 *M*:

$$(0.10 - 4.2 \times 10^{-3})\ M \simeq 0.10\ M$$

We can now insert the equilibrium concentrations into the expression for K_a:

$$K_a = \frac{(4.2 \times 10^{-3})(4.2 \times 10^{-3})}{0.10} = 1.8 \times 10^{-4}$$

Check: The magnitude of our answer is reasonable because K_a for a weak acid is usually between 10^{-3} and 10^{-10}.

(b) The percentage of acid that ionizes is given by the concentration of H^+ or CHO_2^- at equilibrium, divided by the initial acid concentration, multiplied by 100%:

$$\text{Percent ionization} = \frac{[H^+]_{equilibrium}}{[HCHO_2]_{initial}} \times 100\% = \frac{4.2 \times 10^{-3}}{0.10} \times 100\% = 4.2\%$$

PRACTICE EXERCISE

Niacin, one of the B vitamins, has the following molecular structure:

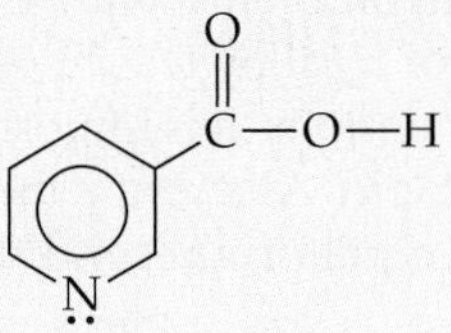

A 0.020 *M* solution of niacin has a pH of 3.26. **(a)** What percentage of the acid is ionized in this solution? **(b)** What is the acid-dissociation constant, K_a, for niacin?
Answers: **(a)** 2.7%; **(b)** 1.5×10^{-5}

Using K_a to Calculate pH

Knowing the value of K_a and the initial concentration of the weak acid, we can calculate the concentration of $H^+(aq)$ in a solution of a weak acid. Let's calculate the pH of a 0.30 *M* solution of acetic acid ($HC_2H_3O_2$), the weak acid responsible for the characteristic odor and acidity of vinegar, at 25°C.

Our *first* step is to write the ionization equilibrium for acetic acid:

$$HC_2H_3O_2(aq) \rightleftharpoons H^+(aq) + C_2H_3O_2^-(aq) \qquad [16.28]$$

According to the structural formula of acetic acid, shown in Table 16.2, the hydrogen that ionizes is the one attached to an oxygen atom. We write this hydrogen separate from the others in the formula to emphasize that only this one hydrogen is readily ionized.

The *second* step is to write the equilibrium-constant expression and the value for the equilibrium constant. From Table 16.2 we have $K_a = 1.8 \times 10^{-5}$. Thus, we can write the following:

$$K_a = \frac{[H^+][C_2H_3O_2^-]}{[HC_2H_3O_2]} = 1.8 \times 10^{-5} \qquad [16.29]$$

As the *third* step, we need to express the concentrations that are involved in the equilibrium reaction. This can be done with a little accounting, as described in Sample Exercise 16.10. Because we want to find the equilibrium value for $[H^+]$, let's call this quantity x. The concentration of acetic acid before any of it ionizes is 0.30 *M*. The chemical equation tells us that for each molecule of $HC_2H_3O_2$ that ionizes, one $H^+(aq)$ and one $C_2H_2O_2^-(aq)$ are formed. Consequently, if x moles per liter of $H^+(aq)$ form at equilibrium, x moles per liter of $C_2H_3O_2^-(aq)$ must also form, and x moles per liter of $HC_2H_3O_2$ must be ionized. This gives rise to the following table with the equilibrium concentrations shown on the last line:

	$HC_2H_3O_2(aq)$ $\rightleftharpoons$	$H^+(aq)$	+ $C_2H_3O_2^-(aq)$
Initial	0.30 *M*	0	0
Change	$-x$ *M*	$+x$ *M*	$+x$ *M*
Equilibrium	$(0.30 - x)$ *M*	x *M*	x *M*

As the *fourth* step of the problem, we need to substitute the equilibrium concentrations into the equilibrium-constant expression. The substitutions give the following equation:

$$K_a = \frac{[H^+][C_2H_3O_2^-]}{[HC_2H_3O_2]} = \frac{(x)(x)}{0.30 - x} = 1.8 \times 10^{-5} \qquad [16.30]$$

This expression leads to a quadratic equation in x, which we can solve by using an equation-solving calculator or by using the quadratic formula. We can also simplify the problem, however, by noting that the value of K_a is quite small. As a result, we anticipate that the equilibrium will lie far to the left and that x will be very small compared to the initial concentration of acetic acid. Thus, we will *assume* that x is negligible compared to 0.30, so that $0.30 - x$ is essentially equal to 0.30.

$$0.30 - x \simeq 0.30$$

As we will see, we can (and should!) check the validity of this assumption when we finish the problem. By using this assumption, Equation 16.30 now becomes

$$K_a = \frac{x^2}{0.30} = 1.8 \times 10^{-5}$$

Solving for x, we have

$$x^2 = (0.30)(1.8 \times 10^{-5}) = 5.4 \times 10^{-6}$$

$$x = \sqrt{5.4 \times 10^{-6}} = 2.3 \times 10^{-3}$$

$$[H^+] = x = 2.3 \times 10^{-3}\,M$$

$$pH = -\log(2.3 \times 10^{-3}) = 2.64$$

We should now go back and check the validity of our simplifying assumption that $0.30 - x \simeq 0.30$. The value of x we determined is so small that, for this number of significant figures, the assumption is entirely valid. We are thus satisfied that the assumption was a reasonable one to make. Because x represents the moles per liter of acetic acid that ionize, we see that, in this particular case, less than 1% of the acetic acid molecules ionize:

$$\text{Percent ionization of } HC_2H_3O_2 = \frac{0.0023\,M}{0.30\,M} \times 100\% = 0.77\%$$

As a general rule, if the quantity x is more than about 5% of the initial value, it is better to use the quadratic formula. You should always check the validity of any simplifying assumptions after you have finished solving a problem.

Finally, we can compare the pH value of this weak acid to a solution of a strong acid of the same concentration. The pH of the 0.30 M solution of acetic acid is 2.64. By comparison, the pH of a 0.30 M solution of a strong acid such as HCl is $-\log(0.30) = 0.52$. As expected, the pH of a solution of a weak acid is higher than that of a solution of a strong acid of the same molarity.

SAMPLE EXERCISE 16.11

Calculate the pH of a 0.20 M solution of HCN. (Refer to Table 16.2 or Appendix D for the value of K_a.)

Solution

Analyze and Plan: We are given the molarity of a weak acid and are asked for the pH. From Table 16.2, K_a for HCN is 4.9×10^{-10}. We proceed as in the example just worked in the text, writing the chemical equation and constructing a table of initial and equilibrium concentrations in which the equilibrium concentration of H^+ is our unknown.

Solve: Writing both the chemical equation for the ionization reaction that forms $H^+(aq)$ and the equilibrium-constant (K_a) expression for the reaction:

$$HCN(aq) \rightleftharpoons H^+(aq) + CN^-(aq)$$

$$K_a = \frac{[H^+][CN^-]}{[HCN]} = 4.9 \times 10^{-10}$$

Next we tabulate the concentrations of the species involved in the equilibrium reaction, letting $x = [H^+]$ at equilibrium:

$$HCN(aq) \rightleftharpoons H^+(aq) + CN^-(aq)$$

	$HCN(aq)$	$H^+(aq)$	$CN^-(aq)$
Initial	0.20 M	0	0
Change	$-x$ M	$+x$ M	$+x$ M
Equilibrium	$(0.20 - x)$ M	x M	x M

Substituting the equilibrium concentrations from the table into the equilibrium-constant expression yields

$$K_a = \frac{(x)(x)}{0.20 - x} = 4.9 \times 10^{-10}$$

We next make the simplifying approximation that x, the amount of acid that dissociates, is small compared with the initial concentration of acid; that is, $0.20 - x \simeq 0.20$. Thus,

$$\frac{x^2}{0.20} = 4.9 \times 10^{-10}$$

Solving for x, we have

$$x^2 = (0.20)(4.9 \times 10^{-10}) = 0.98 \times 10^{-10}$$

$$x = \sqrt{0.98 \times 10^{-10}} = 9.9 \times 10^{-6}\,M = [H^+]$$

9.9×10^{-6} is much smaller than 5% of 0.20, the initial HCN concentration. Our simplifying approximation is therefore appropriate. We now calculate the pH of the solution:

$$pH = -\log[H^+] = -\log(9.9 \times 10^{-6}) = 5.00$$

PRACTICE EXERCISE

The K_a for niacin (Practice Exercise 16.10) is 1.5×10^{-5}. What is the pH of a 0.010 M solution of niacin?
Answer: 3.41

(a)

(b)

▲ **Figure 16.8** Demonstration of the relative rates of reaction of two acid solutions of the same concentration with Mg metal. (a) The flask on the left contains 1 M $HC_2H_3O_2$; the one on the right contains 1 M HCl. Each balloon contains the same amount of magnesium metal. (b) When the Mg metal is dropped into the acid, H_2 gas is formed. The rate of H_2 formation is higher for the 1 M HCl solution on the right as evidenced by more gas in the balloon.

The result obtained in Sample Exercise 16.11 is typical of the behavior of weak acids; the concentration of $H^+(aq)$ is only a small fraction of the concentration of the acid in solution. Those properties of the acid solution that relate directly to the concentration of $H^+(aq)$, such as electrical conductivity and rate of reaction with an active metal, are much less evident for a solution of a weak acid than for a solution of a strong acid. Figure 16.8 ◀ presents an experiment that demonstrates the difference in concentration of $H^+(aq)$ in weak and strong acid solutions of the same concentration. The rate of reaction with the metal is much faster for the solution of a strong acid.

To determine the pH of a solution of a weak acid, you might think that the percent ionization of the acid would be simpler to use than the acid-dissociation constant. However, the percent ionization at a particular temperature depends not only on the identity of the acid but also on its concentration. As shown in Figure 16.9 ▶, the percent ionization of a weak acid decreases as its concentration increases. This fact is further demonstrated in Sample Exercise 16.12.

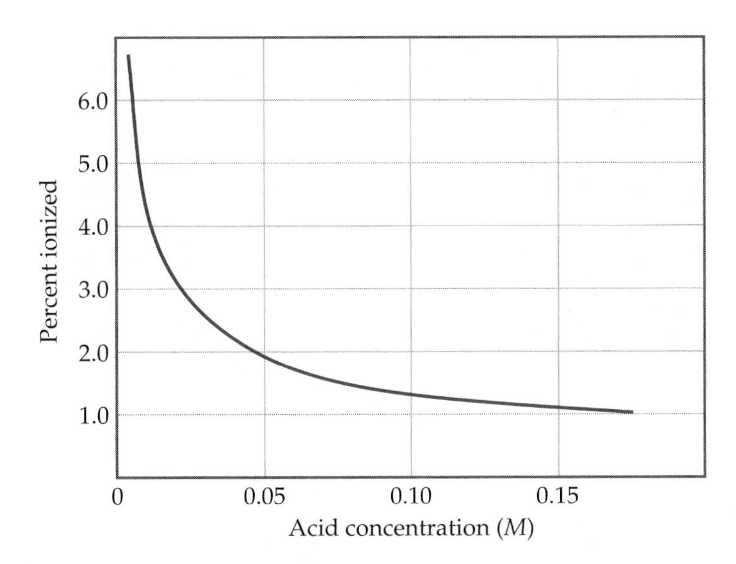

◀ **Figure 16.9** The percent ionization of a weak acid decreases with increasing concentration. The data shown are for acetic acid.

SAMPLE EXERCISE 16.12

Calculate the percentage of HF molecules ionized in **(a)** a 0.10 *M* HF solution; **(b)** a 0.010 *M* HF solution.

Solution

Analyze: We are asked to calculate the percent ionization of two HF solutions of different concentration.

Plan: We approach this problem as we would previous equilibrium problems. We begin by writing the chemical equation for the equilibrium and tabulating the known and unknown concentrations of all species. We then substitute the equilibrium concentrations into the equilibrium-constant expression and solve for the unknown concentration, that of H^+.

Solve: (a) The equilibrium reaction and equilibrium concentrations are as follows:

$$HF(aq) \rightleftharpoons H^+(aq) + F^-(aq)$$

	$HF(aq)$	$H^+(aq)$	$F^-(aq)$
Initial	0.10 M	0	0
Change	$-x$ M	$+x$ M	$+x$ M
Equilibrium	$(0.10 - x)$ M	x M	x M

The equilibrium-constant expression is

$$K_a = \frac{[H^+][F^-]}{[HF]} = \frac{(x)(x)}{0.10 - x} = 6.8 \times 10^{-4}$$

When we try solving this equation using the approximation $0.10 - x = 0.10$ (that is, by neglecting the concentration of acid that ionizes in comparison with the initial concentration), we obtain

$$x = 8.2 \times 10^{-3}\ M$$

Because this value is greater than 5% of 0.10 *M*, we should work the problem without the approximation, using an equation-solving calculator or the quadratic formula. Rearranging our equation and writing it in standard quadratic form, we have

$$x^2 = (0.10 - x)(6.8 \times 10^{-4})$$
$$= 6.8 \times 10^{-5} - (6.8 \times 10^{-4})x$$
$$x^2 + (6.8 \times 10^{-4})x - 6.8 \times 10^{-5} = 0$$

This equation can be solved using the standard quadratic formula.

$$x = \frac{-b \pm \sqrt{b^2 - 4ac}}{2a}$$

Substituting the appropriate numbers gives

$$x = \frac{-6.8 \times 10^{-4} \pm \sqrt{(6.8 \times 10^{-4})^2 + 4(6.8 \times 10^{-5})}}{2}$$
$$= \frac{-6.8 \times 10^{-4} \pm 1.6 \times 10^{-2}}{2}$$

Of the two solutions, only the one that gives a positive value for x is chemically reasonable. Thus,

$$x = [H^+] = [F^-] = 7.9 \times 10^{-3}\,M$$

From our result we can calculate the percent of molecules ionized:

$$\text{Percent ionization of HF} = \frac{\text{concentration ionized}}{\text{original concentration}} \times 100\%$$

$$= \frac{7.9 \times 10^{-3}\,M}{0.10\,M} \times 100\% = 7.9\%$$

(b) Proceeding similarly for the 0.010 M solution, we have

$$\frac{x^2}{0.010 - x} = 6.8 \times 10^{-4}$$

Solving the resultant quadratic expression, we obtain

$$x = [H^+] = [F^-] = 2.3 \times 10^{-3}\,M$$

The percentage of molecules ionized is

$$\frac{0.0023}{0.010} \times 100\% = 23\%$$

Notice that in diluting the solution by a factor of 10, the percentage of molecules ionized increases by a factor of 3. This result is in accord with what we see in Figure 16.9. It is also what we would expect from Le Châtelier's principle. (Section 15.6) There are more "particles" or reaction components on the right side of the equation than on the left. Dilution causes the reaction to shift in the direction of the larger number of particles because this counters the effect of the decreasing concentration of particles.

PRACTICE EXERCISE

Calculate the percentage of niacin molecules ionized in **(a)** the solution in Practice Exercise 16.11; **(b)** a $1.0 \times 10^{-3}\,M$ solution of niacin.
Answers: **(a)** 3.9%; **(b)** 12%

Polyprotic Acids

Many acids have more than one ionizable H atom. These acids are known as **polyprotic acids.** For example, each of the H atoms in sulfurous acid (H_2SO_3) can ionize in successive steps:

$$H_2SO_3(aq) \rightleftharpoons H^+(aq) + HSO_3^-(aq) \qquad K_{a1} = 1.7 \times 10^{-2} \qquad [16.31]$$

$$HSO_3^-(aq) \rightleftharpoons H^+(aq) + SO_3^{2-}(aq) \qquad K_{a2} = 6.4 \times 10^{-8} \qquad [16.32]$$

The acid-dissociation constants for these equilibria are labeled K_{a1} and K_{a2}. The numbers on the constants refer to the particular proton of the acid that is ionizing. Thus, K_{a2} always refers to the equilibrium involving removal of the second proton of a polyprotic acid.

In the preceding example K_{a2} is much smaller than K_{a1}. On the basis of electrostatic attractions, we would expect a positively charged proton to be lost more readily from the neutral H_2SO_3 molecule than from the negatively charged HSO_3^- ion. This observation is general: *It is always easier to remove the first proton from a polyprotic acid than the second.* Similarly, for an acid with three ionizable protons, it is easier to remove the second proton than the third. Thus, the K_a values become successively smaller as successive protons are removed.

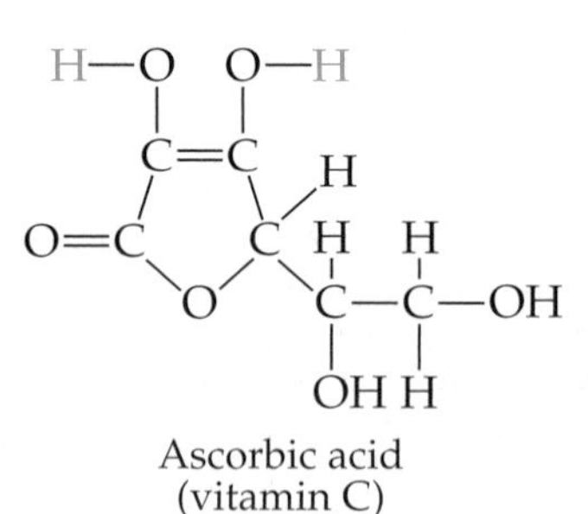

Ascorbic acid (vitamin C)

Citric acid

The acid-dissociation constants for a few common polyprotic acids are listed in Table 16.3 ▶. A more complete list is given in Appendix D. The structures for ascorbic and citric acids are shown in the margin. Notice that the K_a values for successive losses of protons from these acids usually differ by a factor of at least 10^3. Notice also that the value of K_{a1} for sulfuric acid is listed simply as "large." Sulfuric acid is a strong acid with respect to the removal of the first proton. Thus, the reaction for the first ionization step lies completely to the right:

$$H_2SO_4(aq) \longrightarrow H^+(aq) + HSO_4^-(aq) \qquad \text{(complete ionization)}$$

HSO_4^-, on the other hand, is a weak acid for which $K_{a2} = 1.2 \times 10^{-2}$.

Because K_{a1} is so much larger than subsequent dissociation constants for these polyprotic acids, almost all of the $H^+(aq)$ in the solution comes from the first ionization reaction. As long as successive K_a values differ by a factor of 10^3 or more, it is possible to obtain a satisfactory estimate of the pH of polyprotic acid solutions by considering only K_{a1}.

TABLE 16.3 Acid-Dissociation Constants of Some Common Polyprotic Acids

Name	Formula	K_{a1}	K_{a2}	K_{a3}
Ascorbic	$H_2C_6H_6O_6$	8.0×10^{-5}	1.6×10^{-12}	
Carbonic	H_2CO_3	4.3×10^{-7}	5.6×10^{-11}	
Citric	$H_3C_6H_5O_7$	7.4×10^{-4}	1.7×10^{-5}	4.0×10^{-7}
Oxalic	$H_2C_2O_4$	5.9×10^{-2}	6.4×10^{-5}	
Phosphoric	H_3PO_4	7.5×10^{-3}	6.2×10^{-8}	4.2×10^{-13}
Sulfurous	H_2SO_3	1.7×10^{-2}	6.4×10^{-8}	
Sulfuric	H_2SO_4	Large	1.2×10^{-2}	
Tartaric	$H_2C_4H_4O_6$	1.0×10^{-3}	4.6×10^{-5}	

SAMPLE EXERCISE 16.13

The solubility of CO_2 in pure water at 25°C and 0.1 atm pressure is 0.0037 *M*. The common practice is to assume that all of the dissolved CO_2 is in the form of carbonic acid (H_2CO_3), which is produced by reaction between the CO_2 and H_2O:

$$CO_2(aq) + H_2O(l) \rightleftharpoons H_2CO_3(aq)$$

What is the pH of a 0.0037 *M* solution of H_2CO_3?

Solution

Analyze: We are asked to determine the pH of a 0.0037 *M* solution of a polyprotic acid.

Plan: H_2CO_3 is a diprotic acid; the two acid-dissociation constants, K_{a1} and K_{a2} (Table 16.3), differ by more than a factor of 10^3. Consequently, the pH can be determined by considering only K_{a1}, thereby treating the acid as if it were a monoprotic acid.

Solve: Proceeding as in Sample Exercises 16.11 and 16.12, we can write the equilibrium reaction and equilibrium concentrations as follows:

	$H_2CO_3(aq)$ $\rightleftharpoons$	$H^+(aq)$ +	$HCO_3^-(aq)$
Initial	$0.0037\ M$	0	0
Change	$-x\ M$	$+x\ M$	$+x\ M$
Equilibrium	$(0.0037 - x)\ M$	$x\ M$	$x\ M$

The equilibrium-constant expression is as follows:

$$K_{a1} = \frac{[H^+][HCO_3^-]}{[H_2CO_3]} = \frac{(x)(x)}{0.0037 - x} = 4.3 \times 10^{-7}$$

Solving this equation using an equation-solving calculator, we get

$$x = 4.0 \times 10^{-5}\ M$$

Alternatively, because K_{a1} is small, we can make the simplifying approximation that x is small, so that

$$0.0037 - x \simeq 0.0037$$

Thus,

$$\frac{(x)(x)}{0.0037} = 4.3 \times 10^{-7}$$

Solving for x, we have

$$x^2 = (0.0037)(4.3 \times 10^{-7}) = 1.6 \times 10^{-9}$$

$$x = [H^+] = [HCO_3^-] = \sqrt{1.6 \times 10^{-9}} = 4.0 \times 10^{-5}\ M$$

The small value of x indicates that our simplifying assumption was justified. The pH is therefore

$$pH = -\log[H^+] = -\log(4.0 \times 10^{-5}) = 4.40$$

Comment: If we were asked to solve for $[CO_3^{2-}]$, we would need to use K_{a2}. Let's illustrate that calculation. Using the values of $[HCO_3^-]$ and $[H^+]$ calculated above, and setting $[CO_3^{2-}] = y$, we have the following initial and equilibrium concentration values:

	$HCO_3^-(aq)$ $\rightleftharpoons$	$H^+(aq)$ +	$CO_3^{2-}(aq)$
Initial	$4.0 \times 10^{-5}\ M$	$4.0 \times 10^{-5}\ M$	0
Change	$-y\ M$	$+y\ M$	$+y\ M$
Equilibrium	$(4.0 \times 10^{-5} - y)\ M$	$(4.0 \times 10^{-5} + y)\ M$	$y\ M$

Assuming that y is small compared to 4.0×10^{-5}, we have

$$K_{a2} = \frac{[H^+][CO_3^{2-}]}{[HCO_3^-]} = \frac{(4.0 \times 10^{-5})(y)}{4.0 \times 10^{-5}} = 5.6 \times 10^{-11}$$

$$y = 5.6 \times 10^{-11}\ M = [CO_3^{2-}]$$

The value calculated for y is indeed very small compared to 4.0×10^{-5}, showing that our assumption was justified. It also shows that the ionization of HCO_3^- is negligible compared to that of H_2CO_3, as far as production of H^+ is concerned. However, it is the *only* source of CO_3^{2-}, which has a very low concentration in the solution. Our calculations thus tell us that in a solution of carbon dioxide in water most of the CO_2 is in the form of CO_2 or H_2CO_3, a small fraction ionizes to form H^+ and HCO_3^-, and an even smaller fraction ionizes to give CO_3^{2-}.

PRACTICE EXERCISE

Calculate the pH and concentration of oxalate ion, $[C_2O_4^{2-}]$, in a 0.020 M solution of oxalic acid ($H_2C_2O_4$) (see Table 16.3).
Answer: pH = 1.80; $[C_2O_4^{2-}] = 6.4 \times 10^{-5}$ M

16.7 Weak Bases

Many substances behave as weak bases in water. Weak bases react with water, abstracting protons from H_2O, thereby forming the conjugate acid of the base and OH^- ions.

$$B(aq) + H_2O \rightleftharpoons HB^+ + OH^-(aq) \qquad [16.33]$$

The most commonly encountered weak base is ammonia.

$$NH_3(aq) + H_2O(l) \rightleftharpoons NH_4^+(aq) + OH^-(aq) \qquad [16.34]$$

The equilibrium-constant expression for this reaction can be written as

$$K_b = \frac{[NH_4^+][OH^-]}{[NH_3]} \qquad [16.35]$$

Water is the solvent, so it is omitted from the equilibrium-constant expression.

As with K_w and K_a, the subscript "b" denotes that this equilibrium constant refers to a particular type of reaction, namely the ionization of a weak base in water. The constant K_b is called the **base-dissociation constant.** *The constant K_b always refers to the equilibrium in which a base reacts with H_2O to form the corresponding conjugate acid and OH^-.* Table 16.4 ▼ lists the names, formulas, Lewis structures,

TABLE 16.4 Some Weak Bases and Their Aqueous Solution Equilibria

Base	Lewis Structure	Conjugate Acid	Equilibrium Reaction	K_b
Ammonia (NH_3)	H—N̈—H (with H on N)	NH_4^+	$NH_3 + H_2O \rightleftharpoons NH_4^+ + OH^-$	1.8×10^{-5}
Pyridine (C_5H_5N)	(ring)N:	$C_5H_5NH^+$	$C_5H_5N + H_2O \rightleftharpoons C_5H_5NH^+ + OH^-$	1.7×10^{-9}
Hydroxylamine (H_2NOH)	H—N̈—Ö̤H (with H on N)	H_3NOH^+	$H_2NOH + H_2O \rightleftharpoons H_3NOH^+ + OH^-$	1.1×10^{-8}
Methylamine (NH_2CH_3)	H—N̈—CH_3 (with H on N)	$NH_3CH_3^+$	$NH_2CH_3 + H_2O \rightleftharpoons NH_3CH_3^+ + OH^-$	4.4×10^{-4}
Hydrosulfide ion (HS^-)	[H—S̤̈:]$^-$	H_2S	$HS^- + H_2O \rightleftharpoons H_2S + OH^-$	1.8×10^{-7}
Carbonate ion (CO_3^{2-})	[:Ö: / C / :Ö̤ and Ö̤:]$^{2-}$	HCO_3^-	$CO_3^{2-} + H_2O \rightleftharpoons HCO_3^- + OH^-$	1.8×10^{-4}
Hypochlorite ion (ClO^-)	[:C̤̈l—Ö̤:]$^-$	HClO	$ClO^- + H_2O \rightleftharpoons HClO + OH^-$	3.3×10^{-7}

equilibrium reactions, and values of K_b for several weak bases in water. Appendix D includes a more extensive list. These bases contain one or more lone pairs of electrons because a lone pair is necessary to form the bond with H^+. Notice that in the neutral molecules in Table 16.4 the lone pairs are on nitrogen atoms. The other bases listed are anions derived from weak acids.

SAMPLE EXERCISE 16.14

Calculate the concentration of OH^- in a 0.15 *M* solution of NH_3.

Solution

Analyze: We are given the concentration of a weak base and are asked to determine the concentration of OH^-.

Plan: We will use essentially the same procedure here as used in solving problems involving the ionization of weak acids, that is, we write the chemical equation and tabulate initial and equilibrium concentrations.

Solve: We first write the ionization reaction and the corresponding equilibrium-constant (K_b) expression.

$$NH_3(aq) + H_2O(l) \rightleftharpoons NH_4^+(aq) + OH^-(aq)$$

$$K_b = \frac{[NH_4^+][OH^-]}{[NH_3]} = 1.8 \times 10^{-5}$$

We then tabulate the equilibrium concentrations involved in the equilibrium.

	$NH_3(aq)$ +	$H_2O(l)$ $\rightleftharpoons$	$NH_4^+(aq)$ +	$OH^-(aq)$
Initial	0.15 *M*	—	0	0
Change	$-x$ *M*	—	$+x$ *M*	$+x$ *M*
Equilibrium	$(0.15 - x)$ *M*	—	x *M*	x *M*

(We ignore the concentration of H_2O because it is not involved in the equilibrium-constant expression.) Inserting these quantities into the equilibrium-constant expression gives the following:

$$K_b = \frac{[NH_4^+][OH^-]}{[NH_3]} = \frac{(x)(x)}{0.15 - x} = 1.8 \times 10^{-5}$$

Because K_b is small, we can neglect the small amount of NH_3 that reacts with water, as compared to the total NH_3 concentration; that is, we can neglect x relative to 0.15 *M*. Then we have

$$\frac{x^2}{0.15} = 1.8 \times 10^{-5}$$

$$x^2 = (0.15)(1.8 \times 10^{-5}) = 2.7 \times 10^{-6}$$

$$x = [NH_4^+] = [OH^-] = \sqrt{2.7 \times 10^{-6}} = 1.6 \times 10^{-3}\ M$$

Check: The value obtained for x is only about 1% of the NH_3 concentration, 0.15 *M*. Therefore, neglecting x relative to 0.15 was justified.

PRACTICE EXERCISE

Which of the following compounds should produce the highest pH as a 0.05 *M* solution: pyridine, methylamine, or nitrous acid?

Answer: methylamine

Types of Weak Bases

How can we recognize from a chemical formula whether a molecule or ion is able to behave as a weak base? Weak bases fall into two general categories. The first category contains neutral substances that have an atom with a nonbonding pair of electrons that can serve as a proton acceptor. Most of these bases, including all of the uncharged bases listed in Table 16.4, contain a nitrogen atom. These substances include ammonia and a related class of compounds called **amines**. In organic amines, one or more of the N—H bonds in NH_3 is replaced with a bond between N and C. Thus, the replacement of one N—H bond in NH_3 with a N—CH_3 bond gives methylamine, NH_2CH_3 (usually written CH_3NH_2). Like NH_3, amines can abstract a proton from a water molecule by forming an additional N—H bond, as shown here for methylamine:

$$H-\ddot{N}(H)-CH_3(aq) + H_2O(l) \rightleftharpoons \left[H-N(H)(H)-CH_3 \right]^+(aq) + OH^-(aq) \qquad [16.36]$$

The chemical formula for the conjugate acid of methylamine is usually written $CH_3NH_3^+$.

The second general category of weak bases consists of the anions of weak acids. In an aqueous solution of sodium hypochlorite (NaClO), for example, NaClO dissolves in water to give Na^+ and ClO^- ions. The Na^+ ion is always a spectator ion in acid-base reactions. (Section 4.3) The ClO^- ion, however, is the conjugate base of a weak acid, hypochlorous acid. Consequently, the ClO^- ion acts as a weak base in water:

$$ClO^-(aq) + H_2O(l) \rightleftharpoons HClO(aq) + OH^-(aq) \qquad K_b = 3.33 \times 10^{-7} \qquad [16.37]$$

SAMPLE EXERCISE 16.15

A solution is made by adding solid sodium hypochlorite (NaClO) to enough water to make 2.00 L of solution. If the solution has a pH of 10.50, how many moles of NaClO were added to the water?

Solution

Analyze and Plan: We are given the pH of a 2.00-L solution of NaClO and must calculate the number of moles of NaClO needed to raise the pH to 10.50. NaClO is an ionic compound consisting of Na^+ and ClO^- ions. As such, it is a strong electrolyte that completely dissociates in solution into Na^+, which is a spectator ion, and ClO^- ion, which is a weak base with $K_b = 3.33 \times 10^{-7}$ (Equation 16.37). From the pH we can determine the equilibrium concentration of OH^-. We can then construct a table of initial and equilibrium concentrations in which the initial concentration of ClO^- is our unknown.

Solve: We can calculate $[OH^-]$ by using either Equation 16.16 or Equation 16.19; we will use the latter method here:

$$pOH = 14.00 - pH = 14.00 - 10.50 = 3.50$$

$$[OH^-] = 10^{-3.50} = 3.16 \times 10^{-4}\ M$$

This concentration is high enough that we can assume that Equation 16.37 is the only source of OH^-; that is, we can neglect any OH^- produced by the autoionization of H_2O. We now assume a value of x for the initial concentration of ClO^- and solve the equilibrium problem in the usual way:

	$ClO^-(aq)$	+ $H_2O(l)$ ⇌	$HClO(aq)$	+ $OH^-(aq)$
Initial	$x\ M$	—	0	0
Change	$-3.16 \times 10^{-4}\ M$	—	$+3.16 \times 10^{-4}\ M$	$+3.16 \times 10^{-4}\ M$
Final	$(x - 3.16 \times 10^{-4})M$	—	$3.16 \times 10^{-4}\ M$	$3.16 \times 10^{-4}\ M$

We now use the expression for the base-dissociation constant to solve for x:

$$K_b = \frac{[HClO][OH^-]}{[ClO^-]} = \frac{(3.16 \times 10^{-4})^2}{x - 3.16 \times 10^{-4}} = 3.33 \times 10^{-7}$$

Thus,

$$x = \frac{(3.16 \times 10^{-4})^2}{3.33 \times 10^{-7}} + (3.16 \times 10^{-4}) = 0.30\ M$$

We say that the solution is 0.30 M in NaClO, even though some of the ClO^- ions have reacted with water. Because the solution is 0.30 M in NaClO and the total volume of solution is 2.00 L, 0.60 mol of NaClO is the amount of the salt that was added to the water.

PRACTICE EXERCISE

A solution of NH_3 in water has a pH of 10.50. What is the molarity of the solution?
Answer: 0.0056 M

16.8 Relationship Between K_a and K_b

We've seen in a qualitative way that the stronger acids have the weaker conjugate bases. To see if we can find a corresponding *quantitative* relationship, let's consider the NH_4^+ and NH_3 conjugate acid-base pair. Each of these species reacts with water:

$$NH_4^+(aq) \rightleftharpoons NH_3(aq) + H^+(aq) \quad [16.38]$$

$$NH_3(aq) + H_2O(l) \rightleftharpoons NH_4^+(aq) + OH^-(aq) \quad [16.39]$$

Each of these equilibria is expressed by a characteristic dissociation constant:

$$K_a = \frac{[NH_3][H^+]}{[NH_4^+]} \qquad K_b = \frac{[NH_4^+][OH^-]}{[NH_3]}$$

When Equations 16.38 and 16.39 are added together, the NH_4^+ and NH_3 species cancel and we are left with just the autoionization of water.

$$\begin{array}{rcl} NH_4^+(aq) & \rightleftharpoons & NH_3(aq) + H^+(aq) \\ NH_3(aq) + H_2O(l) & \rightleftharpoons & NH_4^+(aq) + OH^-(aq) \\ \hline H_2O(l) & \rightleftharpoons & H^+(aq) + OH^-(aq) \end{array}$$

Recall that when two equations are added to give a third, the equilibrium constant associated with the third equation equals the product of the equilibrium constants for the two equations added together. (Section 15.2)

Applying this rule to our present example, when we multiply K_a and K_b, we obtain the following:

$$K_a \times K_b = \left(\frac{[NH_3][H^+]}{[NH_4^+]}\right)\left(\frac{[NH_4^+][OH^-]}{[NH_3]}\right)$$

$$= [H^+][OH^-] = K_w$$

Thus, the result of multiplying K_a times K_b is just the ion-product constant for water, K_w (Equation 16.16). This is what we would expect, moreover, because adding Equations 16.38 and 16.39 gave us the autoionization equilibrium for water, for which the equilibrium constant is K_w.

This relationship is so important that it should receive special attention: *The product of the acid-dissociation constant for an acid and the base-dissociation constant for its conjugate base is the ion-product constant for water.*

$$K_a \times K_b = K_w \quad [16.40]$$

As the strength of an acid increases (larger K_a), the strength of its conjugate base must decrease (smaller K_b,) so that the product $K_a \times K_b$ equals 1.0×10^{-14} at 25°C. The K_a and K_b data in Table 16.5 ▼ demonstrate this relationship.

TABLE 16.5 Some Conjugate Acid-Base Pairs

Acid	K_a	Base	K_b
HNO_3	(Strong acid)	NO_3^-	(Negligible basicity)
HF	6.8×10^{-4}	F^-	1.5×10^{-11}
$HC_2H_3O_2$	1.8×10^{-5}	$C_2H_3O_2^-$	5.6×10^{-10}
H_2CO_3	4.3×10^{-7}	HCO_3^-	2.3×10^{-8}
NH_4^+	5.6×10^{-10}	NH_3	1.8×10^{-5}
HCO_3^-	5.6×10^{-11}	CO_3^{2-}	1.8×10^{-4}
OH^-	(Negligible acidity)	O^{2-}	(Strong base)

Chemistry at Work Amines and Amine Hydrochlorides

Many amines with low molecular weights have unpleasant "fishy" odors. Amines and NH_3 are produced by the anaerobic (absence of O_2) decomposition of dead animal or plant matter. Two such amines with very disagreeable odors are $H_2N(CH_2)_4NH_2$, known as *putrescine*, and $H_2N(CH_2)_5NH_2$, known as *cadaverine*.

Many drugs, including quinine, codeine, caffeine, and amphetamine (Benzedrine[TM]), are amines. Like other amines, these substances are weak bases; the amine nitrogen is readily protonated upon treatment with an acid. The resulting products are called *acid salts*. If we use A as the abbreviation for an amine, the acid salt formed by reaction with hydrochloric acid can be written as AH^+Cl^-. It is also sometimes written as $A \cdot HCl$ and referred to as a hydrochloride. Amphetamine hydrochloride, for example, is the acid salt formed by treating amphetamine with HCl:

$$C_6H_5-CH_2-\underset{\displaystyle CH_3}{\underset{|}{CH}}-\ddot{N}H_2(aq) + HCl(aq) \longrightarrow$$

Amphetamine

$$C_6H_5-CH_2-\underset{\displaystyle CH_3}{\underset{|}{CH}}-NH_3^+Cl^-(aq)$$

Amphetamine hydrochloride

Such acid salts are much less volatile, more stable, and generally more water-soluble than the corresponding neutral amines. Many drugs that are amines are sold and administered as acid salts. Some examples of over-the-counter medications that contain amine hydrochlorides as active ingredients are shown in Figure 16.10 ▼.

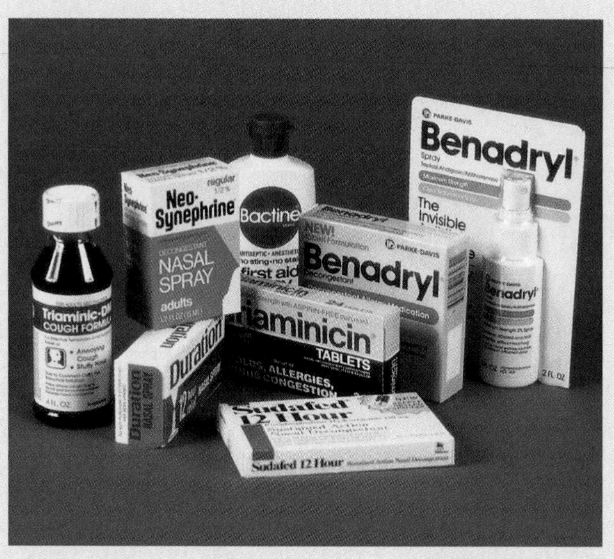

▲ **Figure 16.10** Some over-the-counter medications in which an amine hydrochloride is a major active ingredient.

By using Equation 16.40, we can calculate K_b for any weak base if we know K_a for its conjugate acid. Similarly, we can calculate K_a for a weak acid if we know K_b for its conjugate base. As a practical consequence, ionization constants are often listed for only one member of a conjugate acid-base pair. For example, Appendix D does not contain K_b values for the anions of weak acids because these can be readily calculated from the tabulated K_a values for their conjugate acids.

If you look up the values for acid- or base-dissociation constants in a chemistry handbook, you may find them expressed as pK_a or pK_b (that is, as $-\log K_a$ or $-\log K_b$). (Section 16.4) Equation 16.40 can be written in terms of pK_a and pK_b by taking the negative log of both sides.

$$pK_a + pK_b = pK_w = 14.00 \quad \text{at } 25°C \qquad [16.41]$$

SAMPLE EXERCISE 16.16

Calculate **(a)** the base-dissociation constant, K_b, for the fluoride ion (F^-); **(b)** the acid-dissociation constant, K_a, for the ammonium ion (NH_4^+).

Solution

Analyze: We are asked to determine dissociation constants for F^-, the conjugate base of HF, and NH_4^+, the conjugate acid of NH_3.

Plan: Although neither F^- nor NH_4^+ appears in the tables, we can find the tabulated values for ionization constants for HF and NH_3, and use the relationship between K_a and K_b to calculate the ionization constants for each of the conjugates.

Solve: (a) K_a for the weak acid, HF, is given in Table 16.2 and Appendix D as $K_a = 6.8 \times 10^{-4}$. We can use Equation 16.40 to calculate K_b for the conjugate base, F^-:

$$K_b = \frac{K_w}{K_a} = \frac{1.0 \times 10^{-14}}{6.8 \times 10^{-4}} = 1.5 \times 10^{-11}$$

(b) K_b for NH_3 is listed in Table 16.4 and in Appendix D as $K_b = 1.8 \times 10^{-5}$. Using Equation 16.40, we can calculate K_a for the conjugate acid, NH_4^+:

$$K_a = \frac{K_w}{K_b} = \frac{1.0 \times 10^{-14}}{1.8 \times 10^{-5}} = 5.6 \times 10^{-10}$$

PRACTICE EXERCISE

(a) Which of the following anions has the largest base-dissociation constant: NO_2^-, PO_4^{3-}, or N_3^-? **(b)** The base quinoline has the following structure:

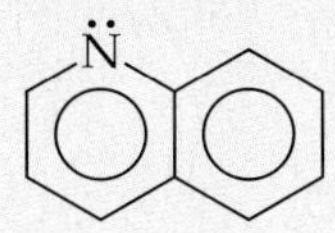

Its conjugate acid is listed in handbooks as having a pK_a of 4.90. What is the base-dissociation constant for quinoline?
Answers: **(a)** PO_4^{3-} ($K_b = 2.4 \times 10^{-2}$); **(b)** 7.9×10^{-10}

16.9 Acid-Base Properties of Salt Solutions

Even before you began this chapter, you were undoubtedly aware of many substances that are acidic, such as HNO_3, HCl, and H_2SO_4, and others that are basic, such as NaOH and NH_3. However, our recent discussions have indicated that ions can also exhibit acidic or basic properties. For example, we calculated K_a for NH_4^+ and K_b for F^- in Sample Exercise 16.16. Such behavior implies that salt solutions can be acidic or basic. Before proceeding with further discussions of acids and bases, let's examine the way dissolved salts can affect pH.

We can assume that when salts dissolve in water, they are completely dissociated; nearly all salts are strong electrolytes. Consequently, the acid-base properties of salt solutions are due to the behavior of their constituent cations and anions. Many ions are able to react with water to generate $H^+(aq)$ or $OH^-(aq)$. This type of reaction is often called **hydrolysis**. The pH of an aqueous salt solution can be predicted qualitatively by considering the ions of which the salt is composed.

An Anion's Ability to React with Water

In general, an anion, X^-, in solution can be considered the conjugate base of an acid. For example, Cl^- is the conjugate base of HCl and $C_2H_3O_2^-$ is the conjugate base of $HC_2H_3O_2$. Whether an anion reacts with water to produce hydroxide depends upon the strength of the acid to which it is conjugate. To identify the acid and assess its strength, we can simply add a proton to the anion's formula:

$$X^- \text{ plus a proton gives HX}$$

If the acid determined in this way is one of the strong acids listed at the beginning of Section 16.5, then the anion in question will have a negligible tendency to abstract protons from water (Section 16.2) and the following equilibrium will lie entirely to the left:

$$X^-(aq) + H_2O(l) \rightleftharpoons HX(aq) + OH^-(aq) \qquad [16.42]$$

Consequently, the anion X^- will not affect the pH of the solution. The presence of Cl^- in an aqueous solution, for example, does not result in the production of any OH^- and does not affect the pH.

Conversely, if HX is *not* one of the seven strong acids, then it is a weak acid. In this case the conjugate base X^- will react to a small extent with water to produce the weak acid and hydroxide ions, thus causing the pH to be higher (more basic) than it otherwise would be. Acetate ion ($C_2H_3O_2^-$), for example, being the conjugate base of a weak acid, reacts with water to produce acetic acid and hydroxide ions, thereby increasing the pH of the solution.*

$$C_2H_3O_2^-(aq) + H_2O(l) \rightleftharpoons HC_2H_3O_2(aq) + OH^-(aq) \qquad [16.43]$$

Anions that still have ionizable protons, such as HSO_3^-, are amphoteric: They can act as either acids or bases (Sample Exercise 16.2). Their behavior toward water will be determined by the relative magnitudes of K_a and K_b for the ion, as shown in Sample Exercise 16.17.

A Cation's Ability to React with Water

Polyatomic cations whose formulas contain one or more protons can be considered the conjugate acids of weak bases. NH_4^+, for example, is the conjugate acid of the weak base NH_3. Recall from Section 16.2 the inverse relationship between the strength of a base and that of its conjugate acid. Thus, a cation such as NH_4^+ will donate a proton to water, producing hydronium ions and lowering the pH:

$$NH_4^+(aq) + H_2O(l) \rightleftharpoons NH_3(aq) + H_3O^+(aq) \qquad [16.44]$$

Most metal ions can also react with water to decrease the pH of an aqueous solution. The mechanism by which metal ions produce acidic solutions is described in Section 16.11. However, ions of alkali metals and of the heavier alkaline earth metals do not react with water and therefore do not affect pH. Note that these exceptions are the cations found in the strong bases. (Section 16.5)

Combined Effect of Cation and Anion in Solution

If an aqueous salt solution contains an anion that does not react with water and a cation that does not react with water, we expect the pH to be neutral. If the solution contains an anion that reacts with water to produce hydroxide and a cation that does not react with water, we expect the pH to be basic. If the solution contains a cation that reacts with water to produce hydronium and an anion that does not react with water, we expect the pH to be acidic. Finally, a solution may contain an anion and a cation *both* capable of reacting with water. In this case, both hydroxide and hydronium will be produced. Whether the solution is basic, neutral, or acidic will depend upon the relative abilities of the ions to react with water.

To summarize:

1. An anion that is the conjugate base of a strong acid, for example, Br^-, will not affect the pH of a solution.
2. An anion that is the conjugate base of a weak acid, for example, CN^-, will cause an increase in pH.
3. A cation that is the conjugate acid of a weak base, for example, $CH_3NH_3^+$, will cause a decrease in pH.
4. With the exception of ions of group 1A and heavier members of group 2A (Ca^{2+}, Sr^{2+}, and Ba^{2+}), metal ions will cause a decrease in pH.
5. When a solution contains both the conjugate base of a weak acid and the conjugate acid of a weak base, the ion with the largest ionization constant will have the greatest influence on the pH.

Figure 16.11 ▶ demonstrates the influence of several salts on pH.

* These rules apply to what are called normal salts. These salts are those that contain no ionizable protons on the anion. The pH of an acid salt (such as $NaHCO_3$ or NaH_2PO_4) is affected not only by the hydrolysis of the anion but also by its acid dissociation, as shown in Sample Exercise 16.17.

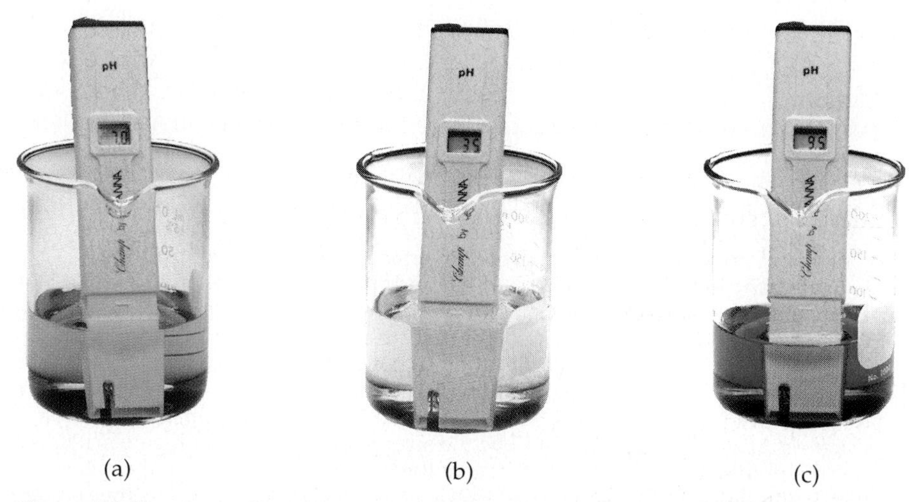

▲ **Figure 16.11** Depending on the ions involved, salt solutions can be neutral, acidic, or basic. These three solutions contain the acid-base indicator bromthymol blue. (a) The NaCl solution is neutral (pH = 7.0); (b) the NH_4Cl solution is acidic (pH = 3.5); (c) the NaClO solution is basic (pH = 9.5).

SAMPLE EXERCISE 16.17

Predict whether the salt Na_2HPO_4 will form an acidic or basic solution on dissolving in water.

Solution

Analyze and Plan: We are asked to predict whether a solution of Na_2HPO_4 will be acidic or basic. Because Na_2HPO_4 is an ionic compound, we divide it into its component ions, Na^+ and HPO_4^{2-}, and consider whether each is acidic or basic. Because Na^+ is the cation of a strong base, NaOH, we know that Na^+ has no influence on pH. It is merely a spectator ion in acid-base chemistry. Thus, our analysis of whether the solution is acidic or basic must focus on the behavior of the HPO_4^{2-} ion. We need to consider the fact that HPO_4^{2-} can act as either an acid or a base.

$$HPO_4^{2-}(aq) \rightleftharpoons H^+(aq) + PO_4^{3-}(aq) \qquad [16.45]$$

$$HPO_4^{2-}(aq) + H_2O \rightleftharpoons H_2PO_4^-(aq) + OH^-(aq) \qquad [16.46]$$

The reaction with the larger ionization constant will determine whether the solution is acidic or basic.

Solve: The value of K_a for Equation 16.45, as shown in Table 16.3, is 4.2×10^{-13}. We must calculate the value of K_b for Equation 16.46 from the value of K_a for its conjugate acid, $H_2PO_4^-$. We make use of the relationship shown in Equation 16.40.

$$K_a \times K_b = K_w$$

We want to know K_b for the base HPO_4^{2-}, knowing the value of K_a for the conjugate acid $H_2PO_4^-$:

$$K_b(HPO_4^{2-}) \times K_a(H_2PO_4^-) = K_w = 1.0 \times 10^{-14}$$

Because K_a for $H_2PO_4^-$ is 6.2×10^{-8} (Table 16.3), we calculate K_b for HPO_4^{2-} to be 1.6×10^{-7}. This is more than 10^5 times larger than K_a for HPO_4^{2-}; thus the reaction shown in Equation 16.46 predominates over that in Equation 16.45, and the solution will be basic.

PRACTICE EXERCISE

Predict whether the dipotassium salt of citric acid ($K_2HC_6H_5O_7$) will form an acidic or basic solution in water (see Table 16.3 for data).
Answer: acidic

SAMPLE EXERCISE 16.18

List the following solutions in order of increasing acidity (decreasing pH): (i) 0.1 *M* $Ba(C_2H_3O_2)_2$; (ii) 0.1 *M* NH_4Cl; (iii) 0.1 *M* NH_3CH_3Br; (iv) 0.1 *M* KNO_3.

Solution

Analyze: We are asked to arrange a series of salt solutions from least acidic to most acidic.

Plan: We can determine whether the pH of a solution is acidic, basic, or neutral by identifying the ions in solution, and by assessing how each ion will affect the pH.

Solve: Solution (i) contains barium ions and acetate ions. Ba^{2+}, is the cation of the strong base $Ba(OH)_2$ and will therefore not affect the pH. The anion, $C_2H_3O_2^-$, is the conjugate base of the weak acid $HC_2H_3O_2$ and will hydrolyze to produce OH^- ions, thereby making the solution basic. Solutions (ii) and (iii) both contain cations that are conjugate acids of weak bases and anions that are conjugate bases of strong acids. Both solutions will therefore be acidic. Solution (i) contains NH_4^+, which is the conjugate acid of NH_3 ($K_b = 1.8 \times 10^{-5}$). Solution (iii) contains $NH_3CH_3^+$, which is the conjugate acid of NH_2CH_3 ($K_b = 4.4 \times 10^{-4}$). Because NH_3 has the smaller K_b and is the weaker of the two bases, NH_4^+ will be the stronger of the two conjugate acids. Solution (ii) will therefore be the more acidic of the two. Solution (iv) contains the K^+ ion, which is the cation of the strong base KOH, and the NO_3^- ion, which is the conjugate base of the strong acid HNO_3. Neither of the ions in solution (iv) will react with water to any appreciable extent, making the solution neutral. Thus, the order of increasing acidity is 0.1 *M* $Ba(C_2H_3O_2)_2$ < 0.1 *M* KNO_3 < 0.1 *M* NH_3CH_3Br < 0.1 *M* NH_4Cl.

PRACTICE EXERCISE

In each of the following, indicate which salt will form the more acidic (or less basic) 0.010 *M* solution: **(a)** $NaNO_3$, $Fe(NO_3)_3$; **(b)** KBr, KBrO; **(c)** CH_3NH_3Cl, $BaCl_2$ **(d)** NH_4NO_2, NH_4NO_3.

Answers: **(a)** $Fe(NO_3)_3$; **(b)** KBr; **(c)** CH_3NH_3Cl; **(d)** NH_4NO_3

16.10 Acid-Base Behavior and Chemical Structure

When a substance is dissolved in water, it may behave as an acid, behave as a base, or exhibit no acid-base properties. How does the chemical structure of a substance determine which of these behaviors is exhibited by the substance? For example, why do some substances that contain OH groups behave as bases, releasing OH^- ions into solution, whereas others behave as acids, ionizing to release H^+ ions? Why are some acids stronger than others? In this section we will discuss briefly the effects of chemical structure on acid-base behavior.

Factors That Affect Acid Strength

A molecule containing H will transfer a proton only if the H—X bond is polarized in the following way:

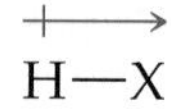

In ionic hydrides such as NaH, the reverse is true; the H atom possesses a negative charge and behaves as a proton acceptor (Equation 16.23). Nonpolar H—X bonds, such as the H—C bond in CH_4, produce neither acidic nor basic aqueous solutions.

A second factor that helps determine whether a molecule containing an H—X bond will donate a proton is the strength of the bond. Very strong bonds are less easily dissociated than weaker ones. This factor is important, for example, in the case of the hydrogen halides. The H—F bond is the most polar H—X bond. You therefore might expect that HF would be a very strong acid if the first factor were all that mattered. However, the energy required to dissociate HF into

H and F atoms is much higher than it is for the other hydrogen halides, as shown in Table 8.4. As a result, HF is a weak acid, whereas all the other hydrogen halides are strong acids in water.

A third factor that affects the ease with which a hydrogen atom ionizes from HX is the stability of the conjugate base, X^-. In general, the greater the stability of the conjugate base, the stronger is the acid. The strength of an acid is often a combination of all three factors: the polarity of the H—X bond, the strength of the H—X bond, and the stability of the conjugate base, X^-.

Binary Acids

In general, the H—X bond strength is the most important factor determining acid strength among the binary acids (those containing hydrogen and just one other element) in which X is in the same *group* in the periodic table. The strength of an H—X bond tends to decrease as the element X increases in size. As a result, the bond strength decreases, and the acidity increases down a group. Thus, HCl is a stronger acid than HF, and H_2S is a stronger acid than H_2O.

Bond strengths change less moving across a row in the periodic table than they do down a group. As a result, bond polarity is the major factor determining acidity for binary acids in the same *row*. Thus, acidity increases as the electronegativity of the element X increases, as it generally does moving from left to right in a row. For example, the acidity of the second-row elements varies in the following order: $CH_4 < NH_3 \ll H_2O < HF$. Because the C—H bond is essentially nonpolar, CH_4 shows no tendency to form H^+ and CH_3^- ions. Although the N—H bond is polar, NH_3 has a nonbonding pair of electrons on the nitrogen atom that dominates its chemistry, so NH_3 acts as a base rather than as an acid. The periodic trends in the acid strengths of binary compounds of hydrogen and the nonmetals of periods 2 and 3 are summarized in Figure 16.12 ▼.

Oxyacids

Many common acids, such as sulfuric acid, contain one or more O—H bonds:

```
        :Ö:
         |
   H—Ö—S—Ö—H
         |
        :Ö:
```

Acids in which OH groups and possibly additional oxygen atoms are bound to a central atom are called **oxyacids**. The OH group is also present in bases. What factors determine whether an OH group will behave as a base or as an acid?

	GROUP 4A	5A	6A	7A
Period 2	CH_4 No acid or base properties	NH_3 Weak base	H_2O ---	HF Weak acid
Period 3	SiH_4 No acid or base properties	PH_3 Weak base	H_2S Weak acid	HCl Strong acid

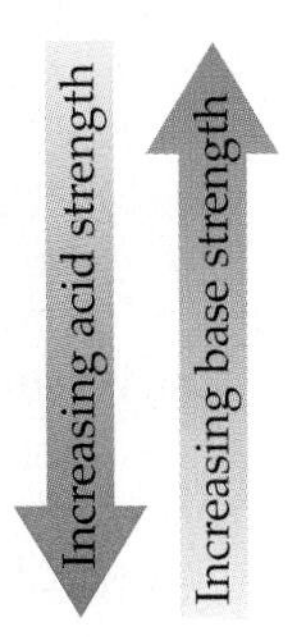

Increasing acid strength →

← Increasing base strength

◀ **Figure 16.12** Acid-base properties of the binary compounds of hydrogen and nonmetals of periods 2 and 3.

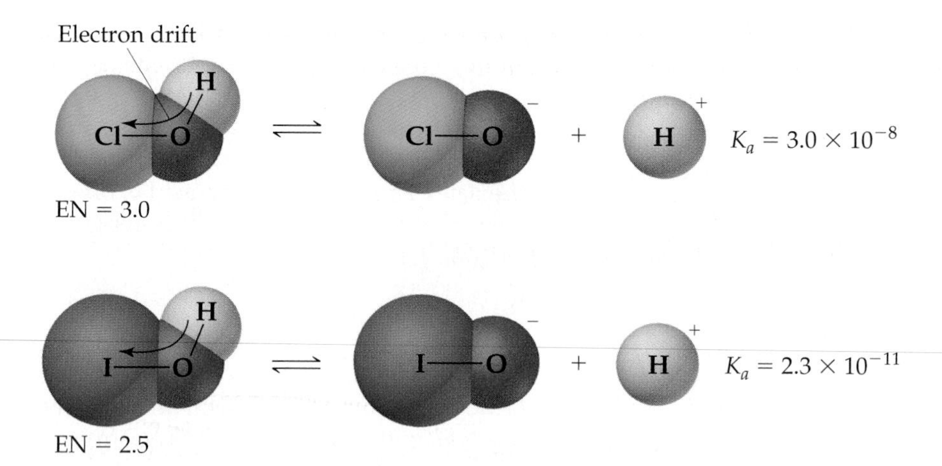

▶ **Figure 16.13** As the electronegativity of the atom attached to an OH group increases, the ease with which the hydrogen atom is ionized increases. The drift of electrons toward the electronegative atom further polarizes the O—H bond, which favors ionization. In addition, the electronegative atom will help stabilize the conjugate base, which also leads to a stronger acid. Because Cl is more electronegative than I, HClO is a stronger acid than HIO.

Let's consider an OH group bound to some atom Y, which might in turn have other groups attached to it:

$$\rangle\!\!-\text{Y—O—H}$$

At one extreme, Y might be a metal, such as Na, K, or Mg. Because of their low electronegativities, the pair of electrons shared between Y and O is completely transferred to oxygen, and an ionic compound containing OH^- is formed. Such compounds are therefore sources of OH^- ions and behave as bases.

When Y is a nonmetal, the bond to O is covalent and the substance does not readily lose OH^-. Instead, these compounds are either acidic or neutral. As a general rule, as the electronegativity of Y increases, so will the acidity of the substance. This happens for two reasons: First, the O—H bond becomes more polar, thereby favoring loss of H^+ (Figure 16.13 ▲). Second, because the conjugate base is usually an anion, its stability generally increases as the electronegativity of Y increases.

Many oxyacids contain additional oxygen atoms bonded to the central atom Y. The additional electronegative oxygen atoms pull electron density from the O—H bond, further increasing its polarity. Increasing the number of oxygen atoms also helps stabilize the conjugate base by increasing its ability to "spread out" its negative charge. Thus, the strength of an acid will increase as additional electronegative atoms bond to the central atom Y.

We can summarize these ideas as two simple rules that relate the acid strength of oxyacids to the electronegativity of Y and to the number of groups attached to Y.

1. *For oxyacids that have the same number of OH groups and the same number of O atoms, acid strength increases with increasing electronegativity of the central atom Y.* For example, the strength of the hypohalous acids, which have the structure H—O—Y, increases as the electronegativity of Y increases (Table 16.6 ◀).
2. *For oxyacids that have the same central atom Y, acid strength increases as the number of oxygen atoms attached to Y increases.* For example, the strength of the oxyacids of chlorine steadily increases from hypochlorous acid (HClO) to perchloric acid ($HClO_4$):

TABLE 16.6 Electronegativity Values (EN) of Y and Acid-Dissociation Constants (K_a) of the Hypohalous Acids, H—O—Y

Acid	EN of Y	K_a
HClO	3.0	3.0×10^{-8}
HBrO	2.8	2.5×10^{-9}
HIO	2.5	2.3×10^{-11}

Hypochlorous	Chlorous	Chloric	Perchloric
H—O—Cl:	H—O—Cl—O:	H—O—Cl(—O:)—O:	H—O—Cl(—O:)(—O:)—O:
$K_a = 3.0 \times 10^{-8}$	$K_a = 1.1 \times 10^{-2}$	Strong acid	Strong acid

Increasing acid strength →

Because the oxidation number of the central atom increases as the number of attached O atoms increases, this correlation can be stated in an equivalent way: In a series of oxyacids, the acidity increases as the oxidation number of the central atom increases.

SAMPLE EXERCISE 16.19

Arrange the compounds in each of the following series in order of increasing acid strength: **(a)** AsH_3, HI, NaH, H_2O; **(b)** H_2SeO_3, H_2SeO_4, H_2O.

Solution

Analyze: We are asked to arrange two sets of compounds in order from weakest acid to strongest acid.

Plan: For the binary acids in part (a), we will consider the electronegativities of As, I, Na, and O, respectively. For the oxyacids in part (b), we will consider the number of oxygen atoms bonded to the central atom and the similarities between the Se-containing compounds and some more familiar acids.

Solve: (a) The elements from the left side of the periodic table form the most basic binary hydrogen compounds because the hydrogen in these compounds carries a negative charge. Thus NaH should be the most basic compound on the list. Because arsenic is less electronegative than oxygen, we might expect that AsH_3 would be a weak base toward water. That is also what we would predict by an extension of the trends shown in Figure 16.13. Further, we expect that the binary hydrogen compounds of the halogens, as the most electronegative element in each period, will be acidic relative to water. In fact, HI is one of the strong acids in water. Thus the order of increasing acidity is $NaH < AsH_3 < H_2O < HI$.

(b) The acidity of oxyacids increases as the number of oxygen atoms bonded to the central atom increases. Thus, H_2SeO_4 will be a stronger acid than H_2SeO_3; in fact, the Se atom in H_2SeO_4 is in its maximum positive oxidation state, and so we expect it to be a comparatively strong acid, much like H_2SO_4. H_2SeO_3 is an oxyacid of a nonmetal that is similar to H_2SO_3. As such, we expect that H_2SeO_3 is able to donate a proton to H_2O, indicating that H_2SeO_3 is a stronger acid than H_2O. Thus, the order of increasing acidity is $H_2O < H_2SeO_3 < H_2SeO_4$.

PRACTICE EXERCISE

In each of the following pairs choose the compound that leads to the more acidic (or less basic) solution: **(a)** HBr, HF; **(b)** PH_3, H_2S; **(c)** HNO_2, HNO_3; **(d)** H_2SO_3, H_2SeO_3.
Answers: **(a)** HBr; **(b)** H_2S; **(c)** HNO_3; **(b)** H_2SO_3

Carboxylic Acids

Another large group of acids is illustrated by acetic acid ($HC_2H_3O_2$):

```
     H  :O:
     |  ||
  H—C—C—Ö—H
     |     ..
     H
```

The portion of the structure shown in blue is called the *carboxyl group*, which is often written as COOH. Thus, the chemical formula of acetic acid is often written as CH_3COOH, where only the hydrogen atom in the carboxyl group can be ionized. Acids that contain a carboxyl group are called **carboxylic acids**, and they form the largest category of organic acids. Formic acid and benzoic acid, whose structures are drawn in the margin, are further examples of this large and important category of acids.

```
   :O:
   ||
H—C—Ö—H
       ..
```

Formic acid

```
       :O:
       ||
(C6H5)—C—Ö—H
           ..
```

Benzoic acid

Acetic acid (CH_3COOH) is a weak acid ($K_a = 1.8 \times 10^{-5}$). Methanol (CH_3OH), on the other hand, is not an acid in water. Two factors contribute to the acidic behavior of carboxylic acids. First, the additional oxygen atom attached to the carboxyl group carbon draws electron density from the O—H bond, increasing its polarity and helping to stabilize the conjugate base. Second, the conjugate base of a carboxylic acid (a *carboxylate anion*) can exhibit resonance

(Section 8.6), which contributes further to the stability of the anion by spreading the negative charge over several atoms:

$$H_3C{-}C({=}\ddot{O}{:}){-}\ddot{O}{:}^- \xleftrightarrow{\text{resonance}} H_3C{-}C({-}\ddot{O}{:}^-){=}\ddot{O}$$

The acid strength of carboxylic acids also increases as the number of electronegative atoms in the acid increases. For example, trifluoroacetic acid (CF_3COOH) has $K_a = 5.0 \times 10^{-1}$; the replacement of three hydrogen atoms of acetic acid with more electronegative fluorine atoms leads to a large increase in acid strength.

16.11 Lewis Acids and Bases

For a substance to be a proton acceptor (a Brønsted–Lowry base), it must have an unshared pair of electrons for binding the proton. NH_3, for example, acts as a proton acceptor. Using Lewis structures, we can write the reaction between H^+ and NH_3 as follows:

$$H^+ + {:}NH_3 \longrightarrow [NH_4]^+$$

G. N. Lewis was the first to notice this aspect of acid-base reactions. He proposed a definition of acid and base that emphasizes the shared electron pair: A **Lewis acid** is an electron-pair acceptor, and a **Lewis base** is an electron-pair donor.

ANIMATION
Lewis Acid-Base Theory

Every base that we have discussed thus far—whether it be OH^-, H_2O, an amine, or an anion—is an electron-pair donor. Everything that is a base in the Brønsted–Lowry sense (a proton acceptor) is also a base in the Lewis sense (an electron-pair donor). In the Lewis theory, however, a base can donate its electron pair to something other than H^+. The Lewis definition therefore greatly increases the number of species that can be considered acids; H^+ is a Lewis acid, but not the only one. For example, consider the reaction between NH_3 and BF_3. This reaction occurs because BF_3 has a vacant orbital in its valence shell. (Section 8.7) It therefore acts as an electron-pair acceptor (a Lewis acid) toward NH_3, which donates the electron pair:

$$\underset{\text{Base}}{H_3N{:}} + \underset{\text{Acid}}{BF_3} \longrightarrow H_3N{-}BF_3$$

Our emphasis throughout this chapter has been on water as the solvent and on the proton as the source of acidic properties. In such cases we find the Brønsted–Lowry definition of acids and bases to be the most useful. In fact, when we speak of a substance as being acidic or basic, we are usually thinking of aqueous solutions and using these terms in the Arrhenius or Brønsted–Lowry sense. The advantage of the Lewis theory is that it allows us to treat a wider variety of

Chemistry and Life The Amphoteric Behavior of Amino Acids

Amino acids are the building blocks of proteins. The general structure of amino acids is shown here, where different amino acids have different R groups attached to the central carbon atom.

$$H-\ddot{N}(H)-C(R)(H)-C(=\ddot{O}:)-\ddot{O}-H$$

Amine group (basic) Carboxyl group (acidic)

For example, in *glycine*, which is the simplest amino acid, R is a hydrogen atom, whereas in *alanine* R is a CH_3 group.

$$H_2N-C(H)(H)-COOH \qquad H_2N-C(CH_3)(H)-COOH$$

Glycine Alanine

Amino acids contain a carboxyl group and can therefore serve as acids. They also contain an NH_2 group, characteristic of amines (Section 16.7), and thus they can also act as bases. Amino acids, therefore, are amphoteric. For glycine, we might expect that the acid and base reactions with water would be as follows:

Acid: $H_2N-CH_2-COOH(aq) + H_2O(l) \rightleftharpoons$

$$H_2N-CH_2-COO^-(aq) + H_3O^+(aq) \quad [16.47]$$

Base: $H_2N-CH_2-COOH(aq) + H_2O(l) \rightleftharpoons$

$$^+H_3N-CH_2-COOH(aq) + OH^-(aq) \quad [16.48]$$

The pH of a solution of glycine in water is about 6.0, indicating that it is a slightly stronger acid than a base.

The acid-base chemistry of amino acids is somewhat more complicated than shown in Equations 16.47 and 16.48, however. Because the COOH can act as an acid and the NH_2 group can act as a base, amino acids undergo a "self-contained" Brønsted–Lowry acid-base reaction in which the proton of the carboxyl group is transferred to the basic nitrogen atom:

$$H-\ddot{N}(H)-C(H)(H)-C(=O)-OH \rightleftharpoons H-N^+(H)(H)-C(H)(H)-C(=O)-O^- \quad [16.49]$$

proton transfer

Neutral molecule Zwitterion

Although the form of the amino acid on the right side of Equation 16.49 is electrically neutral overall, it has a positively charged end and a negatively charged end. A molecule of this type is called a *zwitterion* (German for "hybrid ion").

Do amino acids exhibit any properties indicating that they behave as zwitterions? If so, they should behave similar to ionic substances. (Section 8.2) Crystalline amino acids (Figure 16.14 ▼) have relatively high melting points, usually above 200°C, which is characteristic of ionic solids. Amino acids are far more soluble in water than in nonpolar solvents. In addition, the dipole moments of amino acids are large, consistent with a large separation of charge in the molecule. Thus, the ability of amino acids to act simultaneously as acids and bases has important effects on their properties.

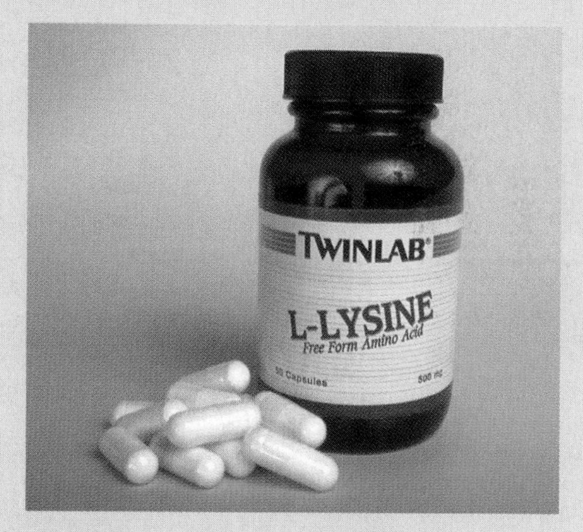

▲ **Figure 16.14** Lysine, one of the amino acids found in proteins, is available as a dietary supplement.

reactions, including those that do not involve proton transfer, as acid-base reactions. To avoid confusion, a substance like BF_3 is rarely called an acid unless it is clear from the context that we are using the term in the sense of the Lewis definition. Instead, substances that function as electron-pair acceptors are referred to explicitly as "Lewis acids."

Lewis acids include molecules that, like BF_3, have an incomplete octet of electrons. In addition, many simple cations can function as Lewis acids. For example, Fe^{3+} interacts strongly with cyanide ions to form the ferricyanide ion, $Fe(CN)_6^{3-}$.

$$Fe^{3+} + 6{:}C{\equiv}N{:}^- \longrightarrow [Fe(C{\equiv}N{:})_6]^{3-}$$

The Fe^{3+} ion has vacant orbitals that accept the electron pairs donated by the CN^- ions; we will learn more in Chapter 24 about just which orbitals are used by the Fe^{3+} ion. The metal ion is highly charged, too, which contributes to the interaction with CN^- ions.

Some compounds with multiple bonds can behave as Lewis acids. For example, the reaction of carbon dioxide with water to form carbonic acid (H_2CO_3) can be pictured as an attack by a water molecule on CO_2, in which the water acts as an electron-pair donor and the CO_2 as an electron-pair acceptor, as shown in the margin. The electron pair of one of the carbon–oxygen double bonds is moved onto the oxygen, leaving a vacant orbital on the carbon that can act as an electron-pair acceptor. We have shown the shift of these electrons with arrows. After forming the initial acid-base product, a proton moves from one oxygen to another, thereby forming carbonic acid. A similar kind of Lewis acid-base reaction takes place when any oxide of a nonmetal dissolves in water to form an acidic solution.

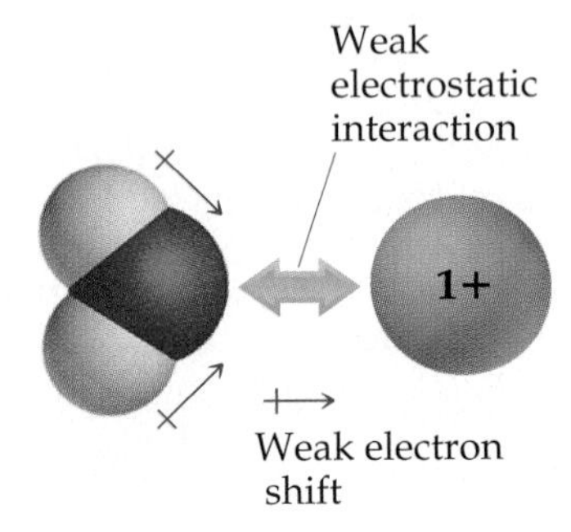

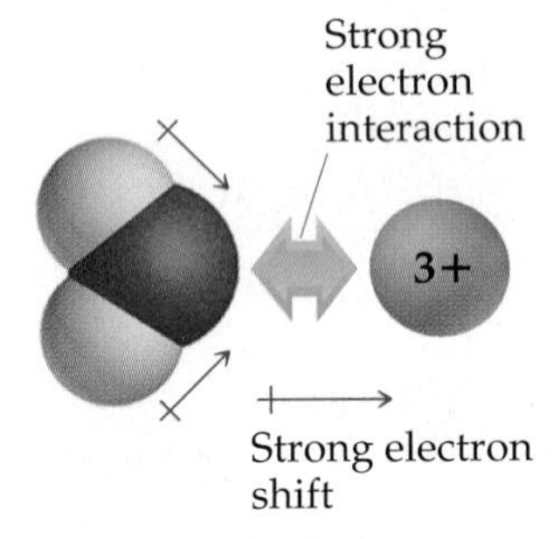

▲ **Figure 16.15** Interaction of a water molecule with a cation of 1+ charge or 3+ charge. The interaction is much stronger with the smaller ion of higher charge, causing that hydrated ion to be more acidic

Hydrolysis of Metal Ions

As we have already seen, most metal ions behave as acids in aqueous solution. (Section 16.9) For example, an aqueous solution of $Cr(NO_3)_3$ is quite acidic. An aqueous solution of $ZnCl_2$ is also acidic, though to a lesser extent. The Lewis concept helps explain the interactions between metal ions and water molecules that give rise to this acidic behavior.

Because metal ions are positively charged, they attract the unshared electron pairs of water molecules. It is primarily this interaction, referred to as *hydration*, that causes salts to dissolve in water. (Section 13.1) The process of hydration can be thought of as a Lewis acid-base interaction in which the metal ion acts as a Lewis acid and the water molecules as Lewis bases. When a water molecule interacts with the positively charged metal ion, electron density is drawn from the oxygen, as illustrated in Figure 16.15 ◀. This flow of electron density causes the O—H bond to become more polarized; as a result, water molecules bound to the metal ion are more acidic than those in the bulk solvent.

The hydrated Fe^{3+} ion, $Fe(H_2O)_6{}^{3+}$, which we usually represent simply as $Fe^{3+}(aq)$, acts as a source of protons:

$$Fe(H_2O)_6{}^{3+}(aq) \rightleftharpoons Fe(H_2O)_5(OH)^{2+}(aq) + H^+(aq) \quad [16.50]$$

The acid-dissociation constant for this hydrolysis reaction has the value $K_a = 2 \times 10^{-3}$, so $Fe^{3+}(aq)$ is a fairly strong acid. Acid-dissociation constants for hydrolysis reactions generally increase with increasing charge and decreasing radius of the ion (Figure 16.15). Thus, the Cu^{2+} ion, which has a smaller charge and a larger radius than Fe^{3+}, forms less acidic solutions than Fe^{3+}: The K_a for $Cu^{2+}(aq)$ is 1×10^{-8}. The acid hydrolysis of a number of salts of metal ions is demonstrated in Figure 16.16 ▶. Note that the Na^+ ion, which is large and has only a 1+ charge (and which we have previously identified as the cation of a strong base), exhibits no acid hydrolysis and yields a neutral solution.

SAMPLE INTEGRATIVE EXERCISE 16: Putting Concepts Together

Phosphorous acid (H_3PO_3) has the following Lewis structure.

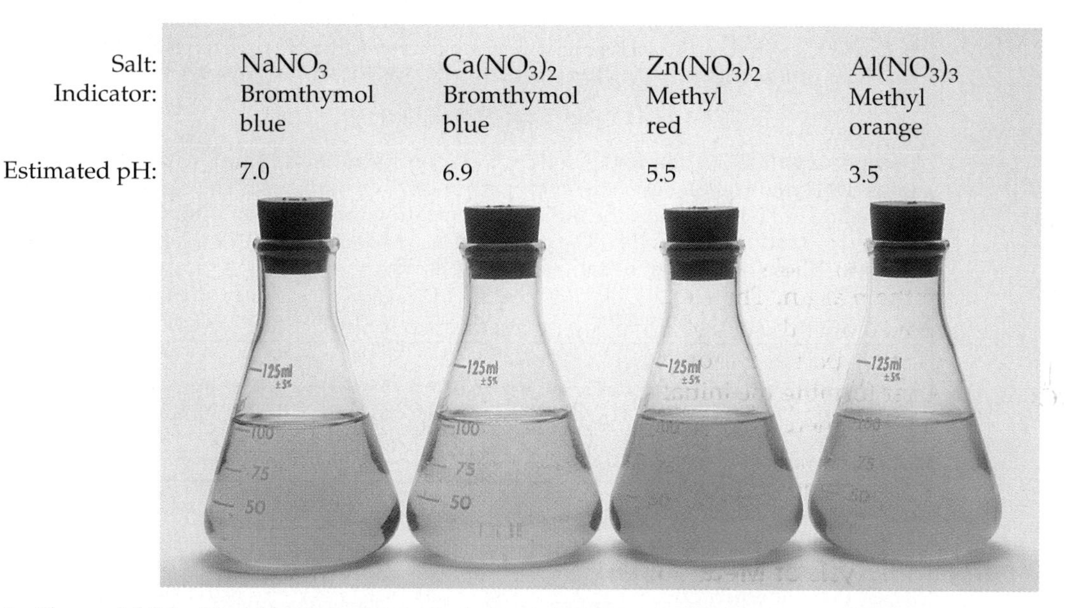

▲ **Figure 16.16** The pH values of 1.0 *M* solutions of a series of nitrate salts, as estimated using acid-base indicators. From left to right, $NaNO_3$, $Ca(NO_3)_2$, $Zn(NO_3)_2$, and $Al(NO_3)_3$.

(a) Explain why H_3PO_3 is diprotic and not triprotic. **(b)** A 25.0-mL sample of a solution of H_3PO_3 is titrated with 0.102 *M* NaOH. It requires 23.3 mL of NaOH to neutralize both acidic protons. What is the molarity of the H_3PO_3 solution? **(c)** This solution has a pH of 1.59. Calculate the percent ionization and K_{a1} for H_3PO_3, assuming that $K_{a1} \gg K_{a2}$. **(d)** How does the osmotic pressure of a 0.050 *M* solution of HCl compare with that of a 0.050 *M* solution of H_3PO_3? Explain.

Solution The problem asks us to explain why there are only two ionizable protons in the H_3PO_3 molecule. Further, we are asked to calculate the molarity of a solution of H_3PO_3, given titration-experiment data. We then need to calculate the percent ionization of the H_3PO_3 solution in part (b). Finally, we are asked to compare the osmotic pressure of a 0.050 *M* solution of H_3PO_3 with that of an HCl solution of the same concentration.

We will use what we have learned about molecular structure and its impact on acidic behavior to answer part (a). We will then use stoichiometry and the relationship between pH and $[H^+]$ to answer parts (b) and (c). Finally, we will consider acid strength in order to compare the colligative properties of the two solutions in part (d).

(a) Acids have polar H—X bonds. From Figure 8.6 we see that the electronegativity of H is 2.1 and that of P is also 2.1. Because the two elements have the same electronegativity, the H—P bond is nonpolar. (Section 8.4) Thus, this H cannot be acidic. The other two H atoms, however, are bonded to O, which has an electronegativity of 3.5. The H—O bonds are therefore polar, with H having a partial positive charge. These two H atoms are consequently acidic.

(b) The chemical equation for the neutralization reaction is

$$H_3PO_3(aq) + 2NaOH(aq) \rightarrow Na_2HPO_3(aq) + H_2O(l)$$

From the definition of molarity, $M = \text{mol/L}$, we see that moles $= M \times L$. (Section 4.6) Thus, the number of moles of NaOH added to the solution is $(0.0233\ L)(0.102\ mol/L) = 2.377 \times 10^{-3}$ mol NaOH. The balanced equation indicates that 2 mol of NaOH is consumed for each mole of H_3PO_3. Thus, the number of moles of H_3PO_3 in the sample is

$$(2.377 \times 10^{-3}\ \text{mol NaOH})\left(\frac{1\ \text{mol}\ H_3PO_3}{2\ \text{mol NaOH}}\right) = 1.189 \times 10^{-3}\ \text{mol}\ H_3PO_3$$

The concentration of the H_3PO_3 solution therefore equals $(1.189 \times 10^{-3}\ \text{mol})/(0.0250\ J) = 0.0475\ M$.

(c) From the pH of the solution, 1.59, we can calculate $[H^+]$ at equilibrium.

$$[H^+] = \text{antilog}(-1.59) = 10^{-1.59} = 0.026\ M\ (2\ \text{significant figures})$$

Because $K_{a1} \gg K_{a2}$, the vast majority of the ions in solution are from the first ionization step of the acid.

$$H_3PO_3(aq) \rightleftharpoons H^+(aq) + H_2PO_3^-(aq)$$

Because one $H_2PO_3^-$ ion forms for each H^+ ion formed, the equilibrium concentrations of H^+ and $H_2PO_3^-$ are equal: $[H^+] = [H_2PO_3^-] = 0.026\ M$. The equilibrium concentration of H_3PO_3 equals the initial concentration minus the amount that ionizes to form H^+ and $H_2PO_3^-$: $[H_3PO_3] = 0.0475\ M - 0.026\ M = 0.022\ M$ (2 significant figures). These results can be tabulated as follows:

	$H_3PO_3(aq)$ $\rightleftharpoons$	$H^+(aq)$ +	$H_2PO_3^-(aq)$
Initial	0.0475 *M*	0	0
Change	−0.026 *M*	+0.026 *M*	+0.026 *M*
Equilibrium	0.022 *M*	0.026 *M*	0.026 *M*

The percent ionization is

$$\text{Percent ionization} = \frac{[H^+]_{\text{equilibrium}}}{[H_3PO_3]_{\text{initial}}} \times 100\% = \frac{0.026\ M}{0.0475\ M} \times 100\% = 55\%$$

The first acid-dissociation constant is

$$K_{a1} = \frac{[H^+][H_2PO_3^-]}{[H_3PO_3]} = \frac{(0.026)(0.026)}{0.022} = 0.030$$

(d) Osmotic pressure is a colligative property and depends on the total concentration of particles in solution. (Section 13.5) Because HCl is a strong acid, a 0.050 *M* solution will contain 0.050 *M* $H^+(aq)$ and 0.050 *M* $Cl^-(aq)$, or a total of 0.100 mol/L of particles. Because H_3PO_3 is a weak acid, it ionizes to a lesser extent than HCl, and hence there are fewer particles in the H_3PO_3 solution. As a result, the H_3PO_3 solution will have the lower osmotic pressure.

Summary and Key Terms

Section 16.1 Acids and bases were first recognized by the properties of their aqueous solutions. For example, acids turn litmus red, whereas bases turn litmus blue. Arrhenius recognized that the properties of acidic solutions are due to $H^+(aq)$ ions and those of basic solutions are due to $OH^-(aq)$ ions.

Section 16.2 The Brønsted–Lowry concept of acids and bases is more general than the Arrhenius concept and emphasizes the transfer of a proton (H^+) from an acid to a base. The H^+ ion, which is merely a proton with no surrounding valence electrons, is strongly bound to water. For this reason, the **hydronium ion**, $H_3O^+(aq)$, is often used to represent the predominant form of H^+ in water instead of the simpler $H^+(aq)$.

A **Brønsted–Lowry acid** is a substance that donates a proton to another substance; a **Brønsted–Lowry base** is a substance that accepts a proton from another substance. Water is an **amphoteric** substance, one that can function as either a Brønsted–Lowry acid or base, depending on the substance with which it reacts.

The **conjugate base** of a Brønsted–Lowry acid is the species that remains when a proton is removed from the acid. The **conjugate acid** of a Brønsted–Lowry base is the species formed by adding a proton to the base. Together, an acid and its conjugate base (or a base and its conjugate acid) are called a **conjugate acid-base pair**.

The acid-base strengths of conjugate acid-base pairs are related: The stronger an acid, the weaker its conjugate base; the weaker an acid, the stronger its conjugate base. In every acid-base reaction, the position of the equilibrium favors the transfer of the proton from the stronger acid to the stronger base.

Section 16.3 Water ionizes to a slight degree, forming $H^+(aq)$ and $OH^-(aq)$. The extent of this **autoionization** is expressed by the **ion-product constant** for water:

$$K_w = [H^+][OH^-] = 1.0 \times 10^{-14}\ (25°C)$$

This relationship describes both pure water and aqueous solutions. The K_w expression indicates that the product of $[H^+]$ and $[OH^-]$ is a constant. Thus, as $[H^+]$ increases, $[OH^-]$

decreases. Acidic solutions are those that contain more $H^+(aq)$ than $OH^-(aq)$; basic solutions contain more $OH^-(aq)$ than $H^+(aq)$.

Section 16.4 The concentration of $H^+(aq)$ can be expressed in terms of **pH**: $pH = -\log [H^+]$. At 25°C the pH of a neutral solution is 7.00, whereas the pH of an acidic solution is below 7.00, and the pH of a basic solution is above 7.00. The pX notation is also used to represent the negative log of other small quantities, as in pOH and pK_w. The pH of a solution can be measured using a pH meter, or it can be estimated using acid-base indicators.

Section 16.5 Strong acids are strong electrolytes, ionizing completely in aqueous solution. The common strong acids are HCl, HBr, HI, HNO_3, $HClO_3$, $HClO_4$, and H_2SO_4. The conjugate bases of strong acids have negligible basicity.

Common strong bases are the ionic hydroxides of alkali metals and the heavy alkaline earth metals. The cations of strong bases have negligible acidity.

Section 16.6 Weak acids are weak electrolytes; only part of the molecules exist in solution in ionized form. The extent of ionization is expressed by the **acid-dissociation constant**, K_a, which is the equilibrium constant for the reaction $HA(aq) \rightleftharpoons H^+(aq) + A^-(aq)$, which can also be written $HA(aq) + H_2O(l) \rightleftharpoons H_3O^+(aq) + A^-(aq)$. The larger the value of K_a, the stronger the acid. The concentration of a weak acid and its K_a value can be used to calculate the pH of a solution.

Polyprotic acids, such as H_2SO_3, have more than one ionizable proton. These acids have acid-dissociation constants that decrease in magnitude in the order $K_{a1} > K_{a2} > K_{a3}$. Because nearly all the $H^+(aq)$ in a polyprotic acid solution comes from the first dissociation step, the pH can usually be estimated satisfactorily by considering only K_{a1}.

Sections 16.7 and 16.8 Weak bases include NH_3, **amines**, and the anions of weak acids. The extent to which a weak base reacts with water to generate the corresponding conjugate acid and OH^- is measured by the **base-dissociation constant**, K_b. This is the equilibrium constant for the reaction $B(aq) + H_2O(l) \rightleftharpoons HB^+(aq) + OH^-(aq)$, where B is the base.

The relationship between the strength of an acid and the strength of its conjugate base is expressed quantitatively by the equation $K_a \times K_b = K_w$, where K_a and K_b are dissociation constants for conjugate acid-base pairs.

Section 16.9 The acid-base properties of salts can be ascribed to the behavior of their respective cations and anions. The reaction of ions with water, with a resultant change in pH, is called **hydrolysis**. The cations of the alkali metals and the alkaline earth metals and the anions of strong acids do not undergo hydrolysis. They are always spectator ions in acid-base chemistry.

Section 16.10 The tendency of a substance to show acidic or basic characteristics in water can be correlated with its chemical structure. Acid character requires the presence of a highly polar H—X bond. Acidity is also favored when the H—X bond is weak and when the X^- ion is very stable.

For **oxyacids** with the same number of OH groups and the same number of O atoms, acid strength increases with increasing electronegativity of the central atom. For oxyacids with the same central atom, acid strength increases as the number of oxygen atoms attached to the central atom increases. The structures of **carboxylic acids**, which are organic acids containing the COOH group, also helps us to understand their acidity.

Section 16.11 The Lewis concept of acids and bases emphasizes the shared electron pair rather than the proton. A **Lewis acid** is an electron-pair acceptor, and a **Lewis base** is an electron-pair donor. The Lewis concept is more general than the Brønsted–Lowry concept because it can apply to cases in which the acid is some substance other than H^+. The Lewis concept helps to explain why many hydrated metal cations form acidic aqueous solutions. The acidity of these cations generally increases as their charge increases and as the size of the metal ion decreases.

Exercises

Arrhenius and Brønsted–Lowry Acids and Bases

16.1 Although HCl and H_2SO_4 have very different properties as pure substances, their aqueous solutions possess many common properties. List some general properties of these solutions, and explain their common behavior in terms of the species present.

16.2 Although pure NaOH and CaO have very different properties, their aqueous solutions possess many common properties. List some general properties of these solutions, and explain their common behavior in terms of the species present.

16.3 **(a)** What is the difference between the Arrhenius and the Brønsted–Lowry definitions of an acid? **(b)** $NH_3(g)$ and $HCl(g)$ react to form the ionic solid $NH_4Cl(s)$ (Figure 16.3). Which substance is the Brønsted–Lowry acid in this reaction? Which is the Brønsted–Lowry base?

16.4 **(a)** What is the difference between the Arrhenius and the Brønsted–Lowry definitions of a base? **(b)** When ammonia is dissolved in water, it behaves both as an Arrhenius base and as a Brønsted–Lowry base. Explain.

16.5 Give the conjugate base of the following Brønsted–Lowry acids: **(a)** H_2SO_3; **(b)** $HC_2H_3O_2$; **(c)** $H_2AsO_4^-$; **(d)** NH_4^+.

16.6 Give the conjugate acid of the following Brønsted–Lowry bases: **(a)** $HAsO_4^{2-}$; **(b)** CH_3NH_2; **(c)** SO_4^{2-}; **(d)** $H_2PO_4^-$.

16.7 Designate the Brønsted–Lowry acid and the Brønsted–Lowry base on the left side of each of the following

equations, and also designate the conjugate acid and conjugate base on the right side:
(a) $NH_4^+(aq) + CN^-(aq) \rightleftharpoons HCN(aq) + NH_3(aq)$
(b) $(CH_3)_3N(aq) + H_2O(l) \rightleftharpoons (CH_3)_3NH^+(aq) + OH^-(aq)$
(c) $HCHO_2(aq) + PO_4^{3-}(aq) \rightleftharpoons CHO_2^-(aq) + HPO_4^{2-}(aq)$

16.8 Designate the Brønsted–Lowry acid and the Brønsted–Lowry base on the left side of each equation, and also designate the conjugate acid and conjugate base on the right side.
(a) $CHO_2^-(aq) + H_2O(l) \rightleftharpoons HCHO_2(aq) + OH^-(aq)$
(b) $HSO_4^-(aq) + HCO_3^-(aq) \rightleftharpoons SO_4^{2-}(aq) + H_2CO_3(aq)$
(c) $HSO_3^-(aq) + H_3O^+(aq) \rightleftharpoons H_2SO_3(aq) + H_2O(l)$

16.9 **(a)** The hydrogen oxalate ion ($HC_2O_4^-$) is amphoteric. Write a balanced chemical equation showing how it acts as an acid toward water and another equation showing how it acts as a base toward water. **(b)** What is the conjugate acid of $HC_2O_4^-$? What is its conjugate base?

16.10 **(a)** Write an equation for the reaction in which $H_2C_6O_5H_7^-(aq)$ acts as a base in $H_2O(l)$. **(b)** Write an equation for the reaction in which $H_2C_6O_5H_7^-(aq)$ acts as an acid in $H_2O(l)$. **(c)** What is the conjugate acid of $H_2C_6O_5H_7^-$? What is its conjugate base?

16.11 Label each of the following as being a strong acid, a weak acid, or a species with negligible acidity. In each case write the formula of its conjugate base: **(a)** HNO_2; **(b)** H_2SO_4; **(c)** HPO_4^{2-}; **(d)** CH_4; **(e)** $CH_3NH_3^+$ (an ion related to NH_4^+).

16.12 Label each of the following as being a strong base, a weak base, or a species with negligible basicity. In each case write the formula of its conjugate acid: **(a)** $C_2H_3O_2^-$; **(b)** HCO_3^-; **(c)** O^{2-}; **(d)** Cl^-; **(e)** NH_3.

16.13 **(a)** Which of the following is the stronger Brønsted–Lowry acid, HBrO or HBr? **(b)** Which is the stronger Brønsted–Lowry base, F^- or Cl^-? Briefly explain your choices.

16.14 **(a)** Which of the following is the stronger Brønsted–Lowry acid, HNO_3 or HNO_2? **(b)** Which is the stronger Brønsted–Lowry base, NH_3 or H_2O? Briefly explain your choices.

16.15 Predict the products of the following acid-base reactions, and also predict whether the equilibrium lies to the left or to the right of the equation:
(a) $HCO_3^-(aq) + F^-(aq) \rightleftharpoons$
(b) $O^{2-}(aq) + H_2O(l) \rightleftharpoons$
(c) $HC_2H_3O_2(aq) + HS^-(aq) \rightleftharpoons$

16.16 Predict the products of the following acid-base reactions, and also predict whether the equilibrium lies to the left or to the right of the equation:
(a) $Cl^-(aq) + H_3O^+(aq) \rightleftharpoons$
(b) $HNO_2(aq) + H_2O(l) \rightleftharpoons$
(c) $NO_3^-(aq) + H_2O(l) \rightleftharpoons$

Autoionization of Water

16.17 **(a)** What is meant by the term *autoionization*? **(b)** Explain why pure water is a poor conductor of electricity. **(c)** You are told that an aqueous solution is acidic. What does this statement mean?

16.18 **(a)** Write a chemical equation that illustrates the autoionization of water. **(b)** Write the expression for the ion-product constant for water, K_w. Why is $[H_2O]$ absent from this expression? **(c)** A solution is described as basic. What is meant by this statement?

16.19 Calculate $[H^+]$ for each of the following solutions, and indicate whether the solution is acidic, basic, or neutral: **(a)** $[OH^-] = 0.00005\ M$; **(b)** $[OH^-] = 3.2 \times 10^{-9}\ M$; **(c)** a solution in which $[OH^-]$ is 100 times greater than $[H^+]$.

16.20 Calculate $[OH^-]$ for each of the following solutions, and indicate whether the solution is acidic, basic, or neutral: **(a)** $[H^+] = 0.0041\ M$; **(b)** $[H^+] = 3.5 \times 10^{-9}\ M$; **(c)** a solution in which $[H^+]$ is ten times greater than $[OH^-]$.

16.21 At the freezing point of water (0°C), $K_w = 1.2 \times 10^{-15}$. Calculate $[H^+]$ and $[OH^-]$ for a neutral solution at this temperature.

16.22 Deuterium oxide (D_2O, where D is deuterium, the hydrogen-2 isotope) has an ion-product constant, K_w, of 8.9×10^{-16} at 20°C. Calculate $[D^+]$ and $[OD^-]$ for pure (neutral) D_2O at this temperature.

The pH Scale

16.23 By what factor does $[H^+]$ change for a pH change of **(a)** 2.00 units; **(b)** 0.50 units?

16.24 Consider two solutions, solution A and solution B. $[H^+]$ in solution A is 500 times greater than that in solution B. What is the difference in the pH values of the two solutions?

16.25 **(a)** If NaOH is added to water, how does $[H^+]$ change? How does pH change? **(b)** Use the pH benchmarks in Figure 16.5 to estimate the pH of a solution with $[H^+] = 0.00003\ M$. Is the solution acidic or basic? **(c)** If pH = 7.8, first estimate and then calculate the molar concentrations of $H^+(aq)$ and $OH^-(aq)$ in the solution.

16.26 **(a)** If HNO_3 is added to water, how does $[OH^-]$ change? How does pH change? **(b)** Use the pH benchmarks in Figure 16.5 to estimate the pH of a solution with $[OH^-] = 0.014\ M$. Is the solution acidic or basic? **(c)** If pH = 6.6, first estimate and then calculate the molar concentrations of $H^+(aq)$ and $OH^-(aq)$ in the solution.

16.27 Complete the following table by calculating the missing entries and indicating whether the solution is acidic or basic.

$[H^+]$	$[OH^-]$	pH	pOH	acidic or basic?
$7.5 \times 10^{-3}\ M$				
	$3.6 \times 10^{-10}\ M$			
		8.25		
			5.70	

16.28 Complete the following table by calculating the missing entries. In each case indicate whether the solution is acidic or basic.

pH	pOH	$[H^+]$	$[OH^-]$	acidic or basic?
6.21				
	10.13			
		$3.5 \times 10^{-3}\ M$		
			$5.6 \times 10^{-4}\ M$	

16.29 The average pH of normal arterial blood is 7.40. At normal body temperature (37°C), $K_w = 2.4 \times 10^{-14}$. Calculate $[H^+]$ and $[OH^-]$ for blood at this temperature.

16.30 Carbon dioxide in the atmosphere dissolves in raindrops to produce carbonic acid (H_2CO_3), causing the pH of clean, unpolluted rain to range from about 5.2 to 5.6. What are the ranges of $[H^+]$ and $[OH^-]$ in the raindrops?

Strong Acids and Bases

16.31 **(a)** What is a strong acid? **(b)** A solution is labeled 0.500 *M* HCl. What is $[H^+]$ for the solution? **(c)** Which of the following are strong acids: HF, HCl, HBr, HI?

16.32 **(a)** What is a strong base? **(b)** A solution is labeled 0.125 *M* $Sr(OH)_2$. What is $[OH^-]$ for the solution? **(c)** Is the following statement true or false? Because $Mg(OH)_2$ is not very soluble, it cannot be a strong base. Explain.

16.33 Calculate the pH of each of the following strong acid solutions: **(a)** $8.5 \times 10^{-3}\ M$ HBr; **(b)** 1.52 g of HNO_3 in 575 mL of solution; **(c)** 5.00 mL of 0.250 *M* $HClO_4$ diluted to 50.0 mL; **(d)** a solution formed by mixing 10.0 mL of 0.100 *M* HBr with 20.0 mL of 0.200 *M* HCl.

16.34 Calculate the pH of each of the following strong acid solutions: **(a)** 0.0575 *M* HNO_3; **(b)** 0.723 g of $HClO_4$ in 2.00 L of solution; **(c)** 5.00 mL of 1.00 *M* HCl diluted to 0.750 L; **(d)** a mixture formed by adding 50.0 mL of 0.020 *M* HCl to 125 mL of 0.010 *M* HI.

16.35 Calculate $[OH^-]$ and pH for **(a)** $1.5 \times 10^{-3}\ M$ $Sr(OH)_2$; **(b)** 2.250 g of LiOH in 250.0 mL of solution; **(c)** 1.00 mL of 0.175 *M* NaOH diluted to 2.00 L; **(d)** a solution formed by adding 5.00 mL of 0.105 *M* KOH to 15.0 mL of $9.5 \times 10^{-2}\ M$ $Ca(OH)_2$.

16.36 Calculate $[OH^-]$ and pH for each of the following strong base solutions: **(a)** 0.0050 *M* KOH; **(b)** 2.055 g of KOH in 500.0 mL of solution; **(c)** 10.0 mL of 0.250 *M* $Ca(OH)_2$ diluted to 500.0 mL; **(d)** a solution formed by mixing 10.0 mL of 0.015 *M* $Ba(OH)_2$ with 30.0 mL of $7.5 \times 10^{-3}\ M$ NaOH.

16.37 Calculate the concentration of an aqueous solution of NaOH that has a pH of 11.50.

16.38 Calculate the concentration of an aqueous solution of $Ca(OH)_2$ that has a pH of 12.00.

[16.39] Calculate the pH of a solution made by adding 15.00 g of sodium hydride (NaH) to enough water to make 2.500 L of solution.

[16.40] Calculate the pH of a solution made by adding 2.50 g of lithium oxide (Li_2O) to enough water to make 1.200 L of solution.

Weak Acids

16.41 Write the chemical equation and the K_a expression for the ionization of each of the following acids in aqueous solution. First show the reaction with $H^+(aq)$ as a product and then with the hydronium ion: **(a)** $HBrO_2$; **(b)** $HC_3H_5O_2$.

16.42 Write the chemical equation and the K_a expression for the acid dissociation of each of the following acids in aqueous solution. First show the reaction with $H^+(aq)$ as a product and then with the hydronium ion: **(a)** HC_6H_5O; **(b)** HCO_3^-.

16.43 Lactic acid ($HC_3H_5O_3$) has one acidic hydrogen. A 0.10 *M* solution of lactic acid has a pH of 2.44. Calculate K_a.

16.44 Phenylacetic acid ($HC_8H_7O_2$) is one of the substances that accumulates in the blood of people with phenylketonuria, an inherited disorder that can cause mental retardation or even death. A 0.085 *M* solution of $HC_8H_7O_2$ is found to have a pH of 2.68. Calculate the K_a value for this acid.

16.45 A 0.200 *M* solution of a weak acid HA is 9.4% ionized. Using this information, calculate $[H^+]$, $[A^-]$, [HA], and K_a for HA.

16.46 A 0.100 *M* solution of chloroacetic acid ($ClCH_2COOH$) is 11.0% ionized. Using this information, calculate $[ClCH_2COO^-]$, $[H^+]$, $[ClCH_2COOH]$, and K_a for chloroacetic acid.

16.47 A particular sample of vinegar has a pH of 2.90. Assuming that acetic acid is the only acid that vinegar contains ($K_a = 1.8 \times 10^{-5}$), calculate the concentration of acetic acid in the vinegar.

16.48 How many moles of HF ($K_a = 6.8 \times 10^{-4}$) must be present in 0.500 L to form a solution with a pH of 2.70?

16.49 The acid-dissociation constant for benzoic acid ($HC_7H_5O_2$) is 6.3×10^{-5}. Calculate the equilibrium concentrations of H_3O^+, $C_7H_5O_2^-$, and $HC_7H_5O_2$ in the solution if the initial concentration of $HC_7H_5O_2$ is 0.050 *M*.

16.50 The acid-dissociation constant for hypochlorous acid (HClO) is 3.0×10^{-8}. Calculate the concentrations of H_3O^+, ClO^-, and HClO at equilibrium if the initial concentration of HClO is 0.0075 *M*.

16.51 Calculate the pH of each of the following solutions (K_a and K_b values are given in Appendix D): **(a)** 0.095 *M* propionic acid ($HC_3H_5O_2$); **(b)** 0.100 *M* hydrogen chromate ion ($HCrO_4^-$); **(c)** 0.120 *M* pyridine (C_5H_5N).

16.52 Determine the pH of each of the following solutions (K_a and K_b values are given in Appendix D): **(a)** 0.125 M hypochlorous acid; **(b)** 0.0085 M phenol; **(c)** 0.095 M hydroxylamine.

16.53 Saccharin, a sugar substitute, is a weak acid with $pK_a = 2.32$ at 25°C. It ionizes in aqueous solution as follows:

$$HNC_7H_4SO_3(aq) \rightleftharpoons H^+(aq) + NC_7H_4SO_3^-(aq)$$

What is the pH of a 0.10 M solution of this substance?

16.54 The active ingredient in aspirin is acetylsalicylic acid ($HC_9H_7O_4$), a monoprotic acid with $K_a = 3.3 \times 10^{-4}$ at 25°C. What is the pH of a solution obtained by dissolving two extra-strength aspirin tablets, containing 500 mg of acetylsalicylic acid each, in 250 mL of water?

16.55 Calculate the percent ionization of hydrazoic acid (HN_3) in solutions of each of the following concentrations (K_a is given in Appendix D): **(a)** 0.400 M; **(b)** 0.100 M; **(c)** 0.0400 M.

16.56 Calculate the percent ionization of $HCrO_4^-$ in solutions of each of the following concentrations (K_a is given in Appendix D): **(a)** 0.250 M; **(b)** 0.0800 M; **(c)** 0.0200 M.

[16.57] Show that for a weak acid, the percent ionization should vary as the inverse square root of the acid concentration.

[16.58] For solutions of a weak acid, a graph of pH versus the log of the initial acid concentration should be a straight line. What is the magnitude of the slope of that line?

[16.59] Citric acid, which is present in citrus fruits, is a triprotic acid (Table 16.3). Calculate the pH and the citrate ion ($C_6H_5O_7^{3-}$) concentration for a 0.050 M solution of citric acid. Explain any approximations or assumptions that you make in your calculations.

[16.60] Tartaric acid is found in many fruits, including grapes. It is partly responsible for the dry texture of certain wines. Calculate the pH and the tartarate ion ($C_4H_4O_6^{2-}$) concentration for a 0.250 M solution of tartaric acid, for which the acid-dissociation constants are listed in Table 16.3. Explain any approximations or assumptions that you make in your calculation.

Weak Bases

16.61 What is the essential structural feature of all Brønsted–Lowry bases?

16.62 What are two kinds of molecules or ions that commonly function as weak bases?

16.63 Write the chemical equation and the K_b expression for the ionization of each of the following bases in aqueous solution: **(a)** dimethylamine, $(CH_3)_2NH$; **(b)** carbonate ion, CO_3^{2-}; **(c)** formate ion, CHO_2^-.

16.64 Write the chemical equation and the K_b expression for the reaction of each of the following bases with water: **(a)** propylamine, $C_3H_7NH_2$; **(b)** monohydrogen phosphate ion, HPO_4^{2-}; **(c)** benzoate ion, $C_6H_5CO_2^-$.

16.65 Calculate the molar concentration of OH^- ions in a 0.075 M solution of ethylamine ($C_2H_5NH_2$) ($K_b = 6.4 \times 10^{-4}$). Calculate the pH of this solution.

16.66 Calculate the molar concentration of OH^- ions in a 1.15 M solution of hypobromite ion (BrO^-; $K_b = 4.0 \times 10^{-6}$). What is the pH of this solution?

16.67 Ephedrine, a central nervous system stimulant, is used in nasal sprays as a decongestant. This compound is a weak organic base:

$$C_{10}H_{15}ON(aq) + H_2O(l) \rightleftharpoons C_{10}H_{15}ONH^+(aq) + OH^-(aq)$$

A 0.035 M solution of ephedrine has a pH of 11.33. **(a)** What are the equilibrium concentrations of $C_{10}H_{15}ON$, $C_{10}H_{15}ONH^+$, and OH^-? **(b)** Calculate K_b for ephedrine.

16.68 Codeine ($C_{18}H_{21}NO_3$) is a weak organic base. A 5.0×10^{-3} M solution of codeine has a pH of 9.95. Calculate the value of K_b for this substance. What is the pK_b for this base?

The K_a–K_b Relationship; Acid-Base Properties of Salts

16.69 Although the acid-dissociation constant for phenol (C_6H_5OH) is listed in Appendix D, the base-dissociation constant for the phenolate ion ($C_6H_5O^-$) is not. **(a)** Explain why it is not necessary to list both K_a for phenol and K_b for the phenolate ion. **(b)** Calculate the K_b for the phenolate ion. **(c)** Is the phenolate ion a weaker or stronger base than ammonia?

16.70 We can calculate K_b for the carbonate ion if we know the K_a values of carbonic acid (H_2CO_3). **(a)** Is K_{a1} or K_{a2} of carbonic acid used to calculate K_b for the carbonate ion? Explain. **(b)** Calculate K_b for the carbonate ion. **(c)** Is the carbonate ion a weaker or stronger base than ammonia?

16.71 **(a)** Given that K_a for acetic acid is 1.8×10^{-5} and that for hypochlorous acid is 3.0×10^{-8}, which is the stronger acid? **(b)** Which is the stronger base, the acetate ion or the hypochlorite ion? **(c)** Calculate K_b values for $C_2H_3O_2^-$ and ClO^-.

16.72 **(a)** Given that K_b for ammonia is 1.8×10^{-5} and that for hydroxylamine is 1.1×10^{-8}, which is the stronger base? **(b)** Which is the stronger acid, the ammonium ion or the hydroxylammonium ion? **(c)** Calculate K_a values for NH_4^+ and H_3NOH^+.

16.73 Using data from Appendix D, calculate $[OH^-]$ and pH for each of the following solutions: **(a)** 0.10 M NaCN; **(b)** 0.080 M Na_2CO_3; **(c)** a mixture that is 0.10 M in $NaNO_2$ and 0.20 M in $Ca(NO_2)_2$.

16.74 Using data from Appendix D, calculate $[OH^-]$ and pH for each of the following solutions: **(a)** 0.036 M NaF; **(b)** 0.127 M Na_2S; **(c)** a mixture that is 0.035 M in $NaC_2H_3O_2$ and 0.055 M in $Ba(C_2H_3O_2)_2$.

16.75 Predict whether aqueous solutions of the following compounds are acidic, basic, or neutral: **(a)** NH_4Br; **(b)** $FeCl_3$; **(c)** Na_2CO_3; **(d)** $KClO_4$; **(e)** $NaHC_2O_4$.

16.76 Predict whether aqueous solutions of the following substances are acidic, basic, or neutral: **(a)** CsBr; **(b)** $Al(NO_3)_3$; **(c)** KCN; **(d)** $[CH_3NH_3]Cl$; **(e)** $KHSO_4$.

16.77 An unknown salt is either NaF, NaCl, or NaOCl. When 0.050 mol of the salt is dissolved in water to form 0.500 L of solution, the pH of the solution is 8.08. What is the identity of the salt?

16.78 An unknown salt is either KBr, NH_4Cl, KCN, or K_2CO_3. If a 0.100 *M* solution of the salt is neutral, what is the identity of the salt?

16.79 Sorbic acid ($HC_6H_7O_2$) is a weak monoprotic acid with $K_a = 1.7 \times 10^{-5}$. Its salt (potassium sorbate) is added to cheese to inhibit the formation of mold. What is the pH of a solution containing 11.25 g of potassium sorbate in 1.75 L of solution?

16.80 Trisodium phosphate (Na_3PO_4) is available in hardware stores as TSP and is used as a cleaning agent. The label on a box of TSP warns that the substance is very basic (caustic or alkaline). What is the pH of a solution containing 50.0 g of TSP in a liter of solution?

Acid-Base Character and Chemical Structure

16.81 How does the acid strength of an oxyacid depend on **(a)** the electronegativity of the central atom; **(b)** the number of nonprotonated oxygen atoms in the molecule?

16.82 **(a)** How does the strength of an acid vary with the polarity and strength of the H—X bond? **(b)** How does the acidity of the binary acid of an element vary as a function of the electronegativity of the element? How does this relate to the position of the element in the periodic table?

16.83 Explain the following observations: **(a)** HNO_3 is a stronger acid than HNO_2; **(b)** H_2S is a stronger acid than H_2O; **(c)** H_2SO_4 is a stronger acid than HSO_4^-; **(d)** H_2SO_4 is a stronger acid than H_2SeO_4; **(e)** CCl_3COOH is a stronger acid than CH_3COOH.

16.84 Explain the following observations: **(a)** HCl is a stronger acid than H_2S; **(b)** H_3PO_4 is a stronger acid than H_3AsO_4; **(c)** $HBrO_3$ is a stronger acid than $HBrO_2$; **(d)** $H_2C_2O_4$ is a stronger acid than $HC_2O_4^-$; **(e)** benzoic acid (C_6H_5COOH) is a stronger acid than phenol (C_6H_5OH).

16.85 Based on their compositions and structures and on conjugate acid-base relationships, select the stronger base in each of the following pairs: **(a)** BrO^- or ClO^-; **(b)** BrO^- or BrO_2^-; **(c)** HPO_4^{2-} or $H_2PO_4^-$.

16.86 Based on their compositions and structures and on conjugate acid-base relationships, select the stronger base in each of the following pairs: **(a)** NO_3^- or NO_2^-; **(b)** PO_4^{3-} or AsO_4^{3-}; **(c)** HCO_3^- or CO_3^{2-}.

16.87 Indicate whether each of the following statements is true or false. For each statement that is false, correct the statement so that it is true. **(a)** In general, the acidity of binary acids increases from left to right in a given row of the periodic table. **(b)** In a series of acids that have the same central atom, acid strength increases with the number of hydrogen atoms bonded to the central atom. **(c)** Hydrotelluric acid (H_2Te) is a stronger acid than H_2S because Te is more electronegative than S.

16.88 Indicate whether each of the following statements is true or false. For each statement that is false, correct the statement so that it is true. **(a)** Acid strength in a series of H—X molecules increases with increasing size of X. **(b)** For acids of the same general structure but differing electronegativities of the central atoms, acid strength decreases with increasing electronegativity of the central atom. **(c)** The strongest acid known is HF because fluorine is the most electronegative element.

Lewis Acids and Bases

16.89 If a substance is an Arrhenius base, is it necessarily a Brønsted–Lowry base? Is it necessarily a Lewis base? Explain.

16.90 If a substance is a Lewis acid, is it necessarily a Brønsted–Lowry acid? Is it necessarily an Arrhenius acid? Explain.

16.91 Identify the Lewis acid and Lewis base among the reactants in each of the following reactions:
(a) $Fe(ClO_4)_3(s) + 6H_2O(l) \rightleftharpoons Fe(H_2O)_6^{3+}(aq) + 3ClO_4^-(aq)$
(b) $CN^-(aq) + H_2O(l) \rightleftharpoons HCN(aq) + OH^-(aq)$
(c) $(CH_3)_3N(g) + BF_3(g) \rightleftharpoons (CH_3)_3NBF_3(s)$
(d) $HIO(lq) + NH_2^-(lq) \rightleftharpoons NH_3(lq) + IO^-(lq)$
(*lq* denotes liquid ammonia as solvent)

16.92 Identify the Lewis acid and Lewis base in each of the following reactions:
(a) $HNO_2(aq) + OH^-(aq) \rightleftharpoons NO_2^-(aq) + H_2O(l)$
(b) $FeBr_3(s) + Br^-(aq) \rightleftharpoons FeBr_4^-(aq)$
(c) $Zn^{2+}(aq) + 4NH_3(aq) \rightleftharpoons Zn(NH_3)_4^{2+}(aq)$
(d) $SO_2(g) + H_2O(l) \rightleftharpoons H_2SO_3(aq)$

16.93 Predict which member of each pair produces the more acidic aqueous solution: **(a)** K^+ or Cu^{2+}; **(b)** Fe^{2+} or Fe^{3+}; **(c)** Al^{3+} or Ga^{3+}. Explain.

16.94 Which member of each pair produces the more acidic aqueous solution: **(a)** $ZnBr_2$ or $CdCl_2$; **(b)** CuCl or $Cu(NO_3)_2$; **(c)** $Ca(NO_3)_2$ or $NiBr_2$? Explain.

Additional Exercises

16.95 Indicate whether each of the following statements is correct or incorrect. For those that are incorrect, explain why they are wrong.
(a) Every Brønsted–Lowry acid is also a Lewis acid.
(b) Every Lewis acid is also a Brønsted–Lowry acid.
(c) Conjugate acids of weak bases produce more acidic solutions than conjugate acids of strong bases.
(d) K^+ ion is acidic in water because it causes hydrating water molecules to become more acidic.
(e) The percent ionization of a weak acid in water increases as the concentration of acid decreases.

16.96 Indicate whether each of the following statements is correct or incorrect. For those that are incorrect, explain why they are wrong.
(a) Every Arrhenius base is also a Brønsted–Lowry base.
(b) Every Brønsted–Lowry acid is a Lewis base.
(c) Conjugate bases of strong acids produce more basic solutions than conjugate bases of weak acids.
(d) Al^{3+} ion is acidic in water because it causes hydrating water molecules to become more acidic.
(e) The percent ionization of a weak base in water increases as the concentration of base increases.

16.97 Hemoglobin plays a part in a series of equilibria involving protonation-deprotonation and oxygenation-deoxygenation. The overall reaction is approximately as follows:

$$HbH^+(aq) + O_2(aq) \rightleftharpoons HbO_2(aq) + H^+(aq)$$

where Hb stands for hemoglobin, and HbO_2 for oxyhemoglobin. **(a)** The concentration of O_2 is higher in the lungs and lower in the tissues. What effect does high $[O_2]$ have on the position of this equilibrium? **(b)** The normal pH of blood is 7.4. Is the blood acidic, basic, or neutral? **(c)** If the blood pH is lowered by the presence of large amounts of acidic metabolism products, a condition known as acidosis results. What effect does lowering blood pH have on the ability of hemoglobin to transport O_2?

16.98 What is the pH of a solution that is 2.5×10^{-9} *M* in NaOH?

16.99 Which of the following solutions has the highest pH? **(a)** a 0.1 *M* solution of a strong acid or a 0.1 *M* solution of a weak acid; **(b)** a 0.1 *M* solution of an acid with $K_a = 2 \times 10^{-3}$ or one with $K_a = 8 \times 10^{-6}$; **(c)** a 0.1 *M* solution of a base with $pK_b = 4.5$ or one with $pK_b = 6.5$.

16.100 The hydrogen phthalate ion ($HC_8H_5O_4^-$) is a weak monoprotic acid. When 525 mg of potassium hydrogen phthalate is dissolved in enough water to form 250 mL of solution, the pH of the solution is 4.24. **(a)** Calculate K_a for this acid. **(b)** Calculate the percent ionization of the acid.

[16.101] A hypothetical acid H_2X is both a strong acid and a diprotic acid. **(a)** Calculate the pH of a 0.050 *M* solution of H_2X, assuming that only one proton ionizes per acid molecule. **(b)** Calculate the pH of the solution from part (a), now assuming that both protons of each acid molecule completely ionize. **(c)** In an experiment it is observed that the pH of a 0.050 *M* solution of H_2X is 1.27. Comment on the relative acid strengths of H_2X and HX^-. **(d)** Would a solution of the salt NaHX be acidic, basic, or neutral? Explain.

16.102 Arrange the following 0.10 *M* solutions in order of increasing acidity (decreasing pH): (i) NH_4NO_3; (ii) $NaNO_3$; (iii) $NH_4C_2H_3O_2$; (iv) NaF; (v) $NaC_2H_3O_2$.

[16.103] What are the concentrations of H^+, $H_2PO_4^-$, HPO_4^{2-}, and PO_4^{3-} in a 0.0250 *M* solution of H_3PO_4?

[16.104] Many moderately large organic molecules containing basic nitrogen atoms are not very soluble in water as neutral molecules, but they are frequently much more soluble as their acid salts. Assuming that pH in the stomach is 2.5, indicate whether each of the following compounds would be present in the stomach as the neutral base or in the protonated form: nicotine, $K_b = 7 \times 10^{-7}$; caffeine, $K_b = 4 \times 10^{-14}$; strychnine, $K_b = 1 \times 10^{-6}$; quinine, $K_b = 1.1 \times 10^{-6}$.

[16.105] The amino acid glycine ($H_2N{-}CH_2{-}COOH$) can participate in the following equilibria in water:

$$H_2N{-}CH_2{-}COOH + H_2O \rightleftharpoons H_2N{-}CH_2{-}COO^- + H_3O^+ \quad K_a = 4.3 \times 10^{-3}$$

$$H_2N{-}CH_2{-}COOH + H_2O \rightleftharpoons {}^+H_3N{-}CH_2{-}COOH + OH^- \quad K_b = 6.0 \times 10^{-5}$$

(a) Use the values of K_a and K_b to estimate the equilibrium constant for the intramolecular proton transfer to form a zwitterion:

$$H_2N{-}CH_2{-}COOH \rightleftharpoons {}^+H_3N{-}CH_2{-}COO^-$$

What assumptions did you need to make? **(b)** What is the pH of a 0.050 *M* aqueous solution of glycine? **(c)** What would be the predominant form of glycine in a solution with pH 13? With pH 1?

[16.106] The Lewis structure for acetic acid is shown in Table 16.2. Replacing hydrogen atoms on the carbon with chlorine atoms causes an increase in acidity, as follows:

Acid	Formula	K_a (25°C)
Acetic	CH_3COOH	1.8×10^{-5}
Chloroacetic	$CH_2ClCOOH$	1.4×10^{-3}
Dichloroacetic	$CHCl_2COOH$	3.3×10^{-2}
Trichloroacetic	CCl_3COOH	2×10^{-1}

Using Lewis structures as the basis of your discussion, explain the observed trend in acidities in the series. Calculate the pH of a 0.010 *M* solution of each acid.

Integrative Exercises

16.107 Calculate the number of $H^+(aq)$ ions in 1.0 mL of pure water at 25°C.

16.108 The volume of an adult's stomach ranges from about 50 mL when empty to 1 L when full. If its volume is 400 mL and its contents have a pH of 2, how many moles of H^+ does it contain? Assuming that all the H^+ comes from HCl, how many grams of sodium hydrogen carbonate will totally neutralize the stomach acid?

16.109 Atmospheric CO_2 levels have risen by nearly 20% over the past 40 years from 315 ppm to 375 ppm. **(a)** Given that the average pH of clean, unpolluted rain today is 5.4, determine the pH of unpolluted rain 40 years ago. Assume that carbonic acid (H_2CO_3) formed by the reaction of CO_2 and water is the only factor influencing pH.

$$CO_2(g) + H_2O(l) \rightleftharpoons H_2CO_3(aq)$$

(b) What volume of CO_2 at 25°C and 1.0 atm is dissolved in a 20.0-L bucket of today's rainwater?

[16.110] In many reactions the addition of $AlCl_3$ produces the same effect as the addition of H^+. **(a)** Draw a Lewis structure for $AlCl_3$ in which no atoms carry formal charges, and determine its structure using the VSEPR method. **(b)** What characteristic is notable about the structure in part (a) that helps us understand the acidic character of $AlCl_3$? **(c)** Predict the result of the reaction between $AlCl_3$ and NH_3 in a solvent that does not participate as a reactant. **(d)** Which acid-base theory is most suitable for discussing the similarities between $AlCl_3$ and H^+?

[16.111] What is the boiling point of a 0.10 *M* solution of $NaHSO_4$ if the solution has a density of 1.002 g/mL?

[16.112] Cocaine is a weak organic base whose molecular formula is $C_{17}H_{21}NO_4$. An aqueous solution of cocaine was found to have a pH of 8.53 and an osmotic pressure of 52.7 torr at 15°C. Calculate K_b for cocaine.

[16.113] The iodate ion is reduced by sulfite according to the following reaction:

$$IO_3^-(aq) + 3SO_3^{2-}(aq) \rightarrow I^-(aq) + 3SO_4^{2-}(aq)$$

The rate of this reaction is found to be first order in IO_3^-, first order in SO_3^{2-}, and first order in H^+. **(a)** Write the rate law for the reaction. **(b)** By what factor will the rate of the reaction change if the pH is lowered from 5.00 to 3.50? Does the reaction proceed faster or slower at the lower pH? **(c)** By using the concepts discussed in Section 14.6, explain how the reaction can be pH-dependent even though H^+ does not appear in the overall reaction.

[16.114] **(a)** Using dissociation constants from Appendix D, determine the value for the equilibrium constant for each of the following reactions. (Remember that when reactions are added, the corresponding equilibrium constants are multiplied.)

(i) $HCO_3^-(aq) + OH^-(aq) \rightleftharpoons CO_3^{2-}(aq) + H_2O(l)$

(ii) $NH_4^+(aq) + CO_3^{2-}(aq) \rightleftharpoons NH_3(aq) + HCO_3^-(aq)$

(b) We usually use single arrows for reactions when the forward reaction is appreciable (*K* much greater than 1) or when products escape from the system, so that equilibrium is never established. If we follow this convention, which of these equilibria might be written with a single arrow?

[16.115] Lactic acid, $CH_3CH(OH)COOH$, received its name because it is present in sour milk as a product of bacterial action. It is also responsible for the soreness in muscles after vigorous exercise. **(a)** The pK_a of lactic acid is 3.85. Compare this with the value for propionic acid (CH_3CH_2COOH, $pK_a = 4.89$), and explain the difference. **(b)** Calculate the lactate ion concentration in a 0.050 *M* solution of lactic acid. **(c)** When a solution of sodium lactate, $(CH_3CH(OH)COO)Na$, is mixed with an aqueous copper(II) solution, it is possible to obtain a solid salt of copper(II) lactate as a blue-green hydrate, $(CH_3CH(OH)COO)_2Cu \cdot xH_2O$. Elemental analysis of the solid tells us that the solid is 22.9% Cu and 26.0% C by mass. What is the value for *x* in the formula for the hydrate? **(d)** The acid-dissociation constant for the $Cu^{2+}(aq)$ ion is 1.0×10^{-8}. Based on this value and the acid-dissociation constant of lactic acid, predict whether a solution of copper(II) lactate will be acidic, basic, or neutral. Explain your answer.

eMedia Exercises

16.116 You can measure the pH in the **Acids and Bases** simulation (*eChapter 16.4*) for aqueous solutions of 13 different compounds. **(a)** List the compounds available in the simulation, and identify each as a strong acid, a weak acid, a strong base, or a weak base. **(b)** For each compound, measure and record the pH of a 0.05 *M* solution.

16.117 **(a)** Using data from the **Acids and Bases** simulation (*eChapter 16.4*), calculate the K_a of HNO_2. **(b)** Determine the percent ionization of HNO_2 at 2.0 *M*, 0.20 *M*, 0.020 *M*, and 0.0020 *M* concentrations. **(c)** Explain the trend in percent ionization using Le Châtelier's principle.

16.118 The **Introduction to Aqueous Acids** animation (*eChapter 16.5*) illustrates the ionization in water of two different strong acids and one weak acid. **(a)** Given that all three ionize to produce hydrogen ion in water, what is it about the behavior of the weak acid that makes it different from the strong acids? **(b)** What is the consequence of this difference in terms of pH?

16.119 The **Introduction to Aqueous Bases** animation (*eChapter 16.5*) illustrates the ionization of a weak base. **(a)** Write the equation that corresponds to the ionization of ammonia in water. **(b)** In the animation the ammonia molecule ionizes, producing aqueous hydroxide. What is it about the behavior of ammonia in water that makes ammonia a *weak* base?

16.120 The K_b of ammonia is 1.8×10^{-5}. **(a)** Calculate the concentration of aqueous ammonia that would have a pH of 8.5. Use the **Acids and Bases** simulation (*eChapter 16.4*) to check your answer. **(b)** What is the percent ionization of ammonia at this concentration?

16.121 Use the **Equilibrium Constant** simulation (*eChapter 16.6*) to experiment with the ionization of a weak acid. Choose the $HA \rightleftharpoons H^+ + A^-$ reaction, and enter the K_a of acetic acid (1.8×10^{-5}). **(a)** Enter a starting HA concentration of 0.1 *M*, and determine the equilibrium concentration of H^+. **(b)** By what percentage did the concentration of HA change? **(c)** By what percentage did the concentration of H^+ change? **(d)** Calculate pH and pOH of the 0.1 *M* solution of acetic acid.

Chapter 17

Additional Aspects of Aqueous Equilibria

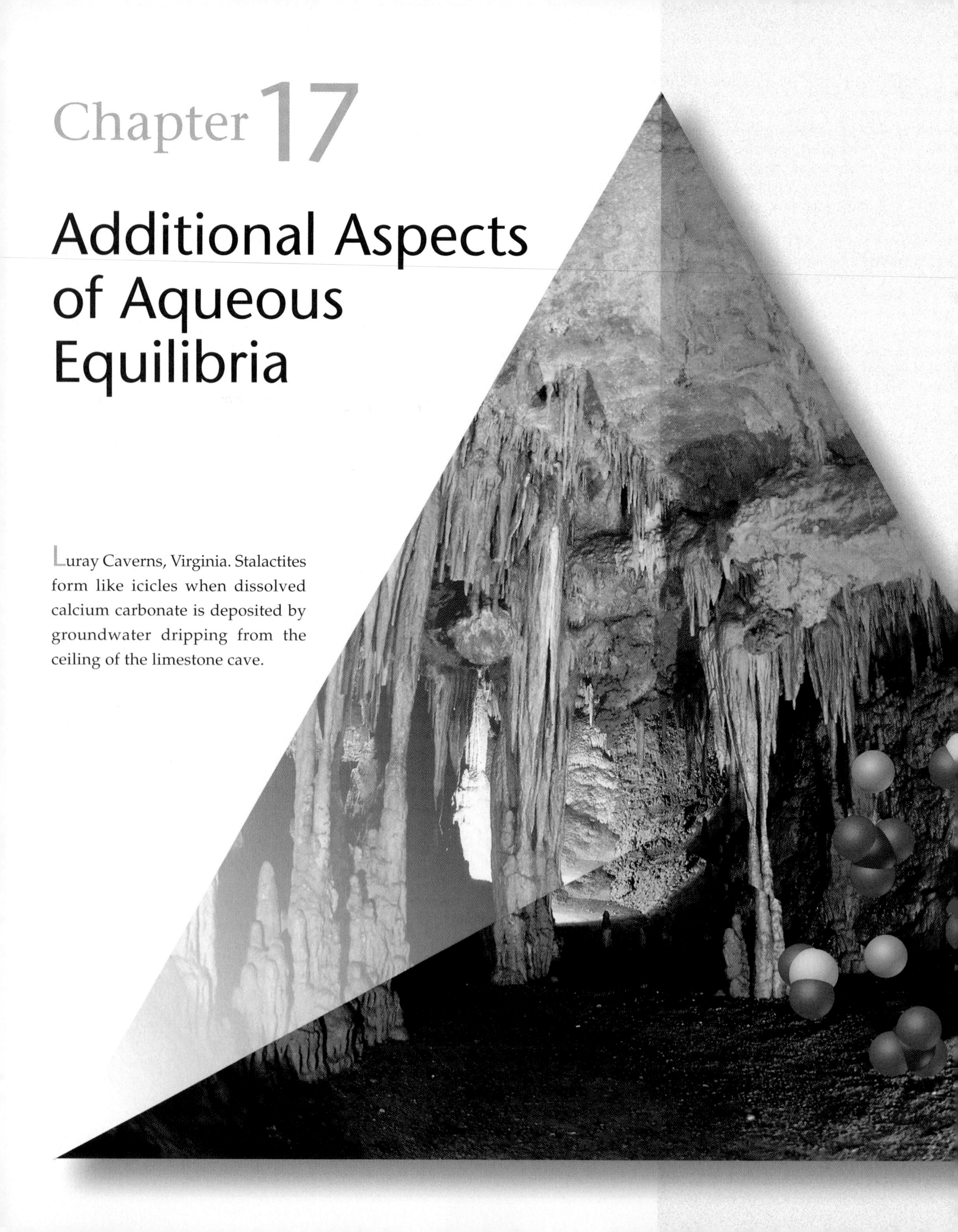

Luray Caverns, Virginia. Stalactites form like icicles when dissolved calcium carbonate is deposited by groundwater dripping from the ceiling of the limestone cave.

WATER IS THE most common and most important solvent on Earth. In a sense, it is the solvent of life. It is difficult to imagine how living matter in all its complexity could exist with any liquid other than water as the solvent. Water occupies its position of importance not only because of its abundance but also because of its exceptional ability to dissolve a wide variety of substances. Aqueous solutions encountered in nature, such as biological fluids and seawater, contain many solutes. Consequently, many equilibria take place simultaneously in these solutions.

In this chapter we take a step toward understanding such complex solutions by looking first at further applications of acid-base equilibria. We then broaden our discussion to include two additional types of aqueous equilibria, those involving slightly soluble salts and those involving the formation of metal complexes in solution.

▶ What's Ahead ◀

- We begin by considering a specific example of Le Châtelier's principle known as the *common-ion effect*.
- We then consider the composition of *buffered solutions*, or *buffers*, and learn how they resist pH change upon the addition of small amounts of strong acid or strong base.
- We continue by examining acid-base titration in detail, and we explore how to determine pH at any point in an acid-base titration.
- Next we learn how to use equilibrium constants known as *solubility-product constants* to determine to what extent a sparingly soluble salt will dissolve in water, and we investigate some of the factors that affect solubility.
- Continuing the discussion of solubility equilibria, we learn how to precipitate ions selectively.
- The chapter concludes with an explanation of how the principles of solubility and complexation equilibria can be used to identify ions qualitatively in solution.

17.1 The Common-Ion Effect

In Chapter 16 we examined the equilibrium concentrations of ions in solutions containing a weak acid or a weak base. We now consider solutions that contain not only a weak acid, such as acetic acid ($HC_2H_3O_2$), but also a soluble salt of that acid, such as $NaC_2H_3O_2$. What happens when $NaC_2H_3O_2$ is added to a solution of $HC_2H_3O_2$? Because $C_2H_3O_2^-$ is a weak base, the pH of the solution increases; that is, $[H^+]$ decreases. It is instructive, however, to view this effect from the perspective of Le Châtelier's principle. (Section 15.6)

Like most salts, $NaC_2H_3O_2$ is a strong electrolyte. Consequently, it dissociates completely in aqueous solution to form Na^+ and $C_2H_3O_2^-$ ions. In contrast, $HC_2H_3O_2$ is a weak electrolyte that ionizes as follows:

$$HC_2H_3O_2(aq) \rightleftharpoons H^+(aq) + C_2H_3O_2^-(aq) \qquad [17.1]$$

The addition of $C_2H_3O_2^-$, from $NaC_2H_3O_2$, causes this equilibrium to shift to the left, thereby decreasing the equilibrium concentration of $H^+(aq)$.

$$HC_2H_3O_2(aq) \rightleftharpoons H^+(aq) + C_2H_3O_2^-(aq)$$

Addition of $C_2H_3O_2^-$ shifts equilibrium, reducing $[H^+]$

The dissociation of the weak acid $HC_2H_3O_2$ decreases when we add the strong electrolyte $NaC_2H_3O_2$, which has an ion in common with it. We can generalize this observation, which we call the **common-ion effect**: *The extent of ionization of a weak electrolyte is decreased by adding to the solution a strong electrolyte that has an ion in common with the weak electrolyte.* Sample Exercises 17.1 and 17.2 illustrate how equilibrium concentrations may be calculated when a solution contains a mixture of a weak electrolyte and a strong electrolyte that have a common ion. The procedures are similar to those encountered for weak acids and weak bases in Chapter 16.

SAMPLE EXERCISE 17.1

What is the pH of a solution made by adding 0.30 mol of acetic acid ($HC_2H_3O_2$) and 0.30 mol of sodium acetate ($NaC_2H_3O_2$) to enough water to make 1.0 L of solution?

Solution

Analyze: We are asked to determine the pH of a solution containing equal concentrations of acetic acid and a salt containing the acetate ion.

Plan: In any problem in which we must determine the pH of a solution containing a mixture of solutes, it is helpful to proceed by a series of logical steps.

First, identify the major species in solution and consider their acidity or basicity. Because $HC_2H_3O_2$ is a weak electrolyte and $NaC_2H_3O_2$ is a strong electrolyte, the major species in the solution are $HC_2H_3O_2$ (a weak acid), Na^+ (which is neither acidic nor basic), and $C_2H_3O_2^-$ (which is the conjugate base of $HC_2H_3O_2$).

Second, identify the important equilibrium reaction. The pH of the solution will be controlled by the dissociation equilibrium of $HC_2H_3O_2$, which involves both $HC_2H_3O_2$ and $C_2H_3O_2^-$.

$$HC_2H_3O_2(aq) \rightleftharpoons H^+(aq) + C_2H_3O_2^-(aq)$$

(We have written the equilibrium using $H^+(aq)$ rather than $H_3O^+(aq)$, but both representations of the hydrated hydrogen ion are equally valid.)

Because $NaC_2H_3O_2$ was added to the solution, the values of $[H^+]$ and $[C_2H_3O_2^-]$ are not the same. The Na^+ ion is merely a spectator ion and will have no influence on the pH. (Section 16.9)

Third, calculate the initial and equilibrium concentrations of each of the species that participates in the equilibrium. We can tabulate the concentrations much as we have done in solving other equilibrium problems. (Section 15.5)

$$HC_2H_3O_2(aq) \rightleftharpoons H^+(aq) + C_2H_3O_2^-(aq)$$

	$HC_2H_3O_2(aq)$	$H^+(aq)$	$C_2H_3O_2^-(aq)$
Initial	0.30 M	0	0.30 M
Change	$-x$ M	$+x$ M	$+x$ M
Equilibrium	$(0.30 - x)$ M	x M	$(0.30 + x)$ M

The equilibrium concentration of $C_2H_3O_2^-$ (the common ion) is the initial concentration due to the $NaC_2H_3O_2$ (0.30 M) plus the change in concentration (x) due to the ionization of $HC_2H_3O_2$.

The equilibrium-constant expression is

$$K_a = 1.8 \times 10^{-5} = \frac{[H^+][C_2H_3O_2^-]}{[HC_2H_3O_2]}$$

(The dissociation constant for $HC_2H_3O_2$ at 25°C is from Appendix D; addition of $NaC_2H_3O_2$ does not change the value of this constant.)

Solve: Substituting the equilibrium-constant concentrations into the equilibrium expression gives

$$K_a = 1.8 \times 10^{-5} = \frac{x(0.30 + x)}{0.30 - x}$$

Because K_a is small, we assume that x is small compared to the original concentrations of $HC_2H_3O_2$ and $C_2H_3O_2^-$ (0.30 M each). We can therefore simplify our equation before solving for x.

$$K_a = 1.8 \times 10^{-5} = \frac{x(0.30)}{0.30}$$

$$x = 1.8 \times 10^{-5}\ M = [H^+]$$

The resulting value of x is indeed small relative to 0.30, justifying the approximation made in simplifying the problem.

Fourth, calculate the pH from the equilibrium concentration of $H^+(aq)$.

$$pH = -\log(1.8 \times 10^{-5}) = 4.74$$

In Section 16.6 we calculated that a 0.30 M solution of $HC_2H_3O_2$ has a pH of 2.64, corresponding to $[H^+] = 2.3 \times 10^{-3}\ M$. Thus, the addition of $NaC_2H_3O_2$ has substantially decreased $[H^+]$, as would be expected from Le Châtelier's principle.

PRACTICE EXERCISE

Calculate the pH of a solution containing 0.085 M nitrous acid, (HNO_2; $K_a = 4.5 \times 10^{-4}$), and 0.10 M potassium nitrite (KNO_2).

Answer: 3.42

SAMPLE EXERCISE 17.2

Calculate the fluoride ion concentration and pH of a solution that is 0.20 M in HF and 0.10 M in HCl.

Solution

Analyze: We are asked to determine the concentration of fluoride ion and the pH in a solution containing the weak acid HF and the strong acid HCl.

Plan: Because HF is a weak acid and HCl is a strong acid, the major species in solution are HF, H^+, and Cl^-. The problem asks for $[F^-]$, which is formed by ionization of HF. Thus, the important equilibrium is

$$HF(aq) \rightleftharpoons H^+(aq) + F^-(aq)$$

The common ion in this problem is the hydronium ion. Now we can tabulate the initial and equilibrium concentrations of each species involved in this equilibrium (the Cl^- is merely a spectator ion).

	$HF(aq)$ $\rightleftharpoons$	$H^+(aq)$ +	$F^-(aq)$
Initial	0.20 M	0.10 M	0
Change	$-x$ M	$+x$ M	$+x$ M
Equilibrium	$(0.20 - x)$ M	$(0.10 + x)$ M	x M

The equilibrium constant for the ionization of HF, from Appendix D, is 6.8×10^{-4}.
Solve: Substituting the equilibrium-constant concentrations into the equilibrium expression gives

$$K_a = 6.8 \times 10^{-4} = \frac{[H^+][F^-]}{[HF]} = \frac{(0.10 + x)(x)}{0.20 - x}$$

If we assume that x is small relative to 0.10 or 0.20 M, this expression simplifies to give

$$\frac{(0.10)(x)}{0.20} = 6.8 \times 10^{-4}$$

$$x = \frac{0.20}{0.10}(6.8 \times 10^{-4}) = 1.4 \times 10^{-3}\ M = [F^-]$$

This F^- concentration is substantially smaller than it would be in a 0.20 M solution of HF with no added HCl. The common ion, H^+, suppresses the ionization of HF. The concentration of $H^+(aq)$ is

$$[H^+] = (0.10 + x)\ M \simeq 0.10\ M$$

Thus, pH = 1.00. Notice that for all practical purposes, $[H^+]$ is due entirely to the HCl; the HF makes a negligible contribution by comparison.

PRACTICE EXERCISE

Calculate the formate ion concentration and pH of a solution that is 0.050 M in formic acid ($HCHO_2$; $K_a = 1.8 \times 10^{-4}$) and 0.10 M in HNO_3.
Answer: $[CHO_2^-] = 9.0 \times 10^{-5}$; pH = 1.00

Sample Exercises 17.1 and 17.2 both involve weak acids. The ionization of a weak base is also decreased by the addition of a common ion. For example, the addition of NH_4^+ (as from the strong electrolyte NH_4Cl) causes the base-dissociation equilibrium of NH_3 to shift to the left, decreasing the equilibrium concentration of OH^- and lowering the pH.

$$NH_3(aq) + H_2O(l) \rightleftharpoons NH_4^+(aq) + OH^-(aq) \qquad [17.2]$$

Addition of NH_4^+ shifts equilibrium, reducing $[OH^-]$

17.2 Buffered Solutions

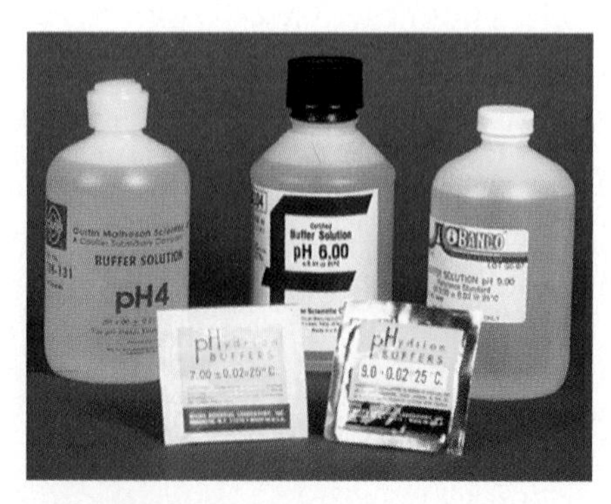

▲ **Figure 17.1** Prepackaged buffer solutions and ingredients for forming buffer solutions of predetermined pH.

Solutions such as those discussed in Section 17.1, which contain a weak conjugate acid-base pair, can resist drastic changes in pH upon the addition of small amounts of strong acid or strong base. These solutions are called **buffered solutions** (or merely **buffers**). Human blood, for example, is a complex aqueous mixture with a pH buffered at about 7.4 (see the "Chemistry and Life" box near the end of this section). Much of the chemical behavior of seawater is determined by its pH, buffered at about 8.1 to 8.3 near the surface. Buffered solutions find many important applications in the laboratory and in medicine (Figure 17.1 ◀).

Composition and Action of Buffered Solutions

A buffer resists changes in pH because it contains both an acidic species to neutralize OH^- ions and a basic one to neutralize H^+ ions. The acidic and basic species that make up the buffer, however, must not consume each other through a neutralization reaction. These requirements are fulfilled by a weak acid-base conjugate pair such as $HC_2H_3O_2$–$C_2H_3O_2^-$ or NH_4^+–NH_3. Thus, buffers are often prepared by mixing a weak acid or a weak base with a salt of that acid or base. The $HC_2H_3O_2$–$C_2H_3O_2^-$ buffer can be prepared, for example, by adding $NaC_2H_3O_2$ to a solution of $HC_2H_3O_2$; the NH_4^+–NH_3 buffer can be prepared by adding NH_4Cl to a solution of NH_3. By choosing appropriate components and adjusting their relative concentrations, we can buffer a solution at virtually any pH.

To understand better how a buffer works, let's consider a buffer composed of a weak acid (HX) and one of its salts (MX, where M^+ could be Na^+, K^+, or another cation). The acid-dissociation equilibrium in this buffered solution involves both the acid and its conjugate base.

$$HX(aq) \rightleftharpoons H^+(aq) + X^-(aq) \qquad [17.3]$$

The corresponding acid-dissociation-constant expression is

$$K_a = \frac{[H^+][X^-]}{[HX]} \qquad [17.4]$$

Solving this expression for $[H^+]$, we have

$$[H^+] = K_a \frac{[HX]}{[X^-]} \qquad [17.5]$$

We see from this expression that $[H^+]$, and thus the pH, is determined by two factors: the value of K_a for the weak-acid component of the buffer and the ratio of the concentrations of the conjugate acid-base pair, $[HX]/[X^-]$.

If OH^- ions are added to the buffered solution, they react with the acid component of the buffer to produce water and the base component (X^-).

$$OH^-(aq) + HX(aq) \longrightarrow H_2O(l) + X^-(aq) \qquad [17.6]$$

This reaction causes [HX] to decrease and $[X^-]$ to increase. As long as the amounts of HX and X^- in the buffer are large compared to the amount of OH^- added, however, the ratio $[HX]/[X^-]$ doesn't change much, and thus the change in pH is small. A specific example of such a buffer, the HF/F^- buffer, is shown in Figure 17.2 ▼.

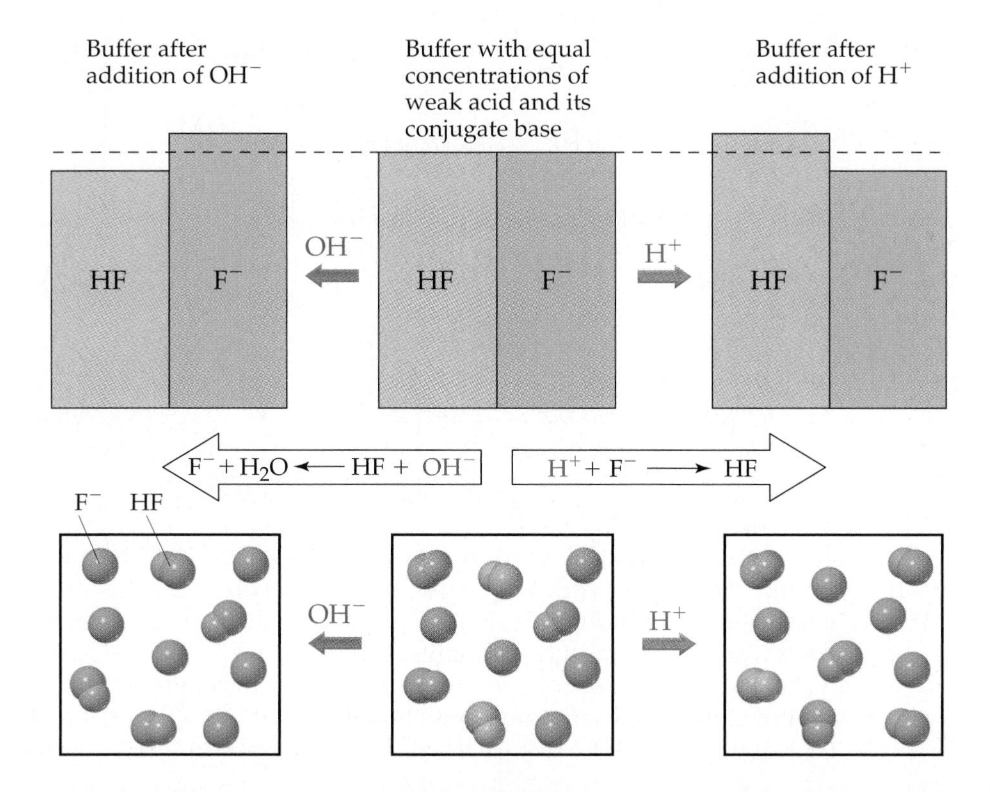

◀ **Figure 17.2** A buffer consisting of a mixture of the weak acid HF and its conjugate base F^-. When a small portion of OH^- is added to the buffer (left), it reacts with HF, decreasing [HF] and increasing $[F^-]$ in the buffer. Conversely, when a small portion of H^+ is added to the buffer (right), it reacts with F^-, decreasing $[F^-]$ and increasing [HF] in the buffer. Because pH depends on the ratio of F^- to HF, the resulting pH change is small.

If H^+ ions are added, they react with the base component of the buffer.

$$H^+(aq) + X^-(aq) \longrightarrow HX(aq) \qquad [17.7]$$

This reaction can also be represented using H_3O^+:

$$H_3O^+(aq) + X^-(aq) \longrightarrow HX(aq) + H_2O(l)$$

Using either equation, the reaction causes $[X^-]$ to decrease and $[HX]$ to increase. As long as the change in the ratio $[HX]/[X^-]$ is small, the change in pH will be small.

Figure 17.2 shows a buffer consisting of equal concentrations of hydrofluoric acid and fluoride ion (center). The addition of OH^- (left) reduces [HF] and increases $[F^-]$. The addition of H^+ (right) reduces $[F^-]$ and increases [HF].

Buffers most effectively resist a change in pH in *either* direction when the concentrations of weak acid and conjugate base are about the same. From Equation 17.5 we see that when the concentrations of weak acid and conjugate base are equal, $[H^+] = K_a$ and therefore $pH = pK_a$. For this reason, we usually try to select a buffer whose acid form has a pK_a close to the desired pH.

Buffer Capacity and pH

Two important characteristics of a buffer are its capacity and its pH. **Buffer capacity** is the amount of acid or base the buffer can neutralize before the pH begins to change to an appreciable degree. The buffer capacity depends on the amount of acid and base from which the buffer is made. The pH of the buffer depends on the K_a for the acid and on the relative concentrations of the acid and base that comprise the buffer. According to Equation 17.5, for example, $[H^+]$ for a 1-L solution that is 1 M in $HC_2H_3O_2$ and 1 M in $NaC_2H_3O_2$ will be the same as for a 1-L solution that is 0.1 M in $HC_2H_3O_2$ and 0.1 M in $NaC_2H_3O_2$. The first solution has a greater buffering capacity, however, because it contains more $HC_2H_3O_2$ and $C_2H_3O_2^-$. The greater the amounts of the conjugate acid-base pair, the more resistant the ratio of their concentrations, and hence the pH, is to change.

Because conjugate acid-base pairs share a common ion, we can use the same procedures to calculate the pH of a buffer that we used to treat the common-ion effect (see Sample Exercise 17.1). However, an alternate approach is sometimes taken that is based on an equation derived from Equation 17.5. Taking the negative log of both sides of Equation 17.5, we have

$$-\log [H^+] = -\log\left(K_a \frac{[HX]}{[X^-]}\right) = -\log K_a - \log\frac{[HX]}{[X^-]}$$

Because $-\log [H^+] = pH$ and $-\log K_a = pK_a$, we have

$$pH = pK_a - \log\frac{[HX]}{[X^-]} = pK_a + \log\frac{[X^-]}{[HX]} \qquad [17.8]$$

In general,

$$pH = pK_a + \log\frac{[\text{base}]}{[\text{acid}]} \qquad [17.9]$$

where [acid] and [base] refer to the equilibrium concentrations of the conjugate acid-base pair. Note that when [base] = [acid], $pH = pK_a$.

Equation 17.9 is known as the **Henderson–Hasselbalch equation**. Biologists, biochemists, and others who work frequently with buffers often use this equation to calculate the pH of buffers. In doing equilibrium calculations, we have seen that we can normally neglect the amounts of the acid and base of the buffer that ionize. Therefore, we can usually use the starting concentrations of the acid and base components of the buffer directly in Equation 17.9.

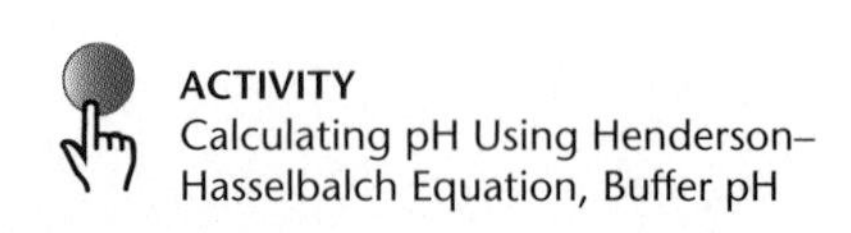

ACTIVITY
Calculating pH Using Henderson–Hasselbalch Equation, Buffer pH

SAMPLE EXERCISE 17.3

What is the pH of a buffer that is 0.12 M in lactic acid ($HC_3H_5O_3$) and 0.10 M in sodium lactate? For lactic acid, $K_a = 1.4 \times 10^{-4}$.

Solution

Analyze: We are asked to calculate the pH of a buffer containing lactic acid and its conjugate base, the lactate ion.

Plan: We will first determine the pH using the method described in Section 17.1. The major species in solution are $HC_3H_5O_3$, Na^+, and $C_3H_5O_3^-$.

The pH will be controlled by the acid-dissociation equilibrium of lactic acid. The initial and equilibrium concentrations of the species involved in this equilibrium are

$$HC_3H_5O_3(aq) \rightleftharpoons H^+(aq) + C_3H_5O_3^-(aq)$$

	$HC_3H_5O_3(aq)$	$H^+(aq)$	$C_3H_5O_3^-(aq)$
Initial	0.12 M	0	0.10 M
Change	$-x$ M	$+x$ M	$+x$ M
Equilibrium	$(0.12 - x)$ M	x M	$(0.10 + x)$ M

Solve: The equilibrium concentrations are governed by the equilibrium expression:

$$K_a = 1.4 \times 10^{-4} = \frac{[H^+][C_3H_5O_3^-]}{[HC_3H_5O_3]} = \frac{x(0.10 + x)}{(0.12 - x)}$$

Because K_a is small and a common ion is present, we expect x to be small relative to either 0.12 or 0.10 M. Thus, our equation can be simplified to give

$$K_a = 1.4 \times 10^{-4} = \frac{x(0.10)}{0.12}$$

Solving for x gives a value that justifies our approximation:

$$[H^+] = x = \left(\frac{0.12}{0.10}\right)(1.4 \times 10^{-4}) = 1.7 \times 10^{-4}\ M$$

$$pH = -\log(1.7 \times 10^{-4}) = 3.77$$

Alternatively, we could have used the Henderson–Hasselbalch equation to calculate pH directly:

$$pH = pK_a + \log\left(\frac{[\text{base}]}{[\text{acid}]}\right) = 3.85 + \log\left(\frac{0.10}{0.12}\right)$$

$$= 3.85 + (-0.08) = 3.77$$

PRACTICE EXERCISE

Calculate the pH of a buffer composed of 0.12 M benzoic acid and 0.20 M sodium benzoate. (Refer to Appendix D.)
Answer: 4.42

SAMPLE EXERCISE 17.4

How many moles of NH_4Cl must be added to 2.0 L of 0.10 M NH_3 to form a buffer whose pH is 9.00? (Assume that the addition of NH_4Cl does not change the volume of the solution.)

Solution

Analyze: Here we are asked to determine the amount of ammonium ion required to prepare a buffer of a specific pH.

Plan: The major species in the solution will be NH_4^+, Cl^-, and NH_3. Of these, the Cl^- ion is a spectator (it is the conjugate base of a strong acid). Thus, the NH_4^+–NH_3 conjugate acid-base pair will determine the pH of the buffer solution. The equilibrium relationship between NH_4^+ and NH_3 is given by the base-dissociation constant for NH_3:

$$NH_3(aq) + H_2O(l) \rightleftharpoons NH_4^+(aq) + OH^-(aq) \qquad K_b = \frac{[NH_4^+][OH^-]}{[NH_3]} = 1.8 \times 10^{-5}$$

Because K_b is small and the common ion NH_4^+ is present, the equilibrium concentration of NH_3 will essentially equal its initial concentration:

$$[NH_3] = 0.10\ M$$

We obtain $[OH^-]$ from the pH:

$$pOH = 14.00 - pH = 14.00 - 9.00 = 5.00$$

and so

$$[OH^-] = 1.0 \times 10^{-5}\ M$$

Solve: We now use the expression for K_b to obtain $[NH_4^+]$.

$$[NH_4^+] = K_b\frac{[NH_3]}{[OH^-]} = (1.8 \times 10^{-5})\frac{(0.10\ M)}{(1.0 \times 10^{-5}\ M)} = 0.18\ M$$

Thus, in order for the solution to have pH = 9.00, $[NH_4^+]$ must equal 0.18 *M*. The number of moles of NH_4Cl needed is given by the product of the volume of the solution and its molarity.

$$(2.0\ L)(0.18\ mol\ NH_4Cl/L) = 0.36\ mol\ NH_4Cl$$

Comment: Because NH_4^+ and NH_3 are a conjugate acid-base pair, we could use the Henderson–Hasselbalch equation (Equation 17.9) to solve this problem. To do so requires first using Equation 16.41 to calculate pK_a for NH_4^+ from the value of pK_b for NH_3. We suggest you try this approach to convince yourself that you can use the Henderson–Hasselbalch equation for buffers for which you are given K_b for the conjugate base rather than K_a for the conjugate acid.

PRACTICE EXERCISE

Calculate the concentration of sodium benzoate that must be present in a 0.20 *M* solution of benzoic acid ($HC_7H_5O_2$) to produce a pH of 4.00.
Answer: 0.13 *M*

Addition of Strong Acids or Bases to Buffers

Let's now consider in a more quantitative way the response of a buffered solution to the addition of a strong acid or base. In solving these problems, it is important to understand that reactions between strong acids and weak bases proceed essentially to completion, as do those between strong bases and weak acids. Thus, as long as we do not exceed the buffering capacity of the buffer, we can assume that the strong acid or strong base is completely consumed by reaction with the buffer.

Consider a buffer that contains a weak acid HX and its conjugate base X^-. When a strong acid is added to this buffer, the added H^+ is consumed by X^- to produce HX; thus, [HX] increases and $[X^-]$ decreases. When a strong base is added to the buffer, the added OH^- is consumed by HX to produce X^-; in this case [HX] decreases and $[X^-]$ increases.

To calculate how the pH of the buffer responds to the addition of a strong acid or a strong base, we follow the strategy outlined in Figure 17.3 ▼:

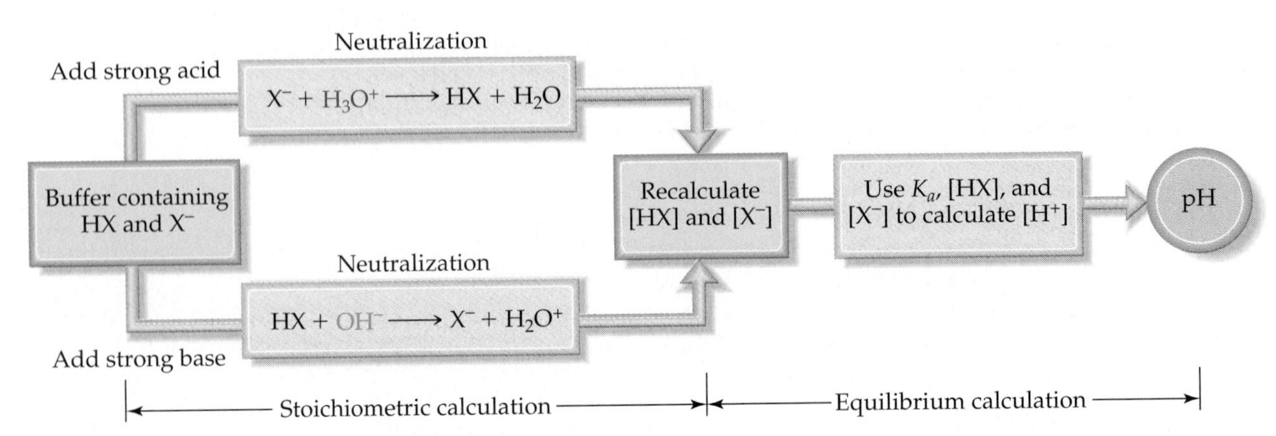

▲ **Figure 17.3** Outline of the procedure used to calculate the pH of a buffer after the addition of strong acid or strong base. As long as the amount of added acid or base does not exceed the buffer capacity, the Henderson–Hasselbalch equation, Equation 17.9, can be used for the equilibrium calculation.

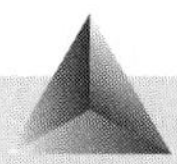

Chemistry and Life Blood as a Buffered Solution

Many of the chemical reactions that occur in living systems are extremely sensitive to pH. Many of the enzymes that catalyze important biochemical reactions, for example, are effective only within a narrow pH range. For this reason the human body maintains a remarkably intricate system of buffers, both within tissue cells and in the fluids that transport cells. Blood, the fluid that transports oxygen to all parts of the body (Figure 17.4 ▶), is one of the most prominent examples of the importance of buffers in living beings.

Human blood is slightly basic with a normal pH of 7.35 to 7.45. Any deviation from this normal pH range can have extremely disruptive effects on the stability of cell membranes, the structures of proteins, and the activities of enzymes. Death may result if the blood pH falls below 6.8 or rises above 7.8. When the pH falls below 7.35, the condition is called *acidosis*; when it rises above 7.45, the condition is called *alkalosis*. Acidosis is the more common tendency because ordinary metabolism generates several acids within the body.

The major buffer system that is used to control the pH of blood is the *carbonic acid–bicarbonate buffer system*. Carbonic acid (H_2CO_3) and bicarbonate ion (HCO_3^-) are a conjugate acid-base pair. In addition, carbonic acid can decompose into carbon dioxide gas and water. The important equilibria in this buffer system are

$$H^+(aq) + HCO_3^-(aq) \rightleftharpoons H_2CO_3(aq) \rightleftharpoons H_2O(l) + CO_2(g) \quad [17.10]$$

Several aspects of these equilibria are notable. First, although carbonic acid is a diprotic acid, the carbonate ion (CO_3^{2-}) is unimportant in this system. Second, one of the components of this equilibrium, CO_2, is a gas, which provides a mechanism for the body to adjust the equilibria. Removal of CO_2 via exhalation shifts the equilibria to the right, consuming H^+ ions. Third, the buffer system in blood operates at a pH of 7.4, which is fairly far removed from the pK_{a1} value of H_2CO_3 (6.1 at physiological temperatures). In order for the buffer to have a pH of 7.4, the ratio [base]/[acid] must have a value of about 20. In normal blood plasma the concentrations of HCO_3^- and H_2CO_3 are about 0.024 *M* and 0.0012 *M*, respectively. As a consequence, the buffer has a high capacity to neutralize additional acid, but only a low capacity to neutralize additional base.

The principal organs that regulate the pH of the carbonic acid–bicarbonate buffer system are the lungs and kidneys. Some of the receptors in the brain are sensitive to the concentrations of H^+ and CO_2 in bodily fluids. When the concentration of CO_2 rises, the equilibria in Equation 17.10 shift to the left, which leads to the formation of more H^+. The receptors trigger a reflex to breathe faster and deeper, increasing the rate of elimination of CO_2 from the lungs and shifts the equilibria back to the right. The kidneys absorb or release H^+ and HCO_3^-; much of the excess acid leaves the body in urine, which normally has a pH of 5.0 to 7.0.

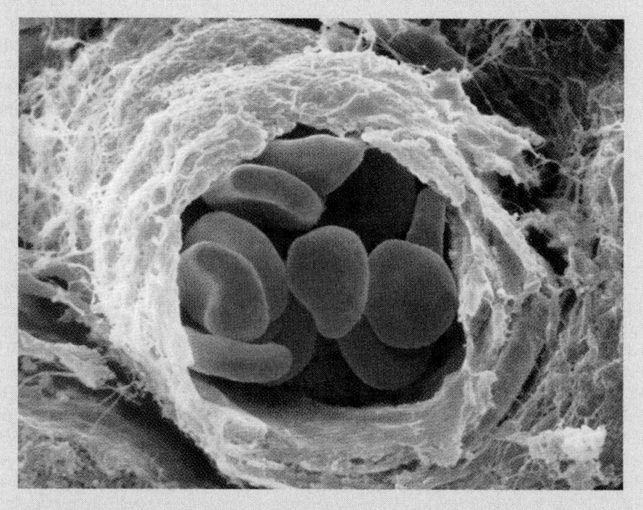

▲ **Figure 17.4** A scanning electromicrograph of a group of red blood cells traveling through a small branch of an artery. Blood is a buffered solution whose pH is maintained between 7.35 and 7.45.

The regulation of the pH of blood plasma relates directly to the effective transport of O_2 to bodily tissues. Oxygen is carried by the protein hemoglobin, which is found in red blood cells. Hemoglobin (Hb) reversibly binds both H^+ and O_2. These two substances compete for the Hb, which can be represented approximately by the following equilibrium:

$$HbH^+ + O_2 \rightleftharpoons HbO_2 + H^+ \quad [17.11]$$

Oxygen enters the blood through the lungs, where it passes into the red blood cells and binds to Hb. When the blood reaches tissue in which the concentration of O_2 is low, the equilibrium in Equation 17.11 shifts to the left and O_2 is released. An increase in H^+ ion concentration (decrease in blood pH) also shifts this equilibrium to the left, as does increasing temperature.

During periods of strenuous exertion, three factors work together to ensure the delivery of O_2 to active tissues: (1) As O_2 is consumed, the equilibrium in Equation 17.11 shifts to the left according to Le Châtelier's principle. (2) Exertion raises the temperature of the body, also shifting the equilibrium to the left. (3) Large amounts of CO_2 are produced by metabolism, which shifts the equilibrium in Equation 17.10 to the left, thus decreasing the pH. Other acids, such as lactic acid, are also produced during strenuous exertion as tissues become starved for oxygen. The decrease in pH shifts the hemoglobin equilibrium to the left, delivering more O_2. In addition, the decrease in pH stimulates an increase in the rate of breathing, which furnishes more O_2 and eliminates CO_2. Without this elaborate arrangement, the O_2 in tissues would be rapidly depleted, making further activity impossible.

1. Consider the acid-base neutralization reaction, and determine its effect on [HX] and $[X^-]$. This stage of the procedure is a *stoichiometry calculation*.
2. Use K_a and the new concentrations of [HX] and $[X^-]$ from step 1 to calculate $[H^+]$. This second stage of the procedure is a standard *equilibrium calculation* and is most easily done using the Henderson–Hasselbalch equation.

The complete procedure is illustrated in Sample Exercise 17.5.

SAMPLE EXERCISE 17.5

A buffer is made by adding 0.300 mol $HC_2H_3O_2$ and 0.300 mol $NaC_2H_3O_2$ to enough water to make 1.00 L of solution. The pH of the buffer is 4.74 (Sample Exercise 17.1). **(a)** Calculate the pH of this solution after 0.020 mol of NaOH is added, and, for comparison, **(b)** calculate the pH that would result if 0.020 mol of NaOH were added to 1.00 L of pure water (neglect any volume changes).

Solution

Analyze: We are asked to determine the pH of a buffer after addition of a small amount of strong base, and to compare the pH change to the pH that would result if we were to add the same amount of strong base to pure water.

Plan: (a) Solving this problem involves the two steps outlined in Figure 17.3.

Stoichiometric Calculation: We assume that the OH^- provided by NaOH is completely consumed by $HC_2H_3O_2$, the weak-acid component of the buffer. A useful convention for this type of calculation is to write the number of moles of each species prior to the reaction above the equation and the number of moles of each species after reaction below the equation. Prior to the reaction in which the added hydroxide is consumed by acetic acid, there are 0.300 mol each of acetic acid and acetate ion, and 0.020 mol of hydroxide ion.

$$\begin{array}{lccccc} \text{Before reaction:} & 0.300\text{ mol} & & 0.020\text{ mol} & & & & 0.300\text{ mol} \\ & HC_2H_3O_2(aq) & + & OH^-(aq) & \longrightarrow & H_2O(l) & + & C_2H_3O_2^-(aq) \end{array}$$

Because the amount of OH^- added is smaller than the amount of $HC_2H_3O_2$, all the added OH^- will be consumed. An equal amount of $HC_2H_3O_2$ will be consumed, and the same amount of $C_2H_3O_2^-$ will be produced. We write these new, post-reaction amounts below the equation.

$$\begin{array}{lccccccc} \text{Before reaction:} & 0.300\text{ mol} & & 0.020\text{ mol} & & & & 0.300\text{ mol} \\ & HC_2H_3O_2(aq) & + & OH^-(aq) & \longrightarrow & H_2O(l) & + & C_2H_3O_2^-(aq) \\ \text{After reaction:} & 0.280\text{ mol} & & 0\text{ mol} & & & & 0.320\text{ mol} \end{array}$$

Equilibrium Calculation: We now turn our attention to the equilibrium that will determine the pH of the buffer, namely the ionization of acetic acid.

$$HC_2H_3O_2(aq) \rightleftharpoons H^+(aq) + C_2H_3O_2^-(aq)$$

Solve: Using the new quantities of $HC_2H_3O_2$ and $C_2H_3O_2^-$, we can determine the pH using the Henderson–Hasselbalch equation.

$$pH = 4.74 + \log\frac{0.320\text{ mol}}{0.280\text{ mol}} = 4.80$$

Note that we can use mole amounts in place of concentrations in the Henderson–Hasselbalch equation.

Comment: If 0.020 mol of H^+ were added to the buffer, we would proceed in a similar way to calculate the resulting pH of the buffer. In this case, the pH decreases by 0.06 units giving pH = 4.69, as shown in the figure below.

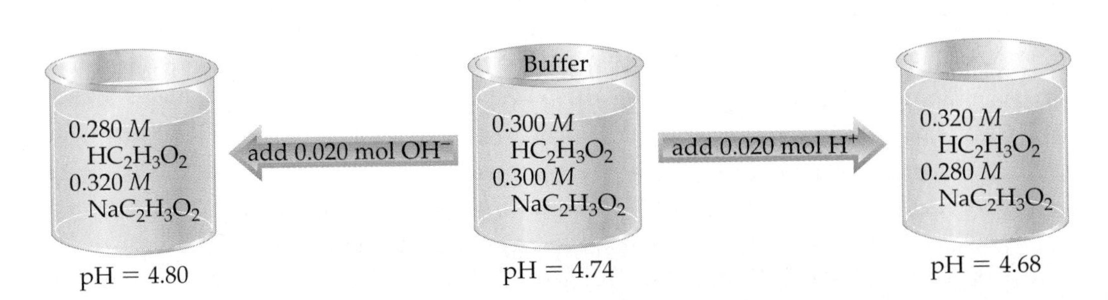

(b) To determine the pH of a solution made by adding 0.020 mol of NaOH to 1.00 L of pure water, we can first determine pOH using Equation 16.18 and subtract from 14.

$$pH = 14 - (-\log 0.020) = 12.30$$

Note that although the small amount of NaOH is enough to change the pH of water significantly, the pH of the buffer changes very little.

PRACTICE EXERCISE

Determine **(a)** the pH of the original buffer described in Sample Exercise 17.5 after the addition of 0.020 mol HCl, and **(b)** the pH of the solution that would result from the addition of 0.020 mol HCl to 1.00 L of pure water.
Answers: **(a)** 4.68; **(b)** 1.70

17.3 Acid-Base Titrations

In Section 4.6 we briefly described *titrations*. In an acid-base titration, a solution containing a known concentration of base is slowly added to an acid (or the acid is added to the base). Acid-base indicators can be used to signal the *equivalence point* of a titration (the point at which stoichiometrically equivalent quantities of acid and base have been brought together). Alternatively, a pH meter can be used to monitor the progress of the reaction producing a **pH titration curve**, a graph of the pH as a function of the volume of the added titrant. The shape of the titration curve makes it possible to determine the equivalence point in the titration. The titration curve can also be used to select suitable indicators and to determine the K_a of the weak acid or the K_b of the weak base being titrated.

A typical apparatus for measuring pH during a titration is illustrated in Figure 17.5 ▶. The titrant is added to the solution from a buret, and the pH is continually monitored using a pH meter. To understand why titration curves have certain characteristic shapes, we will examine the curves for three kinds of titrations: (1) strong acid–strong base; (2) weak acid–strong base; and (3) polyprotic acid–strong base. We will also briefly consider how these curves relate to those involving weak bases.

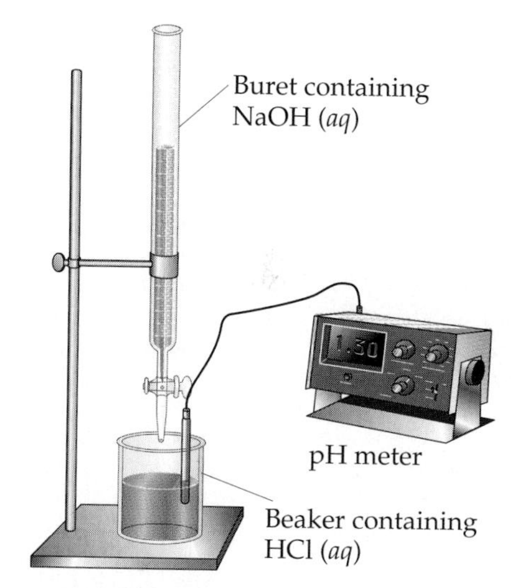

▲ **Figure 17.5** A typical setup for using a pH meter to measure data for a titration curve. In this case a standard solution of NaOH (the titrant) is added by buret to a solution of HCl that is to be titrated. The solution is stirred during the titration to ensure uniform composition.

Strong Acid–Strong Base Titrations

The titration curve produced when a strong base is added to a strong acid has the general shape shown in Figure 17.6 ▶. This curve depicts the pH change that occurs as 0.100 *M* NaOH is added to 50.0 mL of 0.100 *M* HCl. The pH can be calculated at various stages of the titration. To help understand these calculations, we can divide the curve into four regions:

1. *The initial pH*: The pH of the solution before the addition of any base is determined by the initial concentration of the strong acid. For a solution of 0.100 *M* HCl, $[H^+] = 0.100\ M$ and hence $pH = -\log(0.100) = 1.000$. Thus, the initial pH is low.
2. *Between the initial pH and the equivalence point*: As NaOH is added, the pH increases slowly at first and then rapidly in the vicinity of the equivalence point. The pH of the solution before the equivalence point is determined by the concentration of acid that has not yet been neutralized. This calculation is illustrated in Sample Exercise 17.6(a).
3. *The equivalence point*: At the equivalence point an equal number of moles of the NaOH and HCl have reacted, leaving only a solution of their salt, NaCl. The pH of the solution is 7.00 because the cation of a strong base (in this case Na^+) and the anion of a strong acid (in this case Cl^-) do not hydrolyze and therefore have no appreciable effect on pH. ∞ (Section 16.9)
4. *After the equivalence point*: The pH of the solution after the equivalence point is determined by the concentration of the excess NaOH in the solution. This calculation is illustrated in Sample Exercise 17.6(b).

▶ **Figure 17.6** The pH curve for titration of a solution of 50.0 mL of 0.100 *M* of a strong acid with a solution of 0.100 *M* of a strong base, in this case, HCl and NaOH.

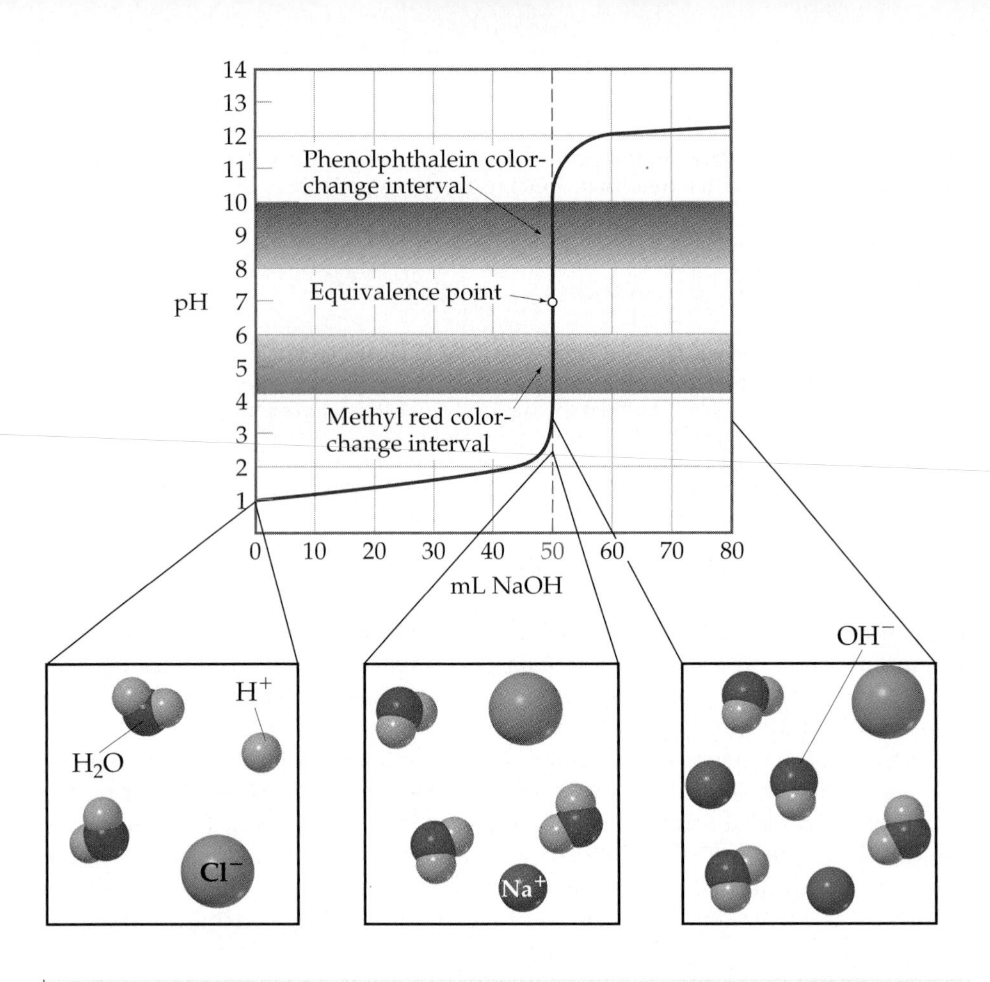

SAMPLE EXERCISE 17.6

Calculate the pH when the following quantities of 0.100 *M* NaOH solution have been added to 50.0 mL of 0.100 *M* HCl solution: **(a)** 49.0 mL; **(b)** 51.0 mL.

Solution

Analyze: We are asked to calculate the pH at two points in the titration of a strong acid with a strong base. The first point is just prior to the equivalence point, so we expect the pH to be determined by the small amount of strong acid that has not yet been neutralized. The second point is just after the equivalence point, so we expect this pH to be determined by the small amount of excess strong base.

(a) Plan: As the NaOH solution is added to the HCl solution, $H^+(aq)$ reacts with $OH^-(aq)$ to form H_2O. Both Na^+ and Cl^- are spectator ions, having negligible effect on the pH. In order to determine the pH of the solution, we must first determine how many moles of H^+ were originally present and how many moles of OH^- were added. We can then calculate how many moles of each ion remain after the neutralization reaction. In order to calculate $[H^+]$, and hence pH, we must also remember that the volume of the solution increases as we add titrant, thus diluting the concentration of all solutes present.

The number of moles of H^+ in the original HCl solution is given by the product of the volume of the solution (50.0 mL = 0.0500 L) and its molarity (0.100 *M*).

$$(0.0500 \text{ L soln})\left(\frac{0.100 \text{ mol H}^+}{1 \text{ L soln}}\right) = 5.00 \times 10^{-3} \text{ mol H}^+$$

Likewise, the number of moles of OH^- in 49.0 mL of 0.100 *M* NaOH is

$$(0.0490 \text{ L soln})\left(\frac{0.100 \text{ mol OH}^-}{1 \text{ L soln}}\right) = 4.90 \times 10^{-3} \text{ mol OH}^-$$

Because we have not yet reached the equivalence point, there are more moles of H^+ present than OH^-. Each mole of OH^- will react with one mole of H^+. Using the convention introduced in Sample Exercise 17.5,

Before reaction:	5.00×10^{-3} mol		4.90×10^{-3} mol	
	$H^+(aq)$	+	$OH^-(aq)$	$\longrightarrow H_2O(l)$
After reaction:	0.10×10^{-3} mol		0.00 mol	

Solve: During the course of the titration the volume of the reaction mixture increases as the NaOH solution is added to the HCl solution. Thus, at this point in the titration the solution has a volume of 50.0 mL + 49.0 mL = 99.0 mL. (We assume that the total volume is the sum of the volumes of the acid and base solutions.) Thus, the concentration of $H^+(aq)$ is

$$[H^+] = \frac{\text{moles } H^+(aq)}{\text{liters soln}} = \frac{0.10 \times 10^{-3}\text{ mol}}{0.09900\text{ L}} = 1.0 \times 10^{-3}\,M$$

The corresponding pH equals $-\log(1.0 \times 10^{-3}) = 3.00$.

(b) Plan: We proceed in the same way as we did in part (a), except we are now past the equivalence point and have more OH^- in the solution than H^+. As before, the initial number of moles of each reactant is determined from their volumes and concentrations. The reactant present in smaller stoichiometic amount (the limiting reactant) is consumed completely, leaving an excess this time of hydroxide ion.

Before reaction:	5.00×10^{-3} mol		5.10×10^{-3} mol		
	$H^+(aq)$	+	$OH^-(aq)$	$\longrightarrow$	$H_2O(l)$
After reaction:	0.0 mol		0.10×10^{-3} mol		

Solve: In this case the total volume of the solution is 50.0 mL + 51.0 mL = 101.0 mL = 0.1010 L. Hence, the concentration of $OH^-(aq)$ in the solution is

$$[OH^-] = \frac{\text{moles } OH^-(aq)}{\text{liters soln}} = \frac{0.10 \times 10^{-3}\text{ mol}}{0.1010\text{ L}} = 1.0 \times 10^{-3}\,M$$

Thus, the pOH of the solution equals $-\log(1.0 \times 10^{-3}) = 3.00$, and the pH equals $14.00 - \text{pOH} = 14.00 - 3.00 = 11.00$.

PRACTICE EXERCISE

Calculate the pH when the following quantities of 0.10 M HNO_3 have been added to 25.0 mL of 0.10 M KOH solution: **(a)** 24.9 mL; **(b)** 25.1 mL.
Answers: **(a)** 10.30; **(b)** 3.70

Optimally, an indicator would change color at the equivalence point in a titration. In practice, however, that is unnecessary. The pH changes very rapidly near the equivalence point, and in this region merely a drop of titrant can change the pH by several units. Thus, an indicator beginning and ending its color change anywhere on this rapid-rise portion of the titration curve will give a sufficiently accurate measure of the volume of titrant needed to reach the equivalence point. The point in a titration where the indicator changes color is called the *end point* to distinguish it from the actual equivalence point that it closely approximates.

In Figure 17.6 we see that the pH changes very rapidly from about 4 to about 10 near the equivalence point. Consequently, an indicator for this strong acid–strong base titration can change color anywhere in this range. Most strong acid–strong base titrations are carried out using phenolphthalein as an indicator (Figure 4.19) because it dramatically changes color in this range. From Figure 16.7 we see that phenolphthalein changes color from pH 8.3 to 10.0. Several other indicators would also be satisfactory, including methyl red, which changes color from pH 4.2 to 6.0 (Figure 17.7 ▶).

Titration of a solution of a strong base with a solution of a strong acid would yield an analogous curve of pH versus added acid. In this case, however, the pH would be high at the outset of the titration and low at its completion, as shown in Figure 17.8 ▶.

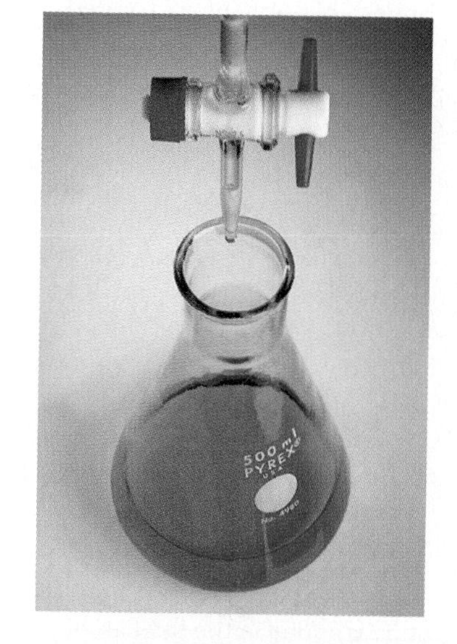

(a)

(b)

▲ **Figure 17.7** Change in appearance of a solution containing methyl red indicator in the pH range 4.2 to 6.3. The characteristic acid color is shown in (a), and the characteristic basic color in (b).

Weak Acid–Strong Base Titrations

The curve for the titration of a weak acid by a strong base is very similar in shape to that for the titration of a strong acid by a strong base. Consider, for example, the titration curve for the titration of 50.0 mL of 0.100 M acetic acid ($HC_2H_3O_2$) with

0.100 M NaOH shown in Figure 17.9 ▼. We can calculate the pH at points along this curve using principles we have discussed earlier. As in the case of the titration of a strong acid by a strong base, we can divide the curve into four regions:

1. *The initial pH*: This pH is just the pH of the 0.100 M $HC_2H_3O_2$. We performed calculations of this kind in Section 16.6. The calculated pH of 0.100 M $HC_2H_3O_2$ is 2.89.
2. *Between the initial pH and the equivalence point*: To determine pH in this range, we must consider the neutralization of the acid.

$$HC_2H_3O_2(aq) + OH^-(aq) \longrightarrow C_2H_3O_2^-(aq) + H_2O(l) \qquad [17.12]$$

Prior to reaching the equivalence point, part of the $HC_2H_3O_2$ is neutralized to form $C_2H_3O_2^-$. Thus, the solution contains a mixture of $HC_2H_3O_2$ and $C_2H_3O_2^-$.

The approach we take in calculating the pH in this region of the titration curve involves two main steps. First, we consider the neutralization reaction between $HC_2H_3O_2$ and OH^- to determine the concentrations of $HC_2H_3O_2$ and $C_2H_3O_2^-$ in the solution. Next, we calculate the pH of this buffer pair using procedures developed in Sections 17.1 and 17.2. The general procedure is diagrammed in Figure 17.10 ▶ and illustrated in Sample Exercise 17.7.

3. *The equivalence point*: The equivalence point is reached after adding 50.0 mL of 0.100 M NaOH to the 50.0 mL of 0.100 M $HC_2H_3O_2$. At this point the 5.00×10^{-3} mol of NaOH completely reacts with the 5.00×10^{-3} mol of $HC_2H_3O_2$ to form 5.00×10^{-3} mol of their salt, $NaC_2H_3O_2$. The Na^+ ion of this salt has no significant effect on the pH. The $C_2H_3O_2^-$ ion, however, is a weak base, and the pH at the equivalence point is therefore greater than 7. Indeed, the pH at the equivalence point is always above 7 in a weak acid–strong base titration because the anion of the salt formed is a weak base.

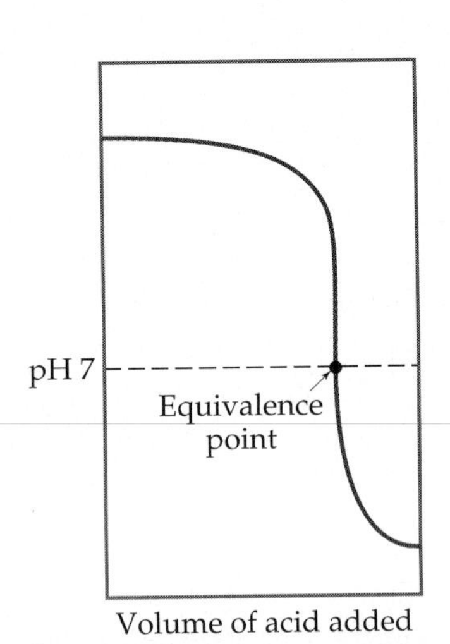

▲ **Figure 17.8** The shape of a pH curve for titration of a strong base with a strong acid.

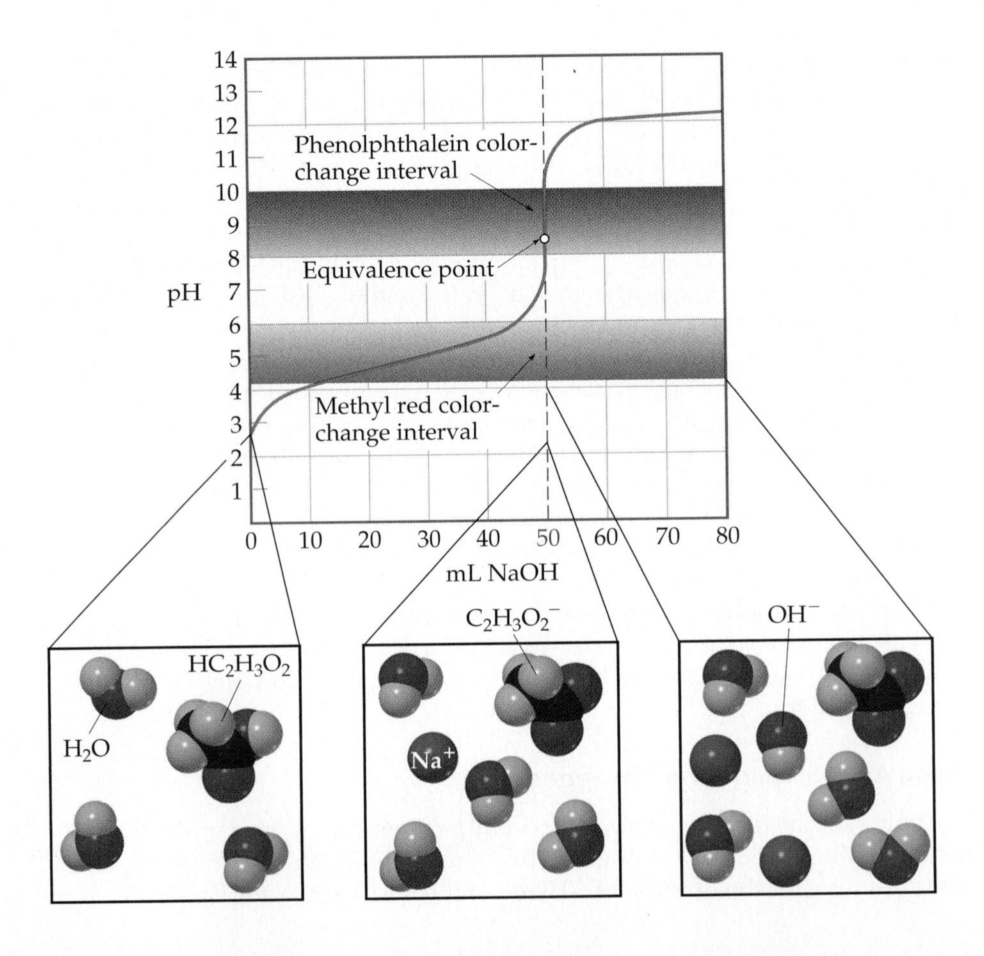

▶ **Figure 17.9** The line shows the variation in pH as 0.100 M NaOH solution is added in the titration of 50.0 mL of 0.100 M acetic acid solution.

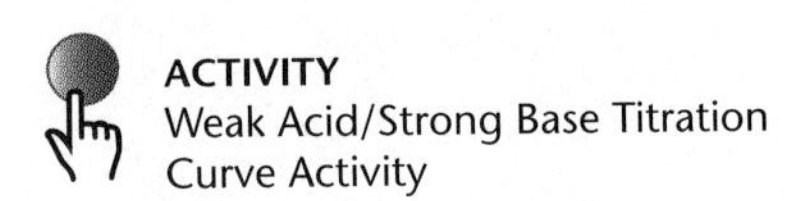

ACTIVITY
Weak Acid/Strong Base Titration Curve Activity

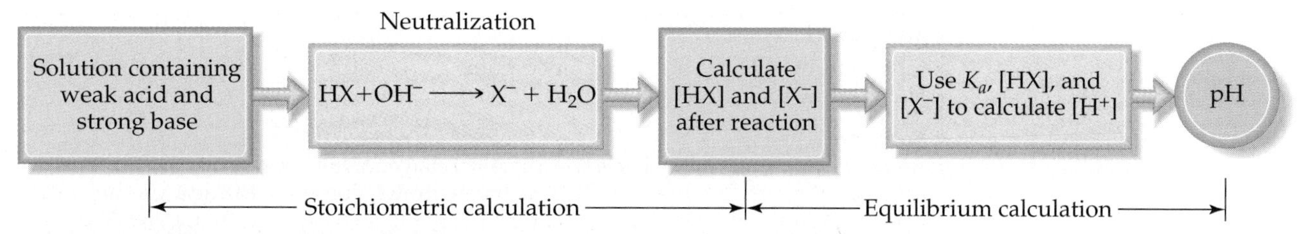

▲ **Figure 17.10** Outline of the procedure used to calculate the pH of a mixture in which a weak acid has been partially neutralized by strong base. An analogous procedure can be used for the addition of strong acid to a weak base.

4. *After the equivalence point*: In this region of the titration curve, $[OH^-]$ from the reaction of $C_2H_3O_2^-$ with water is negligible compared to $[OH^-]$ from the excess NaOH. Thus, the pH is determined by the concentration of OH^- from the excess NaOH. The method for calculating pH in this region is therefore like that for the strong acid–strong base titration illustrated in Sample Exercise 17.6(b). Thus, the addition of 51.0 mL of 0.100 *M* NaOH to 50.0 mL of either 0.100 *M* HCl or 0.100 *M* $HC_2H_3O_2$ yields the same pH, 11.00. Notice in Figures 17.6 and 17.9 that the titration curves for the titrations of both the strong acid and the weak acid are the same after the equivalence point.

SAMPLE EXERCISE 17.7

Calculate the pH of the solution formed when 45.0 mL of 0.100 *M* NaOH is added to 50.0 mL of 0.100 *M* $HC_2H_3O_2$ ($K_a = 1.8 \times 10^{-5}$).

Solution

Analyze: We are asked to calculate the pH prior to the equivalence point of the titration of a weak acid with a strong base.

Plan: We first must determine the number of moles of weak acid and strong base that have been combined. This will tell us how much of the weak acid's conjugate base has been produced, and we can solve for pH using the equilibrium-constant expression.

Solve: *Stoichiometric Calculation*: The product of the volume and concentration of each solution gives the number of moles of each reactant present before the neutralization:

$$(0.0450\ \text{L soln})\left(\frac{0.100\ \text{mol NaOH}}{1\ \text{L soln}}\right) = 4.50 \times 10^{-3}\ \text{mol NaOH}$$

$$(0.0500\ \text{L soln})\left(\frac{0.100\ \text{mol HC}_2\text{H}_3\text{O}_2}{1\ \text{L soln}}\right) = 5.00 \times 10^{-3}\ \text{mol HC}_2\text{H}_3\text{O}_2$$

The 4.50×10^{-3} mol of NaOH consumes 4.50×10^{-3} mol of $HC_2H_3O_2$:

	$HC_2H_3O_2^+(aq)$ +	$OH^-(aq)$ ⟶	$C_2H_3O_2^-(aq)$ + $H_2O(l)$
Before reaction:	5.00×10^{-3} mol	4.50×10^{-3} mol	0.0 mol
After reaction:	0.50×10^{-3} mol	0.0 mol	4.50×10^{-3} mol

The total volume of the solution is

$$45.0\ \text{mL} + 50.0\ \text{mL} = 95.0\ \text{mL} = 0.0950\ \text{L}$$

The resulting molarities of $HC_2H_3O_2$ and $C_2H_3O_2^-$ after the reaction are therefore

$$[HC_2H_3O_2] = \frac{0.50 \times 10^{-3}\ \text{mol}}{0.0950\ \text{L}} = 0.0053\ M$$

$$[C_2H_3O_2^-] = \frac{4.50 \times 10^{-3}\ \text{mol}}{0.0950\ \text{L}} = 0.0474\ M$$

Equilibrium Calculation: The equilibrium between $HC_2H_3O_2$ and $C_2H_3O_2^-$ must obey the equilibrium-constant expression for $HC_2H_3O_2$:

$$K_a = \frac{[H^+][C_2H_3O_2^-]}{[HC_2H_3O_2]} = 1.8 \times 10^{-5}$$

Solving for $[H^+]$ gives

$$[H^+] = K_a \times \frac{[HC_2H_3O_2]}{[C_2H_3O_2^-]} = (1.8 \times 10^{-5}) \times \left(\frac{0.0053}{0.0474}\right) = 2.0 \times 10^{-6}\ M$$

$$\text{pH} = -\log(2.0 \times 10^{-6}) = 5.70$$

Comment: We could have solved for pH equally well using the Henderson-Hasselbalch equation.

PRACTICE EXERCISE

(a) Calculate the pH in the solution formed by adding 10.0 mL of 0.050 *M* NaOH to 40.0 mL of 0.0250 *M* benzoic acid ($HC_7H_5O_2$, $K_a = 6.3 \times 10^{-5}$). **(b)** Calculate the pH in the solution formed by adding 10.0 mL of 0.100 *M* HCl to 20.0 mL of 0.100 *M* NH_3.

Answers: **(a)** 4.20; **(b)** 9.26

SAMPLE EXERCISE 17.8

Calculate the pH at the equivalence point in the titration of 50.0 mL of 0.100 *M* $HC_2H_3O_2$ with 0.100 *M* NaOH.

Solution

Analyze: We are asked to determine the pH at the equivalence point of the titration of a weak acid with a strong base. Because the neutralization of a weak acid produces the corresponding conjugate base, we expect the pH to be basic at the equivalence point.

Plan: We should first determine how many moles of acetic acid there are initially. This will tell us how many moles of acetate ion there will be in solution at the equivalence point. We then must determine the final volume of the resulting solution, and the concentration of acetate ion. From this point this is simply a weak-base equilibrium problem like those in Section 16.7.

Solve: The acetic acid solution contains 5.00×10^{-3} mol $HC_2H_3O_2$. Hence 5.00×10^{-3} mol of $C_2H_3O_2^-$ is formed. The volume of this salt solution is the sum of the volumes of the acid and base, 50.0 mL + 50.0 mL = 100.0 mL = 0.1000 L. Thus, the concentration of $C_2H_3O_2^-$ is

$$[C_2H_3O_2^-] = \frac{5.00 \times 10^{-3}\text{ mol}}{0.1000\text{ L}} = 0.0500\ M$$

The $C_2H_3O_2^-$ ion is a weak base.

$$C_2H_3O_2^-(aq) + H_2O(l) \rightleftharpoons HC_2H_3O_2(aq) + OH^-(aq)$$

The K_b for $C_2H_3O_2^-$ can be calculated from the K_a value of its conjugate acid, $K_b = K_w/K_a = (1.0 \times 10^{-14})/(1.8 \times 10^{-5}) = 5.6 \times 10^{-10}$. Using the K_b expression, we have

$$K_b = \frac{[HC_2H_3O_2^-][OH^-]}{[C_2H_3O_2^-]} = \frac{(x)(x)}{0.0500 - x} = 5.6 \times 10^{-10}$$

Making the approximation that $0.0500 - x \simeq 0.0500$, and then solving for x, we have $x = [OH^-] = 5.3 \times 10^{-6} M$, which gives pOH = 5.28 and pH = 8.72.

PRACTICE EXERCISE

Calculate the pH at the equivalence point when **(a)** 40.0 mL of 0.025 *M* benzoic acid ($HC_7H_5O_2$, $K_a = 6.3 \times 10^{-5}$) is titrated with 0.050 *M* NaOH; **(b)** 40.0 mL of 0.100 *M* NH_3 is titrated with 0.100 *M* HCl

Answers: **(a)** 8.21; **(b)** 5.28

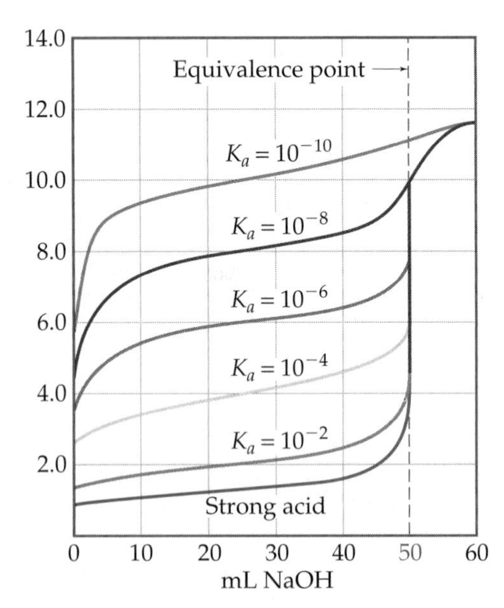

▲ **Figure 17.11** Influence of acid strength on the shape of the curve for titration with NaOH. Each curve represents titration of 50.0 mL of 0.10 *M* acid with 0.10 *M* NaOH.

The pH titration curves for weak acid–strong base titrations differ from those for strong acid–strong base titrations in three noteworthy ways:

1. The solution of the weak acid has a higher initial pH than a solution of a strong acid of the same concentration.
2. The pH change at the rapid-rise portion of the curve near the equivalence point is smaller for the weak acid than it is for the strong acid.
3. The pH at the equivalence point is above 7.00 for the weak acid–strong base titration.

To illustrate these differences further, consider the family of titration curves shown in Figure 17.11 ◀. As expected, the initial pH of the weak acid solutions is always higher than that of the strong acid solution of the same concentration. Notice also that the pH change near the equivalence point becomes less marked as the acid becomes weaker (that is, as K_a becomes smaller). Finally, the pH at the equivalence point steadily increases as K_a decreases.

Because the pH change near the equivalence point becomes smaller as K_a decreases, the choice of indicator for a weak acid–strong base titration is more critical than it is for a strong acid–strong base titration. When 0.100 *M* $HC_2H_3O_2$ ($K_a = 1.8 \times 10^{-5}$) is titrated with 0.100 *M* NaOH, for example, as shown in Figure 17.9, the pH increases rapidly only over the pH range of about 7 to 10. Phenolphthalein is therefore an ideal indicator because it changes color from pH 8.3 to 10.0, close to the pH at the equivalence point. Methyl red is

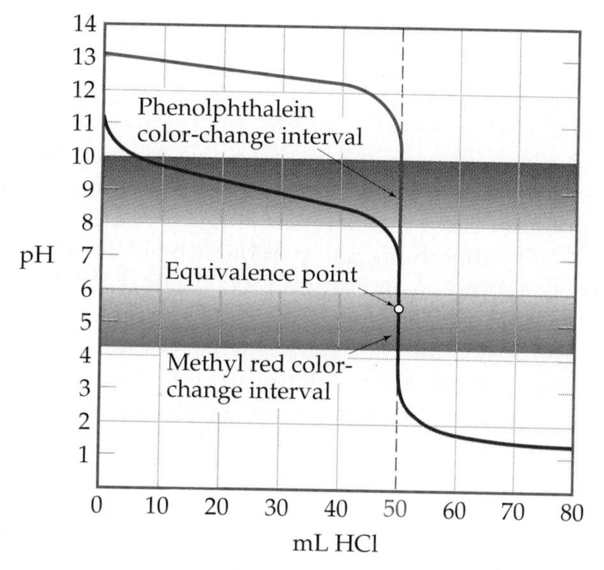

◀ **Figure 17.12** The blue line shows pH versus volume of added HCl in the titration of 50.0 mL of 0.10 *M* ammonia with 0.10 *M* HCl. The red line segment shows the graph of pH versus added acid for the titration of 0.10 *M* NaOH.

ACTIVITY
Weak Base/Strong Acid Titration Curve Activity.

a poor choice, however, because its color change occurs from 4.2 to 6.0, which begins well before the equivalence point is reached.

Titration of a weak base (such as 0.100 *M* NH_3) with a strong acid solution (such as 0.100 *M* HCl) leads to the titration curve shown in Figure 17.12 ▲. In this particular example the equivalence point occurs at pH 5.28. Thus, methyl red would be an ideal indicator, but phenolphthalein would be a poor choice.

Titrations of Polyprotic Acids

When weak acids contain more than one ionizable H atom, as in phosphorous acid (H_3PO_3), reaction with OH^- occurs in a series of steps. Neutralization of H_3PO_3 proceeds in two stages. (Sample Integrative Exercise 16)

$$H_3PO_3(aq) + OH^-(aq) \longrightarrow H_2PO_3^-(aq) + H_2O(l) \quad [17.13]$$

$$H_2PO_3^-(aq) + OH^-(aq) \longrightarrow HPO_3^{2-}(aq) + H_2O(l) \quad [17.14]$$

When the neutralization steps of a polyprotic acid or polybasic base are sufficiently separated, the substance exhibits a titration curve with multiple equivalence points. Figure 17.13 ▼ shows the two distinct equivalence points in the titration curve for the H_3PO_3-$H_2PO_3^-$-HPO_3^{2-} system.

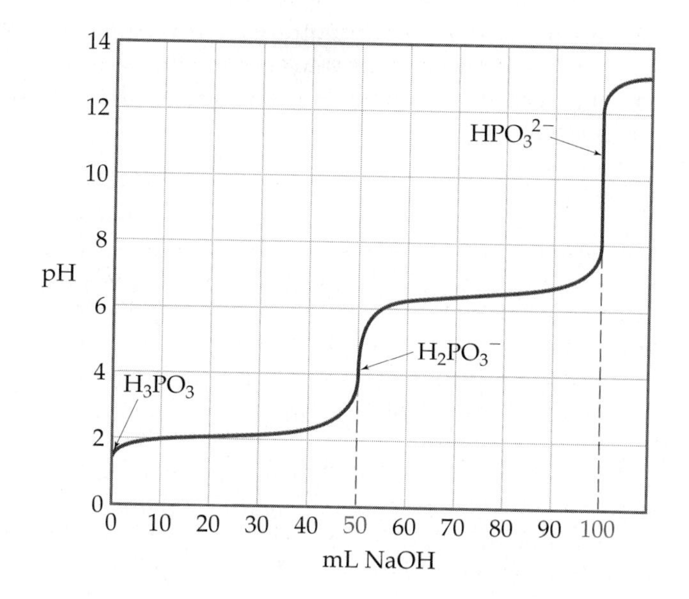

◀ **Figure 17.13** Titration curve for the reaction of 50.0 mL of 0.10 *M* H_3PO_3 with 0.10 *M* NaOH.

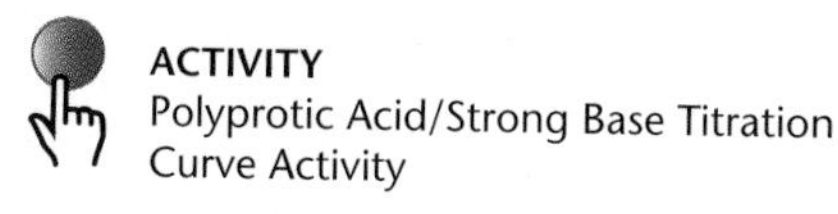

ACTIVITY
Polyprotic Acid/Strong Base Titration Curve Activity

17.4 Solubility Equilibria

The equilibria that we have considered thus far in this chapter have involved acids and bases. Furthermore, they have been homogeneous; that is, all the species have been in the same phase. Through the rest of this chapter we will consider the equilibria involved in the dissolution or precipitation of ionic compounds. These reactions are heterogeneous.

The dissolving and precipitating of compounds are phenomena that occur both within us and around us. Tooth enamel dissolves in acidic solutions, for example, causing tooth decay. The precipitation of certain salts in our kidneys produces kidney stones. The waters of the earth contain salts dissolved as water passes over and through the ground. Precipitation of $CaCO_3$ from groundwater is responsible for the formation of stalactites and stalagmites within limestone caves such as the one shown in the chapter-opening photo.

In our earlier discussion of precipitation reactions we considered some general rules for predicting the solubility of common salts in water. (Section 4.2) These rules give us a qualitative sense of whether a compound will have a low or high solubility in water. By considering solubility equilibria, in contrast, we can make quantitative predictions about the amount of a given compound that will dissolve. We can also use these equilibria to analyze the factors that affect solubility.

The Solubility-Product Constant, K_{sp}

ACTIVITY
K_{sp} Activity

Recall that a *saturated solution* is one in which the solution is in contact with undissolved solute. (Section 13.2) Consider, for example, a saturated aqueous solution of $BaSO_4$ that is in contact with solid $BaSO_4$. Because the solid is an ionic compound, it is a strong electrolyte and yields $Ba^{2+}(aq)$ and $SO_4^{2-}(aq)$ ions upon dissolving. The following equilibrium is readily established between the undissolved solid and hydrated ions in solution:

$$BaSO_4(s) \rightleftharpoons Ba^{2+}(aq) + SO_4^{2-}(aq) \qquad [17.15]$$

As with any equilibrium, the extent to which this dissolution reaction occurs is expressed by the magnitude of its equilibrium constant. Because this equilibrium equation describes the dissolution of a solid, the equilibrium constant indicates how soluble the solid is in water and is referred to as the **solubility-product constant** (or simply the **solubility product**). It is denoted K_{sp}, where *sp* stands for solubility product. The equilibrium-constant expression for this process is written according to the same rules as those that apply to any equilibrium-constant expression. That is, the concentration terms of the products are multiplied together, and each is raised to the power of its stoichiometric coefficient in the balanced chemical equation, and these are divided by the concentration terms of the reactants multiplied together, and each raised to the power of its stoichiometric coefficient. Solids, liquids, and solvents do not appear in the equilibrium-constant expressions for heterogeneous equilibria (Section 15.3), however, so *the solubility product equals the product of the concentration of the ions involved in the equilibrium, each raised to the power of its coefficient in the equilibrium equation.* Thus, the solubility-product expression for the equilibrium expressed in Equation 17.15 is

$$K_{sp} = [Ba^{2+}][SO_4^{2-}] \qquad [17.16]$$

Even though $[BaSO_4]$ is excluded from the equilibrium-constant expression, some undissolved $BaSO_4(s)$ must be present in order for the system to be at equilibrium.

In general, the solubility-product constant (K_{sp}) is the equilibrium constant for the equilibrium that exists between a solid ionic solute and its ions in a saturated aqueous solution. The values of K_{sp} at 25°C for many ionic solids are tabulated in

Appendix D. The value of K_{sp} for $BaSO_4$ is 1.1×10^{-10}, a very small number, indicating that only a very small amount of the solid will dissolve in water.

SAMPLE EXERCISE 17.9

Write the expression for the solubility-product constant for CaF_2, and look up the corresponding K_{sp} value in Appendix D.

Solution

Analyze and Plan: We are asked to write an equilibrium-constant expression for the process by which CaF_2 dissolves in water. We apply the same rules for writing any equilibrium-constant expression, making sure to exclude the solid reactant from the expression. We assume that the compound dissociates completely into its component ions.

$$CaF_2(s) \rightleftharpoons Ca^{2+}(aq) + 2F^-(aq)$$

Solve: Following the italicized rule stated previously, the expression for K_{sp} is

$$K_{sp} = [Ca^{2+}][F^-]^2$$

In Appendix D we see that this K_{sp} has a value of 3.9×10^{-11}.

PRACTICE EXERCISE

Give the solubility-product-constant expressions and the values of the solubility-product constants (from Appendix D) for the following compounds: **(a)** barium carbonate; **(b)** silver sulfate.

Answers: **(a)** $K_{sp} = [Ba^{2+}][CO_3^{2-}] = 5.0 \times 10^{-9}$; **(b)** $K_{sp} = [Ag^+]^2[SO_4^{2-}] = 1.5 \times 10^{-5}$

Solubility and K_{sp}

It is important to distinguish carefully between solubility and the solubility-product constant. The solubility of a substance is the quantity that dissolves to form a saturated solution. (Section 13.2) Solubility is often expressed as grams of solute per liter of solution (g/L). The molar solubility is the number of moles of the solute that dissolve in forming a liter of saturated solution of the solute (mol/L). The solubility-product constant (K_{sp}) is the equilibrium constant for the equilibrium between an ionic solid and its saturated solution.

The solubility of a substance can change considerably as the concentrations of other solutes change. The solubility of $Mg(OH)_2$, for example, depends highly on pH. The solubility is also affected by the concentrations of other ions in solution, especially Mg^{2+}. In contrast, the solubility-product constant, K_{sp}, has only one value for a given solute at any specific temperature.*

In principle it is possible to use the K_{sp} value of a salt to calculate solubility under a variety of conditions. In practice, great care must be taken in doing so for the reasons indicated in "A Closer Look" on the limitations of solubility products in Section 17.5. Agreement between measured solubility and that calculated from K_{sp} is usually best for salts whose ions have low charges (1+ and 1−) and do not hydrolyze. Figure 17.14 ▼ summarizes the relationships among various expressions of solubility and K_{sp}.

* This is strictly true only for very dilute solutions. The values of equilibrium constants are somewhat altered when the total concentration of ionic substances in water is increased. However, we will ignore these effects, which are taken into consideration only for work that requires exceptional accuracy.

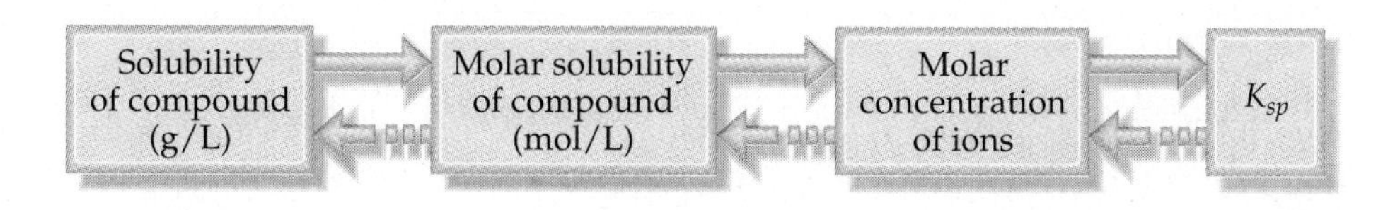

◀ **Figure 17.14** Outline of steps involved in interconverting solubility and K_{sp}.

SAMPLE EXERCISE 17.10

Solid silver chromate is added to pure water at 25°C. Some of the solid remains undissolved at the bottom of the flask. The mixture is stirred for several days to ensure that equilibrium is achieved between the undissolved $Ag_2CrO_4(s)$ and the solution. Analysis of the equilibrated solution shows that its silver ion concentration is $1.3 \times 10^{-4} M$. Assuming that Ag_2CrO_4 dissociates completely in water and that there are no other important equilibria involving the Ag^+ or CrO_4^{2-} ions in the solution, calculate K_{sp} for this compound.

Solution

Analyze: We are given the equilibrium concentration of silver ion in a saturated solution of silver chromate. From this we are asked to determine the value of the solubility-product constant for the dissolution of silver chromate. The equilibrium equation and the expression for K_{sp} are

$$Ag_2CrO_4(s) \rightleftharpoons 2Ag^+(aq) + CrO_4^{2-}(aq) \qquad K_{sp} = [Ag^+]^2[CrO_4^{2-}]$$

Plan: We know that at equilibrium $[Ag^+] = 1.3 \times 10^{-4} M$. All the Ag^+ and CrO_4^{2-} ions in the solution come from the Ag_2CrO_4 that dissolves. From the chemical formula of silver chromate, we know that there must be 2 Ag^+ ions in solution for each CrO_4^{2-} ion in solution. Consequently, the concentration of CrO_4^{2-} is half the concentration of Ag^+.

$$[CrO_4^{2-}] = \left(\frac{1.3 \times 10^{-4} \text{ mol } Ag^+}{\text{L}}\right)\left(\frac{1 \text{ mol } CrO_4^{2-}}{2 \text{ mol } Ag^+}\right) = 6.5 \times 10^{-5} M$$

Solve: We can now calculate the value of K_{sp}.

$$K_{sp} = [Ag^+]^2[CrO_4^{2-}] = (1.3 \times 10^{-4})^2(6.5 \times 10^{-5}) = 1.1 \times 10^{-12}$$

This value agrees well with the one given in Appendix D, 1.2×10^{-12}.

PRACTICE EXERCISE

A saturated solution of $Mg(OH)_2$ in contact with undissolved solid is prepared at 25°C. The pH of the solution is found to be 10.17. Assuming that $Mg(OH)_2$ dissociates completely in water and that there are no other simultaneous equilibria involving the Mg^{2+} or OH^- ions in the solution, calculate K_{sp} for this compound.

Answer: 1.6×10^{-12}

SAMPLE EXERCISE 17.11

The K_{sp} for CaF_2 is 3.9×10^{-11} at 25°C. Assuming that CaF_2 dissociates completely upon dissolving and that there are no other important equilibria affecting its solubility, calculate the solubility of CaF_2 in grams per liter.

Solution

Analyze: We are given K_{sp} for CaF_2 and are asked to determine solubility. Recall that the *solubility* of a substance is the quantity that can dissolve in solvent, whereas the *solubility-product constant*, K_{sp}, is an equilibrium constant.

Plan: We can approach this problem by using our standard techniques for solving equilibrium problems. Assume initially that none of the salt has dissolved, and then allow x moles/liter of CaF_2 to dissociate completely when equilibrium is achieved.

	$CaF_2(s)$ $\rightleftharpoons$	$Ca^{2+}(aq)$ +	$2F^-(aq)$
Initial	—	0	0
Change	—	$+x$ M	$+2x$ M
Equilibrium	—	x M	$2x$ M

The stoichiometry of the equilibrium dictates that $2x$ moles/liter of F^- are produced for each x moles/liter of CaF_2 that dissolve.

Solve: We now use the expression for K_{sp} and substitute the equilibrium concentrations to solve for the value of x.

$$K_{sp} = [Ca^{2+}][F^-]^2 = (x)(2x)^2 = 4x^3 = 3.9 \times 10^{-11}$$

$$x = \sqrt[3]{\frac{3.9 \times 10^{-11}}{4}} = 2.1 \times 10^{-4} M$$

(Remember that $\sqrt[3]{y} = y^{1/3}$; to calculate the cube root of a number, you can use the y^x function on your calculator, with $x = \frac{1}{3}$.) Thus, the molar solubility of CaF_2 is 2.1×10^{-4} mol/L. The mass of CaF_2 that dissolves in water to form a liter of solution is

$$\left(\frac{2.1 \times 10^{-4} \text{ mol } CaF_2}{1 \text{ L soln}}\right)\left(\frac{78.1 \text{ g } CaF_2}{1 \text{ mol } CaF_2}\right) = 1.6 \times 10^{-2} \text{ g } CaF_2/\text{L soln}$$

Comment: Because F^- is the anion of a weak acid, you might expect that the hydrolysis of the ion would affect the solubility of CaF_2. The basicity of F^- is so small ($K_b = 1.5 \times 10^{-11}$), however, that the hydrolysis occurs to only a slight extent and does not significantly influence the solubility. The reported solubility is 0.017 g/L at 25°C, in good agreement with our calculation.

PRACTICE EXERCISE

The K_{sp} for LaF_3 is 2×10^{-19}. What is the solubility of LaF_3 in water in moles per liter?
Answer: 9.28×10^{-6} mol/L

17.5 Factors That Affect Solubility

The solubility of a substance is affected not only by temperature but also by the presence of other solutes. The presence of an acid, for example, can have a major influence on the solubility of a substance. In Section 17.4 we considered the dissolving of ionic compounds in pure water. In this section we examine three factors that affect the solubility of ionic compounds: the presence of common ions, the pH of the solution, and the presence of complexing agents. We will also examine the phenomenon of amphoterism, which is related to the effects of both pH and complexing agents.

Common-Ion Effect

The presence of either $Ca^{2+}(aq)$ or $F^-(aq)$ in a solution reduces the solubility of CaF_2, shifting the solubility equilibrium of CaF_2 to the left.

$$CaF_2(s) \rightleftharpoons Ca^{2+}(aq) + 2F^-(aq)$$

Addition of Ca^{2+} or F^- shifts equilibrium, reducing solubility

This reduction in solubility is another application of the common-ion effect. (Section 17.1) In general, *the solubility of a slightly soluble salt is decreased by the presence of a second solute that furnishes a common ion.* Figure 17.15 ▶ shows how the solubility of CaF_2 decreases as NaF is added to the solution. Sample Exercise 17.12 shows how the K_{sp} can be used to calculate the solubility of a slightly soluble salt in the presence of a common ion.

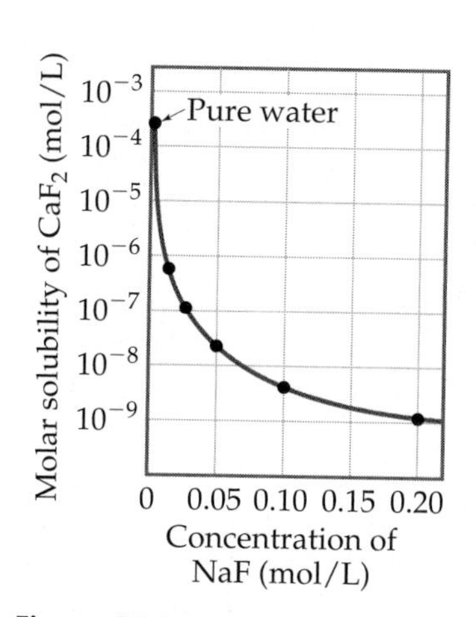

▲ **Figure 17.15** The effect of the concentration of NaF on the solubility of CaF_2 demonstrates the common-ion effect. Notice that the solubility of CaF_2 is on a logarithmic scale.

A Closer Look Limitations of Solubility Products

The concentrations of ions calculated from K_{sp} sometimes deviate appreciably from those found experimentally. In part, these deviations are due to electrostatic interactions between ions in solution, which can lead to ion pairs. (See "A Closer Look" on colligative properties of electrolyte solutions in Section 13.5.) These interactions increase in magnitude both as the concentrations of the ions increase and as their charges increase. The solubility calculated from K_{sp} tends to be low unless it is corrected to account for these interactions between ions. Chemists have developed procedures for correcting for these "ionic-strength" or "ionic-activity" effects, and these procedures are examined in more advanced chemistry courses. As an example of the effect of these interionic interactions, consider $CaCO_3$ (calcite) whose solubility product, $K_{sp} = 4.5 \times 10^{-9}$, gives a calculated solubility of 6.7×10^{-5} mol/L. Making corrections for the interionic interactions in the solution yields a higher solubility, 7.3×10^{-5} mol/L. The reported solubility, however, is twice as high (1.4×10^{-4} mol/L), so there must be one or more additional factors involved.

Another common source of error in calculating ion concentrations from K_{sp} is ignoring other equilibria that occur simultaneously in the solution. It is also possible, for example, that acidbase or complex-ion equilibria take place simultaneously with solubility equilibria. In particular, both basic anions and cations with high charge-to-size ratios undergo hydrolysis reactions that can measurably increase the solubilities of their salts. For example, $CaCO_3$ contains the basic carbonate ion ($K_b = 1.8 \times 10^{-4}$), which hydrolyzes in water: $CO_3^{2-}(aq) + H_2O(l) \rightleftharpoons HCO_3^-(aq) + OH^-(aq)$. If we consider both the effect of the interionic interactions in the solution and the effect of the simultaneous solubility and hydrolysis equilibria, we calculate a solubility of 1.4×10^{-4} mol/L, in agreement with the measured value.

Finally, we generally assume that ionic compounds dissociate completely into their component ions when they dissolve. This assumption is not always valid. When MgF_2 dissolves, for example, it yields not only Mg^{2+} and F^- ions but also MgF^+ ions in solution. Thus, we see that calculating solubility using K_{sp} can be more complicated than it first appears and it requires considerable knowledge of the equilibria occurring in solution.

SAMPLE EXERCISE 17.12

Calculate the molar solubility of CaF_2 at 25°C in a solution that is **(a)** 0.010 M in $Ca(NO_3)_2$; **(b)** 0.010 M in NaF.

Solution

Analyze and Plan: We are asked to determine the solubility of CaF_2 in the presence of two different strong electrolytes, each of which contains an ion common to CaF_2. As in Sample Exercise 17.11, the solubility product at 25°C is

$$K_{sp} = [Ca^{2+}][F^-]^2 = 3.9 \times 10^{-11}$$

The value of K_{sp} is unchanged by the presence of additional solutes. Because of the common-ion effect, however, the solubility of the salt will decrease in the presence of common ions.

(a) We can again use our standard equilibrium techniques. In this instance, however, the initial concentration of Ca^{2+} is 0.010 M because of the dissolved $Ca(NO_3)_2$:

	$CaF_2(s) \rightleftharpoons$	$Ca^{2+}(aq)$ +	$2F^-(aq)$
Initial	—	0.010 M	0
Change	—	$+x$ M	$+2x$ M
Equilibrium	—	$(0.010 + x)$ M	$2x$ M

Solve: Substituting into the solubility-product expression gives

$$K_{sp} = 3.9 \times 10^{-11} = [Ca^{2+}][F^-]^2 = (0.010 + x)(2x)^2$$

This would be a messy problem to solve exactly, but fortunately it is possible to simplify matters greatly. Even without the common-ion effect, the solubility of CaF_2 is very small. Assume that the 0.010 M concentration of Ca^{2+} from $Ca(NO_3)_2$ is very much greater than the small additional concentration resulting from the solubility of CaF_2; that is, x is small compared to 0.010 M, and $0.010 + x \simeq 0.010$. We then have

$$3.9 \times 10^{-11} = (0.010)(2x)^2$$

$$x^2 = \frac{3.9 \times 10^{-11}}{4(0.010)} = 9.8 \times 10^{-10}$$

$$x = \sqrt{9.8 \times 10^{-10}} = 3.1 \times 10^{-5} M$$

The very small value for x validates the simplifying assumption we have made. Our calculation indicates that 3.1×10^{-5} mol of solid CaF_2 dissolves per liter of the 0.010 M $Ca(NO_3)_2$ solution.

(b) In this case the common ion is F^-, and at equilibrium we have

$$[Ca^{2+}] = x \quad \text{and} \quad [F^-] = 0.010 + 2x$$

Assuming that $2x$ is small compared to 0.010 M (that is, $0.010 + 2x \approx 0.010$), we have

$$3.9 \times 10^{-11} = x(0.010)^2$$

$$x = \frac{3.9 \times 10^{-11}}{(0.010)^2} = 3.9 \times 10^{-7}\,M$$

Thus, 3.9×10^{-7} mol of solid CaF_2 should dissolve per liter of 0.010 M NaF solution.

Comment: If you compare the results in parts (a) and (b), you will see that although both Ca^{2+} and F^- reduce the solubility of CaF_2, their effects are not the same. The effect of F^- is more pronounced than that of Ca^{2+} because $[F^-]$ appears to the second power in the K_{sp} expression for CaF_2, whereas $[Ca^{2+}]$ appears to the first power.

PRACTICE EXERCISE

The value for K_{sp} for manganese(II) hydroxide, $Mn(OH)_2$, is 1.6×10^{-13}. Calculate the molar solubility of $Mn(OH)_2$ in a solution that contains 0.020 M NaOH.

Answer: $4.0 \times 10^{-10}\,M$

Solubility and pH

The solubility of any substance whose anion is basic will be affected to some extent by the pH of the solution. Consider $Mg(OH)_2$, for example, for which the solubility equilibrium is

$$Mg(OH)_2(s) \rightleftharpoons Mg^{2+}(aq) + 2OH^-(aq) \qquad K_{sp} = 1.8 \times 10^{-11} \qquad [17.17]$$

A saturated solution of $Mg(OH)_2$ has a calculated pH of 10.52 and contains $[Mg^{2+}] = 1.7 \times 10^{-4}M$. Now suppose that solid $Mg(OH)_2$ is equilibrated with a solution buffered at a more acidic pH of 9.0. The pOH, therefore, is 5.0, so $[OH^-] = 1.0 \times 10^{-5}$. Inserting this value for $[OH^-]$ into the solubility-product expression, we have

$$K_{sp} = [Mg^{2+}][OH^-]^2 = 1.8 \times 10^{-11}$$

$$[Mg^{2+}](1.0 \times 10^{-5})^2 = 1.8 \times 10^{-11}$$

$$[Mg^{2+}] = \frac{1.8 \times 10^{-11}}{(1.0 \times 10^{-5})^2} = 0.18\,M$$

Thus, $Mg(OH)_2$ dissolves in the solution until $[Mg^{2+}] = 0.18\,M$. It is apparent that $Mg(OH)_2$ is quite soluble in this solution. If the concentration of OH^- were reduced even further by making the solution more acidic, the Mg^{2+} concentration would have to increase to maintain the equilibrium condition. Thus, a sample of $Mg(OH)_2$ will dissolve completely if sufficient acid is added (Figure 17.16 ▶).

The solubility of almost any ionic compound is affected if the solution is made sufficiently acidic or basic. The effects are very noticeable, however, only when one or both ions involved are at least moderately acidic or basic. The metal hydroxides, such as $Mg(OH)_2$, are examples of compounds containing a strongly basic ion, the hydroxide ion.

As we have seen, the solubility of $Mg(OH)_2$ greatly increases as the acidity of the solution increases. The solubility of CaF_2 increases as the solution becomes more acidic, too, because the F^- ion is a weak base; it is the conjugate base of the weak acid HF. As a result, the solubility equilibrium of CaF_2 is shifted to the right as the concentration of F^- ions is reduced by protonation to form HF. Thus, the solution process can be understood in terms of two consecutive reactions.

$$CaF_2(s) \rightleftharpoons Ca^{2+}(aq) + 2F^-(aq) \qquad [17.18]$$

$$F^-(aq) + H^+(aq) \rightleftharpoons HF(aq) \qquad [17.19]$$

MOVIE
Dissolution of $Mg(OH)_2$ by Acid, Precipitation Reactions

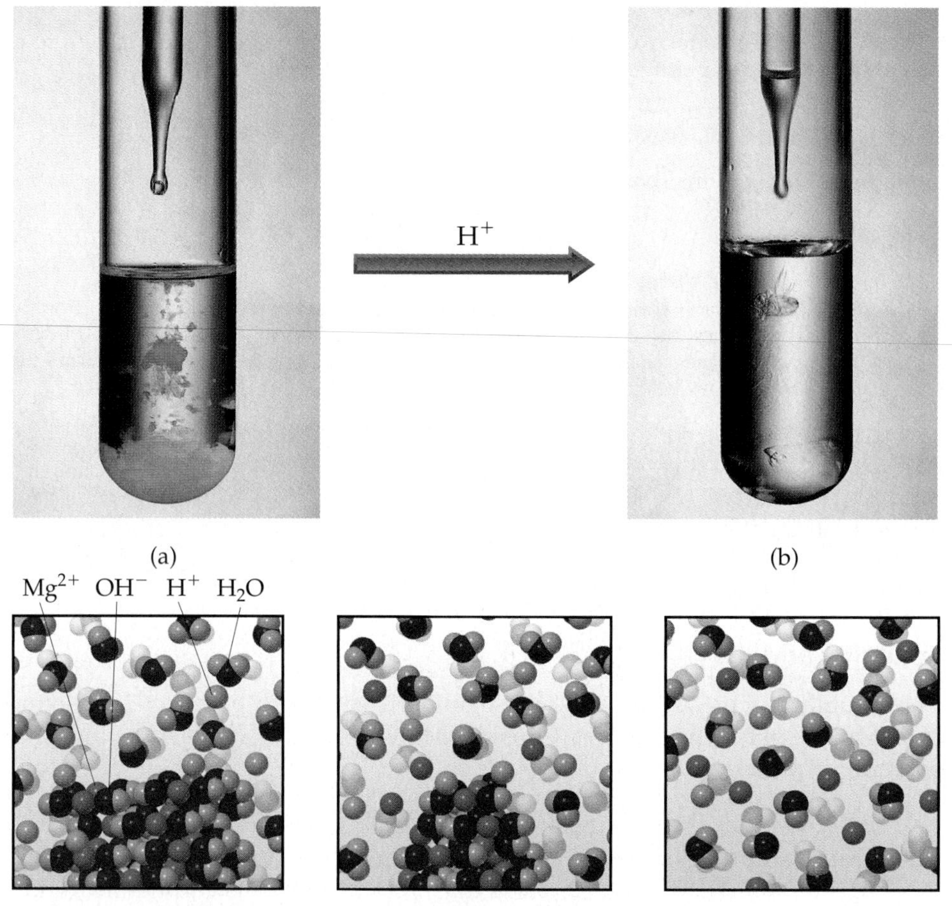

▲ **Figure 17.16** (a) A precipitate of $Mg(OH)_2(s)$. (b) The precipitate dissolves upon addition of acid. The molecular art shows the dissolving of the $Mg(OH)_2$ by H^+ ions. (The anions accompanying the acid have been omitted to simplify the art.)

Chemistry and Life Sinkholes

A principal cause of sinkholes is the dissolution of limestone, which is calcium carbonate, by groundwater. Although $CaCO_3$ has a relatively small solubility-product constant, it is quite soluble in the presence of acid.

$$CaCO_3(s) \rightleftharpoons Ca^{2+}(aq) + CO_3^{2-}(aq) \qquad K_{sp} = 4.5 \times 10^{-9}$$

Rainwater is naturally acidic, with a pH range of 5 to 6, and can become more acidic when it comes into contact with decaying plant matter. Because carbonate ion is the conjugate base of the weak acid, hydrogen carbonate ion (HCO_3^-), it readily combines with hydrogen ion.

$$CO_3^{2-}(aq) + H^+(aq) \longrightarrow HCO_3^-(aq)$$

The consumption of carbonate ion shifts the dissolution equilibrium to the right, thus increasing the solubility of $CaCO_3$. This can have profound consequences in areas where the terrain consists of porous calcium carbonate bedrock covered by a relatively thin layer of clay and/or topsoil. As acidic water percolates through and gradually dissolves the limestone, it creates underground voids. A sinkhole results when the overlying ground can no longer be supported by the remaining bedrock and collapses into the underground cavity [Figure 17.17(a) ▶]. Sinkholes are one of a variety of geologic features known as *karst* landforms. Other karst landforms, also caused by the dissolution of bedrock by groundwater, include caves and underground streams. The sudden formation of large sinkholes can pose a serious threat to life and property [Figure 17.17(b)]. The existence of deep sinkholes also increases the risk of contamination of the aquifer.

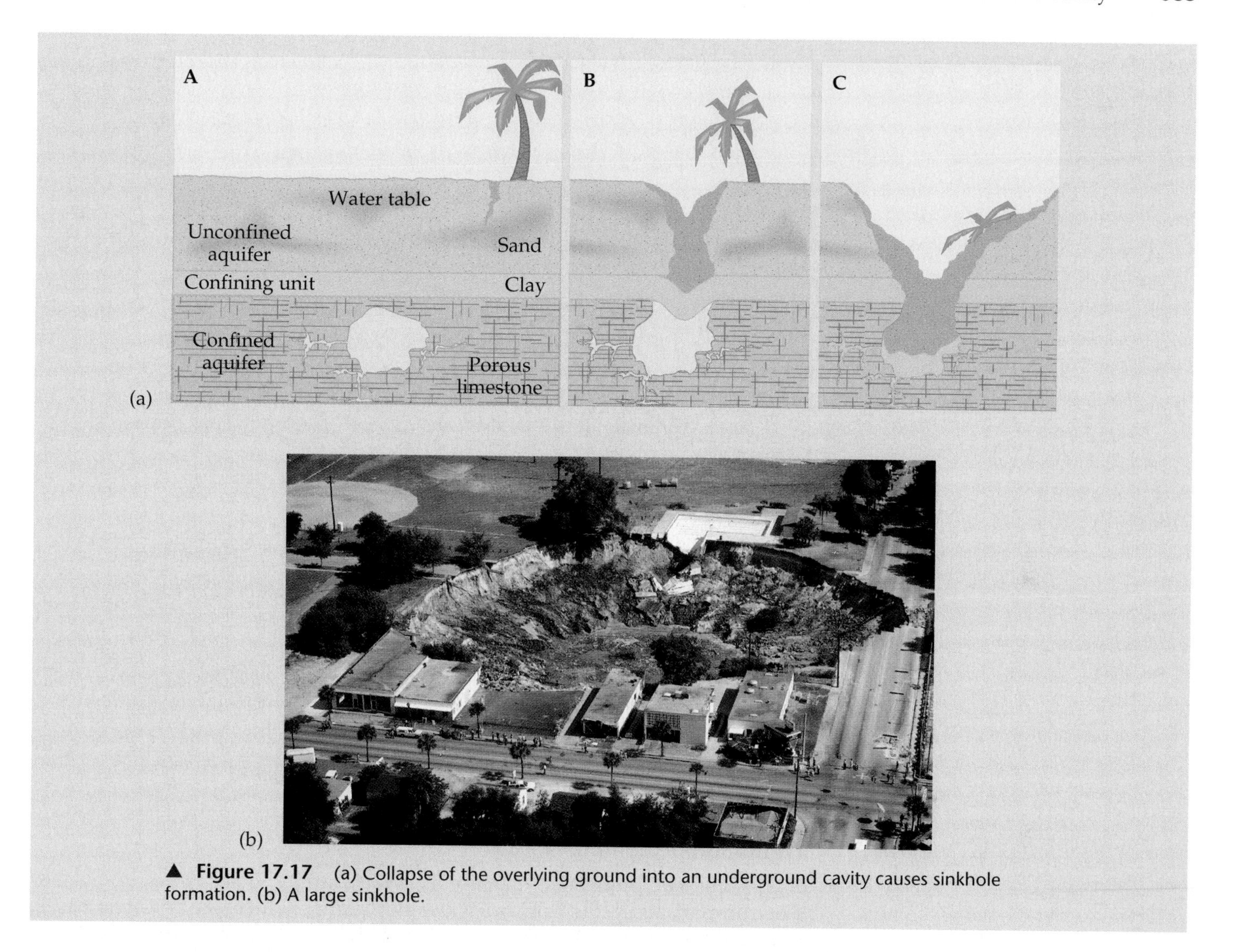

▲ **Figure 17.17** (a) Collapse of the overlying ground into an underground cavity causes sinkhole formation. (b) A large sinkhole.

The equation for the overall process is

$$CaF_2(s) + 2H^+(aq) \rightleftharpoons Ca^{2+}(aq) + 2HF(aq) \quad [17.20]$$

Figure 17.18 ▶ shows how the solubility of CaF_2 changes with pH.

Other salts that contain basic anions, such as CO_3^{2-}, PO_4^{3-}, CN^-, or S^{2-}, behave similarly. These examples illustrate a general rule: *The solubility of slightly soluble salts containing basic anions increases as* $[H^+]$ *increases (as pH is lowered).* The more basic the anion, the more the solubility is influenced by pH. Salts with anions of negligible basicity (the anions of strong acids) are unaffected by pH changes.

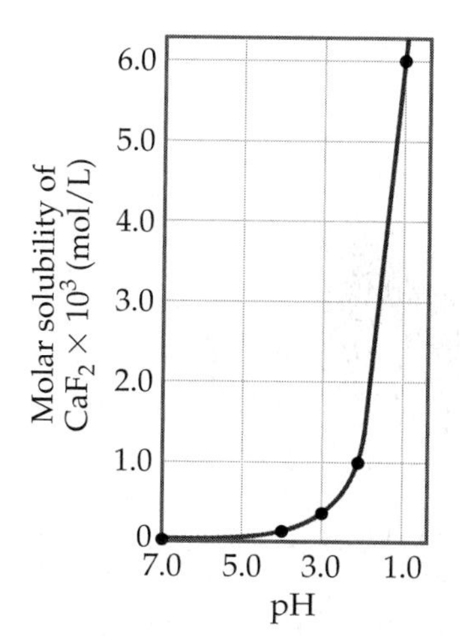

▲ **Figure 17.18** The effect of pH on the solubility of CaF_2. The pH scale is given with acidity increasing to the right. Notice that the vertical scale has been multiplied by 10^3.

SAMPLE EXERCISE 17.13

Which of the following substances will be more soluble in acidic solution than in basic solution: **(a)** $Ni(OH)_2(s)$; **(b)** $CaCO_3(s)$; **(c)** $BaF_2(s)$; **(d)** $AgCl(s)$?

Solution

Analyze and Plan: The problem lists four sparingly soluble salts, and we are asked to determine which will be more soluble at low pH than at high pH. In order to do this, we must determine which of the salts dissociates to produce a basic anion that will be consumed to an appreciable extent by H^+ ion.

Solve: (a) $Ni(OH)_2(s)$ will be more soluble in acidic solution because of the basicity of OH^-; the H^+ ion reacts with the OH^- ion, forming water.

$$Ni(OH)_2(s) \rightleftharpoons Ni^{2+}(aq) + 2OH^-(aq)$$
$$2OH^-(aq) + 2H^+(aq) \rightleftharpoons 2H_2O(l)$$
$$\text{Overall: } Ni(OH)_2(s) + 2H^+(aq) \rightleftharpoons Ni^{2+}(aq) + 2H_2O(l)$$

(b) Similarly, $CaCO_3(s)$ dissolves in acid solutions because CO_3^{2-} is a basic anion.

$$CaCO_3(s) \rightleftharpoons Ca^{2+}(aq) + CO_3^{2-}(aq)$$
$$CO_3^{2-}(aq) + 2H^+(aq) \rightleftharpoons H_2CO_3(aq)$$
$$H_2CO_3(aq) \longrightarrow CO_2(g) + H_2O(l)$$
$$\text{Overall: } CaCO_3(s) + 2H^+(aq) \longrightarrow Ca^{2+}(aq) + CO_2(g) + H_2O(l)$$

(c) The solubility of BaF_2 is also enhanced by lowering the pH, because F^- is a basic anion.

$$BaF_2(s) \rightleftharpoons Ba^{2+}(aq) + 2F^-(aq)$$
$$2F^-(aq) + 2H^+(aq) \rightleftharpoons 2HF(aq)$$
$$\text{Overall: } BaF_2(s) + 2H^+(aq) \longrightarrow Ba^{2+}(aq) + 2HF(aq)$$

(d) The solubility of AgCl is unaffected by changes in pH because Cl^- is the anion of a strong acid and therefore has negligible basicity.

PRACTICE EXERCISE

Write the net ionic equation for the reaction of the following copper(II) compounds with acid: **(a)** CuS; **(b)** $Cu(N_3)_2$.

Answers: **(a)** $CuS(s) + H^+(aq) \rightleftharpoons Cu^{2+}(aq) + HS^-(aq)$;
(b) $Cu(N_3)_2(s) + 2H^+(aq) \rightleftharpoons Cu^{2+}(aq) + 2HN_3(aq)$

Formation of Complex Ions

A characteristic property of metal ions is their ability to act as Lewis acids, or electron-pair acceptors, toward water molecules, which act as Lewis bases, or electron-pair donors. (Section 16.11) Lewis bases other than water can also interact with metal ions, particularly with transition-metal ions. Such interactions can dramatically affect the solubility of a metal salt. AgCl, for example, which has $K_{sp} = 1.8 \times 10^{-10}$, will dissolve in the presence of aqueous ammonia because Ag^+ interacts with the Lewis base NH_3, as shown in Figure 17.19 ▶. This process can be viewed as the sum of two reactions, the dissolution of AgCl and the Lewis acid–base interaction between Ag^+ and NH_3.

$$AgCl(s) \rightleftharpoons Ag^+(aq) + Cl^-(aq) \quad [17.21]$$
$$Ag^+(aq) + 2NH_3(aq) \rightleftharpoons Ag(NH_3)_2^+(aq) \quad [17.22]$$
$$\text{Overall: } AgCl(s) + 2NH_3(aq) \rightleftharpoons Ag(NH_3)_2^+(aq) + Cl^-(aq) \quad [17.23]$$

The presence of NH_3 drives the top reaction, the dissolution of AgCl, to the right as $Ag^+(aq)$ is consumed to form $Ag(NH_3)_2^+$.

For a Lewis base such as NH_3 to increase the solubility of a metal salt, it must be able to interact more strongly with the metal ion than water does. The NH_3 must displace solvating H_2O molecules (Sections 13.1 and 16.11) in order to form $Ag(NH_3)_2^+$:

$$Ag^+(aq) + 2NH_3(aq) \rightleftharpoons Ag(NH_3)_2^+(aq) \quad [17.24]$$

An assembly of a metal ion and the Lewis bases bonded to it, such as $Ag(NH_3)_2^+$, is called a **complex ion**. The stability of a complex ion in aqueous

◀ **Figure 17.19** A saturated solution of AgCl in contact with solid AgCl. When concentrated ammonia is added, Ag^+ ions are consumed in the formation of the complex ion $Ag(NH_3)_2^+$. Removal of Ag^+ ions from the solution pulls the dissolution equilibrium to the right, as shown in Equation 17.24, causing AgCl to dissolve. Addition of sufficient ammonia results in complete dissolution of the AgCl solid. The molecular art shows the AgCl solid being dissolved by addition of NH_3.

solution can be judged by the size of the equilibrium constant for its formation from the hydrated metal ion. For example, the equilibrium constant for formation of $Ag(NH_3)_2^+$ (Equation 17.24) is 1.7×10^7:

$$K_f = \frac{[Ag(NH_3)_2^+]}{[Ag^+][NH_3]^2} = 1.7 \times 10^7 \qquad [17.25]$$

The equilibrium constant for this kind of reaction is called a **formation constant**, K_f. The formation constants for several complex ions are listed in Table 17.1 ▼.

TABLE 17.1 Formation Constants for Some Metal Complex Ions in Water at 25°C

Complex Ion	K_f	Equilibrium Equation
$Ag(NH_3)_2^+$	1.7×10^7	$Ag^+(aq) + 2NH_3(aq) \rightleftharpoons Ag(NH_3)_2^+(aq)$
$Ag(CN)_2^-$	1×10^{21}	$Ag^+(aq) + 2CN^-(aq) \rightleftharpoons Ag(CN)_2^-(aq)$
$Ag(S_2O_3)_2^{3-}$	2.9×10^{13}	$Ag^+(aq) + 2S_2O_3^{2-}(aq) \rightleftharpoons Ag(S_2O_3)_2^{3-}(aq)$
$CdBr_4^{2-}$	5×10^3	$Cd^{2+}(aq) + 4Br^-(aq) \rightleftharpoons CdBr_4^{2-}(aq)$
$Cr(OH)_4^-$	8×10^{29}	$Cr^{3+}(aq) + 4OH^- \rightleftharpoons Cr(OH)_4^-(aq)$
$Co(SCN)_4^{2-}$	1×10^3	$Co^{2+}(aq) + 4SCN^-(aq) \rightleftharpoons Co(SCN)_4^{2-}(aq)$
$Cu(NH_3)_4^{2+}$	5×10^{12}	$Cu^{2+}(aq) + 4NH_3(aq) \rightleftharpoons Cu(NH_3)_4^{2+}(aq)$
$Cu(CN)_4^{2-}$	1×10^{25}	$Cu^{2+}(aq) + 4CN^-(aq) \rightleftharpoons Cu(CN)_4^{2-}(aq)$
$Ni(NH_3)_6^{2+}$	1.2×10^9	$Ni^{2+}(aq) + 6NH_3(aq) \rightleftharpoons Ni(NH_3)_6^{2+}(aq)$
$Fe(CN)_6^{4-}$	1×10^{35}	$Fe^{2+}(aq) + 6CN^-(aq) \rightleftharpoons Fe(CN)_6^{4-}(aq)$
$Fe(CN)_6^{3-}$	1×10^{42}	$Fe^{3+}(aq) + 6CN^-(aq) \rightleftharpoons Fe(CN)_6^{3-}(aq)$

SAMPLE EXERCISE 17.14

Calculate the concentration of Ag^+ present in solution at equilibrium when concentrated ammonia is added to a 0.010 M solution of $AgNO_3$ to give an equilibrium concentration of $[NH_3] = 0.20\ M$. Neglect the small volume change that occurs when NH_3 is added.

Solution

Analyze and Plan: We are asked to determine what concentration of aqueous silver ion will remain uncombined when the ammonia concentration is brought to 0.20 M in a solution originally 0.010 M in $AgNO_3$. Because K_f for the formation of $Ag(NH_3)_2^+$ is quite large, we will begin with the assumption that essentially all the Ag^+ is converted to $Ag(NH_3)_2^+$, in accordance with Equation 17.24, and approach the problem as though we are concerned with the *dissociation* of $Ag(NH_3)_2^+$ rather than its *formation*. In order to facilitate this approach, we will need to reverse the equation to represent the formation of Ag^+ and NH_3 from $Ag(NH_3)_2^+$ and also make the corresponding change to the equilibrium constant.

$$Ag(NH_3)_2^+(aq) \rightleftharpoons Ag^+(aq) + 2NH_3(aq)$$

$$\frac{1}{K_f} = \frac{1}{1.7 \times 10^7} = 5.9 \times 10^{-8}$$

If $[Ag^+]$ is 0.010 M initially, then $[Ag(NH_3)_2^+]$ will be 0.010 M following addition of the NH_3. We now construct a table to solve this equilibrium problem. Note that the NH_3 concentration given in the problem is an *equilibrium* concentration rather than an initial concentration.

Solve:

	$Ag(NH_3)_2^+(aq)$ $\rightleftharpoons$	$Ag^+(aq)$ +	$2NH_3(aq)$
Initial	0.010 M	0 M	
Change	$-x\ M$	$+x\ M$	
Equilibrium	$0.010 - x\ M$	$x\ M$	0.20 M

Because the concentration of Ag^+ is very small, we can ignore x in comparison with 0.010. Thus, $0.010 - x \approx 0.010\ M$. Substituting these values into the equilibrium-constant expression for the dissociation of $Ag(NH_3)_2^+$, we obtain

$$\frac{[Ag^+][NH_3]^2}{[Ag(NH_3)_2^+]} = \frac{(x)(0.20)^2}{0.010} = 5.9 \times 10^{-8}$$

Solving for x, we obtain $x = 1.5 \times 10^{-8} M = [Ag^+]$. Thus, formation of the $Ag(NH_3)_2^+$ complex drastically reduces the concentration of free Ag^+ ion in solution.

PRACTICE EXERCISE

Calculate $[Cr^{3+}]$ in equilibrium with $Cr(OH)_4^-$ when 0.010 mol of $Cr(NO_3)_3$ is dissolved in a liter of solution buffered at pH 10.0.
Answer: $1 \times 10^{-16} M$

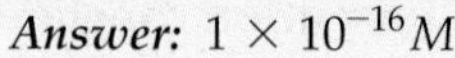

Chemistry and Life Tooth Decay and Fluoridation

Tooth enamel consists mainly of a mineral called hydroxyapatite, $Ca_{10}(PO_4)_6(OH)_2$. It is the hardest substance in the body. Tooth cavities are caused when acids dissolve tooth enamel.

$$Ca_{10}(PO_4)_6(OH)_2(s) + 8H^+(aq) \longrightarrow 10Ca^{2+}(aq) + 6HPO_4^{2-}(aq) + 2H_2O(l)$$

The resultant Ca^{2+} and HPO_4^{2-} ions diffuse out of the tooth enamel and are washed away by saliva. The acids that attack the hydroxyapatite are formed by the action of specific bacteria on sugars and other carbohydrates present in the plaque adhering to the teeth.

Fluoride ion, present in drinking water, toothpaste, and other sources, can react with hydroxyapatite to form fluoroapatite, $Ca_{10}(PO_4)_6F_2$. This mineral, in which F^- has replaced OH^-, is much more resistant to attack by acids because the fluoride ion is a much weaker Brønsted–Lowry base than the hydroxide ion.

Because the fluoride ion is so effective in preventing cavities, it is added to the public water supply in many places to give a concentration of 1 mg/L (1 ppm). The compound added may be NaF or Na_2SiF_6. Na_2SiF_6 reacts with water to release fluoride ions by the following reaction:

$$SiF_6^{2-}(aq) + 2H_2O(l) \longrightarrow 6F^-(aq) + 4H^+(aq) + SiO_2(s)$$

About 80% of all toothpastes now sold in the United States contain fluoride compounds, usually at the level of 0.1% fluoride by mass. The most common compounds in toothpastes are sodium fluoride (NaF), sodium monofluorophosphate (Na_2PO_3F), and stannous fluoride (SnF_2).

The general rule is that the solubility of metal salts increases in the presence of suitable Lewis bases, such as NH_3, CN^-, or OH^-, if the metal forms a complex with the base. The ability of metal ions to form complexes is an extremely important aspect of their chemistry. In Chapter 24 we will take a much closer look at complex ions. In that chapter and others we will see applications of complex ions to areas such as biochemistry, metallurgy, and photography.

Amphoterism

Some metal hydroxides and oxides that are relatively insoluble in neutral water dissolve in strongly acidic and strongly basic solutions. These substances are soluble in strong acids and bases because they themselves are capable of behaving as either an acid or base; they are *amphoteric*. (Section 16.2) Amphoteric substances include the hydroxides and oxides of Al^{3+}, Cr^{3+}, Zn^{2+}, and Sn^{2+}.

These species dissolve in acidic solutions because they contain basic anions. What makes amphoteric oxides and hydroxides special, though, is that they also dissolve in strongly basic solutions (Figure 17.20 ▼). This behavior results from the formation of complex anions containing several (typically four) hydroxides bound to the metal ion.

$$Al(OH)_3(s) + OH^-(aq) \rightleftharpoons Al(OH)_4^-(aq) \qquad [17.26]$$

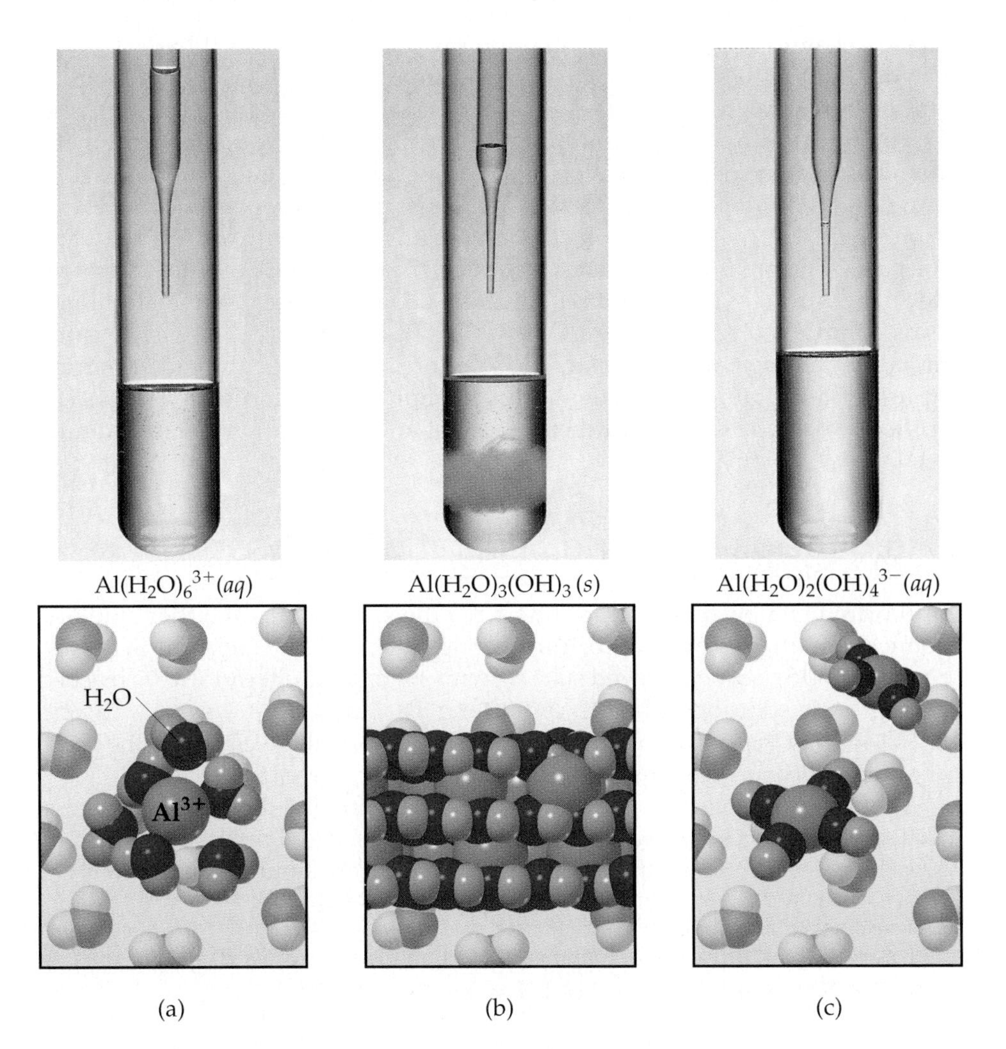

◀ **Figure 17.20** As NaOH is added to a solution of Al^{3+} (a), a precipitate of $Al(OH)_3$ forms (b). As more NaOH is added, the $Al(OH)_3$ dissolves (c), demonstrating the amphoterism of the $Al(OH)_3$.

Amphoterism is often explained by the behavior of the water molecules that surround the metal ion and that are bonded to it by Lewis acid–base interactions. (Section. 16.11) For example, $Al^{3+}(aq)$ is more accurately represented as $Al(H_2O)_6{}^{3+}(aq)$ because six water molecules are bonded to the Al^{3+} in aqueous solution. Recall from Section 16.11 that this hydrated ion is a weak acid. As a strong base is added, $Al(H_2O)_6{}^{3+}$ loses protons in a stepwise fashion, eventually forming neutral and water-insoluble $Al(H_2O)_3(OH)_3$. This substance then dissolves upon removal of an additional proton to form the anion $Al(H_2O)_2(OH)_4{}^-$. The reactions that occur are as follows:

$$Al(H_2O)_6{}^{3+}(aq) + OH^-(aq) \rightleftharpoons Al(H_2O)_5(OH)^{2+}(aq) + H_2O(l)$$

$$Al(H_2O)_5(OH)^{2+}(aq) + OH^-(aq) \rightleftharpoons Al(H_2O)_4(OH)_2{}^+(aq) + H_2O(l)$$

$$Al(H_2O)_4(OH)_2{}^+(aq) + OH^-(aq) \rightleftharpoons Al(H_2O)_3(OH)_3(s) + H_2O(l)$$

$$Al(H_2O)_3(OH)_3(s) + OH^-(aq) \rightleftharpoons Al(H_2O)_2(OH)_4{}^-(aq) + H_2O(l)$$

Removing additional protons is possible, but each successive reaction occurs less readily than the one before. As the charge on the ion becomes more negative, it becomes increasingly difficult to remove a positively charged proton. Addition of an acid reverses these reactions. The proton adds in a stepwise fashion to convert the OH^- groups to H_2O, eventually re-forming $Al(H_2O)_6{}^{3+}$. The common practice is to simplify the equations for these reactions by excluding the bound H_2O molecules. Thus, we usually write Al^{3+} instead of $Al(H_2O)_6{}^{3+}$, $Al(OH)_3$ instead of $Al(H_2O)_3(OH)_3$, $Al(OH)_4{}^-$ instead of$Al(H_2O)_2(OH)_4{}^-$, and so forth.

The extent to which an insoluble metal hydroxide reacts with either acid or base varies with the particular metal ion involved. Many metal hydroxides—such as $Ca(OH)_2$, $Fe(OH)_2$, and $Fe(OH)_3$—are capable of dissolving in acidic solution but do not react with excess base. These hydroxides are not amphoteric.

The purification of aluminum ore in the manufacture of aluminum metal provides an interesting application of the property of amphoterism. As we have seen, $Al(OH)_3$ is amphoteric, whereas $Fe(OH)_3$ is not. Aluminum occurs in large quantities as the ore *bauxite*, which is essentially Al_2O_3 with additional water molecules. The ore is contaminated with Fe_2O_3 as an impurity. When bauxite is added to a strongly basic solution, the Al_2O_3 dissolves because the aluminum forms complex ions, such as $Al(OH)_4{}^-$. The Fe_2O_3 impurity, however, is not amphoteric and remains as a solid. The solution is filtered, getting rid of the iron impurity. Aluminum hydroxide is then precipitated by addition of acid. The purified hydroxide receives further treatment and eventually yields aluminum metal. (Section 23.3)

17.6 Precipitation and Separation of Ions

Equilibrium can be achieved starting with the substances on either side of a chemical equation. The equilibrium among $BaSO_4(s)$, $Ba^{2+}(aq)$, and $SO_4{}^{2-}(aq)$ (Equation 17.15) can be achieved starting with solid $BaSO_4$. It can also be reached starting with solutions of salts containing Ba^{2+} and $SO_4{}^{2-}$, say $BaCl_2$ and Na_2SO_4. When these two solutions are mixed, $BaSO_4$ will precipitate if the product of the ion concentrations, $Q = [Ba^{2+}][SO_4{}^{2-}]$, is greater than K_{sp}.

The use of the reaction quotient, Q, to determine the direction in which a reaction must proceed to reach equilibrium was discussed earlier. (Section 15.5) The possible relationships between Q and K_{sp} are summarized as follows:

If $Q > K_{sp}$, precipitation occurs until $Q = K_{sp}$.
If $Q = K_{sp}$, equilibrium exists (saturated solution).
If $Q < K_{sp}$, solid dissolves until $Q = K_{sp}$.

SAMPLE EXERCISE 17.15

Will a precipitate form when 0.10 L of $8.0 \times 10^{-3} M$ $Pb(NO_3)_2$ is added to 0.40 L of $5.0 \times 10^{-3} M$ Na_2SO_4?

Solution

Analyze: The problem asks us to determine whether or not a precipitate will form when two salt solutions are combined.
Plan: We should determine the concentrations of all ions immediately upon mixing of the solutions and compare the value of the reaction quotient, Q, to the solubility-product constant for any potentially insoluble product. The possible metathesis products are $PbSO_4$ and $NaNO_3$. Sodium salts are quite soluble; $PbSO_4$ has a K_{sp} of 6.3×10^{-7} (Appendix D), however, and will precipitate if the Pb^{2+} and SO_4^{2-} ion concentrations are high enough for Q to exceed K_{sp} for the salt.

Solve: When the two solutions are mixed, the total volume becomes 0.10 L + 0.40 L = 0.50 L. The number of moles of Pb^{2+} in 0.10 L of $3.0 \times 10^{-3} M$ $Pb(NO_3)_2$ is

$$(0.10\ \text{L})\left(8.0 \times 10^{-3}\ \frac{\text{mol}}{\text{L}}\right) = 8.0 \times 10^{-4}\ \text{mol}$$

The concentration of Pb^{2+} in the 0.50-L mixture is therefore

$$[Pb^{2+}] = \frac{8.0 \times 10^{-4}\ \text{mol}}{0.50\ \text{L}} = 1.6 \times 10^{-3} M$$

The number of moles of SO_4^{2-} is

$$(0.40\ \text{L})\left(5.0 \times 10^{-3}\ \frac{\text{mol}}{\text{L}}\right) = 2.0 \times 10^{-3}\ \text{mol}$$

Therefore, $[SO_4^{2-}]$ in the 0.50-L mixture is

$$[SO_4^{2-}] = \frac{2.0 \times 10^{-3}\ \text{mol}}{0.50\ \text{L}} = 4.0 \times 10^{-3} M$$

We then have

$$Q = [Pb^{2+}][SO_4^{2-}] = (1.6 \times 10^{-3})(4.0 \times 10^{-3}) = 6.4 \times 10^{-6}$$

Because $Q > K_{sp}$, $PbSO_4$ will precipitate.

PRACTICE EXERCISE

Will a precipitate form when 0.050 L of $2.0 \times 10^{-2} M$ NaF is mixed with 0.010 L of $1.0 \times 10^{-2} M$ $Ca(NO_3)_2$?
Answer: Yes, because $Q = 4.6 \times 10^{-8}$ is larger than $K_{sp} = 3.9 \times 10^{-11}$.

Selective Precipitation of Ions

Ions can be separated from each other based on the solubilities of their salts. Consider a solution containing both Ag^+ and Cu^{2+}. If HCl is added to the solution, AgCl ($K_{sp} = 1.8 \times 10^{-10}$) precipitates, while Cu^{2+} remains in solution because $CuCl_2$ is soluble. Separation of ions in an aqueous solution by using a reagent that forms a precipitate with one or a few of the ions is called *selective precipitation.*

SAMPLE EXERCISE 17.16

A solution contains $1.0 \times 10^{-2} M$ Ag^+ and $2.0 \times 10^{-2} M$ Pb^{2+}. When Cl^- is added to the solution, both AgCl ($K_{sp} = 1.8 \times 10^{-10}$) and $PbCl_2$ ($K_{sp} = 1.7 \times 10^{-5}$)precipitate from the solution. What concentration of Cl^- is necessary to begin the precipitation of each salt? Which salt precipitates first?

Solution

Analyze: We are asked to determine the concentration of Cl^- necessary to begin the precipitation from a solution containing Ag^+ and Pb^{2+} ions, and to predict which metal chloride will begin to precipitate first.
Plan: We are given K_{sp} values for the two possible precipitates. Using these and the metal ion concentrations, we can calculate what concentration of Cl^- ion would be necessary to begin precipitation of each. The salt requiring the lower Cl^- ion concentration will precipitate first.
Solve: For AgCl we have

$$K_{sp} = [Ag^+][Cl^-] = 1.8 \times 10^{-10}$$

Because $[Ag^+] = 1.0 \times 10^{-2}\,M$, the greatest concentration of Cl^- that can be present without causing precipitation of AgCl can be calculated from the K_{sp} expression.

$$K_{sp} = [1.0 \times 10^{-2}][Cl^-] = 1.8 \times 10^{-10}$$

$$[Cl^-] = \frac{1.8 \times 10^{-10}}{1.0 \times 10^{-2}} = 1.8 \times 10^{-8}\,M$$

Any Cl^- in excess of this very small concentration will cause AgCl to precipitate from solution. Proceeding similarly for $PbCl_2$, we have

$$K_{sp} = [Pb^{2+}][Cl^-]^2 = 1.7 \times 10^{-5}$$

$$[2.0 \times 10^{-2}][Cl^-]^2 = 1.7 \times 10^{-5}$$

$$[Cl^-]^2 = \frac{1.7 \times 10^{-5}}{2.0 \times 10^{-2}} = 8.5 \times 10^{-4}$$

$$[Cl^-] = \sqrt{8.5 \times 10^{-4}} = 2.9 \times 10^{-2}\,M$$

Thus, a concentration of Cl^- in excess of $2.9 \times 10^{-2}\,M$ will cause $PbCl_2$ to precipitate.

Comparing the concentrations of Cl^- required to precipitate each salt, we see that as Cl^- is added to the solution, AgCl will precipitate first because it requires a much smaller concentration of Cl^-. Thus, Ag^+ can be separated from Pb^{2+} by slowly adding Cl^- so $[Cl^-]$ is between $1.8 \times 10^{-8}\,M$ and $2.9 \times 10^{-2}\,M$.

PRACTICE EXERCISE

A solution consists of 0.050 *M* Mg^{2+} and 0.020 *M* Cu^{2+}. Which ion will precipitate first as OH^- is added to the solution? What concentration of OH^- is necessary to begin the precipitation of each cation? [$K_{sp} = 1.8 \times 10^{-11}$ for $Mg(OH)_2$ and $K_{sp} = 2.2 \times 10^{-20}$ for $Cu(OH)_2$.]

Answer: $Cu(OH)_2$ precipitates first. $Cu(OH)_2$ begins to precipitate when $[OH^-]$ exceeds $1.0 \times 10^{-9}\,M$; $Mg(OH)_2$ begins to precipitate when $[OH^-]$ exceeds $1.9 \times 10^{-5}\,M$.

Sulfide ion is often used to separate metal ions because the solubilities of sulfide salts span a wide range and depend greatly on the pH of the solution. Cu^{2+} and Zn^{2+}, for example, can be separated by bubbling H_2S gas through a properly acidified solution. Because CuS ($K_{sp} = 6 \times 10^{-37}$) is less soluble than ZnS ($K_{sp} = 2 \times 10^{-25}$), CuS precipitates from an acidified solution (pH = 1) while ZnS does not (Figure 17.21 ▶):

$$Cu^{2+}(aq) + H_2S(aq) \rightleftharpoons CuS(s) + 2H^+(aq) \qquad [17.27]$$

The CuS can be separated from the Zn^{2+} solution by filtration. The CuS can then be dissolved by using a high concentration of H^+, shifting the equilibrium shown in Equation 17.27 to the left.

17.7 Qualitative Analysis for Metallic Elements

In this chapter we have seen several examples of equilibria involving metal ions in aqueous solution. In this final section we look briefly at how solubility equilibria and complex-ion formation can be used to detect the presence of particular metal ions in solution. Before the development of modern analytical instrumentation, it was necessary to analyze mixtures of metals in a sample by so-called wet chemical methods. For example, a metallic sample that might contain several metallic elements was dissolved in a concentrated acid solution. This solution was then tested in a systematic way for the presence of various metal ions.

Qualitative analysis determines only the presence or absence of a particular metal ion, whereas **quantitative analysis** determines how much of a given substance is present. Wet methods of qualitative analysis have become less

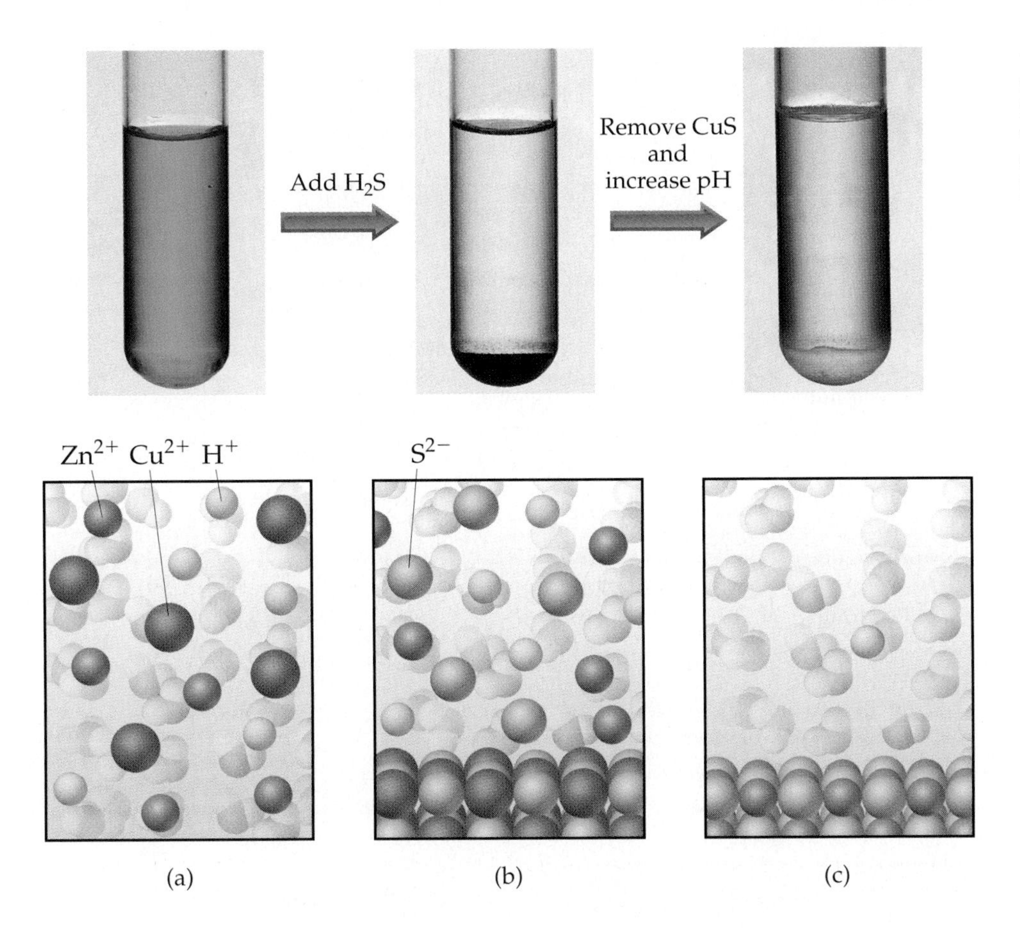

◀ **Figure 17.21** (a) Solution containing $Zn^{2+}(aq)$ and $Cu^{2+}(aq)$. (b) When H_2S is added to a solution whose pH exceeds 0.6, CuS precipitates. (c) After CuS is removed, the pH is increased, allowing ZnS to precipitate.

important as a means of analysis. They are frequently used in general chemistry laboratory programs, however, to illustrate equilibria, to teach the properties of common metal ions in solution, and to develop laboratory skills. Typically, such analyses proceed in three stages. (1) The ions are separated into broad groups on the basis of solubility properties. (2) The individual ions within each group are then separated by selectively dissolving members in the group. (3) The ions are then identified by means of specific tests.

A scheme in general use divides the common cations into five groups, as shown in Figure 17.22 ▶. The order of addition of reagents is important. The most selective separations—those that involve the smallest number of ions—are carried out first. The reactions that are used must proceed so far toward completion that any concentration of cations remaining in the solution is too small to interfere with subsequent tests. Let's take a closer look at each of these five groups of cations, briefly examining the logic used in this qualitative analysis scheme.

1. *Insoluble chlorides:* Of the common metal ions, only Ag^+, Hg_2^{2+}, and Pb^{2+} form insoluble chlorides. When dilute HCl is added to a mixture of cations, therefore, only AgCl, Hg_2Cl_2, and $PbCl_2$ will precipitate, leaving the other cations in solution. The absence of a precipitate indicates that the starting solution contains no Ag^+, Hg_2^{2+}, and Pb^{2+}.
2. *Acid-insoluble sulfides:* After any insoluble chlorides have been removed, the remaining solution, now acidic, is treated with H_2S. Only the most insoluble metal sulfides—CuS, Bi_2S_3, CdS, PbS, HgS, As_2S_3, Sb_2S_3, and SnS_2—can precipitate. (Note the very small values of K_{sp} for some of these sulfides in Appendix D.) Those metal ions whose sulfides are somewhat more soluble—for example, ZnS or NiS—remain in solution.

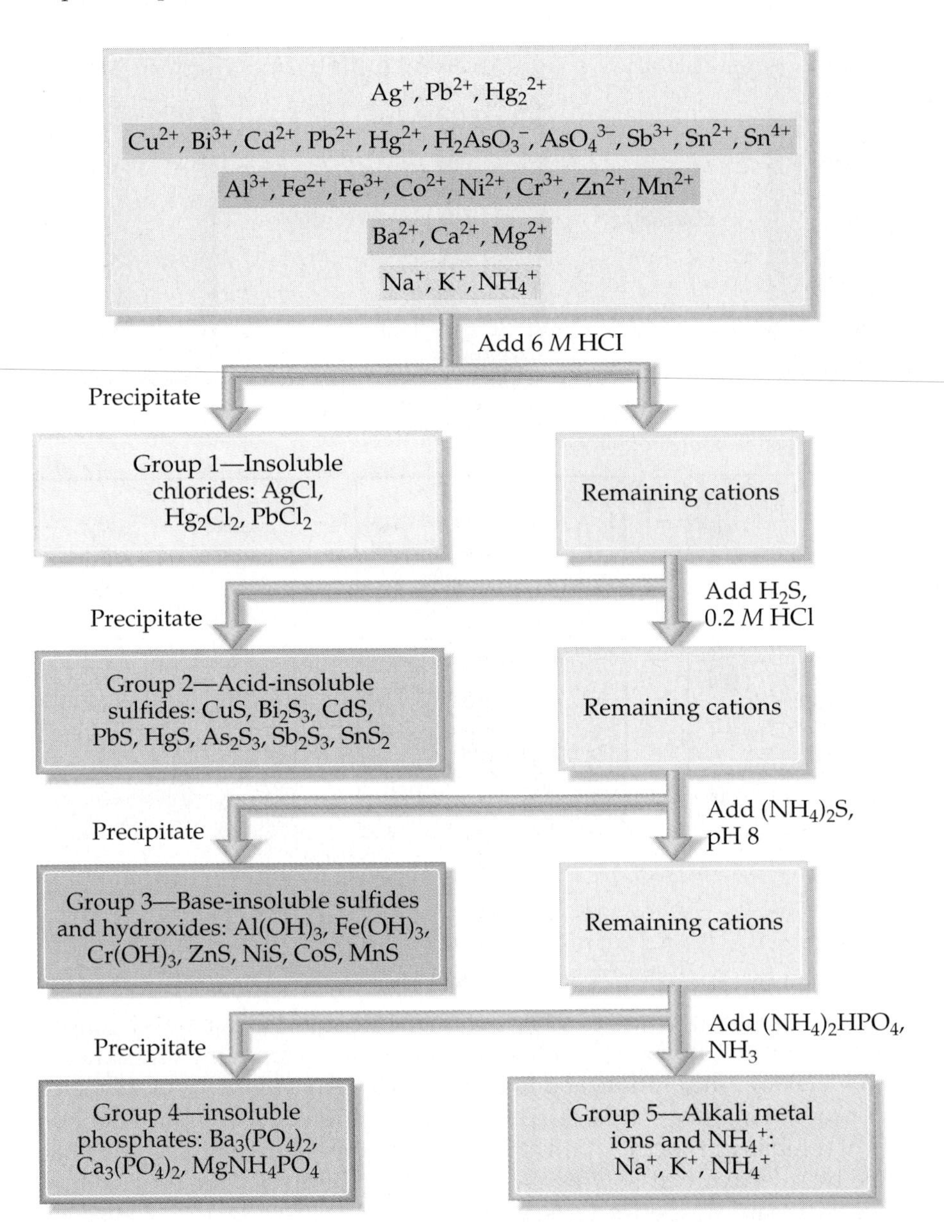

▶ **Figure 17.22** Qualitative analysis scheme for separating cations into groups.

3. *Base-insoluble sulfides and hydroxides:* After the solution is filtered to remove any acid-insoluble sulfides, the remaining solution is made slightly basic, and $(NH_4)_2S$ is added. In basic solutions the concentration of S^{2-} is higher than in acidic solutions. Thus, the ion products for many of the more soluble sulfides are made to exceed their K_{sp} values and precipitation occurs. The metal ions precipitated at this stage are Al^{3+}, Cr^{3+}, Fe^{3+}, Zn^{2+}, Ni^{2+}, Co^{2+}, and Mn^{2+}. (Actually, the Al^{3+}, Fe^{3+}, and Cr^{3+} ions do not form insoluble sulfides; instead they are precipitated as insoluble hydroxides at the same time.)
4. *Insoluble phosphates:* At this point the solution contains only metal ions from periodic table groups 1A and 2A. Adding $(NH_4)_2HPO_4$ to a basic solution precipitates the group 2A elements Mg^{2+}, Ca^{2+}, Sr^{2+}, and Ba^{2+} because these metals form insoluble phosphates.
5. *The alkali metal ions and NH_4^+:* The ions that remain after removing the insoluble phosphates form a small group. We can test for each ion individually. A flame test can be used to determine the presence of K^+, for example, because the flame turns a characteristic violet color if K^+ is present.

Additional separation and testing is necessary to determine which ions are present within each of the groups. Consider, for example, the ions of the insoluble chloride group. The precipitate containing the metal chlorides is boiled in water. The $PbCl_2$ is relatively soluble in hot water, whereas AgCl and Hg_2Cl_2 are not. The hot solution is filtered, and a solution of Na_2CrO_4 is added to the filtrate. If Pb^{2+} is present, a yellow precipitate of $PbCrO_4$ forms. The test for Ag^+ consists of treating the metal chloride precipitate with dilute ammonia. Only Ag^+ forms an ammonia complex. If AgCl is present in the precipitate, it will dissolve in the ammonia solution.

$$AgCl(s) + 2NH_3(aq) \rightleftharpoons Ag(NH_3)_2^+(aq) + Cl^-(aq) \qquad [17.28]$$

After treatment with ammonia, the solution is filtered and the filtrate made acidic by adding nitric acid. The nitric acid removes ammonia from solution by forming NH_4^+, thus releasing Ag^+, which re-forms the AgCl precipitate.

$$Ag(NH_3)_2^+(aq) + Cl^-(aq) + 2H^+(aq) \rightleftharpoons AgCl(s) + 2NH_4^+(aq) \qquad [17.29]$$

The analyses for individual ions in the acid-insoluble and base-insoluble sulfides are a bit more complex, but the same general principles are involved. The detailed procedures for carrying out such analyses are given in many laboratory manuals.

SAMPLE INTEGRATIVE EXERCISE 17: Putting Concepts Together

A sample of 1.25 L of HCl gas at 21°C and 0.950 atm is bubbled through 0.500 L of 0.150 M NH_3 solution. Assuming that all of the HCl dissolves and that the volume of the solution remains 0.500 L, calculate the pH of the resulting solution.

Solution The number of moles of HCl gas is calculated from the ideal-gas law.

$$n = \frac{PV}{RT} = \frac{(0.950 \text{ atm})(1.25 \text{ L})}{(0.0821 \text{ L-atm/mol-K})(294 \text{ K})} = 0.0492 \text{ mol HCl}$$

The number of moles of NH_3 in the solution is given by the product of the volume of the solution and its concentration.

$$\text{mol } NH_3 = (0.500 \text{ L})(0.150 \text{ mol } NH_3/\text{L}) = 0.0750 \text{ mol } NH_3$$

The acid HCl and base NH_3 react, transferring a proton from HCl to NH_3, producing NH_4^+ and Cl^- ions.

$$HCl(g) + NH_3(aq) \longrightarrow NH_4^+(aq) + Cl^-(aq)$$

To determine the pH of the solution, we first calculate the amount of each reactant and each product present at the completion of the reaction.

	$HCl(g)$	+ $NH_3(aq)$	$\longrightarrow$ $NH_4^+(aq)$	+ $Cl^-(aq)$
Before reaction:	0.0492 mol	0.0750 mol	0 mol	0 mol
After reaction:	0 mol	0.0258 mol	0.0492 mol	0.0492 mol

Thus, the reaction produces a solution containing a mixture of NH_3, NH_4^+, and Cl^-. The NH_3 is a weak base ($K_b = 1.8 \times 10^{-5}$), NH_4^+ is its conjugate acid, and Cl^- is neither acidic nor basic. Consequently, the pH depends on $[NH_3]$ and $[NH_4^+]$.

$$[NH_3] = \frac{0.0258 \text{ mol } NH_3}{0.500 \text{ L soln}} = 0.0516\ M$$

$$[NH_4^+] = \frac{0.0492 \text{ mol } NH_4^+}{0.500 \text{ L soln}} = 0.0984\ M$$

We can calculate the pH using either K_b for NH_3 or K_a for NH_4^+. Using the K_b expression, we have

	$NH_3(aq)$	+ $H_2O(l)$ $\rightleftharpoons$	$NH_4^+(aq)$	+ $OH^-(aq)$
Initial	0.0516 M	—	0.0984 M	0
Change	$-x\ M$	—	$+x\ M$	$+x\ M$
Equilibrium	$(0.0516 - x)\ M$	—	$(0.0984 + x)\ M$	$x\ M$

$$K_b = \frac{[NH_4^+][OH^-]}{[NH_3]} = \frac{(0.0984 + x)(x)}{(0.0516 - x)} \simeq \frac{(0.0984)x}{0.0516} = 1.8 \times 10^{-5}$$

$$x = [OH^-] = \frac{(0.0516)(1.8 \times 10^{-5})}{0.0984} = 9.4 \times 10^{-6}\,M$$

Hence, $pOH = -\log(9.4 \times 10^{-6}) = 5.03$ and $pH = 14.00 - pOH = 14.00 - 5.03 = 8.97$

Summary and Key Terms

Section 17.1 In this chapter we've considered several types of important equilibria that occur in aqueous solution. Our primary emphasis has been on acid-base equilibria in solutions containing two or more solutes and on solubility equilibria. The dissociation of a weak acid or weak base is repressed by the presence of a strong electrolyte that provides an ion common to the equilibrium. This phenomenon is called the **common-ion effect**.

Section 17.2 A particularly important type of acid-base mixture is that of a weak conjugate acid-base pair. Such mixtures function as **buffered solutions (buffers)**. Addition of small amounts of a strong acid or a strong base to a buffered solution causes only small changes in pH because the buffer reacts with the added acid or base. (Strong acid–strong base, strong acid–weak base, and weak acid–strong base reactions proceed essentially to completion.) Buffered solutions are usually prepared from a weak acid and a salt of that acid or from a weak base and a salt of that base. Two important characteristics of a buffered solution are its **buffer capacity** and its pH. The pH can be calculated using K_a or K_b. The relationship between pH, pK_a, and the concentrations of an acid and its conjugate base can be expressed by the **Henderson–Hasselbalch equation**: $pH = pK_a + \log\frac{[base]}{[acid]}$.

Section 17.3 The plot of the pH of an acid (or base) as a function of the volume of added base (or acid) is called a **pH titration curve**. Titration curves aid in selecting a proper pH indicator for an acid-base titration. The titration curve of a strong acid–strong base titration exhibits a large change in pH in the immediate vicinity of the equivalence point; at the equivalence point for this titration, $pH = 7$. For strong acid–weak base or weak acid–strong base titrations, the pH change in the vicinity of the equivalence point is not as large. Furthermore, the pH at the equivalence point is not 7 in either of these cases. Rather, it is the pH of the salt solution that results from the neutralization reaction. It is possible to calculate the pH at any point of the titration curve by first considering the stoichiometry of the reaction between the acid and base and then examining equilibria involving remaining solute species.

Section 17.4 The equilibrium between a solid compound and its ions in solution provides an example of heterogeneous equilibrium. The **solubility-product constant** (or simply the **solubility product**), K_{sp}, is an equilibrium constant that expresses quantitatively the extent to which the compound dissolves. The K_{sp} can be used to calculate the solubility of an ionic compound, and the solubility can be used to calculate K_{sp}.

Section 17.5 Several experimental factors, including temperature, affect the solubilities of ionic compounds in water. The solubility of a slightly soluble ionic compound is decreased by the presence of a second solute that furnishes a common ion (the common-ion effect). The solubility of compounds containing basic anions increases as the solution is made more acidic (as pH decreases). Salts with anions of negligible basicity (the anions of strong acids) are unaffected by pH changes.

The solubility of metal salts is also affected by the presence of certain Lewis bases that react with metal ions to form stable **complex ions**. Complex-ion formation in aqueous solution involves the displacement by Lewis bases (such as NH_3 and CN^-) of water molecules attached to the metal ion. The extent to which such complex formation occurs is expressed quantitatively by the **formation constant** for the complex ion. Amphoteric metal hydroxides are those slightly soluble metal hydroxides that dissolve on addition of either acid or base. Acid-base reactions involving the OH^- or H_2O groups bound to the metal ions give rise to the amphoterism.

Section 17.6 Comparison of the ion product, Q, with the value of K_{sp} can be used to judge whether a precipitate will form when solutions are mixed or whether a slightly soluble salt will dissolve under various conditions. Precipitates form when $Q > K_{sp}$. Ions can be separated from each other based on the solubilities of their salts.

Section 17.7 Metallic elements vary a great deal in the solubilities of their salts, in their acid-base behavior, and in their tendencies to form complex ions. These differences can be used to separate and detect the presence of metal ions in mixtures. **Qualitative analysis** determines the presence or absence of species in a sample, whereas **quantitative analysis** determines how much of each species is present. The qualitative analysis of metal ions in solution can be carried out by separating the ions into groups on the basis of precipitation reactions and then analyzing each group for individual metal ions.

Exercises

Common-Ion Effect

17.1 **(a)** What is the common-ion effect? **(b)** Give an example of a salt that can decrease the ionization of HNO_2 in solution.

17.2 **(a)** Consider the equilibrium $B(aq) + H_2O(l) \rightleftharpoons HB^+(aq) + OH^-(aq)$. Using Le Châtelier's principle, explain the effect of the presence of a salt of HB^+ on the ionization of B. **(b)** Give an example of a salt that can decrease the ionization of NH_3 in solution.

17.3 Does the pH increase, decrease, or remain the same when each of the following is added? **(a)** $NaNO_2$ to a solution of HNO_2; **(b)** $(CH_3NH_3)Cl$ to a solution of CH_3NH_2; **(c)** sodium formate to a solution of formic acid; **(d)** potassium bromide to a solution of hydrobromic acid; **(e)** HCl to a solution of $NaC_2H_3O_2$.

17.4 Indicate whether the pH increases, decreases, or remains the same when each of the following is added: **(a)** $Ca(C_7H_5O_2)_2$ to a solution of $HC_7H_5O_2$; **(b)** pyridinium nitrate, $(C_5H_5NH)(NO_3)$, to a solution of pyridine, C_5H_5N; **(c)** ammonia to a solution of hydrochloric acid; **(d)** sodium hydrogen carbonate to a solution of carbonic acid; **(e)** $NaClO_4$ to a solution of NaOH.

17.5 Using information from Appendix D, calculate the pH and the propionate ion concentration, $[C_3H_5O_2^-]$, of a solution that is 0.060 *M* in potassium propionate ($KC_3H_5O_2$) and 0.085 *M* in propionic acid ($HC_3H_5O_2$).

17.6 Using information from Appendix D, calculate the pH and the trimethylammonium ion concentration of a solution that is 0.075 *M* in trimethylamine, $(CH_3)_3N$, and 0.10 *M* in trimethylammonium chloride, $(CH_3)_3NHCl$.

17.7 Calculate the pH of the following solutions: **(a)** 0.160 *M* in sodium formate ($NaCHO_2$) and 0.260 *M* in formic acid ($HCHO_2$); **(b)** 0.210 *M* in pyridine (C_5H_5N) and 0.350 *M* in pyridinium chloride (C_5H_5NHCl).

17.8 Calculate the pH of the following: **(a)** a solution made by combining 50.0 mL of 0.15 *M* acetic acid and 50.0 mL of 0.20 *M* sodium acetate; **(b)** a solution made by combining 125 mL of 0.050 *M* hydrofluoric acid with 50.0 mL of 0.10 *M* sodium fluoride.

17.9 **(a)** Calculate the percent ionization of 0.0075 *M* butanoic acid ($K_a = 1.5 \times 10^{-5}$). **(b)** Calculate the percent ionization of 0.0075 *M* butanoic acid in a solution containing 0.085 *M* sodium butanoate.

17.10 **(a)** Calculate the percent ionization of 0.085 *M* lactic acid ($K_a = 1.4 \times 10^{-4}$). **(b)** Calculate the percent ionization of 0.085 *M* lactic acid in a solution containing 0.050 *M* sodium lactate.

Buffers

17.11 Explain why a mixture of $HC_2H_3O_2$ and $NaC_2H_3O_2$ can act as a buffer while a mixture of HCl and NaCl cannot.

17.12 Explain why a mixture of HCl and NaF can act as a buffer but a mixture of HF and NaCl cannot.

17.13 **(a)** Calculate the pH of a buffer that is 0.12 *M* in lactic acid and 0.11 *M* in sodium lactate. **(b)** Calculate the pH of a buffer formed by mixing 85 mL of 0.13 *M* lactic acid with 95 mL of 0.15 *M* sodium lactate.

17.14 **(a)** Calculate the pH of a buffer that is 0.100 *M* in $NaHCO_3$ and 0.125 *M* in Na_2CO_3. **(b)** Calculate the pH of a solution formed by mixing 55 mL of 0.20 *M* $NaHCO_3$ with 65 mL of 0.15 *M* Na_2CO_3.

17.15 A buffer is prepared by adding 20.0 g of acetic acid ($HC_2H_3O_2$) and 20.0 g of sodium acetate ($NaC_2H_3O_2$) to enough water to form 2.00 L of solution. **(a)** Determine the pH of the buffer. **(b)** Write the complete ionic equation for the reaction that occurs when a few drops of hydrochloric acid are added to the buffer. **(c)** Write the complete ionic equation for the reaction that occurs when a few drops of sodium hydroxide solution are added to the buffer.

17.16 A buffer is prepared by adding 5.0 g of ammonia (NH_3) and 20.0 g of ammonium chloride (NH_4Cl) to enough water to form 2.50 L of solution. **(a)** What is the pH of this buffer? **(b)** Write the complete ionic equation for the reaction that occurs when a few drops of nitric acid are added to the buffer. **(c)** Write the complete ionic equation for the reaction that occurs when a few drops of potassium hydroxide solution are added to the buffer.

17.17 How many moles of sodium hypobromite (NaBrO) should be added to 1.00 L of 0.050 *M* hypobromous acid (HBrO) to form a buffer solution of pH 9.15? Assume that no volume change occurs when the NaBrO is added.

17.18 How many grams of sodium lactate ($NaC_3H_5O_3$) should be added to 1.00 L of 0.150 *M* lactic acid ($HC_3H_5O_3$) to form a buffer solution with pH 2.90? Assume that no volume change occurs when the $NaC_3H_5O_3$ is added.

17.19 A buffer solution contains 0.10 mol of acetic acid and 0.13 mol of sodium acetate in 1.00 L. **(a)** What is the pH of this buffer? **(b)** What is the pH of the buffer after the addition of 0.02 mol of KOH? **(c)** What is the pH of the buffer after the addition of 0.02 mol of HNO_3?

17.20 A buffer solution contains 0.12 mol of propionic acid ($HC_3H_5O_2$) and 0.10 mol of sodium propionate ($NaC_3H_5O_2$) in 1.50 L. **(a)** What is the pH of this buffer? **(b)** What is the pH of the buffer after the addition of 0.01 mol of NaOH? **(c)** What is the pH of the buffer after the addition of 0.01 mol of HI?

17.21 **(a)** What is the ratio of HCO_3^- to H_2CO_3 in blood of pH 7.4? **(b)** What is the ratio of HCO_3^- to H_2CO_3 in an exhausted marathon runner whose blood pH is 7.1?

17.22 A buffer, consisting of $H_2PO_4^-$ and HPO_4^{2-}, helps control the pH of physiological fluids. Many carbonated soft drinks also use this buffer system. What is the pH of a soft drink in which the major buffer ingredients are 6.5 g of NaH_2PO_4 and 8.0 g of Na_2HPO_4 per 355 mL of solution?

Acid-Base Titrations

17.23 The accompanying graph shows the titration curves for two monoprotic acids. **(a)** Which curve is that of a strong acid? **(b)** What is the approximate pH at the equivalence point of each titration? **(c)** How do the original concentrations of the two acids compare if 40.0 mL of each is titrated to the equivalence point with the same volume of 0.100 M base?

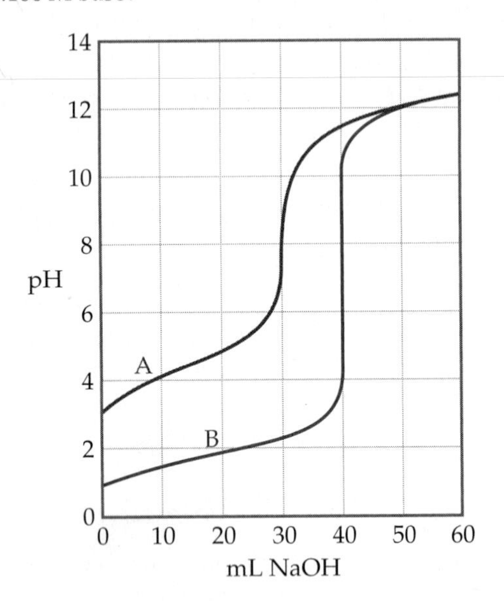

17.24 How does titration of a strong, monoprotic acid with a strong base differ from titration of a weak, monoprotic acid with a strong base with respect to the following: **(a)** quantity of base required to reach the equivalence point; **(b)** pH at the beginning of the titration; **(c)** pH at the equivalence point; **(d)** pH after addition of a slight excess of base; **(e)** choice of indicator for determining the equivalence point?

17.25 Two monoprotic acids, both 0.100 M in concentration, are titrated with 0.100 M NaOH. The pH at the equivalence point for HX is 8.8 and that for HY is 7.9. **(a)** Which is the weaker acid? **(b)** Which indicators in Figure 16.7 could be used to titrate each of these acids?

17.26 Assume that 30.0 mL of a 0.10 M solution of a weak base B that accepts one proton is titrated with a 0.10 M solution of the monoprotic strong acid HX. **(a)** How many moles of HX have been added at the equivalence point? **(b)** What is the predominant form of B at the equivalence point? **(c)** What factor determines the pH at the equivalence point? **(d)** Which indicator, phenolphthalein or methyl red, is likely to be the better choice for this titration?

17.27 How many milliliters of 0.0850 M NaOH are required to titrate each of the following solutions to the equivalence point: **(a)** 40.0 mL of 0.0900 M HNO_3; **(b)** 35.0 mL of 0.0720 M HBr; **(c)** 50.0 mL of a solution that contains 1.85 g of HCl per liter?

17.28 How many milliliters of 0.105 M HCl are needed to titrate each of the following solutions to the equivalence point: **(a)** 55.0 mL of 0.0950 M NaOH; **(b)** 23.5 mL of 0.117 M KOH; **(c)** 125.0 mL of a solution that contains 1.35 g of NaOH per liter?

17.29 A 20.0-mL sample of 0.200 M HBr solution is titrated with 0.200 M NaOH solution. Calculate the pH of the solution after the following volumes of base have been added: **(a)** 15.0 mL; **(b)** 19.9 mL; **(c)** 20.0 mL; **(d)** 20.1 mL; **(e)** 35.0 mL.

17.30 A 30.0-mL sample of 0.200 M KOH is titrated with 0.150 M $HClO_4$ solution. Calculate the pH after the following volumes of acid have been added: **(a)** 30.0 mL; **(b)** 39.5 mL; **(c)** 39.9 mL; **(d)** 40.0 mL; **(e)** 40.1 mL.

17.31 A 35.0-mL sample of 0.150 M acetic acid ($HC_2H_3O_2$) is titrated with 0.150 M NaOH solution. Calculate the pH after the following volumes of base have been added: **(a)** 0 mL; **(b)** 17.5 mL; **(c)** 34.5 mL; **(d)** 35.0 mL; **(e)** 35.5 mL; **(f)** 50.0 mL.

17.32 Consider the titration of 30.0 mL of 0.030 M NH_3 with 0.025 M HCl. Calculate the pH after the following volumes of titrant have been added: **(a)** 0 mL; **(b)** 10.0 mL; **(c)** 20.0 mL; **(d)** 35.0 mL; **(e)** 36.0 mL; **(f)** 37.0 mL.

17.33 Calculate the pH at the equivalence point for titrating 0.200 M solutions of each of the following bases with 0.200 M HBr: **(a)** sodium hydroxide (NaOH); **(b)** hydroxylamine (NH_2OH); **(c)** aniline ($C_6H_5NH_2$).

17.34 Calculate the pH at the equivalence point in titrating 0.100 M solutions of each of the following with 0.080 M NaOH: **(a)** hydrobromic acid (HBr); **(b)** lactic acid ($HC_3H_5O_3$); **(c)** sodium hydrogen chromate ($NaHCrO_4$).

Solubility Equilibria and Factors Affecting Solubility

17.35 **(a)** Why is the concentration of undissolved solid not explicitly included in the expression for the solubility-product constant? **(b)** Write the expression for the solubility-product constant for each of the following strong electrolytes: AgI, $SrSO_4$, $Fe(OH)_2$, and Hg_2Br_2.

17.36 **(a)** Explain the difference between solubility and solubility-product constant. **(b)** Write the expression for the solubility-product constant for each of the following ionic compounds: $MnCO_3$, $Hg(OH)_2$, and $Cu_3(PO_4)_2$.

17.37 **(a)** If the molar solubility of CaF_2 at 35°C is 1.24×10^{-3} mol/L, what is K_{sp} at this temperature? **(b)** It is found that 1.1×10^{-2} g of SrF_2 dissolves per 100 mL of aqueous solution at 25°C. Calculate the solubility product for SrF_2. **(c)** The K_{sp} of $Ba(IO_3)_2$ at 25°C is 6.0×10^{-10}. What is the molar solubility of $Ba(IO_3)_2$?

17.38 **(a)** The molar solubility of $PbBr_2$ at 25°C is 1.0×10^{-2} mol/L. Calculate K_{sp}. **(b)** If 0.0490 g of $AgIO_3$ dissolves per liter of solution, calculate the solubility-product constant. **(c)** Using the appropriate K_{sp} value from Appendix D, calculate the solubility of $Cu(OH)_2$ in grams per liter of solution.

17.39 A 1.00-L solution saturated at 25°C with calcium oxalate (CaC_2O_4) contains 0.0061 g of CaC_2O_4. Calculate the solubility-product constant for this salt at 25°C.

17.40 A 1.00-L solution saturated at 25°C with lead(II) iodide contains 0.54 g of PbI_2. Calculate the solubility-product constant for this salt at 25°C.

17.41 Using Appendix D, calculate the molar solubility of AgBr in **(a)** pure water; **(b)** 3.0×10^{-2} *M* $AgNO_3$ solution; **(c)** 0.10 *M* NaBr solution.

17.42 Calculate the solubility of LaF_3 in grams per liter in **(a)** pure water; **(b)** 0.025 *M* KF solution; **(c)** 0.150 *M* $LaCl_3$ solution.

17.43 Calculate the solubility of $Mn(OH)_2$ in grams per liter when buffered at pH **(a)** 7.0; **(b)** 9.5; **(c)** 11.8.

17.44 Calculate the molar solubility of $Fe(OH)_2$ when buffered at pH **(a)** 7.0; **(b)** 10.0; **(c)** 12.0.

17.45 Which of the following salts will be substantially more soluble in acidic solution than in pure water: **(a)** $ZnCO_3$; **(b)** ZnS; **(c)** BiI_3; **(d)** AgCN; **(e)** $Ba_3(PO_4)_2$?

17.46 For each of the following slightly soluble salts, write the net ionic equation, if any, for reaction with acid: **(a)** MnS; **(b)** PbF_2; **(c)** $AuCl_3$; **(d)** $Hg_2C_2O_4$; **(e)** CuBr.

17.47 From the value of K_f listed in Table 17.1, calculate the concentration of Cu^{2+} in 1.0 L of a solution that contains a total of 1×10^{-3} mol of copper(II) ion and that is 0.10 *M* in NH_3.

17.48 To what final concentration of NH_3 must a solution be adjusted to just dissolve 0.020 mol of NiC_2O_4 ($K_{sp} = 4 \times 10^{-10}$) in 1.0 L of solution? (*Hint*: You can neglect the hydrolysis of $C_2O_4^{2-}$ because the solution will be quite basic.)

17.49 By using the values of K_{sp} for AgI and K_f for $Ag(CN)_2^-$, calculate the equilibrium constant for the following reaction:

$$AgI(s) + 2CN^-(aq) \rightleftharpoons Ag(CN)_2^-(aq) + I^-(aq)$$

17.50 Using the value of K_{sp} for Ag_2S, K_{a1} and K_{a2} for H_2S, and $K_f = 1.1 \times 10^5$ for $AgCl_2^-$, calculate the equilibrium constant for the following reaction:

$$Ag_2S(s) + 4Cl^-(aq) + 2H^+(aq) \rightleftharpoons 2AgCl_2^-(aq) + H_2S(aq)$$

Precipitation; Qualitative Analysis

17.51 **(a)** Will $Ca(OH)_2$ precipitate from solution if the pH of a 0.050 *M* solution of $CaCl_2$ is adjusted to 8.0? **(b)** Will Ag_2SO_4 precipitate when 100 mL of 0.050 *M* $AgNO_3$ is mixed with 10 mL of 5.0×10^{-2} *M* Na_2SO_4 solution?

17.52 **(a)** Will $Co(OH)_2$ precipitate from solution if the pH of a 0.020 *M* solution of $Co(NO_3)_2$ is adjusted to 8.5? **(b)** Will $AgIO_3$ precipitate when 100 mL of 0.010 *M* $AgNO_3$ is mixed with 10 mL of 0.015 *M* $NaIO_3$? (K_{sp} of $AgIO_3$ is 3.1×10^{-8}.)

17.53 Calculate the minimum pH needed to precipitate $Mn(OH)_2$ so completely that the concentration of Mn^{2+} is less than 1 μg per liter [1 part per billion (ppb)].

17.54 Suppose that a 20-mL sample of a solution is to be tested for Cl^- ion by addition of 1 drop (0.2 mL) of 0.10 *M* $AgNO_3$. What is the minimum number of grams of Cl^- that must be present in order for AgCl(*s*) to form?

17.55 A solution contains 2.0×10^{-4} *M* Ag^+ and 1.5×10^{-3} *M* Pb^{2+}. If NaI is added, will AgI ($K_{sp} = 8.3 \times 10^{-17}$) or PbI_2 ($K_{sp} = 7.9 \times 10^{-9}$) precipitate first? Specify the concentration of I^- needed to begin precipitation.

17.56 A solution of Na_2SO_4 is added dropwise to a solution that is 0.0150 *M* in Ba^{2+} and 0.0150 *M* in Sr^{2+}. **(a)** What concentration of SO_4^{2-} is necessary to begin precipitation? (Neglect volume changes. $BaSO_4$: $K_{sp} = 1.1 \times 10^{-10}$; $SrSO_4$: $K_{sp} = 3.2 \times 10^{-7}$.) **(b)** Which cation precipitates first? **(c)** What is the concentration of SO_4^{2-} when the second cation begins to precipitate?

17.57 A solution containing an unknown number of metal ions is treated with dilute HCl; no precipitate forms. The pH is adjusted to about 1, and H_2S is bubbled through. Again, no precipitate forms. The pH of the solution is then adjusted to about 8. Again, H_2S is bubbled through. This time a precipitate forms. The filtrate from this solution is treated with $(NH_4)_2HPO_4$. No precipitate forms. Which metal ions discussed in Section 17.7 are possibly present? Which are definitely absent within the limits of these tests?

17.58 An unknown solid is entirely soluble in water. On addition of dilute HCl, a precipitate forms. After the precipitate is filtered off, the pH is adjusted to about 1 and H_2S is bubbled in; a precipitate again forms. After filtering off this precipitate, the pH is adjusted to 8 and H_2S is again added; no precipitate forms. No precipitate forms upon addition of $(NH_4)_2HPO_4$. The remaining solution shows a yellow color in a flame test. Based on these observations, which of the following compounds might be present, which are definitely present, and which are definitely absent: CdS, $Pb(NO_3)_2$, HgO, $ZnSO_4$, $Cd(NO_3)_2$, and Na_2SO_4?

17.59 In the course of various qualitative analysis procedures, the following mixtures are encountered: **(a)** Zn^{2+} and Cd^{2+}; **(b)** $Cr(OH)_3$ and $Fe(OH)_3$; **(c)** Mg^{2+} and K^+; **(d)** Ag^+ and Mn^{2+}. Suggest how each mixture might be separated.

17.60 Suggest how the cations in each of the following solution mixtures can be separated: **(a)** Na^+ and Cd^{2+}; **(b)** Cu^{2+} and Mg^{2+}; **(c)** Pb^{2+} and Al^{3+}; **(d)** Ag^+ and Hg^{2+}.

17.61 **(a)** Precipitation of the group 4 cations (Figure 17.22) requires a basic medium. Why is this so? **(b)** What is the most significant difference between the sulfides precipitated in group 2 and those precipitated in group 3? **(c)** Suggest a procedure that would serve to redissolve the group 3 cations following their precipitation.

17.62 A student who is in a great hurry to finish his laboratory work decides that his qualitative analysis unknown contains a metal ion from the insoluble phosphate group, group 4 (Figure 17.22). He therefore tests his sample directly with $(NH_4)_2HPO_4$, skipping earlier tests for the metal ions in groups 1, 2, and 3. He observes a precipitate and concludes that a metal ion from group 4 is indeed present. Why is this possibly an erroneous conclusion?

Additional Exercises

17.63 Furoic acid ($HC_5H_3O_3$) has a K_a value of 6.76×10^{-4} at 25°C. Calculate the pH at 25°C of **(a)** a solution formed by adding 35.0 g of furoic acid and 30.0 g of sodium furoate ($NaC_5H_3O_3$) to enough water to form 0.250 L of solution; **(b)** a solution formed by mixing 30.0 mL of 0.250 *M* $HC_5H_3O_3$ and 20.0 mL of 0.22 *M* $NaC_5H_3O_3$ and diluting the total volume to 125 mL; **(c)** a solution prepared by adding 50.0 mL of 1.65 *M* NaOH solution to 0.500 L of 0.0850 *M* $HC_5H_3O_3$.

17.64 A certain organic compound that is used as an indicator for acid-base reactions exists in aqueous solution as equal concentrations of the acid form, HB, and the base form, B^-, at a pH of 7.80. What is the pK_a for the acid form of this indicator, HB?

17.65 The acid-base indicator bromcresol green is a weak acid. The yellow acid and blue base forms of the indicator are present in equal concentrations in a solution when the pH is 4.68. What is the pK_a for bromcresol green?

17.66 Equal quantities of 0.010 *M* solutions of an acid HA and a base B are mixed. The pH of the resulting solution is 9.2. **(a)** Write the equilibrium equation and equilibrium-constant expression for the reaction between HA and B. **(b)** If K_a for HA is 8.0×10^{-5}, what is the value of the equilibrium constant for the reaction between HA and B? **(c)** What is the value of K_b for B?

17.67 Two buffers are prepared by adding an equal number of moles of formic acid ($HCHO_2$) and sodium formate ($NaCHO_2$) to enough water to make 1.00 L of solution. Buffer A is prepared using 1.00 mol each of formic acid and sodium formate. Buffer B is prepared by using 0.010 mol of each. **(a)** Calculate the pH of each buffer, and explain why they are equal. **(b)** Which buffer will have the greater buffer capacity? Explain. **(c)** Calculate the change in pH for each buffer upon the addition of 1.0 mL of 1.00 *M* HCl. **(d)** Calculate the change in pH for each buffer upon the addition of 10 mL of 1.00 *M* HCl. **(e)** Discuss your answers for parts (c) and (d) in light of your response to part (b).

17.68 Suppose you need to prepare a buffer at pH 8.6. Using Appendix D, select at least two different acid-base pairs that would be appropriate. Describe the composition of the buffers.

17.69 A biochemist needs 750 mL of an acetic acid–sodium acetate buffer with pH 4.50. Solid sodium acetate ($NaC_2H_3O_2$) and glacial acetic acid ($HC_2H_3O_2$) are available. Glacial acetic acid is 99% $HC_2H_3O_2$ by mass and has a density of 1.05 g/mL. If the buffer is to be 0.20 *M* in $HC_2H_3O_2$, how many grams of $NaC_2H_3O_2$ and how many milliliters of glacial acetic acid must be used?

17.70 A sample of 0.2140 g of an unknown monoprotic acid was dissolved in 25.0 mL of water and titrated with 0.0950 *M* NaOH. The acid required 27.4 mL of base to reach the equivalence point. **(a)** What is the molar mass of the acid? **(b)** After 15.0 mL of base had been added in the titration, the pH was found to be 6.50. What is the K_a for the unknown acid?

17.71 Show that the pH at the halfway point of a titration of a weak acid with a strong base (where the volume of added base is half of that needed to reach the equivalence point) is equal to pK_a for the acid.

17.72 Potassium hydrogen phthalate, often abbreviated KHP, can be obtained in high purity and is used to determine the concentrations of solutions of strong bases. Strong bases react with the hydrogen phthalate ion as follows:

$$HP^-(aq) + OH^-(aq) \longrightarrow H_2O(l) + P^{2-}(aq)$$

The molar mass of KHP is 204.2 g/mol and K_a for the HP^- ion is 3.1×10^{-6}. **(a)** If a titration experiment begins with 0.4885 g of KHP and has a final volume of about 100 mL, which indicator from Figure 16.7 would be most appropriate? **(b)** If the titration required 38.55 mL of NaOH solution to reach the end point, what is the concentration of the NaOH solution?

17.73 If 40.00 mL of 0.100 *M* Na_2CO_3 is titrated with 0.100 *M* HCl, calculate **(a)** the pH at the start of the titration; **(b)** the volume of HCl required to reach the first equivalence point and the predominant species present at this point; **(c)** the volume of HCl required to reach the second equivalence point and the predominant species present at this point; **(d)** the pH at the second equivalence point.

17.74 A hypothetical weak acid, HA, was combined with NaOH in the following proportions: 0.20 mol of HA, 0.080 mol of NaOH. The mixture was diluted to a total volume of 1.0 L, and the pH measured. **(a)** If pH = 4.80, what is the pK_a of the acid? **(b)** How many additional moles of NaOH should be added to the solution to increase the pH to 5.00?

[17.75] What is the pH of a solution made by mixing 0.30 mol NaOH, 0.25 mol Na_2HPO_4, and 0.20 mol H_3PO_4 with water and diluting to 1.00 L?

17.76 You have 1.0 *M* solutions of H_3PO_4 and NaOH. Describe how you would prepare a buffered solution of pH 7.20 with the highest possible buffering capacity from these reagents. Describe the composition of the buffer.

[17.77] Suppose you want to do a physiological experiment that calls for a pH 6.5 buffer. You find that the organism with which you are working is not sensitive to the weak acid H_2X ($K_{a1} = 2 \times 10^{-2}$; $K_{a2} = 5.0 \times 10^{-7}$), or its sodium salts. You have available a 1.0 *M* solution of this acid and a 1.0 *M* solution of NaOH. How much of the NaOH solution should be added to 1.0 L of the acid to give a buffer at pH 6.50? (Ignore any volume change.)

[17.78] How many microliters of 1.000 *M* NaOH solution must be added to 25.00 mL of a 0.1000 *M* solution of lactic acid ($HC_3H_5O_3$) to produce a buffer with pH = 3.75?

17.79 For each pair of compounds, use K_{sp} values to determine which has the greater molar solubility: **(a)** CdS or CuS; **(b)** $PbCO_3$ or $BaCrO_4$; **(c)** $Ni(OH)_2$ or $NiCO_3$; **(d)** AgI or Ag_2SO_4.

17.80 A saturated solution of $Mg(OH)_2$ in water has a pH of 10.38. Estimate the K_{sp} for this compound.

17.81 What concentration of Ca^{2+} remains in solution after CaF_2 has been precipitated from a solution that is 0.20 *M* in F^- at 25°C?

17.82 The solubility-product constant for barium permanganate, $Ba(MnO_4)_2$, is 2.5×10^{-10}. Suppose that solid $Ba(MnO_4)_2$ is in equilibrium with a solution of $KMnO_4$. What concentration of $KMnO_4$ is required to establish a concentration of 2.0×10^{-8} *M* for the Ba^{2+} ion in solution?

17.83 Calculate the ratio of $[Ca^{2+}]$ to $[Fe^{2+}]$ in a lake in which the water is in equilibrium with deposits of both $CaCO_3$ and $FeCO_3$, assuming that the water is slightly basic and that the hydrolysis of the carbonate ion can therefore be ignored.

[17.84] The solubility products of $PbSO_4$ and $SrSO_4$ are 6.3×10^{-7} and 3.2×10^{-7}, respectively. What are the values of $[SO_4{}^{2-}]$, $[Pb^{2+}]$, and $[Sr^{2+}]$ in a solution at equilibrium with both substances?

[17.85] What pH buffer solution is needed to give a Mg^{2+} concentration of 3.0×10^{-2} *M* in equilibrium with solid magnesium oxalate?

[17.86] The value of K_{sp} for $Mg_3(AsO_4)_2$ is 2.1×10^{-20}. The AsO_4^{3-} ion is derived from the weak acid H_3AsO_4 ($pK_{a1} = 2.22$; $pK_{a2} = 6.98$; $pK_{a3} = 11.50$). When asked to calculate the molar solubility of $Mg_3(AsO_4)_2$ in water, a student used the K_{sp} expression and assumed that $[Mg^{2+}] = 1.5[AsO_4^{3-}]$. Why was this a mistake?

[17.87] The solubility product for $Zn(OH)_2$ is 3.0×10^{-16}. The formation constant for the hydroxo complex, $Zn(OH)_4^{2-}$, is 4.6×10^{17}. What concentration of OH^- is required to dissolve 0.015 mol of $Zn(OH)_2$ in a liter of solution?

Integrative Exercises

17.88 **(a)** Write the net ionic equation for the reaction that occurs when a solution of hydrochloric acid (HCl) is mixed with a solution of sodium formate ($NaCHO_2$). **(b)** Calculate the equilibrium constant for this reaction. **(c)** Calculate the equilibrium concentrations of Na^+, Cl^-, H^+, CHO_2^-, and $HCHO_2$ when 50.0 mL of 0.15 *M* HCl is mixed with 50.0 mL of 0.15 *M* $NaCHO_2$.

17.89 **(a)** A 0.1044-g sample of an unknown monoprotic acid requires 22.10 mL of 0.0500 *M* NaOH to reach the end point. What is the molecular weight of the unknown? **(b)** As the acid is titrated, the pH of the solution after the addition of 11.05 mL of the base is 4.89. What is the K_a for the acid? **(c)** Using Appendix D, suggest the identity of the acid. Do both the molecular weight and K_a value agree with your choice?

17.90 A sample of 7.5 L of NH_3 gas at 22°C and 735 torr is bubbled into a 0.50-L solution of 0.40 *M* HCl. Assuming that all the NH_3 dissolves and that the volume of the solution remains 0.50 L, calculate the pH of the resulting solution.

17.91 Aspirin has the following structural formula:

At body temperature (37°C), K_a for aspirin equals 3×10^{-5}. If two aspirin tablets, each having a mass of 325 mg, are dissolved in a full stomach whose volume is 1 L and whose pH is 2, what percent of the aspirin is in the form of neutral molecules?

17.92 What is the pH at 25°C of water saturated with CO_2 at a partial pressure of 1.10 atm? The Henry's law constant for CO_2 at 25°C is 3.1×10^{-2} mol/L-atm. The CO_2 is an acidic oxide, reacting with H_2O to form H_2CO_3.

17.93 The osmotic pressure of a saturated solution of strontium sulfate at 25°C is 21 torr. What is the solubility product of this salt at 25°C?

17.94 A concentration of 10–100 parts per billion (by mass) of Ag^+ is an effective disinfectant in swimming pools. However, if the concentration exceeds this range, the Ag^+ can cause adverse health effects. One way to maintain an appropriate concentration of Ag^+ is to add a slightly soluble salt to the pool. Using K_{sp} values from Appendix D, calculate the equilibrium concentration of Ag^+ in parts per billion that would exist in equilibrium with **(a)** AgCl; **(b)** AgBr; **(c)** AgI.

[17.95] Fluoridation of drinking water is employed in many places to aid in the prevention of dental caries. Typically the F^- ion concentration is adjusted to about 1 ppb. Some water supplies are also "hard"; that is, they contain certain cations such as Ca^{2+} that interfere with the action of soap. Consider a case where the concentration of Ca^{2+} is 8 ppb. Could a precipitate of CaF_2 form under these conditions? (Make any necessary approximations.)

eMedia Exercises

17.96 The **Common-Ion Effect** movie (*eChapter 17.1*) shows the solubility of an iodide salt being reduced by the addition of sodium iodide. **(a)** How would you expect the solubility of the salt to be affected by the addition of a strong acid? **(b)** How would you expect it to be affected by the addition of a strong base? Explain your reasoning for both answers.

17.97 **(a)** Using data from the **Calculating pH Using the Henderson–Hasselbalch Equation** activity (*eChapter 17.2*), determine the pK_a of benzoic acid ($C_6H_5CO_2H$). **(b)** Using the pK_a that you determined in part (a), calculate the pH of a 0.0015 *M* aqueous solution of benzoic acid. **(c)** Will the pH of the solution in part (b) change if it is diluted with water to twice its original volume? Explain. **(d)** Will the pH of a buffer change when it is diluted with water to twice its original volume? If not, explain why not.

17.98 In Exercise 17.15 you calculated the pH of a buffer prepared by adding 20.0 g each of acetic acid and sodium acetate to enough water to make a 2.00-L solution. **(a)** Use the **Calculating pH Using the Henderson–Hasselbalch Equation** simulation (*eChapter 17.2*) to prepare this buffer and verify your answer to Exercise 17.15. **(b)** Describe what is meant by the term *buffer capacity*. **(c)** Of which could you add more to the above buffer without causing a drastic change in pH, strong acid or strong base?

17.99 The **Dissolution of $Mg(OH)_2$ by Acid** animation (*eChapter 17.5*) shows how the relatively insoluble solid, $Mg(OH)_2$, can be made more soluble in water by the addition of acid. **(a)** Write the net ionic equation for the process by which magnesium hydroxide dissolves in neutral water. **(b)** Write the net ionic equation for the combination of two hydronium ions with two hydroxide ions. **(c)** Show that the net ionic equations in parts (a) and (b) add to give the overall net ionic equation shown in the animation. **(d)** Calculate the equilibrium constant for the process represented by the overall net ionic equation (i.e., dissolution of magnesium hydroxide in aqueous acid). **(e)** Determine the solubility of magnesium hydroxide in 0.010 *M* HCl.

Chapter 18

Chemistry of the Environment

Polar stratospheric clouds form over Antartica in the winter months. By removing NO_2 from the atmosphere, they cause reduced formation of $ClONO_2$ and increased concentration of ClO, a major catalyst for reduction of stratospheric ozone.

What's Ahead

- In this chapter we will introduce the nature of Earth's atmosphere and hydrosphere and indicate some of the ways in which human activities have altered these vital components of our environment.
- We'll study the temperature and pressure profiles of Earth's atmosphere, and its chemical composition.
- The upper regions of the atmosphere, where the pressure is very low, absorb a great deal of high-energy radiation from the Sun through *photoionization* and *photodissociation* reactions. By filtering out high-energy radiation, these processes make it possible for life as we know it to exist on Earth.
- Ozone in the *stratosphere* also acts as a filter of high-energy ultraviolet light. Human activities have depleted the ozone layer by introducing chemicals into the stratosphere that perturb the natural cycle of ozone formation and decomposition. Notable among these are the *chlorofluorocarbons* (CFCs).
- The lowest region of the atmosphere, the *troposphere*, is the region in which we live. Many minor constituents of the troposphere affect air quality and the acidity of rainfall. The concentrations of many of these minor constituents, including those that cause *acid rain* and *photochemical smog*, have been increased by human activities.
- Carbon dioxide is a particularly important minor constituent of the atmosphere because it acts as a "greenhouse" gas; that is, it causes a warming of Earth's atmosphere. The combustion of fossil fuels (coal, oil, and natural gas) is projected to cause a doubling of atmospheric CO_2 concentration over the preindustrial level by about 2050. This doubling is expected to cause significant warming of the atmosphere, with attendant climatic changes.
- Almost all the water on Earth is in the world ocean. Obtaining freshwater for human consumption, industrial processes, and agricultural use from seawater is energy-intensive. We count on freshwater sources to supply most of our needs, but these sources often require treatment to render them usable.
- *Green chemistry* is an international initiative to make all industrial products, processes and chemical reactions compatible with a sustainable society and environment. We'll examine some reactions and processes in which the goals of green chemistry can be advanced.

IN 1992 REPRESENTATIVES of 172 countries met in Rio de Janeiro, Brazil, for the United Nations Conference on Environment and Development—a conference that became known as the Earth Summit. Five years later, in December 1997, representatives of 130 nations met in Kyoto, Japan, to discuss the impact of human activities on global warming. Out of that meeting came an initiative to work toward a global treaty that would, among other things, spell out actions to be taken to reduce emissions of gases that cause global warming. In July 2001 in Bonn, Germany, 178 nations signed a treaty based on the so-called Kyoto Protocols.* These efforts to address environmental concerns at the international level indicate that many of the most urgent environmental problems are global in nature.

The economic growth of both developed and underdeveloped nations depends critically on chemical processes. These range from treatment of water supplies to industrial processes, some of which produce products or byproducts that are harmful to the environment. We are now in a position to apply the principles we have learned in earlier chapters to an understanding of these processes. In this chapter we consider some aspects of the chemistry of our environment, focusing on Earth's *atmosphere* and the aqueous environment, called the *hydrosphere*.

Both the atmosphere and hydrosphere of our planet make life as we know it possible. Thus, management of the environment so as to maintain and enhance the quality of life is one of the most important concerns of our time. Our daily decisions as consumers mirror those of the leaders meeting in Bonn and similar international meetings: We must weigh the costs versus the benefits of our actions. Unfortunately, the environmental impacts of our decisions are often very subtle and not immediately evident.

* The United States, almost alone among nations, refused to sign the treaty.

18.1 Earth's Atmosphere

Because most of us have never been very far from Earth's surface, we tend to take for granted the many ways in which the atmosphere determines the environment in which we live. In this section we will examine some of the important characteristics of our planet's atmosphere.

The temperature of the atmosphere varies in a complex manner as a function of altitude, as shown in Figure 18.1(a) ▼. The atmosphere is divided into four regions based on this temperature profile. Just above the surface, in the **troposphere**, the temperature normally decreases with increasing altitude, reaching a minimum of about 215 K at about 12 km. Nearly all of us live out our entire lives in the troposphere. Howling winds and soft breezes, rain, sunny skies—all that we normally think of as "weather"—occur in this region. Commercial jet aircraft typically fly about 10 km (33,000 ft) above Earth, an altitude that approaches the upper limit of the troposphere, which we call the *tropopause*.

Above the tropopause the temperature increases with altitude, reaching a maximum of about 275 K at about 50 km. This region is called the **stratosphere**. Beyond the stratosphere are the *mesosphere* and the *thermosphere*. Notice in Figure 18.1 that the temperature extremes that form the boundaries for each region are denoted by the suffix *-pause*. The boundaries are important because gases mix across them relatively slowly. For example, pollutant gases generated in the troposphere find their way into the stratosphere very slowly.

Unlike the temperature changes that occur in the atmosphere, the pressure of the atmosphere decreases in a regular way with increasing elevation, as shown in Figure 18.1(b). Atmospheric pressure drops off much more rapidly at lower elevations than at higher ones because of the atmosphere's compressibility. Thus, the pressure decreases from an average value of 760 torr (101 kPa) at sea level to 2.3×10^{-3} torr

▶ **Figure 18.1** (a) Temperature variations in the atmosphere at altitudes below 110 km. (b) Variations in atmospheric pressure with altitude. At 80 km the pressure is approximately 0.01 torr.

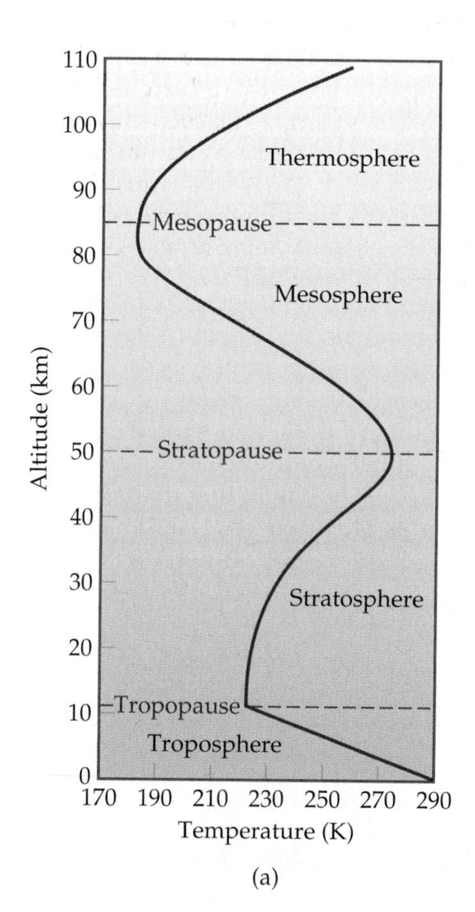

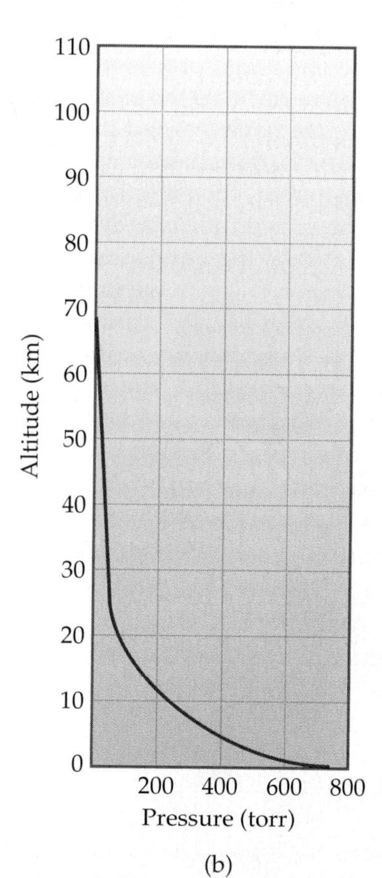

(3.1×10^{-4} kPa) at 100 km, to only 1.0×10^{-6} torr (1.3×10^{-7} kPa) at 200 km. The troposphere and stratosphere together account for 99.9% of the mass of the atmosphere, with 75% of the mass in the troposphere.

▲ **Figure 18.2** The aurora borealis, or northern lights. This luminous display in the northern sky is produced by collisions of high-speed electrons and protons from the Sun with air molecules. The charged particles are channeled toward the polar regions by Earth's magnetic field.

Composition of the Atmosphere

The atmosphere is an extremely complex system. Its temperature and pressure change over a wide range with altitude, as we have just seen. The atmosphere is bombarded by radiation and energetic particles from the Sun. This barrage of energy has profound chemical effects, especially on the outer reaches of the atmosphere (Figure 18.2 ▶). In addition, because of Earth's gravitational field, lighter atoms and molecules tend to rise to the top. As a result of all these factors, the composition of the atmosphere is not uniform.

Table 18.1 ▼ shows the composition by mole fraction of dry air near sea level. Although traces of many substances are present, N_2 and O_2 make up about 99% of the entire atmosphere. The noble gases and CO_2 make up most of the remainder.

When speaking of trace constituents of substances, we commonly use *parts per million* (ppm) as the unit of concentration. When applied to substances in aqueous solution, parts per million refers to grams of the substance per million grams of solution. ∞ (Section 13.2) When dealing with gases, however, one part per million refers to one part by *volume* in 1 million volume units of the whole. Because volume (V) is proportional to the amount (n) of gas via the ideal-gas equation ($PV = nRT$), volume fraction and mole fraction are the same. Thus, 1 ppm of a trace constituent of the atmosphere amounts to 1 mol of that constituent in 1 million moles of total gas; that is, the concentration in ppm is equal to the mole fraction times 10^6. Table 18.1 lists the mole fraction of CO_2 in the atmosphere as 0.000375. Its concentration in ppm is $0.000375 \times 10^6 = 375$ ppm.

Before we consider the chemical processes that occur in the atmosphere, let's review some of the important chemical properties of the two major components of the atmosphere, N_2 and O_2. Recall that the N_2 molecule possesses a triple bond between the nitrogen atoms. ∞ (Section 8.3) This very strong bond is largely responsible for the very low reactivity of N_2, which undergoes reaction only under extreme conditions. The bond energy in O_2 (495 kJ/mol) is much lower than that for N_2 (941 kJ/mol), so O_2 is much more reactive than N_2. Oxygen reacts with many substances to form oxides. The oxides of nonmetals, for example SO_2, usually form acidic solutions when dissolved in water. The oxides of active metals, for example CaO, form basic solutions when dissolved in water. ∞ (Section 7.6)

TABLE 18.1 Composition of Dry Air near Sea Level

Component*	Content (mole fraction)	Molar Mass
Nitrogen	0.78084	28.013
Oxygen	0.20948	31.998
Argon	0.00934	39.948
Carbon dioxide	0.000375	44.0099
Neon	0.00001818	20.183
Helium	0.00000524	4.003
Methane	0.000002	16.043
Krypton	0.00000114	83.80
Hydrogen	0.0000005	2.0159
Nitrous oxide	0.0000005	44.0128
Xenon	0.000000087	131.30

*Ozone, sulfur dioxide, nitrogen dioxide, ammonia, and carbon monoxide are present as trace gases in variable amounts.

SAMPLE EXERCISE 18.1

What is the concentration (in parts per million) of water vapor in a sample of air if the partial pressure of water is 0.80 torr and the total pressure of the air is 735 torr?

Solution

Analyze and Plan: We are given the partial pressure of water vapor and the total pressure of an air sample. Recall that the partial pressure of a given component of a mixture of gases is given by the product of its mole fraction and the total pressure of the mixture. (Section 10.6)

$$P_{H_2O} = X_{H_2O}P_t$$

Solve: Solving for the mole fraction of water vapor in the mixture, X_{H_2O}, gives

$$X_{H_2O} = \frac{P_{H_2O}}{P_t} = \frac{0.80 \text{ torr}}{735 \text{ torr}} = 0.0011$$

The concentration in ppm is the mole fraction times 10^6:

$$0.0011 \times 10^6 = 1100 \text{ ppm}$$

PRACTICE EXERCISE

The concentration of CO in a sample of air is found to be 4.3 ppm. What is the partial pressure of the CO if the total air pressure is 695 torr?

Answer: 3.0×10^{-3} torr

18.2 Outer Regions of the Atmosphere

Although the outer portion of the atmosphere, beyond the stratosphere, contains only a small fraction of the atmospheric mass, it forms the outer defense against the hail of radiation and high-energy particles that continually bombard Earth. As this occurs, the molecules and atoms of the upper atmosphere undergo chemical changes.

Photodissociation

The Sun emits radiant energy over a wide range of wavelengths. The shorter-wavelength, higher-energy radiations in the ultraviolet range of the spectrum are sufficiently energetic to cause chemical changes. Recall that electromagnetic radiation can be pictured as a stream of photons. (Section 6.2) The energy of each photon is given by the relationship $E = h\nu$, where h is Planck's constant and ν is the frequency of the radiation. For a chemical change to occur when radiation falls on Earth's atmosphere, two conditions must be met. First, there must be photons with energy sufficient to accomplish whatever chemical process is being considered. Second, molecules must absorb these photons. When these requirements are met, the energy of the photons is converted into some other form of energy within the molecule.

The rupture of a chemical bond resulting from absorption of a photon by a molecule is called **photodissociation**. Photodissociation does not form ions. The bond cleavage leaves half the bonding electrons with each of the two atoms forming two neutral particles.

One of the most important processes occurring in the upper atmosphere above about 120-km elevation is the photodissociation of the oxygen molecule:

$$O_2(g) + h\nu \longrightarrow 2O(g) \qquad [18.1]$$

The minimum energy required to cause this change is determined by the dissociation energy of O_2, 495 kJ/mol. In Sample Exercise 18.2 we calculate the longest wavelength photon having sufficient energy to dissociate the O_2 molecule.

SAMPLE EXERCISE 18.2

What is the maximum wavelength of light (in nanometers) that has enough energy per photon to dissociate the O_2 molecule?

Solution

Analyze and plan: We are asked to determine the wavelength of a photon that has just sufficient energy to break the O—O bond in O_2. We first need to calculate the energy required to break the O—O bond in one molecule, then find the wavelength of a photon of this energy.

Solve: The dissociation energy of O_2 is 495 kJ/mol. Using this value, we can calculate the amount of energy needed to break the bond in a single O_2 molecule:

$$\left(495 \times 10^3 \frac{\text{J}}{\text{mol}}\right)\left(\frac{1 \text{ mol}}{6.022 \times 10^{23} \text{ molecule}}\right) = 8.22 \times 10^{-19} \frac{\text{J}}{\text{molecule}}$$

We next use the Planck relation, $E = h\nu$, to calculate the frequency, ν, of a photon that has this amount of energy:

$$\nu = \frac{E}{h} = \frac{8.22 \times 10^{-19} \text{ J}}{6.626 \times 10^{-34} \text{ J-s}} = 1.24 \times 10^{15} \text{ s}^{-1}$$

Finally, we use the relationship between the frequency and wavelength of light (Section 6.1) to calculate the wavelength of the light:

$$\lambda = \frac{c}{\nu} = \left(\frac{3.00 \times 10^8 \text{ m/s}}{1.24 \times 10^{15} \text{ /s}}\right)\left(\frac{10^9 \text{ nm}}{1 \text{ m}}\right) = 242 \text{ nm}$$

Thus, ultraviolet light of wavelength 242 nm has sufficient energy per photon to photodissociate an O_2 molecule. Because photon energy increases as wavelength *decreases*, any photon of wavelength *shorter* than 242 nm will have sufficient energy to dissociate O_2.

PRACTICE EXERCISE

The bond energy in N_2 is 941 kJ/mol (Table 8.4). What is the longest wavelength photon that has sufficient energy to dissociate N_2?
Answer: 127 nm

Fortunately for us, O_2 absorbs much of the high-energy, short-wavelength radiation from the solar spectrum before it reaches the lower atmosphere. As it does, atomic oxygen, O, is formed. At higher elevations the dissociation of O_2 is very extensive. At 400 km, for example, only 1% of the oxygen is in the form of O_2; the other 99% is atomic oxygen. At 130 km, O_2 and O are just about equally abundant. Below 130 km, O_2 is more abundant than O.

The bond-dissociation energy of N_2 is very high (Table 8.4). As shown in Practice Exercise 18.2, only photons of very short wavelength possess sufficient energy to dissociate N_2. Furthermore, N_2 does not readily absorb photons, even when they possess sufficient energy. As a result, very little atomic nitrogen is formed in the upper atmosphere by dissociation of N_2.

Photoionization

In 1901 Guglielmo Marconi received a radio signal in St. John's, Newfoundland, that had been transmitted from Land's End, England, some 2900 km away. Because radio waves were thought to travel in straight lines, it had been assumed that radio communication over large distances on Earth would be impossible. Marconi's successful experiment suggested that Earth's atmosphere in some way substantially affects radio-wave propagation. His discovery led to intensive study of the upper atmosphere. In about 1924 the existence of electrons in the upper atmosphere was established by experimental studies.

For each electron present in the upper atmosphere, there must be a corresponding positively charged ion. The electrons in the upper atmosphere result mainly from the **photoionization** of molecules, caused by solar radiation. Photoionization occurs when a molecule absorbs radiation, and the absorbed energy causes loss of an electron. For photoionization to occur, therefore, the molecule must absorb a photon, and the photon must have enough energy to remove an electron.

TABLE 18.2 Ionization Processes, Ionization Energies, and Maximum Wavelengths Capable of Causing Ionization

Process	Ionization Energy (kJ/mol)	λ_{max} (nm)
$N_2 + h\nu \longrightarrow N_2^+ + e^-$	1495	80.1
$O_2 + h\nu \longrightarrow O_2^+ + e^-$	1205	99.3
$O + h\nu \longrightarrow O^+ + e^-$	1313	91.2
$NO + h\nu \longrightarrow NO^+ + e^-$	890	134.5

Some of the more important ionization processes occurring in the upper atmosphere above about 90 km are shown in Table 18.2 ▲, together with the ionization energies and λ_{max}, the maximum wavelength of a photon capable of causing ionization. Photons with energies sufficient to cause ionization have wavelengths in the high-energy region of the ultraviolet. These wavelengths are completely filtered out of the radiation reaching Earth because they are absorbed by the upper atmosphere.

18.3 Ozone in the Upper Atmosphere

While N_2, O_2, and O absorb photons with wavelengths shorter than 240 nm, ozone is the key absorber of photons with wavelengths of 240 to 310 nm. Let's consider how ozone forms in the upper atmosphere and how it absorbs photons.

Below an altitude of 90 km, most of the short-wavelength radiation capable of photoionization has been absorbed. Radiation capable of dissociating the O_2 molecule is sufficiently intense, however, for photodissociation of O_2 (Equation 18.1) to remain important down to an altitude of 30 km. In the region between 30 and 90 km, the concentration of O_2 is much greater than that of atomic oxygen. Therefore, the O atoms that form in this region undergo frequent collisions with O_2 molecules, resulting in the formation of ozone, O_3:

$$O(g) + O_2(g) \longrightarrow O_3^*(g) \qquad [18.2]$$

The asterisk over the O_3 denotes that the ozone molecule contains an excess of energy. The reaction of O with O_2 to form O_3^* releases 105 kJ/mol. This energy must be transferred away from the O_3^* molecule in a very short time, or it will simply fly apart again into O_2 and O—a decomposition that is the reverse of the process by which O_3^* is formed.

An energy-rich O_3^* molecule can release its excess energy by colliding with another atom or molecule and transferring some of the excess energy to it. Let's represent the atom or molecule with which O_3^* collides as M. (Usually M is N_2 or O_2 because these are the most abundant molecules.) The formation of O_3^* and the transfer of excess energy to M are summarized by the following equations:

$$O(g) + O_2(g) \rightleftharpoons O_3^*(g) \qquad [18.3]$$

$$O_3^*(g) + M(g) \longrightarrow O_3(g) + M^*(g) \qquad [18.4]$$

$$O(g) + O_2(g) + M(g) \longrightarrow O_3(g) + M^*(g) \quad \text{(net)} \qquad [18.5]$$

The rate at which O_3 forms according to Equations 18.3 and 18.4 depends on two factors that vary in opposite directions with increasing altitude. First, the formation of O_3^*, according to Equation 18.3, depends on the presence of O atoms. At low altitudes most of the radiation energetic enough to dissociate O_2 has been absorbed; thus, the formation of O is favored at higher altitudes. Second, both Equations 18.3 and 18.4 depend on molecular collisions. The concentration of molecules is greater at low altitudes, however, so the frequency of collisions between

O and O_2 (Equation 18.3) and between O_3^* and M (Equation 18.4) are both greater at lower altitudes. Because these processes vary with altitude in opposite directions, the highest rate of O_3 formation occurs in a band at an altitude of about 50 km, near the stratopause [Figure 18.1(a)]. Overall, roughly 90% of Earth's ozone is found in the stratosphere, between the altitudes of 10 and 50 km.

Once formed, the ozone molecule does not last long. Ozone is capable of absorbing solar radiation, which decomposes it back into O_2 and O. Because only 105 kJ/mol is required for this process, photons of wavelength shorter than 1140 nm are sufficiently energetic to photodissociate O_3. The strongest and most important absorptions, however, are of photons from 200 to 310 nm. If it were not for the layer of ozone in the stratosphere, these high-energy photons would penetrate to Earth's surface. Plant and animal life as we know it could not survive in the presence of this high-energy radiation. The "ozone shield" is therefore essential for our continued well-being. The ozone molecules that form this essential shield against radiation represent only a tiny fraction of the oxygen atoms present in the stratosphere, however, because they are continually destroyed even as they are formed.

The photodecomposition of ozone reverses the reaction that forms it. We thus have a cyclic process of ozone formation and decomposition, summarized as follows:

$$O_2(g) + h\nu \longrightarrow O(g) + O(g)$$

$$O(g) + O_2(g) + M(g) \longrightarrow O_3(g) + M^*(g) \quad \text{(heat released)}$$

$$O_3(g) + h\nu \longrightarrow O_2(g) + O(g)$$

$$O(g) + O(g) + M(g) \longrightarrow O_2(g) + M^*(g) \quad \text{(heat released)}$$

The first and third processes are photochemical; they use a solar photon to initiate a chemical reaction. The second and fourth processes are exothermic chemical reactions. The net result of all four processes is a cycle in which solar radiant energy is converted into thermal energy. The ozone cycle in the stratosphere is responsible for the rise in temperature that reaches its maximum at the stratopause, as illustrated in Figure 18.1.

The scheme described for the formation and decomposition of ozone molecules accounts for some, but not all, of the facts about the ozone layer. Many chemical reactions occur that involve substances other than just oxygen. In addition, the effects of turbulence and winds that mix up the stratosphere must be considered. A very complicated picture results. The overall result of ozone formation and removal reactions, coupled with atmospheric turbulence and other factors, is to produce an ozone profile in the upper atmosphere as shown in Figure 18.3 ▼.

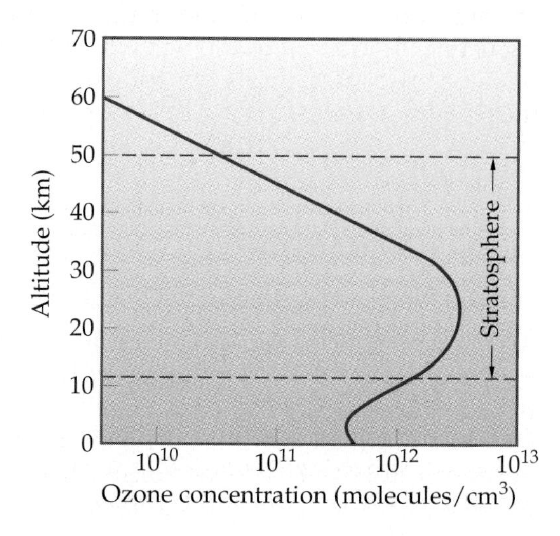

◀ **Figure 18.3** Variation in ozone concentration in the atmosphere, as a function of altitude.

Depletion of the Ozone Layer

ANIMATION
Catalytic Destruction of Stratospheric Ozone, CFCs and Stratospheric Ozone

In 1995 the Nobel Prize for chemistry was awarded to F. Sherwood Rowland, Mario Molina, and Paul Crutzen for their studies of ozone depletion in the stratosphere. In 1970 Crutzen showed that naturally occurring nitrogen oxides catalytically destroy ozone. Rowland and Molina recognized in 1974 that chlorine from **chlorofluorocarbons** (CFCs) may deplete the ozone layer that protects Earth's surface from damaging ultraviolet radiation. These substances, principally $CFCl_3$ (Freon-11™) and CF_2Cl_2 (Freon-12™), have been widely used as propellants in spray cans, as refrigerant and air-conditioner gases, and as foaming agents for plastics. They are virtually unreactive in the lower atmosphere. Furthermore, they are relatively insoluble in water and are therefore not removed from the atmosphere by rainfall or by dissolution in the oceans. Unfortunately, the lack of reactivity that makes them commercially useful also allows them to survive in the atmosphere and to diffuse eventually into the stratosphere. It is estimated that several million tons of chlorofluorocarbons are now present in the atmosphere.

As CFCs diffuse into the stratosphere, they are exposed to high-energy radiation, which can cause photodissociation. The C—Cl bonds are considerably weaker than the C—F bonds (Table 8.4). As a result, free chlorine atoms are formed readily in the presence of light with wavelengths in the range of 190 to 225 nm, as shown in the following equation for Freon-12™:

$$CF_2Cl_2(g) + h\nu \longrightarrow CF_2Cl(g) + Cl(g) \qquad [18.6]$$

Calculations suggest that chlorine atom formation occurs at the greatest rate at an altitude of about 30 km.

Atomic chlorine reacts rapidly with ozone to form chlorine monoxide (ClO) and molecular oxygen (O_2):

$$Cl(g) + O_3(g) \longrightarrow ClO(g) + O_2(g) \qquad [18.7]$$

Equation 18.7 follows a second-order rate law with a very large rate constant:

$$\text{Rate} = k\,[Cl][O_3] \qquad k = 7.2 \times 10^9\ M^{-1}\,s^{-1} \text{ at } 298\ K \qquad [18.8]$$

Under certain conditions the ClO generated in Equation 18.7 can react to regenerate free Cl atoms. One way that this can happen is by photodissociation of the ClO:

$$ClO(g) + h\nu \longrightarrow Cl(g) + O(g) \qquad [18.9]$$

The Cl atoms generated in Equation 18.9 can react with more O_3, according to Equation 18.7. These two equations form a cycle for the Cl atom–catalyzed decomposition of O_3 to O_2 as we see when we add the equations in the following way:

$$\begin{aligned}
2Cl(g) + 2O_3(g) &\longrightarrow 2ClO(g) + 2O_2(g)\\
2ClO(g) + h\nu &\longrightarrow 2Cl(g) + 2O(g)\\
O(g) + O(g) &\longrightarrow O_2(g)\\
\hline
2Cl(g) + 2O_3(g) + 2ClO(g) + 2O(g) &\longrightarrow 2Cl(g) + 2ClO(g) + 3O_2(g) + 2O(g)
\end{aligned}$$

The equation can be simplified by eliminating like species from each side of the equation to give

$$2O_3(g) \xrightarrow{Cl} 3O_2(g) \qquad [18.10]$$

Because the rate of Equation 18.7 increases linearly with [Cl], the rate at which ozone is destroyed increases as the quantity of Cl atoms increases. Thus, the greater the amount of CFCs that diffuse into the stratosphere, the faster the destruction of the ozone layer. Rates of diffusion of molecules from the troposphere into the stratosphere are slow. Nevertheless, a thinning of the ozone layer over the South Pole has already been observed, particularly during the months of September and October (Figure 18.4 ▶). Scientists have also found evidence that the North Pole suffers a similar, but less pronounced, ozone loss during late winter. There is also increasing evidence of some depletion at lower latitudes.

Because of the environmental problems associated with CFCs, steps have been taken to limit their manufacture and use. A major step was the signing of the Montreal Protocol on Substances that Deplete the Ozone Layer in 1987, in which participating nations agreed to reduce CFC production. More stringent limits were set in 1992, when representatives of approximately 100 nations agreed to ban the production and use of CFCs by 1996. Nevertheless, because CFCs are unreactive and because they diffuse so slowly into the stratosphere, scientists estimate that ozone depletion will continue for many years to come.

What substances will replace CFCs? At this time the main alternatives are hydrofluorocarbons, compounds in which C—H bonds replace the C—Cl bonds of CFCs. One such compound in current use is CH_2FCF_3, known as HFC-134a. Changing from CFCs to alternatives such as HFCs is costly. For example, the cost to change the air-conditioning equipment in commercial buildings in the United States to make it compatible with CFC replacements is estimated to be about 2 billion dollars. Furthermore, the present CFC substitutes are somewhat less efficient as refrigerants, requiring slightly more energy to use, which will also increase costs to consumers. Nevertheless, scientists and policymakers have concluded that the costs are necessary to protect our environment.

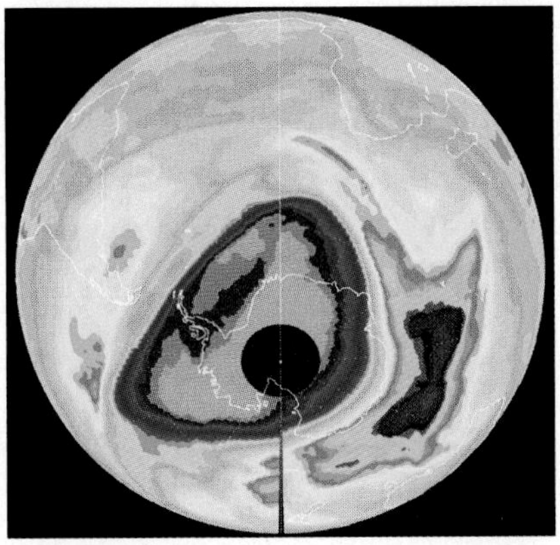

▲ **Figure 18.4** Map of the total ozone present in the Southern Hemisphere, taken September 16, 2000, from an orbiting satellite. The different colors represent different ozone concentrations. The center area, which is over Antarctica, is the area of lowest ozone concentration.

18.4 Chemistry of the Troposphere

The troposphere consists primarily of N_2 and O_2, which together comprise 99% of Earth's atmosphere at sea level (Table 18.1). Other gases, although present only at very low concentrations, can have major effects on our environment. Table 18.3 ▼ lists the major sources and typical concentrations of some of the important minor constituents of the troposphere. Many of these substances occur to only a slight extent in the natural environment, but exhibit much higher concentrations in certain areas as a result of human activities. In this section we will discuss the most important characteristics of a few of these substances and their chemical roles as air pollutants. As we will see, most form as either a direct or an indirect result of our widespread use of combustion reactions.

TABLE 18.3 Sources and Typical Concentrations of Some Minor Atmospheric Constituents

Minor Constituent	Sources	Typical Concentrations
Carbon dioxide, CO_2	Decomposition of organic matter; release from the oceans; fossil-fuel combustion	375 ppm throughout the troposphere
Carbon monoxide, CO	Decomposition of organic matter; industrial processes; fossil-fuel combustion	0.05 ppm in unpolluted air; 1–50 ppm in urban traffic areas
Methane, CH_4	Decomposition of organic matter; natural-gas seepage	1.77 ppm throughout the troposphere
Nitric oxide, NO	Electrical discharges; internal combustion engines; combustion of organic matter	0.01 ppm in unpolluted air; 0.2 ppm in smog
Ozone, O_3	Electrical discharges; diffusion from the stratosphere; photochemical smog	0 to 0.01 ppm in unpolluted air; 0.5 ppm in photochemical smog
Sulfur dioxide, SO_2	Volcanic gases; forest fires; bacterial action; fossil-fuel combustion; industrial processes	0 to 0.01 ppm in unpolluted air; 0.1–2 ppm in polluted urban environment

TABLE 18.4 Median Concentrations of Atmospheric Pollutants in a Typical Urban Atmosphere

Pollutant	Concentration (ppm)
Carbon monoxide	10
Hydrocarbons	3
Sulfur dioxide	0.08
Nitrogen oxides	0.05
Total oxidants (ozone and others)	0.02

MOVIE
Carbon Dioxide Behaves as an Acid in Water

Sulfur Compounds and Acid Rain

Sulfur-containing compounds are present to some extent in the natural, unpolluted atmosphere. They originate in the bacterial decay of organic matter, in volcanic gases, and from other sources listed in Table 18.3. The concentration of sulfur-containing compounds in the atmosphere resulting from natural sources is very small compared with the concentrations built up in urban and industrial environments as a result of human activities. Sulfur compounds, chiefly sulfur dioxide, SO_2, are among the most unpleasant and harmful of the common pollutant gases. Table 18.4 ◀ shows the concentrations of several pollutant gases in a *typical* urban environment (not one that is particularly affected by smog). According to these data, the level of sulfur dioxide is 0.08 ppm or higher about half the time. This concentration is considerably lower than that of other pollutants, notably carbon monoxide. Nevertheless, SO_2 is regarded as the most serious health hazard among the pollutants shown, especially for people with respiratory difficulties.

Combustion of coal and oil accounts for about 80% of the total SO_2 released in the United States. The extent to which SO_2 emissions are a problem in the burning of coal and oil depends on the level of their sulfur concentration. Some oil, such as that from the Middle East, is relatively low in sulfur, whereas other oil, such as that from Venezuela, has a higher sulfur content. Because of concern about SO_2 pollution, low-sulfur oil is in greater demand and is thus more expensive.

Coals also vary in their sulfur content. Much of the coal from east of the Mississippi is relatively high in sulfur content, up to 6% by mass. Much of the coal from the western states has a lower sulfur content. This coal, however, also has a lower heat content per unit mass of coal, so the difference in sulfur content per unit of heat produced is not as large as is often assumed.

More than 30 million tons of SO_2 is released into the atmosphere in the United States each year. (By comparison, the eruption of Mount Pinatubo in the Philippines in 1991 spewed 15 million to 30 million tons of SO_2 into the atmosphere.) Sulfur dioxide itself is harmful to both human health and property; furthermore, atmospheric SO_2 can be oxidized to SO_3 by any of several different pathways (such as reaction with O_2 or O_3). When SO_3 dissolves in water, it produces sulfuric acid, H_2SO_4:

$$SO_3(g) + H_2O(l) \longrightarrow H_2SO_4(aq)$$

Many of the environmental effects ascribed to SO_2 are actually due to H_2SO_4.

The presence of SO_2 in the atmosphere and the sulfuric acid that it produces result in the phenomenon of **acid rain**. (Nitrogen oxides, which form nitric acid, are also major contributors to acid rain.) Uncontaminated rainwater is naturally acidic and generally has a pH value of about 5.6. The primary source of this natural acidity is CO_2, which reacts with water to form carbonic acid, H_2CO_3. Acid rain, however, is more acidic than normal rainwater and typically has a pH value of about 4. This acidity has affected many lakes in northern Europe, northern United States, and Canada, reducing fish populations and affecting other parts of the ecological network within the lakes and surrounding forests.

The pH of most natural waters containing living organisms is between 6.5 and 8.5. At pH levels below 4.0 all vertebrates, most invertebrates, and many microorganisms are destroyed. The lakes that are most susceptible to damage are those with low concentrations of basic ions, such as HCO_3^-, that buffer them against changes in pH. Over 300 lakes in New York State contain no fish, and 140 lakes in Ontario, Canada, are devoid of life. The acid rain that appears to have killed the organisms in these lakes originates hundreds of kilometers upwind in the Ohio Valley and Great Lakes regions.

Because acids react with metals and with carbonates, acid rain is corrosive both to metals and to stone building materials. Marble and limestone, for example, whose major constituent is $CaCO_3$, are readily attacked by acid rain (Figure 18.5 ▶). Billions of dollars each year are lost as a result of corrosion due to SO_2 pollution.

One way to reduce the quantity of SO_2 released into the environment is to remove sulfur from coal and oil before it is burned. Unfortunately, this is currently too expensive to be technologically feasible. However, several methods have been developed for removing SO_2 from the gases formed when coal and oil are combusted. Powdered limestone ($CaCO_3$), for example, can be injected into the furnace of a power plant, where it decomposes into lime (CaO) and carbon dioxide:

$$CaCO_3(s) \longrightarrow CaO(s) + CO_2(g)$$

The CaO then reacts with SO_2 to form calcium sulfite:

$$CaO(s) + SO_2(g) \longrightarrow CaSO_3(s)$$

The solid particles of $CaSO_3$ as well as much of the unreacted SO_2 can be removed from the furnace gas by passing it through an aqueous suspension of lime (Figure 18.6 ▼). Not all the SO_2 is removed, however, and, given the enormous quantities of coal and oil burned worldwide, pollution by SO_2 will probably remain a problem for some time.

(a)

(b)

▲ **Figure 18.5** (a) This statue at the Field Museum in Chicago shows the effects of corrosion from acid rain and atmospheric pollutants. (b) The same statue after restoration.

Carbon Monoxide

Carbon monoxide is formed by the incomplete combustion of carbon-containing material, such as fossil fuels. In terms of total mass, CO is the most abundant of all the pollutant gases. The level of CO present in unpolluted air is low, probably on the order of 0.05 ppm. The estimated total amount of CO in the atmosphere is about 5.2×10^{14} g. In the United States alone, roughly 1×10^{14} g of CO is produced each year, about two thirds of which comes from automobiles.

Carbon monoxide is a relatively unreactive molecule and consequently poses no direct threat to vegetation or materials. It does affect humans, however. It has the unusual ability to bind very strongly to **hemoglobin**, the iron-containing

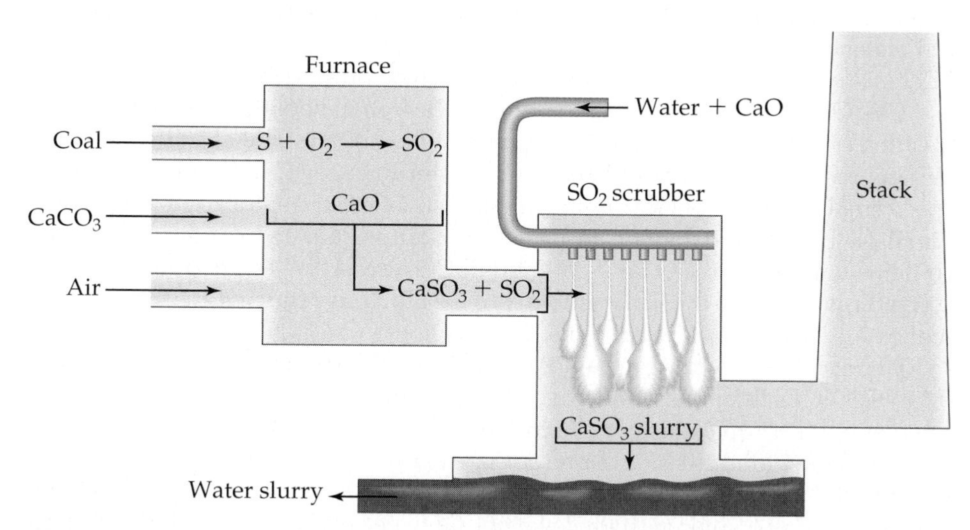

▲ **Figure 18.6** Common method for removing SO_2 from combusted fuel. Powdered limestone decomposes into CaO, which reacts with SO_2 to form $CaSO_3$. The $CaSO_3$ and any unreacted SO_2 enter a purification chamber called a scrubber, where a shower of CaO and water converts the remaining SO_2 into $CaSO_3$ and precipitates the $CaSO_3$ into a watery residue called a slurry.

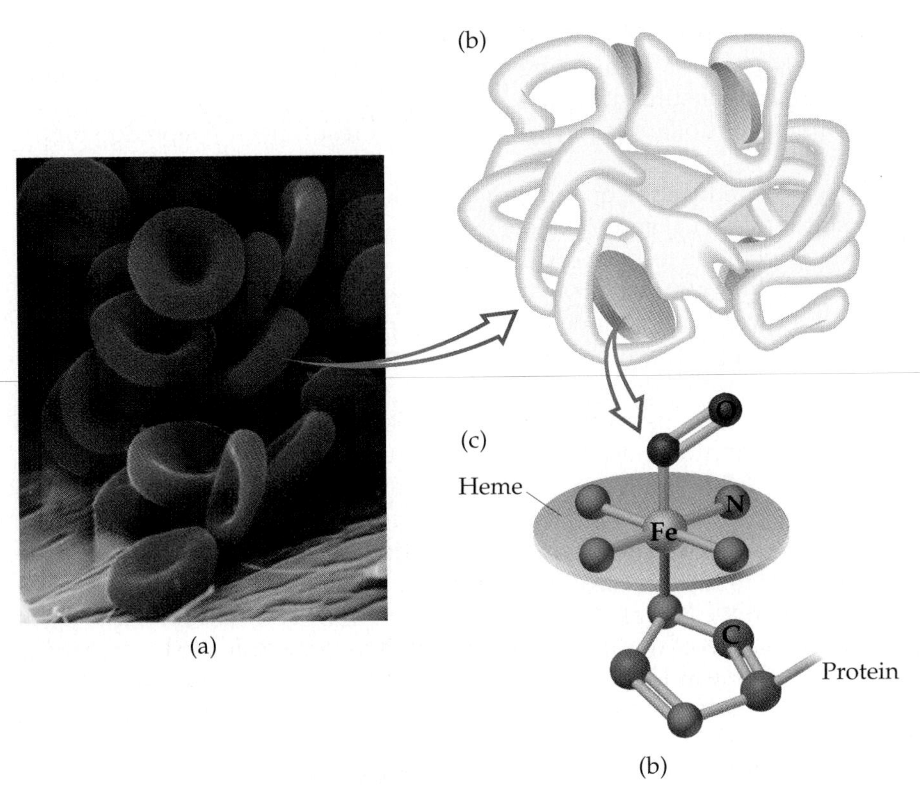

▲ **Figure 18.7** Red blood cells (a) contain hemoglobin (b). The hemoglobin contains four heme units, each of which can bind an O_2 molecule (c). When exposed to CO, the heme binds CO in preference to O_2.

protein in red blood cells [Figure 18.7(a) ▲] that transports oxygen in blood. Hemoglobin consists of four protein chains loosely held together in a cluster [Figure 18.7(b)]. Each chain has a heme molecule within its folds. The structure of heme is shown in Figure 18.7(c). Note that iron is situated in the center of a plane of four nitrogen atoms. A hemoglobin molecule in the lungs picks up an O_2 molecule, which reacts with the iron atom to form a species called *oxyhemoglobin*. As the blood circulates, the oxygen molecule is released in tissues as needed for cell metabolism, that is, for the chemical processes occurring in the cell. (See the "Chemistry and Life" box on blood as a buffered solution in Section 17.2.)

Like O_2, CO also binds very strongly to the iron in hemoglobin. The complex is called *carboxyhemoglobin* and is represented as COHb. The affinity of human hemoglobin for CO is about 210 times greater than for O_2. As a result, a relatively small quantity of CO can inactivate a substantial fraction of the hemoglobin in the blood for oxygen transport. For example, a person breathing air that contains only 0.1% of CO takes in enough CO after a few hours of breathing to convert up to 60% of the hemoglobin into COHb, thus reducing the blood's normal oxygen-carrying capacity by 60%.

Under normal conditions a nonsmoker breathing unpolluted air has about 0.3 to 0.5% COHb in the bloodstream. This amount arises mainly from the production of small quantities of CO in the course of normal body chemistry and from the small amount of CO present in clean air. Exposure to higher concentrations of CO causes the COHb level to increase, which in turn leaves fewer Hb sites to which O_2 can bind. If the level of COHb becomes too high, oxygen transport is effectively shut down and death occurs. Because CO is colorless and odorless, CO poisoning occurs with very little warning. Improperly ventilated combustion devices, such as kerosene lanterns and stoves, thus pose a potential health hazard (Figure 18.8 ◀).

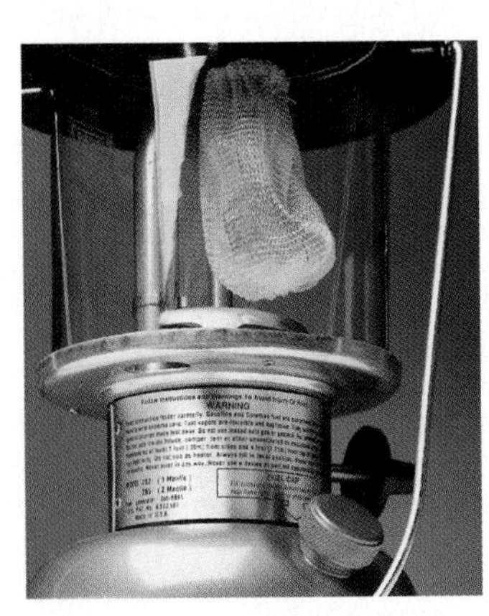

▲ **Figure 18.8** Kerosene lamps and stoves have warning labels concerning use in enclosed spaces, such as an indoor room. Incomplete combustion can produce colorless, odorless carbon monoxide, CO, which is toxic.

Nitrogen Oxides and Photochemical Smog

Nitrogen oxides are primary components of smog, a phenomenon with which city dwellers are all too familiar. The term *smog* refers to a particularly unpleasant condition of pollution in certain urban environments that occurs when weather conditions produce a relatively stagnant air mass. The smog made famous by Los Angeles, but now common in many other urban areas as well, is more accurately described as **photochemical smog** because photochemical processes play a major role in its formation (Figure 18.9 ▶).

▲ **Figure 18.9** Photochemical smog. Smog is produced largely by the action of sunlight on automobile exhaust gases.

Nitric oxide, NO, forms in small quantities in the cylinders of internal combustion engines by the direct combination of nitrogen and oxygen:

$$N_2(g) + O_2(g) \rightleftharpoons 2NO(g) \qquad \Delta H = 180.8 \text{ kJ} \qquad [18.11]$$

As noted in the "Chemistry at Work" box in Section 15.6, the equilibrium constant K for this reaction increases from about 10^{-15} at 300 K (near room temperature) to about 0.05 at 2400 K (approximately the temperature in the cylinder of an engine during combustion). Thus, the reaction is more favorable at higher temperatures. Before the installation of pollution-control devices, typical emission levels of NO_x were 4 g/mi. (The x is either 1 or 2 because both NO and NO_2 are formed, although NO predominates.) Present auto emission standards call for NO_x emission levels of less than 0.4 g/mi, but this is scheduled to be reduced to only 0.07 g/mi by 2004. Table 18.5 ▼ summarizes the federal standards for hydrocarbons and NO_x emissions since 1975 as well as the more restrictive standards enforced in California.

In air NO is rapidly oxidized to nitrogen dioxide (NO_2):

$$2NO(g) + O_2(g) \rightleftharpoons 2NO_2(g) \qquad \Delta H = -113.1 \text{ kJ} \qquad [18.12]$$

The equilibrium constant for this reaction decreases from about 10^{12} at 300 K to about 10^{-5} at 2400 K. The photodissociation of NO_2 initiates the reactions associated with photochemical smog. The dissociation of NO_2 into NO and O requires 304 kJ/mol, which corresponds to a photon wavelength of 393 nm. In sunlight, therefore, NO_2 undergoes dissociation to NO and O:

$$NO_2(g) + h\nu \longrightarrow NO(g) + O(g) \qquad [18.13]$$

The atomic oxygen formed undergoes several possible reactions, one of which gives ozone, as described earlier:

$$O(g) + O_2 + M(g) \longrightarrow O_3(g) + M^*(g) \qquad [18.14]$$

Ozone is a key component of photochemical smog. Although it is an essential UV screen in the upper atmosphere, it is an undesirable pollutant in the troposphere. It is extremely reactive and toxic, and breathing air that contains appreciable amounts of ozone can be especially dangerous for asthma sufferers, exercisers, and the elderly. We therefore have two ozone problems: excessive amounts in many urban environments, where it is harmful, and depletion in the stratosphere, where it is vital.

TABLE 18.5 National Tailpipe Emission Standards*

Year	Hydrocarbons (g/mi)	Nitrogen Oxides (g/mi)
1975	1.5 (0.9)	3.1 (2.0)
1980	0.41 (0.41)	2.0 (1.0)
1985	0.41 (0.41)	1.0 (0.4)
1990	0.41 (0.41)	1.0 (0.4)
1995	0.25 (0.25)	0.4 (0.4)
2004		0.07 (0.05)

*California standards in parentheses

In addition to nitrogen oxides and carbon monoxide, an automobile engine also emits unburned *hydrocarbons* as pollutants. These organic compounds, which are composed entirely of carbon and hydrogen, are the principal components of gasoline (Section 25.1). A typical engine without effective emission controls emits about 10 to 15 g of these compounds per mile. Current standards require that hydrocarbon emissions be less than 0.25 g/mi.

Reduction or elimination of smog requires that the essential ingredients for its formation be removed from automobile exhaust. Catalytic converters are designed to drastically reduce the levels of NO_x and hydrocarbons, two of the major ingredients of smog (see the "Chemistry at Work" box in Section 14.6). However, emission-control systems are notably unsuccessful in poorly maintained automobiles.

Water Vapor, Carbon Dioxide, and Climate

We have seen how the atmosphere makes life as we know it possible on Earth by screening out harmful short-wavelength radiation. In addition, the atmosphere is essential in maintaining a reasonably uniform and moderate temperature on the surface of the planet. The two atmospheric components of greatest importance in maintaining Earth's surface temperature are carbon dioxide and water.

Earth is in overall thermal balance with its surroundings. This means that Earth radiates energy into space at a rate equal to the rate at which it absorbs energy from the Sun. The Sun has a temperature of about 6000 K. As seen from outer space, Earth is relatively cold, with a temperature of about 254 K. The distribution of wavelengths in the radiation emitted from an object is determined by its temperature. Why does Earth, viewed from outside its atmosphere, appear so much colder than the temperature we usually experience at its surface? The troposphere, transparent to visible light, is not transparent to infrared radiation. Figure 18.10 ▼ shows the distribution of radiation from Earth's surface and the wavelengths absorbed by atmospheric water vapor and carbon dioxide. According to the graph, these atmospheric gases absorb much of the outgoing radiation from Earth's surface. In doing so, they help to maintain a livably uniform temperature at the surface by holding in, as it were, the infrared radiation from the surface, which we feel as heat. The influence of H_2O, CO_2, and certain other atmospheric gases on Earth's temperature is often called the *greenhouse effect* (see the "Chemistry at Work" box in Section 3.7).

The partial pressure of water vapor in the atmosphere varies greatly from place to place and time to time, but it is generally highest near Earth's surface and drops off very sharply with increased elevation. Because water vapor absorbs infrared radiation so strongly, it plays the major role in maintaining the atmospheric temperature at night, when the surface is emitting radiation into space and

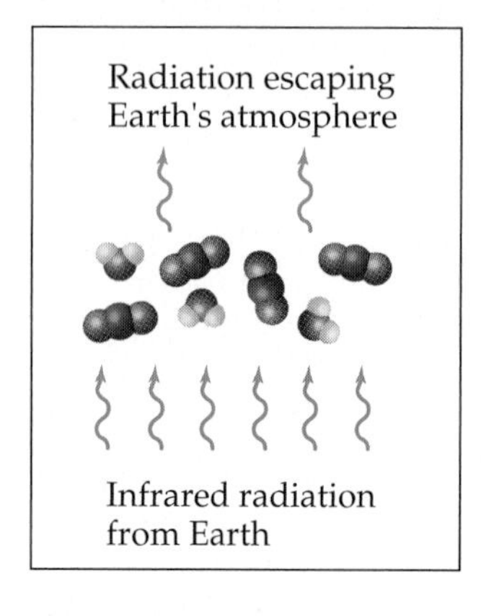

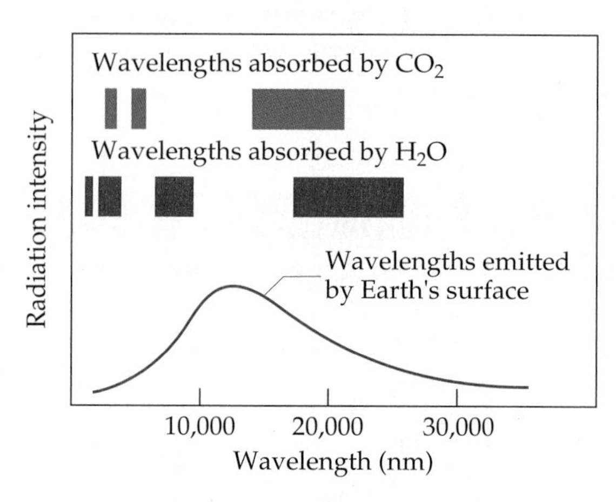

▶ **Figure 18.10** (a) Carbon dioxide and water absorb certain wavelengths of infrared radiation, which helps keep energy from escaping from Earth's surface. (b) The distribution of the wavelengths absorbed by CO_2 and H_2O compared to the wavelengths emitted by Earth's surface.

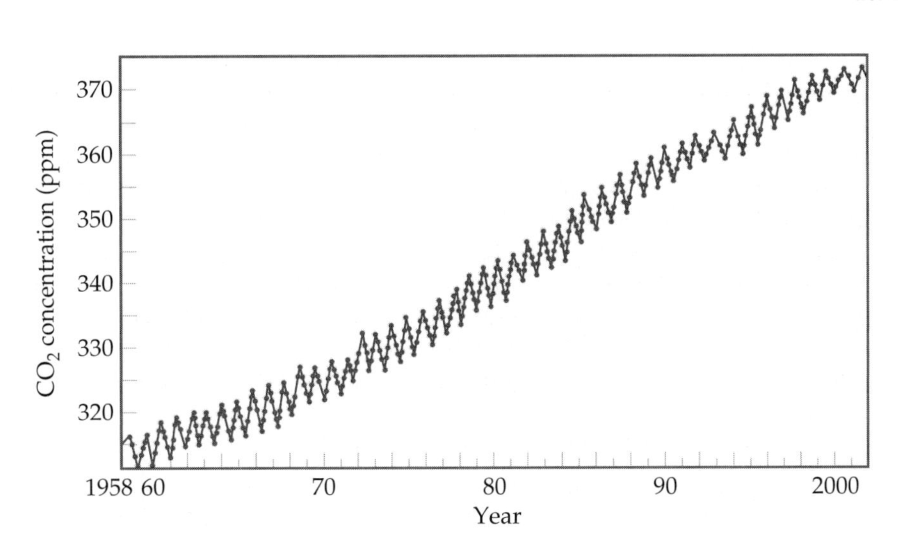

◀ **Figure 18.11** The concentration of atmospheric CO_2 has risen more than 15% since the late 1950s. These data were recorded at the Mauna Loa Observatory in Hawaii by monitoring the absorption of infrared radiation. The sawtooth shape of the graph is due to regular seasonal variations in CO_2 concentration for each year.

not receiving energy from the Sun. In very dry desert climates, where the water-vapor concentration is unusually low, it may be extremely hot during the day, but very cold at night. In the absence of an extensive layer of water vapor to absorb and then radiate part of the infrared radiation back to Earth, the surface loses this radiation into space and cools off very rapidly.

Carbon dioxide plays a secondary, but very important, role in maintaining the surface temperature. The worldwide combustion of fossil fuels, principally coal and oil, on a prodigious scale in the modern era has sharply increased the carbon dioxide level of the atmosphere. Measurements carried out over several decades show that the CO_2 concentration in the atmosphere is steadily increasing (Figure 18.11 ▲). Indeed, the CO_2 level has increased by 30% to over 375 ppm

A Closer Look Methane as a Greenhouse Gas

Although CO_2 receives most of the attention, other gases in total make an equal contribution to the greenhouse effect. Chief among these is methane, CH_4. Each methane molecule has about 25 times the greenhouse effect of a CO_2 molecule. Studies of atmospheric gas trapped long ago in the Greenland and Antarctic ice sheets show that the concentration of methane in the atmosphere has increased during the industrial age, from preindustrial values in the range of 0.3-0.7 ppm to the present value of about 1.8 ppm.

Methane is formed in biological processes that occur in low-oxygen environments. So-called anaerobic bacteria, which flourish in swamps and landfills, near the roots of rice plants, and in the digestive systems of cows and other ruminant animals, produce methane (Figure 18.12 ▶). It also leaks into the atmosphere during natural-gas extraction and transport (see "Chemistry at Work" box, Section 10.5). It is estimated that about two thirds of present-day methane emissions, which are increasing by about 1% per year, are related to human activities.

Methane has a half-life in the atmosphere of about 10 years, whereas CO_2 is much longer-lived. This might at first seem a good thing, but there are indirect effects to consider. Some methane is oxidized in the stratosphere, producing water vapor, a powerful greenhouse gas that is otherwise virtually absent from the stratosphere. In the troposphere methane is attacked by reactive species such as OH radicals, or nitrogen oxides, eventually producing other greenhouse gases such as O_3. It has been estimated that the climate-changing effects of CH_4 are at least one-third those of CO_2, and perhaps even half as large. Given this large contribution, important reductions of the greenhouse effect could be achieved by reducing methane emissions or capturing the emissions for use as a fuel.

▲ **Figure 18.12** Ruminant animals such as cows and sheep produce methane in their digestive systems. In Australia sheep and cattle produce about 14% of the country's total greenhouse emissions.

since preindustrial times. A consensus is emerging among scientists that this increase is already perturbing Earth's climate and that it may be responsible for the observed increase in the average global air temperature of 0.3°C to 0.6°C over the past century.

On the basis of present and expected future rates of fossil-fuel use, the atmospheric CO_2 level is expected to double from its present level sometime between 2050 and 2100. Computer models predict that this increase will result in an average global temperature increase of 1°C to 3°C. Major changes in global climate could result from a temperature change of this magnitude. Because so many factors go into determining climate, we cannot predict with certainty what changes will occur. Clearly, however, humanity has acquired the potential, by changing the concentration of CO_2 and other heat-trapping gases in the atmosphere, to substantially alter the climate of the planet.

18.5 The World Ocean

Water is the most common liquid on Earth. It covers 72% of Earth's surface and is essential to life. Our bodies are about 65% water by mass. Because of extensive hydrogen bonding, water has unusually high melting and boiling points and a high heat capacity. (Section 11.2) Its highly polar character is responsible for its exceptional ability to dissolve a wide range of ionic and polar-covalent substances. Many reactions occur in water, including reactions in which H_2O itself is a reactant. Recall, for example, that H_2O can participate in acid-base reactions as either a proton donor or a proton acceptor. (Section 16.4) In Chapter 20 we will see that H_2O can also participate in oxidation-reduction reactions as either a donor or an acceptor of electrons. All these properties play a role in our environment.

Seawater

The vast layer of salty water that covers so much of the planet is connected and is generally constant in composition. For this reason, oceanographers speak of a world ocean rather than of the separate oceans we learn about in geography books. The world ocean is huge. Its volume is 1.35×10^9 km^3. Almost all the water on Earth, 97.2%, is in the world ocean (Figure 18.13 ▼). Of the remaining 2.8%, 2.1% is in the form of ice caps and glaciers. All the freshwater—in lakes, rivers, and groundwater—amounts to only 0.6%. Most of the remaining 0.1% is in brackish (salty) water, such as that in the Great Salt Lake in Utah.

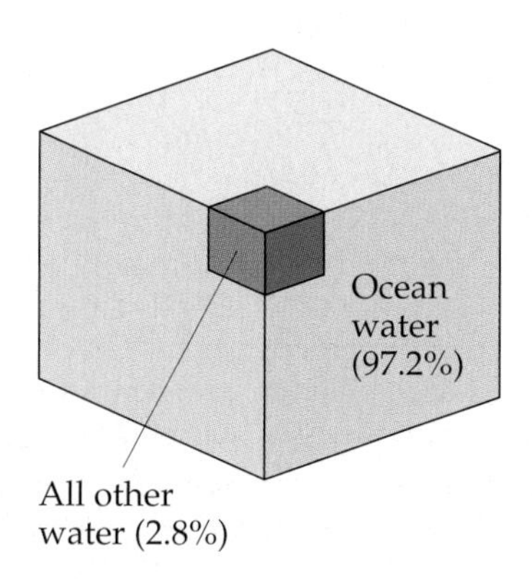

▶ **Figure 18.13** Most water on Earth is in the oceans.

TABLE 18.6 Ionic Constituents of Seawater Present in Concentrations Greater than 0.001 g/kg (1 ppm)

Ionic Constituent	g/kg Seawater	Concentration (M)
Chloride, Cl^-	19.35	0.55
Sodium, Na^+	10.76	0.47
Sulfate, SO_4^{2-}	2.71	0.028
Magnesium, Mg^{2+}	1.29	0.054
Calcium, Ca^{2+}	0.412	0.010
Potassium, K^+	0.40	0.010
Carbon dioxide*	0.106	2.3×10^{-3}
Bromide, Br^-	0.067	8.3×10^{-4}
Boric acid, H_3BO_3	0.027	4.3×10^{-4}
Strontium, Sr^{2+}	0.0079	9.1×10^{-5}
Fluoride, F^-	0.0013	7.0×10^{-5}

*CO_2 is present in seawater as HCO_3^- and CO_3^{2-}.

Seawater is often referred to as saline water. The **salinity** of seawater is the mass in grams of dry salts present in 1 kg of seawater. In the world ocean the salinity averages about 35. To put it another way, seawater contains about 3.5% dissolved salts by mass. The list of elements present in seawater is very long. Most, however, are present only in very low concentrations. Table 18.6 ▲ lists the 11 ionic species that are most abundant in seawater.

The sea is so vast that if a substance is present in seawater to the extent of only 1 part per billion (ppb, that is, 1×10^{-6} g per kilogram of water), there is still 5×10^9 kg of it in the world ocean. Nevertheless, the ocean is rarely used as a source of raw materials because the cost of extracting the desired substances is too high. Only three substances are obtained from seawater in commercially important amounts: sodium chloride, bromine, and magnesium.

Desalination

Because of its high salt content, seawater is unfit for human consumption and for most of the uses to which we put water. In the United States the salt content of municipal water supplies is restricted by health codes to no more than about 500 ppm. This amount is much lower than the 3.5% dissolved salts present in seawater and the 0.5% or so present in brackish water found underground in some regions. The removal of salts from seawater or brackish water to make the water usable is called **desalination.**

Water can be separated from dissolved salts by *distillation* (described in the "Closer Look" box in Section 13.5) because water is a volatile substance and the salts are nonvolatile. The principle of distillation is simple enough, but carrying out the process on a large scale presents many problems. As water is distilled from a vessel containing seawater, for example, the salts become more and more concentrated and eventually precipitate out.

Seawater can also be desalinated using **reverse osmosis**. Recall that osmosis is the net movement of solvent molecules, but not solute molecules, through a semipermeable membrane. (Section 13.5) In osmosis the solvent passes from the more dilute solution into the more concentrated one. However, if sufficient external pressure is applied, osmosis can be stopped and, at still higher pressures, reversed. When this occurs, solvent passes from the more concentrated into the more dilute solution. In a modern reverse-osmosis facility tiny hollow fibers are

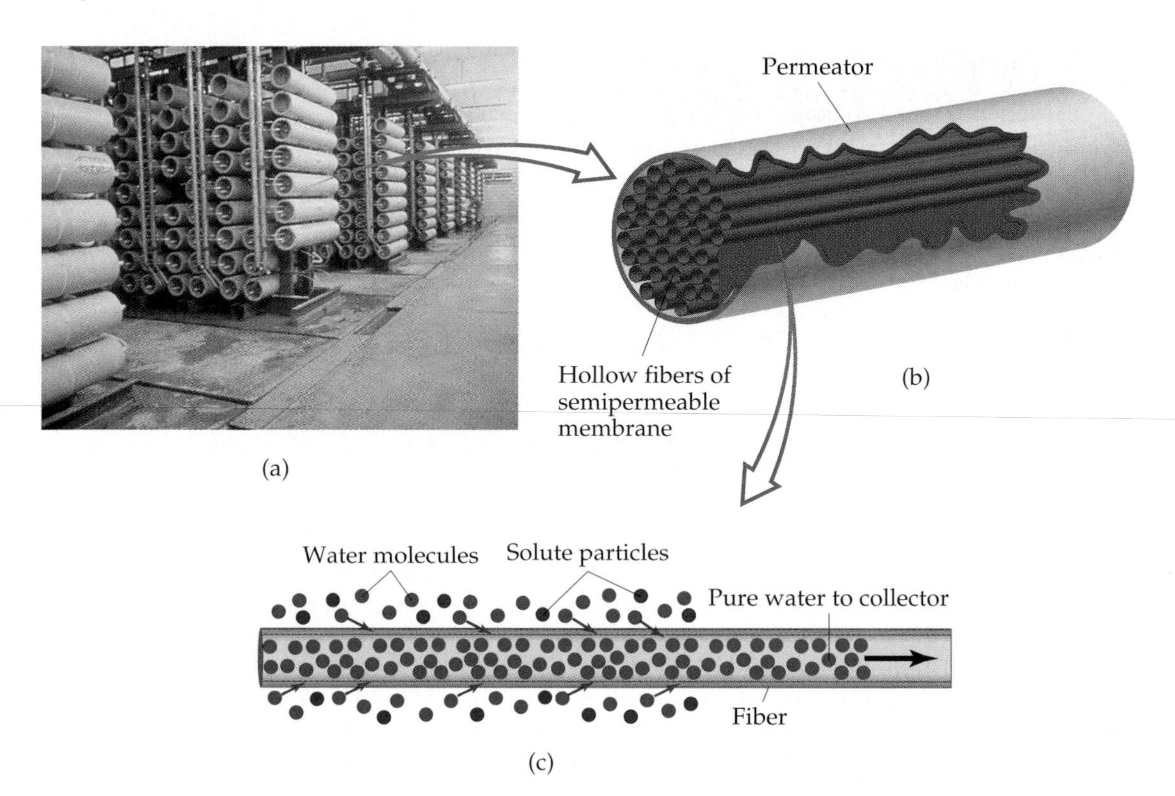

▲ **Figure 18.14** (a) A room inside a reverse-osmosis desalination plant. (b) Each cylinder shown in (a) is called a permeator and contains several million tiny hollow fibers. (c) When seawater is introduced under pressure into a permeator, water passes through the fiber wall into the fibers and is thereby separated from the ions of the salt.

used as the semipermeable membrane. Water is introduced under pressure into the fibers, and desalinated water is recovered as illustrated in Figure 18.14 ▲.

The world's largest desalination plant is located in Jubail, Saudi Arabia. This plant provides 50% of that country's drinking water by using reverse osmosis to desalinate seawater from the Persian Gulf. Such plants are becoming increasingly common in the United States. In 1992, for example, the city of Santa Barbara, California, opened a reverse-osmosis plant that can produce 8 million gallons of drinking water a day. Small-scale, manually operated reverse-osmosis desalinators are now also available for use in camping, traveling, and at sea (Figure 18.15 ◀).

▲ **Figure 18.15** Survivor-35 Katadyn North America Model MROD-35-LS hand-operated water desalinator that works by reverse osmosis. It can produce 4.5 L (1.2 gal) of pure water from seawater in an hour.

18.6 Freshwater

An adult needs about 2 liters of water per day for drinking. In the United States our daily use of water per person far exceeds this subsistence level, amounting to an average of about 300 L/day for personal consumption and hygiene. We use about 8 L/person for cooking and drinking, about 120 L for cleaning (bathing, laundering, and housecleaning), 80 L for flushing toilets, and 80 L for lawns. We use far greater quantities indirectly in agriculture and industry to produce food and other items. For example, about 1×10^5 L of water is used in the manufacture of 1000 kg of steel, about the quantity of steel in an average automobile.

The total amount of freshwater on Earth is not a very large fraction of the total water present. Indeed, freshwater is one of our most precious resources. Freshwater forms by evaporation from the oceans and the land. The water vapor that accumulates in the atmosphere is transported by global atmospheric circulation, eventually returning to Earth as rain and snow.

As rain falls and as water runs off the land on its way to the oceans, it dissolves a variety of cations (mainly Na^+, K^+ Mg^{2+}, Ca^{2+}, and Fe^{2+}), anions (mainly Cl^-, SO_4^{2-}, and HCO_3^-), and dissolved gases (principally O_2, N_2, and CO_2). As we use water, it becomes laden with additional dissolved material, including the wastes of human society. As our population and output of environmental pollutants increase, we find that we must spend ever-increasing amounts of money and resources to guarantee a supply of freshwater.

Dissolved Oxygen and Water Quality

The amount of dissolved O_2 in water is an important indicator of the quality of water. Water fully saturated with air at 1 atm and 20°C contains about 9 ppm of O_2. Oxygen is necessary for fish and much other aquatic life. Cold-water fish require that the water contain at least 5 ppm of dissolved oxygen for survival. Aerobic bacteria consume dissolved oxygen in order to oxidize organic materials and so meet their energy requirements. The organic material that the bacteria are able to oxidize is said to be **biodegradable**. This oxidation occurs by a complex set of chemical reactions, and the organic material disappears gradually.

Excessive quantities of biodegradable organic materials in water are detrimental because they deplete the water of the oxygen necessary to sustain normal animal life. Typical sources of these biodegradable materials, which are called *oxygen-demanding wastes*, include sewage, industrial wastes from food-processing plants and paper mills, and effluent (liquid waste) from meatpacking plants.

In the presence of oxygen the carbon, hydrogen, nitrogen, sulfur, and phosphorus in biodegradable material end up mainly as CO_2, HCO_3^-, H_2O, NO_3^-, SO_4^{2-}, and phosphates. The formation of these oxidation products sometimes reduces the amount of dissolved oxygen to the point where aerobic bacteria can no longer survive. Anaerobic bacteria then take over the decomposition process, forming CH_4, NH_3, H_2S, PH_3, and other products, several of which contribute to the offensive odors of some polluted waters.

Plant nutrients, particularly nitrogen and phosphorus, contribute to water pollution by stimulating excessive growth of aquatic plants. The most visible results of excessive plant growth are floating algae and murky water. More significantly, however, as plant growth becomes excessive, the amount of dead and decaying plant matter increases rapidly, a process called *eutrophication* (Figure 18.16 ▼). Decaying plants consume O_2 as they are biodegraded, leading to the depletion of

◀ **Figure 18.16** The growth of algae and duckweed in this pond is due to agricultural wastes. The wastes feed the growth of algae and weeds, which deplete the oxygen in the water, a process called eutrophication. A eutrophic lake cannot support fish.

oxygen in the water. Without sufficient supplies of oxygen, the water, in turn, cannot sustain any form of animal life. The most important sources of nitrogen and phosphorus compounds in water are domestic sewage (phosphate-containing detergents and nitrogen-containing body wastes), runoff from agricultural land (fertilizers containing both nitrogen and phosphorus), and runoff from livestock areas (animal wastes containing nitrogen).

Treatment of Municipal Water Supplies

The water needed for domestic uses, agriculture, and industrial processes is taken from naturally occurring lakes, rivers, and underground sources or from reservoirs. Much of the water that finds its way into municipal water systems is "used" water; it has already passed through one or more sewage systems or industrial plants. Consequently, this water must be treated before it is distributed to our faucets. Municipal water treatment usually involves five steps: coarse filtration, sedimentation, sand filtration, aeration, and sterilization. Figure 18.17 ▼ shows a typical treatment process.

After coarse filtration through a screen, the water is allowed to stand in large settling tanks in which finely divided sand and other minute particles can settle out. To aid in removing very small particles, the water may first be made slightly basic by adding CaO. Then $Al_2(SO_4)_3$ is added. The aluminum sulfate reacts with OH^- ions to form a spongy, gelatinous precipitate of $Al(OH)_3$ ($K_{sp} = 1.3 \times 10^{-33}$). This precipitate settles slowly, carrying suspended particles down with it, thereby removing nearly all finely divided matter and most bacteria. The water is then filtered through a sand bed. Following filtration, the water may be sprayed into the air to hasten the oxidation of dissolved organic substances.

The final stage of the operation normally involves treating the water with a chemical agent to ensure the destruction of bacteria. Ozone is most effective, but it must be generated at the place where it is used. Chlorine, Cl_2, is therefore more convenient. Chlorine can be shipped in tanks as a liquefied gas and dispensed from the tanks through a metering device directly into the water supply. The amount used depends on the presence of other substances with which the chlorine might react and on the concentrations of bacteria and viruses to be removed. The sterilizing action of chlorine is probably due not to Cl_2 itself, but to hypochlorous acid, which forms when chlorine reacts with water:

$$Cl_2(aq) + H_2O(l) \longrightarrow HClO(aq) + H^+(aq) + Cl^-(aq) \qquad [18.15]$$

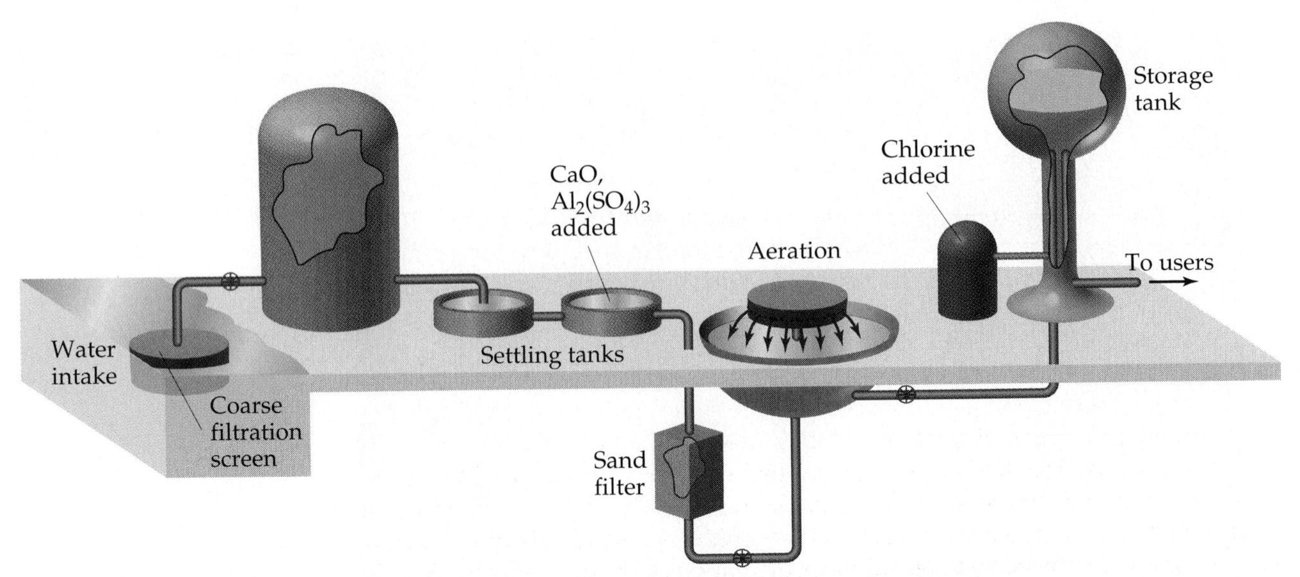

▲ **Figure 18.17** Common steps in treating water for a public water system.

A Closer Look Water Softening

Water containing a relatively high concentration of Ca^{2+}, Mg^{2+}, and other divalent cations is called **hard water**. Although the presence of these ions is generally not a health threat, they can make water unsuitable for some household and industrial uses. For example, these ions react with soaps to form an insoluble soap scum, the stuff of bathtub rings. In addition, mineral deposits may form when water containing these ions is heated. When water containing calcium ions and bicarbonate ions is heated, some carbon dioxide is driven off. As a result, the solution becomes less acidic and insoluble calcium carbonate forms:

$$Ca^{2+}(aq) + 2HCO_3^-(aq) \longrightarrow CaCO_3(s) + CO_2(g) + H_2O(l)$$

The solid $CaCO_3$ coats the surface of hot-water systems and tea kettles, thereby reducing heating efficiency. These deposits, called *scale*, can be especially serious in boilers where water is heated under pressure in pipes running through a furnace. Formation of scale reduces the efficiency of heat transfer and reduces the flow of water through pipes (Figure 18.18 ▼).

The removal of the ions that cause hard water is called water *softening*. Not all municipal water supplies require water softening. In those that do, the water is generally taken from underground sources in which it has had considerable contact with limestone, $CaCO_3$, and other minerals containing Ca^{2+}, Mg^{2+}, and Fe^{2+}. The **lime-soda process** is used for large-scale municipal water-softening operations. The water is treated with lime, CaO [or slaked lime, $Ca(OH)_2$], and soda ash, Na_2CO_3. These chemicals precipitate Ca^{2+} as $CaCO_3$ ($K_{sp} = 4.5 \times 10^{-9}$) and Mg^{2+} as $Mg(OH)_2$ ($K_{sp} = 1.6 \times 10^{-12}$):

$$Ca^{2+}(aq) + CO_3^{2-}(aq) \longrightarrow CaCO_3(s)$$

$$Mg^{2+}(aq) + 2OH^-(aq) \longrightarrow Mg(OH)_2(s)$$

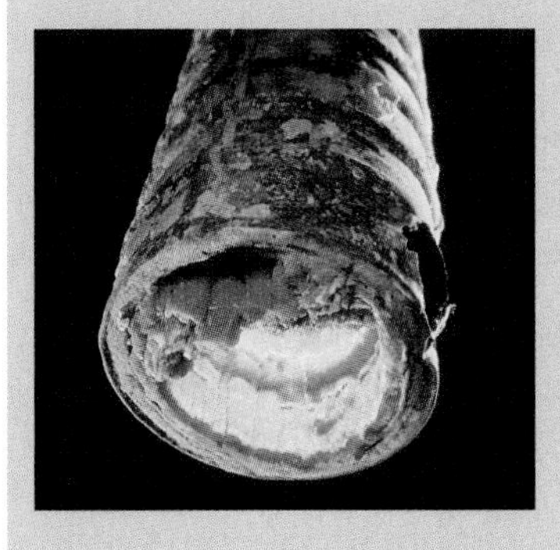

◀ **Figure 18.18** A section of water pipe that has been coated on the inside with $CaCO_3$ and other insoluble salts deposited from hard water.

Ion exchange is a typical household method for water softening. In this procedure the hard water is passed through a bed of an ion-exchange resin: plastic beads with covalently bound anion groups such as $—COO^-$ or $—SO_3^-$. These negatively charged groups have Na^+ ions attached to balance their charges. The Ca^{2+} ions and other cations in the hard water are attracted to the anionic groups and displace the lower-charged Na^+ ions into the water. Thus, one type of ion is exchanged for another. To maintain charge balance, 2 Na^+ ions enter the water for each Ca^{2+} removed. If we represent the resin with its anionic site as $R—COO^-$, we can write the equation for the process as follows:

$$2Na(R—COO)(s) + Ca^{2+}(aq) \rightleftharpoons Ca(R—COO)_2(s) + 2Na^+(aq)$$

Water softened in this way contains an increased concentration of Na^+ ions. Although Na^+ ions do not form precipitates or cause other problems associated with hard-water cations, individuals concerned with their sodium intake, such as those with high blood pressure (hypertension), should avoid drinking water softened in this way.

When all the available Na^+ ions have been displaced from the ion-exchange resin, the resin is regenerated by flushing it with a concentrated solution of NaCl. Homeowners can do this by charging their units with large amounts of NaCl(*s*), which can be purchased at most grocery stores. The high concentration of Na^+ forces the equilibrium shown in the preceding equation to shift to the left, causing the Na^+ ions to displace the hard-water cations, which are flushed down the drain.

18.7 Green Chemistry

As the human population has grown to its present number of about 6 billion, we have severely impacted our natural environment. In the quest for food and shelter, we have stripped entire regions of the original wildlife and vegetation. Modern agriculture and manufacturing have generated many substances harmful to the environment. Mining operations, for example, bring to the surface minerals that, with rainfall, produce highly acidic runoff that poisons streams. Industrial operations produce wastes that pollute the air and groundwater.

There is a growing realization that if humankind is to thrive in the future, we must create a sustainable society; that is, one in which processes are in balance with the planet's natural processes, in which no toxic materials are released to the environment, and in which our needs are met with renewable resources. Finally, all of this must be accomplished with the expenditure of the least possible amount of energy.

Although chemical industry is but a small part of the whole, chemical processes are involved in nearly all aspects of modern life. Chemistry is therefore at the heart of efforts to accomplish these goals. The **green chemistry** initiative promotes the design and application of chemical products and processes that are compatible with human health and that preserve the environment. Some of the major principles that govern green chemistry are the following:

- It is better to prevent waste than to treat or clean it up after it has been created.
- In synthesizing new substances, the method employed should generate as little waste product as possible. Those substances that are generated should possess little or no toxicity to human health and the environment.
- Chemical processes should be designed to be as energy-efficient as possible, avoiding high temperatures and pressures.
- Catalysts that permit the use of common and safe reagents should be employed whenever possible.
- The raw materials for chemical processes should be renewable feedstocks whenever technically and economically feasible.
- Auxiliary substances, such as solvents, should be eliminated or made as innocuous as possible.

Let's consider some of the areas in which green chemistry can operate to improve environmental quality.

Solvents and Reagents

A major area of concern in chemical processes is the use of volatile organic compounds as solvents for reactions. The solvent is not generally consumed in the reaction, but there are unavoidable releases to the atmosphere even in the most carefully controlled processes. Further, the solvent may be toxic or may decompose at least to some extent during reaction, thus creating wastes. The use of supercritical fluids ("Chemistry at Work" box, Section 11.4) represents a way to replace the conventional solvent by CO_2, a nontoxic gas already present in the atmosphere that can be recycled. Du Pont chemical company, for example, has invested in a production facility to make polytetrafluoroethylene, $-[CF_2CF_2]_n-$ (Teflon™) and copolymers with tetrafluoroethylene in liquid or supercritical CO_2. In this case the CO_2 replaces chlorofluorocarbon solvents, which, aside from their costs, have harmful effects on Earth's ozone layer (Section 18.3).

As a further example, *para*-xylene is oxidized to form terephthalic acid, which in turn is used to make polyethylene terephthalate (PET) plastic and polyester fiber (Section 12.2, Table 12.1):

$$CH_3-C_6H_4-CH_3 + 3O_2 \xrightarrow[\text{catalyst}]{190°C,\ 20\ atm} HO-\overset{O}{\overset{\|}{C}}-C_6H_4-\overset{O}{\overset{\|}{C}}-OH + 3H_2O$$

para-Xylene Terephthalic acid

This commercial process requires pressurization and a relatively high temperature. The catalyst is a manganese/cobalt mixture, oxygen is the oxidizing agent, and the solvent is acetic acid (CH_3COOH). A group at Nottingham University in England has developed an alternative route that employs supercritical water as solvent (Table 11.5) and hydrogen peroxide as the oxidant. This alternative process has several potential advantages, most particularly the elimination of acetic acid as solvent and the use of an innocuous oxidizing agent.

Whether it can successfully replace the existing commercial process depends on many factors, which will require further research.

Another environmentally friendly substance that is a promising candidate as a reagent or solvent is dimethyl carbonate, which has polar character and a relatively low boiling point (90°C). It could replace less environmentally friendly substances such as dimethyl sulfate and methyl halides as a reagent that supplies the methyl (CH_3) group in reactions:

$CH_3—O—S(=O)_2—O—CH_3$ Dimethyl sulfate

$CH_3—Cl$ Methyl chloride

$CH_3—O—C(=O)—O—CH_3$ Dimethyl carbonate

▲ **Figure 18.19** These CD's are formed from Lexan™ polycarbonate.

It might also be used in place of a reagent such as phosgene, Cl—CO—Cl. Not only is phosgene itself toxic, the production of phosgene forms CCl_4 as an undesirable by-product:

$$CO(g) + Cl_2(g) \longrightarrow COCl_2(g) + CCl_4(g) \text{ (by-product)}$$

Phosgene is widely used as a reagent in commercially important reactions, such as formation of polycarbonate plastics (Figure 18.19 ▶):

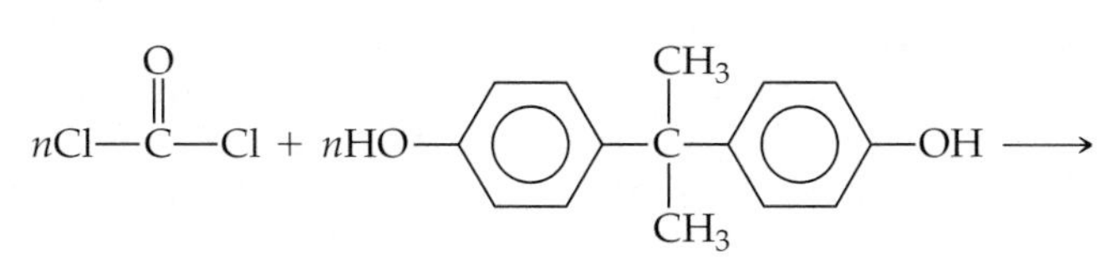

—[C(=O)—O—C6H4—C(CH3)2—C6H4—O]n—C(=O)—O— + 2nHCl

Lexan™ polycarbonate

If dimethyl carbonate could be substituted for phosgene in such reactions, the by product of the reaction would be methanol, CH_3OH, rather than HCl.

Other Processes

Many processes that are important in modern society use chemicals not found in nature. Let's briefly examine two of these, dry cleaning and the coating of auto bodies to prevent corrosion, and consider alternatives being developed to reduce harmful environmental impacts.

Dry cleaning of clothing typically uses a chlorinated organic solvent such as tetrachloroethylene ($Cl_2C{=}CCl_2$), which may cause cancer in humans. The widespread use of this and related solvents in dry cleaning, metal cleaning, and other industrial processes has contaminated groundwater in some areas. Alternative dry-cleaning methods that employ liquid CO_2, along with special cleaning agents, are being successfully commercialized (Figure 18.20 ▶).

▲ **Figure 18.20** This dry-cleaning apparatus employs liquid CO_2 as solvent.

The metal bodies of cars are coated extensively during manufacture to prevent corrosion. One of the key steps is the electrodeposition of a layer of metal

▶ **Figure 18.21** An automobile body receives a corrosion protection coating containing yttrium in place of lead.

ions that creates an interface between the vehicle body and polymeric coatings that serve as the undercoat for painting. In the past, lead has been a metal of choice for inclusion in the electrodeposition mixture. Lead is highly toxic, however, so its use in other paints and coatings has been virtually eliminated. The PPG Industries, Inc., a major producer of automotive coatings, has developed a relatively nontoxic yttrium hydroxide alternative to lead (Figure 18.21 ▲). When this coating is subsequently heated, the hydroxide is converted to the oxide, producing an insoluble ceramic-like coating.

Water Purification

Access to clean water is essential to the workings of a stable, thriving society. We have seen in the previous section that disinfection of water is an important step in water treatment for human consumption. Water disinfection is one of the greatest public health innovations in human history. It has dramatically decreased the incidences of water-borne bacterial diseases such as cholera and typhus. But this great benefit comes at a price.

In 1974 scientists in both Europe and the United States discovered that chlorination of water produces a group of by-products that had previously gone undetected. These by-products are called *trihalomethanes* (THMs) because all have a single carbon atom and three halogen atoms: $CHCl_3$, $CHCl_2Br$, $CHClBr_2$, and $CHBr_3$. These and many other chlorine- and bromine-containing organic substances are produced by the reaction of aqueous chlorine with organic materials present in nearly all natural waters, as well as with substances that are by-products of human activity. Recall that chlorine dissolves in water to form HOCl, which is the active oxidizing agent (Section 7.8):

$$Cl_2(g) + H_2O(l) \longrightarrow HOCl(aq) + HCl(aq) \quad [18.16]$$

HOCl in turn reacts with organic substances to form the THMs. Bromine enters through the reaction of HOCl with dissolved bromide ion:

$$HOCl(aq) + Br^-(aq) \longrightarrow HOBr(aq) + Cl^-(aq) \quad [18.17]$$

$HOBr(aq)$ halogenates organic substances analogously to $HOCl(aq)$.

Some THMs and other halogenated organic substances are suspected carcinogens, others interfere with the body's endocrine system. As a result, the

World Health Organization and the EPA have placed concentration limits of 100–200 $\mu g/L$ (100–200 ppb) on the total quantity of such substances in drinking water. The goal is to reduce the levels of THMs and related substances in the drinking water supply while preserving the antibacterial effectiveness of the water treatment. In some cases simply lowering the concentration of chlorine may provide adequate disinfection while reducing the concentrations of THMs formed. Alternative oxidizing agents, such as ozone (O_3) or chlorine dioxide (ClO_2), produce less of the halogenated substances, but they have their own disadvantages. Each is capable of oxidizing aqueous bromide, as shown, for example, for ozone:

$$O_3(aq) + Br^-(aq) + H_2O(l) \longrightarrow HOBr(aq) + O_2(aq) + OH^-(aq) \quad [18.18]$$

$$HOBr(aq) + 2O_3(aq) \longrightarrow BrO_3^-(aq) + 2O_2(aq) + H^+(aq) \quad [18.19]$$

As we have seen, HOBr(aq) is capable of reacting with dissolved organic substances to form halogenated organic compounds. Furthermore, bromate ion has been shown to cause cancer in animal tests.

There seem to be no completely satisfactory alternatives to chlorination at present. The risks of cancer from THMs and related substances in municipal water are very low, however, compared to the risks of cholera, typhus, and gastrointestinal disorders from untreated water. When the water supply is cleaner to begin with, less disinfectant is needed; thus, the danger of contamination through disinfection is reduced. Once the THMs are formed, their concentrations in the water supply can be reduced by aeration because the THMs are more volatile than water. Alternatively, they can be removed by adsorption onto activated charcoal or other adsorbents, although these are costly procedures.

SAMPLE INTEGRATIVE EXERCISE 18: Putting Ideas Together

(a) Acids from acid rain or other sources are no threat to lakes in areas where the rock is limestone (calcium carbonate), which can neutralize the excess acid. Where the rock is granite, however, no such neutralization occurs. How does the limestone neutralize the acid? **(b)** Acidic water can be treated with basic substances to increase the pH, although such a procedure is usually only a temporary cure. Calculate the minimum mass of lime, CaO, needed to adjust the pH of a small lake (4.0×10^9 L) from 5.0 to 6.5. Why might more lime be needed?

Solution (a) The carbonate ion, which is the anion of a weak acid, is basic. (Sections 16.2 and 16.7) Thus, the carbonate ion, CO_3^{2-}, reacts with $H^+(aq)$. If the concentration of $H^+(aq)$ is small, the major product is the bicarbonate ion, HCO_3^-. If the concentration of $H^+(aq)$ is higher, however, H_2CO_3 forms and decomposes to CO_2 and H_2O. (Section 4.3)

(b) The initial and final concentrations of $H^+(aq)$ in the lake are obtained from their pH values:

$$[H^+]_{initial} = 10^{-5.0} = 1 \times 10^{-5}\ M \text{ and } [H^+]_{final} = 10^{-6.5} = 3 \times 10^{-7}\ M$$

Using the volume of the lake, we can calculate the number of moles of $H^+(aq)$ at both pH values:

$$(1 \times 10^{-5}\ \text{mol/L})(4.0 \times 10^9\ \text{L}) = 4 \times 10^4\ \text{mol}$$

$$(3 \times 10^{-7}\ \text{mol/L})(4.0 \times 10^9\ \text{L}) = 1 \times 10^3\ \text{mol}$$

Hence the change in the amount of $H^+(aq)$ is $4 \times 10^4\ \text{mol} - 1 \times 10^3\ \text{mol} \approx 4 \times 10^4\ \text{mol}$.

Let's assume that all the acid in the lake is completely ionized, so that only the free $H^+(aq)$ measured by the pH needs to be neutralized. We will need to neutralize at least that much acid, although there may be a great deal more acid in the lake than that.

The oxide ion of CaO is very basic. (Section 16.5) In the neutralization reaction one mole of O^{2-} reacts with two moles of H^+ to form H_2O. Thus, 2.9×10^4 mol of H^+ requires the following number of grams of CaO:

$$(4 \times 10^4 \text{ mol H}^+)\left(\frac{1 \text{ mol CaO}}{2 \text{ mol H}^+}\right)\left(\frac{56.1 \text{ CaO}}{1 \text{ mol CaO}}\right) = 1 \times 10^6 \text{ g CaO.}$$

This amounts to slightly more than a ton of CaO. That would not be very costly because CaO is an inexpensive base, selling for less than $100 per ton when purchased in large quantity. The amount of CaO calculated above, however, is the very minimum amount needed because there are likely to be weak acids in the water that must also be neutralized. This liming procedure has been used to adjust the pH of some small lakes to bring their pH into the range necessary for fish to live. The lake in our example would be about a half mile long, having about the same distance across and an average depth of 20 ft.

Summary and Key Terms

Sections 18.1 and 18.2 In these sections we examined the physical and chemical properties of Earth's atmosphere. The complex temperature variations in the atmosphere give rise to four regions, each with characteristic properties. The lowest of these regions, the **troposphere**, extends from the surface up to an altitude of about 12 km. Above the troposphere, in order of increasing altitude, are the **stratosphere**, mesosphere, and thermosphere. In the upper reaches of the atmosphere only the simplest chemical species can survive the bombardment of highly energetic particles and radiation from the Sun. The average molecular weight of the atmosphere at high elevations is lower than that at Earth's surface because the lightest atoms and molecules diffuse upward and because of **photodissociation**, which is the breaking of bonds in molecules due to the absorption of light. Absorption of radiation may also lead to the formation of ions via **photoionization**.

Section 18.3 Ozone is produced in the upper atmosphere from the reaction of atomic oxygen with O_2. Ozone is itself decomposed by absorption of a photon or by reaction with an active species such as NO. **Chlorofluorocarbons** can undergo photodissociation in the stratosphere, introducing atomic chlorine, which is capable of catalytically destroying ozone. A marked reduction in the ozone level in the upper atmosphere would have serious adverse consequences because the ozone layer filters out certain wavelengths of ultraviolet light that are not removed by any other atmospheric component.

Section 18.4 In the troposphere the chemistry of trace atmospheric components is of major importance. Many of these minor components are pollutants. Sulfur dioxide is one of the more noxious and prevalent examples. It is oxidized in air to form sulfur trioxide, which, upon dissolving in water, forms sulfuric acid. The oxides of sulfur are major contributors to **acid rain**. One method of preventing the escape of SO_2 from industrial operations is to react the SO_2 with CaO to form calcium sulfite ($CaSO_3$).

Carbon monoxide is found in high concentrations in automobile engine exhaust and in cigarette smoke. CO is a health hazard because it can form a strong bond with **hemoglobin** and thus reduce the capacity for blood to transfer oxygen from the lungs.

Photochemical smog is a complex mixture of components in which both nitrogen oxides and ozone play important roles. Smog components are generated mainly in automobile engines, and smog control consists largely of controlling auto emissions.

Carbon dioxide and water vapor are the major components of the atmosphere that strongly absorb infrared radiation. CO_2 and H_2O are therefore critical in maintaining Earth's temperature. The concentrations of CO_2 and other so-called "greenhouse gases" in the atmosphere are thus important in determining worldwide climate. As a result of the extensive combustion of fossil fuels (coal, oil, and natural gas), the carbon dioxide level of the atmosphere is steadily increasing.

Section 18.5 Seawater contains about 3.5% by mass of dissolved salts and is described as having a **salinity** of 35. Because most of the world's water is in the oceans, humans may eventually look to the seas for freshwater. **Desalination** is the removal of dissolved salts from seawater or brackish water to make it fit for human consumption. Desalination may be accomplished by distillation or by **reverse osmosis**.

Section 18.6 Freshwater contains many dissolved substances, including dissolved oxygen, which is necessary for fish and other aquatic life. Substances that are decomposed by bacteria are said to be **biodegradable**. Because the oxidation of biodegradable substances by aerobic bacteria consumes dissolved oxygen, these substances are called oxygen-demanding wastes. The presence of an excess amount of oxygen-demanding wastes in water can deplete the dissolved oxygen sufficiently to kill fish and produce offensive odors. Plant nutrients can contribute to the problem by stimulating the growth of plants that become oxygen-demanding wastes when they die.

The water available from freshwater sources may require treatment before it can be used domestically. The several steps usually used in municipal water treatment include coarse filtration, sedimentation, sand filtration, aeration, sterilization, and sometimes water softening. Water softening is required when the water contains ions such as Mg^{2+} and Ca^{2+}, which react with soap to form soap scum. Water containing such ions is called **hard water**. The **lime-soda process**, which involves adding CaO and Na_2CO_3 to hard water, is sometimes used for large-scale municipal water softening. Individual homes usually rely on **ion exchange**, a process by which hard-water ions are exchanged for Na^+ ions.

Section 18.7 The **green chemistry** initiative promotes the design and application of chemical products and processes that are compatible with human health and that preserve the environment. The areas in which the principles of green chemistry can operate to improve environmental quality include choices of solvents and reagents for chemical reactions, development of alternative processes, and improvements in existing systems and practices.

Exercises

Earth's Atmosphere

18.1 **(a)** What is the primary basis for the division of the atmosphere into different regions? **(b)** Name the regions of the atmosphere, indicating the altitude interval for each one.

18.2 **(a)** What name is given to the boundary between the troposphere and stratosphere? **(b)** How are the boundaries between the regions of the atmosphere determined? **(c)** Explain why the stratosphere, which is more than 20 mi thick, has a smaller total mass than the troposphere, which is less than 10 mi thick.

18.3 The concentration of ozone in Mexico City has been measured at 0.37 ppm. Calculate the partial pressure of ozone at this concentration if the atmospheric pressure is 650 torr.

18.4 From the data in Table 18.1 calculate the partial pressures of carbon dioxide and argon when the total atmospheric pressure is 98.6 kPa.

18.5 The estimated average concentration of carbon monoxide in air in the United States in 1991 was 6.0 ppm. Calculate the number of CO molecules in 1.0 L of this air at a pressure of 745 torr and a temperature of 17°C.

18.6 **(a)** From the data in Table 18.1, what is the concentration of neon in the atmosphere in ppm? **(b)** What is the concentration of neon in the atmosphere in molecules per L, assuming an atmospheric pressure of 743 torr and a temperature of 295°C?

The Upper Atmosphere: Ozone

18.7 The dissociation energy of a carbon–bromine bond is typically about 210 kJ/mol. What is the maximum wavelength of photons that can cause C—Br bond dissociation?

18.8 In CF_3Cl the C—Cl bond-dissociation energy is 339 kJ/mol. In CCl_4 the C—Cl bond-dissociation energy is 293 kJ/mol. What is the range of wavelengths of photons that can cause C—Cl bond rupture in one molecule but not in the other?

18.9 Use the energy requirements to explain why photodissociation of oxygen is more important than photoionization of oxygen at altitudes below about 90 km.

18.10 Give two reasons why photodissociation of N_2 is a relatively unimportant process compared to photodissociation of O_2.

18.11 **(a)** Why is the temperature of the stratosphere higher near the stratopause than near the tropopause? **(b)** Explain how ozone is formed in the stratosphere.

18.12 **(a)** What is the principal mechanism by which oxygen atoms are created at 120-km elevation? **(b)** Why do oxygen atoms exist longer at a 120-km altitude than at 50-km altitude? **(c)** At Earth's surface, what is the biological significance of the stratospheric ozone layer?

18.13 What is a hydrofluorocarbon? Why are these compounds potentially less harmful to the ozone layer than CFCs?

18.14 Draw the Lewis structure for the chlorofluorocarbon CFC-11, $CFCl_3$. What chemical characteristics of this substance allow it to effectively deplete stratospheric ozone?

18.15 **(a)** Why is the fluorine present in chlorofluorocarbons not also involved in depletion of the ozone layer? **(b)** What are the chemical forms in which chlorine exists in the stratosphere following cleavage of the carbon–chlorine bond?

18.16 Would you expect the substance $CFBr_3$ to be effective in depleting the ozone layer, assuming that it is present in the stratosphere? Explain.

Chemistry of the Troposphere

18.17 What are the major adverse health effects of each of the following pollutants: **(a)** CO; **(b)** SO_2; **(c)** O_3?

18.18 Compare typical concentrations of CO, SO_2, and NO in unpolluted air (Table 18.3) and urban air (Table 18.4), and indicate in each case at least one possible source of the higher values in Table 18.4.

18.19 For each of the following gases, make a list of known or possible naturally occurring sources: **(a)** CH_4; **(b)** SO_2; **(c)** NO; **(d)** CO.

18.20 Why is rainwater naturally acidic, even in the absence of polluting gases such as SO_3?

18.21 **(a)** Write a chemical equation that describes the attack of acid rain on limestone, $CaCO_3$. **(b)** If a limestone sculpture were treated to form a surface layer of calcium sulfate, would this help to slow down the effects of acid rain? Explain.

18.22 The first stage in corrosion of iron in the atmosphere is oxidation to Fe^{2+}. **(a)** Write a balanced chemical equation to show the reaction of iron with acid rain. **(b)** Would you expect the same sort of reaction to occur with a silver surface? Explain.

18.23 Alcohol-based fuels for automobiles lead to the production of formaldehyde (CH_2O) in exhaust gases. Formaldehyde undergoes photodissociation, which contributes to photochemical smog:

$$CH_2O + h\nu \longrightarrow CHO + H$$

The maximum wavelength of light that can cause this reaction is 335 nm. **(a)** In what part of the electromagnetic spectrum is light with this wavelength found? **(b)** What is the maximum strength of a bond, in kJ/mol, that can be broken by absorption of a photon of 335-nm light? **(c)** Compare your answer from part (b) to the appropriate value from Table 8.4. What do you conclude about the C—H bond energy in formaldehyde?

18.24 An important reaction in the formation of photochemical smog is the photodissociation of NO_2:

$$NO_2 + h\nu \longrightarrow NO(g) + O(g)$$

The maximum wavelength of light that can cause this reaction is 420 nm. **(a)** In what part of the electromagnetic spectrum is light with this wavelength found? **(b)** What is the maximum strength of a bond, in kJ/mol, that can be broken by absorption of a photon of 420-nm light?

18.25 Explain why increasing concentrations of CO_2 in the atmosphere affect the quantity of energy leaving Earth, but do not affect the quantity entering from the Sun.

18.26 **(a)** With respect to absorption of radiant energy, what distinguishes a greenhouse gas from a non-greenhouse gas? **(b)** CH_4 is a greenhouse gas, but Ar is not. How might the molecular structure of CH_4 explain why it is a greenhouse gas?

The World Ocean

18.27 What is the molarity of Na^+ in a solution of NaCl whose salinity is 5.3 if the solution has a density of 1.03 g/mL?

18.28 Phosphorus is present in seawater to the extent of 0.07 ppm by mass. If the phosphorus is present as phosphate, PO_4^{3-}, calculate the corresponding molar concentration of phosphate.

18.29 A first-stage recovery of magnesium from seawater is precipitation of $Mg(OH)_2$ with CaO:

$$Mg^{2+}(aq) + CaO(s) + H_2O(l) \longrightarrow Mg(OH)_2(s) + Ca^{2+}(aq)$$

What mass of CaO is needed to precipitate 5.0×10^6 g of $Mg(OH)_2$?

18.30 Assuming a 10% efficiency of recovery, how many liters of seawater must be processed to obtain 10^8 kg of bromine in a commercial production process, assuming the bromide ion concentration listed Table 18.6?

[18.31] Suppose that one wishes to use reverse osmosis to reduce the salt content of brackish water containing 0.22 *M* total salt concentration to a value of 0.01 *M*, thus rendering it usable for human consumption. What is the minimum pressure that needs to be applied in the permeators (Figure 18.14) to achieve this goal, assuming that the operation occurs at 298 K? [*Hint*: Refer to Section 13.5]

[18.32] Assume that a portable reverse-osmosis apparatus such as that shown in Figure 18.15 operates on seawater, whose concentrations of constituent ions are listed in Table 18.6, and that the desalinated water output has an effective molarity of about 0.02 *M*. What minimum pressure must be applied by hand-pumping at 305 K to cause reverse osmosis to occur? [*Hint*: Refer to Section 13.5.]

Freshwater

18.33 List the common products formed when an organic material containing the elements carbon, hydrogen, oxygen, sulfur, and nitrogen decomposes **(a)** under aerobic conditions; **(b)** under anaerobic conditions.

18.34 **(a)** Explain why the concentration of dissolved oxygen in freshwater is an important indicator of the quality of the water. **(b)** How is the solubility of oxygen in water affected by increasing temperature?

18.35 The following organic anion is found in most detergents:

$$H_3C-(CH_2)_9-C(H)(CH_3)-C_6H_4-SO_3^-$$

Assume that the anion undergoes aerobic decomposition in the following manner:

$$2C_{18}H_{29}SO_3^-(aq) + 51O_2(aq) \longrightarrow 36CO_2(aq) + 28H_2O(l) + 2H^+(aq) + 2SO_4^{2-}(aq)$$

What is the total mass of O_2 required to biodegrade 1.0 g of this substance?

18.36 The average daily mass of O_2 taken up by sewage discharged in the United States is 59 g per person. How many liters of water at 9 ppm O_2 are totally depleted of oxygen in 1 day by a population of 85,000 people?

18.37 Write a balanced chemical equation to describe how magnesium ions are removed in water treatment by the addition of slaked lime, $Ca(OH)_2$.

18.38 **(a)** Which of the following ionic species is, or could be, responsible for hardness in a water supply: Ca^{2+}; K^+; Mg^{2+}; Fe^{2+}; Na^+? **(b)** What properties of an ion determine whether it will contribute to water hardness?

18.39 How many moles of $Ca(OH)_2$ and Na_2CO_3 should be added to soften 1.0×10^3 L of water in which $[Ca^{2+}] = 5.0 \times 10^{-4}$ *M* and $[HCO_3^-] = 7.0 \times 10^{-4}$ *M*?

18.40 The concentration of Ca^{2+} in a particular water supply is 5.7×10^{-3} *M*. The concentration of bicarbonate ion, HCO_3^-, in the same water is 1.7×10^{-3} *M*. What masses of $Ca(OH)_2$ and Na_2CO_3 must be added to 5.0×10^7 L of this water to reduce the level of Ca^{2+} to 20% of its original level?

18.41 What is the function served by addition of $Al_2(SO_4)_3$ to slightly basic water in the course of water treatment?

18.42 Ferrous sulfate ($FeSO_4$) is often used as a coagulant in water purification. The iron(II) salt is dissolved in the water to be purified, then oxidized to the iron(III) state by dissolved oxygen, at which time gelatinous $Fe(OH)_3$ forms, assuming the pH is above about 6. Write balanced chemical equations for the oxidation of Fe^{2+} to Fe^{3+} by dissolved oxygen, and for the formation of $Fe(OH)_3(s)$ by reaction of $Fe^{3+}(aq)$ with $HCO_3^-(aq)$.

Green Chemistry

18.43 One of the principles of green chemistry is that it is better to prevent formation of wastes than to clean them up once they are formed. How, if at all, does this principle relate to energy efficiency?

18.44 Discuss how catalysts can make processes more energy-efficient.

18.45 Explain how the use of dimethylcarbonate in place of phosgene is consistent with the first principle of green chemistry, that it is better to avoid waste production than to find ways to clean it up after it is created.

18.46 The Baeyer–Villiger reaction is a classic organic oxidation reaction for converting ketones to lactones, as in the following example:

Ketone + 3-Chloroperbenzoic acid ⟶ Lactone + 3-Chlorobenzoic acid

The reaction is used in the manufacture of plastics and pharmaceuticals. 3-Chloroperbenzoic acid is somewhat shock-sensitive, however, and prone to explode. Secondly, 3-Chlorobenzoic acid is a waste product. An alternative process being developed uses hydrogen peroxide and a catalyst consisting of tin deposited within a solid support. The catalyst is readily recovered from the reaction mixture. **(a)** What would you expect to be the other product of oxidation of the ketone to lactone by hydrogen peroxide? **(b)** What principles of green chemistry are addressed by use of the proposed process?

Additional Exercises

18.47 A friend of yours has seen each of the following items in newspaper articles and would like an explanation: **(a)** acid rain; **(b)** greenhouse gas; **(c)** photochemical smog; **(d)** ozone depletion. Give a brief explanation of each term and identify one or two of the chemicals associated with each.

18.48 Suppose that on another planet the atmosphere consists of 17% Ar, 38% CH_4, and 45% O_2. What is the average molar mass at the surface? What is the average molar mass at an altitude at which all the O_2 is photodissociated?

18.49 If an average O_3 molecule "lives" only 100–200 seconds in the stratosphere before undergoing dissociation, how can O_3 offer any protection from ultraviolet radiation?

18.50 Show how Equations 18.7 and 18.9 can be added to give Equation 18.10. (You may need to multiply one of the reactions by a factor to have them add properly.)

18.51 *Halons* are fluorocarbons that contain bromine, such as $CBrF_3$. They are used extensively as foaming agents for fighting fires. Like CFCs, halons are very unreactive and ultimately can diffuse into the stratosphere. **(a)** Based on the data in Table 8.4, would you expect photodissociation of Br atoms to occur in the stratosphere? **(b)** Propose a mechanism by which the presence of halons in the stratosphere could lead to the depletion of stratospheric ozone.

[18.52] The *hydroxyl radical*, OH, is formed at low altitudes via the reaction of excited oxygen atoms with water:

$$O^*(g) + H_2O(g) \longrightarrow 2OH(g)$$

Once produced, the hydroxyl radical is very reactive. Explain why each of the following series of reactions affects the pollution in the troposphere:

(a) $OH + NO_2 \longrightarrow HNO_3$

(b) $OH + CO + O_2 \longrightarrow CO_2 + OOH$

$OOH + NO \longrightarrow OH + NO_2$

(c) $OH + CH_4 \longrightarrow H_2O + CH_3$
$CH_3 + O_2 \longrightarrow OOCH_3$
$OOCH_3 + NO \longrightarrow OCH_3 + NO_2$

18.53 Explain, using Le Châtelier's principle, why the equilibrium constant for the formation of NO from N_2 and O_2 increases with increasing temperature, whereas the equilibrium constant for the formation of NO_2 from NO and O_2 decreases with increasing temperature.

18.54 The affinity of carbon monoxide for hemoglobin is about 210 times that of O_2. Assume a person is inhaling air that contains 112 ppm of CO. If all the hemoglobin leaving the lungs carries either oxygen or CO, calculate the fraction in the form of carboxyhemoglobin.

18.55 Natural gas consists primarily of methane, $CH_4(g)$. **(a)** Write a balanced chemical equation for the complete combustion of methane to produce $CO_2(g)$ as the only carbon-containing product. **(b)** Write a balanced chemical equation for the incomplete combustion of methane to produce $CO(g)$ as the only carbon-containing product. **(c)** At 25°C and 1.0 atm pressure, what is the minimum quantity of dry air needed to combust 1.0 L of $CH_4(g)$ completely to $CO_2(g)$? Natural gas has been used in conjunction with a catalyst to remove NO and NO_2 from an industrial gas stream. What are the likely products of the reactions of CH_4 with NO and NO_2 under such conditions?

18.56 One of the possible consequences of global warming is an increase in the temperature of ocean water. The oceans serve as a "sink" for CO_2 by dissolving large amounts of it. **(a)** How would the solubility of CO_2 in the oceans be affected by an increase in the temperature of the water? **(b)** Discuss the implications of your answer to part (a) for the problem of global warming.

18.57 The solar energy striking Earth averages 169 watts per square meter. The energy radiated from Earth's surface averages 390 watts per square meter. Comparing these numbers, one might expect that the planet would cool quickly, yet it does not. Why not?

18.58 Write balanced chemical equations for each of the following reactions: **(a)** The nitric oxide molecule undergoes photodissociation in the upper atmosphere. **(b)** The nitric oxide molecule undergoes photoionization in the upper atmosphere. **(c)** Nitric oxide undergoes oxidation by ozone in the stratosphere. **(d)** Nitrogen dioxide dissolves in water to form nitric acid and nitric oxide.

18.59 **(a)** Explain why $Mg(OH)_2$ precipitates when CO_3^{2-} ion is added to a solution containing Mg^{2+}. **(b)** Will $Mg(OH)_2$ precipitate when 4.0 g of Na_2CO_3 is added to 1.00 L of a solution containing 125 ppm of Mg^{2+}?

18.60 Describe some of the pros and cons of using chlorine rather than ozone as the disinfecting agent for municipal water supplies.

18.61 Current pulp-bleaching technology in the paper industry uses chlorine as the oxidizing agent. From a green chemistry perspective, given what you have read in this chapter, what are the shortcomings of such a process?

[18.62] It has recently been pointed out that there may be increased amounts of NO in the troposphere as compared with the past because of massive use of nitrogen-containing compounds in fertilizers. Assuming that NO can eventually diffuse into the stratosphere, how might it affect the conditions of life on Earth? Using the index to this text, look up the chemistry of nitrogen oxides. What chemical pathways might NO in the troposphere follow?

Integrative Exercises

18.63 The estimated average concentration of NO_2 in air in the United States in 1994 was 0.021 ppm. **(a)** Calculate the partial pressure of the NO_2 in a sample of this air when the atmospheric pressure is 745 torr (99.1 kPa). **(b)** How many molecules of NO_2 are present under these conditions at 20°C in a room that measures $15 \times 14 \times 8$ ft?

[18.64] In 1986 Georgia Power company's electrical power plant in Taylorsville, Georgia, burned 8,376,726 tons of coal, a national record at that time. **(a)** Assuming that the coal was 83% carbon and 2.5% sulfur and that combustion was complete, calculate the number of tons of carbon dioxide and sulfur dioxide produced by the plant during the year. **(b)** If 55% of the SO_2 could be removed by reaction with powdered CaO to form $CaSO_3$, how many tons of $CaSO_3$ would be produced?

18.65 The water supply for a midwestern city contains the following impurities: coarse sand; finely divided particulates; nitrate ion; trihalomethanes; dissolved phosphorus in the form of phosphates; potentially harmful bacterial strains; dissolved organic substances. Which of the following processes or agents, if any, is effective in removing each of these impurities: coarse sand filtration; activated carbon filtration; aeration; ozonization; precipitation with aluminum hydroxide?

18.66 The concentration of H_2O in the stratosphere is about 5 ppm. It undergoes photodissociation as follows:

$$H_2O(g) \longrightarrow H(g) + OH(g)$$

(a) Using Table 8.4, calculate the wavelength required to cause this dissociation.

(b) The hydroxyl radicals, OH, can react with ozone, giving the following reactions:

$$OH(g) + O_3(g) \longrightarrow HO_2(g) + O_2(g)$$
$$HO_2(g) + O(g) \longrightarrow OH(g) + O_2(g)$$

What overall reaction results from these two elementary reactions? What is the catalyst in the overall reaction? Explain.

18.67 The standard enthalpies of formation of ClO and ClO_2 are 101 and 102 kJ/mol, respectively. Using these data and the thermodynamic data in Appendix C, calculate the overall enthalpy change for each step in the following catalytic cycle:

$$ClO(g) + O_3(g) \longrightarrow ClO_2(g) + O_2(g)$$
$$ClO_2(g) + O(g) \longrightarrow ClO(g) + O_2(g)$$

What is the enthalpy change for the overall reaction that results from these two steps?

18.68 The main reason that distillation is a costly method for purifying water is the high energy required to heat and vaporize water. **(a)** Using the density, specific heat, and heat of vaporization of water from Appendix B, calculate the amount of energy required to vaporize 1.00 gallon of water beginning with water at 20°C. **(b)** If the energy is provided by electricity costing \$0.085/kwh, calculate its cost. **(c)** If distilled water sells in the grocery store for \$1.26 per gallon, what percentage of the sales price is represented by the cost of the energy?

[18.69] A reaction that contributes to the depletion of ozone in the stratosphere is the direct reaction of oxygen atoms with ozone:

$$O(g) + O_3(g) \longrightarrow 2O_2(g)$$

At 298 K the rate constant for this reaction is $4.8 \times 10^5/M/s$. **(a)** Based on the units of the rate constant, write the likely rate law for this reaction. **(b)** Would you expect this reaction to occur via a single elementary process? Explain why or why not. **(c)** From the magnitude of the rate constant, would you expect the activation energy of this reaction to be large or small? Explain. **(d)** Use ΔH_f° values from Appendix C to estimate the enthalpy change for this reaction. Would this reaction raise or lower the temperature of the stratosphere?

[18.70] Assume that an equilibrium mixture of N_2, O_2, and NO is achieved at a temperature of 2400 K:

$$N_2(g) + O_2(g) \rightleftharpoons 2NO(g) \quad K_{eq} = 0.05$$

If the original reaction mixture consists of air at sea level and at 1.0 atm in a 1.0-L vessel, what is the partial pressure of NO at equilibrium at 2400 K? What is the concentration of NO in ppm?

18.71 Nitrogen dioxide (NO_2) is the only important gaseous species in the lower atmosphere that absorbs visible light. **(a)** Write the Lewis structure(s) for NO_2. **(b)** How does this structure account for the fact that NO_2 dimerizes to form N_2O_4? Based on what you can find about this dimerization reaction in the text, would you expect to find the NO_2 that forms in an urban environment to be in the form of dimer? Explain. **(c)** What would you expect as products, if any, for the reaction of NO_2 with CO? **(d)** Would you expect NO_2 generated in an urban environment to migrate to the stratosphere? Explain.

[18.72] If the pH of a 1.0-in. rainfall over 1500 mi^2 is 2.5, how many kilograms of H_2SO_4 are present, assuming that it is the only acid contributing to the pH?

[18.73] The Henry's law constant for CO_2 in water at 25°C is $3.1 \times 10^{-2} M/atm$. **(a)** What is the solubility of CO_2 in water at this temperature if the solution is in contact with air at normal atmospheric pressure? **(b)** Assume that all of this CO_2 is in the form of H_2CO_3 produced by the reaction between CO_2 and H_2O:

$$CO_2(aq) + H_2O(l) \longrightarrow H_2CO_3(aq)$$

What is the pH of this solution?

[18.74] The precipitation of $Al(OH)_3$ ($K_{sp} = 1.3 \times 10^{-33}$) is sometimes used to purify water. **(a)** Estimate the pH at which precipitation of $Al(OH)_3$ will begin if 2.0 lb of $Al_2(SO_4)_3$ is added to 1000 gal of water. **(b)** Approximately how many pounds of CaO must be added to the water to achieve this pH?

eMedia Exercises

18.75 The formation of ozone molecules in the stratosphere is shown in the **Stratospheric Ozone** movie (*eChapter 18.2*). **(a)** Describe this process, including equations where appropriate. **(b)** Explain why the process described in part (a) does not occur to the same extent in the troposphere as it does in the stratosphere.

18.76 The **Stratospheric Ozone** movie (*eChapter 18.2*) shows two ways in which an ozone molecule can dissociate. **(a)** Describe both of these processes. **(b)** Which of these processes protects Earth from harmful radiation?

18.77 The **Catalytic Destruction of Stratospheric Ozone** movie (*eChapter 18.3*) shows how NO lowers the concentration of ozone by catalyzing its decomposition. Based on the movie, describe how you think a higher NO_2 concentration in the stratosphere would affect ozone concentration. Explain your reasoning, and use equations where necessary.

18.78 The **CFCs and Stratospheric Ozone** movie (*eChapter 18.3*) portrays chlorofluorocarbons as very stable molecules. **(a)** If CFCs are so stable, why are they considered to be a serious problem in the upper atmosphere? **(b)** If the countries of the world were to stop producing CFCs altogether, their effects in the upper atmosphere would continue for decades. Explain why this is so.

18.79 In Exercise 18.20, you explained why rainwater is acidic even in the absence of pollution, such as SO_3 gas, and the movie **Carbon Dioxide Behaves as an Acid in Water** (*eChapter 18.4*) shows how CO_2 acidifies water. Compare the typical atmospheric concentrations (Table 18.3) of CO_2 and SO_2, which is oxidized to SO_3 in the troposphere. Which one would you expect to have the greater influence on the pH of rainwater? Explain.

Chapter 19

Chemical Thermodynamics

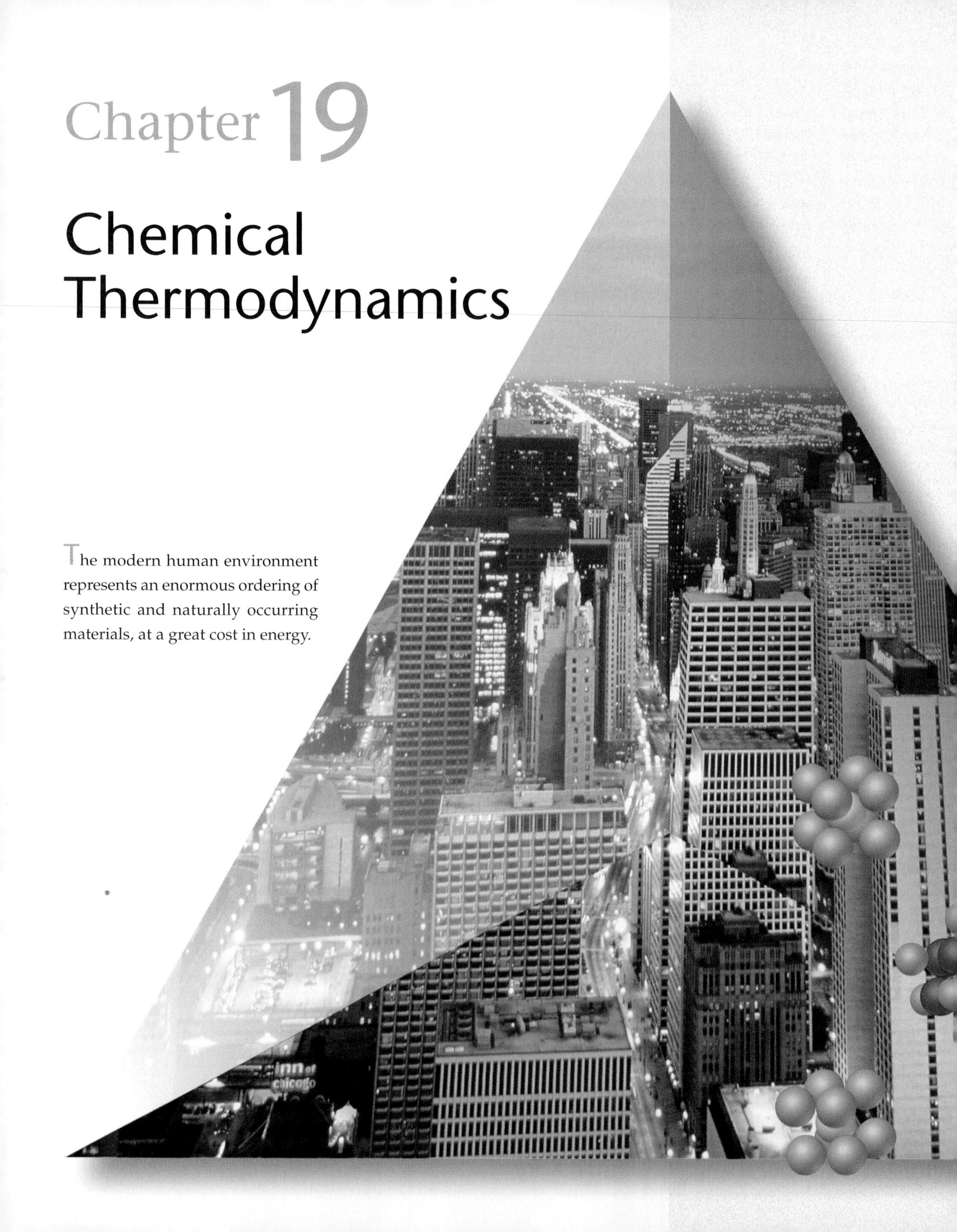

The modern human environment represents an enormous ordering of synthetic and naturally occurring materials, at a great cost in energy.

WE HAVE ALREADY looked at two of the most important questions that chemists ask when designing and understanding chemical reactions: First, how rapidly does the reaction progress? Second, how far toward completion does the reaction proceed? The first question is addressed by a study of the reaction rate, which we discussed in Chapter 14. The second question involves the equilibrium constant, which was the focus of Chapter 15. Let's briefly review how these concepts are related.

In Chapter 14 we learned that the rates of reactions are largely controlled by a factor related to *energy;* namely, the activation energy of the reaction. (Section 14.5) The lower the activation energy, the faster a reaction proceeds. In Chapter 15 we saw that equilibrium depends on the rates of the forward and reverse reactions: Equilibrium is reached when the opposing reactions occur at equal rates. (Section 15.1) Because reaction rates are closely tied to energy, it is logical that equilibrium should also depends in some way on energy. In this chapter we will see how chemical equilibrium is related to the energies of the reactants and products. To do so, we will take a deeper look at *chemical thermodynamics*, the area of chemistry that explores energy relationships.

We first encountered thermodynamics in Chapter 5, where we discussed the nature of energy, the first law of thermodynamics, and the concept of enthalpy. Recall that, for a reaction that occurs at constant pressure, the enthalpy change equals the heat transferred between the system and its surroundings. (Section 5.3) As we discussed in the "Strategies in Chemistry" box in Section 5.4, the enthalpy change of a reaction is an important guide as to whether the reaction is likely to proceed. However, we also pointed out that enthalpy is not the *only* factor that governs whether the reactants or products of a reaction are more favored. For example, ice melts at room temperature even though the process is endothermic. Similarly, $NH_4NO_3(s)$ readily dissolves in water even though ΔH_{soln} is positive. (Section 13.1) Clearly, the sign of the enthalpy change alone is not enough to tell us whether a reaction will proceed.

In this chapter we will address several aspects of chemical thermodynamics. We will see that in addition to enthalpy, we must consider the change in the *randomness* or *disorder* that accompanies a chemical reaction. We alluded to this notion in Section 13.1. Finally, we will learn how to combine the enthalpy change of a reaction with the change in randomness to define a new type of energy that relates directly to equilibrium. We begin by introducing a new aspect to our discussion of thermodynamics, namely the idea of *spontaneous* processes.

▶ What's Ahead ◀

- In this chapter we will put a more formal stamp on our understanding that changes that occur in nature have a directional character: They move *spontaneously* in one direction but not in the reverse direction.
- The thermodynamic function, *entropy*, is a state function (like enthalpy and internal energy), which may be thought of as a measure of disorder or randomness.
- The *second law of thermodynamics* tells us that in any spontaneous process the net entropy of the universe (system plus surroundings) increases.
- The *third law of thermodynamics* states that the entropy of a perfect crystalline solid at 0 K is zero. From this reference point the absolute entropies of pure substances at temperatures above absolute zero can be calculated from experimental data.
- The *free energy* (or *Gibbs free energy*) is a measure of how far removed a system is from equilibrium. It measures the maximum amount of useful work obtainable from a given process and yields information on the direction in which a chemical system will proceed spontaneously.
- The standard free-energy change for a chemical reaction can be used to calculate the equilibrium constant for the process.

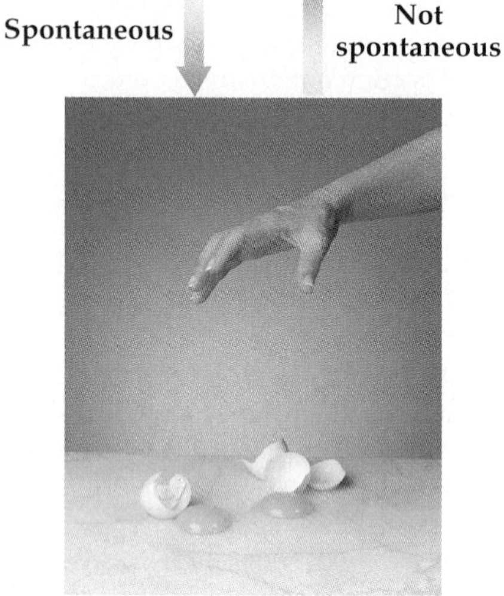

▲ **Figure 19.1** A spontaneous process. This sequence of photographs has an inherent direction: We recognize that the photo of the broken eggs was taken after the photo of the unbroken eggs. The process is spontaneous in one direction, and not spontaneous in the reverse direction.

▲ **Figure 19.2** The combustion of natural gas (methane) in air, producing carbon dioxide and water, is a spontaneous process, once initiated.

19.1 Spontaneous Processes

In Chapter 5 we discussed the first law of thermodynamics, which states that *energy is conserved.* ∞ (Section 5.2) In other words, energy is neither created nor destroyed in any process, whether it be the falling of a brick, the melting of an ice cube, or the combustion of gasoline.* Energy can be transferred between the system and the surroundings or can be converted from one form to another, but the total energy remains constant. We expressed the first law of thermodynamics mathematically as $\Delta E = q + w$, where ΔE is the change in the internal energy of a system, q is the heat absorbed by the system from the surroundings, and w is the work done on the system by the surroundings.

The first law helps us to balance the books, so to speak, on the heat exchanged and the work done in a particular process or reaction. However, because energy is conserved, we can't use the change in energy itself as a criterion for whether the process is favored to occur; anything we do to lower the energy of the system will raise the energy of the surroundings, and vice versa. Nevertheless, our experience tells us that certain processes will *always* occur, even though the energy of the universe is conserved. For instance, water placed in a freezer turns into ice. A shiny nail left outdoors will eventually rust. If you touch a hot object, heat is transferred to you. For all of these processes, energy is conserved, as it must be according to the first law of thermodynamics. And yet they occur without any ongoing outside intervention; such processes are said to be **spontaneous**.

A spontaneous process has a definite direction in which it occurs. For example, if you drop an egg above a hard surface, it will fall and break on impact. (Figure 19.1 ◀) Now imagine that you were to see a video clip in which a broken egg rises from the floor and reassembles itself. You would conclude that the video is running in reverse—broken eggs simply do not magically rise and reassemble themselves! An egg falling and breaking is spontaneous. An example of a chemical process with definite direction is the combustion of natural gas (methane). Once lit, natural gas burns on a stovetop, producing carbon dioxide and water (Figure 19.2 ◀). Carbon dioxide and water do not spontaneously combine to reform methane. The combustion of methane is spontaneous. Neither reverse process, the un-breaking of an egg or the combination of carbon dioxide and water to reform methane, is spontaneous, even though energy is conserved in both the forward and reverse directions. Clearly, there is something other than the internal energy that determines whether a process is spontaneous.

The spontaneity of a process can depend on temperature. Consider, for example, the endothermic process of melting ice under atmospheric pressure. When $T > 0°C$, ice melts spontaneously; the reverse process, liquid water turning into ice, is not spontaneous at these temperatures. However, when $T < 0°C$, the opposite is true. Liquid water converts into ice spontaneously, and the conversion of ice into water is *not* spontaneous (Figure 19.3 ▶). What happens at $T = 0°C$, the normal melting point of water? At the normal melting point of a substance, the solid and liquid phases are in equilibrium. ∞ (Section 11.6) At this particular temperature the two phases are interconverting at the same rate, and there is no preferred direction for the process: Both the forward and reverse processes occur with equal preference, and the process is not spontaneous in either direction.

* The first law as stated here does not apply to nuclear reactions, such as those occurring in the stars or in nuclear reactors, in which mass and energy are interconverted. We discuss nuclear processes in more detail in Chapter 21.

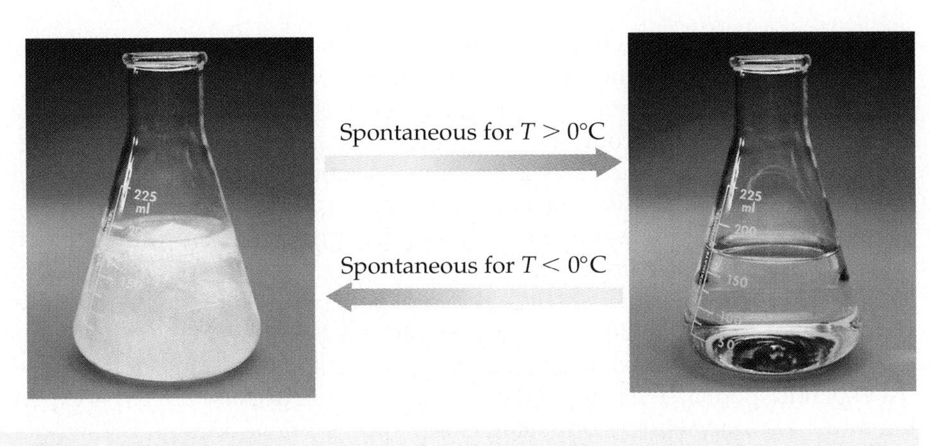

◀ **Figure 19.3** The spontaneity of a process can depend on the temperature. At $T > 0°C$ ice melts spontaneously to liquid water. At $T < 0°C$ the reverse process, water freezing to ice, is spontaneous. At $T = 0°C$ the two states are in equilibrium and neither conversion occurs spontaneously.

SAMPLE EXERCISE 19.1

Predict whether the following processes are spontaneous as described, are spontaneous in the reverse direction, or are in equilibrium: **(a)** When a piece of metal heated to 150°C is added to water at 40°C, the water gets hotter. **(b)** Water at room temperature decomposes into $H_2(g)$ and $O_2(g)$. **(c)** Benzene vapor, $C_6H_6(g)$, at a pressure of 1 atm condenses to liquid benzene at the normal boiling point of benzene, 80.1°C.

Solution

Analyze and Plan: We are asked to judge whether each process will proceed spontaneously in the direction indicated, in the reverse direction, or neither.

Solve: (a) This process is spontaneous. Whenever two objects of different temperature are brought into contact, heat is transferred from the hotter object to the colder one. (Section 5.1) In this instance heat is transferred from the hot metal to the cooler water. The final temperature, after the metal and water achieve the same temperature (thermal equilibrium), is a value between the initial temperatures of the metal and the water. **(b)** Experience tells us that this process is not spontaneous; rather, the *reverse* process—the reaction of H_2 and O_2 to form H_2O—is spontaneous once initiated by a spark or flame (Figure 5.12). **(c)** By definition, the normal boiling point is the temperature at which the vapor with a pressure of 1 atm is in equilibrium with the liquid. Thus, this is an equilibrium situation. Neither the condensation of benzene vapor nor the reverse process is spontaneous. If the temperature were less than 80.1°C, the condensation of benzene vapor would be spontaneous.

PRACTICE EXERCISE

Under 1 atm pressure $CO_2(s)$ (Dry Ice™) sublimes at −78°C. Is the transformation of $CO_2(s)$ to $CO_2(g)$ a spontaneous process at −100°C and 1 atm pressure?
Answer: No, the reverse process is spontaneous at this temperature.

Reversible and Irreversible Processes

To understand better why certain processes are spontaneous, we need to consider more closely the ways in which the state of a system can change. Recall that quantities such as temperature, internal energy, and enthalpy are *state functions*, properties that define a state and do not depend on how the system arrives at that state. (Section 5.2) We also saw that the heat transferred between the system and the surroundings (q) and the work done by or on the system (w) are *not* state functions; the values of q and w depend on the specific path taken from one state to another.

A **reversible process** is a special way in which the state of a system can change. In a reversible process, the change in the system is made in such a way that the system can be restored to its original state by *exactly* reversing the change. In other words, we can completely reverse the change in the system with no net change in either the system or the surroundings. As an example of a reversible process, let's consider again the interconversion of ice and water that is shown in Figure 19.3. At 1 atm pressure, ice and liquid water are in equilibrium with one another at 0°C. Now imagine that we melt 1 mol of ice at 0°C, 1 atm, to form 1 mol of liquid water at 0°C, 1 atm. We can achieve this change by adding a certain amount of heat to the system from the surroundings: $q = \Delta H_{fus}$. If we

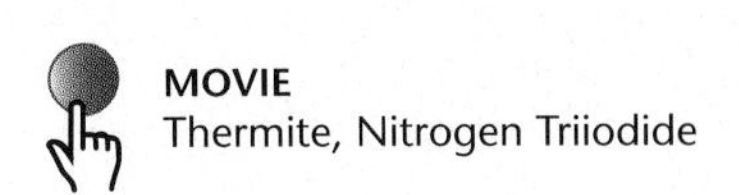

want to return the system to its original state (ice at 0°C), we can simply reverse the procedure by removing the same amount of heat, ΔH_{fus}, *from* the system *to* the surroundings.* After we reverse the change, it is as if nothing has happened; both the system and surroundings are exactly as they were at the beginning. It is important to realize that there is only one specific value of q for any *reversible isothermal* path between two states of a system.

An **irreversible process** is one that cannot simply be reversed to restore the system and its surroundings to their original state. When a system changes by an irreversible process, it must take a different path (with different values of q and w) to get back to its original state. For example, imagine a gas in the cylinder-and-piston arrangement shown in Figure 19.4 ▶. When the partition is removed, the gas expands spontaneously to fill the evacuated space. Because the gas is expanding into a vacuum with no opposing external pressure, it does no P-V work on the surroundings ($w = 0$). (Section 5.3) We can use the piston to compress the gas back to its original state. But doing so requires that the surroundings do work on the system ($w > 0$). In other words, the path to restore the system to its original state requires a different value of w (and, by the first law of thermody-

* In actuality, for the process described to be truly reversible, we would have to add and remove the heat infinitely slowly. All reversible processes occur infinitely slowly; thus, no process that we can observe is truly reversible. We could make the melting of ice at 0°C very nearly reversible by adding the heat very slowly.

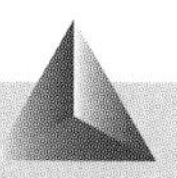

A Closer Look Reversibility and Work

If a process is spontaneous, it can be used to accomplish work, as we do when we burn gasoline in our car engines. Can all the energy change associated with burning gasoline be used to do work, or is there some limit? Scientists and engineers have studied the operation of engines with this very practical question in mind.

The amount of work we can extract from any spontaneous process depends on the manner in which it is conducted. To illustrate this idea, let's first consider the simple arrangement shown in the accompanying figure on the next page. A weight M resting on a platform is our system. For simplicity, let's assume that the rope and platforms are weightless and the pulley is frictionless. If there is no counterbalancing weight m on the right, the platform bearing weight M will fall to the ground without accomplishing any work at all on the surroundings. The potential energy it possessed is all converted to kinetic energy, which in turn is converted to heat at the time of impact.

Now suppose that the counterbalancing weight m is the same as the weight on the platform, M. If M moves from the height h to the ground, it will cause m to be lifted from the ground to height h. This means that the system (that is, M) will have done work on the surroundings that is exactly equal to the loss of its own potential energy in moving from height h to the ground. In this case none of the energy initially possessed by M is converted into heat. But can this process actually occur? If m exactly equals M, there will be no spontaneous movement of M. We can imagine, however, an infinitesimally small force being applied ever so slowly to move the weights. The resultant hypothetical process is completely *reversible*. The process occurs in such a way that the system can be restored to its original state by *exactly* reversing the change.

A reversible change produces the maximum amount of work that can be achieved by the system on the surroundings ($w_{rev} = w_{max}$). Although the process is hypothetical, it gives us a goal by which we can judge the efficiency with which a device is able to convert energy to work.

We can apply these ideas to other processes, such as the expansion of a gas. Gases can do work when they expand, by pushing against an external pressure. Imagine an ideal gas expanding while maintaining a constant temperature. (A process that occurs at constant temperature is said to be *isothermal*.)

What might a reversible, isothermal expansion of an ideal gas be like? It will occur only if the external pressure acting on the piston exactly balances the pressure exerted by the gas. Under these conditions, the piston will not move unless we imagine the external pressure being reduced infinitely slowly, allowing the pressure of the confined gas to readjust to maintain a balance in the two pressures. This gradual, infinitely slow process in which the external pressure and internal pressure are always in equilibrium is reversible. If we reverse the process and compress the gas in the same infinitely slow manner, we are able to return the gas to its original volume. The complete cycle of expansion and compression in this hypothetical process, moreover, is accomplished without any net change to the surroundings. *A process is reversible in this thermodynamic sense not merely because the system can be returned to its original condition, but because the overall cycle can be accomplished without any associated change in the surroundings.* Furthermore, any process that is not reversible is irreversible. Because real processes can at best only approximate the slow, ever-in-equilibrium change associated with reversible processes, all real processes are irreversible to some extent. Thus, *any spontaneous process will be irreversible*, and even if we return the system to the original condition, the surroundings will remain changed.

There is yet one more important conclusion to draw from studies of reversible processes. In Section 5.2 we noted that

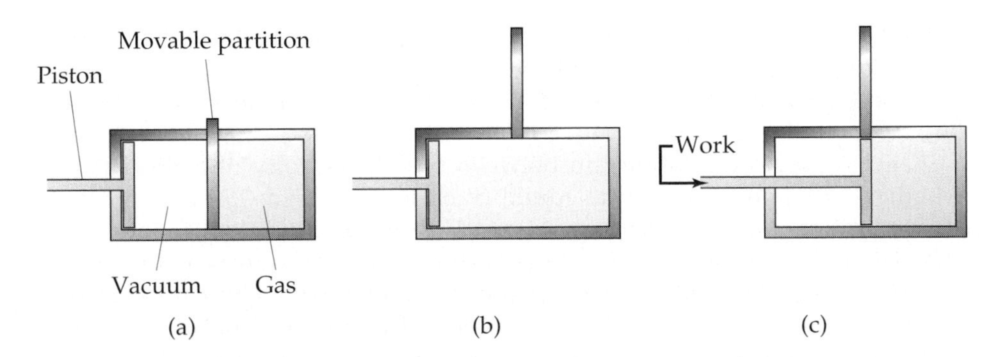

◀ **Figure 19.4** Restoring the system to its original state after an irreversible process changes the surroundings. In (a) the gas is confined to the right half of the cylinder by a partition. When the partition is removed (b), the gas spontaneously (irreversibly) expands to fill the whole cylinder. No work is done by the system during this expansion. In (c) we can use the piston to compress the gas back to its original state. Doing so requires that the surroundings do work on the system, which changes the surroundings forever.

namics, a different value of q) than did the path by which the system was first changed. The expansion of a gas into a vacuum is an irreversible process. Likewise, ice melting at room temperature and water freezing at −30°C are irreversible processes.

The fact that the system must take a different path back to its starting point is not the only irreversible aspect of an irreversible process. We noted before that when a reversible process is reversed, both the system *and* the surroundings are returned to their original conditions. This is not the case for an irreversible process. Although we can restore the system to its original state by choosing a different path, the surroundings will have changed from their original condition.

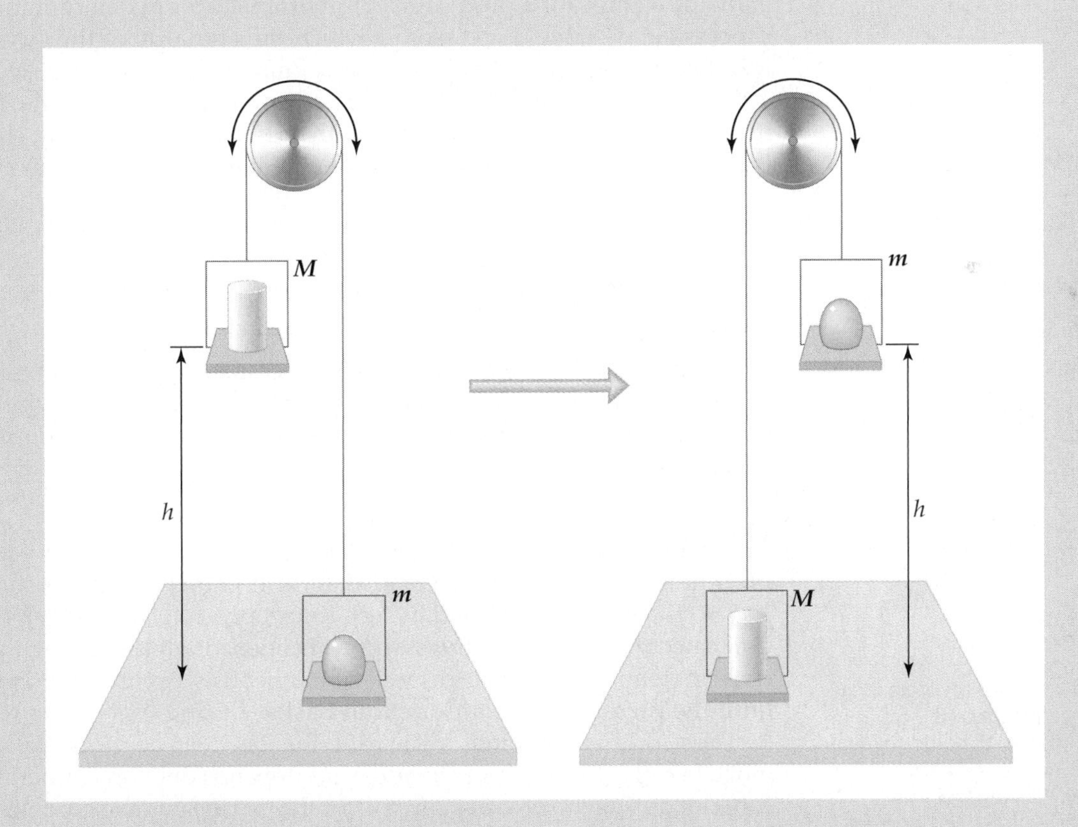

neither heat nor work are functions of state, but functions of path. (That is, their magnitudes depend on how the process is performed.) The maximum work that can be accomplished (that is, w_{rev}), however, has a unique value for any given change in the system. In a similar way, q_{rev}, the heat gained or lost by the system if the process were conducted reversibly, also has a unique value for any given process. So every process, whether conducted reversibly or not, has associated with it benchmark quanties, w_{rev} and q_{rev}, that depend only on the initial and final states of the system. Thus, w_{rev} and q_{rev} are state functions. (This does not contradict our earlier statement that q and w are not state functions, because we have specified that w_{rev} and q_{rev} are the values for a particular, unique pathway, the reversible one.)

There is nothing we can do to prevent these changes to the surroundings—they are an unavoidable result of irreversible processes. In Section 19.2 we will see that these necessary changes in the surroundings are an important aspect of the second law of thermodynamics.

There is a close relationship between the reversibility of a process and whether it is spontaneous or at equilibrium. Recall Figure 19.3, in which we showed the spontaneous melting of ice at $T > 0°C$ and the spontaneous freezing of liquid water at $T < 0°C$. Both of these processes are irreversible. At $T = 0°C$ ice and water are in equilibrium, and they can convert back and forth reversibly. These observations are examples of two very important concepts regarding reversible and irreversible processes:

1. Whenever a chemical system is in equilibrium, reactants and products can interconvert *reversibly*.
2. In any spontaneous process, the path between reactants and products is *irreversible*.

Finally, it is important to realize that just because a process is spontaneous does not necessarily mean that it will occur at an observable rate. A spontaneous reaction can be very fast, as in the case of acid-base neutralization, or very slow, as in the rusting of iron. Thermodynamics can tell us the *direction* and *extent* of a reaction, but it tells us nothing about its *speed*.

What factors make a process spontaneous? In Chapter 5 we saw that the enthalpy change for a process is an important factor in determining whether the process is favorable. Processes in which the enthalpy of the system decreases (exothermic processes) tend to occur spontaneously. However, we will see that considering only the enthalpy change of a process is not enough. The spontaneity of a process also depends on how the disorder of the system changes during the process. In Section 19.2 we consider the matter of disorder in greater detail.

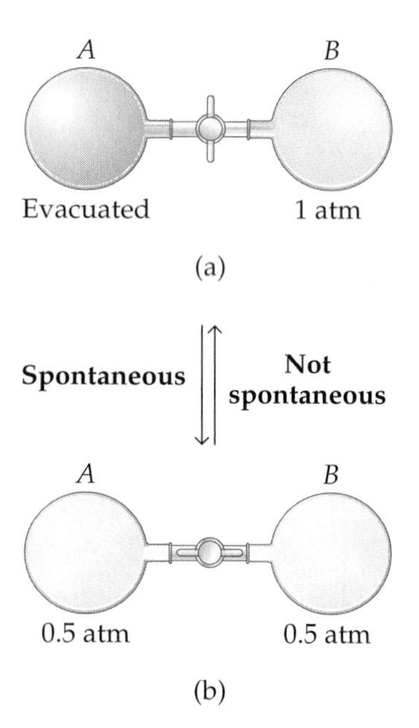

▲ **Figure 19.5** Expansion of an ideal gas into an evacuated space is spontaneous. In (a) flask *B* holds an ideal gas at 1 atm pressure and flask *A* is evacuated. In (b) the stopcock connecting the flasks has been opened. The ideal gas expands to occupy both flasks *A* and *B* at a pressure of 0.5 atm. The reverse process is not spontaneous.

19.2 Entropy and the Second Law of Thermodynamics

We have now seen several examples of processes that occur spontaneously, some exothermic and some endothermic. In order to understand why spontaneous processes occur, let's consider the expansion of a gas into a vacuum at the molecular level.

The Spontaneous Expansion of a Gas

Imagine an ideal gas confined to a 1-L flask at 1 atm pressure, as shown in Figure 19.5 ◀. The flask is connected by a closed stopcock to another 1-L flask, which is evacuated. Now suppose the stopcock is opened while keeping the system at a constant temperature. The gas will spontaneously expand into the second flask until the pressure is 0.5 atm in both flasks. During this constant-temperature (*isothermal*) expansion into a vacuum, the gas does no work ($w = 0$). Furthermore, because the energy of an ideal gas depends only on temperature, which is constant during the process, $\Delta E = 0$ for the expansion.* Nevertheless, the process is spontaneous. The reverse process, in which the gas that is evenly distributed between the two flasks spontaneously moves entirely into one of the flasks, is inconceivable. This reverse process, however, would also involve no heat

* We learned in the "Closer Look" box in Section 5.3 that PV has the units of energy. The product PV is the internal energy of an ideal gas. For an ideal gas, $PV = nRT$. If T is constant, the expression on the right-hand side of the equation is constant. The energy of the gas, therefore, is constant, so for a change occurring at constant temperature, ΔE is zero.

transferred or work done. Like the spontaneous expansion of the gas, it would also have $\Delta E = 0$. (Remember that ΔE for the reverse process is the negative of ΔE for the forward process: $\Delta E_{reverse} = -\Delta E_{forward} = 0$.) Clearly, some factor other than heat or work is important in making the expansion of the gas spontaneous.

We can get a sense of what makes the expansion of a gas spontaneous by envisioning the gas as a collection of particles in constant motion, as we did in discussing the kinetic-molecular theory of gases. (Section 10.7) When the stopcock is opened, we can view the expansion of the gas as the ultimate result of the gas molecules moving randomly throughout the larger volume. Let's look at this idea a little more closely. Suppose we were able to track two of the gas molecules as they undergo their motion. Before the stopcock is opened, both molecules are confined to the right-hand flask, as shown in Figure 19.6(a) ▼. After the stopcock is opened, the molecules travel randomly throughout the entire apparatus. Thus, as shown in Figure 19.6(b), there are four possible arrangements in which we find the two molecules. Because of the random motion of the molecules, each of these four arrangements is equally likely. Note that now only one of the four arrangements corresponds to both molecules' being in the right-hand flask; the probability is $\frac{1}{2}$ that each molecule is in the right-hand flask, so the probability that *both* molecules are in the right-hand flask is $\left(\frac{1}{2}\right)^2 = \frac{1}{4}$.

If we were to apply the same analysis to *three* gas molecules, the probability that all three are in the right-hand flask at the same time is $\left(\frac{1}{2}\right)^3 = \frac{1}{8}$. Now let's consider a *mole* of gas. The probability that all the molecules are in the right-hand flask at the same time is $\left(\frac{1}{2}\right)^N$, where $N = 6.02 \times 10^{23}$. This is a staggeringly small number! Thus, there is essentially zero likelihood that all the gas molecules will be found in the right-hand flask at the same time. This analysis of the microscopic behavior of the gas molecules leads to the expected macroscopic behavior: The gas spontaneously expands to fill both the left and right flasks, and it will not spontaneously all go back into the right-hand flask.

The gas expands because of the tendency for the molecules to "spread out" among the different arrangements that they can take. Before the stopcock is opened, there is only one possible distribution of the molecules: They are all in the right-hand flask. When the stopcock is opened, the arrangement in which all the molecules are in the right-hand flask is but one of an extremely large number of possible arrangements. The most probable arrangements by far are those in which there are essentially equal numbers of molecules in each flask. When the gas spreads throughout the entire apparatus, any given molecule could be in either flask rather than confined to the right-hand flask. We say, therefore, that the arrangement of gas molecules becomes more random or disordered than it was when the gas molecules were entirely in the right-hand flask. As we have seen before, *processes in which the disorder of the system increases tend to occur spontaneously*. (Section 13.1) We will see shortly that this idea is the basis of the second law of thermodynamics.

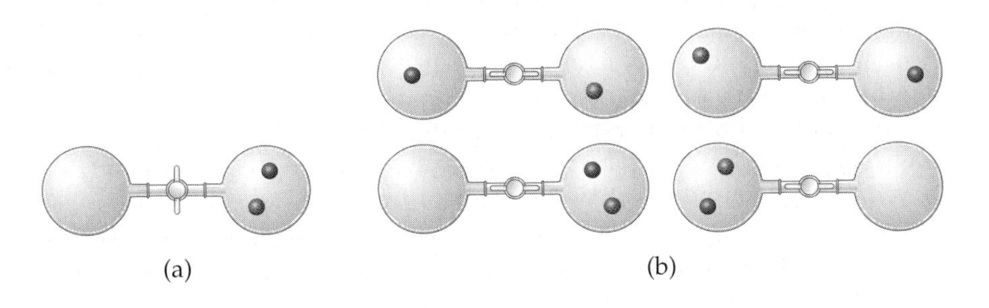

◀ **Figure 19.6** Possible locations of two of the gas molecules involved in the expansion in Figure 19.5. The two molecules are colored red and blue to keep track of them. (a) Before the stopcock is opened, both molecules are in the right-hand flask. (b) After the stopcock is opened, there are four possible arrangements of the two molecules between the two flasks. Only one of the four arrangements corresponds to both molecules' being in the right-hand flask. The greater number of possible arrangements corresponds to greater disorder of the system.

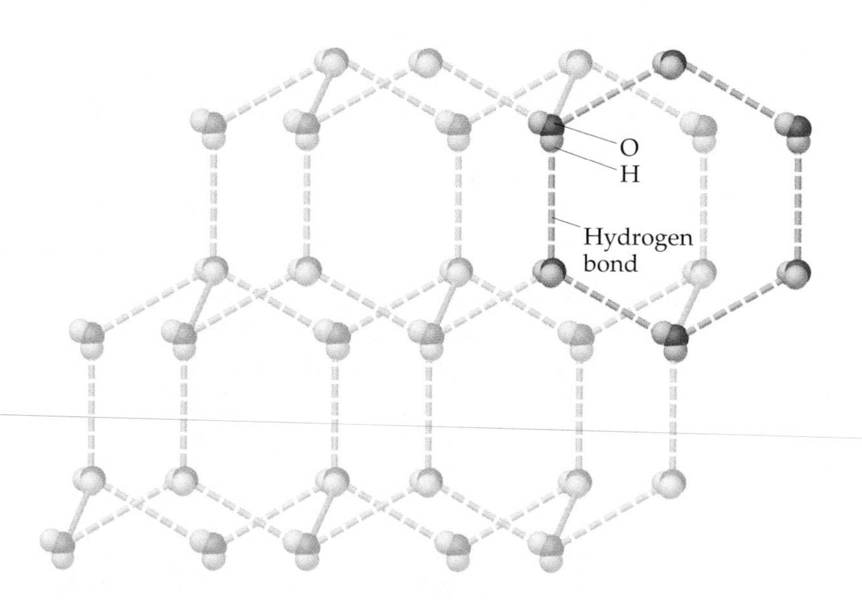

▶ **Figure 19.7** Structure of ice, a well-ordered system.

Entropy

The isothermal expansion of a gas is spontaneous because of the increase in the randomness or disorder of the gas molecules upon expansion. Let's extend this concept to two other spontaneous processes that we have discussed. First, ice melts spontaneously at temperatures above its melting point even though it is an endothermic process. (Section 11.6) Second, salts such as $NH_4NO_3(s)$ and $KCl(s)$ readily dissolve in $H_2O(l)$ even though $\Delta H_{soln} > 0$. (Section 13.1) On the basis of enthalpy, both of these processes are unfavorable. But each shares something in common with the expansion of a gas: In each of these processes the products are in a more random or disordered state than the reactants.

Let's first consider the melting of ice. The molecules of water that make up an ice crystal are held rigidly in place in the ice crystal lattice (Figure 19.7 ▲). When the ice melts, the water molecules are free to move about with respect to one another and to tumble around. Thus, in liquid water the individual water molecules are more randomly distributed than in the solid. The well-ordered solid structure is replaced by the much more disordered liquid structure.

A similar situation applies to a KCl crystal when it dissolves in water. In solid KCl the K^+ and Cl^- ions are in a highly ordered, crystalline state. (Section 11.7) When the solid dissolves, the ions are free to move about in the water. They are in a much more random and disordered state than before. At the same time, however, water molecules are held around the ions as water of hydration, as shown in Figure 19.8 ▶. (Section 13.1) These water molecules are in a *more* ordered state than before because they are now confined to the immediate environment of the ions. Therefore, the dissolving of a salt involves both disordering *and* ordering processes. The disordering processes are usually dominant, and so the overall effect is an increase in disorder of the system upon dissolving most salts in water.

As these examples illustrate, the change in disorder along with the change in energy affects the spontaneity of chemical processes. The disorder is expressed by a thermodynamic quantity called **entropy**, given the symbol S. *The more disordered or random a system, the larger its entropy.* Like internal energy and enthalpy, entropy is a state function. (Section 5.2) The change in entropy of a system, $\Delta S = S_{final} - S_{initial}$, depends only on the initial and final states of the system and not on the particular pathway by which the system changes. A positive value of ΔS indicates that the final state is more disordered than the initial state. Thus, when a gas expands into a larger volume, as in Figure 19.5, its entropy increases and ΔS is a positive number. Likewise, when ice melts, the system becomes more

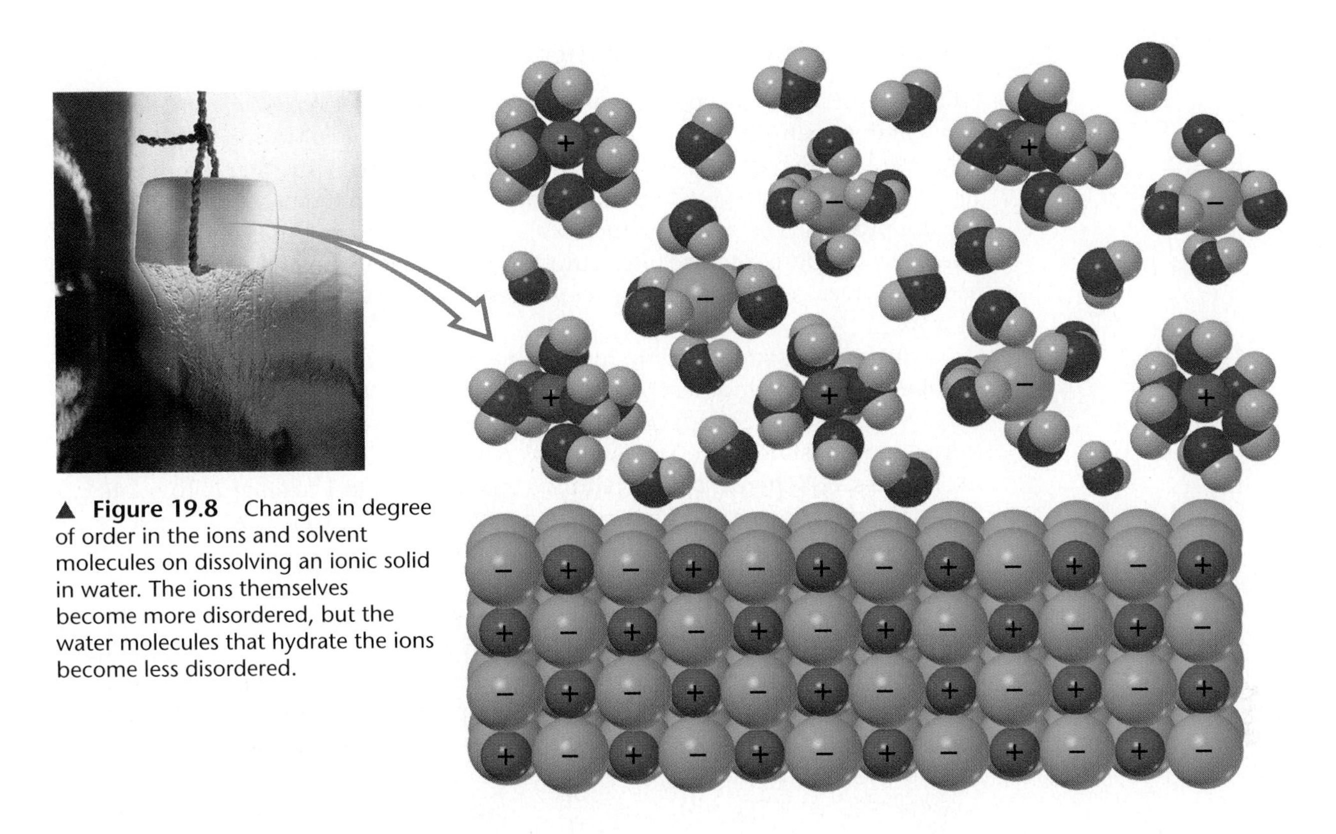

▲ **Figure 19.8** Changes in degree of order in the ions and solvent molecules on dissolving an ionic solid in water. The ions themselves become more disordered, but the water molecules that hydrate the ions become less disordered.

disordered and $\Delta S > 0$. A negative value of ΔS indicates that the final state is more ordered or less random than the initial state.

SAMPLE EXERCISE 19.2

By considering the disorder in the reactants and products, predict whether ΔS is positive or negative for each of the following processes:

(a) $H_2O(l) \longrightarrow H_2O(g)$
(b) $Ag^+(aq) + Cl^-(aq) \longrightarrow AgCl(s)$
(c) $4Fe(s) + 3O_2(g) \longrightarrow 2Fe_2O_3(s)$

Solution

Analyze: We are given three equations and asked to predict the sign of ΔS for these processes.

Plan: The sign of ΔS will be positive if there is an increase in disorder and negative if there is a decrease. Thus, we need to evaluate each equation to determine whether there is an increase or decrease in disorder or randomness.

Solve: (a) The evaporation of a liquid is accompanied by a large increase in volume. One mole of water (18 g) occupies about 18 mL as a liquid and 22.4 L as a gas at STP. Because the molecules are distributed throughout a much larger volume in the gaseous state than in the liquid state, an increase in disorder accompanies vaporization. Therefore, ΔS is positive.

(b) In this process the ions that are free to move about in the larger volume of the solution form a solid in which the ions are confined to highly ordered positions. Thus, there is a decrease in disorder, and ΔS is negative.

(c) The particles of a solid are much more highly ordered and confined to specific locations than are the molecules of a gas. Because a gas is converted into part of a solid product, disorder decreases and ΔS is negative.

PRACTICE EXERCISE

Indicate whether each of the following reactions produces an increase or decrease in the entropy of the system:

(a) $CO_2(s) \longrightarrow CO_2(g)$
(b) $CaO(s) + CO_2(g) \longrightarrow CaCO_3(s)$

Answers: **(a)** increase; **(b)** decrease

Relating Entropy to Heat Transfer and Temperature

At this point in our discussion entropy might strike you as a somewhat abstract concept that seems unconnected to the other thermodynamic quantities we have discussed. However, we can relate the change in entropy to other familiar quantities; in particular, the change in entropy is related to the heat transferred during the process. The way in which we calculate ΔS from other thermodynamic quantities involves the use of calculus and is generally beyond the scope of this text. However, there are some special cases in which we can determine the value of ΔS.

Entropy is a state function. In order to understand the relationship between change in entropy and heat, we need to remember that heat is *not* a state function. (Section 5.2) Suppose a system undergoes a process in which it changes from an initial state (state 1) to a final state (state 2). The heat transferred during the process, q, depends on the path we take from state 1 to state 2. To relate ΔS to heat, we need to consider a particular type of path from state 1 to state 2,* namely a *reversible* path. In Section 19.1 we discussed reversible processes, such as the melting of ice at 0°C. As we noted then, there is only one specific value of q for any reversible path between two states. We will denote the heat transferred along such a path as q_{rev}, where the subscript "rev" reminds us that the path between the states is reversible.

For a process that occurs at constant temperature, the entropy change of the system is the value of q_{rev} divided by the absolute temperature:

$$\Delta S_{sys} = \frac{q_{rev}}{T} \quad \text{(constant } T\text{)} \qquad [19.1]$$

Although there are many possible paths that could take the system from state 1 to state 2, there is only one possible value of q_{rev}. Because q_{rev} and T are both path-independent, S is a state function. Thus, Equation 19.1 applies to *any* isothermal process between states, not just those that are reversible.

An example of a reversible process that occurs at constant temperature is a phase change at the temperature at which the phases are in equilibrium, such as water boiling at 100°C. Let's calculate the entropy change, ΔS_{vap}, when 1 mol of water is converted into 1 mol of steam at 1 atm pressure. The amount of heat transferred to the system during this process, q_{rev}, is the heat of vaporization ΔH_{vap}, and the temperature at which the process is reversible is the normal boiling point, T_b. (Section 11.4) For H_2O, $\Delta H_{vap} = +40.67$ kJ/mol and $T_b = 100°C = 373$ K. Thus,

$$\Delta S_{vap} = \frac{\Delta H_{vap}}{T_b} = \frac{(1 \text{ mol})(+40.67 \text{ kJ/mol})(1000 \text{ J/1 kJ})}{(373 \text{ K})} = +109 \text{ J/K}$$

Notice that the change in entropy is positive; the molecules in $H_2O(g)$ have more disorder than those in $H_2O(l)$. The units for ΔS, J/K, are energy divided by temperature, which we expect from Equation 19.1.

SAMPLE EXERCISE 19.3

The element mercury, Hg, is a silvery liquid at room temperature. The normal freezing point of mercury is −38.9°C, and its molar enthalpy of fusion is $\Delta H_{fus} = 2.29$ kJ/mol. What is the entropy change of the system when 50.0 g of Hg(*l*) freezes at the normal freezing point?

Solution

Analyze: We first recognize that Hg(*l*) and Hg(*s*) are in equilibrium at the normal freezing point. As such, the freezing of Hg(*l*) at its freezing point is a reversible process,

* The special relationship between ΔS and the heat transferred in a reversible process is similar to the special relationship between ΔH and the heat transferred at constant pressure. (Section 5.3)

analogous to the boiling of water at its normal boiling point. We next realize that freezing is an *exothermic* process; heat is transferred from the system to the surroundings when a liquid freezes ($q < 0$). The enthalpy of fusion is defined for melting. Because freezing is the reverse process of melting, the enthalpy change that accompanies the freezing of 1 mol of Hg is $-\Delta H_{fus} = -2.29$ kJ/mol.

Plan: We can use $-\Delta H_{fus}$ and the atomic weight of Hg to calculate q for freezing 50.0 g of Hg:

$$q = (-2.29 \text{ kJ/mol Hg})\left(\frac{1000 \text{ J}}{1 \text{ kJ}}\right)\left(\frac{1 \text{ mol Hg}}{200.59 \text{ g Hg}}\right)(50.0 \text{ g Hg}) = -571 \text{ J}$$

Freezing at the normal freezing point is a reversible process, so we can use this value of q as q_{rev} in Equation 19.1. We must first, however, convert the temperature to K: $-38.9°\text{C} = (-38.9 + 273.15) \text{ K} = 234.3 \text{ K}$.

Solve: We can now calculate the value of ΔS_{sys}:

$$\Delta S_{sys} = \frac{q_{rev}}{T} = (-571 \text{ J})/(234.3 \text{ K}) = -2.44 \text{ J/K}$$

Check: The entropy change is negative, implying that the system becomes more ordered during the process. This makes sense because solids are more ordered (lower entropy) than are liquids. Notice also that the magnitude of this entropy change is much smaller than that for the conversion of a mole of $H_2O(l)$ to $H_2O(g)$, which we calculated earlier. Gases are far more disordered than liquids or solids. The magnitudes of entropy changes for processes that involve the conversion of a solid or liquid into a gas (or vice versa) are generally larger than for processes that involve only solids and liquids.

PRACTICE EXERCISE

The normal boiling point of ethanol, C_2H_5OH, is 78.3°C (see Figure 11.12), and its molar enthalpy of vaporization is 38.56 kJ/mol. What is the change in entropy when 68.3 g of $C_2H_5OH(g)$ at 1 atm pressure condenses to liquid at the normal boiling point?
Answer: −163 J/K.

The Second Law of Thermodynamics

With the introduction of the concept of entropy, we are now in a position to discuss why certain processes are spontaneous. The law that expresses the notion that there is an inherent direction in which processes occur is called the **second law of thermodynamics.**

In the context of chemistry, the second law usually is expressed in terms of entropy. We must consider both the change in entropy of the system *and* the

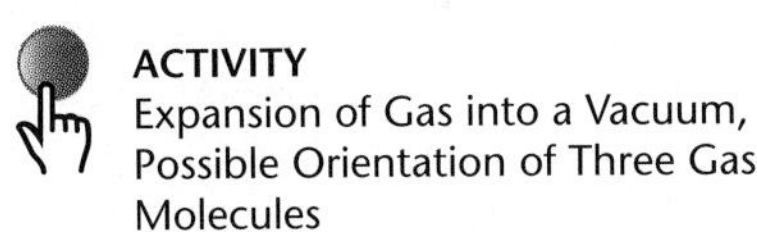

ACTIVITY
Expansion of Gas into a Vacuum, Possible Orientation of Three Gas Molecules

A Closer Look The Entropy Change of Expansion

The expansion of the gas illustrated in Figure 19.5 causes an increase in entropy. To illustrate how this entropy change can be calculated, we can employ Equation 19.1, which applies to the expansion of an ideal gas. For an isothermal expansion, the reversible work is given by*

$$w_{rev} = -nRT \ln \frac{V_2}{V_1}$$

Recall that for this process, $\Delta E = 0$. Thus using Equation 5.5, we see that $q_{rev} = -w_{rev}$. Then, using Equation 19.1,

$$\Delta S_{sys} = \frac{nRT \ln \frac{V_2}{V_1}}{T} = nR \ln \frac{V_2}{V_1}$$

For 1.00 L of an ideal gas at 1.00 atm pressure, assuming the temperature is 0°C, $n = 4.46 \times 10^{-2}$ mol. Thus, choosing the appropriate units for R (Table 10.2), we have

$$\Delta S_{sys} = (4.46 \times 10^{-2} \text{ mol})\left(\frac{8.314 \text{ J}}{\text{mol-K}}\right)(\ln 2) = 0.26 \text{ J/K}$$

This increase in entropy is a measure of the increased randomness, or molecular disorder, in the system due to the expansion. The second law of thermodynamics, therefore, is really about probability: *Under the influence of motional energy, a system tends to assume the state of maximum probability.*

* This formula for reversible work of expansion requires calculus for its derivation. It is not necessary for you to remember it.

change in entropy of the surroundings. The total change in entropy, called *the change in the entropy of the universe* and denoted ΔS_{univ}, is the sum of the changes in entropy of the system, ΔS_{sys}, and of the surroundings, ΔS_{surr}:

$$\Delta S_{univ} = \Delta S_{sys} + \Delta S_{surr} \quad [19.2]$$

In terms of ΔS_{univ}, the second law can be expressed as follows: In any reversible process, $\Delta S_{univ} = 0$, whereas in any irreversible (spontaneous) process, $\Delta S_{univ} > 0$. We can summarize these statements with the following two equations:

$$\textit{Reversible process:} \quad \Delta S_{univ} = \Delta S_{sys} + \Delta S_{surr} = 0$$
$$\textit{Irreversible process:} \quad \Delta S_{univ} = \Delta S_{sys} + \Delta S_{surr} > 0 \quad [19.3]$$

Thus, *the entropy of the universe increases in any spontaneous process.* Unlike energy, entropy is *not* conserved; S_{univ} is continually increasing.

To understand the implications of the second law of thermodynamics, think about straightening your desk—filing away papers, putting pens and pencils in their places, and so forth. If the desk is singled out as the system, the entropy of the system *decreases* during the process: The desk is more highly ordered than it was before it was straightened. But in order to achieve this greater order for the system, you, as part of the surroundings, must metabolize food in order to do the work of straightening up. During the process you will also generate heat, which is released into the air around you. These effects lead to an *increase* in the entropy of the surroundings. The second law demands that the increase in the entropy of the surroundings *must* be greater than the decrease in the entropy of the system, so that $\Delta S_{univ} > 0$.

Chemical systems exhibit the same behavior. That is, the entropy of the universe must increase during a spontaneous process even if the entropy of the system decreases. For example, let's consider the spontaneous oxidation of iron to $Fe_2O_3(s)$ (Figure 19.9 ◀):

$$4Fe(s) + 3O_2(g) \longrightarrow 2Fe_2O_3(s) \quad [19.4]$$

As discussed in Sample Exercise 19.2, this spontaneous process results in a decrease in the entropy of the system. Because the reaction is exothermic, however, the entropy of the surroundings increases as the heat evolved increases the thermal motion of surrounding molecules. We have not yet considered how to *calculate* the change in entropy of the surroundings; nevertheless, the second law tells us that $\Delta S_{sys} + \Delta S_{surr} > 0$. No process that produces order (a decrease in entropy) in a system can proceed without producing an equal or even greater disorder (an increase in entropy) in its surroundings.

A special circumstance of the second law concerns the entropy change in an **isolated system**, one that doesn't exchange energy or matter with its surroundings. For example, when a gas expands under the conditions shown in Figure 19.5, there is no exchange of heat, work, or matter with the surroundings; the gas is an isolated system. Any process that occurs in an isolated system leaves the surroundings completely unchanged. Therefore, because S is a state function, $\Delta S_{surr} = 0$ for such a process. Thus, for the special case of an isolated system, the second law becomes:

$$\textit{Reversible process, isolated system:} \quad \Delta S_{sys} = 0$$
$$\textit{Irreversible process, isolated system:} \quad \Delta S_{sys} > 0 \quad [19.5]$$

Spontaneous

▲ **Figure 19.9** The rusting of iron, such as this nail, is a spontaneous process. The reaction, Equation 19.4, leads to a decrease in the entropy of the system—that is, ΔS_{sys} is negative. Nevertheless, the second law demands that ΔS_{univ} must be positive, which means that ΔS_{surr} is a positive number of greater magnitude than ΔS_{sys}.

SAMPLE EXERCISE 19.4

Consider the reversible melting of 1 mol of ice in a large, isothermal water bath at 0°C and 1 atm pressure. The enthalpy of fusion of ice is 6.01 kJ/mol. Calculate the entropy change in the system and in the surroundings, and the overall change in entropy of the universe for this process.

Analyze and Plan: We are given ΔH_{fusion} for ice, and are asked to calculate entropy changes for the system, the surroundings, and the universe for the melting of 1 mol of ice at 0°C and 1 atm pressure. Because the ice and water bath are at the same temperature, the process described is theoretically reversible. Under these conditions slow, steady additions of infinitesimal amounts of heat will cause the ice to melt without changing the temperature of the system. Because the process involves the melting of *1 mol* of ice, the amount of heat absorbed by the ice equals the molar enthalpy of fusion, 6.01 kJ. In accord with the *first law of thermodynamics*, an equal amount of heat must be given off by the surroundings. We can employ Equation 19.1 to calculate the entropy changes.

Solve: The entropy change associated with melting 1 mol of ice is

$$\Delta S_{sys} = \frac{q_{rev}}{T} = \frac{\Delta H_{fusion}}{T} = \left(\frac{6.01 \text{ kJ/mol}}{273 \text{ K}}\right)\left(\frac{1000 \text{ J}}{1 \text{ kJ}}\right) = 22.0 \text{ J/mol-K}$$

The entropy change of the surroundings, ΔS_{surr}, has the same magnitude but opposite sign, because heat flows from the surroundings to the system. The net entropy change in the universe is therefore zero, as we would expect for a reversible process.

PRACTICE EXERCISE

The molar enthalpy of vaporization for liquid bromine is 30.71 kJ/mol. Calculate the entropy change in the system, the surroundings, and the universe for the reversible vaporization of 1 mol of liquid bromine (Br_2) at its normal boiling point (59°C).
Answer: 93 J/K-mol

Chemistry and Life Entropy and Life

Living systems are highly organized. A ginkgo leaf, for example, reveals the beautiful patterns of form and color shown in Figure 19.10(a) ▼. Animal systems, such as ourselves, are incredibly complex structures in which a host of substances come together in organized ways to form cells, tissue, organs and so on. These various components must operate in synchrony for the organism as a whole to be viable. If even one key system strays far from its optimal state, the organism as a whole may die.

To make a living system from its component molecules, such as a ginkgo leaf from sugar molecules, cellulose and the other substances present in the leaf, requires a very large reduction in entropy. It would seem, then, that living systems might violate the second law of thermodynamics. They seem spontaneously to become more, not less, organized as they develop. To get the full picture, however, we must take into account the surroundings.

We know that a system can be made more organized—that is, made to move toward lower entropy—if we do work on it. When we do work on a gas, for example, by compressing it isothermally, the entropy of the gas is lowered. The energy for the work done is provided by the surroundings, and in the process the net entropy change in the universe is positive. The striking thing about living systems is that they are organized to spontaneously recruit the energy from their surroundings. Some single-celled organisms, called *autotrophs*, capture energy from sunlight and store it in molecules such as sugars and fats [Figure 19.10(b)]. Others, called *heterotrophs*, absorb food molecules from their surroundings and break them down in their digestive systems to provide needed energy. Whatever their mode of existence, however, living systems gain their order at the expense of the surroundings. Each cell exists at the expense of an increase in the entropy of the universe.

(a) 

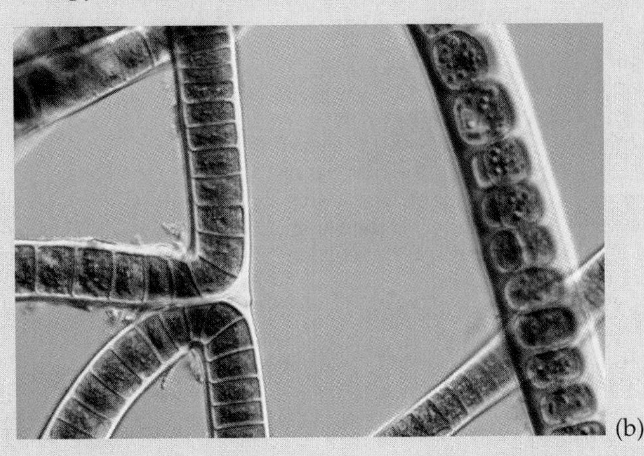 (b)

▲ **Figure 19.10** (a) This ginkgo leaf represents a highly organized living system. (b) Cyanobacteria absorb light and utilize the energy to synthesize the substances needed for growth.

Throughout most of the remainder of this chapter, we will focus mainly on the systems that we encounter, rather than on their surroundings. To simplify the notation, we will usually refer to the entropy change of the system merely as ΔS rather than explicitly indicating ΔS_{sys}.

19.3 The Molecular Interpretation of Entropy

In Section 19.2 we introduced entropy as a state function. We saw that the entropy of a system is an indicator of its randomness or disorder. As chemists, we are usually interested in relating the macroscopic observations that we make on the system to the microscopic description of a system in terms of atoms and molecules. In this section we will look more closely at how the structure and behavior of molecules affect their entropy. We will also learn about the third law of thermodynamics, which is concerned with the entropy of substances at absolute zero.

We have already seen that the expansion of a gas, which is a macroscopic observation, can be explained by looking at the microscopic behavior of the gas molecules. The entropy of the system increases ($\Delta S > 0$) as the gas molecules spread out in a larger volume. Similarly, phase changes from solid to liquid or from liquid to gas also lead to an increased entropy of the system.

Other processes lead to a decrease in the entropy of the system. For example, condensing a gas or freezing a liquid results in an increase in the order of the system; consequently, the entropy decreases for these processes ($\Delta S < 0$). A reaction that leads to a decrease in the number of gaseous molecules generally leads to a decrease in entropy. For example, consider the reaction between nitric oxide gas and oxygen gas to form nitrogen dioxide gas:

$$2NO(g) + O_2(g) \longrightarrow 2NO_2(g) \qquad [19.6]$$

The entropy change for this reaction is negative ($\Delta S < 0$) because three molecules of gas react to form two molecules of gas (Figure 19.11 ▼).

How can we relate these changes in entropy to changes at the molecular level? For the reaction in Equation 19.6, the formation of new N—O bonds imposes more order on the system; the fact that the atoms of the system are more "tied up" in the products than in the reactants leads to a decrease in the entropy of the system. The formation of the new bonds decreases the number of *degrees of freedom*, or forms of motion, available to the atoms; that is, the atoms are less free to move in random fashion because of the formation of new bonds. In general, the greater the number of degrees of freedom of a system, the greater its entropy.

The degrees of freedom of molecules are associated with three different types of motion for the molecule. The entire molecule can move in one direction, as in the

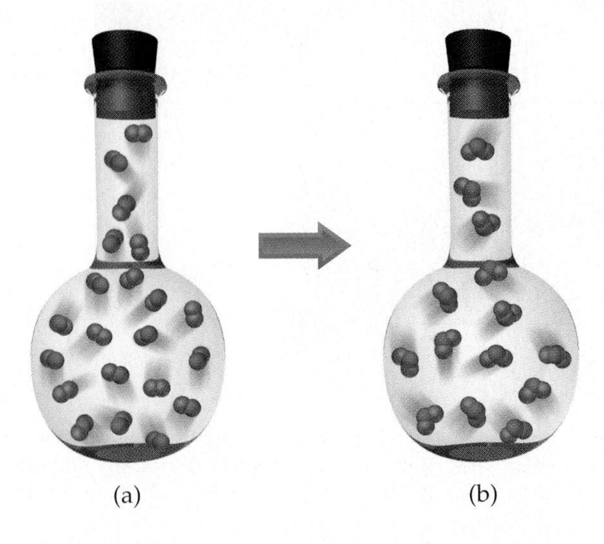

▶ **Figure 19.11** A decrease in the number of gaseous molecules leads to a decrease in the entropy of the system. When the $NO(g)$ and $O_2(g)$ in (a) react to form the $NO_2(g)$ in (b), the number of gaseous molecules decreases. The atoms have fewer degrees of freedom because new N—O bonds form, and the entropy decreases.

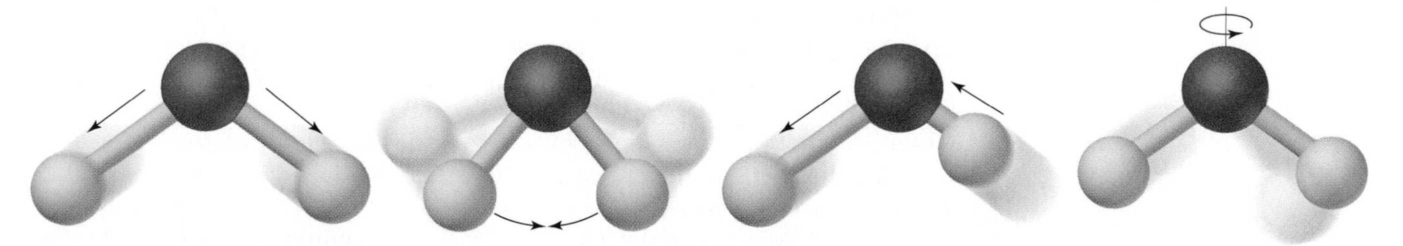

▲ **Figure 19.12** Examples of vibrational and rotational motion, illustrated for the water molecule. Vibrational motions involve periodic displacements of the atoms with respect to one another. Rotational motions involve the spinning of a molecule about an axis.

movements of gas molecules. We call such movement **translational motion**. The molecules in a gas have more translational motion than those in a liquid, which in turn have more translational motion than the molecules in a solid. The atoms within a molecule may also undergo **vibrational motion**, in which they move periodically toward and away from one another, much as a tuning fork vibrates about its equilibrium shape. In addition, molecules may possess **rotational motion**, as though they were spinning like tops. Figure 19.12 ▲ shows the vibrational motions and one of the rotational motions possible for the water molecule. These forms of motion are ways in which the molecule can store energy. The greater the energy that is stored in translational, vibrational, or rotational motion, the greater the entropy.

If we decrease the thermal energy of a system by lowering the temperature, the energy stored in translational, vibrational, and rotational forms of motion decreases. As less energy is stored, the entropy of the system decreases. If we keep lowering the temperature, do we reach a state in which these motions are essentially shut down, a point of perfect order? This question is addressed by the **third law of thermodynamics**, which states that *the entropy of a pure crystalline substance at absolute zero is zero:* $S(0\ \text{K}) = 0$.

Figure 19.13 ▼ shows schematically a pure crystalline solid. At absolute zero all the units of the lattice are without thermal motion in their lattice sites. The condition of $S = 0$ corresponds to perfect order. If this arrangement could be achieved, the individual atoms and molecules would be in a *perfect* crystalline lattice, as well

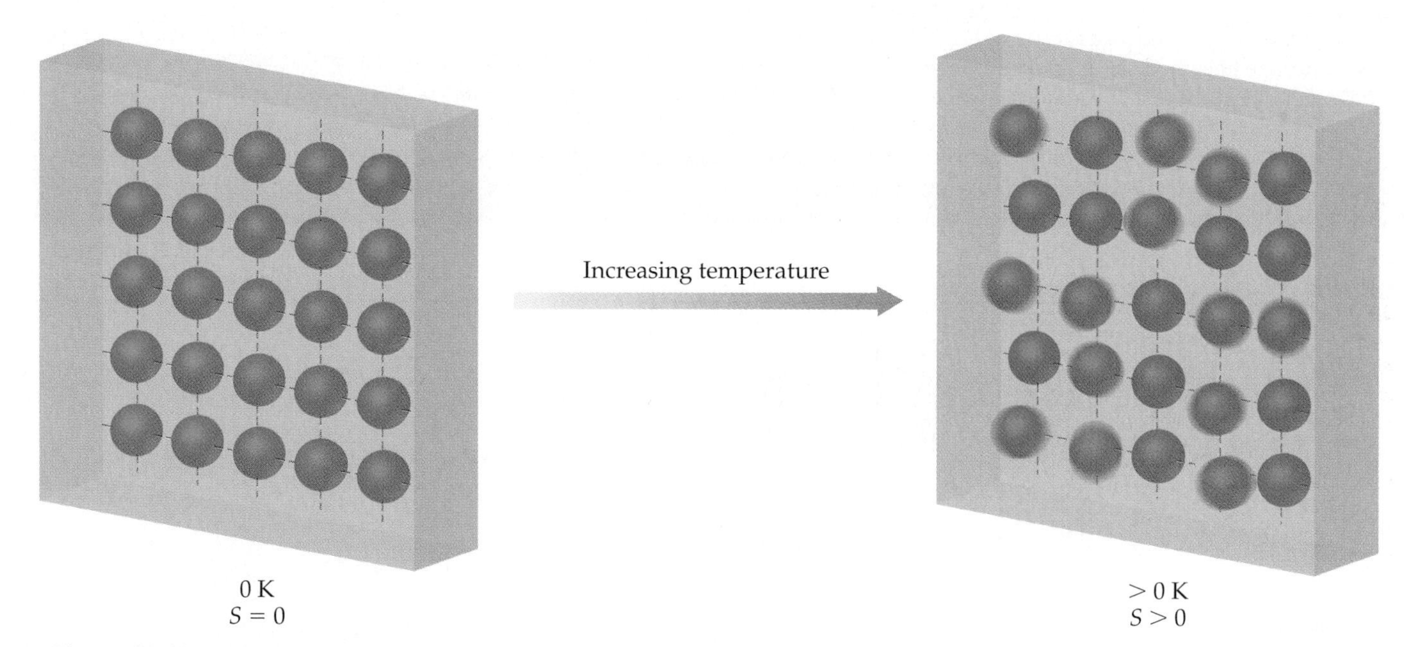

▲ **Figure 19.13** A perfectly ordered crystalline solid at and above 0 K. At absolute zero all lattice units are in their lattice sites, devoid of thermal motion. As temperature increases, the atoms or molecules gain energy and exhibit increasing vibrational motion.

defined in position as they can be. As the temperature is increased from absolute zero, the atoms or molecules in the crystal gain energy in the form of vibrational motion about their lattice positions. Thus, the degrees of freedom of the crystal increase. The entropy of the lattice therefore increases with temperature because vibrational motion causes the atoms or molecules to be less ordered.

What happens to the entropy of the substance as we continue to heat it? Figure 19.14 ▶ is a plot of how the entropy of a typical substance varies with temperature. We see that the entropy of the solid continues to increase steadily with increasing temperature up to the melting point of the solid. When the solid melts, the atoms or molecules are free to move about the entire volume of the substance. The added degrees of freedom for the individual molecules greatly increase the entropy of the substance. We therefore see a sharp increase in the entropy at the melting point. After all the solid has melted to liquid, the temperature again increases and with it, the entropy.

At the boiling point of the liquid another abrupt increase in entropy occurs. We can understand this increase as resulting from the increased volume in which the molecules may be found. The increase in volume means an increase in randomness. As the gas is heated, the entropy increases steadily as more energy is stored in the translational motion of the gas molecules. At higher temperatures the distribution of molecular speeds is spread out toward higher values (Figure 10.18). The expansion of the range of speeds of the gas molecules leads to the increased kinetic energy and increased disorder, and hence increased entropy.

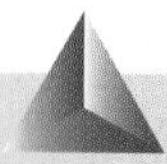

A Closer Look Entropy, Disorder, and Ludwig Boltzmann

The entropy of a system is related to its disorder. The fact that we can assign a definite value to the entropy of a system implies that disorder can somehow be quantified. The quantitative relationship of entropy to disorder was first established by the Austrian physicist Ludwig Boltzmann (1844–1906). Boltzmann reasoned that the disorder of a particular state of a system, and thus its entropy, is related to the number of possible arrangements of molecules in the state.

We can illustrate Boltzmann's idea by using the poker hands shown in Table 19.1 ▼. The probability that a poker hand will contain five *specific* cards is the same, regardless of which five cards are specified. Thus, there is an equal probability of dealing either of the specific hands shown in Table 19.1. However, the first hand, a royal flush (the ten through ace of a single suit), strikes us as much more highly ordered than the second hand, a "nothing." The reason for this is clear if we compare the number of arrangements of five cards that correspond to a royal flush to the number corresponding to a nothing. There are only 4 poker hands that are in the "state" of a royal flush; in contrast, there are over 1.3 million nothing hands. The nothing state has a higher degree of disorder than the royal-flush state because there are so many more arrangements of cards that correspond to the nothing state.

We used similar reasoning when we discussed the isothermal expansion of a gas (Figure 19.5). When the stopcock is opened, there are more possible arrangements of the

TABLE 19.1 A Comparison of the Number of Combinations that Can Lead to a Royal Flush and to a "Nothing" Hand in Poker

Hand	State	Number of Hands that Lead to This State
	Royal flush	4
	"Nothing"	1,302,540

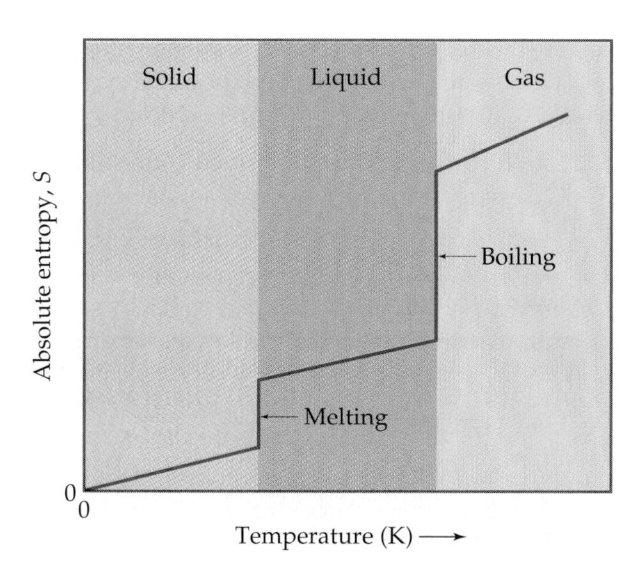

◀ **Figure 19.14** Entropy changes that occur as the temperature of a crystalline solid is increased from absolute zero.

Entropy generally increases with increasing temperature. Further, the entropies of the phases of a given substance follow the order $S_{solid} < S_{liquid} < S_{gas}$. This ordering fits in nicely with our picture of the relative degrees of disorder in solids, liquids, and gases. (Section 11.1)

gas molecules, as depicted in Figure 19.6. Thus, the randomness is greater when the stopcock is opened. Likewise, there are more possible arrangements of the H_2O molecules in liquid water than in ice (Figure 1.4). Hence, liquid water is more disordered than ice.

Boltzmann showed that the entropy of an isolated system equals a constant times the natural logarithm of the number of possible arrangements of atoms or molecules in the system:

$$S = k \ln W \qquad [19.7]$$

W is the number of possible arrangements in the system, and k is a constant known as *Boltzmann's constant*. Boltzmann's constant is the "atomic equivalent" of the gas constant, R. It equals R (usually expressed in joules) divided by Avogadro's number:

$$k = \frac{R}{N} = \frac{8.31 \text{ J/mol-K}}{6.02 \times 10^{23} \text{ mol}^{-1}} = 1.38 \times 10^{-23} \text{ J/K}$$

A pure crystalline substance at absolute zero is assumed to have only one arrangement of atoms or molecules; that is, $W = 1$. Thus, Equation 19.7 is consistent with the third law of thermodynamics: If $W = 1$, then $S = k \ln 1 = 0$. At any temperature above absolute zero, the atoms acquire energy, more arrangements become possible, and so $W > 1$ and $S > 0$.

Boltzmann made many other significant contributions to science, particularly in the area of *statistical mechanics*, which is the derivation of bulk thermodynamic properties for large collections of atoms or molecules by using the laws of probability. For example, the molecular speed distributions shown in Figure 10.19 are derived by using statistical mechanics; such plots are known as *Maxwell–Boltzmann distributions*.

Unfortunately, Boltzmann's life had a tragic ending. He strongly believed in the existence of atoms, which, as strange as it seems to us now, was a controversial viewpoint in physics at the beginning of the twentieth century. In poor health and unable to endure continual intellectual attacks on his beliefs, Boltzmann committed suicide on September 5, 1906. Ironically, it was only a few years later that the work of Thomson, Millikan, and Rutherford led to acceptance of the nuclear atom model. (Section 2.2) Although Boltzmann made many contributions to science, the connection between entropy and disorder is arguably his greatest, and it is inscribed on his gravestone (Figure 19.15 ▼).

▲ **Figure 19.15** Ludwig Boltzmann's gravestone in Vienna is inscribed with his famous relationship between the entropy of a state and the number of arrangements available in the state (in Boltzmann's time, "log" was used to represent the natural logarithm).

We can extend these observations about the entropy changes of a pure substance to provide some generalizations about the expected entropy changes for chemical reactions. In general, the entropy is expected to increase for processes in which

1. Liquids or solutions are formed from solids.
2. Gases are formed from either solids or liquids.
3. The number of molecules of gas increases during a chemical reaction.

SAMPLE EXERCISE 19.5

Choose the sample of matter that has greater entropy in each pair, and explain your choice: **(a)** 1 mol of NaCl(*s*) or 1 mol of HCl(*g*) at 25°C; **(b)** 2 mol of HCl(*g*) or 1 mol of HCl(*g*) at 25°C; **(c)** 1 mol of HCl(*g*) or 1 mol of Ar(*g*) at 25°C; **(d)** 1 mol of $N_2(s)$ at 24 K or 1 mol of $N_2(g)$ at 298 K.

Solution

Analyze and Plan: We need to select the system in each pair that has the greater entropy. To do this we examine the state of the system and the complexity of the molecules it comprises.

Solve: **(a)** Gaseous HCl has the higher entropy because gases are more disordered than solids. **(b)** The sample containing 2 mol of HCl has twice the entropy of the sample containing 1 mol. **(c)** The HCl sample has the higher entropy because the HCl molecule is capable of storing energy in more ways than is Ar. HCl molecules can rotate and vibrate; Ar atoms cannot. **(d)** The gaseous N_2 sample has the higher entropy because gases are more disordered than solids.

PRACTICE EXERCISE

Choose the substance with the greater entropy in each case: **(a)** 1 mol of $H_2(g)$ at STP or 1 mol of $H_2(g)$ at 100°C and 0.5 atm; **(b)** 1 mol of $H_2O(s)$ at 0°C or 1 mol of $H_2O(l)$ at 25°C; **(c)** 1 mol of $H_2(g)$ at STP or 1 mol of $SO_2(g)$ at STP; **(d)** 1 mol of $N_2O_4(g)$ at STP or 2 mol of $NO_2(g)$ at STP.

Answers: **(a)** 1 mol of $H_2(g)$ at 100°C and 0.5 atm; **(b)** 1 mol of $H_2O(l)$ at 25°C; **(c)** 1 mol of $SO_2(g)$ at STP; **(d)** 2 mol of $NO_2(g)$ at STP

SAMPLE EXERCISE 19.6

Predict whether the entropy change of the system in each of the following isothermal reactions is positive or negative.

(a) $CaCO_3(s) \longrightarrow CaO(s) + CO_2(g)$
(b) $N_2(g) + 3H_2(g) \longrightarrow 2NH_3(g)$
(c) $N_2(g) + O_2(g) \longrightarrow 2NO(g)$

Solution

Analyze and Plan: We are asked to determine the direction of entropy change in each chemical process. In comparing the entropies of reactants and products, we look to see which have the greater number of moles of gas and, if relevant, which are composed of more complex molecules.

Solve: **(a)** The entropy change here is positive because a solid is converted into a solid and a gas. Gaseous substances generally possess more entropy than solids, and so whenever the products contain more moles of gas than the reactants, the entropy change is probably positive. **(b)** The entropy change in formation of NH_3 from N_2 and H_2 is negative because there are fewer moles of gas in the product than in the reactants. **(c)** This represents a case in which the entropy change will be small because the same number of moles of gas is involved in the reactants and in the product. The sign of ΔS is impossible to predict based on our discussions thus far, but we can predict that ΔS will be close to zero.

PRACTICE EXERCISE

Predict whether ΔS is positive or negative in each of the following processes:

(a) $HCl(g) + NH_3(g) \longrightarrow NH_4Cl(s)$
(b) $2SO_2(g) + O_2(g) \longrightarrow 2SO_3(g)$
(c) cooling nitrogen gas from 20°C to −50°C

Answers: **(a)** negative; **(b)** negative; **(c)** negative

19.4 Entropy Changes in Chemical Reactions

In Section 5.5 we discussed how calorimetry can be used to measure ΔH for chemical reactions. No comparable, easy method exists for measuring ΔS for a reaction. By using experimental measurements of the variation of heat capacity with temperature, however, we can determine the absolute entropy, S, for many substances at any temperature. (The theory and the methods used for these measurements and calculations are beyond the scope of this text.) Absolute entropies are based on the reference point of zero entropy for perfect crystalline solids at 0 K (the third law). Entropies are usually tabulated as molar quantities, in units of joules per mole-Kelvin (J/mol-K).

The molar entropy values of substances in their standard states are known as **standard molar entropies** and are denoted $S°$. The standard state for any substance is defined as the pure substance at 1 atm pressure.* Table 19.2 ▶ lists the values of $S°$ for several substances at 298 K; Appendix C gives a more extensive list.

We can make several observations about the $S°$ values in Table 19.2:

1. Unlike enthalpies of formation, the standard molar entropies of elements at the reference temperature of 298 K are *not* zero.
2. The standard molar entropies of gases are greater than those of liquids and solids, consistent with our interpretation of experimental observations, as represented in Figure 19.14.
3. Standard molar entropies generally increase with increasing molar mass. [Compare Li(*s*), Na(*s*) and K(*s*).]
4. Standard molar entropies generally increase with an increasing number of atoms in the formula of a substance.

These last two observations are consistent with our discussion of molecular motion in Section 19.3. In general, the number and significance of the vibrational degrees of freedom of molecules increase with increasing mass and increasing number of atoms.

The entropy change in a chemical reaction is given by the sum of the entropies of the products less the sum of the entropies of the reactants:

$$\Delta S° = \sum nS°(\text{products}) - \sum mS°(\text{reactants}) \qquad [19.8]$$

As in Equation 5.31, the coefficients n and m are the coefficients in the chemical equation, as illustrated in Sample Exercise 19.7.

TABLE 19.2 Standard Molar Entropies of Selected Substances at 298 K

Substance	$S°$, J/mol-K
Gases	
$H_2(g)$	130.6
$N_2(g)$	191.5
$O_2(g)$	205.0
$H_2O(g)$	188.8
$NH_3(g)$	192.5
$CH_3OH(g)$	237.6
$C_6H_6(g)$	269.2
Liquids	
$H_2O(l)$	69.9
$CH_3OH(l)$	126.8
$C_6H_6(l)$	172.8
Solids	
Li(*s*)	29.1
Na(*s*)	51.4
K(*s*)	64.7
Fe(*s*)	27.2
$FeCl_3(s)$	142.3
NaCl(*s*)	72.3

SAMPLE EXERCISE 19.7

Calculate $\Delta S°$ for the synthesis of ammonia from $N_2(g)$ and $H_2(g)$ at 298 K:

$$N_2(g) + 3H_2(g) \longrightarrow 2NH_3(g)$$

Solution

Analyze and Plan: We are asked to calculate the entropy change for the synthesis of ammonia from its constituent elements. Standard molar entropy values for the reactants and the product are given in Table 19.2.

Solve: Using Equation 19.8 we have

$$\Delta S° = 2S°(NH_3) - [S°(N_2) + 3S°(H_2)]$$

* The standard pressure used in thermodynamics is actually no longer 1 atm, but is now based on the SI unit for pressure, the pascal (Pa). The standard pressure is 10^5 Pa, a quantity known as a *bar*: 1 bar = 10^5 Pa = 0.987 atm. Because 1 bar differs from 1 atm by only 1.3 %, we will continue to refer to the standard pressure as 1 atm.

Substituting the appropriate $S°$ values from Table 19.2 yields

$$\Delta S° = (2 \text{ mol})(192.5 \text{ J/mol-K}) - [(1 \text{ mol})(191.5 \text{ J/mol-K}) + (3 \text{ mol})(130.6 \text{ J/mol-K})]$$

$$= -198.3 \text{ J/K}$$

The value for $\Delta S°$ is negative, in agreement with our qualitative prediction in Sample Exercise 19.6(b).

PRACTICE EXERCISE

Using the standard entropies in Appendix C, calculate the standard entropy change, $\Delta S°$, for the following reaction at 298 K:

$$Al_2O_3(s) + 3H_2(g) \longrightarrow 2Al(s) + 3H_2O(g)$$

Answer: 180.39 J/K

Entropy Changes in the Surroundings

Tabulated absolute entropy values can be used to calculate the standard entropy change occurring in a system, such as a chemical reaction, as just described. But what about the entropy change occurring in the surroundings? We should recognize that the surroundings serve essentially as a large, constant-temperature heat source (or heat sink). The change in entropy of the surroundings will depend on how much heat is absorbed or given off by the system. For an isothermal process, the entropy change of the surroundings is given by

$$\Delta S_{surr} = \frac{-q_{sys}}{T} \qquad [19.9]$$

For a reaction occurring at constant pressure, q_{sys} is simply the enthalpy change for the reaction, ΔH. For the reaction in Sample Exercise 19.7, the formation of ammonia from $H_2(g)$ and $N_2(g)$ at 298 K, q_{sys} is the enthalpy change for reaction under standard conditions, $\Delta H°$. (Section 5.7) Using the procedures described in Section 5.7, we have

$$\Delta H°_{rxn} = 2\Delta H°_f[NH_3(g)] - 3\Delta H°_f[H_2(g)] - \Delta H°_f[N_2(g)]$$

$$= 2(-46.19 \text{ kJ}) - 3(0 \text{ kJ}) - (0 \text{ kJ}) = -92.38 \text{ kJ}$$

Thus, at 298 K the formation of ammonia from $H_2(g)$ and $N_2(g)$ is exothermic. Absorption of the heat given off by the system results in an increase in the entropy of the surroundings:

$$\Delta S_{surr} = \frac{92.38 \text{ kJ}}{298 \text{ K}} = 0.310 \text{ kJ/K} = 310 \text{ J/K}$$

Notice that the magnitude of the entropy gained by the surroundings (310 J/K) is greater than that lost by the system (198.4 J/K, as calculated in Sample Exercise 19.7):

$$\Delta S_{univ} = \Delta S_{sys} + \Delta S_{surr} = -198.4 \text{ J/K} + 310 \text{ J/K} = 112 \text{ J/K}$$

Because ΔS_{univ} is positive for any spontaneous reaction, this calculation indicates that when $NH_3(g)$, $H_2(g)$, and $N_2(g)$ are together in their standard states (each at 1 atm pressure), the reaction system will move spontaneously toward formation of $NH_3(g)$. Keep in mind that while the thermodynamic calculations indicate that formation of ammonia is spontaneous, they do not tell us anything about the rate at which ammonia is formed. Establishing equilibrium in this system within a reasonable period of time requires a catalyst, as discussed in Section 15.6.

19.5 Gibbs Free Energy

We have seen examples of endothermic processes that are spontaneous, such as the dissolution of ammonium nitrate in water (Section 13.1). We learned in our discussion of the solution process that the driving force for a spontaneous, endothermic process is an increase in disorder, or entropy, of the system. However, we have also encountered processes that are spontaneous and yet proceed with a *decrease* in the entropy of the system, such as the highly exothermic formation of sodium chloride from its constituent elements. (Section 8.2) Spontaneous processes that result in a decrease in the system's entropy are always exothermic. Thus, the spontaneity of a reaction seems to involve two thermodynamic concepts, enthalpy and entropy. There should be a way to use ΔH and ΔS to predict whether a given reaction occurring at constant temperature and pressure will be spontaneous. The means for doing so was first developed by the American mathematician J. Willard Gibbs (1839–1903). Gibbs (Figure 19.16 ▶) proposed a new state function, now called the **Gibbs free energy** (or just **free energy**). The Gibbs free energy, G, of a state is defined as

$$G = H - TS \qquad [19.10]$$

where T is the absolute temperature. For a process occurring at constant temperature, the change in free energy of the system, ΔG, is given by the expression

$$\Delta G = \Delta H - T\Delta S \qquad [19.11]$$

To see how the function G relates to reaction spontaneity, recall that for a reaction occurring at constant temperature and pressure

$$\Delta S_{\text{univ}} = \Delta S_{\text{sys}} + \Delta S_{\text{surr}} = \Delta S_{\text{sys}} + \left(\frac{-\Delta H_{\text{sys}}}{T}\right)$$

Multiplying both sides by $(-T)$ gives us

$$-T\Delta S_{\text{univ}} = \Delta H_{\text{sys}} - T\Delta S_{\text{sys}} \qquad [19.12]$$

Comparing Equation 19.12 with Equation 19.11, we see that the free-energy change in a process occurring at constant temperature and pressure, ΔG, is equal to $-T\Delta S_{\text{univ}}$. We know that for spontaneous processes, ΔS_{univ} is positive. Thus, the sign of ΔG provides us with extremely valuable information about the spontaneity of processes that occur at constant temperature and pressure. If both T and P are constant, the relationship between the sign of ΔG and the spontaneity of a reaction is as follows:

1. If ΔG is negative, the reaction is spontaneous in the forward direction.
2. If ΔG is zero, the reaction is at equilibrium.
3. If ΔG is positive, the reaction in the forward direction is nonspontaneous; work must be supplied from the surroundings to make it occur. However, the reverse reaction will be spontaneous.

An analogy is often drawn between the free-energy change during a spontaneous reaction and the potential-energy change when a boulder rolls down a hill. Potential energy in a gravitational field "drives" the boulder until it reaches a state of minimum potential energy in the valley [Figure 19.17(a) ▶]. Similarly, the free energy of a chemical system decreases until it reaches a minimum value [Figure 19.17(b)]. When this minimum is reached, a state of equilibrium exists. *In any spontaneous process at constant temperature and pressure, the free energy always decreases.*

As a specific illustration of these ideas, let's return to the Haber process for the synthesis of ammonia from nitrogen and hydrogen, which we discussed extensively in Chapter 15:

$$N_2(g) + 3H_2(g) \rightleftharpoons 2NH_3(g)$$

▲ **Figure 19.16** Josiah Willard Gibbs (1839–1903) was the first person to be awarded a Ph.D. in science from an American university (Yale, 1863). From 1871 until his death, he held the chair of mathematical physics at Yale. Gibbs developed much of the theoretical foundation that led to the development of chemical thermodynamics.

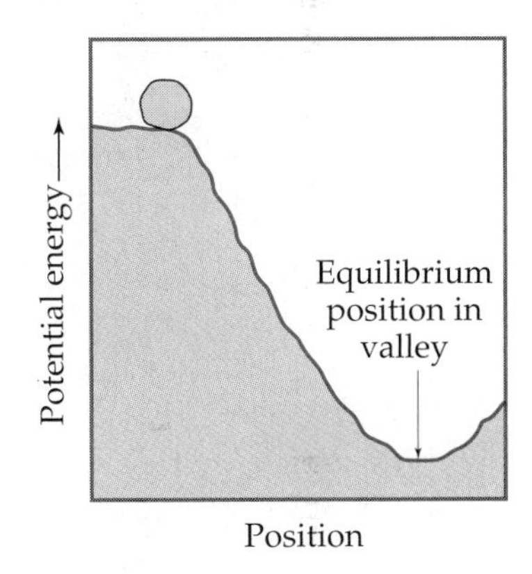

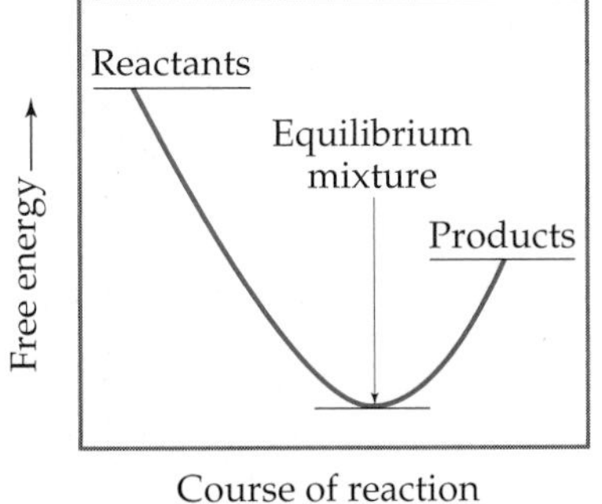

▲ **Figure 19.17** Analogy between the potential-energy change of a boulder rolling down a hill (a) and the free-energy change in a spontaneous reaction (b). The equilibrium position in (a) is given by the minimum potential energy available to the system. The equilibrium position in (b) is given by the minimum free energy available to the system.

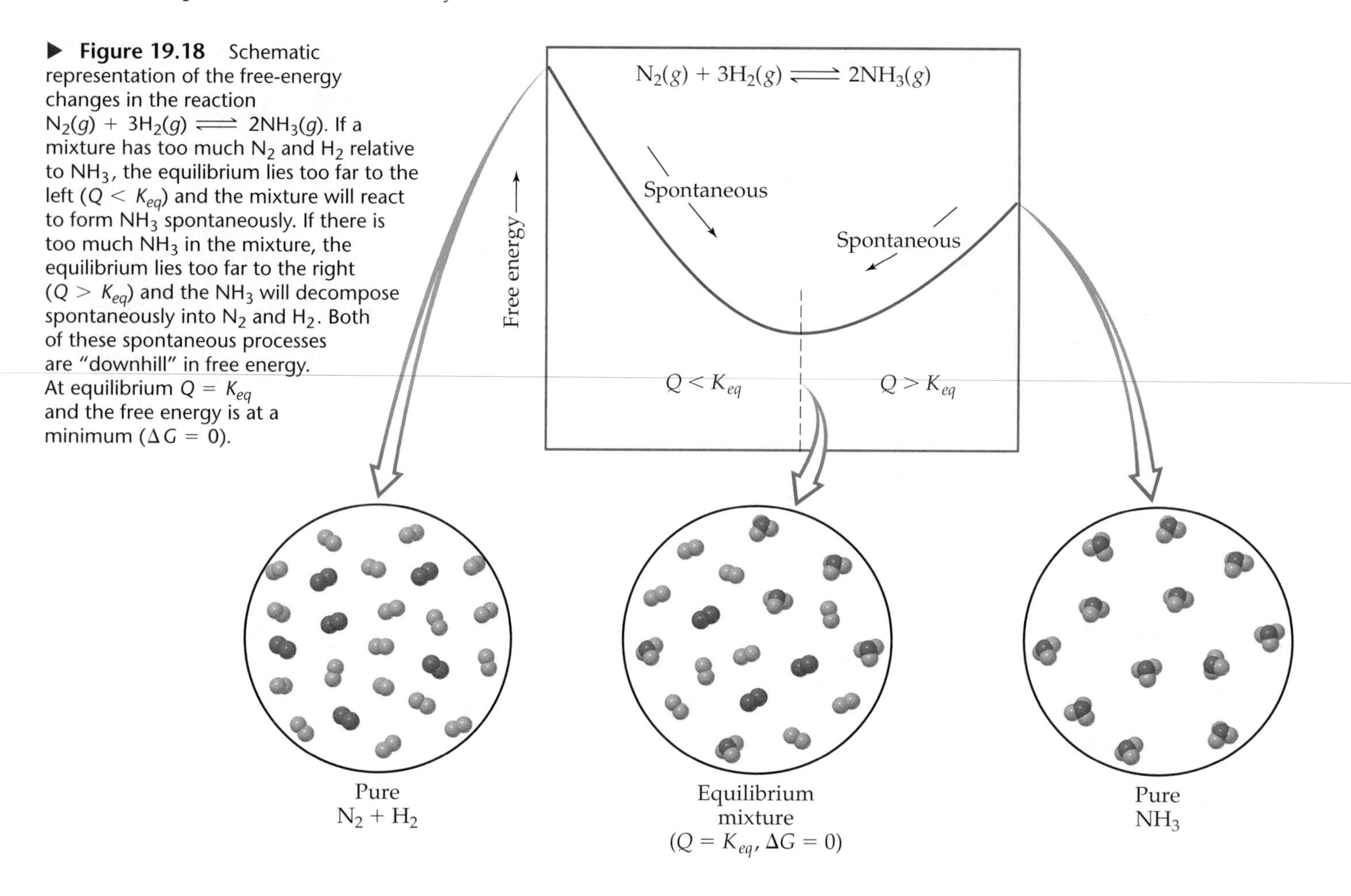

▶ **Figure 19.18** Schematic representation of the free-energy changes in the reaction $N_2(g) + 3H_2(g) \rightleftharpoons 2NH_3(g)$. If a mixture has too much N_2 and H_2 relative to NH_3, the equilibrium lies too far to the left ($Q < K_{eq}$) and the mixture will react to form NH_3 spontaneously. If there is too much NH_3 in the mixture, the equilibrium lies too far to the right ($Q > K_{eq}$) and the NH_3 will decompose spontaneously into N_2 and H_2. Both of these spontaneous processes are "downhill" in free energy. At equilibrium $Q = K_{eq}$ and the free energy is at a minimum ($\Delta G = 0$).

Imagine that we have a reaction vessel that allows us to maintain a constant temperature and pressure and that we have a catalyst that allows the reaction to proceed at a reasonable rate. What will happen if we charge the vessel with a certain number of moles of N_2 and three times that number of moles of H_2? As we saw in Figure 15.6(a), the N_2 and H_2 will react spontaneously to form NH_3 until equilibrium is achieved. Similarly, Figure 15.6(b) demonstrates that if we charged the vessel with pure NH_3, it will decompose spontaneously to form N_2 and H_2 until equilibrium is reached. In each case the free energy of the system is lowered on the way to equilibrium which represents a minimum in the free energy. We illustrate these cases in Figure 19.18 ▲.

This is a good time to remind ourselves of the significance of the reaction quotient, Q, for a system that is not at equilibrium. ∞ (Section 15.5) Recall that when $Q < K_{eq}$, there is an excess of reactants relative to products. The reaction will proceed spontaneously in the forward direction to reach equilibrium. When $Q > K_{eq}$, the reaction will proceed spontaneously in the reverse direction. At equilibrium $Q = K_{eq}$. We have illustrated these points in Figure 19.18. In Section 19.7 we will see how to use the value of Q to calculate the value of ΔG for systems that are not at equilibrium.

Standard Free-Energy Changes

Free energy is a state function, like enthalpy. We can tabulate **standard free energies of formation** for substances, just as we can tabulate standard enthalpies of formation. ∞ (Section 5.7) It is important to remember that standard values for these functions imply a particular set of conditions, or standard states. The standard state for gaseous substances is 1 atm pressure. For solid substances, the standard state is the pure solid; for liquids, the pure liquid. For substances in solution, the standard state is normally a concentration of 1 *M*. (In very accurate work it may be necessary to make certain corrections, but we need not worry about these.) The temperature usually chosen for purposes of tabulating data is 25°C, but we will calculate ΔG° at other temperatures as well. Just as for the standard heats of formation, the free energies of elements in their standard states are set to zero. This arbitrary choice of a reference point has no effect on the quantity in which we are really interested, namely, the *difference* in free energy between reactants and products. The rules about standard states are summarized in Table 19.3 ▶. A listing of standard free energies of formation, denoted ΔG_f°, appears in Appendix C.

The standard free energies of formation are useful in calculating the *standard free-energy change* for chemical processes. The procedure is analogous to the calculation of ΔH° (Equation 5.31) and ΔS° (Equation 19.8):

$$\Delta G^\circ = \sum n\Delta G_f^\circ(\text{products}) - \sum m\Delta G_f^\circ(\text{reactants}) \qquad [19.13]$$

TABLE 19.3 Conventions Used in Establishing Standard Free Energies

State of Matter	Standard State
Solid	Pure solid
Liquid	Pure liquid
Gas	1 atm pressure
Solution	1 *M* concentration
Elements	Standard free energy of formation of an element in its standard state is defined as zero

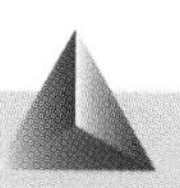

A Closer Look What's "Free" About Free Energy?

The Gibbs free energy is a remarkable thermodynamic quantity. Because so many chemical reactions are carried out under conditions of near-constant pressure and temperature, chemists, biochemists, and engineers use the sign and magnitude of ΔG as exceptionally useful tools in the design and implementation of chemical and biochemical reactions. We will see examples of the usefulness of ΔG throughout the remainder of this chapter and this text.

There are two common questions that often arise when one first learns about the Gibbs free energy: Why does the sign of ΔG tell us about the spontaneity of reactions? What is "free" about free energy? We will address these two questions here by using concepts discussed in Chapter 5 and earlier in this chapter.

In Section 19.2 we saw that the second law of thermodynamics governs the spontaneity of processes. In order to apply the second law (Equation 19.3), however, we must determine ΔS_{univ}, which is often difficult to evaluate. By using the Gibbs free energy under conditions of constant temperature and pressure, we can relate ΔS_{univ} to quantities that depend only on changes in the system, namely ΔH and ΔS (as before, if we do not put a subscript on these quantities, we are referring to the system). Our first step in seeing this relationship is to recall Equation 19.1, which states that at constant temperature ΔS equals the amount of heat that would be transferred to the system in a reversible process, divided by the temperature:

$$\Delta S = q_{rev}/T \quad (\text{constant } T)$$

Remember that because entropy is a state function, this is the entropy change regardless of whether the system changes reversibly or irreversibly. Similarly, if T is constant, the entropy change of the surroundings is given by the heat transferred to the surroundings, q_{surr}, divided by the temperature. Because heat transferred *to* the surroundings must be transferred *from* the system, it follows that $q_{surr} = -q_{sys}$. Combining these ideas allows us to relate ΔS_{surr} to q_{sys}:

$$\Delta S_{surr} = q_{surr}/T = -q_{sys}/T \quad (\text{constant } T) \qquad [19.14]$$

If P is also constant, $q_{sys} = q_P = \Delta H$ (Equation 5.10), so

$$\Delta S_{surr} = -q_{sys}/T = -\Delta H/T \quad (\text{constant } T, P) \qquad [19.15]$$

We can now use Equation 19.3 to calculate ΔS_{univ} in terms of ΔS and ΔH:

$$\Delta S_{univ} = \Delta S + \Delta S_{surr} = \Delta S - \Delta H/T \quad (\text{constant } T, P) \qquad [19.16]$$

Thus, under conditions of constant temperature and pressure, the second law (Equation 19.3) becomes:

Reversible process: $\Delta S - \Delta H/T = 0$

Irreversible process: $\Delta S - \Delta H/T > 0$

$$(\text{constant } T, P) \qquad [19.17]$$

Now we can see the relationship between ΔG and the second law. If we multiply the preceding equations by $-T$ and rearrange, we reach the following conclusion:

Reversible process: $\Delta G = \Delta H - T\Delta S = 0$

Irreversible process: $\Delta G = \Delta H - T\Delta S < 0$

$$(\text{constant } T, P) \qquad [19.18]$$

We see that we can use the sign of ΔG to conclude whether a reaction is spontaneous, nonspontaneous, or at equilibrium. The magnitude of ΔG is also significant. A reaction for which ΔG is large and negative, such as the burning of gasoline, is much more capable of doing work on the surroundings than is a reaction for which ΔG is small and negative, such as ice melting at room temperature. In fact, thermodynamics tells us that *the change in free energy for a process, ΔG, equals the maximum useful work that can be done by the system on its surroundings in a spontaneous process occurring at constant temperature and pressure:*

$$w_{max} = \Delta G \qquad [19.19]$$

This relationship explains why ΔG is called the *free* energy. It is the portion of the energy change of a spontaneous reaction that is free to do useful work. The remainder of the energy enters the environment as heat.

What use can be made of this standard free-energy change for a chemical reaction? The quantity ΔG° tells us whether a mixture of reactants and products, each present under standard conditions, would spontaneously react in the forward direction to produce more products ($\Delta G^\circ < 0$) or in the reverse direction to form more reactants ($\Delta G^\circ > 0$). Because ΔG_f° values are readily available for a large number of substances, the standard free-energy change is easy to calculate for many reactions of interest.

For processes that are not spontaneous ($\Delta G > 0$), the free-energy change is a measure of the *minimum* amount of work that must be done to cause the process to occur. In actual cases we always need to do more than this theoretical minimum amount because of the inefficiencies in the way the changes occur.

SAMPLE EXERCISE 19.8

(a) By using data from Appendix C, calculate the standard free-energy change for the following reaction at 298 K:

$$P_4(g) + 6Cl_2(g) \longrightarrow 4PCl_3(g)$$

(b) What is ΔG° for the reverse of the above reaction?

Solution

Analyze and Plan: We are asked to calculate the free-energy change for the indicated reaction, and then to determine the free-energy change of its reverse. To do this we look up the free-energy values for the products and reactants, multiply the molar quantities by the coefficients in the balanced equation, and subtract the total for the reactants from that for the products.

Solve: $Cl_2(g)$ is in its standard state, so ΔG_f° is zero for this reactant. $P_4(g)$, however, is not in its standard state, so ΔG_f° is not zero for this reactant. From the balanced equation and using Appendix C, we have:

$$\begin{aligned}\Delta G^\circ_{rxn} &= 4\Delta G_f^\circ[PCl_3(g)] - \Delta G_f^\circ[P_4(g)] - 6\Delta G_f^\circ[Cl_2(g)] \\ &= 4(-269.6 \text{ kJ/mol}) - (24.4) - 0 \\ &= -1102.8 \text{ kJ/mol}\end{aligned}$$

The fact that ΔG° is negative tells us that a mixture of $P_4(g)$, $Cl_2(g)$, and $PCl_3(g)$ at 25°C, each present at a partial pressure of 1 atm, would react spontaneously in the forward direction to form more PCl_3. Remember, however, that the value of ΔG° tells us nothing about the rate at which the reaction occurs.

(b) Remember that $\Delta G = G(\text{products}) - G(\text{reactants})$. If we reverse the reaction, we reverse the roles of the reactants and products. Thus, reversing the reaction changes the sign of ΔG, just as reversing the reaction changes the sign of ΔH. (Section 5.4) Hence, using the result from part (a):

$$4PCl_3(g) \longrightarrow P_4(g) + 6Cl_2(g) \qquad \Delta G^\circ = +1102.8 \text{ kJ}$$

PRACTICE EXERCISE

By using data from Appendix C, calculate ΔG° at 298 K for the combustion of methane: $CH_4(g) + 2O_2(g) \longrightarrow CO_2(g) + 2H_2O(g)$.

Answer: -800.7 kJ

SAMPLE EXERCISE 19.9

In Section 5.7 we used Hess's law to calculate ΔH° for the combustion of propane gas at 298 K (see Figure 5.22):

$$C_3H_8(g) + 5O_2(g) \longrightarrow 3CO_2(g) + 4H_2O(l) \qquad \Delta H^\circ = -2220 \text{ kJ}$$

(a) *Without using data from Appendix C*, predict whether ΔG° for this reaction is more negative or less negative than ΔH°. **(b)** Use data from Appendix C to calculate the standard free-energy change for the reaction at 298 K. Is your prediction from part (a) correct?

Solution

Analyze and Plan: In part (a) we must predict the value for $\Delta G°$ relative to that for $\Delta H°$ on the basis of the balanced equation for the reaction. In part (b) we must calculate the value for $\Delta G°$ and compare with our qualitative prediction. The free-energy change incorporates both the change in enthalpy and the change in entropy for the reaction (Equation 19.11), so under standard conditions:

$$\Delta G° = \Delta H° - T\Delta S°$$

To determine whether $\Delta G°$ is more negative or less negative than $\Delta H°$, we need to determine the sign of the term $T\Delta S°$. T is the absolute temperature, 298 K, so it is a positive number. We can predict the sign of $\Delta S°$ by looking at the reaction.

Solve: (a) We see that the reactants consist of 6 mol of gas, and the products consist of 3 mol of gas and 4 mol of liquid. Thus, the number of moles of gas has decreased significantly during the reaction. By using the general rules we discussed in Section 19.3, we would expect a decrease in the number of gas molecules to lead to a decrease in the entropy of the system—the products are less disordered than the reactants. We therefore expect $\Delta S°$ and $T\Delta S°$ to be negative numbers. Because we are subtracting $T\Delta S°$, which is a negative number, we would predict that $\Delta G°$ is *less negative* than $\Delta H°$.

(b) Using Equation 19.13 and values from Appendix C, we can calculate the value of $\Delta G°$:

$$\Delta G° = 3\Delta G_f°[CO_2(g)] + 4\Delta G_f°[H_2O(l)] - \Delta G_f°[C_3H_8(g)] - 5\Delta G_f°[O_2(g)]$$

$$= 3\text{ mol}(-394.4\text{ kJ/mol}) + 4\text{ mol}(-237.13\text{ kJ/mol}) -$$

$$1\text{ mol}(-23.47\text{ kJ/mol}) - 5\text{ mol}(0\text{ kJ/mol}) = -2108\text{ kJ}$$

Notice that we have been careful to use the value of $\Delta G_f°$ for $H_2O(l)$; as in the calculation of ΔH values, the phases of the reactants and products are important. As we predicted, $\Delta G°$ is less negative than $\Delta H°$ because of the decrease in entropy during the reaction.

PRACTICE EXERCISE

Consider the combustion of propane to form $CO_2(g)$ and $H_2O(g)$ at 298 K: $C_3H_8(g) + 5O_2(g) \longrightarrow 3CO_2(g) + 4H_2O(g)$. Would you expect $\Delta G°$ to be more negative or less negative than $\Delta H°$?

Answer: more negative

19.6 Free Energy and Temperature

We have seen that tabulations of $\Delta G_f°$, such as those in Appendix C, make it possible to calculate $\Delta G°$ for reactions at the standard temperature of 25°C. However, we are often interested in examining reactions at other temperatures. How is the change in free energy affected by the change in temperature? Let's look again at Equation 19.11:

$$\Delta G = \Delta H - T\Delta S = \underbrace{\Delta H}_{\text{enthalpy term}} + \underbrace{(-T\Delta S)}_{\text{entropy term}}$$

Notice that we have written the expression for ΔG as a sum of two contributions, an enthalpy term, ΔH, and an entropy term, $-T\Delta S$. Because the value of $-T\Delta S$ depends directly on the absolute temperature T means that ΔG will vary with temperature. T is a positive number at all temperatures other than absolute zero. We know that the enthalpy term, ΔH, can be positive or negative. The entropy term, $-T\Delta S$, can also be positive or negative. When ΔS is positive, which means the final state is more disordered than the initial state, the term $-T\Delta S$ is negative. When ΔS is negative, the term $-T\Delta S$ is positive.

The sign of ΔG, which tells us whether a process is spontaneous, will depend on the signs and magnitudes of ΔH and $-T\Delta S$. When both ΔH and $-T\Delta S$ are negative, ΔG will always be negative and the process will be spontaneous at all

TABLE 19.4 Effect of Temperature on the Spontaneity of Reactions

ΔH	ΔS	$-T\Delta S$	$\Delta G = \Delta H - T\Delta S$	Reaction Characteristics	Example
−	+	−	Always negative	Spontaneous at all temperatures	$2O_3(g) \longrightarrow 3O_2(g)$
+	−	+	Always positive	Nonspontaneous at all temperatures; reverse reaction always spontaneous	$3O_2(g) \longrightarrow 2O_3(g)$
−	−	+	Negative at low T; positive at high T	Spontaneous at low T; becomes nonspontaneous at high T	$H_2O(l) \longrightarrow H_2O(s)$
+	+	−	Positive at low T; negative at high T	Nonspontaneous at low T; becomes spontaneous at high T	$H_2O(s) \longrightarrow H_2O(l)$

temperatures. Likewise, when both ΔH and $-T\Delta S$ are positive, ΔG will always be positive and the process will be nonspontaneous at all temperatures (the reverse process will be spontaneous at all temperatures). When ΔH and $-T\Delta S$ have opposite signs, however, the sign of ΔG will depend on the magnitudes of these two terms. In these instances temperature is an important consideration. Generally, ΔH and ΔS change very little with temperature. However, the value of T directly affects the magnitude of $-T\Delta S$. As the temperature increases, the magnitude of the term $-T\Delta S$ increases and it will become relatively more important in determining the sign and magnitude of ΔG.

For example, let's consider once more the melting of ice to liquid water at 1 atm pressure (Figure 19.3):

$$H_2O(s) \longrightarrow H_2O(l) \qquad \Delta H > 0, \Delta S > 0$$

This process is endothermic, which means that ΔH is positive. We also know that the entropy increases during this process, so ΔS is positive and $-T\Delta S$ is negative. At temperatures below 0°C the magnitude of ΔH is greater than that of $-T\Delta S$. Hence, the positive enthalpy term dominates, leading to a positive value for ΔG. The positive value of ΔG means that the melting of ice is not spontaneous at $T < 0°C$; rather, the reverse process, the freezing of liquid water into ice, is spontaneous at these temperatures. What happens at temperatures greater than 0°C? As the temperature increases, so does the magnitude of the entropy term $-T\Delta S$. When $T > 0°C$, the magnitude of $-T\Delta S$ is greater than the magnitude of ΔH. At these temperatures the negative entropy term dominates, which leads to a negative value for ΔG. The negative value of ΔG tells us that the melting of ice is spontaneous at $T > 0°C$. At the normal melting point of water, $T = 0°C$, the two phases are in equilibrium. Recall that $\Delta G = 0$ at equilibrium; at $T = 0°C$, ΔH and $-T\Delta S$ are equal in magnitude and opposite in sign, so they cancel one another and give $\Delta G = 0$.

The possible situations for the relative signs of ΔH and ΔS are given in Table 19.4 ▲, along with examples of each. By applying the concepts we have developed for predicting entropy changes, we often can predict how ΔG will change with temperature.

Our discussion of the temperature dependence of ΔG is also relevant to standard free-energy changes. As we saw in Sample Exercise 19.9, under standard conditions, Equation 19.11 becomes:

$$\Delta G° = \Delta H° - T\Delta S° \qquad [19.20]$$

We can readily calculate the values of $\Delta H°$ and $\Delta S°$ at 298 K from the data tabulated in Appendix C. If we assume that the values of $\Delta H°$ and $\Delta S°$ do not change with temperature, we can use Equation 19.20 to estimate the value of $\Delta G°$ at temperatures other than 298 K. We illustrate this procedure in Sample Exercise 19.10.

SAMPLE EXERCISE 19.10

The Haber process for the production of ammonia involves the following equilibrium:

$$N_2(g) + 3H_2(g) \rightleftharpoons 2NH_3(g)$$

Assume that ΔH° and ΔS° for this reaction do not change with temperature. **(a)** Predict the direction in which ΔG° for this reaction changes with increasing temperature. **(b)** Calculate the values of ΔG° for the reaction at 25°C and 500°C.

Solution

Analyze and Plan: In part (a) we are asked to predict the direction in which ΔG° for the ammonia synthesis reaction changes as temperature increases. To do this we need to determine the sign of ΔS for the reaction. In part (b) we need to determine ΔG° for the reaction at two different temperatures.

Solve: (a) Equation 19.20 tells us that ΔG° is the sum of the enthalpy term ΔH° and the entropy term $-T\Delta S^\circ$. The temperature dependence of ΔG° comes from the entropy term. We expect ΔS° for this reaction to be negative because the number of moles of gas is smaller in the products. Because ΔS° is negative the term $-T\Delta S^\circ$ is positive and grows larger with increasing temperature. As a result, ΔG° becomes less negative (or more positive) with increasing temperature. Thus, the driving force for the production of NH_3 becomes smaller with increasing temperature.

(b) We can readily calculate ΔH° and ΔS° for the reaction by using the data in Appendix C. In fact, we have already done so! We calculated the value of ΔH° in Sample Exercise 15.13 (Section 15.6), and the value of ΔS° was determined in Sample Exercise 19.7: $\Delta H^\circ = -92.38$ kJ and $\Delta S^\circ = -198.4$ J/K. If we assume that these values don't change with temperature, we can calculate ΔG° at any temperature by using Equation 19.20. At $T = 298$ K we have:

$$\Delta G = -92.38 \text{ kJ} - (298 \text{ K})(-198.4 \text{ J/K})\left(\frac{1 \text{ kJ}}{1000 \text{ J}}\right)$$

$$= -92.38 \text{ kJ} + 59.1 \text{ kJ} = -33.3 \text{ kJ}$$

At $T = 500 + 273 = 773$ K we have

$$\Delta G^\circ = -92.38 \text{ kJ} - (773 \text{ K})\left(-198.4 \frac{\text{J}}{\text{K}}\right)\left(\frac{1 \text{ kJ}}{1000 \text{ J}}\right)$$

$$= -92.38 \text{ kJ} + 153 \text{ kJ} = 61 \text{ kJ}$$

Notice that we have been careful to convert $-T\Delta S^\circ$ into units of kJ so that it can be added to ΔH°, which has units of kJ.

Comment: Increasing the temperature from 298 K to 773 K changes ΔG° from -33.3 kJ to $+61$ kJ. Of course, the result at 773 K depends on the assumption that ΔH° and ΔS° do not change with temperature. In fact, these values do change slightly with temperature. Nevertheless, the result at 773 K should be a reasonable approximation. The positive increase in ΔG° with increasing T agrees with our prediction in part (a) of this exercise. Our result indicates that a mixture of $N_2(g)$, $H_2(g)$, and $NH_3(g)$, each present at a partial pressure of 1 atm, will react spontaneously at 298 K to form more $NH_3(g)$. In contrast, at 773 K the positive value of ΔG° tells us that the reverse reaction is spontaneous. Thus, when the mixture of three gases, each at a partial pressure of 1 atm is heated to 773 K, some of the $NH_3(g)$ spontaneously decomposes into $N_2(g)$ and $H_2(g)$.

ACTIVITY
Gibbs Free Energy

PRACTICE EXERCISE

(a) Using standard enthalpies of formation and standard entropies in Appendix C, calculate ΔH° and ΔS° at 298 K for the following reaction: $2SO_2(g) + O_2(g) \longrightarrow 2SO_3(g)$.
(b) Using the values obtained in part (a), estimate ΔG° at 400 K.
Answers: **(a)** $\Delta H^\circ = -196.6$ kJ, $\Delta S^\circ = -189.6$ J/K; **(b)** $\Delta G^\circ = -120.8$ kJ

ANIMATION
Air Bags

MOVIE
Formation of Water

19.7 Free Energy and the Equilibrium Constant

In Section 19.5 we saw a special relationship between ΔG and equilibrium: For a system at equilibrium, $\Delta G = 0$. We have also seen how we can use tabulated thermodynamic data, such as those in Appendix C, to calculate values of the standard free-energy change, ΔG°. In this final section of this chapter, we will learn two more ways in which we can use free energy as a powerful tool in our analysis of chemical reactions. First, we will learn how to use the value of ΔG° to calculate the value of ΔG under *nonstandard* conditions. Second, we will see how we can directly relate the value of ΔG° for a reaction to the value of the equilibrium constant for the reaction.

The set of standard conditions for which ΔG° values pertain are given in Table 19.3. Most chemical reactions occur under nonstandard conditions. For any chemical process the general relationship between the standard free-energy change, ΔG°, and the free-energy change under any other conditions, ΔG, is given by the following expression:

$$\Delta G = \Delta G^\circ + RT \ln Q \qquad [19.21]$$

In this equation R is the ideal-gas constant, 8.314 J/mol-K; T is the absolute temperature; and Q is the reaction quotient that corresponds to the particular reaction mixture of interest. (Section 15.5) Recall that the expression for Q is identical to the equilibrium-constant expression except that the reactants and products need not necessarily be at equilibrium.

Under standard conditions the concentrations of all the reactants and products are equal to 1. Thus, under standard conditions $Q = 1$ and therefore $\ln Q = 0$. We see that Equation 19.21 therefore reduces to $\Delta G = \Delta G^\circ$ under standard conditions, as it should.

SAMPLE EXERCISE 19.11

As we saw in Section 11.5, the *normal boiling point* is the temperature at which a pure liquid is in equilibrium with its vapor at a pressure of 1 atm. **(a)** Write the chemical equation that defines the normal boiling point of liquid carbon tetrachloride, $CCl_4(l)$. **(b)** What is the value of ΔG° for the equilibrium in part (a)? **(c)** Use thermodynamic data in Appendix C and Equation 19.20 to estimate the normal boiling point of CCl_4.

Solution

Analyze and Plan: (a) We must write a chemical equation that describes the physical equilibrium between liquid and gaseous CCl_4 at the normal boiling point. **(b)** We must determine the value of ΔG° for CCl_4 in equilibrium with its vapor at the normal boiling point. **(c)** We are asked to estimate the normal boiling point of CCl_4 based on available thermodynamic data.

Solve: (a) The normal boiling point of CCl_4 is the temperature at which pure liquid CCl_4 is in equilibrium with its vapor at a pressure of 1 atm:

$$CCl_4(l) \rightleftharpoons CCl_4(g; 1\ \text{atm})$$

(b) At equilibrium, $\Delta G = 0$. In any normal boiling-point equilibrium, both the liquid and the vapor are in their standard states (Table 19.3). As a consequence, $Q = 1$, $\ln Q = 0$, and $\Delta G = \Delta G^\circ$ for this process. Thus, we conclude that $\Delta G^\circ = 0$ for the equilibrium involved in the normal boiling point of any liquid. We would also find that $\Delta G^\circ = 0$ for the equilibria relevant to normal melting points and normal sublimation points of solids.

(c) Combining Equation 19.20 with the result from part (b), we see the following equality at the normal boiling point T_b of $CCl_4(l)$ or any other pure liquid:

$$\Delta G^\circ = \Delta H^\circ - T_b \Delta S^\circ = 0$$

Solving the equation for T_b, we obtain $T_b = \Delta H^\circ/\Delta S^\circ$. Strictly speaking, we would need the values of ΔH° and ΔS° for the equilibrium between $CCl_4(l)$ and $CCl_4(g)$ at the normal boiling point to do this calculation. However, we can *estimate* the boiling point by using the values of ΔH° and ΔS° for CCl_4 at 298 K, which we can obtain from the data in Appendix C and Equations 5.31 and 19.8:

$$\Delta H^\circ = (1\ \text{mol})(-106.7\ \text{kJ/mol}) - (1\ \text{mol})(-139.3\ \text{kJ/mol}) = +32.6\ \text{kJ}$$

$$\Delta S^\circ = (1\ \text{mol})(309.4\ \text{J/mol-K}) - (1\ \text{mol})(214.4\ \text{J/mol-K}) = +95.0\ \text{J/K}$$

Notice that, as expected, the process is endothermic ($\Delta H > 0$) and produces more disorder ($\Delta S > 0$). We can now use these values to estimate T_b for $CCl_4(l)$:

$$T_b = \frac{\Delta H^\circ}{\Delta S^\circ} = \left(\frac{32.6\ \text{kJ}}{95.0\ \text{J/K}}\right)\left(\frac{1000\ \text{J}}{1\ \text{kJ}}\right) = 343\ \text{K} = 70^\circ\text{C}$$

Note also that we have used the conversion factor between J and kJ to make sure that the units of ΔH° and ΔS° match.

Check: The experimental normal boiling point of $CCl_4(l)$ is 76.5°C. The small deviation of our estimate from the experimental value is due to the assumption that ΔH° and ΔS° do not change with temperature.

PRACTICE EXERCISE

Use data in Appendix C to estimate the normal boiling point, in K, for elemental bromine, $Br_2(l)$. (The experimental value is given in Table 11.3.)
Answer: 330 K

When the concentrations of reactants and products are nonstandard, we must calculate the value of Q in order to determine the value of ΔG. We illustrate how this is done in Sample Exercise 19.12.

SAMPLE EXERCISE 19.12

We will continue to explore the Haber process for the synthesis of ammonia:

$$N_2(g) + 3H_2(g) \rightleftharpoons 2NH_3(g)$$

Calculate ΔG at 298 K for a reaction mixture that consists of 1.0 atm N_2, 3.0 atm H_2, and 0.50 atm NH_3.

Solution

Analyze and Plan: We are asked to calculate ΔG *under nonstandard conditions.* To use Equation 19.21, we first need to calculate the value of the reaction quotient Q for the specified partial pressures of the gases.

Solve:

$$Q = \frac{P_{NH_3}{}^2}{P_{N_2}P_{H_2}{}^3} = \frac{(0.50)^2}{(1.0)(3.0)^3} = 9.3 \times 10^{-3}$$

In Sample Exercise 19.10 we calculated ΔG° for this reaction: $\Delta G^\circ = -33.3$ kJ. We will have to make some changes to the units of this quantity in applying Equation 19.21, however. In order for the units to match correctly, we will use ΔG° in units of kJ/mol. We will use "per mole" to mean "per mole of the reaction as written." Thus, we will use $\Delta G^\circ = -33.3$ kJ/mol, which implies per mole of N_2, per three moles of H_2, and per two moles of NH_3. We can now use Equation 19.21 to calculate ΔG for these nonstandard conditions:

$$\Delta G = \Delta G^\circ + RT \ln Q$$

$$= (-33.3 \text{ kJ/mol}) + (8.314 \text{ J/mol-K})(298 \text{ K})(1 \text{ kJ}/1000 \text{ J}) \ln (9.3 \times 10^{-3})$$

$$= (-33.3 \text{ kJ/mol}) + (-11.6 \text{ kJ/mol}) = -44.9 \text{ kJ/mol}$$

We see that ΔG becomes more negative, changing from −33.3 kJ/mol to −44.9 kJ/mol, as the pressures of N_2, H_2, and NH_3 are changed from 1.0 atm each (standard conditions, ΔG°) to 1.0 atm, 3.0 atm, and 0.50 atm, respectively. The larger negative value for ΔG indicates a larger "driving force" to produce NH_3. We would have made the same prediction on the basis of Le Châtelier's principle. (Section 15.6) Relative to standard conditions, we have increased the pressure of a reactant (H_2) and decreased the pressure of the product (NH_3). Le Châtelier's principle predicts that both of these changes should shift the reaction more to the product side, thereby forming more NH_3.

PRACTICE EXERCISE

Calculate ΔG at 298 K for the reaction of nitrogen and hydrogen to form ammonia if the reaction mixture consists of 0.50 atm N_2, 0.75 atm H_2, and 2.0 atm NH_3.
Answer: −26.0 kJ/mol

We can now use Equation 19.21 to derive the relationship between ΔG° and the equilibrium constant K_{eq}. At equilibrium, $\Delta G = 0$. Further, recall that the reaction quotient Q equals the equilibrium constant K_{eq} when the system is at equilibrium. Thus, at equilibrium, Equation 19.21 transforms as follows:

$$\Delta G = \Delta G^\circ + RT \ln Q$$

$$0 = \Delta G^\circ + RT \ln K_{eq}$$

$$\Delta G^\circ = -RT \ln K_{eq} \qquad [19.22]$$

From Equation 19.22 we can see that if $\Delta G°$ is negative, then $\ln K_{eq}$ must be positive. A positive value for $\ln K_{eq}$ means $K_{eq} > 1$. Therefore, the more negative $\Delta G°$ is, the larger the equilibrium constant, K_{eq}. Conversely, if $\Delta G°$ is positive, then $\ln K_{eq}$ is negative, which means that $K_{eq} < 1$. To summarize:

$$\Delta G°\text{, negative:} \qquad K_{eq} > 1$$

$$\Delta G°\text{, zero:} \qquad K_{eq} = 1$$

$$\Delta G°\text{, positive:} \qquad K_{eq} < 1$$

Equation 19.22 also allows us to calculate the value of K_{eq} if we know the value of $\Delta G°$. If we solve the equation for K_{eq}, we obtain

$$K_{eq} = e^{-\Delta G°/RT} \qquad [19.23]$$

As we pointed out in Sample Exercise 19.10, some care is necessary in the choice of units. We will express $\Delta G°$ in kJ/mol. For the reactants and products in the equilibrium-constant expression, we use the following conventions: Gas pressures are given in atm; solution concentrations are given in moles per liter (molarity); and solids, liquids, and solvents do not appear in the expression. (Section 15.3) We illustrate the use of Equation 19.23 in Sample Exercise 19.13.

SAMPLE EXERCISE 19.13

Use standard free energies of formation to calculate the equilibrium constant K_{eq} at 25°C for the reaction involved in the Haber process:

$$N_2(g) + 3H_2(g) \rightleftharpoons 2NH_3(g)$$

Solution

Analyze and Plan: We are required to use Equation 19.23 to evaluate the equilibrium constant for formation of $NH_3(g)$ from $H_2(g)$ and $N_2(g)$. The expression for K_{eq} for this reaction is written as

$$K_{eq} = \frac{P_{NH_3}{}^2}{P_{N_2}P_{H_2}{}^3}$$

in which the pressures of the gases are expressed in atmospheres. The standard free-energy change for the reaction is calculated in Sample Exercise 19.10: $\Delta G° = -33.3$ kJ/mol $= -33{,}300$ J/mol (remember that we use kJ/mol or J/mol as the units of $\Delta G°$ when using Equations 19.14, 19.15, or 19.16).

Solve: We can use this value to calculate $-\Delta G°/RT$, the exponent in Equation 19.23:

$$\frac{-\Delta G°}{RT} = \frac{-(-33{,}300\ \text{J/mol})}{[(8.314\ \text{J/mol-K})(298\ \text{K})]} = 13.4$$

We insert this value into Equation 19.22 to obtain K_{eq}:

$$K_{eq} = e^{-\Delta G°/RT} = e^{13.4} = 7 \times 10^5$$

Comment: This is a large equilibrium constant, which indicates that the product, NH_3, is greatly favored in the equilibrium mixture at 25°C. The equilibrium constants for temperatures in the range 300°C to 600°C, given in Table 15.2, are much smaller than the value at 25°C. Clearly, a low-temperature equilibrium favors the production of ammonia more than a high-temperature one. Nevertheless, the Haber process is carried out at high temperatures because the reaction is extremely slow at room temperature.
Remember: Thermodynamics can tell us the direction and extent of a reaction, but tells us nothing about the rate at which it will occur. If a catalyst were found that would permit the reaction to proceed at a rapid rate at room temperature, high pressures would not be needed to force the equilibrium toward NH_3.

PRACTICE EXERCISE

Use data from Appendix C to calculate the standard free-energy change, $\Delta G°$, and the equilibrium constant, K_{eq}, at 298 K for the following reaction: $H_2(g) + Br_2(l) \rightleftharpoons 2HBr(g)$.
Answer: -106.4 kJ/mol; 5×10^{18}

Chemistry and Life Driving Nonspontaneous Reactions

Many desirable chemical reactions, including a large number that are central to living systems, are nonspontaneous as written. For example, consider the extraction of copper metal from the mineral *chalcocite*, which contains Cu_2S. The decomposition of Cu_2S to its elements is nonspontaneous:

$$Cu_2S(s) \longrightarrow 2Cu(s) + S(s) \qquad \Delta G^\circ = +86.2 \text{ kJ}$$

Because ΔG° is very positive, we cannot obtain Cu(*s*) directly via this reaction. Instead, we must find some way to "do work" on the reaction in order to force it to occur as we wish. We can do this by coupling the reaction to another one so that the overall reaction *is* spontaneous. For example, we can envision the S(*s*) reacting with $O_2(g)$ to form $SO_2(g)$:

$$S(s) + O_2(g) \longrightarrow SO_2(g) \qquad \Delta G^\circ = -300.4 \text{ kJ}$$

By coupling these reactions together, we can extract much of the copper metal via a spontaneous reaction:

$$Cu_2S(s) + O_2(g) \longrightarrow 2Cu(s) + SO_2(g)$$

$$\Delta G^\circ = (+86.2 \text{ kJ}) + (-300.4 \text{ kJ}) = -214.2 \text{ kJ}$$

In essence, we have used the spontaneous reaction of S(*s*) with $O_2(g)$ to provide the free energy needed to extract the copper metal from the mineral.

Biological systems employ the same principle of using spontaneous reactions to drive nonspontaneous ones. Many of the biochemical reactions that are essential for the formation and maintenance of highly ordered biological structures are not spontaneous. These necessary reactions are made to occur by coupling them with spontaneous reactions that release energy. The metabolism of food is the usual source of the free energy needed to do the work of maintaining biological systems. For example, complete oxidation of the sugar *glucose*, $C_6H_{12}O_6$, to CO_2 and H_2O yields substantial free energy:

$$C_6H_{12}O_6(s) + 6O_2(g) \longrightarrow 6CO_2(g) + 6H_2O(l)$$

$$\Delta G^\circ = -2880 \text{ kJ}$$

This energy can be used to drive nonspontaneous reactions in the body. However, a means is necessary to transport the energy released by glucose metabolism to the reactions that require energy. One way, shown in Figure 19.19 ▼, involves the interconversion of adenosine triphosphate (ATP) and adenosine diphosphate (ADP), molecules that are related to the building blocks of nucleic acids. The conversion of ATP to ADP releases free energy ($\Delta G^\circ = -30.5$ kJ) that can be used to drive other reactions.

In the human body the metabolism of glucose occurs via a complex series of reactions, most of which release free energy. The free energy released during these steps is used in part to reconvert lower-energy ADP back to higher-energy ATP. Thus, the ATP-ADP interconversions are used to store energy during metabolism and to release it as needed to drive nonspontaneous reactions in the body. If you take a course in biochemistry, you will have the opportunity to learn more about the remarkable sequence of reactions used to transport free energy throughout the human body.

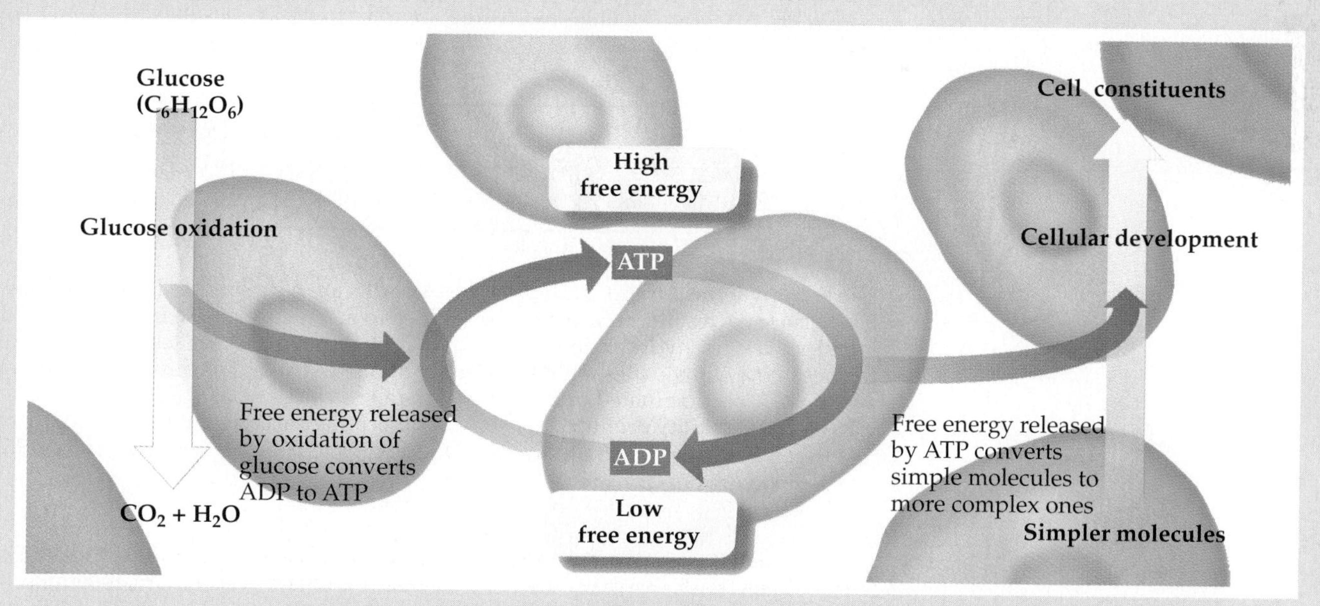

▲ **Figure 19.19** Schematic representation of part of the free-energy changes that occur in cell metabolism. The oxidation of glucose to CO_2 and H_2O produces free energy. This released free energy is used to convert ADP into the more energetic ATP. The ATP is then used, as needed, as an energy source to convert simple molecules into more complex cell constituents. When ATP releases its free energy, it is converted back into ADP.

SAMPLE INTEGRATIVE EXERCISE 19: Putting Concepts Together

Consider the simple salts NaCl(*s*) and AgCl(*s*). We will examine the equilibria in which these salts dissolve in water to form aqueous solutions of ions:

$$NaCl(s) \rightleftharpoons Na^+(aq) + Cl^-(aq)$$

$$AgCl(s) \rightleftharpoons Ag^+(aq) + Cl^-(aq)$$

(a) Calculate the value of $\Delta G°$ at 298 K for each of the preceding reactions. **(b)** The two values from part (a) are very different. Is this difference primarily due to the enthalpy term or the entropy term of the standard free-energy change? **(c)** Use the values of $\Delta G°$ to calculate the K_{sp} values for the two salts at 298 K. **(d)** Sodium chloride is considered a soluble salt, whereas silver chloride is considered insoluble. Are these descriptions consistent with the answers to part (c)? **(e)** How will $\Delta G°$ for the solution process of these salts change with increasing T? What effect should this change have on the solubility of the salts?

Solution **(a)** We will use Equation 19.13 along with ΔG_f° values from Appendix C to calculate the ΔG_{soln}° values for each equilibrium. (As we did in Section 13.1, we use the subscript "soln" to indicate that these are thermodynamic quantities for the formation of a solution.) We find:

$$\Delta G_{soln}^\circ(NaCl) = (-261.9\ kJ/mol) + (-131.2\ kJ/mol) - (-384.0\ kJ/mol)$$
$$= -9.1\ kJ/mol$$

$$\Delta G_{soln}^\circ(AgCl) = (+77.11\ kJ/mol) + (-131.2\ kJ/mol) - (-109.70\ kJ/mol)$$
$$= +55.6\ kJ/mol$$

(b) We can write ΔG_{soln}° as the sum of an enthalpy term, ΔH_{soln}°, and an entropy term, $-T\Delta S_{soln}^\circ$: $\Delta G_{soln}^\circ = \Delta H_{soln}^\circ + (-T\Delta S_{soln}^\circ)$. We can calculate the values of ΔH_{soln}° and ΔS_{soln}° by using Equations 5.31 and 19.8. We can then calculate $-T\Delta S_{soln}^\circ$ at $T = 298$ K. All these calculations are now familiar to us. The results are summarized in the following table:

Salt	ΔH_{soln}°	ΔS_{soln}°	$-T\Delta S_{soln}^\circ$
NaCl	+3.6 kJ/mol	+43.2 J/mol-K	−12.9 kJ/mol
AgCl	+65.7 kJ/mol	+34.3 J/mol-K	−10.2 kJ/mol

The entropy terms for the solution of the two salts are very similar. That seems sensible because each solution process should lead to a similar increase in disorder as the salt dissolves into hydrated ions. (Section 13.1) In contrast, we see a very large difference in the enthalpy term for the solution of the two salts. The difference in the values of ΔG_{soln}° is dominated by the difference in the values of ΔH_{soln}°.

(c) The solubility product, K_{sp}, is the equilibrium constant for the solution process. (Section 17.4) As such, we can relate K_{sp} directly to ΔG_{soln}° by using Equation 19.23:

$$K_{sp} = e^{-\Delta G_{soln}^\circ/RT}$$

We can calculate the K_{sp} values in the same way we applied Equation 19.23 in Sample Exercise 19.13. We use the ΔG_{soln}° values we obtained in part (a), remembering to convert them from kJ/mol to J/mol:

$$\text{NaCl:}\quad K_{sp} = [Na^+(aq)][Cl^-(aq)] = e^{-(-9100)/[(8.314)(298)]} = e^{+3.7} = 40$$

$$\text{AgCl:}\quad K_{sp} = [Ag^+(aq)][Cl^-(aq)] = e^{-(+55600)/[(8.314)(298)]} = e^{-22.4} = 1.9 \times 10^{-10}$$

The value calculated for the K_{sp} of AgCl is very close to that listed in Appendix D.

(d) A soluble salt is one that dissolves appreciably in water. (Section 4.2) The K_{sp} value for NaCl is greater than 1, indicating that NaCl dissolves to a great extent. The K_{sp} value for AgCl is very small, indicating that very little dissolves in water. Silver chloride should indeed be considered an insoluble salt.

(e) As we expect, the solution process has a positive value of ΔS for both salts (see the table on the preceding page). As such, the entropy term of the free-energy change, $-T\Delta S^\circ_{soln}$, is negative. If we assume that ΔH°_{soln} and ΔS°_{soln} do not change much with temperature, then an increase in T will serve to make ΔG°_{soln} more negative. Thus, the driving force for the dissolving salts will increase with increasing T, and we therefore expect the solubility of the salts to increase with increasing T. In Figure 13.17 we see that the solubility of NaCl (and that of nearly any salt) increases with increasing temperature. (Section 13.3)

Summary and Key Terms

Introduction and Section 19.1 In this chapter we examined some of the aspects of chemical thermodynamics, the area of chemistry that explores energy relationships. Most reactions and chemical processes have an inherent directionality: They are **spontaneous** in one direction and not spontaneous in the reverse direction. The spontaneity of a process is related to the thermodynamic path the system takes from the initial state to the final state. In a **reversible process**, such as the melting and freezing of ice at 0° C, the system can go back and forth between states along the same path. In an **irreversible process** the system can't return to its original state along the same path. Any spontaneous process is irreversible.

Section 19.2 The spontaneous nature of processes is related to a thermodynamic state function called **entropy**. Entropy, denoted S, is related to randomness or disorder; the greater the disorder, the greater the entropy. A process that increases the randomness of the system, such as the expansion of a gas, leads to a positive value of ΔS. For a process that occurs at constant temperature, the entropy change of the system is given by the heat absorbed by the system along a reversible path, divided by the temperature: $\Delta S = q_{rev}/T$. The way entropy controls the spontaneity of processes is given by the **second law of thermodynamics**, which governs the change in the entropy of the universe, $\Delta S_{univ} = \Delta S_{sys} + \Delta S_{surr}$. The second law states that in a reversible process $\Delta S_{univ} = 0$; in an irreversible (spontaneous) process $\Delta S_{univ} > 0$. Entropy values are usually expressed in units of joules per kelvin, J/K.

An **isolated system** is one that does not exchange energy or matter with its surroundings. For an isolated system the second law requires that ΔS_{sys} is zero for a reversible process and greater than zero for an irreversible process.

Sections 19.3 and 19.4 Entropy changes in a chemical system are associated with an increase in the number of ways the particles of the system can be arranged in space. Molecules can change their arrangements by moving in a number of different ways. In **translational motion** the entire molecule moves in space. Molecules can also undergo **vibrational motion**, in which the atoms of the molecule move toward and away from one another in periodic fashion, and **rotational motion**, in which the entire molecule spins like a top. These types of motion, and therefore the entropy of the system, decrease with decreasing temperature. The **third law of thermodynamics** states that the entropy of a pure crystalline solid at 0 K is zero.

The third law allows us to assign entropy values for substances at different temperatures. Under standard conditions the entropy of a mole of a substance is called its **standard molar entropy**, denoted S°. From tabulated values of S° we can calculate the entropy change for any process under standard conditions.

Section 19.5 The **Gibbs free energy** (or just **free energy**), G, is a thermodynamic state function that combines the two state functions enthalpy and entropy: $G = H - TS$. For processes that occur at constant temperature, $\Delta G = \Delta H - T\Delta S$. For a process or reaction occurring at constant temperature and pressure, the sign of ΔG relates to the spontaneity of the process. When ΔG is negative, the process is spontaneous. When ΔG is positive, the process is nonspontaneous; the reverse process is spontaneous. At equilibrium the process is reversible and ΔG is zero. The free energy is also a measure of the maximum useful work that can be performed by a system in a spontaneous process.

The standard free-energy change, ΔG°, for any process can be calculated from tabulations of **standard free**

energies of formation, ΔG_f°, which are defined in a fashion analogous to standard enthalpies of formation, ΔH_f° The value of ΔG_f°, for a pure element in its standard state is defined to be zero.

Sections 19.6 and 19.7 The values of ΔH and ΔS generally do not vary much with temperature. As a consequence, the dependence of ΔG with temperature is governed mainly by the value of T in the expression $\Delta G = \Delta H - T\Delta S$. The entropy term $-T\Delta S$ has the greater effect on the temperature dependence of ΔG, and hence on the spontaneity of the process. For example, a process for which $\Delta H > 0$ and $\Delta S > 0$, such as the melting of ice, can be nonspontaneous ($\Delta G > 0$) at low temperatures and spontaneous ($\Delta G < 0$) at higher temperatures.

Under nonstandard conditions, ΔG is related to ΔG° and the value of the reaction quotient, Q: $\Delta G = \Delta G^\circ + RT \ln Q$. At equilibrium ($\Delta G = 0$, $Q = K_{eq}$), $\Delta G^\circ = -RT \ln K_{eq}$. Thus, the standard free-energy change is directly related to the equilibrium constant for the reaction. This relationship can be used to explain the temperature dependence of equilibrium constants.

Exercises

Spontaneous Processes

19.1 Which of the following processes are spontaneous and which are nonspontaneous: **(a)** the melting of ice cubes at −5°C and 1 atm pressure; **(b)** dissolution of sugar in a cup of hot coffee; **(c)** the reaction of nitrogen atoms to form N_2 molecules at 25°C and 1 atm; **(d)** alignment of iron filings in a magnetic field; **(e)** formation of CH_4 and O_2 molecules from CO_2 and H_2O at room temperature and 1 atm of pressure?

19.2 Which of the following processes are spontaneous: **(a)** spreading of the fragrance of perfume through a room; **(b)** separating a mixture of N_2 and O_2 into two separate samples, one that is pure N_2 and one that is pure O_2; **(c)** the bursting of an inflated balloon; **(d)** the reaction of sodium metal with chlorine gas to form sodium chloride; **(e)** the dissolution of $HCl(g)$ in water to form concentrated hydrochloric acid?

19.3 **(a)** Give two examples of endothermic processes that are spontaneous. **(b)** Give an example of a process that is spontaneous at one temperature but nonspontaneous at a different temperature.

19.4 **(a)** A nineteenth-century chemist, Marcellin Berthelot, suggested that all chemical processes that proceed spontaneously are exothermic. Is this correct? If you think not, offer some counterexamples.

19.5 **(a)** Consider the vaporization of liquid water to steam at a pressure of 1 atm. **(a)** Is this process endothermic or exothermic? **(b)** In what temperature range is it a spontaneous process? **(c)** In what temperature range is it a nonspontaneous process? **(d)** At what temperature are the two phases in equilibrium?

19.6 The normal freezing point of 1-propanol (C_3H_8O, see Figure 2.28) is −127°C. **(a)** Is the freezing of 1-propanol an endothermic or exothermic process? **(b)** In what temperature range is the freezing of 1-propanol a spontaneous process? **(c)** In what temperature range is it a nonspontaneous process? **(d)** Is there any temperature at which liquid and solid 1-propanol are in equilibrium? Explain.

19.7 **(a)** What is special about a *reversible* process? **(b)** Suppose a reversible process is reversed, restoring the system to its original state. What can be said about the surroundings after the process is reversed? **(c)** Under what circumstances will the vaporization of water to steam be a reversible process?

19.8 **(a)** What is meant by calling a process *irreversible*? **(b)** After an irreversible process the system is restored to its original state. What can be said about the condition of the surroundings after the system is restored to its original state? **(c)** Under what conditions will the condensation of a liquid be an irreversible process?

19.9 Consider a process in which an ideal gas changes from state 1 to state 2 in such a way that its temperature changes from 300 K to 200 K. Does the change in ΔE depend on the particular pathway taken to carry out this change of state? Explain.

19.10 A system goes from state 1 to state 2 and back to state 1. **(a)** What is the relationship between the value of ΔE for going from state 1 to state 2 to that for going from state 2 back to state 1? **(b)** Without further information, can you conclude anything about the amount of heat transferred to the system as it goes from state 1 to state 2 as compared to that upon going from state 2 back to state 1? **(c)** Suppose the changes in state are reversible processes. Can you conclude anything about the work done by the system upon going from state 1 to state 2 as compared to that upon going from state 2 back to state 1?

19.11 Consider a system consisting of an ice cube. If the ice cube melts reversibly at 0°C, is ΔE zero for the process? Explain.

19.12 Consider what happens when a sample of the explosive TNT (see Chemistry at Work, Section 8.8) is detonated. **(a)** Is the detonation a spontaneous process? **(b)** What is the sign of q for this process? **(c)** Can you determine whether w is positive, negative, or zero for the process? Explain. **(d)** Can you determine the sign of ΔE for the process? Explain.

Entropy and the Second Law of Thermodynamics

19.13 For the isothermal expansion of a gas into a vacuum, $\Delta E = 0$, $q = 0$, and $w = 0$. **(a)** Is this a spontaneous process? **(b)** Explain why no work is done by the system during this process. **(c)** In thermodynamics, what is the "driving force" for the expansion of the gas?

19.14 Explain why it is possible to consider the heat gained or lost from a system in a reversible process as a state function, whereas q is normally not considered a state function.

19.15 Suppose that four gas molecules are placed in the right-hand flask of the apparatus in Figure 19.5. The left-hand flask is evacuated. **(a)** By analogy to Figure 19.6, after the stopcock is opened, how many different arrangements of the molecules are possible? **(b)** How many of the arrangements correspond to all the molecules' being in the right-hand flask? **(c)** How does the observation in part (b) explain the spontaneous expansion of the gas?

19.16 Suppose we have a dual flask system with eight molecules in it. The specific molecules are numbered 1 through 8. Two of the possible arrangements of the molecules are shown in the figure that follows:

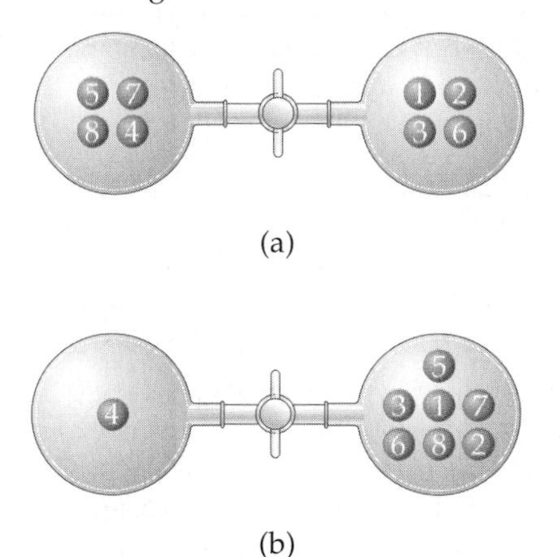

(a)

(b)

(a) Which of these two arrangements, if either, is more probable than the other? **(b)** Is your answer to part (a) consistent with our expectation that the gas molecules, on average, distribute themselves equally between the two flasks? Explain.

19.17 **(a)** What is *entropy*? **(b)** During a chemical process the system becomes more ordered. What is the sign of the change in the entropy of the system for the process? **(c)** Does ΔS for a process depend on the path taken from the initial to the final state of the system? Explain.

19.18 **(a)** Give an example of a process in which the entropy of the system decreases. **(b)** What is the sign of ΔS for the process? **(c)** What is the significance of the statement that entropy is a state function?

19.19 **(a)** What do you expect for the sign of ΔS in a chemical reaction in which two moles of gaseous reactants are converted to three moles of gaseous products? **(b)** For which of the processes in Exercise 19.2 does the entropy of the system increase?

19.20 **(a)** In a chemical reaction two gases combine to form a solid. What do you expect for the sign of ΔS? **(b)** For which of the processes in Exercise 19.1 does the entropy of the system increase?

19.21 How does the entropy of the system change when the following occur: **(a)** a solid melts; **(b)** a liquid vaporizes; **(c)** a solid dissolves in water; **(d)** a gas liquefies?

19.22 Why is the increase in entropy of the system greater for the vaporization of a substance than for its melting?

19.23 The normal boiling point of methanol (CH_3OH) is 64.7°C, and its molar enthalpy of vaporization is $\Delta H_{vap} =$ 71.8 kJ/mol. **(a)** When $CH_3OH(l)$ boils at its normal boiling point, does its entropy increase or decrease? **(b)** Calculate the value of ΔS when 1.00 mol of $CH_3OH(l)$ is vaporized at 64.7°C.

19.24 The element cesium (Cs) freezes at 28.4°C, and its molar enthalpy of fusion is $\Delta H_{fus} = 2.09$ kJ/mol **(a)** When molten cesium solidifies to Cs(s) at its normal melting point, is ΔS positive or negative? **(b)** Calculate the value of ΔS when 15.0 g of Cs(l) solidifies at 28.4°C.

19.25 **(a)** Express the second law of thermodynamics in words. **(b)** If the entropy of the system increases during a reversible process, what can you say about the entropy change of the surroundings? **(c)** In a certain spontaneous process the system undergoes an entropy change, $\Delta S = 42$ J/K. What can you conclude about ΔS_{surr}?

19.26 Express the second law of thermodynamics as a mathematical equation. **(b)** In a particular spontaneous process the entropy of the system decreases. What can you conclude about the sign and magnitude of ΔS_{surr}? **(c)** During a certain reversible process the surroundings undergo an entropy change, $\Delta S_{surr} = -78$ J/K. What is the entropy change of the system for this process?

The Molecular Interpretation of Entropy

19.27 **(a)** State the third law of thermodynamics. **(b)** Distinguish between translational motion, vibrational motion, and rotational motion of a molecule. **(c)** Illustrate these three kinds of motion with sketches for the HCl molecule.

19.28 **(a)** You are told that the entropy of a certain system is zero. What do you know about the system? **(b)** The energy of a gas is increased by heating it. Using CO_2 as an example, illustrate the different ways in which additional energy can be distributed among the molecules of the gas.

19.29 For each of the following pairs, choose the substance with the higher entropy per mole at a given temperature: **(a)** Ar(l) or Ar(g); **(b)** He(g) at 3 atm pressure or He(g) at 1.5 atm pressure; **(c)** 1 mol of Ne(g) in 15.0 L or 1 mol of Ne(g) in 1.50 L; **(d)** $CO_2(g)$ or $CO_2(s)$.

19.30 For each of the following pairs, indicate which substance possesses the larger standard entropy: **(a)** 1 mol of $P_4(g)$ at 300°C, 0.01 atm, or 1 mol of $As_4(g)$ at 300°C, 0.01 atm; **(b)** 1 mol of $H_2O(g)$ at 100°C, 1 atm, or 1 mol of $H_2O(l)$ at 100°C, 1 atm; **(c)** 0.5 mol of $N_2(g)$ at 298 K, 20-L volume, or 0.5 mol $CH_4(g)$ at 298 K, 20-L volume; **(d)** 100 g $Na_2SO_4(s)$ at 30°C or 100 g $Na_2SO_4(aq)$ at 30°C.

19.31 Predict the sign of the entropy change of the system for each of the following reactions:
(a) $2SO_2(g) + O_2(g) \longrightarrow 2SO_3(g)$
(b) $Ba(OH)_2(s) \longrightarrow BaO(s) + H_2O(g)$
(c) $CO(g) + 2H_2(g) \longrightarrow CH_3OH(l)$
(d) $FeCl_2(s) + H_2(g) \longrightarrow Fe(s) + 2HCl(g)$

19.32 Predict the sign of ΔS_{sys} for each of the following processes: **(a)** molten Fe solidifies; **(b)** LiCl(*s*) is formed from Li(*s*) and $Cl_2(g)$; **(c)** zinc metal dissolves in hydrochloric acid, forming $ZnCl_2(aq)$ and $H_2(g)$; **(d)** silver bromide precipitates upon mixing $AgNO_3(aq)$ and KBr(*aq*).

19.33 Use Appendix C to compare the standard enthalpies at 25°C for the following pairs of substances. In each case explain the difference in the entropy values. **(a)** Sc(*s*) and Sc(*g*); **(b)** $NH_3(g)$ and $NH_3(aq)$; **(c)** 1 mol $P_4(g)$ and 2 mol $P_2(g)$; **(d)** C(graphite) and C(diamond).

19.34 Using Appendix C, compare the standard entropies at 25°C for the following pairs of substances. For each pair, explain the difference in the entropy values. **(a)** CuO(*s*) and $Cu_2O(s)$; **(b)** 1 mol $N_2O_4(g)$ and 2 mol $NO_2(g)$; **(c)** $CH_3OH(g)$ and $CH_3OH(l)$; **(d)** 1 mol PbO(*s*) plus 1 mol $CO_2(g)$ and 1 mol $PbCO_3(s)$.

19.35 Use Appendix C to compare the absolute entropies of the following gaseous hydrocarbons: methane (CH_4), ethane (C_2H_6), propane (C_3H_8), and butane (C_4H_{10}). What do you conclude about the trend in $S°$ as the number of carbon atoms increases?

[19.36] The absolute entropies at 298 K for certain of the group 4A elements are as follows: C(*s*, diamond) = 2.43 J/mol-K; Si(*s*) = 18.81 J/mol-K; Ge(*s*) = 31.09 J/mol-K; and Sn(*s*) = 51.18 J/mol-K. All but Sn have the diamond structure. How do you account for the trend in the $S°$ values?

19.37 Using $S°$ values from Appendix C, calculate $\Delta S°$ values for the following reactions. In each case account for the sign of $\Delta S°$:
(a) $C_2H_4(g) + H_2(g) \longrightarrow C_2H_6(g)$
(b) $N_2O_4(g) \longrightarrow 2NO_2(g)$
(c) $Be(OH)_2(s) \longrightarrow BeO(s) + H_2O(g)$
(d) $2CH_3OH(g) + 3O_2(g) \longrightarrow 2CO_2(g) + 4H_2O(g)$

19.38 Calculate $\Delta S°$ values for the following reactions by using tabulated $S°$ values from Appendix C. In each case explain the sign of $\Delta S°$:
(a) $N_2H_4(g) + H_2(g) \longrightarrow 2NH_3(g)$
(b) $Al(s) + 3Cl_2(g) \longrightarrow 2AlCl_3(s)$
(c) $Mg(OH)_2(s) + 2HCl(g) \longrightarrow MgCl_2(s) + 2H_2O(l)$
(d) $2CH_4(g) \longrightarrow C_2H_6(g) + H_2(g)$

Gibbs Free Energy

19.39 **(a)** For a process that occurs at constant temperature, express the change in Gibbs free energy in terms of changes in the enthalpy and entropy of the system. **(b)** For a certain process that occurs at constant *T* and *P*, the value of ΔG is positive. What can you conclude? **(c)** What is the relationship between ΔG for a process and the rate at which it occurs?

19.40 **(a)** What is the meaning of the standard free-energy change, $\Delta G°$, as compared with ΔG? **(b)** For any process that occurs at constant temperature and pressure, what is the significance of $\Delta G = 0$? **(c)** For a certain process, ΔG is large and negative. Does this mean that the process necessarily occurs rapidly?

19.41 For a certain chemical reaction, $\Delta H° = -35.4$ kJ and $\Delta S° = -85.5$ J/K. **(a)** Is the reaction exothermic or endothermic? **(b)** Does the reaction lead to an increase or decrease in the disorder of the system? **(c)** Calculate $\Delta G°$ for the reaction at 298 K. **(d)** Is the reaction spontaneous at 298 K?

19.42 A certain reaction has $\Delta H° = -19.5$ kJ and $\Delta S° = +42.7$ J/K. **(a)** Is the reaction exothermic or endothermic? **(b)** Does the reaction lead to an increase or decrease in the disorder of the system? **(c)** Calculate $\Delta G°$ for the reaction at 298 K. **(d)** Is the reaction spontaneous at 298 K?

19.43 Using data in Appendix C, calculate $\Delta H°$, $\Delta S°$, and $\Delta G°$ at 298 K for each of the following reactions. In each case show that $\Delta G° = \Delta H° - T\Delta S°$.
(a) $H_2(g) + F_2(g) \longrightarrow 2HF(g)$
(b) $C(s, \text{graphite}) + 2Cl_2(g) \longrightarrow CCl_4(g)$
(c) $2PCl_3(g) + O_2(g) \longrightarrow 2POCl_3(g)$
(d) $2CH_3OH(g) + H_2(g) \longrightarrow C_2H_6(g) + 2H_2O(g)$

19.44 Use data in Appendix C to calculate $\Delta H°$, $\Delta S°$, and $\Delta G°$ at 25°C for each of the following reactions. In each case show that $\Delta G° = \Delta H° - T\Delta S°$.
(a) $Ni(s) + Cl_2(g) \longrightarrow NiCl_2(s)$
(b) $CaCO_3(s, \text{calcite}) \longrightarrow CaO(s) + CO_2(g)$
(c) $P_4O_{10}(s) + 6H_2O(l) \longrightarrow 4H_3PO_4(aq)$
(d) $2CH_3OH(l) + 3O_2(g) \longrightarrow 2CO_2(g) + 4H_2O(l)$

19.45 Using data from Appendix C, calculate $\Delta G°$ for the following reactions. Indicate whether each reaction is spontaneous under standard conditions.
(a) $2SO_2(g) + O_2(g) \longrightarrow 2SO_3(g)$
(b) $NO_2(g) + N_2O(g) \longrightarrow 3NO(g)$
(c) $6Cl_2(g) + 2Fe_2O_3(s) \longrightarrow 4FeCl_3(s) + 3O_2(g)$
(d) $SO_2(g) + 2H_2(g) \longrightarrow S(s) + 2H_2O(g)$

19.46 Using data from Appendix C, calculate the change in Gibbs free energy for each of the following reactions. In each case indicate whether the reaction is spontaneous under standard conditions.
(a) $H_2(g) + Cl_2(g) \longrightarrow 2HCl(g)$
(b) $MgCl_2(s) + H_2O(l) \longrightarrow MgO(s) + 2HCl(g)$
(c) $2NH_3(g) \longrightarrow N_2H_4(g) + H_2(g)$
(d) $2NOCl(g) \longrightarrow 2NO(g) + Cl_2(g)$

19.47 Cyclohexane (C_6H_{12}) is a liquid hydrocarbon at room temperature. **(a)** Write a balanced equation for the combustion of $C_6H_{12}(l)$ to form $CO_2(g)$ and $H_2O(l)$. **(b)** Without using thermochemical data, predict whether $\Delta G°$ for this reaction is more negative or less negative than $\Delta H°$.

19.48 Sulfur dioxide reacts with strontium oxide as follows:

$$SO_2(g) + SrO(s) \longrightarrow SrSO_3(s)$$

(a) Without using thermochemical data, predict whether $\Delta G°$ for this reaction is more negative or less negative than ΔH°. **(b)** If you had only standard enthalpy data for this reaction, how would you go about making a rough estimate of the value of ΔG° at 298 K, using data from Appendix C on other substances?

19.49 Classify each of the following reactions as one of the four possible types summarized in Table 19.4:
(a) $N_2(g) + 3F_2(g) \longrightarrow 2NF_3(g)$
$\Delta H^\circ = -249$ kJ; $\Delta S = -278$ J/K
(b) $N_2(g) + 3Cl_2(g) \longrightarrow 2NCl_3(g)$
$\Delta H^\circ = 460$ kJ; $\Delta S^\circ = -275$ J/K
(c) $N_2F_4(g) \longrightarrow 2NF_2(g)$ $\Delta H^\circ = 85$ kJ; $\Delta S^\circ = 198$ J/K

19.50 From the values given for ΔH° and ΔS°, calculate ΔG° for each of the following reactions at 298 K. If the reaction is not spontaneous under standard conditions at 298 K, at what temperature (if any) would the reaction become spontaneous?
(a) $2PbS(s) + 3O_2(g) \longrightarrow 2PbO(s) + 2SO_2(g)$
$\Delta H^\circ = -844$ kJ; $\Delta S^\circ = -165$ J/K
(b) $2POCl_3(g) \longrightarrow 2PCl_3(g) + O_2(g)$
$\Delta H^\circ = 572$ kJ; $\Delta S^\circ = 179$ J/K

19.51 A particular reaction is spontaneous at 450 K. The enthalpy change for the reaction is +34.5 kJ. What can you conclude about the sign and magnitude of ΔS for the reaction?

19.52 A certain reaction is nonspontaneous at −25°C. The entropy change for the reaction is 95 J/K. What can you conclude about the sign and magnitude of ΔH?

19.53 For a particular reaction, $\Delta H = -32$ kJ and $\Delta S = -98$ J/K. Assume that ΔH and ΔS do not vary with temperature. **(a)** At what temperature will the reaction have $\Delta G = 0$? **(b)** If T is increased from that in part (a), will the reaction be spontaneous or nonspontaneous?

19.54 Reactions in which a substance decomposes by losing CO_2 are called *decarboxylation* reactions. The decarboxylation of acetic acid proceeds as follows:

$$CH_3COOH(l) \longrightarrow CH_4(g) + CO_2(g)$$

By using data from Appendix C, calculate the minimum temperature at which this process will be spontaneous under standard conditions. Assume that ΔH° and ΔS° do not vary with temperature.

19.55 Consider the following reaction between oxides of nitrogen:

$$NO_2(g) + N_2O(g) \longrightarrow 3NO(g)$$

(a) Use data in Appendix C to predict how ΔG° for the reaction varies with increasing temperature. **(b)** Calculate ΔG° at 800 K, assuming that ΔH° and ΔS° do not change with temperature. Under standard conditions is the reaction spontaneous at 800 K? **(c)** Calculate ΔG° at 1000 K. Is the reaction spontaneous under standard conditions at this temperature?

19.56 Methanol (CH_3OH) can be made by the controlled oxidation of methane:

$$CH_4(g) + \tfrac{1}{2}O_2(g) \longrightarrow CH_3OH(g)$$

(a) Use data in Appendix C to calculate ΔH° and ΔS° for this reaction. **(b)** How is ΔG° for the reaction expected to vary with increasing temperature? **(c)** Calculate ΔG° at 298 K. Under standard conditions, is the reaction spontaneous at this temperature? **(d)** Is there a temperature at which the reaction would be at equilibrium under standard conditions and that is low enough so that the compounds involved are likely to be stable?

19.57 **(a)** Use data in Appendix C to estimate the boiling point of benzene, $C_6H_6(l)$. **(b)** Use a reference source, such as the *CRC Handbook of Chemistry and Physics*, to find the experimental boiling point of benzene. How do you explain any deviation between your answer in part (a) and the experimental value?

[19.58] **(a)** Using data in Appendix C, estimate the temperature at which the free energy change for the transformation from $I_2(s)$ to $I_2(g)$ is zero. What assumptions must you make in arriving at this estimate? **(b)** Use a reference source, such as WebElements (http://www.webelements.com/), to find the experimental melting and boiling points of I_2. **(c)** Which of the values in part (b) is closer to the value you obtained in part (a)? Can you explain why this is so?

19.59 Acetylene gas, $C_2H_2(g)$, is used in welding. **(a)** Write a balanced equation for the combustion of acetylene gas to $CO_2(g)$ and $H_2O(l)$. **(b)** How much heat is produced in burning a mole of C_2H_2 under standard conditions if both reactants and products are brought to 298 K? **(c)** What is the maximum amount of useful work that can be accomplished under standard conditions by this reaction?

19.60 **(a)** How much heat is produced in burning a mole of ethylene (C_2H_4) under standard conditions if reactants and products are brought to 298 K and $H_2O(l)$ is formed? **(b)** What is the maximum amount of useful work that can be accomplished under standard conditions by this system?

Free Energy and Equilibrium

19.61 Explain qualitatively how ΔG changes for each of the following reactions as the partial pressure of O_2 is increased:
(a) $2CO(g) + O_2(g) \longrightarrow 2CO_2(g)$
(b) $2H_2O_2(l) \longrightarrow 2H_2O(l) + O_2(g)$
(c) $2KClO_3(s) \longrightarrow 2KCl(s) + 3O_2(g)$

19.62 Indicate whether ΔG increases, decreases, or does not change when the partial pressure of H_2 is increased in each of the following reactions:
(a) $N_2(g) + 3H_2(g) \longrightarrow 2NH_3(g)$
(b) $2HBr(g) \longrightarrow H_2(g) + Br_2(g)$
(c) $2H_2(g) + C_2H_2(g) \longrightarrow C_2H_6(g)$

19.63 Consider the reaction $2NO_2(g) \longrightarrow N_2O_4(g)$. **(a)** Using data from Appendix C, calculate ΔG° at 298 K. **(b)** Calculate ΔG at 298 K if the partial pressures of NO_2 and N_2O_4 are 0.40 atm and 1.60 atm, respectively.

19.64 Consider the reaction $H_2(g) + F_2(g) \longrightarrow 2HF(g)$. **(a)** Using data from Appendix C, calculate ΔG° at 298 K. **(b)** Calculate ΔG at 298 K if the reaction mixture consists of 8.0 atm of H_2, 4.5 atm of F_2, and 0.36 atm of HF.

19.65 Use data from Appendix C to calculate K_{eq} at 298 K for each of the following reactions:
(a) $H_2(g) + I_2(g) \rightleftharpoons 2HI(g)$
(b) $C_2H_5OH(g) \rightleftharpoons C_2H_4(g) + H_2O(g)$
(c) $3C_2H_2(g) \rightleftharpoons C_6H_6(g)$

19.66 Write the equilibrium-constant expression and calculate the value of the equilibrium constant for each of the following reactions at 298 K, using data from Appendix C:
(a) $NaHCO_3(s) \rightleftharpoons NaOH(s) + CO_2(g)$
(b) $2HBr(g) + Cl_2(g) \rightleftharpoons 2HCl(g) + Br_2(g)$
(c) $2SO_2(g) + O_2(g) \rightleftharpoons 2SO_3(g)$

19.67 Consider the decomposition of barium carbonate:

$$BaCO_3(s) \rightleftharpoons BaO(s) + CO_2(g)$$

Using data from Appendix C, calculate the equilibrium pressure of CO_2 at **(a)** 298 K and **(b)** 1100 K.

19.68 Consider the following reaction:

$$PbCO_3(s) \rightleftharpoons PbO(s) + CO_2(g)$$

Using data in Appendix C, calculate the equilibrium pressure of CO_2 in the system at **(a)** 120°C and **(b)** 480°C.

19.69 The value of K_a for nitrous acid (HNO_2) at 25°C is given in Appendix D. **(a)** Write the chemical equation for the equilibrium that corresponds to K_a. **(b)** By using the value of K_a, calculate ΔG° for the dissociation of nitrous acid in aqueous solution. **(c)** What is the value of ΔG at equilibrium? **(d)** What is the value of ΔG when $[H^+] = 5.0 \times 10^{-2}\ M$, $[NO_2^-] = 6.0 \times 10^{-4}\ M$, and $[HNO_2] = 0.20\ M$?

19.70 The K_b for methylamine (CH_3NH_2) at 25°C is given in Appendix D. **(a)** Write the chemical equation for the equilibrium that corresponds to K_b. **(b)** By using the value of K_b, calculate ΔG° for the equilibrium in part (a). **(c)** What is the value of ΔG at equilibrium? **(d)** What is the value of ΔG when $[H^+] = 1.5 \times 10^{-8}\ M$, $[CH_3NH_3^+] =$ $[H^+] = 1.5 \times 10^{-8}\ M$, $[CH_3NH_3^+] = 5.5 \times 10^{-4}\ M$, and $[CH_3NH_2] = 0.120\ M$?

Additional Exercises

19.71 Indicate whether each of the following statements is true or false. If it is false, correct it. **(a)** The feasibility of manufacturing NH_3 from N_2 and H_2 depends entirely on the value of ΔH for the process $N_2(g) + 3H_2(g) \longrightarrow 2NH_3(g)$. **(b)** The reaction of $H_2(g)$ with $Cl_2(g)$ to form $HCl(g)$ is a spontaneous process. **(c)** A spontaneous process can in principle be conducted reversibly. **(d)** Spontaneous processes in general require that work be done to force them to proceed. **(e)** Spontaneous processes are those that are exothermic and that lead to a higher degree of order in the system.

19.72 Suppose an ideal gas is compressed to half its original volume at a constant temperature of 300 K. What can you say about **(a)** the change in internal energy of the gas, and **(b)** the change in entropy of the gas?

19.73 For each of the following processes, indicate whether the signs of ΔS and ΔH are expected to be positive, negative, or about zero. **(a)** A solid sublimes. **(b)** The temperature of a sample of Co(s) is lowered from 60°C to 25°C. **(c)** Ethyl alcohol evaporates from a beaker. **(d)** A diatomic molecule dissociates into atoms. **(e)** A piece of charcoal is combusted to form $CO_2(g)$ and $H_2O(g)$.

19.74 The reaction $2Mg(s) + O_2(g) \longrightarrow 2MgO(s)$ is highly spontaneous and has a negative value for ΔS°. The second law of thermodynamics states that in any spontaneous process there is always an increase in the entropy of the universe. Is there an inconsistency between the above reaction and the second law?

19.75 **(a)** What is an *isolated* system? **(b)** An isolated system undergoes a change in state. What can be said about the values of ΔE, q, and w? **(c)** Express the second law of thermodynamics mathematically for an isolated system.

19.76 Propanol (C_3H_7OH) melts at −126.5°C and boils at 97.4°C. Draw a qualitative sketch of how the absolute entropy changes as propanol vapor at 150°C and 1 atm is cooled to solid propanol at −150°C and 1 atm.

19.77 Cyclopropane and propylene are isomers of C_3H_6. Based on the molecular structures shown, which of these isomers would you expect to have the higher absolute entropy at 25°C?

Cyclopropane Propylene

[19.78] Three of the forms of elemental carbon are graphite, diamond, and buckminsterfullerene. The absolute entropies at 298 K for graphite and diamond are listed in Appendix C. **(a)** Account for the difference in the S° values of graphite and diamond in light of their structures (Figure 11.41). **(b)** What would you expect for the S° value of buckminsterfullerene (Figure 11.43) relative to the values for graphite and diamond? Explain.

19.79 For the majority of the compounds listed in Appendix C, the value of ΔG_f° is more positive (or less negative) than the value of ΔH_f°. **(a)** Explain this observation, using $NH_3(g)$, $CCl_4(l)$, and $KNO_3(s)$ as examples. **(b)** An exception to this observation is $CO(g)$. Explain the trend in the ΔH_f° and ΔG_f° values for this molecule.

19.80 Consider the following three reactions:
(i) $2RbCl(s) + 3O_2(g) \longrightarrow 2RbClO_3(s)$
(ii) $C_2H_2(g) + 4Cl_2(g) \longrightarrow 2CCl_4(l) + H_2(g)$
(iii) $TiCl_4(l) + 2H_2O(l) \longrightarrow TiO_2(s) + 4HCl(aq)$
(a) For each of the reactions, use data in Appendix C to calculate ΔH°, ΔG°, and ΔS° at 25°C. **(b)** Which of these reactions are spontaneous under standard conditions at 25°C? **(c)** For each of the reactions, predict the manner in which the change in free energy varies with an increase in temperature.

19.81 Using the data in Appendix C and given the pressures listed, calculate ΔG for each of the following reactions:
(a) $N_2(g) + 3H_2(g) \longrightarrow 2NH_3(g)$
$P_{N_2} = 2.6$ atm, $P_{H_2} = 5.9$ atm, $P_{NH_3} = 1.2$ atm
(b) $2N_2H_4(g) + 2NO_2(g) \longrightarrow 3N_2(g) + 4H_2O(g)$
$P_{N_2H_4} = P_{NO_2} = 5.0 \times 10^{-2}$ atm, $P_{N_2} = 0.5$ atm, $P_{H_2O} = 0.3$ atm
(c) $N_2H_4(g) \longrightarrow N_2(g) + 2H_2(g)$
$P_{N_2H_4} = 0.5$ atm, $P_{N_2} = 1.5$ atm, $P_{H_2} = 2.5$ atm

19.82 **(a)** For each of the following reactions, predict the sign of $\Delta H°$ and $\Delta S°$ and discuss briefly how these factors determine the magnitude of K_{eq}. **(b)** Based on your general chemical knowledge, predict which of these reactions will have $K_{eq} > 1$. **(c)** In each case indicate whether K_{eq} should increase or decrease with increasing temperature.
(i) $2Mg(s) + O_2(g) \rightleftharpoons 2MgO(s)$
(ii) $2KI(s) \rightleftharpoons 2K(g) + I_2(g)$
(iii) $Na_2(g) \rightleftharpoons 2Na(g)$
(iv) $2V_2O_5(s) \rightleftharpoons 4V(s) + 5O_2(g)$

19.83 Acetic acid can be manufactured by combining methanol with carbon monoxide, an example of a *carbonylation* reaction:

$$CH_3OH(l) + CO(g) \longrightarrow CH_3COOH(l)$$

(a) Calculate the equilibrium constant for the reaction at 25°C. **(b)** Industrially, this reaction is run at temperatures above 25°C. Will an increase in temperature produce an increase or decrease in the mole fraction of acetic acid at equilibrium? Why are elevated temperatures used? **(c)** At what temperature will this reaction have an equilibrium constant equal to 1? (You may assume that $\Delta H°$ and $\Delta S°$ are temperature-independent, and you may ignore any phase changes that might occur.)

19.84 The oxidation of glucose ($C_6H_{12}O_6$) in body tissue produces CO_2 and H_2O. In contrast, anaerobic decomposition, which occurs during fermentation, produces ethanol (C_2H_5OH) and CO_2. **(a)** Using data given in Appendix C, compare the equilibrium constants for the following reactions:

$$C_6H_{12}O_6(s) + 6O_2(g) \rightleftharpoons 6CO_2(g) + 6H_2O(l)$$

$$C_6H_{12}O_6(s) \rightleftharpoons 2C_2H_5OH(l) + 2CO_2(g)$$

(b) Compare the maximum work that can be obtained from these processes under standard conditions.

[19.85] The conversion of natural gas, which is mostly methane, into products that contain two or more carbon atoms, such as ethane (C_2H_6), is a very important industrial chemical process. In principle, methane can be converted into ethane and hydrogen:

$$2CH_4(g) \longrightarrow C_2H_6(g) + H_2(g)$$

In practice, this reaction is carried out in the presence of oxygen:

$$2CH_4(g) + \tfrac{1}{2}O_2(g) \longrightarrow C_2H_6(g) + H_2O(g)$$

(a) Using the data in Appendix C, calculate K_{eq} for these reactions at 25°C and 500°C. **(b)** Is the difference in $\Delta G°$ for the two reactions due primarily to the enthalpy term (ΔH) or the entropy term ($-T\Delta S$)? **(c)** Explain how the preceding reactions are an example of driving a nonspontaneous reaction, as discussed in the "Chemistry and Life" box in Section 19.7. **(d)** The reaction of CH_4 and O_2 to form C_2H_6 and H_2O must be carried out carefully to avoid a competing reaction. What is the most likely competing reaction?

[19.86] Cells use the hydrolysis of adenosine triphosphate (ATP) as a source of energy (Figure 19.19). The conversion of ATP to ADP has a standard free-energy change of −30.5 kJ/mol. If all the free energy from the metabolism of glucose,

$$C_6H_{12}O_6(s) + 6O_2(g) \longrightarrow 6CO_2(g) + 6H_2O(l)$$

goes into the conversion of ADP to ATP, how many moles of ATP can be produced for each mole of glucose?

[19.87] The potassium-ion concentration in blood plasma is about 5.0×10^{-3} *M*, whereas the concentration in muscle-cell fluid is much greater (0.15 *M*). The plasma and intracellular fluid are separated by the cell membrane, which we assume is permeable only to K^+. **(a)** What is ΔG for the transfer of 1 mol of K^+ from blood plasma to the cellular fluid at body temperature (37°C)? **(b)** What is the minimum amount of work that must be used to transfer this K^+?

[19.88] The relationship between the temperature of a reaction, its standard enthalpy change, and the equilibrium constant at that temperature can be expressed as the following linear equation:

$$\ln K_{eq} = \frac{-\Delta H°}{RT} + \text{constant}$$

(a) Explain how this equation can be used to determine $\Delta H°$ experimentally from the equilibrium constants at several different temperatures. **(b)** Derive the preceding equation using relationships given in this chapter. To what is the constant equal?

[19.89] We saw in the "Closer Look" box in Section 19.2 that the entropy change for the isothermal expansion or compression of an ideal gas is given by

$$\Delta S = nR \ln \frac{V_2}{V_1} \qquad \text{(ideal gas, constant } T\text{)}$$

where V_1 and V_2 are the initial and final volumes of the gas, respectively. **(a)** With reference to Equation 19.1, what must be the expression for q_{rev} when a gas undergoes an isothermal expansion or compression? **(b)** Calculate the entropy change when 0.50 mol of an ideal gas expands at constant temperature from an initial volume of 10.0 L to a final volume of 75.0 L. **(c)** Is the sign of the entropy change in part (b) consistent with your expectations? **(d)** A sample of 8.5 mol of an ideal gas is compressed isothermally to a final volume that is one-eighth its original volume. Calculate the entropy change.

[19.90] One way to derive the equation in Exercise 19.89 depends on the observation that, at constant *T*, the number of ways, *W*, of arranging *m* ideal-gas particles in a volume *V* is proportional to the volume raised to the *m* power:

$$W \propto V^m$$

Use this relationship and Boltzmann's relationship between entropy and number of arrangements (Equation 19.7) to derive the equation for the entropy change for the isothermal expansion or compression of *n* moles of an ideal gas.

[19.91] An ad in an auto magazine trumpets a new device that saves on gas. It is called the "Entropy Converter." The ad reads, "Attach the Entropy Converter to your carburetor and see amazing improvement in your gas mileage! Did you know that increases in entropy when the fuel is burned help to drive your car? The Entropy Converter automatically lowers the entropy of the gas molecules as they enter the carburetor. Then, when they are burned in the engine, the entropy increases more than ever, yielding much more energy to drive your car!" Indicate whether you think this device might work. If not, explain why.

Integrative Exercises

19.92 Most liquids follow *Trouton's rule*, which states that the molar entropy of vaporization lies in the range of 83 to 93 J/mol-K. The normal boiling points and enthalpies of vaporization of several organic liquids are as follows:

Substance	Normal Boiling Point (°C)	ΔH_{vap}(kJ/mol)
Acetone, $(CH_3)_2CO$	56.1	29.1
Dimethyl ether, $(CH_3)_2O$	−24.8	21.5
Ethanol, C_2H_5OH	78.4	38.6
Octane, C_8H_{18}	125.6	34.4
Pyridine, C_5H_5N	115.3	35.1

(a) Calculate ΔS_{vap} for each of the liquids. Do all of the liquids obey Trouton's rule? **(b)** With reference to intermolecular forces (Section 11.2), can you explain any exceptions to the rule? **(c)** Would you expect water to obey Trouton's rule? By using data in Appendix B, check the accuracy of your conclusion. **(d)** Chlorobenzene (C_6H_5Cl) boils at 131.8°C. Use Trouton's rule to estimate ΔH_{vap} for this substance.

19.93 Consider the polymerization of ethylene to polyethylene. (Section 12.2) **(a)** What would you predict for the sign of the entropy change during polymerization (ΔS_{poly})? Explain your reasoning. **(b)** The polymerization of ethylene is a spontaneous process at room temperature. What can you conclude about the enthalpy change during polymerization (ΔH_{poly})? **(c)** Use average bond enthalpies (Table 8.4) to estimate the value of ΔH_{poly} per ethylene monomer added. **(d)** Polyethylene is an *addition polymer*. By comparison, Nylon 6.6 is a *condensation polymer*. How would you expect ΔS_{poly} for a condensation polymer to compare to that for an addition polymer? Explain.

19.94 In chemical kinetics the *entropy of activation* is the entropy change for the process in which the reactants reach the activated complex. The entropy of activation for bimolecular processes is usually negative. Explain this observation with reference to Figure 14.13.

19.95 The following processes were all discussed in Chapter 18, "Chemistry of the Environment." Estimate whether the entropy of the system increases or decreases during each process: **(a)** photodissociation of $O_2(g)$; **(b)** formation of ozone from oxygen molecules and oxygen atoms; **(c)** diffusion of CFCs into the stratosphere; **(d)** desalination of water by reverse osmosis.

19.96 Carbon disulfide (CS_2) is a toxic, highly flammable substance. The following thermodynamic data are available for $CS_2(l)$ and $CS_2(g)$ at 298 K:

	ΔH_f°(kJ/mol)	ΔG_f°(kJ/mol)
$CS_2(l)$	89.7	65.3
$CS_2(g)$	117.4	67.2

(a) Draw the Lewis structure of the molecule. What do you predict for the bond order of the C—S bonds? **(b)** Use the VSEPR method to predict the structure of the CS_2 molecule. **(c)** Liquid CS_2 burns in O_2 with a blue flame, forming $CO_2(g)$ and $SO_2(g)$. Write a balanced equation for this reaction. **(d)** Using the data in the preceding table and in Appendix C, calculate ΔH° and ΔG° for the reaction in part (c). Is the reaction exothermic? Is it spontaneous at 298 K? **(e)** Use the data in the preceding table to calculate ΔS° at 298 K for the vaporization of $CS_2(l)$. Is the sign of ΔS° as you would expect for a vaporization? **(f)** Using data in the preceding table and your answer to part (e), estimate the boiling point of $CS_2(l)$. Do you predict that the substance will be a liquid or a gas at 298 K and 1 atm?

[19.97] The following data compare the standard enthalpies and free energies of formation of some crystalline ionic substances and 1 *m* aqueous solutions of the substances:

Substance	ΔH_f°(kJ/mol)	ΔG_f°(kJ/mol)
$AgNO_3(s)$	−124.4	−33.4
$AgNO_3(aq, 1\ m)$	−101.7	−34.2
$MgSO_4(s)$	−1283.7	−1169.6
$MgSO_4(aq, 1\ m)$	−1374.8	−1198.4

(a) Write the formation reaction for $AgNO_3(s)$. Based on this reaction, do you expect the entropy of the system to increase or decrease upon the formation of $AgNO_3(s)$? **(b)** Use ΔH_f° and ΔG_f° of $AgNO_3(s)$ to determine the entropy change upon formation of the substance. Is your answer consistent with your reasoning in part (a)? **(c)** Is dissolving $AgNO_3$ in water an exothermic or endothermic process? What about dissolving $MgSO_4$ in water? **(d)** For both $AgNO_3$ and $MgSO_4$, use the data to calculate the entropy change when the solid is dissolved in water. **(e)** Discuss the results from part (d) with reference to material presented in this chapter and in the second "Closer Look" box in Section 13.5.

[19.98] Consider the following equilibrium:

$$N_2O_4(g) \rightleftharpoons 2NO_2(g)$$

Thermodynamic data on these gases are given in Appendix C. You may assume that ΔH° and ΔS° do not vary with temperature. **(a)** At what temperature will an equilibrium mixture contain equal amounts of the two gases? **(b)** At what temperature will an equilibrium mixture of 1 atm total pressure contain twice as much NO_2 as N_2O_4? **(c)** At what temperature will an equilibrium mixture of 10 atm total pressure contain twice as much NO_2 as N_2O_4? **(d)** Rationalize the results from parts (b) and (c) by using Le Châtelier's principle. (Section 15.6)

[19.99] The reaction

$$SO_2(g) + 2H_2S(g) \rightleftharpoons 3S(s) + 2H_2O(g)$$

is the basis of a suggested method for removal of SO_2 from power-plant stack gases. The standard free energy of each substance is given in Appendix C. **(a)** What is the equilibrium constant for the reaction at 298 K? **(b)** In principle, is this reaction a feasible method of removing SO_2? **(c)** If $P_{SO_2} = P_{H_2S}$ and the vapor pressure of water is 25 torr, calculate the equilibrium SO_2 pressure in the system at 298 K. **(d)** Would you expect the process to be more or less effective at higher temperatures?

19.100 When most elastomeric polymers (e.g., a rubber band) are stretched, the molecules become more ordered, as illustrated here:

Suppose you stretch a rubber band. **(a)** Do you expect the entropy of the system to increase or decrease? **(b)** If the rubber band were stretched isothermally, would heat need to be absorbed or emitted to maintain constant temperature?

eMedia Exercises

19.101 The **Gibbs Law of Thermodynamics** simulation (*eChapter 19.5*) allows you to explore the relationship between spontaneity and temperature. **(a)** Of the five reactions available in the simulation, which are spontaneous at all temperatures? **(b)** Which are nonspontaneous at all temperatures? **(c)** Is it possible for a reaction with a positive enthalpy change and a negative entropy change to be spontaneous? Explain.

19.102 Using information from the **Gibbs Law of Thermodynamics** simulation (*eChapter 19.5*), determine **(a)** the temperature below which the reaction $N_2O_4(g) \longrightarrow 2NO_2(g)$ will occur spontaneously; **(b)** the temperature below which the reaction $CaCO_3(s) \longrightarrow CaO(s) + CO_2(g)$ will occur spontaneously; **(c)** the temperature above which the reaction $Si(s) + 2Cl_2(g) \longrightarrow SiCl_4(l)$ will occur spontaneously.

19.103 The **Enthalpy of Solution** simulation (*eChapter 13.1*) lets you measure temperature changes for the combination of several different compounds with water. **(a)** Using data from this simulation, determine the signs of $\Delta G°$, $\Delta H°$, and $\Delta S°$ for the dissolution of ammonium nitrate. **(b)** Is the dissolution of ammonium nitrate spontaneous at all temperatures? Explain.

19.104 Watch the **Air Bags** movie again (*eChapter 19.5*), and answer the following questions. **(a)** What are the signs of $\Delta G°$, $\Delta H°$, and $\Delta S°$ for the decomposition of sodium azide? **(b)** Is there a temperature at which the decomposition of sodium azide is not spontaneous? Explain.

19.105 When a balloon filled with hydrogen is ignited, hydrogen and oxygen react explosively to form water. The **Formation of Water** movie (*eChapter 19.5*) states, "The equilibrium lies far to the right." **(a)** Explain what this statement means in terms of products and reactants. **(b)** Explain the relationship between the magnitude of an equilibrium constant and the sign of $\Delta G°$.

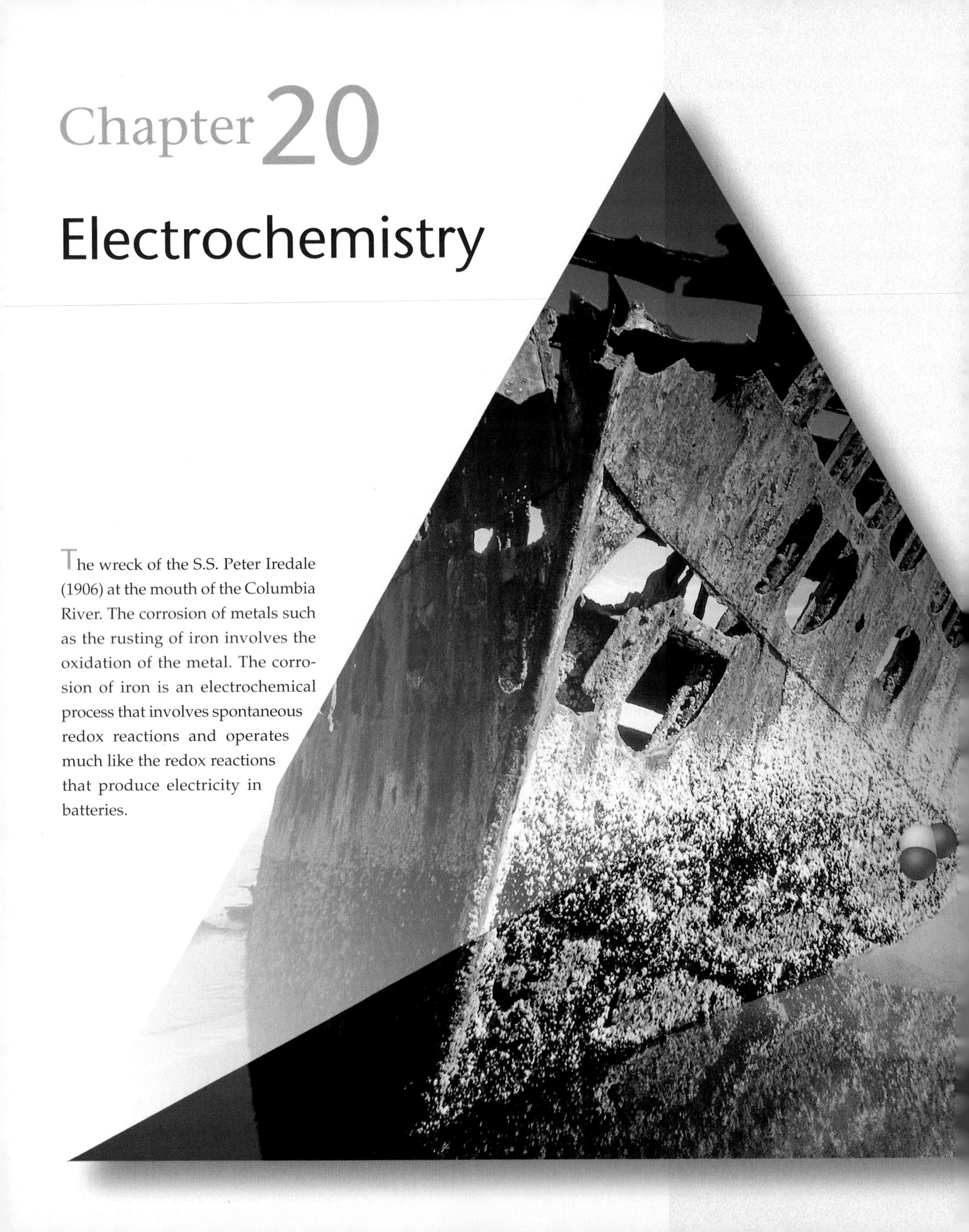

Chapter 20

Electrochemistry

The wreck of the S.S. Peter Iredale (1906) at the mouth of the Columbia River. The corrosion of metals such as the rusting of iron involves the oxidation of the metal. The corrosion of iron is an electrochemical process that involves spontaneous redox reactions and operates much like the redox reactions that produce electricity in batteries.

OXIDATION-REDUCTION (REDOX) reactions are among the most common and important chemical reactions. They are involved in a wide variety of important processes including the rusting of iron, the manufacture and action of bleaches, and the respiration of animals. As we discussed earlier, *oxidation* refers to the loss of electrons. Conversely, *reduction* refers to the gain of electrons. (Section 4.4) Thus, oxidation-reduction reactions occur when electrons are transferred from the atom that is oxidized to the atom that is reduced. When zinc metal is added to a strong acid (Figure 20.1 ▶), for example, electrons are transferred from zinc atoms (zinc is oxidized) to hydrogen ions (hydrogen is reduced):

$$Zn(s) + 2H^+(aq) \longrightarrow Zn^{2+}(aq) + H_2(g) \qquad [20.1]$$

The transfer of electrons that occurs in the reaction in Figure 20.1 produces energy in the form of heat; the reaction is thermodynamically "downhill" and proceeds spontaneously. The transfer of electrons that occurs during oxidation-reduction reactions can also be used to produce energy in the form of electricity. In other instances we use electrical energy to make certain nonspontaneous chemical processes occur. **Electrochemistry** is the study of the relationships between electricity and chemical reactions. Our discussion of electrochemistry will provide insight into such diverse topics as the construction of batteries, the spontaneity of reactions, the corrosion of metals, and electroplating. We will begin by reviewing how oxidation numbers are used to identify redox reactions.

▶ What's Ahead ◀

- After a brief review of *oxidation-reduction (redox) reactions*, we consider how to balance redox equations.
- We next consider *voltaic cells*, those that utilize spontaneous redox reactions to produce electricity. The electrode where oxidation occurs is the *anode*, and that where reduction occurs is the *cathode*.
- One of the important characteristics of a voltaic cell is its *emf* or voltage, which is the difference in the electrical potentials at the two electrodes.
- Electrode potentials, which are tabulated for reduction reactions under standard conditions (*standard reduction potentials*), are used to calculate the voltages of cells, to determine the relative strengths of oxidizing agents and reducing agents, and to predict whether or not particular redox reactions are spontaneous.
- The voltage associated with cells that run under nonstandard conditions can be calculated using standard voltages and the Nernst equation.
- We describe concentration cells and batteries, which are both based on voltaic cells.
- We then discuss *corrosion*, a spontaneous electrochemical process involving metals.
- Finally, we focus on nonspontaneous redox reactions, examining *electrolytic cells*, which use electricity; and we use this discussion to examine the relationship between the quantity of current flowing through a cell and the amount of products obtained.

▲ **Figure 20.1** The addition of zinc metal to hydrochloric acid leads to a spontaneous oxidation-reduction reaction: Zinc metal is oxidized to $Zn^{2+}(aq)$, and $H^+(aq)$ is reduced to $H_2(g)$, which produces the vigorous bubbling.

20.1 Oxidation-Reduction Reactions

How do we determine whether a given chemical reaction is an oxidation-reduction reaction? We can do so by keeping track of the oxidation numbers of all the elements involved in the reaction. This procedure tells us which elements (if any) are changing oxidation state. For example, the reaction in Equation 20.1 can be written

$$\underset{0}{Zn}(s) + 2\underset{+1}{H^+}(aq) \longrightarrow \underset{+2}{Zn^{2+}}(aq) + \underset{0}{H_2}(g) \qquad [20.2]$$

ANIMATION
Oxidation-Reduction Reactions Part I
Oxidation-Reduction Reactions Part II

By writing the oxidation number of each element above or below the equation, we can see the oxidation state changes that occur: The oxidation state of Zn changes from 0 to +2, and that of H changes from +1 to 0.

In a reaction such as Equation 20.2, a clear transfer of electrons occurs. Zinc loses electrons as $Zn(s)$ is converted to $Zn^{2+}(aq)$, and hydrogen gains electrons as $H^+(aq)$ is turned into $H_2(g)$. In other reactions the oxidation states change, but we can't say that any substance literally gains or loses electrons. For example, consider the combustion of hydrogen gas:

$$2\underset{0}{H_2}(g) + \underset{0}{O_2}(g) \longrightarrow 2\underset{+1}{H_2}\underset{-2}{O}(g) \qquad [20.3]$$

Hydrogen has been oxidized from the 0 to the +1 oxidation state, and oxygen has been reduced from the 0 to the −2 oxidation state. Therefore, Equation 20.3 is an oxidation-reduction reaction. Water is not an ionic substance, however, so there is not a complete transfer of electrons from hydrogen to oxygen as water is formed. Using oxidation numbers, therefore, is a convenient form of "bookkeeping," but *you should not generally equate the oxidation state of an atom with its actual charge in a chemical compound.* (A Closer Look: Oxidation Numbers, Formal Charges, and Partial Charges, Section 8.5)

In any redox reaction, both oxidation and reduction must occur. In other words, if one substance is oxidized, then another must be reduced. The substance that makes it possible for another substance to be oxidized is called the **oxidizing agent**, or **oxidant**. The oxidizing agent removes electrons from another substance by acquiring them itself; thus, the oxidizing agent is itself reduced. Similarly, a **reducing agent**, or **reductant**, is a substance that gives up electrons, thereby causing another substance to be reduced. The reducing agent is oxidized in the process. In Equation 20.2, $H^+(aq)$ is the oxidizing agent, and $Zn(s)$ is the reducing agent.

SAMPLE EXERCISE 20.1

The nickel-cadmium (nicad) battery, a rechargeable "dry cell" used in battery-operated devices, uses the following redox reaction to generate electricity:

$$Cd(s) + NiO_2(s) + 2H_2O(l) \longrightarrow Cd(OH)_2(s) + Ni(OH)_2(s)$$

Identify the substances that are oxidized and reduced, and indicate which are oxidizing agents and which are reducing agents.

Solution

Analyze: We are given a redox equation and asked to identify the substance oxidized and the substance reduced and to label one as the oxidizing agent and the other as the reducing agent.

Plan: First, we assign oxidation numbers to all the atoms in the reaction, and determine the elements that are changing oxidation number. Second, we apply the definitions of oxidation and reduction.

Solve:

$$\underset{(0)}{Cd}(s) + \underset{(+4)}{Ni}\underset{(-2)}{O_2}(s) + 2\underset{(+1)}{H_2}\underset{(-2)}{O}(l) \longrightarrow \underset{(+2)}{Cd}(\underset{(-2)}{O}\underset{(+1)}{H})_2(s) + \underset{(+2)}{Ni}(\underset{(-2)}{O}\underset{(+1)}{H})_2(s)$$

Cd increases in oxidation number from 0 to +2, and Ni decreases from +4 to +2. Because the Cd atom increases in oxidation number, it is oxidized (loses electrons), and it therefore serves as the reducing agent. The Ni atom decreases in oxidation number as NiO_2 is converted into $Ni(OH)_2$. Thus, NiO_2 is reduced (gains electrons), and it therefore serves as the oxidizing agent.

Comment: A common mnemonic for remembering oxidation and reduction is "LEO the lion says GER": *l*osing *e*lectrons is *o*xidation; *g*aining *e*lectrons is *r*eduction.

PRACTICE EXERCISE

Identify the oxidizing and reducing agents in the following oxidation-reduction equation:

$$2H_2O(l) + Al(s) + MnO_4^-(aq) \longrightarrow Al(OH)_4^-(aq) + MnO_2(s)$$

Answer: $Al(s)$ is the reducing agent; $MnO_4^-(aq)$ is the oxidizing agent.

20.2 Balancing Oxidation-Reduction Equations

Whenever we balance a chemical equation, we must obey the law of conservation of mass: The amount of each element must be the same on both sides of the equation. As we balance oxidation-reduction reactions, there is an additional requirement: The gains and losses of electrons must be balanced. In other words, if a substance loses a certain number of electrons during a reaction, then another substance must gain that same number of electrons. In many simple chemical reactions, such as Equation 20.2, balancing the electrons is handled "automatically"; we can balance the equation without explicitly considering the transfer of electrons. Many redox reactions are more complex than Equation 20.2, however, and cannot be balanced easily without taking into account the number of electrons lost and gained in the course of the reaction. In this section we examine a systematic procedure for balancing redox equations.

Half-Reactions

Although oxidation and reduction must take place simultaneously, it is often convenient to consider them as separate processes. For example, the oxidation of Sn^{2+} by Fe^{3+}

$$Sn^{2+}(aq) + 2Fe^{3+}(aq) \longrightarrow Sn^{4+}(aq) + 2Fe^{2+}(aq)$$

can be considered to consist of two processes: (1) the oxidation of Sn^{2+} (Equation 20.4) and (2) the reduction of Fe^{3+} (Equation 20.5).

$$\text{Oxidation:} \quad Sn^{2+}(aq) \longrightarrow Sn^{4+}(aq) + 2e^- \quad [20.4]$$

$$\text{Reduction:} \quad 2Fe^{3+}(aq) + 2e^- \longrightarrow 2Fe^{2+}(aq) \quad [20.5]$$

Notice that in the oxidation process electrons are shown as products, whereas in the reduction process they are shown as reactants.

Equations that show either oxidation or reduction alone, as in Equations 20.4 and 20.5, are called **half-reactions**. In the overall redox reaction the number of electrons lost in the oxidation half-reaction must equal the number of electrons gained in the reduction half-reaction. When this condition is met and each half-reaction is balanced, the electrons on each side cancel when the two half-reactions are added to give the overall balanced oxidation-reduction equation.

Balancing Equations by the Method of Half-Reactions

The use of half-reactions provides a general method for balancing oxidation-reduction equations. As an example, let's consider the reaction that occurs between permanganate ion (MnO_4^-) and oxalate ion ($C_2O_4^{2-}$) in acidic aqueous solutions. When MnO_4^- is added to an acidified solution of $C_2O_4^{2-}$, the deep purple color of the MnO_4^- ion fades, as illustrated in Figure 20.2 ▼. Bubbles of CO_2 form, and the solution takes on the pale pink color of Mn^{2+}. We can therefore write the unbalanced equation as follows:

$$MnO_4^-(aq) + C_2O_4^{2-}(aq) \longrightarrow Mn^{2+}(aq) + CO_2(aq) \quad [20.6]$$

(a)

(b)

(c)

▶ **Figure 20.2** Titration of an acidic solution of $Na_2C_2O_4$ with $KMnO_4(aq)$. (a) As the reaction proceeds, deep purple MnO_4^- is rapidly reduced to extremely pale pink Mn^{2+} by $C_2O_4^{2-}$. (b) When all the $C_2O_4^{2-}$ is consumed, the purple color of MnO_4^- persists. The end point corresponds to the faintest discernible purple color in the solution. (c) Beyond the end point, the solution becomes deep purple because of excess MnO_4^-.

Experiments also show that H^+ is consumed and H_2O is produced in the reaction. We will see that these facts can be deduced in the course of balancing the equation.

To complete and balance Equation 20.6 by the method of half-reactions, we begin with the unbalanced reaction and write two incomplete half-reactions, one involving the oxidant and the other involving the reductant.

$$MnO_4^-(aq) \longrightarrow Mn^{2+}(aq)$$
$$C_2O_4^{2-}(aq) \longrightarrow CO_2(g)$$

We have not explicitly stated which substance is oxidized and which is reduced. This information emerges as we balance the half-reactions.

We can now complete and balance the half-reactions separately. First, the atoms undergoing oxidation or reduction are balanced by adding coefficients on one side or the other as necessary. Then, the remaining elements are balanced in the same way. If the reaction occurs in acidic aqueous solution, H^+ and H_2O can be added either to reactants or to products to balance hydrogen and oxygen. Similarly, in basic solution the equation can be completed using OH^- and H_2O. These species are in large supply in the respective solutions, and their formation as products or their use as reactants can easily go undetected experimentally. In the permanganate half-reaction we already have one manganese atom on each side of the equation. However, we have four oxygens on the left and none on the right side; four H_2O molecules are needed among the products to balance the four oxygen atoms in MnO_4^-.

$$MnO_4^-(aq) \longrightarrow Mn^{2+}(aq) + 4H_2O(l)$$

The eight hydrogen atoms that this introduces among the products can then be balanced by adding 8 H^+ to the reactants.

$$8H^+(aq) + MnO_4^-(aq) \longrightarrow Mn^{2+}(aq) + 4H_2O(l)$$

There are now equal numbers of each type of atom on both sides of the equation, but the charge still needs to be balanced. The total charge of the reactants is $8(1+) + (1-) = 7+$, and that of the products is $(2+) + 4(0) = 2+$. To balance the charge, five electrons are added to the reactant side.*

$$5e^- + 8H^+(aq) + MnO_4^-(aq) \longrightarrow Mn^{2+}(aq) + 4H_2O(l)$$

For the oxalate half-reaction, mass balance requires the production of two CO_2 molecules for each oxalate ion that reacts:

$$C_2O_4^{2-}(aq) \longrightarrow 2CO_2(g)$$

Mass is now balanced. We can balance the charge by adding two electrons to the products, giving a balanced half-reaction.

$$C_2O_4^{2-}(aq) \longrightarrow 2CO_2(g) + 2e^-$$

Now that we have two balanced half-reactions, we need to multiply each by an appropriate factor so that the number of electrons gained in one half-reaction equals the number of electrons lost in the other. The half-reactions are then added to give the overall balanced equation. In our example the MnO_4^- half-reaction must be multiplied by 2, and the $C_2O_4^{2-}$ half-reaction must be multiplied by 5 so that the same number of electrons (10) appears on both sides of the equation:

$$10e^- + 16H^+(aq) + 2MnO_4^-(aq) \longrightarrow 2Mn^{2+}(aq) + 8H_2O(l)$$
$$5C_2O_4^{2-}(aq) \longrightarrow 10CO_2(g) + 10e^-$$
$$\overline{16H^+(aq) + 2MnO_4^-(aq) + 5C_2O_4^{2-}(aq) \longrightarrow 2Mn^{2+}(aq) + 8H_2O(l) + 10CO_2(g)}$$

---

* Although the oxidation numbers of the elements need not be used in balancing a half-reaction by this method, oxidation numbers can be used as a check. In this example MnO_4^- contains manganese in a +7 oxidation state. Because manganese changes from a +7 to a +2 oxidation state, it must gain five electrons, as we have just concluded.

The balanced equation is the sum of the balanced half-reactions. Note that the electrons on the reactant and product sides of the equation cancel each other.

We can check the balanced equation by counting atoms and charges: There are 16 H, 2 Mn, 28 O, 10 C, and a net charge of 4+ on both sides of the equation, confirming that the equation is correctly balanced.

We can summarize the procedure for balancing a redox reaction that occurs in acidic solution as follows:

ACTIVITY
Balancing Redox Equation in Acid

1. Divide the equation into two incomplete half-reactions, one for oxidation and the other for reduction.
2. Balance each half-reaction.
   - **(a)** First, balance the elements other than H and O.
   - **(b)** Next, balance the O atoms by adding H_2O.
   - **(c)** Then, balance the H atoms by adding H^+.
   - **(d)** Finally, balance the charge by adding e^- to the side with the greater overall positive charge.
3. Multiply each half-reaction by an integer so that the number of electrons lost in one half-reaction equals the number gained in the other.
4. Add the two half-reactions and simplify where possible by canceling species appearing on both sides of the equation.
5. Check the equation to make sure that there are the same number of atoms of each kind and the same total charge on both sides.

SAMPLE EXERCISE 20.2

Complete and balance the following equation by the method of half-reactions:

$$Cr_2O_7^{2-}(aq) + Cl^-(aq) \longrightarrow Cr^{3+}(aq) + Cl_2(g) \qquad \text{(acidic solution)}$$

Solution

Analyze: We are given a partial redox equation for a reaction occurring in acidic solution and asked to balance it.

Plan: We use the five steps summarized in the text preceding this sample exercise.

Solve: First, we divide the equation into two half-reactions.

$$Cr_2O_7^{2-}(aq) \longrightarrow Cr^{3+}(aq)$$

$$Cl^-(aq) \longrightarrow Cl_2(g)$$

Second, we balance each half-reaction. In the first half-reaction the presence of $Cr_2O_7^{2-}$ among the reactants requires two Cr^{3+} among the products. The 7 oxygen atoms in $Cr_2O_7^{2-}$ are balanced by adding 7 H_2O to the products. The 14 hydrogen atoms in 7 H_2O are then balanced by adding 14 H^+ to the reactants:

$$14H^+(aq) + Cr_2O_7^{2-}(aq) \longrightarrow 2Cr^{3+}(aq) + 7H_2O(l)$$

Charge is balanced by adding electrons to the left side of the equation so that the total charge is the same on both sides.

$$6e^- + 14H^+(aq) + Cr_2O_7^{2-}(aq) \longrightarrow 2Cr^{3+}(aq) + 7H_2O(l)$$

In the second half-reaction, 2 Cl^- are required to balance one Cl_2.

$$2Cl^-(aq) \longrightarrow Cl_2(g)$$

We add two electrons to the right sides to attain charge balance.

$$2Cl^-(aq) \longrightarrow Cl_2(g) + 2e^-$$

Third, we equalize the number of electrons transferred in the two half-reactions. To do so, we multiply the second half-reaction by 3 so that the number of electrons gained in the first half-reaction (6) equals the number lost in the second, allowing the electrons to cancel when the half-reactions are added.

Fourth, the equations are added to give the balanced equation:

$$14H^+(aq) + Cr_2O_7^{2-}(aq) + 6Cl^-(aq) \longrightarrow 2Cr^{3+}(aq) + 7H_2O(l) + 3Cl_2(g)$$

Check: There are equal numbers of atoms of each kind on both sides of the equation (14 H, 2 Cr, 7 O, 6 Cl). In addition, the charge is the same on both sides (6+). Thus, the equation is correctly balanced.

PRACTICE EXERCISE

Complete and balance the following oxidation-reduction equations using the method of half-reactions. Both reactions occur in acidic solution.
(a) $Cu(s) + NO_3^-(aq) \longrightarrow Cu^{2+}(aq) + NO_2(g)$
(b) $Mn^{2+}(aq) + NaBiO_3(s) \longrightarrow Bi^{3+}(aq) + MnO_4^-(aq)$
Answers: **(a)** $Cu(s) + 4H^+(aq) + 2NO_3^-(aq) \longrightarrow Cu^{2+}(aq) + 2NO_2(g) + 2H_2O(l)$
(b) $2Mn^{2+}(aq) + 5NaBiO_3(s) + 14H^+(aq) \longrightarrow$
$2MnO_4^-(aq) + 5Bi^{3+}(aq) + 5Na^+(aq) + 7H_2O(l)$

Balancing Equations for Reactions Occurring in Basic Solution

If a redox reaction occurs in basic solution, the equation must be completed by using OH^- and H_2O rather than H^+ and H_2O. The half-reactions can be balanced initially as if they occurred in acidic solution. The H^+ ions can then be "neutralized" by adding an equal number of OH^- ions to both sides of the equation and canceling, where appropriate, the resulting water molecules. This procedure is shown in Sample Exercise 20.3.

SAMPLE EXERCISE 20.3

Complete and balance the following equation:

$$CN^-(aq) + MnO_4^-(aq) \longrightarrow CNO^-(aq) + MnO_2(s) \quad \text{(basic solution)}$$

Solution

Analyze: We are given an incomplete redox equation and asked to complete and balance it, assuming that the reaction occurs in basic solution.

Plan: We go through the first four steps as if the reaction were occurring in acidic solution. We then add the appropriate number of OH^- ions to each side of the equation, combining H^+ and OH^- to form H_2O. We complete the process by simplifying the equation.

Solve: First, we write the incomplete and unbalanced half-reactions:

$$CN^-(aq) \longrightarrow CNO^-(aq)$$
$$MnO_4^-(aq) \longrightarrow MnO_2(s)$$

Second, we initially balance each half-reaction as if it took place in acidic solution. The resultant balanced half-reactions are

$$CN^-(aq) + H_2O(l) \longrightarrow CNO^-(aq) + 2H^+(aq) + 2e^-$$
$$3e^- + 4H^+(aq) + MnO_4^-(aq) \longrightarrow MnO_2(s) + 2H_2O(l)$$

Third, we multiply the top equation by 3 and the bottom one by 2 to equalize electron loss and gain in the two half-reactions:

$$3CN^-(aq) + 3H_2O(l) \longrightarrow 3CNO^-(aq) + 6H^+(aq) + 6e^-$$
$$6e^- + 8H^+(aq) + 2MnO_4^-(aq) \longrightarrow 2MnO_2(s) + 2H_2O(l)$$

Fourth, the half-reactions are added:

$$6e^- + 8H^+(aq) + 3CN^-(aq) + 3H_2O(l) + 2MnO_4^-(aq) \longrightarrow 3CNO^-(aq) + 6H^+(aq) + 2MnO_2(s) + 2H_2O(l) + 6e^-$$

Simplifing the equation gives:

$$2H^+(aq) + 3CN^-(aq) + 2Mn^{2+}(aq) \longrightarrow 3CNO^-(aq) + 2MnO_2(s) + H_2O(aq)$$

Fifth, because H^+ does not exist in any appreciable concentration in basic solution, we remove it from the equation by adding an appropriate amount of OH^- to both sides of the equation to neutralize the $2H^+$. The $2OH^-$ and $2H^+$ form $2H_2O$:

$$[2OH^-(aq) + 2H^+(aq)] + 3CN^-(aq) + 2MnO_4^-(aq) \longrightarrow 3CNO^-(aq) + 2MnO_2(s) + H_2O(l) + 2OH^-(aq)$$
$$2H_2O(l) + 3CN^-(aq) + 2MnO_4^-(aq) \longrightarrow 3CNO^-(aq) + 2MnO_2(s) + H_2O(l) + 2OH^-(aq)$$

The half-reaction can be simplified because H_2O occurs on both sides of the equation. The simplified equation is

$$H_2O(l) + 3CN^-(aq) + 2MnO_4^-(aq) \longrightarrow 3CNO^-(aq) + 2MnO_2(s) + 2OH^-(aq)$$

Check: The result is checked by counting atoms and charges:

There are 3 C, 3 N, 2 H, 9 O, 2 Mn, and a charge of 5− on both sides of the equation.

PRACTICE EXERCISE

Complete and balance the following equations for oxidation-reduction reactions that occur in basic solution:
(a) $NO_2^-(aq) + Al(s) \longrightarrow NH_3(aq) + Al(OH)_4^-(aq)$
(b) $Cr(OH)_3(s) + ClO^-(aq) \longrightarrow CrO_4^{2-}(aq) + Cl_2(g)$
Answers: **(a)** $NO_2^-(aq) + 2Al(s) + 5H_2O(l) + OH^-(aq) \longrightarrow NH_3(aq) + 2Al(OH)_4^-(aq)$
(b) $2Cr(OH)_3(s) + 6ClO^-(aq) \longrightarrow 2CrO_4^{2-}(aq) + 3Cl_2(g) + 2OH^-(aq) + 2H_2O(l)$

20.3 Voltaic Cells

The energy released in a spontaneous redox reaction can be used to perform electrical work. This task is accomplished through a **voltaic** (or **galvanic**) **cell**, a device in which the transfer of electrons takes place through an external pathway rather than directly between reactants.

One such spontaneous reaction occurs when a strip of zinc is placed in contact with a solution containing Cu^{2+}. As the reaction proceeds, the blue color of $Cu^{2+}(aq)$ ions fades, and copper metal deposits on the zinc. At the same time, the zinc begins to dissolve. These transformations are shown in Figure 20.3 ▼ and are summarized by Equation 20.7:

$$Zn(s) + Cu^{2+}(aq) \longrightarrow Zn^{2+}(aq) + Cu(s) \qquad [20.7]$$

Figure 20.4 ▶ shows a voltaic cell that uses the redox reaction between Zn and Cu^{2+} given in Equation 20.7. Although the setup shown in Figure 20.4 is more complex than that in Figure 20.3, the reaction is the same in both cases. The significant difference between the two setups is that the Zn metal and $Cu^{2+}(aq)$ are not in direct contact in the voltaic cell. Instead, the Zn metal is placed in contact with $Zn^{2+}(aq)$ in one compartment of the cell, and Cu metal is placed in contact with $Cu^{2+}(aq)$ in another compartment. Consequently, the reduction of the Cu^{2+} can occur only by a flow of electrons through an external circuit, namely the wire that connects the Zn and Cu strips.

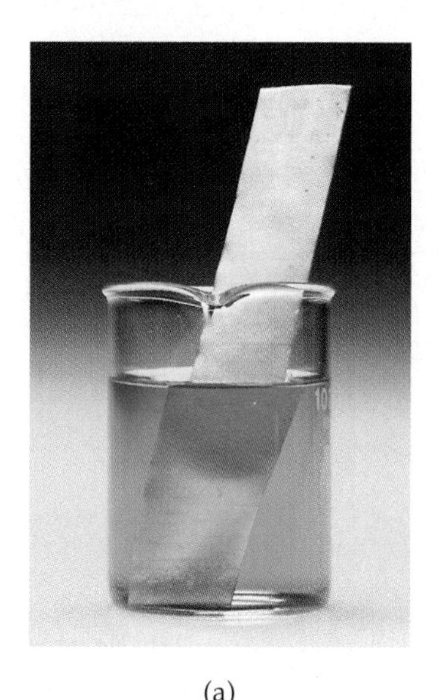
(a)

(b)

▲ **Figure 20.3** (a) A strip of zinc is placed in a solution of copper(II) sulfate. (b) Electrons are transferred from the zinc to the Cu^{2+} ion, forming Zn^{2+} ions and $Cu(s)$. As the reaction proceeds, the zinc dissolves, the blue color due to $Cu^{2+}(aq)$ fades, and copper metal (the dark material on the zinc strip and on the bottom of the beaker) is deposited.

The two solid metals that are connected by the external circuit are called *electrodes*. By definition, the electrode at which oxidation occurs is called the **anode**; the electrode at which reduction occurs is called the **cathode**.* Each of the two compartments of the voltaic cell is called a *half-cell*. One half-cell is the site of the oxidation half-reaction and the other is the site of the reduction half-reaction. In our present example Zn is oxidized and Cu^{2+} is reduced:

Anode (oxidation half-reaction) $\quad Zn(s) \longrightarrow Zn^{2+}(aq) + 2e^-$

Cathode (reduction half-reaction) $\quad Cu^{2+}(aq) + 2e^- \longrightarrow Cu(s)$

Electrons become available as zinc metal is oxidized at the anode. They flow through the external circuit to the cathode, where they are consumed as $Cu^{2+}(aq)$ is reduced. Because Zn(*s*) is oxidized in the cell, the zinc electrode loses mass, and the concentration of the Zn^{2+} solution increases as the cell operates. Similarly, the Cu electrode gains mass, and the Cu^{2+} solution becomes less concentrated as Cu^{2+} is reduced to Cu(*s*).

For a voltaic cell to work, the solutions in the two half-cells must remain electrically neutral. As Zn is oxidized in the anode compartment, Zn^{2+} ions enter the solution. Thus, there must be some means for positive ions to migrate out of the anode compartment or for negative ions to migrate in to keep the solution electrically neutral. Similarly, the reduction of Cu^{2+} at the cathode removes positive charge from the solution, leaving an excess of negative charge in that half-cell. Thus, positive ions must migrate into the compartment or negative ions must migrate out.

In Figure 20.4, a porous glass disc separating the two compartments allows a migration of ions that maintains the electrical neutrality of the solutions. In Figure 20.5 ▼, a *salt bridge* serves this purpose. A salt bridge consists of a U-shaped tube that contains an electrolyte solution, such as $NaNO_3(aq)$, whose ions will not react with other ions in the cell or with the electrode materials. The electrolyte is often incorporated into a gel so that the electrolyte solution does not run out when the U-tube is inverted. As oxidation and reduction proceed at the electrodes, ions from the salt bridge migrate to neutralize charge in the cell compartments.

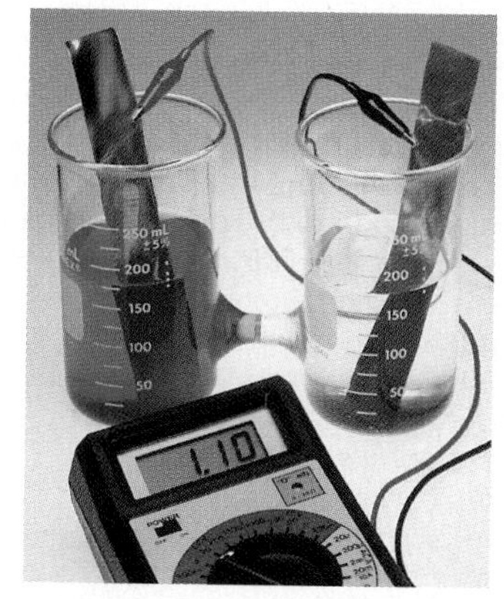

▲ **Figure 20.4** A voltaic cell based on the reaction in Equation 20.7. The left compartment contains 1 *M* $CuSO_4$ and a copper electrode. The one on the right contains 1 *M* $ZnSO_4$ and a zinc electrode. The solutions are connected by a porous glass disc, which permits contact of the two solutions. The metal electrodes are connected through a voltmeter, which reads the potential of the cell, 1.10 V.

* To help remember these definitions, note that *anode* and *oxidation* both begin with a vowel, and *cathode* and *reduction* both begin with a consonant.

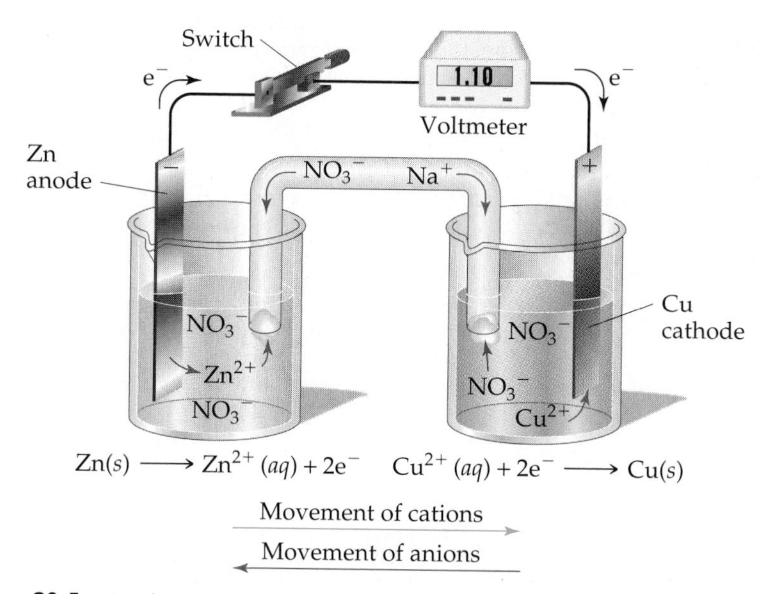

▲ **Figure 20.5** A voltaic cell that uses a salt bridge to complete the electrical circuit.

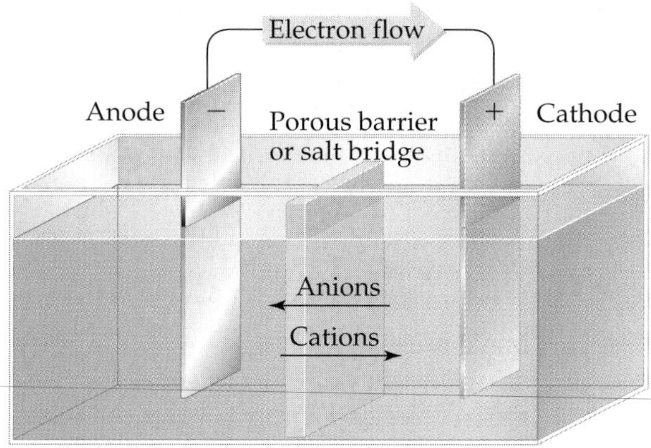

▶ **Figure 20.6** A summary of the terminology used to describe voltaic cells. Oxidation occurs at the anode; reduction occurs at the cathode. The electrons flow spontaneously from the negative anode to the positive cathode. The electrical circuit is completed by the movement of ions in solution. Anions move toward the anode, whereas cations move toward the cathode. The cell compartments can be separated by either a porous glass barrier (as in Figure 20.4) or by a salt bridge (as in Figure 20.5).

Whatever means is used to allow ions to migrate between half-cells, *anions always migrate toward the anode and cations toward the cathode.* In fact, no measurable electron flow will occur between electrodes unless a means is provided for ions to migrate through the solution from one electrode compartment to another, thereby completing the circuit.

Figure 20.6 ▲ summarizes the relationships among the anode, the cathode, the chemical process occurring in a voltaic cell, the direction of migration of ions in solution, and the motion of electrons between electrodes in the external circuit. Notice in particular that *in any voltaic cell the electrons flow from the anode through the external circuit to the cathode.* Because the negatively charged electrons flow from the anode to the cathode, the anode in a voltaic cell is labeled with a negative sign and the cathode with a positive sign; we can envision the electrons as being attracted to the positive cathode from the negative anode through the external circuit.*

SAMPLE EXERCISE 20.4

The following oxidation-reduction reaction is spontaneous:

$$Cr_2O_7^{2-}(aq) + 14H^+(aq) + 6I^-(aq) \longrightarrow 2Cr^{3+}(aq) + 3I_2(s) + 7H_2O(l)$$

A solution containing $K_2Cr_2O_7$ and H_2SO_4 is poured into one beaker, and a solution of KI is poured into another. A salt bridge is used to join the beakers. A metallic conductor that will not react with either solution (such as platinum foil) is suspended in each solution, and the two conductors are connected with wires through a voltmeter or some other device to detect an electric current. The resultant voltaic cell generates an electric current. Indicate the reaction occurring at the anode, the reaction at the cathode, the direction of electron and ion migrations, and the signs of the electrodes.

Solution

Analyze: We are given the equation for a spontaneous reaction and a description of how the cell is constructed. We are asked to write the half-reactions occurring at the anode and at the cathode, as well as the directions of electron and ion movements and the signs assigned to the electrodes.

Plan: Our first step is to divide the chemical equation into half-reactions so that we can identify the oxidation and the reduction processes. We then use the definitions of the anode and cathode and the other terminology summarized in Figure 20.5.

Solve: In one half-reaction, $Cr_2O_7^{2-}(aq)$ is converted into $Cr^{3+}(aq)$. Starting with these ions and then completing and balancing the half-reaction, we have

$$Cr_2O_7^{2-}(aq) + 14H^+(aq) + 6e^- \longrightarrow 2Cr^{3+}(aq) + 7H_2O(l)$$

* Although the anode and cathode are labeled with − and + signs respectively, you should not interpret the labels as charges on the electrodes. The labels simply tell us the electrode at which the electrons are released to the external circuit (the anode) and received from the external circuit (the cathode). The actual charges on the electrodes are essentially zero.

In the other half-reaction, $I^-(aq)$ is converted to $I_2(s)$:

$$6I^-(aq) \longrightarrow 3I_2(s) + 6e^-$$

Now we can use the summary in Figure 20.6 to help us describe the voltaic cell. The first half-reaction is the reduction process (electrons shown on the reactant side of the equation), and by definition, this process occurs at the cathode. The second half-reaction is the oxidation (electrons shown on the product side of the equation), which occurs at the anode. The I^- ions are the source of electrons, and the $Cr_2O_7^{2-}$ ions accept the electrons. Hence, the electrons flow through the external circuit from the electrode immersed in the KI solution (the anode) to the electrode immersed in the $K_2Cr_2O_7$-H_2SO_4 solution (the cathode). The electrodes themselves do not react in any way; they merely provide a means of transferring electrons from or to the solutions. The cations move through the solutions toward the cathode, and the anions move toward the anode. The anode (from which the electrons move) is the negative electrode, and the cathode (toward which the electrons move) is the positive electrode.

PRACTICE EXERCISE

The two half-reactions in a voltaic cell are

$$Zn(s) \longrightarrow Zn^{2+}(aq) + 2e^-$$

$$ClO_3^-(aq) + 6H^+(aq) + 6e^- \longrightarrow Cl^-(aq) + 3H_2O(l)$$

(a) Indicate which reaction occurs at the anode and which at the cathode. **(b)** Which electrode is consumed in the cell reaction? **(c)** Which electrode is positive?
Answers: **(a)** The first reaction occurs at the anode, the second reaction at the cathode. **(b)** The anode (Zn) is consumed in the cell reaction. **(c)** The cathode is positive.

A Molecular View of Electrode Processes

To better understand the relationship between voltaic cells and spontaneous redox reactions, let's look at what happens at the atomic or molecular level. The actual processes involved in the transfer of electrons are quite complex; nevertheless, we can learn much by examining these processes in a simplified way.

Let's first consider the spontaneous redox reaction between Zn(*s*) and $Cu^{2+}(aq)$, illustrated in Figure 20.3. During the reaction Zn(*s*) is oxidized to $Zn^{2+}(aq)$ and $Cu^{2+}(aq)$ is reduced to Cu(*s*). Figure 20.7 ▼ shows a schematic diagram of how these processes occur at the atomic level. We can envision a Cu^{2+} ion coming into contact with the strip of Zn metal, as in Figure 20.7(a). Two electrons are transferred directly from a Zn atom to the Cu^{2+} ion, leading to a Zn^{2+} ion and a Cu atom. The Zn^{2+} ion migrates away into the aqueous solution while the Cu atom remains deposited on the metal strip [Figure 20.7(b)]. As the reaction proceeds, we produce more and more Cu(*s*) and deplete the $Cu^{2+}(aq)$, as we saw in Figure 20.3(b).

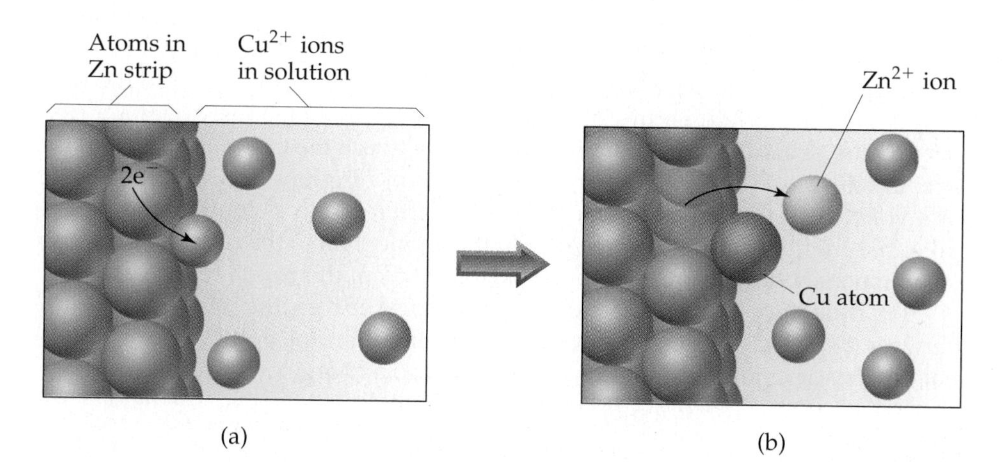

◀ **Figure 20.7** Depiction of the reaction between Zn(*s*) and $Cu^{2+}(aq)$ at the atomic level. The water molecules and anions in the solution are not shown. (a) A Cu^{2+} ion comes in contact with the surface of the Zn strip and gains two electrons from a Zn atom; the Cu^{2+} ion is reduced, and the Zn atom is oxidized. (b) The resulting Zn^{2+} ion enters the solution, and the Cu atom remains deposited on the strip.

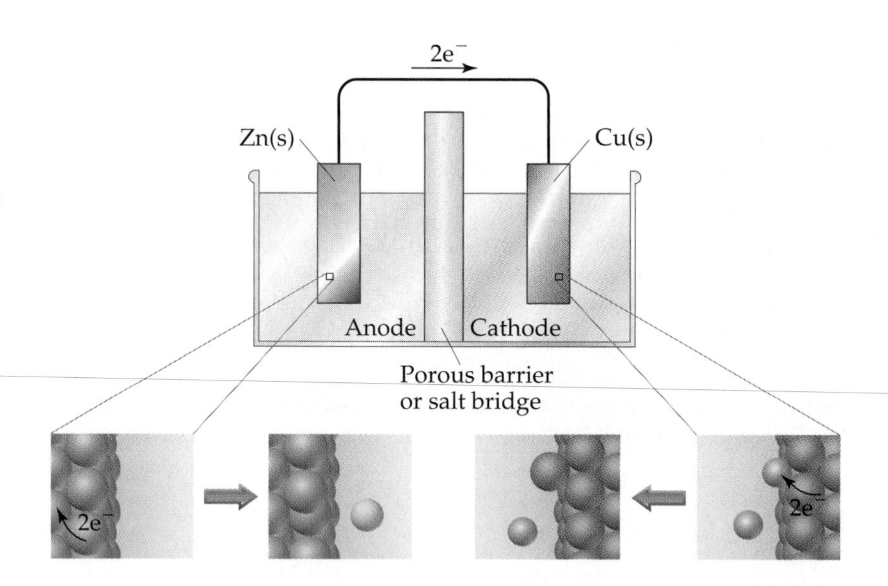

▶ **Figure 20.8** Depiction of the voltaic cell in Figure 20.5 at the atomic level. At the anode a Zn atom loses two electrons and becomes a Zn^{2+} ion; the Zn atom is oxidized. The electrons travel through the external circuit to the cathode. At the cathode a Cu^{2+} ion gains the two electrons, forming a Cu atom; the Cu^{2+} ion is reduced. Ions migrate through the porous barrier to maintain charge balance between the compartments.

The voltaic cell in Figure 20.5 is also based on the oxidation of Zn(*s*) and the reduction of $Cu^{2+}(aq)$. In this case, however, the electrons are not transferred directly between the reacting species. Figure 20.8 ▲ shows qualitatively what happens at each of the electrodes of the cell. At the surface of the anode, a Zn atom "loses" two electrons and becomes a $Zn^{2+}(aq)$ ion in the anode compartment. We envision the two electrons traveling from the anode through the wire to the cathode. At the surface of the cathode, the two electrons reduce a Cu^{2+} ion to a Cu atom, which is deposited on the cathode. As we noted earlier, the flow of electrons from the anode to the cathode is possible only if ions are transferred through the salt bridge to maintain overall charge balance for each of the two compartments.

The redox reaction between Zn and Cu^{2+} is spontaneous regardless of whether they react directly or in the separate compartments of a voltaic cell. In each case the overall reaction is the same—only the path by which the electrons are transferred from the Zn atom to a Cu^{2+} ion is different. In Section 20.4 we will examine *why* this reaction is spontaneous.

20.4 Cell EMF

Why do electrons transfer spontaneously from a Zn atom to a Cu^{2+} ion, either directly as in the reaction of Figure 20.3 or through an external circuit as in the voltaic cell of Figure 20.5? In this section we will examine the "driving force" that pushes the electrons through an external circuit in a voltaic cell.

The chemical processes that constitute any voltaic cell are spontaneous in the way in which we described spontaneous processes in Chapter 19. In a simple sense, we can compare the electron flow caused by a voltaic cell to the flow of water in a waterfall (Figure 20.9 ▶). Water flows spontaneously over a waterfall because of a difference in potential energy between the top of the falls and the stream below. ∞ (Section 5.1) In a similar fashion, electrons flow from the anode of a voltaic cell to the cathode because of a difference in potential energy. The potential energy of electrons is higher in the anode than in the cathode, and they spontaneously flow through an external circuit from the anode to the cathode.

The difference in potential energy per electrical charge (the *potential difference*) between two electrodes is measured in units of volts. One volt (V) is the potential difference required to impart 1 J of energy to a charge of 1 coulomb (C).

$$1\ \text{V} = 1\frac{\text{J}}{\text{C}}$$

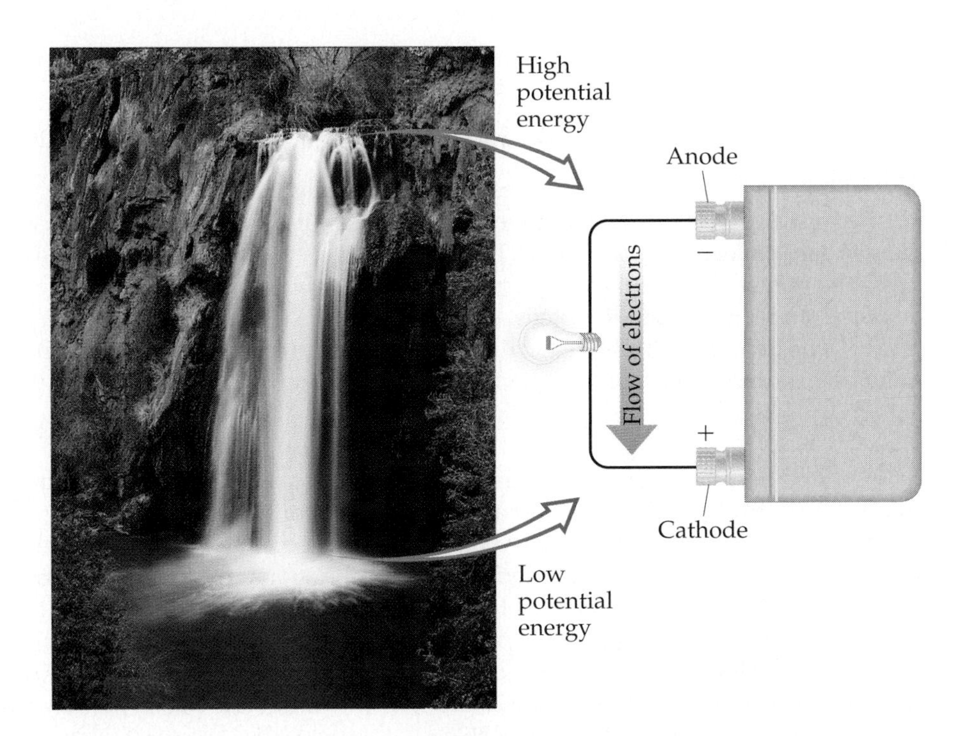

◀ **Figure 20.9** The flow of electrons from the anode to the cathode of a voltaic cell can be likened to the flow of water over a waterfall. Water flows over the waterfall because its potential energy is lower at the bottom of the falls than at the top. Likewise, if there is an electrical connection between the anode and cathode of a voltaic cell, electrons flow from the anode to the cathode in order to lower their potential energy.

The potential difference between the two electrodes of a voltaic cell provides the driving force that pushes electrons through the external circuit. Therefore, we call this potential difference the **electromotive** ("causing electron motion") **force**, or **emf**. The emf of a cell, denoted E_{cell}, is also called the **cell potential**. Because E_{cell} is measured in volts, we often refer to it as the *cell voltage*. For any cell reaction that proceeds spontaneously, such as that in a voltaic cell, the cell potential will be *positive*.

The emf of a particular voltaic cell depends on the specific reactions that occur at the cathode and anode, the concentrations of reactants and products, and the temperature, which we will assume to be 25°C unless otherwise noted. In this section we will focus on cells that are operated at 25°C under *standard conditions*. Recall from Section 19.5 that standard conditions include 1 *M* concentrations for reactants and products in solution and 1 atm pressure for those that are gases (Table 19.3). Under standard conditions the emf is called the **standard emf** or the **standard cell potential** and is denoted E°_{cell}. For the Zn-Cu voltaic cell in Figure 20.5, for example, the standard cell potential at 25°C is +1.10V.

$$Zn(s) + Cu^{2+}(aq, 1\,M) \longrightarrow Zn^{2+}(aq, 1\,M) + Cu(s) \qquad E^\circ_{cell} = +1.10\text{ V}$$

Recall that the superscript ° indicates standard-state conditions. (Section 5.7)

Standard Reduction (Half-Cell) Potentials

The emf or cell potential of a voltaic cell depends on the particular cathode and anode half-cells involved. We could, in principle, tabulate the standard cell potentials for all possible cathode/anode combinations. However, it is not necessary to undertake this arduous task. Rather, we can assign a standard potential to each individual half-cell, and then use these half-cell potentials to determine E°_{cell}.

The cell potential is the difference between two electrode potentials, one associated with the cathode and the other associated with the anode. By convention, the potential associated with each electrode is chosen to be the potential for reduction to occur at that electrode. Thus, standard electrode potentials are tabulated for reduction reactions; they are **standard reduction potentials**, denoted E°_{red}. The cell potential, E°_{cell}, is given by the standard reduction potential of the cathode

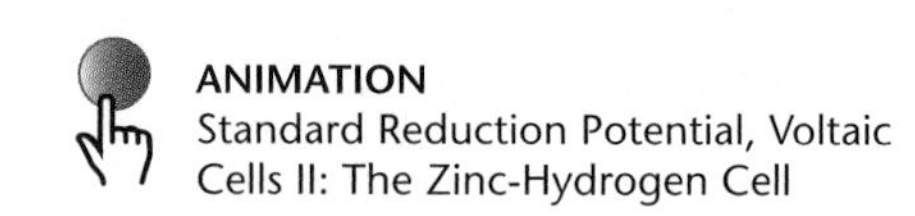
ANIMATION
Standard Reduction Potential, Voltaic Cells II: The Zinc-Hydrogen Cell

reaction, E°_{red}(cathode), *minus* the standard reduction potential of the anode reaction, E°_{red}(anode):

$$E^\circ_{cell} = E^\circ_{red}(\text{cathode}) - E^\circ_{red}(\text{anode}) \quad [20.8]$$

We will discuss Equation 20.8 in greater detail shortly.

Because every voltaic cell involves two half-cells, it is not possible to measure the standard reduction potential of a half-reaction directly. If we assign a standard reduction potential to a certain reference half-reaction, however, we can then determine the standard reduction potentials of other half-reactions relative to that reference. The reference half-reaction is the reduction of $H^+(aq)$ to $H_2(g)$ under standard conditions, which is assigned a standard reduction potential of exactly 0 V.

$$2H^+(aq, 1\ M) + 2e^- \longrightarrow H_2(g, 1\ \text{atm}) \qquad E^\circ_{red} = 0\ \text{V} \quad [20.9]$$

ANIMATION
Standard Reduction Potential, Voltaic Cells II: The Zinc-Hydrogen Cell

An electrode designed to produce this half-reaction is called a **standard hydrogen electrode** (SHE). An SHE consists of a platinum wire connected to a piece of platinum foil covered with finely divided platinum that serves as an inert surface for the reaction. The electrode is encased in a glass tube so that hydrogen gas under standard conditions (1 atm) can bubble over the platinum, and the solution contains $H^+(aq)$ under standard (1 *M*) conditions (Figure 20.10 ▼).

Figure 20.11 ▶ shows a voltaic cell using an SHE and a standard Zn^{2+}/Zn electrode. The spontaneous reaction is the one shown in Figure 20.1, namely the oxidation of Zn and the reduction of H^+.

$$Zn(s) + 2H^+(aq) \longrightarrow Zn^{2+}(aq) + H_2(g)$$

Notice that the Zn^{2+}/Zn electrode is the anode and the SHE is the cathode, and that the cell voltage is +0.76 V. By using the defined standard reduction potential of $H^+(E^\circ_{red} = 0)$ and Equation 20.8, we can determine the standard reduction potential for the Zn^{2+}/Zn half-reaction:

$$E^\circ_{cell} = E^\circ_{red}(\text{cathode}) - E^\circ_{red}(\text{anode})$$
$$+0.76\ \text{V} = 0\ \text{V} - E^\circ_{red}(\text{anode})$$
$$E^\circ_{red}(\text{anode}) = -0.76\ \text{V}$$

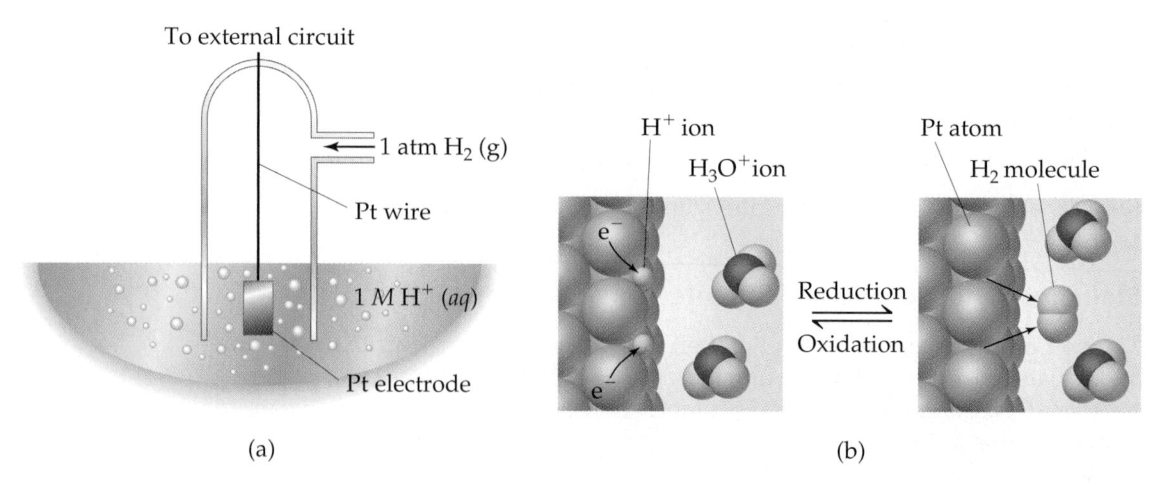

▲ **Figure 20.10** The standard hydrogen electrode (SHE) is used as a reference electrode. (a) An SHE consists of an electrode with finely divided Pt in contact with $H_2(g)$ at 1 atm pressure and an acidic solution with $[H^+] = 1\ M$. (b) Molecular depiction of the processes that occur at the SHE. When the SHE is the cathode of a cell, two H^+ ions each accept an electron from the Pt electrode and are reduced to H atoms. The H atoms bond together to form H_2. When the SHE is the anode of a cell, the reverse process occurs: An H_2 molecule at the electrode surface loses two electrons and is oxidized to H^+. The H^+ ions in solution are hydrated. (Section 16.2)

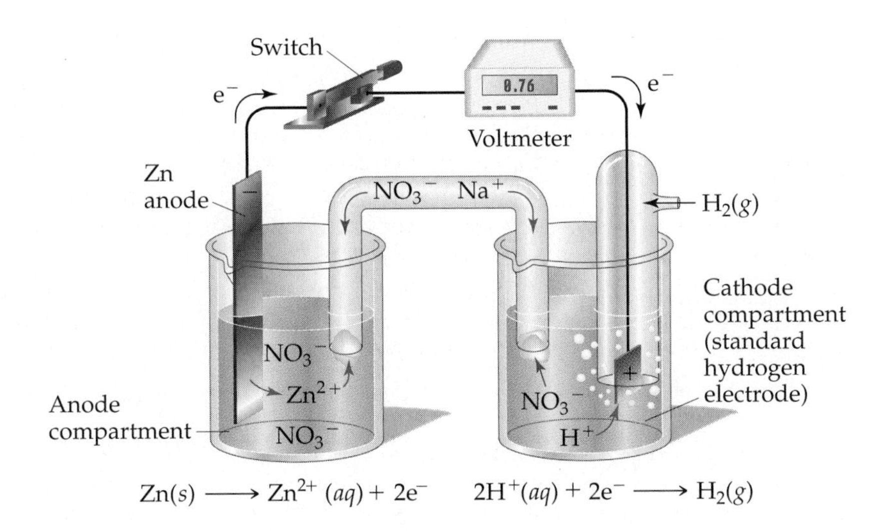

◀ **Figure 20.11** A voltaic cell using a standard hydrogen electrode.

Thus, a standard reduction potential of −0.76 V can be assigned to the reduction of Zn^{2+} to Zn.

$$Zn^{2+}(aq, 1\ M) + 2e^- \longrightarrow Zn(s) \qquad E^\circ_{red} = -0.76\ V$$

We write the reaction as a reduction even though it is "running in reverse," as an oxidation, in the cell in Figure 20.11. *Whenever we assign a potential to a half-reaction, we write the reaction as a reduction.*

The standard reduction potentials for other half-reactions can be established from other cell potentials in a fashion analogous to that used for the Zn^{2+}/Zn half-reaction. Table 20.1 ▼ lists some standard reduction potentials; a more complete list is found in Appendix E. These standard reduction potentials, often called *half-cell potentials*, can be combined to calculate the emfs of a large variety of voltaic cells.

TABLE 20.1 Standard Reduction Potentials in Water at 25°C

Potential (V)	Reduction Half-Reaction
+2.87	$F_2(g) + 2e^- \longrightarrow 2F^-(aq)$
+1.51	$MnO_4^-(aq) + 8H^+(aq) + 5e^- \longrightarrow Mn^{2+}(aq) + 4H_2O(l)$
+1.36	$Cl_2(g) + 2e^- \longrightarrow 2Cl^-(aq)$
+1.33	$Cr_2O_7^{2-}(aq) + 14H^+(aq) + 6e^- \longrightarrow 2Cr^{3+}(aq) + 7H_2O(l)$
+1.23	$O_2(g) + 4H^+(aq) + 4e^- \longrightarrow 2H_2O(l)$
+1.06	$Br_2(l) + 2e^- \longrightarrow 2Br^-(aq)$
+0.96	$NO_3^-(aq) + 4H^+(aq) + 3e^- \longrightarrow NO(g) + 2H_2O(l)$
+0.80	$Ag^+(aq) + e^- \longrightarrow Ag(s)$
+0.77	$Fe^{3+}(aq) + e^- \longrightarrow Fe^{2+}(aq)$
+0.68	$O_2(g) + 2H^+(aq) + 2e^- \longrightarrow H_2O_2(aq)$
+0.59	$MnO_4^-(aq) + 2H_2O(l) + 3e^- \longrightarrow MnO_2(s) + 4OH^-(aq)$
+0.54	$I_2(s) + 2e^- \longrightarrow 2I^-(aq)$
+0.40	$O_2(g) + 2H_2O(l) + 4e^- \longrightarrow 4OH^-(aq)$
+0.34	$Cu^{2+}(aq) + 2e^- \longrightarrow Cu(s)$
0	$2H^+(aq) + 2e^- \longrightarrow H_2(g)$
−0.28	$Ni^{2+}(aq) + 2e^- \longrightarrow Ni(s)$
−0.44	$Fe^{2+}(aq) + 2e^- \longrightarrow Fe(s)$
−0.76	$Zn^{2+}(aq) + 2e^- \longrightarrow Zn(s)$
−0.83	$2H_2O(l) + 2e^- \longrightarrow H_2(g) + 2OH^-(aq)$
−1.66	$Al^{3+}(aq) + 3e^- \longrightarrow Al(s)$
−2.71	$Na^+(aq) + e^- \longrightarrow Na(s)$
−3.05	$Li^+(aq) + e^- \longrightarrow Li(s)$

Because electrical potential measures potential energy per electrical charge, standard reduction potentials are intensive properties. Thus, *changing the stoichiometric coefficient in a half-reaction does not affect the value of the standard reduction potential*. For example, E°_{red} for the reduction of 2 mol of Zn^{2+} is the same as that for the reduction of 1 mol of Zn^{2+}:

$$2Zn^{2+}(aq, 1\,M) + 4e^- \longrightarrow 2Zn(s) \qquad E^\circ_{red} = -0.76\text{ V}$$

SAMPLE EXERCISE 20.5

For the Zn-Cu^{2+} voltaic cell shown in Figure 20.5, we have

$$Zn(s) + Cu^{2+}(aq, 1\,M) \longrightarrow Zn^{2+}(aq, 1\,M) + Cu(s) \qquad E^\circ_{cell} = 1.10\text{ V}$$

Given that the standard reduction potential of Zn^{2+} is −0.76 V, calculate the E°_{red} for the reduction of Cu^{2+} to Cu.

$$Cu^{2+}(aq, 1\,M) + 2e^- \longrightarrow Cu(s)$$

Solution

Analyze: We are given E°_{cell} and E°_{red} for Zn^{2+} and asked to calculate E°_{red} for Cu^{2+}.

Plan: In the voltaic cell, Zn is oxidized and is therefore the anode. Thus, the given E°_{red} for Zn^{2+} is E°_{red} (anode). Because Cu^{2+} is reduced, it is in the cathode half-cell. Thus, the unknown reduction potential for Cu^{2+} is E°_{red} (cathode). Knowing E°_{cell} and E°_{red} (anode), we can use Equation 20.8 to solve for E°_{red} (cathode).

Solve:

$$E^\circ_{cell} = E^\circ_{red}(\text{cathode}) - E^\circ_{red}(\text{anode})$$

$$1.10\text{ V} = E^\circ_{red}(\text{cathode}) - (-0.76\text{ V})$$

$$E^\circ_{red}(\text{cathode}) = 1.10\text{ V} - 0.76\text{ V} = 0.34\text{ V}$$

Check: This standard reduction potential agrees with the one listed in Table 20.1.

Comment: The standard reduction potential for Cu^{2+} can be represented as $E^\circ_{Cu^{2+}} = 0.34$ V and that for Zn^{2+} as $E^\circ_{Zn^{2+}} = -0.76$ V. The subscript identifies the ion that is reduced in the reduction half-reaction.

PRACTICE EXERCISE

A voltaic cell is based on the following half-reactions:

$$In^+(aq) \longrightarrow In^{3+}(aq) + 2e^-$$

$$Br_2(l) + 2e^- \longrightarrow 2Br^-(aq)$$

The standard emf for this cell is 1.46 V. Using the data in Table 20.1, calculate E°_{red} for the reduction of In^{3+} to In^+.

Answer: −0.40 V

SAMPLE EXERCISE 20.6

Using the standard reduction potentials listed in Table 20.1, calculate the standard emf for the voltaic cell described in Sample Exercise 20.4, which is based on the following reaction:

$$Cr_2O_7^{2-}(aq) + 14H^+(aq) + 6I^-(aq) \longrightarrow 2Cr^{3+}(aq) + 3I_2(s) + 7H_2O(l)$$

Solution

Analyze: We are given the equation for a redox reaction and asked to use data in Table 20.1 to calculate the standard emf (standard potential) for the associated voltaic cell.

Plan: Our first step is to identify the half-reactions that occur at the cathode and the anode, which we did in Sample Exercise 20.4. Then we can use data from Table 20.1 and Equation 20.8 to calculate the standard emf.

Solve: The half-reactions are

$$\text{Cathode: } Cr_2O_7^{2-}(aq) + 14H^+(aq) + 6e^- \longrightarrow 2Cr^{3+}(aq) + 7H_2O(l)$$

$$\text{Anode: } 6I^-(aq) \longrightarrow 3I_2(s) + 6e^-$$

According to Table 20.1, the standard reduction potential for the reduction of $Cr_2O_7^{2-}$ to Cr^{3+} is +1.33 V, and the standard reduction potential for the reduction of I_2 to I^- (the reverse of the oxidation half-reaction) is +0.54 V. We then use these values in Equation 20.8.

$$E^\circ_{cell} = E^\circ_{red}(\text{cathode}) - E^\circ_{red}(\text{anode}) = 1.33\text{ V} - 0.54\text{ V} = 0.79\text{ V}$$

Although the iodide half-reaction at the anode must be multiplied by 3 in order to obtain a balanced equation for the reaction, the value of E°_{red} is *not* multiplied by 3. As we have noted, the standard reduction potential is an intensive property, so it is independent of the specific stoichiometric coefficients.

Check: The cell potential, 0.79 V, is a positive number. As noted earlier, a voltaic cell must have a positive emf in order to operate.

PRACTICE EXERCISE

Using data in Table 20.1, calculate the standard emf for a cell that employs the following overall cell reaction:

$$2Al(s) + 3I_2(s) \longrightarrow 2Al^{3+}(aq) + 6I^-(aq)$$

Answer: +2.20 V

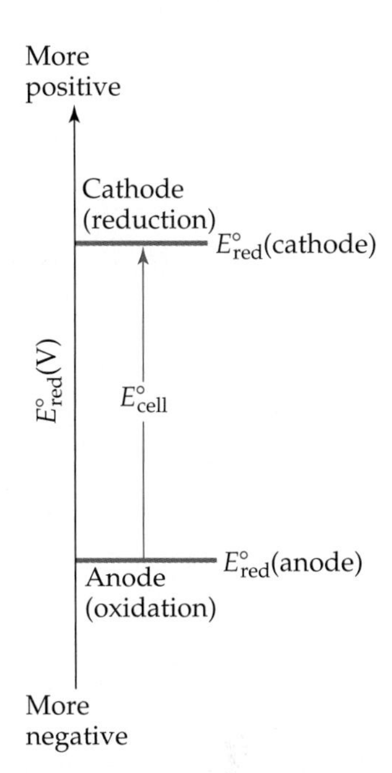

▲ **Figure 20.12** The standard cell potential of a voltaic cell measures the difference in the standard reduction potentials of the cathode and the anode reactions: $E^\circ_{cell} = E^\circ_{red}(\text{cathode}) - E^\circ_{red}(\text{anode})$. In a voltaic cell the cathode reaction is always the one that has the more positive (or less negative) value for E°_{red}.

We are now in a position to discuss Equation 20.8 more fully. For each of the half-cells in a voltaic cell, the standard reduction potential provides a measure of the driving force for reduction to occur: *The more positive the value of* E°_{red} *the greater the driving force for reduction.* In any voltaic cell the reaction at the cathode has a more positive value of E°_{red} than does the reaction at the anode. In essence, the greater driving force of the cathode half-reaction is used to force the anode reaction to occur "in reverse," as an oxidation.

Equation 20.8 tells us that the cell potential, E°_{cell}, is the difference between the standard reduction potential of the cathode half-reaction, E°_{red} (cathode), and that of the anode half-reaction, E°_{red} (anode). We can interpret E°_{cell} as the "net driving force" that pushes the electrons through an external circuit. Equation 20.8 is illustrated graphically in Figure 20.12 ▶, in which the standard reduction potentials are shown on a scale, with more positive E°_{red} values higher on the scale (as in Table 20.1). In any voltaic cell the cathode half-reaction is the one higher on the scale, and the difference between the two standard reduction potentials is the cell potential. Figure 20.13 ▶ shows the specific values of E°_{red} for the Zn-Cu voltaic cell in Figure 20.5.

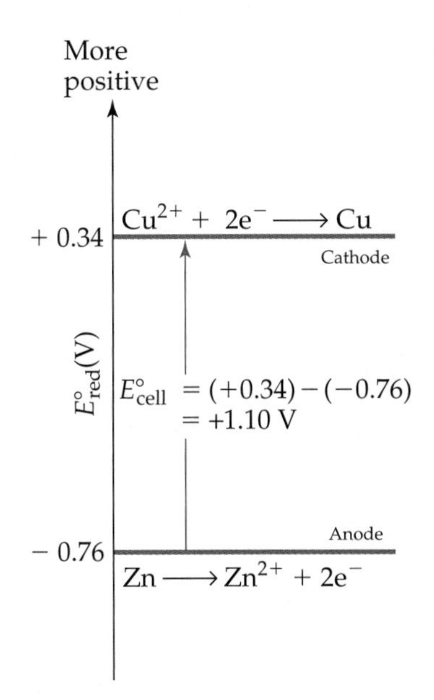

▲ **Figure 20.13** The half-cell potentials for the voltaic cell in Figure 20.5, diagrammed in the style of Figure 20.12.

SAMPLE EXERCISE 20.7

A voltaic cell is based on the following two standard half-reactions:

$$Cd^{2+}(aq) + 2e^- \longrightarrow Cd(s)$$

$$Sn^{2+}(aq) + 2e^- \longrightarrow Sn(s)$$

By using the data in Appendix E, determine **(a)** the half-reactions that occur at the cathode and the anode, and **(b)** the standard cell potential.

Solution

Analyze: We have to look up E°_{red} for two half-reactions and use these values to predict the cathode and anode of the cell and to calculate its standard cell potential, E°_{cell}.

Plan: The cathode will have the reduction with the most positive E°_{red} value. The anode will have the less positive E°_{red}. To write the half-reaction at the anode, we reverse the half-reaction written for the reduction.

Solve: According to Appendix E, $E^\circ_{red}(Cd^{2+}/Cd) = -0.403$ V and $E^\circ_{red}(Sn^{2+}/Sn) = -0.136$ V. The standard reduction potential for Sn^{2+} is more positive (less negative) than that for Cd^{2+}; hence, the reduction of Sn^{2+} is the reaction that occurs at the cathode.

$$\text{Cathode:}\quad Sn^{2+}(aq) + 2e^- \longrightarrow Sn(s)$$

The anode reaction therefore is the loss of electrons by Cd.

$$\text{Anode:} \quad Cd(s) \longrightarrow Cd^{2+}(aq) + 2e^-$$

(b) The cell potential is given by Equation 20.8.

$$E^\circ_{cell} = E^\circ_{red}(\text{cathode}) - E^\circ_{red}(\text{anode}) = (-0.136\ V) - (-0.403\ V) = 0.267\ V$$

Notice that it is unimportant that the E°_{red} values of both half-reactions are negative; the negative values merely indicate how these reductions compare to the reference reaction, the reduction of $H^+(aq)$.

Check: The cell potential is positive, as it must be for a voltaic cell.

PRACTICE EXERCISE

A voltaic cell is based on a Co^{2+}/Co half-cell and an AgCl/Ag half-cell. **(a)** What reaction occurs at the anode? **(b)** What is the standard cell potential?
Answers: **(a)** $Co \longrightarrow Co^{2+} + 2e^-$; **(b)** +0.499 V

Oxidizing and Reducing Agents

We have thus far used the tabulation of standard reduction potentials to examine voltaic cells. We can also use E°_{red} values to understand aqueous reaction chemistry. Recall, for example, the reaction between Zn(*s*) and $Cu^{2+}(aq)$ shown in Figure 20.3.

$$Zn(s) + Cu^{2+}(aq) \longrightarrow Zn^{2+}(aq) + Cu(s)$$

Zinc metal is oxidized and $Cu^{2+}(aq)$ is reduced in this reaction. These substances are in direct contact, however, so we are not producing usable electrical work; the direct contact essentially "short-circuits" the cell. Nevertheless, the driving force for the reaction is the same as that in a voltaic cell, as in Figure 20.5. Because the E°_{red} value for the reduction of Cu^{2+} (0.34 V) is more positive than the E°_{red} value for the reduction of Zn^{2+}(−0.76 V), the reduction of $Cu^{2+}(aq)$ by Zn(*s*) is a spontaneous process.

We can generalize the relationship between the value of E°_{red} and the spontaneity of redox reactions: *The more positive the E°_{red} value for a half-reaction, the greater the tendency for the reactant of the half-reaction to be reduced and, therefore, to oxidize another species.* In Table 20.1, for example, F_2 is the most easily reduced species, so it is the strongest oxidizing agent listed.

$$F_2(g) + 2e^- \longrightarrow 2F^-(aq) \qquad E^\circ_{red} = 2.87\ V$$

Among the most frequently used oxidizing agents are the halogens, O_2, and oxyanions such as MnO_4^-, $Cr_2O_7^{2-}$, and NO_3^-, whose central atoms have high positive oxidation states. According to Table 20.1, all these species undergo reduction with large positive values of E°_{red}.

Lithium ion (Li^+) is the most difficult species to reduce and is therefore the poorest oxidizing agent:

$$Li^+(aq) + e^- \longrightarrow Li(s) \qquad E^\circ_{red} = -3.05\ V$$

Because Li^+ is so difficult to reduce, the reverse reaction, the oxidation of Li(*s*) to $Li^+(aq)$, is a highly favorable reaction. *The half-reaction with the smallest reduction potential is most easily reversed as an oxidation.* Thus, lithium metal has a great tendency to transfer electrons to other species. In water, Li is the strongest reducing agent among the substances listed in Table 20.1.

Commonly used reducing agents include H_2 and the active metals such as the alkali metals and the alkaline earth metals. Other metals whose cations have negative E°_{red} values—Zn and Fe, for example—are also used as reducing agents. Solutions of reducing agents are difficult to store for extended periods because of the ubiquitous presence of O_2, a good oxidizing agent. For example, developer solutions used in photography are mild reducing agents; they have only a limited shelf life because they are readily oxidized by O_2 from the air.

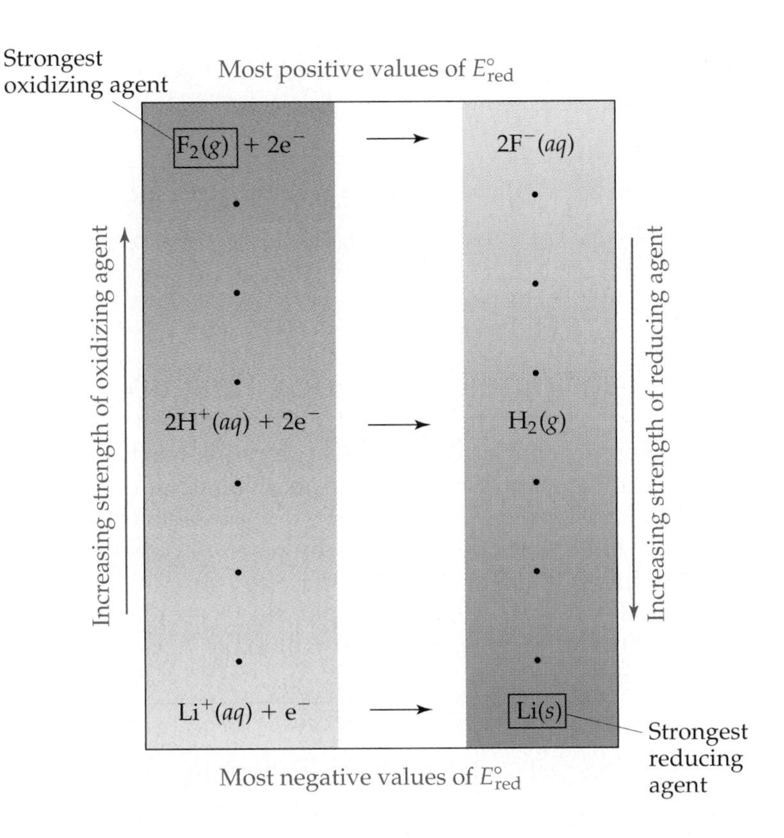

◀ **Figure 20.14** The standard reduction potentials, E°_{red}, listed in Table 20.1 are related to the ability of substances to serve as oxidizing or reducing agents. Species on the left side of the half-reactions can act as oxidizing agents, and those on the right side can serve as reducing agents. As E°_{red} becomes more positive, the oxidizing strength of the species on the left increases. As E°_{red} becomes more negative, the reducing strength of the species on the right increases.

The list of E°_{red} values in Table 20.1 orders the ability of substances to act as oxidizing or reducing agents and is summarized in Figure 20.14 ▲. The substances that are most readily reduced (strong oxidizing agents) are the reactants on the top left of the table. Their products, on the top right of the table, are oxidized with difficulty (weak reducing agents). The substances on the bottom left of the table are reduced with difficulty, but their products are readily oxidized. This inverse relationship between oxidizing and reducing strength is similar to the inverse relationship between the strengths of conjugate acids and bases (Figure 16.4).

To help you remember the relationships between the strengths of oxidizing and reducing agents, recall the very exothermic reaction between sodium metal and chlorine gas to form sodium chloride (Figure 8.2). In this reaction $Cl_2(g)$ is reduced (it serves as a strong oxidizing agent) and Na(*s*) is oxidized (it serves as a strong reducing agent). The products of this reaction—Na^+ and Cl^- ions—are very weak oxidizing and reducing agents, respectively.

SAMPLE EXERCISE 20.8

Using Table 20.1, rank the following ions in order of increasing strength as oxidizing agents: $NO_3^-(aq)$, $Ag^+(aq)$, $Cr_2O_7^{2-}(aq)$.

Solution

Analyze: We are given several ions and asked to rank their abilities to act as oxidizing agents.

Plan: The more readily an ion is reduced (the more positive its E°_{red} value), the stronger it is as an oxidizing agent.

Solve: From Table 20.1, we have

$$NO_3^-(aq) + 4H^+(aq) + 3e^- \longrightarrow NO(g) + 2H_2O(l) \qquad E^\circ_{red} = +0.96\ V$$

$$Ag^+(aq) + e^- \longrightarrow Ag(s) \qquad E^\circ_{red} = +0.80\ V$$

$$Cr_2O_7^{2-}(aq) + 14H^+(aq) + 6e^- \longrightarrow 2Cr^{3+}(aq) + 7H_2O(l) \qquad E^\circ_{red} = +1.33\ V$$

Because the standard reduction potential of $Cr_2O_7^{2-}$ is the most positive, $Cr_2O_7^{2-}$ is the strongest oxidizing agent of the three. The rank order is $Ag^+ < NO_3^- < Cr_2O_7^{2-}$.

PRACTICE EXERCISE

Using Table 20.1, rank the following species from the strongest to the weakest reducing agent: $I^-(aq)$, $Fe(s)$, $Al(s)$.

Answer: $Al(s) > Fe(s) > I^-(aq)$

20.5 Spontaneity of Redox Reactions

We have observed that voltaic cells use redox reactions that proceed spontaneously. Any reaction that can occur in a voltaic cell to produce a positive emf must be spontaneous. Consequently, it is possible to decide whether a redox reaction will be spontaneous by using half-cell potentials to calculate the emf associated with it.

The following discussion will pertain to general redox reactions, not just reactions in voltaic cells. Thus, we will make Equation 20.8 more general by writing it as

$$E° = E°_{red}(\text{reduction process}) - E°_{red}(\text{oxidation process}) \qquad [20.10]$$

In modifying Equation 20.8, we have dropped the subscript "cell" to indicate that the calculated emf does not necessarily refer to a voltaic cell. Similarly, we have generalized the standard reduction potentials on the right side of the equation by referring to the reduction and oxidation processes, rather than the cathode and the anode. We can now make a general statement about the spontaneity of a reaction and its associated emf, E: *A positive value of* E *indicates a spontaneous process, and a negative value of* E *indicates a nonspontaneous one.* We will use E to represent the emf under nonstandard conditions, and $E°$ to indicate the standard emf.

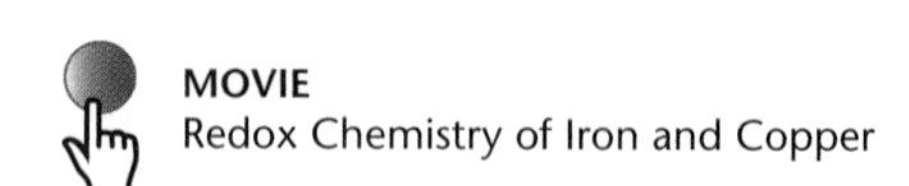

SAMPLE EXERCISE 20.9

Using standard reduction potentials (Table 20.1), determine whether the following reactions are spontaneous under standard conditions:

(a) $Cu(s) + 2H^+(aq) \longrightarrow Cu^{2+}(aq) + H_2(g)$

(b) $Cl_2(g) + 2I^-(aq) \longrightarrow 2Cl^-(aq) + I_2(s)$

Solution

Analyze and Plan: To determine whether a redox reaction is spontaneous under standard conditions, we first need to write its reduction and oxidation half-reactions. We can then use the standard reduction potentials and Equation 20.10 to calculate the standard emf, $E°$, for the reaction. Finally, the sign of $E°$, if positive, indicates that the reaction is spontaneous.

Solve: (a) In this reaction Cu is oxidized to Cu^{2+} and H^+ is reduced to H_2. The corresponding half-reactions and associated standard reduction potentials are

Reduction: $2H^+(aq) + 2e^- \longrightarrow H_2(g) \qquad E°_{red} = 0\text{ V}$

Oxidation: $Cu(s) \longrightarrow Cu^{2+}(aq) + 2e^- \qquad E°_{red} = +0.34\text{ V}$

Notice that for the oxidation process we use the standard reduction potential from Table 20.1 for the reduction of Cu^{2+} to Cu. We now calculate $E°$ by using Equation 20.10.

$$E° = E°_{red}(\text{reduction process}) - E°_{red}(\text{oxidation process})$$
$$= (0\text{ V}) - (0.34\text{ V}) = -0.34\text{ V}$$

Because the value of $E°$ is negative, the reaction is not spontaneous in the direction written. Copper metal does not react with acids in this fashion. The reverse reaction, however, *is* spontaneous: Cu^{2+} can be reduced by H_2.

(b) We follow a procedure analogous to that in (a):

Reduction: $Cl_2(g) + 2e^- \longrightarrow 2Cl^-(aq)$ $E^\circ_{red} = +1.36$ V

Oxidation: $2I^-(aq) \longrightarrow I_2(s) + 2e^-$ $E^\circ_{red} = +0.54$ V

In this case $E^\circ = (1.36 \text{ V}) - (0.54 \text{ V}) = +0.82$ V. Because the value of E° is positive, this reaction is spontaneous and could be used to build a voltaic cell.

PRACTICE EXERCISE

Using the standard reduction potentials listed in Appendix E, determine which of the following reactions are spontaneous under standard conditions:
(a) $I_2(s) + 5Cu^{2+}(aq) + 6H_2O(l) \longrightarrow 2IO_3^-(aq) + 5Cu(s) + 12H^+(aq)$
(b) $Hg^{2+}(aq) + 2I^-(aq) \longrightarrow Hg(l) + I_2(s)$
(c) $H_2SO_3(aq) + 2Mn(s) + 4H^+(aq) \longrightarrow S(s) + 2Mn^{2+}(aq) + 3H_2O(l)$
Answer: Reactions (b) and (c) are spontaneous.

We can use standard reduction potentials to understand the activity series of metals. (Section 4.4) Recall that any metal in the activity series will be oxidized by the ions of any metal below it. We can now recognize the origin of this rule on the basis of standard reduction potentials. The activity series, tabulated in Table 4.5, consists of the oxidation reactions of the metals, ordered from the strongest reducing agent at the top to the weakest reducing agent at the bottom. (Thus, the ordering is "flipped over" relative to that in Table 20.1.) For example, nickel lies above silver in the activity series. We therefore expect nickel to displace silver, according to the following net reaction:

$$Ni(s) + 2Ag^+(aq) \longrightarrow Ni^{2+}(aq) + 2Ag(s)$$

In this reaction Ni is oxidized and Ag^+ is reduced. Therefore, using data from Table 20.1, the standard emf for the reaction is

$$\begin{aligned} E^\circ &= E^\circ_{red}(Ag^+/Ag) - E^\circ_{red}(Ni^{2+}/Ni) \\ &= (+0.80 \text{ V}) - (-0.28 \text{ V}) = +1.08 \text{ V} \end{aligned}$$

The positive value of E° indicates that the displacement of silver by nickel is a spontaneous process. Remember that although the silver half-reaction is multiplied by two, the reduction potential is not.

EMF and Free-Energy Change

The change in Gibbs free energy, ΔG, is a measure of the spontaneity of a process that occurs at constant temperature and pressure. (Section 19.5) Because the emf, E, of a redox reaction indicates whether the reaction is spontaneous, the emf and the free-energy change are related by Equation 20.11.

$$\Delta G = -nFE \quad [20.11]$$

In this equation n is a positive number without units that represents the number of electrons transferred in the reaction. The constant F is called *Faraday's constant*, named after Michael Faraday (Figure 20.15 ▶). Faraday's constant is the quantity of electrical charge on 1 mol of electrons. This quantity of charge is called a **faraday** (F).

$$1 F = 96{,}500 \text{ C/mol} = 96{,}500 \text{ J/V-mol}$$

▲ **Figure 20.15** Michael Faraday (1791–1867) was born in England, a child of a poor blacksmith. At the age of 14 he was apprenticed to a bookbinder who gave Faraday time to read and to attend lectures. In 1812 he became an assistant in Humphry Davy's laboratory at the Royal Institution. He succeeded Davy as the most famous and influential scientist in England, making an amazing number of important discoveries, including his formation of the quantitative relationships between electrical current and the extent of chemical reaction in electrochemical cells.

The units of ΔG calculated by using Equation 20.11 are J/mol; as in Equation 19.21, we use "per mole" to mean "per mole of reaction as written." (Section 19.7)

Both n and F are positive numbers. Thus, a positive value of E in Equation 20.11 leads to a negative value of ΔG. Remember: *A positive value of* E *and a negative value of* ΔG *both indicate that a reaction is spontaneous.* When the reactants and products are all in their standard states, Equation 20.11 can be modified to relate ΔG° and E°.

$$\Delta G^\circ = -nFE^\circ \quad [20.12]$$

SAMPLE EXERCISE 20.10

(a) Use the standard reduction potentials listed in Table 20.1 to calculate the standard free-energy change, $\Delta G°$, for the following reaction:

$$4Ag(s) + O_2(g) + 4H^+(aq) \longrightarrow 4Ag^+(aq) + 2H_2O(l)$$

(b) Suppose the reaction in part (a) were written as

$$2Ag(s) + \frac{1}{2}O_2(g) + 2H^+(aq) \longrightarrow 2Ag^+(aq) + H_2O(l)$$

What are the values of $E°$ and $\Delta G°$ when the reaction is written in this way?

Solution

Analyze: We are asked to determine $\Delta G°$ for a redox reaction using standard reduction potentials.

Plan: We use the data in Table 20.1 and Equation 20.10 to determine $E°$ for the reaction and then use $E°$ in Equation 20.12 to calculate $\Delta G°$.

Solve: (a) We first calculate $E°$ by breaking the equation into two half-reactions, as we did in Sample Exercise 20.9, and then obtain $E°_{red}$ values from Table 20.1 (or Appendix E):

Reduction:	$O_2(g) + 4H^+(aq) + 4e^- \longrightarrow 2H_2O(l)$	$E°_{red} = +1.23\text{ V}$
Oxidation:	$4Ag(s) \longrightarrow 4Ag^+(aq) + 4e^-$	$E°_{red} = +0.80\text{ V}$

Even though the second half-reaction has 4Ag, we use the $E°_{red}$ value directly from Table 20.1 because emf is an intensive property.

Using Equation 20.10, we have

$$E° = (1.23\text{V}) - (0.80\text{ V}) = 0.43\text{ V}$$

The half-reactions show the transfer of 4 electrons. Thus, for this reaction $n = 4$. We now use Equation 20.12 to calculate $\Delta G°$:

$$\begin{aligned}\Delta G° &= -nFE° \\ &= -(4)(96{,}500\text{ J/V-mol})(+0.43)\text{ V} \\ &= -1.7 \times 10^5\text{ J/mol} = -170\text{ kJ/mol}\end{aligned}$$

The positive value of $E°$ leads to a negative value of $\Delta G°$.

(b) The overall equation is the same as that in part (a), multiplied by $\frac{1}{2}$. The half-reactions are

Reduction	$\frac{1}{2}O_2(g) + 2H^+(aq) + 2e^- \longrightarrow H_2O(l)$	$E°_{red} = +1.23\text{ V}$
Oxidation:	$2Ag(s) \longrightarrow 2Ag^+(aq) + 2e^-$	$E°_{red} = +0.80\text{ V}$

The values of $E°_{red}$ are the same as they were in part (a); they are not changed by multiplying the half-reactions by $\frac{1}{2}$. Thus, $E°$ has the same value as in part (a): $E° = +0.43$ V. Notice, though, that the value of n has changed to $n = 2$, which is $\frac{1}{2}$ the value in part (a). Thus, $\Delta G°$ is half as large as in part (a).

$$\Delta G° = -(2)(96{,}500\text{ J/V-mol})(+0.43\text{ V}) = -83\text{ kJ/mol}.$$

Remember: $E°$ is an *intensive* quantity, so multiplying a chemical equation by a certain factor will not affect the value of $E°$. Multiplying an equation will change the value of n, however, and hence the value of $\Delta G°$. The change in free energy, in units of kJ/mol of reaction as written, is an *extensive* quantity.

PRACTICE EXERCISE

Consider the following reaction:

$$3Ni^{2+}(aq) + 2Cr(OH)_3(s) + 10OH^-(aq) \longrightarrow 3Ni(s) + 2CrO_4^{2-}(aq) + 8H_2O(l)$$

(a) What is the value of n for this reaction? **(b)** Use the data in Appendix E to calculate $\Delta G°$ for this reaction.

Answers: **(a)** 6; **(b)** +87 kJ/mol

20.6 Effect of Concentration on Cell EMF

We have seen how to calculate the emf of a cell when the reactants and products are under standard conditions. As a voltaic cell is discharged, however, the reactants of the reaction are consumed and the products are generated, so the concentrations of these substances change. The emf progressively drops until $E = 0$, at which point we say the cell is "dead." At that point the concentrations of the reactants and products cease to change; they are at equilibrium. In this section we will examine how the cell emf depends on the concentrations of the reactants and products of the cell reaction. The emf generated under nonstandard conditions can be calculated by using an equation first derived by Walther Nernst (1864–1941), a German chemist who established many of the theoretical foundations of electrochemistry.

The Nernst Equation

The dependence of the cell emf on concentration can be obtained from the dependence of the free-energy change on concentration. (Section 19.7) Recall that the free-energy change, ΔG, is related to the standard free-energy change, ΔG°:

$$\Delta G = \Delta G^\circ + RT \ln Q \qquad [20.13]$$

The quantity Q is the reaction quotient, which has the form of the equilibrium-constant expression except that the concentrations are those that exist in the reaction mixture at a given moment. (Section 15.5)

Substituting $\Delta G = -nFE$ (Equation 20.11) into Equation 20.13 gives

$$-nFE = -nFE^\circ + RT \ln Q$$

Solving this equation for E gives the **Nernst equation**:

$$E = E^\circ - \frac{RT}{nF} \ln Q \qquad [20.14]$$

This equation is customarily expressed in terms of common (base 10) logarithms:

$$E = E^\circ - \frac{2.303\ RT}{nF} \log Q \qquad [20.15]$$

At $T = 298$ K the quantity $2.303RT/F$ equals 0.0592 V, so the equation simplifies to

$$E = E^\circ - \frac{0.0592\ \text{V}}{n} \log Q \qquad (T = 298\ \text{K}) \qquad [20.16]$$

We can use this equation to find the emf produced by a cell under nonstandard conditions or to determine the concentration of a reactant or product by measuring the emf of the cell.

To show how Equation 20.16 might be used, consider the following reaction, which we have discussed earlier:

$$\text{Zn}(s) + \text{Cu}^{2+}(aq) \longrightarrow \text{Zn}^{2+}(aq) + \text{Cu}(s)$$

In this case $n = 2$ (two electrons are transferred from Zn to Cu^{2+}) and the standard emf is +1.10 V. Thus, at 298 K the Nernst equation gives

$$E = 1.10\ \text{V} - \frac{0.0592\ \text{V}}{2} \log \frac{[\text{Zn}^{2+}]}{[\text{Cu}^{2+}]} \qquad [20.17]$$

Recall that pure solids are excluded from the expression for Q. (Section 15.5) According to Equation 20.17, the emf increases as $[Cu^{2+}]$ increases and as $[Zn^{2+}]$ decreases. For example, when $[Cu^{2+}]$ is 5.0 M and $[Zn^{2+}]$ is 0.050 M, we have

$$E = 1.10\ \text{V} - \frac{0.0592\ \text{V}}{2} \log\left(\frac{0.050}{5.0}\right)$$

$$= 1.10\ \text{V} - \frac{0.0592\ \text{V}}{2}(-2.00) = 1.16\ \text{V}$$

Thus, increasing the concentration of the reactant (Cu^{2+}) and decreasing the concentration of the product (Zn^{2+}) relative to standard conditions increases the emf of the cell ($E = +1.16$ V) relative to standard conditions ($E° = +1.10$ V). We could have anticipated this result by applying Le Châtelier's principle. (Section 15.6)

In general, if the concentrations of reactants increase relative to those of products, the emf increases. Conversely, if the concentrations of products increase relative to reactants, the emf decreases. As a voltaic cell operates, reactants are converted into products, which increases the value of Q and decreases the emf.

SAMPLE EXERCISE 20.11

Calculate the emf at 298 K generated by the cell described in Sample Exercise 20.4 when $[Cr_2O_7^{2-}] = 2.0\ M$, $[H^+] = 1.0\ M$, $[I^-] = 1.0\ M$, and $[Cr^{3+}] = 1.0 \times 10^{-5}\ M$.

$$Cr_2O_7^{2-}(aq) + 14H^+(aq) + 6I^-(aq) \longrightarrow 2Cr^{3+}(aq) + 3I_2(s) + 7H_2O(l)$$

Solution

Analyze: We are given a chemical equation for a voltaic cell and the concentrations of reactants and products under which it operates. We are asked to calculate the emf of the cell under these nonstandard conditions.

Plan: To calculate the emf of a cell under nonstandard conditions, we use the Nernst equation, Equation 20.16.

Solve: We first calculate $E°$ for the cell from standard reduction potentials (Table 20.1 or Appendix E). The standard emf for this reaction was calculated in Sample Exercise 20.6: $E° = 0.79$ V. As you will see if you refer back to that exercise, the balanced equation shows six electrons transferred from reducing agent to oxidizing agent, so $n = 6$. The reaction quotient, Q, is

$$Q = \frac{[Cr^{3+}]^2}{[Cr_2O_7^{2-}][H^+]^{14}[I^-]^6} = \frac{(1.0 \times 10^{-5})^2}{(2.0)(1.0)^{14}(1.0)^6} = 5.0 \times 10^{-11}$$

Using Equation 20.16, we have

$$E = 0.79\text{ V} - \frac{0.0592\text{ V}}{6}\log(5.0 \times 10^{-11})$$

$$= 0.79\text{ V} - \frac{0.0592\text{ V}}{6}(-10.30)$$

$$= 0.79\text{ V} + 0.10\text{ V} = 0.89\text{ V}$$

Check: This result is qualitatively what we expect: Because the concentration of $Cr_2O_7^{2-}$ (a reactant) is greater than 1 M and the concentration of Cr^{3+} (a product) is less than 1 M, the emf is greater than $E°$. Q is about 10^{-10}, so log Q is about -10. Thus, the correction to $E°$ is about $0.06 \times (10)/6$ which is 0.1, in agreement with the more detailed calculation.

PRACTICE EXERCISE

Calculate the emf generated by the cell described in the practice exercise accompanying Sample Exercise 20.6 when $[Al^{3+}] = 4.0 \times 10^{-3} M$ and $[I^-] = 0.010\ M$.
Answer: $E = +2.36$ V

SAMPLE EXERCISE 20.12

If the voltage of a Zn–H^+ cell (like that in Figure 20.11) is 0.45 V at 25°C when $[Zn^{2+}] = 1.0\ M$ and $P_{H_2} = 1.0$ atm, what is the concentration of H^+?

Solution

Analyze: We are given a description of a voltaic cell, its emf, and the concentrations of all reactants and products except H^+, which we are asked to calculate.

Plan: First we write the chemical equation for the cell and use standard reduction potentials to calculate $E°$. We then solve the Nernst equation for Q, and finally use Q to solve for $[H^+]$.

Solve: The cell reaction is

$$Zn(s) + 2H^+(aq) \longrightarrow Zn^{2+}(aq) + H_2(g)$$

The standard emf is

$$E° = E°_{red}(\text{reduction}) - E°_{red}(\text{oxidation})$$
$$= 0\text{ V} - (-0.76\text{ V}) = +0.76\text{ V}$$

Because each Zn atom loses 2 electrons,

$$n = 2$$

Using the Nernst equation (Equation 20.16), we have

$$0.45\text{ V} = 0.76\text{ V} - \frac{0.0592\text{ V}}{2}\log Q$$

$$\log Q = (0.76\text{ V} - 0.45\text{ V})\left(\frac{2}{0.0592\text{ V}}\right) = 10.47$$

$$Q = 10^{10.47} = 3.0 \times 10^{10}$$

Q has the form of the equilibrium constant for the reaction

$$Q = \frac{[Zn^{2+}]P_{H_2}}{[H^+]^2} = \frac{(1.0)(1.0)}{[H^+]^2} = 3.0 \times 10^{10}$$

Solving for $[H^+]$, we have

$$[H^+]^2 = \frac{1.0}{3.0 \times 10^{10}} = 3.3 \times 10^{-11}$$

$$[H^+] = \sqrt{3.3 \times 10^{-11}} = 5.8 \times 10^{-6}\ M$$

Comment: A voltaic cell whose cell reaction involves H^+ can be used to measure $[H^+]$ or pH. A pH meter is a specially designed voltaic cell with a voltmeter calibrated to read pH directly. ∞ (Section 16.4)

PRACTICE EXERCISE

What is the pH of the solution in the cathode compartment of the cell pictured in Figure 20.11 when $P_{H_2} = 1.0$ atm, $[Zn^{2+}]$ in the anode compartment is 0.10 M, and cell emf is 0.542 V?
Answer: pH = 4.19

Concentration Cells

In each of the voltaic cells that we have looked at thus far, the reactive species at the anode has been different from the one at the cathode. Cell emf depends on concentration, however, so a voltaic cell can be constructed using the *same* species in both the anode and cathode compartments as long as the concentrations are different. A cell based solely on the emf generated because of a difference in a concentration is called a **concentration cell**.

A diagram of a concentration cell is shown in Figure 20.16(a) ▼. One compartment consists of a strip of nickel metal immersed in a 1.00 M solution of

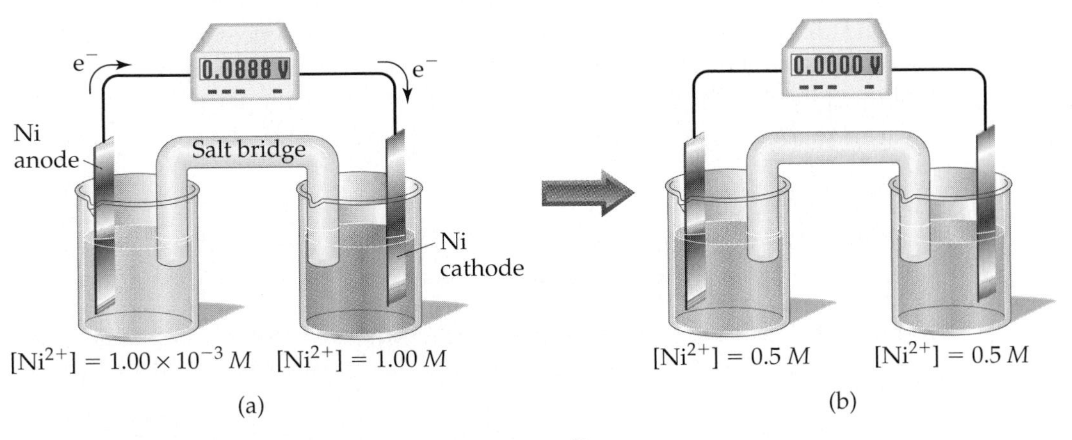

▲ **Figure 20.16** A concentration cell based on the Ni^{2+}-Ni cell reaction. In (a) the concentrations of $Ni^{2+}(aq)$ in the two compartments are unequal, and the cell generates an electrical current. The cell operates until the concentrations of $Ni^{2+}(aq)$ in the two compartments become equal, (b) at which point the cell has reached equilibrium and is "dead."

$Ni^{2+}(aq)$. The other compartment also has an Ni(s) electrode, but it is immersed in a 1.00×10^{-3} M solution of $Ni^{2+}(aq)$. The two compartments are connected by a salt bridge and by an external wire with a voltmeter. The half-cell reactions are the reverse of one another.

$$\text{Anode:} \quad Ni(s) \longrightarrow Ni^{2+}(aq) + 2e^- \qquad E^\circ_{red} = -0.28 \text{ V}$$

$$\text{Cathode:} \quad Ni^{2+}(aq) + 2e^- \longrightarrow Ni(s) \qquad E^\circ_{red} = -0.28 \text{ V}$$

Although the *standard* emf for this cell is zero, $E^\circ_{cell} = E^\circ_{red}(\text{cathode}) - E^\circ_{red}(\text{anode}) = (-0.28 \text{ V}) - (-0.28 \text{ V}) = 0 \text{ V}$, the cell operates under *nonstandard* conditions because the concentration of $Ni^{2+}(aq)$ is different in the two compartments. In fact, the cell will operate until the concentrations of Ni^{2+} in both compartments are equal. Oxidation of Ni(s) occurs in the half-cell containing the more dilute solution, thereby increasing the concentration of $Ni^{2+}(aq)$. It is therefore the anode compartment of the cell. Reduction of $Ni^{2+}(aq)$ occurs in the half-cell containing the more concentrated solution, thereby decreasing the concentration of $Ni^{2+}(aq)$, making it the cathode compartment. The *overall* cell reaction is therefore

$$\text{Anode:} \quad Ni(s) \longrightarrow Ni^{2+}(aq;\ \text{dilute}) + 2e^-$$

$$\text{Cathode:} \quad Ni^{2+}(aq;\ \text{concentrated}) + 2e^- \longrightarrow Ni(s)$$

$$\text{Overall:} \quad Ni^{2+}(aq;\ \text{concentrated}) \longrightarrow Ni^{2+}(aq;\ \text{dilute})$$

We can calculate the emf of a concentration cell by using the Nernst equation. For this particular cell we see that $n = 2$. The expression for the reaction quotient for the overall reaction is $Q = [Ni^{2+}]_{dilute}/[Ni^{2+}]_{concentrated}$. Thus, the emf at 298 K is

$$\begin{aligned} E &= E^\circ - \frac{0.0592}{n} \log Q \\ &= 0 - \frac{0.0592}{2} \log \frac{[Ni^{2+}]_{dilute}}{[Ni^{2+}]_{concentrated}} = -\frac{0.0592}{2} \log \frac{1.00 \times 10^{-3}\ M}{1.00\ M} \\ &= +0.0888 \text{ V} \end{aligned}$$

This concentration cell generates an emf of nearly 0.09 V even though $E^\circ = 0$. The difference in concentration provides the driving force for the cell. When the concentrations in the two compartments become the same, the value of $Q = 1$ and $E = 0$.

The idea of generating a potential by a difference in concentration is the basis for the operation of pH meters (Figure 16.6). It is also a critical aspect in the regulation of the heartbeat in mammals, as is discussed in the "Chemistry and Life" box in this section.

SAMPLE EXERCISE 20.13

A voltaic cell is constructed with two hydrogen electrodes. Electrode 1 has $P_{H_2} = 1.00$ atm and an unknown concentration of $H^+(aq)$. Electrode 2 is a standard hydrogen electrode ($[H^+] = 1.00\ M$, $P_{H_2} = 1.00$ atm). At 298 K the measured cell voltage is 0.211 V, and the electrical current is observed to flow from electrode 1 through the external circuit to electrode 2. Calculate $[H^+]$ for the solution at electrode 1. What is its pH?

Solution

Analyze: We are given the voltage of a concentration cell and the direction in which the current flows. We also have the concentrations of all reactants and products except for $[H^+]$ in half-cell 1, which is our unknown.

Plan: We can use the Nernst equation to determine Q and then use Q to calculate the unknown concentration. Because this is a concentration cell, $E^\circ_{cell} = 0$ V.
Solve: Using the Nernst equation, we have

$$0.211 \text{ V} = 0 - \frac{0.0592 \text{ V}}{2} \log Q$$

$$\log Q = -(0.211\text{V})\left(\frac{2}{0.0592 \text{ V}}\right) = -7.13$$

$$Q = 10^{-7.13} = 7.4 \times 10^{-8}$$

Because current flows from electrode 1 to electrode 2, electrode 1 is the anode of the cell and electrode 2 is the cathode. The electrode reactions are therefore as follows, with the concentration of $H^+(aq)$ in electrode 1 represented with the unknown x:

Electrode 1:	$H_2(g; 1.00 \text{ atm}) \longrightarrow 2H^+(aq; x\ M) + 2e^-$	$E^\circ_{red} = 0$ V
Electrode 2:	$2H^+(aq; 1.00\ M) + 2e^- \longrightarrow H_2(g; 1.00 \text{ atm})$	$E^\circ_{red} = 0$ V

Thus,

$$Q = \frac{[H^+(\text{electrode 1})]^2 P_{H_2}(\text{electrode 2})}{[H^+(\text{electrode 2})]^2 P_{H_2}(\text{electrode 1})}$$

$$= \frac{x^2(1.00)}{(1.00)^2(1.00)} = x^2 = 7.4 \times 10^{-8}$$

$$x = \sqrt{7.4 \times 10^{-8}} = 2.7 \times 10^{-4}$$

At electrode 1, therefore, $[H^+] = 2.7 \times 10^{-4}\ M$, and the pH of the solution is

$$\text{pH} = -\log[H^+] = -\log(2.7 \times 10^{-4}) = 3.57$$

Comment: The concentration of H^+ at electrode 1 is lower than that in electrode 2, which is why electrode 1 is the anode of the cell: The oxidation of H_2 to $H^+(aq)$ increases $[H^+]$ at electrode 1.

PRACTICE EXERCISE

A concentration cell is constructed with two $Zn(s)$-$Zn^{2+}(aq)$ half-cells. The first half-cell has $[Zn^{2+}] = 1.35\ M$, and the second half-cell has $[Zn^{2+}] = 3.75 \times 10^{-4}\ M$. **(a)** Which half-cell is the anode of the cell? **(b)** What is the emf of the cell?
Answers: **(a)** the second half-cell; **(b)** 0.105 V

Cell EMF and Chemical Equilibrium

The Nernst equation helps us understand why the emf of a voltaic cell drops as it discharges: As reactants are converted to products, the value of Q increases, so the value of E decreases, eventually reaching $E = 0$. Because $\Delta G = -nFE$ (Equation 20.11), it follows that $\Delta G = 0$ when $E = 0$. Recall that a system is at equilibrium when $\Delta G = 0$. (Section 19.7) Thus, when $E = 0$, the cell reaction has reached equilibrium, and no net reaction is occurring in the voltaic cell.

At equilibrium the reaction quotient equals the equilibrium constant: $Q = K_{eq}$ at equilibrium. (Section 15.5) Substituting $E = 0$ and $Q = K_{eq}$ into the Nernst equation (Equation 20.14) gives

$$0 = E^\circ - \frac{RT}{nF} \ln K_{eq}$$

At 298 K this equation simplifies to

$$0 = E^\circ - \frac{0.0592}{n} \log K_{eq} \qquad (T = 298 \text{ K})$$

which can be rearranged to give

$$\log K_{eq} = \frac{nE^\circ}{0.0592} \qquad (T = 298 \text{ K}) \qquad [20.18]$$

Thus, the equilibrium constant for a redox reaction can be obtained from the value of the standard emf for the reaction.

Chemistry and Life Heartbeats and Electrocardiography

The human heart is a marvel of efficiency and dependability. In a typical day an adult's heart pumps more than 7000 L of blood through the circulatory system, usually with no maintenance required beyond a sensible diet and lifestyle. We generally think of the heart as a mechanical device, a muscle that circulates blood via regularly spaced muscular contractions. However, more than two centuries ago, two pioneers in electricity, Luigi Galvani (1729–1787) and Alessandro Volta (1745–1827), discovered that the contractions of the heart are controlled by electrical phenomena, as are nerve impulses throughout the body. The pulses of electricity that cause the heart to beat result from a remarkable combination of electrochemistry and the properties of semipermeable membranes. (Section 13.5)

Cell walls are membranes with variable permeability with respect to a number of physiologically important ions (especially Na^+, K^+, and Ca^{2+}). The concentrations of these ions are different for the fluids inside the cells (the *intracellular fluid*, or ICF) and outside the cells (the *extracellular fluid*, or ECF). In cardiac muscle cells, for example, the concentrations of K^+ in the ICF and ECF are typically about 135 millimolar (mM) and 4 mM, respectively. For Na^+, however, the concentration difference between the ICF and ECF is opposite that for K^+; typically, $[Na^+]_{ICF} = 10$ mM and $[Na^+]_{ECF} = 145$ mM.

The cell membrane is initially permeable to K^+ ions, but much less so to Na^+ and Ca^{2+}. The difference in concentration of K^+ ions between the ICF and ECF generates a concentration cell: Even though the same ions are present on both sides of the membrane, there is a potential difference between the two fluids that we can calculate using the Nernst equation with $E^\circ = 0$. At physiological temperature (37°C) the potential in millivolts for moving K^+ from the ECF to the ICF is

$$E = E^\circ - \frac{2.30RT}{nF} \log \frac{[K^+]_{ICF}}{[K^+]_{ECF}}$$

$$= 0 - (61.5 \text{ mV}) \log\left(\frac{135 \text{ m}M}{4 \text{ m}M}\right) = -94 \text{ mV}$$

In essence, the interior of the cell and the ECF together serve as a voltaic cell. The negative sign for the potential indicates that work is required to move K^+ into the intracellular fluid.

Changes in the relative concentrations of the ions in the ECF and ICF lead to changes in the emf of the voltaic cell. The cells of the heart that govern the rate of heart contraction are called the *pacemaker cells*. The membranes of the cells regulate the concentrations of ions in the ICF, allowing them to change in a systematic way. The concentration changes cause the emf to change in a cyclic fashion, as shown in Figure 20.17 ▲. The emf cycle determines the rate at which the heart beats. If the pacemaker cells malfunction because of disease or injury, an artificial pacemaker can be surgically implanted. The artificial pacemaker is a small battery that generates the electrical pulses needed to trigger the contractions of the heart.

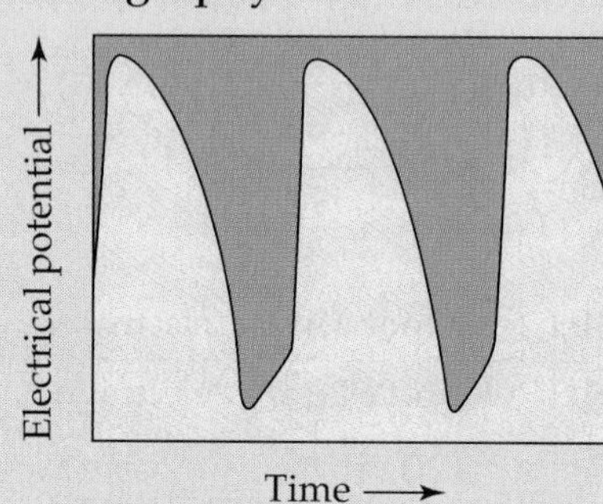

◀ **Figure 20.17** Variation of the electrical potential caused by changes of ion concentrations in the pacemaker cells of the heart.

In the late 1800s scientists discovered that the electrical impulses that cause the contraction of the heart muscle are strong enough to be detected at the surface of the body. This observation formed the basis for *electrocardiography*, noninvasive monitoring of the heart by using a complex array of electrodes on the skin to measure voltage changes during heartbeats. A typical electrocardiogram is shown in Figure 20.18 ▼. It is quite striking that, although the heart's major function is the *mechanical* pumping of blood, it is most easily monitored by using the *electrical* impulses generated by tiny voltaic cells.

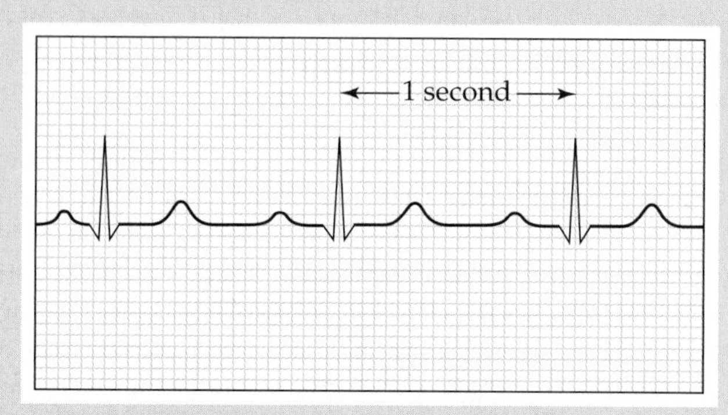

▲ **Figure 20.18** An electrocardiogram (ECG) printout records the electrical events monitored by electrodes attached to the body surface. The horizontal axis is time, and the vertical displacement is the emf.

SAMPLE EXERCISE 20.14

Using standard reduction potentials from Appendix E, calculate the equilibrium constant for the oxidation of Fe^{2+} by O_2 in acidic solution.

$$O_2(g) + 4H^+(aq) + 4Fe^{2+}(aq) \longrightarrow 4Fe^{3+}(aq) + 2H_2O(l)$$

Solution

Analyze: We are given a redox equation and asked to use standard reduction potentials to calculate the equilibrium constant, K_{eq}.

Plan: K_{eq} can be calculated from $E°$ using Equation 20.18. Thus, we must first determine $E°$ from the $E°_{red}$ values for its half-reactions.

Solve: The two half-reactions and their standard reduction potentials from Appendix E are as follows:

Reduction: $O_2(g) + 4H^+(aq) + 4e^- \longrightarrow 2H_2O(l)$ $\quad E°_{red} = +1.23$

Oxidation: $4Fe^{2+}(aq) \longrightarrow 4Fe^{3+}(aq) + 4e^-$ $\quad E°_{red} = +0.77$

Thus,

$$E° = (1.23\ V) - (0.77\ V) = 0.46\ V, \quad \text{and} \quad n = 4$$

Using Equation 20.18, we have

$$\log K_{eq} = \frac{nE°}{0.0592\ V} = \frac{4(0.46\ V)}{0.0592\ V} = 31$$

$$K_{eq} = 1 \times 10^{31}$$

Comment: The large magnitude of K_{eq} indicates that Fe^{2+} ions are unstable in acidic solutions in the presence of O_2 (unless a suitable reducing agent is present).

PRACTICE EXERCISE

Using standard reduction potentials (Appendix E), calculate the equilibrium constant at 25°C for the reaction

$$Br_2(l) + 2Cl^-(aq) \longrightarrow Cl_2(g) + 2Br^-(aq)$$

Answer: $K_{eq} = 1.2 \times 10^{-10}$

20.7 Batteries

A **battery** is a portable, self-contained electrochemical power source that consists of one or more voltaic cells (Figure 20.19 ▼). For example, the common 1.5-V batteries used to power flashlights and many consumer electronic devices are single voltaic cells. Greater voltages can be achieved by using multiple voltaic cells in a single battery, as is the case in 12-V automotive batteries. When cells are connected in series (with the cathode of one attached to the anode of another), the battery produces a voltage that is the sum of the emfs of the individual cells. Higher emfs can also be achieved by using multiple batteries in series (Figure 20.20 ▶). The electrodes of batteries are marked following the convention of Figure 20.6—the cathode is labeled with a plus sign and the anode with a minus sign.

▲ **Figure 20.19** Batteries are voltaic cells that serve as portable sources of electricity. Batteries vary markedly in size and in the electrochemical reaction used to generate electricity.

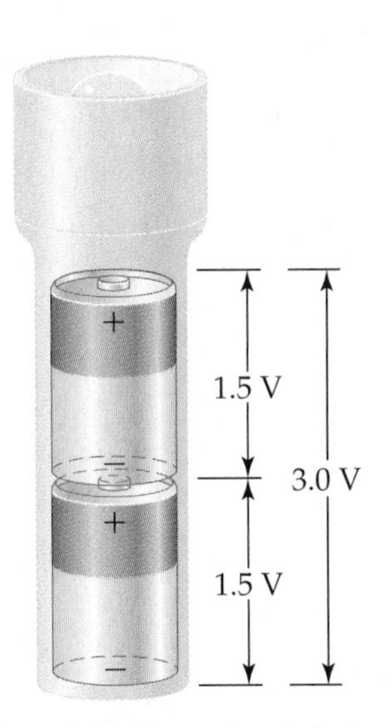

▲ **Figure 20.20** When batteries are connected in series, as in most flashlights, the total emf is the sum of the individual emfs.

Although any spontaneous redox reaction can serve as the basis for a voltaic cell, making a commercial battery that has specific performance characteristics can require considerable ingenuity. The substances that are oxidized at the anode and reduced by the cathode determine the emf of a battery, and the usable life of the battery depends on the quantities of these substances packaged in the battery. Usually the anode and the cathode compartments are separated by a barrier analogous to the porous barrier of Figure 20.6.

Different applications require batteries with different properties. The battery required to start a car, for example, must be capable of delivering a large electrical current for a short time period. The battery that powers a heart pacemaker, on the other hand, must be very small and capable of delivering a small but steady current over an extended time period. Some batteries are *primary* cells, meaning that they can't be recharged. A primary cell must be discarded or recycled after its emf drops to zero. A *secondary* cell can be recharged from an external power source after its emf has dropped.

In this section we will briefly discuss some common batteries. As we do so, notice how the principles we have discussed so far in this chapter help us understand these important sources of portable electrical energy.

Lead-Acid Battery

A 12-V lead-acid automotive battery consists of six voltaic cells in series, each producing 2 V. The cathode of each cell consists of lead dioxide (PbO_2) packed on a metal grid. The anode of each cell is composed of lead. Both electrodes are immersed in sulfuric acid. The electrode reactions that occur during discharge are as follows:

$$\text{Cathode:}\quad PbO_2(s) + HSO_4^-(aq) + 3H^+(aq) + 2e^- \longrightarrow PbSO_4(s) + 2H_2O(l)$$

$$\text{Anode:}\quad Pb(s) + HSO_4^-(aq) \longrightarrow PbSO_4(s) + H^+(aq) + 2e^-$$

$$PbO_2(s) + Pb(s) + 2HSO_4^-(aq) + 2H^+(aq) \longrightarrow 2PbSO_4(s) + 2H_2O(l) \qquad [20.19]$$

The standard cell potential can be obtained from the standard reduction potentials in Appendix E:

$$E^\circ_{cell} = E^\circ_{red}(\text{cathode}) - E^\circ_{red}(\text{anode}) = (+1.685\text{ V}) - (-0.356\text{ V}) = +2.041\text{ V}$$

The reactants Pb and PbO_2 serve as the electrodes. Because the reactants are solids, there is no need to separate the cell into anode and cathode compartments; the Pb and PbO_2 cannot come into direct physical contact unless one electrode plate touches another. To keep the electrodes from touching, wood or glass-fiber spacers are placed between them (Figure 20.21 ▶).

Using a reaction whose reactants and products are solids has another benefit. Because solids are excluded from the reaction quotient Q, the relative amounts of $Pb(s)$, $PbO_2(s)$, and $PbSO_4(s)$ have no effect on the emf of the lead storage battery, helping the battery maintain a relatively constant emf during discharge. The emf does vary somewhat with use because the concentration of H_2SO_4 varies with the extent of cell discharge. As Equation 20.19 indicates, H_2SO_4 is consumed during the discharge.

One advantage of a lead-acid battery is that it can be recharged. During recharging an external source of energy is used to reverse the direction of Equation 20.19, regenerating $Pb(s)$ and $PbO_2(s)$.

$$2PbSO_4(s) + 2H_2O(l) \longrightarrow PbO_2(s) + Pb(s) + 2HSO_4^-(aq) + 2H^+(aq)$$

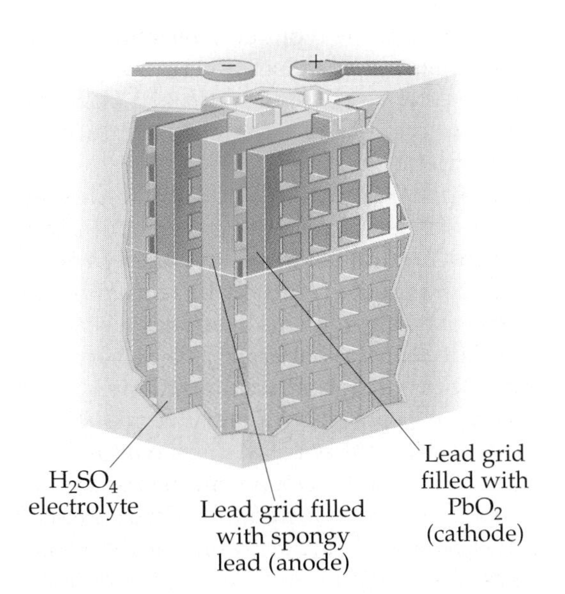

◀ **Figure 20.21** Schematic cutaway drawing of a portion of a 12-V automotive lead-acid battery. Each anode/cathode pair of electrodes produces a potential of about 2 V. Six pairs of electrodes are connected in series, producing the desired battery voltage.

In an automobile the energy necessary for recharging the battery is provided by a generator driven by the engine. Recharging is possible because $PbSO_4$ formed during discharge adheres to the electrodes. As the external source forces electrons from one electrode to another, the $PbSO_4$ is converted to Pb at one electrode and to PbO_2 at the other.

Alkaline Battery

The most common primary (nonrechargeable) battery is the alkaline battery. More than 10^{10} alkaline batteries are produced annually. The anode of this battery consists of powdered zinc metal immobilized in a gel in contact with a concentrated solution of KOH (hence the name *alkaline* battery). The cathode is a mixture of $MnO_2(s)$ and graphite, separated from the anode by a porous fabric. The battery is sealed in a steel can to reduce the risk of leakage of the concentrated KOH. A schematic view of an alkaline battery is shown in Figure 20.22 ▶. The cell reactions are complex, but can be approximately represented as follows:

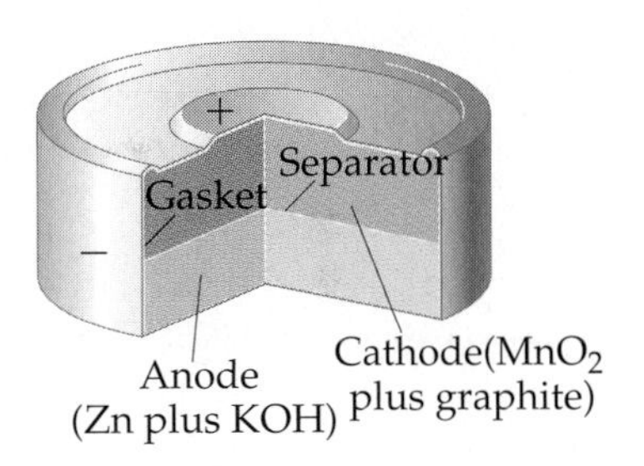

▲ **Figure 20.22** Cutaway view of a miniature alkaline battery.

$$\text{Cathode:} \quad 2MnO_2(s) + 2H_2O(l) + 2e^- \longrightarrow 2MnO(OH)(s) + 2OH^-(aq)$$

$$\text{Anode:} \quad Zn(s) + 2OH^-(aq) \longrightarrow Zn(OH)_2(s) + 2e^-$$

The emf of an alkaline battery is 1.55 V at room temperature. The alkaline battery provides far superior performance over the older "dry cells" that were also based on MnO_2 and Zn as the electrochemically active substances.

Nickel-Cadmium, Nickel-Metal-Hydride, and Lithium-Ion Batteries

The tremendous growth in high-power-demand portable electronic devices, such as cellular phones, notebook computers, and video recorders, has increased the demand for lightweight, readily recharged batteries. One of the most common rechargeable batteries is the nickel-cadmium (nicad) battery. During discharge, cadmium metal is oxidized at the anode of the battery while nickel oxyhydroxide [NiO(OH)(s)] is reduced at the cathode.

$$\text{Cathode:} \quad 2NiO(OH)(s) + 2H_2O(l) + 2e^- \longrightarrow 2Ni(OH)_2(s) + 2OH^-(aq)$$

$$\text{Anode:} \quad Cd(s) + 2OH^-(aq) \longrightarrow Cd(OH)_2(s) + 2e^-$$

As in the lead-acid battery, the solid reaction products adhere to the electrodes, which permits the electrode reactions to be reversed during charging. A single nicad voltaic cell has an emf of 1.30 V. Nicad battery packs typically contain three or more cells in series to produce the higher emfs needed by most electronic devices.

There are drawbacks to nickel-cadmium batteries. Cadmium is a toxic heavy metal. Its use increases the weight of batteries and provides an environmental hazard—roughly 1.5 billion nickel-cadmium batteries are produced annually, and these must eventually be recycled as they lose their ability to be recharged. Some of these problems have been alleviated by the development of nickel-metal-hydride (NiMH) batteries. The cathode reaction of NiMH batteries is the same as that for nickel-cadmium batteries, but the anode reaction is very different. The anode consists of a metal *alloy*, such as $ZrNi_2$, that has the ability to absorb hydrogen atoms (we will discuss alloys in Section 23.6). During the oxidation at the anode, the hydrogen atoms lose electrons, and the resultant H^+ ions react with OH^- ions to form H_2O, a process that is reversed during charging.

The newest rechargeable battery to receive large use in consumer electronic devices is the lithium-ion (Li-ion) battery. Because lithium is a very light element, Li-ion batteries achieve a greater *energy density*—the amount of energy stored per unit mass—than nickel-based batteries. The technology of Li-ion batteries is very different from that of the other batteries we have described here, and it is based on the ability of Li^+ ions to be inserted into and removed from certain layered solids. For example, Li^+ ions can be inserted reversibly into layers of graphite (Figure 11.41). In most commercial cells, one electrode is graphite or some other carbon-based material, and the other is usually made of lithium cobalt oxide ($LiCoO_2$). When charged, cobalt ions are oxidized and Li^+ ions migrate into the graphite. During discharge the Li^+ ions spontaneously migrate from the graphite anode to the cathode, enabling current to flow through the external circuit.

Fuel Cells

The thermal energy released by the combustion of fuels can be converted to electrical energy. The heat may convert water to steam, which drives a turbine that in turn drives the generator. Typically, a maximum of only 40% of the energy from combustion is converted to electricity; the remainder is lost as heat. The direct production of electricity from fuels by a voltaic cell could, in principle, yield a higher rate of conversion of the chemical energy of the reaction. Voltaic cells that perform this conversion using conventional fuels, such as H_2 and CH_4, are called **fuel cells**. Strictly speaking, fuel cells are *not* batteries because they are not self-contained systems.

The most promising fuel-cell system involves the reaction of $H_2(g)$ and $O_2(g)$ to form $H_2O(l)$ as the only product. These cells generate electricity twice as efficiently as the best internal combustion engine. Under basic conditions the electrode reactions in the hydrogen fuel cell are

$$\text{Cathode:} \quad 4e^- + O_2(g) + 2H_2O(l) \longrightarrow 4OH^-(aq)$$

$$\text{Anode:} \quad 2H_2(g) + 4OH^-(aq) \longrightarrow 4H_2O(l) + 4e^-$$

$$2H_2(g) + O_2(g) \longrightarrow 2H_2O(l)$$

The standard emf of an H_2-O_2 fuel cell is +1.23 V, reflecting the large driving force for the reaction of H_2 and O_2 to form H_2O.

Until recently, fuel cells were impractical because they required high operating temperatures in order to allow the cell reaction to occur at an appreciable rate. Newly developed semipermeable membranes and catalysts allow H_2-O_2

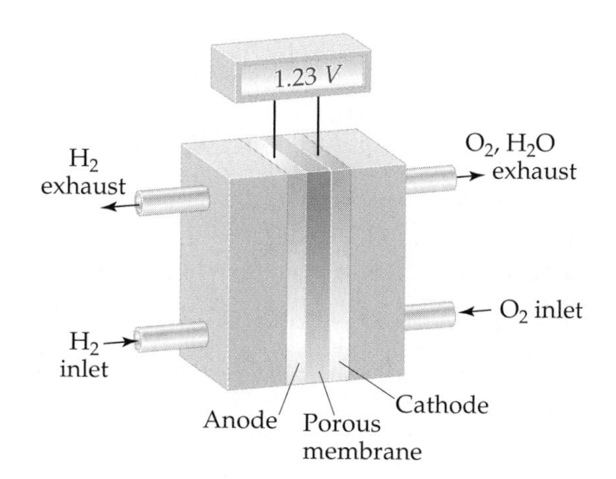

◀ **Figure 20.23** A low-temperature H_2-O_2 fuel cell. The porous membrane allows the H^+ ions generated by the oxidation of H_2 at the anode to migrate to the cathode, where H_2O is formed.

fuel cells to operate at temperatures below 100°C. A schematic drawing of a low-temperature H_2-O_2 fuel cell is shown in Figure 20.23 ▲. This technology is the basis for pollution-free fuel cell–powered vehicles, such as those discussed in the "Chemistry at Work" box in Section 1.4. Currently a great deal of research is going into improving fuel cells. Much effort is being directed toward developing fuel cells that use conventional fuels such as hydrocarbons and alcohols, which are not as difficult to handle and distribute as hydrogen gas.

20.8 Corrosion

Batteries are examples of how spontaneous redox reactions can be used productively. In this section we will examine the undesirable redox reactions that lead to the **corrosion** of metals. Corrosion reactions are spontaneous redox reactions in which a metal is attacked by some substance in its environment and converted to an unwanted compound.

For nearly all metals, oxidation is a thermodynamically favorable process in air at room temperature. When the oxidation process is not inhibited in some way, it can be very destructive. Oxidation can also form an insulating protective oxide layer, however, that prevents further reaction of the underlying metal. On the basis of the standard reduction potential for Al^{3+}, for example, we would expect aluminum metal to be very readily oxidized. The many aluminum soft-drink and beer cans that litter the environment are ample evidence, however, that aluminum undergoes only very slow chemical corrosion. The exceptional stability of this active metal in air is due to the formation of a thin protective coat of oxide—a hydrated form of Al_2O_3—on the surface of the metal. The oxide coat is impermeable to O_2 or H_2O and so protects the underlying metal from further corrosion. Magnesium metal is similarly protected. Some metal alloys, such as stainless steel, likewise form protective impervious oxide coats.

Corrosion of Iron

The rusting of iron (Figure 20.24 ▶) is a familiar corrosion process that carries a significant economic impact. It is estimated that up to 20% of the iron produced annually in the United States is used to replace iron objects that have been discarded because of rust damage.

▲ **Figure 20.24** Corrosion of iron in a shipyard. The corrosion of iron is an electrochemical process of great economic importance. The annual cost of metallic corrosion in the United States is estimated to be $70 billion.

The rusting of iron requires both oxygen and water. Other factors—such as the pH of the solution, the presence of salts, contact with metals more difficult to oxidize than iron, and stress on the iron—can accelerate rusting.

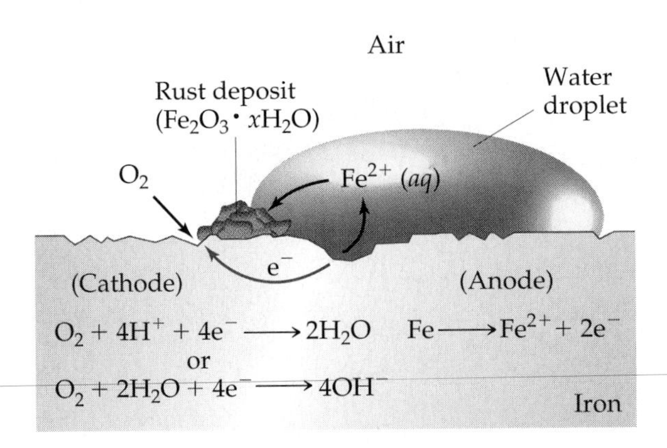

▶ **Figure 20.25** Corrosion of iron in contact with water.

The corrosion of iron is electrochemical in nature. Not only does the corrosion process involve oxidation and reduction, the metal itself conducts electricity. Thus, electrons can move through the metal from a region where oxidation occurs to another region where reduction occurs, as in voltaic cells.

Because the standard reduction potential for the reduction of $Fe^{2+}(aq)$ is less positive than that for the reduction of O_2, Fe(*s*) can be oxidized by $O_2(g)$.

$$\text{Cathode:} \quad O_2(g) + 4H^+(aq) + 4e^- \longrightarrow 2H_2O(l) \qquad E^\circ_{red} = 1.23\text{ V}$$

$$\text{Anode:} \quad Fe(s) \longrightarrow Fe^{2+}(aq) + 2e^- \qquad E^\circ_{red} = -0.44\text{ V}$$

A portion of the iron can serve as an anode at which the oxidation of Fe to Fe^{2+} occurs. The electrons produced migrate through the metal to another portion of the surface that serves as the cathode, at which O_2 is reduced. The reduction of O_2 requires H^+, so lowering the concentration of H^+ (increasing the pH) makes the reduction of O_2 less favorable. Iron in contact with a solution whose pH is greater than 9 does not corrode.

The Fe^{2+} formed at the anode is eventually oxidized further to Fe^{3+}, which forms the hydrated iron(III) oxide known as rust*.

$$4Fe^{2+}(aq) + O_2(g) + 4H_2O(l) + 2xH_2O(l) \longrightarrow 2Fe_2O_3 \cdot xH_2O(s) + 8H^+(aq)$$

Because the cathode is generally the area having the largest supply of O_2, rust often deposits there. If you look closely at a shovel after it has stood outside in the moist air with wet dirt adhered to its blade, you may notice that pitting has occurred under the dirt but that rust has formed elsewhere, where O_2 is more readily available. The corrosion process is summarized in Figure 20.25 ▲.

The enhanced corrosion caused by the presence of salts is usually evident on autos in areas where roads are heavily salted during winter. Like a salt bridge in a voltaic cell, the ions of the salt provide the electrolyte necessary to complete the electrical circuit.

Preventing the Corrosion of Iron

Iron is often covered with a coat of paint or another metal such as tin or zinc to protect its surface against corrosion. Covering the surface with paint or tin is simply a means of preventing oxygen and water from reaching the iron surface. If the coating is broken and the iron is exposed to oxygen and water, corrosion will begin.

---

* Frequently, metal compounds obtained from aqueous solution have water associated with them. For example, copper(II) sulfate crystallizes from water with 5 mol of water per mole of $CuSO_4$. We represent this formula as $CuSO_4 \cdot 5H_2O$. Such compounds are called hydrates. (Section 13.1) Rust is a hydrate of iron(III) oxide with a variable amount of water of hydration. We represent this variable water content by writing the formula as $Fe_2O_3 \cdot xH_2O$.

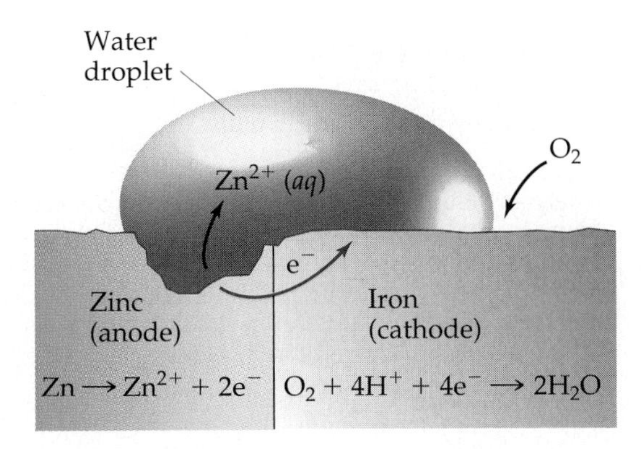

◀ **Figure 20.26** Cathodic protection of iron in contact with zinc.

Galvanized iron, which is iron coated with a thin layer of zinc, uses the principles of electrochemistry to protect the iron from corrosion even after the surface coat is broken. The standard reduction potentials for iron and zinc are

$$Fe^{2+}(aq) + 2e^- \longrightarrow Fe(s) \qquad E^\circ_{red} = -0.44 \text{ V}$$

$$Zn^{2+}(aq) + 2e^- \longrightarrow Zn(s) \qquad E^\circ_{red} = -0.76 \text{ V}$$

Because the E°_{red} value for the reduction of Fe^{2+} is less negative (more positive) than that for the reduction of Zn^{2+}, Fe^{2+} is easier to reduce than Zn^{2+}. Conversely, Zn(*s*) is easier to oxidize than Fe(*s*). Thus, even if the zinc coating is broken and the galvanized iron is exposed to oxygen and water, the zinc, which is most easily oxidized, serves as the anode and is corroded instead of the iron. The iron serves as the cathode at which O_2 is reduced, as shown in Figure 20.26 ▲.

Protecting a metal from corrosion by making it the cathode in an electrochemical cell is known as **cathodic protection**. The metal that is oxidized while protecting the cathode is called the *sacrificial anode*. Underground pipelines are often protected against corrosion by making the pipeline the cathode of a voltaic cell. Pieces of an active metal such as magnesium are buried along the pipeline and connected to it by wire, as shown in Figure 20.27 ▼. In moist soil, where corrosion can occur, the active metal serves as the anode, and the pipe experiences cathodic protection.

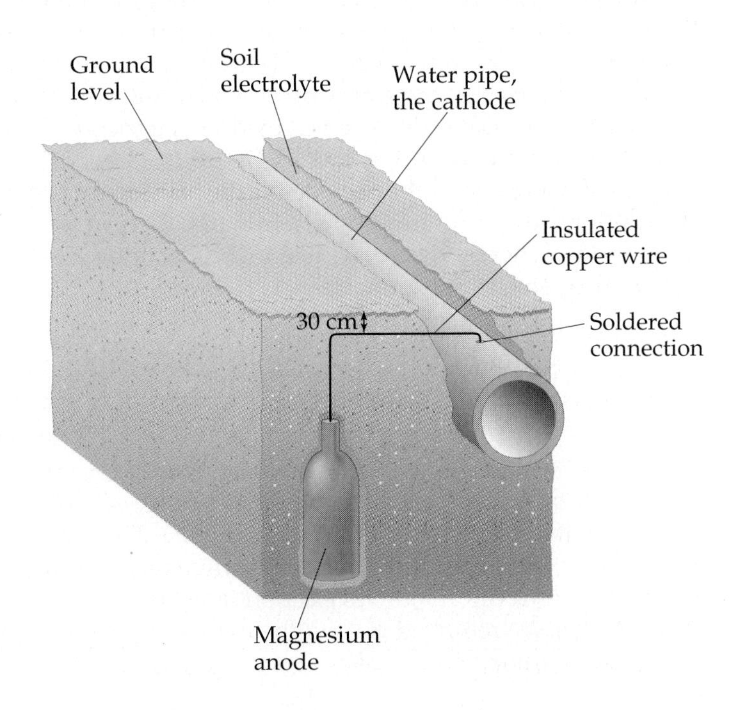

◀ **Figure 20.27** Cathodic protection of an iron water pipe. The magnesium anode is surrounded by a mixture of gypsum, sodium sulfate, and clay to promote conductivity of ions. The pipe, in effect, is the cathode of a voltaic cell.

SAMPLE EXERCISE 20.15

Predict the nature of the corrosion that would take place if an iron gutter were nailed to a house using aluminum nails.

Solution

Analyze: We are asked to describe how the corrosion occurs when Fe is in contact with Al.
Plan: A voltaic cell can be formed at the point of contact of the two metals. The metal that is more easily oxidized will serve as the anode and thereby undergo corrosion. The other metal will serve as the cathode. To determine which metal is oxidized most readily, we must compare their standard reduction potentials from Table 20.1 or Appendix E.
Solve: The standard reduction potentials of Fe and Al are as follows:

$$\text{Cathode:} \quad Fe^{2+}(aq) + 2e^- \longrightarrow Fe(s) \qquad E^\circ_{red} = -0.44\ V$$

$$\text{Anode:} \quad Al(s) \longrightarrow Al^{3+}(aq) + 3e^- \qquad E^\circ_{red} = -1.66\ V$$

Fe will be the cathode because its E°_{red} value is less negative than that of Al. (That is, Fe^{2+} is reduced more readily than Al^{3+}, so Fe is harder to oxidize than Al.) The iron gutter will thus be protected against corrosion in the vicinity of the aluminum nail because the iron serves as the cathode during the electrochemical corrosion. The nail will corrode, however, eventually leaving the gutter on the ground!

PRACTICE EXERCISE

Based on the standard reduction potentials in Table 20.1, which of the following metals could provide cathodic protection to iron: Al, Cu, Ni, Zn?
Answer: Al, Zn

20.9 Electrolysis

Voltaic cells are based on spontaneous oxidation-reduction reactions. Conversely, it is possible to use electrical energy to cause nonspontaneous redox reactions to occur. For example, electricity can be used to decompose molten sodium chloride into its component elements:

$$2NaCl(l) \longrightarrow 2Na(l) + Cl_2(g)$$

Such processes, which are driven by an outside source of electrical energy, are called **electrolysis reactions** and take place in **electrolytic cells**.

An electrolytic cell consists of two electrodes in a molten salt or a solution. A battery or some other source of direct electrical current acts as an electron pump, pushing electrons into one electrode and pulling them from the other. Just as in voltaic cells, the electrode at which the reduction occurs is called the cathode, and the electrode at which oxidation occurs is called the anode. In the electrolysis of molten NaCl, shown in Figure 20.28 ▶, Na^+ ions pick up electrons and are reduced to Na at the cathode. As the Na^+ ions near the cathode are depleted, additional Na^+ ions migrate in. Similarly, there is net movement of Cl^- ions to the anode, where they are oxidized. The electrode reactions for the electrolysis of molten NaCl are summarized as follows:

$$\text{Cathode:} \quad 2Na^+(l) + 2e^- \longrightarrow 2Na(l)$$

$$\text{Anode:} \quad 2Cl^-(l) \longrightarrow Cl_2(g) + 2e^-$$

$$2Na^+(l) + 2Cl^-(l) \longrightarrow 2Na(l) + Cl_2(g)$$

Notice the manner in which the voltage source is connected to the electrodes in Figure 20.28. In a voltaic cell (or any other source of direct current), the electrons move from the negative terminal (Figure 20.6). Thus, the electrode of the electrolytic cell that is connected to the negative terminal of the voltage source is the cathode of the cell; it receives electrons that are used to reduce a substance. The electrons that are removed during the oxidation process at the anode travel to the positive terminal of the voltage source, thus completing the circuit of the cell.

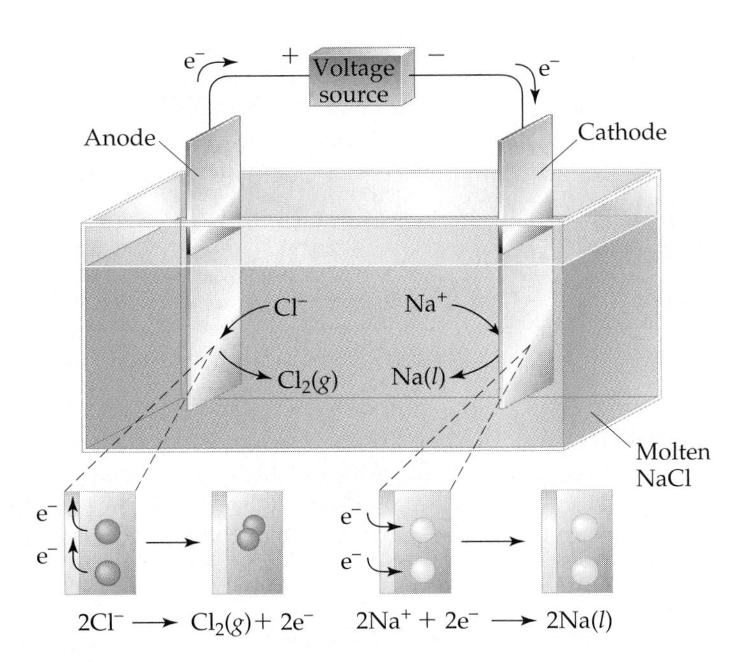

◀ **Figure 20.28** Electrolysis of molten sodium chloride. Cl^- ions are oxidized to $Cl_2(g)$ at the anode, and Na^+ ions are reduced to $Na(l)$ at the cathode.

The electrolysis of molten salts is an important industrial process for the production of active metals such as sodium and aluminum. We have more to say about them in Chapter 23, when we discuss how ores are refined into metals.

Electrolysis of Aqueous Solutions

ANIMATION
Electrolysis of Water

Because of the high melting points of ionic substances, the electrolysis of molten salts requires very high temperatures. (Section 11.8) Do we obtain the same products if we electrolyze the aqueous solution of a salt instead of the molten salt? The electrolysis of an aqueous solution is complicated by the presence of water, because we must consider whether the water is oxidized (to form O_2) or reduced (to form H_2) rather than the ions of the salt.

Suppose we place an aqueous solution of NaF in an electrolytic cell. The possible reactants in the cell are Na^+, F^-, and H_2O. Both Na^+ and H_2O can be reduced, but F^- cannot because the fluoride ion cannot gain further electrons. Thus, the possible reactions at the cathode are

$$Na^+(aq) + e^- \longrightarrow Na(s) \qquad E^\circ_{red} = -2.71\text{ V}$$

$$2H_2O(l) + 2e^- \longrightarrow H_2(g) + 2OH^-(aq) \qquad E^\circ_{red} = -0.83\text{ V}$$

Recall that the more positive (or less negative) the value of E°_{red}, the more favorable the reduction. (Section 20.4) Thus, the reduction of H_2O to H_2 is far more favorable than the reduction of Na^+ to Na. Hydrogen gas is produced at the cathode.

Either F^- or H_2O must be oxidized at the anode because Na^+ cannot lose further electrons.

$$2F^-(aq) \longrightarrow F_2(g) + 2e^- \qquad E^\circ_{red} = +2.87\text{ V}$$

$$2H_2O(l) \longrightarrow O_2(g) + 4H^+(aq) + 4e^- \qquad E^\circ_{red} = +1.23\text{ V}$$

Because oxidation is the reverse of reduction, the more negative (or less positive) the value of E°_{red}, the more favorable the oxidation. Thus, it is far easier to oxidize H_2O than to oxidize F^-. In fact, it is easier still to oxidize the $OH^-(aq)$ that is produced at the cathode.

$$4OH^-(aq) \longrightarrow O_2(g) + 2H_2O(l) + 4e^- \qquad E^\circ_{red} = +0.40\text{ V}$$

Regardless of whether H_2O or OH^- is oxidized, $O_2(g)$ is produced at the anode in preference to $F_2(g)$. Thus, the electrolysis of NaF(*aq*) leads only to the reduction and oxidation of H_2O. The NaF(*aq*) serves only as an electrolyte to allow the electricity to be conducted through the electrolytic cell.

We can use Equation 20.8 to calculate the minimum emf required for the preceding electrolysis.

$$\text{Cathode:}\quad 4H_2O(l) + 4e^- \longrightarrow 2H_2(g) + 4OH^-(aq) \qquad E^\circ_{red} = -0.83\ V$$

$$\text{Anode:}\quad 4OH^-(aq) \longrightarrow O_2(g) + 2H_2O(l) + 4e^- \qquad E^\circ_{red} = +0.40\ V$$

$$2H_2O(l) \longrightarrow 2H_2(g) + O_2(g) \qquad E^\circ_{red} = -1.23\ V$$

Notice that $E^\circ_{cell} = E^\circ_{red}(\text{cathode}) - E^\circ_{red}(\text{anode})$ is negative, so the process is not spontaneous and must be driven by an outside source of energy.

The electrolysis of NaCl(*aq*) leads to a somewhat unexpected result. At the cathode H_2O is reduced to H_2 as with NaF(*aq*). The possible reactions at the anode are the oxidation of $H_2O(l)$ or the oxidation of $Cl^-(aq)$.

$$2H_2O(l) \longrightarrow O_2(g) + 4H^+(aq) + 4e^- \qquad E^\circ_{red} = +1.23\ V$$

$$2Cl^-(aq) \longrightarrow Cl_2(g) + 2e^- \qquad E^\circ_{red} = +1.36\ V$$

Based on the E°_{red} values, we would expect H_2O to be oxidized in preference to Cl^-. Experiments show, however, that Cl^- is usually oxidized rather than H_2O. This counterintuitive result occurs because of the kinetics of the electrode process—in essence, even though the oxidation of H_2O is *thermodynamically* favored, the activation energy for the oxidation of Cl^- is lower, so it is *kinetically* favored. (Section 14.5) The observed cell reactions for the electrolysis of NaCl(*aq*) are summarized as follows:

$$\text{Cathode:}\quad 2H_2O(l) + 2e^- \longrightarrow H_2(g) + 2OH^-(aq)$$

$$\text{Anode:}\quad 2Cl^-(aq) \longrightarrow Cl_2(g) + 2e^-$$

$$2H_2O(l) + 2Cl^-(aq) \longrightarrow H_2(g) + Cl_2(g) + 2OH^-(aq)$$

This electrolytic process is industrially significant because the products—H_2, Cl_2, and NaOH—are important commercial substances.

SAMPLE EXERCISE 20.16

Electrolysis of AgF(*aq*) in an acidic solution leads to the formation of silver metal and oxygen gas. **(a)** Write the half-reaction that occurs at each electrode. **(b)** Calculate the minimum external emf required for this process under standard conditions.

Solution

Analyze: We are told that the electrolysis of an aqueous solution of AgF produces Ag and O_2, and asked to write the half-reactions and to calculate the minimum emf required for the process.

Plan: When an aqueous solution of an ionic compound is electrolyzed, the possible reactants are H_2O and the ions of the solute (in this case Ag^+ and F^-). Because the products are Ag and O_2, the reactants must be Ag^+ and H_2O. Writing the half-reactions for these processes reveals which is oxidation and which is reduction and therefore which occurs at the anode and which at the cathode. The minimum emf is found by using Equation 20.8 to calculate the standard cell potential.

Solve: (a) The cathode is the electrode where reduction occurs. Because $Ag^+(aq)$ is being reduced to Ag(*s*), the half-reaction is the following:

$$Ag^+(aq) + e^- \longrightarrow Ag(s) \qquad E^\circ_{red} = +0.799\ V$$

The standard reduction potential for the reduction of Ag^+ is more positive than for the reduction of either $H_2O(l)$ to $H_2(g)$ ($E^\circ_{red} = -0.83$ V) or $H^+(aq)$ to $H_2(g)$ ($E^\circ_{red} = 0.0$ V). The more positive the value of E°_{red}, the more favorable the reduction. Thus, Ag^+ is the most favorable species for reduction in the solution.

The possible anode half-reactions are the oxidation of F^- to F_2 or the oxidation of H_2O to O_2. (Because the solution is acidic, the concentration of OH^- is expected to be small, so we won't consider the oxidation of OH^-.) The problem states that $O_2(g)$ is produced, so the anode reaction is

$$2H_2O(l) \longrightarrow O_2(g) + 4H^+(aq) + 4e^- \qquad E^\circ_{red} = +1.23\text{ V}$$

As noted previously in our discussion of the electrolysis of NaF(*aq*), the oxidation of H_2O is more favorable than the oxidation of F^-.

(b) The standard cell emf is

$$E^\circ_{cell} = E^\circ_{red}(\text{cathode}) - E^\circ_{red}(\text{anode}) = (+0.799\text{ V}) - (+1.23\text{ V}) = -0.43\text{ V}$$

Because the cell emf is negative, an external emf of at least +0.43 V must be provided to force the electrolysis reaction to occur.

PRACTICE EXERCISE

The electrolysis of $CuCl_2(aq)$ produces Cu(*s*) and $Cl_2(g)$. What is the minimum external emf needed to drive this electrolysis under standard conditions?

Answer: +1.02 V

Electrolysis with Active Electrodes

Thus far in our discussion of electrolysis, we have encountered only electrodes that were *inert*; they did not undergo reaction but merely served as the surface where oxidation and reduction occurred. Several practical applications of electrochemistry, however, are based on *active* electrodes—electrodes that participate in the electrolysis process. *Electroplating*, for example, uses electrolysis to deposit a thin layer of one metal on another metal in order to improve beauty or resistance to corrosion (Figure 20.29 ▼). We can illustrate the principles of electrolysis with active electrodes by describing how to electroplate nickel on a piece of steel.

Figure 20.30 ▶ illustrates the electrolytic cell for our electroplating experiment. The anode of the cell is a strip of nickel metal and the cathode is the piece of steel that will be electroplated. The electrodes are immersed in a solution of $NiSO_4(aq)$. What happens at the electrodes when the external voltage source is turned on? Reduction will occur at the cathode. The standard reduction potential of Ni^{2+}(E°_{red} = −0.28 V) is less negative than that of H_2O (E°_{red} = −0.83 V), so Ni^{2+} will be preferentially reduced at the cathode.

At the anode we need to consider which substances can be oxidized. For the $NiSO_4(aq)$ solution, only the H_2O solvent is readily oxidized because neither Na^+ nor the $SO_4{}^{2-}$ can be oxidized (both already have their elements in their highest possible oxidation state). The Ni atoms in the anode, however, can undergo oxidation. Thus, the two possible processes are:

$$2H_2O(l) \longrightarrow O_2(g) + 4H^+(aq) + 4e^- \qquad E^\circ_{red} = +1.23\text{ V}$$

$$Ni(s) \longrightarrow Ni^{2+}(aq) + 2e^- \qquad E^\circ_{red} = -0.28\text{ V}$$

(a)

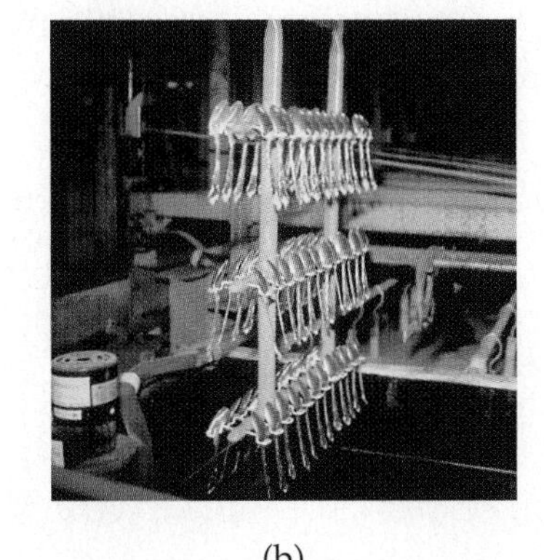

(b)

▲ **Figure 20.29** Electroplating of silverware. (a) The silverware is being withdrawn from the electroplating bath. (b) The polished final product.

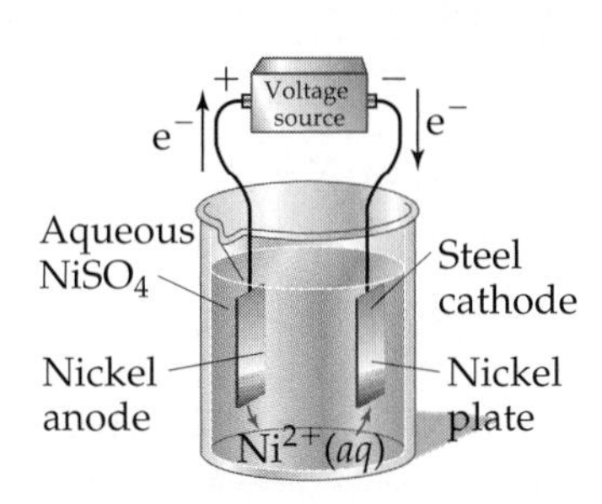

▲ **Figure 20.30** Electrolytic cell with an active metal electrode. Nickel dissolves from the anode to form $Ni^{2+}(aq)$. At the cathode $Ni^{2+}(aq)$ is reduced and forms a nickel "plate" on the cathode.

Because this is an oxidation, the half-reaction with the more negative value of E°_{red} is favored. We therefore expect Ni(*s*) to be oxidized at the anode. We can summarize the electrode reactions as follows:

Cathode (steel strip): $Ni^{2+}(aq) + 2e^- \longrightarrow Ni(s)$ $\quad E^\circ_{red} = -0.28$ V

Anode (nickel strip): $Ni(s) \longrightarrow Ni^{2+}(aq) + 2e^-$ $\quad E^\circ_{red} = -0.28$ V

If we look at the overall reaction, it appears as if nothing has been accomplished. During the electrolysis, however, we are transferring Ni atoms from the Ni anode to the steel cathode, plating the steel electrode with a thin layer of nickel atoms. The standard emf for the overall reaction is $E^\circ_{cell} = E^\circ_{red}(\text{cathode}) - E^\circ_{red}(\text{anode}) = 0$ V. Only a small emf is needed to provide the "push" to transfer the nickel atoms from one electrode to the other. In Chapter 23 we will explore further the utility of electrolysis with active electrodes as a means of purifying crude metals.

Quantitative Aspects of Electrolysis

The stoichiometry of a half-reaction shows how many electrons are needed to achieve an electrolytic process. For example, the reduction of Na^+ to Na is a one-electron process:

$$Na^+ + e^- \longrightarrow Na$$

Thus, 1 mol of electrons will plate out 1 mol of Na metal, 2 mol of electrons will plate out 2 mol of Na metal, and so forth. Similarly, 2 mol of electrons are required to produce 1 mol of copper from Cu^{2+}, and 3 mol of electrons are required to produce 1 mol of aluminum from Al^{3+}.

$$Cu^{2+} + 2e^- \longrightarrow Cu$$

$$Al^{3+} + 3e^- \longrightarrow Al$$

For any half-reaction, the amount of a substance that is reduced or oxidized in an electrolytic cell is directly proportional to the number of electrons passed into the cell.

The quantity of charge passing through an electrical circuit, such as that in an electrolytic cell, is generally measured in *coulombs*. As noted in Section 20.5, the charge on 1 mol of electrons is 96,500 C (1 faraday).

$$1\ F = 96{,}500\ \text{C/mol e}^-$$

A coulomb is the quantity of charge passing a point in a circuit in 1 s when the current is 1 ampere (A).* Therefore, the number of coulombs passing through a cell can be obtained by multiplying the amperage and the elapsed time in seconds.

$$\text{Coulombs} = \text{amperes} \times \text{seconds} \qquad [20.20]$$

ACTIVITY
Electrolysis

Figure 20.31 ▼ shows how the quantities of substances produced or consumed in electrolysis are related to the quantity of electrical charge that is used. The same relationships can also be applied to voltaic cells.

* Conversely, current is the rate of flow of electricity. An ampere (often referred to merely as an amp) is the current associated with the flow of 1 C past a point each second.

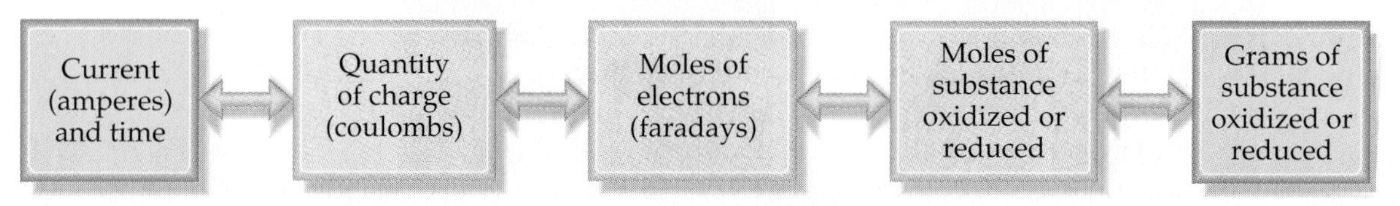

▲ **Figure 20.31** The steps relating the quantity of electrical charge used in electrolysis to the amounts of substances oxidized or reduced.

SAMPLE EXERCISE 20.17

Calculate the number of grams of aluminum produced in 1.00 hr by the electrolysis of molten $AlCl_3$ if the electrical current is 10.0 A.

Solution

Analyze: We are told that $AlCl_3$ is electrolyzed to form Al and asked to calculate the number of grams of Al produced in 1.00 hr with 10.0 A.

Plan: Figure 20.31 provides a road map of the problem. First, the product of the amperage and the time in seconds gives the number of coulombs of electrical charge being used (Equation 20.20). Second, the coulombs can be converted to faradays ($1\ F = 96{,}500\ C/mol\ e^-$), which tells us the number of moles of electrons being supplied. Third, reduction of 1 mol of Al^{3+} to Al requires 3 mol of electrons. Hence we can use the number of moles of electrons to calculate the number of moles of Al metal it produces. Finally, we convert moles of Al into grams.

Solve: First, we calculate the coulombs of electrical charge that are passed into the electrolytic cell:

$$\text{Coulombs} = \text{amperes} \times \text{seconds}$$

$$= (10.0\ A)(1.00\ hr)\left(\frac{3600\ s}{1\ hr}\right)\left(\frac{1\ C}{1\ A\text{-}s}\right) = 3.60 \times 10^4\ C$$

Second, we calculate the number of moles of electrons (the number of faradays of electrical charge) that pass into the cell:

$$\text{Moles } e^- = (3.60 \times 10^4\ C)\left(\frac{1\ mol\ e^-}{96{,}500\ C}\right) = 0.373\ mol\ e^-$$

Third, we relate the number of moles of electrons to the number of moles of aluminum being formed, using the half-reaction for the reduction of Al^{3+}:

$$Al^{3+} + 3e^- \longrightarrow Al$$

Thus, 3 mol of electrons (3 *F* of electrical charge) are required to form 1 mol of Al:

$$\text{Moles Al} = (0.373\ mol\ e^-)\left(\frac{1\ mol\ Al}{3\ mol\ e^-}\right) = 0.124\ mol\ Al$$

Finally, we convert moles to grams:

$$\text{Grams Al} = (0.124\ mol\ Al)\left(\frac{27.0\ g\ Al}{1\ mol\ Al}\right) = 3.36\ g\ Al$$

Because each step involves a multiplication by a new factor, the steps can be combined into a single sequence of factors:

$$\text{Grams Al} = (3.60 \times 10^4\ C)\left(\frac{1\ mol\ e^-}{96{,}500\ C}\right)\left(\frac{1\ mol\ Al}{3\ mol\ e^-}\right)\left(\frac{27.0\ g\ Al}{1\ mol\ Al}\right) = 3.36\ g\ Al$$

PRACTICE EXERCISE

(a) The half-reaction for formation of magnesium metal upon electrolysis of molten $MgCl_2$ is $Mg^{2+} + 2e^- \longrightarrow Mg$. Calculate the mass of magnesium formed upon passage of a current of 60.0 A for a period of 4.00×10^3 s. **(b)** How many seconds would be required to produce 50.0 g of Mg from $MgCl_2$ if the current is 100.0 A?

Answers: **(a)** 30.2 g of Mg; **(b)** 3.97×10^3 s

Electrical Work

We have already seen that a positive value of E is associated with a negative value for the free-energy change and, thus, with a spontaneous process. We also know that for any spontaneous process, ΔG is a measure of the maximum useful work, w_{max}, that can be extracted from the process: $\Delta G = w_{max}$. (Section 19.5) Because $\Delta G = -nFE$, the maximum useful electrical work obtainable from a voltaic cell is

$$w_{max} = -nFE \qquad [20.21]$$

The cell emf, E, is a positive number for a voltaic cell, so w_{max} will be a negative number for a voltaic cell. Work done *by* a system *on* its surroundings is indicated by a negative sign for w. (Section 5.2) Thus, the negative value for w_{max} means that a voltaic cell does work on its surroundings.

In an electrolytic cell we use an external source of energy to bring about a nonspontaneous electrochemical process. In this case, ΔG is positive and E_{cell} is negative. To force the process to occur, we need to apply an external potential, E_{ext}, which must be larger in magnitude than E_{cell}: $E_{ext} > -E_{cell}$. For example, if a nonspontaneous process has $E_{cell} = -0.9$ V, then the external potential E_{ext} must be greater than 0.9 V in order for the process to occur.

When an external potential E_{ext} is applied to a cell, the surroundings are doing work on the system. The amount of work performed is given by

$$w = nFE_{ext} \qquad [20.22]$$

Unlike Equation 20.21, there is no minus sign in Equation 20.22. The work calculated in Equation 20.22 will be a positive number because the surroundings are doing work on the system. The quantity n in Equation 20.22 is the number of moles of electrons forced into the system by the external potential. The product $n \times F$ is the total electrical charge supplied to the system by the external source of electricity.

Electrical work can be expressed in energy units of watts times time. The **watt** (W) is a unit of electrical power (that is, the rate of energy expenditure).

$$1\ \text{W} = 1\ \text{J/s}$$

Thus, a watt-second is a joule. The unit employed by electric utilities is the kilowatt-hour (kWh), which equals 3.6×10^6 J.

$$1\ \text{kWh} = (1000\ \text{W})(1\ \text{hr})\left(\frac{3600\ \text{s}}{1\ \text{hr}}\right)\left(\frac{1\ \text{J/s}}{1\ \text{W}}\right) = 3.6 \times 10^6\ \text{J} \qquad [20.23]$$

Using these considerations, we can calculate the maximum work obtainable from the voltaic cells and the minimum work required to bring about desired electrolysis reactions.

SAMPLE EXERCISE 20.18

Calculate the number of kilowatt-hours of electricity required to produce 1.0×10^3 kg of aluminum by electrolysis of Al^{3+} if the applied emf is 4.50 V.

Solution

Analyze: We are given the mass of Al produced from Al^{3+} and the applied voltage and asked to calculate the energy, in kilowatt-hours, required for the reduction.

Plan: From the mass of Al, we can calculate first the number of moles of Al, then the number of coulombs required to obtain that mass. We can then use Equation 20.22 ($w = nFE_{ext}$), where nF is the total charge in coulombs and E_{ext} is the applied potential, 4.50 V.

Solve: First, we need to calculate nF, the number of coulombs required.

$$\text{Coulombs} = (1.00 \times 10^3\ \text{kg Al})\left(\frac{1000\ \text{g Al}}{1\ \text{kg Al}}\right)\left(\frac{1\ \text{mol Al}}{27.0\ \text{g Al}}\right)\left(\frac{3\ F}{1\ \text{mol Al}}\right)\left(\frac{96{,}500\ \text{C}}{1\ F}\right)$$

$$= 1.07 \times 10^{10}\ \text{C}$$

We can now employ Equation 20.22 to calculate w. In doing so, we must apply several conversion factors, including Equation 20.23, which gives the conversion between kilowatt-hours and joules:

$$\text{Kilowatt-hours} = (1.07 \times 10^{10}\text{C})(4.50\ \text{V})\left(\frac{1\ \text{J}}{1\ \text{C-V}}\right)\left(\frac{1\ \text{kWh}}{3.6 \times 10^6\ \text{J}}\right)$$

$$= 1.34 \times 10^4\ \text{kWh}$$

Comment: This quantity of energy does not include the energy used to mine, transport, and process the aluminum ore, and to keep the electrolysis bath molten during electrolysis. A typical electrolytic cell used to reduce aluminum ore to aluminum metal is only 40% efficient, with 60% of the electrical energy being dissipated as heat. It therefore requires on the order of 33 kWh of electricity to produce 1 kg of aluminum. The aluminum industry consumes about 2% of the electrical energy generated in the United States. Because this is used mainly to reduce aluminum, recycling this metal saves large quantities of energy.

PRACTICE EXERCISE

Calculate the number of kilowatt-hours of electricity required to produce 1.00 kg of Mg from electrolysis of molten $MgCl_2$ if the applied emf is 5.00 V. Assume that the process is 100% efficient.
Answer: 11.0 kWh

SAMPLE INTEGRATIVE EXERCISE 20: Putting Concepts Together

The K_{sp} at 298 K for iron(II) fluoride is 2.4×10^{-6}. **(a)** Write a half-reaction that gives the likely products of the two-electron reduction of $FeF_2(s)$ in water. **(b)** Use the K_{sp} value and the standard reduction potential of $Fe^{2+}(aq)$ to calculate the standard reduction potential for the half-reaction in part (a). **(c)** Rationalize the difference in the reduction potential for the half-reaction in part (a) with that for $Fe^{2+}(aq)$.

Solution **(a)** Iron(II) fluoride is an ionic substance that consists of Fe^{2+} and F^- ions. We are asked to predict where two electrons could be added to FeF_2. We can't envision adding the electrons to the F^- ions to form F^{2-}, so it seems likely that we could reduce the Fe^{2+} ions to $Fe(s)$. We therefore predict the following half-reaction:

$$FeF_2(s) + 2e^- \longrightarrow Fe(s) + 2F^-(aq)$$

In general, the reduction of an ionic substance forms the reduced form of the cation of the salt. We saw an example of this type of reaction in reverse for the anode reaction during the discharge of the lead-acid battery: $Pb(s) + SO_4^{2-}(aq) \longrightarrow PbSO_4(s) + 2e^-$.

(b) The K_{sp} value refers to the following equilibrium (Section 17.4):

$$FeF_2(s) \rightleftharpoons Fe^{2+}(aq) + 2F^-(aq) \qquad K_{sp} = [Fe^{2+}][F^-]^2 = 2.4 \times 10^{-6}$$

To relate this value to the standard reduction potential in part (a), we first need to recognize that we can write the equilibrium reaction as the sum of two half-reactions. The reduction step is the reaction from part (a), and the oxidation step is the oxidation of $Fe(s)$ to $Fe^{2+}(aq)$.

$$\begin{array}{llll} \text{Reduction:} & FeF_2(s) + 2e^- \longrightarrow Fe(s) + 2F^-(aq) & & E^\circ_{red} = ? \\ \text{Oxidation:} & Fe(s) \longrightarrow Fe^{2+}(aq) + 2e^- & & E^\circ_{red} = -0.440\text{ V} \\ \hline & FeF_2(s) \longrightarrow Fe^{2+}(aq) + 2F^-(aq) & & \end{array}$$

Notice that we have written expressions for the E°_{red} values for the two half-reactions. The value for the reduction step is what we seek; the value for the oxidation step is obtained from Appendix E.

Second, we can relate the K_{sp} value to the standard emf, E°, by using Equation 20.18. According to the half-reactions, $n = 2$, so

$$\log K_{sp} = \frac{2E^\circ}{0.0592}$$

Solving for E°:

$$E^\circ = \tfrac{1}{2}(0.0592)(\log K_{sp}) = \tfrac{1}{2}(0.0592)(\log 2.4 \times 10^{-6})$$
$$= \tfrac{1}{2}(0.0592)(-5.62) = -0.166\text{ V}$$

Thus, the very small value of K_{sp} corresponds to a negative value for E°.

Finally, we can calculate the value of E° from the E°_{red} values for the half-reactions by using Equation 20.10. By doing so, we can solve for the E°_{red} value for the reduction of FeF_2:

$$E^\circ = E^\circ_{red}(\text{reduction process}) - E^\circ_{red}(\text{oxidation process})$$
$$-0.166\text{ V} = E^\circ_{red}(\text{reduction process}) - (-0.440\text{ V})$$
$$E^\circ_{red}(\text{reduction process}) = (-0.440\text{ V}) + (-0.166\text{ V}) = -0.606\text{ V}$$

(c) The standard reduction potential for $FeF_2(-0.606V)$ is more negative than that for $Fe^{2+}(-0.440V)$, so the reduction of FeF_2 is the less favorable process. When FeF_2 is reduced, we not only reduce the Fe^{2+} ions, but also break up the ionic solid. The latter process requires overcoming the lattice energy of FeF_2. (Section 8.2) Because this additional energy must be overcome, the reduction of FeF_2 is less favorable than the reduction of Fe^{2+}.

Summary and Key Terms

Introduction and Section 20.1 In this chapter we have focused on **electrochemistry**, the branch of chemistry that relates electricity and chemical reactions. Electrochemistry involves oxidation-reduction reactions, also called redox reactions. These reactions involve a change in the oxidation state of one or more elements. In every oxidation-reduction reaction one substance is oxidized (its oxidation state increases) and one substance is reduced (its oxidation state decreases). The substance that is oxidized is referred to as a **reducing agent**, or **reductant**, because it causes the reduction of some other substance. Similarly, the substance that is reduced is referred to as an **oxidizing agent**, or **oxidant**, because it causes the oxidation of some other substance.

Section 20.2 An oxidization-reduction reaction can be balanced by dividing the reaction into two **half-reactions**, one for the oxidation and one for the reduction. A half-reaction is a balanced chemical equation that includes electrons. In oxidation half-reactions the electrons are on the product (right) side of the reaction; we can envision that these electrons are transferred from a substance when it is oxidized. In reduction half-reactions the electrons are on the reactant (left) side of the reaction. Each half-reaction is balanced separately, and the two are brought together with proper coefficients to balance the electrons on each side of the equation.

Section 20.3 A **voltaic** (or **galvanic**) **cell** uses a spontaneous oxidation-reduction reaction to generate electricity. In a voltaic cell the oxidation and reduction half-reactions often occur in separate compartments. Each compartment has a solid surface called an electrode, where the half-reaction occurs. The electrode where oxidation occurs is called the **anode**; reduction occurs at the **cathode**. The electrons released at the anode flow through an external circuit (where they do electrical work) to the cathode. Electrical neutrality in the solution is maintained by the migration of ions between the two compartments through a device such as a salt bridge.

Section 20.4 A voltaic cell generates an **electromotive force (emf)** that moves the electrons from the anode to the cathode through the external circuit. The origin of emf is a difference in the electrical potential energy of the two electrodes in the cell. The emf of a cell is called its **cell potential**, E_{cell}, and is measured in volts. The cell potential under standard conditions is called the **standard emf** or the **standard cell potential** and is denoted E°_{cell}.

A **standard reduction potential**, E°_{red}, can be assigned for an individual half-reaction. This is achieved by comparing the potential of the half-reaction to that of the **standard hydrogen electrode** (SHE), which is defined to have $E^{\circ}_{red} = 0$ V and is based on the following half-reaction:

$$2H^+(aq, 1\,M) + 2e^- \longrightarrow H_2(g, 1\text{ atm}) \qquad E^{\circ}_{red} = 0\text{ V}$$

The standard cell potential of a voltaic cell is the difference between the standard reduction potentials of the half-reactions that occur at the cathode and the anode: $E^{\circ}_{cell} = E^{\circ}_{red}(\text{cathode}) - E^{\circ}_{red}(\text{anode})$. The value of E°_{cell} is positive for a voltaic cell.

For a reduction half-reaction, E°_{red} is a measure of the tendency of the reduction to occur; the more positive the value for E°_{red}, the greater the tendency of the substance to be reduced. Thus, E°_{red} provides a measure of the oxidizing strength of a substance. Fluorine (F_2) has the most positive value for E°_{red} and is the strongest oxidizing agent. Substances that are strong oxidizing agents produce products that are weak reducing agents, and vice versa.

Section 20.5 The emf, E, is related to the change in Gibbs free energy, ΔG: $\Delta G = -nFE$, where n is the number of electrons transferred during the redox process and F is a unit called the **faraday**. The faraday is the amount of charge on 1 mol of electrons: $1F = 96{,}500$ C/mol. Because E is related to ΔG, the sign of E indicates whether a redox process is spontaneous: $E > 0$ indicates a spontaneous process, and $E < 0$ indicates a nonspontaneous one.

Section 20.6 The emf of a redox reaction varies with temperature and with the concentrations of reactants and products. The **Nernst equation** relates the emf under nonstandard conditions to the standard emf and the reaction quotient Q:

$$E = E^{\circ} - (RT/nF)\ln Q = E^{\circ} - (0.0592/n)\log Q$$

The factor 0.0592 is valid when $T = 298$ K. A **concentration cell** is a voltaic cell in which the same half-reaction occurs at both the anode and cathode but with different concentrations of reactants in each compartment.

At equilibrium, $Q = K_{eq}$ and $E = 0$. The standard emf is therefore related to the equilibrium constant. At $T = 298$ K, the relation is

$$\log K_{eq} = nE^{\circ}/0.0592$$

Section 20.7 A **battery** is a self-contained electrochemical power source that contains one or more voltaic cells. Batteries are based on a variety of different redox reactions. Several common batteries were discussed. The lead-acid battery, the nickel-cadmium battery, the nickel-metal-hydride battery, and the lithium-ion battery are examples of rechargeable batteries. The common alkaline dry cell is not rechargeable. **Fuel cells** are voltaic cells that utilize redox reactions involving conventional fuels, such as H_2 and CH_4.

Section 20.8 Electrochemical principles help us understand **corrosion**, undesirable redox reactions in which a metal is attacked by some substance in its environment.

The corrosion of iron into rust is caused by the presence of water and oxygen, and it is accelerated by the presence of electrolytes, such as road salt. The protection of a metal by putting it in contact with another metal that more readily undergoes oxidation is called **cathodic protection**. Galvanized iron, for example, is coated with a thin layer of zinc; because zinc is oxidized more readily than iron, the zinc serves as a sacrificial anode in the redox reaction.

Section 20.9 An **electrolysis reaction**, which is carried out in an **electrolytic cell**, employs an external source of electricity to drive a nonspontaneous electrochemical reaction. The negative terminal of the external source is connected to the cathode of the cell, and the positive terminal to the anode. The current-carrying medium within an electrolytic cell may be either a molten salt or an electrolyte solution. The products of electrolysis can generally be predicted by comparing the reduction potentials associated with possible oxidation and reduction processes. The electrodes in an electrolytic cell can be active, meaning that the electrode can be involved in the electrolysis reaction. Active electrodes are important in electroplating and in metallurgical processes.

The quantity of substances formed during electrolysis can be calculated by considering the number of electrons involved in the redox reaction and the amount of electrical charge that passes into the cell. The maximum amount of electrical work produced by a voltaic cell is given by the product of the total charge delivered, nF, and the emf, E: $w_{max} = -nFE$. The work performed in an electrolysis is given by $w = nFE_{ext}$, where E_{ext} is the applied external potential. The **watt** is a unit of power: 1 W = 1 J/s. Electrical work is often measured in kilowatt-hours.

Exercises

Oxidation-Reduction Reactions

20.1 **(a)** What is meant by the term *oxidation*? **(b)** On which side of an oxidation half-reaction do the electrons appear? **(c)** What is meant by the term *oxidant*?

20.2 **(a)** What is meant by the term *reduction*? **(b)** On which side of a reduction half-reaction do the electrons appear? **(c)** What is meant by the term *reductant*?

20.3 In each of the following balanced oxidation-reduction equations, identify those elements that undergo changes in oxidation number and indicate the magnitude of the change in each case.
(a) $I_2O_5(s) + 5CO(g) \longrightarrow I_2(s) + 5CO_2(g)$
(b) $2Hg^{2+}(aq) + N_2H_4(aq) \longrightarrow 2Hg(l) + N_2(g) + 4H^+(aq)$
(c) $3H_2S(aq) + 2H^+(aq) + 2NO_3^-(aq) \longrightarrow 3S(s) + 2NO(g) + 4H_2O(l)$
(d) $Ba^{2+}(aq) + 2OH^-(aq) + H_2O_2(aq) + 2ClO_2(aq) \longrightarrow Ba(ClO_2)_2(s) + 2H_2O(l) + O_2(g)$

20.4 Indicate whether the following balanced equations involve oxidation-reduction. If they do, identify the elements that undergo changes in oxidation number.
(a) $PBr_3(l) + 3H_2O(l) \longrightarrow H_3PO_3(aq) + 3HBr(aq)$
(b) $NaI(aq) + 3HOCl(aq) \longrightarrow NaIO_3(aq) + 3HCl(aq)$
(c) $3SO_2(g) + 2HNO_3(aq) + 2H_2O(l) \longrightarrow 3H_2SO_4(aq) + 2NO(g)$
(d) $2H_2SO_4(aq) + 2NaBr(s) \longrightarrow Br_2(l) + SO_2(g) + Na_2SO_4(aq) + 2H_2O(l)$

20.5 At 900°C titanium tetrachloride vapor reacts with molten magnesium to form solid titanium metal and molten magnesium chloride. **(a)** Write a balanced equation for this reaction. **(b)** Which substance is the reductant, and which is the oxidant?

20.6 Hydrazine (N_2H_4) and dinitrogen tetroxide (N_2O_4) form a self-igniting mixture that has been used as a rocket propellant. The reaction products are N_2 and H_2O. **(a)** Write a balanced chemical equation for this reaction. **(b)** Which substance serves as the reducing agent, and which as the oxidizing agent?

20.7 Complete and balance the following half-reactions. In each case indicate whether oxidation or reduction occurs.
(a) $Sn^{2+}(aq) \longrightarrow Sn^{4+}(aq)$
(b) $TiO_2(s) \longrightarrow Ti^{2+}(aq)$ (acidic solution)
(c) $ClO_3^-(aq) \longrightarrow Cl^-(aq)$ (acidic solution)
(d) $OH^-(aq) \longrightarrow O_2(g)$ (basic solution)
(e) $SO_3^{2-}(aq) \longrightarrow SO_4^{2-}(aq)$ (basic solution)

20.8 Complete and balance the following half-reactions. In each case indicate whether oxidation or reduction occurs.
(a) $Mo^{3+}(aq) \longrightarrow Mo(s)$
(b) $H_2SO_3(aq) \longrightarrow SO_4^{2-}(aq)$ (acidic solution)
(c) $NO_3^-(aq) \longrightarrow NO(g)$ (acidic solution)
(d) $Mn^{2+}(aq) \longrightarrow MnO_2(s)$ (basic solution)
(e) $Cr(OH)_3(s) \longrightarrow CrO_4^{2-}(aq)$ (basic solution)

20.9 Complete and balance the following equations, and identify the oxidizing and reducing agents:
(a) $Cr_2O_7^{2-}(aq) + I^-(aq) \longrightarrow Cr^{3+}(aq) + IO_3^-(aq)$ (acidic solution)
(b) $MnO_4^-(aq) + CH_3OH(aq) \longrightarrow Mn^{2+}(aq) + HCO_2H(aq)$ (acidic solution)
(c) $I_2(s) + OCl^-(aq) \longrightarrow IO_3^-(aq) + Cl^-(aq)$ (acidic solution)
(d) $As_2O_3(s) + NO_3^-(aq) \longrightarrow H_3AsO_4(aq) + N_2O_3(aq)$ (acidic solution)
(e) $MnO_4^-(aq) + Br^-(aq) \longrightarrow MnO_2(s) + BrO_3^-(aq)$ (basic solution)
(f) $Pb(OH)_4^{2-}(aq) + ClO^-(aq) \longrightarrow PbO_2(s) + Cl^-(aq)$ (basic solution)

20.10 Complete and balance the following equations, and identify the oxidizing and reducing agents:
(a) $NO_2^-(aq) + Cr_2O_7^{2-}(aq) \longrightarrow Cr^{3+}(aq) + NO_3^-(aq)$ (acidic solution)
(b) $As(s) + ClO_3^-(aq) \longrightarrow H_3AsO_3(aq) + HClO(aq)$ (acidic solution)

(c) $Cr_2O_7^{2-}(aq) + CH_3OH(aq) \longrightarrow HCO_2H(aq) + Cr^{3+}(aq)$ (acidic solution)
(d) $MnO_4^-(aq) + Cl^-(aq) \longrightarrow Mn^{2+}(aq) + Cl_2(aq)$ (acidic solution)
(e) $H_2O_2(aq) + ClO_2(aq) \longrightarrow ClO_2^-(aq) + O_2(g)$ (basic solution)
(f) $H_2O_2(aq) + Cl_2O_7(aq) \longrightarrow ClO_2^-(aq) + O_2(g)$ (basic solution)

Voltaic Cells; Cell Potential

20.11 (a) What are the similarities and differences between Figure 20.3 and Figure 20.4? (b) Why are Na^+ ions drawn into the cathode compartment as the voltaic cell shown in Figure 20.5 operates?

20.12 (a) What is the role of the porous glass disc shown in Figure 20.4? (b) Why do NO_3^- ions migrate into the anode compartment as the voltaic cell shown in Figure 20.5 operates?

20.13 A voltaic cell similar to that shown in Figure 20.5 is constructed. One electrode compartment consists of a silver strip placed in a solution of $AgNO_3$, and the other has an iron strip placed in a solution of $FeCl_2$. The overall cell reaction is

$$Fe(s) + 2Ag^+(aq) \longrightarrow Fe^{2+}(aq) + 2Ag(s)$$

(a) Write the half-reactions that occur in the two electrode compartments. (b) Which electrode is the anode, and which is the cathode? (c) Indicate the signs of the electrodes. (d) Do electrons flow from the silver electrode to the iron electrode or from the iron to the silver? (e) In which directions do the cations and anions migrate through the solution?

20.14 A voltaic cell similar to that shown in Figure 20.5 is constructed. One electrode compartment consists of an aluminum strip placed in a solution of $Al(NO_3)_3$, and the other has a nickel strip placed in a solution of $NiSO_4$. The overall cell reaction is

$$2Al(s) + 3Ni^{2+}(aq) \longrightarrow 2Al^{3+}(aq) + 3Ni(s)$$

(a) Write the half-reactions that occur in the two electrode compartments. (b) Which electrode is the anode, and which is the cathode? (c) Indicate the signs of the electrodes. (d) Do electrons flow from the aluminum electrode to the nickel electrode or from the nickel to the aluminum? (e) In which directions do the cations and anions migrate through the solution?

20.15 (a) What is meant by the term *electromotive force*? (b) What is the definition of the *volt*? (c) What is meant by the term *cell potential*?

20.16 (a) Which electrode in Figure 20.4 corresponds to the higher potential energy for the electrons? (b) What are the units for electrical potential? How does this unit relate to energy expressed in joules? (c) What is special about a *standard* cell potential?

20.17 (a) Write the half-reaction that occurs at a hydrogen electrode when it serves as the cathode of a voltaic cell. (b) What is *standard* about the standard hydrogen electrode? (c) What is the role of the platinum foil in a standard hydrogen electrode?

20.18 (a) Write the half-reaction that occurs at a hydrogen electrode when it serves as the anode of a voltaic cell. (b) The platinum electrode in an SHE is specially prepared to have a large surface area. Propose a reason that this is done. (c) Sketch a standard hydrogen electrode.

20.19 (a) What is a *standard reduction potential*? (b) What is the standard reduction potential of a standard hydrogen electrode? (c) Based on the standard reduction potentials listed in Appendix E, which is the more favorable process: the reduction of $Ag^+(aq)$ to $Ag(s)$ or the reduction of $Sn^{2+}(aq)$ to $Sn(s)$?

20.20 (a) Why is it impossible to measure the standard reduction potential of a single half-reaction? (b) Describe how the standard reduction potential of a half-reaction can be determined. (c) By using data in Appendix E, determine which is the more unfavorable reduction: $Cd^{2+}(aq)$ to $Cd(s)$ or $Ca^{2+}(aq)$ to $Ca(s)$.

20.21 A voltaic cell that uses the reaction

$$Tl^{3+}(aq) + 2Cr^{2+}(aq) \longrightarrow Tl^+(aq) + 2Cr^{3+}(aq)$$

has a measured standard cell potential of +1.19 V. (a) Write the two half-cell reactions. (b) By using data from Appendix E, determine E°_{red} for the reduction of $Tl^{3+}(aq)$ to $Tl^+(aq)$. (c) Sketch the voltaic cell, label the anode and cathode, and indicate the direction of electron flow.

20.22 A voltaic cell that uses the reaction

$$PdCl_4^{2-}(aq) + Cd(s) \longrightarrow Pd(s) + 4Cl^-(aq) + Cd^{2+}(aq)$$

has a measured standard cell potential of +1.03V. (a) Write the two half-cell reactions. (b) By using data from Appendix E, determine E°_{red} for the reaction involving Pd. (c) Sketch the voltaic cell, label the anode and cathode, and indicate the direction of electron flow.

20.23 Using standard reduction potentials (Appendix E), calculate the standard emf for each of the following reactions:
(a) $Cl_2(g) + 2I^-(aq) \longrightarrow 2Cl^-(aq) + I_2(s)$
(b) $Ni(s) + 2Ce^{4+}(aq) \longrightarrow Ni^{2+}(aq) + 2Ce^{3+}(aq)$
(c) $Fe(s) + 2Fe^{3+}(aq) \longrightarrow 3Fe^{2+}(aq)$
(d) $2Al^{3+}(aq) + 3Ca(s) \longrightarrow 2Al(s) + 3Ca^{2+}(aq)$

20.24 Using data in Appendix E, calculate the standard emf for each of the following reactions:
(a) $H_2(g) + F_2(g) \longrightarrow 2H^+(aq) + 2F^-(aq)$
(b) $Cu(s) + Ba^{2+}(aq) \longrightarrow Cu^{2+}(aq) + Ba(s)$
(c) $3Fe^{2+}(aq) \longrightarrow Fe(s) + 2Fe^{3+}(aq)$
(d) $Hg_2^{2+}(aq) + 2Cu^+(aq) \longrightarrow 2Hg(l) + 2Cu^{2+}(aq)$

20.25 The standard reduction potentials of the following half-reactions are given in Appendix E:

$$Ag^+(aq) + e^- \longrightarrow Ag(s)$$
$$Cu^{2+}(aq) + 2e^- \longrightarrow Cu(s)$$
$$Ni^{2+}(aq) + 2e^- \longrightarrow Ni(s)$$
$$Cr^{3+}(aq) + 3e^- \longrightarrow Cr(s)$$

(a) Determine which combination of these half-cell reactions leads to the cell reaction with the largest positive cell emf, and calculate the value. (b) Determine which combination of these half-cell reactions leads to the cell reaction with the smallest positive cell emf, and calculate the value.

20.26 Given the following half-reactions and associated standard reduction potentials:

$AuBr_4^-(aq) + 3e^- \longrightarrow Au(s) + 4Br^-(aq)$ $\quad E^\circ_{red} = -0.858\ V$

$Eu^{3+}(aq) + e^- \longrightarrow Eu^{2+}(aq)$ $\quad E^\circ_{red} = -0.43\ V$

$IO^-(aq) + H_2O(l) + 2e^- \longrightarrow I^-(aq) + 2OH^-(aq)$ $\quad E^\circ_{red} = +0.49\ V$

$Sn^{2+}(aq) + 2e^- \longrightarrow Sn(s)$ $\quad E^\circ_{red} = -0.14\ V$

(a) Write the cell reaction for the combination of these half-cell reactions that leads to the largest positive cell emf, and calculate the value. **(b)** Write the cell reaction for the combination of half-cell reactions that leads to the smallest positive cell emf, and calculate that value.

20.27 The half-reactions in a voltaic cell are the following (or their reverse):

$$Sn^{4+}(aq) + 2e^- \longrightarrow Sn^{2+}(aq)$$
$$MnO_4^-(aq) + 8H^+(aq) + 5e^- \longrightarrow Mn^{2+}(aq) + 4H_2O(l)$$

(a) By referring to Appendix E, select the reduction process that is more favorable. **(b)** What reaction occurs at the cathode of the cell? **(c)** What reaction occurs at the anode? **(d)** Write a balanced equation for the overall cell reaction. **(e)** What is the standard cell potential?

20.28 A voltaic cell similar to the one shown in Figure 20.11 is constructed. One electrode compartment has an aluminum strip in contact with a solution of $Al(NO_3)_3$, and the other is a standard hydrogen electrode. **(a)** By referring to Table 20.1 or Appendix E, write the half-reactions involved and determine which electrode is the anode and which is the cathode. **(b)** Will the aluminum strip gain or lose mass as the cell operates? **(c)** Write a balanced equation for the overall cell reaction. **(d)** What is the standard emf of the cell?

20.29 A 1 *M* solution of $Cu(NO_3)_2$ is placed in a beaker with a strip of Cu metal. A 1 *M* solution of $SnSO_4$ is placed in a second beaker with a strip of Sn metal. The two beakers are connected by a salt bridge, and the two metal electrodes are linked by wires to a voltmeter. **(a)** Which electrode serves as the anode, and which as the cathode? **(b)** Which electrode gains mass and which loses mass as the cell reaction proceeds? **(c)** Write the equation for the overall cell reaction. **(d)** What is the emf generated by the cell under standard conditions?

20.30 A voltaic cell consists of a strip of lead metal in a solution of $Pb(NO_3)_2$ in one beaker, and in the other beaker a platinum electrode is immersed in a NaCl solution, with Cl_2 gas bubbled around the electrode. The two beakers are connected with a salt bridge. **(a)** Which electrode serves as the anode, and which as the cathode? **(b)** Does the Pb electrode gain or lose mass as the cell reaction proceeds? **(c)** Write the equation for the overall cell reaction. **(d)** What is the emf generated by the cell under standard conditions?

Oxidizing and Reducing Agents; Spontaneity

20.31 **(a)** For a strong reductant, would you expect E°_{red} to be positive or negative? **(b)** Are reducing agents found on the left or right side of reduction half-reactions?

20.32 **(a)** Would you expect to find a strong oxidant near the top or the bottom of the substances listed in Table 20.1? **(b)** Would you expect to find an oxidant on the left or on the right side of a reduction half-reaction?

20.33 From each of the following pairs of substances, use data in Appendix E to choose the one that is the stronger oxidizing agent:
(a) $Cl_2(g)$ or $Br_2(l)$
(b) $Ni^{2+}(aq)$ or $Cd^{2+}(aq)$
(c) $BrO_3^-(aq)$ or $IO_3^-(aq)$
(d) $H_2O_2(aq)$ or $O_3(g)$

20.34 From each of the following pairs of substances, use data in Appendix E to choose the one that is the stronger reducing agent:
(a) Fe(*s*) or Mg(*s*)
(b) Ca(*s*) or Al(*s*)
(c) H_2(*g*, acidic solution) or $H_2S(g)$
(d) $H_2SO_3(aq)$ or $H_2C_2O_4(aq)$

20.35 By using the data in Appendix E, determine whether each of the following substances is likely to serve as an oxidant or a reductant: **(a)** $Cl_2(g)$; **(b)** MnO_4^-(*aq*, acidic solution); **(c)** Ba(*s*); **(d)** Zn (*s*).

20.36 Is each of the following substances likely to serve as an oxidant or a reductant: **(a)** Na(*s*); **(b)** $O_3(g)$; **(c)** $Ce^{3+}(aq)$; **(d)** $Sn^{2+}(aq)$?

20.37 **(a)** Assuming standard conditions, arrange the following in order of increasing strength as oxidizing agents in acidic solution: $Cr_2O_7^{2-}$, H_2O_2, Cu^{2+}, Cl_2, O_2. **(b)** Arrange the following in order of increasing strength as reducing agents in acidic solution: Zn, I^-, Sn^{2+}, H_2O_2, Al.

20.38 Based on the data in Appendix E, **(a)** which of the following is the strongest oxidizing agent, and which is the weakest in acidic solution: Ce^{4+}, Br_2, H_2O_2, Zn? **(b)** Which of the following is the strongest reducing agent, and which is the weakest in acidic solution: F^-, Zn, $N_2H_5^+$, I_2, NO?

20.39 The standard reduction potential for the reduction of $Eu^{3+}(aq)$ to $Eu^{2+}(aq)$ is −0.43V. Using Appendix E, which of the following substances is capable of reducing $Eu^{3+}(aq)$ to $Eu^{2+}(aq)$ under standard conditions: Al, Co, H_2O_2, $N_2H_5^+$, $H_2C_2O_4$?

20.40 The standard reduction potential for the reduction of $RuO_4^-(aq)$ to $RuO_4^{2-}(aq)$ is +0.59 V. By using Appendix E, which of the following substances can oxidize $RuO_4^{2-}(aq)$ to $RuO_4^-(aq)$ under standard conditions: $Cr_2O_7^{2-}(aq)$, $ClO^-(aq)$, $Pb^{2+}(aq)$, $I_2(s)$, $Ni^{2+}(aq)$?

20.41 **(a)** What is the relationship between the emf of a reaction and its spontaneity? **(b)** Which of the reactions in Exercise 20.23 are spontaneous under standard conditions? **(c)** What is ΔG° at 298 K for each of the reactions in Exercise 20.23?

20.42 **(a)** What is the relationship between the emf of a reaction and the change in Gibbs free energy? **(b)** Which of the reactions in Exercise 20.24 are spontaneous under standard conditions? **(c)** Calculate the standard free-energy change at 25°C for each of the reactions in Exercise 20.24.

20.43 Given the following reduction half-reactions:

$Fe^{3+}(aq) + e^- \longrightarrow Fe^{2+}(aq)$ $E^\circ_{red} = +0.77\ V$

$S_2O_6^{2-}(aq) + 4H^+(aq) + 2e^- \longrightarrow 2H_2SO_3(aq)$ $E^\circ_{red} = +0.60\ V$

$N_2O(aq) + 2H^+(aq) + 2e^- \longrightarrow N_2(g) + H_2O(l)$ $E^\circ_{red} = -1.77\ V$

$VO_2^+(aq) + 2H^+(aq) + e^- \longrightarrow VO^{2+}(aq) + H_2O(l)$ $E^\circ_{red} = +1.00\ V$

(a) Write balanced chemical equations for the oxidation of $Fe^{2+}(aq)$ by $S_2O_6^{2-}(aq)$, by $N_2O(aq)$, and by $VO_2^+(aq)$. **(b)** Calculate ΔG° for each reaction at 298 K.

20.44 For each of the following reactions, write a balanced equation, calculate the emf, and calculate ΔG° at 298 K. **(a)** Aqueous iodide ion is oxidized to $I_2(s)$ by $Hg_2^{2+}(aq)$. **(b)** In acidic solution, copper(I) ion is oxidized to copper(II) ion by nitrate ion. **(c)** In basic solution, $Cr(OH)_3(s)$ is oxidized to $CrO_4^{2-}(aq)$ by $ClO^-(aq)$.

EMF and Concentration

20.45 **(a)** Under what circumstances is the Nernst equation applicable? **(b)** What is the numerical value of the reaction quotient, Q, under standard conditions? **(c)** What happens to the emf of a cell if the concentrations of the reactants are increased?

20.46 **(a)** A voltaic cell is constructed with all reactants and products in their standard states. Will this condition hold as the cell operates? Explain. **(b)** What is the significance of the factor "0.0592 V" in the Nernst equation? **(c)** What happens to the emf of a cell if the concentrations of the products are increased?

20.47 What is the effect on the emf of the cell shown in Figure 20.11 of each of the following changes? **(a)** The pressure of the H_2 gas is increased in the cathode compartment. **(b)** Zinc nitrate is added to the anode compartment. **(c)** Sodium hydroxide is added to the cathode compartment, decreasing $[H^+]$. **(d)** The area of the anode is doubled.

20.48 A voltaic cell utilizes the following reaction:

$$Al(s) + 3Ag^+(aq) \longrightarrow Al^{3+}(aq) + 3Ag(s)$$

What is the effect on the cell emf of each of the following changes? **(a)** Water is added to the anode compartment, diluting the solution. **(b)** The size of the aluminum electrode is increased. **(c)** A solution of $AgNO_3$ is added to the cathode compartment, increasing the quantity of Ag^+ but not changing its concentration. **(d)** HCl is added to the $AgNO_3$ solution, precipitating some of the Ag^+ as AgCl.

20.49 A voltaic cell is constructed that uses the following reaction and operates at 298 K:

$$Zn(s) + Ni^{2+}(aq) \longrightarrow Zn^{2+}(aq) + Ni(s)$$

(a) What is the emf of this cell under standard conditions? **(b)** What is the emf of this cell when $[Ni^{2+}] = 3.00\ M$ and $[Zn^{2+}] = 0.100\ M$? **(c)** What is the emf of the cell when $[Ni^{2+}] = 0.200\ M$ and $[Zn^{2+}] = 0.900\ M$?

20.50 A voltaic cell utilizes the following reaction and operates at 298 K:

$$3Ce^{4+}(aq) + Cr(s) \longrightarrow 3Ce^{3+}(aq) + Cr^{3+}(aq)$$

(a) What is the emf of this cell under standard conditions? **(b)** What is the emf of this cell when $[Ce^{4+}] = 2.0\ M$, $[Ce^{3+}] = 0.010\ M$, and $[Cr^{3+}] = 0.010\ M$? **(c)** What is the emf of the cell when $[Ce^{4+}] = 0.35\ M$, $[Ce^{3+}] = 0.85\ M$, and $[Cr^{3+}] = 1.2\ M$?

20.51 A voltaic cell utilizes the following reaction and operates at 298 K:

$$4Fe^{2+}(aq) + O_2(g) + 4H^+(aq) \longrightarrow 4Fe^{3+}(aq) + 2H_2O(l)$$

(a) What is the emf of this cell under standard conditions? **(b)** What is the emf of this cell when $[Fe^{2+}] = 3.0\ M$, $[Fe^{3+}] = 0.010\ M$, $P_{O_2} = 0.50$ atm, and the pH of the solution in the cathode compartment is 3.00?

20.52 A voltaic cell utilizes the following reaction and operates at 20°C:

$$2Fe^{3+}(aq) + H_2(g) \longrightarrow 2Fe^{2+}(aq) + 2H^+(aq)$$

(a) What is the emf of this cell under standard conditions? **(b)** What is the emf for this cell when $[Fe^{3+}] = 1.50\ M$, $P_{H_2} = 0.50$ atm, $[Fe^{2+}] = 0.0010\ M$, and the pH in both compartments is 5.00.

20.53 A voltaic cell is constructed with two Zn^{2+}-Zn electrodes. The two cell compartments have $[Zn^{2+}] = 5.00\ M$ and $[Zn^{2+}] = 1.00 \times 10^{-2}\ M$, respectively. **(a)** Which electrode is the anode of the cell? **(b)** What is the standard emf of the cell? **(c)** What is the cell emf for the concentrations given? **(d)** For each electrode, predict whether $[Zn^{2+}]$ will increase, decrease, or stay the same as the cell operates.

20.54 A voltaic cell is constructed with two silver–silver chloride electrodes, each of which is based on the following half-reaction:

$$AgCl(s) + e^- \longrightarrow Ag(s) + Cl^-(aq)$$

The two cell compartments have $[Cl^-] = 0.0150\ M$ and $[Cl^-] = 2.55\ M$, respectively. **(a)** Which electrode is the cathode of the cell? **(b)** What is the standard emf of the cell? **(c)** What is the cell emf for the concentrations given? **(d)** For each electrode, predict whether $[Cl^-]$ will increase, decrease, or stay the same as the cell operates.

20.55 The cell in Figure 20.11 could be used to provide a measure of the pH in the cathode compartment. Calculate the pH of the cathode compartment solution if the cell emf at 298 K is measured to be +0.684 V when $[Zn^{2+}] = 0.30\ M$ and $P_{H_2} = 0.90$ atm.

20.56 A voltaic cell is constructed that is based on the following reaction:

$$Sn^{2+}(aq) + Pb(s) \longrightarrow Sn(s) + Pb^{2+}(aq)$$

(a) If the concentration of Sn^{2+} in the cathode compartment is 1.00 M and the cell generates an emf of +0.22 V, what is the concentration of Pb^{2+} in the anode compartment? **(b)** If the anode compartment contains $[SO_4^{2-}] = 1.00\ M$ in equilibrium with $PbSO_4(s)$, what is the K_{sp} of $PbSO_4$?

20.57 Using the standard reduction potentials listed in Appendix E, calculate the equilibrium constant for each of the following reactions at 298 K:
(a) $Fe(s) + Ni^{2+}(aq) \longrightarrow Fe^{2+}(aq) + Ni(s)$
(b) $Co(s) + 2H^+(aq) \longrightarrow Co^{2+}(aq) + H_2(g)$

(c) $10Br^-(aq) + 2MnO_4^-(aq) + 16H^+(aq) \longrightarrow 2Mn^{2+}(aq) + 8H_2O(l) + 5Br_2(l)$

20.58 Using the standard reduction potentials listed in Appendix E, calculate the equilibrium constant for each of the following reactions at 298 K:
(a) $2VO_2^+(aq) + 4H^+(aq) + 2Ag(s) \longrightarrow 2VO^{2+}(aq) + 2H_2O(l) + 2Ag^+(aq)$
(b) $3Ce^{4+}(aq) + Bi(s) + H_2O(l) \longrightarrow 3Ce^{3+}(aq) + BiO^+(aq) + 2H^+(aq)$
(c) $N_2H_5^+(aq) + 4Fe(CN)_6^{3-}(aq) \longrightarrow N_2(g) + 5H^+(aq) + 4Fe(CN)_6^{4-}(aq)$

20.59 A cell has a standard emf of +0.177 V at 298 K. What is the value of the equilibrium constant for the cell reaction **(a)** if $n = 1$? **(b)** if $n = 2$? **(c)** if $n = 3$?

20.60 At 298 K a cell reaction has a standard emf of +0.17 V. The equilibrium constant for the cell reaction is 5.5×10^5. What is the value of n for the cell reaction?

Batteries; Corrosion

20.61 **(a)** What is a *battery*? **(b)** What is the difference between a *primary* battery and a *secondary* battery? **(c)** A certain application requires a portable 7.5-V power source. Will it be possible to use a single battery based on a single voltaic cell as the power source? Explain.

20.62 **(a)** What happens to the emf of a battery as it is used? Why does this happen? **(b)** The AA-size and D-size alkaline batteries are both 1.5-V batteries that are based on the same electrode reactions. What is the major difference between the two batteries? What performance feature is most affected by this difference?

20.63 During a period of discharge of a lead-acid battery, 382 g of Pb from the anode is converted into $PbSO_4(s)$. What mass of $PbO_2(s)$ is reduced at the cathode during this same period?

20.64 During the discharge of an alkaline battery, 12.9 g of Zn are consumed at the anode of the battery. What mass of MnO_2 is reduced at the cathode during this discharge?

20.65 Heart pacemakers are often powered by lithium–silver chromate "button" batteries. The overall cell reaction is:

$$2Li(s) + Ag_2CrO_4(s) \longrightarrow Li_2CrO_4(s) + 2Ag(s)$$

(a) Lithium metal is the reactant at one of the electrodes of the battery. Is it the anode or the cathode? **(b)** Choose the two half-reactions from Appendix E that *most closely approximate* the reactions that occur in the battery. What standard emf would be generated by a voltaic cell based on these half-reactions? **(c)** The battery generates an emf of +3.5 V. How close is this value to the one calculated in part (b)?

20.66 Mercuric oxide dry-cell batteries are often used where a high-energy density is required, such as in watches and cameras. The two half-cell reactions that occur in the battery are

$$HgO(s) + H_2O(l) + 2e^- \longrightarrow Hg(l) + 2OH^-(aq)$$
$$Zn(s) + 2OH^-(aq) \longrightarrow ZnO(s) + H_2O(l) + 2e^-$$

(a) Write the overall cell reaction. **(b)** The value of E°_{red} for the cathode reaction is +0.098 V. The overall cell potential is +1.35 V. Assuming that both half-cells operate under standard conditions, what is the standard reduction potential for the anode reaction? **(c)** Why is the potential of the anode reaction different than would be expected if the reaction occurred in an acidic medium?

20.67 **(a)** Suppose that an alkaline battery were manufactured using cadmium metal rather than zinc. What effect would this have on the cell emf? **(b)** What environmental advantage is provided by the use of nickel-metal-hydride batteries over the nickel-cadmium batteries?

20.68 **(a)** The nonrechargeable lithium batteries used for photography use lithium metal as the anode. What advantages might be realized by using lithium rather than zinc, cadmium, lead, or nickel? **(b)** The rechargeable lithium-ion battery does not use lithium metal as an electrode material. Nevertheless, it still has a substantial advantage over nickel-based batteries. Why is this the case?

20.69 **(a)** Write the anode and cathode reactions that cause the corrosion of iron metal to aqueous iron(II). **(b)** Write the balanced half-reactions involved in the air oxidation of $Fe^{2+}(aq)$ to $Fe_2O_3 \cdot 3H_2O$.

20.70 **(a)** Based on standard reduction potentials, would you expect copper metal to oxidize under standard conditions in the presence of oxygen and hydrogen ions? **(b)** When the Statue of Liberty was refurbished, Teflon spacers were placed between the iron skeleton and the copper metal on the surface of the statue. What role do these spacers play?

20.71 **(a)** How does the zinc coating in galvanized iron provide protection for the underlying iron? **(b)** Why is the protection provided by zinc called *cathodic protection*?

20.72 **(a)** Magnesium metal is used as a sacrificial anode to protect underground pipes from corrosion. What is a "sacrificial anode"? **(b)** An iron object is plated with a coating of cobalt to protect against corrosion. Does the cobalt protect iron by cathodic protection? Explain.

Electrolysis; Electrical Work

20.73 **(a)** What is *electrolysis*? **(b)** Are electrolysis reactions thermodynamically spontaneous? Explain. **(c)** What process occurs at the anode in the electrolysis of molten NaCl?

20.74 **(a)** What is an *electrolytic cell*? **(b)** The negative terminal of a voltage source is connected to an electrode of an electrolytic cell. Is the electrode the anode or the cathode of the cell? Explain. **(c)** The electrolysis of water is often done with a small amount of sulfuric acid added to the water. What is the role of the sulfuric acid?

20.75 **(a)** Why are different products obtained when molten $MgCl_2$ and aqueous $MgCl_2$ are electrolyzed with inert electrodes? **(b)** Predict the products in each case. **(c)** What is the minimum emf required in each case?

20.76 **(a)** What are the expected half-reactions at each electrode upon electrolysis of molten $AlBr_3$ using inert electrodes?

(b) What are the expected half-reactions upon electrolyzing aqueous $AlBr_3$? (c) What is the minimum emf required in each case?

20.77 Sketch a cell for the electrolysis of aqueous $CuCl_2$ using inert electrodes. Indicate the directions in which the ions and electrons move. Give the electrode reactions, and label the anode and cathode, indicating which is connected to which terminal of the voltage source.

20.78 Sketch a cell for the electrolysis of aqueous HBr using copper electrodes. Give the electrode reactions, labeling the anode and cathode. Calculate the minimum applied voltage required for electrolysis to occur, assuming standard conditions.

20.79 (a) A $Cr^{3+}(aq)$ solution is electrolyzed using a current of 7.75 A. What mass of Cr(s) is plated out after 1.50 days? (b) What amperage is required to plate out 0.250 mol Cr from a Cr^{3+} solution in a period of 8.00 hr?

20.80 Metallic magnesium can be made by the electrolysis of molten $MgCl_2$. (a) What mass of Mg is formed by passing a current of 5.25 A through molten $MgCl_2$ for 2.50 days? (b) How many minutes are needed to plate out 10.00 g Mg from molten $MgCl_2$ using 3.50 A current?

20.81 (a) In the electrolysis of aqueous NaCl, how many liters of $Cl_2(g)$ (at STP) are generated by a current of 15.5 A for a period of 75.0 min? (b) How many moles of NaOH(*aq*) are formed in the solution in this period?

20.82 (a) How many seconds does it take to produce 5.0 L of H_2 measured at 725 torr and 23°C by electrolysis of water using a current of 1.5 A? (b) How many grams of O_2 are produced at the same time?

20.83 A voltaic cell is based on the following reaction:

$$Sn(s) + I_2(s) \longrightarrow Sn^{2+}(aq) + 2I^-(aq)$$

Under standard conditions, what is the maximum electrical work, in joules, that the cell can accomplish if 0.850 mol of Sn is consumed?

20.84 Consider the voltaic cell illustrated in Figure 20.5, which is based on the cell reaction

$$Zn(s) + Cu^{2+}(aq) \longrightarrow Zn^{2+}(aq) + Cu(s)$$

Under standard conditions, what is the maximum electrical work, in joules, that the cell can accomplish if 50.0 g of copper are plated out?

20.85 (a) Calculate the mass of Li formed by electrolysis of molten LiCl by a current of 7.5×10^4 A flowing for a period of 24 hr. Assume the electrolytic cell is 85% efficient. (b) What is the energy requirement for this electrolysis per mole of Li formed if the applied emf is +7.5 V?

20.86 Elemental calcium is produced by the electrolysis of molten $CaCl_2$. (a) What mass of calcium can be produced by this process if a current of 6.5×10^3 A is applied for 48 hr? Assume that the electrolytic cell is 68% efficient. (b) What is the total energy requirement for this electrolysis if the applied emf is +5.00 V?

Additional Exercises

20.87 A *disproportionation* reaction is an oxidation-reduction reaction in which the same substance is oxidized and reduced. Complete and balance the following disproportionation reactions:

(a) $MnO_4^{2-}(aq) \longrightarrow MnO_4^-(aq) + MnO_2(s)$ (acidic solution)

(b) $H_2SO_3(aq) \longrightarrow S(s) + HSO_4^-(aq)$ (acidic solution)

(c) $Cl_2(aq) \longrightarrow Cl^-(aq) + ClO^-(aq)$ (basic solution)

20.88 The following oxidation-reduction reaction in acidic solution is spontaneous:

$$5Fe^{2+}(aq) + MnO_4^-(aq) + 8H^+(aq) \longrightarrow 5Fe^{3+}(aq) + Mn^{2+}(aq) + 4H_2O(l)$$

A solution containing $KMnO_4$ and H_2SO_4 is poured into one beaker, and a solution of $FeSO_4$ is poured into another. A salt bridge is used to join the beakers. A platinum foil is placed in each solution, and the two solutions are connected by a wire that passes through a voltmeter. (a) Sketch the cell, indicating the anode and the cathode, the direction of electron movement through the external circuit, and the direction of ion migrations through the solutions. (b) Sketch the process that occurs at the atomic level at the surface of the anode. (c) Calculate the emf of the cell under standard conditions. (d) Calculate the emf of the cell at 298 K when the concentrations are the following: pH = 0.0, $[Fe^{2+}] = 0.10\ M$, $[MnO_4^-] = 1.50\ M$, $[Fe^{3+}] = 2.5 \times 10^{-4}\ M$, $[Mn^{2+}] = 0.010\ M$.

20.89 A common shorthand way to represent a voltaic cell is to list its components as follows:

anode|anode solution||cathode solution|cathode

A double vertical line represents a salt bridge or a porous barrier. A single vertical line represents a change in phase, such as from solid to solution. (a) Write the half-reactions and overall cell reaction represented by $Fe|Fe^{2+}||Ag^+|Ag$; sketch the cell. (b) Write the half-reactions and overall cell reaction represented by $Zn|Zn^{2+}||H^+|H_2$; sketch the cell. (c) Using the notation just described, represent a cell based on the following reaction:

$$ClO_3^-(aq) + 3Cu(s) + 6H^+(aq) \longrightarrow Cl^-(aq) + 3Cu^{2+}(aq) + 3H_2O(l)$$

Pt is used as an inert electrode in contact with the ClO_3^- and Cl^-. Sketch the cell.

20.90 A voltaic cell is constructed from two half-cells. The first contains a Cd(*s*) electrode immersed in 1 *M* $Cd^{2+}(aq)$ solution. The other contains an Rh(*s*) electrode in 1 *M* $Rh^{3+}(aq)$ solution. The overall cell potential is +1.20 V, and as the cell operates, the concentration of the $Rh^{3+}(aq)$ solution decreases and the mass of the Rh electrode increases. (a) Write a balanced equation for the overall cell reaction. (b) Which electrode is the anode, and which is the cathode? (c) What is the standard reduction potential for the reduction of $Rh^{3+}(aq)$ to Rh(*s*)? (d) What is the value of $\Delta G°$ for the cell reaction?

20.91 Predict whether the following reactions will be spontaneous in acidic solution under standard conditions:

(a) oxidation of Sn to Sn^{2+} by I_2 (to form I^-); (b) reduction of Ni^{2+} to Ni by I^- (to form I_2); (c) reduction of Ce^{4+} to Ce^{3+} by H_2O_2; (d) reduction of Cu^{2+} to Cu by Sn^{2+} (to form Sn^{4+}).

[20.92] Gold exists in two common positive oxidation states, +1 and +3. The standard reduction potentials for these oxidation states are

$$Au^+(aq) + e^- \longrightarrow Au(s) \qquad E^\circ_{red} = +1.69 \text{ V}$$
$$Au^{3+}(aq) + 3e^- \longrightarrow Au(s) \qquad E^\circ_{red} = +1.50 \text{ V}$$

(a) Can you use these data to explain why gold does not tarnish in the air? **(b)** Suggest several substances that should be strong enough oxidizing agents to oxidize gold metal. **(c)** Would $Au^+(aq)$ disproportionate (see Exercise 20.87) spontaneously into $Au^{3+}(aq)$ and Au(*s*)? **(d)** Based on your answers to parts (b) and (c), predict the result of reacting gold metal with fluorine gas.

20.93 A voltaic cell is constructed from an $Ni^{2+}(aq)$-Ni(*s*) half-cell and an $Ag^+(aq)$-Ag(*s*) half-cell. The initial concentration of $Ni^{2+}(aq)$ in the Ni^{2+}-Ni half-cell is $[Ni^{2+}] = 0.0100\ M$. The initial cell voltage is +1.12 V. **(a)** By using data in Table 20.1, calculate the standard emf of this voltaic cell. **(b)** Will the concentration of $Ni^{2+}(aq)$ increase or decrease as the cell operates? **(c)** What is the initial concentration of $Ag^+(aq)$ in the Ag^+-Ag half-cell?

[20.94] A voltaic cell is constructed that uses the following half-cell reactions:

$$Cu^+(aq) + e^- \longrightarrow Cu(s)$$
$$I_2(s) + 2e^- \longrightarrow 2I^-(aq)$$

The cell is operated at 298 K with $[Cu^+] = 2.5\ M$ and $[I^-] = 3.5\ M$. **(a)** Determine E for the cell at these concentrations. **(b)** Which electrode is the anode of the cell? **(c)** Is the answer to part (b) the same as it would be if the cell were operated under standard conditions? **(d)** If $[Cu^+]$ were equal to 1.4 *M*, at what concentration of I^- would the cell have zero potential?

20.95 By using the standard reduction potentials in Appendix E, calculate the equilibrium constant at 298 K for the disproportionation of copper(I) ion.

$$2Cu^+(aq) \longrightarrow Cu(s) + Cu^{2+}(aq)$$

20.96 The common rectangular 9-V alkaline batteries that are used in many portable devices are based on the same electrode reactions as the 1.5-V D-size alkaline batteries. What can you conclude about the internal construction of the 9-V battery relative to the 1.5-V battery?

20.97 **(a)** Write the reactions for the discharge and charge of a nickel-cadmium rechargeable battery. **(b)** Given the following reduction potentials, calculate the standard emf of the cell:

$$Cd(OH)_2(s) + 2e^- \longrightarrow Cd(s) + 2OH^-(aq) \qquad E^\circ_{red} = -0.76 \text{ V}$$
$$NiO(OH)(s) + H_2O(l) + e^- \longrightarrow Ni(OH)_2(s) + OH^-(aq) \qquad E^\circ_{red} = +0.49 \text{ V}$$

(c) A typical nicad voltaic cell generates an emf of +1.30 V. Why is there a difference between this value and the one you calculated in part (b)?

20.98 Iron is often coated with a thin layer of tin to prevent corrosion. **(a)** Use electrode potentials to determine whether tin provides cathodic protection for the iron. **(b)** What do you suppose happens if the tin coating on iron is broken in the presence of air and water?

20.99 If you were going to apply a small potential to a steel ship resting in the water as a means of inhibiting corrosion, would you apply a negative or a positive charge? Explain.

20.100 The following quotation is taken from an article dealing with corrosion of electronic materials: "Sulfur dioxide, its acidic oxidation products, and moisture are well established as the principal causes of outdoor corrosion of many metals." Using Ni as an example, explain why the factors cited affect the rate of corrosion. Write chemical equations to illustrate your points. [*Note*: NiO(*s*) is soluble in acidic solution.]

20.101 Two wires from a battery are tested with a piece of filter paper moistened with NaCl solution containing phenolphthalein, an acid-base indicator that is colorless in acid and pink in base. When the wires touch the paper about an inch apart, the rightmost wire produces a pink coloration on the filter paper and the leftmost produces none. Which wire is connected to the positive terminal of the battery? Explain.

[20.102] **(a)** How many coulombs are required to plate a layer of chromium metal 0.25 mm thick on an auto bumper with a total area of 0.32 m^2 from a solution containing CrO_4^{2-}? The density of chromium metal is 7.20 g/cm^3. **(b)** What current flow is required for this electroplating if the bumper is to be plated in 10.0 s? **(c)** If the external source has an emf of +6.0 V and the electrolytic cell is 65% efficient, how much electrical power is expended to electroplate the bumper?

20.103 The element indium is to be obtained by electrolysis of a molten halide of the element. Passing a current of 3.20 A for a period of 40.0 min forms 4.57 g of In. What is the oxidation state of indium in the halide melt?

20.104 **(a)** What is the maximum amount of work that a 6-V lead-acid battery of a golf cart can accomplish if it is rated at 300 A-hr? **(b)** List some of the reasons why this amount of work is never realized.

[20.105] Some years ago a unique proposal was made to raise the *Titanic*. The plan involved placing pontoons within the ship using a surface-controlled submarine-type vessel. The pontoons would contain cathodes and would be filled with hydrogen gas formed by the electrolysis of water. It has been estimated that it would require about 7×10^8 mol of H_2 to provide the buoyancy to lift the ship [*Journal of Chemical Education*, **50**, 61, (1973)]. **(a)** How many coulombs of electrical charge would be required? **(b)** What is the minimum voltage required to generate H_2 and O_2 if the pressure on the gases at the depth of the wreckage (2 mi) is 300 atm? **(c)** What is the minimum electrical energy required to raise the *Titanic* by electrolysis? **(d)** What is the minimum cost of the electrical energy required to generate the necessary H_2 if the electricity costs 23¢ per kilowatt-hour to generate at the site?

Integrative Exercises

20.106 Copper dissolves in concentrated nitric acid with the evolution of NO(g), which is subsequently oxidized to $NO_2(g)$ in air (Figure 1.11). In contrast, copper does not dissolve in concentrated hydrochloric acid. Explain these observations by using standard reduction potentials from Table 20.1.

20.107 Under standard conditions, the following reaction is spontaneous at 25°C:

$$O_2(g) + 4H^+(aq) + 4Br^-(aq) \longrightarrow 2H_2O(l) + 2Br_2(l)$$

Will the reaction be spontaneous if $[H^+]$ is adjusted using a buffer composed of 0.10 M benzoic acid ($HC_7H_5O_2$) and 0.12 M sodium benzoate ($NaC_7H_5O_2$)?

20.108 Consider the general oxidation of a species A in solution: $A \longrightarrow A^+ + e^-$. The term "oxidation potential" is sometimes used to describe the ease with which species A is oxidized—the easier a species is to oxidize, the greater its oxidation potential. **(a)** What is the relationship between the standard oxidation potential of A and the standard reduction potential of A^+? **(b)** Which of the metals listed in Table 4.5 has the highest standard oxidation potential? Which has the lowest? **(c)** For a series of substances, the trend in oxidation potential is often related to the trend in the first ionization energy. Explain why this relationship is sensible.

[20.109] As seen in "A Closer Look" box in Section 4.4, gold metal dissolves in aqua regia, a mixture of concentrated hydrochloric and concentrated nitric acids. The following standard reduction potentials are important in gold chemistry:

$$Au^{3+}(aq) + 3e^- \longrightarrow Au(s) \qquad E^\circ_{red} = +1.498 \text{ V}$$

$$AuCl_4^-(aq) + 3e^- \longrightarrow Au(s) + 4Cl^-(aq) \qquad E^\circ_{red} = +1.002 \text{ V}$$

(a) Use half-reactions to write a balanced equation for the reaction of Au and nitric acid to produce Au^{3+} and NO(g), and calculate the standard emf of this reaction. Is this reaction spontaneous? **(b)** Use half-reactions to write a balanced equation for the reaction of Au and hydrochloric acid to produce $AuCl_4^-(aq)$ and $H_2(g)$, and calculate the standard emf of this reaction. Is this reaction spontaneous? **(c)** Use half-reactions to write a balanced equation for the reaction of Au and aqua regia to produce $AuCl_4^-(aq)$ and NO(g), and calculate the standard emf of this reaction. Is this reaction spontaneous under standard conditions? **(d)** By using the Nernst equation, explain why aqua regia made from *concentrated* hydrochloric and nitric acids is able to dissolve gold.

20.110 A voltaic cell is based on $Ag^+(aq)/Ag(s)$ and $Fe^{3+}(aq)/Fe^{2+}(aq)$ half-cells. **(a)** What is the standard emf of the cell? **(b)** Which reaction occurs at the cathode, and which at the anode of the cell? **(c)** Use S° values in Appendix C and the relationship between cell potential and free-energy change to predict whether the standard cell potential increases or decreases when the temperature is raised above 25°C.

20.111 Hydrogen gas has the potential as a clean fuel in reaction with oxygen. The relevant reaction is

$$2H_2(g) + O_2(g) \longrightarrow 2H_2O(l)$$

Consider two possible ways of utilizing this reaction as an electrical energy source: (i) Hydrogen and oxygen gases are combusted and used to drive a generator, much as coal is currently used in the electric power industry; (ii) hydrogen and oxygen gases are used to generate electricity directly by using fuel cells that operate at 85°C. **(a)** Use data in Appendix C to calculate ΔH° and ΔS° for the above reaction. We will assume that these values do not change appreciably with temperature. **(b)** Based on the values from part (a), what trend would you expect for the magnitude of ΔG for the above reaction as the temperature increases? **(c)** What is the significance of the change in the magnitude of ΔG with temperature with respect to the utility of hydrogen as a fuel (recall Equation 19.19)? **(d)** Based on the analysis here, would it be more efficient to use the combustion method or the fuel-cell method to generate electrical energy from hydrogen?

20.112 Cytochrome, a complicated molecule that we will represent as $CyFe^{2+}$, reacts with the air we breathe to supply energy required to synthesize adenosine triphosphate (ATP). The body uses ATP as an energy source to drive other reactions. (Section 19.7) At pH 7.0 the following reduction potentials pertain to this oxidation of $CyFe^{2+}$:

$$O_2(g) + 4H^+(aq) + 4e^- \longrightarrow 2H_2O(l) \qquad E^\circ_{red} = +0.82 \text{ V}$$

$$CyFe^{3+}(aq) + e^- \longrightarrow CyFe^{2+}(aq) \qquad E^\circ_{red} = +0.22 \text{ V}$$

(a) What is ΔG for the oxidation of $CyFe^{2+}$ by air? **(b)** If the synthesis of 1.00 mol of ATP from adenosine diphosphate (ADP) requires a ΔG of 37.7 kJ, how many moles of ATP are synthesized per mole of O_2?

[20.113] The standard potential for the reduction of AgSCN(s) is +0.0895 V.

$$AgSCN(s) + e^- \longrightarrow Ag(s) + SCN^-(aq)$$

Using this value and the electrode potential for $Ag^+(aq)$, calculate the K_{sp} for AgSCN.

[20.114] The K_{sp} value for PbS(s) is 8.0×10^{-28}. By using this value together with an electrode potential from Appendix E, determine the value of the standard reduction potential for the reaction

$$PbS(s) + 2e^- \longrightarrow Pb(s) + S^{2-}(aq)$$

[20.115] A pH meter (Figure 16.6) employs a voltaic cell for which the cell potential is very sensitive to pH. A simple (but impractical) pH meter can be constructed by using two hydrogen electrodes: one standard hydrogen electrode (Figure 20.10) and a hydrogen electrode (with 1 atm pressure of H_2 gas) dipped into the solution of unknown pH. The two half-cells are connected by a salt bridge or porous glass disk. **(a)** Sketch the cell described above. **(b)** Write the half-cell reactions for the cell, and calculate the standard emf. **(c)** What is the pH of the solution in the half-cell that has the standard hydrogen electrode? **(d)** What is the cell emf when the pH of the unknown solution is 5.0? **(e)** How precise would a voltmeter have to be in order to detect a change in the pH of 0.01 pH units?

20.116 If 0.500 L of a 0.600 *M* $SnSO_4(aq)$ solution is electrolyzed for a period of 25.00 min using a current of 4.50 A and inert electrodes, what is the final concentration of each ion remaining in the solution? (Assume that the volume of the solution does not change.)

[20.117] A student designs an ammeter (a device to measure electrical current) that is based on the electrolysis of water into hydrogen and oxygen gases. When electrical current of unknown magnitude is run through the device for 2.00 min, 12.3 mL of water-saturated $H_2(g)$ are collected. The temperature of the system is 25.5°C, and the atmospheric pressure is 768 torr. What is the magnitude of the current in A?

eMedia Exercises

20.118 The oxidation of zinc by molecular oxygen is shown in the **Oxidation-Reduction Reactions—Part I** movie (*eChapter 20.1*). **(a)** Write the equation for the reaction between zinc and oxygen. **(b)** Assign oxidation states to each element on each side of the equation. **(c)** Is it possible for oxidation to occur without oxygen? Explain.

20.119 In Exercise 20.25 you chose from a list of half-reactions to construct voltaic cells of specified, relative emf values. The reaction shown in the **Formation of Silver Crystals** movie (*eChapter 20.2*) can be represented with two of the half-reactions listed in Exercise 20.25. **(a)** Write and balance the equation for the reaction in the movie. **(b)** Which species is the oxidizing agent and which is the reducing agent in this reaction?

20.120 The second part of the **Oxidation-Reduction Reactions—Part II** movie (*eChapter 20.1*) illustrates the reaction that occurs when zinc metal is placed in an aqueous solution containing Cu^{2+} ions. The **Voltaic Cells I** movie (*eChapter 20.3*) uses the same reaction for the construction of a voltaic cell. **(a)** Explain why the reaction as depicted in the **Oxidation-Reduction Reactions—Part II** movie cannot be used to generate a voltage. **(b)** Sketch a voltaic cell based on the Zn-Cu reaction. Label and give the purpose of each component of the cell.

20.121 In the **Redox Chemistry of Iron and Copper** movie (*eChapter 20.5*) an iron nail is placed in an aqueous solution of copper sulfate. **(a)** Write the half-reaction for the formation of the copper coating on the nail. **(b)** Is there another half-reaction occurring in the solution? If so, what is it? **(c)** What is the sign of ΔG for the overall process that occurs?

20.122 Chromium metal can be electroplated onto a copper electrode from an aqueous solution of $H_2Cr_2O_7$, as shown in the **Electroplating** movie (*eChapter 20.9*). **(a)** Write the half-reaction for the reduction of dichromate ion to chromium metal. **(b)** What current would have to be applied for 15 minutes in order to plate out 0.75 gram of chromium metal from a solution of $H_2Cr_2O_7$?

Chapter 21

Nuclear Chemistry

In this facility, intense gamma rays from cobalt-60 are used to treat various kinds of cancer, by focusing the ionizing radiation on the cancerous cells.

AS WE HAVE progressed through this text, our focus has been on chemical reactions, specifically reactions in which electrons play a dominant role. In this chapter we consider *nuclear reactions*, changes in matter originating in the nucleus of an atom. When nuclei change spontaneously, emitting radiation, they are said to be **radioactive**. As we will see, there are other kinds of nuclear reactions as well. *Nuclear chemistry* is the study of nuclear reactions and their uses in chemistry.

Nuclear chemistry affects our lives in a variety of ways. Radioactive elements are widely used in medicine as diagnostic tools and as a means of treatment, especially for cancer (Figure 21.1 ▶). They are also used to help determine the mechanisms of chemical reactions, to trace the movement of atoms in biological systems, and to date important historical artifacts. Nuclear reactions are used both to generate electricity and to create weapons of massive destruction. Although the growth of commercial nuclear power has slowed in the United States, it still accounts for about 20% of the total electricity generated. The use of nuclear energy and the disposal of nuclear wastes, however, are extremely controversial social and political issues. Because these topics evoke such a strong emotional reaction, it is often difficult to sift fact from opinion to make rational decisions. It is imperative, therefore, that we have some understanding of nuclear reactions and the uses of radioactive substances.

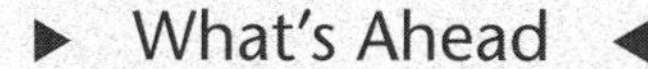

▶ What's Ahead ◀

- In this chapter we'll learn how to describe nuclear reactions by equations analogous to chemical equations, in which the nuclear charges and masses of reactants and products are in balance.
- We'll find that radioactive nuclei decay most commonly by emission of *alpha, beta,* or *gamma radiations*.
- Nuclear stability is determined to a large extent by the *neutron-to-proton ratio*. For stable nuclei, this ratio increases with increasing atomic number, as the number of neutrons increasingly exceeds the number of protons.
- All nuclei with 84 or more protons are radioactive. Heavy nuclei gain stability by a series of nuclear disintegrations leading to a stable nucleus.
- *Nuclear transmutations* are nuclear reactions that are induced by bombardment of a nucleus by a neutron or accelerated charged particle, such as an alpha particle.
- We will learn that radioisotope decays are first-order kinetic processes that exhibit characteristic half-lives. The decays of radioisotopes can be used to determine the ages of ancient artifacts and geological formations.
- Energy changes in nuclear reactions are related to mass changes via Einstein's famous equation, $E = mc^2$. The *binding energies of nuclei* are reflected in the difference between the nuclear mass and the sum of the masses of the nucleons of which they are composed.
- In a *nuclear fission* reaction, a heavy nucleus splits under nuclear bombardment to form two or more product nuclei, with release of energy. This type of nuclear reaction is the energy source for nuclear power plants.
- *Nuclear fusion* results from the fusion of two light nuclei to form a more stable, heavier one.
- The radiation emitted in nuclear reactions has the potential to cause damage to biological materials. Alpha, beta, and gamma rays cause cell damage in animals that can in turn lead to cancer and other illnesses.

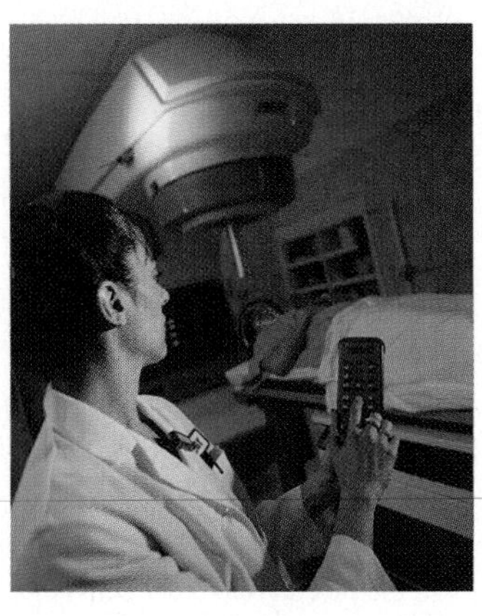

▲ **Figure 21.1** Radiation from radioisotopes such as cobalt-60 or from other sources of high-energy radiation is used in treating cancer.

21.1 Radioactivity

In order to understand nuclear reactions, we must review and develop some ideas introduced in Section 2.3. First, recall that two subatomic particles reside in the nucleus, the *proton* and the *neutron*. We will refer to these particles as **nucleons**. Recall also that all atoms of a given element have the same number of protons; this number is the element's *atomic number*. The atoms of a given element can have different numbers of neutrons, however, so they can have different *mass numbers*; the mass number is the total number of nucleons in the nucleus. Atoms with the same atomic number but different mass numbers are known as *isotopes*.

The different isotopes of an element are distinguished by their mass numbers. For example, the three naturally occurring isotopes of uranium are uranium-234, uranium-235, and uranium-238, where the numerical suffixes represent the mass numbers. These isotopes are also labeled, using chemical symbols, as $^{234}_{92}U$, $^{235}_{92}U$, and $^{238}_{92}U$. The superscript is the mass number; the subscript is the atomic number.

Different isotopes have different natural abundances. For example, 99.3% of naturally occurring uranium is uranium-238, 0.7% is uranium-235, and only a trace is uranium-234. Different nuclei also have different stabilities. Indeed, the nuclear properties of an atom depend on the number of protons and neutrons in its nucleus. Recall that a *nuclide* is a nucleus with a specified number of protons and neutrons. (Section 2.3) Nuclei that are radioactive are called **radionuclides**, and atoms containing these nuclei are called **radioisotopes**.

Nuclear Equations

The vast majority of nuclei found in nature are stable and remain intact indefinitely. Radionuclides, however, are unstable and spontaneously emit particles and electromagnetic radiation. Emission of radiation is one of the ways in which an unstable nucleus is transformed into a more stable one with less energy. The emitted radiation is the carrier of the excess energy. Uranium-238, for example, is radioactive, undergoing a nuclear reaction in which helium-4 nuclei are spontaneously emitted. The helium-4 particles are known as **alpha particles**, and a stream of these particles is called *alpha radiation*. When a uranium-238 nucleus loses an alpha particle, the remaining fragment has an atomic number of 90 and a mass number of 234. It is therefore a thorium-234 nucleus. We represent this reaction by the following *nuclear equation*:

$$^{238}_{92}U \longrightarrow ^{234}_{90}Th + ^{4}_{2}He \qquad [21.1]$$

When a nucleus spontaneously decomposes in this way, it is said to have decayed, or to have undergone *radioactive decay*. Because an alpha particle is involved in this reaction, scientists also describe the process as alpha decay.

In Equation 21.1 the sum of the mass numbers is the same on both sides of the equation $(238 = 234 + 4)$. Likewise, the sum of the atomic numbers on both sides of the equation is equal $(92 = 90 + 2)$. Mass numbers and atomic numbers must be balanced in all nuclear equations.

The radioactive properties of the nucleus are essentially independent of the state of chemical combination of the atom. In writing nuclear equations, therefore, we are not concerned with the chemical form of the atom in which the nucleus resides. It makes no difference whether we are dealing with the atom in the form of an element or of one of its compounds.

SAMPLE EXERCISE 21.1

What product is formed when radium-226 undergoes alpha decay?

Solution

Analyze and Plan: We are asked to determine the nucleus that results when radium-226 loses an alpha particle. We can best do this by writing a balanced nuclear reaction for the process.

Solve: The periodic table or a list of elements shows that radium has an atomic number of 88. The complete chemical symbol for radium-226 is therefore $^{226}_{88}\text{Ra}$. An alpha particle is a helium-4 nucleus, and so its symbol is $^{4}_{2}\text{He}$ (sometimes written as $^{4}_{2}\alpha$). The alpha particle is a product of the nuclear reaction, and so the equation is of the form

$$^{226}_{88}\text{Ra} \longrightarrow {}^{A}_{Z}\text{X} + {}^{4}_{2}\text{He}$$

where A is the mass number of the product nucleus and Z is its atomic number. Mass numbers and atomic numbers must balance, so $226 = A + 4$ and $88 = Z + 2$. Hence, $A = 222$ and $Z = 86$. Again, from the periodic table, the element with $Z = 86$ is radon (Rn). The product, therefore, is $^{222}_{86}\text{Rn}$, and the nuclear equation is

$$^{226}_{88}\text{Ra} \longrightarrow {}^{222}_{86}\text{Rn} + {}^{4}_{2}\text{He}$$

PRACTICE EXERCISE

What element undergoes alpha decay to form lead-208?
Answer: $^{212}_{84}\text{Po}$

Types of Radioactive Decay

The three most common kinds of radioactive decay are alpha (α), beta (β), and gamma (γ) radiation. Table 21.1 ▼ summarizes some of the important properties of these kinds of radiation. As we have just discussed, alpha radiation consists of a stream of helium-4 nuclei known as alpha particles, which we denote as $^{4}_{2}\text{He}$ or $^{4}_{2}\alpha$.

TABLE 21.1 Properties of Alpha, Beta, and Gamma Radiation

	Type of Radiation		
Property	α	β	γ
Charge	2+	1−	0
Mass	6.64×10^{-24} g	9.11×10^{-28} g	0
Relative penetrating power	1	100	10,000
Nature of radiation	$^{4}_{2}\text{He}$ nuclei	Electrons	High-energy photons

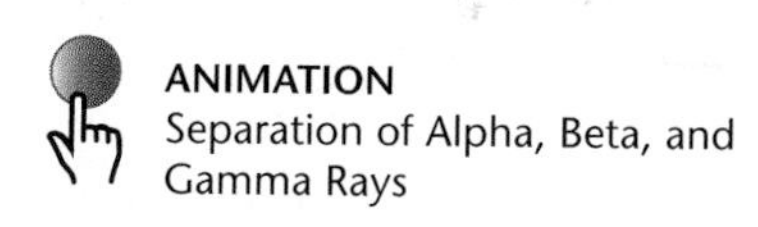
ANIMATION
Separation of Alpha, Beta, and Gamma Rays

Beta radiation consists of streams of **beta particles**, which are high-speed electrons emitted by an unstable nucleus. Beta particles are represented in nuclear equations by the symbol $^{0}_{-1}\text{e}$ or sometimes $^{0}_{-1}\beta$. The superscript zero indicates that the mass of the electron is exceedingly small compared to the mass of a nucleon. The subscript −1 represents the negative charge of the particle, which is opposite that of the proton. Iodine-131 is an isotope that undergoes decay by beta emission:

$$^{131}_{53}\text{I} \longrightarrow {}^{131}_{54}\text{Xe} + {}^{0}_{-1}\text{e} \qquad [21.2]$$

In Equation 21.2 beta decay causes the atomic number to increase from 53 to 54. Beta emission is equivalent to the conversion of a neutron ($^{1}_{0}\text{n}$) to a proton ($^{1}_{1}\text{p}$ or $^{1}_{1}\text{H}$), thereby increasing the atomic number by 1:

$$^{1}_{0}\text{n} \longrightarrow {}^{1}_{1}\text{p} + {}^{0}_{-1}\text{e} \qquad [21.3]$$

Just because an electron is ejected from the nucleus, however, we should not think that the nucleus is composed of these particles, any more than we consider a match to be composed of sparks simply because it gives them off when struck. The electron comes into being only when the nucleus undergoes a nuclear reaction.

Gamma radiation (or gamma rays) consists of high-energy photons (that is, electromagnetic radiation of very short wavelength). Gamma radiation changes neither the atomic number nor the mass number of a nucleus and is represented as $^{0}_{0}\gamma$, or merely γ. It almost always accompanies other radioactive emission because it represents the energy lost when the remaining nucleons reorganize into more stable arrangements. Generally, the gamma rays are not shown when writing nuclear equations.

Two other types of radioactive decay are positron emission and electron capture. A **positron** is a particle that has the same mass as an electron, but an opposite charge.* The positron is represented as ${}^{0}_{1}e$. The isotope carbon-11 decays by positron emission:

$$ {}^{11}_{6}C \longrightarrow {}^{11}_{5}B + {}^{0}_{1}e \qquad [21.4]$$

Positron emission causes the atomic number to decrease from 6 to 5. The emission of a positron has the effect of converting a proton to a neutron, thereby decreasing the atomic number of the nucleus by 1:

$$ {}^{1}_{1}p \longrightarrow {}^{1}_{0}n + {}^{0}_{1}e \qquad [21.5]$$

Electron capture is the capture by the nucleus of an electron from the electron cloud surrounding the nucleus. Rubidium-81 undergoes decay in this fashion, as shown in Equation 21.6:

$$ {}^{81}_{37}Rb + {}^{0}_{-1}e \text{ (orbital electron)} \longrightarrow {}^{81}_{36}Kr \qquad [21.6]$$

Because the electron is consumed rather than formed in the process, it is shown on the reactant side of the equation. Electron capture, like positron emission, has the effect of converting a proton to a neutron:

$$ {}^{1}_{1}p + {}^{0}_{-1}e \longrightarrow {}^{1}_{0}n \qquad [21.7]$$

Table 21.2 ◀ summarizes the symbols used to represent the various elementary particles commonly encountered in nuclear reactions.

TABLE 21.2 Common Particles in Radioactive Decay and Nuclear Transformations

Particle	Symbol
Neutron	${}^{1}_{0}n$
Proton	${}^{1}_{1}H$ or ${}^{1}_{1}p$
Electron	${}^{0}_{-1}e$
Alpha particle	${}^{4}_{2}He$ or ${}^{4}_{2}\alpha$
Beta particle	${}^{0}_{-1}e$ or ${}^{0}_{-1}\beta$
Positron	${}^{0}_{1}e$

SAMPLE EXERCISE 21.2

Write nuclear equations for the following processes: **(a)** mercury-201 undergoes electron capture; **(b)** thorium-231 decays to form protactinium-231.

Solution

Analyze and Plan: We must write balanced nuclear equations in which the masses and charges of reactants and products are equal. We can begin by writing the complete chemical symbols for the nuclei and decay particles that are given in the problem.

Solve: (a) The information given in the question can be summarized as

$$ {}^{201}_{80}Hg + {}^{0}_{-1}e \longrightarrow {}^{A}_{Z}X $$

Because the mass numbers must have the same sum on both sides of the equation, $201 + 0 = A$. Thus, the product nucleus must have a mass number of 201. Similarly, balancing the atomic numbers gives $80 - 1 = Z$. Thus, the atomic number of the product nucleus must be 79, which identifies it as gold (Au):

$$ {}^{201}_{80}Hg + {}^{0}_{-1}e \longrightarrow {}^{201}_{79}Au $$

(b) In this case we must determine what type of particle is emitted in the course of the radioactive decay:

$$ {}^{231}_{90}Th \longrightarrow {}^{231}_{91}Pa + {}^{A}_{Z}X $$

From $231 = 231 + A$ and $90 = 91 + Z$, we deduce that $A = 0$ and $Z = -1$. According to Table 21.2, the particle with these characteristics is the beta particle (electron). We therefore write the following:

$$ {}^{231}_{90}Th \longrightarrow {}^{231}_{91}Pa + {}^{0}_{-1}e $$

PRACTICE EXERCISE

Write a balanced nuclear equation for the reaction in which oxygen-15 undergoes positron emission.

Answer: ${}^{15}_{8}O \longrightarrow {}^{15}_{7}N + {}^{0}_{1}e$

* The positron has a very short life because it is annihilated when it collides with an electron, producing gamma rays: ${}^{0}_{1}e + {}^{0}_{-1}e \longrightarrow 2\,{}^{0}_{0}\gamma$.

21.2 Patterns of Nuclear Stability

The stability of a particular nucleus depends on a variety of factors, and no single rule allows us to predict whether a particular nucleus is radioactive and how it might decay. There are, however, several empirical observations that will help you predict the stability of a nucleus.

Neutron-to-Proton Ratio

Because like charges repel each other, it may seem surprising that a large number of protons can reside within the small volume of the nucleus. At close distances, however, a strong force of attraction, called the *strong nuclear force*, exists between nucleons. Neutrons are intimately involved in this attractive force. All nuclei with two or more protons contain neutrons. The more protons packed in the nucleus, the more neutrons are needed to bind the nucleus together. Stable nuclei with low atomic numbers (up to about 20) have approximately equal numbers of neutrons and protons. For nuclei with higher atomic numbers, the number of neutrons exceeds the number of protons. Indeed, the number of neutrons necessary to create a stable nucleus increases more rapidly than the number of protons, as shown in Figure 21.2 ▼. Thus, the neutron-to-proton ratios of stable nuclei increase with increasing atomic number.

The colored band in Figure 21.2 is the area within which all stable nuclei are found and is known as the *belt of stability*. The belt of stability ends at element 83 (bismuth). *All nuclei with 84 or more protons (atomic number* $\geq$ *84) are radioactive.* For example, all isotopes of uranium, atomic number 92, are radioactive.

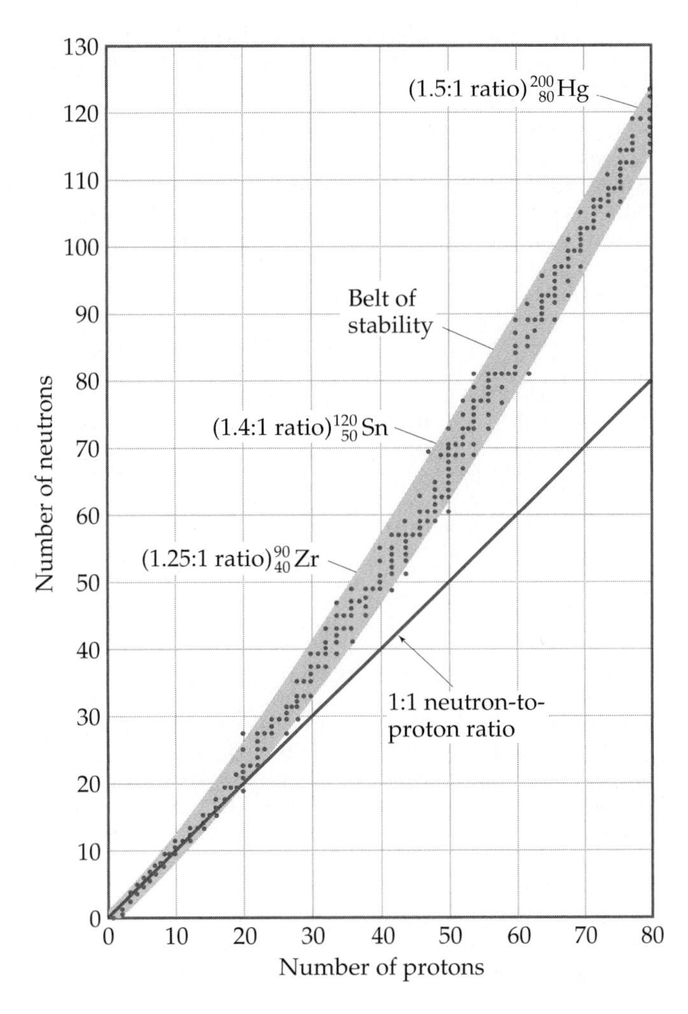

◀ **Figure 21.2** Plot of the number of neutrons versus the number of protons in stable nuclei. As the atomic number increases, the neutron-to-proton ratio of the stable nuclei increases. The stable nuclei are located in the shaded area of the graph known as the belt of stability. The majority of radioactive nuclei occur outside this belt.

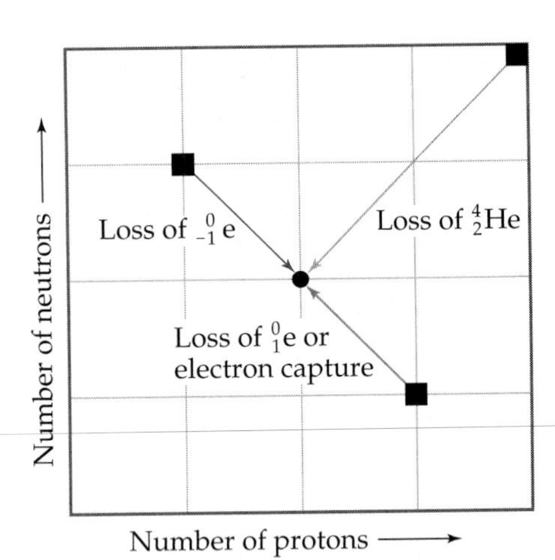

▲ **Figure 21.3** Results of alpha emission ($^{4}_{2}He$), beta emission ($^{0}_{-1}e$), positron emission ($^{0}_{1}e$), and electron capture on the number of protons and neutrons in a nucleus. Moving from left to right or from bottom to top, each square represents an additional proton or neutron, respectively. Moving in the reverse direction indicates the loss of a proton or neutron.

The type of radioactive decay that a particular radionuclide undergoes depends to a large extent on its neutron-to-proton ratio compared to those of nearby nuclei within the belt of stability. We can envision three general situations:

1. *Nuclei above the belt of stability (high neutron-to-proton ratios).* These neutron-rich nuclei can lower their ratio and move toward the belt of stability by emitting a beta particle. Beta emission decreases the number of neutrons and increases the number of protons in a nucleus, as shown in Equation 21.3.
2. *Nuclei below the belt of stability (low neutron-to-proton ratios).* These proton-rich nuclei can increase their ratio by either positron emission or electron capture. Both kinds of decay increase the number of neutrons and decrease the number of protons, as shown in Equations 21.5 and 21.7. Positron emission is more common than electron capture among the lighter nuclei; however, electron capture becomes increasingly common as nuclear charge increases.
3. *Nuclei with atomic numbers ≥ 84.* These heavy nuclei, which lie beyond the upper right edge of the band of stability, tend to undergo alpha emission. Emission of an alpha particle decreases both the number of neutrons and the number of protons by 2, moving the nucleus diagonally toward the belt of stability.

These three situations are summarized in Figure 21.3 ◀.

SAMPLE EXERCISE 21.3

Predict the mode of decay of **(a)** carbon-14; **(b)** xenon-118.

Solution

Analyze and Plan: We are asked to predict the modes of decay of two nuclei. To do this we must calculate the neutron-to-proton ratios and compare the values with those for nuclei that lie within the belt of stability shown in Figure 21.2.

Solve: (a) Carbon has an atomic number of 6. Thus, carbon-14 has 6 protons and $14 - 6 = 8$ neutrons, giving it a neutron-to-proton ratio of $\frac{8}{6} = 1.3$. Elements with low atomic numbers normally have stable nuclei with approximately equal numbers of neutrons and protons. Thus, carbon-14 has a high neutron-to-proton ratio, and we expect that it will decay by emitting a beta particle:

$$^{14}_{6}C \longrightarrow ^{0}_{-1}e + ^{14}_{7}N$$

This is indeed the mode of decay observed for carbon-14.

(b) Xenon has an atomic number of 54. Thus, xenon-118 has 54 protons and $118 - 54 = 64$ neutrons, giving it a neutron-to-proton ratio of $\frac{64}{54} = 1.2$. According to Figure 21.2, stable nuclei in this region of the belt of stability have higher neutron-to-proton ratios than xenon-118. The nucleus can increase this ratio by either positron emission or electron capture:

$$^{118}_{54}Xe \longrightarrow ^{0}_{1}e + ^{118}_{53}I$$

$$^{118}_{54}Xe + ^{0}_{-1}e \longrightarrow ^{118}_{53}I$$

In this case both modes of decay are observed.

Comment: Keep in mind that our guidelines don't always work. For example, thorium-233, $^{233}_{90}Th$, which we might expect to undergo alpha decay, actually undergoes beta decay. Furthermore, a few radioactive nuclei actually lie within the belt of stability. Both $^{146}_{60}Nd$ and $^{148}_{60}Nd$, for example, are stable and lie in the belt of stability. $^{147}_{60}Nd$, however, which lies between them, is radioactive.

PRACTICE EXERCISE

Predict the mode of decay of **(a)** plutonium-239; **(b)** indium-120.

Answers: **(a)** α decay; **(b)** β decay

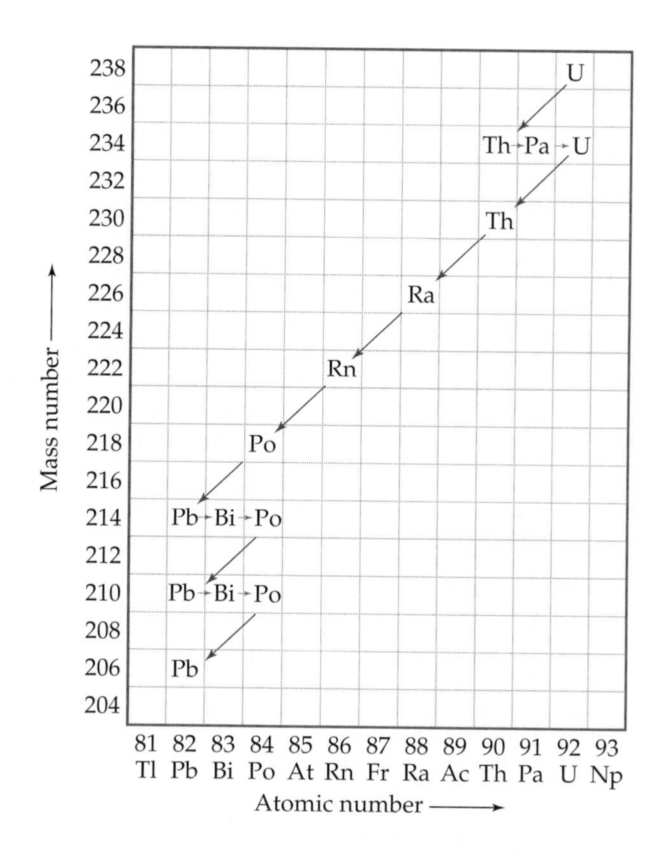

◀ **Figure 21.4** Nuclear disintegration series for uranium-238. The ${}^{238}_{92}U$ nucleus decays to ${}^{234}_{90}Th$. Subsequent decay processes eventually form the stable ${}^{206}_{82}Pb$ nucleus. Each blue arrow corresponds to the loss of an alpha particle; each red arrow corresponds to the loss of a beta particle.

ACTIVITY
Uranium-238 Decay Series

Radioactive Series

Some nuclei, like uranium-238, cannot gain stability by a single emission. Consequently, a series of successive emissions occurs. As shown in Figure 21.4 ▲, uranium-238 decays to thorium-234, which is radioactive and decays to protactinium-234. This nucleus is also unstable and subsequently decays. Such successive reactions continue until a stable nucleus, lead-206, is formed. A series of nuclear reactions that begins with an unstable nucleus and terminates with a stable one is known as a **radioactive series**, or a **nuclear disintegration series**. Three such series occur in nature. In addition to the series that begins with uranium-238 and terminates with lead-206, there is one that begins with uranium-235 and ends with lead-207, and one that begins with thorium-232 and ends with lead-208.

Further Observations

Two further observations can help you predict nuclear stability:

- Nuclei with 2, 8, 20, 28, 50, or 82 protons, or 2, 8, 20, 28, 50, 82, or 126 neutrons, are generally more stable than nuclei that do not contain these numbers of nucleons. These numbers of protons and neutrons are called **magic numbers**.
- Nuclei with even numbers of both protons and neutrons are generally more stable than those with odd numbers of nucleons, as shown in Table 21.3 ▶.

These observations can be understood in terms of the *shell model of the nucleus*, in which nucleons are described as residing in shells analogous to the shell structure for electrons in atoms. Just as certain numbers of electrons (2, 8, 18, 36, 54, and 86) correspond to stable closed-shell electron configurations, so also the magic numbers of nucleons represent closed shells in nuclei. As an example of the stability of nuclei with magic numbers of nucleons, note that the radioactive series depicted in Figure 21.4 ends with formation of the stable ${}^{206}_{82}Pb$ nucleus, which has a magic number of protons (82).

TABLE 21.3 The Number of Stable Isotopes with Even and Odd Numbers of Protons and Neutrons

Number of Stable Isotopes	Protons	Neutrons
157	Even	Even
53	Even	Odd
50	Odd	Even
5	Odd	Odd

Evidence also suggests that pairs of protons and pairs of neutrons have a special stability, analogous to the pairs of electrons in molecules. Thus, stable nuclei with an even number of protons and an even number of neutrons are far more numerous that those with odd numbers (Table 21.3).

SAMPLE EXERCISE 21.4

Which of the following nuclei are especially stable: ${}^{4}_{2}He$, ${}^{40}_{20}Ca$, ${}^{98}_{43}Tc$?

Solution

Analyze and Plan: We are asked to identify especially stable nuclei. To do this we look to see whether the numbers of protons and neutrons correspond to magic numbers.

Solve: The ${}^{4}_{2}He$ nucleus (the alpha particle) has a magic number of both protons (2) and neutrons (2) and is very stable. The ${}^{40}_{20}Ca$ nucleus also has a magic number of both protons (20) and neutrons (20) and is especially stable.

The ${}^{98}_{43}Tc$ nucleus does not have a magic number of either protons or neutrons. In fact, it has an odd number of both protons (43) and neutrons (55). There are very few stable nuclei with odd numbers of both protons and neutrons. Indeed, technetium-98 is radioactive.

PRACTICE EXERCISE

Which of the following nuclei would you expect to exhibit a special stability: ${}^{118}_{50}Sn$, ${}^{210}_{85}At$, ${}^{208}_{82}Pb$?

Answer: ${}^{118}_{50}Sn$, ${}^{208}_{82}Pb$

21.3 Nuclear Transmutations

Thus far we have examined nuclear reactions in which a nucleus spontaneously decays. A nucleus can also change identity if it is struck by a neutron or by another nucleus. Nuclear reactions that are induced in this way are known as **nuclear transmutations**.

The first conversion of one nucleus into another was performed in 1919 by Ernest Rutherford. He succeeded in converting nitrogen-14 into oxygen-17, plus a proton, using the high-velocity alpha particles emitted by radium. The reaction is

$$ {}^{14}_{7}N + {}^{4}_{2}He \longrightarrow {}^{17}_{8}O + {}^{1}_{1}H \qquad [21.8] $$

This reaction demonstrated that nuclear reactions can be induced by striking nuclei with particles such as alpha particles. Such reactions made it possible to synthesize hundreds of radioisotopes in the laboratory.

Nuclear transmutations are sometimes represented by listing, in order, the target nucleus, the bombarding particle, the ejected particle, and the product nucleus. Written in this fashion, Equation 21.8 is ${}^{14}_{7}N(\alpha, p){}^{17}_{8}O$. The alpha particle, proton, and neutron are abbreviated as α, p, and n, respectively.

SAMPLE EXERCISE 21.5

Write the balanced nuclear equation for the process summarized as ${}^{27}_{13}Al(n, \alpha){}^{24}_{11}Na$.

Solution

Analyze and Plan: We must go from the condensed descriptive form of the nuclear reaction to the balanced nuclear equation. This amounts to writing n and α showing their associated subcripts and superscripts.

Solve: The n is the abbreviation for a neutron (${}^{1}_{0}n$), and α represents an alpha particle (${}^{4}_{2}He$). The neutron is the bombarding particle, and the alpha particle is a product. Therefore, the nuclear equation is

$$ {}^{27}_{13}Al + {}^{1}_{0}n \longrightarrow {}^{24}_{11}Na + {}^{4}_{2}He $$

PRACTICE EXERCISE

Using a shorthand notation, write the nuclear reaction

$${}^{16}_{8}\text{O} + {}^{1}_{1}\text{H} \longrightarrow {}^{13}_{7}\text{N} + {}^{4}_{2}\text{He}$$

Answer: ${}^{16}_{8}\text{O}(\text{p}, \alpha){}^{13}_{7}\text{N}$

Using Charged Particles

Charged particles, such as alpha particles, must be moving very fast in order to overcome the electrostatic repulsion between them and the target nucleus. The higher the nuclear charge on either the projectile or the target, the faster the projectile must be moving to bring about a nuclear reaction. Many methods have been devised to accelerate charged particles, using strong magnetic and electrostatic fields. These **particle accelerators**, popularly called "atom smashers," bear such names as *cyclotron* and *synchrotron*. The cyclotron is illustrated in Figure 21.5 ▶. The hollow D-shaped electrodes are called "dees." The projectile particles are introduced into a vacuum chamber within the cyclotron. The particles are then accelerated by making the dees alternately positively and negatively charged. Magnets placed above and below the dees keep the particles moving in a spiral path until they are finally deflected out of the cyclotron and emerge to strike a target substance. Particle accelerators have been used mainly to synthesize heavy elements and to investigate the fundamental structure of matter. Figure 21.6 ▼ shows an aerial view of Fermilab, the National Accelerator Laboratory near Chicago.

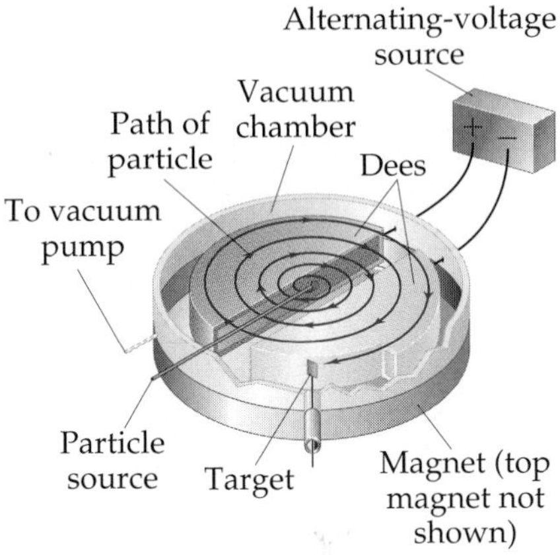

▲ **Figure 21.5** Schematic drawing of a cyclotron. Charged particles are accelerated around the ring by applying alternating voltage to the dees.

Using Neutrons

Most synthetic isotopes used in quantity in medicine and scientific research are made using neutrons as projectiles. Because neutrons are neutral, they are not repelled by the nucleus. Consequently, they do not need to be accelerated, as do charged particles, in order to cause nuclear reactions. (Indeed, they cannot be accelerated.) The necessary neutrons are produced by the reactions that occur in nuclear reactors. Cobalt-60, for example, used in radiation therapy for cancer, is produced by neutron capture. Iron-58 is placed in a nuclear reactor, where it is bombarded by neutrons. The following sequence of reactions takes place:

$${}^{58}_{26}\text{Fe} + {}^{1}_{0}\text{n} \longrightarrow {}^{59}_{26}\text{Fe} \qquad [21.9]$$

$${}^{59}_{26}\text{Fe} \longrightarrow {}^{59}_{27}\text{Co} + {}^{0}_{-1}\text{e} \qquad [21.10]$$

$${}^{59}_{27}\text{Co} + {}^{1}_{0}\text{n} \longrightarrow {}^{60}_{27}\text{Co} \qquad [21.11]$$

Transuranium Elements

Artificial transmutations have been used to produce the elements with atomic number above 92. These are known as the **transuranium elements** because they occur immediately following uranium in the periodic table. Elements 93

◀ **Figure 21.6** An aerial view of the Fermi National Accelerator Laboratory at Batavia, Illinois. Particles are accelerated to very high energies by circulating them through magnets in the ring, which has a circumference of 6.3 km.

(neptunium, Np) and 94 (plutonium, Pu) were first discovered in 1940. They were produced by bombarding uranium-238 with neutrons:

$$^{238}_{92}U + ^{1}_{0}n \longrightarrow ^{239}_{92}U \longrightarrow ^{239}_{93}Np + ^{0}_{-1}e \quad [21.12]$$

$$^{239}_{93}Np \longrightarrow ^{239}_{94}Pu + ^{0}_{-1}e \quad [21.13]$$

Elements with larger atomic numbers are normally formed in small quantities in particle accelerators. Curium-242, for example, is formed when a plutonium-239 target is struck with accelerated alpha particles:

$$^{239}_{94}Pu + ^{4}_{2}He \longrightarrow ^{242}_{96}Cm + ^{1}_{0}n \quad [21.14]$$

In 1994 a team of European scientists synthesized element 111 by bombarding a bismuth target for several days with a beam of nickel atoms:

$$^{209}_{83}Bi + ^{64}_{28}Ni \longrightarrow ^{272}_{111}X + ^{1}_{0}n$$

Amazingly, their discovery was based on the detection of only three atoms of the new element. The nuclei are very short-lived, and they undergo alpha decay within milliseconds of their synthesis. The same group of scientists also reported the synthesis of element 112 in 1996. As of this writing, names and symbols have not yet been chosen for these new elements.

21.4 Rates of Radioactive Decay

Why are some radioisotopes, such as uranium-238, found in nature, whereas others are not and must by synthesized? To answer this question you need to realize that different nuclei undergo radioactive decay at different rates. Many radioisotopes decay essentially completely in a matter of seconds or less, so we do not find them in nature. Uranium-238, on the other hand, decays very slowly; therefore, despite its instability, we can still observe it in nature. An important characteristic of a radioisotope is its rate of radioactive decay.

Radioactive decay is a first-order kinetic process. Recall that a first-order process has a characteristic **half-life,** which is the time required for half of any given quantity of a substance to react. (Section 14.4) The rates of decay of nuclei are commonly discussed in terms of their half-lives. Each isotope has its own characteristic half-life. For example, the half-life of strontium-90 is 28.8 yr. If we started with 10.0 g of strontium-90, only 5.0 g of that isotope would remain after 28.8 yr, 2.5 g would remain after another 28.8 yr, and so on. Strontium-90 decays to yttrium-90, as shown in Equation 21.15:

$$^{90}_{38}Sr \longrightarrow ^{90}_{39}Y + ^{0}_{-1}e \quad [21.15]$$

The loss of strontium-90 as a function of time is shown in Figure 21.7 ▼.

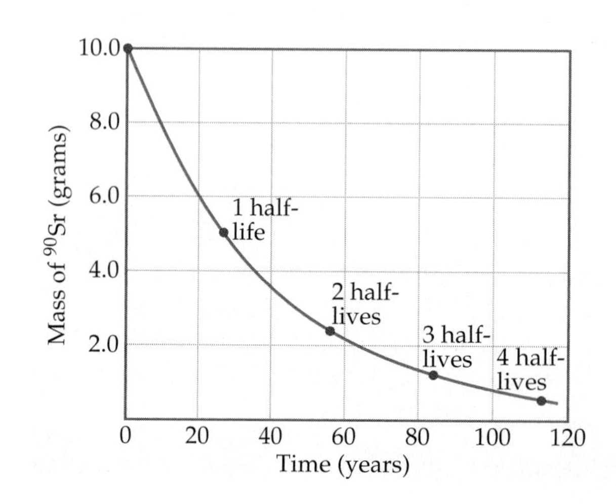

▶ **Figure 21.7** Decay of a 10.0-g sample of $^{90}_{38}Sr$ ($t_{1/2}$ = 28.8 yr).

TABLE 21.4 The Half-lives and Type of Decay for Several Radioisotopes

	Isotope	Half-life (yr)	Type of Decay
Natural radioisotopes	${}^{238}_{92}U$	4.5×10^9	Alpha
	${}^{235}_{92}U$	7.0×10^8	Alpha
	${}^{232}_{90}Th$	1.4×10^{10}	Alpha
	${}^{40}_{19}K$	1.3×10^9	Beta
	${}^{14}_{6}C$	5715	Beta
Synthetic radioisotopes	${}^{239}_{94}Pu$	24,000	Alpha
	${}^{137}_{55}Cs$	30	Beta
	${}^{90}_{38}Sr$	28.8	Beta
	${}^{131}_{53}I$	0.022	Beta

Half-lives as short as millionths of a second and as long as billions of years are known. The half-lives of some radioisotopes are listed in Table 21.4 ▲. One important feature of half-lives for nuclear decay is that they are unaffected by external conditions such as temperature, pressure, or state of chemical combination. Unlike toxic chemicals, therefore, radioactive atoms cannot be rendered harmless by chemical reaction or by any other practical treatment. At this point we can do nothing but allow these nuclei to lose radioactivity at their characteristic rates. In the meantime we must take precautions to isolate radioisotopes because of the damage radiation can cause.

SAMPLE EXERCISE 21.6

The half-life of cobalt-60 is 5.3 yr. How much of a 1.000-mg sample of cobalt-60 is left after a 15.9-yr period?

Solution

Analyze and Plan: We are given the half-life for cobalt-60 and asked to calculate the amount of cobalt-60 remaining from an initial 1.000-mg sample after a 15.9-yr period. To do this we utilize the constant half-life characteristic of a first-order decay process.

Solve: A period of 15.9 yr is three half-lives for cobalt-60. At the end of one half-life 0.500 mg of cobalt-60 remains, 0.250 mg at the end of two half-lives, and 0.125 mg at the end of three half-lives.

PRACTICE EXERCISE

Carbon-11, used in medical imaging, has a half-life of 20.4 min. The carbon-11 nuclides are formed and then incorporated into a desired compound. The resulting sample is injected into a patient, and the medical image is obtained. The entire process takes five half-lives. What percentage of the original carbon-11 remains at this time?

Answer: 3.12%

▲ **Figure 21.8** The shroud of Turin, a linen cloth over 4 m long, bears a faint image of a man, seen here in a photographic negative of the image. The shroud has been alleged to have been the burial shroud of Jesus Christ. Numerous tests have been performed on fragments of the cloth to determine its origin and age. Scientists using radiocarbon dating have concluded that the linen was made between A.D. 1260 and 1390.

Dating

Because the half-life of any particular nuclide is constant, the half-life can serve as a nuclear clock to determine the ages of different objects. Carbon-14, for example, has been used to determine the age of organic materials (Figure 21.8 ▶). The procedure is based on the formation of carbon-14 by neutron capture in the upper atmosphere:

$$ {}^{14}_{7}N + {}^{1}_{0}n \longrightarrow {}^{14}_{6}C + {}^{1}_{1}p \qquad [21.16]$$

This reaction provides a small but reasonably constant source of carbon-14. The carbon-14 is radioactive, undergoing beta decay with a half-life of 5715 yr:

$$^{14}_{6}C \longrightarrow\ ^{14}_{7}N + \ ^{0}_{-1}e \qquad [21.17]$$

In using radiocarbon dating, we generally assume that the ratio of carbon-14 to carbon-12 in the atmosphere has been constant for at least 50,000 yr. The carbon-14 is incorporated into carbon dioxide, which is in turn incorporated, through photosynthesis, into more complex carbon-containing molecules within plants. When the plants are eaten by animals, the carbon-14 becomes incorporated within them. Because a living plant or animal has a constant intake of carbon compounds, it is able to maintain a ratio of carbon-14 to carbon-12 that is identical with that of the atmosphere. Once the organism dies, however, it no longer ingests carbon compounds to replenish the carbon-14 that is lost through radioactive decay. The ratio of carbon-14 to carbon-12 therefore decreases. By measuring this ratio and contrasting it to that of the atmosphere, we can estimate the age of an object. For example, if the ratio diminishes to half that of the atmosphere, we can conclude that the object is one half-life, or 5715 yr, old. This method cannot be used to date objects older than about 50,000 yr. After this length of time the radioactivity is too low to be measured accurately.

The radiocarbon-dating technique has been checked by comparing the ages of trees determined by counting their rings and by radiocarbon analysis. As a tree grows, it adds a ring each year. In the old growth the carbon-14 decays, while the concentration of carbon-12 remains constant. The two dating methods agree to within about 10%. Most of the wood used in these tests was from California bristlecone pines, which reach ages up to 2000 yr. By using trees that died at a known time thousands of years ago, it is possible to make comparisons back to about 5000 B.C.

Other isotopes can be similarly used to date other types of objects. For example, it takes 4.5×10^9 yr for half of a sample of uranium-238 to decay to lead-206. The age of rocks containing uranium can therefore be determined by measuring the ratio of lead-206 to uranium-238. If the lead-206 had somehow become incorporated into the rock by normal chemical processes instead of by radioactive decay, the rock would also contain large amounts of the more abundant isotope lead-208. In the absence of large amounts of this "geonormal" isotope of lead, it is assumed that all of the lead-206 was at one time uranium-238.

The oldest rocks found on Earth are approximately 3×10^9 yr old. This age indicates that Earth's crust has been solid for at least this length of time. Scientists estimate that it required 1 to 1.5×10^9 yr for Earth to cool and its surface to become solid. This places the age of Earth at 4.0 to 4.5×10^9 (about 4.5 billion) yr.

Calculations Based on Half-life

So far our discussion has been mainly qualitative. We now consider the topic of half-lives from a more quantitative point of view. This approach enables us to answer questions of the following types: How do we determine the half-life of uranium-238? Similarly, how do we quantitatively determine the age of an object?

Radioactive decay is a first-order kinetic process. Its rate, therefore, is proportional to the number of radioactive nuclei N in the sample:

$$\text{Rate} = kN \qquad [21.18]$$

The first-order rate constant, k, is called the *decay constant*. The rate at which a sample decays is called its **activity**, and it is often expressed as the number of disintegrations observed per unit time. The **becquerel** (Bq) is the SI unit for expressing the activity of a particular radiation source (that is, the rate at which nuclear disintegrations are occurring). A becquerel is defined as one nuclear disintegration per second. An older, but still widely used, unit of activity is the **curie** (Ci), defined as 3.7×10^{10} disintegrations per second, which is the rate of decay of 1 g of radium. Thus, a 4.0-mCi sample of cobalt-60 undergoes $(4.0 \times 10^{-3}) \times (3.7 \times 10^{10}$ disintegrations per second$) = 1.5 \times 10^{8}$ disintegrations per second and has an activity of 1.5×10^{8} Bq. As a radioactive sample decays, the amount of radiation emanating from the sample decays as well. For example, the half-life of cobalt-60 is 5.26 yr. The 4.0-mCi sample of cobalt-60 would, after 5.26 yr have a radiation activity of 2.0 mCi, or 7.5×10^{7} Bq.

As we saw in Section 14.4, a first-order rate law can be transformed into the following equation:

$$\ln \frac{N_t}{N_0} = -kt \qquad [21.19]$$

In this equation t is the time interval of decay, k is the decay constant, N_0 is the initial number of nuclei (at time zero), and N_t is the number remaining after the time interval. Both the mass of a particular radioisotope and its activity are proportional to the number of radioactive nuclei. Thus, either the ratio of the mass at any time t to the mass at time $t = 0$ or the ratio of the activities at time t and $t = 0$ can be substituted for N_t/N_0 in Equation 21.19.

From Equation 21.19 we can obtain the relationship between the decay constant, k, and half-life, $t_{1/2}$. (Section 14.4)

$$k = \frac{0.693}{t_{1/2}} \qquad [21.20]$$

Thus, if we know the value of either the decay constant or the half-life, then we can calculate the value of the other.

SAMPLE EXERCISE 21.7

A rock contains 0.257 mg of lead-206 for every milligram of uranium-238. The half-life for the decay of uranium-238 to lead-206 is 4.5×10^{9} yr. How old is the rock?

Solution

Analyze and Plan: We're told that a rock sample has a certain amount of lead-206 for every unit weight of uranium–238 and asked to estimate the age of the rock. Presumably the lead-206 is due entirely to radioactive decay of uranium-228, to form lead-206, with a known half-life. To apply first-order kinetics expressions (Equations 21.19 and 21.20) to calculate the time elapsed since the rock was formed, we need first to calculate how much initial uranium-238 there was for every 1 milligram that remains today.

Solve: Let's assume that the rock contains 1.000 mg of uranium-238 at present. The amount of uranium-238 in the rock when it was first formed therefore equals 1.000 mg plus the quantity that decayed to lead-206. We obtain the latter quantity by multiplying the present mass of lead-206 by the ratio of the mass number of uranium to that of lead, into which it has decayed. The total original $^{238}_{92}U$ was thus

$$\text{Original } ^{238}_{92}U = 1.000 \text{ mg} + \frac{238}{206}(0.257 \text{ mg})$$
$$= 1.297 \text{ mg}$$

Using Equation 21.20, we can calculate the decay constant for the process from its half-life:

$$k = \frac{0.693}{4.5 \times 10^{9} \text{ yr}} = 1.5 \times 10^{-10} \text{ yr}^{-1}$$

Rearranging Equation 21.19 to solve for time, t, and substituting known quantities gives

$$t = -\frac{1}{k}\ln\frac{N_t}{N_0} = -\frac{1}{1.5 \times 10^{-10} \text{ yr}^{-1}}\ln\frac{1.000}{1.297} = 1.7 \times 10^{9} \text{ yr}$$

PRACTICE EXERCISE

A wooden object from an archeological site is subjected to radiocarbon dating. The activity of the sample due to ^{14}C is measured to be 11.6 disintegrations per second. The activity of a carbon sample of equal mass from fresh wood is 15.2 disintegrations per second. The half-life of ^{14}C is 5715 yr. What is the age of the archeological sample?

Answer: 2230 yr

SAMPLE EXERCISE 21.8

If we start with 1.000 g of strontium-90, 0.953 g will remain after 2.00 yr. **(a)** What is the half-life of strontium-90? **(b)** How much strontium-90 will remain after 5.00 yr? **(c)** What is the initial activity of the sample in Bq and in Ci?

Solution (a) Analyze and Plan: We are asked to calculate a half-life, $t_{1/2}$, based on data that tell us how much of a radioactive nucleus has decayed in a given period of time ($N_0 = 1.000$ g, $N_t = 0.953$ g, and $t = 2.00$ yr). We do this by first calculating the rate constant for the decay, k, then using that to compute $t_{1/2}$.

Solve: Equation 21.19 is solved for the decay constant, k, and then Equation 21.20 is used to calculate half-life, $t_{1/2}$:

$$k = -\frac{1}{t}\ln\frac{N_t}{N_0} = -\frac{1}{2.00\text{ yr}}\ln\frac{0.953\text{ g}}{1.000\text{ g}}$$

$$= -\frac{1}{2.00\text{ yr}}(-0.0481) = 0.0241\text{ yr}^{-1}$$

$$t_{1/2} = \frac{0.693}{k} = \frac{0.693}{0.0241\text{ yr}^{-1}} = 28.8\text{ yr}$$

(b) Analyze and Plan: We are asked to calculate the amount of a radionuclide remaining after a given period of time, N_t, using the initial quantity, N_0, and the rate constant for decay, k, calculated in part (a).

Solve: Again using Equation 21.19, with $k = 0.0241\text{ yr}^{-1}$, we have

$$\ln\frac{N_t}{N_0} = -kt = -(0.0241\text{ yr}^{-1})(5.00\text{ yr}) = -0.120$$

N_t/N_0 is calculated from $\ln(N_t/N_0) = -0.120$ using the e^x or INV LN function of a calculator:

$$\frac{N_t}{N_0} = e^{-0.120} = 0.887$$

Because $N_0 = 1.000$ g, we have

$$N_t = (0.887)N_0 = (0.887)(1.000\text{ g}) = 0.887\text{ g}$$

(c) Analyze and Plan: We are asked to calculate the activity of the sample in becquerels and curies. To do this we must calculate the number of disintegrations per second per atom, then multiply by the number of atoms in the sample.

Solve: The number of disintegrations per atom per second is given by the rate constant k.

$$k = \left(\frac{0.0241}{\text{yr}}\right)\left(\frac{1\text{ yr}}{365\text{ d}}\right)\left(\frac{1\text{ d}}{24\text{ hr}}\right)\left(\frac{1\text{ hr}}{3600\text{ s}}\right) = 7.64 \times 10^{-10}\text{ s}^{-1}$$

To obtain the total number of disintegrations per second, we calculate the number of atoms in the sample and multiply this by k, where we express k as the number of disintegrations per atom per second:

$$(1.00\text{ g }^{90}\text{Sr})\left(\frac{1\text{ mol }^{90}\text{Sr}}{90\text{ g }^{90}\text{Sr}}\right)\left(\frac{6.02 \times 10^{23}\text{ atoms Sr}}{1\text{ mol }^{90}\text{Sr}}\right) = 6.7 \times 10^{21}\text{ atoms }^{90}\text{Sr}$$

$$\text{Total disintegrations/s} = \left(\frac{7.64 \times 10^{-10}\text{ disintegrations}}{\text{atom}\cdot\text{s}}\right)(6.7 \times 10^{21}\text{ atoms})$$

$$= 5.1 \times 10^{12}\text{ disintegrations/s}$$

Because a Bq is one disintegration per second, the activity is just 5.11×10^{12} Bq. The activity in Ci is given by

$$(5.1 \times 10^{12}\text{ disintegrations/s})\left(\frac{1\text{ Ci}}{3.7 \times 10^{10}\text{ disintegrations/s}}\right) = 1.4 \times 10^{2}\text{ Ci}$$

We have used only two significant figures in products of these calculations because we don't know the atomic weight of ^{90}Sr to more than two significant figures without looking it up in a special source.

PRACTICE EXERCISE

A sample to be used for medical imaging is labeled with ^{18}F, which has a half-life of 110 min. What percentage of the original activity in the sample remains after 300 min?
Answer: 15.1%

21.5 Detection of Radioactivity

A variety of methods have been devised to detect emissions from radioactive substances. Becquerel discovered radioactivity because of the effect of radiation on photographic plates. Photographic plates and film have long been used to detect radioactivity. The radiation affects photographic film in much the same way as X rays do. With care, film can be used to give a quantitative measure of activity. The greater the extent of exposure to radiation, the darker the area of the developed negative. People who work with radioactive substances carry film badges to record the extent of their exposure to radiation (Figure 21.9 ▶).

Radioactivity can also be detected and measured using a device known as a **Geiger counter**. The operation of a Geiger counter is based on the ionization of matter caused by radiation. The ions and electrons produced by the ionizing radiation permit conduction of an electrical current. The basic design of a Geiger counter is shown in Figure 21.10 ▼. It consists of a metal tube filled with gas. The cylinder has a "window" made of material that can be penetrated by alpha, beta, or gamma rays. In the center of the tube is a wire. The wire is connected to one terminal of a source of direct current, and the metal cylinder is attached to the other terminal. Current flows between the wire and metal cylinder whenever ions are produced by entering radiation. The current pulse created when radiation enters the tube is amplified; each pulse is counted as a measure of the amount of radiation.

Certain substances that are electronically excited by radiation can also be used to detect and measure radiation. For example, some substances excited by radiation give off light as electrons return to their lower-energy states. These substances are called *phosphors*. Different substances respond to different particles. Zinc sulfide, for example, responds to alpha particles. An instrument called a **scintillation counter** (Figure 21.11 ▶) is used to detect and measure radiation, based on the tiny flashes of light produced when radiation strikes a suitable phosphor. The flashes are magnified electronically and counted to measure the amount of radiation.

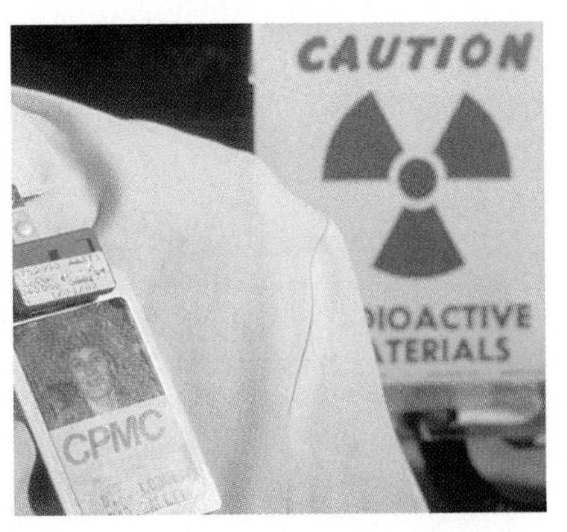

▲ **Figure 21.9** Badge dosimeter. Badges such as the one on this worker's lapel are used to monitor the extent to which the individual has been exposed to high-energy radiation. The radiation dose is determined from the extent of fogging of the film in the dosimeter. Monitoring the radiation in this way helps prevent overexposure for people whose jobs require them to use radioactive materials or X rays.

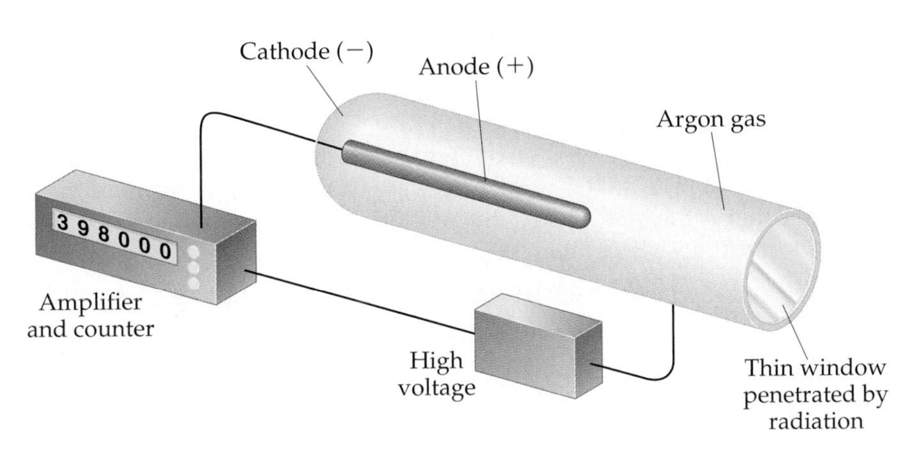

▲ **Figure 21.10** Schematic representation of a Geiger counter.

Radiotracers

Because radioisotopes can be detected so readily, they can be used to follow an element through its chemical reactions. The incorporation of carbon atoms from CO_2 into glucose in photosynthesis, for example, has been studied using CO_2 containing carbon-14:

$$6\,{}^{14}CO_2 + 6H_2O \xrightarrow[\text{chlorophyll}]{\text{sunlight}} {}^{14}C_6H_{12}O_6 + 6O_2 \quad [21.21]$$

The CO_2 is said to be labeled with the carbon-14. Detection devices such as scintillation counters follow the carbon-14 as it moves from the CO_2 through the various intermediate compounds to glucose.

The use of radioisotopes is possible because all isotopes of an element have essentially identical chemical properties. When a small quantity of a radioisotope is mixed with the naturally occurring stable isotopes of the same element, all the

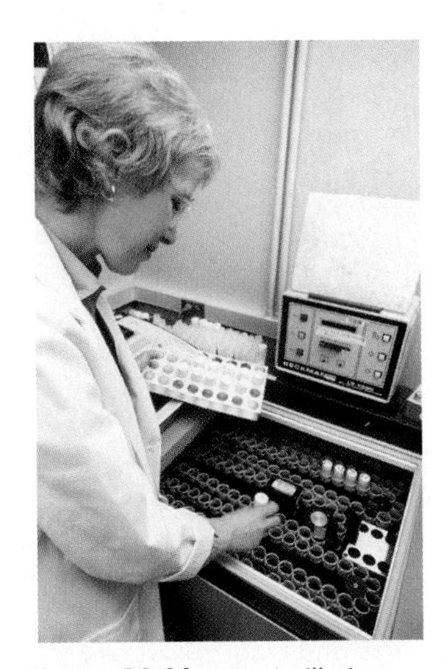

▲ **Figure 21.11** A scintillation counter, used to quantitatively measure the level of radiation.

isotopes go through the same reactions together. The element's path is revealed by the radioactivity of the radioisotope. Because the radioisotope can be used to trace the path of the element, it is called a **radiotracer**.

21.6 Energy Changes in Nuclear Reactions

The energies associated with nuclear reactions can be considered with the aid of Einstein's famous equation relating mass and energy:

$$E = mc^2 \qquad [21.22]$$

In this equation E stands for energy, m for mass, and c for the speed of light, 3.00×10^8 m/s. This equation states that the mass and energy of an object are proportional. If a system loses mass, it loses energy (exothermic); if it gains mass, it gains energy (endothermic). Because the proportionality constant in the equation, c^2, is such a large number, even small changes in mass are accompanied by large changes in energy.

The mass changes in chemical reactions are too small to detect easily. For example, the mass change associated with the combustion of a mole of CH_4 (an exothermic process) is -9.9×10^{-9} g. Because the mass change is so small, it is possible to treat chemical reactions as though mass is conserved.

The mass changes and the associated energy changes in nuclear reactions are much greater than those in chemical reactions. The mass change accompanying the radioactive decay of a mole of uranium-238, for example, is 50,000 times greater than that for the combustion of a mole of CH_4. Let's examine the energy change for this nuclear reaction:

$${}^{238}_{92}\text{U} \longrightarrow {}^{234}_{90}\text{Th} + {}^{4}_{2}\text{He}$$

The nuclei in this reaction have the following masses: ${}^{238}_{98}\text{U}$, 238.0003 amu; ${}^{234}_{90}\text{Th}$, 233.9942 amu; and ${}^{4}_{2}\text{He}$, 4.0015 amu. The mass change, Δm, is the total mass of the products minus the total mass of the reactants. The mass change for the decay of a *mole* of uranium-238 can then be expressed in grams:

$$233.9942 \text{ g} + 4.0015 \text{ g} - 238.0003 \text{ g} = -0.0046 \text{ g}$$

The fact that the system has lost mass indicates that the process is exothermic. All spontaneous nuclear reactions are exothermic.

The energy change per mole associated with this reaction can be calculated using Einstein's equations:

$$\begin{aligned}\Delta E &= \Delta(mc^2) = c^2\Delta m \\ &= (2.9979 \times 10^8 \text{ m/s})^2(-0.0046 \text{ g})\left(\frac{1 \text{ kg}}{1000 \text{ g}}\right) \\ &= -4.1 \times 10^{11}\frac{\text{kg-m}^2}{\text{s}^2} = -4.1 \times 10^{11} \text{ J}\end{aligned}$$

Notice that Δm is converted to kilograms, the SI unit of mass, to obtain ΔE in joules, the SI unit for energy.

SAMPLE EXERCISE 21.9

How much energy is lost or gained when a mole of cobalt-60 undergoes beta decay: ${}^{60}_{27}\text{Co} \longrightarrow {}^{0}_{-1}\text{e} + {}^{60}_{28}\text{Ni}$? The mass of the ${}^{60}_{27}\text{Co}$ atom is 59.933819 amu, and that of a ${}^{60}_{28}\text{Ni}$ atom is 59.930788 amu.

Solution

Analyze and Plan: We are asked to calculate the energy change in a nuclear reaction. To do this we must calculate the mass change in the process. We are given atomic masses, but we need the masses of the nuclei in the reaction. We calculate these by taking account of the masses of the electrons that contribute to the atomic masses.

Solve: A $^{60}_{27}Co$ atom has 27 electrons. The mass of an electron is 5.4858×10^{-4} amu. (See the list of fundamental constants in the back inside cover). We subtract the mass of the 27 electrons from the mass of the $^{60}_{27}Co$ *atom* to find the mass of the $^{60}_{27}Co$ *nucleus*:

$$59.933819 \text{ amu} - (27)(5.4858 \times 10^{-4} \text{ amu}) = 59.919007 \text{ amu (or 59.919007 g/mol)}$$

Likewise, for $^{60}_{28}Ni$, the mass of the nucleus is

$$59.930788 \text{ amu} - (28)(5.4858 \times 10^{-4} \text{ amu}) = 59.915428 \text{ amu (or 59.915428 g/mol)}$$

The mass change in the nuclear reaction is the total mass of the products minus the mass of the reactant:

$$\Delta m = \text{mass of electron} + \text{mass } ^{60}_{28}\text{Ni nucleus} - \text{mass of } ^{60}_{27}\text{Co nucleus}$$
$$= 0.00054858 \text{ amu} + 59.915428 \text{ amu} - 59.919007 \text{ amu}$$
$$= -0.003031 \text{ amu}$$

Thus, when a mole of cobalt-60 decays,

$$\Delta m = -0.003031 \text{ g}$$

Because the mass decreases ($\Delta m < 0$), energy is released ($\Delta E < 0$). The quantity of energy released *per mole* of cobalt-60 is calculated using Equation 21.22:

$$\Delta E = c^2 \Delta m$$
$$= (2.9979 \times 10^8 \text{m/s})^2(-0.003031 \text{ g})\left(\frac{1 \text{ kg}}{1000 \text{ g}}\right)$$
$$= -2.724 \times 10^{11} \frac{\text{kg-m}^2}{\text{s}^2} = 2.724 \times 10^{11} \text{ J}$$

PRACTICE EXERCISE

Positron emission from ^{11}C,

$$^{11}_{6}C \longrightarrow {}^{11}_{5}B + {}^{0}_{1}e$$

occurs with release of 2.87×10^{11} J per mole of ^{11}C. What is the mass change per mole of ^{11}C in this nuclear reaction?

Answer: -3.19×10^{-3} g

Nuclear Binding Energies

Scientists discovered in the 1930s that the masses of nuclei are always less than the masses of the individual nucleons of which they are composed. For example, the helium-4 nucleus has a mass of 4.00150 amu. The mass of a proton is 1.00728 amu, and that of a neutron is 1.00866 amu. Consequently, two protons and two neutrons have a total mass of 4.03188 amu:

$$\text{Mass of two protons} = 2(1.00728 \text{ amu}) = 2.01456 \text{ amu}$$
$$\text{Mass of two neutrons} = 2(1.00866 \text{ amu}) = 2.01732 \text{ amu}$$
$$\text{Total mass} = 4.03188 \text{ amu}$$

The mass of the individual nucleons is 0.03038 amu greater than that of the helium-4 nucleus:

$$\text{Mass of two protons and two neutrons} = 4.03188 \text{ amu}$$
$$\text{Mass of } ^{4}_{2}\text{He nucleus} = 4.00150 \text{ amu}$$
$$\text{Mass difference} = 0.03038 \text{ amu}$$

The mass difference between a nucleus and its constituent nucleons is called the **mass defect**. The origin of the mass defect is readily understood if we consider that energy must be added to a nucleus in order to break it into separated protons and neutrons:

$$\text{Energy} + {}^{4}_{2}\text{He} \longrightarrow 2\,{}^{1}_{1}\text{p} + 2\,{}^{1}_{0}\text{n} \qquad [21.23]$$

Chemistry and Life Medical Applications of Radiotracers

Radiotracers have found wide use as diagnostic tools in medicine. Table 21.5 ▶ lists some of the radiotracers and their uses. These radioisotopes are incorporated into a compound that is administered to the patient, usually intravenously. The diagnostic use of these isotopes is based upon the ability of the radioactive compound to localize and concentrate in the organ or tissue under investigation. Iodine-131, for example, has been used to test the activity of the thyroid gland. This gland is the only important user of iodine in the body. The patient drinks a solution of NaI containing iodine-131. Only a very small amount is used so that the patient does not receive a harmful dose of radioactivity. A Geiger counter placed close to the thyroid, in the neck region, determines the ability of the thyroid to take up the iodine. A normal thyroid will absorb about 12% of the iodine within a few hours.

The medical applications of radiotracers are further illustrated by positron emission tomography (PET). PET is used for clinical diagnosis of many diseases. In this method compounds containing radionuclides that decay by positron emission are injected into a patient. These compounds are chosen to enable researchers to monitor blood flow, oxygen and glucose metabolic rates, and other biological functions. Some of the most interesting work involves the study of the brain, which depends on glucose for most of its energy. Changes in how this sugar is metabolized or used by the brain may signal a disease such as cancer, epilepsy, Parkinson's disease, or schizophrenia.

TABLE 21.5 Some Radionuclides Used as Radiotracers

Nuclide	Half-life	Area of the Body Studied
Iodine-131	8.04 days	Thyroid
Iron-59	44.5 days	Red blood cells
Phosphorus-32	14.3 days	Eyes, liver, tumors
Technetium-99	6.0 hours	Heart, bones, liver, and lungs
Sodium-24	14.8 hours	Circulatory system

The compound to be detected in the patient must be labeled with a radionuclide that is a positron emitter. The most widely used nuclides are carbon-11 (half-life 20.4 min), fluorine-18 (half-life 110 min), oxygen-15 (half-life 2 min), and nitrogen-13 (half-life 10 min). Glucose, for example, can be labeled with ^{11}C. Because the half-lives of positron emitters are so short, the chemist must quickly incorporate the radionuclide into the sugar (or other appropriate) molecule and inject the compound immediately. The patient is placed in an elaborate instrument [Figure 21.12(a) ▼] that measures the positron emission and constructs a computer-based image of the organ in which the emitting compound is localized. The nature of this image [Figure 21.12(b)] provides clues to the presence of disease or other abnormality and helps medical researchers understand how a particular disease affects the functioning of the brain.

(a) (b)

▲ **Figure 21.12** (a) In positron emission tomography (PET) a patient is injected with a solution of a radiolabeled compound that quickly moves to the brain. Radioactive nuclei within the compound emit positrons. The PET instrument measures the positron emissions and develops a three-dimensional image of the brain. (b) PET images of the human brain showing areas active in obsessive-compulsive behavior. Each view represents a different cross section of the brain. The red and yellow areas are the active areas, as indicated by blood flow detected by the radioactive tracer.

TABLE 21.6 Mass Defects and Binding Energies for Three Nuclei

Nucleus	Mass of Nucleus (amu)	Mass of Individual Nucleons (amu)	Mass Defect (amu)	Binding Energy (J)	Binding Energy per Nucleon (J)
${}^{4}_{2}He$	4.00150	4.03188	0.03038	4.53×10^{-12}	1.13×10^{-12}
${}^{56}_{26}Fe$	55.92068	56.44914	0.52846	7.90×10^{-11}	1.41×10^{-12}
${}^{238}_{92}U$	238.00031	239.93451	1.93420	2.89×10^{-10}	1.21×10^{-12}

The addition of energy to a system must be accompanied by a proportional increase in mass. The mass change for the conversion of helium-4 into separated nucleons is $\Delta m = 0.03038$ amu, as shown in these calculations. The energy required for this process is calculated as follows:

$$\Delta E = c^2 \Delta m$$

$$= (2.9979 \times 10^8 \text{ m/s})^2 (0.03038 \text{ amu}) \left(\frac{1 \text{ g}}{6.022 \times 10^{23} \text{ amu}} \right) \left(\frac{1 \text{ kg}}{1000 \text{ g}} \right)$$

$$= 4.534 \times 10^{-12} \text{ J}$$

The energy required to separate a nucleus into its individual nucleons is called the **nuclear binding energy**. The larger the binding energy, the more stable is the nucleus toward decomposition. The nuclear binding energies of helium-4 and two other nuclei (iron-56 and uranium-238) are compared in Table 21.6 ▲. The binding energies per nucleon (that is, the binding energy of each nucleus divided by the total number of nucleons in that nucleus) are also compared in the table.

The binding energies per nucleon can be used to compare the stabilities of different combinations of nucleons (such as 2 protons and 2 neutrons arranged either as ${}^{4}_{2}He$ or $2\,{}^{2}_{1}H$.) Figure 21.13 ▼ shows the binding energy per nucleon plotted against mass number. The binding energy per nucleon at first increases in magnitude as the mass number increases, reaching about 1.4×10^{-12} J for nuclei whose mass numbers are in the vicinity of iron-56. It then decreases slowly to about 1.2×10^{-12} J for very heavy nuclei. This trend indicates that nuclei of intermediate mass numbers are more tightly bound (and therefore more stable) than those with either smaller or larger mass numbers. This trend has two significant consequences: First, heavy nuclei gain stability and therefore give off energy if they are

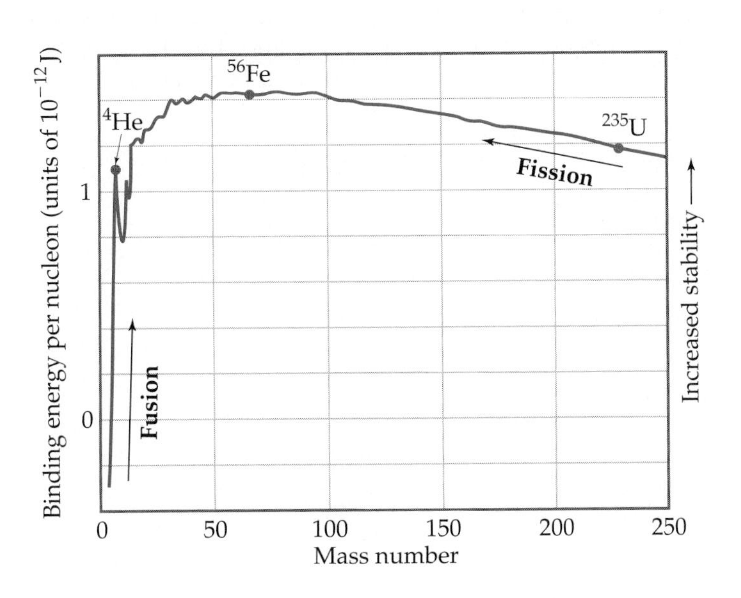

◀ **Figure 21.13** The average binding energy per nucleon increases to a maximum at a mass number of 50 to 60 and decreases slowly thereafter. As a result of these trends, fusion of light nuclei and fission of heavy nuclei are exothermic processes.

fragmented into two mid-sized nuclei. This process, known as **fission**, is used to generate energy in nuclear power plants. Second, even greater amounts of energy are released if very light nuclei are combined or fused together to give more massive nuclei. This **fusion** process is the essential energy-producing process in the Sun. We will look more closely at fission and fusion in Sections 21.7 and 21.8.

21.7 Nuclear Fission

According to our discussion of the energy changes in nuclear reactions (Section 21.6), both the splitting of heavy nuclei (fission) and the union of light nuclei (fusion) are exothermic processes. Commercial nuclear power plants and the most common forms of nuclear weaponry depend on the process of nuclear fission for their operation. The first nuclear fission to be discovered was that of uranium-235. This nucleus, as well as those of uranium-233 and plutonium-239, undergoes fission when struck by a slow-moving neutron.* This induced fission process is illustrated in Figure 21.14 ▼. A heavy nucleus can split in many different ways. Two different ways that the uranium-235 nucleus splits are shown in Equations 21.24 and 21.25:

$$ {}^{1}_{0}\text{n} + {}^{235}_{92}\text{U} \longrightarrow {}^{137}_{52}\text{Te} + {}^{97}_{40}\text{Zr} + 2\,{}^{1}_{0}\text{n} \qquad [21.24]$$

$$ {}^{1}_{0}\text{n} + {}^{235}_{92}\text{U} \longrightarrow {}^{142}_{56}\text{Ba} + {}^{91}_{36}\text{Kr} + 3\,{}^{1}_{0}\text{n} \qquad [21.25]$$

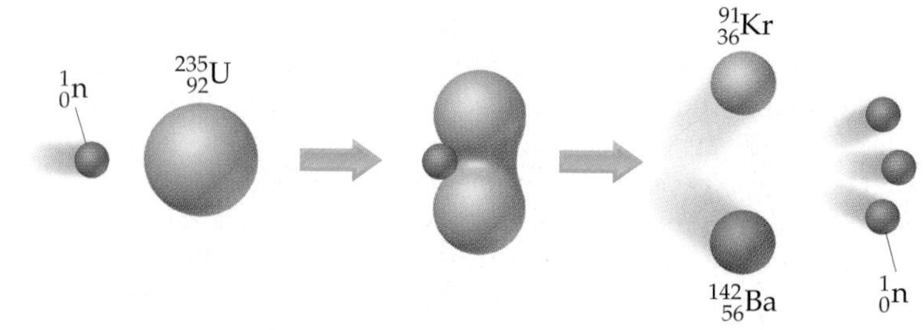

▶ **Figure 21.14** Schematic representation of the fission of uranium-235 showing one of its many fission patterns. In this process 3.5×10^{-11} J of energy is produced per ^{235}U nucleus.

More than 200 different isotopes of 35 different elements have been found among the fission products of uranium-235. Most of them are radioactive.

On the average, 2.4 neutrons are produced by every fission of uranium-235. If one fission produces 2 neutrons, these 2 neutrons can cause two fissions. The 4 neutrons thereby released can produce four fissions, and so forth, as shown in Figure 21.15 ▼. The number of fissions and the energy released quickly escalate,

* Other heavy nuclei can be induced to undergo fission. However, these three are the only ones of practical importance.

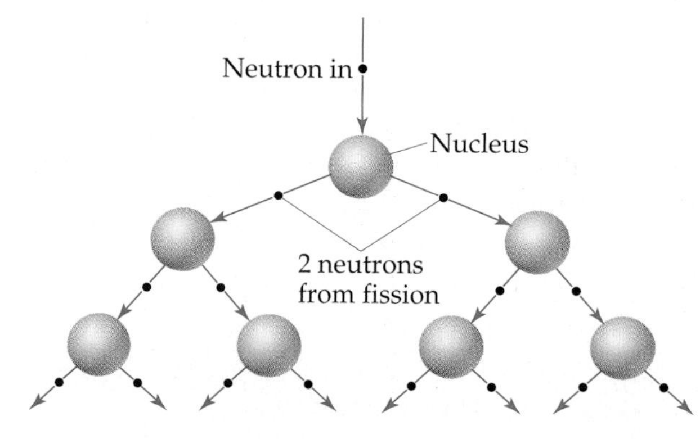

▶ **Figure 21.15** Chain fission reaction in which each fission produces two neutrons. The process leads to an accelerating rate of fission, with the number of fissions potentially doubling at each stage.

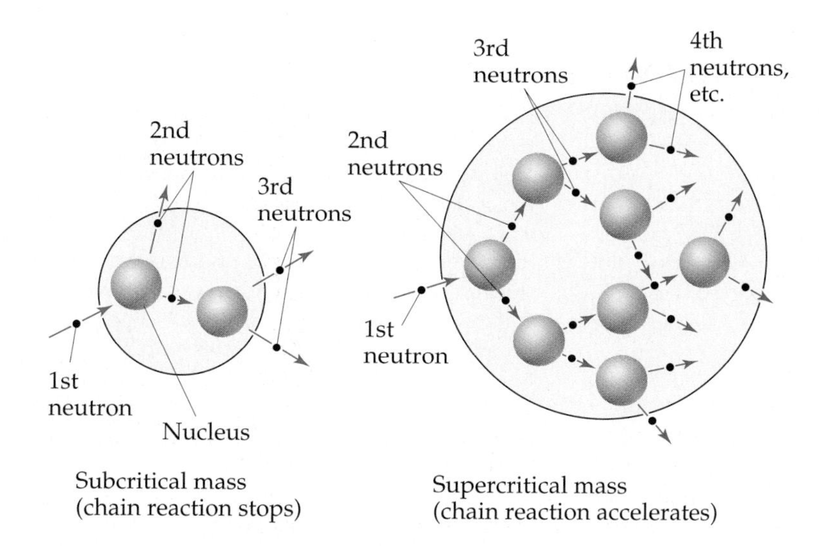

◀ **Figure 21.16** The chain reaction in a subcritical mass soon stops because neutrons are lost from the mass without causing fission. As the size of the mass increases, fewer neutrons are able to escape. In a supercritical mass, the chain reaction is able to accelerate.

and if the process is unchecked, the result is a violent explosion. Reactions that multiply in this fashion are called **chain reactions**.

In order for a fission chain reaction to occur, the sample of fissionable material must have a certain minimum mass. Otherwise, neutrons escape from the sample before they have the opportunity to strike other nuclei and cause additional fission. The chain stops if enough neutrons are lost. The amount of fissionable material large enough to maintain the chain reaction with a constant rate of fission is called the **critical mass**. When a critical mass of material is present, one neutron on average from each fission is subsequently effective in producing another fission. The critical mass of uranium-235 is about 1 kg. If more than a critical mass of fissionable material is present, very few neutrons escape. The chain reaction thus multiplies the number of fissions, which can lead to a nuclear explosion. A mass in excess of a critical mass is referred to as a **supercritical mass**. The effect of mass on a fission reaction is illustrated in Figure 21.16 ▲.

Figure 21.17 ▶ shows a schematic diagram of the first atomic bomb used in warfare, the bomb that was dropped on Hiroshima, Japan, on August 6, 1945. To trigger a fission reaction, two subcritical masses of uranium-235 are slammed together using chemical explosives. The combined masses of the uranium form a supercritical mass, which leads to a rapid, uncontrolled chain reaction and, ultimately, a nuclear explosion. The energy released by the bomb dropped on Hiroshima was equivalent to that of 20,000 tons of TNT (it therefore is called a *20-kiloton* bomb). Unfortunately, the basic design of a fission-based atomic bomb is quite simple. The fissionable materials are potentially available to any nation with a nuclear reactor. This simplicity has resulted in the proliferation of atomic weapons.

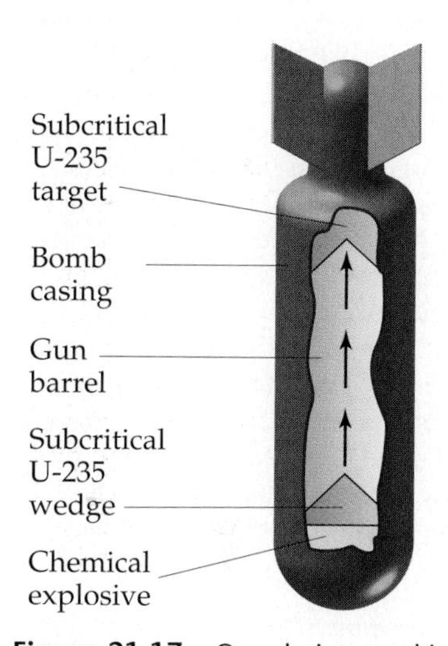

▲ **Figure 21.17** One design used in atomic bombs. A conventional explosive is used to bring two subcritical masses together to form a supercritical mass.

Nuclear Reactors

Nuclear fission produces the energy generated by nuclear power plants. The "fuel" of the nuclear reactor is a fissionable substance, such as uranium-235. Typically, uranium is enriched to about 3% uranium-235 and then used in the form of UO_2 pellets. These enriched uranium pellets are encased in zirconium or stainless steel tubes. Rods composed of materials such as cadmium or boron control the fission process by absorbing neutrons. These *control rods* regulate the flux of neutrons to keep the reaction chain self-sustaining, while preventing the reactor core from overheating.*

* The reactor core cannot reach supercritical levels and explode with the violence of an atomic bomb because the concentration of uranium-235 is too low. However, if the core overheats, sufficient damage might be done to release radioactive materials into the environment.

A Closer Look The Dawning of the Nuclear Age

The fission of uranium-235 was first achieved in the late 1930s by Enrico Fermi and his colleagues in Rome, and shortly thereafter by Otto Hahn and his coworkers in Berlin. Both groups were trying to produce transuranium elements. In 1938 Hahn identified barium among his reaction products. He was puzzled by this observation and questioned the identification because the presence of barium was so unexpected. He sent a detailed letter describing his experiments to Lise Meitner, a former coworker. Meitner had been forced to leave Germany because of the anti-Semitism of the Third Reich and had settled in Sweden. She surmised that Hahn's experiment indicated a new nuclear process was occurring in which the uranium-235 split. She called this process *nuclear fission*.

Meitner passed word of this discovery to her nephew, Otto Frisch, a physicist working at Niels Bohr's institute in Copenhagen. He repeated the experiment, verifying Hahn's observations and finding that tremendous energies were involved. In January 1939 Meitner and Frisch published a short article describing this new reaction. In March 1939 Leo Szilard and Walter Zinn at Columbia University discovered that more neutrons are produced than are used in each fission. As we have seen, this allows a chain reaction process to occur.

News of these discoveries and an awareness of their potential use in explosive devices spread rapidly within the scientific community. Several scientists finally persuaded Albert Einstein, the most famous physicist of the time, to write a letter to President Roosevelt explaining the implications of these discoveries. Einstein's letter, written in August 1939, outlined the possible military applications of nuclear fission and emphasized the danger that weapons based on fission would pose if they were to be developed by the Nazis. Roosevelt judged it imperative that the United States investigate the possibility of such weapons. Late in 1941 the decision was made to build a bomb based on the fission reaction. An enormous research project, known as the "Manhattan Project," began.

On December 2, 1942, the first artificial self-sustaining nuclear fission chain reaction was achieved in an abandoned squash court at the University of Chicago (Figure 21.18 ▼). This accomplishment led to the development of the first atomic bomb, at Los Alamos National Laboratory in New Mexico in July 1945. In August 1945 the United States dropped atomic bombs on two Japanese cities, Hiroshima and Nagasaki. The nuclear age had arrived.

▲ **Figure 21.18** The first self-sustaining nuclear fission reactor was built on a squash court at the University of Chicago. The painting depicts the scene in which the scientists witnessed the reactor as it became self-sustaining on December 2, 1942.

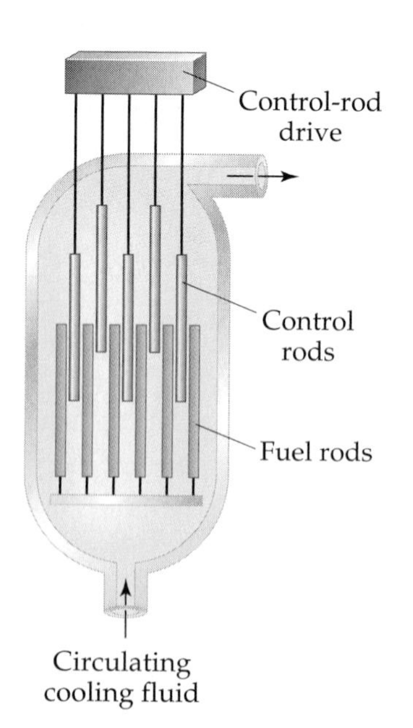

▲ **Figure 21.19** Reactor core showing fuel elements, control rods, and cooling fluid.

The reactor is started by a neutron-emitting source; it is stopped by inserting the control rods more deeply into the reactor core, the site of the fission (Figure 21.19 ◀). The reactor core also contains a *moderator*, which acts to slow down neutrons so that they can be captured more readily by the fuel. A *cooling liquid* circulates through the reactor core to carry off the heat generated by the nuclear fission. The cooling liquid can also serve as the neutron moderator.

The design of a nuclear power plant is basically the same as that of a power plant that burns fossil fuel (except that the burner is replaced by a reactor core). In both instances steam is used to drive a turbine connected to an electrical generator. The steam must be condensed; therefore, additional cooling water, generally obtained from a large source such as a river or lake, is needed. The nuclear power plant design shown in Figure 21.20 ▶ is currently the most popular. The primary coolant, which passes through the core, is in a closed system. Other coolants never pass through the reactor core at all. This lessens the chance that radioactive products could escape the core. Additionally, the reactor is surrounded by a reinforced concrete shell to shield personnel and nearby residents from radiation, and to protect the reactor from external forces.

Fission products accumulate as the reactor operates. These products decrease the efficiency of the reactor by capturing neutrons. The reactor must be stopped periodically so that the nuclear fuel can be replaced or reprocessed. When the fuel

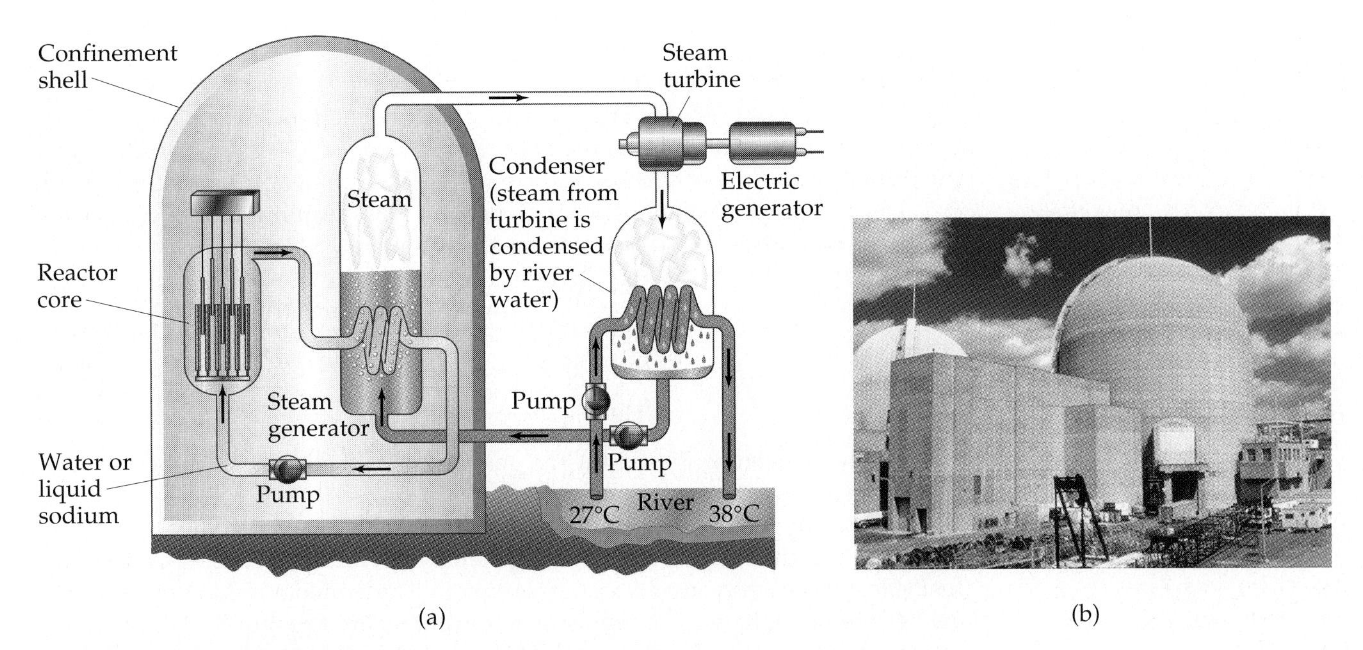

▲ **Figure 21.20** (a) The basic design of a nuclear power plant. Heat produced by the reactor core is carried by a cooling fluid such as water or liquid sodium to a steam generator. The steam so produced is used to drive an electrical generator. (b) A nuclear power plant in Salem, New Jersey. Notice the dome-shaped concrete containment shell.

rods are removed from the reactor, they are initially very radioactive. It was originally intended that they be stored for several months in pools at the reactor site to allow decay of short-lived radioactive nuclei. They were then to be transported in shielded containers to reprocessing plants where the fuel would be separated from the fission products. Reprocessing plants have been plagued with operational difficulties, however, and there is intense opposition to the transport of nuclear wastes on the nation's highways. Even if the transportation difficulties could be overcome, the high level of radioactivity of the spent fuel makes reprocessing a hazardous operation. At present, the spent fuel rods are simply being kept in storage at reactor sites.

Storage poses a major problem because the fission products are extremely radioactive. It is estimated that 20 half-lives are required for their radioactivity to reach levels acceptable for biological exposure. Based on the 28.8-yr half-life of strontium-90, one of the longer-lived and most dangerous of the products, the wastes must be stored for 600 yr. If plutonium-239 is not removed, storage must be for longer periods because plutonium-239 has a half-life of 24,000 yr. It is otherwise advantageous, however, to remove plutonium-239 because it can be used as a fissionable fuel.

A considerable amount of research is being devoted to disposal of radioactive wastes. At present, the most attractive possibilities appear to be formation of glass, ceramic, or synthetic rock from the wastes, as a means of immobilizing them. These solid materials would then be placed in containers of high corrosion resistance and durability and buried deep underground. The United States is currently investigating Yucca Mountain in Nevada as a possible disposal site. Because the radioactivity will persist for a long time, there must be assurances that the solids and their containers will not crack from the heat generated by nuclear decay, allowing radioactivity to find its way into underground water supplies.

21.8 Nuclear Fusion

Recall from Section 21.6 that energy is produced when light nuclei are fused into heavier ones. Reactions of this type are responsible for the energy produced by the Sun. Spectroscopic studies indicate that the Sun is composed of 73% H, 26% He, and only 1% of all other elements, by mass. Among the several fusion processes that are believed to occur are the following:

$$^{1}_{1}H + ^{1}_{1}H \longrightarrow ^{2}_{1}H + ^{0}_{1}e \quad [21.26]$$

$$^{1}_{1}H + ^{2}_{1}H \longrightarrow ^{3}_{2}He \quad [21.27]$$

$$^{3}_{2}He + ^{3}_{2}He \longrightarrow ^{4}_{2}He + 2^{1}_{1}H \quad [21.28]$$

$$^{3}_{2}He + ^{1}_{1}H \longrightarrow ^{4}_{2}He + ^{0}_{1}e \quad [21.29]$$

Theories have been proposed for the generation of the other elements through fusion processes.

Fusion is appealing as an energy source because of the availability of light isotopes and because fusion products are generally not radioactive. Despite this fact, fusion is not presently used to generate energy. The problem is that high energies are needed to overcome the repulsion between nuclei. The required energies are achieved by high temperatures. Fusion reactions are therefore also known as **thermonuclear reactions**. The lowest temperature required for any fusion is that needed to fuse deuterium ($^{2}_{1}H$) and tritium ($^{3}_{1}H$), shown in Equation 21.30. This reaction requires a temperature of about 40,000,000 K:

$$^{2}_{1}H + ^{3}_{1}H \longrightarrow ^{4}_{2}He + ^{1}_{0}n \quad [21.30]$$

Such high temperatures have been achieved by using an atomic bomb to initiate the fusion process. This is done in the thermonuclear, or hydrogen, bomb. This approach is unacceptable, however, for controlled power generation.

Numerous problems must be overcome before fusion becomes a practical energy source. In addition to the high temperatures necessary to initiate the reaction, there is the problem of confining the reaction. No known structural material is able to withstand the enormous temperatures necessary for fusion. Research has centered on the use of an apparatus called a *tokamak*, which uses strong magnetic fields to contain and to heat the reaction (Figure 21.21 ▼).

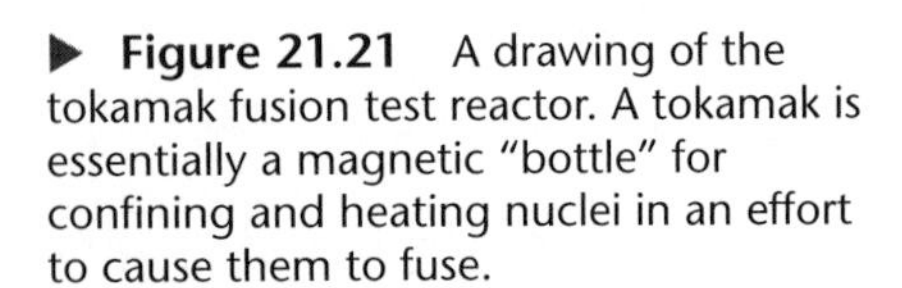

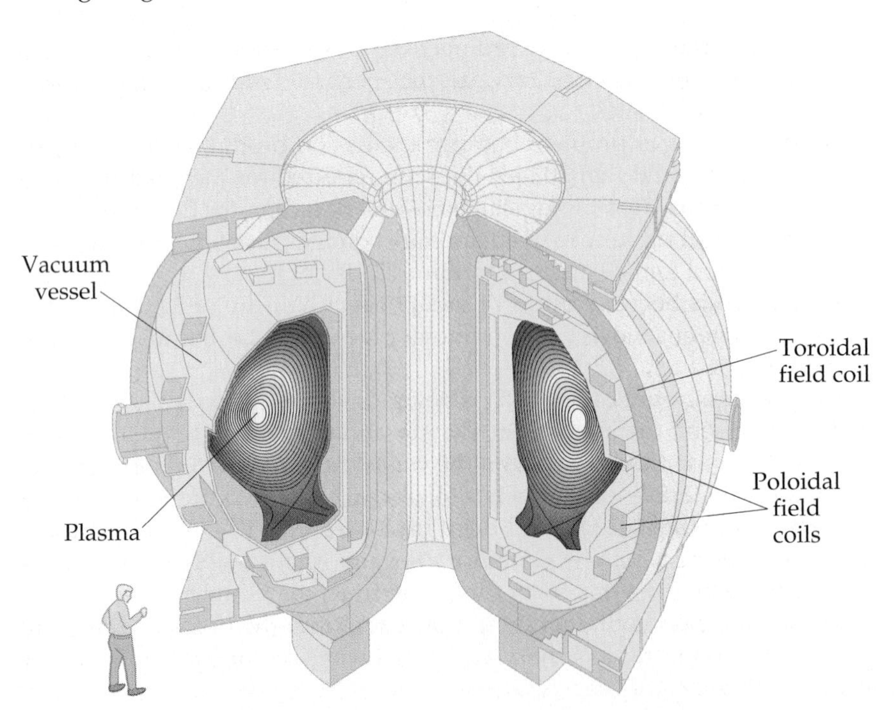

▶ **Figure 21.21** A drawing of the tokamak fusion test reactor. A tokamak is essentially a magnetic "bottle" for confining and heating nuclei in an effort to cause them to fuse.

Temperatures of nearly 3,000,000 K have been achieved in a tokamak, but this is not yet hot enough to initiate continuous fusion. Much research has also been directed at the use of powerful lasers to generate the necessary temperatures.

21.9 Biological Effects of Radiation

We are continually bombarded by radiation from both natural and artificial sources. We are exposed to infrared, ultraviolet, and visible radiation from the Sun, for example, in addition to radio waves from radio and television stations, microwaves from microwave ovens, and X rays from various medical procedures. We are also exposed to radioactivity from the soil and other natural materials. Understanding the different energies of these various kinds of radiation is necessary to understand their different effects on matter.

When matter absorbs radiation, the energy of the radiation can cause either excitation or ionization of the matter. Excitation occurs when the absorbed radiation excites electrons to higher energy states or increases the motion of molecules, causing them to move, vibrate, or rotate. Ionization occurs when the radiation removes an electron from an atom or molecule. In general, radiation that causes ionization, called **ionizing radiation**, is far more harmful to biological systems than radiation that does not cause ionization. The latter, called **non-ionizing radiation**, is generally of lower energy, such as radiofrequency electromagnetic radiation, (Section 6.2) or slow-moving neutrons. Most living tissue contains at least 70% water by mass. When living tissue is irradiated, most of the energy of the radiation is absorbed by water molecules. Thus, it is common to define ionizing radiation as radiation that can ionize water, a process requiring a minimum energy of 1216 kJ/mol. Alpha, beta, and gamma rays (as well as X rays and higher-energy ultraviolet radiation) possess energies in excess of this quantity and are therefore forms of ionizing radiation.

When ionizing radiation passes through living tissue, electrons are removed from water molecules, forming highly reactive H_2O^+ ions. An H_2O^+ ion can react with another water molecule to form an H_3O^+ ion and a neutral OH molecule:

$$H_2O^+ + H_2O \longrightarrow H_3O^+ + OH \qquad [21.31]$$

The unstable and highly reactive OH molecule is a **free radical**, a substance with one or more unpaired electrons, as seen in the Lewis structure for this molecule shown here, $\cdot\ddot{\underset{\cdot\cdot}{O}}—H$. The presence of the unpaired electron is often emphasized by writing the species with a single dot, $\cdot OH$. In cells and tissues, such particles can attack a host of surrounding biomolecules to produce new free radicals, which, in turn, attack yet other compounds. Thus, the formation of a single free radical can initiate a large number of chemical reactions that are ultimately able to disrupt the normal operations of cells.

The damage produced by radiation depends on the activity and energy of the radiation, the length of exposure, and whether the source is inside or outside the body. Gamma rays are particularly harmful outside the body, because they penetrate human tissue very effectively, just as X rays do. Consequently, their damage is not limited to the skin. In contrast, most alpha rays are stopped by skin, and beta rays are able to penetrate only about 1 cm beyond the surface of the skin (Figure 21.22 ▶). Neither is as dangerous as gamma rays, therefore, unless the radiation source somehow enters the body. Within the body, alpha rays are particularly dangerous because they transfer their energy efficiently to the surrounding tissue, initiating considerable damage.

In general, the tissues that show the greatest damage from radiation are those that reproduce at a rapid rate, such as bone marrow, blood-forming tissues, and lymph nodes. The principal effect of extended exposure to low doses of radiation

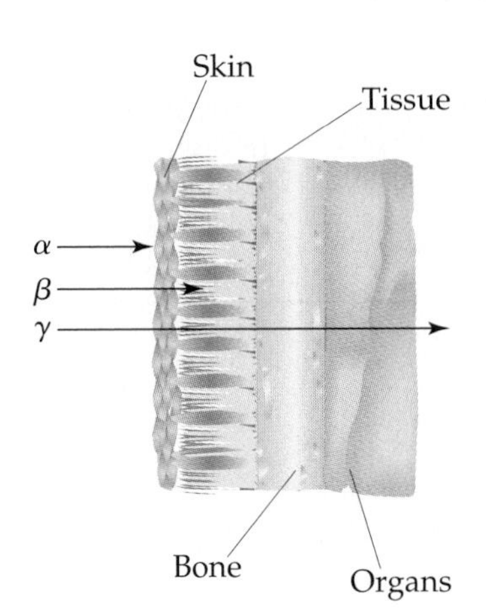

▲ **Figure 21.22** The relative penetrating abilities of alpha, beta, and gamma radiation.

is to cause cancer. Cancer is caused by damage to the growth-regulation mechanism of cells, inducing cells to reproduce in an uncontrolled manner. Leukemia, which is characterized by excessive growth of white blood cells, is probably the major cancer problem associated with radiation.

In light of the biological effects of radiation, it is important to determine whether any levels of exposure are safe. Unfortunately, we are hampered in our attempts to set realistic standards because we don't fully understand the effects of long-term exposure to radiation. Scientists concerned with setting health standards have used the hypothesis that the effects of radiation are proportional to exposure, even down to low doses. Any amount of radiation is assumed to cause some finite risk of injury, and the effects of high dosage rates are extrapolated to those of lower ones. Other scientists believe, however, that there is a threshold below which there are no radiation risks. Until scientific evidence enables us to settle the matter with some confidence, it is safer to assume that even low levels of radiation present some danger.

Radiation Doses

Two units commonly used to measure the amount of exposure to radiation are the *gray* and the *rad*. The **gray** (Gy), which is the SI unit of absorbed dose, corresponds to the absorption of 1 J of energy per kilogram of tissue. The **rad** (*r*adiation *a*bsorbed *d*ose) corresponds to the absorption of 1×10^{-2} J of energy per kilogram of tissue. Thus, 1 Gy = 100 rads. The rad is the unit most often used in medicine.

Not all forms of radiation harm biological materials with the same efficiency. A rad of alpha radiation, for example, can produce more damage than a rad of beta radiation. To correct for these differences, the radiation dose is multiplied by a factor that measures the relative biological damage caused by the radiation. This multiplication factor is known as the *relative biological effectiveness* of the radiation, abbreviated *RBE*. The RBE is approximately 1 for gamma and beta radiation, and 10 for alpha radiation. The exact value of the RBE varies with dose rate, total dose, and the type of tissue affected. The product of the radiation dose in rads and the RBE of the radiation gives the effective dosage in units of **rem** (*r*oentgen *e*quivalent for *m*an):

$$\text{Number of rems} = (\text{number of rads})(\text{RBE}) \qquad [21.32]$$

The SI unit for effective dosage is the sievert (Sv), obtained by multiplying the RBE times the SI unit for radiation dose, the gray; hence, 1 Sv = 100 rem. The rem is the unit of radiation damage that is usually used in medicine.

The effects of short-term exposures to radiation appear in Table 21.7 ◀. An exposure of 600 rem is fatal to most humans. To put this number in perspective, a typical dental X ray entails an exposure of about 0.5 mrem. The average exposure for a person in one year due to all natural sources of ionizing radiation (called *background radiation*) is about 360 mrem.

TABLE 21.7 Effects of Short-Term Exposures to Radiation

Dose (rem)	Effect
0 to 25	No detectable clinical effects
25 to 50	Slight, temporary decrease in white blood cell counts
100 to 200	Nausea; marked decrease in white blood cells
500	Death of half the exposed population within 30 days after exposure

Radon

The radioactive noble gas radon has been much publicized in recent years as a potential risk to health. Radon-222 is a product of the nuclear disintegration series of uranium-238 (Figure 21.4) and is continually generated as uranium in rocks and soil decays. As Figure 21.23 ▶ indicates, radon exposure is estimated to account for more than half the 360-mrem average annual exposure to ionizing radiation.

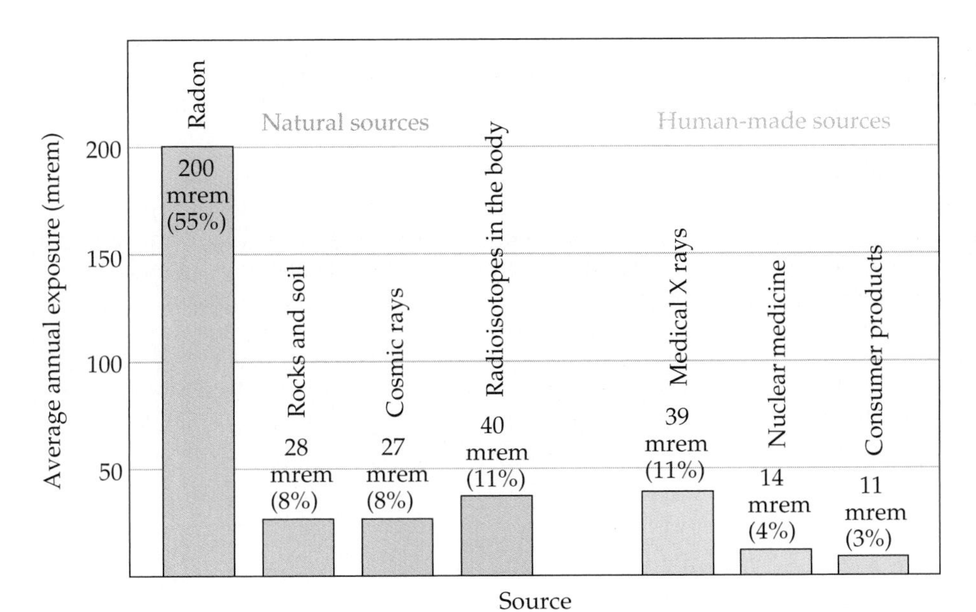

◀ **Figure 21.23** A graph of the sources of the average annual exposure of the U.S. population to high-energy radiation. The total average annual exposure is 360 mrem. (Data from "Ionizing Radiation Exposure of the Population of the United States," Report 93, 1987, National Council on Radiation Protection.)

Chemistry and Life Radiation Therapy

High-energy radiation poses a health hazard because of the damage it does to cells. Healthy cells are either destroyed or damaged by radiation, leading to physiological disorders. Radiation can also destroy *unhealthy* cells, however, including cancerous cells. All cancers are characterized by the runaway growth of abnormal cells. This growth can produce masses of abnormal tissue, called *malignant tumors*. Malignant tumors can be caused by the exposure of healthy cells to high-energy radiation. Somewhat paradoxically, however, malignant tumors can be destroyed by exposing them to the same radiation because rapidly reproducing cells are very susceptible to radiation damage. Thus, cancerous cells are more susceptible to destruction by radiation than healthy ones, allowing radiation to be used effectively in the treatment of cancer. As early as 1904, physicians attempted to use the radiation emitted by radioactive substances to treat tumors by destroying the mass of unhealthy tissue. The treatment of disease by high-energy radiation is called *radiation therapy*.

Many different radionuclides are currently used in radiation therapy. Some of the more commonly used ones are listed in Table 21.8 ▼, along with their half-lives. Most of the half-lives are quite short, meaning that these radioisotopes emit a great deal of radiation in a short period of time (Figure 21.24 ▶).

TABLE 21.8 Some Radioisotopes Used in Radiation Therapy

Isotope	Half-life	Isotope	Half-life
^{32}P	14.3 days	^{137}Cs	30 yr
^{60}Co	5.26 yr	^{192}Ir	74.2 days
^{90}Sr	28.8 yr	^{198}Au	2.7 days
^{125}I	60.25 days	^{222}Rn	3.82 days
^{131}I	8.04 days	^{226}Ra	1600 yr

▲ **Figure 21.24** Vials that contain a salt of cesium-137, which is used in radiation therapy. The blue glow is from the radioactivity of the cesium. In Goiânia, Brazil, in 1987 a cylinder containing cesium-137 was left in an abandoned medical clinic. It was discovered by unsuspecting townspeople, who were fascinated by the strange blue glow. The results were tragic: Four people died from radiation exposure, and 249 others were contaminated.

The radiation source used in radiation therapy may be inside or outside the body. In almost all cases radiation therapy is designed to use the high-energy gamma radiation emitted by radioisotopes. Alpha and beta radiation, which are not as penetrating as gamma radiation, can be blocked by appropriate packaging. For example, ^{192}Ir is often administered as "seeds" consisting of a core of radioactive isotope coated with 0.1 mm of platinum metal. The platinum coating stops the alpha and beta rays, but the gamma rays penetrate it readily. The radioactive seeds can be surgically implanted in a tumor. In other cases human physiology allows the radioisotope to be ingested. For example, most of the iodine in the human body ends up in the thyroid gland (see the "Chemistry and Life" box in Section 21.6), so thyroid cancer can be treated by using large doses of ^{131}I. Radiation therapy on deep organs, where a surgical implant is impractical, often uses a ^{60}Co "gun" outside the body to shoot a beam of gamma rays at the tumor. Particle accelerators are also used as an external source of high-energy radiation for radiation therapy.

Because gamma radiation is so strongly penetrating, it is nearly impossible to avoid damaging healthy cells during radiation therapy. Most cancer patients undergoing radiation treatments experience unpleasant and dangerous side effects such as fatigue, nausea, hair loss, a weakened immune system, and even death. In many cases, therefore, radiation therapy is used only if other cancer treatments, such as *chemotherapy* (the treatment of cancer with powerful drugs), are unsuccessful. Nevertheless, radiation therapy is one of the major weapons we have in the fight against cancer.

The interplay between the chemical and nuclear properties of radon makes it a health hazard. Because radon is a noble gas, it is extremely unreactive and is therefore free to escape from the ground without chemically reacting along the way. It is readily inhaled and exhaled with no direct chemical effects. The half-life of ^{222}Rn, however, is only 3.82 days. It decays by losing an alpha particle into a radioisotope of polonium:

$$^{222}_{86}\text{Rn} \longrightarrow {}^{218}_{84}\text{Po} + {}^{4}_{2}\text{He} \qquad [21.33]$$

Because radon has such a short half-life and alpha particles have a high RBE, inhaled radon is considered a probable cause of lung cancer. Even worse, however, is the decay product, polonium-218, which is an alpha-emitting solid that has an even shorter half-life (3.11 min) than radon-222:

$$^{218}_{84}\text{Po} \longrightarrow {}^{214}_{82}\text{Pb} + {}^{4}_{2}\text{He} \qquad [21.34]$$

▲ **Figure 21.25** Radon home-testing kit. Kits such as this are available for measuring radon levels in the home.

The atoms of polonium-218 can become trapped in the lungs, where they continually bathe the delicate tissue with harmful alpha radiation. The resulting damage is estimated to result in as many as 10% of lung cancer deaths.

The U.S. Environmental Protection Agency (EPA) has recommended that radon-222 levels not exceed 4 pCi per liter of air in homes. Homes located in areas where the natural uranium content of the soil is high often have levels much greater than that. As a result of public awareness, radon-testing kits are readily available in many parts of the country (Figure 21.25 ◀).

SAMPLE INTEGRATIVE EXERCISE 21: Putting Concepts Together

Potassium ion is present in foods and is an essential nutrient in the human body. One of the naturally occurring isotopes of potassium, potassium-40, is radioactive. Potassium-40 has a natural abundance of 0.0117% and a half-life of $t_{1/2} = 1.28 \times 10^9$ yr. It undergoes radioactive decay in three ways: 98.2% is by electron capture, 1.35% is by beta emission, and 0.49% is by positron emission. **(a)** Why should we expect ^{40}K to be radioactive? **(b)** Write the nuclear equations for the three modes by which ^{40}K decays. **(c)** How many $^{40}K^+$ ions are present in 1.00 g of KCl? **(d)** How long does it take for 1.00% of the ^{40}K in a sample to undergo radioactive decay?

Solution (a) The ^{40}K nucleus contains 19 protons and 21 neutrons. There are very few stable nuclei with odd numbers of both protons and neutrons. (Section 21.2)

(b) Electron capture is capture of an inner-shell electron ($_{-1}^{0}e$) by the nucleus:

$$_{19}^{40}K + _{-1}^{0}e \longrightarrow _{18}^{40}Ar$$

Beta emission is loss of a beta particle ($_{-1}^{0}e$) by the nucleus:

$$_{19}^{40}K \longrightarrow _{20}^{40}Ca + _{-1}^{0}e$$

Positron emission is loss of $_{1}^{0}e$ by the nucleus:

$$_{19}^{40}K \longrightarrow _{18}^{40}Ar + _{1}^{0}e$$

(c) The total number of K^+ ions in the sample is

$$(1.00 \text{ g KCl})\left(\frac{1 \text{ mol KCl}}{74.55 \text{ g KCl}}\right)\left(\frac{1 \text{ mol K}^+}{1 \text{ mol KCl}}\right)\left(\frac{6.022 \times 10^{23} \text{ K}^+}{1 \text{ mol K}^+}\right) = 8.08 \times 10^{21} \text{ K}^+ \text{ ions}$$

Of these, 0.0117% are $^{40}K^+$ ions:

$$(8.08 \times 10^{21} \text{ K}^+ \text{ ions})\left(\frac{0.0117\ ^{40}\text{K}^+ \text{ ions}}{100 \text{ K}^+ \text{ ions}}\right) = 9.45 \times 10^{17}\ ^{40}\text{K}^+ \text{ ions}$$

(d) The decay constant (the rate constant) for the radioactive decay can be calculated from the half-life using Equation 21.20:

$$k = \frac{0.693}{t_{1/2}} = \frac{0.693}{1.28 \times 10^9 \text{ yr}} = (5.41 \times 10^{-10})/\text{yr}$$

The rate equation, Equation 21.19, then allows us to calculate the time required:

$$\ln\frac{N_t}{N_0} = -kt$$

$$\ln\frac{99}{100} = -((5.41 \times 10^{-10})/\text{yr})t$$

$$-0.01005 = -((5.41 \times 10^{-10})/\text{yr})t$$

$$t = \frac{-0.01005}{(-5.41 \times 10^{-10})/\text{yr}} = 1.86 \times 10^7 \text{yr}$$

That is, it would take 18.6 million years for just 1.00% of the ^{40}K in a sample to decay.

Summary and Key Terms

Introduction and Section 21.1 The nucleus contains protons and neutrons, both of which are called **nucleons**. Nuclei that are **radioactive** spontaneously emit radiation. These radioactive nuclei are called **radionuclides**, and the atoms containing them are called **radioisotopes**. When a radionuclide decomposes, it is said to undergo radioactive decay. In nuclear equations, reactant and product nuclei are represented by giving their mass numbers and atomic numbers, as well as their chemical symbol. The totals of the mass numbers on both sides of the equation are equal; the totals of the atomic numbers on both sides are also equal. There are five common kinds of radioactive decay: emission of **alpha particles** ($_{2}^{4}He$), emission of **beta particles** ($_{-1}^{0}e$), **positron** emission, **electron capture**, and emission of **gamma radiation** ($_{0}^{0}\gamma$).

Section 21.2 The neutron-to-proton ratio is an important factor determining nuclear stability. By comparing a nuclide's neutron-to-proton ratio with those in the band of stability, we can predict the mode of radioactive decay. In general, neutron-rich nuclei tend to emit beta particles; proton-rich nuclei tend to either emit positrons or undergo electron capture; and heavy nuclei tend to emit alpha particles. The presence of **magic numbers** of nucleons and an even number of protons and neutrons also help determine the stability of a nucleus. A nuclide may undergo a series of decay steps before a stable nuclide forms. This series of steps is called a **radioactive series**, or a **nuclear disintegration series**.

Section 21.3 **Nuclear transmutations**, induced conversions of one nucleus into another, can be brought about by bombarding nuclei with either charged particles or neutrons. **Particle accelerators** increase the kinetic energies of positively charged particles, allowing these particles to overcome their electrostatic repulsion by the nucleus.

Nuclear transmutations are used to produce the **transuranium elements**, those elements with atomic numbers greater than that of uranium.

Sections 21.4 and 21.5 The SI unit for the activity of a radioactive source is the **becquerel** (Bq), defined as one nuclear disintegration per second. A related unit, the **curie** (Ci), corresponds to 3.7×10^{10} disintegrations per second. Nuclear decay is a first-order process. The decay rate (**activity**) is therefore proportional to the number of radioactive nuclei, and radionuclides have constant **half-lives**. Some can therefore be used to date objects; ^{14}C, for example, is used to date certain organic objects.

Geiger counters and **scintillation counters** count the emissions from radioactive samples. The ease of detection of radioisotopes also permits their use as **radiotracers**, to follow elements through their reactions.

Section 21.6 The energy produced in nuclear reactions is accompanied by measurable losses of mass in accordance with Einstein's relationship, $\Delta E = c^2 \Delta m$. The difference in mass between nuclei and the nucleons of which they are composed is known as the **mass defect**. The mass defect of a nuclide makes it possible to calculate its **nuclear binding energy**, the energy required to separate the nucleus into individual nucleons. Energy is produced when heavy nuclei split (**fission**) and when light nuclei fuse (**fusion**).

Sections 21.7 and 21.8 Uranium-235, uranium-233, and plutonium-239 undergo fission when they capture a neutron. The resulting nuclear reaction is a nuclear **chain reaction**, a reaction in which the neutrons produced in one fission cause further fission reactions. A reaction that maintains a constant rate is said to be critical, and the mass necessary to maintain this constant rate is called a **critical mass**. A mass in excess of the critical mass is termed a **supercritical mass**. In nuclear reactors the fission is controlled to generate a constant power. The reactor core consists of fissionable fuel, control rods, a moderator, and cooling fluid. The nuclear power plant resembles a conventional power plant except that the reactor core replaces the fuel burner. There is concern about the disposal of highly radioactive nuclear wastes that are generated in nuclear power plants.

Nuclear fusion requires high temperatures because nuclei must have large kinetic energies to overcome their mutual repulsions. They are therefore called **thermonuclear reactions**. It is not yet possible to generate a controlled fusion process.

Section 21.9 **Ionizing radiation** is energetic enough to remove an electron from a water molecule; radiation with less energy is called **nonionizing radiation**. Ionizing radiation generates **free radicals**, reactive substances with one or more unpaired electrons. The effects of long-term exposure to low levels of radiation are not completely understood, but it is usually assumed that the extent of biological damage varies in direct proportion to the level of exposure.

The amount of energy deposited in biological tissue by radiation is called the radiation dose and is measured in units of grays or rads. One **gray** (Gy) corresponds to a dose of 1 J/kg of tissue. The **rad** is a smaller unit; 100 rads = 1 Gy. The effective dose, which measures the biological damage created by the deposited energy, is measured in units of rems or sieverts (Sv). The **rem** is obtained by multiplying the number of rads by the relative biological effectiveness (RBE); 100 rem = 1 Sv.

Exercises

Radioactivity

21.1 Indicate the number of protons and neutrons in the following nuclei: **(a)** $^{55}_{25}Mn$ **(b)** ^{201}Hg; **(c)** potassium-39.

21.2 Indicate the number of protons and neutrons in the following nuclei: **(a)** $^{126}_{55}Cs$; **(b)** ^{119}Sn; **(c)** barium-141.

21.3 Give the symbol for each of the following particles: **(a)** proton; **(b)** positron; **(c)** beta particle.

21.4 Give the symbol for each of the following particles: **(a)** neutron; **(b)** electron; **(c)** alpha particle.

21.5 Write balanced nuclear equations for the following processes: **(a)** bismuth-214 undergoes beta decay; **(b)** gold-195 undergoes electron capture; **(c)** potassium-38 undergoes positron emission; **(d)** plutonium-242 emits alpha radiation.

21.6 Write balanced nuclear equations for the following transformations: **(a)** neodymium-141 undergoes electron capture; **(b)** gold-201 decays to a mercury isotope; **(c)** selenium-81 undergoes beta decay; **(d)** strontium-83 decays by positron emission.

21.7 Decay of which nucleus will lead to the following products: **(a)** bismuth-211 by beta decay; **(b)** chromium-50 by positron emission; **(c)** tantalum-179 by electron capture; **(d)** radium-226 by alpha decay?

21.8 What particle is produced during the following decay processes: **(a)** sodium-24 decays to magnesium-24; **(b)** mercury-188 decays to gold-188; **(c)** iodine-122 decays to xenon-122; **(d)** plutonium-242 decays to uranium-238?

21.9 The naturally occurring radioactive decay series that begins with $^{235}_{92}U$ stops with formation of the stable $^{207}_{82}Pb$ nucleus. The decays proceed through a series of alpha-particle and beta-particle emissions. How many of each type of emission are involved in this series?

21.10 A radioactive decay series that begins with $^{232}_{90}Th$ ends with formation of the stable nuclide $^{208}_{82}Pb$. How many alpha-particle emissions and how many beta-particle emissions are involved in the sequence of radioactive decays?

Nuclear Stability

21.11 Predict the type of radioactive decay process for the following radionuclides: **(a)** $^{8}_{5}B$; **(b)** $^{68}_{29}Cu$ **(c)** neptunium-241; **(d)** chlorine-39.

21.12 Each of the following nuclei undergoes either beta or positron emission. Predict the type of emission for each: **(a)** $^{66}_{32}Ge$ **(b)** $^{105}_{45}Rh$ **(c)** iodine-137; **(d)** cerium-133.

21.13 Indicate whether each of the following nuclides lies within the belt of stability in Figure 21.2: **(a)** neon-24; **(b)** chlorine-32; **(c)** tin-108; **(d)** polonium-216. For any that do not, describe a nuclear decay process that would alter the neutron-to-proton ratio in the direction of increased stability.

21.14 Indicate whether each of the following nuclides lies within the belt of stability in Figure 21.2: **(a)** $^{79}_{35}Br$; **(b)** $^{94}_{43}Tc$; **(c)** $^{76}_{36}Kr$; **(d)** $^{113}_{48}Cd$. For any that do not, describe a nuclear decay process that would alter the neutron-to-proton ratio in the direction of increased stability.

21.15 One of the nuclides in each of the following pairs is radioactive. Predict which is radioactive and which is stable: **(a)** $^{39}_{19}K$ and $^{40}_{19}K$; **(b)** ^{209}Bi and ^{208}Bi, **(c)** magnesium-25 and neon-24. Explain.

21.16 In each of the following pairs, which nuclide would you expect to be the more abundant in nature: **(a)** $^{115}_{48}Cd$ or $^{112}_{48}Cd$; **(b)** $^{30}_{13}Al$ or $^{27}_{13}Al$ **(c)** palladium-106 or palladium-113; **(d)** xenon-128 or cesium-128? Justify your choices.

21.17 Which of the following nuclides have magic numbers of both protons and neutrons: **(a)** helium-4; **(b)** carbon-12; **(c)** calcium-40; **(d)** nickel-58; **(e)** lead-208?

21.18 Tin-112 is a stable nuclide, but indium-112 is radioactive, with a half-life of only 14 min. How can we explain this difference in nuclear stability:

21.19 Which of the following nuclides of group 6A elements would you expect to be radioactive: **(a)** $^{14}_{8}O$, $^{32}_{16}S$, $^{78}_{34}Se$, $^{115}_{52}Te$, or $^{208}_{84}Po$? Justify your choices.

21.20 Which of the following nuclides would you expect to be radioactive: **(a)** $^{62}_{28}Ni$; **(b)** $^{58}_{29}Cu$; **(c)** $^{108}_{47}Ag$; **(d)** tungsten-184; **(e)** polonium-206? Justify your choices.

Nuclear Transmutations

21.21 Why are nuclear transmutations involving neutrons generally easier to accomplish than those involving protons or alpha particles?

21.22 Rutherford was able to carry out the first nuclear transmutation reactions by bombarding nitrogen-14 nuclei with alpha particles. In the famous experiment on scattering of alpha particles by gold foil (Section 2.2), however, a nuclear transmutation reaction did not occur. What is the difference between the two experiments? What would one need to do to carry out a successful nuclear transmutation reaction involving gold nuclei and alpha particles?

21.23 Complete and balance the following nuclear equations by supplying the missing particle:

(a) $^{32}_{16}S + ^{1}_{0}n \longrightarrow ^{1}_{1}p + ?$

(b) $^{7}_{4}Be + ^{0}_{-1}e$(orbital electron) $\longrightarrow ?$

(c) $? \longrightarrow ^{187}_{76}Os + ^{0}_{-1}e$

(d) $^{98}_{42}Mo + ^{2}_{1}H \longrightarrow ^{1}_{0}n + ?$

(e) $^{235}_{92}U + ^{1}_{0}n \longrightarrow ^{135}_{54}Xe + 2^{1}_{0}n + ?$

21.24 Complete and balance the following nuclear equations by supplying the missing particle:

(a) $^{252}_{98}Cf + ^{10}_{5}B \longrightarrow 3^{1}_{0}n + ?$

(b) $^{2}_{1}H + ^{3}_{2}He \longrightarrow ^{4}_{2}He + ?$

(c) $^{1}_{1}H + ^{11}_{5}B \longrightarrow 3?$

(d) $^{122}_{53}I \longrightarrow ^{122}_{54}Xe + ?$

(e) $^{59}_{26}Fe \longrightarrow ^{0}_{-1}e + ?$

21.25 Write balanced equations for each of the following nuclear reactions: **(a)** $^{238}_{92}U(n,\gamma)^{239}_{92}U$; **(b)** $^{14}_{7}N(p,\alpha)^{11}_{6}C$; **(c)** $^{18}_{8}O(n,\beta)^{19}_{9}F$.

21.26 Write balanced equations for **(a)** $^{238}_{92}U(\alpha,n)^{241}_{94}Pu$; **(b)** $^{14}_{7}N(\alpha,p)^{17}_{8}O$; **(c)** $^{59}_{26}Fe(\alpha,\beta)^{63}_{29}Cu$.

Rates of Radioactive Decay

21.27 Harmful chemicals are often destroyed by chemical treatment. An acid, for example, can be neutralized by a base. Why can't chemical treatment be used to destroy the radioactive products produced in a nuclear reactor?

21.28 It has been suggested that strontium-90 (generated by nuclear testing) deposited in the hot desert will undergo radioactive decay more rapidly because it will be exposed to much higher average temperatures. Is this a reasonable suggestion?

21.29 The half-life of tritium (hydrogen-3) is 12.3 yr. If 48.0 mg of tritium is released from a nuclear power plant during the course of an accident, what mass of this nuclide will remain after 12.3 yr? After 49.2 yr?

21.30 It takes 5.2 minutes for a 1.000-g sample of ^{210}Fr to decay to 0.250 g. What is the half-life of ^{210}Fr?

21.31 A sample of curium-243 was prepared. After 1.00 yr the activity of the sample had declined from 3012 disintegrations per second to 2921 disintegrations per second. What is the half-life of the decay process?

21.32 A sample of zinc-72 has an initial activity of 2310 counts per minute on a device that measures the level of radioactivity. After 120 hours the activity has declined to 457 counts per minute. What is the half-life for decay of zinc-72?

21.33 How much time is required for a 5.75-mg sample of ^{51}Cr to decay to 1.50 mg if it has a half-life of 27.8 days?

21.34 Cobalt-60 has a half-life of 5.26 yr. The cobalt-60 in a radiotherapy unit must be replaced when its radioactivity falls to 75% of the original sample. If the original sample was purchased in August 2000, when will it be necessary to replace the cobalt-60?

[21.35] Radium-226, which undergoes alpha decay, has a half-life of 1600 yr. **(a)** How many alpha particles are emitted in 1.0 min by a 5.0-mg sample of ^{226}Ra? **(b)** What is the activity of the sample in mCi?

[21.36] Cobalt-60, which undergoes beta decay, has a half-life of 5.26 yr. **(a)** How many beta particles are emitted in 45.5 s by a 2.44-mg sample of ^{60}Co? **(b)** What is the activity of the sample in Bq?

21.37 A wooden artifact from a Chinese temple has a ^{14}C activity of 24.9 counts per minute as compared with an activity of 32.5 counts per minute for a standard of zero age. From the half-life for ^{14}C decay, 5715 yr, determine the age of the artifact.

21.38 The cloth shroud from around a mummy is found to have a ^{14}C activity of 8.9 disintegrations per minute per gram of carbon as compared with living organisms that undergo 15.2 disintegrations per minute per gram of carbon. From the half-life for ^{14}C decay, 5715 yr, calculate the age of the shroud.

21.39 The half-life for the process $^{238}U \longrightarrow {}^{206}Pb$ is 4.5×10^9 yr. A mineral sample contains 50.0 mg of ^{238}U and 14.0 mg of ^{206}Pb. What is the age of the mineral?

21.40 Potassium-40 decays to argon-40 with a half-life of 1.27×10^9 yr. What is the age of a rock in which the mass ratio of ^{40}Ar to ^{40}K is 3.6?

Energy Changes

21.41 The combustion of one mole of graphite releases 393.5 kJ of energy. What is the mass change that accompanies the loss of this energy?

21.42 An analytical laboratory balance typically measures mass to the nearest 0.1 mg. What energy change would accompany the loss of 0.1 mg in mass?

21.43 How much energy must be supplied to break a single sodium-23 nucleus into separated protons and neutrons if the nucleus has a mass of 22.983733 amu? How much energy is required per mole of this nucleus?

21.44 How much energy must be supplied to break a single ^{21}Ne nucleus into separated protons and neutrons if the nucleus has a mass of 20.98846 amu? What is the nuclear binding energy for 1 mol of ^{21}Ne?

21.45 Calculate the binding energy per nucleon for the following nuclei: **(a)** $^{12}_{6}C$ (nuclear mass, 11.996708 amu); **(b)** ^{37}Cl (nuclear mass, 36.956576 amu); **(c)** barium-137 (atomic mass, 136.905812 amu).

21.46 Calculate the binding energy per nucleon for the following nuclei: **(a)** $^{14}_{7}N$ (nuclear mass, 13.999234 amu); **(b)** ^{48}Ti (nuclear mass, 47.935878 amu); **(c)** mercury-201 (atomic mass, 200.970277 amu).

21.47 The solar radiation falling on Earth amounts to 1.07×10^{16} kJ/min. **(a)** What is the mass equivalence of the solar energy falling on Earth in a 24-hr period? **(b)** If the energy released in the reaction

$$^{235}U + {}^{1}_{0}n \longrightarrow {}^{141}_{56}Ba + {}^{92}_{36}Kr + 3\,{}^{1}_{0}n$$

(^{235}U nuclear mass, 234.9935 amu; ^{141}Ba nuclear mass, 140.8833 amu, ^{92}Kr nuclear mass, 91.9021 amu) is taken as typical of that occurring in a nuclear reactor, what mass of uranium-235 is required to equal 0.10% of the solar energy that falls on Earth in 1.0 day?

21.48 Based on the following atomic mass values—^{1}H, 1.00782 amu; ^{2}H, 2.01410 amu; ^{3}H, 3.01605 amu; ^{3}He, 3.01603 amu; ^{4}He, 4.00260 amu—and the mass of the neutron given in the text, calculate the energy released per mole in each of the following nuclear reactions, all of which are possibilities for a controlled fusion process:

(a) $^{2}_{1}H + {}^{3}_{1}H \longrightarrow {}^{4}_{2}He + {}^{1}_{0}n$
(b) $^{2}_{1}H + {}^{2}_{1}H \longrightarrow {}^{3}_{2}He + {}^{1}_{0}n$
(c) $^{2}_{1}H + {}^{3}_{2}He \longrightarrow {}^{4}_{2}He + {}^{1}_{1}H$

21.49 Which of the following nuclei is likely to have the largest mass defect per nucleon: **(a)** ^{59}Co; **(b)** ^{11}B; **(c)** ^{118}Sn; **(d)** ^{243}Cm? Explain your answer.

21.50 Based on Figure 21.13, explain why energy is released in the course of the fission of heavy nuclei.

Effects and Uses of Radioisotopes

21.51 Explain how you might use radioactive ^{59}Fe (a beta emitter with $t_{1/2} = 44.5$ days) to determine the extent to which rabbits are able to convert a particular iron compound in their diet into blood hemoglobin, which contains iron atoms.

21.52 Chlorine-36 is a convenient radiotracer. It is a weak beta emitter, with $t_{1/2} = 3 \times 10^5$ yr. Describe how you would use this radiotracer to carry out each of the following experiments. **(a)** Determine whether trichloroacetic acid, CCl_3COOH, undergoes any ionization of its chlorines as chloride ion in aqueous solution. **(b)** Demonstrate that the equilibrium between dissolved $BaCl_2$ and solid $BaCl_2$ in a saturated solution is a dynamic process. **(c)** Determine the effects of soil pH on the uptake of chloride ion from the soil by soybeans.

21.53 Explain the function of the following components of a nuclear reactor: **(a)** control rods; **(b)** moderator.

21.54 Explain the following terms that apply to fission reactions: **(a)** chain reaction; **(b)** critical mass.

21.55 Complete and balance the nuclear equations for the following fission reactions:
(a) ${}^{235}_{92}U + {}^{1}_{0}n \longrightarrow {}^{160}_{62}Sm + {}^{72}_{30}Zn + \_\_\_\_ {}^{1}_{0}n$
(b) ${}^{239}_{94}Pu + {}^{1}_{0}n \longrightarrow {}^{144}_{58}Ce + \_\_\_\_ + 2{}^{1}_{0}n$

21.56 Complete and balance the nuclear equations for the following fission or fusion reactions:
(a) ${}^{2}_{1}H + {}^{2}_{1}H \longrightarrow {}^{3}_{2}He + \_\_\_\_$
(b) ${}^{233}_{92}U + {}^{1}_{0}n \longrightarrow {}^{133}_{51}Sb + {}^{98}_{41}Nb + \_\_\_\_ {}^{1}_{0}n$

21.57 A portion of the Sun's energy comes from the reaction

$$4{}^{1}_{1}H \longrightarrow {}^{4}_{2}He + 2{}^{0}_{1}e$$

This reaction requires a temperature of about 10^6 to 10^7 K. Why is such a high temperature required?

[21.58] The spent fuel rods from a fission reactor are much more intensely radioactive than the original fuel rods. **(a)** What does this tell you about the products of the fission process in relationship to the belt of stability, Figure 21.2? **(b)** Given that only two or three neutrons are released per fission event, and knowing that the nucleus undergoing fission has a neutron-to-proton ratio characteristic of a heavy nucleus, what sorts of decay would you expect to be dominant among the fission products?

21.59 Why is $\cdot OH$ more dangerous to an organism than OH^-?

21.60 Use Lewis structures to represent the reactants and products in Equation 21.31. Why is the H_2O^+ ion a free radical species?

21.61 A laboratory rat is exposed to an alpha-radiation source whose activity is 8.7 mCi. **(a)** What is the activity of the radiation in disintegrations per second? In becquerels? **(b)** The rat has a mass of 250 g and is exposed to the radiation for 2.0 s, absorbing 65% of the emitted alpha particles, each having an energy of 9.12×10^{-13} J. Calculate the absorbed dose in millirads and grays. **(c)** If the RBE of the radiation is 9.5, calculate the effective absorbed dose in mrem and Sv.

21.62 A 65-kg person is accidentally exposed for 116 s to a 21-mCi source of beta radiation coming from a sample of ${}^{90}Sr$. **(a)** What is the activity of the radiation source in disintegrations per second? In becquerels? **(b)** Each beta particle has an energy of 8.75×10^{-14} J, and 6.5% of the radiation is absorbed by the person. Assuming that the absorbed radiation is spread over the person's entire body, calculate the absorbed dose in rads and in grays. **(c)** If the RBE of the beta particles is 1.0, what is the effective dose in mrem and in sieverts? **(d)** Based on Figure 21.23, how does the magnitude of this dose of radiation compare with average background radiation?

Additional Exercises

21.63 Radon-222 decays to a stable nucleus by a series of three alpha emissions and two beta emissions. What is the stable nucleus that is formed?

21.64 A free neutron is unstable and decays into a proton with a half-life of 10.4 min. **(a)** What other particle forms? **(b)** Why don't neutrons in atomic nuclei decay at the same rate?

21.65 The 13 known nuclides of zinc range from ${}^{60}Zn$ to ${}^{72}Zn$. The naturally occurring nuclides have mass numbers 64, 66, 67, 68, and 70. What mode or modes of decay would you expect for the least massive radioactive nuclides of zinc? What mode for the most massive nuclides?

21.66 Chlorine has two stable nuclides, ${}^{35}Cl$ and ${}^{37}Cl$. In contrast, ${}^{36}Cl$ is a radioactive nuclide that decays by beta emission. **(a)** What is the product of decay of ${}^{36}Cl$? **(b)** Based on the empirical rules about nuclear stability, explain why the nucleus of ${}^{36}Cl$ is less stable than either ${}^{35}Cl$ or ${}^{37}Cl$.

21.67 Nuclear scientists have synthesized approximately 1600 nuclei not known in nature. More might be discovered with heavy-ion bombardment using high-energy particle accelerators. Complete and balance the following reactions, which involve heavy-ion bombardments:
(a) ${}^{6}_{3}Li + {}^{56}_{28}Ni \longrightarrow ?$
(b) ${}^{40}_{20}Ca + {}^{248}_{96}Cm \longrightarrow ?$
(c) ${}^{88}_{38}Sr + {}^{84}_{36}Kr \longrightarrow {}^{116}_{46}Pd + ?$
(d) ${}^{40}_{20}Ca + {}^{238}_{92}U \longrightarrow {}^{70}_{30}Zn + 4{}^{1}_{0}n + 2?$

[21.68] Radon-212 is an alpha emitter with a half-life of 25 min. How many alpha particles are emitted in 1.0 s from a 1.0-pg sample of this nuclide? What is the activity of this sample in curies?

[21.69] The synthetic radioisotope technetium-99, which decays by beta emission, is the most widely used isotope in nuclear medicine. The following data were collected on a sample of ${}^{99}Tc$:

Disintegrations per Minute	Time (hr)
180	0
130	2.5
104	5.0
77	7.5
59	10.0
46	12.5
24	17.5

Make a graph of these data similar to Figure 21.7, and determine the half-life. (You may wish to make a graph of the natural log of the disintegration rate versus time; a little rearranging of Equation 21.19 will produce an equation for a linear relation between ln N_t and t; from the slope you can obtain k.)

[21.70] According to current regulations, the maximum permissible dose of strontium-90 in the body of an adult is $1 \mu Ci$ (1×10^{-6} Ci). Using the relationship rate $= kN$, calculate the number of atoms of strontium-90 to which this dose corresponds. To what mass of strontium-90 does this correspond ($t_{1/2}$ for strontium-90 is 28.8 yr)?

[21.71] Suppose you had a detection device that could count every decay from a radioactive sample of plutonium-239 ($t_{1/2}$ is 24,000 yr). How many counts per second would you obtain from a sample containing 0.500 g of plutonium-239? (*Hint*: Look at Equations 21.19 and 21.20.)

21.72 Methyl acetate (CH_3COOCH_3) is formed by the reaction of acetic acid with methyl alcohol. If the methyl alcohol is labeled with oxygen-18, the oxygen-18 ends up in the methyl acetate:

$$CH_3\overset{\overset{\displaystyle O}{\|}}{C}OH + H^{18}OCH_3 \longrightarrow CH_3\overset{\overset{\displaystyle O}{\|}}{C}{}^{18}OCH_3 + H_2O$$

Do the C—OH bond of the acid and the O—H bond of the alcohol break in the reaction, or do the O—H bond of the acid and the C—OH bond of the alcohol break? Explain.

21.73 An experiment was designed to determine whether an aquatic plant absorbed iodide ion from water. Iodine-131 ($t_{1/2} = 8.04$ days) was added as a tracer, in the form of iodide ion, to a tank containing the plants. The initial activity of a 1.00-μL sample of the water was 175 counts per minute. After 32 days the level of activity in a 1.00-μL sample was 12.2 counts per minute. Did the plants absorb iodide from the water?

[21.74] A 26.00-g sample of water containing tritium, ${}^3_1H$, emits 1.50×10^3 beta particles per second. Tritium is a weak beta emitter, with a half-life of 12.3 yr. What fraction of all the hydrogen in the water sample is tritium? (*Hint*: Use Equations 21.19 and 21.20.)

21.75 The nuclear masses of ${}^7Be$, ${}^9Be$, and ${}^{10}Be$ are 7.0147, 9.0100, and 10.0113 amu, respectively. Which of these nuclei has the largest binding energy per nucleon?

21.76 The Sun radiates energy into space at the rate of 3.9×10^{26} J/s. **(a)** Calculate the rate of mass loss from the Sun in kg/s. **(b)** How does this mass loss arise?

[21.77] The average energy released in the fission of a single uranium-235 nucleus is about 3×10^{-11} J. If the conversion of this energy to electricity in a nuclear power plant is 40% efficient, what mass of uranium-235 undergoes fission in a year in a plant that produces 1000 MW (megawatts)? Recall that a watt is 1 J/s.

[21.78] Tests on human subjects in Boston in 1965 and 1966, following the era of atomic bomb testing, revealed average quantities of about 2 pCi of plutonium radioactivity in the average person. How many disintegrations per second does this level of activity imply? If each alpha particle deposits 8×10^{-13} J of energy and if the average person weighs 75 kg, calculate the number of rads and rems of radiation in 1 yr from such a level of plutonium.

Integrative Exercises

21.79 A 49.5-mg sample of sodium perchlorate contains radioactive chlorine-36 (whose atomic mass is 36.0 amu). If 31.0% of the chlorine atoms in the sample are chlorine-36 and the remainder are naturally occurring nonradioactive chlorine atoms, how many disintegrations per second are produced by this sample? The half-life of chlorine-36 is 3.0×10^5 yr.

21.80 Calculate the mass of propane, $C_3H_8(g)$, that must be burned in air to evolve the same quantity of energy as produced by the fusion of 1.0 g of hydrogen in the following fusion reaction:

$$4\,{}^1_1H \longrightarrow {}^4_2He + 2\,{}^0_1e$$

Assume that all the products of the combustion of C_3H_8 are in their gas phases. Use data from Exercise 21.48, Appendix C, and the inside covers of the text.

21.81 A sample of an alpha emitter having an activity of 0.18 Ci is stored in a 15.0-mL sealed container at 22°C for 235 days. **(a)** How many alpha particles are formed during this time? **(b)** Assuming that each alpha particle is converted to a helium atom, what is the partial pressure of helium gas in the container after this 235-day period?

[21.82] Charcoal samples from Stonehenge in England were burned in O_2, and the resultant CO_2 gas bubbled into a solution of $Ca(OH)_2$ (limewater), resulting in the precipitation of $CaCO_3$. The $CaCO_3$ was removed by filtration and dried. A 788-mg sample of the $CaCO_3$ had a radioactivity of 1.5×10^{-2} Bq due to carbon-14. By comparison, living organisms undergo 15.3 disintegrations per minute per gram of carbon. Using the half-life of carbon-14, 5715 yr, calculate the age of the charcoal sample.

[21.83] Using the ionization energy of water, 1216 kJ/mol, calculate the longest wavelength of electromagnetic radiation that is classified as ionizing radiation.

[21.84] When a positron is annihilated by combination with an electron, two photons of equal energy result. What is the wavelength of these photons? Are they gamma ray photons?

[21.85] A 25.0-mL sample of 0.050 *M* barium nitrate solution was mixed with 25.0 mL of 0.050 *M* sodium sulfate solution labeled with radioactive sulfur-35. The activity of the initial sodium sulfate solution was 1.22×10^6 Bq/mL. After the resultant precipitate was removed by filtration, the remaining filtrate was found to have an activity of 250 Bq/mL. **(a)** Write a balanced chemical equation for the reaction that occurred. **(b)** Calculate the K_{sp} for the precipitate under the conditions of the experiment.

eMedia Exercises

21.86 The **Separation of Alpha, Beta, and Gamma Rays** movie (*eChapter 21.1*) shows how an electric field deflects alpha and beta emissions. Predict the product nuclei if each of the following were to decay by either α or β decay: **(a)** ^{235}U; **(b)** ^{209}Bi; **(c)** ^{108}Sn; **(d)** ^{214}Pb.

21.87 The half-life of ^{238}U is 4.5×10^9 yr. **(a)** Use the **Radioactive Decay** simulation (*eChapter 21.4*) to determine the half-life of ^{235}U. **(b)** Which of the two isotopes is more abundant in nature? In light of the two half-lives, does the relative abundance of the two isotopes make sense? Explain.

21.88 Use the **Radioactive Decay** simulation (*eChapter 21.4*) to determine **(a)** the half-life of ^{232}Th; **(b)** the rate constant for decay of ^{232}Th; **(c)** the mass of ^{232}Th remaining after 6.8 billion years. (Starting mass: 20.00 kg)

21.89 In Exercise 21.40 you determined the rate constant for the decay of ^{40}K to ^{40}Ar in order to calculate the age of a rock. **(a)** Using that rate constant, calculate the mass of ^{40}K remaining after 2.4 billion years. (Starting mass: 20.00 kg) Check your answer with the **Radioactive Decay** simulation (*eChapter 21.4*). **(b)** By what process does ^{40}K decay to ^{40}Ar?

21.90 Naturally occurring uranium consists primarily of ^{238}U, with a very small amount of ^{235}U. For the uranium to be fissionable, the concentration of ^{235}U must be enriched by converting all of the uranium into the volatile compound, UF_6. The UF_6 is warmed to a temperature at which it is a gas, and the two uranium isotopes are separated (by a modified effusion process) on the basis of molecular speed. Recall from Chapter 10 that, at any particular temperature, gas molecules have the same average kinetic energy—heavier molecules move more slowly than lighter molecules. Review this concept with the **Gas Phase Boltzmann Distribution** activity (*eChapter 10.7*). Compare the molecular masses and average molecular speeds of He and N_2. **(a)** By what factor do their molecular masses differ? **(b)** By what factor do their average molecular speeds differ? **(c)** What is the relationship between the factors in (a) and (b)? **(d)** By how much do the masses of $^{238}UF_6$ and $^{235}UF_6$ differ? **(e)** By how much do you expect their average molecular speeds to differ? **(f)** Based on your answer to (e), comment on the probable difficulty and expense associated with the enrichment of uranium.

Chapter 22

Chemistry of the Nonmetals

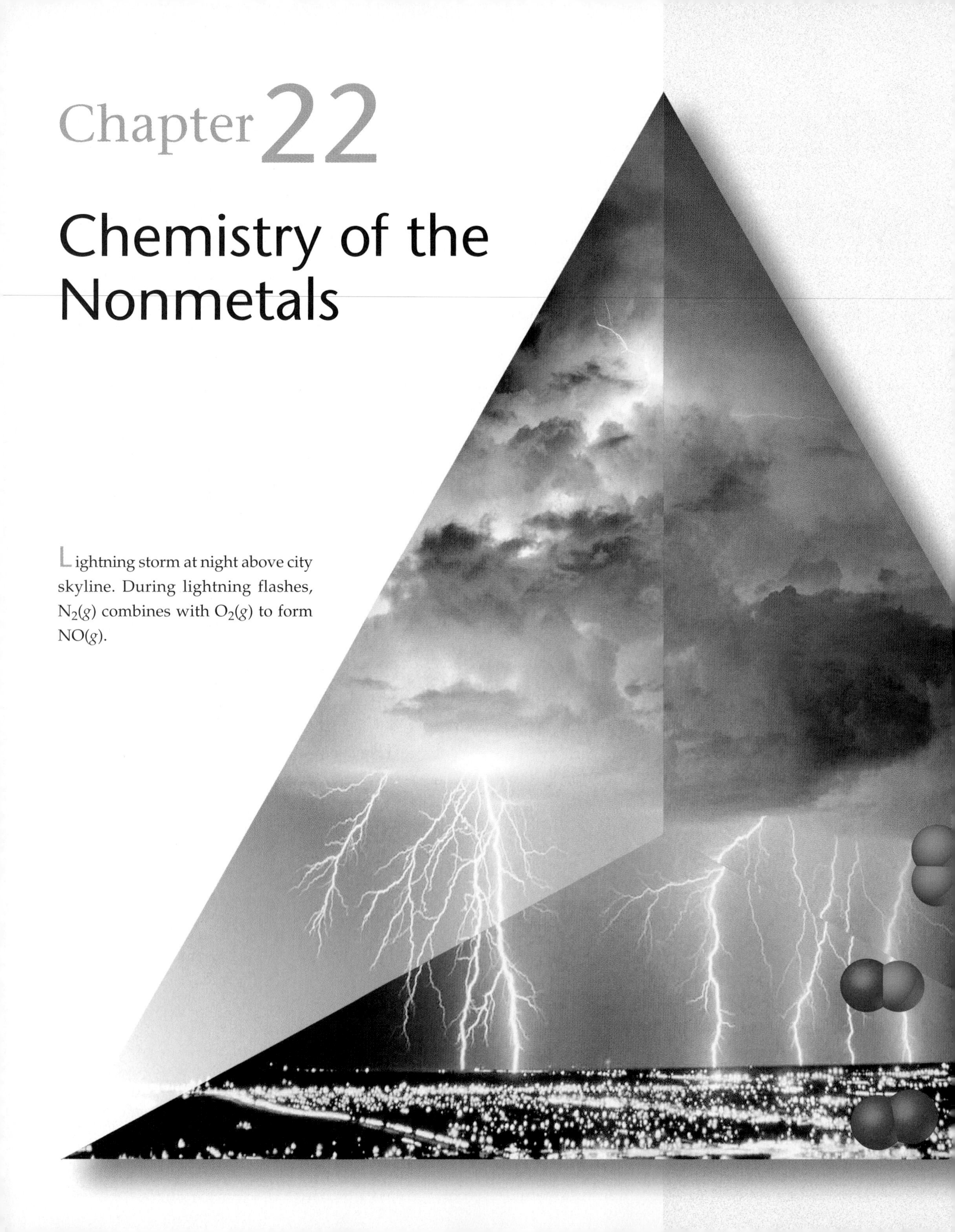

Lightning storm at night above city skyline. During lightning flashes, $N_2(g)$ combines with $O_2(g)$ to form $NO(g)$.

IN THE PREVIOUS chapters of this book we have discussed chemical principles, such as the laws of thermodynamics, the formation of chemical bonds, the behavior of different phases of matter, the factors that influence reaction rates and equilibria, and so forth. In the course of explaining these principles, we have described the chemical and physical properties of many substances. We have done little, however, to examine the elements and their compounds in a systematic fashion. This aspect of chemistry, referred to as *descriptive chemistry*, is the subject of the next several chapters.

We will take a panoramic view of the descriptive chemistry of the nonmetallic elements in this chapter, starting with hydrogen and then progressing, group by group, from right to left in the periodic table. As we examine the elements, we will consider how they occur in nature, how they are isolated from their sources, and how they are used. As we do so, we will encounter the chemistry of the most commercially important compounds of the elements. We will emphasize hydrogen, oxygen, nitrogen, and carbon. These four nonmetals form many commercially important compounds, and they account for 99% of the atoms required by living cells. We will discuss further aspects of the chemistry of these elements when we consider organic and biological chemistry in Chapter 25.

As you study descriptive chemistry, it is important to look for trends rather than trying to memorize all the facts presented. The periodic table is your most valuable tool in this task.

▶ What's Ahead ◀

- We begin with a review of general periodic trends and types of chemical reactions, which will help us to focus on general patterns of behavior as we examine each family in the periodic table.
- The first nonmetal we consider is hydrogen, an element that forms compounds with most other nonmetals.
- Next, we consider the noble gases, the elements of group 8A, which exhibit very limited chemical reactivity (the fluorides and oxides of Xe are the most numerous).
- The halogens, group 7A, have a rich and important chemistry, and F and Cl in particular are commercially very important.
- Oxygen is the most abundant element in both Earth's crust and the human body, and its chemistry includes oxide and peroxide compounds.
- Of the other members of group 6A (S, Se, Te, and Po), sulfur is the most important.
- Nitrogen is a key component of our atmosphere and forms compounds in which its oxidation number ranges from −3 to +5, including such important compounds as NH_3 and HNO_3.
- Of the other members of group 5A (P, As, Sb, and Bi), phosphorus is most important.
- The chemistries of carbon and silicon are the most significant within group 4A, with carbon having many important inorganic and organic compounds, and silicon oxides being the framework substances in Earth's crust.

22.1 General Concepts: Periodic Trends and Chemical Reactions

Recall that we can classify elements as metals, metalloids, and nonmetals. (Section 7.6) Except for hydrogen, which is a special case, the nonmetals occupy the upper right portion of the periodic table. This division of elements relates nicely to trends in the properties of the elements as summarized in Figure 22.1 ▼. Electronegativity, for example, increases as we move left to right across a row of the table, and it decreases as we move down a particular group. The nonmetals thus have higher electronegativities than the metals. This difference leads to the formation of ionic solids in reactions between metals and nonmetals. (Sections 7.6, 8.2, and 8.4) In contrast, compounds formed between nonmetals are molecular substances and are often gases, liquids, or volatile solids at room temperature. (Sections 7.8 and 8.4)

Among the nonmetals, we have seen that the chemistry exhibited by the first member of a group can differ in several important ways from that of subsequent members. For example, nonmetals in the third row and below can accommodate more than eight electrons in their valence shells. (Section 8.7) Another important difference is that the first element in any group can more readily form π bonds than members lower in the group. This trend is due, in part, to atomic size. Small atoms are able to approach each other more closely. As a result, the sideways overlap of p orbitals, which results in the formation of π bonds, is more effective for the first element in each group (Figure 22.2 ▼). More effective overlap means stronger π bonds, as is reflected in the bond enthalpies of their multiple bonds. (Section 8.8) For example, the difference in the bond enthalpies of C—C and C=C bonds is about 270 kJ/mol (Table 8.4); this value reflects the "strength" of a carbon–carbon π bond. By comparison, the strength of a silicon–silicon π bond is only about 100 kJ/mol, significantly lower than that for carbon. As we shall see, pi bonds are particularly important in the chemistry of carbon, nitrogen, and oxygen, which frequently form double bonds. The elements in rows 3, 4, 5, and 6 of the periodic table, on the other hand, have a tendency to form only single bonds.

ACTIVITY
Carbon-Silicon Sigma Overlap

The ability of an atom to form π bonds is one of the important factors in determining the structures of the nonmetals and their compounds. Compare, for example, the elemental forms of carbon and silicon. Carbon has three crystalline allotropes: diamond, graphite, and buckminsterfullerene. (Section 11.8)

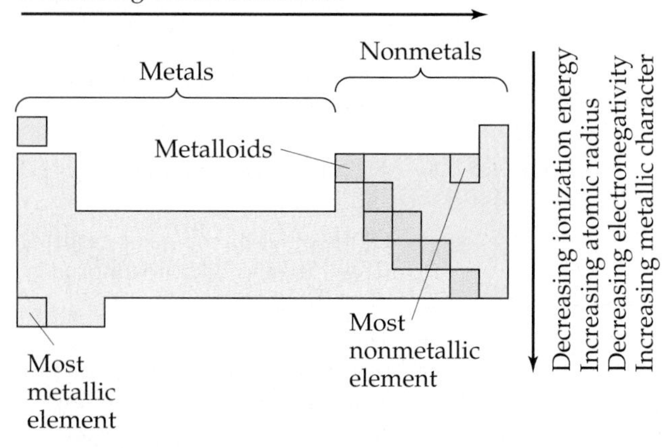

▲ **Figure 22.1** Trends in key properties of the elements as a function of position in the periodic table.

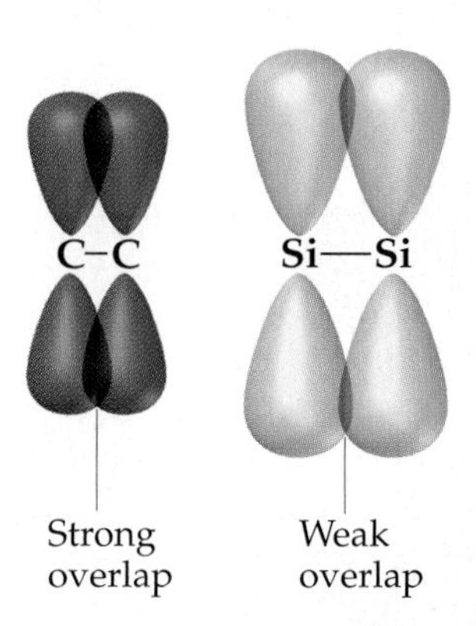

▲ **Figure 22.2** Comparison of π-bond formation by sideways overlap of p orbitals between two carbon atoms and between two silicon atoms. The distance between nuclei increases as we move from carbon to silicon. The p orbitals do not overlap as effectively between two silicon atoms because of this greater separation.

Diamond is a covalent-network solid that has C—C σ bonds but no π bonds. Graphite and buckminsterfullerene have π bonds that result from the sideways overlap of *p* orbitals. Elemental silicon exists only as a diamond-like covalent-network solid with σ bonds; silicon exhibits no form that is analogous to either graphite or buckminsterfullerene, apparently because Si—Si π bonds are weak.

We likewise see significant differences in the dioxides of carbon and silicon (Figure 22.3 ▶). CO_2 is a molecular substance with C—O double bonds, whereas SiO_2 contains no double bonds. SiO_2 is a covalent-network solid in which four oxygen atoms are bonded to each silicon atom by single bonds, forming an extended structure that has the empirical formula SiO_2.

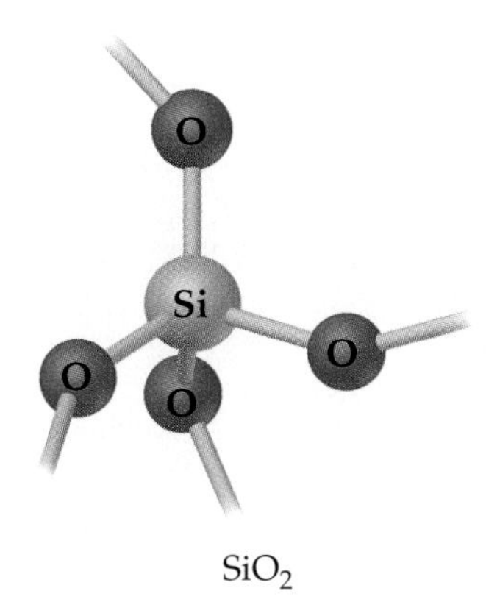

SiO_2

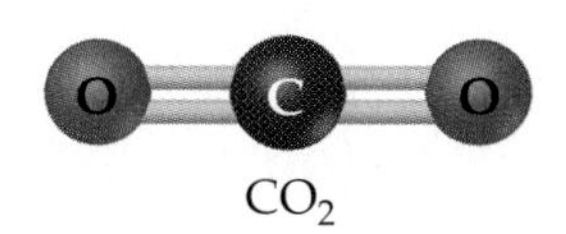

CO_2

▲ **Figure 22.3** Comparison of the structures of CO_2 and SiO_2; CO_2 has double bonds, whereas SiO_2 has only single bonds.

SAMPLE EXERCISE 22.1

Consider the elements Li, K, N, P, and Ne. From this list select the element that **(a)** is the most electronegative; **(b)** has the greatest metallic character; **(c)** can bond to more than four surrounding atoms in a molecule; **(d)** forms π bonds most readily.

Solution

Analyze: We are given a list of elements and asked to predict several properties that can be related to periodic trends.

Plan: We can use the preceding discussion, particularly the summary in Figure 22.1, to guide us to the answers. Thus, we need first to locate each element in the periodic table.

Solve: (a) Electronegativity increases as we proceed toward the upper right portion of the periodic table, excluding the noble gases. Thus, nitrogen (N) is the most electronegative element among those listed. **(b)** Metallic character correlates inversely with electronegativity—the less electronegative an element, the greater its metallic character. The element with the greatest metallic character is therefore potassium (K), which is closest to the lower left corner of the periodic table. **(c)** Nonmetals tend to form molecular compounds, so we can narrow our choice to the three nonmetals on the list: N, P, and Ne. In order to form more than four bonds, an element must be able to expand its valence shell to allow more than an octet of electrons around it. Valence-shell expansion occurs for elements in the third row of the periodic table and below; nitrogen and neon are both in the second row and will not undergo valence-shell expansion. Thus, the answer is phosphorus (P). **(d)** Nonmetals in the second row form π bonds more readily than elements in the third row and below. There are no compounds known that contain covalent bonds to the noble gas Ne. Thus, the other second-row element, N, is the element from the list that forms π bonds most readily.

PRACTICE EXERCISE

Consider the elements Be, C, Cl, Sb, and Cs. Select the element that **(a)** has the lowest electronegativity; **(b)** has the greatest nonmetallic character; **(c)** is most likely to participate in extensive π bonding; **(d)** is most likely to be a metalloid.

Answers: **(a)** Cs; **(b)** Cl; **(c)** C; **(d)** Sb

Chemical Reactions

We will present a large number of chemical reactions in this and later chapters. You will find it helpful to observe general trends in the patterns of reactivity. We have already encountered several general categories of reactions: combustion reactions (Section 3.2), metathesis reactions (Section 4.2), Brønsted–Lowry acid-base (proton-transfer) reactions (Section 16.2), Lewis acid-base reactions (Section 16.11), and redox reactions (Section 20.1). Because O_2 and H_2O are abundant in our environment, it is particularly important to consider the possible reactions of these substances with other compounds. About one third of the reactions discussed in this chapter involve either O_2 (oxidation or combustion reactions) or H_2O (especially proton-transfer reactions).

In combustion reactions with O_2, hydrogen-containing compounds produce H_2O. Carbon-containing ones produce CO_2 (unless the amount of O_2 is insufficient, in which case CO or even C can form). Nitrogen-containing compounds tend to form N_2, although NO can form in special cases. The following reactions illustrate these generalizations:

$$2CH_3OH(l) + 3O_2(g) \longrightarrow 2CO_2(g) + 4H_2O(g) \quad [22.1]$$

$$4CH_3NH_2(g) + 9O_2(g) \longrightarrow 4CO_2(g) + 10H_2O(g) + 2N_2(g) \quad [22.2]$$

The formation of H_2O, CO_2, and N_2 reflects the high thermodynamic stabilities of these substances, which are indicated by the large bond energies for the O—H, C=O, and N≡N bonds that they contain (463, 799, and 941 kJ/mol, respectively). (Section 8.8)

When dealing with proton-transfer reactions, remember that the weaker a Brønsted–Lowry acid, the stronger its conjugate base. (Section 16.2) For example, H_2, OH^-, NH_3, and CH_4 are exceedingly weak proton donors that have *no* tendency to act as acids in water. Thus, species formed from them by removing one or more protons (such as H^-, O^{2-}, and NH_2^-) are extremely strong bases. All react readily with water, removing protons from H_2O to form OH^-. The following reactions are illustrative:

$$CH_3^-(aq) + H_2O(l) \longrightarrow CH_4(g) + OH^-(aq) \quad [22.3]$$

$$N^{3-}(aq) + 3H_2O(l) \longrightarrow NH_3(aq) + 3OH^-(aq) \quad [22.4]$$

Substances that are stronger proton donors than H_2O, such as HCl, H_2SO_4, $HC_2H_3O_2$, and other acids, also react readily with basic anions.

SAMPLE EXERCISE 22.2

Predict the products formed in each of the following reactions, and write a balanced equation:

(a) $CH_3NHNH_2(g) + O_2(g) \longrightarrow$
(b) $Mg_3P_2(s) + H_2O(l) \longrightarrow$
(c) $NaCN(s) + HCl(aq) \longrightarrow$

Solution

Analyze: We are given the reactants for three chemical equations and asked to predict the products and then balance the equations.
Plan: We need to examine the reactants to see if there is a possible reaction type that we might recognize. In (a) the carbon compound is reacting with O_2, which suggests a combustion reaction. In (b) water reacts with an ionic compound. The anion, P^{3-}, is a strong base and H_2O is able to act as an acid, so the reactants suggest an acid-base (proton-transfer) reaction. In (c) we have an ionic compound and a strong acid. Again a proton-transfer reaction is suggested.
Solve: **(a)** Based on the elemental composition of the carbon compound, this combustion reaction should produce CO_2, H_2O, and N_2:

$$2CH_3NHNH_2(g) + 5O_2(g) \longrightarrow 2CO_2(g) + 6H_2O(g) + 2N_2(g)$$

(b) Mg_3P_2 is ionic, consisting of Mg^{2+} and P^{3-} ions. The P^{3-} ion, like N^{3-}, has a strong affinity for protons and reacts with H_2O to form OH^- and PH_3 (PH^{2-}, PH_2^-, and PH_3 are all exceedingly weak proton donors).

$$Mg_3P_2(s) + 6H_2O(l) \longrightarrow 2PH_3(g) + 3Mg(OH)_2(s)$$

$Mg(OH)_2$ has low solubility in water and will precipitate.

(c) NaCN consists of Na^+ and CN^- ions. The CN^- ion is basic (HCN is a weak acid). Thus, CN^- reacts with protons to form its conjugate acid.

$$NaCN(s) + HCl(aq) \longrightarrow HCN(aq) + NaCl(aq)$$

HCN has limited solubility in water and escapes as a gas. HCN is *extremely* toxic; in fact, this reaction is used to produce the lethal gas in gas chambers.

PRACTICE EXERCISE

Write a balanced equation for the reaction of solid sodium hydride with water.
Answer: $NaH(s) + H_2O(l) \longrightarrow NaOH(aq) + H_2(g)$

22.2 Hydrogen

The English chemist Henry Cavendish (1731–1810) first isolated pure hydrogen. Because the element produces water when burned in air, the French chemist Lavoisier gave it the name *hydrogen*, which means "water producer" (Greek: *hydro*, water; *gennao*, to produce).

Hydrogen is the most abundant element in the universe. It is the nuclear fuel consumed by our Sun and other stars to produce energy. (Section 21.8) Although about 70% of the universe is composed of hydrogen, it constitutes only 0.87% of Earth's mass. Most of the hydrogen on our planet is found associated with oxygen. Water, which is 11% hydrogen by mass, is the most abundant hydrogen compound. Hydrogen is also an important part of petroleum, cellulose, starch, fats, alcohols, acids, and a wide variety of other materials.

Isotopes of Hydrogen

The most common isotope of hydrogen, ${}^{1}_{1}H$, has a nucleus consisting of a single proton. This isotope, sometimes referred to as **protium**,* comprises 99.9844% of naturally occurring hydrogen.

Two other isotopes are known: ${}^{2}_{1}H$, whose nucleus contains a proton and a neutron, and ${}^{3}_{1}H$, whose nucleus contains a proton and two neutrons (Figure 22.4 ▶). The ${}^{2}_{1}H$ isotope, called **deuterium**, comprises 0.0156% of naturally occurring hydrogen. It is not radioactive. Deuterium is often given the symbol D in chemical formulas, as in D_2O (deuterium oxide), which is known as *heavy water*.

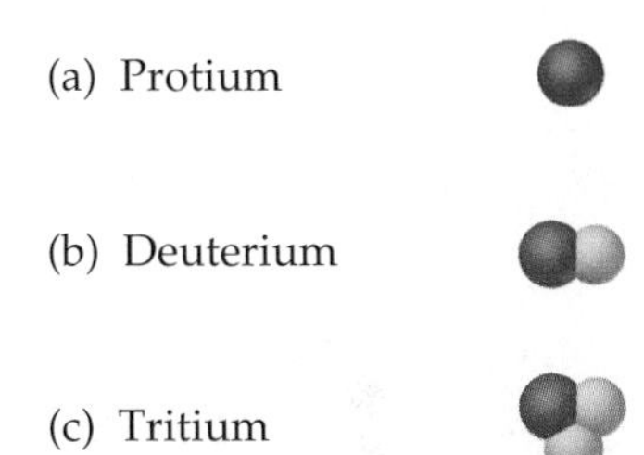

▲ **Figure 22.4** A depiction of the nuclei of the three isotopes of hydrogen. (a) Protium, ${}^{1}_{1}H$, has a single proton (depicted as a red sphere) in its nucleus. (b) Deuterium, ${}^{2}_{1}H$, has a proton and a neutron (depicted as a gray sphere). (c) Tritium, ${}^{3}_{1}H$, has a proton and two neutrons.

Because an atom of deuterium is about twice as massive as an atom of protium, the properties of deuterium-containing substances vary somewhat from those of the "normal" protium-containing analogs. For example, the normal melting and boiling points of D_2O are 3.81°C and 101.42°C, respectively, whereas they are 0.00°C and 100.00°C for H_2O. Not surprisingly, the density of D_2O at 25°C (1.104 g/mL) is greater than that of H_2O (0.997 g/mL). Replacing protium with deuterium (a process called *deuteration*) can also have a profound effect on the rates of reactions, a phenomenon called a *kinetic-isotope effect*. In fact, heavy water can be obtained by the electrolysis of ordinary water because D_2O undergoes electrolysis at a slower rate and therefore becomes concentrated during the electrolysis.

The third isotope, ${}^{3}_{1}H$, is known as **tritium**. It is radioactive, with a half-life of 12.3 yr.

$$ {}^{3}_{1}H \longrightarrow {}^{3}_{2}He + {}^{\ 0}_{-1}e \qquad t_{1/2} = 12.3 \text{ yr} \qquad [22.5] $$

Tritium is formed continuously in the upper atmosphere in nuclear reactions induced by cosmic rays; because of its short half-life, however, only trace quantities exist naturally. The isotope can be synthesized in nuclear reactors by neutron bombardment of lithium-6.

$$ {}^{6}_{3}Li + {}^{1}_{0}n \longrightarrow {}^{3}_{1}H + {}^{4}_{2}He \qquad [22.6] $$

Deuterium and tritium have proved valuable in studying the reactions of compounds containing hydrogen. A compound can be "labeled" by replacing one or more ordinary hydrogen atoms at specific locations within a molecule with deuterium or tritium. By comparing the location of the label in the reactants with that in the products, the mechanism of the reaction can often be inferred. When methyl alcohol (CH_3OH) is placed in D_2O, for example, the H atom of the O—H bond exchanges rapidly with the D atoms in D_2O, forming

* Giving unique names to isotopes is limited to hydrogen. Because of the proportionally large differences in their masses, the isotopes of H show appreciably more differences in their chemical and physical properties than isotopes of heavier elements.

CH_3OD. The H atoms of the CH_3 group do not exchange. This experiment demonstrates the kinetic stability of C—H bonds and reveals the speed at which the O—H bond in the molecule breaks and re-forms.

Properties of Hydrogen

Hydrogen is the only element that is not a member of any family in the periodic table. Because of its $1s^1$ electron configuration, it is generally placed above lithium in the periodic table. However, it is definitely *not* an alkali metal. It forms a positive ion much less readily than any alkali metal; the ionization energy of the hydrogen atom is 1312 kJ/mol, whereas that of lithium is 520 kJ/mol.

Hydrogen is sometimes placed above the halogens in the periodic table because the hydrogen atom can pick up one electron to form the *hydride ion*, H^-, which has the same electron configuration as helium. The electron affinity of hydrogen ($E = -73$ kJ/mol), however, is not as large as that of any halogen; the electron affinity of fluorine is −328 kJ/mol, and that of iodine is −295 kJ/mol. ∞ (Section 7.5) In general, hydrogen shows no closer resemblance to the halogens than it does to the alkali metals.

Elemental hydrogen exists at room temperature as a colorless, odorless, tasteless gas composed of diatomic molecules. We can call H_2 dihydrogen, but it is more commonly referred to as molecular hydrogen or simply hydrogen. Because H_2 is nonpolar and has only two electrons, attractive forces between molecules are extremely weak. As a result, the melting point (−259°C) and boiling point (−253°C) of H_2 are very low.

The H—H bond enthalpy (436 kJ/mol) is high for a single bond (Table 8.4). By comparison, the Cl—Cl bond enthalpy is only 242 kJ/mol. Because H_2 has a strong bond, most reactions of H_2 are slow at room temperature. However, the molecule is readily activated by heat, irradiation, or catalysis. The activation process generally produces hydrogen atoms, which are very reactive. Once H_2 is activated, it reacts rapidly and exothermically with a wide variety of substances.

Hydrogen forms strong covalent bonds with many elements, including oxygen; the O—H bond enthalpy is 463 kJ/mol. The formation of the strong O—H bond makes hydrogen an effective reducing agent for many metal oxides. When H_2 is passed over heated CuO, for example, copper is produced.

$$CuO(s) + H_2(g) \longrightarrow Cu(s) + H_2O(g) \qquad [22.7]$$

When H_2 is ignited in air, a vigorous reaction occurs, forming H_2O.

$$2H_2(g) + O_2(g) \longrightarrow 2H_2O(g) \qquad \Delta H^\circ = -483.6 \text{ kJ} \qquad [22.8]$$

Air containing as little as 4% H_2 (by volume) is potentially explosive. Combustion of hydrogen-oxygen mixtures is commonly used in liquid-fuel rocket engines such as those of the space shuttles. The hydrogen and oxygen are stored at low temperatures in liquid form. The destruction of the space shuttle *Challenger* in 1986 was because of the explosion of its hydrogen and oxygen fuel tanks caused by the malfunction of a solid-fuel booster rocket.

Preparation of Hydrogen

When a small quantity of H_2 is needed in the laboratory, it is usually obtained by the reaction between an active metal such as zinc and a dilute strong acid such as HCl or H_2SO_4.

$$Zn(s) + 2H^+(aq) \longrightarrow Zn^{2+}(aq) + H_2(g) \qquad [22.9]$$

Because H_2 is quite insoluble in water, it can be collected by displacement of water, as shown in Figure 22.5 ▶.

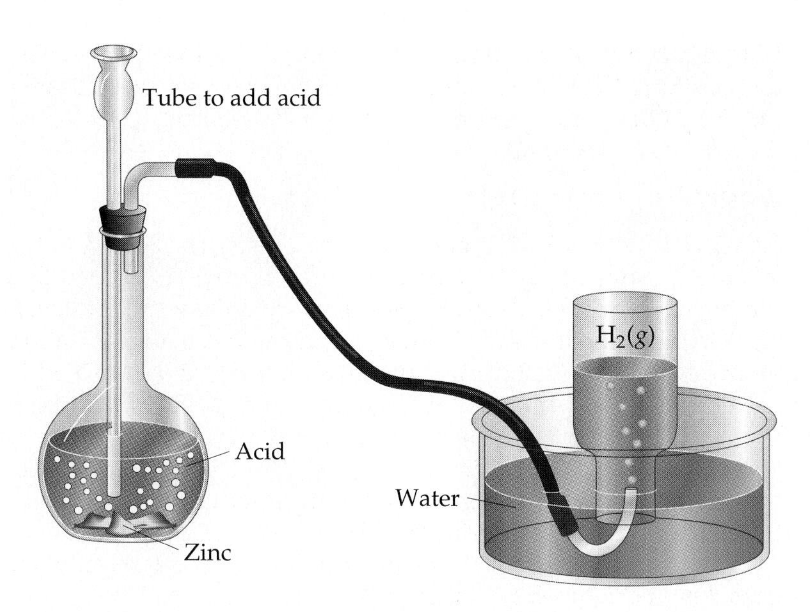

◀ **Figure 22.5** Apparatus commonly used in the laboratory for preparation of hydrogen.

Large quantities of H_2 are produced by reacting methane (CH_4, the principal component of natural gas) with steam at 1100°C. We can view this process as involving the following reactions:

$$CH_4(g) + H_2O(g) \longrightarrow CO(g) + 3H_2(g) \quad [22.10]$$

$$CO(g) + H_2O(g) \longrightarrow CO_2(g) + H_2(g) \quad [22.11]$$

When heated to about 1000°C, carbon also reacts with steam to produce a mixture of H_2 and CO gases.

$$C(s) + H_2O(g) \longrightarrow H_2(g) + CO(g) \quad [22.12]$$

This mixture, known as *water gas*, is used as an industrial fuel.

Simple electrolysis of water consumes too much energy and is consequently too costly to be used commercially to produce H_2. However, H_2 is produced as a by-product in the electrolysis of brine (NaCl) solutions in the course of Cl_2 and NaOH manufacture:

$$2NaCl(aq) + 2H_2O(l) \xrightarrow{\text{electrolysis}} H_2(g) + Cl_2(g) + 2NaOH(aq) \quad [22.13]$$

Uses of Hydrogen

Hydrogen is a commercially important substance: About 2×10^8 kg (200,000 tons) is produced annually in the United States. Over two thirds of the H_2 produced is used to synthesize ammonia by the Haber process. ∞ (Section 15.1) Hydrogen is also used to manufacture methanol (CH_3OH) via the catalytic reaction of CO and H_2 at high pressure and temperature.

$$CO(g) + 2H_2(g) \longrightarrow CH_3OH(g) \quad [22.14]$$

Binary Hydrogen Compounds

Hydrogen reacts with other elements to form compounds of three general types: (1) ionic hydrides, (2) metallic hydrides, and (3) molecular hydrides.

The **ionic hydrides** are formed by the alkali metals and by the heavier alkaline earths (Ca, Sr, and Ba). These active metals are much less electronegative than hydrogen. Consequently, hydrogen acquires electrons from them to form hydride ions (H^-), as shown here:

$$2Li(s) + H_2(g) \longrightarrow 2LiH(s) \quad [22.15]$$

$$Ca(s) + H_2(g) \longrightarrow CaH_2(s) \quad [22.16]$$

The ionic hydrides are high-melting solids (LiH melts at 680°C).

▶ **Figure 22.6** The reaction of CaH_2 with water is vigorous and exothermic. The reddish purple color is due to phenolphthalein added to the water, which indicates the formation of OH^- ions. The bubbling is due to the formation of H_2 gas.

(a)

(b)

The hydride ion is very basic and reacts readily with compounds having even weakly acidic protons to form H_2. For example, H^- reacts readily with H_2O.

$$H^-(aq) + H_2O(l) \longrightarrow H_2(g) + OH^-(aq) \qquad [22.17]$$

Ionic hydrides can therefore be used as convenient (although expensive) sources of H_2. Calcium hydride (CaH_2) is sold commercially and is used to inflate life rafts, weather balloons, and the like, where a simple, compact means of generating H_2 is desired. The reaction of CaH_2 with H_2O is shown in Figure 22.6 ▲.

The reaction between H^- and H_2O (Equation 22.17) is not only an acid-base reaction but also a redox reaction. The H^- ion, therefore, is a good base *and* a good reducing agent. In fact, hydrides are able to reduce O_2 to OH^-:

$$2NaH(s) + O_2(g) \longrightarrow 2NaOH(s) \qquad [22.18]$$

Thus, hydrides are normally stored in an environment that is free of both moisture and air.

Metallic hydrides are formed when hydrogen reacts with transition metals. These compounds are so named because they retain their metallic conductivity and other metallic properties. In many metallic hydrides, the ratio of metal atoms to hydrogen atoms is not fixed or in small whole numbers. The composition can vary within a range, depending on the conditions of synthesis. TiH_2 can be produced, for example, but preparations usually yield $TiH_{1.8}$, which has about 10% less hydrogen than TiH_2. These nonstoichiometric metallic hydrides are sometimes called *interstitial hydrides*. They may be considered to be solutions of hydrogen atoms in the metal, with the hydrogen atoms occupying the holes of interstices between metal atoms in the solid lattice. This is an oversimplification, however, because there is evidence for chemical interaction between metal and hydrogen.

The **molecular hydrides**, formed by nonmetals and semimetals, are either gases or liquids under standard conditions. The simple molecular hydrides are listed in Figure 22.7 ◀, together with their standard free energies of formation, ΔG_f°. In each family the thermal stability (measured by ΔG_f°) decreases as we move down the family. (Recall that the more stable a compound is with respect to its elements under standard conditions, the more negative ΔG_f° is). We will discuss the molecular hydrides further in the course of examining the other nonmetallic elements.

4A	5A	6A	7A
$CH_4(g)$ −50.8	$NH_3(g)$ −16.7	$H_2O(l)$ −237	$HF(g)$ −271
$SiH_4(g)$ +56.9	$PH_3(g)$ +18.2	$H_2S(g)$ −33.0	$HCl(g)$ −95.3
$GeH_4(g)$ +117	$AsH_3(g)$ +111	$H_2Se(g)$ +71	$HBr(g)$ −53.2
	$SbH_3(g)$ +187	$H_2Te(g)$ +138	$HI(g)$ +1.30

▲ **Figure 22.7** Standard free energies of formation (kJ/mol) of molecular hydrides.

22.3 Group 8A: The Noble Gases

The elements of group 8A are chemically unreactive. Indeed, most of our references to these elements have been in relation to their physical properties, as when we discussed intermolecular forces. (Section 11.2) The relative inertness of these elements is due to the presence of a completed octet of valence-shell electrons (except He, which has a filled 1*s* shell). The stability of such an arrangement is reflected in the high ionization energies of the group 8A elements. (Section 7.4)

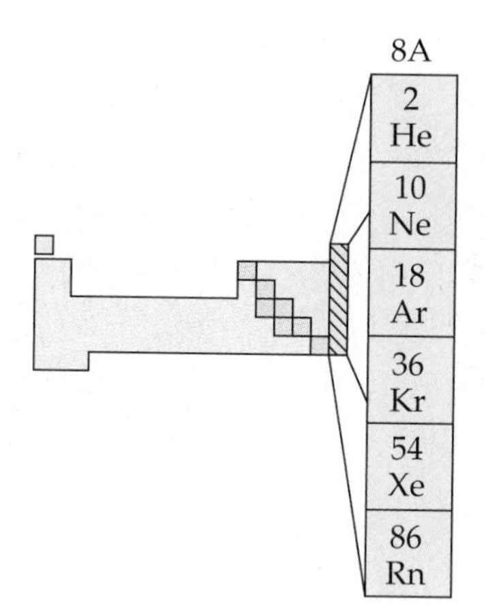

The group 8A elements are all gases at room temperature. They are components of Earth's atmosphere, except for radon, which exists only as a short-lived radioisotope. (Section 21.9) Only argon is relatively abundant (Table 18.1). Neon, argon, krypton, and xenon are recovered from liquid air by distillation. Argon is used as a blanketing atmosphere in electric lightbulbs. The gas conducts heat away from the filament but does not react with it. It is also used as a protective atmosphere to prevent oxidation in welding and certain high-temperature metallurgical processes. Neon is used in electric signs; the gas is caused to radiate by passing an electric discharge through the tube. (Section 6.3)

Helium is, in many ways, the most important of the noble gases. Liquid helium is used as a coolant to conduct experiments at very low temperatures. Helium boils at 4.2 K under 1 atm pressure, the lowest boiling point of any substance. Fortunately, helium is found in relatively high concentrations in many natural-gas wells. Some of this helium is separated to meet current demands, and some is kept for later use.

Noble-Gas Compounds

Because the noble gases are exceedingly stable, they will undergo reaction only under rigorous conditions. Furthermore, we might expect that the heavier noble gases would be most likely to form compounds because their ionization energies are lower (Figure 7.10). A lower ionization energy suggests the possibility of sharing an electron with another atom, leading to a chemical bond. In addition, because the group 8A elements (except helium) already contain eight electrons in their valence shell, formation of covalent bonds will require an expanded valence shell. Valence-shell expansion occurs most readily with larger atoms. (Section 8.7)

The first noble-gas compound was prepared in 1962 by Neil Bartlett while he was on the faculty of the University of British Columbia. His work caused a sensation because it undercut the belief that the noble-gas elements were truly chemically inert. Bartlett initially worked with xenon in combination with fluorine, the element we would expect to be most reactive. Since that time chemists have prepared several xenon compounds of fluorine and oxygen. Some properties of these substances are listed in Table 22.1 ▼. The three fluorides (XeF_2, XeF_4,

TABLE 22.1 Properties of Xenon Compounds

Compound	Oxidation State of Xe	Melting Point (°C)	ΔH_f° (kJ/mol)[a]
XeF_2	+2	129	−109(*g*)
XeF_4	+4	117	−218(*g*)
XeF_6	+6	49	−298(*g*)
$XeOF_4$	+6	−41 to −28	+146(*l*)
XeO_3	+6	—[b]	+402(*s*)
XeO_2F_2	+6	31	+145(*s*)
XeO_4	+8	—[c]	—

[a] At 25°C, for the compound in the state indicated.
[b] A solid; decomposes at 40°C.
[c] A solid; decomposes at −40°C.

and XeF_6) are made by direct reaction of the elements. By varying the ratio of reactants and altering reaction conditions, one or the other of the three compounds can be obtained. The oxygen-containing compounds are formed when the fluorides are reacted with water, as in Equations 22.19 and 22.20:

$$XeF_6(s) + H_2O(l) \longrightarrow XeOF_4(l) + 2HF(g) \quad [22.19]$$

$$XeF_6(s) + 3H_2O(l) \longrightarrow XeO_3(aq) + 6HF(aq) \quad [22.20]$$

SAMPLE EXERCISE 22.3

Use the VSEPR model to predict the structure of XeF_4.

Solution

Analyze and Plan: To predict the structure, we must first write the Lewis structure for the molecule. We then count the number of electron pairs (domains) around the central Xe atom and use that number and the number of bonds to predict the geometry, as discussed in Section 9.2.

Solve: The total number of valence-shell electrons involved is 36 (8 from xenon and 7 from each of the four fluorines). This leads to the Lewis structure shown in Figure 22.8(a) ◀. Xe has 12 electrons in its valence shell, so we expect an octahedral disposition of 6 electron pairs. Two of these are nonbonded pairs. Because nonbonded pairs have a larger volume requirement than bonded pairs (Section 9.2), it is reasonable to expect these nonbonded pairs to be opposite one another. The expected structure is square planar, as shown in Figure 22.8(b). The experimentally determined structure agrees with this prediction.

PRACTICE EXERCISE

Describe the electron-pair geometry and the molecular geometry of XeF_2.

Answer: trigonal bipyramidal; linear

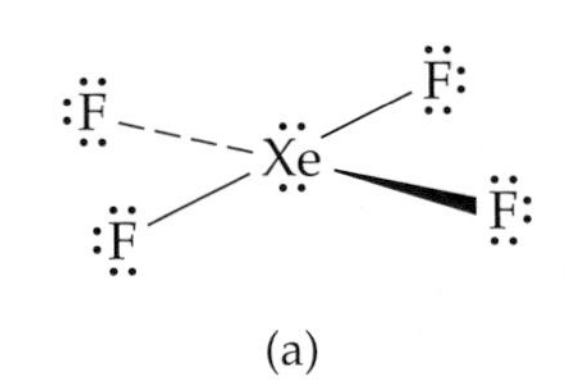

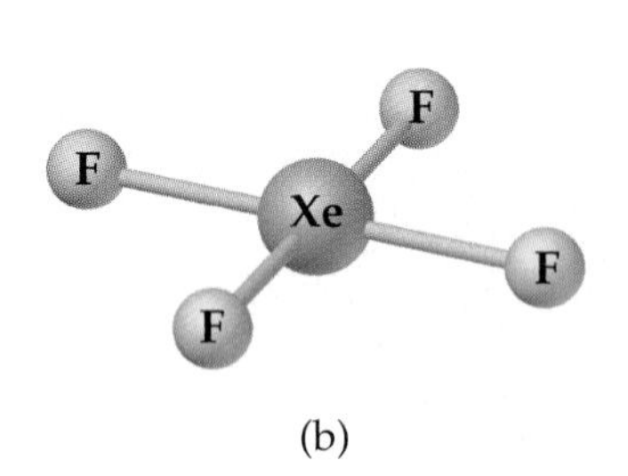

▲ **Figure 22.8** (a) Electron-pair and (b) molecular geometries of XeF_4.

The enthalpies of formation of the xenon fluorides are negative (Table 22.1), which suggests that these compounds should be reasonably stable. This is indeed found to be the case. They are, however, powerful fluorinating agents and must be handled in containers that do not readily react to form fluorides. The enthalpies of formation of the oxyfluorides and oxides of xenon, on the other hand, are positive, so these compounds are quite unstable.

The other noble-gas elements form compounds much less readily than xenon. Only one binary krypton compound, KrF_2, is known with certainty, and it decomposes to its elements at −10°C.

22.4 Group 7A: The Halogens

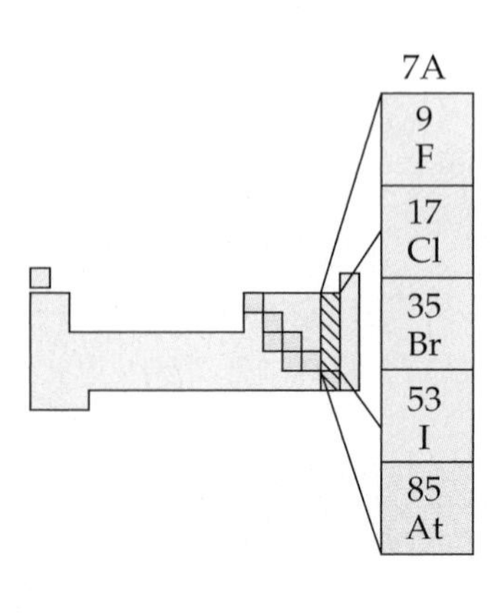

The elements of group 7A, the halogens, have outer electron configurations of ns^2np^5, where n ranges from 2 through 6. The halogens have large negative electron affinities (Section 7.5), and they most often achieve a noble-gas configuration by gaining an electron, which results in a −1 oxidation state. Fluorine, being the most electronegative element, exists in compounds only in the −1 state. The other halogens also exhibit positive oxidation states up to +7 in combination with more electronegative atoms such as O. In the positive oxidation states the halogens tend to be good oxidizing agents, readily accepting electrons.

Chlorine, bromine, and iodine are found as the halides in seawater and in salt deposits. The concentration of iodine in these sources is very small, but it is concentrated by certain seaweeds. When they are harvested, dried, and burned, iodine can be extracted from the ashes. Fluorine occurs in the minerals fluorspar

TABLE 22.2 Some Properties of the Halogens

Property	F	Cl	Br	I
Atomic radius (Å)	0.71	0.99	1.14	1.33
Ionic radius, X^- (Å)	1.33	1.81	1.96	2.20
First ionization energy (kJ/mol)	1681	1251	1140	1008
Electron affinity (kJ/mol)	−328	−349	−325	−295
Electronegativity	4.0	3.0	2.8	2.5
X—X single-bond enthalpy (kJ/mol)	155	242	193	151
Reduction potential (V): $\frac{1}{2}X_2(aq) + e^- \longrightarrow X^-(aq)$	2.87	1.36	1.07	0.54

MOVIE
Physical Properties of the Halogens, Formation of Sodium Chloride

(CaF_2), cryolite (Na_3AlF_6), and fluorapatite [$Ca_5(PO_4)_3F$].* Only fluorspar is an important commercial source of fluorine.

All isotopes of astatine are radioactive. The longest lived isotope is astatine-210, which has a half-life of 8.1 hr and decays mainly by electron capture. Because astatine is so unstable to nuclear decay, very little is known about its chemistry.

Properties and Preparation of the Halogens

Some of the properties of the halogens are summarized in Table 22.2 ▲. Most of the properties vary in a regular fashion as we go from fluorine to iodine. The electronegativity steadily decreases, for example, from 4.0 for fluorine to 2.5 for iodine. The halogens have the highest electronegativities in each horizontal row of the periodic table.

Under ordinary conditions the halogens exist as diatomic molecules. The molecules are held together in the solid and liquid states by London dispersion forces. (Section 11.2) Because I_2 is the largest and most polarizable of the halogen molecules, the intermolecular forces between I_2 molecules are the strongest. Thus, I_2 has the highest melting point and boiling point. At room temperature and 1 atm pressure, I_2 is a solid, Br_2 is a liquid, and Cl_2 and F_2 are gases. Chlorine readily liquefies upon compression at room temperature and is normally stored and handled in liquid form in steel containers.

The comparatively low bond enthalpy in F_2 (155 kJ/mol) accounts in part for the extreme reactivity of elemental fluorine. Because of its high reactivity, F_2 is very difficult to work with. Certain metals, such as copper and nickel, can be used to contain F_2 because their surfaces form a protective coating of metal fluoride. Chlorine and the heavier halogens are also reactive, although less so than fluorine. They combine directly with most elements except the noble gases.

Because of their high electronegativities, the halogens tend to gain electrons from other substances and thereby serve as oxidizing agents. The oxidizing ability of the halogens, which is indicated by their standard reduction potentials, decreases going down the group. As a result, a given halogen is able to oxidize the anions of the halogens below it in the group. For example, Cl_2 will oxidize Br^- and I^-, but not F^-, as seen in Figure 22.9 ▶.

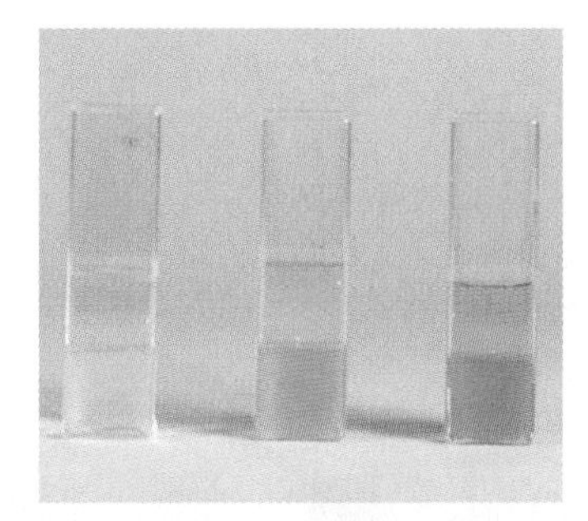

▲ **Figure 22.9** Aqueous solutions of NaF, NaBr, and NaI (from left to right) to which Cl_2 has been added. Each solution is in contact with carbon tetrachloride (CCl_4), which forms the lower layer in each container. The halogens are more soluble in CCl_4 than in H_2O. The F^- ion in the NaF solution (left) does not react with Cl_2; both the aqueous and CCl_4 layers remain colorless. The Br^- ion (center) is oxidized by Cl_2 to form Br_2, producing a yellow aqueous layer and an orange CCl_4 layer. The I^- ion (right) is oxidized to I_2, producing an amber aqueous layer and a violet CCl_4 layer.

* Minerals are solid substances that occur in nature. They are usually known by their common names rather than by their chemical names. What we know as *rock* is merely an aggregate of different kinds of minerals.

SAMPLE EXERCISE 22.4

Write the balanced equation for the reaction, if any, that occurs between **(a)** $I^-(aq)$ and $Br_2(l)$; **(b)** $Cl^-(aq)$ and $I_2(s)$.

Solution

Analyze: We are asked to determine whether a reaction occurs when a particular halide and halogen are combined.

Plan: A given halogen is able to reduce anions of the halogens below it in the periodic table. Thus, the smaller (lower atomic number) halogen will end up as the halide ion. If the halogen with smaller atomic number is already the halide, there will be no reaction. Thus, the key to determining whether or not a reaction will occur is locating the elements in the periodic table.

Solve: (a) Br_2 is able to oxidize (remove electrons from) the anions of the halogens below it in the periodic table. Thus, it will oxidize I^-.

$$2I^-(aq) + Br_2(l) \longrightarrow I_2(s) + 2Br^-(aq)$$

(b) Cl^- is the anion of a halogen above iodine in the periodic table. Thus, I_2 cannot oxidize Cl^-; there is no reaction.

PRACTICE EXERCISE

Write the balanced chemical equation for the reaction that occurs between $Br^-(aq)$ and $Cl_2(aq)$.

Answer: $2Br^-(aq) + Cl_2(aq) \longrightarrow Br_2(l) + 2Cl^-(aq)$

Notice in Table 22.2 that the reduction potential of F_2 is exceptionally high. Fluorine gas readily oxidizes water:

$$F_2(aq) + H_2O(l) \longrightarrow 2HF(aq) + \tfrac{1}{2}O_2(g) \qquad E° = 1.80 \text{ V} \qquad [22.21]$$

Fluorine cannot be prepared by electrolytic oxidation of aqueous solutions of fluoride salts because water itself is oxidized more readily than F^-. (Section 20.9) In practice, the element is formed by electrolytic oxidation of a solution of KF in anhydrous HF. The KF reacts with HF to form a salt, $K^+ HF_2^-$, which acts as the current carrier in the liquid. (The HF_2^- ion is stable because of very strong hydrogen bonding). The overall cell reaction is

$$2KHF_2(l) \longrightarrow H_2(g) + F_2(g) + 2KF(l) \qquad [22.22]$$

Chlorine is produced mainly by electrolysis of either molten or aqueous sodium chloride, as described in Sections 20.9 and 23.4. Both bromine and iodine are obtained commercially from brines containing the halide ions by oxidation with Cl_2.

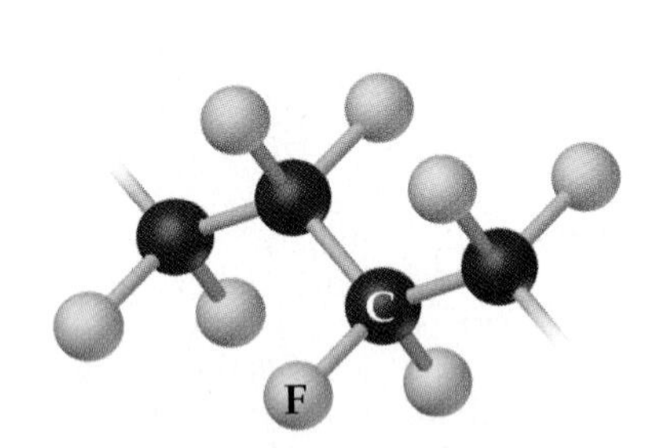

▲ **Figure 22.10** Structure of Teflon™, a fluorocarbon polymer. Teflon™ is an analog of polyethylene (Section 12.2), in which the H atoms are replaced with F atoms.

Uses of the Halogens

Fluorine is an important industrial chemical. It is used, for example, to prepare fluorocarbons—very stable carbon-fluorine compounds used as refrigerants, lubricants, and plastics. Teflon™ (Figure 22.10 ◀) is a polymeric fluorocarbon noted for its high thermal stability and lack of chemical reactivity. (The "Chemistry at Work" box in Section 25.4 describes the discovery of this interesting and important material.)

Chlorine is by far the most commercially important halogen. About 1.2×10^{10} kg (13 million tons) of Cl_2 is produced annually in the United States. In addition, hydrogen chloride production is about 4.4×10^9 kg (4.8 million tons) annually. About half of this chlorine finds its way eventually into the manufacture of chlorine-containing organic compounds such as vinyl chloride (C_2H_3Cl), used in making polyvinyl chloride (PVC) plastics. (Section 12.2) Much of the remainder is used as a bleaching agent in the paper and textile industries. When Cl_2 dissolves in cold dilute base, it disproportionates into Cl^- and hypochlorite, ClO^-.

$$Cl_2(aq) + 2OH^-(aq) \rightleftharpoons Cl^-(aq) + ClO^-(aq) + H_2O(l) \qquad [22.23]$$

Sodium hypochlorite (NaClO) is the active ingredient in many liquid bleaches. Chlorine is also used in water treatment to oxidize and thereby destroy bacteria. (Section 18.6)

Neither bromine nor iodine is as widely used as fluorine and chlorine. Bromine, however, is needed for the silver bromide used in photographic film. A common use of iodine is as KI in table salt. Iodized salt (Figure 22.11 ▶) provides the small amount of iodine necessary in our diets; it is essential for the formation of thyroxin, a hormone secreted by the thyroid gland. Lack of iodine in the diet results in an enlarged thyroid gland, a condition called *goiter*.

▲ **Figure 22.11** Iodized salt contains 0.02% KI by mass.

The Hydrogen Halides

All of the halogens form stable diatomic molecules with hydrogen. Aqueous solutions of HCl, HBr, and HI are strong acids.

The hydrogen halides can be formed by direct reaction of the elements. The most important means of preparing them, however, is by reacting a salt of the halide with a strong nonvolatile acid. Hydrogen fluoride and hydrogen chloride are prepared in this manner by reaction of an inexpensive, readily available salt with concentrated sulfuric acid.

$$CaF_2(s) + H_2SO_4(l) \xrightarrow{\Delta} 2HF(g) + CaSO_4(s) \quad [22.24]$$

$$NaCl(s) + H_2SO_4(l) \xrightarrow{\Delta} HCl(g) + NaHSO_4(s) \quad [22.25]$$

Neither hydrogen bromide nor hydrogen iodide can be prepared by analogous reactions of salts with H_2SO_4 because H_2SO_4 oxidizes Br^- and I^- (Figure 22.12 ▶). This difference in reactivity reflects the greater ease of oxidation of Br^- and I^- relative to F^- and Cl^-. These undesirable oxidations are avoided by using a nonvolatile acid, such as H_3PO_4, that is a poorer oxidizing agent than H_2SO_4.

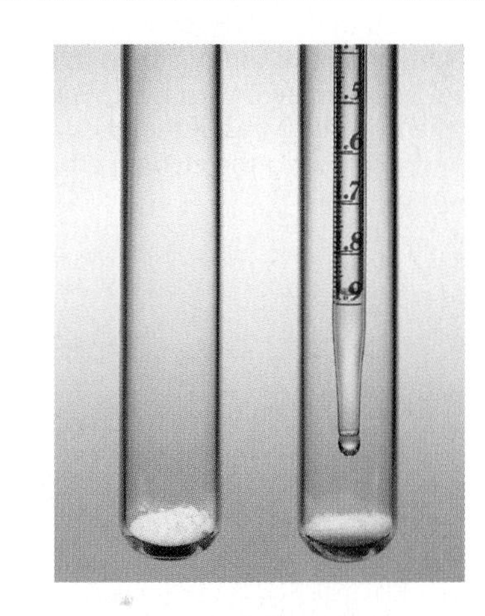

(a)

(b)

▲ **Figure 22.12** (a) Sodium iodide in the left test tube and sodium bromide on the right. Sulfuric acid is in the pipet. (b) Addition of sulfuric acid to the test tubes oxidizes sodium iodide to form the dark-colored iodine on the left. Sodium bromide is oxidized to the yellow-brown bromine on the right. When more concentrated, bromine has a reddish brown color.

SAMPLE EXERCISE 22.5

Write a balanced equation for the formation of hydrogen bromide gas from the reaction of solid sodium bromide with phosphoric acid.

Solution

Analyze: We are asked to write a balanced equation for the reaction between NaBr and H_3PO_4 to form HBr and another product.

Plan: As in Equations 22.24 and 22.25, a metathesis reaction takes place. Let's assume that only one of the hydrogens of H_3PO_4 undergoes reaction. (The actual number depends on the reaction conditions.) Thus, the remaining $H_2PO_4^-$ ion will be associated with the Na^+ ion as NaH_2PO_4 among the products of the equation.

Solve: The balanced equation is

$$NaBr(s) + H_3PO_4(aq) \longrightarrow NaH_2PO_4(s) + HBr(g)$$

PRACTICE EXERCISE

Write the balanced equation for the preparation of HI from NaI and H_3PO_4.

Answer: $NaI(s) + H_3PO_4(l) \longrightarrow NaH_2PO_4(s) + HI(g)$

The hydrogen halides form hydrohalic acid solutions when dissolved in water. These solutions exhibit the characteristic properties of acids, such as reactions with active metals to produce hydrogen gas. (Section 4.4) Hydrofluoric acid also reacts readily with silica (SiO_2) and with various silicates to form hexafluorosilicic acid (H_2SiF_6), as in these examples:

$$SiO_2(s) + 6HF(aq) \longrightarrow H_2SiF_6(aq) + 2H_2O(l) \quad [22.26]$$

$$CaSiO_3(s) + 8HF(aq) \longrightarrow H_2SiF_6(aq) + CaF_2(s) + 3H_2O(l) \quad [22.27]$$

▲ **Figure 22.13** Etched or frosted glass. Designs such as this are produced by first coating the glass with wax. The wax is then removed in the areas to be etched. When treated with hydrofluoric acid, the exposed areas of the glass are attacked, producing the etching effect.

Glass consists mostly of silicate structures (Section 22.10), and these reactions allow HF to etch or frost glass (Figure 22.13 ◀). It is also the reason that HF is stored in wax or plastic containers rather than glass.

Interhalogen Compounds

Because the halogens exist as diatomic molecules, diatomic molecules of two different halogen atoms exist. These compounds are the simplest examples of **interhalogens**, compounds formed between two different halogen elements such as ClF and IF_5.

With one exception, the higher interhalogen compounds have a central Cl, Br, or I atom surrounded by 3, 5, or 7 fluorine atoms. The large size of the iodine atom allows the formation of IF_3, IF_5, and IF_7, in which the oxidation state of I is +3, +5, and +7, respectively. With a central bromine atom, which is smaller than an iodine atom, only BrF_3 and BrF_5 can be formed. Chlorine, which is smaller still, can form ClF_3 and, with difficulty, ClF_5. The only higher interhalogen compound that does not have outer F atoms is ICl_3; the large size of the I atom can accommodate three Cl atoms, whereas Br is not large enough to allow $BrCl_3$ to form.

SAMPLE EXERCISE 22.6

Use the VSEPR model to describe the molecular geometry of BrF_3.

Solution

Analyze and Plan: To predict the geometry of BrF_3, we must first write its Lewis structure. We then count the number of electron domains around the central Br atom to determine the electron-domain geometry and then look at how many bonding electron pairs it has, to determine how to change the electron-domain geometry to the molecular one.

Solve: The Lewis structure of BrF_3 has three bonding pairs of electrons and two nonbonding pairs around the central Br atom. According to the VSEPR model, these five pairs of electrons are disposed about the central atom at the vertices of a trigonal bipyramid (Table 9.3). Because the nonbonding pairs require a larger space, they are placed in the equatorial plane of the trigonal bipyramid:

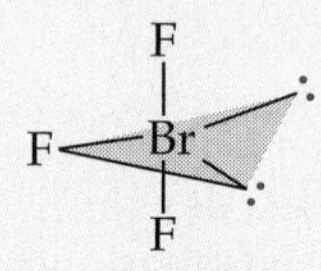

Because the unshared pairs push the bonding pairs back a little, the molecule has the shape of a bent T.

PRACTICE EXERCISE

Use the VSEPR model to predict the molecular geometry of IF_5.
Answer: square pyramidal

Because the interhalogen compounds contain a halogen atom in a positive oxidation state, they are exceedingly reactive. They invariably are powerful oxidizing agents. When the compound acts as an oxidant, the oxidation state of the central halogen atom is decreased to a more preferable value (usually 0 or −1) as in the following example:

$$2CoCl_2(s) + 2ClF_3(g) \longrightarrow 2CoF_3(s) + 3Cl_2(g) \quad [22.28]$$

Oxyacids and Oxyanions

Table 22.3 ▶ summarizes the formulas of the known oxyacids of the halogens and the way they are named.* (Section 2.8) The acid strengths of the oxyacids increase with increasing oxidation state of the central halogen atom. (Section 16.10)

* Fluorine forms one oxyacid, HOF. Because the electronegativity of fluorine is greater than that of oxygen, we must consider fluorine to be in a −1 oxidation state and oxygen to be in the 0 oxidation state in this compound.

TABLE 22.3 The Oxyacids of the Halogens

Oxidation State of Halogen	Formula of Acid: Cl	Br	I	Acid Name
+1	HClO	HBrO	HIO	*Hypo*hal*ous* acid
+3	$HClO_2$	—	—	Hal*ous* acid
+5	$HClO_3$	$HBrO_3$	HIO_3	Hal*ic* acid
+7	$HClO_4$	$HBrO_4$	HIO_4, H_5IO_6	*Per*hal*ic* acid

All the oxyacids are strong oxidizing agents. The oxyanions, formed on removal of H^+ from the oxyacids, are generally more stable than the oxyacids. Hypochlorite salts are used as bleaches and disinfectants because of the powerful oxidizing capabilities of the ClO^- ion. Sodium chlorite, is used as a bleaching agent. Chlorate salts are similarly very reactive. For example, potassium chlorate is used to make matches and fireworks.

Perchloric acid and its salts are the most stable of the oxyacids and oxyanions. Dilute solutions of perchloric acid are quite safe, and most perchlorate salts are stable except when heated with organic materials. When heated, perchlorates can become vigorous, even violent, oxidizers. Considerable caution should be exercised, therefore, when handling these substances, and it is crucial to avoid contact between perchlorates and readily oxidized material such as active metals and combustible organic compounds. The use of ammonium perchlorate (NH_4ClO_4) as the oxidizer in the solid booster rockets for the space shuttle demonstrates the oxidizing power of perchlorates. The solid propellant contains a mixture of NH_4ClO_4 and powdered aluminum, the reducing agent. Each shuttle launch requires about 6×10^5 kg (700 tons) of NH_4ClO_4 (Figure 22.14 ▶).

▲ **Figure 22.14** Launch of the space shuttle *Columbia* from the Kennedy Space Center.

There are two oxyacids that have iodine in the +7 oxidation state. These periodic acids are HIO_4 (called metaperiodic acid) and H_5IO_6 (called paraperiodic acid). The two forms exist in equilibrium in aqueous solution.

$$H_5IO_6(aq) \rightleftharpoons H^+(aq) + IO_4^-(aq) + 2H_2O(l) \qquad K_{eq} = 0.015 \qquad [22.29]$$

HIO_4 is a strong acid, and H_5IO_6 is a weak one; the first two acid-dissociation constants for H_5IO_6 are $K_{a1} = 2.8 \times 10^{-2}$ and $K_{a2} = 4.9 \times 10^{-9}$. The structure of H_5IO_6 is given in Figure 22.15 ▶. The large size of the iodine atom allows it to accommodate six surrounding oxygen atoms. The smaller halogens do not form acids of this type.

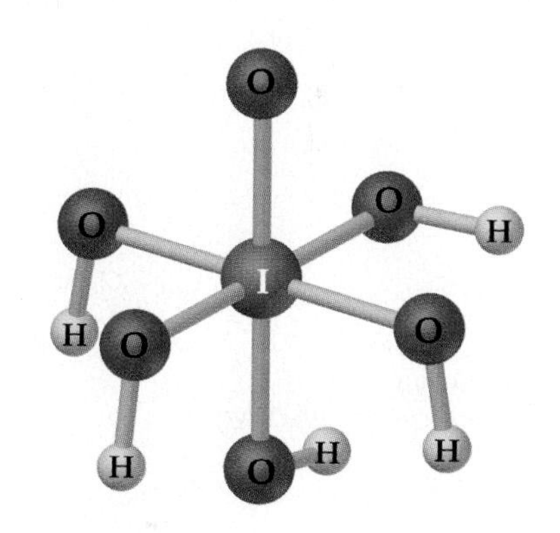

▲ **Figure 22.15** Paraperiodic acid (H_5IO_6).

22.5 Oxygen

By the middle of the seventeenth century, scientists recognized that air contained a component associated with burning and breathing. That component was not isolated until 1774, however, when Joseph Priestley (Figure 22.16 ▶) discovered oxygen. Lavoisier subsequently named the element *oxygen*, meaning "acid former."

Oxygen is found in combination with other elements in a great variety of compounds. Indeed, oxygen is the most abundant element by mass both in Earth's crust and in the human body. It is the oxidizing agent for the metabolism of our foods and is crucial to human life.

▼ **Figure 22.16** Joseph Priestley (1733–1804). Priestley became interested in chemistry at the age of 39. Because Priestley lived next door to a brewery from which he could obtain carbon dioxide, his studies focused on this gas first and were later extended to other gases. Because he was suspected of sympathizing with the American and French Revolutions, his church, home, and laboratory in Birmingham, England, were burned by a mob in 1791. Priestley had to flee in disguise. He eventually emigrated to the United States in 1794, where he lived his remaining years in relative seclusion in Pennsylvania.

Properties of Oxygen

Oxygen has two allotropes, O_2 and O_3. When we speak of molecular oxygen or simply oxygen, it is usually understood that we are speaking of *dioxygen* (O_2), the normal form of the element; O_3 is called *ozone*.

At room temperature dioxygen is a colorless and odorless gas. It condenses to a liquid at −183°C and freezes at −218°C. It is only slightly soluble in water, but its presence in water is essential to marine life.

The electron configuration of the oxygen atom is [He]$2s^2 2p^4$. Thus, oxygen can complete its octet of electrons either by picking up two electrons to form the oxide ion (O^{2-}), or by sharing two electrons. In its covalent compounds it tends to form two bonds: either as two single bonds, as in H_2O, or as a double bond, as in formaldehyde ($H_2C{=}O$). The O_2 molecule itself contains a double bond. (Section 9.8)

The bond in O_2 is very strong (the bond enthalpy is 495 kJ/mol). Oxygen also forms strong bonds with many other elements. Consequently, many oxygen-containing compounds are thermodynamically more stable than O_2. In the absence of a catalyst, however, most reactions of O_2 have high activation energies and thus require high temperatures to proceed at a suitable rate. Once a sufficiently exothermic reaction begins, however, it may accelerate rapidly, producing a reaction of explosive violence.

Preparation of Oxygen

Nearly all commercial oxygen is obtained from air. The normal boiling point of O_2 is −183°C, whereas that of N_2, the other principal component of air, is −196°C. Thus, when air is liquefied and then allowed to warm, the N_2 boils off, leaving liquid O_2 contaminated mainly by small amounts of N_2 and Ar.

A common laboratory method for preparing O_2 is the thermal decomposition of potassium chlorate ($KClO_3$) with manganese dioxide (MnO_2) added as a catalyst:

MOVIE
Reactions with Oxygen

$$2KClO_3(s) \xrightarrow{MnO_2} 2KCl(s) + 3O_2(g) \qquad [22.30]$$

Like H_2, O_2 can be collected by displacement of water because of its relatively low solubility (Figure 22.5).

Much of the O_2 in the atmosphere is replenished through the process of photosynthesis, in which green plants use the energy of sunlight to generate O_2 from atmospheric CO_2. Photosynthesis, therefore, not only regenerates O_2, but also uses up CO_2.

▲ **Figure 22.17** Welding with an oxyacetylene torch. The heat of combustion of acetylene is exceptionally high, thus giving rise to a very high flame temperature.

Uses of Oxygen

Oxygen is one of the most widely used industrial chemicals, ranking behind only sulfuric acid (H_2SO_4) and nitrogen (N_2). About 2.5×10^{10} kg (28 million tons) of O_2 is used annually in the United States. Oxygen can be shipped and stored either as a liquid or in steel containers as a compressed gas. About 70% of the O_2 output, however, is generated where it is needed.

Oxygen is by far the most widely used oxidizing agent. Over half of the O_2 produced is used in the steel industry, mainly to remove impurities from steel. It is also used to bleach pulp and paper. (Oxidation of colored compounds often gives colorless products.) In medicine, oxygen eases breathing difficulties. It is also used together with acetylene (C_2H_2) in oxyacetylene welding (Figure 22.17 ◀). The reaction between C_2H_2 and O_2 is highly exothermic, producing temperatures in excess of 3000°C:

$$2C_2H_2(g) + 5O_2(g) \longrightarrow 4CO_2(g) + 2H_2O(g) \qquad \Delta H^\circ = -2510 \text{ kJ} \qquad [22.31]$$

Ozone

Ozone is a pale-blue poisonous gas with a sharp, irritating odor. Most people can detect about 0.01 ppm in air. Exposure to 0.1 to 1 ppm produces headaches, burning eyes, and irritation to the respiratory passages.

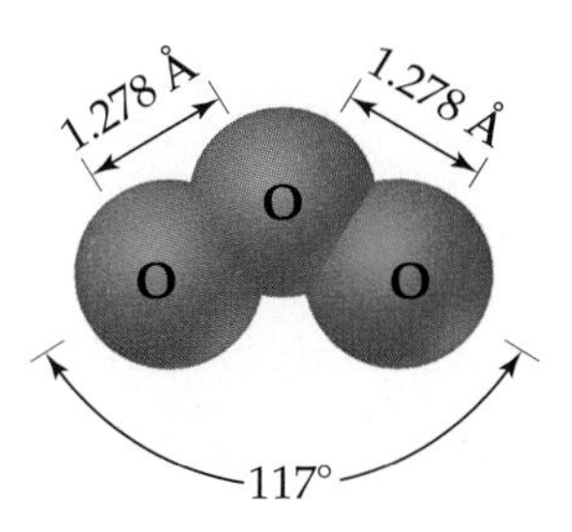

▲ **Figure 22.18** Structure of the ozone molecule.

3-D MODEL
Ozone

The structure of the O_3 molecule is shown in Figure 22.18 ▶. The molecule possesses a π bond that is delocalized over the three oxygen atoms. (Section 8.6) The molecule dissociates readily, forming reactive oxygen atoms:

$$O_3(g) \longrightarrow O_2(g) + O(g) \qquad \Delta H^\circ = 105 \text{ kJ} \qquad [22.32]$$

Ozone is a stronger oxidizing agent than dioxygen. One measure of this oxidizing power is the high standard reduction potential of O_3, compared to that of O_2.

$$O_3(g) + 2H^+(aq) + 2e^- \longrightarrow O_2(g) + H_2O(l) \qquad E^\circ = 2.07 \text{ V} \qquad [22.33]$$

$$O_2(g) + 4H^+(aq) + 4e^- \longrightarrow 2H_2O(l) \qquad E^\circ = 1.23 \text{ V} \qquad [22.34]$$

Ozone forms oxides with many elements under conditions where O_2 will not react; indeed, it oxidizes all the common metals except gold and platinum.

Ozone can be prepared by passing electricity through dry O_2 in an apparatus such as that shown in Figure 22.19 ▼:

$$3O_2(g) \xrightarrow{\text{electricity}} 2O_3(g) \qquad \Delta H^\circ = 285 \text{ kJ} \qquad [22.35]$$

Ozone cannot be stored for long, except at low temperature, because it readily decomposes to O_2. The decomposition is catalyzed by certain metals, such as Ag, Pt, and Pd, and by many transition-metal oxides.

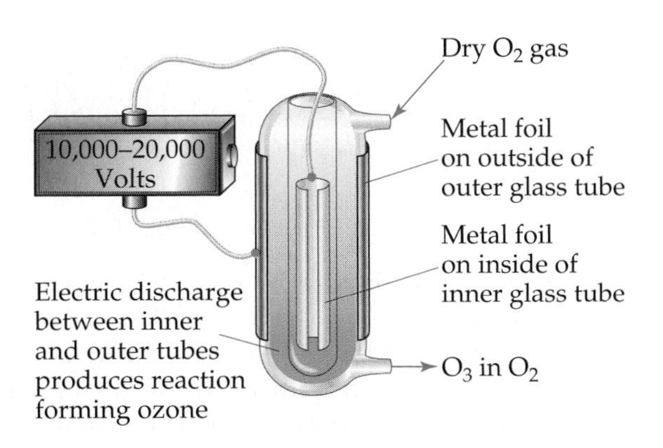

◀ **Figure 22.19** Apparatus for producing ozone from O_2.

SAMPLES EXERCISE 22.7

Using ΔG_f° for ozone from Appendix C, calculate the equilibrium constant, K_{eq}, for Equation 22.35 at 298.0 K.

Solution

Analyze: We are asked to calculate the equilibrium constant for the formation of O_3 from O_2 (Equation 22.35) given the temperature and ΔG_f°.

Plan: The relationship between the standard free-energy change, ΔG_f°, for a reaction and the equilibrium constant for the reaction was given in Section 19.7, Equation 19.22.

Solve: From Appendix C we have $\Delta G_f^\circ(O_3) = 163.4 \text{ kJ/mol}$

Thus, for Equation 22.35, $\Delta G^\circ = (2 \text{ mol } O_3)(163.4 \text{ kJ/mol } O_3) = 326.8 \text{ kJ}$

From Equation 19.22, we have $\Delta G^\circ = -RT \ln K_{eq}$

Thus,

$$\ln K_{eq} = \frac{-\Delta G^\circ}{RT} = \frac{-326.8 \times 10^3 \text{ J}}{(8.314 \text{ J/K-mol})(298.0 \text{ K})} = -131.9$$

$$K_{eq} = e^{-131.9} = 5 \times 10^{-58}$$

Comment: In spite of the unfavorable equilibrium constant, ozone can be prepared from O_2 as described in the preceding text. The unfavorable free energy of formation is overcome by energy from the electrical discharge, and O_3 is removed before the reverse reaction can occur, so a nonequilibrium mixture results.

PRACTICE EXERCISE

Using the data in Appendix C, calculate ΔG° and the equilibrium constant (K_{eq}) for Equation 22.32 at 298.0 K.

Answer: $\Delta G^\circ = 66.7 \text{ kJ}$; $K_{eq} = 2 \times 10^{-12}$

At the present time the uses of ozone as an industrial chemical are relatively limited. Ozone is sometimes used to treat domestic water in place of chlorine. Like Cl_2, it kills bacteria and oxidizes organic compounds. The largest use of ozone, however, is in the preparation of pharmaceuticals, synthetic lubricants, and other commercially useful organic compounds, where O_3 is used to sever carbon–carbon double bonds.

Ozone is an important component of the upper atmosphere, where it screens out ultraviolet radiation. In this way ozone protects Earth from the effects of these high-energy rays. For this reason, depletion of stratospheric ozone is a major scientific concern. (Section 18.3) In the lower atmosphere, however, ozone is considered an air pollutant. It is a major constituent of smog. (Section 18.4) Because of its oxidizing power, it damages living systems and structural materials, especially rubber.

Oxides

The electronegativity of oxygen is second only to that of fluorine. As a result, oxygen exhibits negative oxidation states in all compounds except those with fluorine, OF_2 and O_2F_2. The −2 oxidation state is by far the most common. Compounds in this oxidation state are called *oxides*.

ANIMATION
Periodic Trends: Acid-Base Behavior of Oxides

MOVIE
Carbon Dioxide Behaves as an Acid in Water

Nonmetals form covalent oxides. Most of these oxides are simple molecules with low melting and boiling points. SiO_2 and B_2O_3, however, have polymeric structures. (Sections 22.10 and 22.11) Most nonmetal oxides combine with water to give oxyacids. Sulfur dioxide (SO_2), for example, dissolves in water to give sulfurous acid (H_2SO_3):

$$SO_2(g) + H_2O(l) \longrightarrow H_2SO_3(aq) \quad [22.36]$$

This reaction and that of SO_3 with H_2O to form H_2SO_4 are largely responsible for acid rain. (Section 18.4) The analogous reaction of CO_2 with H_2O to form carbonic acid (H_2CO_3) causes the acidity of carbonated water.

Oxides that react with water to form acids are called **acidic anhydrides** (anhydride means "without water") or **acidic oxides**. A few nonmetal oxides, especially ones with the nonmetal in a low oxidation state—such as N_2O, NO, and CO—do not react with water and are not acidic anhydrides.

Most metal oxides are ionic compounds. Those ionic oxides that dissolve in water react to form hydroxides and are consequently called **basic anhydrides** or **basic oxides**. Barium oxide (BaO), for example, reacts with water to form barium hydroxide [$Ba(OH)_2$].

$$BaO(s) + H_2O(l) \longrightarrow Ba(OH)_2(aq) \quad [22.37]$$

▲ **Figure 22.20** Barium oxide (BaO), the white solid at the bottom of the container, reacts with water to produce barium hydroxide [$Ba(OH)_2$]. The reddish purple color of the solution is caused by phenolphthalein and indicates the presence of OH^- ions in solution.

This reaction is shown in Figure 22.20 ◀. These kinds of reactions are due to the high basicity of the O^{2-} ion and its virtually complete hydrolysis in water.

$$O^{2-}(aq) + H_2O(l) \longrightarrow 2OH^-(aq) \quad [22.38]$$

Even those ionic oxides that are water insoluble tend to dissolve in strong acids. Iron(III) oxide, for example, dissolves in acids:

$$Fe_2O_3(s) + 6H^+(aq) \longrightarrow 2Fe^{3+}(aq) + 3H_2O(l) \quad [22.39]$$

This reaction is used to remove rust ($Fe_2O_3 \cdot nH_2O$) from iron or steel before a protective coat of zinc or tin is applied.

Oxides that can exhibit both acidic and basic characters are said to be *amphoteric*. (Section 16.2) If a metal forms more than one oxide, the basic character of the oxide decreases as the oxidation state of the metal increases.

Oxide Compound	State of Oxidation Cr	Nature of Oxide
CrO	+2	Basic
Cr_2O_3	+3	Amphoteric
CrO_3	+6	Acidic

Peroxides and Superoxides

Compounds containing O—O bonds and oxygen in an oxidation state of −1 are called *peroxides*. Oxygen has an oxidation state of $-\frac{1}{2}$ in O_2^-, which is called the *superoxide* ion. The most active metals (K, Rb, and Cs) react with O_2 to give superoxides (KO_2, RbO_2, and CsO_2). Their active neighbors in the periodic table (Na, Ca, Sr, and Ba) react with O_2, producing peroxides (Na_2O_2, CaO_2, SrO_2, and BaO_2). Less active metals and nonmetals produce normal oxides. (Section 7.8)

When superoxides dissolve in water, O_2 is produced:

$$4KO_2(s) + 2H_2O(l) \longrightarrow 4K^+(aq) + 4OH^-(aq) + 3O_2(g) \qquad [22.40]$$

▲ **Figure 22.21** A self-contained breathing apparatus is used by firefighters and rescue workers. The source of oxygen is the reaction between potassium superoxide (KO_2) and water in the breath.

Because of this reaction, potassium superoxide (KO_2) is used as an oxygen source in masks worn for rescue work (Figure 22.21 ▶). Moisture in the breath causes the compound to decompose to form O_2 and KOH. The KOH so formed removes CO_2 from the exhaled breath:

$$2OH^-(aq) + CO_2(g) \longrightarrow H_2O(l) + CO_3^{2-}(aq) \qquad [22.41]$$

Hydrogen peroxide (H_2O_2) is the most familiar and commercially important peroxide. The structure of H_2O_2 is shown in Figure 22.22 ▼. Pure hydrogen peroxide is a clear, syrupy liquid with a density of 1.47 g/cm^3 at 0°C. It melts at −0.4°C, and its normal boiling point is 151°C. These properties are characteristic of a highly polar, strongly hydrogen-bonded liquid such as water. Concentrated hydrogen peroxide is a dangerously reactive substance because the decomposition to form water and oxygen gas is very exothermic.

$$2H_2O_2(l) \longrightarrow 2H_2O(l) + O_2(g) \qquad \Delta H^\circ = -196.1 \text{ kJ} \qquad [22.42]$$

Hydrogen peroxide is marketed as a chemical reagent in aqueous solutions of up to about 30% by mass. A solution containing about 3% H_2O_2 by mass is sold in drugstores and used as a mild antiseptic; somewhat more concentrated solutions are employed to bleach fabrics.

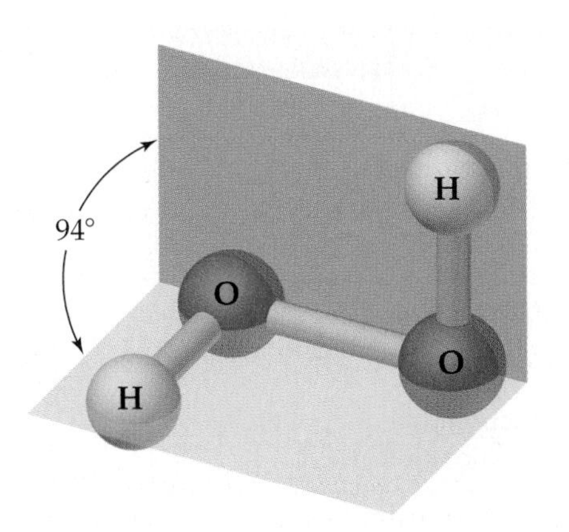

◀ **Figure 22.22** The molecular structure of hydrogen peroxide (H_2O_2). Note that the atoms of the molecule do not lie in a single plane.

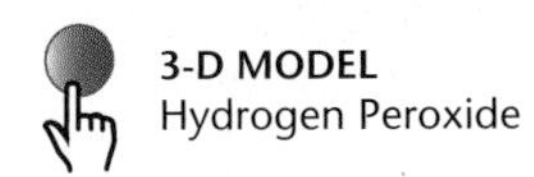

The peroxide ion is also a by-product of metabolism that results from the reduction of molecular oxygen (O_2). The body disposes of this reactive species with enzymes such as peroxidase and catalase.

Hydrogen peroxide can act as either an oxidizing or a reducing agent. The reduction half-reactions in acid solution are

$$2H^+(aq) + H_2O_2(aq) + 2e^- \longrightarrow 2H_2O(l) \qquad E° = 1.78 \text{ V} \qquad [22.43]$$

$$O_2(g) + 2H^+(aq) + 2e^- \longrightarrow H_2O_2(aq) \qquad E° = 0.68 \text{ V} \qquad [22.44]$$

The combination of these two half-reactions leads to the **disproportionation** of H_2O_2 into H_2O and O_2, shown in Equation 22.43. This reaction is strongly favored ($E° = 1.78 \text{ V} - 0.68 \text{ V} = +1.10 \text{ V}$). Disproportionation occurs when an element is simultaneously oxidized and reduced.

22.6 The Other Group 6A Elements: S, Se, Te, and Po

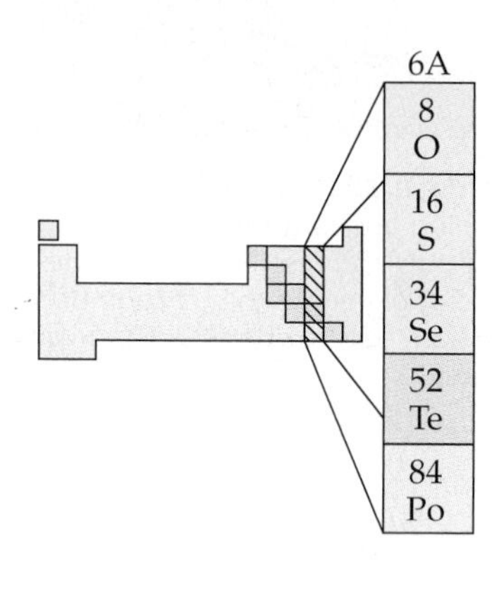

In addition to oxygen, the other group 6A elements are sulfur, selenium, tellurium, and polonium. In this section we will survey the properties of the group as a whole and then examine the chemistry of sulfur, selenium, and tellurium. We will not say much about polonium, which has no stable isotopes and is found only in minute quantities in radium-containing minerals.

General Characteristics of the Group 6A Elements

The group 6A elements possess the general outer-electron configuration ns^2np^4, where n has values ranging from 2 through 6. Thus, these elements may attain a noble-gas electron configuration by the addition of two electrons, which results in a -2 oxidation state. Because the group 6A elements are nonmetals, this is a common oxidation state. Except for oxygen, however, the group 6A elements are also commonly found in positive oxidation states up to $+6$, and they can have expanded valence shells. Thus, compounds such as SF_6, SeF_6, and TeF_6 occur in which the central atom is in the $+6$ oxidation state with more than an octet of valence electrons.

Table 22.4 ▼ summarizes some of the more important properties of the atoms of the group 6A elements. In most of the properties listed in Table 22.4, we see a regular variation as a function of increasing atomic number. For example, atomic and ionic radii increase and ionization energies decrease, as expected, as we move down the family.

TABLE 22.4 Some Properties of the Group 6A Elements

Property	O	S	Se	Te
Atomic radius (Å)	0.73	1.04	1.17	1.43
X^{2-} ionic radius (Å)	1.40	1.84	1.98	2.21
First ionization energy (kJ/mol)	1314	1000	941	869
Electron affinity (kJ/mol)	−141	−200	−195	−190
Electronegativity	3.5	2.5	2.4	2.1
X—X single-bond enthalpy (kJ/mol)	146*	266	172	126
Reduction potential to H_2X in acidic solution (V)	1.23	0.14	−0.40	−0.72

*Based on O—O bond energy in H_2O_2.

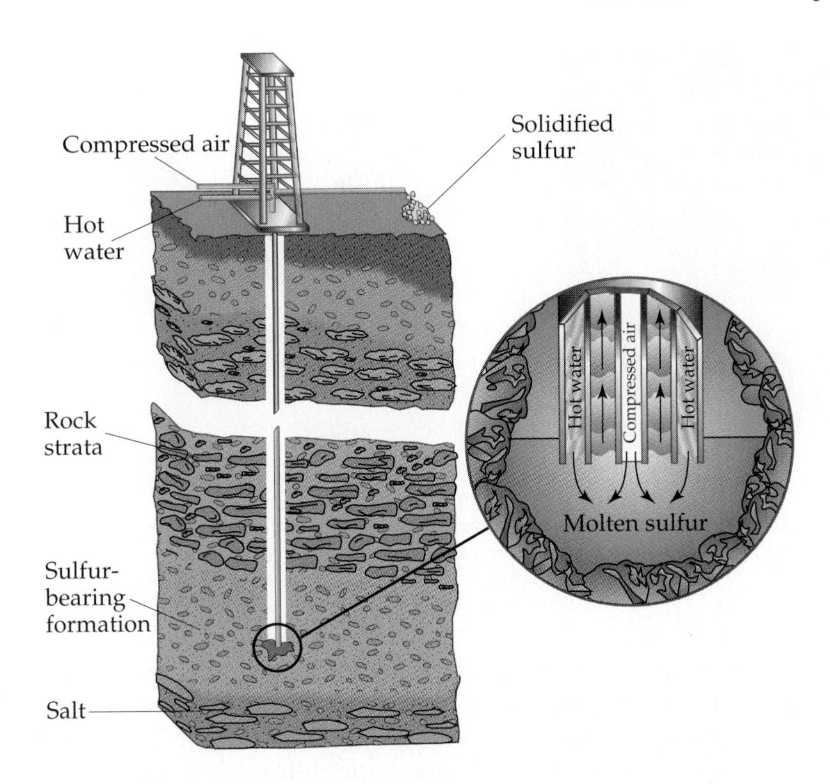

◀ **Figure 22.23** Mining sulfur by the Frasch process. The process is named after Herman Frasch, who invented it in the early 1890s. The process is particularly useful for recovering sulfur from deposits located under quicksand or water.

Occurrences and Preparation of S, Se, and Te

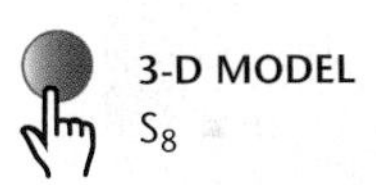

Large underground deposits are the principal source of elemental sulfur. The *Frasch process*, illustrated in Figure 22.23 ▲, is used to obtain the element from these deposits. The method is based on the low melting point and low density of sulfur. Superheated water is forced into the deposit, where it melts the sulfur. Compressed air then forces the molten sulfur up a pipe to the surface above, where the sulfur cools and solidifies.

Sulfur also occurs widely as sulfide and sulfate minerals. Its presence as a minor component of coal and petroleum poses a major problem. Combustion of these "unclean" fuels leads to serious sulfur oxide pollution. (Section 18.4) Much effort has been directed at removing this sulfur, and these efforts have increased the availability of sulfur. The sale of this sulfur helps to partially offset the costs of the desulfurizing processes and equipment.

Selenium and tellurium occur in rare minerals such as Cu_2Se, PbSe, Ag_2Se, Cu_2Te, PbTe, Ag_2Te, and Au_2Te. They also occur as minor constituents in sulfide ores of copper, iron, nickel, and lead.

Properties and Uses of Sulfur, Selenium, and Tellurium

As we normally encounter it, sulfur is yellow, tasteless, and nearly odorless. It is insoluble in water and exists in several allotropic forms. The thermodynamically stable form at room temperature is rhombic sulfur, which consists of puckered S_8 rings, as shown in Figure 22.24 ▶. When heated above its melting point (113°C), sulfur undergoes a variety of changes. The molten sulfur first contains S_8 molecules and is fluid because the rings readily slip over one another. Further heating of this straw-colored liquid causes the rings to break; the fragments then join to form very long molecules that can become entangled. The sulfur consequently becomes highly viscous. This change is marked by a color change to

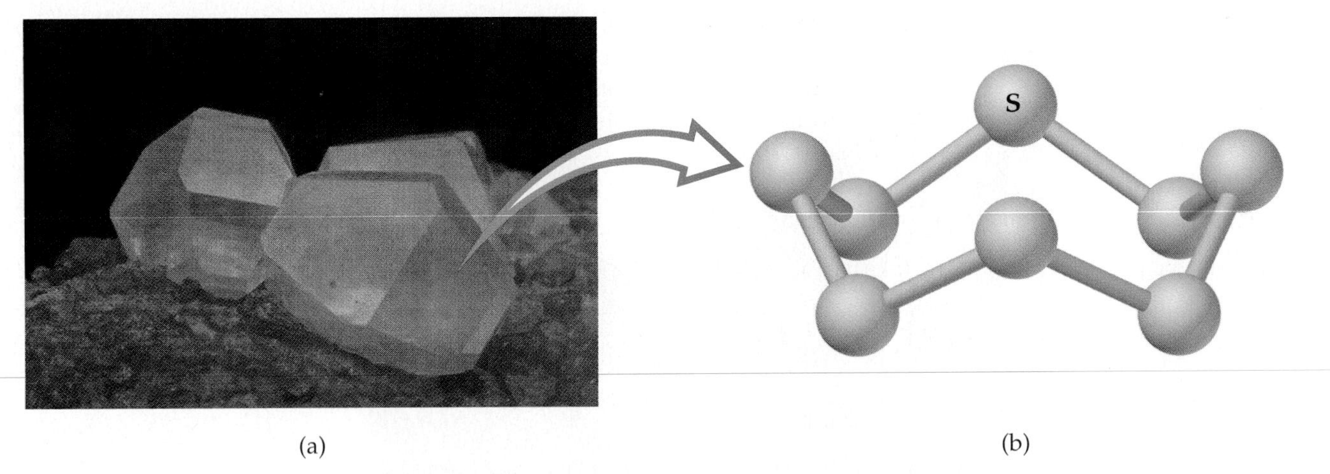

▲ **Figure 22.24** The common yellow crystalline form of rhombic sulfur consists of S_8 molecules. These molecules are puckered rings of S atoms.

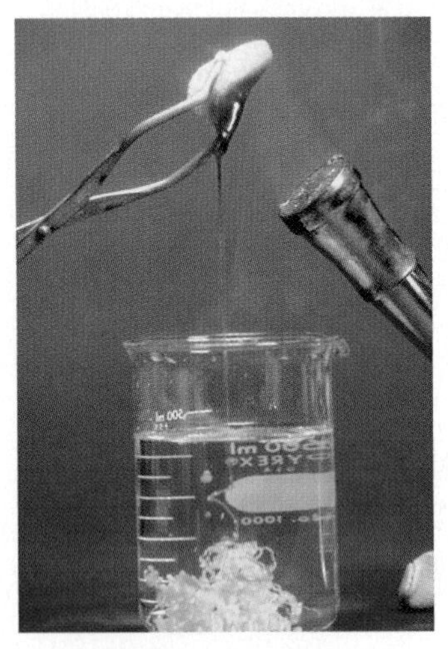

▲ **Figure 22.25** When sulfur is heated above its melting point (113°C), it becomes dark and viscous. Here the liquid is shown falling into cold water, where it again solidifies.

dark reddish brown (Figure 22.25 ◀). Further heating breaks the chains, and the viscosity again decreases.

Most of the 1.4×10^{10} kg (15 million tons) of sulfur produced in the United States each year is used to manufacture sulfuric acid. Sulfur is also used to vulcanize rubber, a process that toughens rubber by introducing cross-linking between polymer chains. (Section 12.2)

The most stable allotropes of both selenium and tellurium are crystalline substances containing helical chains of atoms, as illustrated in Figure 22.26 ◀. Each atom of the chain is close to atoms in adjacent chains, and it appears that some sharing of electron pairs between these atoms occurs.

The electrical conductivity of selenium is very low in the dark, but increases greatly upon exposure to light. This property of the element is utilized in photoelectric cells and light meters. Photocopiers also depend on the photoconductivity of selenium. Photocopy machines contain a belt or drum coated with a film of selenium. This drum is electrostatically charged and then exposed to light reflected from the image being photocopied. The electric charge drains from the selenium, where the selenium is made conductive by exposure to light. A black powder (the "toner") sticks only to the areas that remain charged. The photocopy is made when the toner is transferred to a sheet of plain paper, which is heated to fuse the toner to the paper.

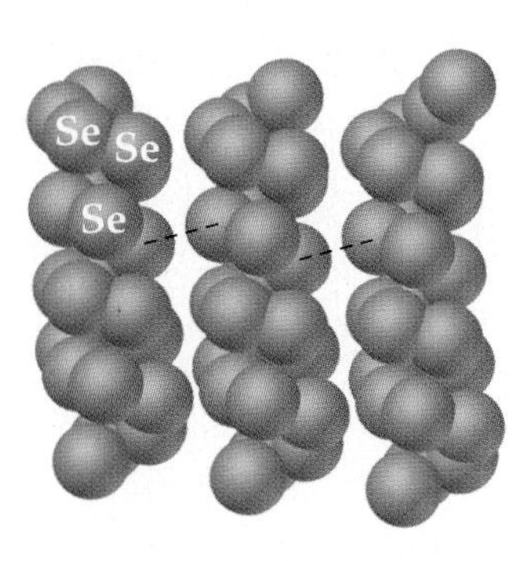

▲ **Figure 22.26** Portion of the structure of crystalline selenium. The dashed lines represent weak bonding interactions between atoms in adjacent chains. Tellurium has the same structure.

Sulfides

Sulfur forms compounds by direct combination with many elements. When the element is less electronegative than sulfur, *sulfides*, which contain S^{2-}, form. Iron(II) sulfide (FeS) forms, for example, by direct combination of iron and sulfur. Many metallic elements are found in the form of sulfide ores, such as PbS (galena) and HgS (cinnabar). A series of related ores containing the disulfide ion, S_2^{2-} (analogous to the peroxide ion), are known as *pyrites*. Iron pyrite, FeS_2, occurs as golden yellow cubic crystals (Figure 22.27 ▶). Because it has been occasionally mistaken for gold by miners, it is often called "fool's gold."

One of the most important sulfides is hydrogen sulfide (H_2S). This substance is not normally produced by direct union of the elements because it is unstable at elevated temperature and decomposes into the elements. It is normally prepared by action of dilute acid on iron(II) sulfide.

$$FeS(s) + 2H^+(aq) \longrightarrow H_2S(aq) + Fe^{2+}(aq) \qquad [22.45]$$

One of hydrogen sulfide's most readily recognized properties is its odor; H_2S is largely responsible for the offensive odor of rotten eggs. Hydrogen sulfide is actually quite toxic. Fortunately, our noses are able to detect H_2S in extremely low, nontoxic concentrations. Sulfur-containing organic molecules, which are similarly odoriferous, are added to natural gas to give it a detectable odor.

▲ **Figure 22.27** Iron pyrite (FeS_2) is also known as fool's gold because its color has fooled people into thinking it was gold. Gold is much more dense and much softer than iron pyrite.

Oxides, Oxyacids, and Oxyanions of Sulfur

Sulfur dioxide is formed when sulfur is combusted in air; it has a choking odor and is poisonous. The gas is particularly toxic to lower organisms, such as fungi, so it is used to sterilize dried fruit and wine. At 1 atm pressure and room temperature SO_2 dissolves in water to produce a solution of about 1.6 *M* concentration. The SO_2 solution is acidic, and we describe it as sulfurous acid (H_2SO_3). Sulfurous acid is a diprotic acid:

$$H_2SO_3(aq) \rightleftharpoons H^+(aq) + HSO_3^-(aq) \qquad K_{a1} = 1.7 \times 10^{-2}\ (25°C) \qquad [22.46]$$

$$HSO_3^-(aq) \rightleftharpoons H^+(aq) + SO_3^{2-}(aq) \qquad K_{a2} = 6.4 \times 10^{-8}\ (25°C) \qquad [22.47]$$

Salts of SO_3^{2-} (sulfites) and HSO_3^- (hydrogen sulfites or bisulfites) are well known. Small quantities of Na_2SO_3 or $NaHSO_3$ are used as food additives to prevent bacterial spoilage. Because some people are extremely allergic to sulfites, all food products with sulfites must now carry a warning disclosing their presence.

Although combustion of sulfur in air produces mainly SO_2, small amounts of SO_3 are also formed. The reaction produces mainly SO_2 because the activation-energy barrier for further oxidation to SO_3 is very high unless the reaction is catalyzed. Sulfur trioxide is of great commercial importance because it is the anhydride of sulfuric acid. In the manufacture of sulfuric acid, SO_2 is first obtained by burning sulfur. The SO_2 is then oxidized to SO_3, using a catalyst such as V_2O_5 or platinum. The SO_3 is dissolved in H_2SO_4 because it does not dissolve quickly in water (Equation 22.48). The $H_2S_2O_7$ formed in this reaction, called pyrosulfuric acid, is then added to water to form H_2SO_4, as shown in Equation 22.49.

$$SO_3(g) + H_2SO_4(l) \longrightarrow H_2S_2O_7(l) \qquad [22.48]$$

$$H_2S_2O_7(l) + H_2O(l) \longrightarrow 2H_2SO_4(l) \qquad [22.49]$$

Commercial sulfuric acid is 98% H_2SO_4. It is a dense, colorless, oily liquid that boils at 340°C. Sulfuric acid has many useful properties: It is a strong acid, a good dehydrating agent, and a moderately good oxidizing agent. Its dehydrating ability is demonstrated in Figure 22.28 ▼.

Year after year, the production of sulfuric acid is the largest of any chemical produced in the United States. About 4.0×10^{10} kg (44 million tons) are

(a)

(b)

(c)

◀ **Figure 22.28** The reaction between sucrose ($C_{12}H_{22}O_{11}$) and concentrated sulfuric acid. Sucrose is a carbohydrate, containing two H atoms for each O atom. Sulfuric acid, which is an excellent dehydrating agent, removes H_2O from the sucrose to form carbon, the black mass remaining at the end of the reaction.

MOVIE
Dehydration of Sugar

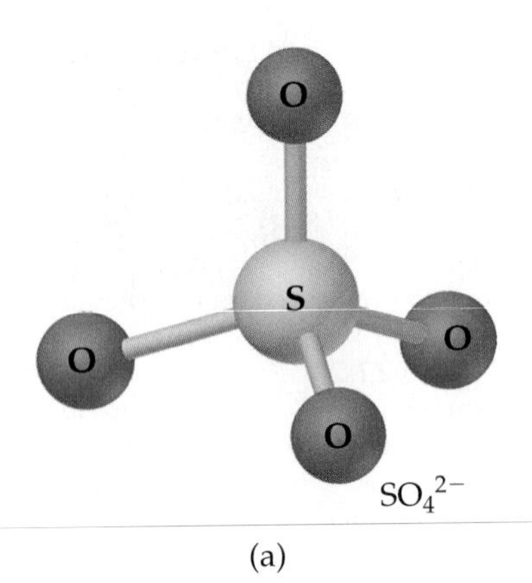

(a)

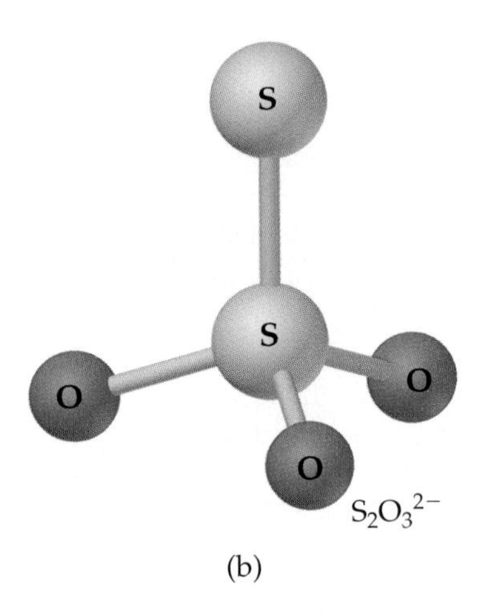

(b)

▲ **Figure 22.29** Comparison of the structures of (a) the sulfate ion (SO_4^{2-}) and (b) the thiosulfate ion ($S_2O_3^{2-}$).

produced annually in this country. Sulfuric acid is employed in some way in almost all manufacturing. Consequently, its consumption is considered a measure of industrial activity.

Sulfuric acid is classified as a strong acid, but only the first hydrogen is completely ionized in aqueous solution. The second hydrogen ionizes only partially.

$$H_2SO_4(aq) \longrightarrow H^+(aq) + HSO_4^-(aq)$$

$$HSO_4^-(aq) \rightleftharpoons H^+(aq) + SO_4^{2-}(aq) \qquad K_a = 1.1 \times 10^{-2}$$

Consequently, sulfuric acid forms two series of compounds: sulfates and bisulfates (or hydrogen sulfates). Bisulfate salts are common components of the "dry acids" used for adjusting the pH of swimming pools and hot tubs; they are also components of many toilet bowl cleaners.

The thiosulfate ion ($S_2O_3^{2-}$) is related to the sulfate ion and is formed by boiling an alkaline solution of SO_3^{2-} with elemental sulfur.

$$8SO_3^{2-}(aq) + S_8(s) \longrightarrow 8S_2O_3^{2-}(aq) \qquad [22.50]$$

The term *thio* indicates substitution of sulfur for oxygen. The structures of the sulfate and thiosulfate ions are compared in Figure 22.29 ◀. When acidified, the thiosulfate ion decomposes to form sulfur and H_2SO_3.

The pentahydrated salt of sodium thiosulfate ($Na_2S_2O_3 \cdot 5H_2O$), known as "hypo," is used in photography. Photographic film consists of a suspension of microcrystals of AgBr in gelatin. When exposed to light, some AgBr decomposes, forming very small grains of silver. When the film is treated with a mild reducing agent (the "developer"), Ag^+ ions in AgBr near the silver grains are reduced, forming an image of black metallic silver. The film is then treated with sodium thiosulfate solution to remove the unexposed AgBr. The thiosulfate ion reacts with AgBr to form a soluble silver thiosulfate complex.

$$AgBr(s) + 2S_2O_3^{2-}(aq) \rightleftharpoons Ag(S_2O_3)_2^{3-}(aq) + Br^-(aq) \qquad [22.51]$$

This step in the process is called "fixing." Thiosulfate ion is also used in quantitative analysis as a reducing agent for iodine:

$$2S_2O_3^{2-}(aq) + I_2(s) \longrightarrow 2I^-(aq) + S_4O_6^{2-}(aq) \qquad [22.52]$$

22.7 Nitrogen

Nitrogen was discovered in 1772 by the Scottish botanist Daniel Rutherford. He found that when a mouse was enclosed in a sealed jar, the animal quickly consumed the life-sustaining component of air (oxygen) and died. When the "fixed air" (CO_2) in the container was removed, a "noxious air" remained that would not sustain combustion or life. That gas is known to us now as nitrogen.

Nitrogen constitutes 78% by volume of Earth's atmosphere, where it occurs as N_2 molecules. Although nitrogen is a key element in living organisms, compounds of nitrogen are not abundant in Earth's crust. The major natural deposits of nitrogen compounds are those of KNO_3 (saltpeter) in India and $NaNO_3$ (Chile saltpeter) in Chile and other desert regions of South America.

Properties of Nitrogen

Nitrogen is a colorless, odorless, and tasteless gas composed of N_2 molecules. Its melting point is −210°C, and its normal boiling point is −196°C.

The N_2 molecule is very unreactive because of the strong triple bond between nitrogen atoms (the N≡N bond enthalpy is 941 kJ/mol, nearly twice that for the bond in O_2; see Table 8.4). When substances burn in air, they normally react with O_2 but not with N_2. When magnesium burns in air, however,

$$NO_3^- \xrightarrow{+0.79\ V} NO_2 \xrightarrow{+1.12\ V} HNO_2 \xrightarrow{+1.00\ V} NO \xrightarrow{+1.59\ V} N_2O \xrightarrow{+1.77\ V} N_2 \xrightarrow{+0.27\ V} NH_4^+$$

$NO_3^- \rightarrow NO$: +0.96 V

$NO_3^- \rightarrow N_2$: +1.25 V

◀ **Figure 22.30** Standard reduction potentials in acid solution for some common nitrogen-containing compounds. Reduction of NO_3^- to NO_2 in acid solution, for example, has a standard electrode potential of 0.79 V (leftmost entry). You should be able to balance this half-reaction using the techniques discussed in Section 20.2.

it also reacts with N_2 to form magnesium nitride (Mg_3N_2). A similar reaction occurs with lithium, forming Li_3N.

$$3Mg(s) + N_2(g) \longrightarrow Mg_3N_2(s) \qquad [22.53]$$

The nitride ion is a strong Brønsted–Lowry base. It reacts with water to form ammonia (NH_3), as in the following reaction:

$$Mg_3N_2(s) + 6H_2O(l) \longrightarrow 2NH_3(aq) + 3Mg(OH)_2(s) \qquad [22.54]$$

The electron configuration of the nitrogen atom is $[He]2s^22p^3$. The element exhibits all formal oxidation states from +5 to −3, as shown in Table 22.5 ▶. The +5, 0, and −3 oxidation states are the most common and generally the most stable of these. Because nitrogen is more electronegative than all elements except fluorine, oxygen, and chlorine, it exhibits positive oxidation states only in combination with these three elements.

TABLE 22.5 Oxidation States of Nitrogen

Oxidation State	Examples
+5	N_2O_5, HNO_3, NO_3^-
+4	NO_2, N_2O_4
+3	HNO_2, NO_2^-, NF_3
+2	NO
+1	N_2O, $H_2N_2O_2$, $N_2O_2^{2-}$, HNF_2
0	N_2
−1	NH_2OH, NH_2F
−2	N_2H_4
−3	NH_3, NH_4^+, NH_2^-

Figure 22.30 ▲ summarizes the standard reduction potentials for interconversion of several common nitrogen species. The potentials in the diagram are large and positive, which indicates that the nitrogen oxides and oxyanions shown are strong oxidizing agents.

Preparation and Uses of Nitrogen

Elemental nitrogen is obtained in commercial quantities by fractional distillation of liquid air. About 3.6×10^{10} kg (40 million tons) of N_2 is produced annually in the United States.

Because of its low reactivity, large quantities of N_2 are used as an inert gaseous blanket to exclude O_2 during the processing and packaging of foods, the manufacture of chemicals, the fabrication of metals, and the production of electronic devices. Liquid N_2 is employed as a coolant to freeze foods rapidly.

The largest use of N_2 is in the manufacture of nitrogen-containing fertilizers, which provide a source of *fixed* nitrogen. We have previously discussed nitrogen fixation in the "Chemistry and Life" box in Section 14.7 and in the "Chemistry at Work" box in Section 15.1. Our starting point in fixing nitrogen is the manufacture of ammonia via the Haber process. (Section 15.1) The ammonia can then be converted into a variety of useful, simple nitrogen-containing species, as shown in Figure 22.31 ▼. Many of the reactions along this chain of conversion are discussed in more detail later in this section.

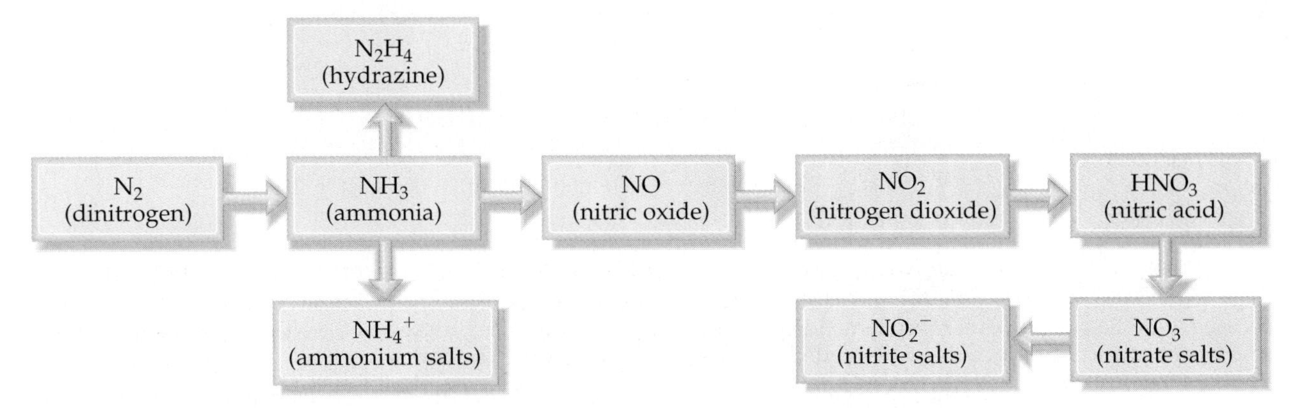

▲ **Figure 22.31** Sequence of conversion of N_2 into common nitrogen compounds.

Hydrogen Compounds of Nitrogen

Ammonia is one of the most important compounds of nitrogen. It is a colorless toxic gas that has a characteristic irritating odor. As we have noted in previous discussions, the NH_3 molecule is basic ($K_b = 1.8 \times 10^{-5}$). (Section 16.7)

In the laboratory NH_3 can be prepared by the action of NaOH on an ammonium salt. The NH_4^+ ion, which is the conjugate acid of NH_3, transfers a proton to OH^-. The resultant NH_3 is volatile and is driven from the solution by mild heating.

$$NH_4Cl(aq) + NaOH(aq) \longrightarrow NH_3(g) + H_2O(l) + NaCl(aq) \quad [22.55]$$

Commercial production of NH_3 is achieved by the Haber process.

$$N_2(g) + 3H_2(g) \longrightarrow 2NH_3(g) \quad [22.56]$$

About 1.6×10^{10} kg (18 million tons) of ammonia is produced annually in the United States. About 75% is used for fertilizer.

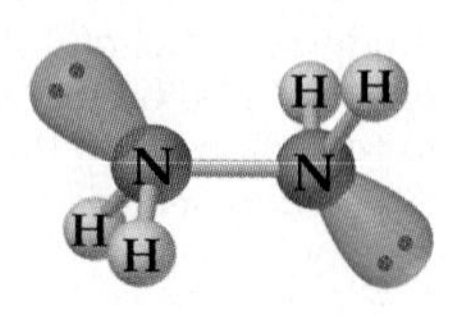

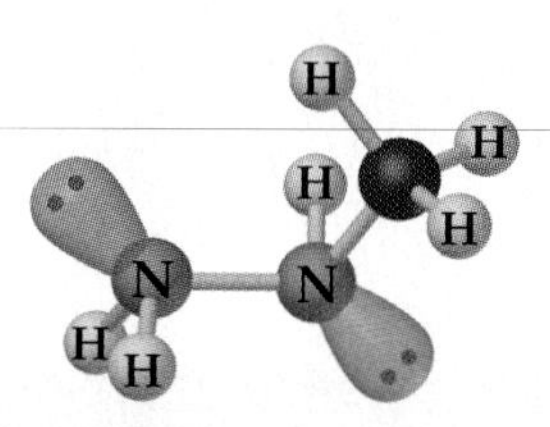

▲ **Figure 22.32** Structures of hydrazine (N_2H_4) and methylhydrazine (CH_3NHNH_2).

Hydrazine (N_2H_4) bears the same relationship to ammonia that hydrogen peroxide does to water. As shown in Figure 22.32 ◀, the hydrazine molecule contains an N—N single bond. Hydrazine is quite poisonous. It can be prepared by the reaction of ammonia with hypochlorite ion (OCl^-) in aqueous solution.

$$2NH_3(aq) + OCl^-(aq) \longrightarrow N_2H_4(aq) + Cl^-(aq) + H_2O(l) \quad [22.57]$$

The reaction is complex, involving several intermediates, including chloramine (NH_2Cl). The poisonous NH_2Cl bubbles out of solution when household ammonia and chlorine bleach (which contains OCl^-) are mixed. This reaction is one reason for the frequently cited warning not to mix bleach and household ammonia.

Pure hydrazine is a colorless, oily liquid that explodes on heating in air. It can be handled safely in aqueous solution, where it behaves as a weak base ($K_b = 1.3 \times 10^{-6}$). The compound is a strong and versatile reducing agent. The major use of hydrazine and compounds related to it, such as methylhydrazine (Figure 22.32), is as rocket fuels.

SAMPLE EXERCISE 22.8

Hydroxylamine (NH_2OH) reduces copper(II) to the free metal in acid solutions. Write a balanced equation for the reaction, assuming that N_2 is the oxidation product.

Solution

Analyze and Plan: We are told that the reaction of NH_2OH and Cu^{2+} produces N_2 and Cu. We are asked to write a balanced equation for the reaction. Because the Cu and N change oxidation numbers during the reaction, this is a redox reaction, and it can be balanced by the method of half-reactions discussed in Section 20.2.

Solve: The unbalanced and incomplete half-reactions are

$$Cu^{2+}(aq) \longrightarrow Cu(s)$$
$$NH_2OH(aq) \longrightarrow N_2(g)$$

Balancing these equations as described in Section 20.2 gives

$$Cu^{2+}(aq) + 2e^- \longrightarrow Cu(s)$$
$$2NH_2OH(aq) \longrightarrow N_2(g) + 2H_2O(l) + 2H^+(aq) + 2e^-$$

Adding these half-reactions gives the balanced equation:

$$Cu^{2+}(aq) + 2NH_2OH(aq) \longrightarrow Cu(s) + N_2(g) + 2H_2O(l) + 2H^+(aq)$$

PRACTICE EXERCISE

(a) In power plants, hydrazine is used to prevent corrosion of the metal parts of steam boilers by the O_2 dissolved in the water. The hydrazine reacts with O_2 in water to give N_2 and H_2O. Write a balanced equation for this reaction. **(b)** Methylhydrazine, $N_2H_3CH_3(l)$, is used with the oxidizer dinitrogen tetroxide, $N_2O_4(l)$, to power the steering rockets of the space shuttle orbiter. The reaction of these two substances produces N_2, CO_2, and H_2O. Write a balanced equation for this reaction.

Answers: **(a)** $N_2H_4(aq) + O_2(aq) \longrightarrow N_2(g) + 2H_2O(l)$; **(b)** $5N_2O_4(l) + 4N_2H_3CH_3(l) \longrightarrow 9N_2(g) + 4CO_2(g) + 12H_2O(g)$

(a)

(b)

(c)

◀ **Figure 22.33** (a) Nitric oxide (NO) can be prepared by the reaction of copper with 6 *M* nitric acid. In this photo a jar containing 6 *M* HNO_3 has been inverted over some pieces of copper. Colorless NO, which is only slightly soluble in water, is collected in the jar. The blue color of the solution is due to the presence of Cu^{2+} ions. (b) Colorless NO gas, collected as shown on the left. (c) When the stopper is removed from the jar of NO, the NO reacts with oxygen in the air to form yellow-brown NO_2.

Oxides and Oxyacids of Nitrogen

Nitrogen forms three common oxides: N_2O (nitrous oxide), NO (nitric oxide), and NO_2 (nitrogen dioxide). It also forms two unstable oxides that we will not discuss, N_2O_3 (dinitrogen trioxide) and N_2O_5 (dinitrogen pentoxide).

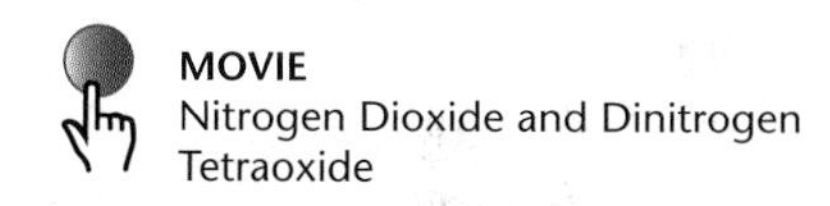

Nitrous oxide (N_2O) is also known as laughing gas because a person becomes somewhat giddy after inhaling only a small amount of it. This colorless gas was the first substance used as a general anesthetic. It is used as the compressed gas propellant in several aerosols and foams, such as in whipped cream. It can be prepared in the laboratory by carefully heating ammonium nitrate to about 200°C.

$$NH_4NO_3(s) \xrightarrow{\Delta} N_2O(g) + 2H_2O(g) \qquad [22.58]$$

Nitric oxide (NO) is also a colorless gas, but, unlike N_2O, it is slightly toxic. It can be prepared in the laboratory by reduction of dilute nitric acid, using copper or iron as a reducing agent, as shown in Figure 22.33 ▲.

$$3Cu(s) + 2NO_3^-(aq) + 8H^+(aq) \longrightarrow 3Cu^{2+}(aq) + 2NO(g) + 4H_2O(l) \qquad [22.59]$$

It is also produced by direct reaction of N_2 and O_2 at high temperatures. This reaction is a significant source of nitrogen oxide air pollutants. (Section 18.4) The direct combination of N_2 and O_2 is not used for commercial production of NO, however, because the yield is low; the equilibrium constant K_{eq} at 2400 K is only 0.05 (see the "Chemistry at Work" box in Section 15.6).

The commercial route to NO (and hence to other oxygen-containing compounds of nitrogen) is via the catalytic oxidation of NH_3.

$$4NH_3(g) + 5O_2(g) \xrightarrow[850°C]{\text{Pt catalyst}} 4NO(g) + 6H_2O(g) \qquad [22.60]$$

The catalytic conversion of NH_3 to NO is the first step in a three-step process known as the **Ostwald process**, by which NH_3 is converted commercially into nitric acid (HNO_3) (Figure 22.34 ▶). Nitric oxide reacts readily with O_2, forming NO_2 when exposed to air (see Figure 22.33).

$$2NO(g) + O_2(g) \longrightarrow 2NO_2(g) \qquad [22.61]$$

When dissolved in water, NO_2 forms nitric acid.

$$3NO_2(g) + H_2O(l) \longrightarrow 2H^+(aq) + 2NO_3^-(aq) + NO(g) \qquad [22.62]$$

Nitrogen is both oxidized and reduced in this reaction, so it disproportionates. The reduction product NO can be converted back into NO_2 by exposure to air and thereafter dissolved in water to prepare more HNO_3.

Recently NO has been found to be an important neurotransmitter in the human body. It causes the muscles that line blood vessels to relax, thus allowing an increased passage of blood.

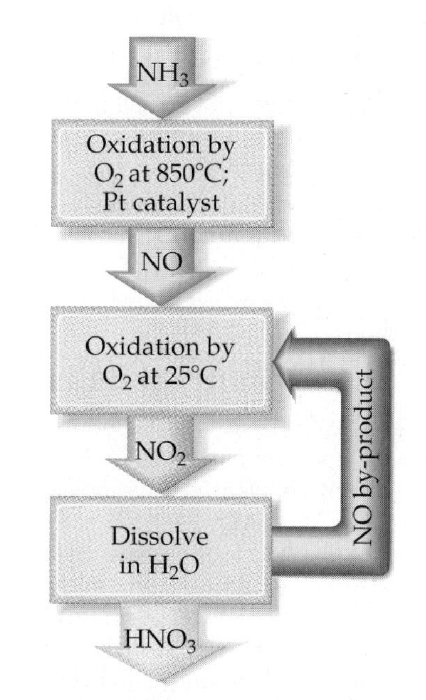

▲ **Figure 22.34** The Ostwald process for converting NH_3 to HNO_3.

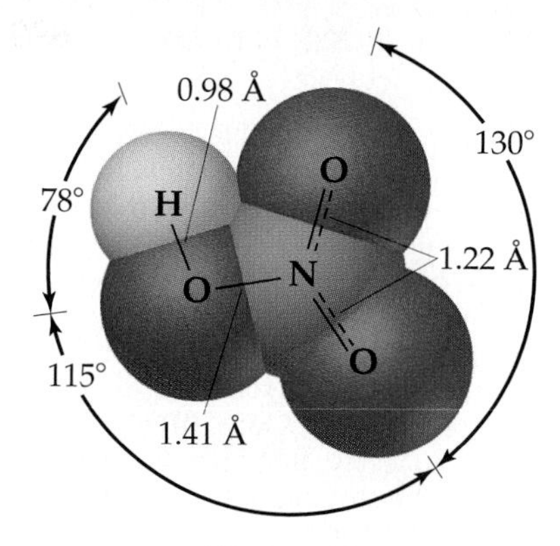

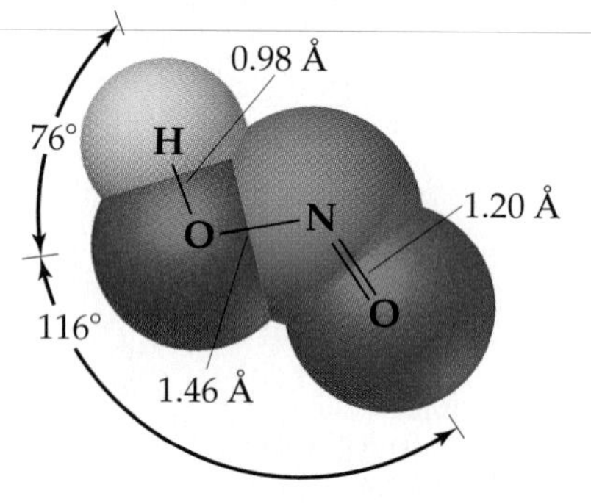

▲ **Figure 22.35** Structures of nitric acid and nitrous acid.

Nitrogen dioxide (NO_2) is a yellow-brown gas (Figure 22.33). Like NO, it is a major constituent of smog. (Section 18.4) It is poisonous and has a choking odor. As discussed in the Introduction of Chapter 15, NO_2 and N_2O_4 exist in equilibrium (Figures 15.1 and 15.2):

$$2NO_2(g) \rightleftharpoons N_2O_4(g) \qquad \Delta H^\circ = -58 \text{ kJ} \qquad [22.63]$$

The two common oxyacids of nitrogen are nitric acid (HNO_3) and nitrous acid (HNO_2) (Figure 22.35 ◀). *Nitric acid* is a colorless, corrosive liquid. Nitric acid solutions often take on a slightly yellow color (Figure 22.36 ◀) as a result of small amounts of NO_2 formed by photochemical decomposition:

$$4HNO_3(aq) \xrightarrow{h\nu} 4NO_2(g) + O_2(g) + 2H_2O(l) \qquad [22.64]$$

Nitric acid is a strong acid. It is also a powerful oxidizing agent, as the following standard reduction potentials indicate:

$$NO_3^-(aq) + 2H^+(aq) + e^- \longrightarrow NO_2(g) + H_2O(l) \qquad E^\circ = +0.79 \text{ V} \qquad [22.65]$$

$$NO_3^-(aq) + 4H^+(aq) + 3e^- \longrightarrow NO(g) + 2H_2O(l) \qquad E^\circ = +0.96 \text{ V} \qquad [22.66]$$

Concentrated nitric acid will attack and oxidize most metals, except Au, Pt, Rh, and Ir.

About 8×10^9 kg (9 million tons) of nitric acid is produced annually in the United States. Its largest use is in the manufacture of NH_4NO_3 for fertilizers, which accounts for about 80% of that produced. HNO_3 is also used in the production of plastics, drugs, and explosives.

Among the explosives made from nitric acid are nitroglycerin, trinitrotoluene (TNT), and nitrocellulose. The reaction of nitric acid with glycerin to form nitroglycerin is shown in Equation 22.67.

$$\begin{array}{c} H \\ | \\ H-C-OH \\ | \\ H-C-OH \\ | \\ H-C-OH \\ | \\ H \end{array} + 3HNO_3 \longrightarrow \begin{array}{c} H \\ | \\ H-CONO_2 \\ | \\ H-CONO_2 \\ | \\ H-CONO_2 \\ | \\ H \end{array} + 3H_2O \qquad [22.67]$$

▲ **Figure 22.36** Colorless nitric acid solution (left) becomes yellow upon standing in sunlight (right).

The following reaction occurs when nitroglycerin explodes:

$$4C_3H_5N_3O_9(l) \longrightarrow 6N_2(g) + 12CO_2(g) + 10H_2O(g) + O_2(g) \qquad [22.68]$$

All the products of this reaction contain very strong bonds. As a result, the reaction is very exothermic. Furthermore, a tremendous amount of gaseous products forms from the liquid. The sudden formation of these gases, together with their expansion resulting from the heat generated by the reaction, produces the explosion. (See the "Chemistry at Work" box in Section 8.8.)

Chemistry and Life Nitrites in Food

Nitrite salts are used as a food additive in cured meats such as bacon, hot dogs, and ham. The nitrite ion serves two functions as an additive. First, it retards spoilage by inhibiting the growth of bacteria, especially *Clostridium botulinum*, which produces the potentially fatal food poisoning known as *botulism*. It also preserves the appetizing flavor and red color of the meat. Debate over the continued use of nitrites in cured meat products arises because HNO_2 (formed when NO_2^- reacts with stomach acid) can react with amino acids to form compounds known as *nitrosamines* (Figure 22.37 ▶). These reactions are thought to occur in the gastrointestinal tract and can also occur at high temperatures, such as those that occur during frying. Nitrosamines have been shown to produce cancer in laboratory animals, causing the U.S. Food and Drug Administration to reduce the limits of allowable concentrations of NO_2^- in foods. Food biochemists have been exploring nitrite-free forms of preserving meats.

◀ **Figure 22.37** General structure of a nitrosamine. The symbol R represents an organic group, such as methyl (CH_3) or ethyl (C_2H_5). Different nitrosamines have different R groups.

Nitrous acid (HNO_2) (Figure 22.35) is considerably less stable than HNO_3 and tends to disproportionate into NO and HNO_3. It is normally made by action of a strong acid, such as H_2SO_4, on a cold solution of a nitrite salt, such as $NaNO_2$. Nitrous acid is a weak acid ($K_a = 4.5 \times 10^{-4}$).

22.8 The Other Group 5A Elements: P, As, Sb, and Bi

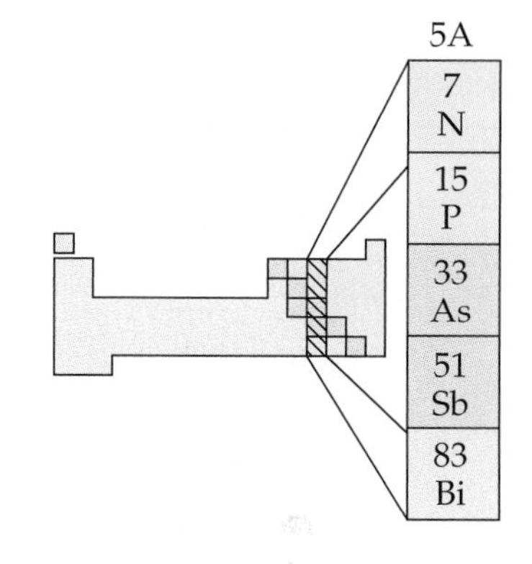

Nitrogen is the most important of the group 5A elements. Of the other elements in this group—phosphorus, arsenic, antimony, and bismuth—phosphorus has a central role in several aspects of biochemistry and environmental chemistry. In this section we will explore the chemistry of these other group 5A elements, with an emphasis on the chemistry of phosphorus.

General Characteristics of the Group 5A Elements

The group 5A elements possess the outer-shell electron configuration ns^2np^3, where n has values ranging from 2 to 6. A noble-gas configuration results from the addition of three electrons to form the -3 oxidation state. Ionic compounds containing X^{3-} ions are not common, however, except for salts of the more active metals, such as in Li_3N. More commonly, the group 5A element acquires an octet of electrons via covalent bonding. The oxidation number may range from -3 to $+5$, depending on the nature and number of the atoms to which the group 5A element is bonded.

Because of its lower electronegativity, phosphorus is found more frequently in positive oxidation states than is nitrogen. Furthermore, compounds in which phosphorus has the $+5$ oxidation state are not as strongly oxidizing as the corresponding compounds of nitrogen. Conversely, compounds in which phosphorus has a -3 oxidation state are much stronger reducing agents than are the corresponding compounds of nitrogen.

Some of the important properties of the group 5A elements are listed in Table 22.6 ▼. The general pattern that emerges from these data is similar to what we have seen before with other groups: Size and metallic character increase as atomic number increases within the group.

The variation in properties among the elements of group 5A is more striking than that seen in groups 6A and 7A. Nitrogen at the one extreme exists as a gaseous diatomic molecule; it is clearly nonmetallic in character. At the other extreme, bismuth is a reddish white, metallic-looking substance that has most of the characteristics of a metal.

The values listed for X—X single-bond enthalpies are not very reliable because it is difficult to obtain such data from thermochemical experiments. However, there is no doubt about the general trend: a low value for the N—N single bond, an increase

TABLE 22.6 Properties of the Group 5A Elements

Property	N	P	As	Sb	Bi
Atomic radius (Å)	0.75	1.10	1.21	1.41	1.55
First ionization energy (kJ/mol)	1402	1012	947	834	703
Electron affinity (kJ/mol)	>0	−72	−78	−103	−91
Electronegativity	3.0	2.1	2.0	1.9	1.9
X—X single-bond enthalpy (kJ/mol)*	163	200	150	120	—
X≡X triple-bond enthalpy (kJ/mol)	941	490	380	295	192

*Approximate values only.

at phosphorus, and then a gradual decline to arsenic and antimony. From observations of the elements in the gas phase, it is possible to estimate the X≡X triple-bond enthalpies, as listed in Table 22.6. Here we see a trend that is different from that for the X—X single bond. Nitrogen forms a much stronger triple bond than the other elements, and there is a steady decline in the triple-bond enthalpy down through the group. These data help us to appreciate why nitrogen alone of the group 5A elements exists as a diatomic molecule in its stable state at 25°C. All the other elements exist in structural forms with single bonds between the atoms.

Occurrence, Isolation, and Properties of Phosphorus

Phosphorus occurs mainly in the form of phosphate minerals. The principal source of phosphorus is phosphate rock, which contains phosphate mainly as $Ca_3(PO_4)_2$. The element is produced commercially by reduction of calcium phosphate with carbon in the presence of SiO_2:

$$2Ca_3(PO_4)_2(s) + 6SiO_2(s) + 10C(s) \xrightarrow{1500°C} P_4(g) + 6CaSiO_3(l) + 10CO(g) \quad [22.69]$$

The phosphorus produced in this fashion is the allotrope known as white phosphorus. This form distills from the reaction mixture as the reaction proceeds.

White phosphorus consists of P_4 tetrahedra (Figure 22.38 ◀). The 60° bond angles in P_4 are unusually small for molecules, so there is much strain in the bonding, which is consistent with the high reactivity of white phosphorus. This allotrope bursts spontaneously into flames if exposed to air. It is a white waxlike solid that melts at 44.2°C and boils at 280°C. When heated in the absence of air to about 400°C, it is converted to a more stable allotrope known as red phosphorus. This form does not ignite on contact with air. It is also considerably less poisonous than the white form. Both allotropes are shown in Figure 22.39 ◀. We will denote elemental phosphorus as simply P(*s*).

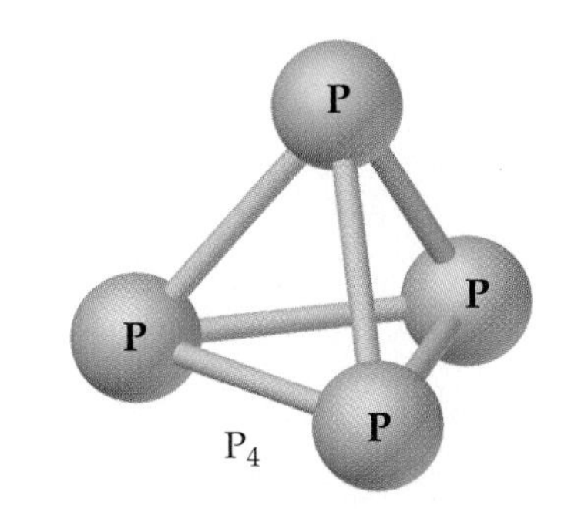

▲ **Figure 22.38** Tetrahedral structure of the P_4 molecule of white phosphorus.

Phosphorus Halides

Phosphorus forms a wide range of compounds with the halogens, the most important of which are the trihalides and pentahalides. Phosphorus trichloride (PCl_3) is commercially the most significant of these compounds and is used to prepare a wide variety of products, including soaps, detergents, plastics, and insecticides.

Phosphorus chlorides, bromides, and iodides can be made by the direct oxidation of elemental phosphorus with the elemental halogen. PCl_3, for example, which is a liquid at room temperature, is made by passing a stream of dry chlorine gas over white or red phosphorus.

$$2P(s) + 3Cl_2(g) \longrightarrow 2PCl_3(l) \quad [22.70]$$

If excess chlorine gas is present, an equilibrium is established between PCl_3 and PCl_5.

$$PCl_3(l) + Cl_2(g) \rightleftharpoons PCl_5(s) \quad [22.71]$$

Because F_2 is such a strong oxidant, the direct reaction of phosphorus with F_2 usually produces PF_5, in which phosphorus is in its most positive oxidation state.

$$2P(s) + 5F_2(g) \longrightarrow 2PF_5(g) \quad [22.72]$$

The phosphorus halides hydrolyze on contact with water. The reactions occur readily, and most of the phosphorus halides fume in air as a result of reaction with water vapor. In the presence of excess water the products are the corresponding phosphorus oxyacid and hydrogen halide.

$$PBr_3(l) + 3H_2O(l) \longrightarrow H_3PO_3(aq) + 3HBr(aq) \quad [22.73]$$

$$PCl_5(l) + 4H_2O(l) \longrightarrow H_3PO_4(aq) + 5HCl(aq) \quad [22.74]$$

▲ **Figure 22.39** White and red allotropes of phosphorus. White phosphorus is very reactive and is normally stored under water to protect it from oxygen. Red phosphorus is much less reactive than white phosphorus, and it is not necessary to store it under water.

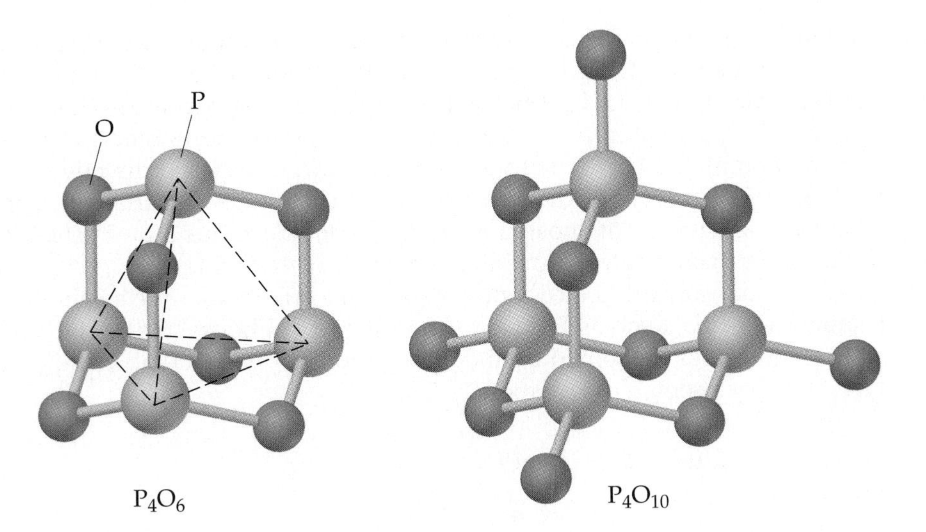

Figure 22.40 Structures of P_4O_6 and P_4O_{10}.

Oxy Compounds of Phosphorus

Probably the most significant compounds of phosphorus are those in which the element is combined in some way with oxygen. Phosphorus(III) oxide (P_4O_6) is obtained by allowing white phosphorus to oxidize in a limited supply of oxygen. When oxidation takes place in the presence of excess oxygen, phosphorus(V) oxide (P_4O_{10}) forms. This compound is also readily formed by oxidation of P_4O_6. These two oxides represent the two most common oxidation states for phosphorus, +3 and +5. The structural relationship between P_4O_6 and P_4O_{10} is shown in Figure 22.40 ▲. Notice the resemblance these molecules have to the P_4 molecule, shown in Figure 22.38; all three substances have a P_4 core.

SAMPLE EXERCISE 22.9

The reactive chemicals on the tip of a "strike anywhere" match are usually P_4S_3 and an oxidizing agent such as $KClO_3$. When the match is struck on a rough surface, the heat generated by the friction ignites the P_4S_3, and the oxidizing agent brings about rapid combustion. The products of the combustion of P_4S_3 are P_4O_{10} and SO_2. Calculate the standard enthalpy change for the combustion of P_4S_3 in air, given the following standard enthalpies of formation: P_4S_3 (−154.4 kJ/mol); P_4O_{10} (−2940 kJ/mol); SO_2 (−296.9 kJ/mol).

Solution

Analyze: We are given the reactants (P_4S_3 and O_2 from air) and the products (P_4O_{10} and SO_2) for a reaction together with their standard enthalpies of formation and asked to calculate the standard enthalpy change for the reaction.

Plan: We first need a balanced chemical equation for the reaction. The enthalpy change for the reaction is then equal to the enthalpies of formation of products minus those of reactants (Equation 5.28, Section 5.7). We also need to recall that the standard enthalpy of formation of any element in its standard state is zero. Thus, $\Delta H_f^\circ(O_2) = 0$.

Solve: The chemical equation for the combustion is

$$P_4S_3(s) + 8O_2(g) \longrightarrow P_4O_{10}(s) + 3SO_2(g)$$

Thus, we can write

$$\begin{aligned}\Delta H^\circ &= \Delta H_f^\circ(P_4O_{10}) + 3\Delta H_f^\circ(SO_2) - \Delta H_f^\circ(P_4S_3) - 8\Delta H_f^\circ(O_2)\\ &= -2940\text{ kJ} + 3(-296.9)\text{ kJ} - (-154.4\text{ kJ}) - 8(0)\\ &= -3676\text{ kJ}\end{aligned}$$

Comment: The reaction is strongly exothermic, making it evident why P_4S_3 is used on match tips.

PRACTICE EXERCISE

Write the balanced equation for the reaction of P_4O_{10} with water, and calculate ΔH° for this reaction using data from Appendix C.
Answer: $P_4O_{10}(s) + 6H_2O(l) \longrightarrow 4H_3PO_4(aq)$; −498.0 kJ

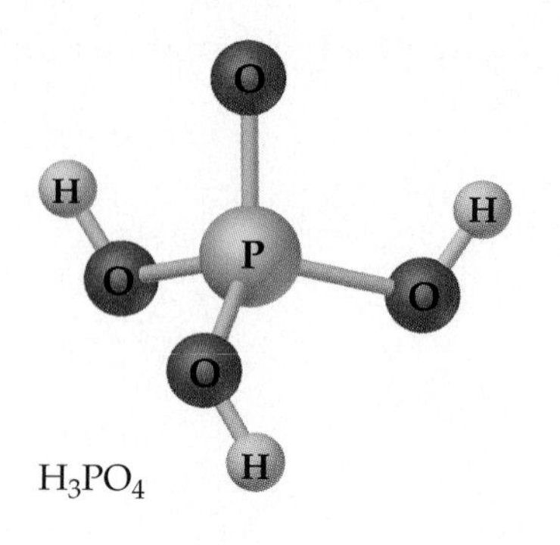

H_3PO_4

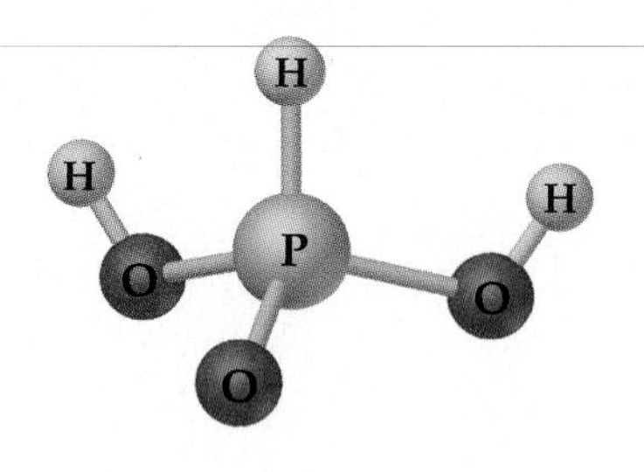

H_3PO_3

▲ **Figure 22.41** Structures of H_3PO_4 and H_3PO_3.

Phosphorus(V) oxide is the anhydride of phosphoric acid (H_3PO_4), a weak triprotic acid. In fact, P_4O_{10} has a very high affinity for water and is consequently used as a drying agent. Phosphorus(III) oxide is the anhydride of phosphorous acid (H_3PO_3), a weak diprotic acid.* The structures of H_3PO_4 and H_3PO_3 are shown in Figure 22.41 ◀. The hydrogen atom that is attached directly to phosphorus in H_3PO_3 is not acidic, because the P—H bond is essentially nonpolar.

One characteristic of phosphoric and phosphorous acids is their tendency to undergo condensation reactions when heated. A *condensation reaction* is one in which two or more molecules combine to form a larger molecule by eliminating a small molecule, such as H_2O. (Section 12.2) The reaction in which two H_3PO_4 molecules are joined by the elimination of one H_2O molecule to form $H_4P_2O_7$ is represented in Equation 22.75.

$$(HO)_2P(=O)\text{—O—H} + \text{H—O—}P(=O)(OH)_2 \longrightarrow (HO)_2P(=O)\text{—O—}P(=O)(OH)_2 + H_2O \quad [22.75]$$

These atoms are eliminated as H_2O

Further condensation produces phosphates with an empirical formula of HPO_3.

$$nH_3PO_4 \longrightarrow (HPO_3)_n + nH_2O \quad [22.76]$$

Two phosphates with this empirical formula, one cyclic and the other polymeric, are shown in Figure 22.42 ◀. The three acids H_3PO_4, $H_4P_2O_7$, and $(HPO_3)_n$ all contain phosphorus in the +5 oxidation state, and all are therefore called phosphoric acids. To differentiate them, the prefixes *ortho-*, *pyro-*, and *meta-* are used: H_3PO_4 is orthophosphoric acid, $H_4P_2O_7$ is pyrophosphoric acid, and $(HPO_3)_n$ is metaphosphoric acid.

$(HPO_3)_3$
Trimetaphosphoric acid

Repeating unit from which empirical formula is obtained

$$\text{—O—}P(=O)(OH)\text{—O—}P(=O)(OH)\text{—O—}P(=O)(OH)\text{—O—}$$

$(HPO_3)_n$
Polymetaphosphoric acid

▲ **Figure 22.42** Structures of trimetaphosphoric acid and polymetaphosphoric acid.

Phosphoric acid and its salts find their most important uses in detergents and fertilizers. The phosphates in detergents are often in the form of sodium tripolyphosphate ($Na_5P_3O_{10}$). A typical detergent formulation contains 47% phosphate, 16% bleaches, perfumes, and abrasives, and 37% linear alkylsulfonate (LAS) surfactant (shown as follows):

$$CH_3\text{—}(CH_2)_9\text{—}CH(CH_3)\text{—}C_6H_4\text{—}SO_2\text{—}O^-Na^+$$

(We have used the notation for the benzene ring, as described in Section 8.6.) The phosphate ions form bonds with metal ions that contribute to the hardness of water. This keeps the metal ions from interfering with the action of the

* Note that the element phosphor*us* (FOS · for · us) has a *-us* suffix, whereas phosphor*ous* (fos · FOR · us) acid has an *-ous* suffix.

surfactant. The phosphates also keep the pH above 7 and thus prevent the surfactant molecules from becoming protonated (gaining an H^+ ion).

Most mined phosphate rock is converted to fertilizers. The $Ca_3(PO_4)_2$ in phosphate rock is insoluble ($K_{sp} = 2.0 \times 10^{-29}$). It is converted to a soluble form for use in fertilizers by treating the phosphate rock with sulfuric or phosphoric acid.

$$Ca_3(PO_4)_2(s) + 3H_2SO_4(aq) \longrightarrow 3CaSO_4(s) + 2H_3PO_4(aq) \quad [22.77]$$

$$Ca_3(PO_4)_2(s) + 4H_3PO_4(aq) \longrightarrow 3Ca^{2+}(aq) + 6H_2PO_4^-(aq) \quad [22.78]$$

The mixture formed when ground phosphate rock is treated with sulfuric acid and then dried and pulverized is known as superphosphate. The $CaSO_4$ formed in this process is of little use in soil except when deficiencies in calcium or sulfur exist. It also dilutes the phosphorus, which is the nutrient of interest. If the phosphate rock is treated with phosphoric acid, the product contains no $CaSO_4$ and has a higher percentage of phosphorus. This product is known as triple superphosphate. Although the solubility of $Ca(H_2PO_4)_2$ allows it to be assimilated by plants, it also allows it to be washed from the soil and into bodies of water, thereby contributing to water pollution. (Section 18.6)

Phosphorus compounds are important in biological systems. The element occurs in phosphate groups in RNA and DNA, the molecules responsible for the control of protein biosynthesis and transmission of genetic information. (Section 25.11) It also occurs in adenosine triphosphate (ATP), which stores energy within biological cells.

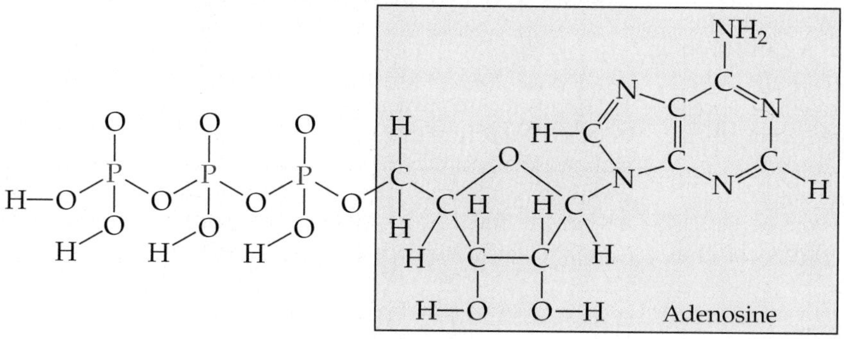

The P—O—P bond of the end phosphate group is broken by hydrolysis with water, forming adenosine diphosphate (ADP). This reaction releases 33 kJ of energy.

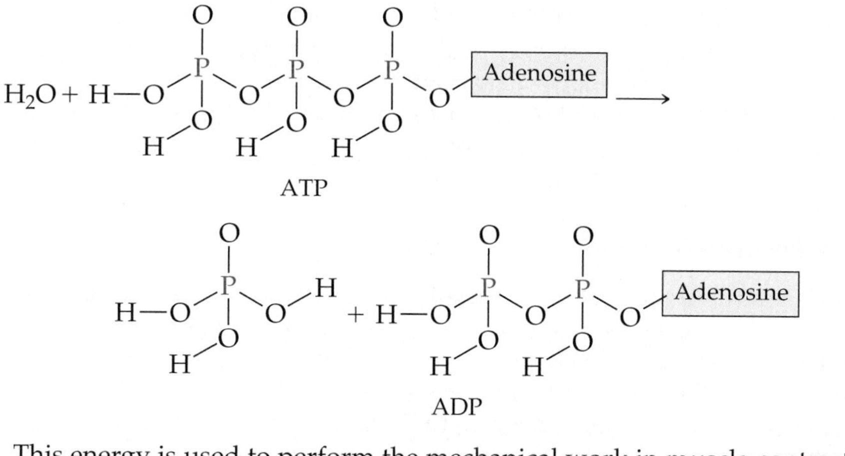

[22.79]

This energy is used to perform the mechanical work in muscle contraction and in many other biochemical reactions (Figure 19.19).

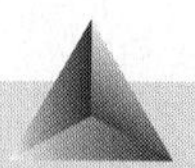

Chemistry and Life Arsenic in Drinking Water

In 2001 the Environmental Protection Agency (EPA) issued a rule reducing the standard for arsenic in public water supplies from 50 ppb (equivalent to 50 $\mu g/L$) to 10 ppb, effective by 2006. Most regions of the United States tend to have low to moderate (2–10 ppb) groundwater arsenic levels (Figure 22.43 ▶). The Western region tends to have higher levels, coming mainly from natural geological sources in the area.

In water the most common forms of arsenic are the arsenate ion and its protonated hydrogen anions (AsO_4^{3-}, $HAsO_4^{2-}$, and $H_2AsO_4^-$), and the arsenite ion and its protonated forms (AsO_3^{3-}, $HAsO_3^{2-}$, $H_2AsO_3^-$, and H_3AsO_3). These species are collectively referred to by the oxidation number of the arsenic as arsenic(V) and arsenic(III), respectively. Arsenic(V) is more prevalent in oxygen-rich (aerobic) surface waters, whereas arsenic(III) is more likely to occur in oxygen-poor (anaerobic) groundwaters. In the pH range from 4 to 10, the arsenic(V) is present primarily as $HAsO_4^{2-}$ and $H_2AsO_4^-$, and the arsenic(III) is present primarily as the neutral acid, H_3AsO_3.

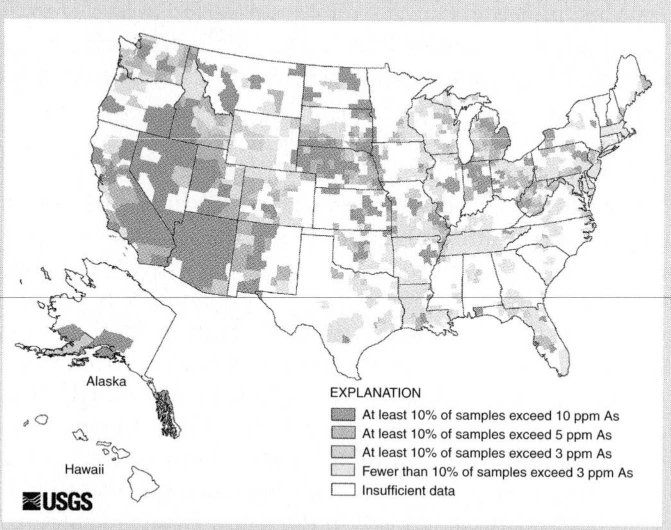

▲ **Figure 22.43** Counties in which at least 10% of the groundwater samples exceeded 10 ppm As are indicated by the darkest color on the scale. As the color of the scale becomes lighter, the scale moves from 10 ppm to 5 ppm to 3 ppm and then cases where fewer than 10% of samples exceed 3 ppm. The white areas are those for which there were insufficient data.

One of the challenges in determining the health effects of arsenic in drinking waters is the different chemistry of the arsenic(V) and arsenic(III), as well as the different concentrations required for physiological responses in different individuals. Statistical studies correlating arsenic levels with the occurrence of disease, however, indicate a lung and bladder cancer risk arising from even low levels of arsenic. A 2001 report from the National Research Council suggests, for example, that people who consume water with 3 ppb arsenic daily have about a 1 in 1000 risk of developing these forms of cancer during their lifetime. At 10 ppb, the risk is approximately 3 in 1000.

The current technologies for removing arsenic perform most effectively when treating arsenic in the form of arsenic(V), so water treatment strategies require preoxidation of the drinking water. Once in the form of arsenic(V), there are a number of possible removal strategies. For example, $Fe_2(SO_4)_3$ could be added to precipitate $FeAsO_4$, which is then removed by filtration. Small utilities in areas where arsenic occurs naturally in groundwater fear that the costs of reducing arsenic even to the 10-ppb level will force them out of business, leaving households dependent on untreated well water.

22.9 Carbon

Carbon constitutes only 0.027% of Earth's crust, so it is not an abundant element. Although some carbon occurs in elemental form as graphite and diamond, most is found in combined form. Over half occurs in carbonate compounds, such as $CaCO_3$. Carbon is also found in coal, petroleum, and natural gas. The importance of the element stems in large part from its occurrence in all living organisms: Life as we know it is based on carbon compounds. In this section we will take a brief look at carbon and its most common inorganic compounds. We will discuss organic chemistry in Chapter 25.

Elemental Forms of Carbon

Carbon exists in three crystalline forms: graphite, diamond, and fullerenes. ∞ (Section 11.8) *Graphite* is a soft, black, slippery solid that has a metallic luster and conducts electricity. It consists of parallel sheets of carbon atoms; the sheets are held together by London forces [Figure 11.41(b)].

Diamond is a clear, hard solid in which the carbon atoms form a covalent network [Figure 11.41(a)]. Diamond is denser than graphite ($d = 2.25\ g/cm^3$ for graphite; $d = 3.51\ g/cm^3$ for diamond). At very high pressures and temperatures (on the order of 100,000 atm at 3000°C) graphite converts to diamond (Figure 22.44 ▶). About 3×10^4 kg of industrial-grade diamonds are synthesized each year, mainly for use in cutting, grinding, and polishing tools.

Fullerenes are molecular forms of carbon that were discovered in the mid-1980s (see "A Closer Look" in Section 11.8). Fullerenes consist of individual molecules like C_{60} and C_{70}. The C_{60} molecules resemble soccer balls (Figure 11.43). The chemical properties of these substances are currently being explored by numerous research groups.

Carbon also exists in three common microcrystalline or amorphous forms of graphite. **Carbon black** is formed when hydrocarbons such as methane are heated in a very limited supply of oxygen.

$$CH_4(g) + O_2(g) \longrightarrow C(s) + 2H_2O(g) \quad [22.80]$$

It is used as a pigment in black inks; large amounts are also used in making automobile tires. **Charcoal** is formed when wood is heated strongly in the absence of air. Charcoal has a very open structure, giving it an enormous surface area per unit mass. Activated charcoal, a pulverized form whose surface is cleaned by heating with steam, is widely used to adsorb molecules. It is used in filters to remove offensive odors from air and colored or bad-tasting impurities from water. **Coke** is an impure form of carbon formed when coal is heated strongly in the absence of air. It is widely used as a reducing agent in metallurgical operations. ∞ (Section 23.2)

▲ **Figure 22.44** Graphite and a synthetic diamond prepared from graphite. Most synthetic diamonds lack the size, color, and clarity of natural diamonds and are therefore not used in jewelry.

Oxides of Carbon

Carbon forms two principal oxides: carbon monoxide (CO) and carbon dioxide (CO_2). *Carbon monoxide* is formed when carbon or hydrocarbons are burned in a limited supply of oxygen.

$$2C(s) + O_2(g) \longrightarrow 2CO(g) \quad [22.81]$$

It is a colorless, odorless, and tasteless gas (mp = −199°C; bp = −192°C). It is toxic because it can bind to hemoglobin and thus interfere with oxygen transport. ∞ (Section 18.4) Low-level poisoning results in headaches and drowsiness; high-level poisoning can cause death. Carbon monoxide is produced by automobile engines, and it is a major air pollutant.

Carbon monoxide is unusual in that it has a lone pair of electrons on carbon: :C≡O:. It is also isoelectronic with N_2, so you might imagine that CO would be equally unreactive. Moreover, both substances have high bond energies (1072 kJ/mol for C≡O and 941 kJ/mol for N≡N). Because of the lower nuclear charge on carbon (compared with either N or O), however, the lone pair on carbon is not held as strongly as that on N or O. Consequently, CO is better able to function as an electron-pair donor (Lewis base) than N_2. It forms a wide variety of covalent compounds, known as metal carbonyls, with transition metals. $Ni(CO)_4$, for example, is a volatile, toxic compound that is formed by simply warming metallic nickel in the presence of CO. The formation of metal carbonyls is the first step in the transition-metal catalysis of a variety of reactions of CO.

Carbon monoxide has several commercial uses. Because it burns readily, forming CO_2, it is employed as a fuel.

$$2CO(g) + O_2(g) \longrightarrow 2CO_2(g) \qquad \Delta H^\circ = -566 \text{ kJ} \quad [22.82]$$

It is also an important reducing agent, widely used in metallurgical operations to reduce metal oxides, such as the iron oxides in blast furnaces.

$$Fe_3O_4(s) + 4CO(g) \longrightarrow 3Fe(s) + 4CO_2(g) \quad [22.83]$$

This reaction is discussed in greater detail in Section 23.2. Carbon monoxide is also used in the preparation of several organic compounds. In Section 22.2 we saw that it can be combined catalytically with H_2 to manufacture methanol (CH_3OH) (Equation 22.14).

Chemistry at Work Carbon Fibers and Composites

The properties of graphite are anisotropic; that is, they differ in different directions through the solid. Along the carbon planes, graphite possesses great strength because of the number and strength of the carbon–carbon bonds in this direction. The bonds between planes are relatively weak, however, making graphite weak in that direction.

Fibers of graphite can be prepared in which the carbon planes are aligned to varying extents parallel to the fiber axis. These fibers are also lightweight (density of about 2 g/cm^3) and chemically quite unreactive. The oriented fibers are made by first slowly pyrolyzing (decomposing by action of heat) organic fibers at about 150°C to 300°C. These fibers are then heated to about 2500°C to graphitize them (convert amorphous carbon to graphite). Stretching the fiber during pyrolysis helps orient the graphite planes parallel to the fiber axis. More amorphous carbon fibers are formed by pyrolysis of organic fibers at lower temperatures (1200°C to 1400°C). These amorphous materials, commonly called *carbon fibers*, are the type most commonly used in commercial materials.

Composite materials that take advantage of the strength, stability, and low density of carbon fibers are widely used. Composites are combinations of two or more materials. These materials are present as separate phases and are combined to form structures that take advantage of certain desirable properties of each component. In carbon composites the graphite fibers are often woven into a fabric that is embedded in a matrix that binds them into a solid structure. The fibers transmit loads evenly throughout the matrix. The finished composite is thus stronger than any one of its components.

Epoxy systems are useful matrices because of their excellent adherence. They are used widely in a number of applications, including high-performance graphite sports equipment such as tennis racquets, golf clubs, and, most recently, bicycle frames (Figure 22.45 ▼). Epoxy systems can be used only when the temperature remains below 150°C. More heat-resistant resins are required for many aerospace applications, where carbon composites now find wide use.

▲ **Figure 22.45** Carbon composites are used extensively in aerospace and automotive applications, and in sporting goods. This high-performance (and high-priced!) off-road bike has a carbon-fiber-composite frame, which makes it very lightweight and helps it absorb road shocks.

Carbon dioxide is produced when carbon-containing substances are burned in excess oxygen.

$$C(s) + O_2(g) \longrightarrow CO_2(g) \qquad [22.84]$$

$$C_2H_5OH(l) + 3O_2(g) \longrightarrow 2CO_2(g) + 3H_2O(g) \qquad [22.85]$$

It is also produced when many carbonates are heated.

$$CaCO_3(s) \longrightarrow CaO(s) + CO_2(g) \qquad [22.86]$$

Large quantities are also obtained as a by-product of the fermentation of sugar during the production of ethanol.

$$\underset{\text{Glucose}}{C_6H_{12}O_6(aq)} \xrightarrow{\text{yeast}} \underset{\text{Ethanol}}{2C_2H_5OH(aq)} + 2CO_2(g) \qquad [22.87]$$

In the laboratory, CO_2 is normally produced by the action of acids on carbonates, as shown in Figure 22.46 ▶:

$$CO_3^{2-}(aq) + 2H^+(aq) \longrightarrow CO_2(g) + H_2O(l) \quad [22.88]$$

Carbon dioxide is a colorless and odorless gas. It is a minor component of Earth's atmosphere but a major contributor to the so-called greenhouse effect. (Section 18.4) Although it is not toxic, high concentrations increase respiration rate and can cause suffocation. It is readily liquefied by compression. When cooled at atmospheric pressure, however, it condenses as a solid rather than as a liquid. The solid sublimes at atmospheric pressure at −78°C. This property makes solid CO_2 valuable as a refrigerant that is always free of the liquid form. Solid CO_2 is known as *Dry Ice*™. About half of the CO_2 consumed annually is used for refrigeration. The other major use is in the production of carbonated beverages. Large quantities are also used to manufacture *washing soda* ($Na_2CO_3 \cdot 10H_2O$) and *baking soda* ($NaHCO_3$). Baking soda is so named because the following reaction occurs in baking:

▲ **Figure 22.46** Solid $CaCO_3$ reacts with a solution of hydrochloric acid to produce CO_2 gas, seen here as the bubbles in the beaker.

$$NaHCO_3(s) + H^+(aq) \longrightarrow Na^+(aq) + CO_2(g) + H_2O(l) \quad [22.89]$$

The $H^+(aq)$ is provided by vinegar, sour milk, or the hydrolysis of certain salts. The bubbles of CO_2 that form are trapped in the dough, causing it to rise. Washing soda is used to precipitate metal ions that interfere with the cleansing action of soap.

Carbonic Acid and Carbonates

Carbon dioxide is moderately soluble in H_2O at atmospheric pressure. The resultant solutions are moderately acidic due to the formation of carbonic acid (H_2CO_3).

$$CO_2(aq) + H_2O(l) \rightleftharpoons H_2CO_3(aq) \quad [22.90]$$

Carbonic acid is a weak diprotic acid. Its acidic character causes carbonated beverages to have a sharp, slightly acidic taste.

Although carbonic acid cannot be isolated as a pure compound, hydrogen carbonates (bicarbonates) and carbonates can be obtained by neutralizing carbonic acid solutions. Partial neutralization produces HCO_3^-, and complete neutralization gives CO_3^{2-}.

The HCO_3^- ion is a stronger base than acid ($K_b = 2.3 \times 10^{-8}$; $K_a = 5.6 \times 10^{-11}$). Consequently, aqueous solutions of HCO_3^- are weakly alkaline.

$$HCO_3^-(aq) + H_2O(l) \rightleftharpoons H_2CO_3(aq) + OH^-(aq) \quad [22.91]$$

The carbonate ion is much more strongly basic ($K_b = 1.8 \times 10^{-4}$).

$$CO_3^{2-}(aq) + H_2O(l) \rightleftharpoons HCO_3^-(aq) + OH^-(aq) \quad [22.92]$$

Minerals containing the carbonate ion are plentiful. The principal carbonate minerals are calcite ($CaCO_3$), magnesite ($MgCO_3$), dolomite [$MgCa(CO_3)_2$], and siderite ($FeCO_3$). Calcite is the principal mineral in limestone rock, large deposits of which occur in many parts of the world. It is also the main constituent of marble, chalk, pearls, coral reefs, and the shells of marine animals such as clams and oysters. Although $CaCO_3$ has low solubility in pure water, it dissolves readily in acidic solutions with evolution of CO_2.

$$CaCO_3(s) + 2H^+(aq) \rightleftharpoons Ca^{2+}(aq) + H_2O(l) + CO_2(g) \quad [22.93]$$

Because water containing CO_2 is slightly acidic (Equation 22.90), $CaCO_3$ dissolves slowly in this medium:

$$CaCO_3(s) + H_2O(l) + CO_2(g) \longrightarrow Ca^{2+}(aq) + 2HCO_3^-(aq) \quad [22.94]$$

▶ **Figure 22.47** Carlsbad Caverns, New Mexico.

This reaction occurs when surface waters move underground through limestone deposits. It is the principal way that Ca^{2+} enters groundwater, producing "hard water." (Section 18.6) If the limestone deposit is deep enough underground, the dissolution of the limestone produces a cave. Two well-known limestone caves are Mammoth Cave in Kentucky and Carlsbad Caverns in New Mexico (Figure 22.47 ▲).

One of the most important reactions of $CaCO_3$ is its decomposition into CaO and CO_2 at elevated temperatures, given earlier in Equation 22.86. About 2.0×10^{10} kg (22 million tons) of calcium oxide, known as lime or quicklime, is produced in the United States annually. Because calcium oxide reacts with water to form $Ca(OH)_2$, it is an important commercial base. It is also important in making mortar, which is a mixture of sand, water, and CaO used in construction to bind bricks, blocks, and rocks together. Calcium oxide reacts with water and CO_2 to form $CaCO_3$, which binds the sand in the mortar.

$$CaO(s) + H_2O(l) \rightleftharpoons Ca^{2+}(aq) + 2OH^-(aq) \qquad [22.95]$$

$$Ca^{2+}(aq) + 2OH^-(aq) + CO_2(aq) \longrightarrow CaCO_3(s) + H_2O(l) \qquad [22.96]$$

Carbides

The binary compounds of carbon with metals, metalloids, and certain nonmetals are called **carbides**. There are three types: ionic, interstitial, and covalent. The ionic carbides are formed by the more active metals. The most common ionic carbides contain the *acetylide* ion (C_2^{2-}). This ion is isoelectronic with N_2, and its Lewis structure, $[:C{\equiv}C:]^{2-}$, has a carbon–carbon triple bond. The most important ionic carbide is calcium carbide (CaC_2), which is produced by the reduction of CaO with carbon at high temperature:

$$2CaO(s) + 5C(s) \longrightarrow 2CaC_2(s) + CO_2(g) \qquad [22.97]$$

The carbide ion is a very strong base that reacts with water to form acetylene ($H{-}C{\equiv}C{-}H$), as in the following reaction:

$$CaC_2(s) + 2H_2O(l) \longrightarrow Ca(OH)_2(aq) + C_2H_2(g) \qquad [22.98]$$

Calcium carbide is therefore a convenient solid source of acetylene, which is used in welding (Figure 22.17).

Interstitial carbides are formed by many transition metals. The carbon atoms occupy open spaces (interstices) between metal atoms in a manner analogous to the interstitial hydrides. (Section 22.2) Tungsten carbide, for example, is very hard and heat-resistant, and is thus used to make cutting tools.

Covalent carbides are formed by boron and silicon. Silicon carbide (SiC), known as *Carborundum*™, is used as an abrasive and in cutting tools. Almost as hard as diamond, SiC has a diamond-like structure with alternating Si and C atoms.

Other Inorganic Compounds of Carbon

Hydrogen cyanide, HCN (Figure 22.48 ▶), is an extremely toxic gas that has the odor of bitter almonds. It is produced by the reaction of a cyanide salt, such as NaCN, with an acid [see Sample Exercise 22.2(c)].

Aqueous solutions of HCN are known as hydrocyanic acid. Neutralization with a base, such as NaOH, produces cyanide salts, such as NaCN. Cyanides find use in the manufacture of several well-known plastics, including nylon and Orlon™. The CN^- ion forms very stable complexes with most transition metals. (Section 17.5) The toxic action of CN^- is caused by its combination with iron(III) in cytochrome oxidase, a key enzyme that is involved in respiration.

Carbon disulfide, CS_2 (Figure 22.48), is an important industrial solvent for waxes, greases, celluloses, and other nonpolar substances. It is a colorless volatile liquid (bp 46.3°C). The vapor is very poisonous and highly flammable. The compound is formed by direct reaction of carbon and sulfur at high temperature.

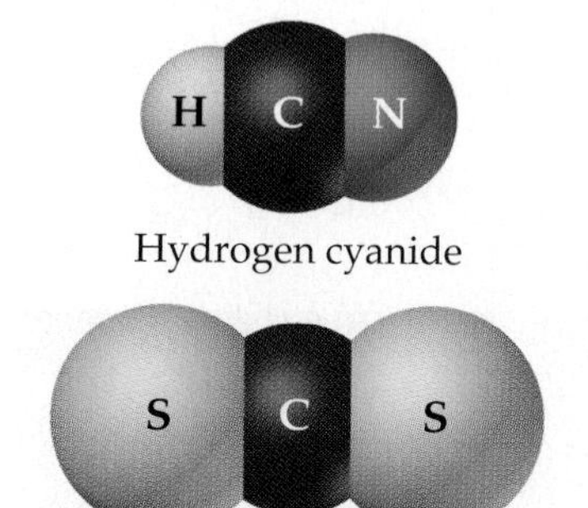

▲ **Figure 22.48** Structures of hydrogen cyanide and carbon disulfide.

22.10 The Other Group 4A Elements: Si, Ge, Sn, and Pb

The other elements of group 4A, in addition to carbon, are silicon, germanium, tin, and lead. The general trend from nonmetallic to metallic character as we go down a family is strikingly evident in group 4A. Carbon is a nonmetal; silicon and germanium are metalloids; tin and lead are metals. In this section we will consider a few general characteristics of group 4A and then look more thoroughly at silicon.

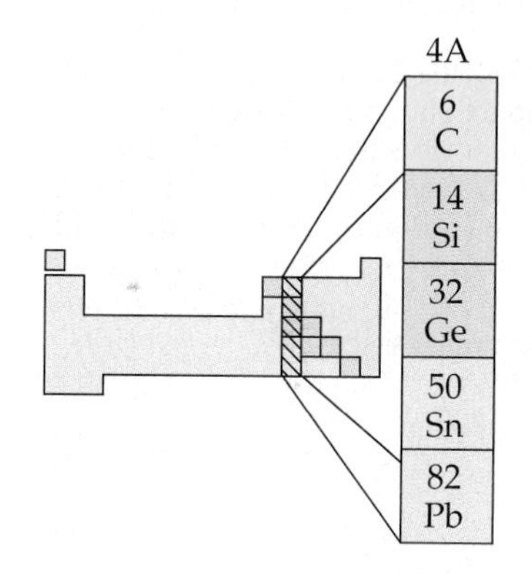

General Characteristics of the Group 4A Elements

Some properties of the group 4A elements are given in Table 22.7 ▼. The elements possess the outer-shell electron configuration ns^2np^2. The electronegativities of the elements are generally low; carbides that formally contain C^{4-} ions are observed only in the case of a few compounds of carbon with very active metals. Formation of 4+ ions by electron loss is not observed for any of these elements; the ionization energies are too high. The +2 oxidation state is found in the chemistry of germanium, tin, and lead, however, and it is the principal oxidation state for lead. The vast majority of the compounds of the group 4A elements are covalently bonded. Carbon forms a maximum of four bonds. The other members of the family are able to form higher coordination numbers through valence-shell expansion.

TABLE 22.7 Some Properties of the Group 4A Elements

Property	C	Si	Ge	Sn	Pb
Atomic radius (Å)	0.77	1.17	1.22	1.40	1.46
First ionization energy (kJ/mol)	1086	786	762	709	716
Electronegativity	2.5	1.8	1.8	1.8	1.9
X—X single-bond enthalpy (kJ/mol)	348	226	188	151	—

Carbon differs from the other group 4A elements in its pronounced ability to form multiple bonds both with itself and with other nonmetals, especially N, O, and S. The origin of this behavior was considered earlier. (Section 22.1)

Table 22.7 shows that the strength of a bond between two atoms of a given element decreases as we go down group 4A. Carbon–carbon bonds are quite strong. Carbon, therefore, has a striking ability to form compounds in which carbon atoms are bonded to one another in extended chains and rings, which accounts for the large number of organic compounds that exist. Other elements, especially those in the vicinity of carbon in the periodic table, can also form chains and rings, but these bonds are far less important in the chemistries of these other elements. The Si—Si bond strength (226 kJ/mol), for example, is much smaller than the Si—O bond strength (386 kJ/mol). As a result, the chemistry of silicon is dominated by the formation of Si—O bonds, and Si—Si bonds play a rather minor role.

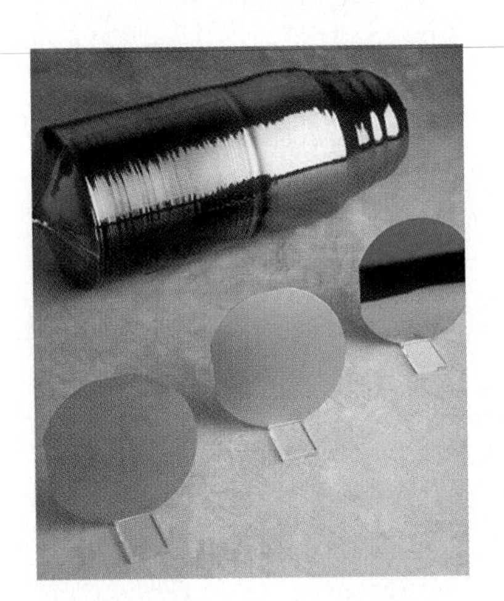

▲ **Figure 22.49** Elemental silicon. To prepare electronic devices, silicon powder is melted, drawn into a single crystal, (top) by zone refining. Wafers of silicon (bottom), cut from the crystal, are subsequently treated by a series of elegant techniques to produce various electronic devices.

Occurrence and Preparation of Silicon

Silicon is the second most abundant element, after oxygen, in Earth's crust. It occurs in SiO_2 and in an enormous variety of silicate minerals. The element is obtained by the reduction of molten silicon dioxide with carbon at high temperature.

$$SiO_2(l) + 2C(s) \longrightarrow Si(l) + 2CO(g) \qquad [22.99]$$

Elemental silicon has a diamond-type structure [see Figure 11.41(a)]. Crystalline silicon is a gray metallic-looking solid that melts at 1410°C (Figure 22.49 ◀). The element is a semiconductor (Section 23.5) and is thus used in making transistors and solar cells. To be used as a semiconductor, it must be extremely pure, possessing less than 10^{-7}% (1 ppb) impurities. One method of purification is to treat the element with Cl_2 to form $SiCl_4$. The $SiCl_4$ is a volatile liquid that is purified by fractional distillation and then converted back to elemental silicon by reduction with H_2:

$$SiCl_4(g) + 2H_2(g) \longrightarrow Si(s) + 4HCl(g) \qquad [22.100]$$

The element can be further purified by the process of zone refining. In the zone-refining process, a heated coil is passed slowly along a silicon rod, as shown in Figure 22.50 ◀. A narrow band of the element is thereby melted. As the molten area is swept slowly along the length of the rod, the impurities concentrate in the molten region, following it to the end of the rod. The purified top portion of the rod is retained for manufacture of electronic devices.

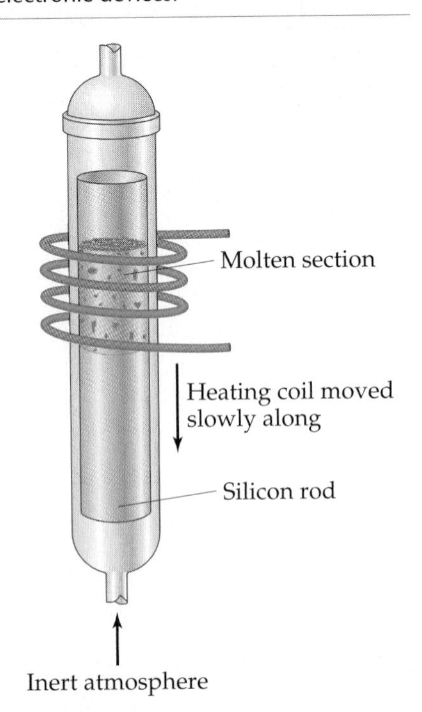

▲ **Figure 22.50** Zone-refining apparatus.

Silicates

Silicon dioxide and other compounds that contain silicon and oxygen comprise over 90% of Earth's crust. **Silicates** are compounds in which a silicon atom is surrounded in a tetrahedral fashion by four oxygens, as shown in Figure 22.51 ▶. In silicates, silicon is found in its most common oxidation state, +4. The simple SiO_4^{4-} ion, which is known as the orthosilicate ion, is found in very few silicate minerals. We can view the silicate tetrahedra, however, as "building blocks" that are used to build mineral structures. The individual tetrahedra are linked together by a common oxygen atom that serves as a vertex of both tetrahedra.

We can link two silicate tetrahedra together, for example, by sharing one oxygen atom, as shown in Figure 22.52 ▶. The resultant structure, called the *disilicate* ion, has two Si atoms and seven O atoms. Si and O are in the +4 and −2 oxidation states, respectively, in all silicates, so the overall charge of the ion must be consistent with these oxidation states. Thus, the charge on Si_2O_7 is $(2)(+4) + (7)(-2) = -6$; it is the $Si_2O_7^{6-}$ ion. The mineral *thortveitite* ($Sc_2Si_2O_7$) contains $Si_2O_7^{6-}$ ions.

In most silicate minerals a large number of silicate tetrahedra are linked together to form chains, sheets, or three-dimensional structures. We can connect

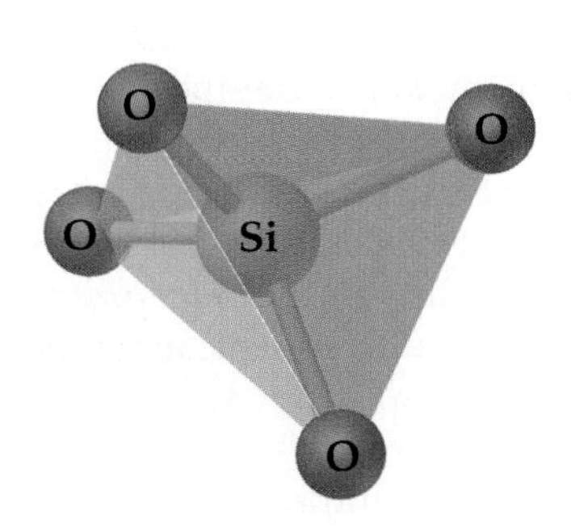

▲ **Figure 22.51** Structure of the SiO_4 tetrahedron of the SiO_4^{4-} ion. This ion is found in several minerals, such as zircon ($ZrSiO_4$).

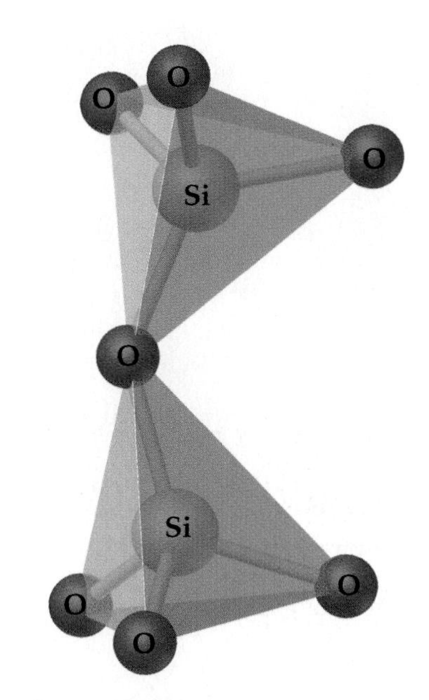

▲ **Figure 22.52** Geometrical structure of the $Si_2O_7^{6-}$ ion, which is formed by the sharing of an oxygen atom by two silicon atoms. This ion occurs in several minerals, such as hardystonite [$Ca_2Zn(Si_2O_7)$].

two vertices of each tetrahedron to two other tetrahedra, for example, leading to an infinite chain with an ···O—Si—O—Si··· backbone. This structure, called a single-strand silicate chain, is represented in Figure 22.53(a) ▼. As shown, this chain can be viewed as repeating units of the $Si_2O_6^{4-}$ ion or, in terms of its simplest formula, SiO_3^{2-}. The mineral *enstatite* ($MgSiO_3$) consists of rows of single-strand silicate chains with Mg^{2+} ions between the strands to balance charge.

In Figure 22.53(b) each silicate tetrahedron is linked to three others, forming an infinite two-dimensional sheet structure. The simplest formula of this infinite sheet is $Si_2O_5^{2-}$. The mineral *talc*, also known as talcum powder, has the formula $Mg_3(Si_2O_5)_2(OH)_2$ and is based on this sheet structure. The Mg^{2+} and OH^- ions lie between the silicate sheets. The slippery feel of talcum powder is due to the silicate sheets sliding relative to one another, much like the sheets of carbon atoms slide in graphite, giving graphite its lubricating properties. (Section 11.8)

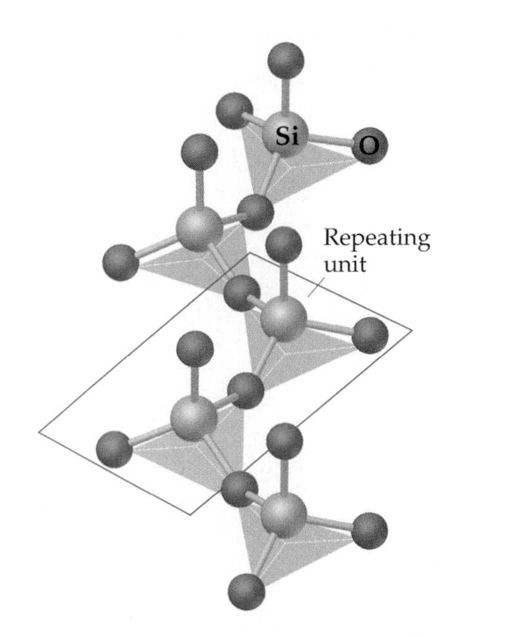

Single-strand silicate chain, $Si_2O_6^{4-}$

(a)

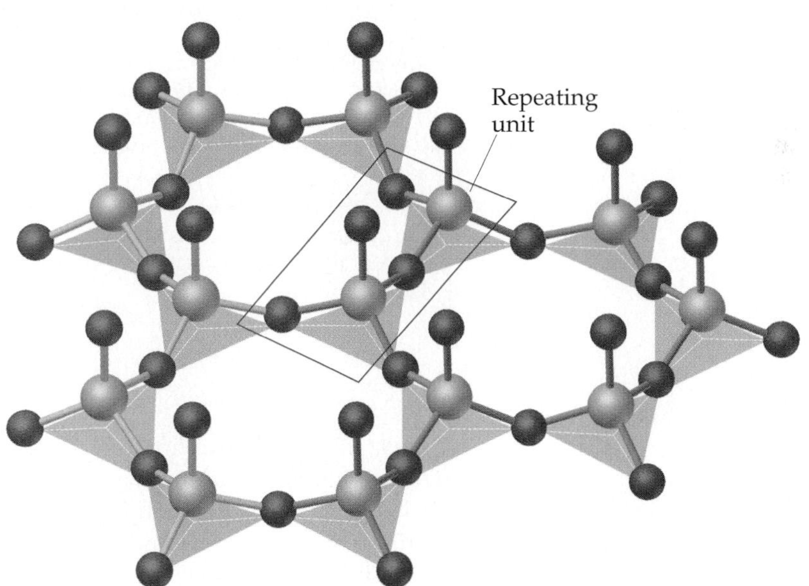

Two-dimensional silicate sheet, $Si_2O_5^{2-}$

(b)

▲ **Figure 22.53** Silicate structures consist of tetrahedra linked together through their vertices. Tetrahedra are linked through a shared oxygen atom. (a) Representation of an infinite single-strand silicate chain. Each tetrahedron is linked to two others. The box shows the repeating unit of the chain, which is similar to the unit cell of solids (Section 11.7); the chain can be viewed as an infinite number of repeating units, laid side by side. The repeating unit has a formula of $Si_2O_6^{4-}$, or as a simplest formula, SiO_3^{2-}. (b) Representation of a two-dimensional sheet structure. Each tetrahedron is linked to three others. The repeating unit of the sheet has the formula $Si_2O_5^{2-}$.

▲ **Figure 22.54** A sample of serpentine asbestos. Note the fibrous character of this silicate mineral.

Asbestos is a general term applied to a group of fibrous silicate minerals. These minerals possess chainlike arrangements of the silicate tetrahedra or sheet structures in which the sheets are formed into rolls. The result is that the minerals have a fibrous character, as shown in Figure 22.54 ◄. Asbestos minerals have been widely used as thermal insulation, especially in high-temperature applications, because of the great chemical stability of the silicate structure. In addition, the fibers can be woven into asbestos cloth, which can be used for fireproof curtains and other applications. However, the fibrous structure of asbestos minerals poses a health risk. Tiny asbestos fibers readily penetrate soft tissues, such as the lungs, where they can cause diseases, including cancer. The use of asbestos as a common building material has therefore been discontinued.

When all four vertices of each SiO_4 tetrahedron are linked to other tetrahedra, the structure extends in three dimensions. This linking of the tetrahedra forms quartz (SiO_2), which was depicted two-dimensionally in Figure 11.30(a). Because the structure is locked together in a three-dimensional array much like diamond [Figure 11.41(a)], quartz is harder than strand- or sheet-type silicates.

SAMPLE EXERCISE 22.10

The mineral *chrysotile* is a noncarcinogenic asbestos mineral that is based on the sheet structure shown in Figure 22.53(b). In addition to silicate tetrahedra, the mineral contains Mg^{2+} and OH^- ions. Analysis of the mineral shows that there are 1.5 Mg atoms per Si atom. What is the simplest formula for chrysotile?

Solution

Analyze: A mineral is described that has a sheet silicate structure with Mg^{2+} and OH^- ions to balance charge and 1.5 Mg for each 1 Si. We are asked to write the chemical formula for the mineral.

Plan: As shown in Figure 22.53(b), the silicate sheet structure is based on the $Si_2O_5{}^{2-}$ ion. We first add Mg^{2+} to give the proper Mg/Si ratio. We then add OH^- ions to obtain a neutral compound.

Solve: The observation that the Mg : Si ratio equals 1.5 is consistent with three Mg^{2+} ions per $Si_2O_5{}^{2-}$ ion. The addition of three Mg^{2+} ions would make $Mg_3(Si_2O_5)^{4+}$. In order to achieve charge balance in the mineral, there must be four OH^- ions per $Si_2O_5{}^{2-}$ ion. Thus, the simplest formula of chrysotile is $Mg_3(Si_2O_5)(OH)_4$.

PRACTICE EXERCISE

The cyclosilicate ion consists of three silicate tetrahedra linked together in a ring. The ion contains three Si atoms and nine O atoms. What is the overall charge on the ion?
Answer: 6−

Glass

Quartz melts at approximately 1600°C, forming a tacky liquid. In the course of melting, many silicon–oxygen bonds are broken. When the liquid is rapidly cooled, silicon–oxygen bonds are re-formed before the atoms are able to arrange themselves in a regular fashion. An amorphous solid, known as quartz glass or silica glass, results (see Figure 11.30). Many different substances can be added to SiO_2 to cause it to melt at a lower temperature. The common **glass** used in windows and bottles is known as soda-lime glass. It contains CaO and Na_2O in addition to SiO_2 from sand. The CaO and Na_2O are produced by heating two inexpensive chemicals, limestone ($CaCO_3$) and soda ash (Na_2CO_3). These carbonates decompose at elevated temperatures:

$$CaCO_3(s) \longrightarrow CaO(s) + CO_2(g) \qquad [22.101]$$

$$Na_2CO_3(s) \longrightarrow Na_2O(s) + CO_2(g) \qquad [22.102]$$

Other substances can be added to soda-lime glass to produce color or to change the properties of the glass in various ways. The addition of CoO, for example, produces the deep blue color of "cobalt glass." Replacing Na_2O by K_2O results in a harder glass that has a higher melting point. Replacing CaO by PbO results in a denser "lead crystal" glass with a higher refractive index. Lead crystal is used for decorative glassware; the higher refractive index gives this glass a particularly sparkling appearance. Addition of nonmetal oxides, such a B_2O_3 and P_4O_{10}, which form network structures related to the silicates, also changes the properties of the glass. Adding B_2O_3 creates a glass with a higher melting point and a greater ability to withstand temperature changes. Such glasses, sold commercially under trade names such as Pyrex® and Kimax®, are used where resistance to thermal shock is important, such as in laboratory glassware or coffeemakers.

Silicones

Silicones consist of O—Si—O chains in which the remaining bonding positions on each silicon are occupied by organic groups such as CH_3.

```
     H3C  CH3    H3C  CH3    H3C  CH3
        \/          \/          \/
. . .   Si          Si          Si        . . .
    \  /  \        /  \        /  \      /
     O      O            O            O
```

Depending on the length of the chain and the degree of cross-linking between chains, silicones can be either oils or rubberlike materials. Silicones are nontoxic and have good stability toward heat, light, oxygen, and water. They are used commercially in a wide variety of products, including lubricants, car polishes, sealants, and gaskets. They are also used for waterproofing fabrics. When applied to a fabric, the oxygen atoms form hydrogen bonds with the molecules on the surface of the fabric. The hydrophobic (water-repelling) organic groups of the silicone are then left pointing away from the surface as a barrier.

22.11 Boron

At this point boron is the only other element left to consider in our survey of the nonmetals. Boron is the only element of group 3A that can be considered nonmetallic. The element has an extended network structure. Its melting point (2300°C) is intermediate between that of carbon (3550°C) and that of silicon (1410°C). The electron configuration of boron is $[He]2s^22p^1$.

Numerous molecules contain only boron and hydrogen, a family of compounds called **boranes**. The simplest borane is BH_3. This molecule contains only six valence electrons and is therefore an exception to the octet rule. (Section 8.7) As a result, BH_3 reacts with itself to form *diborane* (B_2H_6). This reaction can be viewed as a Lewis acid-base reaction (Section 16.11), in which one B—H bonding pair of electrons in each BH_3 molecule is donated to the other. As a result, diborane is an unusual molecule in which hydrogen atoms appear to form two bonds (Figure 22.55 ▶).

Sharing hydrogen atoms between the two boron atoms compensates somewhat for the deficiency in valence electrons around each boron. Nevertheless, diborane is an extremely reactive molecule that is spontaneously flammable in air. The reaction of B_2H_6 with O_2 is extremely exothermic.

$$B_2H_6(g) + 3O_2(g) \longrightarrow B_2O_3(s) + 3H_2O(g) \qquad \Delta H^\circ = -2030 \text{ kJ} \qquad [22.103]$$

Other boranes, such as pentaborane(9) (B_5H_9), are also very reactive. Decaborane ($B_{10}H_{14}$) is stable in air at room temperature, but it undergoes a very exothermic reaction with O_2 at higher temperatures. Boranes have been explored as solid fuels for rockets.

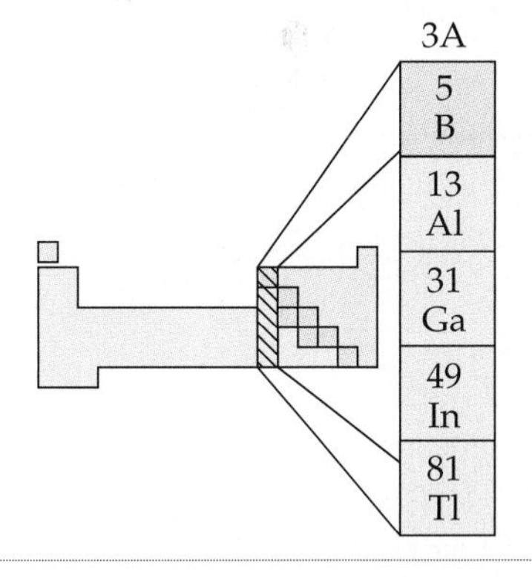

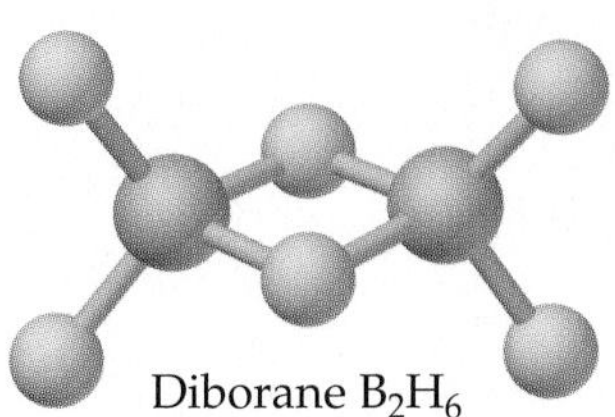

▲ **Figure 22.55** The structure of diborane (B_2H_6). Two of the H atoms bridge between the two B atoms, giving a planar B_2H_2 core to the molecule. Two of the remaining H atoms lie on either side of the B_2H_2 core, giving a nearly tetrahedral bonding environment about the B atoms.

Boron and hydrogen also form a series of anions, called *borane anions*. Salts of the borohydride ion (BH_4^-) are widely used as reducing agents. This ion is isoelectronic with CH_4 and NH_4^+. The lower charge of the central atom in BH_4^- means that the hydrogens of the BH_4^- are "hydridic"; that is, they carry a partial negative charge. Thus, it is not surprising that borohydrides are good reducing agents. Sodium borohydride ($NaBH_4$) is a commonly used reducing agent for certain organic compounds, and you will encounter it again should you take a course in organic chemistry.

The only important oxide of boron is boric oxide (B_2O_3). This substance is the anhydride of boric acid, which we may write as H_3BO_3 or $B(OH)_3$. Boric acid is so weak an acid ($K_a = 5.8 \times 10^{-10}$) that solutions of H_3BO_3 are used as an eyewash. Upon heating, boric acid loses water by a condensation reaction similar to that described for phosphorus in Section 22.8:

$$4H_3BO_3(s) \longrightarrow H_2B_4O_7(s) + 5H_2O(g) \qquad [22.104]$$

The diprotic acid $H_2B_4O_7$ is called tetraboric acid. The hydrated sodium salt, $Na_2B_4O_7 \cdot 10H_2O$, called borax, occurs in dry lake deposits in California and can also be readily prepared from other borate minerals. Solutions of borax are alkaline, and the substance is used in various laundry and cleaning products.

SAMPLE INTEGRATIVE EXERCISE 22: Putting Concepts Together

The interhalogen compound BrF_3 is a volatile, straw-colored liquid. The compound exhibits appreciable electrical conductivity due to autoionization.

$$2BrF_3(l) \rightleftharpoons BrF_2^+(solv) + BrF_4^-(solv)$$

(a) What are the molecular structures of the BrF_2^+ and BrF_4^- ions? **(b)** The electrical conductivity of BrF_3 decreases with increasing temperature. Is the autoionization process exothermic or endothermic? **(c)** One chemical characteristic of BrF_3 is that it acts as a Lewis acid toward fluoride ions. What do we expect will happen when KBr is dissolved in BrF_3?

Solution (a) The BrF_2^+ ion has a total of $7 + 2(7) - 1 = 20$ valence-shell electrons. The Lewis structure for the ion is

$$\left[:\ddot{\underset{..}{F}}-\ddot{\underset{..}{Br}}-\ddot{\underset{..}{F}}:\right]^+$$

Because there are four electron-pair domains around the central Br atom, the resulting electron-pair geometry is tetrahedral. (Section 9.2) Because two of these domains are occupied by bonding pairs of electrons, the molecular geometry is nonlinear.

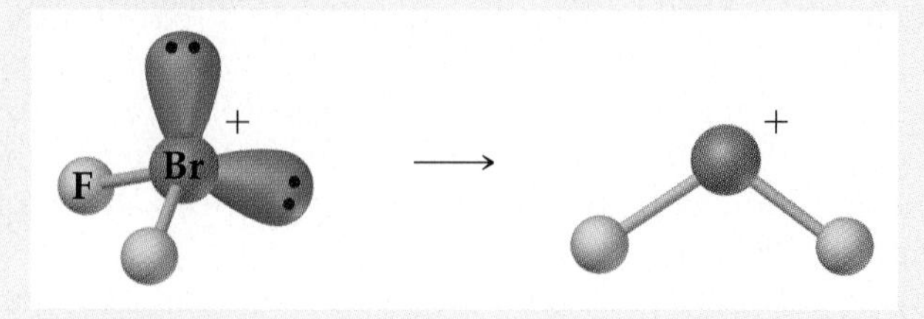

The BrF_4^- ion has a total of $7 + 4(7) + 1 = 36$ electrons, leading to the following Lewis structure.

$$\left[\begin{array}{c} :\ddot{F}\!: \\ :\ddot{\underset{..}{F}}-\underset{..}{\ddot{Br}}-\ddot{\underset{..}{F}}: \\ :\ddot{F}\!: \end{array}\right]^-$$

Because there are six electron-pair domains around the central Br atom in this ion, the electron-pair geometry is octahedral. The two nonbonding pairs of electrons are located opposite each other on the octahedron, leading to a square-planar molecular geometry.

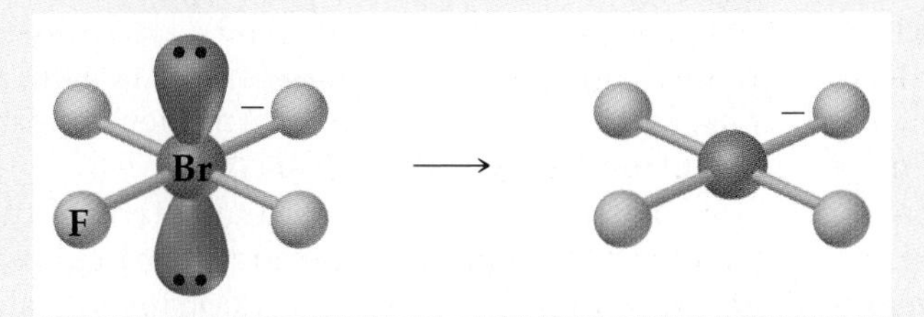

(c) The observation that conductivity decreases as temperature increases indicates that there are fewer ions present in the solution at the higher temperature. Thus, increasing the temperature causes the equilibrium to shift to the left. According to Le Châtelier's principle, this shift indicates that the reaction is exothermic as it proceeds from left to right. (Section 15.6)

(d) A Lewis acid is an electron-pair acceptor. (Section 16.11) The fluoride ion has four valence-shell electron pairs and can act as a Lewis base (an electron-pair donor). Thus, we can envision the following reaction occurring:

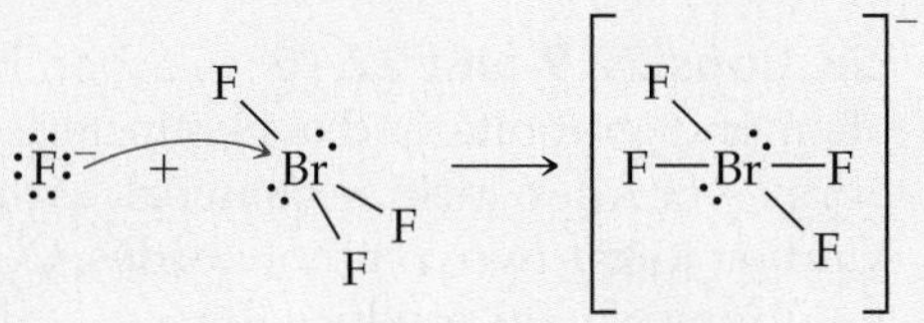

$$F^- + BrF_3 \longrightarrow BrF_4^-$$

Summary and Key Terms

Introduction and Section 22.1 The periodic table is useful for organizing and remembering the descriptive chemistry of the elements. Among elements of a given group, size increases with increasing atomic number, and electronegativity and ionization energy decrease. Nonmetallic character parallels electronegativity, so the most nonmetallic elements are found in the upper right portion of the periodic table. Among the nonmetallic elements, the first member of each group differs dramatically from the other members; it forms a maximum of four bonds to other atoms and exhibits a much greater tendency to form π bonds than the heavier elements in its group.

Because O_2 and H_2O are abundant in our world, we focus on two important and general reaction types as we discuss the descriptive chemistry of the nonmetals: oxidation by O_2 and proton-transfer reactions involving H_2O or aqueous solutions.

Section 22.2 Hydrogen has three isotopes: **protium** (^1_1H), **deuterium** (^2_1H), and **tritium** (^3_1H). Hydrogen is not a member of any particular periodic group, although it is usually placed above lithium. The hydrogen atom can either lose an electron, forming H^+, or gain one, forming H^- (the hydride ion). Because the H—H bond is relatively strong, H_2 is fairly unreactive unless activated by heat or a catalyst. Hydrogen forms a very strong bond to oxygen, so the reactions of H_2 with oxygen-containing compounds usually lead to the formation of H_2O. Because the bonds in CO and CO_2 are even stronger than the O—H bond, the reaction of H_2O with carbon or certain organic compounds leads to the formation of H_2. The $H^+(aq)$ ion is able to oxidize many metals, leading to metal ions and the formation of $H_2(g)$. The electrolysis of water also forms $H_2(g)$.

The binary compounds of hydrogen are of three general types: **ionic hydrides** (formed by active metals), **metallic hydrides** (formed by transition metals), and **molecular hydrides** (formed by nonmetals). The ionic hydrides contain the H^- ion; because this ion is extremely basic, ionic hydrides react with H_2O to form H_2 and OH^-.

Sections 22.3 and 22.4 The noble gases (group 8A) exhibit a very limited chemical reactivity because of the exceptional stability of their electron configurations. The xenon fluorides and oxides and KrF_2 are the best established compounds of the noble gases.

The halogens (group 7A) occur as diatomic molecules. They have the highest electronegativities of the elements in each row of the periodic table. All except fluorine exhibit oxidation states varying from -1 to $+7$. Fluorine is the most electronegative element, so it is restricted to the oxidation states 0 and -1. The oxidizing power of the element (the tendency to form the -1 oxidation state) decreases as we proceed down the group. The hydrogen halides are among the most useful compounds of these elements; these gases dissolve in water to form the hydrohalic acids, such as $HCl(aq)$. Hydrofluoric acid reacts with silica and is therefore used to etch glass. The **interhalogens** are compounds formed between two different halogen elements. Chlorine, bromine, and iodine form a series of oxyacids, in which the halogen atom is in a positive oxidation state. These compounds and their associated oxyanions are strong oxidizing agents.

Sections 22.5 and 22.6 Oxygen has two allotropes, O_2 and O_3 (ozone). Ozone is unstable compared to O_2, and it is a stronger oxidizing agent than O_2. Most reactions of O_2 lead to oxides, compounds in which oxygen is in the -2 oxidation state. The soluble oxides of nonmetals generally produce acidic aqueous solutions; they are called **acidic anhydrides** or **acidic oxides.** In contrast, soluble metal oxides produce basic solutions and are called **basic anhydrides** or **basic oxides.** Many metal oxides that are insoluble in water dissolve in acid, accompanied by the formation of H_2O. Peroxides contain O—O bonds and oxygen in the -1 oxidation state. Peroxides are unstable, decomposing to O_2 and oxides. In such reactions peroxides are simultaneously oxidized and reduced, a process called **disproportionation.** Superoxides contain the O_2^- ion in which oxygen is in the $-\frac{1}{2}$ oxidation state.

Sulfur is the most important of the other group 6A elements. It has several allotropic forms; the most stable one at room temperature consists of S_8 rings. Sulfur forms two oxides, SO_2 and SO_3, and both are important atmospheric pollutants. Sulfur trioxide is the anhydride of sulfuric acid, the most important sulfur compound and the most-produced industrial chemical. Sulfuric acid is a strong acid and a good dehydrating agent. Sulfur forms several oxyanions as well, including the SO_3^{2-} (sulfite), SO_4^{2-} (sulfate), and $S_2O_3^{2-}$ (thiosulfate) ions. Sulfur is found combined with many metals as a sulfide, in which sulfur is in the -2 oxidation state. These compounds often react with acids to form hydrogen sulfide (H_2S), which smells like rotten eggs.

Sections 22.7 and 22.8 Nitrogen is found in the atmosphere as N_2 molecules. Molecular nitrogen is chemically very stable because of the strong N≡N bond. Molecular nitrogen can be converted into ammonia via the Haber process; once the ammonia is made, it can be converted into a variety of different compounds that exhibit nitrogen oxidation states ranging from -3 to $+5$. The most important industrial conversion of ammonia is the **Ostwald process**, in which ammonia is oxidized to nitric acid (HNO_3). Nitrogen has three important oxides: nitrous oxide (N_2O), nitric oxide (NO), and nitrogen dioxide (NO_2). Nitrous acid (HNO_2) is a weak acid; its conjugate base is the nitrite ion (NO_2^-). Another important nitrogen compound is hydrazine (N_2H_4).

Phosphorus is the most important of the remaining group 5A elements. It occurs in nature in phosphate minerals. Phosphorus has several allotropes, including white phosphorus, which consists of P_4 tetrahedra. In reaction with the halogens, phosphorus forms trihalides (PX_3) and pentahalides (PX_5). These compounds undergo hydrolysis to produce an oxyacid of phosphorus and HX. Phosphorus forms two oxides, P_4O_6 and P_4O_{10}. Their corresponding acids, phosphorous acid and phosphoric acid, undergo condensation reactions when heated. Phosphorus compounds are important in biochemistry and as fertilizers.

Sections 22.9 and 22.10 Carbon has three allotropes: diamond, graphite, and buckminsterfullerene. Amorphous forms of carbon include **charcoal, carbon black**, and **coke.** Carbon forms two common oxides, CO and CO_2. Aqueous solutions of CO_2 produce the weak diprotic acid carbonic acid (H_2CO_3), which is the parent acid of hydrogen carbonate and carbonate salts. Binary compounds of carbon are called **carbides.** Carbides may be ionic, interstitial, or covalent. Calcium carbide (CaC_2) contains the strongly basic acetylide ion (C_2^{2-}), which reacts with water to form acetylene. Other important carbon compounds include hydrogen cyanide (HCN) and its corresponding cyanide salts, and carbon disulfide (CS_2). Carbon also forms a vast number of organic compounds, which are discussed in Chapter 25.

The other group 4A elements show great diversity in physical and chemical properties. Silicon, the second most abundant element, is a semiconductor. It reacts with Cl_2 to form $SiCl_4$, a liquid at room temperature. Silicon forms strong Si—O bonds and therefore occurs in a variety of silicate minerals. **Silicates** consist of SiO_4 tetrahedra, linked together at their vertices to form chains, sheets, or three-dimensional structures. The most common three-dimensional silicate is quartz (SiO_2). **Glass** is an amorphous (noncrystalline) form of SiO_2. Silicones contain O—Si—O chains with organic groups bonded to the Si atoms. Like silicon, germanium is a metalloid; tin and lead are metallic.

Section 22.11 Boron is the only group 3A element that is a nonmetal. It forms a variety of compounds with hydrogen, called boron hydrides or **boranes.** Diborane (B_2H_6) has an unusual structure with two hydrogen atoms that bridge between the two boron atoms. Boranes react with oxygen to form boric oxide (B_2O_3), in which boron is in the $+3$ oxidation state. Boric oxide is the anhydride of boric acid (H_3BO_3). Boric acid readily undergoes condensation reactions.

Exercises

Periodic Trends and Chemical Reactions

22.1 Identify each of the following elements as a metal, nonmetal, or metalloid: **(a)** antimony; **(b)** strontium; **(c)** cerium; **(d)** selenium; **(e)** rhodium; **(f)** krypton.

22.2 Identify each of the following elements as a metal, nonmetal, or metalloid: **(a)** rhenium; **(b)** arsenic; **(c)** argon; **(d)** zirconium; **(e)** tellurium; **(f)** gallium.

22.3 Consider the elements Li, K, Cl, C, Ne, and Ar. From this list select the element that **(a)** is most electronegative; **(b)** has the greatest metallic character; **(c)** most readily forms a positive ion; **(d)** has the smallest atomic radius; **(e)** forms π bonds most readily.

22.4 Consider the elements O, Ba, Co, Be, Br, and Se. From this list select the element that **(a)** is most electronegative; **(b)** exhibits a maximum oxidation state of +7; **(c)** loses an electron most readily; **(d)** forms π bonds most readily; **(e)** is a transition metal.

22.5 Explain the following observations: **(a)** The highest fluoride compound formed by nitrogen is NF_3, whereas phosphorus readily forms PF_5. **(b)** Although CO is a well-known compound, SiO doesn't exist under ordinary conditions. **(c)** AsH_3 is a stronger reducing agent than NH_3.

22.6 Explain the following observations: **(a)** HNO_3 is a stronger oxidizing agent than H_3PO_4. **(b)** Silicon can form an ion with six fluorine atoms, SiF_6^{2-}, whereas carbon is able to bond to a maximum of four, CF_4. **(c)** There are three compounds formed by carbon and hydrogen that contain two carbon atoms each (C_2H_2, C_2H_4, and C_2H_6), whereas silicon forms only one analogous compound (Si_2H_6).

22.7 Complete and balance the following equations:
(a) $LiN_3(s) + H_2O(l) \longrightarrow$
(b) $C_3H_7OH(l) + O_2(g) \longrightarrow$
(c) $NiO(s) + C(s) \longrightarrow$
(d) $AlP(s) + H_2O(l) \longrightarrow$
(e) $Na_2S(s) + HCl(aq) \longrightarrow$

22.8 Complete and balance the following equations:
(a) $NaOCH_3(s) + H_2O(l) \longrightarrow$
(b) $CuO(s) + HNO_3(aq) \longrightarrow$
(c) $WO_3(s) + H_2(g) \longrightarrow$
(d) $NH_2OH(l) + O_2(g) \longrightarrow$
(e) $Al_4C_3(s) + H_2O(l) \longrightarrow$

Hydrogen, the Noble Gases, and the Halogens

22.9 **(a)** Give the names and chemical symbols for the three isotopes of hydrogen. **(b)** List the isotopes in order of decreasing natural abundance.

22.10 Which isotope of hydrogen is radioactive? Write the nuclear equation for the radioactive decay of this isotope.

22.11 Why is hydrogen often placed in either group 1A or 7A of the periodic table?

22.12 Why are the properties of hydrogen different from those of either the group 1A or 7A elements?

22.13 Give a balanced equation for the preparation of H_2 using **(a)** Mg and an acid; **(b)** carbon and steam; **(c)** methane and steam.

22.14 List **(a)** three commercial means of producing H_2; **(b)** three industrial uses of H_2.

22.15 Complete and balance the following equations:
(a) $NaH(s) + H_2O(l) \longrightarrow$
(b) $Fe(s) + H_2SO_4(aq) \longrightarrow$
(c) $H_2(g) + Br_2(g) \longrightarrow$
(d) $Na(l) + H_2(g) \longrightarrow$
(e) $PbO(s) + H_2(g) \longrightarrow$

22.16 Write balanced equations for each of the following reactions (some of these are analogous to reactions shown in the chapter). **(a)** Aluminum metal reacts with acids to form hydrogen gas. **(b)** Steam reacts with magnesium metal to give magnesium oxide and hydrogen. **(c)** Manganese(IV) oxide is reduced to manganese(II) oxide by hydrogen gas. **(d)** Calcium hydride reacts with water to generate hydrogen gas.

22.17 Identify the following hydrides as ionic, metallic, or molecular: **(a)** B_2H_6; **(b)** RbH; **(c)** $Th_4H_{1.5}$.

22.18 Identify the following hydrides as ionic, metallic, or molecular: **(a)** BaH_2; **(b)** H_2Te; **(c)** $TiH_{1.7}$.

22.19 Why does xenon form stable compounds with fluorine, whereas argon does not?

22.20 Why were the noble gases the last family of elements to be discovered?

22.21 Write the chemical formula for each of the following compounds, and indicate the oxidation state of the halogen or noble-gas atom in each: **(a)** bromate ion; **(b)** hydroiodic acid; **(c)** bromine trifluoride; **(d)** sodium hypochlorite; **(e)** perchloric acid; **(f)** xenon tetrafluoride.

22.22 Write the chemical formula for each of the following, and indicate the oxidation state of the halogen or noble-gas atom in each; **(a)** calcium hypobromite; **(b)** bromic acid; **(c)** xenon trioxide; **(d)** perchlorate ion; **(e)** iodous acid; **(f)** iodine pentafluoride.

22.23 Name the following compounds: **(a)** $KClO_3$; **(b)** $Ca(IO_3)_2$; **(c)** $AlCl_3$; **(d)** $HBrO_3$; **(e)** H_5IO_6; **(f)** XeF_4.

22.24 Name the following compounds: **(a)** $Fe(ClO_3)_3$; **(b)** $HClO_2$; **(c)** XeF_6; **(d)** BrF_5; **(e)** $XeOF_4$; **(f)** HIO_3 (named as an acid).

22.25 Explain each of the following observations: **(a)** At room temperature I_2 is a solid, Br_2 is a liquid, and Cl_2 and F_2 are both gases. **(b)** F_2 cannot be prepared by electrolytic oxidation of aqueous F^- solutions. **(c)** The boiling point of HF is much higher than those of the other hydrogen halides. **(d)** The halogens decrease in oxidizing power in the order $F_2 > Cl_2 > Br_2 > I_2$.

22.26 Explain the following observations: **(a)** for a given oxidation state, the acid strength of the oxyacid in aqueous solution decreases in the order chlorine > bromine > iodine. **(b)** Hydrofluoric acid cannot be stored in glass bottles. **(c)** HI cannot be prepared by treating NaI with sulfuric acid. **(d)** The interhalogen ICl_3 is known, but $BrCl_3$ is not.

22.27 Write balanced equations for each of the following reactions (some of which are analogous but not identical to reactions shown in this chapter): **(a)** Bromine forms hypobromite ion on addition to aqueous base. **(b)** Chlorine reacts with an aqueous solution of sodium iodide.

22.28 Write balanced chemical equations for each of the following reactions (some of which are analogous but not identical to reactions shown in this chapter): **(a)** Hydrogen bromide is produced upon heating calcium bromide with phosphoric acid. **(b)** Aqueous hydrogen fluoride reacts with solid calcium carbonate, forming water-insoluble calcium fluoride.

22.29 Predict the geometric structures of the following: **(a)** ICl_4^-; **(b)** ClO_3^-; **(c)** H_5IO_6; **(d)** XeF_2.

22.30 The interhalogen compound $BrF_3(l)$ reacts with antimony(V) fluoride to form the salt $(BrF_2^+)(SbF_6^-)$. Write the Lewis structure for both the cation and anion in this substance, and describe the likely structure of each.

Oxygen and the Group 6A Elements

22.31 **(a)** List three industrial uses of O_2. **(b)** List two industrial uses of O_3.

22.32 Give the structure of ozone. Explain why the O—O bond length in ozone (1.28 Å) is longer than that in O_2 (1.21 Å).

22.33 Complete and balance the following equations:
(a) $CaO(s) + H_2O(l) \longrightarrow$
(b) $Al_2O_3(s) + H^+(aq) \longrightarrow$
(c) $Na_2O_2(s) + H_2O(l) \longrightarrow$
(d) $N_2O_3(g) + H_2O(l) \longrightarrow$
(e) $KO_2(s) + H_2O(l) \longrightarrow$
(f) $NO(g) + O_3(g) \longrightarrow$

22.34 Write balanced equations for each of the following reactions. **(a)** When mercury(II) oxide is heated, it decomposes to form O_2 and mercury metal. **(b)** When copper(II) nitrate is heated strongly, it decomposes to form copper(II) oxide, nitrogen dioxide, and oxygen. **(c)** Lead(II) sulfide, $PbS(s)$, reacts with ozone to form $PbSO_4(s)$ and $O_2(g)$. **(d)** When heated in air, $ZnS(s)$ is converted to ZnO. **(e)** Potassium peroxide reacts with $CO_2(g)$ to give potassium carbonate and O_2.

22.35 Predict whether each of the following oxides is acidic, basic, amphoteric, or neutral: **(a)** CO; **(b)** CO_2; **(c)** CaO; **(d)** Al_2O_3.

22.36 Select the more acidic member of each of the following pairs: **(a)** Mn_2O_7 and MnO_2; **(b)** SnO and SnO_2; **(c)** SO_2 and SO_3; **(d)** SiO_2 and SO_2; **(e)** Ga_2O_3 and In_2O_3; **(f)** SO_2 and SeO_2.

22.37 Write the chemical formula for each of the following compounds, and indicate the oxidation state of the group 6A element in each: **(a)** selenium trioxide; **(b)** sodium thiosulfate; **(c)** sulfur tetrafluoride; **(d)** hydrogen sulfide; **(e)** sulfurous acid.

22.38 Write the chemical formula for each of the following compounds, and indicate the oxidation state of the group 6A element in each: **(a)** selenous acid; **(b)** potassium hydrogen sulfite; **(c)** hydrogen telluride; **(d)** carbon disulfide; **(e)** calcium sulfate.

22.39 In aqueous solution, hydrogen sulfide reduces **(a)** Fe^{3+} to Fe^{2+}; **(b)** Br_2 to Br^-; **(c)** MnO_4^- to Mn^{2+}; **(d)** HNO_3 to NO_2. In all cases, under appropriate conditions, the product is elemental sulfur. Write a balanced net ionic equation for each reaction.

22.40 An aqueous solution of SO_2 reduces **(a)** aqueous $KMnO_4$ to $MnSO_4(s)$; **(b)** acidic aqueous $K_2Cr_2O_7$ to aqueous Cr^{3+}; **(c)** aqueous $Hg_2(NO_3)_2$ to mercury metal. Write balanced equations for these reactions.

22.41 Write the Lewis structure for each of the following species and indicate the structure of each: **(a)** SeO_3^{2-}; **(b)** S_2Cl_2; **(c)** chlorosulfonic acid, HSO_3Cl (chlorine is bonded to sulfur).

22.42 The SF_5^- ion is formed when $SF_4(g)$ reacts with fluoride salts containing large cations, such as $CsF(s)$. Draw the Lewis structures for SF_4 and SF_5^-, and predict the molecular structure of each.

22.43 Write a balanced equation for each of the following reactions: **(a)** Sulfur dioxide reacts with water. **(b)** Solid zinc sulfide reacts with hydrochloric acid. **(c)** Elemental sulfur reacts with sulfite ion to form thiosulfate. **(d)** Sulfur trioxide is dissolved in sulfuric acid.

22.44 Write a balanced equation for each of the following reactions. (You may have to guess at one or more of the reaction products, but you should be able to make a reasonable guess, based on your study of this chapter.) **(a)** Hydrogen selenide can be prepared by reaction of an aqueous acid solution on aluminum selenide. **(b)** Sodium thiosulfate is used to remove excess Cl_2 from chlorine-bleached fabrics. The thiosulfate ion forms SO_4^{2-} and elemental sulfur, while Cl_2 is reduced to Cl^-.

Nitrogen and the Group 5A Elements

22.45 Write the chemical formula for each of the following compounds, and indicate the oxidation state of nitrogen in each: **(a)** sodium nitrite; **(b)** ammonia; **(c)** nitrous oxide; **(d)** sodium cyanide; **(e)** nitric acid; **(f)** nitrogen dioxide.

22.46 Write the chemical formula for each of the following compounds, and indicate the oxidation state of nitrogen in each: **(a)** nitrous acid; **(b)** hydrazine; **(c)** potassium cyanide; **(d)** sodium nitrate; **(e)** ammonium chloride; **(f)** lithium nitride.

22.47 Write the Lewis structure for each of the following species, and describe its geometry: **(a)** NH_4^+; **(b)** HNO_3; **(c)** N_2O; **(d)** NO_2.

22.48 Write the Lewis structure for each of the following species, and describe its geometry: **(a)** HNO_2; **(b)** N_3^-; **(c)** $N_2H_5^+$; **(d)** NO_3^-.

22.49 Complete and balance the following equations:
(a) $Mg_3N_2(s) + H_2O(l) \longrightarrow$
(b) $NO(g) + O_2(g) \longrightarrow$
(c) $N_2O_5(g) + H_2O(l) \longrightarrow$
(d) $NH_3(aq) + H^+(aq) \longrightarrow$
(e) $N_2H_4(l) + O_2(g) \longrightarrow$

22.50 Write balanced net ionic equations for each of the following reactions: **(a)** Dilute nitric acid reacts with zinc metal with formation of nitrous oxide. **(b)** Concentrated nitric acid reacts with sulfur with formation of nitrogen dioxide. **(c)** Concentrated nitric acid oxidizes sulfur dioxide with formation of nitric oxide. **(d)** Hydrazine is burned in excess fluorine gas, forming NF_3. **(e)** Hydrazine reduces CrO_4^{2-} to $Cr(OH)_4^-$ in base (hydrazine is oxidized to N_2).

22.51 Write complete balanced half-reactions for **(a)** reduction of nitrate ion to N_2 in acidic solution; **(b)** oxidation of

NH_4^+ to N_2 in acidic solution. What is the standard reduction potential in each case? (See Figure 22.30.)

22.52 Write complete balanced half-reactions for **(a)** reduction of nitrate ion to NO in acidic solution; **(b)** oxidation of HNO_2 to NO_2 in acidic solution. What is the standard reduction potential in each case? (See Figure 22.30.)

22.53 Write formulas for the following compounds, and indicate the oxidation state of the group 5A element in each: **(a)** orthophosphoric acid; **(b)** arsenous acid; **(c)** antimony(III) sulfide; **(d)** calcium dihydrogen phosphate; **(e)** potassium phosphide.

22.54 Write formulas for the following compounds, and indicate the oxidation state of the group 5A element in each: **(a)** phosphorous acid; **(b)** pyrophosphoric acid; **(c)** antimony trichloride; **(d)** magnesium arsenide; **(e)** diphosphorus pentoxide.

22.55 Account for the following observations: **(a)** Phosphorus forms a pentachloride, but nitrogen does not. **(b)** H_3PO_2 is a monoprotic acid. **(c)** Phosphonium salts, such as PH_4Cl, can be formed under anhydrous conditions, but they can't be made in aqueous solution. **(d)** White phosphorus is extremely reactive.

22.56 Account for the following observations: **(a)** H_3PO_3 is a diprotic acid. **(b)** Nitric acid is a strong acid, whereas phosphoric acid is weak. **(c)** Phosphate rock is ineffective as a phosphate fertilizer. **(d)** Phosphorus does not exist at room temperature as diatomic molecules, but nitrogen does. **(e)** Solutions of Na_3PO_4 are quite basic.

22.57 Write a balanced equation for each of the following reactions: **(a)** preparation of white phosphorus from calcium phosphate; **(b)** hydrolysis of PCl_3; **(c)** preparation of PCl_3 from P_4.

22.58 Write a balanced equation for each of the following reactions: **(a)** hydrolysis of PCl_5; **(b)** dehydration of orthophosphoric acid to form pyrophosphoric acid; **(c)** reaction of P_4O_{10} with water.

Carbon, the Other Group 4A Elements, and Boron

22.59 Give the chemical formulas for **(a)** hydrocyanic acid; **(b)** Carborundum™; **(c)** calcium carbonate; **(d)** calcium acetylide.

22.60 Give the chemical formulas for **(a)** carbonic acid; **(b)** sodium cyanide; **(c)** potassium hydrogen carbonate; **(d)** acetylene.

22.61 Write the Lewis structure of each of the following species: **(a)** CN^-; **(b)** CO; **(c)** C_2^{2-}; **(d)** CS_2; **(e)** CO_2; **(f)** CO_3^{2-}.

22.62 Indicate the geometry and the type of hybrid orbitals used by each carbon atom in the following species: **(a)** $CH_3C{\equiv}CH$; **(b)** NaCN; **(c)** CS_2; **(d)** C_2H_6.

22.63 Complete and balance the following equations:

(a) $ZnCO_3(s) \xrightarrow{\Delta}$

(b) $BaC_2(s) + H_2O(l) \longrightarrow$

(c) $C_2H_4(g) + O_2(g) \longrightarrow$

(d) $CH_3OH(l) + O_2(g) \longrightarrow$

(e) $NaCN(s) + HCl(aq) \longrightarrow$

22.64 Complete and balance the following equations:

(a) $CO_2(g) + OH^-(aq) \longrightarrow$

(b) $NaHCO_3(s) + H^+(aq) \longrightarrow$

(c) $CaO(s) + C(s) \xrightarrow{\Delta}$

(d) $C(s) + H_2O(g) \xrightarrow{\Delta}$

(e) $CuO(s) + CO(g) \longrightarrow$

22.65 Write a balanced equation for each of the following reactions: **(a)** Hydrogen cyanide is formed commercially by passing a mixture of methane, ammonia, and air over a catalyst at 800°C. Water is a by-product of the reaction. **(b)** Baking soda reacts with acids to produce carbon dioxide gas. **(c)** When barium carbonate reacts in air with sulfur dioxide, barium sulfate and carbon dioxide form.

22.66 Write a balanced equation for each of the following reactions: **(a)** Burning magnesium metal in a carbon dioxide atmosphere reduces the CO_2 to carbon. **(b)** In photosynthesis, solar energy is used to produce glucose ($C_6H_{12}O_6$) and O_2 out of carbon dioxide and water. **(c)** When carbonate salts dissolve in water, they produce basic solutions.

22.67 Write the formulas for the following compounds, and indicate the oxidation state of the group 4A element or of boron in each: **(a)** boric acid; **(b)** silicon tetrabromide; **(c)** lead(II) chloride; **(d)** sodium tetraborate decahydrate (borax); **(e)** boric oxide.

22.68 Write the formulas for the following compounds, and indicate the oxidation state of the group 4A element or of boron in each: **(a)** silicon dioxide; **(b)** germanium tetrachloride; **(c)** sodium borohydride; **(d)** stannous chloride; **(e)** diborane.

22.69 Select the member of group 4A that best fits each of the following descriptions: **(a)** forms the most acidic oxide; **(b)** is most commonly found in the +2 oxidation state; **(c)** is a component of sand.

22.70 Select the member of group 4A that best fits each of the following descriptions: **(a)** forms chains to the greatest extent; **(b)** forms the most basic oxide; **(c)** is a metalloid that can form 2+ ions.

22.71 What empirical formula and unit charge are associated with each of the following structural types: **(a)** isolated SiO_4 tetrahedra; **(b)** a chain structure of SiO_4 tetrahedra joined at the corners to adjacent units; **(c)** a structure consisting of tetrahedra joined at the corners to form a six-membered ring of alternating Si and O atoms?

22.72 Two silicate anions are known in which the linking of the tetrahedra forms a closed ring. One of these cyclic silicate anions contains three silicate tetrahedra, linked into a ring. The other contains six silicate tetrahedra. **(a)** Sketch these cyclic silicate anions. **(b)** Determine the formula and charge of each of the anions.

22.73 **(a)** How does the structure of diborane (B_2H_6) differ from that of ethane (C_2H_6)? **(b)** By using concepts discussed in Chapter 8, explain why diborane adopts the geometry that it does. **(c)** What is the significance of the statement that the hydrogen atoms in diborane are described as hydridic?

22.74 Write a balanced equation for each of the following reactions: **(a)** Diborane reacts with water to form boric acid and molecular hydrogen. **(b)** Upon heating, boric acid undergoes a condensation reaction to form tetraboric acid. **(c)** Boron oxide dissolves in water to give a solution of boric acid.

Additional Exercises

22.75 In your own words, define the following terms: **(a)** isotope; **(b)** allotrope; **(c)** disproportionation; **(d)** interhalogen; **(e)** Frasch process; **(f)** Ostwald process; **(g)** condensation reaction.

22.76 **(a)** How many grams of H_2 can be stored in 10.0 lb of the alloy FeTi if the hydride $FeTiH_2$ is formed? **(b)** What volume does this quantity of H_2 occupy at STP?

22.77 Starting with D_2O, suggest preparations of **(a)** ND_3; **(b)** D_2SO_4; **(c)** NaOD; **(d)** DNO_3; **(e)** C_2D_2; **(f)** DCN.

22.78 Although the ClO_4^- and IO_4^- ions have been known for a long time, BrO_4^- was not synthesized until 1965. The ion was synthesized by oxidizing the bromate ion with xenon difluoride, producing xenon, hydrofluoric acid, and the perbromate ion. Write the balanced equation for this reaction.

22.79 Which of the following substances will burn in oxygen: SiH_4; SiO_2; CO; CO_2; Mg; CaO? Why won't some of these substances burn in oxygen?

22.80 Write a balanced equation for the reaction of each of the following compounds with water; **(a)** $SO_2(g)$; **(b)** $Cl_2O(g)$; **(c)** $Na_2O(s)$; **(d)** $BaC_2(s)$; **(e)** $RbO_2(s)$; **(f)** $Mg_3N_2(s)$; **(g)** $Na_2O_2(s)$; **(h)** NaH(*s*).

22.81 What is the anhydride for each of the following acids: **(a)** H_2SO_4; **(b)** $HClO_3$; **(c)** HNO_2; **(d)** H_2CO_3; **(e)** H_3PO_4?

22.82 Elemental sulfur is capable of reacting under suitable conditions with Fe, F_2, O_2, or H_2. Write balanced equations to describe these reactions. In which reactions is sulfur acting as a reducing agent and in which as an oxidizing agent?

22.83 A sulfuric acid plant produces a considerable amount of heat. This heat is used to generate electricity, which helps reduce operating costs. The synthesis of H_2SO_4 consists of three main chemical processes: (1) oxidation of S to SO_2; (2) oxidation of SO_2 to SO_3; (3) the dissolving of SO_3 in H_2SO_4 and its reaction with water to form H_2SO_4. If the third process produces 130 kJ/mol, how much heat is produced in preparing a mole of H_2SO_4 from a mole of S? How much heat is produced in preparing a ton of H_2SO_4?

22.84 **(a)** What is the oxidation state of P in PO_4^{3-} and of N in NO_3^-? **(b)** Why doesn't N form a stable NO_4^{3-} ion analogous to P?

22.85 **(a)** What structural feature do the molecules P_4, P_4O_6, and P_4O_{10} have in common? What is the common structural feature of all the acids containing phosphorus(V)? **(b)** Sodium trimetaphosphate ($Na_3P_3O_9$) and sodium tetrametaphosphate ($Na_4P_4O_{12}$) are used as water-softening agents. They contain cyclic $P_3O_9^{3-}$ and $P_4O_{12}^{4-}$ ions, respectively. Propose reasonable structures for these ions.

[22.86] **(a)** Calculate the P—P distance in both P_4O_6 and P_4O_{10} from the following data: the P—O—P bond angle for P_4O_6 is 127.5°, while that for P_4O_{10} is 124.5°. The P—O distance (to bridging oxygens) is 1.65 Å in P_4O_6 and 1.60 Å in P_4O_{10} **(b)** Rationalize the relative P—P bond distances in the two compounds.

22.87 Ultrapure germanium, like silicon, is used in semiconductors. Germanium of "ordinary" purity is prepared by the high-temperature reduction of GeO_2 with carbon. The Ge is converted to $GeCl_4$ by treatment with Cl_2 and then purified by distillation; $GeCl_4$ is then hydrolyzed in water to GeO_2 and reduced to the elemental form with H_2. The element is then zone refined. Write a balanced chemical equation for each of the chemical transformations in the course of forming ultrapure Ge from GeO_2.

22.88 Complete and balance the following equations:
(a) $MnO_4^-(aq) + H_2O_2(aq) + H^+(aq) \longrightarrow$
(b) $Fe^{2+}(aq) + H_2O_2(aq) \longrightarrow$
(c) $I^-(aq) + H_2O_2(aq) + H^+(aq) \longrightarrow$
(d) $MnO_2(s) + H_2O_2(aq) + H^+(aq) \longrightarrow$
(e) $I^-(aq) + O_3(g) \longrightarrow I_2(s) + O_2(g) + OH^-(aq)$

22.89 Hydrogen peroxide is capable of oxidizing **(a)** K_2S to S; **(b)** SO_2 to SO_4^{2-}; **(c)** NO_2^- to NO_3^-; **(d)** As_2O_3 to AsO_4^{3-}; **(e)** Fe^{2+} to Fe^{3+}. Write a balanced net ionic equation for each of these redox reactions.

22.90 Complete and balance the following equations:
(a) $Li_3N(s) + H_2O(l) \longrightarrow$
(b) $NH_3(aq) + H_2O(l) \longrightarrow$
(c) $NO_2(g) + H_2O(l) \longrightarrow$
(d) $2NO_2(g) \longrightarrow$
(e) $NH_3(g) + O_2(g) \xrightarrow{\text{catalyst}}$
(f) $CO(g) + O_2(g) \longrightarrow$
(g) $H_2CO_3(aq) \xrightarrow{\Delta}$
(h) $Ni(s) + CO(g) \longrightarrow$
(i) $CS_2(g) + O_2(g) \longrightarrow$
(j) $CaO(s) + SO_2(g) \longrightarrow$
(k) $Na(s) + H_2O(l) \longrightarrow$
(l) $CH_4(g) + H_2O(g) \xrightarrow{\Delta}$
(m) $LiH(s) + H_2O(l) \longrightarrow$
(n) $Fe_2O_3(s) + 3H_2(g) \longrightarrow$

Integrative Exercises

22.91 What pressure of gas is formed when 0.500 g of XeO_3 decomposes completely to the free elements at 30°C in a 1.00-L volume?

[22.92] Using the thermochemical data in Table 22.1 and Appendix C, calculate the average Xe—F bond enthalpies in XeF_2, XeF_4, and XeF_6, respectively. What is the significance of the trend in these quantities?

22.93 Hydrogen gas has a higher fuel value than natural gas on a mass basis but not on a volume basis. Thus, hydrogen is not competitive with natural gas as a fuel transported long distances through pipelines. Calculate the heat of combustion of H_2 and CH_4 (the principal component of natural gas) **(a)** per mole of each; **(b)** per gram of each; **(c)** per cubic meter of each at STP. Assume $H_2O(l)$ as a product.

22.94 The solubility of Cl_2 in 100 g of water at STP is 310 cm^3. Assume that this quantity of Cl_2 is dissolved and equilibrated as follows:

$$Cl_2(aq) + H_2O(l) \rightleftharpoons Cl^-(aq) + HClO(aq) + H^+(aq)$$

If the equilibrium constant for this reaction is 4.7×10^{-4}, calculate the equilibrium concentration of HClO formed.

22.95 The dissolved oxygen present in any highly pressurized, high-temperature steam boiler can be extremely corrosive to its metal parts. Hydrazine, which is completely miscible with water, can be added to remove oxygen by reacting with it to form nitrogen and water. **(a)** Write the balanced equation for the reaction between gaseous hydrazine and oxygen. **(b)** Calculate the enthalpy change accompanying this reaction. **(c)** Oxygen in air dissolves in water to the extent of 9.1 ppm at 20°C at sea level. How many grams of hydrazine are required to react with all the oxygen in 3.0×10^4 L (the volume of a small swimming pool) under these conditions?

22.96 One method proposed for removing SO_2 from the flue gases of power plants involves reaction with aqueous H_2S. Elemental sulfur is the product. **(a)** Write a balanced chemical equation for the reaction. **(b)** What volume of H_2S at 27°C and 740 torr would be required to remove the SO_2 formed by burning 1.0 ton of coal containing 3.5% S by mass? **(c)** What mass of elemental sulfur is produced? Assume that all reactions are 100% efficient.

22.97 The maximum allowable concentration of $H_2S(g)$ in air is 20 mg per kilogram of air (20 ppm by mass). How many grams of FeS would be required to react with hydrochloric acid to produce this concentration at 1.00 atm and 25°C in an average room measuring 2.7 m × 4.3 m × 4.3 m? (Under these conditions, the average molar mass of air is 29.0 g/mol.)

22.98 The standard heats of formation of $H_2O(g)$, $H_2S(g)$, $H_2Se(g)$, and $H_2Te(g)$ are −241.8, −20.17, +29.7, and +99.6 kJ/mol, respectively. The enthalpies necessary to convert the elements in their standard states to 1 mol of gaseous atoms are 248, 277, 227, and 197 kJ/mol of atoms for O, S, Se, and Te, respectively. The enthalpy for dissociation of H_2 is 436 kJ/mol. Calculate the average H—O, H—S, H—Se, and H—Te bond enthalpies, and comment on their trend.

22.99 When bromine is extracted from seawater, the seawater is first brought to a pH of 3.5 and then treated with Cl_2. Assume that we start with 1.00×10^3 L of seawater whose density is 1.03 g/mL and whose pH is 7.0 and which contains 67 ppm Br^-. Calculate the number of grams of H_2SO_4 needed to adjust the pH and the number of grams of Cl_2 needed to give a 15% excess of the amount needed to react with the Br^- to obtain Br_2.

22.100 Hydrazine has been employed as a reducing agent for metals. Using standard reduction potentials, predict whether the following metals can be reduced to the metallic state by hydrazine under standard conditions in acidic solution: **(a)** Fe^{2+}; **(b)** Sn^{2+}; **(c)** Cu^{2+}; **(d)** Ag^+; **(e)** Cr^{3+}.

22.101 If the lunar lander on the Apollo moon missions used 4.0 tons of dimethylhydrazine, $(CH_3)_2NNH_2$, as fuel, how many tons of N_2O_4 oxidizer were required to react with it? (The reaction produces N_2, CO_2, and H_2O.)

22.102 Both dimethylhydrazine, $(CH_3)_2NNH_2$, and methylhydrazine, CH_3NHNH_2, have been used as rocket fuels. When dinitrogen tetroxide (N_2O_4) is used as the oxidizer, the products are H_2O, CO_2, and N_2. If the thrust of the rocket depends on the volume of the products produced, which of the substituted hydrazines produces a greater thrust per gram total mass of oxidizer plus fuel? [Assume that both fuels generate the same temperature and that $H_2O(g)$ is formed.]

22.103 Carbon forms an unusual, unstable oxide of formula C_3O_2, called carbon suboxide. Carbon suboxide is made by using P_2O_5 to dehydrate the dicarboxylic acid called malonic acid, which has the formula HOOC—CH_2—COOH. **(a)** Write a balanced reaction for the production of carbon suboxide from malonic acid. **(b)** Suggest a Lewis structure for C_3O_2. [*Hint:* The Lewis structure of malonic acid suggests which atoms are connected to which.] **(c)** By using the information in Table 8.3, predict the C—C and C—O bond lengths in C_3O_2. **(d)** Sketch the Lewis structure of a product that could result by the addition of 2 mol of H_2 to 1 mol of C_3O_2.

22.104 Boron nitride has a graphite-like structure with B—N bond distances of 1.45 Å within sheets and a separation of 3.30 Å between sheets. At high temperatures the BN assumes a diamond-like form that is harder than diamond. Rationalize the similarity between BN and elemental carbon.

Media Exercises

22.105 When a balloon filled with hydrogen is ignited, hydrogen and oxygen react explosively to form water, as shown in the **Formation of Water** movie (*eChapter 22.2*). **(a)** Is the formation of water an oxidation-reduction reaction? **(b)** If your answer to part (a) is yes, does hydrogen act as an oxidizing agent or as a reducing agent? **(c)** Neither air nor pure hydrogen is flammable, but mixtures of the two are flammable—even explosive. Explain how the explosion takes place with the hydrogen contained inside a balloon.

22.106 The **Physical Properties of the Halogens** movie (*eChapter 22.4*) shows three of the halogens and tells what the physical state of each is at room temperature. Given that all of the halogens are nonpolar, diatomic molecules, explain using intermolecular forces why the three halogens exist in three different states at room temperature.

22.107 **(a)** Watch the **Carbon Dioxide Behaves as an Acid in Water** movie (*eChapter 22.5*), and write the equation for the formation of carbonic acid from carbon dioxide and water. **(b)** Explain what is meant by the term *acid anhydride*. **(c)** What is the acid anhydride of sulfuric acid?

22.108 Watch the **Reactions with Oxygen** movie (*eChapter 22.5*), and compare the combustion of sulfur with that of white phosphorus. **(a)** Based on your observations, which reaction do you think has the higher activation energy? Explain. **(b)** For each of these two reactions, draw an energy profile like the one in Figure 14.15.

22.109 The **Nitrogen Dioxide and Dinitrogen Tetroxide** movie (*eChapter 22.7*) shows three sealed test tubes containing mixtures of NO_2 and N_2O_4. **(a)** Write an equation corresponding to the equilibrium between these two species. **(b)** Based on the movie, which species predominates at high temperature and which predominates at low temperature? **(c)** Does the equation you wrote for part (a) correspond to an endothermic reaction or an exothermic reaction? Explain your reasoning.

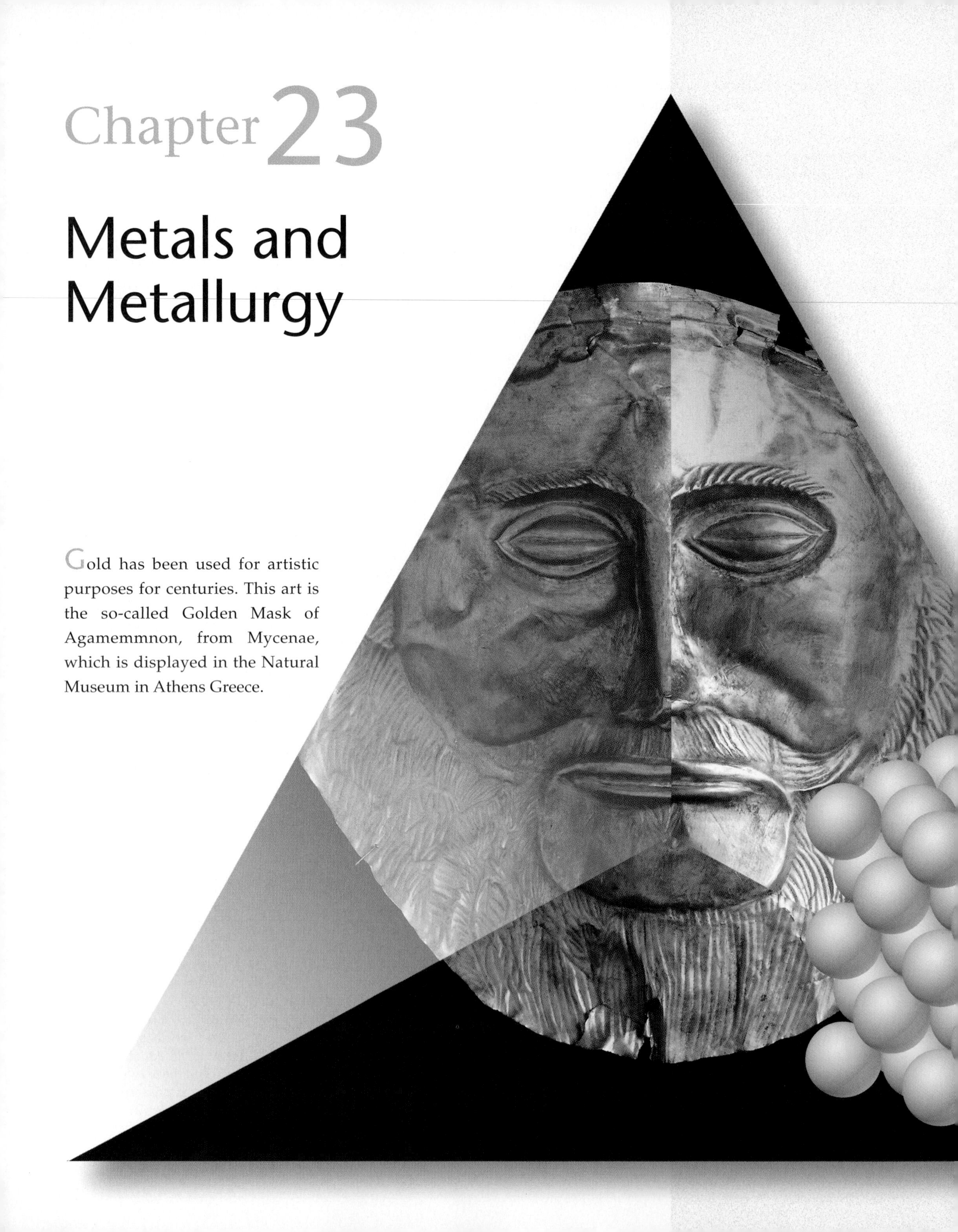

Chapter 23

Metals and Metallurgy

Gold has been used for artistic purposes for centuries. This art is the so-called Golden Mask of Agamemmnon, from Mycenae, which is displayed in the Natural Museum in Athens Greece.

IN CHAPTER 22 we examined the chemistry of nonmetallic elements. In this chapter we turn our attention to metals. Metals have played a major role in the development of civilization. Early history is often divided into the Stone Age, the Bronze Age, and the Iron Age, based on the compositions of the tools used in each era. Modern societies rely on a large variety of metals for making tools, machines, and other items. Chemists and other scientists have found uses for even the least abundant metals as they search for materials to meet evolving technological needs. To illustrate this point, Figure 23.1 ▶ shows the approximate composition of a high-performance jet engine. Notice that iron, long the dominant metal of technology, is not even present to a significant extent.

In this chapter we will consider the chemical forms in which metallic elements occur in nature and the means by which we obtain metals from these sources. We will also examine the bonding in solids and see how metals and mixtures of metals, called *alloys*, are employed in modern technology. Finally, we will look specifically at the properties of transition metals. As we will see, metals have a varied and interesting chemistry.

▶ What's Ahead ◀

- We begin by examining the occurrence of metals in the *lithosphere*, and by considering a few of their common *minerals*.
- *Metallurgy* is the technology of extracting metals from their natural sources and preparing them for use.
- *Pyrometallurgy*, metallurgy at high temperatures, is considered with particular attention to iron.
- *Hydrometallurgy*, the extraction of metals using aqueous solutions, is discussed with particular attention to aluminum.
- *Electrometallurgy*, which utilizes electrochemistry to reduce and refine metals, is examined with attention to sodium, aluminum, and copper.
- The physical properties of metals are reviewed and then accounted for in terms of two models for *metallic bonding*—the electron-sea model and the molecular-orbital model.
- Different kinds of *alloys*—solution alloys, heterogeneous alloys, and intermetallic compounds—are described.
- General characteristics of *transition metals* are discussed, examining their physical properties, electron configurations, oxidation states, and magnetic properties.
- Finally, the chemistry of a few transition metals (Cr, Fe, and Cu) is examined.

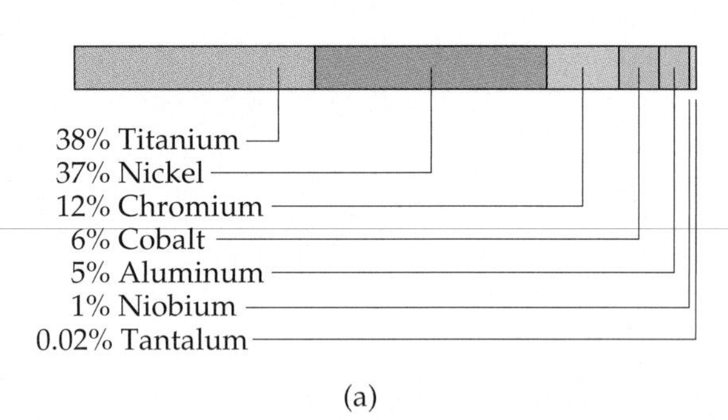

▲ **Figure 23.1** (a) Metallic elements employed in construction of a jet engine. (b) A modern jet engine.

23.1 Occurrence and Distribution of Metals

The portion of our environment that constitutes the solid earth beneath our feet is called the **lithosphere**. The lithosphere provides most of the materials we use to feed, clothe, shelter, and entertain ourselves. Although the bulk of Earth is solid, we have access to only a small region near the surface. Whereas the Earth's radius is 6370 km, the deepest mine extends only about 4 km into Earth.

Many of the metals that are most useful to us are not especially abundant in that portion of the lithosphere to which we have ready access. Consequently, the occurrence and distribution of *concentrated* deposits of these elements often play a role in international politics as nations compete for access to these materials. Deposits that contain metals in economically exploitable quantities are known as **ores**. Usually, the compounds or elements that we desire must be separated from a large quantity of unwanted material and then chemically processed to make them useful. About 2.3×10^4 kg (25 tons) of materials are extracted from the lithosphere and processed annually to support each person in our country. Because the richest sources of many substances are becoming exhausted, it may be necessary in the future to process larger volumes of lower-quality raw materials. Consequently, extraction of the compounds and elements we need could cost more in both energy and environmental impact.

Minerals

With the exception of gold and the platinum-group metals (Ru, Rh, Pd, Os, Ir, and Pt), most metallic elements are found in nature in solid inorganic compounds called **minerals**. Table 23.1 ▶ lists the principal mineral sources of several common metals, three of which are shown in Figure 23.2 ▶. Notice that minerals are identified by common names rather than by chemical names. Names of minerals are usually based on the locations where they were discovered, the person who discovered them, or some characteristic such as color. The name *malachite*, for example, comes from the Greek word *malache*, the name of a type of tree whose leaves are the color of the mineral.

Commercially, the most important sources of metals are oxide, sulfide, and carbonate minerals. Silicate minerals (Section 22.10) are very abundant, but they are generally difficult to concentrate and reduce. Therefore, most silicates are not economical sources of metals.

▲ **Figure 23.2** Three common minerals: (a) chalcopyrite, (b) rutile (in a matrix of quartz), and (c) cinnabar.

TABLE 23.1 Principal Mineral Sources of Some Common Metals

Metal	Mineral	Composition
Aluminum	Bauxite	Al_2O_3
Chromium	Chromite	$FeCr_2O_4$
Copper	Chalcocite	Cu_2S
	Chalcopyrite	$CuFeS_2$
	Malachite	$Cu_2CO_3(OH)_2$
Iron	Hematite	Fe_2O_3
	Magnetite	Fe_3O_4
Lead	Galena	PbS
Manganese	Pyrolusite	MnO_2
Mercury	Cinnabar	HgS
Molybdenum	Molybdenite	MoS_2
Tin	Cassiterite	SnO_2
Titanium	Rutile	TiO_2
	Ilmenite	$FeTiO_3$
Zinc	Sphalerite	ZnS

ACTIVITY
Mineral Sources of Mineral Ores

Metallurgy

Metallurgy is the science and technology of extracting metals from their natural sources and preparing them for practical use. It usually involves several steps: (1) mining, (2) concentrating the ore or otherwise preparing it for further treatment, (3) reducing the ore to obtain the free metal, (4) refining or purifying the metal, and (5) mixing the metal with other elements to modify its properties. This last process produces an *alloy*, a metallic material that is composed of two or more elements (Section 23.6).

After being mined, an ore is usually crushed and ground and then treated to concentrate the desired metal. The concentration stage relies on differences in the properties of the mineral and the undesired material that accompanies it, which is called *gangue* (pronounced "gang"). Crude iron ore, for example, is enriched and formed into pellets (Figure 23.3 ▶).

▲ **Figure 23.3** At this open pit iron ore mine in Michigan's Upper Peninsula, crude ore is concentrated and formed into pellets suitable for shipment.

After an ore is concentrated, a variety of chemical processes are used to obtain the metal in suitable purity. In Sections 23.2–23.4 we will examine some of the most common metallurgical processes. You will see that these techniques depend on many of the basic concepts that were discussed earlier in the text.

23.2 Pyrometallurgy

A large number of metallurgical processes utilize high temperatures to alter the mineral chemically and to ultimately reduce it to the free metal. The use of heat to alter or reduce the mineral is called **pyrometallurgy**. (*Pyro* means "at high temperature.")

Calcination is the heating of an ore to bring about its decomposition and the elimination of a volatile product. The volatile product could be, for example, CO_2 or H_2O. Carbonates are often calcined to drive off CO_2, forming the metal oxide. For example,

$$PbCO_3(s) \xrightarrow{\Delta} PbO(s) + CO_2(g) \qquad [23.1]$$

Most carbonates decompose reasonably rapidly at temperatures in the range of 400°C to 500°C, although $CaCO_3$ requires a temperature of about 1000°C. Most hydrated minerals lose H_2O at temperatures on the order of 100°C to 300°C.

Roasting is a thermal treatment that causes chemical reactions between the ore and the furnace atmosphere. Roasting may lead to oxidation or reduction and may be accompanied by calcination. An important roasting process is the oxidation of sulfide ores, in which the metal is converted to the oxide, as in the following examples:

$$2ZnS(s) + 3O_2(g) \longrightarrow 2ZnO(s) + 2SO_2(g) \qquad [23.2]$$

$$2MoS_2(s) + 7O_2(g) \longrightarrow 2MoO_3(s) + 4SO_2(g) \qquad [23.3]$$

The sulfide ore of a less active metal, such as mercury, can be roasted to the free metal:

$$HgS(s) + O_2(g) \longrightarrow Hg(g) + SO_2(g) \qquad [23.4]$$

In many instances the free metal also can be obtained by using a reducing atmosphere during roasting. Carbon monoxide provides such an atmosphere, and it is frequently used to reduce metal oxides:

$$PbO(s) + CO(g) \longrightarrow Pb(l) + CO_2(g) \qquad [23.5]$$

This method of reduction is not always feasible, however, especially with active metals, which are difficult to reduce.

Smelting is a melting process in which the materials formed in the course of chemical reactions separate into two or more layers. Smelting often involves a roasting stage in the same furnace. Two of the important types of layers formed in smelters are molten metal and slag. The molten metal may consist almost entirely of a single metal, or it may be a solution of two or more metals.

Slag consists mainly of molten silicate minerals, with aluminates, phosphates, and other ionic compounds as constituents. A slag is formed when a basic metal oxide such as CaO reacts at high temperatures with molten silica (SiO_2):

$$CaO(l) + SiO_2(l) \longrightarrow CaSiO_3(l) \qquad [23.6]$$

Pyrometallurgical operations may involve not only the concentration and reduction of a mineral, but also the refining of the metal. **Refining** is the treatment of a crude, relatively impure metal product from a metallurgical process to improve its purity and to define its composition better. Sometimes the goal of the refining process is to obtain the metal itself in pure form. The goal may also be to produce a mixture with a well-defined composition, however, as in the production of steels from crude iron.

The Pyrometallurgy of Iron

The most important pyrometallurgical operation is the reduction of iron. Iron occurs in many different minerals, but the most important sources are two iron oxide minerals—hematite (Fe_2O_3) and magnetite (Fe_3O_4). As the higher-grade deposits of these minerals have become depleted, lower-grade ores have been tapped. *Taconite*, which consists of fine-grained silica with variable ratios of hematite and magnetite, has increased in importance as a source of iron from the great Mesabi Range lying west of Lake Superior.

The reduction of iron oxides can be accomplished in a *blast furnace* such as the one illustrated in Figure 23.4 ▼. A blast furnace is essentially a huge chemical reactor capable of continuous operation. The largest furnaces are over 60 m high and 14 m wide. When operating at full capacity, they produce up to 10,000 tons of iron per day.

The blast furnace is charged at the top with a mixture of iron ore, coke, and limestone. Coke is coal that has been heated in the absence of air to drive off volatile components. It is about 85 to 90% carbon. Coke serves as the fuel, producing heat as it is burned in the lower part of the furnace. It is also the source of the reducing gases CO and H_2. Limestone ($CaCO_3$) serves as the source of the basic oxide CaO, which reacts with silicates and other components of the ore to form slag. Air, which enters the blast furnace at the bottom after preheating, is also an important raw material; it is required for combustion of the coke. Production of 1 kg of crude iron, called *pig iron*, requires about 2 kg of ore, 1 kg of coke, 0.3 kg of limestone, and 1.5 kg of air.

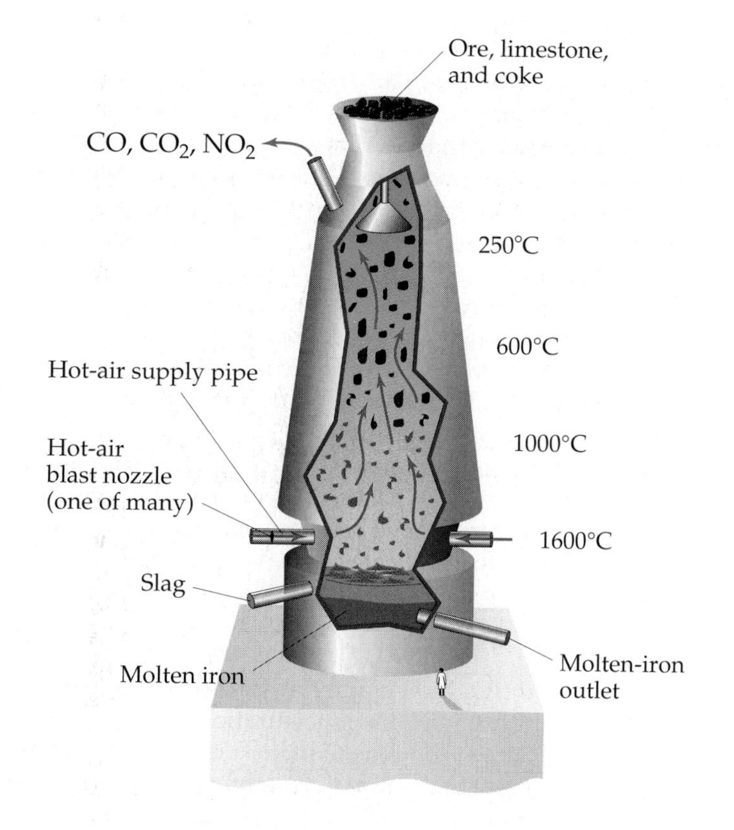

◀ **Figure 23.4** A blast furnace used for reduction of iron ore. Notice the increasing temperatures as the materials pass downward through the furnace.

In the furnace oxygen reacts with the carbon in the coke to form carbon monoxide:

$$2C(s) + O_2(g) \longrightarrow 2CO(g) \qquad \Delta H = -221 \text{ kJ} \qquad [23.7]$$

Water vapor present in the air also reacts with carbon, forming both carbon monoxide and hydrogen:

$$C(s) + H_2O(g) \longrightarrow CO(g) + H_2(g) \qquad \Delta H = +131 \text{ kJ} \qquad [23.8]$$

The reaction of coke with oxygen is exothermic and provides heat for furnace operation, whereas its reaction with water vapor is endothermic. Addition of water vapor to the air thus provides a means of controlling furnace temperature.

In the upper part of the furnace, limestone decomposes to form CaO and CO_2. Here, also, the iron oxides are reduced by CO and H_2. For example, the important reactions for Fe_3O_4 are

$$Fe_3O_4(s) + 4CO(g) \longrightarrow 3Fe(s) + 4CO_2(g) \qquad \Delta H = -15 \text{ kJ} \qquad [23.9]$$

$$Fe_3O_4(s) + 4H_2(g) \longrightarrow 3Fe(s) + 4H_2O(g) \qquad \Delta H = +150 \text{ kJ} \qquad [23.10]$$

Reduction of other elements present in the ore also occurs in the hottest parts of the furnace, where carbon is the major reducing agent.

Molten iron collects at the base of the furnace, as shown in Figure 23.4. It is overlaid with a layer of molten slag formed by the reaction of CaO with the silica present in the ore (Equation 23.6). The layer of slag over the molten iron helps to protect it from reaction with the incoming air. Periodically the furnace is tapped to drain off slag and molten iron. The iron produced in the furnace may be cast into solid ingots. Most, however, is used directly in the manufacture of steel. For this purpose, it is transported, while still liquid, to the steelmaking shop (Figure 23.5 ◀). The production of pig iron using blast furnaces has decreased in recent years because of alternative reduction processes and the increased use of iron scrap in steelmaking. Blast furnaces nevertheless remain a significant means of reducing iron oxides.

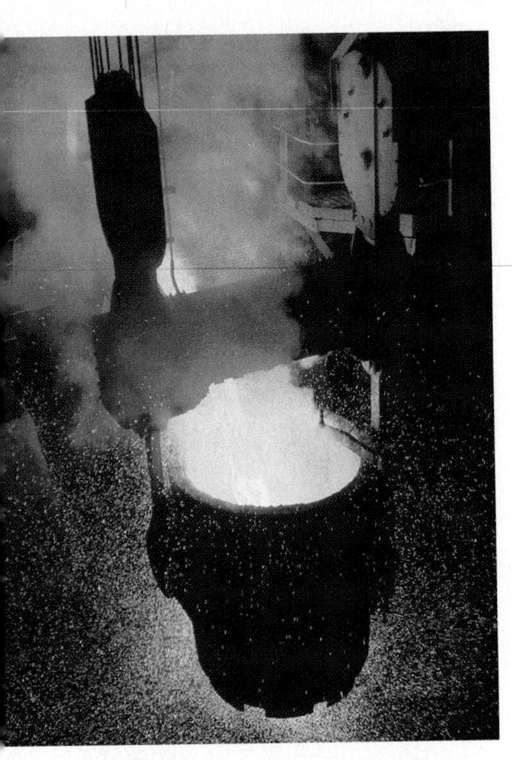

▲ **Figure 23.5** Molten iron being poured for transport to a basic oxygen converter. Steelmakers convert iron to steel by adding scrap steel and other metals as alloying agents.

Formation of Steel

Steel is an alloy of iron. The production of iron from its ore is a chemical reduction process that results in a crude iron containing many undesired impurities. Iron from a blast furnace typically contains 0.6 to 1.2% silicon, 0.4 to 2.0% manganese, and lesser amounts of phosphorus and sulfur. In addition, there is considerable dissolved carbon. In the production of steel these impurity elements are removed by oxidation in a vessel called a *converter*. In modern steelmaking, the oxidizing agent is either pure O_2 or O_2 diluted with argon. Air cannot be used directly as the source of O_2 because N_2 reacts with the molten iron to form iron nitride, which causes the steel to become brittle.

A cross-sectional view of one converter design appears in Figure 23.6 ▶. In this converter O_2, diluted with argon, is blown directly into the molten metal. The oxygen reacts exothermically with carbon, silicon, and many metal impurities, reducing the concentrations of these elements in the iron. Carbon and sulfur are expelled as CO and SO_2 gases, respectively. Silicon is oxidized to SiO_2 and adds to whatever slag may have been present initially in the melt. Metal oxides react with the SiO_2 to form silicates. The presence of a basic slag is also important for removal of phosphorus:

$$3CaO(l) + P_2O_5(l) \longrightarrow Ca_3(PO_4)_2(l) \qquad [23.11]$$

Nearly all of the O_2 blown into the converter is consumed in the oxidation reactions. By monitoring the O_2 concentration in the gas coming from the converter, it is possible to tell when the oxidation is essentially complete. Oxidation of the impurities present in the iron normally requires about 20 min. When the desired

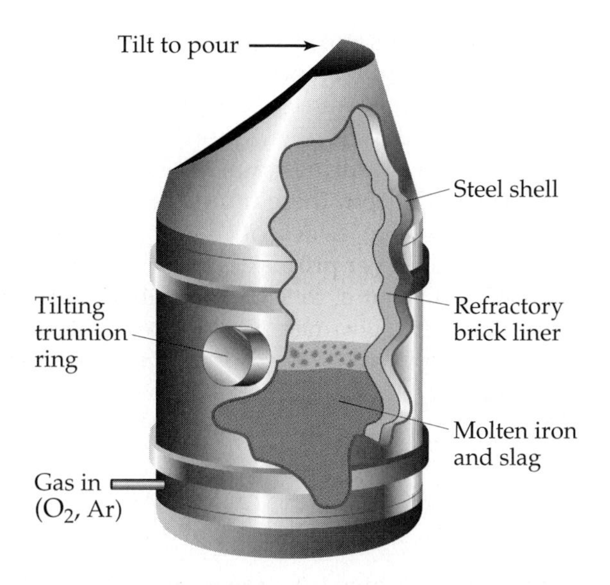

◀ **Figure 23.6** A converter for refining of iron. A mixture of oxygen and argon is blown through the molten iron and slag. The heat generated by the oxidation of impurities maintains the mixture in a molten state. When the desired composition is attained, the converter is tilted to pour out its contents.

composition is attained, the contents of the converter are dumped into a large ladle. To produce steels with various kinds of properties, alloying elements are added as the ladle is being filled. The still-molten mixture is then poured into molds, where it solidifies.

23.3 Hydrometallurgy

Pyrometallurgical operations require large quantities of energy and are often a source of atmospheric pollution, especially by sulfur dioxide. For some metals, other techniques have been developed in which the metal is extracted from its ore by use of aqueous reactions. These processes are called **hydrometallurgy** (*hydro* means "water").

The most important hydrometallurgical process is **leaching**, in which the desired metal-containing compound is selectively dissolved. If the compound is water soluble, water by itself is a suitable leaching agent. More commonly the agent is an aqueous solution of an acid, a base, or a salt. Often the dissolving process involves formation of a complex ion. (Section 17.5) As an example, we can consider the leaching of gold.

As noted in the "Closer Look" box in Section 4.4, gold metal is often found relatively pure in nature. As concentrated deposits of elemental gold have been depleted, lower-grade sources have become more important. Gold from low-grade ores can be concentrated by placing the crushed ore on large concrete slabs and spraying a solution of NaCN over it. In the presence of CN^- and air, the gold is oxidized, forming the stable $Au(CN)_2^-$ ion, which is soluble in water:

$$4Au(s) + 8CN^-(aq) + O_2(g) + 2H_2O(l) \longrightarrow 4Au(CN)_2^-(aq) + 4OH^-(aq) \quad [23.12]$$

After a metal ion is selectively leached from an ore, it is precipitated from solution as the free metal or as an insoluble ionic compound. Gold, for example, is obtained from its cyanide complex by reduction with zinc powder:

$$2Au(CN)_2^-(aq) + Zn(s) \longrightarrow Zn(CN)_4^{2-}(aq) + 2Au(s) \quad [23.13]$$

The Hydrometallurgy of Aluminum

Among metals, aluminum is second only to iron in commercial use. World production of aluminum is about 1.5×10^{10} kg (16 million tons) per year. The most useful ore of aluminum is *bauxite*, in which Al is present as hydrated oxides, $Al_2O_3 \cdot xH_2O$. The value of x varies, depending on the particular mineral present. Because bauxite deposits in the United States are limited, most of the ore used in the production of aluminum must be imported.

The major impurities found in bauxite are SiO_2 and Fe_2O_3. It is essential to separate Al_2O_3 from these impurities before the metal is recovered by electrochemical reduction, as described in Section 23.4. The process used to purify bauxite, called the **Bayer process**, is a hydrometallurgical procedure. The ore is first crushed and ground, then digested in a concentrated aqueous NaOH solution, about 30% NaOH by mass, at a temperature in the range of 150°C to 230°C. Sufficient pressure, up to 30 atm, is maintained to prevent boiling. The Al_2O_3 dissolves in this solution, forming the complex aluminate ion, $Al(OH)_4^-$:

$$Al_2O_3 \cdot H_2O(s) + 2H_2O(l) + 2OH^-(aq) \longrightarrow 2Al(OH)_4^-(aq) \qquad [23.14]$$

The iron(III) oxides do not dissolve in the strongly basic solution. This difference in the behavior of the aluminum and iron compounds arises because Al^{3+} is amphoteric, whereas Fe^{3+} is not. (Section 17.5) Thus, the aluminate solution can be separated from the iron-containing solids by filtration. The pH of the solution is then lowered, causing the aluminum hydroxide to precipitate.

After the aluminum hydroxide precipitate has been filtered, it is calcined in preparation for electroreduction to the metal. The solution recovered from the filtration is reconcentrated so that it can be used again. This task is accomplished by heating to evaporate water from the solution, a procedure that requires much energy and is the most costly part of the Bayer process.

23.4 Electrometallurgy

Many processes that are used to reduce metal ores or refine metals are based on electrolysis. (Section 20.9) Collectively these processes are referred to as **electrometallurgy**. Electrometallurgical procedures can be broadly differentiated according to whether they involve electrolysis of a molten salt or of an aqueous solution.

Electrolytic methods are important for obtaining the more active metals, such as sodium, magnesium, and aluminum. These metals cannot be obtained from aqueous solution because water is more easily reduced than the metal ions. The standard reduction potentials of water under both acidic and basic conditions are more positive than those of $Na^+ (E^\circ_{red} = -2.71\ V)$, $Mg^{2+} (E^\circ_{red} = -2.37\ V)$, and $Al^{3+} (E^\circ_{red} = -1.66\ V)$:

$$2H^+(aq) + 2e^- \longrightarrow H_2(g) \qquad E^\circ_{red} = 0.00\ V \qquad [23.15]$$

$$2H_2O(l) + 2e^- \longrightarrow H_2(g) + 2OH^-(aq) \qquad E^\circ_{red} = -0.83\ V \qquad [23.16]$$

To form such metals by electrochemical reduction, therefore, we must employ a molten-salt medium in which the metal ion of interest is the most readily reduced species.

Electrometallurgy of Sodium

In the commercial preparation of sodium, molten NaCl is electrolyzed in a specially designed cell called the **Downs cell**, illustrated in Figure 23.7 ▶. Calcium chloride ($CaCl_2$) is added to lower the melting point of the molten NaCl from the normal melting point of 804°C to around 600°C. The $Na(l)$ and $Cl_2(g)$ produced in the electrolysis are kept from coming in contact and reforming NaCl. In addition, the Na must be prevented from contact with oxygen because the metal would quickly oxidize under the high-temperature conditions of the cell reaction.

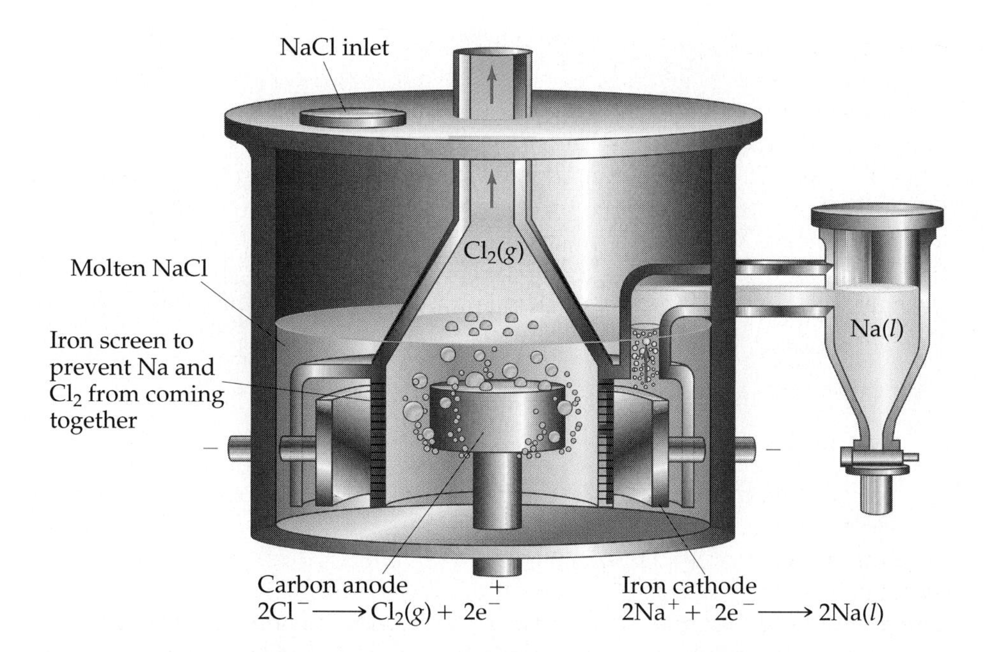

◀ **Figure 23.7** The Downs cell in the commercial production of so

Electrometallurgy of Aluminum

In Section 23.3 we discussed the Bayer process, in which bauxite is concentrated to produce aluminum hydroxide. When this concentrate is calcined at temperatures in excess of 1000°C, anhydrous aluminum oxide (Al_2O_3) is formed. Anhydrous aluminum oxide melts at over 2000°C. This is too high to permit its use as a molten medium for electrolytic formation of free aluminum. The electrolytic process commercially used to produce aluminum is known as the **Hall process**, named after its inventor, Charles M. Hall (see the "Closer Look" box in this section). The purified Al_2O_3 is dissolved in molten cryolite (Na_3AlF_6) which has a melting point of 1012°C and is an effective conductor of electric current. A schematic diagram of the electrolysis cell is shown in Figure 23.8 ▼. Graphite rods are employed as anodes and are consumed in the electrolysis process. The electrode reactions are as follows:

$$\text{Anode:} \quad C(s) + 2O^{2-}(l) \longrightarrow CO_2(g) + 4e^- \qquad [23.17]$$

$$\text{Cathode:} \quad 3e^- + Al^{3+}(l) \longrightarrow Al(l) \qquad [23.18]$$

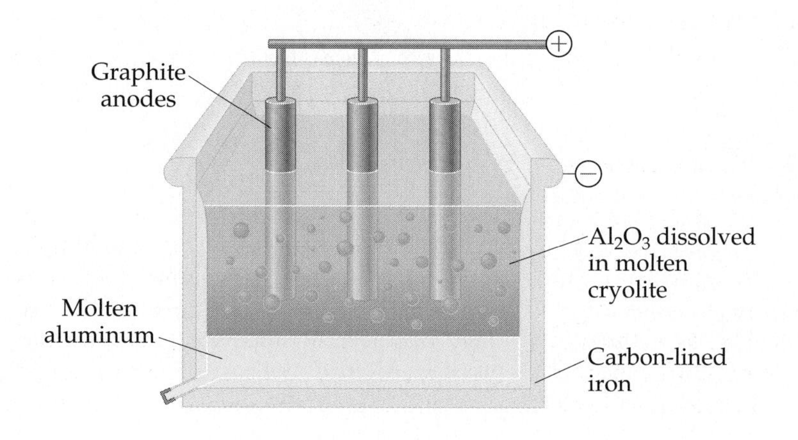

◀ **Figure 23.8** Typical Hall process electrolysis cell used to form aluminum metal through reduction. Because molten aluminum is denser than the mixture of cryolite (Na_3AlF_6) and Al_2O_3, the metal collects at the bottom of the cell.

3.9 The quantities of olite, graphite, and energy produce 1000 kg of .

4000 kg bauxite (~50% Al_2O_3)

1900 kg Al_2O_3

70 kg cryolite
450 kg C anodes

Electrolytic cell

56×10^9 J energy
(4.5 V, 10^5 A)

1000 kg Al

ACTIVITY
Electrolysis

The amounts of raw materials and energy required to produce 1000 kg of aluminum metal from bauxite by this procedure are summarized in Figure 23.9 ▲.

Electrorefining of Copper

Copper is widely used to make electrical wiring and in other applications that utilize its high electrical conductivity. Crude copper, which is usually obtained by pyrometallurgical methods, is not suitable to serve in electrical applications because impurities greatly reduce the metal's conductivity.

Purification of copper is achieved by electrolysis, as illustrated in Figure 23.11 ▶. Large slabs of crude copper serve as the anodes in the cell, and thin sheets of pure copper serve as the cathodes. The electrolyte consists of an acidic solution of $CuSO_4$. Application of a suitable voltage to the electrodes causes oxidation of copper metal at the anode and reduction of Cu^{2+} to form copper metal at the cathode. This strategy can be used because copper is both oxidized and

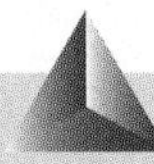

A Closer Look Charles M. Hall

Charles M. Hall (Figure 23.10 ▶) began to work on the problem of reducing aluminum in about 1885 after he had learned from a professor of the difficulty of reducing ores of very active metals. Prior to the development of his electrolytic process, aluminum was obtained by a chemical reduction using sodium or potassium as the reducing agent. Because the procedure was very costly, aluminum metal was very expensive. As late as 1852, the cost of aluminum was \$545 per pound, far greater than the cost of gold. During the Paris Exposition in 1855 aluminum was exhibited as a rare metal, even though it is the third most abundant element in Earth's crust.

Hall, who was 21 years old when he began his research, utilized handmade and borrowed equipment in his studies, and used a woodshed near his home as his laboratory. In about a year's time he was able to solve the problem of reducing aluminum. His procedure consisted of finding an ionic compound that could be melted to form a conducting medium that would dissolve Al_2O_3, but would not interfere in the electrolysis reactions. The relatively rare mineral cryolite (Na_3AlF_6), found in Greenland, met these criteria. Ironically, Paul Héroult, who was the same age as Hall, made the same discovery in France at about the same time. As a result of the research of Hall and Héroult, large-scale production of aluminum became commercially feasible, and aluminum became a common and familiar metal.

◀ **Figure 23.10** Charles M. Hall (1863–1914) as a young man.

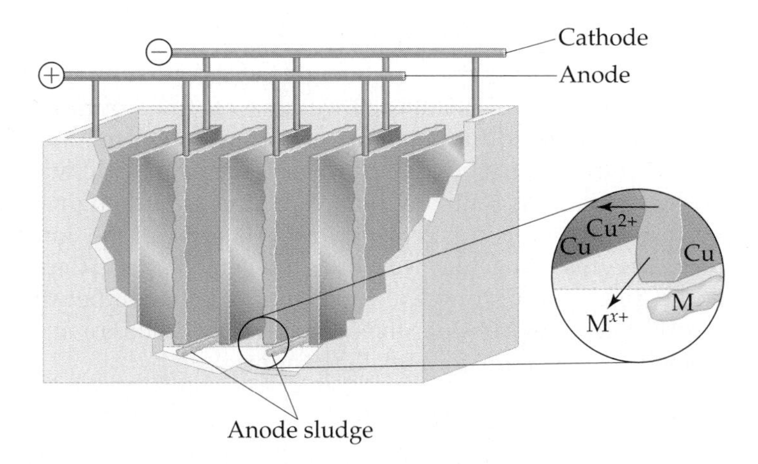

◀ **Figure 23.11** Electrolysis cell for refining of copper. As the anodes dissolve away, the cathodes on which the pure metal is deposited grow in size.

reduced more readily than water. The relative ease of reduction of Cu^{2+} and H_2O is seen by comparing their standard reduction potentials:

$$Cu^{2+}(aq) + 2e^- \longrightarrow Cu(s) \qquad E^\circ_{red} = +0.34\ V \qquad [23.19]$$

$$2H_2O(l) + 2e^- \longrightarrow H_2(g) + 2OH^-(aq) \qquad E^\circ_{red} = -0.83\ V \qquad [23.20]$$

The impurities in the copper anode include lead, zinc, nickel, arsenic, selenium, tellurium, and several precious metals including gold and silver. Metallic impurities that are more active than copper are readily oxidized at the anode, but do not plate out at the cathode because their reduction potentials are more negative than that for Cu^{2+}. Less active metals, however, are not oxidized at the anode. Instead they collect below the anode as a sludge that is collected and processed to recover the valuable metals. The anode sludges from copper-refining cells provide one fourth of U.S. silver production and about one eighth of U.S. gold production.

SAMPLE EXERCISE 23.1

Nickel is one of the chief impurities in the crude copper that is subjected to electrorefining. What happens to this nickel in the course of the electrolytic process?

Solution

Analyze: We are asked to predict whether nickel can be oxidized at the anode and reduced at the cathode during the electrorefining of copper.

Plan: We need to compare the standard reduction potentials of Ni^{2+} and Cu^{2+}. The more negative the reduction potential, the less readily the ion is reduced but the more readily the metal itself is oxidized. (Section 20.4)

Solve: The standard reduction potential for Ni^{2+} is more negative than that for Cu^{2+}:

$$Ni^{2+}(aq) + 2e^- \longrightarrow Ni(s) \qquad E^\circ_{red} = -0.28\ V$$

$$Cu^{2+}(aq) + 2e^- \longrightarrow Cu(s) \qquad E^\circ_{red} = +0.34\ V$$

As a result, nickel is more readily oxidized than copper, assuming standard conditions. Although we do not have standard conditions in the electrolytic cell, we nevertheless expect that nickel is preferentially oxidized at the anode. Because the reduction of Ni^{2+} occurs less readily than the reduction of Cu^{2+}, the Ni^{2+} accumulates in the electrolyte solution, while the Cu^{2+} is reduced at the cathode. After a time it is necessary to recycle the electrolyte solution to remove the accumulated metal ion impurities, such as Ni^{2+}.

PRACTICE EXERCISE

Zinc is another common impurity in copper. Using standard reduction potentials, determine whether zinc will accumulate in the anode sludge or in the electrolytic solution during the electrorefining of copper.

Answer: It is found in the electrolytic solution because the standard reduction potential of Zn^{2+} is more negative than that of Cu^{2+}.

23.5 Metallic Bonding

In our discussion of metallurgy we have confined ourselves to discussing the methods employed for obtaining metals in pure form. Metallurgy is also concerned with understanding the properties of metals and with developing useful new materials. As with any branch of science and engineering, our ability to make advances is coupled to our understanding of the fundamental properties of the systems with which we work. At several places in the text we have referred to the differences in physical and chemical behavior between metals and nonmetals. Let's now consider the distinctive properties of metals and relate these properties to a model for metallic bonding.

Physical Properties of Metals

You have probably held a length of copper wire or an iron bolt. Perhaps you have even seen the surface of a freshly cut piece of sodium metal. These substances, although distinct from one another, share certain similarities that enable us to classify them as metallic. A clean metal surface has a characteristic luster. In addition metals that we can handle with bare hands have a characteristic cold feeling related to their high heat conductivity. Metals also have high electrical conductivities; electrical current flows easily through them. Current flow occurs without any displacement of atoms within the metal structure and is due to the flow of electrons within the metal. The heat conductivity of a metal usually parallels its electrical conductivity. Silver and copper, for example, which possess the highest electrical conductivities among the elements, also possess the highest heat conductivities. This observation suggests that the two types of conductivity have the same origin in metals, which we will soon discuss.

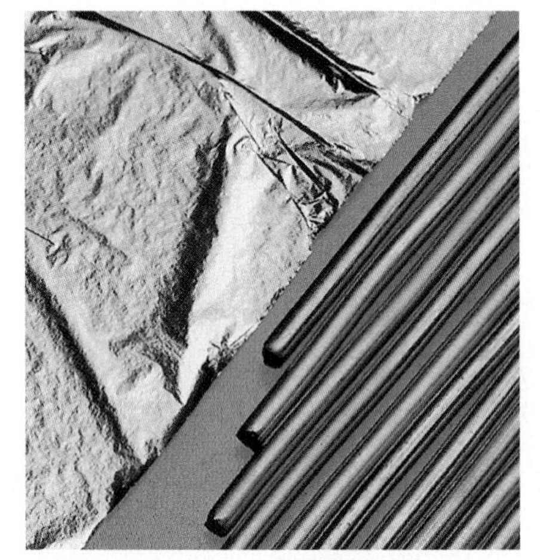

▲ **Figure 23.12** Gold leaf (left) and copper wire (right) demonstrate the characteristic malleability and ductility of metals, respectively.

Most metals are *malleable*, which means that they can be hammered into thin sheets, and *ductile*, which means that they can be drawn into wires (Figure 23.12 ◀). These properties indicate that the atoms are capable of slipping with respect to one another. Ionic solids or crystals of most covalent compounds do not exhibit such behavior. These types of solids are typically brittle and fracture easily. Consider, for example, the difference between dropping an ice cube and a block of aluminum metal onto a concrete floor.

Most metals form solid structures in which the atoms are arranged as close-packed spheres. Copper, for example, possesses a cubic close-packed structure in which each copper atom is in contact with 12 other copper atoms. ∞ (Section 11.7) The number of valence-shell electrons available for bond formation is insufficient for a copper atom to form an electron-pair bond to each of its neighbors. If each atom is to share its bonding electrons with all its neighbors, these electrons must be able to move from one bonding region to another.

Electron-Sea Model for Metallic Bonding

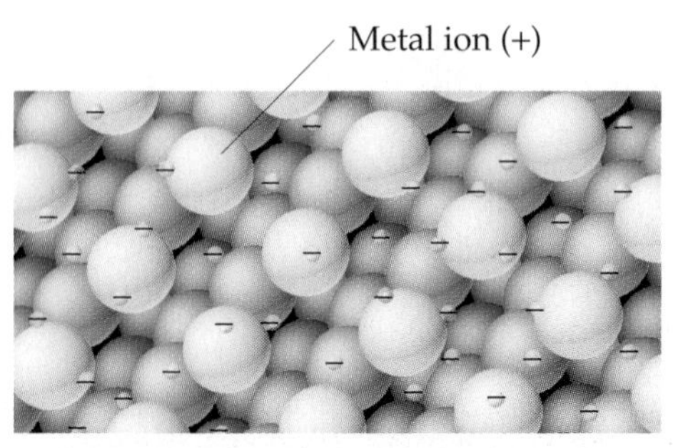

▲ **Figure 23.13** Schematic illustration of the electron-sea model of the electronic structure of metals. Each sphere is a positively charged metal ion.

One very simple model that accounts for some of the most important characteristics of metals is the *electron-sea model*. In this model the metal is pictured as an array of metal cations in a "sea" of valence electrons, as illustrated in Figure 23.13 ◀. The electrons are confined to the metal by electrostatic attractions to the cations, and they are uniformly distributed throughout the structure. The electrons are mobile, however, and no individual electron is confined to any particular metal ion. When a metal wire is connected to the terminals of a battery, electrons flow through the metal toward the positive terminal and into the metal from the battery at the negative terminal. The high heat conductivity of metals is also accounted for by the mobility of the electrons, which permits ready transfer of kinetic energy throughout the solid. The ability of metals to deform (their malleability and ductility) can be explained by the fact that metal atoms form bonds to many neighbors. Changes in the positions of the atoms brought about in reshaping the metal are partly accommodated by a redistribution of electrons.

TABLE 23.2 Melting Points of Selected Transition Metals

	Group 3B	Group 6B	Group 8B
Metal	Sc	Cr	Ni
Melting point (°C)	1541	1857	1455
Metal	Y	Mo	Pd
Melting point (°C)	1522	2617	1554
Metal	La	W	Pt
Melting point (°C)	918	3410	1772

The electron-sea model, however, does not adequately explain all properties. According to the model, for example, the strength of bonding between metal atoms should increase as the number of valence electrons increases, resulting in a corresponding increase in melting points. The group 6B metals (Cr, Mo, W), however, which are in the center of the transition metals, have the highest melting points in their respective periods. The melting points on either side of the center are lower (Table 23.2 ▲), which implies that the strength of metallic bonding first increases with increasing number of electrons and then decreases. Similar trends are seen in other physical properties of the metals, such as the heat of fusion, hardness, and boiling point.

In order to explain some of the physical properties of metals, we need a more refined model than the electron-sea model to describe metallic bonding. We obtain a better model by applying the concepts of molecular-orbital theory to metals. (Sections 9.7 and 9.8)

Molecular-Orbital Model for Metals

In considering the structures of molecules such as benzene, we saw that in some cases electrons are delocalized, or distributed, over several atoms. (Section 11.8) The bonding in metals can be thought of in a similar way. The valence atomic orbitals on one metal atom overlap with those on several nearest neighbors, which in turn overlap with atomic orbitals on still other atoms.

We saw in Section 9.7 that the overlap of atomic orbitals leads to the formation of molecular orbitals. The number of molecular orbitals equals the number of atomic orbitals that overlap. In a metal the number of atomic orbitals that interact or overlap is very large. Thus, the number of molecular orbitals is also very large. Figure 23.14 ▼ shows schematically what happens as increasing numbers of metal atoms come together to form molecular orbitals. As overlap of atomic orbitals occurs, bonding and antibonding molecular-orbital combinations are formed. The energies of these molecular orbitals lie at closely spaced intervals in the energy range between the highest- and lowest-energy orbitals. Consequently, interaction of all the valence atomic orbitals of each metal atom with the orbitals of adjacent metal atoms gives rise to a huge number of molecular orbitals that extend over the entire metal structure. The energy separations between these metal orbitals are so tiny that for all practical purposes we may think of the orbitals as forming a continuous *band* of allowed energy states, referred to as an *energy band*, as shown in Figure 23.14.

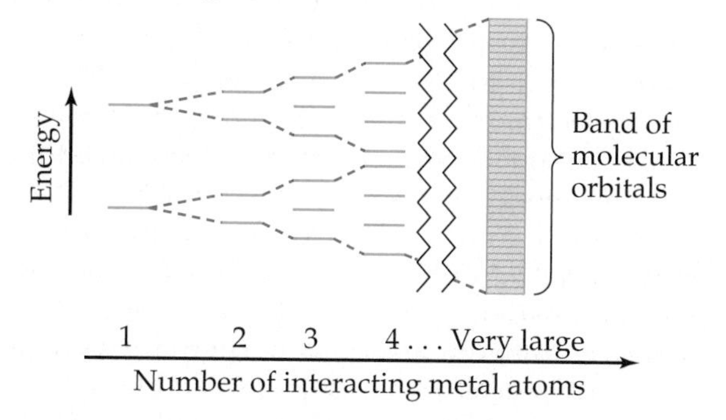

◀ **Figure 23.14** Schematic illustration of how the number of molecular orbitals increases and their energy spacing decreases as the number of interacting atoms increases. In metals these interactions form a nearly continuous *band* of molecular orbitals that are delocalized throughout the metal lattice. The number of electrons available does not completely fill these orbitals.

A Closer Look Insulators and Semiconductors

A solid exhibits metallic character because it has a partially filled energy band, as shown in Figure 23.15 ▼(a); there are more molecular orbitals in the band than are needed to accommodate all the bonding electrons of the structure. Thus, an excited electron may easily move to the nearby higher orbital. In some solids, however, such as diamond, the electrons completely fill the allowed levels in a band. When we apply molecular-orbital theory to diamond, therefore, we find that the bands of allowed energies are as shown in Figure 23.15(b). The carbon 2*s* and 2*p* atomic orbitals combine to form two energy bands, each of which accommodates four electrons per carbon atom. One of these is completely filled with electrons. The other is completely empty. There is a large energy gap between the two bands. Because there is no readily available vacant orbital into which the highest-energy electrons can move under the influence of an applied electrical potential, diamond is a poor electrical conductor. Solids in which the energy bands are either completely filled or completely empty are electrical *insulators*.

Silicon and germanium have electronic structures like diamond. However, the energy gap between the filled and empty bands decreases as we move from carbon (diamond) to silicon to germanium, as seen in Table 23.3 ▼. For silicon and germanium, the energy gap is small enough that at ordinary temperatures a few electrons have sufficient energy to jump from the filled band (called the *valence band*) to the empty band (called the *conduction band*). As a result, there are some empty orbitals in the valence band, permitting electrical conductivity. The electrons in the upper energy band also serve as carriers of electrical current. Therefore, silicon and germanium are *semiconductors*, solids with electrical conductivities between those of metals and those of insulators. Other substances, such as gallium arsenide (GaAs), also behave as semiconductors.

The electrical conductivity of a semiconductor or insulator can be modified by adding small amounts of other substances. This process, called *doping*, causes the solid to have either too few or too many electrons to fill the valence band. Consider what happens to silicon when a small amount of phos-

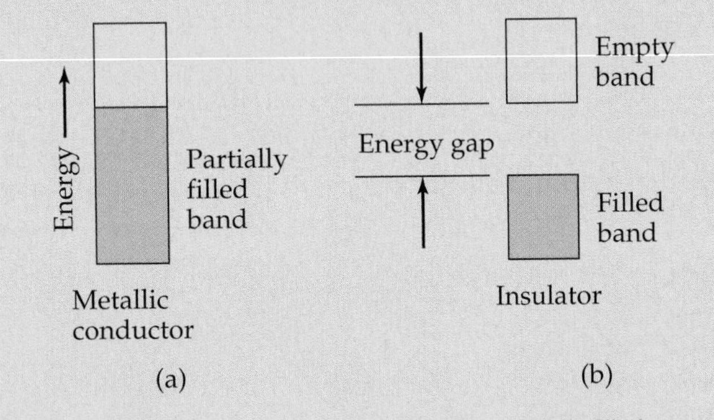

▲ **Figure 23.15** Metallic conductors have partially filled energy bands, as shown in (a). Insulators have filled and empty energy bands, as in (b).

TABLE 23.3 Energy Gaps in Diamond, Silicon, and Germanium

Element	Energy Gap (kJ/mol)
C	502
Si	100
Ge	67

The electrons available for metallic bonding do not completely fill the available molecular orbitals; we can think of the energy band as a partially filled container for electrons. The incomplete filling of the energy band gives rise to characteristic metallic properties. The electrons in orbitals near the top of the occupied levels require very little energy input to be "promoted" to still higher energy orbitals, which are unoccupied. Under the influence of any source of excitation, such as an applied electrical potential or an input of thermal energy, electrons move into previously vacant levels and are thus freed to move through the lattice, giving rise to electrical and thermal conductivity.

Trends in properties of transition metals, such as the melting point (Table 23.2), can be readily explained by the molecular-orbital model. Recall the molecular-orbital description of second-period diatomic molecules. (Section 9.8) Half of the molecular orbitals were bonding, and half were antibonding. As we proceed across the period, the bond order generally increases until N_2, at which point it begins to decrease. This trend occurs because N_2 possesses the right number of electrons to completely fill the bonding molecular orbitals while leaving the higher-energy antibonding molecular orbitals empty.

The energy states that lead to the band for transition metals can likewise be divided roughly into two types: lower-energy states that result from metal–metal bonding interactions, and those at higher energy that result from metal–metal antibonding interactions. The group 6B metals (Cr, Mo, W) possess the correct number of electrons to fill the portion of the energy band that results phorus or another element from group 5A is added. The phosphorus atoms substitute for silicon atoms at random sites in the structure. Phosphorus, however, possesses five valence-shell electrons per atom, as compared to four for silicon. There is no room for these extra electrons in the valence band. They must therefore occupy the conduction band, as illustrated in Figure 23.16 ▼. These higher-energy electrons have access to many vacant orbitals within the energy band they occupy and serve as carriers of electrical current [Figure 23.16(b)]. Silicon doped with phosphorus in this manner is called an *n-type* semiconductor, because this doping introduces extra *n*egative charges (electrons) into the system.

If the silicon is doped instead with a group 3A element, such as gallium, the Ga atoms that substitute for silicon have one electron too few to meet their bonding requirements to the four neighboring silicon atoms. The valence band is thus incompletely filled, as illustrated in Figure 23.16(c). Under the influence of an applied field, electrons can move from occupied molecular orbitals to the few that are vacant in the valence band. A semiconductor formed by doping silicon with a group 3A element is called a *p-type* semiconductor because this doping creates electron vacancies that can be thought of as *p*ositive holes in the system. The modern electronics industry is based on integrated circuitry formed from silicon or germanium doped with various elements to create the desired electronic characteristics (Figure 23.17 ▼).

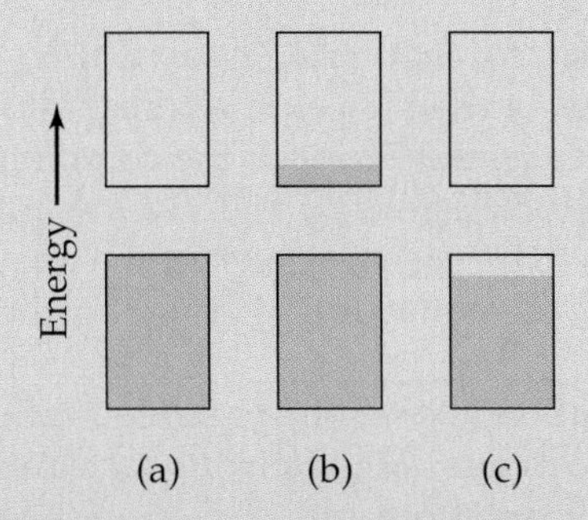

▲ **Figure 23.16** Effect of doping on the occupancy of the allowed energy levels in silicon. (a) *Pure silicon.* The valence-shell electrons just fill the lower-energy allowed energy band. (b) *Silicon doped with phosphorus.* Excess electrons occupy the lowest-energy orbitals in the higher-energy band of allowed energies. These electrons are capable of conducting current. (c) *Silicon doped with gallium.* There are not quite enough electrons to fully occupy the orbitals of the lower-energy allowed band. The presence of vacant orbitals in this band allows current to flow.

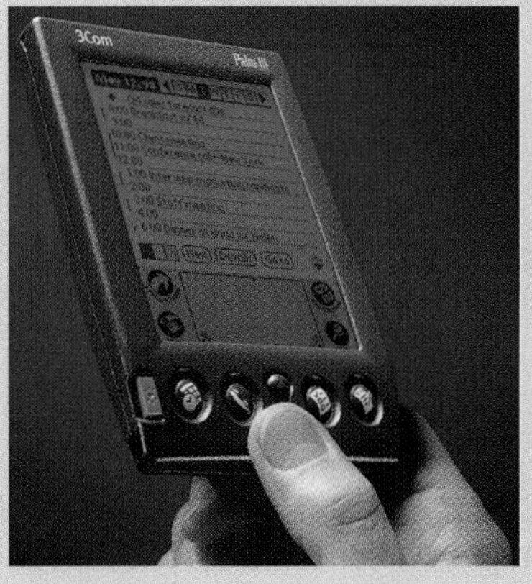

◀ **Figure 23.17** Semiconductors permit tremendous miniaturization of electronic devices, as illustrated by this hand-held computer.

from metal–metal bonding interactions and to leave the metal–metal antibonding orbitals empty. Metals with a smaller number of electrons than the group 6B metals have fewer metal–metal bonding orbitals occupied. Metals with a greater number of electrons than the group 6B metals have more metal–metal antibonding orbitals occupied. In each case the metal–metal bonding should be weaker than that of the group 6B metals, consistent with the trends in melting point and other properties. Factors other than the number of electrons (such as atomic radius, nuclear charge, and the particular packing structure of the metal) also play a role in determining the properties of metals.

This molecular-orbital model of metallic bonding (or *band theory*, as it is also called) is not so different in some respects from the electron-sea model. In both models the electrons are free to move about in the solid. The molecular-orbital model is more quantitative than the simple electron-sea model, however, so many properties of metals can be accounted for by quantum mechanical calculations using molecular-orbital theory.

23.6 Alloys

An **alloy** is a material that contains more than one element and has the characteristic properties of metals. The alloying of metals is of great importance because it is one of the primary ways of modifying the properties of pure metallic elements.

TABLE 23.4 Some Common Alloys

Primary Element	Name of Alloy	Composition by Mass	Properties	Uses
Bismuth	Wood's metal	50% Bi, 25% Pb, 12.5% Sn, 12.5% Cd	Low melting point (70°C)	Fuse plugs, automatic sprinklers
Copper	Yellow brass	67% Cu, 33% Zn	Ductile, takes polish	Hardware items
Iron	Stainless steel	80.6% Fe, 0.4% C, 18% Cr, 1% Ni	Resists corrosion	Tableware
Lead	Plumber's solder	67% Pb, 33% Sn	Low melting point (275°C)	Soldering joints
Silver	Sterling silver	92.5% Ag, 7.5% Cu	Bright surface	Tableware
	Dental amalgam	70% Ag, 18% Sn, 10% Cu, 2% Hg	Easily worked	Dental fillings

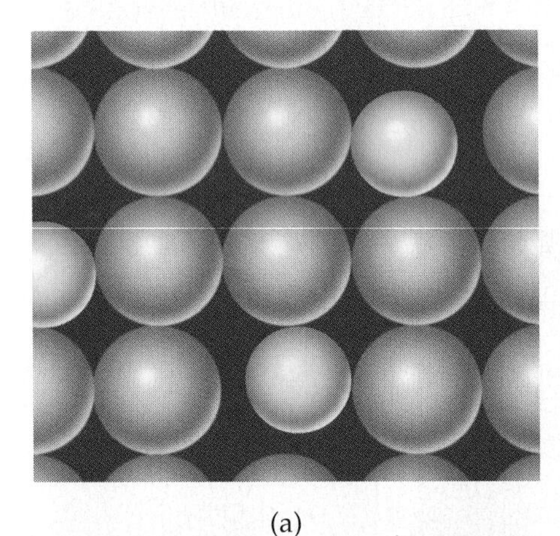

(a)

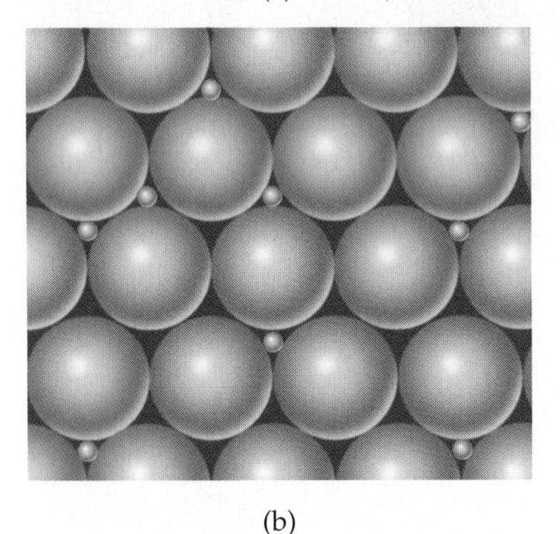

(b)

▲ **Figure 23.18** (a) Substitutional and (b) interstitial alloys. The blue spheres represent host metal; the yellow spheres represent the other components of the alloy.

Nearly all the common uses of iron, for example, involve alloy compositions. Pure gold, moreover, is too soft to be used in jewelry, whereas alloys of gold and copper are quite hard. Pure gold is termed 24 carat; the common alloy used in jewelry is 14 carat, meaning that it is 58% gold $\left(\frac{14}{24} \times 100\%\right)$. A gold alloy of this composition has suitable hardness to be used in jewelry. The alloy can be either yellow or white, depending on the elements added. Some further examples of alloys are given in Table 23.4 ▲.

Alloys can be classified as solution alloys, heterogeneous alloys, and intermetallic compounds. **Solution alloys** are homogeneous mixtures in which the components are dispersed randomly and uniformly. Atoms of the solute can take positions normally occupied by a solvent atom, thereby forming a *substitutional alloy*, or they can occupy interstitial positions, positions in the "holes" between the solvent atoms, thereby forming an *interstitial alloy*. These types are diagrammed in Figure 23.18 ◀.

Substitutional alloys are formed when the two metallic components have similar atomic radii and chemical-bonding characteristics. For example, silver and gold form such an alloy over the entire range of possible compositions. When two metals differ in radii by more than about 15%, solubility is more limited.

For an interstitial alloy to form, the component present in the interstitial positions between the solvent atoms must have a much smaller covalent radius than the solvent atoms. Typically, an interstitial element is a nonmetal that bonds to neighboring atoms. The presence of the extra bonds provided by the interstitial component causes the metal lattice to become harder, stronger, and less ductile. Steel, for example, is an alloy of iron that contains up to 3% carbon. Steel is much harder and stronger than pure iron. *Mild steels* contain less than 0.2% carbon; they are malleable and ductile and are used to make cables, nails, and chains. *Medium steels* contain 0.2 to 0.6% carbon; they are tougher than mild steels and are used to make girders and rails. *High-carbon steel*, used in cutlery, tools, and springs, contains 0.6 to 1.5% carbon. In all these cases other elements may be added to form *alloy steels*. Vanadium and chromium may be added to impart strength and to increase resistance to fatigue and corrosion. For example, a rail steel used in Sweden on lines bearing heavy ore carriers contains 0.7% carbon, 1% chromium, and 0.1% vanadium.

One of the most important iron alloys is stainless steel, which contains about 0.4% carbon, 18% chromium, and 1% nickel. The chromium is obtained by carbon reduction of chromite ($FeCr_2O_4$) in an electric furnace. The product of the reduction is *ferrochrome* ($FeCr_2$), which is then added in the appropriate amount to molten iron that comes from the converter to achieve the desired steel composition. The ratio of elements present in the steel may vary over a wide range, imparting a variety of specific physical and chemical properties to the materials.

In **heterogeneous alloy** the components are not dispersed uniformly. In the form of steel known as pearlite, for example, two distinct phases—essentially pure iron and the compound Fe_3C, known as cementite—are present in alternating layers. In general, the properties of heterogeneous alloys depend not only on the composition but also on the manner in which the solid is formed from the molten mixture. Rapid cooling leads to distinctly different properties than are obtained by slow cooling.

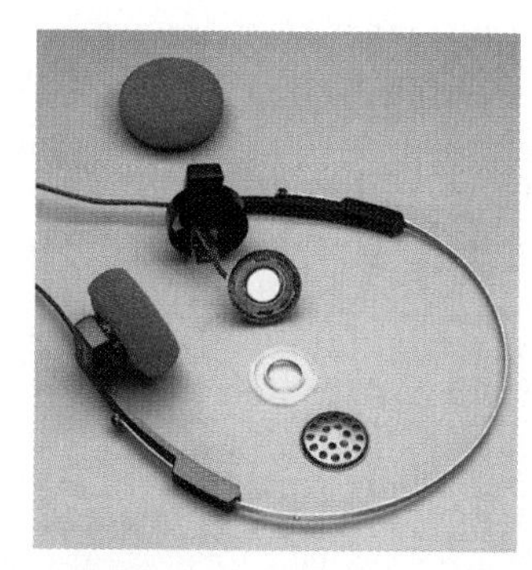

▲ **Figure 23.19** Interior of a lightweight audio headset. The assembly can be made small because of the very strong magnetism of the Co_5Sm alloy used.

Intermetallic Compounds

Intermetallic compounds are homogeneous alloys that have definite properties and compositions. For example, copper and aluminum form a compound, $CuAl_2$, known as duraluminum. Intermetallic compounds play many important roles in modern society. The intermetallic compound Ni_3Al is a major component of jet aircraft engines because of its strength and low density. Razor blades are often coated with Cr_3Pt, which adds hardness, allowing the blade to stay sharp longer. The compound Co_5Sm is used in the permanent magnets in lightweight headsets (Figure 23.19 ▶) because of its high magnetic strength per unit weight.

A Closer Look Shape Memory Alloys

In 1961 a naval engineer, William J. Buechler, made an unexpected and fortunate discovery. In searching for the best metal to use in missile nose cones, he tested many metal alloys. One of them, an intermetallic compound of nickel and titanium, NiTi, behaved very oddly. When he struck the cold metal, the sound was a dull thud. When he struck the metal at a higher temperature, however, it resonated like a bell. Mr. Buechler knew that the way in which sound propagates in a metal is related to its metallic structure. Clearly, the structure of the NiTi alloy had changed as it went from cold to warm. As it turned out, he had discovered an alloy that has shape memory.

Metals and metal alloys consist of many tiny crystalline areas (*crystallites*). When a metal is formed into a certain shape at high temperature, the crystallites are forced into a particular arrangement with respect to one another. Upon cooling a normal metal, the crystallites are "locked" in place by the bonds between them. When the metal is subsequently bent, the resulting stresses are sometimes elastic, as in a spring. Often, however, the metal merely deforms (e.g., when we bend a nail or crumple a sheet of aluminum foil). In these cases bending weakens the bonds that tie the crystallites together, and upon repeated flexing, the metal breaks apart.

In a shape memory alloy, the atoms can exist in two different bonding arrangements, representing two different solid-state phases. ∞ (Section 11.7) The higher-temperature phase has strong, fixed bonds between the atoms in the crystallites. By contrast, the lower-temperature phase is quite flexible with respect to the arrangements between the atoms. Thus, when the metal is distorted at low temperature, the stresses of the distortion are absorbed *within* the crystallites, by changes in the atomic lattice. In the higher-temperature phase, however, the atomic lattice is stiff, and stresses due to bending are absorbed by the bonds between crystallites, as in a normal metal.

To see how a shape memory metal behaves, suppose we bend a bar of NiTi alloy into a semicircle [Figure 23.20(a) ▶] and then heat it to about 500°C. We then cool the metal to below the transition temperature for the phase change to the low-temperature, flexible form. Although the cold metal remains in the semicircular form, as in Figure 23.20(b), it is now quite flexible and can readily be straightened or bent into another shape. When the metal is subsequently warmed and passes through the phase change to the "stiff" phase, it "remembers" its original shape and immediately returns to it as shown in Figure 23.20(c).

There are many uses for such shape memory alloys. The curved shape in a dental brace, for example, can be formed at high temperature into the curve that the teeth are desired to follow. Then at low temperature, where the metal is flexible, it can be shaped to fit the mouth of the wearer of the braces. When the brace is inserted in the mouth and warms to body temperature, the metal passes into the stiff phase and exerts a force against the teeth as it attempts to return to its original shape. Other uses for shape memory metals include heat-actuated shutoff valves in industrial process lines, which need no outside power source. Inserted into the face of a golf club, shape memory metals are said to impart more spin to the ball and greater control of the club.

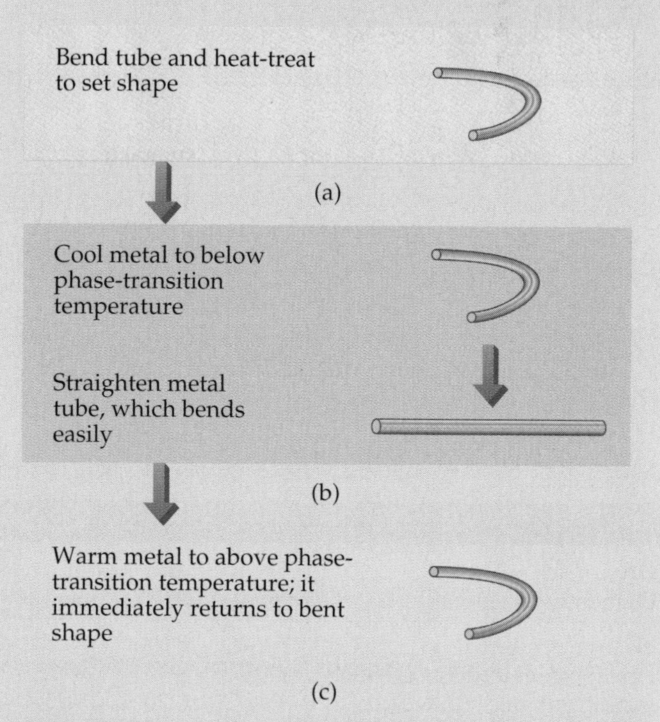
Bend tube and heat-treat to set shape
(a)
Cool metal to below phase-transition temperature
Straighten metal tube, which bends easily
(b)
Warm metal to above phase-transition temperature; it immediately returns to bent shape
(c)

▲ **Figure 23.20** Illustration of the behavior of a shape-memory alloy.

These examples illustrate some unusual ratios of combining elements. Nothing we have yet discussed in this text would lead us to predict such compositions. Among the many fundamental problems that remain unresolved in chemistry is that of developing a good theoretical model to predict the stoichiometries of intermetallic compounds.

23.7 Transition Metals

Many of the most important metals of modern society are transition metals. Transition metals, which occupy the *d* block of the periodic table (Figure 23.21 ▼), include such familiar elements as chromium, iron, nickel, and copper. They also include less familiar elements that have come to play important roles in modern technology, such as those in the high-performance jet engine pictured in Figure 23.1. In this section we consider some of the physical and chemical properties of transition metals.

Physical Properties

Several physical properties of the elements of the first transition series are listed in Table 23.5 ▼. Some of these properties, such as ionization energy and atomic radius, are characteristic of isolated atoms. Others, including density and melting point, are characteristic of the bulk solid metal.

The atomic properties vary in similar ways across each series. Notice, for example, that the bonding atomic radii of the transition metals shown in Figure 23.22 ▶ exhibit the same pattern in all three series. The trend in atomic radii is complex because it is the product of several factors, some of which work in opposite directions. In general, we would expect the atomic radius to decrease steadily as we proceed from left to right across the transition series because of the

▼ Figure 23.21 Transition metals are those elements that occupy the *d* block of the periodic table.

3B	4B	5B	6B	7B	8B	8B	8B	1B	2B
21 Sc	22 Ti	23 V	24 Cr	25 Mn	26 Fe	27 Co	28 Ni	29 Cu	30 Zn
39 Y	40 Zr	41 Nb	42 Mo	43 Tc	44 Ru	45 Rh	46 Pd	47 Ag	48 Cd
71 Lu	72 Hf	73 Ta	74 W	75 Re	76 Os	77 Ir	78 Pt	79 Au	80 Hg

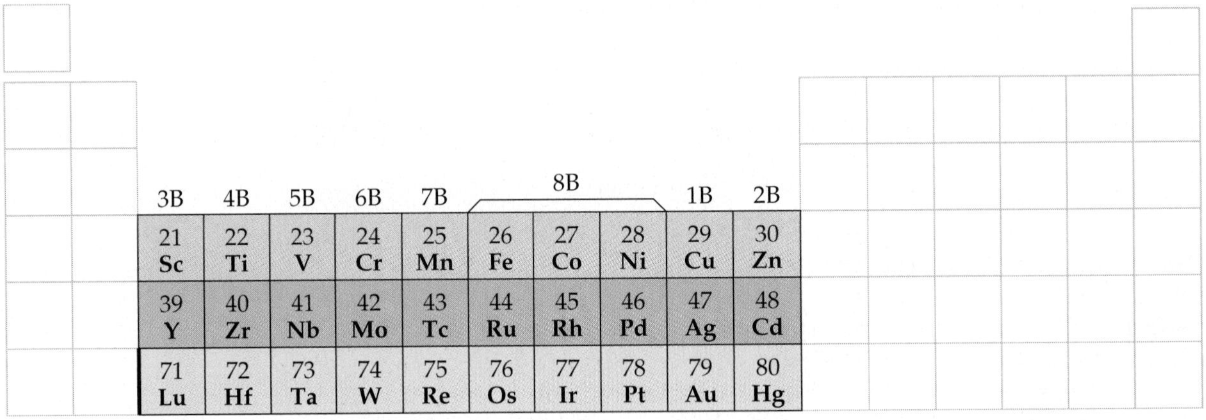

TABLE 23.5 Properties of the First Transition-Series Elements

Group:	3B	4B	5B	6B	7B	8B			1B	2B
Element:	Sc	Ti	V	Cr	Mn	Fe	Co	Ni	Cu	Zn
Electron configuration	$3d^1 4s^2$	$3d^2 4s^2$	$3d^3 4s^2$	$3d^5 4s^1$	$3d^5 4s^2$	$3d^6 4s^2$	$3d^7 4s^2$	$3d^8 4s^2$	$3d^{10} 4s^1$	$3d^{10} 4s^2$
First ionization energy (kJ/mol)	631	658	650	653	717	759	758	737	745	906
Bonding atomic radius (Å)	1.44	1.36	1.25	1.27	1.39	1.25	1.26	1.21	1.38	1.31
Density (g/cm^3)	3.0	4.5	6.1	7.9	7.2	7.9	8.7	8.9	8.9	7.1
Melting point (°C)	1541	1660	1917	1857	1244	1537	1494	1455	1084	420

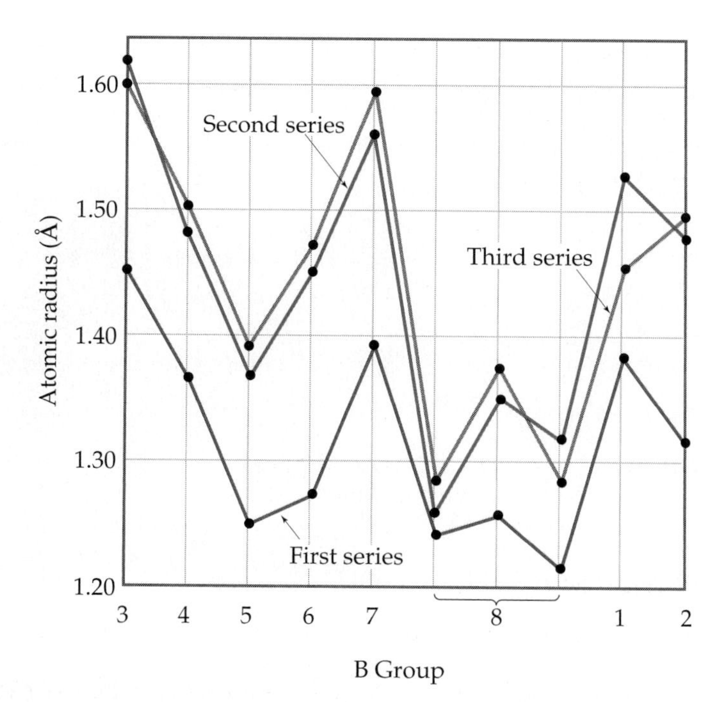

◀ **Figure 23.22** Variation in atomic radius of transition metals as a function of the periodic table group number.

increasing effective nuclear charge experienced by the valence electrons. Indeed, for groups 3, 4, and 5, this is the trend observed. As the number of *d* electrons increases, however, not all of them are employed in bonding. Nonbonding electrons exert repulsive effects that cause increased bond distances, and we see their effects in the maximum that occurs at group 7, and in the increase seen as we move past the group 8 elements. The bonding atomic radius is an empirical quantity that is especially difficult to define for elements such as the transition metals, which can exist in various oxidation states. Nevertheless, comparisons of the variations from one series to another are valid.

The incomplete screening of the nuclear charge by added electrons produces an interesting and important effect in the third transition-metal series. In general, the bonding atomic radius increases as we move down in a family because of the increasing principal quantum number of the outer-shell electrons. ∞ (Section 7.3) Once we move beyond the group 3 elements, however, the second and third transition-series elements have virtually the same bonding atomic radii. In group 5, for example, tantalum has virtually the same radius as niobium. This effect has its origin in the lanthanide series, the elements with atomic numbers 57 through 70, which occur between Ba and Lu (Figure 23.21). The filling of 4*f* orbitals through the lanthanide elements causes a steady increase in the effective nuclear charge, producing a contraction in size, called the **lanthanide contraction**. This contraction just offsets the increase we would expect as we go from the second to the third series. Thus, the second- and third-series transition metals in each group have about the same radii all the way across the series. As a consequence, the second- and third-series metals in a given group have great similarity in their chemical properties. For example, the chemical properties of zirconium and hafnium are remarkably similar. They always occur together in nature, and they are very difficult to separate.

Electron Configurations and Oxidation States

Transition metals owe their location in the periodic table to the filling of the *d* subshells. When these metals are oxidized, however, they lose their outer *s* electrons before they lose electrons from the *d* subshell. ∞ (Section 7.4) The electron

▶ **Figure 23.23** Salts of transition-metal ions and their solutions. From left to right: Mn^{2+}, Fe^{2+}, Co^{2+}, Ni^{2+}, Cu^{2+}, and Zn^{2+}.

configuration of Fe is $[Ar]3d^64s^2$, for example, whereas that of Fe^{2+} is $[Ar]3d^6$. Formation of Fe^{3+} requires loss of one $3d$ electron, giving $[Ar]3d^5$. Most transition-metal ions contain partially occupied d subshells. The existence of these d electrons is partly responsible for several characteristics of transition metals:

1. They often exhibit more than one stable oxidation state.
2. Many of their compounds are colored, as shown in Figure 23.23 ▲. (We will discuss the origin of these colors in Chapter 24.)
3. Transition metals and their compounds exhibit interesting and important magnetic properties.

Figure 23.24 ▼ summarizes the common nonzero oxidation states for the first transition series. The oxidation states shown as large circles are those most frequently encountered either in solution or in solid compounds. The ones shown as small circles are less common. Notice that Sc occurs only in the +3 oxidation state and Zn occurs only in the +2 oxidation state. The other metals, however, exhibit a variety of oxidation states. For example, Mn is commonly found in solution in the +2 (Mn^{2+}) and +7 (MnO_4^-) oxidation states. In the solid state the +4 oxidation state (as in MnO_2) is common. The +3, +5, and +6 oxidation states are less common.

The +2 oxidation state, which commonly occurs for nearly all of these metals, is due to the loss of their two outer 4*s* electrons. This oxidation state is found for all these elements except Sc, where the 3+ ion with an [Ar] configuration is particularly stable.

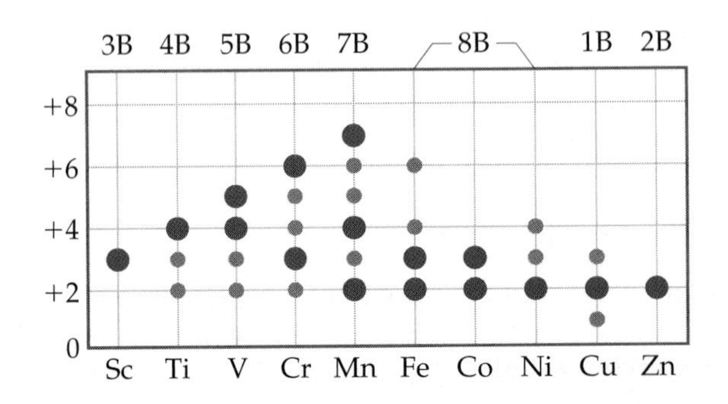

▶ **Figure 23.24** Nonzero oxidation states of the first transition series. The most common oxidation states are indicated by the larger circles.

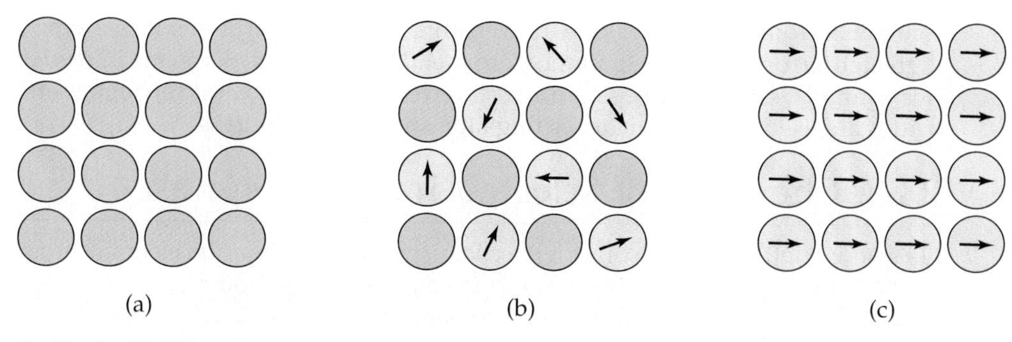

▲ **Figure 23.25** Types of magnetic behavior. (a) Diamagnetic: no centers (atoms or ions) with magnetic moments. (b) Simple paramagnetic: centers with magnetic moments not aligned unless the substance is in a magnetic field. (c) Ferromagnetic: coupled centers aligned in a common direction.

Oxidation states above +2 are due to successive losses of 3*d* electrons. From Sc through Mn the maximum oxidation state increases from +3 to +7, equaling in each case the total number of 4*s* plus 3*d* electrons in the atom. Thus, manganese has a maximum oxidation state of $2 + 5 = +7$. As we move to the right beyond the Mn in the first transition series, the maximum oxidation state decreases. In the second and third transition series the maximum oxidation state is +8, which is achieved in RuO_4 and OsO_4. In general, the maximum oxidation states are found only when the metals are combined with the most electronegative elements, especially O, F, and, possibly, Cl.

Magnetism

The magnetic properties of transition metals and their compounds are both interesting and important. Measurements of magnetic properties provide information about chemical bonding. In addition, many important uses are made of magnetic properties in modern technology.

An electron possesses a "spin" that gives it a magnetic moment, causing it to behave like a tiny magnet. (Section 9.8) Figure 23.25(a) ▲ represents a diamagnetic solid, one in which all the electrons in the solid are paired. When a diamagnetic substance is placed in a magnetic field, the motions of the electrons cause the substance to be very weakly repelled by the magnet.

When an atom or ion possesses one or more unpaired electrons, the substance is *paramagnetic*. (Section 9.8) In a paramagnetic solid the unpaired electrons on the atoms or ions of the solid are not influenced by the electrons on adjacent atoms or ions. The magnetic moments on the individual atoms or ions are randomly oriented, as shown in Figure 23.25(b). When placed in a magnetic field, however, the magnetic moments become aligned roughly parallel to one another, producing a net attractive interaction with the magnet. Thus, a paramagnetic substance is drawn into a magnetic field.

You are probably much more familiar with the magnetic behavior of simple iron magnets (Figure 23.26 ▶), a much stronger form of magnetism called **ferromagnetism**. Ferromagnetism arises when the unpaired electrons of the atoms or ions in a solid are influenced by the orientations of the electrons of their neighbors. The most stable (lowest-energy) arrangement results when the spins of electrons on adjacent atoms or ions are aligned in the same direction, as shown in Figure 23.25(c). When a ferromagnetic solid is placed in a magnetic field, the electrons tend to align strongly along the magnetic field. The attraction for the magnetic field that results may be as much as 1 million times stronger than that for a simple paramagnetic substance. When the external magnetic field is removed, the interactions between the electrons cause the solid as a whole to maintain a magnetic moment. We then refer to it as a *permanent magnet*. The most

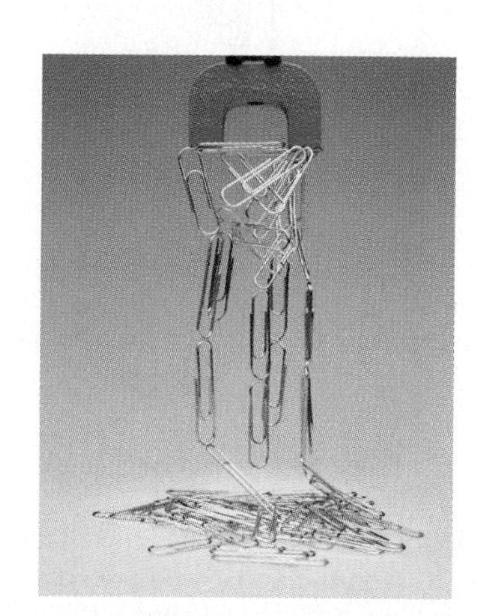

▲ **Figure 23.26** Permanent magnets are made from ferromagnetic materials.

common ferromagnetic solids are the elements Fe, Co, and Ni. Many alloys exhibit greater ferromagnetism than the pure metals themselves. Some metal oxides (for example, CrO_2 and Fe_3O_4) are also ferromagnetic. Several ferromagnetic oxides are used in magnetic recording tape and computer disks.

23.8 Chemistry of Selected Transition Metals

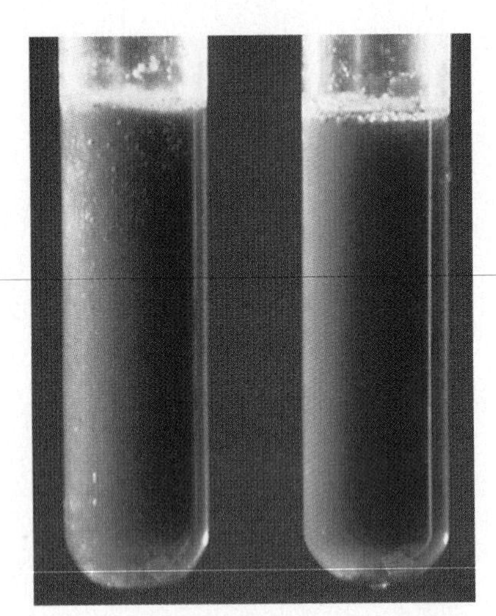

▲ **Figure 23.27** The tube on the left contains the violet hydrated chromium(III) ion, $Cr(H_2O)_6{}^{3+}$. The tube on the right contains the green $[(H_2O)_4Cr(OH)_2Cr(H_2O)_4]^{4+}$ ion.

Let's now briefly consider some of the chemistry of three common elements from the first transition series: chromium, iron, and copper. As you read this material, look for the trends that illustrate the generalizations outlined earlier.

Chromium

Chromium dissolves slowly in dilute hydrochloric or sulfuric acid, liberating hydrogen. In the absence of air, the reaction results in the formation of a sky blue solution of the chromium(II) or chromous ion:

$$Cr(s) + 2H^+(aq) \longrightarrow Cr^{2+}(aq) + 2H_2(g) \quad [23.21]$$

In the presence of air, the chromium(II) ion is rapidly oxidized by O_2 to form the chromium(III) ion. The reaction produces the green $[(H_2O)_4Cr(OH)_2Cr(H_2O)_4]^{4+}$ ion (Figure 23.27 ◀, right). In a strongly acidic solution, this ion reacts slowly with H^+ ions to form the violet $[Cr(H_2O)_6]^{3+}$ ion (Figure 23.27, left), which is often represented simply as $Cr^{3+}(aq)$. The overall reaction in acid solution is often given simply as shown in Equation 23.22.

$$4Cr^{2+}(aq) + O_2(g) + 4H^+(aq) \longrightarrow 4Cr^{3+}(aq) + 2H_2O(l) \quad [23.22]$$

Chromium is frequently encountered in aqueous solution in the +6 oxidation state. In basic solution the yellow chromate ion ($CrO_4{}^{2-}$) is the most stable. In acidic solution the dichromate ion ($Cr_2O_7{}^{2-}$) is formed:

$$CrO_4{}^{2-}(aq) + H^+(aq) \rightleftharpoons HCrO_4{}^-(aq) \quad [23.23]$$

$$2HCrO_4{}^-(aq) \rightleftharpoons Cr_2O_7{}^{2-}(aq) + H_2O(l) \quad [23.24]$$

MOVIE
Redox Chemistry of Iron and Copper

Equation 23.24 is a condensation reaction in which water is split out from two $HCrO_4{}^-$ ions. Similar reactions occur among the oxyanions of other elements, such as phosphorus. (Section 22.8) The equilibrium between the dichromate and chromate ions is readily observable because $CrO_4{}^{2-}$ is bright yellow and $Cr_2O_7{}^{2-}$ is deep orange, as seen in Figure 23.28 ◀. The dichromate ion in acidic solution is a strong oxidizing agent, as evidenced by its large, positive reduction potential. By contrast, the chromate ion in basic solution is not a particularly strong oxidizing agent.

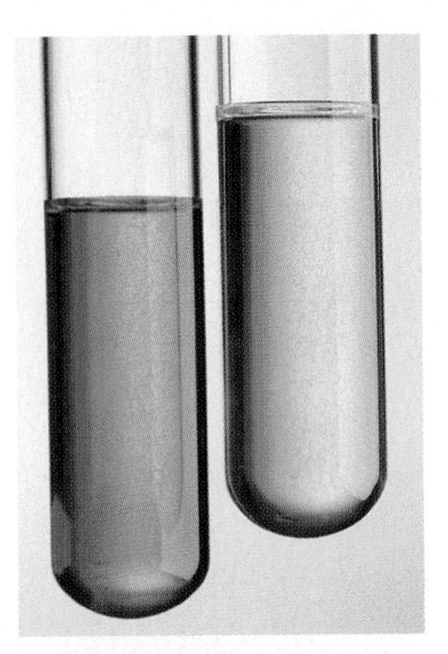

▲ **Figure 23.28** Sodium chromate, Na_2CrO_4 (on the right), and potassium dichromate, $K_2Cr_2O_7$ (on the left), illustrate the difference in color of the chromate and dichromate ions.

Iron

We have already discussed the metallurgy of iron in considerable detail in Section 23.2. Here we consider some of its important aqueous solution chemistry. Iron exists in aqueous solution in either the +2 (ferrous) or +3 (ferric) oxidation states. It often appears in natural waters because these waters come in contact with deposits of $FeCO_3$ ($K_{sp} = 3.2 \times 10^{-11}$). Dissolved CO_2 in the water can then help dissolve the mineral:

$$FeCO_3(s) + CO_2(aq) + H_2O(l) \longrightarrow Fe^{2+}(aq) + 2HCO_3{}^-(aq) \quad [23.25]$$

The dissolved Fe^{2+}, together with Ca^{2+} and Mg^{2+}, contributes to water hardness. (Section 18.6)

The standard reduction potentials in Appendix E reveal much about the kind of chemical behavior we should expect to observe for iron. The potential for reduction from the +2 state to the metal is negative; however, the reduction from the +3 to the +2 state is positive. Iron, therefore, should react with nonoxidizing acids such as dilute sulfuric acid or acetic acid to form $Fe^{2+}(aq)$, as indeed it does. In the presence of air, however, $Fe^{2+}(aq)$ tends to be oxidized to $Fe^{3+}(aq)$, as shown by the positive standard emf for Equation 23.26:

$$4Fe^{2+}(aq) + O_2(g) + 4H^+(aq) \longrightarrow 4Fe^{3+}(aq) + 2H_2O(l) \quad E° = +0.46 \text{ V} \quad [23.26]$$

You may have seen instances in which water dripping from a faucet or other outlet has left a brown stain (see Figure 23.29 ▶). The brown color is due to insoluble iron(III) oxide, formed by oxidation of iron(II) present in the water:

$$4Fe^{2+}(aq) + 8HCO_3^-(aq) + O_2(g) \longrightarrow 2Fe_2O_3(s) + 8CO_2(g) + 4H_2O(l) \quad [23.27]$$

▲ **Figure 23.29** The presence of dissolved iron salts in a water supply leads to stains from deposits of Fe_2O_3.

When iron metal reacts with an oxidizing acid such as warm, dilute nitric acid, $Fe^{3+}(aq)$ is formed directly:

$$Fe(s) + NO_3^-(aq) + 4H^+(aq) \longrightarrow Fe^{3+}(aq) + NO(g) + 2H_2O(l) \quad [23.28]$$

In the +3 oxidation state iron is soluble in acidic solution as the hydrated ion, $Fe(H_2O)_6^{3+}$. However, this ion hydrolyzes readily (Section 16.11):

$$Fe(H_2O)_6^{3+}(aq) \rightleftharpoons Fe(H_2O)_5(OH)^{2+}(aq) + H^+(aq) \quad [23.29]$$

When an acidic solution of iron(III) is made more basic, a gelatinous red-brown precipitate, most accurately described as a hydrous oxide, $Fe_2O_3 \cdot nH_2O$, is formed (Figure 23.30 ▶). In this formulation n represents an indefinite number of water molecules, depending on the precise conditions of the precipitation. Usually, the precipitate that forms is represented merely as $Fe(OH)_3$. The solubility of $Fe(OH)_3$ is very low ($K_{sp} = 4 \times 10^{-38}$). It dissolves in strongly acidic solution but not in basic solution. The fact that it does *not* dissolve in basic solution is the basis of the Bayer process, in which aluminum is separated from impurities, primarily iron(III). ∞ (Section 23.3)

▲ **Figure 23.30** Addition of a NaOH solution to an aqueous solution of Fe^{3+} causes $Fe(OH)_3$ to precipitate.

Copper

In its aqueous solution chemistry, copper exhibits two oxidation states: +1 (cuprous) and +2 (cupric). In the +1 oxidation state copper possesses a $3d^{10}$ electron configuration. Salts of Cu^+ are often water insoluble and are mostly white in color. In solution the Cu^+ ion readily disproportionates:

$$2Cu^+(aq) \longrightarrow Cu^{2+}(aq) + Cu(s) \quad K_{eq} = 1.2 \times 10^6 \quad [23.30]$$

Because of this reaction and because copper(I) is readily oxidized to copper(II) under most solution conditions, the +2 oxidation state is by far the more common.

Many salts of Cu^{2+}, including $Cu(NO_3)_2$, $CuSO_4$, and $CuCl_2$, are water soluble. Copper sulfate pentahydrate ($CuSO_4 \cdot 5H_2O$), a widely used salt, has four water molecules bound to the copper ion and a fifth held to the SO_4^{2-} ion by hydrogen bonding. The salt is blue. (It is often called *blue vitriol*; see Figure 23.31 ▶.) Aqueous solutions of Cu^{2+}, in which the copper ion is coordinated by water molecules, are also blue. Among the insoluble compounds of copper(II) is $Cu(OH)_2$, which is formed when NaOH is added to an aqueous Cu^{2+} solution (Figure 23.32 ▶). This blue compound readily loses water on heating to form black copper(II) oxide:

▲ **Figure 23.31** Crystals of copper(II) sulfate pentahydrate, $CuSO_4 \cdot 5H_2O$.

$$Cu(OH)_2(s) \longrightarrow CuO(s) + H_2O(l) \quad [23.31]$$

CuS is one of the least soluble copper(II) compounds ($K_{sp} = 6.3 \times 10^{-36}$). This black substance does not dissolve in NaOH, NH_3, or nonoxidizing acids such as HCl. It does dissolve in HNO_3, however, which oxidizes the sulfide to sulfur:

$$3CuS(s) + 8H^+(aq) + 2NO_3^-(aq) \longrightarrow 3Cu^{2+}(aq) + 3S(s) + 2NO(g) + 4H_2O(l) \quad [23.32]$$

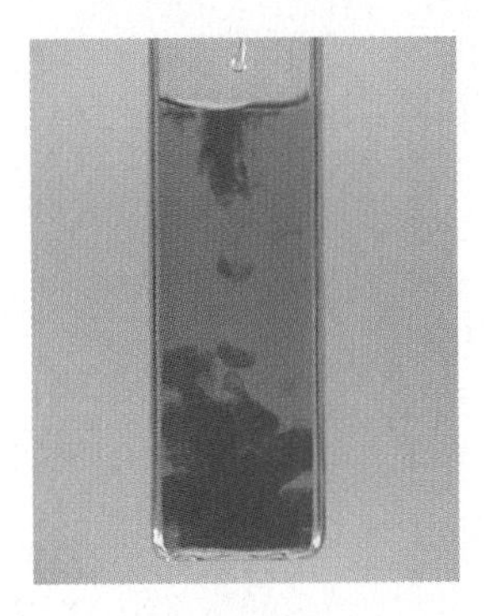

▲ **Figure 23.32** Addition of a NaOH solution to an aqueous solution of Cu^{2+} causes $Cu(OH)_2$ to precipitate.

$CuSO_4$ is often added to water to stop algae or fungal growth, and other copper preparations are used to spray or dust plants to protect them from lower organisms and insects. Copper compounds are not generally toxic to human beings, except in massive quantities. Our daily diet normally includes from 2 to 5 mg of copper.

SAMPLE INTEGRATIVE EXERCISE 23: Putting Concepts Together

The most important commercial ore of chromium is *chromite* ($FeCr_2O_4$). **(a)** What is the most reasonable assignment of oxidation states to Fe and Cr in this ore? **(b)** Chromite can be reduced in an electric arc furnace (which provides the required heat) using coke (carbon). Write a balanced chemical equation for this reduction, which forms ferrochrome ($FeCr_2$). **(c)** Two of the major forms of chromium in the +6 oxidation state are CrO_4^{2-} and $Cr_2O_7^{2-}$. Draw Lewis structures for these species (*Hint*: You may find it helpful to consider the Lewis structures of nonmetallic anions of the same formula.) **(d)** Chromium metal is used in alloys (e.g., stainless steel) and in electroplating, but the metal itself is not widely used, in part because it is not ductile at ordinary temperatures. From what we have learned in this chapter about metallic bonding and properties, suggest why chromium is less ductile than most metals.

Solution **(a)** Because each oxygen has an oxidation number of −2, the four oxygens represent a total of −8. If the metals have whole-number oxidation numbers, our choices are Fe = +4 and Cr = +2, or Fe = +2 and Cr = +3. The latter choice seems the more reasonable because a +4 oxidation number for iron is unusual. (Although one alternative would be that Fe is +3 and the two Cr have different oxidation states of +2 and +3, the properties of chromite indicate that the two Cr have the same oxidation number.)

(b) The balanced equation is

$$2C(s) + FeCr_2O_4(s) \longrightarrow FeCr_2(s) + 2CO_2(g)$$

(c) We expect that in CrO_4^{2-}, the Cr will be surrounded tetrahedrally by four oxygens. The electron configuration of the Cr atom is $[Ar]3d^5 4s^1$, giving it six electrons that can be used in bonding, much like the S atom in SO_4^{2-}. These six electrons must be shared with four O atoms, each of which has six valence-shell electrons. In addition, the ion has a 2− charge. Thus we have a total of 6 + 4(6) + 2 = 32 valence electrons to place in the Lewis structure. Putting one electron pair in each Cr—O bond and adding unshared electron pairs to the oxygens, we require precisely 32 electrons to achieve an octet around each atom:

In $Cr_2O_7^{2-}$ the structure is analogous to that of the diphosphate ion ($P_2O_7^{4-}$), which we discussed in Section 22.8. We can think of the $Cr_2O_7^{2-}$ ion as formed by a condensation reaction as shown in Equation 23.24.

(d) Recall that chromium, with six electrons available for bonding, has relatively strong metallic bonding among the metals of the transition series, as evidenced by its high melting point (Table 23.2). This means that distortions of the metallic lattice of the sort that occur when metals are drawn into wires will require more energy than for other metals with weaker metallic bonding.

Summary and Key Terms

Section 23.1 The metallic elements are extracted from the **lithosphere**, the uppermost solid portion of our planet. Metallic elements occur in nature in **minerals**, which are solid inorganic substances found in various deposits, or **ores**. The desired components of an ore must be separated from the undesired components, called *gangue*. **Metallurgy** is concerned with obtaining metals from these sources and with understanding and modifying the properties of metals.

Section 23.2 **Pyrometallurgy** is the use of heat to bring about chemical reactions that convert an ore from one chemical form to another. In **calcination** an ore is heated to drive off a volatile substance, as in heating a carbonate ore to drive off CO_2. In **roasting**, the ore is heated under conditions that bring about reaction with the furnace atmosphere. For example, sulfide ores might be heated to oxidize sulfur to SO_2. In a **smelting** operation two or more layers of mutually insoluble materials form in the furnace. One layer consists of molten metal, and the other layer (**slag**) is composed of molten silicate minerals and other ionic materials such as phosphates.

Iron, the most important metal in modern society, is obtained from its oxide ores by reduction in a blast furnace. The reducing agent is carbon, in the form of coke. Limestone ($CaCO_3$) is added to react with the silicates present in the crude ore to form slag. The raw iron from the blast furnace, called pig iron, is usually taken directly to a converter, where **refining** takes place to form various kinds of steel. In the converter the molten iron reacts with pure oxygen to oxidize impurity elements.

Section 23.3 **Hydrometallurgy** is the use of chemical processes occurring in aqueous solution to separate a mineral from its ore or one particular element from others. In **leaching**, an ore is treated with an aqueous reagent to dissolve one component selectively. In the **Bayer process** aluminum is selectively dissolved from bauxite by treatment with concentrated NaOH solution.

Section 23.4 **Electrometallurgy** is the use of electrolytic methods to prepare or purify a metallic element. Sodium is prepared by electrolysis of molten NaCl in a **Downs cell**. Aluminum is obtained in the **Hall process** by electrolysis of Al_2O_3 in molten cryolite (Na_3AlF_6). Copper is purified by electrolysis of aqueous copper sulfate solution using anodes composed of impure copper.

Section 23.5 The properties of metals can be accounted for in a qualitative way by the electron-sea model, in which the electrons are visualized as being free to move throughout the metal structure. In the molecular-orbital model the valence atomic orbitals of the metal atoms interact to form an energy band that is incompletely filled by valence electrons. The orbitals that constitute the energy band are delocalized over the atoms of the metal, and their energies are closely spaced. Because the energy differences between orbitals in the band are so small, promoting electrons to higher-energy orbitals requires very little energy. This gives rise to high electrical and thermal conductivity, as well as other characteristic metallic properties. In an insulator, on the other hand, all the orbitals of a band are completely filled and there is a large energy gap between the filled and the next unfilled band.

Section 23.6 **Alloys** are materials that possess characteristic metallic properties and are composed of more than one element. Usually, one or more metallic elements are major components. **Solution alloys** are homogeneous alloys in which the components are distributed uniformly throughout. In **heterogeneous alloys** the components are not distributed uniformly; instead, two or more distinct phases with characteristic compositions are present. **Intermetallic compounds** are homogeneous alloys that have definite properties and compositions.

Sections 23.7 and 23.8 Transition metals are characterized by incomplete filling of the *d* orbitals. The presence of *d* electrons in transition elements leads to multiple oxidation states. As we proceed through a given series of transition metals, the effective nuclear charge for the valence electrons increases slowly. As a result, the later transition elements in a given row tend to adopt lower oxidation states and have slightly smaller ionic radii. Although the atomic and ionic radii increase in the second series as compared to the first, the elements of the second and third series are similar with respect to these and other properties. This similarity is due to the **lanthanide contraction**. Lanthanide elements, atomic numbers 57 through 70, display an increase in effective nuclear charge that compensates for the increase in principal quantum number in the third series.

The presence of unpaired electrons in valence orbitals leads to interesting magnetic behavior in transition metals and their compounds. In **ferromagnetic** substances the unpaired electron spins on atoms in a solid are affected by those on neighboring atoms. In a magnetic field the spins become aligned along the direction of the magnetic field. When the magnetic field is removed, this orientation remains, giving the solid a magnetic moment as observed in permanent magnets.

In this chapter we also briefly considered some chemistry of three of the common transition metals: chromium, iron, and copper.

Exercises

Metallurgy

23.1 Two of the most heavily utilized metals are aluminum and iron. What are the most important natural sources of these elements? In what oxidation state is each metal found in nature?

23.2 **(a)** Pyrolusite (MnO_2) is a commercially important mineral of manganese. What is the oxidation state of Mn in this metal? **(b)** Name some reagents that might be used to reduce this ore to the metal.

23.3 Explain in your own words what is meant by the statement, "This ore consists of a small concentration of chalcopyrite together with considerable gangue."

23.4 What is meant by the following terms: **(a)** calcination; **(b)** leaching; **(c)** smelting; **(d)** slag?

23.5 Complete and balance each of the following equations:

(a) $PbS(s) + O_2(g) \xrightarrow{\Delta}$

(b) $PbCO_3(s) \xrightarrow{\Delta}$

(c) $WO_3(s) + H_2(g) \xrightarrow{\Delta}$

(d) $ZnO(s) + CO(g) \xrightarrow{\Delta}$

23.6 Complete and balance each of the following equations:

(a) $CdS(s) + O_2(g) \xrightarrow{\Delta}$

(b) $CoCO_3(s) \xrightarrow{\Delta}$

(c) $Cr_2O_3(s) + Na(l) \longrightarrow$

(d) $VCl_3(g) + K(l) \longrightarrow$

(e) $BaO(s) + P_2O_5(l) \longrightarrow$

23.7 A sample containing $PbSO_4$ is to be refined to Pb metal via calcination, followed by roasting. **(a)** What volatile product would you expect to be produced by calcination? **(b)** Propose an appropriate atmosphere to accompany the roasting. **(c)** Write balanced chemical equations for the two steps.

23.8 Suppose a metallurgist wanted to use cobalt(II) carbonate as a source of cobalt metal. **(a)** What products would you expect from calcination of this substance? **(b)** With what reagent could you react the product of the calcination in a roasting operation to form Co metal? **(c)** Write balanced chemical equations for the processes discussed in parts (a) and (b).

23.9 Write balanced chemical equations for the reduction of FeO and Fe_2O_3 by H_2 and by CO.

23.10 What is the major reducing agent in the reduction of iron ore in a blast furnace? Write a balanced chemical equation for the reduction process.

23.11 What role does each of the following materials play in the chemical processes that occur in a blast furnace: **(a)** air; **(b)** limestone; **(c)** coke; **(d)** water? Write balanced chemical equations to illustrate your answers.

23.12 **(a)** In the basic oxygen process for steel formation, what reactions cause the temperature in the converter to increase? **(b)** Write balanced chemical equations for the oxidation of carbon, sulfur, and silicon in the converter.

23.13 **(a)** Why is the Bayer process a necessary step in the production of aluminum metal? **(b)** What difference in chemical properties is used in the Bayer process to separate Al_2O_3 from Fe_2O_3?

23.14 What roles are played by O_2 and by CN^- in the leaching of gold from low-grade ores?

23.15 Describe how electrometallurgy could be employed to purify crude cobalt metal. Describe the compositions of the electrodes and electrolyte, and write out all electrode reactions.

23.16 The element tin is generally recovered from deposits of the ore cassiterite (SnO_2). The oxide is reduced with carbon, and the crude metal is purified by electrolysis. Write balanced chemical equations for the reduction process and for the electrode reactions in the electrolysis. (Assume that an acidic solution of $SnSO_4$ is employed as an electrolyte in the electrolysis.)

Metals and Alloys

23.17 Sodium is a highly malleable substance, whereas sodium chloride is not. Explain this difference in properties.

23.18 Silicon has the same crystal structure as diamond (Figure 11.41). Based on this fact, do you think silicon is likely to exhibit metallic properties or not? Explain your answer.

23.19 Silver has the highest electrical and thermal conductivities of any metal. How does the electron-sea model account for these conductivities?

23.20 **(a)** Compare the electronic structures of chromium and selenium. In what respects are they similar, and in what respects do they differ? **(b)** Chromium is a metal, and selenium is a nonmetal. What factors are important in determining this difference in properties?

23.21 The densities of the elements K, Ca, Sc, and Ti are 0.86, 1.5, 3.2, and 4.5 g/cm^3, respectively. What factors are likely to be of major importance in determining this variation?

23.22 The heat of atomization, which is the enthalpy change for the process $M(s) \longrightarrow M(g)$, where $M(g)$ is the atomic form of the metal, varies in the first transition row as follows:

Element	Ca	Sc	Ti	V
ΔH_{atom} (kJ/mol)	178	378	471	515

Use a model for metallic bonding to account for this variation.

23.23 According to band theory, how do insulators differ from conductors? How do semiconductors differ from conductors?

23.24 Which would you expect to be a better conductor of electricity, germanium or germanium doped with arsenic? Explain, using the molecular-orbital model.

23.25 Tin exists in two allotropic forms: Gray tin has a diamond structure, and white tin has a close-packed structure.

Which of these allotropic forms would you expect to be more metallic in character? Explain why the electrical conductivity of white tin is much greater than that of gray tin. Which form would you expect to have the longer Sn—Sn bond distance?

[23.26] As we learned in Chapter 11, graphite is a good conductor of electricity in the dimension of the layer structure [Figure 11.41(b)]. Use the molecular-orbital model for metals to explain the conductivity of graphite.

23.27 Define the term *alloy*. Distinguish among solution alloys, heterogeneous alloys, and intermetallic compounds.

23.28 Distinguish between substitutional and interstitial alloys. What conditions favor formation of substitutional alloys?

Transition Metals

23.29 Which of the following properties are better considered characteristic of the free isolated atoms, and which are characteristic of the bulk metal: **(a)** electrical conductivity; **(b)** first ionization energy; **(c)** atomic radius; **(d)** melting point; **(e)** heat of vaporization; **(f)** electron affinity?

23.30 Which of the following species would you expect to possess metallic properties: **(a)** $TiCl_4$; **(b)** NiCo alloy; **(c)** W; **(d)** Ge; **(e)** Hg_2^{2+}? Explain in each case.

23.31 What is meant by the term lanthanide contraction? What properties of the transition elements are affected by the lanthanide contraction?

23.32 Zirconium and hafnium are the group 4B elements in the second and third transition series. The atomic radii of these elements are virtually the same (Figure 23.22). Explain this similarity in atomic radius.

23.33 Write the formula for the fluoride corresponding to the highest expected oxidation state for each of the following: **(a)** Sc; **(b)** Co; **(c)** Zn.

23.34 Write the formula for the oxide corresponding to the highest expected oxidation state for each of the following: **(a)** Cd; **(b)** W; **(c)** Nb.

23.35 Why does chromium exhibit several oxidation states in its compounds, whereas aluminum exhibits only the +3 oxidation state?

23.36 The element vanadium exhibits multiple oxidation states in its compounds, including +2. The compound VCl_2 is known, whereas $ScCl_2$ is unknown. Use electron configurations and effective nuclear charges to account for this difference in behavior.

23.37 Write the expected electron configuration for each of the following ions: **(a)** Cr^{3+}; **(b)** Au^{3+}; **(c)** Ru^{2+}; **(d)** Cu^{+}; **(e)** Mn^{4+}; **(f)** Ir^{+}.

23.38 What is the expected electron configuration for each of the following ions: **(a)** Ti^{2+}; **(b)** Co^{3+}; **(c)** Pd^{2+}; **(d)** Mo^{3+}; **(e)** Ru^{3+}; **(f)** Ni^{4+}?

23.39 Which would you expect to be more easily oxidized, Ti^{2+} or Ni^{2+}?

23.40 Which would you expect to be the stronger reducing agent, Cr^{2+} or Fe^{2+}?

23.41 How does the presence of air affect the relative stabilities of ferrous and ferric ions?

23.42 Give the chemical formulas and colors of the chromate and dichromate ions. Which of these is more stable in acidic solution?

23.43 Write balanced chemical equations for the reaction between iron and **(a)** hydrochloric acid; **(b)** nitric acid.

23.44 MnO_2 reacts with aqueous HCl to yield $MnCl_2(aq)$ and chlorine gas. **(a)** Write a balanced chemical equation for the reaction. **(b)** Is this an oxidation-reduction reaction? If yes, identify the oxidizing and reducing agents.

23.45 On the atomic level, what distinguishes a paramagnetic material from a diamagnetic one? How does each behave in a magnetic field?

23.46 **(a)** What characteristics of a ferromagnetic material distinguish it from one that is paramagnetic? **(b)** What type of interaction must occur in the solid to bring about ferromagnetic behavior? **(c)** Must a substance contain iron to be ferromagnetic? Explain.

Additional Exercises

23.47 Write a chemical equation for the reaction that occurs when PbS is roasted in air. Why might a sulfuric acid plant be located near a plant that roasts sulfide ores?

23.48 Explain why aluminum, magnesium, and sodium metals are obtained by electrolysis instead of by reduction with chemical reducing agents.

23.49 Make a list of the chemical reducing agents employed in the production of metals, as described in this chapter. For each of them, identify a metal that can be formed using that reducing agent.

23.50 Write balanced chemical equations for each of the following verbal descriptions: **(a)** Vanadium oxytrichloride ($VOCl_3$) is formed by the reaction of vanadium(III) chloride with oxygen. **(b)** Niobium(V) oxide is reduced to the metal with hydrogen gas. **(c)** Iron(III) ion in aqueous solution is reduced to iron(II) ion in the presence of zinc dust. **(d)** Niobium(V) chloride reacts with water to yield crystals of niobic acid ($HNbO_3$).

23.51 Write a balanced chemical equation to correspond to each of the following verbal descriptions: **(a)** NiO(*s*) can be solubilized by leaching with aqueous sulfuric acid. **(b)** After concentration, an ore containing the mineral carrollite ($CuCo_2S_4$) is leached with aqueous sulfuric acid to produce a solution containing copper ions and cobalt ions. **(c)** Titanium dioxide is treated with chlorine in the presence of carbon as a reducing agent to form $TiCl_4$. **(d)** Under oxygen pressure ZnS(*s*) reacts at 150°C with aqueous sulfuric acid to form soluble zinc sulfate, with deposition of elemental sulfur.

23.52 The crude copper that is subjected to electrorefining contains selenium and tellurium as impurities. Describe the probable fate of these elements during electrorefining, and relate your answer to the positions of these elements in the periodic table.

23.53 Why is the +2 oxidation state common among the transition metals? Why do so many transition metals exhibit a variety of oxidation states?

[23.54] Write balanced chemical equations that correspond to the steps in the following brief account of the metallurgy of molybdenum: Molybdenum occurs primarily as the sulfide, MoS_2. On boiling with concentrated nitric acid, a white residue of MoO_3 is obtained. This is an acidic oxide; when it is dissolved in hot, excess concentrated ammonia, ammonium molybdate crystallizes on cooling. On heating ammonium molybdate, white MoO_3 is obtained. On further heating to 1200°C in hydrogen, a gray powder of metallic molybdenum is obtained.

23.55 Distinguish between **(a)** a substitutional alloy and an intermetallic compound; **(b)** a paramagnetic and a diamagnetic substance; **(c)** a semiconductor and an insulator; **(d)** metallic conduction and electrolytic conduction of electricity.

23.56 Pure silicon is a very poor conductor of electricity. Titanium, which also possesses four valence-shell electrons, is a metallic conductor. Explain the difference.

[23.57] The thermodynamic stabilities of the three complexes $Zn(H_2O)_4{}^{2+}$, $Zn(NH_3)_4{}^{2+}$, and $Zn(CN)_4{}^{2-}$ increase from the H_2O to the NH_3 to the CN^- complex. How do you expect the reduction potentials of these three complexes to compare?

23.58 Why do paramagnetic substances experience such a weak attraction to a magnet compared with ferromagnetic substances?

23.59 Indicate whether each of the following compounds is expected to be diamagnetic or paramagnetic, and give a reason for your answer in each case: **(a)** $NbCl_5$; **(b)** $CrCl_2$ **(c)** CuCl; **(d)** RuO_4; **(e)** $NiCl_2$.

[23.60] Associated with every ferromagnetic solid is a temperature known as its Curie temperature. When heated above its Curie temperature, the substance no longer exhibits ferromagnetism, but rather becomes paramagnetic. Use the kinetic-molecular theory of solids to explain this observation.

23.61 Write balanced chemical equations for each of the following reactions characteristic of elemental manganese: **(a)** It reacts with aqueous HNO_3 to form a solution of manganese(II) nitrate. **(b)** When solid manganese(II) nitrate is heated to 450 K, it decomposes to MnO_2. **(c)** When MnO_2 is heated to 700 K, it decomposes to Mn_3O_4. **(d)** When solid $MnCl_2$ is reacted with $F_2(g)$, it forms MnF_3 (one of the products is ClF_3).

23.62 Predict what happens in the following cases: **(a)** Freshly precipitated $Fe(OH)_2$ is exposed to air. **(b)** KOH is added to an aqueous solution of copper(II) nitrate. **(c)** Sodium hydroxide is added to an aqueous solution of potassium dichromate.

[23.63] Based on the chemistry described in this chapter and others, propose balanced chemical equations for the following sequence of reactions involving nickel: **(a)** The ore millerite, which contains NiS, is roasted in an atmosphere of oxygen to produce an oxide. **(b)** The oxide is reduced to the metal using coke. **(c)** Dissolving the metal in hydrochloric acid produces a green solution. **(d)** Adding excess sodium hydroxide to the solution causes a gelatinous green material to precipitate. **(e)** Upon heating, the green material loses water and yields a green powder.

[23.64] Indicate whether each of the following solids is likely to be an insulator, a metallic conductor, an n-type semiconductor, or a p-type semiconductor: **(a)** TiO_2; **(b)** Ge doped with In; **(c)** Cu_3;Al **(d)** Pd; **(e)** SiC; **(f)** Bi.

Integrative Exercises

23.65 If a blast furnace uses Fe_2O_3 to produce 9.00×10^3 tons of Fe each day, what is the minimum quantity of carbon required in the furnace, assuming that the actual reducing agent is actually the carbon monoxide?

23.66 **(a)** A charge of 3.3×10^6 kg of material containing 27% Cu_2S and 13% FeS is added to a converter and oxidized. What mass of $SO_2(g)$ is formed? **(b)** What is the molar ratio of Cu to Fe in the resulting mixture of oxides? **(c)** What are the likely formulas of the oxides formed in the oxidation reactions, assuming an excess of oxygen? **(d)** Write balanced equations representing each of the oxidation reactions.

23.67 Using the concepts discussed in Chapter 13, indicate why the molten metal and slag phases formed in the blast furnace shown in Figure 23.4 are immiscible.

23.68 In an electrolytic process nickel sulfide is oxidized in a two-step reaction:

$$Ni_3S_2(s) \longrightarrow Ni^{2+}(aq) + 2NiS(s) + 2e^-$$

$$NiS(s) \longrightarrow Ni^{2+}(aq) + S(s) + 2e^-$$

What mass of Ni^{2+} is produced in solution by passing a current of 67 A for a period of 11.0 hr, assuming the cell is 90% efficient?

[23.69] **(a)** Using the data in Appendix C, estimate the free-energy change for the following reaction at 1200°C:

$$Si(s) + 2MnO(s) \longrightarrow SiO_2(s) + 2Mn(s)$$

(b) What does this value tell you about the feasibility of carrying out this reaction at 1200°C?

[23.70] **(a)** In the converter employed in steel formation (Figure 23.6), oxygen gas is blown at high temperature directly into a container of molten iron. Iron is converted to rust on exposure to air at room temperature, but the iron is not extensively oxidized in the converter. Explain why this is so. **(b)** The oxygen introduced into the converter reacts with various impurities, particularly with carbon, phosphorus, sulfur, silicon, and impurity metals. What are the products of these reactions, and where do they end up in the process?

23.71 Copper(I) is an uncommon oxidation state in aqueous acidic solution because it disproportionates into Cu^{2+} and Cu. Use data from Appendix E to calculate the equilibrium constant for the reaction $2Cu^+(aq) \rightleftharpoons Cu^{2+}(aq) + Cu(s)$.

23.72 The reduction of metal oxides is often accomplished using carbon monoxide as a reducing agent. Carbon (coke) and carbon dioxide are usually present, leading to the following reaction:

$$C(s) + CO_2(g) \rightleftharpoons 2CO(g)$$

Using data from Appendix C, calculate the equilibrium constant for this reaction at 298 K and at 2000 K, assuming that the enthalpies and entropies do not depend upon temperature.

23.73 Magnesium is obtained by electrolysis of molten $MgCl_2$. **(a)** Why isn't an aqueous solution of $MgCl_2$ used in the electrolysis? **(b)** Several cells are connected in parallel by very large copper buses that convey current to the cells. Assuming that the cells are 96% efficient in producing the desired products in electrolysis, what mass of Mg is formed by passing a current of 97,000 A for a period of 24 hr?

23.74 Vanadium(V) fluoride is a colorless substance that melts at 19.5°C and boils at 48.3°C. Vanadium(III) fluoride, on the other hand, is yellow-green in color and melts at 800°C. **(a)** Suggest a structure and bonding for VF_5 that accounts for its melting and boiling points. Can you identify a compound of a nonmetallic element that probably has the same structure? **(b)** VF_3 is prepared by the action of HF on heated VCl_3. Write a balanced equation for this reaction. **(c)** While VF_5 is a known compound, the other vanadium(V) halides are unknown. Suggest why these compounds might be unstable. (*Hint*: The reasons might have to do with both size and electronic factors.)

23.75 The galvanizing of iron sheet can be carried out electrolytically using a bath containing a zinc sulfate solution. The sheet is made the cathode, and a graphite anode is used. Calculate the cost of the electricity required to deposit a 0.49-mm layer of zinc on both sides of an iron sheet 2.0 m wide and 80 m long if the current is 30 A, the voltage is 3.5 V, and the energy efficiency of the process is 90%. Assume the cost of electricity is $0.082 per kilowatt hour. The density of zinc is 7.1 g/cm^3.

[23.76] Silver is found as Ag_2S in the ore argentite. **(a)** By using data in Table 17.2 and Appendix D.3, determine the equilibrium constant for the cyanidation of Ag_2S to $Ag(CN)_2^-$. **(b)** Based on your answer to part (a), would you consider cyanidation to be a practical means of leaching silver from argentite ore? **(c)** Silver is also found as AgCl in the ore horn silver. Would it be feasible to use cyanidation as a leaching process for this ore?

[23.77] The heats of atomization, ΔH_{atom}, in kJ/mol, of the first transition series of elements are as follows:

Element	Ca	Sc	Ti	V	Cr	Mn	Fe	Co	Ni	Cu
ΔH_{atom}	178	378	471	515	397	281	415	426	431	338

(a) Write an equation for the process involved in atomization, and describe the electronic and structural changes that occur. **(b)** ΔH_{atom} varies irregularly in the series following V. How can this be accounted for, at least in part, using the electronic configurations of the gaseous atoms? (*Hint*: Recall the discussions of Sections 6.8 and 6.9.)

eMedia Exercises

23.78 The exothermic reduction of iron by aluminum is shown in the **Thermite** movie (*eChapter 23.2*). **(a)** Determine the oxidation states of iron and aluminum in the reactants and products. **(b)** What is the purpose of adding potassium chlorate, sugar, and concentrated sulfuric acid at the beginning of the experiment? What principle of kinetics does this illustrate?

23.79 In the **Electroplating** movie (*eChapter 23.8*), chromium metal is reduced at the cathode of an electrolytic cell. The source of chromium ion to be reduced is aqueous chromic acid. **(a)** Using the information in the table, explain why sodium metal cannot be produced from an aqueous solution of sodium chloride.

Half-Reaction	E°_{red} **(V)**
$Cl_2(g) + 2e^- \longrightarrow 2Cl^-(aq)$	1.359
$Na^+(aq) + e^- \longrightarrow Na(s)$	−2.71
$2H^+(aq) + 2e^- \longrightarrow H_2(g)$	0.000
$2H_2O(l) + 2e^- \longrightarrow H_2(g) + 2OH^-(aq)$	−0.83
$Ca^{2+}(aq) + 2e^- \longrightarrow Ca(s)$	−2.87

The molten salt that is electrolyzed to produce sodium metal is actually a mixture of sodium chloride and calcium chloride. **(b)** Explain why another compound would be added to the sodium chloride. **(c)** Would the presence of $CaCl_2$ in the melt result in the production of impure sodium? Use data from the table to explain your reasoning.

23.80 Use the band theory **Metallic Bonding** activity (*eChapter 23.5*) to determine the melting points of two or more metals in the same transition series. **(a)** What appears to be the relationship between the melting point and the relative numbers of electrons in bonding and antibonding orbitals? **(b)** Explain why the electron-sea theory of metal bonding would not predict the observed trend in melting points of transition metals.

23.81 In the **Redox Chemistry of Iron and Copper** movie (*eChapter 23.8*), copper from a copper sulfate solution is reduced and forms a coating of copper metal on an iron nail. Based on this observation, predict what would happen to an iron impurity in crude copper during the electrolytic refinement of copper. Would the iron be oxidized and become part of the electrolyte solution in the cell, or would it become part of the anode sludge? Explain.

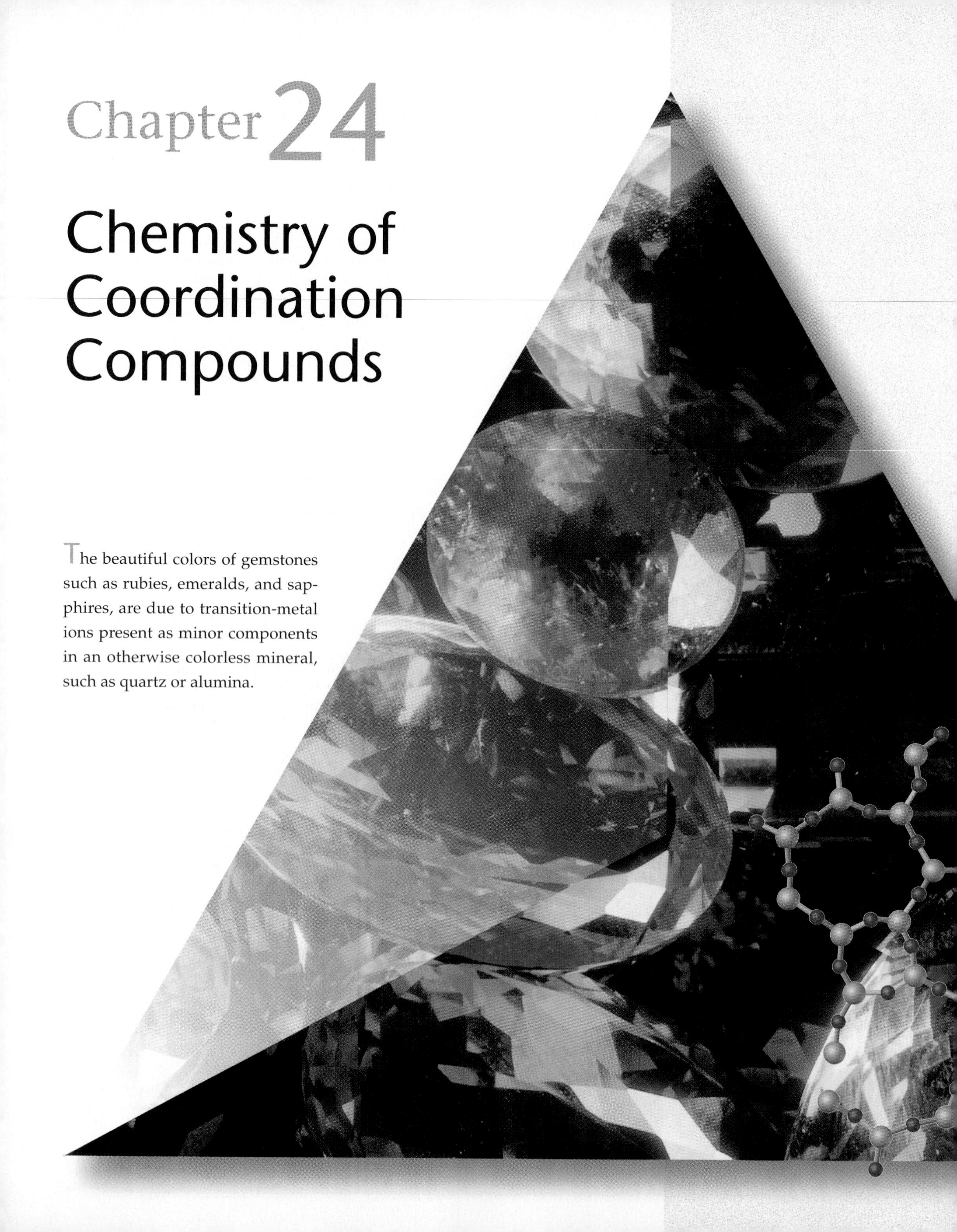

Chapter 24

Chemistry of Coordination Compounds

The beautiful colors of gemstones such as rubies, emeralds, and sapphires, are due to transition-metal ions present as minor components in an otherwise colorless mineral, such as quartz or alumina.

THE COLORS ASSOCIATED with chemistry are not only beautiful—they are also informative, providing insights into the structure and bonding of matter. Compounds of the transition metals constitute an important group of colored substances. Some of them are used in paint pigments; others produce the colors in glass and precious gems. Why do these compounds have color, and why do these colors change as the ions or molecules bonded to the metal change? The chemistry that we explore in this chapter will help us to answer these questions.

In earlier chapters we have seen that metal ions can function as Lewis acids, forming covalent bonds with a variety of molecules and ions that function as Lewis bases. (Section 16.11) We have encountered many ions and compounds that result from such interactions. We discussed $[Fe(H_2O)_6]^{3+}$ and $[Ag(NH_3)_2]^+$, for example, in our coverage of equilibria in Sections 16.11 and 17.5. Hemoglobin, an important iron compound that is responsible for the oxygen-carrying capacity of blood. (Sections 13.6 and 18.4) In Section 23.3 we saw that hydrometallurgy depends on the formation of species such as $[Au(CN)_2]^-$. In this chapter we will focus on the rich and important chemistry associated with such complex assemblies of metals surrounded by molecules and ions. Metal compounds of this kind are called *coordination compounds*.

▸ What's Ahead ◂

- We begin by introducing the concepts of *metal complexes* and *ligands*, and providing a brief history of the development of *coordination chemistry*.
- Next we examine some of the common geometries exhibited by coordination complexes for different *coordination numbers*.
- Our discussion then turns to *polydentate ligands*, which are ligands with more than one *donor atom*, and to some of their special properties, including their important roles in biological systems.
- We introduce the *nomenclature* used to name coordination compounds.
- Coordination compounds exhibit *isomerism*, in which two compounds have the same composition but different structures. The types of isomerism exhibited by coordination compounds are described, including *structural isomers*, *geometric isomers*, and *optical isomers*, which are two isomers of a compound that are mirror images of one another.
- We will discuss the basic notions of *color* and *magnetism* in coordination compounds.
- In order to explain some of the interesting spectral and magnetic properties of coordination compounds, we present the *crystal-field theory*.

24.1 Metal Complexes

Species such as $[Ag(NH_3)_2]^+$ that are assemblies of a central metal ion bonded to a group of surrounding molecules or ions are called **metal complexes** or merely *complexes*. If the complex carries a net charge, it is generally called a *complex ion*. (Section 17.5) Compounds that contain complexes are known as **coordination compounds**. Most of the coordination compounds that we will examine contain transition-metal ions, although ions of other metals can form complexes as well.

The molecules or ions that surround the metal ion in a complex are known as **ligands** (from the Latin word *ligare*, meaning "to bind"). There are two NH_3 ligands bonded to Ag^+ in $[Ag(NH_3)_2]^+$. Ligands are normally either anions or polar molecules. Every ligand has at least one unshared pair of valence electrons, as illustrated in the following examples:

$$:\ddot{O}(H)-H \qquad :N(H)_2-H \qquad :\ddot{Cl}:^- \qquad :C\equiv N:^-$$

In forming a complex, the ligands are said to *coordinate* to the metal.

The Development of Coordination Chemistry: Werner's Theory

Because compounds of the transition metals exhibit beautiful colors, the chemistry of these elements greatly fascinated chemists even before the periodic table was introduced. In the late 1700s through the 1800s many coordination compounds were isolated and studied. These compounds showed properties that seemed puzzling in light of the bonding theories at the time. Table 24.1 ▼, for example, lists a series of compounds that result from the reaction of cobalt(III) chloride with ammonia. These compounds have strikingly different colors. Even the last two listed, which were both formulated as $CoCl_3 \cdot 4NH_3$, have different colors.

All the compounds in Table 24.1 are strong electrolytes (Section 4.1), but they yield different numbers of ions when dissolved in water. For example, dissolving $CoCl_3 \cdot 6NH_3$ in water yields four ions per formula unit, whereas $CoCl_3 \cdot 5NH_3$ yields only three ions per formula unit. Furthermore, the reaction of the compounds with excess aqueous silver nitrate leads to the precipitation of variable amounts of AgCl(*s*); the precipitation of AgCl(*s*) in this way is often used to test for the number of "free" Cl^- ions in an ionic compound. When $CoCl_3 \cdot 6NH_3$ is treated with excess $AgNO_3(aq)$, three moles of AgCl(*s*) are produced per mole of complex, so all three Cl^- ions in the formula can react to form AgCl(*s*). By contrast, when $CoCl_3 \cdot 5NH_3$ is treated with $AgNO_3(aq)$ in an analogous fashion, only two moles of AgCl(*s*) precipitate per mole of complex; one of the Cl^- ions in the compound does not react to form AgCl(*s*). These results are summarized in Table 24.1.

TABLE 24.1 Properties of Some Ammonia Complexes of Cobalt(III)

Original Formulation	Color	Ions per Formula Unit	"Free" Cl^- Ions per Formula Unit	Modern Formulation
$CoCl_3 \cdot 6NH_3$	Orange	4	3	$[Co(NH_3)_6]Cl_3$
$CoCl_3 \cdot 5NH_3$	Purple	3	2	$[Co(NH_3)_5Cl]Cl_2$
$CoCl_3 \cdot 4NH_3$	Green	2	1	*trans*-$[Co(NH_3)_4Cl_2]Cl$
$CoCl_3 \cdot 4NH_3$	Violet	2	1	*cis*-$[Co(NH_3)_4Cl_2]Cl$

In 1893 the Swiss chemist Alfred Werner (1866–1919) proposed a theory that successfully explained the observations in Table 24.1, and it became the basis for understanding coordination chemistry. Werner proposed that metal ions exhibit both "primary" and "secondary" valences. The primary valence is the oxidation state of the metal, which for the complexes in Table 24.1 is +3. (Section 4.4) The secondary valence is the number of atoms directly bonded to the metal ion, which is also called the **coordination number**. For these cobalt complexes, Werner deduced a coordination number of six with the ligands in an octahedral arrangement (Figure 9.9) around the Co ion.

Werner's theory provided a beautiful explanation for the results in Table 24.1. The NH_3 molecules in the complexes are ligands that are bonded to the Co ion; if there are fewer than six NH_3 molecules, the remaining ligands are Cl^- ions. The central metal and the ligands bound to it constitute the **coordination sphere** of the complex. In writing the chemical formula for a coordination compound, Werner suggested using square brackets to set off the groups within the coordination sphere from other parts of the compound. He therefore proposed that $CoCl_3 \cdot 6NH_3$ and $CoCl_3 \cdot 5NH_3$ are better written as $[Co(NH_3)_6]Cl_3$ and $[Co(NH_3)_5Cl]Cl_2$, respectively. He further proposed that the chloride ions that are part of the coordination sphere are bound so tightly that they do not become freed up when the complex is dissolved in water. Thus, dissolving $[Co(NH_3)_5Cl]Cl_2$ in water produces a $[Co(NH_3)_5Cl]^{2+}$ ion and two Cl^- ions; only the two "free" Cl^- ions are able to react with $Ag^+(aq)$ to form $AgCl(s)$.

Werner's ideas also explained why there are two distinctly different forms of $CoCl_3 \cdot 4NH_3$. Using Werner's postulates, we formulate the compound as $[Co(NH_3)_4Cl_2]Cl$. As shown in Figure 24.1 ▶, there are two different ways to arrange the ligands in the $[Co(NH_3)_4Cl_2]^+$ complex, called the *cis* and *trans* forms. In *cis*-$[Co(NH_3)_4Cl_2]^+$ the two chloride ligands occupy adjacent vertices of the octahedral arrangement. In *trans*-$[Co(NH_3)_4Cl_2]^+$ the chlorides are opposite one another. As seen in Table 24.1, the difference in these arrangements causes the complexes to have different colors.

The insight into the bonding in coordination compounds that Werner provided is even more remarkable when we realize that his theory predated Lewis's ideas of covalent bonding by more than 20 years! Because of his tremendous contributions to coordination chemistry, Werner was awarded the 1913 Nobel Prize in chemistry.

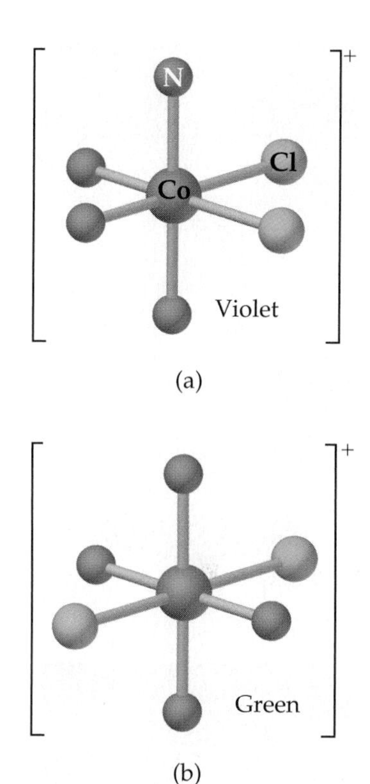

▲ **Figure 24.1** The two forms (isomers) of the complex $[Co(NH_3)_4Cl_2]^+$. In (a) *cis*-$[Co(NH_3)_4Cl_2]^+$ the two Cl ligands occupy adjacent vertices of the octahedron, whereas in (b) *trans*-$[Co(NH_3)_4Cl_2]^+$ they are opposite one another. (The blue spheres represent the coordinated NH_3 ligands.)

SAMPLE EXERCISE 24.1

Palladium(II) tends to form complexes with a coordination number of 4. One such compound was originally formulated as $PdCl_2 \cdot 3NH_3$. **(a)** Suggest the appropriate coordination compound formulation for this compound. **(b)** Suppose an aqueous solution of the compound is treated with excess $AgNO_3(aq)$. How many moles of $AgCl(s)$ are formed per mole of $PdCl_2 \cdot 3NH_3$?

Solution **(a) Analyze and Plan:** We are given the coordination number of Pd(II) and the other groups in the compound. To write the formula correctly, we need to determine what ligands are attached to Pd(II) in the compound.

Solve: By analogy to the ammonia complexes of cobalt(III), we might expect that the three NH_3 groups of $PdCl_2 \cdot 3NH_3$ serve as ligands attached to the Pd(II) ion. The fourth ligand around Pd(II) is one of the chloride ions. The second chloride ion is not a ligand; it serves only as an anion in this ionic compound. We conclude that the correct formulation is $[Pd(NH_3)_3Cl]Cl$.

(b) We expect that the chloride ion that serves as a ligand will not be precipitated as $AgCl(s)$ following the reaction with $AgNO_3(aq)$. Thus, only the single "free" Cl^- can react. We therefore expect to produce one mole of $AgCl(s)$ per mole of complex. The balanced equation is the following:

$$[Pd(NH_3)_3Cl]Cl(aq) + AgNO_3(aq) \longrightarrow [Pd(NH_3)_3Cl]NO_3(aq) + AgCl(s)$$

This is a metathesis reaction (Section 4.2) in which one of the cations is the $[Pd(NH_3)_3Cl]^+$ complex ion.

PRACTICE EXERCISE

Predict the number of ions produced per formula unit in an aqueous solution of $CoCl_2 \cdot 6H_2O$.

Answer: three (the complex ion and two chloride ions)

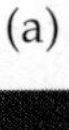

(a)

(b)

▲ **Figure 24.2** When an aqueous solution of NH_4SCN is added to an aqueous solution of Fe^{3+}, the intensely colored $[Fe(H_2O)_5NCS]^{2+}$ ion is formed.

The Metal-Ligand Bond

The bond between a ligand and a metal ion is an example of an interaction between a Lewis base and a Lewis acid. (Section 16.11) Because the ligands have unshared pairs of electrons, they can function as Lewis bases (electron-pair donors). Metal ions (particularly transition-metal ions) have empty valence orbitals, so they can act as Lewis acids (electron-pair acceptors). We can picture the bond between the metal ion and ligand as the result of their sharing a pair of electrons that was initially on the ligand:

$$Ag^+(aq) + 2\!:\!\underset{\displaystyle H}{\overset{\displaystyle H}{N}}\!-\!H(aq) \longrightarrow \left[H\!-\!\underset{\displaystyle H}{\overset{\displaystyle H}{N}}\!:\!Ag\!:\!\underset{\displaystyle H}{\overset{\displaystyle H}{N}}\!-\!H \right]^+(aq) \quad [24.1]$$

The formation of metal-ligand bonds can markedly alter the properties we observe for the metal ion. A metal complex is a distinct chemical species that has physical and chemical properties different from the metal ion and the ligands from which it is formed. Complexes, for example, may have colors that differ dramatically from those of their component metal ions and ligands. Figure 24.2 ◀ shows the color change that occurs when aqueous solutions of SCN^- and Fe^{3+} are mixed, forming $[Fe(H_2O)_5NCS]^{2+}$.

Complex formation can also significantly change other properties of metal ions, such as their ease of oxidation or reduction. Ag^+, for example, is readily reduced in water.

$$Ag^+(aq) + e^- \longrightarrow Ag(s) \qquad E^\circ = +0.799\ V \quad [24.2]$$

In contrast, the $[Ag(CN)_2]^-$ ion is not so easily reduced because complexation by CN^- ions stabilizes silver in the +1 oxidation state.

$$[Ag(CN)_2]^-(aq) + e^- \longrightarrow Ag(s) + 2CN^-(aq) \qquad E^\circ = -0.31\ V \quad [24.3]$$

Hydrated metal ions are actually complex ions in which the ligand is water. Thus, $Fe^{3+}(aq)$ consists largely of $[Fe(H_2O)_6]^{3+}$. (Section 16.11) Complex ions form in aqueous solutions from reactions in which ligands such as NH_3, SCN^-, and CN^- replace H_2O molecules in the coordination sphere of the metal ion.

Charges, Coordination Numbers, and Geometries

The charge of a complex is the sum of the charges on the central metal and on its surrounding ligands. In $[Cu(NH_3)_4]SO_4$ we can deduce the charge on the complex if we first recognize that SO_4 represents the sulfate ion and therefore has a 2− charge. Because the compound is neutral, the complex ion must have a 2+ charge, $[Cu(NH_3)_4]^{2+}$. We can then use the charge of the complex ion to deduce the oxidation number of copper. Because the NH_3 ligands are neutral molecules, the oxidation number of copper must be +2.

$$+2 + 4(0) = +2$$
$$\downarrow \quad \downarrow \quad \swarrow$$
$$[Cu(NH_3)_4]^{2+}$$

SAMPLE EXERCISE 24.2

What is the oxidation number of the central metal in $[Rh(NH_3)_5Cl](NO_3)_2$?

Solution

Analyze and Plan: In order to determine the oxidation number of the Rh atom, we need to figure out what charges are contributed by the other groups in the substance. The overall charge is zero, so the oxidation number of the metal must balance the charge due to the rest of the compound.

Solve: The NO_3 group is the nitrate anion, which has a 1− charge, NO_3^-. The NH_3 ligands are neutral and the Cl is a coordinated chloride ion, which has a 1− charge, Cl^-. The sum of all the charges must be zero.

$$x + 5(0) + (-1) + 2(-1) = 0$$

$$[Rh(NH_3)_5Cl](NO_3)_2$$

The oxidation number of rhodium, x, must therefore be +3.

PRACTICE EXERCISE

What is the charge of the complex formed by a platinum(II) metal ion surrounded by two ammonia molecules and two bromide ions?

Answer: zero

SAMPLE EXERCISE 24.3

A complex ion contains a chromium(III) bound to four water molecules and two chloride ions. What is its formula?

Solution The oxidation state of the metal is +3, water is neutral, and chloride has a 1− charge:

$$+3 + 4(0) + 2(-1) = +1$$

$$Cr(H_2O)_4Cl_2$$

The charge on the ion is 1+, $[Cr(H_2O)_4Cl_2]^+$.

PRACTICE EXERCISE

Write the formula for the complex described in the Practice Exercise accompanying Sample Exercise 24.2.

Answer: $[Pt(NH_3)_2Br_2]$

Recall that the number of atoms directly bonded to the metal atom in a complex is called the *coordination number*. The atom of the ligand bound directly to the metal is called the **donor atom**. Nitrogen, for example, is the donor atom in the $[Ag(NH_3)_2]^+$ complex shown in Equation 24.1. The silver ion in $[Ag(NH_3)_2]^+$ has a coordination number of 2, whereas each cobalt ion in the Co(III) complexes in Table 24.1 has a coordination number of 6.

Some metal ions exhibit constant coordination numbers. The coordination number of chromium(III) and cobalt(III) is invariably 6, for example, and that of platinum(II) is always 4. The coordination numbers of most metal ions vary with the ligand, however. The most common coordination numbers are 4 and 6.

The coordination number of a metal ion is often influenced by the relative sizes of the metal ion and the surrounding ligands. As the ligand gets larger, fewer can coordinate to the metal ion. Thus, iron(III) is able to coordinate to six fluorides in $[FeF_6]^{3-}$ but coordinates to only four chlorides in $[FeCl_4]^-$. Ligands that transfer substantial negative charge to the metal also produce reduced coordination numbers. For example, six neutral ammonia molecules can coordinate to nickel(II), forming $[Ni(NH_3)_6]^{2+}$, but only four negatively charged cyanide ions can coordinate, forming $[Ni(CN)_4]^{2-}$.

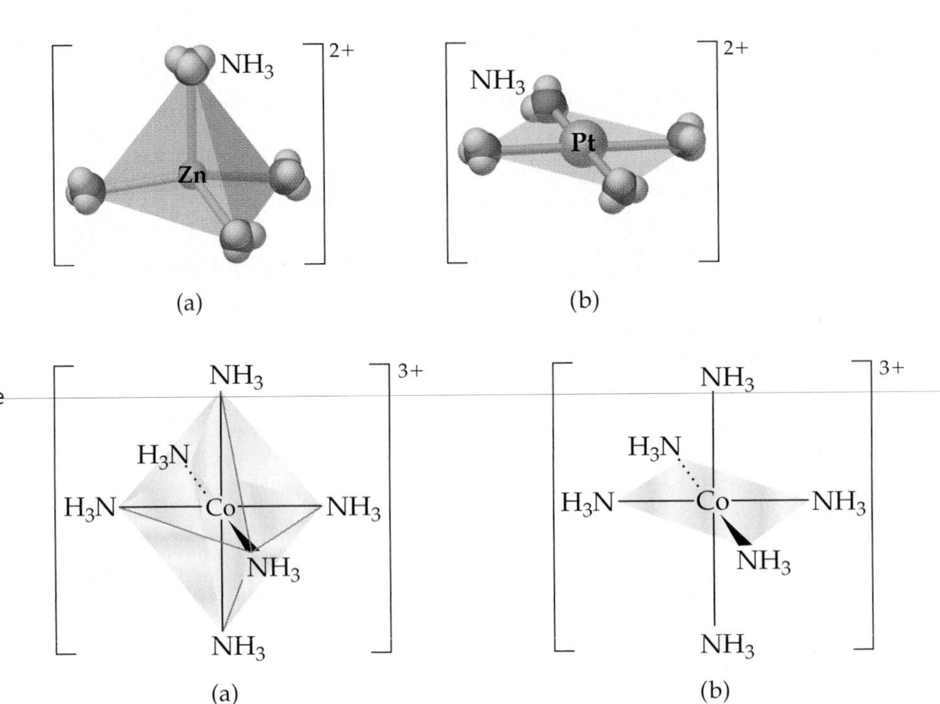

▶ **Figure 24.3** Structures of (a) $[Zn(NH_3)_4]^{2+}$ and (b) $[Pt(NH_3)_4]^{2+}$, illustrating the tetrahedral and square-planar geometries, respectively. These are the two common geometries for complexes in which the metal ion has a coordination number of 4.

▶ **Figure 24.4** Two representations of an octahedral coordination sphere, the common geometric arrangement for complexes in which the metal ion has a coordination number of 6.

Four-coordinate complexes have two common geometries—tetrahedral and square planar—as shown in Figure 24.3 ▲. The tetrahedral geometry is the more common of the two and is especially common among nontransition metals. The square-planar geometry is characteristic of transition-metal ions with eight *d* electrons in the valence shell, such as platinum(II) and gold(III).

The vast majority of 6-coordinate complexes have an octahedral geometry, as shown in Figure 24.4(a) ▲. The octahedron is often represented as a planar square with ligands above and below the plane, as in Figure 24.4(b). Recall, however, that all positions on an octahedron are geometrically equivalent. (Section 9.2)

3-D MODEL
Ethalyenediamine

24.2 Ligands with More than One Donor Atom

The ligands that we have discussed so far, such as NH_3 and Cl^-, are called **monodentate ligands** (from the Latin, meaning "one-toothed"). These ligands possess a single donor atom and are able to occupy only one site in a coordination sphere. Some ligands have two or more donor atoms that can simultaneously coordinate to a metal ion, thereby occupying two or more coordination sites. They are called **polydentate ligands** ("many-toothed"). Because they appear to grasp the metal between two or more donor atoms, polydentate ligands are also known as **chelating agents** (from the Greek word *chele*, "claw"). One such ligand is *ethylenediamine*.

$$H_2\ddot{N}-CH_2-CH_2-\ddot{N}H_2$$

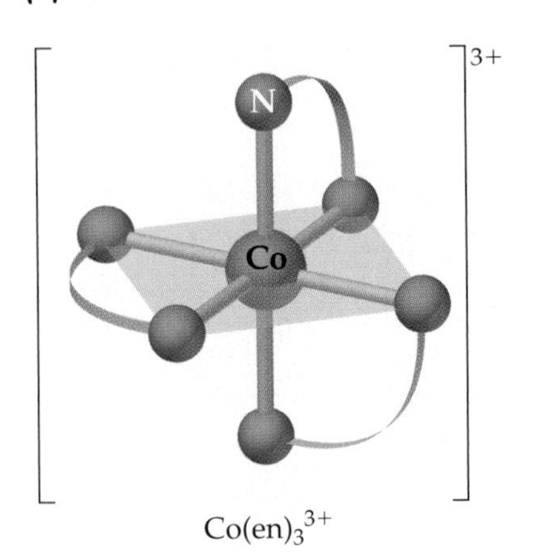

$Co(en)_3{}^{3+}$

▲ **Figure 24.5** The $[Co(en)_3]^{3+}$ ion, showing how each bidentate ethylenediamine ligand is able to occupy two positions in the coordination sphere.

Ethylenediamine, which is abbreviated en, has two nitrogen atoms (shown in color) that have unshared pairs of electrons. These donor atoms are sufficiently far apart that the ligand can wrap around a metal ion with the two nitrogen atoms simultaneously bonding to the metal in adjacent positions. The $[Co(en)_3]^{3+}$ ion, which contains three ethylenediamine ligands in the octahedral coordination sphere of cobalt(III), is shown in Figure 24.5 ◀. Notice that the ethylenediamine has been written in a shorthand notation as two nitrogen atoms connected by an

3-D MODEL
Oxalate Ion

Oxalate ion

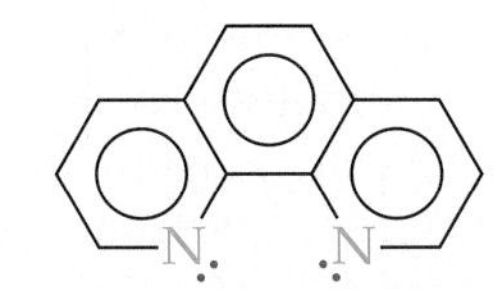

Ortho-phenanthroline
(*o*-phen)

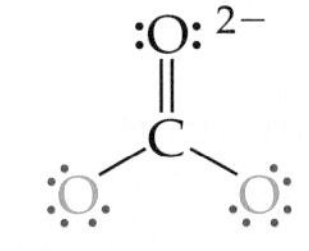

Carbonate ion

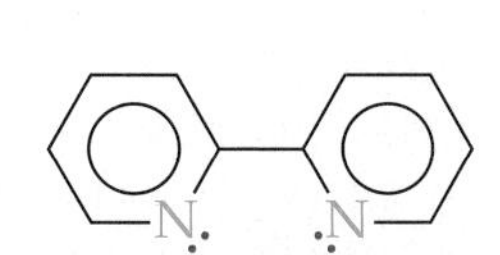

Bipyridine
(bipy)

▲ **Figure 24.6** Structures of some bidentate ligands. The coordinating atoms are shown in blue.

arc. Ethylenediamine is a **bidentate ligand** ("two-toothed" ligand) because it can occupy two coordination sites. The structures of several other bidentate ligands are shown in Figure 24.6 ▲.

The ethylenediaminetetraacetate ion, abbreviated $[EDTA]^{4-}$, is an important polydentate ligand that has six donor atoms.

$[EDTA]^{4-}$

It can wrap around a metal ion using all six of these donor atoms, as shown in Figure 24.7 ▶, although it sometimes binds to a metal using only five of its six donor atoms.

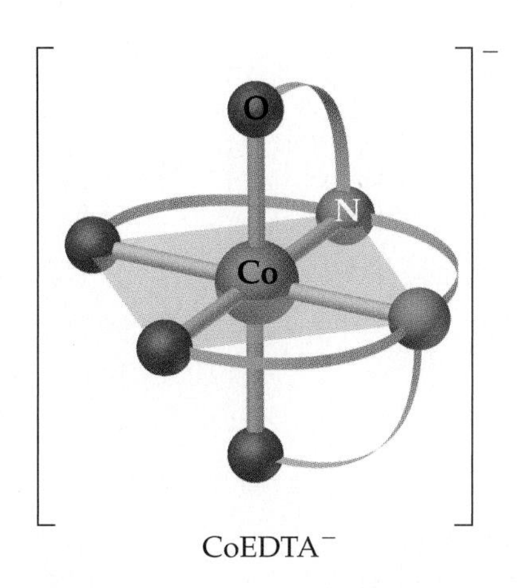

$CoEDTA^{-}$

▲ **Figure 24.7** The $[CoEDTA]^{-}$ ion, showing how the ethylenediaminetetraacetate ion is able to wrap around a metal ion, occupying six positions in the coordination sphere.

In general, chelating ligands form more stable complexes than do related monodentate ligands. The formation constants for $[Ni(NH_3)_6]^{2+}$ and $[Ni(en)_3]^{2+}$, shown in Equations 24.4 and 24.5, illustrate this observation.

$$[Ni(H_2O)_6]^{2+}(aq) + 6NH_3(aq) \rightleftharpoons [Ni(NH_3)_6]^{2+}(aq) + 6H_2O(l) \qquad K_f = 1.2 \times 10^9 \qquad [24.4]$$

$$[Ni(H_2O)_6]^{2+}(aq) + 3en(aq) \rightleftharpoons [Ni(en)_3]^{2+}(aq) + 6H_2O(l) \qquad K_f = 6.8 \times 10^{17} \qquad [24.5]$$

Although the donor atom is nitrogen in both instances, $[Ni(en)_3]^{2+}$ has a formation constant that is more than 10^8 times larger than that of $[Ni(NH_3)_6]^{2+}$. The generally larger formation constants for polydentate ligands as compared with the corresponding monodentate ligands is known as the **chelate effect**. We examine the origin of this effect in greater detail in "A Closer Look" in this section.

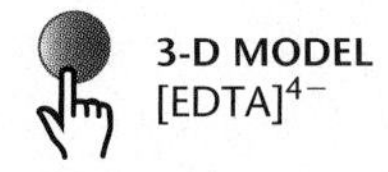

Chelating agents are often used to prevent one or more of the customary reactions of a metal ion without actually removing it from solution. For example, a metal ion that interferes with a chemical analysis can often be complexed and its interference thereby removed. In a sense, the chelating agent hides the metal ion. For this reason, scientists sometimes refer to these ligands as sequestering agents. (The word *sequester* means to remove, set apart, or separate.)

Phosphates such as sodium tripolyphosphate, shown here, are used to complex or sequester metal ions in hard water so these ions cannot interfere with the action of soap or detergents: (Section 18.6)

$$Na_5\left[O{-}P(=O)(O){-}O{-}P(=O)(O){-}O{-}P(=O)(O){-}O\right]$$

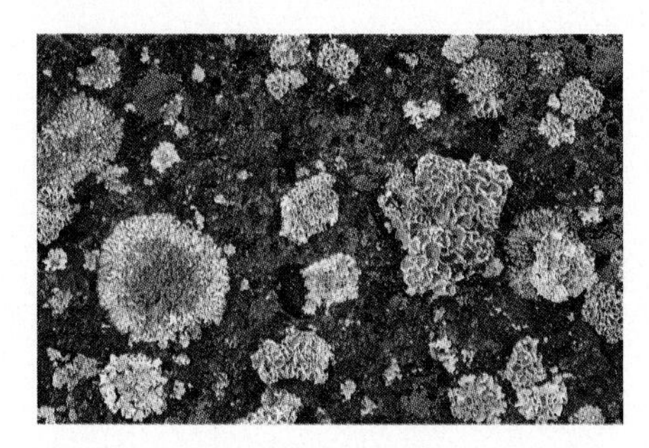

▲ **Figure 24.8** Lichens growing on a rock surface. Lichens obtain the nutrients needed for growth from a variety of sources. Using chelating agents, they are able to extract needed metallic elements from the rocks on which they grow.

Chelating agents such as EDTA are used in consumer products, including many prepared foods such as salad dressings and frozen desserts, to complex trace metal ions that catalyze decomposition reactions. Chelating agents are used in medicine to remove metal ions such as Hg^{2+}, Pb^{2+}, and Cd^{2+}, which are detrimental to health. One method of treating lead poisoning is to administer $Na_2[Ca(EDTA)]$. The EDTA chelates the lead, allowing it to be removed from the body via urine. Chelating agents are also quite common in nature. Mosses and lichens secrete chelating agents to capture metal ions from the rocks they inhabit (Figure 24.8 ◀).

Metals and Chelates in Living Systems

Ten of the 29 elements known to be necessary for human life are transition metals. ("Chemistry and Life," Section 2.7) These ten—V, Cr, Mn, Fe, Co, Ni, Cu, Zn, Mo, and Cd—owe their roles in living systems mainly to their ability to form complexes with a variety of donor groups present in biological systems. Metal ions are integral parts of many enzymes, which are the body's catalysts. (Section 14.7)

Although our bodies require only small quantities of metals, deficiencies can lead to serious illness. A deficiency of manganese, for example, can lead to convulsive disorders. Some epilepsy patients have been helped by the addition of manganese to their diets.

Among the most important chelating agents in nature are those derived from the *porphine* molecule, which is shown in Figure 24.10 ▶. This molecule can coordinate to a metal using the four nitrogen atoms as donors. Upon coordination to a metal, the two H atoms shown bonded to nitrogen are displaced. Complexes derived from porphine are called **porphyrins**. Different porphyrins contain different metal ions and have different substituent groups attached to the carbon atoms at the ligand's periphery. Two of the most important porphyrin or porphyrin-like compounds are *heme*, which contains Fe(II), and *chlorophyll*, which contains Mg(II).

A Closer Look Entropy and the Chelate Effect

When we examined thermodynamics more closely in Chapter 19, we learned that many chemical processes are driven by positive changes in the entropy of the system. (Section 19.3) The special stability associated with the formation of chelates, called the *chelate effect*, can also be explained by looking at the entropy changes that occur when polydentate ligands bind to a metal ion. To understand this effect better, let's look at some reactions in which two H_2O ligands of the square-planar Cu(II) complex $[Cu(H_2O)_4]^{2+}$ are replaced by other ligands. First, let's consider replacing the H_2O ligands with NH_3 ligands at 27°C to form $[Cu(H_2O)_2(NH_3)_2]^{2+}$, the structure of which is shown in Figure 24.9(a) ▶.

$$[Cu(H_2O)_4]^{2+}(aq) + 2NH_3(aq) \rightleftharpoons [Cu(H_2O)_2(NH_3)_2]^{2+}(aq) + 2H_2O(l)$$

$$\Delta H^\circ = -46\ \text{kJ};\quad \Delta S^\circ = -8.4\ \text{J/K};\quad \Delta G^\circ = -43\ \text{kJ}$$

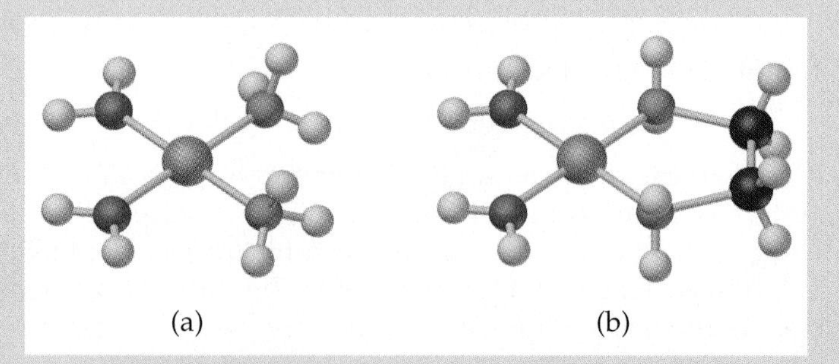

▲ **Figure 24.9** Ball-and-stick representations of the square-planar complexes (a) $[Cu(H_2O)_2(NH_3)_2]^{2+}$ and (b) $[Cu(H_2O)_2(en)]^{2+}$. The red spheres represent the H_2O ligands, and the blue spheres represent the NH_3 or en ligands.

The thermodynamic data provide us with information about the relative abilities of H_2O and NH_3 to serve as ligands in these systems. In general, NH_3 binds more tightly to metal ions than H_2O, so these kinds of substitution reactions are exothermic ($\Delta H < 0$). The stronger bonding of the NH_3 ligands also causes $[Cu(H_2O)_2(NH_3)_2]^{2+}$ to be more rigid, which is probably the reason that the entropy change for the reaction is slightly negative. By using Equation 19.18, we can use the value of ΔG° to calculate the equilibrium constant of the reaction at 27°C. The resulting value, $K_{eq} = 3.1 \times 10^7$, tells us that the equilibrium lies far to the right, favoring substitution of H_2O by NH_3. For this equilibrium, the change in the enthalpy is large and negative enough to overcome the negative change in the entropy.

How does this situation change if instead of two NH_3 ligands we use a single bidentate ethylenediamine (en) ligand and

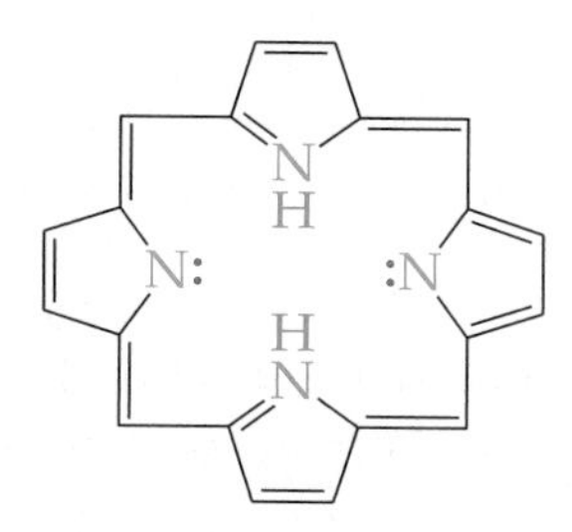

▲ **Figure 24.10** Structure of the porphine molecule. This molecule forms a tetradentate ligand with the loss of the two protons bound to nitrogen atoms. Porphine is the basic component of porphyrins, complexes that play a variety of important roles in nature.

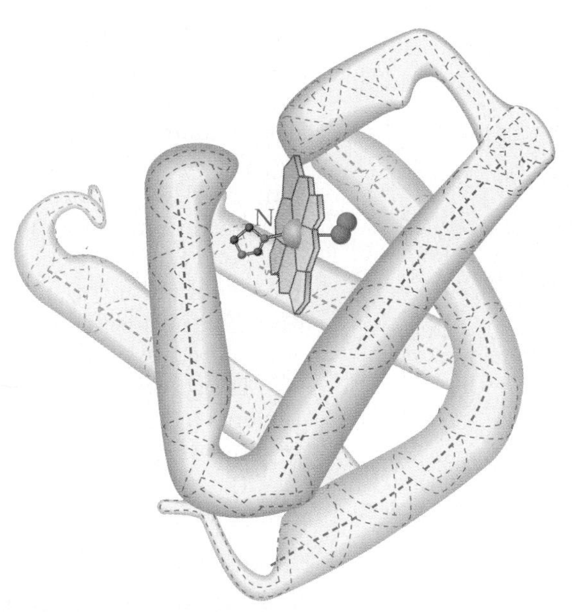

▲ **Figure 24.11** A schematic structure of myoglobin, a protein that stores oxygen in cells. Myoglobin has a molecular weight of about 18,000 amu and contains one heme unit, symbolized by the red disk. The heme unit is bound to the protein through a nitrogen-containing ligand, represented by the blue N on the left. In the oxygenated form an O_2 molecule is coordinated to the heme group, as shown. The three-dimensional structure of the protein chain is represented by the continuous purple cylinder. The helical sections are denoted by the dashed lines. The protein wraps around to make a kind of pocket for the heme group.

Figure 24.11 ▲ shows a schematic structure of myoglobin, a protein that contains one heme group. Myoglobin is a *globular protein*, one that folds into a compact, roughly spherical shape. Globular proteins are generally soluble in water and are mobile within cells. Myoglobin is found in the cells of skeletal muscle,

form $[Cu(H_2O)_2(en)]^{2+}$ [Figure 24.9(b)]? The equilibrium reaction and thermodynamic data are

$$[Cu(H_2O)_4]^{2+}(aq) + en(aq) \rightleftharpoons [Cu(H_2O)_2(en)]^{2+}(aq) + 2H_2O(l)$$

$$\Delta H^\circ = -54 \text{ kJ}; \quad \Delta S^\circ = +23 \text{ J/K}; \quad \Delta G^\circ = -61 \text{ kJ}$$

The en ligand binds slightly more strongly to a Cu^{2+} ion than two NH_3 ligands, so the enthalpy change on forming $[Cu(H_2O)_2(en)]^{2+}$ is slightly more negative than for $[Cu(H_2O)_2(NH_3)_2]^{2+}$. There is a big difference in the entropy change, however. Whereas the entropy change for forming $[Cu(H_2O)_2(NH_3)_2]^{2+}$ is negative, the entropy change for forming $[Cu(H_2O)_2(en)]^{2+}$ is positive, indicating a greater degree of disorder. We can explain this observation by using the concepts we discussed in Section 19.4. Because a single en ligand occupies two coordination sites, two molecules of H_2O are released upon binding one en ligand. Thus, there are three molecules on the right side of the equation, whereas there are only two on the left side, all of which are part of the same aqueous solution. The greater number of molecules on the right leads to the positive entropy change for the equilibrium. The slightly more negative value of ΔH° coupled with the positive entropy change leads to a much more negative value of ΔG° and a correspondingly larger equilibrium constant: $K_{eq} = 4.2 \times 10^{10}$.

We can combine the earlier equations to show that the formation of $[Cu(H_2O)_2(en)]^{2+}$ is thermodynamically preferred over the formation of $[Cu(H_2O)_2(NH_3)_2]^{2+}$. If we add the second reaction to the reverse of the first reaction, we obtain

$$[Cu(H_2O)_2(NH_3)_2]^{2+}(aq) + en(aq) \rightleftharpoons [Cu(H_2O)_2(en)]^{2+}(aq) + 2NH_3(aq)$$

The thermochemical data for this equilibrium reaction can be obtained from those given earlier.

$$\Delta H^\circ = (-54 \text{ kJ}) - (-46 \text{ kJ}) = -8 \text{ kJ}$$

$$\Delta S^\circ = (+23 \text{ J/K}) - (-8.4 \text{ J/K}) = +31 \text{ J/K}$$

$$\Delta G^\circ = (-61 \text{ kJ}) - (-43 \text{ kJ}) = -18 \text{ kJ}$$

Notice that at 27°C (300 K), the entropic contribution ($-T\Delta S^\circ$) to the free-energy change is negative and greater in magnitude than the enthalpic contribution (ΔH°). The resulting value of K_{eq} for this reaction, 1.4×10^3, shows that the formation of the chelate complex is much more favorable.

The chelate effect is important in biochemistry and molecular biology. The additional thermodynamic stabilization provided by entropic effects helps to stabilize biological metal-chelate complexes, such as porphyrins, and can allow changes in the oxidation state of the metal ion while retaining the structural integrity of the complex.

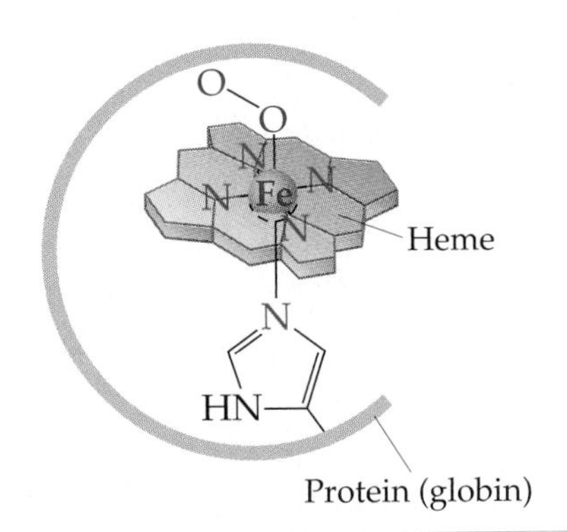

▲ **Figure 24.12** Schematic representation of oxymyoglobin or oxyhemoglobin. The iron is bound to four nitrogen atoms of the porphyrin, to a nitrogen from the surrounding protein, and to an O_2 molecule.

particularly in seals, whales, and porpoises. It stores oxygen in cells until it is needed for metabolic activities. Hemoglobin, the protein that transports oxygen in human blood, is made up of four heme-containing subunits, each of which is very similar to myoglobin.

The coordination environment of the iron in myoglobin and hemoglobin is illustrated schematically in Figure 24.12 ◀. The iron is coordinated to the four nitrogen atoms of the porphyrin and to a nitrogen atom from the protein chain. The sixth position around the iron is occupied either by O_2 (in oxyhemoglobin, the bright red form) or by water (in deoxyhemoglobin, the purplish red form). The oxy form is shown in Figure 24.12. Some substances, such as CO, are poisonous because they bind to iron more strongly than does O_2. (Section 18.4)

The **chlorophylls**, which are porphyrins that contain Mg(II), are the key components in the conversion of solar energy into forms that can be used by living organisms. This process, called **photosynthesis**, occurs in the leaves of green plants. In photosynthesis, carbon dioxide and water are converted to carbohydrate, with the release of oxygen.

$$6CO_2 + 6H_2O \xrightarrow{48\,h\nu} C_6H_{12}O_6 + 6O_2 \quad [24.6]$$

The product of this reaction is the sugar glucose, $C_6H_{12}O_6$, which serves as a fuel in biological systems. (Section 5.8) The formation of one mole of glucose requires the absorption of 48 mol of photons from sunlight or other sources of light. The photons are absorbed by chlorophyll-containing pigments in the leaves of plants. The structure of the most abundant chlorophyll, called chlorophyll *a*, is shown in Figure 24.13 ◀.

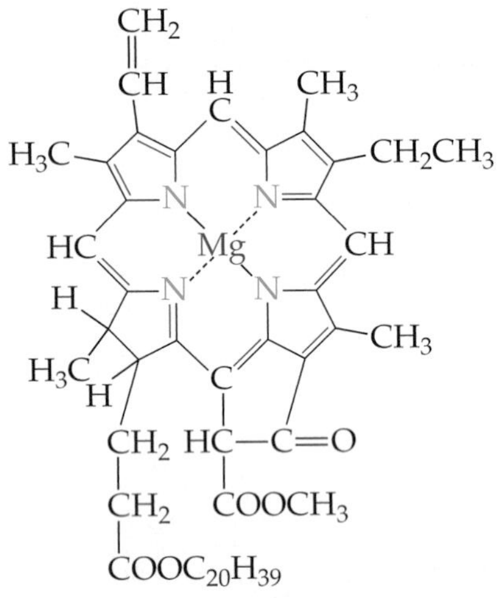

▲ **Figure 24.13** Structure of chlorophyll *a*. All chlorophyll molecules are essentially alike; they differ only in details of the side chains.

Chlorophylls contain a Mg^{2+} ion bound to four nitrogen atoms arranged around the metal in a planar array. The nitrogen atoms are part of a porphine-like ring (Figure 24.10). The series of alternating, or *conjugated*, double bonds in the ring surrounding the metal ion is similar to ones found in many organic dyes. ("Chemistry at Work," Section 9.8) This system of conjugated double bonds makes it possible for chlorophyll to absorb light strongly in the visible region of the spectrum. Figure 24.14 ▼ compares the absorption spectrum of chlorophyll to the distribution of visible solar energy at Earth's surface. Chlorophyll is green because it absorbs red light (maximum absorption at 655 nm) and blue light (maximum absorption at 430 nm) and transmits green light.

The solar energy absorbed by chlorophyll is converted by a complex series of steps into chemical energy. This stored energy is then used to drive the reaction in Equation 24.6 to the right, a direction in which it is highly endothermic. Plant photosynthesis is nature's solar-energy-conversion machine; all living systems on Earth depend on it for continued existence (Figure 24.15 ▼).

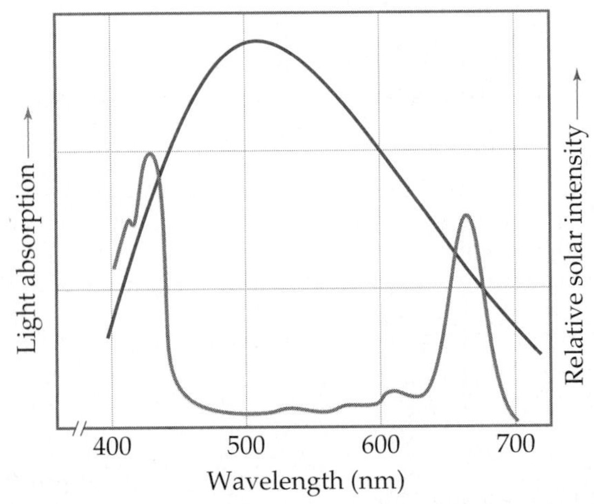

▲ **Figure 24.14** Absorption spectrum of chlorophyll (green curve), in comparison with the solar radiation at ground level (red curve).

▲ **Figure 24.15** The absorption and conversion of solar energy that occurs in leaves provides the energy necessary to drive all the living processes of the plant, including growth.

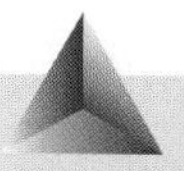

Chemistry and Life The Battle for Iron in Living Systems

Although iron is the fourth most abundant element in Earth's crust, living systems have difficulty assimilating enough iron to satisfy their needs. Consequently, iron-deficiency anemia is a common problem in humans. In plants, chlorosis, an iron deficiency that results in yellowing of leaves, is also commonplace. Living systems have difficulty assimilating iron because most iron compounds in nature have a very low solubility in water. Microorganisms have adapted to this problem by secreting an iron-binding compound, called a *siderophore*, that forms an extremely stable water-soluble complex with iron(III). One such complex is called *ferrichrome*; its structure is shown in Figure 24.16 ▼. The iron-binding strength of a siderophore is so great that it can extract iron from Pyrex™ glassware, and it readily solubilizes the iron in iron oxides.

The overall charge of ferrichrome is zero, which makes it possible for the complex to pass through the rather hydrophobic walls of cells. When a dilute solution of ferrichrome is added to a cell suspension, iron is found entirely within the cells in an hour. When ferrichrome enters the cell, the iron is removed through an enzyme-catalyzed reaction that reduces the iron to iron(II). Iron in the lower oxidation state is not strongly complexed by the siderophore. Microorganisms thus acquire iron by excreting a siderophore into their immediate environment and then taking the resulting iron complex into the cell. The overall process is illustrated in Figure 24.17 ▶.

In humans, iron is assimilated from food in the intestine. A protein called *transferrin* binds iron and transports it across the intestinal wall to distribute it to other tissues in the body. The normal adult carries a total of about 4 g of iron. At any one time, about 3 g, or 75%, of this iron is in the blood, mostly in the form of hemoglobin. Most of the remainder is carried by transferrin.

A bacterium that infects the blood requires a source of iron if it is to grow and reproduce. The bacterium excretes a siderophore into the blood to compete with transferrin for the iron it holds. The formation constants for iron binding are about the same for transferrin and siderophores. The more iron available to the bacterium, the more rapidly it can reproduce, and, thus, the more harm it can do. Several years ago, New Zealand clinics regularly gave iron supplements to infants soon after birth. However, the incidence of certain bacterial infections was eight times higher in treated than in untreated infants. Presumably, the presence of more iron in the blood than absolutely necessary makes it easier for bacteria to obtain the iron needed for growth and reproduction.

In the United States it is common medical practice to supplement infant formula with iron sometime during the first year of life because human milk is virtually devoid of iron. Given what is now known about iron metabolism by bacteria, many research workers in nutrition believe that iron supplementation is not generally justified or wise.

For bacteria to continue to multiply in the bloodstream, they must synthesize new supplies of siderophores. Synthesis of siderophores in bacteria slows, however, as the temperature is increased above the normal body temperature of 37°C, and it stops completely at 40°C. This suggests that fever in the presence of an invading microbe is a mechanism used by the body to deprive bacteria of iron.

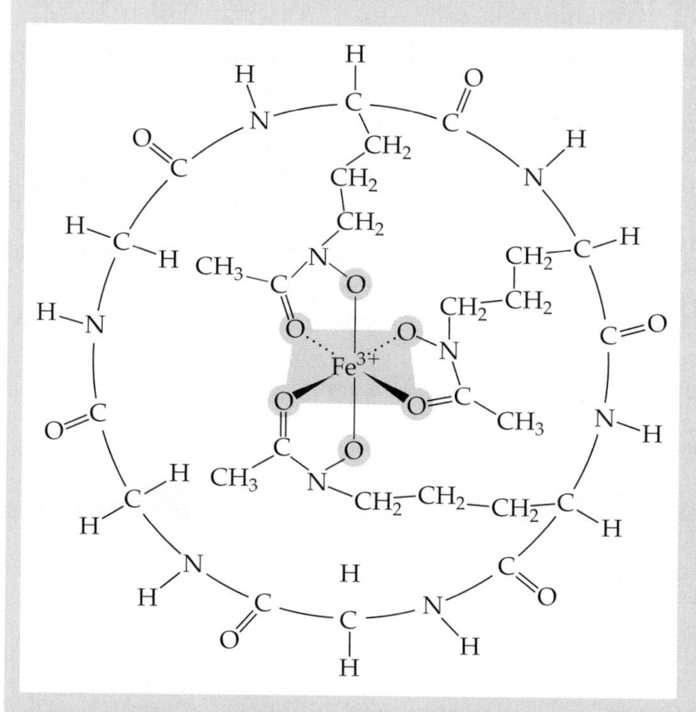

▲ **Figure 24.16** The structure of ferrichrome. In this complex an Fe^{3+} ion is coordinated by six oxygen atoms. The complex is very stable; it has a formation constant of about 10^{30}. The overall charge of the complex is zero.

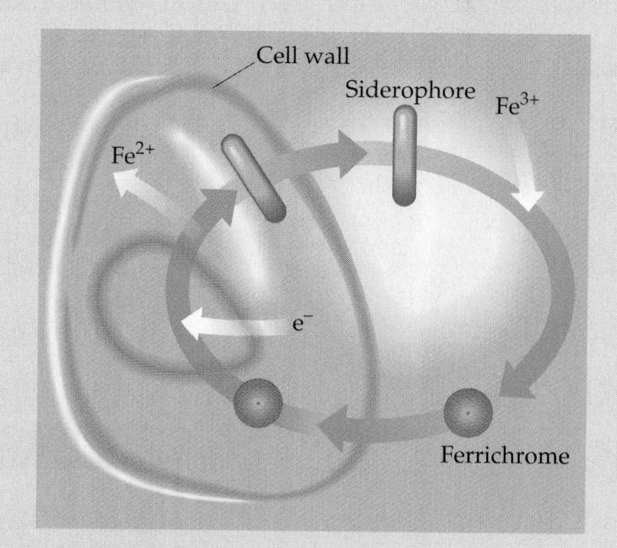

▲ **Figure 24.17** The iron-transport system of a bacterial cell. The iron-binding ligand, called a siderophore, is synthesized inside the cell and excreted into the surrounding medium. It reacts with Fe^{3+} ion to form ferrichrome, which is then absorbed by the cell. Inside the cell the ferrichrome is reduced, forming Fe^{2+}, which is not tightly bound by the siderophore. Having released the iron for use in the cell, the siderophore may be recycled back into the medium.

24.3 Nomenclature of Coordination Chemistry

When complexes were first discovered and few were known, they were named after the chemist who originally prepared them. A few of these names persist; for example, the dark red substance $NH_4[Cr(NH_3)_2(NCS)_4]$ is still known as Reinecke's salt. Once the structures of complexes were more fully understood, it became possible to name them in a more systematic manner. Let's consider two examples:

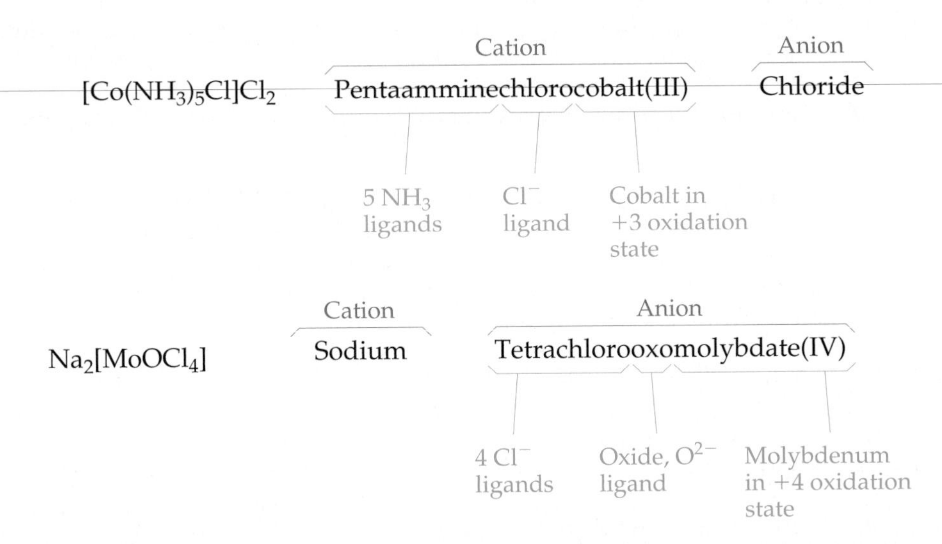

These examples illustrate how coordination compounds are named. The rules that govern naming of this class of substances are as follows:

1. *In naming salts, the name of the cation is given before the name of the anion.* Thus, in $[Co(NH_3)_5Cl]Cl_2$ we name the $[Co(NH_3)_5Cl]^{2+}$ cation and then Cl^-.
2. *Within a complex ion or molecule, the ligands are named before the metal. Ligands are listed in alphabetical order, regardless of charge on the ligand. Prefixes that give the number of ligands are not considered part of the ligand name in determining alphabetical order.* Thus, in the $[Co(NH_3)_5Cl]^{2+}$ ion we name the ammonia ligands first, then the chloride, then the metal: pentaamminechlorocobalt(III). In writing the formula, however, the metal is listed first.
3. *The names of the anionic ligands end in the letter* o, *whereas neutral ones ordinarily bear the name of the molecules.* Some common ligands and their names are listed in Table 24.2 ▼. Special names are given to H_2O (aqua), NH_3 (ammine), and CO (carbonyl). For example, $[Fe(CN)_2(NH_3)_2(H_2O)_2]^+$ would be named as diamminediaquadicyanoiron(III) ion.

TABLE 24.2 Some Common Ligands

Ligand	Name in Complexes	Ligand	Name in Complexes
Azide, N_3^-	Azido	Oxalate, $C_2O_4^{2-}$	Oxalato
Bromide, Br^-	Bromo	Oxide, O^{2-}	Oxo
Chloride, Cl^-	Chloro	Ammonia, NH_3	Ammine
Cyanide, CN^-	Cyano	Carbon monoxide, CO	Carbonyl
Fluoride, F^-	Fluoro	Ethylenediamine, en	Ethylenediamine
Hydroxide, OH^-	Hydroxo	Pyridine, C_5H_5N	Pyridine
Carbonate, CO_3^{2-}	Carbonato	Water, H_2O	Aqua

4. *Greek prefixes* (di-, tri-, tetra-, penta-, and hexa-) *are used to indicate the number of each kind of ligand when more than one is present. If the ligand itself contains a prefix of this kind* (for example, ethylenediamine), *alternate prefixes are used* (bis-, tris-, tetrakis-, pentakis-, hexakis-) *and the ligand name is placed in parentheses.* For example, the name for $[Co(en)_3]Br_3$ is tris(ethylenediamine)-cobalt(III) bromide.
5. *If the complex is an anion, its name ends in* -ate. The compound $K_4[Fe(CN)_6]$ is named potassium hexacyanoferrate(II), for example, and the ion $[CoCl_4]^{2-}$ is named tetrachlorocobaltate(II) ion.
6. *The oxidation number of the metal is given in parentheses in Roman numerals following the name of the metal.*

The following substances and their names demonstrate the application of these rules:

$[Ni(NH_3)_6]Br_2$	hexaamminenickel(II) bromide
$[Co(en)_2(H_2O)(CN)]Cl_2$	aquacyanobis(ethylenediamine)cobalt(III) chloride
$Na_2[MoOCl_4]$	sodium tetrachlorooxomolybdate(IV)

SAMPLE EXERCISE 24.4

Name the following compounds: **(a)** $[Cr(H_2O)_4Cl_2]Cl$; **(b)** $K_4[Ni(CN)_4]$.

Solution

Analyze and Plan: To name the complexes, we need to determine the ligands in the complexes and their names, and the oxidation state of the metal ion. We then put the information together following the rules listed previously.

Solve: **(a)** There are four water molecules, which are indicated as tetraaqua, and two chloride ions, indicated as dichloro. The oxidation state of Cr is +3.

$$+3 + 4(0) + 2(-1) + (-1) = 0$$
$$[Cr(H_2O)_4Cl_2]Cl$$

Thus, we have chromium(III). Finally, the anion is chloride. Putting these parts together, the name of the compound is

tetraaquadichlorochromium(III) chloride

(b) The complex has four cyanide ions, CN^-, which we indicate as tetracyano. The oxidation state of the nickel is zero.

$$4(+1) + 0 + 4(-1) = 0$$
$$K_4[Ni(CN)_4]$$

Because the complex is an anion, the metal is indicated as nickelate(0). Putting these parts together and naming the cation first, we have

potassium tetracyanonickelate(0)

PRACTICE EXERCISE

Name the following compounds: **(a)** $[Mo(NH_3)_3Br_3]NO_3$; **(b)** $(NH_4)_2[CuBr_4]$. **(c)** Write the formula for sodium diaquadioxalatoruthenate(III).

Answers: **(a)** triamminetribromomolybdenum(IV) nitrate; **(b)** ammonium tetrabromocuprate(II) **(c)** $Na[Ru(H_2O)_2(C_2O_4)_2]$

24.4 Isomerism

When two or more compounds have the same composition but a different arrangement of atoms, we call them **isomers**. Isomerism—the existence of isomers—is a characteristic feature of both organic and inorganic compounds. Although isomers are composed of the same collection of atoms, they usually differ in one or more physical or chemical properties such as color, solubility, or rate of reaction with some reagent. We will consider two main kinds of isomers in

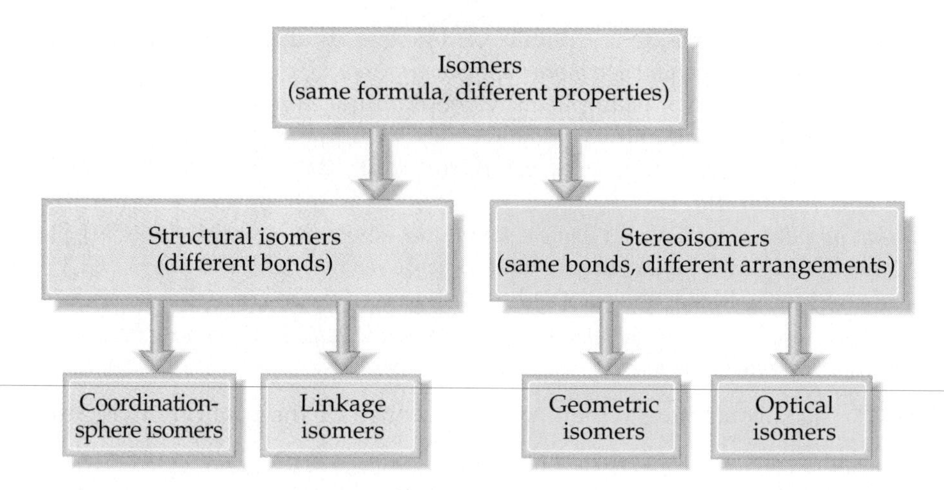

▶ **Figure 24.18** Forms of isomerism in coordination compounds.

coordination compounds: **structural isomers** (which have different bonds) and **stereoisomers** (which have the same bonds but different spatial arrangements of the bonds). Each of these classes also has subclasses, as shown in Figure 24.18 ▲.

Structural Isomerism

Many different types of structural isomerism are known in coordination chemistry. Figure 24.18 gives two examples: linkage isomerism and coordination-sphere isomerism. **Linkage isomerism** is a relatively rare but interesting type that arises when a particular ligand is capable of coordinating to a metal in two different ways. The nitrite ion, NO_2^-, for example, can coordinate through either a nitrogen or an oxygen atom, as shown in Figure 24.19 ▼. When it coordinates through the nitrogen atom, the NO_2^- ligand is called *nitro*; when it coordinates through the oxygen atom, it is called *nitrito* and is generally written ONO^-. The isomers shown in Figure 24.19 have different properties. The N-bonded isomer is yellow, for example, whereas the O-bonded isomer is red. Another ligand capable of coordinating through either of two donor atoms is thiocyanate, SCN^-, whose potential donor atoms are N and S.

Coordination-sphere isomers differ in the ligands that are directly bonded to the metal, as opposed to being outside the coordination sphere in the solid lattice. For example, $CrCl_3(H_2O)_6$ exists in three common forms: $[Cr(H_2O)_6]Cl_3$ (a violet compound), $[Cr(H_2O)_5Cl]Cl_2 \cdot H_2O$ (a green compound), and $[Cr(H_2O)_4Cl_2]Cl \cdot 2H_2O$ (also a green compound). In the two green compounds the water has been displaced from the coordination sphere by chloride ions and occupies a site in the crystal lattice.

Stereoisomerism

Stereoisomerism is the most important form of isomerism. **Stereoisomers** have the same chemical bonds but different spatial arrangements. In the square-planar complex $[Pt(NH_3)_2Cl_2]$, for example, the chloro ligands can be either adjacent to

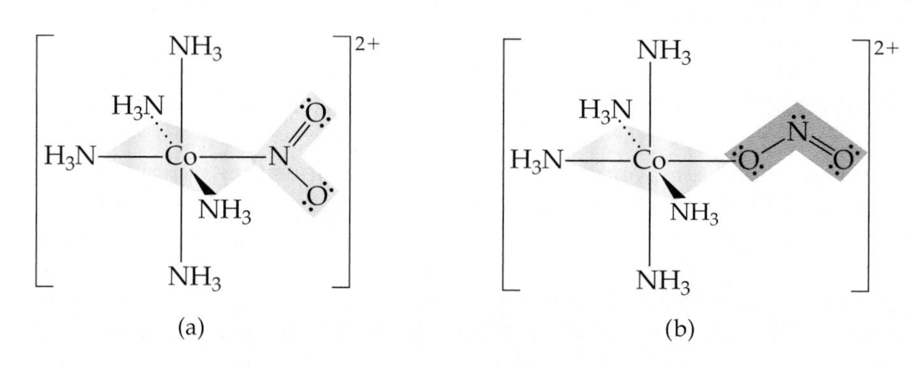

▶ **Figure 24.19** (a) Yellow N-bound, and (b) red O-bound isomers of $[Co(NH_3)_5NO_2]^{2+}$.

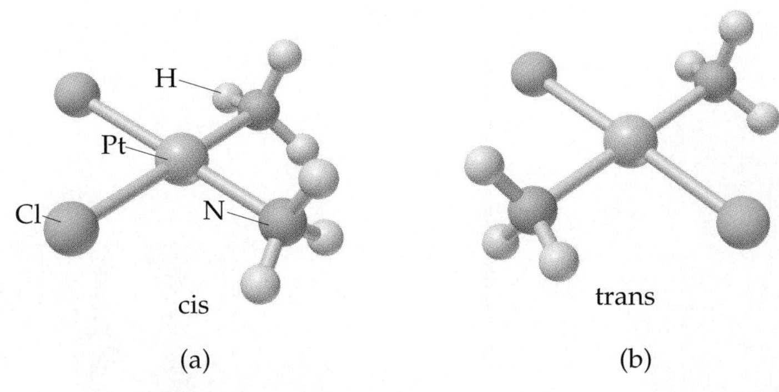

◀ **Figure 24.20** (a) *Cis* and (b) *trans* geometric isomers of the square-planar $[Pt(NH_3)_2Cl_2]$.

or opposite each other, as illustrated in Figure 24.20 ▲. This particular form of isomerism, in which the arrangement of the constituent atoms is different though the same bonds are present, is called **geometric isomerism**. The isomer in Figure 24.20(a), with like ligands in adjacent positions, is called the *cis* isomer. The isomer in Figure 24.20(b), with like ligands across from one another, is called the *trans* isomer. Geometric isomers generally have different properties, such as colors, solubilities, melting points, and boiling points. They may also have markedly different chemical reactivities. For example, *cis*-$[Pt(NH_3)_2Cl_2]$, also called *cisplatin*, is effective in the treatment of testicular, ovarian, and certain other cancers, whereas the *trans* isomer is ineffective.

Geometric isomerism is possible also in octahedral complexes when two or more different ligands are present. The *cis* and *trans* isomers of the tetraamminedichlorocobalt(III) ion were shown in Figure 24.1. As noted in Section 24.1 and Table 24.1, these two isomers have different colors. Their salts also possess different solubilities in water.

Because all the corners of a tetrahedron are adjacent to one another, *cis-trans* isomerism is not observed in tetrahedral complexes.

SAMPLE EXERCISE 24.5

The Lewis structure of the CO molecule indicates that the molecule has a lone pair on the C atom and one on the O atom (:C≡O:). When CO binds to a transition-metal atom, it nearly always does so by using the lone pair on the C atom. How many geometric isomers are there for tetracarbonyldichloroiron(II)?

Solution

Analyze and Plan: Given the name of a complex, we need to determine the chemical formula, propose a likely geometry, and then determine the number of isomers.

Solve: The name indicates that the complex has four carbonyl (CO) ligands and two chloro (Cl^-) ligands, so its formula is $Fe(CO)_4Cl_2$. The complex therefore has a coordination number of 6, and we can assume that it has an octahedral geometry. Like $[Co(NH_3)_4Cl_2]^+$ (Figure 24.1), it has four ligands of one type and two of another. Consequently, it possesses two isomers: one with the Cl^- ligands across the metal from each other (*trans*-$Fe(CO)_4Cl_2$) and one with the Cl^- ligands adjacent (*cis*-$Fe(CO)_4Cl_2$).

In principle, the CO ligand could exhibit linkage isomerism by binding to a metal atom via the lone pair on the O atom. When bonded this way, a CO ligand is called an *isocarbonyl* ligand. Metal isocarbonyl complexes are extremely rare, and we do not normally have to consider the possibility that CO will bind in this way.

Comment: In general, the number of isomers of a complex can be determined by making a series of drawings of the structure with ligands in different locations. It is easy, however, to overestimate the number of geometric isomers. Sometimes different orientations of a single isomer are incorrectly thought to be different isomers. If two structures can be rotated so that they are equivalent, they are not isomers of each other. The problem of identifying isomers is compounded by the difficulty we often have in visualizing three-dimensional molecules from their two-dimensional representations. It is sometimes easier to determine the number of isomers if we use three-dimensional models.

PRACTICE EXERCISE

How many isomers exist for square-planar $[Pt(NH_3)_2ClBr]$?
Answer: two

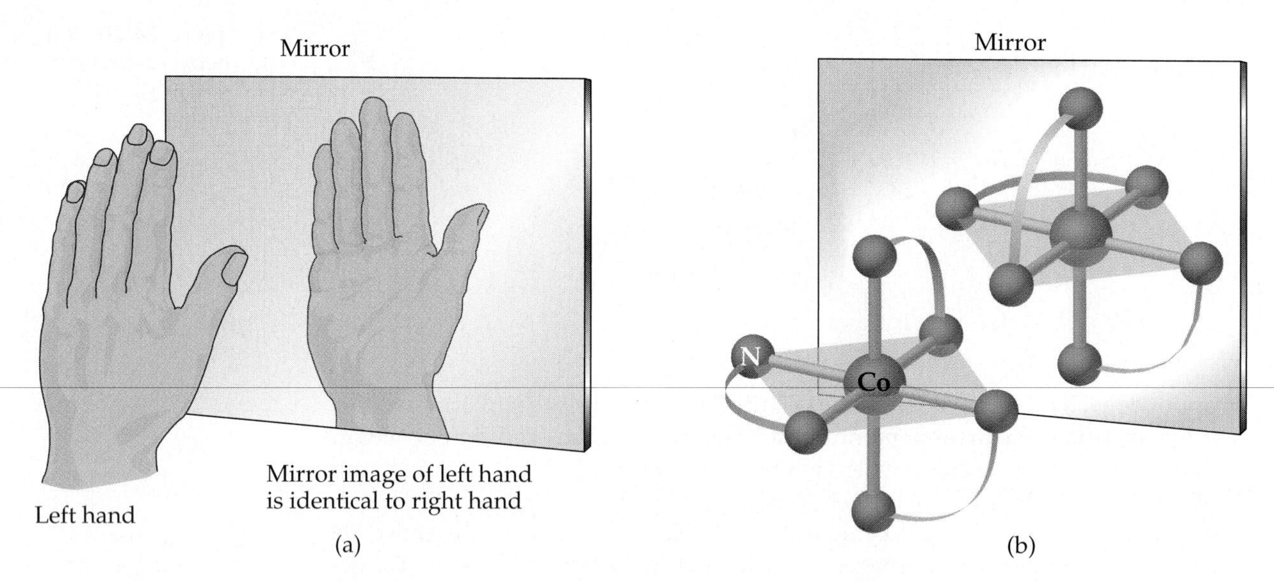

▲ **Figure 24.21** Just as our hands are nonsuperimposable mirror images of each other (a), so too are optical isomers such as the two optical isomers of $[Co(en)_3]^{3+}$ (b).

A second type of stereoisomerism is known as **optical isomerism**. Optical isomers, called **enantiomers**, are mirror images that cannot be superimposed on each other. They bear the same resemblance to each other that your left hand bears to your right hand. If you look at your left hand in a mirror, the image is identical to your right hand [Figure 24.21(a) ▲]. No matter how hard you try, however, you cannot superimpose your two hands on one another. An example of a complex that exhibits this type of isomerism is the $[Co(en)_3]^{3+}$ ion. Figure 24.21(b) shows the two enantiomers of $[Co(en)_3]^{3+}$ and their mirror-image relationship to each other. Just as there is no way that we can twist or turn our right hand to make it look identical to our left, so also there is no way to rotate one of these enantiomers to make it identical to the other. Molecules or ions that are not superimposable on their mirror image are said to be **chiral** (pronounced KY-rul). Enzymes are among the most important chiral molecules and, as noted in Section 24.2, many enzymes contain complexed metal ions. A molecule need not contain a metal atom to be chiral, however; in Section 25.7, you will see that many organic molecules, including some of those that are important in biochemistry, are chiral.

SAMPLE EXERCISE 24.6

Does either *cis*- or *trans*-$[Co(en)_2Cl_2]^+$ have optical isomers?

Solution

Analyze and Plan: The en ligand is a bidentate ligand, so this complex is six-coordinate. We need to determine the structures of the isomers and whether each of them has a nonsuperimposable mirror image.

Solve: By sketching an octahedral structure (see, for example, Figure 24.5), you should draw out both the *cis* and *trans* isomers of $[Co(en)_2Cl_2]^+$, then their mirror images. Note that the mirror image of the *trans* isomer, in which the Cl ligands are opposite one another, is identical to the original. Consequently *trans*-$[Co(en)_2Cl_2]^+$ does not exhibit optical isomerism. The mirror image of *cis*-$[Co(en)_2Cl_2]^+$, however, is different from the original, so there are optical isomers (enantiomers) for this complex: *cis*-$[Co(en)_2Cl_2]^+$ is a chiral complex.

PRACTICE EXERCISE

Does the square-planar complex ion $[Pt(NH_3)(N_3)ClBr]^-$ have optical isomers?
Answer: no

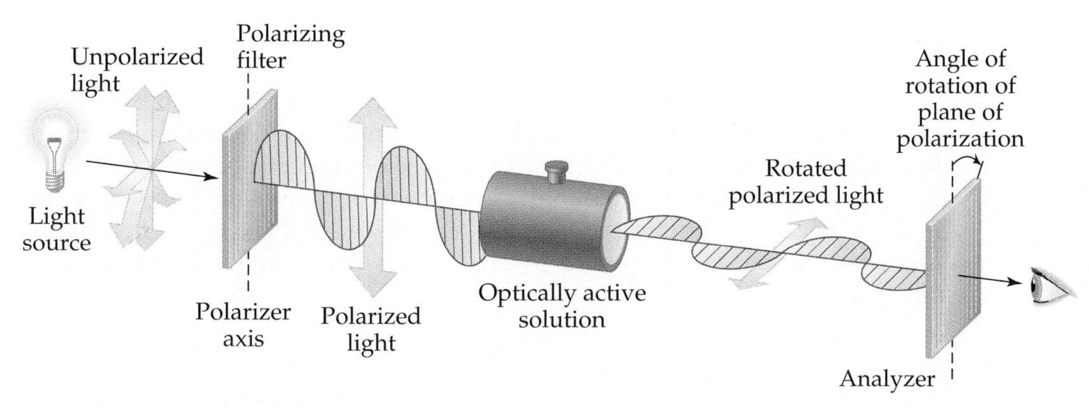

▲ **Figure 24.22** Effect of an optically active solution on the plane of polarization of plane-polarized light. The unpolarized light is passed through a polarizer. The resultant polarized light thereafter passes through a solution containing a dextrorotatory optical isomer. As a result, the plane of polarization of the light is rotated to the right relative to an observer looking toward the light source.

ANIMATION
Optical Activity

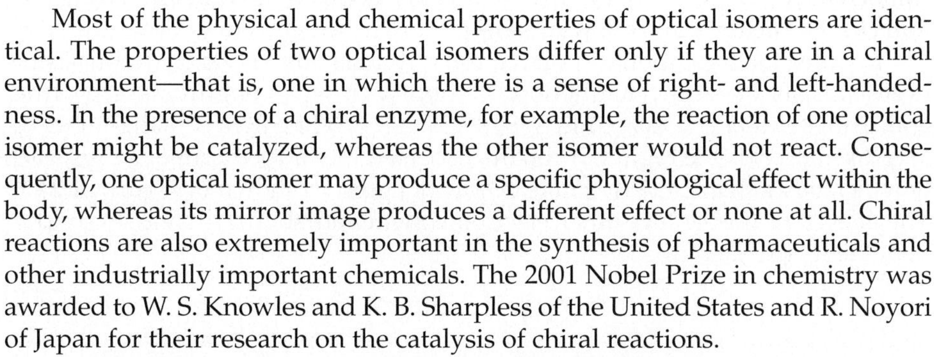

Most of the physical and chemical properties of optical isomers are identical. The properties of two optical isomers differ only if they are in a chiral environment—that is, one in which there is a sense of right- and left-handedness. In the presence of a chiral enzyme, for example, the reaction of one optical isomer might be catalyzed, whereas the other isomer would not react. Consequently, one optical isomer may produce a specific physiological effect within the body, whereas its mirror image produces a different effect or none at all. Chiral reactions are also extremely important in the synthesis of pharmaceuticals and other industrially important chemicals. The 2001 Nobel Prize in chemistry was awarded to W. S. Knowles and K. B. Sharpless of the United States and R. Noyori of Japan for their research on the catalysis of chiral reactions.

Optical isomers are usually distinguished from each other by their interaction with plane-polarized light. If light is polarized—for example, by passage through a sheet of Polaroid™ film—the light waves are vibrating in a single plane, as shown in Figure 24.22 ▲. If the polarized light is passed through a solution containing one optical isomer, the plane of polarization is rotated either to the right (clockwise) or to the left (counterclockwise). The isomer that rotates the plane of polarization to the right is **dextrorotatory**; it is labeled the dextro, or *d*, isomer (Latin *dexter*, "right"). Its mirror image rotates the plane of polarization to the left; it is **levorotatory** and is labeled the levo, or *l*, isomer (Latin *laevus*, "left"). The isomer of $[Co(en)_3]^{3+}$ on the left in Figure 24.21(b) is found experimentally to be the *l* isomer of this ion. Its mirror image is the *d* isomer. Because of their effect on plane-polarized light, chiral molecules are said to be **optically active**.

When a substance with optical isomers is prepared in the laboratory, the chemical environment during the synthesis is not usually chiral. Consequently, equal amounts of the two isomers are obtained; the mixture is said to be **racemic**. A racemic mixture will not rotate polarized light because the rotatory effects of the two isomers cancel each other. In order to separate the isomers from the racemic mixture, the isomers must be placed in a chiral environment. For example, one optical isomer of the chiral tartrate anion,* $C_4H_4O_6{}^{2-}$, can be used to separate a racemic mixture of $[Co(en)_3]Cl_3$. If *d*-tartrate is added to an aqueous solution of $[Co(en)_3]Cl_3$, *d*-$[Co(en)_3]$ (*d*-$C_4H_4O_6$)Cl will precipitate, leaving *l*-$[Co(en)_3]^{3+}$ in solution.

---

* When sodium ammonium tartrate, $NaNH_4C_4H_4O_6$, is crystallized from solution, the two optical isomers form separate crystals whose shapes are mirror images of each other. In 1848 Louis Pasteur achieved the first separation of a racemic mixture into optical isomers in an unusual way: Under a microscope, he separated by hand the "right-handed" crystals of this compound from the "left-handed" ones.

24.5 Color and Magnetism

Studies of the colors and magnetic properties of transition-metal complexes have played an important role in the development of modern models for metal-ligand bonding. We have discussed the various types of magnetic behavior in Section 23.7, and we have also discussed the interaction of radiant energy with matter in Section 6.3. Let's briefly examine the significance of these two properties for transition-metal complexes before we try to develop a model for metal-ligand bonding.

Color

In Figure 23.23 we saw the diverse range of colors exhibited by salts of transition-metal ions and their aqueous solutions. (Section 23.7) In these examples the coordination sphere about the metal is occupied by water molecules. In general, the color of a complex depends on the particular element, its oxidation state, and the ligands bound to the metal. Figure 24.23 ▼ shows how the pale blue color characteristic of $[Cu(H_2O)_4]^{2+}$ changes to a deep blue color as NH_3 ligands replace H_2O ligands to form $[Cu(NH_3)_4]^{2+}$.

In order for a compound to have color, it must absorb visible light. Visible light consists of electromagnetic radiation with wavelengths ranging from approximately 400 nm to 700 nm. (Section 6.1) White light contains all wavelengths in this visible region. It can be dispersed into a spectrum of colors, each of which has a characteristic range of wavelengths, as shown in Figure 24.24 ▼. The energy of this or any other electromagnetic radiation is inversely proportional to its wavelength: (Section 6.2)

$$E = h\nu = h(c/\lambda) \quad [24.7]$$

▶ **Figure 24.23** An aqueous solution of $CuSO_4$ is pale blue because of $[Cu(H_2O)_4]^{2+}$ (left). When $NH_3(aq)$ is added (middle and right), the deep blue $[Cu(NH_3)_4]^{2+}$ ion forms.

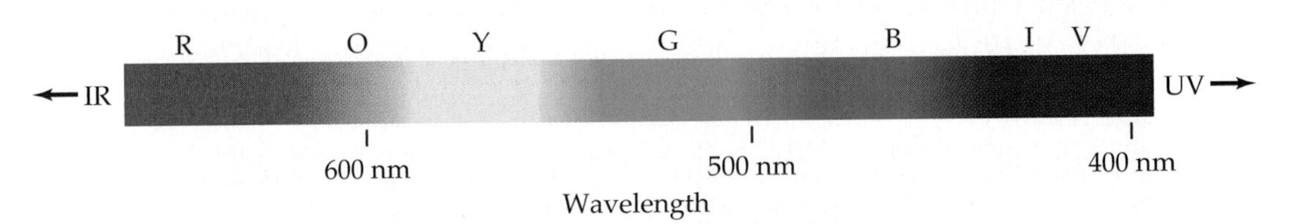

▲ **Figure 24.24** Visible spectrum showing the relation between color and wavelength.

A compound will absorb visible radiation when that radiation possesses the energy needed to move an electron from its lowest energy (ground) state to some excited state. (Section 6.3) Thus, the particular energies of radiation that a substance absorbs dictate the colors that it exhibits.

When a sample absorbs visible light, the color we perceive is the sum of the remaining colors that are reflected or transmitted by an object and strike our eyes. An opaque object reflects light, whereas a transparent one transmits it. If an object absorbs all wavelengths of visible light, none reaches our eyes from that object. Consequently, it appears black. If it absorbs no visible light, it is white or colorless. If it absorbs all but orange, the material appears orange. We also perceive an orange color, however, when visible light of all colors except blue strikes our eyes. Orange and blue are **complementary colors**; the removal of blue from white light makes the light look orange, and vice versa. Thus, an object has a particular color for one of two reasons: (1) It reflects or transmits light of that color; (2) it absorbs light of the complementary color. Complementary colors can be determined using an artist's color wheel, shown in Figure 24.25 ▶. The wheel shows the colors of the visible spectrum, from red to violet. Complementary colors, such as orange and blue, appear as wedges opposite each other on the wheel.

ACTIVITY
Color Wheel Activity

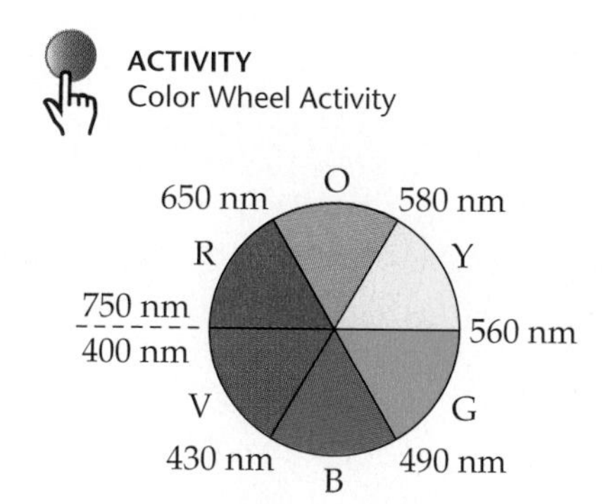

▲ **Figure 24.25** Artist's color wheel, showing the colors that are complementary to one another and the wavelength range of each color.

SAMPLE EXERCISE 24.7

The complex ion *trans*-$[Co(NH_3)_4Cl_2]^+$ absorbs light primarily in the red region of the visible spectrum (the most intense absorption is at 680 nm). What is the color of the complex?

Solution Because the complex absorbs red light, its color will be complementary to red. From Figure 24.25, we see that green is complementary to red, so the complex appears green. As noted in Table 24.1, this green complex was one of those that helped Werner establish his theory of coordination. The other geometric isomer of this complex, *cis*-$[Co(NH_3)_4Cl_2]^+$, absorbs yellow light and therefore appears violet.

PRACTICE EXERCISE

The $[Cr(H_2O)_6]^{2+}$ ion has an absorption band at about 630 nm. Which of the following colors—sky blue, yellow, green, or deep red—is most likely to describe this ion?
Answer: sky blue

The amount of light absorbed by a sample as a function of wavelength is known as its **absorption spectrum**. The visible absorption spectrum of a transparent sample can be determined as shown in Figure 24.26 ▼. The spectrum of $[Ti(H_2O)_6]^{3+}$, which we will discuss in Section 24.6, is shown in Figure 24.27 ▶. The absorption maximum of $[Ti(H_2O)_6]^{3+}$ is at about 500 nm. Because the sample absorbs most strongly in the green and yellow regions of the visible spectrum, it appears red-violet.

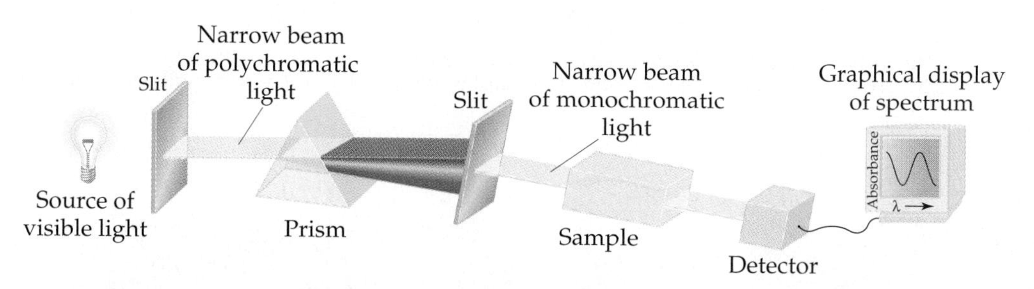

▲ **Figure 24.26** Experimental determination of the absorption spectrum of a solution. The prism is rotated so that different wavelengths of light pass through the sample. The detector measures the amount of light reaching it, and this information can be displayed as the absorption at each wavelength.

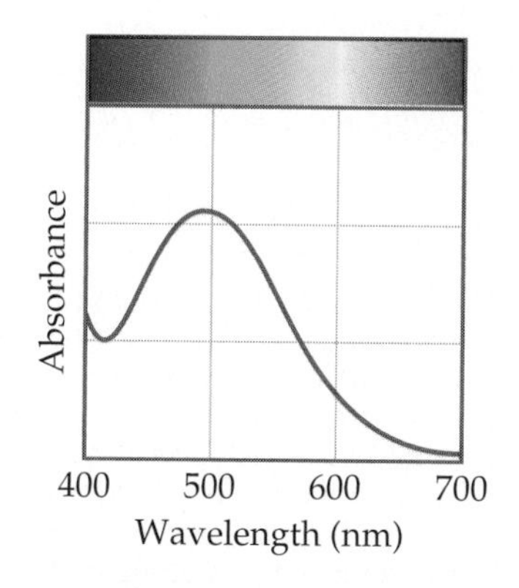

▲ **Figure 24.27** Visible absorption spectrum of the $[Ti(H_2O)_6]^{3+}$ ion.

Magnetism

Many transition-metal complexes exhibit simple paramagnetism, as described in Sections 9.8 and 23.7. In such compounds the individual metal ions possess some number of unpaired electrons. It is possible to determine the number of unpaired electrons per metal ion from the degree of paramagnetism. The experiments reveal some interesting comparisons. Compounds of the complex ion $[Co(CN)_6]^{3-}$ have no unpaired electrons, for example, but compounds of the $[CoF_6]^{3-}$ ion have four unpaired electrons per metal ion. Both complexes contain Co(III) with a $3d^6$ electron configuration. Clearly, there is a major difference in the ways in which the electrons are arranged in the metal orbitals in these two cases. Any successful bonding theory must explain this difference, and we shall present such a theory in Section 24.6.

24.6 Crystal-Field Theory

Scientists have long recognized that many of the magnetic properties and colors of transition-metal complexes are related to the presence of *d* electrons in metal orbitals. In this section we will consider a model for bonding in transition-metal complexes, called the **crystal-field theory**, that accounts for many of the observed properties of these substances.*

The ability of a metal ion to attract ligands such as water around itself is a Lewis acid–base interaction. (Section 16.11) The base—that is, the ligand—donates a pair of electrons into a suitable empty orbital on the metal, as shown in Figure 24.28 ▼. Much of the attractive interaction between the metal ion and the surrounding ligands is due, however, to the electrostatic forces between the positive charge on the metal and negative charges on the ligands. If the ligand is ionic, as in the case of Cl^- or SCN^-, the electrostatic interaction occurs between the positive charge on the metal center and the negative charge on the ligand. When the ligand is neutral, as in the case of H_2O or NH_3, the negative ends of these polar molecules, which contain an unshared electron pair, are directed toward the metal. In this case the attractive interaction is of the ion-dipole type. (Section 11.2) In either case the result is the same: The ligands are attracted strongly toward the metal center. Because of the electrostatic attraction between the positive metal ion and the electrons of the ligands, the assembly of the metal ion and the ligands is lower in energy than the fully separated charges.

Although the positive metal ion is attracted to the electrons in the ligands, the *d* electrons on the metal ion feel a repulsion from the ligands (negative charges repel one another). Let's examine this effect more closely, and particularly for the case in which the ligands form an octahedral array around the metal ion. For the crystal-field model, we consider the ligands to be negative points of charge that repel the electrons in the *d* orbitals. Figure 24.29 ▶ shows the effects of these point charges on the energies of the *d* orbitals in two steps. In the first step, the *average* energy of the *d* orbitals is raised by the presence of the point charges.

* The name *crystal field* arose because the theory was first developed to explain the properties of solid, crystalline materials, such as ruby. The same theoretical model applies to complexes in solution.

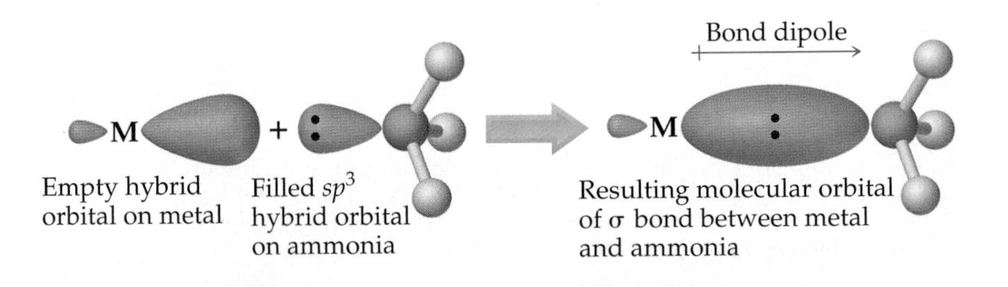

▶ **Figure 24.28** Representation of the metal-ligand bond in a complex as a Lewis acid–base interaction. The ligand, which acts as a Lewis base, donates charge to the metal via a metal hybrid orbital. The bond that results is strongly polar, with some covalent character. It is often sufficient to assume that the metal-ligand interaction is entirely electrostatic in character, as is done in the crystal-field model.

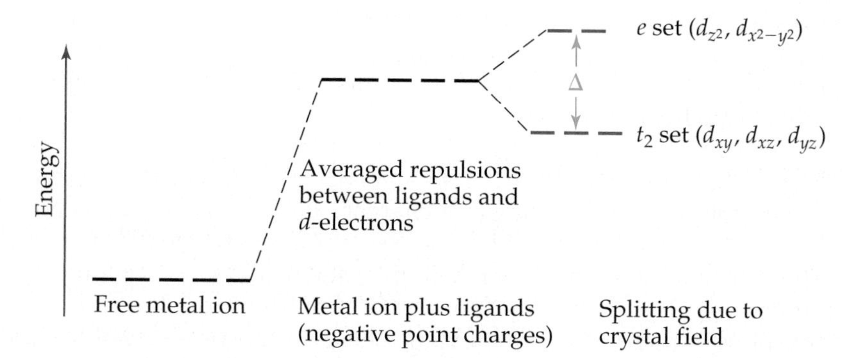

◀ **Figure 24.29** Effects of an octahedral crystal field on the energies of the five *d* orbitals of a transition-metal ion. On the left the energies of the *d* orbitals of a free ion are shown. When negative charges are brought up to the ion, the average energy of the *d* orbitals increases (center). On the right the splitting of the *d* orbitals due to the octahedral field is shown. Because the repulsion felt by the d_{z^2} and $d_{x^2-y^2}$ orbitals is greater than that felt by the d_{xy} d_{xz}, and d_{yz}, orbitals, the five *d* orbitals split into a lower-energy set of three (the t_2 set) and a higher-energy set of two (the *e* set).

Hence, all five *d* orbitals are raised in energy by the same amount. In the second step we consider what happens to the energies of the individual *d* orbitals when the ligands sit in an octahedral arrangement. In a 6-coordinate octahedral complex we can envision the ligands approaching along the *x*-, *y*-, and *z*-axes, as shown in Figure 24.30(a) ▼; this arrangement is called an *octahedral crystal field*.

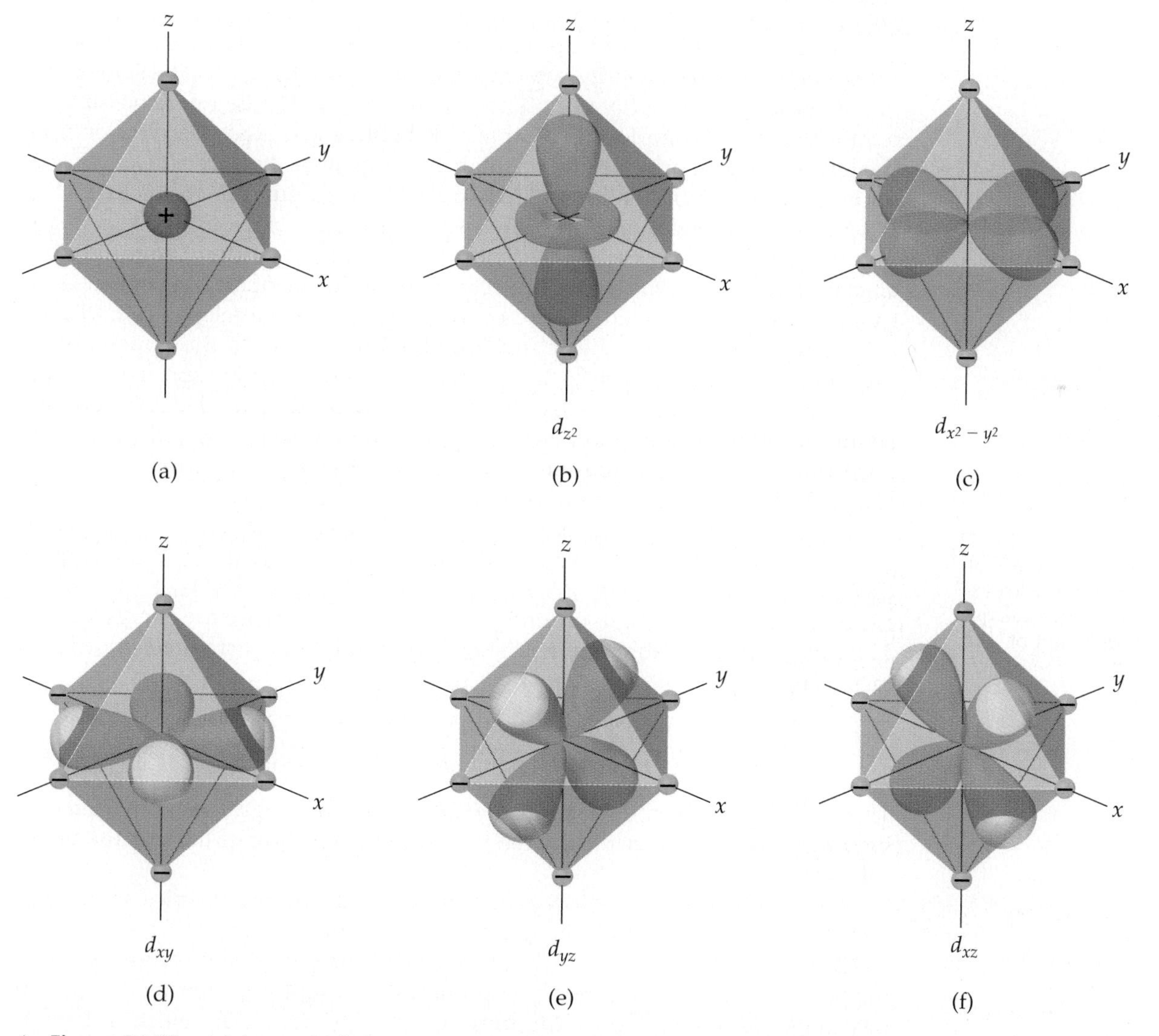

▲ **Figure 24.30** (a) An octahedral array of negative charges approaching a metal ion. (b–f) The orientations of the *d* orbitals relative to the negative charges. Notice that the lobes of the d_{z^2} and $d_{x^2-y^2}$ orbitals (b and c) point toward the charges, whereas the lobes of the d_{xy}, d_{yz}, and d_{xz} orbitals (d–f) point between the charges.

Because the d orbitals on the metal ion have different shapes, they do not all have the same energy under the influence of the crystal field. To see why, we must consider the shapes of the d orbitals and how their lobes are oriented relative to the ligands.

Figure 24.30(b–f) shows the five d orbitals in an octahedral crystal field. Notice that the d_{z^2} and $d_{x^2-y^2}$ orbitals have lobes directed *along* the x-, y-, and z-axes pointing toward the point charges, whereas the d_{xy}, d_{xz}, and d_{yz} orbitals have lobes that are directed *between* the axes along which the charges approach. The high symmetry of the octahedral crystal field dictates that the d_{z^2} and $d_{x^2-y^2}$ orbitals experience the same amount of repulsion from the crystal field. Those two orbitals therefore have the same energy in the presence of the crystal field. Likewise, the d_{xy}, d_{xz}, and d_{yz} orbitals experience exactly the same repulsion, so those three orbitals stay at the same energy. Because their lobes point right at the negative charges, the d_{z^2} and $d_{x^2-y^2}$ orbitals experience stronger repulsions than those in the d_{xy}, d_{xz}, and d_{yz} orbitals. As a result, an energy separation, or splitting, occurs between the three lower-energy d orbitals (called the t_2 set of orbitals) and the two higher-energy ones (called the e set),* as shown on the right side of Figure 24.29. The energy gap between the two sets of d orbitals is labeled Δ, a quantity that is often called the *crystal-field splitting energy*.

The crystal-field model helps us account for the observed colors in transition-metal complexes. The energy gap between the d orbitals, Δ, is of the same order of magnitude as the energy of a photon of visible light. It is therefore possible for a transition-metal complex to absorb visible light, which excites an electron from the lower-energy d orbitals into the higher-energy ones. In the $[Ti(H_2O)_6]^{3+}$ ion, for example, the Ti(III) ion has an $[Ar]3d^1$ electron configuration (recall that when determining the electron configurations of transition-metal ions, we remove the s electrons first). (Section 7.4) Ti(III) is thus called a "d^1 ion." In the ground state of $[Ti(H_2O)_6]^{3+}$, the single $3d$ electron resides in one of the three lower-energy orbitals in the t_2 set. Absorption of light with a wavelength of 495 nm (242 kJ/mol) excites the $3d$ electron from the lower t_2 set to the upper e set of orbitals, as shown in Figure 24.31 ◀, generating the absorption spectrum shown in Figure 24.27. Because this transition involves exciting an electron from one set of d orbitals to the other, we call it a ***d-d* transition**. As noted earlier, the absorption of visible radiation that produces this d-d transition causes the $[Ti(H_2O)_6]^{3+}$ ion to appear purple.

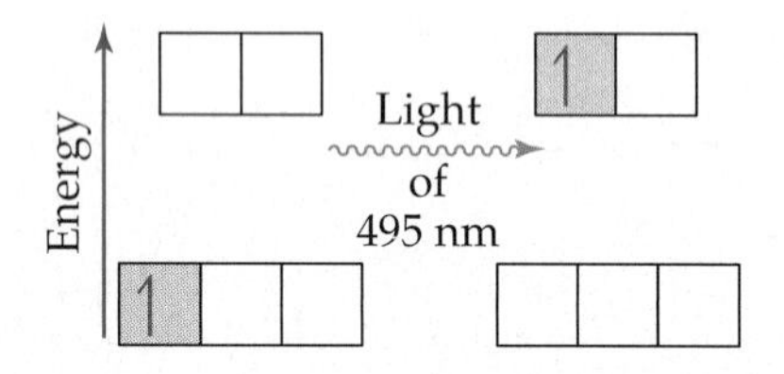

▲ **Figure 24.31** The $3d$ electron of $[Ti(H_2O)_6]^{3+}$ is excited from the lower-energy d orbitals to the higher-energy ones when irradiated with light of 495-nm wavelength.

The magnitude of the energy gap, Δ, and consequently the color of a complex depend on both the metal and the surrounding ligands. For example, $[Fe(H_2O)_6]^{3+}$ is light violet, $[Cr(H_2O)_6]^{3+}$is violet, and $[Cr(NH_3)_6]^{3+}$ is yellow. Ligands can be arranged in order of their abilities to increase the energy gap, Δ. The following is an abbreviated list of common ligands arranged in order of increasing Δ:

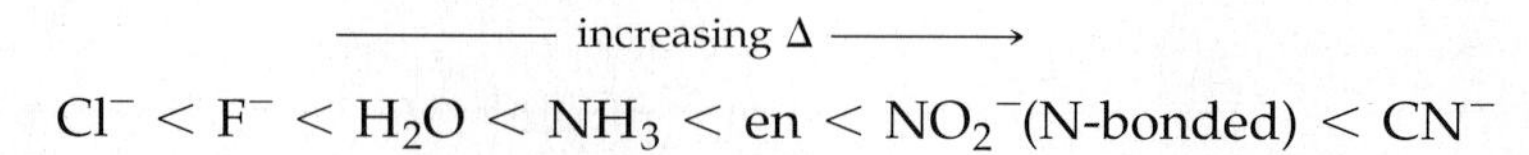

$$Cl^- < F^- < H_2O < NH_3 < en < NO_2^-\text{(N-bonded)} < CN^-$$

This list is known as the **spectrochemical series**. The magnitude of Δ increases by roughly a factor of two from the far left to the far right of the spectrochemical series.

Ligands that lie on the low-Δ end of the spectrochemical series are termed *weak-field ligands*; those on the high-Δ end are termed *strong-field ligands*. Figure 24.32 ▶ shows schematically what happens to the crystal-field splitting when the ligand is varied in a series of chromium(III) complexes. Because a Cr atom has an $[Ar]3d^5 4s^1$ electron configuration, Cr^{3+} has an $[Ar]3d^3$ electron configuration; Cr(III), therefore, is a d^3 ion. Consistent with Hund's rule, the three $3d$

* The labels t_2 for the d_{xy}, d_{xz}, and d_{yz} orbitals and e for the d_{z^2} and $d_{x^2-y^2}$ orbitals come from the application of a branch of mathematics called *group theory* to crystal-field theory. Group theory can be used to analyze the effects of symmetry on molecular properties.

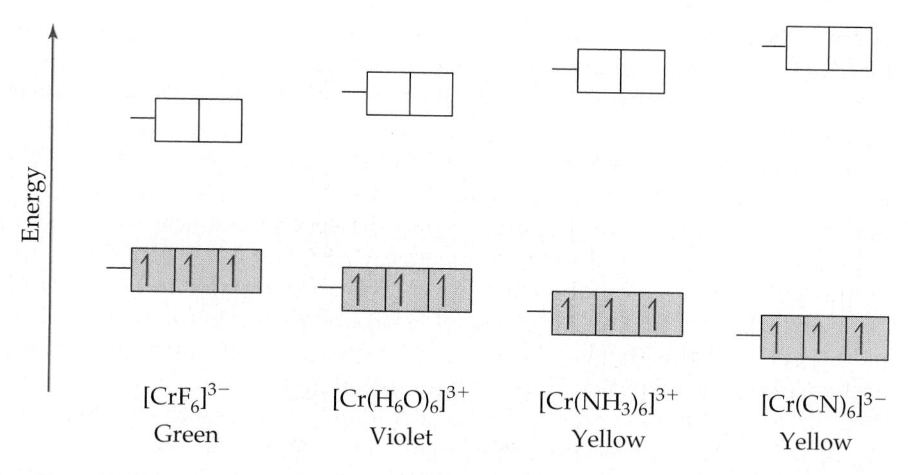

◀ **Figure 24.32** Crystal-field splitting in a series of octahedral chromium(III) complexes.

ACTIVITY
D-Orbital Population in Cobalt Complexes, Crystal Field Theory: Chromium Complexes

electrons occupy the t_2 set of orbitals, with one electron in each and all the spins the same. (Section 6.8) As the field exerted by the six surrounding ligands increases, the splitting of the metal *d* orbitals increases. Because the absorption spectrum is related to this energy separation, these complexes vary in color.

SAMPLE EXERCISE 24.8

Which of the following complexes of Ti^{3+} exhibits the shortest wavelength absorption in the visible spectrum: $[Ti(H_2O)_6]^{3+}$, $[Ti(en)_3]^{3+}$, or $[TiCl_6]^{3-}$?

Solution Each of these ions is an octahedral Ti(III) complex. Ti(III) is a d^1 ion, so we anticipate that the absorption is due to a *d-d* transition in which the 3*d* electron is excited from the lower t_2 set to the higher *e* set. The wavelength of the absorption is determined by the magnitude of the splitting Δ. The larger the splitting, the shorter the wavelength of the absorption. Of the three ligands involved—H_2O, en, and Cl^-—we see that ethylenediamine (en) is highest in the spectrochemical series and will therefore cause the largest splitting of the t_2 and *e* sets of orbitals. Thus, the complex with the shortest wavelength absorption is $[Ti(en)_3]^{3+}$.

PRACTICE EXERCISE

The absorption spectrum of $[Ti(NCS)_6]^{3-}$ shows a band that lies intermediate in wavelength between those for $[TiCl_6]^{3-}$ and $[TiF_6]^{3-}$. What can we conclude about the place of NCS^- in the spectrochemical series?
Answer: It lies between Cl^- and F^-; that is, $Cl^- < NCS^- < F^-$.

Electron Configurations in Octahedral Complexes

The crystal-field model also helps us understand the magnetic properties and some important chemical properties of transition-metal ions. From Hund's rule, we expect that electrons will always occupy the lowest-energy vacant orbitals first and that they will occupy a set of degenerate orbitals one at a time with their spins parallel. (Section 6.8) Thus, if we have a d^1, d^2, or d^3 octahedral complex, the electrons will go into the lower-energy t_2 set of orbitals, with their spins parallel, as shown in Figure 24.33 ▼. When a fourth electron must be added, a problem arises. If the electron is added to a lower-energy t_2 orbital, an energy gain of magnitude Δ is realized, as compared with placing the electron in a higher-energy *e* orbital. There is a penalty for doing this, however, because the electron must now be paired up with the electron already occupying the orbital. The energy required to do this, relative to putting it in another orbital with parallel spin, is called the **spin-pairing energy**. The spin-pairing energy arises from the greater

Ti^{3+}, a d^1 ion V^{3+}, a d^2 ion Cr^{3+}, a d^3 ion

1   1 1   1 1 1

◀ **Figure 24.33** Electron configurations associated with one, two, and three electrons in the 3*d* orbitals in octahedral complexes.

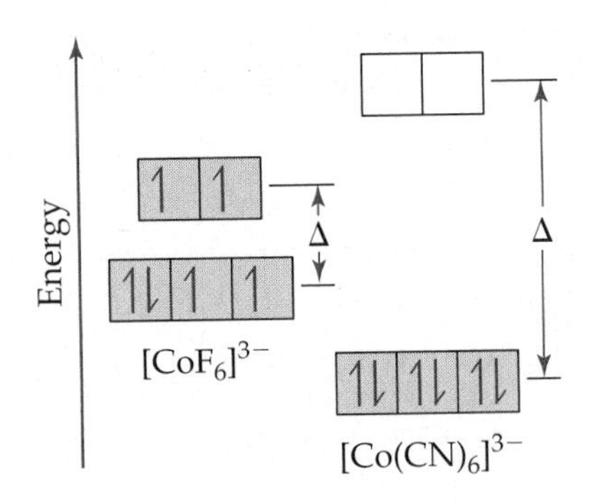

▲ **Figure 24.34** Population of d orbitals in the high-spin $[CoF_6]^{3-}$ ion (small Δ) and low-spin $[Co(CN)_6]^{3-}$ ion (large Δ).

electrostatic repulsion of two electrons that share an orbital as compared with two that are in different orbitals with the same electron spin.

The ligands that surround the metal ion and the charge on the metal ion often play major roles in determining which of the two electronic arrangements arises. In both the $[CoF_6]^{3-}$ and $[Co(CN)_6]^{3-}$ ions, the ligands have a 1− charge. The F^- ion, however, is on the low end of the spectrochemical series, so it is a weak-field ligand. The CN^- ion is on the high end of the spectrochemical series, so it is a strong-field ligand. It produces a larger energy gap than the F^- ion. The splittings of the d-orbital energies in these two complexes are compared in Figure 24.34 ◀.

Cobalt(III) has an [Ar]$3d^6$ electron configuration, so these are both d^6 complexes. Let's imagine that we add these six electrons one at a time to the d orbitals of the CoF_6^{3-} ion. The first three will go into the lower-energy t_2 orbitals with their spins parallel. The fourth electron could go into one of the t_2 orbitals, pairing up with one of those already present. Doing so would result in an energy gain of Δ as compared with putting the electron in one of the higher-energy e orbitals. However, it would cost energy in an amount equal to the spin-pairing energy. Because F^- is a weak-field ligand, Δ is small, and the more stable arrangement is the one in which the electron is placed in one of the e orbitals. Similarly, the fifth electron we add goes into the other e orbital. With all of the orbitals containing at least one electron, the sixth must be paired up, and it goes into a lower-energy t_2 orbital; we end up with four of the electrons in the t_2 set of orbitals and two electrons in the e set. In the case of the $[Co(CN)_6]^{3-}$ complex, the crystal-field splitting is much larger. The spin-pairing energy is smaller than Δ, so all six electrons are paired in the t_2 orbitals, as illustrated in Figure 24.34.

The $[CoF_6]^{3-}$ complex is a **high-spin complex**; that is, the electrons are arranged so that they remain unpaired as much as possible. The $[Co(CN)_6]^{3-}$ ion, on the other hand, is a **low-spin complex**; that is, the electrons are arranged so that they remain paired as much as possible. These two different electronic arrangements can be readily distinguished by measuring the magnetic properties of the complex, as described earlier. The absorption spectrum also shows characteristic features that indicate the electronic arrangement.

SAMPLE EXERCISE 24.9

Predict the number of unpaired electrons in 6-coordinate high-spin and low-spin complexes of Fe^{3+}.

Solution Fe^{3+} is a d^5 ion. In a high-spin complex, all five of these electrons are unpaired, with three in the t_2 orbitals and two in the e orbitals. In a low-spin complex, all five electrons reside in the t_2 set of d orbitals, so there is one unpaired electron.

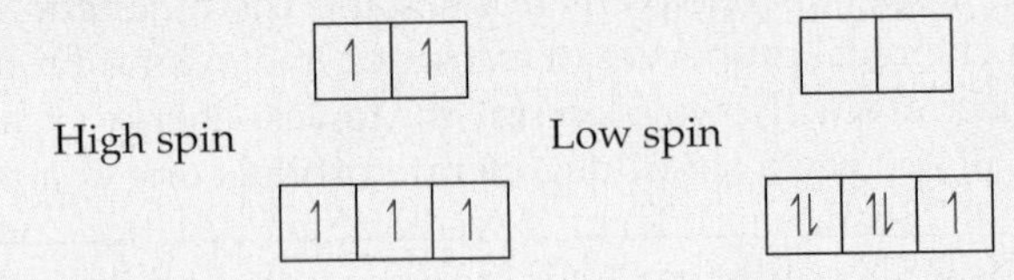

PRACTICE EXERCISE

For which d electron configurations in octahedral complexes is it possible to distinguish between high-spin and low-spin arrangements?
Answer: d^4, d^5, d^6, d^7

Tetrahedral and Square-Planar Complexes

Thus far we have considered the crystal-field model only for complexes having an octahedral geometry. When there are only four ligands about the metal, the geometry is generally tetrahedral, except for the special case of metal ions with a d^8 electron configuration, which we will discuss in a moment. The crystal-field splitting of the metal d orbitals in tetrahedral complexes differs from that in octahedral complexes. Four equivalent ligands can interact with a central metal ion most effectively by approaching along the vertices of a tetrahedron. It turns out—and this is not easy to explain in just a few sentences—that the splitting of the metal d orbitals in a tetrahedral crystal is just the opposite of that for the octahedral case. That is, the three metal d orbitals in the t_2 set are raised in energy, and the two orbitals in the e set are lowered, as illustrated in Figure 24.35 ▶. Because there are only four ligands instead of six, as in the octahedral case, the crystal-field splitting is much smaller for tetrahedral complexes. Calculations show that for the same metal ion and ligand set, the crystal-field splitting for a tetrahedral complex is only four-ninths as large as for the octahedral complex. For this reason all tetrahedral complexes are high spin; the crystal field is never large enough to overcome the spin-pairing energies.

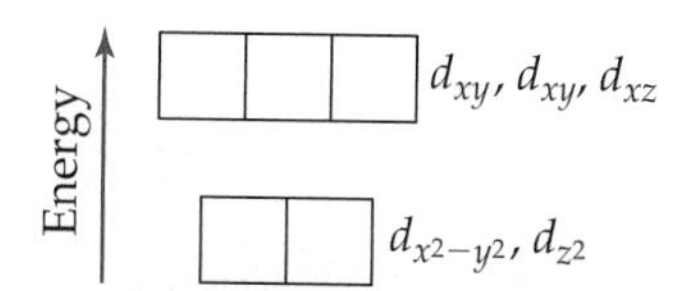

▲ **Figure 24.35** Energies of the d orbitals in a tetrahedral crystal field.

Square-planar complexes, in which four ligands are arranged about the metal ion in a plane, can be envisioned as formed by removing two ligands from along the vertical z-axis of the octahedral complex. The changes that occur in the energy levels of the d orbitals are illustrated in Figure 24.36 ▶. Note in particular that the d_{z^2} orbital is now considerably lower in energy than the $d_{x^2-y^2}$ orbital because the ligands along the vertical z-axis have been removed.

Square-planar complexes are characteristic of metal ions with a d^8 electron configuration. They are nearly always low spin; that is, the eight d electrons are spin-paired to form a diamagnetic complex. Such an electronic arrangement is particularly common among the ions of heavier metals, such as Pd^{2+}, Pt^{2+}, Ir^+, and Au^{3+}.

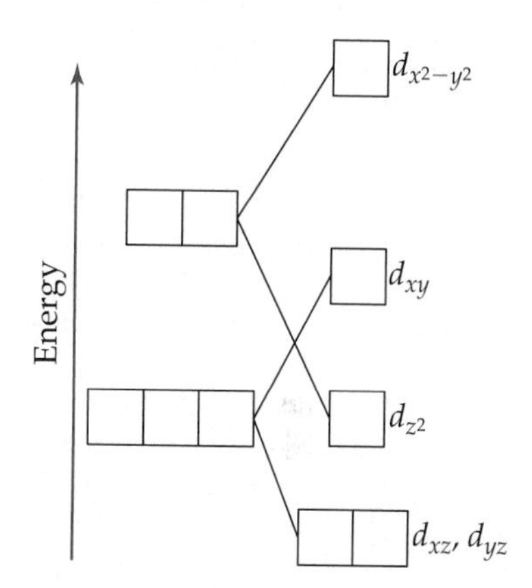

▲ **Figure 24.36** Effect on the relative energies of the d orbitals caused by removing the two negative charges from the z-axis of an octahedral complex. When the charges are completely removed, the square-planar geometry results.

SAMPLE EXERCISE 24.10

Four-coordinate nickel(II) complexes exhibit both square-planar and tetrahedral geometries. The tetrahedral ones, such as $[NiCl_4]^{2-}$, are paramagnetic; the square-planar ones, such as $[Ni(CN)_4]^{2-}$, are diamagnetic. Show how the d electrons of nickel(II) populate the d orbitals in the appropriate crystal-field splitting diagram in each case.

Solution Nickel(II) has an electron configuration of $[Ar]3d^8$. The population of the d electrons in the two geometries is given as follows:

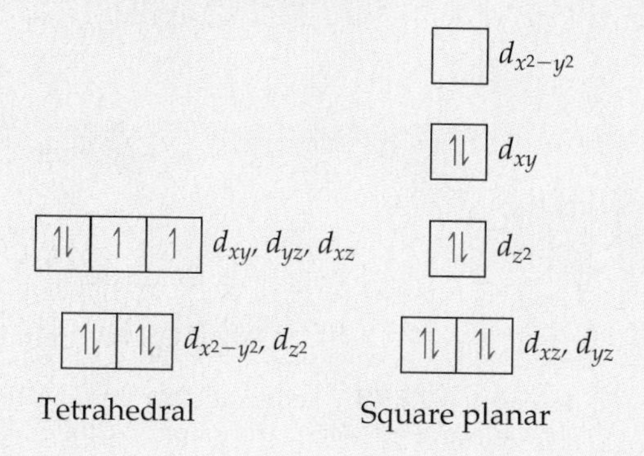

PRACTICE EXERCISE

How many unpaired electrons do you predict for the tetrahedral $[CoCl_4]^{2-}$ ion?
Answer: three

We have seen that the crystal-field model provides a basis for explaining many features of transition-metal complexes. In fact, it can be used to explain many observations in addition to those we have discussed. Many lines of evidence show, however, that the bonding between transition-metal ions and ligands must have some covalent character. Molecular-orbital theory (Sections 9.7 and 9.8) can also be used to describe the bonding in complexes, although the application of molecular-orbital theory to coordination compounds is beyond the scope of our discussion. The crystal-field model, although not entirely accurate in all details, provides an adequate and useful first description of the electronic structure of complexes.

A Closer Look Charge-Transfer Color

▲ **Figure 24.37** From left to right, $KMnO_4$, K_2CrO_4, and $KClO_4$. $KMnO_4$ and K_2CrO_4 are intensely colored due to ligand-to-metal charge transfer (LMCT) transitions in the MnO_4^- and CrO_4^{2-} anions. There are no valence *d* orbitals on Cl, so the charge-transfer transition for ClO_4^- requires ultraviolet light and $KClO_4$ is white.

In the laboratory portion of your course, you have probably seen many colorful compounds of the transition metals. Many of these exhibit color because of *d-d* transitions, in which visible light excites electrons from one *d* orbital to another. There are other colorful transition-metal complexes, however, that derive their color from a rather different type of excitation involving the *d* orbitals. Two such common substances are the deep violet permanganate ion (MnO_4^-) and the bright yellow chromate ion (CrO_4^{2-}), salts of which are shown in Figure 24.37 ▲. Both MnO_4^- and CrO_4^{2-} are tetrahedral complexes.

The permanganate ion strongly absorbs visible light with a maximum absorption at a wavelength of 565 nm. The strong absorption in the yellow portion of the visible spectrum is responsible for the violet appearance of salts and solutions of the ion (violet is the complementary color to yellow). What is happening during this absorption? The MnO_4^- ion is a complex of Mn(VII), which has a d^0 electron configuration. As such, the absorption in the complex cannot be due to a *d-d* transition because there are no *d* electrons to excite! That does not mean, however, that the *d* orbitals are not involved in the transition. The excitation in the MnO_4^- ion is due to a *charge-transfer transition*, in which an electron on one of the oxygen ligands is excited into a vacant *d* orbital on the Mn atom (Figure 24.38 ▶). In essence, an electron is transferred from a ligand to the metal, so this transition is called a *ligand-to-metal charge-transfer (LMCT) transition*. An LMCT transition is also responsible for the color of the CrO_4^{2-}, which is a d^0 Cr(VI) complex. Also shown in Figure 24.37 is a salt of the perchlorate ion (ClO_4^-). Like MnO_4^-, the ClO_4^- is tetrahedral and has its central atom in the +7 oxidation state. However, because the Cl atom doesn't have low-lying *d* orbitals, exciting an electron requires a more energetic photon than for MnO_4^-. The first absorption for ClO_4^- is in the ultraviolet portion of the spectrum, so all of the visible light is transmitted and the salt appears white.

Other complexes exhibit charge-transfer excitations in which an electron from the metal atom is excited to an empty orbital on a ligand. Such an excitation is called a *metal-to-ligand charge-transfer (MLCT) transition.*

Charge-transfer transitions are generally more intense than *d-d* transitions. Many metal-containing pigments used for oil painting, such as cadmium yellow (CdS), chrome yellow ($PbCrO_4$), and red ochre (Fe_2O_3), have intense colors because of charge-transfer transitions.

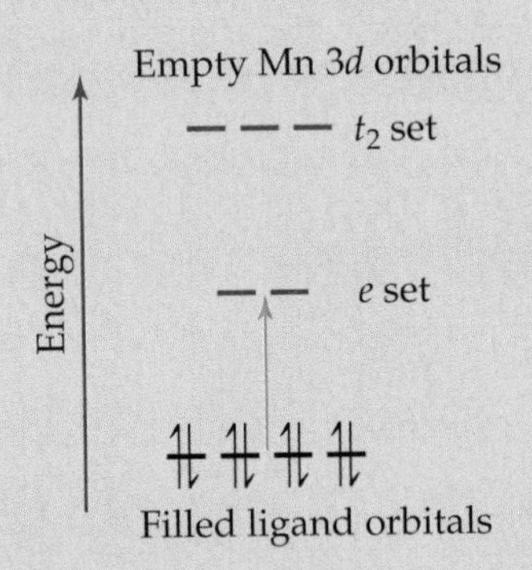

▲ **Figure 24.38** Schematic diagram of the ligand-to-metal charge transfer (LMCT) transition in MnO_4^-. As shown by the blue arrow, an electron is excited from a nonbonding pair on O into one of the empty *d* orbitals on Mn.

SAMPLE INTEGRATIVE EXERCISE 24: Putting Concepts Together

The oxalate ion has the Lewis structure shown in Figure 24.6. **(a)** Show the geometrical structure of the complex formed by coordination of oxalate to cobalt(II), forming $[Co(C_2O_4)(H_2O)_4]$. **(b)** Write the formula for the salt formed upon coordination of three oxalate ions to Co(II), assuming that the charge-balancing cation is Na^+. **(c)** Sketch all of the possible geometric isomers for the cobalt complex formed in part (b). Are any of these isomers chiral? Explain. **(d)** The equilibrium constant for the formation of the cobalt(II) complex produced by coordination of three oxalate anions, as in part (b), is 5.0×10^9. By comparison, the formation constant for formation of the cobalt(II) complex with three molecules of *ortho*-phenanthroline (Figure 24.6) is 9×10^{19}. From these results, what conclusions can you draw regarding the relative Lewis base properties of the two ligands toward cobalt(II)? **(e)** Using the approach described in Sample

Exercise 17.14, calculate the concentration of free aqueous Co(II) ion in a solution initially containing 0.040 *M* oxalate ion and 0.0010 *M* $Co^{2+}(aq)$.

Solution **(a)** The complex formed by coordination of one oxalate ion is octahedral (see figure in margin).

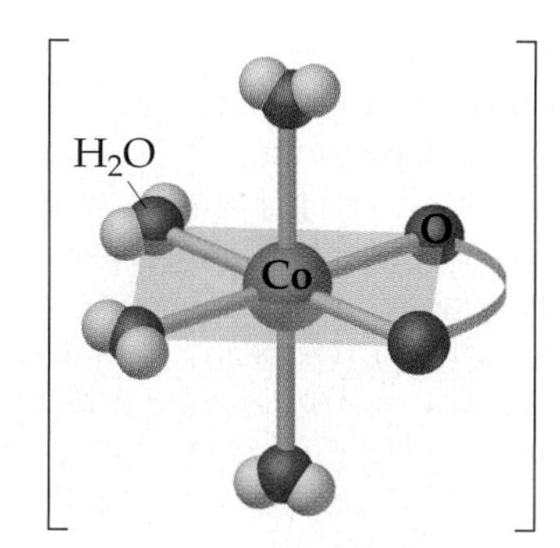

(b) Because the oxalate ion has a charge of 2−, the net charge of a complex with three oxalate anions and one Co^{2+} ion is 4−. Therefore, the coordination compound has the formula $Na_4[Co(C_2O_4)_3]$. **(c)** There is only one geometric isomer. The complex is chiral, however, in the same way as the $[Co(en)_3]^{3+}$ complex, shown in Figure 24.21(b). These two mirror images are not superimposable, so there are two enantiomers.

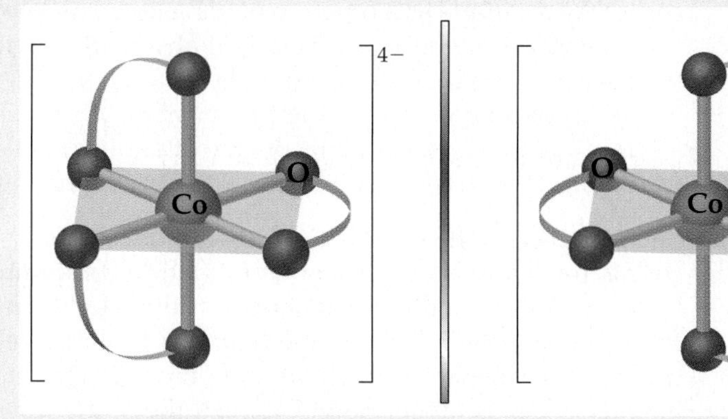

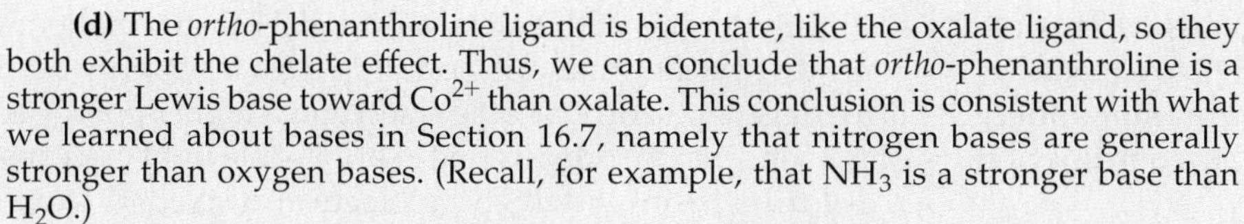

(d) The *ortho*-phenanthroline ligand is bidentate, like the oxalate ligand, so they both exhibit the chelate effect. Thus, we can conclude that *ortho*-phenanthroline is a stronger Lewis base toward Co^{2+} than oxalate. This conclusion is consistent with what we learned about bases in Section 16.7, namely that nitrogen bases are generally stronger than oxygen bases. (Recall, for example, that NH_3 is a stronger base than H_2O.)

(e) The equilibrium we must consider involves three moles of oxalate ion (represented as Ox^{2-}).

$$Co^{2+}(aq) + 3Ox^{2-}(aq) \rightleftharpoons [Co(Ox)_3]^{4-}(aq)$$

The formation constant expression is

$$K_f = \frac{[[Co(Ox)_3]^{4-}]}{[Co^{2+}][Ox^{2-}]^3}$$

Because K_f is so large, we can assume that essentially all of the Co^{2+} is converted to the oxalato complex. Under that assumption, the final concentration of $[Co(Ox)_3]^{3-}$ is 0.0010 *M* and that of oxalate ion is $[Ox^{2-}] = (0.040) - 3(0.0010) = 0.037\ M$ (three Ox^{2-} ions react with each Co^{2+} ion). We then have

$$[Co^{2+}] = xM,\ [Ox^{2-}] \cong 0.037\ M,\ [[Co(Ox)_3]^{4-}] \cong 0.0010\ M.$$

Inserting these values into the equilibrium-constant expression, we have

$$K_f = \frac{(0.0010)}{x(0.037)} = 5 \times 10^9$$

Solving for x, we obtain 4×10^{-9} *M*. From this we can see that the oxalate has complexed all but a tiny fraction of the Co^{2+} present in solution.

Summary and Key Terms

Section 24.1 **Coordination compounds** are substances that contain **metal complexes**; metal complexes contain metal ions bonded to several surrounding anions or molecules known as **ligands**. The metal ion and its ligands comprise the **coordination sphere** of the complex. The atom of the ligand that bonds to the metal ion is the **donor atom**. The number of donor atoms attached to the metal ion is the **coordination number** of the metal ion. The most common coordination numbers are 4 and 6; the most common coordination geometries are tetrahedral, square planar, and octahedral.

Sections 24.2 and 24.3 Ligands that occupy only one site in a coordination sphere are called **monodentate ligands**. If a ligand has several donor atoms that can coordinate simultaneously to the metal ion, it is a **polydentate**

ligand and is also referred to as a **chelating agent**. Two common examples are ethylenediamine, denoted en, which is a **bidentate ligand**, and the ethylenediaminetetraacetate ion, $[EDTA]^{4-}$, which has six potential donor atoms. In general, chelating agents form more stable complexes than do related monodentate ligands, an observation known as the **chelate effect**. Many biologically important molecules, such as the **porphyrins**, are complexes of chelating agents. A related group of plant pigments known as **chlorophylls** are important in **photosynthesis**, the process by which plants use solar energy to convert CO_2 and H_2O into carbohydrates.

As in the nomenclature of other inorganic compounds, specific rules are followed for naming coordination compounds. In general, the number and type of ligands attached to the metal ion are specified, as is the oxidation state of the metal ion.

Section 24.4 **Isomers** are compounds with the same composition but different arrangements of atoms and therefore different properties. **Structural isomers** differ in the bonding arrangements of the ligands. One simple form of structural isomerism, known as **linkage isomerism**, occurs when a ligand is capable of coordinating to a metal ion through either of two donor atoms. **Coordination-sphere isomers** contain different ligands in the coordination sphere.

Stereoisomers are isomers with the same chemical bonding arrangements but different spatial arrangements of ligands. The most common forms of stereoisomerism are **geometric isomerism** and **optical isomerism**. Geometric isomers differ from one another in the relative locations of donor atoms in the coordination sphere; the most common are *cis-trans* isomers. Optical isomers are nonsuperimposable mirror images of one another. Geometric isomers differ from one another in their chemical and physical properties; optical isomers or **enantiomers** are **chiral**, however, meaning that they have a specific "handedness" and they differ only in the presence of a chiral environment. Optical isomers can be distinguished from one another by their interactions with plane-polarized light; solutions of one isomer rotate the plane of polarization to the right (**dextrorotatory**), and solutions of its mirror image rotate the plane to the left (**levorotatory**). Chiral molecules, therefore, are **optically active**. A 50–50 mixture of two optical isomers does not rotate plane-polarized light and is said to be **racemic**.

Section 24.5 Studies of the colors and magnetic properties of transition-metal complexes have played important roles in the formulation of bonding theories for these compounds. A substance has a particular color because it either (1) reflects or transmits light of that color, or (2) absorbs light of the **complementary color**. The amount of light absorbed by a sample as a function of wavelength is known as its **absorption spectrum**. The light absorbed provides the energy to excite electrons to higher-energy states.

It is possible to determine the number of unpaired electrons in a complex from its degree of paramagnetism. Compounds with no unpaired electrons are diamagnetic.

Section 24.6 The **crystal-field theory** successfully accounts for many properties of coordination compounds, including their color and magnetism. In this model the interaction between metal ion and ligand is viewed as electrostatic. Because some *d* orbitals are pointing right at the ligands whereas others point between them, the ligands split the energies of the metal *d* orbitals. For an octahedral complex, the *d* orbitals are split into a lower-energy set of three degenerate orbitals (the t_2 set) and a higher-energy set of two degenerate orbitals (the *e* set). Visible light can cause a ***d-d* transition**, in which an electron is excited from a lower-energy *d* orbital to a higher-energy *d* orbital. The **spectrochemical series** lists ligands in order of their ability to split the *d*-orbital energies in octahedral complexes.

Strong-field ligands create a splitting of *d*-orbital energies that is large enough to overcome the **spin-pairing energy**. The *d* electrons then preferentially pair up in the lower-energy orbitals, producing a **low-spin complex**. When the ligands exert a weak crystal field, the splitting of the *d* orbitals is small. The electrons then occupy the higher-energy *d* orbitals in preference to pairing up in the lower-energy set, producing a **high-spin complex**.

The crystal-field model also applies to tetrahedral and square-planar complexes, which leads to different *d*-orbital splitting patterns. In a tetrahedral crystal field, the splitting of the *d* orbitals is exactly opposite that of the octahedral case. The splitting by a tetrahedral crystal field is much smaller than that of an octahedral crystal field, so tetrahedral complexes are always high-spin complexes.

Exercises

Introduction to Metal Complexes

24.1 **(a)** Define the italicized phrases in the following sentence: A *metal complex* is found in which the *coordination number* is six, with four H_2O and two NH_3 *ligands*. **(b)** Explain why the formation of a metal-ligand bond is an example of a Lewis acid–base interaction.

24.2 **(a)** What is the difference between Werner's concepts of *primary valence* and *secondary valence*? What terms do we now use for these concepts? **(b)** Why can the NH_3 molecule serve as a ligand but the BH_3 molecule cannot?

24.3 A complex is written as $NiBr_2 \cdot 6NH_3$. **(a)** What is the oxidation state of the Ni atom in this complex? **(b)** What is the likely coordination number for the complex? **(c)** If the complex is treated with excess $AgNO_3(aq)$, how many moles of AgBr will precipitate per mole of complex?

24.4 A certain complex of metal M is formulated as $MCl_3 \cdot 3H_2O$. The coordination number of the complex is not known, but is expected to be four or six. **(a)** Would reaction of the complex with $AgNO_3(aq)$ provide information about the coordination number? **(b)** Would conductivity measurements provide information about the coordination number?

24.5 Indicate the coordination number of the metal and the oxidation number of the metal in each of the following complexes:
(a) $Na_2[CdCl_4]$ **(b)** $K_2[MoOCl_4]$
(c) $[Co(NH_3)_4Cl_2]Cl$ **(d)** $[Ni(CN)_5]^{3-}$
(e) $K_3[V(C_2O_4)_3]$ **(f)** $[Zn(en)_2]Br_2$

24.6 Indicate the coordination number of the metal and the oxidation number of the metal in each of the following complexes:
(a) $K_3[Co(CN)_6]$ **(b)** $[Mn(H_2O)_5Br]^+$
(c) $[Pt(NH_3)_3Br_3]Br$ **(d)** $[Co(C_2O_4)(NH_3)_4]$
(e) $[V(H_2O)_4(SCN)_2]^+$ **(f)** $[Mo(en)_2F_2]NO_3$

24.7 Determine the number and type of each donor atom in each of the complexes in Exercise 24.5.

24.8 What are the number and types of donor atoms in each of the complexes in Exercise 24.6?

Polydentate Ligands; Nomenclature

24.9 **(a)** What is the difference between a monodentate ligand and a bidentate ligand? **(b)** How many bidentate ligands are necessary to fill the coordination sphere of a six-coordinate complex? **(c)** You are told that a certain molecule can serve as a tridentate ligand. Based on this statement, what do you know about the molecule?

24.10 For each of the following polydentate ligands, determine (i) the maximum number of coordination sites that the ligand can occupy on a single metal ion, and (ii) the number and type of donor atoms in the ligand: **(a)** ethylenediamine (en); **(b)** bipyridine (bipy); **(c)** the oxalate anion ($C_2O_4^{2-}$); **(d)** the 2− ion of the porphine molecule (Figure 24.10); **(e)** $[EDTA]^{4-}$.

24.11 Polydentate ligands can vary in the number of coordination positions they occupy. In each of the following, identify the polydentate ligand present, and indicate the probable number of coordination positions it occupies:
(a) $[Co(NH_3)_4(o\text{-phen})]Cl_3$ **(b)** $[Cr(C_2O_4)(H_2O)_4]Br$
(c) $[Cr(EDTA)(H_2O)]^-$ **(d)** $[Zn(en)_2](ClO_4)_2$

24.12 Indicate the likely coordination number of the metal in each of the following complexes:
(a) $[Cd(en)Cl_2]$ **(b)** $[Hg(bipy)Br_2]$
(c) $[Co(o\text{-phen})_2Cl_2]NO_3$ **(d)** $[Ce(EDTA)]$

24.13 **(a)** What is meant by the term *chelate effect*? **(b)** What thermodynamic factor is generally responsible for the chelate effect? **(c)** Why are polydentate ligands often called *sequestering agents*?

24.14 *Pyridine* (C_5H_5N), abbreviated py, is the following molecule:

(a) Is pyridine a monodentate or polydentate ligand? **(b)** Consider the following equilibrium reaction:

$$[Ru(py)_4(bipy)]^{2+} + 2py \rightleftharpoons [Ru(py)_6]^{2+} + bipy$$

What would you predict for the magnitude of the equilibrium constant for this equilibrium? Explain the basis for your answer.

24.15 Write the formula for each of the following compounds, being sure to use brackets to indicate the coordination sphere:
(a) hexaamminechromium(III) nitrate
(b) tetraamminecarbonatocobalt(III) sulfate
(c) dichlorobis(ethylenediamine)platinum(IV) bromide
(d) potassium diaquatetrabromovanadate(III)
(e) bis(ethylenediamine)zinc(II) tetraiodomercurate(II)

24.16 Write the formula for each of the following compounds, being sure to use brackets to indicate the coordination sphere:
(a) pentaaquabromomanganese(III) sulfate
(b) tris(bipyridine)ruthenium(II) nitrate
(c) dichlorobis(*ortho*-phenanthroline)iron(III) perchlorate
(d) sodium tetrabromo(ethylenediamine)cobaltate(III)
(e) hexaaminenickel(II) tris(oxalato)chromate(III)

24.17 Write the names of the following compounds using the standard nomenclature rules for coordination complexes:
(a) $[Rh(NH_3)_4Cl_2]Cl$ **(b)** $K_2[TiCl_6]$
(c) $MoOCl_4$ **(d)** $[Pt(H_2O)_4(C_2O_4)]Br_2$

24.18 Write names for the following coordination compounds:
(a) $[Nb(en)Cl_3]SO_4$ **(b)** $Mo(CO)_3(C_5H_5N)_3$
(c) $NH_4[AuCl_4]$ **(d)** $[Ir(NH_3)_4(H_2O)_2](NO_3)_3$

Isomerism

24.19 By writing formulas or drawing structures related to any one of the following complexes, illustrate **(a)** geometric isomerism; **(b)** linkage isomerism; **(c)** optical isomerism; **(d)** coordination-sphere isomerism. The complexes are:

$[Co(NH_3)_4Br_2]Cl$; $[Pd(NH_3)_2(ONO)_2]$; *cis*-$[V(en)_2Cl_2]^+$.

24.20 **(a)** Draw the two linkage isomers of $[Co(NH_3)_5SCN]^{2+}$. **(b)** Draw the two geometric isomers of $[Co(NH_3)_3Cl_3]^{2+}$. **(c)** Two compounds with the formula $Co(NH_3)_5ClBr$ can be prepared. Use structural formulas to show how they differ. What kind of isomerism does this illustrate?

24.21 A four-coordinate complex MA_2B_2 is prepared and found to have two different isomers. Is it possible to determine from this information whether the complex is square planar or tetrahedral? If so, which is it?

24.22 Consider an octahedral complex MA_3B_3. How many geometric isomers are expected for this compound? Will any of the isomers be optically active? If so, which ones?

24.23 Draw the *cis* and *trans* isomers of the $[Co(en)_2(NH_3)Cl]^{2+}$ ion. Are either or both of these isomers chiral? If so, draw the two enantiomers.

24.24 Draw the distinct geometric isomers of $[Rh(bipy)(NH_3)_3Br]^{2+}$. Do any of these geometric isomers have optical isomers? If so, identify them and draw the structures of their enantiomers.

24.25 Sketch all the possible stereoisomers of **(a)** tetrahedral $[Cd(H_2O)_2Cl_2]$; **(b)** square-planar $[IrCl_2(PH_3)_2]^-$; **(c)** octahedral $[Fe(o\text{-phen})_2Cl_2]^+$.

24.26 Sketch all the possible stereoisomers of **(a)** tetrahedral $[Zn(en)(CN)_2]$; **(b)** octahedral $[CoBr_2Cl_2(en)]^-$; **(c)** square-planar $[Pd(en)Cl(SCN)]$.

Color; Magnetism; Crystal-Field Theory

24.27 **(a)** To the closest 100 nm, what are the largest and smallest wavelengths of visible light? **(b)** What is meant by the term *complementary color*? **(c)** What is the significance of complementary colors in understanding the colors of metal complexes?

24.28 **(a)** A complex absorbs light with wavelength of 530 nm. Do you expect it to have color? **(b)** A solution of a compound appears green. Does this observation necessarily mean that all colors of visible light other than green are absorbed by the solution? Explain. **(c)** What information is usually presented in a *visible absorption spectrum* of a compound?

24.29 What is the observed color of a coordination compound that absorbs radiation of wavelength 580 nm?

24.30 Notice in Figure 13.7 that an aqueous solution of $NiCl_2$ is green. Assuming that the green color is due to a single absorption band, sketch the shape of the absorption curve for the solution, analogous to that for $[Ti(H_2O)_6]^{3+}$ in Figure 24.27.

24.31 In crystal-field theory, ligands are modeled with point negative charges. What is the basis of this assumption, and how does it relate to the nature of metal-ligand bonds?

24.32 Explain why the d_{xy}, d_{xz}, and d_{yz} orbitals lie lower in energy than the d_{z^2} and $d_{x^2-y^2}$ orbitals in the presence of an octahedral arrangement of ligands about the central metal ion.

24.33 **(a)** Sketch a diagram that shows the definition of the *crystal-field splitting energy* (Δ) for an octahedral crystal field. **(b)** What is the relationship between the magnitude of Δ and the energy of the *d-d* transition for a d^1 complex? **(c)** What is the *spectrochemical series*?

24.34 As shown in Figure 24.27, the *d-d* transition of $[Ti(H_2O)_6]^{3+}$ produces an absorption maximum at a wavelength of 500 nm. **(a)** What is the magnitude of Δ for $[Ti(H_2O)_6]^{3+}$ in kJ/mol? **(b)** How would the magnitude of Δ change if the H_2O ligands in $[Ti(H_2O)_6]^{3+}$ were replaced with NH_3 ligands?

24.35 Explain why many cyano complexes of divalent transition-metal ions are yellow, whereas many aqua complexes of these ions are blue or green.

24.36 The $[Ni(H_2O)_6]^{2+}$ ion is green, whereas the $[Ni(NH_3)_6]^{2+}$ ion is purple. Predict the predominant color of light absorbed by each ion. Which ion absorbs light with the shorter wavelength? Do your conclusions agree with the spectrochemical series?

24.37 Give the number of *d* electrons associated with the central metal ion in each of the following complexes: **(a)** $[Ru(en)_3]Cl_3$; **(b)** $K_2[Cu(CN)_4]$; **(c)** $Na_3[Co(NO_2)_6]$; **(d)** $[Mo(EDTA)]ClO_4$; **(e)** $K_3[ReCl_6]$.

24.38 Give the number of *d* electrons associated with the central metal ion in each of the following complexes: **(a)** $K_3[Fe(CN)_6]$; **(b)** $[Mn(H_2O)_6](NO_3)_2$; **(c)** $Na_2[CoCl_4]$; **(d)** $Na[Ag(CN)_2]$; **(e)** $[Sr(EDTA)]^{2-}$.

24.39 For each of the following metals, write the electronic configuration of the atom and its 3+ ion: **(a)** Mn; **(b)** Ru; **(c)** Rh. Draw the crystal-field energy-level diagram for the *d* orbitals of an octahedral complex, and show the placement of the *d* electrons for each 3+ ion, assuming a strong-field complex. How many unpaired electrons are there in each case?

24.40 For each of the following metals, write the electronic configuration of the atom and its 2+ ion: **(a)** Ru; **(b)** Mo; **(c)** Co. Draw the crystal-field energy-level diagram for the *d* orbitals of an octahedral complex, and show the placement of the *d* electrons for each 2+ ion, assuming a weak-field complex. How many unpaired electrons are there in each case?

24.41 Draw the crystal-field energy-level diagrams, and show the placement of *d* electrons for each of the following: **(a)** $[Cr(H_2O)_6]^{2+}$ (four unpaired electrons); **(b)** $[Mn(H_2O)_6]^{2+}$ (high spin); **(c)** $[Ru(NH_3)_5H_2O]^{2+}$ (low spin); **(d)** $[IrCl_6]^{2-}$ (low spin); **(e)** $[Cr(en)_3]^{3+}$; **(f)** $[NiF_6]^{4-}$.

24.42 Draw the crystal-field energy-level diagrams, and show the placement of electrons for the following complexes: **(a)** $[VCl_6]^{3-}$; **(b)** $[FeF_6]^{3-}$ (a high-spin complex); **(c)** $[Ru(bipy)_3]^{3+}$ (a low-spin complex); **(d)** $[NiCl_4]^{2-}$ (tetrahedral); **(e)** $[PtBr_6]^{2-}$; **(f)** $[Ti(en)_3]^{2+}$.

24.43 The complex $[Mn(NH_3)_6]^{2+}$ contains five unpaired electrons. Sketch the energy-level diagram for the *d* orbitals, and indicate the placement of electrons for this complex ion. Is the ion a high-spin or a low-spin complex?

24.44 The ion $[Fe(CN)_6]^{3-}$ has one unpaired electron, whereas $[Fe(NCS)_6]^{3-}$ has five unpaired electrons. From these results, what can you conclude about whether each complex is high spin or low spin? What can you say about the placement of NCS^- in the spectrochemical series?

Additional Exercises

24.45 Give one or more examples of each of the following:
(a) An octahedral complex containing two bidentate and two monodentate ligands
(b) A complex with coordination number 4
(c) A high-spin and a low-spin complex of the same metal ion
(d) A ligand that is capable of linkage isomerism
(e) A complex ion that exhibits geometric isomerism
(f) A complex that exhibits optical isomerism but not geometric isomerism

24.46 Based on the molar conductance values listed here for the series of platinum(IV) complexes, write the formula for each complex so as to show which ligands are in the coordination sphere of the metal. By way of example, the molar conductances of NaCl and $BaCl_2$ are 107 ohm^{-1} and 197 ohm^{-1}, respectively.

Complex	Molar Conductance (ohm^{-1})* of 0.050 *M* Solution
$Pt(NH_3)_6Cl_4$	523
$Pt(NH_3)_4Cl_4$	228
$Pt(NH_3)_3Cl_4$	97
$Pt(NH_3)_2Cl_4$	0
$KPt(NH_3)Cl_5$	108

*The ohm is a unit of resistance; conductance is the inverse of resistance.

24.47 **(a)** A compound with formula $RuCl_3 \cdot 5H_2O$ is dissolved in water, forming a solution that is approximately the same color as the solid. Immediately after forming the solution, the addition of excess $AgNO_3(aq)$ forms 2 mol of solid AgCl per mole of complex. Write the formula for the compound, showing which ligands are likely to be present in the coordination sphere. **(b)** After a solution of

$RuCl_3 \cdot 5H_2O$ has stood for about a year, addition of $AgNO_3(aq)$ precipitates 3 mol of AgCl per mole of complex. What has happened in the ensuing time?

24.48 Sketch the structure of the complex in each of the following compounds:
(a) *cis*-$[Co(NH_3)_4(H_2O)_2](NO_3)_2$
(b) $Na_2[Ru(H_2O)Cl_5]$
(c) *trans*-$NH_4[Co(C_2O_4)_2(H_2O)_2]$
(d) *cis*-$[Ru(en)_2Cl_2]$

24.49 **(a)** Give the full name for each of the compounds in Exercise 24.48. **(b)** Will any of the complexes be optically active? Explain.

24.50 The molecule *dimethylphosphinoethane* $[(CH_3)_2PCH_2CH_2P(CH_3)_2$, which is abbreviated dmpe] is used as a ligand for some complexes that serve as catalysts. A complex that contains this ligand is $Mo(CO)_4(dmpe)$. **(a)** Draw a Lewis structure of dmpe, and determine whether it can serve as a polydentate ligand. **(b)** Determine the oxidation state of Mo in$Mo(CO)_4(dmpe)$. **(c)** Sketch the structure of $Mo(CO)_4(dmpe)$ and determine whether it can have multiple isomers.

24.51 Although the *cis* configuration is known for $[Pt(en)Cl_2]$, no *trans* form is known. **(a)** Explain why the *trans* compound is not possible. **(b)** Suggest what type of ligand would be required to form a *trans*-bidentate coordination to a metal atom.

[24.52] The acetylacetone ion forms very stable complexes with many metallic ions. It acts as a bidentate ligand, coordinating to the metal at two adjacent positions. Suppose that one of the CH_3 groups of the ligand is replaced by a CF_3 group, as shown:

Trifluoromethyl acetylacetonate (tfac)

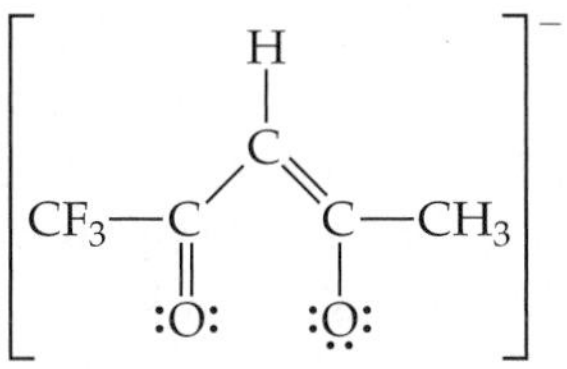

$$\left[CF_3-\underset{\underset{:O:}{\|}}{C}=\overset{\overset{H}{|}}{C}-\underset{\underset{:\ddot{O}:}{|}}{C}-CH_3 \right]^-$$

Sketch all possible isomers for the complex with three tfac ligands on cobalt(III). (You can use the symbol ●‿○ to represent the ligand.)

24.53 Give brief statements about the relevance of the following complexes in living systems: **(a)** hemoglobin; **(b)** chlorophylls; **(c)** siderophores.

24.54 Write balanced chemical equations to represent the following observations. (In some instances the complex involved has been discussed previously in the text.) **(a)** Solid silver chloride dissolves in an excess of aqueous ammonia. **(b)** The green complex $[Cr(en)_2Cl_2]Cl$, on treatment with water over a long time, converts to a brown-orange complex. Reaction of $AgNO_3$ with a solution of the product precipitates 3 mol of AgCl per mole of Cr present. (Write two chemical equations.) **(c)** When an NaOH solution is added to a solution of $Zn(NO_3)_2$, a precipitate forms. Addition of excess NaOH solution causes the precipitate to dissolve. (Write two chemical equations.) **(d)** A pink solution of $Co(NO_3)_2$ turns deep blue on addition of concentrated hydrochloric acid.

24.55 Some metal complexes have a coordination number of 5. One such complex is $Fe(CO)_5$, which adopts a *trigonal bipyramidal* geometry (see Figure 9.8). **(a)** Write the name for $Fe(CO)_5$, using the nomenclature rules for coordination compounds. **(b)** What is the oxidation state of Fe in this compound? **(c)** Suppose one of the CO ligands is replaced with a CN^- ligand, forming $[Fe(CO)_4(CN)]^-$. How many geometric isomers would you predict this complex to have?

24.56 What properties of the ligand determine the size of the splitting of the *d*-orbital energies in the presence of an octahedral arrangement of ligands about a central transition-metal ion? Explain.

24.57 Which of the following objects is chiral? **(a)** a left shoe; **(b)** a slice of bread; **(c)** a wood screw; **(d)** a molecular model of $Zn(en)Cl_2$; **(e)** a typical golf club.

24.58 The complexes $[V(H_2O)_6]^{3+}$ and $[VF_6]^{3-}$ are both known. **(a)** Draw the *d*-orbital energy-level diagram for V(III) octahedral complexes. **(b)** What gives rise to the colors of these complexes? **(c)** Which of the two complexes would you expect to absorb light of higher energy? Explain.

[24.59] One of the more famous species in coordination chemistry is the Creutz–Taube complex,

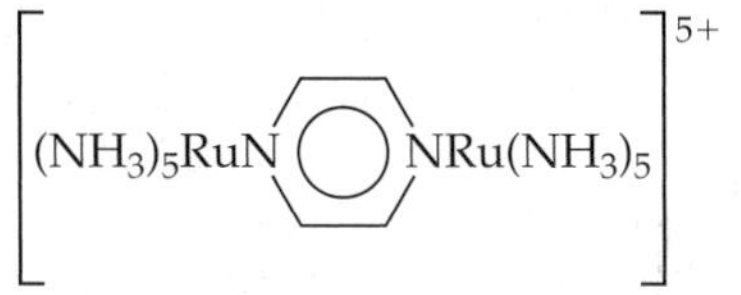

$$\left[(NH_3)_5Ru\text{N}\bigcirc\text{NRu}(NH_3)_5 \right]^{5+}$$

It is named for the two scientists who discovered it and initially studied its properties. The central ligand is pyrazine, a planar six-membered ring with nitrogens at opposite sides. **(a)** How can you account for the fact that the complex, which has only neutral ligands, has an odd overall charge? **(b)** The metal is in a low-spin configuration in both cases. Assuming octahedral coordination, draw the *d*-orbital energy-level diagram for each metal. **(c)** In many experiments, the two metal ions appear to be in exactly equivalent states. Can you think of a reason that this might appear to be so, recognizing that electrons move very rapidly compared to nuclei?

24.60 Solutions of $[Co(NH_3)_6]^{2+}$, $[Co(H_2O)_6]^{2+}$ (both octahedral), and $[CoCl_4]^{2-}$ (tetrahedral) are colored. One is pink, one is blue, and one is yellow. Based on the spectrochemical series and remembering that the energy splitting in tetrahedral complexes is normally much less than that in octahedral ones, assign a color to each complex.

24.61 Oxyhemoglobin, with an O_2 bound to iron, is a low-spin Fe(II) complex; deoxyhemoglobin, without the O_2 molecule, is a high-spin complex. How many unpaired electrons are centered on the metal ion in each case? Explain in a general way why the two forms of hemoglobin have different colors (hemoglobin is red, whereas deoxyhemoglobin has a bluish cast).

24.62 Sketch two sets of *x*- and *y*-axes with point negative charges equidistant from the origin along the $\pm x$- and $\pm y$-axes. On one set of axes, sketch a d_{xy} orbital. On the other, sketch a $d_{x^2-y^2}$ orbital. Use your sketches to explain why the $d_{x^2-y^2}$ orbital is higher in energy than the d_{xy} orbital in a square-planar crystal field.

24.63 In each of the following pairs of complexes, which would you expect to absorb at the longer wavelength: **(a)** $[FeF_6]^{4-}$ or $[Fe(NH_3)_6]^{2+}$; **(b)** $[V(H_2O)_6]^{2+}$ or $[Cr(H_2O)_6]^{3+}$; **(c)** $[Co(NH_3)_6]^{2+}$ or $[CoCl_4]^{2-}$? Explain your reasoning in each case.

[24.64] Consider the tetrahedral anions ${VO_4}^{3-}$ (orthovanadate ion), ${CrO_4}^{2-}$ (chromate ion), and ${MnO_4}^-$ (permanganate ion). **(a)** These anions are *isoelectronic*. What is meant by this statement? **(b)** Would you expect these anions to exhibit *d-d* transitions? Explain. **(c)** As mentioned in "A Closer Look" on charge-transfer color, the violet color of MnO_4^- is due to a *ligand-to-metal charge transfer* (LMCT) transition. What is meant by this term? **(d)** The LMCT transition in MnO_4^- occurs at a wavelength of 565 nm. The CrO_4^{2-} ion is yellow. Is the wavelength of the LMCT transition for chromate larger or smaller than that for MnO_4^-? Explain. **(e)** The VO_4^{3-} ion is colorless. Is this observation consistent with the wavelengths of the LMCT transitions in MnO_4^- and CrO_4^{2-}?

[24.65] The red color of ruby is due to the presence of Cr(III) ions at octahedral sites in the close-packed oxide lattice of Al_2O_3. Draw the crystal-field-splitting diagram for Cr(III) in this environment. Suppose that the ruby crystal is subjected to high pressure. What do you predict for the variation in the wavelength of absorption of the ruby as a function of pressure? Explain.

24.66 In 2001, chemists at SUNY-Stonybrook succeeded in synthesizing the complex *trans*-$[Fe(CN)_4(CO)_2]^{2-}$, which could be a model of complexes that may have played a role in the origin of life. **(a)** Sketch the structure of the complex. **(b)** The complex is isolated as a sodium salt. Write the complete name of this salt. **(c)** What is the oxidation state of Fe in this complex? How many *d* electrons are associated with the Fe in this complex? **(d)** Would you expect this complex to be high-spin or low-spin? Explain.

[24.67] When Alfred Werner was developing the field of coordination chemistry, it was argued by some that the optical activity he observed in the chiral complexes he had prepared was because of the presence of carbon atoms in the molecule. To disprove this argument, Werner synthesized a chiral complex of cobalt that had no carbon atoms in it, and he was able to resolve it into its enantiomers. Design a cobalt(III) complex that would be chiral if it could be synthesized and that contains no carbon atoms. (It may not be possible to synthesize the complex you design, but we won't worry about that for now.)

24.68 Many trace metal ions exist in the bloodstream as complexes with amino acids or small peptides. The anion of the amino acid glycine, symbol gly, is capable of acting as a bidentate ligand, coordinating to the metal through nitrogen and $-O^-$ atoms.

$$H_2NCH_2\overset{\overset{\displaystyle O}{\|}}{C}-O^-$$

How many isomers are possible for **(a)** $[Zn(gly)_2]$ (tetrahedral); **(b)** $[Pt(gly)_2]$ (square-planar); **(c)** $[Co(gly)_3]$ (octahedral)? Sketch all possible isomers. Use N‿O to represent the ligand.

[24.69] Suppose that a transition-metal ion were in a lattice in which it was in contact with just two nearby anions, located on opposite sides of the metal. Diagram the splitting of the metal *d* orbitals that would result from such a crystal field. Assuming a strong field, how many unpaired electrons would you expect for a metal ion with six *d* electrons? (*Hint:* Consider the linear axis to be the *z*-axis.)

Integrative Exercises

24.70 Metallic elements are essential components of many important enzymes operating within our bodies. *Carbonic anhydrase*, which contains Zn^{2+}, is responsible for rapidly interconverting dissolved CO_2 and bicarbonate ion, HCO_3^-. The zinc in carbonic anhydrase is coordinated by three nitrogen-containing groups and a water molecule. The enzyme's action depends on the fact that the coordinated water molecule is more acidic than the bulk solvent molecules. Explain this fact in terms of Lewis acid–base theory (Section 16.11).

24.71 Two different compounds have the formulation $CoBr(SO_4)\cdot 5NH_3$. Compound A is dark violet and compound B is red-violet. When compound A is treated with $AgNO_3(aq)$, no reaction occurs, whereas compound B reacts with $AgNO_3(aq)$ to form a white precipitate. When compound A is treated with $BaCl_2(aq)$, a white precipitate is formed, whereas compound B has no reaction with $BaCl_2(aq)$. **(a)** Is Co in the same oxidation state in these complexes? **(b)** Explain the reactivity of compounds A and B with $AgNO_3(aq)$ and $BaCl_2(aq)$. **(c)** Are compounds A and B isomers of one another? If so, which category from Figure 24.18 best describes the isomerism observed for these complexes? **(d)** Would compounds A and B be expected to be strong electrolytes, weak electrolytes, or nonelectrolytes?

[24.72] The molecule *methylamine* (CH_3NH_2) can act as a monodentate ligand. The following are equilibrium reactions and the thermochemical data at 298 K for reactions of methylamine and en with $Cd^{2+}(aq)$:

$$Cd^{2+}(aq) + 4CH_3NH_2(aq) \rightleftharpoons [Cd(CH_3NH_2)_4]^{2+}(aq)$$
$$\Delta H^\circ = -57.3\text{ kJ};\ \Delta S^\circ = -67.3\text{ J/K};\ \Delta G^\circ = -37.2\text{ kJ}$$
$$Cd^{2+}(aq) + 2en(aq) \rightleftharpoons [Cd(en)_2]^{2+}(aq)$$
$$\Delta H^\circ = -56.5\text{ kJ};\ \Delta S^\circ = +14.1\text{ J/K};\ \Delta G^\circ = -60.7\text{ kJ}$$

(a) Calculate ΔG° and the equilibrium constant K_{eq} for the following *ligand exchange* reaction:

$$[Cd(CH_3NH_2)_4]^{2+}(aq) + 2en(aq) \rightleftharpoons [Cd(en)_2]^{2+}(aq) + 4CH_3NH_2(aq)$$

(b) Based on the value of K_{eq} in part (a), what would you conclude about this reaction? What concept is demonstrated? **(c)** Determine the magnitudes of the enthalpic (ΔH°) and the entropic ($-T\Delta S^\circ$) contributions to ΔG° for the ligand exchange reaction. Explain the relative magnitudes. **(d)** Based on information in this exercise and in "A Closer Look" box on the chelate effect, predict the sign of ΔH° for the following hypothetical reaction:

$$[Cd(CH_3NH_2)_4]^{2+}(aq) + 4NH_3(aq) \rightleftharpoons [Cd(NH_3)_4]^{2+}(aq) + 4CH_3NH_2(aq)$$

24.73 A palladium complex formed from a solution containing bromide ion and pyridine, C_5H_5N (a good electron-pair donor), is found on elemental analysis to contain 37.6% bromine, 28.3% carbon, 6.60% nitrogen, and 2.37% hydrogen by mass. The compound is slightly soluble in several organic solvents; its solutions in water or alcohol do not conduct electricity. It is found experimentally to have a zero dipole moment. Write the chemical formula, and indicate its probable structure.

24.74 A manganese complex formed from a solution containing potassium bromide and oxalate ion is purified and analyzed. It contains 10.0% Mn, 28.6% potassium, 8.8% carbon, and 29.2% bromine by mass. The remainder of the compound is oxygen. An aqueous solution of the complex has about the same electrical conductivity as an equimolar solution of $K_4[Fe(CN)_6]$. Write the formula of the compound, using brackets to denote the manganese and its coordination sphere.

24.75 **(a)** In early studies it was observed that when the complex $[Co(NH_3)_4Br_2]Br$ was placed in water, the electrical conductivity of a 0.05 *M* solution changed from an initial value of 191 ohm^{-1} to a final value of 374 ohm^{-1} over a period of an hour or so. Suggest an explanation for the observed results. (See Exercise 24.46 for relevant comparison data.) **(b)** Write a balanced chemical equation to describe the reaction. **(c)** A 500-mL solution is made up by dissolving 3.87 g of the complex. As soon as the solution is formed, and before any change in conductivity has occurred, a 25.00-mL portion of the solution is titrated with 0.0100 *M* $AgNO_3$ solution. What volume of $AgNO_3$ solution do you expect to be required to precipitate the free $Br^-(aq)$? **(d)** Based on the response you gave to part (b), what volume of $AgNO_3$ solution would be required to titrate a fresh 25.00-mL sample of $[Co(NH_3)_4Br_2]Br$ after all conductivity changes have occurred?

24.76 The total concentration of Ca^{2+} and Mg^{2+} in a sample of hard water was determined by titrating a 0.100-L sample of the water with a solution of $EDTA^{4-}$. The $EDTA^{4-}$ chelates the two cations:

$$Mg^{2+} + [EDTA]^{4-} \longrightarrow [Mg(EDTA)]^{2-}$$
$$Ca^{2+} + [EDTA]^{4-} \longrightarrow [Ca(EDTA)]^{2-}$$

It requires 31.5 mL of 0.0104 *M* $[EDTA]^{4-}$ solution to reach the end point in the titration. A second 0.100-L sample was then treated with sulfate ion to precipitate Ca^{2+} as calcium sulfate. The Mg^{2+} was then titrated with 18.7 mL of 0.0104 *M* $[EDTA]^{4-}$. Calculate the concentrations of Mg^{2+} and Ca^{2+} in the hard water in mg/L.

24.77 The value of Δ for the $[CrF_6]^{3-}$ complex is 182 kJ/mol. Calculate the expected wavelength of the absorption corresponding to promotion of an electron from the lower-energy to the higher-energy *d*-orbital set in this complex. Should the complex absorb in the visible range? (You may need to review Sample Exercise 6.3; remember to divide by Avogadro's number.)

[24.78] A Cu electrode is immersed in a solution that is 1.00 *M* in $[Cu(NH_3)_4]^{2+}$ and 1.00 *M* in NH_3. When the cathode is a standard hydrogen electrode, the emf of the cell is found to be +0.08 V. What is the formation constant for $[Cu(NH_3)_4]^{2+}$?

[24.79] The complex $[Ru(EDTA)(H_2O)]^-$ undergoes substitution reactions with several ligands, replacing the water molecule with the ligand.

$$[Ru(EDTA)(H_2O)]^- + L \longrightarrow [Ru(EDTA)L]^- + H_2O$$

The rate constants for several ligands are as follows:

Ligand, L	$k\ (M^{-1}s^{-1})$
Pyridine	6.3×10^3
SCN^-	2.7×10^2
CH_3CN	3.0×10

(a) One possible mechanism for this substitution reaction is that the water molecule dissociates from the complex in the rate-determining step, and then the ligand L fills the void in a rapid second step. A second possible mechanism is that L approaches the complex, begins to form a new bond to the metal, and displaces the water molecule, all in a single concerted step. Which of these two mechanisms is more consistent with the data? Explain. **(b)** What do the results suggest about the relative basicities of the three ligands toward Ru(III)? **(c)** Assuming that the complexes are all low spin, how many unpaired electrons are in each?

eMedia Exercises

24.80 After watching the **Isomerism** movie (*eChapter 24.3*), answer the following questions. **(a)** Which of the three geometries can exhibit *cis-trans* isomerism? Draw structures of examples to support your answer. **(b)** Which of the three geometries can exhibit optical isomerism? Again, draw structures to support your answer.

24.81 *Cis* and *trans* isomers of tetraamminedichlorocobalt(III) ion are shown in the **Isomerism** movie (*eChapter 24.3*). **(a)** Is either of these isomers chiral? If so, which one? **(b)** For the following, tell whether *cis-trans* isomerism is possible and whether or not each isomer is chiral: $[Co(NH_3)_2Cl_4]^-$; $[Co(NH_3)_5Cl]^{2+}$; $[Cr(C_2O_4)_2(H_2O)_2]^-$; and $[Cr(en)_2Br_2]^+$. **(c)** Is it possible for an octahedral complex to be incapable of exhibiting *cis-trans* isomerism and yet still be chiral? Explain.

24.82 The **Optical Activity** movie (*eChapter 24.3*) illustrates the interaction of plane-polarized light with a solution containing a single enantiomer of a chiral compound. **(a)** What is meant by the term *enantiomer*? **(b)** In what ways do enantiomers differ in their physical and chemical properties? **(c)** What is a racemic mixture? **(d)** How does a solution containing a racemic mixture interact with plane-polarized light?

24.83 Enantiomers are designated either *d* or *l*, depending on whether they rotate the plane of plane-polarized light clockwise or counterclockwise, respectively. Watch the **Optical Activity** movie (*eChapter 24.3*). **(a)** Draw both enantiomers for each of the following optically active complexes: *cis*-$[Cr(en)_2Cl_2]^+$; $[Co(en)_3]^{3+}$; *cis*-$[Cr(C_2O_4)_2(H_2O)_2]^-$. **(b)** Which of the following solutions would you expect to rotate the plane of plane-polarized light? An equimolar mixture of *l-cis*-$[Cr(en)_2Cl_2]^+$ and *d-cis*-$[Cr(en)_2Cl_2]^+$; an equimolar mixture of *d-cis*-$[Cr(en)_2Cl_2]^+$ and *trans*-$[Cr(en)_2Cl_2]^+$; an equimolar mixture of *d*-$[Co(en)_3]^{3+}$ and *d-cis*-$[Cr(en)_2Cl_2]^+$. Explain your predictions.

Chapter 25

The Chemistry of Life: Organic and Biological Chemistry

The genetic information that provides both the similarities and the differences between these two tigers is transferred from one generation to the next by the replication of molecules of deoxyribonucleic acid, or DNA.

THE ELEMENT CARBON forms a vast number of compounds. Over 16 million carbon-containing compounds are known, and about 90% of the new compounds synthesized each year contain carbon. The study of carbon compounds constitutes a separate branch of chemistry known as **organic chemistry**. This term arose from the eighteenth-century belief that organic compounds could be formed only by living systems. This idea was disproved in 1828 by the German chemist Friedrich Wöhler when he synthesized urea (H_2NCONH_2), an organic substance found in the urine of mammals, by heating ammonium cyanate (NH_4OCN), an inorganic substance.

The notion that organic chemicals and living organisms are connected is certainly true in one sense: Life as we know it could not exist without a vast array of complex, biologically important organic molecules. The study of the chemistry of living species is called *biological chemistry*, or **biochemistry**.

In this final chapter we present a brief view of some of the elementary aspects of organic chemistry and biochemistry. Many of you will study these subjects in greater detail by taking additional courses devoted entirely to these topics. As you read the materials that follow, you will notice that many of the concepts important for understanding the fundamentals of organic chemistry and biochemistry have been developed in earlier chapters.

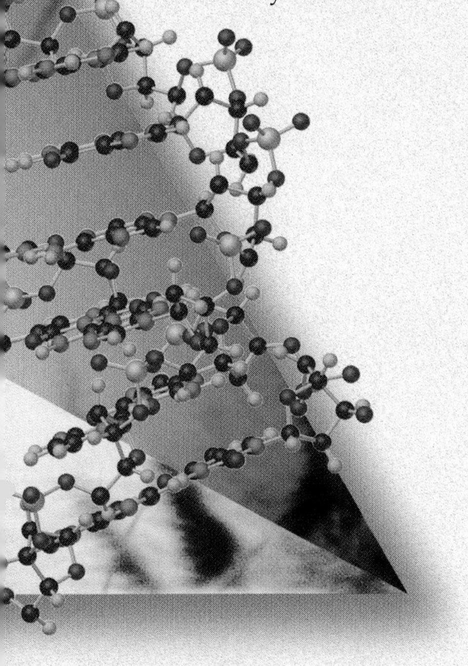

▶ What's Ahead ◀

- We begin with a brief review of the structures and reactivity of organic compounds.
- We then consider the *hydrocarbons*, compounds containing only C and H.
- There are several classes of hydrocarbons. Those in which all bonds are single bonds are called *alkanes*. Those with one or more C=C bonds are called *alkenes*, whereas those with one or more C≡C bonds are called *alkynes*. *Aromatic* hydrocarbons have at least one planar ring with delocalized π electrons.
- One reason for the huge number of organic compounds is the existence of *isomers*, compounds with identical compositions whose molecules have different structures.
- A central organizing principle of organic chemistry is the *functional group*, a group of atoms at which most of the compound's chemical reactions occur. Several functional groups, including the alcohol group, are discussed.
- We conclude the chapter by considering several classes of biochemically important molecules—*proteins*, *carbohydrates*, and *nucleic acids*.

25.1 Some General Characteristics of Organic Molecules

What is it about carbon that leads to the tremendous diversity in its compounds and allows them to play such crucial roles in biology and industry? Let's consider some general features of organic molecules, and as we do, let's review some principles that we learned in earlier chapters.

The Structures of Organic Molecules

The three-dimensional structures of organic and biochemical molecules play an essential role in determining their physical and chemical behaviors. Because carbon has four valence electrons ([He]$2s^2 2p^2$), it forms four bonds in virtually all its compounds. When all four bonds are single bonds, the electron pairs are disposed in a tetrahedral arrangement. (Section 9.2) In the hybridization model the carbon 2*s* and 2*p* orbitals are then sp^3 hybridized. (Section 9.5) When there is one double bond, the arrangement is trigonal planar (sp^2 hybridization). With two double bonds or a triple bond, it is linear (*sp* hybridization). Examples are shown in Figure 25.1 ▼.

C—H bonds occur in almost every organic molecule. Because the valence shell of H can hold only two electrons, hydrogen forms only one covalent bond. As a result, hydrogen atoms are always located on the surface of organic molecules, as in the propane molecule:

```
     H   H   H
     |   |   |
 H — C — C — C — H
     |   |   |
     H   H   H
```

The C—C bonds form the backbone or skeleton of the molecule while the H atoms are on the surface or "skin" of the molecule.

The bonding arrangements about individual atoms are important in determining overall molecular shape. In turn, the overall shapes of organic and biochemical molecules are also important in determining how they will react with other molecules, and how rapidly. They also determine important physical properties.

The Stabilities of Organic Substances

In Section 8.8 we learned about the average strengths of various chemical bonds, including those that are characteristic of organic molecules, such as C—H, C—C, C—N, C—O, and C=O bonds. Carbon forms strong bonds with a variety of elements, especially H, O, N, and the halogens. Carbon also has an exceptional ability to bond to itself, forming a variety of molecules with

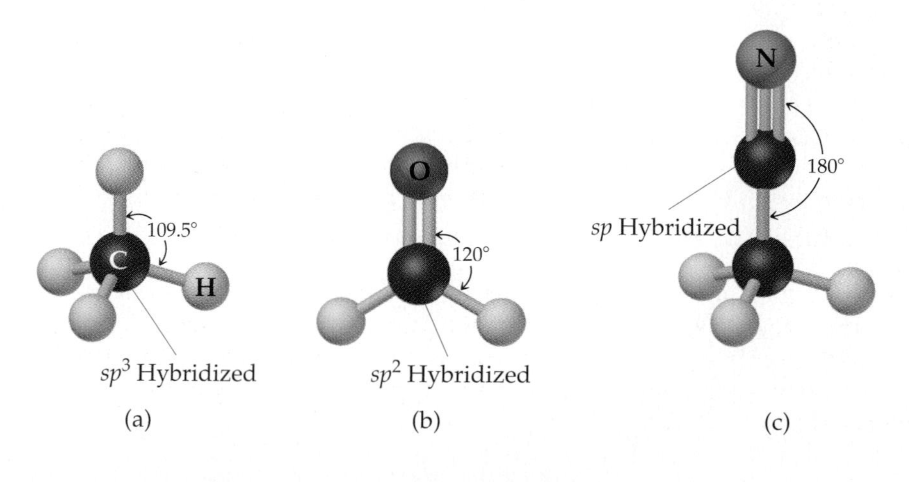

▶ **Figure 25.1** Molecular models showing the three common geometries around carbon: (a) tetrahedral in methane (CH_4), where the carbon is bonded to four other atoms; (b) trigonal planar in formaldehyde (CH_2O), where the carbon is bonded to three other atoms; and (c) linear in acetonitrile (CH_3CN), where the top carbon is bonded to two atoms.

chains or rings of carbon atoms. As we saw in Chapter 8, double bonds are generally stronger than single bonds, and triple bonds are stronger than double bonds. Increasing bond strength with bond order is accompanied by a shortening of the bond. Thus, carbon–carbon bond lengths decrease in the order $C{-}C > C{=}C > C{\equiv}C$.

We know from calorimetric measurements that reaction of a simple organic substance such as methane (CH_4) with oxygen is highly exothermic. (Sections 5.6, 5.7, and 5.8) Indeed, the combustion of methane (natural gas) keeps many of our homes warm during the winter months! Although the reactions of most organic compounds with oxygen are exothermic, great numbers of them are stable indefinitely at room temperature in the presence of air because the activation energy required for combustion to begin is large.

Most reactions with low or moderate activation barriers begin when a region of high electron density on one molecule encounters a region of low electron density on another molecule. The regions of high electron density may be due to the presence of a multiple bond or to the more electronegative atom in a polar bond. Because of their strength and lack of polarity, C—C single bonds are relatively unreactive. C—H bonds are also largely unreactive for the same reasons. The C—H bond is nearly nonpolar because the electronegativities of C (2.5) and H (2.1) are close. To better understand the implications of these facts, consider ethanol:

```
    H   H
    |   |
H — C — C — O — H
    |   |
    H   H
```

The differences in the electronegativity values of C (2.5) and O (3.5) and those of O and H (2.1) indicate that the C—O and O—H bonds are quite polar. Thus, the chemical reactions of ethanol involve these bonds. A group of atoms such as the C—O—H group, which determines how an organic molecule functions or reacts, is called a **functional group**. The functional group is the center of reactivity in an organic molecule.

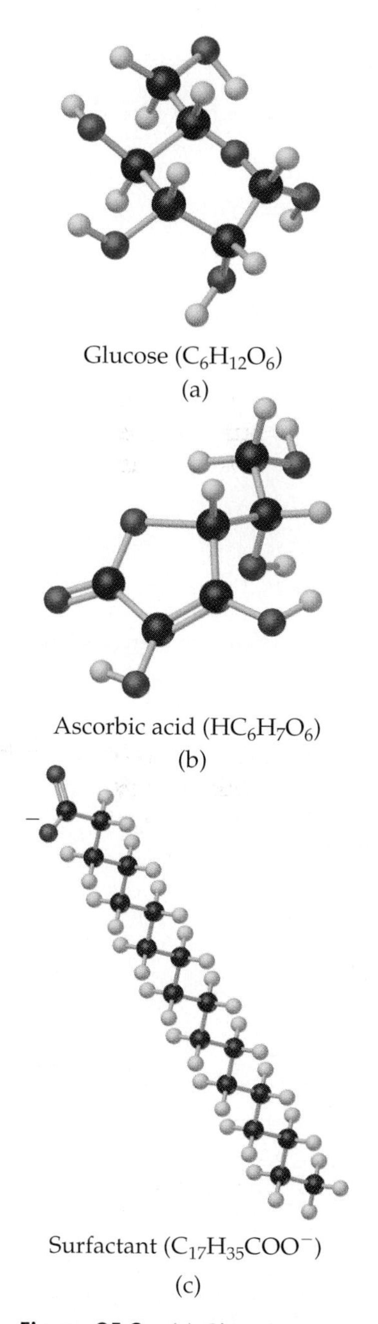

▲ **Figure 25.2** (a) Glucose ($C_6H_{12}O_6$), a simple sugar; (b) ascorbic acid ($HC_6H_7O_6$), known as vitamin C; (c) the sterate ion ($C_{17}H_{35}COO^-$), an ion that functions as a surfactant. (To allow the figure to fit the allocated space, the surfactant is drawn to a different relative scale than the models of glucose and ascorbic acid.)

Solubility and Acid-Base Properties of Organic Substances

In most organic substances the most prevalent bonds are carbon–carbon and carbon–hydrogen, which have low polarity. For this reason, the overall polarity of organic molecules is often low. They are generally soluble in nonpolar solvents and not very soluble in water. (Section 13.3) Molecules that are soluble in polar solvents such as water are those that have polar groups on the surface of the molecule, such as found in glucose [Figure 25.2(a)] ▶ or ascorbic acid [vitamin C, Figure 25.2(b)]. Surfactant organic molecules have a long, nonpolar part that extends into a nonpolar medium, and a polar, ionic "head group" that extends into a polar medium such as water [Figure 25.2(c)]. (Section 13.6) This type of structure is found in many biochemically important substances, as well as soaps and detergents.

Many organic substances contain acidic or basic groups. The most important acidic substances are the carboxylic acids, which bear the functional group $-COOH$. (Section 16.10) The most important basic substances are amines, which bear the $-NH_2$, $-NHR$, or $-NR_2$ groups, where R is an organic group consisting of some combination of C—C and C—H bonds, such as $-CH_3$ or $-C_2H_5$. (Section 16.7)

As you read this chapter, you will find many concept links to related materials in earlier chapters, many of them to the sections just discussed. *We strongly encourage you to follow these links and review the earlier materials.* Doing so will definitely enhance your understanding and appreciation of organic chemistry and biochemistry.

25.2 Introduction to Hydrocarbons

Because the compounds of carbon are so numerous, it is convenient to organize them into families that exhibit structural similarities. The simplest class of organic compounds is the *hydrocarbons*, compounds composed only of carbon and hydrogen. The key structural feature of hydrocarbons (and of most other organic substances) is the presence of stable carbon–carbon bonds. Carbon is the only element capable of forming stable, extended chains of atoms bonded through single, double, or triple bonds.

Hydrocarbons can be divided into four general types, depending on the kinds of carbon–carbon bonds in their molecules. Figure 25.3 ▼ shows an example of each of the four types: alkanes, alkenes, alkynes, and aromatic hydrocarbons. In these hydrocarbons, as well as in other organic compounds, each C atom invariably has four bonds (four single bonds, two single bonds and one double bond, or one single bond and one triple bond).

Alkanes are hydrocarbons that contain only single bonds, as in ethane (C_2H_6). Because alkanes contain the largest possible number of hydrogen atoms per carbon atom, they are called *saturated hydrocarbons*. **Alkenes**, also known as olefins, are hydrocarbons that contain a C=C double bond, as in ethylene (C_2H_4). **Alkynes** contain a C≡C triple bond, as in acetylene (C_2H_2). In **aromatic hydrocarbons** the carbon atoms are connected in a planar ring structure, joined by both σ and π bonds between carbon atoms. Benzene (C_6H_6) is the best-known example of an aromatic hydrocarbon. Alkenes,

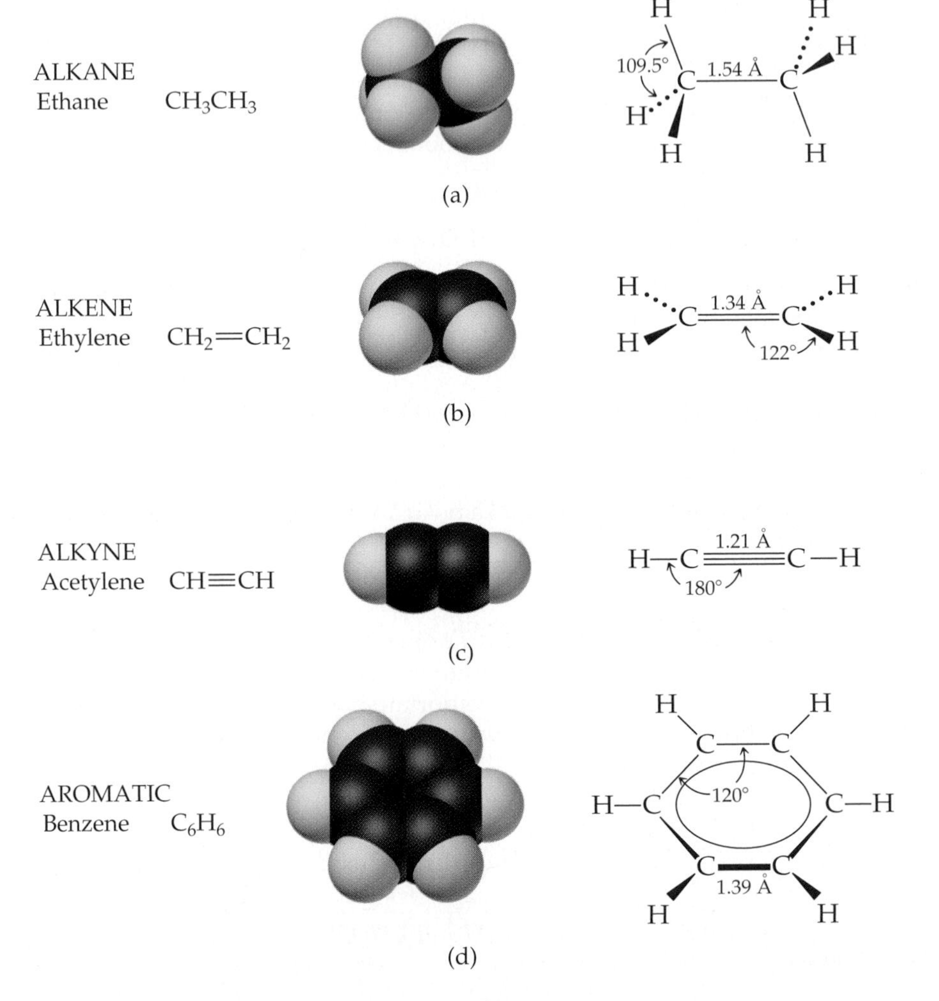

▶ **Figure 25.3** Names, geometrical structures, and molecular formulas for examples of each type of hydrocarbon.

3-D MODEL
Ethane, Ethylene, Acetylene, Benzene

alkynes, and aromatic hydrocarbons are called *unsaturated hydrocarbons* because they contain less hydrogen than an alkane having the same number of carbon atoms.

The members of these different classes of hydrocarbons exhibit different chemical behaviors, as we will see shortly. Their physical properties, however, are similar in many ways. Because carbon and hydrogen do not differ greatly in electronegativity (2.5 for carbon, 2.1 for hydrogen), hydrocarbon molecules are relatively nonpolar. Thus, they are almost completely insoluble in water, but they dissolve readily in other nonpolar solvents. Furthermore, their melting points and boiling points are determined by London dispersion forces. Hence, hydrocarbons tend to become less volatile with increasing molar mass. (Section 11.2) As a result, hydrocarbons of very low molecular weight, such as C_2H_6 (bp = −89°C), are gases at room temperature; those of moderate molecular weight, such as C_6H_{14} (bp = 69°C), are liquids; and those of high molecular weight, such as docosane ($C_{22}H_{46}$; mp = 44°C), are solids.

25.3 Alkanes

Table 25.1 ▼ lists several of the simplest alkanes. Many of these substances are familiar because they are used so widely. Methane is a major component of natural gas and is used for home heating and in gas stoves and hot-water heaters. Propane is the major component of bottled gas used for home heating and cooking in areas where natural gas is not available. Butane is used in disposable lighters and in fuel canisters for gas camping stoves and lanterns. Alkanes with from 5 to 12 carbon atoms per molecule are found in gasoline.

The formulas for the alkanes given in Table 25.1 are written in a notation called *condensed structural formulas*. This notation reveals the way in which atoms are bonded to one another but does not require drawing in all the bonds. For example, the Lewis structure and the condensed structural formulas for butane (C_4H_{10}) are

$$\begin{array}{ccccccccc} & & H & & H & & H & & H & \\ & & | & & | & & | & & | & \\ H & - & C & - & C & - & C & - & C & - & H \\ & & | & & | & & | & & | & \\ & & H & & H & & H & & H & \end{array}$$

$$H_3C-CH_2-CH_2-CH_3$$

or

$$CH_3CH_2CH_2CH_3$$

TABLE 25.1 First Several Members of the Straight-Chain Alkane Series

Molecular Formula	Condensed Structural Formula	Name	Boiling Point (°C)
CH_4	CH_4	Methane	−161
C_2H_6	CH_3CH_3	Ethane	−89
C_3H_8	$CH_3CH_2CH_3$	Propane	−44
C_4H_{10}	$CH_3CH_2CH_2CH_3$	Butane	−0.5
C_5H_{12}	$CH_3CH_2CH_2CH_2CH_3$	Pentane	36
C_6H_{14}	$CH_3CH_2CH_2CH_2CH_2CH_3$	Hexane	68
C_7H_{16}	$CH_3CH_2CH_2CH_2CH_2CH_2CH_3$	Heptane	98
C_8H_{18}	$CH_3CH_2CH_2CH_2CH_2CH_2CH_2CH_3$	Octane	125
C_9H_{20}	$CH_3CH_2CH_2CH_2CH_2CH_2CH_2CH_2CH_3$	Nonane	151
$C_{10}H_{22}$	$CH_3CH_2CH_2CH_2CH_2CH_2CH_2CH_2CH_2CH_3$	Decane	174

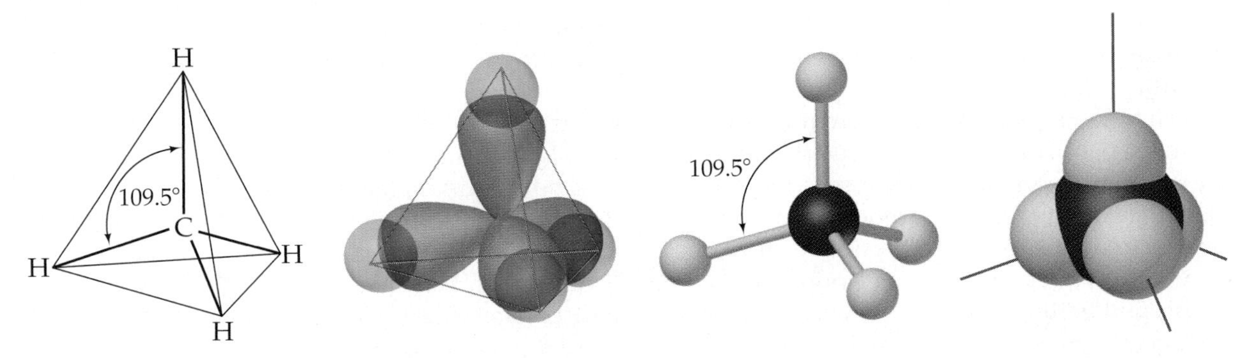

▲ **Figure 25.4** Representations of the three-dimensional arrangement of bonds about carbon in methane.

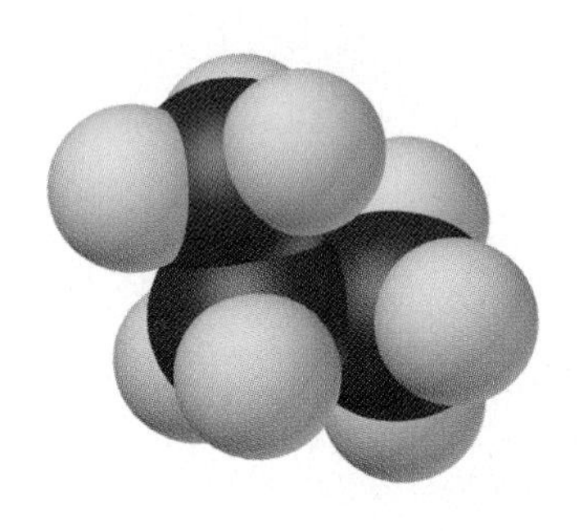

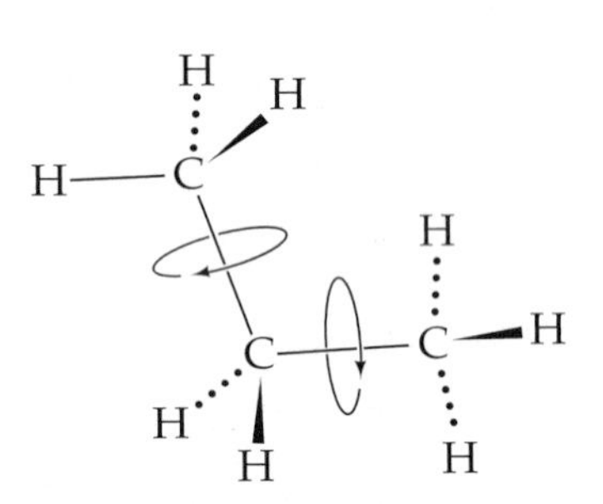

▲ **Figure 25.5** Three-dimensional models for propane (C_3H_8), showing rotations about the carbon–carbon single bonds.

We will frequently use either Lewis structures or condensed structural formulas to represent organic compounds. Notice that each carbon atom in an alkane has four single bonds, whereas each hydrogen atom forms one single bond. Notice also that each succeeding compound in the series listed in Table 25.1 has an additional CH_2 unit.

Structures of Alkanes

The Lewis structures and condensed structural formulas for alkanes do not tell us anything about the three-dimensional structures of these substances. According to the VSEPR model, the geometry about each carbon atom in an alkane is tetrahedral; that is, the four groups attached to each carbon are located at the vertices of a tetrahedron. (Section 9.2) The three-dimensional structures can be represented as shown for methane in Figure 25.4 ▲. The bonding may be described as involving sp^3 hybridized orbitals on the carbon. (Section 9.5)

Rotation about a carbon–carbon single bond is relatively easy, and it occurs very rapidly at room temperature. To visualize such rotation, imagine grasping the top left methyl group in Figure 25.5 ◀, which shows the structure of propane, and spinning it relative to the rest of the structure. Because motion of this sort occurs very rapidly in alkanes, a long-chain alkane molecule is constantly undergoing motions that cause it to change its shape, something like a length of chain that is being shaken.

Structural Isomers

The alkanes listed in Table 25.1 are called *straight-chain hydrocarbons* because all the carbon atoms are joined in a continuous chain. Alkanes consisting of four or more carbon atoms can also form *branched chains* called *branched-chain hydrocarbons*. Figure 25.6 ▶ shows the condensed formulas and space-filling models for all the possible structures of alkanes containing four and five carbon atoms. There are two ways that four carbon atoms can be joined to give C_4H_{10}: as a straight chain (left) or a branched chain (right). For alkanes with five carbon atoms (C_5H_{12}), there are three different arrangements.

Compounds with the same molecular formula but with different bonding arrangements (and hence different structures) are called **structural isomers**. The structural isomers of a given alkane differ slightly from one another in physical properties. Note the melting and boiling points of the isomers of butane and pentane, given in Figure 25.6. The number of possible structural isomers increases rapidly with the number of carbon atoms in the alkane. There are 18 possible isomers having the molecular formula C_8H_{18}, for example, and 75 possible isomers with the molecular formula $C_{10}H_{22}$.

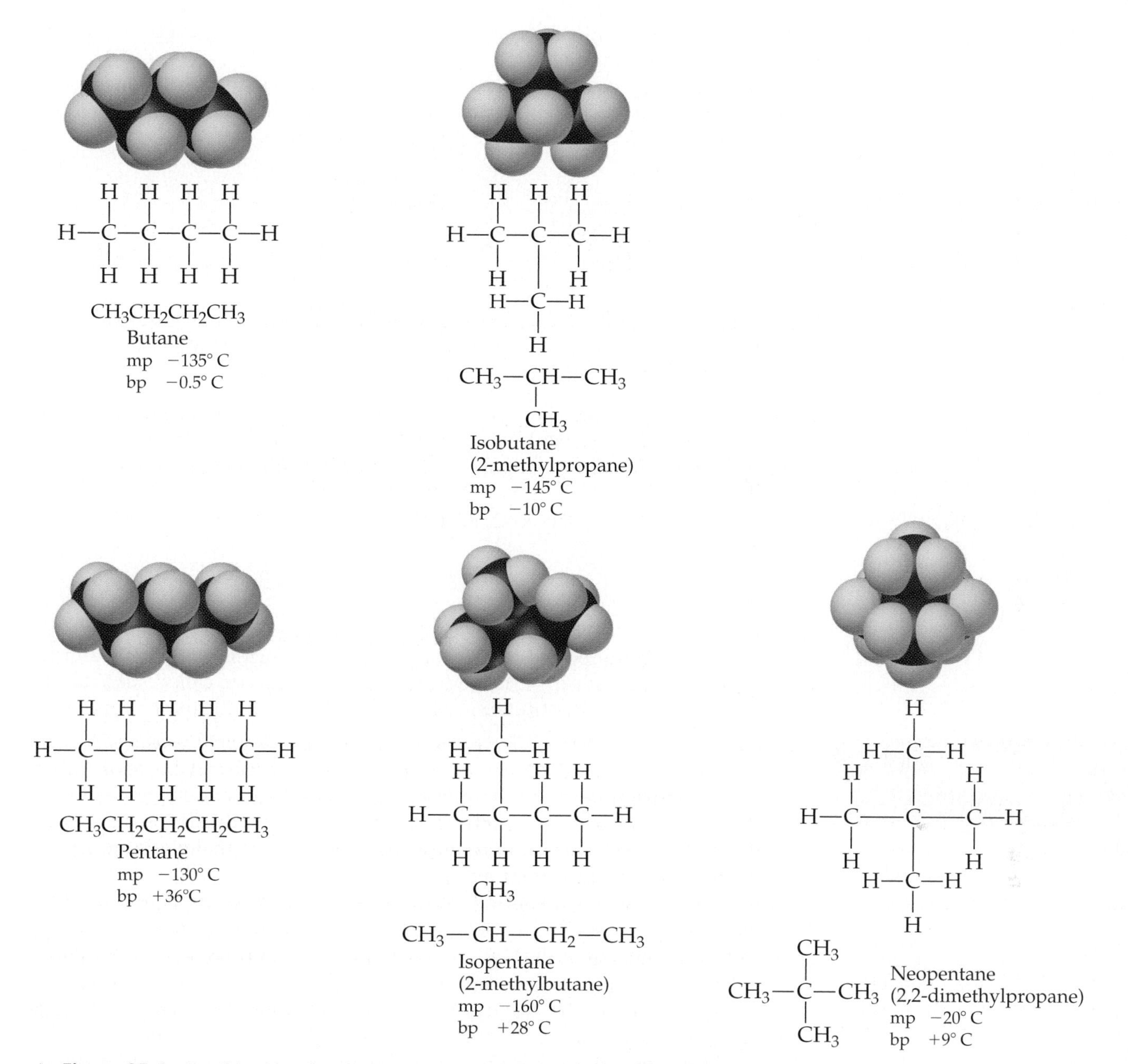

▲ **Figure 25.6** Possible structures, names, and melting and boiling points of alkanes of formula C_4H_{10} and C_5H_{12}.

Nomenclature of Alkanes

The first names given to the structural isomers shown in Figure 25.6 are the so-called common names. The isomer in which one CH_3 group is branched off the major chain is labeled the *iso-* isomer (for example, isobutane). As the number of isomers grows, however, it becomes impossible to find a suitable prefix to denote each isomer. The need for a systematic means of naming organic compounds was recognized early in the history of organic chemistry. In 1892 an organization called the International Union of Chemistry met in Geneva, Switzerland, to formulate rules for systematic naming of organic substances. Since that time the task of updating the rules for naming compounds has fallen to the International Union of Pure and Applied Chemistry (IUPAC). Chemists everywhere, regardless of their nationality or political affiliation, subscribe to a common system for naming compounds.

The IUPAC names for the isomers of butane and pentane are the ones given in parentheses for each compound in Figure 25.6. These names as well as those of other organic compounds have three parts to them:

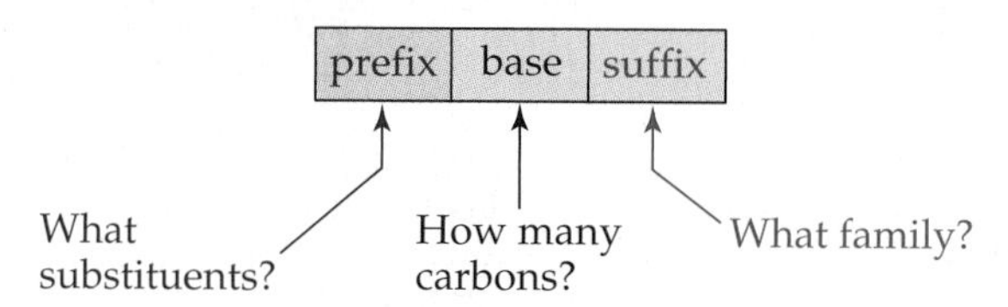

The following steps summarize the procedures used to arrive at the names of alkanes, which all have names ending with the suffix *-ane*. We use a similar approach to write the names of other organic compounds.

1. *Find the longest continuous chain of carbon atoms, and use the name of this chain (Table 25.1) as the base name of the compound.* The longest chain may not always be written in a straight line, as seen in the following example:

   CH_3—$\overset{2}{C}H\overset{1}{C}H_3$
   |
   $\underset{3}{C}H_2$—$\underset{4}{C}H_2$—$\underset{5}{C}H_2$—$\underset{6}{C}H_3$

   2-Methyl*hexane*

   Because this compound has a chain of six C atoms, it is named as a substituted hexane. Groups attached to the main chain are called *substituents* because they are substituted in place of an H atom on the main chain.

2. *Number the carbon atoms in the longest chain, beginning with the end of the chain that is nearest to a substituent.* In our example we number the C atoms from the upper right because that places the CH_3 substituent on the second C atom of the chain; if we number from the lower right, the CH_3 would be on the fifth C atom. The chain is numbered from the end that gives the lowest number for the substituent position.

3. *Name and give the location of each substituent group.* A substituent group that is formed by removing an H atom from an alkane is called an **alkyl group**. Alkyl groups are named by replacing the *-ane* ending of the alkane name with *-yl*. The methyl group (CH_3), for example, is derived from methane (CH_4). Likewise, the ethyl group (C_2H_5) is derived from ethane (C_2H_6). Table 25.2 ◀ lists several common alkyl groups. The name 2-methylhexane indicates the presence of a methyl (CH_3) group on the second carbon atom of a hexane (six-carbon) chain.

TABLE 25.2 Condensed Structural Formulas and Common Names for Several Alkyl Groups

Group	Name
CH_3—	Methyl
CH_3CH_2—	Ethyl
$CH_3CH_2CH_2$—	Propyl
$CH_3CH_2CH_2CH_2$—	Butyl
$HC(CH_3)_2$— (HC— with CH_3 above and CH_3 below)	Isopropyl
CH_3—C— (with CH_3 above and CH_3 below)	*t*-Butyl

4. *When two or more substituents are present, list them in alphabetical order.* When there are two or more of the same substituent, the number of substituents of that type is indicated by a prefix: *di-* (two), *tri-* (three), *tetra-* (four), *penta-* (five), and so forth. Notice how the following example is named:

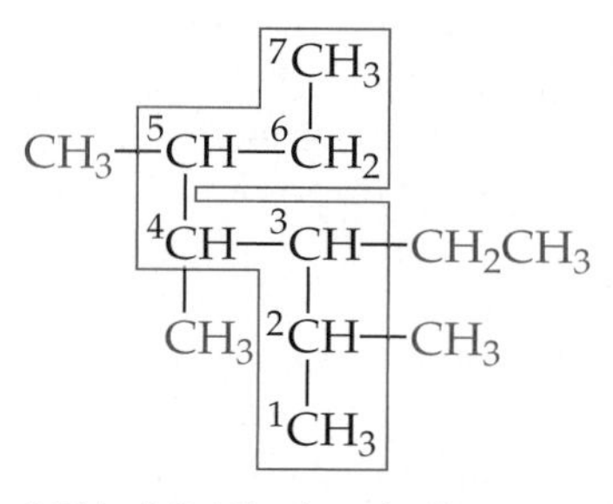

3-Ethyl-2,4,5-trimethylheptane

SAMPLE EXERCISE 25.1

Give the systematic name for the following alkane:

$$\begin{array}{l} CH_3{-}CH_2{-}CH{-}CH_3 \\ \qquad\quad | \\ CH_3{-}CH{-}CH_2 \\ \qquad\qquad | \\ \quad CH_3{-}CH_2 \end{array}$$

Solution

Analyze: We are given the structural formula of an alkane and asked to give its name.

Plan: Because the hydrocarbon is an alkane, its name ends in *-ane*. The name of the parent hydrocarbon is based on the longest continuous chain of carbon atoms, as summarized in Table 25.1. Branches are alkyl groups, named after the number of carbon atoms in the branch and located by counting C atoms along the longest continuous chain.

Solve: The longest continuous chain of C atoms extends from the upper left CH_3 group to the lower left CH_3 group and is seven carbon atoms long:

$$\begin{array}{l} {}^1CH_3{-}{}^2CH_2{-}{}^3CH{-}CH_3 \\ \qquad\qquad | \\ \quad CH_3{-}{}^4CH{-}{}^5CH_2 \\ \qquad\qquad\quad | \\ \qquad {}^7CH_3{-}{}^6CH_3 \end{array}$$

The parent compound is thus heptane. There are two CH_3 (methyl) groups that branch off the main chain. Hence this compound is a dimethylheptane. To specify the location of the two methyl groups, we must number the carbon atoms from the end that gives the lowest number possible to the carbons bearing side chains. This means that we should start numbering with the upper left carbon. There is a methyl group on carbon 3, and one on carbon 4. The compound is thus 3,4-dimethylheptane.

PRACTICE EXERCISE

Name the following alkane:

$$\begin{array}{l} \qquad CH_3{-}CH{-}CH_3 \\ \qquad\qquad\quad | \\ CH_3{-}CH{-}CH_2 \\ \qquad\quad | \\ \qquad CH_3 \end{array}$$

Answer: 2,4-dimethylpentane

SAMPLE EXERCISE 25.2

Write the condensed structural formula for 3-ethyl-2-methylpentane.

Solution

Analyze: We are given the systematic name for a hydrocarbon and asked to write its structural formula.

Plan: Because the compound's name ends in *-ane*, it is an alkane, meaning that all the carbon–carbon bonds are single bonds. The parent hydrocarbon is pentane, indicating five C atoms (Table 25.1). There are two alkyl groups specified, an ethyl group (two carbon atoms, C_2H_5) and a methyl group (one carbon atom, CH_3). Counting from left to right along the five-carbon chain, the ethyl group will be attached to the third C atom and the methyl group will be attached to the second C atom.

Solve: We begin by writing a string of five C atoms attached to each other by single bonds. These represent the backbone of the parent pentane chain:

$$C{-}C{-}C{-}C{-}C$$

We next place a methyl group on the second carbon, and an ethyl group on the third carbon atom of the chain. Hydrogens are then added to all the other carbon atoms to make the four bonds to each carbon, giving the following condensed structure:

$$\begin{array}{l} \qquad\quad CH_3 \\ \qquad\qquad | \\ CH_3{-}CH{-}CH{-}CH_2{-}CH_3 \\ \qquad\qquad\qquad | \\ \qquad\qquad\quad CH_2CH_3 \end{array}$$

The formula can be written even more concisely as $CH_3CH(CH_3)CH(C_2H_5)CH_2CH_3$

In this formula the branching alkyl groups are indicated in parentheses.

PRACTICE EXERCISE

Write the condensed structural formula for 2,3-dimethylhexane.

Answer: $CH_3CH(CH_3)—CH(CH_3)CH_2CH_2CH_3$ or $CH_3CH(CH_3)CH(CH_3)CH_2CH_2CH_3$

Cycloalkanes

Alkanes can form not only branched chains, but also rings or cycles. Alkanes with this form of structure are called **cycloalkanes**. Figure 25.7 ▶ illustrates a few cycloalkanes. Cycloalkane structures are sometimes drawn as simple polygons in which each corner of the polygon represents a CH_2 group. This method of representation is similar to that used for benzene rings. (Section 8.6) In the case of aromatic structures each corner represents a CH group.

Chemistry at Work Gasoline

Petroleum, or crude oil, is a complex mixture of organic compounds, mainly hydrocarbons, with smaller quantities of other organic compounds containing nitrogen, oxygen, or sulfur. The tremendous demand for petroleum to meet the world's energy needs has led to the tapping of oil wells in such forbidding places as the North Sea and northern Alaska.

The usual first step in the *refining*, or processing, of petroleum is to separate it into fractions on the basis of boiling point. The fractions commonly taken are shown in Table 25.3 ▼. Because gasoline is the most commercially important of these fractions, various processes are used to maximize its yield.

Gasoline is a mixture of volatile hydrocarbons containing varying amounts of aromatic hydrocarbons in addition to alkanes. In an automobile engine, a mixture of air and gasoline vapor is compressed by a piston and then ignited by a spark plug. The burning of the gasoline should create a strong, smooth expansion of gas, forcing the piston outward and imparting force along the driveshaft of the engine. If the gas burns too rapidly, the piston receives a single hard slam rather than a strong, smooth push. The result is a "knocking" or "pinging" sound and a reduction in the efficiency with which energy produced by the combustion is converted to work.

The *octane number* of a gasoline is a measure of its resistance to knocking. Gasolines with high octane numbers burn more smoothly and are thus more effective fuels (Figure 25.8 ▶). Branched alkanes and aromatic hydrocarbons have higher octane numbers than the straight-chain alkanes. The octane number of gasoline is obtained by comparing its knocking characteristics with those of "isooctane" (2,2,4-trimethylpentane) and heptane. Isooctane is assigned an octane number of 100, whereas heptane is assigned 0. Gasoline with the same knocking characteristics as a mixture of 90% isooctane and 10% heptane is rated as 90 octane.

The gasoline obtained directly from fractionation of petroleum (called *straight-run* gasoline) contains mainly straight-chain hydrocarbons and has an octane number around 50. It is therefore subjected to a process called *cracking*, which converts the straight-chain alkanes into more desirable branched-chain ones (Figure 25.9 ▶). Cracking is also used to produce aromatic hydrocarbons and to convert some of the less volatile kerosene and fuel-oil fraction into compounds with lower molecular weights that are suitable for use as automobile fuel. In the cracking process the hydrocarbons are mixed with a catalyst and heated to 400°C to 500°C. The catalysts used are naturally occurring clay

TABLE 25.3 Hydrocarbon Fractions from Petroleum

Fraction	Size Range of Molecules	Boiling-Point Range (°C)	Uses
Gas	C_1 to C_5	−160 to 30	Gaseous fuel, production of H_2
Straight-run gasoline	C_5 to C_{12}	30 to 200	Motor fuel
Kerosene, fuel oil	C_{12} to C_{18}	180 to 400	Diesel fuel, furnace fuel, cracking
Lubricants	C_{16} and up	350 and up	Lubricants
Paraffins	C_{20} and up	Low-melting solids	Candles, matches
Asphalt	C_{36} and up	Gummy residues	Surfacing roads

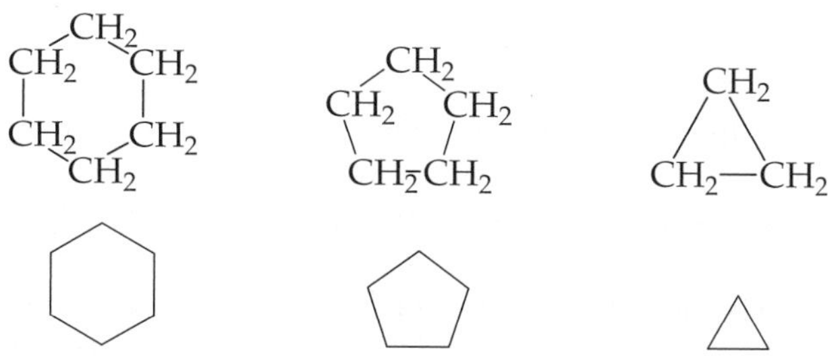

$$\begin{array}{c} \text{CH}_2 \\ \text{CH}_2 \quad \text{CH}_2 \\ | \qquad | \\ \text{CH}_2 \quad \text{CH}_2 \\ \text{CH}_2 \end{array} \qquad \begin{array}{c} \text{CH}_2 \\ \text{CH}_2 \quad \text{CH}_2 \\ \text{CH}_2\text{—CH}_2 \end{array} \qquad \begin{array}{c} \text{CH}_2 \\ \text{CH}_2\text{—CH}_2 \end{array}$$

Cyclohexane Cyclopentane Cyclopropane

◀ **Figure 25.7** Condensed structural formulas for three cycloalkanes.

Carbon rings containing fewer than five carbon atoms are strained because the C—C—C bond angle in the smaller rings must be less than the 109.5° tetrahedral angle. The amount of strain increases as the rings get smaller. In cyclopropane, which has the shape of an equilateral triangle, the angle is only 60°; this molecule is therefore much more reactive than propane, its straight-chain analog.

Cycloalkanes, particularly the small-ring compounds, sometimes behave chemically like unsaurated hydrocarbons, which we will discuss shortly. The general formula for cycloalkanes, C_nH_{2n}, differs from the general formula for straight-chain alkanes, C_nH_{2n+2}.

minerals or synthetic Al_2O_3–SiO_2 mixtures. In addition to forming molecules more suitable for gasoline, cracking results in the formation of hydrocarbons of lower molecular weight, such as ethylene and propene. These substances are used in a variety of reactions to form plastics and other chemicals.

The octane rating of gasoline is also increased by adding certain compounds called *antiknock agents* or octane enhancers. Until the mid-1970s the principal antiknock agent was tetraethyl lead, $(C_2H_5)_4Pb$. It is no longer used, however, because of the environmental hazards of lead and because it poisons catalytic converters. ("Chemistry at Work," Section 14.7) Aromatic compounds such as toluene ($C_6H_5CH_3$) and oxygenated hydrocarbons such as ethanol (CH_3CH_2OH) and methyl *t*-butyl ether (MTBE, shown below) are now generally used as antiknock agents.

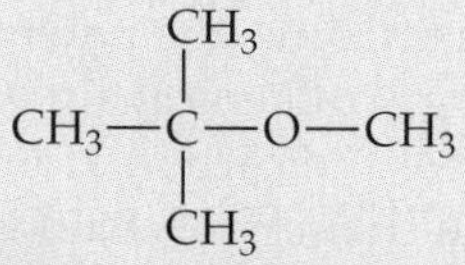

$$\begin{array}{c} \text{CH}_3 \\ | \\ \text{CH}_3\text{—C—O—CH}_3 \\ | \\ \text{CH}_3 \end{array}$$

Use of MTBE may soon be banned, however, because it finds its way into drinking-water supplies from spills and leaking storage tanks, giving the water a bad smell and taste and perhaps producing adverse health effects.

◀ **Figure 25.8** The octane rating of gasoline measures its resistance to knocking when burned in an engine. The octane rating of this gasoline is 89, as shown on the face of the pump.

▲ **Figure 25.9** Petroleum is separated into fractions by distillation and is subjected to catalytic cracking in a refinery, as shown here.

Reactions of Alkanes

Because they contain only C—C and C—H bonds, most alkanes are relatively unreactive. At room temperature, for example, they do not react with acids, bases, or strong oxidizing agents, and they are not even attacked by boiling nitric acid. Their low chemical reactivity is due primarily to the strength and lack of polarity of C—C and C—H bonds.

Alkanes are not completely inert, however. One of their most commercially important reactions is *combustion* in air, which is the basis of their use as fuels. (Section 3.2) For example, the complete combustion of ethane proceeds as follows:

$$2C_2H_6(g) + 7O_2(g) \longrightarrow 4CO_2(g) + 6H_2O(l) \qquad \Delta H = -2855 \text{ kJ}$$

In the following sections we will see that hydrocarbons can be modified to impart greater reactivity by introducing unsaturation into the carbon–carbon framework and by attaching other reactive groups to the hydrocarbon backbone.

25.4 Unsaturated Hydrocarbons

The presence of one or more multiple bonds makes the structure and reactivity of unsaturated hydrocarbons significantly different from those of alkanes.

Alkenes

Alkenes are unsaturated hydrocarbons that contain a C=C bond. The simplest alkene is $CH_2{=}CH_2$, called ethene (IUPAC) or ethylene. Ethylene is a plant hormone. It plays important roles in seed germination and fruit ripening. The next member of the series is $CH_3{-}CH{=}CH_2$, called propene or propylene. For alkenes with four or more carbon atoms, several isomers exist for each molecular formula. For example, there are four isomers of C_4H_8, as shown in Figure 25.10 ▼. Notice both their structures and their names.

The names of alkenes are based on the longest continuous chain of carbon atoms that contains the double bond. The name given to the chain is obtained from the name of the corresponding alkane (Table 25.1) by changing the ending from *-ane* to *-ene*. The compound on the left in Figure 25.10, for example, has a double bond as part of a three-carbon chain; thus, the parent alkene is propene.

The location of the double bond along an alkene chain is indicated by a prefix number that designates the number of the carbon atom that is part of the double bond and is nearest an end of the chain. The chain is always numbered from the end that brings us to the double bond sooner and hence gives the smallest-numbered prefix. In propene the only possible location for the double bond is between the first and second carbons; thus, a prefix indicating its loca-

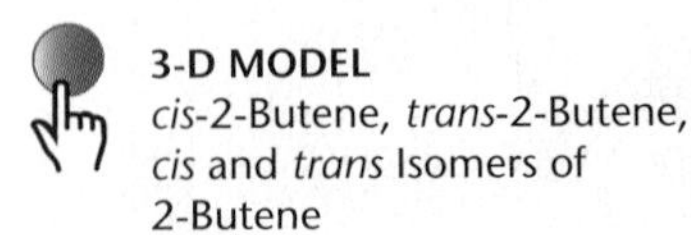

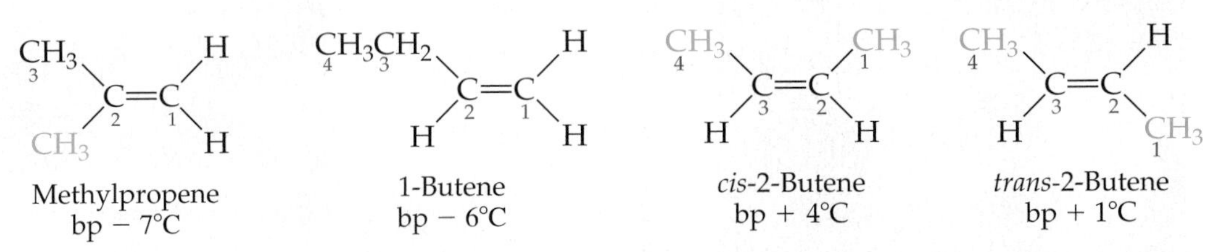

▲ **Figure 25.10** Structures, names, and boiling points of alkenes with molecular formula C_4H_8.

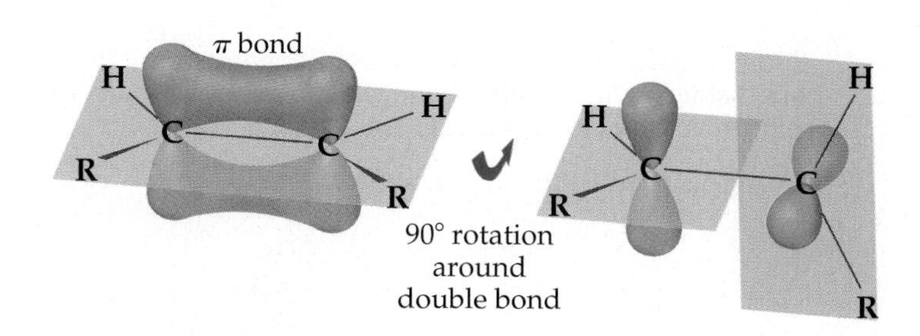

◀ **Figure 25.11** Schematic illustration of rotation about a carbon–carbon double bond in an alkene. The overlap of the *p* orbitals that form the π bond is lost in the rotation. For this reason, rotation about carbon–carbon double bonds does not occur readily.

tion is unnecessary. For butene (Figure 25.10) there are two possible positions for the double bond, either after the first carbon (1-butene) or after the second carbon (2-butene).

If a substance contains two or more double bonds, each is located by a numerical prefix. The ending of the name is altered to identify the number of double bonds: diene (two), triene (three), and so forth. For example, $CH_2{=}CH{-}CH_2{-}CH{=}CH_2$ is 1,4-pentadiene.

The two isomers on the right in Figure 25.10 differ in the relative locations of their terminal methyl groups. These two compounds are **geometric isomers**, compounds that have the same molecular formula and the same groups bonded to one another, but differ in the spatial arrangement of these groups. (Section 24.4) In the *cis* isomer the two methyl groups are on the same side of the double bond, whereas in the *trans* isomer they are on opposite sides. Geometric isomers possess distinct physical properties and often differ significantly in their chemical behavior.

Geometric isomerism in alkenes arises because, unlike the C—C bond, the C=C bond resists twisting. Recall that the double bond between two carbon atoms consists of a σ and a π bond. (Section 9.6) Figure 25.11 ▲ shows a *cis* alkene. The carbon–carbon bond axis and the bonds to the hydrogen atoms and to the alkyl groups (designated R) are all in a plane. The *p* orbitals that overlap sideways to form the π bond are perpendicular to the molecular plane. As Figure 25.11 shows, rotation around the carbon–carbon double bond requires the π bond to be broken, a process that requires considerable energy (about 250 kJ/mol). While rotation about a double bond doesn't occur easily, rotation about a double bond is a key process in the chemistry of vision. ("Chemistry and Life," Section 9.6)

SAMPLE EXERCISE 25.3

Draw all the isomers of pentene, C_5H_{10}. (Consider only those with an unbranched hydrocarbon chain.)

Solution

Analyze: We are asked to draw all the isomers (both structural and geometric) for an alkene with a five-carbon chain.

Plan: Because the compound is named pentene and not pentadiene or pentatriene, we know that the five-carbon chain contains only one carbon–carbon double bond. Thus, we can begin by first placing the double bond in various locations along the chain, remembering that the chain can be numbered from either end. After finding the different distinct locations for the double bond, we can consider whether the molecule can have *cis* and *trans* isomers.

Solve: There can be a double bond after either the first carbon (1-pentene) or second carbon (2-pentene). These are the only two possibilities because the chain can be numbered from either end. (Thus, what we might erroneously call 4-pentene is actually 1-pentene, as seen by numbering the carbon chain from the other end.)

Because the first C atom in 1-pentene is bonded to two H atoms, there are no *cis-trans* isomers. On the other hand, there are *cis* and *trans* isomers for 2-pentene. Thus, the three isomers for pentene are

$CH_3—CH_2—CH_2—CH{=}CH_2$

1-Pentene

$CH_3—CH_2$ and CH_3 on the same side of C=C, H and H below

cis-2-Pentene

$CH_3—CH_2$ and H above C=C, H and CH_3 below (opposite sides)

trans-2-Pentene

(You should convince yourself that *cis*- or *trans*-3-pentene is identical to *cis*- or *trans*-2-pentene, respectively.)

PRACTICE EXERCISE

How many straight-chain isomers are there of hexene, C_6H_{12}?
Answer: 5 (1-hexene, *cis*-2-hexene, *trans*-2-hexene, *cis*-3-hexene, *trans*-3-hexene)

Alkynes

Alkynes are unsaturated hydrocarbons containing one or more C≡C bonds. The simplest alkyne is acetylene (C_2H_2), a highly reactive molecule. When acetylene is burned in a stream of oxygen in an oxyacetylene torch, the flame reaches about 3200 K. The oxyacetylene torch is widely used in welding, which requires high temperatures. Alkynes in general are highly reactive molecules. Because of their higher reactivity, they are not as widely distributed in nature as alkenes; alkynes, however, are important intermediates in many industrial processes.

Alkynes are named by identifying the longest continuous chain in the molecule containing the triple bond and modifying the ending of the name as listed in Table 25.1 from *-ane* to *-yne*, as shown in Sample Exercise 25.4.

SAMPLE EXERCISE 25.4

Name the following compounds:

(a) $CH_3CH_2CH_2—CH(CH_3)$ and CH_3 on the same side of C=C, H and H below

(b) $CH_3CH_2CH_2CH(CH_2CH_2CH_3)—C{\equiv}CH$

Solution

Analyze: We are given the structural formulas for two compounds, the first an alkene and the second an alkyne, and asked to name the compounds.
Plan: In each case the name is based on the number of carbon atoms in the longest continuous carbon chain that contains the multiple bond. In the case of the alkene, care must be taken to indicate whether *cis-trans* isomerism is possible and, if so, which isomer is given.
Solve: (a) The longest continuous chain of carbons that contains the double bond is seven in length. The parent compound is therefore heptene. Because the double bond begins at carbon 2 (numbering from the end closest to the double bond), the parent hydrocarbon chain is named 2-heptene. A methyl group is found at carbon atom 4. Thus, the compound is 4-methyl-2-heptene. The geometrical configuration at the double bond is *cis* (i.e., the alkyl groups are bonded to the double bond on the same side). Thus, the full name is 4-methyl-*cis*-2-heptene.

(b) The longest continuous chain of carbon atoms containing the triple bond is six, so this compound is a derivative of hexyne. The triple bond comes after the first carbon (numbering from the right), making it a derivative of 1-hexyne. The branch from the hexyne chain contains three carbon atoms, making it a propyl group. Because it is located on the third carbon atom of the hexyne chain, the molecule is 3-propyl-1-hexyne.

PRACTICE EXERCISE

Draw the condensed structural formula for 4-methyl-2-pentyne.

Answer: $CH_3-C\equiv C-\underset{\displaystyle CH_3}{\underset{|}{CH}}-CH_3$

Addition Reactions of Alkenes and Alkynes

ACTIVITY
Addition Reactions of Alkenes

The presence of carbon–carbon double or triple bonds in hydrocarbons markedly increases their chemical reactivity. The most characteristic reactions of alkenes and alkynes are **addition reactions**, in which a reactant is added to the two atoms that form the multiple bond. A simple example is the addition of a halogen such as Br_2 to ethylene:

$$H_2C{=}CH_2 + Br_2 \longrightarrow \underset{\displaystyle Br}{\underset{|}{H_2C}}-\underset{\displaystyle Br}{\underset{|}{CH_2}} \qquad [25.1]$$

The pair of electrons that form the π bond in ethylene is uncoupled and is used to form two new bonds to the two bromine atoms. The σ bond between the carbon atoms is retained.

Addition of H_2 to an alkene converts it to an alkane:

$$CH_3CH{=}CHCH_3 + H_2 \xrightarrow{\text{Ni, 500°C}} CH_3CH_2CH_2CH_3 \qquad [25.2]$$

The reaction between an alkene and H_2, referred to as *hydrogenation*, does not occur readily under ordinary conditions of temperature and pressure. One reason for the lack of reactivity of H_2 toward alkenes is the high bond enthalpy of the H_2 bond. To promote the reaction, it is necessary to use a catalyst that assists in rupturing the H—H bond. The most widely used catalysts are finely divided metals on which H_2 is adsorbed. (Section 14.7)

Hydrogen halides and water can also add to the double bond of alkenes, as illustrated by the following reactions of ethylene:

$$CH_2{=}CH_2 + HBr \longrightarrow CH_3CH_2Br \qquad [25.3]$$

$$CH_2{=}CH_2 + H_2O \xrightarrow{H_2SO_4} CH_3CH_2OH \qquad [25.4]$$

The addition of water is catalyzed by a strong acid, such as H_2SO_4.

The addition reactions of alkynes resemble those of alkenes, as shown in the following examples:

$$CH_3C\equiv CCH_3 + Cl_2 \longrightarrow \begin{matrix} Cl & & CH_3 \\ & C{=}C & \\ CH_3 & & Cl \end{matrix} \qquad [25.5]$$

2-Butyne — *trans*-2,3-Dichloro-2-butene

$$CH_3C\equiv CCH_3 + 2Cl_2 \longrightarrow CH_3-\underset{\displaystyle Cl}{\overset{\displaystyle Cl}{\overset{|}{\underset{|}{C}}}}-\underset{\displaystyle Cl}{\overset{\displaystyle Cl}{\overset{|}{\underset{|}{C}}}}-CH_3 \qquad [25.6]$$

2-Butyne — 2,2,3,3-Tetrachlorobutane

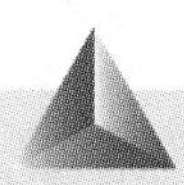

Chemistry at Work The Accidental Discovery of Teflon™

A *polymer* is a material with a high molecular weight that is formed when simple molecules called monomers join together in long chains. (Section 12.2) Polymers may be either natural or synthetic in origin. Later in this chapter we will see several natural polymers, such as proteins and starch. In Section 12.2 we discussed several synthetic polymers, such as polyethylene and nylon. Another synthetic polymer is Teflon™, which was discovered quite by accident.

In 1938 a scientist at DuPont named Roy J. Plunkett made a rather curious observation: A tank of gaseous *tetrafluoroethylene* ($CF_2{=}CF_2$) that was supposed to be full seemed to have no gas in it. Rather than discard the tank, Plunkett decided to cut it open. He found that the inside of the tank was coated with a waxy white substance that was remarkably unreactive toward even the most corrosive reagents. The compound was formed by the *addition polymerization* (Section 12.2) of tetrafluoroethylene:

$$n CF_2{=}CF_2 \xrightarrow{\text{polymerization}} -\!\!\left(CF_2-CF_2 \right)_n\!\!-$$

As it turned out, the properties of Teflon™ were ideal for an immediate and important application in the development of the first atomic bomb. Uranium hexafluoride (UF_6), which was used to separate fissionable ^{235}U by gaseous diffusion (see the "Chemistry at Work" box in Section 10.8), is an extremely corrosive material. Teflon™ was used as a gasket material in the gaseous diffusion plant. It is now used in a variety of applications, from nonstick cookware to space suits.

Plunkett's desire to know more about something that just didn't seem right is a wonderful example of how natural scientific curiosity can lead to remarkable discoveries. If you wish to read about more such accidental discoveries, we recommend Royston M. Roberts, *Serendipity: Accidental Discoveries in Science*, John Wiley and Sons, 1989.

SAMPLE EXERCISE 25.5

Write the structural formula for the product of the hydrogenation of 3-methyl-1-pentene.

Solution

Analyze: We are asked to predict the compound formed when a particular alkene undergoes hydrogenation (reaction with H_2).

Plan: To determine the structural formula of the reaction product, we must first write the structural formula or Lewis structure of the reactant. In the hydrogenation of the alkene, H_2 adds to the double bond, producing an alkane. (That is, each carbon atom of the double bond forms a bond to an H atom and the double bond is converted to a single bond.)

Solve: The name of the starting compound tells us that we have a chain of five carbon atoms with a double bond at one end (position 1) and a methyl group on the third carbon from that end (position 3):

$$CH_2{=}CH-\underset{}{\overset{\displaystyle CH_3 \atop |}{CH}}-CH_2-CH_3$$

Hydrogenation—the addition of two H atoms to the carbons of the double bond—leads to the following alkane:

$$CH_3-CH_2-\overset{\displaystyle CH_3 \atop |}{CH}-CH_2-CH_3$$

Comment: The longest chain in the product alkane has five carbon atoms; its name is therefore 3-methylpentane

PRACTICE EXERCISE

Addition of HCl to an alkene forms 2-chloropropane. What is the alkene?
Answer: propene

Mechanism of Addition Reactions

As our understanding of chemistry has grown, chemists have been able to advance from simply cataloging reactions known to occur to explaining *how* they occur. An explanation of how a reaction proceeds is called a *mechanism*. (Section 14.6)

In Equation 25.3 we considered the addition reaction between HBr and an alkene. This reaction is thought to proceed in two steps. In the first step, which is rate determining [(Section 14.6)], the HBr molecule attacks the electron-rich double bond, transferring a proton to one of the two alkene carbons. In the reaction of 2-butene with HBr, for example, the first step proceeds as follows:

$$CH_3CH{=}CHCH_3 + HBr \longrightarrow \left[\begin{array}{c} \overset{\delta+}{CH_3CH}{\cdots}CHCH_3 \\ \vdots \\ H \\ \vdots \\ Br^{\delta-} \end{array}\right] \longrightarrow CH_3\overset{+}{C}H{-}CH_2CH_3 + Br^- \quad [25.7]$$

The pair of electrons that formed the π bond between the carbon atoms in the alkane is used to form the new C—H bond.

The second step, involving the addition of Br^- to the positively charged carbon, is faster:

$$CH_3\overset{+}{C}H{-}CH_2CH_3 + Br^- \longrightarrow \left[\begin{array}{c} CH_3\overset{\delta+}{C}H{-}CH_2CH_3 \\ \vdots \\ Br^{\delta-} \end{array}\right] \longrightarrow \begin{array}{c} CH_3CHCH_2CH_3 \\ | \\ Br \end{array} \quad [25.8]$$

In this reaction the bromide ion (Br^-) donates a pair of electrons to the carbon, forming the new C—Br bond.

Because the first rate-determining step in the reaction involves both the alkene and acid, the rate law for the reaction is second order, first order each in alkene and HBr:

$$\text{Rate} = -\frac{\Delta[CH_3CH{=}CHCH_3]}{\Delta t} = k[CH_3CH{=}CHCH_3][HBr] \quad [25.9]$$

The energy profile for the reaction is shown in Figure 25.12 ▼. The first energy maximum represents the transition state in the first step of the mechanism. The second maximum represents the transition state for the second step. Notice that there is an energy minimum between the first and second steps of the reaction. This energy minimum corresponds to the energies of the intermediate species, $CH_3\underset{+}{C}H{-}CH_2CH_3$ and Br^-.

To show electron movement in reactions like these, chemists often use curved arrows, which point in the direction of the electron flow (from a negative charge

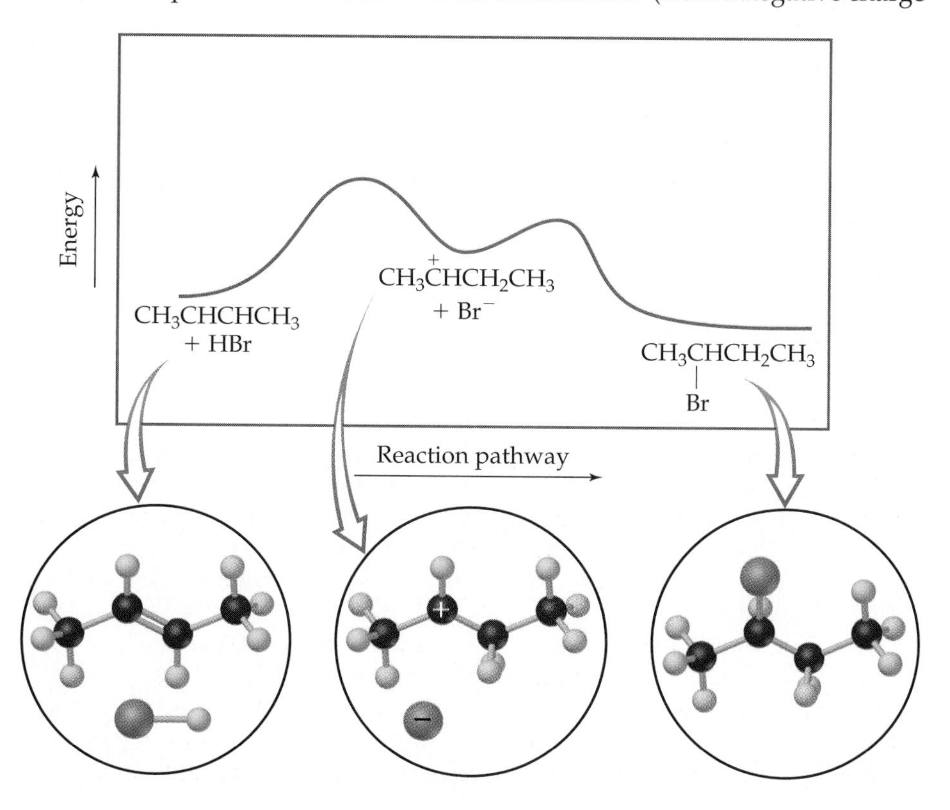

◀ **Figure 25.12** Energy profile for addition of HBr to 2-butene, $CH_3CH{=}CHCH_3$.

toward a positive charge). For the addition of HBr to 2-butene, the shifts in electron positions are shown as follows:

$$CH_3CH{=}CHCH_3 + H{-}\ddot{\underset{..}{Br}}{:} \xrightarrow{\text{slow}} CH_3{-}\overset{H}{\underset{+}{C}}{-}\overset{H}{\underset{H}{C}}{-}CH_3 + {:}\ddot{\underset{..}{Br}}{:}^- \xrightarrow{\text{fast}} CH_3{-}\overset{H}{\underset{:\underset{..}{Br}:}{C}}{-}\overset{H}{\underset{H}{C}}{-}CH_3$$

Aromatic Hydrocarbons

Aromatic hydrocarbons are members of a large and important class of hydrocarbons. The simplest member of the series is benzene (C_6H_6) whose structure is shown in Figure 25.3. The planar, highly symmetrical structure of benzene, with its 120° bond angles, suggests a high degree of unsaturation. You might therefore expect benzene to resemble the unsaturated hydrocarbons and to be highly reactive. The chemical behavior of benzene, however, is unlike that of alkenes or alkynes. Benzene and the other aromatic hydrocarbons are much more stable than alkenes and alkynes because the π electrons are delocalized in the π orbitals. (Section 9.6)

Each aromatic ring system is given a common name as shown in Figure 25.13 ◀. The aromatic rings are represented by hexagons with a circle inscribed inside to denote aromatic character. Each corner represent a carbon atom. Each carbon is bound to three other atoms—either three carbons or two carbons and a hydrogen. The hydrogen atoms are not shown.

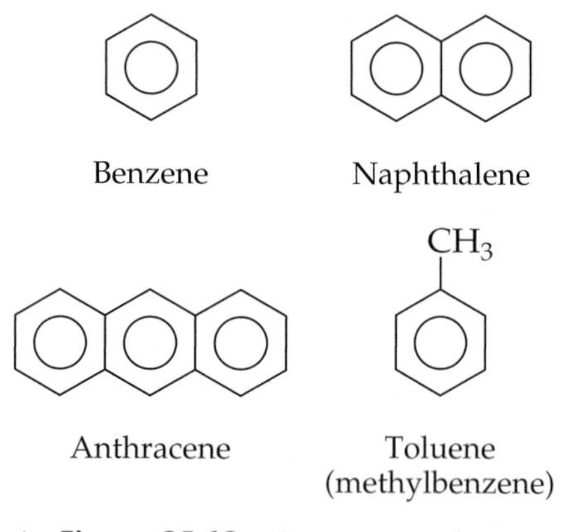

▲ **Figure 25.13** Structures and names of several aromatic compounds.

Although aromatic hydrocarbons are unsaturated, they do not readily undergo addition reactions. The delocalized π bonding causes aromatic compounds to behave quite differently from alkenes and alkynes. Benzene, for example, does not add Cl_2 or Br_2 to its double bonds under ordinary conditions. In contrast, aromatic hydrocarbons undergo **substitution reactions** relatively easily. In a substitution reaction one atom of a molecule is removed and replaced (substituted) by another atom or group of atoms. When benzene is warmed in a mixture of nitric and sulfuric acids, for example, hydrogen is replaced by the nitro group, NO_2:

$$\text{(benzene)} + HNO_3 \xrightarrow{H_2SO_4} \text{(benzene)}{-}NO_2 + H_2O \qquad [25.10]$$

More vigorous treatment results in substitution of a second nitro group into the molecule:

$$\text{(benzene)}{-}NO_2 + HNO_3 \xrightarrow{H_2SO_4} \text{(benzene, 1,3-)}(NO_2)_2 + H_2O \qquad [25.11]$$

There are three possible isomers of benzene with two nitro groups attached. These three isomers are named *ortho-*, *meta-*, and *para-*dinitrobenzene:

ortho-Dinitrobenzene	*meta*-Dinitrobenzene	*para*-Dinitrobenzene
mp 118°C	mp 90°C	mp 174°C

A Closer Look Aromatic Stabilization

We can estimate the stabilization of the π electrons in benzene by comparing the energy required to add hydrogen to benzene to form a saturated compound with the energy required to hydrogenate certain alkenes. The hydrogenation of benzene to form cyclohexane can be represented as

$$C_6H_6 + 3H_2 \longrightarrow C_6H_{12} \qquad \Delta H° = -208 \text{ kJ/mol}$$

The enthalpy change in this reaction is −208 kJ/mol. The heat of hydrogenation of the cyclic alkene cyclohexene is −120 kJ/mol:

$$C_6H_{10} + H_2 \longrightarrow C_6H_{12} \qquad \Delta H° = -120 \text{ kJ/mol}$$

Similarly, the heat released on hydrogenating 1,4-cyclohexadiene is −232 kJ/mol:

$$C_6H_8 + 2H_2 \longrightarrow C_6H_{12}$$

1,4-Cyclohexadiene $\Delta H° = -232$ kJ/mol

From these last two reactions, it appears that the heat of hydrogenating each double bond is roughly 116 kJ/mol for each bond. There is the equivalent of three double bonds in benzene. We might expect, therefore, that the heat of hydrogenating benzene would be about three times −116, or −348 kJ/mol, if benzene behaved as though it were "cyclohexatriene"; that is, if it behaved as though it had three isolated double bonds in a ring. Instead, the heat released is 140 kJ less than this, indicating that benzene is more stable than would be expected for three double bonds. The difference of 140 kJ/mol between the "expected" heat of hydrogenation, −348 kJ/mol, and the observed heat of hydrogenation, −208 kJ/mol, is due to stabilization of the π electrons through delocalization in the π orbitals that extend around the ring in this aromatic compound.

Mainly the *meta* isomer is formed in the reaction of nitric acid with nitrobenzene.

The bromination of benzene, which is carried out using $FeBr_3$ as a catalyst, is another substitution reaction:

$$C_6H_6 + Br_2 \xrightarrow{FeBr_3} C_6H_5Br + HBr \qquad [25.12]$$

In a similar reaction, called the *Friedel-Crafts reaction*, alkyl groups can be substituted onto an aromatic ring by reaction of an alkyl halide with an aromatic compound in the presence of $AlCl_3$ as a catalyst:

$$C_6H_6 + CH_3CH_2Cl \xrightarrow{AlCl_3} C_6H_5CH_2CH_3 + HCl \qquad [25.13]$$

25.5 Functional Groups: Alcohols and Ethers

The reactivity of organic compounds can be attributed to particular atoms or groups of atoms within the molecules. A site of reactivity in an organic molecule is called a *functional group* because it controls how the molecule behaves or functions. (Section 25.1) As we have seen, the presence of C=C double bonds or C≡C triple bonds in a hydrocarbon markedly increases its reactivity. Furthermore, these functional groups each undergo characteristic reactions. Each distinct kind of functional group often undergoes the same kinds of reactions in every molecule, regardless of the size and complexity of the molecule. Thus, the chemistry of an organic molecule is largely determined by the functional groups it contains.

Table 25.4 ▶ lists the most common functional groups and gives examples of each. Notice that in addition to C=C double bonds or C≡C triple bonds, there are also many functional groups that contain elements other than just C and H. Many of the functional groups contain other nonmetals such as O and N.

We can think of organic molecules as being composed of functional groups that are bonded to one or more alkyl groups. The alkyl groups, which are made

TABLE 25.4 Common Functional Groups in Organic Compounds

Functional Group	Type of Compound	Suffix or Prefix	Example	Systematic Name (common name)
$>C=C<$	Alkene	*ene*	$H_2C=CH_2$	Ethene (Ethylene)
$-C \equiv C-$	Alkyne	*-yne*	$H-C \equiv C-H$	Ethyne (Acetylene)
$-\overset{\vert}{\underset{\vert}{C}}-\ddot{O}-H$	Alcohol	*-ol*	$H_3C-\ddot{O}-H$	Methanol (Methyl alcohol)
$-\overset{\vert}{\underset{\vert}{C}}-\ddot{O}-\overset{\vert}{\underset{\vert}{C}}-$	Ether	*ether*	$H_3C-\ddot{O}-CH_3$	Dimethyl ether
$-\overset{\vert}{\underset{\vert}{C}}-\ddot{X}:$ (X = halogen)	Haloalkane	*halo-*	$H_3C-\ddot{Cl}:$	Chloromethane (Methyl chloride)
$-\overset{\vert}{\underset{\vert}{C}}-\underset{\vert}{\ddot{N}}-$	Amine	*-amine*	$H_3C-CH_2-\ddot{N}H_2$	Ethylamine
$-\overset{:O:}{\overset{\Vert}{C}}-H$	Aldehyde	*-al*	$H_3C-\overset{:O:}{\overset{\Vert}{C}}-H$	Ethanal (Acetaldehyde)
$-\overset{\vert}{\underset{\vert}{C}}-\overset{:O:}{\overset{\Vert}{C}}-\overset{\vert}{\underset{\vert}{C}}-$	Ketone	*-one*	$H_3C-\overset{:O:}{\overset{\Vert}{C}}-CH_3$	Propanone (Acetone)
$-\overset{:O:}{\overset{\Vert}{C}}-\ddot{O}-H$	Carboxylic acid	*-oic acid*	$H_3C-\overset{:O:}{\overset{\Vert}{C}}-\ddot{O}-H$	Ethanoic acid (Acetic acid)
$-\overset{:O:}{\overset{\Vert}{C}}-\ddot{O}-\overset{\vert}{\underset{\vert}{C}}-$	Ester	*-oate*	$H_3C-\overset{:O:}{\overset{\Vert}{C}}-\ddot{O}-CH_3$	Methyl ethanoate (Methyl acetate)
$-\overset{:O:}{\overset{\Vert}{C}}-\underset{\vert}{\ddot{N}}-$	Amide	*-amide*	$H_3C-\overset{:O:}{\overset{\Vert}{C}}-\ddot{N}H_2$	Ethanamide (Acetamide)

of C—C and C—H single bonds, are the less reactive portions of the organic molecules. In describing general features of organic compounds, chemists often use the designation R to represent any alkyl group: methyl, ethyl, propyl, and so on. Alkanes, for example, which contain no functional group, are represented as

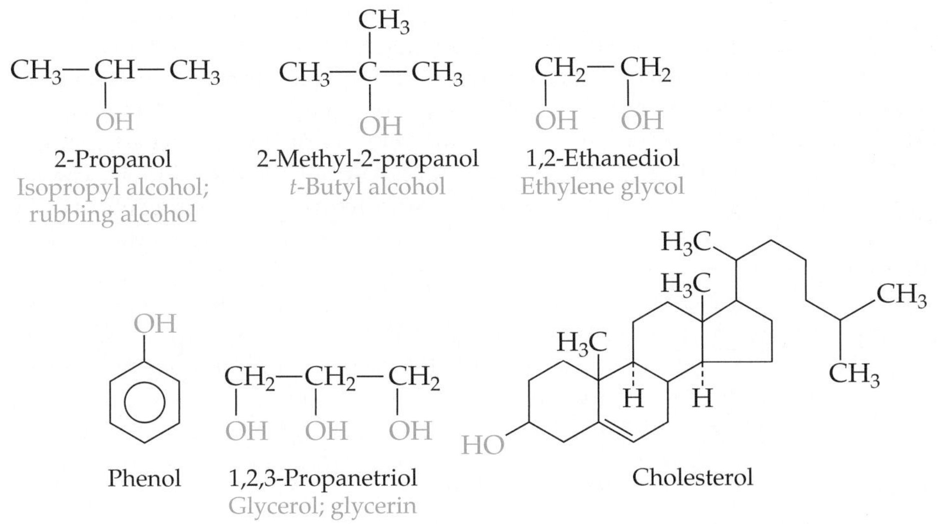

$CH_3—CH(OH)—CH_3$
2-Propanol
Isopropyl alcohol; rubbing alcohol

$CH_3—C(CH_3)(OH)—CH_3$
2-Methyl-2-propanol
t-Butyl alcohol

$CH_2(OH)—CH_2(OH)$
1,2-Ethanediol
Ethylene glycol

OH
Phenol

$CH_2(OH)—CH_2(OH)—CH_2(OH)$
1,2,3-Propanetriol
Glycerol; glycerin

H_3C H_3C CH_3 H_3C H H CH_3 HO
Cholesterol

◀ **Figure 25.14** Structural formulas of several important alcohols. Their common names are given in blue.

R—H. Alcohols, which contain the O—H, or alcohol functional group, are represented as R—OH. If two or more different alkyl groups are present in a molecule, we will designate them as R, R′, R″, and so forth. In this section we examine the structure and chemical properties of two functional groups, alcohols and ethers. In Section 25.6 we consider some additional functional groups that contain C=O bonds.

Alcohols (R—OH)

Alcohols are hydrocarbon derivatives in which one or more hydrogens of a parent hydrocarbon have been replaced by a *hydroxyl* or *alcohol* functional group, OH. Figure 25.14 ▲ shows the structural formulas and names of several alcohols. Note that the name for an alcohol ends in *-ol*. The simple alcohols are named by changing the last letter in the name of the corresponding alkane to *-ol*—for example, ethan*e* becomes ethan*ol*. Where necessary, the location of the OH group is designated by an appropriate numeral prefix that indicates the number of the carbon atom bearing the OH group, as shown in the examples in Figure 25.14.

The O—H bond is polar, so alcohols are much more soluble in polar solvents such as water than are hydrocarbons. The OH functional group can also participate in hydrogen bonding. As a result, the boiling points of alcohols are much higher than those of their parent alkanes.

Figure 25.15 ▶ shows several familiar commercial products that consist entirely or in large part of an organic alcohol. Let's consider how some of the more important alcohols are formed and used.

The simplest alcohol, methanol (methyl alcohol), has many important industrial uses and is produced on a large scale. Carbon monoxide and hydrogen are heated together under pressure in the presence of a metal oxide catalyst:

$$CO(g) + 2H_2(g) \xrightarrow[400°C]{200-300 \text{ atm}} CH_3OH(g) \qquad [25.14]$$

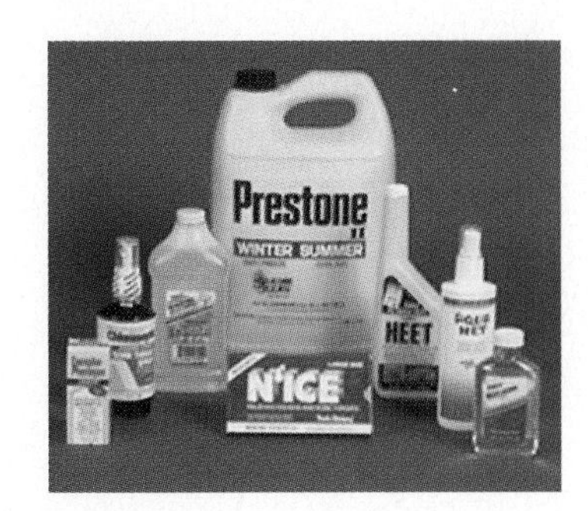

▲ **Figure 25.15** Some commercial products that are composed entirely or mainly of alcohols.

Because methanol has a very high octane rating as an automobile fuel, it is used as a gasoline additive and as a fuel in its own right.

Ethanol (ethyl alcohol, C_2H_5OH) is a product of the fermentation of carbohydrates such as sugar and starch. In the absence of air, yeast cells convert carbohydrates into a mixture of ethanol and CO_2, as shown in Equation 25.15. In the process, yeast derives energy necessary for growth:

$$C_6H_{12}O_6(aq) \xrightarrow{\text{yeast}} 2C_2H_5OH(aq) + 2CO_2(g) \qquad [25.15]$$

This reaction is carried out under carefully controlled conditions to produce beer, wine, and other beverages in which ethanol is the active ingredient.

Many polyhydroxyl alcohols (those containing more than one OH group) are known. The simplest of these is 1,2-ethanediol (ethylene glycol, $HOCH_2CH_2OH$). This substance is the major ingredient in automobile antifreeze. Another common polyhydroxyl alcohol is 1,2,3-propanetriol (glycerol, $HOCH_2CH(OH)CH_2OH$). It is a viscous liquid that dissolves readily in water and is widely used as a skin softener in cosmetic preparations. It is also used in foods and candies to keep them moist.

Phenol is the simplest compound with an OH group attached to an aromatic ring. One of the most striking effects of the aromatic group is the greatly increased acidity of the OH group. Phenol is about 1 million times more acidic in water than a typical nonaromatic alcohol such as ethanol. Even so, it is not a very strong acid ($K_a = 1.3 \times 10^{-10}$). Phenol is used industrially to make several kinds of plastics and dyes. It is also used as a topical anesthetic in many sore throat sprays.

Cholesterol, shown in Figure 25.14, is a biochemically important alcohol. The OH group forms only a small component of this rather large molecule, so cholesterol is only slightly soluble in water (0.26 g per 100 mL of H_2O). Cholesterol is a normal component of our bodies; when present in excessive amounts, however, it may precipitate from solution. It precipitates in the gallbladder to form crystalline lumps called *gallstones*. It may also precipitate against the walls of veins and arteries and thus contribute to high blood pressure and other cardiovascular problems. The amount of cholesterol in our blood is determined not only by how much cholesterol we eat but also by total dietary intake. There is evidence that excessive caloric intake leads the body to synthesize excessive cholesterol.

Ethers (R—O—R′)

Compounds in which two hydrocarbon groups are bonded to one oxygen are called **ethers**. Ethers can be formed from two molecules of alcohol by splitting out a molecule of water. The reaction is thus a dehydration; it is catalyzed by sulfuric acid, which takes up water to remove it from the system:

$$CH_3CH_2{-}OH + H{-}OCH_2CH_3 \xrightarrow{H_2SO_4} CH_3CH_2{-}O{-}CH_2CH_3 + H_2O \qquad [25.16]$$

A reaction in which water is split out from two substances is called a *condensation reaction*. (Sections 12.2 and 22.8)

Ethers are used as solvents. Both diethyl ether and the cyclic ether tetrahydrofuran are common solvents for organic reactions:

$CH_3CH_2{-}O{-}CH_2CH_3$

Diethyl ether

$CH_2{-}CH_2$ | | CH_2 CH_2 (ring closed through O)

Tetrahydrofuran (THF)

25.6 Compounds with a Carbonyl Group

Several of the functional groups listed in Table 25.4 contain a C=O double bond. This particular group of atoms is called a **carbonyl group**. The carbonyl group, together with the atoms that are attached to the carbon of the carbonyl group, defines several important functional groups that we consider in this section.

Aldehydes $\left(R-\overset{\overset{\displaystyle O}{\|}}{C}-H\right)$ and Ketones $\left(R-\overset{\overset{\displaystyle O}{\|}}{C}-R'\right)$

In **aldehydes** the carbonyl group has at least one hydrogen atom attached, as in the following examples:

$$H-\overset{\overset{\displaystyle O}{\|}}{C}-H \qquad CH_3-\overset{\overset{\displaystyle O}{\|}}{C}-H$$

Methanal
Formaldehyde

Ethanal
Acetaldehyde

In **ketones** the carbonyl group occurs at the interior of a carbon chain and is therefore flanked by carbon atoms:

$$CH_3-\overset{\overset{\displaystyle O}{\|}}{C}-CH_3 \qquad CH_3-\overset{\overset{\displaystyle O}{\|}}{C}-CH_2CH_3$$

Propanone
Acetone

2-Butanone
Methyl ethyl ketone

Aldehydes and ketones can be prepared by carefully controlled oxidation of alcohols. It is fairly easy to oxidize alcohols. Complete oxidation results in formation of CO_2 and H_2O, as in the burning of methanol:

$$CH_3OH(g) + \tfrac{3}{2}O_2(g) \longrightarrow CO_2(g) + 2H_2O(g)$$

Controlled partial oxidation to form other organic substances, such as aldehydes and ketones, is carried out by using various oxidizing agents such as air, hydrogen peroxide (H_2O_2), ozone (O_3), and potassium dichromate ($K_2Cr_2O_7$).

Many compounds found in nature possess an aldehyde or ketone functional group. Vanilla and cinnamon flavorings are naturally occurring aldehydes. The ketones carvone and camphor impart the characteristic flavors of spearmint leaves and caraway seeds.

Ketones are less reactive than aldehydes and are used extensively as solvents. Acetone, which boils at 56°C, is the most widely used ketone. The carbonyl functional group imparts polarity to the solvent. Acetone is completely miscible with water, yet it dissolves a wide range of organic substances. 2-Butanone ($CH_3COCH_2CH_3$), which boils at 80°C, is also used industrially as a solvent.

Carboxylic Acids $\left(R-\overset{\overset{\displaystyle O}{\|}}{C}-OH\right)$

We first discussed carboxylic acids in Section 16.10. **Carboxylic acids** contain the *carboxyl* functional group, which is often written as COOH. These weak acids are widely distributed in nature and are commonly used in consumer products [Figure 25.16(a) ▶]. They are also important in the manufacture of polymers used

(a)

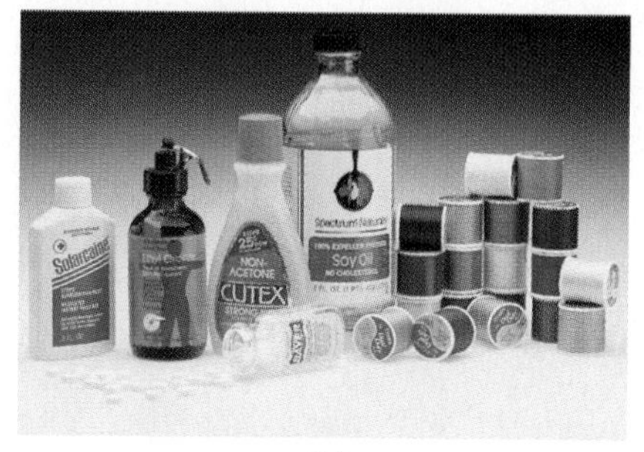

(b)

▶ **Figure 25.16** Carboxylic acids and esters are components of many household items: (a) Spinach and some cleaners contain oxalic acid; vinegar contains acetic acid; vitamin C is ascorbic acid; citrus fruits contain citric acid; and aspirin is acetylsalicylic acid (also an ester). (b) Many sunburn lotions contain benzocaine (an ester); some nail polish remover is ethyl acetate; vegetable oils, polyester thread, and aspirin are also esters.

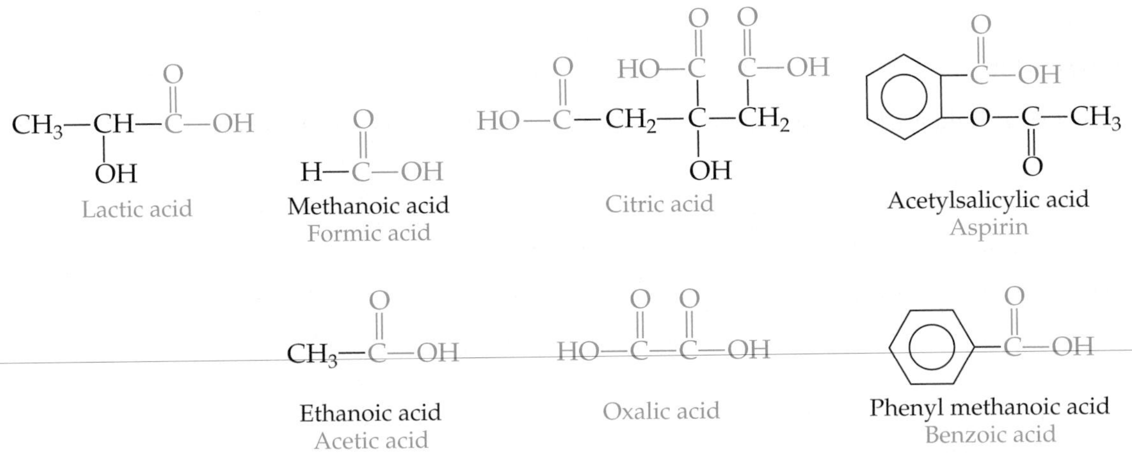

Lactic acid: $CH_3-CH(OH)-C(=O)-OH$

Methanoic acid (Formic acid): $H-C(=O)-OH$

Citric acid: $HO-C(=O)-CH_2-C(OH)(COOH)-CH_2-C(=O)-OH$

Acetylsalicylic acid (Aspirin): $C_6H_4(-C(=O)-OH)(-O-C(=O)-CH_3)$

Ethanoic acid (Acetic acid): $CH_3-C(=O)-OH$

Oxalic acid: $HO-C(=O)-C(=O)-OH$

Phenyl methanoic acid (Benzoic acid): $C_6H_5-C(=O)-OH$

▲ **Figure 25.17** Structural formulas of several common carboxylic acids. The IUPAC names are given for the monocarboxylic acids, but they are generally referred to by their common names.

to make fibers, films, and paints. Figure 25.17 ▲ shows the structural formulas of several carboxylic acids. Notice that oxalic acid and citric acid contain two and three carboxyl groups, respectively. The common names of many carboxylic acids are based on their historical origins. Formic acid, for example, was first prepared by extraction from ants; its name is derived from the Latin word *formica*, meaning "ant."

Carboxylic acids can be produced by oxidation of alcohols in which the OH group is attached to a CH_2 group, such as ethanol or propanol. Under appropriate conditions the corresponding aldehyde may be isolated as the first product of oxidation. These transformations are shown for ethanol in the following equations, in which (O) represents an oxidant that can provide oxygen atoms (such as $K_2Cr_2O_7$):

$$\underset{\text{Ethanol}}{CH_3CH_2OH} + (O) \longrightarrow \underset{\text{Acetaldehyde}}{CH_3\overset{O}{\overset{\|}{C}}H} + H_2O \qquad [25.17]$$

$$\underset{\text{Acetaldehyde}}{CH_3\overset{O}{\overset{\|}{C}}H} + (O) \longrightarrow \underset{\text{Acetic acid}}{CH_3\overset{O}{\overset{\|}{C}}OH} \qquad [25.18]$$

The air oxidation of ethanol to acetic acid is responsible for causing wines to turn sour, producing vinegar.

Acetic acid can also be produced by the reaction of methanol with carbon monoxide in the presence of a rhodium catalyst:

$$CH_3OH + CO \xrightarrow{\text{catalyst}} CH_3-\overset{O}{\overset{\|}{C}}-OH \qquad [25.19]$$

This reaction involves, in effect, the insertion of a carbon monoxide molecule between the CH_3 and OH groups. A reaction of this kind is called *carbonylation*.

Esters $(R-\overset{O}{\overset{\|}{C}}-O-R')$

Carboxylic acids can undergo condensation reactions with alcohols to form **esters**:

$$\underset{\text{Acetic acid}}{CH_3-\overset{O}{\overset{\|}{C}}-OH} + \underset{\text{Ethanol}}{HO-CH_2CH_3} \longrightarrow \underset{\text{Ethyl acetate}}{CH_3-\overset{O}{\overset{\|}{C}}-O-CH_2CH_3} + H_2O \quad [25.20]$$

Esters are compounds in which the H atom of a carboxylic acid is replaced by a hydrocarbon group:

$$-\overset{O}{\overset{\|}{C}}-O-\overset{|}{\underset{|}{C}}-$$

Figure 25.16(b) shows some common esters. Esters are named by using first the group from which the alcohol is derived and then the group from which the acid is derived.

Esters generally have very pleasant odors. They are largely responsible for the pleasant aromas of fruit. Pentyl acetate ($CH_3COOCH_2CH_2CH_2CH_2CH_3$), for example, is responsible for the odor of bananas.

When esters are treated with an acid or a base in aqueous solution, they are hydrolyzed; that is, the molecule is split into its alcohol and acid components:

$$\underset{\text{Methyl propionate}}{CH_3CH_2-\overset{O}{\overset{\|}{C}}-O-CH_3} + Na^+ + OH^- \longrightarrow$$

$$\underset{\text{Sodium propionate}}{CH_3CH_2-\overset{O}{\overset{\|}{C}}-O^-} + Na^+ + \underset{\text{Methanol}}{CH_3OH} \quad [25.21]$$

In this example the hydrolysis was carried out in basic medium. The products of the reaction are the sodium salt of the carboxylic acid and the alcohol.

The hydrolysis of an ester in the presence of a base is called **saponification**, a term that comes from the Latin word for soap (*sapon*). Naturally occurring esters include fats and oils. In the soap-making process an animal fat or a vegetable oil is boiled with a strong base, usually NaOH. The resultant soap consists of a mixture of sodium salts of long-chain carboxylic acids (called fatty acids), which form during the saponification reaction (Figure 25.18 ▼).

◀ **Figure 25.18** Saponification of fats and oils has long been used to make soap. This etching shows a step in the soap-making process during the mid-nineteenth century.

SAMPLE EXERCISE 25.6

In a basic aqueous solution, esters react with hydroxide ion to form the salt of the carboxylic acid and the alcohol from which the ester is constituted. Name each of the following esters, and indicate the products of their reaction with aqueous base.

(a) $C_6H_5-C(=O)-OCH_2CH_3$ **(b)** $CH_3CH_2CH_2-C(=O)-O-C_6H_5$

Solution

Analyze: We are given two esters and asked to name them and to predict the products formed when they undergo hydrolysis (split into an alcohol and carboxylate ion) in basic solution.

Plan: Esters are formed by the condensation reaction between an alcohol and a carboxylic acid. To name an ester we must analyze its structure and determine the identities of the alcohol and acid from which it is formed. We can identify the alcohol by adding an OH to the alkyl group attached to the O atom of the carboxyl (COO) group. We can identify the acid by adding an H group to the O atom of the carboxyl group. The first part of an ester name indicates the alcohol portion, and the second indicates the acid portion. The name conforms to how the ester undergoes hydrolysis in base, reacting with base to form an alcohol and a carboxylate anion.

Solve: (a) This ester is derived from ethanol (CH_3CH_2OH) and benzoic acid (C_6H_5COOH). Its name is therefore ethyl benzoate. The net ionic equation for reaction of ethyl benzoate with hydroxide ion is as follows:

$$C_6H_5-C(=O)-OCH_2CH_3(aq) + OH^-(aq) \longrightarrow C_6H_5-C(=O)-O^-(aq) + HOCH_2CH_3(aq)$$

The products are benzoate ion and ethanol.

(b) This ester is derived from phenol (C_6H_5OH) and butyric acid ($CH_3CH_2CH_2COOH$). The residue from the phenol is called the phenyl group. The ester is therefore named phenyl butyrate. The net ionic equation for the reaction of phenyl butyrate with hydroxide ion is as follows:

$$CH_3CH_2CH_2C(=O)-O-C_6H_5(aq) + OH^-(aq) \longrightarrow CH_3CH_2CH_2C(=O)-O^-(aq) + HO-C_6H_5(aq)$$

The products are butyrate ion and phenol.

PRACTICE EXERCISE

Write the structural formula for the ester formed from propyl alcohol and propionic acid.

Answer: $CH_3CH_2C(=O)-O-CH_2CH_2CH_3$

Amines and Amides

Amines are organic bases. (Section 16.7) They have the general formula R_3N, where R may be H or a hydrocarbon group, as in the following examples:

$CH_3CH_2NH_2$	$(CH_3)_3N$	$C_6H_5—NH_2$
Ethylamine	Trimethylamine	Phenylamine Aniline

Amines containing a hydrogen bonded to nitrogen can undergo condensation reactions with carboxylic acids to form **amides**:

$$CH_3\overset{O}{\overset{\|}{C}}—OH + H—N(CH_3)_2 \longrightarrow CH_3\overset{O}{\overset{\|}{C}}—N(CH_3)_2 + H_2O \qquad [25.22]$$

We may consider the amide functional group to be derived from a carboxylic acid with an NR_2 group replacing the OH of the acid, as in these additional examples:

$CH_3\overset{O}{\overset{\|}{C}}—NH_2$	$C_6H_5—\overset{O}{\overset{\|}{C}}—NH_2$
Ethanamide Acetamide	Phenylmethanamide Benzamide

The amide linkage

$$R—\overset{O}{\overset{\|}{C}}—\underset{\underset{H}{|}}{N}—R'$$

where R and R′ are organic groups, is the key functional group in the structures of proteins, as we will see in Section 25.9.

25.7 Chirality in Organic Chemistry

A molecule possessing a nonsuperimposable mirror image is termed **chiral** (Greek *cheir*, "hand"). (Section 24.4) *Compounds containing carbon atoms with four different attached groups are inherently chiral.* A carbon atom with four different attached groups is called a *chiral center*. For example, the structural formula of 2-bromopentane is

$$CH_3—\underset{\underset{H}{|}}{\overset{\overset{Br}{|}}{C}}—CH_2CH_2CH_3$$

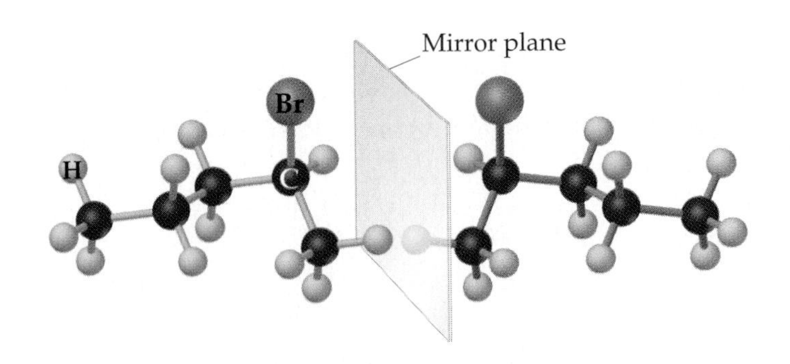

▶ **Figure 25.19** The two enantiomeric forms of 2-bromopentane. The mirror-image isomers are not superimposable.

All four groups attached to the second carbon are different, making that carbon a chiral center. Figure 25.19 ▲ illustrates the two nonsuperimposable mirror images of this molecule. If you imagine trying to move the left-hand molecule to the right and turning it in every possible way, you will conclude that it cannot be superimposed on the right-hand molecule. Nonsuperimposable mirror images are called optical isomers or *enantiomers*. (Section 24.4) Organic chemists use the labels *R* and *S* to distinguish the two forms. We need not go into the rules for deciding on the labels.

Enantiomers, such as the two forms of 2-bromopentane shown in Figure 25.19, have identical physical properties, such as melting and boiling points, and identical chemical properties when they react with nonchiral reagents. Only in a chiral environment do they exhibit different behaviors. One of the more interesting properties of chiral substances is that their solutions may rotate the plane of polarized light, as explained in Section 24.4.

Chirality is very common in organic substances. It is not often observed, however, because when a chiral substance is synthesized in a typical chemical reaction, the two enantiomeric species are formed in precisely the same quantity. The resulting mixture of isomers is called a *racemic* mixture. A racemic mixture of enantiomers does not rotate the plane of polarized light because the two forms rotate the light to equal extents in opposite directions. (Section 24.4)

Naturally occurring substances often are found as just one enantiomer. An example is tartaric acid, which has the structural formula*

```
    COOH
     |
  H—C—OH
     |
  H—C—OH
     |
    COOH
```

This compound has not one, but two chiral centers. Both the inner carbon atoms are attached to four different groups. Tartaric acid is found in nature as the free acid, as a salt of calcium or potassium, in fruit extracts, and especially as crystals deposited during wine fermentation. The naturally occurring form is optically active in solution. Tartaric acid is typical; many biologically important molecules are chiral. In Section 25.9 we will examine the amino acids, all of which (except for glycine) are chiral and found in nature as just one of the enantiomers.

Many drugs of importance in human medicine are chiral substances. When a drug is administered as a racemic mixture, it often turns out that only one of the enantiomers has beneficial results. The other is either inert or nearly so, or may even have a harmful effect. For this reason, a great deal of attention has been given in recent years to methods for synthesizing the desired enantiomer of chiral drugs. Synthesizing just one of the enantiomers of a chiral substance can be

* Louis Pasteur discovered chirality while he was studying crystalline samples of salts of tartaric acid.

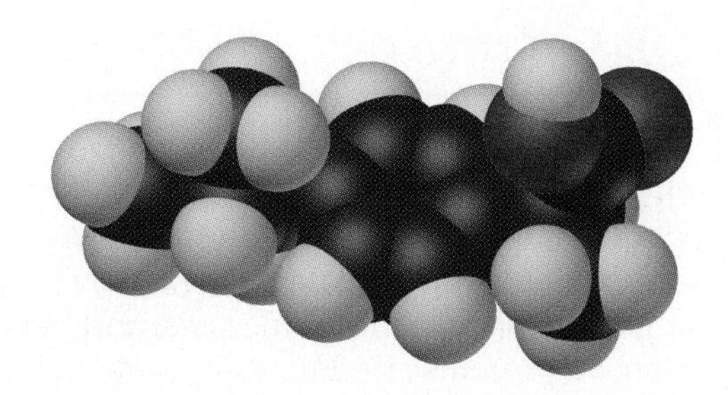

◀ **Figure 25.20** Structural formula and space-filling molecular model of *S*-ibuprofen.

very difficult and costly, but the rewards are worth the effort. Worldwide sales of single-enantiomer drugs amount to over $125 billion annually! As an example, the nonsteroidal analgesic ibuprofen (marketed under the trade names Advil™, Motrin™, and Nuprin™) is a chiral molecule that is sold as the racemic mixture. A preparation consisting of just the most active enantiomer, however, relieves pain and reduces inflammation much more rapidly than the racemic mixture. Approval to market the single-enantiomer version of this drug, *S*-ibuprofen (Figure 25.20 ▲), is being sought from the Food and Drug Administration.

25.8 Introduction to Biochemistry

The several types of organic functional groups discussed in Sections 25.5 and 25.6 generate a vast array of molecules with very specific chemical reactivities. Nowhere is this specificity more apparent than in *biochemistry*—the chemistry of living organisms.

Before we discuss specific biochemical molecules, we can make some general observations. As we will see, many biologically important molecules are quite large. The synthesis of these molecules is a remarkable aspect of biochemistry, one that places large demands on the chemical systems in living organisms. Organisms build biomolecules from much smaller and simpler substances that are readily available in the biosphere. The synthesis of large molecules requires energy because most of the reactions are endothermic. The ultimate source of this energy is the Sun. Mammals and other animals have essentially no capacity for using solar energy directly; rather they depend on plant photosynthesis to supply the bulk of their energy needs. ∞ (Section 24.2)

In addition to requiring large amounts of energy, living organisms are highly organized. The complexity of all the substances that make up even the simplest single-cell organisms and the relationships among all the many chemical processes that occur are truly amazing. In thermodynamic terms, living systems are very low in entropy as compared to the raw materials from which they are formed. Thus, the chemistry of living systems must continually resist the tendency toward increased entropy that is required by the second law of thermodynamics. ∞ ("Chemistry and Life," Section 19.2) Maintaining a high degree of order places additional energetic requirements on organisms.

We have introduced you to some important biochemical applications of fundamental chemical ideas in the "Chemistry and Life" essays that appear throughout this text. A complete summary of the topics covered and their locations in the text are included in the table of contents. The remainder of this chapter will serve as only a brief introduction to other aspects of biochemistry. Nevertheless, you will see some patterns emerging. Hydrogen bonding (Section 11.2), for example, is critical to the function of many biochemical systems and the geometry of molecules can govern their biological importance and activity. Many of the large molecules in living systems are polymers (Section 12.2) of much smaller molecules. These **biopolymers** can be classified into three broad categories: proteins, polysaccharides (carbohydrates), and nucleic acids. We discuss these classes of biopolymers in Sections 25.9, 25.10, and 25.11, respectively.

25.9 Proteins

Proteins are macromolecular substances present in all living cells. About 50% of your body's dry weight is protein. Proteins serve as the major structural components in animal tissues; they are a key part of skin, nails, cartilage, and muscles. Other proteins catalyze reactions, transport oxygen, serve as hormones to regulate specific body processes, and perform other tasks. Whatever their function, all proteins are chemically similar, being composed of the same basic building blocks, called **amino acids**.

Amino Acids

The building blocks of all proteins are α-amino acids, which are substances in which the amino group is located on the carbon atom immediately adjacent to the carboxylic acid group. Thus, there is always one carbon atom between the amino group and the carboxylic acid group. The general formula for an α-amino acid is represented in the following ways:

$$H_2N-\underset{R}{\overset{H}{C}}-\overset{O}{\overset{\|}{C}}-OH \quad \text{or} \quad {}^{+}H_3N-\underset{R}{\overset{H}{C}}-\overset{O}{\overset{\|}{C}}-O^-$$

The doubly ionized form, called the zwitterion, usually predominates at near-neutral values of pH. This form results from the transfer of a proton from the carboxylic acid group to the basic amine group. ("Chemistry and Life," Section 16.10)

Amino acids differ from one another in the nature of their R groups. Figure 25.21 ▼ shows the structural formulas of several of the 22 amino acids found in proteins.* Our bodies use 20 of these and can synthesize 10 of them in sufficient

▶ **Figure 25.21** Condensed structural formulas for several amino acids, with the three-letter abbreviation for each acid. The acids are shown in the zwitterionic form in which they exist in water at near-neutral pH.

$$H-\underset{H}{\overset{H_3N^+}{C}}-\overset{O}{\overset{\|}{C}}-O^-$$
Glycine (Gly)

$$CH_3-\underset{H}{\overset{H_3N^+}{C}}-\overset{O}{\overset{\|}{C}}-O^-$$
Alanine (Ala)

$$CH_3-\underset{CH_3}{CH}-\underset{H}{\overset{H_3N^+}{C}}-\overset{O}{\overset{\|}{C}}-O^-$$
Valine (Val)

$$C_6H_5-CH_2-\underset{H}{\overset{H_3N^+}{C}}-\overset{O}{\overset{\|}{C}}-O^-$$
Phenylalanine (Phe)

$$HO-\overset{O}{\overset{\|}{C}}-CH_2-\underset{H}{\overset{H_3N^+}{C}}-\overset{O}{\overset{\|}{C}}-O^-$$
Aspartic acid (Asp)

$$HO-\overset{O}{\overset{\|}{C}}-CH_2-CH_2-\underset{H}{\overset{H_3N^+}{C}}-\overset{O}{\overset{\|}{C}}-O^-$$
Glutamic acid (Glu)

$$HO-CH_2-\underset{H}{\overset{H_3N^+}{C}}-\overset{O}{\overset{\|}{C}}-O^-$$
Serine (Ser)

$$HS-CH_2-\underset{H}{\overset{H_3N^+}{C}}-\overset{O}{\overset{\|}{C}}-O^-$$
Cysteine (Cys)

$$H_2N-(CH_2)_4-\underset{H}{\overset{H_3N^+}{C}}-\overset{O}{\overset{\|}{C}}-O^-$$
Lysine (Lys)

* The 22nd amino acid was discovered in 2002 by researdners at Ohio State University.

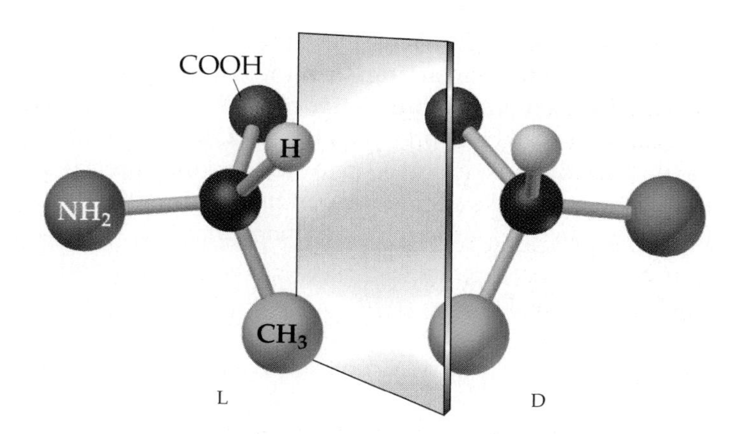

◀ **Figure 25.22** Alanine is chiral and therefore has two enantiomers, which are nonsuperimposable mirror images of each other.

amounts for our needs. The other ten must be ingested and are called *essential amino acids* because they are necessary components of our diet.

The α-carbon atom of the amino acids, which bears both the ammonium and carboxylate groups, has four different groups attached to it.* The amino acids are thus chiral. Figure 25.22 ▲ shows the two enantiomers of the amino acid alanine. For historical reasons, the two enantiomeric forms of amino acids are often distinguished by the labels D- (from the Latin *dexter*, "right") and L- (from the Latin *laevus*, "left"). All the amino acids normally found in proteins are "left-handed"; that is, all have the L- configuration at the carbon center (except for glycine, which is not chiral). Only amino acids with this specific configuration at the chiral carbon center form proteins in living organisms.

Polypeptides and Proteins

Amino acids are linked together into proteins by amide groups, one of the functional groups introduced in Section 25.6:

$$\mathrm{R{-}\overset{\overset{\displaystyle O}{\|}}{C}{-}\underset{\underset{\displaystyle H}{|}}{N}{-}R}$$

Each of these amide groups is called a **peptide bond** when it is formed by amino acids. A peptide bond is formed by a condensation reaction between the carboxyl group of one amino acid and the amino group of another amino acid. Alanine and glycine, for example, can react to form the dipeptide glycylalanine:

$$\underset{\text{Glycine (Gly)}}{\mathrm{H_3\overset{+}{N}{-}\underset{\underset{\displaystyle H}{|}}{\overset{\overset{\displaystyle H}{|}}{C}}{-}\overset{\overset{\displaystyle O}{\|}}{C}{-}O^-}} + \underset{\text{Alanine (Ala)}}{\mathrm{H{-}\underset{\underset{\displaystyle H}{|}}{\overset{\overset{\displaystyle H}{|}}{\overset{+}{N}}}{-}\underset{\underset{\displaystyle CH_3}{|}}{\overset{\overset{\displaystyle H}{|}}{C}}{-}\overset{\overset{\displaystyle O}{\|}}{C}{-}O^-}} \longrightarrow$$

$$\underset{\text{Glycylalanine (Gly-Ala)}}{\mathrm{H_3\overset{+}{N}{-}\underset{\underset{\displaystyle H}{|}}{\overset{\overset{\displaystyle H}{|}}{C}}{-}\overset{\overset{\displaystyle O}{\|}}{C}{-}\underset{\underset{\displaystyle H}{|}}{N}{-}\underset{\underset{\displaystyle CH_3}{|}}{\overset{\overset{\displaystyle H}{|}}{C}}{-}\overset{\overset{\displaystyle O}{\|}}{C}{-}O^-}} + H_2O \qquad [25.23]$$

* The sole exception is glycine, for which R = H. For this amino acid, there are two H atoms on the α-carbon atom.

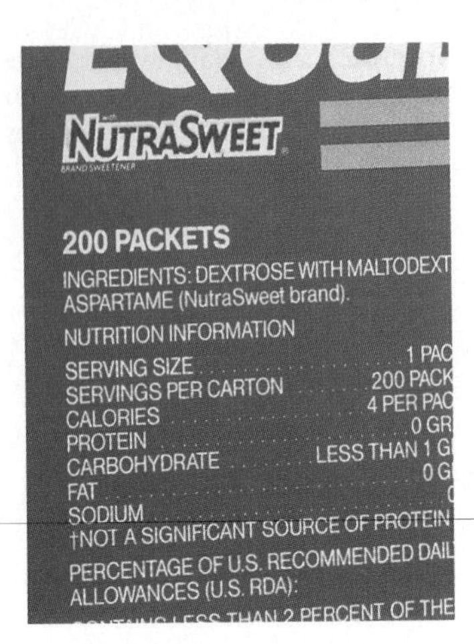

▲ **Figure 25.23** The artificial sweetener aspartame is the methyl ester of a dipeptide.

The amino acid that furnishes the carboxyl group for peptide-bond formation is named first, with a *-yl* ending; then the amino acid furnishing the amino group is named. Based on the three-letter codes for the amino acids from Figure 25.22, glycylalanine can be abbreviated Gly-Ala. In this notation it is understood that the unreacted amino group is on the left and the unreacted carboxyl group on the right. The artificial sweetener *aspartame* (Figure 25.23 ◀) is the methyl ester of the dipeptide of aspartic acid and phenylalanine:

```
       H  O     H  O
       |  ||    |  ||
  H2N—C—C—N—C—C—O—CH3
       |     |  |
      CH2    H  CH2
       |        |
      C—OH     C6H5 (benzene ring)
      ||
      O
```

SAMPLE EXERCISE 25.7

Draw the structural formula for alanylglycylserine.

Solution

Analyze: We are given the name of a substance with peptide bonds and asked to write its structural formula.

Plan: The name of this substance suggests that three amino acids—alanine, glycine, and serine—have been linked together, forming a *tripeptide*. Note that the ending *-yl* has been added to each amino acid except for the last one, serine. By convention, the first-named amino acid (alanine, in this case) has a free amino group and the last-named one (serine) has a free carboxyl group. Thus, we can construct the structural formula of the tripeptide from its amino acid building blocks (Figure 25.21).

Solve: We first combine the carboxyl group of alanine with the amino group of glycine to form a peptide bond and then the carboxyl group of glycine with the amino group of serine to form another peptide group. The resulting tripeptide consists of three "building blocks" connected by peptide bonds:

```
Amino group ————————————→ Carboxyl group

  +     H  O  |   H  O  |    H  O
        |  || |   |  || |    |  ||
   H3N—C—C—|N—C—C—|N—C—C—O⁻
        |     |  |  |  |   |  |
       CH3    |H  H |  H   CH2OH

       Ala       Gly         Ser
```

We can abbreviate this tripeptide as Ala-Gly-Ser.

PRACTICE EXERCISE

Name the dipeptide that has the following structure, and give its abbreviation:

```
  +      H   O     H  O
         |   ||    |  ||
   H3N—C—C—N—C—C—O⁻
         |     |  |
     HOCH2     H  CH2
                  |
                  COOH
```

Answer: serylaspartic acid; Ser-Asp

Polypeptides are formed when a large number of amino acids are linked together by peptide bonds. Proteins are linear (i.e., unbranched) polypeptide molecules with molecular weights ranging from about 6000 to over 50 million amu. Because 20 different amino acids are linked together in proteins and because proteins consist of hundreds of amino acids, the number of possible arrangements of amino acids within proteins is virtually limitless.

Protein Structure

The arrangement, or sequence, of amino acids along a protein chain is called its **primary structure**. The primary structure gives the protein its unique identity. A change in even one amino acid can alter the biochemical characteristics of the protein. For example, sickle-cell anemia is a genetic disorder resulting from a single replacement in a protein chain in hemoglobin. The chain that is affected contains 146 amino acids. The substitution of a single amino acid with a hydrocarbon side chain for one that has an acidic functional group in the side chain alters the solubility properties of the hemoglobin, and normal blood flow is impeded. ("Chemistry and Life," Section 13.6)

Proteins in living organisms are not simply long, flexible chains with random shapes. Rather, the chains coil or stretch in particular ways. The **secondary structure** of a protein refers to how segments of the protein chain are oriented in a regular pattern.

One of the most important and common secondary structure arrangements is the ***α*-helix**, first proposed by Linus Pauling and R. B. Corey. The helix arrangement is shown in schematic form in Figure 25.24 ▼. Imagine winding a long protein chain in a helical fashion around a long cylinder. The helix is held in position

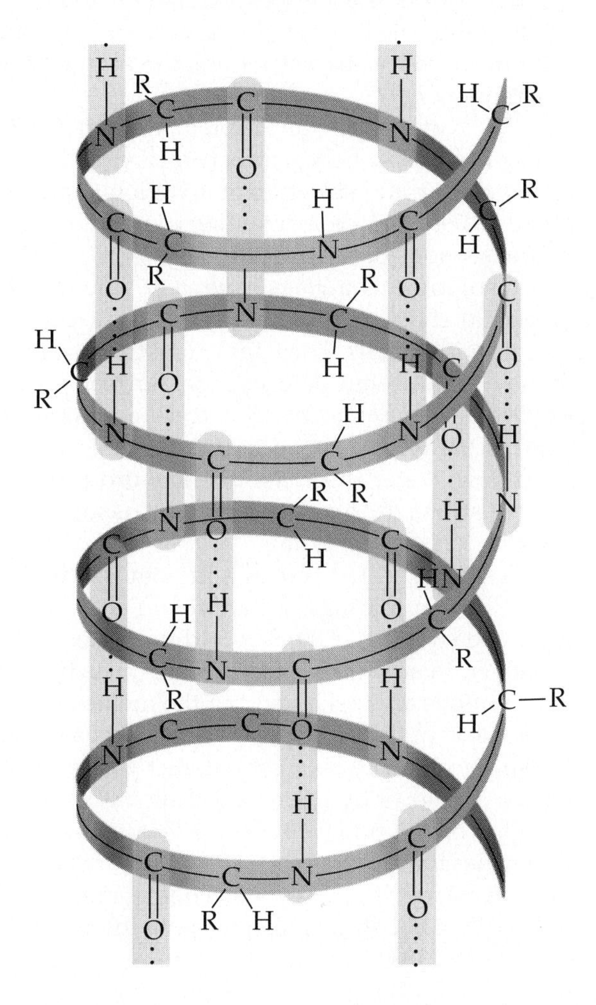

◀ **Figure 25.24** *α*-Helix structure for a protein. The symbol R represents any one of the several side chains shown in Figure 25.21.

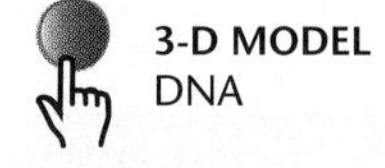

Chemistry and Life The Origins of Chirality in Living Systems

The presence of a "handedness" in the molecules that make up living systems is a key feature of life on Earth. The insistence of nature on just one chiral form in the molecules of life is called *homochirality*. How did the dominance of the L- amino acids arise? Why are the naturally occurring helices of protein and DNA, which we will discuss in Section 25.11, all right-turn helices? Homochirality could have arisen by chance in the course of evolution or because it was "seeded" in some way at the beginnings of life. One theory is that chirality was introduced early in Earth's evolutionary history, through seeding by chiral amino acids that fell on the planet from outer space.

Examination of the Murchison meteorite, which fell to Earth during this century, has revealed the presence of amino acids. For some of the acids, there appears to be a surplus of the L- form. One theory proposes that the chiral amino acids could have been synthesized in interstellar space by the action of circularly polarized starlight.* Astronomers in Australia have recently observed circular polarization in the infrared light from a region of intense star birth in the Orion nebula. These workers calculate that a similar degree of circular polarization could be present in the visible and ultraviolet light from this source. Light that has the energy required to break chemical bonds, if circularly polarized, could give rise to new chiral molecules with a preference for one enantiomer over the other. Perhaps the homochirality we observe on Earth today arose, through a process of amplification and refinement in the course of evolutionary development, from molecules that were formed in interstellar space when the planet was young.

* Circularly polarized light is like plane-polarized light, as shown in Figure 24.22, except that the plane continuously rotates either to the left or to the right. Thus, in a sense, circularly polarized light is chiral.

by hydrogen-bond interactions between N—H bonds and the oxygens of nearby carbonyl groups. The pitch of the helix and the diameter of the cylinder must be such that (1) no bond angles are strained and (2) the N—H and C=O functional groups on adjacent turns are in proper position for hydrogen bonding. An arrangement of this kind is possible for some amino acids along the chain, but not for others. Large protein molecules may contain segments of the chain that have the α-helical arrangement interspersed with sections in which the chain is in a random coil.

The overall shape of a protein, determined by all the bends, kinks, and sections of rodlike α-helical structure, is called the **tertiary structure**. Figure 24.11 shows the tertiary structure of myoglobin, a protein with a molecular weight of about 18,000 amu and containing one heme group. (Section 24.2) Notice the helical sections of the protein.

Myoglobin is a *globular protein*, one that folds into a compact, roughly spherical shape. Globular proteins are generally soluble in water and are mobile within cells. They have nonstructural functions, such as combating the invasion of foreign objects, transporting and storing oxygen, and acting as catalysts. The *fibrous proteins* form a second class of proteins. In these substances the long coils align themselves in a more or less parallel fashion to form long, water-insoluble fibers. Fibrous proteins provide structural integrity and strength to many kinds of tissue and are the main components of muscle, tendons, and hair.

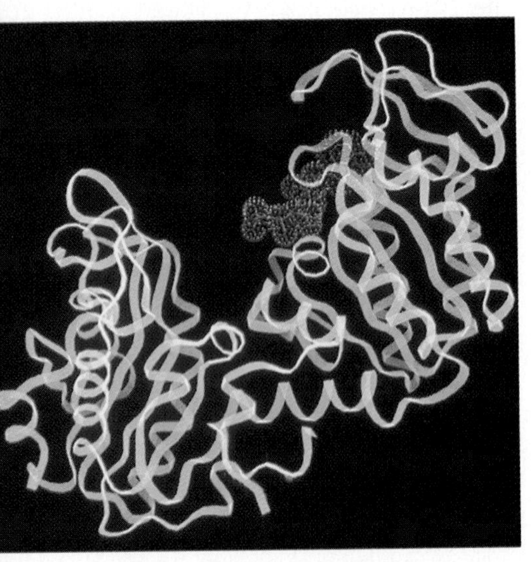

▲ **Figure 25.25** A computer-generated structure of an enzyme showing the carbon backbone as a green ribbon. The substrate (violet) is shown at the active site.

The tertiary structure of a protein is maintained by many different interactions. Certain foldings of the protein chain lead to lower-energy (more stable) arrangements than do other folding patterns. For example, a globular protein dissolved in aqueous solution folds in such a way that the nonpolar hydrocarbon portions are tucked within the molecule, away from the polar water molecules. Most of the more polar acidic and basic side chains, however, project into the solution where they can interact with water molecules through ion-dipole, dipole-dipole, or hydrogen-bonding interactions.

One of the most important classes of proteins is *enzymes*, large protein molecules that serve as catalysts. (Section 14.7) Enzymes usually catalyze only very specific reactions. Their tertiary structure generally dictates that only certain substrate molecules can interact with the active site of the enzyme (Figure 25.25 ◀).

25.10 Carbohydrates

Carbohydrates are an important class of naturally occurring substances found in both plant and animal matter. The name **carbohydrate** (hydrate of carbon) comes from the empirical formulas for most substances in this class, which can be written as $C_x(H_2O)_y$. For example, **glucose**, the most abundant carbohydrate, has the molecular formula $C_6H_{12}O_6$, or $C_6(H_2O)_6$. Carbohydrates are not really hydrates of carbon; rather, they are polyhydroxy aldehydes and ketones. Glucose, for example, is a six-carbon aldehyde sugar, whereas *fructose*, the sugar that occurs widely in fruit, is a six-carbon ketone sugar (Figure 25.26 ▶).

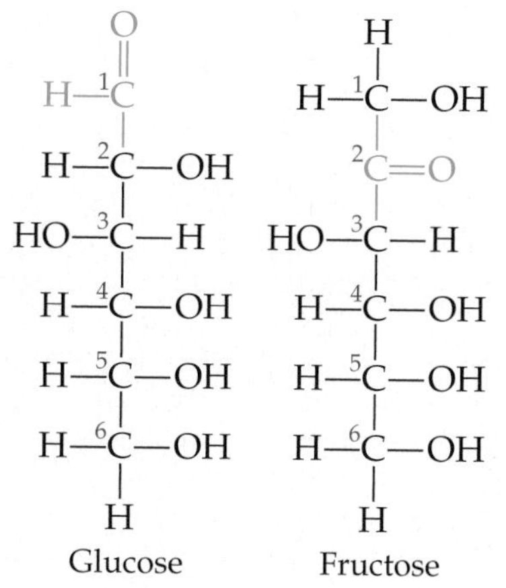

Glucose, having both alcohol and aldehyde functional groups and having a reasonably long and flexible backbone, can react with itself to form a six-member-ring structure, as shown in Figure 25.27 ▼. In fact, only a small percentage of the glucose molecules are in the open-chain form in aqueous solution. Although the ring is often drawn as if it were planar, the molecules are actually nonplanar because of the tetrahedral bond angles around the C and O atoms of the ring.

Figure 25.27 shows that the ring structure of glucose can have two different relative orientations. In the α form the OH group on carbon 1 and the CH_2OH group on carbon 5 point in opposite directions. In the β form they point in the same direction. Although the difference between the α and β forms might seem small, it has enormous biological consequences. As we will soon see, this one small change in structure accounts for the vast difference in properties between starch and cellulose.

Fructose can cyclize to form either five- or six-member rings. The five-member ring forms when the OH group on carbon 5 reacts with the carbonyl group on carbon 2:

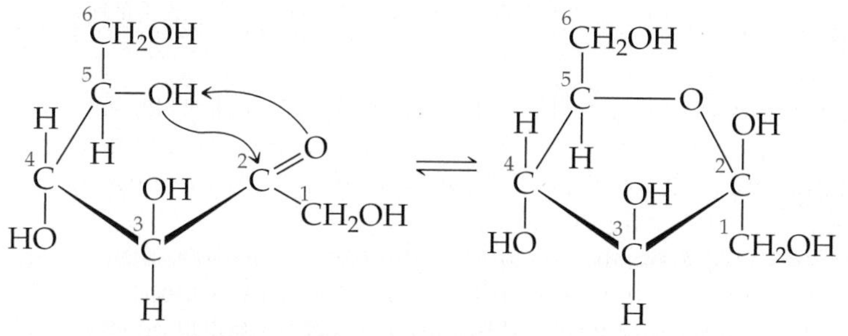

[25.24]

The six-member ring results from the reaction between the OH group on carbon 5 and the carbonyl group on carbon 1.

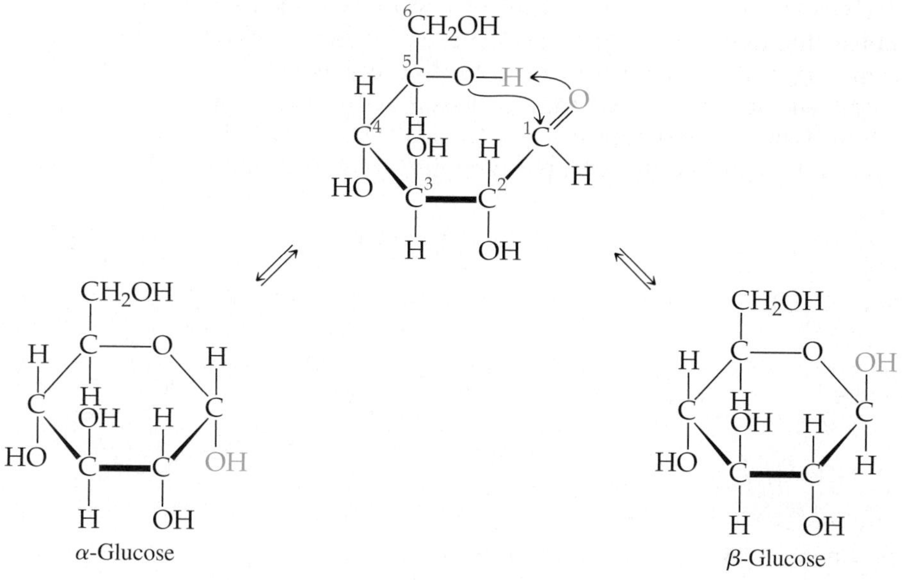

α-Glucose

β-Glucose

▲ **Figure 25.26** Linear structures of glucose and fructose.

Glucose

Fructose

◀ **Figure 25.27** Glucose reacts with itself to form two different six-member-ring structures, designated α and β.

SAMPLE EXERCISE 25.8

How many chiral carbon atoms are there in the open-chain form of glucose (Figure 25.26)?

Solution

Analyze: We are given the structure of glucose and asked to determine the number of chiral carbons in the molecule.

Plan: A chiral carbon has four different groups attached (Section 25.7).

Solve: The carbon atoms numbered 2, 3, 4, and 5 each have four different groups attached to them, as indicated here:

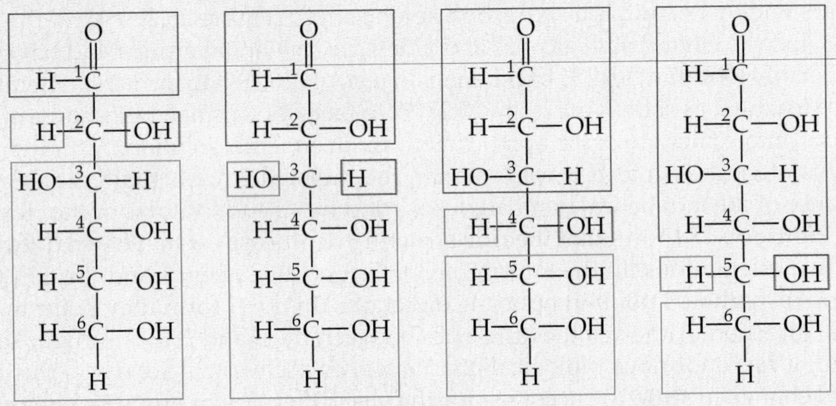

Carbon atoms 1 and 6 have only three different substituents on them. Thus, there are four chiral carbon atoms in the glucose molecule.

PRACTICE EXERCISE

How many chiral carbon atoms are there in the open-chain form of fructose (Figure 25.26)?

Answer: three

Disaccharides

Both glucose and fructose are examples of **monosaccharides**, simple sugars that can't be broken into smaller molecules by hydrolysis with aqueous acids. Two monosaccharide units can be linked together by a condensation reaction to form a *disaccharide*. The structures of two common disaccharides, *sucrose* (table sugar) and *lactose* (milk sugar), are shown in Figure 25.28 ▼.

The word *sugar* makes us think of sweetness. All sugars are sweet, but they differ in the degree of sweetness we perceive when we taste them. Sucrose is about six times sweeter than lactose, slightly sweeter than glucose, but only about half as sweet as fructose. Disaccharides can be reacted with water (hydrolyzed) in the presence of an acid catalyst to form monosaccharides. When sucrose is hydrolyzed, the mixture of glucose and fructose that forms, called

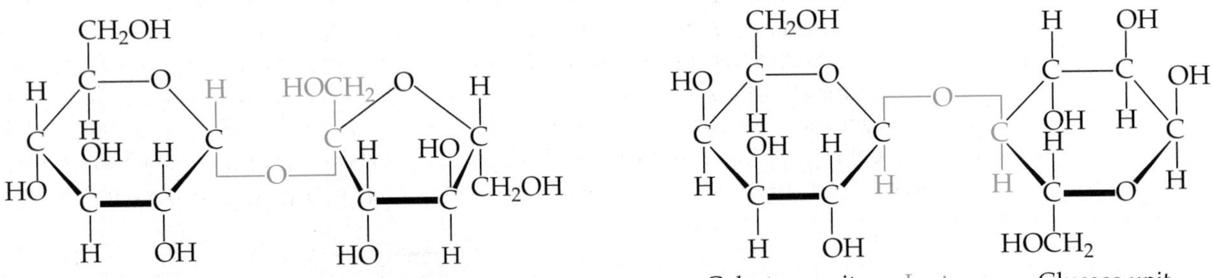

Glucose unit — Sucrose — Fructose unit

Galactose unit — Lactose — Glucose unit

▲ **Figure 25.28** The structures of two disaccharides, sucrose (left) and lactose (right).

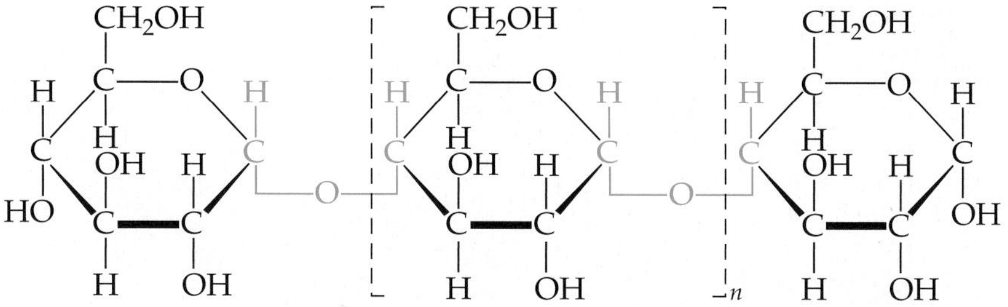

▲ **Figure 25.29** Structure of a starch molecule. The molecule consists of many units of the kind enclosed in brackets, joined by linkages of the α form. (That is, the C—O bonds on the linking carbons are on the opposite side of the ring from the CH_2OH groups.)

*invert sugar,** is sweeter to the taste than the original sucrose. The sweet syrup present in canned fruits and candies is largely invert sugar formed from hydrolysis of added sucrose.

Polysaccharides

Polysaccharides are made up of many monosaccharide units joined together by a bonding arrangement similar to those shown for the disaccharides in Figure 25.28. The most important polysaccharides are starch, glycogen, and cellulose, which are formed from repeating glucose units.

Starch is not a pure substance. The term refers to a group of polysaccharides found in plants. Starches serve as a major method of food storage in plant seeds and tubers. Corn, potatoes, wheat, and rice all contain substantial amounts of starch. These plant products serve as major sources of needed food energy for humans. Enzymes within the digestive system catalyze the hydrolysis of starch to glucose.

Some starch molecules are unbranched chains, whereas others are branched. Figure 25.29 ▲ illustrates an unbranched starch structure. Notice, in particular, that the glucose units are in the α form (that is, the bridging oxygen atom is opposite the CH_2OH groups).

Glycogen is a starchlike substance synthesized in the body. Glycogen molecules vary in molecular weight from about 5000 to more than 5 million amu. Glycogen acts as a kind of energy bank in the body. It is concentrated in the muscles and liver. In muscles it serves as an immediate source of energy; in the liver it serves as a storage place for glucose and helps to maintain a constant glucose level in the blood.

Cellulose forms the major structural unit of plants. Wood is about 50% cellulose; cotton fibers are almost entirely cellulose. Cellulose consists of an unbranched chain of glucose units, with molecular weights averaging more than 500,000 amu. The structure of cellulose is shown in Figure 25.30 ▼. At first glance

* The term *invert sugar* comes from the fact that rotation of the plane of polarized light by the glucose-fructose mixture is in the opposite direction, or inverted, from that of the sucrose solution.

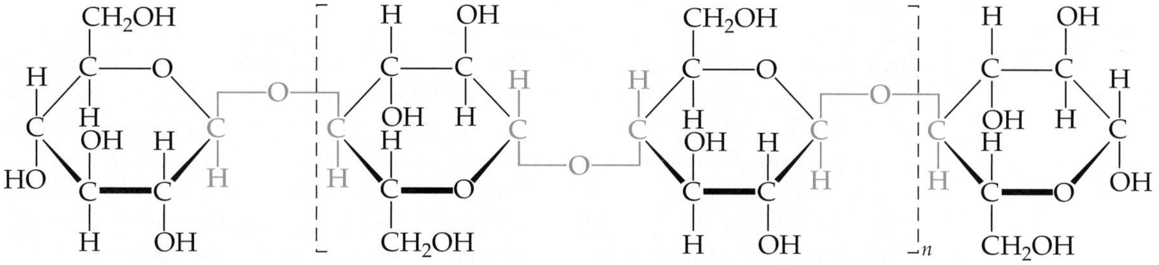

▲ **Figure 25.30** Structure of cellulose. Like starch, cellulose is a polymer. The repeating unit is shown between brackets. The linkage in cellulose is of the β form, different from that in starch (see Figure 25.29).

▶ Figure 25.31 Structures of starch (a) and cellulose (b). These representations show the geometrical arrangements of bonds about each carbon atom. The glucose rings are oriented differently with respect to one another in the two structures.

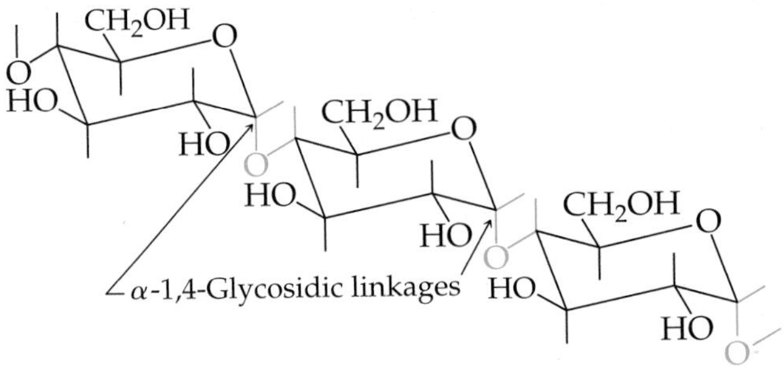

this structure looks very similar to that of starch. In cellulose, however, the glucose units are in the β form (i.e., the bridging oxygen atom is on the same side as the CH_2OH groups).

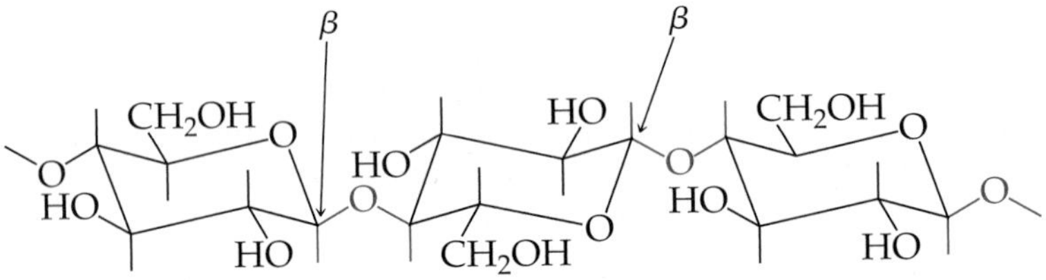

The distinction between starch and cellulose is made clearer when we examine their structures in a more realistic three-dimensional representation, as shown in Figure 25.31 ▲. The individual glucose units have different relationships to one another in the two structures. Because of this fundamental difference, enzymes that readily hydrolyze starches do not hydrolyze cellulose. Thus, you might eat a pound of cellulose and receive no caloric value from it whatsoever, even though the heat of combustion per unit weight is essentially the same for both cellulose and starch. A pound of starch, in contrast, would represent a substantial caloric intake. The difference is that the starch is hydrolyzed to glucose, which is eventually oxidized with release of energy. Cellulose, however, is not readily hydrolyzed by enzymes present in the body, and so it passes through the digestive system relatively unchanged. Many bacteria contain enzymes, called cellulases, that hydrolyze cellulose. These bacteria are present in the digestive systems of grazing animals, such as cattle, that use cellulose for food.

25.11 Nucleic Acids

Nucleic acids are a class of biopolymers that are the chemical carriers of an organism's genetic information. **Deoxyribonucleic acids (DNA)** are huge molecules (Figure 1.2c) whose molecular weights may range from 6 million to 16 million amu. **Ribonucleic acids (RNA)** are smaller molecules, with molecular weights in the range of 20,000 to 40,000 amu. Whereas DNA is found primarily in the nucleus of the cell, RNA is found mostly outside the nucleus in the *cytoplasm*, the nonnuclear material enclosed within the cell membrane. The DNA stores the genetic information of the cell and controls the production of proteins. The RNA carries the information stored by DNA out of the nucleus of the cell into the cytoplasm, where the information can be used in protein synthesis.

The monomers of nucleic acids, called **nucleotides**, are formed from the following units:

1. A phosphoric acid molecule, H_3PO_4
2. A five-carbon sugar
3. A nitrogen-containing organic base

The sugar component of RNA is *ribose*, whereas that in DNA is *deoxyribose*.

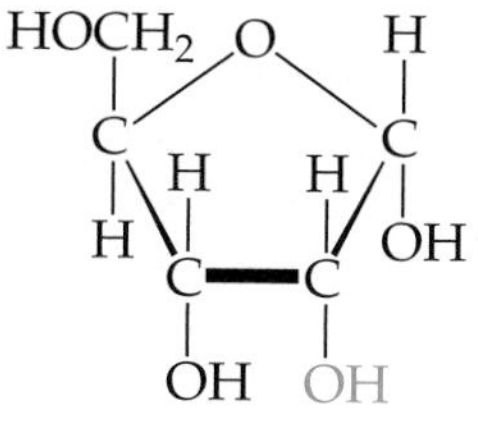

Ribose

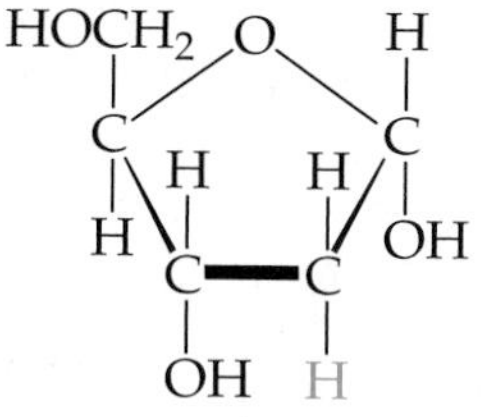

Deoxyribose

*Deoxy*ribose differs from ribose only in having one fewer oxygen atom at carbon 2.

The following nitrogen bases are found in DNA and RNA:

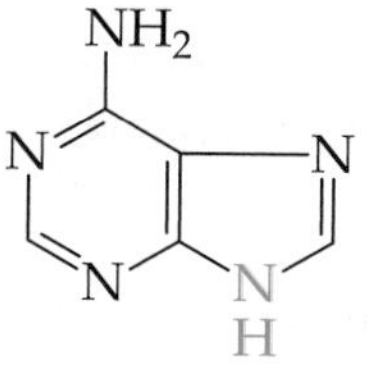

Adenine (A) DNA RNA

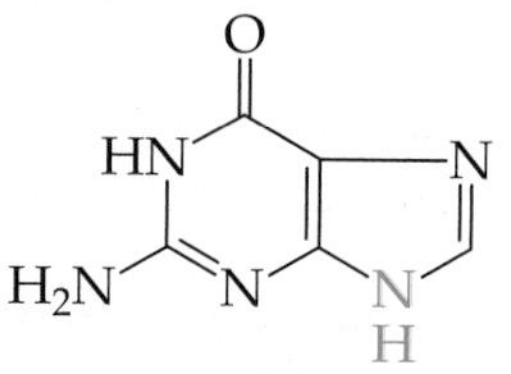

Guanine (G) DNA RNA

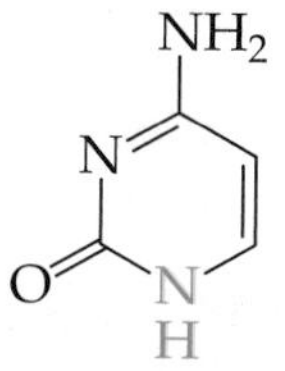

Cytosine (C) DNA RNA

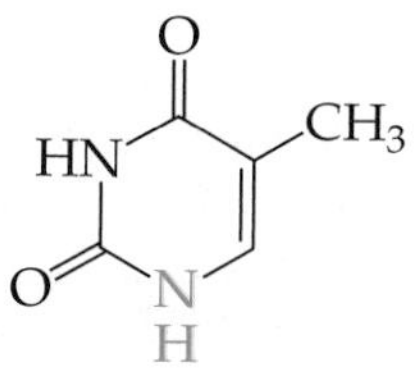

Thymine (T) DNA

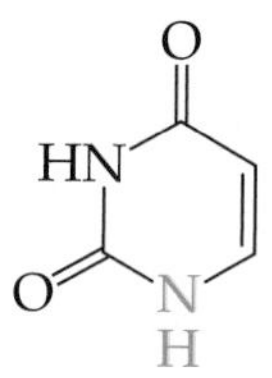

Uracil (U) RNA

The base is attached to a ribose or deoxyribose molecule through a bond to the nitrogen atom shown in color. An example of a nucleotide in which the base is adenine and the sugar is deoxyribose is shown in Figure 25.32 ▼.

Nucleic acids are polynucleotides formed by condensation reactions between an OH group of the phosphoric acid unit on one nucleotide and an OH group of

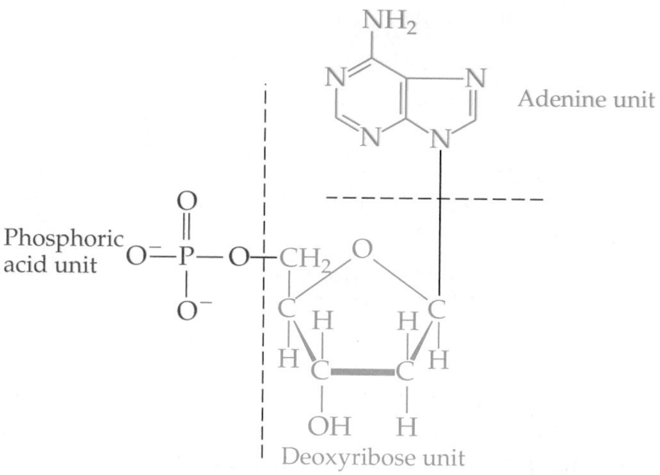

Phosphoric acid unit

Adenine unit

Deoxyribose unit

◀ **Figure 25.32** Structure of deoxyadenylic acid, a nucleotide formed from phosphoric acid, deoxyribose, and an organic base, adenine.

the sugar of another nucleotide. Figure 25.33 ◀ shows a portion of the polymeric chain of a DNA molecule.

DNA molecules consist of two deoxyribonucleic acid chains or strands that are wound together in the form of a **double helix**, as shown in Figure 25.34 ▼. The drawing on the right [Figure 25.34(b)] has been simplified to show the essential features of the structure. The sugar and phosphate groups form the backbone of each strand. The bases (represented by the letters T, A, C, and G) are attached to the sugars. The two strands are held together by attractions between the bases in one strand and those in the other strand. These attractions involve both London dispersion interactions and hydrogen bonds. (Section 11.2) As shown in Figure 25.35 ▶, the structures of thymine (T) and adenine (A) make them perfect partners for hydrogen bonding. Likewise, cytosine (C) and guanine (G) form ideal hydrogen-bonding partners. In the double-helix structure, therefore, each thymine on one strand is opposite an adenine on the other strand. Likewise, each cytosine is opposite a guanine. The double-helix structure with complementary bases on the two strands is the key to understanding how DNA functions.

The two strands of DNA unwind during cell division, and new complementary strands are constructed on the unraveling strands (Figure 25.36 ▶). This process results in two identical double-helix DNA structures, each containing one strand from the original structure and one newly synthesized strand. This replication process allows genetic information to be transmitted when cells divide. The structure of DNA is also the key to understanding protein synthesis, the means by which viruses infect cells, and many other problems of central importance to modern biology. These themes are beyond the scope of this book. If you take courses in the life sciences, however, you will learn a good deal about such matters.

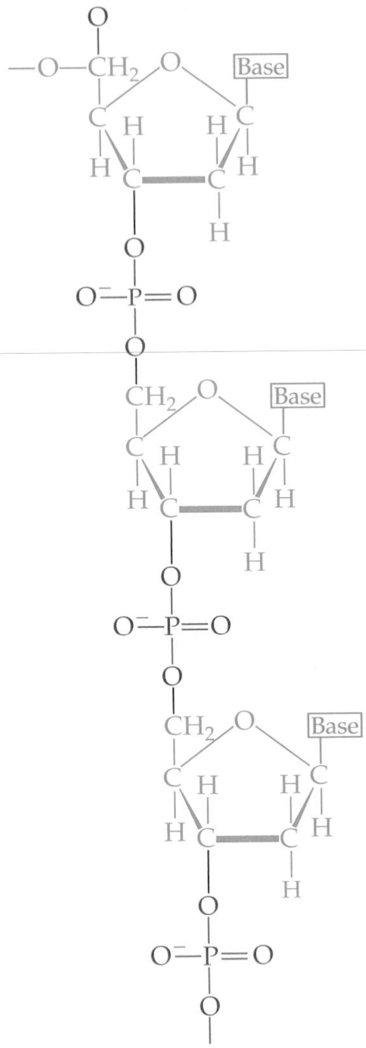

▲ **Figure 25.33** Structure of a polynucleotide. Because the sugar in each nucleotide is deoxyribose, this polynucleotide is of the form found in DNA.

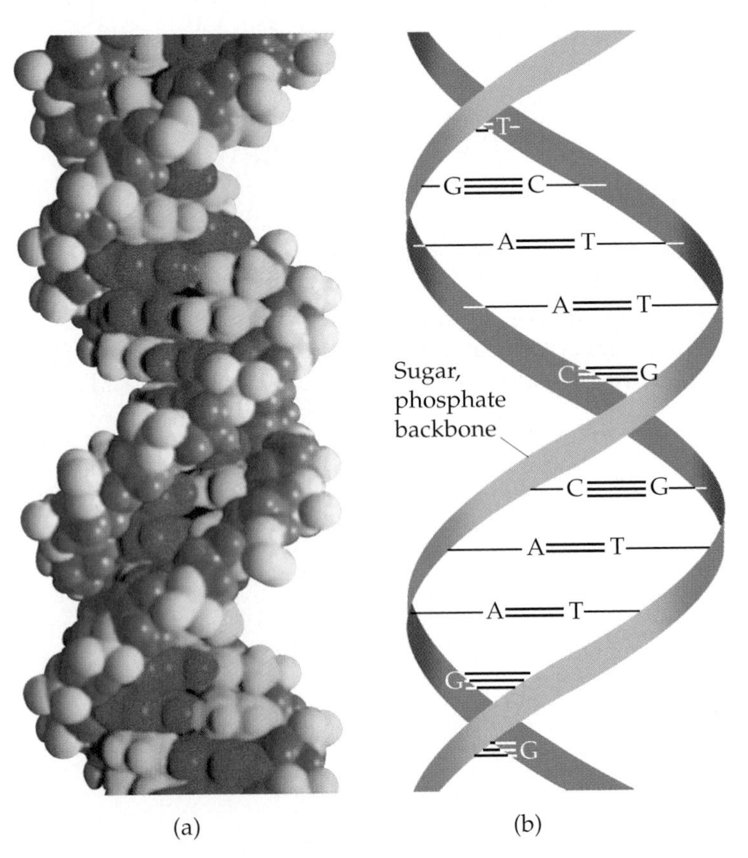

▲ **Figure 25.34** (a) A computer-generated model of a DNA double helix. The dark-blue and light-blue atoms represent the sugar-phosphate chains that wrap around the outside. Inside the chains are the bases, shown in red and yellow-green. (b) A schematic illustration of the double helix showing the hydrogen-bond interactions between complementary base pairs.

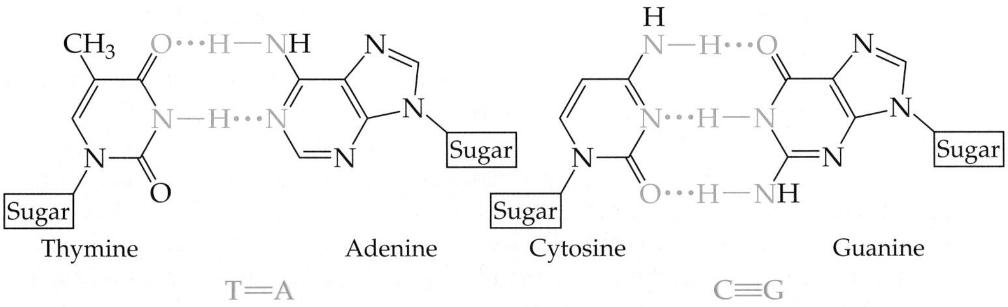

Thymine Adenine Cytosine Guanine

T=A C≡G

▲ **Figure 25.35** Hydrogen bonding between complementary base pairs. The hydrogen bonds shown here are responsible for formation of the double-stranded helical structure of DNA, as shown in Figure 25.34(b).

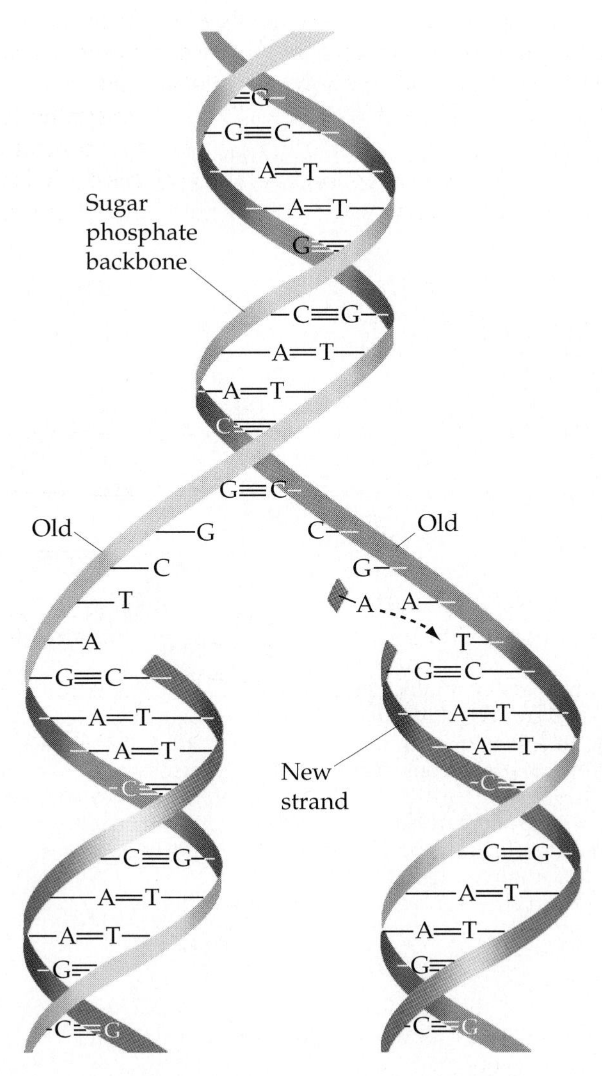

▲ **Figure 25.36** A schematic representation of DNA replication. The original DNA double helix partially unwinds, and new nucleotides line up on each strand in complementary fashion. Hydrogen bonds help align the new nucleotides with the original DNA chain. When the new nucleotides are joined by condensation reactions, two identical double-helix DNA molecules result.

SAMPLE INTEGRATIVE EXERCISE 25: Putting Concepts Together

Pyruvic acid has the following structure:

$$CH_3-\overset{\overset{\displaystyle O}{\|}}{C}-\overset{\overset{\displaystyle O}{\|}}{C}-OH$$

It is formed in the body from carbohydrate metabolism. In the muscle it is reduced to lactic acid in the course of exertion. The acid-dissociation constant for pyruvic acid is 3.2×10^{-3}. **(a)** Why does pyruvic acid have a higher acid-dissociation constant than acetic acid? **(b)** Would you expect pyruvic acid to exist primarily as the neutral acid or as dissociated ions in muscle tissue, assuming a pH of about 7.4 and an acid concentration of $2 \times 10^{-4}\ M$? **(c)** What would you predict for the solubility properties of pyruvic acid? Explain. **(d)** What is the hybridization of each carbon atom in pyruvic acid? **(e)** Assuming H atoms as the reducing agent, write a balanced chemical equation for the reduction of pyruvic acid to lactic acid (Figure 25.17). (Although H atoms don't exist as such in biochemical systems, biochemical reducing agents deliver hydrogen for such reductions.)

Solution **(a)** The acid ionization constant for pyruvic acid should be somewhat greater than that of acetic acid because the carbonyl function on the α-carbon atom exerts an electron-withdrawing effect on the carboxylic acid group. In the C—O—H bond system the electrons are shifted from hydrogen, facilitating loss of the hydrogen as a proton. (Section 16.10)

(b) To determine the extent of ionization, we first set up the ionization equilibrium and equilibrium-constant expression. Using HPv as the symbol for the acid, we have

$$HPv \rightleftharpoons H^+ + Pv^-$$

$$K_a = \frac{[H^+][Pv^-]}{[HPv]} = 3.2 \times 10^{-3}$$

Let $[Pv^-] = x$. Then the concentration of undissociated acid is $2 \times 10^{-4} - x$. The concentration of $[H^+]$ is fixed at 4.0×10^{-8} (the antilog of the pH value). Substituting, we obtain

$$3.2 \times 10^{-3} = \frac{[4.0 \times 10^{-8}][x]}{[2 \times 10^{-4} - x]}$$

Solving for x, we obtain $x[3.2 \times 10^{-3} + 4.0 \times 10^{-8}] = 6.4 \times 10^{-7}$.

The second term in the brackets is negligible compared to the first, so $x = [Pv^-] = 6.4 \times 10^{-7}/3.2 \times 10^{-3} = 2 \times 10^{-4}\ M$. This is the initial concentration of acid, which means that essentially all the acid has dissociated. We might have expected this result because the acid is quite dilute and the acid-dissociation constant is fairly high.

(c) Pyruvic acid should be quite soluble in water because it has polar functional groups and a small hydrocarbon component. It is miscible with water, ethanol, and diethyl ether.

(d) The methyl group carbon has sp^3 hybridization. The carbon carrying the carbonyl group has sp^2 hybridization because of the double bond to oxygen. Similarly, the carboxylic acid carbon is sp^2 hybridized.

(e) The balanced chemical equation for this reaction is

$$CH_3\overset{\overset{\displaystyle O}{\|}}{C}COOH + 2(H) \longrightarrow CH_3\underset{\underset{\displaystyle H}{|}}{\overset{\overset{\displaystyle OH}{|}}{C}}COOH$$

Essentially, the ketonic functional group has been reduced to an alcohol.

Strategies in Chemistry What Now?

If you are reading this box, you have made it to the end of our text. We congratulate you on the tenacity and dedication that you have exhibited to make it this far!

As an epilogue, we offer the ultimate study strategy in the form of a question: What do you plan to do with the knowledge of chemistry that you have gained thus far in your studies? Many of you will enroll in additional courses in chemistry as part of your required curriculum. For others this will be the last formal course in chemistry that you will take. Regardless of the career path you plan to take—whether it's chemistry, one of the biomedical fields, engineering, the liberal arts, or whatever—we hope that this text has increased your appreciation of the chemistry in the world around you. If you pay attention, you will be aware of encounters with chemistry on a daily basis, from food and pharmaceutical labels to gasoline pumps to sports equipment to news reports.

We have also tried to give you a sense of the dynamic nature of chemistry. Chemistry is constantly changing. Research chemists synthesize new compounds, develop new reactions, uncover chemical properties that were previously unknown, find new applications for known compounds, and refine theories. You may wish to participate in the fascinating venture of chemical research by taking part in an undergraduate research program. Given all the answers that chemists seem to have, you may be surprised at the large number of questions that they still find to ask.

Finally, we hope you have enjoyed using this textbook. We certainly enjoyed putting so many of our thoughts about chemistry on paper. We truly believe it to be the central science, one that benefits all who learn about it and from it.

Summary and Key Terms

Introduction and Section 25.1 This chapter introduces **organic chemistry**, which is the study of carbon compounds (typically compounds containing carbon–carbon bonds), and **biochemistry**, which is the study of the chemistry of living organisms. We have encountered many aspects of organic chemistry in earlier chapters. Carbon forms four bonds in its stable compounds. The C—C single bonds and the C—H bonds tend to have low reactivity. Those bonds that have a high electron density (such as multiple bonds or bonds with an atom of high electronegativity) tend to be the sites of reactivity in an organic compound. These sites of reactivity are called **functional groups**.

Section 25.2 The simplest types of organic compounds are hydrocarbons, those composed of only carbon and hydrogen. There are four major kinds of hydrocarbons: alkanes, alkenes, alkynes, and aromatic hydrocarbons. **Alkanes** are composed of only C—H and C—C single bonds. **Alkenes** contain one or more carbon–carbon double bonds. **Alkynes** contain one or more carbon–carbon triple bonds. **Aromatic hydrocarbons** contain cyclic arrangements of carbon atoms bonded through both σ and delocalized π bonds. Alkanes are saturated hydrocarbons; the others are unsaturated.

Section 25.3 Alkanes may form straight-chain, branched-chain, and cyclic arrangements. Isomers are substances that possess the same molecular formula, but differ in the arrangements of atoms. In **structural isomers** the bonding arrangements of the atoms differ. Different isomers are given different systematic names. The naming of hydrocarbons is based on the longest continuous chain of carbon atoms in the structure. The locations of **alkyl groups**, which branch off the chain, are specified by numbering along the carbon chain. Alkanes with ring structures are called **cycloalkanes**. Alkanes are relatively unreactive. They do, however, undergo combustion in air, and their chief use is as sources of heat energy produced by combustion.

Section 25.4 The names of alkenes and alkynes are based on the longest continuous chain of carbon atoms that contains the multiple bond, and the location of the multiple bond is specified by a numerical prefix. Alkenes exhibit not only structural isomerism but geometric (*cis-trans*) isomerism as well. In **geometric isomers** the bonds are the same, but the molecules have different geometries. Geometric isomerism is possible in alkenes because rotation about the C=C double bond is restricted.

Alkenes and alkynes readily undergo **addition reactions** to the carbon–carbon multiple bonds. Additions of acids, such as HBr, proceed via a rate-determining step in which a proton is transferred to one of the alkene or alkyne carbon atoms. Addition reactions are difficult to carry out with aromatic hydrocarbons, but **substitution reactions** are easily accomplished in the presence of catalysts.

Sections 25.5 and 25.6 The chemistry of organic compounds is dominated by the nature of their functional groups. The functional groups we have considered are

R—O—H (Alcohol)

R—C(=O)—H (Aldehyde)

>C=C< (Alkene)

—C≡C— (Alkyne)

R—C(=O)—N< (Amide)

R—N(—R′ (or H))—R″ (or H) (Amine)

R—C(=O)—O—H (Carboxylic acid)

R—C(=O)—O—R′ (Ester)

R—O—R′ (Ether)

R—C(=O)—R′ (Ketone)

R, R′, and R″ represent hydrocarbon groups—for example, methyl (CH_3) or phenyl (C_6H_5).

Alcohols are hydrocarbon derivatives containing one or more OH groups. **Ethers** are formed by a condensation reaction of two molecules of alcohol. Several functional groups contain the **carbonyl (C=O) group**, including **aldehydes, ketones, carboxylic acids, esters,** and **amides.** Aldehydes and ketones can be produced by oxidation of certain alcohols. Further oxidation of the aldehydes produces carboxylic acids. Carboxylic acids can form esters by a condensation reaction with alcohols, or they can form amides by a condensation reaction with amines. Esters undergo hydrolysis (**saponification**) in the presence of strong bases.

Section 25.7 Molecules that possess nonsuperimposable mirror images are termed **chiral**. The two nonsuperimposable forms of a chiral molecule are called *enantiomers*. In carbon compounds a chiral center is created when all four groups bonded to a central carbon atom are different, as in 2-bromobutane. Many of the molecules occurring in living systems, such as the amino acids, are chiral and exist in nature in only one enantiomeric form. Many drugs of importance in human medicine are chiral, and the enantiomers may produce very different biochemical effects. For this reason, synthesis of only the effective isomers of chiral drugs has become a high priority.

Sections 25.8 and 25.9 Many of the molecules that are essential for life are large natural polymers that are constructed from smaller molecules called monomers. Three of these **biopolymers** were considered in this chapter: proteins, polysaccharides (carbohydrates), and nucleic acids.

Proteins are polymers of **amino acids**. They are the major structural materials in animal systems. All naturally occurring proteins are formed from 20 amino acids. The amino acids are linked by **peptide bonds**. A **polypeptide** is a polymer formed by linking many amino acids by peptide bonds.

Amino acids are chiral substances. Usually only one of the enantiomers is found to be biologically active. Protein structure is determined by the sequence of amino acids in the chain (its **primary structure**), the coiling or stretching of the chain (its **secondary structure**), and the overall shape of the complete molecule (its **tertiary structure**). One of the most important secondary structure arrangements is the **alpha (α) helix**.

Section 25.10 **Carbohydrates**, which are polyhydroxy aldehydes and ketones, are the major structural constituents of plants and are a source of energy in both plants and animals. **Glucose** is the most common **monosaccharide** or simple sugar. Two monosaccharides can be linked together by means of a condensation reaction to form a disaccharide. **Polysaccharides** are complex carbohydrates made up of many monosaccharide units joined together. The three most important polysaccharides are **starch**, which is found in plants; **glycogen**, which is found in mammals; and **cellulose**, which is also found in plants.

Section 25.11 **Nucleic acids** are biopolymers that carry the genetic information necessary for cell reproduction; they also control cell development through control of protein synthesis. The building blocks of these biopolymers are **nucleotides**. There are two types of nucleic acids, the **ribonucleic acids (RNA)** and the **deoxyribonucleic acids (DNA)**. These substances consist of a polymeric backbone of alternating phosphate and ribose or deoxyribose sugar groups, with organic bases attached to the sugar molecules. The DNA polymer is a double-stranded helix (a **double helix**) held together by hydrogen bonding between matching organic bases situated across from one another on the two strands. The hydrogen bonding between specific base pairs is the key to gene replication and protein synthesis.

Exercises

Introduction to Organic Compounds; Hydrocarbons

25.1 Predict the ideal values for the bond angles about each carbon atom in the propanal molecule. Indicate the hybridization of orbitals for each carbon.

$$CH_3CH_2\overset{\overset{\displaystyle O}{\|}}{C}H$$

25.2 Identify the carbon atom(s) in the structure shown that has (have) each of the following hybridizations: **(a)** sp^3; **(b)** sp; **(c)** sp^2.

$$N{\equiv}C{-}CH_2{-}CH_2{-}CH{=}CH{-}\underset{\underset{\displaystyle \overset{|}{H}}{\displaystyle C{=}O}}{\overset{}{\underset{|}{C}HOH}}$$

25.3 List five elements that are commonly found in organic compounds. Which ones are more electronegative than carbon?

25.4 Which of the following bonds in an organic compound would you expect to be reactive? C—C, C—O, C—H, C—Cl. Explain.

25.5 **(a)** What is the difference between a straight-chain and branched-chain alkane? **(b)** What is the difference between an alkane and an alkyl group? **(c)** Why are alkanes said to be saturated?

25.6 What structural features help us identify a compound as **(a)** an alkane; **(b)** a cylcoalkane; **(c)** an alkene; **(d)** an alkyne; **(e)** a saturated hydrocarbon; **(f)** an aromatic hydrocarbon?

25.7 Give the molecular formula of a hydrocarbon containing five carbon atoms that is **(a)** an alkane; **(b)** a cycloalkane;

(c) an alkene; **(d)** an alkyne. Which are saturated and which are unsaturated hydrocarbons?

25.8 Give the molecular formula of a cyclic alkane, a cyclic alkene, a linear alkyne, and an aromatic hydrocarbon that in each case contains six carbon atoms. Which are saturated and which are unsaturated hydrocarbons?

25.9 Give the general formula for any dialkene; that is, a straight-chain hydrocarbon with two double bonds along the chain.

25.10 Give the general formula for any cyclic alkene; that is, a cyclic hydrocarbon with one double bond.

25.11 Draw all the possible noncyclic structural isomers of C_5H_{10}. Name each compound.

25.12 Write the condensed structural formulas for as many alkenes and alkynes as you can think of that have the molecular formula C_6H_{10}.

25.13 All the structures that follow have the same molecular formula, C_8H_{18}. Which structures are the same molecule? (*Hint:* One way to do this is to determine the chemical name for each.)

CH_3
(a) $CH_3CCH_2CHCH_3$
CH_3 CH_3

CH_3 CH_3
(b) $CH_3CHCHCH_2$
CH_2
CH_3

CH_3
(c) $CH_3CHCHCH_3$
$CHCH_3$
CH_3

CH_3CHCH_3
(d) $CH_3CHCHCH_3$
CH_3

25.14 Which of the structures that follow represent the same substance? (See the hint in Exercise 25.13.)

(a) $CH_3CH_2CHCH_2CH_3$
CH_2
CH_3

CH_3
(b) $CH_3CH_2CHCHCH_3$
CH_3

CH_3
(c) $CH_3CH_2CCH_2CH_3$
CH_3

(d) $CH_3CHCH_2CH_3$
CH_3CH
CH_3

25.15 What are the approximate bond angles **(a)** about carbon in an alkane; **(b)** about a doubly bonded carbon atom in an alkene; **(c)** about a triply bonded carbon atom in an alkyne?

25.16 What are the characteristic hybrid orbitals employed by **(a)** carbon in an alkane; **(b)** carbon in a double bond in an alkene; **(c)** carbon in the benzene ring; **(d)** carbon in a triple bond in an alkyne?

25.17 Draw the structural formula or give the name, as appropriate, for the following:

CH_3 H H H H
(a) H—C—C—C—C—C—H
H H H CH_3 H

CH_3
(b) $CH_3CH_2CH_2CH_2CH_2CH_2CCH_2CHCH_3$
CH_2 CH_3
CH_3

(c) 3-methylhexane
(d) 4-ethyl-2,2-dimethyloctane
(e) methylcyclohexane

25.18 Draw the structural formula or give the name, as appropriate, for the following:

CH_3CH_2 CH_2CH_3
(a) CH_3CCH_2CH
CH_3 CH_3

CH_3
(b) $CH_3CH_2CH_2CCH_3$
$CH_3CHCH_2CH_3$

(c) 2,5-dimethylnonane
(d) 3-ethyl-4,4-dimethylheptane
(e) 1-ethyl-4-methylcyclohexane

25.19 Name the following compounds:

(a) CH_3CHCH_3
$CHCH_2CH_2CH_2CH_3$
CH_3

(b) CH_3CH_2(H)C=C(H)$CH_2CHCH_2CH_3$ (with CH_3 on the CH)

(c) Br–(benzene ring)–Br (para)

CH_2CH_3
(d) $HC{\equiv}CCH_2CCH_3$
CH_3

(e) (cyclobutane)—CH_3

25.20 Name the following compounds:

(a) CH_2—CH_2
Cl—CH HC—Cl
CH_2—CH_2

(b) $HC{\equiv}C—CH_2—Cl$

(c) *trans*-$CH_3CH{=}CHCH_2CH_2CH_3$

CH_3
(d) CH_3CH_2—C—CH_2Cl
(benzene ring)

(e) *cis*-$CH_2{=}CH—CH{=}CH—CH_2Cl$

25.21 Why is geometric isomerism possible for alkenes, but not for alkanes and alkynes?

25.22 Using butene as an example, distinguish between structural and geometric isomers.

25.23 Indicate whether each of the following molecules is capable of geometrical (*cis-trans*) isomerism. For those that are, draw the structures of each: **(a)** 1,1-dichloro-1-butene; **(b)** 2,4-dichloro-2-butene; **(c)** 1,4-dichlorobenzene; **(d)** 4,5-dimethyl-2-pentyne.

25.24 Draw all the distinct geometric isomers of 2,4-hexadiene.

25.25 What is the octane number of a mixture of 35% heptane and 65% isooctane?

25.26 Describe two ways in which the octane number of a gasoline consisting of alkanes can be increased.

Reactions of Hydrocarbons

25.27 **(a)** What is the difference between a substitution reaction and an addition reaction? Which one is commonly observed with alkenes, and which one with aromatic hydrocarbons? **(b)** Using condensed structural formulas, write the balanced equation for the addition reaction of 2,3-dimethyl-2-butene with Br_2. **(c)** Write a balanced chemical equation for the substitution reaction of Cl_2 with *p*-dichlorobenzene in the presence of $FeCl_3$ as a catalyst.

25.28 Using condensed structural formulas, write a balanced chemical equation for each of the following reactions: **(a)** hydrogenation of cyclohexene; **(b)** addition of H_2O to *trans*-2-pentene using H_2SO_4 as a catalyst (two products); **(c)** reaction of 2-chloropropane with benzene in the presence of $AlCl_3$.

25.29 **(a)** When cyclopropane is treated with HI, 1-iodopropane is formed. A similar type of reaction does not occur with cyclopentane or cyclohexane. How do you account for the reactivity of cyclopropane? **(b)** Suggest a method of preparing ethylbenzene, starting with benzene and ethylene as the only organic reagents.

25.30 **(a)** One test for the presence of an alkene is to add a small amount of bromine and look for the disappearance of the brown color. This test does not work for detecting the presence of an aromatic hydrocarbon. Explain. **(b)** Write a series of reactions leading to *para*-bromoethylbenzene, beginning with benzene and using other reagents as needed. What isomeric side products might also be formed?

25.31 Describe the intermediate that is thought to form in the addition of a hydrogen halide to an alkene, using cyclohexene as the alkene in your description.

25.32 The rate law for addition of Br_2 to an alkene is first order in Br_2 and first order in alkene. Does this fact prove that the mechanism of addition of Br_2 to an alkene proceeds in the same manner as for addition of HBr? Explain.

25.33 The molar heat of combustion of gaseous cyclopropane is -2089 kJ/mol; that for gaseous cyclopentane is -3317 kJ/mol. Calculate the heat of combustion per CH_2 group in the two cases, and account for the difference.

25.34 The heat of combustion of decahydronaphthalene, ($C_{10}H_{18}$) is -6286 kJ/mol. The heat of combustion of naphthalene ($C_{10}H_8$) is -5157 kJ/mol. (In both cases $CO_2(g)$ and $H_2O(l)$ are the products.) Using these data and data in Appendix C, calculate the heat of hydrogenation of naphthalene. Does this value provide any evidence for aromatic character in naphthalene?

Functional Groups and Chirality

25.35 Identify the functional groups in each of the following compounds:

(a) $CH_3\underset{\underset{O}{\|}}{C}CH_2CH_3$ **(b)** $CH_3\underset{\underset{OH}{|}}{C}{=}O$ **(c)** $\underset{\underset{OH}{|}}{CH_2}CH_2CH_3$

(d) $CH_3O\underset{\underset{O}{\|}}{C}CH_2CH_3$ **(e)** $H_2N\underset{\underset{O}{\|}}{C}CH_3$ **(f)** $CH_3CH_2NHCH_3$

25.36 Identify the functional groups in each of the following compounds:

(a) $HC{\equiv}C{-}CH_2{-}\overset{\overset{O}{\|}}{C}{-}H$

(b) 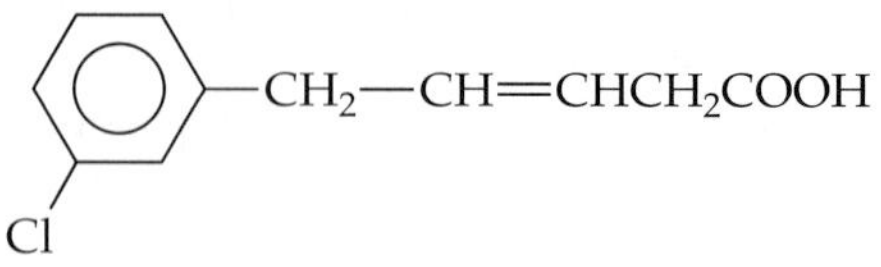 (benzene ring, Cl)$-CH_2-CH{=}CHCH_2COOH$

(c) 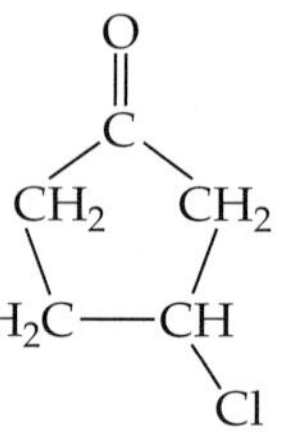 Cyclic: $\overset{O}{\|}$ C; CH_2 CH_2; $H_2C{-}CH$; Cl

(d) $CH_3CH{=}CH\overset{\overset{O}{\|}}{C}{-}OCH_2CH_3$

(e) 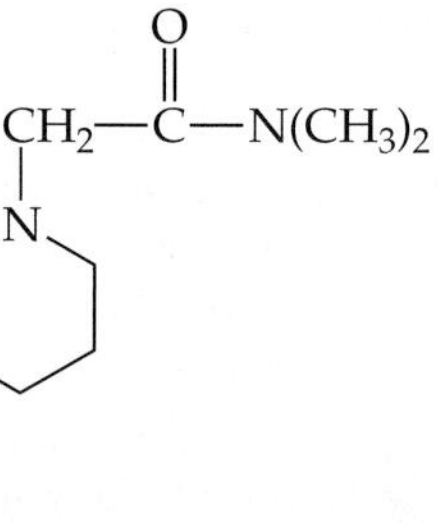 $\underset{\underset{N}{|}}{CH_2}{-}\overset{\overset{O}{\|}}{C}{-}N(CH_3)_2$ (N in a six-membered ring)

25.37 Give the structural formula for **(a)** an aldehyde that is an isomer of acetone; **(b)** an ether that is an isomer of 1-propanol.

25.38 **(a)** Give the empirical formula and structural formula for a cyclic ether containing four carbon atoms in the ring. **(b)** Write the structural formula for a straight-chain compound that is a structural isomer of your answer to part (a).

25.39 The IUPAC name for a carboxylic acid is based on the name of the hydrocarbon with the same number of carbon atoms. The ending *-oic* is appended, as in ethanoic acid, which is the IUPAC name for acetic acid,

$$CH_3\overset{\overset{\large O}{\|}}{C}OH$$

Give the IUPAC name for each of the following acids:

(a) $H\overset{\overset{O}{\|}}{C}OH$

(b) $CH_3CH_2CH_2\overset{\overset{O}{\|}}{C}OH$

(c) $CH_3CH_2\underset{\underset{CH_3}{|}}{C}HCH_2\overset{\overset{O}{\|}}{C}OH$

25.40 Aldehydes and ketones can be named in a systematic way by counting the number of carbon atoms (including the carbonyl carbon) that they contain. The name of the aldehyde or ketone is based on the hydrocarbon with the same number of carbon atoms. The ending *-al*, for aldehyde, or *-one*, for ketone, is added as appropriate. Draw the structural formulas for the following aldehydes or ketones: **(a)** propanal; **(b)** 2-pentanone; **(c)** 3-methyl-2-butanone; **(d)** 2-methylbutanal.

25.41 Draw the condensed structure of the compounds formed by condensation reactions between **(a)** benzoic acid and ethanol; **(b)** ethanoic acid and methylamine; **(c)** acetic acid and phenol. Name the compound in each case.

25.42 Draw the condensed structures of the esters formed from **(a)** butanoic acid and methanol; **(b)** benzoic acid and 2-propanol; **(c)** propanoic acid and dimethylamine. Name the compound in each case.

25.43 Write a balanced chemical equation using condensed structural formulas for the saponification (base hydrolysis) of **(a)** methyl propionate; **(b)** phenyl acetate.

25.44 Write a balanced chemical equation using condensed structural formulas for **(a)** the formation of propyl acetate from the appropriate acid and alcohol; **(b)** the saponification (base hydrolysis) of methyl benzoate.

25.45 Write the condensed structural formula for each of the following compounds: **(a)** 2-butanol; **(b)** 1,2-ethanediol; **(c)** methyl formate; **(d)** diethyl ketone; **(e)** diethyl ether.

25.46 Write the condensed structural formula for each of the following compounds: **(a)** 3,3-dichlorobutyraldehyde; **(b)** methyl phenyl ketone; **(c)** *para*-bromobenzoic acid; **(d)** methyl-*trans*-2-butenyl ether; **(e)** *N,N*-dimethylbenzamide.

25.47 **(a)** Draw the structure for 2-bromo-2-chloro-3-methylpentane, and indicate any chiral carbons in the molecule. **(b)** Does 3-chloro-3-methylhexane have optical isomers? Why or why not?

25.48 **(a)** Identify the compounds in Exercise 25.18 that have chiral carbon atoms. **(b)** Do any of the compounds in Exercise 25.20 have optical isomers?

Proteins

25.49 **(a)** What is an α-amino acid? **(b)** How do amino acids react to form proteins?

25.50 What properties of the side chains (R groups) of amino acids affect their behavior? Give examples to illustrate your reply.

25.51 Draw the two possible dipeptides formed by condensation reactions between glycine and valine.

25.52 Write a chemical equation for the formation of alanylserine from the constituent amino acids.

25.53 **(a)** Draw the condensed structure of the tripeptide Ala-Glu-Lys. **(b)** How many different tripeptides can be made from the amino acids serine and phenylalanine? Give the abbreviations for each of these tripeptides, using the three-letter codes for amino acids.

25.54 **(a)** What amino acids would be obtained by hydrolysis of the following tripeptide?

$$H_2N\underset{\underset{(CH_3)_2CH}{|}}{C}H\overset{\overset{O}{\|}}{C}NH\underset{\underset{H_2COH}{|}}{C}H\overset{\overset{O}{\|}}{C}NH\underset{\underset{H_2CCH_2\underset{\underset{O}{\|}}{C}OH}{|}}{C}H\overset{\overset{O}{\|}}{C}OH$$

(b) How many different tripeptides can be made from the amino acids glycine, serine, and glutamic acid? Give the abbreviation for each of these tripeptides, using the three-letter codes for amino acids.

25.55 Describe the primary, secondary, and tertiary structures of proteins.

25.56 Describe the role of hydrogen bonding in determining the α-helix structure of a protein.

Carbohydrates

25.57 In your own words, define the following terms: **(a)** carbohydrate; **(b)** monosaccharide; **(c)** disaccharide.

25.58 What is the difference between α-glucose and β-glucose? Show the condensation of two glucose molecules to form a disaccharide with an α linkage; with a β linkage.

25.59 The structural formula for the linear form of galactose is as follows:

```
        O
        ‖
        CH
        |
    H—C—OH
        |
   HO—C—H
        |
   HO—C—H
        |
    H—C—OH
        |
       CH2OH
```

(a) How many chiral carbons are present in the molecule? **(b)** Draw the structure of the six-member-ring form of this sugar.

25.60 The structural formula for the linear form of D-mannose is as follows:

```
        O
        ‖
        CH
        |
   HO—C—H
        |
   HO—C—H
        |
    H—C—OH
        |
    H—C—OH
        |
       CH2OH
```

(a) How many chiral carbons are present in the molecule? **(b)** Draw the structure of the six-member-ring form of this sugar.

25.61 What is the empirical formula of glycogen? What is the unit that forms the basis of the glycogen polymer? What form of linkage joins these monomeric units?

25.62 What is the empirical formula of cellulose? What is the unit that forms the basis of the cellulose polymer? What form of linkage joins these monomeric units?

Nucleic Acids

25.63 Describe a nucleotide. Draw the structural formula for deoxycytidine monophosphate, analogous to deoxyadenylic acid, in which cytosine is the organic base.

25.64 A nucleoside consists of an organic base of the kind shown in Section 25.11, bound to ribose or deoxyribose. Draw the structure for deoxyguanosine, formed from guanine and deoxyribose.

25.65 Write a balanced chemical equation using condensed formulas for the condensation reaction between a mole of deoxyribose and a mole of doubly ionized phosphoric acid, HPO_4^{2-}.

25.66 A nucleotide undergoes hydrolysis under neutral conditions to yield 1 mol of $H_2PO_4^-$ and an organic product. The same starting material undergoes hydrolysis under acidic conditions to yield thymidine and ribose-monophosphate. Draw the structure of the unknown substance.

25.67 When samples of double-stranded DNA are analyzed, the quantity of adenine present equals that of thymine. Similarly, the quantity of guanine equals that of cytosine. Explain the significance of these observations.

25.68 Imagine a single DNA strand containing a section with the following base sequence: A, C, T, C, G, A. What is the base sequence of the complementary strand?

Additional Exercises

25.69 Draw the condensed structural formulas for two molecules with the formula C_3H_4O.

25.70 How many structural isomers are there for a five-member carbon chain with one double bond? For a six-member carbon chain with two double bonds?

25.71 There are no known stable cyclic compounds with ring sizes of seven or less that have an alkyne linkage in the ring. Why is this? Could a ring with a larger number of carbon atoms accommodate an alkyne linkage? Explain how you would use ball-and-stick molecular models to try to answer this question.

25.72 Draw the combined Lewis and condensed structural formulas for the *cis* and *trans* isomers of 2-pentene. Can cyclopentene exhibit *cis-trans* isomerism? Explain.

25.73 Why do alkenes but not alkanes or alkynes, exhibit *cis-trans* isomerism?

25.74 Explain why *trans*-1,2-dichloroethene has no dipole moment, whereas *cis*-1,2-dichloroethene has a dipole moment.

25.75 Write the structural formulas for as many alcohols as you can think of that have empirical formula C_3H_6O.

25.76 How many molecules of HBr would you expect to react readily with each molecule of styrene?

C_6H_5—CH=CH_2

[25.77] Dinitromethane, $(NO_2)_2CH_2$, is a dangerously reactive substance that decomposes readily on warming. On the other hand, dichloromethane is relatively unreactive. Why is the nitro compound so reactive compared to the chloro compound? (*Hint:* Consider the oxidation numbers of the atoms involved and the possible products of decomposition.)

25.78 Identify each of the functional groups in the following molecules:

(a) $CH_2{=}CH{-}O{-}CH{=}CH_2$ (an anesthetic)

(b) 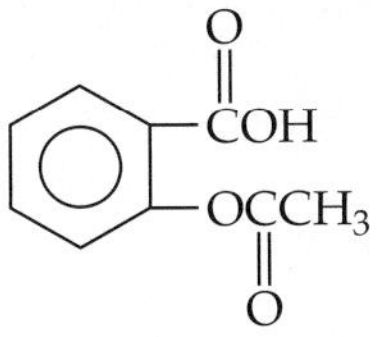 (acetylsalicylic acid, aspirin) [benzene ring bearing $\overset{O}{\overset{\|}{C}}OH$ and $O\overset{\|}{\underset{O}{C}}CH_3$]

(c) 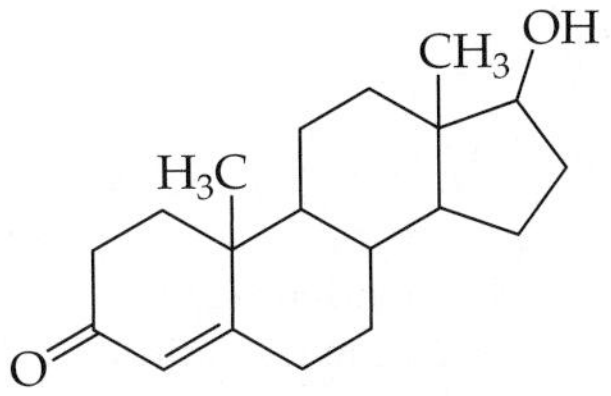 (testosterone, a male sex hormone) [steroid skeleton with CH_3, OH, H_3C, and O]

25.79 Write a condensed structural formula for each of the following: **(a)** an acid with the formula $C_4H_8O_2$; **(b)** a cyclic ketone with the formula C_5H_8O; **(c)** a dihydroxy compound with formula $C_3H_8O_2$; **(d)** a cyclic ester with formula $C_5H_8O_2$.

25.80 Although carboxylic acids and alcohols both contain an —OH group, one is acidic in water and the other is not. Explain the difference.

25.81 Give the condensed formulas for the carboxylic acid and the alcohol from which each of the following esters is formed:

(a) 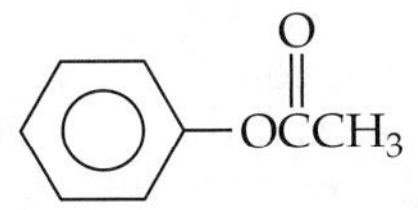 [benzene ring]—$O\overset{O}{\overset{\|}{C}}CH_3$ **(b)** 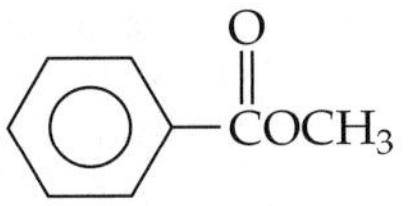 [benzene ring]—$\overset{O}{\overset{\|}{C}}OCH_3$

[25.82] Indole smells rather terrible in high concentrations, but has a pleasant floral-like odor when highly diluted. It has the following structure:

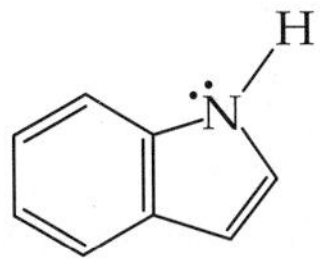

[indole structure with N–H]

Indole is a planar molecule. The nitrogen is a very weak base, with a K_b of 2×10^{-12}. Explain how this information indicates that the indole molecule is aromatic in character.

25.83 Locate the chiral carbon atoms, if any, in each of the following substances:

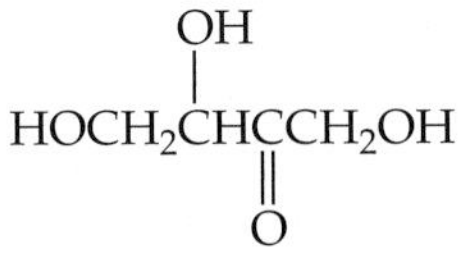

(a) 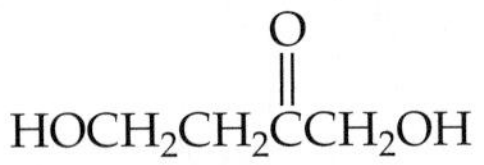 $HOCH_2CH_2\overset{O}{\overset{\|}{C}}CH_2OH$ **(b)** $HOCH_2\overset{OH}{\overset{|}{C}H}\underset{O}{\underset{\|}{C}}CH_2OH$

(c) $HO\overset{O}{\overset{\|}{C}}\underset{NH_2}{\underset{|}{C}H}\overset{CH_3}{\overset{|}{C}H}C_2H_5$

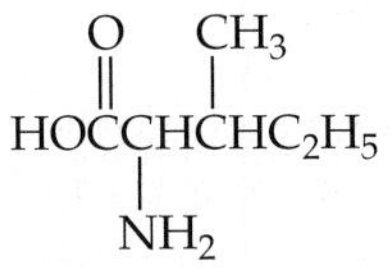

25.84 Draw the condensed structural formula of each of the following tripeptides: **(a)** Val-Gly-Asp; **(b)** Phe-Ser-Ala.

25.85 Glutathione is a tripeptide found in most living cells. Partial hydrolysis yields Cys-Gly and Glu-Cys. What structures are possible for glutathione?

25.86 Starch, glycogen, and cellulose are all polymers of glucose. What are the structural differences among them?

25.87 Monosaccharides can be categorized in terms of the number of carbon atoms (pentoses have five carbons and hexoses have six carbons) and according to whether they contain an aldehyde (*aldo-* prefix, as in aldopentose) or ketone group (*keto-* prefix, as in ketopentose). Classify glucose and fructose (Figure 25.26) in this way.

25.88 Write a complementary nucleic acid strand for the following strand, using the concept of complementary base pairing: GGTACT.

Integrative Exercises

25.89 Explain why the boiling point of ethanol (78°C) is much higher than that of its isomer, dimethyl ether (−25°C), and why the boiling point of CH_2F_2 (−52°C) is far above that of CF_4 (−128°C).

[25.90] An unknown organic compound is found on elemental analysis to contain 68.1% carbon, 13.7% hydrogen, and 18.2% oxygen by mass. It is slightly soluble in water. Upon careful oxidation it is converted into a compound that behaves chemically like a ketone and contains 69.7% carbon, 11.7% hydrogen, and 18.6% oxygen by mass. Indicate two or more reasonable structures for the unknown.

[25.91] An organic compound is analyzed and found to contain 66.7% carbon, 11.2% hydrogen, and 22.2% oxygen by mass. The compound boils at 79.6°C. At 100°C and 0.970 atm the vapor has a density of 2.28 g/L. The compound has a carbonyl group and cannot be oxidized to a carboxylic acid. Suggest a structure for the compound.

[25.92] An unknown substance is found to contain only carbon and hydrogen. It is a liquid that boils at 49°C at 1 atm pressure. Upon analysis it is found to contain 85.7% carbon and 14.3% hydrogen by mass. At 100°C and 735 torr the vapor of this unknown has a density of 2.21 g/L. When it is dissolved in hexane solution and bromine water is added, no reaction occurs. What is the identity of the unknown compound?

25.93 The standard free energy of formation of solid glycine is -369 kJ/mol, whereas that of solid glycylglycine is -488 kJ/mol. What is $\Delta G°$ for the condensation of glycine to form glycylglycine?

25.94 One of the most important molecules in biochemical systems is adenosine triphosphate (ATP), for which the structure is

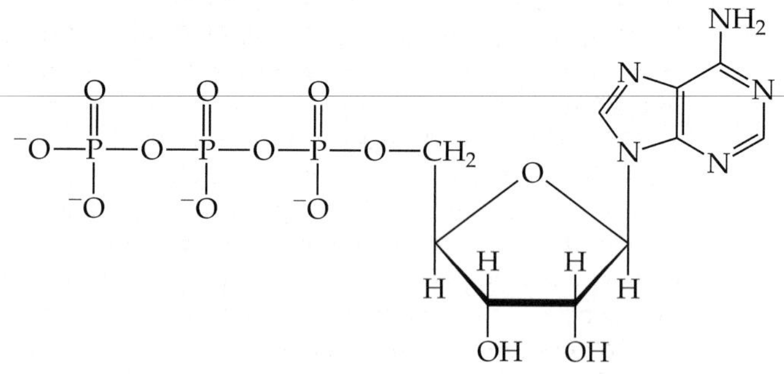

ATP

ATP is the principal carrier of biochemical energy. It is considered an energy-rich compound because the hydrolysis of ATP to yield adenosine diphosphate (ADP) and inorganic phosphate is spontaneous under aqueous biochemical conditions. **(a)** Write a balanced equation for the reaction of ATP with water to yield ADP and inorganic phosphate ion. [*Hint:* Hydrolysis reactions are just the reverse of condensation reactions (Section 22.8).] **(b)** What would you expect for the sign of the free-energy change for this reaction? **(c)** ADP can undergo further hydrolysis. What would you expect for the product of that reaction?

25.95 A typical amino acid with one amino group and one carboxylic acid group such as alanine (Figure 25.21) can exist in water in several ionic forms. **(a)** Suggest the forms of the amino acid at low pH and at high pH. **(b)** Amino acids are reported to have two pK_a values, one in the range of 2 to 3 and the other in the range of 9 to 10. Alanine, for example, has pK_a values of about 2.3 and 9.6. Using species such as acetic acid and ammonia as models, suggest the origin of the two pK_a values.

[25.96] The protein ribonuclease A in its native, or most stable, form is folded into a compact globular shape. **(a)** Does the native form have a lower or higher free energy than the denatured form, in which the protein is an extended chain? **(b)** What is the sign of the entropy change in going from the denatured to the folded form? **(c)** In the folded form the ribonuclease A has four —S—S— bonds that bridge between parts of the chain, as shown in the accompanying figure. What effect do you predict that these four linkages have on the free energy and entropy of the folded form as compared with a hypothetical folded structure that does not possess the four —S—S— linkages? Explain. **(d)** A gentle reducing agent converts the four —S—S— linkages to eight —S—H bonds. What effect do you predict this would have on the free energy and entropy of the protein?

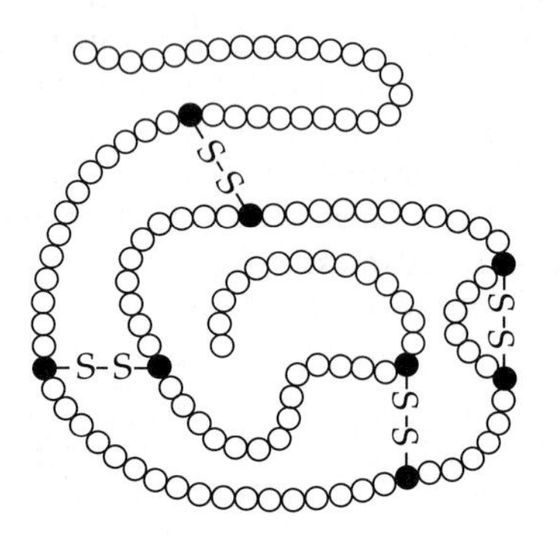

Native ribonuclease A

[25.97] The monoanion of adenosine monophosphate (AMP) is an intermediate in phosphate metabolism:

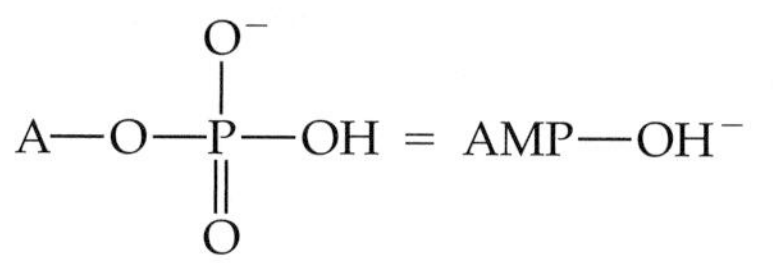

$$A{-}O{-}\overset{O^-}{\underset{O}{P}}{-}OH = AMP{-}OH^-$$

where A = adenosine. If the pK_a for this anion is 7.21, what is the ratio of [AMP—OH$^-$] to [AMP—O^{2-}] in blood at pH 7.0?

eMedia Exercises

25.98 Use the **Boiling Point** activity (*eChapter 25.2*) to plot the boiling points of the first six straight-chain alkanes (methane, ethane, propane, butane, pentane, and hexane), and their corresponding alcohols. **(a)** For a given number of carbons, which has the higher boiling point, the alkane or the alcohol? Use intermolecular forces to explain this observation. **(b)** Does the difference in boiling point for an alkane and its corresponding alcohol get smaller or larger as the number of carbons increases? **(c)** Using your observation from part (b), describe how the relative importance of hydrogen bonding and London dispersion forces changes with increasing length of a carbon chain. Which type of interaction contributes more significantly to the *overall* intermolecular forces for small molecules, and which for large molecules?

25.99 Plot the boiling points of at least five different alcohols and their corresponding amines (Example: *tert*-butanol and *tert*-butylamine) using the **Boiling Point** activity (*eChapter 25.2*). **(a)** Based on the boiling points, which functional group (alcohol or amine) appears to contribute more significantly to hydrogen bonding? Explain. **(b)** Draw Lewis structures of both alcohol and amine functional groups. Use the Lewis structures to support the conclusion that one functional group exhibits more significant hydrogen bonding than the other.

25.100 Watch the **Chirality** movie (*eChapter 25.7*) and answer the following question.What conditions are required for a carbon in an organic molecule to be chiral?

25.101 The **Optical Activity** movie (*eChapter 25.7*) illustrates the behavior of optically active molecules. **(a)** What conditions are necessary for an organic molecule to be optically active? **(b)** What is a *racemic mixture*, and why does it not rotate plane-polarized light? **(c)** Do *cis* and *trans* isomers of dichloroethene exhibit optical activity? Explain.

25.102 Exercise 25.53(b) asks for the number of and abbreviations for the tripeptides that can be made from the amino acids serine and phenylalanine. Using information from the **Proteins and Amino Acids** movie (*eChapter 25.9*) and the table of structural formulas of amino acids in Figure 25.21, draw the structures for each of the tripeptides in your answer to Exercise 25.53(b).

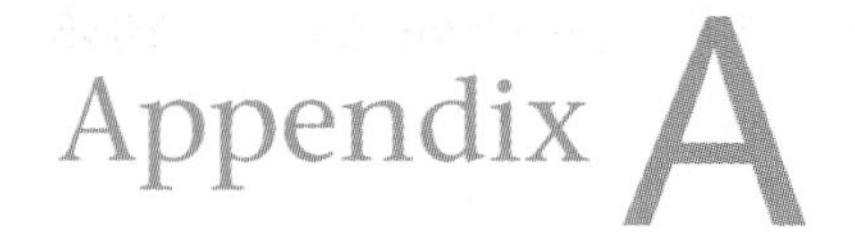

Mathematical Operations

A.1 Exponential Notation

The numbers used in chemistry are often either extremely large or extremely small. Such numbers are conveniently expressed in the form

$$N \times 10^n$$

where N is a number between 1 and 10, and n is the exponent. Some examples of this *exponential notation*, which is also called *scientific notation*, follow.

1,200,000 is 1.2×10^6 (read "one point two times ten to the sixth power")

0.000604 is 6.04×10^{-4} (read "six point zero four times ten to the negative fourth power")

A positive exponent, as in the first example, tells us how many times a number must be multiplied by 10 to give the long form of the number:

$$1.2 \times 10^6 = 1.2 \times 10 \times 10 \times 10 \times 10 \times 10 \times 10 \quad \text{(six tens)}$$
$$= 1{,}200{,}000$$

It is also convenient to think of the positive exponent as the number of places the decimal point must be moved to the *left* to obtain a number greater than 1 and less than 10: If we begin with 3450 and move the decimal point three places to the left, we end up with 3.45×10^3.

In a related fashion, a negative exponent tells us how many times we must divide a number by 10 to give the long form of the number.

$$6.04 \times 10^{-4} = \frac{6.04}{10 \times 10 \times 10 \times 10} = 0.000604$$

It is convenient to think of the negative exponent as the number of places the decimal point must be moved to the *right* to obtain a number greater than 1 but less than 10: If we begin with 0.0048 and move the decimal point three places to the right, we end up with 4.8×10^{-3}.

In the system of exponential notation, with each shift of the decimal point one place to the right, the exponent *decreases* by 1:

$$4.8 \times 10^{-3} = 48 \times 10^{-4}$$

Similarly, with each shift of the decimal point one place to the left, the exponent *increases* by 1:

$$4.8 \times 10^{-3} = 0.48 \times 10^{-2}$$

Many scientific calculators have a key labeled EXP or EE, which is used to enter numbers in exponential notation. To enter the number 5.8×10^3 on such a calculator, the key sequence is

[5] [·] [8] [EXP] (or [EE]) [3]

On some calculators the display will show 5.8, then a space, followed by 03, the exponent. On other calculators, a small 10 is shown with an exponent 3.

To enter a negative exponent, use the key labeled +/−. For example, to enter the number 8.6×10^{-5}, the key sequence is

8	·	6	EXP	+/−	5

When entering a number in exponential notation, do not key in the 10 if your calculator has the EXP or EE button.

In working with exponents, it is important to recall that $10^0 = 1$. The following rules are useful for carrying exponents through calculations.

1. **Addition and Subtraction** In order to add or subtract numbers expressed in exponential notation, the powers of 10 must be the same.

$$\begin{aligned}(5.22 \times 10^4) + (3.21 \times 10^2) &= (522 \times 10^2) + (3.21 \times 10^2)\\ &= 525 \times 10^2 \qquad \text{(3 significant figures)}\\ &= 5.25 \times 10^4\end{aligned}$$

$$\begin{aligned}(6.25 \times 10^{-2}) - (5.77 \times 10^{-3}) &= (6.25 \times 10^{-2}) - (0.577 \times 10^{-2})\\ &= 5.67 \times 10^{-2} \qquad \text{(3 significant figures)}\end{aligned}$$

When you use a calculator to add or subtract, you need not be concerned with having numbers with the same exponents, because the calculator automatically takes care of this matter.

2. **Multiplication and Division** When numbers expressed in exponential notation are multiplied, the exponents are added; when numbers expressed in exponential notation are divided, the exponent of the denominator is subtracted from the exponent of the numerator.

$$\begin{aligned}(5.4 \times 10^2)(2.1 \times 10^3) &= (5.4)(2.1) \times 10^{2+3}\\ &= 11 \times 10^5\\ &= 1.1 \times 10^6\end{aligned}$$

$$(1.2 \times 10^5)(3.22 \times 10^{-3}) = (1.2)(3.22) \times 10^{5-3} = 3.9 \times 10^2$$

$$\frac{3.2 \times 10^5}{6.5 \times 10^2} = \frac{3.2}{6.5} \times 10^{5-2} = 0.49 \times 10^3 = 4.9 \times 10^2$$

$$\frac{5.7 \times 10^7}{8.5 \times 10^{-2}} = \frac{5.7}{8.5} \times 10^{7-(-2)} = 0.67 \times 10^9 = 6.7 \times 10^8$$

3. **Powers and Roots** When numbers expressed in exponential notation are raised to a power, the exponents are multiplied by the power. When the roots of numbers expressed in exponential notation are taken, the exponents are divided by the root.

$$\begin{aligned}(1.2 \times 10^5)^3 &= (1.2)^3 \times 10^{5\times 3}\\ &= 1.7 \times 10^{15}\\ \sqrt[3]{2.5 \times 10^6} &= \sqrt[3]{2.5} \times 10^{6/3}\\ &= 1.3 \times 10^2\end{aligned}$$

Scientific calculators usually have keys labeled x^2 and $\sqrt{x}$ for squaring and taking the square root of a number, respectively. To take higher powers or roots, many calculators have y^x and $\sqrt[x]{y}$ (or INV y^x) keys. For example, to perform the operation $\sqrt[3]{7.5 \times 10^{-4}}$ on such a calculator, you would key in 7.5×10^{-4}, press the $\sqrt[x]{y}$ key (or the INV and then the y^x keys), enter the root, 3, and finally press =. The result is 9.1×10^{-2}.

SAMPLE EXERCISE 1

Perform each of the following operations, using your calculator where possible:
(a) Write the number 0.0054 in standard exponential notation
(b) $(5.0 \times 10^{-2}) + (4.7 \times 10^{-3})$
(c) $(5.98 \times 10^{12})(2.77 \times 10^{-5})$
(d) $\sqrt[4]{1.75 \times 10^{-12}}$

Solution (a) Because we move the decimal three places to the right to convert 0.0054 to 5.4, the exponent is −3:

$$5.4 \times 10^{-3}$$

Scientific calculators are generally able to convert numbers to exponential notation using one or two keystrokes. Consult your instruction manual to see how this operation is accomplished on your calculator.

(b) To add these numbers longhand, we must convert them to the same exponent.

$$(5.0 \times 10^{-2}) + (0.47 \times 10^{-2}) = (5.0 + 0.47) \times 10^{-2} = 5.5 \times 10^{-2}$$

(Note that the result has only two significant figures.) To perform this operation on a calculator, we enter the first number, strike the + key, then enter the second number and strike the = key.

(c) Performing this operation longhand, we have

$$(5.98 \times 2.77) \times 10^{12-5} = 16.6 \times 10^{7} = 1.66 \times 10^{8}$$

On a scientific calculator, we enter 5.98×10^{12}, press the × key, enter 2.77×10^{-5}, and press the = key.

(d) To perform this operation on a calculator, we enter the number, press the $\sqrt[x]{y}$ key (or the INV and y^x keys), enter 4, and press the = key. The result is 1.15×10^{-3}.

PRACTICE EXERCISE

Perform the following operations: **(a)** Write 67,000 in exponential notation, showing two significant figures; **(b)** $(3.378 \times 10^{-3}) - (4.97 \times 10^{-5})$; **(c)** $(1.84 \times 10^{15})/(7.45 \times 10^{-2})$; **(d)** $(6.67 \times 10^{-8})^3$.
Answers: **(a)** 6.7×10^4; **(b)** 3.328×10^{-3}; **(c)** 2.47×10^{16}; **(d)** 2.97×10^{-22}

A.2 Logarithms

Common Logarithms

The common, or base-10, logarithm (abbreviated log) of any number is the power to which 10 must be raised to equal the number. For example, the common logarithm of 1000 (written log 1000) is 3 because raising 10 to the third power gives 1000.

$$10^3 = 1000, \text{ therefore, } \log 1000 = 3$$

Further examples are

$$\log 10^5 = 5$$
$$\log 1 = 0 \quad (\text{Remember that } 10^0 = 1)$$
$$\log 10^{-2} = -2$$

In these examples the common logarithm can be obtained by inspection. However, it is not possible to obtain the logarithm of a number such as 31.25 by inspection. The logarithm of 31.25 is the number x that satisfies the following relationship:

$$10^x = 31.25$$

Most electronic calculators have a key labeled LOG that can be used to obtain logarithms. For example, we can obtain the value of log 31.25 by entering 31.25 and pressing the LOG key. We obtain the following result:

$$\log 31.25 = 1.4949$$

Notice that 31.25 is greater than 10 (10^1) and less than 100 (10^2). The value for log 31.25 is accordingly between log 10 and log 100, that is, between 1 and 2.

Significant Figures and Common Logarithms

For the common logarithm of a measured quantity, the number of digits after the decimal point equals the number of significant figures in the original number. For example, if 23.5 is a measured quantity (three significant figures), then log 23.5 = 1.371 (three significant figures after the decimal point).

Antilogarithms

The process of determining the number that corresponds to a certain logarithm is known as obtaining an *antilogarithm*. It is the reverse of taking a logarithm. For example, we saw above that log 23.5 = 1.371. This means that the antilogarithm of 1.371 equals 23.5.

$$\log 23.5 = 1.371$$

$$\text{antilog } 1.371 = 23.5$$

The process of taking the antilog of a number is the same as raising 10 to a power equal to that number.

$$\text{antilog } 1.371 = 10^{1.371} = 23.5$$

Many calculators have a key labeled 10^x that allows you to obtain antilogs directly. On others, it will be necessary to press a key labeled INV (for *inverse*), followed by the LOG key.

Natural Logarithms

Logarithms based on the number e are called natural, or base e, logarithms (abbreviated ln). The natural log of a number is the power to which e (which has the value 2.71828...) must be raised to equal the number. For example, the natural log of 10 equals 2.303.

$$e^{2.303} = 10, \text{ therefore } \ln 10 = 2.303$$

Your calculator probably has a key labeled LN that allows you to obtain natural logarithms. For example, to obtain the natural log of 46.8, you enter 46.8 and press the LN key.

$$\ln 46.8 = 3.846$$

The natural antilog of a number is e raised to a power equal to that number. If your calculator can calculate natural logs, it will also be able to calculate natural antilogs. On some calculators, there is a key labeled e^x that allows you to calculate natural antilogs directly; on others, it will be necessary to first press the INV key followed by the LN key. For example, the natural antilog of 1.679 is given by

$$\text{Natural antilog } 1.679 = e^{1.679} = 5.36$$

The relation between common and natural logarithms is as follows:

$$\ln a = 2.303 \log a$$

Notice that the factor relating the two, 2.303, is the natural log of 10, which we calculated above.

Mathematical Operations Using Logarithms

Because logarithms are exponents, mathematical operations involving logarithms follow the rules for the use of exponents. For example, the product of z^a and z^b (where z is any number) is given by

$$z^a \cdot z^b = z^{(a+b)}$$

Similarly, the logarithm (either common or natural) of a product equals the *sum* of the logs of the individual numbers.

$$\log ab = \log a + \log b \qquad \ln ab = \ln a + \ln b$$

For the log of a quotient,

$$\log (a/b) = \log a - \log b \qquad \ln (a/b) = \ln a - \ln b$$

Using the properties of exponents, we can also derive the rules for the logarithm of a number raised to a certain power.

$$\log a^n = n \log a \qquad \ln a^n = n \ln a$$

$$\log a^{1/n} = (1/n) \log a \qquad \ln a^{1/n} = (1/n) \ln a$$

pH Problems

One of the most frequent uses for common logarithms in general chemistry is in working pH problems. The pH is defined as $-\log [H^+]$, where $[H^+]$ is the hydrogen ion concentration of a solution (Section 16.4). The following sample exercise illustrates this application.

SAMPLE EXERCISE 2

(a) What is the pH of a solution whose hydrogen ion concentration is 0.015 *M*?
(b) If the pH of a solution is 3.80, what is its hydrogen ion concentration?

Solution **(a)** We are given the value of $[H^+]$. We use the LOG key of our calculator to calculate the value of $\log [H^+]$. The pH is obtained by changing the sign of the value obtained. (Be sure to change the sign *after* taking the logarithm.)

$$[H^+] = 0.015$$

$$\log [H^+] = -1.82 \qquad \text{(2 significant figures)}$$

$$pH = -(-1.82) = 1.82$$

(b) To obtain the hydrogen ion concentration when given the pH, we must take the antilog of $-$pH.

$$pH = -\log [H^+] = 3.80$$

$$\log [H^+] = -3.80$$

$$[H^+] = \text{antilog}\,(-3.80) = 10^{-3.80} = 1.6 \times 10^{-4}\,M$$

PRACTICE EXERCISE

Perform the following operations: **(a)** $\log (2.5 \times 10^{-5})$; **(b)** ln 32.7; **(c)** antilog -3.47; **(d)** $e^{-1.89}$.
Answers: **(a)** -4.60; **(b)** 3.487; **(c)** 3.4×10^{-4}; **(d)** 1.5×10^{-1}

A.3 Quadratic Equations

An algebraic equation of the form $ax^2 + bx + c = 0$ is called a *quadratic equation*. The two solutions to such an equation are given by the quadratic formula:

$$x = \frac{-b \pm \sqrt{b^2 - 4ac}}{2a}$$

SAMPLE EXERCISE 3

Find the values of x that satisfy the equation $2x^2 + 4x = 1$.

Solution To solve the given equation for x, we must first put it in the form

$$ax^2 + bx + c = 0$$

and then use the quadratic formula. If

$$2x^2 + 4x = 1$$

then

$$2x^2 + 4x - 1 = 0$$

Using the quadratic formula, where $a = 2$, $b = 4$, and $c = -1$, we have

$$x = \frac{-4 \pm \sqrt{(4)(4) - 4(2)(-1)}}{2(2)}$$

$$= \frac{-4 \pm \sqrt{16 + 8}}{4} = \frac{-4 \pm \sqrt{24}}{4} = \frac{-4 \pm 4.899}{4}$$

The two solutions are

$$x = \frac{0.899}{4} = 0.225 \quad \text{and} \quad x = \frac{-8.899}{4} = -2.225$$

Often in chemical problems the negative solution has no physical meaning, and only the positive answer is used.

A.4 Graphs

Often the clearest way to represent the interrelationship between two variables is to graph them. Usually, the variable that is being experimentally varied, called the *independent variable*, is shown along the horizontal axis (x-axis). The variable that responds to the change in the independent variable, called the *dependent variable*, is then shown along the vertical axis (y-axis). For example, consider an experiment in which we vary the temperature of an enclosed gas and measure its pressure. The independent variable is temperature, and the dependent variable is pressure. The data shown in Table 1 ▶ can be obtained by means of this experiment. These data are shown graphically in Figure 1 ▼. The relationship between temperature and pressure is linear. The equation for any straight-line graph has the form

TABLE 1 Interrelation Between Pressure and Temperature

Temperature (°C)	Pressure (atm)
20.0	0.120
30.0	0.124
40.0	0.128
50.0	0.132

$$y = mx + b$$

where m is the slope of the line, and b is the intercept with the y-axis. In the case of Figure 1, we could say that the relationship between temperature and pressure takes the form

$$P = mT + b$$

where P is pressure in atm and T is temperature in °C. As shown in Figure 1, the slope is 4.10×10^{-4} atm/°C, and the intercept—the point where the line crosses the y-axis—is 0.112 atm. Therefore, the equation for the line is

$$P = \left(4.10 \times 10^{-4} \frac{\text{atm}}{°\text{C}}\right)T + 0.112 \text{ atm}$$

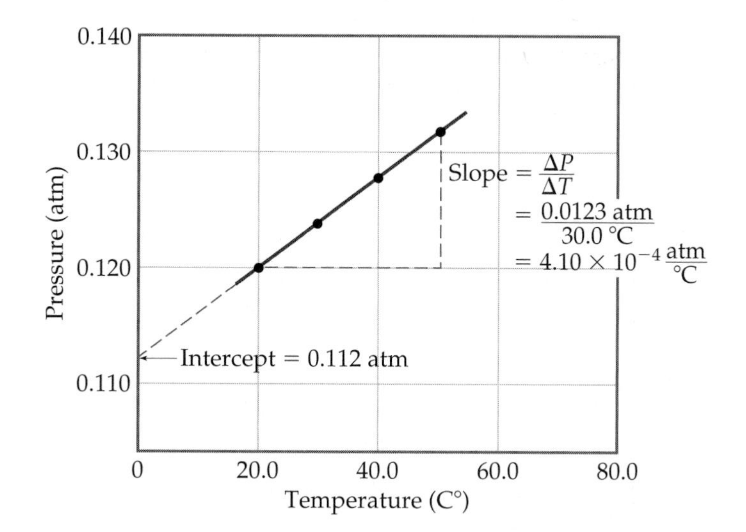

◀ **Figure 1**

Appendix B

Properties of Water

Density: 0.99987 g/mL at 0°C
1.00000 g/mL at 4°C
0.99707 g/mL at 25°C
0.95838 g/mL at 100°C

Heat of fusion: 6.008 kJ/mol at 0°C

Heat of vaporization: 44.94 kJ/mol at 0°C
44.02 kJ/mol at 25°C
40.67 kJ/mol at 100°C

Ion-product constant, K_w: 1.14×10^{-15} at 0°C
1.01×10^{-14} at 25°C
5.47×10^{-14} at 50°C

Specific heat: Ice (−3°C)—2.092 J/g-K
Water at 14.5°C—4.184 J/g-K
Steam (100°C)—1.841 J/g-K

Vapor Pressure (torr)

T(°C)	*P*	*T*(°C)	*P*	*T*(°C)	*P*	*T*(°C)	*P*
0	4.58	21	18.65	35	42.2	92	567.0
5	6.54	22	19.83	40	55.3	94	610.9
10	9.21	23	21.07	45	71.9	96	657.6
12	10.52	24	22.38	50	92.5	98	707.3
14	11.99	25	23.76	55	118.0	100	760.0
16	13.63	26	25.21	60	149.4	102	815.9
17	14.53	27	26.74	65	187.5	104	875.1
18	15.48	28	28.35	70	233.7	106	937.9
19	16.48	29	30.04	80	355.1	108	1004.4
20	17.54	30	31.82	90	525.8	110	1074.6

Appendix C

Thermodynamic Quantities for Selected Substances at 298.15 K (25°C)

Substance	ΔH_f° (kJ/mol)	ΔG_f° (kJ/mol)	S° (J/mol-K)
Aluminum			
Al(*s*)	0	0	28.32
$AlCl_3(s)$	−705.6	−630.0	109.3
$Al_2O_3(s)$	−1669.8	−1576.5	51.00
Barium			
Ba(*s*)	0	0	63.2
$BaCO_3(s)$	−1216.3	−1137.6	112.1
BaO(*s*)	−553.5	−525.1	70.42
Beryllium			
Be(*s*)	0	0	9.44
BeO(*s*)	−608.4	−579.1	13.77
$Be(OH)_2(s)$	−905.8	−817.9	50.21
Bromine			
Br(*g*)	111.8	82.38	174.9
$Br^-(aq)$	−120.9	−102.8	80.71
$Br_2(g)$	30.71	3.14	245.3
$Br_2(l)$	0	0	152.3
HBr(*g*)	−36.23	−53.22	198.49
Calcium			
Ca(*g*)	179.3	145.5	154.8
Ca(*s*)	0	0	41.4
$CaCO_3$ (*s*, calcite)	−1207.1	−1128.76	92.88
$CaCl_2(s)$	−795.8	−748.1	104.6
$CaF_2(s)$	−1219.6	−1167.3	68.87
CaO(*s*)	−635.5	−604.17	39.75
$Ca(OH)_2(s)$	−986.2	−898.5	83.4
$CaSO_4(s)$	−1434.0	−1321.8	106.7
Carbon			
C(*g*)	718.4	672.9	158.0
C(*s*, diamond)	1.88	2.84	2.43
C(*s*, graphite)	0	0	5.69
$CCl_4(g)$	−106.7	−64.0	309.4
$CCl_4(l)$	−139.3	−68.6	214.4
$CF_4(g)$	−679.9	−635.1	262.3
$CH_4(g)$	−74.8	−50.8	186.3
$C_2H_2(g)$	226.7	209.2	200.8
$C_2H_4(g)$	52.30	68.11	219.4
$C_2H_6(g)$	−84.68	−32.89	229.5
$C_3H_8(g)$	−103.85	−23.47	269.9
$C_4H_{10}(g)$	−124.73	−15.71	310.0
$C_4H_{10}(l)$	−147.6	−15.0	231.0
$C_6H_6(g)$	82.9	129.7	269.2
$C_6H_6(l)$	49.0	124.5	172.8
$CH_3OH(g)$	−201.2	−161.9	237.6
$CH_3OH(l)$	−238.6	−166.23	126.8
$C_2H_5OH(g)$	−235.1	−168.5	282.7
$C_2H_5OH(l)$	−277.7	−174.76	160.7
$C_6H_{12}O_6(s)$	−1273.02	−910.4	212.1
CO(*g*)	−110.5	−137.2	197.9
$CO_2(g)$	−393.5	−394.4	213.6
$HC_2H_3O_2(l)$	−487.0	−392.4	159.8
Cesium			
Cs(*g*)	76.50	49.53	175.6
Cs(*l*)	2.09	0.03	92.07
Cs(*s*)	0	0	85.15
CsCl(*s*)	−442.8	−414.4	101.2
Chlorine			
Cl(*g*)	121.7	105.7	165.2
$Cl^-(aq)$	−167.2	−131.2	56.5
$Cl_2(g)$	0	0	222.96
HCl(*aq*)	−167.2	−131.2	56.5
HCl(*g*)	−92.30	−95.27	186.69
Chromium			
Cr(*g*)	397.5	352.6	174.2
Cr(*s*)	0	0	23.6
$Cr_2O_3(s)$	−1139.7	−1058.1	81.2
Cobalt			
Co(*g*)	439	393	179
Co(*s*)	0	0	28.4
Copper			
Cu(*g*)	338.4	298.6	166.3
Cu(*s*)	0	0	33.30
$CuCl_2(s)$	−205.9	−161.7	108.1
CuO(*s*)	−156.1	−128.3	42.59
$Cu_2O(s)$	−170.7	−147.9	92.36
Fluorine			
F(*g*)	80.0	61.9	158.7
$F^-(aq)$	−332.6	−278.8	−13.8
$F_2(g)$	0	0	202.7
HF(*g*)	−268.61	−270.70	173.51

Substance	ΔH_f° (kJ/mol)	ΔG_f° (kJ/mol)	S° (J/mol-K)
Hydrogen			
$H(g)$	217.94	203.26	114.60
$H^+(aq)$	0	0	0
$H^+(g)$	1536.2	1517.0	108.9
$H_2(g)$	0	0	130.58
Iodine			
$I(g)$	106.60	70.16	180.66
$I^-(aq)$	−55.19	−51.57	111.3
$I_2(g)$	62.25	19.37	260.57
$I_2(s)$	0	0	116.73
$HI(g)$	25.94	1.30	206.3
Iron			
$Fe(g)$	415.5	369.8	180.5
$Fe(s)$	0	0	27.15
$Fe^{2+}(aq)$	−87.86	−84.93	113.4
$Fe^{3+}(aq)$	−47.69	−10.54	293.3
$FeCl_2(s)$	−341.8	−302.3	117.9
$FeCl_3(s)$	−400	−334	142.3
$FeO(s)$	−271.9	−255.2	60.75
$Fe_2O_3(s)$	−822.16	−740.98	89.96
$Fe_3O_4(s)$	−1117.1	−1014.2	146.4
$FeS_2(s)$	−171.5	−160.1	52.92
Lead			
$Pb(s)$	0	0	68.85
$PbBr_2(s)$	−277.4	−260.7	161
$PbCO_3(s)$	−699.1	−625.5	131.0
$Pb(NO_3)_2(aq)$	−421.3	−246.9	303.3
$Pb(NO_3)_2(s)$	−451.9	—	—
$PbO(s)$	−217.3	−187.9	68.70
Lithium			
$Li(g)$	159.3	126.6	138.8
$Li(s)$	0	0	29.09
$Li^+(aq)$	−278.5	−273.4	12.2
$Li^+(g)$	685.7	648.5	133.0
$LiCl(s)$	−408.3	−384.0	59.30
Magnesium			
$Mg(g)$	147.1	112.5	148.6
$Mg(s)$	0	0	32.51
$MgCl_2(s)$	−641.6	−592.1	89.6
$MgO(s)$	−601.8	−569.6	26.8
$Mg(OH)_2(s)$	−924.7	−833.7	63.24
Manganese			
$Mn(g)$	280.7	238.5	173.6
$Mn(s)$	0	0	32.0
$MnO(s)$	−385.2	−362.9	59.7
$MnO_2(s)$	−519.6	−464.8	53.14
$MnO_4^-(aq)$	−541.4	−447.2	191.2
Mercury			
$Hg(g)$	60.83	31.76	174.89
$Hg(l)$	0	0	77.40
$HgCl_2(s)$	−230.1	−184.0	144.5
$Hg_2Cl_2(s)$	−264.9	−210.5	192.5
Nickel			
$Ni(g)$	429.7	384.5	182.1
$Ni(s)$	0	0	29.9
$NiCl_2(s)$	−305.3	−259.0	97.65
$NiO(s)$	−239.7	−211.7	37.99
Nitrogen			
$N(g)$	472.7	455.5	153.3
$N_2(g)$	0	0	191.50
$NH_3(aq)$	−80.29	−26.50	111.3
$NH_3(g)$	−46.19	−16.66	192.5
$NH_4^+(aq)$	−132.5	−79.31	113.4
$N_2H_4(g)$	95.40	159.4	238.5
$NH_4CN(s)$	0.0	—	—
$NH_4Cl(s)$	−314.4	−203.0	94.6
$NH_4NO_3(s)$	−365.6	−184.0	151
$NO(g)$	90.37	86.71	210.62
$NO_2(g)$	33.84	51.84	240.45
$N_2O(g)$	81.6	103.59	220.0
$N_2O_4(g)$	9.66	98.28	304.3
$NOCl(g)$	52.6	66.3	264
$HNO_3(aq)$	−206.6	−110.5	146
$HNO_3(g)$	−134.3	−73.94	266.4
Oxygen			
$O(g)$	247.5	230.1	161.0
$O_2(g)$	0	0	205.0
$O_3(g)$	142.3	163.4	237.6
$OH^-(aq)$	−230.0	−157.3	−10.7
$H_2O(g)$	−241.82	−228.57	188.83
$H_2O(l)$	−285.83	−237.13	69.91
$H_2O_2(g)$	−136.10	−105.48	232.9
$H_2O_2(l)$	−187.8	−120.4	109.6
Phosphorus			
$P(g)$	316.4	280.0	163.2
$P_2(g)$	144.3	103.7	218.1
$P_4(g)$	58.9	24.4	280
$P_4(s, \text{red})$	−17.46	−12.03	22.85
$P_4(s, \text{white})$	0	0	41.08
$PCl_3(g)$	−288.07	−269.6	311.7
$PCl_3(l)$	−319.6	−272.4	217
$PF_5(g)$	−1594.4	−1520.7	300.8
$PH_3(g)$	5.4	13.4	210.2
$P_4O_6(s)$	−1640.1	—	—
$P_4O_{10}(s)$	−2940.1	−2675.2	228.9
$POCl_3(g)$	−542.2	−502.5	325
$POCl_3(l)$	−597.0	−520.9	222
$H_3PO_4(aq)$	−1288.3	−1142.6	158.2
Potassium			
$K(g)$	89.99	61.17	160.2
$K(s)$	0	0	64.67
$KCl(s)$	−435.9	−408.3	82.7

Substance	ΔH_f° (kJ/mol)	ΔG_f° (kJ/mol)	S° (J/mol-K)
$KClO_3(s)$	−391.2	−289.9	143.0
$KClO_3(aq)$	−349.5	−284.9	265.7
$K_2CO_3(s)$	−1150.18	−1064.58	155.44
$KNO_3(s)$	−492.70	−393.13	132.9
$K_2O(s)$	−363.2	−322.1	94.14
$KO_2(s)$	−284.5	−240.6	122.5
$K_2O_2(s)$	−495.8	−429.8	113.0
$KOH(s)$	−424.7	−378.9	78.91
$KOH(aq)$	−482.4	−440.5	91.6
Rubidium			
$Rb(g)$	85.8	55.8	170.0
$Rb(s)$	0	0	76.78
$RbCl(s)$	−430.5	−412.0	92
$RbClO_3(s)$	−392.4	−292.0	152
Scandium			
$Sc(g)$	377.8	336.1	174.7
$Sc(s)$	0	0	34.6
Selenium			
$H_2Se(g)$	29.7	15.9	219.0
Silicon			
$Si(g)$	368.2	323.9	167.8
$Si(s)$	0	0	18.7
$SiC(s)$	−73.22	−70.85	16.61
$SiCl_4(l)$	−640.1	−572.8	239.3
$SiO_2(s, \text{quartz})$	−910.9	−856.5	41.84
Silver			
$Ag(s)$	0	0	42.55
$Ag^+(aq)$	105.90	77.11	73.93
$AgCl(s)$	−127.0	−109.70	96.11
$Ag_2O(s)$	−31.05	−11.20	121.3
$AgNO_3(s)$	−124.4	−33.41	140.9
Sodium			
$Na(g)$	107.7	77.3	153.7
$Na(s)$	0	0	51.45
$Na^+(aq)$	−240.1	−261.9	59.0
$Na^+(g)$	609.3	574.3	148.0
$NaBr(aq)$	−360.6	−364.7	141.00

Substance	ΔH_f° (kJ/mol)	ΔG_f° (kJ/mol)	S° (J/mol-K)
$NaBr(s)$	−361.4	−349.3	86.82
$Na_2CO_3(s)$	−1130.9	−1047.7	136.0
$NaCl(aq)$	−407.1	−393.0	115.5
$NaCl(g)$	−181.4	−201.3	229.8
$NaCl(s)$	−410.9	−384.0	72.33
$NaHCO_3(s)$	−947.7	−851.8	102.1
$NaNO_3(aq)$	−446.2	−372.4	207
$NaNO_3(s)$	−467.9	−367.0	116.5
$NaOH(aq)$	−469.6	−419.2	49.8
$NaOH(s)$	−425.6	−379.5	64.46
Strontium			
$SrO(s)$	−592.0	−561.9	54.9
$Sr(g)$	164.4	110.0	164.6
Sulfur			
$S(s, \text{rhombic})$	0	0	31.88
$S_8(g)$	102.3	49.7	430.9
$SO_2(g)$	−296.9	−300.4	248.5
$SO_3(g)$	−395.2	−370.4	256.2
$SO_4^{2-}(aq)$	−909.3	−744.5	20.1
$SOCl_2(l)$	−245.6	—	—
$H_2S(g)$	−20.17	−33.01	205.6
$H_2SO_4(aq)$	−909.3	−744.5	20.1
$H_2SO_4(l)$	−814.0	−689.9	156.1
Titanium			
$Ti(g)$	468	422	180.3
$Ti(s)$	0	0	30.76
$TiCl_4(g)$	−763.2	−726.8	354.9
$TiCl_4(l)$	−804.2	−728.1	221.9
$TiO_2(s)$	−944.7	−889.4	50.29
Vanadium			
$V(g)$	514.2	453.1	182.2
$V(s)$	0	0	28.9
Zinc			
$Zn(g)$	130.7	95.2	160.9
$Zn(s)$	0	0	41.63
$ZnCl_2(s)$	−415.1	−369.4	111.5
$ZnO(s)$	−348.0	−318.2	43.9

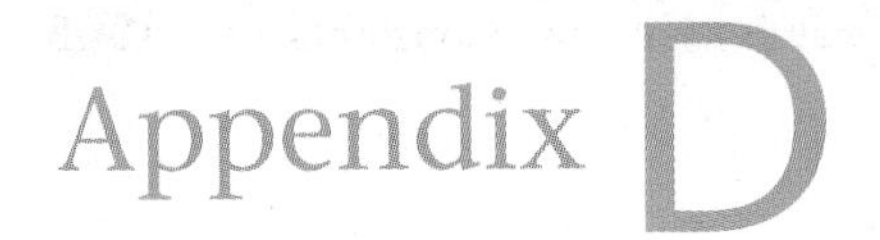

Aqueous-Equilibrium Constants

TABLE 1 Dissociation Constants for Acids at 25°C

Name	Formula	K_{a1}	K_{a2}	K_{a3}
Acetic	$HC_2H_3O_2$	1.8×10^{-5}		
Arsenic	H_3AsO_4	5.6×10^{-3}	1.0×10^{-7}	3.0×10^{-12}
Arsenous	H_3AsO_3	5.1×10^{-10}		
Ascorbic	$HC_6H_7O_6$	8.0×10^{-5}	1.6×10^{-12}	
Benzoic	$HC_7H_5O_2$	6.3×10^{-5}		
Boric	H_3BO_3	5.8×10^{-10}		
Butanoic	$HC_4H_7O_2$	1.5×10^{-5}		
Carbonic	H_2CO_3	4.3×10^{-7}	5.6×10^{-11}	
Chloroacetic	$HC_2H_2O_2Cl$	1.4×10^{-3}		
Chlorous	$HClO_2$	1.1×10^{-2}		
Citric	$H_3C_6H_5O_7$	7.4×10^{-4}	1.7×10^{-5}	4.0×10^{-7}
Cyanic	$HCNO$	3.5×10^{-4}		
Formic	$HCHO_2$	1.8×10^{-4}		
Hydroazoic	HN_3	1.9×10^{-5}		
Hydrocyanic	HCN	4.9×10^{-10}		
Hydrofluoric	HF	6.8×10^{-4}		
Hydrogen chromate ion	$HCrO_4^-$	3.0×10^{-7}		
Hydrogen peroxide	H_2O_2	2.4×10^{-12}		
Hydrogen selenate ion	$HSeO_4$	2.2×10^{-2}		
Hydrosulfuric acid	H_2S	9.5×10^{-8}	1×10^{-19}	
Hypobromous	$HBrO$	2.5×10^{-9}		
Hypochlorous	$HClO$	3.0×10^{-8}		
Hypoiodous	HIO	2.3×10^{-11}		
Iodic	HIO_3	1.7×10^{-1}		
Lactic	$HC_3H_5O_3$	1.4×10^{-4}		
Malonic	$H_2C_3H_2O_4$	1.5×10^{-3}	2.0×10^{-6}	
Nitrous	HNO_2	4.5×10^{-4}		
Oxalic	$H_2C_2O_4$	5.9×10^{-2}	6.4×10^{-5}	
Paraperiodic	H_5IO_6	2.8×10^{-2}	5.3×10^{-9}	
Phenol	HC_6H_5O	1.3×10^{-10}		
Phosphoric	H_3PO_4	7.5×10^{-3}	6.2×10^{-8}	4.2×10^{-13}
Propionic	$HC_3H_5O_2$	1.3×10^{-5}		
Pyrophosphoric	$H_4P_2O_7$	3.0×10^{-2}	4.4×10^{-3}	
Selenous	H_2SeO_3	2.3×10^{-3}	5.3×10^{-9}	
Sulfuric	H_2SO_4	Strong acid	1.2×10^{-2}	
Sulfurous	H_2SO_3	1.7×10^{-2}	6.4×10^{-8}	
Tartaric	$H_2C_4H_4O_6$	1.0×10^{-3}	4.6×10^{-5}	

TABLE 2 Dissociation Constants for Bases at 25°C

Name	Formula	K_b
Ammonia	NH_3	1.8×10^{-5}
Aniline	$C_6H_5NH_2$	4.3×10^{-10}
Dimethylamine	$(CH_3)_2NH$	5.4×10^{-4}
Ethylamine	$C_2H_5NH_2$	6.4×10^{-4}
Hydrazine	H_2NNH_2	1.3×10^{-6}
Hydroxylamine	$HONH_2$	1.1×10^{-8}
Methylamine	CH_3NH_2	4.4×10^{-4}
Pyridine	C_5H_5N	1.7×10^{-9}
Trimethylamine	$(CH_3)_3N$	6.4×10^{-5}

TABLE 3 Solubility-Product Constants for Compounds at 25°C

Name	Formula	K_{sp}	Name	Formula	K_{sp}
Barium carbonate	$BaCO_3$	5.0×10^{-9}	Lead(II) fluoride	PbF_2	3.6×10^{-8}
Barium chromate	$BaCrO_4$	2.1×10^{-10}	Lead(II) sulfate	$PbSO_4$	6.3×10^{-7}
Barium fluoride	BaF_2	1.7×10^{-6}	Lead(II) sulfide*	PbS	3×10^{-28}
Barium oxalate	BaC_2O_4	1.6×10^{-6}	Magnesium hydroxide	$Mg(OH)_2$	1.6×10^{-12}
Barium sulfate	$BaSO_4$	1.1×10^{-10}	Magnesium carbonate	$MgCO_3$	3.5×10^{-8}
Cadmium carbonate	$CdCO_3$	1.8×10^{-14}	Magnesium oxalate	MgC_2O_4	8.6×10^{-5}
Cadmium hydroxide	$Cd(OH)_2$	2.5×10^{-14}	Manganese(II) carbonate	$MnCO_3$	5.0×10^{-10}
Cadmium sulfide*	CdS	8×10^{-28}	Manganese(II) hydroxide	$Mn(OH)_2$	1.6×10^{-13}
Calcium carbonate (calcite)	$CaCO_3$	4.5×10^{-9}	Manganese(II) sulfide*	MnS	2×10^{-53}
Calcium chromate	$CaCrO_4$	7.1×10^{-4}	Mercury(I) chloride	Hg_2Cl_2	1.2×10^{-18}
Calcium fluoride	CaF_2	3.9×10^{-11}	Mercury(I) iodide	Hg_2I_2	1.1×10^{-28}
Calcium hydroxide	$Ca(OH)_2$	6.5×10^{-6}	Mercury(II) sulfide*	HgS	2×10^{-53}
Calcium phosphate	$Ca_3(PO_4)_2$	2.0×10^{-29}	Nickel(II) carbonate	$NiCO_3$	1.3×10^{-7}
Calcium sulfate	$CaSO_4$	2.4×10^{-5}	Nickel(II) hydroxide	$Ni(OH)_2$	6.0×10^{-16}
Chromium(III) hydroxide	$Cr(OH)_3$	1.6×10^{-30}	Nickel(II) sulfide*	NiS	3×10^{-20}
Cobalt(II) carbonate	$CoCO_3$	1.0×10^{-10}	Silver bromate	$AgBrO_3$	5.5×10^{-5}
Cobalt(II) hydroxide	$Co(OH)_2$	1.3×10^{-15}	Silver bromide	$AgBr$	5.0×10^{-13}
Cobalt(II) sulfide*	CoS	5×10^{-22}	Silver carbonate	Ag_2CO_3	8.1×10^{-12}
Copper(I) bromide	$CuBr$	5.3×10^{-9}	Silver chloride	$AgCl$	1.8×10^{-10}
Copper(II) carbonate	$CuCO_3$	2.3×10^{-10}	Silver chromate	Ag_2CrO_4	1.2×10^{-12}
Copper(II) hydroxide	$Cu(OH)_2$	4.8×10^{-20}	Silver iodide	AgI	8.3×10^{-17}
Copper(II) sulfide*	CuS	6×10^{-37}	Silver sulfate	Ag_2SO_4	1.5×10^{-5}
Iron(II) carbonate	$FeCO_3$	2.1×10^{-11}	Silver sulfide*	Ag_2S	6×10^{-51}
Iron(II) hydroxide	$Fe(OH)_2$	7.9×10^{-16}	Strontium carbonate	$SrCO_3$	9.3×10^{-10}
Lanthanum fluoride	LaF_3	2×10^{-19}	Tin(II) sulfide*	SnS	1×10^{-26}
Lanthanum iodate	$La(IO_3)_3$	6.1×10^{-12}	Zinc carbonate	$ZnCO_3$	1.0×10^{-10}
Lead(II) carbonate	$PbCO_3$	7.4×10^{-14}	Zinc hydroxide	$Zn(OH)_2$	3.0×10^{-16}
Lead(II) chloride	$PbCl_2$	1.7×10^{-5}	Zinc oxalate	ZnC_2O_4	2.7×10^{-8}
Lead(II) chromate	$PbCrO_4$	2.8×10^{-13}	Zinc sulfide*	ZnS	2×10^{-25}

*For a solubility equilibrium of the type $MS(s) + H_2O(l) \rightleftharpoons M^{2+}(aq) + HS^-(aq) + OH^-(aq)$

Standard Reduction Potentials at 25°C

Half-Reaction	$E°$ (V)
$Ag^+(aq) + e^- \longrightarrow Ag(s)$	+0.799
$AgBr(s) + e^- \longrightarrow Ag(s) + Br^-(aq)$	+0.095
$AgCl(s) + e^- \longrightarrow Ag(s) + Cl^-(aq)$	+0.222
$Ag(CN)_2^-(aq) + e^- \longrightarrow Ag(s) + 2CN^-(aq)$	−0.31
$Ag_2CrO_4(s) + 2e^- \longrightarrow 2Ag(s) + CrO_4^{2-}(aq)$	+0.446
$AgI(s) + e^- \longrightarrow Ag(s) + I^-(aq)$	−0.151
$Ag(S_2O_3)_2^{3-} + e^- \longrightarrow Ag(s) + 2S_2O_3^{2-}(aq)$	+0.01
$Al^{3+}(aq) + 3e^- \longrightarrow Al(s)$	−1.66
$H_3AsO_4(aq) + 2H^+(aq) + 2e^- \longrightarrow H_3AsO_3(aq) + H_2O(l)$	+0.559
$Ba^{2+}(aq) + 2e^- \longrightarrow Ba(s)$	−2.90
$BiO^+(aq) + 2H^+(aq) + 3e^- \longrightarrow Bi(s) + H_2O(l)$	+0.32
$Br_2(l) + 2e^- \longrightarrow 2Br^-(aq)$	+1.065
$BrO_3^-(aq) + 6H^+(aq) + 5e^- \longrightarrow Br_2(l) + 3H_2O(l)$	+1.52
$2CO_2(g) + 2H^+(aq) + 2e^- \longrightarrow H_2C_2O_4(aq)$	−0.49
$Ca^{2+}(aq) + 2e^- \longrightarrow Ca(s)$	−2.87
$Cd^{2+}(aq) + 2e^- \longrightarrow Cd(s)$	−0.403
$Ce^{4+}(aq) + e^- \longrightarrow Ce^{3+}(aq)$	+1.61
$Cl_2(g) + 2e^- \longrightarrow 2Cl^-(aq)$	+1.359
$HClO(aq) + H^+(aq) + e^- \longrightarrow Cl_2(g) + H_2O(l)$	+1.63
$ClO^-(aq) + H_2O(l) + 2e^- \longrightarrow Cl^-(aq) + 2OH^-(aq)$	+0.89
$ClO_3^-(aq) + 6H^+(aq) + 5e^- \longrightarrow Cl_2(g) + 3H_2O(l)$	+1.47
$Co^{2+}(aq) + 2e^- \longrightarrow Co(s)$	−0.277
$Co^{3+}(aq) + e^- \longrightarrow Co^{2+}(aq)$	+1.842
$Cr^{3+}(aq) + 3e^- \longrightarrow Cr(s)$	−0.74
$Cr^{3+}(aq) + e^- \longrightarrow Cr^{2+}(aq)$	−0.41
$Cr_2O_7^{2-}(aq) + 14H^+(aq) + 6e^- \longrightarrow 2Cr^{3+}(aq) + 7H_2O(l)$	+1.33
$CrO_4^{2-}(aq) + 4H_2O(l) + 3e^- \longrightarrow Cr(OH)_3(s) + 5OH^-(aq)$	−0.13
$Cu^{2+}(aq) + 2e^- \longrightarrow Cu(s)$	+0.337
$Cu^{2+}(aq) + e^- \longrightarrow Cu^+(aq)$	+0.153
$Cu^+(aq) + e^- \longrightarrow Cu(s)$	+0.521
$CuI(s) + e^- \longrightarrow Cu(s) + I^-(aq)$	−0.185
$F_2(g) + 2e^- \longrightarrow 2F^-(aq)$	+2.87
$Fe^{2+}(aq) + 2e^- \longrightarrow Fe(s)$	−0.440
$Fe^{3+}(aq) + e^- \longrightarrow Fe^{2+}(aq)$	+0.771
$Fe(CN)_6^{3-}(aq) + e^- \longrightarrow Fe(CN)_6^{4-}(aq)$	+0.36
$2H^+(aq) + 2e^- \longrightarrow H_2(g)$	0.000
$2H_2O(l) + 2e^- \longrightarrow H_2(g) + 2OH^-(aq)$	−0.83

Half-Reaction	$E°$ (V)
$HO_2^-(aq) + H_2O(l) + 2e^- \longrightarrow 3OH^-(aq)$	+0.88
$H_2O_2(aq) + 2H^+(aq) + 2e^- \longrightarrow 2H_2O(l)$	+1.776
$Hg_2^{2+}(aq) + 2e^- \longrightarrow 2Hg(l)$	+0.789
$2Hg^{2+}(aq) + 2e^- \longrightarrow Hg_2^{2+}(aq)$	+0.920
$Hg^{2+}(aq) + 2e^- \longrightarrow Hg(l)$	+0.854
$I_2(s) + 2e^- \longrightarrow 2I^-(aq)$	+0.536
$IO_3^-(aq) + 6H^+(aq) + 5e^- \longrightarrow I_2(s) + 3H_2O(l)$	+1.195
$K^+(aq) + e^- \longrightarrow K(s)$	−2.925
$Li^+(aq) + e^- \longrightarrow Li(s)$	−3.05
$Mg^{2+}(aq) + 2e^- \longrightarrow Mg(s)$	−2.37
$Mn^{2+}(aq) + 2e^- \longrightarrow Mn(s)$	−1.18
$MnO_2(s) + 4H^+(aq) + 2e^- \longrightarrow Mn^{2+}(aq) + 2H_2O(l)$	+1.23
$MnO_4^-(aq) + 8H^+(aq) + 5e^- \longrightarrow Mn^{2+}(aq) + 4H_2O(l)$	+1.51
$MnO_4^-(aq) + 2H_2O(l) + 3e^- \longrightarrow MnO_2(s) + 4OH^-(aq)$	+0.59
$HNO_2(aq) + H^+(aq) + e^- \longrightarrow NO(g) + H_2O(l)$	+1.00
$N_2(g) + 4H_2O(l) + 4e^- \longrightarrow 4OH^-(aq) + N_2H_4(aq)$	−1.16
$N_2(g) + 5H^+(aq) + 4e^- \longrightarrow N_2H_5^+(aq)$	−0.23
$NO_3^-(aq) + 4H^+(aq) + 3e^- \longrightarrow NO(g) + 2H_2O(l)$	+0.96
$Na^+(aq) + e^- \longrightarrow Na(s)$	−2.71
$Ni^{2+}(aq) + 2e^- \longrightarrow Ni(s)$	−0.28
$O_2(g) + 4H^+(aq) + 4e^- \longrightarrow 2H_2O(l)$	+1.23
$O_2(g) + 2H_2O(l) + 4e^- \longrightarrow 4OH^-(aq)$	+0.40
$O_2(g) + 2H^+(aq) + 2e^- \longrightarrow H_2O_2(aq)$	+0.68
$O_3(g) + 2H^+(aq) + 2e^- \longrightarrow O_2(g) + H_2O(l)$	+2.07
$Pb^{2+}(aq) + 2e^- \longrightarrow Pb(s)$	−0.126
$PbO_2(s) + HSO_4^-(aq) + 3H^+(aq) + 2e^- \longrightarrow PbSO_4(s) + 2H_2O(l)$	+1.685
$PbSO_4(s) + H^+(aq) + 2e^- \longrightarrow Pb(s) + HSO_4^-(aq)$	−0.356
$PtCl_4^{2-}(aq) + 2e^- \longrightarrow Pt(s) + 4Cl^-(aq)$	+0.73
$S(s) + 2H^+(aq) + 2e^- \longrightarrow H_2S(g)$	+0.141
$H_2SO_3(aq) + 4H^+(aq) + 4e^- \longrightarrow S(s) + 3H_2O(l)$	+0.45
$HSO_4^-(aq) + 3H^+(aq) + 2e^- \longrightarrow H_2SO_3(aq) + H_2O(l)$	+0.17
$Sn^{2+}(aq) + 2e^- \longrightarrow Sn(s)$	−0.136
$Sn^{4+}(aq) + 2e^- \longrightarrow Sn^{2+}(aq)$	+0.154
$VO_2^+(aq) + 2H^+(aq) + e^- \longrightarrow VO^{2+}(aq) + H_2O(l)$	+1.00
$Zn^{2+}(aq) + 2e^- \longrightarrow Zn(s)$	−0.763

Answers to Selected Exercises

Chapter 1

1.1 (a) Heterogeneous mixture (b) homogeneous mixture (c) pure substance (d) homogeneous mixture. **1.3** (a) Al (b) Na (c) Br (d) Cu (e) Si (f) N (g) Mg (h) He **1.5** (a) Hydrogen (b) magnesium (c) lead (d) silicon (e) fluorine (f) tin (g) manganese (h) arsenic **1.7** C is a compound; it contains carbon and oxygen. A is a compound; it contains at least carbon and oxygen. B is not defined by the data given; it is probably a compound because few elements exist as white solids.

1.9

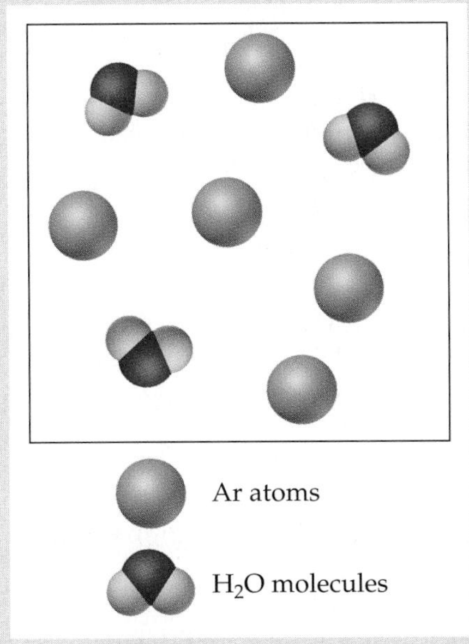

1.11 Physical properties: silvery white; lustrous; melting point = 649°C; boiling point = 1105°C; density at 20°C = 1.738 g/mL; pounded into sheets; drawn into wires; good conductor. Chemical properties: burns in air; reacts with Cl_2. **1.13** (a) Chemical (b) physical (c) physical (d) chemical (e) chemical **1.15** First heat the liquid to 100°C to evaporate the water. If there is a residue, measure the physical properties of the residue, such as color, density, and melting point. If the properties match those of NaCl, the water contained dissolved table salt. If the properties don't match, the residue is a different dissolved solid. If there is no residue, no dissolved solid is present. **1.17** (a) 1×10^{-1} (b) 1×10^{-2} (c) 1×10^{-15} (d) 1×10^{-6} (e) 1×10^{6} (f) 1×10^{3} (g) 1×10^{-9} (h) 1×10^{-3} (i) 1×10^{-12} **1.19** (a) 2.55×10^{-2} g (b) 0.40 nm (c) 575 μm **1.21** (a) Time (b) density (c) length (d) area (e) temperature (f) volume (g) temperature **1.23** (a) 1.59 g/cm^3. Carbon tetrachloride, 1.59 g/mL, is more dense than water, 1.00 g/mL; carbon tetrachloride will sink rather than float on water. (b) 1.609 kg (c) 50.35 mL **1.25** (a) Calculated density = 0.86 g/mL. The substance is probably toluene, density = 0.866 g/mL. (b) 40.4 mL ethylene glycol (c) 1.11×10^3 g nickel **1.27** 4.6×10^{-8} m; 46 nm **1.29** (a) 17°C (b) 422.1°F (c) 506 K (d) 108°F (e) 1644 K **1.31** Exact: (c), (d), and (f) **1.33** 7.5 cm. There are two significant figures in this measurement; the number of cm can be read precisely, but there is some estimating (uncertainty) required to read tenths of a centimeter. **1.35** (a) 4 (b) 3 (c) 4 (d) 3 (e) 5 **1.37** (a) 3.002×10^2 (b) 4.565×10^5 (c) 6.543×10^{-3} (d) 9.578×10^{-4} (e) 5.078×10^4 (f) -3.500×10^{-2} **1.39** (a) 27.04 (b) −8.0 (c) 1.84×10^{-3} (d) 7.66×10^{-4} **1.41** Arrange conversion factors so that the starting units cancel and the new units remain in the appropriate place, either numerator or denominator. **1.43** (a) 76 mL (b) 50 nm (c) 6.88×10^{-4} s (d) 1.55 g/L (e) 6.151×10^{-3} L/s **1.45** (a) 4.32×10^5 s (b) 88.5 m (c) \$0.499/L (d) 46.6 km/hr (e) 1.420 L/s (f) 707.9 cm^3 **1.47** (a) 1.2×10^2 L (b) 4×10^2 mg (c) 9.64 km/L (d) 26 mL/g **1.49** 52 kg air **1.51** 467 ft **1.53** Use the kg as a unit for comparison.
5 lb potatoes < 2.5 kg; 5 kg sugar = 5 kg;
1 gal = 4 qt ≈ 4 L ≈ 4 kg. The order of mass from lightest to heaviest is 5 lb potatoes < 1 gal water < 5 kg sugar.
1.55 Composition is the contents of a substance; structure is the arrangement of these contents. **1.58** 8.47 g O; the *law of constant composition* **1.61** 27.1 K; −411.0°F **1.64** Al has the largest diameter, 1.92 cm; Pb has the smallest, 1.19 cm. Note that Pb and Ag, with similar densities, have similar diameters; Al, with a much smaller density, has a much larger diameter. **1.66** (a) 1.05×10^{13} g NaOH (b) 4.94×10^{-3} km^3 **1.69** Freezing point of H_2O = 5.50°G **1.71** (a) 3.9×10^8 m (b) 5.8×10^5 s **1.74** (a) 2.98×10^3 cm^3 (b) 0.0482 m^3 (c) 655 kg Hg **1.76** (a) 61.5% Au (b) 15 carat gold **1.79** Carbon tetrachloride: 1.5940 g/cm^3; hexane: 0.6603 g/cm^3; benzene: 0.87654 g/cm^3; methylene iodide: 3.3254 g/cm^3. Only methylene iodide will separate the two granular solids.

Chapter 2

2.1 Postulate 4 of the atomic theory states that the relative number and kinds of atoms in a compound are constant, regardless of the source. Therefore, 1.0 g of pure water should always contain the same relative amounts of hydrogen and oxygen, no matter where or how the sample is obtained. **2.3** (a) 0.5711 g O/1 g N; 1.142 g O/1 g N; 2.284 g O/1 g N; 2.855 g O/1 g N (b) The numbers in part (a) obey the *law of multiple proportions*. Multiple proportions arise because atoms are the indivisible entities combining, as stated in Dalton's atomic theory. **2.5** (1) Electric and magnetic fields deflected the rays in the same way they would deflect negatively charged particles. (2) A metal plate exposed to cathode rays acquired a negative charge. **2.7** (a) In Millikan's oil-drop experiment, X rays interact with gaseous atoms or molecules in the chamber, forming positive ions and free electrons. The free electrons are then able to recombine with the ions or cling to the oil drops. (b) If the positive plate were lower than the negative plate, the oil drops "coated" with negatively charged electrons would be attracted to the positively charged plate and would descend much more quickly. (c) The more times a measurement is repeated, the better the chance of detecting and compensating for experimental errors. Millikan wanted to demonstrate the validity of his result via its reproducibility. **2.9** (a) Because γ rays are not deflected by the electric field, they carry no charge. (b) If α and β rays are deflected in opposite directions in an electric field, they must have opposite electrical charges. **2.11** (a) 0.19 nm; 1.9×10^2 or 190 pm (b) 2.6×10^6 Kr atoms (c) 2.9×10^{-23} cm^3 **2.13** (a) proton, neutron, electron (b) proton = +1, neutron = 0, electron = −1 (c) The neutron is most massive, the electron least massive. (The neutron and proton have very similar masses). **2.15** (a) ^{28}Si: 14 p, 14 n, 14 e (b) ^{60}Ni: 28 p, 32 n, 28 e (c) ^{85}Rb: 37 p, 48 n, 37 e (d) ^{128}Xe: 54 p, 74 n, 54 e (e) ^{195}Pt: 78 p, 117 n, 78 e (f) ^{238}U: 92 p, 146 n, 92 e

2.17

Symbol	^{52}Cr	^{75}As	^{40}Ca	^{222}Rn	^{193}Ir
Protons	24	33	20	86	77
Neutrons	28	42	20	136	116
Electrons	24	33	20	86	77
Mass no.	52	75	40	222	193

2.19 (a) $^{179}_{72}Hf$ (b) $^{40}_{18}Ar$ (c) $^{4}_{2}He$ (d) $^{115}_{49}In$ (e) $^{28}_{14}Si$ **2.21** (a) $^{12}_{6}C$ (b) Atomic weights are average atomic masses, the sum of the mass of each naturally occurring isotope of an element times its fractional abundance. Each Cl atom will have the mass of one of the naturally occurring isotopes, while the "atomic weight" is an average value. **2.23** 207 amu **2.25** (a) In Thomson's cathode-ray experiments and in mass spectrometry a stream of charged particles is passed through the poles of a magnet. The charged particles are deflected by the magnetic field according to their mass and charge. (b) The x-axis label is atomic weight, and the y-axis label is signal intensity. (c) Uncharged particles are not deflected in a magnetic field. The effect of the magnetic field on charged moving particles is the basis of their separation by mass. **2.27** (a) average atomic mass = 24.31 amu (b)

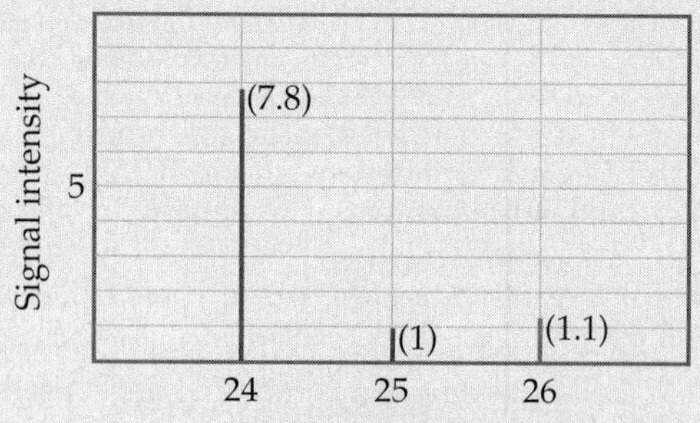

2.29 (a) Ag (metal) (b) He (nonmetal) (c) P (nonmetal) (d) Cd (metal) (e) Ca (metal) (f) Br (nonmetal) (g) As (metalloid) **2.31** (a) K, alkali metals (metal) (b) I, halogens (nonmetal) (c) Mg, alkaline earth metals (metal) (d) Ar, noble gases (nonmetal) (e) S, chalcogens (nonmetal) **2.33** An empirical formula shows the simplest ratio of the different atoms in a molecule. A molecular formula shows the exact number and kinds of atoms in a molecule. A structural formula shows how these atoms are arranged. **2.35** (a) molecular: C_6H_6; empirical: CH (b) molecular: S_iCl_4; empirical: S_iCl_4 **2.37** (a) 6 (b) 6 (c) 12

2.39 (a)

```
             H       H
             |       |
C2H6O,   H—C—O—C—H
             |       |
             H       H
```

(b)

```
             H   H
             |   |
C2H6O,   H—C—C—O—H
             |   |
             H   H
```

(c)

```
           H
           |
CH4O,   H—C—O—H
           |
           H
```

(d) PF_3,

```
F—P—F
  |
  F
```

2.41 (a) $AlBr_3$ (b) C_4H_5 (c) C_2H_4O (d) P_2O_5 (e) C_3H_2Cl (f) BNH_2 **2.43** (a) Al^{3+} (b) Ca^{2+} (c) S^{2-} (d) I^- (e) Cs^+ **2.45** (a) GaF_3, gallium(III) fluoride (b) LiH, lithium hydride (c) AlI_3, aluminum iodide (d) K_2S, potassium sulfide **2.47** (a) $CaBr_2$ (b) NH_4Cl (c) $Al(C_2H_3O_2)_3$ (d) K_2SO_4 (e) $Mg_3(PO_4)_2$ **2.49** Molecular: (a) B_2H_6 (b) CH_3OH (f) NOCl (g) NF_3. Ionic: (c) $LiNO_3$ (d) Sc_2O_3 (e) CsBr (h) Ag_2SO_4 **2.51** (a) ClO_2^- (b) Cl^- (c) ClO_3^- (d) ClO_4^- (e) ClO^- **2.53** (a) Aluminum fluoride (b) iron(II) hydroxide (ferrous hydroxide) (c) copper(II) nitrate (cupric nitrate) (d) barium perchlorate (e) lithium phosphate (f) mercury(I) sulfide (mercurous sulfide) (g) calcium acetate (h) chromium(III) carbonate (chromic carbonate) (i) potassium chromate (j) ammonium sulfate **2.55** (a) Cu_2O (b) K_2O_2 (c) $Al(OH)_3$ (d) $Zn(NO_3)_2$ (e) Hg_2Br_2 (f) $Fe_2(CO_3)_3$ (g) NaBrO **2.57** (a) Bromic acid (b) hydrobromic acid (c) phosphoric acid (d) HClO (e) HIO_3 (f) H_2SO_3 **2.59** (a) Sulfur hexafluoride (b) iodine pentafluoride (c) xenon trioxide (d) N_2O_4 (e) HCN (f) P_4S_6 **2.61** (a) $ZnCO_3$, ZnO, CO_2 (b) HF, SiO_2, SiF_4,H_2O (c) SO_2, H_2O, H_2SO_3 (d) H_3P (or PH_3) (e) $HClO_4$, Cd, $Cd(ClO_4)_2$ (f) VBr_3 **2.63** (a) A hydrocarbon is a compound composed of the elements hydrogen and carbon only. (b) All alkanes are hydrocarbons, but compounds other than alkanes can also be hydrocarbons.

(c)

```
    H   H
    |   |
H—C—C—H
    |   |
    H   H
```

(d)

```
    H   H   H   H
    |   |   |   |
H—C—C—C—C—H
    |   |   |   |
    H   H   H   H
```

Molecular: C_4H_{10}
Empirical: C_2H_5

2.65 (a) *Functional groups* are groups of specific atoms that are constant from one molecule to the next. (b) —OH

(c)

```
    H   H   H   H
    |   |   |   |
H—C—C—C—C—OH
    |   |   |   |
    H   H   H   H
```

2.69 Radioactivity is the spontaneous emission of radiation from a substance. Becquerel's discovery showed that atoms could decay, or degrade, *implying* that they are not indivisible. However, it wasn't until Rutherford and others characterized the nature of radioactive emissions that the full significance of the discovery was apparent. **2.72** (a) 2 protons, 1 neutron, 2 electrons (b) Tritium, ^3H, is more massive. (c) A precision of 1×10^{-27} g would be required to differentiate between ^3H and ^3He. **2.76** (a) $^{16}_{8}O$, $^{17}_{8}O$, $^{18}_{8}O$ (b) All isotopes are atoms of the same element, oxygen, with the same atomic number, 8 protons in the nucleus and 8 electrons. We expect their electron arrangements to be the same and their chemical properties to be very similar. Each has a different number of neutrons, a different mass number, and a different atomic mass. **2.78** (a) The 68.926 amu isotope has 31 protons, 38 neutrons, and the symbol $^{69}_{31}Ga$. The 70.926 amu isotope has 31 protons, 40 neutrons, and the symbol $^{71}_{31}Ga$. (b) ^{69}Ga = 60.3%, ^{71}Ga = 39.7% **2.81** (a) 5 significant figures (b) 0.05444% **2.84** (a) $^{266}_{106}Sg$ has 106 protons, 160 neutrons, and 106 electrons (b) Sg is in group 6B (or 6) and immediately below tungsten, W. We expect the chemical properties of Sg to most closely resemble those of W. **2.87** (a) nickel(II) oxide, 2+ (b) manganese(IV) oxide, 4+ (c) chromium(III) oxide, 3+ (d) molybdenum(VI) oxide, 6+ **2.90** (a) sodium chloride (b) sodium bicarbonate (or sodium hydrogen carbonate) (c) sodium hypochlorite (d) sodium hydroxide (e) ammonium carbonate

(f) calcium sulfate **2.94** (a) CH (b) No. Benzene is not an alkane because alkanes are hydrocarbons with all single bonds. (c) The molecular formula is C_6H_6O or C_6H_5OH. The structural formula is

Chapter 3

3.1 (a) Conservation of mass (b) Subscripts in chemical formulas should not be changed when balancing equations, because changing the subscript changes the identity of the compound (*law of constant composition*). (c) (*g*), (*l*), (*s*), (*aq*) **3.3** Equation (a) best fits the diagram. **3.5** (a) $2SO_2(g) + O_2(g) \longrightarrow 2SO_3(g)$
(b) $P_2O_5(s) + 3H_2O(l) \longrightarrow 2H_3PO_4(aq)$
(c) $CH_4(g) + 4Cl_2(g) \longrightarrow CCl_4(l) + 4HCl(g)$
(d) $Al_4C_3(s) + 12H_2O(l) \longrightarrow 4Al(OH)_3(s) + 3CH_4(g)$
(e) $C_4H_{10}O(l) + 6O_2(g) \longrightarrow 4CO_2(g) + 5H_2O(l)$
(f) $2Fe(OH)_3(s) + 3H_2SO_4(aq) \longrightarrow Fe_2(SO_4)_3(aq) + 6H_2O(l)$
(g) $Mg_3N_2(s) + 4H_2SO_4(aq) \longrightarrow 3MgSO_4(aq) + (NH_4)_2SO_4(aq)$
3.7 (a) $CaC_2(s) + 2H_2O(l) \longrightarrow Ca(OH)_2(aq) + C_2H_2(g)$
(b) $2KClO_3(s) \xrightarrow{\Delta} 2KCl(s) + 3O_2(g)$
(c) $Zn(s) + H_2SO_4(aq) \longrightarrow ZnSO_4(aq) + H_2(g)$
(d) $PCl_3(l) + 3H_2O(l) \longrightarrow H_3PO_3(aq) + 3HCl(aq)$
(e) $3H_2S(g) + 2Fe(OH)_3(s) \longrightarrow Fe_2S_3(s) + 6H_2O(g)$
3.9 (a) Determine the formula by balancing the positive and negative charges in the ionic product. All ionic compounds are solids. $2Na(s) + Br_2(l) \longrightarrow 2NaBr(s)$ (b) The second reactant is $O_2(g)$. The products are $CO_2(g)$ and $H_2O(l)$.
$2C_6H_6(l) + 15O_2(g) \longrightarrow 12CO_2(g) + 6H_2O(l)$
3.11 (a) $Mg(s) + Cl_2(g) \longrightarrow MgCl_2(s)$
(b) $Ni(OH)_2(s) \longrightarrow NiO(s) + HO_2(g)$
(c) $C_8H_8(l) + 10O_2(g) \longrightarrow 8CO_2(g) + 4H_2O(l)$
(d) $2C_5H_{12}O(l) + 15O_2(g) \longrightarrow 10CO_2(g) + 12H_2O(l)$
3.13 (a) $2Al(s) + 3Cl_2(g) \longrightarrow 2AlCl_3(s)$ combination
(b) $C_2H_4(g) + 3O_2(g) \longrightarrow 2CO_2(g) + 2H_2O(l)$ combustion
(c) $6Li(s) + N_2(g) \longrightarrow 2Li_3N(s)$ combination
(d) $PbCO_3(s) \longrightarrow PbO(s) + CO_2(g)$ decomposition
(e) $C_7H_8O_2(l) + 8O_2(g) \longrightarrow 7CO_2(g) + 4H_2O(l)$ combustion.
3.15 (a) 34.1 amu (b) 118.7 amu (c) 142.3 amu (d) 132.1 amu (e) 212.3 amu (f) 159.6 amu (g) 222.5 amu **3.17** (a) 49.9% (b) 45.0% (c) 43.2% (d) 67.6% (e) 60.0% **3.19** (a) 79.2% (b) 63.2% (c) 64.6% **3.21** (a) 6.022×10^{23} (b) The formula weight of a substance in amu has the same numerical value as the molar mass expressed in grams. **3.23** 23 g Na contains 1 mol of atoms; 0.5 mol H_2O contains 1.5 mol atoms; 6.0×10^{23} N_2 molecules contains 2 mol atoms
3.25 4.4×10^{24} kg. One mole of shot-put balls weighs 0.73 times as much as Earth. **3.27** (a) 72.8 g CaH_2 (b) 0.0219 mol $Mg(NO_3)_2$ (c) 1.48×10^{23} CH_3OH molecules (d) 3.52×10^{24} H atoms **3.29** (a) 0.856 g $Al_2(SO_4)_3$ (b) 1.69×10^{-3} mol Cl^- (c) 0.248 g $C_8H_{10}N_4O_2$ (d) 387 g cholesterol/mol
3.31 (a) molar mass = 162.3 g (b) 3.08×10^{-5} mol allicin (c) 1.86×10^{19} allicin molecules (d) 3.71×10^{19} S atoms
3.33 (a) 1.15×10^{21} H atoms (b) 9.62×10^{19} $C_6H_{12}O_6$ molecules (c) 1.60×10^{-4} mol $C_6H_{12}O_6$ (d) 0.0287 g $C_6H_{12}O_6$
3.35 3.28×10^{-8} mol C_2H_3Cl/L; 1.97×10^{16} molecules/L
3.37 (a) NO_2 (b) No, because we have no way of knowing whether the empirical and molecular formulas are the same. NO_2 represents the simplest ratio of atoms in a molecule, but not the only possible molecular formula. **3.39** (a) C_2H_6O (b) Fe_2O_3 (c) CH_2O **3.41** (a) $CSCl_2$ (b) C_3OF_6 (c) Na_3AlF_6

3.43 (a) C_6H_{12} (b) NH_2Cl **3.45** (a) empirical formula, $C_4H_5N_2O$; molecular formula, $C_8H_{10}N_4O_2$ (b) empirical formula and molecular formula, $NaC_5H_8O_4N$ **3.47** (a) C_7H_8 (b) The empirical and molecular formulas are $C_{10}H_{20}O$.
3.49 $x = 10$; $Na_2CO_3 \cdot 10\,H_2O$ **3.51** If the equation is not balanced, the mole ratios derived from the coefficients will be incorrect and lead to erroneous calculated amounts of products. **3.53** 4.0 mol CH_4 can produce 4.0 mol CO and 12.0 mol H_2. **3.55** (a) 2.4 mol HF (b) 5.25 g NaF (c) 0.610 g Na_2SiO_3
3.57 (a) $Al_2S_3(s) + 6H_2O(l) \longrightarrow 2Al(OH)_3(s) + 3H_2S(g)$ (b) 10.9 g $Al(OH)_3$ **3.59** (a) 3.75 mol N_2 (b) 9.28 g NaN_3 (c) 548 g NaN_3 **3.61** (a) 5.50×10^{-3} mol Al (b) 1.47 g $AlBr_3$
3.63 (a) The *limiting reactant* determines the maximum number of product moles resulting from a chemical reaction; any other reactant is an *excess reactant*. (b) The limiting reactant regulates the amount of products because it is completely used up during the reaction; no more product can be made when one of the reactants is unavailable.

3.65 N_2 = , NH_3 = .

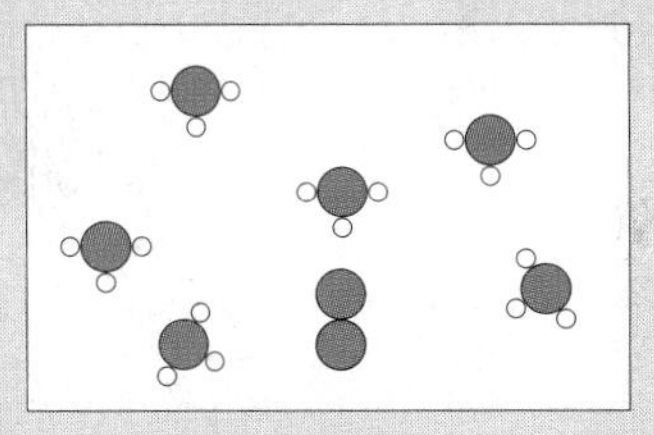

$N_2 + 3H_2 \longrightarrow 2NH_3$. Eight N atoms (4 N_2 molecules) require 24 H atoms (12 H_2 molecules) for complete reaction. Only 9 H_2 molecules are available, so H_2 is the limiting reactant. Nine H_2 molecules (18 H atoms) determine that 6 NH_3 molecules are produced. One N_2 molecule is in excess.
3.67 (a) 2125 bicycles (b) 630 frames left over, 130 handlebars left over (c) the wheels **3.69** NaOH is the limiting reactant; 0.850 mol Na_2CO_3 can be produced; 0.15 mol CO_2 remain.
3.71 (a) $NaHCO_3$ is the limiting reactant. (b) 0.524 g CO_2 (c) 0.238 g citric acid remain **3.73** 0.00 g $AgNO_3$ (limiting reactant), 4.32 g Na_2CO_3, 5.68 g Ag_2CO_3, 3.50 g $NaNO_3$
3.75 (a) The theoretical yield is 60.3 g C_6H_5Br. (b) 94.0% yield **3.77** 6.73 g Li_3N actual yield
3.79 (a) $C_4H_8O_2(l) + 5O_2(g) \longrightarrow 4CO_2(g) + 4H_2O(l)$
(b) $Cu(OH)_2(s) \longrightarrow CuO(s) + H_2O(g)$
(c) $Zn(s) + Cl_2(g) \longrightarrow ZnCl_2(s)$ **3.81** (a) 0.0208 mol C, 1.25×10^{22} C atoms (b) 2.77×10^{-3} mol $C_9H_8O_4$, 1.67×10^{21} $C_9H_8O_4$ molecules **3.83** (a) mass = 4.6638×10^{-19} g Si (b) volume = 2.0×10^{-19} cm^3 (c) edge length = 5.9×10^{-7} cm(= 5.9 nm) **3.85** (a) The empirical formula is $C_{10}H_{18}O$. (b) The molecular formula is $C_{10}H_{18}O$. **3.87** C_6H_5Cl **3.90** (a) 7.6×10^{-5} mol NaI (b) 3.44×10^{-3} g NaI **3.92** 1.1 kg H_2O **3.95** 10.2 g $KClO_3$, 20.0 g $KHCO_3$, 13.8 g K_2CO_3, 56.0 g KCl **3.98** 1.57×10^{24} O atoms **3.100** 52 kg CO_2 **3.102** (a) $S(s) + O_2(g) \longrightarrow SO_2(g)$; $SO_2(g) + CaO(s) \longrightarrow CaSO_3(s)$ (b) 1.7×10^5 kg $CaSO_3$/day

Chapter 4

4.1 Tap water contains enough dissolved electrolytes to complete a circuit between an electrical appliance and our body, producing a shock. **4.3** When CH_3OH dissolves, the neutral CH_3OH molecules that are dispersed throughout the solution do not carry charge and the solution is nonconducting. When $HC_2H_3O_2$ dissolves, a few molecules ionize to form $H^+(aq)$ and $C_2H_3O_2^-(aq)$. These few ions carry some charge and the solution is weakly conducting.
4.5 (a) $ZnCl_2(aq) \longrightarrow Zn^{2+}(aq) + 2Cl^-(aq)$
(b) $HNO_3(aq) \longrightarrow H^+(aq) + NO_3^-(aq)$ **4.7** AX is a

nonelectrolyte, AY is a weak electrolyte, and AZ is a strong electrolyte. **4.9** $HCHO_2$ molecules, H^+ ions, and CHO_2^- ions; $HCHO_2(aq) \rightleftharpoons H^+(aq) + CHO_2^-(aq)$ **4.11** (a) Soluble (b) insoluble (c) soluble (d) insoluble (e) soluble
4.13 (a) $Na_2CO_3(aq) + 2AgNO_3(aq) \longrightarrow Ag_2CO_3(s) + 2NaNO_3(aq)$ (b) No precipitate
(c) $FeSO_4(aq) + Pb(NO_3)_2(aq) \longrightarrow PbSO_4(s) + Fe(NO_3)_2(aq)$
4.15 (a) $2Na^+(aq) + CO_3^{2-}(aq) + Mg^{2+}(aq) + SO_4^{2-}(aq) \longrightarrow MgCO_3(s) + 2Na^+(aq) + SO_4^{2-}(aq)$
$Mg^{2+}(aq) + CO_3^{2-}(aq) \longrightarrow MgCO_3(s)$
(b) $Pb^{2+}(aq) + 2NO_3^-(aq) + 2Na^+(aq) + S^{2-}(aq) \longrightarrow PbS(s) + 2Na^+(aq) + 2NO_3^-(aq)$
$Pb^{2+}(aq) + S^{2-}(aq) \longrightarrow PbS(s)$
(c) $6NH_4^+(aq) + 2PO_4^{3-}(aq) + 3Ca^{2+}(aq) + 6Cl^-(aq) \longrightarrow Ca_3(PO_4)_2(s) + 6NH_4^+(aq) + 6Cl^-(aq)$
$3Ca^{2+}(aq) + 2PO_4^{3-}(aq) \longrightarrow Ca_3(PO_4)_2(s)$
4.17 The solution must contain Ba^{2+}. It could contain K^+ and Ba^{2+} together, but since we are dealing with a single salt, we assume that only Ba^{2+} is present. **4.19** The solution that forms a precipitate with $H_2SO_4(aq)$ is $Pb(NO_3)_2(aq)$; the other one is $Mg(NO_3)_2(aq)$. **4.21** (a) A *monoprotic acid* has one ionizable (acidic) H, whereas a *diprotic acid* has two. (b) A *strong acid* is completely ionized in aqueous solution, whereas only a fraction of *weak acid* molecules are ionized. (c) An *acid* is an H^+ donor, and a *base* is an H^+ acceptor. **4.23** (a) strong acid (b) weak acid (c) weak base (d) strong base **4.25** (a) acid, mixture of ions and molecules (weak electrolyte) (b) none of the above, entirely molecules (nonelectrolyte) (c) salt, entirely ions (strong electrolyte) (d) base, entirely ions (strong electrolyte) **4.27** (a) H_2SO_3, weak electrolyte (b) C_2H_5OH, nonelectrolyte (c) NH_3, weak electrolyte (d) $KClO_3$, strong electrolyte (e) $Cu(NO_3)_2$, strong electrolyte
4.29 (a) $2HBr(aq) + Ca(OH)_2(aq) \longrightarrow CaBr_2(aq) + 2H_2O(l)$
$H^+(aq) + OH^-(aq) \longrightarrow H_2O(l)$
(b) $Cu(OH)_2(s) + 2HClO_4(aq) \longrightarrow Cu(ClO_4)_2(aq) + 2H_2O(l)$
$Cu(OH)_2(s) + 2H^+(aq) \longrightarrow 2H_2O(l) + Cu^{2+}(aq)$
(c) $Al(OH)_3(s) + 3HNO_3(aq) \longrightarrow Al(NO_3)_3(aq) + 3H_2O(l)$
$Al(OH)_3(s) + 3H^+(aq) \longrightarrow 3H_2O(l) + Al^{3+}(aq)$
4.31 (a) $CdS(s) + H_2SO_4(aq) \longrightarrow CdSO_4(aq) + H_2S(g)$
$CdS(s) + 2H^+(aq) \longrightarrow H_2S(g) + Cd^{2+}(aq)$
(b) $MgCO_3(s) + 2HClO_4(aq) \longrightarrow Mg(ClO_4)_2(aq) + H_2O(l) + CO_2(g)$;
$MgCO_3(s) + 2H^+(aq) \longrightarrow H_2O(l) + CO_2(g) + Mg^{2+}(aq)$
4.33 (a) $FeO(s) + 2H^+(aq) \longrightarrow H_2O(l) + Fe^{2+}(aq)$
(b) $NiO(s) + 2H^+(aq) \longrightarrow H_2O(l) + Ni^{2+}(aq)$ **4.35** (a) In terms of electron transfer, *oxidation* is the loss of electrons by a substance, and *reduction* is the gain of electrons (LEO says GER). (b) Relative to oxidation numbers, when a substance is oxidized, its oxidation number increases. When a substance is reduced, its oxidation number decreases. **4.37** The most easily oxidized metals are near the bottom of groups on the left side of the chart, especially groups 1A and 2A. The least easily oxidized metals are on the lower right of the transition metals, particularly those near the bottom of groups 8B and 1B.
4.39 (a) +6 (b) +4 (c) +7 (d) +1 (e) 0 (f) −1
4.41 (a) $Ni \longrightarrow Ni^{2+}$, Ni is oxidized; $Cl_2 \longrightarrow 2Cl^-$, Cl is reduced (b) $Fe^{2+} \longrightarrow Fe$, Fe is reduced; $Al \longrightarrow Al^{3+}$, Al is oxidized (c) $Cl_2 \longrightarrow 2Cl^-$, Cl is reduced; $2I^- \longrightarrow I_2$, I is oxidized (d) $S^{2-} \longrightarrow SO_4^{2-}$, S is oxidized; $H_2O_2 \longrightarrow H_2O$; O is reduced **4.43** (a) $Mn(s) + H_2SO_4(aq) \longrightarrow MnSO_4(aq) + H_2(g)$; $Mn(s) + 2H^+(aq) \longrightarrow Mn^{2+}(aq) + H_2(g)$
(b) $2Cr(s) + 6HBr(aq) \longrightarrow 2CrBr_3(aq) + 3H_2(g)$
$2Cr(s) + 6H^+(aq) \longrightarrow 2Cr^{3+}(aq) + 3H_2(g)$
(c) $Sn(s) + 2HCl(aq) \longrightarrow SnCl_2(aq) + H_2(g)$
$Sn(s) + 2H^+(aq) \longrightarrow Sn^{2+}(aq) + H_2(g)$
(d) $2Al(s) + 6HCHO_2(aq) \longrightarrow 2Al(CHO_2)_3(aq) + 3H_2(g)$
$2Al(s) + 6HCHO_2(aq) \longrightarrow 2Al^{3+}(aq) + 6CHO_2^-(aq) + 3H_2(g)$
4.45 (a) $2Al(s) + 3NiCl_2(aq) \longrightarrow 2AlCl_3(aq) + 3Ni(s)$ (b) no reaction (c) $2Cr(s) + 3NiSO_4(aq) \longrightarrow Cr_2(SO_4)_3(aq) + 3Ni(s)$
(d) $Mn(s) + 2HBr(aq) \longrightarrow MnBr_2(aq) + H_2(g)$
(e) $H_2(g) + CuCl_2(aq) \longrightarrow Cu(s) + 2HCl(aq)$
4.47 (a) i. $Zn(s) + Cd^{2+}(aq) \longrightarrow Cd(s) + Zn^{2+}(aq)$;
ii. $Cd(s) + Ni^{2+}(aq) \longrightarrow Ni(s) + Cd^{2+}(aq)$ (b) Cd is between Zn and Ni on the activity series. (c) Place an iron strip in $CdCl_2(aq)$. If Cd(*s*) is deposited, Cd is less active than Fe; if there is no reaction, Cd is more active than Fe. Do the same test with Co if Cd is less active than Fe or with Cr if Cd is more active than Fe. **4.49** (a) Intensive; the *ratio* of amount of solute to total amount of solution is the same, regardless of how much solution is present. (b) The term *0.50 mol HCl* defines an amount (~18 g) of the pure substance HCl. The term 0.50 *M* HCl is a ratio; it indicates that there are 0.50 mol of HCl solute in 1.0 liter of solution. **4.51** (a) 0.0863 *M* NH_4Cl (b) 0.0770 mol HNO_3 (c) 83.3 mL of 1.50 *M* KOH
4.53 (a) 4.46 g KBr (b) 0.145 *M* $Ca(NO_3)_2$ (c) 20.3 mL of 1.50 *M* Na_3PO_4 **4.55** (a) 0.15 *M* K_2CrO_4 has the highest K^+ concentration. (b) 30.0 mL of 0.15 *M* K_2CrO_4 has more K^+ ions.
4.57 (a) 0.14 *M* Na^+, 0.14 *M* OH^- (b) 0.25 *M* Ca^{2+}, 0.50 *M* Br^- (c) 0.25 *M* CH_3OH (d) 0.067 *M* K^+, 0.067 *M* ClO_3^-, 0.13 *M* Na^+, 0.067 *M* SO_4^{2-} **4.59** (a) 1.69 mL 14.8 *M* NH_3 (b) 0.592 *M* NH_3 **4.61** (a) Add 6.42 g $C_{12}H_{22}O_{11}$ to a 125-mL volumetric flask, dissolve in a small volume of water, and add water to the mark on the neck of the flask. Agitate thoroughly to ensure total mixing. (b) Thoroughly rinse, clean, and fill a 50-mL buret with the 1.50 *M* $C_{12}H_{22}O_{11}$. Dispense 26.7 mL of this solution into a 400-mL volumetric container, add water to the mark, and mix thoroughly. **4.63** 1.398 *M* $HC_2H_3O_2$
4.65 0.117 g NaCl **4.67** (a) 38.0 mL of 0.115 *M* $HClO_4$ (b) 769 mL of 0.128 *M* HCl (c) 0.408 *M* $AgNO_3$ (d) 0.275 g KOH **4.69** 27 g $NaHCO_3$ **4.71** 1.22×10^{-2} *M* $Ca(OH)_2$ solution; the solubility of $Ca(OH)_2$ is 0.0904 g in 100 mL solution.
4.73 (a) $NiSO_4(aq) + 2KOH(aq) \longrightarrow Ni(OH)_2(s) + K_2SO_4(aq)$ (b) $Ni(OH)_2$ (c) KOH is the limiting reactant.
(d) 0.927 g $Ni(OH)_2$ (e) 0.0667 *M* $Ni^{2+}(aq)$, 0.0667 *M* $K^+(aq)$, 0.100 *M* $SO_4^{2-}(aq)$ **4.75** 91.40% $Mg(OH)_2$ **4.77** The precipitate is CdS(*s*). $Na^+(aq)$ and $NO_3^-(aq)$ are spectator ions and remain in solution, along with any excess reactant ions. The net ionic equation is $Cd^{2+}(aq) + S^{2-}(aq) \longrightarrow CdS(s)$.
4.80 (a) $Al(OH)_3(s) + 3H^+(aq) \longrightarrow Al^{3+}(aq) + 3H_2O(l)$
(b) $Mg(OH)_2(s) + 2H^+(aq) \longrightarrow Mg^{2+}(aq) + 2H_2O(l)$
(c) $MgCO_3(s) + 2H^+(aq) \longrightarrow Mg^{2+}(aq) + H_2O(l) + CO_2(g)$
(d) $NaAl(CO_3)(OH)_2(s) + 4H^+(aq) \longrightarrow Na^+(aq) + Al^{3+}(aq) + 3H_2O(l) + CO_2(g)$
(e) $CaCO_3(s) + 2H^+(aq) \longrightarrow Ca^{2+}(aq) + H_2O(l) + CO_2(g)$
4.83 (a) No reaction (b) $Zn(s) + Pb^{2+}(aq) \longrightarrow Zn^{2+}(aq) + Pb(s)$ (c) no reaction (d) $Zn(s) + Fe^{2+}(aq) \longrightarrow Zn^{2+}(aq) + Fe(s)$ (e) $Zn(s) + Cu^{2+}(aq) \longrightarrow Zn^{2+}(aq) + Cu(s)$ (f) no reaction
4.86 1.70 *M* KBr **4.89** 30 mol Na^+ **4.91** 0.368 *M* H_2O_2
4.93 1.81×10^{19} Na^+ ions **4.96** 5.1×10^3 kg Na_2CO_3
4.99 0.233 *M* Cl^- **4.102** (a) +5 (b) silver arsenate (c) 5.22% As

Chapter 5

5.1 An object can possess energy by virtue of its motion or position. Kinetic energy depends on the mass of the object and its velocity. Potential energy depends on the position of the object relative to the body with which it interacts. **5.3** (a) 84 J (b) 20 cal (c) As the ball hits the sand, its speed (and hence its kinetic energy) drops to zero. Most of the kinetic energy is transferred to the sand, which deforms when the ball lands. Some energy is released as heat through friction between the ball and the sand. **5.5** 1 Btu = 1054 J **5.7** 2.1×10^3 kcal **5.9** As the pellet rises against the force of gravity, kinetic energy imparted by the air gun is changed to potential energy. When all kinetic energy has been transferred to potential energy (or lost as heat through friction), the pellet stops rising and falls to Earth. In principle, if enough kinetic energy could be imparted to the pellet, it could escape the force of gravity and move into space. For an air gun and a pellet, this is practically impossible. **5.11** (a) The *system* is the well-defined part of the universe whose energy changes are being studied. (b) A closed system can exchange heat but not mass with its surroundings. **5.13** (a) *Work* is a force applied over a distance. (b) The amount of work done is the magnitude of the force times the distance over which it is applied. $w = F \times d$. **5.15** (a) Gravity; work is done because the force of gravity is opposed and the pencil is lifted. (b) Mechanical force; work is done because the force of the coiled spring is opposed as the spring is compressed over a distance. **5.17** (a) In any chemical or physical change, energy can be neither created nor destroyed; energy is conserved. (b) The *internal energy (E)* of a system is the sum of all the kinetic and potential energies of the system components. (c) Internal energy increases when work is done on the system and when heat is transferred to the system. **5.19** (a) $\Delta E = -152$ kJ, exothermic (b) $\Delta E = +0.75$ kJ, endothermic (c) $\Delta E = +14.0$ kJ, endothermic **5.21** (a) System (iii) is endothermic. (b) $\Delta E < 0$ for system (iii). (c) $\Delta E > 0$ for systems (i) and (ii). **5.23** (a) Since little or no work is done by the system in case (2), the gas will absorb most of the energy as heat; the case (2) gas will have the higher temperature. (b) In case (2) $w \approx 0$ and $q \approx 100$ J. In case (1) a significant amount of energy will be used to do work on the surroundings ($-w$), but some will be absorbed as heat ($+q$). (c) ΔE is greater for case (2) because the entire 100 J increases the internal energy of the system, rather than a part of the energy doing work on the surroundings. **5.25** (a) A *state function* is a property that depends only on the physical state (pressure, temperature, etc.) of the system, not on the route used to get to the current state. (b) Internal energy *is* a state function; work *is not* a state function. (c) Temperature is a state function; regardless of how hot or cold the sample has been, the temperature depends only on its present condition. **5.27** (a) For the many processes that occur at constant atmospheric pressure, the enthalpy change is a meaningful measure of the energy change associated with the process. (b) Only under conditions of constant pressure is ΔH for a process equal to the heat transferred during the process. (c) The process is exothermic. **5.29** (a) $HC_2H_3O_2(l) + 2O_2(g) \longrightarrow 2H_2O(l) + 2CO_2(g)$, $\Delta H = -871.7$ kJ

(b) $HC_2H_3O_2(l) + 2O_2(g)$

$\Delta H = -871.7$ kJ

$2\,H_2O(l) + 2CO_2(g)$

5.31 The reactant, $2Cl(g)$, has the higher enthalpy. **5.33** (a) Exothermic (b) −59 kJ heat transferred (c) 6.43 g MgO produced (d) +112 kJ heat absorbed **5.35** (a) −35.4 kJ (b) −0.759 kJ (c) +12.3 J **5.37** At constant pressure, $\Delta E = \Delta H - P\Delta V$. The values of either P and ΔV or T and Δn must be known to calculate ΔE from ΔH. **5.39** $\Delta E = -125$ kJ, $\Delta H = -89$ kJ **5.41** (a) $\Delta H = +726.5$ kJ (b) $\Delta H = -1453$ kJ (c) The exothermic forward reaction is more likely to be thermodynamically favored. (d) Vaporization is endothermic. If the product were $H_2O(g)$, the reaction would be more endothermic and would have a smaller negative ΔH. **5.43** (a) J/°C or J/K (b) J/g-°C or J/g-K **5.45** (a) 4.184 J/g-K (b) 774 J/°C (c) 904 kJ **5.47** 3.47×10^4 J **5.49** $\Delta H = -45.7$ kJ/mol NaOH **5.51** $\Delta E_{rxn} = -25.5$ kJ/g $C_6H_4O_2$ or -2.75×10^3 kJ/mol $C_6H_4O_2$ **5.53** (a) Heat capacity of the complete calorimeter = 14.4 kJ/°C (b) 5.40°C **5.55** If a reaction can be described as a series of steps, ΔH for the reaction is the sum of the enthalpy changes for each step. As long as we can describe a route where ΔH for each step is known, ΔH for any process can be calculated. **5.57** (a) $\Delta H = +90$ kJ

(b)

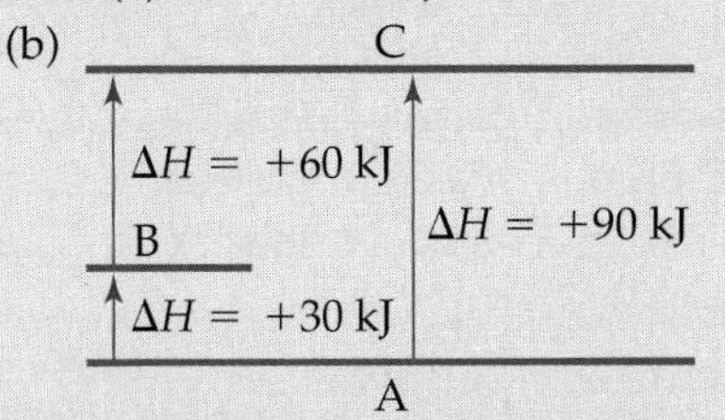

The process of A forming C can be described as A forming B and B forming C. **5.59** $\Delta H = -1300.0$ kJ **5.61** $\Delta H = -2.49 \times 10^3$ kJ **5.63** (a) *Standard conditions* for enthalpy changes are $P = 1$ atm and some common temperature, usually 298 K. (b) *Enthalpy of formation* is the enthalpy change that occurs when a compound is formed from its component elements. (c) *Standard enthalpy of formation* ΔH_f° is the enthalpy change that accompanies formation of one mole of a substance from elements in their standard states. **5.65** Yes, it would still be possible to have tables of standard enthalpies of formation like Table 5.3. Standard enthalpies of formation are the net enthalpy difference between a compound and its component elements in their standard states. Regardless of the value of the enthalpy of formation of the elements, the magnitude of the difference in enthalpies should be the same (assuming the same reaction stoichiometry). **5.67** (a) $\frac{1}{2}N_2(g) + \frac{3}{2}H_2(g) \longrightarrow NH_3(g)$, $\Delta H_f^\circ = -46.19$ kJ (b) $\frac{1}{8}S_8(s) + O_2(g) \longrightarrow SO_2(g)$, $\Delta H_f^\circ = -296.9$ kJ (c) $Rb(s) + \frac{1}{2}Cl_2(g) + \frac{3}{2}O_2(g) \longrightarrow RbClO_3(s)$, $\Delta H_f^\circ = -392.4$ kJ (d) $N_2(g) + 2H_2(g) + \frac{3}{2}O_2(g) \longrightarrow NH_4NO_3(s)$, $\Delta H_f^\circ = -365.6$ kJ **5.69** $\Delta H_{rxn}^\circ = -847.6$ kJ **5.71** (a) $\Delta H_{rxn}^\circ = -196.6$ kJ (b) $\Delta H_{rxn}^\circ = 37.1$ kJ (c) $\Delta H_{rxn}^\circ = -556.7$ kJ (d) $\Delta H_{rxn}^\circ = -68.3$ kJ **5.73** $\Delta H_f^\circ = -248$ kJ **5.75** $\Delta H_f^\circ = -924.8$ kJ **5.77** (a) $C_8H_{18}(l) + \frac{25}{2}O_2(g) \longrightarrow 8CO_2(g) + 9H_2O(g)$, $\Delta H = -5069$ kJ (b) $8C(s, gr) + 9H_2(g) \longrightarrow C_8H_{18}(l)$ (c) $\Delta H_f^\circ = -255$ kJ **5.79** (a) *Fuel value* is the amount of heat produced when 1 gram of a substance (fuel) is combusted. (b) Glucose, $C_6H_{12}O_6$, is *blood sugar*. It is important because glucose is the fuel that is carried by blood to cells and combusted to produce energy in the body. (c) 5 g of fat **5.81** 104 or 1×10^2 Cal/serving **5.83** 59.7 Cal **5.85** (a) $\Delta H_{comb} = -1850$ kJ/mol C_3H_4, -1926 kJ/mol C_3H_6, -2044 kJ/mol C_3H_8 (b) $\Delta H_{comb} = -4.61 \times 10^4$ kJ/kg C_3H_4, -4.58×10^4 kJ/kg C_3H_6, -4.635×10^4 kJ/kg C_3H_8 (c) These three substances yield nearly identical quantities of heat per unit mass, but propane is marginally higher than the other two. **5.87** (a) 469.4 m/s (b) 5.124×10^{-21} J (c) 3.086 kJ/mol **5.90** The spontaneous air bag reaction is probably exothermic, with $-\Delta H$ and thus $-q$. When the bag inflates, work is done by the system, so the sign of w is also negative. **5.93** (a) $q = 0$, $w > 0$, $\Delta E > 0$ (b) The sign of q is negative. The changes in state described in both cases are identical. ΔE is the same in both cases, but the distribution of energy transferred as either work or heat is different in the two scenarios.

5.96 1.8×10^4 or 18,000 bricks
5.100 (a, b) $CH_4(g) + O_2(g) \longrightarrow C(s) + 2H_2O(l)$, $\Delta H° = -496.9$ kJ; $CH_4(g) + \frac{3}{2}O_2(g) \longrightarrow CO(g) + 2H_2O(l)$, $\Delta H° = -607.4$ kJ; $CH_4(g) + 2O_2(g) \longrightarrow CO_2(g) + 2H_2O(l)$, $\Delta H° = -890.4$ kJ (c) Assuming that $O_2(g)$ is present in excess, the reaction that produces $CO_2(g)$ has the most negative ΔH per mole of CH_4 burned and therefore the most thermodynamically stable products. **5.103** 1,3 butadiene: (a) $\Delta H = -2543.4$ kJ/mol C_4H_6 (b) 47 kJ/g (c) 11.18% H. 1-butene: (a) $\Delta H = -2718.5$ kJ/mol C_4H_8 (b) 48 kJ/g (c) 14.37% H. *n*-butane: (a) $\Delta H = -2878.5$ kJ/mol C_4H_{10} (b) 50 kJ/g (c) 17.34% H. (d) As the mass % H increases, the fuel value (kJ/g) of the hydrocarbon increases, given the same number of C atoms. A graph of the data suggests that mass % H and fuel value are directly proportional when the number of C atoms is constant. **5.107** (a) 1.479×10^{-18} J/molecule (b) 1×10^{-15} J/photon. The X ray has approximately 1000 times more energy than is produced by the combustion of 1 molecule of $CH_4(g)$. **5.111** (a) 3.18 g Cu (b) $Cu(OH)_2$ (c) $CuSO_4(aq) + 2KOH(aq) \longrightarrow Cu(OH)_2(s) + K_2SO_4(aq)$, $Cu^{2+}(aq) + 2OH^-(aq) \longrightarrow Cu(OH)_2(s)$ (d) $\Delta H = -52$ kJ

Chapter 6

6.1 (a) Meters (b) 1/seconds (c) meters/second **6.3** (a) True (b) False. The frequency of radiation decreases as the wavelength increases. (c) False. Ultraviolet light has shorter wavelengths than visible light. (d) False. Electromagnetic radiation and sound waves travel at different speeds. **6.5** wavelength of X rays < ultraviolet < green light < red light < infrared < radio waves **6.7** (a) 6.63×10^{20} s^{-1} (b) 1.18×10^{-8} m (c) neither is visible (d) 2.25×10^6 m **6.9** 6.88×10^{14} s^{-1}; blue **6.11** (a) *Quantization* means that energy can only be absorbed or emitted in specific amounts or multiples of these amounts. This minimum amount of energy is equal to a constant times the frequency of the radiation absorbed or emitted; $E = h\nu$. (b) In everyday activities, macroscopic objects such as our bodies gain and lose total amounts of energy much larger than a single quantum $h\nu$. The gain or loss of the relatively minuscule quantum of energy is unnoticed. **6.13** (a) 2.45×10^{-19} J (b) 1.80×10^{-20} J (c) 25.3 nm; ultraviolet **6.15** (a) $\lambda = 3.3\ \mu m$, $E = 6.0 \times 10^{-20}$ J; $\lambda = 0.154$ nm, $E = 1.29 \times 10^{-15}$ J (b) The 3.3 μm photon is in the infrared and the 0.154 nm photon is in the X-ray region; the X-ray photon has the greater energy.
6.17 (a) 6.11×10^{-19} J/photon (b) 368 kJ/mol (c) 1.64×10^{15} photons **6.19** 8.1×10^{16} photons/s
6.21 (a) $E_{min} = 7.22 \times 10^{-19}$ J (b) $\lambda = 275$ nm (c) $E_{120} = 1.66 \times 10^{-18}$ J. The excess energy of the 120 nm photon is converted into the kinetic energy of the emitted electron. $E_k = 9.3 \times 10^{-19}$ J/electron. **6.23** When applied to atoms, the notion of quantized energies means that only certain values of ΔE are allowed. These are represented by the lines in the emission spectra of excited atoms. **6.25** (a) Emitted (b) absorbed (c) emitted **6.27** $E_2 = -5.45 \times 10^{-19}$ J; $E_6 = -0.606 \times 10^{-19}$ J; $\Delta E = 4.84 \times 10^{-19}$ J; $\lambda = 410$ nm; visible, violet.
6.29 (a) Only lines with $n_f = 2$ represent ΔE values and wavelengths that lie in the visible portion of the spectrum. Lines with $n_f = 1$ have shorter wavelengths and lines with $n_f > 2$ have longer wavelengths than visible radiation. (b) $n_i = 3, n_f = 2$; $\lambda = 6.56 \times 10^{-7}$ m; this is the red line at 656 nm. $n_i = 4, n_f = 2$; $\lambda = 4.86 \times 10^{-7}$ m; this is the blue line at 486 nm. $n_i = 5, n_f = 2$; $\lambda = 4.34 \times 10^{-7}$ m; this is the violet line at 434 nm. **6.31** (a) Ultraviolet region (b) $n_i = 6, n_f = 1$ **6.33** (a) $\lambda = 5.6 \times 10^{-37}$ m (b) $\lambda = 2.65 \times 10^{-34}$ m (c) $\lambda = 2.3 \times 10^{-13}$ m

6.35 4.14×10^3 m/s **6.37** (a) $\Delta x \geq 4 \times 10^{-27}$ m (b) $\Delta x \geq 3 \times 10^{-10}$ m **6.39** The Bohr model states with 100% certainty that the electron in hydrogen can be found 0.53 Å from the nucleus. The quantum mechanical model is a statistical model that states the probability of finding the electron in certain regions around the nucleus. While 0.53 Å is the radius with highest probability, that probability is always less than 100%. **6.41** (a) $n = 4, l = 3, 2, 1, 0$ (b) $l = 2, m_l = -2, -1, 0, 1, 2$
6.43 (a) $3p$: $n = 3, l = 1$ (b) $2s$: $n = 2, l = 0$ (c) $4f$: $n = 4, l = 3$ (d) $5d$: $n = 5, l = 2$ **6.45** (a) impossible, $1p$ (b) possible (c) possible (d) impossible, $2d$

6.47

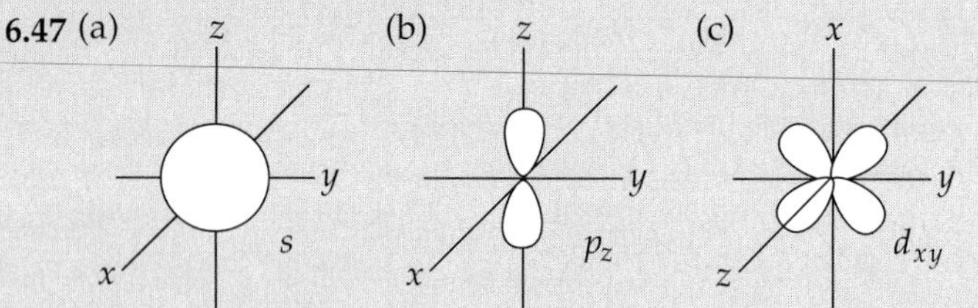

6.49 (a) The hydrogen atom 1*s* and 2*s* orbitals have the same overall spherical shape, but the 2*s* orbital has a larger radial extension and one more node than the 1*s* orbital. (b) A single 2*p* orbital is directional in that its electron density is concentrated along one of the three Cartesian axes of the atom. The $d_{x^2-y^2}$ orbital has electron density along both the *x*- and *y*-axes, while the p_x orbital has density only along the *x*-axis. (c) The average distance of an electron from the nucleus in a 3*s* orbital is greater than for an electron in a 2*s* orbital. (d) $1s < 2p < 3d < 4f < 6s$ **6.51** (a) In the hydrogen atom, orbitals with the same principle quantum number, *n*, have the same energy. (b) In a many-electron atom, for a given *n* value, orbital energy increases with increasing *l* value: $s < p < d < f$
6.53 (a) $+\frac{1}{2}, -\frac{1}{2}$ (b) a magnet with a strong inhomogeneous magnetic field (c) they must have different m_s values; the Pauli exclusion principle **6.55** (a) 10 (b) 2 (c) 6 (d) 14
6.57 (a) Each box represents an orbital. (b) Electron spin is represented by the direction of the half-arrows. (c) No. In Be, there are no electrons in subshells that have degenerate orbitals, so Hund's rule is not used. **6.59** (a) Cs, [Xe]$6s^1$ (b) Ni, [Ar]$4s^2 3d^8$ (c) Se, [Ar]$4s^2 3d^{10} 4p^4$ (d) Cd, [Kr]$5s^2 4d^{10}$ (e) Ac, [Rn]$7s^2 6d^1$ (f) Pb, [Xe]$6s^2 4f^{14} 5d^{10} 6p^2$

6.61

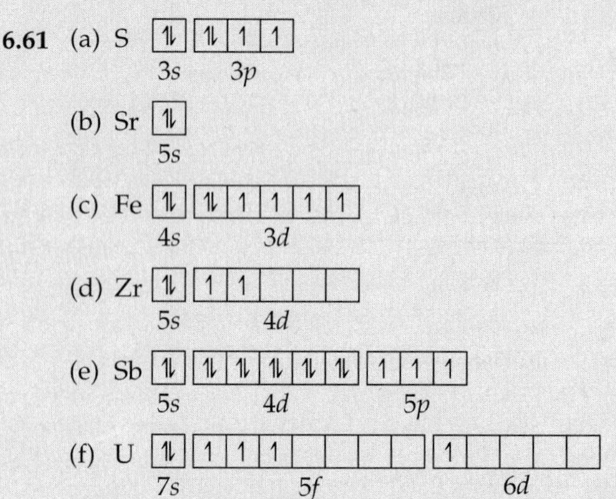

(a) 2 unpaired electrons (b) 0 unpaired electrons (c) 4 unpaired electrons (d) 2 unpaired electrons (e) 3 unpaired electrons (f) 4 unpaired electrons **6.63** (a) Mg (b) Al (c) Cr (d) Te **6.65** (a) The fifth electron would fill the 2*p* subshell before the 3*s*. (b) Either the core is [He], or the outer electron configuration should be $3s^2 3p^3$. (c) The 3*p* subshell would fill before the $3d$. **6.67** (a) $\lambda_A = 3.6 \times 10^{-8}$ m, $\lambda_B = 8.0 \times 10^{-8}$ m (b) $\nu_A = 8.4 \times 10^{15}\ s^{-1}$, $\nu_B = 3.7 \times 10^{15}\ s^{-1}$ (c) A, ultraviolet; B, ultraviolet **6.69** 46.7 min **6.71** 1.6×10^{18} photons **6.73** 3.6×10^6 photons/s, 1.3×10^{-12} J/s **6.75** (a) Radiation from the Sun is a continuous spectrum. When gaseous atoms in the Sun's atmosphere are exposed to this radiation, the electrons in these atoms change from their ground state to one of several allowed excited states. Thus, the dark lines are the wavelengths that correspond to allowed energy changes in atoms of the solar atmosphere. The continuous background is all other wavelengths of solar radiation. (b) The scientist should record the absorption spectrum of pure neon or other elements of interest. The black lines should appear at the same wavelengths regardless of the source of neon. **6.77** $v = 1.02 \times 10^7$ m/s **6.79** (a) l (b) n and l (c) m_s (d) m_l **6.81** (a) 1 (b) 3 (c) 5 (d) 9 **6.83** (a) The xy plane, where $z = 0$ (b) the yz and xz planes, where $x = 0$ and $y = 0$ (c) the planes that bisect the x and y axes and contain the z axis, where $x^2 - y^2 = 0$ **6.85** Mt, [Rn] $7s^25f^{14}6d^7$ **6.87** 1.7×10^{28} photons

Chapter 7

7.1 Mendeleev placed elements with similar chemical and physical properties within a family or column of the table. For undiscovered elements, he left blanks. He predicted properties for the "blanks" based on properties of other elements in the family and on either side. **7.3** (a) *Effective nuclear charge*, Z_{eff}, is a representation of the average electrical field experienced by a single electron. It is the average environment created by the nucleus and the other electrons in the molecule, expressed as a net positive charge at the nucleus. (b) Going from left to right across a period, effective nuclear charge increases. **7.5** (a) K, 1+ (b) Br, 7+ **7.7** The $n = 3$ electrons in Kr experience a greater effective nuclear charge and thus have a greater probability of being closer to the nucleus. **7.9** Atomic radii are determined by distances between atoms in various situations. Bonding radii are calculated from the internuclear separation of two atoms joined by a chemical bond. Nonbonding radii are calculated from the internuclear separation between two gaseous atoms that collide and move apart but do not bond. **7.11** 1.44 Å **7.13** From the sum of the atomic radii, As—I = 2.52 Å. This is very close to the experimental value of 2.55 Å. **7.15** (a) Decrease (b) increase (c) F < S < P < As **7.17** (a) Be < Mg < Ca (b) Br < Ge < Ga (c) Si < Al < Tl **7.19** (a) Electrostatic repulsions are reduced by removing an electron from a neutral atom, effective nuclear charge increases, and the cation is smaller. (b) The additional electrostatic repulsion produced by adding an electron to a neutral atom decreases the effective nuclear charge experienced by the valence electrons, and increases the size of the anion. (c) Going down a column, valence electrons are further from the nucleus, and they experience greater shielding by core electrons. The greater radial extent of the valence electrons outweighs the increase in Z. **7.21** The blue sphere is a metal; its size decreases on reaction, characteristic of the change in radius when a metal atom forms a cation. The red sphere is a nonmetal; its size increases on reaction, characteristic of the change in radius when a nonmetal atom forms an anion. **7.23** (a) An *isoelectronic* series is a group of atoms or ions that have the same number of electrons. (b) (i) Cl^-: Ar (ii) Se^{2-}: Kr (iii) Mg^{2+}: Ne **7.25** (a) Since the number of electrons in an isoelectronic series is the same, repulsion and shielding effects are usually similar for the different particles. As Z increases, the valence electrons are more strongly attracted to the nucleus and the size of the particle decreases. (b) A $2p$ electron in Na^+ **7.27** (a) Se < Se^{2-} < Te^{2-} (b) Co^{3+} < Fe^{3+} < Fe^{2+} (c) Ti^{4+} < Sc^{3+} < Ca (d) Be^{2+} < Na^+ < Ne **7.29** $Te(g) \longrightarrow Te^+(g) + e^-$; $Te^+(g) \longrightarrow Te^{2+}(g) + e^-$; $Te^{2+}(g) \longrightarrow Te^{3+}(g) + e^-$ **7.31** (a) According to Coulomb's law, the energy of an electron in an atom is negative. In order to increase the energy of the electron and remove it from the atom, energy must be added to the atom. Ionization energy, ΔE for this process, is positive. (b) F has a greater first ionization energy than O because F has a greater Z_{eff} and the outer electrons in both elements are approximately the same distance from the nucleus. (c) The second ionization energy of an element is greater than the first because more energy is required to overcome the larger Z_{eff} of the 1+ cation than that of the neutral atom. **7.33** (a) The smaller the atom, the larger its first ionization energy (of the nonradioactive elements). (b) He has the largest, and Cs the smallest first ionization energy. **7.35** (a) Ne (b) Mg (c) Cr (d) Br (e) Ge **7.37** (a) Sb^{3+}, [Kr]$5s^24d^{10}$ (b) Ga^+, [Ar]$4s^23d^{10}$ (c) P^{3-}, [Ne]$3s^23p^6$ or [AR] (d) Cr^{3+}, [Ar]$3d^3$ (e) Zn^{2+}, [Ar]$3d^{10}$ (f) Ag^+, [Kr]$4d^{10}$ **7.39** (a) Co^{2+}, [Ar]$3d^7$, 3 unpaired electrons (b) In^+, [Kr]$5s^24d^{10}$, 0 unpaired electrons **7.41** Ionization energy: $Se(g) \longrightarrow Se^+(g) + e^-$; [Ar]$4s^23d^{10}4p^4 \longrightarrow$ [Ar]$4s^23d^{10}4p^3$; electron affinity: $Se(g) + e^- \longrightarrow Se^-(g)$; [Ar]$4s^23d^{10}4p^4 \longrightarrow$ [Ar]$4s^23d^{10}4p^5$ **7.43** $Li + 1e^- \longrightarrow Li^-$; [He]$2s^1 \longrightarrow$ [He]$2s^2$; $Be + 1e^- \longrightarrow Be^-$; [He]$2s^2 \longrightarrow$ [He]$2s^22p^1$. Adding an electron to Li completes the $2s$ subshell. The added electron experiences essentially the same effective nuclear charge as the other valence electron, there is an overall stabilization and ΔE is negative. An extra electron in Be would occupy the higher energy $2p$ subshell. This electron is shielded from the full nuclear charge by the $2s$ electrons and does not experience a stabilization in energy; ΔE is positive. **7.45** The smaller the first ionization energy of an element, the greater the metallic character of that element. **7.47** (a) Li (b) Na (c) Sn (d) Al **7.49** Ionic: MgO, Li_2O, Y_2O_3; molecular: SO_2, P_2O_5, N_2O, XeO_3. Ionic compounds are formed by combining a metal and a nonmetal; molecular compounds are formed by two or more nonmetals. **7.51** (a) An acidic oxide dissolved in water produces an acidic solution; a basic oxide dissolved in water produces a basic solution. (b) Oxides of nonmetals, such as SO_3, are acidic; oxides of metals, such as CaO, are basic.
7.53 (a) $BaO(s) + H_2O(l) \longrightarrow Ba(OH)_2(aq)$
(b) $FeO(s) + 2HClO_4(aq) \longrightarrow Fe(ClO_4)_2(aq) + H_2O(l)$
(c) $SO_3(g) + H_2O(l) \longrightarrow H_2SO_4(aq)$
(d) $CO_2(g) + 2NaOH(aq) \longrightarrow Na_2CO_3(aq) + H_2O(l)$
7.55 (a) Na, [Ne]$3s^1$; Mg, [Ne]$3s^2$ (b) When forming ions, both adopt the stable configuration of Ne; Na loses one electron and Mg two electrons to achieve this configuration. (c) The effective nuclear charge of Mg is greater, so its ionization energy is greater. (d) Mg is less reactive because it has a higher ionization energy. (e) The atomic radius of Mg is smaller because the effective nuclear charge is greater. **7.57** (a) Ca is more reactive because it has a lower ionization energy than Mg. (b) K is more reactive because it has a lower ionization energy than Ca.
7.59 (a) $2K(s) + Cl_2(g) \longrightarrow 2KCl(s)$
(b) $SrO(s) + H_2O(l) \longrightarrow Sr(OH)_2(aq)$
(c) $4Li(s) + O_2(g) \longrightarrow 2Li_2O(s)$
(d) $2Na(s) + S(l) \longrightarrow Na_2S(s)$ **7.61** H, $1s^1$; Li, [He] $2s^1$; F, [He] $2s^22p^5$. Like Li, H has only one valence electron, and its most common oxidation number is +1. Like F, H needs only one electron to adopt the stable electron configuration of the nearest noble gas; both H and F can exist in the −1 oxidation state. **7.63** (a) F, [He] $2s^22p^5$; Cl, [Ne] $3s^23p^5$ (b) F and Cl are in the same group, and both adopt a 1− ionic charge. (c) The

$2p$ valence electrons in F are closer to the nucleus and more tightly held than are the $3p$ electrons of Cl, so the ionization energy of F is greater. (d) The high ionization energy of F coupled with a relatively large exothermic electron affinity makes it more reactive than Cl toward H_2O. (e) While F has approximately the same effective nuclear charge as Cl, its small atomic radius gives rise to large repulsions when an extra electron is added, so the overall electron affinity of F is less exothermic than that of Cl. (f) The $2p$ valence electrons in F are closer to the nucleus so the atomic radius is smaller than that of Cl. **7.65** Under ambient conditions, the group 8A elements are all gases that are extremely unreactive, so the name "inert gases" seemed appropriate. It is inappropriate because both Xe and Kr were found to react with substances having a strong tendency to remove electrons, such as F_2. **7.67** (a) $2O_3(g) \longrightarrow 3O_2(g)$ (b) $Xe(g) + F_2(g) \longrightarrow XeF_2(g)$; $Xe(g) + 2F_2(g) \longrightarrow XeF_4(s)$; $Xe(g) + 3F_2(g) \longrightarrow XeF_6(s)$ (c) $S(s) + H_2(g) \longrightarrow H_2S(g)$ (d) $2F_2(g) + 2H_2O(l) \longrightarrow 4HF(aq) + O_2(g)$ **7.69** (a) Te has more metallic character and is a better electrical conductor. (b) At room temperature, oxygen molecules are diatomic and exist in the gas phase. Sulfur molecules are 8-membered rings and exist in the solid state. (c) Chlorine is generally more reactive than bromine because Cl atoms have a greater (more exothermic) electron affinity than Br atoms. **7.71** Up to $Z = 83$, there are three instances where atomic weights are reversed relative to atomic numbers: Ar and K; Co and Ni; Te and I. In each case the most abundant isotope of the element with the larger atomic number has one more proton, but fewer neutrons than the element with the smaller atomic number. The smaller number of neutrons causes the element with the larger Z to have a smaller than expected atomic weight. **7.73** (a) Na (b) Si^{3+} (c) The greater the effective nuclear charge experienced by a valence electron, the larger the ionization energy for that electron. According to Table 7.2, I_1 for Na is 496 kJ/mol. I_4 for Si is 4360 kJ/mol. **7.76** (a) Mo–F distance = 2.16 Å (b) S–F distance = 1.73 Å (c) Cl–F distance = 1.70 Å **7.79** The completed $4f$ subshell in Hf leads to a much larger change in Z and Z_{eff} going from Zr to Hf than in going from Y to La. This larger increase in Z_{eff} going from Zr to Hf leads to a smaller increase in atomic radius than in going from Y to La. **7.82** Ionization energy of F^-: $F^-(g) \longrightarrow F(g) + 1e^-$; electron affinity of F: $F(g) + 1e^- \longrightarrow F^-(g)$. The two processes are the reverse of each other. The energies are equal in magnitude but opposite in sign. $I_1(F^-) = -E(F)$.

7.84 O, $[He]2s^22p^4$ — 2s: [↑↓] 2p: [↑↓][↑][↑]

O^{2-}, $[He]2s^22p^6 = [Ne]$ — 2s: [↑↓] 2p: [↑↓][↑↓][↑↓]

O^{3-}, $[Ne]3s^1$ — The third electron would be added to the 3s orbital, which is farther from the nucleus and more strongly shielded by the [Ne] core. The overall attraction of this 3s electron for the oxygen nucleus is not large enough for O^{3-} to be a stable particle.

7.86 (a) The group 2B metals have complete $(n-1)d$ subshells. An additional electron would occupy an np subshell and be substantially shielded by both ns and $(n-1)d$ electrons. This is not a lower energy state than the neutral atom and a free electron. (b) Group 1B elements have the generic electron configuration $ns^1(n-1)d^{10}$. An additional electron would complete the ns subshell and experience repulsion with the other ns electron. Going down the group, size of the ns subshell increases and repulsion effects decrease, so effective nuclear charge increases and electron affinities become more negative. **7.89** $O_2 < Br_2 < K < Mg$. O_2 and Br_2 are nonpolar nonmetals. O_2, with the much lower molar mass, should have the lower melting point. K and Mg are metallic solids with higher melting points than the two nonmetals. Since alkaline earth metals are typically harder, more dense, and higher melting than alkali metals, Mg should have the highest melting point of the group. This order of melting points is confirmed by data in Tables 7.4, 7.5, 7.6, and 7.7. **7.91** Ionization energy increases upon moving one place to the right in a horizontal row of the table, and decreases moving one place down in a family. Similarly, atomic size decreases in moving one place to the right and increases in moving downward. Thus, two elements such as Li and Mg that are diagonally related tend to have similar ionization energies and atomic sizes, which gives rise to some similarities in chemical behavior. **7.94** Chlorine and bromine are much closer in ionization energy and electron affinity to carbon than they are to the metals. Carbon has a much greater tendency than a metal to keep its own electrons and at least some attraction for the electrons of other elements. Carbon is unlikely to form a simple cation, so compounds of carbon and the halogens are molecular, rather than ionic. **7.96** (a) Li, $[He]2s^1$; $Z_{eff} \approx 1+$. (b) $I_1 \approx 5.45 \times 10^{-19}$ J/atom ≈ 328 kJ/mol (c) The estimated value of 328 kJ/mol is less than the Table 7.4 value of 520 kJ/mol. Our estimate for Z_{eff} was a lower limit; the [He] core electrons do not perfectly shield the $2s$ electron from the nuclear charge. (d) Based on the experimental ionization energy, $Z_{eff} = 1.26$. This value is greater than the estimate from part (a), which is consistent with the explanation in part (c). **7.99** (a) Mg_3N_2 (b) $Mg_3N_2(s) + 3H_2O(l) \longrightarrow 3MgO(s) + 2NH_3(g)$; the driving force is the production of $NH_3(g)$ (c) 17% Mg_3N_2 (d) $3Mg(s) + 2NH_3(g) \longrightarrow Mg_3N_2(s) + 3H_2(g)$. NH_3 is the limiting reactant and 0.46 g H_2 are formed. (e) $\Delta H^\circ_{rxn} = -368.70$ kJ

Chapter 8

8.1 (a) Valence electrons are those that take part in chemical bonding. This usually means the electrons beyond the core noble-gas configuration of the atom, although it is sometimes only the outer-shell electrons. (b) A nitrogen atom has 5 valence electrons. (c) The atom (Si) has 4 valence electrons. **8.3** P, $1s^22s^22p^63s^23p^3$. A $3s$ electron is a valence electron; a $2s$ (or $1s$) electron is a nonvalence electron. The $3s$ valence electron is involved in chemical bonding, while the $2s$ or $1s$ nonvalence electron is not.

8.5 (a) $\dot{\text{Ca}}\cdot$ (b) $:\ddot{\text{P}}\cdot$ (c) $:\ddot{\underset{..}{\text{Ne}}}:$ (d) $\cdot\dot{\text{B}}\cdot$

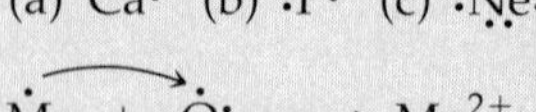

8.7 $\text{Mg} + \cdot\ddot{\text{O}}: \longrightarrow Mg^{2+} + [:\ddot{\underset{..}{\text{O}}}:]^{2-}$

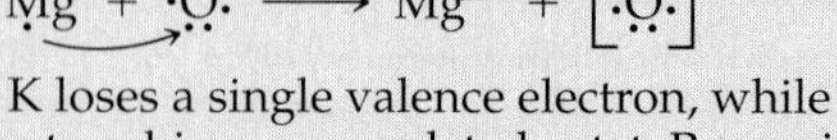

8.9 K loses a single valence electron, while Ca loses two electrons to achieve a completed octet. Removing one electron from the core of either K^+ or Ca^{2+} would be energetically unfavorable because the core electrons are stabilized by a strong electrostatic attraction for the nucleus. Even a large lattice energy is not enough to promote removal of a core electron. **8.11** (a) AlF_3 (b) K_2S (c) Y_2O_3 (d) Mg_3N_2 **8.13** (a) Sr^{2+}, [Kr], noble-gas configuration (b) Ti^{2+}, $[Ar]3d^2$ (c) Se^{2-}, $[Ar]4s^23d^{10}4p^6 = [Kr]$, noble-gas configuration (d) Ni^{2+}, $[Ar]3d^8$ (e) Br^-, $[Ar]4s^23d^{10}4p^6 = [Kr]$, noble-gas configuration (f) Mn^{3+}, $[Ar]3d^4$ **8.15** (a) *Lattice energy* is the energy required to totally separate one mole of solid ionic compound into its gaseous ions. (b) The magnitude of the lattice energy depends on the magnitudes of the charges of the two ions, their radii and the arrangement of ions in the lattice. **8.17** KF, 808 kJ/mol; CaO, 3414 kJ/mol; ScN, 7547 kJ/mol The interionic distances in the three compounds are similar. For compounds with similar ionic separations, the lattice energies should be related as the product of the charges of the ions. The lattice energies above are approximately related as 1:4:9. Slight variations are due to the small differences in ionic separations. **8.19** Since the ionic charges are the same in the two compounds, the KBr and CsCl separations must be approximately equal. **8.21** The large attractive energy between oppositely charged Ca^{2+} and O^{2-} more than compensates for the energy required to form Ca^{2+} and O^{2-} from the neutral atoms. **8.23** The lattice energy of RbCl(*s*) is +692 kJ/mol. This value is smaller than the lattice energy for NaCl because Rb^+ has a larger ionic radius than Na^+ and therefore cannot approach Cl^- as closely as Na^+ can. **8.25** (a) A *covalent bond* is the bond formed when two atoms share one or more pairs of electrons. (b) Any simple compound whose component atoms are nonmetals, such as H_2, SO_2, and CCl_4, are molecular and have covalent bonds between atoms. (c) Covalent because it is a gas at room temperature and below.

8.27 :C̤̈l· + :C̤̈l· + :C̤̈l· + ·Ṅ: ⟶ :C̤̈l—N(—Cl:)(—Cl:)

8.29 (a) :Ö=Ö: (b) A double bond is required because there are not enough electrons to satisfy the octet rule with single bonds and unshared pairs. (c) The greater the number of shared electron pairs between two atoms, the shorter the distance between the atoms. An O=O double bond is shorter than an O—O single bond. **8.31** (a) *Electronegativity* is the ability of an atom in a molecule to attract electrons to itself. (b) The range of electronegativities on the Pauling scale is 0.7–4.0. (c) Fluorine is the most electronegative element. (d) Cesium is the least electronegative element that is not radioactive. **8.33** (a) S (b) C (c) As (d) Mg **8.35** The bonds in (a), (b), and (d) are polar. The more electronegative element in each polar bond is: (a) O (b) F (d) O **8.37** (a) A polar molecule has a measurable dipole moment while a nonpolar molecule has a zero net dipole moment. (b) Yes. If X and Y have different electronegativities, the electron density around the more electronegative atom will be greater, producing a charge separation or dipole in the molecule. (c) The dipole moment, μ, is the product of the magnitude of the separated charges, Q, and the distance between them, r. $\mu = Qr$ **8.39** The calculated charge on H and F is 0.41 e. **8.41** (a) MnO_2, ionic (b) P_2S_3, covalent (c) CoO, ionic (d) copper(I) sulfide, ionic (e) chlorine trifluoride, covalent (f) vanadium(V) fluoride, ionic

8.43 (a) H—Si(H)(H)—H (b) :C≡O: (c) :F̤̈—S̤̈—F̤̈: (d) :Ö—S(:Ö:)(:Ö—H)—Ö—H (e) [:Ö—C̤̈l—Ö:]⁻ (f) H—N̈(H)—Ö—H

8.45 (a) [:N≡O:]⁺ (formal charges 0, +1) (b) :Ö: (−1) on P (+1), with three :C̤̈l: (0) (c) [Cl (+3) bonded to four :Ö: (−1 each)]⁻ (d) Cl (+2) bonded to three O (−1 each) and one O—H (0, 0)

8.47 (a) [Ö=N̈—Ö:]⁻ ⟷ [:Ö—N̈=Ö]⁻ (b) O_3 is isoelectronic with NO_2^-; both have 18 valence electrons. (c) Since each N—O bond has partial double-bond character, the N—O bond length in NO_2^- should be shorter than in species with formal N—O single bonds. **8.49** The more electron pairs shared by two atoms, the shorter the bond. Thus, the C—O bond lengths vary in the order $CO < CO_2 < CO_3^{2-}$. **8.51** (a) Two equally valid Lewis structures can be drawn for benzene.

[Two Kekulé structures of benzene, C_6H_6, with ⟷ between them]

The concept of resonance dictates that the true description of bonding is some hybrid or blend of the two Lewis structures. The most obvious blend of these two resonance structures is a molecule with six equivalent C—C bonds with equal lengths. (b) In order for the six C—C bonds in benzene to be equivalent, each must have some double-bond character. That is, more than one pair but fewer-than two pairs of electrons are involved in each C—C bond. This model predicts a uniform C—C bond length that is shorter than a single bond but longer than a double bond. **8.53** (a) The *octet rule* states that atoms will gain, lose, or share electrons until they are surrounded by eight valence electrons. (b) The octet rule applies to the individual ions in an ionic compound. For example, in $MgCl_2$, Mg loses 2 e^- to become Mg^{2+} with the electron configuration of Ne. Each Cl atom gains one electron to form Cl^- with the electron configuration of Ar. **8.55** The most common exceptions to the octet rule are molecules with more than eight electrons around one or more atoms. **8.57** (a) CO_3^{2-} has three resonance structures, and all obey the octet rule.

[Three resonance structures of CO_3^{2-}, each with one C=O double bond, connected by ⟷]

(b) H—B(H)—H, 6 electrons around B (c) [:Ï—Ï—Ï:]⁻, 10 electrons around central I (d) :F̤̈—Ge(:F̤̈:)(:F̤̈:)—F̤̈: (e) [AsF_6]⁻, 12 electrons around As

8.59 (a) :C̈l—Be—C̈l:; this structure violates the octet rule.
(formal charges: 0, 0, 0)

(b) C̈l═Be═C̈l ⟷ :C̈l—Be≡Cl ⟷ Cl≡Be—C̈l:
(formal charges: +1 −2 +1; 0 −2 +2; +2 −2 0)

(c) Since formal charges are minimized on the structure that violates the octet rule, this form is probably most important. **8.61** (a) $\Delta H = -304$ kJ (b) $\Delta H = -82$ kJ (c) $\Delta H = -467$ kJ **8.63** (a) −288 kJ (b) −116 kJ (c) −1299 kJ **8.65** (a) Exothermic (b) The ΔH calculated from bond enthalpies (−97 kJ) is slightly more exothermic (more negative) than that obtained using ΔH_f° values (−92.38 kJ). **8.67** The average Ti—Cl bond enthalpy is 430 kJ/mol. **8.69** (a) group 4A (b) group 2A (c) group 5A **8.71** $E = -8.65 \times 10^{-19}$ J; on a molar basis $E = -521$ kJ. The absolute value is less than the lattice energy, 808 kJ/mol. The difference represents the added energy of putting all the K^+F^- ion pairs together in a three-dimensional array. **8.73** (a) 779 kJ/mol (b) 627 kJ/mol (c) 2195 kJ/mol **8.76** (b) H_2S and (c) NO_2^- contain polar bonds. **8.79** (a) +1 (b) −1 (c) +1 (assuming the odd electron is on N) (d) 0 (e) +3 **8.81** (a) In the leftmost Lewis structure the more electronegative oxygen atom has the negative formal charge (−1), so this structure is likely to be most important. (b) The longer than typical N≡N length and the shorter than typical N═O length indicate that the middle and right structures, with less favorable formal charges, do contribute to the overall structure. This physical data indicates that while formal charge can be used to predict which resonance form will be more important to the observed structure, the influence of minor contributors on the true structure cannot be ignored. **8.83** ΔH is +42 kJ for the first reaction and −200 kJ for the second. The latter is much more favorable because the formation of 2 mol of O—H bonds is more exothermic than the formation of 1 mol of H—H bonds. **8.85** (a) $\Delta H = 7.85$ kJ/g $C_3H_5N_3O_9$ (b) $4C_7H_5N_3O_6(s) \longrightarrow 6N_2(g) + 7CO_2(g) + 10H_2O(g) + 21C(s)$ **8.88** (a) Ti^{2+}, $[Ar]3d^2$; Ca, $[Ar]4s^2$. The 2 valence electrons in Ti^{2+} and Ca are in different principal quantum levels and different subshells. (b) In Ca the 4*s* is lower in energy than the 3*d*, while in Ti^{2+} the 3*d* is lower in energy than the 4*s*. (c) There is only one 4*s* orbital, so the 2 valence electrons in Ca are paired; there are 5 degenerate 3*d* orbitals, so the 2 valence electrons in Ti^{2+} are unpaired. **8.90** The "second electron affinity" of O is +750 kJ **8.95** (a) $\Delta H = 1551$ kJ (b) $\Delta H = 1394$ kJ (c) $\Delta H = 1353$ kJ **8.97** (a) Br—Br, $D(g) = 193$ kJ/mol, $D(l) = 223.6$ kJ/mol (b) C—Cl, $D(g) = 328$ kJ/mol, $D(l) = 336.1$ kJ/mol (c) O—O, $D(g) = 146$ kJ/mol, $D(l) = 192.7$ kJ/mol (d) Average bond enthalpy in the liquid phase is the sum of the enthalpy of vaporization for the molecule and the gas phase bond-dissociation enthalpies, divided by the number of bonds dissociated. This is greater than the gas phase bond-dissociation enthalpy owing to the contribution from the enthalpy of vaporization.

Chapter 9

9.1 Yes. The only possible bond angles in this arrangement are 120° angles. **9.3** (a) An *electron domain* is a region in a molecule where electrons are most likely to be found. (b) Like the balloons in Figure 9.5, each electron domain occupies a finite volume of space, so they also adopt an arrangement where repulsions are minimized. **9.5** (a) Trigonal planar (b) tetrahedral (c) trigonal bipyramidal (d) octahedral **9.7** The electron-domain geometry indicated by VSEPR describes the arrangement of all bonding and nonbonding electron domains. The molecular geometry describes just the atomic positions. In NH_3 there are 4 electron domains around nitrogen, so the electron-domain geometry is tetrahedral. Because there are 3 bonding and 1 nonbonding domains, the molecular geometry is trigonal pyramidal. **9.9** (a) Tetrahedral, tetrahedral (b) trigonal bipyramidal, T-shaped (c) octahedral, square pyramidal **9.11** (a) Tetrahedral, trigonal pyramidal (b) linear, linear (c) linear, linear (d) tetrahedral, trigonal pyramidal (e) trigonal bipyramidal, seesaw (f) octahedral, square planar **9.13** (a) i, trigonal planar; ii, tetrahedral; iii, trigonal bipyramidal (b) i, 0; ii, 1; iii, 2 (c) N and P (d) Cl (or Br or I). This T-shaped molecular geometry arises from a trigonal bipyramidal electron-domain geometry with 2 nonbonding domains. Assuming each F atom has 3 nonbonding domains and forms only single bonds with A, A must have 7 valence electrons and be in or below the third row of the periodic table to produce these electron-domain and molecular geometries. **9.15** (a) 1–109°, 2–109° (b) 3–109°, 4–109° (c) 5–180° (d) 6–120°, 7–109°, 8–109° **9.17** (a) Although both ions have 4 bonding electron domains, the 6 total domains around Br require octahedral domain geometry and square-planar molecular geometry, while the 4 total domains about B lead to tetrahedral domain and molecular geometry. (b) CF_4 will have bond angles closest to the value predicted by VSEPR because there are no nonbonding electron domains around C. In SF_4 the single nonbonding domain will occupy more space, "push back" the bonding domains, and lead to bond angles that are nonideal. **9.19** Each species has 4 electron domains, but the number of nonbonding domains decreases from 2 to 0, going from NH_2^- to NH_4^+. Since nonbonding domains occupy more space than bonding domains, the bond angles expand as the number of nonbonding domains decreases. **9.21** Yes. The dipole moment vector points along the O—S—O angle bisector with the negative end of the dipole pointing away from the S atom. **9.23** (a) In Exercise 9.13, molecules (ii) and (iii) will have nonzero dipole moments. Molecule (i) has no nonbonding electron pairs on A, and the 3 A—F bond dipoles are oriented so that they cancel. Molecules (ii) and (iii) have nonbonding electron pairs on A and their bond dipoles do not cancel. (b) In Exercise 9.14, molecules (i) and (ii) have a zero dipole moment. **9.25** CO, NCl_3, and SF_2 are polar. **9.27** The middle isomer has a zero net dipole moment. **9.29** (a) *Orbital overlap* occurs when valence atomic orbitals on two adjacent atoms share the same region of space. (b) In valence-bond theory, orbital overlap allows two bonding electrons to mutually occupy the space between the bonded nuclei. (c) Valence-bond theory is a combination of the atomic orbital concept and the Lewis model of electron-pair bonding. **9.31** (a) *sp*, 180° (b) sp^3, 109° (c) sp^2, 120° (d) sp^3d^2, 90° and 180° (e) sp^3d, 90°, 120°, and 180°

9.33 [:Ö—S̈—Ö: with :Ö: below S]$^{2-}$ 4 electron domains around S; tetrahedral electron-domain geometry; trigonal-pyramidal molecular geometry; sp^3 hybrid orbitals; "ideal" O—S—O angle ~107° (The nonbonding electron domain will reduce the tetrahedral angles somewhat.)

9.35 (a) B, $[He]2s^22p^1$. One 2*s* electron is promoted to an empty 2*p* orbital. The 2*s* and two 2*p* orbitals that each contain one electron are hybridized to form three equivalent hybrid orbitals in a trigonal planar arrangement. (b) sp^2

(c)

B

(d) A single 2*p* orbital is unhybridized. It lies perpendicular to the trigonal plane of the sp^2 hybrid orbitals.

9.37 (a) sp^2 (b) sp^3 (c) sp (d) sp^3d (e) sp^3d^2

9.39 (a) σ (b) π

(c) A σ bond is generally stronger than a π bond because there is more extensive orbital overlap.

9.41 (a)

```
    H              O
    |              ‖
H—C—H              C
    |            /   \
    H           H     H
```

(b) sp^3, sp^2 (c) The C atom in CH_4 is sp^3 hybridized; there are no unhybridized p orbitals available for the π overlap required by multiple bonds. In CH_2O the C atom is sp^2 hybridized, with one p atomic orbital available to form the π overlap in the C=O double bond. **9.43** (a) 24 valence electrons (b) 18 valence electrons form σ bonds (c) 2 valence electrons form π bonds (d) 4 valence electrons are nonbonding (e) The central C atom is sp^2 hybridized **9.45** (a) ~109° about the leftmost C, sp^3, ~120° about the right-hand C, sp^2 (b) The doubly bonded O can be viewed as sp^2, and the other as sp^3; the nitrogen is sp^3 with approximately 109° bond angles. (c) nine σ bonds, one π bond **9.47** (a) In a localized π bond, the electron density is concentrated between the two atoms forming the bond. In a delocalized π bond, the electron density is spread over all the atoms that contribute p orbitals to the network. (b) The existence of more than one resonance form is a good indication that a molecule will have delocalized π bonding. (c) delocalized **9.49** (a) Both atomic and molecular orbitals have a characteristic energy and shape; each can hold a maximum of two electrons. Atomic orbitals are localized, and their energies are the result of interactions between the subatomic particles in a single atom. Molecular orbitals can be delocalized and their energies are influenced by interactions between electrons on several atoms. (b) There is a net lowering in energy that accompanies bond formation, because the electrons in H_2 are strongly attracted to both H nuclei. (c) 2

9.51 (a)

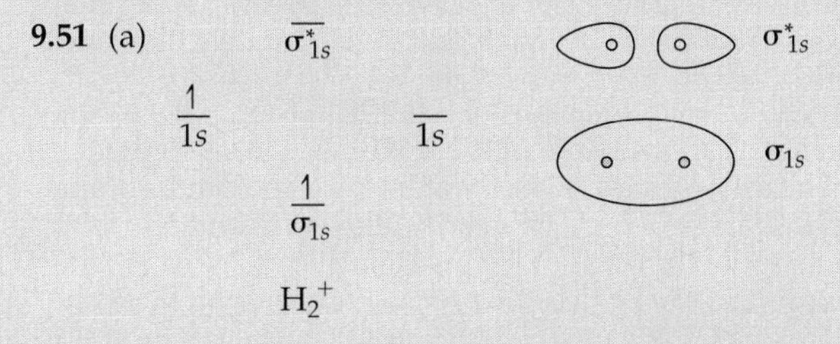

(b) There is one electron in H_2^+. (c) σ_{1s}^1 (d) BO $= \frac{1}{2}$ (e) Yes. If the single electron in H_2^+ is excited to the σ_{1s}^* orbital, its energy is higher than the energy of an H 1s atomic orbital and H_2^+ will decompose into a hydrogen atom and a hydrogen ion.

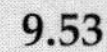

9.53

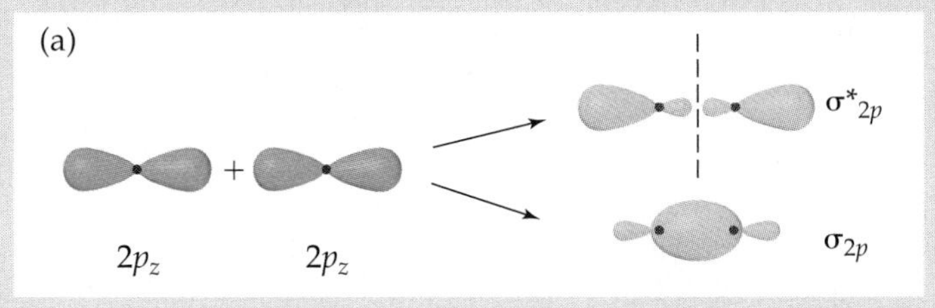

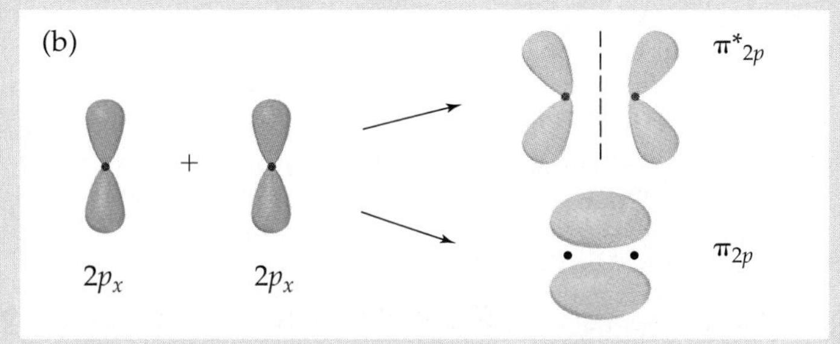

(c) $\sigma_{2p} < \pi_{2p} < \pi_{2p}^* < \sigma_{2p}^*$ **9.55** (a) When comparing the same two bonded atoms, bond order and bond energy are directly related, while bond order and bond length are inversely related. When comparing different bonded nuclei, there are no simple relationships. (b) Be_2 is not expected to exist; it has a bond order of zero and is not energetically favored over isolated Be atoms. Be_2^+ has a bond order of 0.5 and is slightly lower in energy than isolated Be atoms. It will probably exist under special experimental conditions, but be unstable.
9.57 (a, b) Substances with no unpaired electrons are weakly repelled by a magnetic field. This property is called *diamagnetism*. (c) O_2^{2-}, Be_2^{2+}
9.59 (a) B_2^+, $\sigma_{2s}^2\sigma_{2s}^{*2}\pi_{2p}^1$, increase (b) Li_2^+, $\sigma_{1s}^2\sigma_{1s}^{*2}\sigma_{2s}^1$, increase (c) N_2^+, $\sigma_{2s}^2\sigma_{2s}^{*2}\pi_{2p}^4\sigma_{2p}^1$, increase (d) Ne_2^{2+}, $\sigma_{2s}^2\sigma_{2s}^{*2}\pi_{2p}^4\pi_{2s}^{*4}$, decrease
9.61 CN, $\sigma_{2s}^2\sigma_{2s}^{*2}\pi_{2p}^4\sigma_{2p}^1$, bond order = 2.5, paramagnetic; CN^+, $\sigma_{2s}^2\sigma_{2s}^{*2}\pi_{2p}^4$, bond order = 2.0, diamagnetic; CN^-, $\sigma_{2s}^2\sigma_{2s}^{*2}\pi_{2p}^4\sigma_{2p}^2$, bond order = 3.0, diamagnetic
9.63 (a) 3s, 3p_x, 3p_y, 3p_z (b) π_{3p} (c) 2 (d) If the MO diagram for P_2 is similar to that of N_2, P_2 will have no unpaired electrons and be diamagnetic. **9.65** (a) bent (b) tetrahedral (c) bent (d) T-shaped (e) linear **9.67** SiF_4 is tetrahedral, SF_4 is seesaw, XeF_4 is square planar. The shapes are different because the number of nonbonding electron domains is different in each molecule, even though all have four bonding electron domains. Bond angles and thus molecular shape are determined by the total number of electron domains. **9.69** (a) Two sigma, two pi (b) two sigma, two pi (c) three sigma, one pi (d) four sigma, one pi **9.72** The compound on the right has a nonzero dipole moment.

9.74

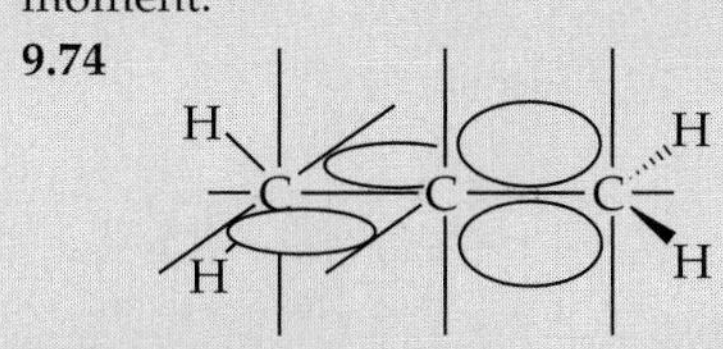

(a) The molecule is nonplanar. (b) Allene has no dipole moment. (c) The bonding in allene would not be described as delocalized. The π electron clouds of the two adjacent C=C are mutually perpendicular, so there is no overlap and no delocalization of π electrons.

9.77 (a) O=O—O: ⟷ :O—O=O

To accommodate the delocalized π bonding indicated in the resonance structures above, all O atoms must be sp^2 hybridized. (b) For the resonance structure on the left, both sigma bonds are formed by overlap of sp^2 hybrid orbitals, the π bond is formed by overlap of atomic p orbitals, one of the nonbonded pairs on the right terminal O atom is in a p atomic orbital, and the remaining 5 nonbonded pairs are in sp^2 hybrid orbitals. (c) Only unhybridized p atomic orbitals can be used to form a delocalized π system. (d) The delocalized π system then contains 4 electrons, 2 from the π bond and 2 from the nonbonded pair in the p orbital. **9.79** N_2^{2-} and O_2^{2-} are likely to be stable species, F_2^{2-} is not. **9.82** (a) HNO_2 (b) $\ddot{O}=\ddot{N}-\ddot{O}-H$ (c) The geometry around N is trigonal planar. (d) sp^2 hybridization around N (e) three sigma, one pi **9.87** From bond-dissociation enthalpies, $\Delta H = 5364$ kJ; according to Hess's law, $\Delta H° = 5535$ kJ. The difference in the two results, 171 kJ, is due to the resonance stabilization in benzene. The amount of energy actually required to decompose 1 mol of $C_6H_6(g)$ is greater than the sum of the localized bond enthalpies.

Chapter 10

10.1 (a) A gas is much less dense than a liquid. (b) A gas is much more compressible than a liquid. (c) All mixtures of gases are homogenous. Similar liquid molecules form homogeneous mixtures, while very dissimilar molecules form heterogeneous mixtures. **10.3** (a) $F = m \times a$. The forces they exert on the floor are exactly equal. (b) $P = F/A$. The person standing on one foot applies this force over a smaller area and thus exerts a greater pressure on the floor. **10.5** (a) 10.3 m (b) 2.0 atm **10.7** (a) The tube can have any cross-sectional area. (b) At equilibrium the force of gravity per unit area acting on the mercury column at the level of the outside mercury is not equal to the force of gravity per unit area acting on the atmosphere. (c) The column of mercury is held up by the pressure of the atmosphere applied to the exterior pool of mercury. **10.9** (a) 0.349 atm (b) 265 mm Hg (c) 3.53×10^4 Pa (d) 0.353 bar **10.11** (a) $P = 773.4$ torr (b) The pressure in Chicago is greater than standard atmospheric pressure, and so it makes sense to classify this weather system as a "high-pressure system." **10.13** 1.7×10^3 kPa **10.15** (i) 0.29 atm (ii) 1.063 atm (iii) 0.136 atm

10.17 (a) $V_2 = \frac{5}{3} V_1$

300 K, V_1 → 500 K, V_2

(b) $V_2 = \frac{1}{2} V_1$

1 atm, V_1 → 2 atm, V_2

10.19 (a) 5.39 L (b) 15.2 L **10.21** (a) If equal volumes of gases at the same temperature and pressure contain equal numbers of molecules and molecules react in the ratios of small whole numbers, it follows that the volumes of reacting gases are in the ratios of small whole numbers. (b) Since the two gases are at the same temperature and pressure, the ratio of the numbers of atoms is the same as the ratio of volumes. There are 1.5 times as many Xe atoms as Ne atoms. **10.23** (a) $PV = nRT$; P in atmospheres, V in liters, n in moles, T in kelvins. (b) An *ideal gas* exhibits pressure, volume, and temperature relationships described by the equation $PV = nRT$. **10.25** Flask A contains the gas with $\mathcal{M} = 30$ g/mol, and flask B contains the gas with $\mathcal{M} = 60$ g/mol. **10.27** (a) 42.1 L (b) 32.5 K (c) 3.96 atm (d) 0.320 mol **10.29** 1.7×10^4 kg H_2 **10.31** (a) 91 atm (b) 2.3×10^2 L **10.33** (a) 39.7 g Cl_2 (b) 12.5 L (c) 377 K (d) 2.53 atm **10.35** (a) $n = 2 \times 10^{-4}$ mol O_2 (b) The roach needs 8×10^{-3} mol O_2 in 48 hr, more than 100% of the O_2 in the jar. **10.37** For gas samples at the same conditions, molar mass determines density. Of the three gases listed, (c) Cl_2 has the largest molar mass. **10.39** (c) Because the helium atoms are of lower mass than the average air molecule, the helium gas is less dense than air. The balloon thus weighs less than the air displaced by its volume. **10.41** (a) $d = 1.77$ g/L (b) $\mathcal{M} = 80.1$ g/mol **10.43** $\mathcal{M} = 89.4$ g/mol **10.45** 3.5×10^{-9} g Mg **10.47** 2.94×10^3 L NH_3 **10.49** 0.402 g Zn **10.51** (a) When the stopcock is opened, the volume occupied by $N_2(g)$ increases from 2.0 L to 5.0 L. P of $N_2 = 0.40$ atm (b) When the gases mix, the volume of $O_2(g)$ increases from 3.0 L to 5.0 L. P of $O_2 = 1.2$ atm (c) $P_t = 1.6$ atm **10.53** (a) P of He = 1.88 atm, P of Ne = 1.10 atm, P of Ar = 0.360 atm (b) $P_t = 3.34$ atm **10.55** P of $N_2 = 0.98$ atm, P of $O_2 = 0.39$ atm, P of $CO_2 = 0.20$ atm **10.57** 2.5 mole % O_2 **10.59** $P_t = 2.70$ atm **10.61** (a) Increase in temperature at constant volume or decrease in volume or increase in pressure (b) decrease in temperature (c) increase in volume, decrease in pressure (d) increase in temperature **10.63** The fact that gases are readily compressible supports the assumption that most of the volume of a gas sample is empty space. **10.65** (a) Vessel A has more molecules. (b) The density of CO is 1.25 g/L and the density of SO_2 is 1.33 g/L vessel B has more mass. (c) The average kinetic energy of the molecules in vessel B is higher. (d) $u_A/u_B = 1.46$. The molecules in vessel A have the greater rms speed. **10.67** (a) In order of increasing speed: $CO_2 \approx N_2O < F_2 < HF < H_2$ (b) $u_{H_2} = 1.92 \times 10^3$ m/s, $u_{CO_2} = 4.12 \times 10^2$ m/s **10.69** The order of increasing rate of effusion is: $^2H^{37}Cl < {}^1H^{37}Cl < {}^2H^{35}Cl < {}^1H^{35}Cl$ **10.71** As_4S_6 **10.73** (a) Non-ideal-gas behavior is observed at very high pressures and low temperatures. (b) The real volumes of gas molecules and attractive intermolecular forces between molecules cause gases to behave nonideally. **10.75** According to the ideal-gas law, the ratio PV/RT should be constant for a given gas sample at all combinations of pressure, volume, and temperature. If this ratio changes with increasing pressure, the gas sample is not behaving ideally. **10.77** Ar ($a = 1.34$, $b = 0.0322$) will behave more like an ideal gas than CO_2 ($a = 3.59$, $b = 0.0427$) at high pressures. **10.79** (a) $P = 0.917$ atm (b) $P = 0.896$ atm **10.81** Over time, the gases will mix perfectly. Each bulb will contain 4 blue and 3 red atoms. **10.83** 3.3 mm^3 **10.86** 5.4×10^3 g O_2 **10.90** (a) $NH_3(g)$ will remain after reaction. (b) $P = 0.957$ atm **10.92** Oxygen is 70.1 mole % of the mixture. **10.95** Only item (b) is satisfactory. Item (c) would not have supported a column of Hg because it is open at both ends. Item (d) is not tall enough to support a nearly 760-mm Hg column. Items (a) and (e) are inappropriate for the same reason: They don't have a uniform cross-sectional area. **10.98** (a) As a gas is compressed at constant temperature, the number of intermolecular collisions increases. Intermolecular attraction causes some of these collisions to be inelastic, which amplifies the deviation from ideal behavior. (b) As the temperature of a gas is increased at constant volume, a larger fraction of the molecules has sufficient kinetic energy to overcome intermolecular attractions and the effect of intermolecular attraction becomes less significant. **10.101** $\Delta H = -1.1 \times 10^{14}$ kJ (assuming $H_2O(l)$ is a product) **10.105** (a) The partial pressure of IF_5 is 0.515 atm. (b) The mole fraction of IF_5 is 0.544.

Chapter 11

11.1 (a) Solid < liquid < gas (b) gas < liquid < solid **11.3** In the liquid and solid states the particles are touching and there is very little empty space, so the volumes occupied by a unit mass are very similar and the densities are similar. In the gas phase the molecules are far apart, so a unit mass occupies a much greater volume than in the liquid or solid, and the density of the gas phase is much less. **11.5** As the temperature of a substance increases, the average kinetic energy of the particles increases. As the average kinetic energy increases, more particles are able to overcome intermolecular attractive forces and move to a less ordered state, from solid to liquid to gas. **11.7** (a) London dispersion forces (b) dipole-dipole forces (c) dipole-dipole forces and in certain cases hydrogen bonding **11.9** (a) Nonpolar covalent molecule; London dispersion forces only (b) polar covalent molecule with O—H bonds; hydrogen bonding, dipole-dipole forces and London dispersion forces (c) polar covalent molecule; dipole-dipole and London dispersion forces (but not hydrogen bonding) **11.11** (a) *Polarizability* is the ease with which the charge distribution in a molecule can be distorted to produce a transient dipole. (b) Te is the most polarizable because its valence electrons are farthest from the nucleus and least tightly held. (c) In order of increasing polarizability: $CH_4 < SiH_4 < SiCl_4 < GeCl_4 < GeBr_4$ (d) The magnitude of London dispersion forces and thus the boiling points of molecules increase as polarizability increases. The order of increasing boiling points is the order of increasing polarizability given in (c). **11.13** (a) H_2S (b) CO_2 (c) CCl_4 **11.15** Rodlike butane molecules and spherical 2-methylpropane molecules both experience dispersion forces. The larger contact surface between butane molecules produces a higher boiling point. **11.17** CH_3NH_2 and CH_3OH. Molecules with N—H, O—H and F—H bonds form hydrogen bonds with like molecules. **11.19** (a) HF has the higher boiling point because hydrogen bonding is stronger than dipole-dipole forces. (b) $CHBr_3$ has the higher boiling point because it has the higher molar mass, indicating greater polarizability and stronger dispersion forces. (c) ICl has the higher boiling point because the molecules have similar molar masses (hence similar dispersion forces), but ICl is polar giving it dipole-dipole forces that are absent for the nonpolar Br_2. **11.21** High surface tension, high boiling point, high specific heat. **11.23** (a) Viscosities and surface tensions of liquids both increase as intermolecular forces become stronger. (b) Surface tension and viscosity decrease as temperature and average kinetic energy of molecules increase. **11.25** (a) $CHBr_3$ has a higher molar mass, is more polarizable and has stronger dispersion forces, so the surface tension is greater. (b) As temperature increases, the viscosity of the oil decreases because the average kinetic energies of the molecules increase. (c) Adhesive forces between polar water and nonpolar car wax are weak, so the large surface tension of water draws the liquid into the shape with the smallest surface area, a sphere. **11.27** Endothermic: melting, vaporization, sublimation; exothermic: condensation, freezing, deposition. **11.29** Melting does not require separation of molecules, so the energy requirement is smaller than for vaporization, where molecules must be separated. **11.31** 2.2×10^3 g H_2O **11.33** 105 kJ **11.35** (a) The *critical pressure* is the pressure required to cause liquefaction at the critical temperature. (b) As the force of attraction between molecules increases, the critical temperature of the compound increases. (c) All the gases in Table 11.5 can be liquefied at the temperature of liquid nitrogen, given sufficient pressure. **11.37** (a) No effect (b) No effect (c) Vapor pressure decreases with increasing intermolecular attractive forces because fewer molecules have sufficient kinetic energy to overcome the attractive forces and escape to the vapor phase. (d) Vapor pressure increases with increasing temperature because average kinetic energies of molecules increase. **11.39** $CBr_4 < CHBr_3 < CH_2Br_2 < CH_2Cl_2 < CH_3Cl < CH_4$. The trend is dominated by dispersion forces even though four of the molecules are polar. The order of increasing volatility is the order of increasing vapor pressure, decreasing molar mass, and decreasing strength of dispersion forces. **11.41** (a) The temperature of the water in the two pans is the same. (b) Vapor pressure does not depend on either volume or surface area of the liquid. At the same temperature, the vapor pressures of water in the two containers are the same. **11.43** (a) Approximately 17°C (b) approximately 26°C **11.45** (a) 79°C (b) The vapor pressure of diethyl ether at 12°C is approximately 325 torr, less than the atmospheric pressure of 340 torr. If an open-end manometer was used, the arm open to the atmosphere would be lower than the arm open to the container. **11.47** The liquid-gas line of a phase diagram ends at the critical point, the temperature and pressure beyond which the gas and liquid phases are indistinguishable. **11.49** (a) $H_2O(g)$ will condense to $H_2O(s)$ at approximately 4 mm Hg; at a higher pressure, perhaps 5 atm or so, $H_2O(s)$ will melt to form $H_2O(l)$. (b) At 100°C and 0.50 atm, water is in the vapor phase. As it cools, water vapor condenses to the liquid at approximately 82°C, the temperature where the vapor pressure of liquid water is 0.50 atm. Further cooling results in freezing at approximately 0°C. The freezing point of water increases with decreasing pressure, so at 0.50 atm, the freezing temperature is very slightly above 0°C. **11.51** (a)

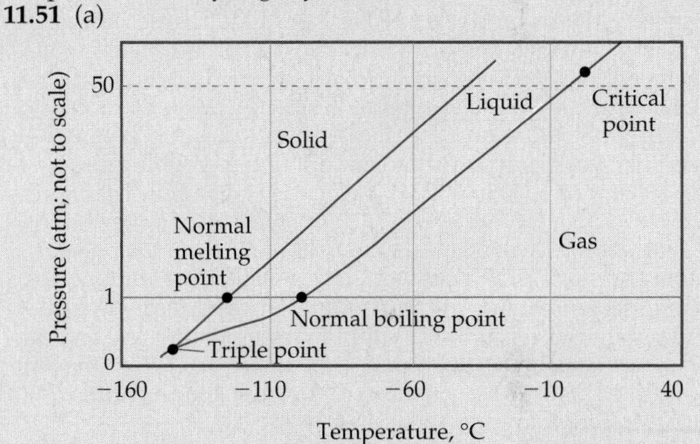

(b) Xe(*s*) is denser than Xe(*l*) because the solid-liquid line on the phase diagram is normal. (c) Cooling Xe(*g*) at 100 torr will cause deposition of the solid because 100 torr is below the pressure of the triple point. **11.53** In a crystalline solid, the component particles are arranged in an ordered repeating pattern. In an amorphous solid, there is no orderly structure. **11.55** The unit cell is the building block of the crystal lattice. When repeated in three dimensions, it produces the crystal lattice. It is a parallelepiped with characteristic distances and angles. Unit cells can be primitive or centered. **11.57** The large difference in melting points is due to the very different forces imposing atomic order in the solid state. Much more kinetic energy is required to disrupt the delocalized metallic bonding in gold than to overcome the relatively weak London dispersion forces in Xe. **11.59** (a) $r = 1.355$ Å (b) density $= 22.67$ g/cm^3 **11.61** Atomic weight $= 55.8$ g/mol **11.63** (a) 12 (b) 6 (c) 8 **11.65** $a = 6.13$ Å **11.67** (a) The U^{4+} ions in UO_2 are represented by the smaller spheres in Figure 11.42(c). The ratio of large spheres to small spheres matches the O^{2-} to U^{4+} ratio in the chemical formula, so the small ones must represent U^{4+}. (b) density $= 10.97$ g/cm^3 **11.69** (a) Hydrogen bonding, dipole-dipole forces, London dispersion forces (b) covalent chemical bonds (c) ionic bonds (d) metallic bonds **11.71** In molecular solids, relatively weak intermolecular forces bind the molecules in the lattice, so relatively little energy is required to disrupt these forces. In covalent-network solids, covalent bonds join atoms into an extended network. Melting or deforming a covalent-network solid means breaking covalent bonds, which requires a large amount of energy. **11.73** Because of its relatively high melting point and properties as a conducting solution, the solid must be ionic. **11.75** (a) B, covalent-network lattice like C(s), versus weak dispersion forces in BF_3 (b) NaCl, ionic versus metallic bonding (c) TiO_2, higher charge on O^{2-} than on Cl^- (d) MgF_2, higher charge on Mg^{2+} than on Na^+ **11.78** (a) SO_2, IF, HBr (b) CH_3NH_2, HCOOH **11.80** (a) The *cis* isomer has stronger dipole-dipole forces and the higher boiling point. (b) Since the nonpolar *trans* isomer with weaker intermolecular forces has the higher melting point, it must pack more efficiently in the solid state. **11.83** (a) Decrease (b) increase (c) increase (d) increase (e) increase (f) increase (g) increase **11.86** The two O—H groups in ethylene glycol are involved in many hydrogen bonding interactions, leading to its high boiling point and viscosity, relative to pentane, which experiences only dispersion forces. **11.88** The vacuum pump reduces the pressure of the atmosphere above the water until atmospheric pressure equals the vapor pressure of water and the water boils. Boiling is an endothermic process, and the temperature drops if the system is not able to absorb heat from the surroundings fast enough. As the temperature of the water decreases, the water freezes. **11.93** (a) 1 atom (b) 2 atoms (c) 4 atoms **11.95** The most effective diffraction occurs when distances between layers of atoms in the crystal are similar to the wavelength of the light being diffracted. Molybdenum X rays of 0.71 Å are on the same order of magnitude as interlayer distances in crystals and are diffracted. Visible light, 400–700 nm or 4000 to 7000 Å, is too long to be diffracted effectively. **11.100** (a) The pressure in the tank must be greater than atmospheric pressure. As long as some liquid is present, the gas pressure in the tank will be constant. (b) If butane gas escapes from the tank, butane liquid will vaporize (evaporate) to maintain the equilibrium vapor pressure. Vaporization is an endothermic process, the butane will absorb heat from the surroundings, and the temperature of the tank and the liquid butane will decrease. (c) 56.8 kJ; $V = 67.9$ L **11.103** P(benzene vapor) $= 98.6$ torr.

Chapter 12

12.1 Both an ordinary liquid and a nematic liquid crystal phase are fluids; they are converted directly to the solid phase upon cooling. The nematic phase is cloudy and more viscous than an ordinary liquid. Upon heating, the nematic phase is converted to an ordinary liquid. **12.3** In the solid state the relative orientation of the molecules is fixed and repeating in all three dimensions. When a substance changes to the nematic liquid crystalline phase, the molecules remain aligned in one dimension; translational motion is allowed, but rotational motion is restricted. Transformation to the isotropic liquid phase destroys the one-dimensional order, resulting in free translational and rotational motion. **12.5** The presence of polar groups or nonbonded electron pairs leads to relatively strong dipole-dipole interactions between molecules. These are a significant part of the orienting forces necessary for liquid crystal formation. **12.7** In the nematic phase there is one-dimensional order, while in a smectic phase, there is two-dimensional order. In a smectic phase, the long directions of the molecules and the ends of the molecules are aligned. **12.9** A nematic phase is composed of sheets of molecules aligned along their lengths, with no additional order within the sheet or between sheets. A cholesteric phase also contains this kind of sheet, but with some ordering between sheets. **12.11** n-Decane does not have a sufficiently high chain length or molecular mass to be considered a polymer.

12.13 $CH_3-C(=O)-O-H + H-O-CH_2-CH_3 \longrightarrow$ (Acetic acid; Ethanol)

$CH_3-C(=O)-O-CH_2CH_3 + H_2O$ (Ethyl acetate)

If a dicarboxylic acid and a dialcohol are combined, there is the potential for propagation of the polymer chain at both ends of both monomers.

12.15 (a) $H_2C=CHCl$ (H, H on one carbon; H, Cl on the other)

(b) $H_2N-CH_2-CH_2-CH_2-CH_2-CH_2-CH_2-NH_2$ + $HO-C(=O)-CH_2-CH_2-CH_2-CH_2-C(=O)-OH$

(c) $HO-CH_2-CH_2-OH$ + $HO-C(=O)-C_6H_4-C(=O)-OH$

12.17 HOOC–C_6H_4–COOH and H_2N–C_6H_4–NH_2

12.19 $nH_2N-(CH_2)_3C(=O)-O-H + H-N(H)-(CH_2)_3C(=O)-OH \longrightarrow$

$\left[NH-(CH_2)_3C(=O)-NH-(CH_2)_3C(=O)\right]_n$

12.21 Flexibility of molecular chains causes flexibility of the bulk polymer. Flexibility is enhanced by molecular features that inhibit order, such as branching, and diminished by features that encourage order, such as cross-linking or delocalized π-electron density. *Cross-linking*, the formation of chemical bonds between polymer chains, reduces flexibility of the molecular chains, increases the hardness of the material, and decreases the chemical reactivity of the polymer. **12.23** The function of the polymer determines whether high molecular masses and high degree of crystallinity are desirable properties. If the polymer will be used as a flexible wrapping or fiber, rigidity is an undesirable property. **12.25** Is the neoprene biocompatible? Does it provoke inflammatory reactions? Does neoprene meet the physical requirements of a flexible lead? Will it remain resistant to degradation and maintain elasticity? Can neoprene be prepared in sufficiently pure form so that it can be classified as *medical grade*? **12.27** Current vascular graft materials cannot be lined with cells similar to those in the native artery. The body detects that the graft is "foreign," and platelets attach to the inside surfaces, causing blood clots. The inside surfaces of future vascular implants need to accommodate a lining of cells that do not attract or attach to platelets. **12.29** In order for skin cells in a

culture medium to develop into synthetic skin, a mechanical matrix that holds the cells in contact with one another must be present. The matrix must be strong, biocompatible, and biodegradable. It probably has polar functional groups that form hydrogen bonds with biomolecules in the tissue cells.
12.31 Ceramics are not readily recyclable because of their extremely high melting points and rigid ionic or covalent-network structures. **12.33** Very small, uniformly sized and shaped particles are required for the production of a strong ceramic object by sintering. Upon heating to initiate condensation reactions, the more uniform the particle size and the greater the total surface area of the solid, the more chemical bonds are formed and the stronger the ceramic object. **12.35** Steel reinforcing rods are added to concrete to resist stress applied along the long direction of the rod. By analogy, the shape of the reinforcing material in the ceramic composite should be rodlike, with a length much greater than its diameter. Rods can be oriented in many directions, so that the brittle material, concrete or ceramic composite, is strengthened in all directions. **12.37** Each Si is bound to four C atoms, and each C is bound to four Si atoms in a tetrahedral arrangement, producing an extended three-dimensional network. The extended three-dimensional nature of the structure produces the exceptional hardness, and the covalent character of the bonding network provides the great thermal stability. **12.39** A superconducting material offers no resistance to the flow of electrical current. Superconductive materials could transmit electricity with much greater efficiency than current carriers. **12.41** The sharp drop in resistivity of MgB_2 near 39 K is the superconducting transition temperature, T_c. **12.43** It is difficult to mold ceramic superconductors into useful shapes such as wires; these wires are fragile at best; the amount of current per cross-sectional area that can be carried by these wires is limited; superconducting ceramics require very low temperatures that render them impractical for widespread use. **12.45** Adhesion is due to attractive intermolecular forces. These include ion-dipole, dipole-dipole, dispersion forces, and hydrogen bonding between substances with like bonding characteristics. **12.47** The coating in Figure 12.31 is a metallic film that reflects most of the incident sunlight. The exclusion of sunlight from the interior of the building reduces glare and cooling load. The opacity of the film provides privacy. **12.49** A dipole moment roughly parallel to the long dimension of the molecule would cause the molecules to reorient when an electric field is applied perpendicular to the usual direction of molecular orientation. **12.52** At the temperature where a substance changes from the solid to the liquid crystalline phase, kinetic energy sufficient to overcome most of the long-range order in the solid has been supplied. A relatively small increase in temperature is required to overcome the remaining aligning forces and produce an isotropic liquid. **12.54** At low Antarctic temperatures, the liquid crystalline phase is closer to its freezing point. The molecules have less kinetic energy due to temperature, and the applied voltage may not be sufficient to overcome orienting forces among the ends of molecules. If some or all of the molecules do not rotate when the voltage is applied, the display will not function properly.

12.58 (a) $\left[-Si(CH_3)_2- \right]_n \xrightarrow{400°C} \left[-SiH(CH_3)-CH_2- \right]_n$

$\left[-SiH(CH_3)-CH_2- \right]_n \xrightarrow{1200°C} [SiC]_n + CH_4(g) + H_2(g)$

(b) $2NbBr_5(g) + 5H_2(g) \longrightarrow 2Nb(s) + 10HBr(g)$

(c) $SiCl_4(l) + 4C_2H_5OH(l) \longrightarrow Si(OC_2H_5)_4(s) + 4HCl(g)$

(d) $n\ C_6H_5-CH{=}CH_2 \longrightarrow \left[-CH(C_6H_5)-CH_2- \right]_n$

12.62 (a) $SiO_2(s)$ (b) $TiO_2(s)$ (c) $Ge(s)$
12.64 (a) $\Delta H = -82$ kJ/mol C_2H_4
(b) $\Delta H = -14$ kJ/mol (of either reactant) (c) $\Delta H = 0$ kJ
12.68 (a) $x = 0.22$ (b) Hg and Cu both have more than one stable oxidation state. If different ions in the solid lattice have different charges, the average charge is a noninteger value. Ca and Ba are stable only in the +2 oxidation state and are unlikely to have noninteger average charge. (c) Ba^{2+} is largest; Cu^{2+} is smallest.

Chapter 13

13.1 If the magnitude of ΔH_3 is small relative to the magnitude of ΔH_1, ΔH_{soln} will be large and endothermic (energetically unfavorable) and not much solute will dissolve.
13.3 (a) Dispersion (b) ion-dipole (c) hydrogen bonding (d) dipole-dipole **13.5** (a) ΔH_1 (b) ΔH_3 **13.7** (a) Since the solute and solvent experience very similar London dispersion forces, the energy required to separate them individually and the energy released when they are mixed are approximately equal. $\Delta H_1 + \Delta H_2 \approx -\Delta H_3$. Thus, ΔH_{soln} is nearly zero. (b) Since no strong intermolecular forces prevent the molecules from mixing, they do so spontaneously because of the increase in disorder. **13.9** (a) Supersaturated (b) Add a seed crystal. A seed crystal provides a nucleus of already preligned molecules, so that ordering of the dissolved particles (crystallization) is more facile. **13.11** (a) Unsaturated (b) saturated (c) saturated (d) unsaturated **13.13** The liquids water and glycerol form homogenous mixtures (solutions) regardless of the relative amounts of the two components. Glycerol has an —OH group on each C atom in the molecule. This structure facilitates strong hydrogen bonding similar to that in water. **13.15** As n increases, water solubility decreases and hexane solubility increases. **13.17** (a) $CaCl_2$ (b) C_6H_5OH **13.19** (a) A sealed container is required to maintain a partial pressure of $CO_2(g)$ greater than 1 atm above the beverage. (b) Since the solubility of gases increases with decreasing temperature, some $CO_2(g)$ will remain dissolved in the beverage if it is kept cool.
13.21 $C_{He} = 5.6 \times 10^{-4}\ M$, $C_{N_2} = 9.0 \times 10^{-4}\ M$
13.23 (a) 2.57% Na_2SO_4 by mass (b) 6.56 ppm Ag
13.25 (a) $X_{CH_3OH} = 0.017$ (b) 3.0% CH_3OH by mass (c) 0.96 m CH_3OH **13.27** (a) 0.283 M $Mg(NO_3)_2$ (b) 1.12 M $LiClO_4 \cdot 3H_2O$ (c) 0.350 M HNO_3 **13.29** (a) 7.27 m C_6H_6 (b) 0.285 m NaCl **13.31** (a) 43.01% H_2SO_4 by mass (b) $X_{H_2SO_4} = 0.122$ (c) 7.69 m H_2SO_4 (d) 5.827 M H_2SO_4
13.33 (a) $X_{CH_3OH} = 0.177$ (b) 5.23 m CH_3OH (c) 3.53 M CH_3OH **13.35** (a) 6.38×10^{-2} mol $CaBr_2$ (b) 7.50×10^{-3} mol KCl (c) 6.94×10^{-3} mol $C_6H_{12}O_6$
13.37 (a) Weigh out 1.3 g KBr, dissolve in water, dilute with stirring to 0.75 L. (b) Weigh out 2.62 g of KBr, dissolve it in 122 g of H_2O to make 125 g of 0.180 m solution. (c) Weigh 244 g KBr and dissolve in enough H_2O to make 1.85 L solution. (d) Weigh 10.1 g KBr, dissolve it in a small amount of water, and dilute to 0.568 L. **13.39** 15 M NH_3
13.41 (a) 31.9% $C_3H_6(OH)_2$ by mass (b) 6.17 m $C_3H_6(OH)_2$
13.43 freezing point depression, $\Delta T_f = K_f(m)$; boiling point elevation, $\Delta T_b = K_b(m)$; osmotic pressure, $\pi = MRT$; vapor

pressure lowering, $P_A = X_A P_A^\circ$ **13.45** (a) An ideal solution is a solution that obeys Raoult's law. (b) The experimental vapor pressure, 67 mm Hg, is less than the value predicted by Raoult's law, 74.5 mm Hg, for an ideal solution. The solution is not ideal. **13.47** (a) $P_{H_2O} = 186.0$ torr (b) 192 g $C_3H_8O_2$ **13.49** (a) $X_{Eth} = 0.2812$ (b) $P_{soln} = 238$ torr (c) X_{Eth} in vapor = 0.472 **13.51** (a) Because NaCl is a strong electrolyte, one mole of NaCl produces twice as many dissolved particles as one mole of the molecular solute $C_6H_{12}O_6$. Boiling-point elevation is directly related to total moles of dissolved particles, so 0.10 *m* NaCl has the higher boiling point.
(b) 0.10 *m* NaCl: $\Delta T_b = 0.101°C$, $T_b = 100.1°C$;
0.10 *m* $C_6H_{12}O_6$: $\Delta T_b = 0.051°C$, $T_b = 100.1°C$
13.53 0.030 *m* phenol > 0.040 *m* glycerin = 0.020 *m* KBr
13.55 (a) $T_f = -115.3°C$, $T_b = 78.8°C$ (b) $T_f = -67.9°C$, $T_b = 64.6°C$ (c) $T_f = -0.91°C$, $T_b = 100.3°C$
13.57 $\pi = 0.0271$ atm **13.59** $\mathcal{M} = 1.8 \times 10^2$ g/mol adrenaline **13.61** $\mathcal{M} = 1.39 \times 10^4$ g/mol lysozyme **13.63** (a) $i = 2.76$ (b) The more concentrated the solution, the greater the ion-pairing and the smaller the measured value of *i*. **13.65** (a) In the gaseous state, particles are far apart and intermolecular attractive forces are small. When two gases combine, all terms in Equation 13.1 are essentially zero and the mixture is always homogeneous. (b) To determine whether Faraday's dispersion is a true solution or a colloid, shine a beam of light on it. If light is scattered, the dispersion is a colloid.
13.67 (a) Hydrophobic (b) hydrophilic (c) hydrophobic
13.69 Electrostatic repulsions between groups at the surface of the dispersed particles inhibit coalescence. Hydrophilic colloids can be coagulated by adding electrolytes, and some colloids can be coagulated by heating. **13.71** The periphery of the BHT molecule is mostly hydrocarbon-like groups, such as $-CH_3$. The one $-OH$ group is rather buried inside and probably does little to enhance solubility in water. Thus, BHT is more likely to be soluble in the nonpolar hydrocarbon hexane, C_6H_{14}, than in polar water. **13.73** (a) 1×10^{-4} *M* (b) 60 mm Hg **13.76** A solution that is 12 ppm KCl has the higher molarity of K^+ ions. **13.79** (a) 0.16 *m* Na (b) 2.0 *M* Na (c) Clearly, molality and molarity are not the same for this amalgam. Only in the instance that one kg solvent and the mass of one liter of solution are nearly equal do the two concentration units have similar values. **13.82** (a) $T_f = -0.6°C$ (b) $T_f = -0.4°C$ **13.85** (a) $K_b = 2.34°C/m$ (b) $\mathcal{M} = 2.6 \times 10^2$ g/mol **13.89** (a) CF_4, 1.7×10^{-4} *m*; $CClF_3$, 9×10^{-4} *m*; CCl_2F_2, 2.3×10^{-3} *m*; $CHClF_2$, 3.5×10^{-2} *m* (b) Molality and molarity are numerically similar when kilograms solvent and liters solution are nearly equal. This is true when solutions are dilute and when the density of the solvent is nearly 1 g/mL, as in this problem. (c) Water is a polar solvent; the solubility of solutes increases as their polarity increases. Nonpolar CF_4 has the lowest solubility and the most polar fluorocarbon, $CHClF_2$, has the greatest solubility in H_2O. (d) The Henry's law constant for $CHClF_2$ is 3.5×10^{-2} mol/L-atm. This value is greater than the Henry's law constant for $N_2(g)$ because $N_2(g)$ is nonpolar and of lower molecular mass than $CHClF_2$.
13.93 (a) cation (*g*) + anion (*g*) + solvent

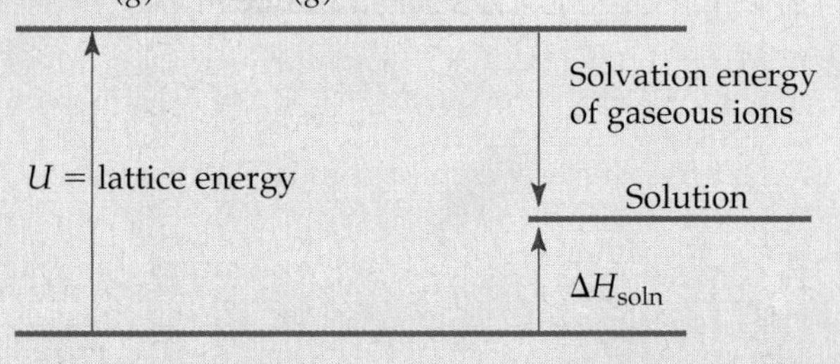

(b) Lattice energy (*U*) is inversely related to the distance between ions, so salts with large cations like $(CH_3)_4N^+$ have smaller lattice energies than salts with simple cations like Na^+. Also, the $-CH_3$ groups in the large cation are capable of dispersion interactions with nonpolar groups of the solvent molecules, resulting in a more negative solvation energy of the gaseous ions. Overall, for salts with larger cations, lattice energy is smaller (less positive), the solvation energy of the gaseous ions is more negative, and ΔH_{soln} is less endothermic. These salts are more soluble in polar nonaqueous solvents.
13.96 The empirical and molecular formula is $C_{12}H_{16}N_2O_3$.

Chapter 14

14.1 (a) *Reaction rate* is the change in the amount of products or reactants in a given amount of time. (b) Rates depend on concentration of reactants, surface area of reactants, temperature, and presence of catalyst. (c) The stoichiometry of the reaction (mole ratios of reactants and products) must be known to relate rate of disappearance of reactants to rate of appearance of products.

14.3

Time (min)	Mol A	(a) Mol B	[A] (mol/L)	Δ[A] (mol/L)	(b) Rate (*M*/s)
0	0.065	0.000	0.65		
10	0.051	0.014	0.51	−0.14	2.3×10^{-4}
20	0.042	0.023	0.42	−0.09	1.5×10^{-4}
30	0.036	0.029	0.36	−0.06	1.0×10^{-4}
40	0.031	0.034	0.31	−0.05	0.8×10^{-4}

(c) $\Delta[B]_{avg}/\Delta t = 1.3 \times 10^{-4}$ *M*/s

14.5

Time (s)	Time Interval (s)	Concentration (*M*)	Δ*M*	Rate (*M*/s)
0		0.0165		
2,000	2,000	0.0110	−0.0055	28×10^{-7}
5,000	3,000	0.00591	−0.0051	17×10^{-7}
8,000	3,000	0.00314	−0.00277	9.23×10^{-7}
12,000	4,000	0.00137	−0.00177	4.43×10^{-7}
15,000	3,000	0.00074	−0.00063	2.1×10^{-7}

14.7 From the slopes of the tangents to the graph, the rates are -1.2×10^{-6} *M*/s at 5000 s, -5.8×10^{-7} *M*/s at 8000 s.
14.9 (a) $-\Delta[H_2O_2]/\Delta t = \Delta[H_2]/\Delta t = \Delta[O_2]/\Delta t$
(b) $-\frac{1}{2}\Delta[N_2O]/\Delta t = \frac{1}{2}\Delta[N_2]/\Delta t = \Delta[O_2]/\Delta t$
(c) $-\Delta[N_2]/\Delta t = -\frac{1}{3}\Delta[H_2]/\Delta t = \frac{1}{2}\Delta[NH_3]/\Delta t$
14.11 (a) $-\Delta[O_2]/\Delta t = 0.43$ mol/s; $\Delta[H_2O]/\Delta t = 0.85$ mol/s (b) P_{total} decreases by 12 torr/min **14.13** (a) If [A] doubles, the rate will increase by a factor of four; the rate constant, *k*, is unchanged. Rate is proportional to $[A]^2$, so when the value of [A] doubles, rate changes by 2^2, or 4. The rate constant, *k*, is the proportionality constant that does not change unless the temperature changes. (b) The reaction is second order in A, first order in B, and third order overall. (c) units of $k = M^{-2}\,s^{-1}$
14.15 (a) Rate = $k[N_2O_5]$ (b) Rate = 1.16×10^{-4} *M*/s (c) When the concentration of N_2O_5 doubles, the rate of the reaction doubles. **14.17** (a, b) $k = 1.7 \times 10^2\ M^{-1}\,s^{-1}$ (c) If $[OH^-]$ is tripled, the rate triples. **14.19** (a) $x = 1$ (b) $x = 3$ (c) $x = 0$. The rate does not depend on [A].
14.21 (a) Rate = $k[OCl^-][I^-]$ (b) $k = 6.0 \times 10^9\ M^{-1}\,s^{-1}$ (c) Rate = 3.0×10^3 *M*/s **14.23** (a) Rate = $k[NO]^2[O_2]$

(b, c) $k_{avg} = 7.11 \times 10^3 \, M^{-2} s^{-1}$ **14.25** (a) Rate $= k[NO]^2[Br_2]$ (b) $k_{avg} = 1.2 \times 10^4 \, M^{-2} s^{-1}$ (c) $\frac{1}{2}\Delta[NOBr]/\Delta t = -\Delta[Br_2]/\Delta t$ (d) $-\Delta[Br_2]/\Delta t = 8.4 \, M/s$ **14.27** (a) $[A]_0$ is the molar concentration of reactant A at time 0. $[A]_t$ is the molar concentration of reactant A at time t. $t_{1/2}$ is the time required to reduce $[A]_0$ by a factor of 2. k is the rate constant for a particular reaction. (b) A graph of ln[A] versus time yields a straight line for a first-order reaction. **14.29** (a) $k = 3.0 \times 10^{-6} \, s^{-1}$ (b) $t_{1/2} = 3.2 \times 10^4$ s **14.31** (a) $P_{65} = 20$ torr (b) $t = 51$ s **14.33** Plot ln $P_{SO_2Cl_2}$ versus time, $k = -\text{slope} = 2.19 \times 10^{-5} \, s^{-1}$ **14.35** (a) The plot of 1/[A] versus time is linear, so the reaction is second order in [A]. (b) $k = 0.040 \, M^{-1} \text{min}^{-1}$ (c) $t_{1/2} = 38$ min **14.37** (a) The plot of $1/[NO_2]$ versus time is linear, so the reaction is second order in NO_2. (b) $k = \text{slope} = 10 \, M^{-1} s^{-1}$ **14.39** (a) The central idea of the *collision model* is that molecules must collide to react. (b) The energy of the collision and the orientation of the molecules when they collide determine whether a reaction will occur. (c) At a higher temperature, there are more total collisions and each collision is more energetic. **14.41** $f = 4.94 \times 10^{-2}$. At 400 K, approximately 1 out of 20 molecules has this kinetic energy.

14.43 (a)

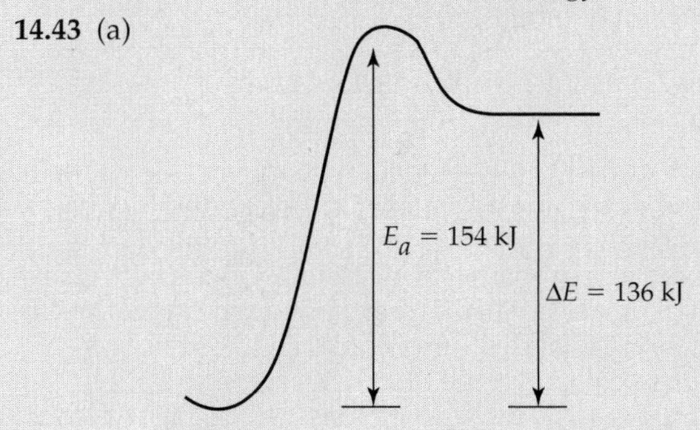

(b) E_a (reverse) = 18 kJ/mol **14.45** Reaction (b) is fastest and reaction (c) is slowest. **14.47** (a) $k = 1.1 \, s^{-1}$ (b) $k = 4.9 \, s^{-1}$ **14.49** A plot of ln k versus $1/T$ has a slope of -5.71×10^3; $E_a = -R(\text{slope}) = 47.5$ kJ/mol. **14.51** The reaction will occur 88 times faster at 50°C, assuming equal initial concentrations. **14.53** (a) An *elementary step* is a process that occurs in a single event; the order is given by the coefficients in the balanced equation for the step. (b) A *unimolecular* elementary step involves only one reactant molecule; a *bimolecular* elementary step involves two reactant molecules. (c) A *reaction mechanism* is a series of elementary steps that describe how an overall reaction occurs and explain the experimentally determined rate law. **14.55** (a) unimolecular, rate $= k[Cl_2]$ (b) bimolecular, rate $= k[OCl^-][H_2O]$ (c) bimolecular, rate $= k[NO][Cl_2]$ **14.57** There is one intermediate, B, and two transition states. The B $\longrightarrow$ C step is faster, and the overall reaction A $\longrightarrow$ C is exothermic.
14.59 (a) $H_2(g) + 2ICl(g) \longrightarrow I_2(g) + 2HCl(g)$ (b) HI is the intermediate. (c) first step: rate $= k[H_2][ICl]$; second step: rate $= k[HI][ICl]$ (d) If the first step is slow, the observed rate law is rate $= k[H_2][ICl]$. **14.61** (a) Rate $= k[NO][Cl_2]$ (b) The second step must be slow relative to the first step. **14.63** (a) A catalyst increases the rate of reaction by decreasing the activation energy, E_a, or increasing the frequency factor, A. (b) A *homogeneous catalyst* is in the same phase as the reactants, while a *hetereogeneous catalyst* is in a different phase.
14.65 (a) Multiply the coefficients in the first reaction by two and sum. (b) $NO_2(g)$ is a catalyst because it is consumed and then reproduced in the reaction sequence. (c) This is a homogeneous catalysis. **14.67** Use of chemically stable supports makes it possible to obtain very large surface areas per unit mass of the precious metal catalyst because the metal can be deposited in a very thin, even monomolecular, layer on the surface of the support. **14.69** To put two D atoms on a single carbon, it is necessary that one of the already existing C—H bonds in ethylene be broken while the molecule is adsorbed, so that the H atom moves off as an adsorbed atom and is replaced by a D atom. This requires a larger activation energy than simply adsorbing C_2H_4 and adding one D atom to each carbon.
14.71 (a) Living organisms operate efficiently in a very narrow temperature range; the role of enzymes as homogeneous catalysts that speed up desirable reactions without heating and undesirable side effects is crucial for biological systems.
(b) *catalase*: $2H_2O_2 \longrightarrow 2H_2O + O_2$; *nitrogenase*: $N_2 \longrightarrow 2NH_3$(nitrogen fixation) **14.73** (a) The catalyzed reaction is approximately 10,000,000 times faster at 25°C.
(b) The catalyzed reaction is 180,000 times faster at 125°C.
14.75 $\Delta[Cl^-]/\Delta t = 7.8 \times 10^{-7} \, M/s$
14.78 (a) Rate $= k[HgCl_2][C_2O_4^{2-}]^2$ (b) $k = 8.7 \times 10^{-3} \, M^{-2} s^{-1}$ (c) Rate $= 1.0 \times 10^{-5} \, M/s$ **14.80** (a) $k = 4.28 \times 10^{-4} \, s^{-1}$ (b) [urea] $= 0.059 \, M$ (c) $t_{1/2} = 1.62 \times 10^3$ s **14.84** A plot of ln K versus $1/T$ is linear with slope $= -1.751 \times 10^4$. $E_a = -(\text{slope})R = 1.5 \times 10^2$ kJ/mol.
14.87 (a) $Cl_2(g) + CHCl_3(g) \longrightarrow HCl(g) + CCl_4(g)$ (b) $Cl(g)$, $CCl_3(g)$ (c) step 1, unimolecular; step 2, bimolecular; step 3, bimolecular (d) Step 2, the slow step, is rate determining. (e) Rate $= k[CHCl_3][Cl_2]^{1/2}$ **14.89** *Enzyme*: carbonic anhydrase; *substrate*: carbonic acid (H_2CO_3); *turnover number*: 1×10^7 molecules/s. **14.91** partial pressure of $O_2 = 0.402$ atm **14.93** (a) Use an open-end manometer, a clock, a ruler, and a constant-temperature bath. Load the flask with HCl(*aq*), and read the height of the Hg in both arms of the manometer. Quickly add Zn(*s*) to the flask, and record time = 0 when the Zn(*s*) contacts the acid. Record the height of the Hg in one arm of the manometer at convenient time intervals such as 5 sec. Calculate the pressure of $H_2(g)$ at each time. Since $P = (n/V)RT$, $\Delta P/\Delta t$ at constant temperature is an acceptable measure of reaction rate. (b) Keep the amount of Zn(*s*) constant, and vary the concentration of HCl(*aq*) to determine the reaction order for H^+ and Cl^-. Keep the concentration of HCl(*aq*) constant, and vary the amount of Zn(*s*) to determine the order for Zn(*s*). Combine this information to write the rate law. (c) $-\Delta[H^+]/\Delta t = 2\Delta[H_2]/\Delta t$; the rate of disappearance of H^+ is twice the rate of appearance of $H_2(g)$. $[H_2] = P/RT$ (d) By changing the temperature of the constant-temperature bath, measure the rate data at several temperatures and calculate the rate constant k at these temperatures. Plot ln k versus $1/T$; the slope of the line is $-E_a/R$. (e) Measure rate data at constant temperature, HCl concentration, and mass of Zn(*s*), varying only the form of the Zn(*s*). Compare the rate of reaction for metal strips and granules. **14.96** Changes in temperature change the kinetic energy of the various groups on the enzyme and their tendency to form intermolecular associations or break free from them. At temperatures above the temperature of maximum activity, sufficient kinetic energy has been imparted so that the three-dimensional structure of the enzyme is destroyed. This is the process of *denaturation*. In the lock-and-key model of enzyme action, the active site is the specific location in the enzyme where the reaction takes place. The precise geometry of the active site both accommodates and activates the substrate. When an enzyme is denatured, its activity is destroyed because the active site has collapsed.

Chapter 15

15.1 Yes. In the fourth and fifth boxes, the relative amounts (concentrations) of A and B are constant. Although the reaction is ongoing, the rates of A $\longrightarrow$ B and B $\longrightarrow$ A are equal.

15.3 (a) $K_{eq} = 2.8 \times 10^{-2}$ (b) Since $k_f < k_r$, in order for the two rates to be equal, [A] must be greater than [B]. **15.5** (a) The *law of mass action* expresses the relationship between the concentrations of reactants and products at equilibrium for any reaction. $K_{eq} = P_{NOBr_2}/P_{NO} \times P_{Br_2}$ (b) The *equilibrium-constant expression* is an algebraic equation where the variables are the equilibrium concentrations of the reactants and products for a specific chemical reaction. The *equilibrium constant* is a number; it is the ratio calculated from the equilibrium expression for a particular chemical reaction. (c) Introduce a known quantity of $NOBr_2(g)$ into a vessel of known volume at constant (known) temperature. After equilibrium has been established, measure the total pressure in the flask. Using an equilibrium table, calculate equilibrium pressures, concentrations and the value of K_{eq}. **15.7** (a) $K_{eq} = P_{N_2O} \times P_{NO_2}/P^3_{NO}$; homogeneous (b) $K_{eq} = P_{CS_2} \times P^4_{H_2}/P_{CH_4} \times P^2_{H_2S}$; homogeneous (c) $K_{eq} = P^4_{CO}/P_{Ni(CO)_4}$; heterogeneous (d) $K_{eq} = [H^+][F^-]/[HF]$; homogeneous (e) $K_{eq} = [Ag^+]^2/[Zn^{2+}]$; heterogeneous **15.9** (a) mostly reactants (b) mostly products **15.11** (a) $K_{eq} = 4.2 \times 10^2$ (b) The equilibrium favors SO_3 at this temperature. **15.13** (a) $K_{eq} = 8.93$ (b) $K_{eq} = 1.25 \times 10^{-2}$ (c) $K_{eq} = 79.7$ **15.15** (a) $K_{eq} = 0.10$ (b) $K_{eq} = 1.2 \times 10^{-4}$ (c) $K_{eq} = 8.3 \times 10^3$ **15.17** (a) $K_{eq} = [Hg]^4 P_{O_2}/[Hg_2O]^2$ (b) The molar concentration, the ratio of moles of a substance to volume occupied by the substance, is a constant for pure solids and liquids. (c) $K_{eq} = P_{O_2}$ **15.19** $K_{eq} = 1.84 \times 10^{-2}$ **15.21** $K_{eq} = 51$ **15.23** (a) $P_{H_2} = 0.299$ atm, $P_{N_2} = 0.466$ atm, $P_{H_2O} = 3.394$ atm (to the proper significant figures, these values are 0.3 atm, 0.5 atm, and 3.4 atm) (b) $K_{eq} = 25.65$ or 3×10^1 (one significant figure) **15.25** (a) $P_{CO_2} = 3.87$, $P_{H_2} = 1.82$, $P_{CO} = 0.23$ (b) $K_{eq} = 0.11$ **15.27** (a) A *reaction quotient* is the result of a general set of concentrations whereas the equilibrium constant requires equilibrium concentrations. (b) to the right (c) The concentrations used to calculate Q must be equilibrium concentrations. **15.29** (a) $Q = 3.35 \times 10^{-7}$; the reaction will proceed to the left. (b) $Q = 1.68 \times 10^{-10}$; the reaction will proceed to the right. (c) $Q = 2.19 \times 10^{-10}$; the mixture is at equilibrium. **15.31** $P_{Cl_2} = 4.98$ atm **15.33** (a) $P_{Br_2} = 0.980$ atm, $P_{Br} = 0.361$ atm; 0.451 g Br (b) $P_{HI} = 4.7$ atm, 21 g HI **15.35** $P_{NO} = 0.43$ atm, $P_{N_2} = P_{O_2} = 18$ atm **15.37** The equilibrium pressure of $Br_2(g)$ is 0.416 atm. **15.39** (a) $P_{PH_3} = P_{BCl_3} = 0.233$ atm (b) A bit more than 0.608 g PH_3BCl_3 is needed. **15.41** $P_{IBr} = 15.5$ atm, $P_{I_2} = P_{Br_2} = 0.926$ atm **15.43** (a) Shift equilibrium to the right (b) decrease the value of K (c) shift equilibrium to the left (d) no effect (e) no effect (f) shift equilibrium to the right **15.45** (a) No effect (b) no effect (c) increase equilibrium constant (d) no effect **15.47** (a) $\Delta H° = -155.7$ kJ (b) The reaction is exothermic, so the equilibrium constant will decrease with increasing temperature. (c) Δn does not equal zero, so a change in volume at constant temperature will affect the fraction of products in the equilibrium mixture. **15.49** (a) $K_{eq} = 1.5 \times 10^{-39}$ (b) Reactants are much more plentiful than products at equilibrium. **15.51** $K_{eq} = 18$ **15.54** (a) $P_t = 0.74$ atm (b) $K_{eq} = 0.40$ **15.57** $P_{PH_3} = 6.8 \times 10^{-2}$ atm **15.60** (a) $Q = 1.7$; $Q > K_{eq}$; the reaction will shift to the left. (b) $Q = 1.7$; $Q > K_{eq}$; the reaction will shift to the left. (c) $Q = 5.8 \times 10^{-3}$; $Q < K_{eq}$; the reaction mixture will shift to the right. **15.63** The maximum value of Q is 4.1; reduction will occur. **15.66** $P_{H_2} = P_{I_2} = 1.48$ atm; $P_{HI} = 10.22$ atm **15.69** The patent claim is false. A catalyst does not alter the position of equilibrium in a system, only the rate of approach to the equilibrium condition. **15.70** (a) (i) $K_{eq} = [Na^+]/[Ag^+]$ (ii) $K_{eq} = [Hg^{2+}]^3/[Al^{3+}]^2$ (iii) $K_{eq} = [Zn^{2+}]P_{H_2}/[H_2]^2$ (b) Using data from Table 4.5: (i) Ag^+ is far below Na, so the reaction will proceed to the right and K_{eq} will be large. (ii) Al^{3+} is above Hg, so the reaction will not proceed to the right and K_{eq} will be small. (iii) H^+ is below Zn, so the reaction will proceed to the right and K_{eq} will be large. (c) $K_{eq} < 1$ for this reaction, so Cd is below Fe. **15.72** (a) At equilibrium, the forward and reverse reactions occur at equal rates. (b) Reactants are favored at equilibrium. (c) A catalyst lowers the activation energy for both the forward and reverse reactions. (d) The ratio of the rate constants remains unchanged. (e) The value of K_{eq} will increase with increasing temperature. **15.75** At 850°C, $K_{eq} = 14.1$; at 950°C, $K_{eq} = 78.8$; at 1050°C, $K_{eq} = 2.7 \times 10^2$; at 1200°C, $K_{eq} = 1.7 \times 10^3$. Because K_{eq} increases with increasing temperature, the reaction is endothermic.

Chapter 16

16.1 Solutions of HCl and H_2SO_4 conduct electricity, taste sour, turn litmus paper red (are acidic), neutralize solutions of bases, and react with active metals to form $H_2(g)$. HCl and H_2SO_4 solutions have these properties in common because both compounds are strong acids. That is, they both dissociate completely in H_2O to form $H^+(aq)$ and an anion. (HSO_4^- is not completely dissociated, but the first dissociation step for H_2SO_4 is complete.) The presence of ions enables the solutions to conduct electricity; the presence of $H^+(aq)$ in excess of 1×10^{-7} *M* accounts for all other properties listed. **16.3** (a) The Arrhenius definition of an acid is confined to aqueous solution; the Brønsted–Lowry definition applies to any physical state. (b) HCl is the Brønsted–Lowry acid; NH_3 is the Brønsted–Lowry base. **16.5** (a) HSO_3^- (b) $C_2H_3O_2^-$ (c) $HAsO_4^{2-}$ (d) NH_3

16.7

B–L Acid	+	B–L Base	⇌	Conjugate Acid	+	Conjugate Base
(a) $NH_4^+(aq)$		$CN^-(aq)$		$HCN(aq)$		$NH_3(aq)$
(b) $H_2O(l)$		$(CH_3)_3N(aq)$		$(CH_3)_3NH^+(aq)$		$OH^-(aq)$
(c) $HCHO_2(aq)$		$PO_4^{3-}(aq)$		$HPO_4^{2-}(aq)$		$CHO_2^-(aq)$

16.9 (a) Acid: $HC_2O_4^-(aq) + H_2O(l) \rightleftharpoons C_2O_4^{2-}(aq) + H_3O^+(aq)$ Base: $HC_2O_4^-(aq) + H_2O(l) \rightleftharpoons H_2C_2O_4(aq) + OH^-(aq)$ (b) $H_2C_2O_4$ is the conjugate acid of $HC_2O_4^-$. $C_2O_4^{2-}$ is the conjugate base of $HC_2O_4^-$. **16.11** (a) weak, NO_2^- (b) strong, HSO_4^- (c) weak, PO_4^{3-} (d) negligible, CH_3^- (e) weak, CH_3NH_2. **16.13** (a) HBr. It is one of the seven strong acids. (b) F^-. HCl is a stronger acid than HF, so F^- is the stronger conjugate base. **16.15** (a) $HF(aq) + CO_3^{2-}(aq)$, the equilibrium lies to the left (b) $OH^-(aq) + OH^-(aq)$, the equilibrium lies to the right (c) $H_2S(aq) + C_2H_3O_2^-(aq)$, the equilibrium lies to the right. **16.17** (a) *Autoionization* is the ionization of a neutral molecule into an anion and a cation. The equilibrium expression for the autoionization of water is

$H_2O(l) \rightleftharpoons H^+(aq) + OH^-(aq)$. (b) Pure water is a poor conductor of electricity because it contains very few ions. (c) If a solution is acidic, it contains more H^+ than OH^-. **16.19** (a) $[H^+] = 2 \times 10^{-10}$ *M*, basic (b) $[H^+] = 3.1 \times 10^{-6}$ *M*, acidic (c) $[H^+] = 1.0 \times 10^{-8}$ *M*, basic **16.21** $[H^+] = [OH^-] = 3.5 \times 10^{-8}$ *M* **16.23** (a) $[H^+]$ changes by a factor of 100. (b) $[H^+]$ changes by a factor of 3.2 **16.25** (a) $[H^+]$ decreases, pH increases (b) The pH is between 4 and 5. By calculation, pH = 4.5; the solution is acidic. (c) pH = 7.8 is between pH 7 and pH 8, closer to pH = 8. A good estimate is 3×10^{-8} M H^+ and 7×10^{-7} *M* OH^-. By calculation, $[H^+] = 2 \times 10^{-8}$ *M* H^+; $[OH^-] = 6 \times 10^{-7}$ *M* OH^-

16.27

$[H^+]$	$[OH^-]$	pH	pOH	Acidic or Basic
7.5×10^{-3} *M*	1.3×10^{-12} *M*	2.12	11.88	acidic
2.8×10^{-5} *M*	3.6×10^{-10} *M*	4.56	9.44	acidic
5.6×10^{-9} *M*	1.8×10^{-6} *M*	8.25	5.75	basic
5.0×10^{-9} *M*	2.0×10^{-6} *M*	8.30	5.70	basic

16.29 $[H^+] = 4.0 \times 10^{-8}$ *M*, $[OH^-] = 6.0 \times 10^{-7}$ *M* **16.31** (a) A *strong* acid is completely dissociated into ions in aqueous solution. (b) $[H^+] = 0.500$ *M* (c) HCl, HBr, HI **16.33** (a) $[H^+] = 8.5 \times 10^{-3}$ *M*, pH = 2.07 (b) $[H^+] = 0.0419$ *M*, pH = 1.377 (c) $[H^+] = 0.0250$ *M*, pH = 1.602 (d) $[H^+] = 0.167$ *M*, pH = 0.778 **16.35** (a) $[OH^-] = 3.0 \times 10^{-3}$ *M*, pH = 11.48 (b) $[OH^-] = 0.3758$ *M*, pH = 13.5750 (c) $[OH^-] = 8.75 \times 10^{-5}$ *M*, pH = 9.942 (d) $[OH^-] = 0.17$ *M*, pH = 13.23 **16.37** 3.2×10^{-3} *M* NaOH **16.39** pH = 13.400 **16.41** (a) $HBrO_2(aq) \rightleftharpoons H^+(aq) + BrO_2^-(aq)$, $K_a = [H^+][BrO_2^-]/[HBrO_2]$; $HBrO_2(aq) + H_2O(l) \rightleftharpoons H_3O^+(aq) + BrO_2^-(aq)$, $K_a = [H_3O^+][BrO_2^-]/[HBrO_2]$ (b) $HC_3H_5O_2(aq) \rightleftharpoons H^+(aq) + C_3H_5O_2^-(aq)$, $K_a = [H^+][C_3H_5O_2^-]/[HC_3H_5O_2]$; $HC_3H_5O_2(aq) + H_2O(l) \rightleftharpoons H_3O^+(aq) + C_3H_5O_2^-(aq)$, $K_a = [H_3O^+][C_3H_5O_2^-]/[HC_3H_5O_2]$ **16.43** $K_a = 1.4 \times 10^{-4}$ **16.45** $[H^+] = [X^-] = 0.019$ *M*, $[HX] = 0.181$ *M*, $K_a = 2.0 \times 10^{-3}$ **16.47** 0.089 *M* $HC_2H_3O_2$ **16.49** $[H^+] = [C_7H_5O_2^-] = 1.8 \times 10^{-3}$ *M*, $[HC_7H_5O_2] = 0.048$ *M* **16.51** (a) $[H^+] = 1.1 \times 10^{-3}$ *M*, pH = 2.95 (b) $[H^+] = 1.7 \times 10^{-4}$ *M*, pH = 3.76 (c) $[OH^-] = 1.4 \times 10^{-5}$ *M*, pH = 9.15 **16.53** $[H^+] = 2.0 \times 10^{-2}$ *M*, pH = 1.71 **16.55** (a) $[H^+] = 2.8 \times 10^{-3}$ *M*, 0.69% ionization (b) $[H^+] = 1.4 \times 10^{-3}$ *M*, 1.4% ionization (c) $[H^+] = 8.7 \times 10^{-4}$ *M*, 2.2% ionization **16.57** $HX(aq) \rightleftharpoons H^+(aq) + X^-(aq)$; $K_a = [H^+][X^-]/[HX]$. Assume that the percent of acid that ionizes is small. Let $[H^+] = [X^-] = y$. $K_a = y^2/[HX]$; $y = K_a^{1/2}/[HX]^{1/2}$. Percent ionization = y/[HX] × 100. Substituting for y, percent ionization = $100\ K_a^{1/2}[HX]^{1/2}/[HX]$ or $100\ K_a^{1/2}/[HX]^{1/2}$. That is, percent ionization varies inversely as the square root of the concentration of HX. **16.59** $[H^+] = 5.72 \times 10^{-3}$ *M*, pH = 2.24, $[HC_6H_5O_7^{3-}] = 1.2 \times 10^{-9}$ *M*. The approximation that the first ionization is less than 5% of the total acid concentration is not valid; the quadratic equation must be solved. The $[H^+]$ produced from the second and third ionizations is small with respect to that present from the first step; the second and third ionizations can be neglected when calculating the $[H^+]$ and pH. **16.61** All Brønsted–Lowry bases contain at least one nonbonded (lone) pair of electrons to attract H^+. **16.63** (a) $(CH_3)_2NH(aq) + H_2O(l) \rightleftharpoons (CH_3)_2NH_2^+(aq) + OH^-(aq)$; $K_b = [(CH_3)_2NH_2^+][OH^-]/[(CH_3)_2NH]$ (b) $CO_3^{2-}(aq) + H_2O(l) \rightleftharpoons HCO_3^-(aq) + OH^-(aq)$; $K_b = [HCO_3^-][OH^-]/[CO_3^{2-}]$ (c) $CHO_2^-(aq) + H_2O(l) \rightleftharpoons HCHO_2(aq) + OH^-(aq)$; $K_b = [HCHO_2][OH^-]/[CHO_2^-]$ **16.65** From the quadratic formula, $[OH^-] = 6.6 \times 10^{-3}$ *M*, pH = 11.82. **16.67** (a) $[C_{10}H_{15}ON] = 0.033$ *M*, $[C_{10}H_{15}ONH^+] = [OH^-] = 2.1 \times 10^{-3}$ *M* (b) $K_b = 1.4 \times 10^{-4}$ **16.69** (a) For a conjugate acid/conjugate base pair such as $C_6H_5OH/C_6H_5O^-$, K_b for the conjugate base can always be calculated from K_a for the conjugate acid, so a separate list of K_b values is not necessary. (b) $K_b = 7.7 \times 10^{-5}$ (c) Phenolate is a stronger base than NH_3. **16.71** (a) Acetic acid is stronger. (b) Hypochlorite ion is the stronger base. (c) K_b for $C_2H_3O_2^- = 5.6 \times 10^{-10}$, K_b for $ClO^- = 3.3 \times 10^{-7}$ **16.73** (a) $[OH^-] = 1.4 \times 10^{-3}$ *M*, pH = 11.15 (b) $[OH^-] = 3.8 \times 10^{-3}$ *M*, pH = 11.58 (c) $[NO_2^-] = 0.50$ *M*, $[OH^-] = 3.3 \times 10^{-6}$ *M*, pH = 8.52 **16.75** (a) Acidic (b) acidic (c) basic (d) neutral (e) acidic **16.77** The unknown is NaF. **16.79** $[OH^-] = 5.0 \times 10^{-6}$ *M*, pH = 8.70 **16.81** (a) As the electronegativity of the central atom (X) increases, the strength of the oxyacid increases. (b) As the number of nonprotonated oxygen atoms in the molecule increases, the strength of the oxyacid increases. **16.83** (a) HNO_3 is a stronger acid because it has one more nonprotonated oxygen atom and thus a higher oxidation number on N. (b) For binary hydrides, acid strength increases going down a family, so H_2S is a stronger acid than H_2O. (c) H_2SO_4 is a stronger acid because H^+ is much more tightly held by the anion HSO_4^-. (d) For oxyacids, the greater the electronegativity of the central atom, the stronger the acid, so H_2SO_4 is the stronger acid. (e) CCl_3COOH is stronger because the electronegative Cl atoms withdraw electron density from other parts of the molecule, which weakens the O—H bond and makes H^+ easier to remove. **16.85** (a) BrO^- (b) BrO^- (c) HPO_4^{2-} **16.87** (a) True (b) False. In a series of acids that have the same central atom, acid strength increases with the number of nonprotonated oxygen atoms bonded to the central atom. (c) False. H_2Te is a stronger acid than H_2S because the H—Te bond is longer, weaker, and more easily dissociated than the H—S bond. **16.89** Yes. The Arrhenius definition of a base, an $OH^-(aq)$ donor, is most restrictive; the Brønsted definition, an H^+ acceptor, is more general; and the Lewis definition, and electron-pair donor, is most general. Any substance that fits the narrow Arrhenius definition will fit the broader Brønsted and Lewis definitions. **16.91** (a) Acid, $Fe(ClO_4)_3$ or Fe^{3+}; base, H_2O (b) Acid, H_2O; base, CN^- (c) Acid, BF_3; base, $(CH_3)_3N$ (d) Acid, HIO; base, NH_2^- **16.93** (a) Cu^{2+}, higher cation charge (b) Fe^{3+}, higher cation charge (c) Al^{3+}, smaller cation radius, same charge **16.95** (a) Correct. (b) Incorrect. A Brønsted acid must have ionizable hydrogen. Lewis acids are electron-pair acceptors, but need not have ionizable hydrogen. (c) Correct. (d) Incorrect. K^+ is a negligible Lewis acid because its relatively large ionic radius and low positive charge render it a poor attractor of electron pairs. (e) Correct. **16.98** Assume *T* = 25°C. For acid

or base solute concentrations less than 1×10^{-6} M, consider the autoionization of water as a source of $H^+(aq)$ or $OH^-(aq)$. For this solution, $[H^+] = 9.9 \times 10^{-8}$ M, pH = 7.01.
16.99 (a) A 0.1 M solution of a weak acid (b) the acid with $K_a = 8 \times 10^{-6}$ (c) the base with $pK_b = 4.5$
16.101 (a) $[H^+] = 0.050$ M, pH = 1.30 (b) $[H^+] = 0.10$ M, pH = 1.00 (c) The pH assuming no ionization of HX^- is 1.30, so HX^- is not completely ionized; H_2X, which is completely ionized, is a stronger acid than HX^-. (d) Since H_2X is a strong acid, HX^- has no tendency to act like a base. HX^- does act like a weak acid, so a solution of NaHX would be acidic.
16.104 For all the compounds except caffeine, $[BH^+]/[B] > 1$ and the protonated form dominates. Caffeine, a very weak base, exists as the neutral base. **16.107** 6.0×10^{13} H^+ ions
16.109 (a) To the precision of the reported data, the pH of rainwater 40 years ago was 5.4, no different from the pH today. With extra significant figures, $[H^+] = 3.63 \times 10^{-6}$ M, pH = 5.440. (b) A 20.0-L bucket of today's rainwater contains 0.02 L (with extra significant figures, 0.200 L) of dissolved CO_2.
16.112 [cocaine] $= 2.93 \times 10^{-3}$ M, $[OH^-] = 3.4 \times 10^{-6}$ M, $K_b = 3.9 \times 10^{-9}$ **16.114** (a) K(i) $= 5.6 \times 10^3$, K(ii) = 10 (b) Both (i) and (ii) have $K > 1$, so both could be written with a single arrow.

Chapter 17

17.1 (a) The extent of dissociation of a weak electrolyte is decreased when a strong electrolyte containing an ion in common with the weak electrolyte is added to it. (b) $NaNO_2$
17.3 (a) pH increases (b) pH decreases (c) pH increases (d) no change (e) pH decreases
17.5 $[H^+] = 1.8 \times 10^{-5}$ M, pH = 4.73, $[C_3H_5O_2^-] = 6.0 \times 10^{-2}$ M
17.7 (a) pH = 3.53 (b) pH = 5.01 **17.9** (a) 4.5% ionization (b) 0.018% ionization **17.11** In a mixture of $HC_2H_3O_2$ and $NaC_2H_3O_2$, $HC_2H_3O_2$ reacts with added base and $C_2H_3O_2^-$ combines with added acid, leaving $[H^+]$ relatively unchanged. Although HCl and Cl^- are a conjugate acid-base pair, Cl^- has no tendency to combine with added acid to form undissociated HCl. Any added acid simply increases $[H^+]$ in an HCl-NaCl mixture. **17.13** (a) pH = 3.82 (b) pH = 3.96
17.15 (a) pH = 4.60
(b) $Na^+(aq) + C_2H_3O_2^-(aq) + H^+(aq) + Cl^-(aq) \longrightarrow HC_2H_3O_2(aq) + Na^+(aq) + Cl^-(aq)$
(c) $HC_2H_3O_2(aq) + Na^+(aq) + OH^-(aq) \longrightarrow C_2H_3O_2^-(aq) + H_2O(l) + Na^+(aq)$
17.17 0.18 mol NaBrO **17.19** (a) pH = 4.86 (b) pH = 5.0 (c) pH = 4.71 **17.21** (a) $[HCO_3^-]/[H_2CO_3] = 11$ (b) $[HCO_3^-]/[H_2CO_3] = 5.4$ **17.23** (a) Curve B (b) pH at the approximate equivalence point of curve A = 8.0, pH at the approximate equivalence point of curve B = 7.0 (c) For equal volumes of A and B, the concentration of acid B is greater, since it requires a larger volume of base to reach the equivalence point. **17.25** (a) HX is weaker. The higher the pH at the equivalence point, the stronger the conjugate base (X^-) and the weaker the conjugate acid (HX). (b) Phenolphthalein, which changes color in the pH 8–10 range, is perfect for HX and probably appropriate for HY. **17.27** (a) 42.4 mL NaOH soln (b) 29.6 mL NaOH soln (c) 29.8 mL NaOH soln
17.29 (a) pH = 1.54 (b) pH = 3.30 (c) pH = 7.00 (d) pH = 10.69 (e) pH = 12.74 **17.31** (a) pH = 2.78 (b) pH = 4.74 (c) pH = 6.58 (d) pH = 8.81 (e) pH = 11.03
(f) pH = 12.42 **17.33** (a) pH = 7.00 (b) $[HONH_3^+] = 0.100$ M, pH = 3.52 (c) $[C_6H_5NH_3^+] = 0.100$ M, pH = 2.82 **17.35** (a) The concentration of undissolved solid does not appear in the solubility product expression because it is constant as long as there is solid present. (b) $K_{sp} = [Ag^+][I^-]$; $K_{sp} = [Sr^{2+}][SO_4^{2-}]$; $K_{sp} = [Fe^{2+}][OH^-]^2$; $K_{sp} = [Hg_2^{2+}][Br^-]^2$
17.37 (a) $K_{sp} = 7.63 \times 10^{-9}$ (b) $K_{sp} = 2.7 \times 10^{-9}$ (c) 5.3×10^{-4} mol $Ba(IO_3)_2$/L **17.39** $K_{sp} = 2.3 \times 10^{-9}$
17.41 (a) 7.1×10^{-7} mol AgBr/L (b) 1.7×10^{-11} mol AgBr/L (c) 5.0×10^{-12} mol AgBr/L **17.43** (a) 1.4×10^3 g $Mn(OH)_2$/L (b) 1.4×10^{-2} g/L (c) 3.6×10^{-7} g/L **17.45** More soluble in acid: (a) $ZnCO_3$ (b) ZnS (d) AgCN (e) $Ba_3(PO_4)_2$
17.47 $[Cu^{2+}] = 2 \times 10^{-12}$ M **17.49** $K = K_{sp} \times K_f = 8 \times 10^4$
17.51 (a) $Q < K_{sp}$; no $Ca(OH)_2$ precipitates (a) $Q < K_{sp}$; no Ag_2SO_4 precipitates **17.53** pH = 13.0 **17.55** AgI will precipitate first, at $[I^-] = 4.2 \times 10^{-13}$ M. **17.57** The first two experiments eliminate group 1 and 2 ions (Figure 17.22). The absence of insoluble carbonate precipitates in the filtrate from the third experiment rules out group 4 ions. The ions that might be in the sample are those from group 3, Al^{3+}, Fe^{2+}, Zn^{2+}, Cr^{3+}, Ni^{2+}, Co^{2+}, or Mn^{2+}, and from group 5, NH_4^+, Na^+, or K^+. **17.59** (a) Make the solution acidic with 0.5 M HCl; saturate with H_2S. CdS will precipitate; ZnS will not. (b) Add excess base; $Fe(OH)_3(s)$ precipitates, but Cr^{3+} forms the soluble complex $Cr(OH)_4^-$. (c) Add $(NH_4)_2HPO_4$; Mg^{2+} precipitates as $MgNH_4PO_4$; K^+ remains soluble. (d) Add 6 M HCl; precipitate Ag^+ as $AgCl(s)$. **17.61** (a) Base is required to increase $[PO_4^{3-}]$ so that the solubility product of the metal phosphates of interest is exceeded and the phosphate salts precipitate. (b) K_{sp} for the cations in group 3 is much larger, and so to exceed K_{sp}, a higher $[S^{2-}]$ is required. (c) They should all redissolve in strongly acidic solution.
17.63 (a) pH = 3.025 (b) pH = 2.938 (c) pH = 12.862
17.65 $pK_a = 4.68$ **17.68** HOBr and NaOBr in a mole ratio of 1 to 1. $H_2NNH_3^+$ and H_2NNH_2 in a mole ratio of 1 to 3.1.
17.70 (a) $\mathcal{M} = 82.2$ g/mol (b) $K_a = 3.8 \times 10^{-7}$
17.72 (a) The pH at the equivalence point is 8.94; either phenolphthalein or thymol blue is appropriate. Phenolphthalein is usually chosen because the colorless-to-pink end point is easier to see. (b) 0.06206 M NaOH. **17.75** The resulting 1.00-L solution is a buffer containing 0.10 mol $H_2PO_4^-$ and 0.35 mol HPO_4^{2-}. pH = 7.75 **17.78** 1.1×10^3 μL of 1.0 M NaOH **17.81** $[Ca^{2+}]$ remaining $= 9.8 \times 10^{-10}$ M
17.84 $[SO_4^{2-}] = 9.7 \times 10^{-4}$ M, $[Pb^{2+}] = 6.5 \times 10^{-4}$ M, $[Sr^{2+}] = 3.3 \times 10^{-4}$ M **17.87** $[OH^-]$ must be greater than or equal to 1.0×10^{-2} M.
17.88 (a) $H^+(aq) + CHO_2^-(aq) \rightleftharpoons HCHO_2(aq)$ (b) $K = 5.6 \times 10^3$ (c) $[Na^+] = [Cl^-] = 0.075$ M, $[H^+] = [HCHO_2^-] = 3.7 \times 10^{-3}$ M, $[HCHO_2] = 0.071$ M
17.90 $[NH_4^+] = 0.10$ M, $[NH_3] = 0.050$ M, pH = 8.95
17.93 $[Sr^{2+}] = [SO_4^{2-}] = 5.7 \times 10^{-4}$ M, $K_{sp} = 3.2 \times 10^{-7}$

Chapter 18

18.1 (a) Its temperature profile (b) troposphere, 0 to 12 km; stratosphere, 12 to 50 km; mesosphere, 50 to 85 km; thermosphere, 85 to 110 km **18.3** The partial pressure of O_3 is 2.4×10^{-4} torr. **18.5** 1.5×10^{17} CO molecules
18.7 570 nm **18.9** Photoionization of O_2 requires 1205 kJ/mol.

Photodissociation requires only 495 kJ/mol. At lower elevations, high-energy short-wavelength solar radiation has already been absorbed. Below 90 km, the increased concentration of O_2 and the availability of longer wavelength radiation cause the photodissociation process to dominate.
18.11 (a) The rate of ozone formation, an exothermic process, is highest at about 50 km, near the stratopause. The heat generated by this formation reaction causes the temperature to be higher near the stratopause than the lower altitude tropopause. (b) The first step in the formation of O_3 is the photodissociation of O_2 to form two O atoms. Then, an O atom and an O_2 molecule collide to form $O_3{}^*$, a species with excess energy. If a carrier molecule such as N_2 or O_2 collides with $O_3{}^*$ and removes the excess energy, O_3 is formed. **18.13** A *hydrofluorocarbon* is a compound that contains hydrogen, fluorine, and carbon; it contains hydrogen in place of chlorine. HFCs are potentially less harmful than CFCs because photodissociation does not produce Cl atoms, which catalyze the destruction of ozone.
18.15 (a) The C—F bond requires more energy for dissociation than the C—Cl bond and is not readily cleaved by the available wavelengths of UV light. (b) Chlorine is present as chlorine atoms and chlorine oxide molecules, Cl and ClO, respectively. **18.17** (a) CO binds with hemoglobin in the blood to block O_2 transport to the cells; people with CO poisoning suffocate from lack of O_2. (b) SO_2 is very corrosive to tissue and contributes to respiratory disease, especially for people with other respiratory problems. It is also a major source of acid rain, which damages forests and wildlife in natural waters. (c) O_3 is extremely reactive and toxic because of its strong oxidizing ability. The products of its reactions with other atmospheric pollutants cause eye irritation and breathing difficulties. **18.19** (a) Methane, CH_4, arises from decomposition of organic matter by certain microorganisms; it also escapes from underground gas deposits. (b) SO_2 is released in volcanic gases and also is produced by bacterial action on decomposing vegetable and animal matter. (c) Nitric oxide, NO, results from oxidation of decomposing organic matter and is formed in lightning flashes. (d) CO is a possible product of some vegetable matter decay.
18.21 (a) $H_2SO_4(aq) + CaCO_3(s) \longrightarrow CaSO_4(s) + H_2O(l) + CO_2(g)$ (b) The $CaSO_4(s)$ would be much less reactive with acidic solution, since it would require a strongly acidic solution to shift the relevant equilibrium to the right.
$CaSO_4(s) + 2H^+(aq) \rightleftharpoons Ca^{2+}(aq) + 2HSO_4{}^-(aq)$
18.23 (a) ultraviolet (b) 357 kJ/mol (c) The average C—H bond energy from Table 8.4 is 413 kJ/mol. The C—H bond energy in CH_2O, 357 kJ/mol, is less than the "average" C—H bond energy. **18.25** Incoming and outgoing energies are in different regions of the electromagnetic spectrum. CO_2 is transparent to incoming visible radiation but absorbs outgoing infrared radiation. **18.27** 0.093 M Na^+ **18.29** 4.8×10^6 g CaO
18.31 The minimum pressure required to initiate reverse osmosis is greater than 5.1 atm. **18.33** (a) $CO_2(g)$, $HCO_3{}^-$, $H_2O(l)$, $SO_4{}^{2-}$, $NO_3{}^-$, $HPO_4{}^{2-}$, $H_2PO_4{}^-$ (b) $CH_4(g)$, $H_2S(g)$, $NH_3(g)$, $PH_3(g)$ **18.35** 2.5 g O_2
18.37 $Mg^{2+}(aq) + Ca(OH)_2(s) \longrightarrow Mg(OH)_2(s) + Ca^{2+}(aq)$
18.39 0.35 mol $Ca(OH)_2$, 0.15 mol Na_2CO_3 **18.41** $Al_2(SO_4)_3$ reacts with OH^- to form $Al(OH)_3(s)$, a gelatinous precipitate that occludes fine particles and bacteria present in the water. The $Al(OH)_3(s)$ settles slowly, removing the undesirable particulate matter. **18.43** Production of any form of energy requires a fuel and generates waste products. A more energy-efficient device or process uses less energy, which requires less fuel and generates fewer waste products. **18.45** Use of dimethylcarbonate in place of phosgene as a carbonyl source condenses methanol, CH_3OH, rather than HCl. Methanol is much less toxic than HCl, and it has the potential to be a second useful product, rather than waste. **18.48** Average molar mass at the surface = 27 g/mol. Average molar mass at the altitude where all O_2 has photodissociated = 19 g/mol.
18.50 $2[18.7] + [18.9] = 2Cl(g) + 2O_3(g) + 2ClO(g) \longrightarrow$
$2ClO(g) + 3O_2(g) + 2Cl(g) = 2O_3(g) \xrightarrow{Cl} 3O_2(g) = [18.10]$.
18.53 The formation of NO(g) is endothermic, so K_{eq} increases with increasing temperature. The oxidation of NO(g) to $NO_2(g)$ is exothermic, so the value of K_{eq} decreases with increasing temperature. **18.56** (a) The solubility of $CO_2(g)$ in the ocean would decrease if the temperature of the ocean increased. (b) If the solubility of $CO_2(g)$ in the ocean decreased because of global warming, more $CO_2(g)$ would be released into the atmosphere, perpetuating a cycle of increasing temperature and concomitant release of $CO_2(g)$ from the ocean.
18.59 (a) $CO_3{}^{2-}$ hydrolyzes in aqueous solution to produce OH^-. If $OH^-(aq)$ is sufficient to exceed K_{sp} for $Mg(OH)_2$, the solid will precipitate. (b) $Q = 3.5 \times 10^{-8}$; $Q > K_{sp}$, so $Mg(OH)_2$ will precipitate. **18.63** (a) $P_{NO_2} = 1.6 \times 10^{-5}$ torr (b) 2×10^{19} NO_2 molecules **18.66** (a) $\lambda = 258$ nm (b) $O_3(g) + O(g) \longrightarrow 2O_2(g)$. Because it is consumed and then reproduced, OH(g) is the catalyst in the overall reaction, another pathway for the destruction of ozone. **18.70** $P_{NO} = 0.08$ atm; concentration of NO = 8×10^4 ppm **18.73** (a) 1.1×10^{-5} M (b) by solving the quadratic formula, $[H^+] = 2.0 \times 10^{-6}$ M, pH = 5.71

Chapter 19

19.1 Spontaneous: b, c, d; nonspontaneous: a, e
19.3 (a) $NH_4NO_3(s)$ dissolves in water, as in a chemical cold pack. Naphthalene (moth balls) sublimes at room temperature. (b) Melting of a solid is spontaneous above its melting point but nonspontaneous below its melting point. **19.5** (a) endothermic (b) at or above 100°C (c) below 100°C (d) at 100°C
19.7 (a) For a *reversible* process, the forward and reverse changes occur by the same path. There is only one reversible pathway for a specified set of conditions. Work can be realized only from a reversible process. (b) There is no net change in the surroundings. (c) The vaporization of water to steam is reversible if it occurs at the boiling temperature of water for a specified external (atmospheric) pressure. **19.9** No. ΔE is a state function. $\Delta E = q + w$; q and w are not state functions. Their values do depend on path, but their sum, ΔE, does not.
19.11 We know that melting is a process that increases the energy of the system even though there is no change in temperature. ΔE is not zero for the process. **19.13** (a) Yes. (b) $w = -P_{ext}\Delta V$. Since the gas expands into a vacuum, $P_{ext} = 0$ and $w = 0$. (c) The driving force for this expansion is the increase in disorder of the system. **19.15** (a) 16 total arrangements (b) only one arrangement (c) The gas will spontaneously adopt the state with maximum disorder, the state with the most possible arrangements for the molecules. **19.17** (a) Entropy is related to disorder. (b) Negative. (c) No. ΔS is a state function, so it is independent of path. **19.19** (a) ΔS is positive. (b) ΔS is positive for Exercise 19.2 (a) and (c). **19.21** S increases in (a), (b),

and (c); S decreases in (d) **19.23** (a) ΔS increases (b) 213 J/K **19.25** (a) For a spontaneous process, the entropy of the universe increases; for a reversible process, the entropy of the universe does not change. (b) For a reversible process, if the entropy of the system increases, the entropy of the surroundings must decrease by the same amount. (c) For a spontaneous process, the entropy of the universe must increase, so the entropy of the surroundings must decrease by less than 42 J/K. **19.27** (a) The entropy of a pure crystalline substance at absolute zero is zero. (b) In *translational* motion the entire molecule moves in a single direction; in *rotational* motion the molecule rotates or spins around a fixed axis. In *vibrational* motion the bonds within a molecule stretch and bend, but the average position of the atoms does not change.

(c) H—Cl $\longrightarrow$ H—Cl
Translational

O—C—O
Rotational

H—Cl $\longleftrightarrow$ H——Cl $\longleftrightarrow$ H—Cl
Vibrational

19.29 (a) Ar(g) (b) He(g) at 1.5 atm (c) 1 mol of Ne(g) in 15.0 L (d) $CO_2(g)$ **19.31** (a) $\Delta S < 0$ (b) $\Delta S > 0$ (c) $\Delta S < 0$ (d) $\Delta S > 0$ **19.33** (a) Sc(s), 34.6 J/mol-K; Sc(g), 174.7 J/mol-K. In general, the gas phase of a substance has a larger $S°$ than the solid phase because of the greater volume and motional freedom of the molecules. (b) $NH_3(g)$, 192.5 J/mol-K; $NH_3(aq)$, 111.3 J/mol-K. Molecules in the gas phase have more motional freedom than molecules in solution. (c) 1 mol of $P_4(g)$, 280 J/K; 2 mol of $P_2(g)$, 2(218.1) = 436.2 J/K. More particles have a greater number of arrangements. (d) C (diamond), 2.43 J/mol-K; C (graphite) 5.69 J/mol-K. The internal entropy in graphite is greater because there is translational freedom among planar sheets of C atoms, while there is very little freedom within the covalent-network diamond lattice. (d) 1 mol of $H_2(g)$, 130.58 J/K; 2 mol of H(g), 229.20 J/K. More particles have a greater number of degrees of freedom. **19.35** The $S°$ value for each of the hydrocarbons is $CH_4(g)$, 186.3 J/mol-K; $C_2H_6(g)$, 229.5 J/mol-K; $C_3H_8(g)$, 269.9 J/mol-K; $C_4H_{10}(g)$, 310.0 J/mol-K. As the number of C atoms increases, the increased structural complexity leads to more motional degrees of freedom and $S°$ of the hydrocarbon increases. **19.37** (a) $\Delta S° = -120.5$ J/K. $\Delta S°$ is negative because there are fewer moles of gas in the products. (b) $\Delta S° = +176.6$ J/K. $\Delta S°$ is positive because there are more moles of gas in the products. (c) $\Delta S° = +152.39$ J/K. $\Delta S°$ is positive because the product contains more total particles and more moles of gas. (d) $\Delta S° = +92.3$ J/K. $\Delta S°$ is positive because there are more moles of gas in the products. **19.39** (a) $\Delta G = \Delta H - T\Delta S$ (b) If ΔG is positive, the process is nonspontaneous, but the reverse process is spontaneous. (c) There is no relationship between ΔG and rate of reaction. **19.41** (a) Exothermic (b) $\Delta S°$ is negative; the reaction leads to a decrease in disorder. (c) $\Delta G° = -9.9$ kJ (d) If all reactants and products are present in their standard states, the reaction is spontaneous at this temperature. **19.43** (a) $\Delta H° = -537.22$ kJ, $\Delta S° = 13.7$ J/K, $\Delta G° = -541.40$ kJ, $\Delta G° = \Delta H° - T\Delta S° = -541.31$ kJ (b) $\Delta H° = -106.7$ kJ, $\Delta S° = -142.2$ J/K, $\Delta G° = -64.0$ kJ, $\Delta G° = \Delta H° - T\Delta S° = -64.3$ kJ (c) $\Delta H° = -508.3$ kJ, $\Delta S° = -178$ J/K, $\Delta G° = -465.8$ kJ, $\Delta G° = \Delta H° - T\Delta S° = -455.1$ kJ. The discrepancy in $\Delta G°$ values is due to experimental uncertainties in the tabulated thermodynamic data. (d) $\Delta H° = -165.9$ kJ, $\Delta S° = 1.4$ J/K, $\Delta G° = 166.2$ kJ, $\Delta G° = \Delta H° - T\Delta S° = -166.3$ kJ **19.45** (a) $\Delta G° = -140.0$ kJ, spontaneous (b) $\Delta G° = +104.70$ kJ, nonspontaneous (c) $\Delta G° = +146$ kJ, nonspontaneous (d) $\Delta G° = -156.7$ kJ, spontaneous **19.47** (a) $C_6H_{12}(l) + 12O_2(g) \longrightarrow 6CO_2(g) + 12H_2O(l)$ (b) Because $\Delta S°$ is negative, $\Delta G°$is less negative than $\Delta H°$. **19.49** (a) The forward reaction is spontaneous at low temperatures, but becomes nonspontaneous at higher temperatures. (b) The reaction is nonspontaneous in the forward direction at all temperatures. (c) The forward reaction is nonspontaneous at low temperatures, but becomes spontaneous at higher temperatures. (d) The reaction becomes spontaneous in the forward direction at very high temperatures. **19.51** $\Delta S > +76.7$ J/K **19.53** (a) $T = 330$ K (b) nonspontaneous **19.55** (a) $\Delta H° = 155.7$ kJ,$\Delta S° = 171.4$ J/K. Since $\Delta S°$ is positive, $\Delta G°$ becomes more negative with increasing temperature. (b) $\Delta G° = 19$ kJ. The reaction is not spontaneous under standard conditions at 800 K. (c) $\Delta G° = -15.7$ kJ. The reaction is spontaneous under standard conditions at 1000 K. **19.57** (a) $T_b = 79$°C (b) From the *Handbook of Chemistry and Physics*, 74th Edition, $T_b = 80.1$°C. The values are remarkably close; the small difference is due to deviation from ideal behavior by $C_6H_6(g)$ and experimental uncertainty in the boiling point measurement and the thermodynamic data. **19.59** (a) $C_2H_2(g) + \frac{5}{2}O_2(g) \longrightarrow 2CO_2(g) + H_2O(l)$ (b) -1299.5 kJ of heat produced/mol C_2H_2 burned (c) $w_{max} = -1235.1$ kJ/mol C_2H_2 **19.61** (a) ΔG becomes more negative. (b) ΔG becomes more positive. (c) ΔG becomes more positive. **19.63** (a) $\Delta G° = -5.40$ kJ (b) $\Delta G = 0.30$ kJ **19.65** (a) $\Delta G° = -16.77$ kJ, $K_p = 870$ (b) $\Delta G° = 8.0$ kJ, $K_p = 0.04$ (c) $\Delta G° = -497.9$ kJ, $K_p = 2 \times 10^{87}$ **19.67** $\Delta H° = 269.3$ kJ, $\Delta S° = 0.1719$ kJ/K (a) $P_{CO_2} = 6.0 \times 10^{-39}$ atm (b) $P_{CO_2} = 1.6 \times 10^{-4}$ atm **19.69** (a) $HNO_2(aq) \rightleftharpoons H^+(aq) + NO_2^-(aq)$ (b) $\Delta G° = 19.1$ kJ (c) $G = 0$ at equilibrium (d) $\Delta G = -2.72$ kJ **19.73** (a) $\Delta H > 0$, $\Delta S > 0$ (b) $\Delta H < 0$, $\Delta S < 0$ (c) $\Delta H > 0$, $\Delta S > 0$ (d) $\Delta H > 0$, $\Delta S > 0$ (e) $\Delta H < 0$, $\Delta S > 0$ **19.77** Propylene will have a higher $S°$ at 25°C. In propylene, there is free rotation around the C—C single bond; this greater motional freedom leads to a higher absolute entropy. **19.81** (a) $\Delta G° = -33.32$ kJ, $\Delta G = -47.98$ kJ (b) $\Delta G° = -1336.8$ kJ, $\Delta G = -1324.2$ kJ (c) $\Delta G° = -159.4$ kJ, $\Delta G = -152.1$ kJ **19.83** (a) $K_{eq} = 4 \times 10^{15}$ (b) An increase in temperature will decrease the mole fraction of CH_3COOH at equilibrium. Elevated temperatures must be used to increase the speed of the reaction. (c) $K_{eq} = 1$ at 836 K or 563°C. **19.86** $\Delta G° = -2878.8$ kJ, 94.4 mol ATP/mol glucose **19.89** (a) $q_{rev} = nRT \ln(V_2/V_1)$ (b) $\Delta S = 8.4$ J/K (c) When a gas expands, there are more possible arrangements for the particles, and entropy increases. The positive sign for ΔS in part (b) is consistent with this prediction. (d) $\Delta S = -1.5 \times 10^2$ J/K **19.93** (a) The polymerization of ethylene reduces the number of particles in the system, so ΔS_{poly} is expected to be negative. (b) If the reaction is spontaneous and the entropy of the system decreases, the enthalpy of polymerization must be exothermic. (c) $\Delta H_{poly} = 1.36 \times 10^{-19}$ J/C_2H_4 monomer (d) In terms of structure, a condensation polymer imposes more order on the monomer(s) than an addition polymer. But, condensation polymerization does not lead to a reduction in the number of particles in the system, so ΔS_{poly} will be less negative than for addition polymerization. **19.96** (a) $\ddot{S}=C=\ddot{S}$ The C—S bond order is approximately 2. (b) 2e$^-$ domains around C, linear e$^-$ domain geometry, linear molecular structure (c) $CS_2(l) + 3O_2(g) \longrightarrow CO_2(g) + 2SO_2(g)$ (d) $\Delta H° = -1077.0$ kJ; $\Delta G° = -1060.5$ kJ. The reaction is exothermic ($-\Delta H°$) and spontaneous ($-\Delta G°$) at 298 K. (e) $\Delta S_{vap} = 86.6$ J/K. ΔS_{vap} is always positive because the gas phase occupies a greater volume and has more motional freedom and a larger absolute entropy than the liquid. (f) $T_b = 320$ K $= 47°$C. CS_2 is a liquid at 298 K, 1 atm. **19.99** (a) $K_{eq} = 8 \times 10^{15}$ (b) The process is feasible in principle. However, use of $H_2S(g)$ produces a severe safety hazard. (c) $P_{SO_2} = 5 \times 10^{-7}$ atm (d) The process will be less effective at elevated temperatures.

Chapter 20

20.1 (a) *Oxidation* is the loss of electrons. (b) Electrons appear on the products' side (right side). (c) The *oxidant* is the reactant that is reduced. **20.3** (a) I, +5 to 0; C, +2 to +4 (b) Hg, +2 to 0; N, −2 to 0 (c) N, +5 to +2; S, −2 to 0 (d) Cl, +4 to +3; O, −1 to 0 **20.5** (a) $TiCl_4(g) + 2Mg(l) \longrightarrow Ti(s) + 2MgCl_2(l)$ (b) Mg(*l*) is the reductant; $TiCl_4(g)$ is the oxidant. **20.7** (a) $Sn^{2+}(aq) \longrightarrow Sn^{4+}(aq) + 2e^-$, oxidation (b) $TiO_2(s) + 4H^+(aq) + 2e^- \longrightarrow Ti^{2+}(aq) + 2H_2O(l)$, reduction (c) $ClO_3^-(aq) + 6H^+(aq) + 6e^- \longrightarrow Cl^-(aq) + 3H_2O(l)$, reduction (d) $4OH^-(aq) \longrightarrow O_2(g) + 2H_2O(l) + 4e^-$, oxidation (e) $SO_3^{2-}(aq) + 2OH^-(aq) \longrightarrow SO_4^{2-}(aq) + H_2O(l) + 2e^-$, oxidation **20.9** (a) $Cr_2O_7^{2-}(aq) + I^-(aq) + 8H^+(aq) \longrightarrow 2Cr^{3+}(aq) + IO_3^-(aq) + 4H_2O(l)$; oxidizing agent, $Cr_2O_7^{2-}$; reducing agent, I^- (b) $4MnO_4^-(aq) + 5CH_3OH(aq) + 12H^+(aq) \longrightarrow 4Mn^{2+}(aq) + 5HCO_2H(aq) + 11H_2O(l)$; oxidizing agent, MnO_4^- reducing agent, CH_3OH (c) $I_2(s) + 5OCl^-(aq) + H_2O(l) \longrightarrow 2IO_3^-(aq) + 5Cl^-(aq) + 2H^+(aq)$; oxidizing agent, OCl^-; reducing agent, I_2 (d) $As_2O_3(s) + 2NO_3^-(aq) + 2H_2O(l) + 2H^+(aq) \longrightarrow 2H_3AsO_4(aq) + N_2O_3(aq)$; oxidizing agent, NO_3^-; reducing agent, As_2O_3 (e) $2MnO_4^-(aq) + Br^-(aq) + H_2O(l) \longrightarrow 2MnO_2(s) + BrO_3^-(aq) + 2OH^-(aq)$; oxidizing agent, MnO_4^-; reducing agent, Br^- (f) $Pb(OH)_4^{2-}(aq) + ClO^-(aq) \longrightarrow PbO_2(s) + Cl^-(aq) + 2OH^-(aq) + H_2O(l)$; oxidizing agent, ClO^-; reducing agent, $Pb(OH)_4^{2-}$ **20.11** (a) The reaction $Cu^{2+}(aq) + Zn(s) \longrightarrow Cu(s) + Zn^{2+}(aq)$ is occurring in both Figures. In Figure 20.3 the reactants are in contact, while in Figure 20.4 the oxidation half-reaction and reduction half-reaction are occurring in separate compartments. In Figure 20.3 the flow of electrons cannot be isolated or utilized; in Figure 20.4 electrical current is isolated and flows through the voltmeter. (b) Na^+ cations are drawn into the cathode compartment to maintain charge balance as Cu^{2+} ions are removed. **20.13** (a) $Ag^+(aq) + 1e^- \longrightarrow Ag(s)$; $Fe(s) \longrightarrow Fe^{2+}(aq) + 2e^-$ (b) Fe(*s*) is the anode, Ag(*s*) is the cathode. (c) Fe(*s*) is negative; Ag(*s*) is positive. (d) Electrons flow from the Fe electrode (−) toward the Ag electrode (+). (e) Cations migrate toward the Ag(*s*) cathode; anions migrate toward the Fe(*s*) anode. **20.15** *Electromotive force*, emf, is the potential energy difference between an electron at the anode and an electron at the cathode of a voltaic cell. (b) One *volt* is the potential energy difference required to impart 1 J of energy to a charge of 1 coulomb. (c) *Cell potential*, E_{cell}, is the emf of an electrochemical cell. **20.17** (a) $2H^+(aq) + 2e^- \longrightarrow H_2(g)$ (b) A *standard* hydrogen electrode, SHE, has components that are at standard conditions, 1 *M* $H^+(aq)$ and $H_2(g)$ at 1 atm. (c) The platinum foil in a SHE serves as an inert electron carrier and a solid reaction surface. **20.19** (a) A *standard reduction potential* is the relative potential of a reduction half-reaction measured at standard conditions. (b) $E°_{red} = 0$ (c) The reduction of $Ag^+(aq)$ to Ag(*s*) is much more energetically favorable. **20.21** (a) $Cr^{2+}(aq) \longrightarrow Cr^{3+}(aq) + e^-$; $Tl^{3+}(aq) + 2e^- \longrightarrow Tl^+(aq)$ (b) $E°_{red} = 0.78$ V

(c)

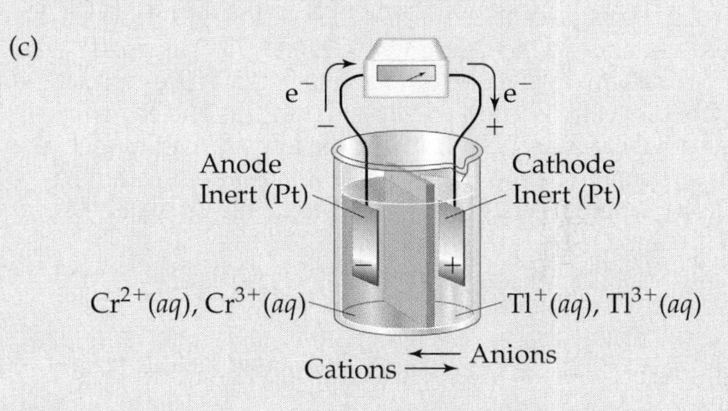

20.23 (a) $E° = 0.823$ V (b) $E° = 1.89$ V (c) $E° = 1.211$ V (d) $E° = 1.21$ V **20.25** (a) $3Ag^+(aq) + Cr(s) \longrightarrow 3Ag(s) + Cr^{3+}(aq)$, $E° = 1.54$ V (b) Two of the combinations have essentially equal $E°$ values: $2Ag^+(aq) + Cu(s) \longrightarrow 2Ag(s) + Cu^{2+}(aq)$, $E° = 0.462$ V; $3Ni^{2+}(aq) + 2Cr(s) \longrightarrow 3Ni(s) + 2Cr^{3+}(aq)$, $E° = 0.46$ V **20.27** (a) $MnO_4^-(aq) + 8H^+(aq) + 5e^- \longrightarrow Mn^{2+}(aq) + 4H_2O(l)$, $E°_{red} = 1.51$ V (b) The half-reaction in part (a) (c) $Sn^{2+}(aq) \longrightarrow Sn^{4+}(aq) + 2e^-$ (d) $5Sn^{2+}(aq) + 2MnO_4^-(aq) + 16H^+(aq) \longrightarrow 5Sn^{4+}(aq) + 2Mn^{2+}(aq) + 8H_2O(l)$ (e) $E° = 1.36$ V **20.29** (a) Anode, Sn(*s*); cathode, Cu(*s*). (b) The copper electrode gains mass as Cu is plated out, and the tin electode looses mass as Sn is oxidized. (c) $Cu^{2+}(aq) + Sn(s) \longrightarrow Cu(s) + Sn^{2+}(aq)$. (d) $E° = 0.473$ V. **20.31** (a) Negative (b) right **20.33** (a) $Cl_2(g)$ (b) $Ni^{2+}(aq)$ (c) $BrO_3^-(aq)$ (d) $O_3(g)$ **20.35** (a) $Cl_2(aq)$, strong oxidant (b) $MnO_4^-(aq)$, acidic, strong oxidant (c) Ba(*s*), strong reductant (d) Zn(*s*), reductant **20.37** (a) $Cu^{2+}(aq) < O_2(g) < Cr_2O_7^{2-}(aq) < Cl_2(g) < H_2O_2(aq)$ (b) $H_2O_2(aq) < I^-(aq) < Sn^{2+}(aq) < Zn(s) < Al(s)$ **20.39** Al and $H_2C_2O_4$ **20.41** (a) The more positive the emf of a reaction, the more spontaneous the reaction. (b) Reactions (a), (b), and (c) in Exercise 20.23 are spontaneous. (c) 20.23 (a) $\Delta G° = -159$ kJ; 20.23 (b) $\Delta G° = -365$ kJ; 20.23 (c) $\Delta G° = -234$ kJ; 20.23 (d) $\Delta G° = -701$ kJ **20.43** (a) $2Fe^{2+}(aq) + S_2O_6^{2-}(aq) + 4H^+(aq) \longrightarrow 2Fe^{3+}(aq) + 2H_2SO_3(aq)$; $2Fe^{2+}(aq) + N_2O(aq) + 2H^+(aq) \longrightarrow 2Fe^{3+}(aq) + N_2(g) + H_2O(l)$; $Fe^{2+}(aq) + VO_2^+(aq) + 2H^+(aq) \longrightarrow Fe^{3+}(aq) + VO^{2+}(aq) + H_2O(l)$ (b) $E° = -0.17$ V, $\Delta G° = 33$ kJ; $E° = -2.54$ V, $\Delta G° = 4.90 \times 10^2$ kJ; $E° = 0.23$ V, $\Delta G° = -22$ kJ **20.45** (a) The *Nernst equation* is applicable when the components of an electrochemical cell are at nonstandard conditions. (b) $Q = 1$ (c) Q decreases and E increases **20.47** (a) E decreases (b) E decreases (c) E decreases (d) no effect **20.49** (a) $E° = 0.48$ V (b) $E = 0.53$ V (c) $E = 0.46$ V **20.51** (a) $E° = 0.46$ V (b) $E = 0.42$ V **20.53** (a) The compartment with $[Zn^{2+}] = 1.00 \times 10^{-2}$ M is the anode. (b) $E° = 0$ (c) $E = 0.0799$ V (d) In the anode compartment $[Zn^{2+}]$ increases; in the cathode compartment $[Zn^{2+}]$ decreases **20.55** $E° = 0.763$ V, pH = 1.6 **20.57** (a) $E° = 0.16$ V, $K = 2.54 \times 10^5 = 3 \times 10^5$ (b) $E° = 0.277$ V, $K_{eq} = 2.3 \times 10^9$ (c) $E° = 0.44$ V, $K_{eq} = 10^{74}$ **20.59** (a) $K_{eq} = 9.8 \times 10^2$ (b) $K_{eq} = 9.5 \times 10^5$ (c) $K_{eq} = 9.3 \times 10^8$ **20.61** (a) A *battery* is a portable, self-contained electrochemical power source composed of one or more voltaic cells. (b) A *primary* battery is not rechargeable, while a *secondary* battery can be recharged. (c) No. No single voltaic cell is capable of producing 7.5 V. Three 2.5 V voltaic cells connected in series would produce the desired voltage. **20.63** 441 g PbO_2 **20.65** (a) The anode (b) $E° = 3.50$ V (c) The emf of the battery, 3.5 V, is exactly the cell potential calculated in part (b). **20.67** (a) The cell emf will have a smaller value. (b) NiMH batteries use an alloy such as $ZrNi_2$ as the anode material. This eliminates the use and disposal problems associated with Cd, a toxic heavy metal. **20.69** (a) anode: $Fe(s) \longrightarrow Fe^{2+}(aq) + 2e^-$; cathode: $O_2(g) + 4H^+(aq) + 4e^- \longrightarrow 2H_2O(l)$ (b) $2Fe^{2+}(aq) + 3H_2O(l) + 3H_2O(l) \longrightarrow Fe_2O_3 \cdot 3H_2O(s) + 6H^+(aq) + 2e^-$; $O_2(g) + 4H^+(aq) + 4e^- \longrightarrow 2H_2O(l)$ **20.71** (a) Zn^{2+} has a more negative reduction potential than Fe^{2+}. If both Zn and Fe are exposed to O_2, Zn will be oxidized and Fe will not; Zn acts as a sacrificial anode. (b) Zn protects Fe by making it the cathode in the electrochemical process; this is called *cathodic protection*. **20.73** (a) *Electrolysis* is an electrochemical process driven by an outside energy source. (b) By definition, electrolysis reactions are nonspontaneous. (c) $2Cl^-(l) \longrightarrow Cl_2(g) + 2e^-$ **20.75** (a) The products are different because in aqueous electrolysis water is reduced in preference to Mg^{2+}. (b) $MgCl_2(l) \longrightarrow Mg(l) + Cl_2(g)$; $2Cl^-(aq) + 2H_2O(l) \longrightarrow Cl_2(g) + H_2(g) + OH^-(aq)$ (c) Mg^{2+} is reduced, $E° = -3.73$ V; H_2O is reduced, $E° = -2.19$ V **20.77** Cl^- is oxidized in preference to water because production of Cl_2 is kinetically favored.

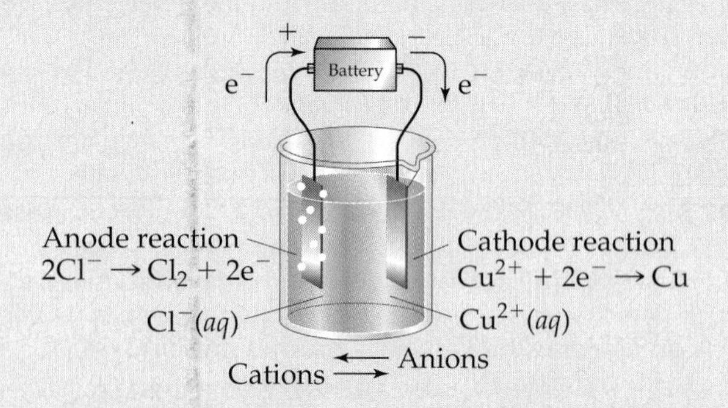

20.79 (a) 180 g Cr(s) (b) 2.51 A **20.81** (a) 10.5 L $Cl_2(g)$ (b) 0.940 mol NaOH **20.83** $w_{max} = -110$ kJ **20.85** (a) 4.0×10^5 g Li (b) 0.24 kWh/mol Li **20.87** (a) $3MnO_4^{2-}(aq) + 4H^+(aq) \longrightarrow 2MnO_4^-(aq) + MnO_2(s) + 2H_2O(l)$ (b) $3H_2SO_3(aq) \longrightarrow S(s) + 2HSO_4^-(aq) + 2H^+(aq) + H_2O(l)$ (c) $Cl_2(aq) + 2OH^-(aq) \longrightarrow Cl^-(aq) + ClO^-(aq) + H_2O(l)$

20.90 (a) $2Rh^{3+}(aq) + 3Cd(s) \longrightarrow 2Rh(s) + 3Cd^{2+}(aq)$ (b) Cd(s) is the anode, and Rh(s) is the cathode. (c) $E°_{red} = 0.80$ V (d) $\Delta G° = -695$ kJ **20.93** (a) $E° = 1.08$ V (b) $[Ni^{2+}]$ increases as the cell operates. (c) $[Ag^+] = 0.474 = 0.5$ M **20.95** $E° = 0.368$ V, $K_{eq} = 1.6 \times 10^6$ **20.97** (a) In discharge, $Cd(s) + 2NiO(OH)(s) + 2H_2O(l) \longrightarrow Cd(OH)_2(s) + 2Ni(OH)_2(s)$. In charging, the reverse reaction occurs. (b) $E° = 1.25$ V (c) 1.25 V is the standard cell potential, $E°$. The concentrations of reactants and products inside the battery are adjusted so that the cell output is greater than $E°$. **20.100** Water (moisture) provides an ion-transport medium that completes the voltaic cell and facilitates corrosion. Corrosion occurs more readily in acid solution because O_2 has a more positive reduction potential in the presence of $H^+(aq)$. SO_2 and its oxidation products dissolve in water to produce acidic solutions, which encourage corrosion. The anodic and cathodic reactions for the corrosion of Ni are $Ni(s) \longrightarrow Ni^{2+}(aq) + 2e^-$, $E°_{red} = -0.28$ V; $O_2(g) + 4H^+(aq) + 4e^- \longrightarrow 2H_2O(l)$, $E°_{red} = 1.23$ V. NiO(s), formed by the dry air oxidation of Ni, serves as a coating to protect against further corrosion. However, NiO dissolves in acidic solutions, which exposes Ni(s) to further wet corrosion. $NiO(s) + 2H^+(aq) \longrightarrow Ni^{2+}(aq) + H_2O(l)$ **20.103** Indium is in the +2 oxidation state in the molten halide. **20.107** The sign of E is negative, so the reaction is not spontaneous in this buffer. **20.110** (a) $E° = 0.028$ V (b) anode: $Fe^{2+}(aq) \longrightarrow Fe^{3+}(aq) + 1e^-$; cathode: $Ag^+(aq) + e^- \longrightarrow Ag(s)$ (c) $\Delta S° = 148.5$ J. Since $\Delta S°$ is positive, $\Delta G°$ will become more negative and $E°$ will become more positive as temperature is increased. **20.113** $K_{sp} = 1 \times 10^{-12}$ **20.116** $[Sn^{2+}] = 0.530$ M, $[H^+] = 0.140$ M, $[SO_4^{2-}]$ remains unchanged.

Chapter 21

21.1 (a) 25 protons, 30 neutrons (b) 80 protons, 121 neutrons (c) 19 protons, 20 neutrons **21.3** (a) ${}^1_1p$ (b) ${}^0_1e$ (c) ${}^0_{-1}\beta$ **21.5** (a) ${}^{214}_{83}Bi \longrightarrow {}^{214}_{84}Po + {}^0_{-1}e$ (b) ${}^{195}_{79}Au + {}^0_{-1}e$ (orbital electron) $\longrightarrow {}^{195}_{78}Pt$ (c) ${}^{38}_{19}K \longrightarrow {}^{38}_{18}Ar + {}^0_1e$ (d) ${}^{242}_{94}Pu \longrightarrow {}^{238}_{92}U + {}^4_2He$ **21.7** (a) ${}^{211}_{82}Pb \longrightarrow {}^{211}_{83}Bi + {}^0_{-1}\beta$ (b) ${}^{50}_{25}Mn \longrightarrow {}^{50}_{24}Cr + {}^0_1e$ (c) ${}^{179}_{74}W + {}^0_{-1}e \longrightarrow {}^{179}_{73}Ta$ (d) ${}^{230}_{90}Th \longrightarrow {}^{226}_{88}Ra + {}^4_2He$ **21.9** 7 alpha emissions, 4 beta emissions **21.11** (a) Positron emission (for low atomic numbers, positron emission is more common than electron capture) (b) beta emission (c) beta emission (d) beta emission **21.13** (a) No (low neutron/proton ratio, could be a positron emitter or undergo orbital electron capture) (b) no (high atomic number, alpha emitter) (c) no (high neutron/proton ratio, beta emitter) (d) no (low neutron/proton ratio, could be a positron emitter or undergo orbital electron capture). **21.15** (a) Stable: ${}^{39}_{19}K$, 20 neutrons is a magic number (b) stable: ${}^{209}_{83}Bi$, 126 neutrons is a magic number (c) stable: ${}^{25}_{12}Mg$; ${}^{24}_{10}Ne$ has a much higher neutron/proton ratio **21.17** (a) ${}^4_2He$ (c) ${}^{40}_{20}Ca$ (e) ${}^{126}_{82}Pb$ **21.19** Radioactive: ${}^{14}_{8}O$, ${}^{115}_{52}Te$—low neutron/proton ratio; ${}^{208}_{84}Po$—atomic number ≥ 84 **21.21** Protons and alpha particles are positively charged and must be moving very fast to overcome electrostatic forces which would repel them from the target nucleus. Neutrons are electrically neutral and not repelled by the nucleus. **21.23** (a) ${}^{32}_{16}S + {}^1_0n \longrightarrow {}^1_1p + {}^{32}_{15}P$

(b) $^{7}_{4}Be + ^{0}_{-1}e$ (orbital electron) $\longrightarrow ^{7}_{3}Li$
(c) $^{187}_{75}Re \longrightarrow ^{187}_{76}Os + ^{0}_{-1}e$ (d) $^{98}_{42}Mo + ^{2}_{1}H \longrightarrow ^{1}_{0}n + ^{99}_{43}Tc$
(e) $^{235}_{92}U + ^{1}_{0}n \longrightarrow ^{135}_{54}Xe + ^{99}_{38}Sr + 2^{1}_{0}n$
21.25 (a) $^{238}_{92}U + ^{1}_{0}n \longrightarrow ^{239}_{92}U + ^{0}_{0}\gamma$
(b) $^{14}_{7}N + ^{1}_{1}H \longrightarrow ^{11}_{6}C + ^{4}_{2}He$ (c) $^{18}_{8}O + ^{1}_{0}n \longrightarrow ^{19}_{9}F + ^{0}_{-1}e$
21.27 The energy changes involved in chemical reactions are much too small to allow us to alter nuclear properties via chemical processes. The nuclei that are formed in a nuclear reaction will continue to emit radioactivity regardless of any chemical changes we bring to bear. **21.29** 24.0 mg tritium remain after 12.3 yr, 3.0 mg after 49.2 yr
21.31 $k = 0.0307\ yr^{-1}$, $t_{1/2} = 22.6$ yr **21.33** $k = 0.0249\ d^{-1}$, $t = 53.9$ d **21.35** (a) 4.1×10^{-12} g ^{226}Ra decays in 1.0 min, 1.1×10^{10} alpha particles emitted (b) 4.9 mCi
21.37 $k = 1.21 \times 10^{-4}\ yr^{-1}$, $t = 2.20 \times 10^{3}$ yr
21.39 $k = 1.5 \times 10^{-10}\ yr^{-1}$; the original rock contained 66.2 mg ^{238}U and is 1.8×10^{9} yr old. **21.41** $\Delta m = 4.378 \times 10^{-9}$ g
21.43 $\Delta m = 0.200287$ amu, $\Delta E = 2.98912 \times 10^{-11}$ J/^{23}Na nucleus required, 1.80009×10^{13} J/mol ^{23}Na
21.45 (a) mass defect = 0.098940 amu,
binding energy/nucleon = 1.2305×10^{-12} J
(b) mass defect = 0.340423 amu,
binding energy/nucleon = 1.37312×10^{-12} J
(c) mass defect = 1.234249 amu,
binding energy/nucleon = 1.344536×10^{-12} J
21.47 (a) 1.71×10^{5} kg/d (b) 2.1×10^{8} g ^{235}U **21.49** (a) ^{59}Co; it has the largest binding energy per nucleon, and binding energy gives rise to mass defect. **21.51** The ^{59}Fe is incorporated into the diet component and fed to the rabbits. Blood samples are removed from the animals, the red blood cells separated, and the radioactivity of the sample measured. If the iron in the dietary compounds has been incorporated into blood hemoglobin, the blood cell sample should show beta emission. Samples can be taken at various times to determine the rate of iron uptake, rate of loss of iron from the blood, and so forth. **21.53** (a) *Control rods* control neutron flux so that there are enough neutrons to sustain the chain reaction but not so many that the core overheats. (b) A *moderator* slows neutrons so that they are more easily captured by fissioning nuclei.
21.55 (a) $4^{1}_{0}n$ (b) $^{94}_{36}Kr$ **21.57** The extremely high temperature is required to overcome the electrostatic charge repulsions between the nuclei so that they can come together to react.
21.59 ·OH is a free radical; it can react with almost any particle (atom, molecule, ion) to acquire an electron and become OH^-. This often starts a disruptive chain of reactions, each producing a different free radical. Hydroxide ion, OH^-, on the other hand, will be attracted to cations or the positive end of a polar molecule. The acid-base reactions of OH^- are usually much less disruptive to the organism than the chain of redox reactions initiated by ·OH radical. **21.61** (a) 3.2×10^{8} dis/s, 3.2×10^{8} Bq (b) 1.5×10^{2} mrad, 1.5×10^{-3} Gy
(c) 1.5×10^{3} mrem, 1.5×10^{-2} Sv **21.63** $^{210}_{82}Pb$ **21.65** The most massive radionuclides will have the highest neutron/proton ratios, so they will decay by a process that lowers this ratio, beta emission. The least massive nuclides will decay by a process that increases the neutron/proton ratio, positron emission or orbital electron capture.
21.68 1.3×10^{6} alpha particles/s; 3.5×10^{-5} Ci
21.70 3.7×10^{4} disintegrations/s; 4.8×10^{13} ^{90}Sr nuclei or 7.2×10^{-9} g ^{90}Sr **21.73** If there is no depletion of iodide due to plant uptake, the calculated activity after 32 days is 11.1 counts/min. The observed activity, 12.2 counts/min, is actually higher than this; we can assume that the plants did not absorb iodide. **21.75** ^{7}Be, 8.612×10^{-13} J/nucleon; ^{9}Be, 1.035×10^{-12} J/nucleon; ^{10}Be: 1.042×10^{-12} J/nucleon. The binding energies/nucleon for ^{9}Be and ^{10}Be are very similar; that for ^{10}Be is slightly higher. **21.77** 2×10^{5} g U/yr
21.79 7.55×10^{19} ^{36}Cl nuclei, $k = 7.3 \times 10^{-14}\ s^{-1}$, 5.5×10^{6} dis/s **21.81** (a) 1.4×10^{17} alpha particles (b) $P_{He} = 0.28$ torr

Chapter 22

22.1 Metals: (b) Sr, (c) Ce, (e) Rh; nonmetals: (d) Se, (f) Kr; metalloid: (a) Sb **22.3** (a) Cl (b) K (c) K in the gas phase (lowest ionization energy), Li in aqueous solution (most positive $E°$ value) (d) Ne; Ne and Ar are difficult to compare because they do not form compounds and their radii are not measured in the same way as other elements. **22.5** (a) N is too small a central atom to fit five fluorine atoms, and it does not have available *d* orbitals, which can help accommodate more than eight electrons. (b) Si does not readily form π bonds, which are necessary to satisfy the octet rule for both atoms in the molecule. (c) As has a lower electronegativity than N; that is, it more readily gives up electrons to an acceptor and is more easily oxidized.
22.7 (a) $LiN_3(s) + H_2O(l) \longrightarrow HN_3(aq) + LiOH(aq)$
(b) $2C_3H_7OH(l) + 9O_2(g) \longrightarrow 6CO_2(g) + 8H_2O(l)$
(c) $NiO(s) + C(s) \longrightarrow CO(g) + Ni(s)$ or
$2NiO(s) + C(s) \longrightarrow CO_2(g) + 2Ni(s)$
(d) $AlP(s) + 3H_2O(l) \longrightarrow PH_3(g) + Al(OH)_3(s)$
(e) $Na_2S(s) + 2HCl(aq) \longrightarrow H_2S(g) + 2NaCl(aq)$
22.9 (a) $^{1}_{1}H$, protium; $^{2}_{1}H$, deuterium; $^{3}_{1}H$, tritium
(b) in order of decreasing natural abundance: protium > deuterium > tritium **22.11** Like other elements in group 1A, hydrogen has only one valence electron. Like other elements in group 7A, hydrogen needs only one electron to complete its valence shell.
22.13 (a) $Mg(s) + 2H^+(aq) \longrightarrow Mg^{2+}(aq) + H_2(g)$
(b) $C(s) + H_2O(g) \xrightarrow{1000°C} CO(g) + H_2(g)$
(c) $CH_4(g) + H_2O(g) \xrightarrow{1100°C} CO(g) + 3H_2(g)$
22.15 (a) $NaH(s) + H_2O(l) \longrightarrow NaOH(aq) + H_2(g)$
(b) $Fe(s) + H_2SO_4(aq) \longrightarrow Fe^{2+}(aq) + H_2(g) + SO_4^{2-}(aq)$
(c) $H_2(g) + Br_2(g) \longrightarrow 2HBr(g)$ (d) $Na(l) + H_2(g) \longrightarrow 2NaH(s)$
(e) $PbO(s) + H_2(g) \xrightarrow{\Delta} Pb(s) + H_2O(g)$
22.17 (a) molecular (b) ionic (c) metallic **22.19** Xenon has a lower ionization energy than argon; because the valence electrons are not as strongly attracted to the nucleus, they are more readily promoted to a state in which the atom can form bonds with fluorine. Also, Xe is larger and can more easily accommodate an expanded octet of electrons. **22.21** (a) BrO_3^-, +5 (b) HI, −1 (c) BrF_3, Br, +3; F, −1 (d) NaOCl, +1 (e) $HClO_4$, +7 (f) XeF_4, +4 **22.23** (a) potassium chlorate (b) calcium iodate (c) aluminum chloride (d) bromic acid (e) paraperiodic acid (f) xenon tetrafluoride **22.25** (a) Van der Waals intermolecular attractive forces increase with increasing number of electrons in the atoms. (b) F_2 reacts with water.
$F_2(g) + H_2O(l) \longrightarrow 2HF(g) + O_2(g)$. That is, fluorine is too strong an oxidizing agent to exist in water. (c) HF has extensive hydrogen bonding. (d) Oxidizing power is related to

electronegativity. Electronegativity and oxidizing power decrease in the order given.
22.27 (a) $Br_2(l) + 2OH^-(aq) \longrightarrow BrO^-(aq) + Br^-(aq) + H_2O(l)$ (b) $Cl_2(g) + 3I^-(aq) \longrightarrow I_3^-(aq) + 2Cl^-(aq)$
22.29 (a) square planar (b) trigonal pyramidal (c) octahedral about the central iodine (d) linear **22.31** (a) As an oxidizing agent in steelmaking; to bleach pulp and paper; in oxyacetylene torches; in medicine to assist in breathing (b) synthesis of pharmaceuticals, lubricants, and other organic compounds where C=C bonds are cleaved; in water treatment
22.33 (a) $CaO(s) + H_2O(l) \longrightarrow Ca^{2+}(aq) + 2OH^-(aq)$
(b) $Al_2O_3(s) + 6H^+(aq) \longrightarrow 2Al^{3+}(aq) + 3H_2O(l)$
(c) $Na_2O_2(s) + 2H_2O(l) \longrightarrow 2Na^+(aq) + 2OH^-(aq) + H_2O_2(aq)$
(d) $N_2O_3(g) + H_2O(l) \longrightarrow 2HNO_2(aq)$
(e) $2KO_2(s) + 2H_2O(l) \longrightarrow 2K^+(aq) + 2OH^-(aq) + O_2(g) + H_2O_2(aq)$
(f) $NO(g) + O_3(g) \longrightarrow NO_2(g) + O_2(g)$ **22.35** (a) Neutral (b) acidic (c) basic (d) amphoteric **22.37** (a) SeO_3, +6 (b) $Na_2S_2O_3$, +2 (c) SF_4, +4 (d) H_2S, −2 (e) H_2SO_3, +4
22.39 (a) $2Fe^{3+}(aq) + H_2S(aq) \longrightarrow 2Fe^{2+}(aq) + S(s) + 2H^+(aq)$
(b) $Br_2(l) + H_2S(aq) \longrightarrow 2Br^-(aq) + S(s) + 2H^+(aq)$
(c) $2MnO_4^-(aq) + 6H^+(aq) + 5H_2S(aq) \longrightarrow 2Mn^{2+}(aq) + 5S(s) + 8H_2O(l)$
(d) $2NO_3^-(aq) + H_2S(aq) + 2H^+(aq) \longrightarrow 2NO_2(aq) + S(s) + 2H_2O(l)$

22.41 (a) [:Ö—Se—Ö: with :Ö: below Se]$^{2-}$

Trigonal pyramidal

(b) :Cl̈—S̈—S̈—C̈l:

Bent (free rotation around S—S bond)

(c) :Ö: above S; :Ö—S—C̈l:; :Ö—H below S

Tetrahedral (around S)

22.43 (a) $SO_2(s) + H_2O(l) \longrightarrow H_2SO_3(aq) \rightleftharpoons H^+(aq) + HSO_3^-(aq)$
(b) $ZnS(s) + 2HCl(aq) \longrightarrow ZnCl_2(aq) + H_2S(g)$
(c) $8SO_3^{2-}(aq) + S_8(s) \longrightarrow 8S_2O_3^{2-}(aq)$
(d) $SO_3(aq) + H_2SO_4(l) \longrightarrow H_2S_2O_7(l)$ **22.45** (a) $NaNO_2$, +3 (b) NH_3, −3 (c) N_2O, +1 (d) NaCN, −3 (e) HNO_3, +5 (f) NO_2, +4

22.47 (a) [H—N—H with H above and H below N]$^+$

Tetrahedral

(b) :Ö—N(=O:)—Ö—H ⟷ Ö=N(—Ö:)—Ö—H ⟷ Ö:—N(—Ö:)=Ö—H

The geometry around nitrogen is trigonal planar, but the hydrogen atom is not required to lie in this plane. The third resonance form makes a much smaller contribution to the structure than the first two.

(c) :N̈=N=Ö: ⟷ :N≡N—Ö: ⟷ :N̈—N≡O:

The molecule is linear. Again, the third resonance form makes less contribution to the structure because of the high formal charges involved.

(d) Ö=N̈—Ö: ⟷ :Ö—N̈=Ö:

The molecule is bent (nonlinear).

22.49 (a) $Mg_3N_2(s) + 6H_2O(l) \longrightarrow 3Mg(OH)_2(s) + 2NH_3(aq)$
(b) $2NO(g) + O_2(g) \longrightarrow 2NO_2(g)$
(c) $N_2O_5(g) + H_2O(l) \longrightarrow 2H^+(aq) + 2NO_3^-(aq)$
(d) $NH_3(aq) + H^+(aq) \longrightarrow NH_4^+(aq)$
(e) $N_2H_4(l) + O_2(g) \longrightarrow N_2(g) + 2H_2O(g)$
22.51 (a) $2NO_3^-(aq) + 12H^+(aq) + 10e^- \longrightarrow N_2(g) + 6H_2O(l)$, $E^\circ_{red} = +1.25$ V (b) $2NH_4^+(aq) \longrightarrow N_2(g) + 8H^+(aq) + 6e^-$, $E^\circ_{red} = 0.27$ V **22.53** (a) H_3PO_4, +5 (b) H_3AsO_3, +3 (c) Sb_2S_3, +3 (d) $Ca(H_2PO_4)_2$, +5 (e) K_3P, −3
22.55 (a) Phosphorus is a larger atom than nitrogen, and P has energetically available 3*d* orbitals, which participate in the bonding, but nitrogen does not. (b) Only one of the three hydrogens in H_3PO_2 is bonded to oxygen. The other two are bonded directly to phosphorus and are not easily ionized. (c) PH_3 is a weaker base than H_2O so any attempt to add H^+ to PH_3 in the presence of H_2O causes protonation of H_2O. (d) Because of the severely strained bond angles in P_4 molecules, white phosphorus is highly reactive.
22.57 (a) $2Ca_3(PO_4)_2(s) + 6SiO_2(s) + 10C(s) \longrightarrow P_4(g) + 6CaSiO_3(l) + 10CO_2(g)$
(b) $3H_2O(l) + PCl_3(l) \longrightarrow H_3PO_3(aq) + 3H^+(aq) + 3Cl^-(aq)$
(c) $6Cl_2(g) + P_4(s) \longrightarrow 4PCl_3(l)$ **22.59** (a) HCN (b) SiC (c) $CaCO_3$ (d) CaC_2

22.61 (a) [:C≡N:]$^-$ (b) :C≡O: (c) [:C≡C:]$^{2-}$ (d) S̈=C=S̈ (e) Ö=C=Ö (f) [:Ö: above C; :O̤ and :Ö: below C]$^{2-}$

one of three equivalent resonance structures

22.63 (a) $ZnCO_3(s) \xrightarrow{\Delta} ZnO(s) + CO_2(g)$
(b) $BaC_2(s) + 2H_2O(l) \longrightarrow Ba^{2+}(aq) + 2OH^-(aq) + C_2H_2(g)$
(c) $C_2H_4(g) + 3O_2(g) \longrightarrow 2CO_2(g) + 2H_2O(g)$
(d) $2CH_3OH(l) + 3O_2(g) \longrightarrow 2CO_2(g) + 4H_2O(g)$
(e) $NaCN(s) + H^+(aq) \longrightarrow Na^+(aq) + HCN(g)$
22.65 (a) $2CH_4(g) + 2NH_3(g) + 3O_2(g) \xrightarrow[cat]{800^\circ C} 2HCN(g) + 6H_2O(g)$
(b) $NaHCO_3(s) + H^+(aq) \longrightarrow CO_2(g) + H_2O(l) + Na^+(aq)$
(c) $2BaCO_3(s) + O_2(g) + 2SO_2(g) \longrightarrow 2BaSO_4(s) + 2CO_2(g)$
22.67 (a) H_3BO_3, +3 (b) $SiBr_4$, +4 (c) $PbCl_2$, +2 or $PbCl_4$, +4 (d) $Na_2B_4O_7 \cdot 10H_2O$, +3 (e) B_2O_3, +3 **22.69** (a) Carbon (b) lead (c) silicon **22.71** (a) SiO_4^{4-} (b) SiO_3^{2-} (c) SiO_3^{2-}
22.73 (a) Diborane has bridging H atoms linking the two B atoms. The structure of ethane has the C atoms bound directly, with no bridging atoms. (b) B_2H_6 is an electron-deficient molecule. The 6 valence electron pairs are all involved in B—H sigma bonding, so the only way to satisfy the octet rule at B is to have the bridging H atoms shown in Figure 22.55. (c) The term *hydridic* indicates that the H atoms in B_2H_6 have more

than the usual amount of electron density for a covalently bound H atom. **22.76** (a) 88.2 g H_2 (b) 980 L **22.79** SiH_4, CO, Mg **22.81** (a) SO_3 (b) Cl_2O_5 (c) N_2O_3 (d) CO_2 (e) P_2O_5 **22.84** (a) PO_4^{3-}, +5; NO_3^-, +5 (b) The Lewis structure for NO_4^{3-} would be

```
        ..       3-
     [ :O: ]
   ..   |   ..
  [:O — N — O:]
   ..   |   ..
     [ :O: ]
        ..
```

The formal charge on N is +1 and on each O atom is −1. The four electronegative oxygen atoms withdraw electron density, leaving the nitrogen deficient. Since N can form a maximum of four bonds, it cannot form a π bond with one or more of the O atoms to regain electron density, as the P atom in PO_4^{3-} does. Also the short N—O distance would lead to a tight tetrahedron of O atoms subject to steric repulsion.
22.87 $GeO_2(s) + C(s) \longrightarrow Ge(l) + CO_2(g)$
$Ge(l) + 2Cl_2(g) \longrightarrow GeCl_4(l)$
$GeCl_4(l) + 2H_2O(l) \longrightarrow GeO_2(s) + 4HCl(g)$
$GeO_2(s) + 2H_2(g) \longrightarrow Ge(s) + 2H_2O(l)$
22.91 $P = 0.174$ atm **22.93** (a) $\Delta H_{comb} = -285.83$ kJ/mol H_2, −890.4 kJ/mol CH_4 (b) $\Delta H_{comb} = -141.79$ kJ/g H_2, −55.50 kJ/g CH_4 (c) $\Delta H_{comb} = -1.276 \times 10^4$ kJ/m^3 H_2, 3.975×10^4 kJ/m^3 CH_4
22.96 (a) $SO_2(g) + 2H_2S(s) \longrightarrow 3S(s) + 2H_2O$ or $8SO_2(g) + 16H_2S(g) \longrightarrow 3S_8(s) + 16H_2O(g)$
(b) 2.0×10^3 mol = 5.01×10^4 L $H_2S(g)$
(c) 9.5×10^4 g S = 210 lb S per ton of coal combusted.
22.98 D(H—O) = 463 kJ, D(H—S) = 367 kJ, D(H—Se) = 316 kJ, D(H—Te) = 266 kJ. The H—X bond energy decreases steadily in the series, probably due to the increasing size of the orbital from X with which the hydrogen 1*s* orbital must overlap. **22.101** 12 tons N_2O_4

Chapter 23

23.1 Iron: hematite, Fe_2O_3; magnetite, Fe_3O_4. Aluminum: bauxite, $Al_2O_3 \cdot xH_2O$. In ores, iron is present as the 3+ ion or as both the 2+ and 3+ ions as in magnetite. Aluminum is always present in the +3 oxidation state. **23.3** An ore consists of a little bit of the stuff we want (chalcopyrite, $CuFeS_2$) and lots of other junk (gangue).

23.5 (a) $2PbS(s) + 3O_2(s) \xrightarrow{\Delta} 2PbO(s) + 2SO_2(g)$
(b) $PbCO_3(s) \xrightarrow{\Delta} PbO(s) + CO_2(g)$
(c) $WO_3(s) + 3H_2(g) \xrightarrow{\Delta} W(s) + 3H_2O(g)$
(d) $ZnO(s) + CO(g) \xrightarrow{\Delta} Zn(l) + CO_2(g)$

23.7 (a) $SO_3(g)$ (b) $CO(g)$ provides a reducing environment for the transformation of Pb^{2+} to Pb.
(c) $PbSO_4(s) \longrightarrow PbO(s) + SO_3(g)$;
$PbO(s) + CO(g) \longrightarrow Pb(s) + CO_2(g)$
23.9 $FeO(s) + H_2(g) \longrightarrow Fe(s) + H_2O(g)$;
$FeO(s) + CO(g) \longrightarrow Fe(s) + CO_2(g)$;
$Fe_2O_3(s) + 3H_2(g) \longrightarrow 2Fe(s) + 3H_2O(g)$;
$Fe_2O_3(s) + 3CO(g) \longrightarrow 2Fe(s) + 3CO_2(g)$ **23.11** (a) Air serves mainly to oxidize coke to CO; this exothermic reaction also provides heat for the furnace: $2C(s) + O_2(g) \longrightarrow 2CO(g)$, $\Delta H = -221$ kJ. (b) Limestone, $CaCO_3$, is the source of basic oxide for slag formation: $CaCO_3(s) \xrightarrow{\Delta} CaO(s) + CO_2(g)$; $CaO(l) + SiO_2(l) \longrightarrow CaSiO_3(l)$. (c) Coke is the fuel for the blast furnace and the source of CO, the major reducing agent in the furnace. $2C(s) + O_2(g) \longrightarrow 2CO(g)$;
$4CO(g) + Fe_3O_4(s) \longrightarrow 4CO_2(g) + 3Fe(l)$ (d) Water acts as a source of hydrogen and as a means of controlling temperature. $C(s) + H_2O(g) \longrightarrow CO(g) + H_2(g)$, $\Delta H = +131$ kJ
23.13 (a) The Bayer process is necessary to separate the iron-containing solids from bauxite before electroreduction.
(b) Because it is amphoteric, Al^{3+} reacts with excess OH^- to form the soluble complex ion $Al(OH)_4^-$, while the Fe^{3+} solids cannot. This allows separation of the unwanted iron-containing solids by filtration. **23.15** To purify crude cobalt electrochemically, use an electrolysis cell in which the crude metal is the anode, a thin sheet of pure cobalt is the cathode, and the electrolyte is an aqueous solution of a soluble cobalt salt such as $CoSO_4 \cdot 7H_2O$. Reduction of water doesn't occur, because of kinetic effects. Anode reaction:
$Co(s) \longrightarrow Co^{2+}(aq) + 2e^-$; cathode reaction:
$Co^{2+}(aq) + 2e^- \longrightarrow Co(s)$. **23.17** Sodium is metallic; each atom is bonded to many others. When the metal lattice is distorted, many bonds remain intact. In NaCl the ionic forces are strong, and the ions are arranged in very regular arrays. The ionic forces tend to be broken along certain cleavage planes in the solid, and the substance does not tolerate much distortion before cleaving. **23.19** In the electron-sea model, electrons move about in the metallic lattice, while silver atoms remain more or less fixed in position. Under the influence of an applied potential, the electrons are free to move throughout the structure, giving rise to thermal and electrical conductivity.
23.21 The variation in densities indicates shorter metal-metal bond distances moving from left to right in the series. It seems that the extent of metal-metal bonding increases in the series and that all the valence electrons in these elements are involved in metallic bonding. **23.23** According to band theory, an *insulator* has energy bands that are either completely filled or completely empty, with a large energy gap between the full and empty bands. A *conductor* has partially filled energy bands. A *semiconductor* has a filled or partially filled energy band separated by a small energy gap from an empty or nearly empty band. **23.25** White tin is more metallic in character; it has a higher conductivity and a longer Sn—Sn distance (3.02 Å) than gray tin (2.81 Å). **23.27** An *alloy* contains atoms of more than one element and has the properties of a metal. In a *solution alloy* the components are randomly dispersed. In a *heterogeneous alloy* the components are not evenly dispersed and can be distinguished at a macroscopic level. In an *intermetallic compound* the components have interacted to form a compound substance, as in Cu_3As. **23.29** Isolated atoms: (b), (c), and (f); bulk metal: (a), (d), and (e) **23.31** The *lanthanide contraction* is the name given to the decrease in atomic size due to the buildup in effective nuclear charge as we move through the lanthanides (elements 57 to 70) and beyond them. The lanthanide contraction affects size-related properties such as ionization energy, electron affinity, and density.
23.33 (a) ScF_3 (b) CoF_3 (c) ZnF_2 **23.35** Chromium, $[Ar]4s^1 3d^5$, has six valence electrons, some or all of which can be involved in bonding, leading to multiple stable oxidation states. Al, $[Ne]3s^2 3p^1$, has only three valence electrons, which are all lost or shared during bonding, producing the +3 state exclusively. **23.37** (a) Cr^{3+}, $[Ar]3d^3$ (b) Au^{3+}, $[Xe]4f^{14}5d^8$

(c) Ru^{2+}, $[Kr]4d^6$ (d) Cu^+, $[Ar]3d^{10}$ (e) Mn^{4+}, $[Ar]3d^3$ (f) Ir^+: $[Xe]6s^14f^{14}5d^7$ **23.39** Ti^{2+} **23.41** Fe^{2+} is a reducing agent that is readily oxidized to Fe^{3+} in the presence of O_2 from air. **23.43** (a) $Fe(s) + 2HCl(aq) \longrightarrow FeCl_2(aq) + H_2(g)$
(b) $Fe(s) + 4HNO_3(g) \longrightarrow Fe(NO_3)_3(aq) + NO(g) + 2H_2O(l)$
23.45 The unpaired electrons in a paramagnetic material cause it to be weakly attracted into a magnetic field. A diamagnetic material, where all electrons are paired, is very weakly repelled by a magnetic field. **23.47** $PbS(s) + O_2(g) \longrightarrow Pb(l) + SO_2(g)$. $SO_2(g)$ is a product of roasting sulfide ores. In an oxygen-rich environment, $SO_2(g)$ is oxidized to $SO_3(g)$, which dissolves in $H_2O(l)$ to form sulfuric acid, $H_2SO_4(aq)$. A sulfuric acid plant near a roasting plant would provide a means for disposing of hazardous $SO_2(g)$ that would also generate a profit. **23.49** $CO(g)$: $Pb(s)$; $H_2(g)$: $Fe(s)$; $Zn(s)$: $Au(s)$
23.52 Because selenium and tellurium are both nonmetals, we expect them to be difficult to oxidize. Both Se and Te are likely to accumulate as the free elements in the so-called anode slime, along with noble metals that are not oxidized.
23.55 (a) Intermetallic compounds have a definite stoichiometry and properties, while substitutional alloys have a range of compositions. Both are homogeneous solution alloys. (b) A paramagnetic substance has unpaired electrons and is attracted into a magnetic field. A diamagnetic substance has only paired electrons and is weakly repelled by a magnetic field. (c) Insulators have a filled valence band, with a large energy gap between the valence and the conduction band. Semiconductors have a filled valence band but a smaller band gap, so that some electrons can move to the conduction band. (d) In metallic conduction, metal atoms are stationary while a few valence electrons act as charge carriers. In electrolytic conduction mobile ions carry charge throughout the liquid. **23.57** $E°$ will become more negative as the stability (K_f value) of the complex increases. **23.60** In a ferromagnetic solid the spins of all unpaired electrons are parallel. As the temperature of the solid increases, the increased kinetic energy of the atoms overcomes the force aligning the electron spins. The substance becomes paramagnetic; it still has unpaired electrons, but their spins are no longer aligned.
23.63 (a) $2NiS(s) + 3O_2(g) \longrightarrow 2NiO(s) + 2SO_2(g)$
(b) $2C(s) + O_2(g) \longrightarrow 2CO(g)$;
$C(s) + H_2O(g) \longrightarrow CO(g) + H_2(g)$;
$NiO(s) + CO(g) \longrightarrow Ni(s) + CO_2(g)$;
$NiO(s) + H_2(g) \longrightarrow Ni(s) + H_2O(g)$
(c) $Ni(s) + 2HCl(aq) \longrightarrow NiCl_2(aq) + H_2(g)$
(d) $NiCl_2(aq) + 2NaOH(aq) \longrightarrow Ni(OH)_2(s) + 2NaCl(aq)$
(e) $Ni(OH)_2(s) \longrightarrow NiO(s) + H_2O(g)$ **23.65** 2.63×10^6 kg or 2.90×10^3 ton **23.68** 7.3×10^2 g $Ni^{2+}(aq)$ **23.70** (a) At the conditions in the converter, the H^+ necessary for oxidation of Fe in air is not present. (b) The products of these reactions are $CO_2(g)$, $SO_2(g)$, $P_2O_5(l)$, SiO_2, and $M_xO_y(l)$. CO_2 and SO_2 escape as gases. P_2O_5 reacts with $CaO(l)$ to form $Ca_3(PO_4)_2(l)$, which is removed with the basic slag layer. SiO_2 and metal oxides can combine to form other silicates; all are removed with the basic slag layer. **23.73** (a) The standard reduction potential for $H_2O(l)$ is much greater than that of $Mg^{2+}(aq)$ (−0.83 V vs. −2.37 V). In aqueous solution $H_2O(l)$ would be preferentially reduced and no $Mg(s)$ would be obtained.
(b) 1.0×10^3 kg Mg **23.76** (a) $K_{eq} = 6 \times 10^{-9}$ (b) The equilibrium constant is much less than one; the process is not practical. (c) $K_{eq} = 2 \times 10^{11}$. Since $K_{eq} \gg 1$ for this process, it is potentially useful. However, the magnitude of K_{eq} says nothing about the rate of reaction. It could require heat, a catalyst, or both to be practical.

Chapter 24

24.1 (a) A *metal complex* consists of a central metal ion bonded to a number of surrounding molecules or ions. The number of bonds formed by the central metal ion is the *coordination number*. The surrounding molecules or ions are the *ligands*.
(b) Metal ions, by virtue of their positive charge and empty *d*, *s*, and *p* orbitals, act as electron-pair acceptors, or Lewis acids. Ligands, which have at least one unshared electron pair, act as electron-pair donors, or Lewis acids. **24.3** (a) +2 (b) 6
(c) 2 mol $AgBr(s)$ will precipitate per mole of complex.
24.5 (a) Coordination number = 4, oxidation number = +2
(b) 5, +4 (c) 6, +3 (d) 5, +2 (e) 6, +3 (f) 4, +2 **24.7** (a) $4Cl^-$ (b) $4Cl^-$, $1O^{2-}$ (c) 4N, $2Cl^-$ (d) 5C (e) 6O (f) 4N **24.9** (a) A monodendate ligand binds to a metal via one atom, a bidendate ligand through two atoms. (b) Three bidendate ligands fill the coordination sphere of a six-coordinate complex. (c) A tridendate ligand has at least three atoms with unshared electron pairs in the correct orientation to simultaneously bind to one or more metal ions. **24.11** (a) *Ortho*-phenanthroline, *o*-phen, is bidentate (b) oxalate, $C_2O_4^{2-}$, is bidentate (c) ethylenediaminetetraacetate, EDTA, is hexadentate (d) ethylenediamine, en, is bidentate **24.13** (b) The increase in entropy, $+\Delta S$, associated with the substitution of a chelating ligand for two or more monodentate ligands generally gives rise to the *chelate effect*. Chemical reactions with $+\Delta S$ tend to be spontaneous, have negative ΔG and large positive values of K_{eq}.
24.15 (a) $[Cr(NH_3)_6](NO_3)_3$ (b) $[Co(NH_3)_4CO_3]_2SO_4$
(c) $[Pt(en)_2Cl_2]Br_2$ (d) $K[V(H_2O)_2Br_4]$ (e) $[Zn(en)_2][HgI_4]$
24.17 (a) tetraaminedichlororhodium(III) chloride (b) potassium hexachlorotitanate(IV) (c) tetrachlorooxomolybdenum(VI) (d) tetraaqua(oxalato)platinum(IV) bromide

24.19 (a)

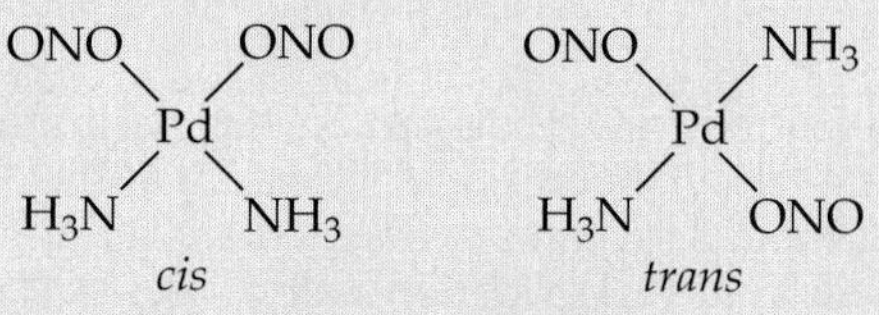

ONO ONO
Pd
H_3N NH_3
cis

ONO NH_3
Pd
H_3N ONO
trans

(b) $[Pd(NH_3)_2(ONO)_2]$, $[Pd(NH_3)_2(NO_2)_2]$

(c)

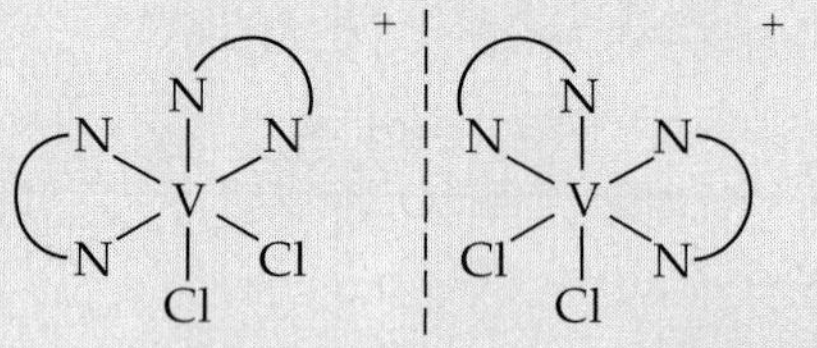

+
N N N
V
N Cl
Cl

+
N N N
V
Cl N
Cl

(d) $[Co(NH_3)_4Br_2]Cl$, $[Co(NH_3)_4BrCl]Br$ **24.21** Yes. No structural or stereoisomers are possible for a tetrahedral complex of the form MA_2B_2. The complex must be square planar with *cis* and *trans* geometric isomers. **24.23** In the *trans* isomer the Cl—Co—NH_3 bond angle is 180°; in the *cis* isomer this bond angle is 90°. The *cis* isomer is chiral. **24.25** (a) one isomer (b) *trans* and *cis* isomers with 180° and 90° Cl—Ir—Cl angles, respectively (c) *trans* and *cis* isomers with 180° and 90° Cl—Fe—Cl angles, respectively. The *cis* isomer is optically active. **24.27** (a) Visible light has wavelengths between 400 and 700 nm. (b) *Complementary* colors are opposite each other on an artist's color wheel. (c) A colored metal complex absorbs visible light of its complementary color. **24.29** Blue to blue-violet **24.31** Most of the attraction between a metal ion and a ligand is electrostatic. Whether the interaction is ion-ion or ion-dipole, the ligand is strongly attracted to the metal center and can be modeled as a point negative charge.

24.33 (a)

$d_{x^2-y^2}, d_{z^2}$

Δ

d_{xy}, d_{xz}, d_{yz}

(b) The magnitude of Δ and the energy of the *d-d* transition for a d^1 complex are equal. (c) The spectrochemical series is an ordering of ligands according to their ability to increase the energy gap Δ. **24.35** A yellow color is due to absorption of light around 400 to 430 nm, a blue color to absorption near 620 nm. The shorter wavelength corresponds to a higher-energy electron transition and larger Δ value. Cyanide is a stronger-field ligand, and its complexes are expected to have larger Δ values than aqua complexes. **24.37** (a) Ru^{3+}, d^5 (b) Cu^{2+}, d^9 (c) Co^{3+}, d^6 (d) Mo^{5+}, d^1 (e) Re^{3+}, d^4 **24.39** (a) $[Ar]4s^23d^5$, $[Ar]3d^4$, 2 unpaired electrons (b) $[Kr]5s^14d^7$, $[Kr]4d^5$, 1 unpaired electron (c) $[Kr]5s^14d^8$, $[Kr]4d^6$, 0 unpaired electrons **24.41** All complexes in this exercise are six-coordinate octahedral.

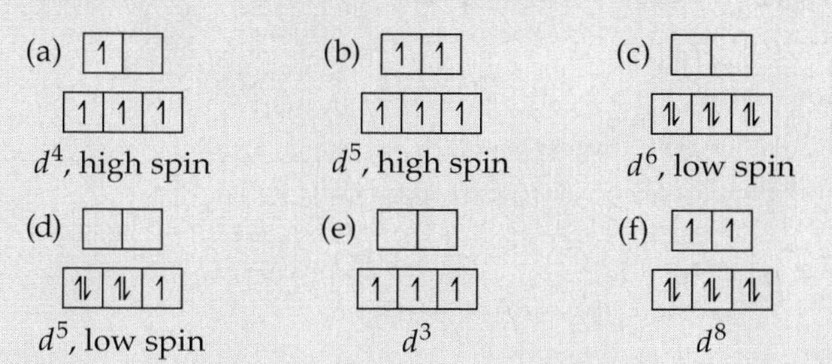

24.43

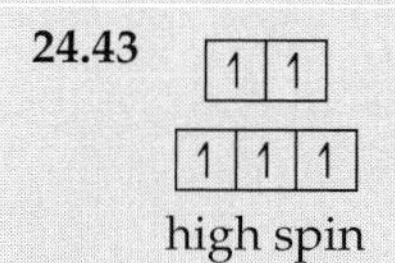

24.45 (a) $[Ni(en)_2Cl_2]$ (b) $K_2[Ni(CN)_4]$ (c) $[CoF_6]^{3-}$, high spin; $[Co(NH_3)_6]^{3+}$, low spin (d) thiocyanate, SCN^- or NCS^- (e) $[Co(en)_2Cl_2]Cl$ (f) $[Co(en)_3]Cl_3$

24.48 (a) $[Co(NH_3)_4(H_2O)_2]^{2+}$ (structure: Co bonded to 2 H_2O and 4 NH_3)

Octahedral

(b) $[RuCl_5(H_2O)]^{2-}$ (structure: Ru bonded to H_2O and 5 Cl)

Octahedral

(c) (structure: Co bonded to 2 H_2O and two bidentate O–O ligands)$^-$

Octahedral

(d) (structure: Ru bonded to 2 Cl and two bidentate N–N ligands)

Octahedral

24.51 (a) In a square-planar complex, if one pair of ligands is *trans*, the remaining two coordination sites are also *trans* to each other. The bidentate ethylenediamine ligand is too short to occupy *trans* coordination sites, so the *trans* isomer of $[Pt(en)Cl_2]$ is unknown. (b) The minimum steric requirement for a *trans*-bidentate ligand is a medium-length chain between the two coordinating atoms that will occupy the *trans* positions. A polydentate ligand such as EDTA is much more likely to occupy *trans* positions because it locks the metal ion in place with multiple coordination sites and shields the metal ion from competing ligands present in the solution.

24.54 (a) $AgCl(s) + 2NH_3(aq) \longrightarrow [Ag(NH_3)_2]^+(aq) + Cl^-(aq)$
(b) $[Cr(en)_2Cl_2]Cl(aq) + 2H_2O(l) \longrightarrow [Cr(en)_2(H_2O)_2]^{3+}(aq) + 3Cl^-(aq)$; $3Ag^+(aq) + 3Cl^-(aq) \longrightarrow 3AgCl(s)$
(c) $Zn(NO_3)_2(aq) + 2NaOH(aq) \longrightarrow Zn(OH)_2(s) + 2NaNO_3(aq)$; $Zn(OH)_2(s) + 2NaOH(aq) \longrightarrow [Zn(OH)_4]^{2-}(aq) + 2Na^+(aq)$
(d) $Co^{2+}(aq) + 4Cl^-(aq) \longrightarrow [CoCl_4]^{2-}(aq)$ **24.57** (a) left shoe (b) wood screw (c) a typical golf club **24.60** $[Co(NH_3)_6]^{3+}$, yellow; $[Co(H_2O)_6]^{2+}$, pink; $[CoCl_4]^{2-}$, blue **24.63** (a) $[FeF_6]^{4-}$. F^- is a weak-field ligand that imposes a smaller Δ and longer λ for the complex ion. (b) $[V(H_2O)_6]^{2+}$. V^{2+} has a lower charge, so the interaction with the ligand will produce a weaker field, a smaller Δ, and a longer absorbed wavelength. (c) $[CoCl_4]^{2-}$. Cl^- is a weak-field ligand that imposes a smaller Δ and a longer λ for the complex ion.

24.66 (a) $[Fe(CN)_4(CO)_2]^{2-}$ (structure: Fe bonded to 4 CN and 2 CO, CO groups *trans*)

(b) sodium dicarbonyltetracyanoferrate(II) (c) +2, 6 *d* electrons (d) We expect the complex to be low spin. Cyanide (and carbonyl) are high on the spectrochemical series, which means the complex will have a large Δ splitting, characteristic of low-spin complexes. **24.71** (a) Co is in the same +3 oxidation state in both complexes. (b) The ions that form precipitates with $AgNO_3(aq)$ and $BaCl_2(aq)$ are outside the coordination sphere. The dark violet compound A forms a precipitate with $BaCl_2(aq)$ but not $AgNO_3(aq)$, so it has SO_4^{2-} outside the coordination sphere and coordinated Br^-, $[Co(NH_3)_5Br]SO_4$. The red-violet compound B forms a precipitate with $AgNO_3(aq)$ but not $BaCl_2(aq)$, so it has Br^- outside the coordination sphere and coordinated SO_4^{2-}, $[Co(NH_3)_5SO_4]Br$. (c) Compounds A and B are coordination-sphere isomers. (d) Compounds A and B are both strong electrolytes. **24.74** $K_4[Mn(ox)_2Br_2]$ **24.76** 47.3 mg Mg^{2+}/L, 53.4 mg Ca^{2+}/L

Chapter 25

25.1 The carbon atoms with two and three hydrogens both have tetrahedral electron-domain geometry, 109° bond angles, and sp^3 hybridization. The carbon atom with one hydrogen has trigonal-planar electron-domain geometry, 120° bond angles, and sp^2 hybridization. **25.3** Carbon, hydrogen, oxygen, nitrogen, sulfur, and phosphorus. Oxygen and nitrogen are more electronegative than carbon; sulfur has the same electronegativity as carbon. **25.5** (a) A *straight-chain hydrocarbon* has all carbon atoms connected in a continuous chain. A *branched-chain hydrocarbon* has a branch; at least one carbon atom is bound to three or more carbon atoms. (b) An *alkane* is a complete molecule composed of carbon and hydrogen in which all bonds are σ bonds. An *alkyl group* is a substituent formed by removing a hydrogen atom from an alkane. **25.7** (a) C_5H_{12} (b) C_5H_{10} (c) C_5H_{10} (d) C_5H_8; saturated: (a), (b); unsaturated: (c), (d) **25.9** C_nH_{2n-2}

25.11

$CH_3-CH_2-CH_2-CH=CH_2$
1-Pentene

$CH_3-CH_2-CH=CH-CH_3$
2-Pentene

$CH_2=CH-CH(CH_3)-CH_3$
3-Methyl-1-butene

$CH_2=C(CH_3)-CH_2-CH_3$
2-Methyl-1-butene

$CH_3-C(CH_3)=CH-CH_3$
2-Methyl-2-butene

25.13 (c) and (d) are the same molecule. **25.15** (a) 109° (b) 120° (c) 180° **25.17** (a) 2-methylhexane (b) 4-ethyl-2,4-dimethyldecane

(c) $CH_3-CH_2-CH_2-CH(CH_3)-CH_2-CH_3$

(d) $CH_3-CH_2-CH_2-CH_2-CH(CH_2CH_3)-CH_2-C(CH_3)_2-CH_3$

(e) [cyclohexane ring with H atoms and one CH_3 substituent]; [ring: CH, CH_3, CH, CH_2, CH_2, CH_2, CH_2]

25.19 2,3-dimethylheptane (b) *cis*-6-methyl-3-octene (c) *para*-dibromobenzene (d) 4,4-dimethyl-1-hexyne (e) methylcyclobutane **25.21** Geometric isomerism in alkenes is the result of restricted rotation about the double bond. In alkanes bonding sites are interchangeable by free rotation about the C—C single bonds. In alkynes there is only one additional bonding site on a triply bound carbon, so no isomerism results. **25.23** (a) no

(b) $(H)(ClH_2C)C=C(Cl)(CH_3)$; $(ClH_2C)(H)C=C(Cl)(CH_3)$

(c) no (d) no **25.25** 65 **25.27** (a) An addition reaction is the addition of some reagent to the two atoms that form a multiple bond. In a substitution reaction one atom or group of atoms replaces another atom. Alkenes typically undergo addition, while aromatic hydrocarbons usually undergo substitution.

(b) $CH_3-C(CH_3)=C(CH_3)-CH_3 + Br_2 \longrightarrow CH_3-CBr(CH_3)-CBr(CH_3)-CH_3$

(c) $C_6H_4Cl_2$ (para-dichlorobenzene) $+ Cl_2 \xrightarrow{FeCl_3} C_6H_3Cl_3 + HCl$

25.29 (a) The 60° C—C—C angles in the cyclopropane ring cause strain that provides a driving force for reactions that result in ring opening. There is no comparable strain in the five- or six-membered rings.
(b) $C_2H_4(g) + HBr(g) \longrightarrow CH_3CH_2Br(l)$;
$C_6H_6(l) + CH_3CH_2Br(l) \xrightarrow{AlCl_3} C_6H_5CH_2CH_3(l) + HBr(g)$
25.31 The intermediate is a carbocation. One of the C atoms formerly involved in the π bond is now bound to a second H atom from the hydrogen halide. The other C atom formerly involved in the π bond carries a full positive charge and forms only three sigma bonds, two to adjacent C atoms and one to H. **25.33** ΔH_{comb}/mol CH_2 for cyclopropane = 693.3 kJ, for cyclopentane = 663.4 kJ. $\Delta H_{comb}/CH_2$ group for cyclopropane is greater because C_3H_6 contains a strained ring. When combustion occurs, the strain is relieved and the stored energy is released. **25.35** (a) Ketone (b) carboxylic acid (c) alcohol (d) ester (e) amide (f) amine

25.37 (a) Propionaldehyde (or propanal):

$H-CH_2-CH_2-C(=O)H$

(b) ethylmethyl ether:

$H-CH_2-CH_2-O-CH_2-H$

25.39 (a) Methanoic acid (b) butanoic acid (c) 3-methylpentanoic acid

25.41 (a) $CH_3CH_2O-C(=O)-C_6H_5$
Ethylbenzoate

(b) $CH_3N(H)-C(=O)CH_3$
N-methylethanamide or N-methylacetamide

(c) $C_6H_5-O-C(=O)CH_3$
Phenylacetate

25.43

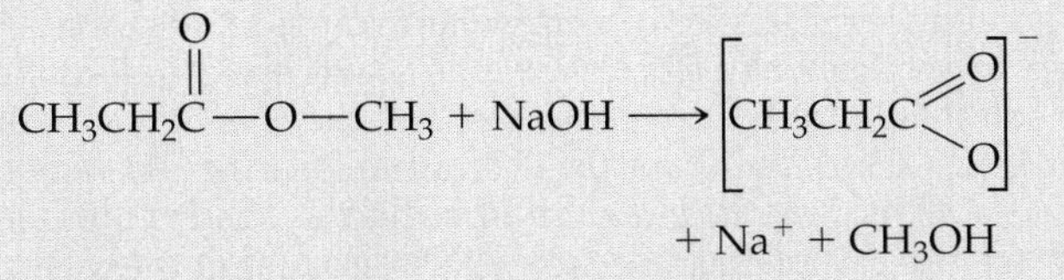

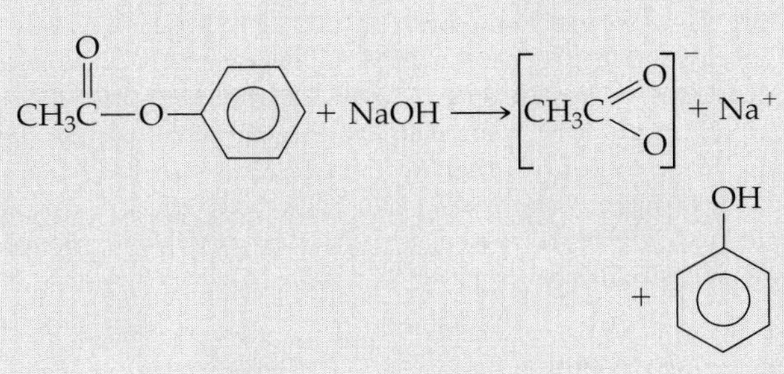

(a) $CH_3CH_2C(=O)—O—CH_3 + NaOH \longrightarrow [CH_3CH_2CO_2]^- + Na^+ + CH_3OH$

(b) $CH_3C(=O)—O—C_6H_5 + NaOH \longrightarrow [CH_3CO_2]^- + Na^+ + C_6H_5OH$

25.45 (a) 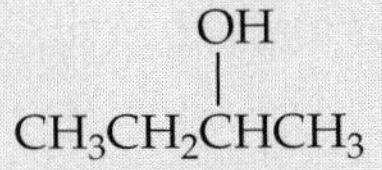 $CH_3CH_2CH(OH)CH_3$ (b) $HOCH_2CH_2OH$

(c) 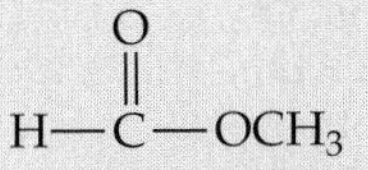 $H—C(=O)—OCH_3$ (d) 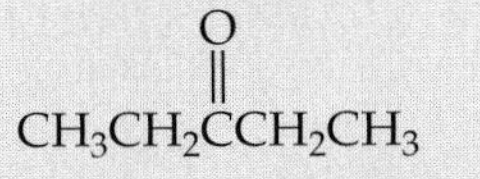 $CH_3CH_2C(=O)CH_2CH_3$

(e) $CH_3CH_2OCH_2CH_3$

25.47 (a) 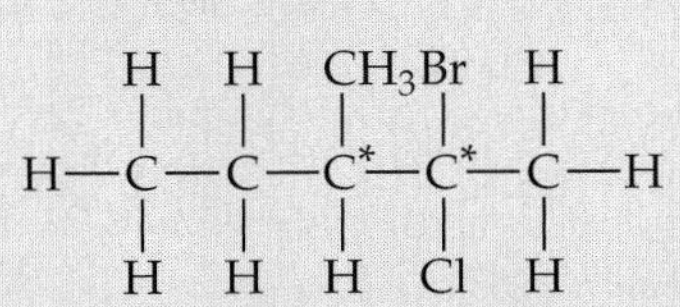 $H—CH_2—CH_2—C^*H(CH_3)—C^*HBrCl—CH_2—H$ ... as drawn: H—C(H)(H)—C(H)(H)—C*(CH₃)(H)—C*(Br)(Cl)—C(H)(H)—H *chiral C atoms

(b) 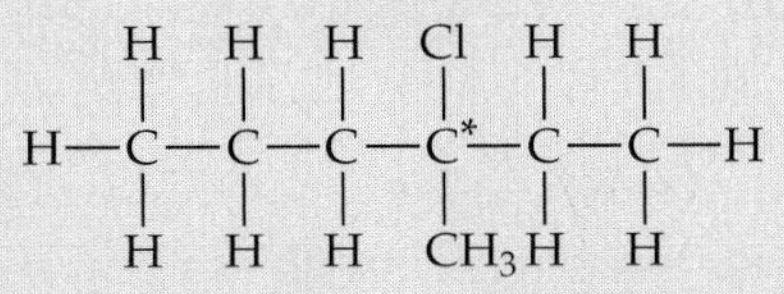 H—C(H)(H)—C(H)(H)—C(H)(H)—C*(Cl)(CH₃)—C(H)(H)—C(H)(H)—H *chiral C atoms

Yes, the molecule has optical isomers. The chiral carbon atom is attached to chloro, methyl, ethyl, and propyl groups. **25.49** (a) An α-amino acid contains an NH_2 function on the carbon adjacent to the carboxcylic acid function. (b) In protein formation, amino acids undergo a condensation reaction between the amino group of one molecule and the carboxylic acid group of another to form the amide linkage. **25.51** Two peptides are possible: $H_2NCH_2CONHCH(CH(CH_3)_2)COOH$ (glycylvaline) and $H_2NCH(CH(CH_3)_2)CONHCH_2COOH$ (valylglycine).

25.53 (a) 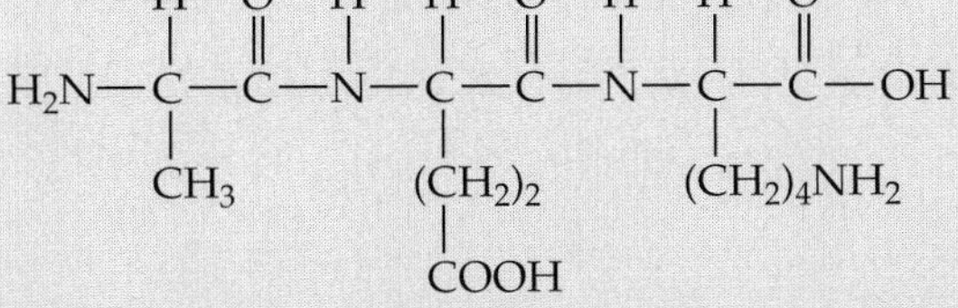 $H_2N—CH(CH_3)—C(=O)—NH—CH((CH_2)_2COOH)—C(=O)—NH—CH((CH_2)_4NH_2)—C(=O)—OH$

(b) Eight: Ser-Ser-Ser; Ser-Ser-Phe; Ser-Phe-Ser; Phe-Ser-Ser; Ser-Phe-Phe; Phe-Ser-Phe; Phe-Phe-Ser; Phe-Phe-Phe
25.55 The *primary structure* of a protein refers to the sequence of amino acids in the chain. The *secondary structure* is the configuration (helical, folded, open) of the protein chain. The *tertiary structure* is the overall shape of the protein determined by the way the segments fold together **25.57** (a) Carbohydrates, or sugars, are polyhydroxyaldehydes or ketones composed of carbon, hydrogen, and oxygen. They are derived primarily from plants and are a major food source for animals. (b) A monosaccharide is a simple sugar molecule that cannot be decomposed into smaller sugar molecules by hydrolysis. (c) A disaccharide is a carbohydrate composed of two simple sugar units. Hydrolysis breaks the disaccharides into two monosaccharides. **25.59** (a) In the linear form of galactose, the aldehydic carbon is C1. Carbon atoms 2, 3, 4, and 5 are chiral because they each carry four different groups. (b) Both the β form (shown here) and the α form (OH on carbon 1 on same side of ring as OH on carbon 2) are possible.

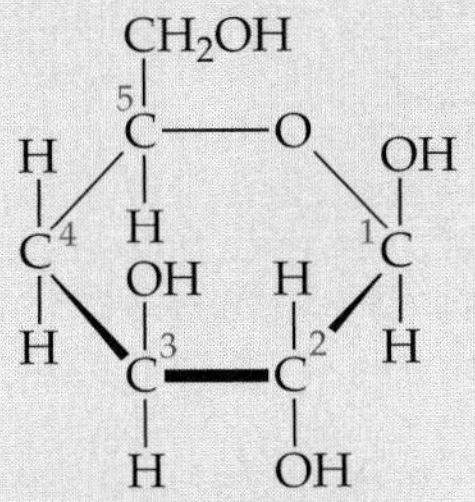

[Structure: galactose ring — CH_2OH on C5; C1 bearing OH and H; C2 bearing H and OH; C3 bearing OH and H; C4 bearing H and H]

Galactose

25.61 The empirical formula of glycogen is $C_6H_{10}O_5$. The six-membered ring form of glucose is the unit that forms the basis of glycogen. The monomeric glucose units are joined by α linkages. **25.63** A nucleotide consists of a nitrogen-containing aromatic compound, a sugar in the furanose (five-membered) ring form, and a phosphoric acid group. The structure of deoxycytidine monophosphate is

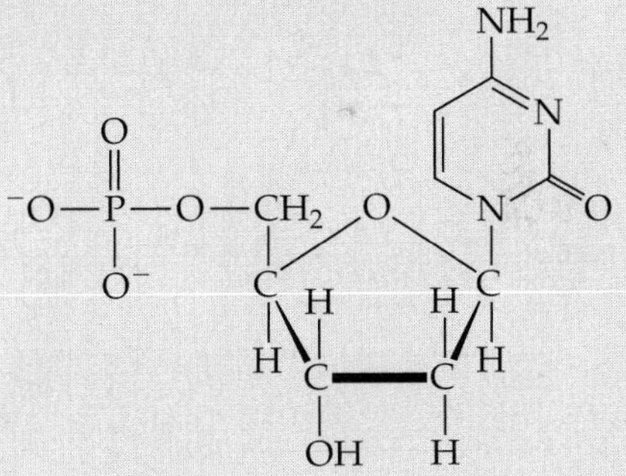

[Structure: $^-O—P(=O)(O^-)—O—CH_2$ attached to deoxyribose ring (OH on C3, H on C2) bonded to cytosine (NH_2, N, C=O)]

25.65 $C_4H_7O_3CH_2OH + HPO_4^{2-} \longrightarrow C_4H_7O_3CH_2—O—PO_3^{2-} + H_2O$
25.67 In the helical structure for DNA, the strands of the polynucleotides are held together by hydrogen-bonding interactions between particular pairs of bases. Adenine and thymine are one base pair, and guanine and cytosine are another. For each adenine in one strand, there is a cytosine in the other, so that total adenine equals total thymine, and total guanine equals total cytosine.

25.69

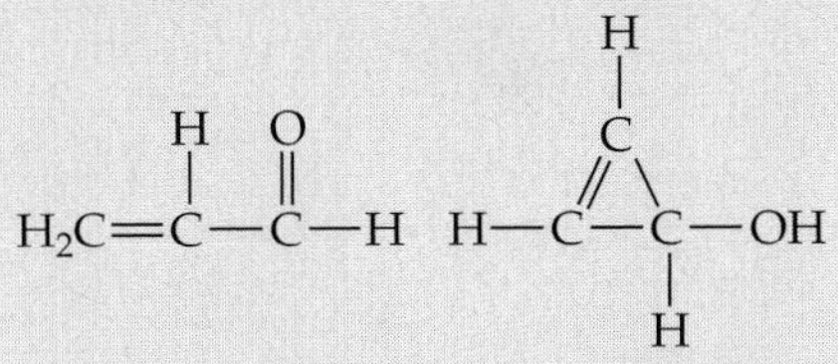

$H_2C{=}CH—C(=O)—H$ 　 H—C(H)—C(H)(—OH) (three-membered ring with C=C: cyclopropenol)

$H—C{\equiv}C—CH_2OH$

25.72

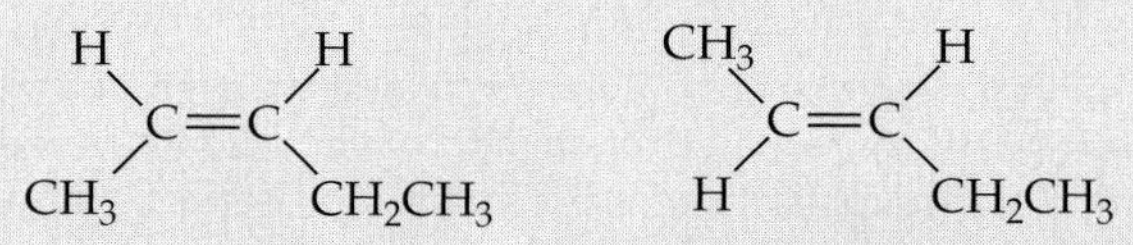

$$\underset{cis}{\overset{H\quad\quad H}{\underset{CH_3\quad\quad CH_2CH_3}{C{=}C}}}\qquad\qquad \underset{trans}{\overset{CH_3\quad\quad H}{\underset{H\quad\quad CH_2CH_3}{C{=}C}}}$$

Cyclopentene does not show *cis-trans* isomerism because the existence of the ring demands that the C—C bonds be *cis* to one another.

25.75 $H_2C{=}CH{-}CH_2OH$ $\quad$ $H_2C{-}C{-}H$ (cyclopropane ring with CH_2 bridging, OH on the C)

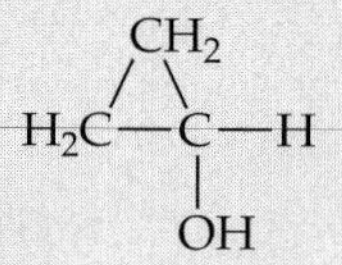

(The —OH group cannot be attached to an alkene carbon atom; these molecules are called "vinyl alcohols" and are unstable.) **25.80** The difference between an alcoholic hydrogen and a carboxylic acid hydrogen lies in the carbon to which the —OH is attached. In a carboxylic acid the electronegative carbonyl oxygen withdraws electron density from the O—H bond, rendering the bond more polar and the H more ionizable. In an alcohol no electronegative atoms are bound to the carbon that holds the —OH group, and the H is tightly bound to the O. **25.83** (a) None (b) The carbon bearing the secondary —OH has four different groups attached, and is thus chiral. (c) The carbon bearing the $—NH_2$ group and the carbon bearing the CH_3 group are both chiral. **25.85** Glu-Cys-Gly is the only possible structure. **25.87** Glucose is an aldohexose, and fructose is a ketohexose. **25.89** Ethanol contains —O—H bonds, which form strong intermolecular hydrogen bonds, while dimethyl ether experiences only weak dipole-dipole and dispersion forces. Polar CH_2F_2 experiences dipole-dipole and dispersion forces, while nonpolar CF_4 experiences only dispersion forces.

25.91 $CH_3\overset{O}{\overset{\|}{C}}CH_2CH_3$

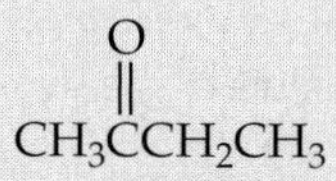

25.93 $\Delta G = -36$ kJ **25.97** $[AMPOH^-]/[AMPO^{2-}] = 0.65$

Glossary

absorption spectrum The amount of light absorbed by a sample as a function of wavelength. (Section 24.5)

accuracy A measure of how closely individual measurements agree with the correct value. (Section 1.5)

acid A substance that is able to donate a H^+ ion (a proton) and hence increases the concentration of $H^+(aq)$ when it dissolves in water. (Section 4.3)

acid-dissociation constant (K_a) An equilibrium constant that expresses the extent to which an acid transfers a proton to solvent water. (Section 16.6)

acidic anhydride (acidic oxide) An oxide that forms an acid when added to water; soluble nonmetal oxides are acidic anhydrides. (Section 22.5)

acidic oxide (acidic anhydride) An oxide that either reacts with a base to form a salt or with water to form an acid. (Section 22.5)

acid rain Rainwater that has become excessively acidic because of absorption of pollutant oxides, notably SO_3, produced by human activities. (Section 18.4)

actinide element Element in which the $5f$ orbitals are only partially occupied. (Section 6.8)

activated complex (transition state) The particular arrangement of atoms found at the top of the potential-energy barrier as a reaction proceeds from reactants to products. (Section 14.5)

activation energy (E_a) The minimum energy needed for reaction; the height of the energy barrier to formation of products. (Section 14.5)

active site Specific site on a heterogeneous catalyst or an enzyme where catalysis occurs. (Section 14.7)

activity The decay rate of a radioactive material, generally expressed as the number of disintegrations per unit time. (Section 21.4)

activity series A list of metals in order of decreasing ease of oxidation. (Section 4.4)

addition polymerization Polymerization that occurs through coupling of monomers with one another, with no other products formed in the reaction. (Section 12.2)

addition reaction A reaction in which a reagent adds to the two carbon atoms of a carbon–carbon multiple bond. (Section 25.4)

adsorption The binding of molecules to a surface. (Section 14.7)

alcohol An organic compound obtained by substituting a hydroxyl group ($-OH$) for a hydrogen on a hydrocarbon. (Sections 2.9 and 25.5)

aldehyde An organic compound that contains a carbonyl group to which at least one hydrogen atom is attached. (Section 25.6)

alkali metals Members of group 1A in the periodic table. (Section 7.7)

alkaline earth metals Members of group 2A in the periodic table. (Section 7.7)

alkanes Compounds of carbon and hydrogen containing only carbon–carbon single bonds. (Sections 2.9 and 25.2)

alkenes Hydrocarbons containing one or more carbon–carbon double bonds. (Section 25.2)

alkyl group A group that is formed by removing a hydrogen atom from an alkane. (Section 25.3)

alkynes Hydrocarbons containing one or more carbon–carbon triple bonds. (Section 25.2)

alloy A substance that has the characteristic properties of a metal and contains more than one element. Often there is one principal metallic component, with other elements present in smaller amounts. Alloys may be homogeneous or heterogeneous in nature. (Section 23.6)

alpha (α) helix A protein structure in which the protein is coiled in the form of a helix, with hydrogen bonds between $C{=}O$ and $N{-}H$ groups on adjacent turns. (Section 25.9)

alpha particles Particles that are identical to helium-4 nuclei, consisting of two protons and two neutrons, symbol ${}^4_2He$ or ${}^4_2\alpha$. (Section 21.1)

amide An organic compound that has an NR_2 group attached to a carbonyl. (Section 25.6)

amine A compound that has the general formula R_3N, where R may be H or a hydrocarbon group. (Section 16.7)

amino acid A carboxylic acid that contains an amino ($-NH_2$) group attached to the carbon atom adjacent to the carboxylic acid ($-COOH$) functional group. (Section 25.9)

amorphous solid A solid whose molecular arrangement lacks a regular, long-range pattern. (Section 11.7)

amphoteric Capable of behaving as either an acid or a base. (Section 16.2)

angstrom A common non-SI unit of length, denoted Å, that is used to measure atomic dimensions: $1 \text{Å} = 10^{-10}$ m. (Section 2.3)

anion A negatively charged ion. (Section 2.7)

anode An electrode at which oxidation occurs. (Section 20.3)

antibonding molecular orbital A molecular orbital in which electron density is concentrated outside the region between the two nuclei of bonded atoms. Such orbitals, designated as σ^* or π^*, are less stable (of higher energy) than bonding molecular orbitals. (Section 9.7)

aqueous solution A solution in which water is the solvent. (Chapter 4: Introduction)

aromatic hydrocarbons Hydrocarbon compounds that contain a planar, cyclic arrangement of carbon atoms linked by both σ and delocalized π bonds. (Section 25.2)

Arrhenius equation An equation that relates the rate constant for a reaction to the frequency factor, A, the activation energy, E_a, and the temperature, T: $k = Ae^{-E_a/RT}$. In its logarithmic form, it is written: $\ln k = -E_a/RT + \ln A$. (Section 14.5)

atmosphere (atm) A unit of pressure equal to 760 torr; 1 atm = 101.325 kPa. (Section 10.2)

atom The smallest representative particle of an element. (Sections 1.1 and 2.1)

atomic mass unit (amu) A unit based on the value of exactly 12 amu for the mass of the isotope of carbon that has six protons and six neutrons in the nucleus. (Sections 2.3 and 3.3)

atomic number The number of protons in the nucleus of an atom of an element. (Section 2.3)

atomic radius An estimate of the size of an atom. See **bonding atomic radius**. (Section 7.3)

atomic weight The average mass of the atoms of an element in atomic mass units (amu); it is numerically equal to the mass in grams of 1 mol of the element. (Section 2.4)

autoionization The process whereby water spontaneously forms low concentrations of $H^+(aq)$ and $OH^-(aq)$ ions by proton transfer from one water molecule to another. (Section 16.3)

Avogadro's hypothesis A statement that equal volumes of gases at the same temperature and pressure contain equal numbers of molecules. (Section 10.3)

Avogadro's law A statement that the volume of a gas maintained at constant temperature and pressure is directly proportional to the quantity of the gas. (Section 10.3)

Avogadro's number The number of ${}^{12}C$ atoms in exactly 12 g of ${}^{12}C$; it equals 6.022×10^{23}. (Section 3.4)

bar A unit of pressure equal to 10^5 Pa. (Section 10.2)

base A substance that is an H^+ acceptor; a base produces an excess of $OH^-(aq)$ ions when it dissolves in water. (Section 4.3)

base-dissociation constant (K_b) An equilibrium constant that expresses the extent to which a base reacts with solvent water, accepting a proton and forming $OH^-(aq)$. (Section 16.7)

basic anhydride (basic oxide) An oxide that forms a base when added to water; soluble metal oxides are basic anhydrides. (Section 22.5)

basic oxide (basic anhydride) An oxide that either reacts with water to form a base or reacts with an acid to form a salt and water. (Section 22.5)

battery A self-contained electrochemical power source that contains one or more voltaic cells. (Section 20.7)

Bayer process A hydrometallurgical procedure for purifying bauxite in the recovery of aluminum from bauxite-containing ores. (Section 23.3)

becquerel The SI unit of radioactivity. It corresponds to one nuclear disintegration per second. (Section 21.4)

Beer's law The light absorbed by a substance (A) equals the product of its molar absorptivity constant (a), the path length through which the light passes (b), and the molar concentration of the substance (c): $A = abc$. (Section 14.2)

beta particles Energetic electrons emitted from the nucleus, symbol ${}_{-1}^{0}e$. (Section 21.1)

bidentate ligand A ligand in which two coordinating atoms are bound to a metal. (Section 24.2)

bimolecular reaction An elementary reaction that involves two molecules. (Section 14.6)

biochemistry The study of the chemistry of living systems. (Chapter 25: Introduction)

biocompatible Any substance or material that is compatible with living systems. (Section 12.3)

biodegradable Organic material that bacteria are able to oxidize. (Section 18.6)

biomaterial Any material that has a biomedical application. (Section 12.3)

biopolymer A polymeric molecule of high molecular weight found in living systems. The three major classes of biopolymer are proteins, carbohydrates, and nucleic acids. (Section 25.8)

body-centered cubic cell A cubic unit cell in which the lattice points occur at the corners and at the center. (Section 11.7)

bomb calorimeter A device for measuring the heat evolved in the combustion of a substance under constant-volume conditions. (Section 5.5)

bond angles The angles made by the lines joining the nuclei of the atoms in a molecule. (Section 9.1)

bond dipole The dipole moment due to the two atoms of a covalent bond. (Section 9.3)

bond enthalpy The enthalpy change, ΔH, required to break a particular bond when the substance is in the gas phase. (Section 8.8)

bonding atomic radius The radius of an atom as defined by the distances separating it from other atoms to which it is chemically bonded. (Section 7.3)

bonding molecular orbital A molecular orbital in which the electron density is concentrated in the internuclear region. The energy of a bonding molecular orbital is lower than the energy of the separate atomic orbitals from which it forms. (Section 9.7)

bonding pair In a Lewis structure a pair of electrons that is shared by two atoms. (Section 9.2)

bond length The distance between the centers of two bonded atoms. (Section 8.8)

bond order The number of bonding electron pairs shared between two atoms, less the number of antibonding electron pairs: bond order = (number of bonding electrons − number of antibonding electrons). (Section 9.7)

bond polarity A measure of how equally the electrons are shared between the two atoms in a chemical bond. (Section 8.4)

boranes Covalent hydrides of boron. (Section 22.11)

Born–Haber cycle A thermodynamic cycle based on Hess's law that relates the lattice energy of an ionic substance to its enthalpy of formation and to other measurable quantities. (Section 8.2)

Boyle's law A law stating that at constant temperature, the product of the volume and pressure of a given amount of gas is a constant. (Section 10.3)

Brønsted–Lowry acid A substance (molecule or ion) that acts as a proton donor. (Section 16.2)

Brønsted–Lowry base A substance (molecule or ion) that acts as a proton acceptor. (Section 16.2)

buffer capacity The amount of acid or base a buffer can neutralize before the pH begins to change appreciably. (Section 17.2)

buffered solution (buffer) A solution that undergoes a limited change in pH upon addition of a small amount of acid or base. (Section 17.2)

calcination The heating of an ore to bring about its decomposition and the elimination of a volatile product. For example, a carbonate ore might be calcined to drive off CO_2. (Section 23.2)

calorie A unit of energy, it is the amount of energy needed to raise the temperature of 1 g of water by 1°C, from 14.5°C to 15.5°C. A related unit is the joule: 1 cal = 4.184 J. (Section 5.1)

calorimeter An apparatus that measures the evolution of heat. (Section 5.5)

calorimetry The experimental measurement of heat produced in chemical and physical processes. (Section 5.5)

capillary action The process by which a liquid rises in a tube because of a combination of adhesion to the walls of the tube and cohesion between liquid particles. (Section 11.3)

carbide A binary compound of carbon with a metal or metalloid. (Section 22.9)

carbohydrates A class of substances formed from polyhydroxy aldehydes or ketones. (Section 25.10)

carbon black A microcrystalline form of carbon. (Section 22.9)

carbonyl group The C=O double bond, a characteristic feature of several organic functional groups, such as ketones and aldehydes. (Section 25.6)

carboxylic acid A compound that contains the —COOH functional group. (Sections 16.10 and 25.6)

catalyst A substance that changes the speed of a chemical reaction without itself undergoing a permanent chemical change in the process. (Section 14.7)

cathode An electrode at which reduction occurs. (Section 20.3)

cathode rays Streams of electrons that are produced when a high voltage is applied to electrodes in an evacuated tube. (Section 2.2)

cathodic protection A means of protecting a metal against corrosion by making it the cathode in a voltaic cell. This can be achieved by attaching a more easily oxidized metal, which serves as an anode, to the metal to be protected. (Section 20.8)

cation A positively charged ion. (Section 2.7)

cell potential A measure of the driving force, or "electrical pressure," for an electrochemical reaction; it is measured in volts: 1 V = 1 J/C. Also called electromotive force. (Section 20.4)

cellulose A polysaccharide of glucose; it is the major structural element in plant matter. (Section 25.10)

Celsius scale A temperature scale on which water freezes at 0° and boils at 100° at sea level. (Section 1.4)

ceramic A solid inorganic material, either crystalline (oxides, carbides, silicates) or amorphous (glasses). Most ceramics melt at high temperatures. (Section 12.4)

chain reaction A series of reactions in which one reaction initiates the next. (Section 21.7)

changes of state Transformations of matter from one state to a different one, for example, from a gas to a liquid. (Section 1.3)

charcoal A form of carbon produced when wood is heated strongly in a deficiency of air. (Section 22.9)

Charles's law A law stating that at constant pressure, the volume of a given quantity of gas is proportional to absolute temperature. (Section 10.3)

chelate effect The generally larger formation constants for polydentate ligands as compared with the corresponding monodentate ligands. (Section 24.2)

chelating agent A polydentate ligand that is capable of occupying two or more sites in the coordination sphere. (Section 24.2)

chemical bond A strong attractive force that exists between atoms in a molecule. (Section 8.1)

chemical changes Processes in which one or more substances are converted into other substances; also called **chemical reactions**. (Section 1.3)

chemical equation A representation of a chemical reaction using the chemical formulas of the reactants and products; a balanced chemical equation contains equal numbers of atoms of each element on both sides of the equation. (Section 3.1)

chemical equilibrium A state of dynamic balance in which the rate of formation of the products of a reaction from the reactants equals the rate of formation of the reactants from the products; at equilibrium the concentrations of the reactants and products remain constant. (Section 4.1; Chapter 15: Introduction.)

chemical formula A notation that uses chemical symbols with numerical subscripts to convey the relative proportions of atoms of the different elements in a substance. (Section 2.6)

chemical kinetics The area of chemistry concerned with the speeds, or rates, at which chemical reactions occur. (Chapter 14: Introduction)

chemical nomenclature The rules used in naming substances. (Section 2.8)

chemical properties Properties that describe a substance's composition and its reactivity; how the substance reacts or changes into other substances. (Section 1.3)

chemical reactions Processes in which one or more substances are converted into other substances; also called **chemical changes**. (Section 1.3)

chemical-vapor deposition A method for forming thin films in which a substance is deposited on a surface and then undergoes some form of chemical reaction to form the film. (Section 12.6)

chemistry The scientific discipline that treats the composition, properties, and transformations of matter. (Chapter 1: Introduction)

chiral A term describing a molecule or an ion that cannot be superimposed on its mirror image. (Sections 24.4 and 25.7)

chlorofluorocarbons Compounds composed entirely of chlorine, fluorine, and carbon. (Section 18.3)

chlorophyll A plant pigment that plays a major role in conversion of solar energy to chemical energy in photosynthesis. (Section 24.2)

cholesteric liquid crystalline phase A liquid crystal formed from flat, disc-shaped molecules that align through a stacking of the molecular discs. (Section 12.1)

coal A naturally occurring solid containing hydrocarbons of high molecular weight, as well as compounds containing sulfur, oxygen, and nitrogen. (Section 5.8)

coke An impure form of carbon, formed when coal is heated strongly in the absence of air. (Section 22.9)

colligative properties Those properties of a solvent (vapor-pressure lowering, freezing-point lowering, boiling-point elevation, osmotic pressure) that depend on the total concentration of solute particles present. (Section 13.5)

collision model A theory based on the idea that molecules must collide to react; it explains the factors influencing reaction rates in terms of the frequency of collisions, the number of collisions with energies exceeding the activation energy, and the probability that the collisions occur with suitable orientations. (Section 14.5)

colloidal dispersions (colloids) Mixtures containing particles larger than normal solutes but small enough to remain suspended in the dispersing medium. (Section 13.6)

combination reaction A chemical reaction in which two or more substances combine to form a single product. (Section 3.2)

combustion reaction A chemical reaction that proceeds with evolution of heat and usually also a flame; most combustion involves reaction with oxygen, as in the burning of a match. (Section 3.2)

common-ion effect A shift of an equilibrium induced by an ion common to the equilibrium. For example, added Na_2SO_4 decreases the solubility of the slightly soluble salt $BaSO_4$, or added $NaC_2H_3O_2$ decreases the percent ionization of $HC_2H_3O_2$. (Section 17.1)

complementary colors Colors that, when mixed in proper proportions, appear white or colorless. (Section 24.5)

complete ionic equation A chemical equation in which dissolved strong electrolytes (such as dissolved ionic compounds) are written as separate ions. (Section 4.2)

complex ion (complex) An assembly of a metal ion and the Lewis bases (ligands) bonded to it. (Sections 17.5 and 24.1)

composite A complex solid mixture of two or more components. One component is usually present in much greater quantity than the others and acts as the primary host matrix for the other components. (Section 12.4)

compound A substance composed of two or more elements united chemically in definite proportions. (Section 1.2)

concentration The quantity of solute present in a given quantity of solvent or solution. (Section 4.5)

concentration cell A voltaic cell containing the same electrolyte and the same electrode materials in both the anode and cathode compartments. The emf of the cell is derived from a difference in the concentrations of the same electrolyte solutions in the compartments. (Section 20.6)

condensation polymerization Polymerization in which molecules are joined together through condensation reactions. (Section 12.2)

condensation reaction A chemical reaction in which a small molecule (such as a molecule of water) is split out from between two reacting molecules, as for example, between an organic acid and an amine function:

$$-\underset{\underset{\displaystyle O}{\|}}{C}-O-H + H-\underset{\underset{\displaystyle H}{|}}{N}- \longrightarrow -\underset{\underset{\displaystyle O}{\|}}{C}-\underset{\underset{\displaystyle H}{|}}{N}- + H_2O$$

(Section 12.2)

conjugate acid A substance formed by addition of a proton to a Brønsted–Lowry base. (Section 16.2)

conjugate acid-base pair An acid and a base, such as H_2O and OH^-, that differ only in the presence or absence of a proton. (Section 16.2)

conjugate base A substance formed by the loss of a proton from a Brønsted–Lowry acid. (Section 16.2)

continuous spectrum A spectrum that contains radiation distributed over all wavelengths. (Section 6.3)

conversion factor A ratio relating the same quantity in two systems of units that is used to convert the units of measurement. (Section 1.6)

coordination compound or complex A compound containing a metal ion bonded to a group of surrounding molecules or ions that act as ligands. (Section 24.1)

coordination number The number of adjacent atoms to which an atom is directly bonded. In a complex, the coordination number of the metal ion is the number of donor atoms to which it is bonded. (Sections 11.7 and 24.1)

coordination sphere The metal ion and its surrounding ligands. (Section 24.1)

coordination-sphere isomers Structural isomers of coordination compounds in which the ligands within the coordination sphere differ. (Section 24.4)

copolymer A complex polymer resulting from the polymerization of two or more monomers. (Section 12.2)

core electrons The electrons that are not in the outermost shell of an atom. (Section 6.8)

corrosion The process by which a metal is oxidized by substances in its environment. (Section 20.8)

covalent bond A bond formed between two or more atoms by a sharing of electrons. (Section 8.1)

covalent-network solids Solids in which the units that make up the three-dimensional network are joined by covalent bonds. (Section 11.8)

critical mass The amount of fissionable material necessary to maintain a chain reaction. (Section 21.7)

critical pressure The pressure at which a gas at its critical temperature is converted to a liquid state. (Section 11.4)

critical temperature The highest temperature at which it is possible to convert the gaseous form of a substance to a liquid. The critical temperature increases with an increase in the magnitude of intermolecular forces. (Section 11.4)

cross-linking The formation of bonds between polymer chains. (Section 12.2)

crystal-field theory A theory that accounts for the colors and the magnetic and other properties of transition-metal complexes in terms of the splitting of the energies of metal ion *d* orbitals by the electrostatic interaction with the ligands. (Section 24.6)

crystal lattice An imaginary network of points on which the repeating unit of the structure of a solid (the contents of the unit cell) may be imagined to be laid down so that the structure of the crystal is obtained. Each point represents an identical environment in the crystal. (Section 11.7)

crystalline solid (crystal) A solid whose internal arrangement of atoms, molecules, or ions shows a regular repetition in any direction through the solid. (Section 11.7)

crystallinity A measure of the extent of crystalline character (order) in a polymer. (Section 12.2)

crystallization The process in which a dissolved solute comes out of solution and forms a crystalline solid. (Section 13.2)

cubic close packing A close-packing arrangement in which the atoms of the third layer of a solid are not directly over those in the first layer. (Section 11.7)

curie A measure of radioactivity: 1 curie = 3.7×10^{10} nuclear disintegrations per second. (Section 21.4)

cycloalkanes Saturated hydrocarbons of general formula C_nH_{2n} in which the carbon atoms form a closed ring. (Section 25.3)

Dalton's law of partial pressures A law stating that the total pressure of a mixture of gases is the sum of the pressures that each gas would exert if it were present alone. (Section 10.6)

***d-d* transition** The transition of an electron from a lower-energy *d* orbital to a higher-energy orbital. (Section 24.6)

decomposition reaction A chemical reaction in which a single compound reacts to give two or more products. (Section 3.2)

degenerate Having the same energy (in several orbitals). (Section 6.7)

delocalized electrons Electrons that are spread over a number of atoms in a molecule rather than localized between a pair of atoms. (Section 9.6)

density The ratio of an object's mass to its volume. (Section 1.4)

deoxyribonucleic acid (DNA) A polynucleotide in which the sugar component is deoxyribose. (Section 25.11)

desalination The removal of salts from seawater, brine, or brackish water to make it fit for human consumption. (Section 18.5)

deuterium The isotope of hydrogen whose nucleus contains a proton and a neutron: ${}^2_1H$. (Section 22.2)

dextrorotatory, or merely dextro or *d* A term used to label a chiral molecule that rotates the plane of polarization of plane-polarized light to the right (clockwise). (Section 24.4)

diamagnetism A type of magnetism that causes a substance with no unpaired electrons to be weakly repelled from a magnetic field. (Section 9.8)

diatomic molecule A molecule composed of only two atoms. (Section 2.6)

diffusion The spreading of one substance through another. (Section 10.8)

dilution The process of preparing a less concentrated solution from a more concentrated one by adding solvent. (Section 4.5)

dimensional analysis A method of problem solving in which units are carried through all calculations. Dimensional analysis ensures that the final answer of a calculation has the desired units. (Section 1.6)

dipole A molecule with one end having a slight negative charge and the other end having a slight positive charge; a polar molecule. (Section 8.4)

dipole-dipole force The force that exists between polar molecules. (Section 11.2)

dipole moment A measure of the separation between the positive and negative charges in polar molecules. (Section 8.4)

displacement reaction A reaction in which an element reacts with a compound, displacing an element from it. (Section 4.4)

disproportionation A reaction in which a species undergoes simultaneous oxidation and reduction [as in $N_2O_3(g) \longrightarrow NO(g) + NO_2(g)$.] (Section 22.5)

donor atom The atom of a ligand that bonds to the metal. (Section 24.1)

double bond A covalent bond involving two electron pairs. (Section 8.3)

double helix The structure for DNA that involves the winding of two DNA polynucleotide chains together in a helical arrangement. The two strands of the double helix are complementary in that the organic bases on the two strands are paired for optimal hydrogen bond interaction. (Section 25.11)

Downs cell A cell used to obtain sodium metal by electrolysis of molten NaCl. (Section 23.4)

dynamic equilibrium A state of balance in which opposing processes occur at the same rate. (Section 11.5)

effective nuclear charge The net positive charge experienced by an electron in a many-electron atom; this charge is not the full nuclear charge because there is some shielding of the nucleus by the other electrons in the atom. (Section 7.2)

effusion The escape of a gas through an orifice or hole. (Section 10.8)

elastomer A material that can undergo a substantial change in shape via stretching, bending, or compression and return to its original shape upon release of the distorting force. (Section 12.2)

electrochemistry The branch of chemistry that deals with the relationships between electricity and chemical reactions. (Chapter 20: Introduction)

electrolysis reaction A reaction in which a nonspontaneous redox reaction is brought about by the passage of current under a sufficient external electrical potential. The devices in which electrolysis reactions occur are called electrolytic cells. (Section 20.9)

electrolyte A solute that produces ions in solution; an electrolytic solution conducts an electric current. (Section 4.1)

electrolytic cell A device in which a nonspontaneous oxidation-reduction reaction is caused to occur by passage of current under a sufficient external electrical potential. (Section 20.9)

electromagnetic radiation (radiant energy) A form of energy that has wave characteristics and that propagates through a vacuum at the characteristic speed of 3.00×10^8 m/s. (Section 6.1)

electrometallurgy The use of electrolysis to reduce or refine metals. (Section 23.4)

electromotive force (emf) A measure of the driving force, or *electrical pressure*, for the completion of an electrochemical reaction. Electromotive force is measured in volts: 1 V = 1 J/C. Also called the cell potential. (Section 20.4)

electron A negatively charged subatomic particle found outside the atomic nucleus; it is a part of all atoms. An electron has a mass 1/1836 times that of a proton. (Section 2.3)

electron affinity The energy change that occurs when an electron is added to a gaseous atom or ion. (Section 7.5)

electron capture A mode of radioactive decay in which an inner-shell orbital electron is captured by the nucleus. (Section 21.1)

electron configuration A particular arrangement of electrons in the orbitals of an atom. (Section 6.8)

electron density The probability of finding an electron at any particular point in an atom; this probability is equal to ψ^2, the square of the wave function. (Section 6.5)

electron domain In the VSEPR model, regions about a central atom in which electrons are likely to be found. (Section 9.2)

electron-domain geometry The three-dimensional arrangement of the electron domains around an atom according to the VSEPR model. (Section 9.2)

electronegativity A measure of the ability of an atom that is bonded to another atom to attract electrons to itself. (Section 8.4)

electronic charge The negative charge carried by an electron; it has a magnitude of 1.602×10^{-19}C. (Section 2.3)

electronic structure The arrangement of electrons of an atom or molecule. (Chapter 6: Introduction)

electron shell A collection of orbitals that have the same value of *n*. For example, the orbitals with $n = 3$ (the 3*s*, 3*p*, and 3*d* orbitals) comprise the third shell. (Section 6.5)

electron spin A property of the electron that makes it behave as though it were a tiny magnet. The electron behaves as if it were spinning on its axis; electron spin is quantized. (Section 6.7)

element A substance that cannot be separated into simpler substances by chemical means. (Sections 1.1 and 1.2)

elementary steps Processes in a chemical reaction that occur in a single event or step. (Section 14.6)

empirical formula (simplest formula) A chemical formula that shows the kinds of atoms and their relative numbers in a substance. (Section 2.6)

enantiomers Two mirror-image molecules of a chiral substance. The enantiomers are nonsuperimposable. (Section 24.4)

endothermic process A process in which a system absorbs heat from its surroundings. (Section 5.2)

energy The ability to do work or to transfer heat. (Section 5.1)

energy-level diagram A diagram that shows the energies of molecular orbitals relative to the atomic orbitals from which they are derived. Also called a **molecular-orbital diagram**. (Section 9.7)

enthalpy A quantity defined by the relationship $H = E + PV$; the enthalpy change, ΔH, for a reaction that occurs at constant pressure is the heat evolved or absorbed in the reaction: $\Delta H = q_p$. (Section 5.3)

enthalpy (heat) **of formation** The enthalpy change that accompanies the formation of a substance from the most stable forms of its component elements. (Section 5.7)

enthalpy of reaction The enthalpy change associated with a chemical reaction. (Section 5.4)

entropy A thermodynamic function associated with the number of different equivalent energy states or spatial arrangements in which a system may be found. It is a thermodynamic state function, which means that once we specify the conditions for a system—that is, the temperature, pressure, and so on—the entropy is defined. (Sections 13.1 and 19.2)

enzyme A protein molecule that acts to catalyze specific biochemical reactions. (Section 14.7)

equilibrium constant The numerical value of the equilibrium-constant expression for a system at equilibrium. The equilibrium constant is denoted by K_{eq}. (Section 15.2)

equilibrium-constant expression The expression that describes the relationship among the concentrations (or partial pressures) of the substances present in a system at equilibrium. The numerator is obtained by multiplying the concentrations of the substances on the product side of the equation, each raised to a power equal to its coefficient in the chemical equation. The denominator similarly contains the concentrations of the substances on the reactant side of the equation. (Section 15.2)

equivalence point The point in a titration at which the added solute reacts completely with the solute present in the solution. (Section 4.6)

ester An organic compound that has an OR group attached to a carbonyl; it is the product of a reaction between a carboxylic acid and an alcohol. (Section 25.6)

ether A compound in which two hydrocarbon groups are bonded to one oxygen. (Section 25.5)

exchange (metathesis) reaction A reaction between compounds that when written as a molecular equation appears to involve the exchange of ions between the two reactants. (Section 4.2)

excited state A higher energy state than the ground state. (Section 6.3)

exothermic process A process in which a system releases heat to its surroundings. (Section 5.2)

extensive property A property that depends on the amount of material considered; for example, mass or volume. (Section 1.3)

face-centered cubic cell A cubic unit cell that has lattice points at each corner as well as at the center of each face. (Section 11.7)

faraday A unit of charge that equals the total charge of 1 mol of electrons: $1\ F = 96{,}500\ C$. (Section 20.5)

***f*-block metals** Lanthanide and actinide elements in which the $4f$ or $5f$ orbitals are partially occupied. (Section 6.9)

ferromagnetism The ability of some substances to become permanently magnetized. (Section 23.7)

first law of thermodynamics A statement of our experience that energy is conserved in any process. We can express the law in many ways. One of the more useful expressions is that the change in internal energy, ΔE, of a system in any process is equal to the heat, q, added to the system, plus the work, w, done on the system by its surroundings: $\Delta E = q + w$. (Section 5.2)

first-order reaction A reaction in which the reaction rate is proportional to the concentration of a single reactant, raised to the first power. (Section 14.4)

fission The splitting of a large nucleus into two smaller ones. (Section 21.6)

force A push or a pull. (Section 5.1)

formal charge The number of valence electrons in an isolated atom minus the number of electrons assigned to the atom in the Lewis structure. (Section 8.5)

formation constant For a metal ion complex, the equilibrium constant for formation of the complex from the metal ion and base species present in solution. It is a measure of the tendency of the complex to form. (Section 17.5)

formula weight The mass of the collection of atoms represented by a chemical formula. For example, the formula weight of NO_2 (46.0 amu) is the sum of the masses of one nitrogen atom and two oxygen atoms. (Section 3.3)

fossil fuels Coal, oil, and natural gas, which are presently our major sources of energy. (Section 5.8)

free energy (Gibbs free energy, G) A thermodynamic state function that gives a criterion for spontaneous change in terms of enthalpy and entropy: $G = H - TS$. (Section 19.6)

free radical A substance with one or more unpaired electrons. (Section 21.9)

frequency The number of times per second that one complete wavelength passes a given point. (Section 6.1)

frequency factor (A) A term in the Arrhenius equation that is related to the frequency of collision and the probability that the collisions are favorably oriented for reaction. (Section 14.5)

fuel cell A voltaic cell that utilizes the oxidation of a conventional fuel, such as H_2 or CH_4, in the cell reaction. (Section 20.7)

fuel value The energy released when 1 g of a substance is combusted. (Section 5.8)

functional group An atom or group of atoms that imparts characteristic chemical properties to an organic compound. (Section 25.1)

fusion The joining of two light nuclei to form a more massive one. (Section 21.6)

galvanic cell See voltaic cell. (Section 20.3)

gamma radiation Energetic electromagnetic radiation emanating from the nucleus of a radioactive atom. (Section 21.1)

gas Matter that has no fixed volume or shape; it conforms to the volume and shape of its container. (Section 1.2)

gas constant (R) The constant of proportionality in the ideal-gas equation. (Section 10.4)

Geiger counter A device that can detect and measure radioactivity. (Section 21.5)

geometric isomers Compounds with the same type and number of atoms and the same chemical bonds but different spatial arrangements of these atoms and bonds. (Sections 24.4 and 25.4)

Gibbs free energy A thermodynamic state function that combines enthalpy and entropy, in the form $G = H - TS$. For a change occurring at constant temperature and pressure, the change in free energy is $\Delta G = \Delta H - T\Delta S$. (Section 19.6)

glass An amorphous solid formed by fusion of SiO_2, CaO, and Na_2O. Other oxides may also be used to form glasses with differing characteristics. (Section 22.10)

glucose A polyhydroxy aldehyde whose formula is $CH_2OH(CHOH)_4CHO$; it is the most important of the monosaccharides. (Section 25.10)

glycogen The general name given to a group of polysaccharides of glucose that are synthesized in mammals and used to store energy from carbohydrates. (Section 25.10)

Graham's law A law stating that the rate of effusion of a gas is inversely proportional to the square root of its molecular weight. (Section 10.8)

gray (Gy) The SI unit for radiation dose corresponding to the absorption of 1 J of energy per kilogram of tissue; 1 Gy = 100 rads. (Section 21.9)

green chemistry chemistry that promotes the design and application of chemical products and processes that are compatible with human health and that preserve the environment. (Section 18.7)

ground state The lowest-energy, or most stable, state. (Section 6.3)

group Elements that are in the same column of the periodic table; elements within the same group or family exhibit similarities in their chemical behavior. (Section 2.5)

Haber process The catalyst system and conditions of temperature and pressure developed by Fritz Haber and coworkers for the formation of NH_3 from H_2 and N_2. (Section 15.1)

half-life The time required for the concentration of a reactant substance to decrease to half its initial value; the time required for half of a sample of a particular radioisotope to decay. (Sections 14.4 and 21.4)

half-reaction An equation for either an oxidation or a reduction that explicitly shows the electrons involved [for example, $Zn^{2+}(aq) + 2e^- \longrightarrow Zn(s)$]. (Section 20.2)

Hall process A process used to obtain aluminum by electrolysis of Al_2O_3 dissolved in molten cryolite, Na_3AlF_6. (Section 23.4)

halogens Members of group 7A in the periodic table. (Section 7.8)

hard water Water that contains appreciable concentrations of Ca^{2+} and Mg^{2+}; these ions react with soaps to form an insoluble material. (Section 18.6)

heat The flow of energy from a body at higher temperature to one at lower temperature when they are placed in thermal contact. (Section 5.1)

heat capacity The quantity of heat required to raise the temperature of a sample of matter by 1°C (or 1 K). (Section 5.5)

heat of fusion The enthalpy change, ΔH, for melting a solid. (Section 11.4)

heat of vaporization The enthalpy change, ΔH, for vaporizing a liquid. (Section 11.4)

hemoglobin An iron-containing protein responsible for oxygen transport in the blood. (Section 18.4)

Henderson–Hasselbalch equation The relationship among the pH, pK_a, and the concentrations of acid and conjugate base in an aqueous solution: $\text{pH} = \text{p}K_a + \log \frac{[\text{acid}]}{[\text{base}]}$. (Section 17.2)

Henry's law A law stating that the concentration of a gas in a solution, C_g, is proportional to the pressure of gas over the solution: $C_g = kP_g$. (Section 13.3)

Hess's law The heat evolved in a given process can be expressed as the sum of the heats of several processes that, when added, yield the process of interest. (Section 5.6)

heterogeneous alloy An alloy in which the components are not distributed uniformly; instead, two or more distinct phases with characteristic compositions are present. (Section 23.6)

heterogeneous catalyst A catalyst that is in a different phase from that of the reactant substances. (Section 14.7)

heterogeneous equilibrium The equilibrium established between substances in two or more different phases, for example, between a gas and a solid or between a solid and a liquid. (Section 15.3)

hexagonal close packing A close-packing arrangement in which the atoms of the third layer of a solid lie directly over those in the first layer. (Section 11.7)

high-spin complex A complex whose electrons populate the *d* orbitals to give the maximum number of unpaired electrons. (Section 24.6)

high-temperature superconductivity The "frictionless" flow of electrical current (superconductivity) at temperatures above 30 K. (Section 12.5)

homogeneous catalyst A catalyst that is in the same phase as the reactant substances. (Section 14.7)

homogeneous equilibrium The equilibrium established between reactant and product substances that are all in the same phase. (Section 15.3)

Hund's rule A rule stating that electrons occupy degenerate orbitals in such a way as to maximize the number of electrons with the same spin. In other words, each orbital has one electron placed in it before pairing of electrons in orbitals occurs. Note that this rule applies only to orbitals that are degenerate, which means that they have the same energy. (Section 6.8)

hybridization The mixing of different types of atomic orbitals to produce a set of equivalent hybrid orbitals. (Section 9.5)

hybrid orbital An orbital that results from the mixing of different kinds of atomic orbitals on the same atom. For example, an sp^3 hybrid results from the mixing, or hybridizing, of one *s* orbital and three *p* orbitals. (Section 9.5)

hydration Solvation when the solvent is water. (Section 13.1)

hydride ion An ion formed by the addition of an electron to a hydrogen atom: H^-. (Section 7.7)

hydrocarbons Compounds composed of only carbon and hydrogen. (Section 2.9)

hydrogen bonding Bonding that results from intermolecular attractions between molecules containing hydrogen bonded to an electronegative element. The most important examples involve oxygen, nitrogen, or fluorine. (Section 11.2)

hydrolysis A reaction with water. When a cation or anion reacts with water, it changes the pH. (Section 16.9)

hydrometallurgy Aqueous chemical processes for recovery of a metal from an ore. (Section 23.3)

hydronium ion (H_3O^+) The predominant form of the proton in aqueous solution. (Section 16.2)

hydrophilic Water attracting. (Section 13.6)

hydrophobic Water repelling. (Section 13.6)

hypothesis A tentative explanation of a series of observations or of a natural law. (Section 1.3)

ideal gas A hypothetical gas whose pressure, volume, and temperature behavior is completely described by the ideal-gas equation. (Section 10.4)

ideal-gas equation An equation of state for gases that embodies Boyle's law, Charles's law, and Avogadro's hypothesis in the form $PV = nRT$. (Section 10.4)

ideal solution A solution that obeys Raoult's law. (Section 13.5)

immiscible liquids Liquids that do not mix. (Section 13.3)

indicator A substance added to a solution to indicate by a color change the point at which the added solute has reacted with all the solute present in solution. (Section 4.6)

instantaneous rate The reaction rate at a particular time as opposed to the average rate over an interval of time. (Section 14.2)

intensive property A property that is independent of the amount of material considered; for example, density. (Section 1.3)

interhalogens Compounds formed between two different halogen elements. Examples include IBr and BrF_3. (Section 22.4)

intermediate A substance formed in one elementary step of a multistep mechanism and consumed in another; it is neither a reactant nor an ultimate product of the overall reaction. (Section 14.6)

intermetallic compound A homogeneous alloy with definite properties and composition. Intermetallic compounds are stoichiometric compounds, but their compositions are not readily explained in terms of ordinary chemical bonding theory. (Section 23.6)

intermolecular forces The short-range attractive forces operating between the particles that make up the units of a liquid or solid substance. These same forces also cause gases to liquefy or solidify at low temperatures and high pressures. (Chapter 11: Introduction)

internal energy The total energy possessed by a system. When a system undergoes a change, the change in internal energy, ΔE, is defined as the heat, q, added to the system, plus the work, w, done on the system by its surroundings: $\Delta E = q + w$. (Section 5.2)

ion Electrically charged atom or group of atoms (polyatomic ion); ions can be positively or negatively charged, depending on whether electrons are lost (positive) or gained (negative) by the atoms. (Section 2.7)

ion-dipole force The force that exists between an ion and a neutral polar molecule that possesses a permanent dipole moment. (Section 11.2)

ion exchange A process in which ions in solution are exchanged for other ions held on the surface of an ion-exchange resin; the exchange of a hard-water cation such as Ca^{2+} for a soft-water cation such as Na^+ is used to soften water. (Section 18.6)

ionic bond A bond between oppositely charged ions. The ions are formed from atoms by transfer of one or more electrons. (Section 8.1)

ionic compound A compound composed of cations and anions. (Section 2.7)

ionic hydrides Compounds formed when hydrogen reacts with alkali metals and also the heavier alkaline earths (Ca, Sr, and Ba); these compounds contain the hydride ion, H^-. (Section 22.2)

ionic solids Solids that are composed of ions. (Section 11.8)

ionization energy The energy required to remove an electron from a gaseous atom when the atom is in its ground state. (Section 7.4)

ionizing radiation Radiation that has sufficient energy to remove an electron from a molecule, thereby ionizing it. (Section 21.9)

ion-product constant For water, K_w is the product of the aquated hydrogen ion and hydroxide ion concentrations: $[H^+][OH^-] = K_w = 1.0 \times 10^{-14}$ at 25°C. (Section 16.3)

irreversible process A process that is not reversible, and as a result some of its potential for accomplishing work is dissipated as heat. Any spontaneous process is irreversible in practice. (Section 19.1)

isoelectronic series A series of atoms, ions, or molecules having the same number of electrons. (Section 7.3)

isolated system A system that does not exchange energy or matter with its surroundings. (Section 19.2)

isomers Compounds whose molecules have the same overall composition but different structures. (Section 24.4)

isotopes Atoms of the same element containing different numbers of neutrons and therefore having different masses. (Section 2.3)

joule (J) The SI unit of energy, 1 kg-m^2/s^2. A related unit is the calorie: 4.184 J = cal. (Section 5.1)

Kelvin scale The absolute temperature scale; the SI unit for temperature is the kelvin. Zero on the Kelvin scale corresponds to −273.15°C; therefore, K = °C + 273.15. (Section 1.4)

ketone A compound in which the carbonyl group occurs at the interior of a carbon chain and is therefore flanked by carbon atoms. (Section 25.6)

kinetic energy The energy that an object possesses by virtue of its motion. (Section 5.1)

kinetic-molecular theory A set of assumptions about the nature of gases. These assumptions, when translated into mathematical form, yield the ideal-gas equation. (Section 10.7)

lanthanide contraction The gradual decrease in atomic and ionic radii with increasing atomic number among the lanthanide elements, atomic numbers 57 through 70. The decrease arises because of a gradual increase in effective nuclear charge through the lanthanide series. (Section 23.7)

lanthanide (rare earth) **element** Element in which the 4*f* subshell is only partially occupied. (Section 6.8)

lattice energy The energy required to separate completely the ions in an ionic solid. (Section 8.2)

law of conservation of mass A scientific law stating that the total mass of the products of a chemical reaction is the same as the total mass of the reactants, so that mass remains constant during the reaction. (Section 3.1)

law of constant composition A law that states that the elemental composition of a pure compound is always the same, regardless of its source; also called the **law of definite proportions**. (Section 1.2)

law of definite proportions A law that states that the elemental composition of a pure substance is always the same, regardless of its source; also called the **law of constant composition**. (Section 1.2)

law of mass action The rules by which the equilibrium constant is expressed in terms of the concentrations of reactants and products, in accordance with the balanced chemical equation for the reaction. (Section 15.2)

leaching The selective dissolution of a desired mineral by passing an aqueous reagent solution through an ore. (Section 23.3)

Le Châtelier's principle A principle stating that when we disturb a system at chemical equilibrium, the relative concentrations of reactants and products shift so as to partially undo the effects of the disturbance. (Section 15.6)

levorotatory, or merely levo or *l* A term used to label a chiral molecule that rotates the plane of polarization of plane-polarized light to the left (counterclockwise). (Section 24.4)

Lewis acid An electron-pair acceptor. (Section 16.11)

Lewis base An electron-pair donor. (Section 16.11)

Lewis structure A representation of covalent bonding in a molecule that is drawn using Lewis symbols. Shared electron pairs are shown as lines, and unshared electron pairs are shown as pairs of dots. Only the valence-shell electrons are shown. (Section 8.3)

Lewis symbol (electron-dot symbol) The chemical symbol for an element with a dot for each valence electron. (Section 8.1)

ligand An ion or molecule that coordinates to a metal atom or to a metal ion to form a complex. (Section 24.1)

lime-soda process A method for removal of Mg^{2+} and Ca^{2+} ions from water to reduce water hardness. The substances added to the water are "lime," CaO [or "slaked lime," $Ca(OH)_2$], and "soda ash," Na_2CO_3, in amounts determined by the concentrations of the undesired ions. (Section 18.6)

limiting reactant (limiting reagent) The reactant present in the smallest stoichiometric quantity in a mixture of reactants; the amount of product that can form is limited by the complete consumption of the limiting reactant. (Section 3.7)

line spectrum A spectrum that contains radiation at only certain specific wavelengths. (Section 6.3)

linkage isomers Structural isomers of coordination compounds in which a ligand differs in its mode of attachment to a metal ion. (Section 24.4)

liquid Matter that has a distinct volume but no specific shape. (Section 1.2)

liquid crystal A substance that exhibits one or more partially ordered liquid phases above the melting point of the solid form. By contrast, in nonliquid crystalline substances the liquid phase that forms upon melting is completely unordered. (Section 12.1)

lithosphere That portion of our environment consisting of the solid Earth. (Section 23.1)

lock-and-key model A model of enzyme action in which the substrate molecule is pictured as fitting rather specifically into the active site on the enzyme. It is assumed that in being bound to the active site, the substrate is somehow activated for reaction. (Section 14.7)

London dispersion forces Intermolecular forces resulting from attractions between induced dipoles. (Section 11.2)

low-spin complex A complex that has the lowest possible number of unpaired electrons. (Section 24.6)

magic numbers Numbers of protons and neutrons that result in very stable nuclei. (Section 21.2)

main-group elements Elements in the *s* and *p* blocks of the periodic table. (Section 6.9)

mass A measure of the amount of material in an object. It measures the resistance of an object to being moved. In SI units, mass is measured in kilograms. (Section 1.4)

mass defect The difference between the mass of a nucleus and the total masses of the individual nucleons that it contains. (Section 21.6)

mass number The sum of the number of protons and neutrons in the nucleus of a particular atom. (Section 2.3)

mass percentage The number of grams of solute in each 100 g of solution. (Section 13.4)

mass spectrometer An instrument used to measure the precise masses and relative amounts of atomic and molecular ions. (Section 2.4)

matter Anything that occupies space and has mass; the physical material of the universe. (Section 1.1)

matter waves The term used to describe the wave characteristics of a particle. (Section 6.4)

mean free path The average distance traveled by a gas molecule between collisions. (Section 10.8)

metal complex (complex ion or complex) An assembly of a metal ion and the Lewis bases bonded to it. (Section 24.1)

metallic bond Bonding in which the bonding electrons are relatively free to move throughout the three-dimensional structure. (Section 8.1)

metallic character The extent to which an element exhibits the physical and chemical properties characteristic of metals, for example, luster, malleability, ductility, and good thermal and electrical conductivity. (Section 7.6)

metallic elements (metals) Elements that are usually solids at room temperature, exhibit high electrical and heat conductivity, and appear lustrous. Most of the elements in the periodic table are metals. (Section 2.5)

metallic hydrides Compounds formed when hydrogen reacts with transition metals; these compounds contain the hydride ion, H^-. (Section 22.2)

metallic solids Solids that are composed of metal atoms. (Section 11.8)

metalloids Elements that lie along the diagonal line separating the metals from the nonmetals in the periodic table; the properties of metalloids are intermediate between those of metals and nonmetals. (Section 2.5)

metallurgy The science of extracting metals from their natural sources by a combination of chemical and physical processes. It is also concerned with the properties and structures of metals and alloys. (Section 23.1)

metathesis (exchange) reaction A reaction in which two substances react through an exchange of their component ions: $AX + BY \longrightarrow AY + BX$. Precipitation and acid-base neutralization reactions are examples of metathesis reactions. (Section 4.2)

metric system A system of measurement used in science and in most countries. The meter and the gram are examples of metric units. (Section 1.4)

mineral A solid, inorganic substance occurring in nature, such as calcium carbonate, which occurs as calcite. (Section 23.1)

miscible Liquids that mix in all proportions. (Section 13.3)

mixture A combination of two or more substances in which each substance retains its own chemical identity. (Section 1.2)

molal boiling-point-elevation constant (K_b) A constant characteristic of a particular solvent that gives the change in boiling point as a function of solution molality: $\Delta T_b = K_b m$. (Section 13.5)

molal freezing-point-depression constant (K_f) A constant characteristic of a particular solvent that gives the change in freezing point as a function of solution molality: $\Delta T_f = K_f m$. (Section 13.5)

molality The concentration of a solution expressed as moles of solute per kilogram of solvent; abbreviated m. (Section 13.4)

molar heat capacity The heat required to raise the temperature of 1 mol of a substance by 1°C. (Section 5.5)

molarity The concentration of a solution expressed as moles of solute per liter of solution; abbreviated M. (Section 4.5)

molar mass The mass of 1 mol of a substance in grams; it is numerically equal to the formula weight in atomic mass units. (Section 3.4)

mole A collection of Avogadro's number (6.022×10^{23}) of objects; for example, a mole of H_2O is 6.022×10^{23} H_2O molecules. (Section 3.4)

molecular compound A compound that consists of molecules. (Section 2.6)

molecular equation A chemical equation in which the formula for each substance is written without regard for whether it is an electrolyte or a nonelectrolyte. (Section 4.2)

molecular formula A chemical formula that indicates the actual number of atoms of each element in one molecule of a substance. (Section 2.6)

molecular geometry The arrangement in space of the atoms of a molecule. (Section 9.2)

molecular hydrides Compounds formed when hydrogen reacts with nonmetals and metalloids. (Section 22.2)

molecularity The number of molecules that participate as reactants in an elementary reaction. (Section 14.6)

molecular orbital An allowed state for an electron in a molecule. According to molecular orbital theory, a molecular orbital is entirely analogous to an atomic orbital, which is an allowed state for an electron in an atom. A molecular orbital may be classified as σ or π, depending on the disposition of electron density with respect to the internuclear axis. (Section 9.7)

molecular orbital diagram A diagram that shows the energies of molecular orbitals relative to the atomic orbitals from which they are derived; also called an **energy-level diagram**. (Section 9.7)

molecular-orbital theory A theory that accounts for the allowed states for electrons in molecules. (Section 9.7)

molecular solids Solids that are composed of molecules. (Section 11.8)

molecular weight The mass of the collection of atoms represented by the chemical formula for a molecule. (Section 3.3)

molecule A chemical combination of two or more atoms. (Sections 1.1 and 2.6)

mole fraction The ratio of the number of moles of one component of a mixture to the total moles of all components; abbreviated X, with a subscript to identify the component. (Section 10.6)

momentum The product of the mass, m, and velocity, v, of a particle. (Section 6.4)

monodentate ligand A ligand that binds to the metal ion via a single donor atom. It occupies one position in the coordination sphere. (Section 24.2)

monomers Molecules with low molecular weights, which can be joined together (polymerized) to form a polymer. (Section 12.2)

monosaccharide A simple sugar, most commonly containing six carbon atoms. The joining together of monosaccharide units by condensation reactions results in formation of polysaccharides. (Section 25.10)

multiple bonding Bonding involving two or more electron pairs. (Section 8.3)

natural gas A naturally occurring mixture of gaseous hydrocarbon compounds composed of hydrogen and carbon. (Section 5.8)

nematic liquid crystalline phase A liquid crystal in which the molecules are aligned in the same general direction, along their long axes, but in which the ends of the molecules are not aligned. (Section 12.1)

Nernst equation An equation that relates the cell emf, E, to the standard emf, $E°$, and the reaction quotient, $Q: E = E° - 2.30RT/nF \log Q$. (Section 20.6)

net ionic equation A chemical equation for a solution reaction in which soluble strong electrolytes are written as ions and spectator ions are omitted. (Section 4.2)

neutralization reaction A reaction in which an acid and a base react in stoichiometrically equivalent amounts; the neutralization reaction between an acid and a metal hydroxide produces water and a salt. (Section 4.3)

neutron An electrically neutral particle found in the nucleus of an atom; it has approximately the same mass as a proton. (Section 2.3)

noble gases Members of group 8A in the periodic table. (Section 7.8)

node A locus of points in an atom at which the electron density is zero. For example, the node in a 2s orbital is a spherical surface. (Section 6.6)

nonbonding pair In a Lewis structure a pair of electrons assigned completely to one atom; also called a lone pair. (Section 9.2)

nonelectrolyte A substance that does not ionize in water and consequently gives a nonconducting solution. (Section 4.1)

nonionizing radiation Radiation that does not have sufficient energy to remove an electron from a molecule. (Section 21.9)

nonmetallic elements (nonmetals) Elements in the upper right corner of the periodic table; nonmetals differ from metals in their physical and chemical properties. (Section 2.5)

nonpolar covalent bond A covalent bond in which the electrons are shared equally. (Section 8.4)

normal boiling point The boiling point at 1 atm pressure. (Section 11.5)

normal melting point The melting point at 1 atm pressure. (Section 11.6)

nuclear binding energy The energy required to decompose an atomic nucleus into its component protons and neutrons. (Section 21.6)

nuclear disintegration series A series of nuclear reactions that begins with an unstable nucleus and terminates with a stable one. Also called a **radioactive series.** (Section 21.2)

nuclear transmutation A conversion of one kind of nucleus to another. (Section 21.3)

nucleic acids Polymers of high molecular weight that carry genetic information and control protein synthesis. (Section 25.11)

nucleon A particle found in the nucleus of an atom. (Section 21.1)

nucleotide Compounds formed from a molecule of phosphoric acid, a sugar molecule, and an organic nitrogen base. Nucelotides form linear polymers called DNA and RNA, which are involved in protein synthesis and cell reproduction. (Section 25.11)

nucleus The very small, very dense, positively charged portion of an atom; it is composed of protons and neutrons. (Section 2.2)

nuclide A nucleus of a specific isotope of an element. (Section 2.3)

octet rule A rule stating that bonded atoms tend to possess or share a total of eight valence-shell electrons. (Section 8.1)

optical isomers Stereoisomers in which the two forms of the compound are nonsuperimposable mirror images. (Section 24.4)

optically active Possessing the ability to rotate the plane of polarized light. (Section 24.4)

orbital An allowed energy state of an electron in the quantum mechanical model of the atom; the term orbital is also used to describe the spatial distribution of the electron. An orbital is defined by the values of three quantum numbers: n, l, and m_l. (Section 6.5)

ore A source of a desired element or mineral, usually accompanied by large quantities of other materials such as sand and clay. (Section 23.1)

organic chemistry The study of carbon-containing compounds, typically containing carbon–carbon bonds. (Section 2.9 and Chapter 25: Introduction)

osmosis The net movement of solvent through a semipermeable membrane toward the solution with greater solute concentration. (Section 13.5)

osmotic pressure The pressure that must be applied to a solution to stop osmosis from pure solvent into the solution. (Section 13.5)

Ostwald process An industrial process used to make nitric acid from ammonia. The NH_3 is catalytically oxidized by O_2 to form NO; NO in air is oxidized to NO_2; HNO_3 is formed in a disproportionation reaction when NO_2 dissolves in water. (Section 22.7)

overall reaction order The sum of the reaction orders of all the reactants appearing in the rate expression. (Section 14.3)

overlap The extent to which atomic orbitals on different atoms share the same region of space. When the overlap between two orbitals is large, a strong bond may be formed. (Section 9.4)

oxidation A process in which a substance loses one or more electrons. (Section 4.4)

oxidation number (oxidation state) A positive or negative whole number assigned to an element in a molecule or ion on the basis of a set of formal rules; to some degree it reflects the positive or negative character of that atom. (Section 4.4)

oxidation-reduction reaction A chemical reaction in which the oxidation states of certain atoms change. (Chapter 20: Introduction)

oxidizing agent, or **oxidant** The substance that is reduced and thereby causes the oxidation of some other substance in an oxidation-reduction reaction. (Section 20.1)

oxyacid A compound in which one or more OH groups, and possibly additional oxygen atoms, are bonded to a central atom. (Section 16.10)

oxyanion A polyatomic ion that contains one or more oxygen atoms. (Section 2.8)

ozone The name given to O_3, an allotrope of oxygen. (Section 7.8)

paramagnetism A property that a substance possesses if it contains one or more unpaired electrons. A paramagnetic substance is drawn into a magnetic field. (Section 9.8)

partial pressure The pressure exerted by a particular gas in a mixture. (Section 10.6)

particle accelerator A device that uses strong magnetic and electrostatic fields to accelerate charged particles. (Section 21.3)

parts per billion (ppb) The concentration of a solution in grams of solute per 10^9 (billion) grams of solution; equals micrograms of solute per liter of solution for aqueous solutions. (Section 13.4)

parts per million (ppm) The concentration of a solution in grams of solute per 10^6 (million) grams of solution; equals milligrams of solute per liter of solution for aqueous solutions. (Section 13.4)

pascal (Pa) The SI unit of pressure: $1\ \text{Pa} = \text{N/m}^2$. (Section 10.2)

Pauli exclusion principle A rule stating that no two electrons in an atom may have the same four quantum numbers (n, l, m_l, and m_s). As a consequence of this principle, there can be no more than two electrons in any one atomic orbital. (Section 6.7)

peptide bond A bond formed between two amino acids. (Section 25.9)

percent yield The ratio of the actual (experimental) yield of a product to its theoretical (calculated) yield, multiplied by 100. (Section 3.7)

periodic table The arrangement of elements in order of increasing atomic number, with elements having similar properties placed in vertical columns. (Section 2.5)

petroleum A naturally occurring combustible liquid composed of hundreds of hydrocarbons and other organic compounds. (Section 5.8)

pH The negative log in base 10 of the aquated hydrogen ion concentration: $\text{pH} = -\log[\text{H}^+]$. (Section 16.4)

pH titration curve See **titration curve.** (Section 17.3).

phase change The conversion of a substance from one state of matter to another. The phase changes we consider are melting and freezing (solid ↔ liquid), sublimation and deposition (solid ↔ gas), and vaporization and condensation (liquid ↔ gas). (Section 11.4)

phase diagram A graphic representation of the equilibria among the solid, liquid, and gaseous phases of a substance as a function of temperature and pressure. (Section 11.6)

photochemical smog A complex mixture of undesirable substances produced by the action of sunlight on an urban atmosphere polluted with automobile emissions. The major starting ingredients are nitrogen oxides and organic substances, notably olefins and aldehydes. (Section 18.4)

photodissociation The breaking of a molecule into two or more neutral fragments as a result of absorption of light. (Section 18.2)

photoionization The removal of an electron from an atom or molecule by absorption of light. (Section 18.2)

photon The smallest increment (a quantum) of radiant energy; a photon of light with frequency v has an energy equal to hv. (Section 6.2)

photosynthesis The process that occurs in plant leaves by which light energy is used to convert carbon dioxide and water to carbohydrates and oxygen. (Section 24.2)

physical changes changes (such as a phase change) that occur with no change in chemical composition. (Section 1.3)

physical properties Properties that can be measured without changing the composition of a substance, for example, color and freezing point. (Section 1.3)

pi (π) bond A covalent bond in which electron density is concentrated above and below the line joining the bonded atoms. (Section 9.6)

pi (π) molecular orbital A molecular orbital that concentrates the electron density on opposite sides of a line that passes through the nuclei. (Section 9.8)

Planck's constant (h) The constant that relates the energy and frequency of a photon, $E = hv$. Its value is 6.626×10^{-34} J-s. (Section 6.2)

plastic A material that can be formed into particular shapes by application of heat and pressure. (Section 12.2)

polar covalent bond A covalent bond in which the electrons are not shared equally. (Section 8.4)

polarizability The ease with which the electron cloud of an atom or a molecule is distorted by an outside influence, thereby inducing a dipole moment. (Section 11.2)

polar molecule A molecule that possesses a nonzero dipole moment. (Section 8.4)

polyatomic ion An electrically charged group of two or more atoms. (Section 2.7)

polydentate ligand A ligand in which two or more donor atoms can coordinate to the same metal ion. (Section 24.2)

polymer A large molecule of high molecular mass, formed by the joining together, or polymerization, of a large number of molecules of low molecular mass. The individual molecules forming the polymer are called monomers. (Section 12.2)

polypeptide A polymer of amino acids that has a molecular weight of less than 10,000. (Section 25.9)

polyprotic acid A substance capable of ionizing more than one proton in water; H_2SO_4 is an example. (Section 16.6)

polysaccharide A substance made up of several monosaccharide units joined together. (Section 25.10)

porphyrin A complex derived from the porphine molecule. (Section 24.2)

positron A particle with the same mass as an electron but with a positive charge, symbol ${}^{0}_{1}\text{e}$. (Section 21.1)

potential energy The energy that an object possesses as a result of its composition or its position with respect to another object. (Section 5.1)

precipitate An insoluble substance that forms in, and separates from, a solution. (Section 4.2)

precipitation reaction A reaction that occurs between substances in solution in which one of the products is insoluble. (Section 4.2)

precision The closeness of agreement among several measurements of the same quantity; the reproducibility of a measurement. (Section 1.5)

pressure A measure of the force exerted on a unit area. In chemistry, pressure is often expressed in units of atmospheres (atm) or torr: 760 torr = 1 atm; in SI units pressure is expressed in pascals (Pa). (Section 10.2)

primary structure The sequence of amino acids along a protein chain. (Section 25.9)

primitive cubic cell A cubic unit cell in which the lattice points are at the corners only. (Section 11.7)

probability density (ψ^2) A value that represents the probability that an electron will be found at a given point in space. (Section 6.5)

product A substance produced in a chemical reaction; it appears to the right of the arrow in a chemical equation. (Section 3.1)

protein A biopolymer formed from amino acids. (Section 25.9)

protium The most common isotope of hydrogen. (Section 22.2)

proton A positively charged subatomic particle found in the nucleus of an atom. (Section 2.3)

pure substance Matter that has a fixed composition and distinct properties. (Section 1.2)

pyrometallurgy A process in which heat converts a mineral in an ore from one chemical form to another and eventually to the free metal. (Section 23.2)

qualitative analysis The determination of the presence or absence of a particular substance in a mixture. (Section 17.7)

quantitative analysis The determination of the amount of a given substance that is present in a sample. (Section 17.7)

quantum The smallest increment of radiant energy that may be absorbed or emitted; the magnitude of radiant energy is $h\nu$. (Section 6.2)

racemic mixture A mixture of equal amounts of the dextrorotatory and levorotatory forms of a chiral molecule. A racemic mixture will not rotate polarized light. (Section 24.4)

rad A measure of the energy absorbed from radiation by tissue or other biological material; 1 rad = transfer of 1×10^{-2} J of energy per kilogram of material. (Section 21.9)

radioactive series A series of nuclear reactions that begins with an unstable nucleus and terminates with a stable one. Also called **nuclear disintegration series**. (Section 21.2)

radioactivity The spontaneous disintegration of an unstable atomic nucleus with accompanying emission of radiation. (Section 2.2; Chapter 21: Introduction)

radioisotope An isotope that is radioactive; that is, it is undergoing nuclear changes with emission of radiation. (Section 21.1)

radionuclide A radioactive nuclide. (Section 21.1)

radiotracer A radioisotope that can be used to trace the path of an element. (Section 21.5)

Raoult's law A law stating that the partial pressure of a solvent over a solution, P_A, is given by the vapor pressure of the pure solvent, P_A°, times the mole fraction of a solvent in the solution, $X_A: P_A = X_A P_A^\circ$. (Section 13.5)

rate constant A constant of proportionality between the reaction rate and the concentrations of reactants that appear in the rate law. (Section 14.3)

rate-determining step The slowest elementary step in a reaction mechanism. (Section 14.6)

rate law An equation that relates the reaction rate to the concentrations of reactants (and sometimes of products also). (Section 14.3)

reactant A starting substance in a chemical reaction; it appears to the left of the arrow in a chemical equation. (Section 3.1)

reaction mechanism A detailed picture, or model, of how the reaction occurs; that is, the order in which bonds are broken and formed, and the changes in relative positions of the atoms as the reaction proceeds. (Section 14.6)

reaction order The power to which the concentration of a reactant is raised in a rate law. (Section 14.3)

reaction quotient (Q) The value that is obtained when concentrations of reactants and products are inserted into the equilibrium expression. If the concentrations are equilibrium concentrations, $Q = K$; otherwise, $Q \neq K$. (Section 15.5)

reaction rate The decrease in concentration of a reactant or the increase in concentration of a product with time. (Section 14.2)

redox (oxidation-reduction) reaction A reaction in which certain atoms undergo changes in oxidation states. The substance increasing in oxidation state is oxidized; the substance decreasing in oxidation state is reduced. (Chapter 20; Introduction)

reducing agent, or **reductant** The substance that is oxidized and thereby causes the reduction of some other substance in an oxidation-reduction reaction. (Section 20.1)

reduction A process in which a substance gains one or more electrons. (Section 4.4)

refining The process of converting an impure form of a metal into a more usable substance of well-defined composition. For example, crude pig iron from the blast furnace is refined in a converter to produce steels of desired compositions. (Section 23.2)

rem A measure of the biological damage caused by radiation; rems = rads × RBE. (Section 21.9)

renewable energy Energy such as solar energy, wind energy, and hydroelectric energy that is from essentially inexhaustible sources. (Section 5.8)

representative (main-group) element Element in which the s and p orbitals are partially occupied. (Section 6.9)

resonance structures (resonance forms) Individual Lewis structures in cases where two or more Lewis structures are equally good descriptions of a single molecule. The resonance structures in such an instance are "averaged" to give a correct description of the real molecule. (Section 8.6)

reverse osmosis The process by which water molecules move under high pressure through a semipermeable membrane from the more concentrated to the less concentrated solution. (Section 18.5)

reversible process A process that can go back and forth between states along exactly the same path; a system at equilibrium is reversible because it can be reversed by an infinitesimal modification of a variable such as temperature. (Section 19.1)

ribonucleic acid (RNA) A polynucleotide in which ribose is the sugar component. (Section 25.11)

roasting Thermal treatment of an ore to bring about chemical reactions involving the furnace atmosphere. For example, a sulfide ore might be roasted in air to form a metal oxide and SO_2. (Section 23.2)

root-mean-square (rms) speed (μ) The square root of the average of the squared speeds of the gas molecules in a gas sample. (Section 10.7)

rotational motion Movement of a molecule as though it is spinning like a top. (Section 19.3)

salinity A measure of the salt content of seawater, brine, or brackish water. It is equal to the mass in grams of dissolved salts present in 1 kg of seawater. (Section 18.5)

salt An ionic compound formed by replacing one or more H^+ of an acid by other cations. (Section 4.3)

saponification Hydrolysis of an ester in the presence of a base. (Section 25.6)

saturated solution A solution in which undissolved solute and dissolved solute are in equilibrium. (Section 13.2)

scientific law A concise verbal statement or a mathematical equation that summarizes a broad variety of observations and experiences. (Section 1.3)

scientific method The general process of advancing scientific knowledge by making experimental observations and by formulating laws, hypotheses, and theories. (Section 1.3)

scintillation counter An instrument that is used to detect and measure radiation by the fluorescence it produces in a fluorescing medium. (Section 21.5)

secondary structure The manner in which a protein is coiled or stretched. (Section 25.9)

second law of thermodynamics A statement of our experience that there is a direction to the way events occur in nature. When a process occurs spontaneously in one direction, it is nonspontaneous in the reverse direction. It is possible to state the second law in many different forms, but they all relate back to the same idea about spontaneity. One of the most common statements found in chemical contexts is that in any spontaneous process the entropy of the universe increases. (Section 19.2)

second-order reaction A reaction in which the overall reaction order (the sum of the concentration-term exponents) in the rate law is 2. (Section 14.4)

sigma (σ) bond A covalent bond in which electron density is concentrated along the internuclear axis. (Section 9.6)

sigma (σ) molecular orbital A molecular orbital that centers the electron density about an imaginary line passing through two nuclei. (Section 9.7)

significant figures The digits that indicate the precision with which a measurement is made; all digits of a measured quantity are significant, including the last digit, which is uncertain. (Section 1.5)

silicates Compounds containing silicon and oxygen, structurally based on SiO_4 tetrahedra. (Section 22.10)

single bond A covalent bond involving one electron pair. (Section 8.3)

SI units The preferred metric units for use in science. (Section 1.4)

slag A mixture of molten silicate minerals. Slags may be acidic or basic, according to the acidity or basicity of the oxide added to silica. (Section 23.2)

smectic liquid crystalline phase A liquid crystal in which the molecules are aligned along their long axes and arranged in sheets, with the ends of the molecules aligned. There are several different kinds of smectic phases. (Section 12.1)

smelting A melting process in which the materials formed in the course of the chemical reactions that occur separate into two or more layers. For example, the layers might be slag and molten metal. (Section 23.2)

sol-gel process A process in which extremely small particles (0.003 to 0.1 μm in diameter) of uniform size are produced in a series of chemical steps followed by controlled heating. (Section 12.4)

solid Matter that has both a definite shape and a definite volume. (Section 1.2)

solubility The amount of a substance that dissolves in a given quantity of solvent at a given temperature to form a saturated solution. (Sections 4.2 and 13.2)

solubility-product constant (solubility product) (K_{sp}) An equilibrium constant related to the equilibrium between a solid salt and its ions in solution. It provides a quantitative measure of the solubility of a slightly soluble salt. (Section 17.4)

solute A substance dissolved in a solvent to form a solution; it is normally the component of a solution present in the smaller amount. (Section 4.1)

solution A mixture of substances that has a uniform composition; a homogeneous mixture. (Section 1.2)

solution alloy A homogeneous alloy, with the components distributed uniformly throughout. (Section 23.6)

solvation The clustering of solvent molecules around a solute particle. (Section 13.1)

solvent The dissolving medium of a solution; it is normally the component of a solution present in the greater amount. (Section 4.1)

specific heat The heat capacity of 1 g of a substance; the heat required to raise the temperature of 1 g of a substance by 1°C. (Section 5.5)

spectator ions Ions that go through a reaction unchanged and that appear on both sides of the complete ionic equation. (Section 4.2)

spectrochemical series A list of ligands arranged in order of their abilities to split the *d*-orbital energies (using the terminology of the crystal-field model). (Section 24.6)

spectrum The distribution among various wavelengths of the radiant energy emitted or absorbed by an object. (Section 6.3)

spin magnetic quantum number (m_s) A quantum number associated with the electron spin; it may have values of $+\frac{1}{2}$ or $-\frac{1}{2}$. (Section 6.7)

spin-pairing energy The energy required to pair an electron with another electron occupying an orbital. (Section 24.6)

spontaneous process A process that is capable of proceeding in a given direction, as written or described, without needing to be driven by an outside source of energy. A process may be spontaneous even though it is very slow. (Section 19.1)

sputtering A method for forming thin films in which the material that is to form the film is made the cathode in a high-voltage gaseous discharge of an inert gas. (Section 12.6)

standard atmospheric pressure Defined as 760 torr or, in SI units, 101.325 kPa. (Section 10.2)

standard emf, also called the **standard cell potential** ($E°$) The emf of a cell when all reagents are at standard conditions. (Section 20.4)

standard enthalpy of formation (ΔH_f°) The change in enthalpy that accompanies the formation of 1 mol of a substance from its elements, with all substances in their standard states. (Section 5.7)

standard enthalpy (ΔH°) The enthalpy change when all reactants and products are at their standard states. (Section 5.7)

standard free energy of formation (ΔG_f°) The change in free energy associated with the formation of a substance from its elements under standard conditions. (Section 19.6)

standard hydrogen electrode An electrode based on the half-reaction $2H^+(1\,M) + 2e^- \longrightarrow H_2(1\text{ atm})$. The standard electrode potential of the standard hydrogen electrode is defined as 0 V. (Section 20.4)

standard molar entropy ($S°$) The entropy value for a mole of a substance in its standard state. (Section 19.4)

standard reduction potential (E_{red}°) The potential of a reduction half-reaction under standard conditions, measured relative to the standard hydrogen electrode. A standard reduction potential is also called a **standard electrode potential**. (Section 20.4)

standard solution A solution of known concentration. (Section 4.6)

standard temperature and pressure (STP) Defined as 0°C and 1 atm pressure; frequently used as reference conditions for a gas. (Section 10.4)

starch The general name given to a group of polysaccharides that acts as energy-storage substances in plants. (Section 25.10)

state function A property of a system that is determined by the state or condition of the system and not by how it got to that state; its value is fixed when temperature, pressure, composition, and physical form are specified; *P, V, T, E,* and *H* are state functions. (Section 5.2)

states of matter The three forms that matter can assume: solid, liquid, and gas. (Section 1.2)

stereoisomers Compounds possessing the same formula and bonding arrangement but differing in the spatial arrangements of the atoms. (Section 24.4)

stoichiometry The relationships among the quantities of reactants and products involved in chemical reactions. (Chapter 3: Introduction)

stratosphere The region of the atmosphere directly above the troposphere. (Section 18.1)

strong acid An acid that ionizes completely in water. (Section 4.3)

strong base A base that ionizes completely in water. (Section 4.3)

strong electrolyte A substance that is completely ionized in solution, for example, strong acids, strong bases, and most salts. (Section 4.1)

structural formula A formula that shows not only the number and kinds of atoms in the molecule but also the arrangement of the atoms. (Section 2.6)

structural isomers Compounds possessing the same formula but differing in the bonding arrangements of the atoms. (Section 24.4)

subatomic particles Particles such as protons, neutrons, and electrons that are smaller than an atom. (Section 2.2)

subshell One or more orbitals with the same set of quantum numbers *n* and *l*. For example, we speak of the 2*p* subshell ($n = 2, l = 1$), which is composed of three orbitals ($2p_x$, $2p_y$, and $2p_z$). (Section 6.5)

substrate A substance that undergoes a reaction at the active site in an enzyme. (Section 14.7)

substitution reactions Reactions in which one atom (or group of atoms) replaces another atom (or group) within a molecule; substitution reactions are typical for alkanes and aromatic hydrocarbons. (Section 25.4)

superconducting ceramic A complex metal oxide that undergoes a transition to a superconducting state at a low temperature. (Section 12.5)

superconducting transition temperature (T_c) The temperature below which a substance exhibits superconductivity. (Section 12.5)

superconductivity The "frictionless" flow of electrons that occurs when a substance loses all resistance to the flow of electrical current. (Section 12.5)

supercritical mass An amount of fissionable material larger than the critical mass. (Section 21.7)

supersaturated solutions Solutions containing more solute than a saturated solution. (Section 13.2)

surface tension The intermolecular, cohesive attraction that causes a liquid to minimize its surface area. (Section 11.3)

surroundings In thermodynamics everything that lies outside the system that we study. (Section 5.1)

system In thermodynamics the portion of the universe that we single out for study. We must be careful to state exactly what the system contains and what transfers of energy it may have with its surroundings. (Section 5.1)

termolecular reaction An elementary reaction that involves three molecules. (Section 14.6)

tertiary structure The overall shape of a large protein, specifically, the manner in which sections of the protein fold back upon themselves or intertwine. (Section 25.9)

theoretical yield The quantity of product that is calculated to form when all of the limiting reagent reacts. (Section 3.7)

theory A tested model or explanation of the general principles of certain phenomena. (Section 1.3)

thermochemistry The relationship between chemical reactions and energy changes. (Chapter 5: Introduction)

thermodynamics The study of energy and its transformation. (Chapter 5: Introduction)

thermonuclear reaction Another name for fusion reactions; reactions in which two light nuclei are joined to form a more massive one. (Section 21.8)

thermoplastic A polymeric material that can be readily reshaped by application of heat and pressure. (Section 12.2)

thermosetting plastic A plastic that, once formed in a particular mold, is not readily reshaped by application of heat and pressure. (Section 12.2)

thin film A film deposited on an underlying substrate to provide decoration or protection from chemical attack or to enhance a desirable property, such as reflectivity, electrical conductivity, color, or hardness. (Section 12.6)

third law of thermodynamics A law stating that the entropy of a pure, crystalline solid at absolute zero temperature is zero: $S(0\text{ K}) = 0$. (Section 19.3)

titration The process of reacting a solution of unknown concentration with one of known concentration (a standard solution). (Section 4.6)

titration curve A graph of pH as a function of added titrant. (Section 17.3)

torr A unit of pressure (1 torr = 1 mm Hg). (Section 10.2)

transition elements (transition metals) Elements in which the *d* orbitals are partially occupied. (Section 6.8)

transition state (activated complex) The particular arrangement of reactant and product molecules at the point of maximum energy in the rate-determining step of a reaction. (Section 14.5)

translational motion Movement in which an entire molecule moves in a definite direction. (Section 19.3)

transuranium elements Elements that follow uranium in the periodic table. (Section 21.3)

triple bond A covalent bond involving three electron pairs. (Section 8.3)

triple point The temperature at which solid, liquid, and gas phases coexist in equilibrium. (Section 11.6)

tritium The isotope of hydrogen whose nucleus contains a proton and two neutrons. (Section 22.2)

troposphere The region of Earth's atmosphere extending from the surface to about 12 km altitude. (Section 18.1)

Tyndall effect The scattering of a beam of visible light by the particles in a colloidal dispersion. (Section 13.6)

uncertainty principle A principle stating there is an inherent uncertainty in the precision with which we can simultaneously specify the position and momentum of a particle. This uncertainty is significant only for extremely small particles, such as electrons. (Section 6.4)

unimolecular reaction An elementary reaction that involves a single molecule. (Section 14.6)

unit cell The smallest portion of a crystal that reproduces the structure of the entire crystal when repeated in different directions in space. It is the repeating unit or "building block" of the crystal lattice. (Section 11.7)

unsaturated solutions Solutions containing less solute than a saturated solution. (Section 13.2)

vacuum deposition A method of forming thin films in which a substance is sublimed at high temperature without decomposition and then deposited on the object to be coated. (Section 12.6)

valence-bond theory A model of chemical bonding in which an electron-pair bond is formed between two atoms by the overlap of orbitals on the two atoms. (Section 9.4)

valence electrons The outermost electrons of an atom; those that occupy orbitals not occupied in the nearest noble-gas element of lower atomic number. The valence electrons are the ones the atom uses in bonding. (Section 6.8)

valence orbitals Orbitals that contain the outer-shell electrons of an atom. (Chapter 7: Introduction)

valence-shell electron-pair repulsion (VSEPR) model A model that accounts for the geometric arrangements of shared and unshared electron pairs around a central atom in terms of the repulsions between electron pairs. (Section 9.2)

van der Waals equation An equation of state for nonideal gases that is based on adding corrections to the ideal-gas equation. The correction terms account for intermolecular forces of attraction and for the volumes occupied by the gas molecules themselves. (Section 10.9)

vapor Gaseous state of any substance that normally exists as a liquid or solid. (Section 10.1)

vapor pressure The pressure exerted by a vapor in equilibrium with its liquid or solid phase. (Section 11.5)

vibrational motion Movement of the atoms within a molecule in which they move periodically toward and away from one another. (Section 19.3)

viscosity A measure of the resistance of fluids to flow. (Section 11.3)

volatile Tending to evaporate readily. (Section 11.5)

voltaic (galvanic) cell A device in which a spontaneous oxidation-reduction reaction occurs with the passage of electrons through an external circuit. (Section 20.3)

vulcanization The process of cross-linking polymer chains in rubber. (Section 12.2)

watt A unit of power; $1\text{ W} = 1\text{ J/s}$. (Section 20.9)

wave function A mathematical description of an allowed energy state (an orbital) for an electron in the quantum mechanical model of the atom; it is usually symbolized by the Greek letter ψ. (Section 6.5)

wavelength The distance between identical points on successive waves. (Section 6.1)

weak acid An acid that only partly ionizes in water. (Section 4.3)

weak base A base that only partly ionizes in water. (Section 4.3)

weak electrolyte A substance that only partly ionizes in solution. (Section 4.1)

work The movement of an object against some force. (Section 5.1)

Photo/Art Credits

Chapter 1: CO01 NASA, Andrew S. Wilson (University of Maryland); Patrick L. Shopbell (Caltech); Chris Simpson (Subaru Telescope); Thaisa Storchi-Bergmann and F.K.B. Barbosa (UFRGS, Brazil); and Martin J. Ward (University of Leicester, U.K.) **1.2a** Andrew Syred/Science Photo Library/Photo Researchers, Inc. **1.2b** Francis G. Mayer/CORBIS **1.2c** G. Murti/Photo Researchers, Inc. **1.3** Richard Megna/Fundamental Photographs **1.4** Dale Wilson/Green Stock/CORBIS **1.7** Charles D. Winters/Photo Researchers, Inc. **1.8a** M. Angelo/CORBIS **1.8b** Richard Megna/Fundamental Photographs **1.11a–c** Donald Clegg and Roxy Wilson Pearson Education/PH College **1.12a–b** Donald Clegg and Roxy Wilson/Pearson Education/PH College **1.14a–c** Richard Megna/Fundamental Photographs **1.16** Richard Megna/Fundamental Photographs **1.17** Australian Postal Service/National Standards Commission **1.21** Tom Pantages/Daimler Chrysler **1.22** NASA/Goddard Space Flight Center

Chapter 2: CO02 Gary Valaskovic/New Objective, Inc. **2.1** CORBIS **2.2** IBMRL Visuals Unlimited **2.3b–c** Richard Megna/Fundamental Photographs **2.6** Radium Institute American/Institute of Physics/Emilio Segre Visual Archives **2.7** G.R. 'Dick' Roberts Photo Library **2.17** Richard Megna/Fundamental Photographs **2.18** Ernest Orlando Lawrence Berkeley National Laboratory, Courtesy AIP Emilio Segre Visual Archives **2.23** Andrew Syred/Science Photo Library/Photo Researchers, Inc. **2.25** Richard Megna/Fundamental Photographs

Chapter 3: CO03 Galen Rowell/Mountain Light Photography Inc. **3.1** Jean-Loup Charmet/Science Photo Library/Photo Researchers, Inc. **3.2** Richard Megna/Fundamental Photographs **p. 77** Dave Carpenter **3.5a–b** Richard Megna/Fundamental Photographs **3.6** Donald Johnston/Getty Images Inc. **3.7** Richard Megna/Fundamental Photographs **3.9** Richard Megna/Fundamental Photographs **p. 108** Carey B. Van Loon **p. 109** Paul Silverman/Fundamental Photographs

Chapter 4: CO04 Paul Silverman/Fundamental Photographs **4.1** Richard Megna/Fundamental Photographs **4.2a–c** Richard Megna/Fundamental Photographs **4.4** Richard Megna/Fundamental Photographs **4.5** Robert Mathena/Fundamental Photographs **4.7** Richard Megna/Fundamental Photographs **4.8a–c** Richard Megna/Fundamental Photographs **4.9** Richard Megna/Fundamental Photographs **4.10** Richard Megna/Fundamental Photographs **4.11** Teri Stratford/Photo Researchers, Inc. **4.13** Richard Megna/Fundamental Photographs **4.14a–c** Peticolas/Megna/Fundamental Photographs **4.15** Erich Lessing/Art Resource, N.Y. **4.16a–d** Donald Clegg and Roxy Wilson/Pearson Education/PH College **4.17a–c** Richard Megna/Fundamental Photographs **4.19a–c** Richard Megna/Fundamental Photographs **p. 149** Richard Megna/Fundamental Photographs

Chapter 5: CO05 Dennis Flaherty/Photo Researchers, Inc. **5.1a** Amoz Eckerson/Visuals Unlimited **5.1b** Tom Pantages **5.8a–b** Richard Megna/Fundamental Photographs **5.14a–b** Donald Clegg and Roxy Wilson/Pearson Education/PH College **5.15** UPI CORBIS **5.20** Gerard Vandystadt/Photo Researchers, Inc. **5.23** Michael Dalton/Fundamental Photographs **5.25** Honda America/Peter Arnold, Inc.

Chapter 6: CO06 Stone/Getty Images, Inc. **6.1** Pal Hermansen/Getty Images Inc. **6.5** AGE/Peter Arnold, Inc. **6.6a** Laura Martin/Visuals Unlimited **6.6b** PhotoDisc, Getty Images, Inc. **6.8** Photograph by Paul Ehrenfest/Courtesy AIP Emilio Segre Visual Archives, Ehrenfest Collection **6.9** Scott Camazine/Photo Researchers, Inc. **6.11a–b** Tom Pantages **6.14** Gelderblom/Eye of Science/Photo Researchers, Inc. **6.15** Photo by Professor Friedrich Hund, Werner Heisenberg Archive, Werner Heisenberg Institut/Max Planck Institut for Physik **6.26** Alfred Pasieka/Science Photo Library/Photo Researchers, Inc.

Chapter 7: CO07 Wildlife Pictures/Peter Arnold, Inc. **7.1** Richard Megna/Fundamental Photographs **7.8** 1ca2: A.E. Eriksson, T.A. Jones, A. Liljas: Refined structure of human anhydrase II at 2.0 Angstrom resolution. Proteins 4 pp. 274 (1988). PDB: H.M. Berman, J. Westbrook, Z. Feng, G. Gilliland, T.N. Bhat, H. Weissig, I.N. Shindyalov, P.E. Bourne: The Protein Data Bank. Nucleic Acids Research, 28 pp. 235–242 (2000). Protein Data Bank (PDB) ID 1ca2 from www.pdb.org **7.13** Peter Marbach/Grant Heilman Photography, Inc. **7.15a–b** Richard Megna/Fundamental Photographs **7.16** Ed Degginger/Color-Pic, Inc. **7.17a–b** Richard Megna/Fundamental Photographs **7.18** Richard Megna/Fundamental Photographs **7.19** Richard Megna/ Fundamental Photographs **7.20** Nigel Cattlin/Holt Studios International/Photo Researchers, Inc. **7.21a–c** Richard Megna/Fundamental Photographs **7.22a–c** H. Eugene LeMay, Jr. **7.23** Adam Woolfitt/CORBIS **7.24** Paparazzi Photography–Dr. Pepper/Seven Up and Jeffrey L. Rodengen, Write Stuff Syndicate, Inc. **7.25** Tom Pantages **7.26** RNHRD NHS Trust/Getty Images Inc. **7.27** Ed Degginger/Color-Pic, Inc. **7.29** Richard Megna/Fundamental Photographs **7.30** Ed Degginger/Color-Pic, Inc.

Chapter 8: CO08 M. Jentoft-Nilsen, F. Hasler, D. Chesters/NASA/GSFC & T. Neilsen/University of Hawaii/CCRT **8.1a–c** Richard Megna/Fundamental Photographs **8.2a–c** Donald Clegg and Roxy Wilson/Pearson Education/PH College **8.12a** Tom Pantages **8.14** CORBIS

Chapter 9: CO09 Omikron/Photo Researchers, Inc. **9.5a–c** Kristen Brochmann/Fundamental Photographs **9.30** Bill Longcore/Photo Researchers, Inc. **9.43** Richard Megna/Fundamental Photographs **9.45a** Phil Degginger/Color-Pic, Inc. **9.45b** Scott Camazine/Photo Researchers, Inc.

Chapter 10: CO10 Getty Images, Inc. **10.4** Andy Levin/Photo Researchers, Inc. **10.5** Roland Seitre/Peter Arnold, Inc. **10.8a–b** Richard Megna/Fundamental Photographs **10.14** Ed Degginger/Color-Pic, Inc. **10.15** Van Bucher/Photo Researchers, Inc. **10.21a–b** Richard Megna/Fundamental Photographs

Chapter 11: CO11 Mark A. Schneider/Photo Researchers, Inc. **11.9** Richard Megna/Fundamental Photographs **p. 414** Bill Watterson/Universal Press Syndicate **11.10** Astrid and Hanns-Frieder Michler/Peter Arnold, Inc. **11.11** Richard Megna/Fundamental Photographs **11.13** Kristen Brochmann/Fundamental Photographs **11.15** Hermann Eisenbeiss/Photo Researchers, Inc. **11.16** Richard Megna/ Fundamental Photographs **11.29a** Dan McCoy/Rainbow **11.29b** Herve Berthoule/ Jacana Scientific Control/Photo Researchers, Inc. **11.29c** Michael Dalton/Fundamental Photographs **11.39** Science Source/Photo Researchers, Inc. **11.43** Phil Degginger/Merck/Color-Pic, Inc. **11.44** Robert L. Whetten/University of California at Los Angeles

Chapter 12: CO12 Onne van der Wal Photography **12.1a–b** Richard Megna/Fundamental Photographs **12.2** Thales Navigation, Inc. **12.6** Richard Megna/Fundamental Photographs **12.8** Richard Megna/Fundamental Photographs **12.11a–b** Richard Megna/Fundamental Photographs **12.14** Tom Pantages **12.15** Leonard Lessin/Peter Arnold, Inc. **12.16** General Electric Global Marketing Communications/Courtesy SMART **12.19** SJM is a registered trademark of St. Jude Medical, Inc./Copyright St. Jude Medical, Inc. 2002/This image is provided courtesy of St. Jude Medical, Inc. All rights reserved. **12.20** Southern Illinois University/Photo Researchers, Inc. **12.21** Advanced Tissue Sciences, Inc. **12.22** Biomet Inc. **12.23** Photo courtesy of Kyocera Industrial Ceramics Corp./Vancouver, WA **12.24** Professor C. Zukoski/Department of Chemical Engineering, University of Illinois/Urbana-Champaign, Illinois **12.25** James L. Amos/ Peter Arnold, Inc. **12.26** David Parker/IMI/Univ. of Birmingham/Science Photo Library/Photo Researchers, Inc. **12.27** Mauro Fermariello/Science Photo Library/ Photo Researchers, Inc. **12.29** John Sohlden/Visuals Unlimited **12.30** © and courtesy Superconductor Technologies Inc. **12.31** Nathan Benn/Woodfin Camp & Associates **12.32** Kristen Brochmann/Fundamental Photographs **12.34** Richard Megna/Fundamental Photographs **12.35** Dean M. Aslam/Department of Electrical and Computer Engineering, Michigan State University.

Chapter 13: CO13 Foto-UNEP-Still Pictures/Peter Arnold, Inc. **13.5** Leonard Lessin/Peter Arnold, Inc. **13.7a–c** Richard Megna/Fundamental Photographs **13.8** Ed Degginger/Color-Pic, Inc. **13.10a–c** Richard Megna/Fundamental Photographs **13.15** Paul Silverman/Fundamental Photographs **13.16** Joan Richardson/Visuals Unlimited **13.21** Grant Heilman/Grant Heilman Photography, Inc. **13.27** Leonard Lessin/Peter Arnold, Inc. **13.28a** E.R. Degginger/Color-Pic, Inc. **13.28b** Gene Rhoden/Visuals Unlimited **13.32** Oliver Meckes & Nicole Ottawa/Photo Researchers, Inc.

Chapter 14: CO14 Milton Heiberg/Photo Researchers, Inc. **14.1a** Michael S./Yamashita CORBIS **14.1b** © S.C. Fried/Photo Researchers, Inc. **14.1c** David N. Davis/Photo Researchers, Inc. **14.2a** Michael Dalton/Fundamental Photographs **14.2b** Richard Megna/Fundamental Photographs **14.11** Richard Megna/Fundamental Photographs **14.16** S.C. Fried/Photo Researchers, Inc. **14.19a–c** Richard Megna/Fundamental Photographs **14.22** Astrid & Hanns-Frieder Michler/Science Photo Library/Photo Researchers, Inc. **14.23** Richard Megna/Fundamental Photographs **14.25a–b** Coordinates by J.C. Cheetham, P.J. Artymiuk, and D.C. Phillips; Images by Laboratory for Molecular Modeling, Pharmacy School, UNC-Chapel Hill **14.27a** Science Photo Library/Photo Researchers, Inc.

Chapter 15: CO15 Richard Megna/Fundamental Photographs **15.2a–c** Richard Megna/Fundamental Photographs **15.5** Ed Degginger/Color-Pic, Inc./Courtesy of Farmland Industries, Inc. **15.14a–c** Richard Megna/Fundamental Photographs **15.15a–c** Richard Megna/ Fundamental Photographs

Chapter 16: CO16 FoodPix/Getty Images Inc. **16.3** Richard Megna/Fundamental Photographs **16.6** Yoav Levy/Phototake NYC **16.8a–b** Donald Clegg and Roxy Wilson/Pearson Education/PH College **16.10** Donald Clegg and Roxy Wilson/Pearson Education/PH College **16.11a-c** Richard Megna/Fundamental Photographs **16.14** Frank LaBua/Pearson Education/PH College **16.16** Tom Pantages

Chapter 17: CO17 Runk/Schoenberger/Grant Heilman Photography, Inc. **17.1** Donald Clegg and Roxy Wilson/Pearson Education/PH College **17.4** Professors P.P. Motta and S. Correr/Science Picture Library/Photo Researchers, Inc. **17.7a–b** Richard Megna/Fundamental Photographs **17.16a–b** Richard Megna/Fundamental Photographs **17.17b** Gerry Davis/Phototake NYC **17.19a–b** Richard Megna/Fundamental Photographs **17.20a–c** Richard Megna/Fundamental Photographs **17.21a–c** Richard Megna/Fundamental Photographs

Chapter 18: CO18 Ken Griffiths/Getty Images, Inc. **18.2** Pekka Parviainen/Science Photo Library/Photo Researchers, Inc. **18.4** NASA Headquarters **18.5a–b** Don & Pat Valenti/Getty Images Inc. **18.7a** Dennis Kunkel/Phototake NYC **18.8** Richard Megna/Fundamental Photographs **18.9** Ulf E. Wallin/Getty Images Inc, **18.12** Australia Picture Library/CORBIS **18.14a** E.I. DuPont de Nemours & Co., Inc. **18.15** © Katadyn North America/Used by permission. **18.16** Robert T. Zappalorti/Photo Researchers, Inc. **18.18** Sheila Terry/Science Photo Library/Photo Researchers, Inc. **18.19** GE Plastics **18.20** Kim Fennema/Visuals Unlimited **18.21** PPG Industries, Inc.

Chapter 19: CO19 Getty Images, Inc. 19.1 Kristen Brochmann/Fundamental Photographs **19.2** PNNL/RDF Visuals Unlimited **19.3** Michael Dalton/Fundamental Photographs **19.8** Richard Megna/Fundamental Photographs **19.9a–b** Richard Megna/Fundamental Photographs **19.10a** Klaus Pavsan/Peter Arnold, Inc. **19.10b** Biophoto Associates/Photo Researchers, Inc. **19.15** Austrian Central Library for Physics, University of Vienna **19.16** Library of Congress

Chapter 20: CO20 Tom Meyers/Getty Images, Inc. **20.1a–b** Richard Megna/ Fundamental Photographs **20.2a–c** Richard Megna/Fundamental Photographs **20.3a–b** Richard Megna/Fundamental Photographs **20.4** Richard Megna/Fundamental Photographs **20.9** Jeff Gnass/Corbis Stock Market **20.15** Burndy Library/Dibner Institute for the History of Science and Technology **20.19** Kaj R. Svenson/Science Photo Library/Photo Researchers, Inc. **20.24** Erich Schrempp/Photo Researchers, Inc. **20.29a–b** Reed Barton/Tom Pantages

Chapter 21: CO21 Science VU/NIH Visuals Unlimited **21.1** Yoav Levy/Phototake NYC **21.6** Fermi National Accelerator Laboratory/Science Photo Library/Photo Researchers, Inc. **21.8** Gianni Tortoli/Science Source/Photo Researchers, Inc. **21.9** Yoav Levy/Phototake NYC **21.11** Kevin Schafer/Peter Arnold, Inc. **21.12a** Hank Morgan/Rainbow **21.12b** Wellcome Dept. of Cognitive Neurology/Science Photo Library/Photo Researchers, Inc. **21.18** Gary Sheahan *Birth of the Atomic Age* Chicago (Illinois); 1957. Historical Society **21.20b** Ed Degginger/Color-Pic, Inc. **21.24** Earl Roberge/Photo Researchers, Inc. **21.25** Richard Megna/Fundamental Photographs

Chapter 22: CO22 Getty Images, Inc. **22.6a–b** Richard Megna/Fundamental Photographs **22.9** Donald Clegg and Roxy Wilson/Pearson Education/PH College **22.11** Paul Silverman/Fundamental Photographs **22.12a–b** Richard Megna/-Fundamental Photographs **22.13** Richard Megna/Fundamental Photographs **22.14** NASA/Johnson Space Center **22.16** The Granger Collection **22.17** John Hill/Getty Images Inc. **22.20** Donald Clegg and Roxy Wilson/Pearson Education/PH College **22.21** Courtesy of DuPont Nomex® **22.24a** Jeffrey A. Scovil **22.25** Lawrence Migdale/Science Source/Photo Researchers, Inc. **22.27** Dan McCoy/Rainbow **22.28** Kristen Brochmann/Fundamental Photographs **22.33a–c** Donald Clegg and Roxy Wilson/Pearson Education/PH College **22.36** Kristen Brochmann/Fundamental Photographs **22.39** Richard Megna/Fundamental Photographs **22.43** USGS (Welch, A.H., Watkins, S.A., Helsel, D.R., and Focazio, M.F., 2000, Arsenic in ground–water resources of the United States: U.S. Geological Survey Fact Sheet 063–00, 4p.) **22.44** General Electric Corporate Research & Development Center **22.45a** Cytec Fiberite Inc./Cytec Industries Inc. **22.45b** John Laptad/Trek Bicycle Corp. **22.46** Richard Megna/Fundamental Photographs **22.47** Chad Ehlers/Getty Images Inc. **22.49** Texas Instruments Incorporated **22.54** National Institute for Occupational Safety & Health

Chapter 23: CO23 Nimtallah/Art Resource, N.Y. (National Archeological Museum, Athens, Greece) **23.1b** Brownie Harris/Corbis Stock Market **23.2a–b** Karl Hartmann/Traudel Sachs Phototake NYC **23.2c** Jeffrey A. Scovil **23.3** The Cleveland-Cliffs Iron Company **23.4b** Science VU/API Visuals Unlimited **23.5** Robin Smith/Getty Images Inc. **23.10** Oberlin College Archives **23.12** Richard Megna/Fundamental Photographs **23.17** 3Com and the 3Com logo are the registered trademarks. Palm V™ and the Palm V™ logo are the trademarks of Palm Computing Inc., 3Com Corporation or its subsidiaries **23.19** Richard Megna/Fundamental Photographs **23.23** Donald Clegg and Roxy Wilson/Pearson Education/PH College **23.26** Michael Dalton/Fundamental Photographs **23.27** Donald Clegg and Roxy Wilson/Pearson Education/PH College **23.28** Richard Megna/Fundamental Photographs **23.29** Alan L. Detrick/Color-Pic, Inc. **23.30** Donald Clegg and Roxy Wilson/Pearson Education/PH College **23.31** Paul Silverman/Fundamental Photographs **23.32** Donald Clegg and Roxy Wilson/Pearson Education/PH College

Chapter 24: CO24 Malcolm S. Kirk/Peter Arnold, Inc. **24.2a–b** Richard Megna/Fundamental Photographs **24.8** Gary C. Will/Visuals Unlimited **24.15** Michael P. Gadomski/Photo Researchers, Inc. **24.23** Richard Megna/Fundamental Photographs **24.37a–c** Tom Pantages

Chapter 25: CO25 Ron Kimball Photography **25.8** Ed Degginger/Color-Pic, Inc. **25.9** Wes Thompson/Corbis Stock Market **25.15** Donald Clegg and Roxy Wilson/Pearson Education/PH College **25.16a–b** Richard Megna/Fundamental Photographs **25.18** Culver Pictures, Inc. **25.23** Richard Megna/Fundamental Photographs **25.25** Oxford Molecular Biophysics Laboratory/Science Photo Library/Photo Researchers, Inc.

Index

COMMON IONS

Positive Ions (Cations)

1+

Ammonium (NH_4^+)
Cesium (Cs^+)
Copper(I) or cuprous (Cu^+)
Hydrogen (H^+)
Lithium (Li^+)
Potassium (K^+)
Silver (Ag^+)
Sodium (Na^+)

2+

Barium (Ba^{2+})
Cadmium (Cd^{2+})
Calcium (Ca^{2+})
Chromium(II) or chromous (Cr^{2+})
Cobalt(II) or cobaltous (Co^{2+})
Copper(II) or cupric (Cu^{2+})
Iron(II) or ferrous (Fe^{2+})
Lead(II) or plumbous (Pb^{2+})
Magnesium (Mg^{2+})
Manganese(II) or manganous (Mn^{2+})
Mercury(I) or mercurous (Hg_2^{2+})
Mercury(II) or mercuric (Hg^{2+})
Strontium (Sr^{2+})
Nickel(II) (Ni^{2+})
Tin(II) or stannous (Sn^{2+})
Zinc (Zn^{2+})

3+

Aluminum (Al^{3+})
Chromium(III) or chromic (Cr^{3+})
Iron(III) or ferric (Fe^{3+})

Negative Ions (Anions)

1−

Acetate ($C_2H_3O_2^-$)
Bromide (Br^-)
Chlorate (ClO_3^-)
Chloride (Cl^-)
Cyanide (CN^-)
Dihydrogen phosphate ($H_2PO_4^-$)
Fluoride (F^-)
Hydride (H^-)
Hydrogen carbonate or bicarbonate (HCO_3^-)
Hydrogen sulfite or bisulfite (HSO_3^-)
Hydroxide (OH^-)
Iodide (I^-)
Nitrate (NO_3^-)
Nitrite (NO_2^-)
Perchlorate (ClO_4^-)
Permanganate (MnO_4^-)
Thiocyanate (SCN^-)

2−

Carbonate (CO_3^{2-})
Chromate (CrO_4^{2-})
Dichromate ($Cr_2O_7^{2-}$)
Hydrogen phosphate (HPO_4^{2-})
Oxide (O^{2-})
Peroxide (O_2^{2-})
Sulfate (SO_4^{2-})
Sulfide (S^{2-})
Sulfite (SO_3^{2-})

3−

Arsenate (AsO_4^{3-})
Phosphate (PO_4^{3-})

FUNDAMENTAL CONSTANTS*

Atomic mass unit	1 amu	$= 1.66053873 \times 10^{-24}$ g
	1 g	$= 6.02214199 \times 10^{23}$ amu
Avogadro's number	N	$= 6.02214199 \times 10^{23}$/mol
Boltzmann's constant	k	$= 1.3806503 \times 10^{-23}$ J/K
Electron charge	e	$= 1.602176462 \times 10^{-19}$ C
Faraday's constant	F	$= 9.64853415 \times 10^{4}$ C/mol
Gas constant	R	$= 0.082058205$ L-atm/mol-K
Mass of electron	m_e	$= 5.485799 \times 10^{-4}$ amu
		$= 9.10938188 \times 10^{-28}$ g
Mass of neutron	m_n	$= 1.0086649$ amu
		$= 1.67492716 \times 10^{-24}$ g
Mass of proton	m_p	$= 1.0072765$ amu
		$= 1.67262158 \times 10^{-24}$ g
Pi	π	$= 3.1415927$
Planck's constant	h	$= 6.62606876 \times 10^{-34}$ J-s
Speed of light	c	$= 2.99792458 \times 10^{8}$ m/s

*Fundamental constants are listed at the National Institute of Standards and Technology website: http://physics.nist.gov/PhysRefData/contents.html

Installation Instructions

Please see the README file for CD-ROM launching instructions.

Minimum System Requirements

PC:
- -Windows 98/Me/2000/NT/XP Pro
- -Pentium II 233 MHz processor
- -64 Mb RAM
- -800 x 600 pixel screen resolution
- -Color monitor running "Thousands of Colors" or higher
- -Audio capable system is recommended
- -4x CD-ROM drive
- -Speakers
- -Netscape Navigator/Communicator 4.75 and higher or Internet Explorer 5.5 and higher.
  **NOTE: Netscape 6 and higher is NOT supported.
- -Macromedia Shockwave 8.5 plug-in (c)1985-2001
- -Chemscape Chime 2.6 SP3 plug-in (c)2001
- -Apple QuickTime 5 plug-in

Macintosh:
- -Power PC with OS 8.1 or greater
- -64 Mb RAM, Virtual Memory Enabled
- -800 x 600 pixel screen resolution
- -Color monitor running "Thousands of Colors" or higher
- -Audio capable system is recommended
- -4x CD-ROM drive
- -Speakers
- -Netscape Navigator/Communicator 4.75 and higher.(Netscape Communicator 4.78 (c)1995-2001 installer is included on CD-ROM).
  **NOTE: Netscape 6 and higher and any version of Microsoft Internet Explorer are NOT supported.
- -Macromedia Shockwave 8.5 plug-in (c)1985-2001
- -Chemscape Chime 2.6 SP3 plug-in (c)2001
- -Apple QuickTime 5 plug-in